曹绪龙

化学驱油论文集

（上册）

曹绪龙　编著

中国海洋大学出版社

·青岛·

图书在版编目(CIP)数据

曹绪龙化学驱油论文集 / 曹绪龙编著. — 青岛 ：中国海洋大学出版社，2023.8

ISBN 978-7-5670-3597-3

Ⅰ. ①曹… Ⅱ. ①曹… Ⅲ. ①石油 — 开采— 文集 Ⅳ. ①TE357.46-53

中国国家版本馆 CIP 数据核字(2023)第 158756 号

CAO XULONG HUAXUE QUYOU LUNWENJI (SHANGCE)

曹绪龙化学驱油论文集(上册)

出版发行	中国海洋大学出版社
社　　址	青岛市香港东路 23 号　　**邮政编码**　266071
网　　址	http://pub.ouc.edu.cn
出 版 人	刘文菁
责任编辑	矫恒鹏
电　　话	0532-85902349
电子信箱	2586345806@qq.com
印　　制	青岛国彩印刷股份有限公司
版　　次	2023 年 8 月第 1 版
印　　次	2023 年 8 月第 1 次印刷
成品尺寸	185 mm×260 mm
印　　张	41.75
字　　数	914 千
印　　数	1～1000
定　　价	598.00 元(上下册)
订购电话	0532-82032573(传真)

发现印装质量问题，请致电 0532-58700166，由印刷厂负责调换。

前　言

提高老油田采收率和新区有效动用是增加经济可采储量和原油产量的主要途径。胜利油田是我国重要的石油工业基地，是一个地质条件复杂、含油层系多、储层种类多、物性变化大、油藏类型多的复式含油盆地，被誉为“石油地质大观园”。不论从油藏类型、油藏条件，还是提高采收率方法来看，依据胜利油藏为对象形成的方法，在全国都具有很高的推广价值。

在石油开采技术不断发展的今天，化学驱作为提高采收率的重要方式，在油田的生产中发挥着越来越重要的作用。胜利油田在20世纪60年代开展化学驱研究，90年代开展矿场试验，经过耐温抗盐聚合物驱理论、耐温抗盐聚合物加合增效理论的研究和发展，建立了高温高盐油藏化学驱理论，形成了聚合物驱、无碱二元复合驱和非均相复合驱技术系列，为油田可持续高质量发展做出了贡献，同时积累了丰富的攻关实践经验。

为进一步推动提高采收率技术的创新与发展，也为了给该领域的科研工作者提供借鉴、学习的机会，本文集选载了曹绪龙及其科研团队著述的154篇文章，以胜利油田为研究对象，全面介绍了胜利油田在化学驱提高采收率方面取得的研究成果。

本文集共分为聚合物及其相互作用、表面活性剂及其相互作用、驱油体系及驱油方法、化学剂浓度分析、驱油实验与数值模拟、矿场应用、论文综述、其他等八个部分，其中，聚合物及其相互作用论文34篇、表面活性剂及其相互作用论文39篇、驱油体系及驱油方法论文38篇、化学剂浓度分析论文3篇、驱油实验与数值模拟论文13篇、矿场应用论文7篇、论文综述3篇、其它文章17篇，内容涉及基础理论研究、驱油剂研发、驱油体系设计、数值模拟研究、矿场试验等诸多方面，是一本专业性强、涉及学科多并具有强烈胜利特色的科技书籍。该书的出版对石油行业的科技工作者和高校从事化学驱研究的师生具有较好的参考和借鉴价值，同时也将对进一步推动我国油田提高采收率发挥积极作用。

本论文集是参与胜利油田化学驱油技术研究的广大科技工作者的集体智慧的结晶，在此向他们表示衷心的感谢！同时也向在本文集编写过程中提供支持与帮助的同志表示谢意。

曹绪龙

2023年6月12日

目录

第一章　聚合物及其相互作用

第二章 表面活性剂及其相互作用

第三章 驱油体系及驱油方法

第四章　化学剂浓度分析

第五章　驱油实验与数值模拟

第六章 矿场应用

第七章 论文综述

第八章 其他

第一章　聚合物及其相互作用

聚合物及其复合体系的热稳定性研究

崔培英　曹绪龙　隋希华　施晓乐

（胜利油田有限公司地质科学研究院）

摘要：针对胜利油区孤岛油田西区油藏条件和现场实际，对不同水解度的聚合物在地层温度下经过长期相互作用引起的黏度、水解度、浓度以及复合体系动态界面张力变化规律进行了室内研究。研究表明，在一定的地层条件下，不同水解度的聚合物经过长期水解均趋于同一水解平衡值，在10%～40%水解度范围内，用初始水解度低的聚合物配制的聚合物及其复合体系的长期热稳定性好。

关键词：胜利油区；聚合物体系；复合体系；热稳定性；水解度

中图分类号：TE345　文献标识码：B　文章编号：1009-9603(2001)03-0058-03

引言

在地层温度条件下，聚合物经过长期热稳定后的黏度、水解度以及浓度变化的趋势是三次采油研究应关注的问题。本文主要针对不同水解度的聚合物开展热稳定性实验研究，确定现场聚合物水解度的最佳适用范围，提高聚合物驱、复合驱体系驱油效率提供实验和理论依据。

1　实验部分

1.1 实验样品

胜利京大公司凝胶聚合物（简称京大聚合物），固含量为27.4%，相对分子质量为867万，水解度为10.7%；胜利长安公司凝胶聚合物（简称长安聚合物），固含量为20.5%，相对分子质量为1 706万，水解度为20.8%；法国产3530S干粉聚合物（简称3530S聚合物），固含量为90.8%，相对分子质量为1 700万，水解度为29.9%；胜利钻井公司干粉聚合物（简称钻井聚合物），固含量为86.2%，相对分子质量为921万，水解度为39.6%。上海助剂厂生产的活性剂BES及北京华能公司生产的活性剂PS样品，所有化学试剂的纯度均不低于化学纯。

实验油样为孤岛西区脱水脱气原油。

实验所用的盐水为模拟孤岛西区注入水,矿化度 6 196 mg/L。

1.2 热稳定性实验

在容量大约为 50 mL 的玻璃安瓿瓶中分别装入配好的 1 500 mg/L 的聚合物溶液、1 500 mg/L 聚合物+1.2%Na_2CO_3+0.2%BES+0.1%PS 的复合体系,经冷冻、抽空、解冻,反复 3 次达到绝氧后再饱和高纯度氮气,火焰封口,置入 70℃恒温箱中,定时取样分析测定。

1.3 参数测定

用 Texas-500 型旋转滴界面张力仪,在 70℃下测量复合体系的动态界面张力。

用 DV-Ⅲ型黏度仪,在水浴温度 70℃条件下测定聚合物及复合体系的黏度。

分别称取 3.33 g 样品,加蒸馏水至 100 g,加入 0.1 mol/L 的 NaOH 溶液 0.5 mL、0.005 mol/L 的 MGC 溶液 5 mL,用 KPVS 滴定并计算聚合物水解度。

采用 Water－2690 型液相色谱仪测定聚合物浓度。

2 结果与讨论

2.1 Na_2CO_3、活性剂对聚合物水解度、黏度的影响

在浓度为 5 000 mg/L 聚合物母液中分别加入 Na_2CO_3、活性剂 PS 和 BES,配制成含有 1 500 mg/L 聚合物、1.2%Na_2CO_3、0.1%PS、0.2%BES 的复合体系,分别测定水解度及黏度(表 1)。结果表明,各组分对聚合物水解度、黏度的影响规律基本一致,其中对水解度较大的钻井聚合物影响较小;对水解度较小的 3530S 聚合物影响较大。

表 1 Na_2CO_3、PS、BES 对聚合物水解度、黏度的影响表

溶液配方	钻井聚合物		3530S 聚合物	
	水解度,%	黏度/mPa·s	水解度,%	黏度/mPa·s
0.15%聚合物	37.2	23.6	30.0	22.9
1.2%Na_2CO_3+0.15%聚合物	39.4	19.9	32.4	15.45
0.1%PS+0.15%聚合物	39.7	20.2	32.4	23.4
0.2%BES+0.15%聚合物	39.7	20.4	32.4	22.5
0.1%PS+0.2%BES+0.15%聚合物	39.8	20.4	37.4	23.6
0.1%PS+1.2%Na_2CO_3+0.15%聚合物	39.7	15.3	36.8	15.9
0.2%BES+1.2%Na_2CO_3+0.15%聚合物	39.7	15.0	35.3	16.1
0.2%BES+0.1%PS+1.2%Na_2CO_3+0.15%聚合物	39.9	14.9	39.6	15.9

就 3530S 聚合物而言,单一 Na_2CO_3 或单一活性剂 PS、BES 的加入,将导致其水解度增大;两种组分分别复配而成的 1.2%Na_2CO_3+0.1%PS,1.2%Na_2CO_3+0.2%

BES、0.1%PS+0.2%BES体系，引起聚合物水解度增加程度大于单一的1.2%Na_2CO_3及单一活性剂所引起聚合物水解度增加的程度，由三种组分复配而成的复合体系，水解度具有最大值。这一结果表明，Na_2CO_3、PS、BES的存在对聚合物水解度的影响具有协同效应，均可提高聚合物水解度。同时还发现，Na_2CO_3是粘度降低的主要因素。

2.2 水解度对聚合物及复合体系热稳定性的影响

2.2.1 水解度的变化规律

地层水矿化度(盐浓度)、pH值及温度是影响聚合物水解的主要因素[1]。一般认为，聚合物的水解度在35%～41%范围内溶液会产生沉淀。用地层水配制的聚合物体系在70℃条件下长期热老化均可导致聚合物水解度的增大。此外，由于大量Na_2CO_3及活性剂的加入，复合体系的pH值一般大于11，初始水解迅速，老化一天后均超过35%。无论是聚合物体系还是复合体系，初始水解度高的聚合物水解度增幅缓慢，初始水解度低的聚合物水解度增幅较大，最终趋于一致，接近40%(图1，图2)。由此可以确定，在温度为70℃、矿化度为6 196 mg/L的孤岛西区注入水条件下，长期热老化后聚合物水解度将达到平衡值40%。

2.2.2 黏度的变化规律

实验表明，在长期热稳定条件下，聚合物体系黏度主要有三方面变化：一是聚合物黏度随时间增长而下降，原因是聚合物分子在水溶液中有解缠作用，当解缠结-缠结达到平衡时，聚合物溶液的黏度趋于稳定；二是聚合物分子水解产生羧酸基，羧酸基之间由于同性电荷相斥使分子膨胀，随水解度增大黏度升高；三是随着时间延长，水解到35%以上，聚合物与地层水中Ca^{2+}、Mg^{2+}产生沉淀，导致溶液黏度下降。其中，水解作用产生沉淀是引起聚合物化学降解的主要原因。

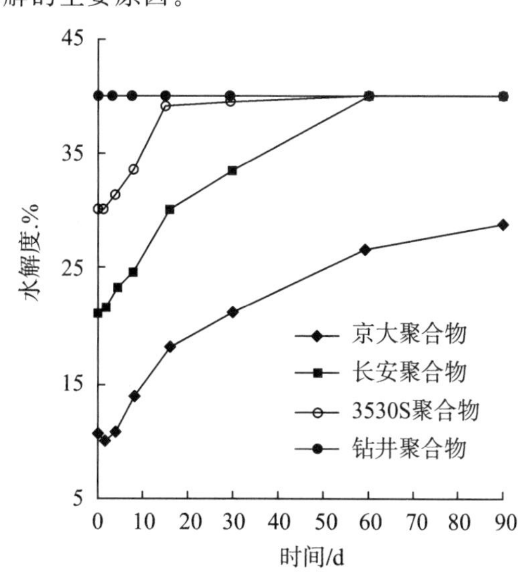

图1　聚合物体系中聚合物水解度变化曲线图

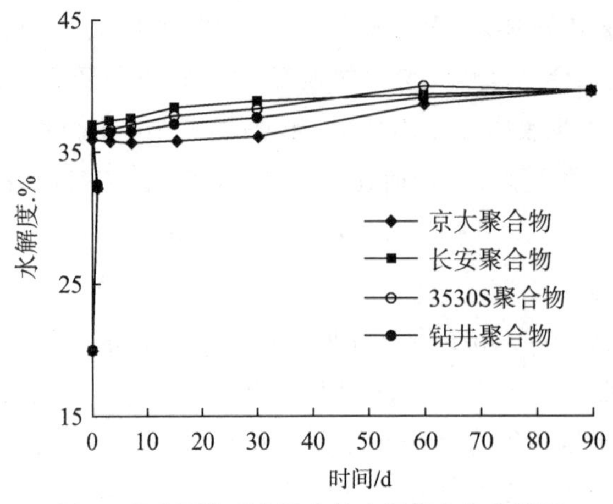

图 2 复合驱体系中聚合物水解度变化曲线图

热稳定初期,低水解度的长安和京大聚合物不断水解(图 3),黏度逐渐增大并趋于稳定,90 d 后,京大聚合物黏度保留率为 198%,长安聚合物黏度保留率为 115%。中水解度的 3530S 聚合物,黏度保留率为 102%,其黏度变化以解缠结-缠结作用和电荷相斥作用为主。水解度高达 39.6%的钻井聚合物,与地层水中 Ca^{2+}、Mg^{2+} 作用产生沉淀,故黏度呈下降趋势,黏度保留率为 34%,以水解作用为主。复合体系中由于大量 Na_2CO_3 及活性剂的加入,黏度较聚合物体系低,初始水解度较高(图 4)。低水解度的聚合物,仍以解缠结-缠结作用和电荷相斥作用为主,虽然初始黏度较聚合物体系的低,但由于水解度比聚合物体系大,黏度上升也大,京大聚合物的黏度保留率为 432%,长安聚合物的黏度保留率为 243%;中高水解度的聚合物以水解作用为主,黏度不断下降,3530S 聚合物黏度保留率为 29.8%,钻井聚合物黏度保留率为 21%。表明低水解度的聚合物可提高聚合物溶液的长期热稳定性,可把高温水解这一不利因素转化为有利因素。

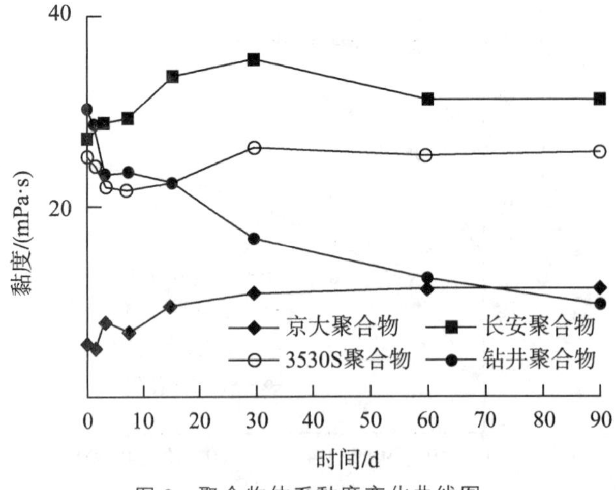

图 3 聚合物体系黏度变化曲线图

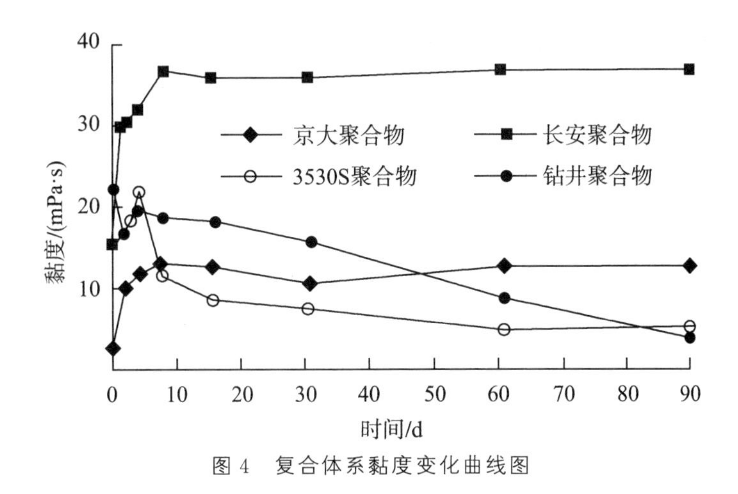

图 4　复合体系黏度变化曲线图

2.2.3 浓度变化规律

热稳定性实验表明，地层水中的离子不会对聚合物体系中聚合物浓度产生影响，所以聚合物体系中聚合物浓度长时间未发生较大变化。复合驱体系中聚合物浓度呈下降趋势（表 2）。实验还发现，聚合物水解度的高低、变化情况与聚合物浓度变化关系不大，但活性剂 PS、BES 中的某些成分可导致聚合物浓度下降。

表 2　聚合物及复合体系中聚合物浓度随时间变化数据表

时间/	聚合物体系中聚合物浓度/(mg/L)				复合体系中聚合物浓度/(mg/L)			
d	京大聚合物	长安聚合物	3530S 聚合物	钻井聚合物	京大聚合物	长安聚合物	3530S 聚合物	钻井聚合物
1	1 319	1 538	1 292	1 568	1 641	1 794	1 527	1 400
3	1 149	1 183	1 252	1 454	1 119	1 363	1 301	1 312
7	1 084	1 382	1 403	1 621	1 120	817	1 288	1 274
15	1 313	1 844	1 297	1 499	1 077	798	1 083	1 157
30	1 307	1 397	1 145	1 586	1 048	755	1 232	1 010
60	1 367	1 271	1 399	1 521	747	584	997	946
90	1 255	1 401	1 298	1 513	645	543	778	938

2.2.4 复合体系界面张力变化

复合体系中的 Na_2CO_3、PS、BES 会导致聚合物水解度的增加，但水解度的差异不会对动态界面力产生较大影响。所以，原油中加入复合体系，动态界面张力变化不大（数量级不变），仅使油水动态界面张力的最低值和稳态值降低。

3　结论

Na_2CO_3、PS、BES 对聚合物水解度的影响存在协同效应，其中 Na_2CO_3 是主要因素。

温度和矿化度（盐浓度）决定了聚合物水解度平衡值的大小。在一定的地层条件下，不同水解度的聚合物经过长期水解均趋于同一平衡值。

聚合物体系中经长期热老化后,水解度增大,浓度变化不大,初始水解度低的聚合物黏度增加,反之亦然。复合体系中聚合物初始水解迅速,经长期热老化后,水解度的增大变缓,并趋于一致;浓度呈下降趋势;动态界面张力变化不大。

长期热稳定性实验发现:低水解度的京大公司、长安公司产聚合物热稳定性较好,黏度稳中有升;中水解度的3530S聚合物热稳定性一般;高水解度的钻井聚合物,黏度损失较大。在10%~40%水解度范围内,低水解度的聚合物比高水解度的聚合物热稳定性好。在现场使用中,为了保证聚合物的增黏作用,应使用水解度较低的聚合物产品。

参考文献

[1] 高树棠,苏树林,杨景纯,等.聚合物驱提高石油采收率[M].北京:石油出版社,1996

Depth profile control is a new technology for improving the oil displacement efficiency in polymer flooding. Compoundion polymer is a new type one that is polymerized by anion, cation and nonion three monomers, and of good adsorptivity and flowability in oil-bearing formations.

Space reticulate-structure colloid of high intensity may come into being after proper crosslinking. Experiment study shows that gel time is longer than 4 days, gel viscosity, breakthrough pressure gradient and water plugging rate are more than 1×10^4 mPa·s,14 MPa/m and 96%respectively for profile control agent of the compound ion polymer. So the polymer is suitable to the depth profile control of polymer flooding. The field experiment presents that output time is prolonged, oil production increased, water cut decreased by a wide margin, concentration of the produced polymer decreased, utilization ratio of the polymer is high and the efect on improving oil recovery is remarkable after profile control by using the compound ion polymer.

Keywords:compound ion polymer, crosslinking, depth profile control, polymer flooding, EORCui Peiying, Cao Xulong, Sui Xihua et al. Study on thermal stability of polymer and its combination system. PGRE, 2001, 8(3):58-60

Laboratory study is done for the change rules of viscosity, degree of hydrolysis, concentration and dynamic interfacial tension of combination system caused by longterm interaction at formation temperature to the polymers with different degree of hydrolysis according to the reservoir and field conditions of western area in Gudao oilfield of Shengli petroliferous province. The research shows that the polymers of different degree of hydrolysis all tended to reach a same balanced value of hydrolysis through longterm hydrolysis at given formation condition. The polymer compounded with low initial degree of hydrolysis and its combination system have good longterm thermal stability at the range of 10%~40% of degree of hydrolysis.

Keywords: S hengli oildom, polymer system, combination system, thermal stability, degree of hydrolysis Hou Jirui, Liu Zhongchun, Xia Huifen et al. Influence of viscoelastic effect in ternary combination system on oil displacement efficiency. PGRE, 2001,8(3):61-64

Rheological behavior of ASP combination system is studied systematically for scale buildup in wellbore and pump block-up caused by high-concentration alkaline agent in the fields of Daqing oilfield. Influence of the alkaline agent on viscoelasticity of ASP combination system is determined. Project comparative tests are made on macroscopic heterogeneity model, natural core and microscopic physical model. The result shows that as the influence factors of oil displacement efficiency viscoelastic efect has more significance than super low interfacial tension; reducing volume of alkaline agent in the combination system properly (interfacial tension should reduce to 0. 01 mN/m) has not influenced oil displacement efficiency obviously. The concept of lowalkaline combination flooding is put forward firstly in this paper and the necessity of super low interfacial tension in the combination flooding is negated. A new thinking to solve the problem of alkaline scale in ASP combination system is found. The limit of capillary criterion theory in the application of the combination flooding is discussed.

Keywords: ASP combination system, rheological property, viscoelastic effect, super low interfacial tension, lowalkaline ternary combination flooding Liu Yijiang, Liu Jisong, Li Xiufu. Application of precrosslinked gel particles to depth profile control. PGRE, 2001,8(3):65-66

Depth profile control in water injector is not only an important technology of secondary oil recovery, but also the technical key of tertiary oil recovery in severe heterogeneous oil reservoirs. Depth profile control technology of precrosslinked gel particles is that polymer crosslinked system formed gel by crosslinking at ground, then gel particles have prepared through the processes of granulation, drying, flouring, screening and so on, at last depth profile control in water injector is carried out. Field experiments proved that the technology has advantages of convenient operation, little influence of reservoir temperature and formation water salinity on profile control agent, no restriction on depth of the profile control, easily storage of the profile control agent and good results in the profile control etc.

Keywords: depth profile control, pre-crosslinked gel, colloidal dispersed gel, heterogeneous oil reservoir, EOR Cai Jingong, Wu Jinlian, Su Haifang et al. Research

本文编辑 高 岩

A study of dilational rheological properties of polymers at interfaces

Cao Xulong[a,b,*] Li Yang[b] Jiang Shengxiang[a] Sun Huanquan[b]
Cagna Alain[c] Dou Lixia[b]

[a] Lanzhou Institute of Chemical Physics, Chinese Academy of Science, Lanzhou 730000, China
[b] Geological Research Institute of Shengli Oilfield, SINOPEC, Dongying 257015, China
[c] I. T. Concept, Parc de Chancdan, 69770 Longessaigne, France
Received 20 May 2003; accepted 16 August 2003

Abstract: Viscoelastic properties of two polymers, partially hydrolyzed polyacrylamide and partially hydrolyzed modified polyacrylamide, widely used in chemical flooding in the petroleum industry, were investigated at three interfaces, water-air, water-dodecane, and water-crude oil, by means of a dilational method provided by I. T. Concept, France, at 85℃. Polymer solutions were prepared in brine with 10 000 mg/L sodium chloride and 2 000 mg/L calcium chloride. It has been shown that the viscoelastic modulus increases with the increment of polymer concentration in the range of 0～1 500 mg/L at the water-air interface. Each polymer shows different viscoelatic behavior at different interfaces. Generally speaking, values of the viscoelastic modulus (E), the real part (E'), and the imaginary part (E'') at the crude oil-water interface for each polymer are lower than at the air-water or water-dodecane interface. The two polymers display different interfacial properties at the same interface. Polymer No. 2 gives more viscous interfaces than polymer No. 1. All the information obtained from this paper will be helpful in understanding the interfacial rheology of ultra-high-molecular-weight polymer solutions.

Keywords: Polymers; Dilational method; Interfacial rheological properties; Dodecane; Crude oil

1 Introduction

Polymers have been widely used in industry for many years. Among their applications, the polymers which have medium-high molecular weight are used in drilling mud; they greatly improve particle suspension properties and mud fluid loss. The very-high-molecular-weight polymers are used in sewage disposal, causing suspended matter to decrease greatly, and the ultra-high-molecular-weight polymers, especially partially hydrolyzed polyacrylamide and its modified products, are used in improving oil recovery in the petroleum industry. It is said that some 8% (OOIP) oil is recovered by polymer flooding after water flooding[1]. Moreover, polymers are extensively used in agriculture and papermaking.

Bulk properties of HPAM solutions such as apparent viscosity and rheology have been reported in many publications[2-5]. Although it is very import to applications in many areas, the interfacial rheology ofpolymer solutions are scarcely reported in literatures. Linear viscoelastic behavior of end-tethered polymer monolayers at the air/water interface was investigated by Luap and Goedel[6]. In their work, the rheological behavior of Langmuir monolayers consisting of polyisoprene chains tethered by one end to the air/water interface was determined by applying a CIR-100 commercial interfacial rheometer from Camtel Ltd., Royston, UK. They used a small-amplitude shear flow and measurements were performed in a two-concentric-ring geometry. Competitive displacement of β-lactoglobulin from the air/water interface by sodium dodecyl sulfate was followed using two surface rheological techniques: shear and dilation. Both surface shear and dilation rheology showed similar trends[7]. Kim et al. utilized a dilation method for interfacial rheology measurement to study dynamic interfacial mechanisms[8]. Dynamic surface properties of asphaltenes and resins at the oil-air interface were reported by Bauget et al., they operated a dynamic drop tensionmeter from I. T. Concept and obtained dilational viscoelastic properties[9]. Benjamins et al. used a modified version of the automatic drop tensionmeter to study viscoelastic properties of triacrylglycerol/water interfaces covered by proteins at 25℃[10]. In this paper, we report experimental studies of dilational rheological properties of two kinds of polymers at three interfaces with a commercial dynamic drop tensionmeter that was equipped with a high-pressure and high-temperature system from I. T. Concept.

2 Experimental section

2.1 Materials

Dodecane was from the No.1 Chemical Reagent Factory of Shanghai, China. Its active content is 99.5%, the density is 0.748 7 g/cm^3, and the refractive index is 1.421 5. Crude oil was obtained from Shengli Oilfield Co. Ltd., SINOPEC. Its viscosity is 42 mPa · s at 50℃ and its density is 0.902 8 g/cm^3. Content of asphaltenes and gum in crude Oil is 25.0% by weight, aromatic hydrocarbon 24.4%, polar material 7.2%, saturated hydrocarbons 43.4%, acid number 2.40 mg KOH/g oil. Polymer No.1 is partially hydrolyzed polyacrylamide from Mitsubishi Corporation, Japan. Its hy-drolysis is 25.0% and active content 90.0%. Its viscosity average molecular weight is about 22×10^6. Polymer No.2 is partially hydrolyzed modified polyacrylamide from Wanquan Oil Chemical Co. Ltd., China, which is 90.8% active content and hydrolysis 23.6%. Its viscosity average molecular weight is 18×10^6. The modified polymer has being called "branched polymer" and alkylene groups were added to the acrylamide molecules. Both polymers were prepared to 5 000 mg/L by dissolving them in brine that contained 100 000 mg/L sodium chloride and 2 000 mg/L calcium chlo-ride.

All experiments were performed at 85℃ and frequency 0.1 Hz.

2.2 Apparatus and principle

The apparatus used in our experiments is from I. T. Concept, France. It is described in Fig.1.

When the surface area of drops is deformed by dilation, the resistance against changes of area can provoke its own elastic and viscous response. Surface dilational modulus in compression and expansion is defined by where γ is the interfacial tension and A is the area of the interface.

$$E = d\gamma / dA / A \tag{1}$$

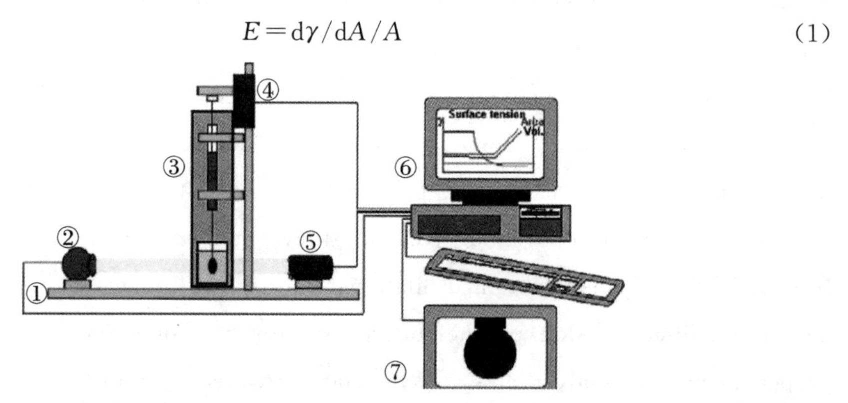

Fig. 1 Schematic of the experimental apparatus: 1, optical bench; 2, light source; 3, syringe driver; 4, motor; 5, camera; 6, computer; 7, video monitor.

In oscillatory experiments, E is a complex quantity where θ is phase angle. The real part $(E') = |E|\cos\theta$ is re-ferred to the elastic component and the imaginary part to the viscous component.

$$E = |E|\cos\theta + i|E|\sin\theta, \tag{2}$$

The modulus E has been extensively investigated both theoretically and experimentally by means of low-amplitude compression/expansion waves at a range of frequencies[10-16].

3 Results and discussion

3.1 Dilational properties of polymers at water-air interface

Variations of viscoelastic modulus (E), real part (E'), and imaginary part (E'') at water-air interfaces versus polymer concentrations in brine for two polymer samples are given in Table 1. Similar trends of dilational rheological properties are observed for the two polymers. The trends are similar to the observation of viscoelastic modulus for several asphaltene concentrations at oil/air interfaces[9]. Viscoelastic modulus (E), real part (E'), and imaginary part (E'') are sharply increased from 1.0, 1.0, 0.2 to 19.6, 15.1, 12.5 mN/m for polymer No.1, while they are rapidly raised to 15.0, 10.6, 10.6 mN/m for polymer No.2, when polymer concentrations are changed from 0 to 750 mg/L. After concentration is over 750 mg/L, viscoelastic modulus (E), real part (E'), and imaginary part (E'') of both polymers are smoothly increased with increment of polymer concentration. Experimental results also showed that rheological properties of polymer No.1 are different from those of polymer No.2. Viscoelastic modulus (E) of polymer No.1 is always larger than that of polymer No.2 at each concentration we have studied, for instance, viscoelastic modulus (E) of polymer No.1 is 24.9 mN/m and viscoelastic modulus (E) of polymer No.2 is 21.8 mN/m when polymer concentration is 1 500 mg/L. Real part (E') and imaginary part (E'') of the two samples give the same trend except that polymer concentration is 1 500 mg/L. From the data, we also find that changes of the real part(E') of the two polymers are less than those of the imaginary part (E'') of the two samples. That means that the real part (E')of polymers plays an important role at air-water interfaces.

3.2 Dilational properties of polymers at water-dodecane interfaces

The experiment results of viscoelastic modulus (E), real part(E'), and imaginary part(E'') at water-dodecane interface versus polymer concentration in brine for two polymer samples are listed in Table 2. It has been found that interfacial rheology at the water-dodecane interface is larger than at the water-air interface when there is no polymer. The trends are again similar for the two polymers, but unlike that mentioned in last paragraph, the maximum values of viscoelastic modulus (E) are when polymer concentration is 1 000 mg/L. They are 18.4, 22.2 mN/m for polymer No.1 and polymer No.2, respectively. Comparing the results of polymer No.1 to polymer No.2, we find that the viscoelastic modulus (E) of polymer No.2 is larger than that of polymer No.1 at any polymer concentration, which is opposed to what is observed at air-water interfaces. It is also found that the imaginary part (E'') of polymer No.1

changes little while the imaginary part (E'') of polymer No. 2 varies when polymer concentration is increased. Because the imaginary part (E'') of polymer No.2 is larger than that of polymer No.1, polymer No.2 provides a more viscous interface.

Table 1 Viscoelastic properties at water-air interfaces for polymer No.1 and polymer No.2 at various concentrations

Polymer	0 mg/L				750 mg/L				1 000 mg/L				1 500 mg/L			
No.	E	E'	E''	IFT	E	E'	E''	IFT	E	E'	E''	IFT	E	E'	E''	IFT
1	1.0	1.0	0.2	61.5	19.6	15.1	12.5	61.5	24.3	20.3	13.3	61.5	24.9	22.1	11.5	61.0
2	1.0	1.0	0.2	61.5	15.0	10.6	10.6	60.8	19.2	14.6	12.4	60.7	21.8	18.2	12.1	60.6

Table 2 Viscoelastic properties at water-dodecane interfaces for polymer No.1 and polymer No.2 at various concentrations

Polymer	0 mg/L				750 mg/L				1 000 mg/L				1 500 mg/L			
No.	E	E'	E''	IFT	E	E'	E''	IFT	E	E'	E''	IFT	E	E'	E''	IFT
1	4.0	3.9	0.9	42.0	14.0	13.5	3.8	41.9	18.4	18.1	3.5	41.5	14.4	12.0	3.9	41.2
2	4.0	3.9	0.9	42.0	14.4	13.2	5.7	41.2	22.2	20.4	8.7	39.8	19.0	17.6	7.1	38.7

Table 3 Viscoelastic properties at water-crude oil interfaces for polymer No.1 and polymer No.2 at various concentrations

Polymer	0 mg/L				750 mg/L				1 000 mg/L				1 500 mg/L			
No.	E	E'	E''	IFT	E	E'	E''	IFT	E	E'	E''	IFT	E	E'	E''	IFT
1	3.1	3.0	0.8	25.3	3.9	3.9	0.4	24.5	3.5	3.3	1.2	24.5	4.9	4.5	1.8	24.3
2	3.1	3.0	0.8	25.3	3.6	3.2	1.1	23.4	3.1	2.6	1.6	23.1	4.3	3.4	2.7	23.0

3.3 Dilational properties of polymers at water-crude oil interfaces

In order to understand dilational rheology of polymers at the water-crude oil interface, some degassed crude oil were obtained from Shengli oilfield and some experiments were performed at the same test conditions of dodecane. The results of two polymer samples are shown in Table 3. When there is no polymer, the interfacial rheology at the crude oilwater is less than at the dodecane-water interface. Interfacial tension between crude oil and brine is 24.2 mN/m but it is 40.4 mN/m between dodecane and brine. Active content in crude oil such as nickel-porphyrin, acid matters, etc., may causes the decrease of interfacial properties. As shown in Table 3, viscoelastic modulus (E), real part (E') and imaginary part (E'') at water-crude oil interfaces do not vary much with the augmentation of polymer concentration. Comparing of polymer No. 1 to polymer No. 2, their viscoelastic moduli (E) are close at different polymer concentrations. The imaginary parts (E'') of two polymers are augmented when polymer concentration goes up; for example, the imaginary part (E'')

of polymer No.1 is from 0.4 to 1.8 mN/m, while that of polymer No.2 is from 1.1 to 2.7 mN/m when polymer concentration is increased from 750 to 1 500 mg/L. Viscoelastic modulus (E) and real part (E') of polymer No.1 are almost higher than that of polymer No.2. Otherwise the imaginary part (E'') of polymer No.1 is less than that of polymer No.2. This indicates that polymer No.2 gives a more viscous interface.

3.4 Comparison of interfacial behavior of two polymers at three interfaces

As mentioned above, different polymers give different interfacial behavior at the same interface and different interfaces make interfacial properties vary for the same polymer. Here is the interfacial rheology compared at the conditions of polymer concentration 750 mg/L, frequency 0.1 Hz, and oscillation after 600 s. Viscoelastic properties of two polymers at the three interfaces, i. e., water-air, water-dodecane, and water-crude oil are shown in Fig. 2. Obviously there are differences of viscoelastic properties among these three interfaces. Generally speaking, viscoelastic properties are lessened along with the decrease of interfacial tension. This is because the two main factors, adsorption properties of the polymer and surface relaxation processes, are expected to depend on the polarity of the interface and the interfacial tension would seem to be a better measure for its polarity. Viscoelstic modulus at water-air interface is about four times higher than at crude oil-water interface. Viscoelastic properties at water-air interfaces are clearly different for two polymers, but they are close at the interfaces of both water-dodecane and crude oil. Comparing the real part (E') and the imaginary part (E'') of polymer No.1 to polymer No.2, it has been found that the real parts (E') of poly-mer No.1 contribute to viscoelastic properties more than those of polymer No. 2; on the other hands, the imaginary parts (E'') of polymer No. 2 contribute to viscoelastic properties more than those of polymer No.1, whether at waterdodecane or water-crude oil interfaces. Therefore polymer No. 2 gives more viscous interfaces than polymer No.1. See Figs.3 and 4.

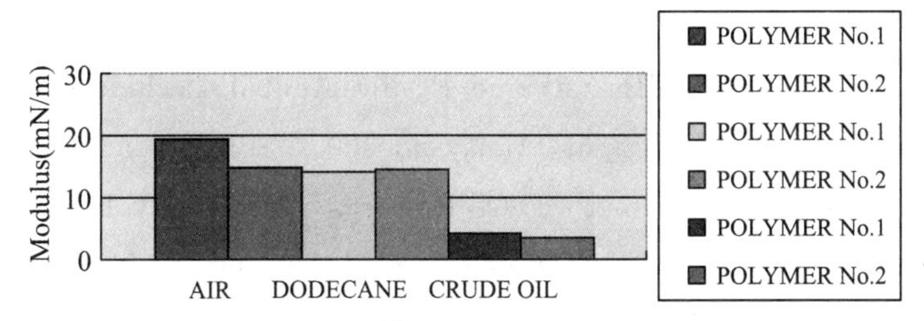

Fig. 2 Comparison of viscoelastic modulus of polymer No.1 at three interfaces of polymer No.2 at concentration 750 mg/L.

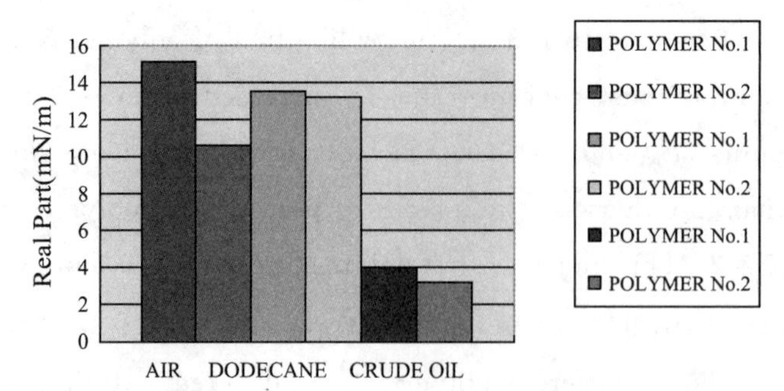

Fig. 3 Elastic modulus of two polymers at three interfaces determined 600 s after formation of the interfaces at concentration 750 mg/L.

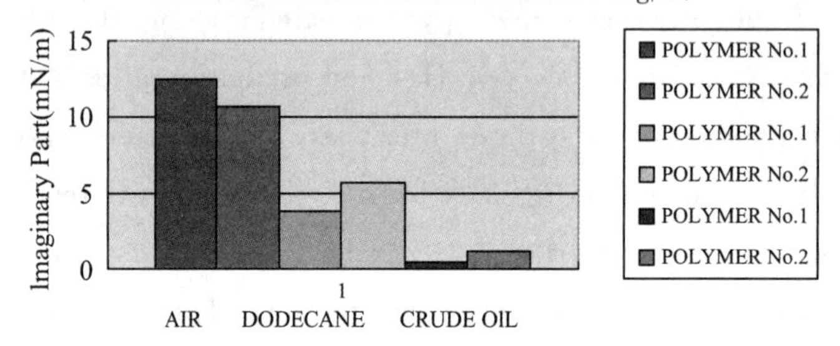

Fig. 4 Imaginary part comparison of polymer No.1 with polymer No.2 at three interfaces at concentration 750 mg/L and frequency 0. 1 Hz.

4 Summary

Experiments on interfacial viscoelastic properties were performed at three interfaces for two polymer samples. From the results we can draw some ideas of interfacial rheology.

Values of viscoelastic modulus at crude oil-water interfaces are lower than at air-water or dodecane-water interfaces for two polymers. Viscoelastic properties are intensively improved by adding polymer. The two polymers have different behavior at the three interfaces; polymer No.2 gives more viscous interfaces.

References

[1] Y. Li, X. Cao, SPE Asia and Pacific Annual Technical Conference and Exhibition, October, Kuala Lumpur, Malaysia, 2000, SPE Paper 68697.

[2] K. S. Sorbie, Polymer-Improved Oil Recovery, Blackie/CRC Press, 1991,37-79.

[3] D. G. Wreath, A Study of Polymer Flooding and Residual Oil Saturation, Thesis, The University of Texas at Austin, 1989,12-20.

[4] M. Wang, Laboratory Investigation of Factors Affecting Residual Oil Saturation by Polymer Flooding, Dissertation, The University of Texas at Austin, 1995,17-27.

[5] C. Wang, X. Cao, SPE Annual Improving Oil Recovery Conference, Tulsa, Oklahoma, 1997, SPE 38321.

[6] C. Luap, W. Goedel, Macromolecules 34(2001)1343-1351.

[7] A. R. Mackie, P. Gunning, P. Wilde, Langmuir 16(2000)8176-8181.

[8] Y. H. Kim, D. T. Wasan, P. J. Breen, Colloids Surf. A 95(1995)235-247.

[9] F. Bauget, D. Langevin, R. Lenormand, J. Colloid Interface Sci. 239(2001) 501-508.

[10] J. Benjamins, A. Cagna, F. H. Lucassen-Reynder, Colloids Surf. A 114 (1996)245-254.

[11] L. Ting, D. T. Wasan, K. Miyano, J. Colloid Interface Sci. 107(1985)345.

[12] J. Lucassen, M. Van de Tempel, J. Colloid Interface Sci. 41(1972)491.

[13] J. M. H. Jassen, H. E. H. Meijer, J. Rheol. 37(1993)597.

[14] A. A. Sonin, A. Bonfillon, D. Langgevin, J. Colloid Interface Sci. 162 (1994)323.

[15] J. J. M. Janssen, A. Boon, W. G. M. Agterof, AIChE J. 40(1994)1929.

[16] D. E. Tambe, M. M. Sharma, J. Colloid Interface Sci. 147(1991)137.

驱油用聚丙烯酰胺溶液界面特性研究

窦立霞[1,3] 曹绪龙[2,3] 江小芳[3] 崔晓红[3] 刘异男[3]

(1 山东大学环境科学与工程学院 山东济南 250000;

2 中国科学院兰州化学物理研究所 甘肃兰州 730000;

3 中石化胜利油田有限公司地质科学研究院 山东东营 257015)

摘要:使用 TRACKR 全自动液滴界面张力仪测量了不同试验条件下十二烷/聚丙烯酰胺溶液界面扩大、缩小时的流变特征及界面张力,开发出了用以表征液膜强度的界面黏弹性 E 的测量方法。在十二烷/聚丙烯酰胺界面形成稳定的过程中界面(扩张)黏弹模量不断增加,并逐渐达到恒定,其中弹性成分所占比例远大于黏性成分即 $E'>E''$。考察的聚合物浓度范围为 50～2 000 mg/L,界面黏弹模量从 17.6 提高到 30.6 mN/m^{-1}。放置 3 d 后的聚合物溶液界面黏弹模量保留率为 82.8%。实验结果表明聚合物浓度变化对界面张力影响不显著。

关键词:聚丙烯酰胺;界面性质;界面扩张流变;界面黏弹模量

聚丙烯酰胺作为一种良好的增黏剂,增加注入水的黏度,降低其流动能力,使油水流度比达到合理值,从而改善波及效率,提高原油采收率。聚合物驱是所有化学驱方法在中国最具潜力的技术,已开始工业化推广,比水驱提高采收率 10%左右[1-3]。虽然,对部分水解聚丙烯酰胺体相溶液性质如流变性、热稳定性、剪切稳定性等研究报道较多[4,5],但对其界面性质的研究报道甚少。界面流变根据界面形变的方式不同,可分为界面剪切流变和界面扩张流变。Clarisse Luap 等人曾用 Camtel 公司(Royston UK)CIR-100 型界面剪切流变仪研究了在空气/水界面上一端连接上聚异戊二烯分子的朗格缪尔(langmiur)单分子层的流变行为[6]。此结果对认识聚合物溶液的界面流变性有所帮助。

界面扩张法是近些年发展起来的研究界面黏弹性的方法[7-9],为了探讨聚合物对油水界面黏弹性的作用,同时为了避免原油活性组分对测量过程中界面性质的影响,本文借助界面扩张法研究了正十二烷与部分水解聚丙烯酰胺溶液的界面特性。

1　实验部分

1.1 仪器和试剂

法国 I. T. CONCEPT 公司的 TRACKR 全自动液滴界面张力仪，IKA-WERKE 数显恒速搅拌器；驱油用部分水解聚丙烯酰胺选用日本三菱公司提供的干粉状样品 Mo-4000，固含量 91.50%，平均分子量 2 000 万，水解度 24%；十二烷，密度 0.748 g/cm^3，北京化工厂生产。

1.2 溶液配制

准确称取 0.4/S(S 为固含量)g 干粉样品，精确至 0.000 1 g，称取(200～0.4/S)g 0.5%NaCl 盐水于烧杯中，调节恒速搅拌器至 400 r/min，使大部分溶液形成旋涡。在 30 s 内将干粉均匀撒在旋涡内壁，调节转速至 700 r/min，搅拌 1 h 配成质量浓度为 2 000 mg/L 的溶液，然后稀释成不同浓度的溶液，待用。

1.3 实验方法和原理

1.3.1 界面张力

将光源、内装聚合物溶液的比色皿和 CCD 照相机成一排放于仪器底座上，调节装有十二烷的注射器和控制马达形成液滴后，液滴剖面通过 CCD 照相机数字转换到电脑上，通过处理在浸入溶液的毛细管顶端形成的十二烷液滴剖面图象，由以下两个方程式计算获得界面张力：

Laplace-Young 方程式：

$$\Delta P = V[(1/R) + (1/R')] \tag{1}$$

液滴在任何水平面的平衡方程：

$$2\pi x V \sin\theta = V(\Delta\rho)g + \pi x^2 \Delta P \tag{2}$$

式中，R，R'为任意曲面 M 的两个曲率半径，若为球面 $R=R'$；ΔP 为弯曲表面 M 上的附加压力，x 为 z 作为纵坐标，x 作为横坐标时，点 M 的横坐标值；θ 为曲面 M 的切线与 x 轴的夹角；Δd 为两种流体的密度差；V 为界面张力，V 为液滴体积，g 为重力加速度。测得去离子水 30℃ 时的表面张力为 71.7 mN/m，与文献值 71.4 mN/m[10] 误差只有 0.4%，证明了该实验方法的准确性。

1.3.2 界面黏弹模量

当滴表面在扩张压缩条件下变形时，便产生阻止表面变化的阻力，从而产生界面黏性和弹性响应。界面黏弹模量 E(mN/m)定义为界面张力与界面面积相对变化的比值：

$$E = dV/(dA/A) = dV/d\ln A \tag{3}$$

当界面面积正弦变化，任何时刻的界面流变可以用 FOURRIER 分析，界面面积形变 dA/A 是输入值，界面张力 γ 是输出结果，分析结果为界面粘弹模量 E，其中实数和虚数部分分别称为存储和损耗模量，分别代表了弹性和黏度部分的贡献。

$$E = |E|e^{je} = E' + iE'' \tag{4}$$

1.3.3 液滴体积的确定

由图1可知,分别取滴体积为12、18、20 μL时测得的十二烷/聚丙烯酰胺(500 mg/L)溶液界面扩张黏弹模量值一致,所以滴体积设定在一定范围内对实验结果无影响。为满足仪器设计原理的要求,液滴体积越大数据越准确;同时保证实验中液滴的稳定,不因体积过大而上浮,以下实验中液滴体积设定为20 μL,此时界面面积约34.75 mm^2,设定振幅为3 μL。

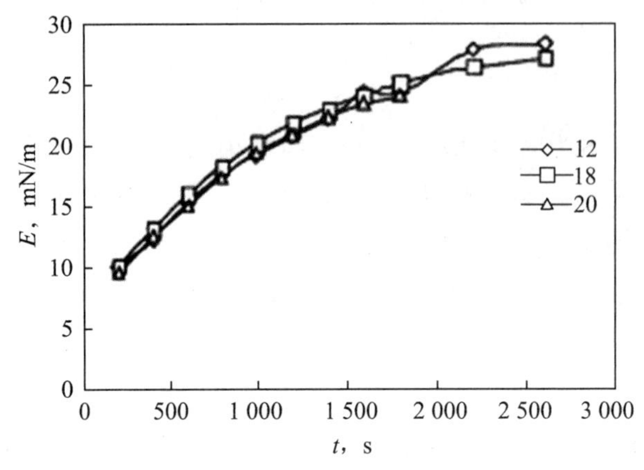

图1 不同液滴体积下十二烷/Mo-4000界面黏弹模量的比较

2 结果与讨论

2.1 界面黏弹模量的平衡

图2为不同浓度的十二烷/聚丙烯酰胺溶液的界面(扩张)黏弹模量随时间的变化曲线。由图2可知,界面黏弹模量随测量时间增长,逐渐增加并在一定时间后趋于平缓。这表明聚合物分子在界面达到吸附平衡需要一定的时间,而且高浓度时,界面黏弹模量达到平衡的较快,而低浓度时较慢。这可能是因为Mo-4000是一种超高分子量的聚合物,其分子线团尺寸不低于0.3 μm[1],具有如此大的水动力学体积使得其在界面的排列分布需要一定时间。而高浓度溶液中分子数目众多,能很快达到吸附平衡状态。

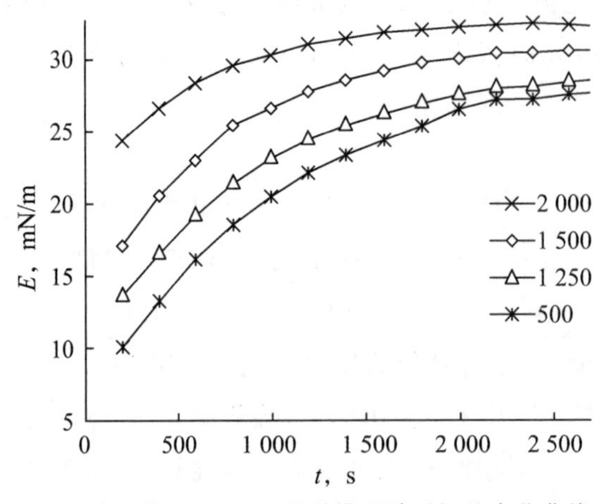

图2 十二烷/Mo-4000黏弹模量随时间的变化曲线

2.2 聚合物浓度的影响

表 1　十二烷/不同浓度聚合物界面参数

浓度/mg/L	E/mN/m^{-1}	$\theta°$	E'	E''	IFT/mN/m^{-1}
0					40.2
50	17.6	8.54	17.5	2.62	39.9
200	17.7	8.80	17.5	2.71	37.6
500	23.7	6.48	23.5	2.67	36.1
1 000	24.4	12.6	23.8	5.30	37.9
1 250	25.4	11.7	24.9	5.17	35.9
1 500	27.9	13.4	27.1	6.48	35.0
1 750	28.1	16.0	27.0	7.72	36.9
2 000	30.6	16.7	29.3	8.80	36.5

由表 1 可知，聚合物没有明显降低或升高界面张力的作用。随聚合物浓度增大，油水界面黏弹模量增加；所有体系中均是存储模量即弹性部分占主导，但浓度增加，损耗模量有不同程度的增加。这是由于聚合物溶液在十二烷液滴界面处形成的吸附层是弹性凝胶体，具有很强的扩张性和可压缩性，所以液滴在做扩张/压缩正弦振荡的过程中，表现出为以弹性为主的特征。聚合物分子链上具有多个链节，而这些链节都可能吸附在界面上，因此聚合物浓度增加时，界面吸附的长链分子会增多，界面吸附层相应加厚，表现为膜强度的增加即黏弹模量增加，最终应有一饱和浓度，有待下一步研究。

2.3 聚合物稳定性的影响

表 2　聚合物稳定性的影响(70℃)

放置时间	η/mPa·s	E/mN·m^{-1}
3 h	25.8	19.8
3 d	23.5	16.4
保留率/%	91.1	82.8

表 2 是 1 500 mg/L 的新鲜和暴氧放置 3 天的聚合物样品表观黏度和黏弹模量的对比。聚合物溶液存在老化现象，热降解、溶解氧氧化降解等因素均会导致聚合物分子链的断裂，结果造成其表观黏度的大幅度下降[11]。同时断链也导致分子量的降低，液滴周围聚合物包围层的厚度就会变薄，可以看到老化后聚合物的界面粘弹模量同样下降。

3　结论

在界面膜稳定过程中，界面黏弹模量随测量时间增加而增加并在一定时间后趋于平衡，浓度不同黏弹性模量 E 达到平衡的时间不同，浓度越大黏弹性模量 E 达到平衡值的时间愈短。Mo-4000(表面非活性物质,对界面张力无影响)随聚合物浓度的增加,界面黏弹模量增加,其中又以弹性成分为主导,即 $E'>E''$。聚合物降解不仅使表观黏度下降,同时降低界面黏弹模量。

Study on Interfacial Characteristics of Polyacrylamide Solutions

Dou Lixia[1,2] Cao Xulong[2,3] Jiang Xiaofang[3] Cui Xiaohong[3]

(1 College of Environmental Science and Engineering, Shandong University, Jinan 250000;

2 Lanzhou Institute of chemical physics, Chinese Academy of Science, lanzhou 730000;

3 Geological Science Research Institute of Sheng li Oilfield Co. Ltd SINOPEC Dongying 257015)

Abstract: Using the TRACKER fully automated drop tensiometer to the interfacial rheology in dilatation or contraction of do decane /polyacry lamide solutions, in different experimental conditions. Ameasurement of the interfacial visco-elastic modulus as a token of the membranes' intensity is founded. During the process of the dodecane/polyacrylamide interface balance, the interfacial visco-elastic modulus increases with increasing time and come to the equation. The elastic component is higher than the viscous component. The interfacial visco-elastic modulus increases with increasing concentration of the polyacrylamide. After aging of the polyacry lamide solutions, The interfacial viscoelastic modulus decreases. The interfacial tension does not chang with polyacrylamide concent ration.

Keywords: Polyacrylamide; Interfacial Characteristics; Interfacial dilational rheology; Interfacial viscoelastic modulus

参考文献

[1] 沈平平,俞稼镛.大幅度提高石油采收率的基础研究[M]北京:石油工业出版社,2001,49:186-191

[2] Yang Li, Xulong Cao. Practice and Knowledge of Polymer flooding in High Temperature and High salinity Reserv oirs of Sheng li Petroliferous Area. SPE paper 68697, SPE Asia and Pacific Annual Technical Conference and Exhibition, Kuala Lumpur, Malaysia, 2000 Oct.

[3] Mengwu Wang. Laboratory investigation of facto rsaffecting residual oil saturation by polymer flooding [D]: 17-27, Disserta tio n, The University of Texas at Austin, 1995 Dec.

[4] Dana George Wreath. A study of polymer flooding and residual oil saturation [D]:12-20, Thesis, The university of Texas at Austin, 1989 Dec.

[5] Kenneth S. Sorbie Polymer-Improved Oil Recovery [M], Blakie and Son Ltd, CRC Press, 1991, 37-79

[6] Clarisse Luap and Werner Goedel. Macromo lecules [J],2001,34:1343-1351

[7] Alan R Mackie, Patrick Gunning, Peter Wilde. Langmuir [J],2000,16:8176-8181

[8] Y. H. Kim, D. T. Wasan, P. J. Breen. Colloids and surfaces A: Physicochemical and Engineering Aspects [J],1995,95:235-247

[9] Fabrice Bauget, Dominique Langevin and Roland Lenormand. J. Colloid Interface Science [J], 2001,239:501-508

[10] 傅献彩,沈文霞,姚天扬. 物理化学 [M]. 北京:高等教育出版社,1992:886

[11] 江小芳,林永红,韩仁功等. 聚合物溶液黏度稳定性影响因素及改进方法[A]. 孙焕泉,王端平,张善文. 胜利油区勘探开发论文集 [C]. 北京:地质出版社,2001:299-303

驱油用聚丙烯酰胺溶液的界面扩张流变特征研究

曹绪龙[1,2] 李 阳[1] 蒋生祥[2] 孙焕泉[1] 窦立霞[1]

(1. 中石化胜利油田有限公司地质科学研究院,山东东营 257015;
2. 中国科学院兰州化学物理研究所,甘肃兰州 730000)

摘要:借助界面扩张流变测量方法,研究了聚合物与正十二烷形成的界面膜特征。主要探讨了液膜稳定过程与界面扩张粘弹模量随测量时间变化的关系、聚合物溶液浓度的影响以及不同聚合物、不同界面膜的流变特征。结果表明,界面粘弹模量的变化反映了膜的吸附平衡过程;正十二烷/聚合物体系膜强度以弹性为主,只是随聚合物浓度增加,粘性模量对膜强度的贡献比例有所上升。聚合物溶液与不同物质所形成的界面具有不同的粘弹性特征,与空气、正十二烷、原油所形成界面的界面扩张粘弹模量 E 的大小次序为 $E_{空气}>E_{正十二烷}>E_{原油}$,即 E 随着两相密度差的减小而减小。

关键词:部分水解聚丙烯酰胺;界面粘弹性;界面扩张方法;正十二烷;原油
中图分类号:O641.3 文献标识码:A

Dilational rheological properties of polyacrylamide solutions at interfaces

Cao Xulong[1,2] Li Yang[1] Jiang Shengxiang[2] Sun Huanquan[1]
Dou Lixia[1]

(1. Geological Science Research Instituteof Shengli Oilfield Company Limited, China National Petrochemical Corporation, Dongying 257015, China;
2. Lanzhou Institute of Chemical Physics, Chinese Academy of Science, Lanzhou 730000, China)

Abstract:The changes of interfacial viscoelasticity along with measuring time and the effect of partially hydrolyzed polyacrylamide(PHPAM) concentration and type viscoelasticity at the interface between PHPAM and dodecane were investigated by means of dilational method. The results show that response of interfacial adsorption process can be made by interfacial viscoelasticity. Elasticity plays an important role for PHPAM/dodcane system even though contribution of viscosity in viscoelasticity goes up along with the increase of PHPAM concentration. Moreover, different viscoelasticity characteristics at different interfaces for PHPAM solution were also found. Among those three interfaces were experimented, and the order of the numerical value of viscoelasticity from large to small is PHPAM/air interface, PHPAM/dodecane interface and PHPAM/crude oil interface.

Keywords: partially hydrolyzed polyacry lamide; interfacial viscoelasticity; dilational method; dodecane; crude oil

乳状液稳定性的主要影响因素之一是界面膜的机械强度,它可用界面压、界面粘弹性(或粘度)和界面膜屈服值等参数予以表征。界面扩张法是近些年发展起来的研究界面粘弹性的方法。Alan R. M ackie 等[1]用界面扩张法研究了空气/水界面上十二烷基磺酸钠对乳酸球蛋白的竞争取代作用,并与表面剪切法进行了对比;Y. H. Kim 等[2]用界面扩张法研究了动态界面机理;Fabrice Bauget 等[3]则用此方法研究了沥青质在油/水界面的动态特征。然而,迄今为止对于聚丙烯酰胺溶液油水膜性质的研究报道很少。于军胜等[4]用自行设计组装的界面粘弹仪,研究了部分水解聚丙烯酰胺和表面活性剂的水溶液与正十二烷形成的油/水界面的膜弹性,Cao Xu-long 等[5]研究了聚丙烯酰胺溶液在不同界面上的界面流变性,此结果对认识驱油过程中聚合物溶液的界面流变性有所帮助。笔者借助界面扩张法研究正十二烷、空气和原油与部分水解聚丙烯酰胺溶液界面的流变特征。

1 实验部分

1.1 材料、试剂与仪器

日本三菱公司生产的商用部分水解聚丙烯酰胺 Mo-4000, 其有效含量 91.50%,相对分子质量 2.2×10^7,水解度 25%,残余单体含量小于 0.1%;高温高盐聚合物 1#,其固含量 90.8%,相对分子质量 1.8×10^7,水解度 23.6%,残余单体含量小于 0.1%,中试样品;上海试剂厂生产的正十二烷,其含量 99.5%,密度 0.748 g/cm³,折光率 1.421 5;10%NaCl+0.2%$CaCl_2$ 实验用盐水,NaCl 与 $CaCl_2$ 为分析纯;取自胜利油田有限公司孤岛采油厂的原油,50℃时粘度 42 mPa·s,密度 0.903 4 g/cm³,其中胶质与沥青质含量 25.0%,芳香烃含量 24.4%,极性物含量 7.2%,饱和烃含量 43.4%。

实验仪器为法国 I. T. CONCEPT 公司生产的 TRACKER 全自动液滴界面张力仪。

1.2 原理

以油为内相,聚合物溶液为外相,在 30℃温度下进行实验,在测试过程中形成水包油型液滴，利用 TRACKER 液滴张力仪测定界面张力与面积的变化，进而求得界面膜的粘弹性模量，即当液滴界面在压缩或扩张条件下变形时，便产生阻止界面变化的阻力，从而产生界面粘性和弹性响应。因此，将界面扩张粘弹模量定义为

$$E=\mathrm{d}\gamma/(\mathrm{d}A/A)=\mathrm{d}\gamma/\mathrm{dln}\ A\ .$$

式中，γ 为界面张力，mN/m；A 为膜面积，mm^2；$\mathrm{d}A$ 为膜面积增量，该参数是对界面张力梯度产生的阻力的量度，也是体系在任其自然时界面张力梯度消失速度的量度。

膜粘弹性越高，膜的机械强度越大，其自修复能力也就越强。在进行压缩和扩张振荡实验时，通过 FOURRIER 变换，E 以复数形式表示为

$$E=|E|\cos\theta+|E|\sin\theta=E'+iE''.$$

式中，θ 为相位角；E 由粘性和弹性两部分组成，实部 E' 为弹性部分，虚部 E'' 为粘性部分。

2 实验结果及其讨论

2.1 液滴形成后的膜稳定过程

界面流变特征能直接用剪切、压缩/扩张方式测量出来。剪切粘弹性是一种对形变的阻抗，是对吸附层的机械力直接测量的结果。扩张流变性是通过测量膜的界面张力(表面压)的变化而得到，是一种对膜抵抗压缩和扩张的测量。实验表明[6]，对于同一单分子层，扩张的界面性质要比相应的剪切性质高出几个数量级。因此，扩张性质是流体力学和界面科学的一个重要参数。

图 1 给出了正十二烷/聚合物 Mo-4000 体系界面张力和界面粘弹模量 E 随时间的变化曲线，显示了液滴形成后膜趋于稳定的整个过程。起初界面上吸附的聚合物分子较少，膜的强度较低，随时间延长，膜逐渐变强，直至稳定；对应着界面粘弹模量逐步增大，并最终达到一平衡值，这时界面膜上两种分子的吸附趋于平衡，而且由于聚合物具有相对分子质量大、粘度高的特点，此过程是典型的扩散控制的界面吸附。同时，随着膜的稳定，界面张力略有下降，后来基本保持不变，由此可以认为液滴稳定过程与界面张力的关系不密切。

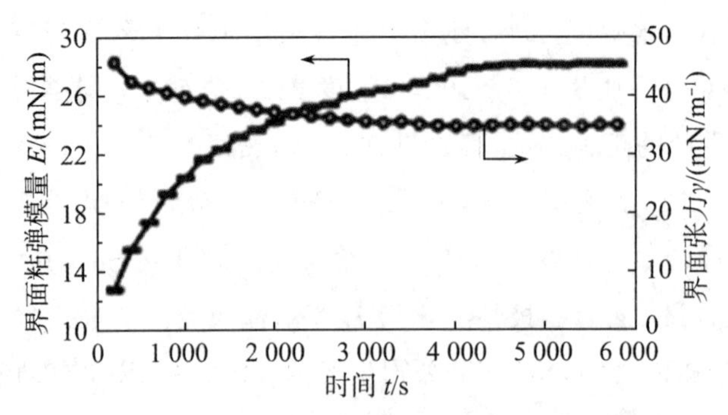

图 1 正十二烷/Mo-4000 体系界面粘弹模量和界面张力随时间的变化曲线

2.2 聚合物浓度对界面膜强度的影响

表1给出了不同浓度的Mo-4000溶液与正十二烷的界面膜特征参数，包括平衡粘弹模量、粘性模量、弹性模量、界面张力、膜寿命以及粘性模量、弹性模量贡献比例。由实验结果可见，随着Mo-4000溶液浓度的增加，界面粘弹模量、粘性模量、弹性模量均增加。由于膜粘弹性的增强意味着膜分子间作用力增大，因此说明聚合物浓度增加，在油/水界面上，聚合物分子间的相互作用力增强。由于Mo-4000溶液属粘弹性流体，与正十二烷形成的界面膜中，以弹性成分为主，随浓度的增加，分子间力作用增强，其贡献比例略有下降，而粘性比例开始上升。

从实验结果还可以看出，正十二烷/聚合物界面膜的寿命较长，原因有两个，一是聚合物溶液具有很强的粘滞力，阻碍了液膜变薄；二是体系具有较高的界面张力，液膜能稳定存在。同时可以看到，聚合物浓度对界面张力基本无影响。

2.3 不同聚合物的界面膜特征

表2给出了浓度为1 000 mg/L的2种聚合物溶液与空气、正十二烷和原油3种物质形成的界面膜特征参数。由实验结果可见，聚合物与不同物质所形成的界面张力和界面粘弹性不同。由于原油中的天然表面活性物质具有较好的界面活性，可明显降低原油和水的界面张力，所以原油/聚合物体系的界面张力最低。总体而言，界面张力和粘弹性随着两相密度差的增大而增大。2种聚合物比较可以看出，虽然1＃聚合物的粘性模量较高，但粘弹模量普遍低，在聚合物驱替过程中容易导致油水界面发生滑动，油膜不易剥离，所以界面粘弹模量也可作为评价聚合物性能的指标之一。

表1　聚合物浓度与膜特征参数的关系

浓度 ρ/(mg/L)	粘弹模量 E/(mN/m)	弹性模量 E'/(mN/m)	粘性模量 E''/(mN/m)	界面张力 γ/(mN/m)	弹性模量贡献比例 E'/E	粘性模量贡献比例 E''/E	膜寿命 τ/s
50	23.8	23.7	2.58	39.1	0.99	0.11	>4 000
200	26.3	26.1	2.90	37.6	0.99	0.11	
500	28.4	28.0	4.81	30.7	0.99	0.17	
1 500	30.4	29.7	6.15	32.6	0.98	0.20	
2 000	32.1	30.9	8.55	35.8	0.96	0.27	

表2　不同聚合物的界面膜特征　　mN·m^{-1}

界面	粘弹模量 E		弹性模量 E'		粘性模量 E''		界面张力 γ	
	Mo-4000	1＃	Mo-4000	1＃	Mo-4000	1＃	Mo-4000	1＃
空气/聚合物	24.3	19.2	20.3	14.6	13.3	12.4	61.5	60.7
正十二烷/聚合物	18.4	17.9	18.1	16.4	3.31	7.12	41.9	39.8
原油/聚合物	17.8	16.9	17.2	14.6	4.58	8.45	25.1	24.2

3 结论

(1) 正十二烷/聚合物体系液膜的稳定过程可以通过界面粘弹模量随时间的变化曲线来表示,是典型的由扩散控制的界面吸附平衡过程。

(2) 正十二烷/聚合物界面膜以弹性为主。随着聚合物浓度增大,界面粘弹模量、粘性模量、弹性模量均有所增加,但粘性模量对粘弹性模量的贡献比例上升。

(3) 聚合物溶液与空气、正十二烷、原油所形成界面的界面粘弹性模量的大小次序为 $E_{空气}>E_{正十二烷}>E_{原油}$。界面粘弹性模量的大小随着两相密度差的减小而减小。可用界面粘弹模量表征不同的聚合物性能。

参考文献

[1] MACKIE Alan R, GUN NING A Patrick, WI LDE Peter J, et al . Competitive displacement of β-lactoglobulin from the air/water interface by sodium dodecyl sulfate [J]. L angmuir, 2000,16(21):8176-8181.

[2] KIMYH, WASANDT, BREENPJ. A study of the dynamic interfacial mechanisms for demulsification of water in oil emulsions [J]. Colloids and Surfaces A: Physicochemical and Engineering A spects,1995,95:235-247 .

[3] BAUGET Fabrice, LANGEVIN Dominique and LENORMAND Roland. Dynamic surface properties of asphaltenes and resins at the oil-air interface [J]. Journal of Colloid and Interface Science,2001,239:501-508.

[4] 于军胜,张嘉云,张金花,等.油水界面的膜弹性与乳状液的稳定性研究 [J]. 化学通报(网络版), 1999(13):C99034. YU Jun-sheng, ZHANG Jia-yun, ZHANG Jin-hua, etal. Study on the relationship betw een interfacial elasticity of oil/ equeus solution and emulsion steability [J]. Chemistry Online,1999(13):C99034.

[5] CAO Xu-long, LI Y ang, JIANG Sheng-xiang, et al. A study of dilational rheological properties of polymers at interfaces [J]. Jour nal of Colloid and Interface Science,2004,270:295-298.

[6] CHANGCH, FRAN SESE J. A n analy sis of the factors affecting dy namic tension measurements with the pulsating bubble surfactometer [J]. Journal of Co lloid and I nterface Science,1994,164:107-113.

本文编辑 李志芬

部分水解聚丙烯酰胺溶液的界面扩张流变特征

曹绪龙[1,2]　李　阳[1]　蒋生祥[2]　孙焕泉[1]　窦立霞[1]　李　英[3]　徐桂英[3]

(1. 中石化胜利油田有限公司　地质科学研究院，山东东营　257015；

2. 中国科学院　兰州化学物理研究所，甘肃兰州　730000；

3. 山东大学　胶体与界面化学实验室，山东济南　250062)

摘要：借助于界面扩张流变测量方法，研究了测量时间对超高分子量聚丙烯酰胺Mo-4000溶液与十二烷形成界面的黏弹性特征；研究了Mo-4000溶液浓度对界面扩张黏弹模量E的影响，探讨了Mo-4000溶液与空气、十二烷、原油所形成界面的E特征. 结果表明，同一浓度的Mo-4000溶液与十二烷形成的E大小随测量时间而变化，初期E较小，随着测量时间的增长而逐渐增大直至达到平衡。在同一测量时间时，Mo-4000溶液浓度愈大E愈大，达到平衡E的时间愈短。不同浓度时，初始E的差异大于平衡E的差异。由于Mo-4000的黏均分子量高达2.2×10^7，故弹性模量E'对E的贡献大，虽然随着溶液浓度的增大，黏性模量E''对E的贡献比例略有上升，但是弹性模量E'对E的贡献仍为主导。此外，Mo-4000溶液与不同物质所形成的不同界面具有不同的黏弹性特征，与空气、十二烷、原油所形成界面的E的大小次序为$E_{空气}>E_{十二烷}>E_{原油}$，即E的大小随两相密度差的减小而减小。

关键词：部分水解聚丙烯酰胺；界面黏弹性；黏弹模量；界面扩张实验方法

中图分类号：O641.3　文献标识码：A

Dilational interfacial properties of PHPAM solutions

Cao Xulong[1,2]　Li Yang[1]　Jiang Shengxiang[2]　Sun Huanquan[1]

Dou Lixia[1]　Li Ying[3]　Xu Guiying[3]

(1. Geological Research Institute of Shengli Oilfield Co. Ltd, SINOPEC, Dongying 257015, Shandong, China;

2. Lanzhou Institute of Chemical Physics, The Chinese Academy of Science, Lanzhou 730000, Gansu, China;

3. Key Laboratory of Colloid and Interface Chemistry, Shandong Univ., Jinan 250100, Shandong, China)

Abstract: The changes of interfacial viscoelasticity with measuring time and the effect of PHPAM concentration on viscoelasticity at the interface between PHPAM Mo-4000 and dodecane were investigated by means of dilational method. Preliminary comparison experiments were also done for different interfaces such as air/PHPAM, dodecane/PHPAM, and crude oil/PHPAM. The results show that interfacial viscoelasticity increases while the testing time lasting with the same PHPAM concentration. The viscoelasticity is smaller at beginning and becoming larger with the time increaseing until the equilibrium state is obtained. The larger the PHPAM concentration, the higher the initial and equilibrium viscoelasticity are at dodecane/PHPAM interface and the shorter the time needed to reach the equilibrium state is. The difference of initial viscoelasticity is larger than that of the equilibrium viscoelasticity for different concentrations. The oscillation experiments show that elastic modulus (E') contributes to viscoelasticity (E) much more than viscous modulus (E''). Although the contribution of E' decreases with the increment of PHPAM concentra-tion, it still plays a main role in the viscoelasticity modulus for the high molecular weight of PHPAM Mo-4000. Moreover, among those three experimented interfaces, viscoelasticity modulus at air/PHPAM surface is the largest one. That at dodecane/PHPAM interface is the second. That at crude oil/PHPAM interface is the smallest. It seems that viscoelasticity between two phases lessens when the difference of density between these two phase decreaseing.

Keywords: partially hydrolyzed polyacrylamide; interfacial viscoelasticity; viscoelasticity modulus; dilational method

0 引言

聚丙烯酰胺在采油、造纸、冶金、环保等众多领域有了越来越多的用途，特别是在化学驱油领域，由于其具有较好的增黏能力、溶解能力、滤过能力以及增加阻力系数和残余阻力系数的能力，从而在矿场得到了广泛的应用。文献[1-3]报道，水驱后进行聚合物驱可增加原油采收率8%以上。虽然，对部分水解聚丙烯酰胺体相溶液性质如流变性、热稳定性、剪切稳定性等研究报道较多[4-5]，但对其界面性质的研究报道甚少。Clarisse Luap等人曾用Camtel公司(Royston UK)CIR-100型界面剪切流变仪研究了在空气/水界面上连接聚异戊二烯分子的朗格缪尔(Langmiur)单分子层的流变行为[6]，研究结果有助于人们认识聚合物溶液的界面流变性。

界面扩张法是近些年发展起来的研究界面黏弹性的方法。Alan R Mackie等人用界面扩张法研究了空气/水界面上十二烷基磺酸钠对乳酸球蛋白的竞争取代作用并与表面剪切法进行了对比[7]。此外，Y. H. Kim等人用界面扩张法研究了动态界面机理[8]，

Fabrice Bauget 等人用此方法研究了沥青质在油水界面的动态特征[9]。本文借助于界面扩张法研究了十二烷、空气和原油与部分水解聚丙烯酰胺溶液界面的流变特征。

1 实验部分

1.1 材料与试剂

十二烷:质量分数 0.995,密度 0.748 g/cm^3,折光率 1.421 5,上海试剂厂生产。部分水解聚丙烯酰胺 Mo-4000:固含量 91.50%,黏均分子量 2.2×10^7,水解度 25%,残余单体含量<0.1%,日本三菱公司生产。NaCl 与 $CaCl_2$ 为分析纯,水为蒸馏水。原油:地层温度 50℃时黏度 42 mPa·s,密度 0.903 4 g/cm^3,其中胶质与沥青质质量分数 0.250,芳香烃质量分数 0.244,极性物质量分数 0.072,饱和烃质量分数 0.434,该样品取自胜利油田有限公司孤岛采油厂。本文所有实验均在 30℃下进行。

1.2 仪器与测量原理

实验仪器为法国 I. T. CONCEPT 公司生产的 TRACKER 界面扩张流变仪。

测量原理:当滴表面在扩张条件下变形时,便产生阻止表面变化的阻力,从而产生界面黏性和弹性响应.在界面面积变化的过程中,界面扩张黏弹模量定义为

$$E=\mathrm{d}\gamma/(\mathrm{d}A/A)=\mathrm{d}\gamma/\mathrm{d}\ln A, \tag{1}$$

式中,γ 为界面张力,A 为界面面积。

当进行扩张压缩振荡实验时,E 以复数形式表示:

$$E=|E|\cos\theta+|E|\sin\theta=E'+iE'', \tag{2}$$

式中,θ 为相位角,E 由黏性和弹性两部分组成:实部 E'为弹性部分,虚部 E''为黏性部分。

2 实验结果与讨论

2.1 不同反应时间测量的界面张力和黏弹模量

图 1 给出了十二烷/部分水解聚丙烯酰胺 Mo-4000 溶液界面张力(IFT)和界面黏弹模量(E)随时间的变化曲线。由图 1 可见,随着时间增长,界面张力逐渐降低,起初变化速度快,1 500 s 后下降渐缓,4 000 s 后基本保持不变。这表明,随着时间的增长十二烷/部分水解聚丙烯酰胺 Mo-4000 溶液界面上两种分子界面上的吸附过程趋向平衡。界面黏弹模量随时间的增长不断增大,起初变化速度快,1 500 s 后变化渐缓,4 000 s 后数值基本保持不变。该特征与 Fabrice Bauget 等人观察到的油/空气界面上沥青质的动态特征相比[9],虽然动态界面张力特征相近,但界面黏弹模量却截然不同。由于部分水解聚丙烯酰胺 Mo-4000 溶液黏度高,本身分子量大,液相中扩散速度小,因此该过程是典型的扩散控制的界面吸附过程。

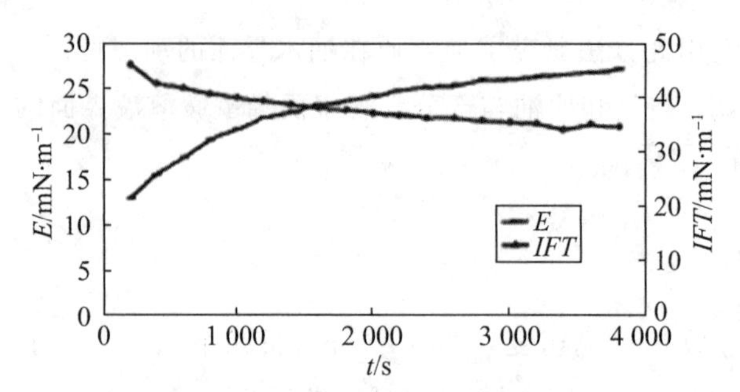

图 1 PHPAM-十二烷的界面黏弹模量和张力随时间的变化曲线

Fig. 1 Viscoelastic modulus and interfacial tension at the PHPAM-dodecane interface for 1 000 mg L PHPAM versus time

依据 Lucassen 和 V an den Tempel 的理论[10]，界面黏弹模量是振荡频率和界面表面活性剂浓度的函数，可用下式表征(为简化公式表达，定义：$\zeta=[\omega_D/4\pi f]^{1/2}$)：

$$E(f,c)=E'(f,c)+iE''(f,c), \tag{3}$$

$$E'(f,c)=E_0(1+\zeta)/(1+2\zeta+2\zeta^2), \tag{4}$$

$$E''(f,c)=E_0\zeta/(1+2\zeta+2\zeta^2). \tag{5}$$

式中，$E(f,c)$为复合黏弹模量(mN/m)，$E'(f,c)$为弹性模量(mN/m)，$E''(f,c)$为黏性模量(mN/m)，E_0 为极限黏弹模量(mN/m)，ω_D 为扩散松弛特征频率(Hz)，f 为振荡频率(Hz)，c 为界面吸附浓度(mol/L)。

上述方程仅能描述实验振荡频率条件下复合黏弹模量的平衡状态。由上述实验结果，可得到在 Mo-4000 质量浓度为 1.00 g/L 时扩散松弛特征频率 ω_D 为 0.78 s^{-1}，极限黏弹模量 E_0 为 28.1 mN/m。

图 2 给出了不同质量浓度条件下，Mo-4000 溶液与十二烷界面的平衡黏弹模量、黏性模量、弹性模量、界面张力、相位角的变化曲线。由图 2 结果可见，随着质量浓度的增加，黏弹模量增加，质量浓度小于 0.10 g/L 时黏弹模量增加较快，大于 0.20 g/L 后黏弹模量增加相对较慢。从实验结果来看，质量浓度越大，黏弹模量越大。对应于 0.10，0.50，0.75，1.50 和 2.00 g/L 质量浓度的黏弹模量分别为 26.8，28.4，29.1，30.4 和 32.1 mN/m。

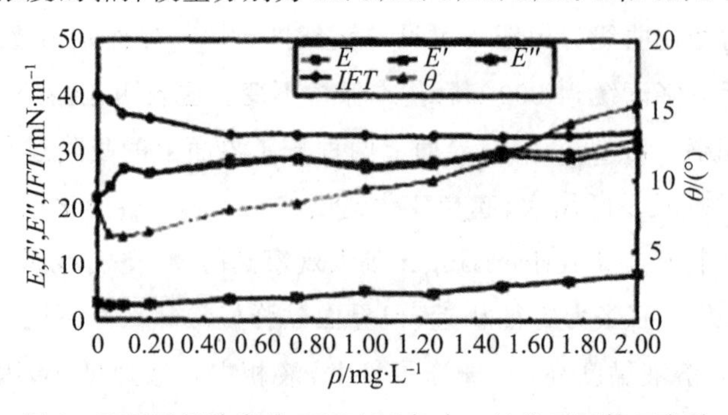

图 2 不同质量浓度的 PHPAM 与十二烷的界面特征参数

Fig. 2 Interfacial characteristic parameters at PHPAM-DODECANE interface for different PHPAM concentrations

2.2 质量浓度对界面黏弹模量的影响

图 3 给出了不同 Mo-4000 质量浓度溶液与十二烷界面的黏弹模量曲线。由实验结果可见,在实验质量浓度范围内,任何质量浓度时,界面黏弹模量都随着时间增长而增大。随着 Mo-4000 溶液质量浓度的增加,界面黏弹模量亦增加;界面黏弹模量起始值差异较大,随着持续时间增长,二者的界面黏弹模量差异减小。就 0.05 g/L 和 2.00 g/L 浓度体系而言,400 s 时后者为前者的 2.84 倍;2 400 s 时后者仅为前者的 1.79 倍。除此之外,从实验结果还可看出,界面黏弹模量随时间的变化因质量浓度而异,高质量浓度时,界面黏弹模量达到平衡较快,而低质量浓度时较慢。从平衡特征来看,0.50 g/L 浓度以上达到平衡较快,初期界面黏弹模量增加快;0.20 mg/L 浓度以下达到平衡较慢,初期界面黏弹模量增加慢。

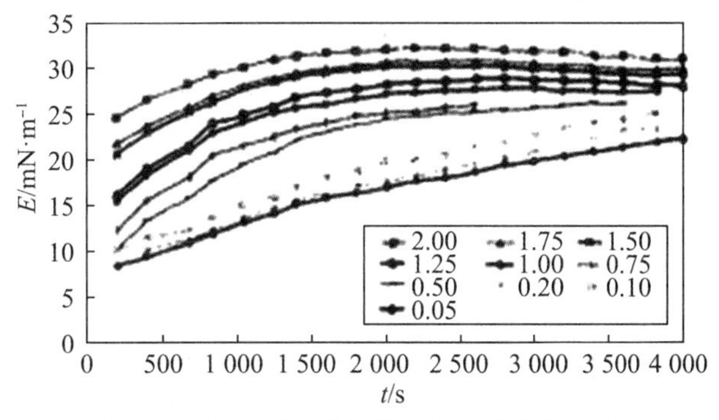

图 3 PHPAM 质量浓度对界面黏弹模量的影响

Fig. 3 The effects of PHPAM concentration on viscoelastic modulus

2.3 黏性模量与弹性模量特征

图 4 给出了不同 Mo-4000 质量浓度溶液与十二烷界面的黏性模量曲线。由实验结果可看出,随着时间的变化,在 1.00 g/L 浓度以下黏性模量基本相当,在该浓度下黏性模量在 400 s 时为 4.96 mN/m,之后基本保持在 5.35 mN/m 左右。在 1.50 g/L 浓度以上黏性模量略有变化。如质量浓度为 2.00 g/L 时,黏性模量由 400 s 时的 9.77 mN/m 下降到 3 600 s 时的 8.14 mN/m,然后略有上升,4 500 s 时为 8.49 mN/m,总的说来,黏性模量随时间变化较小。从图 4 的结果还可看出,随着 Mo-4000 溶液质量浓度的增大,其界面黏性模量增大。对比图 2 可看出,随着质量浓度的增加,相位角增大,因此黏性模量在界面黏弹模量中的贡献比例随质量浓度的增加而增加。

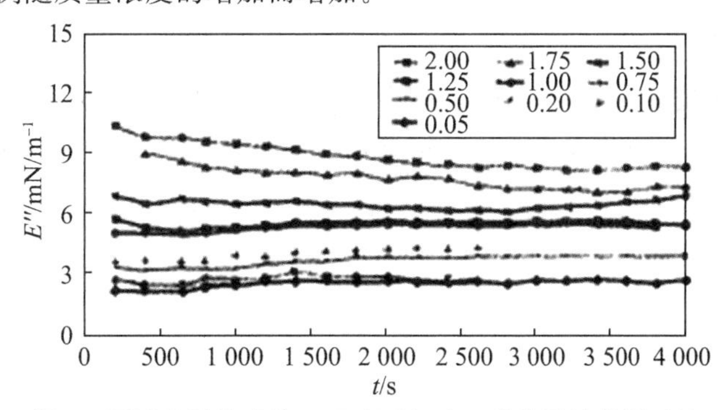

图 4 不同质量浓度的 PHPAM 与十二烷的黏性模量对比

Fig. 4 The viscous modulus of PHPAM-Dodecane interface at different concentrations

从图 2 的实验结果可看出,弹性模量随质量浓度的增加而增加,初期增加速度较快,质量浓度高于 0.20 g/L 之后增加速度较慢。此外,对于超高黏均分子量的 Mo-4000 溶液与十二烷形成的界面,弹性模量在界面黏弹模量中的贡献占主导地位,随质量浓度的增加贡献比例略有下降。

2.4 聚合物 Mo-4000 溶液在不同界面上的黏弹性

表 1 给出了 1.00 g/L 聚合物 Mo-4000 溶液与空气、十二烷和原油 3 种物质形成的界面在平衡状态时的黏弹性测量结果。由实验结果可见,聚丙烯酰胺 Mo-4000 溶液与不同物质形成的界面张力和界面黏弹性不同。总体而言,界面张力随着两相密度差的增大而增大,黏弹性亦随着两相密度差的增大而增大。

表 1 Mo-4000 在不同界面的黏弹性测量结果

Tab. 1 Viscoelasticity of Mo-4000 solution at different interfaces

Interface	E	E'	E''	IFT
Air/PHPAM	24.3	20.3	13.3	61.5
Dodecane/PHPAM	18.4	18.1	3.5	41.9
Oil/PHPAM	3.5	3.3	1.2	24.5

3 结论

(1) 聚丙烯酰胺 Mo-4000 溶液与十二烷界面的黏弹性模量随测量时间的不同而不同,初期 E 随时间的变化相对较大,后期相对较小,直至达到平衡值为止。

(2) 质量浓度对黏弹性模量 E 有明显的影响,质量浓度愈大黏弹性模量 E 愈大,随着测量时间的变化,初期差异明显,平衡时差异相对减小。质量浓度不同黏弹性模量 E 达到平衡所需的时间不同,质量浓度愈小,达到平衡值的时间愈长。

(3) 对所研究的超高分子量聚丙烯酰胺 Mo-4000 溶液与十二烷形成的界面而言,弹性模量 E' 对黏弹性模量 E 的贡献比例大,随着聚丙烯酰胺 Mo-4000 溶液质量浓度的增加,虽然黏性模量 E'' 对黏弹性模量 E 的贡献比例略有上升,但是仍以弹性模量 E' 的贡献比例为主。

(4) 聚丙烯酰胺 Mo-4000 溶液与不同物质所形成的不同界面具有不同的黏弹性特征,其与空气、十二烷、原油所形成界面的界面黏弹性模量 E 的大小次序为 $E_{空气}>E_{十二烷}>E_{原油}$。其界面黏弹性模量 E 的大小随着两相密度差的增大而增大。

参考文献

[1] Li Yang, Cao Xulong. Practice and knowledge of polymer flooding in high temperature and high salinity reservoirs of Shengli Petroliferous Area [R]. Kuala

Lumpur, Malaysia:SPE Asia and Pacific Annual Technical Conference and Exhibition, 2000.

[2] Wang Chenglong, Cao Xulong. Design and application of ASP sy stem to the close well spacing, Gudong Oilfield [R]. Tulsa, Oklahoma:SPE annual improving oil recovery conference,1997.

[3] Wang Mengwu. Laboratory investigation of factors affecting residual oil saturation by polymer flooding [D]. Austin:The University of Texas,1995. 17-27.

[4] Dana George Wreath. A study of polymer flooding and residual oil saturation [D]. Austin:The University of Texas,1989. 12-20.

[5] Kenneth S Sorbie. Polymer-improved oil recovery [M]. Glasgow and London: Blakie and Son Ltd, CRC Press,1991. 37-79.

[6] Clarisse Luap, Werner A Goedel. Linear viscoelastic behavior of end-tethered polymer monolayers at the air water interface [J]. Macromolecules, 2001, 34(5): 1343-1351.

[7] Alan R Mackie, Patrick Gunning, Peter Wilde, et al. Competitive displacement of β-Lactoglobulin from the air water interface by sodium dodecyl sulfate [J]. Langmuir,2000,16(21):8176-8181.

[8] Kim Y H, Wasan D T, Breen P J. A study of dynamic interfacial mechanisms for demulsification of water-in-oil emulsions [J]. Colloids and Surfaces A: Physicochemical and Engineering Aspects,1995,95(2～3):235-247.

[9] Bauget Fabrice, Langevin Dominique, Lenormand Roland. Dynamic surface properties of asphaltenes and resins at the oil-airinterface [J]. Journal of Colloid and Interface Science,2001,239(2):501-508.

[10] J Lucassen, M Van de Tempel. Longitudinal waves on viscoelastic surfaces [J]. Journal of Colloid and Interface Science,1972,41(3):491-498.

本文编辑 胡春霞

Rheological properties of poly(acrylamide-co-sodium acrylate) and poly(acrylamide-co-sodium vinyl sulfonate) solutions

Cao Jie(曹杰)[1] Che Yuju(车玉菊)[1] Cao Xulong(曹绪龙)[2]
Zhang Jichao(张继超)[2] Wang Hongyan(王红艳)[2]
Tan Yebang(谭业邦)[1]

(1. School of Chemistry and Chemical Engineering, Shandong University, Jinan 250100, China;
2. Research Institute of Geology, Shengli Oilfield Company, Dongying 257000, China)

Abstract: Poly (acrylamide-co-sodium acrylate) (PAM/AA-Na) and poly (acrylamide-co-sodium vinyl sulfonate) (PAM/VSS-Na) were prepared by inverse emulsion polymerization. The effects of $CaCl_2$ on PAM/VSS-Na or PAM/VSS-Na aqueous solutions were investigated by steady-flow experiments at 25, 40, 55 and 70℃. The results show that the apparent viscosities of both solutions decrease with addition of $CaCl_2$ or increase of temperature and shear rates. PAM/VSS-Na solution has better performance on the salt tolerance, shear endurance and temperature resistance due to containing sulfonic group in the molecules. Ca^{2+} concentration can affect the viscous activation energy of both solutions and the reason may be that these interactions between Ca^{2+} and also copolymer molecules are related to temperature and competitive in solution. These results may offer the basic data for searching the flooding systems with the ability of temperature resistance, salt tolerance and shear endurance for tertiary oil recovery.

Keywords: polyacrylamide; sodium vinyl sulfonate; rheological properties; salt tolerant; temperature resistant

1 Introduction

Many research groups have focused on the technology of enhanced oil recovery, and the notable progress has been made in chemical flooding, including surfactant flooding, polymer flooding, alkaline-polymer flooding and alkaline-polymer-surfactant flooding[1-6]. The practices show that the rheology and interfacial tension of injected fluid have an important effect on the oil-displacing efficiency in polymer flooding.

At present, one of the most widely used water-soluble polymers is partially hydrolyzed polyacrylamide (HPAM), which controls mobility in the reservoirs by increasing the viscosity of the injected water, and more importantly, by reducing the formation of permeability. However, HPAM has many disadvantages in oil recovery. For example, the viscosity of HPAM aqueous solution exhibits an abrupt reduction in 3% brine[7]. It suffers excessive thermal hydrolysis at high temperatures and as a result may precipitate in the presence of bivalent cations. Substituted or modified monomer, can yield a better product which may be polyacrylamides, or those copolymerized with a suitable thermally stable, salt tolerant or shear endurant[8-9]. It is established that the copolymers of acrylamide with monomers including sulfonic group offer polyelectrolyte behavior in aqueous solutions-characteristics of special interest to tertiary oil recovery[7].

In this work, poly (acrylamide-co-sodium acrylate) (PAM/AA-Na) and poly (acrylamide-co-sodium vinyl sulfonate) (PAM/VSS-Na) were prepared by inverse emulsion polymerization, respectively. The PAM/AA-Na and PAM/VSS-Na solutions with or without $CaCl_2$ were investigated via steady-flow experiments at 25, 40, 55 and 70℃. The aim is to offer the basic data for searching the flooding systems with the ability of temperature resistance, salt tolerance and shear endurance.

2 Experimental

2.1 Materials

Poly (acrylamide co-sodium acrylate)(PAM/AA-Na), which is an intrinsic viscosity of 910 mL/g in 1 mol/L $NaNO_3$ solution, and poly(acrylamide-co-sodium vinyl sulfonate) (PAM/VSS-Na), which is an intrinsic viscosity of 696 mL/g in 1 mol/L $NaNO_3$ solution, were prepared by inverse emulsion polymerization. The emulsions were precipitated by ethanol, and washed by ethanol for six times and mixture of water and ethanol for three times to remove emulsifier and unreacted monomers. The molar ratio of sodium acrylate in PAM/AA-Na is 7.2%, which was determined by conductimetric analysis. And the molar ratio of sodium vinyl sulfonate in PAM/VSS-Na is 6.6%, which was determined by elementary analysis. $CaCl_2$ was purchased from Tianjin Chemical Agents Company, China. The water used was distilled for three times.

2.2 Preparation of sample solutions

Three stock solutions, concentrated PAM/AA-Na, PAM/VSS-Na and $CaCl_2$ solutions were prepared. After stirring for 2 d, PAM/AA-Na and PAM/VSS-Na solution were gently heated at 50℃ to ensure complete dissolution. A series of sample solutions were then prepared at natural pH by mixing different amounts of the stock solutions to obtain desired $CaCl_2$ concentrations between 0 and 3.0 mmol/L. Sample solutions were then stirred for 12 hand left overnight to equilibrate before the viscosity measurements.

2.3 Rheological measurements

The rheological measurements were carried out on a HAAKE RS75 rheometer (Germany) with coaxial cylinder sensor system (Z41 Ti). The temperature was maintained at (25.0 ± 0.1), (40.0 ± 0.1), (55.0 ± 0.1) and (70.0 + 0.1)℃, respectively. For the shear-dependent behavior, the viscosity measurements were carried out at shear rates ranging from 0 to 1 000 s^{-1}.

3 Results and discussion

3.1 Effect of $CaCl_2$ and shear rates on apparent viscosity of both solutions

Fig. 1 shows the steady state shear flow curves of 0.5% PAM/AA-Na and 0.5% PAM/VSS-Na solutions at different $CaCl_2$ concentrations and temperatures. Fig. 2 shows the relative values of apparent viscosity of 0.5% PAM/AA-Na and 0.5% PAM/VSS-Na solutions as a function of $CaCl_2$ concentration and shear rate. From the results, it is found that apparent viscosities (η) of both PAM/AA-Na and PAM/VSS-Na solutions decrease with increasing $CaCl_2$ concentration or temperature. Both solutions exhibit shear-thinning effect at high shear rates and behave as non-Newtonian fluids with increasing shear rates. However, the reduction of apparent viscosity of PAM/VSS-Na solution is lower compared with that of PAM/AA-Na solution as $CaCl_2$: concentration and shear rate increase.

The decrease of the viscosity with shear rate is mainly related to the orientation of macromolecules along the streamline of flow and to the disentanglement of macromolecules with increasing shear force. In general, while in the solution of a good solvent, ionic polymers exist in the maximum possible expanded state to minimize the repulsive interaction between the ionic groups of the same macroion bearing a similar charge.

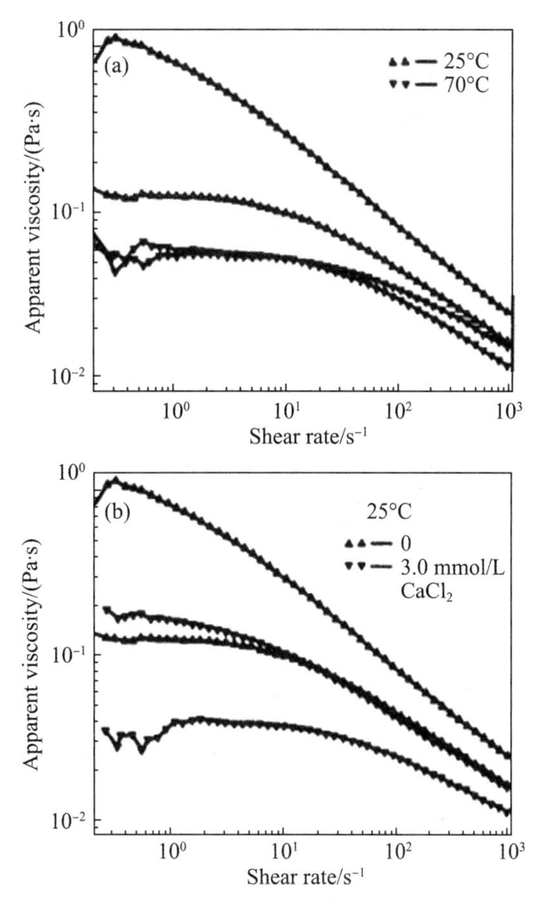

Fig. 1 Steady state shear flow curves of 0.5% PAM/AA-Na(▲,▼) and 0.5% PAM/VSS-Na(△, ▽) solutions at 25℃, 75℃(a) and 0, 3 mmol/L $CaCl_2$ concentrations(b)

And an addition of electrolyte in the ionic polymer solution induces the increase of solution ionic strength and screens the electrostatic charges. Then the macromolecule conformation reduces to the statistical coil conformation. As a consequence, a decrease of the apparent viscosity in copolymer solution containing $CaCl_2$ in comparison with aqueous solution is observed[10].

Because Ca^{2+} can associate with carboxylic group, it facilitates the formation of intra-molecular associations of PAM/AA-Na, which may precipitate in the solution. Sulfonic group has weaker interaction with Ca^{2+} and is larger than carboxylic group, which prevents the degradation of the backbone of copolymer and also makes copolymer to expand more. As a result, the apparent viscosity of PAM/VSS-Na solution is unchanged or increased at lower shear rates and decreases much less at high shear rates compared with that of PAM/AA-Na solution.

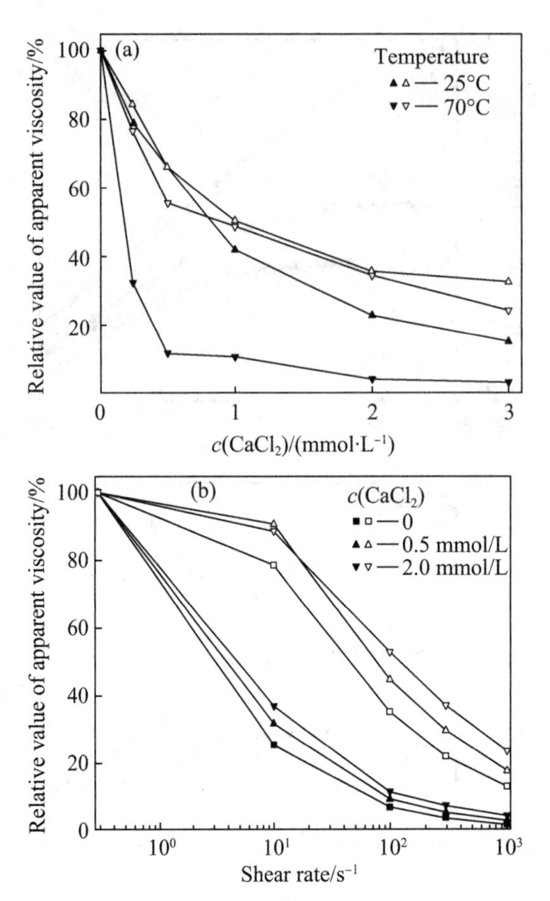

Fig. 2 Relative values of apparent viscosity of 0.5% PAM/AA-Na (▲, ▼) and 0.5% PAM/VSS-Na (△, ▽) solutions at 0.3 s^{-1} as a function of $CaCl_2$ concentration at 25, 70℃ (a) (relative value of each salt free solution is defined as 100%) and relative values of apparent viscosity of 0.5% PAM/AA-Na (■, ▲, ▼) and 0.5% PAM/VSS-Na (□, △, ▽) solutions as function of shear rate at 25℃ at different $CaCl_2$ concentrations(b) (each relative value at 0.3 s^{-1} is defined as 100%)

3.2 Effect of temperature on apparent viscosity

As well known, the increase of temperature can give rise to crimple of PAM/AA-Na molecules due to its dehydrating and destruction of the associational structure[4]. As a consequence, Fig. 3 shows that the viscosities of all solutions decrease at each $CaCl_2$ concentration as temperature increases.

It is reported that the relationship between apparent viscosities of polymer solution and temperatures satisfies Arrhenius equation:

$$\eta = A\exp\left(\frac{\Delta E_\eta}{RT}\right) \tag{1}$$

where η is the apparent viscosity of polymer solution, ΔE_η is the viscous activation energy; R is the gas constant, and T is the absolute temperature. According to this Eqn. (1), $\ln\eta$ and polymer solution should show a straight-line relationship, with slope of $\Delta E_\eta/R$. As a result, viscous activation energy of PAM/AA-Na and PAM/VSS-Na solutions at different $CaCl_2$ concentrations was calculated, as shown in Fig. 4. Viscous activation energy is related to the dependence between viscosity and temperature of polymer solution, and the higher the viscous activation energy is, the more the influence of temperature on the viscosity.

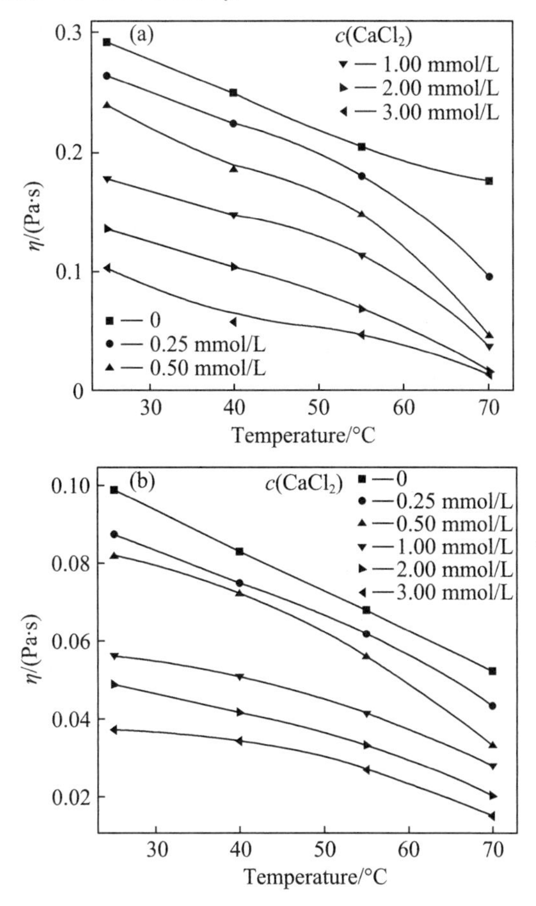

Fig. 3 Viscosity of 0.5% PAM/AA-Na (a) and 0.5% PAM/VSS-Na (b) solutionat 10 s^{-1} as function of temperature at different $CaCl_2$ concentrations

Fig. 4 shows that the viscous activation energy of PAM/VSS-Na is less than that of PAM/AA-Na in $CaCl_2$ solutions at each concentration, which indicates that PAM/VSS-Na and $CaCl_2$ system has better performance on the temperature resistance. However, the viscous activation energy of PAM/VSS-Na is more than that of PAM/AA-Na in salt free solution. The reason may be that sulfonic group is larger than carboxylic group, which indicates that charged groups in PAM/VSS-Na molecules are more crowded than that in PAM/AA-Na molecules when charges are not screened by other salt ions. As a result, PAM/VSS-Na is more sensitive in pure aqueous solution.

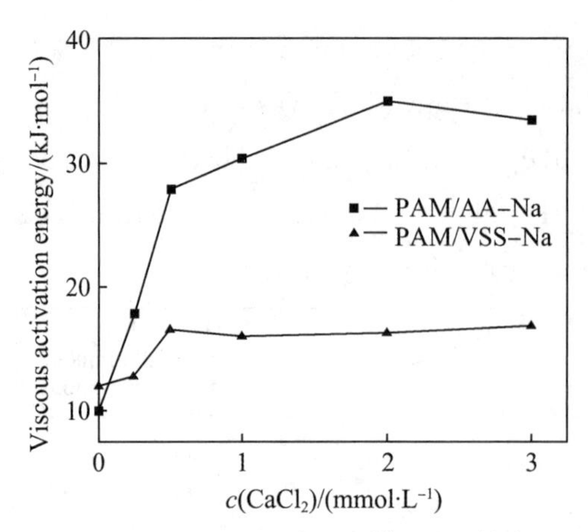

Fig. 4 Changes of viscous activation energies of 0.5% PAM/AA-Na and 0.5% PAM/VSS-Na solutions as function of $CaCl_2$ concentration

Fig. 4 also shows the viscous activation energies of both solutions increase sharply at 0.5 mmol/L $CaCl_2$ concentration. This may indicate the interactions between Ca^{2+} and molecules of PAM/AA-Na or PAM/VSS-Na are related to Ca^{2+} concentration. Because Ca^{2+} could associate with amide group and carboxylic group, and could compress the electrical double layer of charged group, and capture the hydrated shell around polar groups, and screen the electrostatic charges. These interactions may be competitive and related to temperature. As a result, when $CaCl_2$ concentration is less than 0.5 mmol/L, the viscous activation energies of both solutions are lowered because some or all of these interactions may be less owing to lacking enough Ca^{2+} in solutions.

4 Conclusions

1) The apparent viscosities of PAM/AA-Na and PAM/VSS-Na aqueous solutions decrease as Ca^{2+} concentration, temperature and shear rates increase. Both solutions exhibit shear-thinning effect at high shear rates and behave as non-Newtonian fluids when shear rates increase.

2) The reduction of apparent viscosity of PAM/VSS-Na solution is lower compared with that of PAM/AA-Na solution as salt concentration and shear rates increase. which shows that PAM/VSS-Na has better performance on the salt tolerance and shear endurance.

3) The viscous activation energy of PAMA^SS-Na is less than that of PAM/AA-Na in $CaCl_2$ solutions at each concentration, which indicates that PAM/AA-Na is more sensitive to temperature in salt solution. And the viscous activation energy of PAM/VSS-Na is more than that of PAM/AA-Na in pure aqueous solution. This may be because that charged groups in PAM/VSS-Na molecules are more crowded than that in PAM/AA-Na molecules when charges are not screened by other ions.

4) Viscous activation energies of both solutions increase sharply at 0.5 mmol/L

$CaCl_2$ concentration, which is probably because that the interactions between Ca^{2+} and copolymers are competitive and related to temperature.

References

[1] FENG Yu-jun, GRASSL B, BILLON L, KHOUKH A, FRANCOIS J. Effects of NaCl on steady rheological behavior in aqueous solutions of hydrophobically modified polyacrylamide and its partially hydrolyzed analogues prepared by post-modification [J]. Polym lot, 2002,51(10):939-947.

[2] HAN Da-kuang, YANG Cheng-zhi, ZHANG Zheng-qing, LOU Zhu-bong, CHANG Y I. Recent development of enhanced oil recovery in China [J]. J Petrol Sci Eng, 1999,22(3):181-188.

[3] Rosen M J, WANG Hong-zhuang, SHEN Ping-ping, ZHU You-yi. Ultralow interfacial tension for enhanced oil recovery at very low surfactant concentrations [J]. Langmuir, 2005,21(9):3749-3756.

[4] XIN Xia, XU Gui-ying, WU Dan, LI Yi-ming, CAO Xiao-rong. The effect of $CaCl_2$ on the interaction between hydrolyzed polyacrylamide and sodium strarate: Rheological property study [J]. Colloids Surf A,2007,305(1/3):138-144.

[5] GONG Houjiang, XIN Xia, XU Gui-ying, WANG Ya-jing. The dynamic interfacial tension between HPAM/$C_{17}H_{33}$CCKJNa mixed solution and crude oil in the presence of sodium halide [J]. Colloids Surf A,2008,317(1/3):522-527.

[6] YANG M H. The rheological behavior of polyacrylamide solution II: Yield stress [J]. Polym Test,2000,20(6):635-642.

[7] SABHAPONDIT A, BROTHAKUR A, HAQUE 1. Water soluble acrylamidomethyl propane sulfonate (AMPS) copolymer as an enhanced oil recovery chemical [J]. Energy and Fuels,2003,87(12):683-688.

[8] LI Hai-pu, HU Yue-hua, WANG Dian-zuo, XU Jing. Effect of hydroxamic acid polymers on reverse flotation of bauxite [J]. J Cent South Univ Technol,2004,11 (3):291-294.

[9] ZHANG Jian-feng, HU Yue-hua, WANG Dian-zuo. Preparation and determination of hydroximic polyacrylamide [J]. J Cent South Univ Technol,2002,9 (3):177-180.

[10] LEWANDOWSKA K Comparative studies of rheological properties polyacrylamide and partially hydrolyzed polyacrylamide solutions [J]. J Appl Polym Sci,2007,103(4):2235-2241.

(Edited by LONG Huai-zhong)

Kinetics study on the formation of resol with high content of hydroxymethyl group

Gao Pin[1] Zhang Yanfang[1] Huang Guangsu[1] Zheng Jing[1]
Chen Mengmeng[1] Cao Xulong[2] Liu Kun[2] Zhu Yangwen[2]

[1] State Key Lab of Polymer Materials Engineer, College of Polymer Science and Engineering, Sichuan University, Chengdu, China 610065

[2] Geological Research Institute of Shengli Oilfield Co. Ltd, SINOPEC, Dongying, Shandong, China 257015

Abstract: The application of resol as a crosslinking agent with polyacrylamide has become widespread, but it is not clear for forming mechanisms of extraordinarily high activity of the resol. In this research, the major components of the resol and their mass percentage are investigated by Liquid chromatography-mass spectrograph, the mechanism of forming the components is exploited. The evolution of formaldehyde, phenol, and first formed addition products were quantitatively traced by titration and high-perform-ance liquid chromatography, furthermore the kinetics of synthesis is studied based on the analyzing results. The results show that the resol is composed of nine compounds and the sum of percentage for each integral area is more than 99%. In addition, the constant of overall reaction ratek consists of three values of 5.667×10^{-5}, 7.236×10^{-5}, and 23.05×10^{-5} s^{-1}, which means that the reaction can be sep-arated to three stages of hydroxymethylphenol (HMP) pro-ducing stage, dihydroxymethylphenol (DHMP) and trihy-droxymethylphenol (THMP) producing stage and conden-sation stage.

Keywords: resol; hydroxymethyl group; kinetics; crosslinking agent; liquid chromatography-mass spectra

INTRODUCTION

Phenolic-formaldehyde (PF) resins synthesized by phenol and formaldehyde under a basic catalyst have been utilized in a variety fields as electric, construction, automobile, and even aviation and spaceflight industry, owing to their excellent heat-resistance and fire-retardant. The resol, an oligomer derived from the first-step of forming the resin, has been used as adhesive owing to self-crosslinking for long.

The reaction of preparing resol aroused scientists' interests early in the middle of 20th century, but until recent time still a lot of literatures reported some new developments about it. Grenier-Loustalot et al.[1-5] studied the reaction mechanism of phenol, formaldehyde, and the model compounds by highperformance liquid chromatography (HPLC) and ^{13}C NMR spectroscopy. Astarloa-Aierbe et al.[6-10] analyzed the first step of the condensation reaction, taking account of the influence of many parameters such as pH, type of catalyst, formaldehyde/phenol (F/P) molar ratio by HPLC, gas chromatography, and[13] CNMR spectroscopy. With FTIR spectroscopy using multicomponent spectroscopic analysis of a software package QUANT+, Holopainen et al. predicting reliably quantitative results such as the amount of free phenol and the F/P molar ratio.[11] In addition, employing a multiple parameter regression method, Riccardi et al. and Manfredi et al.[12,13] carried out the modelling of the phenolic resol resin.

Recently, an extraordinarily active resol has been widely used in the oil production field as a crosslinking agent because of its good water-solubility and high crosslinking reactivity. The crosslinking reaction occurs between amide groups on polyacrylamide and hydroxymethyl groups on resol. It is evident that the activity is closely related with the amount of hydroxymethyl group in each component, so the higher content of hydroxymethyl group is, the higher crosslinking reactivity will be. Although the gelling property of the resol with polyacrylamide has been studied by many researchers, such as Li et al. and Kong and Song,[14,15] few study on components and forming mechanism of the resol with multi-hydroxymethyl groups has been reported. On the other hand, liquid chromatography-mass spectra (LC-MS) technology is hardly employed for analyzing the composition of resol yet. The LC-MS taking the electrospary ionization (ESI) was an efficient way of qualitative and quantitative analysis, testing under low temperature avoids changing the structure of samples, which often takes place during the vapourization in gas chromatography-mass spectra. The qualitative analyzing method can supply a good deal of precise information for determination on molecular weight of the high reactive oligomer formed in the organic synthesis.[16]

In the present article, the major components of the resol with great content of hydroxymethyl group and their mass percentage were studied. To understand the evolution procedure of the reaction well, the contents of formaldehyde, phenol, and first formed addition products were quantitatively analyzed and the kinetics of the reaction was constructed, gaining some interesting results.

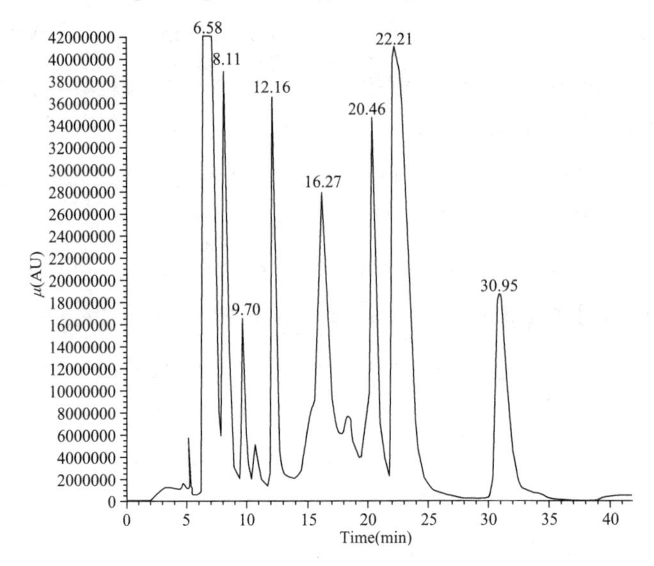

Figure 1 LC-MS chromatogram of resol by UV detector.

EXPERIMENTAL

Materials

Phenol (>99%), formaldehyde (37% aqueous solution), and sodium hydroxide are from Chongqi Maoye, China. Besides, methanol was from Shandong Yuwang, China and water was from Sichuan University.

Synthesis of resol

Resol was synthesized by mixing phenol (P) and formaldehyde (F) in a molar ratio of F/P =2—5, the pH was then adjusted to 9.0 with alkaline catalyst. The mixture was heated to the reaction temperature for 7h and stirred during the reaction. Samples were taken at regular intervals. The time when mixture reached the reaction temperature was defined as zero, and then the reaction was stopped by putting the reactor in a cold water bath (20℃). The samples were kept at −18℃ before test.

Liquid chromatography-mass spectra

LC analysis was performed using a HPLC system (America) coupled to a Finnigan LCQ^{DECA} mass spectrometer, equipped with a Z-spray ESI interface (Waters). Chromatographic separation was achieved with a ThermoQuest C_{18} (5 μm, 250 mm× 4.6 mm, America) column set at 20℃. For the analysis, a mobile phase of methanol/water (v/v) was used an elution of 30%～50% methanol. The flow rate was set at 0.5 mL/min.

High-performance liquid chromatography

Analyses were conducted with a HPLC system consisting of a Waters HPLC equipped with the Waters 515 HPLC pump and 2487 detector. The analytical column was a VP-ODS column (5 μm, 250 mm×4. 6 mm, Shimpack, Japan) set at 20℃. A mobile phase of methanol/water (v/v) was used an elution of 30%~50% methanol. The flow rate was set at 0. 5 mL/ min. The eluate was monitored at 254 nm.

RESULTS AND DISCUSSION

Figures 1 and 2 show the LC-MS and HPLC chromatograms of the same resol sample detected by UV detectors, respectively. Among them, the resol is composed of nine compounds and the sum of percentages for each integral area account for 99%. The t_R of each peak in LC-MS and HPLC has some deviations between both chromatograms because of the different column effectiveness. However, the order and integral areas of the peaks are corresponding in the two chromatograms.

The information from LC-MS test is shown in Table 1. The molecular ion peak and quasi molecular ion peak are strong in LC-MS test owing to the "soft" ionization technology, which can avoid further disintegration of the molecular ion peak, so the molecular weight of the sample is easily ascertained due to the weak peaks of the fragmentions.[16] Figure 3 is the mass spectrum of peak 1, in which the molecular ion peak of m/z = 183 is the base peak, so the molecular weight of the compound corresponding the peak is 184. Combining the mechanism of the addition reaction between phenol and formaldehyde with the information of the test, this compound can be determined to be 2, 4, 6-trihydroxymethyl phenol (2,4,6-THMP).

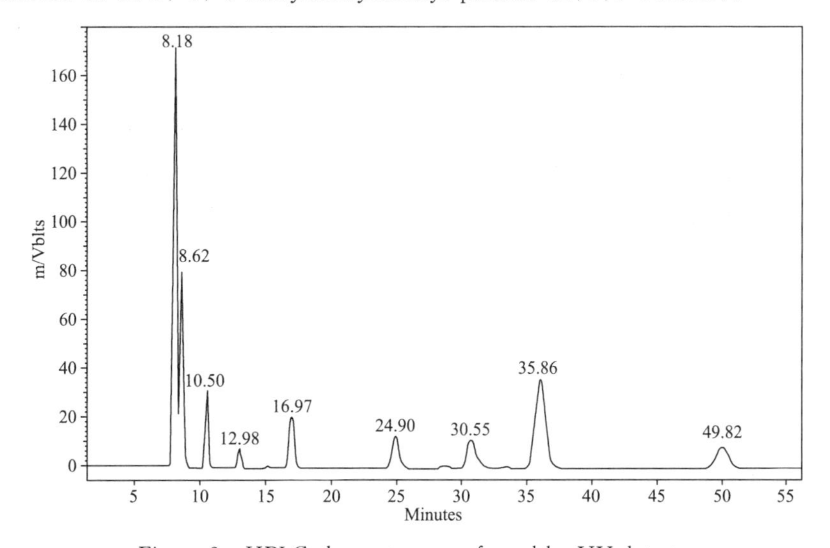

Figure 2 HPLC chromatogram of resol by UV detector

Table 1 LC-MS Information of Resol

Peak no.	t_R (min)	MS information (m/z)
1	6.58	183(M−1), 184
2	11.11	289, 301, 319(M−1), 320
3	12.16	389, 407, 425(M−1), 426
4	16.27	299, 301, 319(M−1), 320
5	20.46	153, 271, 289(M−1), 290
6	22.21	299, 301, 319(M−1), 320
7	30.95	271, 289(M−1), 290

According to the information such as m/z, t_R of the LC-MS test, and different polarity between parahydroxymethyl and ortho-hydroxymethyl, molecular formulas of most compounds are obtained. The MS spectrums of the peaks whose t_Rs are 8.11 and 9.70 are not detected because of the sensitivity and detectability of the detector, but it can be deduced that the compounds corresponding the two peaks are 2, 4-dihydroxymethylphenol and 4-hydroxyme-thylphenol from the former research results.[1-10] In Table 2, the molecular formulas of each components obtained by LC-MS and HPLC information are shown as follows.

Among all the resulting nine components, there is no phenol reactant. It is pronounced that the content of THMP is the highest in the resol, while the sum percentages of integral areas of the dimers occupies about 40% and the tripolymer include only one component whose content is very low. To understand the reaction course distinctly, the kinetics of the reaction with high F/P molar ratio was studied.

Generally speaking, the reaction between phenol and formaldehyde in the alkaline pH range can be divided to two steps: addition reaction of hydroxymethyl groups to the ortho and para free positions of phenol (Fig. 4) and condensation reactions between hydroxymethyl group and one free position in phenol giving rise to methylene bridges or two hydroxymethyl groups forming methylene ether bonds.

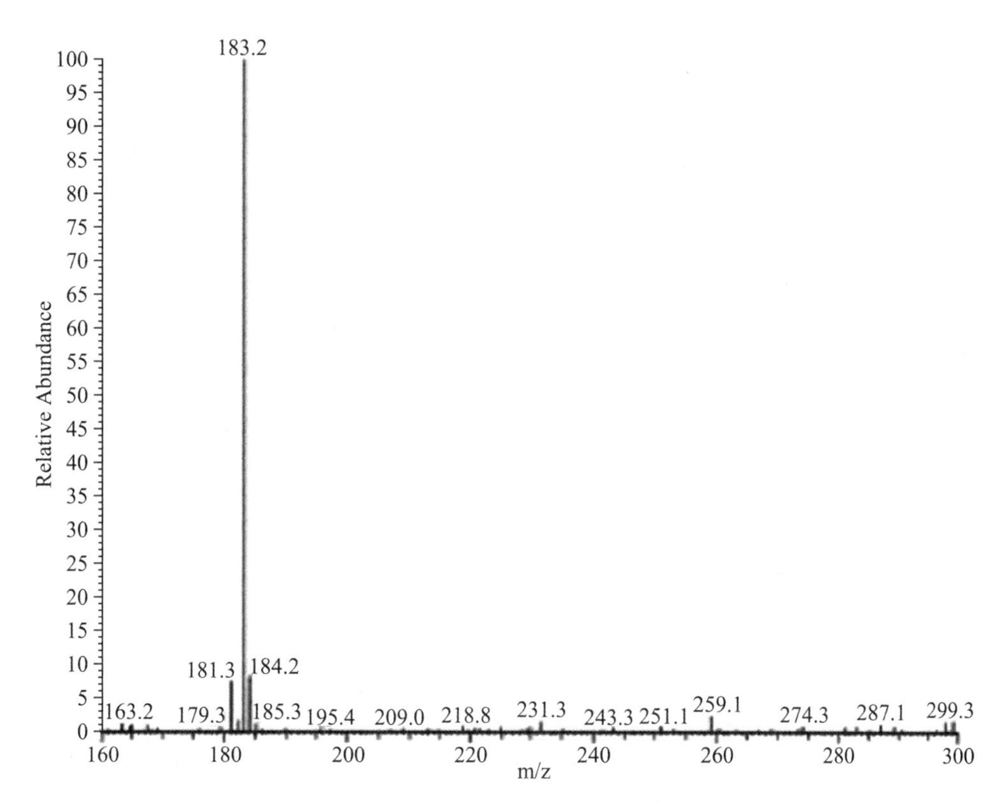

Figure 3 MS chromatogram of peak 1.

Table 2 Molecular Formulas of Components Conjectured by LC-MS and HPLC Information

Molecular formula	t_R(min)	Integral area (%)
OH; HOH_2O—; —CH_2OH; CH_2OH	6.58	31.23
OH; OH; HOH_2C—; —CH_2—; —CH_2OH; CH_2OH; CH_2OH	11.11	1.98
Tripolymer	12.16	6.34
OH; CH_2OH; HOH_2C—; —CH_2—; —OH; CH_2OH; CH_2OH	16.27	5.51
OH; CH_2OH; —CH_2—; —OH; CH_2OH; CH_2OH	20.46	6.38

（续表）

Molecular formula	t_R (min)	Integral area (%)
HOH_2C—CH_2—CH_2OH; HO, CH_2OH, OH, CH_2OH	22.21	20.27
Dimer	30.95	6.04

In the past, most studies about kinetics mechanism of resol synthesis focused on the resol, which has the property of self-setting and the low F/P molar ratio, which is helpful to obtain the self-setting PF with high crosslinking density.[17] In present research, the resol synthesized has excellent crosslinking reactivity with polyacrylamide because of high content of hydroxymethyl group derived from high F/P molar ratio.

The disappearance of the free formaldehyde and free phenol content versus reaction time is depicted in Figure 5. It can be seen that even though the F/P molar ratio is high in the reaction, formaldehyde still reacts sufficiently in the whole synthesis, while concentration of free formaldehyde finally goes down to around 1%, which demonstrates that high F/P ratio has no effect on the reactive extent of formaldehyde. The concentration of free formaldehyde is assumed to be linear decreased in the beginning 100 min. After that, the velocity of consumption is falling down and the concentration of formaldehyde is almost kept at a certain level. The situation of free phenol is similar with that of free formaldehyde, at beginning 150 min the concentration of phenol drops fast and then hardly changed.

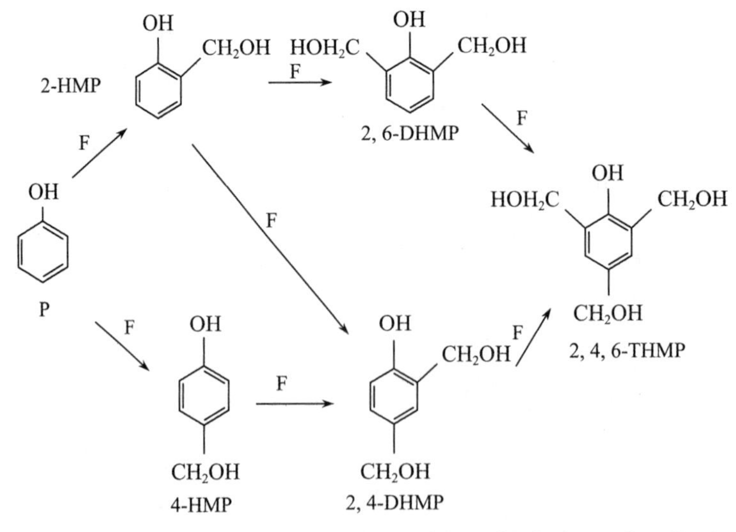

Figure 4 Reaction mechanism for addition of formalde-hyde to phenolic rings.

2-hydroxymethylphenol (2-HMP) and 4-hydroxy-methylphenol (4-HMP) are not only products but also reactants in the synthesis of the resol, the evolutions of them are shown in Figure 6. At the beginning the concentration of 2-HMP is much higher than that of 4-HMP, because in the addition reaction catalyzed with NaOH, the para-position in phenol shows a slightly higher relative reactivity towards formaldehyde than the ortho-position. However, 2-HMP is produced at a higher rate due to the fact that two ortho-positions are available. The ortho/para (o/p) ratio was found to be 1. 7. [18] The concentration of 2-HMP is lower than that of 4-HMP in the end of reaction because 2-HMP is found to be twice as reactive as 4-HMP with respect to either position. [18] Increased reactivity is also found with higher methylolated phenols to be produced, especially remarkable with 2, 6-dihydroxymethylphenol. The ratio variation of the concentration of 2-HMP and 4-HMP (o/p) in this system is shown in Figure 7, which corroborates the para directing characteristic of NaOH discussed above. Though the o/p value is over 2. 0 at the beginning because the lower catalyst concentration is used, [13] it descends quickly as reac-tion time increasing in the most of reaction time the o/p value is less than 1. 8.

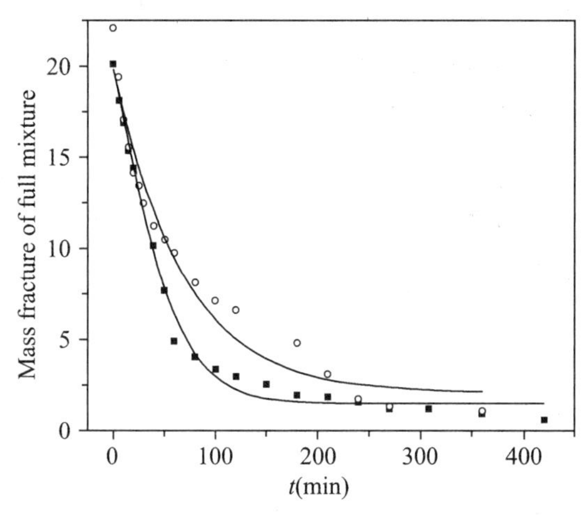

Figure 5　Disappearance of formaldehyde and phenol. ■ formaldehyde, ○ phenol.

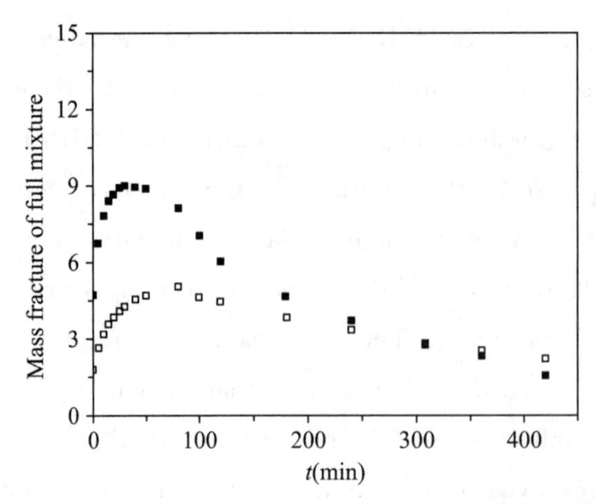

Figure 6　Evolution of HMP by HPLC during synthesis. ■ 2-HMP, □ 4-HMP.

Similarly, both the concentrations of 2, 4-dihydrox-ymethylphenol (2, 4-DHMP) and 2, 6-dihydroxyme-thylphenol (2, 6-DHMP) rise at first and then descend in the procedure of reaction, which means that they all play a role of intermediate products in the reaction system. The evolutions of the two compounds can be illustrated in Figure 8. Because of the coaction of high reactivity of ortho-hydroxymethyl phenol and the effect of para directing, the concentration of 2, 6-DHMP is much lower than that of 2, 4-DHMP except for the beginning of the reaction.

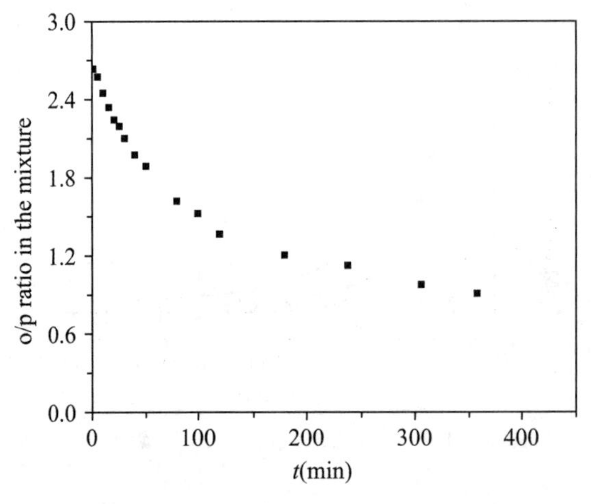

Figure 7　Ortho/para ratio of the resol.

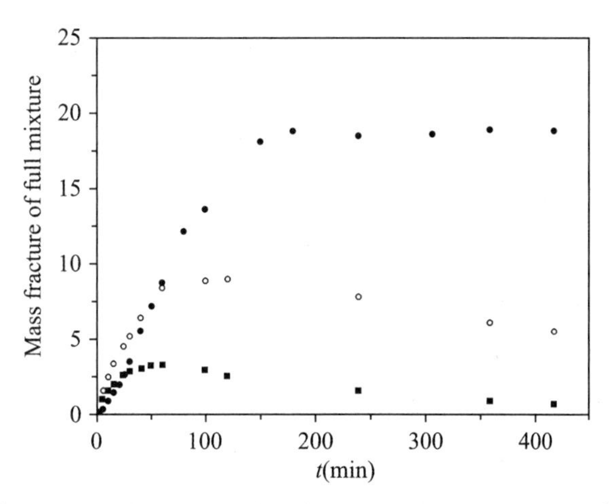

Figure 8 Evolution of DHMP and 2,4, 6-THMP by HPLC during synthesis. ■ 2, 6-HMP, ○ 2,4-HMP,● 2,4,6-THMP.

The concentration of 2, 4, 6-trihydroxymethylphenol (2, 4, 6-THMP) does not take on the downtrend in the whole reaction, which means it continuously keeps at a high level in the reaction system (see Fig. 8). After 150 min of the reaction, the concentration of 2, 4, 6-THMP reaches to the highest value and keeps it to the end of the reaction. This period coincides with the time at which the concentration of free phenol goes down to the lowest value. It means that formaldehyde attacks to the free positions of substituted phenol rapidly to forming 2, 4, 6-THMP whose concentration is the highest in the resol. Consequently there are half reactants transformed to 2, 4, 6-THMP at the end of the reaction due to the high F/P ratio at the beginning of the reaction and the low reaction temperature.

The reaction rate is correlative with concentrations of phenol and formaldehyde because the addition reaction between them belongs to second-order reaction. The rate expression for the formation of the products is

$$dy/dt = k(a-x)(b-y) \quad (1)$$

where a and b are the initial concentrations of formaldehyde and phenol, while x and y are the concentrations of formaldehyde and phenol reacted at time t, respectively. k is the constant of overall reaction rate.

Hypothesizing $y = mx$, substituting the value of y in eq. (1) and integrating, we get

$$kt = \frac{m}{(b-ma)}\ln\left(\frac{a(b-mx)}{b(a-x)}\right) \quad (2)$$

Hypothesizing

$$\gamma = \frac{m}{(b-ma)}\ln\left(\frac{a(b-mx)}{b(a-x)}\right) \quad (3)$$

introduced the data of concentration to the eqs. (2) and (3) and the fitted results show in Figure 9, the slope of it is the value of k.

It can be observed from Figure 9 that γ and t is not simply linear relation in the researched range but instead of three stages of perfect linear relation, while the slopes of the three lines are 5.667×10^{-5}, 7.236×10^{-5}, and 23.05×10^{-5} s^{-1} increased in sequential, an interesting phenomenon. It means that the constant k changes following the composition of the products, and the whole reaction can be expressed by three k_s. The phenomenon could be attributed to the fact that the whole reaction experience three stages of HMP-producing stage, DHMP and THMP-producing stage and condensation stage. The fluctuation of concentration in the reaction system can prove this statement. As shown in Figure 9, the first stage is from zero time to 30 min of the reaction. In this period HMP concentration reaches the highest value (from Fig. 6), it means that HMP plays the role of product, whose concentration is the highest in the whole system. After 30 min, most of HMP starts to occur the addition reaction with formaldehyde, which results in descent of the HMP concentration. The second stage lasts from 30 to 150 min of the reaction. In this stage, HMP in the reaction system react with formaldehyde, DHMP and THMP are formed, the addition reactive rate increase. Compared with the rate of first stage, both concentrations of 2,4-DHMP and 2, 4, 6-THMP arrive to the highest value (see Fig. 8). Here the reaction transit to the last stage from 150 min to the end of reaction. The consumptions of phenol and formaldehyde become lower, but k value increase rapidly, where DHMP and THMP, the major components in the reaction system, condensate to form the dimers and tripolymer. The consumption of THMP and DHMP make the reaction equilibrium shift to the product direction, which induce the increment of the k value.

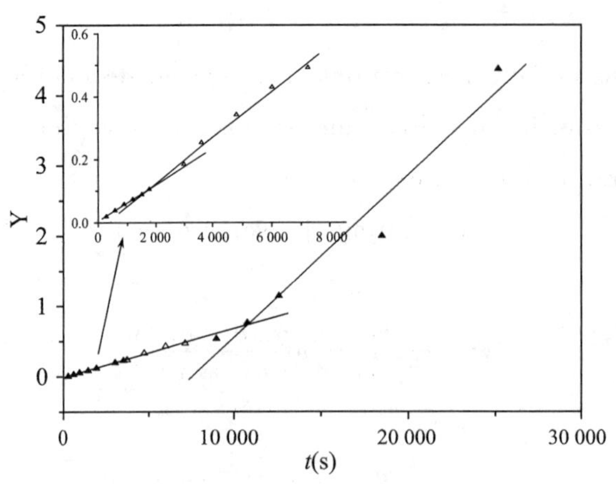

Figure 9 Linear dependence of γ on time.

CONCLUSIONS

A resol with high content of hydroxymethyl has been synthesized; most of the addition products and low molecular weight condensation compounds in it are identified by LC-MS. The graphs obtained by LC-MS can be corresponded with that from HPLC preferably. On the basis of results from the analysis, kinetics of the synthesis is studied. The constant of overall reaction rate k consists of three values of 5.667×10^{-5}, 7.236×10^{-5}, and 23.05×10^{-5} s^{-1}, which means that the reaction can be separated to three stages of HMP-producing stage, DHMP and THMP-producing stage and condensation stage.

References

[1] Grenier-Loustalot, M.-F.; Larroque, S.; Grenier, P.; Leca, J.-P.; Bedel, D. Polymer 1994,35,3046.

[2] Grenier-Loustalot, M.-F.; Larroque, S.; Grande, D.; Grenier, P.; Bedel, D. Polymer 1996,37,1363.

[3] Grenier-Loustalot, M.-F.; Larroque, S.; Grenier, P.; Bedel, D. Polymer 1996,37,939.

[4] Grenier-Loustalot, M.-F.; Larroque, S.; Grenier, P.; Bedel, D. Polymer 1996,37,955.

[5] Grenier-Loustalot, M.-F.; Larroque, S.; Grenier, P.; Bedel, D. Polymer 1996,37,639.

[6] Astarloa-Aierbe, G.; Echeverria, J. M.; Egiburu, J. L.; Ormaetxea, M.; Mondragon, I. Polymer 1998,39,3147.

[7] Astarloa-Aierbe, G.; Echeverria, J. M.; Martin, M. D.; Mondragon, I. Polymer 1998,39,3467.

[8] Astarloa-Aierbe, G.; Echeverria, J. M.; Mondragon, I. Polymer 1999, 40,5873.

[9] Astarloa-Aierbe, G.; Echeverria, J. M.; Vazquez, A.; Mondragon, I. Polymer 2000,41,3311.

[10] Astarloa-Aierbe, G.; Echeverria, J. M.; Riccardi, C. C.; Mondra-gon, I. Polymer 2002,43,2239.

[11] Holopainen, T.; Alvila, L.; Rainio, J.; Pakkanen, T. T. J Appl Polym Sci 1998,69,2175.

[12] Riccardi, C. C.; Astarloa-Aierbe, G.; Echeverria, J. M.; Mondragon, I. Polymer 2002, 43, 1631.

[13] Manfredi, L. B.; Riccardi, C. C.; de la Osa, O.; Vazquez, A. Polym Int 2001, 50, 796.

[14] Li, G.; Xu, J.; Mao, G.; Dai, B.; Zang, S.; Lu, K. Oilfield Chem 2000, 17, 310.

[15] Kong, B.; Song, Z. Acta Petrolei Sinica 2000, 21, 70.

[16] Niessen, W. M. A. J Chromatogr A 1999, 856, 179.

[17] Manfredi, L. B.; de la Osa, O.; Galego Fernndez, N.; Vzquez, A. Polymer 1999, 40, 3867.

[18] Knop, A.; Pilato, L. Phenolic Resins; Springer-Verlag: Berlin, 1985.

一种新型热触变流体的性能研究

韩玉贵　曹绪龙　祝仰文　刘　坤

（中国石化股份胜利油田分公司地质科学研究院，山东东营 257015）

摘要：常规聚合物驱油剂在高温高盐条件下粘度损失严重，严重影响了高温高盐油藏聚合物驱油效果，为了探索具有较高增粘性能的耐温抗盐驱油剂，对一种新型热触变流体的性能进行了室内研究。首先利用质量分数为 2% 的氯化钠水溶液配制热触变流体，然后利用 DV-Ⅲ 旋转粘度计考察了 90℃ 条件下该流体的增粘性能及热稳定性，利用 NCR300 流变仪考察了其流变性能，并记录溶液的外观变化。结果表明，热触变流体初始粘度很低，但随着高温老化时间的延长，粘度逐渐增大，表现出很好的高温增粘性，该热触变流体的粘度受剪切作用影响不大，表现出一定的牛顿流体特性，流体的粘性模量远大于弹性模量。将热触变流体与 HPAM 进行复配，流体既获得了较高的初始粘度，又维持了粘度的长期热稳定性，并且表现出一定的粘弹特性。该类热触变流体在高温高盐油藏通过增大驱替液粘度达到扩大驱替液波及体积的目的，因而在提高原油采收率方面可能具有较好的应用前景。

关键词：热触变流体；常规聚合物；高温老化；粘度；流变性

中图分类号：TE357.43　文献标识码：A　文章编号：1009-9603(2009)02-0061-03

高相对分子质量的部分水解聚丙烯酰胺（HPAM）是化学驱中普遍用作驱油剂的关键高分子材料。HPAM 在淡水中具有较好的增粘行为，但是在盐水中，HPAM 的电负性被盐离子屏蔽，卷曲现象严重，增粘性能下降。当其水解度超过 40% 时，HPAM 与钙镁离子结合产生沉淀。事实上，此类聚合物自身存在剪切变稀、高温分解、高温水解和遇盐降粘的缺点，严重影响了高温高盐油藏聚合物的驱油效果和的使用效率，已经成为当前在高温高盐油藏推广开展聚合物驱油的技术瓶颈。因此，研究开发增粘性能好的耐温抗盐驱油剂迫在眉睫。针对这种现状，耐温抗盐驱油剂的开发研究一直比较活跃[1-5]。

笔者针对加州理工大学开发的一种具有热触变性的流体（TTF）开展了室内研究，对流体在不同高温老化时间的增粘性能、流变性能及作用机理进行了系统的评价与分析，以期探索高温高盐油藏提高采收率的新方法。

1 实验准备

1.1 主要实验药品及仪器

主要实验药品包括:热触变流体样品,主要由A剂和B剂2种成分组成,A剂主要成分为交联聚合物微球,B剂主要成分为一种特殊结构的表面活性剂,通过A剂和B剂相互作用,使流体表现出热触变特性;水解度为22%、相对分子质量约为$2\ 000\times10^4$的HPAM;氯化钠,分析纯;氢氧化钠,分析纯;氮气。

主要实验仪器包括:DV-Ⅲ旋转粘度计、MCR300高温高压流变仪、精度为0.001 g的电子天平。

1.2 实验样品的配制

热触变流体样品称取一定量的质量分数为2%的氯化钠水溶液,通氮除氧30 min,加入A剂搅拌2 h,然后在搅拌的同时慢慢滴入B剂,再搅拌1 h,最后用氢氧化钠溶液将溶液的pH值调至9.0,搅拌30 min,装瓶密封,放置在90℃下的烘箱中老化,然后定期测定该流体的粘度及流变性。

TTF与HPAM复配样品称取一定质量浓度的TTF溶液与一定质量浓度的HPAM溶液混合均匀后,装瓶密封,放置在90℃下的烘箱中老化,然后定期测定该流体的粘度及流变性。

1.3 实验样品的测试条件

测试温度均为75℃,在剪切速率为$7.34\ s^{-1}$的条件下测试粘度;在剪切速率为$0.1\sim400\ s^{-1}$的条件下测试剪切速率对粘度的影响关系;在振荡频率为1 Hz,振荡应力为0.5 Pa的条件下测试流体的粘弹性。

2 热触变流体增粘性能及热老化稳定性评价

测试并观察不同老化时间下的流体粘度和流体外观。实验结果表明(表1),TTF溶液的初始粘度很低,只有2.4 mPa/s,随着高温老化时间的延长,流体粘度逐渐增加,高温老化15 d时其流体粘度增加到23.6 mPa/s,之后流体粘度随高温老化时间的延长略有降低;刚配制的TTF溶液外观为乳白色不透明均匀液体,随着高温老化时间的延长,流体逐渐变成半透明的粘稠液体,最后还有少许沉淀生成。出现上述现象的主要原因是:① A剂聚合物微球随高温老化时间的延长其粒径逐渐变大,致使流体粘度逐渐增大;② B剂会在聚合物微球表面形成一种保护膜,防止聚合物微球之间的相互粘连沉淀,保证了流体的稳定性;③ 老化后期,随着聚合物微球粒径的增大,微球表面活性剂排列变稀,有些微球之间发生粘连沉降,导致流体粘度降低。

表1 热触变流体的粘度随高温老化时间的变化情况

老化时间/d	质量浓度为5 000 mg/L的TTF		TTF与HPAM复配后流体粘度/(mPa/s)
	粘度/(mPa/s)	外观	
0	2.4	乳白色不透明均匀乳液	22.1
3	8.7	粘稠乳白色半透明乳液	59.7
8	16.8	粘稠、半透明	48.1
15	23.6	粘稠、半透明	22.6
26	18.6	粘稠、半透明,有沉淀	18.2

注:TTF与HPAM复配后流体的总有效质量浓度为5 000 mg/L(TTF的质量浓度为4 500 mg/L,HPAM的质量浓度为500 mg/L)

热触变流体在高温条件下粘度不减反增的特点,可以有效地调节水油流度比;A剂逐级封堵坊[6-7]可大大提高驱替液的波及体积;B剂的界面活性,也可不同程度地降低油水界面张力,提高洗油效率。考虑上述3个特点,TTF流体在高温高盐油藏、非均质严重油藏扩大驱替液波及体积、堵水调剖和提高采收率方面可能具有很好的应用效果。

在保证总浓度不变的情况下,热触变流体与HPAM聚合物复配后,流体的初始粘度大大提高,达到了22.1 mPa/s,并且随着高温老化时间的延长流体粘度先增加至59.7 mPa/s,然后逐渐减小,当高温老化26 d时粘度还保持在18.2 mPa/s,与单独TTF溶液粘度基本相当。出现这种现象的原因可能是:线型高相对分子质量的HPAM利用其长的分子链把2个甚至多个聚合物微球串连起来,形成了“临时网状结构”,使其复配后的流体粘度大幅度提高。随着高温老化时间的延长,一方面聚合物微球逐渐膨胀,使流体粘度增加;另一方面由于高温下HPAM降解断链,使HPAM与聚合物微球之间形成的网状结构逐渐被破坏,导致流体粘度降低,通过2方面的相互作用,从而使复配后的流体既实现了较高的初始粘度,又保证了流体粘度的长期稳定性,从而使该流体在近井地带和远井地带都能起到较好的调驱效果。

3 热触变流体的流变性评价

由于驱油剂在地层多孔介质的运移过程中,时刻会受到岩石壁的剪切作用,并且这种剪切作用会对驱油剂的增粘性能造成不同程度的影响,同时夏惠芬等[8-10]研究发现不仅驱油剂的增粘性能影响驱油效率,而且其弹性也会对驱油效率产生不同程度的影响,所以采用流变表征方法研究了不同剪切速率对热触变流体粘度和粘弹特性的影响。

3.1 不同剪切速率对热触变流体粘度的影响

通过对不同高温老化时间热触变流体的流变曲线测试发现(图1,图2),对于单独使用的TTF溶液,随着高温老化时间的延长,粘度增加,并且流体粘度基本不受剪切速率的影响，表现出明显的牛顿流体特性；对于 TTF 与 HPAM 复配流体，老化初期流体粘度随剪切速率的增加逐渐降低，表现出明显的假塑性，随着高温老化时间的延长，流体粘度受剪切速率的影响越来越小，流体假塑性表现得越来越不明显。其原因是：复配后，HPAM 分子链与 TTF 溶液中的聚合物微球之间的相互作用形成一定数量的"临时网状结构"，高温老化初期，随着剪切速率的增大，这种"临时网状结构"逐渐被破坏，导致粘度逐渐降低，流体表现出假塑性；老化后期，随着 HPAM 的降解和分子链断裂，导致流体中"临时网状结构"逐渐消失，流体的假塑性也逐渐减小。

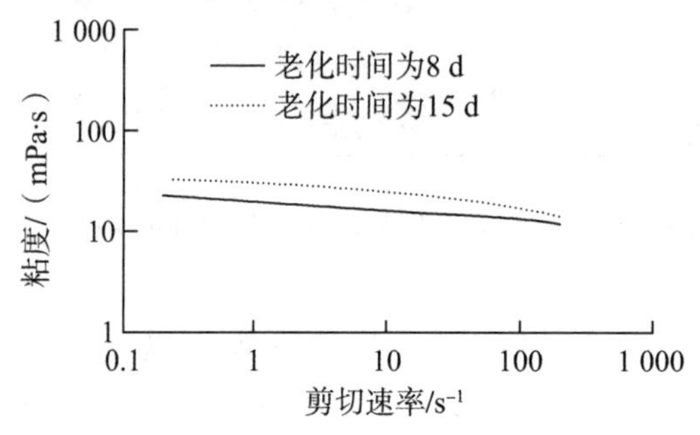

图 1 不同高温老化天数的 TTF 溶液流变性曲线

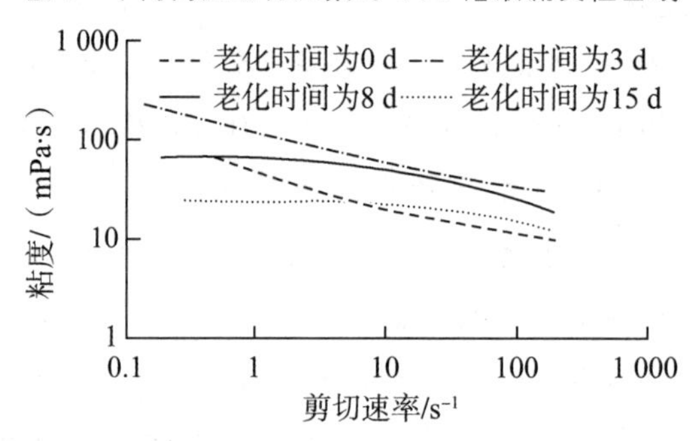

图 2 不同高温老化大数的 TTF 与 HPAM 复配流体流变性曲线

3.2 热触变流体粘弹性测试结果

流体的粘弹性一般通过动态振荡流变测试方法表征，主要测试参数为弹性模量(G')和粘性模量(G'')，弹性模量表征流体中的弹性大小，而粘性模量则表征流体的粘性特征。热触变流体粘弹性测试结果表明(表 2)，当 TTF 溶液单独使用时，高温老化初期，流体只具有很小的粘性，未能测出其弹性模量，而当老化 8 d 时，流体开始表现出一定的粘弹性，并且随着高温老化时间的延长，流体的粘性和弹性都增加；而对于 TTF 与 HPAM 复配后的流体，在高温老化初期就表现出明显的粘弹性，当老化 3 d 时，流体的粘性模量和弹性模量达到最大值，随后流体的粘弹性随老化时间的延长逐渐降低。通过综合比较不同老化时间下流体的粘性模量和弹性模量发现(表 2)，无论单独使用的 TTF 溶液还是 TTF 与 HPAM 复配后溶液，它们的粘性模量都大于弹性模量，因而该类流体表现出明显的粘性。

表 2 热触变流体粘弹性测试结果

老化时间/d	TTF		TTF 与 HPAM 复配后	
	G'/mPa	G''/mPa	G'/mPa	G''/mPa
0			24.3	183
3			89.5	366
8	4.8	180	74.0	310
15	6.2	217	19.5	187

4 结论

通过对热触变流体的室内评价结果发现，随着高温老化时间的延长该类流体的粘度都有不同程度的增加，流体表现出明显的热触变性和较强的耐温性，尤其是 TTF 与 HPAM 复配后，大大提高了流体的初始粘度，同时还表现出明显的热触变性和较好的粘度稳定性。流变测试结果表明，单独使用的 TTF 溶液的粘度受剪切速率影响不大，流体的粘性模量远远大于弹性模量，流体表现出明显的牛顿流体特性；而当 TTF 与 HPAM 复配后，流体具有假塑性和相对较高的弹性模量，且表现出粘弹性流体特征。

参考文献

[1] 曹宝格，李华斌，罗平亚，等.疏水缔合水溶性聚合物 AP4 清水溶液的流变特征 [J].油田化学，2005，22(2)：168-172.

[2] 程杰成，沈兴海，袁士义，等.新型梳形抗盐聚合物的流变性 [J].高分子材料科学与工程，2004，20(4)：119-121.

[3] 赵怀珍，吴肇亮，郑晓宇，等.水溶性交联聚合物微球的制备及性能 [J].精细化工，2002，22(1)：62-65.

[4] 马海霞，林梅钦，李明远.交联聚合物微球体系性质研究 [J].应用化工，2006，35(6)：453-455.

[5] 王涛，孙焕泉，肖建洪，等.孤岛油田东区 1-14 井组聚合物微球技术调驱矿场试验 [J].石油天然气学报，2005，27(6)：779-781.

[6] 王涛，肖建洪，孙焕泉，等.聚合物微球的粒径影响因素及封堵特性 [J].油气地质与采收率，2006，13(4)：80-82.

[7] 孙焕泉，王涛，肖建洪，等.新型聚合物微球逐级深部调剖技术 [J].油气地质与采收率，2006，13(4)：77-79.

[8] 夏惠芬，王德民，刘中春，等.粘弹性聚合物溶液提高微观驱油效率的机理研究 [J].石油学报，2001，22(4)：60-65.

[9] 张宏方,王德民,岳湘安,等.利用聚合物溶液提高驱油效率的实验研究 [J].石油学报,2004,25(2):55-58.

[10] 夏惠芬,王德民,关庆杰,等.聚合物溶液的粘弹性实验 [J].大庆石油学院学报,2002,26(2):105-108.

本文编辑 武云云

HanYugui, CaoXulong, ZhuYangwen et al. Research on the behavior of a new type thermal thixotropic fluid. PGRE, 2009,16(2):61～63

Because the viscosity of conventional polymer oil displacement agent decreases seriously under high temperature and high salt condition, the effect of polymer flooding has been reduced obviously in high temperature and high salt reservoirs. The performance of one new type thermal thixotropic fluid (TTF) was researched. Firstly the TTF solution was made up of 2% NaCl water, then the viscosity and thermostability of TTF solution were tested by DV-Ⅲ Bookfield viscometer under 90℃, and rheological property was tested by NCR300 rheometer and the extrinsic feature was recorded. The results in dicate that the viscosity of the TTF solution gradually increases with aging tim eextended, dis-playing the viscosity increasing behavior in high temperature. The effect of shear rate on the viscosity of TTF solution was weak, expressing definite Newtonian flow character, and the viscosity modulus of TTF solution was far bigger than elasticity modulus. The solution of TTF combined with HPAM was analyzed, test re-sults show that the combination of thermal thixotropic fluid and HPAM can obtain higher fluid viscosity, good thermal stability for long time, and may display definite viscoelastic attribute. This type thermal thixotropic fluid possibly have a good application prospect in high temperature and high salt reservoirs.

Keywords: thermal thixotropic fluid; conventional polymer; high temperature aging; rheological property

HanYugui, Geoscience Research Institute, Shengli Oilfield Com-pany, SINOPEC, Dongying City, Shandong Province, 257015, China

驱油用聚合物溶液的拉伸流变性能

韩玉贵 曹绪龙 宋新旺 赵 华 何冬月

(胜利油田分公司地质科学研究院,山东东营 257015)

摘要:测试不同类型聚合物在不同质量浓度、不同盐度条件下拉伸流变性能。结果表明:随着聚合物溶液质量浓度的增大其拉伸黏度也增大,随着溶液中 NaCl 盐度增加而拉伸黏度减小,但不同类型聚合物拉伸黏度受质量浓度和盐度影响的强弱不同。通过对聚合物拉伸流变性能的测试,可以分析溶液中聚合物分子的存在状态及作用机理,进而指导油田用聚合物的筛选、质量控制及配方设计等;聚合物溶液在长细管岩心中的渗流运移规律明显不同,拉伸流变性能对于聚合物溶液在地层中的渗流运移特性起重要作用。

关键词:拉伸流变;聚合物溶液;黏弹性;多孔介质;长细管岩心

中图分类号:TE341 文献标识码:A 文章编号:1000-1891(2011)02-0041-05

0 引言

早期的聚合物驱油机理认为聚合物的作用是通过增加驱替液流体黏度,降低水油流度比,增大波及体积,从而达到提高原油采收率的目的。张宏方等[1-6]研究提出聚合物驱不仅可以提高波及系数,而且还可以提高水波及域内的驱油效率。提高驱油效率的机理表现为:① 本体黏度可以改善水油流度比,扩大波及体积;② 油水界面黏度是聚合物溶液驱替膜状、孤岛状残余油的主要原因;③ 拉伸黏度使聚合物溶液存在弹性,是驱替盲端残余油及提高地下聚合物有效黏度的主要原因.尽管在这方面的认识尚不统一,但驱油用聚合物溶液流变性的重要程度显而易见。为此,人们研究聚合物溶液流变行为较多。特别关于聚合物溶液剪切流变性研究有较多报道[7-11],曹绪龙、窦立霞等[12-14]对聚合物油水界面流变性能也开展研究,得出超高分子 HPAM 溶液界面黏弹性比疏水缔合型大,且前者黏弹模量大于后者的认识,但到目前为止油田用聚合物溶液拉伸流变性能研究还未见报道。

笔者对不同类型驱油用聚合物的拉伸流变性进行研究,以期探索溶液中聚合物分子的存在状态、分子间作用力强弱及微观驱油机理,指导聚合物驱、二元复合驱中聚合物的筛选、质量控制和配方设计及新型聚合物开发研制等。

1 实验

1.1 材料及仪器

聚合物 HPAM-1,特性黏度为 1 546 mL/g,水解度为 23.3%,山东东营长安化工集团生产,制备过程中分子链上引进一些疏水单体,所以该聚合物在溶液中存在一定的疏水缔合作用;聚合物 HPAM-2,特性黏度为 2 592 mL/g,水解度为 23.5%,日本进口;氯化钠,分析纯,天津市盛大化工销售有限公司生产。

HAAK CaBRE1 拉伸流变仪,美国热电公司生产;MCR300 高温高压流变仪,奥地利安东帕公司生产。

1.2 流变性测试

1.2.1 条件

拉伸流变测试条件:温度为 30℃,测量板直径为 6.0 mm,样品初始高度为 3.0 mm,样品最终高度为 15.0 mm;剪切表观黏度测试条件:温度为 30℃,剪切速率为 7.34 s^{-1}。

1.2.2 过程

分别用蒸馏水配制质量浓度为 500 mg/L,1 000 mg/L,1 500 mg/L,3 000 mg/L 的聚合物溶液,考察质量浓度对不同聚合物流变性的影响;分别用氯化钠质量分数为 0,0.5%,1.0%,1.5%,2.0%溶液配制质量浓度为 1 500 mg/L 聚合物溶液,考察盐度(质量分数)对不同聚合物流变性的影响;分别用十二烷基磺酸钠质量分数为 0,0.05%,0.10%,0.15%,0.20%溶液配制质量浓度为 1 500 mg/L 聚合物溶液,考察表面活性剂对不同聚合物流变性的影响。

1.3 岩心驱替

长细管岩心长度为 16 m,内径为 0.7 mm,均匀分布 5 个测压点,进口、出口各 1 个测压点,渗透率为 20 μm^2,孔隙度为 35%。先用 2 mg/L NaCl 水溶液饱和长细管岩心,然后利用 1 500 mg/L 聚合物溶液以 0.25 mL/min 注入长细管岩心,检测不同测压口的压力,实验温度为 75℃。

2 结果与讨论

2.1 质量浓度

2 种不同质量浓度聚合物拉伸黏度应变曲线见图 1。由图 1 可知,2 种聚合物拉伸黏度随其质量浓度的增加而逐渐增大,但不同聚合物的拉伸黏度随质量浓度变化的幅度不同。分别取应变为 10 时所对应的拉伸黏度为基础,比较质量浓度对不同聚合物拉伸剪切表观黏度影响关系,见图 2。由图 2(a)可知,HPAM-1 聚合物随质量浓度增加其拉伸黏度迅速增大,并且存在一个明显拐点,而 HPAM-2 的只是缓慢增加,并没有明显拐点。

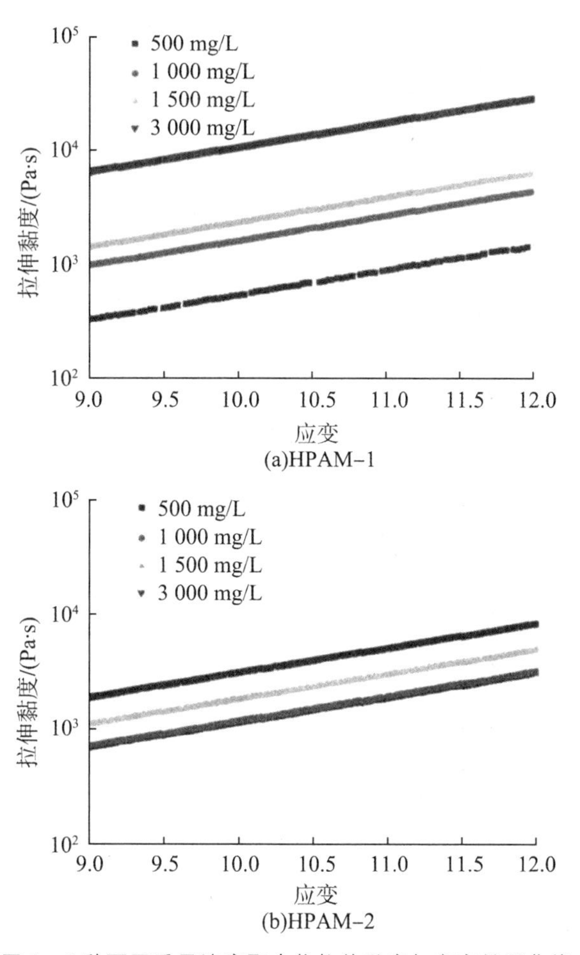

图 1　2 种不同质量浓度聚合物拉伸黏度与应变关系曲线

由图 2(b)可知，随着质量浓度增加 2 种聚合物的剪切表观黏度没有出现明显的拐点，主要是由于溶液中聚合物分子间作用机理和存在状态不同引起的。就溶液间分子间作用力大小而言，一般分子间的疏水缔合作用大于分子之间的缠结作用，并且随着质量浓度增加，聚合物分子间的疏水缔合作用力迅速增大，而分子间的缠结作用没有较大增加，所以在破坏溶液中聚合物网状结构的低剪切速率下，疏水缔合或分子缠结作用而形成的网状结构造成聚合物动力学体积增大，对剪切黏度的贡献是等效的，聚合物相对分子质量越高，分子间越易发生缠结使其水动力学尺寸增大，剪切表观黏度也就越大；涉及破坏溶液中聚合物网状结构的拉伸流变测试时，聚合物分子间作用力更强、网状结构更牢固，其拉伸黏度就越大、弹性越强。

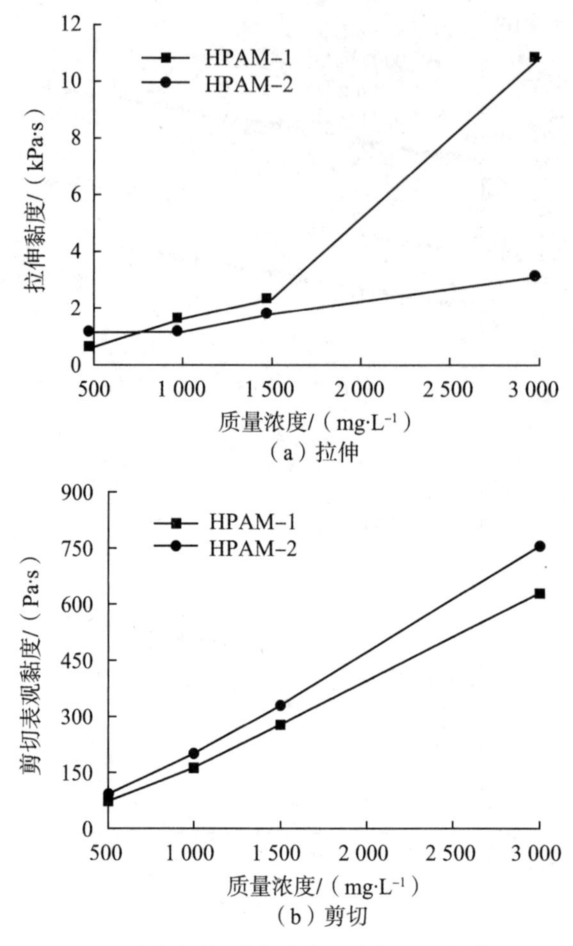

图 2　不同聚合物拉伸、剪切表观黏度与质量浓度关系曲线

通过不同质量浓度聚合物拉伸流变性能测试，一方面可以了解单个聚合物在不同质量浓度条件下弹性大小，进而可以分析溶液中分子存在状态及作用机理，为聚合物驱微观机理解释、驱油配方设计提供直接证据；另一方面可以对不同聚合物类型进行鉴别。由 2 种黏度测试结果证明，通过拉伸流变性能测试可以更全面、客观了解聚合物溶液的性能，为油田用聚合物的筛选、质量控制、新型聚合物开发提供技术指导。

2.2 盐度

盐度对不同聚合物流变性影响的测试结果见图 3。由图 3 可知，随着水中 NaCl 质量分数的增加，2 种聚合物的拉伸黏度和松弛时间减小，但不同聚合物拉伸流变性受盐度影响强弱不同，这也是由聚合物类型不同导致在溶液中分子的存在状态及作用机理不同引起的。对于 HPAM-1 聚合物，溶液中临时动态网状结构的形成是分子间缔合和分子间缠结共同作用的结果，随着溶液中盐度的增加，缠结作用大幅度减少，而分子间缔合作用能够缓冲盐对溶液中临时动态网状结构的破坏。由于分子链的卷曲，分子间缔合的概率逐渐减小，而分子内缔合概率增加，所以在盐度为 1.0％时，是由分子间缔合作用为主向分子内缔合作用为主转化的临界点。

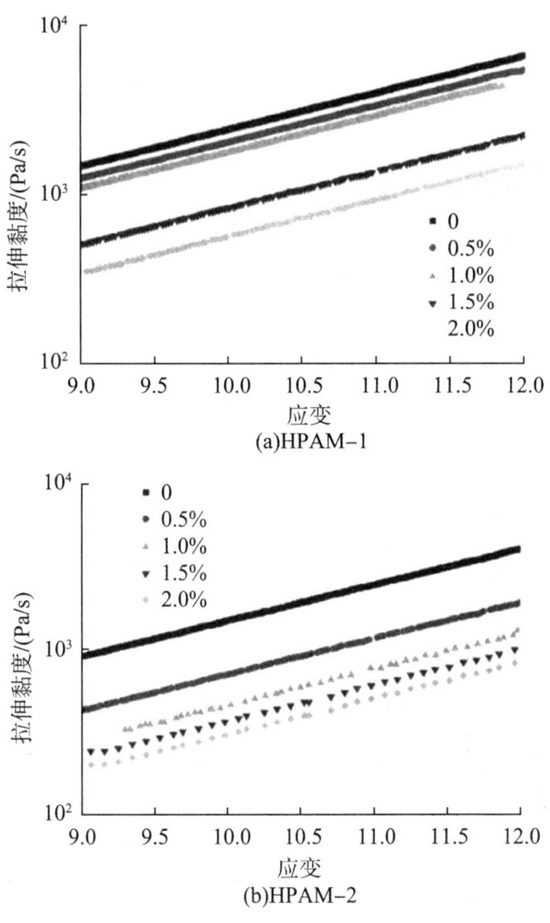

图 3 水溶液中 NaCl 盐度对聚合物拉伸黏度的影响

2 种聚合物拉伸黏度和剪切表观黏度随溶液中盐度变化的影响关系见图 4。由图 4 可知，2 种聚合物的拉伸黏度和剪切表观黏度受溶液中盐度的影响趋势基本相同，只是盐溶液中 HPAM-1 聚合物的拉伸黏度和剪切表观黏度大于 HPAM-2 的，原因是 HPAM-1 聚合物分子间缔合作用能够减小盐对溶液中聚合物临时动态网状结构的破坏。

2.3 注入孔隙体积倍数

HPAM-1 和 HPAM-2 聚合物在长细管岩心运移时的沿程压力变化曲线见图 5。由图 5 可知，通过驱替过程中长细管岩心沿程压力变化可以了解聚合物溶液在多孔介质中的渗流运移规律，HPAM-1 聚合物溶液的注入造成入口(P1)和出口 1(P2)处压力的急剧增加，后面的出口(P3，P4，P5，P6)较低，而 HPAM-2 聚合物溶液注入时，入口、出口 1 至出口 5 处的压力缓慢递减，这种现象的原因也是溶液中聚合物分子存在状态及分子间的作用大小不同造成的。

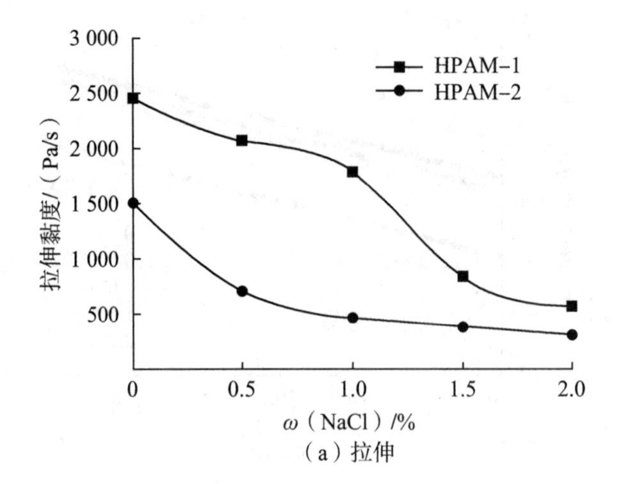

(a)拉伸

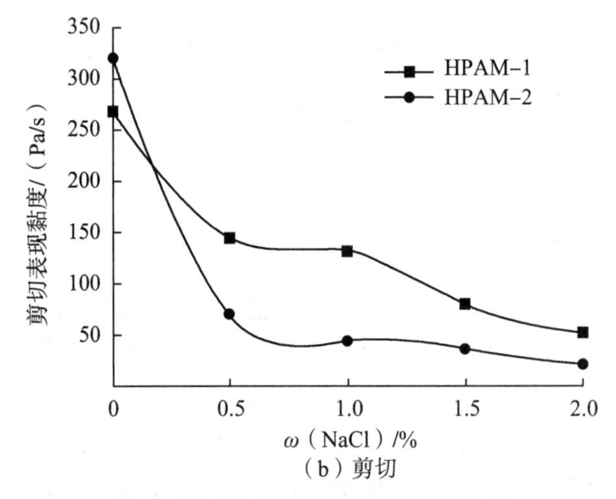

(b)剪切

图 4　不同聚合物黏度与盐度关系曲线

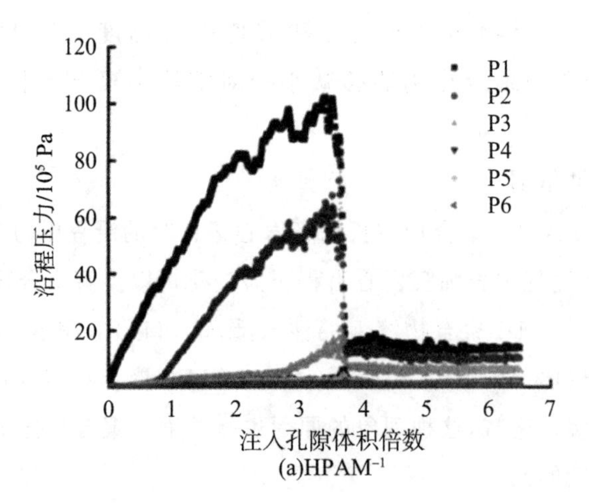

(a)HPAM-1

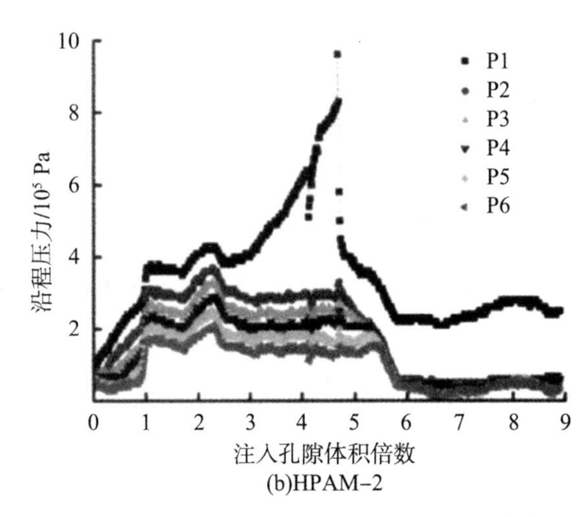

(b)HPAM-2

图 5　2 种聚合物在长细管岩心中运移时的沿程压力变化曲线

3　结束语

通过对 HPAM-1 和 HPAM-2 两种类型聚合物，在不同质量浓度、不同盐度条件下拉伸流变的评价和在长细管岩心中运移渗流规律的研究，发现溶液中聚合物分子间作用力对拉伸黏度起重要作用，因为分子间作用力决定分子间形成的临时交联网状结构的牢固程度，也同时反映其溶液弹性的大小。通过比较聚合物剪切表观黏度测试结果，拉伸流变能够更客观地反映溶液中聚合物分子存在状态及作用大小，同时反映溶液的弹性强弱，而溶液弹性强弱直接影响在多孔介质中的运移情况。通过拉伸流变性能测试可以更全面了解聚合物溶液性能，为微观驱油机理解释、油田用聚合物的筛选、质量控制、新型聚合物开发提供技术指导。

参考文献

[1] 张宏方，王德民，岳湘安，等. 利用聚合物溶液提高驱油效率的实验研究 [J]. 石油学报，2004，25(2)：55-59.

[2] 夏惠芬，王德民，侯吉瑞，等. 聚合物溶液的黏弹性对驱油效率的影响 [J]. 大庆石油学院学报，2002，26(2)：109-12.

[3] 夏惠芬，王德民，刘中春，等. 黏弹性聚合物溶液提高微观驱油效率的机理研究 [J]. 石油学报，2001，22(4)：60-5.

[4] 张宏方，王德民，王立军，等. 聚合物溶液在多孔介质中的渗流规律及其提高驱油效率的机理 [J]. 大庆石油地质与开发，2002，21(4)：57-61.

[5] 王立军. 聚合物溶液黏弹性对提高驱油效率的作用 [D]. 大庆：大庆石油学院，2003.

[6] 张宏方，王德民，岳湘安，等. 不同类型聚合物溶液对采油残余油的作用机理研

究 [J]. 高分子学报,2003(3):321-325.

[7] 曹宝格,罗平亚,李华斌,等. 疏水缔合聚合物溶液黏弹性及流变性研究 [J]. 石油学报,2006,27(1):85-88.

[8] 张星,李兆敏,孙仁远,等. 聚合物流变性实验研究 [J]. 新疆石油地质,2006,27(2):197-199.

[9] 欧阳坚,孙广华,王贵江,等. 耐温抗盐聚合物 TS-45 流变性及驱油效率研究 [J]. 油田化学,2004,21(4):330-333.

[10] 戴玉华,吴飞鹏,李妙贞,等. 新型缔合聚合物 P(AM/POEA)溶液的流变性质 [J]. 高分子材料科学与工程,2005,21(3):121-124.

[11] 程杰成,沈兴海,袁士义,等. 新型梳形抗盐聚合物的流变性 [J]. 高分子材料科学与工程,2004,20(4):119-121.

[12] CaoXulong, LiYang, JiangShengxiang, etal. A study of dilation alrheolo gical proper tiesof poly mersa tinter faces [J]. Journalof ColloidandInterfaceScience, 2004, 270(2):295-298.

[13] 窦立霞,曹绪龙,江小芳,等. 驱油用聚丙烯酰胺溶液界面特性研究 [J]. 胶体与聚合物,2004,22(2):14-6.

[14] 窦立霞,郭龙,王红艳,等. 油/聚合物溶液体系动态界面扩张模量的测量 [J]. 油田化学,2004,21(1):72-74.

Responsive wetting transition on superhydrophobic surfaces with sparsely grafted polymer brushes

Liu Xinjie[ac] Ye Qian[a] Song Xinwang[b] Zhu Yangwen[b] Cao Xulong[b]
Liang Yongmin[a] Zhou Feng*[a]

We demonstrate here that surface wetting transitions and contact angle hysteresis can be significantly altered by manipulating the droplet-surface interaction, which has never been reported before. The dynamic wetting behavior of a pressed water droplet on responsive polymer brushes-modified anodized alumina with pre-modified dilute initiator is shown. The wetting transition between superhydrophobicity and hydrophilicity can or cannot be achieved depending on the responsiveness between droplets of different pH, the concentrations of electrolytes and the environmental temperature and surface grafted stimuli-responsive polymer brushes. The contact angle changes are rather apparent, giving the surface double-faced wetting characteristics. The responsive surface composition regulated wetting will be very useful in understanding wetting theory, and will be helpful experimentally in designing smart surfaces in, for example, microfluidic devices.

Introduction

Wettability is an important property of solid surfaces and is mainly determined by two factors, i. e. the surface topography and chemical composition.[1-12] Superhydrophobic surfaces,[3-9,13-18] with a contact angle (CA) larger than 150°, have attracted much attention due to their practical applications.[19] Superhydrophobic surfaces can usually be achieved by, on the one hand, forming hierarchical micro/nanoscale binary structures, and by using chemical modification with low surface energy materials on the other. Two models, the Wenzel model[1] and the Cassie-Baxter model[2] are most frequently used to account for experimental results. The Wenzel model describes wetting behavior where a liquid droplet on a rough surface is in intimate contact with surface asperities, while the Cassie model describes states where the droplet sits on a solid-air composite surface.[20-25] The key difference between the two regimes is the contact angle hysteresis, i. e. the contact angle hysteresis of the Cassie wetting regime is much lower than that of the Wenzel wetting regime. The underlying trapped air in a Cassie state can be eliminated when a pressure (impact) is applied, resulting in a transition to the Wenzel state; we call this a "metastable" Cassie state. Therefore, the contact mode depends on the way liquid drops are deposited, for example, dragging over a rough surface[26] or being impacted can lead to such a transition.[27] The impacting action of a droplet on a superhydrophobic surface has been studied by several groups.[28-37] Lee et al. confirmed that the impacting action between water droplets and the surface depends on the geometric parameters of the surface and only some certain geometries could result in the robust Cassie regime.[27] Bhushan et al. have developed the criterion where the transition from the Cassie regime to the Wenzel regime was determined by three factors: impact velocity, geometric parameters (length, width and height of patterns, pitch between patterns) of patterned surfaces and liquid property.[29] Generally, the above-mentioned studies are focused on the effect of geometric parameters of surfaces on wetting mode.[38] As will be shown in the present work, surface composition in low fractions might not affect the surface wettability, but the surface wetting transitions and contact angle hysteresis can be significantly altered by manipulating the droplet-surface interaction, which has never been reported before. We demonstrate the dynamic wetting behavior of a pressed water droplet on responsive polymer brushes-modified anodized alumina with premodified dilute initiator.[39] The wetting transition between superhydrophobicity and hydrophilicity can or cannot be achieved depending on the responsiveness between droplets of different pH, the concentrations of electrolytes and the environmental temperature and surface grafted stimuli-responsive polymer brushes. The CA changes are rather apparent, giving the surface double-faced wetting characteristics.

Results and discussion Preparation of surfaces

Scheme 1 gives a general description of the procedures for preparing the superhydrophobic surfaces on anodized aluminium oxide substrates modified with polymer brushes. After the anodized alumina substrates are hydroxylated by oxygen plasma, they are immersed in the mixed silane solution including 3-(trichlorosilyl) propyl-2-bromo-2-methyl propanoate (initiator) and 1H,1 H,2 H,2 H-perfluorooctyl trichlorosilane (PFOTS). The mixed monolayers (including covalent bonding initiator and PFOTS) are formed on the anodized alumina substrates, and this course is shown in Scheme 1a. Then a typical surface-initiated atom transfer radical polymerization (SI-ATRP) of (dimethylamino) ethyl methacrylate (DMAEMA) or N-isopropylacrylamide (NIPAm) is performed, therefore sparse polymer brushes (poly (N-isopropylacrylamide), PNIPAm and poly(dimethylamino) ethyl methacrylate, PDMAEMA) are grafted on the positions where initiators originally existed as shown in Scheme 1 b, c.

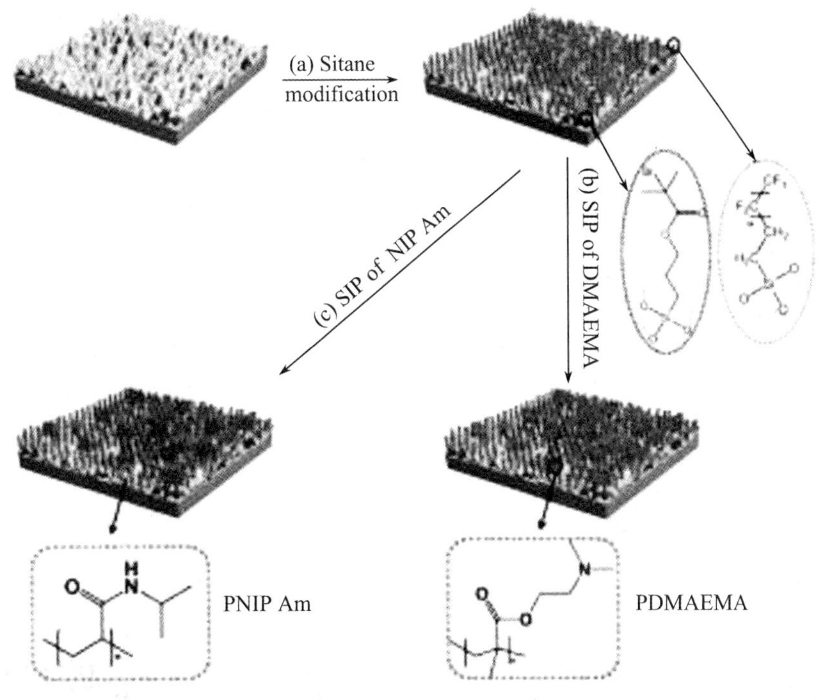

Scheme 1　Schematic illustration of grafting PDMAEMA/ PNIP Am macromolecular anchors on initiator/PFOTS modified anodized alumina substrates.

Fig. 1 a shows a FE-SEM image of the as-prepared anodized alumina surface with nanopore structures. However, the nanopore structures were entirely covered with the PDMAEMA polymer films after the initiator/PFOTS (1 : 5)-modified sample was grafted by PDMAEMA as shown in Fig. 1b,f. With increasing content of PFOTS (In/PF=1 : 10 or 1 : 20), the corresponding grafted PDMAEMA film becomes much thinner than that resulting from higher initiator content (i. e. In/PF=1 : 5), because the competitive growth between initiator and PFOTS leads to a reduction in the initiator assembled on the substrate, and the nanopore structures change from entirely buried (In/ PF=1 : 5 Fig. 1b,f) to partially blocked (Fig. 1c,g) or almost as in the case of the freshly prepared anodized alumina (Fig. 1d,h).

Pressure induced pH dependent wetting transition

Responsive polymer brushes may assume stretched or collapsed conformations when interacting with certain droplets. As shown in Scheme 2, taking PDMAEMA grafted surfaces as one example, if an acid droplet is dropped on the surface, PDMAEMA brushes in the contact area will be protonated, because the amine groups gain protons in acidic conditions and deprotonate in basic conditions (the pK_a of PDMAEMA is 7.5) and will show the stretched state at the molecular level at low pH.40,41 When the droplet is pressed by a superhydrophobic plate (that has low adhesion to the droplet) from above, the droplet will expand, rendering an increased contact area and a greater number of polymer chains stretching off. After release of pressure, the ability of the droplet to round up depends on the equilibrium between the solvation force and the cohesive force of the liquid drop. Macroscopically, the acid droplet may convert from a superhydrophobic/hydrophobic state to a less hydrophobic, even hydrophilic state. On the other hand, if a base droplet is dropped on the substrate, the underlying PDMAEMA will assume a deprotonated collapsed state, so even if the base droplets are pressed as shown in Scheme 2 b, the CAs will change very little and the droplet will return to its original shape. Scheme 2 c, d are the corresponding digital images of acid and base droplets on the PDMAEMA grafted anodized alumina from initiator/ PFOTS= 1 : 8, and the result is in good agreement with the mechanism mentioned in Scheme 2 a, b. In a word, the surface shows double faced wettability to acid droplets, and relatively stable wettability to base droplets.

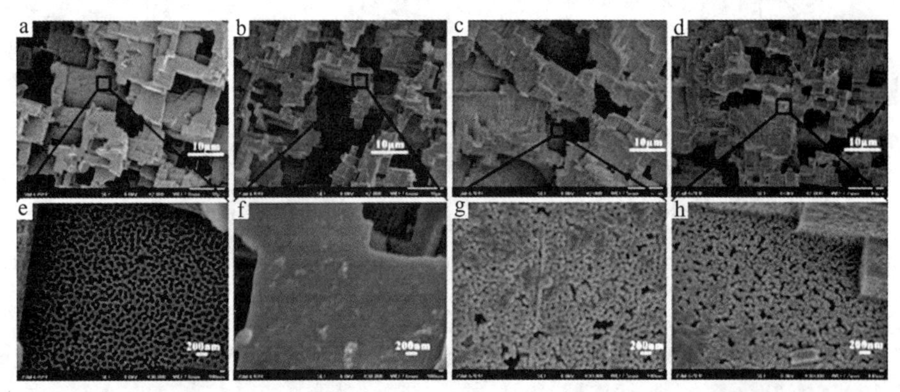

Fig. 1 FE-SEM images of (a) anodized alumina substrate and PDMAEMA grafted Al_2O_3 substrate from (b) initiator/PFOTS=1 : 5, (c) initiator/PFOTS=1 : 10 and (d) initiator/PFOTS=1 : 20; (e), (f), (g) and (h) are the corresponding magnified images respectively.

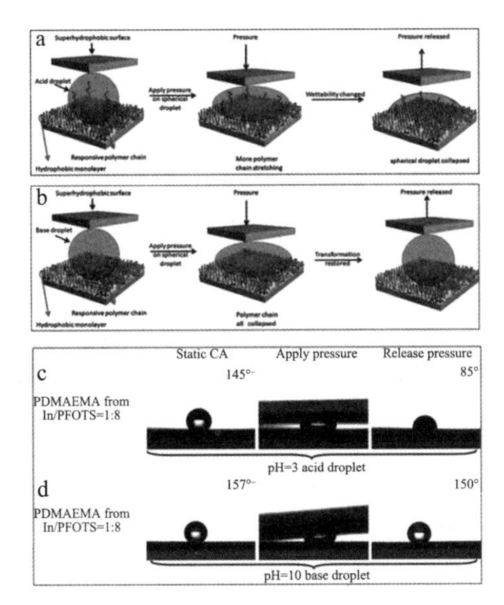

Scheme 2 Schematic illustration of pressure induced (a) acid droplet wettability changes from superhydrophobicity to hydrophilicity and (b) base droplet wettability unchanged on the PDMAEMA macromolecular anchors grafted anodized alumina substrates; (c) and (d) are the corresponding digital images of acid droplet (pH = 3) and base droplet (pH = 10) on the PDMAEMA grafted anodized alumina (from initiator/ PFOTS=1 : 8) before and after application of external pressure.

In order to prove the hypothesis, a detailed investigation was carried out. Firstly, the pressure induced wettability changes of the pH responsive polymer (PDMAEMA) modified sample are studied, and acid droplets (pH=3) and base droplets (pH=10) were used as probing liquids. A certain pressure (lager than 250 Pa) is applied on the pressed droplet, and it can be simply calculated by the Laplace equation ($\Delta P = 4\gamma |\cos\theta|/d, d < 1$ mm, d is the distance between the superhydrophobic plate with ultra-low adhesion and the polymer grafted substrate when the external pressure is applied, γ is the surface tension of the used droplet, and θ, which is close to 180°, is the critical contact angle at the final position when the pressure is applied) . The static contact angle of acid droplets on pure PDMAEMA grafted anodized alumina is about 72°, so the surface is intrinsically hydrophilic, because of the stretched conformation of PDMAEMA chains in an acidic environment. The relatively high CA value is caused by surface roughness. However, if an external pressure was applied on the acid droplet, the acid droplet wouldn't round up again, but spread out and show superhydrophilicity (CA<5°) as shown in Fig.2 a. The result is in agreement with our hypothesis, because the pressing action induces a greater number of polymer chains to change from a collapsed state to a stretched state, with the increased contact area and the greater number of stretched polymer chains helping to capture the droplet. The PDMAEMA grafted sample from mixed monolayers of the initiator/ PFOTS modified sample shows the same trend, but the difference is that the water repellent action of the PFOTS constituent enhances the hydrophobic properties of the sample surface. For the sample from initiator/PFOTS=1 : 5 , the static contact angle of the acid droplet is about 136°, but it decreases to about 77° after the external pressure is applied. With increasing amounts of PFOTS in the bulk assembly solution, such as the sample from initiator/PFOTS=1 : 20 , the contact angle decreases only from 163° to 132° after pressure is applied on acid droplets. From the above mentioned results, it is seen that the maximum difference of CAs before and after applying pressure is about 70° for pure polymer grafted surfaces, and the minimum difference value is about 30° for that grafted from the initiator/PFOTS=1 : 20 surface. The wettability change is rightly induced by interplay between surface chemistry and the droplet environment.

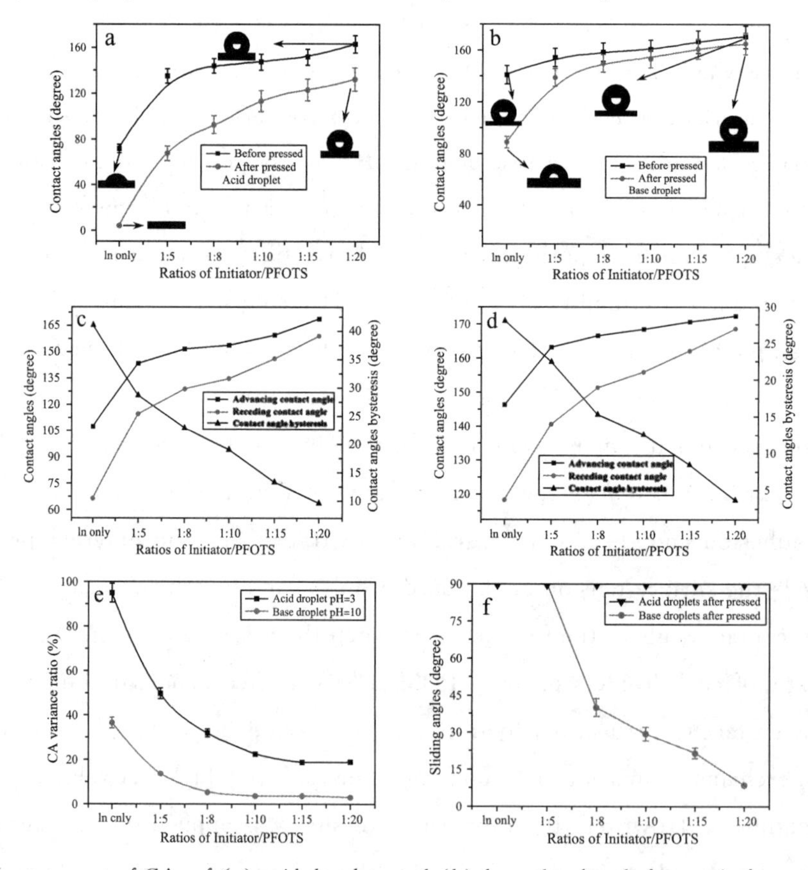

Fig. 2 Measurement of CAs of (a) acid droplets and (b) base droplets before and after pressure was applied; advancing, receding contact angles, and contact angle hysteresis of (c) acid droplets and (d) base droplets before pressing; (e) the corresponding CA variance of acid and base droplets before and after pressure was applied; (f) Sliding angle of acid droplets and base droplets after pressure was applied on PDMAEMA grafted anodized alumina substrates.

To further prove this, the wetting behavior of base droplets under pressure was studied. For the substrate from pure polymer brush-modified sample, the static CA of base droplets is as high as about 141° (Fig. 2 b), because PDMAEMA macromolecules assume a collapsed chain conformation under basic conditions. The static CA would decrease to about 90° after pressure is applied, and this is due to the transition from the Cassie model to the Wenzel model, because the action of pressure physically extrudes the air cushion layer between the base droplet and the solid surface. However, when PFOTS is introduced into the sample, the base droplet becomes more nonwettable than the sample grafted from PDMAEMA only. For samples with a higher fluorine content (Initiator/PFOTS=1：8, 1：10, 1：15 or 1：20), the CA changes of the base droplet before and after pressure are even less than 8°. Therefore, increased fluorine content on the surface leads to a more stable Cassie wetting model, which is in accordance with literature reports.[42] Polymer brushes contribute less to wetting transitions since no strong interaction between polymer and droplet occurs. Fig.2c, d are the corresponding advancing contact angle, receding contact angle and contact angle hysteresis of acid and base droplets before pressing. It can be seen that the hysteresis decreases along with increasing fluorine content and the values for acid droplets are all larger than basic droplets. After pressing, acid droplets will stick onto the surfaces and hysteresis can no longer be measured. The comparison of CA variance of acid and base droplets before and after pressure is applied is summarized in Fig.2e, f. It is seen that the CA variance of acid droplets shows a much larger decrease than that of base droplets. In contrast to acid droplets, base droplets could still slide off the sample surface even with low fluorine contents (beginning from initiator/ PFOTS=1：8) after they are pressed, but the acid droplets could not slide off any sample surfaces. The adhesion of acid droplets and the sliding of base droplets after being pressed proves our hypothesis that PDMAEMA grafted samples possess two totally different wetting behaviors for acid and base droplets under pressure due to the pH response of PDMAEMA chains.

Dye staining is a good way to display the contact between droplet and surface. A typical water-soluble dye (Rhodamine B) mixed in acid and base droplets was used to probe the wetting behavior. Fig.3a shows typical fluorescence images of PDMAEMA grafted samples after contact with droplet and pressing. The yellow color indicates dye staining and the green color shows the non-wetted area. From Fig.3a, we can see that the same sample is much more easily wetted by acid droplets than by base droplets, and the sample becomes much harder to wet by both acid and base droplets with increasing fluorine content. The corresponding percentage of normalized yellow fluorescence distributed in the whole images was summarized in Fig. 3b, and the calculated result is in very good agreement with CA measurement. The above mentioned results provide strong evidence for the double faced wettability of PDMAEMA grafted samples.

Pressure induced temperature dependent wetting transition

A similar philosophy can be extended to other responsive polymer systems that can bring about similar wetting behavior but under different conditions. PNIPAm is a well-known tempera-ture-sensitive polymer with a low critical solution temperature (LCST) of about 32℃. Surfaces coated with PNIPAm switch from hydrophilic below the LCST to hydrophobic above the LCST.[43] We demonstrate here that the PNIPAm grafted sample from different ratios of initiator/PFOTS could also possess double faced wettability by regulating environmental temperature. Fig. 4 shows the corresponding FE-SEM images, and we can see that the grafting of PNIPAm also blocks the nanopore structures to some extent (Fig. 4b), while increasing fluorine content lowers the grafting intensity of PNIPAm (no obvious blocking is found, Fig. 4c). The corresponding advancing contact angle, receding contact angle and contact angle hysteresis of water droplets at 25℃ and 40℃ are summarized in Table 1. It is seen that contact angle hysteresis is much higher at 25℃ than at 40℃ for respective surfaces, indicating virtual interaction between PNIPAm chains and droplets at low temperature, while at 40℃ the interaction becomes trivial.

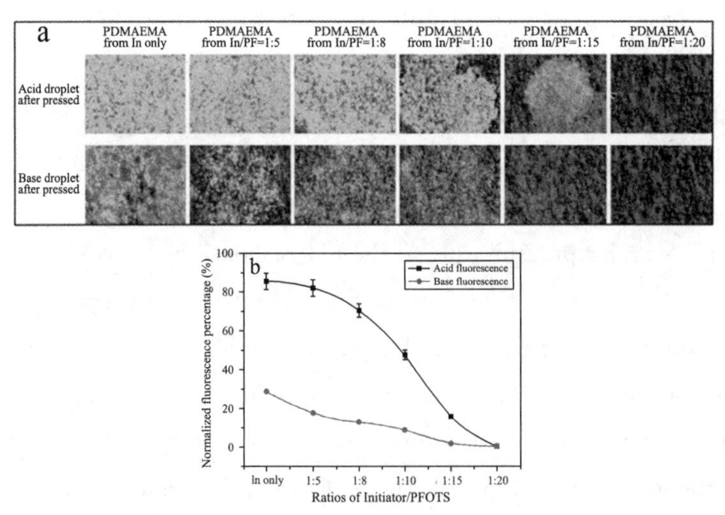

Fig. 3 (a) Fluorescence spectra of acid droplets and base droplets after application of pressure on PDMAEMA grafted anodized alumina substrates; (b) Corresponding fluorescence spectra area pixel percentage as a function of the whole picture pixel (calculated by ImageJ).

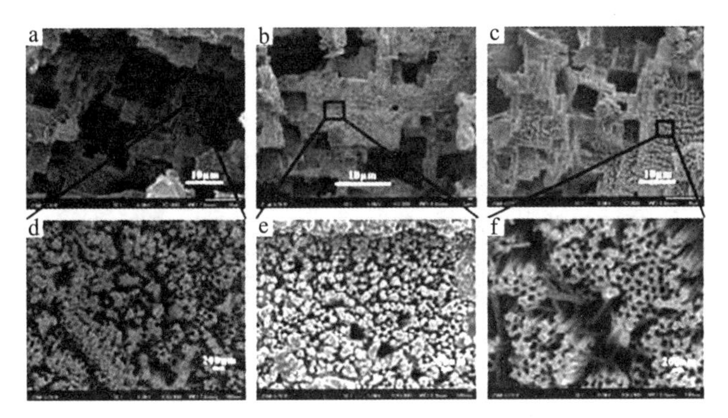

Fig. 4　FE-SEM images of (a) anodized alumina substrate, and PNIPAm grafted Al_2O_3 substrate (b) from initiator/PFOTS= 1 : 10; (c) from initiator/PFOTS=1 : 20; (d), (e), (f) are the respective magnifi ed images.

Fig.5 shows the CA changes of water droplets on the PNIPAm brushes grafted sample under external pressure. Water droplets maintain superhydrophobic states (CA >150°) with the help of pre-modified PFOTS on these PNIPAm grafted samples, while the pure PNIPAm grafted sample without PFOTS shows superhydrophilicity at 25℃ (<LCST), because PNIPAm brushes form intermolecular hydrogen bonds with water molecules. Mixing polymer brushes and PFOTS will change the surface wetting behavior. As is shown in Fig. 5, the surface grafted from initiator/PFOTS=1 : 10 changes from superhydrophobic (CA 153°) to hydrophilic (48°) before and after pressing at 25℃, because more PNIPAm chains will show stretched conformations (more hydrophilic) under the action of external pressure. Increasing the content of PFOTS results in decreased CA changes before and after pressing, 11° and 1° respectively for initiator/PFOTS=1 : 15 or 1 : 20. Pressing droplets on the same PNIPAm grafted samples from different ratios of initiator/ PFOTS at 40℃ (>LCST), CAs of water droplets change only a little and all droplets can round up again. PNIPAm chains maintain a collapsed state above 40℃ and solvation forces will not exist. Thus, even if pressure is applied, water droplets remain superhydrophobic (CA >150°) with low adhesion states (sliding angle smaller than 5°) . In a word, the PNIPAm grafted surface from initiator/ PFOTS=1 : 10 shows obvious double faced wettability below the LCST and grafted polymer chains significantly contribute to the wetting transition.

Table 1 Advancing contact angle (θ_{adv}), receding contact angle (θ_{red}), and contact angle hysteresis (CAH) of PNIPAm grafted samples at 25℃ and 40℃, respectively

PNIPAm grafted sample	25℃			40℃		
	θ_{adv}	θ_{red}	CAH	θ_{adv}	θ_{red}	CAH
From In/PFOTS=1∶10	161.2	129.1	32.1	166.7	154.3	12.4
From In/PFOTS=1∶15	169.0	156.4	12.6	171.5	164.3	7.2
From In/PFOTS=1∶20	170.1	165.4	4.7	174.5	171.2	3.2

Pressure induced electrolyte dependent wetting transition

PDMAEMA can be quaternized to give cationic polymer brushes (Q-PDMAEMA), the solubility of which are determined by the interaction of the counterions with the polyelectrolyte backbone.[44-48] Nitrate counteranions (NO_3^-) lead to hydrophilic polymer chains, whereas Tf_2N^- [$(CF_3SO_2)_2N^-$] shows strong ion-pairing with the quaternary ammonium groups and hence hydrophobic polymer chains. The strong ion-pairing interaction is very stable in pure water and low ionic strength solutions, but can be broken in solutions of high ionic strength, for example 1 M KCl or $NaNO_3$.[46,49] The corresponding advancing contact angle, receding contact angle and contact angle hysteresis of water droplets containing 1 M and 1 mM $NaNO_3$ are summarized in Table 2. It is seen that contact angle hysteresis is much higher for droplets with 1 M $NaNO_3$ than for those with 1 mM $NaNO_3$ for respective surfaces, indicating virtual interaction between Q-PDMAEMA chains with the droplet containing 1 M $NaNO_3$, due to anion exchange at high salt concentration inducing water affinity in the polymer chains, while the anion exchange won't occur when the salt concentration of the solution is low.

Fig. 6a describes the dynamic wetting behavior of 1 mM and 1 M $NaNO_3$ droplets on the Q-PDMAEMA-Tf_2N. CAs of water droplets and 1 mM $NaNO_3$ droplets show a tiny decrease (still larger than 165°), but the CAs of 1 M $NaNO_3$ droplets decrease from about 170° to about 144°, indicating that 1 M $NaNO_3$ droplets wet the substrate much more easily than water droplets and 1 mM $NaNO_3$ droplets due to *on-site* ion exchange interactions. This phenomenon strongly confirms that 1 M $NaNO_3$ droplets carry out more effective anion exchange than 1 mM $NaNO_3$ droplets on the sample surface. Based on this mechanism, we also investigated the wetting behaviors of 1 mM $NaNO_3$ and 1 M $NaNO_3$ droplets on the Q-PDMAEMA (Tf_2N^- as anion) grafted substrates which were previously modified by different ratios of initiator/PFOTS. Fig. 6 b shows the CA variance of 1 mM $NaNO_3$ and 1 M $NaNO_3$ droplets before and after pressing, and the CA variance of 1 M $NaNO_3$ droplets shows a much bigger change than that of 1 mM $NaNO_3$ droplets.

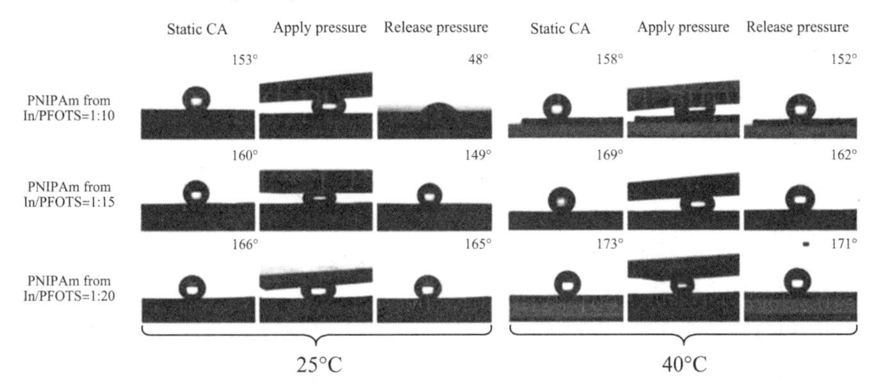

Fig. 5 Corresponding optical images of wettability changes of PNIPAm (from different ratios of In/PFOTS) grafted anodized alumina substrates at 25℃ and at 40℃.

Mechanism analysis

Surface wetting transition from Cassie to Wenzel states can be realized by several means, such as the application of pressure on the droplet, the droplet impacting the surface with a certain speed or by vibration. Extrand[50] indicated that the droplet weight and the contact line density are two key factors influencing the transition that will occur when the droplet weight is larger than the surface tension on the triple line. Lafuma and Quere[25] stated that transition would take place when the contact angle predicted by the Wenzel equation is equal to that predicted by the Cassie equation, and the nature of the transition is the interfacial energy of the Wenzel state equaling that of the Cassie state. Patankar[51] claimed that the wetting state of a droplet depends typically on how the droplet is formed. Nosonovsky and Bhushan[52] suggested that the transition is a dynamic process and is regulated by the surface geometric parameters.[53] The above mentioned wetting transition depends either on the way the droplet is put onto a surface or on surface geometry. It is also possible that surface composition will result in a wetting transition and therefore hysteresis, without affecting surface wetting property. In the present work, we have designed such surfaces with sparsely distributed responsive polymer chains embedded in a superhydrophobic environment. The sparsely grafted polymer brushes do not significantly influence the surface wetting because of low grafting density, but they lead to considerable hysteresis when the droplet interplays with polymers. The effect is more prominent when the contact area is expanded under pressing.

Table 2 Advancing contact angle, receding contact angle, and contact angle hysteresis of 1 M and 1 mM $NaNO_3$ droplets on Q-PDMAEMA grafted samples (Tf_2N^- as anion)

Q-PDMAEMA (Tf_2N^- as anion) grafted sample	1 M $NaNO_3$			1 mM $NaNO_3$		
	θ_{adv}	θ_{red}	CAH	θ_{adv}	θ_{red}	CAH
From In only	150.2	127.3	22.9	156.8	138.5	18.3
From In/PFOTS=1∶5	163.3	141.8	21.5	167.7	155.5	12.2
From In/PFOTS=1∶8	163.9	149.6	14.3	168.4	157.3	11.1
From In/PFOTS=1∶10	165.3	153.7	11.6	171.5	163.3	8.2
From In/PFOTS=1∶15	168.1	160.2	7.9	172.3	167.5	4.8
From In/PFOTS=1∶20	170.5	164.4	6.1	174.3	169.3	5.0

When the liquid droplet is a good solvent for grafted polymers (PDMAEMA with acidic droplets, PNIPAm below the LCST, Q-PDMAEMA with $NaNO_3$ containing droplets, and the polymer chain adopts its stretched state), the polymer under the droplets remains hydrated. The hydration force makes droplets stick to the surface, while surrounding fluorine containing monolayers make the contact line hardly move forward. With the assistance of pressure, the contact line moves forward, allowing more polymer chains to interact with the droplet and to be hydrated, which increases contact angle hysteresis and induces the wetting transition from the Cassie to the Wenzel model as shown in Scheme 3 a. In contrast, if the polymer brushes do not interact with probing liquid droplets (polymer chain shows collapsed conformation), the movement of the contact line of the liquid droplet will not be favored even under pressure, Scheme 3 b.

Conclusions

We have demonstrated that wetting transitions and hysteresis can be closely related to the surface composition at the molecular level. Sparsely distributed grafted responsive polymers on the surface don't significantly change the surface wetting properties, but lead considerably to surfaces with responsive wetting transitions and hysteresis characteristics. When the probing droplet interacts with polymers and they become hydrated, the wetting can be easily realized from the Cassie mode to the Wenzel mode, with the characteristics of high hysteresis and a decrease in contact angle. On the other hand, if the probing droplet doesn't interact with polymers, meaning that polymer will remain in a collapsed state, the droplet will remain in the stable Cassie wetting mode. The responsive surface composition regulated wetting will be very useful in understanding wetting theory, and will be helpful experimentally in designing smart surfaces in, for example, microfluidic devices.

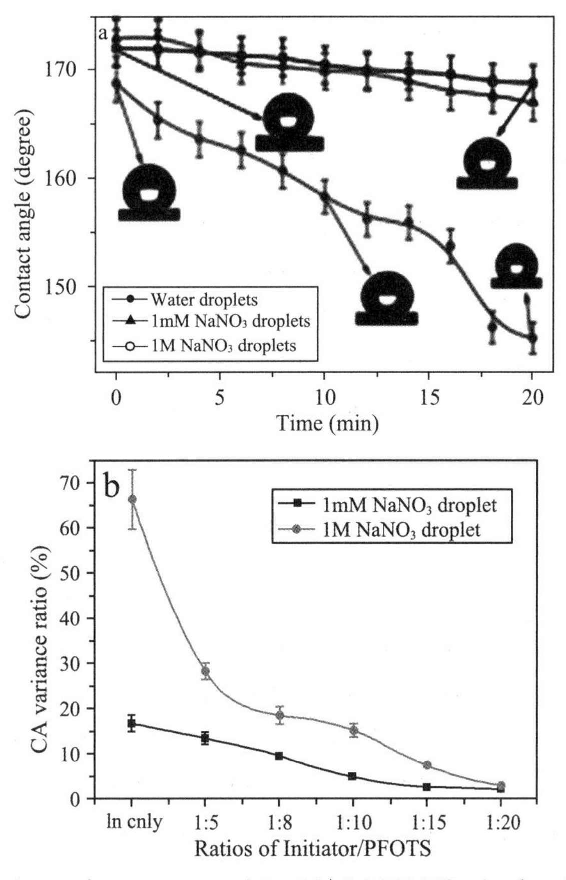

Fig. 6 (a) Time dependence of pure water and 1 mM/ 1 M $NaNO_3$ droplets CAs on Q-PDMAEMA (from Initiator/ PFOTS=1∶5) with Tf_2N^- counterion modified Al_2O_3, (b) the corresponding CA variance of 1 mM and 1 M $NaNO_3$ droplets under the pressure action.

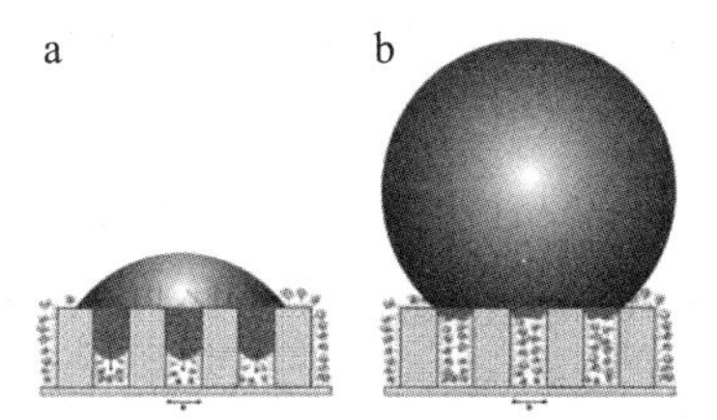

Scheme 3 Schematic illustration of the regulation of surface chemistry on porous anodized alumina substrate (a) stretched state of polymer chain, (b) collapsed state of polymer chain.

Experimental section Preparation of Al_2O_3 substrate and initiator attachment

Al_2O_3 substrates were prepared according to literature procedures.[19] The Al_2O_3 sheets were cleaned by oxygen plasma for 3 min and were immersed in a solution of 50 mL toluene containing 3 μL triethylamine and a mixture of 3-(trichloro-silyl)propyl-2-

bromo-2-methyl propanoate (initiator) and 1H, 1H, 2H, 2H-perfl uorooctyl trichlorosilane (PFOTS) in different molar ratios for 12 h. The samples were taken out of the mixed solution and copiously washed with ultrapure water and hexane, and then blown dry with N_2.

Macromolecular growth

For the polymerization of DMAEMA, 1 mL DMAEMA monomer and 10 mL of 1 : 1 (v/v) $MeOH/H_2O$ mixture were placed in a flask under N_2 flow for 30 min; then 0.060 8 g bipyridyl and 0.030 4 g CuBr were added into a flask and purged with N_2 again; 20 min later the initiator/PFOTS decorated Al_2O_3 sheets were added. The polymerizations were performed at room temperature under N_2. The substrates were taken out of the polymerization solution after one hour and washed with ultrapure water and ethanol. The polymer-grafted samples were further dried under vacuum overnight at 45℃ before further analysis. Quaternization was carried out in 5 mL of iodomethane/25 mL of CH_3NO_2 at room temperature for 24 h, and then the quaternized substrates were immersed in 0.1 M Tf_2NLi solution to carry out anion exchange for 4 h. SI-ATRP of NIPAm was obtained by immersing the initiator/ PFOTS modif i ed sheets into a degassed solution of NIPAm (2.52 g) in a 1 : 1(v/v) mixture of H_2O and CH_3OH (6 mL) containing CuBr (0.032 g) and pentamethyldiethylenetriamine (PMDETA ;0.14 mL) for 10 min at room temperature. After the polymerization, the sheets were rinsed with ultrapure water and absolute ethyl alcohol and then blown dry with N_2.

Characterization

SEM images were obtained on a JSM-6701F fi eld emission scanning electron microscope (FE-SEM) at 5~10 kV. X-ray photoelectron spectra (XPS) were obtained on a multi-functional XPS/AES system (Model PHI-5072, Physical Electronics. Inc., Eden Prairie. MN) by using Mg-Ka radiation (250 W, pass energy of 29.35 eV). The binding energy of C 1s (284.8 eV) was used as the reference. Sessile water droplet contact angle values were acquired at a DSA-100 optical contact angle meter (Kruss Company, Ltd., Germany) at ambient temperature (~22℃). 6 μL deionized water was dropped on the samples using an automatic dispense controller, and the contact angles were determined automatically by using the Laplace-Young fitting algorithm. The average CA values were obtained by measuring the sample at seven different positions of the substrate. The typical fluorescence images were acquired using a fluorescence microscope (BX51, Olympus).

Acknowledgements

This work was supported by the "Hundred Talents Program" of CAS, "973" program (2007CB607601), NSFC (50835009) and State key research project 2008ZX05001-002.

References

[1] R. N. Wenzel, Ind. *Eng. Chem.*, 1936,28:988-994.

[2] A. B. D. Cassie and S. Baxter, *Trans. Faraday Soc.*, 1944,40:0546-0550.

[3] T. Onda, S. Shibuichi, N. Satoh and K. Tsujii, *Langmuir*, 1996,12(9): 2125-2127.

[4] S. Shibuichi, T. Onda, N. Satoh and K. Tsujii, *J. Phys. Chem.*, 1996,100 (50):19512-19517.

[5] W. Chen, A. Y. Fadeev, M. C. Hsieh, D. Oner, J. Youngblood and T. J. McCarthy, *Langmuir*,1999,15(10):3395-3399.

[6] D. Oner and T. J. McCarthy, *Langmuir*,2000,16(20):7777-7782.

[7] L. Feng, S. Li, Y. Li, H. Li, L. Zhang, J. Zhai, Y. Song, B. Liu, L. Jiang and D. Zhu, Adv. Mater., 2002,14(24):1857-1860.

[8] L. Feng, S. H. Li, H. J. Li, J. Zhai, Y. L. Song, L. Jiang and D. B. Zhu, *Angew. Chem., Int. Ed.*,2002,41(7):1221-1223.

[9] H. Y. Erbil, A. L. Demirel, Y. Avci and O. Mert, *Science*, 2003, 299 (5611):1377-1380.

[10] J. Bico, U. Thiele and D. Quere, *Colloids Surf.*, A, 2002, 206 (1-3): 41-46.

[11] S. Herminghaus, *Europhys. Lett.*, 2000,52(2):165-170.

[12] R. Wang, K. Hashimoto, A. Fujishima, M. Chikuni, E. Kojima, A. Kitamura, M. Shimohigoshi and T. Watanabe, *Nature*,1997,388(6641):431-432.

[13] K. K. S. Lau, J. Bico, K. B. K. Teo, M. Chhowalla, G. A. J. Amaratunga, W. I. Milne, G. H. McKinley and K. K. Gleason, *Nano Lett.*, 2003, 3(12):1701-1705.

[14] A. Nakajima, K. Hashimoto and T. Watanabe, *Monatsch. Chem.*,2001, 132(1):31-41.

[15] D. Quere, *Rep. Prog. Phys.*, 2005,68(11):2495-2532.

[16] X. M. Li, D. Reinhoudt and M. Crego-Calama, *Chem. Soc. Rev.*,2007,36

(8):1350-1368.

[17] M. L. Ma and R. M. Hill, *Curr. Opin. Colloid Interface Sci.*,2006,11(4):193-202.

[18] R. Rioboo, M. Voue, A. Vaillant, D. Seveno, J. Conti, A. I. Bondar, D. A. Ivanov and J. De Coninck, *Langmuir*,2008,24(17):9508-9514.

[19] M. Miwa, A. Nakajima, A. Fujishima, K. Hashimoto and T. Watanabe, *Langmuir*,2000,16(13):5754-5760.

[20] A. Marmur, *Langmuir*,2003,19(20):8343-8348.

[21] A. B. D. Cassie, Discuss. *Faraday Soc.*, 1948,3:11-16.

[22] S. Brandon, N. Haimovich, E. Yeger and A. Marmur, J. *Colloid Interface Sci.*, 2003,263(1):237-243.

[23] L. Gao and T. J. McCarthy, *Langmuir*,2006,22(7):2966-2967.

[24] N. A. Patankar, *Langmuir*,2004,20(17):7097-7102.

[25] A. Lafuma and D. Quere, *Nat. Mater.*, 2003,2(7):457-460.

[26] L. Barbieri, E. Wagner and P. Hoffmann, *Langmuir*, 2007, 23(4):1723-1734.

[27] B. He, N. A. Patankar and J. Lee, *Langmuir*,2003,19(12):4999-5003.

[28] R. Rioboo, M. Voue, A. Vaillant and J. De Coninck, *Langmuir*,2008,24(24):14074-14077.

[29] Y. C. Jung and B. Bhushan, *Langmuir*,2008,24(12):6262-6269.

[30] D. Bartolo, F. Bouamrirene, E. Verneuil, A. Buguin, P. Silberzan and S. Moulinet, *Europhys. Lett.*, 2006,74(2):299-305.

[31] M. Callies and D. Quere, *Soft Matter*,2005,1(1):55-61.

[32] M. Reyssat, A. Pepin, F. Marty, Y. Chen and D. Quere, *Europhys. Lett.*, 2006,74(2):306-312.

[33] Y. C. Jung and B. Bhushan, *Langmuir*,2009,25(16):9208-9218.

[34] B. Liu and F. F. Lange, J. *Colloid Interface Sci.*, 2006,298(2):899-909.

[35] D. Richard, C. Clanet and D. Quere, *Nature*,2002,417(6891):811-811.

[36] E. Bormashenko, R. Pogreb, G. Whyman and M. Erlich, *Langmuir*,2007,23(12):6501-6503.

[37] X. Yao, Q. W. Chen, L. Xu, Q. K. Li, Y. L. Song, X. F. Gao, D. Quere and L. Jiang, Adv. *Funct. Mater.*, 2010,20(4):656-662.

[38] M. Nosonovsky and B. Bhushan, *Microelectron. Eng.*, 2007, 84(3):382-386.

[39] T. L. Sun, W. L. Song and L. Jiang, Chem. *Commun.*, 2005:1723-1725.

[40] A. S. Lee, V. Butun, M. Vamvakaki, S. P. Armes, J. A. Pople and A. P. Gast, *Macromolecules*,2002,35(22):8540-8551.

[41] E. S. Gil and S. A. Hudson, *Prog. Polym. Sci.*, 2004, 29 (12): 1173-1222.

[42] Y. K. Lai, C. J. Lin, J. Y. Huang, H. F. Zhuang, L. Sun and T. Nguyen, *Langmuir*,2008,24(8):3867-3873.

[43] T. L. Sun, G. J. Wang, L. Feng, B. Q. Liu, Y. M. Ma, L. Jiang and D. B. Zhu, *Angew. Chem.*, *Int. Ed.*, 2004,43(3):357-360.

[44] O. Azzaroni, A. A. Brown, N. Cheng, A. Wei, A. M. Jonas and W. T. S. Huck, J. *Mater. Chem.*, 2007,17(32):3433-3439.

[45] O. Azzaroni, A. A. Brown and W. T. S. Huck, *Adv. Mater.*, 2007,19(1):151-154.

[46] H. S. Lim, S. G. Lee, D. H. Lee, D. Y. Lee, S. Lee and K. Cho, *Adv. Mater.*, 2008,20(23):4438-4441.

[47] S. Moya, O. Azzaroni, T. Farhan, V. L. Osborne and W. T. S. Huck, *Angew. Chem.*, Int. Ed., 2005,44(29):4578-4581.

[48] B. S. Lee, Y. S. Chi, J. K. Lee, I. S. Choi, C. E. Song, S. K. Namgoong and S. G. Lee, J. *Am. Chem.* Soc., 2004,126(2):480-481.

[49] F. Zhou, H. Y. Hu, B. Yu, V. L. Osborne, W. T. S. Huck and W. M. Liu, *Anal. Chem.*, 2007,79(1):176-182.

[50] C. W. Extrand, *Langmuir*,2002,18(21):7991-7999.

[51] N. A. Patankar, *Langmuir*,2004,20(17):7097-7102.

[52] M. Nosonovsky and B. Bhushan, *Microsyst. Technol.*, 2005, 11 (7): 535-549.

[53] M. Nosonovsky and B. Bhushan, *Nano Lett.*, 2007,7(9):2633-2637.

疏水缔合型聚丙烯酰胺(HAPAM)和常规聚丙烯酰胺(HPAM)的增黏机理

李美蓉[1] 柳 智[2] 曹绪龙[3] 张本艳[3] 张继超[3] 孙方龙[1]

(1. 中国石油大学理学院，山东青岛 266555；
2. 中国石化青岛液化天然气有限责任公司，山东青岛 266400；
3. 胜利油田分公司地质科学研究院三采中心，山东东营 257015)

摘要：疏水缔合型聚丙烯酰胺(HAPAM)与常规聚丙烯酰胺(HPAM)相比具有显著的高效增黏性、耐盐性和理想的耐温稳定性。通过测定HAPAM溶液的电导率和黏度，得到聚合物的临界胶束质量浓度为450 mg/L。通过红外光谱和核磁共振分析，可知HAPAM侧链含有长的烷基链，并且侧链含有双键，这些构成疏水基团，起到疏水缔合作用。扫描电镜观测结果从形态上直接说明了HAPAM和HPAM的增黏和降黏机理，HPAM溶液的表观黏度主是由非结构黏度η_n构成，HAPAM溶液黏度由非结构黏度η_n和结构黏度η_s共同构成。

关键词：常规聚丙烯酰胺；疏水缔合型聚丙烯酰胺；红外光谱；核磁共振；扫描电镜
中图分类号：TE357 文献标识码：A
doi:10. 3969/j. issn. 1001-8719. 2012. 06. 024

Thickening mechanism of hydrophobically associating polyacrylamide and polyacrylamide

Li Meirong[1] Liu Zhi[2] Cao Xulong[3] Zhang Benyan[3]
Zhang Jichao[3] Sun Fanglong[1]

(1. China University of Petroleum, College of Science, Qingdao 266555, China;
2. Qingdao Liquefied Natural Gas Limited Liability Company, SINOPEC, Qingdao 266400, China;
3. Shengl iOilfield Company of Geological Sciences Institute of the three mining centers, Dongying 257015, China)

Abstract: Compared with conventional polyacrylamide (HPAM), hydrophobically associating polyacrylamide (HAPAM) has the performance of increasing viscosity, salt tolerance and temperature stability. By measuring the conductivity and viscosity of the polymer at its various concentrations, the critical micelle concentration of HAPAM of 450 mg/L was obtained. By analyzing the results of IR and NMR, it is confirmed that HAPAM contained long alkyl chains with double bonds, which constitute the hydrophobic group. By the method of scanning electron microscopy, the viscosity mechanisms of HPAM and HAPAM were directly shown. The apparent viscosity of HPAM is mainly composed of non-structure viscosity η_n, while the apparent viscosity of HAPAM is composed of both η_n and structure viscosity η_s.

Keywords: polyacrylamide (HPAM); hydrophobically associating polyacrylamide (HAPAM); IR; NMR; SEM

我国已经进入三次采油阶段,其特点是向油层注入水以外的其他驱油剂,以进一步提高原油的采收率。目前广泛使用的驱油剂有部分水解聚丙烯酰胺(HPAM)、黄原胶以及近几年研制和开发的疏水缔合型聚合物。疏水缔合型聚丙烯酰胺(HAPAM)是目前理想的抗盐、耐温、抗剪切以及高效增黏的新型聚合物,并且具有扩大波及系数、降低油一水界面张力、提高对原油的增溶能力和乳化能力的作用[1-2]。其实质是,在 HPAM 分子链上引入少量疏水基团,使聚合物分子在水溶液中由于静电、氢键或范德华力作用而在分子间产生具有一定强度但又可逆的物理缔合,从而形成巨大的三维网状结构,即使当聚合物溶液浓度较低时,体系仍然有很高的表观黏度,而不像 HPAM 溶液那样,仅靠提高聚合物相对分子质量来实现高效增黏[3]。笔者将 HAPAM 与 HPAM 性能进行比较,并解释二者增黏机理的不同。

1 实验部分

1.1 材料和仪器

常规聚丙烯酰胺,北京恒聚化工集团股份有限公司产品;疏水缔合型聚丙烯酰胺,四川光亚科技股份有限公司产品,型号 ZND;NaCl,分析纯。

Brookfield Engineering Laboratories 公司 Brookfield DV-Ⅱ+Pro 型黏度计;日本日立公司 S-4800 冷场扫描电镜;美国珀金埃尔默公司 Spectrometer One FT-IR 光谱仪;Varian 公司 INOVA-500 型核磁共振仪;山海雷磁创意仪器仪表有限公司 DDS-114 数字电导率仪;常州国华电器有限公司 79-1 型磁力加热搅拌器。

1.2 聚合物溶液黏度和临界胶束浓度的测定

1.2.1 聚合物溶液黏度的测定

参照《SY/T6576.2003 用于提高石油采收率的聚合物评价的推荐作法》的方法配制质量浓度为 5 000 mg/L 的聚合物母液。

用去离子水将聚合物母液稀释至不同浓度,电磁搅拌器搅拌 1 h 后在 60℃下恒温 10 min,然后采用 Brookfield DV-Ⅱ+Pro 型黏度计测定黏度,选择 RV-3 号转子,转速为 120 r/min。

1.2.2 HAPAM 临界胶束浓度的测定

配制不同浓度的 HAPAM 溶液,常温下测定黏度及电导率,黏度和电导率的拐点即为临界胶束浓度。

1.3 聚合物表征

1.3.1 红外光谱

将纯聚合物固体研磨后,取微量粉体进行红外光谱分析。

1.3.2 核磁共振

采用 D_2O 作为溶剂。由于聚合物的相对分子质量比较大,溶液黏度高,为了获得更清晰的谱图,利用超声波促进溶解,同时对其进行降解处理,以降低样品的相对分子质量。

1.3.3 扫描电镜

用去离子水和 3 000 mg/L NaCl 溶液模拟矿化水配制质量浓度不同的聚合物溶液,在 25℃下用磁力搅拌器搅拌约 2 h,然后静置熟化 14 d。用移液管分别取 100 μL 溶液滴在 2 个已经超声波处理的新云母片上,使它尽量铺张成膜,以保持聚合物水化分子的原有结构和形貌。在干燥器中干燥,然后进行喷金。最后用 S-4800 冷场扫描电镜观察聚合物的形态。实验温度 25℃。选择电压 3.0 kV,选择工作距离和放大倍数如扫描电镜照片下方所注[4]。

2 结果与讨论

2.1 HPAM 和 HAPAM 性能比较

不同浓度的常规 HPAM 和 HAPAM 溶液的黏度测定结果示于图 1。不同 NaCl 浓度下,1 500 mg/L HAPAM 和 HPAM 溶液的黏度测定结果示于图 2。不同温度下,1 500 mg/L HPAM 和 HAPAM 溶液的黏度测定结果示于图 3。

从图 1 可以看出,HAPAM 溶液的黏度高于相同浓度下 HPAM 溶液的浓度。从图 2 可以看出,随着 NaCl 浓度的增加,HPAM 溶液的黏度急剧减小,而 HAPAM 溶液的黏度随着盐浓度的增大而升高,说明与 HPAM 相比,HAPAM 具有抗盐性[5]。从图 3 可以看出,随着温度的升高,HPAM 和 HAPAM 溶液的黏度均降低,但在同一温度和相同浓度下,HAPAM 的黏度要高于常规型 HPAM。因此,HAPAM 与 HPAM 相比有明显的黏度优势和很好的抗盐性。

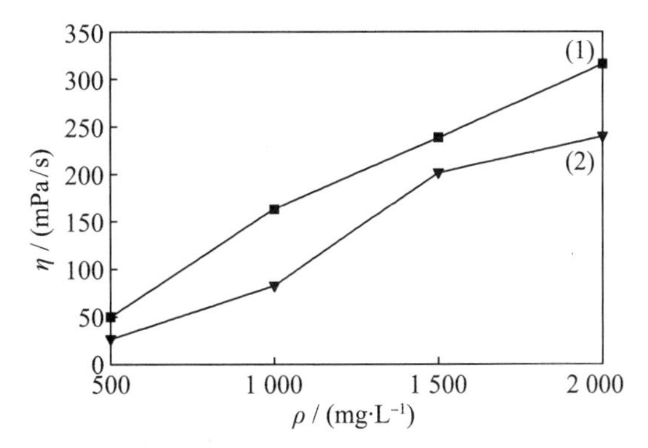

图 1 HPAM 和 HAPAM 溶液的黏度—浓度曲线

Fig. 1 The viscosity-concentration curves of HPAM and HAPAM solutions (1) HAPAM;(2) HPAM

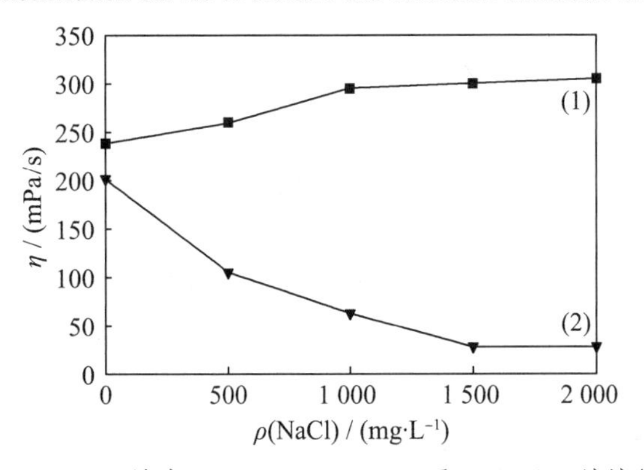

图 2 不同 NaCl 浓度(ρ(NaCl))下 HPAM 和 HAPAM 溶液的黏度

Fig. 2 The viscosity of HPAM and HAPAM solution sunder different ρ(NaCl) (1) HAPAM;(2) HPAM ρ(Polymer)=1 500 mg/L

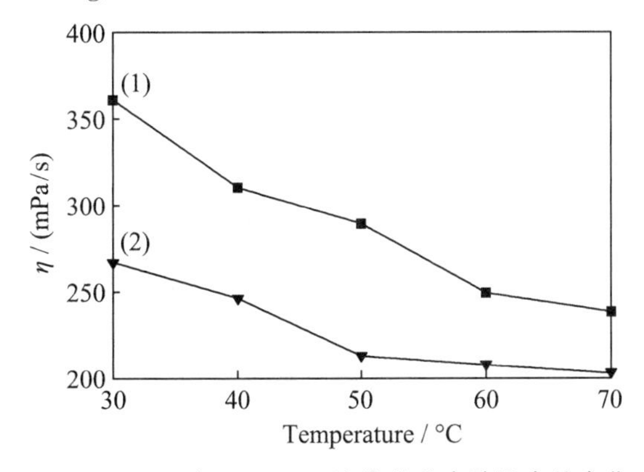

图 3 HPAM 和 HAPAM 溶液的黏度随温度的变化

Fig. 3 The viscosity of HPAM and HAPAM solutions vs temperature (1) HAPAM; (2) HPAM ρ (Polymer)=1 500 mg/L

2.2 HAPAM的临界胶束浓度

测定了不同浓度HAPAM溶液的电导率和黏度,结果示于图4,由此可以得到HAPAM的临界胶束浓度(CMC)。从图4可以看出,HAPAM溶液的电导率和黏度在其质量浓度为450 mg/L处出现拐点,说明其临界胶束质量浓度为450 mg/L。疏水缔合型聚合物是一种亲水性高分子链上带有少量疏水基团的水溶性聚合物,在其溶液浓度高于临界缔合浓度时,分子间发生缔合作用而形成物理交联网络。

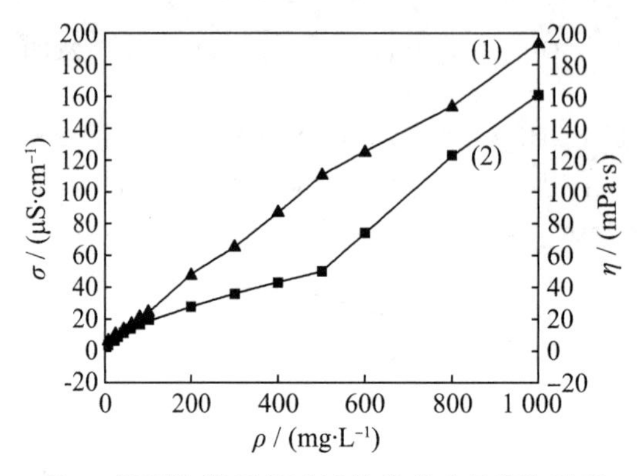

图4 不同浓度HAPAM溶液的电导率和黏度

Fig. 4 The Electrical conductivity and viscosity of HAPAM solutions with different concentrations (1) Electrical conductivity; (2) Viscosity

2.3 HPAM和HAPAM的分子结构表征结果

2.3.1 FT-IR

图5为提纯的HPAM和HAPAM固体样品的FT-IR谱。从图5可以看出,HPAM和HAPAM的FT-IR谱基本相同,在3 420和3 200 cm^{-1}处的峰比较宽,为H_2O中—OH的红外吸收峰,而伯酰胺N—H键的对称伸缩振动峰和不对称伸缩振动峰应该在3 350和3 200 cm^{-1}左右,被—OH峰掩盖。1 660 cm^{-1}峰为C—O键的对称伸缩振动峰,1 410 cm^{-1}峰为C—N键的伸缩振动和N—H键的弯曲振动的偶合峰,以C—N键的伸缩振动为主。2 930 cm^{-1}峰为—CH_2—的不对称伸缩振动峰,2 800 cm^{-1}左右峰为—CH_2—的对称伸缩振动峰。1 450 cm^{-1}峰为—CH_2—CO中的—CH_2—弯曲振动峰。1 320和1 120 cm^{-1}两峰在1 350~1 150 cm^{-1}之间,为仲酰胺C—N键的伸缩振动峰和部分N—H键的振动峰,569 cm^{-1}峰为伯酰胺氨基的面外弯曲振动峰。HAPAM与HPAM的明显差异仅在1 040 cm^{-1}处,该特征峰是由双键产生的烯烃面外弯曲振动,说明与HPAM相比,HAPAM在侧链中的疏水缔合基团含有双键。

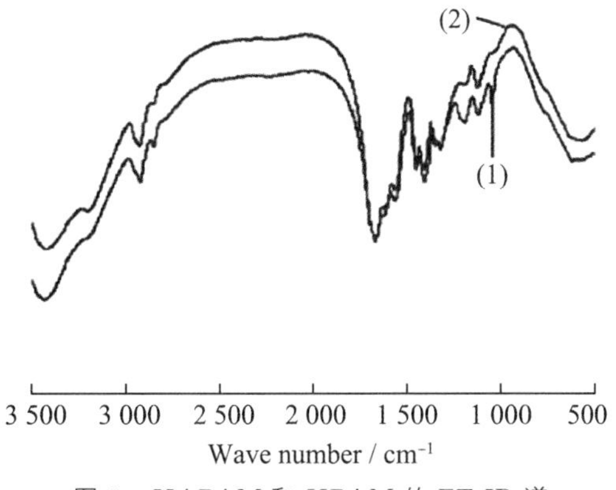

图 5　HAPAM 和 HPAM 的 FT-IR 谱

Fig. 5　FT-IR spectra of HAPAM and HPAM (1) HAPAM; (2) HPAM

2.4 ^{1}H NMR 和^{13}C NMR

图 6 为 HAPAM 的^{1}HNMR 和^{13}CNMR 谱，其各峰位置所代表的官能团结构总结列于表 1 和表 2。由表 1 和表 2 可知，HAPAM 是由 2 个嵌段 $\mathrm{-[-\overset{H_2}{C}-\underset{}{\overset{O=C-NH_2}{\overset{|}{CH}}}-]_x-}$ (A) 和 $\mathrm{-[-\overset{H_2}{C}-\overset{R}{\overset{|}{CH}}-]_y-}$ (B) 共同构成，(B) 中含有长的烷基链 R，并且 R 中含有双键，构成疏水基团，起到疏水缔合作用。

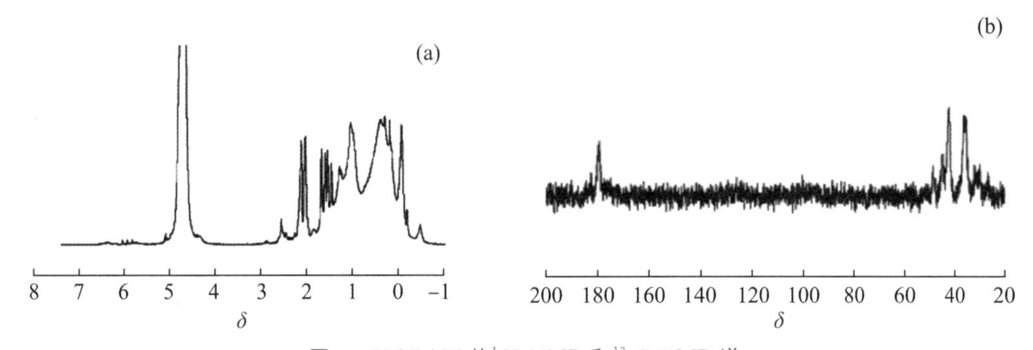

图 6　HAPAM 的^{1}H NMR 和^{13}C NMR 谱

Fig. 6　^{1}H NMR and ^{13}C NMR of HAPAM (a) ^{1}H NMR; (b) ^{13}C NMR

表 1　HAPAM 的^{1}H NMR 谱中化学位移对应的官能团

Table 1　Chemical shifts and relative functional groups in ^{1}H NMR of HAPAM

δ	Functional group
1.457—1.698, 2.042—2.135	CH_2 and CH in $\mathrm{-[-\overset{H_2}{C}-\overset{O=C-NH_2}{\overset{|}{CH}}-]_x-}$ (A)

(续表)

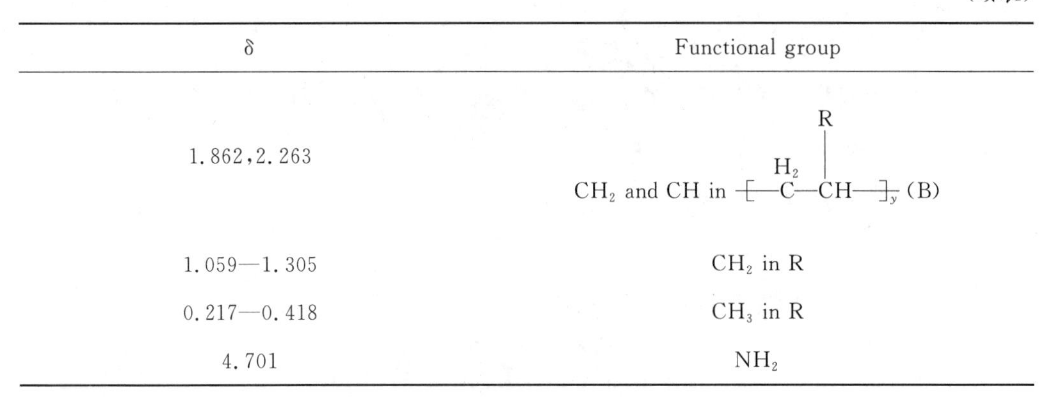

δ	Functional group
1.862,2.263	CH_2 and CH in $-[-\overset{H_2}{C}-\underset{}{\overset{R}{CH}}-]_y-$ (B)
1.059—1.305	CH_2 in R
0.217—0.418	CH_3 in R
4.701	NH_2

表 2 HAPAM 的 13C NMR 中化学位移对应的官能团

Table 2 Chemical shifts and relative functional groups in 13C NMR of HAPAM

δ	Functional group
179.720—182.885	C—Oin (A) C—C in R
35.052—36.140,42.001—45.075	CH_2 and CH in(A and B)
48.888	CH in R
29.931—31.765	CH_2 in R
26.429	CH_3 in R

2.5 HPAM 和 HAPAM 的增黏机理

聚合物溶液的表观黏度由非结构黏度(η_n)和结构黏度(η_s)构成。非结构黏度是指聚合物溶于水后,大分子链在水溶液中水化伸展,同时包裹一定量的水,此时分子的流体力学体积增大,使水溶液黏度提高;结构黏度是指由于大分子链间相互作用(缠绕、缔合、化学键力等),使聚合物溶液整体流体力学体积增大,黏度升高[6]。

图 7 为 1 000 mg/L HPAM 溶液和矿化水配制的 HPAM 溶液的 SEM 照片。从图 7(a)可以看出,HPAM 在溶液中占有的空间体积大,HPAM 分子链长,分子链之间相互缠绕,形成三维网状结构,这种三维网状结构阻碍了水介质的自由运动,致使聚合物的黏度很大。另外,形成这种三维网状结构的原因可从聚合物的化学结构式方面分析,HPAM 分子链中含有一定比例的羧酸根负离子,链节之间有静电斥力,使得卷曲的分子链变得松散舒展,增加了阻碍水介质自由运动的能力,提高了聚合物溶液黏度。观察 HPAM 的表面结构,可以看出 HPAM 分子表面有一层较厚的水化膜,分子链粗细不均匀,链节处有凸起。造成这种形态的原因是 HPAM 分子的亲水基团与水介质之间存在溶剂化作用,HPAM 分子表面形成了较厚的水化膜,束缚了大量水介质,使聚合物黏度增大[7-8]。由此得出,HPAM 分子间只有简单的缠绕,其表观黏度主要由 η_n 构成,η_s 对溶液黏度贡献很小。

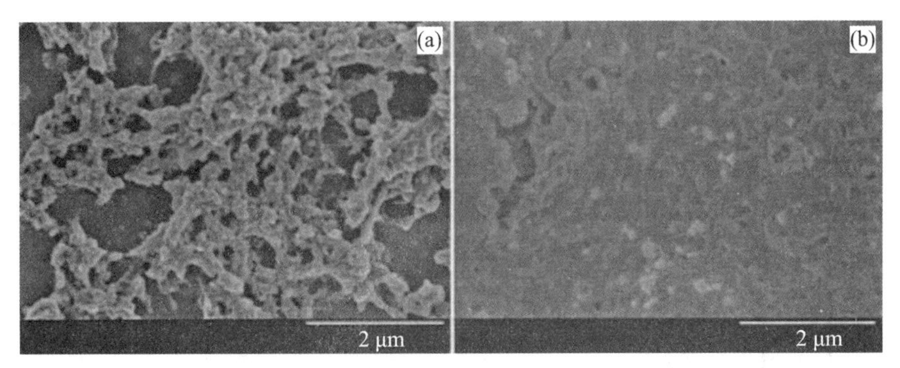

图 7　1 000 mg/L HPAM 溶液和矿化水配制的 HPAM 溶液的扫描电镜照片

Fig. 7　SEM photo of 1 000 mg/L HPAM solution and HPAM solution made up by mineralized water
(a) HPAM solution; (b) HPAM solution made up by mineralized water

从图 7(b)可以看出，在金属离子的作用下，聚合物原来的三维网状结构被破坏，并且不再舒展，变得卷曲。金属离子与羧酸根负离子相互吸引，使分子链节的静电排斥作用减弱，舒展程度降低。对比观察图 7(a)、(b)中 HPAM 的表面结构，可以看出，HPAM 表面发生了明显的变化，表面的溶剂化层在矿化水的作用下，水化层变薄，释放出大量水介质。原因是金属正离子与羧酸根负离子相互作用，降低了 HPAM 分子电荷密度，使分子变得卷曲，同时减弱了极性基团的溶剂化能力，释放大量的“自由水”，从而使黏度大大降低[9]。

图 8 为 HAPAM 溶液浓度达到及大于临界胶束浓度的 SEM 照片。从图 8(a)可以观察到大分子链在水溶液中的水化伸展作用和分子表面的水化膜。HAPAM 正是通过增大其分子的流体力学体积，产生水化膜，使水溶液黏度提高，η_n 对疏水缔合型聚合物溶液表观黏度起着决定性作用。从图 8(b)可以看出，聚合物的形貌类似于一个个的“花朵”，聚合物分子以一个中心为缔合点形成一种交联的形态[10-11]。分子链中的疏水基团之间在静电、氢键或范德华力作用下产生具有一定强度而又可逆的物理缔合，形成巨大的三维网状结构，即分子间缔合，也就是图中的“花蕊”部分；带有亲水基的分子链舒展在水中，在图中为“花瓣”部分。此时，结构黏度 η_s 对疏水缔合型聚合物溶液的表观黏度起着决定性作用。随着聚合物浓度的增加，疏水基团不断增加，缔合强度也随之不断增大，因此溶液黏度随着聚合物浓度的增大而增加。当温度升高时，分子间的物理缔合减弱，黏度降低。

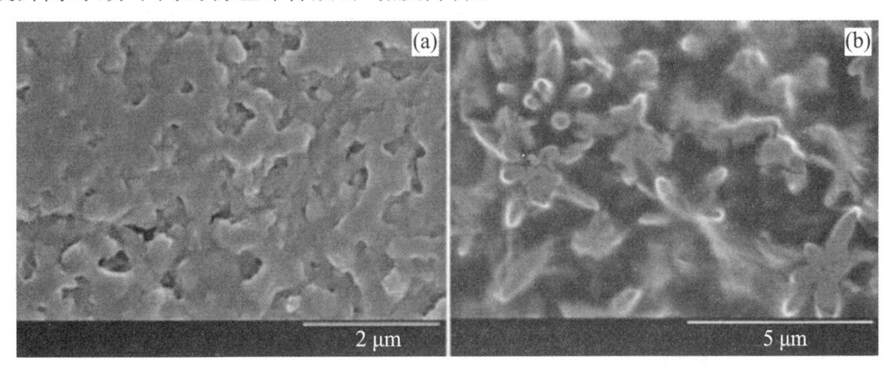

图 8　溶液浓度达到及大于临界胶束浓度的 HAPAM 溶液的扫描电镜照片

Fig. 8　SEM photo of HAPAM solution (a) 50 mg/L; (b) 500 mg/L

图 9 为用矿化水配制的 500 mg/L HAPAM 溶液的 SEM 照片。从图 9 可以看出，HAPAM 的形貌仍是原来花朵状，但在盐的作用下，水化作用明显减弱。原因是电解质的存在会对分子链产生电荷屏蔽作用，并且会压缩分子链的水化膜双电子层，使水化膜变薄，从而分子链伸展程度降低，变得卷曲，流体力学体积减小，体系的 η_n 下降；与此同时，电解质的加入使溶液极性增大，疏水基团逃离作用加大，同时疏水基团的水化膜也变薄，两方面的共同作用使缔合力增大，η_s 随之增大。对于疏水缔合型聚合物，随着电解质浓度增大，η_n 降低，η_s 增加，当 η_n 降低得多 η_s 增加得少，溶液表观黏度下降；当 η_n 降低与 η_s 增加的值相当，溶液表观黏度不变；当 η_n 降低得少 η_s 增加得多，溶液表观黏度增加。因此，疏水缔合型聚合物溶液表观黏度随着电解质浓度增大会出现盐增稠现象，表现出一定的抗盐特性。

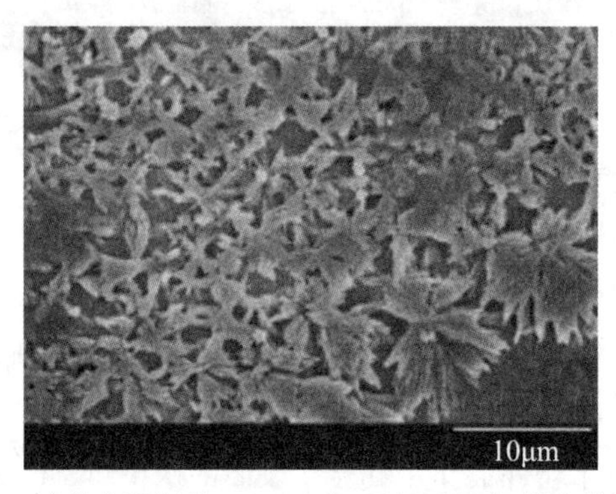

图 9 矿化水配制的 500 mg/L HAPAM 扫描电镜照片

Fig. 9 SEM photo of 500 mg/L HAPAM solution made up by mineralized water

3 结论

(1) 疏水缔合型聚丙烯酰胺(HAPAM)分子中含有长的烷基链，并且侧键中含有双键，构成了疏水基团，起到疏水缔合作用。其临界胶束质量浓度为 450mg/L。

(2) HPAM 溶液的黏度主要由 η_n 构成，随着矿化水的加入，黏度明显降低。而对于 HAPAM 溶液，黏度由 η_n 和 η_s 共同构成，随着盐的加入，如果 η_n 降低得多，η_s 增加得少，溶液表观黏度下降；如果 η_n 降低和 η_s 增加的值相当，溶液表观黏度不变；如果 η_n 降低得少，η_s 增加得多，溶液表观黏度增加。因此，HAPAM 具有明显的黏度优势，并具有很好的抗盐性，更适合油田生产。

参考文献

[1] 陈锡荣，黄凤兴. 驱油用耐温抗盐水溶性聚合物的研究进展 [J]. 石油化工，2009，38，(10)：1132-1136. (CHEN Xirong，HUANG Fengxing. Research development of heat resistant and salt tolerant polymer for enhanced oil recovery [J]. Petrochemical Technology，2009，38，(10)：1132-1136.)

[2] 张红艳,康万利,孟令伟,等.一种驱油用疏水缔合聚丙烯酰胺的乳化性能[J].石油学报(石油加工),2010,26(4):628-634.(ZHANG Hongyan, KANG Wanli,MENG Lingwei, et al. Emulsion characteristics of the hydrophobically associating polyacrylamid used for oilflooding [J]. Acta Petrolei Sinica (Petroleum Processing Section),2010,26(4):628-634.)

[3] 冯玉军,罗传秋,罗平亚,等.疏水缔合水溶性聚丙烯酰胺的溶液结构研究[J].石油学报(石油加工),2001,17(6):39-43.(FENG Yuj un,LUO Chuanqiu, LUO Pingya, et al. Study on characterization of microstructure of hydrophobically associating water-soluble polymer in aqueous media by scanning electron microscopy and environmental scanning electron microscopy [J]. Acta Petrolei Sinica (Petroleum Processing Section),2001,17(6):39-43.)

[4] 朱怀江,赵常青,罗健辉,等.聚合物水化分子的微观结构研究[J].电子显微学报,2005,24(3):205-210.(ZHU Huaijiang, ZHAO Changqing, LUO Jianhui,et al. Study on the microstructure of polymer molecule hydrate [J]. Journal of Chinese Electron Microscopy Society,2005,24(3): 205-210.)

[5] 刘雨文.矿化度对疏水缔合聚合物溶液粘度的影响[J].油气地质与采收率,2003,10(3):62-63.(LIU Yuwen. Influence of salinity on the viscosity of hydrophobic associated polymer solution [J]. Petroleum geology and recovery efficiency,2003,10(3):62-63.)

[6] 贾振福,李早元,钟静霞,等.无机盐种类和浓度对疏水缔合聚合物溶液黏度的影响[J].钻井液与完井液,2007,24(1):53-57.(JIA Zhenfu,LI Zaoyuan, ZHONG jingxia,et al. Inorganic salts and their concentrations: effects on the viscosity of hydrophobically associating polymer solution [J]. Drilling Fluid &Completion Fluid, 2007,24(1) :53-57.)

[7] 康万利,孟令伟,牛井岗,等.矿化度影响HPAM溶液黏度机理[J].高分子材料科学与工程,2006,22(5):175-179(KANG Wanli, MENG Lingwei, NIU Jinggang,et al. Mechanism of the effect of salinity on HPAM solution viscosity [J]. Polymer Materials Science & Engineering,2006,22(5) :175-179.)

[8] 周恩乐,李虹,薄淑琴,等.聚丙烯酰胺的形态结构研究[J].高分子学报,1991,1:51-56.(ZHOU Enle,LI Hong,BO Shuqin,et al. Morphology of polyacrylamide [J]. Acta Polymerica Sinica,1991,1:51-56.)

[9] 包木太,陈庆国,王娜,等.油田污水中聚丙烯酰胺(HPAM)的降解机理研究[J].高分子通报,2008,2:1-9.(BAO Mutai,CHEN Qingguo,WANG Na,et al. Study on the mechanism of HPAM-degradation in the sewage of oilfield [J]. Chinese Polymer Bulletin,2008,2:1-9.)

[10] 蒋留峰,钟传蓉,徐敏,等. 接枝丙烯酰胺共聚物的溶液性能和微结构 [J]. 物理化学学报, 2010,26(3):535-540. (JIANG Liufeng, ZHONG Chuanrong, XU Min, et al. Solution properties and microstructure of branched acrylamide-based copolymers [J]. Acta Phys-Chim Sin, 2010, 26(3):535-540.)

[11] 蒋春勇,段明,方申文,等. 星型疏水缔合聚丙烯酰胺溶液性质的研究 [J]. 石油化工, 2010, 39(2):204-207. (JIANG Chunyong, DUAN Ming, FANG Shenwen, et al. Solution properties of star-shaped hydrophobic-associative acrylamide copolymer [J]. Petrochemical Technology, 2010, 39(2):204-207.)

The effect of microstructure on performance of associative polymer: In solution and porous media

Zhang Peng[a,*] Wang Yefei[a] Yang Yan[a] Zhang Jian[b]
Cao Xulong[c] Song Xinwang[c]

[a] China University of Petroleum (East China), School of Petroleum Engineering, Qingdao 266555, China

[b] CNOOC Research Center, Beijing 100027, China

[c] Geological and Scientifi c Research Institute, SINOPEC Shengli Oilfield Company, Dongying, Shangdong 257015, China

Abstract: Hydrophobically associating polyacrylamide (HAPAM) is considered as a promising candidate of polymer flooding because of its excellent apparent viscosifying capability. However, the effective viscosity and the oil displacement effi ciency of HAPAM are lower than those of conventional polymers at a concentration of 2 000 mg/l. Therefore, the microstructures of two types of polymers were investigated by scanning electron microscopy (SEM) to reveal the relationship between the morphology and properties. The specimens were prepared by accelerated freeze-drying in order to keep polymers with their original morphology without distortion. In aqueous solution, associative polymer exhibited a compact three-dimensional network structure, while only loose network structure was found in conventional polymer. A number of fi laments are attached to the associative polymer skeletons when the polymer concentration is higher than the critical association concentration (CAC). But in the porous media, most of the networks of conventional polymer were larger and more integrated than those of associative polymer. It may be interpreted by absorption loss of HAPAM and spatial confi nement in porous medium.

Keywords: hydrophobically associating polyacrylamide (HAPAM) polymer flooding microstructure effective viscosity scanning electron microscopy (SEM)

1 Introduction

It is well known that polymer flooding is an effective method applied to enhance oil recovery (EOR). It has now become one of the chief means of enhanced oil recovery in Shengli and Daqing oilfields in China (Hou et al., 2009). Currently, partially hydrolyzed polyacrylamide (HPAM) as a conventional polymer is commercially used in the reservoirs of the mild conditions, but not available under harsh conditions. For that reason, a new type of polymer named hydrophobically associating polyacrylamide (HAPAM) has been developed and considered as a promising candidate as thickener in EOR that involves hightemperature and high-salinity reservoir (Gouveia et al., 2008).

HAPAM originates from polyacrylamide by the introduction of a small fraction of hydrophobic groups onto the hydrophilic polymer backbone (Shashkina et al., 2003). The chemical sche-matic of HPAM and HAPAM is shown in Fig. 1. In aqueous solution, above a critical association concentration (CAC), the hydrophobic groups tend to associate together leading to the formation of polymolecular associations, which exhibit excellent viscosifying ability (Feng et al., 2005; Yahaya et al., 2001). For HAPAM, enhancing the salinity is accompanied by an increase in the polarity of the solution, which strengthens the hydrophobic unit interactions leading to higher viscosity (Gao et al., 2005). In addition, Carrageenan solution can also reach high viscosity at high salinities because of the conformation change from random coil to double helix. Moreover, Carrageenans are renewable, nontoxic and not hazardous to the environment (Iglauer et al., 2011).

Past study results concentrated on the apparent viscosity of HAPAM in solution, but a few papers discussed the effective viscosity in porous medium. Compared with apparent viscosity, effective viscosity (μ_{ef}) has more practical value in on-site application, because it truly reflects the viscosity as polymer flows through porous media (Cheng, 1989). It is determined by Eq. (1) (Wang et al., 2001), where RF is the resistance factor, RRF is the residual resistance factor and μw is the water viscosity at test temperature. The RF can be calculated by Eq. (2) by pressure drop of polymer solution (ΔP_P) and brine (ΔP_B) through porous media at the same flow rate. The RFF can be obtained by Eq. (3) based on permeability of brine (k_b) and brine permeability (k_f) after poly-mer injection.

$$\mu_{ef}=\frac{RF}{PFF}\mu_w \tag{1}$$

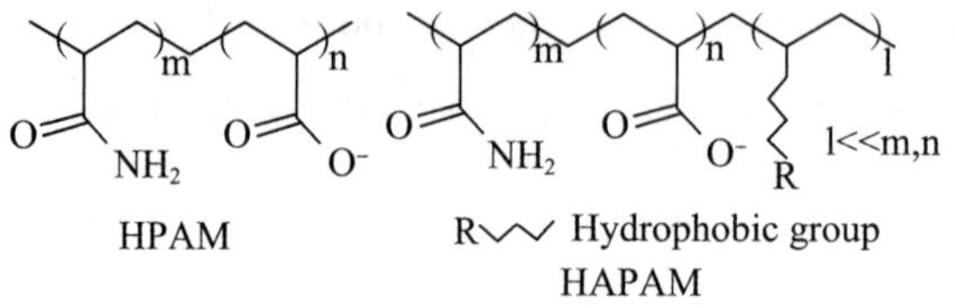

Fig. 1 The chemical schematic of HPAM and HAPAM.

$$RF \equiv \frac{\Delta P_P}{\Delta P_B} \tag{2}$$

$$RFF = \frac{k_b}{k_f} \tag{3}$$

Surprisingly, although the apparent viscosity of HAPAM is much higher than that of HPAM, the effective viscosity and the oil displacement efficiency of HAPAM are lower than HPAM. HPAM outperforms HAPAM in our experiment. To explain such behavior, the microstructures of polymers in solution and porous medium are direct observed by scanning electron microscopy (SEM) in this work. The specimens are prepared by accelerated freeze-drying in order to keep the polymer's original morphology in different circumstances. This process including rapid cooling and ice sublimation, no external force is applied (Feng et al., 2002), so the results obtained authentically reflect the information regarding the polymers in solution or in porous media.

2 Experimental

2.1 Materials

Two polyacrylamide samples were used in this study: conven-tional polyacrylamide (MO4000) and hydrophobically associating polyacrylamide (HNT275), all provided by Geological and Scientific Research Institute of SINOPEC Shengli Oilfield Company. The intrinsic viscosities of MO4000 and HNT275 were 2 213 mL/g and 1 451 mL/g respectively. The synthetic water for all the tests was a brine with $TSD=20,000$ mg/L (deionized water with 1.861 *wt*% sodium chloride and 0.139 *wt*% calcium chloride) and viscosity 1.22 mPa·s at 25℃. Oil sample used in this study was a mixture of dehydrated oil (asphaltenes 7.8%, resins 33.9%, oily constituents 58.3%) from Shengli Oilfield and kerosene, with viscosity of 20 mPa·s at 80℃. The displacement tests were conducted by using artificial cores and sandpacks, where the porosity was approximately 25% and the permeability to water was approximately 1 μm^2.

2.2 Experimental procedures

Apparent viscosities of polymer solutions were measured by using the Brookfield DV-Ⅱ Viscometer at a shear rate of 7.34 s^{-1} and 80℃. Laboratory oildisplacement experiment was carried out as follows. At first, the sandpack (φ25 mm×300 mm) was wet packed with sand of 200～250 mesh from the Jidong oil field. The wet-packed sandpack was flooded with brine, and permeability can be calculated. Then, the sandpack was flooded with the oil sample until water production ceased. Whereafter, brine flooding continued until water cut was greater than 98%. After that, 0.3 pore volume (PV) of polymer slug was injected, followed waterflood until the water cut of efflux reached 98% again. The brine water saturations and residual oil saturations can be calculated by volume equilibrium because the pore volume and volumes of effluent can be measured. More accurate results can be obtained by using the method of Al Mansoori and Pentland (Al Mansoori et al., 2010; Pentland et al., 2010) where conditions permit.

2.3 Lyophilization of polymer solution on mica sheet

Lyophilization is a dehydration process by freezing the mate-rial and then reducing the surrounding pressure to allow the frozen water in the material to sublimate. This makes the polymer to keep the original morphology in water solution.

About 3 μL of polymer solution was transferred to a fresh mica sheet by a micro-syringe. After freezing in liquid nitrogen for 0.5 h, the sample was quickly transferred to a lyophilizer (MELITEK A/S, Alslev, Denmark) in which the condenser temperature was −60℃. The sample was lyophilized for 24 h and the vacuum pressure was 7~8 Pa. All the other procedures were as described above.

2.4 Lyophilization of polymer solution in porous media

Two pore volume (PV) of polymer solution was injected into the artificial core (φ25 mm×70 mm). Next, the core was quickly transferred to a vacuum cup of liquid nitrogen and the vacuum cup was placed in freezer for 24 h at −20℃. Then, the core was put into lyophilizer for 24 h and the vacuum pressure was 7~8 Pa. After freeze-drying, the core was crushed 30~40 mm away from the inlet and core fragments were examined with scanning electronic microscope (SEM, Hitachi S-4800). The accelerating voltage was 3.0 kV.

3 Results and discussion

3.1 In aqueous solution

Fig. 2 shows the apparent viscosity as a function of the polymer concentration in brine with $TSD = 20,000$ mg/l for the two samples we chose in this work. For conventional partially hydrolyzed polymer MO4000, viscosity increased slowly with the increase of concentration without an abrupt change point. But for hydrophobically associating polymer HNT275, apparent viscosity exhibited sharp increase above the characteristic concentration (approximately 1 300 mg/L), which is the critical association concentration (CAC), with increase of concentration.

In order to get better understanding of why HAPAM solution shows stronger viscosity than traditional polymer, the microstructures of polymer solutions were investigated by SEM. Fig. 3a and b respectively shows the SEM micrographs of conventional polymer MO4000 and associative polymer HNT275 at 2 000 mg/L with 300 magnification level. Three-dimensional network structure formed in polymer skeletons both in conventional polymer and associative polymer is clearly observed. But the mashes of associative polymer network are significantly more compact than that of conventional polymer. Fig. 3c is a 2 500× magnification of one region in Fig. 3b. From it, a number of filaments are found attaching to the polymer skeletons, which are nearly non-existent in traditional polymer (Fig. 3a). These filaments have a size of about 0.02～0.08 μm in diameter, which is significantly smaller than the size of polymer skeletons(0.5～2.4 μm). This is likely because of the hydrophobic groups from associative polymer tending for intermolecular association together in aqueous solution, resulting in increase of interaction among the polymer skeletons. As a comparison, we investigated the morphologies when the concentration was below the critical association concentration. As presented in Fig. 3d at 600 mg/L, the mashes of associative polymer were no more compact than that shown in Fig. 3b. The filaments among polymer skeletons shown in Fig. 3e were also less than in Fig. 3c. For associative polymer, the unique thickening ability may be attributed to the compact sizes and distributed filaments in the network.

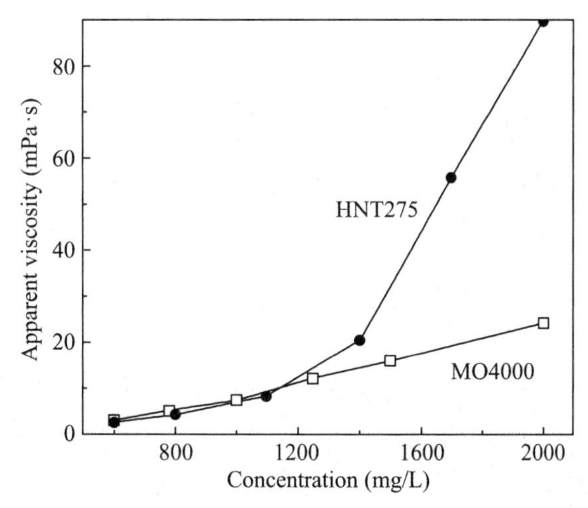

Fig. 2 Apparent viscosity versus concentration for different types of polymer.

3.2 In porous media

Although associative polymer shows the very high apparent viscosity by comparison with conventional polymer, the effective viscosity of the latter is higher than the former in porous media. Similarly, the abilities of HNT275 to increase oil recovery are lower than that of MO4000 as shown in Fig. 4. Compared with new polymer HNT275, the new surfactant classes (Iglauer et al., 2010) can meet the technical requirements as enhanced oil recovery (EOR) agents. Table 1 summarizes the properties of MO4000 and HNT275 at a concentration of 2 000 mg/l.

To make clear the reason for strange behavior of associative polymer, we investigated the microstructure of polymer solution in porous media. Firstly, we choose a low magnifying power of 70× in order to analyze samples with the widest possible field of vision. Fig. 5a and b shows the micrographs of conventional polymer MO4000 and associative polymer HNT275 at 2 000 mg/L in porous media respectively. It can be seen that both polymers still form networks in the pores, but the network scale of MO4000 is larger than that of HNT275. For HNT275, most networks are formed only among a limited number of polymer skeletons. It seems there are not enough polymer skeletons to form a continuous network. The local enlarged SEM photos are shown in Fig. 5c-f. It can be more clearly seen that the networks of conventional polymer are larger and more integrated than those of associative polymer. The sizes of most HNT275 networks are small and obviously have structure defects. Only one relatively complete network is found in whole associative polymer sample as shown in Fig. 5 g. Nevertheless, it is much smaller than MO4000's network as presented in Fig. 5h with the same magnification level (300×). The polymer skeletons of MO4000 have a size of about 0.32～6.13 μm in diameter, which shows little difference from 0.31 to 2.67 μm of HNT275.

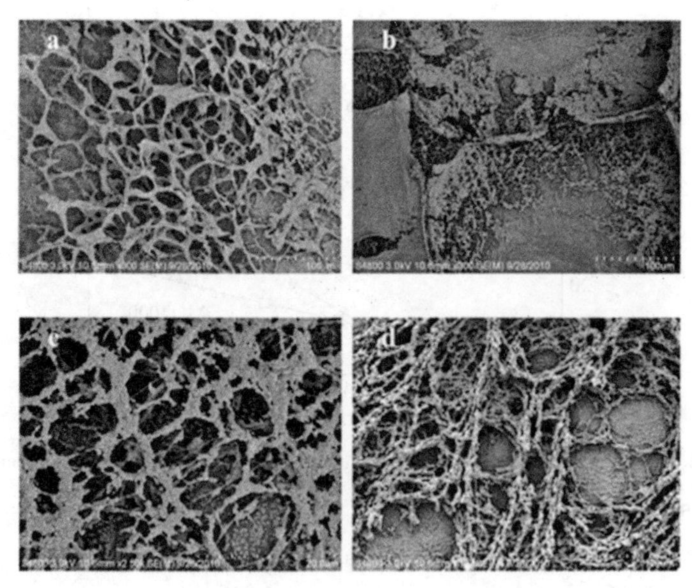

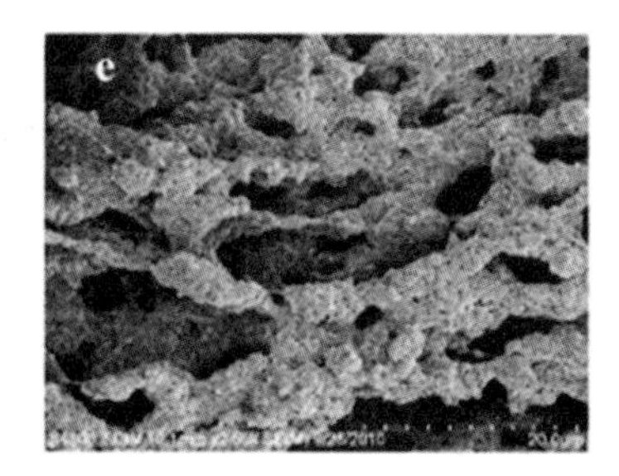

Fig. 3 Comparison of SEM observations for HNT275 and MO4000 in aqueous solution: (a) MO4000, 2 000 mg/L, 300× (b) HNT275, 2 000 mg/L, 300× (c) HNT275, 2 000 mg/L, 2500× (d) HNT275, 600 mg/L,300× (e) HNT275, 600 mg/L, 2500× .

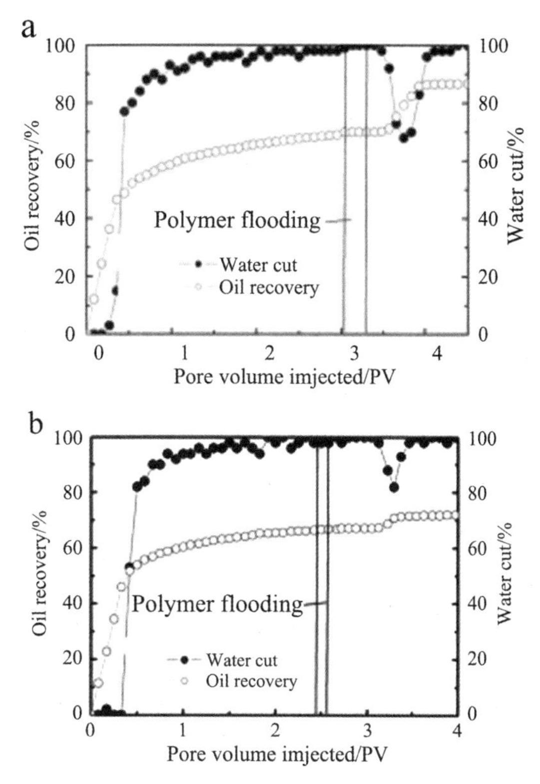

Fig. 4 Oil recovery and water cut curves of two polymer fl oodings at a concentration of 2000 mg/ l under a flow rate of 0.5 mL/min. (a: MO4000; b:HNT275).

The proposed interpretation for the phenomena, supported by the above-mentioned SEM results, is as follows: for associative polymer, polymer chains can adsorb by hydrophobic interaction to form multi-layer sorption because introduction of hydrophobic groups, the adsorbed layer being formed in part from some chains that are not directly in contact with the sand surface (Argillier et al., 1996). Whereas the conventional polymer forms only a 'classical' layer consisting of loops, tails, and trains (Volpert et al., 1998). Therefore, the absorption loss of HNT275 is larger than that of MO4000 in the forepart of the core, causing a reduction in the amount of movable polymer chains in the pores, which produce small and defective network leading to the decrease of effective viscosity. Additionally, the network formed by HNT275 is little more compact than MO4000 in porous medium. Thus associative polymer flows through porous media probably not in the form of associative structure, which also leads to the decrease of effective viscosity. The narrow space in porous may restrict HNT275 to form the associative structure. Generally, the molecular weight of the hydrophobically associating polymer is lower than that of conventional polymer (Wang et al. , 2005). So once the association cannot occur, its viscosifying power may be no better than conventional polymer.

4 Conclusions

1. Associative polymer HNT275 can significantly improve apparent viscosity of brine (*TSD* = 20, 000 mg/L) when the concentration exceeds critical association concentration (approximate 1 300 mg/l), while conventional polymer MO4000 cannot. However, the effective viscosity in porous media and oil displacement efficiency of HNT275 are lower than that of MO4000 at the same concentration of 2 000 mg/L.

2. In aqueous solution, network structures can be found both in the HNT275 and MO4000 solutions. But mashes of HNT275 network are significantly more compact than that of MO4000.

3. In the porous media, most of networks formed by MO4000 skeletons are larger and more integrated than HNT275. For HNT275, most networks are formed by a limited number of polymer skeletons and have structure defect. It may be attributed to the great absorption loss of HNT275 and spatial confinement in porous medium.

Table 1 The properties of MO4000 and HNT275 at a concentration of 2 000 mg/l.

Polymer	[a]μ_a (mPa·s)	*RRF*	*RF*	[b]μ_e (mPa·s)	[c]S_W(%)	[d]S_{OW}(%)	[e]S_{OP}(%)	[f]N_c	Tertiary(%) recovery
MO4000	24.1	5.21	55.3	3.82	78.0	22.0	9.8	0.23	16.6
HNT275	89.8	13.2	44.7	1.22	76.5	23.5	20.0	0.08	4.88

[a] Apparent viscosity.

[b] Effective viscosity.

[c] Water saturation after water flooding.

[d] Oil saturation after water flooding.

[e] Oil saturation after polymer flooding.

[f] Capillary number.

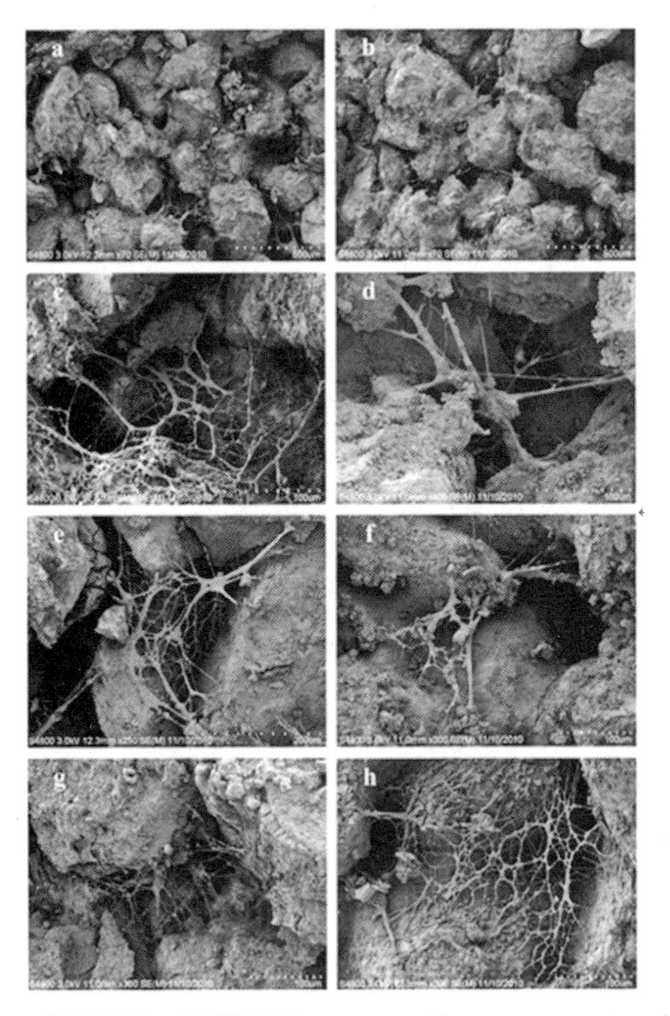

Fig. 5 SEM images of MO4000 and HNT275 in porous media at a concentration of 2 000 mg/l: (a) MO4000, 70×(b) HNT275, 70× (c) MO4000, 450×(d) HNT275, 400×(e) MO4000, 250 V (f) HNT275, 300×(g) HNT275, 300×(h) MO4000, 300×.

Acknowledgments

This work was sponsored by the National S&T Major Project (Grant no.: 2011ZX05024-004) on Chemical flooding for offshore heavy oil field, National 863 Program (Grant no: 2007AA090701) of research on EOR by polymer flooding in Bohai oilfield and "Taishan Scholars" Construction Engineering (no. ts20070704).

References

Al Mansoori, S. K., Itsekiri, E., Iglauer, S., Pentland, C. H., Bijeljic, B., Blunt, M. J., 2010. Measurements of non-wetting phase trapping applied to carbon dioxide storage. Int. J. Greenhouse Gas Control 4,283-288.

Argillier, J. F. , Audibert, A. , Lecourtier, J. , Moan, M. , Rousseau, L. , 1996. Solution and adsorption properties of hydrophobically associating water-soluble polyacrylamides. Colloids Surf. A113,247-257.

Cheng,J. C. , 1989. The flow characteristics of BP16 solution in porous medium. Pet. Geol. Oilfi eld Dev. Daqing (Chinese) 8,47-52.

Feng, Y. , Luo, P. , Luo, C. , Yan, Q. , 2002. Direct visualization of microstructures in hydrophobically modifi ed polyacrylamide aqueous solution by environmental scanning electron microscopy. Polym. Int. 51,931-993.

Feng, Y. J. , Billon, L. , Grassl, B. , Bastiat, G. , Borisov, O. , Franc-ois, J. , 2005. Hydrophobically associating polyacrylamides and their partially hydrolyzed derivatives prepared by post-modifi cation. 2. Properties of non-hydrolyzed polymers in pure water and brine. Polymer 46 ,9283-9295.

Gao, B. J. , Wu, N. , Li, Y. B. , 2005. Interaction between the strong anionic character of strong anions and the hydrophobic association property of hydrophobic blocks in macromolecular chains of a water-soluble copolymer. J. Appl. Polym. Sci. 96,714-722.

Gouveia, L. M. , Paillet, S. , Khoukh, A. , Grassl, B. , Müller, A. J. , 2008. The effect of the ionic strength on the rheological behavior of hydrophobically modifi ed polyacrylamide aqueous solutions mixed with sodium dodecyl sulfate (SDS) or cetyltrimethylammonium p — toluenesulfonate (CTAT). Colloids Surf. A 322, 211-218.

Hou, J. , Li, Z. , Cao, X. , Song, X. , 2009. Integrating genetic algorithm and support vector machine for polymer flooding production performance prediction. J. Pet. Sci. Eng. 68,29-39.

Iglauer, S. , Wu, Y. , Shuler, P. , Tang, Y. , Goddard III, W. A. , 2010. New surfactant classes for enhanced oil recovery and their tertiary oil recovery potential. J. Pet. Sci. Eng. 71,23-29.

Iglauer,S. , Wu,Y. , Shuler, P. , Tang, Y. , Goddard III, W. A. , 2011. Dilute iota-and kappa-Carrageenan solutions with high viscosities in high salinity brines. J. Pet. Sci. Eng. 75,304-311.

Pentland, C. H. , Itsekiri, E. , Al Mansoori, S. K. , Iglauer, S. , Bijeljic, B. , Blunt, M. J. , 2010. Measurement of nonwetting-phase trapping in sandpacks. SPE J. 15,270-277.

Shashkina, Y. A. , Zaroslov, Y. D. , Smirnov, V. A. , Philippova, O. E. , Khokhlov, A. R. , Pryakhina, T. A. , Churochkina, N. A. , 2003. Hydrophobic aggregation in aqueous solutions of hydrophobically modifi ed polyacrylamide in the vicinity of over-lap concentration. Polymer 44,2289-2293.

Volpert, E., Selb, J., Candau, F., Green, N., Argillier, J. F., Audibert, A., 1998. Adsorption of hydrophobically associating polyacrylamides on clay. Langmuir 14,1870-1879.

Wang, J., Luo, P. Y., Zhang, G. Q., 2001. The flow properties of aqueous solution of hydrophobically associating amphoteric polymer NAPs through porous media. Oilfi eld Chem. (Chinese) 18,152-154.

Wang, P. F., Du, M., Li, F. S., 2005. Study of polymerization conditions for synthesiz-ing high molecular weight hydrophobically associating water-soluble polymer. Appl. Chem. Ind. (Chinese) 34,705-707.

Yahaya, G. O., Ahdab, A. A., Ali, S. A., Abu-Sharkh, B. F., Hamad, E. Z., 2001. Solution behavior of hydrophobically associating water-soluble block copolymers of acrylamide and N-benzylacrylamide. Polymer 42,3363-3372.

剪切作用对功能聚合物微观结构性能的影响研究

李美蓉[1] 黄 漫[1] 曲彩霞[1] 曹绪龙[2] 张继超[2] 刘 坤[2]

(1. 中国石油大学(华东)理学院,山东青岛 266580;
2. 中国石化胜利油田地质科学研究院,山东东营 257015)

摘要:梳型聚合物和活性聚合物是目前常用驱油聚合物,其增黏性和黏弹性是评价其驱油能力的重要指标。为考察剪切作用对两种聚合物溶液性能的影响,分别研究了梳型聚合物和活性聚合物溶液经过模拟炮眼剪切前后的宏观和微观性能变化。结果表明,在高速剪切、拉伸应力作用下,梳型聚合物黏度损失率为40.73%,活性聚合物黏度损失率为70.10%;当剪切频率为0.02 Hz时,梳型聚合物界面扩张弹性降低了19.03%,而活性聚合物界面扩张弹性降低了68.03%;相比活性聚合物,梳型聚合物紧密的空间网状结构虽被部分破坏,但仍有疏松的网络结构,且以聚集体的形式紧密地分散在溶液中,通过DLS及AFM测定表明其粒径尺寸稍有变小;可见梳型聚合物抗剪切能力较活性聚合物强。

关键词:梳型聚合物;活性聚合物;炮眼剪切;黏度;界面扩张弹性
中图分类号:O631.1⁺3 文献标识码:A

Effect of shear action on the microcosmic structure and performance of functional polymer used in oil displacement

Li Meirong[1] Huang Man[1] Qu Caixia[1] Cao Xulong[2]
Zhang Jichao[2] Liu Kun[2]

(1. College of Science,China University of Petroleum(East China),Qingdao 266580,China;
2. Geological Scientific Research Institute Shengli Oilfield,SINOPEC,Dongying 257015,China)

Abstract:Comb-shaped polymer and reactive polymer are two kinds of polymers widely used in the tertiary recovery process at present and their viscosity and viscoelasticity are important indexes for evaluating the oil displacement efficiency. To investigate the effect of shear action on the properties of polymer solutions, the macroscopic and microcosmic properties of the comb-shaped polymer and reactive polymer, before and after shearing through the simulated hole cut, were compared. The results showed that at high-speed shearing and tensilestress, the viscosities of the comb-shaped polymer and the reactive polymer are decreased by 40.73% and 70.10%, respectively. With a shear frequency of 0.02 Hz, the interfacial dilatational elasticities of the combtype polymer and the reactive polymer are reduced by 19.03% and 68.03%, respectively. Compared with the reactivepolymer, the comb-shaped polymer still exhibits a loosen network structure and is present as aggregates dispersed in the solution after shearing, though its super-molecular structure is partially destroyed ; the particle size of the comb-shaped polymer, as measured by DLS and AFM, is only slightly reduced. These suggest that the comb-shaped polymer has a superior shear performance to the reactive polymer.

Keywords: comb-shaped polymer; reactive polymer; hole cut; viscosity; interfacial dilatational elasticity

针对目前使用的聚合物普通存在不耐温不抗盐的缺点，开发了耐温抗盐的梳型聚合物和活性聚合物。梳形聚合物是在高分子的侧链同时带亲油基团和亲水基团，由于亲油基团和亲水基团的相互排斥，使得分子内和分子间的卷曲、缠结减少，高分子链在水溶液中排列成梳形。活性聚合物是一种溶于水后吸水溶胀且可变形的高分子。它除了具有一般聚合物的性能外还具有一定的表面活性[1,2]。

驱油聚合物溶液的黏性可以扩大波及系数，从而提高原油的采收率；聚合物的界面扩张黏弹性则可以驱替部分盲孔中的残余油，提高洗油效率，进一步提高采收率[3-5]。功能聚合物溶液在现场注入地下的过程中，将经历一系列的机械剪切，比如注入管线与设备、注聚工艺、完井方式(炮眼)、近井地带的高速剪切降解作用，导致聚合物分子发生断裂，溶液黏度明显下降。近年来，针对炮眼附近的高速剪切和拉伸作用对聚合物溶液宏观性能的影响做了许多研究[6-8]，但对其界面流变性质的研究报道较少。

本实验针对梳型聚合物和活性聚合物，研究炮眼附近剪切作用对聚合物溶液性能的影响，探讨了聚合物溶液的黏度、界面扩张弹性及微观形态的变化，比较两种功能聚合物抗剪切能力，对指导油田开发具有重要意义。

1 实验部分

1.1 材料与试剂

模拟填沙管、乌氏黏度计、中间容器、D-250L 型恒速恒压泵(海安石油科技有限公司)、HW-4B型恒温箱(南通华兴石油仪器有限公司)、Brookfield DV-Ⅱ+Pro型黏度计(美国BROOKFIELD公司)、Zatasizer Nano ZS型高灵敏度动态光散射仪(英国马尔文公司)、S-4800冷场扫描电镜(日本日立公司)、美国DI公司NanoScope Ⅳ a原子力显微镜(美国DI公司)、JMP2000界面膨胀流变仪(上海中辰数字技术设备有限公司)。

活性聚合物、梳型聚合物均由胜利油田地质科学研究院提供,经无水乙醇抽提,真空干燥恒重备用,二次蒸馏水为实验室自制,其他试剂均为分析纯。

1.2 聚合物的增黏性测试

分别将5 000 mg/L的聚合物母液用二次蒸馏水稀释到所需的浓度,用电磁搅拌器搅拌1 h后在25℃下恒温20 min,用BrookfieldDV-Ⅱ+Pro型黏度计测定其黏度,转速为42.22 s^{-1}。

1.3 模拟剪切实验

1.3.1 注入速率

注入速率为10 $m^3/(m \cdot d)$。每米16个射孔,射孔孔眼直径为1 cm,深度为50 cm。计算可得渗流速率为0.046 1 cm/s。实验中渗流速率与实际渗流速率相等,填沙管直径为3.8 cm,计算注入速率30 mL/min。

1.3.2 剪切实验

向模拟装置(4.8 cm×52 cm)充填80~120目磨圆度较好的石英砂,孔隙度22%。按图1所示,将浓度为1 500 mg/L的聚合物溶液以一定的流量通过模拟段,收集产出液并密封保存。

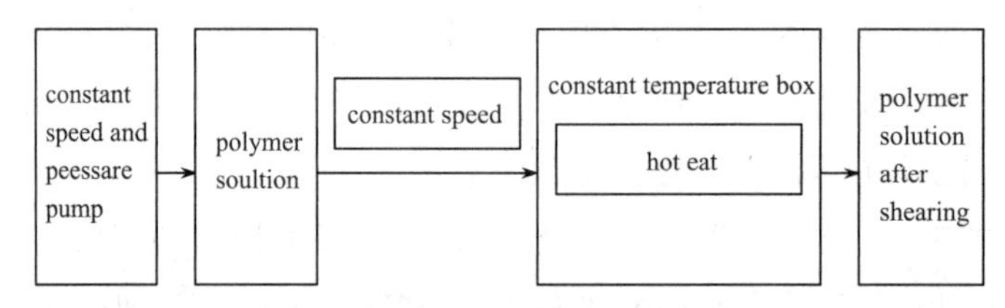

图1 模拟剪切装置流程示意图

Figure 1 Sketch map of process simulation of shear device

1.4 剪切前后功能聚合物界面扩张弹性的测定

Langmuir槽法测定界面扩张黏弹性,将5 000 mg/L的聚合物母液用去离子水稀释至1 500 mg/L聚合物溶液。首先调节滑障到固定的位置,然后挂上吊片将重力清零,必要时用微调按钮。将吊片浸没在溶液中,在恒温循环水浴中静置3 h,改变不同频率,测量剪切前后功能聚合物界面扩张弹性。

2 结果与讨论

2.1 功能聚合物的增黏能力比较

按1.2的实验方法,将5 000 mg/L的聚合物母液用二次蒸馏水稀释到不同的浓度,在25℃测定各浓度梳形聚合物和活性聚合物的黏度,结果见图2。

由图 2 可知,梳型聚合物的黏度比活性聚合物的黏度大。当浓度为 1 500 mg/L 时,梳型聚合物黏度达到 305 mPa·s;而活性聚合物的黏度偏小,黏度只有 182 mPa·s。这是因为梳形聚合物分子主链上带有含极性基团短侧链,侧链之间体积和电性排斥作用使聚合物主链不易卷缩、舒展程度好,呈梳型结构,且分子间的作用力及分子链的缠结作用随着聚合物分子在水溶液中浓度的增加而加强,因而梳形聚合物溶液黏度比较高[9]。

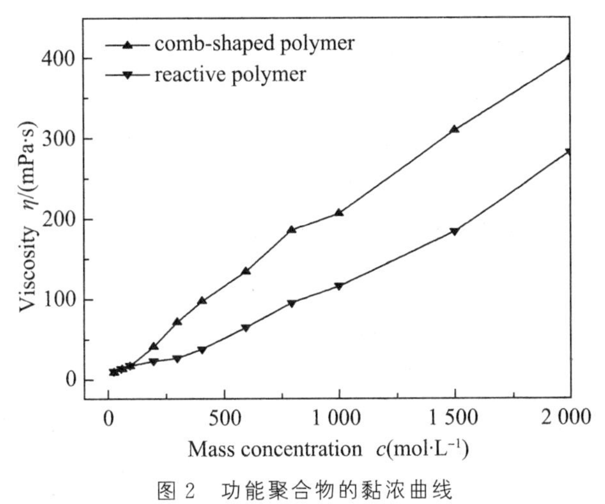

图 2 功能聚合物的黏浓曲线

Figure 2 Relationship between apparent viscosity and concentration of functional polymer

2.2 模拟剪切作用对功能聚合物宏观性能的影响

在高速剪切、拉伸应力作用下不同结构的功能聚合物抗剪切能力不同,分别对活性聚合物和梳型聚合物剪切前后黏度及分子量进行测量,计算黏度损失率,结果见表 1 和表 2。

表 1 不同聚合物在模拟剪切后黏度损失率

Table 1 Viscosity loss of different polymers after simulation shear

Polymer	Viscosity before shearing η/(mPa·s)	Viscosity after shearing η/(mPa·s)	Loss rate of viscosity /%
Reactive polymer	182.30	54.51	70.10
Comb-shaped polymer	305.20	180.89	40.73

表 2 功能聚合物在模拟剪切后相对分子量的变化

Table 2 Relative molecule mass loss of different polymers after simulation shear

Polymer	Relative molecule mass before shearing	Relative molecule mass after shearing	Loss rate of molecule mass%
Reactive polymer	9.79×10^5	5.99×10^5	38.82
Comb-shaped polymer	4.08×10^6	1.73×10^6	57.60

由表 1 可知,梳型聚合物黏度损失率为 40.73%,活性聚合物黏度损失率为 70.10%,黏度损失率是梳型聚合物<活性聚合物。由表 2 可知,分子量降低幅度是梳型聚合物>活性聚合物,这与两种聚合物黏度损失率的顺序并不一致,说明模拟剪切后,聚合物黏度的降低并不是与分子量的变化直接关联,而主要受聚合物缔合微观结构的变化影响。

2.3 剪切作用对功能聚合物微观结构与性能的影响分析

2.3.1 功能聚合物剪切前后表面形貌的改变

聚合物溶液的表观黏度由非结构黏度 η(非)和结构黏度 η(结)构成。非结构黏度是指聚合物溶于水以后,大分子链在水溶液中水化伸展,同时包裹一定水,此时分子的流体力学体积增大,使水溶液表观黏度提高;结构黏度是指由于大分子链间相互作用(缠绕、缔合、化学键力等),使聚合物溶液整体流体力学体积增大,表观黏度升高。用 S-4800 冷场扫描电镜观察两种功能聚合物现场浓度 1 500 mg/L 时模拟剪切前后微观结构。选择电压 5.0 kV,放大倍数为 20 000 倍,实验结果见图 3。

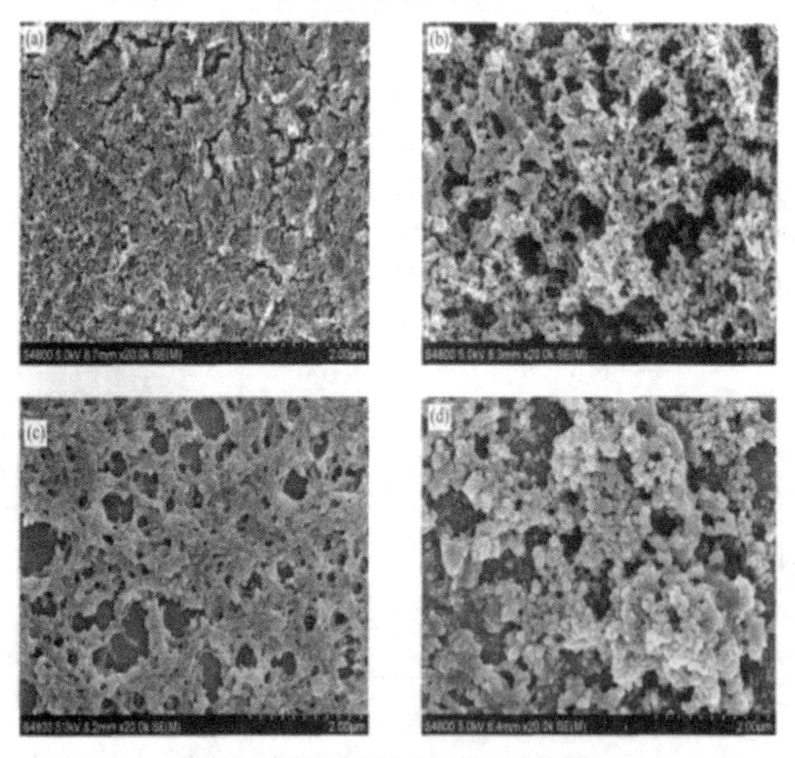

图 3 功能聚合物模拟剪切前后 SEM 照片

Figure 3 SEM images of functional polymer before and after simulations shear (a): reactive polymer SEM before simulations shear;(b): reactive polymer SEM after simulations shear (c): comb-polymer SEM before simulations shear; (d): comb-shaped polymer SEM after simulations shear

由图 3(a)可知,模拟剪切前活性聚合物中多个分子聚集在一起,形成相对独立的聚集体,长分子链相互缠绕,使聚集体之间形成连续空间网络结构,因聚集及缠绕程度均不是很强,骨架较细,故黏度不是很大;由图 3(b)可知,模拟剪切后,炮眼剪切、拉伸作用破坏了活性聚合物溶液长分子链相互缠绕形成的空间网络结构,形成相对独立的小聚集体,因而剪切后聚合物溶液的表观黏度下降明显。

由图 3(c)可知,模拟剪切前梳型聚合物分子链很长,相互缠绕形成较粗的骨架,形成具有不同尺寸孔洞的多层次立体网络结构。这种网络结构既有支撑作用,同时也与水介质之间存在溶剂化作用,在分子表面形成了较厚的水化膜,存在"束缚水",吸附和包裹大量水分子产生形变阻力,网状结构刚性增强,聚合物黏度很大。由图 3(d)知,在高速剪切、拉伸应力作用下,含有极性基团短侧链和较长的分子链可能被切断,分子链舒展程度变差、易卷曲,分子链不规则地缠结在一起,因而剪切后聚合物溶液的表观黏度也有所下降[10]。

2.3.2 功能聚合物模拟剪切前后聚合物粒径分布

用 Zatasizer Nano ZS 型高灵敏度动态光散射仪测定功能聚合物溶液浓度为 500 mg/L 的粒径分布[11]。以 He-Ne 光源,测定功率 10 mW,波长 633 nm,散射光强测定角度为 173°,温度为 25℃,实验结果见图 4。

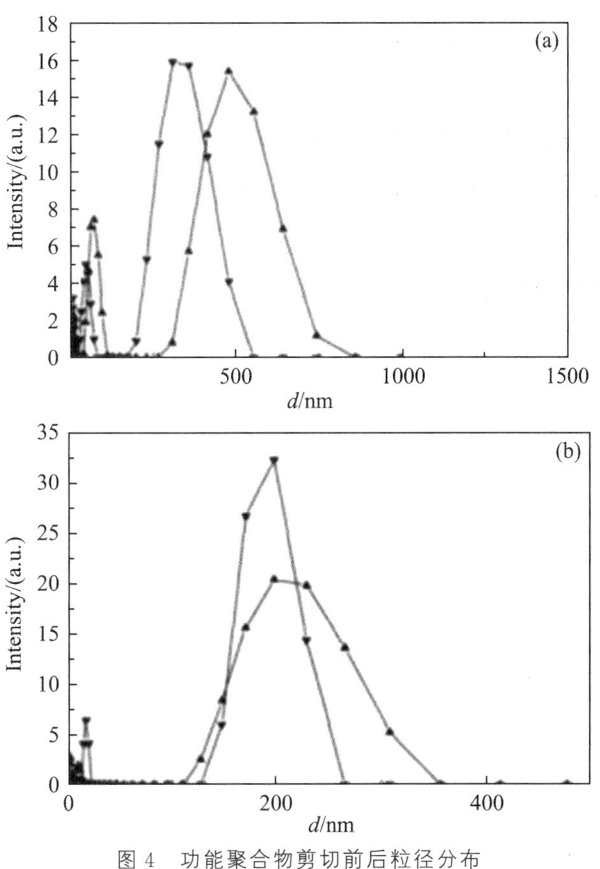

图 4 功能聚合物剪切前后粒径分布

Figure 4 Intensity distribution of functional polymers solutions by DLS before and after simulations shear (a): intensity distribution of reactive polymer; (b): intensity distribution of comb-shaped polymer ▲: before simulations shear; ▼ : after simulations shear

由图 4 可知,模拟剪切后,两种聚合物粒径分布峰均左移,说明模拟剪切的过程中,分子链被切断,分子链之间的缠结作用下降,粒径变小。梳型聚合物中值粒径变化 220～200 nm,而活性聚合物中值粒径变化 500～300 nm;梳型聚合物粒径尺寸变化幅度小,说明梳型聚合物抗剪切能力较强。

2.3.3 功能聚合物模拟剪切前后 AFM 分析

采用 NanoScope Ⅳ a 原子力显微镜,所用探针为商用 Si_3N_4 探针,以轻敲模式在 25℃下观察剪切前后两种功能聚合物溶液成膜后的微观形貌[12],结果见图 5。

由图 5 中活性聚合物溶液剪切前后的 AFM 可以看出,剪切前聚合物形成很多聚集体,在聚集体之间,又以分子间缔合的方式形成网络结构;炮眼剪切后层间的网络结构被破坏,局部分布有零散的小聚集体,通过图 5(a)和(b)的立体图可以看出剪切前后粒径尺寸明显变小,且粒径分析与图 4(a)动态光散射的结果相吻合。

由梳型聚合物溶液剪切前后的 AFM 可以看出,剪切前聚合物溶液中形成了结构比较紧密的聚网络结构;剪切后虽然紧密的空间网状结构被部分破坏,但仍有疏松的网络结构,且以聚集体的形式紧密地分散在溶液中,剪切后粒径尺寸稍有变小,具体见图 5(c)和(d)的立体图,与图 4(b)动态光散射的结果相吻合。

2.3.4 功能聚合物模拟剪切前后界面扩张弹性的测定

曹宝格[13]研究得出,聚合物溶液的弹性是指聚合物溶液在应力作用下溶液变形过程中能量的储存,与溶液内聚合物分子链间作用以及形成的结构有关。界面扩张弹性表征微观界面膜的强度,强度越大,携带出的残余油量越多,驱替效率越高;经过剪切后功能聚合物溶液的结构发生了改变,其界面扩张弹性随频率的变化见图 6。

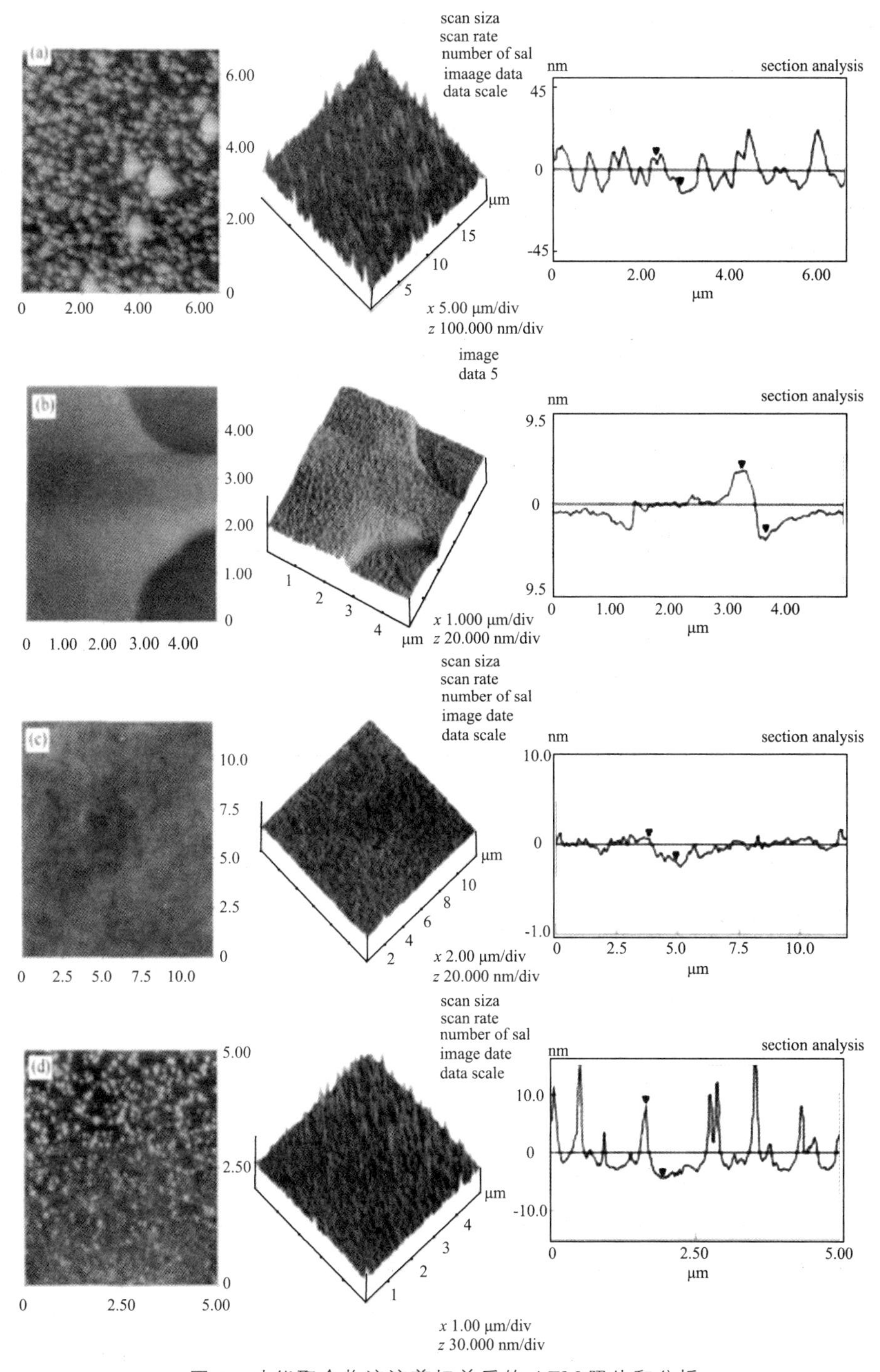

图 5 功能聚合物溶液剪切前后的 AFM 照片和分析

Figure 5 AFM images and section analysis of functional polymers before and after simulations shear (a): AFM images and section analysis of reactive polymer before simulations shear; (b): AFM images and section analysis of reactive polymer after simulations shear; (c): AFM images and section analysis of comb-shaped polymer before simulations shear; (d): AFM images and section analysis of comb-shaped polymer after simulations shear

由图 6 可知,界面扩张弹性随测量频率的增大逐渐增加并在一定时间后趋于平缓,聚合物分子在界面达到吸附平衡需要一定的时间。在高剪切速率下,破坏了功能聚合物分子聚集体间和分子间的缔合结构,分子链发生卷曲,流体力学体积减小,从而使界面扩张黏性降低;相比活性聚合物,梳型聚合物降低的幅度小,当剪切频率为 0.02 Hz 时,梳型聚合物界面扩张弹性降低了 19.03%,而活性聚合物界面扩张弹性降低了 68.03%。这与表观黏度降低的顺序一致,界面扩张弹性再次证明了梳型聚合物抗剪切能力大于活性聚合物。

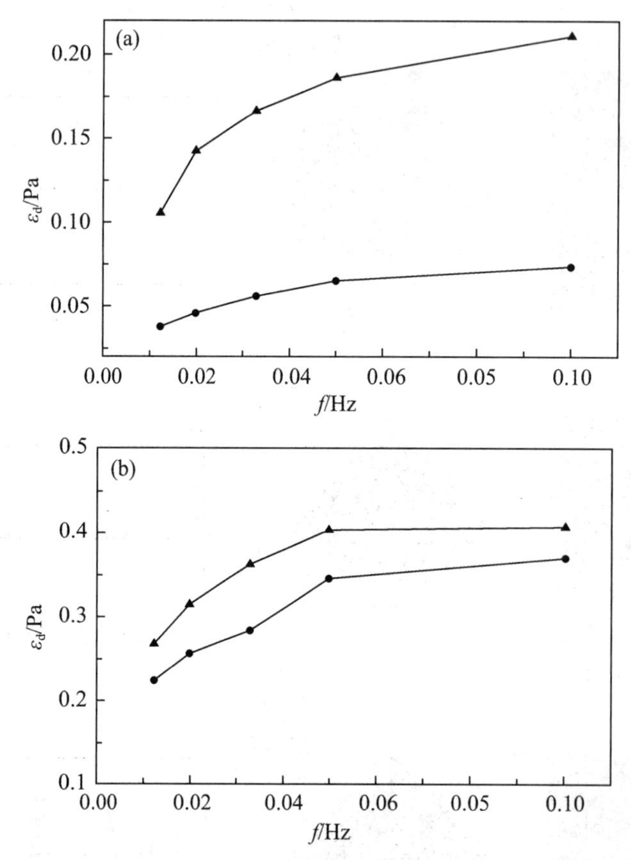

图 6 功能聚合物溶液剪切前后的界面扩张弹性 ε_d 与频率 f 的关系

Figure 6 Relationship betw een interfacial dilatational elasticity and frequency of functional polymers solution before and after simulations shear (a): relationship betw een interfacial dilatational elasticity and frequency of reactive polymer solution; (b): relationship betw een interfacial dilatational elasticity and frequency of comb-shaped polymer solution ●: after simulations shear; ▲: before simulations shear

3 结论

梳型聚合物、活性聚合物为现场驱油用的两种新型功能聚合物。梳型聚合物具有刚性不易卷曲的梳型结构,可形成骨架较粗多层次立体网络结构,增黏效果较强;在高速剪切、拉伸应力作用下，梳型聚合物中含有极性基团短侧链和较长的分子链被切断，分子链舒展程度变差、易卷曲，分子链不规则地缠结在一起，粒径尺寸变化幅度小，因而剪切后聚合物溶液的表观黏度有所下降，界面扩张弹性降低的幅度较小；而活性聚合物溶液长分子链相互缠绕聚合溶液长分子链断离卷曲，形成相对独立的小聚集体，粒径尺寸变化幅度大，因而剪切后聚合物溶液的表观黏度明显下降，界面扩张弹性降低的幅度也较大；梳型聚合物比活性聚合物抗剪切能力强。

参考文献

[1] 梁伟，赵修太，韩有祥，孟凡召，张新慧. 驱油用耐温抗盐聚合物研究进展 [J]. 特种油气藏，2010，17(2)：11-14.

(LIANG Wei，ZHAO Xiu-tai，HAN You-xiang，M ENG Fan-zhao，ZHANG Xin-hui. Research progress on heat and salt resistance polymer flooding special [J]. Oil and Gas Reservoirs，2010，17 (2)：11-14.)

[2] 韩玉贵. 耐温抗盐驱油用化学剂研究进展 [J]. 西南石油大学学报(自然科学版)，2011，33 (3)：150-153.

(HAN Yu-gui. Research progress on heat and salt resistance chemical medicament flooding special [J]. Journal of Southw est Petroleum University, 2011, 33 (3): 150-153.)

[3] 王立军，王德民. 聚合物溶液粘弹性对提高驱油效率的作用 [D]. 大庆：大庆石油学院，2003.

(WANG Li-jun, WANG De-ming. The action that viseoelasticity of Poymer solution on enhancing oil displacement efficiency [D]. 大庆: Daqing Petroleum Institute，2003.)

[4] 夏惠芬，王德民，刘中春，杨清彦. 粘弹性聚合物溶液提高微观驱油效率的机理研究 [J]. 石油学报，2001，22(4)：61-65.

(XIA Hui-fen，WANG De-ming，LIU Zhong -chun，YANG Qing -yan. Research on mechanism of improving microscopic oil displacement efficiency of viscoelastic polymer solution [J]. Acta Petrolei Sinica，2001，22(2)：61-65.)

[5] 张宏方，王德民，岳湘安，王立军. 利用聚合物溶液提高驱油效率的实验研究 [J]. 石油学报，2004，25(2)：55-58.

(ZHANG Hong-fang, WANG De-ming, YUE Xiang-an, WANG Li-jun. Experimental study on enhancing displacement efficiencies using polymer solutions [J]. Acta Petrolei Sinica，2004，25 (2)：55-58.)

[6] 曹绪龙，张磊，程建波. Gemini 表面活性剂对疏水缔合聚丙烯酰胺界面吸附膜

扩张流变性质的影响 [J]. 高等学校化学学报,2010,21(5):998-1002.

(CAO Xu-long,ZHANG Lei,CHENG Jian-bo. Effect of Gemini surfactant on the interfacial dilational properties of hydrophobically modified partly hydrolyzed polyacrylamide [J]. Chemical Research in Chinese Universities. 2010, 31 (5): 998-1002.)

[7] 施雷庭,徐豪飞,叶仲斌,向问陶,周薇,赵文森. 剪切作用对不同聚合物溶液流度控制能力影响研究 [J]. 油田化学,2010,27(2):175-178.

(SHI Lei-ting,XU Hao-fei,YE Zhong-bin,XIANG Wen-tao,ZHOU Wei,ZHAO Wen-shen. The effects of shearing on the mobility control capacity of different polymer solutions [J]. Oilfield Chemistry,2010,27(2):175-178.)

[8] 叶仲斌,彭杨,施雷庭,舒政,陈洪. 多孔介质剪切作用对聚合物溶液粘弹性及驱油效果的影响 [J]. 油气地质与采收率,2008,15(9):59-62.

(YE Zhong -bin, PENG Yang, SHI Lei-ting, SHU Zheng, CHEN Hong. The influence of shear action in porous medium on viscoelasticity and oil displacement efficiency of polymer [J]. Petroleum Geology and Recovery Efficiency,2008,15(9):59-62.)

[9] 陈洪,韩利娟,徐鹏,罗平亚. 疏水改性聚丙烯酰胺的增粘机理研究 [J]. 物理化学学报,2003,19(11):1020～1024.

(CHEN Hong,HAN Li-juan,XU Peng,LUO Ping -ya. The thickening methenism study of hydrophobically modified polycrylamide [J]. Acta Physico -Chimica Sinica, 2003,19 (11):1020-1024.)

[10] 朱怀江,赵常青,罗健辉,魏宝和,杨静波,朱德升,聚合物水化分子的微观结构研究 [J]. 电子显微学报,2005,24(3):205-210.

(ZHU Huai-jiang,ZHAO Chang -qing,LUO Jian-hui,WEI Bao -he,YANG Jing -bo,ZHU De-sheng. Study on the microstructure of polymer molecule hydrate [J]. Journal of Chinese Electron M icroscopy Society,2005,24 (3) : 205-210.)

[11] ZHANG YX, WUFP, LIMZ, WANG E J. Novel approach to synthesizing hydrophobically associating copolymer using template copolymerization: The synthesis and behaviors of acrylamide and 4-(ω-propenoyloxyethoxy) benzoic acid copolymer [J]. J Phys Chem B, 2005,109(47):22250-22255.

[12] 张瑞,叶仲斌,罗平亚. 原子力显微镜在聚合物溶液结构研究中的应用 [J]. 电子显微学报,2010,29(5):475-481.

(ZHANG Rui,YE Zhong-bin,LUO Ping -ya. The atomic force microscopy study on the microstructure of the polymer solution [J]. Journal of Chinese Electron M icroscopy Society,2010,29(5):475-481.)

[13] 曹宝格.驱油用疏水缔合聚合物溶液的流变性及粘弹性试验研究 [D].成都:西南石油学院,2006.

(Cao Baoge. The experiment study on the rheological properties and viscoelastic of hydrophobic associating polymer solution used to displace crude oil [D] . Chengdu: Southwest Petroleum University,2006.)

两类聚合物溶液黏度在岩心渗流方向上的分布研究

宋新旺[1] 吴志伟[2*] 曹绪龙[1] 韩玉贵[1] 岳湘安[2] 张立娟[2]

(中国石化胜利油田地质科学研究院1,东营 257015;

中国石油大学(北京)石油工程教育部重点实验室2,北京 102249)

摘要:为更清晰地认识聚合物溶液在油藏深部黏度的作用,以胜利孤东七区油藏非均质强、储层结构疏松为条件,开展了各种长度下岩心聚合物溶液渗流实验。在岩心出口端取采出液,研究两类聚合物溶液在近井和油藏深部黏度分布特性。结果表明:相同注入速度下,质量浓度为1 500 mg/L的聚合物HJ、B4以及DH5溶液黏度保留率分别为0.817,0.815,0.335,要高于质量浓度为1 800 mg/L的值;相同浓度的聚合物溶液,注入速度越高,黏度略微下降,变化并不明显;部分水解聚丙烯酰胺聚合物HJ和B4在油藏深部(20%～80%)黏度保留率分别为0.642,0.443,比相同质量浓度DH5略高。

关键词:聚合物驱 黏度保留率 局部PV数 注入速度 质量浓度

中图法分类号:TE357.46; 文献标志码:A

随着聚合物驱的广泛应用,黏度成为评价聚合物驱效果的重要指标,聚合物溶液黏度在配制、搅拌、输送、通过泵和阀门、炮眼、地层过程中出现损失[1-3],而随着各大油田相继进入高水期,近井地带的剩余油较少,原油主要集中在远井地带,研究聚合物溶液在远井地带的黏度对认识其在油藏深部提高采收率的作用至关重要。

针对聚合物溶液黏度损失的研究很多,主要采用静态实验、短岩心渗流、采出液循环注入方式研究聚合物溶液黏度受到矿化度、温度、渗透率、机械降解和吸附滞留的影响[4-6],但是采用物理模拟方式研究聚合物溶液黏度损失在渗流方向的分布鲜有报道。现采用三种不同聚合物溶液进行了岩心渗流实验,分析了聚合物类型、质量浓度和注入速度对聚合物溶液黏度的影响,研究了聚合物溶液在渗流方向上的黏度损失分布,对研究聚合物溶液在油藏深部提高采收率幅度具有重要意义。

1 聚合物溶液在短岩心中渗流实验

1.1 实验设备与材料

实验设备与仪器:2PB00C型平流泵,恒温箱,高压中间容器,岩心夹持器(Ø2.5 cm

×10 cm)，压力采集系统，BrookfieldDV-II+黏度计，德国 HAAKE 公司产 RS6000 流变仪。渗流用聚合物：HJ(相对分子质量 2.2×10^7，固含量 91%)，B4(相对分子质量 2.6×10^7，固含量 90.94%)，DH5(相对分子质量 2.0×10^7，固含量 95.23%)，其中 HJ，B4 为不同类别的部分水解聚丙烯酰胺 HPAM，DH5 为疏水缔合聚合物。实验人造岩心：渗透率范围为(1～1.5)μm^2，孔隙度为 27.07%。实验用模拟地层水，其总矿化度和离子浓度见表 1。模拟胜利孤东油藏温度为 68℃。

表 1 地层水各类离子含量及总矿化度 mg/L

离子	含量	离子	含量
Na^++K^+	3 056	SO_4^{2-}	0
Ca^{2+}	223	HCO_3^-	431
Mg^{2+}	60	CO_3^{2-}	41
Cl^-	4 984	总矿化度	8 795

1.2 实验方法

(1) 对烘干岩心抽真空，饱和模拟地层水，计算孔隙体积；

(2) 油藏温度 68℃ 下恒温 12 h 后，用地层矿化度水测渗透率；

(3) 将三种不同的聚合物配制成质量浓度为 1 500 mg/L 和 1 800 mg/L 的溶液，分别以 15 m/d、10 m/d、5 m/d 和 1 m/d 速度注入 10 cm 岩心中，直到各测压点压力达到稳定，并保持围压高于入口端压力(2.0～3.0)MPa，同时隔一定时间取采出液，并测量其黏度变化；

(4) 地层水后续水驱，直到入口端压力稳定，隔一定时间测采出液黏度变化。

1.3 实验结果

1.3.1 聚合物溶液质量浓度对黏度损失的影响

以注入速度 1 m/d 将质量浓度为 1 500 mg/L 和 1 800 mg/L 的三种聚合物溶液注入到 10 cm 长岩心中，不同质量浓度的聚合物溶液的黏度损失不同，黏度损失率随 PV 数变化曲线如图 1 所示。

从图 1 可以看出，黏度损失率随注入量的增加开始降低，之后逐渐增加并平稳在一定值。原因在于开始由于水的稀释、剪切和滞留作用导致黏度迅速下降，随着注入量的增加，聚合物溶液的滞留逐步达到饱和，此时黏度损失主要由剪切引起，注入速度一定时，对聚合物分子的剪切作用趋于稳定。对于三种不同聚合物溶液 HJ、B4 和 DH5 而言，质量浓度 1 500 mg/L 的聚合物溶液黏度保留率分别为 0.817，0.815，0.335，大于质量浓度为 1 800 mg/L 的值，同时质量浓度越高，由吸附滞留带来的黏度损失越小，但对于高质量浓度的聚合物溶液，单位体积内聚合物分子发生缠绕、缔合的几率增大，形成强度较高的网状结构，从而使其结构黏度[7]增加，在经过岩心孔道结构时，孔隙喉道对网状结构的剪切使其变成小分子，网状结构的强度降低，使其黏度保留率降低。从图 1 中还可以看出,对于两种不同类型的聚合物溶液 HPAM 和疏水缔合聚合物,随着注入量的增加,其黏度保留率的平稳值完全不同,相同质量浓度的 HPAM 溶液的黏度保留率明显高于疏水缔合聚合物溶液,主要因为疏水缔合聚合物分子网状结构较牢固[8-10],其注入性较差,大部分聚合物分子主要堆积在岩心的前端,导致进入到岩心的聚合物溶液浓度较低,产出液的黏度较小。

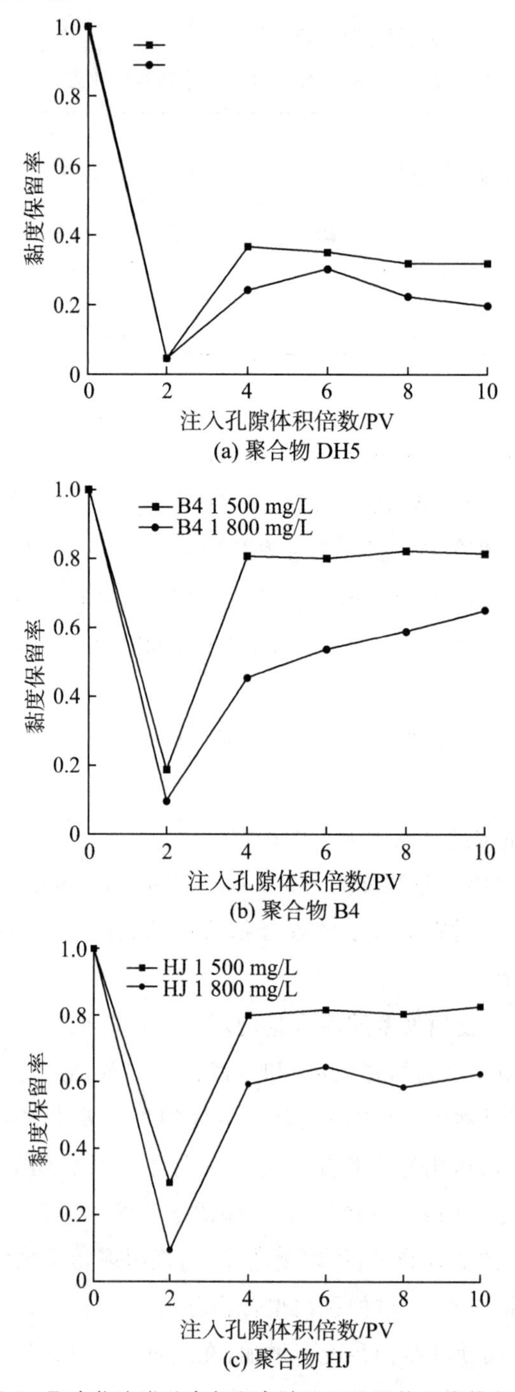

图 1 聚合物溶液黏度保留率随注入孔隙体积倍数变化

1.3.2 注入速度对黏度损失的影响

在 1 m/d、5 m/d、10 m/d、15 m/d 注入速度下向 10 cm 岩心中注入三种质量浓度为 1 500 mg/L 和 1 800 mg/L 的聚合物溶液，通过测试渗流压力达到平稳时采出液黏度，得到黏度保留率与注入速度关系曲线如图 2 所示。

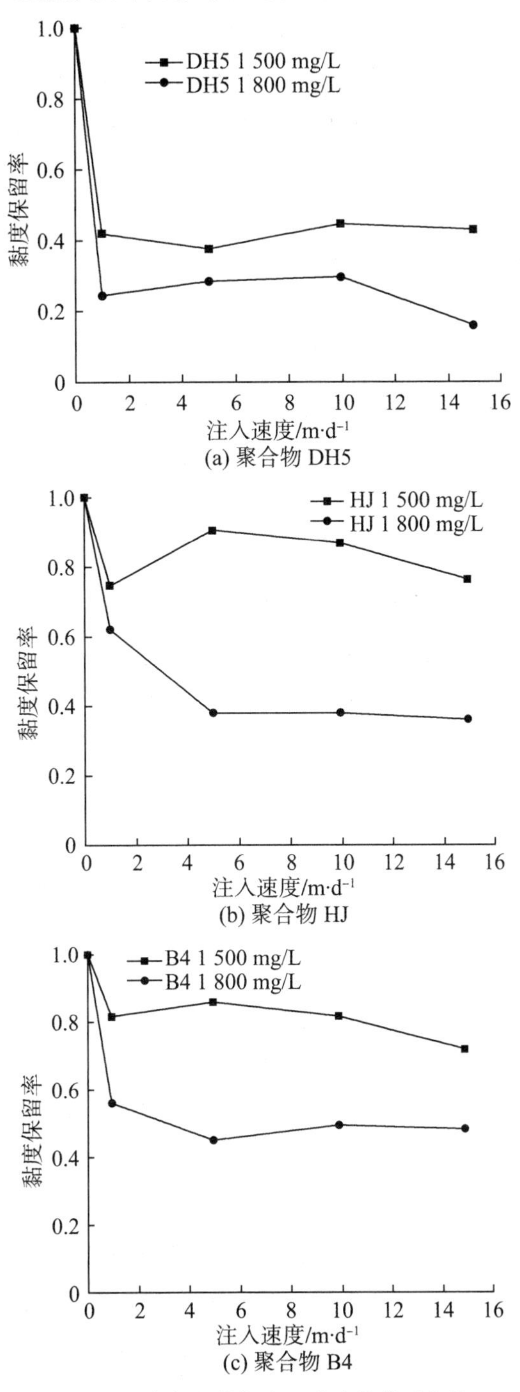

图 2　黏度保留率与注入速度的关系曲线

由图 2 可知,同一类型聚合物溶液,黏度保留率随着注入速度增加整体上是减小的,聚合物溶液表现出剪切稀释性,对三种不同聚合物 DH5、B4 和 HJ 而言,两种不同质量浓度 1 500 mg/L 和 1 800 mg/L 的黏度保留率相差不同,其中 DH5 相差最小,B4 其次,HJ 相差最大,其差值分别为 0.173,0.304,0.382。相同注入速度和相近岩心孔隙结构下,对于质量浓度为 1 500 mg/L 和 1 800 mg/L 的聚合物溶液,质量浓度越低,黏度保留率越大,主要由于质量浓度越高,单位体积内聚合物分子缠绕的几率越大,形成的网状结构越牢固,因此剪切作用越大,对较高质量浓度的聚合物分子而言,剪切造成的黏度损失越大。

2 长岩心中聚合物溶液渗流实验

2.1 实验条件和方法

100 cm 岩心夹持器,其余条件和短岩心渗流实验相同。将质量浓度为 1 500 mg/L 和 1 800 mg/L 的三种聚合物溶液以 5 m/d 的速度分别注到 40 cm、60 cm 和 100 cm 岩心中,隔一段时间测量采出液黏度。将 10 cm、40 cm、60 cm 和 80 cm 岩心的黏度变化折算到 80 cm 的不同位置处黏度变化,因此引入局部 PV 数概念。

2.2 局部 PV 数

矿场应用聚合物驱时,聚合物溶液的用量在 0.25~0.5 PV,该 PV 数是针对整体油藏,对于近井部位,聚合物溶液通过的用量较远井部分要大得多,因此,提出的局部 PV 数是针对某一渗流段通过该段地层的聚合物溶液用量,用 V_r 表示。对于单向渗流油藏而言

$$V_r=\frac{L-L_1}{L_1}V_w \tag{1}$$

式(1)中,L 为渗流的整个岩心长度,m;L_1 为渗流的局部段长度,m,$L>L_1$;V_w,相对整个岩心的 PV 数,即整体 PV 数。

对于径向流油藏而言

$$V_r'=\frac{\pi R^2-\pi r^2}{\pi r^2}V_w'=\left(\frac{R^2}{r^2}-1\right)V_w' \tag{2}$$

式(2)中,R 为注聚整体油藏半径,m;r,渗流段距井口距离,m;V_r',V_w',分别为局部 PV 数和整体油藏 PV 数。

2.3 聚合物溶液的黏度在渗流方向上的分布

对渗流方向上黏度分布的研究,采用长岩心在各测压点取样,会引起岩心内部压力波动从而导致测试结果与实际油藏不符。现采用渗流率和孔隙结构相似的不同长度岩心,将不同岩心的长度折算到油藏不同位置处,测试不同长度岩心出口端采出液黏度即测试油藏不同位置处黏度变化,避免取样带来内部压力扰动。将 10 cm、40 cm、60 cm 和 80 cm 转换为油藏 1/8、1/2、3/4 和 1 处,以 7.34 s^{-1} 的黏度值为例,1 500 mg/L 的聚合

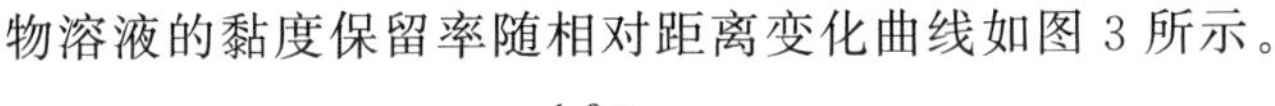
物溶液的黏度保留率随相对距离变化曲线如图 3 所示。

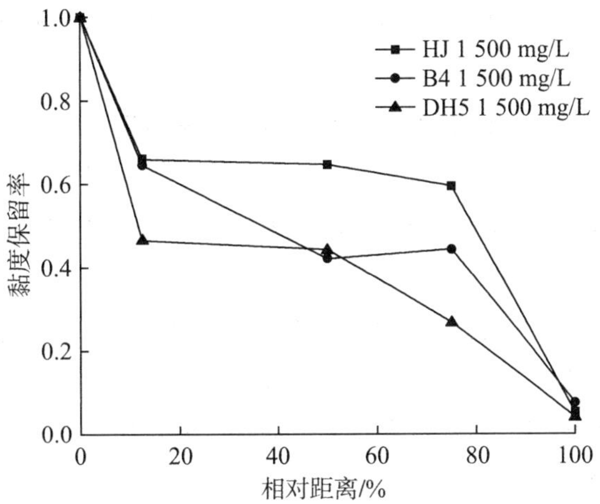

图 3　黏度保留率在渗流方向上的分布曲线

由图 3 可知，对于相同质量浓度的 HJ、B4 和 DH5，疏水缔合聚合物 DH5 在前端注入性差，易堵塞，大部分聚合物溶液滞留在端面，其在油藏深部（20%～80%）黏度保留率仅为 0.438，而 HJ 和 B4 的黏度保留率分别为 0.642 和 0.443，比 DH5 略高，而三种聚合物溶液的黏度损失主要集中在近井地带。通过测试 80 cm 岩心出口端采出液 1PV 时的黏度得出，聚合物溶液渗流前端的黏度急剧下降是由于剪切、吸附滞留和地层水的稀释共同作用。

3　结论

（1）聚合物溶液在渗流过程中，质量浓度对于黏度损失率影响较大，而注入速度对黏度损失率影响较小，因此对于不同油藏条件，在保证良好注入性条件下，应优选出适宜特定油藏条件的质量浓度。

（2）聚合物溶液渗流前端受到地层矿化度水、剪切、滞留作用，造成黏度损失严重，疏水缔合聚合物溶液易在前端造成堵塞，注入性差，黏度保留率低。

（3）聚合物溶液在油藏中黏度损失主要在近井地带，疏水缔合聚合物 DH5 和部分水解聚丙烯酰胺 B4、HJ 在油藏深部的黏度保留率分别为 0.438，0.443 和 0.642。

参考文献

[1] 陈明强，孙志强，王江顺，等. 配注过程中聚合物溶液的黏度损失分析. 西安石油大学学报（自然科学版），2007；22(3)：60-63

[2] 李兆敏，陈辉，黄善波，等. 聚合物动态剪切实验研究. 西安石油大学学报（自然科学版），2009；24(3)：69-72

[3] 邵振波，周吉生，孙刚，等. 部分水解聚丙烯酰胺驱油过程中机械降解研究. 油田

化学,2005;22(1):72-77

[4] 施雷庭,徐豪飞,叶仲斌,等.剪切作用对不同聚合物溶液流度控制能力影响研究.油田化学,2010;27(2):174-178

[5] 王启民,廖广志,牛金刚.聚合物驱油技术的实践与认识.大庆石油地质与开发,1999;18(4):1-5

[6] 袁敏,贾忠伟,袁纯玉.聚合物溶液黏弹性影响因素研究.大庆石油地质与开发,2005;24(5):74-76

[7] 吴文祥,张向宇.不同分子量聚合物溶液在多孔介质中的渗流特性研究.西安石油大学学报(自然科学版),2007;22(2):103-106

[8] 张宏方,王德民,王立军.聚合物溶液在多孔介质中的渗流规律及其提高驱油效率的机理.大庆石油地质与开发,2002;21(8):57-60

[9] 张云宝,任艳滨,卢祥国.几种驱油聚合物在多孔介质中传输能力实验研究.油田化学,2009;26(4):425-428

[10] 岳湘安,侯吉瑞,吕鑫,等.驱油剂界面特性和流变性对石油采收率的综合影响.应用化学,2008;25(8):904-908

Study on viscosity distribution of two kinds of polymer solution along fluids conducting direction

Song Xinwang[1] Wu Zhiwei[2*] Cao Xulong[1] Han Yugui[1]
Yue Xiang'an[2] Zhang Lijuan[2]

(Research Institute of Geology, Shengli Oilfield Company, SINOPEC, Dongying 257015, P. R. China;
MOE Key laboratory of Petroleum Engineering[2] in China University of Petroleum, Beijing 102249, P. R. China)

Abstract: In order to cognize the contribution of polymer solution's viscosity to EOR in deep reservoir, on the reservoir conditions of strong heterogeneity and unconsolidated reservoir structure of seventh region of Gudong oil-field, flow experiments of polymer solution are carried out in the cores with different lengths, produced fluid is taken out from outlet end of different cores to study distribution characteristics of polymer solution's viscosity in near wellbore area and deep reservoir. The experimental results show that, viscosity reserve rate of polymer HJ, B4 and DH5 is respectively 0.817、0.815 and 0.335 with the concentration of 1 500 mg/L, which are much larger than the values of the concentration of 1 800 mg/L. With the injection rate increases, viscosity of the same polymer's concentration declines a little. The viscosity reserve rate of partially hydrolyzed polyacrylamide polymer HJ and B4 is respectively 0.642 and 0.443, both of which are higher than the value of polymer DH5 with the same concentration.

Keywords: polymer flooding viscosity reserve rate partial pore volume injection rate concentration

原子力显微镜与动态光散射研究疏水缔合聚丙烯酰胺微观结构

曲彩霞[1] 李美蓉[1*] 曹绪龙[2] 张继超[2] 陈翠霞[1] 黄 漫[1]

(1. 中国石油大学(华东)理学院,青岛 266580;2. 中国石化胜利油田地质科学研究院,东营 257015)

摘要:驱油用疏水缔合聚丙烯酰胺(HAP)性能优劣由其微观结构决定,优选研究HAP微观结构的方法,方便现场准确筛选功能良好的驱油聚合物。本文通过原子力显微镜(AFM)和扫描电镜(SEM)观察了不同浓度HAP溶液的微观结构,动态光散射(DLS)与AFM分析了粒径分布。结果表明,随着聚合物浓度的增加,DLS测得粒径增大,粒径分布由单峰变为双峰;SEM观察到由无空间网络结构,到相对松散、易被盐破坏空间网络结构,再到不易被盐破坏的完整、致密的网络结构;AFM可以同时得到一定空间结构以及与DLS相符的粒径大小,三种方法综合比较发现,结合三种不同的表征手段能够较好的、较全面的表示HAP在溶液中的微观结构。

关键词:疏水缔合聚丙烯酰胺;微观结构;扫描电镜;原子力显微镜;动态光散射

HAP是指在常规的聚合物主链上引入极少量(摩尔分数小于2%)疏水基团所形成的一类新型聚合物[1]。在三次采油中可以提高水相粘度和降低地层中水的相对渗透率,改善油水前沿的相对流度比,提高驱替体系的波及系数,从而提高原油采收率。HAP溶液获得较高表观粘度以及抗盐性的基本原理是:临界缔合浓度(CAC)之后,不同分子疏水基团之间相互缔合,开始形成聚集体,亲水基长链相互缠绕,开始形成网络结构,随着浓度的增加,分子间缔合增强,骨架变粗,进一步形成一种受盐离子影响较小的空间网络结构[2]。

目前分析聚合物微观结构的方法较多,SEM、AFM和DLS[3]是几种较为常见的分析方法。王等[2]用SEM研究其疏水缔合性质;韩等[4]用动态光散射方法研究了HAP与普通聚丙烯酰胺(HAPM)在水溶液中的行为,找到了HAP和HAPM性能差异的根本原因。陈等[5]利用AFM将HAP与HAPM对比,证实了HAP溶液中网络结构的存在,分析疏水缔合型聚丙烯酰胺的增粘机理。本文利用SEM、AFM和DLS三种方法分析HAP溶液的微观结构,综合比较三种方法的优劣,以筛选出能够全面地反映HAP微观结构的表征手段。

1 实验部分

1.1 试剂与仪器

HAP为胜利油田地质院提供疏水缔合部分水解聚丙烯酰胺(用无水乙醇抽提,真空干燥恒重备用);二次蒸馏水:实验室自制。其他试剂均为分析纯。

采用美国BROOKFIELD公司Brookfield DV-Ⅱ+Pro型粘度计测定聚合物表观粘度;日本日立公司S-4800冷场扫描电镜观察溶液结构;英国马尔文公司Zatasizer Nano ZS型高灵敏度动态光散射仪分析粒径;美国DI公司NanoScope Ⅳa原子力显微镜进行微观分析。

1.2 矿化水的配制

配制矿化度为19 334 mg/L的模拟矿化水,在二次蒸馏水中加入相应的盐。将0.857 2 g六水氯化镁、1.136 g氯化钙、17.34 g氯化钠溶于1 L二次蒸馏水中配制而成。

1.3 粘度的测量

在烧杯中加入二次蒸馏水,开动搅拌器,调节转速至旋涡延伸至容器底部75%高度,匀速加入疏水缔合聚丙酰胺干粉,在200 r/min下连续搅拌3 h至完全溶解,静置24 h后,配成5 000 mg/L的母液。分别将聚合物母液用二次蒸馏水和矿化水稀释到不同浓度,用电磁搅拌器搅拌2 h后在25℃下恒温20 min,剪切速率为42.22 s^{-1},测定其粘度。

1.4 SEM观察溶液结构

用二次蒸馏水和矿化水配制不同质量浓度的聚合物溶液,在室温25℃下,用搅拌器搅拌约2 h,用移液枪分别取100 μL溶液滴在采用超声波处理的新云母片上,使它尽量铺展成膜,尽量保持聚合物水化分子的原有结构和形貌,自然干燥,再进行喷金,最后用S-4800冷场扫描电镜观察聚合物的形态,实验温度为25℃,选择电压5.0 kV。

1.5 DLS分析粒径

用Zatasizer Nano ZS型高灵敏度动态光散射仪测定样品溶液的粒径分布。分析仪采用He-Ne光源,功率10 mW,波长633 nm,散射光强测定角度为173°,测定温度为25℃。

1.6 AFM进行微观分析

移取10 μL聚合物溶液至新鲜剥离的云母片上并尽量铺展使之成膜,室温下干燥。用美国DI公司生产的Nanoscope Ⅳa原子力显微镜,所用探针为商用Si_3N_4探针,以轻敲模式在室温下观察聚合物溶液的微观形貌。

2 结果与讨论

2.1 HAP溶液表观粘度

疏水缔合聚合物溶液的表观粘度主要取决于分子链间的相互作用和分子链(束、团)的粒径。聚合物HAP在蒸馏水和矿化水中的粘浓曲线见图1。

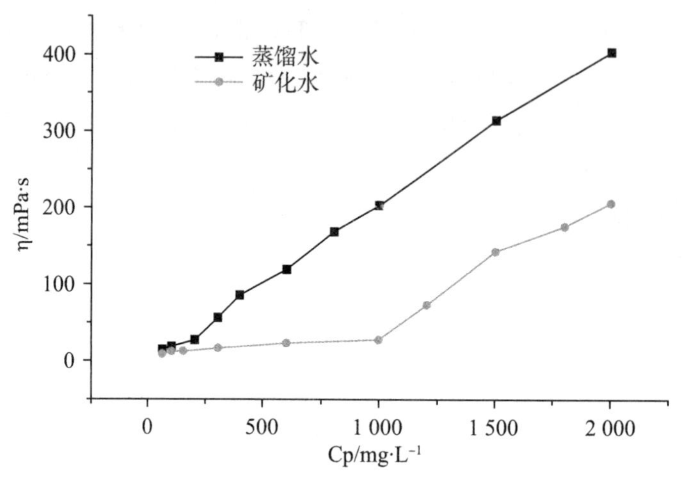

图1 表观粘度与聚合物浓度之间的关系

Figure 1 Relationship between apparent viscosity and concentration of polymer

当HAP的浓度较低时,分子链由于疏水基团的分子内缔合效应而发生卷曲,这时对溶液粘度起作用的主要是聚合物粒径;随着浓度的进一步增加,当聚合物浓度高于其临界缔合浓度后[6],分子间的缔合作用逐渐增强,致使聚合物溶液的粘度显著上升,粘浓曲线以较大斜率变化。由图1知,该聚合物为HAP,它的临界缔合浓度为150 mg·L^{-1}。

矿化度对HAP溶液影响主要有两方面:对静电排斥及水化作用的影响和对疏水缔合作用的影响。一方面,屏蔽聚合物链上羧基离子以及改性引入的磺酸根之间的斥力,使高分子链卷曲收缩,流体力学体积减小,同时,盐离子的去水化作用,使水化膜变薄,粘度降低;另一方面,盐的加入使溶剂的极性增大,HAP分子间缔合增强,导致表观粘度增加,当盐对静电排斥作用的影响大于对疏水缔合作用的影响时,聚合物溶液的表观粘度降低;反之,则聚合物溶液的表观粘度增加。如图1,浓度在500 mg·L^{-1}内,聚合物溶液抗盐性不强,因为在低浓度下,分子间缔合较弱,形成的网络结构较为松散,盐对静电排斥作用的影响大于对疏水缔合作用的影响,使高分子链卷曲收缩,网络结构被破坏,粘度损失率为74.75%。浓度为1 500 mg·L^{-1},聚合物溶液抗盐性增强,因为在高浓度下,分子间缔合较强,形成的网络结构较为致密,虽然盐对静电排斥作用的影响大于对疏水缔合作用的影响,使高分子链卷曲收缩,但高浓度下分子间缔合作用较强,网络结构只有被部分破坏,粘度损失率为54.79%。

2.2 DLS测HAP粒径

由2.1粘度测量知，HAP溶液临界缔合浓度为150 mg·L^{-1}，当浓度为100 mg·L^{-1}，溶液处于临界缔合浓度之前，主要为分子内缔合；当浓度为500 mg·L^{-1}，溶液处于临界缔合浓度之后，主要为分子间缔合；矿化水配制浓度为500 mg·L^{-1}HAP溶液，粘度明显降低。

由图2知，浓度为100 mg·L^{-1}的聚合物溶液，出现单峰，DLS测定的散射光强累计显示，粒径分布于147.7～229.3 nm，可能为单分子聚集体的粒径，表明为分子内缔合产物；浓度为500 mg·L^{-1}的聚合物溶液，出现双峰[8]，粒径分布于61.21～307.6 nm处有一个面积较小的峰，表明有分子内缔合产物，粒径分布于307.6～1 545 nm处有一个面积很大的峰，可能为分子间缔合产生的聚集体的粒径，表明有分子间缔合产物[9,10]；矿化水配制浓度为500 mg·L^{-1}的聚合物溶液，依然有双峰，但双峰均左移，盐离子对分子链产生电荷屏蔽作用，使静电斥力作用减弱，聚合物分子链产生空间卷曲，同时发生去水化，分子链的水化层变薄，相邻水化层间距减小，水化膜粘弹性减小，导致分子线团蜷缩的位阻基本消失，粒径变小，分子间缔合的强度变弱，相应的粘度也会有较大的降低。

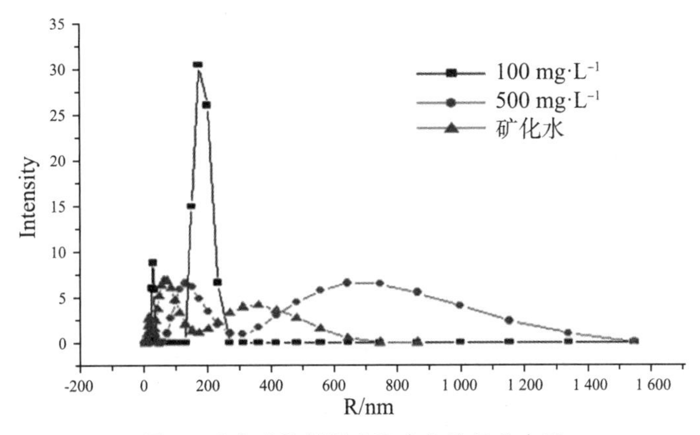

图2　动态光散射测定聚合物粒径分布图

Figure 2　Intensity distribution of HAP solutions by DLS

2.3 冷冻蚀刻SEM与普通SEM观察聚合物溶液微观形态

用液氮速冻法使聚合物溶液快速固化，锁定聚合物水化分子形态，然后在冷冻真空干燥仪中使水分子升华，尽量保持聚合物水化分子的原有结构和形貌，利用SEM观察到聚合物溶液的微观结构为3(a)；自然干燥得到干片，利用SEM观察到聚合物溶液的微观结构为3(b)。

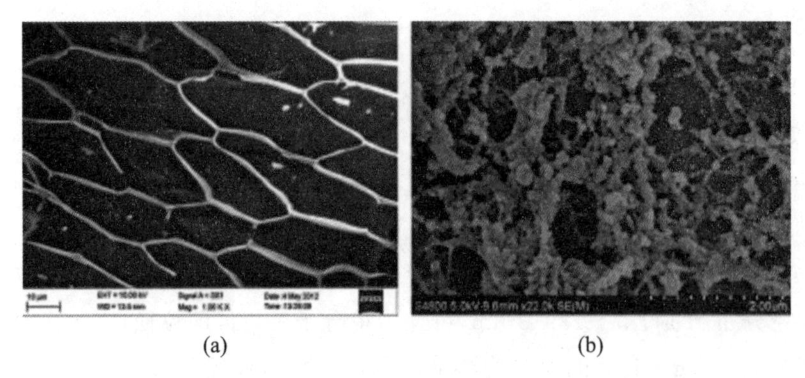

图 3　不同制样方法的 HAP 聚合物 SEM 图

(a) 矿化水配制 500 mg・L^{-1} HAP 聚合物冷冻蚀刻 SEM 图(×10 000)；

(b) 矿化水配制 500 mg・L^{-1} HAP 聚合物 SEM 图(×22 000)

Figure 3　SEM images of HAP polymer at different procedure

(a) 500 mg・L^{-1} SEM of Freeze-etching under mineralized contents(×10 000)；

(b) 500 mg・L^{-1} SEM under mineralized contents(×22 000)

图 3(a)和 3(b)均为矿化水配制浓度 500 mg・L^{-1} 的聚合物溶液，冷冻刻蚀制样，对表面形貌影响较小，可以清楚地看到形成骨架粗细不均匀的空间网络结构[7]；而自然干燥制样过程，浓度变化，使观察的形貌与溶液状态可能存在一定差异，但制样过程尽量将溶液铺展成薄膜，使干燥时间缩短，尽量减少干燥过程对聚合物形貌的影响，然后选择成膜较好的样品进行观察，可以看到松散网络结构，可进行定性解释。

2.4 SEM 观察聚合物溶液微观形态

蒸馏水以及矿化水配制的不同浓度 HAP 溶液的 SEM 见图 3(a)～3(f)。

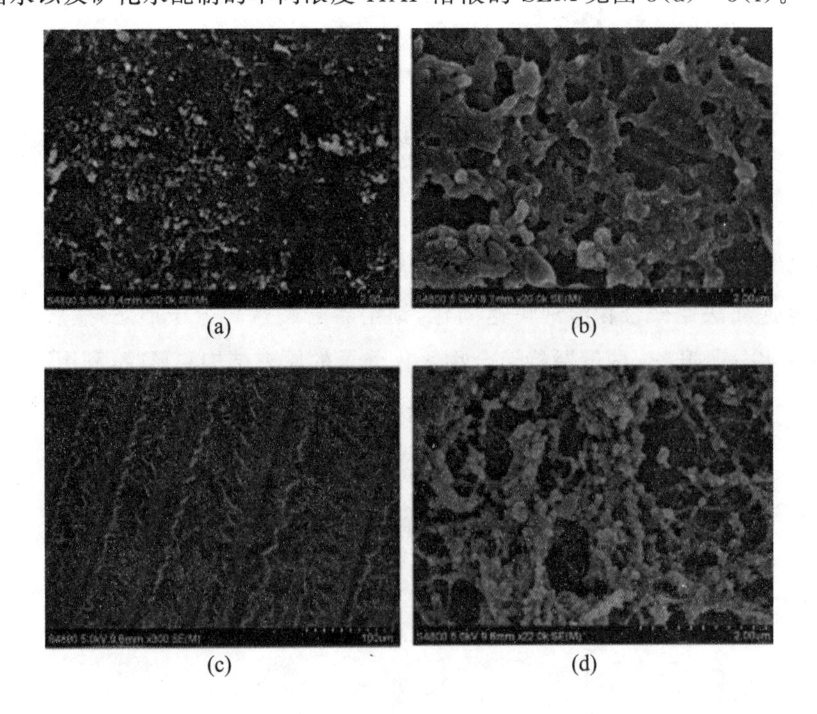

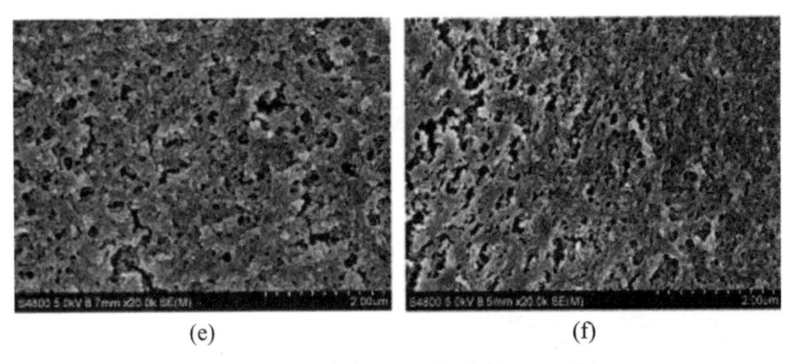

(e) (f)

图 4 不同浓度 HAP 聚合物 SEM 图

(a) 100 mg·L^{-1} SEM 图(×20 000);(b) 500 mg·L^{-1} SEM 图(×20 000);

(c) 矿化水配制 500 mg·L^{-1} SEM 图(×300);(d) 矿化水配制 500 mg·L^{-1} SEM 图(×22 000);

(e) 1 500 mg·L^{-1} SEM 图(×20 000);(f)矿化水配制 1 500 mg·L^{-1} SEM 图(×20 000)

Figure 4 SEM images of polymer at different concentration

(a) 100mg·L^{-1} SEM image(×20 000);

(b) 500 mg·L^{-1} SEM image(×20 000);

(c) 500 mg·L^{-1} SEM image under mineralized contents(×300);

(d) 500 mg·L^{-1} SEM image under mineralized contents(×22 000);

(e) 1 500 mg·L^{-1} SEM image(×20 000);

(f) 1 500 mg·L^{-1} SEM image under mineralized contents(×20 000)

图 4(a)为浓度 100 mg·L^{-1} 即临界浓度之前 SEM 图,聚合物较为松散的分布在空间内,无空间网络结构,可能因聚合物分子链卷曲,主要为分子内缔合,对应溶液粘度很低;图 4(b)为浓度 500 mg·L^{-1} 即临界浓度之后 SEM 图,形成一定的空间网络结构[7],且骨架粗细不均匀,链节处有凸起,出现这种形态的原因是出现分子间缔合作用,疏水基团相互聚集,而聚合物分子链上的亲水基团与水介质之间存在溶剂化作用,将自由水变为束缚水,聚合物分子表面形成了较厚的水化膜,聚合物粘度增加显著;图 4(c)为矿化水配制浓度 500 mg·L^{-1} 的聚合物溶液,聚合物在矿化水中形成了连续的松枝状结构,发现“松枝”主干和细枝都是由很小的颗粒聚集而成,对“松枝”边缘进行 2 μm 范围内的扫描,得到 4(d),可以看出网络结构骨架变细,破坏空间网络结构,主要是盐的加入使溶液极性增大,疏水基团缔合作用增大,但疏水基团的水化膜也变薄,且会对分子链产生电荷屏蔽作用,从而使分子链伸展程度降低,分子链卷曲,粘度也会相应降低;图 4(e)为 1 500 mg·L^{-1} 浓度下 SEM 图,相比 500 mg·L^{-1} 聚合物溶液,形成更完整、致密的多层次网络结构,分子间缔合作用增强,在宏观上表现为表观粘度很大;图 4(f)为矿化水配制浓度 500 mg·L^{-1} 的聚合物溶液,原来致密、均匀的空间网络结构在盐的作用下变的相对不均匀、松散,但为结构破坏不严重,粘度降低也不严重。

2.5 AFM 对聚合物溶液微观形态进行研究

(1) 100 mg・L^{-1} HAP 溶液的 AFM 分析

由图 5(a)AFM 图中可以直观的看出，稀溶液中聚合物不形成空间网络结构，而是零星的分散在溶液中，通过粒径分析知，粒径为 150 nm 左右，高度为 2 nm 以内，与 2.2 动态光散射测得的结果相符，说明在该浓度下确实以分子内缔合为主，呈单分子状态分布在空间中，粘度很低。

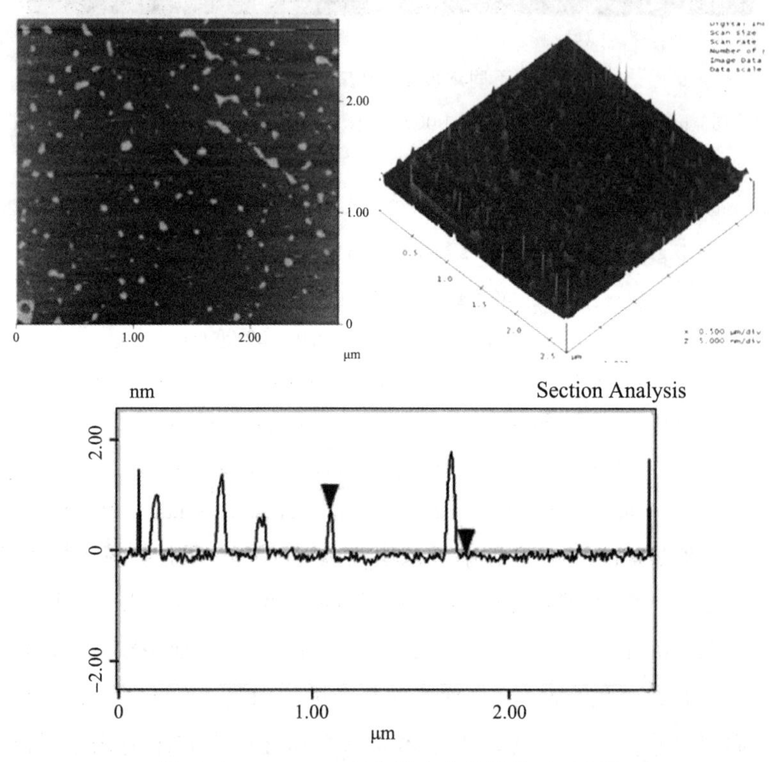

图 5(a)　100 mg・L^{-1} 聚合物溶液的 AFM 图

Figure 5(a)　AFM images and section analysis of 100mg・L^{-1} HAP

(2) 500 mg・L^{-1} HAP 溶液的 AFM 分析

由图 5(b)可以直观的看出，聚合物分子内或分子间缔合结合在一起形成相对独立的聚集体[11]，通过粒径分析知，较浓度 100 mg・L^{-1} 粒径明显增大，聚合物分子链上离子的静电斥力作用，使分子链伸展且水化膜较厚，分子内或分子间缔合形成的聚集粒径为 800 nm 左右，高度为 100 nm 左右，与 2.2 动态光散射测得的结果相符，相应的粘度较大。

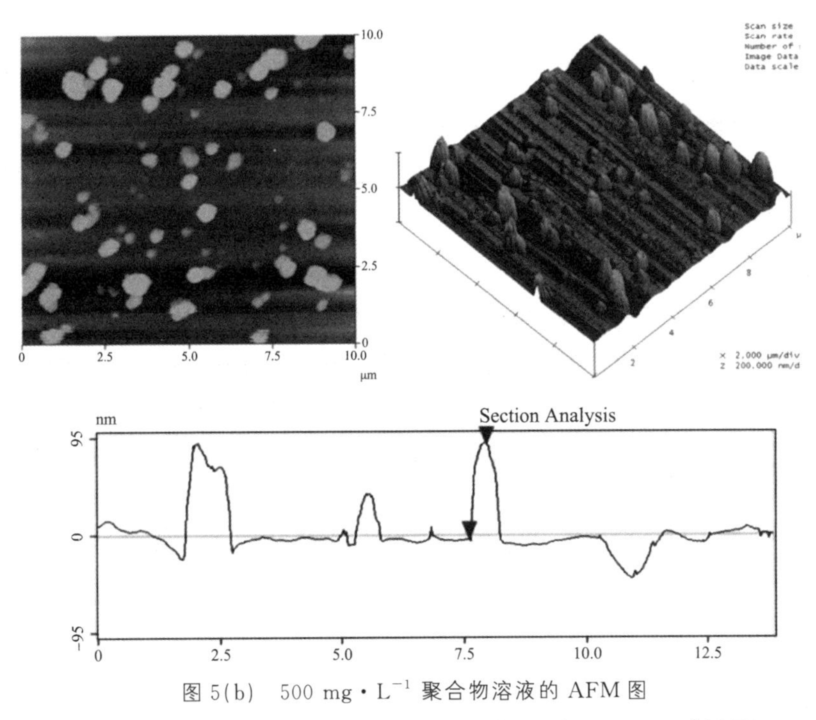

图 5(b) 500 mg·L^{-1} 聚合物溶液的 AFM 图

Figure 5(b) AFM images and section analysis of 500mg·L^{-1} HAP

(3) 矿化水配制 500 mg·L^{-1}HAP 溶液的 AFM 图

由图 5(c)可以直观地看出，在矿化水中聚合物分子内或分子间缔合结合在一起形成相对独立的聚集体，粒径和高度明显变小，原因是盐离子屏蔽聚合物链上羧基离子基团，使高分子链卷曲收缩且产生去水化作用；较二次蒸馏水配制的聚合物溶液，聚集体粒径明显降低，为 200 nm 左右，高度为 30 nm 左右，与 2.2 动态光散射测得的结果相符。

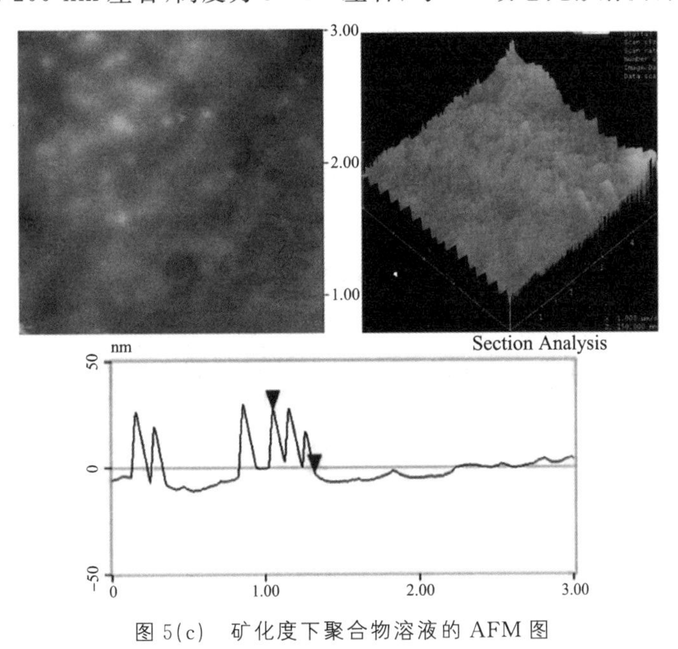

图 5(c) 矿化度下聚合物溶液的 AFM 图

Figure 5(c) AFM images and section analysis of 500 mg·L^{-1} HAP under mineralized contents

3 结论

DLS可测得聚合物溶液中聚集体粒径大小;SEM可观察聚合物溶液干样的空间网络结构;AFM得到与DLS相符粒径大小以及一定的空间结构。说明干燥过程虽对聚合物形貌有一定影响,结合三种不同的表征手段能够较好的、较全面的表示HAP在溶液中的微观结构。

SEM、DLS和AFM研究显示,在临界缔合浓度之前,HAP溶液中形成了许多分散的单个缔合体,较为松散的分布在空间内,粒径较小且为单峰,对应粘度很低;临界缔合浓度之后,开始形成空间网络结构,出现聚集体,粒径变大且为双峰,粘度明显增大,但此时抗盐性不强,在盐离子的影响下,网络结构易被破坏,聚集体粒径明显降低,粘度明显降低;当浓度达到1 500 mg·L^{-1},形成了更完整、致密的多层次的空间网络结构,在宏观上表现为表观粘度很大,此时抗盐性也较强,在盐离子的影响下,网络结构不易被破坏。

参考文献

[1] 钟传蓉,黄荣华,马俊涛.化学世界,2003,(12):660~664.

[2] 王东贤,王琳,宫清涛,廖琳,张路,靳志强,黄玉萍,严峰,赵濉,俞稼镛.感光科学与光化学,2005,3(23):197~202.

[3] 张瑞,叶仲斌,罗平亚.电子显微学报,2010,(5):475~481.

[4] 韩利娟,叶仲斌,陈洪,罗平亚.西南石油学院学报,2006,28(4):71~73.

[5] 陈洪,韩利娟,徐鹏.物理化学学报,2003,19(11):1020~1024.

[6] Aguiar J, Carpena P, Molina-Bolivar J A, Carnero RUIZ C. J COLLOID INTERF SCI,2003,(258):116~122.

[7] 宋春雷,杨青波,张文德,王丕新.应用化学,2009,26(4):1020~1024.

[8] Klucker R,Munch J P,Schosseler F. Macromolecules,1997,(30):3839~3848.

[9] 孙焕泉,张坤玲,陈静,曾胜文,平郑骅,许元泽.高分子学报,2006,(6):79~83.

[10] 段洪东,侯万国,汪庐山,李希明,李伯耿.高分子通报,2002,(2):49~55.

[11] Sun Wei,Long Jun,Xu Zhenghe,Masliyah J H. Langmuir,2008,(24):14015~14021.

Study on microstructure of hydrophobic associating polyacrylamide by atomic force microscopy and dynamic light scattering

Qu Caixia[1] Li Meirong[1*] Cao Xulong[2] Zhang Jichao[2]
Chen Cuixia[1] Huang Man[1]

(1. China University of Petroleum, College of Science, Qingdao 266580, China;
2. Geological Scientific Research Institute Shengli Oilfield, SINOPEC, Dongying 257015, China)

Abstract: The efficiency of hydrophobic associating polyacrylamide (HAP) used for polymer flooding is determined by its microstructure. The study on analysis means of the microstructure can contribute to screening good functional polymer systems for EOR. The atomic force microscopy (AFM) and scanning electron microscope (SEM) are used to observe the microstructures of HAP solutions at different concentrations, while dynamic light scatter (DLS) and AFM are used to analyze particle size distribution. The DLS results show that as the concentration increases particle size is larger and double peaks appear; the relatively loose network with weak salt resistant ability is replaced by dense network structure that is less susceptible to salt. Both the particle size by DLS and the microstructures can be measured by the AFM at the same time. The comparisons of three means at various solution conditions provide comprehensive characterization revealing the relation of structure and properties of HAP.

Keywords: Hydrophobic associating polyacrylamide; Microstructure; EM; AFM; DLS

疏水缔合型和非疏水缔合型驱油聚合物的结构与溶液特征

李美蓉[1] 黄 漫[1] 曲彩霞[1] 曹绪龙[2] 张继超[2] 刘 坤[2]

(1. 中国石油大学理学院,山东 青岛 266580;
2. 中国石化胜利油田分公司地质科学研究院,山东 东营 257015)

摘要:考察三次采油过程中所用的部分水解聚丙烯酰胺(HPAM)和疏水缔合型部分水解聚丙烯酰胺(AHPAM)结构与性质的差异,通过对其结构表征以及对其增黏性、耐温性和抗盐性的测试,结合扫描电镜(SEM)对两种聚合物溶液微观结构形态的观察,讨论疏水基团的存在对聚合物溶液性质的影响。结果表明,相对于HPAM,AHPAM溶液因结构中含有的疏水基之间的相互作用使高分子间进一步形成缔合的网络结构而具有良好的增黏性、耐温性和抗盐性。

关键词:疏水缔合型;聚合物溶液;扫描电镜;疏水基团

中图分类号:TE357.4 文献标志码:A 文章编号:1673-5005(2013)03-0167-05

Structure and solution properties for HPAM and AHPAM used in oil displacement polymer

Li Meirong[1] Huang Man[1] Qu Caixia[1] Cao Xulong[2] Zhang Jichao[2] Liu Kun[2]

(1. College of Science in China University of Petroleum, Qingdao 266580, China;
2. Geological Scientific Research Institute of Shengli Oilfield, SINOPEC, Dongying 257015, China)

Abstract: The structure and properties of partial hydrolyzed polyacrylamide (HPAM) and associating partial hydrolyzed poly-acrylamide(AHPAM) used in the tertiary oil recovery process were compared. Based on their structure characterization, the increasing viscosity characteristics, temperature resistance, salinity resistance were tested. The microstructure morphologies of two kinds of polymer solutions were observed by using scanning electron microscopy. The influence of the hydrophobic groups on the polymer properties was discussed. The results show that compared with HPAM, the property of AHPAM solu-tion is superior in the increasing viscosity characteristics, temperature resistance and salinity resistance due to the structure containing the hydrophobic grouping to form association among super-molecular structures.

Keywords: associating partial hydrolyzed; polymer solution; scanning electron microscopy; hydrophobic groups

聚丙烯酰胺，特别是部分水解聚丙烯酰胺(HPAM)，是聚合物驱油中应用广泛的水溶性高分子，HPAM 中的羧基带电基团之间的静电斥力增大高分子的流体力学半径，从而得到更强的增黏效果，但在高矿化度的环境下，由于离子的屏蔽，静电斥力也大大减弱，这一优点也不显著，高温条件下 HPAM 水解程度加快而失效，因此须对 HPAM 大分子进行改性[1-3]。疏水缔合聚合物因具有的疏水基团之间的相互作用使高分子间进一步形成缔合的网络结构而具有良好的增黏、抗温和耐盐性能[4]。不过这类聚合物因含有疏水基团会导致溶解性能变差，溶解时残留物较多，现场应用时容易堵塞多孔介质。笔者通过红外光谱和核磁共振对两种不同的聚丙烯酰胺的结构进行表征，考察其增黏、抗温、抗盐性能以及微观形态结构，比较两种聚丙烯酰胺的性能。

1 实验

1.1 仪器与试剂

仪器：Brookfield DV-Ⅱ+Pro 型黏度计(美国 BROOKFIELD 公司)；79-1 型磁力加热搅拌器(江苏金坛市金城国胜仪器厂)；数显恒温水浴锅(上海双捷实验设备有限公司)；Spectrometer One FT-IR(Perkin Elmer 公司)；AC-50 核磁共振仪(德国布鲁克公司)；D-250L 型恒速恒压泵(海安石油科技有限公司)；HW-4B 型恒温箱(南通华兴石油仪器有限公司)；S-4800 冷场扫描电镜(日本日立公司生产)。

试剂：聚合物样品由胜利油田地质科学研究院提供；实验用水为二次蒸馏水；其他试剂均为分析纯。

1.2 结构表征方法

红外光谱法：将纯聚合物固体在 100℃下烘干 3 h，研磨碎后，用溴化钾晶片压片制样，用傅里叶变换红外光谱仪分析聚合物的结构。

核磁共振波谱法：将纯聚合物固体在 100℃下烘干 3 h，以 D_2O 为溶剂，用 AC-50 核磁共振仪测定其 ^1H 及 ^{13}C 谱图。

1.3 聚合物的增黏性测试

分别将 5.0 g/L 的聚合物母液用二次蒸馏水稀释到所需的质量浓度，用电磁搅拌器搅拌 1 h 后在 25℃下恒温 20 min,用 Brook fieldDV-Ⅱ＋Pro 型黏度计测定其黏度(RV-3 号转子,120 r/min)。

1.4 聚合物的抗温性测试

分别将 5.0 g/L 的聚合物母液用二次蒸馏水稀释到 1.5 g/L,用电磁搅拌器搅拌 1 h 后分别在 30、40、50、60、70、80℃下恒温 20 min,测定其黏度。

1.5 聚合物的抗盐性测试

分别将 5.0 g/L 的聚合物母液用模拟矿化水稀释到 2.0、1.5、1.0、0.8、0.6、0.4、0.2、0.1、0.06 g/L,用电磁搅拌器搅拌 1 h 后,在 25℃下恒温 20 min,用 Brook field DV-Ⅱ＋Pro 型黏度计测定其在 19.334 g/L 矿化度下的黏度。

1.6 聚合物溶液的微观结构形态

用二次蒸馏水和模拟矿化水配制质量浓度不同的聚合物溶液,在室温 25℃下,用搅拌器搅拌约 2 h,然后静置熟化 14 d,用移液枪分别取 100 μL 溶液滴在采用超声波处理的 2 个新云母片上,使它尽量铺张成膜,尽量保持聚合物水化分子的原有结构和形貌,然后在干燥器中干燥,再进行喷金,最后用 S-4800 冷场扫描电镜观察聚合物的形态。

2 结果分析

2.1 聚合物结构表征

聚合物的红外光谱见图 1,聚合物的核磁共振结果见图 2;其各个峰的位置所代表的官能团结构见表 1。

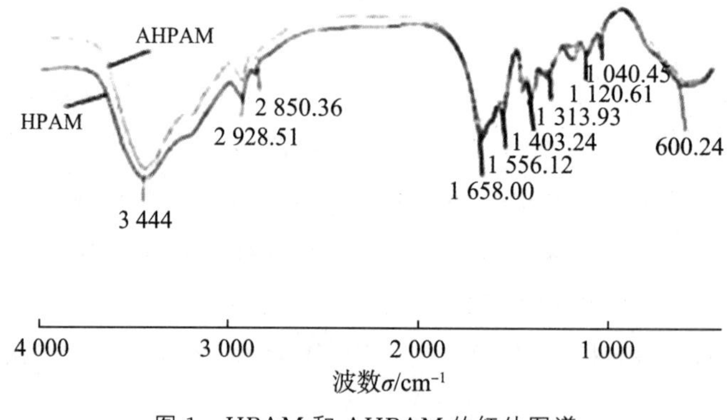

图 1 HPAM 和 AHPAM 的红外图谱

Fig. 1 IR spectrum of HPAM and AHPAM

从图 1 看出,HPAM 和 AHPAM 出峰位置基本相同,在 3 444 cm^{-1} 存在伯酰胺的 N—H 的对称伸缩振动的峰,1 658 cm^{-1} 为 C═O 的对称伸缩振动峰,1400 cm^{-1} 为 C—N 缩振动和 N—H 的弯曲振动的偶合峰,以 C—N 的伸缩振动为主。2 928 cm^{-1} 为 —CH_2— 不对称伸缩振动峰,2 850 cm^{-1} 为 —CH_2— 称伸缩振动峰。1 403 cm^{-1} 为 —CH_2—CO 接近于 —CH_2— 弯曲振动峰。1 313 cm^{-1} 和 1 120 cm^{-1} 为仲酰胺 C—N 键的伸缩振动和部分 N—H 的振动峰,600 cm^{-1} 为酰胺的氨基外面弯动峰。AHPAM 与

HPAM的明显差异在1 040 cm^{-1}，该特征峰是由双键产生的烯烃面外弯曲振动，这说明与HPAM相比，AHPAM在侧链中含有双键这类疏水基团。

表1 化学位移与官能团的关系

Table 1 Relationship between chemical shift and function groups

	化学位移δ	官能团
^{13}C谱	179.720～182.885	C═O
	35.052～36.140和42.001～45.075	分子主链中丙烯酰胺链节—CH_2和—CH
	48.888	分子主链中丙烯酸链节—CH
	29.931～31.765	侧链—CH_2
	26.429	侧链—CH_3
^{1}H谱	1.457～1.698和2.042～2.135	分子主链中丙烯酰胺链节—CH_2和—CH
	1.862和2.263	分子主链中含取代基团链节的—CH_2和—CH
	2.462～2.576	分子主链中丙烯酸链节—CH
	1.059～1.305	侧链—CH_2
	0.217～0.418	侧链—CH_3

从表1可知，AHPAM可知$\delta=5.119$和$\delta=5.828$为C═C上氢的化学位移，$\delta=3.730$为侧链与季铵根原子相连—CH_2的化学位移，$\delta=126.30$为C═C的化学位移，$\delta=57.730$为分子侧链与季铵根原子相连—CH_2的化学位移；AHPAM含有疏水基团C═C和季铵根。并且加硝酸银后，AHPAM溶液产生白色絮状沉淀，也可以验证AHPAM中含有氯离子，也就是季铵盐阳离子。

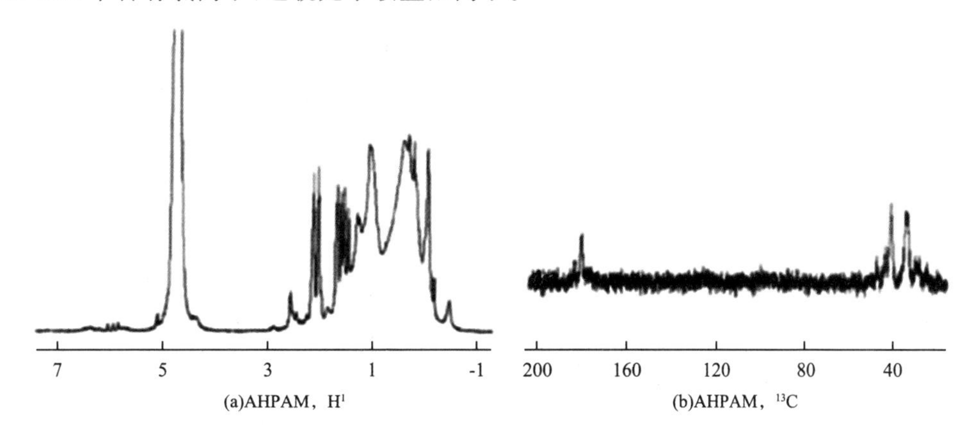

(a)AHPAM，H[1]　　(b)AHPAM，^{13}C

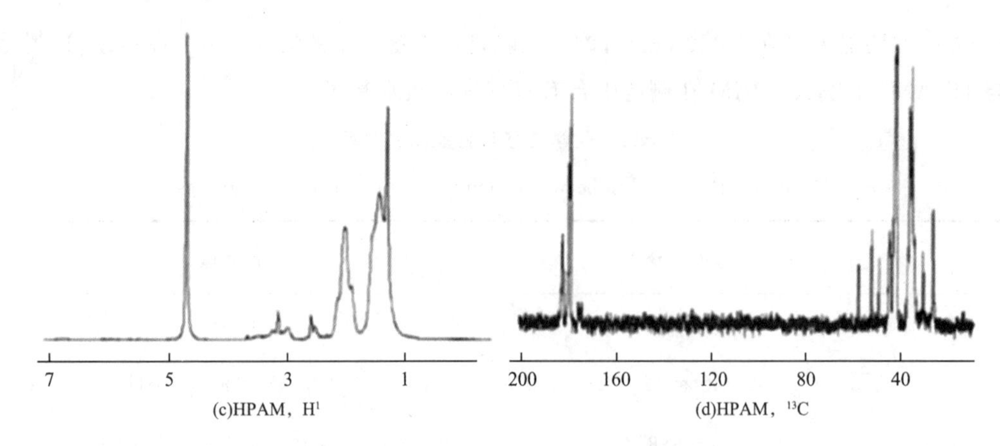

图 2　HPAM 和 AHPAM 的核磁共振图谱

Fig. 2　^{13}C spectrum and ^{1}H spectrum of HPAM and AHPAM

2.2 聚合物性能分析

2.2.1 聚合物的黏浓曲线

分别将 HPAM 和 AHPAM 母液用二次蒸馏水稀释到 2.0、1.5、1.0、0.8、0.6、0.4、0.3、0.2、0.1、0.06 g/L，测量两种聚合物溶液的黏度，结果见图 3。

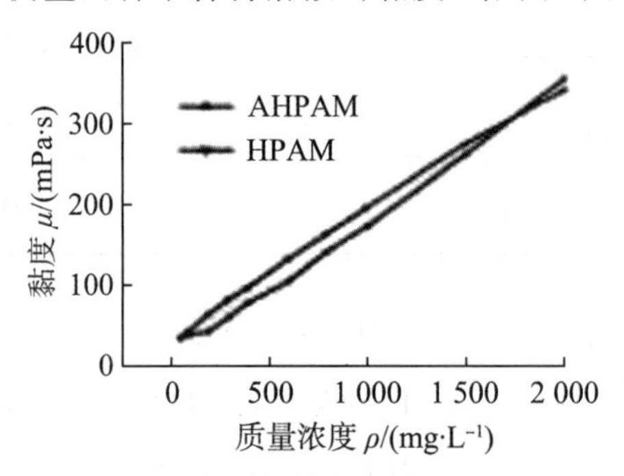

图 3　聚合物的质量浓度与表观黏度的关系

Fig. 3　Relationship between apparent viscosity and concentration of polymer

由图 3 可知，AHPAM 有一质量浓度的转折点即是有临界缔合质量浓度而 HPAM 没有转折质量浓度。临界缔合质量浓度之后溶液的表观黏度大幅度增加，这主要是由于在这个质量浓度点附近，AH-PAM 分子链中的疏水基团结合成一定强度的缔合键，形成了网络结构，且质量浓度越大网络结构越强，宏观上为表观黏度随质量浓度增加急速增大[5-6]。

2.2.2 聚合物的抗温性

AHPAM 有刚性的疏水基团，而 HPAM 没有疏水基团；AHPAM 溶解于水中时，水分子高度有序地围绕在疏水基团周围，使得疏水基团倾向于缔合在一起，以尽可能减少与水分子的接触表面，增加熵值。因此，其缔合是一个“熵驱动”过程，缔合作用随着温度升高而增强。但另一方面，温度升高使水分子热运动加剧，破坏疏水基团周围水分子的“冰山结构”，大大削弱疏水基团间的缔合[7]。两种聚合物的黏度-温度曲线见图 4。

由图 4 知，在相同温度范围内，AHPAM 的黏度大于 HPAM；对于 AHPAM，因结构中有疏水基团，随着温度升高疏水基的缔合作用增强，分子内正负离子间的相互作用减弱，大分子链处于伸展状态，溶液黏度有所提高；但随温度继续升高，疏水基团的热运动加剧，疏水缔合作用被消弱，同时水分子的热运动加剧导致疏水基团周围水分子结构与状态改变，疏水缔合作用进一步消弱，大分子链收缩，黏度下降。所以 AHPAM 溶液的黏度随着温度升高先上升后下降，而 HPAM 因结构中没有疏水基团，不会出现缔合现象；随着温度的升高热运动越来越剧烈，宏观上表现溶液的黏度不断下降。

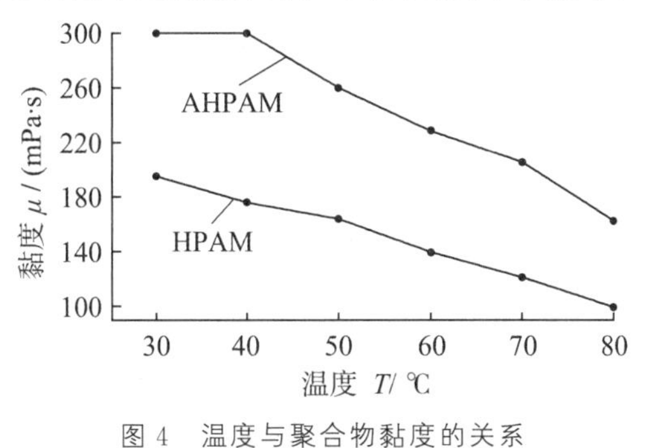

图 4 温度与聚合物黏度的关系

Fig. 4 Relationship between apparent viscosity and temperature

2.2.3 聚合物的抗盐性

矿化度能增强聚合物中链节间的缔合作用[8-9]，其直接原因是随着水溶液中电解质质量浓度的增加，溶剂极性增强，疏水基团间缔合的概率增加。

由图 5 对比分析可知，在相同的矿化度下，AHPAM 黏度下降幅度远小于 HPAM。在相同矿化度下，质量浓度为 1.5 g/L 时，AHPAM 的黏度损失只有 24.4%，而 HPAM 的黏度损失达 88.4%，AHPAM 的耐盐性能明显优于 HPAM。这是因为：一方面，盐的加入使溶剂的极性增大，AHPAM 分子间缔合增多；另一方面 AHPAM 中引入了提高大分子链刚性的基团，可降低 Ca^{2+}、Mg^{2+} 对高分子主链的影响，使得聚合物抗盐性增加。

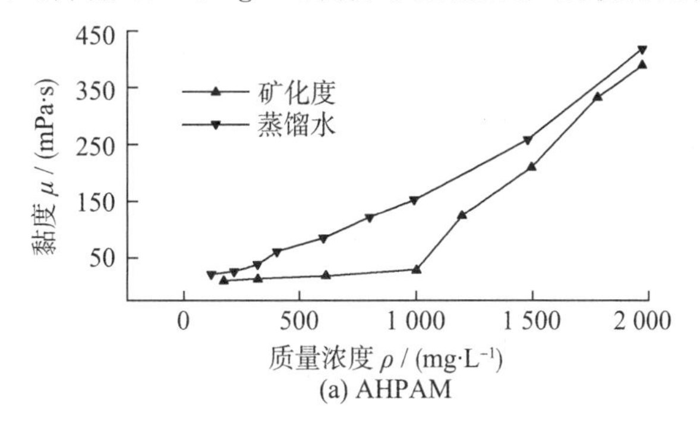

(a) AHPAM

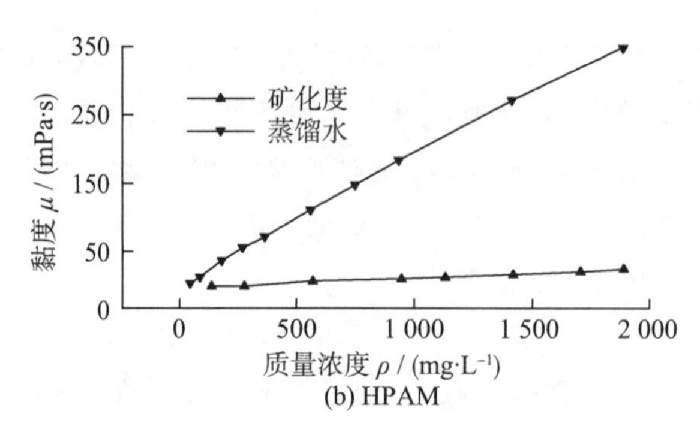

(b) HPAM

图 5 AHPAM 和 HPAM 的抗盐性

Fig. 5 Salinity resistence of AHPAM and HPAM

2.3 聚合物溶液的微观结构形态

2.3.1 相同质量浓度下的 SEM 图像

图 6 为两种聚合物的 SEM 图像。其中,图 6(a)为 0.5 g/L AHPAM 溶液黏度转折前、后各取一滴的 SEM 照片[10-11];图 6(b)分别为 HPAM 溶液 0.5、0.05 g/L 的 SEM 照片。

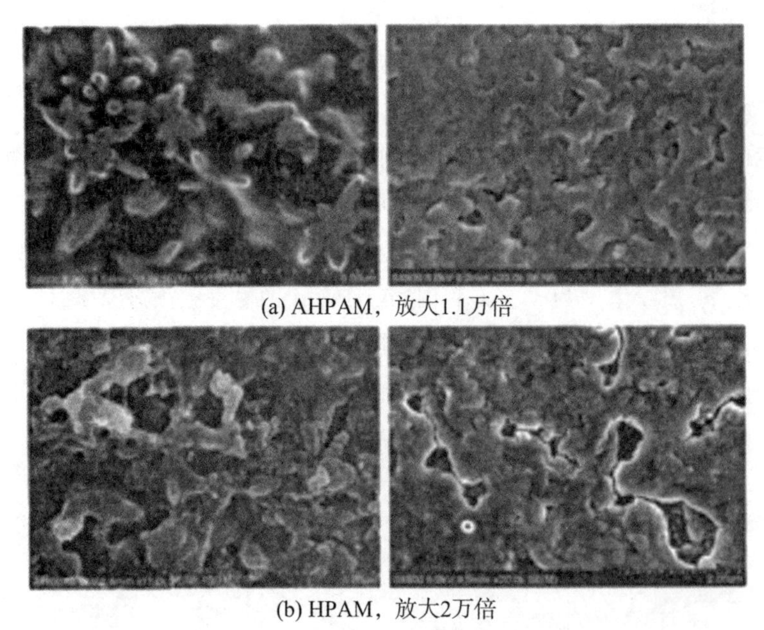
(a) AHPAM，放大1.1万倍

(b) HPAM，放大2万倍

图 6 聚合物在不同浓度下的扫描电镜

Fig. 6 SEM images of polymer for different concentration

由图 6(a)看出,转折后一点,AHPAM 的形貌类似于一个个的"花朵",聚合物以一个中心为缔合点形成一种交联的形态。疏水基团由于憎水作用而发生聚集,使大分子链产生分子间缔合,也就是图中的"花蕊"部分;亲水基团带有亲水基舒展在水中,在图中为"花瓣";AHPAM 表面分子链粗细不均匀,链节处有凸起,造成这种形态的原因是聚合物分子的亲水基团与水介质之间存在溶剂化作用,使分子表面形成了较厚的水化膜,因"束缚水"的存在,使聚合物黏度增大。

图 6(b)看出，HPAM 分子表面有较厚的水化膜，具有明显的溶剂化作用，存在静电斥力作用，静电斥力使分子链变得松散舒展不易卷曲；但 HPAM 微观形态交联效果不好，空间效应在图中都没有表现出来，分子间没有出现空间缔合，溶液黏度较低。

图 6 表明了两种聚合物增黏性差异的原因：对于 AHPAM，结构中有少量疏水基团，在转折点前，聚合物间有很少的缔合，转折点之后，分子的疏水基团缔合在一起，在分子间产生具有一定强度的缔合，从而形成巨大的网状空间结构，因此黏度有明显的改变；对于 HPAM，分子链上不存在疏水基团，仅存在溶剂化作用和静电斥力作用，不存在分子间缔合作用，所以不存在临界缔合浓度。

2.3.2 矿化度的影响

图 7 为两种聚合物在质量浓度 1.5 g/L，矿化度 19.334 g/L 下的 SEM 图像。

从图 7(a)和(b)看出，蒸馏水中 AHPAM 结构中的疏水基团因憎水作用而发生聚集，使大分子链产生分子间缔合，这些聚集体相互交联形成了空间网络体系；盐离子的加入会对分子链产生电荷屏蔽作用，从而分子链伸展程度降低、卷曲，流体力学体积减小，体系黏度下降；电解质的加入使溶液极性增大，疏水基团缔合作用增大，体系黏度上升，两方面的共同作用使聚合物黏度变化不大。

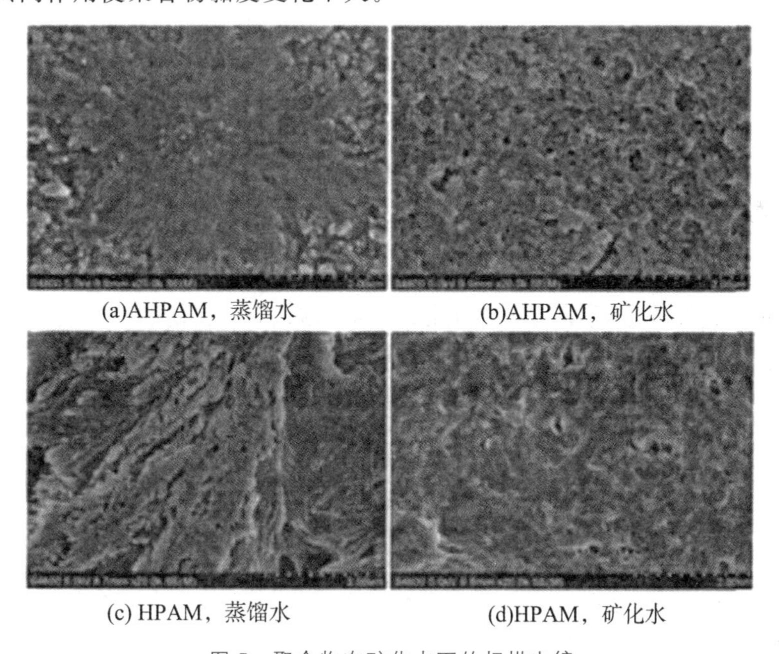

(a)AHPAM，蒸馏水　(b)AHPAM，矿化水

(c) HPAM，蒸馏水　(d)HPAM，矿化水

图 7　聚合物在矿化水下的扫描电镜

Fig. 7　SEM images of polymer with mineralized contents

从图 7(c)和(d)看出，HPAM 在盐的作用下空间网络结构被破坏，空间体积变得卷曲，导致黏度降低。

图 7 表明了两种聚合物抗盐差异的原因：在盐的作用下，不仅有对分子链产生电荷屏蔽作用，从而分子链伸展程度降低、卷曲，而且电解质会压缩分子链的水化膜电子层，使水化膜变薄，体系黏度下降；电解质的加入还使溶液极性增大；AHPAM 分子链上有

少量疏水基团,随溶液极性增大,疏水基团缔合作用增大,使体系黏度上升,此作用与盐对分子链产生电荷屏蔽作用相当,从而体系的黏度下降不大,抗盐性强;HPAM因缺少疏水基团,在盐的作用下,仅有对分子链产生电荷屏蔽作用,从而分子链伸展程度降低、卷曲,并且电解质会压缩分子链的水化膜电子层,使水化膜变薄,体系黏度下降较大,抗盐性差。

综上分析,AHPAM分子链中的疏水基团形成一定强度缔合键,形成了网络结构,且质量浓度越大网络结构越强,宏观上随质量浓度增加表观黏度急速增大。并且疏水基团的存在使得随温度的升高AHPAM溶液黏度先升高后降低,有良好的耐温性;疏水基团之间的缔合作用随电解质的加入而增大,使得总体上AHPAM溶液的黏度变化不大,表现出良好的抗盐性。

参考文献

[1] 孙焕泉,张坤玲,陈静,等.疏水缔合效应对聚丙烯酰胺类水溶液结构和流变性质的影响[J].高分子学报,2006(6):810-814.

SUN Huan-quan,ZHANG Kun-ling,CHEN Jing,et al. Effect of hydrophobic association on structure and rheolog-icai behaviors of polyacyylamide based aqueous solutions [J]. Acta Polymerica Sinica,2006(6):810-814.

[2] 江立鼎,高保娇,孔德轮.丙烯酰胺型阴离子表面活性单体与丙烯酰胺共聚物水溶液的流变特性[J].高分子学报,2007(4):349-354.

JIANG Li-ding,GAO Bao-jiao,KONG De-lun. Reologi-cal behavior of aqueous solutions of copolymers of acryl-amide-type anionic surface-active monomer and acrylam-ide [J]. Acta Polymerica Sinica,2007(4):349-354.

[3] 章云祥,方勤.水溶性高聚物的研究进展[J].功能高分子学报,1997,10(1):103-109.

ZHANG yun-xiang,FANG qin. Advanced development of water-soluble polymer [J]. Journal of Functional Poly-mers,1997,10(1):103-109.

[4] 王东贤,王琳,宫清涛,等.改性聚丙烯酰胺水溶液的疏水缔合性质[J].感光科学与光化学,2005,23(3):197-202.

WANG Dong-xian,WANG Lin. GONG Qing-tao,et al. Study on the hydrophobic association properties of hydro-phobically modified polyacrylamide [J]. Photographic Science and Photochemistry,2005,23(3):197-202.

[5] 冯玉军,罗传秋,罗平亚,等.疏水缔合水溶性聚丙烯酰胺的溶液结构的研究[J].石油学报:石油加工,2001,17(6):39-44.

FENG Yu-jun,LUO Chuan-qiu,LUO Ping-ya,et al. Study on characterization of microstructure of hydrophobi-cally association water-solution polymer in aqueo media by scanning electron microscopy and environmental scanning electron microscopy [J]. Acta Pertolei Sinica (Petroleum Processing Section),2001,17(6):39-44. asp? ouvrage =1184227.

[6] GRUHN V,LAUE R. A heuristic method for detecting problems in business process models [J]. Business ProcessManagementJournal,2010,16(5):806—821.

[7] AWAD A. BPMN-Q : a language to query business processes [C/OL] // Proceedings of EMISA 2007,Nanjing China. 2007:115-128. http://dbis. eprints. uni-ulm. de /205/1/EMISA-Proceedings-Komplett. pdf#page=117.

[8] Business Process Modeling Notation (BPMN) Version 2. 0 [S]. Object Management Group (OMG), Jun 2010. http://www. omg. org/spec/BPMN/ 2. 0/PDF/.

[9] VANHATALO J, VOLZER H, KOEHLER J. The refined process structuretree [J]. Data & Knowledge Engineering,2009,68(9):793-818.

[10] POLYVYANYY A,VANHATALO J,VOLZER H. Sim-plified computation and generalization of the refined process structure tree [M]//Web Services and Formal Methods. 2010,Hoboken,NJ,USA:SpringerBerlin Heidelberg,2011:25-41.

[11] WHITE S A. Process modeling notations and workflow patterns [J]. WorkflowHandbook,2004:265-294.

[12] VAN DER AALST W M P, TER HOFSTEDE A H M. YAWL: yet anotherworkflow language [J]. Information Systems,2005,30(4):245-275.

本文编辑 修荣荣

耐温抗盐缔合聚合物的合成及性能评价

曹绪龙 刘 坤 韩玉贵 何冬月

(中国石化胜利油田分公司地质科学研究院,山东 东营 257015)

摘要:针对胜利油区三类油藏高温高盐的环境,采用胶束聚合法合成了一种耐温抗盐缔合聚合物HAWSP,研究了引发剂、表面活性剂、缔合单体D_iC_8AM物质的量分数及耐温抗盐功能单体AMPS-Na质量分数对HAWSP合成的影响,将其分子结构进行核磁共振氢谱表征,并对其溶液性能进行评价。HAWSP的最佳合成条件为:复合引发剂为Y4,表面活性剂为NS,D_iC_8AM物质的量分数为1.4%~1.6%,AMPS-Na质量分数为15%~17.5%。核磁共振氢谱证实HAWSP由AM,D_iC_8AM和AMPS-Na共聚而成。随着溶液中钙镁离子质量浓度的增加,HAWSP溶液的粘度保留率为55%~60%,而抗盐高分子质量聚合物PAM溶液的仅为30%~40%;在90℃高温下经120 d老化后,质量浓度为1 500 mg/L的HAWSP溶液的粘度为15 mPa·s,而相同质量浓度的PAM溶液在老化时间超过60 d后产生沉淀,HAWSP的耐温抗盐、老化稳定和驱油性能明显优于PAM,原油采收率提高幅度也明显高于PAM,可用作三次采油驱油剂。

关键词:胶束聚合法 缔合聚合物 耐温抗盐性能 老化稳定性能 驱油性能

中图分类号:TE357.431 文献标识码:A 文章编号:1009-9603(2014)02-0010-05

聚合物驱油技术作为油气田提高采收率的主要措施之一,在三次采油领域得到了广泛应用[1-4]。用于三次采油的聚合物包括聚丙烯酰胺、耐温抗盐单体聚合物和疏水缔合聚合物等。然而,随着油藏开采难度的增大,温度和矿化度随之升高,部分驱油用聚合物已不能满足三类油藏开发的要求,主要表现在增粘性、长期稳定性和更恶劣油藏环境下的适应性等方面[5-10]。为此,针对胜利油区三类油藏高温高盐的环境,采用胶束聚合法合成了一种耐温抗盐缔合聚合物HAWSP,该聚合物分子结构中含有缔合单体D_iC_8AM和耐温抗盐功能单体AMPS-Na[11-13],疏水缔合作用和耐温抗盐作用具有明显的协同效应,可使聚合物在高温高矿化度油藏条件下保持优异的增粘性能,同时使聚合物的长期稳定性能得到明显提高;并重点研究了引发剂、表面活性剂、缔合单体物质的量分数和耐温抗盐功能单体质量分数对聚合物溶液性能的影响,以及聚合物溶液的耐温抗盐性能、老化稳定性能和驱油性能。

1 实验器材与方法

1.1 实验器材

实验试剂包括：丙烯酰胺（AM）；2-丙烯酰胺基-甲基丙磺酸（AMPS），工业级，使用时先采用氢氧化钠中和；氢氧化钠、乙二胺四乙酸二钠、氯化钠、氯化镁、氯化钙、过硫酸钾（KPS）、偶氮二异丁咪唑啉盐酸盐（AIBI）、低温氧化还原引发剂（MR）、辅助引发剂（LI），均为分析纯；表面活性剂 SDS，SD-BS，AEO 和 NS，均为分析纯；缔合单体 N，N-二辛基丙烯酰胺（D_iC_8AM），自制；去离子水。

实验仪器包括：Brookfield DV-Ⅲ粘度计；电子天平，精度为±0.001 g；磁力搅拌器；超级恒温水浴，精度为±0.1℃；恒温烘箱，精度为 0.1℃；高纯氮气（体积分数为99.999%）；RFJ 型安瓿瓶封口机；刻度吸管、烧杯等玻璃仪器若干。

1.2 实验方法

HAWSP 的合成 HAWSP 的合成步骤包括：① 在 2 000 mL 烧杯中加入适量的去离子水，按照一定的比例加入 AM，D_iC_8AM，AMPS-Na 和乙二胺四乙酸二钠以及表面活性剂等充分搅匀，然后降温至 5℃，并通高纯氮气除氧 30 min；② 在聚合体系温度为 5℃的条件下，加入一定量的复合引发剂进行绝热聚合，绝热聚合时间为 8 h，同时记录聚合体系内部温度；③ 取出胶体，用剪刀剪成粒径约为 1 mm 的颗粒，按照理论水解度为20%加入氢氧化钠，于 105℃恒温水浴中水解 3 h；④ 将水解后的聚合物胶体颗粒置于105℃的鼓风干燥箱中干燥 2 h，经粉碎和筛分得到白色或微黄色的耐温抗盐缔合聚合物干粉；⑤ 将步骤④得到的聚合物干粉利用无水乙醇洗涤多次，得到白色的耐温抗盐缔合聚合物干粉。

表观粘度测试将耐温抗盐缔合聚合物干粉样品用模拟盐水配制成质量分数为0.5%的母液，模拟盐水总矿化度为 2×10^4 mg/L，其中钙离子质量浓度为 420 mg/L，镁离子质量浓度为 110 mg/L；并用模拟盐水稀释至质量浓度为 1 500 mg/L；采用 Brookfield DV-Ⅲ粘度计，在温度为 90℃、剪切速率为 7.34 s^{-1} 的条件下测定溶液表观粘度。

驱油性能评价实验采用双管并联；岩心长度为 30 cm，直径为 2.5 cm；高、低渗透管渗透率变异系数为 0.2～0.6，平均渗透率为 $1\ 000\times10^{-3}\sim4\ 500\times10^{-3}$ μm^2；原油粘度为 50 mPa·s；实验温度为 85℃，注入速率为 0.46 mL/min。实验步骤包括：① 先水驱至高渗透管含水率达到 100%；② 转注 0.3 倍孔隙体积的聚合物溶液；③ 注聚合物溶液后转水驱，直至含水率达到 98%以上，计算采收率。

2 HAWSP 合成影响因素

2.1 引发剂

根据自由基聚合动力学理论，聚合物分子动力学链长与引发剂浓度的平方根成反比，若要制备高分子质量的聚合物，保持聚合体系低自由基浓度是非常重要的[14]。而多数研究表明,在聚合反应过程中,体系温度不断升高会使引发剂的分解速率明显增加,采用单一引发剂难以合成高分子质量的聚合物,因此丙烯酰胺类聚合物合成中广泛使用复合引发剂[15-18]。复合引发剂包含多种不同类型、不同分解温度的引发剂,随着聚合体系温度的增加,在特定的温度段进行分解产生自由基引发单体聚合,通过合理设计复合引发剂的类型和浓度,可控制聚合体系始终维持较低自由基浓度,使聚合反应平稳发生,可制备高分子质量甚至超高分子质量的聚合物。

设计了 Y1(MR 和 KPS),Y2(MR,AIBI 和 KPS),Y3(MR 和 AIBI),Y4(MR,AIBI 和 LI)共 4 种复合引发剂,由复合引发剂对 HAWSP 聚合体系升温速率的影响(图 1)可以看出,复合引发剂组成对升温速率影响较大。当引发剂中含有 KPS 时,升温速率较快,而采用低温氧化还原引发剂 MR 和水溶性偶氮引发剂 AIBI 组成的复合引发剂时,升温速率明显降低,在此基础上引入一种辅助引发剂 LI 形成的复合引发剂 Y4 升温速率进一步降低,聚合反应更加缓慢平稳,有利于制备具有高效增粘性的耐温抗盐缔合聚合物。

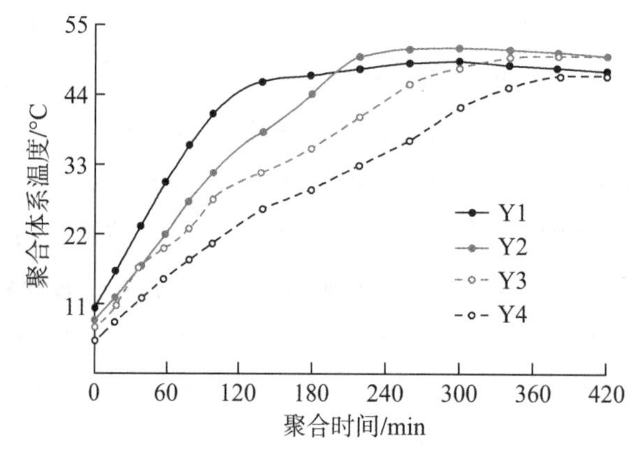

图 1 复合引发剂组成对 HAWSP 聚合体系温度的影响

在 Y1,Y2,Y3 和 Y4 这 4 种复合引发剂条件下制备的聚合物体系的表观粘度分别为 5.2,7.8,15.2 和 25.5 mPa·s,在复合引发剂 Y4 条件下制备的 HAWSP 的增粘性明显优于其他 3 种复合引发剂。因此,Y4 是理想的 HAWSP 合成用复合引发剂。

2.2 表面活性剂

疏水缔合聚合物合成方法较为复杂,这是由于 在合成过程中很难将疏水单体和水溶性单体充分混合。胶束聚合法是公认的制备高效增粘性疏水缔合聚合物最有效的方法[19-20]。胶束聚合法制备的疏水缔合聚合物,疏水链是以微嵌段的形式分布在聚合物大分子主链上,疏水链的这种微嵌段分布大大降低了疏水基团之间形成疏水微区时的熵阻力,使聚合物临界缔合浓度大幅降低,增粘能力加强。

采用胶束聚合法制备耐温抗盐缔合聚合物,需要在聚合体系中加入表面活性剂。表面活性剂的加入对聚合反应的影响主要体现在 3 个方面:① 增溶缔合单体,使其与水溶性单体进行共聚;② 调整疏水微嵌段的结构,控制其在大分子主链上的分布;③ 链转移作用,过多的表面活性剂将大幅降低疏水缔合聚合物的分子质量而使增粘性变差。

由表面活性剂对 HAWSP 表观粘度的影响(图 2)可知,表面活性剂的类型和质量浓度对 HAWSP 的表观粘度具有明显的影响。相对于 SDS,SDBS 和 AEO,表面活性剂 NS 增溶效果最好,链转移作用弱,制备的 HAWSP 增粘性能较好;当 NS 的质量浓度为 400～500 mg/L 时,溶液粘度大于 40 mPa・s。

2.3 D_iC_8AM 物质的量分数

在一定的范围内,随着缔合单体物质的量分数的增大,分子间缔合形成的结构粘度大幅增加,但是在缔合单体物质的量分数过高时,疏水缔合聚合物的分子质量大幅下降,且分子内缔合作用远超过分子间缔合作用,溶液粘度大幅降低,同时还可能导致聚合物不溶解。为此考察了 D_iC_8AM 物质的量分数(疏水单体占总单体的物质的量分数)对 HAWSP 表观粘度的影响。实验结果表明,当 D_iC_8AM 的物质的量分数为 0.4%～1.6%时,随着物质的量分数的增加,溶液粘度增大,这是因为 D_iC_8AM 的物质的量分数增加后,疏水单体在主链上形成的疏水微嵌段越长,更易形成分子间缔合。但是在 D_iC_8AM 的物质的量分数大于 1.6%后,由于疏水单体的链转移作用导致聚合物分子质量降低,分子间缔合作用减弱,溶液粘度大幅度降低。因此,D_iC_8AM 的物质的量分数为 1.4%～1.6%较为合适。

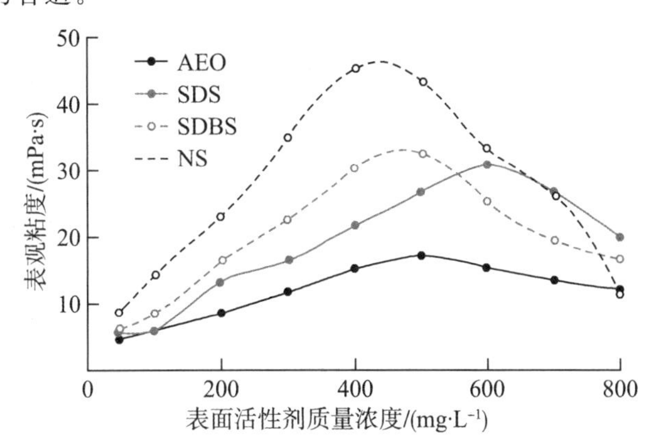

图 2　表面活性剂对 HAWSP 表观粘度的影响

2.4 AMPS-Na 质量分数

引入各种功能单体是改善聚合物耐温抗盐性能的主要途径之一[21]。因此,在疏水缔合聚合物分子中引入耐温抗盐功能单体 AMPS-Na,AMPS-Na 分子结构中具有对盐不敏感的磺酸基团,可使疏水缔合聚合物在高温高盐尤其是高钙镁离子的条件下仍能保持较高的粘度,又由于结构单元中含有庞大的侧基,其共聚物在高温高盐条件下的长期稳定性能将得到明显改善。为此考察了 AMPS-Na 质量分数(AMPS-Na 占总单体的质量分数)对 HAWSP 表观粘度的影响。实验结果表明,在一定范围内,随着 AMPS-Na 质量分数的增加,HAWSP 的表观粘度增大,当 AMPS-Na 的质量分数为 15%～17.5%时,HAWSP 的表观粘度大于 40 mPa・s;而在 AMPS-Na 质量分数超过 17.5%后,HAWSP 的表观粘度降低,这可能是过多的 AMPS-Na 使缔合作用被削弱,造成

HAWSP 分子质量的下降，从而影响了增粘性能。因此，AMPS-Na 的质量分数为 15%～17.5%比较合适。

综上所述，确定了 HAWSP 的主要合成条件为：复合引发剂为 Y4，表面活性剂为 NS，缔合单体 D_iC_8AM 物质的量分数为 1.4%～1.6%，耐温抗盐功能单体 AMPS-Na 质量分数为 15%～17.5%，在此条件下制备了分子质量为 1.05×10^7 g/mol 的耐温抗盐缔合聚合物。

3 HAWSP 分子结构表征

采用 D_2O 为溶剂，利用 ARX400 核磁共振仪(400 MHz，Bruker 公司)对 HAWSP 的分子结构组成进行了核磁共振氢谱分析，并利用端甲基上的氢和主链上亚甲基的氢的峰面积比确定了 HAWSP 的分子结构[22-26]。

根据 HAWSP 的核磁共振氢谱可知：① 化学位移 $\delta=2.021\ 4\times10^{-6}$ 对应主链上的次甲基 Hb，Hd，Hf 和 Hh。② 化学位移 $\delta=3.464\ 3\times10^{-6}$ 对应 AMPS-Na 中的 Hj，证实了含有 AMPS-Na 结构单元。③ 化学位移 $\delta=1.144\ 2\times10^{-6}$ 对应亚甲基单元。通过峰面积可以看出，$\delta=1.144\ 2\times10^{-6}$ 对应亚甲基峰的面积与 $\delta=2.021\ 4\times10^{-6}$ 对应次甲基峰的面积之比大于 2，说明 $\delta=1.144\ 2\times10^{-6}$ 处对应的亚甲基峰不仅含有主链上的亚甲基 Ha，Hc，He 和 Hg，还含有 D_iC_8AM 中的亚甲基单元，证实了 D_iC_8AM 的存在；④ 化学位移 $\delta=0.996\ 4\times10^{-6}$ 对应端甲基单元。通过峰面积可见，$\delta=0.996\ 4\times10^{-6}$ 处对应端甲基峰面积(1.198 3)与 $\delta=2.021\ 4\times10^{-6}$ 处对应次甲基峰面积(3.395 1)之比为 35.295%，这与投料比(优化后物质的量比 AM∶AMPS-Na∶D_iC_8AM＝1.549∶0.073∶0.026)中的端甲基物质的量(由 AMPS-Na 和 D_iC_8AM 中的端甲基组成，为 0.590 7 mol)与次甲基物质的量(1.647 7 mol)的比值(35.85%)非常接近，说明化学位移 $\delta=0.996\ 4\times10^{-6}$ 对应端甲基单元由 AMPS-Na 和 D_iC_8AM 中的端甲基组成，证实了 AMPS-Na 和 D_iC_8AM 的存在。

4 HAWSP 溶液性能评价

以超高分子质量抗盐聚合物 PAM(分子质量为 2.5×10^7 g/mol)为对比，对耐温抗盐缔合聚合物 HAWSP(分子质量为 1.05×10^7 g/mol)的溶液性能进行了评价。

4.1 耐温抗盐性能

评价了质量浓度均为 1 500 mg/L 的 HAWSP 和 PAM 溶液在不同温度和不同钙镁离子质量浓度条件下的表观粘度，其中温度分别为 90 和 95℃，钙镁离子质量浓度为 87～874 mg/L(总矿化度为 3×10^4 mg/L，钙离子与镁离子质量浓度比为 2∶1)。实验结果(图 3)表明，聚合物的分子结构类型与其耐温抗盐性能密切相关，HAWSP 在高温高盐条件下仍能保持较高的粘度，优于 PAM。当溶液中钙镁离子质量浓度从 87 mg/L 增加到 874 mg/L 时，HAWSP 溶液的粘度保留率为 55%～60%，而抗盐高分子质量聚合物 PAM 溶液的仅为 30%～40%，HAWSP 溶液的粘度远远大于同等条件下的 PAM 溶液的粘度，甚至出现了“盐增稠”现象。这是因为 HAWSP 溶液的粘度由结构粘度和非结构粘度组成，其中非结构粘度只与单分子链的流体力学体积有关，而结构粘度主要与疏水基团之间的缔合作用有关，是 HAWSP 粘度的主要组成部分，当矿化度、温度和钙镁离子质量浓度较高时，仍然存在分子间的相互缔合作用，这种分子间相互作用使得 HAWSP 溶液在高温高盐条件下依然保持较高的粘度[27]。

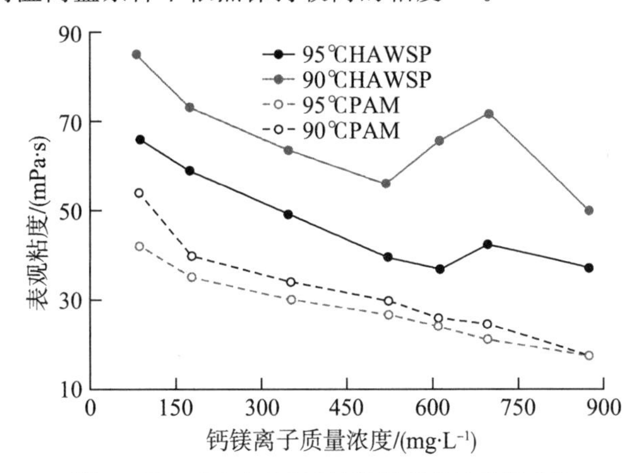

图 3　HAWSP 与 PAM 溶液耐温抗盐性能对比

4.2 老化稳定性能

采用总矿化度为 2×10^4 mg/L(钙镁离子质量浓度分别为 420 和 110 mg/L)的模拟盐水配制质量浓度均为 1 500 mg/L 的 HAWSP 和 PAM 溶液，将其装入 20 mL 的安瓿瓶中，通高纯氮气除氧 10 min 后进行火焰密封，置于 90℃恒温烘箱中进行高温老化，并定期取出测试粘度。实验结果(图 4)表明，HAWSP 的老化稳定性能明显优于 PAM，在 90℃高温下经 120 d 老化后，溶液粘度大于 15 mPa・s，而 PAM 在老化时间超过 60 d 后产生沉淀。缔合单体和耐温抗盐功能单体的协同作用是老化稳定性能优异的原因，耐温抗盐功能单体 AMPS-Na 具有对盐不敏感的磺酸基团和庞大的侧基，且具有抑制酰胺基水解的作用，而缔合单体 D_iC_8AM 能形成分子链间的缔合作用，这些均有利于老化稳定性的提高。

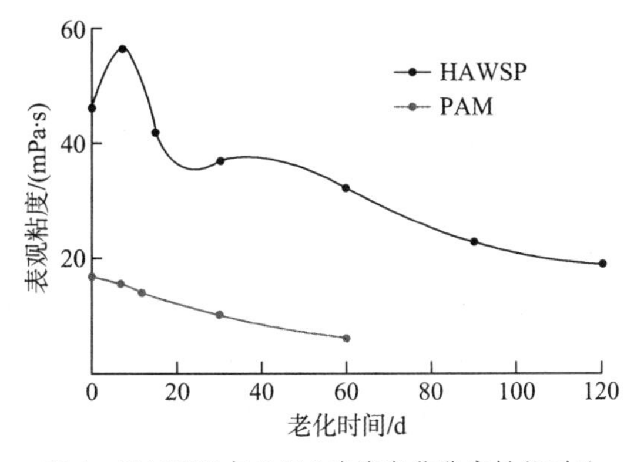

图 4　HAWSP 与 PAM 溶液老化稳定性能对比

4.3 驱油性能

从聚合物驱提高采收率性能评价结果(表1)可以看出,HAWSP提高采收率幅度高于同等条件下的PAM,尤其是在渗透率变异系数为0.5和0.59时,HAWSP比PAM提高了4.8%和4.5%,驱油效果明显。分析认为,HAWSP溶液由于疏水缔合作用形成致密的空间网络结构,表现出更强的粘弹性效应,更有利于提高驱油效率。

表1 聚合物驱提高采收率性能评价结果

聚合物类型	渗透率/10^{-3} μm^2		渗透率	含油饱	水驱理论	最终采	提高采收
	低渗透管	高渗透管	变异系数	和度,%	采收率,%	收率,%	率幅度,%
HAWSP	972	1 495	0.21	80	69.6	83.7	14.1
PAM	980	1 518	0.22	81.4	64	76.1	12.1
HAWSP	1 482	4 436	0.5	77.1	58.5	81.6	23.1
PAM	1 486	4 523	0.51	82.8	49	67.3	18.3
HAWSP	1 820	7 104	0.59	77.1	46	72.6	26.6
PAM	1 611	6 971	0.62	81.7	41.5	63.6	22.1

5 结论

通过优化合成条件,确定了HAWSP聚合物的主要合成条件:复合引发剂为Y4,表面活性剂为NS,D_iC_8AM的物质的量分数为1.4%~1.6%,AMPS-Na质量分数为15%~17.5%,可制备分子质量为1.05×10^7 g/mol的耐温抗盐缔合聚合物。对HAWSP的分子结构进行了核磁共振氢谱表征,证实了HAWSP由AM,D_iC_8AM和AMPS-Na共聚而成。

耐温抗盐缔合聚合物HAWSP的耐温抗盐性、长期稳定性和驱油性能明显优于抗盐高分子质量聚合物PAM,原油采收率提高幅度明显高于PAM,可用作三次采油驱油剂。

参考文献

[1] Khabeev N,Inogamov N. Simulation of micellar-polymer flooding of a layered oil reservoir of nonuniform thickness [J]. Journal of Applied Mechanics and Technical Physics,2008,49(6):985-991.

[2] Samanta A, Bera A, Ojha K, et al. Comparative studies on enhanced oil recovery by alkali-surfactant and polymer flooding [J]. Journal of Petroleum Exploration and Production Technologies, 2012,2(2):67-74.

[3] Zhang Z,Li J,Zhou J. Microscopic roles of “viscoelasticity” in HPMA polymer

flooding for EOR [J]. Transport in Porous Media, 2011,86(1):199－214.

[4] Ramazani S A A, Nourani M, Emadi M, et al. Analytical and experimental study to predict the residual resistance factor on polymer flooding process in fractured medium [J]. Transport in Porous Media,2010,85(3):825－840.

[5] Wyatt K, Pitts M J, Surkalo H, et al. Field chemical flood performance comparison with laboratory displacement in reservoir core [C] . SPE 89385,2004.

[6] 刘剑,张立娟,高伟栋,等. 聚合物溶液在低渗透油层中的适应性实验研究 [J]. 油气地质与采收率,2010,17(3):71-73.

[7] 孙焕泉. 胜利油田三次采油技术的实践与认识 [J]. 石油勘探与开发,2006,33(3):262-266.

[8] 曹瑞波,王晓玲,韩培慧,等. 聚合物驱多段塞交替注入方式及现场应用 [J]. 油气地质与采收率,2012,19(3):71-73.

[9] 袁斌,韩霞. 泡沫复合驱污水配注聚合物溶液粘度损失原因及对策 [J]. 油气地质与采收率,2013,20(2):83-86.

[10] 宋新旺,李哲. 缔合聚合物在多孔介质中的渗流运移特征 [J]. 油气地质与采收率,2012,19(4):50-52.

[11] 姜晨钟,陈新钢,章君,等. 耐温抗盐型聚丙烯酰胺研究进展 [J]. 应用科技,2000,27(8):27-28.

[12] 刘平德,吴肇亮,牛亚斌,等. 耐温抗盐聚合物 KY 的合成及性能影响因素 [J]. 石油钻采工艺,2004,26(4):69-71,85.

[13] 王中华,张辉,黄弘军. 耐温抗盐聚合物驱油剂的设计与合成 [J]. 钻采工艺,1998,21 (6):62-64,67-68.

[14] 潘祖仁. 高分子化学[M]. 3 版. 北京:化学工业出版社,2003.

[15] 陈晓蕾. 共聚法合成阴离子型聚丙烯酰胺石油驱油剂的研究 [D]:西安:西北工业大学,2004.

[16] 张玉凤,沈静,苏文强. 合成工艺对非离子聚丙烯酰胺相对分子质量的影响 [J]. 东北林业大学学报,2006,(5):110-112.

[17] 庞雪君. 丙烯酰胺类聚合物水解度测定方法的改进 [J]. 油气地质与采收率,2010,17(1):71-73.

[18] 马自俊,金日辉. 丙烯酰胺水溶液聚合的几种氧化还原引发体系的研究 [J]. 精细石油化工,1997,(1):41-43.

[19] Hill A, Candau F, Selb J. Properties of hydrophobically associating polyacrylamides:influence of the method of synthesis [J]. Macromolecules, 1993, 26 (17):4521-4532.

[20] Bock J K. Syntactic persistence in language production [J]. Cognitive

Psychology,1986,18(3):355-387.

[21] 薛新生,张健,舒政,等.剪切方式对疏水缔合聚合物溶液性能的影响 [J]. 油气地质与采收率,2013,20(1):59-62.

[22] Volpert E, Selb J, Candau F. Associating behaviour of polyacrylamides hydrophobically modified with dihexylacrylamide [J]. Poly-mer, 1998, 39 (5): 1025-1033.

[23] Volpert E, Selb J, Candau F. Influence of the hydrophobe structure on composition, microstructure, and rheology in associating polyacrylamides prepared by micellar copolymerization [J]. Macromolecules,1996,29(5):1452-1463.

[24] Xue W, Hamley I W, Castelletto V, et al. Synthesis and characterization of hydrophobically modified polyacrylamides and some observations on rheological properties [J]. European Polymer Journal,2004,40(1):47-56.

[25] Regalado E J, Selb J, Candau F. Viscoelastic behavior of semidilute solutions of multisticker polymer chains [J]. Macromolecules,1999,32(25):8 580-8 588.

[26] Castelletto V, Hamley I W, Xue W, et al. Rheological and structural characterization of hydrophobically modified polyacrylamide solutions in the semidilute regime [J]. Macromolecules,2004,37(4):1492-1501.

[27] 薛新生,周薇,向问陶,等.螯合剂 GX 对疏水缔合聚合物溶液粘度保留率的影响 [J]. 油气地质与采收率,2011,18(4): 50-53.

本文编辑 经雅丽

Rheological properties and salt resistance of a hydrophobically associating polyacrylamide

Deng Quanhua[A] Li Haiping[B] Li Ying[A,D] Cao Xulong[C] Yang Yong[C] Song Xinwang[C]

[A]Key Laboratory for Colloid and Interface Chemistry of Education Ministry, Shandong University, Jinan 250100, China.

[B]State Key Laboratory of Crystal Materials, Shandong University, Jinan 250100, China.

[C]Geological Scientific Research Institute, Shengli Oilfield, Dongying 257015, China.

[D]Corresponding author. Email: yingli@sdu.edu.cn

The rheological properties of electrolyte solution of a hydrophobically associating acrylamide-based copolymer (HA-PAM) containing hydrop hobically modified monomer and sodium 2-acryla mido-2-methyl propan esulfonicsul fonate were inves tigated in this paper. The study mainly focussed one ffect sofel ectroly teconcen tration, temperature, and shearrateon the solution rheological properties. HA-PAM exhibited much stronger salt tolerance and shearing resistance than the commonly used partially hydrolyzed polyacrylamide, and has great potential for application in tertiary oil recovery of oilfields with high salinity. The salt resistance mechanism of HA-PAM in solution was investigated by combining molecular simulation and experimental methods. The structure-performance relationship of the salt-resisting polymer may provide useful guidance for design and synthesis of novel water-soluble polymers with high salt resistance.

Introduction

Polymer flooding is one of the most efficient ways to enhance oil recovery (EOR). Partially hydrolyzed polyacryla mide (HPAM) is the most commonly used watersoluble polymer as mobility controller in flooding systems. However, the poor salt tolerance and low shear resistance[1] limit the application of HPAM in oil reservoirs with high salinity and at high temperatures. For exploitation of this kind of reservoirs, it is necessary to develop novel watersoluble polymer systems with high salt and temperature tolerance and shear resistance.

Comb-shaped,[2-5] hydrophobically associating,[6] amphoteric[7] and star-like polymers,[8] polymers containing monomers with temperature and salt-resisting properties,[9] and polyelectrolyte complexes[10] have been synthesized to overcome the shortcomings of HPAM. Among them, hydrophobically associating polymers consisting of hydrophobic side groups, which have attracted the most interest of researchers, can form a transient network possessing unique rheological properties. They can be roughly divided into four types: (Ⅰ) copolymers based on polyacrylamide (PAM) and its derivatives,[11] (Ⅱ) hydrophobically modified poly(vinyl alcohol) (PVA) and poly(ethylene glycol),[12,13] (Ⅲ) hydrophobically modified polysaccharides,[14,15] and (Ⅳ) other copolymers composed of hydrophilic and hydrophobic monomers such as polystyrene-poly(sodium methacrylate) amphiphilic block copolymers and[16] poly(maleic acid/octyl vinyl ether).[9,13,14] These hydrophobically modified water-soluble polymers have achieved commercial acceptance as associative thickeners used in, forexample, cosmetics, paints, and enhancedoilrecovery.[17] Among these hydrophobically modified polymers, researches on the type I polymers have been widely reported in the literature. For example, McCormick et al. synthesized a terpolymercontainingsodium 3-acrylamido-3-methylbutanoate, which possessed strong associative effects and could maintain high viscosity in 0.5M NaCl solution.[18] The hydrophobically associating copolymers containing sulfonic group could exhibit stronger salt resistance.[2,19] Some researches on the salt resistance mechanism of hydrophobically associating copolymers have been reported, but the studies are insufficiently detailed and comprehensive because the corresponding microscopic characterization is lacking.[20,21]

Dissipative particle dynamics (DPD) has been proven to be a useful tool in elucidating mechanisms integrated with theoretical models.[22,23] In DPD models, clusters of atoms are grouped together as single 'bead', enabling direct simulation about the molecular behaviour of the huge systems on long (mesoscopic) length and time scales.

In this work, the rheological properties of a hydrophobically associating acrylamide-based copolymer (HA-PAM) with hydrophobically modified monomer and sodium 2-acrylamido-2-methylpropanesulfonic sulfonate (AMPS) in salt solution were investigated. The effect ofpolymer concentration, electrolyte concentration, temperature, and shear rate on the rheology of the solution was determined. It was found that HA-PAM solutions exhibited high salt and temperature resistance. DPD mesoscopic simulation method was applied to describe the behaviour of the polymer molecules in detail to illustrate the relationship between structure and molecular behaviour of polymers that exhibit salt resistance, and provide a way for the design of novel salt-tolerant polymers for EOR. Molecula dynamics (MD) was also used to give details of polymer molecules in solutions with electrolyte. [24-26]

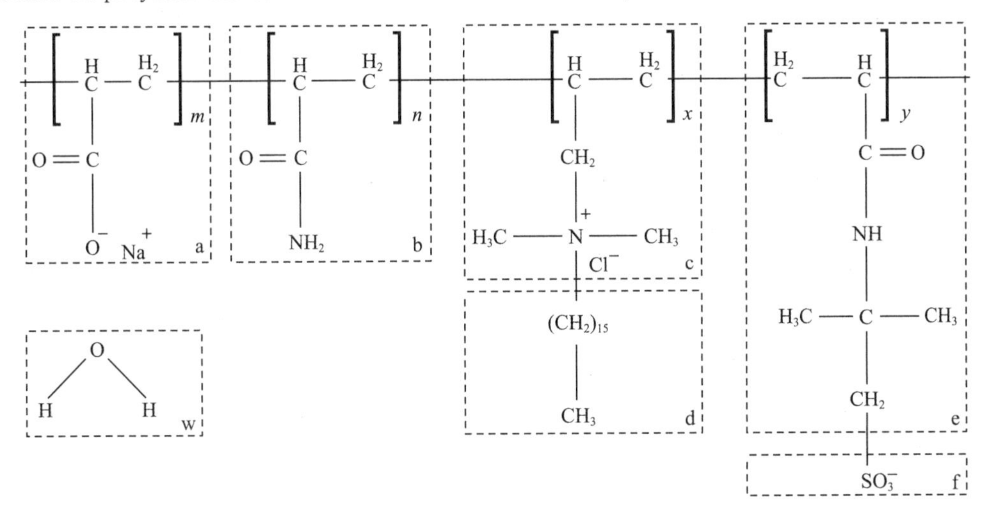

Fig. 1 Chemical structure of HA-PAM. m, n, x, and y correspond to mole percent values of 10, 80, 0. 5, and 9. 5 mol-%, respectively. In DPDsimulation, the segmentsofthe polymermoleculeare representedbybeads, labelledas'a','b','c','d','e', 'f'. The water molecule is represented as'w'.

Experimental

Materials

HA-PAM (Fig. 1) and HPAM usedinthisstudywere suppliedas a gift by Geological Scientific Research Institute of Shengli oil field (China). Their molecular weight sareroughly 3.6×10^5 and 1.7×10^7 $g\cdot mol^{-1}$. Hydrolysis degree of HPAM is 28. 7%. [27] The structure of HA-PAM was characterized by Fourier transform infrared (FT-IR) spectroscopy, and ^{1}HNMR and ^{13}C NMR spectroscopy (Figs S1—S3, Supplementary Material), and corresponds well with the given structure. NaCl and $CaCl_2$ (analytical pure) were purchased from Tianjin Guangcheng Chemical Co. , Ltd (China). Pyrene (97 wt-%), hydrochloric acid, and NaOH (analytical pure) were obtained from Aladdin. Deionized water was used in this work.

Preparation of HA-PAM Solutions

Stock solution of 0. 2 wt-% (w/w) HA-PAM was prepared by dissolving 0. 2 g of solid powder in 100 g H_2O. HA-PAM solu-tions at low concentrations were obtained via dilution of the stock solution. The pH of the HA-PAM solutions was adjusted with

1M HCl and 1M NaOH solutions to 7. 28 干 0. 15. Pyrene was used as fluorescence label. HA-PAM solutions containing pyrene (3.03×10^{-4} g·L^{-1}) were prepared by add-ing 20 mL of 0. 15 g·L^{-1} pyrene ethanol solution to HA-PAM aqueous solution.

Characterization

The structure of HA-PAM was verified using ^{1}HNMR, ^{13}CNMR, and FT-IR spectroscopy. The ^{1}HNMR and ^{13}CNMR spectra were measured in D_2O on a Bruker Avance 300-MHz NMR spectrometer. FT-IR measurements were performed on a Nicolet iS5 Mid Infrared FT-IR spectrometer (Thermofisher, Nicolet iS5, USA).

The molecular weight of HA-PAM was determined through static light scattering measurements on a light scattering instru-ment (Wyatt Technology). The laser was positioned so that the incident beam was vertically polarized. Zimm plots were obtained using Astra software. The polymer was filtered with a0. 45-mmporesizemembrane filterbeforestaticlightscattering measurement. A differential refractive index detector (Optilab-REX) was used to measure the differential refractive indices (dn/dc) of different polymer solutions at 658nm and 258C.

Rheological Measurements

The rheological measurements were carried out on a HAAKE RS75 rheometer (Germany) with a coaxial cylinder sensor system (Z41 Ti). The samples were rested for more than 24 h before measurement to guarantee that no bubbles were present. The experiment temperature was controlled to be 25. 0± 0. 18C, except for the study whereby the effect of temperature was investigated. Before the measurement, all the samples were rested for 30 min after the cylinder reached the measuring position to eliminate the shear influences. In the steady-state shearing experiment, the shear rate ranged from 0. 1 to 10 s^{-1} or 0. 1 to 1000 s^{-1} with a gradient of 0. 5 $(\Delta\tau/\tau)\Delta t\%$ (τ is yield stress) and a maximum waiting time of 20 s at each shear rate step. The stress sweep experiment was carried out from 0. 03 to 10 Pa at a frequency of 0. 5 Hz (3. 14 rads^{-1}) to determine the linear viscoelastic region. The frequency sweep measurement was performed from 0. 05 to 100 rads^{-1} at 0. 04 Pa (within the linear viscoelastic region).

Zero-shear viscosity (η_0) of the solutions was determined by the creep method. The creep measurement was carried out at a constant stress of 0. 05Pa and the variation of solution viscosity with time (t) was recorded. η_0 is the solution viscosity that reaches the equilibrium.

The equilibrium value of dynamic viscosity and modulus were measured within the linear viscoelasticity region (0. 05 Pa) and angular frequency of 0. 2 rads^{-1}. The temperature sweep range was 20—70℃.

Simulation Method

DPD

The simulation system was performed in a 3D box with size of 20×20×20 R_C^3. During the simulation, 20 000 time steps were sufficient for the free energy of this system to asymptoti-cally approach a stable value. Flory-Huggins parameters (x_{ij}) were used to represent the liquid compressibility and mutual solubility. a_{ij} represents the repulsion parameter. The Flory-Huggins parameters (x_{ij}) were obtained by blends, and then transformed into interaction parameters a_{ij} by the following equation: $a_{ij} = a_{ii} \times 3.27 + 25$. The simulated bead density (ρ) was 3.0. Spring constant between different beads was 4.0 according to Groot's work.[28] Root mean square (RMS) end-to-end distance was applied to describe the expansion extent of the polymer chains. The parameters are summarized in Table 1. All simulations were performed using *Materials Studio* 5.0from Accelrys Inc.

Table 1 Bead-bead interaction parameters a_{ij} used in the simulation

Bead	a	b	c	d	e	f	w
a	80						
b	22.9	25					
c	14	29.8	70				
d	75.4	79.1	47	25			
e	29	29.5	27.1	70.9	25		
f	90	45	16	73.3	55.3	100	
w	15	27	28	64.4	31.5	9.6	25

MD

Simulations were performed under canonical ensemble (NVT, constant number of atoms, volume, and temperature),[29] The totalener gyin clud evalen ceandnon-bond interaction terms listed in the following equation: $E_{total} = E_{bonds} + E_{angles} + E_{dihedrals} + E_{cross} + E_{non\text{-}bond}$.

Two polymercha in swith chainleng thsof 200 and 5 000 water molecules were embedded into the cubic cells using Amorphous Cell module. Then, 5, 10, and 75 NaCl were randomly put in the box, respectively, for HA-PAM electrolyte solution and HPAM solution (75 NaCl was used for HPAM solution). The right amount of Cl^- and Nat was added to maintain the charge balance in these systems. The formed simulation cell was 59. 4Å×59. 4Å×59. 4″Å, 59. 45Å×59. 45Å×59. 45Å, 60. 04Å×60. 04Å×60. 04Å, and 59. 304 8Å×59. 304 8Å×59. 304 8Å, respectively. Water was modelled using the simple point charge/extended (SPC/E) model.[30] The temperature was con-trolled at 298K using the Hoover-Nose thermostat.[31] For the long range electrostatic interaction, the Ewald method was used,[32] and the cut-off distance was 9. 5Å. In consideration of the potential functions and atom interaction parameters, a COMPASS force field, the most accurateforcefieldforthecalculation of molecular interactions, was employed in the simulation.[33] All the initial configurations were first minimized for 10 000 steps with the Smart Minimizer method, after which a 1 ns MD run was carried out to equilibrate the system. The run was continued for another 100 ps for data collection. The results were saved every 5 ps. The radial distribution functions (RDF, g(r)) for O^- (—COO^-) of HA-PAM and O of H_2O were used to exhibit variation of the —COO^- electric double layer by increas-ing NaCl amount. Values of g(r) can give information on the appearance frequency of atoms.[34]

Fluorescence Measurement

The fluorescence spectra were recorded on an LS-55 spectrofluorometer (Perkin-Elmer, USA) using a 1. 0 cm quartz cuvette and pyrene was used as a fluorescent probe. The excitation and emission slits were both set at 7. 5 nm. The excitation wave-length was 335 nm. The emission spectra were recorded over a scanning range of 350～650 nm. The analytical wavelengths of the measurement were 373 nm and 384 nm, and I_1 and I_3 rep-resent their intensities, respectively. The ratio of I_1/I_3 was used to estimate the micropolarity.

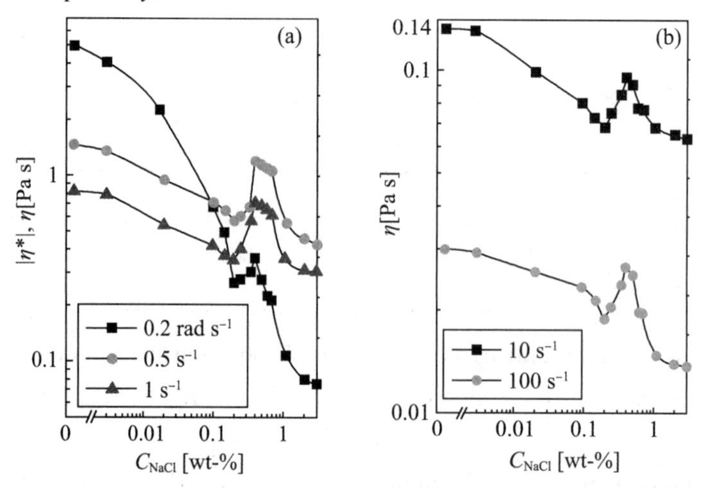

Fig. 2 Effect of NaCl concentration on steady and dynamic viscosity of 0. 1 wt-% HA-PAM solution; T=25℃. (a) Dynamic viscosity at 0. 2 rads^{-1} and steady viscosity at 0. 5 s^{-1} and 1 s^{-1}; (b) steady viscosity at 10 s^{-1} and 100 s^{-1}.

Results and Discussion

Effect of NaCl Concentration on the Rheological Properties of HA-PAM

η_0 obtained via the creep measurement (Fig. S4, Supplementary Material) exhibits a gradual increase with increasing HA-PAM solution concentrations (Fig. S5, Supplementary Material). The overlapping concentration of HA-PAM aqueous solution is determined to be 0.009 5 wt-% (Fig. S5, Supplementary Mate-rial). The tests of HA-PAM solutions in this work were all performed at concentrations higher than 0.009 5 wt-% or in the semi-dilute regime.

Fig. 2 shows that the viscosity of 0.1 wt-% HA-PAM solution initially decreases as NaCl concentration (C_{NaCl}) increases, then increases when C_{NaCl} is higher than 0.2 wt-%; and the maximum viscosity appears at $C_{NaCl} = 0.4$ wt-%, after that, the viscosity decreases. The HA-PAM solutions exhibit enhanced thickening capacity with $C_{NaCl} = 0.2$-0.4 wt-% at various shear rates. The increase of bulk phase viscosity of polyelectrolyte solutions with increasing electrolyte concentrations has been reported widely,[35-37] but the above depicted phenomenon for HA-PAM solution is rarely seen.

The variation of the viscosity of HA-PAM solutions without NaCl or with 0.1wt-%NaCl under shearing is presented in Fig. 3. The curve of HA-PAM solution without NaCl decreased linearly, indicating a typical shear-thinning character, and fitted very well to a power law model (Table S1, Supplementary Material), and the flow behaviour index (n) was 0.208. The curve of HA-PAM solutioncontaining 0.1wt-%NaCl can be divided into three parts (A, B, and C in Fig. 3. The flow behaviour index (n) values of A, B, and C were 0.111, 0.298 and 0.354, respectively. The smaller n value indicated that the degree of shear thinning of HA-PAM solution containing NaCl is stronger at relatively low shear rates ($<0.45\ s^{-1}$). As the shear rate is raised, the n values for curves B and C are larger than that for pure HA-PAM solution, illustrating that the shear-thinning effect gets weaker in salt solutions.

Fig. 4 exhibits the viscosity variation of 0.1 wt-% HA-PAM solution with time when the shear rate was fixed at $0.5\ s^{-1}$. The viscosity of pure HA-PAM solution decreases with time and reaches an equilibrium, finally, showing positive thixotropy.[38] In contrast, for the HA-PAM solution containing 0.1wt-% NaCl, the viscosity increases initially and decreases subsequently, indicating complex thixotropy.[39]

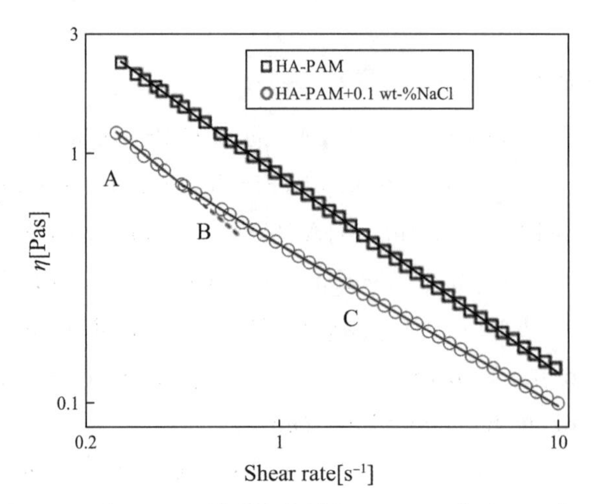

Fig. 3　Variation of viscosity of 0.1 wt-% HA-PAM aqueous solution with and without NaCl as a function of shear rate. A, B, and C represent the three stages of the fitting line.

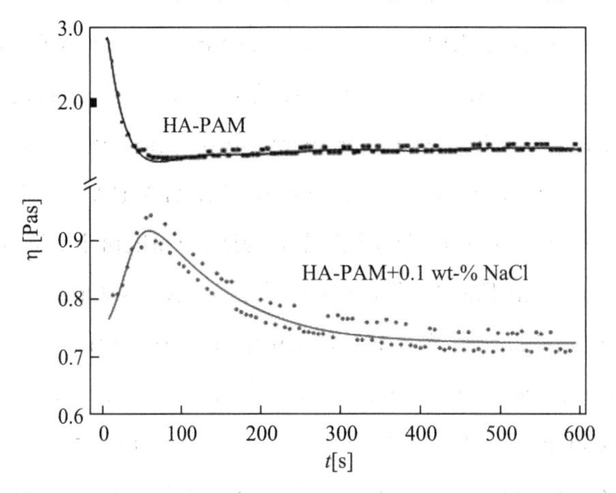

Fig. 4　Change in the viscosity of 0.1 wt-% HA-PAM aqueous solution with and without NaCl as a function of time; the shear rate was maintained at 0.5 s^{-1}.

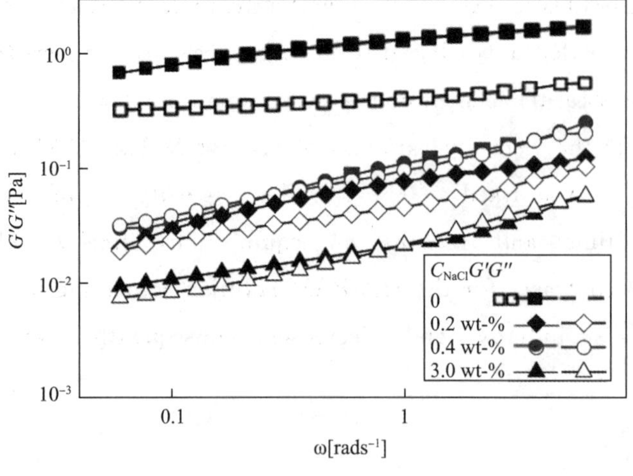

Fig. 5　Variation of the dynamic modulus as a function of angular frequency for 0.1 wt-% HA-PAM solutions with NaCl at concentrations of 0, 0.2, 0.4, and 3.0 wt-% at a shear stress of 0.04 Pa.

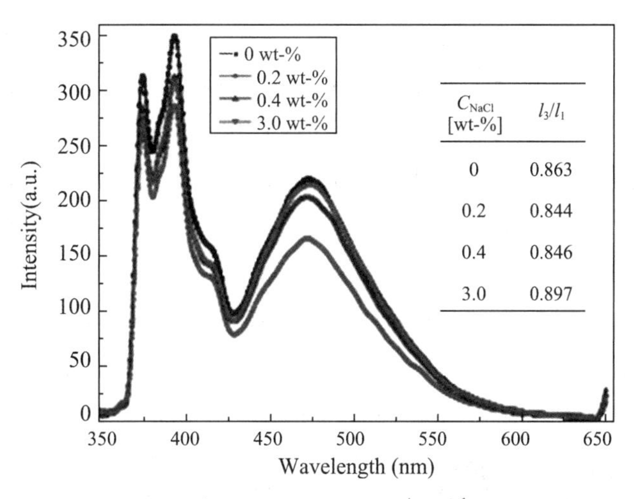

C_{NaCl} [wt-%]	I_3/I_1
0	0.863
0.2	0.844
0.4	0.846
3.0	0.897

Fig. 6 Fluorescence emission spectra of pyrene (3.03×10^{-4} gL^{-1}) in HA-PAM solutions at different NaCl concentrations.

A wide linear viscoelastic region is observed for 0.1 wt-% HA-PAM solutions (Figs S6 and S7, Supplementary Material). Dynamic rheological measurements conducted on HA-PAM aqueous solutions (Fig. 5) revealed that both $G'$and G'' increase monotonously in the entire examined frequency range. In the absence of NaCl, G' is larger than G''over the studied frequency range, indicating that HA-PAM solution is a dominant elastic fluid. As C_{NaCl} is raised, both G' and G'' decrease first, then increase and decrease again, and the phase angle tangent tan(6) increases gradually, reaching almost to 1 when C_{NaCl} is larger than 0.4 wt-%. This suggests the HA-PAM solution transformsfrom a dominant elastic fluid into a typical viscoelastic fluid at high salt concentration. [40]

Molecular Behaviour of HA-PAM Investigated by DPD and Fluorescence

Fig. 6 shows the fluorescence spectra of HA-PAM solutions at different C_{NaCl}. The I_3/I_1 value decreases from 0.863 to 0.844 as C_{NaCl} increases from 0 to 0.2 wt-%, which might arises from the electrolytescreeningeffect. When 0.2 wt-%$<C_{NaCl}<$0.4 wt-%, the I_3/I_1 value increases slightly with increasing C_{NaCl}, indicating the gradual growth of the hydrophobic microdomains induced by the interchain hydrophobic association. When C_{NaCl} increases to 3 wt-%, the increase ofthe I_3/I_1 value may be due to the enhanced intramolecular hydrophobic interaction caused by the aggregation of the curled molecules. The high viscosity of HAPAM solution may result from the hydrophobic interaction, [41] the molecular entanglement, [42] and the increase of the hydrodynamic radius. [43] According to the fluorescence measurements, the hydrophobic interaction plays an important role on the salt resistance.

Fig. 7 shows the DPD snapshots of the polymer molecules in solution under shear, which clearly show the hydrophobic domains formed by aggregation of the hydrophobic groups. The aggregation degree gets larger at higher shear rates. The simulation results agree very well with the experimental results in Figs 3-7, indicating that the hydrophobic interaction between the hydrophobic side chains becomes stronger under shearing. The RMS of the polymer molecules are calculated (Table S2, Supplementary Material). The results show that shear induces molecular stretch. So it could be concluded that shear induces the stretch of the macromolecules that contributes to the increase of the solution viscosity on one hand and causes the increase of intermolecular hydrophobic interaction[44] on the other hand.

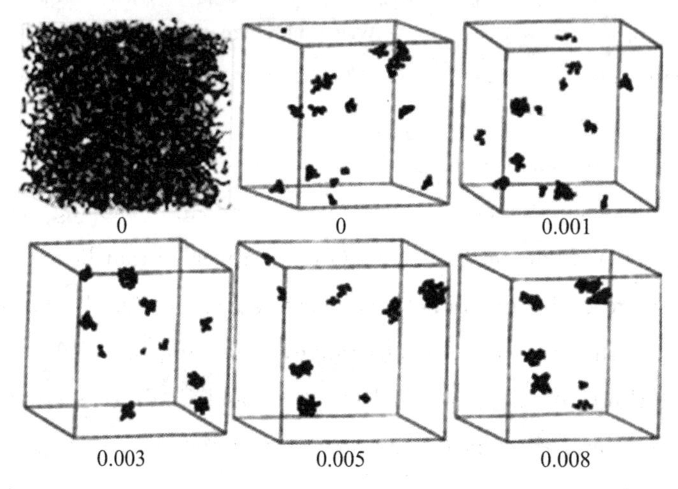

Fig. 7 Structural snapshots of the hydrophobic part of HA-PAM as a function of shear rate (0, 0.001, 0.003, 0.005, and 0.008 s^{-1}). The first image is HA-PAM with all the function groups.

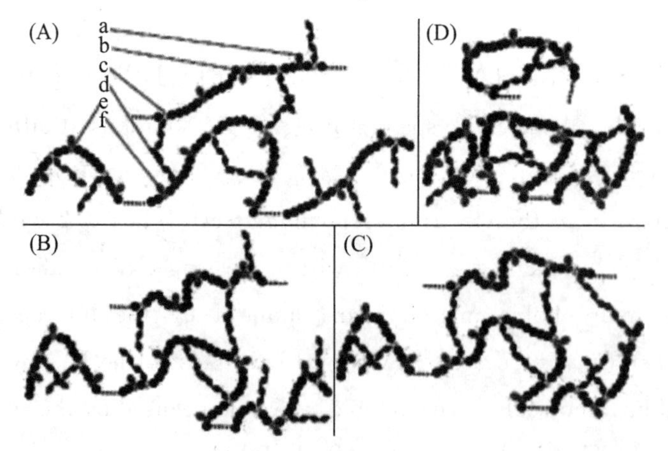

Fig. 8 Schematic illustration of the interaction between HA-PAM mole-cules with increasing electroly teconcen trations. Colour softhe beads are the sameasthosein Fig. 1: a (sodiumacrylate, green); b (acrylamide, darkblue); c+d (hydrophobically associating monomer, yellowt dark yellow); and e+f (sodium 2-acrylamido-2-methylpropanesulfonate, light bluet purple blue). A, B, C, and D represent the HA-PAM chain variation with increasing NaCl.

According to the above results, we can understand the effect of salts on the molecular behaviour of HA-PAM on a micro-scopic level, as shown in Fig. 8. When NaCl was added to the solution, the screening effect of electrolyte leads to the macro-molecular curl [Fig. 8 (A), (B)], the hydrodynamic radius decreases, and intermolecular entanglement is weakened, which causes the decline in the solution viscosity.[45,46] On the other hand, the increase of the salt concentration contributes to the increase of intermolecular hydrophobic association [Fig. 8(C)].[46,47] As shown in Fig. 2(a), the apparent viscosity of 0.1 wt-% HA-PAM solution decreased from 1 462.5 to 570.5 mPa · s with C_{NaCl} increasing from 0 to 0.2 wt-%. This probably results from the screening effect of NaCl. Increase of the NaCl concentration from 0.4 to 3.0 wt-% would decrease the viscosity of HA-PAM solution. This is mainly due to the formation of compact hydrophobically microdomains [Fig. 8(D)].

Effect of $CaCl_2$ Concentration on the Viscosity of HA-PAM Solutions

Fig. 9 shows the effect of $CaCl_2$ concentration (C_{CaCl_2}) on the viscosity of 0.1 wt-% HA-PAM solution. A similar trend of the steady viscosity was observed to that in Fig. 2, but the minimum and maximum viscosity values appear at lower electrolyte con— centrations (C_{CaCl_2}=0.04 wt-% and 0.06 wt-%, respectively).

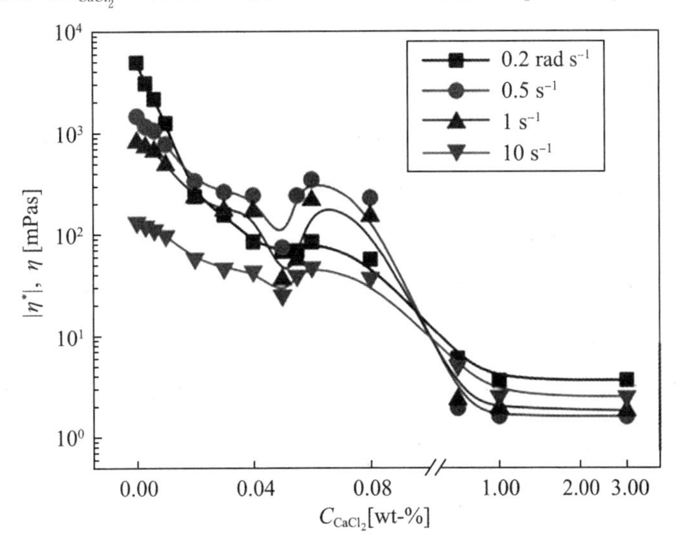

Fig. 9 Variation of viscosity of HA-PAM solution as a function of $CaCl_2$ concentration.

Salt Tolerance of HA-PAM

To assess the salt resistance of HA-PAM solution, the viscosity, dynamic modulus, andviscosityretention of 0.1 wt-% HA-PAM solution are compared with those of 0.1 wt-% HPAM.[27] As shown in Table S3 (Supplementary Material), the chosen elec-trolyte concentrations are 3 wt-% NaCl and 0. 08 wt-% $CaCl_2$. The viscosity (η) at 10 s^{-1}, G' and G'' at 1Hz, and tan(δ) of pure HA-PAM solution are 132. 6 mPas, 1. 17 Pa, 0. 58 Pa, and 0. 49, respectively, whereas the corresponding values of pure HPAM solution are 114. 3 mPas, 0. 42 Pa, 0. 30 Pa, and 0. 73. The 0. 1 wt-% HA-PAM solution exhibits higher viscosity and elas-ticity than 0. 1 wt-% HPAM solution. In the presence of 3 wt-% NaCl, the viscosity, G', and G'' decrease to 63. 1 mPas, 0. 077 Pa, and 0. 073 Pa, respectively for HA-PAM solution, whereas those for HPAM solution are respectively 6. 7 mPas, 2.87×10^{-6} Pa, and 1.50×10^{-3} Pa. The viscosity and modulus of the former are much larger than those of the latter. The HA-PAM solution transforms from a dominant elastic fluid into a viscoelastic fluid (tan(δ) = 0. 94), whereas the HPAM solution transforms into a dominant viscous fluid (tan(δ) = 520. 87). The viscosity retention of HA-PAM solution and HPAM solution is 47. 37 wt-% and 5. 88 wt-%, respectively. The former is much larger than the latter. After the addition of 0. 08 wt-% $CaCl_2$, the viscosity and dynamic modulus of both HA-PAM and HPAM solutions decrease. Those of HPAM solution decline more remarkably, similarly to that with the addition of NaCl (Table S3, Supplementary Material). The viscosity retention of HA-PAM solution, 27. 92 wt-%, is also much larger than that of HPAM solutions, 8. 88 wt. %.

To better understand the mechanism about how the inorganic ions affect the property of polymer molecules, MD simulation method was used to evaluate details of the interaction between the inorganic ions and charged groups of polymers. Fig. 10 shows the RDF for O^- ($-COO^-$) of HA-PAM (or HPAM) and O of H_2O at different NaCl concentrations, and clearly indicates that the electric double layer of $-COO^-$ of HPAM becomes thinner than that of HM-PAM. So the salt tolerance of HA-PAM is better than that of HPAM. HA-PAM would be a good candidate for application in EOR, especially in oilfields with high salinity.

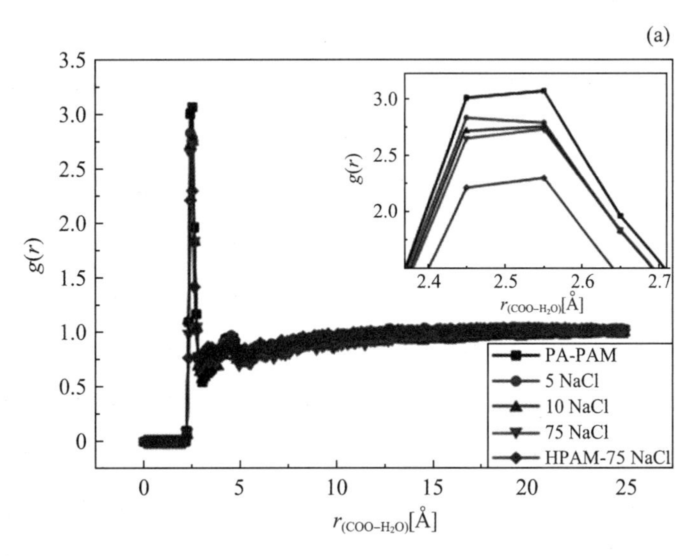

Fig. 10 RDF for O^- (—COO^-) of HA-PAM (or HPAM) and O (H_2O) with different amounts of NaCl at 25℃. The inset image is a high-magnification image of the peak observed in the main graph.

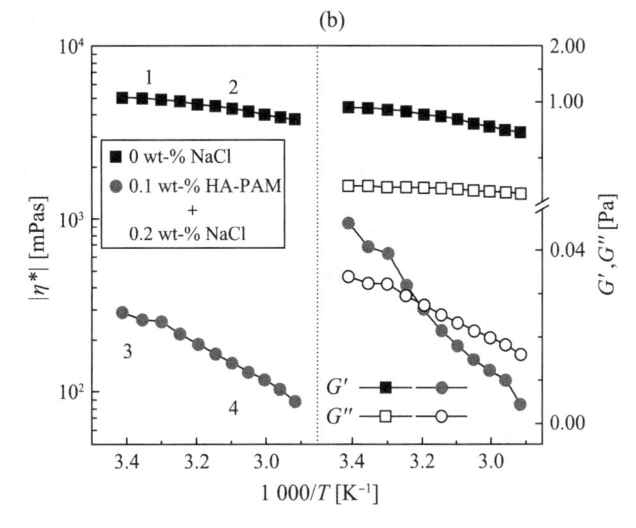

Fig. 11 (a) Dynamic viscosity (0.2 $rads^{-1}$) and (b) modulus of 0.1 wt-% HA-PAM in (■, □) 0 wt-% and 0.2 wt-% NaCl (●, ○) aqueous solutions.

Effect of Temperature on Rheological Properties of HA-PAM Fig. 11 shows the variation of viscosity and dynamic modulus with temperature for HA-PAM solution with C_{NaCl} of 0 and 0.2 wt-%. Both the viscosity and modulus decrease with increasing temperatures over the temperature range studied, and are in agreement with the trends reported in the literature. [48]

Two break points at 35 and 30℃ are observed from the plots of viscosity versus temperature of HA-PAM solutions without and with NaCl, respectively, i. e. the plots disclose two distinct regions (curves 1—4 in Fig. 11). HA-PAM solution containing

NaCl exhibits lower breaking point temperatures. As the temperature is raised, the viscosity decreases more quickly for curves 2 and 4 than for curves 1 and 3. All curves could be well fitted to the Arrhenius equation, $\eta = A\exp(E_a/RT)$, where η, A, E_a, R, and T are the viscosity of the polymer solution, the pre-exponential factor, the Arrhenius activation energy (Tables S4 and S5, Supplementary Material), the gas constant (8.314 J $(K \cdot mol)^{-1}$), and the thermodynamic temperature, respectively.[49] In the case of a network-forming solution, E_a may be suitably termed as the characteristic activation energy of viscous flow per mole of polymer segment.[50] For pure HA-PAM solution, the E_a values are 2.5 kJ mol^{-1} at $T < 35℃$ and 6.1 kJ mol^{-1} at $T > 35℃$. Larger values, i. e. 9.3 kJmol^{-1} at $T < 30℃$ and 22.1 kJmol^{-1} at $T >$ 30℃, are obtained for HA-PAM solution with NaCl. This is probably because the electrolyte screening effect that induced macromolecular curl can cause the increase in the characteristic activation energy of the molecular segment movement, corresponding to higher E_a values.[51] G' is higher than G'' in the tested temperature range for pure HA-PAM solution, indicating a dominant elastic behaviour. For NaCl solution of HA-PAM, G' is higher than G'' at lower temperatures, indicating an elastic behaviour; and G' is smaller than G'' at higher temperatures, indicating a viscous behaviour. This result proves that at higher temperatures, the addition of NaCl would cause obvious decreases in elasticity for HA-PAM solution or that G's more sensitive to temperature on addition of NaCl.

The influence of temperature on the microstructures of solution is complex. Onone hand, an increase in the temperature favours the hydrophobe-hydrophobe association, which is an endothermic process. Extensive intermolecular association leads to a network structure and a large increase in viscosity and dynamic modulus.[52] With the increase of temperature, the hydrophobic parts aggregate gradually, and when the temperature exceeds 65℃, they would be densely packed. On the other hand, at high temperatures, water molecules and polymer chains move faster, and the interchain hydrophobic association would be partially destroyed.[53,54] Meanwhile, the hydrodynamic radius of HA-PAM decreases. Thus, the solution viscosity and dynamic modulus would decline. The nature of the operating effect is dependent on the solution condition. According to the fact that the viscosity of HA-PAM solutions without and with NaCl decreases gradually with increasing temperatures, the latter factors play a decisive role. But the occurrence ofbreaking points at 35 or 30℃ or the slower decrease of viscosity with increasing temperatures for curves 1 and 3 in Fig. 11 indicates that the former factor also plays a role. The break point shifts from 35 to 30℃ after the addition of 0.2 wt-% NaCl, implying that the former factor plays a smaller role with increasing C_{NaCl}. This is probably because the electrolyte screening effect that induced molecular curl is unfavourable to the intermolecular hydrophobe-hydrophobe association.

Conclusions

The rheological properties of HA-PAM solutions were investigated systematically at various solution concentrations, temperatures, and electrolyte concentrations. In the presence of NaCl, the intermolecular hydrophobic interaction becomes stronger and a solution viscosity increase at 0.2 wt-%$<C_{NaCl}<$1.0 wt-% was observed. The HA-PAM solutions exhibited prominent elastic behaviour at low C_{NaCl} and become visco-elastic fluids at high C_{NaCl}. The dynamic viscosity of the solu-tions depended exponentially on the inverse oftemperature, and is attributed to endothermic-driven hydrophobic associating processes and the destruction of the hydration spheres. $CaCl_2$ reduced the solution viscosity more prominently than NaCl. Compared with HPAM solution, the HA-PAM solution exhibits much highersaltresistance. DPD and MD wereused tostudythe mechanism at the molecular level. MD results showed that the compression of the electrical double layers instigated the for-mation of a thinner hydration film, leading to a more collapsed conformation of HA-PAM chain. DPD results showed that suitableshearratescouldstretchouttheHA-PAMchainsfurther and enhance the aggregation between the hydrophobic side chains. All the results indicate that HA-PAM is appropriate for use in EOR of oilfields with high salinity.

Supplementary Material

Information about the characterization of HA-PAM by ^{1}H NMR, ^{13}C NMR, and FT-IR spectroscopy, and rheological data of 0.1 wt-% HPAM and HA-PAM solutions are available on the Journal's website.

Acknowledgement

The National Municipal Science and Technology Project (No.2008ZX05011-002) and National Science Fund of China (No. 21173134) are gratefully acknowledged for financial support.

References

[1] D. A. Z. Wever, F. Picchioni, A. A. Broekhuis, Prog. Polym. Sci. 2011, 36, 1558. doi:10.1016/J.PROGPOLYMSCI.2011.05.006

[2] C. Zhong, L. Jiang, X. Peng, J. Polym. Sci. Pol. Chem. 2010, 48, 1241. doi:10.1002/POLA.23888

[3] D. A. Z. Wever, L. M. Polgar, M. C. Stuart, F. Picchioni, A. A. Broekhuis, Ind. Eng. Chem. Res. 2013, 52, 16993. doi:10.1021/ IE403045Y

[4] D. A. Z. Wever, F. Picchioni, A. A. Broekhuis, Ind. Eng. Chem. Res. 2013, 52, 16352. doi:10.1021/IE402526K

[5] D. A. Z. Wever, F. Picchioni, A. Broekhuis, Eur. Polym. J. 2013, 49, 3289. doi:10.1016/J.EURPOLYMJ.2013.06.036

[6] E. K. Penott-Chang, L. Gouveia, I. J. Fernández, A. J. Müller, A. Díaz-Barrios, A. E. Sáez, Colloid. Surface. A 2007, 295, 99. doi: 10.1016/ J.COLSURFA.2006.08.038

[7] E. E. Kathmann, L. A. White, C. L. McCormick, Polymer 1997, 38, 871. doi:10.1016/S0032-3861(96)00586-1

[8] S. Hietala, P. Mononen, S. Strandman, P. Järvi, M. Torkkeli, K. Jankova, S. Hvilsted, H. Tenhu, Polymer 2007, 48, 4087. doi: 10.1016/J.POLYMER.2007.04.069

[9] A. Sabhapondit, A. Borthakur, I. Haque, Energy Fuels 2003, 17, 683. doi: 10.1021/EF010253T

[10] H. Li, R. Chen, X. Lu, W. Hou, Carbohydr. Polym. 2012, 90, 1330. doi:10.1016/J.CARBPOL.2012.07.001

[11] W. Xue, I. W. Hamley, V. Castelletto, P. D. Olmsted, Eur. Polym. J. 2004, 40, 47. doi:10.1016/J.EURPOLYMJ.2003.09.014

[12] F. Xiong, J. Li, H. Wang, Y. Chen, J. Cheng, J. Zhu, Polymer 2006, 47, 6636. doi:10.1016/J.POLYMER.2006.07.020

[13] G. Yahya, S. Ali, M. Al-Naafa, E. Hamad, J. Appl. Polym. Sci. 1995, 57, 343. doi:10.1002/APP. 1995.070570311

[14] D. Charpentier, G. Mocanu, A. Carpov, S. Chapelle, L. Merle, G. Müller, Carbohydr. Polym. 1997, 33,177. doi:10.1016/S0144-8617 (97)00031-3

[15] C. Esquenet, P. Terech, F. Boué, E. Buhler, Langmuir 2004, 20, 3583. doi:10.1021/LA036395S

[16] P. Raffa, P. Brandenburg, D. A. Z. Wever, A. A. Broekhuis, F. Picchioni,Macromolecules2013, 46,7106. doi:10.1021/MA401453J

[17] S. Wu, R. A. Shanks, G. Bryant, J. Appl. Polym. Sci. 2006, 100, 4348. doi:10.1002/APP.23282

[18] C. L. McCormick, J. C. Middleton, D. F. Cummins, Macromolecules 1992, 25, 1201. doi:10.1021/MA00030A001

[19] Z. Ye,G. Gou, S. Gou, W. Jiang,T. Liu,J. Appl. Polym. Sci. 2013, 128, 2003.

[20] W. Kuang,J. Zhang,R. Li,B. Wu,Y. Tan,Polym. Bull. 2013, 70,3547. doi:10.1007/S00289-013-1039-4

[21] C. Senan, J. Meadows, P. T. Shone, P. A. Williams, Langmuir 1994, 10, 2471. doi:10.1021/LA00019A074

[22] Y. Mei, Y. Han, H. Wang, L. Xie, H. Zhou, J. Surfactants Deterg. 2014, 17, 323. doi:10.1007/S11743-013-1521-X

[23] Q. Li, R. Yuan, L. Ying, J. Appl. Polym. Sci. 2013, 128, 206. doi:10.1002/APP.38169

[24] P. Chen,L. Yao,Y. Liu,J. Luo,G. Zhou,B. Jiang,J. Mol. Model. 2012, 18, 3153. doi:10.1007/S00894-011-1332-9

[25] C. l. Ren, W. D. Tian, I. Szleifer, Y. Q. Ma, Macromolecules 2011, 44, 1719. doi:10.1021/MA1027752

[26] T. Zhao,G. Xu, S. Yuan, Y. Chen,H. Yan,J. Phys. Chem. B2010, 114, 5025. doi:10.1021/JP907438X

[27] H. Li, W. Hou, Food Hydrocoll. 2011, 25, 1547. doi: 10.1016/ J. FOODHYD.2011.01.014

[28] R. D. Groot, Langmuir 2000, 16, 7493. doi:10.1021/LA000010D

[29] J. Jeon, A. V. Dobrynin, J. Phys. Chem. B 2006, 110, 24652. doi:10.1021/JP064288B

[30] H. Yan, X. L. Guo, S. L. Yuan, C. B. Liu, Langmuir 2011, 27, 5762.

doi:10.1021/LA1049869

[31] Q. Liao, A. V. Dobrynin, M. Rubinstein, Macromolecules 2003, 36, 3386. doi:10.1021/MA025995F

[32] H. Yan, S. L. Yuan, G. Y. Xu, C. B. Liu, Langmuir 2010, 26, 10448. doi:10.1021/LA100310W

[33] C. Li, Y. Li, R. Yuan, W. Lv, Langmuir 2013, 29, 5418. doi: 10.1021/ LA4011373

[34] D. Wu, Y. Feng, G. Xu, Y. Chen, X. Cao, Y. Li, Colloid. Surface. A 2007, 299, 117. doi:10.1016/J.COLSURFA.2006.11.031

[35] G. O. Yahya, E. Z. Hamad, Polymer 1995, 36, 3705. doi:10.1016/ 0032-3861(95)93773-F

[36] L. Ye, R. Huang, J. Appl. Polym. Sci. 1999, 74, 211. doi: 10.1002/(SICI)1097-4628(19991003)74:1,211::AID-APP26.3.0.CO;2-T

[37] J. Ma, B. Liang, P. Cui, H. Dai, R. Huang, Polymer 2003, 44, 1281. doi:10.1016/S0032-3861(02)00851-0

[38] S. M. A. Razavi, H. Karazhiyan, Food Hydrocoll. 2009, 23, 908. doi:10.1016/J.FOODHYD.2008.05.010

[39] S. P. Li, W. G. Hou, D. J. Sun, P. Z. Guo, C. X. Jia, J. F. Hu, Langmuir 2003, 19, 3172. doi:10.1021/LA026669W

[40] L. G. Patruyo, A. J. Müller, A. E. Sáez, Polymer 2002, 43, 6481. doi: 10.1016/S0032-3861(02)00598-0

[41] S. L. Cram, H. R. Brown, G. M. Spinks, D. Hourdet, C. Creton, Macromolecules 2005, 38, 2981. doi:10.1021/MA048504V

[42] R. P. Wool, Macromolecules 1993, 26, 1564. doi:10.1021/ MA00059A012

[43] A. R. Saadatabadi, M. Nourani, M. A. Emadi, Iran. Polym. J. 2010, 19, 105.

[44] K. C. Tam, W. K. Ng, R. D. Jenkins, Polymer 2005, 46, 4052. doi:10.1016/J.POLYMER.2005.03.042

[45] S. Biggs, J. Selb, F. Candau, Polymer 1993, 34, 580. doi:10.1016/ 0032-3861(93)90554-N

[46] D. G. Peiffer, Polymer 1990, 31, 2353. doi:10.1016/0032-3861(90) 90324-R

[47] Y. X. Zhang, A. H. Da, G. B. Butler, T. E. Hogen-Esch, J. Polym. Sci. Pol. Chem. 1992, 30, 1383. doi:10.1002/POLA.1992.080300717

[48] J. Wang, L. Li, H. Ke, P. Liu, L. Zheng, X. Guo, S. F. Lincoln, Asia-Pac. J. Chem. Eng. 2009, 4, 537. doi:10.1002/APJ.279

[49] N. Beheshti, A.-L. Kjøniksen, K. Zhu, K. D. Knudsen, B. Nyström, J. Phys. Chem. B 2010, 114, 6273. doi:10.1021/JP100333F

[50] S. P. Patel, G. Ranjan, V. S. Patel, Int. J. Biol. Macromol. 1987, 9, 314. doi:10.1016/0141-8130(87)90001-8

[51] M. V. Badiger, A. Lutz, B. A. Wolf, Polymer 2000, 41, 1377. doi:10.1016/S0032-3861(99)00294-3

[52] J. Desbrieres, Polymer 2004, 45, 3285. doi:10.1016/J.POLYMER.2004.03.032

[53] G. O. Yahaya, A. A. Ahdab, S. A. Ali, B. F. Abu-Sharkh, E. Z. Hamad, Polymer 2001, 42, 3363. doi:10.1016/S0032-3861(00)00711-4

[54] S. Shaikh, S. A. Ali, E. Z. Hamad, B. F. Abu-Sharkh, Polym. Eng. Sci. 1999, 39, 1962. doi:10.1002/PEN.11589

三乙烯四胺对磺化聚丙烯酰胺性能的影响

曹绪龙[1] 胡 岳[2] 宋新旺[1] 祝仰文[1] 韩玉贵[1] 王鲲鹏[2] 陈 湧[2] 刘 育

(1. 胜利油田分公司地质科学研究院,东营 257015;

2. 南开大学化学系,元素有机化学国家重点实验室,天津化学化工协同创新中心,天津 300071)

摘要:研究了水溶液中带有负电荷的磺化聚丙烯酰胺与三乙烯四胺之间的相互作用,考察了三乙烯四胺的引入对磺化聚丙烯酰胺黏度及流变学性质的影响。研究发现,三乙烯四胺的加入可以有效地提高磺化聚丙烯酰胺的黏度、抗剪切性及剪切回复性等性能,为进一步研究其在驱油中的应用提供了理论基础。

关键词:聚丙烯酰胺;三乙烯四胺;非共价相互作用

中图分类号:O632.63 文献标志码:A

以高温高盐油藏为主要研究对象的三次采油技术是目前油田化学领域的研究热点之一。针对该类油藏的特点,聚合物驱油是一种目前广泛采用的提高采收率的方法[1,2]。传统的驱油用聚合物主要是部分水解聚丙烯酰胺[3-7]。在高温高盐油藏中利用聚合物驱油的过程中会遇到诸如聚合物遇盐黏度降低和高温水解等问题,严重影响了该技术的应用[8]。为解决这一问题,目前采用较多的方法是在制备聚丙烯酰胺时引入新的共聚物单体及可抑制酰胺基团水解的基团,从而使部分水解聚丙烯酰胺逐步向疏水缔合型聚丙烯酰胺发展[9-13]。本课题组[14]曾报道带有负电荷的阴离子型聚丙烯酰胺与表面活性剂的相互作用,发现带有负电荷的聚丙烯酰胺与阳离子表面活性剂分子通过静电相互作用,在聚合物链上引入疏水侧链,侧链之间的疏水缔合作用改善了聚丙烯酰胺的抗剪切能力以及剪切回复性。本文研究了阴离子型的磺化聚丙烯酰胺与三乙烯四胺形成的超分子复合体系的黏度和流变性质,重点考察了在水、盐水以及矿化水等不同环境中三乙烯四胺对磺化聚丙烯酰胺黏度、抗剪切性及剪切回复性的影响。与共价修饰不同基团的聚合物体系相比,磺化聚丙烯酰胺与三乙烯四胺复合体系具有易于制备以及剪切回复性能强等诸多优势[15]。同时,有机阳离子寡聚体三乙烯四胺相对于无机强碱具有可降解的特性,考虑到今后聚合物驱油的绿色环保可回收的发展要求,可改变传统碱驱污染地下环境的现状，为石油事业的发展提供了新的思路。

1 实验部分

1.1 试剂与仪器

乙二胺、二乙烯三胺和三乙烯四胺均为分析纯，购自百灵威科技有限公司；去离子水为实验室自制。

DIN 53810 型乌氏黏度计[测试温度为(30±0.1)℃，德国 Schott 公司]；AR1500 型流变仪(美国 TA 公司)；ZETAPALS/BI-200SM 型广角激光散射仪(美国 Brookhaven 公司)。

1.2 实验过程

磺化聚丙烯酰胺(SPAM，分子量 6.2×10^5)参照文献[14]方法制备。SPAM 和三乙烯四胺的结构式见图 1。模拟矿化水的配置：1.47 g $MgCl_2\cdot6H_2O$+1.94 g $CaCl_2$+29.90 g NaCl+993.34 g 水(矿化程度为 32.868 g/L)。

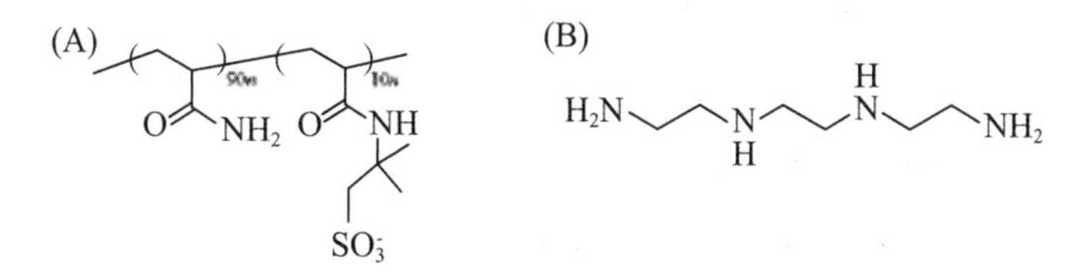

Fig. 1 Structures of SPAM (A) and triethylenetetraamine(B)

2 结果与讨论

2.1 三乙烯四胺对磺化聚丙烯酰胺黏度的影响

考察了纯水、NaCl(1 mol/L)以及矿化水溶液中，固定 SPAM 浓度为 2 mg/mL，加入三乙烯四胺引起的黏度变化。测定方法为将 SPAM 溶液及三乙烯四胺溶液按照 4∶3 的体积比混合，静置 12 h，以不出现絮凝物为溶解标准。在 7.34 s^{-1} 的剪切速率下测得黏度如下：在水、NaCl(1 mol/L)和矿化水中 SPAM 的黏度分别为 0.347，0.024 和 0.021 Pa·s；而在 SPAM/三乙烯四胺复合体系的黏度分别为 0.425，0.055 和 0.051 Pa·s。

无论在纯水或盐水溶液中，SPAM/三乙烯四胺复合体系的黏度均比 SPAM 自身的黏度明显提高。其可能原因是：一方面极性溶剂有利于 SPAM 与三乙烯四胺形成三维网络结构的超分子复合物，从而增加了体系的黏度[16]，并且增溶作用增强了在盐溶液中的溶解性[17]；另一方面，三乙烯四胺较高的正电荷密度屏蔽了盐水溶液中的无机盐阳离子，尤其是钙、镁离子对 SPAM 中磺酸根阴离子的中和与沉淀作用[18]。通过分析 Zeta 电势发现，SPAM 中加入三乙烯四胺后 Zeta 电势发生明显变化。单纯的 SPAM 在水溶液中的电势为−29.37 mV,引入三乙烯四胺后电势变为12.17 mV,说明三乙烯四胺参与了SPAM的组装。通过静电相互作用将阴离子聚合物交联起来,中和了磺化聚丙烯酰胺的负电性,得到了微带正电性的组装体。TEM照片(图2)也证明形成了大尺寸的组装体。此外,考察了SPAM与其他寡聚乙烯二胺如乙二胺、二乙烯三胺等复合体系的黏度,结果表明,乙二胺和二乙烯三胺对SPAM的增黏效果均低于三乙烯四胺。SPAM/三乙烯四胺复合体系在纯水和盐溶液中高于SPAM自身的黏度使得其在油田大规模应用成为可能。

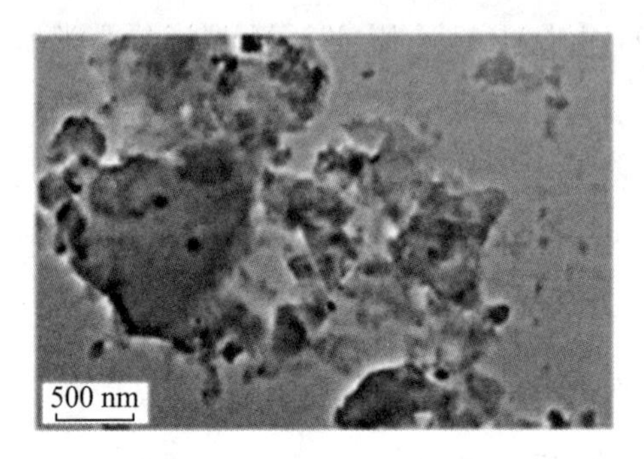

Fig. 2 TEM image of SPAM with triethylenetetraamine

2.2 三乙烯四胺对磺化聚丙烯酰胺流变性质的影响

为考察三乙烯四胺对SPAM流变性能的影响,测试了2 mg/mL的SPAM/三乙烯四胺复合体系(体积比4∶3)的黏度随着剪切速率(0～1 000 s^{-1})的变化情况(图3),并绘制了上述复合体系在200 s^{-1}速率下剪切2 min后,溶液黏度(η)与初始黏度(η_0)之比随时间变化的黏度回复曲线(图4)。

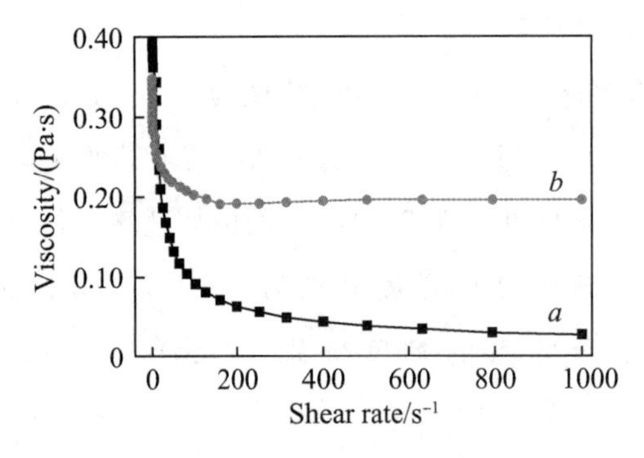

Fig. 3 Effect of shear rate on the apparent viscosity of SPAM (a) and SPAM + triethylenetetraamine (b) in aqueous solution at 45℃

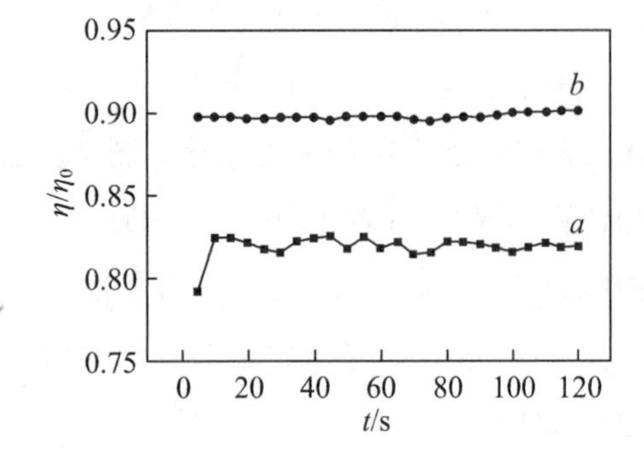

Fig. 4 Recovery of viscosity of aqueous solution of SPAM (a) and SPAM + triethylenetetraamine (b) after shearing for 2 min at 45℃

由图3可见,SPAM/三乙烯四胺复合体系表现出明显高于SPAM自身的抗剪切性。即使在较高的剪切速率(1 000 s^{-1})下,SPAM/三乙烯四胺复合体系的黏度也仅下降40%;

而在相同条件下，SPAM 自身的黏度则下降了 93%。由图 4 可见，SPAM/三乙烯四胺复合体系在剪切停止后其黏度值可以回复到 90%，显示出较好的剪切回复性；而在相同条件下，SPAM 的黏度值仅回复到约 82%。这主要是由 SPAM 与三乙烯四胺之间的弱相互作用所致，如范德华力、氢键和静电相互作用等提高了复合体系的稳定性，使其对外力剪切显示出较好的耐受性。另一方面，当 SPAM/三乙烯四胺复合体系经过高速剪切后，由于范德华力、氢键及静电相互作用等非共价弱相互作用的可逆性，SPAM 与三乙烯四胺重新缔合成为超分子复合体系，剪切时断裂的三维网络结构得以恢复。同时在剪切力作用下体系链状结构更加舒展，使非共价键作用导致的重新缔合过程较为容易，所以复合体系黏度回复较快。进一步的对比实验发现，SPAM 与乙二胺或二乙烯三胺形成的复合体系的抗剪切性和剪切回复性均低于 SPAM/三乙烯四胺复合体系。其可能的原因是在 SPAM 与寡聚乙烯二胺之间的弱相互作用中，静电相互作用起主导作用。乙二胺或二乙烯三胺分子较低的正电荷密度降低了主客体之间的静电相互作用，从而导致 SPAM 与乙二胺或二乙烯三胺形成的复合体系具有相对较低的抗剪切性和剪切回复性。

本文还研究了 SPAM/三乙烯四胺复合体系在盐水溶液和矿化水溶液中的抗剪切能力以及剪切回复性。由图 5 和图 6 可见，在 1 mol/L NaCl 溶液和矿化水溶液中，与单纯的 SPAM 相比，复合体系仍然有很好的流变性能，不仅抗剪切能力明显高于单纯的 SPAM，而且在强剪切力作用后能够很快地回复缔合结构，体现在宏观上的表现就是黏度的恢复能力。推测 NaCl 和其他电解质的加入对聚合物上缔合的有机多胺阳离子产生的双电层起到压缩作用，但还是存在少量的负电性，这种微小的负电性也使聚合物线团尺寸稍微变大[19-21]。Na^+，Ca^{2+} 和 Mg^{2+} 的加入使得疏水集团的逃离更加方便，同时疏水基团的水化膜层变薄。引起表观黏度的增大，同时在盐水中聚丙烯酰胺链内部化学结构的变化导致其链的柔性发生了变化[13,22-24]。

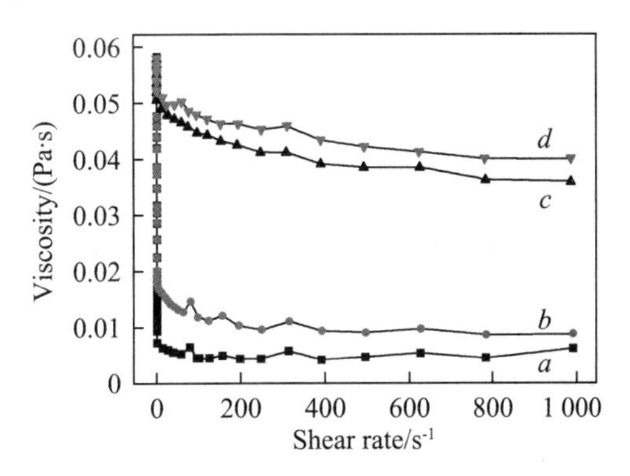

Fig. 5 Effect of shear rate on the apparent viscosity of SPAM+NaCl(a), SPAM+mineralized water (b), SPAM+triethylenetetraamine+NaCl (c) and SPAM+triethylenetetraamine+mineralized water (d) at 45℃

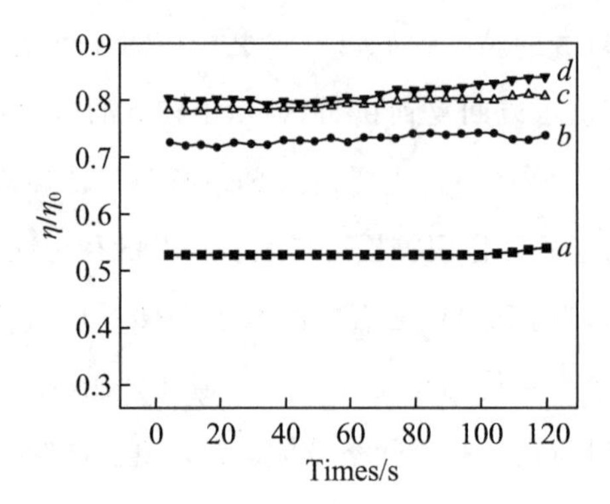

Fig. 6 Recovery of viscosity of SPAM + NaCl (a), SPAM + mineralized water (b), SPAM + triethylenetetraamine+ NaCl (c) and SPAM + triethylenetetraamine + mineralized water (d) after shearing for 2 min at 45℃

3 结论

本实验发现带有正电荷的三乙烯四胺可与带有负电荷的磺化聚丙烯酰胺通过非共价相互作用形成 超分子复合体系,通过超分子作用得到缔合聚合物网络,明显改善了阴离子聚丙烯酰胺的抗剪切能力。静电作用导致的重新缔合较为容易,致使聚合物的黏度恢复能力和剪切回复性得到显著提高。这种超分子缔合体系在盐水和矿化水中仍能保持较好的流变性质,为在油田的实际应用奠定了基础。

参考文献

[1] Song X. W., Petroleum Geology and Recovery Eficiency, 2002, 9 (3), 13-15.(宋新旺.油气地质与采收率,2002,9(3),13-15)

[2] Liu C. T., Li X., Petroleum Geology and Recovery Eficiency, 2012, 19 (1), 66-68.(刘春天,李星.油气地质与采收率,2012,19(1),66-68)

[3] Sun H. Q., Petroleum Exploration and Development, 2006, 33 (3), 262-266.(孙焕泉.石油勘探与开发,2006,33(3),262-266)

[4] Winnik F. M., Regismond S. T. A., Goddard E. D., Langmuir, 1997, 13(1), 111-114.

[5] Feng Y. J., Billon L., Grassl B., Polymer, 2002, 43(7), 2055-2064.

[6] Castelletto V., Hamleg W., Xue W., Macromolecules, 2004, 37 (4), 1497-1501.

[7] Pabon M., Corpart J., Selb J., J. Appl. Polym. Sci., 2004, 91(2), 916-924.

[8] Xu L., Chen Y. Q., Journal of Southwest Petroleum University, 2012, 34(6),

136-140.(徐亮,程玉桥.西南石油大学学报,2012,34(6),136-140)

[9] Castelletto V., Hamleg W., Xue W., Macromolecules,2004,37,1497-1501.

[10] Feng Y.J., Billon L., Grass B., Polymer,2001,43,2055-2064.

[11] Pabon M., Corpart J., Selb J., J. Appl. Polym. Sci., 2004,91(2),916-924.

[12] Cao X.L., Hu Y., Song X.W., Zhu Y.W., Han Y.G., Wang K.P., Guo D.S., Liu Y., Chem. J. Chinese Universities,2014,35(9),2037-2042.(曹旭龙,胡岳,宋新旺,祝仰文,韩玉贵,王鲲鹏,郭东升,刘育.高等学校化学学报,2014,35(9),2037-2042).

[13] Arena G., Casnati A., Contino A., Ungaro R., Tetrahedron Lett., 1997,38,4685-4693.

[14] Cao X.L., Hu Y., Song X.W., Zhu Y.W., Han Y.G., Wang K.P., Guo D.S., Liu Y., Chem. J. Chinese Universities,2015,36(2),395-398.(曹旭龙,胡岳,宋新旺,祝仰文,韩玉贵,王鲲鹏,郭东升,刘育.高等学校化学学报,2015,36(2)395-398).

[15] Wang K.P., Chen Y., Liu Y., Chem. Commun., 2015,51,1647-1649.

[16] Wang K.P., Guo D.S., Zhao H.X., Liu Y., Chem. Eur. J., 2014,20,4023-4031.

[17] Li S.S., Liao M.Y., Jin M.H., Chem. Res. Chinese Universities,2014,30(3),518-520.

[18] Fang Y., Sun G., Bi Y.H., Chem. Res. Chinese Universities,2014,30(5),817-820.

[19] Chen K.L., Zhao Y.H., Yuan X.Y., Chem. Res. Chinese Universities,2014,30(2),339-342.

[20] Liu Li. H., Li X., Cao J., Ling Y.L., Chin. J. Appl. Chem., 2011,28(7),777-784.(刘立华,李鑫,曹菁,令玉林.应用化学,2011,28(7),777-784).

[21] Cao B.G., Li H.B., Luo Y.P., Oil Field Chemistry,2005,22,168-172.

[22] Gao B.J., Guo H.P., Wang J., Macromolecules,2008,41(8),2890-2897.

[23] Li M.R., Liu Z., Cao X.L., Acta Petrolei Sinica,2012,28,1037-1042.

[24] Zhou C., Yang W., Yu Z., Zhou W., Xia Y., Han Z., Wu Q., Polym. Bull., 2011,66(3),407-417.

Influence of triethylenetetraamine on properties of sulfonated polyacrylamide

Cao Xulong[1*]　Hu Yue[2]　Song Xinwang[1]　Zhu Yangwen[1]
Han Yugui[1]　Wang Kunpeng[2]　Chen Yong[2]　Liu Yu[2*]

(1. Geological Scientific Research Institute of Shengli Oilfield Company, SINOPEC, Dongying 257015, China;
2. Department of Chemistry, State Key Laboratory of Elemento-Organic Chemistry, Collaborative Innovation Center of Chemical Science and Engineering (Tianjin), Nankai University, Tianjin 300071, China)

Abstract: The water-soluble negatively charged polyacrylamide was prepared by free radical copolymerization method. The interaction between negatively charged sulfonated polyacrylamide and triethylenetetraamine was investigated in aqueous solution and mineralized water. The results indicated that the addition of triethylenetetraamine can efficiently influence the viscosity and rheological properties of sulfonated polyacrylamide, which provides the theoretical basis for the further application in oil displacement.

Keywords: Polyacrylamide; Triethylenetetraamine; Oil displacement

(Ed. :P,H,F,K)

聚合物分子结构与老化稳定性关系研究

曹绪龙　祝仰文　韩玉贵　刘　坤

中石化胜利油田分公司勘探开发研究院

摘要:在胜利油田三类油藏条件下,系统研究了聚合物分子结构与长期稳定性能的关系。研究表明,疏水缔合聚合物依靠分子链间相互作用而具有优异的长期稳定性能,其次是耐温抗盐单体共聚物,而高分子量聚丙烯酰胺最差;提高缔合单体含量、提高聚合物特性黏数、降低水解度和在分子结构中引入功能单体能明显改善长期稳定性能;当疏水缔合聚合物分子结构中含0.5%(mol)的缔合单体、15%(w)的2-丙烯酰胺基-2-甲基丙磺酸、特性黏数值大于1 400 mL/g和水解度低于20%时,经120天老化后,溶液不会发生相分离,仍具有一定的增黏能力。

关键词:疏水缔合聚合物　三类油藏　分子结构　长期稳定性　相分离

中图分类号:O631.4　文献标志码:A

Study on the relationship of polymer molecular structure and long-termstability

Cao Xulong　Zhu Yangwen　Han Yugui　Liu Kun

(Exploration and Development Research Institute,Sinopec Shengli Oilfeld Company,Dongying 257015,China)

Abstract: The relationship of polymer molecular structure and long-term stability was studied in the third oil reservoirs condition in Shengli Oilfield. The results indicated that hydrophobically associating polymer has best long-term stability performance depending on interaction between molecule chains,then the worse is heat and salt resistance copolymer,and the worst is polyacrylamide with high molecular weight. Through improving hydrophobic monomer content, enhancing polymer molecular weight, reducing the degree of hydrolysis and introducing the function monomer in the molecular structure, the long-term stability performance has been improved obviously. When hydrophobe molar fraction is 0.5%, 2-acrylamido-2-methylpropane sulfonic acid quality score is 15% in the hydrophobically associating polymer molecular structure, the intrinsic viscosity value is more than 1 400 mL/g and the degree of hydrolysis is lower than 20%, the solution will not occur phase separation and still has certain viscosity ability after aging 120 days.

Keywords: hydrophobically associating polymer, the third oil reservoir, molecular structure, long-term stability, phase separation

化学驱作为三次采油的主要技术之一,在国内外油田已经得到了广泛的应用[1-5]。我国的胜利油田从 1992 年开始,对温度和矿化度相对较低的一类和二类油藏开展了大规模的三次采油技术研究,提高采收率效果明显,取得了显著的经济和社会效益。但是,对于温度和矿化度尤其是钙镁离子含量高的三类油藏,目前还未能实施规模化学驱[6-7]。分析认为,制约三类油藏聚合物驱技术发展的主要问题是未有适合此类油藏的聚合物,虽然部分研究人员在三类油藏用耐温抗盐聚合物开发方面做了大量的工作,但在高效性和长效性等方面仍需要进一步提高。

根据化学驱技术对聚合物的要求,老化稳定性作为聚合物现场应用最重要的性能之一,直接影响到聚合物驱提高采收率的能力[8-10]。在水溶性聚合物老化稳定性研究方面,国内外学者在聚合物的水解和降解等方面做了大量的研究。如孔柏岭、余元洲等[11-14]研究了油藏污水中聚丙烯酰胺的老化稳定性能,认为 HPAM 溶液中溶解氧含量是影响长期高温稳定性能的重要因素,溶解氧含量越高,HPAM 溶液的稳定性越差;GuyMuller 等研究了高分子量聚丙烯酰胺溶液的热稳定性,其中水解过程降低了抗二价离子的能力,并强烈依赖于油气藏温度和 pH 值,降解过程导致黏度的降低[15];Ryles 等研究了水溶性聚合物 PAM、还原胶和多糖等在绝氧条件下的长期热稳定性,认为 PAM 降解的主要机理是酰胺基的水解,与二价金属离子相互作用造成黏度的重大损失,最后在极端条件下(高水解度或高二价离子含量)发生相分离,水解速率主要依赖于温度[16]。此外,Yang、Russel 和 Wel-lington 等[17-22]也对高温条件下聚合物的稳定性做了大量研究,然而这些研究基本上是针对某种单一结构的聚合物,而关于不同分子结构聚合物与老化稳定性的关系研究却较少。为此,系统研究了胜利油田三类油藏条件下(温度为 90℃,矿化度约 3.2×10^4 mg/L,其中钙镁离子质量浓度 875 mg/L)聚合物分子结构与长期稳定性能的关系,对开发三类油藏化学驱用聚合物的分子结构设计具有一定的指导作用。

1　试验

1.1 实验仪器

BrookFileld DV-Ⅲ 型黏度计；超级恒温水浴，±0.1℃；电子天平，±0.000 1 g；恒温箱，±0.1℃；恒温水浴，±0.1℃；S212 型恒速搅拌器；HJ-6 多头磁力搅拌器；安瓿瓶密封机。

1.2 试验试剂

不同分子结构的聚合物，自制；高纯氮气，99.999%；NaCl、$CaCl_2$、$MgCl_2$ 均为分析纯；模拟盐水总矿化度 32 529 mg/L，具体组成见表 1。

表 1　模拟盐水离子组成

Table 1　Composition of synthesis brine water

离子类型	Cl^-	Na^+	Ca^{2+}	Mg^{2+}
ρ/(mg·L^{-1})	19 900	11 754	700	175

1.3 评价方法及步骤

(1) 取一定量的粉状聚合物样品，用模拟盐水配制成 5 000 mg/L 的母液，然后用模拟盐水将母液稀释至 1 750 mg/L 的目标液。

(2) 将 1 750 mg/L 的目标液置于安瓿瓶中，通入高纯氮气除氧 10 min 后进行火焰密封，置于 90℃的恒温箱中老化。

(3) 按照不同的老化时间取出目标液，在 Brook-Fileld DV-Ⅲ型黏度计上测试黏度，剪切速率 7.34 s^{-1}，测试温度 90℃。

2　结果与分析

2.1 聚合物类型对长期稳定性的影响

考察了目前常用的 3 种耐温抗盐聚合物(超高分子量聚丙烯酰胺 HPAM(相对分子质量 2 500 万)、疏水缔合聚合物 HAWSP 和耐温抗盐单体共聚物 HST(耐温抗盐单体为 2-丙烯酰胺基-2-甲基丙磺酸)在高温高盐条件下的长期稳定性能，实验结果见表 2。

表 1　聚合物类型对长期稳定性的影响

Table 2　Effect of polymer types on thermal stability

聚合物类型	不同老化时间下的黏度/(mPa·s)					
	0 d	15 d	30 d	60 d	90 d	120 d
HPAM	31.3	12.6	分相	分相	分相	分相
HST	29.5	31.0	22.6	12.4	分相	分相
HAWSP	75.4	86.4	43.6	30.7	11.2	9.7

表 2 说明,聚合物的分子结构与其长期稳定性密切相关。其中,缔合聚合物经 120 天老化后仍未发生相分离,其次是耐温抗盐单体共聚物,经 60 天老化后,黏度已大幅度降低,并在老化时间达到 90 天时发生相分离,而高分子量水解聚丙烯酰胺的长期稳定性能最差,经 30 天老化后溶液发生相分离。

分析认为,耐温抗盐单体共聚物依靠功能单体的抗温抗盐作用和单分子链的水动力学尺寸表现出一定的长期稳定性,而超高分子量 HPAM 仅依靠单分子链的水动力学尺寸,长期稳定性能最差。相比之下,疏水缔合聚合物依靠分子间相互缔合作用比依靠单分子链的水动力学尺寸更具有优势,其长期稳定性明显优于耐温抗盐单体共聚物和超高分子量 HPAM。

2.2 缔合单体含量的影响

疏水缔合聚合物由于疏水缔合作用具有优异的长期稳定性能,而疏水单体含量是影响疏水缔合作用的直接因素。因此,考察了高温高盐条件下,聚合物分子结构中缔合单体摩尔分数对老化稳定性能的影响,实验结果见表 3。

表 3 缔合单体浓度对长期稳定性的影响

Table 3 Effect of associating monomer concentration on thermal stability

缔合单体摩尔分数/%	不同老化时间下的黏度/(mPa·s)					
	0 d	15 d	30 d	60 d	90 d	120 d
0.15	42.5	18.2	14.1	分相	分相	分相
0.30	85.8	26.6	20.5	18.4	分相	分相
0.50	125.4	138.0	125.2	40.2	20.7	17.4
0.75	136.0	142.8	127.4	67.3	45.2	28.8

表 3 表明,在本研究的浓度范围内,随缔合单体摩尔分数的增加,疏水缔合聚合物溶液的长期稳定性提高。当缔合单体的摩尔分数为 0.15%和 0.30%时,分别在老化时间为 60 天和 90 天时发生相分离。而缔合单体的摩尔分数为 0.50%和 0.75%时,经 120 天老化后未发生相分离,溶液黏度值大于 15 mPa·s,老化稳定性能较好。这是因为,疏水单体浓度越高,疏水缔合聚合物分子链间的相互缔合作用越明显,耐温抗盐性能显著,对聚合物溶液的相分离具有一定的抑制作用,因此,长时间老化后溶液黏度保留值较高。

2.3 功能单体的影响

引入各种功能单体是改善聚合物耐温抗盐性能的主要途径之一。本实验在疏水缔合聚合物的分子结构中分别引入耐温抗盐功能单体 AMPS(2-丙烯酰胺基-2-甲基丙磺酸)和 DMAA(N,N-二甲基丙烯酰胺),并研究了其质量分数对疏水缔合聚合物老化稳定性能的影响。实验结果见表 4。

表 4 表明,对于引入功能单体 AMPS 的疏水缔合聚合物,当其质量分数达到 15%和 20%时,经 60 天老化后黏度大于 50 mPa·s,经 120 天老化后仍未发生相分离;而引入

功能单体DMAA的聚合物，当其质量分数达到20%时，经90天老化后溶液黏度为18.8 mPa·s，并在老化时间达到120天时发生相分离，但长期稳定性明显优于DMAA含量为10%和15%的疏水缔合聚合物。

表4 功能单体对长期稳定性的影响

Table 4 Effect of functional monomer on thermal stability

功能单体类型及质量分数	不同老化时间下的溶液黏度/(mPa·s)					
	0 d	15 d	30 d	60 d	90 d	120 d
5%AMPS	46.2	42.2	30.2	15.1	8.7	分相
10%AMPS	87.5	68.1	54.3	28.7	18.9	分相
15%AMPS	192.4	165.0	87.4	51.4	25.4	17.2
20%AMPS	204.2	204.2	98.7	65.2	45.4	21.4
10%DMAA	55.7	34.7	18.2	7.8	分相	分相
15%DMAA	87.3	63.2	40.4	19.7	分相	分相
20%DMAA	85.4	73.2	43.2	23.4	18.8	分相

随疏水缔合聚合物中功能单体AMPS和DMAA含量的增加，长期稳定性能提高。这是因为，功能单体AMPS单元中含有对盐不敏感的磺酸基团及庞大的侧基，而DMAA具有抑制酰胺基水解的作用。因此，在疏水缔合聚合物中引入AMPS和DMAA，可明显改善聚合物在高温高盐条件下的长期稳定性能。

2.4 分子量的影响

聚合物的特性黏数与其长期稳定性能密切相关，考察了不同特性黏数的疏水缔合聚合物在高温高盐条件下的长期稳定性能，实验结果见表5。

表5 疏水缔合聚合物特性黏数对长期稳定性的影响

Table 5 Effect of intrinsic viscosity on thermal stability

$[\eta]/(mL \cdot g^{-1})$	不同老化时间下的溶液黏度/(mPa·s)					
	0 d	15 d	30 d	60 d	90 d	120 d
1 852.4	102.4	112.8	87.4	65.2	34.3	25.4
1 435.5	85.3	90.4	54.6	45.2	21.3	11.6
1 130.5	46.2	44.7	20.4	5.5	分相	分相
930.2	32.5	26.2	6.7	分相	分相	分相
553.7	21.5	13.2	分相	分相	分相	分相

表5表明，随疏水缔合聚合物特性黏数的增加，长期稳定性能增强，当聚合物特性黏数大于1 400 mL/g时，经60天老化后黏度保留率大于50%，经120天老化未发生相分离。当疏水缔合聚合物特性黏数为553.7～1 130.5 mL/g时，在120天的老化时间范围

内均不同程度地发生了相分离,尤其是特性黏数为553.7 mL/g的疏水缔合聚合物,在老化时间仅为30天时就发生相分离。分析认为,疏水缔合聚合物的特性黏数越大,单分子链单分子链上疏水基团数量越多,在溶液中的相对含量越大,疏水基团之间相互接触形成分子间 缔合的能力越强,故长期稳定性越好;另外,随特性黏 数的增加,相同浓度溶液中产生分子链间缠结作用的几率也越大,缠结作用越明显,分子间缔合的几率就越大,故其长期稳定性能更加优异。

2.5 水解度的影响

水解度是化学驱用聚合物的一个重要参数,对聚合物的长期稳定性能具有较大的影响。考察了不同水解度疏水缔合聚合物(该系列疏水缔合聚合物含0.5%(mol)的疏水单体、15%(w)的AMPS、特性黏数值大于1 400 mL/g)在高温高盐条件下的长期稳定性能,实验结果见表6。

表6 疏水缔合聚合物水解度对长期稳定性的影响

Table 6 Effect of hydrolyzed degree on thermal stability

水解度/%	不同老化时间下的溶液黏度/(mPa·s)					
	0 d	15d	30d	60d	90d	120d
1.95	76.5	126.8	187.6	122.4	98.5	76.4
11.25	87.3	141.6	158.4	132.8	76.5	43.2
19.85	112.4	128.0	78.2	43.8	34.2	17.6
29.45	67.8	43.1	26.3	12.3	分相	分相
41.07	53.4	41.5	32.4	分相	分相	分相

表6表明,疏水缔合聚合物的水解度对其长期稳定性有较大的影响。随水解度的增加,聚合物的长期稳定性变差。当聚合物水解度在1.95%~19.85%时,经120天老化后溶液仍未发生相分离,但水解度在29.45%~41.07%时,经60天老化后黏度大幅度降低,甚至产生沉淀。

3 结论

(1) 胜利油田三类油藏高温高盐条件下,疏水缔合聚合物依靠分子链间相互作用而具有优异的长期稳定性能,其次是耐温抗盐单体共聚物,而高分子量水解聚丙烯酰胺最差。

(2) 对于疏水缔合聚合物,提高缔合单体含量、提高分子量、降低水解度和在其分子结构中引入功能单体能明显改善长期稳定性能。

(3) 当疏水聚合物分子结构中缔合单体摩尔分数为0.5%、AMPS质量分数为15%、特性黏数值大于1 400 mL/g和水解度低于20%左右时,在三类油藏条件下经120天老化后,仍具有一定的增黏能力,且不会发生相分离。

参考文献

[1] 王德民，程杰成，吴军政，等. 聚合物驱油技术在大庆油田的应用 [J]. 石油学报，2005，26(1)：74-78.

[2] 薛国勤，孔柏岭，黎锡瑜，等. 河南油田聚合物驱技术 [J]. 石油地质与工程，2009，23(3)：50-52.

[3] 周守为，韩明，向问陶，等. 渤海油田聚合物驱提高采收率技术研究及应用 [J]. 中国海上油气，2006，18(6)：386-389.

[4] PRATAP M，ROY R P，GUPTA R K. Field implementation of polymer EOR technique-a successful experiment in india [J]. Soci-ety of Petroleum Engineers，1997.

[5] DE F，COELHO S L P，BARBOSA L C F. Development and appli-cation of selective polymer Injection to control water production [J]. SPE，1997.

[6] 孙焕泉. 胜利油田三次采油技术的实践与认识 [J]. 石油勘探与开发，2006，33(3)：262-266.

[7] 张爱美. 胜利油区聚合物驱资源分类标准修订及其评价 [J]. 油气地质与采收率，2004，11(5)：68-70.

[8] 吕茂森. 中原油田油藏条件下黄原胶热老化稳定性研究 [J]. 西部探矿工程，2002，14(6)，39-40.

[9] 刘淑芹，吕茂森. 中原油田油藏条件下 AMPS 三元共聚物热老化稳定性研究 [J]. 内蒙古石油化工，2003，29(3)：148-149.

[10] 谭中良. 聚丙烯酰胺高温老化稳定性研究 [J]. 河南石油，1998，(2)：14-16.

[11] 孔柏岭. 长期高温老化对聚丙烯酰胺溶液性能影响的研究 [J]. 油气采收率技术，1996，(4)：7-11.

[12] 余元洲，李宝荣，杨广荣，等. 污水聚丙烯酰胺溶液高温稳定性研究 [J]. 石油勘探与开发，2001，28(1)：66-67.

[13] 孔柏岭. 聚丙烯酰胺的高温水解作用及其选型研究 [J]. 西南石油学院学报，2000，22(1)：66-69.

[14] 孔柏岭，罗九明. 高温油藏条件下聚丙烯酰胺水解反应研究 [J]. 河南石油，1998，25(6)：67-69.

[15] MULLER G. Thermal stability of high-molecular-weight polyac-rylamide aqueous solutions [J]. Polymer Bulletin，1981，5(1)：31-37.

[16] RYLES R G. Chemical stability limits of water-soluble polymers used in oil recovery processes [J]. SPE 13585-PA，1988，3(1)：23-24.

[17] YANG S H，TREIBER L E. Chemical stability of polyacrylamide under

simulated field conditions [J]. SPE 14232-MS,1985.

[18] SHUPE R D. Chemical stability of polyacrylamide polymers [J]. Journal of Petroleum Technology 9299-PA,1981,33(8):1513-1529.

[19] WELLINGTON S L. Biopolymer solution viscosity stabilization-polymer degradation and antioxidant [J]. SPE,1983.

[20] 刘颖. 聚丙烯酰胺的化学降解 [J]. 油气田地面工程,2009,28(5):35-36.

[21] RASHIDI M, BLOKHUS A M, SKAUGE A. Viscosity and retention of sulfonated polyacrylamide polymers at high temperature [J]. Journal of Applied Polymer Science,2011,119(6):3623-3629.

[22] ALAIN ZAITOUN B P. Institut francais du petrole, limiting conditions for the use of hydrolyzed polyacrylamides in brines containing divalent Ions [J]. SPE,1983.

本文编辑 冯学军

对甲氧基苯辛基二甲基烯丙基氯化铵与丙烯酰胺共聚动力学研究

曹绪龙 刘 坤 祝仰文 窦立霞

(胜利油田分公司勘探开发研究院,东营 257015)

摘要:采用定时取样研究了以过硫酸钾-亚硫酸氢钠为引发剂,对甲氧基苯辛基二甲基烯丙基氯化铵(ADMAAC)和丙烯酰胺(AM)在水溶液中的共聚反应动力学,测定了相应的聚合速率方程、聚合表观活化能;采用阴阳离子相互作用测定残余ADMAAC的含量,紫外分光光度法测定残余AM的含量,根据单体投料量和残余量差值,得到低转化率下共聚物的组成,按Kelem-Tudos法得到两单体竞聚率。实验结果表明:聚合反应温度在40℃下,聚合速率方程为:$R_p=K[M]^{1.241}[KPS]^{0.52}[SHS]^{0.55}$;根据Arrhenius经验公式计算出对甲氧基苯辛基二甲基烯丙基氯化铵(ADMAAC)和丙烯酰胺(AM)共聚的表观活化能为73.85 kJ/mol,高于AM水溶液均聚合的活化能E_a(70.32 kJ/mol);两种单体的竞聚率为$r_{ADMAAC}=0.197$、$r_{AM}=4.503$,为ADMAAC-AM共聚合反应控制确定了重要的动力学参数。两单体的竞聚率的积小于1,ADMAAC与AM共聚合行为类型是一种无恒比点的非理想共聚行为,共聚物组成曲线,在对角线下方。

关键词:丙烯酰胺;聚合动力学;聚合速率方程;表观活化能;竞聚率

疏水缔合聚合物(hydrophobically associated water-soluble polymers,HAWSP)指在传统的水溶性聚合物主链上引入极少量疏水基团的一类新型水溶性聚合物[1,2]。在溶液中,由于疏水基团的相互作用,疏水缔合聚合物分子链内和链间存在着缔合作用。当聚合物溶液浓度达到临界缔合浓度(CAC)时,溶液中主要以分子间缔合为主,形成了动态物理交联网络,溶液黏度大幅度提高。季铵盐型阳离子疏水单体因为与丙烯酰胺(AM)的共聚物增粘性能好,耐温抗盐性能强,在油田上得到广泛应用。

目前对于季铵盐型阳离子疏水单体和丙烯酰胺的水溶液聚合制备工艺研究较多,但是对聚合反应动力学的研究少有报道,而这些基础参数是进行聚合物的分子设计、组成及结构与性能控制的基础。有文献报道了烯丙基类单体与丙烯酰胺的动力学研究,其中张伟等[3]研究了过硫酸钾亚硫酸钠氧化还原体一系引发AM与DMDAAC共聚反应动力学行为,在m(AM):m(DMDAAC)=2:1,聚合反应温度为40℃,测定了速率方程和聚合表观活化能。毕可臻等[4]研究了以过硫酸铵、亚硫酸氢钠为引发剂,丙烯酰胺和二甲基二烯丙基氯化铵在水溶液中的共聚反应动力学,测定了单体 AM 与 DMDAAC3 种摩尔比时相应的聚合表观活化能、聚合速率方程和单体竞聚率。Tanaka[5]报道过季铵盐阳离子表面活性单体甲基丙烯酰氧乙基三甲基氯化铵(DMC)与 AM 共聚合反应单体竞聚率的研究,得到单体竞聚率分别是 1.71(DMC)和 0.25(AM)。李万钢[6]研究了季铵盐阳离子表面活性单体甲基丙烯酰氧乙基二甲基苄基氯化铵(DMBAC)与丙烯酰胺(AM)在水溶液中的共聚行为,得到单体竞聚率 $r_1=0.27$(AM),$r_2=2.00$(DMBAC)。但是对于烯丙基类季铵盐阳离子表面活性单体聚合反应动力学研究却鲜有报道。

本研究选取烯丙基类季铵盐阳离子表面活性单体对甲氧基苯辛基二甲基烯丙基氯化铵(ADMAAC)与 AM 共聚,采用逐步称取聚合过程中产物,经过分离洗涤后称量,参照 GB/T 5174-2004[7]测定残余 ADMAAC 含量,采用紫外分光光度法测定残余 AM 含量,研究了 ADMAAC 和 AM 共聚动力学,测定其聚合速率方程、聚合表观活化能和单体竞聚率,认识疏水缔合物的分子结构,为生产工艺提供理论依据,利于生产出适合油田应用的结构产物。

1　实验部分

1.1 主要仪器与试剂

丙烯酰胺(AM),工业级,江西昌九农科化工,使用前经丙酮两次重结晶精制;对甲氧基苯辛基二甲基烯丙基氯化铵(ADMAAC),实验室自制;无水乙醇,分析纯,成都市科龙化工试剂厂;过硫酸钾(KPS),分析纯,成都科龙化工试剂厂;亚硫酸氢钠(SHS),分析纯,成都科龙化工试剂厂;酸性蓝-1,优级纯,北京华迈科生物技术有限责任公司;溴化底米鎓,优级纯,东京化成工业株式会社;实验用水为超纯水,电导率<5 μs/cm,优普纯水机自制;DZF-6050MBE 真空干燥箱,上海精密仪器仪表有限公司;紫外可见分光光度计,UV-2601,北京瑞利分析仪器公司。

1.2 聚合速率方程与表观活化能

将 500 mL 烧杯放置于超级恒温水浴中。向烧杯中加入一定量的 ADMAAC 和 AM 水溶液,通氮气驱氧气 30 min,恒温下加入一定量 KPS-SHS 引发剂,摇匀聚合。分别在 2 min、4 min、6 min、8 min、10 min 取样,倒入快速搅拌的无水乙醇中,并加入阻聚剂羟基苯甲醚终止反应,不断搅拌,将沉淀出来的白色物质剪细,然后再用无水乙醇反复洗涤三次,在真空干燥箱中于 60℃下干燥至恒重,称量所得样品质量以计算转化率,根据转化率-时间关系曲线的斜率得到聚合初始速率(对自由基聚合,在较低转化率 10%以内时[8,9],参与反应的各物种的浓度变化不大,聚合速率基本保持不变):

$$R_P=\frac{-\mathrm{d}[M]}{\mathrm{d}t} \tag{1}$$

溶液聚合中，聚合速率与引发剂[I]、单体浓度[M]关系有以下经验公式：

$$R_p = k[I]^n[M]^m \tag{2}$$

改变单体的浓度(固定 KPS、SHS 浓度)，分别做[M]—t 图，得到不同单体浓度下的聚合速率，以 lgR_P 对 lg[M]作图，通过线性回归求出直线的斜率即可得到单体 M 的反应级数。用同样的方法固定其他两个参数，分别得到氧化剂 KPS、还原剂 SHS 的反应级数。

表观活化能[10]：

根据 Arrhenius 方程

$$k = A\mathrm{e}^{E_a/RT} \tag{3}$$

即

$$R_p = A\mathrm{e}^{E_a/RT}[I]^n[M]^m \tag{4}$$

对上式两边求对数得到：

$$\ln R_p = \frac{\mathrm{E_a}}{RT} + A[I]^n[M]^m \tag{5}$$

固定引发剂和单体的浓度，在一定范围内改变温度，得到 lnR_p 与 $1/T$ 之间的关系，由直线的斜率可求出表观活化能 E_a。

1.3 竞聚率和共聚物组成的测定

将 ADMAAC 与 AM 按一定的摩尔比配制成总浓度为 1.062 8 mol/L 的溶液，通氮气驱氧气 30 min，35℃恒温下加入引发剂[KPS]=4.62×10^{-4} mol/L，[SHS]=4.62×10^{-4} mol/L，摇匀聚合。在反应过程中用玻璃棒蘸取反应溶液浸入无水乙醇中，一旦出现白色沉淀，即倒入快速搅拌的乙醇中，并加入阻聚剂羟基苯甲醚终止反应，不断搅拌，将沉淀出来的白色物质剪细，然后再用无水乙醇反复洗涤三次，收集残余单体溶液并浓缩去除乙醇，用去离子水定容，作为待测液。

然后参照 GB/T 5174-2004 测定阳离子疏水单体 ADMAAC 的残余含量(短碳链非离子型的 AM 对测试结果没有影响)；用可见紫外分光光度计在 245 nm(适当的浓度条件下，该波长处 ADMAAC 没有吸收峰)处测定残余 AM 的吸光度，由 AM 的吸光度-浓度标准曲线得到残余 AM 的含量。根据 ADMAAC 和 AM 的投入量[M_1]、[M_2]减去残余单体含量，计算得到低转化率(10%以内)下共聚物中 ADMAAC 单元和 AM 单元的含量 d[M_1]、d[M_2]，对于二组分共聚体系，采用 Kelem-Tudos[11] 方法算出聚合物中两单体的竞聚率。

$$\eta = \left(r_1 + \frac{r_2}{a}\right)\xi - \frac{r_2}{a} \tag{6}$$

式中，$\eta = \dfrac{G}{a+F}$，$\xi = \dfrac{F}{a+F}$，$a = \sqrt{F_{\min} \times F_{\max}}$，$G = \dfrac{X(Y-1)}{Y}$，$F = \dfrac{X_2}{Y}$，原料中单体摩尔比 $X = \dfrac{[M_1]}{[M_2]}$，低转化率下共聚物组成 $Y = \dfrac{\mathrm{d}[M_1]}{\mathrm{d}[M_2]}$ r_1、r_2 为两种单体的竞聚率。将原料中单

体物质的量和共聚物中聚合单体的组成分别代入式中求得参数。以 ξ 为横坐标，η 为纵坐标作图得一直线，由直线的斜率和截距可分别求得 r_1 和 r_2。

将 r_1 和 r_2 带入二元共聚物组成微分方程[12]：

$$F_1=\frac{r_1 f_1^2+f_1 f_2}{r_1 f_1^2+2 f_1 f_2+r_2 f_2} \tag{7}$$

$$f_2=1-f_1 \tag{8}$$

式中，F_1 为某一瞬间结构单元 M_1 占共聚物总结构单元的摩尔分数，f_1 和 f_2 分别为同一瞬间单体 M_1 和 M_2 占单体混合物的摩尔分数。据此共聚物组成方程可由对共聚物的组成要求确定原料配比，也可从原料投料比预测共聚物的组成。

2　结果与讨论

2.1 ADMAAC 与 AM 共聚合速率方程的测定

2.1.1 单体浓度对聚合速率的影响

当两种单体的物质的量之比恒定时，单体之间的相互影响可以忽略[13]，因此固定单体摩尔比 AM：ADMAAC＝90.74，固定反应温度为 40℃，引发剂[KPS]＝4.62×10^{-4} mol/L，[SHS]＝4.62×10^{-4} mol/L，在 0.633 8 mol/L～1.197 2 mol/L 范围内改变总单体浓度，考察单体浓度对聚合速率 R_p 的影响。以 $\lg R_p$ 对 $\lg[M]$ 作图并进行线性回归，如图 1 所示。

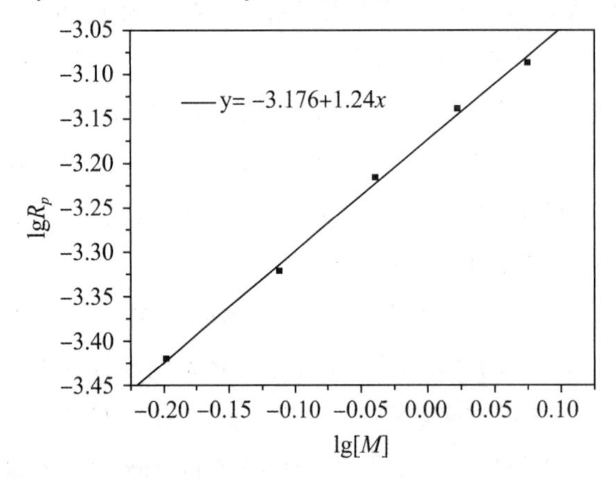

图 1　$\lg R_p$ 与 $\lg[M]$ 关系曲线

Figure 1　Curve of $\ln R_p$ versus $\lg[M]$

聚合反应速率对数与单体的浓度对数呈线性关系，其斜率为 1.24，所以，聚合速率与单体总浓度的关系可表示为 $R_p \propto [M]^{1.24}$。可以看出随着单体浓度的增加，聚合反应速率迅速增加。这是由于单体浓度的增加，增加了单体与自由基作用的机会，减少了自由基的失活。与溶液聚合的级数有一定偏离，对单体浓度级数比理论值大，可能是由于共聚单体间复杂的相互作用，如离子对、极性效应等，使其偏离正常聚合行为[14]。疏水单体 ADMAAC 不仅作为单体参与聚合反应，同时其为长链季铵盐，容易引发向单体的链转移作用，使聚合速率偏离常见 1 级关系。

2.1.2 氧化剂 KPS 浓度对聚合速率的影响

氧化剂的浓度是决定自由基数量的重要因素，而自由基的数量又决定着聚合速率的大小。为此，控制反应温度为 40℃，固定摩尔比 AM ∶ ADMAAC=90.74，$[M]=1.0628$ mol/L，还原剂浓度为 4.62×10^{-4} mol/L，在$(2.77\sim6.47)\times10^{-4}$ mol/L 内，改变氧化剂的浓度进行 ADMAAC 和 AM 的水溶液聚合。以 $\lg R_p$ 与 lg[KPS]作图，斜率即为级数关系，如图 2 所示。

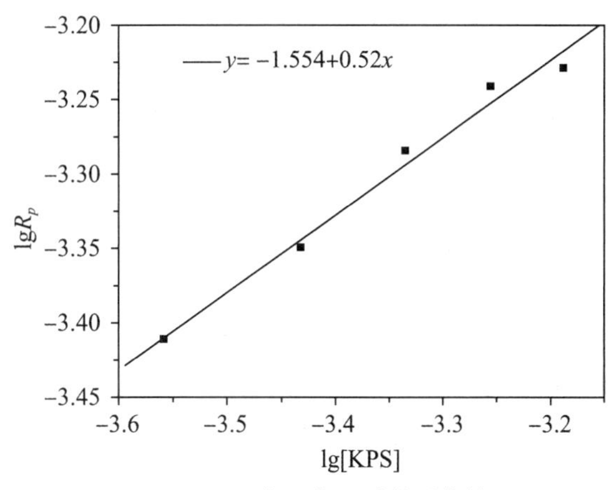

图 2　$\lg R_p$ 与 lg[KPS]关系曲线

Figure 2　Curve of $\ln R_p$ versus lg[KPS]

聚合反应速率对数与 KPS 的浓度对数呈线性关系，其斜率为 0.52，所以，聚合速率与过硫酸钾的关系可表示为 $R_p\propto \text{KPS}^{0.52}$。由此可见，随着氧化剂浓度的增加，产生的自由基数量增大，与单体作用的几率增多，因此聚合反应速度加快。对氧化剂浓度的反应级数略高于 0.5，说明链终止方式主要是双基终止，同时还存在少量的单基终止[15]。

2.1.3 还原剂 SHS 浓度对聚合速率的影响

还原剂的浓度同样是通过影响自由基数量来影响聚合反应速率。实验中选择氧化剂浓度为 4.62×10^{-4} mol/L，其他条件与上述相同，在$(2.77\sim6.47)\times10^{-4}$ mol/L 内，改变还原剂亚硫酸氢钠浓度，进行 ADMAAC 和 AM 的水溶液共聚。以 $\lg R_p$ 与 lg[SHS]作图，斜率即为级数关系，如图 3 所示。

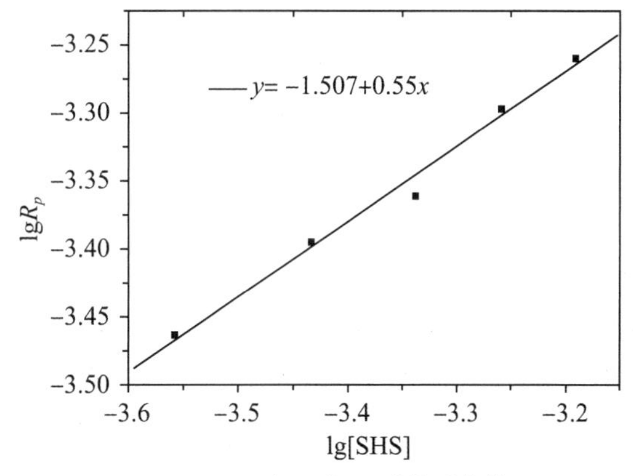

图 3　$\lg R_p$ 与 lg[SHS]关系曲线

Figure 3　Curve of $\ln R_p$ versus lg[SHS]

聚合反应速率对数与 SHS 的浓度对数呈线性关系,其斜率为 0.55,所以,聚合速率与亚硫酸氢钠的关系可表示为 $R_p \propto SHS^{0.55}$。还原剂浓度增大,自由基生成量会因还原剂浓度的增大而减少,但自由基数量的绝对量仍会有所增加,从而反应速率增大。对还原剂浓度的反应级数略高于经典的自由基聚合双基终止反应的数值 0.5,说明聚合反应中同时还存在单基链终止反应,其主要原因是由于表面活性单体 ADMAAC 容易形成自缔合聚集体,其开始引发后,可能形成具有 ADMAAC 嵌段结构的链增长自由基,自由基增长链分子内强烈的疏水缔合作用降低了链增长自由基的扩散能力,而且同类型链增长自由基之间由于正电荷的相互排斥也会降低双基终止的几率,单基终止比例增加,是单基终止和双基终止并存的结果。

由此得到 KPS-SHS 所构成的引发体系的对甲氧基苯辛基二甲基烯丙基氯化铵和丙烯酰胺共聚速率方程近似地可用下式表示:$R_p = K[M]^{1.24}[KPS]^{0.52}[SHS]^{0.55}$。

2.2 ADMAAC 与 AM 共聚合表观活化能测定

固定 AM∶ADMAAC=90.74,[M]=1.062 8 mol/L,引发剂的浓度[KPS]=4.62×10^{-4} mol/L,[SHS]=4.62×10^{-4} mol/L,反应温度分别为 35、40、45、50 和 55℃时,进行 ADMAAC 和 AM 的水溶液共聚,得到不同反应温度下的 R_p,以 $\ln R_p$ 与 $1/T$ 作图并进行线性回归。固定[AM]=1.062 8 mol/L,按照上述方法,进行 AM 的水溶液均聚,得到不同反应温度下的聚合反应速率 R_p,以 $\ln R_p$ 与 $1/T$ 作图并进行线性回归,如图 4 所示。

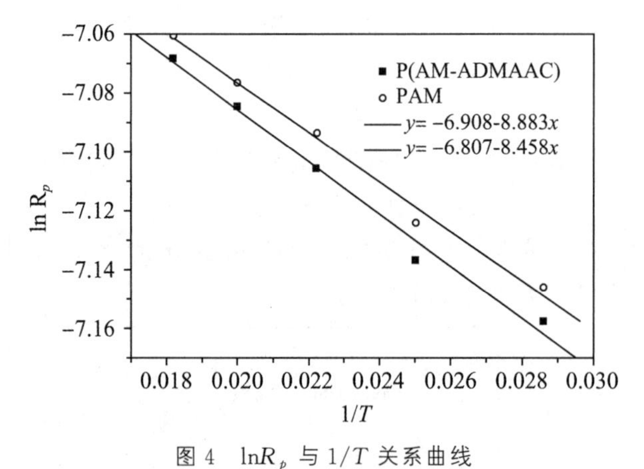

图 4 $\ln R_p$ 与 $1/T$ 关系曲线

Figure 4 Curve of $\ln R_p$ versus $1/T$

如图 4 所示,线性回归斜率分别为 −8.883 和 −8.458,通过计算得到 AM 与 ADMAAC 共聚合的活化能 E_a 为 73.85 kJ/mol,AM 水溶液聚合的活化能 E_a 为 70.32 kJ/mol,AM 与 ADMAAC 共聚合的活化能比 AM 均聚活化能偏高,这是由于 ADMAAC 属于烯丙基类单体,烯丙基类单体具有自阻聚作用,这是由衰减链转移造成的。单体烯丙基连在双键 α 位置上的 C—H 键很弱,而链自由基活泼,结果向单体转移而终止。所形成的烯丙基自由基很稳定,不能再引发,而相互或与其他增长自由基终止,相当于阻聚剂的终止作用。ADMAAC 活性比 AM 低，且空间位阻大，导致链引发和链增长速率降低，进而聚合反应所需的表观活化能升高。

2.3 竞聚率与共聚物组成

测定在转化率≤10%时，ADMAAC 与 AM 在不同摩尔比下共聚物 P(ADMAAC-AM)的组成，并将 ADMAAC 在单体混合物中的摩尔配比 X，ADMAAC 结构单元在共聚物上的摩尔比 Y 及按 K-T 方法计算的参数 η 与 ξ 一并列入表 1。以 η 对 ξ 作图，得共聚体系单体的竞聚率关系见图 5。

从图 5 可见，该共聚体系的参数 η 与 ξ 具有良好的线性相关性。由直线的斜率和截距可求出 P(ADMAAC-AM)体系的竞聚率 $r_{ADMAAC}=0.197$、$r_{AM}=4.503$。可见 ADMAAC 的活性远比 AM 低，烯丙基类阳离子表面活性剂共聚倾向大于均聚倾向，AM 的均聚倾向却大于共聚倾向，因此阳离子单体 ADMAAC 在共聚物中的含量小于投料含量。其主要原因是在于：一方面，ADMAAC 是烯丙基单体，具有较强的自阻聚和链转移作用[16]；ADMAAC 的体积大且带正电荷，由于位阻和电性斥力较大，产生的末端和前末端效应较大，使链自由基与之作用活化能升高，活性显著下降[12]，这两方面的原因都会导致 ADMAAC 活性降低；另一方面，表面活性单体 ADMAAC 在水溶液中表面活性较强，具有一定的自缔合能力，而此处每组实验的 ADMAAC 起始浓度都高于其临界胶束浓度(0.8 mmol/L)[17]，因此，大量的 ADMAAC 在水溶液中缔合形成聚集体。当水溶液中的链增长自由基从水相扩散进入聚集体内引发 ADMAAC 聚合时，由于聚集体内 ADMAAC 浓度较高，类似于本体单体浓度，将有利于阳离子单体均聚合反应的发生，所以在与 AM 的共聚合反应过程中，ADMAAC 的竞聚率要明显高于没有表面活性的烯丙基类单体 DMDAAC[4]($r_{DMDAAC}=0.14$，$r_{AM}=6.11$)，在聚合物链上有形成 ADMAAC 微嵌段结构的可能性。

表 1 P(ADMAAC-AM)共聚物组成以及参数 η 和 ξ

Table 1 The composition of ADMAAC-AM copolymer and the related parameters η and ξ of the K-T method

X=[ADMAAC]/[AM]	转化率/%	Y=d[ADMAAC]/d[AM]	ξ	η
0.005 478	7.063 8	0.001 236	0.206 947	−37.729 589
0.022 280	7.640 2	0.004 681	0.532 704	−23.797 501
0.045 574	9.079 2	0.009 078	0.710 961	−15.458 516
0.057 622	8.765 6	0.013 898	0.719 757	−12.317 446
0.069 979	8.354 9	0.016 538	0.760 948	−10.694 082
0.082 574	9.922 8	0.019 128	0.793 053	−9.420 385

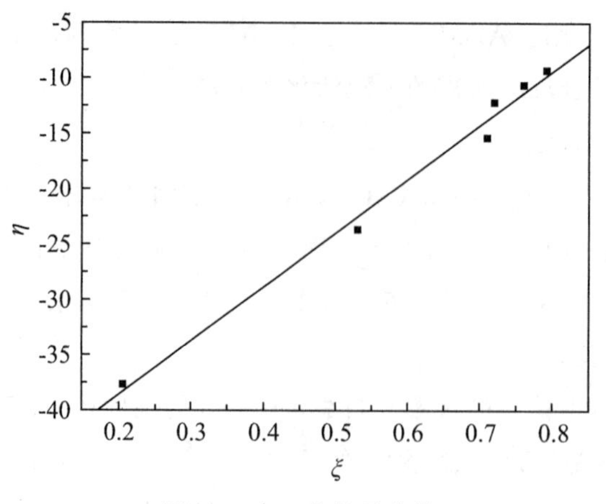

图5 η与ξ的关系曲线

Figure 5 Curve of η versus ξ

将$r_{ADMAAC}=0.197$、$r_{AM}=4.503$带入公式(7)得到聚合物的组成-单体含量的组成曲线，如图6所示，由于以上体系中两单体的竞聚率的积小于1，因此ADMAAC与AM共聚合行为类型是一种无恒比点的非理想共聚行为，共聚物组成曲线，在对角线下方，曲线没有像理想共聚物曲线($r_1 \cdot r_2=1$)表现为在对角线上下对称，且与对角线无交点(恒比共聚点)。

由于活性差异，在共聚过程中，两单体的消耗速度不同，AM消耗得很快，浓度迅速降低，而ADMAAC则消耗较慢，反应混合物中单体比例随反应的进行不断变化，共聚物的组成也随之改变，造成共聚物的结构不均匀。因此，要获得疏水单体含量高且结构均匀的共聚物，除根据竞聚率和共聚物组成微分方程确定单体配比外，还可以采用补加活泼单体法，在共聚过程中，采取先加活性低的ADMAAC单体，再按一定速率陆续补加活性高的单体AM，以保持单体组成恒定，获得组成比较均一的共聚物，利于有针对性的合成适用于不同油田的疏水缔合聚合物。

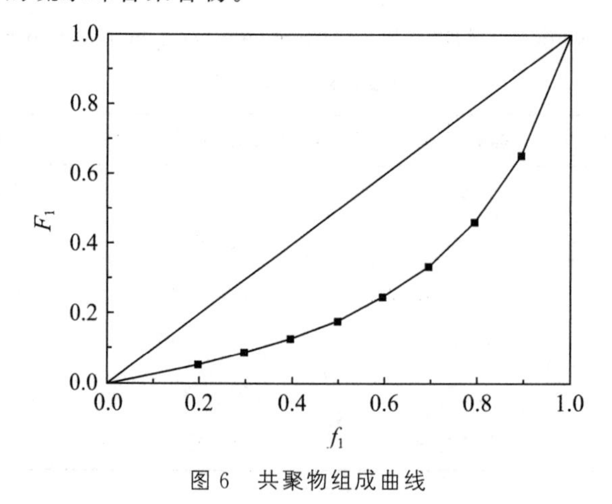

图6 共聚物组成曲线

Figure 6 Composition curve of the copolymers

3 结论

(1) 以过硫酸钾-亚硫酸氢钠为引发剂，反应温度 40℃时，ADMAAC 与 AM 水溶液聚合动力学聚合速率方程为：$R_p=K[M]^{1.241}[KPS]^{0.52}[SHS]^{0.55}$；

(2) ADMAAC 与 AM 共聚的表观活化能为 73.85 kJ/mol，高于 AM 均聚的表观活化能；

(3) 两种单体的竞聚率分别为 $r_{ADMAAC}=0.197$、$r_{AM}=4.503$。表明 ADMAAC 的活性远低于 AM，ADMAAC 与 AM 共聚为非理想共聚。说明 ADMAAC 趋向于形成共聚物，而 AM 更趋向于形成均聚物，在聚合物链上有形成 ADMAAC 微嵌段结构的可能性。

参考文献

[1] Hill A，Candau F，Selb J. Macromolecules，1993，26(17)：4521-4532.

[2] Volpert E，Selb J，Candau F. Macromolecules，1996，29(5)：1452-1463.

[3] 张伟，汪树军，刘红研，潘惠芳. 石油大学学报(自然科学版)，2005，(06)：110-112，118.

[4] 毕可臻，张跃军. 应用化学，2011，(12)：1354-1359.

[5] Tanaka H. J Polym Sci Pol Chem，1986，24(1)：29-36.

[6] 李万刚. 疏水改性阳离子聚丙烯酰胺 P(DMBAC-AM)的研究. 北京：中国科学院研究生院(理化技术研究所)硕士学位论文，2009.

[7] GB/T 5174-2004，表面活性剂洗涤剂阳离子活性物含量的测定. 北京：国家质检总局，2004.

[8] 刘立华，刘汉文，龚竹青. 高分子学报，2007，(2)：183-189.

[9] 高保娇，王旭鹏，李延斌. 高分子学报，2005，(3)：453-457.

[10] 朱明，丁瑶，张静，等. 高分子通报，2014，(2)：173-178.

[11] Braun D，Czerwinski W，Disselhoff G，Tüdös F，Kelen T，Turcsányi B. Makromol Mater Eng，1984，125(1)：161-205.

[12] 潘祖仁. 高分子化学，第 2 版. 北京：化学工业出版社，1997：72.

[13] Ponratnam S，Kapur SL. Makromol Chem Phys，1977，178(4)：1029-1038.

[14] 刘立华. 新型聚季铵盐与聚合硫酸铁复合絮凝剂合成及其基础理论与应用研究. 长沙：中南大学博士学位论文，2004.

[15] Schulz DN，Kaladas JJ，Maurer JJ，Bock J，Pace SJ，Schulz WW. Polymer，1987，28(12)：2110-2115.

[16] 潘祖仁，于在璋. 自由基聚合. 北京：化学工业出版社，1983：214.

[17] 柳建新,郭拥军,祝仰文,杨红萍,杨雪杉,钟金杭,李华兵,罗平亚. 物理化学学报,2012,(7):1757-1763.

Kinetics study on polymerization of p-methoxybenzeneoctyl dimethyl allyl ammonium chloride and acrylamide

Cao Xulong* Liu Kun Zhu Yangwen Dou Lixia

(GeoscienceResearch Institute, ShengliOilfeldCompany, Dongying 257015, China)

Abstract: The kinetics of polymerization of p-methoxybenzeneoctyl dimethyl allyl ammonium chloride (ADMAAC) and acrylamide (AM) initiated by potassium persulfate-sodium bisulfate redoxcomplex in aqueous solution was studied by egular sampling. The polymerization rate equation and apparent activation energy was measured. The content of residual ADMAAC was determined by interaction anion and cation, and the content of residual AM was determined by UV spectrophotometry, and the composition of copolymer at low conversion was obtained by the difference value of input content and residual content of monomer, and the reactivity ratios of monomers in polymerization were obtained by KelenTudos method. The results show that when the polymerization temperature is 40℃, the polymerization rate equation is $R_p = K[M]^{1.241}[KPS]^{0.52}[SHS]^{0.55}$. The apparent activation energy is 73.85 kJ/mol, which Is higher than the activation energy E_a (70.32 kJ/mol) of AM aqueous solution homopolymerization, according to the experience formula of Arrhenius, and the values of reactivity ratios are $r_{ADMAAC} = 0.197$、$r_{AM} = 4.503$. The important kinetic parameters for controlling of ADMAAC-AM copolymerization reaction is determined. The product of reactivity ratio of two monomers is less than 1, the copolymerization behavior type of ADMAAC and AM is a non-ideal copolymerization behavior without azeotropic point, the copolymer composition curve below the diagonal.

Keywords: Acrylamide; Polymerization kinetics; Polymerization rate equation; Apparent activation energy; Reactivity ratios

阴离子聚丙烯酰胺/三乙醇胺超分子体系的表征及性能

祝仰文[1] 刘 歌[2] 曹绪龙[1] 宋新旺[1] 陈 湧[2] 刘 育[2]

(1. 胜利油田分公司勘探开发研究院,东营 257015;
2. 南开大学化学系,元素有机化学国家重点实验室,天津化学化工协同创新中心,天津 300071)

摘要:研究了水溶液和模拟矿化水溶液中带有负电荷的磺化聚丙烯酰胺(SPAM)与三乙醇胺(TEA)之间的相互作用及形成的超分子体系的结构特征,考察了超分子体系的形成对SPAM形貌、流体力学直径、zeta电位、黏度及流变学性质的影响。研究结果表明,超分子体系的形成有利于提高磺化聚丙烯酰胺的黏度、抗剪切性、剪切回复性及抗温耐盐性。

关键词:磺化聚丙烯酰胺;三乙醇胺;超分子体系

中图分类号:O632.63 **文献标志码**:A

聚合物驱油是提高高温高盐油藏采收率的主要方法[1-3]。目前,油田生产中使用的驱油用聚合物主要包括部分水解聚丙烯酰胺[4,5]及黄原胶[6-8]等。这些驱油用聚合物在高温高盐的油藏环境中有可能发生遇盐黏度降低或高温下水解等现象,从而导致驱油效率的降低[9,10]。解决这一问题的方法通常是在制备聚丙烯酰胺时引入新的共聚物单体及可抑制酰胺水解的基团,将部分水解聚丙烯酰胺转化为疏水缔合型聚丙烯酰胺[11-14]。刘育等[15-17]尝试将带有正电荷的表面活性剂或寡聚乙烯二胺与阴离子型聚丙烯酰形成超分子复合体系,通过阴离子型聚丙烯酰胺链上的负电荷与表面活性剂或寡聚乙烯二胺上的正电荷之间的静电相互作用来改善聚丙烯酰胺的抗盐和抗剪切能力。与共价修饰不同基团的聚合物体系相比,阴离子型聚丙烯酰胺与阳离子表面活性剂或寡聚乙烯二胺构筑的超分子复合体系具有易于制备及剪切回复性能强等优势。同时,与无机强碱相比,有机胺分子具有可降解的特性,这不仅符合聚合物驱油技术的绿色、环保、可回收的发展要求,而且能在一定程度上降低传统碱驱技术对环境的污染,从而为石油事业的发展提供新的思路。

本文将具有静电(叔胺基)和氢键(羟基)双重作用位点的三乙醇胺分子与阴离子型聚丙烯酰胺结合,通过阴离子型聚丙烯酰胺链上的磺酸负离子与三乙醇胺上叔胺正离子的静电相互作用及阴离子型聚丙烯酰胺链上的众多氧原子与三乙醇胺上的羟基之间的氢键相互作用的协同贡献构筑超分子体系，并考察了其在水及模拟矿化水等不同环境中的黏度、抗剪切性及剪切回复性等性能。

1 实验部分

1.1 试剂与仪器

三乙醇胺(分析纯，天津市三江赛瑞达科技有限公司)；参考文献[17]方法制备磺化聚丙烯酰胺(SPAM)，分子量 5.2×10^6，$m:n\approx9:1$，磺化率≈10%，结构如图 1 所示；参考文献[16]方法制备模拟矿化水，总矿化度为 32 868 mg/L，$Ca^{2+}+Mg^{2+}$ 总质量浓度为 873 mg/L。

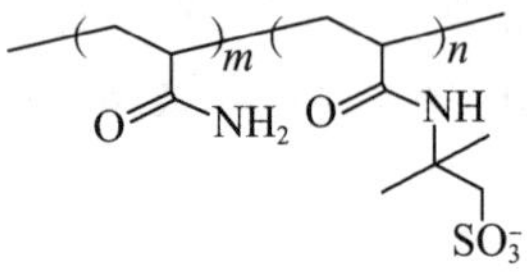

图 1 Structure of SPAM

AR2000 型流变仪(美国 TA 公司)；ZETAPALS/BI-200SM 型广角激光散射仪(美国 Brookhaven 公司)；TecnaiG2F20 型场发射透射电子显微镜(TEM，美国 FEI 公司)。

1.2 实验过程

采用 AR2000 流变仪，在 7.34 s^{-1} 剪切速率下，向浓度为 1 mg/mL 的 SPAM 溶液中逐渐加入三乙醇胺，测定表观黏度变化。

分别在模拟矿化水中配制 SPAM 溶液和 SPAM/三乙醇胺复合溶液，其中 SPAM 浓度为 1 mg/mL，三乙醇胺浓度为 0.8 mg/mL，将样品滴到铜网上自然晾干，观察其微观结构。

分别在模拟矿化水中配制 SPAM 溶液和 SPAM/三乙醇胺复合溶液，SPAM 浓度为 1 mg/mL，三乙醇胺浓度为 0.3 mg/mL。将样品过 800 nm 一次性滤膜，静置 24 h 后，在 25℃下测定其流体力学直径及分布情况。

分别配制 SPAM 溶液和 SPAM/三乙醇胺复合溶液，SPAM 浓度为 1 mg/mL，三乙醇胺浓度为 0.8 mg/mL，在 25℃下测定其 zeta 电势。

分别配制 SPAM 溶液和不同浓度 TEA 的 SPAM/三乙醇胺复合溶液，在应变 20%，温度 25℃下测定其黏弹性能。

分别在水中和模拟矿化水中配制 SPAM 溶液和 SPAM/三乙醇胺复合溶液，SPAM 浓度为 1 mg/mL，三乙醇胺浓度为 0.8 mg/mL，进行抗剪切性能和耐温抗盐性能测试。

2 结果与讨论

2.1 SPAM/三乙醇胺超分子体系的结构表征

通过测定黏度、动态光散射数据和 zeta 电势等对 SPAM/三乙醇胺超分子体系的结构进行表征。固定 SPAM 浓度 1 mg/mL，加入三乙醇胺后引起的黏度变化情况如图 2

所示。SPAM(1 mg/mL)的表观剪切黏度为 81 mPa・s，当加入低浓度的三乙醇胺时，SPAM 黏度变化不大；当加入的三乙醇胺浓度超过 0.4 mg/mL 后，体系的黏度明显增加，说明形成了 SPAM/三乙醇胺超分子体系；加入 0.8 mg/mL 三乙醇胺后，体系的表观黏度达到 205 mPa・s，约为 SPAM 表观黏度的 3.5 倍。这可能是由于 SPAM 与三乙醇胺形成的三维网络结构超分子体系增大了 SPAM 的黏度所致。由图 3 可见，SPAM/三乙醇胺超分子体系显示出一种较为紧凑且有序的形貌特征，并且其中绝大多数 SPAM 聚合物以较为伸展的链状结构存在，纳米线的宽度约为 13 nm。在相同条件下，单纯 SPAM 的 TEM 照片则显示出聚合物链的伸展结构和团聚结构 2 种形式共存的形貌特征，其原因可能是三乙醇胺与 SPAM 之间的弱相互作用拉近了 SPAM 聚合物链之间的距离。此外，三乙醇胺与 SPAM 之间存在大量的氢键，这些氢键具有一定的方向性，从而提高了 SPAM 聚合物链结构的有序性。

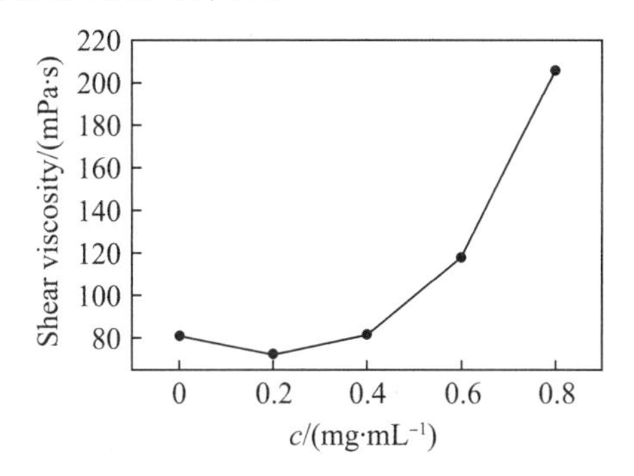

图 2 Effect of TEA concentration on viscosity of SPAM

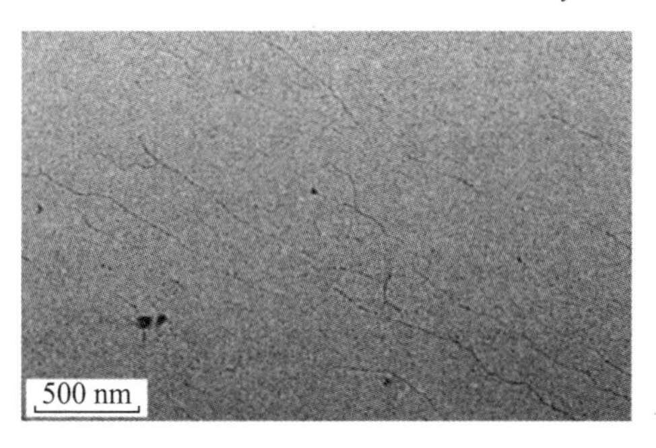

图 3 TEM image of SPAM/TEA supramolecular system

图 4 示出了 SPAM 和 SPAM/三乙醇胺超分子体系的动态光散射测定结果。由图 4 可见，SPAM 的流体力学直径主要分布在 2 个区间：第一个区间以分子内缔合和分子链的相互缠绕为主，其流体力学直径约为 50 nm；第二个区间以分子间缔合为主，其流体力学直径约为 180 nm。而 SPAM/三乙醇胺超分子体系的流体力学直径无论在第一区间(140 nm)还是第二区间(780 nm)均显著增大，而且 SPAM/三乙醇胺超分子体系的流体力学直径在第二区间的增大更明显，说明 SPAM 与三乙醇胺之间的相互作用促进了 SPAM 聚合物分子间的缔合。图 5 给出 SPAM 和 SPAM/三乙醇胺超分子体系的 zeta 电势，可见 SPAM 在加入三乙醇胺后 zeta 电势发生明显变化。由图 5 可见，SPAM 具有较低的 zeta 电势(−31.66 mV)，表明 SPAM 表面带有较多负电荷，容易受到矿化水环境中大量的 Ca^{2+}/

Mg^{2+}的干扰从而导致性能变差。加入三乙醇胺后 *zeta* 电势变为−0.13 mV,说明三乙醇胺通过静电和氢键相互作用将 SPAM 阴离子聚合物交联起来,中和了磺化聚丙烯酰胺的绝大部分负电性,得到了几乎中性的组装体,在一定程度上有利于提高其抗盐性。

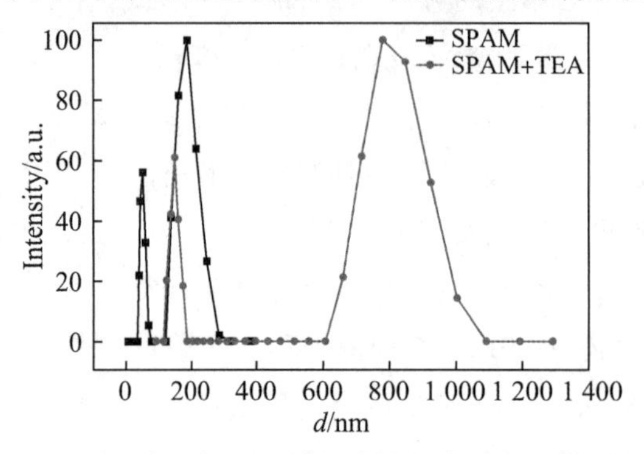

图 4 Hydrodynamicradiusdistribution of SPAM and SPAM /TEA in aqueous solution

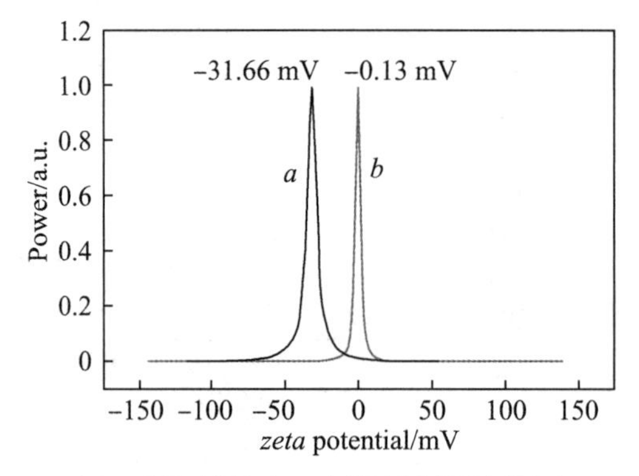

图 5 *Zeta* potentials of SPAM(a) and SPAM/TEA(b) in aqueous solution

2.2 SPAM/三乙醇胺超分子体系的黏弹性

黏弹性是评价聚合物的重要指标之一[18,19],聚合物的黏弹性越好则驱油效率越高[20,21]。实验中测定了 TEA 含量对 SPAM/三乙醇胺超分子体系的黏弹性的影响(见图 6),SPAM 和 SPAM/三乙醇胺超分子体系黏度随剪切速率(0.1~1 000 s^{-1})的变化情况以及在 200 s^{-1} 速率下剪切 5 min 后黏度回复性能(见图 7)。

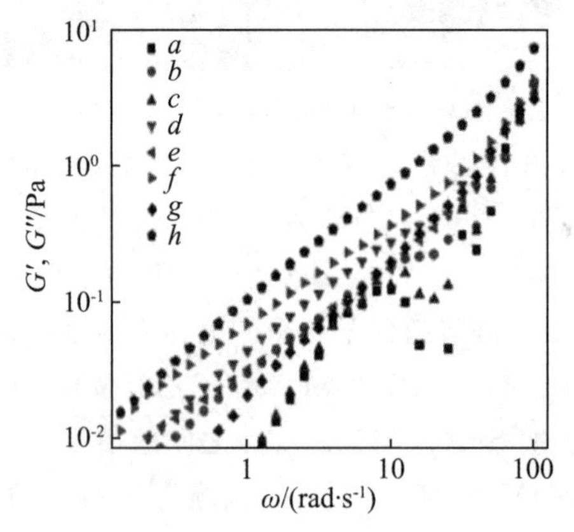

图 6 Effects of TEA concentration on G′(a,c,e,g) and G″ (b,d,f,h) of SPAM in aqueous solution at 25℃ Concentration of SPAM/(mg·mL^{-1}):a,b. 0.2;c,d. 0.4;e,f. 0.6;g,h. 0.8.

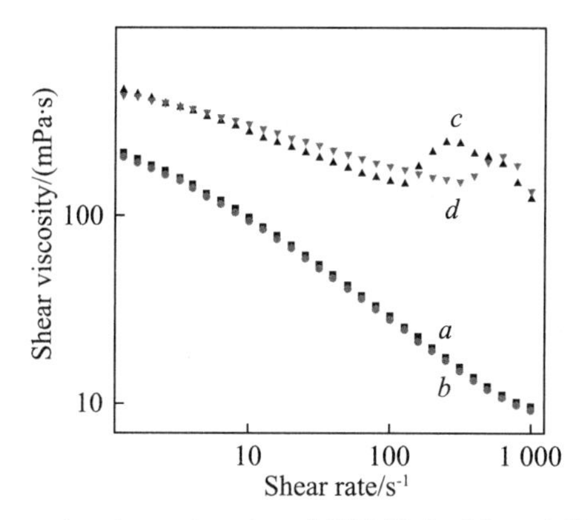

图 7 Effects of shear rate on the sheer viscosity of SPAM (a,b) and SPAM/TEA (c,d) in aqueous solution at 25℃ a,c. Without a pre-shearing; b,d. after a pre-shea-ring at 200 s^{-1} for 5 min.

由图 6 可见,在所检测的角频率范围内,随着 TEA 浓度的增加,SPAM/三乙醇胺超分子体系的弹性模量(G′)和黏性模量(G″)逐渐增大。此外,SPAM/三乙醇胺超分子体系的黏性模量大于弹性模量且二者相差不大,说明超分子体系呈现为接近于半固态的高黏性液体。图 7 示出了 SPAM 和 SPAM/三乙醇胺超分子体系的剪切黏度随剪切速率(0.1～1 000 s^{-1})的变化曲线及以 200 s^{-1} 的速率剪切 5 min 后重复对 SPAM 和 SPAM/三乙醇胺超分子体系进行剪切(速率 0.1～1 000 s^{-1})剪切回复性曲线。SPAM/三乙醇胺超分子体系表现出明显高于 SPAM 的抗剪切性(图 7 谱线 a 和 c)。由图 7 谱线 a 可知,在高剪切速率(1 000 s^{-1})下,SPAM 的黏度下降为 9.5 mPa·s,黏度损失约为 88%;由图 7 谱线 b 可知,相同条件下超分子体系的黏度仅下降到 124.3 mPa·s,黏度损失约为 39%。当以 200 s^{-1} 的速率对单纯的 SPAM 和 SPAM/三乙醇胺超分子体系分别剪切 5 min 后,再重复剪切速率从 0.1～1 000 s^{-1} 的剪切,所得到的剪切曲线 b 和 d 与最初的剪切曲线 a 和 c 基本重合,说明体系具有很好的剪切回复性。这主要是因为 SPAM 与三乙醇胺之间的弱相互作用,如氢键和静电相互作用等的协同作用提高了超分子体系的稳定性,使其对外力剪切显示出较好的耐受性。另一方面,当超分子体系经过剪切后,由于非共价弱相互作用的可逆性,SPAM 与三乙醇胺可以重新组装为超分子体系,使得因受到剪切而被断裂的三维网络结构得以恢复。

2.3 SPAM/三乙醇胺超分子体系的耐温抗盐性能

研究了 SPAM/三乙醇胺超分子体系在高温模拟矿化水溶液中的抗剪切能力及剪切回复性。图 8 显示,在 85℃下模拟矿化水中 SPAM/三乙醇胺超分子体系的表观黏度为 11.9 mPa·s,比相同条件下单纯 SPAM 的表观黏度(1.5 mPa·s)提高了约 8 倍。图 8 剪切曲线 c 表明,随着剪切速率从 10 s^{-1} 逐渐提高至 1 000 s^{-1},SPAM/三乙醇胺超分子体系的黏度几乎不变。另一方面,在以 200 s^{-1} 速率剪切 5 min 后,重复对 SPAM/三乙醇胺超分子体系进行剪切(速率 10～1 000 s^{-1}),所得剪切曲线 d 与最初的剪切曲线 c 依然基本重合。这些现象说明即使在高温矿化水中,SPAM/三乙醇胺超分子体系仍表现出高的抗剪切能力和剪切回复性。推测一方面是因为 SPAM 中磺酸根离子的水化能力强,受温度影响较小;另一方面,SPAM/三乙醇胺超分子体系具有近中性的 zeta 电势,其受矿化水中 Ca^{2+} 和 Mg^{2+} 的影响程度远小于表面带有较多负电荷的 SPAM。

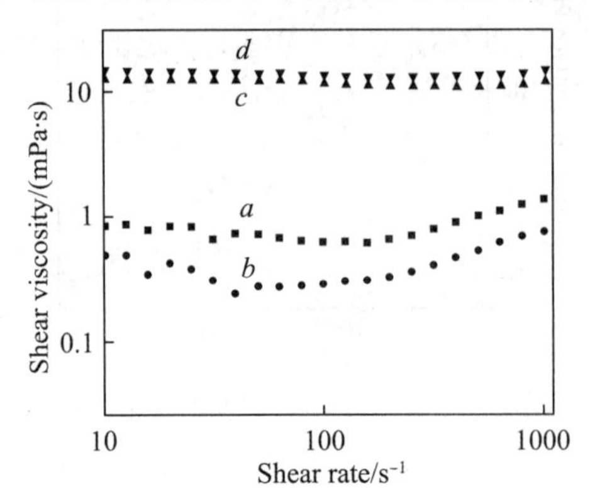

图 8 Effects of shear rate on the apparent viscosity of SPAM(a, b) and SPAM /TEA(c, d) in mineralized water solution at 85℃ a,c. Without a preshearing; b,d. After a preshearing at 200 s^{-1} for 5 min.

3 结论

研究了三乙醇胺与带有负电荷的磺化聚丙烯酰胺通过非共价相互作用形成的超分子体系的结构及 性能。结果表明,通过超分子相互作用得到大尺寸、高黏度超分子体系的方法可明显改善阴离子聚丙烯酰胺在高温、高盐环境中的抗剪切能力和剪切回复性,为超分子化学技术在采油工业中的实际应用奠定了基础。

参考文献

[1] Xu H., Petroleum Drilling Techniques, 2015, 43(2), 78-83.(徐辉。石油钻探技术,2015,43(2),78-83)

[2] Cao X. L., Liu K., Hang Y. G., He D. Y., Petroleum Geology and Recovery Efficiency, 2014, 21(2), 10-14.(曹绪龙,刘坤,韩玉贵,何冬月。油气地质与采收率,2014,21(2),10-14)

[3] Li B., Oilfield Chem., 2014, 31 (2), 278-281.(李彬。油田化学,2014,31(2),278-281)

[4] Pabon M., Corpart J., Selb J., J. Appl. Polym. Sci., 2004, 91(2), 916-924.

[5] Feng Y. J., Billon L., Grassl B., Polym. J., 2002, 43(7), 2055-2064.

[6] Spinelli L. S., Aquino A. S., Lucas E., Almeida A. R., Leal R., Martin A. L., Polym. Eng. Sci., 2008, 48(10), 1885-1891.

[7] Wu Y. X., Feng P. J., Zhong S. Q., Xu F., Ren S., Zhou W., Duan M. F., Xiao J. Y., Drilling and Production Technology, 2009, 32(5), 77-80.(吴月先,冯培基,

钟水清，徐峰，任山，周文，段明峰，肖俊英。钻采工程，2009，32(5)，77-80).

[8] Zhang B. Y.，Sun J. M.，Kang H.，Chemical Engineering of Oil and Gas，1999，28 (1)，49-52.(张伯英，孙景民，康恒. 石油与天然气化工，1999，28(1)，49-52).

[9] Liu Y. Y.，Chen P. K.，Luo J. H.，Zhou G.，Jiang B.，Acta Phys. Chim. Sin.，2010，26 (11)，2907-2914.(刘艳艳，陈攀科，罗健辉，周歌，江波. 物理化学学报，2010，26(11)，2907-2914).

[10] Xu L.，Chen Y. Q.，J. South. Petro. University，2012，34 (6)，136-140.(徐亮，程玉桥。西南] 石油大学学报，2012，34(6)，136-140).

[11Castelletto V.，Hamleg W.，Xue W.，Macromolecules，2004，37 (4)，1497-1501.

[12] Feng Y. J.，Billon L.，Grass B.，Polymer，2001，43(7)，2055-2064.

[13] Pabon M.，Corpart J.，Selb J.，J. Appl. Polym. Sci.，2004，91(2)，916-924.

[14] Hang Y. G.，J . South. Petro. University，2011，33(3)，149-153.(韩玉贵. 西南石油大学学报，2011，33(3)，149-153).

[15]Cao X. L.，Hu Y.，Song X. W.，Zhu Y. W.，Han Y. G.，Wang K. P.，Guo D. S.，Liu Y.，Chem. J. Chinese Universities，2015，36(2)，395-398.(曹旭龙，胡岳，宋新旺，祝仰文，韩玉贵，王鲲鹏，郭东升，刘育. 高等学校化学学报，2015，36(2)，395-398).

[16] Cao X. L.，Hu Y.，Song X. W.，Zhu Y. W.，Han Y. G.，Wang K. P.，Chen Y.，Liu Y.，Chem. J. Chinese Universities，2015，36 (7)，1437-1440.(曹旭龙，胡岳，宋新旺，祝仰文，韩玉贵，王鲲鹏，陈湧，刘育。高等学校化学学报，2015，36(7)，1437-1440).

[17] Cao X. L.，Hu Y.，Song X. W.，Zhu Y. W.，Han Y. G.，Wang K. P.，Guo D. S.，Liu Y.，Chem. J. Chinese Universities，2014，35(9)，2037-2042.(曹旭龙，胡岳，宋新旺，祝仰文，韩玉贵，王鲲鹏，郭东升，刘育. 高等学校化学学报，2014，35(9)，2037-2042).

[18] Hao Y. P.，Blan J. J.，Yang H. L.，Zhang H. L.，Gao G.，Dong L. S.，Chem. Res. Chinese Universities，2016，32(1)，140-148.

[19] Chen G. Y.，Feng B. X.，Zhu K. Y.，Zhao Y. H.，Yuan X. Y.，Chem. Res. Chinese Universities，2015，31(2)，303-307.

[20] Xia H.，Wang D.，Wang G.，Wu J.，Pet. Sci. Technol.，2008，26(4)，398-412.

[21] Zhang Z.，Li J. C.，Zhou J. F.，Transp. Porous. Med.，2011，86(1)，229-244.

Characterization and property of sulfonated polyacrylamide/triethanolamine supramolecular systemt

Zhu Yangwen[1] Liu Ge[2] Cao Xulong[1*] Song Xinwang[1]
Chen Yong[2] Liu Yu[2*]

(1. Research Institute of Exploration and Development, ShengLi Oilfied Company,
China Petro-Chemical Corporation, Dongying 257015, China;
2. Department of Chemistry, State Key Laboratory of Elemento-Organic Chemistry,
Collaborative Innovation Center of Chemical Science and Engineering(Tianjin), Nankai University,
Tianjin 300071, China)

Abstract: Polymer flooding technology is one of the preferred "enhanced oil recovery"(EOR) technologies used for improving oil recovery. Among the various chemical reagents for polymer flooding technology, polyacrylamide has attracted more and more attention. To explore the improving effect of triethanolamine (TEA) on sulfonated polyacrylamide (SPAM), the interaction of negatively charged sulfonated polyacrylamide with triethanolamine and the structure of SPAM/TEA supramolecular system were investigated in aqueous solution and mineralized water by means of transmission electron microscopy (TEM), viscosity, dynamic light scattering (DLS), zeta potential and rheological experiments. The results indicate that, through the cooperative contribution of electrostatic and hydrogen bond interactions, SPAM and TEA can form a large supramolecular assembly with an average diameter of 780 nm measured by DLS. Significantly, the formation of SPAM/TEA supramolecular system can efficiently improve the viscosity and rheological properties of sulfonated polyacrylamide. In addition, the nearly neutral zeta potential of SPAM/TEA supramolecular system can decrease the affect of Ca^{2+} and Mg^{2+} cations in the mineralized water.

Keywords: Sulfonated polyacrylamide; Triethanolamine; Supramolecular system

(Ed. : P,H,W,K)

Interfacial rheological behaviors of inclusion complexes of cyclodextrin and alkanes

Wang Ce[ab]　Cao Xulong[c]　Zhu Yangwen[c]　Xu Zhicheng[b]
Gong Qingtao[b]　Zhang Lei[*b]　Zhang Lu[*b]　Zhao Sui[b]

The transformation of cyclodextrins (CDs) and alkanes from separated monomers to inclusion complexes at the interface is illustrated by analyzing the evolution of interfacial tension along with the variation of interfacial area for an oscillating drop. Amphiphilic intermediates are formed by threading one CD molecule on one alkane molecule at the oil/aqueous interface. After that, the amphiphilic intermediates transform into non-amphiphilic supramolecules which further assemble through hydrogen bonding at the oil/aqueous interface to generate a rigid network. With the accumulation of supramolecules at the interface, microcrystals are formed at the interface. The supramolecules of dodecane@2α-CD grow into microrods which form an unconsolidated shell and gradually cover the drop. However, the microcrystals of dodecane@2β-CD are significantly smaller which fabricate into skin-like films at the interface. The amphiphilic intermediates during the transformation increase the feasibility of self-emulsification and the skin-like films enhance the stability of the emulsion. With these unique properties, CDs can be promising for application in hydrophobic drug delivery, food industry and enhanced oil recovery.

1 Introduction

Host-guest chemistry plays an important role in the construction of supramolecular architectures in aqueous solution and on the interface. As one kind of the most significant host molecules cyclodextrins (CDs) can form stable and specific inclusion complexes with a variety of organic guest molecules.[1,2] CDs are doughnut-like oligosaccharides with a hydrophilic exterior and a hydrophobic cavity.[3,4] Therefore, CDs are able to shelter the hydrophobic parts of guest molecules and form host-guest complexes through hydrophobic interactions, and/or other interactions such as hydrogen bonds.[5] There have been numerous reports about the construction of multifunctional materials and smart/responsive interfaces from CDs and different guests, such as azobenzene,[6-8] adamantane,[9,10] ferrocene,[11] phthalate esters,[12] etc. Surfactants are another type of guests whose hydrophobic moiety can be covered by CDs.[13] As a result, non-amphiphilic inclusion complexes are fabricated and the surface tension and critical micelle concentrations increase.[13-15] Moreover, the inclusion complexes can further self-assemble into non-amphiphilic vesicles and microtubes driven by hydrogen bonding.[16,17]

Apart from that, simple hydrophobic molecules like alkanes can even form inclusion complexes.[18] Zhou CC et al. have found that CDs can assemble with *n*-dodecane to form channel type dimers which further construct vesicles or brick-like vesicles through hydrogen bonding.[19,20] More surprisingly, the inclusion complexes are formed by threading CDs from the aqueous phase on alkane molecules from the emulsion drop surface which grow further into an interfacial film. As a result, CDs have the potential of being used as emulsifiers to replace surfactants in such fields like food and drug delivery where surfactants might have detrimental eff ects.[21,22] Mathapa BG and Paunov VN have found that the inclusion complexes further grow into microcrystals which remain attached on the surface of the emulsion drops and form densely packed layers.[22] Therefore, the adsorption layer of inclusion complexes of CDs and alkanes at the oil/aqueous interface is highly significant for emulsification which is worth a systematic study. Currently, various methods have been used to explore the structure of inclusion complexes and their aggregations in aqueous solution. However, the dynamic process of the transformation from monomers to inclusion complexes is rarely reported. Since the supramolecular interaction between CDs and alkanes takes places at the oil-aqueous interface, a reasonable method is to study the properties of the interface between the CD solution and the oil phase.

Interfacial tension (IFT) reflects the nature of the interfacial interaction and the dynamic changes in the state of interfacial aggregation while the measurement of interfacial rheological properties can provide more accurate information about the structure of the adsorption layers, which has been comprehensively investigated in recent years. For a viscoelastic film, when the adsorption layer is disturbed under periodic compressions and expansions at a given frequency, there will be a response of interfacial tension along with change of area, so interfacial dilational modulus and viscoelasticity can be calculated. Hence, interfacial dilational rheological experiment is widely employed for studying the interfacial behaviors of proteins, phospholipids, polymers, surfactants, and complexes.[23-26] However, the studies of interfacial dilational viscoelasticity of supramolecules are scarcely reported in the literature. The study of the dilational rheology of the interfacial layer between alkanes and the aqueous solution of CDs can not only reflect the strength of the adsorption layer of inclusion complexes of CDs and alkanes, but also reveal the progress of formation of inclusion complexes and how the complexes change when the interface undergoes deformation. Therefore, the interfacial dilational rheological properties are worthy of investigation.

The present article mainly focuses on experimental studies of the dynamic interfacial behavior of alkane-CD supramolecules. We take advantage of the interfacial dilational rheological experiment to track the interfacial tension changes along with the sine periodic oscillation of surface area. We explain how alkanes form surface active agents with CDs at the oil-aqueous interface, and how the amphipathic supramolecules transform into hydrotropic inclusion complexes. The spontaneous formation of biocompatible CD-based inclusion complexes would increase the use of CDs in a wide range of industrial products.

2 Experimental methods

2.1 Materials

The materials α-CD, β-CD and n-dodecane were purchased from Aladdin Industrial Corporation. The solutions of α-CD and β-CD were prepared with ultrapure water with a resistivity of 18.2 MΩ · cm.

2.2 Interfacial tension and interfacial dilatational rheology: pendant drop technique

The interfacial tension and dilatational rheology of the adsorbed layers at the oil-water interfaces were measured by the pendant drop technique using an oscillating pendant drop tensiometer from DataPhysics OCA20 (DataPhysics Company, Germany). A pensile drop of CD solution was injected from a stainless steel needle by means of a microsyringe and maintained vertically in a quartz cuvette cell full of alkane.[27,28] The profile of the oil drop was captured using a charge-coupled device (CCD) camera and transferred to the data acquisition computer, where it was digitized and analyzed by software employing the Laplace equa-tion. As a result, the interfacial tension and interfacial area were obtained at the same time.

To study the interfacial rheology properties, a sinusoidal periodic oscillation was performed on the interfacial area at a frequency of 0.1 Hz. The dilational modulus, which characterizes the interface resistance to the changes of the area which is reflected as the IFT variation, is defined by Gibbs in the following expression:

$$\varepsilon=\frac{d\gamma}{d\ln A} \tag{1}$$

where ε is the dilational modulus, γ is the IFT, and A is the interfacial area.

3 Result and discussion

3.1 Formation process of inclusion complexes of *n*-dodecane and CDs at the interface

Zhou CC et al. have demonstrated the formation of dodecane@2β-CD in aqueous solution with one *n*-dodecane chain threading into two β-CDs using mass analysis with ^1H NMR and ESI-MS.[19] However, it is difficult to determine whether there is the formation of an intermediate before the formation of the final dodecane @ 2β-CD supramolecules. In order to investigate the formation process in this work, the interfacial dilational rheology experiment was carried out which has been proved to be an excellent method to see the interface aggregation and the interfacial behavior. The equilibrium IFTs and interfacial dilational moduli of α-CD and β-CD solutions and *n*-dodecane are shown in Fig. 1.

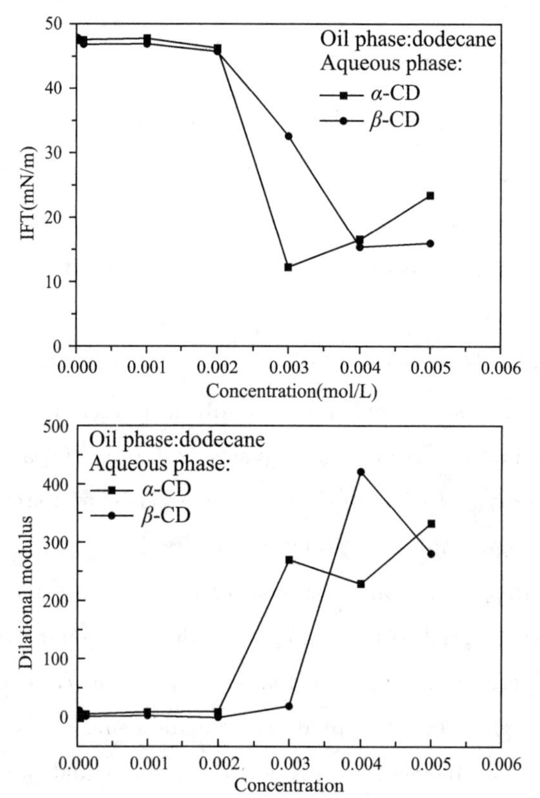

Fig. 1 Equilibrium IFTs and interfacial dilational moduli of the aqueous solution of CDs and *n*-dodecane.

One prominent characteristic of the supramolecules of β-CD and n-dodecane is the outstanding high modulus. As traditional surfactants, the maximal interfacial dilational moduli of pure SDS and C12TAB at the water/kerosene interface are around 15 and 25 $mN \cdot m^{-1}$ at 30℃, respectively. The dilational modulus of the mixture of SDS and C12TAB can increase to 45 $mN \cdot m^{-1}$.[29] Cao C et al. reported that the maximal dilational modulus of lysozyme at the water/decane interface is around 65 $mN \cdot m^{-1}$.[30] Liu XP reported that the maximal dilational modulus of a Gemini surfactant with different charge at the water/decane interface is around 110 $mN \cdot m^{-1}$.[23] Du FP et al. reported that the dilational modulus of a trisiloxane surfactant at the water/ air interface reaches a maximum around 50 $mN \cdot m^{-1}$ at 5×10^{-4} $mol \cdot L^{-1}$, and the concentration scale is similar to the CD concentration studied in this work.[31] As to the supramolecules of CD solution and n-dodecane, the interfacial dilational modulus reaches as high as 419 $mN \cdot m^{-1}$, which indicates that an unusual interaction exists among the molecules at the interface. The original data of how α-CD aqueous-dodecane interfacial tension changes along with interfacial area at various concentrations are displayed in Fig. 2 (A～G). The interfacial area and IFT of a typical surfactant, C12TAB, are also shown in Fig.2(H) as a comparison.

At lower concentration, from 1×10^{-5} $mol \cdot L^{-1}$ to 1×10^{-3} $mol \cdot L^{-1}$, the interfacial dilational modulus of n-dodecane and α-CD solution maintains below 10 $mN \cdot m^{-1}$ because there was almost no interfacial tension response to the deformation of the interfacial area, as shown in Fig. 2(A～C).

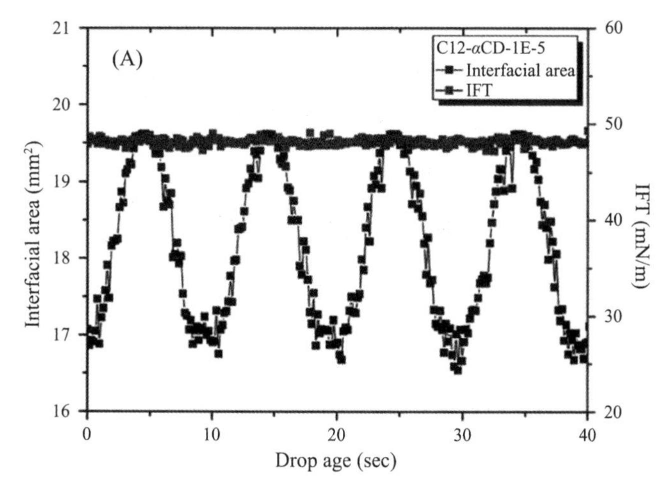

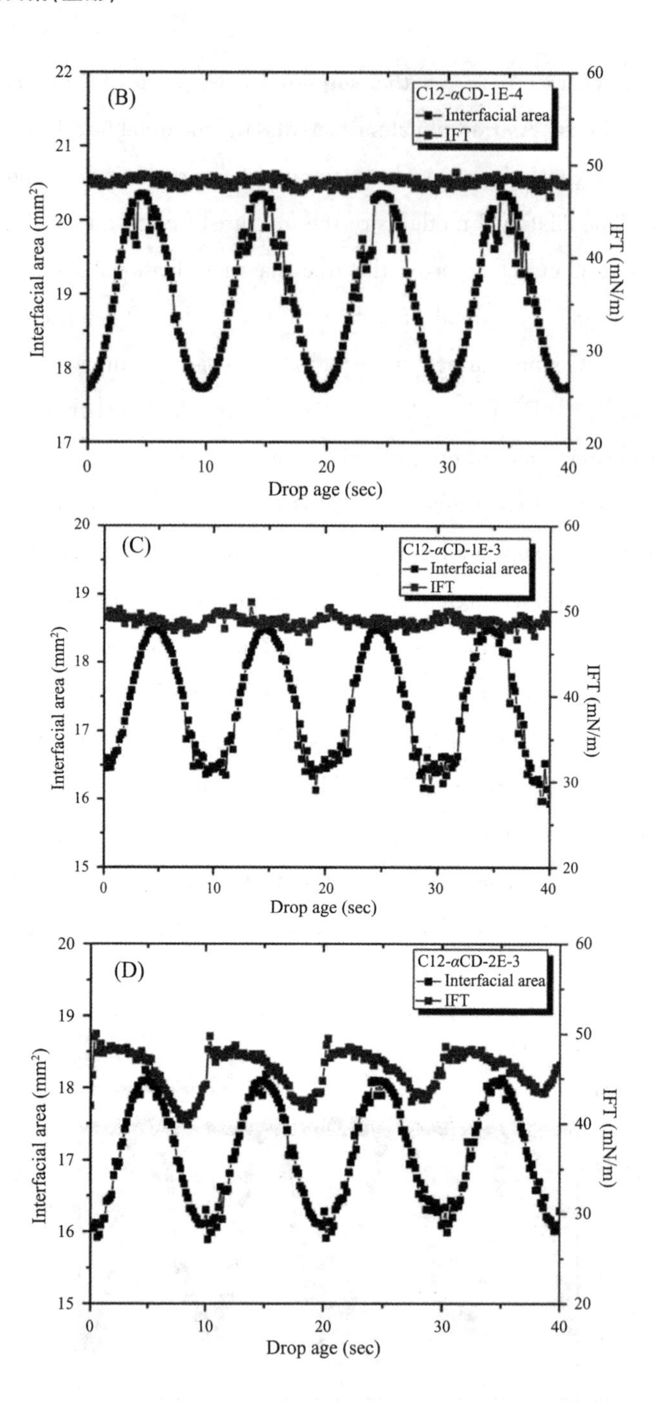
(B)
C12-αCD-1E-4
Interfacial area
IFT
Interfacial area (mm²)
IFT (mN/m)
Drop age (sec)
(C)
C12-αCD-1E-3
Interfacial area
IFT
Interfacial area (mm²)
IFT (mN/m)
Drop age (sec)
(D)
C12-αCD-2E-3
Interfacial area
IFT
Interfacial area (mm²)
IFT (mN/m)
Drop age (sec)

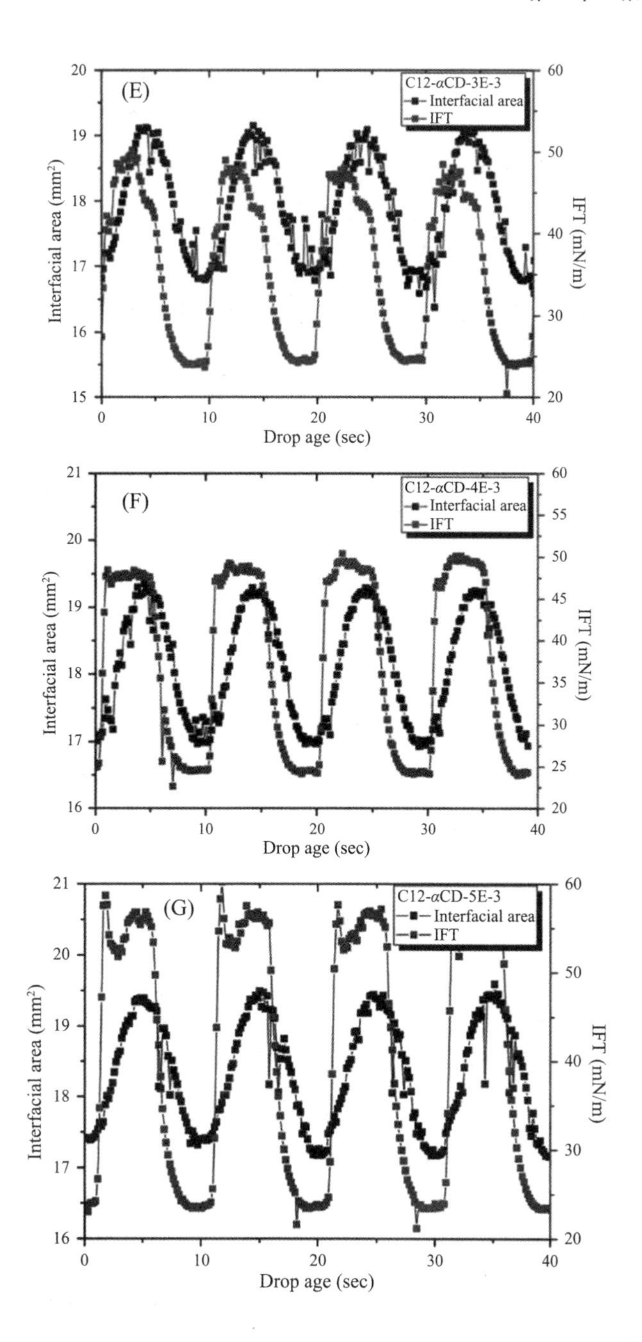
(E)
C12-αCD-3E-3
Interfacial area
IFT
Interfacial area (mm²)
IFT (mN/m)
Drop age (sec)
(F)
C12-αCD-4E-3
Interfacial area
IFT
Interfacial area (mm²)
IFT (mN/m)
Drop age (sec)
(G)
C12-αCD-5E-3
Interfacial area
IFT
Interfacial area (mm²)
IFT (mN/m)
Drop age (sec)

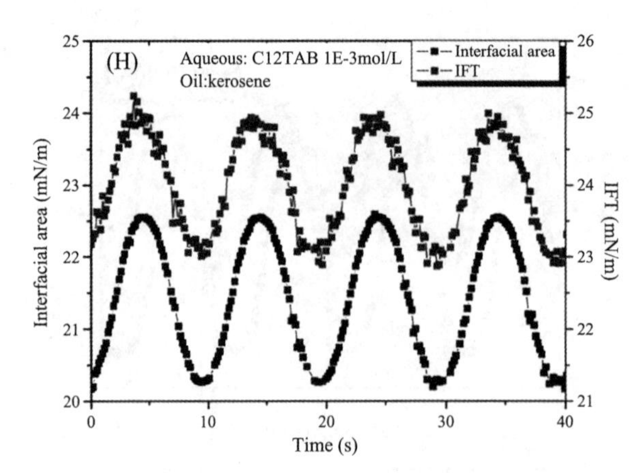

Fig. 2 Interfacial area and IFT of n-dodecane and α-CD aqueous solution at different concentrations (A-G) and that of a typical surfactant (C12TAB) at the aqueous-kerosene interface (H) as a comparison.

This indicates that interfacial active molecules or aggregations at the aqueous-dodecane interface are negligible. However, it is interesting to note that the modulus of α-CD aqueous solution and n-dodecane begins to increase when the concentration increases to 2×10^{-3} mol · L^{-1}. During the compression and expansion of the interface, the minimal IFT reaches as low as 25 mN · m^{-1} which is significantly lower than that of n-dodecane and pure water (around 49.1 mN · m^{-1}). This indicates that amphiphilic supra-molecules are formed at the interface. However, it is obvious that neither n-dodecane nor cyclodextrins have remarkable interfacial activity. Also, in the structure of dodecane@2β-CD, it has been proved in the literature that an n-dodecane molecule is packed up by two β-CDs, which is hence highly hydrophilic and hardly has such strong amphipathy.[19] Thus, it is unlikely to decrease the IFT. Therefore, it can be concluded that the amphiphilic intermediate dodecane@β-CD is formed during the progress of transformation, as shown in Fig. 3.

During the compression-expansion cycles, the interfacial area *versus* the time follows a sinusoidal curve at a given frequency. For typical surfactants, like C12TAB (Fig. 2(H)), the interfacial concentration decreases under expansion, and thus IFT increases along with the interfacial area. The compression of the interfacial area condenses the surfactants at the interface, which decreases IFT. In this progress, the breakage of intermolecular interactions is reversible and would not produce any change in the monolayer structure during the compression-expansion cycle. Thus, when the interfacial area is oscillated harmonically at a given frequency, kinetic IFT response to deformation of the interfacial area exhibits a linear behavior.[32,33] Even though the relaxation processes, such as diffusion exchange of surfactant molecules between the interface and the bulk solution or molecular rearrangements within the layer, may cause a phase difference between the applied area variation and IFT response, the curves of IFT are still sinusoidal periodic symmetric. However, for the interface of cyclodextrin solution and *n*-dodecane, there is strong distortion of the sinusoidal shape of the IFT kinetic curve. During each period, there are flat parts in the IFT curves at the maxima and minima IFT, and IFT can increase dramatically from minima to maxima even under slight area expansion. Unlike the adsorption interfacial layer of traditional surfactants, the interface of CD solution and alkanes experiences complicated transformation because monomers of CDs and alkanes form inclusion complexes through host-guest interaction.[20,22] Previous studies of observable interfacial properties in the nonlinear domain provide valuable information about the two-dimensional microstructure of the interfacial layer, as well as about the evolution of the structure with time.[34,35]

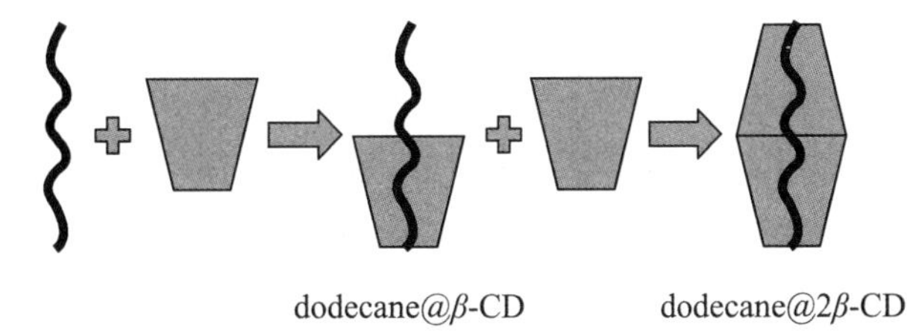

Fig. 3 Scheme of the transition from monomers of CD and *n*-dodecane to supramolecules.

For the primary stage, IFT changes along with time periodi-cally. According to the shape of the curve of IFT and interfacial area, each period can be divided into four stages to reveal the transition of the supramolecule under area deformation. As soon as the contact of aqueous solution and oil phase is established, the amphipathic supramolecule dodecane@α-CD is generated. In Stage 1, with the compression of the interfacial area, the concentration of dodecane@α-CD increases, which leads to a rapid decrease of IFT. As the interfacial area continues to shrink, the amphiphilic intermediate dodecane@α-CD becomes saturated gradually. Then, in Stage 2, IFT keeps constant and a minimal platform appears. Meanwhile, a high concentration gradient between the interface and the bulk is developed during the compression, which may lead to the diffusion of dodecane@α-CD from the interface to the solution. Apart from that, as a result of the dense hydroxyl at the exterior of α-CD, the supramolecules tend to self-assemble into a closely packed layer at the interface through hydrogen bonding and present a low IFT value. A similar distortion has been found in the kinetic surface tension of polyacids and dodecyl trimethyl ammonium bromide. The surface tension decreases with the compression up to a certain value and becomes almost constant after that until the surface expansion takes place when it begins to increase again. These results were explained by the formation of nano-or microparticles in the surface layer at a certain critical surface pressure in the course of compression.[32,36] In Stage 3, the closely packed layer of dodecane@α-CD at the interface is broken by the expansion. With the expansion of the interfacial area, the interfacial concentration of amphiphilic dodecane@α-CD decreases, and thus IFT begins to increase. At the same time, the expansion of the interfacial area provides more spare space for the conformational transition of dodecane@α-CD from an upright position to lying on the interface. This provides a possibility for dodecane@α-CD to transform into dodecane@ 2α-CD which hardly has any interface activity. As a result, IFT increases from 24.8 $mN \cdot m^{-1}$ to 47.1 $mN \cdot m^{-1}$. In Stage 4, with the increase of the interfacial area in a further step, IFT stays at the maximal platform which is almost at the value of pure water and n-dodecane. This indicates that there are few amphiphilic inter-mediates during this stage. The possible reason may be that most amphiphilic intermediates have been transformed into dodecane @ 2α-CDs. Even though the newly generated dodecane@α-CD at the crevices between the dodecane@2α-CD patches would lead to the decrease of IFT, the dilation dilutes the interfacial concentration of dodecane@α-CD at the same time. Therefore, when the two sides reach a balance in this stage, IFT remains almost constant. The transition of the inclusion complexes of α-CD and n-dodecane at the interface under periodical compression and expansion is also illustrated in Fig. 4.

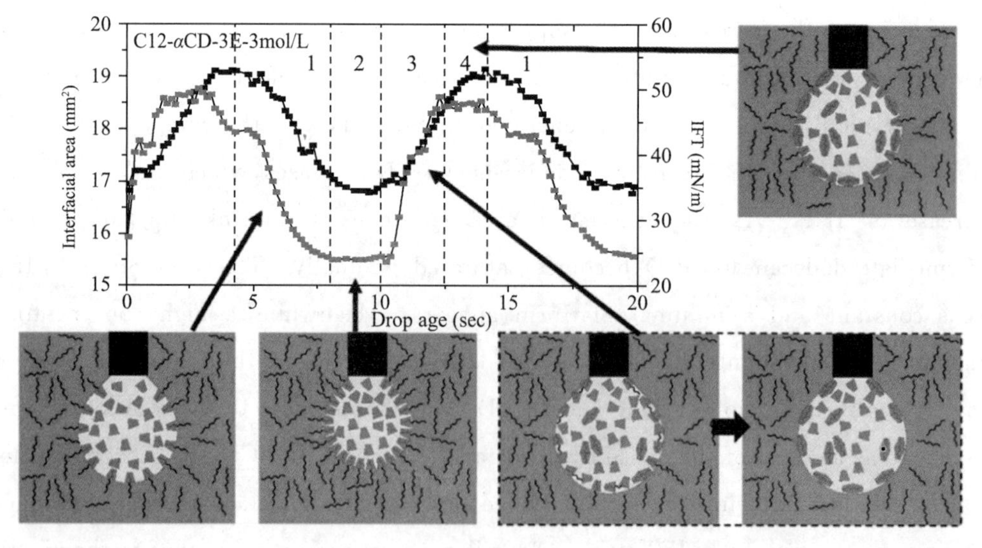

Fig. 4 IFT and interfacial area for α-CD aqueous solution and n-dodecane (the concentration of α-CD is 3×10^{-3} mol L^{-1}, the first 20 s), and the corresponding schematic representation of the transition of the inclusion complexes.

Fig. 5 displays the IFT dynamic periodical curves at different times. It is clear that these curves share a similar shape, which indicates a good reproducibility of the IFT data. The minimal and maximal flats can be observed in each period. However, with the accumulation of dodecane@ 2α-CD at the interface, an increasing area would be occupied by the non-amphiphilic inclusion complexes. Therefore, the concentration of the newly generated amphiphilic intermediate at the interface decreases and the minimal IFT platform increases along with time.

CDs are recognized as a series of cyclic oligosaccharides consisting of 6, 7, or 8 glucose units linked by α-1,4-linkages, named as α-CD, β-CD, or γ-CD, respectively. To explore the influence of the size of CDs on the interfacial properties of CD solution and n-dodecane, IFT measurements of β-CD solution and n-dodecane are performed to compare with that of α-CD. The variation of IFT and interfacial area with time is presented in Fig. 6.

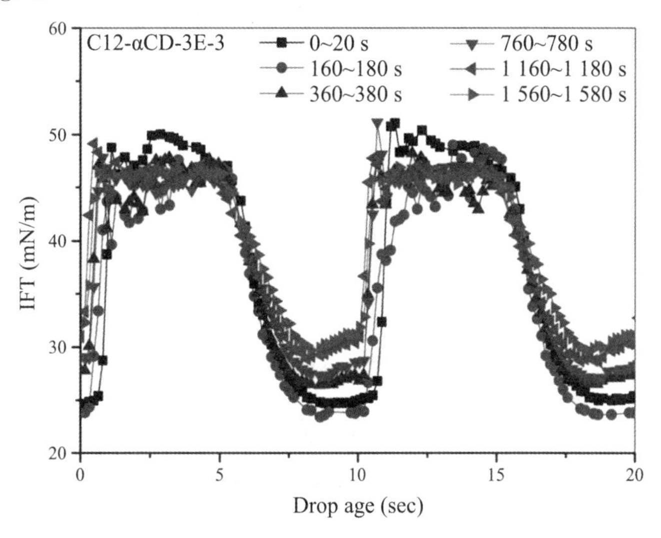

Fig. 5　Dynamic IFT periodical curves at different times, the concentration of α-CD is 3×10^{-3} mol L^{-1}.

The dynamic interfacial tension of β-CD aqueous solution and n-dodecane is similar to that of α-CD. At low concentration, IFT almost keeps constant during the periodical compression and expansion of the drop. At high concentration (4×10^{-3} mol · L^{-1} and 5×10^{-3} mol · L^{-1}), the generation of amphiphilic n-dodecane@β-CD supramolecules with one n-dodecane molecule threading in one β-CD would cause the decrease of IFT. With the compression of the interfacial area and the growth of the supramolecules in number, the interfacial concentration of n-dodecane@β-CD increases which causes the IFT to decrease dramatically to 15.5 mN · m^{-1}. But the expansion of the interfacial area disperses the tightly packed n-dodecane@β-CD and provides sufficient space for dodecane @ β-CD to transform into non-amphiphilic n-dodecane @ 2β-CD. As a consequence, IFT increases to the maximum.

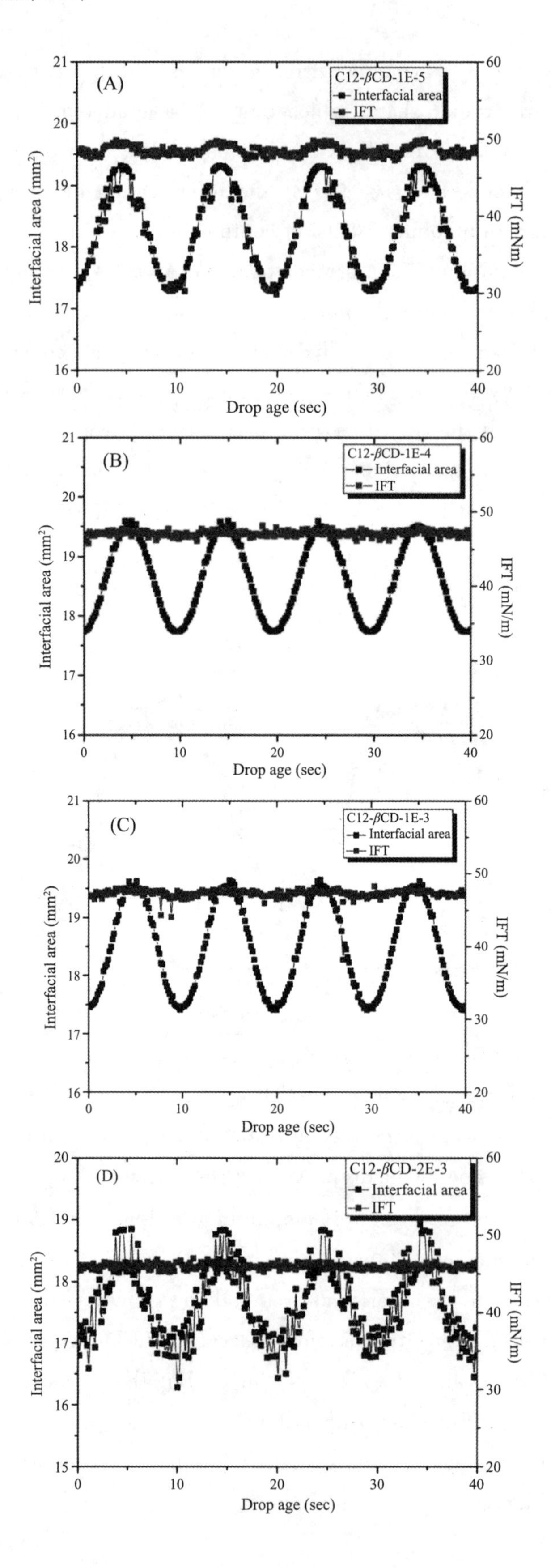
(A)
C12-βCD-1E-5
Interfacial area
IFT
Interfacial area (mm²)
IFT (mNm)
Drop age (sec)
(B)
C12-βCD-1E-4
Interfacial area
IFT
Interfacial area (mm²)
IFT (mN/m)
Drop age (sec)
(C)
C12-βCD-1E-3
Interfacial area
IFT
Interfacial area (mm²)
IFT (mN/m)
Drop age (sec)
(D)
C12-βCD-2E-3
Interfacial area
IFT
Interfacial area (mm²)
IFT (mN/m)
Drop age (sec)

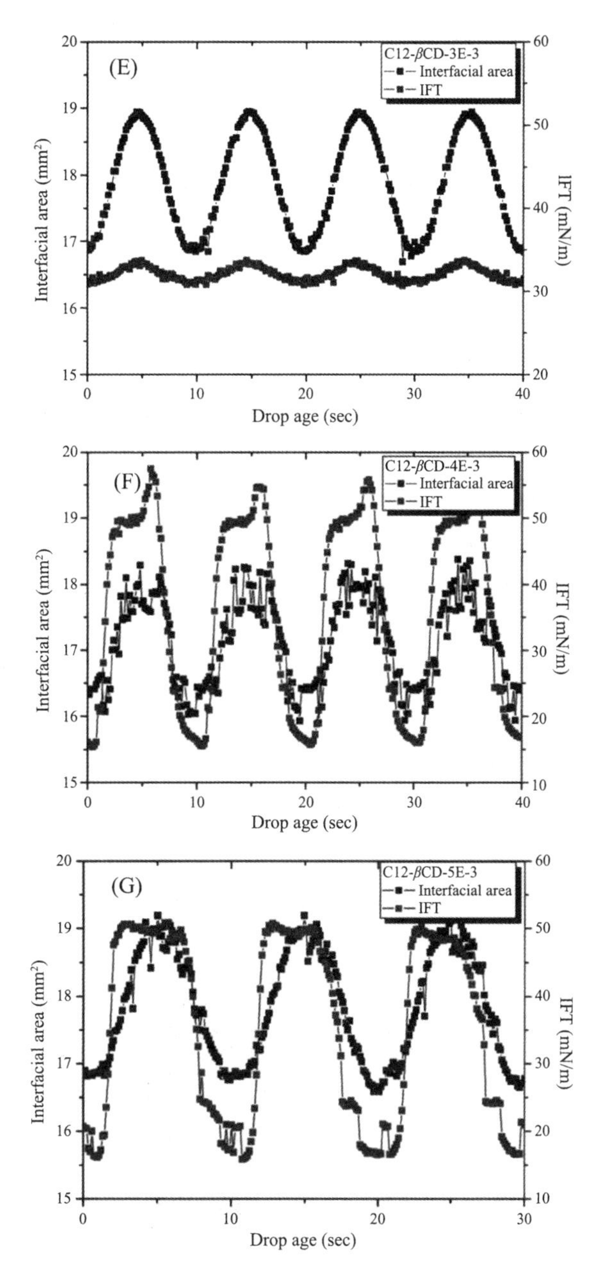

Fig. 6 Interfacial area and IFT of *n*-dodecane and b-CD aqueous solution at diffierent concentrations.

3.2 Supramolecular blocks of CDs and *n*-dodecane

It is interesting that, for both α-CD and β-CD, supramolecular blocks are formed at the interface when the concentration of CDs is over 4×10^{-3} mol · L^{-1}. Mathapa and Vesselin N. Paunov have reported that the inclusion complexes formed by threading CDs from the aqueous phase on *n*-tetradecane or silicone oil molecules from the oil phase grow further into microrods and microplatelets depending on the type of CD used. The microrods from α-CD/tetradecane were longer than 100 μm whereas the microcrystals from β-CD/tetradecane were much shorter(<10 mm).[22] Fortunately, the films formed by these microcrystals at the interface are visualized with the help of a digital microscope in the present work and the captured pictures are shown in Fig. 7. In the case of dodecane@2α-CD, with the accumulation of supramolecules, microcrystals appear at the interface. These microcrystals grow in size and number and gradually cover the interface of the drop. A shall is formed by the crystals of n-dodecane@2α-CD. However, for dodecane@2β-CD, a skin-like film on the drop interface is gradually formed after several periods of oscillation. The presence of such skins becomes visible when the aqueous drop in n-dodecane is shrinking in volume. He Q et al. have reported a similar phenomenon. The adsorption of pure protein or a mixed phospholipid/protein complex also generated skin-like films with network structures by cross-linking at the chloroform/water interface.[24] The driving force of formation of supramolecular films at the n-dodecane/aqueous interface is hydrogen bonding at the exterior of dodecane@2β-CD. Because of intermolecular cross-links of the inclusion complex, a network is formed at the interface. With the accumulation of the inclusion complex, the mono-adsorption layer begins to transform into a multilayer film which will control the shape of the drop. Thus the IFT calculated by the drop profile is no longer authentic.

To reveal the basic constitution unit of the films covering the drop, a 200% expansion has been imposed on the interfacial area of the drop. As can be seen in the photo, the shell of n-dodecane@α-CD is assembled with microrods with a length of 100-200 μm. However, the film of drops of β-CD solution is torn up to irregular pitches, the size of which is below 50 μm.

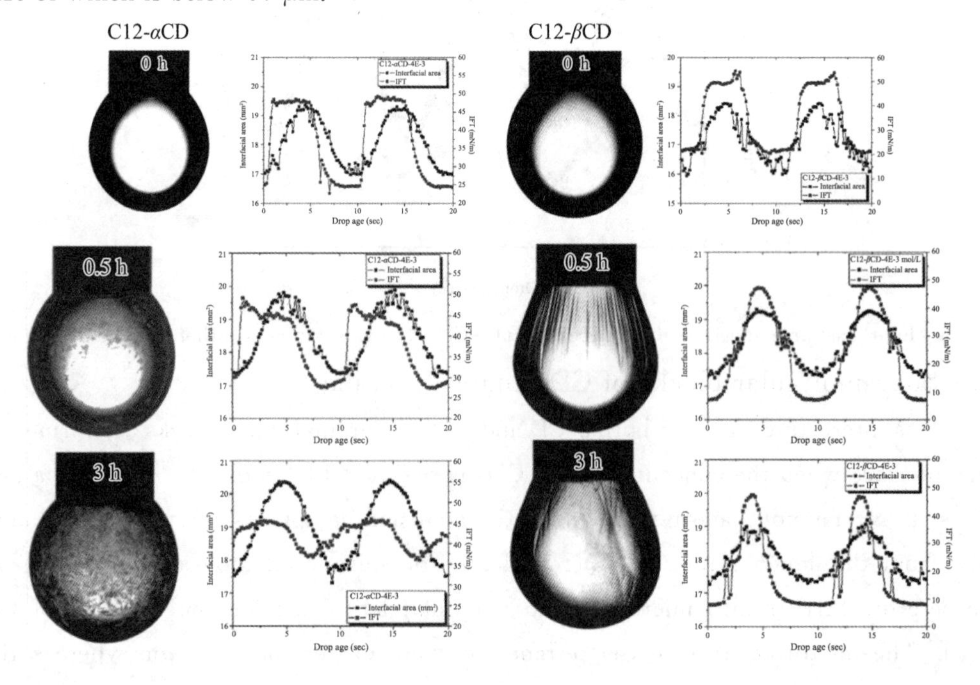

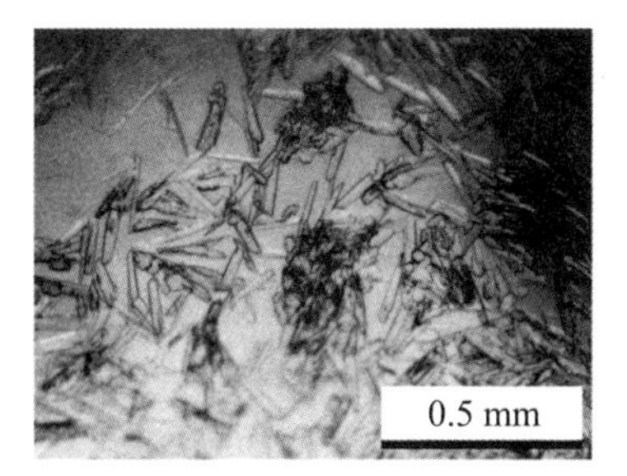

Fig. 7 Films and microcrystals generated by supramolecules at the interface. The concentrations of a-CD and b-CD are 4×10^{-3} mol L^{-1} separately.

4 Conclusion

During the formation of the inclusion complexes of CDs and alkanes at the interface, supramolecules with different interfacial activity are formed. The interfacial activity of the supramolecules will be reflected by the interfacial tension between the aqueous solution and oil. Herein, in the present article, recording the dynamic IFT of an oscillating drop is proposed as a method to study the transformation of supramolecules of CDs and alkanes at the interface.

The transformation route of inclusion complexes of CDs and alkanes is illustrated. Amphiphilic intermediates are formed during this process which causes the decrease of IFT. After that, amphiphilic intermediates convert to non-amphiphilic inclusion complexes which further cross-link at the oil/aqueous interface to generate a rigid network with an extremely high modulus. The supramolecules of dodecane@2α-CD grow into larger microrods and form a shell to cover the drop, while the microcrystals of dodecane@2β-CD are significantly smaller and a skin-like film is formed at the interface.

The *in situ* generation of alkane@CD supramolecules at the oil/water interface expands the application of CDs. Especially, the amphiphilic intermediates during the transformation increase the feasibility of self-emulsification and the skin-like films enhance the stability of the emulsion. These unique properties demonstrate that CDs are promising for application in hydrophobic drug delivery, food industry, enhanced oil recovery, *etc*.

Conflicts of interest

There are no conflicts to declare.

Acknowledgements

The authors acknowledge financial support from the National Science &

Technology Major Project (2016ZX05011-003) and National Natural Science Foundation (21703269) of China.

References

[1] M. D. Yilmaz and J. Huskens, *Soft Matter*, 2012, 8, 11768-11780.

[2] Y. Sun, J. K. Ma, D. M. Tian and H. B. Li, *Chem. Commun.*, 2016, 52, 4602-4612.

[3] A. Harada, J. Li and M. Kamachi, *Nature*, 1992, 356, 325-327.

[4] E. Junquera, G. Tardajos and E. Aicart, *Langmuir*, 1993, 9, 1213-1219.

[5] X. Zhang and C. Wang, *Chem. Soc. Rev.*, 2011, 40, 94-101.

[6] K. Ichimura, S. K. Oh and M. Nakagawa, *Science*, 2000, 288, 1624-1626.

[7] P. B. Wan, Y. G. Jiang, Y. P. Wang, Z. Q. Wang and X. Zhang, *Chem. Commun.*, 2008, 5710-5712.

[8] Y. P. Wang, N. Ma, Z. Q. Wang and X. Zhang, *Angew. Chem.*, *Int. Ed.*, 2007, 46, 2823-2826.

[9] Y. Bai, X. D. Fan, W. Tian, T. T. Liu, H. Yao, Z. Yang, H. T. Zhang and W. B. Zhang, *Polym. Chem.*, 2015, 6, 732-737.

[10] Y. Liu, Y. W. Yang and Y. Chen, *Chem. Commun.*, 2005, 4208-4210.

[11] H. C. Zhang, F. F. Xin, W. An, A. Y. Hao, X. Wang, X. H. Zhao, Z. N. Liu and L. Z. Sun, *Colloids Surf.*, A, 2010, 363, 78-85.

[12] T. Sun, J. Shen, H. Yan, J. C. Hao and A. Y. Hao, *Colloids Surf.*, A, 2012, 414, 41-49.

[13] J. Hernandez-Pascacio, C. Garza, X. Banquy, N. Diaz-Vergara, A. Amigo, S. Ramos, R. Castillo, M. Costas and A. Pineiro, *J. Phys. Chem. B*, 2007, 111, 12625-12630.

[14] A. Pineiro, X. Banquy, S. Perez-Casas, E. Tovar, A. Garcia, A. Villa, A. Amigo, A. E. Mark and M. Costas, *J. Phys. Chem. B*, 2007, 111, 4383-4392.

[15] L. Jiang, Y. Yan, M. Drechsler and J. Huang, *Chem. Commun.*, 2012, 48, 7347-7349.

[16] C. C. Zhou, X. H. Cheng, Y. Yan, J. D. Wang and J. B. Huang, *Langmuir*, 2014, 30, 3381-3386.

[17] C. C. Zhou, X. H. Cheng, Q. Zhao, Y. Yan, J. D. Wang and J. B. Huang, *Langmuir*, 2013, 29, 13175-13182.

[18] N. Szaniszlo, E. Fenyvesi and J. Balla, *J. Inclusion Phenom. Macrocyclic Chem.*, 2005, 53, 241-248.

[19] C. C. Zhou, X. H. Cheng, Q. Zhao, Y. Yan, J. D. Wang and J. B. Huang, *Sci. Rep.*, 2014, 4, 6.

[20] C. C. Zhou, J. B. Huang and Y. Yan, *Soft Matter*, 2016, 12, 1579-1585.

[21] M. Inoue, K. Hashizaki, H. Taguchi and Y. Saito, *J. Oleo Sci.*, 2009, 58, 85-90.

[22] B. G. Mathapa and V. N. Paunov, *Phys. Chem. Chem. Phys.*, 2013, 15, 17903-17914.

[23] X. P. Liu, J. Feng, L. Zhang, S. Zhao and J. Y. Yu, *Acta Phys.-Chim. Sin.*, 2010, 26, 1277-1283.

[24] Q. He, Y. Zhang, G. Lu, R. Miller, H. Mohwald and J. B. Li, *Adv. Colloid Interface Sci.*, 2008, 140, 67-76.

[25] A. Jyoti, R. M. Prokop, J. Li, D. Vollhardt, D. Y. Kwok, R. Miller, H. Mohwald and A. W. Neumann, *Colloids Surf.*, A, 1996, 116, 173-180.

[26] R. Foudazi, S. Qavi, I. Masalova and A. Y. Malkin, *Adv. Colloid Interface Sci.*, 2015, 220, 78-91.

[27] Y. P. Huang, L. Zhang, L. Zhang, L. Luo, S. Zhao and J. Y. Yu, *J. Phys. Chem.* B, 2007, 111, 5640-5647.

[28] L. Zhang, X. C. Wang, Q. T. Gong, L. Zhang, L. Luo, S. Zhao and J. Y. Yu, *J. Colloid Interface Sci.*, 2008, 327, 451-458.

[29] C. Wang, H. B. Fang, Q. T. Gong, Z. C. Xu, Z. Y. Liu, L. Zhang, L. Zhang and S. Zhao, *Energy Fuels*, 2016, 30, 6355-6364.

[30] C. Cao, J. M. Lei, L. Zhang and F. P. Du, *Langmuir*, 2014, 30, 13744-13753.

[31] C. Cao, L. Zhang, X. -X. Zhang and F. -P. Du, *Food Hydrocolloids*, 2013, 30, 456-462.

[32] A. G. Bykov, S. Y. Lin, G. Loglio, R. Miller and B. A. Noskov, *J. Phys. Chem. C*, 2009, 113, 5664-5671.

[33] J. M. RodriguezPatino, C. C. Sanchez, M. R. RodriguezNino and M. C. Fernandez, *Langmuir*, 2001, 17, 4003-4013.

[34] A. G. Bykov, L. Liggieri, B. A. Noskov, P. Pandolfini, F. Ravera and G. Loglio, *Adv. Colloid Interface Sci.*, 2015, 222, 110-118.

[35] A. G. Bykov, B. A. Noskov, G. Loglio, V. V. Lyadinskaya and R. Miller, *Soft Matter*, 2014, 10, 6499-6505.

[36] B. A. Noskov, *Curr. Opin. Colloid Interface Sci.*, 2010, 15, 229-236.

模板聚合法制备 AM/AMPS 共聚物及其性能研究

曹绪龙* 祝仰文 窦立霞

(胜利油田勘探开发研究院,山东 东营 257015)

摘要:分别采用普通溶液聚合法、模板聚合法制备 AM/AMPS 二元共聚物 P(AM-AMPS),采用 K-T 法计算了两种聚合方法中的单体竞聚率。结果表明,采用模板聚合法 AMPS 竞聚率成倍提高。考察了反应条件对模板聚合法制备 P(AM-AMPS)共聚物特性黏数、黏度的影响,确定了最佳合成条件为:模板单体单元与 AMPS 比例为 0.7∶1,AMPS 摩尔百分比为 5%,单体质量分数为 25%~30%,溶液 pH 为 7~8。模板聚合法制备共聚物的耐温抗盐性较普通聚合法明显提高。

关键词:AM/AMPS 模板聚合;聚二甲基二烯丙基氯化铵;竞聚率

中图分类号:TQ317.4;TE39 文献标志码:A

文章编号:0253-4320(2020)S-0102-50

Synthesis of AM-AMPS copolymer by template polymerization process and its solution properties

Cao Xulong* Zhu Yangwen Dou Lixia

(The Exploration and Development Research Institute, Sinopec Shengli Oilfield Branch Company, Dongying 257015, China)

Abstract: Copolymerization of acrylamide and 2-acrylamido-2-methylpropanesulfonic (AMPS) are respectively accomplished by ordinary solution polymerization and template polymerization methods. Reactivity ratios of monomers by two methods are calculated by means of K-T method. The results show that the reactivity ratio of AMPS by template polymerization method is twice that by ordinary solution polymerization. The effects of reaction conditions on the intrinsic viscosity, solubility and apparent viscosity of P(AM-

AMPS) are investigated. The optimum synthesis conditions are determined as follows: the ratio of template unit to AMPS is 0.7 : 1, AMPS content is 15 wt%, the total addition amount of the monomers is in the range of 25%～30%, and the pH of system is 8. The copolymer prepared by template polymerization method exhibits better temperature-resistant and salt-resistant compared to the copolymer prepared by ordinary solution polymerization method.

Keywords: template polymerization of AM/AMPS; polydimethyl diallyl ammonium chloride; reactivity ratio

随着油田的不断开发，开采油藏条件不断恶化，油田开采正逐步进入高温高盐的二类、三类油藏区块。目前常采用的驱油剂聚丙烯酰胺具有耐温抗盐性差的缺点，难以满足该类油藏的应用。在聚合物结构中引入耐温抗盐单体是目前提高聚合物性能常用的方法之一，其中以2-丙烯酰胺-2-甲基丙磺酸（AMPS）单体应用最为广泛[1]，该类单体的磺酸基团上含有S—O键，使S从—OH吸引电子的能力变弱，使其对盐离子的吸引能力也变弱，因而具有良好的抗盐能力；引入AMPS单体还可以提高聚合物对钙镁离子的容忍能力，提高黏度保留率[2-3]。目前研究人员对丙烯酰胺（AM）/AMPS抗盐聚合物的合成与溶液性能做了大量研究，但有关AM/AMPS单体单元微结构对溶液性能影响的研究鲜有报道。

目前合成具有预定结构和序列的嵌段聚合物的常用方法包括活性聚合法、正离子聚合转化法、力化学方法、缩聚法和特殊引发剂法等[4-5]，但都因其合成条件苛刻、单体适应性不佳而有局限。采用模板聚合法可有效控制共聚物组成及序列分布。模板聚合法的原理是将能与单体或生长链通过氢键、静电键合、电子转移、疏水键合、范德华力等相互作用的高分子（模板）预先放入聚合体系，让单体在模板上进行预组装，当聚合体系大分子链自由基遇到模板，则引发模板上预组装的单体，形成一段单体嵌段结构单元，如此反复则可制备嵌段型共聚物[6-7]。张玉玺等[8]采用模板聚合法制备了P（AM—AA）多嵌段共聚物，经过动力学研究表明，采用模板聚合法丙烯酸（AA）单元序列长度明显增加。

本文通过动力学研究获得聚合方法对单体竞聚率的影响，考察了反应条件对模板聚合法制备P（AM—AMPS）共聚物特性黏数、黏度的影响，确定最佳合成条件，并对两种聚合方法制备抗盐聚合物性能进行对比。结果表明，模板聚合法制备抗盐聚合物的耐温抗盐性明显提高。

1 实验部分

1.1 试剂及仪器

聚乙烯醇硫酸钾（PVSK，酯化度98.4%），分析纯，日本和光纯药工业株式会社；聚二甲基二烯丙基氯化铵（PDMDAAC，分子量20 000），工业级，宜兴市清泰净化剂有限公司;AM、AA、AMPS,均为工业级,江西昌九农科;无水乙醇、丙酮、甲苯胺蓝(TBO)、过硫酸铵、亚硫酸氢钠、NaOH、冰醋酸,均为分析纯,成都市科龙化工试剂厂 Brookfield LV DV-Ⅲ黏度计(美国 Brookfield 公司);DZF-6050MBE 真空干燥箱。

1.2 实验方法

1.2.1 竞聚率测试

(1) 竞聚率测试聚合物的制备

普通聚合法:固定单体总浓度为 1.1 mol/L,改变 AM/AMPS 比例,在 30℃条件下加入过硫酸铵－亚硫酸氢钠组成的复合引发体系引发聚合,当聚合体系在乙醇中形成沉淀物时加入 2.5%苯酚阻聚剂终止反应,在无水乙醇中进行沉淀,控制单体转化率＜10%。

模板聚合法:固定单体总浓度为 1.1 mol/L,固定 PDMDAAC 单体单元/AMPS＝1∶1,改变 AM/AMPS 比例,在 30℃条件下加入过硫酸铵－亚硫酸氢钠组成的复合引发体系引发聚合,当聚合体系在乙醇中形成沉淀物时加入 2.5%苯酚阻聚剂终止反应,并在无水乙醇中进行沉淀,控制单体转化率＜10%,将经乙醇沉淀后的聚合物用去离子水溶解成浓度为 5 000 mg/L 的聚合物溶液,将聚合物溶液装入透析袋(截留分子量为 20 000)并置于纯水中进行透析 48 h 以上,此过程中小分子量模板扩散游离出透析袋,完成透析后将透析袋内聚合物溶液经乙醇沉淀､干燥,获得样品。

(2) 聚合物中 AMPS 含量测试

采用胶体滴定法测试 P(AM—AMPS)中 AMPS 含量[9]。取 10 mL 蒸馏水于 250 mL 锥形瓶中,再加入 10 mL 浓度 6×10^{-4} mol/L 的 PDMDAAC(指 DMDAAC 单元的摩尔浓度,后同)溶液,滴加 3 滴 0.1%的甲苯胺蓝指示剂,然后用 1×10^{-3} mol/L 的 PVSK(指单体单元的摩尔浓度,后同)标准溶液滴定,至紫红色终点,所用 PVSK 体积记为 V_0。

用去离子水将共聚物配制成浓度为 100 mg/L 的溶液,取配制好的溶液 10 mL 于 250 mL 锥形瓶中,再加入 10 mL 浓度 6×10^{-4} mol/L 的 PDMDAAC 溶液,搅拌 10 min,滴加 3 滴 0.1%的甲苯胺蓝指示剂,滴定方法同上,所用 PVSK 体积记为 V_1,得到胶体滴定值(V_0-V_1),根据式(1)计算出共聚物中 AMPS 的含量,然后求平均值。

$$w=[(V_0-V_1)C_0\times229]/\mathrm{m}\times100\% \tag{1}$$

式中,w 为共聚物中 AMPS 的含量,%;V_0 为直接滴定 PDMDAAC 消耗的 PVSK 体积,mL;V_1 为反滴定共聚物消耗的 PVSK 体积,mL;C_0 为 PVSK 中单体单元的浓度,mol/L;m 为滴定共聚物试样的质量,g;229 为 AMPS 摩尔质量,g/mol。

(3) 竞聚率常数计算

AM 和 AMPS 的投入量分别为$[M_1]$、$[M_2]$,共聚物中 AM 单元和 AMPS 单元的含量分别为 d$[M_1]$、d$[M_2]$,依据 Kelem-Tudos[10](K—T)法用式(2)算出聚合物中两单体的竞聚率。

$$\eta=(r_1+r_2/\alpha)\xi-r_2/\alpha \tag{2}$$

式中，$\eta=G/(\alpha+F)$；$\xi=F/(\alpha+F)$；$\alpha=\sqrt{F_{min}\times F_{max}}$；$G=[X(Y-1)]/Y$；$F=X^2/Y$；原料中单体摩尔比 $X=[M_1]/[M_2]$；低转化率下共聚物组成 $Y=d[M_1]/d[M_2]$；r_1、r_2 为两种单体的竞聚率。

将原料中单体物质的量和共聚物中聚合单体的组成分别代入式(2)求得参数。以 ξ 为横坐标，η 为纵坐标作图得一直线，由直线的斜率和截距可分别求得 r_1 和 r_2。

1.2.2 不同聚合方法制备的聚合物抗盐耐温性能研究

(1) 聚合物的制备

普通抗盐聚丙烯酰胺的制备：反应容器中按照一定单体总浓度及比例依次加入水、添加剂、AM、AMPS，在 30℃条件下加入过硫酸铵－亚硫酸氢钠组成的复合引发体系引发聚合，反应时间 5～6 h，聚合反应结束后按照水解度 15%加入水解剂 NaOH，在 95℃条件水解 2 h，然后干燥粉碎获得聚合物干粉。

模板聚合法制备 P(AM—AMPS)抗盐聚丙烯酰胺：反应容器中按照一定单体总浓度及比例依次加入水、添加剂、AM、AMPS、PDMDAAC，在 30℃条件下加入过硫酸铵-亚硫酸氢钠组成的复合引发体系引发聚合，反应时间 5～6 h，聚合反应结束后按照水解度 15%加入水解剂 NaOH，在 95℃条件水解 2 h，然后干燥粉碎获得聚合物干粉。将聚合物干粉用去离子水溶解成浓度为 5 000 mg/L 的聚合物溶液，将聚合物溶液装入透析袋(截留分子量为 20 000)并置于纯水中进行透析 48 h 以上，此过程中小分子量模板扩散游离出透析袋，完成透析后将透析袋内聚合物溶液经乙醇沉淀、干燥，获得不含模板的 P(AM—AMPS)抗盐聚丙烯酰胺。

(2) 聚合物抗盐耐温性能测试

依照 GB/T 12005.10-92 测试并计算聚合物特性黏数。将聚合物在总矿化度为 20 000 mg/L(二价离子 500 mg/L)盐水条件溶解成 5 000 mg/L 母液，再采用上述盐水将母液稀释至 2 000 mg/L，采用黏度计在 7.34 s^{-1}、80℃下测试聚合物黏度。

2 结果与讨论

2.1 AM/AMPS 共聚物竞聚率常数测定

2.1.1 共聚物组分测定

依照 1.2.1 竞聚率测试聚合物的制备方法，采用普通聚合法和模板聚合法制备了系列不同 AM/AMPS 比例聚合物，依据 1.2.1 方法测定共聚物中 AMPS 含量，获得普通聚合法和模板聚合法制备的共聚物单元组成，如表 1 所示。

表1　不同聚合方法单体投料与共聚物组成

聚合方法	$X=[M_1]/[M_2]$	转化率/%	$Y=d[M_1]/d[M_2]$
普通聚合法	9∶1	2.44	9.546
	8∶2	3.61	4.213
	7∶3	5.26	2.635
	6∶4	3.68	1.758
	4∶6	6.64	0.952
模板聚合法	9∶1	5.42	8.815
	8∶2	5.68	3.234
	7∶3	4.23	1.659
	6∶4	5.63	0.943
	4∶6	4.21	0.365

2.1.2 竞聚率的计算

依据1.2.1所述方法计算单体竞聚率，见图1。结果表明，采用普通聚合法合成的AM/AMPS共聚物，$r_{AM}=0.98$，$r_{AMPS}=0.546$，两种单体都倾向于共聚，不利于合成长嵌段单元分布聚合物；而采用模板聚合洪合成的AM/AMPS共聚物，$r_{AM}=1.121$，$r_{AMPS}=2.647$，极大地提高了AMPS的竞聚率，使得AMPS单体倾向于自聚形成多AMPS嵌段结构聚合物。

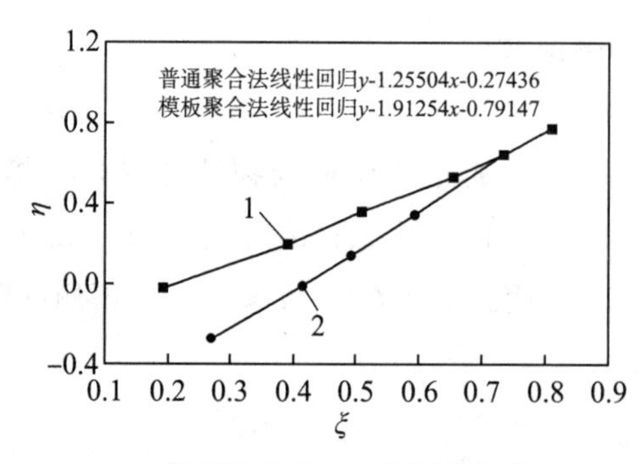

1. 普通聚合法；2. 模板聚合法

图1　K-T法计算不同聚合方法下AM和AMPS竞聚率回归图

2.2 模板聚合法制备AM/AMPS共聚物的影响因素

2.2.1 模板/AMPS比例对聚合物特性黏数、黏度的影响

根据模板聚合的机理，模板含量直接影响AMPS嵌段的结构。当模板含量过低，部分AMPS未在模板上进行“分子组装”，形成AMPS短嵌段；若模板含量过高，AMPS排列于模板上的长度将减少，同样导致AMPS嵌段长度变短，故模板与AMPS存在最佳比例从而达到嵌段长度最大。

固定AMPS摩尔百分比为5%、单体质量分数为25%、溶液pH为7.0、引发剂质量分数为 0.025%，研究模板/AMPS 比例对聚合物特性黏数及黏度的影响，结果如图 2 所示。可以看出，模板具有一定的链转移剂作用，体系中模板含量增加，聚合物特性黏数有所降低，但聚合物溶液黏度先增加后降低，存在最佳模板/AMPS 比例。分析认为，AMPS 嵌段长度增加利于提高聚合物耐温抗盐性能，但模板含量增加不利于聚合物分子量的提高，综合模板/AMPS 比例对聚合物特性黏数、黏度的影响，确定模板/AMPS 最佳比例为 0.7∶1。

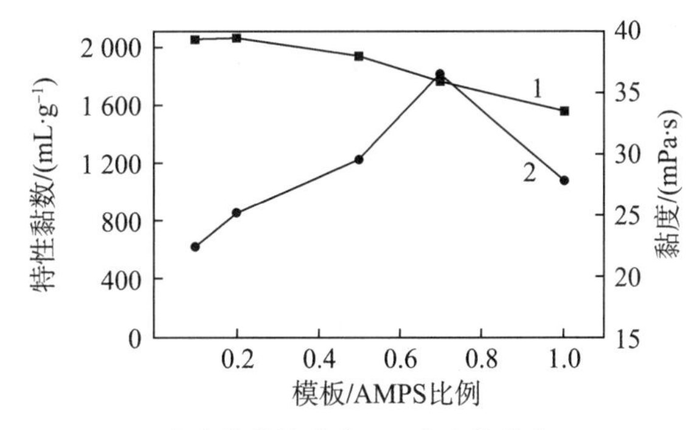

1. 聚合物特性黏数；2. 聚合物黏度

图 2　模板/AMPS 比例与聚合物特性黏数、黏度的关系

2.2.2 AMPS 含量对聚合物特性黏数、黏度的影响

固定模板/AMPS 比例为 0.7∶1、单体质量分数为 25%、溶液 pH 为 7.0、引发剂质量分数为 0.02%，研究 AMPS 含量对聚合物特性黏数与黏度的影响，结果如图 3 所示。可以看出，当 AMPS 摩尔百分比低于 4%时，聚合物特性黏数无明显变化，聚合物黏度明显提高；当 AMPS 摩尔分数大于 5.5%时，聚合物特性黏数开始有所降低，聚合物溶液性能也随之下降，这是因为 AMPS 含有较大侧基，含量过高不利于提高聚合物分子量[11]。综合 AMPS 含量对聚合物特性黏数、溶液黏度的影响，确定 AMPS 最佳含量为摩尔百分比 5%。

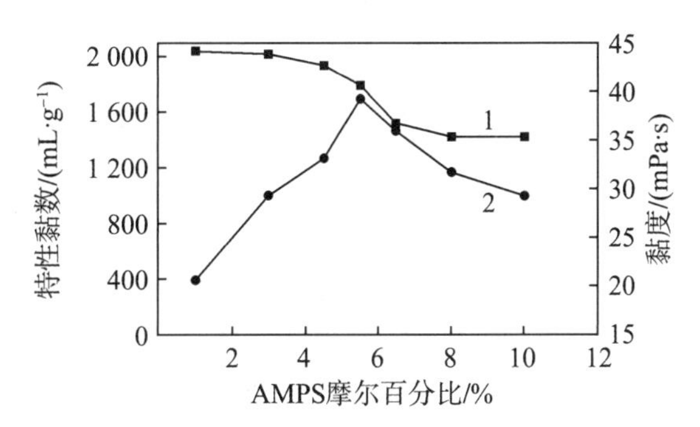

1. 聚合物特性黏数；2. 聚合物黏度

图 3　AMPS 含量与聚合物特性黏数、黏度的关系

2.2.3 单体浓度对聚合物特性黏数、黏度的影响

固定 AMPS 摩尔百分比 5%、模板/AMPS 比例为 0.7∶1、溶液 pH 为 7.0、引发剂质量分数 0.02%，研究单体浓度对聚合物特性黏数、黏度的影响，结果如图 4 所示。可

以看出,随着单体浓度增加,聚合物特性黏数及黏度均先增加后降低,当单体质量分数为25%时聚合物特性黏数达到最大值。这是因为单体浓度增加一方面增加了单体与活性链的碰撞,有利于分子量增加,另一方面浓度过高,聚合体系黏度过大,聚合热不易导出,容易导致聚合体系温度过高,链转移加剧,分子量降低。两个因素使得单体浓度存在最佳值,由于聚合物黏度与聚合物分子量呈正比,通过实验确定单体质量分数为25%~30%最佳。

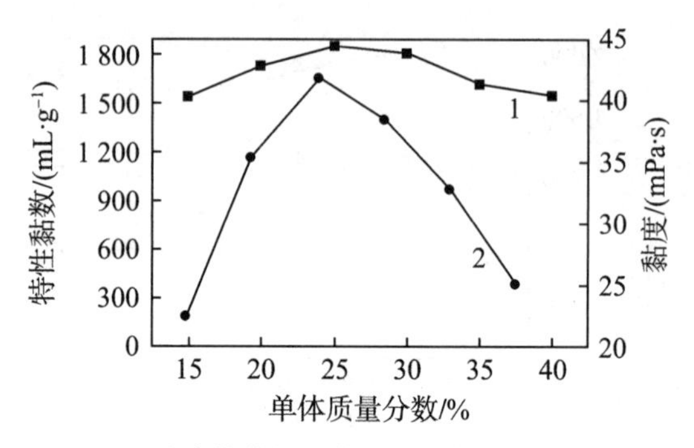

1. 聚合物特性黏数;2. 聚合物黏度

图4 单体浓度与聚合物特性黏数、黏度关系

2.2.4 溶液pH对聚合物特性黏数、黏度的影响

体系的pH对单体活性以及引发剂活性均有影响,适宜的pH利于自由基的缓慢释放,保证聚合平稳进行,有利于提高聚合物分子量。固定AMPS摩尔百分比为5%、模板/AMPS比例为0.7∶1、单体质量分数为26%、引发剂质量分数为0.02%,研究了溶液pH对聚合物特性黏数、黏度的影响,结果如图5所示。可以看出,随着pH增加,聚合物特性黏数与黏度值均先增加后降低。pH主要通过影响聚合物分子量来调节溶液黏度,确定pH为7~8时达到最佳。

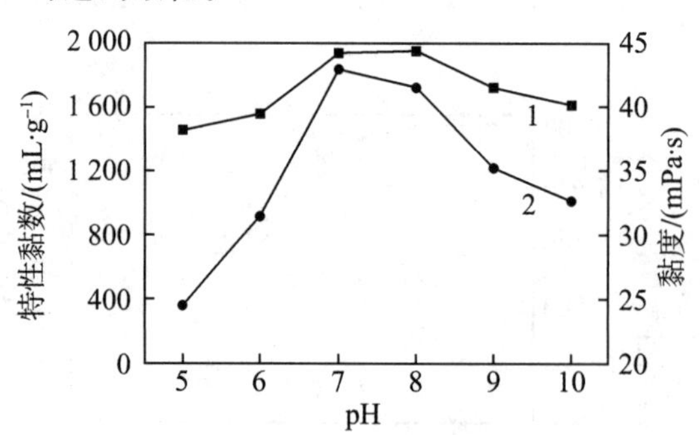

1. 聚合物特性黏数;2. 聚合物黏度

图5 pH与聚合物特性黏数、黏度的关系

2.3 不同聚合方法制备AM/AMPS共聚物的耐温抗盐性能对比

分别采用普通聚合法、模板聚合法制备抗盐聚合物KPY、MPY,其中AM和AMPS摩尔比为95∶5,模板聚合法中模板单体单元/AMPS比例为0.7∶1,经过系列合成优化,获得特性黏数达2 000 mL/g的抗盐聚合物KPY/MPY。将两种不同制备方法获得

的抗盐聚合物进行耐温抗盐性能对比，抗盐性实验中聚合物测试温度 85℃，聚合物浓度 2 000 mg/L；耐温性实验中聚合物溶剂总矿化度为 20 000 mg/L（二价离子 500 mg/L），聚合物浓度 2 000 mg/L，结果见图 6。

实验结果表明，模板聚合法制备的嵌段型抗盐聚合物在耐温、抗盐性能上均明显优于普通聚合法制备的聚合物。通过提高抗盐单体 AMPS 嵌段长度可明显提高聚合物的耐温抗盐性能，这是因为 AMPS 单元在大分子链中呈多嵌段分布，聚集的 AMPS 单元更利于抵抗周围的二价离子，进而提高了溶液的耐温抗盐性能。

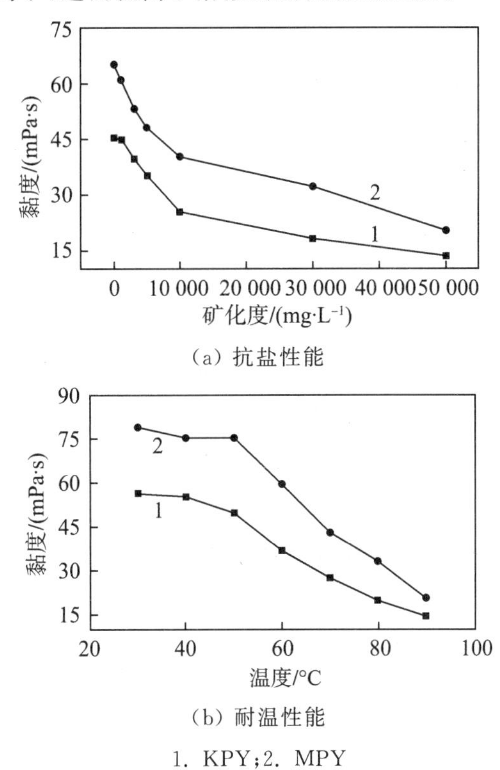

（a）抗盐性能

（b）耐温性能

1. KPY；2. MPY

图 6　不同聚合方法 P(AM-AMPS)耐温抗盐性能对比

3　结论

（1）采用普通聚合法 AM、AMPS 单体竞聚率分别为 $r_{AM}=0.98$，$r_{AMPS}=0.546$；采用模板聚合法 AM、AMPS 单体竞聚率分别为 $r_{AM}=1.121$，$r_{AMPS}=2.647$。采用模板聚合法可明显提高 AMPS 竞聚率，可制备多嵌段 AMPS 分布的抗盐聚合物。

（2）考察反应条件对模板聚合法制备 AM/AMPS 抗盐聚合物特性黏数及黏度的影响，确定了最优合成条件为模板单体单元/AMPS 比例为 0.7∶1，AMPS 摩尔百分比为 5%，单体质量分数为 25%～30%，溶液 pH 为 7～8。

（3）与普通聚合法相比，模板聚合法制备的抗盐聚合物 AMPS 呈嵌段分布，有利于提高聚合物的耐温抗盐性能。

参考文献

[1] 陈锡荣,黄凤兴.驱油用耐温抗盐水溶性聚合物的研究进展[J].石油化工,2009,38(10):1132-1137.

[2] 李奇,蒲万芬,王亚波,等.AM/AMPS共聚物的合成与性质研究[J].应用化工,2012,41(2):300-303,313.

[3] 宋华,翟永刚,丁伟,等.AM/AMPS共聚物的合成与耐温抗盐性能研究[J].化学工业与工程技术,2013,34(5):49-52.

[4] 李子龙,王春浩,杜福胜,等.序列可控聚合[J].高分子通报,2014,(8):13-22.

[5] 罗婕,义建军,胡杰,等.嵌段共聚物合成方法研究进展[J].化工新型材料,2010,38(4):10-12,19.

[6] 李执芬,王身国,习复.模板聚合——一类新的聚合方法[J].化学通报,1981,44(5):1-5.

[7] 沈家瑞.模板聚合的发展和应用[J].化工进展,1985,4(2):20-22.

[8] 张玉玺,吴飞鹏,李妙贞,等.聚烯丙基氯化铵模板存在下丙烯酰胺-丙烯酸共聚合反应动力学研究[J].高分子学报,2004,35(6):889-892.

[9] 刘坤,宋新旺,祝仰文,等.胶体滴定法测定P(AM-co-AMPSNa)中AMPSNa含量[J].应用化工,2015,44(11):2115-2119,2127.

[10] 王孟,刘芝芳.丙烯酰胺-丙烯酸钠共聚合竞聚率的测定[J].南华大学学报(自然科学版),2006,20(1):93-95.

[11] 吕华华.磺酸盐型聚丙烯酰胺研究[D]:青岛:中国石油大学,2008.

大分子交联剂制备颗粒驱油剂及其性能*

曹绪龙

(中国石油化工股份有限公司胜利油田分公司,山东 东营 257015)

摘要:为了解决我国油田开采中出现的非均质性突出、高温高盐等问题,本文突破小分子交联剂制备颗粒驱油剂的技术思路,采用合成的多种新型大分子交联剂制备了系列颗粒型驱油剂,并且对其网络结构、力学性能、耐老化能力以及运移机理进行了研究。研究结果表明,大分子交联剂能够有效提高凝胶的均匀性,赋予了凝胶的优异的力学性能。3种大分子交联剂凝胶的断裂伸长率都在2 000%以上,断裂强度达到0.3 MPa以上,并且90%压缩应变下不破坏。其中MCH0.5-30性能最优,断裂伸长率能够达到2 800%,断裂强度能够达到0.6 MPa。此外其在85℃、30 g/L矿化度的盐水中,老化性能优于小分子交联剂制备的颗粒型驱油剂。大分子交联剂为制备高性能凝胶颗粒驱油剂提供了新的思路,有望在高温高盐的苛刻油藏中得到应用。

关键词:三次采油;水凝胶;黏弹性颗粒;驱油剂

中图分类号:TE357.46:TE39 文献标识码:A

0 前言

目前我国原油对外依存度近70%,严重影响国家能源安全,提高油田采收率具有重要的战略意义。我国东部老油田大部分处于开采中后期阶段,面临着严重的储层非均质性、剩余油高度分散、高温高盐等问题[1]。常规的驱油技术对这类油藏提高采收率的幅度有限,急需开发新型驱油剂和采油新技术[2-5]。对此,胜利油田研发了具有部分支化部分交联结构的黏弹性颗粒驱油剂B-PPG,构筑了黏弹性颗粒驱油剂、聚合物和表面活性剂复配的非均相复合驱体系,在聚合物驱后油藏和高温高盐Ⅲ类油藏提高石油采油率方面取得了显著的降水增油效果[6-8]。为了进一步扩大非均相复合驱技术的应用规模,拓宽应用领域,使非均相复合驱体系适用于高温高盐、强非均质性的苛刻油藏条件,对PPG的性能提出了更高的要求[9-10]。

采用小分子交联剂N,N′-亚甲基双丙烯酰胺(MBA)制备的交联聚丙烯酰胺颗粒驱油剂,虽然能够封堵高渗透率孔道,降低油藏的非均质性,但由于交联点分布不均匀,缺乏有效的能量耗散机制，存在机械性能差、变形能力弱、通过孔喉时容易发生剪切破碎以及不耐高温高盐等问题[11]。本文突破了小分子交联剂制备凝胶颗粒驱油剂的技术思路，采用合成的多种新型大分子交联剂制备了一系列高性能凝胶颗粒驱油剂 B-PPG，通过动态光散射、力学测试、老化试验以及荧光可视渗流渗流实验研究了 B-PPG 的网络结构、机械性能、耐温抗盐能力和在孔隙中的运移规律。

1 实验部分

1.1 材料与仪器

丙烯酰胺(AM)，成都赛乐思科技有限公司厂；过硫酸钾(KPS)，四甲基乙二胺(TMEDA)，小分子联剂 N,N′-亚甲基双丙烯酰胺(MBA)，成都科龙化工试剂厂；氯化钠(NaCl)，氯化钙($CaCl_2$)，六水氯化镁($MgCl_2 \cdot 6H_2O$)，异硫氰酸荧光素，天津市博迪化工有限公司；去离子水，实验室自制；不同结构的种大分子交联剂 TA、MC、MCH，其中 TA 是具有 2～6 个不饱和双键的树枝型交联剂；MC 是含有多个不饱和双键的线性大分子交联剂；MCH 是含有多个不饱和双键的超支化大分子交联剂。

BI-200SM 型动态光散射仪，美国 Brookhaven 公司；Insrtron 5567 型万能材料拉伸机，美国 Instron 公司。

1.2 水凝胶的制备

首先称取一定质量的 AM、交联剂、TMEDA 加入烧杯后用去离子水完全溶解，然后将混合溶液转移到 250 mL 单口烧瓶中，抽气除氧 10 min 后与溶有引发剂 KPS 的溶液混合均匀。最后倒入模具，在 25℃的水浴锅中反应 24 h，即可得到水凝胶。选择不同的交联剂，通过调节变量合成出的水凝胶分别命名为 TAx-y、MCx-y、MCHx-y。x 表示每 100 g 水中交联剂溶液(质量分数为 5%)的质量(g)，y 表示每 100 g 水中单体质量(g)。作为对比，制备小分子交联剂 MBA 摩尔浓度为 5×10^{-3} mol%，单体质量为 30 g 的水凝胶，命名为 SC 凝胶。

1.3 B-PPG 的制备

将模具中的凝胶拿出剪碎后，在 75℃下的烘箱中烘干 4 h 除去多余水份。经过粉碎、筛分后得到一定粒径的 B-PPG 驱油剂颗粒。根据不同交联剂分别将 B-PPG 命名为 $B\text{-}PPG_{SC}$、$B\text{-}PPG_{TA}$、$B\text{-}PPG_{MC}$、$B\text{-}PPG_{MCH}$。

1.4 测试与表征

1.4.1 动态光散射测试

将前驱液直接注入光散射瓶中原位聚合形成水凝胶。使用 BI-200SM 型的动态光散射仪，散射角度设置为 90°，每隔 30 s 旋转样品台得到该点的时间平均光散射强度。重复操作 100 次可以得到样品的总均散射强度。通过仪器自带的软件对归一化时间平均散射光强相关函数($\langle g_{\tau}^{(2)}(\tau)-1\rangle$)进行反拉普拉斯变换，得到水凝胶的归一化特征线宽分布函数 $G(\Gamma)$，以此表征凝胶的网络结构。

1.4.2 力学性能测试

采用 Insrtron 5567 型万能材料拉伸机对凝胶的力学性能进行表征：

(1) 拉伸性能。将样品从模具中取出，用裁刀将样品裁剪成哑铃状（标准为 JISK6251-7）。用镊子将样品夹持在夹具上，控制拉伸速率 100 mm/min 进行拉伸测试，得到样品的应力应变曲线，每个样品重复 5 次。

(2) 压缩性能。将样品将从柱状模具中取出，用圆形裁刀裁成直径约 13 mm、高度约 20 mm 的柱状样品。控制压缩速率 20%/min 进行压缩测试，得样品的压缩曲线，每个样品重复 5 次。

(3) 形变恢复能力。按照上述要求制备好一定规格的样品，在压缩模式下，以 80% 作为最大应变，控制压缩速率 5 mm/min 将样品压缩至最大应变，然后以相同的速率恢复至初始形态。重复该过程 5 次。

(4) 抗缺陷性能测试。按上述要求制备好一定规格的样品后，用剪刀预制一个约为样品宽度 75% 的缺口，控制拉伸速率 100 mm/min，重复该过程 5 次。

1.4.3 耐老化性能测试

将 0.4 g 的 B-PPG 驱油颗粒加入 80 mL 的矿化度为 30 g/L 的盐水中（配方为 1 L 水中加入 27.31 g NaCl、1.11 g $CaCl_2$ 和 3.83 g $MgCl_2 \cdot 6H_2O$），等待 12 h 使其充分溶胀后倒入圆底烧瓶抽气除氧；将抽完气的悬浮液迅速倒入玻璃瓶中，盖上橡胶塞并用胶带缠紧；将玻璃瓶放入老化箱在 85℃ 下进行老化测试，每过一段时间取出测定悬浮液的模量和黏度。

1.4.4 荧光可视微观渗流实验

将 B-PPG 驱油颗粒和异硫氰酸荧光素溶液混合均匀后放置一段时间，待 B-PPG 与荧光分子充分反应后，通过多次洗涤和抽滤后得到染色的 B-PPG。将染色的 B-PPG 和水混合后加入容器中，打开泵将悬浮液以一定的流速灌注入微观渗流模型，然后通过共聚焦显微镜监测 B-PPG 的运移情况。

2 结果与讨论

2.1 凝胶网络的均匀性分析

凝胶网络的均匀性对凝胶的力学强度有着很大的影响，为了验证大分子交联剂在提高凝胶网络均匀性上的作用，运用动态光散射测量了样品在不同位置的时间平均散射强度，结果如图 1 所示。图中虚线与实线之间的差表示时间平均散射强度的波动程度，可以反映凝胶网络结构的不均匀性。可以看到，小分子交联剂制备 SC 凝胶的时间平均散射强度的波动程度要高于大分子交联剂制备的凝胶，这说明大分子交联剂能够有效提高凝胶网络的均匀性。并且随着交联剂含量的增加，凝胶网络的均匀性也逐渐增加。TA 凝胶网络均匀性相比于 SC 更高，可能由于 MBA 和 TA 交联剂官能度的差异。TA 交联剂分子上的双键引发单体，形成链自由基。当链自由基与其他交联剂分子的链自由基双基终止时会形成交联结构，单基终止则会形成支化链结构，其中支化链形成的缠结能够有效降低凝胶网络的不均匀性。而MBA交联剂分子官能度仅为2，难以形成支化链。并且交联剂用量越大，支化链形成的物理缠结越多，网络更加均匀。

2.2 凝胶的松弛行为

为了验证凝胶中存在支化链形成的物理缠结结构，采用动态光散射表征凝胶的松弛行为。图2(a)为SC凝胶的归一化时间平均散射光强相关函数，通过拟合可以得到如图2(b)所示的凝胶松弛时间谱。图3(a)是不同交联剂用量的MC凝胶的归一化时间平均散射光强相关函数函数，通过拟合可以得到如图3(b)所示的凝胶松弛时间谱。从图3(b)可以看到，MC凝胶的松弛时间谱存在两个峰，分别表示快、慢两种松弛模式。其中，快松弛模式的峰强而尖，松弛时间在10^{-5} s数量级，这种松弛行为主要是由于凝胶网络中分子链浓度的波动[12]。由图2可知，SC凝胶同样存在这一松弛行为，并且松弛时间相差不大。慢松弛模式的峰弱而宽，松弛时间在10^{-3} s数量级，这一松弛过程则主要与形成缠结的支化链的扩散有关。凝胶网络中，分子链的扩散会被诸如缠结以及链间的摩擦所阻碍。在大分子交联剂MC凝胶中支化链高度缠结，并且其中的一端形成了化学交联的结构，使得支化链的扩散受到了极大的限制，因此相比于快松弛模式，慢松弛模式的松弛时间更长。此外，如图2所示，SC凝胶不存在这一松弛行为，这从侧面验证了大分子交联剂凝胶中存在支化链形成的物理缠结结构。

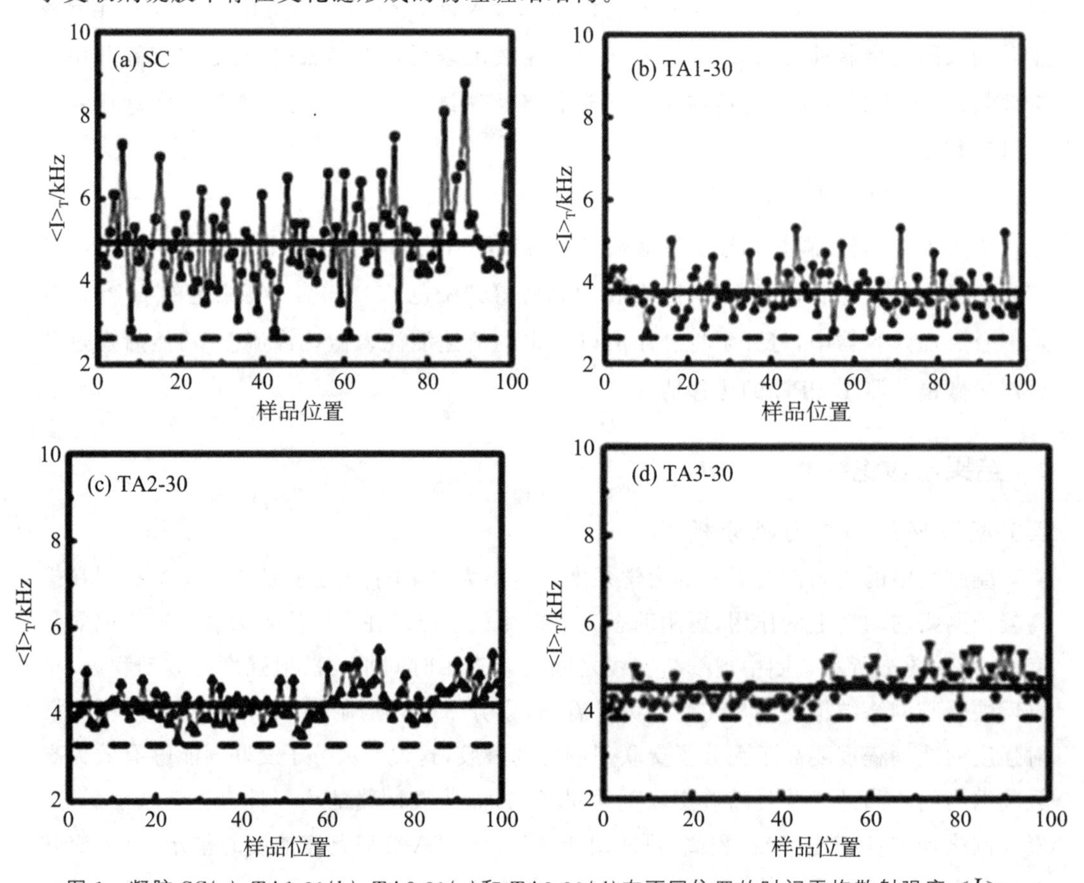

图1　凝胶SC(a)、TA1-30(b)、TA2-30(c)和TA3-30(d)在不同位置的时间平均散射强度$\langle I \rangle_T$

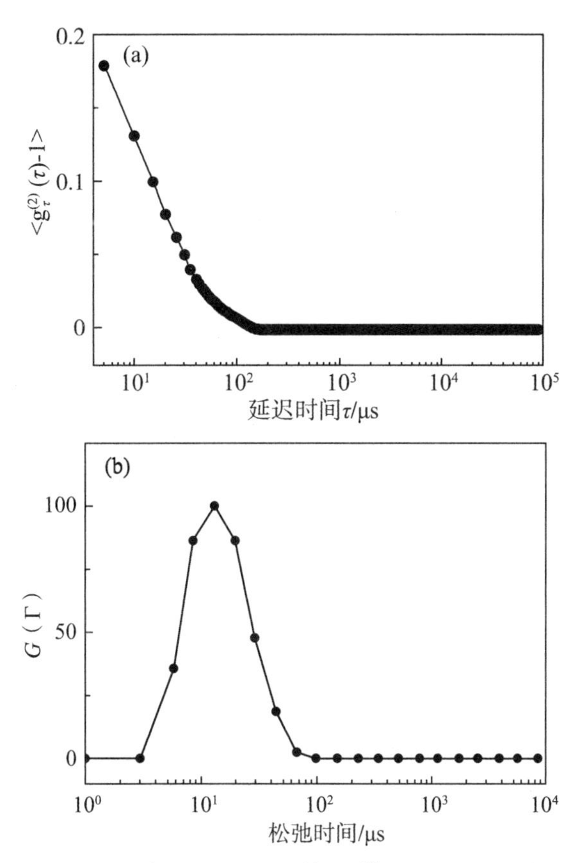

图 2　SC 凝胶的归一化时间平均散射光强相关函数 $<g_\tau^{(2)}(\tau)-1>$ (a) 与归一化特征线宽分布函数 $G(\Gamma)$ (b)

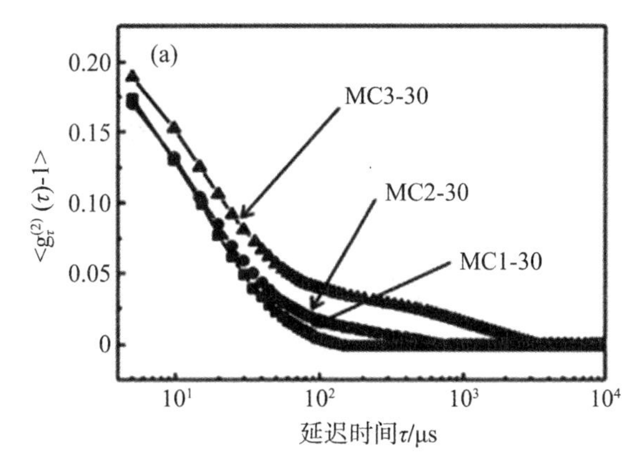

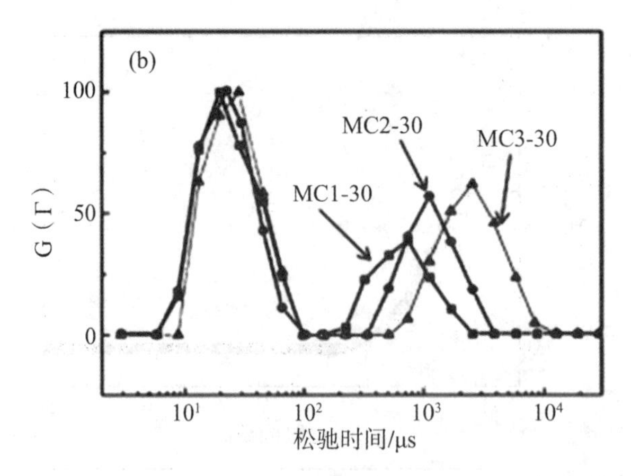

图 3 不同交联剂用量的 MC 凝胶的归一化时间平均散射光强相关函数<$g_{\tau}^{(2)}(\tau)-1$>(a)与归一化特征线宽分布函数 G(Γ)(b)

2.3 凝胶的力学性能

不同大分子交联剂加量下的 TA 凝胶(a)、MC 凝胶(b)和 MCH 凝胶(c)的拉伸应力应变曲线如图 4 所示。由大分子交联剂制备的凝胶具有更加均匀的网络结构以及优异的能量耗散机制,赋予了凝胶优异的力学性能。其中 TA 凝胶随着交联剂用量的增加,力学强度、断裂应变和韧性均先增加后降低。MC 凝胶随着交联剂用量的增加,强度不断提高而断裂应变逐渐降低。MCH 凝胶的强度和伸长率均随着交联剂用量的增加而降低。

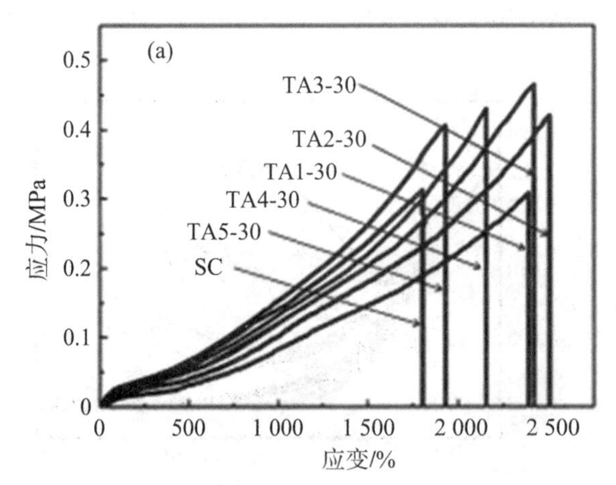

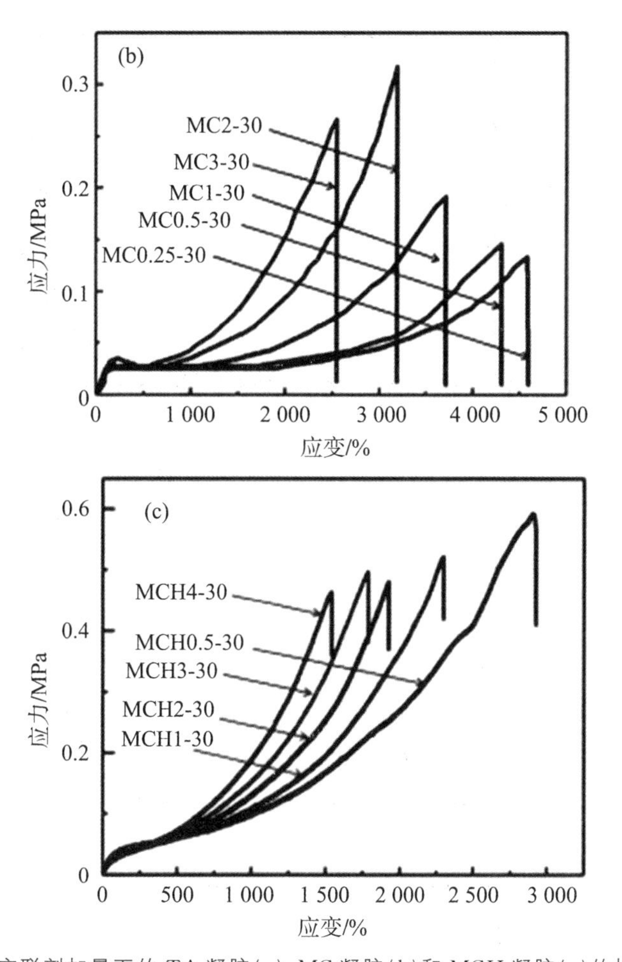

图 4　不同大分子交联剂加量下的 TA 凝胶(a)、MC 凝胶(b)和 MCH 凝胶(c)的拉伸应力应变曲线

3 种大分子交联剂所制备凝胶的断裂伸长率均达到 2 000%以上，其中 MC0.25-30 凝胶的断裂伸长率甚至能达到 4 500%，说明其具有良好的变形能力。并且凝胶的断裂强度不低于 0.3 MPa，其中 MCH 凝胶的断裂强度最高达到了 0.6 MPa，相比于 SC 凝胶力学性能大大提高。大分子交联剂凝胶之所以具有如此优异的力学性能主要是由于支化链形成的物理缠结以及独特的拓扑结构调整这两种能量耗散机制。物理缠结相比于化学交联更弱，当受外力作用时，物理缠结先于化学交联结构被破坏，会耗散一部分能量。此外大分子交联剂可以通过链的运动来调整自身的构象，从而调整交联点的位置，使得应力能够均匀分散，有效地避免了应力集中。这一行为与具有滑环结构的水凝胶极为相似[13]。

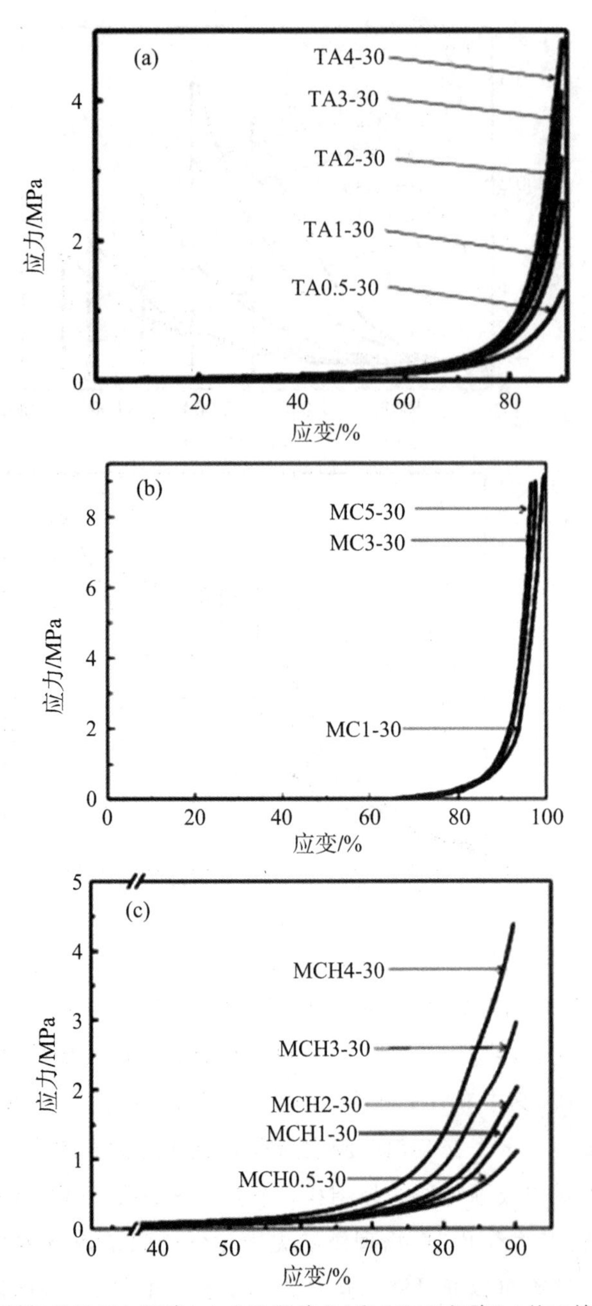

图5　不同交联剂加量下TA凝胶(a)、MC凝胶(b)和MCH凝胶(c)的压缩应力应变曲线

不同大分子交联剂加量下的TA凝胶(a)、MC凝胶(b)和MCH凝胶(c)凝胶的压缩性能如图5所示。随着交联剂的增加，TA凝胶在90%应变时的压缩应力从1.3 MPa逐渐增大到4.9 MPa。而MC凝胶的压缩性能对交联剂用量不敏感，在98%的应变下压缩强度均能达到8.5 MPa以上。随着交联剂用量的增加，MCH凝胶在90%应变下的压缩强度从1 MPa逐渐增加到4.4 MPa。

与拉伸性能相似，大分子交联剂制备的凝胶具有良好的弹性和韧性。当压缩应变达到90%时凝胶依然能够保持完整的形状，不发生破坏，说明其能承受较大的压缩应变。

而且在除去外力之后，变形能够部分恢复。

2.4 凝胶的抗缺陷性能

以力学性能最优异 MC2-30 凝胶为例，进一步研究凝胶的抗缺陷能力。先将样品的一侧剪出一个长为 22.5 mm 的缺口，如图 6(a)所示，然后再对有缺口和无缺口的样品分别进行拉伸，得到试样的力—位移曲线，结果如图 6(b)所示。无缺口试样的位移为 230 mm，有缺口试样断裂应变达到 200 mm，与无缺口试样相差无几，表现出了超强的拉伸性能。说明 MC 凝胶的拉伸性受缺口的影响小。经过计算可以得出 MC 凝胶的断裂能高达 11 676 J/m^2，远高于大多数水凝胶的，与一些韧性弹性体的相似[5,14]。

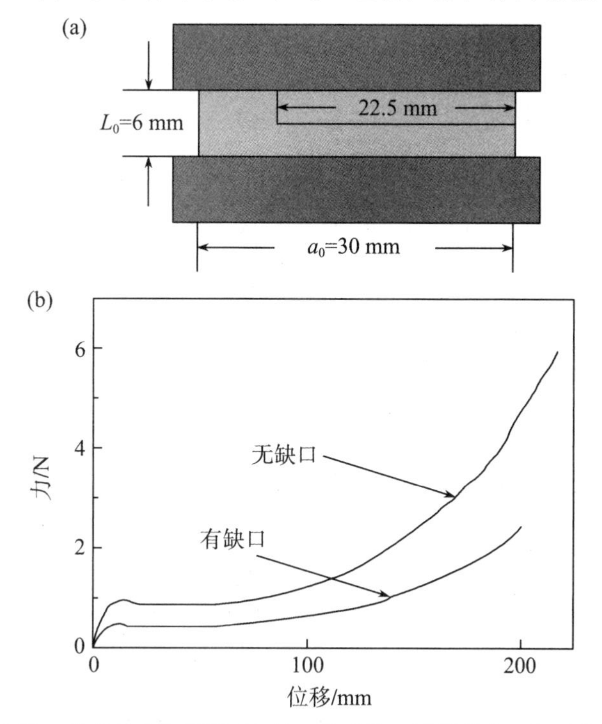

(a) 缺口样品的几何形状(b) 缺口和非缺口试样的力—位移曲线

图 6 MC 凝胶的抗撕裂性能

2.5 B-PPG 的耐老化性能

根据前文，选择对比样 SC 凝胶和力学性能最优异的 TA3-30、MC2-30、MCH0.5-30 凝胶制备的 $B\text{-}PPG_{SC}$(a)、$B\text{-}PPG_{TA}$(b)、$B\text{-}PPG_{MC}$(c)和 $B\text{-}PPG_{MCH}$(d)在盐水中充分溶胀后悬浮液的模量和黏度随老化时间的变化如图 7 所示。4 种 B-PPG 悬浮液的黏度均随着时间的延长呈现先增加后降低的趋势，而模量在整个老化过程中一直处于降低的过程。其中，$B\text{-}PPG_{TA}$ 在老化 21 d 时达到最高黏度 120.3 mPa·s，直到 73 d 后完全降解。而 $B\text{-}PPG_{MC}$ 在 21 d 时最高黏度为 156.5 mPa·s，90 d 时基本完全降解。$B\text{-}PPG_{MCH}$ 在老化 30 d 时黏度最高，达到了 162.8 mPa·s；老化 90 d 后 $B\text{-}PPG_{MCH}$ 才会完全降解。作为对比，$B\text{-}PPG_{SC}$ 悬浮液在老化 15 d 后的黏度最高，为 102 mPa·s，老化 60 d 后 $B\text{-}PPG_{SC}$ 就完全降解，大分子交联剂 B-PPG 表现出了优异的耐老化性能。之所以出现这种现象的原因主要是由于在热的作用下,分子链断裂交联网络开始破坏,形成支化链。更多线性结构的存在使得悬浮液的黏度大大提高。经过进一步老化后,交联结构被完全破坏,B-PPG 变为线性分子,此时悬浮液的黏度达到最高。之后再延长老化时间,线性分子链发生断裂,分子量降低,黏度逐渐降低。大分子交联剂 B-PPG 驱油颗粒在高温、高矿化度的条件下的耐老化性能要明显强于小分子交联剂 B-PPG[15]。

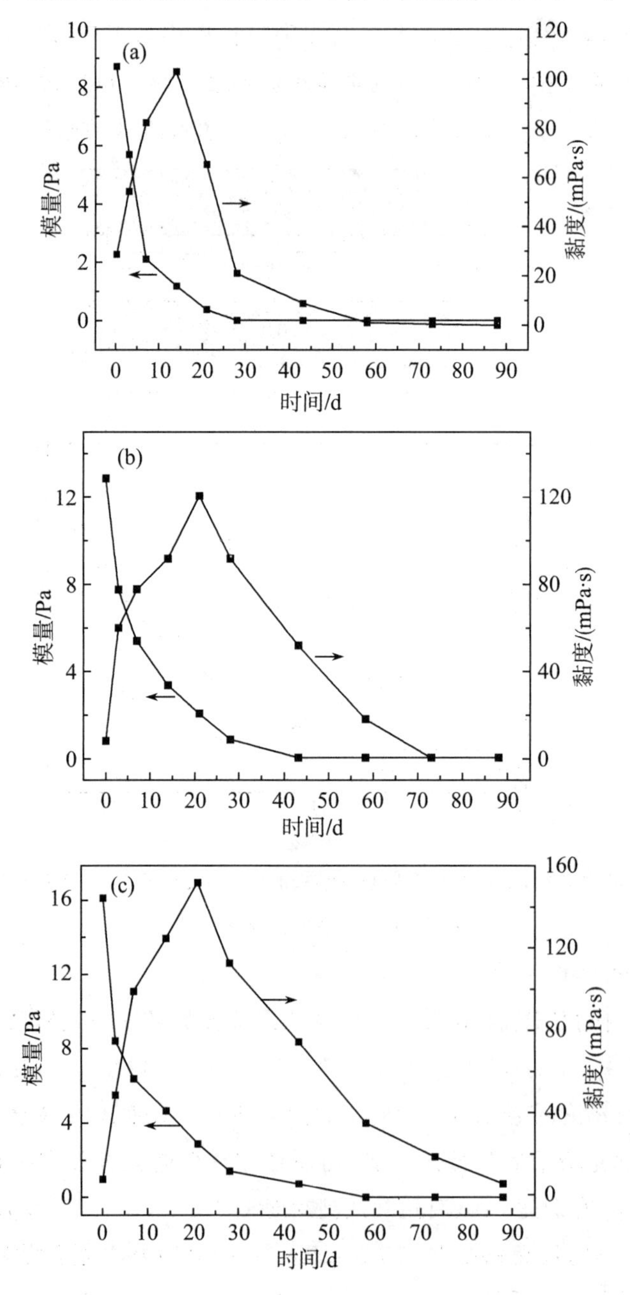

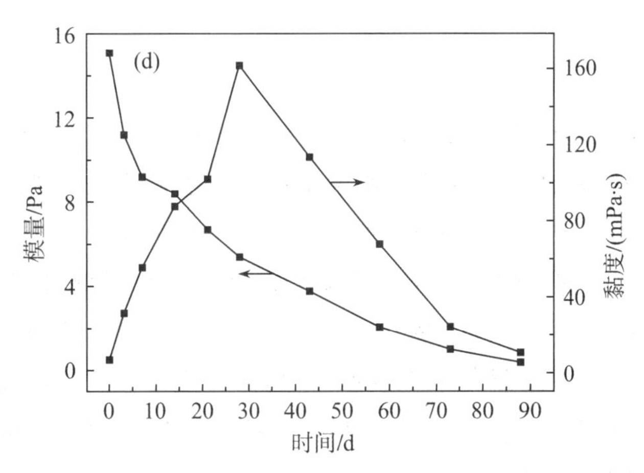

图 7　B-PPG$_{SC}$(a)、B-PPG$_{TA}$(b)、B-PPG$_{MC}$(c)和 B-PPG$_{MCH}$(d)在盐水中的模量和黏度随老化时间的变化(质量浓度 5 g/L)

2.6 荧光可视微观渗流实验分析

为了能够直观地看到 B-PPG 在孔隙中的运移，我们先采用荧光分子对 B-PPG 进行标记，然后使用激光共聚焦显微镜监测 B-PPG 在岩心渗流模型中的运移过程并研究其运移机理，如图 8 所示，从图 8(a)～(e)能观察到 B-PPG 颗粒变形通过孔喉的过程，而在图 8(d)中未观察不到带有绿光的 B-PPG，说明 B-PPG 在入口处出现“真空”，证实了 B-PPG 在孔隙间的运移是间歇的。我们认为当颗粒发生变形通过孔喉时，会使得入口处的压力骤降，须等到 B-PPG 颗粒重新堆积并达到一定的压力后，颗粒才能够再一次通过孔喉。这一现象表明了 B-PPG 颗粒在微观渗流模型中的运移存在“颗粒堆积—颗粒变形—颗粒通过”的过程。

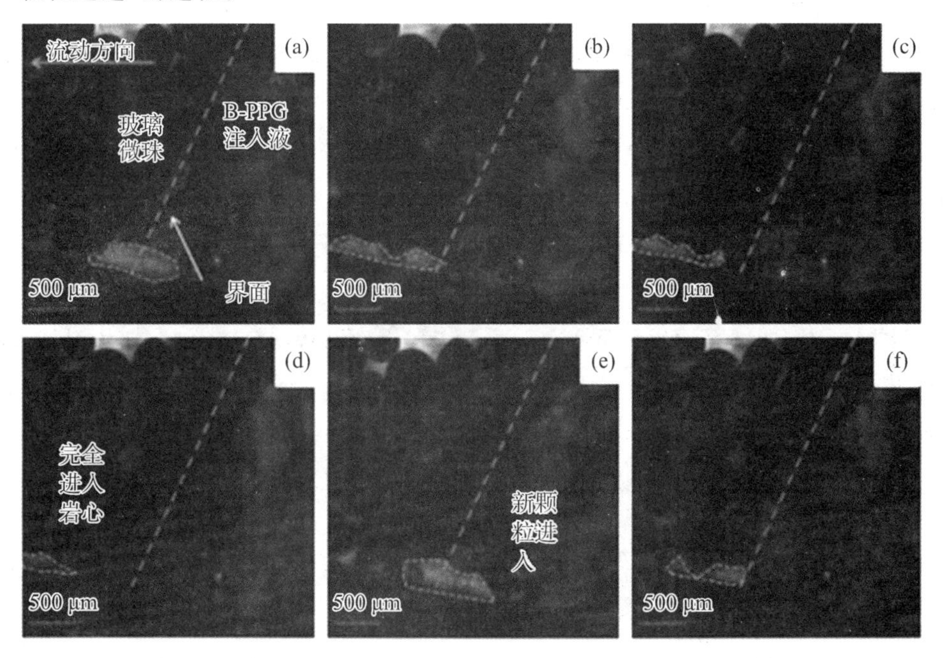

图 8　B-PPG 在渗流模型中的运移过程的图像

3 结论

采用在之前小分子交联剂的基础上合成的多种新的大分子交联剂制备了一系列 B-PPG。与小分子交联剂相比,大分子交联剂能够有效提高 B-PPG 凝胶网络的均匀性。大分子交联剂凝胶存在快、慢两种松弛模式的松弛行为,而小分子交联剂凝胶 SC 仅存在一种快松弛模式的松弛行为。大分子交联剂独特的拓扑结构调整赋予了凝胶优异的能量耗散机制,大大提高了力学性能。此外,B-PPG 颗粒耐老化性能优异,85℃下老化 90 d 依然保持一定的黏度。荧光可视微观渗流试验明确了 B-PPG 的运移机制为"颗粒堆积—颗粒变形—颗粒通过"。大分子交联剂为制备高性能水凝胶提供了新的思路,有望在高温高盐的苛刻油藏中得到应用。

参考文献

[1] 任月嫣. 三次采油技术的现状及未来发展 [J]. 化学工程与装备,2020 (8): 94-95.

[2] 徐江涛. 浅析三次采油用新型聚合物的开发 [J]. 化学工程与装备,2020 (2): 91-92.

[3] 兰海宽,毛艳妮,刘旻,等. 驱油剂聚丙烯酰胺在油田生产中的应用研究 [J]. 化工管理,2019(23):210-211.

[4] 毛佩林. 新型驱油剂聚丙烯酰胺的合成及性能研究 [J]. 橡塑技术与装备,2018,44(18):33-39.

[5] SUN J Y, ZHAO X, ILLEPERUMA W R, et al. Highly stretchable and tough hydrogels [J]. Nature,2012,489(7414):133-136.

[6] 朱勇,姜祖明,李江波,等. 部分交联聚丙烯酰胺驱油剂的渗流、驱油特性及其耐老化性能 [J]. 油田化学,2015,32 (3):360-365.

[7] 姜祖明. 黏弹性颗粒驱油剂的制备与性能 [J]. 塑料工业,2020,48 (4): 148-152.

[8] LI J B, JIANG Z M, WANG Y, et al. Stability, seepage and displacement characteristics of heterogeneous branched-preformed particle gels for enhanced oil recovery [J]. RSC Adv, 2018,8(9):4881-4889.

[9] 任奕,易飞. 海上油田耐温耐盐聚合物驱油剂设计 [J]. 天津化工,2018,32(6): 5-8.

[10] 于志省,夏燕敏,李应成. 耐温抗盐丙烯酰胺系聚合物驱油剂最新研究进展 [J]. 精细化工,2012,29(5):417-424.

[11] 李明远,郑晓宇,林梅钦,等. 交联聚合物溶液深部调驱先导试验 [J]. 石油学

报,2002(6):72-76.

[12] HUANG M, FURUKAWA H, TANAKA Y, et al. Importance of entanglement between first and second components in high-strength double network gels [J]. Macromolecules,2007,40(18):6658-6664.

[13] LIU C,KADONO H,MAYUMI K,et al. Unusual fracture behavior of slide-ring gels with movable cross-links [J]. ACS Macro Lett,2017,6(12):1409-1413.

[14] TAO L S,KUROKAWA T,KURODA S,et al. Physical hydrogels composed of polyampholytes demonstrate high toughness and viscoelasticity [J]. Nat Mater, 2013,12(10):932-937.

[15] 姜祖明,苏智青,黄光速.预交联共聚物驱油剂高温高盐环境下长期耐老化机理研究 [J].油田化学,2010,27(2):166-170.

Synthesis of super high molecular weight Co-polymer of AM/NaA/AMPS by oxidation-reduction and controlled radical polymerization

Ji Yanfeng* Cao Xulong Zhu Yangwen Xu Hui

Exploration and Development Research Institute, Shengli Oilfield, Dongying 257000, Shandong Province, P. R. China;

Abstract: Super high molecular weight co-polymers of AM/NaA/AMPS were prepared by oxidation-reduction [OR-P (AM/NaA/AMPS)] and controlled radical polymerization [CR-P (AM/NaA/AMPS)]. The resulting copolymers were fully characterized, and the reaction conditions for their preparation were optimized. OR-P (AM/NaA/AMPS), CR-P(AM/NaA/AMPS), and conventional partially hydrolyzed polyacrylamide (HPAM) in brine solution were comprehensively characterized by thermogravimetric analysis, scanning electron microscopy, atomic force microscopy, and dynamic light scattering. OR-P(AM/NaA/AMPS) and CR-P(AM/NaA/AMPS) containing AMPS monomer showed better salt resistance, temperature tolerance, and viscosification property than the conventional HPAM polymer, making them more promising for enhanced oil recovery. Through comprehensive comparison and analysis, it was found that OR-P(AM/NaA/AMPS) was more conducive for high-temperature condition due to the existence of xanthone in OR-P(AM/NaA/AMPS). On the other hand, CR-P(AM/NaA/AMPS) was more suitable for high-mineral atmosphere, which could be attributed to its higher intrinsic viscosity.

Keywords: Oxidation-reduction polymerization; Controlled radical polymerization; Elemental composition; Thermogravimetry; Microcosmic aggregation morphology

1 Introduction

Polyacrylamide (PAM) is an important type of hydro-soluble polymer, with flocculating, thickening, drag reduction, and dispersing characteristics. PAM has been used in various areas, ranging from coal and mineral processing, papermaking, water purification, drag reduction, to oil production processes including oil well drilling, reservoir stimulation, water shutoff, profile modification, and particularly tertiary oil recovery. Partially hydrolyzed polyacrylamide (HPAM) is one of the most widely used water-soluble polymers for enhanced oil recovery (EOR).[1-3] In polymer flooding process, HPAM is used to thicken the injected water to mobilize capillary trapped water-flooded oil in the secondary stage, which improves the sweep efficiency to increase the oil recovery factor. However, due to shifting of oil production toward deeper, high-temperature and high-salinity (HTHS) reservoirs, conventional HPAM cannot be used to viscosify the displacing fluid under the HTHS conditions. This is because its viscosity decreases significantly under HTHS conditions due to a charge shielding effect.[4-5] To overcome the limitations of HPAM, a great deal of research has been undertaken on the synthesis of thermally stable and salt-tolerant water soluble polymers.[6-9]

The methods for developing improved PAMs are mainly as follows.[10-13] Firstly, the synthesis of non-associative polymers with structural monomers having temperature stability and salt tolerance properties, and introducing functional structural monomers, such as 2-acrylamide-2-methylpropanesulfonic acid (AMPS), N-vinylpyrrolidone (NVP) and N-vinylamides (NVAs) with desirable properties including restricted hydrolysis, complexing high valence cations, great hydration capacity, and enhancing the rigidity of high molecular chain.[14-17] Secondly, the synthesis of polymers with special interactions, such as hydrophobic associated polymer, amphoteric ionic polymer, and colloidal dispersion gel.[18-19] Thirdly, the synthesis of light cross-linked polymers with increased difficulty in conformational transition, high salt tolerance and increased viscosity due to the presence of cross-linked structure.[20-22] Fourthly, the synthesis of super high molecular weight polymers, which can enhance the thickening capacity by increasing the hydrodynamic volume of the solution of high molecular weight polymer.[23] Among them, the polymer obtained by the second and third methods is prone to a large number of insoluble substances, so the first and second are the main methods for our study.

In this study, two super high molecular weight co-polymers of AM/NaA/AMPS were prepared by oxidation-reduction (OR) and controlled radical (CR) polymerization, respectively. This paper discusses the effects of AMPS on the molecular structure, stretching property and microcosmic morphology of high molecular weight polymers prepared by different methods to provide theoretical and methodological support for the preparation of acrylamide-based polymers.

2 Materials and Methods

2.1 Materials

Acrylamide (AM, 99.5%, Changjiu Agri-Scientific Co. Ltd, Nanchang, China), 2-acrylamide-2-methylpropanesulfonic acid (AMPS, 99.2%, Shandong Lianmeng Chemical Co. Ltd, China), and macroinitiator polyacrylamide-xanthone (PAM-XAN) were prepared in our laboratory using known procedures. Sodium hydroxide, ethylene diamine tetraacetic acid (EDTA), and other solvents were all analytical grade and purchased from Aladdin Chemical Reagent Factory (Shanghai, China). The water used in this study was double distilled using an all-glass apparatus, and nitrogen with a purity of 99.99% was used.

2.2 Synthesis of co-polymers

2.2.1 Synthesis of oxidation-reduction polymer

The terpolymer OR-P (AM/NaA/AMPS) was prepared by the oxidation-reduction copolymerization of AM, sodium acrylate (NaA), and AMPS in pure water, and the pH was adjusted around 9 using NaOH. After 30 min of N_2 purge, a certain amount of initiator was injected into the solution. The polymerization was carried out at 10℃ under N_2 atmosphere for 4 h, and the resulting product was obtained by adding a certain amount of sodium hydroxide into the reaction mixture, followed by freeze-drying, affording the final terpolymer product as a white powder.

The effects of different types of initiator on heating rate, intrinsic viscosity, and apparent viscosity were investigate, the effects of AMPS content were also studied.

2.2.2 Synthesis of controlled radical polymer

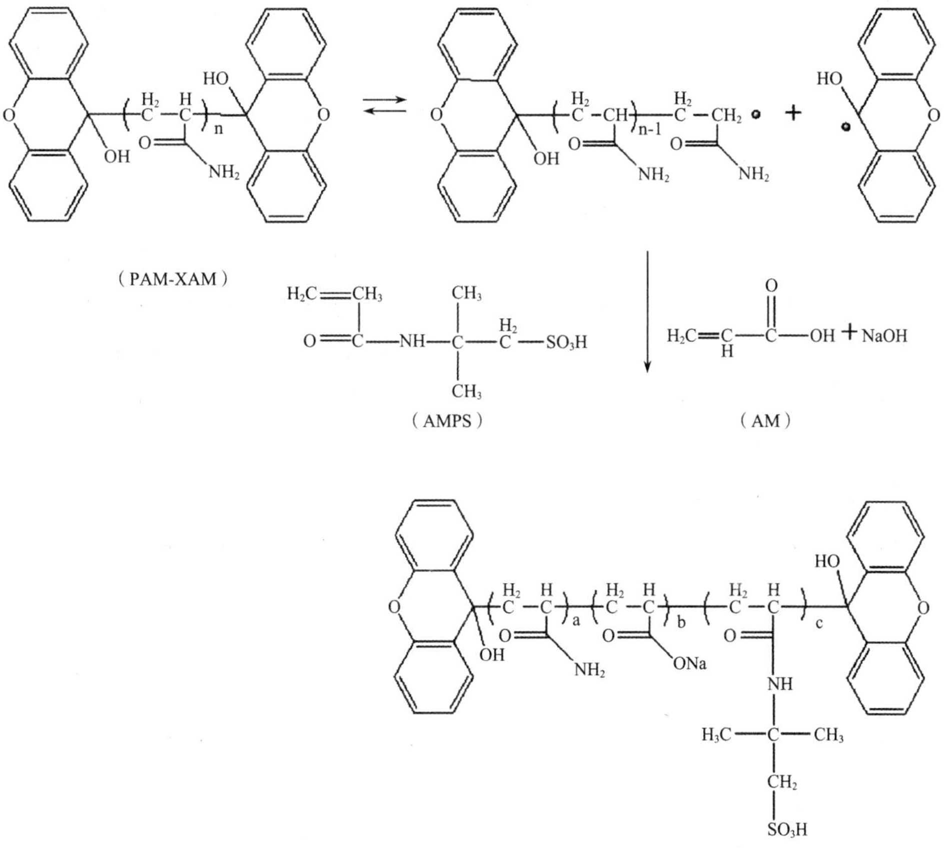

Fig. 1　Synthesis of CR-P(AM/NaA/AMPS)

The terpolymer CR-P (AM/NaA/AMPS) was prepared by controlled radical polymerization of AM and AMPS in pure water, and the pH was adjusted around 9 using NaOH followed by 30 min of N_2 purge. Then, a certain amount of macroinitiator (PAM-XAN) was injected into the solution. The polymerization was carried out at 40℃ under N_2 atmosphere for 6 h. The resulting product was obtained by adding the reaction mixture into a certain amount of sodium hydroxide, followed by freeze-drying, affording the final terpolymer product as a white solid. The synthetic route is shown in Fig. 1.

2.3 Characterizations

2.3.1 Infrared Spectroscopy

The IR spectra of the samples was measured with KBr pellets using a WQF-520A IR Spectrophotometer (Beijing Rayleigh Analytical Instrument Company) in the optical range, 4 500～400 cm^{-1}, by averaging 32 scans at a resolution of 4 cm^{-1}.

2.3.2 Elementary Analysis

The elementary analysis of the AM/AA/NSFM copolymer was carried out using a Vario EL-III elemental analyzer. The content of different element in the copolymer can be obtained by detecting the gases, which are the decomposition products of the copolymer at high temperature.

2.3.3 Thermogravimetric Analysis

The thermal weight loss curves of polymer samples were measured using a TGA-Q500 type thermal analyzer, and the structural changes of the polymers were evaluated according to the thermogravimetric curves. The test conditions of TGA were as follows: Oxygen atmosphere at a flow rate of 100 mL/min; heating rate of 10℃/min; and test temperature range of 25～700℃.

2.3.4 Dynamic Light Scattering

DLS technique has been widely used for determining the hydrodynamic radius of polymers by testing the intensity fluctuation change of sample. The working principle is based on the variation in the intensity of scattered light, which is the diffusion coefficient of the molecule tested by Doppler frequency shift- D_{app}. Using Stokes-Einstein formula, the hydrodynamic radius of particle can be derived as follows.

$$R_{h,app}=k_B T/(6\pi\eta D_{app}) \quad (1)$$

where k_B is the Boltzmann constant, η is the solvent viscosity, and T is the absolute temperature.

DLS spectrum was obtained using a Laser Light Scattering Spectrometer (ALV-5000/E/WIN Multiple Tau Digital Correlator). The sample was centrifuged to remove dust before analysis, and the test temperature was 40℃.

2.3.5 Scanning Electronic Microscopy

SEM images were taken using a JMS-6380LV scanning electron microscope. The polymer solution was frozen by liquid nitrogen and then was made electrically conductive by coating a thin layer of gold in vacuum (approximately 300Å) at 30 W for 30 s. The pictures were taken at an excitation voltage of 10 kV and a magnification of 5 000×.

2.3.6 Atomic Force Microscopy

Atomic force microscopy was used to investigate the surface morphology and thickness of HPAM and the two co-polymers. AFM images were taken using a commercial Nanoscope Ⅲ (Digital Instruments, Santa Barbara, CA) using a Si_3N_4 probe to analyze the apparent morphology of polymer (1 000 mg/L).

3 Results and Discussion

3.1 Optimization of oxidation-reduction polymerization

Table 1 The effect of initiator systems on the molecular weight and viscosity of the polymer

Initiator systems	Heating rate (℃/h)	Intrinsic viscosity (mL/g)	Apparent viscosity (mPa · s)*
Oxidation-reduction initiator	30.0	2 100	7.8
Azo initiator	13.5	2 200	7.8
Oxidation-reduction initiator, Azo initiator	24.5	2 550	11.2
Oxidation-reduction initiator, Azo initiator, assisting agent	8.0	3 350	16.3

* Brine salinity, 32 868 mg/L; Ca^{2+} and Mg^{2+} content, 874 mg/L; Polymer concentration, 1 500 mg/L; Test temperature, 85 ℃.

The experimental results in Table 1 show that the intrinsic and apparent viscosities of OR-P(AM/NaA/AMPS) by using the complex initiation system of OR initiator, Azo initiator, and additive were greater than that by using either OR initiator or Azo initiator alone, while the heating rate of polymerization system was quite different under complex initiation systems. Thus, it can be supposed that the complex initiation system of OR initiator, Azo initiator, and additive can regulate the decomposition rate of initiator, making the polymerization reaction more stable. Therefore, low temperature OR initiator, Azo initiator, and additive were used in this study.

Table 2 The effect of functional monomer AMPS content on the viscosity of polymer

Sample	AMPS content	Apparent viscosity (mPa · s)
1#	20.0%	17.9
2#	15.0%	15.3
3#	12.0%	15.1
4#	10.0%	14.2

Table 2 shows that the apparent viscosity increased with increasing AMPS content. However, AMPS is far more expensive than AM. Therefore, the final AMPS content was chosen to be around 12%~15%.

3.2 Optimization of controlled radical polymer

The intrinsic viscosity of the controlled free radical copolymer products with different AMPS contents was investigated at different temperatures.

Table 3 The intrinsic viscosity of CR-P(AM/NaA/AMPS) prepared under different temperature conditions

AM (g)	AMPS (g)	Reaction temperature (℃)	Intrinsic viscosity (mL/g)
1.8	0.2	55	1 873
1.8	0.2	50	1 976
1.8	0.2	45	crosslinked
1.8	0.2	40	crosslinked
1.6	0.4	55	1 632
1.6	0.4	50	1 818
1.6	0.4	45	1 877
1.6	0.4	40	2 065
1.4	0.6	55	1 223
1.4	0.6	50	1 469
1.4	0.6	45	1 755
1.4	0.6	40	1 826
1.2	0.8	55	997
1.2	0.8	50	1 252
1.2	0.8	45	1 516
1.2	0.8	40	1 644

The experimental results in Table 3 show that at identical monomer concentrations, the intrinsic viscosity of CR-P(AM/NaA/AMPS) increased gradually with decreasing polymerization temperature, indicating that the polymerization reaction was more stable at low temperature. In particular, when the proportion of AM/AMPS (mass ratio 9 : 1) and the polymerization temperature were low, crosslinking reactions were more likely to occur in CR-P(AM/NaA/AMPS), decreasing the solubility. The results showed that the crosslinking reaction was effectively suppressed with increasing AMPS content (AM/AMPS mass ratio changed to 8 : 2, 7 : 3 and 6 : 4). However, with increasing AMPS content, the intrinsic viscosity of the prepared polymers significantly reduced. Based on the crosslinking reactions and the intrinsic viscosity, the optimum values of mass ratio of AM/AMPS and reaction temperature were 8 : 2 and 40℃, respectively.

Table 4　The intrinsic viscosity of CR-P(AM/NaA/AMPS) prepared with different dosages of initiator and monomer content

Macroinitiator (mg)	Monomer content (%)	Intrinsic viscosity (mL/g)
30	25	2 065
25	25	2 182
20	25	2 247
15	25	2 386
10	25	2 518
30	30	2 211
25	30	2 407
20	30	2 596
15	30	2 681
10	30	2 772
30	35	2 170
25	35	2 235
20	35	2 372
15	35	2 470
10	35	2 520

Table 4 illustrates the changes in the intrinsic viscosity with different amounts of initiator and concentration of monomer at a fixed AM/AMPS mass ratio of 8∶2. Table 4 shows that the intrinsic viscosity of co-polymer increased with decreasing concentration of macroinitiator. With increasing monomer percentage (from 25% to 35%), the intrinsic viscosity increased first and then decreased. The optimum dosage of macroinitiator and critical monomer content were selected as follows to obtain the maximum intrinsic viscosity: macroinitiator dosage of 10 mg; and monomer content of 30%.

3.3 Molecular structure and elemental characterization

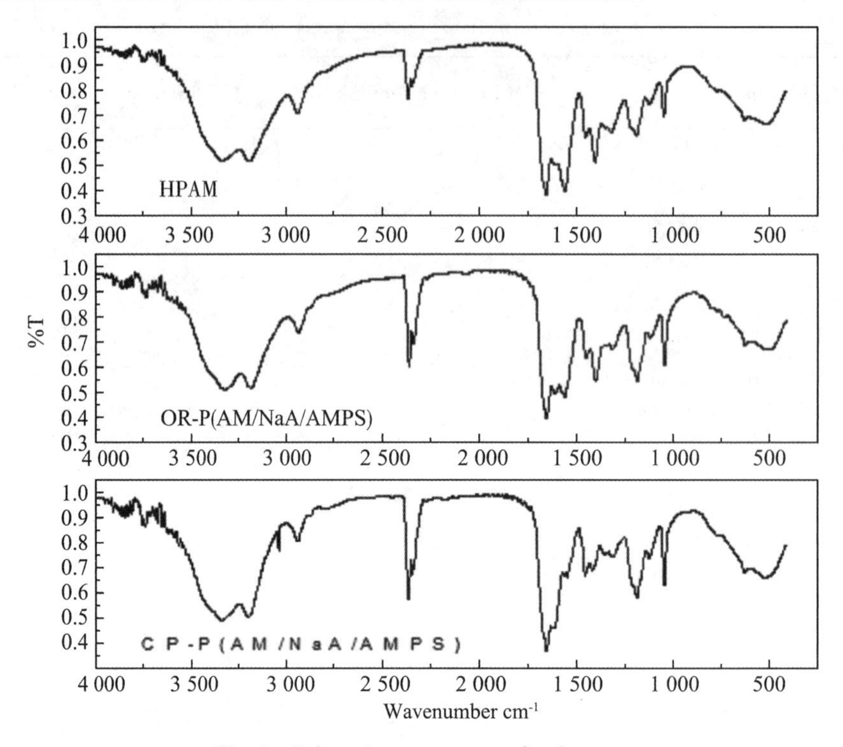

Fig. 2 Infrared spectrogram of polymers

The IR spectra are shown in Fig. 2. The IR spectra of OR-P(AM/NaA/AMPS) indicated that monomer AMPS was copolymerized successfully onto the backbone of PAM chains, as confirmed by the peaks at 1 194 and 1 119 cm^{-1} in the IR spectrum of OR-P(AM/NaA/AMPS), which are the characteristic absorption peaks of sulfonic group due to the S═O stretching vibration of SO_3^{2-} in AMPS. In the IR spectrum of CR-P (AM/NaA/AMPS), the main peaks for the infrared vibration of acrylamide at 3 438 cm^{-1}, 2 929 cm^{-1}, and 1 633 cm^{-1}, and characteristic peak of sulfonic acid at 1 194 cm^{-1} were observed. In addition, the characteristic absorption peak of the benzene ring skeleton at 1 451 cm^{-1} was also observed, indicating the existence of xanthone on the molecular skeleton. According to the IR spectrum of HPAM, the main infrared vibration peaks of acrylamide were observed at 3 438 cm^{-1}, 2 929 cm^{-1}, and 1 633 cm^{-1}, but the infrared vibration peaks of sulfonate group or xanthone were not observed.

Table 5　Analysis and comparison of elements in the co-polymers and HPAM

Element / polymer	C	H	N	O	S
HPAM	38.28	6.55	13.43	29.65	/
OR-P	39.73	6.59	12.51	30.42	1.04
CR-P	38.92	6.62	13.83	28.93	2.01

The different elemental contents in the copolymer and HPAM were obtained by detecting the gases, which are the decomposition products of the copolymer at high temperature. The experimental results are listed in Table 5. The conventional HPAM does not show any significant presence of sulfur, as no AMPS monomer was introduced into HPAM. For the two co-polymers, a considerable percentage of sulfur was observed, confirming that AMPS was copolymerized onto the backbone of PAM chains.

3.4 Analysis and characterization of thermal weight loss of polymer

The thermal weight loss and thermogravimetric curves of the two co-polymers were investigated and compared to that of the conventional HPAM.

Fig. 3 show that the initial weight loss temperature of HPAM was about 204.18℃, while the initial weight loss temperatures of OR-P(AM/NaA/AMPS) and CR-P(AM/NaA/AMPS) were approximately 312.17℃ and 257.20℃, respectively. Compared to HPAM, the TG/TGA data of the two co-polymers not displayed significant weight losses until 400.33℃ and 331.44℃, indicating that the thermal stabilities of two co-polymers were enhanced by the addition of AMPS monomer. Moreover, OR-P(AM/NaA/AMPS) showed better temperature resistance, probably because of higher intrinsic viscosity of OR-P(AM/NaA/AMPS).

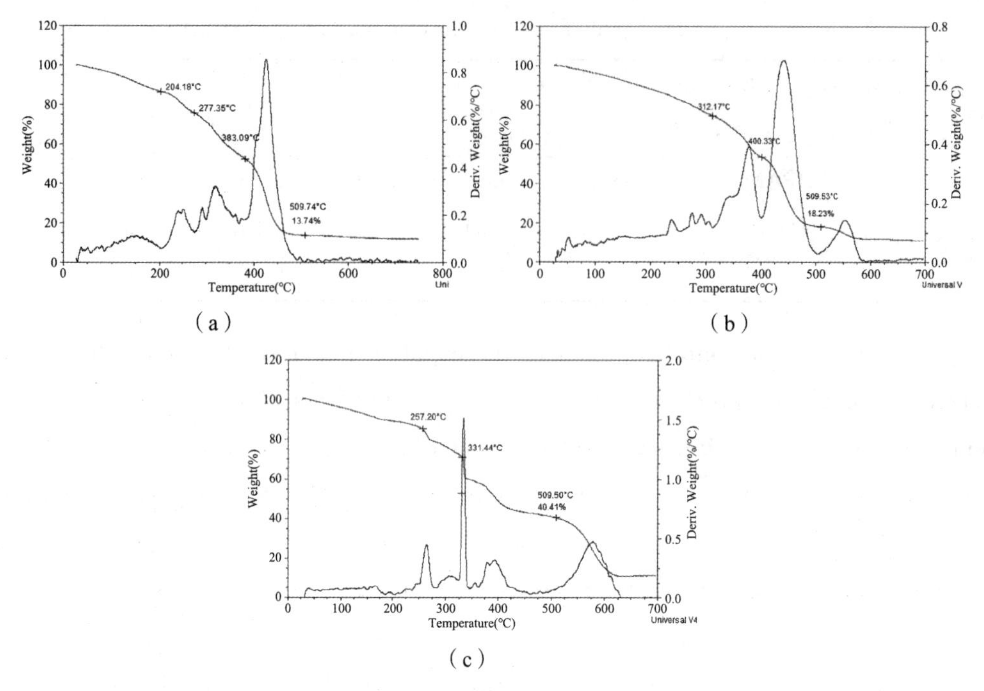

Fig. 3 TG curves of polymer: (a) HPAM; (b) OR-P(AM/NaA/AMPS); (c) CP-P(AM/NaA/AMPS)

3.5 Microcosmic aggregation morphology of polymer solution

The microcosmic aggregation morphologies of OR-P(AM/NaA/AMPS), CR-P(AM/NaA/AMPS) and HPAM were characterized and analyzed by DLS, freeze-etching SEM, and AFM at room temperature.

3.5.1 Study on the hydrodynamic radius of polymer

The hydrodynamic radius of the polymer was investigated by DLS, and the results are shown in Fig. 4. The hydrodynamic radius of the three types of polymers increased with increasing concentration. However, the hydrodynamic radius of both co-polymers was considerably higher than that of conventional HPAM, reflecting the better tackifying ability of the two co-polymers. The average hydrodynamic radius of CR-P(AM/NaA/AMPS) was higher than that of OR-P(AM/NaA/AMPS).

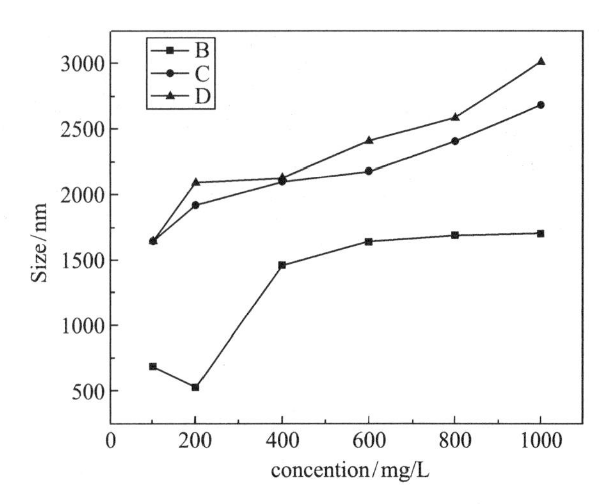

Fig. 4　The change of the average hydrodynamic radius of the polymer with the concentration

3.5.2 Microstructure characterization of polymer solution by Freeze-etching SEM

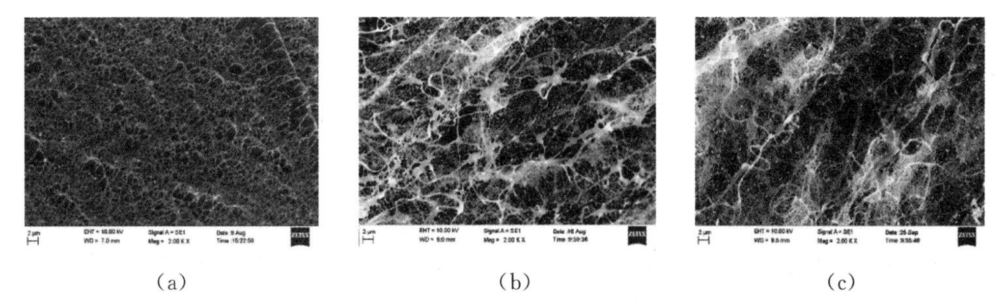

Fig. 5　Microaggregation morphologies of polymer: (a) 0.1% HPAM solution (2 000×); (b) 0.1% OR-P(AM/NaA/AMPS) solution (2 000×); (c) 0.1% CP-P(AM/NaA/AMPS) solution (2 000×).

The microstructures of two co-polymers and HPAM in aqueous solution were investigated by freeze-etching SEM. The polymer solution was prepared in distilled water at a concentration of 0.1% and simulated water at a concentration of 0.2%. The simulated water had a calcium ion content of 874 mg/L. Under the same magnification conditions, the microstructures of the three types of polymer were observed and compared. The SEM images are shown in Fig. 5 and Fig. 6.

Fig. 5 indicated the dense network structure of HPAM, OR-P(AM/NaA/AMPS) and CR-P(AM/NaA/AMPS) in distilled water, and the molecular chains in aqueous solution showed mutual aggregation and winding. Based on this effect, the friction between the molecular chains increased, and the polymer solution showed high viscosity.

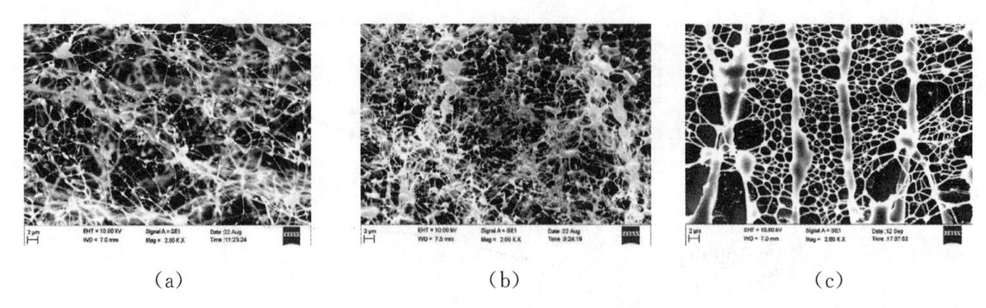

(a) (b) (c)

Fig. 6 Microaggregation morphologies of polymer: (a) 0.2% HPAM solution (2 000×); (b) 0.2% OR-P(AM/NaA/AMPS) solution (2 000×); (c) 0.2% CP-P(AM/NaA/AMPS) solution (2 000×).

As seen from Fig. 6, in the presence of high content of calcium and magnesium ions, the aggregation and entanglement of HPAM and co-polymers solution decreased significantly, and the network structure formed in solution became sparser. This might be attributed to the existence of calcium ions, which can destroy the molecular aggregation and decrease the viscosity of polymer solution. However, the degree of damage to the network structure of the three polymer solutions was quite different. The molecular chains of OR-P(AM/NaA/AMPS) and CR-P(AM/NaA/AMPS) still formed a network between the dense structure to a certain extent, while conventional HPAM solution showed more serious damage. Therefore, the apparent viscosity of co-polymers was also higher. The network structure of CR-P(AM/NaA/AMPS) was more obvious than that of OR-P(AM/NaA/AMPS), resulting in increased salt tolerance, which is consistent with the DLS result.

3.5.3 AFM characterization of the morphology of the polymer aggregates

The micromorphology of polymers was extensively characterized and analyzed by AFM. Fig. 7 shows the AFM images of HPAM with different concentrations. The HPAM solution exhibited a filamentous network structure. When the concentration was high (as shown in Fig. 7a), the fine filament structure overlapped and formed a thin lamellar structure. With decreasing concentration, the overlapped network structure gradually dispersed, forming the structure as shown in Fig. 7b. When the concentration was further reduced to 200×10^{-6} and 100×10^{-6} (Fig. 7c and d), respectively, most of the network nodes disappeared, the mesh structure increased, and the dense network structure became sparse.

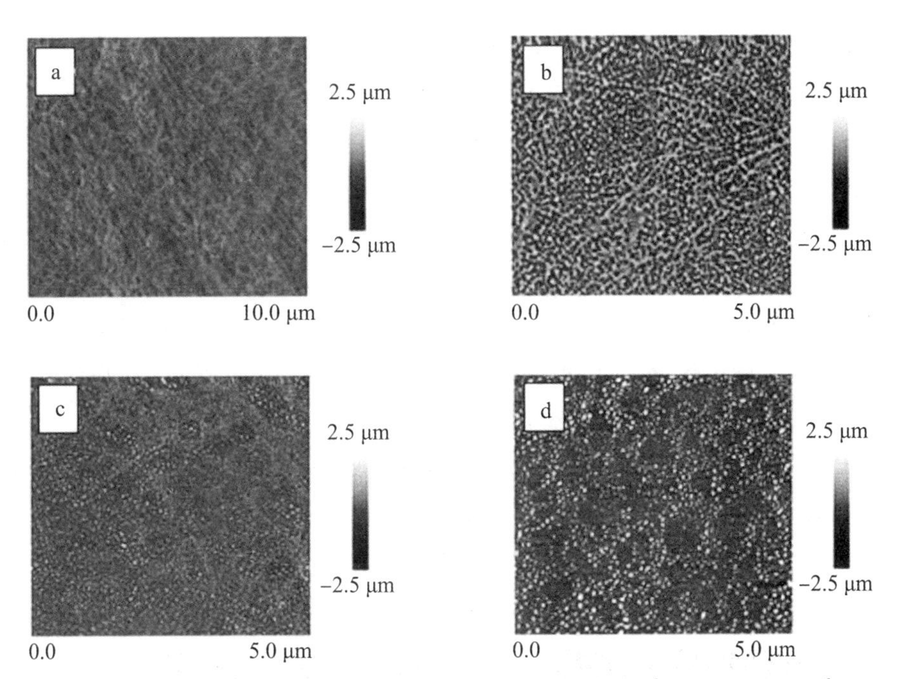

Fig. 7　AFM images of HPAM. The concentration is: (a) $1\ 000\times10^{-6}$; (b) 600×10^{-6}; (c) 200×10^{-6}; (d) 100×10^{-6}

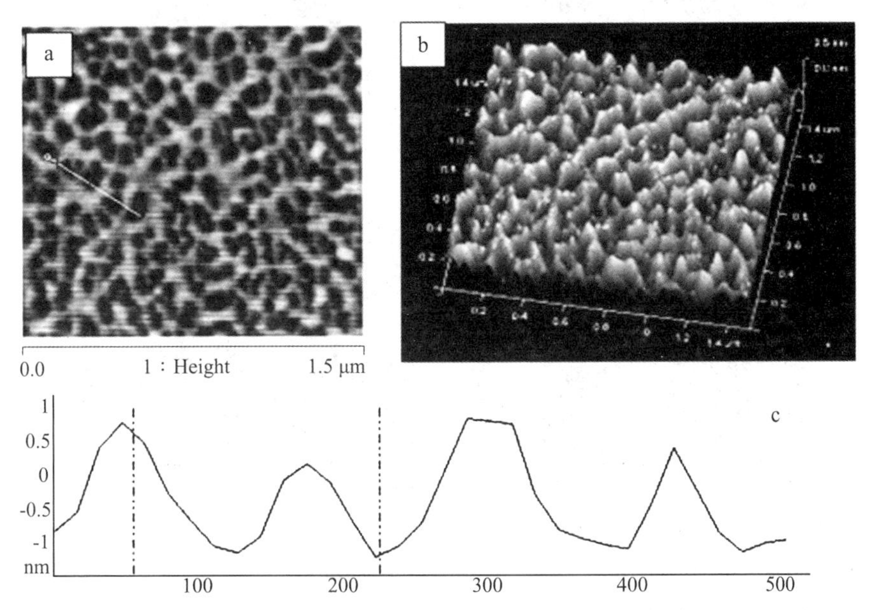

Fig. 8　Profile analysis of HPAM with a concentration of 1000ppm. (a) Ocal enlarged diagram; (b) Three-dimentional stereogram; (c) Height analysis diagram

HPAM solution with a concentration of $1\ 000\times10^{-6}$ was selected for the profile analysis, as shown in Fig. 8. The irregular network structure can be seen from the local enlarged diagram, and the skeleton of the network was not uniform. The height of the skeleton formed by HPAM was not homogeneous, more than 1～2 nm, as obvious from the three-dimensional stereogram. The AFM morphology analysis results showed that increasing the concentration of HPAM led to aggregation. Therefore, the aggregated molecular chains dispersed to form a network structure.

Fig. 9 shows the micrograph of OR-P(AM/NaA/AMPS) in aqueous solution, indicating the disordered network structure of OR-P(AM/NaA/AMPS) was more elongated, superimposed, and assembled on layers. With decreasing concentration, the degree of aggregation decreased, and the long filament structure was broken and gradually disappeared. However, when the concentration was reduced to 200×10^{-6}, the polymer structure still showed the presence of interwoven polymer on the substrate.

The three-dimensional stereogram and height analysis diagram (as shown in Fig. 10) show that the height of the aggregate was mainly between 0.5 and 1 nm, and a part of the multi-layer aggregation reached 1.5 nm. This result indicated that the molecular chain retained its aggregation state, at both low and high concentrations.

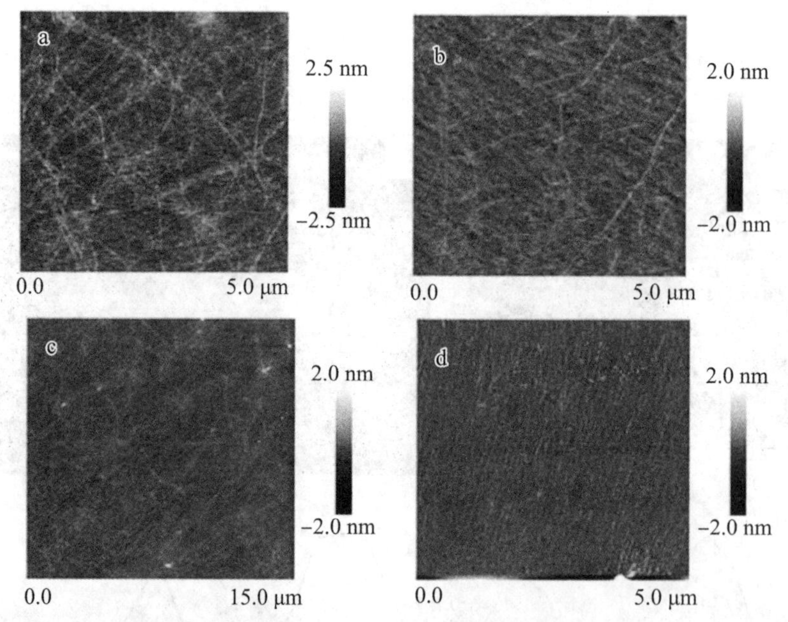

Fig. 9 AFM images of OR-P(AM/NaA/AMPS). The concentration is: (a) $1\,000\times10^{-6}$; (b) 600×10^{-6}; (c) 200×10^{-6}; (d) 100×10^{-6}

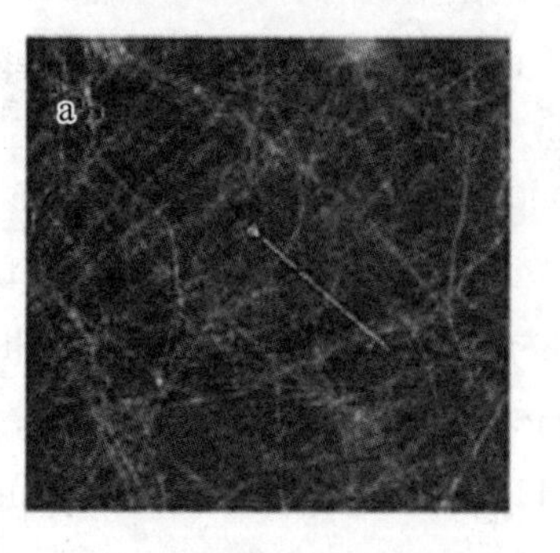

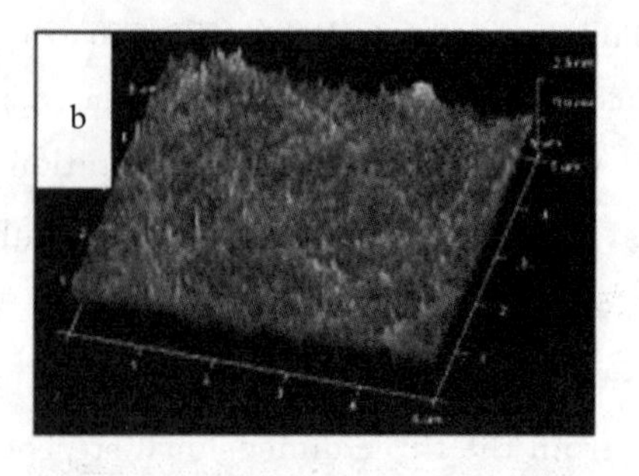

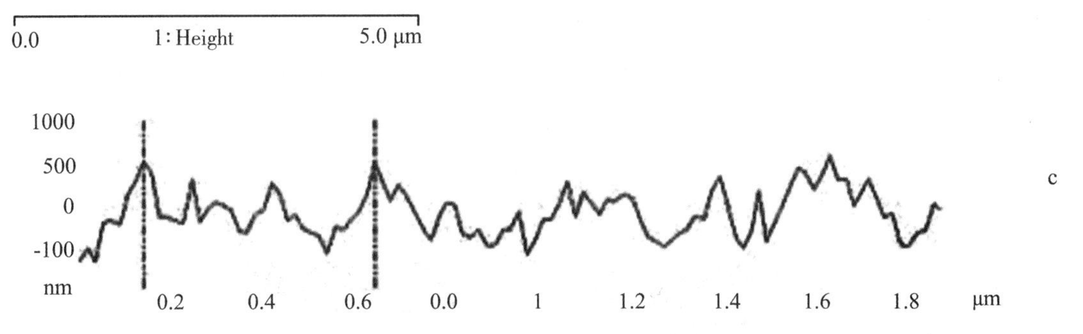

Fig. 10 Profile analysis of OR-P(AM/NaA/AMPS) with a concentration of 1 000 × 10^{-6}. (a) Ocal enlarged diagram; (b) Three-dimentional stereogram; (c) Height analysis diagram

Fig. 11 shows the AFM image of CR-P(AM/NaA/AMPS) in aqueous solution. The structure is similar to that of OR-P(AM/NaA/AMPS). Both copolymers showed filamentary random interlacing nets, and the disordered network structure was slender, stacked together, and assembled. With decreasing concentration, the degree of aggregation decreased, and the long filament structure was broken and gradually disappeared. The interwoven polymer structure was still obtained when the concentration was reduced to 200 × 10^{-6}. The height analysis (Fig. 10) showed that the height was between 0. 5 and 1 nm, which was same as that of OR- P(AM/NaA/AMPS).

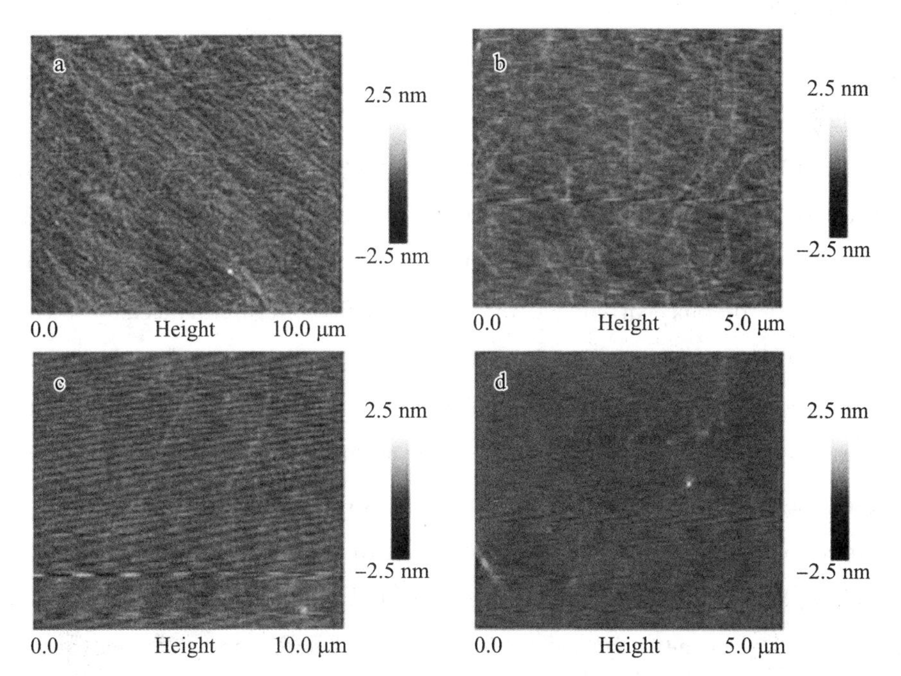

Fig. 11 AFM images of CP-P(AM/NaA/AMPS). The concentration is: (a) 1 000 × 10^{-6}; (b) 600 × 10^{-6}; (c) 200 × 10^{-6}; (d) 100 × 10^{-6}

The AFM characterization results showed that the network structures of OR-P

(AM/NaA/AMPS) and CR-P(AM/NaA/AMPS) (as shown in Fig. 12) can be elongated with increasing density of polymer solution. However, increasing polymer concentration from low to high resulted in aggregation, which caused the chains to disperse and form a network structure. This indicates that the two co-polymer chains can form stronger aggregates than HPAM.

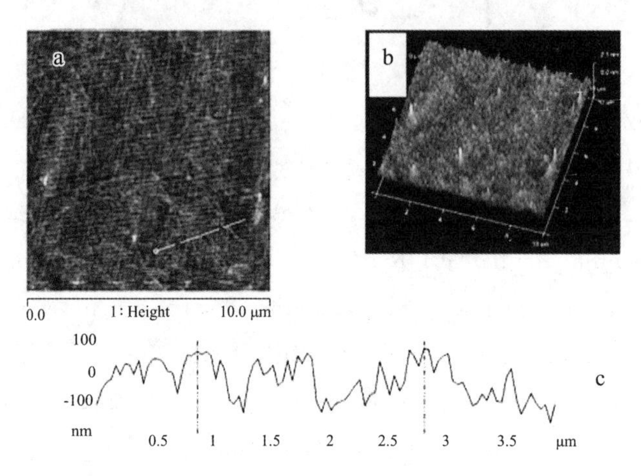

Fig. 12 Profile analysis of CP-P(AM/NaA/AMPS) with a concentration of $1\ 000\times10^{-6}$. (a) Ocal enlarged diagram; (b) Three-dimentional stereogram; (c) Height analysis diagram

4 Conclusions

Two novel temperature-resistant and salt-tolerant co-polymers (OR-P and CR-P) were successfully synthesized by the addition of AMPS monomer via OR and CR polymerization, respectively. The reaction conditions for their preparation were optimized, and the resulting copolymers were fully characterized. The results of TGA and freeze-etching SEM indicated that the temperature resistance OR-P(AM/NaA/AMPS) and salt tolerance of CR-P(AM/NaA/AMPS) were enhanced by introducing functional AMPS monomer. The infrared spectra and elemental analyses provided strong evidence for grafting between AMPS and AM. The DLS results indicated that the two co-polymers had higher hydrodynamic radius. CR-P(AM/NaA/AMPS) had the highest hydrodynamic radius, indicating its higher tackifying ability. The microcosmic aggregation morphologies of the three polymers were investigated by freeze-etching SEM and AFM. The results indicated that all three polymers formed net aggregates, while the net aggregates of OR-P and CR-P were stronger than that of HPAM. The network structure of CR-P(AM/NaA/AMPS) was more obvious than that of OR-P(AM/NaA/AMPS) with increased salt tolerance, which was consistent with the DLS result.

Acknowledgments

This research was supported by the National Science and Technology Major Project (No. 2016ZX05011-003), The Certificate of China Postdoctoral Science Foundation (No. 2016M592241)

References

[1] Thomas S. Enhanced Oil Recovery - An Overview [J]. Oil & Gas Science & Technology, 2007, 63(1):9-19. https://doi.org/10.2516/ogst:2007060

[2] Wever D A Z, Picchioni F, Broekhuis A A. Polymers for enhanced oil recovery: A paradigm for structure-property relationship in aqueous solution [J]. Progress in Polymer Science, 2011, 36(11):1558-1628. https://doi.org/10.1016/j.progpolymsci.2011.05.006

[3] Bao M, Chen Q, Li Y, et al. Biodegradation of partially hydrolyzed polyacrylamide by bacteria isolated from production water after polymer flooding in an oil field [J]. Journal of Hazardous Materials, 2010, 184(1-3):105-110. https://doi.org/10.1016/j.jhazmat.2010.08.011

[4] Gao C. Viscosity of partially hydrolyzed polyacrylamide under shearing and heat [J]. Journal of Petroleum Exploration and Production Technology, 2013. 3(3): 203-206. https://doi.org/10.1007/s13202-013-0051-4

[5] Ladaviere C, Delair T, Domard A, et al. Studies of the thermal stability of maleic anhydride co-polymers in aqueous solution [J]. Polymer Degradation & Stability, 1999, 65(2):231-241. https://doi.org/10.1016/S0141-3910(99)00009-9

[6] Hong C Y, You Y Z, Pan C Y. Synthesis of Water-Soluble Multiwalled Carbon Nanotubes with Grafted Temperature-Responsive Shells by Surface RAFT Polymerization [J]. Chemistry of Materials, 2005, 17(9):2247-2254. https://doi.org/10.1021/cm048054l

[7] Zhang X F, Zhang X, Dai H. Synthesis and solution behaviour of temperature sensitive water soluble polymers bearing alkylaryl hydrophobes [J]. Plastics, Rubber and Composites, 2007, 36(7-8):365-369. https://doi.org/10.1179/174328907X237520

[8] Feng Y J, Wang Z Y, Li F S, et al. Study on solubility of hydrophobically modified polyacryamides [J]. Journal of Southwest Petroleum Institute, 2001, 23(6): 56-59. https://doi.org/10.3863/j.issn.1000-2634.2001.06.17(in Chinese)

[9] Sarsenbekuly B, Kang W, Fan H, et al. Study of salt tolerance and temperature resistance of a hydrophobically modified polyacrylamide based novel

functional polymer for EOR [J]. Colloids & Surfaces A Physicochemical & Engineering Aspects, 2017, 514:91-97. https://doi.org/10.1016/j.colsurfa.2016.10.051

[10] Corlay P, Delamaide E, Petrole I F D U. Status and outlook of the polymer flooding process [J]. Journal of Petroleum Technology, 1996, 48(3): 198-199. https://doi.org/10.1115/1.2792698

[11] Wu Y M, Zhang B Q, Wu T, et al. Properties of the forpolymer ofN-vinylpyrrolidone with itaconic acid, acrylamide and 2-acrylamido-2-methyl-1-propane sulfonic acid as a fluid-loss reducer for drilling fluid at high temperatures [J]. Colloid & Polymer Science, 2001, 279(9):836-842. https://doi.org/10.1007/s003960100494

[12] Zhong C, Jiang L, Peng X. Synthesis and solution behavior of comb-like terpolymers with poly(ethylene oxide) macromonomer [J]. Journal of Polymer Science Part A Polymer Chemistry, 2010, 48(5): 1241-1250. https://doi.org/10.1002/pola.23888

[13] Guo J, Shi X, Yang Z, et al. Synthesis of temperature-resistant and salt-tolerant surfactant SDB-7 and its performance evaluation for Tahe Oilfield flooding (China) [J]. Petroleum Science, 2014, 11(4): 584-589. https://doi.org/10.1007/s12182-014-0375-9 (in Chinese)

[14] Ding C, Ju B, Zhang S. Temperature resistance and salt tolerance of starch derivatives containing sulfonate groups [J]. Starch - St? rke, 2014, 66(3-4):369-375. https://doi.org/10.1002/star.201300127

[15] Song H, Zhang S F, Ma X C, et al. Synthesis and application of starch-graft-poly(AM-co-AMPS) by using a complex initiation system of CS-APS [J]. Carbohydrate Polymers, 2007, 69(1): 189-195. https://doi.org/10.1016/j.carbpol.2006.09.022

[16] Ma J, Cui P, Zhao L, et al. Synthesis and solution behavior of hydrophobic association water-soluble polymers containing arylalkyl group [J]. European Polymer Journal, 2002, 38(8):1627-1633. https://doi.org/10.1016/S0014-3057(02)00034-4

[17] Liu X, Jiang W, Gou S, et al. Synthesis and evaluation of novel water-soluble copolymers based on acrylamide and modular β-cyclodextrin [J]. Carbohydrate Polymers, 2013, 96(1):47-56. https://doi.org/10.1016/j.carbpol.2013.03.053

[18] He Y, Xu Z H, Wu F, et al. Synthesis and characterization of a novel amphiphilic copolymer containing β-cyclodextrin [J]. Colloid and Polymer Science, 2014, 292(7):1725-1733. https://doi.org/10.1007/s00396-014-3235-7

[19] Mao H, Qiu Z, Shen Z, et al. Hydrophobic associated polymer based silica

nanoparticles composite with core-shell structure as a filtrate reducer for drilling fluid at utra-high temperature [J]. Journal of Petroleum Science and Engineering, 2015, 129:1-14. https://doi.org/10.1016/j.petrol.2015.03.003

[20] Bara J E, Hatakeyama E S, Gabriel C J, et al. Synthesis and light gas separations in cross-linked gemini room temperature ionic liquid polymer membranes [J]. Journal of Membrane Science, 2008, 316(1-2):186-191. https://doi.org/10.1016/j.memsci.2007.08.052

[21] KAFOURIS, Demetris, THEMISTOU, et al. Synthesis and characterization of star polymers and cross-linked star polymer model networks with cores based on an asymmetric, hydrolyzable dimethacrylate cross-linker [J]. Chemistry of Materials, 2006, 18(1):119-143. https://doi.org/10.1021/ma0513416

[22] Yang F Y, Liu K Y, Han S Z. Synthesis of super high molecular weight co-polymer of AM/AA/AMPS and its salt by inverse suspension polymerization [J]. Journal of Beijing University of Chemical Technology, 2003, 30(2):5-9. https://doi.org/10.7688/j.issn.1000-1646.2013.02.07

[23] Ye M L, Han D, Shi L H. Studies on determination of molecular weight for ultrahigh molecular weight partially hydrolyzed polyacrylamide [J]. Journal of Applied Polymer Science, 2015, 60(3):317-322. https://doi.org/10.1002/(SICI)1097-4628(19960418)60:33.0.CO;2-O

Preparation and properties of temperature-tolerant viscoelastic particle oil-displacing agent

Cao Xulong

(Shengli Oilfield Company, Sinopec, Dongying, Shandong 257015, P R of China)

Abstract: In order to solve the problems of heterogeneity, high temperature and high salt in the oilfield exploitation, the technical idea of preparing particle oil displacement agents with small molecule crosslinking agents was broken through, and a series of particles type oil-displacing agents were synthesized with a variety of new macromolecular crosslinkers, and the network structure, mechanical properties, aging resistance and migration mechanism was studied. The research results showed that the macromolecular crosslinking agent could effectively improve the uniformity of the gel and endow the gel with excellent mechanical properties. The breaking elongation of the three gels prepared with macromolecular crosslinking agent was above 2 000%, the breaking strength was above 0.3 MPa, and could be compressed to 90% without being damaged. Among them, MCH0.5-30 had the best performance, the breaking elongation could reach up to 2 800%, and the breaking strength could reach up to 0.6 MPa. In addition, it had a high modulus retention rate after aging in brine with a salinity of 30 g/L at 85℃ for 90 days, showing excellent aging resistance. Macromolecular crosslinking agents provide new ideas for the preparation of high-performance oil-displacing agent, which are expected to be applied in harsh oil reservoirs with high temperature and high salinity.

Keywords: tertiary oil recovery; hydrogel; viscoelastic particle; oil-displacing agent

力化学改性木粉纤维制备水凝胶颗粒驱油剂*

孙焕泉[1] 曹绪龙[2] 姜祖明[3] 祝仰文[3] 郭兰磊[3]

(1. 中国石油化工股份有限公司,北京 100728;
2. 中国石油化工股份有限公司胜利油田分公司,山东 东营 257001;
3. 中国石油化工股份有限公司胜利油田分公司勘探开发研究院,山东 东营 257015)

摘要:木粉价廉、来源广,含大量木质纤维素,可生物降解,绿色环保。用木粉复合水凝胶作为驱油剂,可以有效降低生产成本。以丙烯酰胺为原料、力化学改性木粉(PWF-BA)为填充料、过硫酸钾为引发剂,N,N′-亚甲基双丙烯酰胺为交联剂,制备了改性木粉复合水凝胶。研究了复合水凝胶的弹性、黏性、抗剪切能力及在高温高盐条件下的抗老化性能。结果表明,经过化学处理后,木粉中的木质素和半纤维素基本被去除;随球磨时间增加,颗粒尺寸和结晶度显著降低,比表面积增加。当球磨时间为4 h时制得的木粉在聚丙烯酰胺基体中具有良好的分散性和界面相互作用,复合水凝胶的拉伸性能和在85℃盐水中的抗老化性能良好,可在三次采油中作为驱油剂使用。

关键词:木粉;纤维素;水凝胶;力化学改性;驱油剂

中图分类号:TE357.46;TE39 文献标识码:A

石油是一种不可再生的优质能源和基础化工原料,与我国的经济、人民生活、环境等密切相关[1]。目前,我国主要油田已经进入三次采油阶段,其中以聚丙烯酰胺及其衍生物为主的合成类聚合物因具有良好的水溶性和较高的黏度,是应用最广泛、效果最突出的聚合物驱油剂[2]。但当前国际油价持续走低,而且现有的具有良好驱油效果的聚丙烯酰胺驱油剂生产成本较高,不利于我国经济的可持续发展[3-4]。

木粉主要由纤维素、半纤维素和木质素组成。以木粉为代表的生物质资源具有可再生性、来源广泛[5]。虽然木粉具有生物可降解性,但部分木质纤维的降解过程极为漫长,这样会在一个时期内造成对环境的污染以及废弃物的堆积[6]。因此,如何高效地开发和利用这部分资源受到了广泛地关注。由于木质素和半纤维素的保护作用以及纤维素中大量的分子内氢键和分子间氢键,木粉存在刚性大、溶解性差、反应活性低以及与基体材料相容性差等缺点[7-10]。因此,需要对木粉进行改性处理来提高其反应活性和相容性。

本文报道的化学改性木粉复合水凝胶制备驱油剂的方法,可大大降低驱油剂的生产成本,同时高效利用低质木材资源,提高其使用价值。首先,通过碱酸以及力化学法对木粉进行改性处理,提高木粉的反应活性。将改性木粉分散到聚丙烯酰胺凝胶基体中制得复合凝胶,经烘干、粉碎获得颗粒驱油剂,研究了驱油剂的弹性、黏性、抗剪切能力及在高温高盐条件下的抗老化性能。

1 实验部分

1.1 材料与仪器

杨木粉纤维(PWF),74 μm(200 目),工业级,浙江德清林木质纤维公司;丙烯酰胺(AM)、N,N′-亚甲基双丙烯酰胺(MBA),分析纯,成都科龙化工试剂厂;过硫酸钾(KPS)、四甲基乙二胺(TMEDA)、氢氧化钠(NaOH),分析纯,天津市博迪化工有限公司;氯化钠(NaCl)、氯化钙($CaCl_2$)、六水氯化镁($MgCl_2 \cdot 6H_2O$),分析纯,天津市科密欧化学试剂有限公司;去离子水;盐水,矿化度 19 334 mg/L,1 L 去离子水中含 1.11 g $CaCl_2$、3.83 g $MgCl_2 \cdot 6H_2O$、27.31 g NaCl。

Nicolet IS10 红外光谱仪、Apreo S HiVoc 场发射扫描电镜,美国赛默飞世尔科技公司;Malvern Mas-terizer 2000 激光粒度仪,英国马尔文公司;D/MAX-ⅢX 射线衍射仪,荷兰皇家飞利浦公司;TAAR2000ex 旋转流变仪,美国 TA 公司;球磨机,德国飞驰公司;拉伸试验仪,美国 Instron 公司。

1.2 实验方法

1.2.1 木粉的化学处理与球磨改性

首先,将杨木粉在 20%的 NaOH 溶液中浸泡 2 h;接着将碱处理后的木粉用 1 mol/L H_2SO_4 溶液酸化处理 2 h,以除去大部分的木质素和半纤维素,使得被包裹的纤维素羟基裸露出来,提高其反应活性[11]。最后将酸处理的木粉用去离子水洗涤、抽滤直至滤液呈中性。将洗后的木粉置于 80℃真空烘箱中烘干至质量恒定。

通过球磨产生的高速剪切和挤压实现对木粉的进一步力化学改性。将不同球磨时间(0、0.5、1、2、4 h)获得的木粉分别命名为 PWF-BA0h、PWF-BA0.5 h、PWF-BA1 h、PWF-BA2 h 和 PWF-BA4 h。

1.2.2 改性木粉复合凝胶的制备

分别称取 60 g AM、25 μL TMEDA、80 mg KPS 和 6.5 mg MBA 溶于去离子水中,加入 10%(相对于单体的质量分数)经不同球磨时间改性的木粉,然后将混合物转移到 250 mL 烧杯中,在磁力搅拌作用下通氮除氧 15 min。最后将反应液转移到哑铃状模具(拉伸测试)中,于室温下反应 24 h。分别命名为 PWF-BA0h/PAM—PWF-BA4h/PAM 水凝胶。在反应条件和试剂用量不变的条件下,不加木粉作为对比,制备的聚丙烯酰胺水凝胶命名为 PAM 水凝胶。

1.2.3 测试与表征

(1) 红外测试。

用红外光谱仪分析力化学改性木粉 PWF-BA 的化学结构。

(2) 粒径及粒径分布测试。

用激光粒度仪表征经不同球磨时间处理后木粉的粒径及其分布,同时得到比表面积。

(3) X 射线衍射(XRD)测试。

用 X 射线衍射仪分析球磨改性后木粉的结晶度变化。

(4) 机械性能测试。

通过拉伸测试分析复合水凝胶的机械性能。将水凝胶裁剪成 4 mm×1 mm×20 mm 哑铃型样条,以 100 mm/min 的速率用拉伸试验仪将样条拉伸至断裂,记录应力—应变曲线。

(5) 扫描电镜(SEM)测试。

将凝胶样品经冷冻干燥后,在液氮中脆断获得断面,断面喷金后进行 SEM 测试。

(6) 老化性能测试。

在 70℃的真空烘箱中将凝胶样品烘干至恒重,粉碎筛分。将 0.5 g 干凝胶粉末缓慢加入到 100 mL 模拟胜利油田地下油藏环境的盐水溶液中,搅拌 10 min 制备凝胶颗粒悬浮液。凝胶颗粒在室温下溶胀 24 h 后,由于吸水膨胀粒径由 240 μm 变为 290 μm(颗粒的老化性质主要受高温和高盐的影响,常温下的溶胀过程不会造成颗粒的老化)。抽真空去除溶液中的氧气,然后将悬浮液转移到密封瓶中,放入 85℃的烘箱中老化,每隔一段时间取出部分悬浮液滴在旋转流变仪上,测量悬浮液的黏度,剪切速率为 7.34 s^{-1}。

2 结果与讨论

2.1 化学处理前后木粉的结构表征

化学处理前后木粉的红外图谱如图 1 所示。木质素芳香环骨架振动的特征峰位于 1 598、1 507、1 462 cm^{-1},半纤维素中醛酸伸缩振动的特征峰约在 1 743 cm^{-1}。与处理前的曲线相比,这些特征峰在处理后曲线中基本消失,表明经过化学处理后木粉中的木质素和半纤维素基本被除去。因此,认为经过化学处理后的木粉基本由纤维素组成。

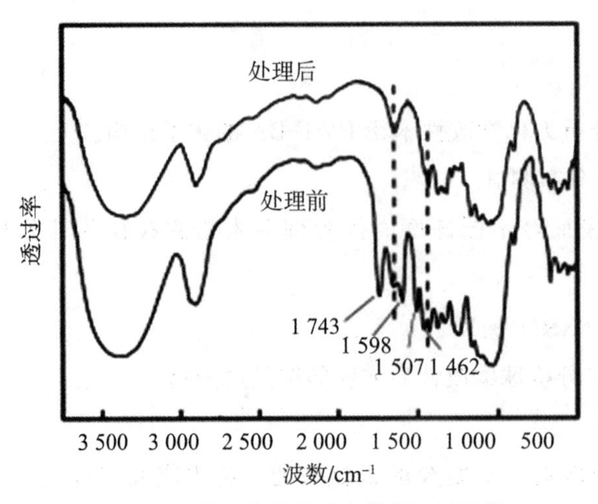

图 1 化学处理前后木粉的红外图谱

2.2 球磨时间对力化学改性木粉的影响

如图 2 所示,经过球磨处理后,木粉的平均粒径逐渐减小,粒径分布的宽度先增大后减小。球磨 0.5 h 后,木粉的平均粒径由 41.89 μm 急剧减小到 23.30 μm,比表面积由 0.24 m^2/g 增至 0.49 m^2/g。随着球磨时间进一步增至 4 h,木粉的平均粒径降至 8.89 μm,比表面积增至 0.99 m^2/g。由图 3 可见,球磨 1 h 导致木粉的平均粒径急剧下降,接近 65%,而进一步球磨 3 h 后导致平均直径进一步下降约 15%。究其原因,可能是大部分的原始木粉颗粒在 1 h 内被研磨成更小的颗粒,只有少量的原始颗粒没有被研磨,导致球磨开始时的粒度分布曲线变宽。随后,较小尺寸的原始颗粒难以被进一步磨细,而尺寸较大的颗粒被磨细,使得颗粒尺寸分布曲线变窄。总的来说,随着球磨时间的增加,PWF-BA 的尺寸明显减小,比表面积逐渐增大。

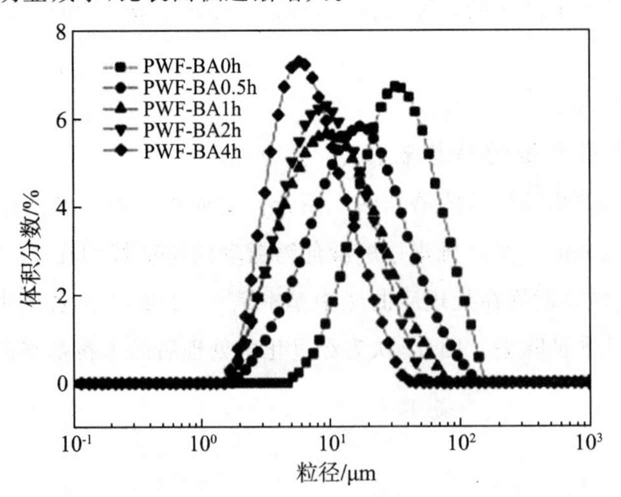

图 2 球磨时间对 PWF-BA 粒径分布的影响

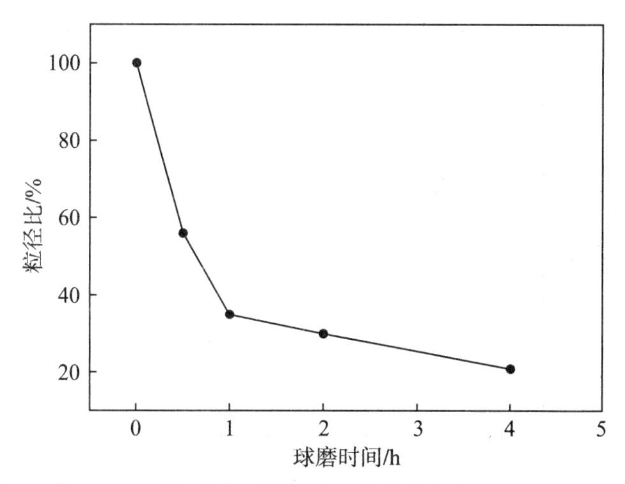

图 3 经过不同时间球磨后木粉的平均粒度与初始平均粒度之比

通过 XRD 分析了随着球磨时间的增加，PWF-BA 的结构变化情况。如图 4 所示，未经球磨处理木粉的衍射曲线(曲线 a)同时呈现纤维素Ⅰ峰和纤维素Ⅱ峰[12-13]。随着球磨时间的增加，PWF-BA 的 XRD 图谱中纤维素Ⅰ峰和纤维素Ⅱ峰强度均明显降低(曲线 b～e)。当木粉经过 4 h 球磨处理后，纤维素Ⅰ峰和纤维素Ⅱ峰几乎完全消失。通过以下方程计算结晶指数(CI)：$CI=(A_T-A_{AM})/A_T$，其中 A_T 为 XRD 图谱曲线下对应的总积分面积，A_{AM} 为力化学改性木粉的非晶区积分面积。通过对木粉球磨 4 h 后的 XRD 曲线(曲线 e)进行高斯拟合，得到非晶区的 XRD 曲线(曲线 f)。由图 5 可见，随着球磨时间的增加，PWF-BA 的结晶指数逐渐下降。球磨时间由 0 h 增至 4 h，PWF-BA 的结晶指数由 71%降至 23%，这是在球磨过程中木粉的结晶区被破坏所致。随着晶体体积的减小，非晶部分增大。由于球磨的作用，大分子链变得更短、更灵活和松散，导致木粉表面的大分子链反应性增加。

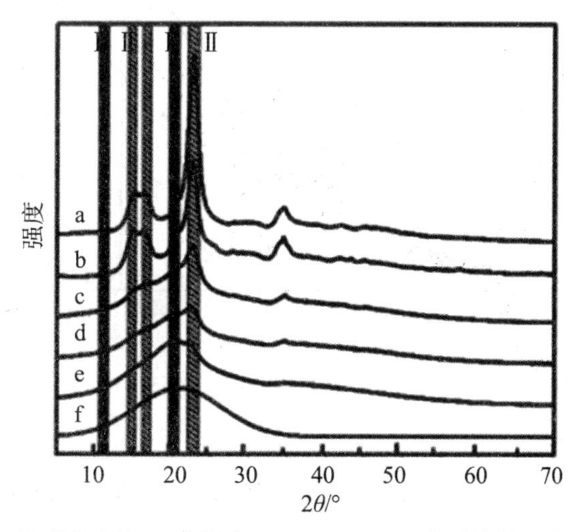

从 a～e 的球磨时间(h)分别为：0、0.5、1、2、4；f 为预计的无定形曲线。

图 4 球磨不同时间的 PWF-BA 的 XRD 图谱

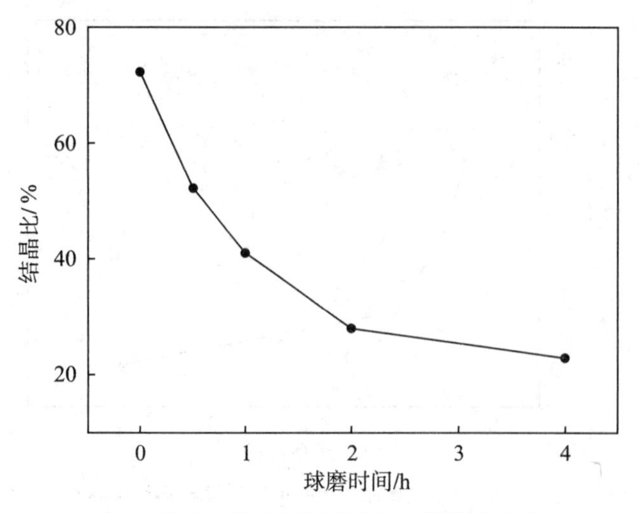

图 5 球磨时间对 PWF-BA 结晶指数的影响

2.3 PWF-BA/PAM 凝胶的机械性能

不同力化学改性时间的 PWF-BA 复合 PAM 水凝胶的应力-应变曲线如图 6 所示。随着球磨时间的增加,复合水凝胶的拉伸强度和断裂伸长率相对于 PWF-BA0h/PAM 样品的拉伸强度和断裂伸长率都有显著提高。在所有样品中,PWF-BA4h/PAM 水凝胶样品具有最大的拉伸强度(280 kPa)和最长的断裂伸长率(1 500%),其力学性能甚至优于不添加任何木粉的 PAM 水凝胶的力学性能。以上结果表明,球磨时间越长,PWF-BA 对 PAM 凝胶的增强效果越好。当球磨时间为 4 h 时,其可以作为有效的增强增韧填料提升 PAM 水凝胶的机械性能。

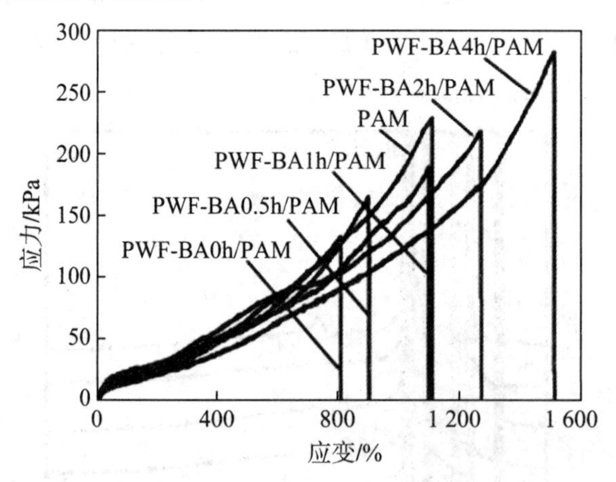

图 6 PAM 凝胶和 PWF-BA/PAM 凝胶的拉伸行为

2.4 复合凝胶的结构

相比于纯的 PAM 凝胶(图 7(a))和 PWF-BA4 h/PAM 凝胶(图 7(c)),PWF-BA0.5 h/PAM 凝胶(图 7(b))的孔径分布较宽,孔隙大小不一。PWF-BA4 h/ PAM 凝胶的孔径相对较小。PWF-BA0.5 h/PAM 凝胶孔径分布较广可能是由于 PWF-BA0.5 h 样品在水凝胶聚合过程中部分聚集在一起,阻碍了活性链的移动,导致交联网络不均匀。

PWF-BA0.5 h/PAM 的不均匀交联网络也是导致其力学性能较差的原因之一。而 PWF-BA4 h 由于粒径小、比表面积大,可以很好地分散在 PAM 基体中。此外,在球磨过程中产生了大量的新鲜纤维表面和新的羟基。与 PWF-BA0.5 h 相比,PWF-BA4 h 的比表面积大,新的羟基基团多,可以与更多的 PAM 链形成氢键作用,形成更致密的网状结构,孔径也更小。这也导致了与 PWF-BA0.5 h 相比,PWF-BA4 h 与基体的界面相互作用更好,如图 7(d)和(e)所示。图 7(d)中 PWF-BA0.5 h 与 PAM 基体界面之间清晰的痕迹表明两相间的相容性较差。相比之下,PWF-BA4 h 由于其较大的比表面积和增加的界面相互作用,可以很好地分散并牢固地嵌入 PAM 基体中(图 7(e))。这可以解释为什么 PWF-BA4 h/PAM 样品的力学性能优于纯 PAM 样品的力学性能。

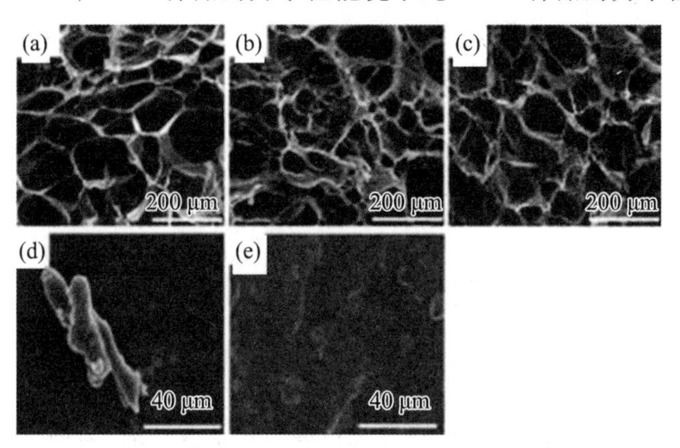

完全溶胀后冻干制备的 PAM 凝胶(a)、PWF-BA0.5 h/PAM 凝胶(b)、PWF-BA4 h/PAM 凝胶(c);未溶胀冻干制备的 PWF-BA0.5 h/PAM 凝胶(d)、PWF-BA4 h/PAM 凝胶(e)。

图 7 PAM 凝胶和 PWF-BA/PAM 凝胶的扫描电镜照片

2.5 凝胶颗粒的老化性能

在前期工作中,已经研究了 PAM 凝胶颗粒在 85℃盐水溶液中的老化机理[14]。在老化初期,交联点之间的链断裂,形成线性支链。由于交联键数的减少,对分子链的限制减少,凝胶的分子链向外扩展,流体力学体积增大,黏度明显提高。同样,对于 PWF-BA/PAM 水凝胶,由于外部支链上的约束比内部支链上的约束小,最初的部分交联也断裂为支链,最外层的支链首先从网络中释放出来。由于缺乏交联网络的限制,交联的 PAM 颗粒增大,流体力学体积增大,导致黏度显著增加,将这一阶段称为第一阶段。经过较长时间的老化,所有交联网络逐渐分解为线性链,随着自由支链的进一步断开,凝胶颗粒的黏度明显降低,导致线性链变短,将这一阶段称为第二阶段。如图 8 所示,纯 PAM 样品的黏度在第 3 d 达到最大值,PWF-BA0.5 h/PAM 凝胶的黏度在第 5 d 达到最大值,而 PWF-BA4 h/PAM 凝胶的黏度在第 9 d 达到最大值。之后这 3 个样品的黏度都呈下降趋势。老化 13 d 后,PWF-BA4h/PAM 凝胶仍保持较高的黏度值,但纯 PAM 的黏度在第 4 d 时即急剧下降,第 13 d 时继续下降至低于初始值。这可能是由于前文提到的 PWF-BA4 h 与 PAM 链之间存在更好的界面相互作用,从而限制了第一阶段的过程,使 PWF-BA4 h/PAM 凝胶具有更好的抗老化性能。总的来说,PWF-BA4 h/PAM 颗粒在高温、高盐条件下具有优异的抗老化性能。此外,PWF-BA4 h/PAM 水凝胶良好的力学性能表明其颗粒经过孔喉后不易破碎。因此,具有良好的机械性能和抗老化性能的 PWF-BA4 h/PAM 复合水凝胶颗粒有望作为驱油剂使用。

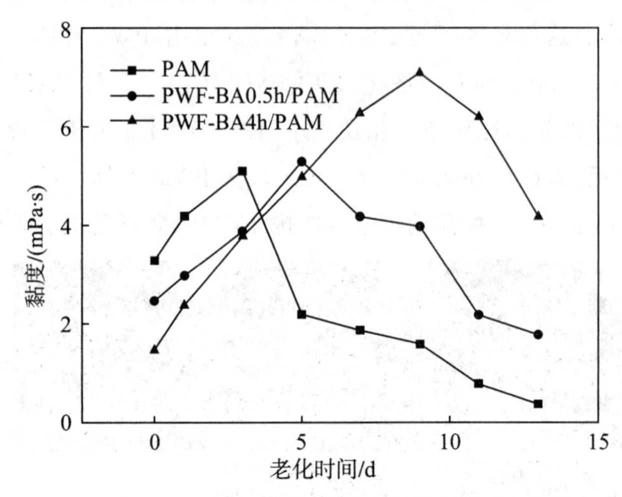

图 8 PAM 凝胶和 PWF-BA/PAM 凝胶在盐水老化过程中的黏度变化

3 结论

将杨木粉(PWF)经碱和酸处理,再通过球磨力化学改性后,与丙烯酰胺、N,N′-亚甲基双丙烯酰胺、过硫酸钾、四甲基乙二胺混合,制得 PWF-BA/PAM 水凝胶。经过化学处理和球磨力化学处理后,PWF 中的木质素和半纤维素基本被去除,颗粒尺寸和结晶度显著降低,比表面积增加。球磨处理 4 h 的力化学改性木粉 PWF-BA4 h 在 PAM 基体中具有良好的分散性和界面相互作用。与纯 PAM 水凝胶相比,PWF-BA4 h/PAM 水凝胶具有更好的拉伸性能和抗老化性能。利用力化学改性木粉 PWF-BA 制备聚丙烯酰胺水凝胶可以减少环境污染和缓解资源短缺,在石油工业中的应用前景良好。

参考文献

[1] 江泽民.对中国能源问题的思考 [J]. 上海交通大学学报,2008(3):345-359.

[2] 万刚,马超,赵林.三次采油用耐温抗盐聚合物的性能评价 [J]. 广东化工,2015,42(16):271-274.

[3] 冈秦麟. 论我国的三次采油技术 [J]. 油气采收率技术. 1998(4):3-9.

[4] 姜祖明,苏智青,黄光速,等.预交联共聚物驱油剂高温高盐环境下长期耐老化机理研究 [J]. 油田化学,2010,27(2):166-170.

[5] 李思远,杨伟,杨鸣波.木塑复合材料挤出成型工艺及性能的研究 [J]. 塑料工业,2003(11):22-24.

[6] 靳玲.无渗透钻井液处理剂合成与表征 [D].成都:四川大学,2007:78.

[7] SANNIGRAHI P,RAGAUSKAS A J,TUSKAN G A. Poplar as a feedstock for biofuels:A review of compositional characteristics [J]. Biofuels,Bioproducts and Biorefining,2010,4(2):209-226.

[8] NISHIYAMA Y,SUGIYAMA J,CHANZY H,et al. Crystal structure and hydrogen bonding system in cellulose I_α from synchrotron X-ray and neutron fiber diffraction [J]. J Am Chem Soc,2003,125(47):14300-14306.

[9] ZHAO Yan, LU Wenjing, CHEN Jiajun, et al. Research progress on hydrothermal dissolution and hydrolysis of lignocellulose and lignocellulosic waste [J]. Front Environ Sci Eng,2014,8(2):151-161.

[10] 滕国敏,张勇,万超瑛,等.木塑复合材料的界面改性方法 [J]. 化工新型材料,2005(5):7-9.

[11] SUN Q,FOSTON M,MENG X,et al. Effect of lignin content on changes occurring in poplar cellulose ultrastructure during dilute acid pretreatment [J]. Biotechnol Biofuels,2014,7(1):150.

[12] OH S Y,YOO D I,SHIN Y,et al. Crystalline structure analysis of cellulose treated with sodium hydroxide and carbon dioxide by means of X-ray diffraction and FTIR spectroscopy [J]. Carbohydr Res,2005,340(15):2376-2391.

[13] OUDIANI A E,CHAABOUNI Y,MSAHLI S,et al. Crystal transition from cellulose I to cellulose II in NaOH treated Agave americana L. fibre [J]. Carbohydr Polym,2011,86(3) :1221-1229.

[14] JIANG Zuming,SU Zhiqing,LI Li,et al. Antiaging mechanism for partly crosslinked polyacrylamide in saline solution under high-temperature and high-salinity conditions [J]. J Macromol Sci B,2013,52(7):113-126.

Preparation of hydrogel particle oil-displacing agent by wood powder fiber with mechanochemical modification

Sun Huanquan[1] Cao Xulong[2] Jiang Zuming[3]
Zhu Yangwen[3] Guo Lanlei[3]

(1. China Petroleum & Chemical Corporation, Beijing 100728, P R of China;
2. Shengli Oilfield Company, Sinopec, Dongying, Shandong 257001, P R of China;
3. Exploration and Development Research Institute, Shengli Oilfield Company, Sinopec, Dongying, Shandong 257015, P R of China)

Abstract: Wood powder is cheap and has a wide range of sources. It contains a large amount of lignocellulose, which is biodegradable and environmentally friendly. The oil-displacement agent compositing wood powder and hydrogel can effectively reduce the cost of production. This work used acrylamide (AM) as monomer, poplar wood fiber mechanochemically modified by ball milling (PWF-BA) as filler, potassium persulfate (KPS) as initiator, N, N'-methylenebisacrylamide (MBA) as crosslinking agent, to synthesize wood powder composite hydrogel. The elasticity, viscosity, shear resistance and anti-aging property of the composite hydrogel under high temperature and high salt conditions were studied. The results showed that after chemical treatment, the lignin and hemicellulose in the wood powder were basically removed. As the milling time increased, the particle size and crystallinity decreased significantly, and the specific surface area increased. When the ball milling time was 4 h, the wood powder obtained had good dispersibility and interfacial interaction in the polyacrylamide matrix. The tensile property of the composite hydrogel and the anti-aging performance in salt water at 85℃ were good. The hydrogel was expected to be used as an oil displacing agent in tertiary recovery.

Keywords: wood powder; fiber; hydrogel; mechanochemical modification; oil-displacing agent

Synthesis and the delayed thickening mechanism of encapsulated polymer for low permeability reservoir production

Gong Jincheng[a] Wang Yanling[a,*] Cao Xulong[b,*]
Yuan Fuqing[b] Ji Yanfeng[b]

[a] School of Petroleum Engineering, China University of Petroleum (East China), Qingdao 266580, China

[b] Exploration and Development Research Institute, Shengli Oilfield, Dongying 257000, China

Abstract: Polymer flooding is a common oil recovery technology. However, the polymer solution is difficult to be injected into low permeability reservoirs due to its high viscosity, making the reservoir reserves unable to be produced. To help tackle this problem, the encapsulated polymers with different shell-core ratios were synthesized with polyurethane as the shell material, and the delayed thickening of polymer latex in the water medium was achieved. The encapsulated polymers were characterized by DLS, SEM, TEM, and TGA to determine the shell-core ratio. The results showed that the encapsulated polymers with the shell-core ratio of 0.25 had a uniform particle size distribution (297.8、531.8 nm), smooth and regular shaped shell, and high thermal degradation temperature (221.6℃). These features facilitated the delayed thickening and stable dispersion of encapsulated polymers in the aqueous medium. When the concentra-tion of encapsulated polymer was 1 500 mg/L, the flocculation of microcapsules in aqueous medium was negligible, and the viscosity of the fluid increased from 1 mPa・s to 12.84 mPa・s after 9 d. In addition, this work elucidated the delayed thickening mechanism of encapsulated polymer, and it contributed to future applications of this material in low permeability reservoirs.

Keywords: Microencapsulation Polyurethane Polymer thickener Shell-core ratio Delayed thickening

1 Introduction

As a basic raw material for the energy and chemical industries, petroleum is becoming more and more in demand as moderniza-tion advances. In the face of continuous consumption of oil resources in medium and high permeability reservoirs, the percentage of reserves in low permeability reservoirs is increasing. Therefore, improving recovery from low permeability reservoirs is particularly important[1-3]. Partially hydrolyzed polyacrylamide (HPAM) is widely used as a chemical oil displacement agent due to its low cost and excellent viscoelasticity after dissolving in water. However, due to its limited temperature and salt resistance, it cannot be used in high-temperature and high-salinity reservoirs[4,5]. Several attempts have been made to increase the adaptability of polymers to high-temperature and high-salinity by introducing temperature and salt-resistant sulfonic acid group ($—SO_3H$) or rigid pyrrolidone (C_5H_9NO) into the main chain of polyacrylamide[6-8]. However, the problems of high injection pressure and viscosity loss of the polymer solution in low permeability reservoirs(<50 人 10^{-3} lm^2) remain unsolved[9]. Given the small pore throat size of low permeability reservoir and the micron hydrodynamic size of polymer molecules in aqueous solution, the pressure required for the polymer solution injection is large[10]. Moreover, the polymer molecular chains are easy to be adsorbed and mechanically captured when moving in the formation, decreasing the solution concentration and its viscosity[9,11]. This not only compromises the expected oil displacement effect but also causes reagent waste.

The unfavorable formation conditions have severely limited the application of polymer flooding. In recent years, encapsulation technologies developed in oilfield-related fields, such as tubing corrosion inhibitors[12], drilling fluid additives[13], and surfactant flooding[14], provide the possibility to solve this problem. Microcapsules are closed micron containers with a shell-core structure, where the core material is encapsulated by the shell material made of natural or synthetic polymers. It can release the core material slowly and can protect the core material from the external environment to a great extent[15]. The selection of the shell material is an important factor affecting its releasing function. Polyurethane (PU) is made by polymerization of isocyanate compounds and diols or polyols at the two-phase interface, and is widely used to prepare various microcapsules because of its good moisture permeability and air permeability. The microcapsules with PU as the shell can not only maintain the original character-istics of the core material, but also have good mechanical perfor-mance and abrasion resistance, whereby ensuring a better slowly-releasing effect of the microcapsule on the core material[16-18]. Among the shell-forming monomers, aliphatic isocyanates such as isophorone diisocyanate (IPDI) and polyether diols such as polyethylene glycol (PEG) are the most commonly used because of their better performance with respect to parameters including low toxicity, toughness, temperature resistance, hydrolysis resistance, and solvent resistance[19-21].

In order to reduce the injection pressure and the migrational viscosity loss, we used microcapsules to encapsulate the polymer latex in this research. By slowly releasing the polymer into the aqueous medium, the initial viscosity of the fluid was reduced and the utilization rate of the polymer was improved. In this study, we used inverse emulsion polymerization to introduce $—SO_3H$ into polyacrylamide molecules to improve the adaptability of polymers to high-temperature and high-salinity reservoirs. In addition, we used in situ polymerization and select IPDI and PEG to prepare encapsulated polymer thickener (ECP) with polyurethane as shell filled with polymer latex. Hence, we prepared ECPs with different shell-core ratios, and characterized the ECPs in detail. We further analyzed the influence of shell-core ratios on the performance of delayed thickening and dispersion stability. In addition, the mechanism of delayed thickening was also revealed.

2 Materials and methodology

2.1 Materials

2-acrylamide-2-methylpropane sulfonic acid (AMPS) is a chemical quality product from Weifang Quanxin Chemical Co., Ltd. Acrylamide (AM), ammonium persulphate (APS), sodium hydrogen sulfite (SHS) and sodium hydroxide (NaOH) are analytically pure reagents purchased from Chengdu Kelong Chemicals Company, China. Isophorone isocyanate (IPDI) is an industrial product purchased from Guangdong Haoyi Chemical Technology Co., Ltd. Polyethylene glycol (PEG 2000) and 1,4-butanediol (BDO) are industrial products obtained from Guangzhou Guangjia Chemical Co., Ltd. Nhexane, sorbitan fatty acid ester (Span80) and sorbitan monooleate polyoxyethylene ether (Tween80) are analytically pure reagents purchased from Sinopharm Chemical Reagent Co., Ltd. The compo-nents of simulated formation water are listed in Table 1.

2.2 Synthesis

2.2.1 Synthesis of polymer latex

Polymer latex with thickening function was synthesized by inverse emulsion polymerization[22]. The inverse emulsion was prepared by mixing a solution of core-forming monomers, AM and AMPS, with a solution of emulsifier in n-hexane. Water-soluble initiators were then used to develop monomer droplets into polymer latex particles [Fig. 1(a)]. The detailed implementation process is as follows:

The AM (12 g) and AMPS (4 g) were dissolved in deionized water (8 g) at 25℃ under magnetic stirring in a vessel, NaOH solution was used to adjust pH to 7.5. Then, the solution was emulsified in n-hexane (14.5 g) containing Span80 (3.75 g) and Tween80 (1.25 g) stirred at a speed of 1 000 rpm for 20 min. After dispersion and stabilization of monomer droplets in organic phase, the inverse emulsion was stirred under N_2 at 400 rpm. Next, APS (0.003 g) and SHS (0.001 5 g) as initiators were added to the inverse emulsion, followed by polymerization at 45℃ for 4 h to obtain polymer latex. Latex particles were obtained by washing the latex with n-hexane repeatedly. The latex particles were dried at 80℃ for 2 h, and then weighed for yield calculation.

2.2.2 Synthesis of encapsulated polymer thickener (ECP)

ECPs were prepared by encapsulating polymer latex particles with PU prepolymer via in situ polymerization[23]. To this end, PU prepolymer was prepared by a solution of shell-forming monomers, IPDI and PEG (IPDI : PEG = 2 : 1, molar ratio), in n-hexane. Then the prepolymer solution was mixed with polymer latex. Due to the existence of hydrophilic polyether and hydrophobic aliphatic ring, the prepolymer can be adsorbed on the interface between the hydrated layer and the organic phase, and form the prepolymer film on the surface of latex particle[22-24]. Finally, the chain extender was used to grow the prepolymer film into the PU shell [Fig. 1(b)].

Shell-core ratio is defined as the mass ratio of shell-forming monomers (IPDI and PEG) to core-forming monomers (AM and AMPS). Four different ECPs ECP-1, ECP-2, ECP-3 and ECP-4 with shell-core ratios of 0.5, 0.375, 0.25 and 0.125, respectively, were prepared to investigate the effect of shell-core ratio on the performance of ECP. The shell-forming monomers (8, 6, 4 and 2 g) were dissolved in n-hexane (5.3 g) at 25℃ with N_2 gas purging before a few drops of catalyst were added to the solution. The solution was then stirred for 4 h while the temperature gradually increased to 60℃ to obtain PU prepolymer solution. After that, the PU prepolymer solution was gently added to the resultant polymer latex in 2.2.1 under mechanical agitation at 1 000 r/min for 20 min to form prepolymer film. Then the chain extender, BDO (4 wt% of shell-forming monomers mass) was gently added to the mixture under N_2, triggering the polymerization at 50℃ for 4 h to obtain the ECP emulsion. Finally, the ECPs were separated using vacuum filtration. The solids were washed with n-hexane, dried at 80℃ for 2 h, and weighed for the yield calculation.

2.3 Characterization

2.3.1 Shell extraction

The synthesized ECPs were dispersed in deionized water, and the suspension was aged at 120℃. The PU shells will crack at this temperature and release latex particles. When the solution viscos-ity reached the maximum, it can be considered that the latex par-ticles had been completely released and dissolved in the water. The undissolved shells were extracted from the suspension after cen-trifugation, and dried at 80℃ for 2 h for chemical structure characterization.

2.3.2 FT-IR

The functional groups of polymer latex, PU shell and ECP were characterized using the KBr tablet method at room temperature with a Fourier Transform Infrared Spectrophotometer (Nicolet is5 ATR-FTIR). The scanning wavenumber range was 650～4 000 cm^{-1}.

Table 1 Components of simulated formation water.

Component	$CaCl_2$	$MgCl_2$	NaCl	Total salinity
Content(mg/L)	1 142.7	404.3	17 787.0	19 334.0

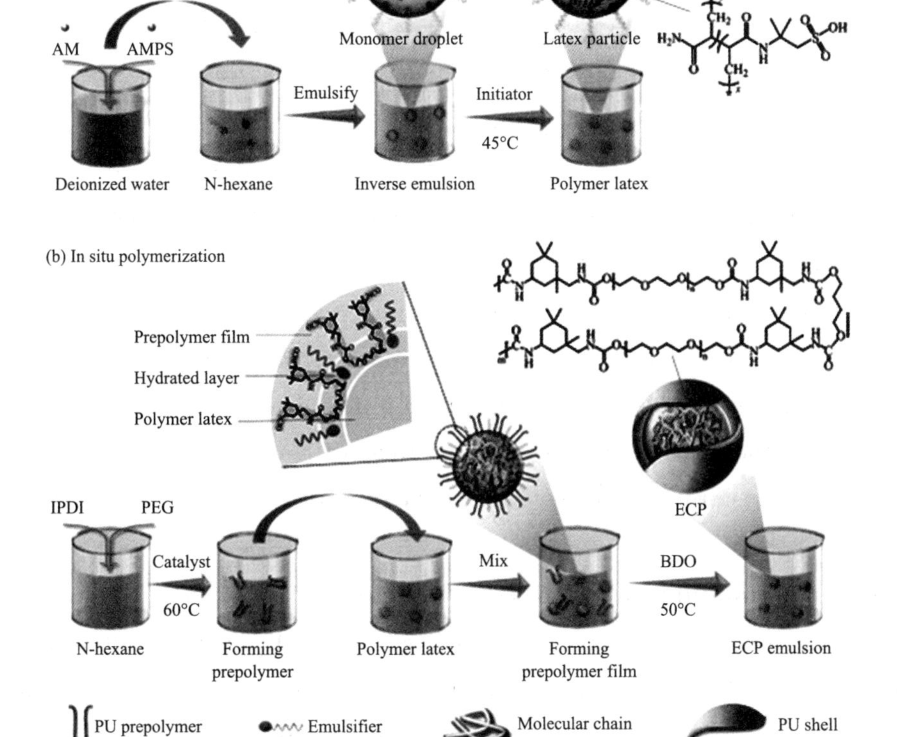

Fig. 1 Graphical synthesis mechanism of ECP.

2.3.3 ^{1}H NMR

The ^{1}H NMR spectrum of the PU shell was measured on a nuclear magnetic resonance spectrometer (Bruker-400 MHz) using $CDCl_3$ as the solvent.

2.3.4 Particle size distribution

Polymer latex and the four different ECP emulsions were diluted with n-hexane to the concentration of 50 mg/L, sonicated until the dispersion was uniform. Particle size distribution was tested by dynamic light scattering (DLS, Malvern, Zetasizer Nano ZS).

2.3.5 Morphology of microcapsules

Surface morphology and shell thickness of the four different ECPs were observed by scanning electron microscope (SEM, JEOL Ltd. JSM-6700F) and transmission electron microscope (TEM, HT-7700). Before the SEM analysis, the samples were dried on silicon wafers and placed under vacuum, flushed with argon, and then gold-sprayed to increase the conductivity of the samples. The sam-ples were placed on the microgrid and tested after infrared drying with an exposure time of 500 ms for the TEM analysis.

2.3.6 Thermogravimetric analysis (TGA)

A thermogravimetric analyzer (TGA, STARe, Mettler Toledo) was used to evaluate the thermostability of latex particles and four ECPs. The temperature range was 30～600℃, the heating rate was 10℃/min, and the N_2 flow rate was 20 mL/min.

2.4 Preparation of ECP suspensions

At room temperature, the ECP emulsion was gently dripped into the simulated formation water under mechanical agitation at 500 r/min for 30 min. The concentration of ECP suspension was 1 500 mg/L.

2.5 Delayed thickening performance of ECP suspensions

The ECP suspensions were transferred into ampoule bottles and blew N_2 for deaeration. The bottles were sealed and placed in the oven at 75℃ for aging. The change rule of the apparent viscosity of the fluid with the aging time was measured using a high-temperature rheometer (MCR301, Anton Paar) at 75℃ and a shear rate of 7.34 s^{-1}.

2.6 Dispersion stability of ECP suspensions

The ECP suspensions were transferred into the test bottles. Turbiscan stability analyzer (Tower, Formulaction) was used to evaluate the stability of the suspensions for 12 h at 75℃. The instrument emitted pulsed near-infrared light ($\lambda=880$ nm). Since the ECP suspensions were opaque fluids, the backscattering detector can detect the backscattered light intensity (BS) every 40 um along the height of the bottle. Therefore, the variation of BS with sample height at different times can be obtained. The expression of BS is shown in equation (1).

$$BS=\sqrt{\frac{3\varphi(1-g)Q_s}{2d}} \tag{1}$$

where φ is the volume fraction of particles, d is the mean diameter of particles, g and Q_s are the optical parameters given by Mie's the-ory. In general, BS is proportional to u, and is inversely proportional to d. However, it is noted that according to Mie's theory, when d is less than 880 nm, BS is proportional to d.

2.7 Delayed thickening mechanism of ECP

The ECP suspension after deaeration was aged at 75℃. The vis-coelasticity of the fluid under different aging times was measured by a rheometer (MCR301, Anton Paar). The vibration frequency f was 0.1～10 Hz and the amplitude c was 10%. Atomic force micro-scopy (AFM, Nanoscope IIIa, Bresso) was used to test the molecular configuration in the fluid under different aging times. The working frequency was 86 kHz and the force constant was 0.12 N/m. 1 mL of sample fluid was added dropwise to the mica sheet and dis-charged simultaneously with the N_2 gas flow.

3 Results and discussions

3.1 Preparation and characterization

The shell-core ratio is the main factor affecting the delayed thickening performance of ECP. Decrease in shell-core ratio can increase the proportion of polymer latex for thickening yet reduce the thickness and strength of the shell, making the shell easier to rupture, and shortening the delayed thickening time. Therefore the shell-core ratio of ECP needs to be determined to balance the thickening ability and release time.

3.1.1 Synthesis

The yields of the synthesized latex particles and ECPs were cal-culated by equations (2) and (3):

$$Y_C(\%)=\frac{W_C}{W_{CM}} \tag{2}$$

$$Y_{ECP}(\%)=\frac{W_{ECP}}{W_{CM}+W_{SM}} \tag{3}$$

where Y_C is the yield of latex particles, Y_{ECP} is the yield of ECP, W_C is the mass of latex particles after being washed and dried, W_{CM} is the dosage of core-forming monomers, W_{ECP} is the mass of ECPs after being washed and dried, and W_{SM} is the dosage of shell-forming monomers.

The yield of latex particles is 97.8%, while all four yields of ECPs are lower than that of polymer latex (Fig. 2). Increasing the shellcore ratio leads to a lower yield. This can be explained that using more prepolymers is not conducive to the formation of prepolymer film due to the increasing viscosity of the organic phase[25]. A lar-ger amount of prepolymers could also make the latex particle sur-face saturated with the adsorbed prepolymers. As such, excessive prepolymers will compromise the yields.

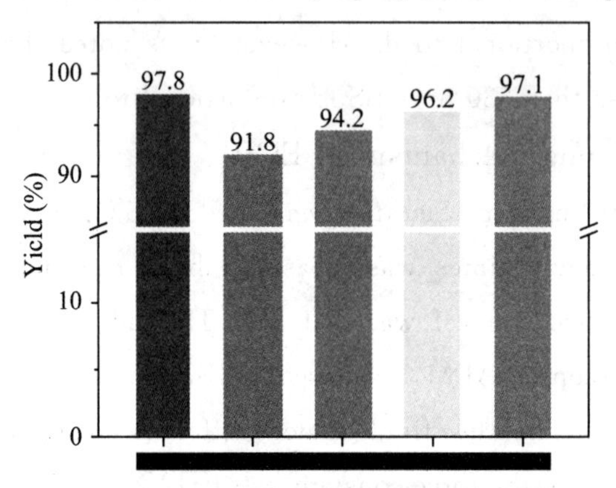

Fig. 2 Yields of polymer latex core and encapsulated polymers.

The solid contents of the synthesized ECP emulsions need to be determined for the preparation of a fluid with a predefined ECP concentration. The actual solid content and theoretical solid con-tent of the ECP emulsion were calculated by equations (4) and (5).

$$S_A(\%)=\frac{W_{ECP}}{E_{\mathrm{Emulsion}}} \tag{4}$$

$$S_T(\%)=\frac{W_{CM}+W_{SM}}{E_{\mathrm{Emulsion}}} \tag{5}$$

where S_A is the actual solid content, S_T is the theoretical solid content, and W_{Emulsion} is the total mass of the ECP emulsion. The decline in yield led to a larger gap between S_A and S_T (Fig. 3).

3.1.2 Chemical structure characterization

FTIR spectroscopy was utilized to characterize the latex particle, PU shell and ECP (Fig. 4a). The FTIR spectrum of latex particle shows the stretching vibration peak at the area of 3 307 cm^{-1}, which is related to N—H bond on-NH_2. The stretching vibration peak at 1 665 cm^{-1} is associated with C═O on the amide group. These characteristic peaks indicate the presence of AM in the molecular chain[26,27]. The peaks at 1 176 cm^{-1} and 1 031 cm^{-1} are caused by the stretching vibration of the O═S═O bond in the —SO_3H group[28,29], which confirms the preparation of latex modified by AMPS. In the spectrum of PU shell, the absorption peaks at 2 923 cm^{-1} and 2 848 cm^{-1} are related to $—CH_3$ in the aliphatic ring and $—CH_2—$ in the polyether chain, respectively. The characteristic peaks at 1 740 cm^{-1}, 1 553 cm^{-1}, and 1 117 cm^{-1} are related to C═O, N—H, and C—O—C in the carbamate bond, respectively[30]. These peaks indicate that the shell contains the PU prepolymer. The absence of NCO labelled peak near 2 260 cm^{-1} indicates the NCO groups in the PU prepolymer react completely with the —OH groups. Because the relative reaction rate between NCO and BDO is 10 times higher than that between NCO and aqueous medium, polyurethane is preferred to polyurea as the shell material[31].

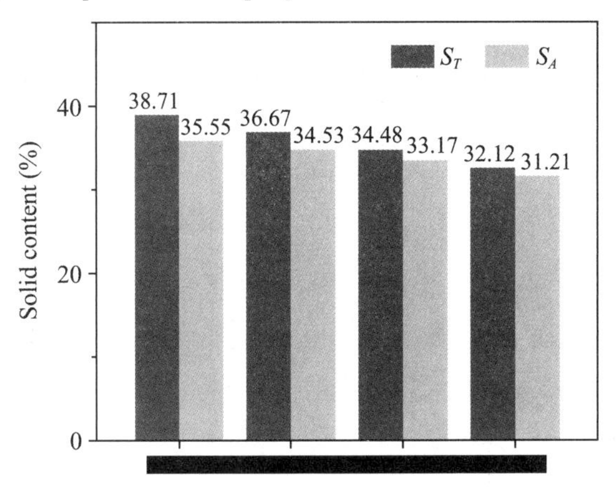

Fig. 3 Solid content of ECP emulsions.

^1H NMR was used to further confirm the shell material as the polyurethane [Fig. 4(b)]. Solvent peaks are at $\delta=2.50\times10^{-6}$ and 3.34×10^{-6}. The peaks at a, b and c (0.86×10^{-6}, 0.84×10^{-6}, 1.24×10^{-6}) are the proton signals of $—CH_3$ and $—CH_2—$, which repre-sent the aliphatic ring structure. The peaks at d, e (2.73×10^{-6}, 2.28×10^{-6}) represent the protons on C attached to urethano. And the peaks at f, g (4.80×10^{-6}, 4.05×10^{-6}) are associated with —NH— of urethano. The large number of $—CH_2—$ on the polyether chain results in a proton peak with a large integrated area at h (3.51×10^{-6}). In addition, two polyol proton peaks with small inte-grated areas at i and j (4.65×10^{-6}, 1.99×10^{-6}) demonstrate the pres-ence of BDO with less dosage. Thus further proving the shell material is polyurethane.

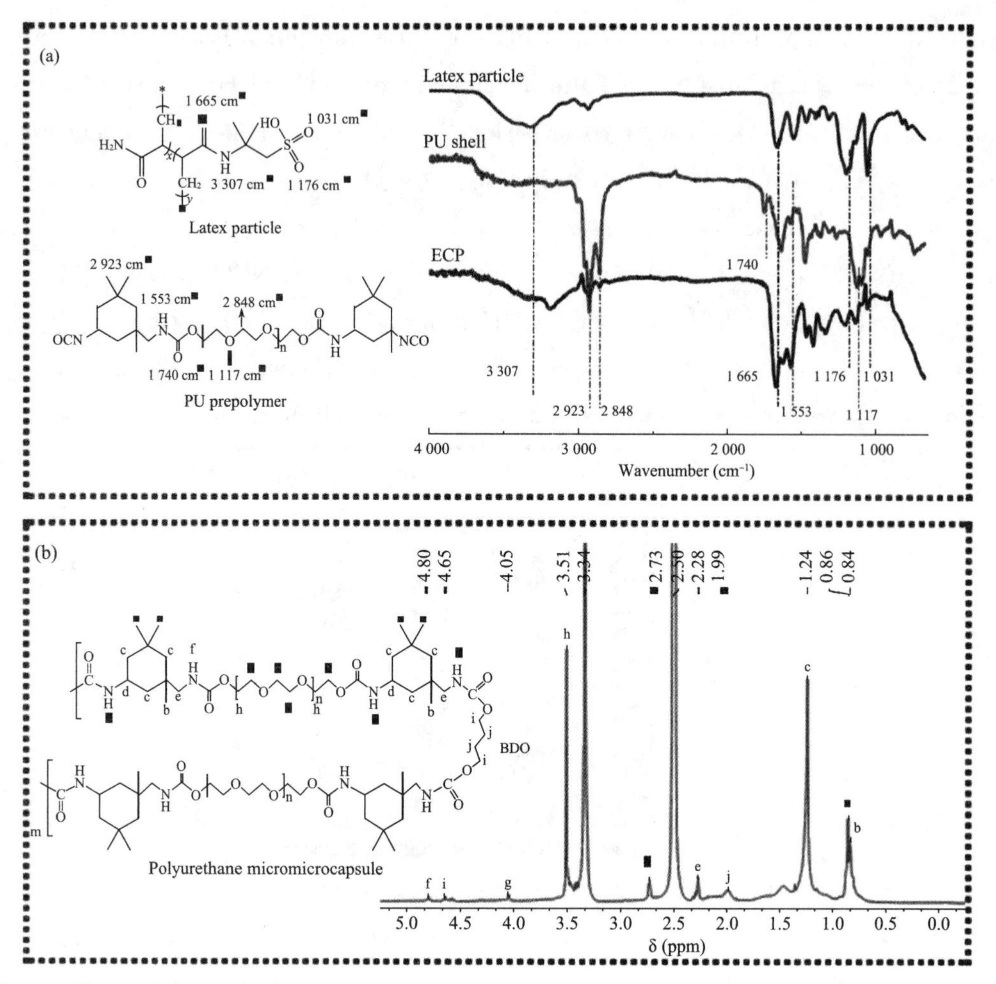

Fig. 4 Chemical structure characterization: (a) FT-IR spectra of latex particle, PU shell and ECP; (b)^{1}H NMR spectrum of PU shell.

In FTIR spectrum of ECP (Fig. 4a), the absorption peaks of the —SO_3H group in the latex particles exist at 1 176 cm^{-1} and 1 031 cm^{-1}, and the absorption peaks of the carbamate bond in the PU shell exist at 1 553 cm^{-1} and 1 117 cm^{-1}. These features confirm that the PU shell successfully encapsulates the latex particle.

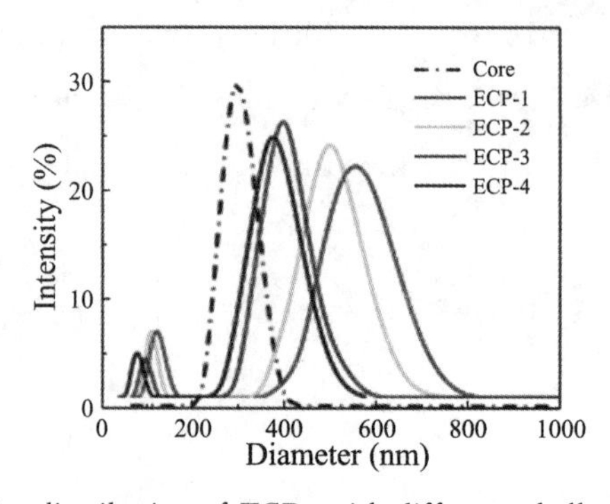

Fig. 5 Size distribution of ECPs with different shell-core ratios.

3.1.3 Size distributions

Size distributions of four ECPs are shown in Fig. 5. The mean thickness of the shell can be calculated from equation (6), where 6_{mean} is the mean shell thickness (nm), $d_{ECP\text{-}mean}$ is the mean size of ECP (nm), and $d_{Core\text{-}mean}$ is the mean size of polymer latex core (nm).

$$6_{mean}\ 1/4\ d_{ECP\text{-}mean} - d_{Core\text{-}mean}$$

ECPs' size increases with a larger shell-core ratio. This is due to the increase of prepolymer dosage can lead to a larger prepolymer film thickness. The mean thickness of the shell increases significantly from 30.1 nm to 105.9 nm with the shellcore ratio increasing from 0.125 to 0.5 (Table 2).

ECP-3 and ECP-4 show a lower uniformity of particle size than ECP-1 and ECP-2, indicating that the increase of shell-core ratio will lead to uneven shells (see detailed discussion in 3.1.4). Small particles can be observed in the size distributions of ECPs, which is due to the aggregation of prepolymers that are not adsorbed on the core surface to form microspheres[32]. However, the number of these self-polymerized microspheres is small due to the low reaction activity between isocyanates. Otherwise, small size and uniform size distribution will be conducive to the stable dispersion of colloidal particles in the fluid[33].

3.1.4 Shell morphology

The morphology of the microcapsule shell is one of the key factors affecting the utilization and delayed thickening effect of the latex core. The spherical microcapsule with a smooth surface and regular shape is preferable, which can improve the encapsulation performance of the microcapsule shell to the core.

SEM and TEM micrographs of the ECPs with different shell-core ratios are shown in Fig. 6. The micrographs of ECP-4 show that the microcapsules are spherical-shaped with wrinkled surfaces. The surfaces of a few microcapsules are even incomplete. This is due to the thin shell with low strength being prone to deform or damage when they are dried at high temperatures. This shell morphology will undoubtedly affect the utilization and delayed thickening effect of latex. Compared with ECP-4, the shell of ECP-3 is more complete and tightly wrapped on the core without wrinkles, and its thickness increases to 44.6 nm. This indicates that increasing the shell thickness can improve the encapsulation performance of microcapsules on the latex core.

Irregularly shaped microcapsules with uneven shell thickness and non-uniform distribution of particle size can be observed in the micrographs of ECP-2 and ECP-1. The larger the shell-core ratio, the more pronounced this phenomenon is, which is consistent with the results of 3. 1. 3. This is because an appropriate amount of prepolymer can be evenly arranged on the surface of the latex core to form a regular film, while an excessive prepolymer will form an irregular arrangement around the latex core through the hydrogen bond between urethano amino (—NH—) and urethano carbonyl (C═O)[34,35] (Fig. 7).

3. 1. 5 Thermogravimetric analysis

TGA was used to evaluate the thermostability of latex core and ECPs with different shell-core ratios. Fig. 8 a shows the TGA and DTG curves of the latex core. The weight loss of the latex core can be divided into five stages. The initial stage (S1) before 117. 5℃ with a weight loss of 4. 18% is corresponding to the volatilization of crystal water in copolymer molecules. The second stage (S2) in the range of 117. 5 写 232. 3℃ with a weight loss of 13. 37% is related to the fracture of acylamino in AM unit. The third stage (S3) occurs with a weight loss of 11. 14% from 232. 2 to 296. 3℃, which can be ascribed to the decomposition of the sul-fonic acid groups in AMPS unit[36]. The thermal degradation stage of the main chains occurs in the range of 296. 3 写 389. 8℃ (S4) with a weight loss of 23. 41%, and the pyrolysis of the main chain destroys the skeleton and removes nitrogen and hydrogen. The last stage (S5) above 389. 8℃ is due to the carbonization of the polymer particles.

The TGA and DTG curves of the ECP-4 are shown in Fig. 8b. The water loss of the first stage (P1) is 9. 38%, which is significantly higher than that of S1. This is because the shell is tightly wrapped on the surface of the latex, making the bound water in the core molecules more difficult to lose during drying[25]. There exists a plateau period (P2) without weight loss in the range of 138. 7～219. 6℃. Such plateau does not appear in the TGA thermograph of the latex core. This indicates that the PU shell has a better temperature resistance, thus it can protect the latex from thermal degradation. Between 219. 6 写 292. 3℃ (P3), the weight loss of 9. 69% is due to the partial cracking of polyols in polyurethane[31] and the decomposition of side groups and branches of latex core. The sharp weight loss of 16. 75% in the range of 292. 3 写 351. 4℃ (P4) corresponds to the thermal degradation of PU shell and latex molecular chains. There are two peaks in the DTG curve at the P4 stage, which may be due to the fact that the shell is lighter than the latex, so the degradation process at the P4 stage shifts from a joint degradation of shell and latex to a single degradation of latex molecular chains. The weight loss above 351. 4℃ is caused by particle carbonization. In contrast to the thermal degradation temperature of 117. 5℃ for the latex core, the thermal degradation of ECP-4 begins at temperatures above 219. 6℃, indicating that the thermostability is greatly improved by coating the latex core with a PU shell.

TGA curves and the weight loss at each stage of the ECPs with different shell-core ratios are shown in Fig. 8c and Fig. 8d, respectively. ECP-1, ECP-2, ECP-3 and ECP-4 have moisture losses of 3. 13%, 4. 76%, 6. 72% and 9. 05% at the P1 stage, respectively. Higher moisture content indicates a higher proportion of inner latex. Correspondingly, the thermal degradation temperature decreases from 228. 3℃ to 219. 6℃ due to the reduction of shell proportion. However, as the shell-core ratio changes, there is no apparent pattern in the weight loss data for the P3, P4 and P5 stages, which is due to the overlap of the degradation temperature of core and shell.

The increase of PU shell thickness can increase the thermal degradation temperature of ECP, which can prolong the delayed thickening time. However, the increase of shell thickness will have a great impact on the core storage, which is detrimental to the thickening effect of latex core.

At a shell-core ratio of 0. 25, the synthesized encapsulated polymer ECP-3 has a high yield, uniform particle size distribution, and regular shell morphology, which facilitates the stable dispersion of the ECP in aqueous media. And it also has satisfactory thermosta-bility and latex core reserve, which are beneficial to the delayed thickening of the ECP. Therefore, it can be presumed that the sus-pension performance of ECP-3 is better in the application.

Table 2 Particle size distribution and shell thickness of the encapsulated polymer.

	Min(nm)	Max(nm)	Mean(nm)	Mean shell thickness(nm)
Core	245. 5	361. 3	290. 3	—
ECP-1	398. 1	710. 6	502. 1	105. 9
ECP-2	379. 9	663. 8	474. 8	92. 3
ECP-3	297. 8	531. 8	382. 1	45. 9
ECP-4	270. 4	482. 8	350. 4	30. 1

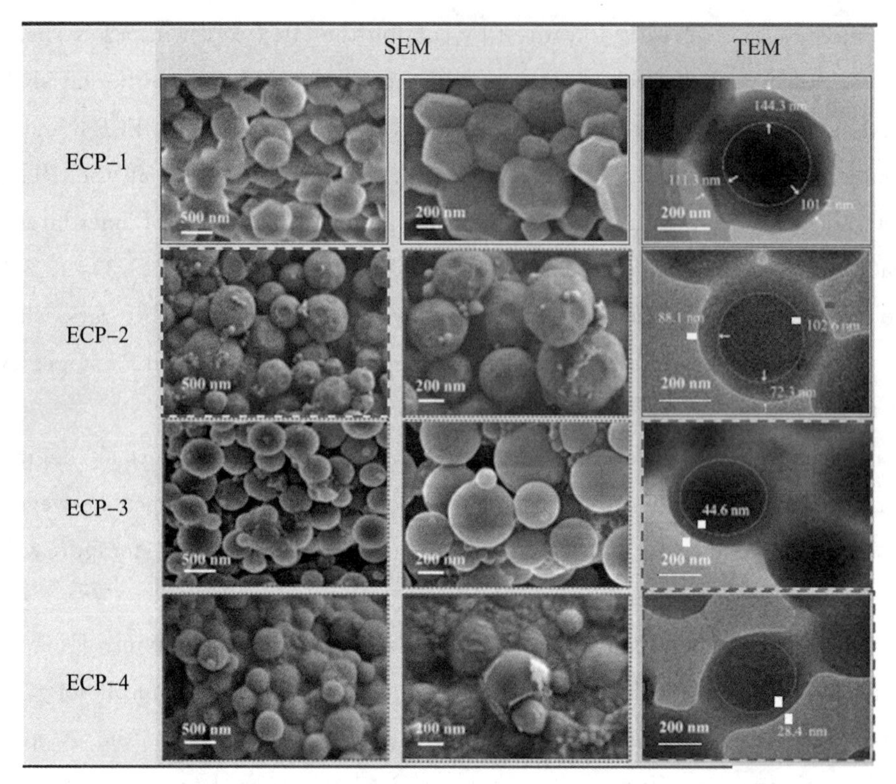

Fig. 6 SEM and TEM micrographs of ECPs with different shell-core ratios.

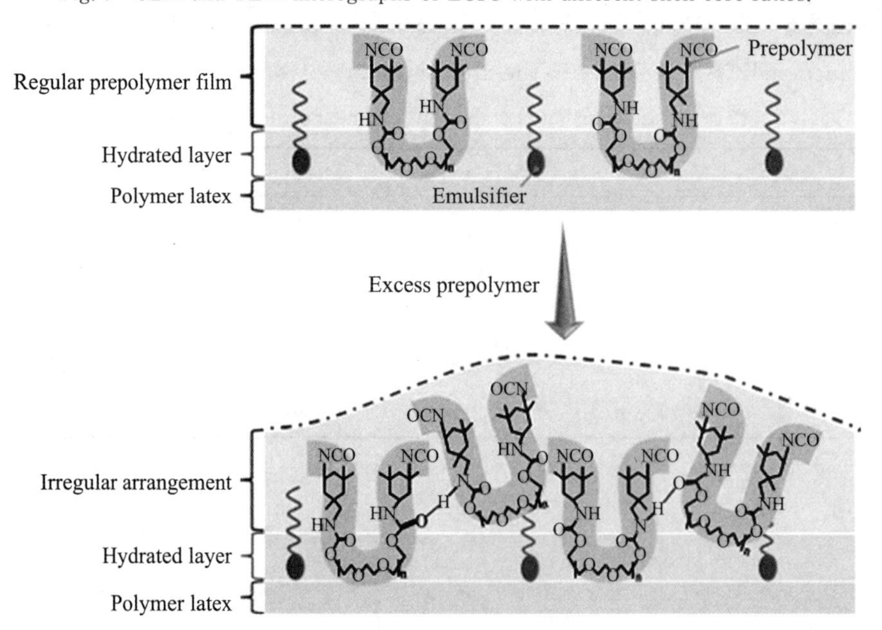

Fig. 7 Schematic of irregular aggregation formed by hydrogen bonding between prepolymers.

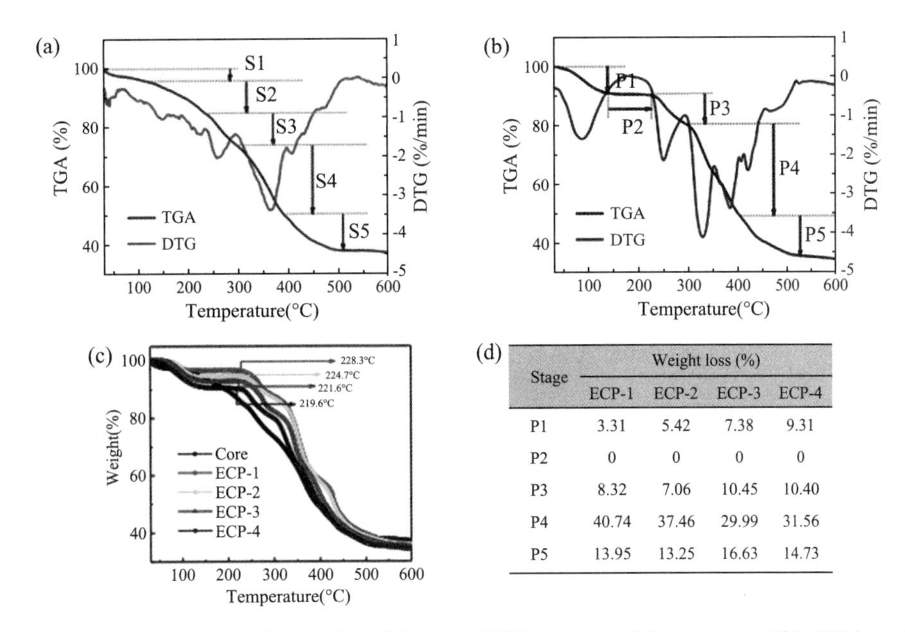

Stage	Weight loss (%)			
	ECP-1	ECP-2	ECP-3	ECP-4
P1	3.31	5.42	7.38	9.31
P2	0	0	0	0
P3	8.32	7.06	10.45	10.40
P4	40.74	37.46	29.99	31.56
P5	13.95	13.25	16.63	14.73

Fig. 8 Thermogravimetric analysis: (a) TGA and DTG curves of latex core; (b) TGA and DTG curves of ECP-4; (c) TGA curves of ECPs with different shell-core ratios; (d) the weight loss of each stage of the ECPs.

3.2 Performance of EPC suspensions

To clarify the effect of shell-core ratio on the performance of ECP, the delayed thickening and dispersion stability of ECP-1, ECP-2, ECP-3, and ECP-4 suspensions were evaluated and discussed.

3.2.1 Delayed thickening performance

To study the effect of shell-core ratio on the delayed thickening performance of ECP, the change rule of the apparent viscosity of the ECP suspensions with the aging time was measured. The viscosity change curves of fluids dispersed with ECP-1, ECP-2, ECP-3, and ECP-4 are shown in Fig. 9a. The changing trend of suspensions' viscosity with time can be divided into two stages: viscosity growth stage and viscosity stability stage. The arrangement of prepolymer on the latex surface will be more dense with the increase of dosage, resulting in a thicker and stronger shell, which is less susceptible to breakage. Thus, as the shell-core ratio increases, the time of the viscosity growth stage increases from 3d to 14d. However, the increase in shell-core ratio leads to the decrease of reserves, thus reducing the maximum viscosity from 13.37 mPa • s to 8.81 mPa • s. The maximum viscosity of the ECP-1 suspension is lower than 10 mPa • s, indicating its thickening effect is poor above 75℃.

The initial suspension has a viscosity of 1 mPas. It shows a strong fluidity when poured outwards, while its flow disappears rapidly without wiredrawing phenomenon when stopping pouring (Fig. 9b). This characteristic would facilitate the fluid injection into the formation.

Once the latex has been released from the capsule and dissolves in the aqueous medium, the fluid is less fluidity due to its higher viscosity. Meanwhile, the wiredrawing phenomenon representing viscoelasticity can be observed (Fig. 9c). In the process of oil displacement, the increase of displacement phase viscosity can improve the water-oil mobility ratio and increase the swept volume. In addition, the displacement phase with good viscoelasticity can effectively displace the residual oil after water displacement, so as to improve the oil displacement efficiency[37].

3.2.2 Dispersion stability performance

Microcapsule particles have a large specific surface area and high surface energy, which tends to cause instability of the suspension. The BS as a function of time for suspensions of ECP with different shell-core ratios is plotted in Fig. 10, which can be used to evaluate their dynamic stability. A slight increase in the BS values is observed at all heights of the sample bottle (Fig. 10b, c, and d). Since the mean diameter of the particles is less than the incident light wavelength (880 nm), the increase in BS value is caused by the increase in particle size after flocculation according to Mie's theory. When the shell-core ratio increases from 0.125 to 0.375, the increment in BS increases from 0.2% to 0.4%, indicating that the increase of microcapsule particle size will accelerate the flocculation rate of particles. This is attributed to the influence of hydrodynamic interaction on particles. According to Einstein Stokes expression[38], with the decrease of particle size, the Brownian diffusion coefficient increases, which makes the particles closer to each other, resulting in the increase of viscous resistance during liquid flow between particles. The increase of viscous resistance will hinder the aggregation process[39].

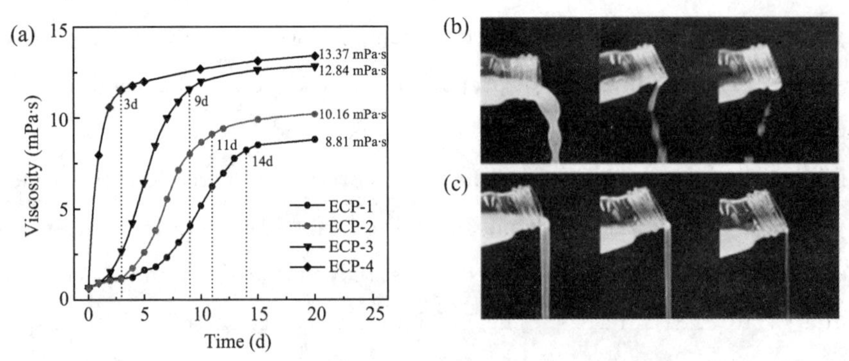

Fig. 9 Delayed thickening performance of ECPs with different shell core ratios: (a) viscosity change curves of suspensions; (b) fluidity of the initial suspension; (c) fluidity of the suspension after thickening.

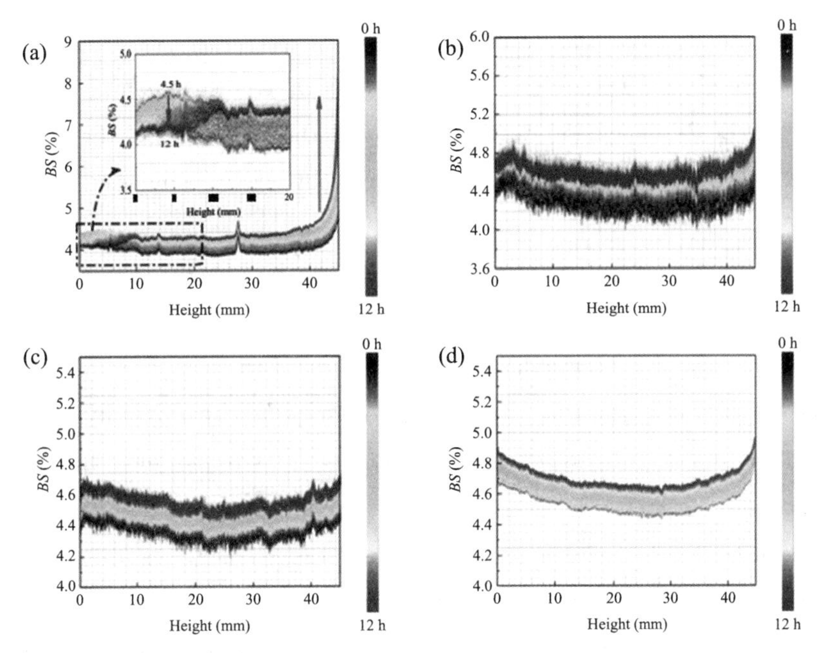

Fig. 10 *BS* spectra of for suspensions of ECP with different shell-core ratios: (a) ECP-1 suspension; (b) ECP-2 suspension; (c) ECP-3 suspension; (d) ECP-4 suspension.

As can be seen from Fig. 10a, the BS at all heights of the sample bottle increases within 4.5 h, indicating that flocculation of the microcapsules occurs. After 4.5 h, the BS below 10 mm begins to decline, while the BS above 40 mm begins to rise obviously, indi-cating that the particle floating occurs, causing the particle volume fraction to fall at the bottom of the suspension and rise at the top of the suspension. During the floating process, the particles continue to aggregate. As a result, the BS continues to increase in the middle of the sample bottle.

The results show that ECP-1 will float up obviously in the suspension, resulting in poor dispersion stability. This is attributed to two reasons. On the one hand, the irregular shell morphology leads to adhesion between the surfaces of the microcapsules when they collide during thermal movement[40,41], thus aggravating the flocculation phenomenon. On the other hand, the PU shell swells slightly in the aqueous medium, causing the structure to become loose and the specific gravity to drop, thus making the floc lighter than water and causing the floc to float under the effect of the density difference.

It can be seen that ECP-3 has good delayed thickening and dispersion stability performance, which is consistent with the characterization results. Hence, the shell-core ratio of 0.25 is preferred.

3.3 Delayed thickening mechanism of ECP

ECP-3 was selected to study the delayed thickening mechanism. The viscoelastic change of the ECP suspension was tested by a rheometer, and the morphological change of the ECP in aqueous media was observed by AFM. The results are shown in Fig. 11.

Uniformly dispersed spherical microcapsules can be observed in the AFM images of the initial suspension at 0d, and there is no network structure in the aqueous medium (Fig. 11 a, b). This indicates that the PU shell has good solvent resistance and preserves the latex particles well. Due to the poor deformability of the polyurethane microcapsules, the dispersion system has no storage modulus (G') which characterizes the elasticity (Fig. 11 c). At high vibration frequencies, frequent collisions between microcapsules generate frictional resistance, resulting in an increase in the loss modulus (G'') characterizing the viscosity in the system[42]. The presence of hydrophilic polyether chains in polyurethanes can attract water molecules into the inner structure of the shell, causing the shell structure to become loose. As a result, the microcapsules swell slightly and the particle size increases to approximately 1 lm.

Due to the poor elasticity of the shell, the swelling behavior will cause the shell to rupture at 2d and release the latex particles (Fig. 11 d). The latex particles form a hydrated layer in the aqueous medium and start to dissolve in the water (Fig. 11 e). Polymer molecular chains dissolve in the aqueous medium and form lowstrength drawn structures, so that changes in G' can be monitored at vibration frequencies below 0.5 Hz. However, the wire-like structure is prone to be destroyed at high vibrational frequencies, resulting in a loss of elasticity of the system (Fig. 11 f).

At 5d, most of the latex particles are dissolved and form a threedimensional network structure in the aqueous medium, giving the system viscoelasticity (Fig. 11 g, h). G'' is always greater than G' in the frequency range of 0.1 to 10 Hz, indicating the system is predominantly viscous, due to the sparse network structure (Fig. 11 i).

As the aging time increases, more latexes are dissolved and the three-dimensional network structure becomes denser (Fig. 11j, k). G'' is larger than G' at frequencies below 2 Hz, indicating that the system is a viscous fluid. Whereas G' is larger than G'' at frequencies above 2 Hz, indicating that the system is an elastic fluid (Fig. 11 l). The better viscoelasticity will facilitate the deformation and migration of the fluid in the porous medium within the reservoir, thus expanding the swept area.

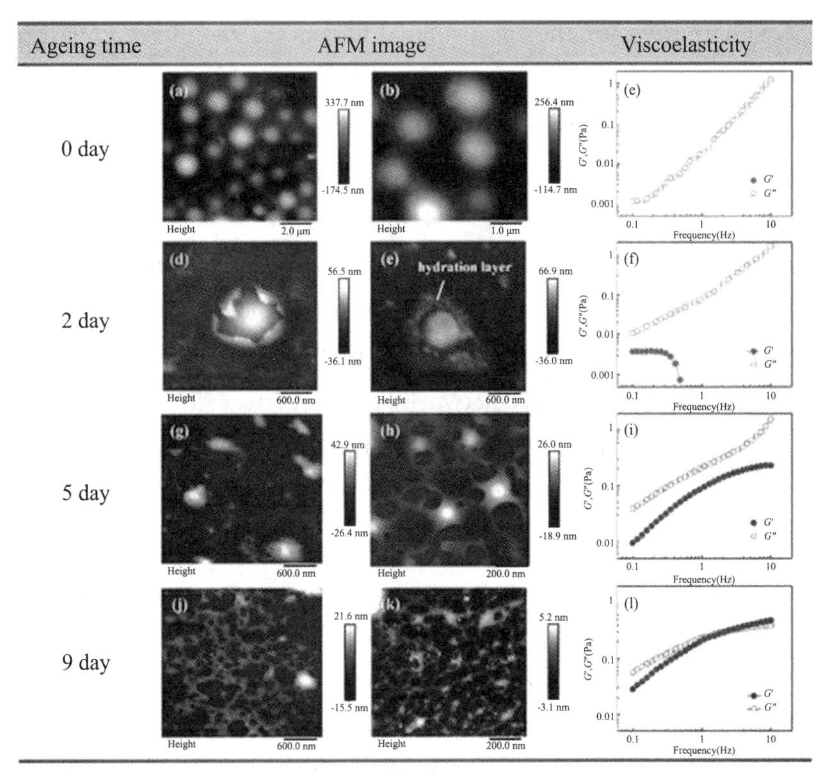

Fig. 11　Morphology and viscoelasticity change of ECP suspension.

The results show that ECP undergoes a process of "shell swelling and rupture-core release-core dissolution" in aqueous media, eventually forming a three-dimensional network structure in water, which transforms the suspension from a low-viscosity fluid to a viscoelastic one.

4　Conclusion

In summary, the PU-based microcapsules with different shellcore ratios filled with latex particles have been synthesized by inverse emulsion polymerization combined with in situ polymerization. The encapsulated polymer can achieve delayed thickening of latex particles in aqueous media, thereby improving the injection capacity and utilization of the polymer in low permeability reservoirs. The increase of shell-core ratio will increase the shell thickness, and improve the integrity and thermostability of the shell. The thermal degradation temperature of the encapsulated polymers with the shell-core ratio of 0.125、0.5 is 219.6、228.3℃, respectively, This feature is conducive to improving the delayed thickening effect of the latex. Excessive prepolymer will lead to uneven particle size distribution and the formation of the irregular shell, accelerate the flocculation of microcapsules in water and lead to unstable dispersion. At the same time, it will also negatively affect the storage of latex in microcapsules. The shellcore ratio of 0.25 is optimized for the encapsulated polymers, they have a uniform particle size distribution, a mean diameter of 382.1 nm and smooth and regular shaped shells. The viscosity of the suspension containing 1 500 mg/L encapsulated polymers increases from 1 mPa・s to 12.84 mPa・s after 9 d. This feature helps to improve the injectability of the fluid into the low permeability reservoir, and improve the oil displacement efficiency in a porous media. The study on the mechanism of delayed thickening shows that due to the hydrophilicity of the polyether chain, the PU shell swells and ruptures in water, and then latex particles are released and dissolved in water, forming a three-dimensional network structure, transforming the low-viscosity suspension to a viscoelastic one. The novel thickening mechanism of encapsulated polymers presents a promising application in the production of low permeability reservoirs.

CRediT authorship contribution statement

Jincheng Gong: Conceptualization, Validation, Formal analysis, Investigation, Writing-original draft, Writing-review & editing, Visualization, Data curation. Yanling Wang: Conceptualization, Supervision, Writing-review & editing. Xulong Cao: Conceptualization, Supervision, Writing-review & editing, Funding acquisi-tion. Fuqing Yuan: Supervision, Writing-review & editing. Yanfeng Ji: Conceptualization, Formal analysis, Investigation, Writing-review & editing, Visualization, Data curation.

Declaration of Competing Interest

The authors declare that they have no known competing financial interests or personal relationships that could have appeared to influence the work reported in this paper.

Acknowledgments

The authors are grateful to the Chemical Group 1 of the Recovery Laboratory, Exploration and Development Research Institute for support. This study is a part of the research project supported by the Joint Funds of the National Natural Science Foundation of China (Grant No. : U21B2070).

References

[1] B. L. Liu, L. Yang, Y. R. Liu, Z. D. Lei, X. B. Chen, K. Wang, J. Li, X.

F. Guo, Q. Yu, Main factors for the development of ultra-low permeability reservoirs, J. Coastal Res. 104(2020)435-439.

[2] JDing, Experimental study on commingled injection in multilayered low. permeability reservoir, Geosyst. Eng. 20(2)(2017)71-80.

[3] T. Song, Q. Feng, T. Schuman, J. Cao, B. J. Bai, A Novel Branched Polymer Gel System with Delayed Gelation Property for Conformance Control, SPE J. 27(01)(2022)105-115.

[4] S. Abdel-Azeim, M. Y. Kanj, Dynamics, aggregation, and interfacial properties of the partially hydrolyzed polyacrylamide polymer for enhanced oil recovery applications: insights from molecular dynamics simulations, Energy Fuels 32(2018) 3335-3343.

[5] A. L. Viken, T. Skauge, P. E. Svendsen, P. A. Time, K. Spildo, Thermothickening and salinity tolerant hydrophobically modified polyacrylamides for polymer flooding, Energy Fuels 32(10)(2018)10421-10427.

[6] Y. Du, Y. Zhu, Y. Ji, H. Xu, H. Zhang, S. Yuan, Effect of salt-resistant monomers on viscosity of modified polymers based on the hydrolyzed poly-acrylamide (HPAM): A molecular dynamics study, J. Mol. Liq. 325(2021)115161, https://doi. org/10. 1016/j. molliq. 2020. 115161.

[7] Y. Bi, W. Li, C. Liu, Z. Xu, X. Jia, Dendrimer-based demulsifiers for polymer flooding oil-in-water emulsions, Energy Fuels 31(5)(2017)5395-5401.

[8] Z. Zhu, W. Kang, B. Sarsenbekuly, H. Yang, C. Dai, R. Yang, H. Fan, Preparation and solution performance for the amphiphilic polymers with different hydrophobic groups, J. Appl. Polym. Sci. 134(20)(2017), https://doi. org/ 10. 1002/app. 44744.

[9] H. Liao, H. Yu, G. Xu, P. Liu, Y. He, Y. Zhou, Polymer-surfactant flooding in low permeability reservoirs: an experimental study, ACS Omega 7(5) (2022)4595-4605.

[10] F. Liu, X. L. Wu, W. S. Zhou, Application of well pattern adjustment for offshore polymer flooding oilfield: a macro-scale and micro-scale study, Chem. Technol. Fuels Oils 56(2020)441-452.

[11] S. Xue, Y. Zhao, C. Zhou, G. Zhang, F. Chen, S. Wang, Improving oil recovery of the heterogeneous low permeability reservoirs by combination of polymer hydrolysis polyacrylamide and two highly biosurfactant-producing bacteria, Sustainability 14(1)(2022) 423, https://doi. org/10. 3390/su14010423.

[12] J. Huang, Y. Zhu, Y. Ma, J. Hu, H. Huang, J. Wei, Q. Yu, pH-

triggered release performance of microcapsule-based inhibitor and its inhibition effect on the reinforcement embedded in mortar, Materials 14(19)(2021) 5517, https://doi.org/10.3390/ma14195517.

[13] M. Ji, S. Liu, H. Xiao, Tribological behaviors of water-based drilling mud with oleic acid-filled microcapsules as lubricant additives for steel-steel contact, Ind. Lubr. Tribol. 72(7)(2020) 835-843.

[14] H. Yu, C. Xue, Y. Qin, Y. Wen, L. Zhang, Y. Li, Preparation and performance of green targeted microcapsules encapsulating surfactants, Colloids Surf. A-Physicochem. Eng. Aspects 623 (2021) 126733, https://doi.org/10.1016/j.colsurfa.2021.126733.

[15] T. Bollhorst, K. Rezwan, M. Maas, Colloidal capsules: nano-and microcapsules with colloidal particle shells, Chem. Soc. Rev. 46(2017) 2091-2126.

[16] H. Fathi Fathabadi, M. Javidi, Self-healing and corrosion performance of polyurethane coating containing polyurethane microcapsules, J. Coat. Technol. Res. 18(5)(2021) 1365-1378.

[17] T. Zheng, K. Chen, W. Y. Chen, B. Wu, Y. Sheng, Y. Xiao, Preparation and characterization of polylactic acid modified polyurethane microcapsules for controlled-release of chlorpyrifos, J. Microencapsul. 36(2019) 62-71.

[18] Y. F. Zhang, J. Z. Liu, J. Li, C. Y. Wang, Q. Ren, Synthesis and storage stability investigation on curing agent microcapsules of imidazole derivatives with aqueous polyurethane as the shell, Polym. Bull. (2022), https://doi.org/10.1007/s00289-021-04063-4.

[19] V. Costa, A. Nohales, P. Félix, C. Guillem, D. Gutiérrez, C. M. Gómez, Structure-property relationships of polycarbonate diol-based polyurethanes as a function of soft segment content and molar mass, J. Appl. Polym. Sci. (2014), https://doi.org/10.1002/app.41704.

[20] R. K. Hedaoo, P. P. Mahulikar, A. B. Chaudhari, S. D. Rajput, V. V. Gite, Fabrication of core-shell novel polyurea microcapsules using isophorone diisocyanate (IPDI) trimer for release system, International Journal of Polymeric Materials and Polymeric, Biomaterials 63(7)(2014) 352-360.

[21] V. H. de Souza Rodrigues, A. Estêvão Carrara, S. S. Rossi, L. Mattos Silva, R. de Cássia Lazzarini Dutra, J. C. N. Dutra, Synthesis, characterization and qualitative assessment of self-healing capacity of PU microcapsules containing TDI and IPDI as a core agent, Mater. Today Commun. 21(2019) 100698, https://doi.org/10.1016/j.mtcomm.2019.100698.

[22] L. Ouyang, L. Wang, F.J. Schork, Synthesis and nucleation mechanism of inverse emulsion polymerization of acrylamide by RAFT polymerization: a comparative study, Polymer 52(1)(2011)63-67.

[23] W. Li, X. Y. Geng, R. Huang, J. P. Wang, N. Wang, X. X. Zhang, Microencapsulated comb-like polymeric solid-solid phase change materials via in-situ polymerization, Polymers 10(172)(2018), https://doi. org/10. 3390/ polym10020172.

[24] D. Saihi, I. Vroman, S. Giraud, S. Bourbigot, Microencapsulation of ammonium phosphate with a polyurethane shell. Part II. Interfacial polymerization technique, React. Funct. Polym. 66(10)(2006)1118-1125.

[25] A. Beglarigale, Y. Seki, N. Y. Demir, H. Yazici, Sodium silicate/ polyurethane microcapsules used for self-healing in cementitious materials: Monomer optimization, characterization, and fracture behavior, Constr. Build. Mater. 162 (2018)57-64.

[26] L. Liu, S. H. Gou, H. C. Zhang, L. H. Zhou, L. Tang, L. Liu, A zwitterionic polymer containing a hydrophobic group: enhanced rheological properties, New J. Chem. 44(2020)9703-9711.

[27] T. Liu, S. Gou, L. Zhou, J. Hao, Y. He, L. Liu, L. Tang, S. Fang, High-viscoelastic graft modified chitosan hydrophobic association polymer for enhanced oil recovery, J. Appl. Polym. Sci. 138(11)(2021)50004, https://doi. org/10. 1002/ app. 50004.

[28] F. Liu, X. Meng, Y. Zhang, L. Ren, F. Nawaz, F.-S. Xiao, Efficient and stable solid acid catalysts synthesized from sulfonation of swelling mesoporous polydivinylbenzenes, J. Catal. 271(1)(2010)52-58.

[29] M. Zhou, R. Yi, Y. Gu, H. Tu, Synthesis and evaluation of a tetra-copolymer for oil displacement, J. Petrol. Sci. Eng. 179(2019)669-674.

[30] M. A. Semsarzadeh, A. H. Navarchian, Effects of NCO/OH ratio and catalyst concentration on structure, temperature resistance, and crosslink density of poly(urethane-isocyanurate), J. Appl. Polym. Sci. 90(2003)963-972.

[31] F. Alizadegan, S. M. Mirabedini, S. Pazokifard, S. Goharshenas Moghadam, R. Farnood, Improving self-healing performance of polyurethane coatings using PUmicrocapsules containing bulky-IPDI-BA and nano-clay, Prog. Org. Coat. 123(2018)350-361.

[32] J. S. Guo, Y. He, D. L. Xie, X. Y. Zhang, Process investigating and modelling for the self-polymerization of toluene diisocyanate (TDI)-based polyurethane prepolymer, J. Mater. Sci. 50(2015)5844-5855.

[33] S. H. Vakili Tahami, Z. Ranjbar, S. Bastani, Aggregation and charging behavior of polydisperse and monodisperse colloidal epoxy-amine adducts, Soft Mater. 11(3)(2013)334-345. Journal of Molecular Liquids 360(2022)119394

[34] C. L. Zhang, J. L. Hu, X. Li, Y. Wu, J. P. Han, Hydrogen-bonding interactions in hard segments of shape memory polyurethane: toluene diisocyanates and 1,6-hexamethylene diisocyanate. A Theoretical and Comparative Study, J. Phys. Chem. A 118(2014)12241-12255.

[35] C. Zhang, J. Hu, Y. Wu, Theoretical studies on hydrogen-bonding interactions in hard segments of shape memory polyurethane-III: isophorone diisocyanate, J. Mol. Struct. 1072(2014)13-19.

[36] A. R. Al Hashmi, R. S. Al Maamari, I. S. Al Shabibi, A. M. Mansoor, A. Zaitoun, H. H. Al Sharji, Rheology and mechanical degradation of high-molecular-weight partially hydrolyzed polyacrylamide during flow through capillaries, J. Petrol. Sci. Eng. 105(2013)100-106.

[37] Y. Du, K. e. Xu, L. Mejia, M. Balhoff, A coreflood-on-a-chip study of viscoelasticity's effect on reducing residual saturation in porous media, Water Resour. Res. 57(8)(2021), https://doi.org/10.1029/2021WR029688.

[38] R. Walser, B. Hess, A. E. Mark, W. F. van Gunsteren, Further investigation on the validity of Stokes-Einstein behaviour at the molecular level, Chem. Phys. Lett. 334(4-6)(2001)337-342.

[39] P. Warszyn'ski, Coupling of hydrodynamic and electric interactions in adsorption of colloidal particles, Adv. Colloid Interface Sci. 84(1-3)(2000)47-142.

[40] G. Van Anders, D. Klotsa, N. K. Ahmed, M. Engel, S. C. Glotzer, Understanding shape entropy through local dense packing, PNAS 111 (2014) E4812-E4821.

[41] A. K. Sharma, V. Thapar, F. A. Escobedo, Solid-phase nucleation free-energy barriers in truncated cubes: interplay of localized orientational order and facet alignment, Soft Matter 14(2018)1996-2005.

[42] N. P. B. Tan, L. H. Keung, W. H. Choi, W. C. Lam, H. N. Leung, Silica-based self-healing microcapsules for selfrepair in concrete, J. Appl. Polym. Sci. 133 (12)(2016), https://doi.org/10.1002/app.43090.

第二章　表面活性剂及其相互作用

不同结构烷基苯磺酸盐油水界面扩张粘弹性质

宋新旺[1,2] 王宜阳[3] 曹绪龙[2] 罗 澜[3] 王 琳[3] 张 路[3,*]
岳湘安[1] 赵 濉[3] 俞稼镛[3]

([1] 中国石油大学(北京),北京 102249;
[2]中国石化胜利油田分公司地质科学研究院,山东东营 257015;
[3]中国科学院理化技术研究所,北京 100080)

摘要:研究了2-甲基-5-(1-庚基辛基)苯磺酸钠、辛基苯磺酸钠和十六烷基苯磺酸钠在正辛烷-水界面上的扩张粘弹性质,考察了链长变化和疏水基支链化对分子界面行为的影响。研究结果表明,链长增加导致分子间相互作用增强,弹性增大;疏水支链在界面上可能由于缠绕和变形产生界面慢弛豫过程,导致较高的扩张模量。

关键词:烷基苯磺酸盐,扩张模量,相角,界面张力弛豫

中图分类号:O647

Dilational viscoelastic properties of sodium alkyl benzene sulfonates with different structures at octane/water interface

Song Xinwang[1,2] Wang Yiyang[3] Cao Xulong[2] Luo Lan[3] Wang Lin[3]
Zhang Lu[3,*] Yue Xiangan[1] Zhao Sui[3] Yu Jiayong[3]

([1]China University of Petroleum (Beijing), Beijing 102249, P. R. China;
[2]Geological Research Institute of Shengli Oilfield Co. Ltd, SINOPEC, Dongying 257015, P. R. China;
[3]Technical Institute of Physics and Chemistry, Chinese Academy ofSciences, Beijing 100080, P. R. China)

Abstract: The dilational viscoelasticity properties of octane-water interface containing three sodium alkyl benzene sulfonates with straight chain (C_8phSO_3Na and $C_{16}phSO_3Na$) and branch chain(7-8-1) respectively were investigated. The influences of alkyl chain length and hydrophobic branched-chain structure on interfacial dilational properties were expounded. It showed that the increase of alkyl chain length resulted in the enhancement of molecular interaction and the increase of interfacial dilational elasticity. The slow relaxation processes were caused by the rearrangement and entanglement of branched-chains at interface and resulted in relative higher interfacial dilational modulus.

Keywords: Sodium alkyl benzene sulfonate, Dilational modulus, Phase angle, Interfacial tension relaxation

表面活性剂分子在液液界面上的吸附状态是引人关注的问题。通过 Gibbs 吸附公式,可以获得界面上表面活性剂分子的浓度,但分子在界面上的相互作用、分子排列方式、界面聚集体等信息仍然难以直接得到。界面扩张流变是通过研究界面平衡张力在不同扰动条件下的反应,从而获得两亲分子界面信息的研究手段[1-9]。实践证明,这是一种常规、简便并且有力的研究方法,国外研究者已经广泛用于两亲分子表面相互作用及其聚集体的研究领域[10-13]。本文研究了不同结构高纯度烷基苯磺酸钠的界面扩张性质。

1 理论基础[14-15]

界面扩张模量定义为界面张力变化相对于相对界面面积变化的比值:

$$\varepsilon=\frac{d\gamma}{d\ln A} \tag{1}$$

对于粘弹性界面,扩张模量可写作复数形式:

$$\varepsilon=\varepsilon_d+i\omega\eta_d \tag{2}$$

其中,ε_d 为扩张弹性,η_d 为扩张粘度。

如果对已达到平衡的表面活性剂吸附膜进行快速扰动,则表面张力衰减曲线可用下式拟合:

$$\Delta\gamma=\sum_{i=1}^{n}\Delta\gamma_i\exp(-\tau_i t) \tag{3}$$

其中,τ_i 是第 i 个过程的特征频率,其倒数 $1/\tau_i=T_i$ 为特征周期;t 是界面扩张驰豫的时间;$\Delta\gamma_i$ 是与第 i 个过程的贡献相关的参数;n 是总的过程的个数。详细理论见文献[1-9]。

2 实验部分

2.1 实验样品及试剂

多取代支链烷基苯磺酸钠 2-甲基-5-(1-庚基辛基)苯磺酸钠(7-8-1)、辛基苯磺酸钠 C_8phSO_3Na 和十六烷基苯磺酸钠 $C_{16}phSO_3Na$ 均为自制,纯度大于 98%[16];正辛烷,天津博迪化学有限公司,分析纯。

2.2 实验方法

界面扩张粘弹性用上海中晨 JMP2000 型界面扩张粘弹性测定仪测定。正辛烷作为油相，水相为用二次蒸馏去离子水配制的不同浓度 7-8-1 和 C_8phSO_3Na 溶液。$C_{16}phSO_3Na$ 为油溶性表面活性剂，实验时配制在正辛烷中。具体实验方法见文献[1-9]。

3　结果与讨论

3.1 浓度对不同结构烷基苯磺酸钠界面扩张模量的影响

多取代支链烷基苯磺酸钠 7-8-1、直链烷基苯磺酸钠 C_8phSO_3Na 和 $C_{16}phSO_3Na$ 溶液的界面扩张模量随浓度的变化趋势见图 1。由图 1 可以看出，三种表面活性剂的扩张模量均随浓度变化各通过一个极大值。

一般而言，在临界胶束浓度之前，表面活性剂体相浓度的增大对界面扩张性质有两方面的影响：一是增大表面活性剂分子的界面浓度，另一方面也增大了从体相向新生成界面通过扩散补充表面活性剂分子的能力。表面活性剂界面浓度的增大会导致界面分子间更强的相互作用和界面形变时更高的界面张力梯度，界面膜的扩张模量增大；而分子从体相向新生成界面的扩散补充有消除界面张力梯度的作用，会降低扩张模量。表面活性剂体相浓度较低时，体相浓度的增加主要体现为对界面吸附量的影响，界面扩张模量随浓度增大而增大；随着浓度进一步增大，从体相向新生界面扩散补充表面活性剂分子的作用占主导地位，模量开始降低。尤其在临界胶束浓度附近，紧挨着界面的亚层上存在胶束，这种扩散补充增强的幅度更大，模量降低的趋势更为明显[14]。

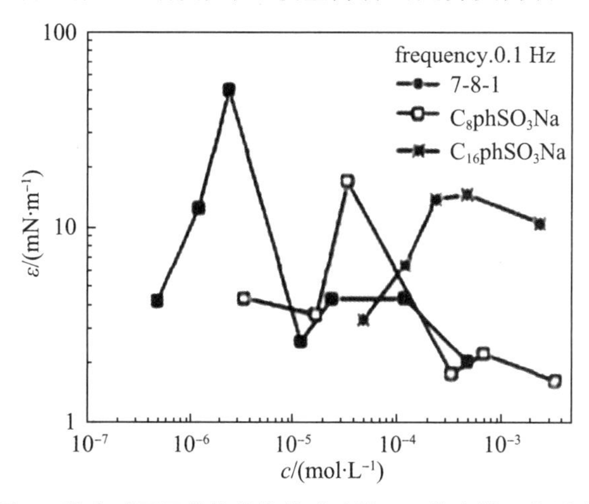

图 1　浓度对不同结构烷基苯磺酸钠界面扩张模量的影响

Fig. 1　Concentration dependency of dilational Modulus for sodium alkyl benzene sulfonates with different structures

实验条件下，7-8-1 的临界胶束浓度为 1.7×10^{-5} mol·L^{-1}，C_8phSO_3Na 的临界胶束浓度为 1.2×10^{-2} mol·L^{-1}，$C_{16}phSO_3Na$ 只能溶解在油相中，在实验浓度范围内没有观察到胶束生成[16]。由于研究的烷基苯磺酸钠均为低分子量表面活性剂，扩散交换能力随体相浓度增大而大大增强，因而在远低于临界胶束浓度时，扩张模量就达到了极大值。从我们前期研究结果看，对于分子量较高的表面活性剂，如破乳剂，由于其扩散较慢,体相浓度增大对扩散交换的增强幅度较小,其扩张模量直到临界胶束浓度附近才开始降低[6]。图 1 中 $C_{16}phSO_3Na$ 的扩张模量最高点的对应浓度比在水相中 C_8phSO_3Na 模量最高点的对应浓度高约一个数量级,这可能是由于 $C_{16}phSO_3Na$ 相对较高的分子量和较强的疏水链间相互作用造成的。

另外,从图 1 还可以看出,7-8-1 在低浓度处有较高的扩张模量,其扩张模量极大值可达 50 $mN \cdot m^{-1}$;而 $C_{16}phSO_3Na$ 的扩张模量在通过极大值后缓慢降低,高浓度时扩张模量较大,表明实验结果与表面活性剂分子的结构密切相关。

3.2 浓度对不同结构烷基苯磺酸钠界面扩张相角的影响

图 2 为三种烷基苯磺酸钠的界面扩张相角随体 相浓度的变化趋势。从图中可以看出,与扩张模量的变化规律相似,相角也随浓度增大通过一个极大值。由于较低浓度条件下 $C_{16}phSO_3Na$ 张力响应较小,无法测量,实验浓度范围内相角已通过极大值。不过,相角并不是在同一浓度处达到极大值:对于支链表面活性剂,相角落后于扩张模量达到极大值;对于直链表面活性剂,相角领先于扩张模量达到极大值。

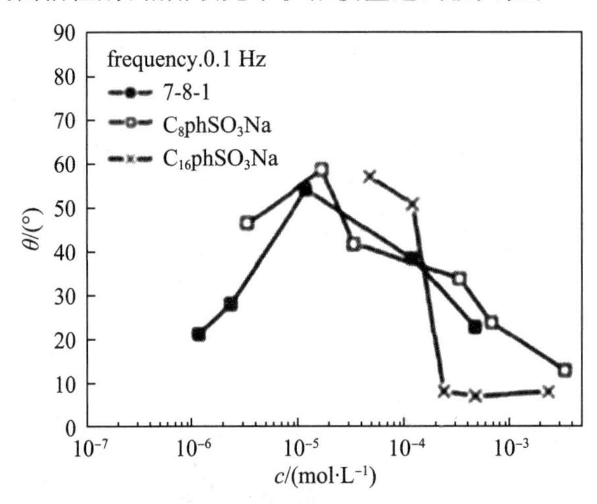

图 2 浓度对不同结构烷基苯磺酸钠界面扩张相角的影响

Fig. 2 Concentration dependency of dilational phase Angle for sodium alkyl benzene sulfonates with different structures

界面扩张模量包括弹性部分和粘性部分,相角反映了粘性部分和弹性部分的比值,是界面膜粘弹特性的定量表征。其中弹性部分又称储能模量,其来源是界面分子因扰动偏离平衡状态导致的能量改变,与分子间相互作用密切相关;粘性部分又称损耗模量,与表面活性剂分子在界面与体相间的交换、界面分子排布方式的改变等弛豫过程相关[14]。值得注意的是,上述弛豫过程将体系偏离平衡态增加的能量部分耗散出去,在贡献给粘度的同时,也可能降低扩张模量。

对于本文研究的直链表面活性剂吸附膜,浓度较低时,扩散交换过程贡献较小,界面膜以弹性为主,相角较低。随浓度增大,扩张模量增大;其中粘性部分增大幅度大于弹性部分,因而相角逐渐增大,界面膜粘性增强。浓度进一步增大,相角通过极大值,说明此时界面分子间相互作用增强明显，界面膜弹性开始增大。比较 C_8phSO_3Na 和 $C_{16}phSO_3Na$ 的相角数据，可以明显看出 $C_{16}phSO_3Na$ 在高浓度时具有较低的相角，这是由于其较长的烷烃链使得界面分子间具有较强的相互作用造成的。同时，这也是图 1 中 $C_{16}phSO_3Na$ 在高浓度处具有较高扩张模量的原因。

对于本文研究的支链表面活性剂吸附膜，相角随浓度变化的极大值落后于扩张模量的极大值，并且其在低浓度处有较高的扩张模量，说明表面活性剂疏水支链对界面扩张性质有着特殊的影响，可能在界面上产生不同于直链分子的特殊弛豫过程。为深入阐明支链化对界面性质的影响，我们进行了界面张力弛豫研究。

表 1 7-8-1 溶液平衡时界面弛豫过程特征时间

Table 1 Interfacial relaxation processes and their characteristic times of 7-8-1 solutions at different concentration

10^6c (7-8-1) / (mol·L^{-1})	$\Delta\gamma_1$ / (mN·m^{-1})	T_1 / s	$\Delta\gamma_2$ / (mN·m^{-1})	T_2 / s	$\Delta\gamma_3$ / (mN·m^{-1})	T_3 / s
0.24	4.8	13.5	0.18	55.6	6.4	232.6
2.24	2.9	7.7	0.14	142.9	2.6	270.3
12	2.2	0.4				
24	2.1	0.5				
48	2.5	0.9				
120	2.2	0.6				
480	2.5	0.8				

cmc(7-8-1)$=1.7\times10^{-5}$ mol·L^{-1}

3.3 不同结构烷基苯磺酸钠的界面弛豫过程研究

界面上吸附分子的行为表现为多种微观弛豫过程。通过分析界面扩张粘弹数据和界面弛豫过程的特征参数，我们可以得知分子间的相互作用以及分子在界面上可能存在的聚集体等信息。对低分子量的表面活性剂分子而言，通常认为存在以下两种类型的弛豫过程：一是分子从体相到界面层的扩散过程；二是分子在界面上的弛豫过程，如分子取向、分子重排等等[14]。

不同结构烷基苯磺酸钠溶液的界面弛豫过程相关参数如表 1～3 所示，表中 T_i 是第 i 个过程的特征时间，$\Delta\gamma_i$ 表征第 i 个过程的相对贡献。由表中可以看出，直链表面活性剂在低浓度时只存在一个快弛豫过程，其特征时间小于 1 s，说明此时体相与界面间的快速交换过程决定界面膜的性质；在高浓度时，由于界面分子间相互作用较强，出现了 10～100 s 的较慢过程，不过快过程的贡献为主。而对于支链表面活性剂 7-8-1，则在低浓度时界面上存在三种弛豫过程，其特征时间从几秒到几百秒；当浓度接近临界胶束浓度时，界面上只存在一个特征时间小于 1 s 的快过程。

表 2 C_8phSO_3Na 溶液平衡时界面弛豫过程特征时间

Table 2 Interfacial relaxation processes and their characteristic times of C_8phSO_3Na solutions at different concentration

10^6 c(C_8phSO_3Na)/(mol·L^{-1})	$\Delta\gamma_1$/(mN·m^{-1})	T_1/s	$\Delta\gamma_2$/(mN·m^{-1})	T_2/s
3.4	1.9	0.3		
17	2.2	0.4		
34	2.4	0.6		
68	3.3	1.1		
170	1.7	1.2		
340	0.22	1.4		
680	0.37	7.1	0.3	90.9

表 3 $C_{16}phSO_3Na$ 溶液平衡时界面弛豫过程特征时间

Table 3 Interfacial relaxation processes and their characteristic times of $C_{16}phSO_3Na$ solutions at different concentrations

10^6 c(C_8phSO_3Na)/(mol·L^{-1})	$\Delta\gamma_1$/(mN·m^{-1})	T_1/s	$\Delta\gamma_2$/(mN·m^{-1})	T_2/s
2.4	1.5	0.4		
12	3.7	0.5		
48	1.6	0.3		
120	2.9	0.4		
240	1.8	13.5	0.42	105.3
480	1.9	3.7	0.85	100

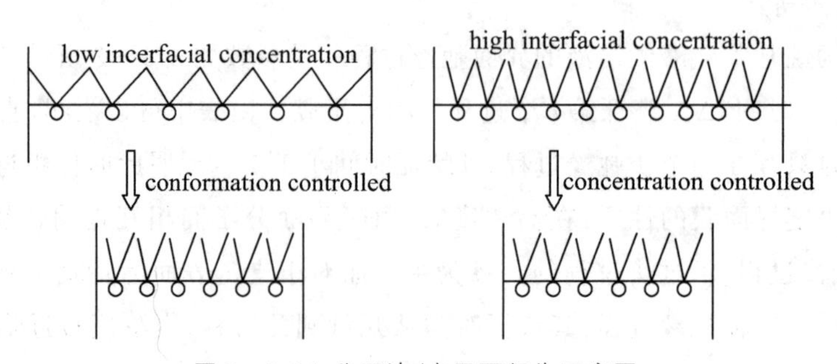

图 3 7-8-1 分子油/水界面行为示意图

Fig. 3 Schematic illustration of the behavior of 7-8-1 Molecule at octane/water interface

对于直链表面活性剂，随浓度增加，分子间相互作用增强，快弛豫过程特征时间变慢。而对于支链表面活性剂 7-8-1 低浓度时独特的界面扩张性质和弛豫过程，我们认为可以归因于其较长的支链间缠绕和变形。7-8-1 分子在界面形变条件下的行为如图 3 所示，浓度较低时，界面压缩条件下 7-8-1 分子舒展支链可能发生缠绕和变形，导致吸附分子取向改变和界面分子重排等慢弛豫过程，产生较高的界面扩张模量。高浓度时，分子以扩散交换为主，其界面扩张性质的变化与直链分子相似。

4 结论

本文研究了多取代支链烷基苯磺酸钠 2-甲基-5-(1-庚基辛基)苯磺酸钠(7-8-1)、辛基苯磺酸钠 C_8phSO_3Na 和十六烷基苯磺酸钠 $C_{16}phSO_3Na$ 在正辛烷-水界面上的扩张粘弹性质，发现三种烷基苯磺酸钠的界面扩张模量和相角均随浓度增加各通过一个极大值，7-8-1 在低浓度处有较高的扩张模量，而 $C_{16}phSO_3Na$ 高浓度时模量较大。界面张力弛豫结果表明，在低浓度时，直链烷基苯磺酸钠以特征时间小于 1 s 的快弛豫过程为主，而支链烷基苯磺酸钠在低浓度时以特征时间高达数百秒的慢弛豫过程为主。疏水支链在界面上的缠绕和变形可能是产生慢弛豫过程的内在原因。

References

[1] Sun, T. L.; Zhang, L.; Wang, Y. Y.; Zhao, S.; Peng, B.; Li, M. Y.; Yu, J. Y. J. Collo id Interface Sci., 2002, 255:241.

[2] Wang, Y. Y.; Zhang, L.; Sun, T. L.; Zhao, S.; Yu, J. Y. J. Collo id Interface Sci., 2003,270:163.

[3] Wang, Y. Y.; Dai, Y. H.; Zhang, L.; Tang, K.; Luo, L.; Gong, Q. T.; Zhao, S.; Li. M. Z.; Wang, E. J.; Yu, J. Y. J. Collo id Interface Sci., 2004,280:76.

[4] Wang, Y. Y.; Dai, Y. H.; Zhang, L.; Luo, L.; Chu, Y. P.; Zhao, S.; Li. M. Z.; Wang, E. J.; Yu, J. Y. Macromolecules,2004,37:2930.

[5] Sun, T. L.; Peng, B.; Xu, Z. M.; Zhang, L.; Zhao, S.; Li, M. Y.; Yu, J. Y. Acta Phys. -Chim. Sin., 2002 , 18 (2):161.[孙涛垒,彭勃,许志明,张路,赵濉,李明远,俞稼镛.物理化学学报(Wu li Huaxue Xuebao),2002,18(2):161]

[6] Wang, Y. Y.; Zhang, L.; Sun, T. L.; Fang, H. B.; Zhao, S.; Yu, J. Y. Acta Phys. -Chim. Sin., 2003,19(4):297.[王宜阳,张路,孙涛垒,方洪波,赵濉,俞稼镛.物理化学学报(Wu li Huaxue Xuebao), 2003,19(4):297]

[7] Wang, Y. Y.; Zhang, L.; Sun, T. L.; Fang, H. B.; Zhao, S.; Yu, J. Y. Acta Phys. -Chim. Sin., 2003,19(5):455.[王宜阳,张路,孙涛垒,方洪波,赵濉,俞稼

镛.物理化学学报(Wu li Huaxue Xuebao),2003,19(5):455]

[8] Wang, Y. Y.; Zhang, L.; Sun, T. L.; Zhao, S.; Yu, J. Y. Chemical Journal ofChinese Universities, 2003,24(11):2044.[王宜阳,张路,孙涛垒,赵濉,俞稼镛.高等学校化学学报(Gaodeng Xuexiao Huaxue Xuebao),2003,24(11):2044]

[9] Sun, T. L.; Zhang, L.; Wang. Y. Y.; Zhao, S.; Yu, J. Y. Chemical Journal of Chinese Universities, 2003,24(12):2243.[孙涛垒,张路,王宜阳,赵濉,俞稼镛.高等学校化学学报(Gaodeng Xuexiao Huaxue Xuebao), 2003,24(12):2243]

[10] Miller, R.; Fainerman, V. B.; Makievski, A. V.; K″agel, J.; Grigoriev, D. O.; Kazakov, V. N.; Sinyachenko, O. V. Advances in Collo id and Interface Science,2000,86:39.

[11] Noskov, B. A.; Akentiev, A. V.; Bilibin, A. Y.; Zorin, I. M.; Miller, R. Advances in Collo id and Interface Science,2003,104:245.

[12] Ivanov, I. B.; Danov, K. D.; Ananthapadmanabhan, K. P.; Lips, A. Advances in Collo id and Interface Science,2005,61:114-115.

[13] Murray, B. S. Current Opin ion in Collo id and Interface Science, 2002, 7:426.

[14] Lucassen-Reynders, E. H. Anionic surfactants: Physical chemistry of surfactant action. Trans. Zhu, B. Y.; Wu, P. Q.; Ding, H. J.; Yang, P. Z. Beijing: China Light Industry Press, 1988:171.[阴离子表面活性剂:表面活性剂作用的物理化学.朱瑶,吴佩强,丁慧君,杨培增,译.北京:中国轻工业出版社,1988:171]

[15] van den Tempel, M.; Lucassen-Reynders, E. H. Advances in Collo id and Interface Science, 1983 , 18:281.

[16] Wang, L. Ph. D. Dissertation. Beijing: Technical Institute of Physics and Chemistry, Chinese Academy of Sciences, 2004.[王琳.博士学位论文.北京:中国科学院理化技术研究所,2004]

Molecular behavior and synergistic effects between sodium dodecylbenzene sulfonate and triton x-100 at oil/water interface

Li Ying[a,*]　He Xiujuan[a]　Cao Xulong[b]　Zhao Guoqing[a]
Tian Xiaoxue[a]　Cui Xiaohong[b]

[a] Key Lab for Colloid and Interface Chemistry of State Education Ministry, Shandong University, Jinan 250100, People's Republic of China

[b] Geological Scientific Research Institute, Shengli Oilfield, DongYing 257015, People's Republic of China

Abstract: Significant synergistic effects between sodium dodecylbenzene sulfonate (SDBS) and nonionic nonylphenol polyethylene oxyether, Triton X-100(TX-100), at the oil/water interface have been investigated by experimental methods and computer simulation. The influences of surfactant concentration, salinity, and the ratio of the two surfactants on the interfacial tension were investigated by conventional interfacial tension methods. A dissipative particle dynamics(DPD) method was used to simulate the adsorption properties of SDBS and TX-100 at the oil/water interface. The experiment and simulation results indicate that ultralow(lower than 10^{-3} mN · m^{-1}) interfacial tension can be obtained at high salinity and very low surfactant concentration. Different distributions of surfactants in the interface and the bulk solution corresponding to the change of salinity have been demonstrated by simulation. Also by computer simulation, we have observed that either SDBS or TX-100 is not distributed uniformly over the interface. Rather, the interfacial layer contains large cavities between SDBS clusters filled with TX-100 clusters. This inhomogeneous distribution helps to enhancing our understanding of the synergistic interaction of the different surfactants. The simulation conclusions are consistent with the experimental results.

Keywords: Synergism effect; Computer simulation; Dissipative particle dynamics; Oil/water interface; Ultralow interfacial tension

1　Introduction

Surfactant adsorption at the oil/water interface, along with the resulting lowering of the interfacial tension, plays an important role in controlling the desired interfacial properties in many practical applications, which range from large-scale industrial operations, such as enhanced oil recovery, to smallscale household uses, such as the stabilization of emulsions in food and cosmetic products. In these applications, different types of combinations, such as anionic/cationic[1,2], nonionic/anionic[3-7], and nonionic/cationic[8,9], are utilized to optimize a certain desired effect, in which the capabilities of a mixture are better than those attainable with the individual components separately, which is known as synergism. Since only a few techniques, such as nonlinear vibrational sum-frequency spectroscopy and second harmonic generation, are available for the investigation of liquid/liquid interfaces, the behavior of surfactants at water/oil interfaces is little investigated at the molecule level, although it is surely important; thus the synergism mechanism is still not clear.

Due to the substantial increase in computational power over the past years, computer simulations have become an important tool for studying such complex interfacial systems[6,10-13]. These kinds of studies allow us to extract more information about dynamical and structural properties of interfacial problems on a molecular level, which are not easy to get from real experiments. Besides, some surfactants are naturally surfacechemically impure and extra purification steps are essential if they are to be studied in detail[14], such as SDBS, researched in the paper; thus the behavior of these surfactants is much lessinvestigated, while computer simulation can avoid this limitation.

In this paper, the oil/water interfacial tension has been measured by conventional methods, and the adsorption layers of SDBS/TX-100 mixed systems were investigated by mesoscopic methods. How the surfactant concentration and salts might affect the monopolizer configuration in the interface has been studied. The synergism mechanism of mixed surfactants has been also analyzed.

2 Experimental

2.1 Materials

Sodium dodecylbenzene sulfonate(SDBS, C. R.) was purchased from Shanghai Chemical Corp(China) and purified; the curve of surface tension to concentration has no low-water mark[15,16]. Tritonx-100(TX-100, C. P.) was purchased from Fluka. NaCl(A. R.) and methylbenzene(A. R.) were purchased from Tianjin Guangcheng Chemical Corp(China). Methylbenzene is used as the oil phase in most experiments if not pointed out otherwise. Freshly distilled water(twice distilled) was used in all solution preparations.

2.2 Experimental methods

The interfacial tension was measured on a TEXAS500 spinning drop interfacial tensiometer at 60±0.1℃. Surfactant solutions were aged overnight.

2.3 Computer simulation method

Dissipative particle dynamics (DPD) is a mesoscopic simulation technique. The simulation strategy is to regard clusters of atoms as single coarse-grained particles or beads. By coarsegraining the atoms of a molecule, it can access a length scale in μm and a time-scale in μs, which is impossible to do with conventional molecular dynamics with an atomistic description. The beads in a DPD simulation move according to the second law of Newton and a certain force. The force consists of soft repulsive conservative forces (F_{ij}^{C}), a pairwise dissipative force (F_{ij}^{D}), and a pairwise random force(F_{ij}^{R}). All three forces tend to 0 when the distance between two particles is larger than the cutoff radius[17-19].

2.4 Model

To studying interfacial properties expediently, oil and water are represented as single beads, respectively. The head group and tail group of the surfactants are represented as two beads, which are connected by a harmonic spring with the spring constant C=4 (in kT units). The names of the surfactants yield information about their structure. For example, SDBS is regarded as a coarse-grained model of HT, and TX-100 is denoted by EC. H and E represent head groups of SDBS and TX-100, respectively, and T and C represent their tail chains. A 20×10×10 box containing a total of 6 000 beads($\rho=3$) is used, and a periodic boundary condition is applied in all three directions. All the simulations are performed using Cerius^2 software on the SGI workstation.

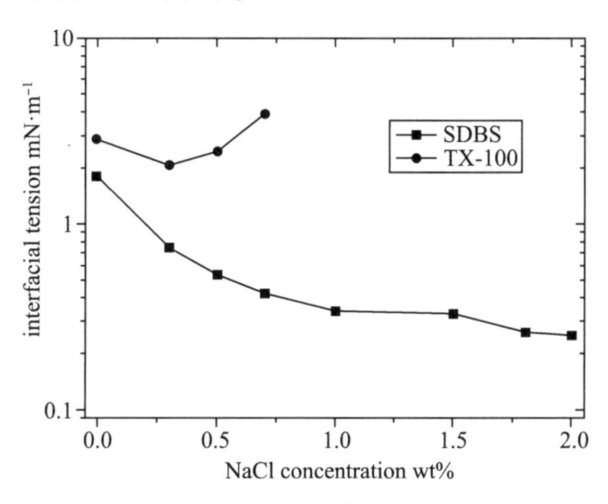

Fig. 1 The methylbenzene/water interfacial tension at different NaCl concentrations. The concentration of surfactant is 0.3 wt%.

3 Results and discussion

3.1 Interfacial behavior for single SDBS and TX-100 surfactant

The oil/water equilibrium interfacial tension, measured between methylbenzene and surfactant solution at different NaCl concentrations, is shown in Fig. 1. The results indicate that the addition of salt makes SDBS more effective in reducing the oil/water interfacial tension; however, when the salinity increases beyond 2.0 wt%, SDBS will salt out. Inorganic salt has little effect on the interface activity of nonionic surfactants. Variation in surfactant properties, such as ionic head groups, due to screening by the addition of salt, can be mimicked in simulation by varying the head-head, head-water repulsion parameters. For ionic surfactants, decreasing the repulsion between head groups corresponds, for example, to adding salt to the solution[20].

It is well known that surfactants in aqueous solutions have two tendencies at the same time: one is to aggregate to form micelles and the other is to adsorb at the interface. Only the latter process corresponds to the equilibrium interface activity of surfactants. Interfacial density d^{int}, obtained by integrating the average number of surfactants at the interface, gives clear information about the two surfactant tendencies[21]. d^{int} increases with surfactant concentration until the interfacial adsorption reaches saturation. After that, additional surfactants cluster together to form a congeries structure in the bulk, and the density in the bulk phase increases, which is useless in decreasing equilibrium interfacial tension. The maximum interfacial density d^{int}_{max} can be considered as a measure of the tendency for surfactant molecules to adsorb at the interface to a certain extent. The larger the d^{int}_{max}, the stronger the tendency to absorb at the interface.

Fig. 2 a shows the effect of decreased head-head repulsion a_{HH} on the interfacial density of SDBS. Since concentrations are sufficient for surfactants to obtain saturation adsorption under any conditions, the highest point of the curve can be considered as the d^{int}_{max}. The maximum of the interfacial density for each a_{HH} assigned in Fig. 2 a is about 0.329, 0.303, 0.287, and 0.260, respectively, which shows that the value of d^{int}_{max} increases as a_{HH} decreases. The addition of salt screened the static repulsion interaction between ionic head groups, which made surfactants prone to adsorb more on the oil/water interface from the water solution.

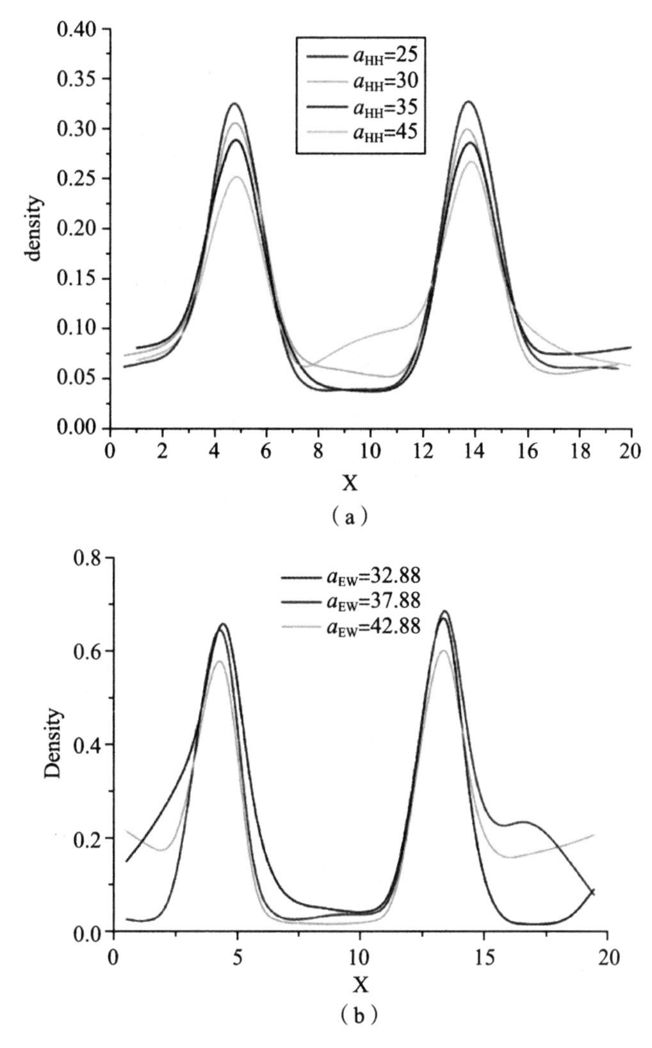

Fig. 2　Effect of the repulsion parameter a_{HH} on the distribution of SDBS(a); effect of the repulsion parameter a_{EW} on that of TX-100 (b). Surfactant concentrations are 0.08 and 0.10 by volume, respectively. The high-density area is the interfacial layer. There are two interface layers because a periodic boundary condition is applied in all three directions. The middle area is water and both sides are oil. The concentrations of SDBS and TX-100 are sufficient to obtain saturation adsorption under any conditions.

Fig. 3 shows a section snapshot of a simulation under the maximum interfacial density of surfactant SDBS($a_{HH}=25$). The head groups of SDBS are not distributed uniformly over the interface. Rather, the interfacial layer contains large cavities filled with water or oil, as the figure shows. This similar observation has been reported on Newton black films[22,23]. Although the addition of salt results in a significant increase in the interfacial density, the interfacial tension can not achieve ultralow because of the cavities on the interface.

Fig. 3 A simulated section snapshot of the SDBS adsorption interface at $d_{max}^{int}=0.329$. Cyan and magenta beads represent head groups and tail chains of SDBS, respectively. Water and oil beads are removed for convenience of observation. For interpretation of the references to color in this figure legend, the reader is referred to the Web version of this article.

In general, inorganic salts are considered to have a weak effect on nonionic surfactants. However, a large amount of inorganic salts would influence the solution properties by their effect on the properties of water, that is, the so-called salting-out effect. In the case of nonionic surfactants with ethylene oxide groups, when salts are introduced, dehydration of the surfactant EO chain would be induced because the ions dissociated from the salts are strongly hydrated and break the hydrogen bonding between ethylene oxide groups and water. Thus, it is suggested that the addition of the salts would make the repulsion parameter between ethylene oxide group and water increase, though not as notably as effect on ionic surfactant head-head repulsion.

The influence of salt on the distribution of TX-100 at the interface can be seen from Fig. 2b. The increase in a_{WE} brings a weak effect on d_{max}^{int} for TX-100 at low salinity, while at high salinity the increase in a_{WE} results in the interfacial density decreasing, as expected and in agreement with experimental results. Moreover, the value of d_{max}^{int} indicates that TX-100 has a stronger tendency to absorb at the interface than SDBS.

It can be concluded that the interfacial tension of each singlesurfactant system cannot achieve an ultralow value no matter what the concentration of salt is, although the surfactant concentration is large enough.

3.2 Interfacial activity and molecular behavior at interfaces for SDBS/TX-100 mixed systems

3.2.1 Adsorption of mixed surfactants at interfaces

The distribution of the mixed surfactants at interfaces corresponding to the total concentration ratio and salinity has been investigated by a simulation method. The ratio of the two surfactants was changed from 0 : 10, 1 : 9, … to 10 : 0(SDBS to TX-100), with total concentration fixed at 0. 1. The results are shown in Fig. 4. The dotted line in the figure is diagonal, representing the pseudo-condition that the distribution of surfactant in the interface is equal to that in the bulk. It is used as reference here.

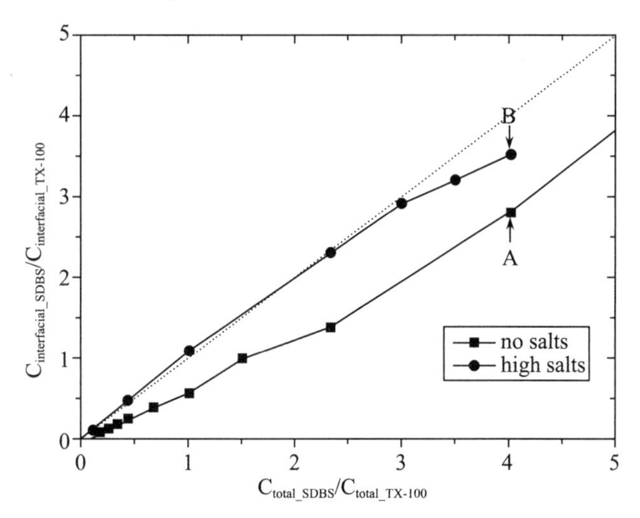

Fig. 4 The distribution of surfactants at the interface and in the total concentration. $C_{total\_SDBS}/C_{total\_TX\text{-}100}$ represents total concentration ratio of SDBS and TX-100, and $C_{interface\_SDBS}/C_{interface\_TX\text{-}100}$ means the ratio of the two surfactants at the interface.

Under the conditions without any salt, nonionic surfactant is interfacially enriched because of its stronger tendency to absorb at the interface than SDBS; thus the interfacial concentration ratio of SDBS to TX-100 is smaller than the total proportion, which yields a curve lower than the diagonal line. At high salinity, the adsorption tendency of SDBS at the oil/water interface has been strengthened, and its bulk concentration becomes very low. The interfacial concentration ratio of SDBS to TX-100 is equal to the total. As SDBS concentration increased to 0. 075, it reached saturated adsorption and additional SDBS molecules began to aggregate in bulk solutions, while TX-100 molecules were still enriched at the interface, thus results in the curve deflecting from the diagonal line. The turning point indicates that SDBS has the largest adsorption at the interface but no excess in the bulk phase. Above that ratio, interfacial density of SDBS stays constant, but TX-100 decreases. The interfacial density of surfactants required to obtain a given interfacial tension reduction reflects its interfacial efficiency[20,24]. From Fig. 1 we can see that SDBS is more efficient at the interface than TX-100. Thus, in order to achieve the lowest interfacial tension, the density of SDBS must be the highest, and also TX-100 adsorbs as much as possible under this condition. So the turning point can be seen as the optimal ratio of mixed surfactants to obtain the lowest interfacial tension, which is about 3 : 1 here.

3.2.2 Effects of surfactant concentration and salinity on interfacial tension in SDBS/TX-100 mixed systems

The effect of surfactant concentration and salinity on interfacial tension in SDBS/TX-100 mixed systems was determined experimentally. It was found that ultralow interfacial tension could be obtained with the mixed surfactant system at the ratio of 7 : 3 to 8 : 2(SDBS to TX-100) at high salinity. The turning point is just in this range. Simulation conclusions are consistent with experimental results.

The effect of the surfactant concentration on the interfacial tension in the SDBS/TX-100 mixed system is shown in Fig. 5, where the weight ratio of the two surfactants was fixed at 8 : 2. The salinity was fixed as 2.0 wt% and the surfactant concentration was changed from 0.01 to 0.5 wt%. It was found that ultralow interfacial tension could be obtained at surfactant concentrations even below 0.03 wt%, the cmc of SDBS. Then adding more surfactants is helpless to decrease the interfacial tension. It was obvious that synergistic interaction between SDBS and TX-100 was the key point for the ultralow interface tension, and it mainly lies at the interfacial density of the surfactant, which was affected by bulk concentration and salinity.

Based on the above results, the total surfactant concentration of the SDBS and TX-100 mixed system was chosen as 0.05 wt% to investigate the effects of salinity on the interfacial activity.

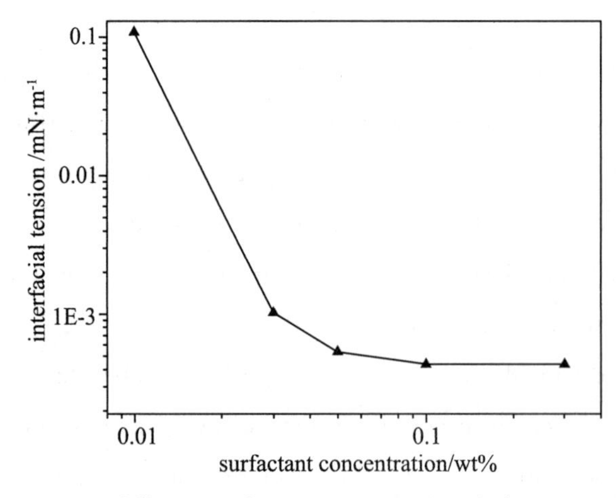

Fig. 5 Interfacial tension at different surfactant concentrations(salt concentration is 2.0 wt%).

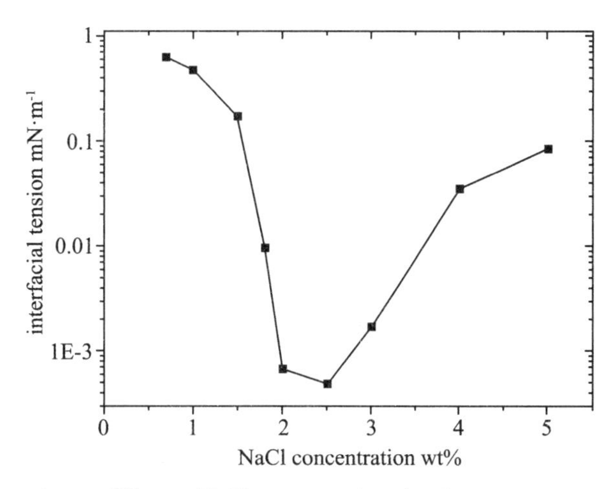

Fig. 6　Interfacial tension at different NaCl concentrations(surfactant concentration is 0.05 wt%).

Fig. 6 shows variation of the interfacial tension as a function of salt concentration at the constant surfactant concentration 0.05 wt%. A sharp decrease in interfacial tension is observed when salinity increases to 1.8 wt%. Then the interfacial tension stays ultralow in the salinity range between 1.8 and 3.0 wt%. Surfactant would salt out at high salinity, resulting in interfacial tension greater than ultralow.

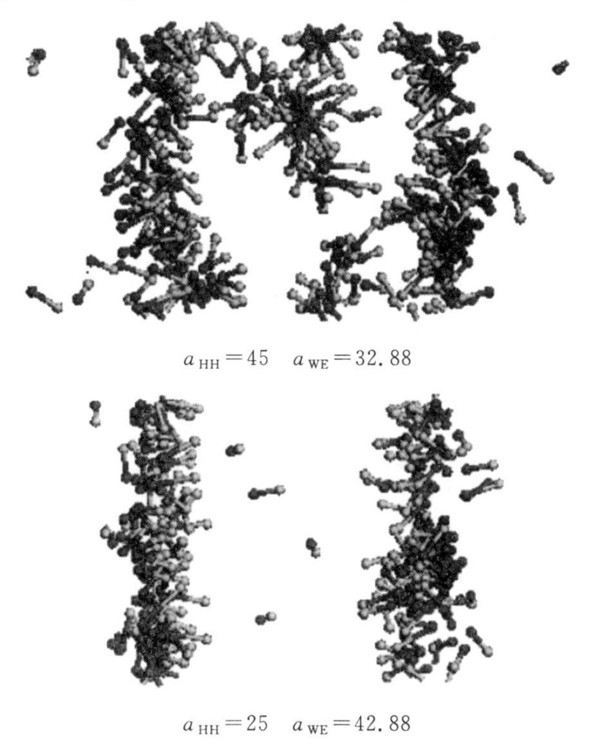

Fig. 7　Effect of adding salt on the distribution of SDBS and TX-100 in the mixed system. Cyan and green beads represent head groups of SDBS and TX-100, respectively, and magenta and red beads represent tail chains. Water and oil beads are removed. For interpretation of the references to color in this figure legend, the reader is referred to the Web version of this article.

The results also show that the mixed surfactants improve the salt resistance of the aqueous solution. The SDBS solution salts out at salinity above 2.0 wt%, but the SDBS/TX-100 mixed solution does not salt out until the salinity is higher than 4.0 wt%. For conventional surfactants[25], the low interfacial tensions usually exist in a narrow salt concentration range known as optimal salinity. Consequently, the appropriate salinity can evidently decrease the interfacial tension of the surfactant mixed system with synergic interaction.

Fig. 7 shows simulation results for the change of distribution of mixed surfactants influenced by salinity at the ratio 8 : 2(SDBS: TX-100). The variety in a_{HH} and a_{WE} makes more and more SDBS molecules move toward and adsorb at the oil/water interface, and thus makes the point A on the lower curve in Fig. 4 rise to the point B on the upper one, and then the interfacial tension decreases.

3.3 The distribution of mixed surfactants at interfaces

According to simulation results, when $a_{HH}=25$ and $a_{WE}=42.88$, the distribution of SDBS and TX-100 on the interface is as shown in Fig. 8. Interestingly, the head groups of SDBS and TX-100 are not distributed uniformly over the interface. The SDBS molecules are placed together with each other and so are the TX-100 molecules. The conclusion is consistent with the result about the behavior of sodium dodecyl sulfate/dodecanol at the air/water interface[12].

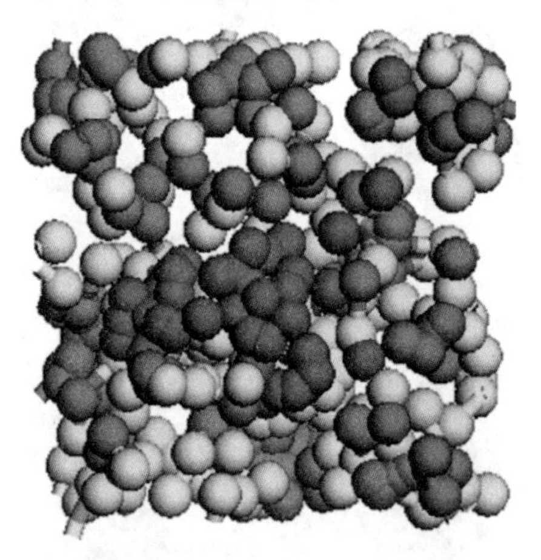

Fig. 8 The distribution of SDBS and TX-100 on the interface. Cyan and green beads represent head groups of SDBS and TX-100, respectively, and magenta and red beads represent tail chains. Water and oil beads are removed. For interpretation of the references to color in this figure legend, the reader is referred to the Web version of this article.

Compared to single SDBS molecular interfacial layers, which consist of large cavities, this kind, as we observed, with TX-100 clusters filling some cavities, are arranged more compactly, thus decreasing the interfacial tension sharply.

Experimentally, another type of nonionic surfactant, Tween-80, has also been used to investigate synergistic effects between SDBS and itself. The interfacial tension of this mixed system can be reduced to 10^{-2} $mN \cdot m^{-1}$ with added salt, which is not as effective as SDBS/TX-100 mixed systems. The difference may lie in the molecular structure distinction between the two nonionic surfactants. Tween-80 shows larger space hampering because of its larger head group(three ethoxy chains) than TX-100. Simulation results show that only a few Tween-80 molecules can insert into the cavities, while most of them aggregate into micelles, which show a weak effect in decreasing interfacial tension. That is, only nonionic surfactants of an opportune size can occupy the cavities formed by ionic surfactants.

4 Conclusions

In this paper, synergistic interaction of SDBS and TX-100 at oil/water interfaces has been investigated by experimental methods and computer simulation. The results show that interfacial tension of single SDBS systems cannot achieve ultralow values because of some cavities in the surfactant interface layer. When TX-100 is introduced, its molecules collocate close to each other in a cluster that filled large cavities formed by SDBS molecules; this system made mixed surfactants arrange more compactly and decreased the interfacial tension to ultralow levels. To achieve the lowest interfacial tension, the density of high-efficiency surfactant, SDBS, must be the highest, which was achieved by adding salts in the paper, and also TX-100 adsorbs at interface as much as possible under this condition. According to interfacial density, the optimal ratio of mixed surfactants to reach lowest interfacial tension at low bulk concentration can be obtained.

The above results make interfacial adsorption tendencies shown in real systems clearer and give us new insight into how the different molecules distribute at the interface, which shows us a direction in which to find synergistic efficiency in real systems.

Acknowledgments

The authors gratefully acknowledge financial support from the National Natural Science Foundation of China(No. 29903006).

Supporting material

The online version of this article contains additional supporting material.

Please visit DOI:10.1016/j.jcis.2006.11.026.

References

[1] L. Chen, J. X. Xiao, K. Ruan, J. M. Ma, Langmuir 18(2002)7250.

[2] S. R. Raghavan, G. Fritz, E. W. Kaler, Langmuir 18(2002)3797.

[3] M. J. Rosen, Z. H. Zhu, T. Gao, J. Colloid Interface Sci. 157(1993)254.

[4] L. Magnus Bergstrolm, J. Phys. Chem. B 109(2005) 12387.

[5] K. Theander, R. J. Pugh, J. Colloid Interface Sci. 267(2003)9.

[6] H. Dominguez, J. Colloid Interface Sci. 274(2004)665.

[7] K. Wojciechowski, J. Buffle, Colloids Surf. A 257(2005)385.

[8] A. Bumajdad, J. Eastoe, J. Colloid Interface Sci. 274(2004)2682.

[9] A. Bumajdad, J. Eastoe, Langmuir 15(1999)5271.

[10] V. B. Fainerman, D. Vollhardt, G. Emrich, J. Phys. Chem. B 105 (2001)4324.

[11] H. Domínguez, J. Phys. Chem. B 106(2002)5915.

[12] H. Domínguez, M. Rivera, Langmuir 21(2005)7257.

[13] Y. Li, G. Xu, Colloids Surf. A 257(2005)385.

[14] J. Eastoe, S. Nave, A. Rankin, A. Paul, A. Downer, Langmuir 16(2000) 4511.

[15] G. X. Zhao, Guo-xi, Phys. Chem. Surfact. [P] (1984)130.

[16] Z. Hou, Z. Li, H. Wang, Oilfield Chem. 16(1999)348.

[17] M. Y. Kuo, H. C. Yang, C. Y. Hua, C. L. Chen, S. Z. Mao, F. Deng, H. U. Wang, Y. R. Du, Chem. Phys. Chem. 5(2004)575.

[18] R. D. Groot, P. B. Warren, J. Chem. Phys. 107(1997)4423.

[19] E. Ryjkina, H. Kuhn, H. Rehage, Angew. Chem. Int. Ed. 41(2002)983.

[20] L. Rekvig, Langmuir 19(2003)8195.

[21] Y. Li, P. Zhang, J. Colloid Interface Sci. 290(2005)275.

[22] F. Bresme, J. Faraudo, Langmuir 20(2004)5127.

[23] F. Bresme, J. Alejandre, J. Chem. Phys. 118(2003)4134.

[24] Y. Li, X. He, Mol. Simul. 31(2005)1027.

[25] S. Nave, J. Eastoe, R. K. Heenan, D. Steyler, I. Grillo, Langmuir 18 (2002)1505.

Effect of inorganic positive ions on the adsorption of surfactant triton x-100 at quartz/solution interface

Shao Yuehua[1] Li Ying[1] Cao Xulong[2] Shen Dazhong[3]
Ma Baomin[1] Wang Hongyan[2]

[1] Key Lab for Colloid and Interface Chemistry of State Education Ministry, Shandong University, Jinan 250100, China;

[2] Geological Scientific Research Institute, Shengli Oilfield, Dongying 257015, China;

[3] School of Chemistry, Chemical Engineering and Material Science, Shandong Normal University, Jinan 250014, China

The electrode-separated piezoelectric sensor (ESPS), an improved setup of quartz crystal microbalance (QCM), has been employed to investigate the adsorption behavior of nonionic surfactant Triton X-100 at the hydrophilic quartz-solution interface in mineralized water medium *in situ*, which contained $CaCl_2$ 0.01 mol · L^{-1}, $MgCl_2$ 0.01 mol · L^{-1}, NaCl 0.35 mol · L^{-1}. In a large scale of surfactant concentration, the effects of Ca^{2+}, Mg^{2+} and Na^+ on the adsorption isotherm and kinetics are obviously different. In aque-ous solution containing NaCl only, adsorption of Triton X-100 on quartz-solution interface is promoted, both adsorption rate and adsorption amount increase. While in mineralized water medium, multivalent positive ions Ca^{2+} and Mg^{2+} are firmly adsorbed on quartz-solution interface, result in the increasing of adsorption rate and adsorption amount at low concentration of surfactant and the peculiar desorption of surfactant at high concentration of Triton X-100. The results got by solution depletion method are in good agreement with which obtained by ESPS. The "bridge" and "separate" effect of inorganic positive ions on the adsorption and desorption mechanism of Triton X-100 at the quartz-solution interface is discussed with molecular dynamics simulations (MD), flame atomic absorption spectrometry (FAAS) and atomic force microscopy (AFM) methods.

quartz crystal microbalance, quartz-solution interface, mineralized water medium, molecular dynamics simulations, Triton X-100

The nature of the adsorption of surfactants on solid sur-faces is of considerable interest as it is important to a number of industrial and technological processes, such as detergency, water treatment, improved oil recovery, ore refinement by flotation, painting cosmetics, agro-chemical formulations and textiles[1]. Exploring the adsorption mechanism of surfactant molecules at the solid-liquid interface is an important way toward mod-eling industrial processes and can give useful clues to the chemistry of the surfactant molecules themselves.

In most of the studies related to surfactant adsorption, solution depletion method was commonly used[2-5] to estimate the adsorption quantity, which provides mainly the information of equilibrium adsorption densities. However, the information of adsorption kinetics is also important for understanding the adsorption process. Besides, although the adsorption mechanism of surfactants from aqueous solutions on various surfaces has been investigated using different techniques, such as quartz crystal microbalance (QCM)[6-8], ellipsometry[9,10], neu-tron reflection[11], small-angle neutron scattering[12], atomic force microscopy[13,14] and molecular dynamics simulations (MD)[15,16], but most of the studies con-cerned in diluted concentration, no more than several times of cmc in most occasions, while practical proc-esses commonly relate to large scale of surfactant con-centration, which has sparely been studied.

Nonionic surfactant Triton X-100 is widely used in many practical processes, especially at the high salinity case because of its excellent salt resistance. But the adsorption of Triton X-100 at high salt concentration, es-pecially with Ca^{2+} and Mg^{2+} presenting has not been studied thoroughly yet.

The QCM device is a useful tool for real time moni-toring the adsorption processes and estimating the ki-netics parameters of surfactants on the solid surface. In this paper, the adsorption isotherm and the adsorption kinetic of Triton X-100 at hydrophilic silica surface has been investigated by *in situ* ESPS, and the kinetics parameters have been obtained. The effect of inorganic positive ions on the adsorption and desorption mecha-nism of Triton X-100 at the quartz-solution interface is discussed with MD, FAAS and AFM methods.

1 Materials and methods

1.1 Materials

Milli-Q water is used to prepare all solutions. All chemicals are of analytical grade or better. The nonionic surfactant Triton X-100 is purchased from Sigma Chemical Co, with > 99% purity. Quartz powder is pur-chased from No. 2 Shanghai Reagent

Manufacture. The specific surface area of quartz powder is determined by the BET adsorption method to be 1.2 $m^2 \cdot g^{-1}$. The glass-ware used for the surfactant solutions is carefully cleaned with bichromate sulfuric acid and then rinsed with excess Milli-Q water.

1.2 Quartz crystal microbalance

The quartz crystal microbalance used is the elec-trode-separated piezoelectric sensor. The change in resonance frequency is directly related to adsorbed amount under certain experimental conditions. Accord-ing to the Sauerbrey relation[17], considering the influ-ence of surface roughness of quartz crystal, a calibration factor of β is added in Eq. (1) to obtain a better estima-tion of adsorption densities in the ESPS method:

$$\Gamma = \Delta m / A = -\beta \Delta F / 2.26 \times 10^{-6} F_0^2. \quad (1)$$

Where Δm is the mass adsorbed onto quartz crystal, A is the geometric area of the sensitive region of the quartz crystal disc. Under our experimental conditions, $\beta = 0.205$, ΔF and F_0 are the frequency shift and the funda-mental frequency of the quartz crystal in Hz, respec-tively. Under our experimental conditions, $F_0 = 5$ MHz, the frequency decrease of 1 Hz corresponds to an ob-served adsorption density of 0.187 $\mu g/cm^2$.

The configuration of the ESPS used is illustrated in Figure 1. A 5 MHz AT-cut quartz crystal disc of 25.4 mm in diameter is used. The quartz disc is fixed to a detection cell made by organic glass with silicon glue.

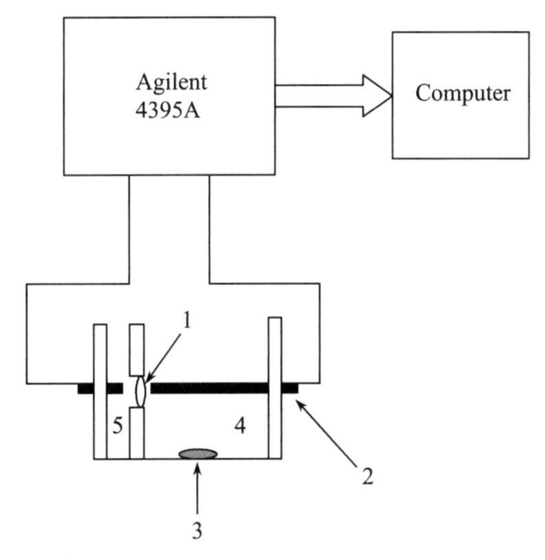

Figure 1 Experimental setup for the monitor adsorption process on the quartz crystal surface by the ESPS method. 1, Quartz crystal; 2, graphite electrode; 3, stirrer; 4, Triton X-100 solution; 5, 1 mol/L NaCl solution.

The two side-by-side graphite electrodes with diame-ter of 6 mm are used as the separated-electrode. The av-erage distance between the separated-electrode and the quartz crystal disc is 6 mm. The side-by-side graphite separated-electrodes are connected to a home-made os-cillator circuit. The volume of the detection cell is 0. 35 mL. The solution is introduced into the detection cell by a peristaltic pump. The oscillating frequencies of the ESPS are recorded by a universal frequency counter (7201, Shijiazhuang No. 4 Radio Factory) connected to a PC computer. The frequency draft of the measurement system is less than 1 Hz in 10 min. The experiments are preformed in room temperature at 298 K.

Before the determination of adsorption, the quartz crystal disc surface is cleaned with NH_4OH : H_2O_2 : H_2O (volume ratio 1 : 1 : 5), then rinsed thoroughly with Milli-Q water then measuring medium. Triton X-100 stock solutions are prepared by the measuring medium used in adsorption experiments, and the measuring me-dium is mineralized water. The conductivity of the stock solution and the measuring medium are adjusted to be the same for each adsorption experiment.

50 mL of measuring medium is added into the detec-tion cell. After 60 s of stabilization under stirring, the stable oscillating frequency of the ESPS, F_1, is recorded as the reference as a function of time with an interval of 15 s for each time point. A certain amount of Triton X-100 stock solution is added into cell, with the injec-tion time being kept constant at approximately 60 s, and then 2. 0 mL of surfactant solution is supplied to the cell. The oscillating frequencies, F_2, are recorded as a func-tion of adsorption time, too. The shifts in the oscillating frequencies, $\Delta F = F_2 - F_1$, are used for the estimation of the adsorbed amount. Each experiment is carried out three times and the results are shown as mean values from all measurements.

1. 3 Molecular dynamics simulation (MD)

The MD simulations numerically integrate Newton's equation of motion such that all the atoms or pseudo-atoms in the system are allowed to evolve in time in response of the forces that act on them. For the present study, the initial set of molecular dynamics simulations considers the evolution of an idealized layer of Triton X-100 surfactants on silica, which consists of a 24-surfactant aggregation structure that is placed 6Å from the negatively charged hydrophilic silica surface, with all the head groups of surfactant molecule in the monolayer are facing towards the silica surface. The system is prepared with 1500 water molecules in a simulation box of dimensions $x = y = 40$Å and z-dimension of 139Å, and hydrocarbon chains of sur-factants extending perpendicular to the xy plane in all-*trans* conformations and every molecule distributing averagely in the xy plane. The water molecules and the silica surfaces are treated in a fully atomistic manner. All simulations are carried out in the NVT ensemble with a time step of 1 fs. The velocity Verlet algorithm[18] is used as the integrator. The temperature is controlled using the Hoover-Nose thermostat with relaxation time of 0. 2 ps[19], and the temperature of the system is kept constant at 298 K by applying velocity rescaling to all the atoms in the system. Ewald summation is used to maintain charge neutrality over the three-dimensional periodic boundary conditions applied in the simulations[18]. At least 1. 5 ns MD production is run to obtain the dynamic information using the trajectory of mole-cules in the simulation box. The results are analyzed from the coordinates which are saved every 0. 2 ps. All the simulations are run in the software Cerius^2 4. 6.

1. 4 Adsorption on quartz powder surface in solution by depletion method

2 g quartz powder was mixed with a 30 mL Triton X-100 solution prepared by the different measuring me-dium until equilibration of the system is reached. After centrifugation, the concentration of Triton X-100 in the clear supernatant solution is determined by a spectral analysis method at the wavelengths of 277 nm. Adsorp-tion densities are estimated from the concentration de-crease.

1. 5 Flame atomic absorption spectrometry (FAAS) method

A Hitachi 180-80 atomic absorption spectrometer (Hi-tachi Ltd. , Tokyo, Japan) was used for the determination of trace amounts of metal ions. The operating parame-ters are shown in Table 1.

Table 1 Operating conditions of FAAS

Elements	Na	Ca	Mg
Wavelength λ(nm)	589. 0	422. 7	285. 2
Slit l (nm)	0. 4	2. 6	2. 6
Lamp current I (mA)	10. 0	7. 5	7. 5
Acetylene flow υ(L · min^{-1})	2. 2	2. 6	2. 0
Air flow υ(L · min^{-1})	9. 4	9. 4	9. 4
Burner height h(mm)	7. 5	7. 5	12. 5

1. 6 Atomic force microscopy (AFM) imaging

The experiments are performed using a Nanoscope Ⅲ (digital instruments) in contact mode with an E-scanner. The imaging method is to use the double layer (or steric) repulsion between the tip and the surface layer and fly the tip over the adsorbed film[13,20,21]. The mica substrate (probing and structure) is freshly cleaned before use with adhesive tape. Silicon nitride N-P cantilevers (digi-tal instruments) with nominal spring constants of 0. 58 and 0. 12 N · m^{-1} are cleaned by UV irradiation for 30 min prior to use. The solution is injected into the cell, sealed with an O-ring, and thermally equilibrated for 15 min to 2 h before imaging. The scan rate, integral gain, and z-deflection range are varied from 8 to 15 Hz, and 0. 4 to 0. 6 nm, respectively (all other gains were set to 0). All experiments are performed at room temperature. All images shown are raw deflection images that have been flattened along the scan lines to remove any tilt from the sample. No other image processing is used. Only 200 nm × 200 nm images are shown below, but similar re-sults are observed in 300 nm × 300 nm and 100 nm × 100 nm images.

2 Results and discussion

2.1 Adsorption isotherm of Triton X-100 on quartz surface

Mean values of the frequency shifts of the ESPS as a result of adsorption of Triton X-100 at hydrophilic quartz surface are determined *in situ* for a variety of concentrations ranging from 0. 004 to 3. 5 mmol · L^{-1} at 298 K, pH 7, and the results are displayed in Figure 2(a). The adsorption isotherm of Triton X-100 on quartz sur-face in pure water is consistent with previous reports[22], while that with Ca^{2+}, Mg^{2+} or Na^{+} presenting is quite different. At low concentration, the adsorbed amounts all increase steeply with the increasing concentration, then reach a plateau in pure water and 0. 4 mol · L^{-1} NaCl bulk solution, *viz.* the total adsorbed amounts reach a plateau and keep constant with the concentration in-creasing after that, though the adsorption amount is lar-ger and get maximum at lower surfactant concentration in the latter medium. While in mineralized water me-dium, the adsorbed amounts of surfactant reach maxi-mum at concentration 1. 2 mmol · L^{-1} and then decrease from 1. 3 mg · m^{-2} to 0. 75 mg · m^{-2}, which is even lower than that in pure water, while concentration increasing from 1. 2 mmol · L^{-1} to 3. 5 mmol · L^{-1}. Solution deple-tion method has also been employed to investigate the adsorption isotherm of Triton X-100 at the hydrophilic quartz-solution interface under the same experimental condition, and the results are displayed in Figure 2 (b), which are in good agreement with that obtained by ESPS. In our opinion, the abnormal adsorption phe-nomenon of Triton X-100 on the quartz surface in min-eralized water medium especially at high concentration should be paid more attention to, for which might be important in practical process.

2.2 Adsorption kinetic of Triton X-100 on quartz surface

Figure 3 shows the time dependence of the adsorption amount of Triton X-100 at quartz surface in a 0.5 mmol • L^{-1} bulk solution in different measuring mediums. The oscillating frequency of the ESPS decreases as the mass load increase. Hence, the adsorption densities can be in situ monitored, and the dynamic adsorption parameters can be estimated from the time responses of the ESPS. Under experimental conditions in the paper, the surfactant is injected just prior to $t=0$ and the adsorption process approaches the equilibrium after 5 min in pure water medium which is quite rapid, while the adsorption processes of Triton X-100 in NaCl solution and mineralized water mediums are even faster, only need 3 and 2 min to approach the adsorption equilibrium, respectively. After the determination, the quartz crystal discs are rinsed thoroughly with surfactantfree measurement medium, and the oscillating frequency can be returned back to its previous baseline (data are not shown in the Figure), which indicated that the adsorption of Triton X-100 onto quartz surface is reversible with respect to the dilution of the bulk phase.

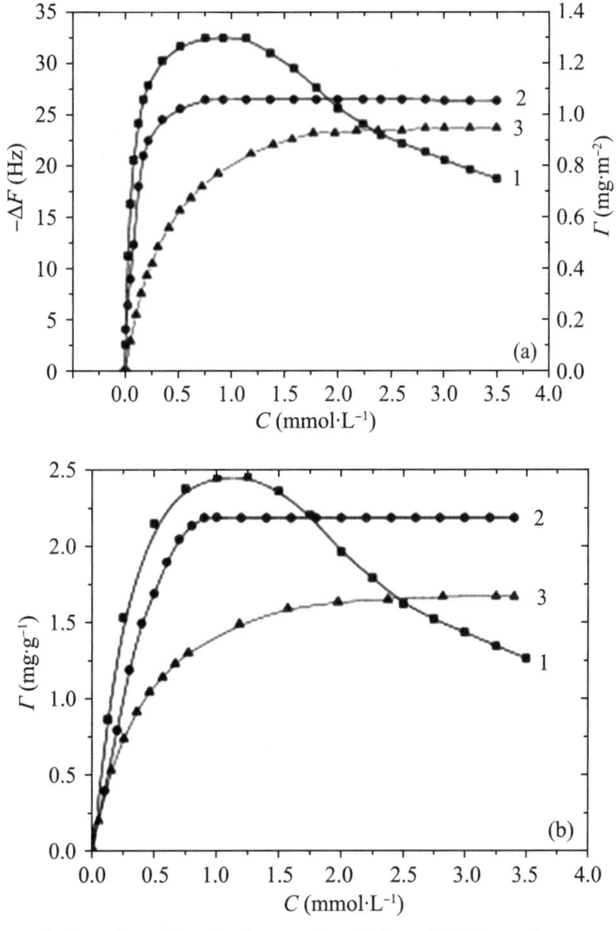

Figure 2　Comparison of the adsorption isotherms for Triton X-100 on the quartz surface. (a) On a quartz crystal disc by ESPS method; (b) on quartz powder surface by solution depletion method. Measuring medium: 1, mineralized water; 2, 0.4 mol • L^{-1} NaCl solution; 3, pure water.

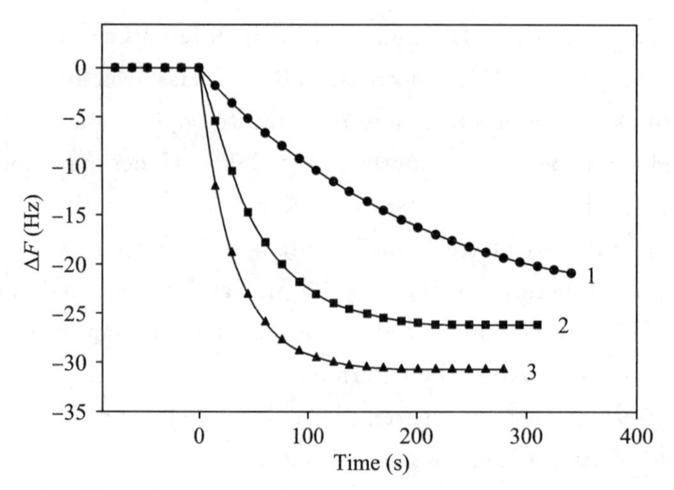

Figure 3 Variation of oscillating frequency shifts of the ESPS with adsorption time of Triton X-100 on a quartz surface with concentration equal to 0. 5 mmol $\cdot$ L^{-1}. The lines are the regression results using eq. (3). Measuring medium: 1, pure water; 2, 0. 4 mol $\cdot$ L^{-1} NaCl solution; 3, min-eralized water.

As shown in Figure 2, the adsorption isotherms of Triton X-100 in pure water, NaCl solution, and mineral-ized water medium are all in the shape of Langmuir type before the maximum point and can be fitted using the Langmuir model in Eq. (2).

$$\Delta F=\frac{\Delta F_{sat}CK}{1+CK}. \tag{2}$$

The response curves of the ESPS, shown in Figure 3, are fitted by Eq. (3)[23]

$$\Delta F=\frac{\Delta F_{sat}CK}{1+CK}[1-\exp[-(k_aC+k_d)t]]. \tag{3}$$

In which k_a and k_d are the rate constants for adsorption and desorption, respectively. K is the adsorption equilib-rium constant, t is the adsorption time, ΔF_{sat} is the oscillating frequency shift corresponding to the density of the saturation adsorption layer and C is the bulk concentration of adsorbed molecule.

The values of the observed adsorption rate constant $k_{obs}=Ck_a+k_d$ are obtained, and the dependence of k_{obs} as a function of C is depicted in Figure 4, of which the good linearity supports the agreement of Langmuir ki-netic model to the adsorption process in the concentrated range discussed above. The values of k_a and k_d corresponds to the slope and intercept of regressed liner of k_{obs} vs C might be got and the results are listed in Table 2. It can be seen that the desorption rate constants k_d vary slightly in different medium, while the adsorption rate constants k_a and adsorption equilibrium constants K increase obviously with inorganic positive ions presenting.

2. 3 Mechanism of effect of inorganic positive ions on adsorption and desorption of Triton X-100 on quartz surface

(1) Molecule dynamic simulation. The simulations predict that the surfactant monolayer adsorbs fairly on the silica. As the system evolves, the surfactants rear-

range into flat hemisphere micelles or monolayer in different solution medium through the applied periodic boundary conditions, as illustrated in Figure 5. The snapshots of the near ultimate states are shown in Figure 6. For the situation in pure water medium, see Figure 6(a), the hemisphere micelles adsorb onto the quartz surface; while in occasions with only NaCl presenting, a surfactant monolayer covers the entire quartz surface, see Figure 6(b); an incompact monolayer covers the entire silica surface in mineralized water medium, too, as shown in Figure 6(c), but is somehow bridged and separated by Ca^{2+} and Mg^{2+} adsorbing onto the quartz surface closely, which is different from Figure 6(b). It should be noticed that few Na^+ is adsorbed onto the quartz surface either in NaCl solution or in mineralized water medium.

Table 2 Adsorption kinetic parameters for Triton X-100 on a quartz surface

Measuring medium	k_a(L · mol^{-1} · s^{-1})	k_d(10^{-3} · s^{-1})	K(10^4 L · mol^{-1})
Mineralized water	141±8.9	3.09±0.03	4.56±0.03
0.4 mol · L^{-1} NaCl solution	108.5±7.4	2.65±0.03	4.09±0.03
Pure water	55.3 ± 7.3	2.52±0.03	2.18±0.03

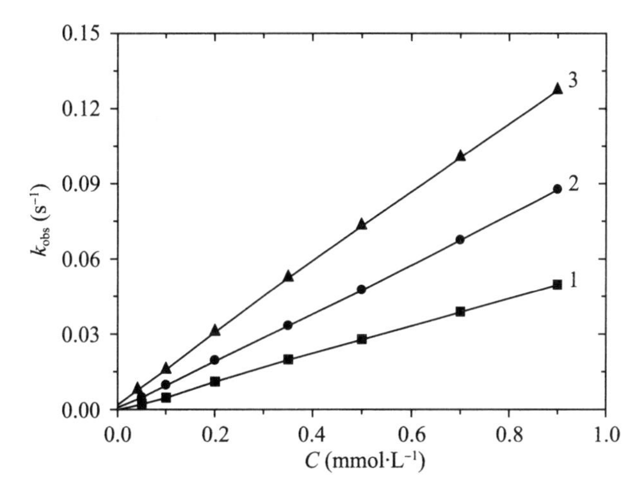

Figure 4 The relationship between k_{obs} and concentration of Triton X-100. Measuring medium: 1, pure water; 2, 0.4 mol · L^{-1} NaCl solution; 3, mineralized water.

(2) Flame atomic absorption spectrometry. The adsorption of positive inorganic ions on the silica surface is verified by FAAS measurement, the results is shown in Table 3. The concentration of Na^+ in solution is almost the same before and after adding quartz powder both in NaCl solution and in mineralized water systems, while the concentration of multivalent positive ions decreased abruptly, which could be concluded to adsorption of these ions on quartz surface. The results given by FAAS measurement accorded well with above MD simulation.

(3) Effect of multivalent positive inorganic ions on the adsorption of Triton X-100

on the silica surface. For a fluid, with no underlying regular structure, the mean square displacement (MSD) should increase lin-eally with simulation time, and the diffusion coefficients of molecule or ions can be obtained from the slope of the MSD verse simulation time, given by the following equation:

$$D = \frac{1}{6N} \lim_{i \to \infty} \frac{d}{dt} \sum_{i=1}^{N} < [r_i(t) - r_i(0)]^2 > \quad (4)$$

where N is the number of diffusive atoms (molecules or ions) in the system. The above differential is approximated by the ratio of MSD and the time difference.

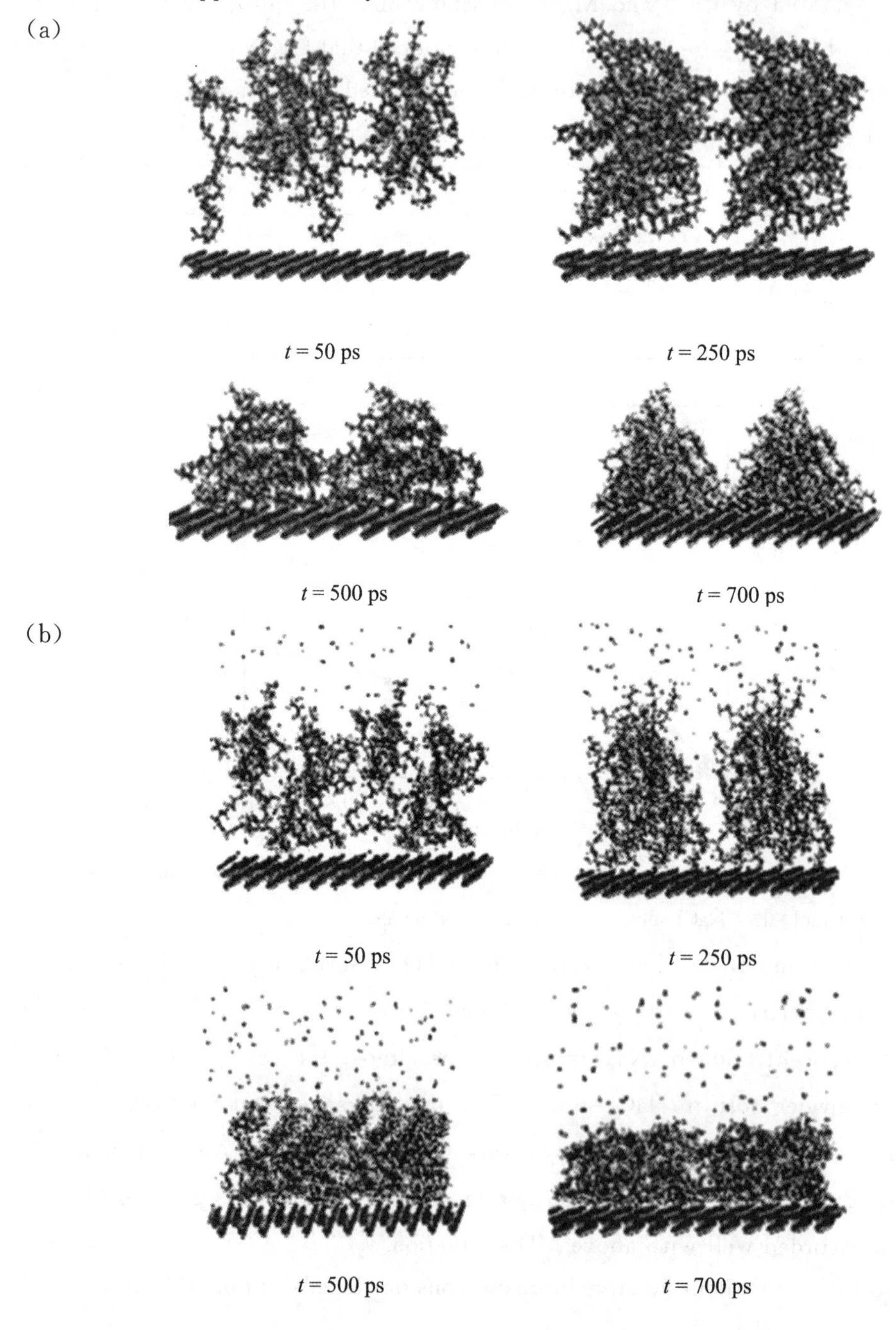

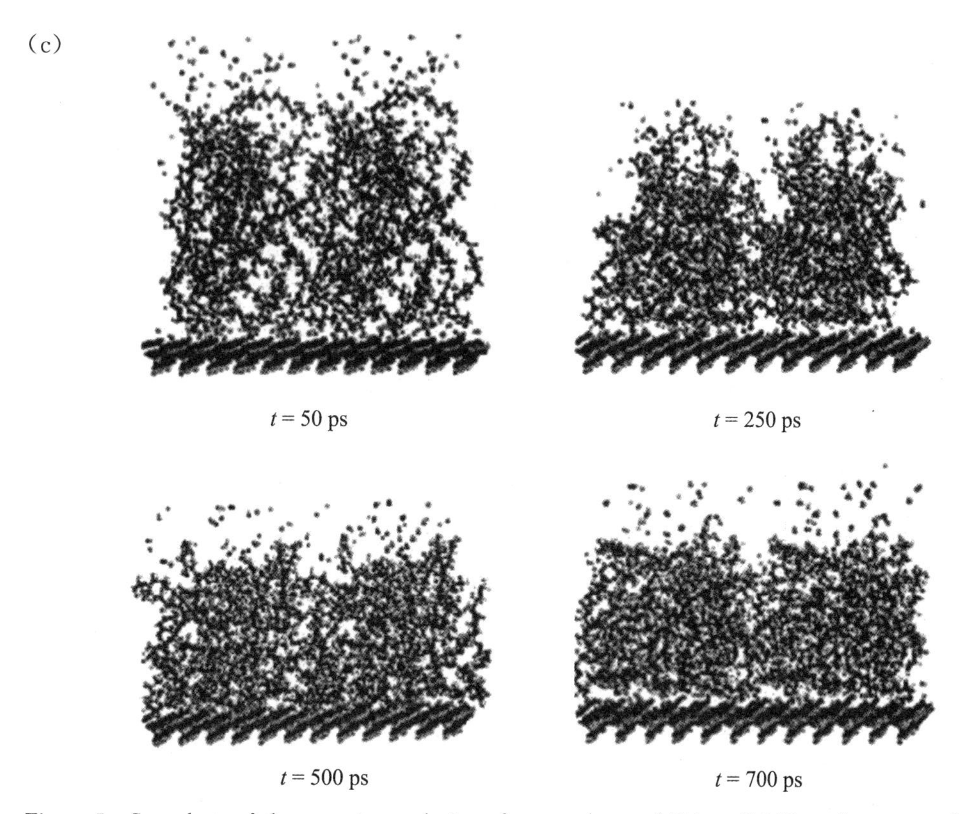

Figure 5 Snapshots of the stepwise evolution of a monolayer of Triton X-100 surfactant on silica with time. Measuring medium: (a) 24-Triton X-100 molecules and 1 500 water molecules; (b) 24-Triton X-100 molecules and 1500 water molecules with 40 Na^+ and 40 Cl^-; (c) 24-Triton X-100 molecules and 1500 water molecules with 50 Na^+, 20 Ca^{2+}, 20 Mg^{2+} and 130 Cl^-. Na^+, Ca^{2+}, Mg^{2+} and Cl^- are represented by blue, green, aqua and cyan, respectively, and the surface atoms are represented by yellow (Si) and red (O) (Colors available in the online version of this article). The water molecules are not shown for clarity. Two cells are repeated in the x-coordinate in order to clearly show the interface. 50 ps: Parts of the head groups of surfactant adsorb onto the nega-tively charged silica surface. The tails bundle together as a result of their hydrophobic interaction with the surrounding water molecules. 250 ps: The struc-ture starts to spread across the surface. Chains of surfactant begin to bunch together to shield themselves from the water. More head groups of surfactant adsorb onto the silica surface. 500 ps: More head groups form a spherical micelle and monolayer micelles structure adsorbed on the silica. 700 ps: This aggregate continues to reorganize itself until it forms a near perfect sphere or monolayer on the silica surface.

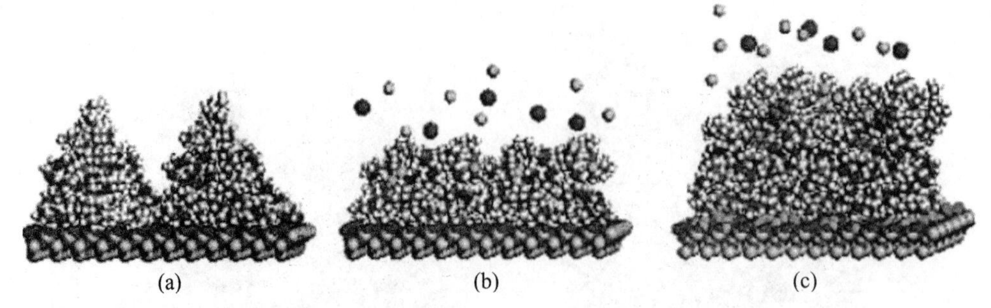

Figure 6 Snapshots of interface aggregation structures formed by Triton X-100 adsorption onto a silica surface in different medi. (a) Pure water; (b) in the aqueous solution with Na^+ and Cl^- presenting; (c) in mineral water medium with Na^+, Ca^{2+}, Mg^{2+} and Cl^- presenting. Water is not shown for clarity. The color scheme and other details are identical to those in Figure 5.

Using Eq. (4), the diffusion coefficients of Triton X-100 molecules are calculated, which are 3.38×10^{-6} $cm^2\cdot s^{-1}$, 4.49×10^{-6} $cm^2\cdot s^{-1}$ and 5.61×10^{-6} $cm^2\cdot s^{-1}$ in pure water medium, NaCl solution and mineralized water medium respectively, indicating that the tendency of surfactant molecules moving to the surface is stronger in mineralized water medium than in NaCl solution and pure water, which are in agreement with that shown in Figure 3. Hence, the adsorption equilibrium constant K and the adsorption rate constants k_a increase obviously with inorganic positive ions presenting is reasonable.

Table 3 Decrease of concentration of inorganic positive ions induced by adsorption on the quartz powder surface in different measuring media

Measuring medium	Ion	$C_0(mol\cdot L^{-1})$	$C_1(mol\cdot L^{-1})$	$C_2(mol\cdot L^{-1})$
0.4 mol/L NaCl solution	Na^+	0.35 ± 0.001	0.3495 ± 0.001	0.0005 ± 0.001
	Na^+	0.35 ± 0.001	0.3496 ± 0.001	0.0004 ± 0.001
Mineralized water	Ca^{2+}	0.01 ± 0.001	0.002 ± 0.001	0.008 ± 0.001
	Mg^{2+}	0.01 ± 0.001	0.004 ± 0.001	0.006 ± 0.001

C_0 and C_1 are the concentration of ions in solution before and after adding quartz powder respectively; C_2 represents the decrease of concentration of ions in the process

Diffusion of inorganic ions has been investigated in the paper, too. The dynamic diffusion process of Ca^{2+}, Mg^{2+} or Na^+ in aqueous solutions are divided into two stages, the first stage describes the diffusion of Ca^{2+}, Mg^{2+} and Na^+ in aqueous solutions before any of them adsorb onto the quartz surface, and the second stage describes the situation after some ions adsorbed onto the quartz surface. The MSD of Ca^{2+}, Mg^{2+} and Na^+ in the two different stages are shown in Figure 7. According to the Eq. (4), the diffusion coefficients of Ca^{2+}, Mg^{2+} and Na^+ in the two stages have been calculated by a linear fit to the corresponding MSD over the simulation time. The relative diffusion coefficients of ions in different stages are listed in Table 4, which show that Ca^{2+}, Mg^{2+} in stage II are obviously less mobile than those in stage I, indicating that some Ca^{2+}, Mg^{2+} adsorb onto the quartz surface firmly by electrostatic interaction as shown in Figure 6(c). The diffusion coefficient of Na^{+} changed little in the two stages and be less than that of Ca^{2+}, Mg^{2+} in stage I, which indicate that the tendency of Na^{+} moving to the silica surface is weaker than that of Ca^{2+} and Mg^{2+}, that match well with the phenomenon few Na^{+} adsorbs onto the quartz surface. These results are also in good agreement with that obtained by FAAS (Table 3).

Multivalent positive ions such as Ca^{2+}, Mg^{2+} adsorb more easily and firmly on the quartz surface than Na^{+}, as a result not only the hydrogen-bonding interaction between the silanol OH group and the ether oxygen groups of Triton X-100 is weakened, but also the adsorption of surfactants aggregation is obstructed because of the space block effect, which result in the decreasing of tendency of adsorption on quartz surface and the desorption of Triton X-100 at high concentration.

Table 4 Diffusion coefficients of Ca^{2+}, Mg^{2+} and Na^{+} in different stages obtained from mean square displacements

Ionic type	Stage Ⅰ $D(10^{-5} cm^2 \cdot s^{-1})$	Stage Ⅱ $D(10^{-5} cm^2 \cdot s^{-1})$
Ca^{2+}	1.99	0.602
Mg^{2+}	1.92	0.79
Na^{+}	1.17	1.10

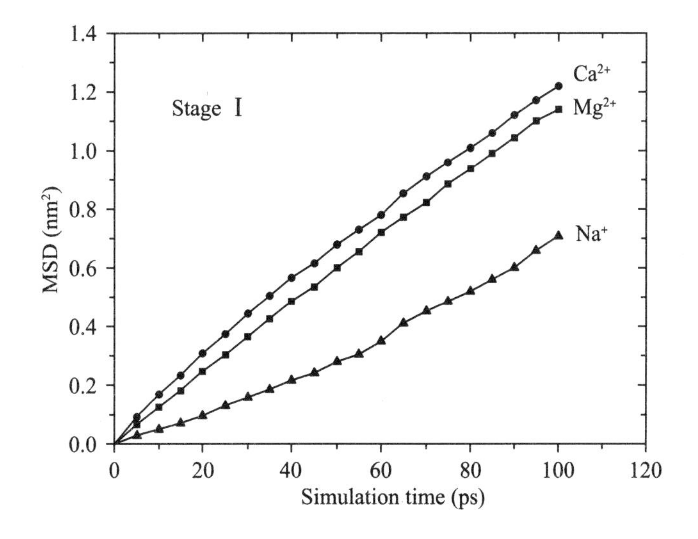

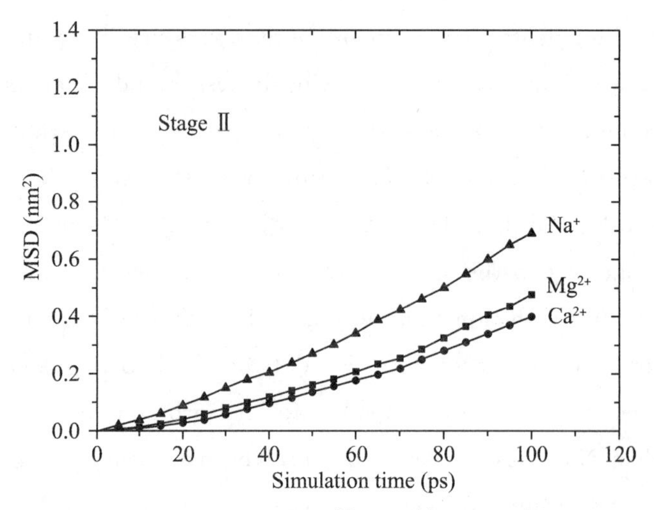

Figure 7 Mean square displacement (MSD) of Ca^{2+}, Mg^{2+} and Na^{+} in different stages in aqueous solutions.

AFM deflection images (200 nm × 200 nm) we have got show adsorbed layer morphologies of Triton X-100 at the mica surface might support the above results primarily based on MD simulation, as shown in Figure 8. At the same concentration of Triton X-100, comparing with that in pure water medium (Figure 8(a)), and in 0. 4 mol · L^{-1} NaCl solution medium (Figure 8(b)), AFM images of Triton X-100 in mineralized water medium (Figure 8(c)) showed that the layer covering the mica surface is thicker than that in Figure 8(a) and 8(b), while there are very obvious nicks, as called "Trailing effect" . The same sample shown in Figure 8(c) was imaged again and the cover layer at the mica surface disappeared. The experiment has been repeated for several times, and the phenomenon are the same. There might be two reasons for "trailing effect", among which are: (ⅰ) viscidity of adsorbed surfactant on mica is relative high and measured substance conglutinated on the AFM tip; (ⅱ) The interaction between the adsorpted surfactant and mica is weak, accordingly result in a loose adsorbed layer of Triton X-100 at the mica surface, which also lead to the "trailing effect". In this experiment, viscidity of experimental system is quite low, and the possibility of the first reason could also be excluded according to the Figure 8(a) and 8(b). So the second reason results in the "trailing effect", which provide an evidence for the "bridge and separate" effect of multivalent inorganic ions on the adsorption of Triton X-100 on the quartz surface.

3 Conclusions

In summary, effect of inorganic positive ions on the adsorption isotherm and kinetics of the nonionic surfactant Triton X-100 on quartzsolution interface is obvious and different. In aqueous solution containing NaCl, adsorption of Triton X-100 on quartz-solution interface is promoted, both adsorption rate and adsorption amount increasing, because of solubility of surfactant in aqueous solution decreasing owing to salt effect and tendency of adsorption on interface increasing. While in mineralized water medium containing Ca^{2+}, Mg^{2+}, these multivalent positive ions adsorb on quartz-solution interface firmly, result in the promotion of adsorption rate and adsorption amount increasing of Triton X-100 at low surfactant concentration and the peculiar desorption of Triton X-100 at high surfactant concentration. The effect of multivalent inorganic ions such as Ca^{2+}, Mg^{2+} on the adsorption of surfactants at surface of hydrophilic elec-tronegative solid should be taken into account in practi-cal processes.

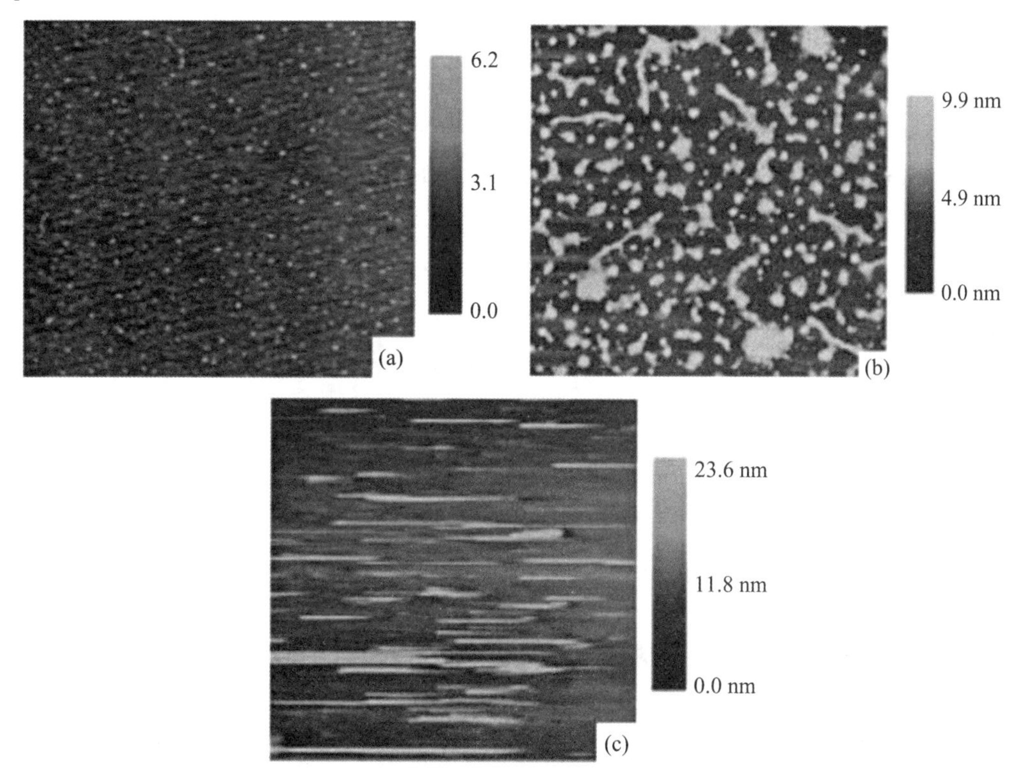

Figure 8 AFM deflection images (200 nm×200 nm) showing adsorbed layer morphologies of Triton X-100 at the mica/aqueous solution interface. (a) 0.2 $mmol \cdot L^{-1}$ Triton X-100; (b) 0.2 mmol/L Triton X-100 + 0.4 $mol \cdot L^{-1}$ NaCl solution; (c) 0.2 $mmol \cdot L^{-1}$ Triton X-100 + mineralized water (TDS=0.4 $mol \cdot L^{-1}$).

References

[1] Scamehorn J F. Phenomena in mixed surfactant systems. ACS Symposium Series 311, Washington, DC: American Chemical Society,1986. 324.

[2] Somasundaran P, Snell E D, Xu Q. Adsorption behavior of alkylarylethoxylated alcohols on silica. J Coll I Sc, 1991, 144(1):165-173.

[3] Desbene P L, Poret F, Treiner C. Adsorption of pure nonionic alkylethoxylated surfactants down to low concentrations at a sil-ica/water interface as determined using a HPLC technique. J Coll I Sc, 1997,190(2):350-356.

[4] Backhaus W K, Klumpp E, Narres H D, Schwuger M J. Adsorption of 2,4-dichlorophenol on montmorillonite and silica: Influence of nonionic surfactants. J Coll I Sc, 2001, 242(1):6-13.

[5] Shen Y H, Preparation of organobentonite using nonionic surfactants. Chemosphere, 2001,44(2):989-995.

[6] Caruso F, Rinia H A, Furlong D N. Gravimetric monitoring of nonionic surfactant adsorption from nonaqueous media onto quartz crystal microbalance electrodes and colloidal silica. Langmuir,1996,12(9):2145-2152.

[7] Cavic B A, Thompson M. Adsorptions of plasma proteins and their elutabilities from a polysiloxane surface studied by an on-line acoustic wave sensor. Anal Chem,2000,72(7):1523-1531.

[8] Hu J T, Yang D L, Kang Q, Shen D Z. Estimation the kinetics parameters for nonspecific adsorption of fibrinogen on quartz surface from the response of an electrodeseparated piezoelectric sensor. Sens Actu B,2003,96(1-2):390-398.

[9] Tiberg F, Johnsonn B, Tang J, Lindman B. Ellipsometry Studies of the self-assembly of nonionic surfactants at the silica-water interface: equilibrium aspects. Langmuir, 1994,10(7):2294-2300.

[10] Brink J, Tiberg F. Adsorption behavior of two binary nonionic sur-factant systems at the silica-water interface. Langmuir, 1996, 12(21):5042-5047.

[11] Penfold J, Staples E J, Tucker I, Thomas R K. Adsorption of mixed cationic and nonionic surfactants at the hydrophilic silicon surface from aqueous solution: the effect of solution composition and concentration. Langmur, 2000,16(23):8879-8883.

[12] Cummins P G. , Staples E, Penfold J. Temperature dependence of the adsorption of hexaethylene glycol monododecyl ether (C12E6) on silica sols. J Phys Chem, 1991,95(15):5902-5905.

[13] Manne S, Gaub H E. Molecular organization of surfactants at solidliquid

interfaces. Science,1995,270(5241):1480-1482.

[14] Patrick H N, Warr G S, Manne S, Aksay I A. Self-assembly structures of nonionic surfactants at graphite/solution interfaces. Langmuir, 1997, 13 (16): 4349-4356.

[15] Goundla S, Steven O N, Preston B M. Molecular dynamics simula-tions of surfactant self-organization at a solid-liquid interface. J Am Chem Soc, 2006,128(3): 848-853.

[16] Kunal S, Patrick C, Mayank J. Morphology and mechanical properties of surfactant aggregates at water-silica interfaces: molecular dynam-ics simulations. Langmuir, 2005,21(12):5337-5342.

[17] Saurbrey G. Verwendung von Schwingquarzen zur Wagung duner Schlichten und zur Mikrowagung. Z Phys, 1959,155:206-222.

[18] Allen M P, Tildesley D J. Computers Simulation of Liquids. Oxford: Clarendon Press,1987:385.

[19] Hoover W G. Canonical dynamics: Equilibrium phase-space distributions. Phys Rev A,1985,31(3):1695-1697.

[20] Manne S, Cleveland J P, Gaub H E, Stucky G D, Hansma P K. Direct visualization of surfactant hemimicelles by force microscopy of the electrical double layer. Langmuir,1994,10(12):4409-4413.

[21] Patrick H N, Warr G G, Manne S. Aksay I A. Surface micellization patterns of quaternary ammonium surfactants on mica. Langmuir,1999,15(5):1685-1692.

[22] Santanu P, Kartic C, A review on experimental studies of surfactant adsorption at the hydrophilic solid-water interface. Adv Coll In Sc, 2004,110(3):75-95.

[23] Shen D Z, Li S H, Li W P, Zhang X L, Wang L Z. Adsorption studies of cationic starch onto quartz surface through an electrode-separated piezoelectric sensor. Microchem J, 2002,71(1):49-55.

孤东二元驱体系中表面活性剂复配增效作用研究及应用

王红艳 曹绪龙 张继超 李秀兰 张爱美

(中国石化胜利油田分公司地质科学研究院,东营 257015)

摘要:报道了孤东PS二元复合驱先导试验中所用胜利石油磺酸盐SLPS与非离子表面活性剂间的复配增效作用研究结果。等摩尔比的磺酸盐AS与非离子剂LS54溶液的动态表面张力能迅速达到平衡且平衡值σ_e低。AS/ LS45混合体系的临界胶束浓度随LS45加入比例的增大不断降低,σ_e值则迅速降低并维持低值。当油相为正辛烷、甲苯、甲苯+正辛烷(1∶2)时,在磺酸盐SDBS中加入非离子剂LS45、TX、Tween、AES后,油水界面张力均大幅降低;在正辛烷中加入甲苯,可以降低与水相间的界面张力。用介观尺寸耗散粒子动力学(DPD)方法模拟,求得SDBS+TX在油水界面的分布密度较SDBS大大增加,导致界面张力大幅下降。按以上原理复配的3 g/LSLPS+1 g/L非离子剂+1.7 g/L聚合物超低界面张力体系,室内物模实验中提高采收率18.1%。在孤乐七区西南$Ng5^4-6^1$层10注16采试验区,从2004年6月起注入SP二元体系(4.5 g/L SLPS +1.5 g/L非离子剂+1.7 g/L聚合物),降水增油效果显著,截止2008年3月,试验区提高采收率5.7%,中心井区提高采收率12.7%且采出程度达到54%。图7表2参12。

关键词:表面活性剂/聚合物二元复合驱油体系;混合表面活性剂;石油磺酸盐;非离子表面活性剂;协同效应;表面张力;界面张力;表面活性剂/聚合物二元复合驱;胜利孤东油田

中图分类号:TE357.46:TE39:O647.2 文献标识码:A

引言

化学复合驱是继聚合物驱后一种更有潜力的三次采油新技术。大庆油田和胜利油田在三元复合驱方面开展了大量的室内研究及矿场试验[1-3]。尽管增油效果明显,但在实施过程中暴露出的结垢、采出液乳化严重[3,4]问题限制了其推广应用。为了探索复合驱技术在胜利油田的应用,开展了无碱体系二元复合驱油体系研究。目前开展的孤东油田七区西

南 $Ng5^4-6^1$ 层二元复合驱先导试验，是我国第一个二元驱矿场试验，其目的是探索聚驱后采用二元复合驱方法提高采收率和大幅度提高一、二类剩余储量采收率的可行性，形成配方设计、方案优化、注采规律等二元复合驱配套技术，为工业化推广提供基础。

孤东油田七区西南 $Ng5^4-6^1$ 层二元复合驱油先导试验中采用的主表面活性剂为胜利石油磺酸盐。石油磺酸盐阴离子极性头间的电性排斥作用较强，疏水链的结构多种多样，在界面上排列不紧密，界面活性不高。选择分子结构适宜、有协同作用的辅助表面活性剂，少量添加即可大大提高石油磺酸盐的界面活性，提高驱油效率。本文针对胜利石油磺酸盐进行了动态行为及与助剂复配协同作用的研究。

1 实验部分

1.1 实验试剂

十二烷基聚氧乙烯聚氧丙烯醚 $C_{12}H_{25}(EO)_4(PO)_5H$(LS45)、$C_{12}H_{25}(EO)_5(PO)_4H$(LS54)均为无色粘稠液体，纯度＞99.95％；十二烷基苯磺酸盐(SDBS)；十六烷基磺酸钠(AS)；十二烷基磺酸钠(SLS)；月桂醇聚环氧乙烷醚硫酸钠(AES)，壬基酚聚环氧乙烷醚(TX，n＝8～9)；聚氧乙烯失水山梨醇单油酸酯 Tween-80。胜利石油磺酸盐(SLPS，胜利中胜环保有限公司生产，平均相对分子质量 400，其中活性物含量 31.4％，活性物中双磺酸盐含量为 14％，未磺化油 25％，无机盐杂质含量 5％)，其余添加剂均为分析试剂。

1.2 实验方法

采用滴体积法在 DVT30 仪上测定动态表面张力，采用 Texas500C 界面张力仪测定界面张力。

2 实验结果与讨论

2.1 磺酸盐与非离子表面活性剂混合体系动态表面张力

具有两亲性的表面活性剂可在空气/水及油/水界面定向排列，形成有序分子层，因此表面过剩是表面活性剂溶液的重要性质，也是其具有表/界面活性的根本原因。而表面活性剂在水溶液中的扩散和吸附等动态行为，对其表面活性的发挥起决定作用，尤其是涉及快速过程时，如稳定分散体系，驱动地层中的原油等，表面活性剂的动态行为起着重要的作用[5,6]。

孤东油田七区西南 $Ng5^4-6^1$ 层二元复合驱油先导试验中使用的主表面活性剂为胜利石油磺酸盐 SLPS。石油磺酸盐是原油或馏分油用发烟硫酸或三氧化硫磺化，然后碱中和得到的混合产物[7]，主要成分是芳烃化合物的单磺酸盐。单一的石油磺酸盐不能使油水界面张力达到超低(10^{-3} mN/m)，需要加入适宜的辅助表面活性剂，提高其界面活性。为了探讨磺酸盐型表面活性剂与非离子表面活性剂混合体系的动态行为和复配协同作用，测定了 LS54 与 AS 混合体系的动态表面张力。表面活性剂总浓度为 1.0×10^{-5} mol/L，采用 7 000 mg/L NaCl 溶液配制，测定温度为 25℃。结果见图 1。

从图 1 可以看出，磺酸盐 AS 在水中的表面张力能很快达到平衡，但表面张力值高(62 mN/m)；非离子表面活性剂 LS54 达到平衡的时间很长，但是最终的表面张力值低(42 mN/m)。而摩尔比 1∶1 SL54+AS 的复配体系既能很快达到平衡，最终表面张力值也明显降低。这说明复配体系扩散速度提高，表面活性增强。此研究结果为二元复合驱中表面活性剂的配方设计提供了技术思路。

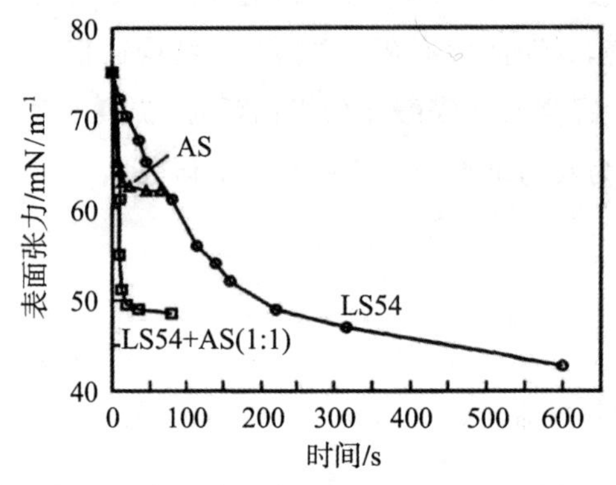

图 1 十六烷基磺酸盐 AS 与非离子表面活性剂 LS54 混合体系的动态表面张力

2.2 磺酸盐与非离子表面活性剂混合体系的临界胶束浓度

表面活性剂混合体系的临界胶束浓度 C_m,c_r 及表面活性剂在胶束和水相之间的分布是影响混合表面活性剂性能的主要方面[8~10]。对于同一类型(阴离子、阳离子、非离子型)混合表面活性剂体系，可以用适用于胶束相的理想溶液理论来预测，但对于不同类型(非离子和离子型)混合表面活性剂体系，尚没有适当的理论进行预测。

考察了不同摩尔比的十六烷基磺酸钠 AS 与十二烷基聚氧乙烯聚氧丙烯醚 LS45 混合体系的临界胶束浓度和平衡表面张力，采用 7 000 mg/L NaCl 溶液配制，测定温度为 25℃，结果见表 1。从表 1 可以看出，加入 LS45 的磺酸盐体系临界胶束浓度迅速下降，加入 5%LS45 的混合体系比单一 AS 体系降低 10 倍以上，而且混合体系的平衡表面张力均低于单一 AS 体系。这说明 AS 和 LS45 形成混合胶束的趋势很大，协同作用很明显。以上研究表明，在磺酸盐表面活性剂中加入少量非离子表面活性剂，在混合胶束形成和表面活性方面即可出现明显的协同作用，使临界胶束浓度大幅度下降，平衡表面张力降低，随着非离子表面活性剂加量的增大，临界胶束浓度进一步降低。

表 1 AS+LS45 混合表面活性剂的临界胶束浓度 C_m,c_r 和平衡表面张力 σ_e(25℃)

摩尔分数		$C_{m,cr}$/(mol/L)	σ_e/(mN/m)
AS	LS45		
1.00	0.00	8.25×10^{-3}	38.0
0.95	0.05	5.53×10^{-4}	32.7
0.90	0.10	4.21×10^{-4}	32.4
0.70	0.30	2.53×10^{-4}	30.7

(续表)

摩尔分数		$C_{m,cr}$/(mol/L)	σ_e/(mN/m)
AS	LS45		
0.50	0.50	8.62×10^{-5}	32.2
0.30	0.70	8.62×10^{-5}	30.6
0.00	1.00	1.47×10^{-5}	38.2

2.3 磺酸盐及复配体系对油相的适应性

测定了各种磺酸盐与不同分子结构的非离子表面活性剂混合水溶液与不同油相间的界面张力的最低值,结果见表 2。采用 7 000 mg/L NaCl 溶液配制,表面活性剂在溶液中的总浓度均为 5 000 mg/L,测定温度为 25℃。表 2 结果表明:加入少量非离子表面活性剂 TX 或 Tween-80 的 SDBS 水溶液,与正辛烷间的界面张力明显降低,表明二者之间有一定的协同作用。以甲苯为油相时,SDBS 与 Tween-80 复配体系的油/水界面张力最小。在实验中发现:随着油相中饱和烷烃含量的增加,界面张力逐渐增大,但只要油相中含有甲苯,混合油相与水的界面张力就低于饱和烷烃与水的界面张力,该现象表明,表面活性剂疏水链与油相分子间的结构相似性对界面张力的降低有利。非离子表面活性剂疏水链含有芳环时,与 SDBS 的复配协同作用最好。

表 2 单一及混合表面活性剂体系与不同油相间的界面张力(25℃)

表面活性剂		界面张力*/(mN/m)		
名称	摩尔比	正辛烷	正辛烷+甲苯**	甲苯
SDBS		1.209	0.456	0.333
SDBS+LS45	9∶1		0.505	
SDBS+TX	8∶2	0.528	0.142	0.227
SDBS+Tween-80	8∶2	0.983	0.321	0.082
SDBS+AES	9∶1		0.418	
SLS		>3	2.460	
TX		0.500		>3

*界面张力最低值。**体积比 2∶1。

2.4 分子模拟研究单一磺酸盐及其复配体系在油水界面的排布

理论研究表明,不同类型和结构的表面活性剂在油水界面共吸附,可由于空间作用力和电性相互作用的改变,形成分子排列更致密更有序的界面膜,使油/水界面张力进一步下降,即存在复配协同作用[11]。

分子模拟技术是从原子水平的相互作用出发,以原子水平的分子模型用计算机模拟分子的结构与行为,进而得到分子体系的各种物理化学性质,以研究结构与性质之间的关系。它既可以模拟分子的静态结构,也可以模拟分子的动态行为。可在分子水平上提供界面的组成和微观结构信息,对确定分子结构与界面性能的关系提供直接证据,具有非常重要的意义[12]。

模拟采用介观尺度耗散粒子动力学(dissipative particles dynamics, DPD)模拟[12]。耗散粒子动力学(DPD)是1992年由Hoogerbrugge和Koelman提出的一种针对复杂流体介观层次的模拟方法。该方法的出发点是积分牛顿运动方程,用一系列的珠子代替体系中的原子簇,利用柔性势能函数进行能量计算,并通过运动方程和三种作用力来描述这些珠子的运动轨迹。油和水用单个珠子描述,表面活性剂分子用两个珠子分别代表头基和尾基,且两珠子之间用谐振子弹力相连,其弹性系数为4.0 KT。模拟步数为20 000步,步长为0.05;模拟盒子的尺寸为:20×10×10,珠子密度为3.0,珠子质量及体系温度均为1.0 DPD单位。

由分子模拟结果(图2)可以看出:界面上的SDBS不可能达到那么致密的排列,即使密度再大,都会有部分空穴存在,其他的SDBS也不可能插入其中,见图2中(a)。当加入TX后,TX的头基不带电,与SDBS的排斥作用较弱,所以能以团簇的形式插入SDBS团簇之间的空穴中,两者起到协同共吸附的作用,见图2中(b、c),界面上表面活性剂的密度大大增加,因而界面张力大大降低。模拟结果与实验结果有很好的对应关系。

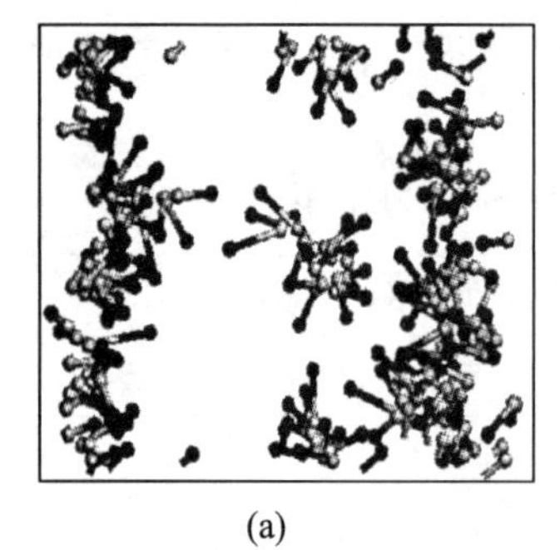

(a)

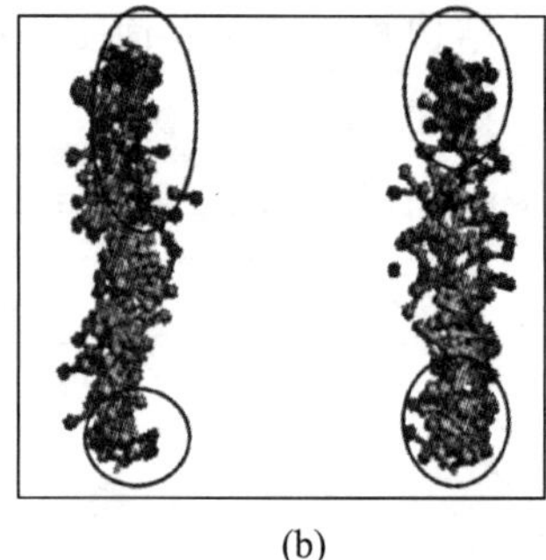

(b)

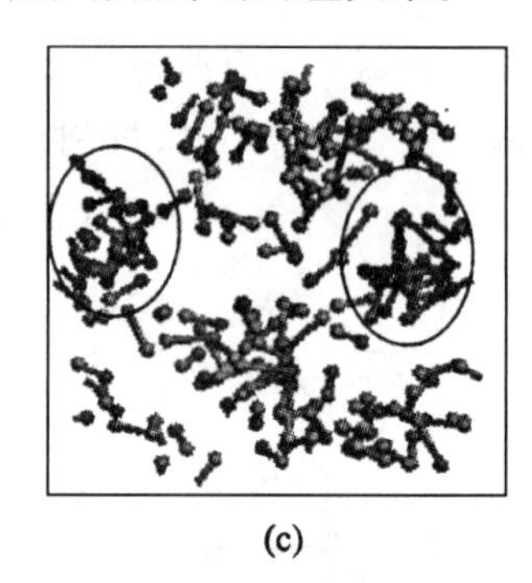

(c)

(a) 单一磺酸盐在界面上的分布形态,黑色代表SDBS的头基,浅灰色代表其尾基;(b) 复配体系在界面上的分布形态,圈中朝向外部的黑色代表TX头基,黑色代表尾基,其他分子朝向内部的代表SDBS头基,朝向外部的代表尾基;(c) 油/水界面上SDBS与TX的共吸附形态,圈中浅灰色代表TX头基,黑色代表尾基,其他分子代表SDBS头基与尾基

图2　分子模拟研究磺酸盐及复配体系在界面的排布形式

3　表面活性剂复配增效作用研究在孤东二元驱先导试验中的应用

二元复合驱先导试验区位于孤东七区西$Ng5^4$-6^1层东南部,含油面积0.94平方千米,地质储量277.5万吨,油层埋深1 261～1 294 m,共有三个含油小层(5^4、5^5、6^1三个小层,其中5^4、5^5发育较好,大片连通)。试验区能代表注水油藏的开采规律;采用常规开发井网,试验结果具有普遍推广意义。试验区油层孔隙度34%,平均渗透率1.320 μm^2,渗透率变异系数0.58,原始含油饱和度72%,地下原油粘度45 mPa·s。原始地层水矿化度为3 152 mg/L,目前产出水矿化度为8 207 mg/L。油层温度68℃,原始地层

压力 12.4 MPa,饱和压力 10.2 MPa。

孤东二元复合驱先导试验区设计井位 26 口,其中,生产井 16 口、注入井 10 口、观察井 3 口,见图 3。注聚前试验区内开生产井 20 口,日产油 64.5 t,日产液 2 360 t,综合含水 98.2%,试验区采出程度 34.4%。

以表面活性剂复配增效作用为基础,根据试验区油水特点及油藏条件,筛选了以非离子表面活性剂为主的助剂 1 号。助剂 1 号的加入使 SLPS 与孤东原油间界面张力大大降低。单一与复配石油磺酸盐体系降低油水界面张力结果比较见图 4。测试条件:配液用水为注入水,矿化度 6 188 mg/L,钙镁总量 70 mg/L;油相为 19-X15 井原油;温度 70℃。表面活性剂总浓度 2 500～6 500 mg/L 之间都能达到超低界面张力(10^{-3} mN/m)。通过配方优化及驱油实验得到的二元复合驱油剂配方为 3 000 mg/L SLPS＋1 000 mg/L 助剂 1 号＋1 700 mg/L 聚合物(北京恒聚Ⅱ型产品,固含量 90%,水解度 23%,1 500 mg/L 聚合物溶液在试验区条件下的黏度为 21.5 mPa·s),加入聚合物后最低的界面张力仍保持 2.0×10^{-3} mN/m。室内物理模拟实验结果表明,此配方提高采收率 18.1%。为减少表面活性剂的吸附与色谱分离,建议矿场注入表面活性剂浓度设计为 4 500 mg/L SLPS＋1 500 mg/L 助剂 1 号。

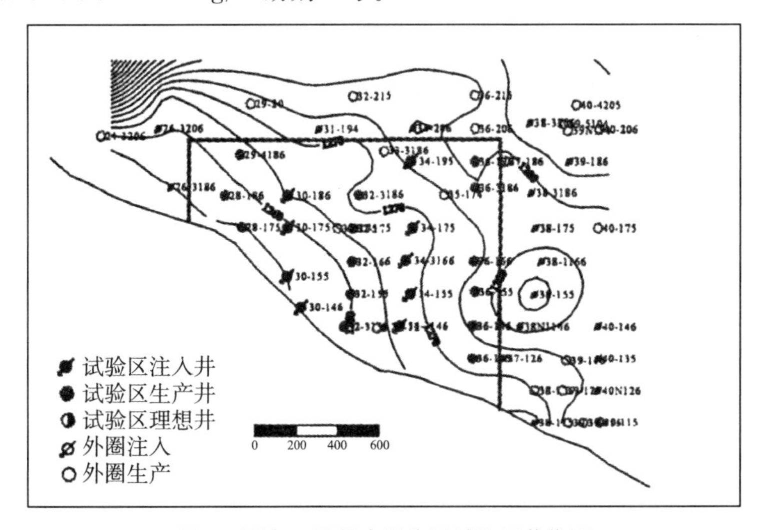

图 3　孤东二元复合驱先导试验区井位图

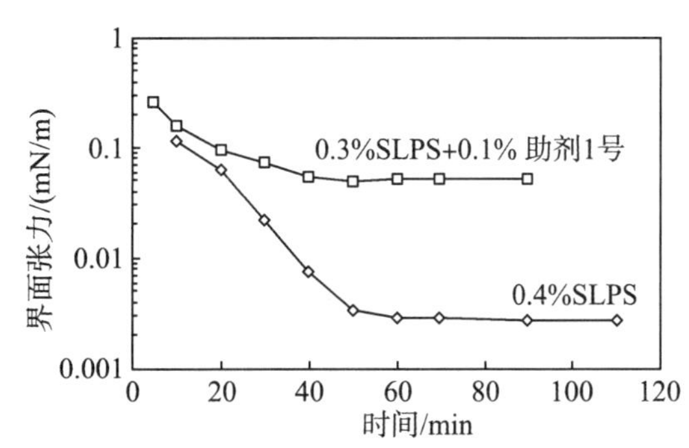

图 4　单一与复配石油磺酸盐体系界面张力测试结果(70℃)

先导试验采用三段塞注入方式,第一段塞为聚合物前置保护段塞,设计注入 0.05 PV(21.84×10^4 m^3)2 000 mg/L 聚合物溶液,连续注 156 天;第二段塞为主体段塞,设计注入 0.3 PV(131.0×10^4 m^3) 1 700 mg/L 聚合物+4 500 mg/L SLPS+1 500 mg/L 助剂 1 号的二元复合驱油剂,需连续注入 936 天;第三段塞为聚合物后保护段塞,设计注入 0.05 PV(21.84×10^4 m^3)1 500 mg/L 的聚合物溶液,连续注入 156 天。

第一段塞从 2003 年 9 月开始注入,实际注入 0.075 PV 后于从 2004 年 6 月开始注主体二元复合驱油段塞,已取得了明显的降水增油效果,见图 5。目前试验区综合含水下降到 89.1%,比见效前的 98.2 t/d 下降了 9.1%。试验区日产油水平上升到 170 t/d,比见效前的 34 t/d 上升了 136 t/d,日产油上升幅度十分明显。截至 2008 年 3 月,试验区已经累积增油 15.8 万吨,提高采收率 5.7%,有 10 口生产井在不同程度上见到降水增油效果。中心井区累增油 8.6 万吨,提高采收率 12.7%,采出程度达到 5%。由于二元驱降水增油效果明显,延长了主段塞的注入量,第三段塞将于 2009 年 1 月开始注入。

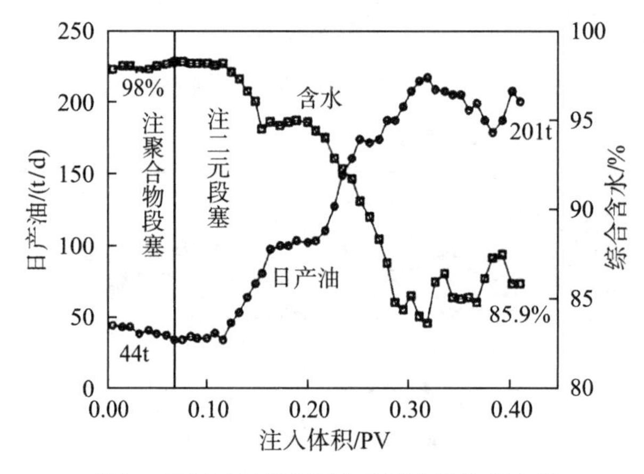

图 5 孤东二元复合驱先导试验区生产曲线

二元复合驱的初步效果说明了方案设计的合理性,同时也验证了配方设计的正确性,为二元复合驱在胜利油田的推广应用奠定了基础。

4 结论

(1) 磺酸盐阴离子表面活性剂与非离子表面活性剂在非平衡态吸附中的协同作用,为以 SLPS 为主剂的二元复合驱配方设计提供了思路。

(2) 在磺酸盐表面活性剂中加入少量非离子表面活性剂,即可使混合表面活性剂的临界胶束浓度大幅度下降,平衡表面张力降低。

(3) 单一的石油磺酸盐 SLPS 不能降低油水界面张力到超低(10^{-3} mN/m),SLPS 与适宜非离子表面活性剂复配的表聚二元复合驱先导试验配方,界面张力达到 2×10^{-3} mN/m,室内提高采收率 18.1%,孤东七区矿场先导试验中已见到明显的降水增油效果。

参考文献

[1] 王宝瑜，曹绪龙，王其伟，等. 孤东小井距三元复合驱现场试验采出液相态变化及组分浓度测定[J]. 油田化学，1994，11(4)：327-330.

[2] 康万利. 大庆油田三元复合驱化学剂作用机理研究 [M]. 北京：石油工业出版社，2001：4-15.

[3] 曹绪龙，孙焕泉，姜颜波，等. 孤岛油田西区复合驱油矿场试验 [J]. 油田化学，2002，19(4)：350-353.

[4] 王红艳，张本艳，张继超，等. 胜利油田三元驱替液用防垢剂研究 [J]. 油田化学，2005，22(3)：252-254.

[5] Cash R L，Cayias J L，Fournier G，et al. The application of low interfacial tension scaling rules to binary hydrocarbon mixtures[J]. J Colloid Interface Sci，1977，59(2)：30-34.

[6] 俞稼镛，宋万超. 化学复合驱基础及进展 [M]. 北京：中国石化出版社，2002：24-31.

[7] 韩冬，沈平平. 表面活性剂驱油原理及应用 [M]. 北京：石油工业出版社，2001：347-351.

[8] Salager J L，Mongan J C. Optimum formuation of surfactant/ water/oil system for minimum interfacial tension or phase behavior，Soc Petrol Engs J，1979，19：107-115.

[9] Myers D. Surfactant Science and Technology [M]. New York：VCH Publishers，2005：6-24.

[10] 沈钟，王果庭. 胶体与表面化学 [M]. 北京：化学工业出版社，2001：347-351.

[11] van der Bogaert R，Joos P. Diffusion-controlled adsorption kinetics for a mixture of surface active agents at the solution-air interface，J Phys Chem，1980，84：190-194.

[12] Dong F L，Li Y，Zhang P. Mesoscopic simulation study on the orientation of surfactants adsorbed at the liquid/liquid interface [J]. Chem Phy Letters，2004，399：215.

Synergistic effect of petroleum sulfonate and non-ionic surfactant and preliminary results of pilot surfactant/polymer flood in gudong oil field

Wang Hongyan Cao Xulong Zhang Jichao Li Xiulan Zhang Aimei

(Geological and Scientific Research Institute, Shengli Oilfield Branch Company, Sinopec, Dongying, Shangdong 257015, P R of China)

Abstract: The results of investigating the synergistic effect of Shengli petroleum sulfonate SLPS and non-ionic surfactants (NISs) in surfactant/polymer (SP) flooding solution for a pilot flood project in Gudong, Shengli, are presented. The dynamic surface tension of solution of sulfonate AS+NIS LS54 mixture of equal molar ratio quickly reaches equilibrium, giving a low equilibrium value. The minimum micelle concentration of AS + LS45 mixture decreases and its equilibrium surface tension decreases fast and then keeps lower values with increasing proportion of LS45 in the surfactant mixture. With n-octane, toluene, and their mixture (2 : 1) used as oil phase and mixed surfactant solution of sulfonate SDBS+NIS LS45, TX, Tween-80 or AES used as water phase, the oil/water interfacial tension (IFT) is significantly decreased to different extent in comparison with plain SDBS solution as water phase while the IFT value between n-octane and a mixed surfactant solution is decreased when toluene is introduced in n-octane. It is shown through mesoscopic simulation using dissipative particles dynamics technique that the distribution of surfactant molecules at oil/water interfaces is much denser for SDBS + TX mixture than that for sole SDBS, which results in significantly reducing IFT values. By using an ultra low IFT system composed of 3 g/L SLPS+1 g/L NIS+1.7 g/L polymer and designed in terms of the concept above described, an enhancement in oil recovery of 18.1% is obtained in a physical modeling experiment. In an SP flood pilot area covering 10 injection and 16 production wells at Ng54-61 pay layers in district 7WS, Gudong, injection of SP flooding solution of 4.5 g/L SLPS+1.5 g/L NIS+1.7 g/L polymer started in June 2004 has resulted in notable reduction in watercut and raise of oil output with an enhancement in oil recovery of 5.7% in whole pilot area and of 12.7% in central area, where 54% of reserves is recovered.

Keywords: surfactant/polymer (SP) combinational flooding solution; mixed surfactant; petroleum sulfonate; nonionic surfactant; synergism; surface tension; interfacial tension; surfactant/polymer (SP) combinational flood; Gudong oil field in Shengli

直链烷基萘磺酸钠的合成及界面活性研究*

王立成[1] 王旭生[1] 宋新旺[2] 曹绪龙[2] 蒋生祥[1*]

(1. 中国科学院兰州化学物理研究所，甘肃兰州 730000；

2. 中国石油化工股份有限公司胜利油田有限公司地质科学研究院，山东东营 257015)

摘要：为研究直链烷基萘磺酸盐结构与界面性能的关系，以α-溴化萘为原料，经交叉耦合反应、磺化反应、中和反应等步骤，合成了5种结构明确的直链烷基萘磺酸钠，烷基链的碳数分别为6、8、10、12、14，产物收率50%～60%。用高效液相色谱、质谱、核磁共振氢谱对其进行了色谱纯度、相对分子质量及结构表征，产物色谱纯度均高于96.0%。测量了各烷基萘磺酸钠的NaCl水溶液与正壬烷的界面张力，证实了临界胶束浓度随烷基碳链的增长而减小。

关键词：直链烷基萘磺酸钠；交叉耦合反应；界面张力；表面活性剂

中图分类号：TQ423 **文献标识码**：A **文章编号**：1003-5214(2009)04-0340-05

Synthesis of long-chain n-alkylnaphthalene sulfonates and study on their interfacial activities

Wang Licheng Wang Xusheng Song Xinwan

Cao Xulong Jiang Shengxiang

(1. Lanzhou Institute of Chemical Physicsof Chinese Academy, Lan zhou 730000, gansu, China;

2. Geolo gical and Scientific Resear chInstitute of Sheng liOilfield

Sinopec, Dongying 257015, Shan dong, China)

Abstract: In order to study the relationship betwen structure and interfacial activiy fivelong-chain n-alkylnaphthalene sulfonates with definite structures were synthesized through three steps including cross-coupling reaction, sulfonation and neutralization, using α-bromonaphthalene as the starting material. The carbon number of the alkyl chain was 6, 8, 10, 12, 14 respectively. The yields of the sulfonates were 50%～60%. HPLC, MS, 1 HNMR were used to characterize the purity, molecular weight, and the structure of the products, of which the purity was all above 96.0%. The interfacial tensions between NaC laqueous solutions of the naphthalene sulfonatesand nonane were measured, showing that the critical micele concentration decreases with the increase of the carbon number of the alky chain.

作为驱油用表面活性剂,石油磺酸盐可以显著降低油水界面张力,且生产原料来源广泛、价格便宜,在三次采油中占有重要位置。石油磺酸盐是由不同结构的烷基芳基磺酸盐组成的复杂混合物,难以分离出单一的组分[1,2]。要搞清楚石油磺酸盐结构与性能的关系,揭示低界面张力形成的机理,十分有必要合成不同结构的高纯度的烷基芳基磺酸盐模型化合物。关于合成高纯度的烷基苯磺酸盐的报道较多,如直链烷基苯磺酸盐[3]、支链烷基苯磺酸盐[4～9]、双尾烷基苯磺酸盐[10,11]及三尾烷基苯磺酸盐[11～13]等;但合成高纯度的烷基萘磺酸盐的报道较少。烷基萘磺酸盐的合成首先是烷基萘中间体的合成。Ronald E 报道了有机锂试剂合成烷基萘等烷基芳烃的方法[14]。KoheiTamao 以膦镍化合物为催化剂,经交叉耦合反应制备了烷基芳烃、烷基烯烃等化合物[15,16]。上述文献均未系统讨论不同长度烷基碳链的烷基萘的合成,已报道的烷基萘的烷基碳数在8以内,较短。谭晓礼通过 Wurtz-Fitig 反应,在金属钠的作用下较系统地合成了己基萘、辛基萘、癸基萘以及相应的磺酸钠盐[17],随后又报道了支链烷基萘磺酸钠的合成[18]。

作者用二氯化[1,3-二(二苯基膦)丙烷]镍(Ⅱ)[NiC$_2$l(dp)]为催化剂,通过 α-溴化萘与溴代烷格氏试剂的交叉耦合反应,合成了己基萘、辛基萘、癸基萘、十二烷基萘以及十四烷基萘;又通过磺化反应、中和反应,合成了相应的磺酸钠盐,合成路线见图1。用高效液相色谱(HPLC)、质谱(MS)、核磁共振氢谱(^{1}HNMR)表征了产物的色谱纯度、相对分子质量及结构;通过界面张力测量,考察了烷基萘磺酸钠的 NaC 水 l 溶液与正壬烷的界面性质。

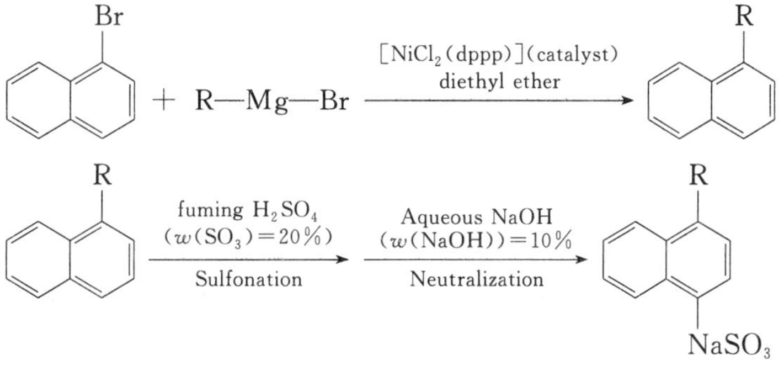

Br + R—Mg—Br → [$NiCl_2$(dppp)](catalyst), diethyl ether → R

R → fuming H_2SO_4 ($w(SO_3)=20\%$), Sulfonation → Aqueous NaOH ($w(NaOH))=10\%$, Neutralization → R, $NaSO_3$

图1　直链烷基萘磺酸钠的合成路线

1 实验部分

1.1 试剂与仪器

α-溴化萘,中国医药(集团)上海化学试剂公司,CP;1-溴正己烷,华东师范大学化工厂,质量分数为98.0%;1-溴正癸烷,上海曙光试剂厂,质量分数为98.0%;1-溴正辛烷、1-溴正十二烷、1-溴正十四烷,上海三友试剂厂,质量分数为98.0%;镁粉,天津市福晨化学试剂厂,质量分数≥99.5%;无水乙醚,天津化学试剂有限公司,质量分数≥99.0%,用前用金属钠干燥;发烟硫酸,北京化工厂,$w(SO_3)=20\%$;氢氧化钠,天津化学试剂有限公司,质量分数≥96.0%;二氯化[1,3-二(二苯基膦)丙烷]镍98%;蒸馏水,实验室自制;无水甲醇,天津市百世化工有限公司,AR;四丁基溴化铵,上海试剂一厂,AR;氯化钠,国药集团化学试剂有限公司,AR。

Aglient 1100型高效液相色谱仪,配二极管阵列检测器,美国Aglient公司(色谱柱为C18柱,规格为150 mm×4.6 mm,粒径5 μm,实验室高压匀浆法装填);Agilent 1100 LC/MSDTrap质谱仪,美国Aglient公司;美国VarianINOVA-400 MHz超导核磁共振仪;TX500C型全量程视频动态界面张力仪,北京盛维基业科技有限公司。

1.2 合成

1.2.1 格氏试剂的合成

将0.60 mol镁粉与100 mL无水乙醚加入到三口烧瓶中;再将0.45 mol溴代烷与100 mL无水乙醚混合均匀,加入恒压滴液漏斗中,逐滴滴加至烧瓶中;电磁搅拌,保持乙醚微沸;滴完后,反应物持续搅拌1 h,结束反应。分别得到正己基溴化镁、正辛基溴化镁、正癸基溴化镁、正十二烷基溴化镁、正十四烷基溴化镁。

1.2.2 直链烷基萘的合成

将0.30 mol α-溴化萘、0.30 g NCi_2l(dp)和150 mL无水乙醚加入三口烧瓶中;将制备的格氏试剂通过恒压滴液漏斗逐滴加入三口烧瓶中,电磁搅拌,保持乙醚微沸,滴完后持续搅拌24 h;反应产物用稀盐酸水解,直至水层澄清;将水层和有机层分开,水层用乙醚萃取3次;将有机层与乙醚层合并,用蒸馏水洗涤3次,无水CaC_2l干燥;过滤去除CaC_2l旋蒸去除乙醚,再减压蒸馏得直链烷基萘。分别为α-正己基萘、α-正辛基萘、α-正癸基萘、α-正十二烷基萘、α-正十四烷基萘。

1.2.3 直链烷基萘磺酸钠的合成

将10 g烷基萘加入三口烧瓶中,滴加发烟硫酸15 mL,电磁搅拌,冰水浴控制温度不超过5℃,滴完后室温搅拌2 h;再滴加20 mL稀盐酸稀释,乙醚萃取;再用氢氧化钠水溶液中和乙醚层至pH=7;旋蒸除去乙醚,溶入无水乙醇,过滤除去无机盐;加入水,用正己烷萃取未磺化的烷基萘;蒸馏除去乙醇、水,得白色烷基萘磺酸钠固体。分别为1-正己基萘-4-磺酸钠(HNS)、1-正辛基萘-4磺-酸钠(ONS)、1-正癸基萘-4-磺酸钠(DNS)、1-正十二烷基萘-4-磺酸钠(DDNS)、1-正十四烷基萘-4磺-酸钠(TDNS)。产物收率50%~60%。

2　结果与讨论

2.1 烷基萘磺酸钠的高效液相色谱表征

（Ⅱ）上海奥普迪诗化学科技有限公司，质量分数色谱条件的选择：以 V(甲醇)：V(水)＝75：25 的混合体系为流动相，内含四丁基溴化铵为离子对添加剂[19,20]，流速为 1.0 mL/min；检测波长 225 nm，烷基萘磺酸钠在 225 nm 处有最大紫外吸收。

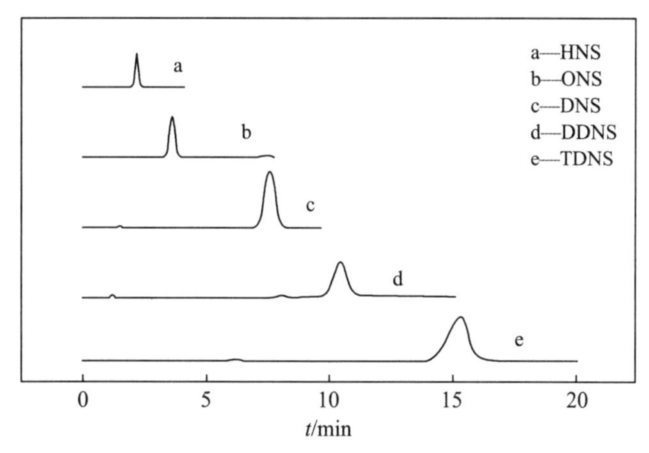

图 2　直链烷基萘磺酸钠的高效液相色谱图

实验证实，随着烷基碳链的增长，烷基萘磺酸钠在碳十八柱上的保留也就增强；流动相中甲醇含量越高，磺酸盐保留越弱；作为离子对试剂的四丁基溴化铵浓度越高，烷基萘磺酸钠保留越强。为使烷基萘磺酸钠有所保留，但又不至于过强，实验选择甲醇体积分数为 25%；正己基萘磺酸钠、正辛基萘磺酸钠、正癸基萘磺酸钠对应的四丁基溴化铵质量浓度为 1 g/L，十二烷基萘磺酸钠与十四烷基萘磺酸钠对应的四丁基溴化铵质量浓度分别为 0.6 g/L 与 0.35 g/L。

从 5 种烷基萘磺酸钠的高效液相色谱图中基本检测不到杂质峰，说明各磺酸盐的纯度很高，采用面积归一化法可知各磺酸盐色谱纯度均大于 96.0%[21]。

2.2 烷基萘磺酸钠的质谱表征

质谱条件：电喷雾电离源 ES,I 负模式检测，雾化气压力 5.52×10^4 Pa 干燥气流速 5 L/min，温度 325℃。

图 3 为 5 种烷基萘磺酸钠的阴离子分子离子峰的 m/Z 值，与烷基萘磺酸钠的相对分子质量减去钠的相对原子质量 $M\text{-}Na^+$ 一致，见表 1。

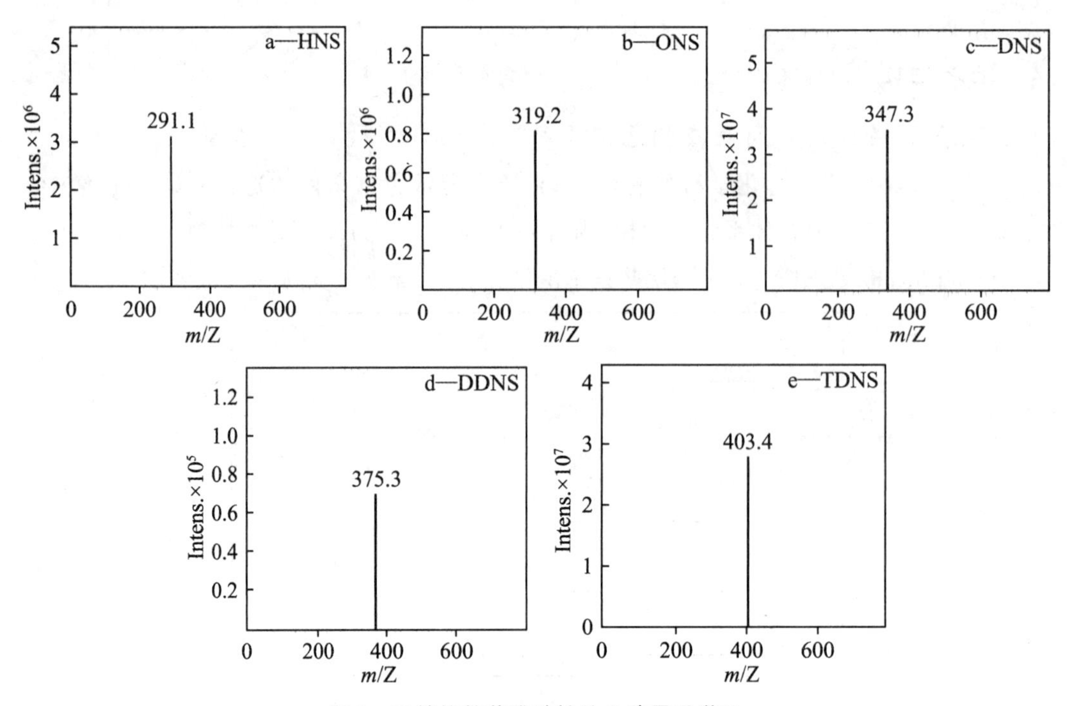

图 3 直链烷基萘磺酸钠的电喷雾质谱图

表 1 直链烷基萘磺酸钠的电喷雾质谱数据

Compounds	m/Z	$M-Na^+$
HNS	291.2	291
ONS	319.2	319
DNS	347.3	347
DDNS	375.3	375
TDNS	403.4	403

2.3 烷基萘磺酸钠的核磁共振氢谱表征

用核磁共振氢谱确定直链烷基萘磺酸钠的结构，溶剂为氘代丙酮 C_3D_6O。对^{1}HNMR 谱中的化学位移分别进行归属，结果如下：

正己基萘磺酸钠：萘环芳氢：9.120～9.097(d,1H)，8.069～8.005(m,2H)，7.496～7.427(m,2H)，7.271～7.253(d,1H)；烷基链脂肪氢：3.087～3.048(,t_2H，naphthy—lCH_2—)，1.738～1.679(m,2H，naphthy—lCH_2—CH_2—)，1.439(m,2H，—CH_2—CH_3)，1.328～1.254〔m,4H，—$(CH_2)_2$—〕，0.893～0.847(,t_3H，—CH_3)。

正辛基萘磺酸钠：萘环芳氢：9.108～9.085(d,1H)，8.069～8.004(m,2H)，7.487～7.444(m,2H)，7.274～7.255(d,1H)；烷基链脂肪氢：3.088～3.050(,t_2H，naphthy—lCH_2—)，1.721～1.701(m,2H，naphthy—lCH_2—CH_2—)，1.437(m,2H，—CH_2—CH_3)，1.330～1.273〔m,8H，—$(CH_2)_4$—〕，0.862(,t_3H，—CH_3)。

正癸基萘磺酸钠：萘环芳氢：9.126～9.105(d,1H)，8.069～8.007(m,2H)，7.468～7.445(m,2H)，7.273～7.255(d,1H)；烷基链脂肪氢：3.087～3.047(,t_2H, naphthy—lCH_2—)，1.720～1.700(m,2H, naphthy—lCH_2—CH_2—)，1.436(m,2H, —CH_2—CH_3)，1.273〔m,12H，—$(CH_2)_6$—〕，0.876～0.846(,t_3H，—CH_3)。

正十二烷基萘磺酸钠：萘环芳氢：9.052(d,1H)，8.040～8.023(m,2H)，7.477(m,2H)，7.271(d,1H)；烷基链脂肪氢：3.066(,t_2H, naphthy—lCH_2—)，1.710(m,2H, naphthy—lCH_2—CH_2—)，1.404(m,2H, —CH_2—CH_3)，1.251～1.100〔m,16H，—$(CH_2)_8$—〕，0.848(,t3H，—CH_3)。

正十四烷基萘磺酸钠：萘环芳氢：9.108(d,1H)，8.006～7.989(m,2H)，7.430(m,2H)，7.243～7.223(d,1H)；烷基链脂肪氢：3.060～3.021(,t_2H, naphthy—lCH_2—)，1.697(m,2H, naphthy—lCH_2—CH_2—)，1.400(m,2H, —CH_2—CH_3)，1.254〔m,20H，—$(CH_2)_{10}$—〕，0.82(,t_3H，—CH_3)。

以正己基萘磺酸钠为例，确定磺酸基的萘环取代位置：经计算可知，化学位移为7.271的氢($H_{\delta 7.271}$)与化学位移为9.120的氢($H_{\delta 9.120}$)相互作用的耦合常数为7.2；化学位移为7.496的氢($H_{\delta 7.271}$)与化学位移为8.069的氢($H_{\delta 8.069}$)相互作用的耦合常数为9.2；由此可知，$H_{\delta 7.271}$与$H_{\delta 9.120}$均处于萘环β位，磺酸钠处于正己基的对位。同理，其他磺酸盐磺酸基的取代位置也是处于烷基的对位。

2.4 烷基萘磺酸钠界面性质的表征

配制系列浓度的直链烷基萘磺酸钠的NaCl水溶液〔w(NaCl)=1%〕，测量各表面活性剂NaCl水溶液与正壬烷[10]的界面张力稳态值，实验温度45℃。以界面张力γ对浓度的对数lgc作图(图4)，得出各直链烷基萘磺酸钠在NaCl水溶液中的临界胶束浓度(CMC)、临界胶束浓度下的界面张力。

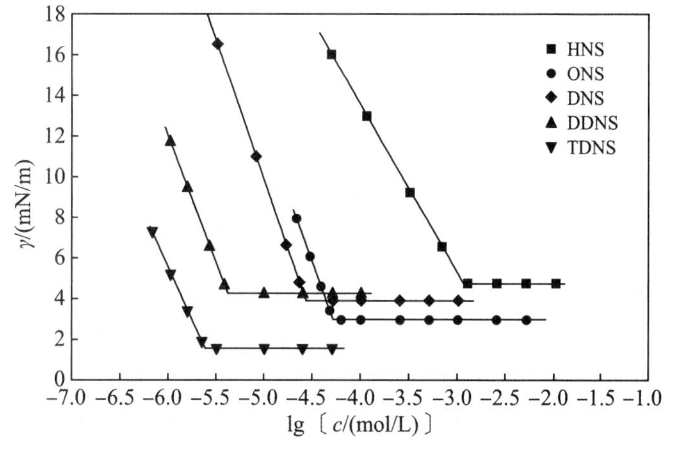

图4 直链烷基萘磺酸钠的NaCl水溶液与正壬烷界面张力的γ-lgc曲线

表2 直链烷基萘磺酸钠 NaCl 水溶液与正壬烷界面张力的计算结果

	Compounds				
	HNS	ONS	DNS	DDNS	TDNS
CMC/(10^5 moL)	109.65	5.13	2.57	0.39	0.24
γ_{cmc}/(mN/m)	4.65	2.97	3.85	4.25	1.53

由表2可以看出，随着直链烷基萘磺酸钠疏水链碳数的增加，CMC依次减小；γ_{CMC}规律性不强，这应该与各烷基萘磺酸钠在油水界面上排布的紧密程度及各磺酸盐的亲水亲油性等性质有关，具体原因有待探讨。

3 结论

作者合成了5种结构明确的直链烷基萘磺酸钠，烷基链的碳数分别为6，8，10，12，14，收率50%～60%。用高效液相色谱、质谱、核磁共振氢谱表征了产物色谱纯度、相对分子质量及结构，产物色谱纯度均高于96%。探讨了不同结构烷基萘磺酸钠的NaCl水溶液与正壬烷的界面性能，随着烷基碳链的增长，直链烷基萘磺酸钠的CMC依次减小。

参考文献

[1] 俞稼镛，宋万超，李之平. 化学复合驱基础及进展[M]. 北京：中国石化出版社，2002:390-394.

[2] 沈平平，俞稼镛. 大幅度提高石油采收率的基础研究 [M]. 北京：石油工业出版社，2001:149-150.

[3] 赵国玺，朱王步瑶. 表面活性剂作用原理[M]. 北京：中国轻工业出版社，2003:698-710.

[4] Yang, JLiZ S, ChengLB, eta. lSynthesis and characterization of mono-isomerical kylbenzene sulfonates [J]. Petroleum Scienceand Technology, 2006, 24:937-984.

[5] 王琳，宫清涛，俞稼镛，等. 支链烷基苯磺酸钠的合成、表征及其结构对表面性质的影响[J]. 石油化工，2004，33(2):104-108.

[6] PaulD B, ChristieHL. New anionic alkylaryl surfactants based on olefin sulfonicacids [J]. JSurfDeter, 2002, 5 (1):39-43.

[7] DoePH, EmaryMEL, WadeW H. Surfactants for producing low interfacial tensionsⅡ:linearal kylbenzene sulfonates with additional alkyl substituents [J]. JAm OilChemistsSo 197 55:505-512.

[8] DoePH, EmaryMEL, WadeW H. Surfactants for producing low

interfacialtension :linearal kylbenzene sulfonates [J]. JAm Oil ChemistsSo1977, 54: 570-577.

[9] GrayFW, GerechtJF, Krem IJ. The preparation of model long chain alkyl benzenesand a studyof their isomeric sulfonation product [J]. JOrgChem, 1955, 20: 511-523.

[10] 宫清涛,王琳,俞稼镛,等.双取代直链烷基苯磺酸钠的合成及其界面活性的研究[J].精细化工,200 22(3):189-208.

[11] DoePH, MahmoudE E, WadeW H. Surfactants for producing low interfacial tension:s Ⅲ. diand trin-alkyl benzene sulfonates [J]. Journal of the American Oil Chemists' Socie 197 55:513-520.

[12] 宫清涛,姜小明,俞稼镛,等.三直链烷基苯磺酸钠的合成及其表面活性的研究[J].石油化工高等学校学报,2005,18(3):20-29.

[13] 姜小明,徐志成,俞稼镛,等.高纯度支链三烷基苯磺酸钠的合成与表面活性[J].精细化工,2004,21 (11):808-811.

[14] RonaldEM, Ei-ichiNegish. iTetrahydrofuranpromotedaryl-alkyl coupling involving organo-lithium reagents [J]. J Org Chem, 1974, 39(23):3452-3453.

[15] KoheiT, KojiS, YoshihisaK. Nicke phosphine complex-catalyzed grignard coulping of alky ,lary, and alkenyl grignard reagents with aryland alkenyl grignard reagents with aryland alkenyl halide: general scope and lmiitations [J]. Buletin of the Chemical Society of Japan, 1976, 49 (7):1958-1969.

[16] KoheiT, KojiS, MakotoK. Selectivecarboncarbonbond formation by cross-coupling of grignard reagents with organic halides catalysis by nicke phosphine complexes [J]. Journal of the American Chemical Society,1972, 94(12):4374-4376.

[17] TanX L, ZhangL, An JY, eta. lSynthesisand studyof the surface properties of long-chain alkyl naphthalene sulfonates [J]. J Surfactants Detergent, 2004, 7(2): 135-169.

[18] 谭晓礼,徐志成,俞稼镛,等.1-(支链烷基)萘和1-(支链烷基)萘-4-磺酸钠的合成[J].精细化工,2004,21(增刊):44-46.

[19] 赵忠奎,李宗石,程侣柏,等.系列烷基甲基萘磺酸盐的高效液相色谱和电喷雾离子化质谱分析[J].分析测试学报,2005,24(3):10-13.

[20] 郭海涛,乔卫红,李宗石,等.系列长链烷基萘磺酸钠的反相离子对高效液相色谱法的研究[J].分析化学研究简报,2003,31(8):985-988.

[21] 赵忠奎,李宗石,乔卫红,等.癸基甲基萘磺酸盐表面活性剂合成及性能研究[J].大连理工大学学报,200 48 (3):318-322.

扩张流变法研究表面活性剂在界面上的聚集行为

曹绪龙[1] 崔晓红[1] 李秀兰[1] 曾胜文[1] 朱艳艳[2] 徐桂英[2*]

([1]中国石化胜利油田分公司地质科学研究院 山东东营 257015;

[2]山东大学胶体与界面化学教育部重点实验室 山东济南 250100)

摘要:近年发展起来的界面流变测定技术在研究界面性质方面具有许多独特之处。本文结合我们的工作,总结了近年来有关该技术在表面活性剂界面聚集行为研究中的应用,讨论了扩张频率、表面活性剂浓度及疏水链长、无机盐和温度对表面扩张流变行为的影响,同时探讨了小分子表面活性剂与高分子表面活性剂表面扩张流变行为的区别以及小分子表面活性剂在气/液界面与液/液表面的扩张流变性的差异。大量研究表明,借助于界面流变性的测定不仅可以研究发生在界面上和界面附近的微观弛豫过程,而且可以探讨界面上超分子聚集体的形成,进而为乳状液和泡沫等分散体系的稳定性提供依据。

关键词:表面活性剂 聚集作用 界面 扩张流变

Study on the aggregation behavior of surfactant at interface by the dilatational rheological methods

Cao Xulong[1] Cui Xiaohong[1] Li Xiulan[1] Zeng Shengwen[1]
Zhu Yanyan[2] Xu Guiying[2*]

([1] Geological Research Institute of Shengli Oilfield Co. Ltd, SINOPEC, Dongying 257015;

[2] Key Laboratory of Colloid and Interface Chemistry (Shandong University), Ministry of Education, Jinan 250100)

Abstract: The applications of dilatational rheological method in studying the aggregation behavior of surfactant at interface are systematically summarized based on our works. The effects of oscillating frequency, concentration and hydrophobic chain of surfactant, electrolyte and temperature on surface dilatational rheology are discussed.

Meanwhile the diference in surface dilatational rheology between low molecular surfactant and macromolecular surfactant has been studied. In addition, the diference of low molecular surfactant on air/water and liquid/liquid interface is investigated. Many studies indicate that these studies not only can provide the microscopic relaxation of the interfacial film, but they also are used to probe the information of super molecular aggregate at interface, which is important to real emulsion and foam formation and stability.

Keywords: Surfactant, Aggregation, Interface, Dilational rheology

表面活性剂在界面上的聚集行为会显著影响泡沫和乳状液等分散体系的性质。因此,有关表面活性剂在界面上聚集行为的研究一直是人们倍感兴趣的课题[1,2]。界面张力法是研究表面活性剂在气/液和液/液界面上聚集行为的经典方法,但是对于有些体系,界面张力值对界面层的变化并不敏感,更难以反映出界面和体相之间的动态性质[3,4]。尤其是对于高分子表面活性物质,表面张力值不能反映其在界面发生的构象或相转变,而这些变化往往对吸附动力学有重要影响,对泡沫和乳状液的稳定起着重要作用[5-7]。界面流变学有望解决上述问题,因此,近年来国外利用界面流变学技术研究表面活性剂在界面上聚集行为的报道大量涌现[8-15],国内在这方面的研究也已受到重视。江龙等[16]曾系统研究过界面膨胀流变特性的测定问题;罗平亚等[17]曾应用均匀设计实验方法研究了聚合物与表面活性剂二元驱油体系的界面流变性;赵濉等[18-27]研究了许多表面活性剂的界面流变特性;汪庐山等[28]研究了泡沫驱用阴离子起泡剂的界面粘弹性及其与泡沫特性的关系,发现随着界面弹性模量增加,泡沫稳定性呈现增强的趋势,弹性变形在泡沫变形中起主导作用,笔者课题组[29-32]通过界面流变性质测定不同表面活性剂与大分子的相互作用,并比较了双子阳离子表面活性剂(Cationic Gemini surfactant)与相应单链表面活性剂的界面聚集行为的差异。

界面流变性是体系的重要特征之一,主要分为界面剪切流变和界面扩张流变[33-36]。实验和理论的研究结果均表明,界面扩张流变参数通常比相应的剪切流变参数大几个数量级[37]。界面扩张粘弹性是流体界面的重要性质,它与乳状液和泡沫的稳定性密切相关,在工农业、生物、制药和日用化工领域起着重要的作用[38-40]。通过扩张流变性测定可研究吸附分子的吸附动力学和界面层的弛豫过程[41]。由于界面扩张流变性依赖于界面上或界面附近存在的微观弛豫过程,因此界面扩张流变参数可以反映界面微观过程的信息,对萃取、抽提、洗涤以及从固体中排液等与界面和表面有关的工业过程也有重要意义[42-46]。尤为重要的是,界面扩张粘弹性质可以用来研究两亲分子在界面上的行为以及流体界面的超分子聚集体。

常用的界面扩张粘弹性测量的方法有[47-53]:振荡气泡法、最大泡压法、Langmuir 槽法等。不同方法的测试频率范围有所区别。Langmuir 槽法用于扩张粘弹性测量的方法又分为:宏观形变法(稳态法)[54]、表面波法[55]和界面张力弛豫法[56]。界面张力弛豫法是通过测量界面发生瞬间形变后的界面张力衰减曲线,经多参数指数拟合,得到界面微观弛豫过程的个数和每个过程的特征弛豫时间等信息,然后通过 Fourier 变换得到界面扩张粘弹性的各种参数。该法的优点是可以得到任意频率下的扩张粘弹性参数。

本文结合笔者课题组的工作,着重介绍界面扩张粘弹性方法研究表面活性剂界面聚集行为的近期研究进展,以期为表面活性剂有序分子膜的研究与应用提供基础数据。

1 表面活性剂在气/液表面上的聚集行为研究

1.1 表面活性剂浓度和频率对扩张模量的影响

界面扩张模量定义为界面张力变化与界面面积相对变化的比值[57-60],即:

$$|\varepsilon| = \frac{d\gamma}{d\ln A} \tag{1}$$

我们利用 Langmuir 槽法对不同浓度表面活性剂水溶液表面上的聚集行为研究发现,所研究体系的扩张模量均随频率增加而增大,且当表面活性剂浓度较低时,扩张频率对扩张模量的影响较小;随表面活性剂浓度增加,扩张模量对扩张频率的依赖性增强;继续增加浓度,扩张频率对扩张模量的影响又变小。但是,不同体系的粘弹性存在一定的差异,其原因可能是不同结构的分子的扩散速率和形成的表面膜微观结构不同。不同表面活性剂水溶液扩张模量随其浓度的变化规律是:随浓度增大,扩张模量总是呈现出先增大后减小的变化趋势。以阳离子双子表面活性剂 1,2-乙烷二(二甲基十二烷基)溴化铵($C_{12}C_2C_{12}$)为例[61],结果示于图 1 中。显然,在其浓度为 2.5×10^{-6} mol/L 处,$C_{12}C_2C_{12}$ 扩张模量达到最大值,即此时膜的弹性最大。一般而言,表面活性剂体相浓度的增大对界面扩张性质有两方面的影响[21]:一是增大表面活性剂的界面浓度,另一方面也增大了从体相向新生成的界面通过扩散补充活性剂分子的能力。表面活性剂界面浓度的增大会导致界面形变时更高的界面张力梯度,因此膜的弹性增大;而活性剂分子从体相向新生成界面的扩散补充则降低了界面张力的梯度,会导致扩张模量的降低。即表面活性剂浓度较低时,界面浓度的增加对弹性和强度的影响占主导地位,膜的弹性和强度随浓度增大而增强,当浓度达到 2.5×10^{-6} mol/L 时,界面上吸附已接近饱和,膜弹性和强度达到最大,继续增加浓度只会增加从体相向新生成的界面扩散补充活性剂分子的趋势,因此膜弹性和强度开始明显减弱。

同样的规律也存在于其它表面活性剂体系,例如:25℃时十二烷基三甲基溴化铵(DTAB)水溶液的表面扩张模量在其浓度为 1.0×10^{-6} mol/L 时达到最大值;十二烷基磺酸钠和十二烷基苯磺酸钠(SDBS)水溶液的表面扩张模量达到最大值所对应的浓度分别为 8.0×10^{-4} 和 5.0×10^{-5} mol/L;而当扩张频率为 0.1 Hz 时,油酸钠浓度在 3.0×10^{-4} mol/L 时扩张模量达到最大值(316 mN/m)。显然,油酸钠的扩张模量远高于其它表面活性剂,这与油酸钠水溶液表面形成较为致密的 $C_{17}H_{33}COO^-$—$C_{17}H_{33}COOH$ 混合吸附膜密切相关。这些结果一方面说明表面活性剂浓度较低时,体系的表面微观弛豫过程以扩散弛豫为主，另一方面也说明表面活性剂在表面和表面附近的微观弛豫过程依赖于其分子结构，而且表明，可以根据需要通过改变表面活性剂的种类和用量来调节界面扩张模量的大小。

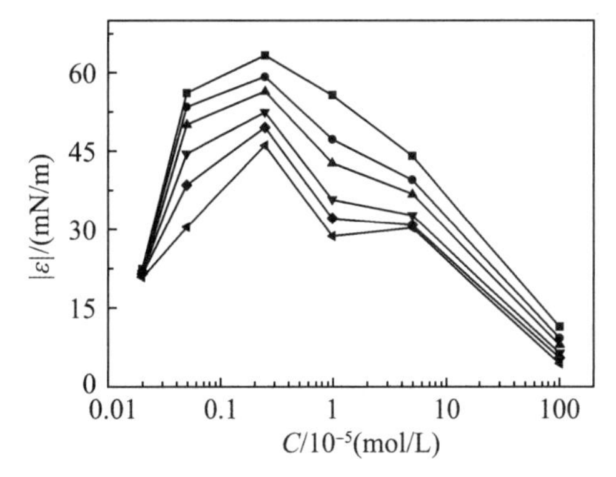

图 1 $C_{12}C_2C_{12}$ 水溶液表面扩张模量随浓度的变化(25℃)[61]

Fig. 1 Interfacial dilational modulus as a function of $C_{12}C_2C_{12}$ concentration at different dilational frequency[61]

■ 0.1 Hz;● 0.05 Hz;▲ 0.033 3 Hz;▼ 0.016 7 Hz;◆ 0.01 Hz;★ 0.005 Hz

1.2 表面活性剂的疏水链长对扩张模量的影响

表面活性剂的疏水链长不同，直接影响其在溶液表面上的聚集行为，因而，扩张粘弹性质会发生显著变化。以具有相同亲水基(三甲基溴化铵)而疏水基分别为十六烷基(CTAB)、十四烷基(TTAB)和十二烷基(DTAB)的阳离子表面活性剂水溶液为例，由其在同一频率下的扩张弹性部分和粘性部分随浓度的变化(图 2)[30]可以看出：在所研究的各种表面活性剂的浓度范围内，CTAB 的扩张弹性值最低；TTAB 的扩张弹性高于 CTAB，扩张弹性值越高说明膜的弹性和强度越大；DTAB 的扩张弹性受浓度的影响较大，较低浓度时扩张弹性值显著高于 CTAB 和 TTAB，但当 DTAB 浓度大于 8 mmol/L 时，其扩张弹性值与 CTAB 接近。不同疏水链长的 3 种表面活性剂的粘性部分的变化趋势在所研究的浓度范围内总体上说 CTAB>TTAB>DTAB。一般认为，粘性部分的大小反映了界面附近各种微观弛豫过程的总和[21]。

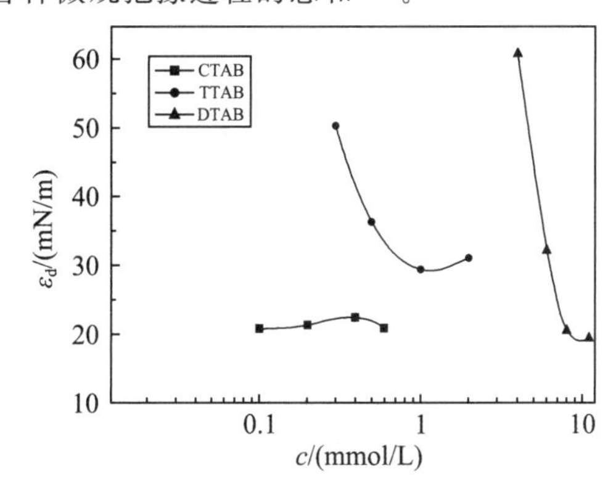

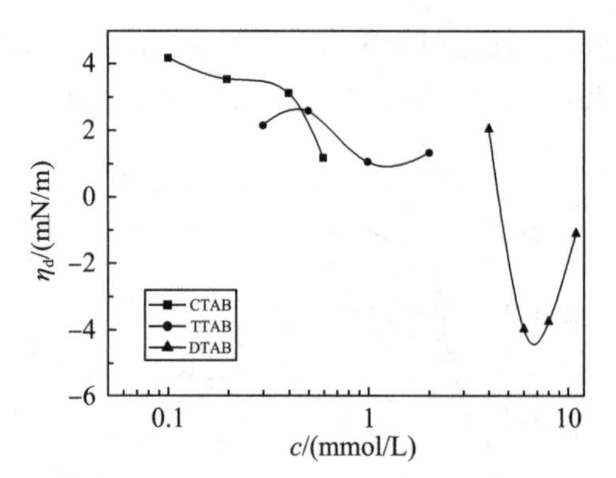

图 2 扩张弹性和扩张粘度随表面活性剂浓度的变化,ω=0.05 Hz[30]

Fig. 2 Dependence of dilational elasticity and viscous component on surfactant concentration[30]

1.3 添加剂对扩张模量的影响

有机醇的存在常引起表面活性剂聚集行为的变化。Wantke 等[62]在频率为 1～500 Hz 范围内,采用振荡气泡法研究了十二烷基硫酸钠(SDS)和十二醇混合体系的界面扩张性质。实验结果表明,纯的十二醇溶液在表面形成不溶性膜,仅表现出弹性行为,而 SDS 溶液则表现出粘弹性行为。当二者混合后,开始混合体系呈现较强的粘性,随时间推移表现出更强的弹性行为,这主要因为测量开始时,SDS 相对十二醇具有较高的表面浓度,随时间增加,界面上的 SDS 分子逐渐被十二醇所替代(图 3)。这与 Vollhardt 得到的结果一致,他们研究了 SDS 和十二醇的动态表面张力,在吸附动力学曲线上发现一个转折点,这表明一级相转变的发生。布鲁斯特角显微镜(BAM)也证实了域的形成。

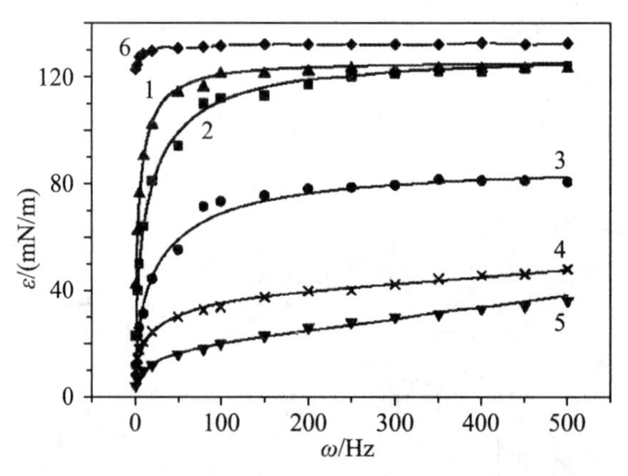

图 3 5×10^{-3} mol/L SDS/十二醇混合体系的界面扩张模量随振荡频率的变化[62]

Fig. 3 The dilational viscoelasticity of mixed 5×10^{-3} mol/L SDS/x mol% dodecanol solutions as a function of dilational frequency[62] 1：0.4 mol%;2：0.1 mol%;3：0.06 mol %;4：0.05 mol%;5：0 mol %;6：纯 1.2×10^{-5} mol/L 十二醇

无机盐能显著影响离子型表面活性剂的聚集行为,从而改变其扩张流变性。以 NaBr 对 $C_{12}C_2C_{12}$ 水溶液扩张模量的影响为例[61](图 4),显然,NaBr 的加入导致体系的扩张模量增大,因为无机盐的加入可以压缩表面活性剂离子头的周围的扩散双电层,削弱离子头之间的静电排斥作用,从而使得表面活性剂更紧密地聚集于表面层中,即表面

层中单位面积上 $C_{12}C_2C_{12}$ 的分子数增加，表面上受到微扰时形成的张力梯度增大，因而扩张模量增大。

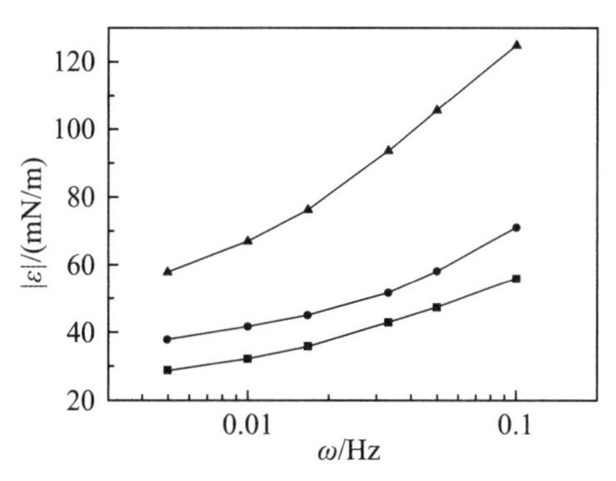

图 4 NaBr 对 $C_{12}C_2C_{12}$ 水溶液扩张模量随扩张频率的影响[61]

Fig. 4 The dilational viscoelasticity of $C_{12}C_2C_{12}$ as a function of dilational frequency in the presence of NaBr[61] $C_{12}C_2C_{12}$ 的浓度：2.5×10^{-6} mol/L ■ 不加 NaBr；● 1×10^{-3} mol/L NaBr；▲ 2×10^{-3} mol/L NaBr

1.4 温度对扩张模量的影响

温度对界面扩张流变性的影响研究表明，扩张模量等参数均随温度的增加而下降。这可能是因为 温度的上升加剧了分子的热运动，此时滑障的扩张压缩速度已小于表面活性剂分子在表面与体相之间的扩散速度，表面上的表面活性剂分子有足够的时间去恢复对平衡后的扰动，即表面层中的表面活性剂分子有足够的时间进行扩散，维持原来的平衡状态，使表面张力变化基本保持不变，因而降低了扩张模量、扩张粘度等参数的数值。但是溶液扩张模量最大值所对应表面活性剂浓度几乎不变，例如，温度在 5～25℃范围内 SDBS 水溶液的表面扩张模量达到最大值所对应的浓度均在 5.0×10^{-5} mol/L附近[63]。

1.5 高分子表面活性剂的界面扩张流变行为

实践中常常利用高分子表面活性剂在气/液表面上的聚集作用来调节泡沫的稳定性，高分子表面活性剂的表面粘弹性对于泡沫基产品的质量起着重要作用。高分子表面活性剂的表面粘弹性与小分子表 面活性剂体系的不同[10, 64-67]。由图 5 示出的扩张频率对星状嵌段聚醚(AP432)水溶液扩张模量的影响可以看出，其扩张模量也随扩张频率的增加而增加，而且当 AP432 浓度较低时，扩张频率对扩张模量的影响较小；随 AP432 浓度增加，扩张模量对扩张频率的依赖性增强直至达到最大值；继续增加浓度，扩张频率对扩张模量的影响又变小。通过比较可以看出，当扩张频率相同时，AP432 扩张模量相对较低[68]。高分子表面活性剂界面粘弹行为特征反映了其单体分子在表面层不同位置之间扩散交换过程的差异，因为在其形成的表面层中可能存在两个微区：紧贴着气相、比较窄的浓度区域(最接近微区，the proximal region)和单体分子的总浓度很低、形成嵌入到液相的尾状(tail)和环状(loop)结构的区域(末端微区，the distal region)[10]。

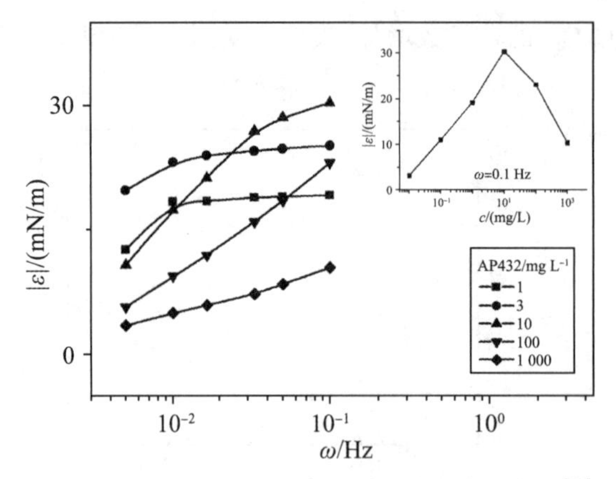

图 5 AP432 溶液的扩张模量随扩张频率的变化[68]

Fig. 5 The variation of dilational viscoelasticity as a function of dilational frequency for AP432 aqueous solutions[68]

对于粘弹性界面，界面应力的变化可以看作弹性和粘性部分的贡献之和，因此，扩张模量也可以用复数形式表示[57-60]：

$$|\varepsilon| = \varepsilon_d + i\omega\eta_d \tag{2}$$

其中，ε_d 和 η_d 分别称为界面扩张弹性和扩张粘度。由于扩张模量中粘性部分的存在，应力与界面面积相对变化之间存在一定的相位差，这种相位差称为扩张粘弹性的相角 θ。

扩张弹性 ε_d 和扩张粘度 η_d 与扩张模量 $|\varepsilon|$、相角 θ 和界面面积正弦变化角频率 ω 的关系可表示为：

$$\varepsilon_d = |\varepsilon| \cos\theta \tag{3}$$

$$\eta_d = (|\varepsilon|/\omega) \sin\theta \tag{4}$$

羟丙基甲基纤维素(HPMC)的扩张弹性和粘性随浓度的变化规律[30]显示，当 HPMC 的浓度为 0.2 g/L 时，体系的扩张弹性达到最大值；若继续增加 HPMC 的浓度，体系的扩张弹性则随之下降。其原因是随 HPMC 浓度增加，界面上的 HPMC 分子数目逐渐增加，但其界面浓度的增加阻碍了部分“尾状”或“环状”结构的 HPMC 分子的完全伸展，故 HPMC 的界面扩张模量随浓度增加而增大，并在 0.2 g/L 处出现一个极大值。若 HPMC 浓度大于 0.2 g/L，由于“最接近区域”和“末端区域”之间的扩散交换加快，使得 HPMC 的界面扩张模量大大降低。

值得注意的是，体系的扩张粘度出现负值现象。这源于体系的相角为负值。我们在研究不同频率下明胶溶液扩张模量随浓度的变化时也发现了负相角现象，而且振荡频率越高，界面膜呈现负相角的可能性越大。这与文献[26,47,64,69-72]报道的扩张粘度出现负值的现象类似。依据这些文献可以认为，正相角意味着响应领先于扰动，这可能是由于界面上分子与体相间的交换引起的，而负相角的出现则可能是界面与体相间的分子交换比较微弱、界面内弛豫过程较强所致，即：当界面层中慢驰豫过程控制界面膜的扩张粘弹性时，如界面膜中存在分子量较高的物质或大的聚集体，得到的相角为负值；而当分子在体相与界面之间的扩散过程控制界面膜的扩张粘弹性质时,得到的相角为正值。研究结果表明,一般表面活性剂浓度较高时,体系容易出现正相角,说明表面活性剂分子在界面和体相之间的扩散交换较快。

2　表面活性剂在液/液界面上聚集行为研究

低分子表面活性剂和高分子表面活性剂在液/液界面上的聚集行为对于乳状液的稳定性至关重要。液/液界面流变性的研究对于食品、医药以及油田开发具有重要意义[73~77]。Murray 等[78]通过界面扩张流变性研究比较了非离子表面活性剂($C_{12}E_6$)在气/液和液/液界面聚集行为的差异;Dicharry 等[79]利用扩张流变方法在不同频率下研究了水/原油(W/O)乳状液的稳定性,发现存在于原油中的沥青质和胶质等天然表面活性剂可以在界面上形成凝胶,而该 W/O 乳状液的稳定性与凝胶的强度和玻璃化转变温度呈正相关。Ivanov 等[12]的研究发现,SDS 在豆油/水界面上的流变参数随其浓度的增大呈现出复杂的变化趋势(图 6)。

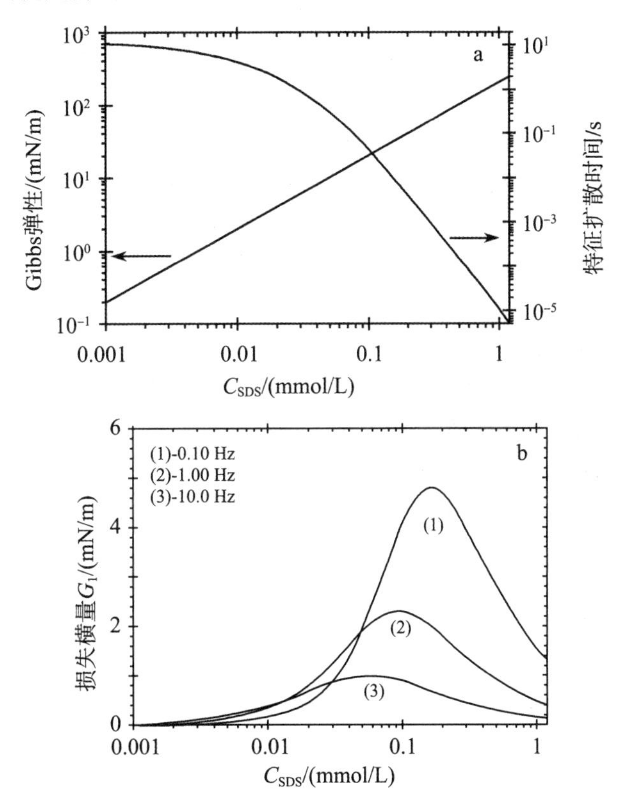

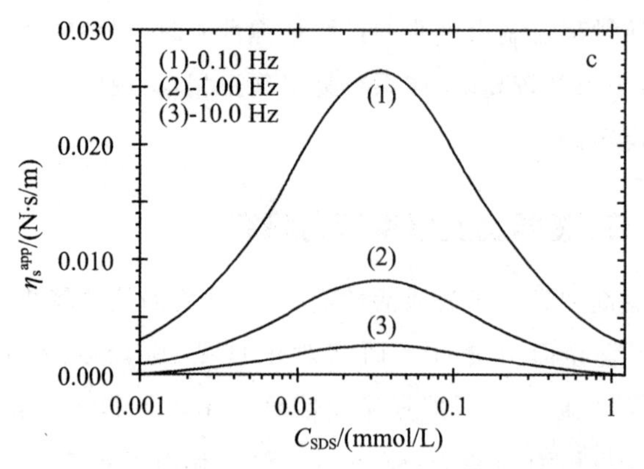

图 6 SDS 在豆油/水界面上的流变参数随其浓度的变化[12]

Fig. 6 Dependence of the Gibbs elasticity, E_G, and the diffusion relaxation time, t_D, (a); the loss modulus, G_i, (b); and the apparent surface viscosity, η_s^{app} (c) on the sodium dodecyl sulfate (SDS) concentration System: soybean oil/water[12] $C_{NaCl}=150\ mmol\cdot L^{-1}$;吸附常数 29.9 L/mmol; $\Gamma_\infty=2.66\ \mu mol/m^2$; $D=5.5\times10^{-10}\ m^2/s$; $\eta_s^{app}=\frac{E_G(2\omega D)^{1/2}}{\omega}$

笔者实验室[29]比较研究了 CTAB 在液体石蜡/水界面与其在气/液表面的的扩张流变差异。结果表明,在气/液表面和液体石蜡/水界面随着扩张频率增加呈现出相同的变化趋势:CTAB 的扩张模量|ε|逐渐增大,而相角 tan θ 逐渐减小,说明弹性部分对吸附层扩张模量有所增加;扩张模量|ε|随 CTAB 浓度增加而单调增大,而相角 tan θ 先略有增加,然后开始下降到接近于零。但在相同 CTAB 浓度下,气/液表面上的扩张模量值大于在油/水界面上的测定值,说明不同的界面,对 CTAB 吸附层的流变性虽然没有本质上的影响,但是数值上有差异。大量的关于表面活性剂在不同界面上构象的研究表明[75,76]:在油/水界面上,由于油分子与表面活性剂分子之间很强的疏水相互作用,表面活性剂分子可以在油/水界面上采取更为直立方式存在,而在气/液表面上由于空气分子与表面活性剂分子间相互作用较弱,所以表面活性剂分子的构象变得更加无序。可以推测,当油/水界面受到压缩或扩张时,只会发生表面活性剂分子在体相和界面上的扩散交换行为,而在气/液界面上,则除了这种扩散交换,还存在表面活性剂分子有序度改变即分子取向的改变,也就是说体系的弛豫行为更强,所以 CTAB 在油/水界面的扩张模量小于在气/液表面之值。

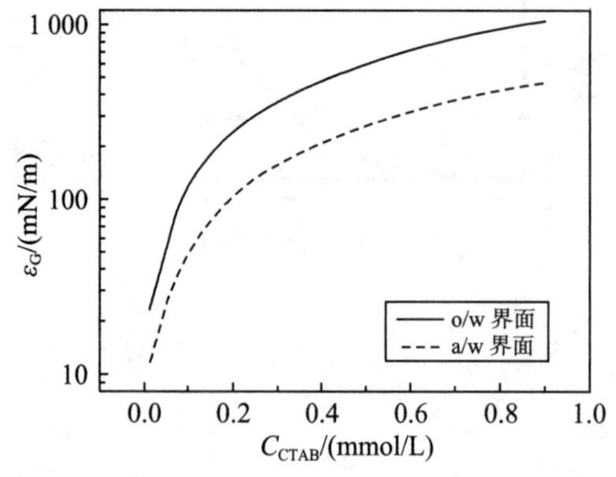

图 7 不同界面上 CTAB 吸附层的 Gibbs 弹性随 CTAB 浓度的变化[29]

Fig. 7 Dependence of Gibbs elasticity (ε_G) on CTAB concentration[29]

根据 Neumann 等[77]的观点，Gibbs 弹性是决定乳状液和泡沫稳定性的重要参数。CTAB 在油/水界面和气/液界面的 Gibbs 弹性计算结果表明，油/水界面 CTAB 吸附层的 Gibbs 弹性高于在气/液界面之值，由于油分子与 CTAB 分子之间存在强的疏水相互作用，因此，在油/水界面上 CTAB 吸附分子之间会插入一部分油分子而占据了一部分空间，导致 CTAB 在油水界面上的吸附量比在气液界面上的吸附量少。对于油/水界面和气/液界面而言，当吸附量发生相同的改变时，对油/水界面张力的影响更加显著，也就是说，油/水界面张力会有更大的变化，所以，油/水界面上 CTAB 吸附层的 Gibbs 弹性大于在气/液界面上之值(图 7)。

研究发现，Gibbs 弹性的变化规律与扩张模量的变化规律不同：随 CTAB 浓度增加，其 Gibbs 弹性逐渐增大，而扩张模量$|\varepsilon|$值却逐渐减小。可能的原因是：(1) Gibbs 弹性值由界面张力结果计算所得，所以只与表面活性剂的体相浓度有函数关系，而扩张模量不仅可以反映出浓度变化引起的界面性质的变化，同时也反映界面面积改变量(α)和改变速度($\acute{a}$)对界面性质的影响[12]。或者说，Gibbs 弹性表征的是吸附膜的平衡态的性质，而扩张模量反应吸附膜的动态性质。(2)确定 Gibbs 弹性所用 Langmuir-Syszkowski 状态方程中忽略了表面活性剂吸附分子在界面上的倾斜角度以及分子的固有可压缩性的影响。在 Fainerman 等[51,80,81]新提出的模型中考虑了固有可压缩性的影响，所得到的 Gibbs 弹性与传统的 Langmuir-Syszkowski 状态方程得出的 Gibbs 弹性存在较大差异。

3 结语

近年来，由于实践的需要以及各种仪器的开发，国内外对扩张流变性质的研究兴趣倍增，不仅已获得了一些规律性知识，而且在基础理论方面也有一定的进展。大量研究表明，通过界面流变性测定研究表面活性剂的界面聚集行为，不仅有助于了解表面活性剂分子在界面上的聚集及其相互作用的机理，而且有助于弄清发生在界面膜内及附近的的弛豫过程，加深对界面膜微观性质的认识，对阐明乳状液和泡沫的稳定性及破乳机理有重要意义。界面膜性质和液膜排液是决定乳状液和泡沫稳定性的关键因素，扩张流变性质是决定乳状液和泡沫界面膜的关键因素。因此这些研究在食品、化妆品和医药生产以及原油乳状液破乳、表面活性剂和聚合物驱油、稠油乳化降粘开采和泡沫驱散等过程中必将发挥重要作用。但是，目前这方面的研究还有一些问题尚未弄情，需要从理论上和技术上进一步深入研究。例如：界面扩张流变研究的方法问题，不同研究方法获得的数据可能相差较大，如何确定方法的适用性；界面扩张流变性参数与界面热力学和动力学性质关系；负相角的出现(即负粘度现象)究竟合理否？如何理解其产生的原因或从技术上如何避免其出现，等等。

现代化生产和生活中涉及表面活性剂、聚合物等两亲分子界面聚集行为的领域愈来愈广泛。界面扩张性质的深入研究必将为开发不同结构的大分子和小分子表面活性剂在驱油、乳化与破乳、发泡与消泡等过程的应用发挥重要作用。

参考文献

[1] M J Rosen. Surfactants and Interfacial Phenomena, Wiley-Interscience, New York, 1978.

[2] Q He, Y Zhang, G Lu et al. Adv. Colloid Interface Sci. , 2008,140:67-76.

[3] B A Noskov, D A Alexandrov, G Loglio et al. Colliods Surf. A, 1999,156:307-313.

[4] C Monteux, R Mangeret, G Laibe et al. Macromolecules, 2006, 39:3408-3414.

[5] C Monteux, G Fuller, V Bergeron. J. Phys. Chem. B, 2004, 108:16473-16482.

[6] C Stubenrauch, R Miller. J. Phys. Chem. B, 2004, 108:6412-6421.

[7] M B J Meinders, T Vliet. Adv. Colloid Interface Sci. , 2004, 108-109:119-126.

[8] R Miller, V B Fainerman, A V Makievskib et al. Adv. Colloid Interface Sci. , 2000,86:39-82.

[9] F Monroy, F Ortega, R G Rubio et al. Adv. Colloid Interface Sci. , 2007, 134-135:175-189.

[10] B A Noskov, A V Akentiev, A Y Bilibin et al. Adv. Colloid Interface Sci. , 2003,104:245-271.

[11] P Koelsch, H Motschmann. Langmuir, 2005,21:6265-6269.

[12] I B Ivanov, K D Danov, K P Ananthapadmanabhan et al. Adv. Colloid Interface Sci. , 2005,114-115:61-92.

[13] B S Murray. Curr. Opin. Colloid Interface Sci. , 2002,7:426-431.

[14] E H Lucassen-Reynders. Surface elasticity and viscosity in compression/dilation, in Anionic surfactants, E H Lucassen-Reynders ed, New York and Basel,1981.

[15] M Van den Tempel, E H Lucassen-Reynders. Adv. Colloid Interface Sci. , 1983,18:281-301.

[16] 江龙,赵丰,唐季安,等. 科学通报,2000,45:2501-2505.

[17] 叶仲斌,刘向君,杨建军,等. 西南石油学院学报,2002,24(5):28-31.

[18] 王宜阳,张路,孙涛垒,等. 物理化学学报,2003,19(5):455-459.

[19] 王宜阳,张路,孙涛垒,等. 高等学校化学学报,2003,24(11):2044-2047.

[20] 孙涛垒,张路,王宜阳,等. 高等学校化学学报,2003,24(12):2243-2247.

[21] Y Wang, L Zhang, T Sun et al. J. Colloid Interface Sci., 2004, 270: 163-170.

[22] Y Y Wang, Y H Dai, L Zhang et al. J. Colloid Interface Sci., 2004,280:76-82.

[23] 宋新旺,王宜阳,曹绪龙,等. 物理化学学报,2006,22:1441-1444.

[24] 王宜阳,张路,李明远,等. 石油学报,2004,20(5):87-93.

[25] T L Sun, L Zhang, Y Y Wang et al. J. Colloid Interface Sci.,2002,255:241-247.

[26] Y Y Wang, Y H Dai, L Zhang et al. Macromolecules, 2004,37:2930-2937.

[27] 孙涛垒,彭勃,许志明,等. 物理化学学报,2002,18:161-165.

[28] 汪庐山,于田田,曹秋芳,等. 油田化学, 2007,24:70-74.

[29] H X Zhang, G Y Xu, D Wu et al. Colloids Surf. A, 2008,317:289-296.

[30] Y M Li, G Y Xu, X Xin et al. Carbohydr. Polym.,2008,72:211-221.

[31] X Xin, G Y Xu, D Wu et al. Colliods Surf. A,2008,322:54-60.

[32] D Wu, Y J Feng, G Y Xu et al. Colliods Surf. A,2007,299:117-123.

[33] H O Lee, T S Jiang, K S Avramidis. J. Colloid Interface Sci.,1991,146:90-122.

[34] G Garofalakis, B S Murray. Langmuir, 2002,18:4765-4774.

[35] B R Blomqvist, T Warnheim, P M Claesson. Langmuir, 2005, 21: 6373-6384.

[36] J T Petkov, T D Gurkov. Langmuir, 2000,16:3703-3711.

[37] A Bonfillon, D Langevin. Langmuir, 1993,9:2172-2177.

[38] 孙涛垒,彭勃,许志明,等. 物理化学学报, 2002,18:161-165.

[39] C Stubenrauch, R Miller. J. Phys. Chem. B, 2004,108:6412-6421.

[40] P Koelsch, H Motschmann. Langmuir, 2005,21:6265-6269.

[41] A Rao, J Kim, R R Thomas. Langmuir, 2005,21:617-621.

[42] T L Sun, L Zhang, Y Y Wang et al. J. Disper. Sci. Tech., 2003,24: 699-707.

[43] P Wilde, A Mackie, F Husband et al. Adv. Colloid Interface Sci., 2004, 108-109:63-71.

[44] Y H Kim, D T Wasan, P J Breen. Colliods Surf. A, 1995,95:235-247.

[45] Y H Kim, D T Wasan. Ind. Eng. Chem. Res., 1996,35:1141-1149.

[46] M B J Meinders, W Kloek, T Vliet. Langmuir, 2001,17:3923-3929.

[47] F Monroy, J G Kahn, D Langevin. Colloids Surf. A, 1998,143:251-260.

[48] V I Kovalchuk, J Krägel, AV Makievski et al. J. Colloid Interface Sci.,

2004,280:498-505.

[49] F Ravera, M Ferrari, E Santini et al. Adv. Colloid Interface Sci. ,2005,117:75-100.

[50] VI Kovalchuk, R M iller, V B Fainerman et al. A dv . Colloid Int erface Sci. ,2005,114:303-312.

[51] C Stubenrauch, V B Fainerman, E V Aksenenko et al. J. Phys. Chem. B,2005,109:1505-1509.

[52] P Erni, P Fischer, E J Windhab. Langmuir, 2005,21:10555-10563.

[53] C M Kausch, Y Kim, V M Russell et al. Langmuir,2003,19:7354-7361.

[54] F V Vader, T F Erkens, M Van den Tempel. Trans. Faraday Soc. ,1964,60:1170-1175.

[55] K Miyano, B M Abrambam, L Ting et al. J. Colloid Interface Sci. ,1983,92:297-302.

[56] T D Karapantsios, M Kostoglou. Colliods Surf. A, 1999,156:49-64.

[57] J Benjamins, A Cagna, E H Lucassen-Reynders. Colloids Surf. A, 1996,114:245-254.

[58] T Fromyr, F K Hansen, A Kotzev et al. Langmuir, 2001,17:5256-5264.

[59] J Lucassen, D Giles. J. Chem. Soc. Faraday Trans. , 1975,71:217-232.

[60] 王宜阳,张路,孙涛垒,等.物理化学学报,2003,19(4):297-301.

[61] 吴丹.山东大学博士学位论文,2007.

[62] K D Wantke, H Fruhner, J Ortegren. Colloids Surf. A,2003,221:185-195.

[63] 程栋,丁洪流,陆维昌,等.华东师范大学学报,2005,4:37-42.

[64] A J Milling, R W Richards, F L Baines et al. Macromolecules, 2001,34:4173-4179.

[65] B A Noskov, A V Akentiev, G Loglio et al. J. Phys. Chem. B, 2000,104:7923-7931.

[66] B A Noskov, A V Akentiev, R Miller. J. Colloid Interface Sci. , 2002,255:417-424.

[67] J W Anseth, A Bialek, R M Hill et al. Langmuir, 2003,19,6349-6356.

[68] X Xin, G Y Xu, Y J Wang et al. Eur. Polym. J. , 2008.

[69] F Monroy, M G Munoz, J E F Rubio et al. J. Phys. Chem. B, 2002,106:5636-5644.

[70] J C Earnshaw, E McCoo. Langmuir, 1995,11:1087-1100.

[71] S K Peace, R W Richards, N Williams. Langmuir,1998,14:667-678.

[72] D M A Buzza, J L Jones, T C B McLeish et al. J. Chem. Phys. ,1998,109:

5008-5024.

[73] D M Sztukowski, H W Yarranton. Langmuir, 2005,21:11651-11658.

[74] K Giribabu, P Ghosh. Chem. Eng. Sci., 2007,62:3057-3067.

[75] M M Knock, G R Bell, E K Hill et al. J. Phys. Chem. B,2003,107:10801-10814.

[76] J Chanda, S Bandyopadhyay. J. Phys. Chem. B, 2006,110:23482-23488.

[77] B Neumann, B Vincent, R Krustev. Langmuir, 2004,20:4336-4344.

[78] B S Murray, A Ventura, C Lallemant. Colliods Surf. A, 1998,143:211-219.

[79] C Dicharry, D Arla, A Sinquin et al. J. Colloid Interface Sci., 2006,297:785-791.

[80] V B Fainerman, E H Lucassen-Reynders. Adv. Colloid Interface Sci., 2002,96:295-323.

[81] V B Fainerman, V I Kovalchuk, E V Aksenenko et al. J. Phys. Chem. B, 2004,108:13700-13705.

阴离子表面活性剂在水溶液中的耐盐机理

赵涛涛[1] 官厚健[1] 徐桂英[1] 曹绪龙[2] 宋新旺[2] 王红艳[2]

(1. 山东大学胶体与界面化学教育部重点实验室,山东济南 250100;

2. 中国石化胜利油田分公司地质科学研究院,山东东营 257015)

摘要:针对驱油用表面活性剂的耐盐性问题,基于国内外 45 篇期刊论文、5 部专著及作者所在研究组近期研究成果,综述了阴离子表面活性剂在水溶液中的耐盐机理,论题包括:前言;① 无机盐对表面活性剂聚集行为的影响;② 无机盐与表面活性剂的相互作用(改变表面活性剂 Kraft 点;改变表面活性剂分子的临界堆积系数,改变表面活性剂在油、水相的分配系数);③ 展望(机理研究;计算模拟的应用)。

关键词:阴离子表面活性剂;水溶液;物化性能;耐盐性;盐/表面活性剂相互作用;作用机理;驱油表面活性剂;综述

中图分类号:TE39:O647.2:O645.13 文献标识码:A

表面活性剂在化学驱油过程中起着重要作用,它能降低油/水界面张力、提高毛管数,从而大幅度提高原油采收率[2]。三次采油技术的发展和油藏条件变化对驱油用表面活性剂提出了更高的要求:具有低油/水界面张力和低吸附损失,抗盐、耐温,与油藏流体配伍,价廉易得[3]。驱油用表面活性剂主要有非离子、阴离子和两性型三大类。阴离子表面活性剂因界面活性高、耐温性好而被广泛使用,但其耐盐性差,不适于多价阳离子含量高的油藏;非离子表面活性剂的耐盐、耐多价阳离子性能好,但在地层中的吸附损失比阴离子表面活性剂高,且不耐高温,价格高;两性表面活性剂大多数都能用于高矿化度、较高温度的油层,且能避免非离子型与阴离子型表面活性剂复配时的色谱分离效应,缺点是价格较高。

为了适应油藏高温高矿化度条件,有关表面活性剂耐盐、耐温性研究一直是备受重视的课题。在国内外许多学者共同努力下,近几年这方面的产品研发和性能研究均取得了较大进展[4-10]。已有的研究大多集中在探讨无机盐对表面活性剂溶液宏观性质如临界胶束浓度($C_{m,cr}$)[11-13]、表面活性剂聚集体大小[14-15]、流变性质[16]等的影响。而有关表面活性剂与无机盐相互作用机理方面的研究报道相对较少。研究表面活性剂的耐盐机理,指导新型耐盐表面活性剂的研制与开发,对于强化采油技术的发展及油田开发具有深远的战略意义和实际意义。本文综述了无机盐对不同阴离子表面活性剂水溶液性能的影响机理及其研究进展，提出了耐盐型表面活性剂应具有的结构特征，以期为耐盐型表面活性剂驱油体系的设计提供基础数据。

1　无机盐对表面活性剂聚集行为的影响

一般而言，在表面活性剂体系中加入无机盐会使其表面活性增强，表现为溶液的表面张力和临界胶束浓度（$C_{m,cr}$）显著降低[117~19]，即适量的无机盐可使表面活性剂降低溶剂表面张力的能力和效率增强[20,21]。另外无机盐还会影响表面张力的时间效应[21]：反离子的价数愈高，溶液表面张力达到平衡所需时间愈短[12]。我们发现水溶液中不含NaCl时，2-乙基-己基琥珀酸酯磺酸钠（简称AOT）的最低表面张力（$\gamma C_{m,cr}$）为27 mN/m，当含有少量NaCl（0.005～0.01 moL）时AOT溶液的$\gamma C_{m,cr}$和$C_{m,cr}$均明显降低，且NaCl浓度愈高，$\gamma C_{m,cr}$和$C_{m,cr}$愈低（图1）[22]。表面活性剂在油/水界面的聚集行为也符合此规律：无机盐能够中和离子型表面活性剂极性基团的电荷、压缩双电层使电荷密度降低，合适的加盐量甚至可使油/水界面张力降至超低水平[23]。

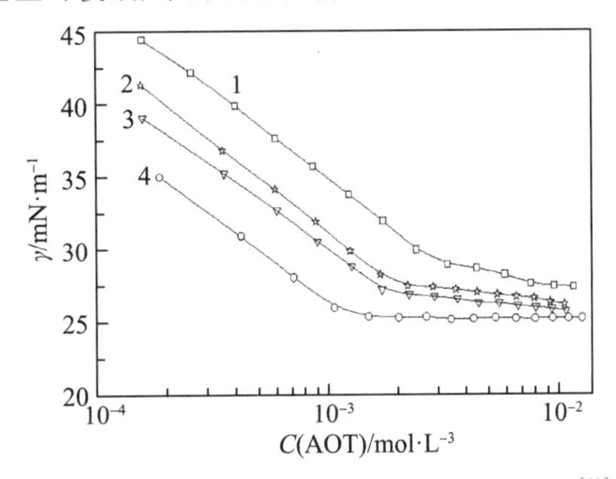

图1　NaCl对AOT溶液表面张力等温线的影响[22]

NaCl浓度（$mol \cdot L^{-1}$）：1—0；2—0.005；3—0.008；4—0.01

无机盐的种类和浓度对不同表面活性剂溶液的$C_{m,cr}$影响规律不同。离子型表面活性剂溶液的$C_{m,cr}$随无机离子加量的增加而降低，这是由于加入的反离子屏蔽了表面电荷，降低了表面电势，从而导致胶束表面Stern层的厚度被压缩。离子型表面活性剂$C_{m,cr}$的对数值与反离子浓度的对数之间有直线关系[20,21]，如式（1）所示。

$$\lg C_{m,cr} = -a\lg C_i + b \tag{1}$$

式中，C_i为反离子的总浓度，单位mol/L，温度恒定时，给定表面活性剂的a和b是常数。例如，AOT水溶液的$C_{m,cr}$与NaCl浓度的关系符合方程（1），常数a和b分别为1.33和－5.10。

离子型表面活性剂在水溶液中形成胶束是其碳氢链的疏水作用和极性基团之间的静电作用的综合结果，疏水作用促进胶束形成，静电排斥力则不利于表面活性剂的聚集。通常情况下，可用式（2）表示两种作用对胶束形成的贡献[24]：

$$RT\ln C_{m,cr} \approx \Delta G_{0m} \tag{2}$$

而 ΔG_{0m} 又可表示为：

$$\Delta G_{0m} \approx \Delta G_m^{hp,0} + \Delta G_m^{el0}$$

式中，ΔG^{hp} 表示疏水部分自由能，ΔG^{el} 表示静电作用对自由能的贡献，当碳氢链相同，反离子分别为一价离子和二价离子时，则有以下结果：

$$RT\ln[C_{m,cr(1)}/C_{m,cr(2)}] = \Delta G^{el}_{c\ m,cr(1)} - \Delta G^{el}_{c\ m,cr(2)} \tag{4}$$

式中，$C_{m,cr(i)}$ 表示反离子价数为 i 时的临界胶束浓度。例如，反离子为 Na^+（十二烷基硫酸钠，简称 SDS)时 $C_{m,cr(1)} \approx 8$ mmol/L，胶束聚集数约为 60，胶束的半径约为 2 nm，表面电荷密度约为 0.19 C/m^2，则每摩尔单体的 $\Delta G_m^{el} = 12.95$ kJ/mol，进而可求得反离子为两价为 1.67 mmol/L(实验测得不同二价十二烷基硫酸的十二烷基硫酸盐的 $C_{m,cr(2)}$ 约盐的 $C_{m,cr(2)}$ 均在 1.66～2.4 mmol/L 范围)[25]。即当反离子价态从 1 变为 2，表面活性剂的 $C_{m,cr}$ 显著降低。说明二价反离子与表面活性剂离子之间结合得更加紧密，与一价反离子相比，解离程度大大减小，表现出非离子表面活性剂的性质。

除了影响界面性质，无机盐对溶液中表面活性剂的聚集体大小和形状也有明显影响[15,19]。Alargova 通过动态光散射研究了高价无机盐对阴离子表面活性剂十二烷基二氧乙烯硫酸钠($CH_3(CH_2)_{11}(OC_2H_4)_2OSO_3N$，a$SDE_2S$)胶束聚集体的影响，发现 Ca^{2+} 和 Al^{3+} 在浓度很低时就能明显改变 SDE_2S 胶束的大小和形状[19]。聚集体尺寸和形状的变化必然影响表面活性剂体系的流变性质[16]，SDE_3S 无-机盐混合体系的零剪切黏度研究表明，固定 SDE_3S 浓度时盐的存在会促进胶束的生长：盐浓度较低时，随盐浓度增大尽管体系中形成的球状胶束逐渐增大，但胶束形状不变，因而溶液黏度变化不大；随无机盐浓度继续增大，体系中形成棒状、虫状胶束，则混合体系的黏度迅速增大；但无机盐浓度超过一定值时会发生盐析作用，棒状、虫状胶束被破坏，溶液中表面活性剂浓度降低，体系黏度又开始降低(图 2)。图 2 的结果还表明，尽管 Ca^{2+} 和 Mg^{2+} 离子荷电量相同，但它们对体系黏度的影响有显著差异，这说明无机离子与阴离子表面活性剂的相互作用不仅仅依赖于静电作用力。

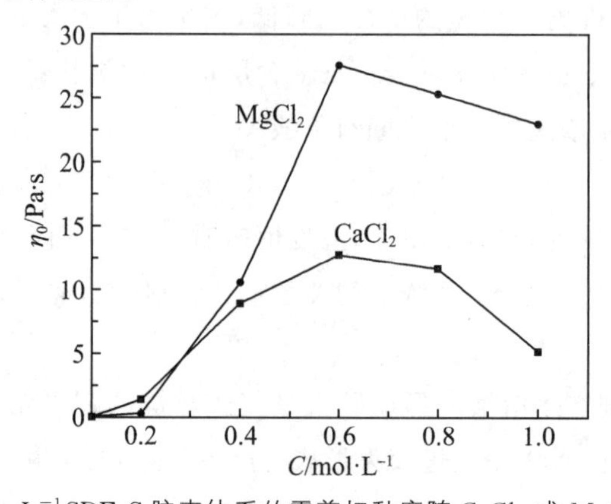

图 2　0.2 mol·L^{-1} SDE_3S 胶束体系的零剪切黏度随 $CaCl_2$ 或 $MgCl_2$ 浓度的变化

表面活性离子与无机盐的相互作用可导致表面活性剂的溶解度下降，甚至沉淀。Somasundaran[26]等研究 SDS 和 Al^{3+} 混合体系时发现，在含有一定浓度 Al^{3+} 的溶液中加入表面活性剂，加入浓度很小时无沉淀生成，加入浓度到达某值后，随着浓度不断增大，体系中沉淀不断生成；而继续增大浓度到一定值后，沉淀会全部溶解。Wanles 考察 SDS-MgC_2l-水体系的相图时也发现存在一个明显的沉淀再—溶解边界[27]。无机盐引起表面活性离子沉淀的最低浓度称为盐的临界浓度(C_s)。C_s 可以作为表面活性剂耐盐性的一个指标，其值与表面活性剂的结构、组成和浓度以及反离子的种类、价数和温度等有关。通过浊度滴定考察表面活性剂在无机盐溶液中的溶解性时发现，溶解度随外加盐浓度增大而下降；若恒定外加盐浓度，当表面活性剂浓度小于 $C_{m,cr}$ 时，溶解度随表面活性剂浓度增大而下降，当浓度接近 $C_{m,cr}$ 时溶解度达最低值；浓度小于 $C_{m,cr}$ 时，二价无机离子 M^{2+} 对阴离子表面活性剂离子(J)的沉淀作用中可能存在以下平衡：

$$nM^{2+} + PJ \rightleftharpoons M_nJ_o \downarrow \tag{5}$$

$$K_S = C_M^n \cdot C_J^P \tag{6}$$

式中，K_s 为 M_nJ_o 的溶度积常数，即表面活性剂 J 的浓度 C_J 愈小，产生沉淀需要的反离子浓度 C_M 愈大。根据相分离模型，表面活性剂的 $C_{m,cr}$ 是其在溶液中以单体形式(分子或离子)存在的最大浓度。因此，可以推断表面活性剂的表面活性愈高，$C_{m,cr}$ 值愈小，产生沉淀需要的无机盐浓度愈大，即表面活性剂的抗盐性愈好。从此意义上讲，表面活性剂的 $C_{m,cr}$ 低者抗盐效果好，表面活性剂同系物相比，碳氢链长者抗盐性好于链短者；复配表面活性剂的抗盐性应好于单一表面活性剂，这已被我们最近的研究所证实：非离子表面活性剂(TX-100)与 SDS 复配体系的 $C_{m,cr}$ 低于 SDS 的，因而 TX-100 的存在能够明显提高 SDS 的抗盐能力(图 3)。

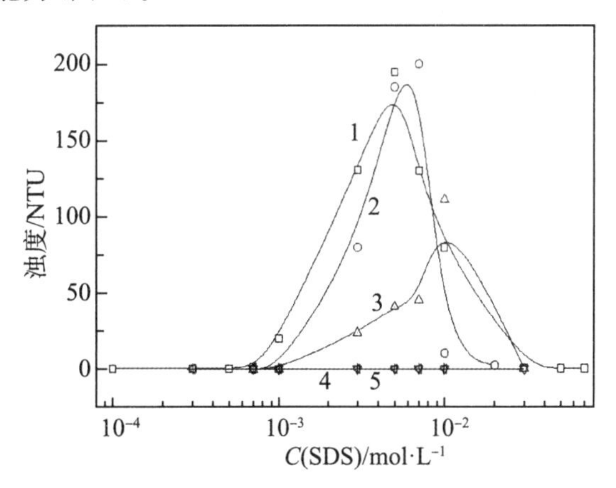

图 3 TX-100 对 SDS 水溶液浊度的影响

[$C(CaCl_2) = 1\times10^{-3}$ mol · L^{-1}]

TX-100 浓度(mol · L^{-1})：1—0；2—5×10^{-4}；3—1×10^{-3}；4—3×10^{-3}；5—1×10^{-2}

自原来的电缩水化层被破坏；区域束缚是指反离子受表面活性离子交叠电场协同作用的吸引而被束缚但当表面活性剂在水溶液的浓度达到 $C_{m,cr}$ 时，体系中形成的胶束等聚集体可将不溶或微溶于水的物质增溶于其中,则沉淀溶解,此时表面活性剂浓度与无机盐浓度之间不存在溶度积关系。

大量的研究结果均表明,无机盐能够显著影响离子型表面活性剂的聚集行为,从而改变体系的性能。而这种影响不仅依赖于表面活性剂的结构、性质和浓度,也与无机盐的本性和浓度密切相关。为了控制表面活性剂的盐效应,必须揭示无机盐与阴离子表面活性剂相互作用的机理,以便更好地调控适合于高矿化度油藏的驱油体系。

2 无机盐与表面活性剂相互作用机理

2.1 无机盐破坏表面活性剂的电性和水化作用

表面活性剂是一种具有两亲结构的分子,在水中的水化作用分为两部分[28]:荷电基团、极性基团的水化和非极性基团的水化。由于离子型表面活性剂荷电基团周围存在很强的静电场,以水的介电常数为 80 计,在水溶液中距一价离子 0.2～0.3 nm 处的电场可达 1×10^{6} V/cm 在如此强的电场下,荷电基团周围会形成三个水化域[28](图 4):在最靠近离子区域,水分子强烈取向、密集排列而形成结构较为规整的水化层,其密度比一般水大,形成时体积缩小,此层水化水称为电缩水化水,或简称水化水;电缩水化层以外的区域为部分失去规整性的水化层,排列较水化水混乱;这层不规整水化层以外的区域为普通状态的自由水。电缩水化水层的厚度与表面活性剂的耐盐性密切相关。有时也将围绕表面活性剂离子基团周围的水化层称为特征水化域,而相邻离子基团相互靠近而产生的水化层称为协同水化域。疏水基的水化是当一个非极性分子溶入水中时,其周围的水分子形成类似冰的结构[21],成为"冰岛"。"冰岛"形成时有两种相反的体积效应同时发生,一是类冰结构形成导致体积增加,二是非极性分子挤入类冰结构内部使体积缩小。通常后一效应大于前者,造成疏水基水化的负体积效应。

无机盐反离子因静电作用而被吸附、束缚在表面活性离子的周围。束缚作用分为定位束缚和区域束缚两类[29～33],定位束缚是指无机反离子与表面活性离子中的一个或一个以上的带电基团直接接触,作用十分强烈,彼此合为一体,以致反离子和基团各自原来的电缩水化层被破坏;区域束缚是指反离子受表面活性离子交叠电场协同作用的吸引而被束缚在表面活性离子周围的协同水化层,在此区域内尽管反离子对表面活性离子也起电性中和作用,但它与表面活性离子没有直接接触,仍然可以自由活动,它们各自的电缩水化层未遭到破坏,因此释放出的水化水要少得多。表面活性剂离子对一价反离子的束缚作用以区域束缚为主,对二价或更高价反离子主要是定位束缚,即一价反离子只能进入协同水化域,二价和更高价反离子则可进入特征水化域,对表面活性剂的去水化作用更强。但影响水化作用的因素不仅是无机反离子的荷电量,还要考虑其半径和水化半径,例如无机反离子对阴离子表面活性剂去水化能力大小顺序为:$Na^{+}<Mg^{2+}<Ca^{2+}<Ba^{2+}$。最容易去水化的基团是 $—PO_3H^{-}$,其次是 $—COO^{-}$,最难的是 $—SO_3^{-}$。在 Ca^{2+}、Sr^{2+} 和 Mg^{2+} 中,Ca^{2+} 对 $—COO^{-}$ 和 $—SO_3^{-}$ 基团的去水化作用最强,Mg^{2+} 最弱。$—COO^{-}$ 和 $—SO_3^{-}$ 基团比较,二价反离子对 $—COO^{-}$ 基的影响

较大。这是由于—SO_3H基团中有两个 S→O 的 p-dπ 配键，增强了 S 从—OH 基吸引电子的能力，使之容易解离，伴随着解离，自由能降低较多，离解产生的—SO_3^- 基团较稳定，对阳离子的吸引力较弱，阳离子较难进入—SO_3^- 的水化层。而—COOH 基团中无 p-dπ 配键，显弱酸性，解离度较小，自由能降低较少，离解产生的—COO^- 基团的稳定性较低，对阳离子的吸引力较强，阳离子容易进入其水化层，因而去水化作用较强[34]。

2.2 无机盐改变表面活性剂 Kraft 点

离子型表面活性剂在温度上升到一定值时其溶解度随温度上升而迅速增大，存在明显的转折点，此温度称作该表面活性剂的 Kraft 点（T_{Kr}）。T_{Kr} 可看作是表面活性剂晶体、胶束和表面活性剂单体三者之间呈现相平衡时的温度[35]。T_{Kr} 所对应的浓度即是表面活性剂在该温度时的 $C_{m,cr}$。无机盐对阴离子表面活性剂溶解度的影响大于其它类型，例如 Ca^{2+} 可与 SDS 形成钙盐（$(C_{12}H_{25}SO_4)_2Ca$(CDS)，SDS 的 $C_{m,cr}$ 和 T_{Kr} 分别是 8.1 mmol/L 和 9℃，而 CDS 的则分别是 1.2 mmoL 和 50℃，即 SDS 水溶液中含有钙盐时室温下表面活性剂会沉淀析出，不能形成胶束等聚集体，因而降低表面张力的能力下降，更不会具有增溶等能力。但若将 SDS 改性为阴-非表面活性剂 $C_{12}H_{25}(OCH_2CH_2)_nSO_4N$，a 则其抗盐能力大大提高，即使形成钙盐，其 T_{Kr} 也不会升高，甚至会降低。表 1 列出不同反离子的十二烷基聚氧乙烯基硫酸盐（$C_{12}H_{25}(OCH_2CH_2)_nO_SO_3$Ml/，z 为反离子的价数）的 $C_{m,cr}$、最低表面张力 γ_{min} 和 T_{Kr}。当烷基硫酸盐分子中的氧乙烯基团数不大于 6 时，表面活性剂的 T_{Kr} 随氧乙烯基团数增多而下降，$C_{12}H_{25}(OCH_2CH_2)_2OSO_3Ca_{1/2}$ 即使在 0.5 moLCaC$_2$l 溶液中，T_{Kr} 也在 0～1℃之间，而且 $C_{m,cr}$ 也随氧乙烯基团数增多而减小。这表明通过改变表面活性剂的分子结构，可同时满足提高表面活性和降低 T_{Kr} 的需要。

表 1　不同反离子 $C_{12}H_{25}(OCH_2CH_2)_nOSO_3Mz$ 的 $C_{m,cr}$、γ_{min} 和 T_{Kr}[36,37]

表面活性剂	$C_{m,cr}$/moL^{-1}(25℃)	T_{Kr}/℃	γ_{min}/mN·m^{-1}
$C_{12}H_{25}OSO_3Ca_{1/2}$	0.002 4(55℃)	50	
$C_{12}H_{25}OCH_2$-$CH_2OSO_3Ca_{1/2}$	0.000 9	15	29.2
$C_{12}H_{25}(OCH_2$-$CH_2)_2OSO_3Ca_{1/2}$	0.000 7	<0	
$C_{12}H_{25}OSO_3Mg_{1/2}$	0.001 8	25	
$C_{12}H_{25}OCH_2$-$CH_2OSO_3Mg_{1/2}$	0.001 0	<0	30.3
$C_{12}H_{25}(OCH_2$-$CH_2)_2OSO_3Mg_{1/2}$		<0	
$C_{12}H_{25}OSO_3Na$	0.008 1	9	37.2
$C_{12}H_{25}OCH_2$-CH_2OSO_3Na	0.004 2	5	37.8
$C_{12}H_{25}(OCH_2$-$CH_2)_2OSO_3Na$	0.003 0	<0	
$C_{12}H_{25}(OCH_2$-$CH_2)_3OSO_3Na$	0.002 8	<0	
$C_{12}H_{25}(OCH_2$-$CH_2)_6OSO_3Na$	0.001 6	<0	

值得注意的是,$C_mH_{2m+1}NHCO(CH_2)nOSO_3M$(缩写为 m-n-M,M=N,a$Ca_{1/2}$)系列表面活性剂,当 n=1 时其钠盐的 T_K 反 r 而比钙盐的高(表 2)[24]。如 25℃时 12-1-$Ca_{1/2}$ 的 $C_{m,cr}$=1.4 mmol/L;35℃时 12-1-Na 的 $C_{m,cr}$=5.2 mmol/L;十二烷基硫酸的钠盐在 35℃时 $C_{m,cr}$=8.3 mmol/L,而钙盐 55℃时则为 2.4 mmol/L。对比表 2 结果可知,较高的 T_{Kr} 总是与较大的 ΔH 相对应,如 SDS 的 ΔH 低于其相应的钙盐,其 T_{Kr} 亦然。水合作用可能是影响焓变的主要因素,对于 m-n-Na(n>1)系列,其水合作用比相应的钙盐强,相转变的 ΔH 较小,故 T_{Kr} 较低。当 n=1 时不能只考虑水合作用的影响。可能是由于分子几何构型的限制,Ca^{2+} 和表面活性阴离子之间不能形成较强的静电作用以使—NH 基团之间形成氢键,即此溶解过程受熵变(ΔS)影响较大。与相关的钠盐相比,m-1-$Ca_{1/2}$ 系列晶体水合状态的结构更不规则,这可能是其 T_{Kr} 较低的主要原因。

表 2 $C_mH_{2m+1}NHCO(CH_2)_nOSO_3M$(M=N,a1/2Ca)及相应表面活性剂的 $C_{m,cr}$、T_{Kr} 及溶剂化焓变(ΔH_s)[24]

表面活性剂	温度/℃	$C_{m,cr}$/m·mol/L^{-1}	T_{kr}/℃	ΔH_s/kJ·mol^{-1}
12-1-Na	32	5.20	25.7	85.7
	50	6.40		
12-1-$Ca_{1/2}$	25	1.40	<0	
	55	1.80		
14-1-Na	50	1.30	35.8	63.4
14-1-$Ca_{1/2}$			22.7	19.5
16-1-Na	50	0.34	47.1	43.3
16-1-$Ca_{1/2}$			42.3	23.4
12-3-Na	35	440	192	467
12-3-$Ca_{1/2}$			55.3	64.1
12-5-Na	50	3.10	40.9	49.5
12-5-$Ca_{1/2}$			65.5	68.8
SDS	35	8.30	9.0	50.0
CDS	55	2.40	50.0	75.0
n-$(CH_2)_{12}OCH_2SO_3Na$	35	4.80	5.0	39.3
n-$(CH_2)_{12}OCH_2SO_3Ca_{1/2}$	25	0.92	15.0	89.4

2.3 无机盐改变表面活性剂分子的临界堆积参数

无机盐可改变表面活性剂分子的堆积参数 P,从而改变表面活性剂体系的微观结构。P 可以用式(7)表示[38]:

$$P=V_C/lA_0 \tag{7}$$

其中,V_c 为表面活性剂疏水尾基体积,A_0 为表面活性剂极性头基所占最小面积,为表面活性剂胶束中尾链最大伸展长度。根据 P 值可以预测形成聚集体的大体形状：P≤1/3 时一般形成球状胶束；1/3≤P≤1/2 时则形成棒状、柱状胶束；1/2≤P≤1 时形成层状相；P≥1 时易形成反胶束。

当体系中不含无机盐且表面活性剂浓度略大于 $C_{m,cr}$ 时，胶束一般以球状结构存在，表面活性剂浓度大于 10 倍 $C_{m,cr}$ 时，胶束一般以不对称状结构存在。当体系中存在无机盐时即使表面活性剂浓度不大于 10 倍 $C_{m,c}$ 也会形成不对称状胶束[39]，这时反离子、表面活性离子与胶束发生强烈的静电相互作用，压缩表面活性剂分子极性头的水化层和胶束的双电层，使表面活性离子极性头在胶束表面所占的面积减小，胶束体积剧增。由于无机阳离子的尺寸相对于表面活性离子的极性头较小，一个表面活性剂分子极性头周围可能吸附几个无机阳离子，因此要考虑无机阳离子的荷电数。对于一价表面活性阴离子，等摩尔的阳离子甚至可以使胶束净电荷为零。

低，就可能形成更大的聚集体，进而从溶液中析出，产生相分离。二价反离子则可与一价表面活性阴离子形成特征离子对，导致表面活性剂分子的堆积参数 P 显著改变，从而影响表面活性剂在界面和体相中的聚集行为。

2.4 无机盐改变表面活性剂在油、水相的分配系数

三次采油过程中，表面活性剂必定与油类物质接触。实践已经证明微乳液与油的界面张力可达超低水平，因而其驱油效率是目前所有驱油体系中最高的。要获得超低界面张力必须考虑三个因素：界面上表面活性剂的浓度、表面电荷密度和表面活性剂在油水相的分配系数 K_{wo}。表面活性剂可使油水界面张力从 40 mN/m 降至约 1 mN/m；若要进一步降低界面张力，必须改变界面电荷密度和调节表面活性剂在油水相中的 K_{wo}。

无机盐在微乳液形成过程中所起作用还依赖于表面活性剂的浓度。当表面活性剂浓度和无机盐含量均低时，表面活性剂在水中形成胶束，少量分布在油相；高盐度条件下表面活性剂在油相分布较多，在水中分布较少，形成反胶束。固定表面活性剂浓度时，随含盐量增加表面活性剂从水相进入油相，在特定含盐度时 $K_{wo}=1$，此时可形成超低界面张力体系。但对于较高浓度表面活性剂的油/水体系，增加无机盐的含量，微乳液可从 WinsorⅠ(WinsorⅢ(WinsorⅠ(图 4)。当达到最佳中相微乳液时，其剩余油相体积等于剩余水相体积。当在油相和水相的表面活性剂 $K_{wo}=1$ 时，则称此无机盐浓度为最佳盐度，此时中相微乳液与油相的界面张力和中相微乳液与水相的界面张力均可达超低水平。用此微乳液作为驱替液，可以获得最高的驱油效率。

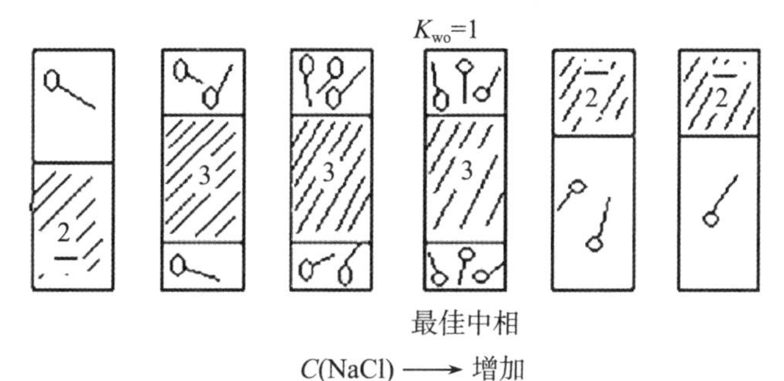

图 4 NaCl 浓度对表面活性剂浓体系相态的影响示意图[40] 2—下相或上相微乳液(WinsorⅠ或Ⅱ)；3—中相微乳液(WinsorⅢ)

通过以上讨论可以得到如下结论。

(1) 少量无机盐的存在有利于提高表面活性剂的界面活性,但无机盐含量高时可能导致表面活性胶束体积增大,表面荷电量减小,静电排斥作用降剂从溶液中沉淀析出,从而丧失降低界面张力和形成胶束的能力。相对而言,磺酸盐表面活性剂抗钙、镁离子的能力好于羧酸盐。

(2) 无机盐对驱油用阴离子表面活性剂的性能影响包括:减弱静电斥力,破坏水化层,甚至形成离子对,导致表面活性剂疏水性增强,进而改变其 Kraft 点、临界堆积参数和在油水相的分配系数。

(3) 不同类型表面活性剂复配体系的 $C_{m,cr}$ 降低,阴/非离子表面活性剂复配体系的抗盐性好于阴离子表面活性剂,一方面钙、镁离子形成离子对所需的单体浓度减小,离子对形成趋势减弱,另一方面形成的胶束对钙、镁盐沉淀具有增溶作用。

烷基硫酸盐或烷基磺酸盐分子中引入氧乙烯基,形成的烷基聚氧乙烯基醚硫酸或磺酸盐表面活性剂不仅 $C_{m,cr}$ 和 T_{Kr} 显著降低,而且其抗钙、镁离子的能力显著提高。

3 展望

关于无机盐与表面活性剂作用的微观机理,已有大量的探索性研究。Poison-Boltzman (PB)理论曾描述靠近带电界面的点电荷分布[41],但用于解释反离子在带电界面附近的分布时,有时与实验结果相矛盾。Biesheuvel 等[42]认为,当一个由阴离子表面活性剂组成的单分子层和含有不同体积、不同电荷数离子的溶液相接触时,体积较小的离子会与体积较大的离子在界面处竞争吸附。例如,从 5 mmol/L/Mg^{2+}(水化半径 0.43 nm)和 5/mmol/l L+i 的溶液中竞争吸附时,L_1^+ 竞争不过 Mg^{2+},扩散层中出现的是大量的 Mg^{2+} 而不是 L_1^+,这与 PB 理论相一致。如果把 L_1^+ 换成体积更小的 C^{S+},则结果不同。当电荷密度最高时(600 mC/m^2),靠近界面区域的二价离子被一价离子(C^{S+})所取代,在扩散层区域只包含一价离子,这一实验结果与 PB 理论相一致[43]。在考察不同的无机盐对带电界面的影响时,不但要考虑无机离子本身的性质,还要考察表面电荷密度[44]。近年来,计算机模拟已广泛用于表面活性剂体系的研究[45~49]。我们用分子动力学模拟方法,通过考察径向分布函数、均方根位移等参数的变化,研究了 NaCl、$CaCl_2$ 和 Na_2SO_4 对 AOT/异辛烷/水体系性能的影响[50]。结果表明 $CaCl_2$ 对 AOT 在界面上的聚集影响最大,使 Na^+ 与 O(H_2O 中氧原子)的径向分布函数峰值发生的变化最大;而 Na_2SO_4 由于离子数较多,使离子强度增大,对聚集的影响也较大,使 N^{a+} 与 O 的径向分布函数峰值有一定增大;NaCl 对 AOT 的影响最小。这几种无机盐都可使 AOT 中 S 原子均方根位移增大,这说明无机盐破坏了亲水头基的水化膜,使得 S 容易移动。最近,我们利用分子动力学方法对比研究了无机盐对直链和支链十六烷基苯磺酸钠(1C16、5C16)的表面聚集行为的影响,发现如下规律:Ca^{2+} 的影响大于 Mg^{2+};二价离子对支链型 5C16 的影响较小,即支链烷基苯磺酸盐的抗钙镁能力好于直链者。这些研究表明,计算机模拟技术可以提供一些微观信息,为解释实验现象提供理论支撑。

参考文献

[1] 赵福麟. 采油用剂 [M] . 东营:石油大学出版社，1997:14 -19.

[2] 彭朴. 采油用表面活性剂[M] . 北京:化学工业出版社，2003:19-22.

[3] 朱友益,沈平平. 三次采油复合驱用表面活性剂:合成、性能及应用[M] . 北京:石油工业出版社,2002:9 -18.

[4] 王业飞,赵福麟. 非离子 -阴离子型表面活性剂的耐盐性能[J]. 油田化学,1999,l6:336-340.

[5] 杜娟,曾宪诚. 金属盐作用下的表面活性剂临界胶束浓度的变化研究[J]. 四川大学学报（自然科学版),2002,39:721-724.

[6] 肖进新,肖寒,蓝亭,等. SDS 及其与十二烷基三乙基溴化铵混合体系在矿化水中的表面活性[J]. 化学学报,2004,62:351 -354.

[7] 唐善彧,杨承志. 用螯合剂抑制表面活性剂驱油过程中石油磺酸盐的沉淀损失[J]. 油田化学,1989,3:237-242.

[8] Dushkin C D, StoichevTL, HorozovT S, etal. Dynamics of foams of ethoxylated ionic surfactant in the presence of micelles and multivalentions [J]. ColoidPolym Sc,i 2003,281:130-142.

[9] Carlson, I Edlund H, Person G, eta. l Competition between monovalent and divalent counter ions in surfactant systems [J]. J Coloid Interface Sc,i 1996,180:598-604.

[10] 陈珍珍,蒋宝源. 用于三次采油的烷基聚氧乙基硫酸盐耐盐表面活性剂性能的研究[J]. 油田化学,1988,5:285-289.

[11] Person CM, Jonson A P, Bergstm M, etal. Testing the Gouy-Chapman theory by means of surface tension measurements for SDS-NaCl Omixtures [J]. JColloid InterfaceSc,i 200,267:151-154.

[12] VakarelskiIU, DushkinC D. Efectof the counter ions on the surface properties of surfactant solutions kinetics of the surface tension and surface potential [J]. Coloids SurfA, 2000,163:177-190.

[13] SantosFK G, NetoE LB, MouraM CPA, etal. Molecular behavior of ionic and nonionic surfactants in saline medium [J]. Coloids SurfA, 2009,333:156-162.

[14] JusufiA, HyninenA P, Hataja M, etal. Electrostatic screening and charge correlation effects in micelization of ionic surfactants [J]. JPhysChemB, 2009,113:6314-6320.

[15] RenoncourtA, VlachyN, Bauduin P, etal. Specific alkali cation effects in the transition from micelles to vesicles through saltaddition [J]. Langmur 2007,23:2376-2381.

[16] MuJH, LiG. Z, JiaX L, etal. Rheological properties and microstructures of anionic micelar solutions in the presence of different in organic salts [J]. JPhysChemB, 2002,106:1 1685-11693.

[17] SantosFKG, NetoELB, MouraM C P A, etal. Molecular behavior of ionic and non ionic surfactants in saline medium [J]. ColoidsSurfA, 2009,333:156-162.

[18] KoelschP, MotschmanH. Varying the counterionsata charged interface [J]. Langmuir 2005,21:3436-3442.

[19] A largovaR, Petkov,J PetsevD, etal. Light scatering study of sodium dodecyl polyoxyethylene-2-sulfonate micelles in the presence of multivalent counter ions [J]. Langmuir 1995,11: 1530-1536.

[20] MyersD. Surfactan Science and Technology, 2ndEd [M]. New York: VCHPublishersIn,c1992:8 1-126.

[21] 赵国玺，朱步瑶.表面活性剂作用原理[M].北京:中国轻工业出版社,2003:263-270.

[22] LuanYuxia, XuGuiyin,g Yuan Shilin,g . Comparative studies of structuraly smiilar surfactant: s sodium bis (2-ethylhexyl) phosphate and sodium bis (2-ethylhexyl) sulfosucinate [J]. Langmur 2002,18:8700-8705.

[23] Gong Houjian, Xin Xia, Xu Guiying, eta. l The dynamic interfacia ltension betwen HPAM/C17 H33 COONa mixed solution and crude oil in the presence of sodium halide [J]. ColoidsSurf A, 2008,317:522 -527.

[24] ZapfA, Beck R, HofmanH. Calcium surfactants: a review [J]. AdvColoidInterfaceSc,i 2003,100-102:349-380.

[25] 宋爱新.新颖有序聚集体结构、性质及其模板效应研究[D].兰州:中国科学院博士学位论文,2005.

[26] SomasundaranP, AnanthapadmanabhanKP, CelikM S. Precpitation-redisolution phenomena in sulfonate-aluminum chloride solutions [J]. Langmuir 1988, 4:1061-1063.

[27] WanlesE ,J DuckerW A. Weak influence of divalentions on anionic surfactant surface-aggregation [J]. Langmuir 1997,13:1463-1474.

[28] 李卓美.高分子泥浆降失水剂的分子结构与其耐盐性能的关系,[J].油田化学，1986,3:103-113.

[29] StrausUP, LeungY P. Volume Changes as a Criterion for Site Binding of Counter ions by Polyelectrolytes [J]. JAmChem So 196 87:1476-1480.

[30] StrausUP, AnderP. Molecular dimendions and interactions of lithium polyphosphate in aqueous lithium bromide solutions [J]. JPhysChem, 1962,66:2235-2239.

[31] StrausUP, SiegelA. Counterion binding by polyelectrolytes VI the binding of magnesium ion by polyphosphates in aqueous electrolyte solutions [J]. JPhysChem, 1963,67:2683-2687.

[32] Lapanje S, RiceSA. On the ionization of polystyrene sulfonic acid [J]. JAmChem So 1961,83:496-497.

[33] KotinL, NagasawaM. A Study of the ionization of polystyrene sulfonic acid by proton magnetic resonance [J]. JAm Chem So,c 196 83:1026-1028.

[34] 李卓美,梁国眉,张雪馨,等.高分子泥浆处理剂耐盐性的研究(盐对处理剂的水化性和溶解性的影响[J].油田化学,1982:171-176.

[35] MoroiY. M iceles theoreticaland aplied aspects [M]. New York:PlenumPres, s 1992:113-128.

[36] ShinodaK, H iralT. Ionic surfactants applicable in the presence of multivalent cation. s physicochemical properties [J]. J Phys Chem,1977,81:1842-1845.

[37] HatoM, ShinodaK. Kraftpoints of calcium and sodium dodecyl poly-(oxyethylene) sulfates and their mixtures [J]. J Phys Chem,1973,77:378-381.

[38] HiemenzPC, RajagopalanR. PrinciplesofColoidand Surface Chemistry3rd ED [M]. NewYork:MarcelDekerInc,1997:367-372.

[39] Lucasen-ReyndersEH. Anionic surfactants Physical Chemistry of Surfactant Action [M]. New York:Marcel DekerInc,1981:57 -85.

[40] 李干佐,郑利强,徐桂英.石油开采中的胶体化学[M].北京:化学工业出版社,2008:248-251.

[41] BuW, Vaknin D, TravesetA. How acurate isPoison-Boltzman theory for monovalent ions near highly charged interfaces [J]. Langmur 2006,22:5673-5681.

[42] BiesheuvelPM, vanSoestbergenM. Counterion volume effects in mixed electrical double layers [J]. JColoid Interface Sc,2007,316:490-499.

[43] ShapovalovVL, BrezesinskiG. Break down of the Gouy-Chapman model for highly charged Langmuir monolayers: counter on size effect [J]. JPhysChemB, 2006, 110:10032-10040.

[44] KrischM ,J DA,uriaR, BrownM A. The efectofan organic surfactanton the liquid-vapor interface of an electrolyte solution [J]. JPhysChemC, 2007, 111: 13497-13509.

[45] BandyopadhyayS, Chanda J. Monolayer of monododecyl diethylene glycol surfactants adsorbed at the air/water interfac: e amolecular dynamics study [J]. Langmur 20019:10443-10448.

[46] Chanda ,J Bandyopadhyay S. Molecular dynamics study of a surfactant

monolaye radsorbed at the wate rinterface [J]. J ChemTheoryCompu, t 2005, 1: 963-971.

[47] KhuranaE, N ielsenSO, KleinM L. Geminisurfactantsatthe air/water interface: A fully atomistic molecular dynamics study [J]. JPhysChemB, 2006, 10: 22136-22142.

[48] PetrovM, M inofarB, VrbkaL. Aqueous ionic and com-plementary zwiterionic soluble surfactant: Molecular dynamics smiulations and sum frequency generation spectroscopy of the surfaces [J]. Langmur 2006, 22: 2498-2505.

[49] PoghosyanA H, ArsenyanL H, Gharabekyan H H. Molecular dynamics study of poly (dialyldmiethylammonium chloride)/sodium dodecyl sulfate/ decanowatersystems [J]. J PhysChem B, 2009, 113: 1303-1310.

[50] Chen Yijian, Xu Guiyin, g Yuan Shiling. Molecular dynamics smiulations of AOT at iso octane/water interface [J]. ColoidsSurfA, 2006, 273: 174-178.

Investigation of salts tolerance of anionic surfactants in aqueous solutions

Zhao Taota Gong Houjian[1] Xu Guiying[1]
Cao Xulong[2] Song Xinwang[2] Wang Hongyan[2]

(1. Key Laboratory of Coloid and Interface Chemistry of Education Ministry in Shandong Universi Jinan, Shangdong250100, PR China;
2. Gelogical Scientific Research Institue Shengli Oilfield Branch Compan,y Sinopec, Dongying Shandong 25701 PR of China)

Abstract: Aimed at unsatisfactory tolerance to inorganic salts of EOR surfactant sandbased on 45 periodical papers 5 monographs, and the results of authors, recent researches, the mechanisms involved in salts tolerance of anionic surfactants (ASs) in aqueous solutions were reviewed and covered following topics: introductory part; effects of in organic salts on aggregation behavior of AS surfactant/salt interactions leading to changes in Kraft point critical packing coefficient and oil/water phase partition coefficient of ASPBT concept in use and computer simulation.

Keywords: s anionic surfactant; aqueous solution; physicochemical properties; tolerance to salt; ssalt/surfactant interaction; functioning mechanisms; surfactants for EOR; review

常规和亲油性石油磺酸盐的组成及界面活性研究

王　帅[1]　王旭生[1]　曹绪龙[2]　宋新旺[2]　刘　霞[1]　蒋生祥[1]

(1. 中国科学院兰州化学物理研究所甘肃省天然药物重点实验室，甘肃兰州 730000；
2. 中国石油化工股份有限公司胜利油田有限公司地质科学研究院，山东东营 257015)

摘要：采用液-液萃取法从胜利油田用石油磺酸盐原样品中分离纯化出常规和亲油性石油磺酸盐样品，对该样品进行了液相色谱分析、电喷雾质谱分析和亲水亲油平衡值测定，并采用旋转滴法考察了它们降低油/水界面张力的能力。结果表明，亲油性较强的石油磺酸盐与常规石油磺酸盐的组成不同，其主要以单磺酸盐的形式存在，而常规石油磺酸盐中单、双磺酸盐含量相当。在界面活性方面，亲油性石油磺酸盐降低界面张力的能力远远低于常规石油磺酸盐，而且其单元体系和二元复配体系均不能达到超低界面张力；而常规石油磺酸盐复配体系可以将界面张力降至 10^{-4} mN/m。同时对石油磺酸盐的组成结构与界面活性之间的相互关系进行了初步探讨。

关键词：石油磺酸盐；界面张力；色谱；质谱

中图分类号：T E357.4　**文献标识码**：A

Composition and interfacial activities of regular and hydrophobic petroleum sulfonates

Wang Shuai[1]　Wang Xusheng[1]　Cao Xulong[2]　Song Xinwang[2]
Liu Xia[1]　Jiang Shengxiang[1]

(1. Key Laboratory f or Natural Medicine of Gansu Province, Lanzhou Institute of Chemical Physics, Chinese Academy of Sciences, Lanzhou Gansu 730000, P. R. China;
2. Geological and Scientific Research Institute of Shengli Oilfield, Sinopec, Dongying Shandong 257015, P. R. China)

Received 10 November 2009; revised 10 March 2010; accepted 19 March 2010

Abstract: Regular and hydro phobic petroleum sulfonates were extracted and purified from the crude samples of petroleum sulfonates used in shengli oilfield by liquid-liquid extraction method. Analysis of these substances using liquid chromatography, mass spectrometry and determination of hydrophile-lipophile balance values were completed . Meanwhile, interfacial tension value s of oil/water system respectively consisted of regular and hydrophobic sulfonates were measured by rotated dropping method. The results show that hydrophobic sulfoantes are different from regular sulfonates in composition, and it mainly consists of mono sulfonates, while the contents of mono and disulfoantes in regular sulfonates are similar. Meanwhile its inter facial activity is much low er than that of regular sulfonates, and it can not obtain ultralow interfacial tension values, not only for unit but also for mixed systems . While mixed systems of regular sulfoantes can decrease interfacial tension values to 10-4 mN/m. Besides, the relationship between the structure of petroleum sulfonates and their interfacial activities was discussed.

Keywords: Petroleum sulfonates; Interfacial tension; Chromatography; Mass spectrometry Corresponding author. Tel.: +86-931-4968206; fax: +86-931-8277088; e-mail: ws777879@163.com

石油磺酸盐是油田中常用的一种阴离子表面活性剂，性能优良的石油磺酸盐具有较高的油水界面活性，通过将原油乳化，降低油水界面张力，增加岩石表面的润湿性，提高原油流动性，来达到驱油目的[1]。由于石油磺酸盐是由石油馏分经磺化而成，其组成和结构随原料油和磺化工艺的不同而差异较大，这种差异直接关联着不同油田和不同采油区的驱油性能[2]。在进行驱油配方研究时，为了更准确地了解石油磺酸盐的驱油性能，往往需要对石油磺酸盐产品进行提纯处理，除去未磺化油和无机盐等组分[3-4]。目前对石油磺酸盐性能的研究工作[5-11]，主要集中在纯化后的石油磺酸盐部分，或未进行分离纯化的石油磺酸盐整体部分。由于石油磺酸盐油溶性的差异，在分离纯化的过程中，产品中亲油性较强的一部分磺酸盐会随着未磺化油同时除去，而对油溶性较强的这部分石油磺酸盐，未见有关研究报道。这部分磺酸盐的具体组成、降低油/水界面张力的能力等信息均未知，在样品处理过程中与未磺化油一起去除是否合理，有待于进一步的考察。因此了解和研究未磺化油中亲油性较强的石油磺酸盐的组成及其界面活性是非常必要的。

本文针对胜利油田二元复合驱油体系的石油磺酸盐产品中的亲油性石油磺酸盐，进行分离提取、分析检测及界面活性方面的研究，分析比较常规和亲油性石油磺酸盐的组成差异及与界面活性之间的构效关系，为生产石油磺酸盐过程中石油馏分的筛选和磺化工艺的改进提供科学信息。

1 实验部分

1.1 仪器与试剂

Agilent 1100 型高效液相色谱仪，配二极管阵列检测器，美国 Agilent 公司；Agilent 1100 M SD Trap 质谱仪，美国 Agilent 公司；TX 500C 型全量程视频动态界面张力仪，北京盛维基业科技有限公司。

试剂：无水乙醇、无水甲醇、异丙醇、正己烷、磷酸二氢钠、氢氧化钠、碳酸钠(分析纯，天津百世化工有限公司)；蒸馏水。

1.2 亲油性石油磺酸盐的提取分离

本文采取液-液萃取法分离纯化石油磺酸盐[3]。首先除去水及易挥发组分、无机盐和未磺化油，得到常规的石油磺酸盐样品，标记为 PS-a。然后对于未磺化油部分，采用体积比为 1∶1 的异丙醇/水萃取，获得亲油性石油磺酸盐样品，标记为 PS-b。

1.3 色谱分析

对于样品 PS-a 和 PS-b，采用液相色谱法对其进行分离分析[12-13]，考察其所含单、双石油磺酸盐的含量差异，色谱分析条件如下：色谱柱，强阴离子 SAX 色谱柱(5 μm，50 mm×4.6 mm I.D.)；流动相，A 甲醇/水(体积比 60∶40)，B 甲醇/磷酸二氢钠(体积比 60∶40)，0～1 min，100%A；1～6 min，50%A；检测波长，280 nm，流速 1.0 mL/min。

1.4 质谱分析

采用电喷雾质谱法，考察样品 PS-a 和 PS-b 所含主要成份的结构信息。质谱分析条件如下：雾化气 10 psi，干燥气 7 L/min，温度 325℃，离子阱负离子检测模式，相对分子质量扫描范围 50～1 000 Da。

1.5 亲水亲油平衡值 HLB 测定

亲水亲油平衡值可以衡量表面活性剂的亲水性或亲油性大小，本文采用乳化法分别对样品 PS-a 和 PS-b 进行 HLB 值测定[14]。实验中，选用煤油为基准油，Tween80 作为标准活性剂，按照下式(1)计算：

$$\mathrm{HLB}=14-b/a \tag{1}$$

其中，b/a 为待测样品溶液和 Tween80 标准溶液的体积比。

1.6 界面张力测定

采用旋转滴法测定所配溶液的界面张力。分别考察了样品 PS-a 和 PS-b 单元体系，以及分别与聚合物(聚丙烯酰胺 PAM，胜利油田提供，相对分子质量 1 700 万，水解度 25%)和无机碱(NaOH、Na_2CO_3)组成的二元体系的界面张力。测定条件如下：温度 70℃，转速 5 000 r/min，矿化度 20 000 mg/L，模拟油为胜利油田胜坨原油。

2 结果与讨论

2.1 色谱分析

样品 PS-a 和 PS-b 的液相色谱分析结果如图 1 所示。

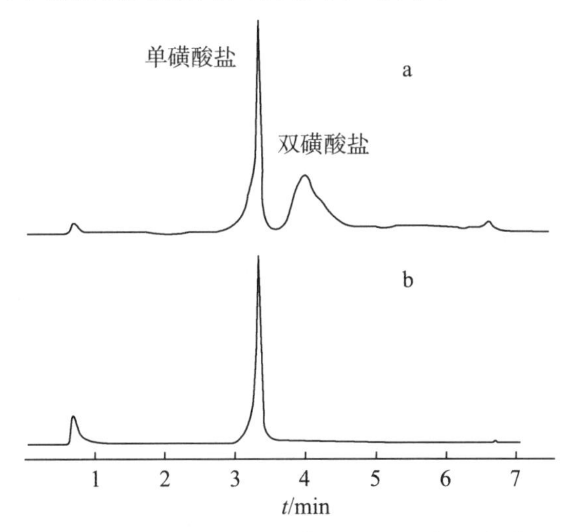

Fig. 1 Chromatograms of PS-a (a)and PS-b (b)

图 1 PS-a(a)和 PS-b(b)色谱分离

从色谱图及色谱数据分析结果得知:PS-a 和 PS-b 均含有单磺酸盐和双磺酸盐,但两者所含单、双磺酸盐的量差别较大;如果以峰面积为评价标准,样品 PS-b 中几乎不含双磺酸盐部分,而样品 PS-a 中单、双磺酸盐所含比例相当,约为 1∶1。

2.2 结构解析

在确定的质谱分析条件下,样品 PS-a 和 PS-b 的质谱图如图 2 所示。

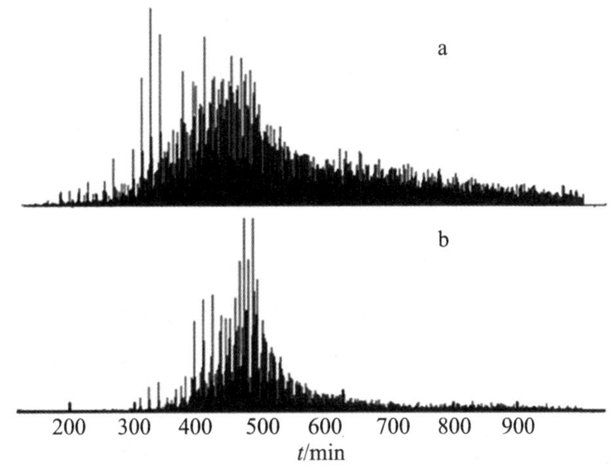

Fig. 2 MS spectrums of PS-a (a)and PS-b (b)

图 2 PS-a(a)和 PS-b(b)质谱图

从质谱图中可以初步得出:PS-b 的相对分子质量分布范围较窄,且丰度较大的分子离子峰较为集中,聚集在 487 min 附近;而 PS-a 的主要分子离子峰较为分散,聚集在 320、459 min 附近。为了更加具体的对两者组成的结构差异进行分析比较,选出各自质谱图中相对丰度最大的 20 个分子离子峰,对其进行结构鉴别,对应的化合物结构信息如图 3 所示。

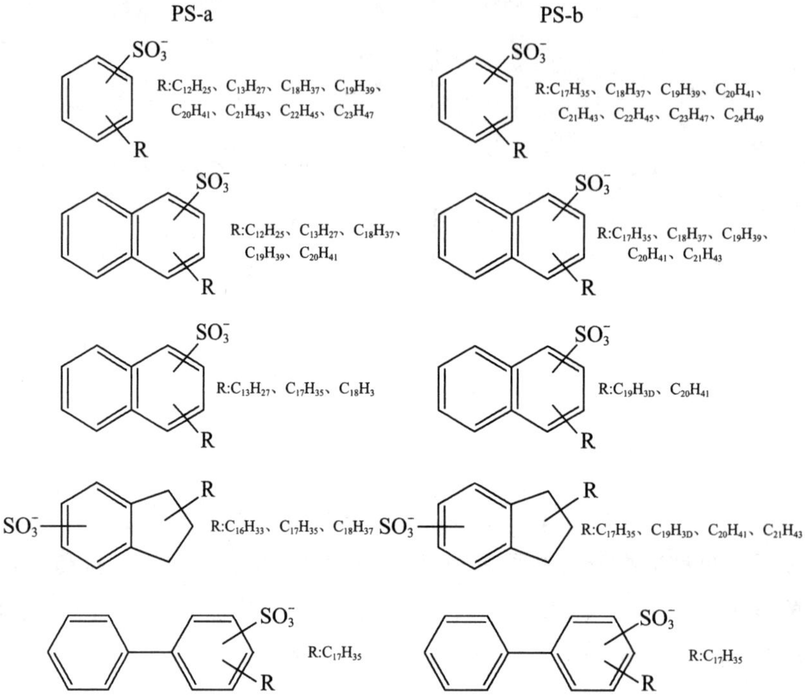

PS-a

SO_3^- R: $C_{12}H_{25}$、$C_{13}H_{27}$、$C_{18}H_{37}$、$C_{19}H_{39}$、$C_{20}H_{41}$、$C_{21}H_{43}$、$C_{22}H_{45}$、$C_{23}H_{47}$

SO_3^- R: $C_{12}H_{25}$、$C_{13}H_{27}$、$C_{18}H_{37}$、$C_{19}H_{39}$、$C_{20}H_{41}$

SO_3^- R: $C_{13}H_{27}$、$C_{17}H_{35}$、$C_{18}H_{3}$

SO_3^- R: $C_{16}H_{33}$、$C_{17}H_{35}$、$C_{18}H_{37}$

SO_3^- R: $C_{17}H_{35}$

PS-b

SO_3^- R: $C_{17}H_{35}$、$C_{18}H_{37}$、$C_{19}H_{39}$、$C_{20}H_{41}$、$C_{21}H_{43}$、$C_{22}H_{45}$、$C_{23}H_{47}$、$C_{24}H_{49}$

SO_3^- R: $C_{17}H_{35}$、$C_{18}H_{37}$、$C_{19}H_{39}$、$C_{20}H_{41}$、$C_{21}H_{43}$

SO_3^- R: $C_{19}H_{3D}$、$C_{20}H_{41}$

SO_3^- R: $C_{17}H_{35}$、$C_{19}H_{3D}$、$C_{20}H_{41}$、$C_{21}H_{43}$

SO_3^- R: $C_{17}H_{35}$

Fig. 3 Molecular structure of main compounds in PS-a and in PS-b

图 3 PS-a 和 PS-b 中主要化合物的分子式结构

对于 PS-b,相对含量较大的 20 个化合物中,按照相对含量大小依次为:烷基萘磺酸盐(C_{17}~C_{21})、烷基苯磺酸盐(C_{17}~C_{24})、烷基茚满磺酸盐(C_{17}~C_{21})、烷基四氢萘磺酸盐(C_{19}~C_{20});而对于 PS-a,按照相对含量大小依次为:烷基苯磺酸盐(C_{11}~C_{13}、C_{18}~C_{23})、烷基萘磺酸盐(C_{12}、C_{18}~C_{20})、烷基茚满磺酸盐(C_{16}~C_{18})、烷基四氢萘磺酸盐(C_{13}、C_{17}~C_{18})。

结果表明:样品 PS-b 中,萘环结构占主要,苯环结构次之,另外支链碳数分布较为集中,如萘环支链碳数分布在 17~21;样品 PS-a 中,苯环结构占主要,萘环结构次之,支链碳数分布较为分散,如苯环支链碳数除分布在 18~23,还分布在 11~13。

2.3 HLB 值测定

在不同体积比的待测/标准表面活性剂配比下,分别测得 PS-a 和 PS-b 的乳化层体积如图 4 所示。

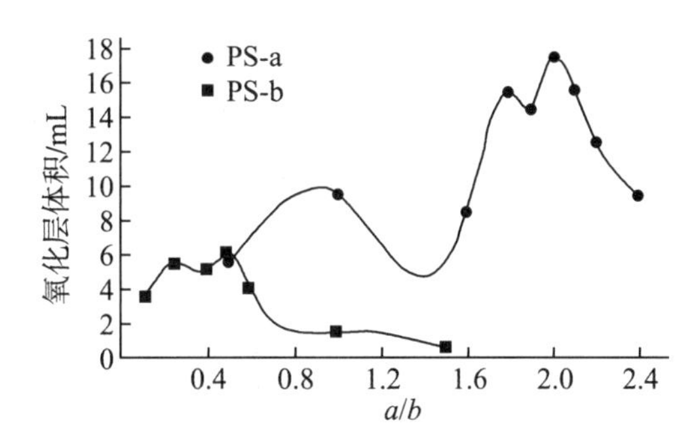

Fig. 4 Experimental results of HLB values of PS-a and PS-b

图 4 PS-a 和 PS-b 的 HLB 值实验结果

根据实验结果，按照乳化层体积最大时的待测/标准表面活性剂配比，计算出 PS-a 和 PS-b 的 HLB 值分别为 13.5 和 12.0。PS-b 的 HLB 值小于 PS-a 的 HLB 值，表明 PS-b 的亲水性比 PS-a 的亲水性小。根据 Davis 法计算 HLB 值的原理[15]，可知：

$$HLB = 7 + \sum(\text{亲水基团常数}) - \sum(\text{疏水基团常数})$$

对于亲水基团数相同的 PS-a 和 PS-b 而言，HLB 值越小，Σ(疏水基团常数)就越大，即疏水基团碳链就越长(参照常见基团亲水及疏水基团常数表[14])。因此，可以推测：PS-b 所含主要组份的相对分子质量较 PS-a 的大，这种结果与质谱分析结果相吻合。另外，从图 4 中可以得知：PS-b 的乳化效果明显低于 PS-a，因此推测，在界面活性方面，PS-b 不如 PS-a 的效果好。

2.4 界面活性

对于 PS-a 和 PS-b 单元体系，测得其界面张力分别为 2.5×10^{-1} mN/m 和 NaN(代表油滴附于旋转壁上，不能有效拉展)。因此根据此实验结果，可知 PS-b 具有非常弱或无界面活性。

对于石油磺酸盐-聚合物复配体系，界面张力测定结果如图 5 所示(实验中 PAM 质量分数分别为 0.1%、0.2%、0.3%，不同浓度下测试结果一致，且实际数值大于此设定值)。

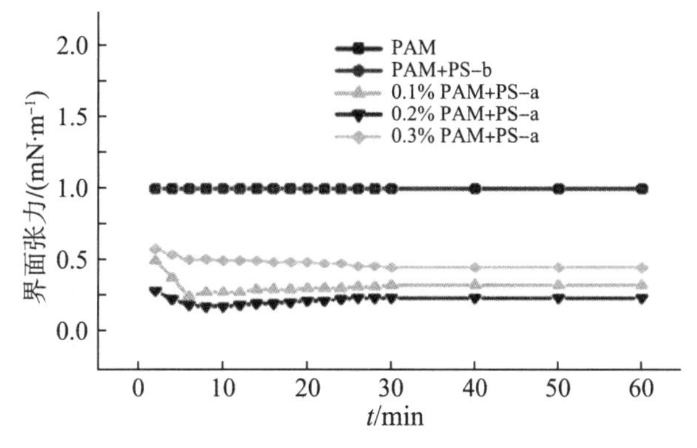

Fig. 5 Interfacial tension values of PS-PAM system

图 5 PS-PA M 二元体系的界面张力

由图 5 结果可知，PAM 单元体系以及 PAM-PS-b 二元体系，均不能降低原油的界面张力；为了便于在图形中对比分析，将其界面张力设定为 1.0mN/m 而 PAM-PS-a 二元体系可以将原油的界面张力维持在 10^{-1} 数量级，数值与 PS-a 单元体系的界面张力数值相近，因此可以得出：无论是 PS 单元体系，还是 PAM-PS 复配体系，PS-b 均不能降低原油的界面张力。

对于石油磺酸盐-无机碱复配体系，界面张力测定结果如图 6 所示。

由图 6 可以看出：PS-b 与 NaOH 复配，原油的界面张力只可以降到 10^{-2} 数量级，而 PS-a 与 NaOH 复配，可以将原油的界面张力降至 10^{-4} 数量级；PS-b 与 Na_2CO_3 复配，原油的界面张力可以降到 10^{-2} 数量级；PS-a 与 Na_2CO_3 复配，可以将原油的界面张力降至 10^{-3} 数量级。综上分析可以得出：PS-b 的界面活性远远低于 PS-a；PS-a 的复配体系可以达到超低界面张力(10^{-3}～10^{-4} mN/m)，而 PS-b 不能。

综上所述，通过对比分析常规和亲油性石油磺酸盐的组成及界面活性，可以得出：亲油性石油磺酸盐中主要化合物的相对分子质量较常规石油磺酸盐的大，且相对分子质量分布范围较窄；其降低油/水界面张力的能力很弱，不能达到超低界面张力的要求。另外，根据样品结构解析及界面活性测试结果，在降低界面张力方面，可以推测(针对胜利油田区块)：苯环结构磺酸盐占主体的样品较萘环结构占主体的样品要好；苯环支链碳数分布要合适，既含有较低碳数如 C_{11}～C_{13}，又含有较高碳数如 C_{18}～C_{23} 的样品效果较好；样品中即含有较高比例单磺酸盐、又含有较高比例双磺酸盐时，降低界面张力效果较好。

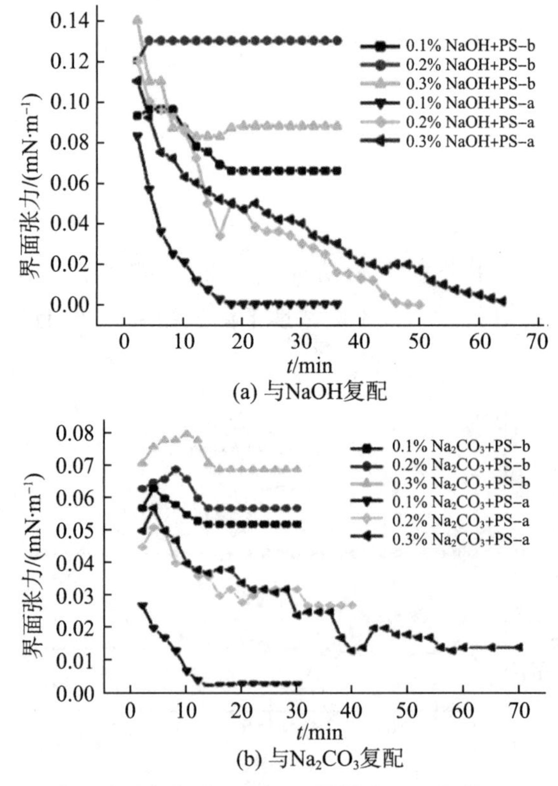

Fig. 6 Interfacial tension values of PS-inorganic base system

图 6 PS-无机碱体系的界面张力

参考文献

[1] 韩冬,沈平平.表面活性剂驱油原理及应用[M].北京:石油工业出版社,2001.

[2] 赵国玺,朱涉瑶.表面活性剂作用原理[M].北京:中国轻工业出版社,2003.

[3] 蒋怀远,饶福焕,蒋宝源.驱油用石油磺酸盐成份分析[J].油田化学,1985,2(1):75-82 .

[4] 程斌,张志军,梁成浩.三次采油用石油磺酸盐的组成和结构分析[J].精细石油化工进展,2004,5(6):14-19 .

[5] 马涛,张晓凤,邵红云,等.驱油用表面活性剂的研究进展[J].精细石油化工,2008,25(4):78-82 .

[6] 陈卫民,陈丽文.两种石油磺酸盐与不同原油间界面张力的探讨[J].日用化学品科学,2008,31(11):27-30.

[7] 王慧云,刘爱芹,温新民.影响油田采出水界面电性质的因素[J].中国石油大学学报:自然科学版,2008,32(3):143-147 .

[8] 王凤清,王玉斗,吴应湘,等.驱油用石油磺酸盐的合成与性能评价[J].中国石油大学学报:自然科学版,2008,32(2):138-141.

[9] 王玉梅.新型石油磺酸盐性能研究[J].油气田地面工程,2009,21(8):21-22 .

[10] 段友智,李阳,范维玉,等.石油磺酸盐组成及其油水体 系界面 张力的关系研究[J].石油化工高等学校学报,2009,22(3):46-50.

[11] 申娜,孙菱翎,刘洁,等.润滑油抽出油制取石油磺酸钠的研究[J].辽宁石油化工大学学报,2008,28(2):22-25.

[12] 蒋生祥,陈立仁,赵明方.多维高效液相色谱法分析石油中的微量石油磺酸盐[J].分析化学,1992,20(6):677-679.

[13] 赵亮,曹绪龙,王红艳,等.离子交换色谱测定原油中的单双石油磺酸盐[J].分析测试学报,2007,26(4):496-499.

[14] 周家华,崔英德.表面活性剂 HLB 值的分析测定与计算 Ⅰ :HLB 值的分析测定[J].精细石油化工,2001,3(2):11-14 .

[15] 周家华,崔英德,吴雅红.表面活性剂 HLB 值的分析测定与计算 Ⅱ :HLB 值的计算[J].精细石油化工,2001,7(4):38-41 .

(Ed .:YY L ,Z)

磺酸盐与非离子表面活性剂协同作用的研究*

王立成[1,2] 曹绪龙[3] 宋新旺[3] 刘淑娟[1] 刘 霞[1] 蒋生祥[1]

(1. 中国科学院兰州化学物理研究所,甘肃兰州 730000;
2. 中国科学院研究生院,北京 100049;
3. 中国石化胜利油田分公司地质科学研究院,山东东营 257015)

摘要:为研究磺酸盐与非离子表面活性剂在降低油水界面张力方面的协同作用,选取了 3 种支链烷基苯磺酸盐与 3 种聚氧乙烯醚非离子表面活性剂,测量了表面活性剂及其复配体系与系列正构烷烃间的界面张力,考察了表面活性剂的亲水亲油性能、复配体积比及 NaCl 浓度对协同作用的影响。结果表明,磺酸盐与非离子表面活性剂复配体系的亲水亲油性能具有加和性,亲油性的磺酸盐 8C18ΦS 与亲水性的非离子表面活性剂复配可以获得亲水亲油性适中的复配体系,使界面张力降低,实现协同作用;而亲水性的磺酸盐 3C10ΦS、7C14ΦS 与亲水性的非离子表面活性剂的复配体系仍然是亲水性的,界面活性低。此外,复配体积比、NaCl 加量也可有效影响复配体系的协同作用。其中,在复配比 2∶1、NaCl 加量 1%时,8C18ΦS 与 LAP-9 复配体系的界面张力最低值为 4.0×10^{-4} mN/m,而 3C10ΦS、7C14ΦS 与 LAP-9 复配体系的界面张力较高,分别位于 0.4、1.0 mN/m 与 0.035、0.13 mN/m 之间。图 7 参 16

关键词:支链烷基苯磺酸盐;非离子表面活性剂;正构烷烃;界面张力;协同作用

中图分类号:TE357.46;O647.2;TQ423 **文献标识码**:A

表面活性剂在三次采油中占有重要位置,其作用机理是通过表面活性剂分子的两亲性将原油与驱替液之间的界面张力降至超低(<0.01 mN/m),进而启动地藏中的残余油,提高原油采收率[1-3]。因此,选择合适的表面活性剂体系,确保超低油水界面张力的形成是关键。

众所周知,表面活性剂与表面活性剂复配可以产生协同作用[2,4],而这种协同作用是降低油水界面张力的有效手段。因此,了解表面活性剂复配体系协同作用的机理,有助于获得理想的界面张力。然而,由于油水界面比较复杂[5],研究表面活性剂复配体系在降低油水界面张力方面的协同作用机理的报道较少。Rosen 等[6-11]推导了一系列公式来研究表面活性剂间的协同作用,指出表面活性剂分子与表面活性剂分子间的相互吸引作用越强，协同作用就越强。张路等[12-13]研究了石油磺酸盐与水溶性表面活性剂间的协同作用，提出协同作用主要受油相性质、表面活性剂的亲水亲油性能以及 NaCl 浓度的影响，并强调协同作用取决于表面活性剂分子与水分子、油分子间的相互作用，而不是表面活性剂分子与表面活性剂分子间的相互作用。李英等[14]利用耗散颗粒动力学模拟了十二烷基苯磺酸钠与非离子表面活性剂 TX-100 在油水界面的协同作用，指出在油水界面表面活性剂分子以团簇的形式存在，非离子表面活性剂团簇插入磺酸盐表面活性剂团簇的空隙之中，在油水紧密排布，产生协同作用，界面张力降低。

本文通过测量支链烷基苯磺酸盐与聚氧乙烯醚非离子表面活性剂及其复配体系同系列正构烷烃的界面张力，研究了磺酸盐与非离子表面活性剂间的协同作用，并从亲水亲油性能的角度阐释了表面活 性剂复配体系产生协同作用的机理，指出了实现协同作用的途径。该工作对三次采油用表面活性剂配 方的筛选具有重要的指导意义。

1 实验部分

1.1 化学试剂

4-(3′-癸基)苯磺酸钠(3C10ΦS)，4-(7′-十四烷基)苯磺酸钠(7C14ΦS)，4-(8′-十八烷基)苯磺酸钠(8C18ΦS)，这 3 种支链烷基苯磺酸盐均为实验室合成[15]，纯度大于 95.0%，结构如图 1 所示；3 种非离子表面活性剂分别为聚氧乙烯(9)月桂醇醚(LAP-9)、聚氧乙烯(10)辛基酚醚(OP-10)与吐温 20(TW-20)，化学纯，国药集团化学试剂有限公司；正己烷至正十四烷系列正构烷烃，色谱纯，默克公司；NaCl，分析纯，国药集团化学试剂有限公司；二次蒸馏水。

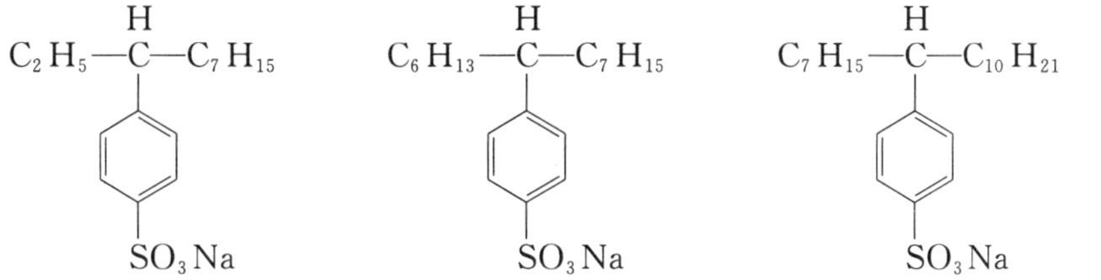

C_2H_5—CH(苯基)—C_7H_{15}，SO_3Na	C_6H_{13}—CH(苯基)—C_7H_{15}，SO_3Na	C_7H_{15}—CH(苯基)—$C_{10}H_{21}$，SO_3Na
3C10ΦS	7C14ΦS	8C18ΦS

图 1 3 种支链烷基苯磺酸盐的结构式

1.2 界面张力的测量

分别配制 3 种支链烷基苯磺酸盐与 3 种非离子表面活性剂的水溶液。其中，表面活性剂的质量分数均为 0.1%，NaCl 加量分别为 1%与 2%。将磺酸盐与非离子表面活性剂以不同的体积比进行复配，用 TX500C 型全量程视频动态界面张力仪(北京盛维基业科技有限公司)测量表面活性剂及其复配体系与系列正构烷烃之间的界面张力。实验温度 70℃，界面张力数据均为稳态平衡值。

2 结果与讨论

2.1 单一表面活性剂的亲水亲油性能

表面活性剂的亲水亲油性能与表面活性剂的界面活性有直接关系。亲水性表面活性剂主要溶解在水中,不能有效富集在油水界面上,界面活性低;亲油性表面活性剂主要溶解在油中,也不能有效富集在油水界面上,界面活性也低;而亲水亲油性能适中的表面活性剂,既可溶解在油中,也可溶解在水中,可以有效富集在油水界面上,使界面张力降低,界面活性高[12]。测量表面活性剂水溶液与系列正构烷烃之间的界面张力,绘制表面活性剂的烷烃扫描曲线,是反映表面活性剂油水界面活性的重要手段。因此,可以根据烷烃扫描曲线,判断表面活性剂的亲水亲油性能。

针对系列不同碳数的烷烃,同一种表面活性剂会表现不同的亲水亲油性能。在低碳数烷烃处,由于油相分子间作用力较弱,表面活性剂分子较易溶入油相,从而表面活性剂在低碳数烷烃中的溶解度较大;而在高碳数烷烃处,油相分子间作用力较强,表面活性剂分子难以溶入油相,故表面活性剂在高碳数烷烃中的溶解度较小。因此,伴随着烷烃碳数(alkane carbon number,ACN)的增加,表面活性剂在油相中的溶解度越来越小。而同一种表面活性剂在水相中的溶解度是一定的,因此,针对系列不同碳数的烷烃,同一种表面活性剂会在某一特定碳数的烷烃处产生最低界面张力,该烷烃的碳数称为该表面活性剂的 $n_{\min}$ 值[12,16]。$n_{\min}$ 值可以反映表面活性剂的亲水亲油性能,$n_{\min}$ 值越低,亲水性越强;$n_{\min}$ 值越高,亲油性越强。为了更具体的表示表面活性剂针对不同烷烃的亲水亲油性能,用 ACN/$n_{\min}$ 表示该表面活性剂针对不同碳数烷烃的亲水亲油性能。ACN/$n_{\min}$ 值大于 1,表面活性剂针对该 ACN 值的烷烃是亲水性的,ACN/$n_{\min}$ 值越大,亲水性越强;ACN/$n_{\min}$ 值小于 1,表面活性剂针对该 ACN 值的烷烃是亲油性的,ACN/$n_{\min}$ 值越小,亲油性越强;ACN/$n_{\min}$ 值等于 1,表面活性剂针对该 ACN 值烷烃的亲水亲油能力适中,界面张力最低,界面活性最高。此外,根据在烷烃扫描范围内是否可以观察到表面活性剂的 $n_{\min}$ 值,烷烃扫描曲线可以分为两类:① 烷烃扫描曲线先降低再升高,在某一特定碳数的烷烃处可以明显观察到最低界面张力和 $n_{\min}$ 值;② 烷烃扫描曲线一直升高或降低,没有最小值,可以延伸曲线推测不同表面活性剂 $n_{\min}$ 值的相对大小。

图 2 为 3 种支链烷基苯磺酸盐、LAP-9 的烷烃扫描曲线,NaCl 质量分数 1%。可以看出,3C10ΦS 的界面张力位于 0.6 至 1.5 mN/m 之间,界面活性较低,且界面张力随着烷烃碳数的增加而升高。通过延伸曲线可以推测其 $n_{\min}$ 值小于 6,ACN/$n_{\min}$ 值大于 1,说明 3C10ΦS 的亲水性很强,这是 3C10ΦS 的亲油基过短所致。7C14ΦS 的界面张力位于 0.02 至 0.1 mN/m 之间,界面活性有所提高,$n_{\min}$ 值有所增加,这与 7C14ΦS 的亲油基碳数增加,亲水性减弱相一致。但整体而言,7C14ΦS 的亲水性仍较强,其 $n_{\min}$ 值小于 6,ACN/$n_{\min}$ 值大于 1。8C18ΦS 的界面张力位于 0.1～0.001mN/m 之间,并随烷烃碳数的增加不断降低,其 $n_{\min}$ 值大于 14,针对扫描烷烃,ACN/$n_{\min}$ 值小于 1,说明 8C18ΦS

是亲油性的。值得一提的是，在高碳数烷烃（正十二烷至正十四烷），8C18ΦS 实现了超低界面张力，界面活性高。这说明对于高碳数烷烃，8C18ΦS 的亲油基碳链长度较为合适，亲水亲油性能接近适中，ACN/n_{min} 值接近 1。非离子表面活性剂 LAP-9 界面活性较低，界面张力位于 0.1～0.4mN/m 之间，且随着烷烃碳数的增加而增加，n_{min} 值小于 6，ACN/n_{min} 值均大于 1，说明 LAP-9 的亲水性较强。根据 LAP-9 的烷烃扫描曲线位于 3C10ΦS 与 7C14ΦS 之间，可以判断 LAP-9 的亲水亲油性能位于 3C10ΦS 与 7C14ΦS 之间。

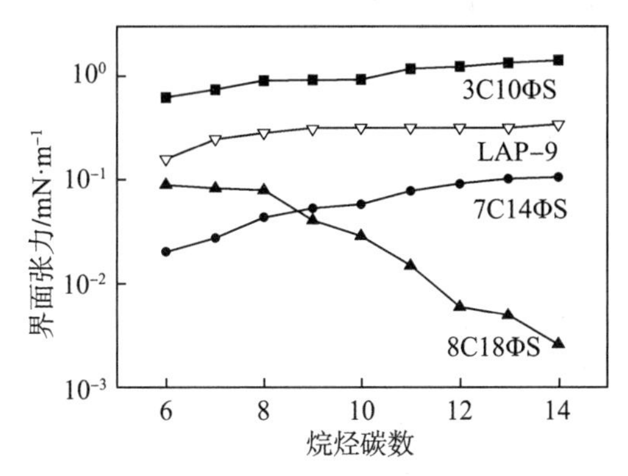

图 2 单一表面活性剂与正构烷烃间的界面张力

简而言之，针对正己烷至正十四烷，3C10ΦS、7C14ΦS 与 LAP-9 均是亲水性的，亲水性顺序为 3C10ΦS＞LAP-9＞7C14ΦS；而 8C18ΦS 是亲油性的，在高碳数烷烃处的亲水亲油性能接近适中，实现了超低界面张力。

2.2 表面活性剂亲水亲油性能对复配体系协同作用的影响

磺酸盐与 LAP-9 复配体系的界面活性见图 3，复配体积比 2：1，NaCl 质量分数 1%。由图 3 可以看出，3C10ΦS 与 LAP-9 复配体系的界面张力位于 0.4～1.0 mN/m 之间，界面活性较低。这是由于亲水性的 3C10ΦS 与亲水性的 LAP-9 复配得到的复配体系仍然是亲水性的，不能有效降低界面张力。此外，复配体系的烷烃扫描曲线位于两种单一表面活性剂之间，证实了复配体系的亲水亲油性位于两种单一表面活性剂之间。图 3 中 7C14ΦS 与 LAP-9 复配体系的界面张力、烷烃扫描曲线也位于两种单一表面活性剂之间，进一步验证了亲水性的磺酸盐与亲水性的非离子表面活性剂复配得到的复配体系仍是亲水性的，且亲水亲油性位于两种单一表面活性剂之间。同时，也说明了磺酸盐与非离子表面活性剂复配，体系的亲水亲油性具有加和性。以磺酸盐 8C18ΦS 为标准，8C18ΦS 与 LAP-9 复配体系的界面张力可以分为两组：一组是在低碳数烷烃（正己烷至正十一烷）处，复配体系的界面张力比 8C18ΦS 的低，即磺酸盐与非离子表面活性剂复配具有协同作用。这是由于在低碳数烷烃处，8C18ΦS 的亲油性较强，而 LAP-9 是亲水性的，二者的复配体系可以获得比较适中的亲水亲油性能，故界面张力降低。另一组是在高碳数烷烃（正十二烷至正十四烷）处，复配体系的界面张力比 8C18ΦS 的高，为负作用。这是由于在高碳数烷烃处，8C18ΦS 的亲水亲油性能接近适中，在此基础上，添加亲水性的 LAP-9 获得的复配体系成为了亲水性的,故界面张力升高。这进一步说明了磺酸盐与非离子表面活性剂复配,体系的亲水亲油性能具有加和性,并证实了降低油水界面张力的协同作用来自复配体系亲水亲油性能的改善,实现协调作用的途径为亲油性的磺酸盐与亲水性的非离子表面活性剂复配。

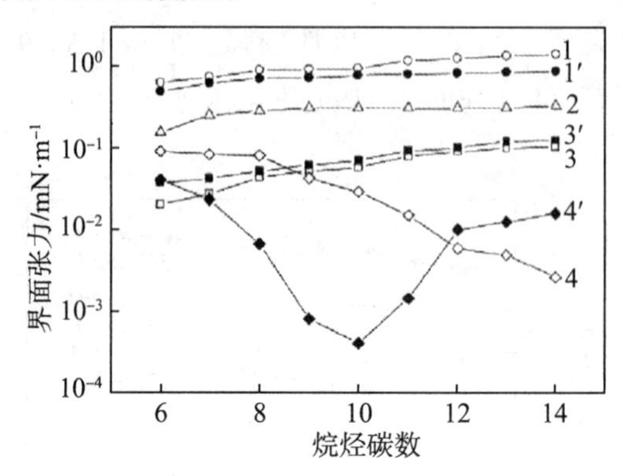

图 3 磺酸盐与 LAP-9 复配体系同正构烷烃间的界面张力(复配体积比 2∶1)

1—3C10ΦS;1′—3C10ΦS、LAP-9 复配体系;2—LAP-9;3—7C14ΦS;3′—7C14ΦS、LAP-9 复配体系;4—8C18ΦS;4′—8C18ΦS、LAP-9 复配体系

2.3 复配比例对复配体系协同作用的影响

复配体积比对 8C18ΦS 与 LAP-9 复配体系同正构烷烃间界面张力的影响见图 4,NaCl 质量分数 1%。当体积比为 4∶1 时,复配体系的界面张力在所有扫描烷烃上均比 8C18ΦS 的低,具有协同作用。原因在于 8C18ΦS 是亲油性的,加入少量的 LAP-9 后,体系的亲油性减弱而向亲水亲油性能适中的方向靠拢,故界面张力降低。当二者的比例为 2∶1 时,在低碳数烷烃处界面张力进一步降低,而在高碳数烷烃处界面张力开始升高,产生了负作用,由协同作用转向负作用的转折点约为 11.8。这是由于随着亲水性 LAP-9 比例的升高,复配体系在高碳数烷烃处已经成为亲水性的,故界面张力升高。当比例调至 1∶1 时,在正己烷至正壬烷处复配体系的界面张力低于单一磺酸盐的界面张力,具有协同作用,而在正癸烷至正十四烷处均为负作用,转折点降至约 9.1。此外,1∶1 复配体系的界面张力随着烷烃碳数的增加一直升高,其 $n_{\min}$ 值小于 6,烷烃扫描曲线与单一磺酸盐 7C14ΦS 的相似,说明 1∶1 复配体系针对所有扫描烷烃均是亲水性的。当复配体积比调至 1∶2 与 1∶4 时,复配体系由协同作用转为负作用的转折点分别降至 8.2 与 6.6 附近,复配体系的烷烃扫描曲线也越来越接近亲水性的 LAP-9,即随着非离子表面活性剂 LAP-9 比例的增加,复配体系的亲水性也在逐渐增强。以上实验结果进一步证实了磺酸盐与非离子表面活性剂复配体系的协同作用源自体系亲水亲油性能的改善。同时,也说明了复配比例可以有效调节复配体系的亲水亲油性能,进而影响体系的协同作用。

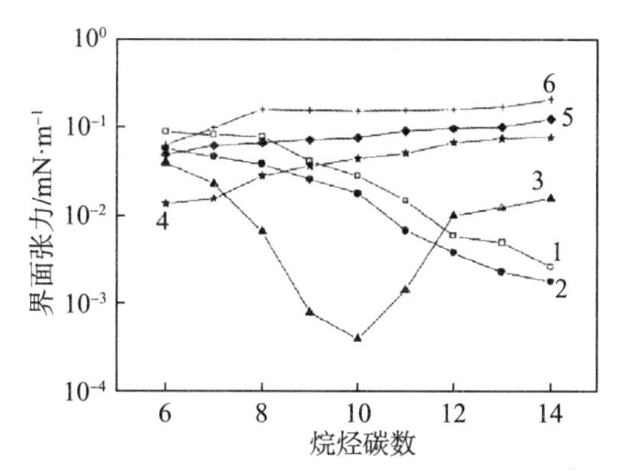

图 4　复配体积比对 8C18ΦS 与 LAP-9 复配体系同正构烷烃间界面张力的影响

8C18ΦS 与 LAP-9 体积比：1—8C18 ΦS；2—4∶1；3—2∶1；4—1∶1；5—1∶2；6—1∶4

2.4 NaCl 加量对复配体系协同作用的影响

在水溶液中，NaCl 的 Na^+ 可以被磺酸盐的亲水基团磺酸根 $-SO_3^-$ 吸附，形成双电层，即 Na^+ 可以屏蔽 $-SO_3^-$ 的负电荷，增强磺酸盐的亲油性。NaCl 浓度越高，磺酸盐的亲油性越强。而非离子表面活性剂亲水基团为聚氧乙烯链，没有电荷，因此其亲水亲油性能基本不受无机盐的影响。两种 NaCl 加量下，8C18ΦS 与 LAP-9 复配体系同正构烷烃间的界面张力见图 5。可以看出，在 2%NaCl 条件下，8C18ΦS 与 LAP-9 复配体系的界面张力均比单一的磺酸盐要低，即协同作用存在于所有扫描烷烃上。这与 1%NaCl 条件下，在低碳数烷烃有协同作用、在高碳数烷烃是负作用有所不同。原因在于在 2% NaCl 条件下，磺酸盐 8C18ΦS 的亲油性增强了，而且针对所有扫描烷烃均具有较强的亲油性，这与在 1%NaCl 条件下，8C18ΦS 在高碳数烷烃处亲油性较弱，亲水亲油性能接近适中有所不同。NaCl 加量 2%时，以 2∶1 的复配体积比往亲油性强的 8C18ΦS 中添加亲水性的 LAP-9 不足以使复配体系成为亲水性的，故不会出现负作用。由此可见，可以通过无机盐调节磺酸盐的亲水亲油性来影响复配体系的协同作用。

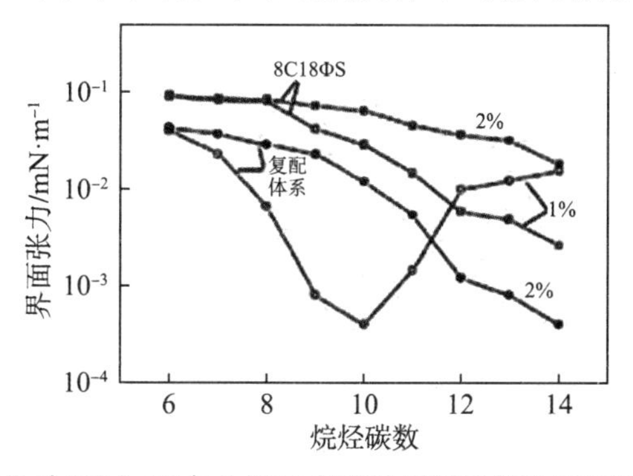

图 5　NaCl 加量对 8C18 ΦS 与 LAP-9 复配体系同正构烷烃间界面张力的影响

数据点为方形的曲线是单一 8C18ΦS，数据点为圆形的曲线是 8C18ΦS 与 LAP-9 复配体系（复配体积比 2∶1），图中数值为 NaCl 加量

2.5 8C18ΦS 与其他非离子表面活性剂的协同作用

8C18ΦS 与非离子表面活性剂 OP-10、TW-20 复配体系同正构烷烃间的界面张力见图 6,NaCl 质量分数 1%。可以看出,与 LAP-9 相似,8C18ΦS 与 OP-10、TW-20 复配体系在合适的复配比下也可以降低油水界面张力,实现协同作用。这是由于 OP-10 与 TW-20 也均为亲水性的非离子表面活性剂。此外,由于 OP-10、TW-20 与 LAP-9 在化学结构上的差异性,也证实了磺酸盐与非离子表面活性剂复配能否产生 41 同作用,并不取决于表面活性剂的化学结构,而主要取决于表面活性剂的亲水亲油性能。

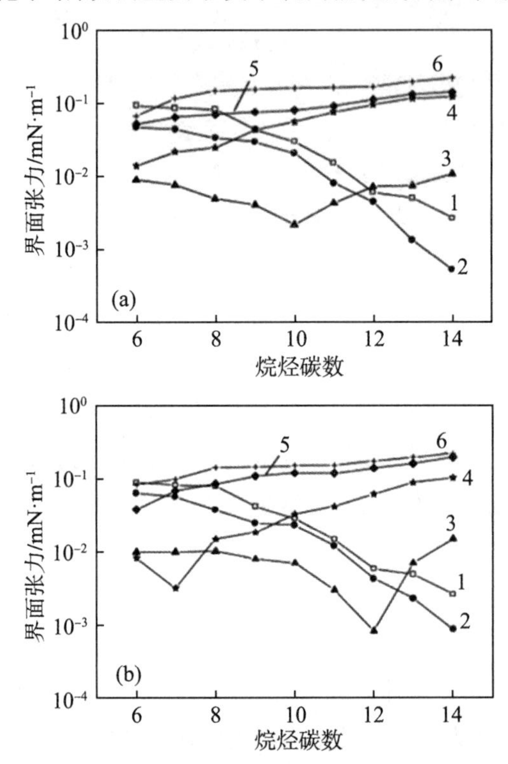

图 6 8C18ΦS 与 OP-10(a)、TW-20 (b)复配体系在不同复配比下同正构烷烃间的界面张力

8C18ΦS 与 OP-10(a)、TW-20(b)体积比:1—8C18ΦS;2—4∶1;3—2∶1;4—1∶1;5—1∶2; 6—1∶4

3 结论

从亲水亲油性能的角度对磺酸盐与非离子表面活性剂复配体系在降低油水界面张力方面的协同作用进行了阐释,指出了复配体系的协同作用源自体系亲水亲油性能的改善。其中,亲油性的磺酸盐与亲水性的非离子表面活性剂复配,可以获得亲水亲油性能适中的复配体系,降低界面张力,实现协同作用。此外,复配比是调节复配体系亲水亲油性能的重要参数,也可以通过 NaCl 调节磺酸盐的亲水亲油性来影响复配体系的协同作用。该结论对三次采油工作中表面活性剂配方体系的筛选具有重要意义。

参考文献

[1] Rosen M J, Wang H Z, Shen P P, et al. Ultralow interfacialtension for enhanced oil recovery at very low surfactant concentrations [J]. Langmuir, 2005,21(9):3749-3756.

[2] Zhao Z K, Bi C G, Qiao W H, et al. Dynamic interfacial tension behavior of the novel surfactant solutions and Daqing crude oil [J]. Colloids Surf A,2007,294(1-3):191-202.

[3] Zhu Y Y, Xu G Y, Gong H J, et al. Production of ultra-low interfacial tension between crude oil and mixed brine solution of Triton X-100 and its oligomer Tyloxapol with cetyltrimethylammonium bromide induced by hydrolyzed polyacrylamide [J]. Colloids Surf A,2009,332(2-3):90-97.

[4] Hou Z S, Li Z P, Wang H Q. The interaction of sodium dodecyl sulfonate and petroleum sulfonate with nonionic surfactants (TritonX-100, Triton X-114) [J]. Colloids Surf A,2000,166(2):243-249.

[5] Han L J, Ye Z B, Chen H, et al. The interfacial tension between cationic gemini surfactant solution and crude oil [J]. J Surfactants Deterg, 2009, 12(3):185-190.

[6] Rosen M J, Murphy D S. Synergism in binary-mixtures of surfactants. 5. 2-phase liquid liquid-systems at low surfactant concentrations [J]. J Colloid Interface Sci,1986,110(1):224-236.

[7] Rosen M J, Murphy D S. Synergism in binary-mixtures of surfactants. 8. Effect of the hydrocarbon in hydrocarbon water-systems [J]. J Colloid Interface Sci, 1989,129(1):208-216.

[8] Murphy D S, Rosen M J. Effect of the nonaqueous phase on interfacial properties of surfactants. 1. Thermodynamics and interfacial properties of a zwitterionic surfactant in hydrocarbon water-systems [J]. J Phys Chem,1988,92(10):2870-2873.

[9] Rosen M J, Murphy D S. Synergism in mixtures containing zwitterionic surfactants [J]. Langmuir,1991,7(11):885-888.

[10] Rosen M J. Predicting synergism in binary mixtures of surfactants [J]. Prog Colloid Polym Sci, 1994,95:39-47.

[11] Rosen M J. Molecular interactions and the quantitative prediction of synergism in mixtures of surfactants [J]. Prog Colloid Polym Sci,1998,109:35-41.

[12] Zhang L, Luo L, Zhao S, et al. Studies of synergism/antagonism for lowering dynamic interfacial tensions in surfactant/alkali/acidic oil systems, part 2:

synergism/antagonism in binary surfactant mixtures [J]. J Colloid Interface Sci，2002，251(1):166-171.

[13] 张路，罗澜，赵濉，等. 油相性质对水相中混合表面活性剂协同效应的影响[J]. 油田化学，2000，17(3):268-271.

[14] Li Y，He X J，Cao X L，et al. Molecualr behavior and synergistic effects between sodium dodecylbenzene sulfonate and Triton X-100 at oil/water interface [J]. J Colloid Interface Sci，2007，307(1):215-220.

[15] 杨捷. 系列烷基苯磺酸盐异构体纯化合物的合成、界面性能及构效关系的研究[D]. 大连理工大学博士学位论文，2005.

[16] Doe P H，El-Emary M，Wade W H. Surfactants for producing low interfacial tensions I：linear alkylbenzene sulfonates [J]. J Am Oil Chem Soc，1977，54(12)：570-577.

Viscosity and interfacial tension study for alkali/surfactant/polymer complex system with blend surfactants

Li Mingyan[1] Lu Xiangguo[1] Sun Gang[2] Xu Dianping[3] Yu Tao[4]

(1. Key Laboratory of Enhanced Oil and Gas Recovery of Education Ministry in Northeast Petroleum University, Daqing, Heilongjiang 163318, P R of China;
2. Exploration and Development Research Institute, Daqing Oilfield Company, Ltd, Petro China, Daqing, Heilongjiang 163712, P R of China;
3. The Fourth Oil Production Factory, Daqing Oilfield Company, Ltd, PetroChina, Daqing, Heilong jiang 163511, P R of China;
4. Daqing Oilfield Refining and Chemical Company, PetroChina, Daqing, Heilongjiang 163000, P R of China)

Abstract: Heavy alkyl benzene sulfonate (DH) and mahogany sulfonate (LH) were typical surfactants in enhancing oil recovery. NaOH/DH/HPAM system and Na_2CO_3/LH/HPAM compound system showed favorable oil-water interfacial activity. In order to investigate the possibility of blending DH and LH, a series of experiments were conducted to evaluate the interfacial tension (IFT) and viscosity of Na_2CO_3/ blending surfactant/polymer system (ASP system) in respects of mass ratio (DH : LH), surfactant concentration and alkali concentration. The results showed that the ratio of DH to LH and the concentration of surfactant had little effect on the viscosity of ASP system. When the ratio of DH to LH was of 1 : 1, the interfacial tension between complex system and crude oil was ultra low which decreased to 8.63×10^{-3} mN/m at 80 minutes. The interfacial tension increased with decreasing dosage of blending surfactant. When the concentration of surfactant was of 0.25%, the interfacial activity was best. The viscosity and interfacial tension of complex system decreased when the concentration of alkali increased with 0.2%—1.0% Na_2CO_3 dosage. The recommended dosage of Na_2CO_3 was of 0.8%—1.0%.

Keywords: heavy alkyl benzene sulfonate; mahogany sulfonate; surfactant; weak alkali/surfactant/polymer system

Efect of electrolytes on interfacial tensions of alkyl ether carboxylate solutions

Liu Ziyu Zhang Lei Cao Xulong Song Xinwang Jin Zhiqiang Zhang Lu Zhao Sui*

Technical Institute of Physics and Chemistry, Chinese Academy of Sciences, Beijing 100190, People's Republic of China

Graduate University of Chinese Academy of Sciences, Beijing 100049, People's Republic of China

Geological Scientific Research Institute of Shengli Oilfield Company Limited, SINOPEC, Dongying 257015, People's Republic of China

Abstract: Alkyl ether carboxylate is one type of surfactant that can produce ultralow interfacial tension (IFT) under high-salinity and high-temperature conditions. In this paper, the influence of counterions on dynamic IFTs of fatty alcohol polyoxyethylene carboxylate ($C_{12}EO_3C$) against alkanes has been studied. The efect of the temperature on the IFT has been investigated. On the basis of our experimental results, one can find that the NaCl concentration has little efect on the IFT, while divalent ions can reduce the IFT to an ultralow value. With the increasing $CaCl_2$ or $MgCl_2$ concentration, dynamic IFT passes through a minimum at a particular salt concentration ("V" shape). Moreover, the stable value of IFT achieves an ultralow value and also passes through a minimum at the same salt concentration. $MgCl_2$ has a stronger tendency to achieve ultralow IFT than that of $CaCl_2$, while the addition of $CaCl_2$ has a stronger tendency to partition surfactant molecules to the oil phase. Ultralow IFT could also be achieved by improving the temperature because of the enhancement of oil solubility of the surfactant. An interfacial model combining two mechanisms, partitioning the surfactant into the oil phase and decreasing the charge repulsive force between interfacial surfactant molecules, responsible for the efect of the electrolyte on dynamic IFT has been provided. All experimental results above can be explained well. Our studies are of great significance in designing ultralow IFT formulation for the reservoir in a high temperature and with high-salinity formation water.

1 INTRODUCTION

It is well-known that the primary requirement needed to mobilize residual oil is a sufficiently low interfacial tension (IFT) to give a capillary number large enough to overcome the capillary forces and allow for the oil to flow during the enhanced oil recovery (EOR) process involving the employment of surfactants. 1 The study of the EOR surfactant, such as petroleum sulfonate and alkylbenzene sulfonates, has been carried out extensively for common reservoirs. However, for high-temperature and high-salinity oil reservoirs, petroleum sulfonate and alkylbenzene sulfonates are easy to be deposited, which results in low oil recovery. Therefore, there is a great importance on the study of surfactants applying to hightemperature and high-salinity oil reservoirs.

Recently, advances including the development of new synthetic surfactants and an increased understanding of the relationship between the surfactant structure and its performance have made it possible to rapidly identify promising high-performance surfactants for EOR. New types of surfactants that possess several hydrophobic and hydrophilic moieties in the same molecule have attracted considerable interest because they exhibit extraordinary surface activity. 2 Fatty alcohol polyoxyethylene ether carboxylate (AEC) and sulfonate (AES) are good new-type surfactants widely studied in surface activity, foaming capability, emulsifying property, and the performance of decreasing the IFT between oil and water. 3-12 The chemical structure of AEC contains two hydrophilic groups ($-CH_2CH_2O-$ and $-CH_2COO-$), and that makes AEC display greater surface activity than the traditional surfactants. Because of the existence of $-CH_2CH_2O-$, AEC have both anionic and non-ionic properties. Witthayapanyanon et al. found that the introduction of additional PO and EO groups in the extended surfactant yielded lower IFT and lower optimum salinity, both of which are desirable in most formulations.[4]

Although many attempts have been made to the IFT behaviors of AES6-8 and foaming properties of AEC,[11,12] the fundamental mechanisms for the lowering of IFT between oil and AEC solutions under high-salinity circumstances have not been recognized clearly until now.

The IFT of surfactant solution depends upon the number of surfactant molecules per unit area at the interface. Surfactant adsorption depends upon their unique amphiphilic properties, which is known as the balance between hydrophobic and hydrophilic forces of the tail and headgroup.[13] Surfactant solubility limits the magnitude to which this approach can be used; that is, as the hydrophobic tail becomes longer, the surfactant eventually loses its water solubility. For a given surfactant, the greater concentration of surfactant molecules at the surface results in the lower surface tension.

Surfactant counterion has a tremendous impact on interfacial properties of ionic surfactant solutions. During the past few decades, there have been numerous studies on the surfactants with counterions or excess inorganic salt. Chan and Shah[14] reported that the partitioning of the surfactant in the oil phase increases as the salinity increases and decreases as the chain length of oil increases. Zhang et al. 15 pointed that the synergism/antagonism for lowering IFT depended upon the factors that can change the surfactant partition coefficients, such as the hydrophilic-lipophilic ability of the surfactant, the salinity, and the alkane carbon number of the oil phase. Chattopadhyay et al. 16 investigates the effect of counterions, such as Li^{+}, Na^{+}, Cs^{+}, and Mg^{2+}, of dodecyl sulfate on interfacial properties in relation to foaming. Yan and Guo17 used molecular dynamics simulations to study the effect of Ca^{2+} ions on the hydration shell of sodium dodecyl carboxylate (SDC) and sodium dodecyl sulfonate (SDSn) monolayers at vapor/liquid interfaces.

The divalent ion in formation water has a significant impact on interfacial intension. However, the effect of the divalent ion on ultralow IFTs has not been thoroughly researched, and different influences between Mg^{2+} and Ca^{2+} ions are little studied. Recently, the interesting molecular mechanism responsible for the effect of NaCl on IFT of alkyl benzene sulfonates has been explored by our group. However, the mechanism for the effect of Mg^{2+} and Ca^{2+} ions has not been carried out because of the precipitation of sulfonates. Because fatty alcohol polyoxyethylene ether carboxylate shows good salt tolerance, in this work, we concentrated on the effects of three ions Na^{+}, Ca^{2+}, Mg^{2+} on the IFT between n-alkanes and surfactant solutions. The optimum electrolyte concentration and the IFT at optimum formulation were discussed. The molecular mechanism for achieving ultralow IFT was also studied for these systems. It has a great important guidance for EOR of high-temperature and high-salinity oil reservoirs.

2 EXPERIMENTAL SECTION

For all of the systems studied, the used surfactant, fatty alcohol polyoxyethylene ether carboxylate ($C_{12}EO_3C$) with a dodecyl and three ethoxyl groups, was synthesized in our laboratory, which has a purity of about 90% (Scheme 1). The oils used in this study were a

Scheme 1. Chemical Structure of $C_{12}EO_3C$

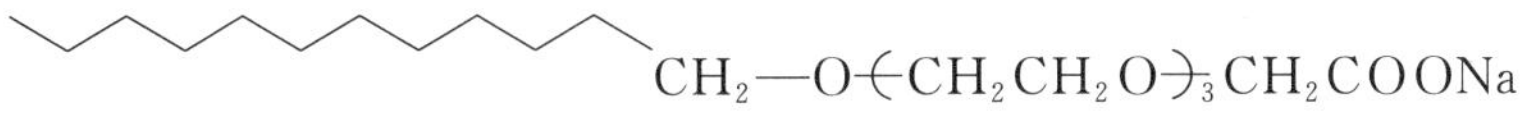

$$CH_2-O\!\left(CH_2CH_2O\right)_3CH_2COONa$$

homologous series of alkanes with chain lengths from C_6 to C_{14} with 99+ mol % purity. All inorganic reagents used were analytical-grade. Double-distilled water was used in the preparation of the aqueous solutions.

The spinning drop technique was employed to measure dynamic IFTs. The standard spinning-drop tensiometer had been modified by the addition of video equipment and an interface to a personal computer. 18 The computer had been fitted with a special video board and a menu-driven image enhancement and analysis program. The video board can "capture" a droplet image for immediate analysis. Analysis usually consists of the measurement of the drop length and drop width.

The volumetric ratio of water/oil in the spinning drop tensiometer is about 200. Samples were assumed to be equilibrated when measured values of IFT remained unchanged for 30 min. The IFT measurements were performed at 30~85±0.5℃.

3 RESULTS AND DISCUSSION

3.1 Efect of the Surfactant Concentration on IFT.

Reduction of IFT depends directly upon the replacement of molecules of solvent at the interface by surfactant molecules.

When the surfactant molecules replace water and oil molecules of the original interface, the interaction across the interface is now between the hydrophilic group of the surfactant and water molecules on one side of the interface and between the hydrophobic group of the surfactant and oil molecules on the other side of the interface.[13] Therefore, the tension across the interface is significantly reduced by the adsorption of surfactant molecules because these interactions are much stronger than the original interaction between the oil and water molecules. Surfactant adsorption depends upon their unique amphiphilic properties, which is known as the balance between hydrophobic and hydrophilic forces of the tail and head groups. Upon equilibration, it is likely that a fraction of surfactant will dissolve in the oil and the rest will dissolve in brine. For a given surfactant, the greater the concentration of surfactant molecules at the surface, the larger the number of surfactant molecules per unit area adsorbed on the interface, which results in a lower IFT. However, the extent of lowering the IFT is related to the properties of the surfactant.[19]

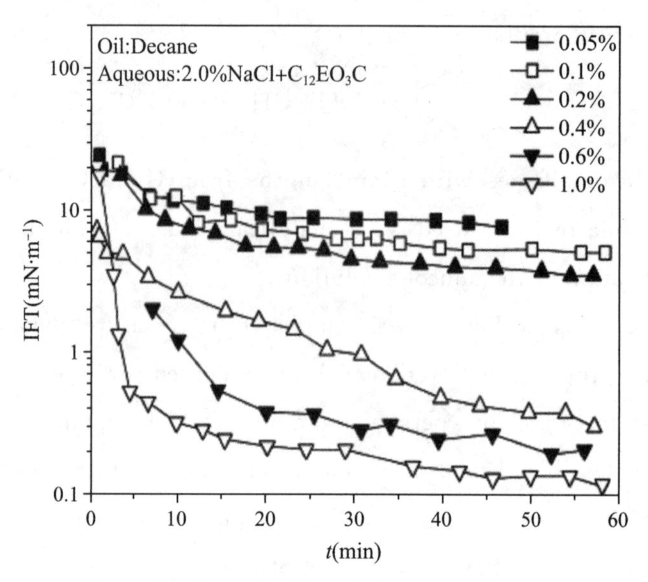

Figure 1 Dynamic IFTs of $C_{12}EO_3C$ solutions.

The efect of the surfactant concentration on the occurrence and detection of transient IFTs between decane and NaCl brine solutions has been investigated. Figure 1 shows the dynamic IFTs of various solutions containing from 0.05 to 1.0 mass% $C_{12}EO_3C$ at 2.0 mass % NaCl.

We can see clearly from Figure 1 a continuous decrease of dynamic IFT to the equilibrium value for all solutions. The time needed for achieving equilibrium becomes shorter with the increasing bulk concentration. At the same time, the stable IFT value at equilibrium decreases when the surfactant concentration increases, and at the concentration of 1 mass%, the IFT reaches 10^{-1} mN/m order of magnitude. These experimental results are in common with EOR surfactants reported in the literature.

3.2 Efect of the hydrophilic-lipophilic Ability on the IFTs of $C_{12}EO_3C$ Solutions.

The partitioning of the surfactant in the alkanes decreases with the increase of the alkane carbon number, and consequently, the surfactant concentration in brine increases. Therefore, at a specific alkane carbon number, where the minimum IFT is observed, the surfactant concentration becomes equal in both the oil and brine phases and the interfacial surfactant concentration is maximal. 14 Each surfactant or surfactant mixture may produce a minimum IFT when measured against a different n-alkane. The alkane carbon number for the minimum IFT is called $n_{\min}$ of this surfactant solution.[20,21]

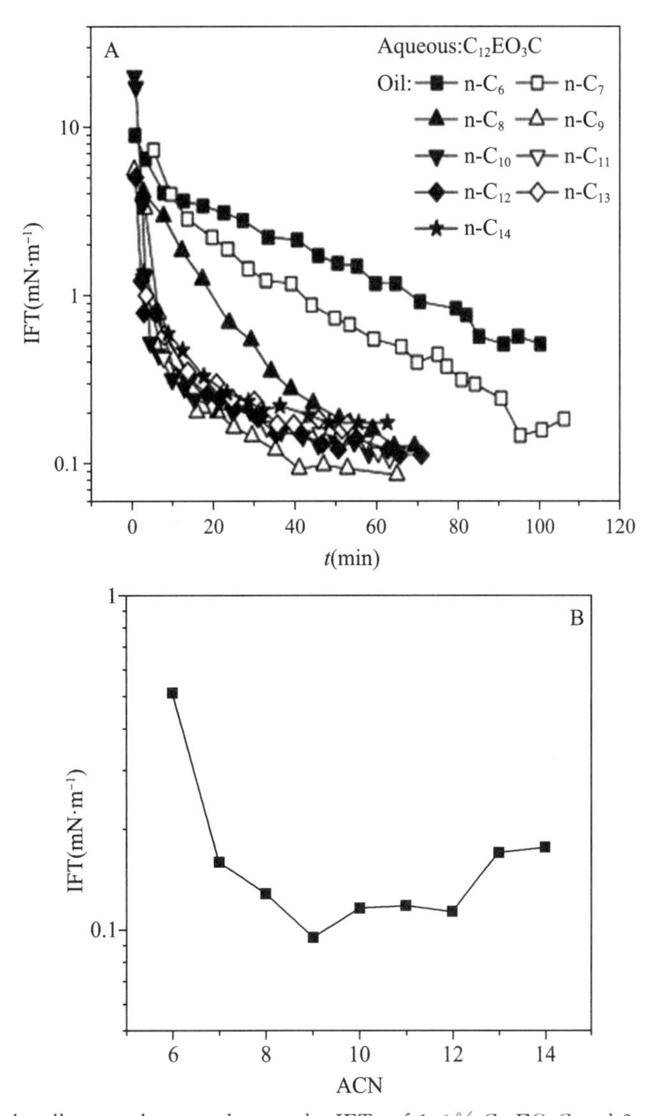

Figure 2. Efect of the alkane carbon number on the IFTs of 1.0% $C_{12}EO_3C$ and 2.0% NaCl solutions.

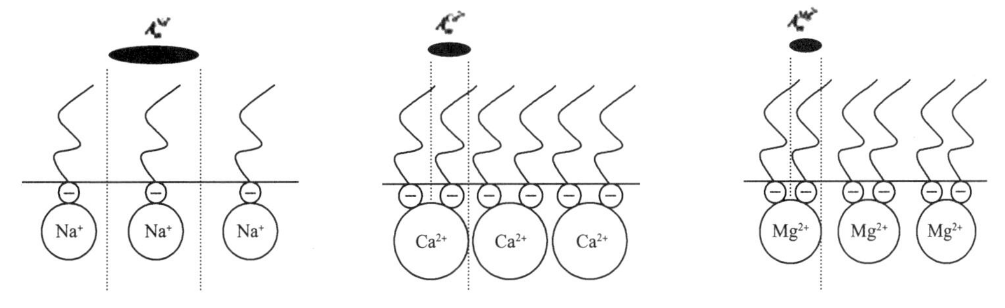

Figure 3. Efect of counterions on molecular packing of the anionic surfactant.

Figure 2A shows the dynamic IFTs for the solution of 1.0 mass% $C_{12}EO_3C$ and 2.0 mass % NaCl against di erent chain lengths of alkane as a function of interfacial age. Figure 2B shows the stable value of dynamic IFT as a function of the alkyl carbon number (ACN).

It is observed from Figure 2A that the time for achieving equilibrium becomes shorter with the increasing ACN. This is because surfactant molecules tend to partition into the oil phase with lower ACN, which results in more time for achieving equilibrium among the oil phase, the aqueous phase, and the interface. We can learn from Figure 2B that the n_{min} value of the surfactant solution is about 9.

3.3 Efect of the Electrolyte on Interfacial Properties of $C_{12}EO_3C$ Solutions.

3.3.1 Efect of the Ionic Type on IFT.

Interfacial properties can be changed with the addition of counterions. The extent of binding of the counterion increases with an increase in the polarizability and valence of counterions and decreases with an increase in its hydrated radius.[22] Thus, in aqueous solution, the extent of binding of the counterion shows $Na^+ < Ca^{2+} < Mg^{2+}$. This can be explained by considering the molecular arrangement at the interface shown in Figure 3 according to the literature.[23]

Because fatty alcohol polyoxyethylene ether carboxylate is an anionic surfactant, this study mainly focuses on the impact of cationic type (Na^+, Ca^{2+}, and Mg^{2+}) on IFT shown in Figure 4. Figure 4A represents the dynamic IFT for the 1.0 mass% $C_{12}EO_3C$ as a function of interfacial age at diferent NaCl concentrations. We can find that all of the dynamic IFTs decrease at first and are kept stable eventually. There is no intensive change for the stable value of dynamic IFT with the increase of the NaCl concentration. Therefore, we can see that the addition of NaCl has a weak efect on partitioning $C_{12}EO_3C$ from the aqueous phase to the oil phase; as a result, the NaCl concentration has little efect on the stable value of IFT in this case.

Panels B and C of Figure 4 show the dynamic IFTs for the 1.0 mass% $C_{12}EO_3C$ as a function of interfacial age at di erent $CaCl_2$ and $MgCl_2$ concentrations, respectively. We can see from the results that the dynamic IFT exhibits a minimum at a particular salt concentration. At a lower surfactant concen-tration, the dynamic IFTs decrease at first and are kept stable eventually. As one increases the concentration of $CaCl_2$/$MgCl_2$, at the critical concentration, the IFT exhibits a minimum. Beyond this concentration, the transient values of IFTs (at several minutes) dropped quickly in any case and there existed an apparent minimum in dynamic IFT curve, followed by a gradual increase to an equilibrium value ("V" shape). In these experiments, the critical concentration of $CaCl_2$ is 0.1 mass % and the critical concentration of $MgCl_2$ is 0.05 mass %.

The IFT is a function of the interfacial surfactant concentration, which depends chiefly upon its rates of adsorption and desorption from the interface. The reason that the dynamic IFT shows a "V" shape curve may be the result of the progress of the surfactant moving from the interface to the oil becoming faster than that adsorbing from the aqueous phase to the interface as interfacial age passes. However, this phenomenon has not occurred with the addition of NaCl.

Figure 4D shows the efect of electrolytes on the stable value of dynamic IFTs as a function of cationic strength. It can be

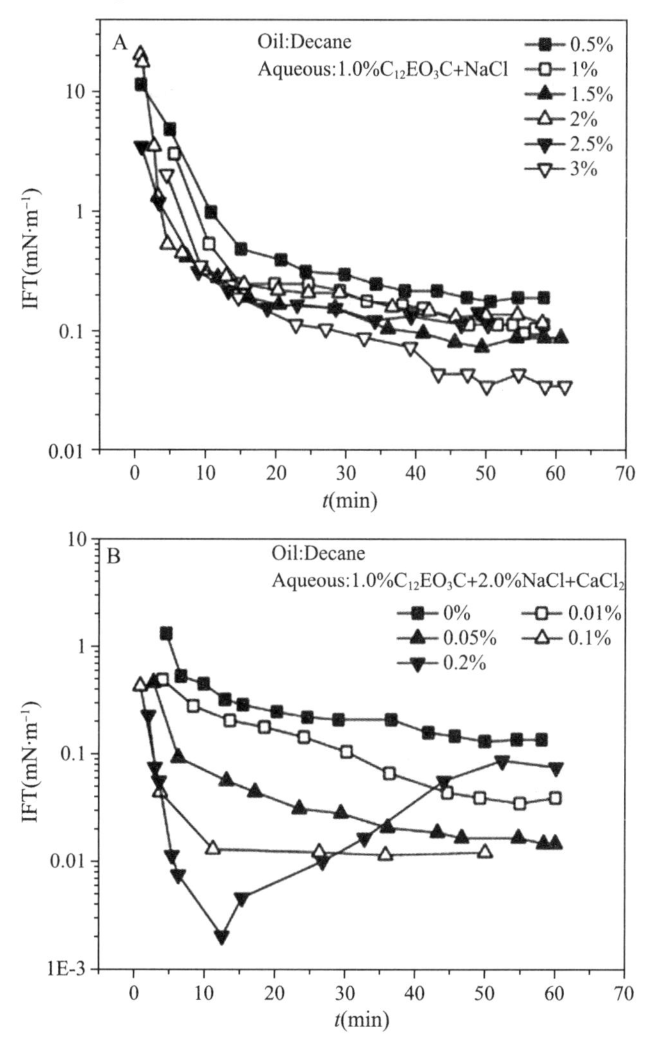

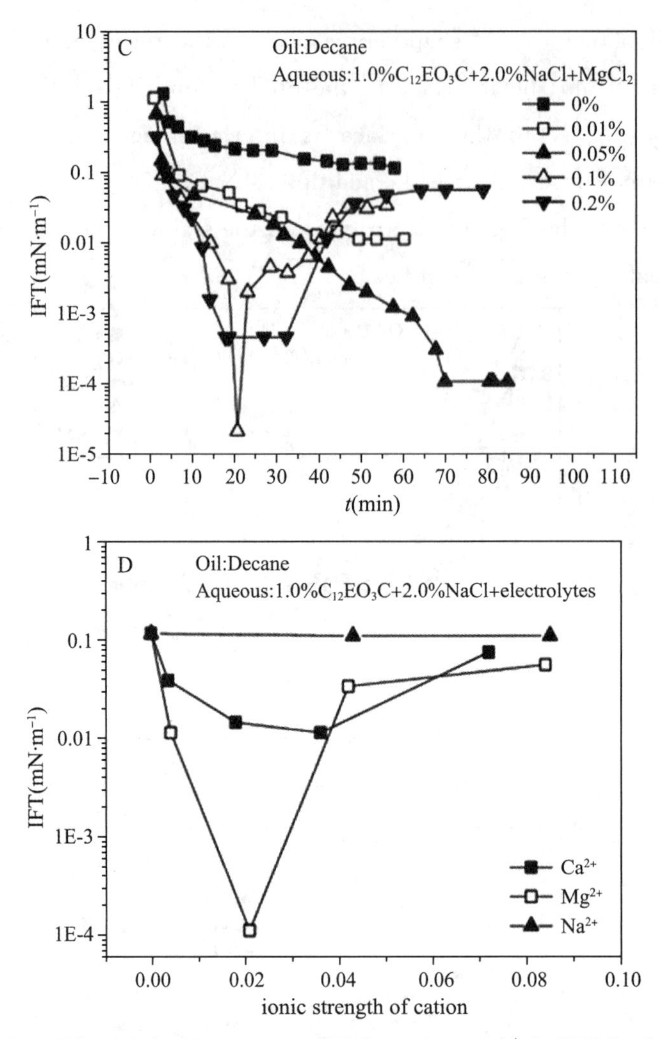

Figure 4 Efect of the ionic type on IFT for 1.0 mass % $C_{12}EO_3C$ solution.

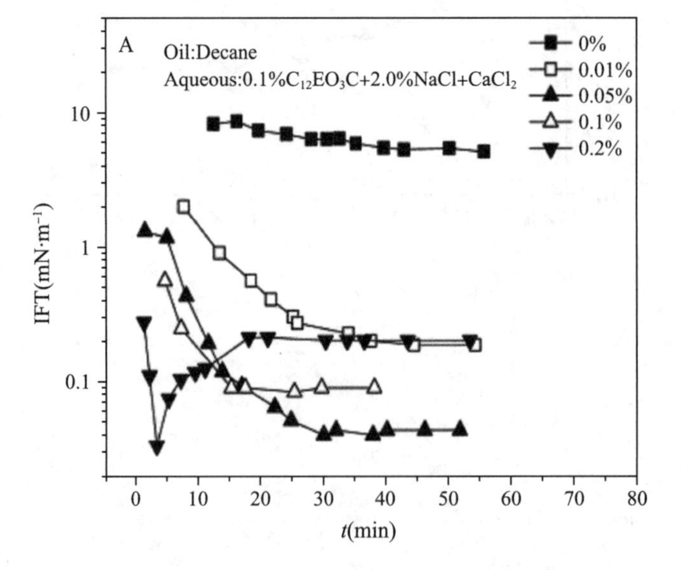

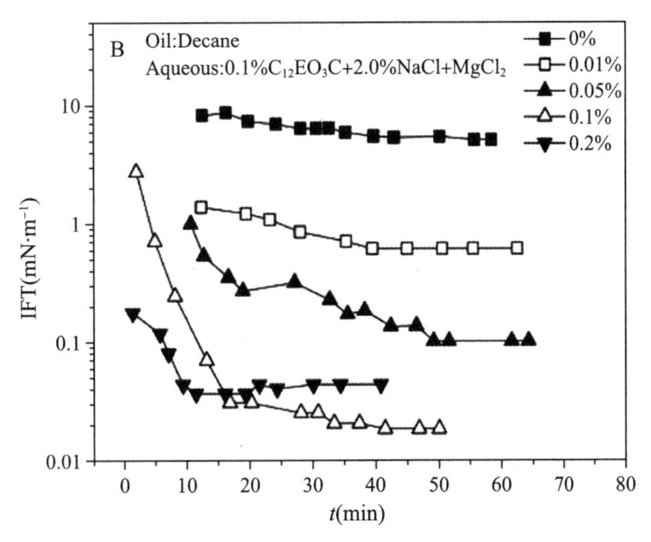

Figure 5 Efect of the divalent ion on IFT for 0.1 mass % $C_{12}EO_3C$ solutions.

clearly seen that the NaCl concentration has little efect on the stable value of IFT. However, for $CaCl_2/MgCl_2$ systems, the IFT exhibits a minimum at a special salt concentration. The efect of Mg^{2+} in our study is greater than that of Ca^{2+}. These experimental results may be related to the extent of binding of the divalent ion to the surfactant. It is interesting to note that the critical concentration for dynamic IFT coincided with the optimum electrolyte concentration for the stable value of IFT. At this special $MgCl_2$ concentration, the IFT reaches 10^{-4} mN/m order of magnitude.

We must point out that the obvious decrease of IFT may be attributed to the compact arrangement of surfactant molecules

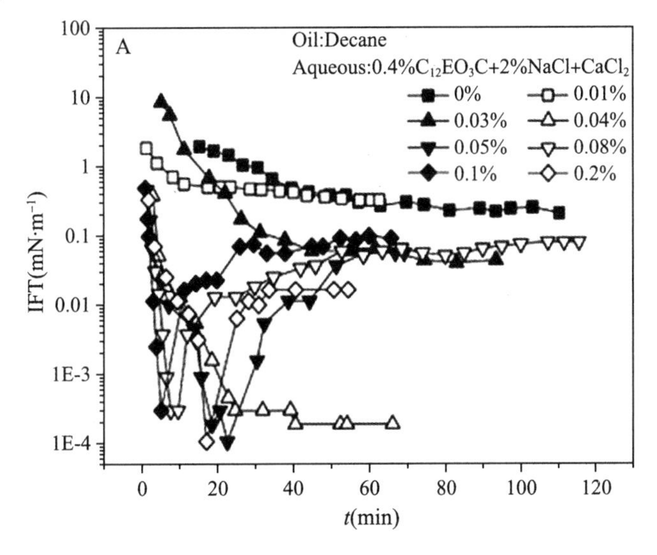

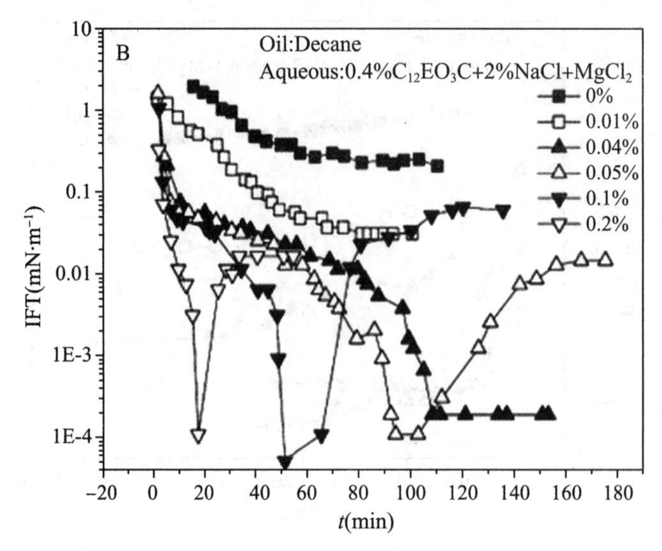

Figure 6 Efect of the divalent ion on IFT for 0.4 mass% $C_{12}EO_3C$.

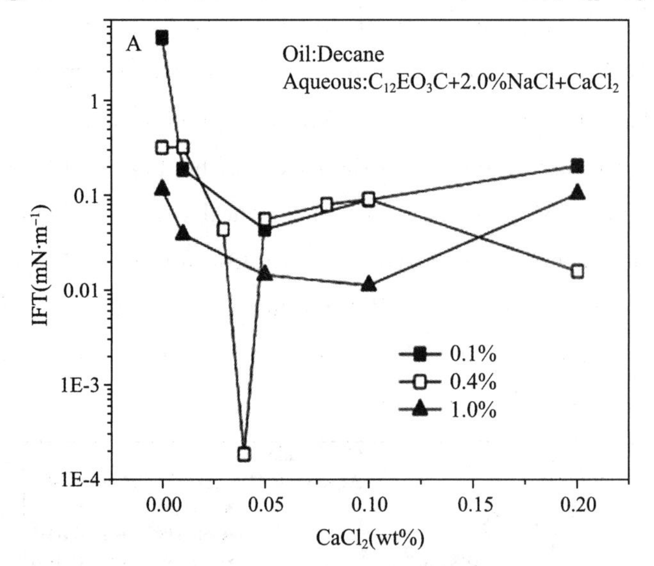

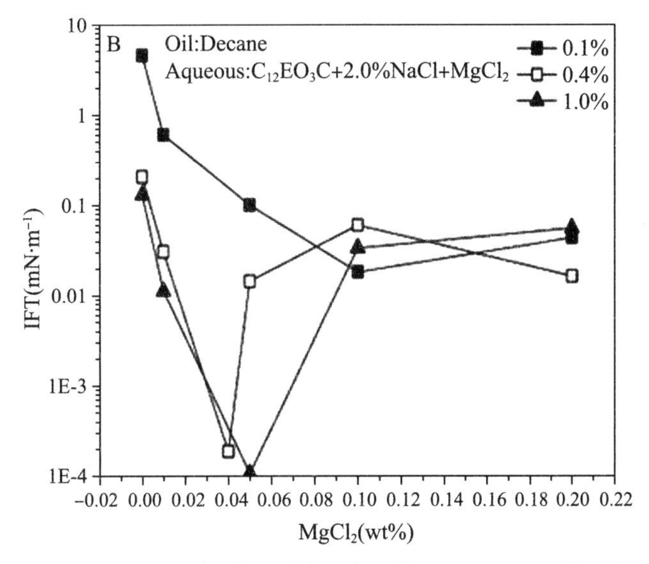

Figure 7　Efect of the divalent ion on the stable value of IFT at various $C_{12}EO_3C$ concentrations.

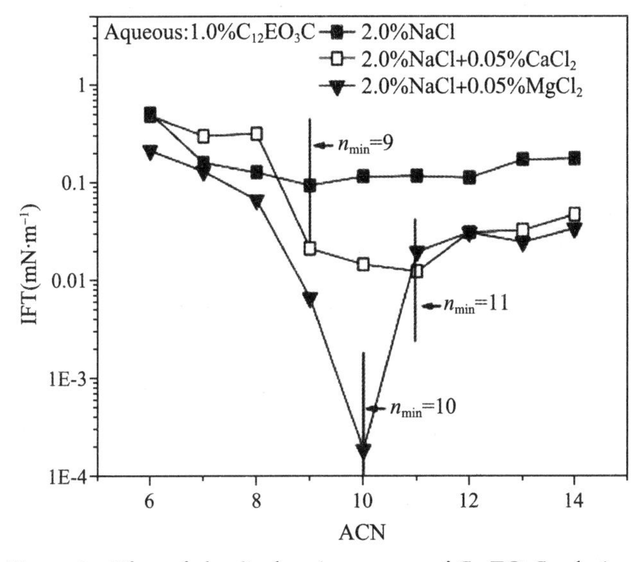

Figure 8　Efect of the divalent ion on n_{min} of $C_{12}EO_3C$ solution.

at the interface by the addition of the divalent ion, as shown in Figure 3. However, the minima in both dynamic IFT curve and stable value versus ionic strength curves cannot be explained well by the binding of counterion to interfacial surfactant molecules. We will discuss the mechanism responsible for the above experimental results in the following part.

3.3.2 Effect of the Divalent Ion on Interfacial Properties at Various $C_{12}EO_3C$ Concentrations.

The efects of divalent ions on interfacial properties at various $C_{12}EO_3C$ concentrations have been investigated. Panels A and B of Figure 5 show the dynamic IFT for 0.1 mass % $C_{12}EO_3C$ solutions as a function of interfacial age at di erent $CaCl_2$ and $MgCl_2$ concentrations, respectively. In this case, the dynamic IFT curves correspond to the 1.0 mass % $C_{12}EO_3C$ systems for $CaCl_2$. However, with the addition of $MgCl_2$, the "V" shape disappears during the experimental $MgCl_2$ concentration range. It seems that Ca^{2+} has more efect on the dynamic IFT behavior than that of Mg^{2+} at 0.1 mass % $C_{12}EO_3C$ concentration.

Figure 6 describes the dynamic IFT for the 0.4 mass % $C_{12}EO_3C$ as a function of interfacial age at di erent $CaCl_2/MgCl_2$ concentrations. We can see from the results that the dynamic IFT curves corresponded to the 1.0 mass % $C_{12}EO_3C$

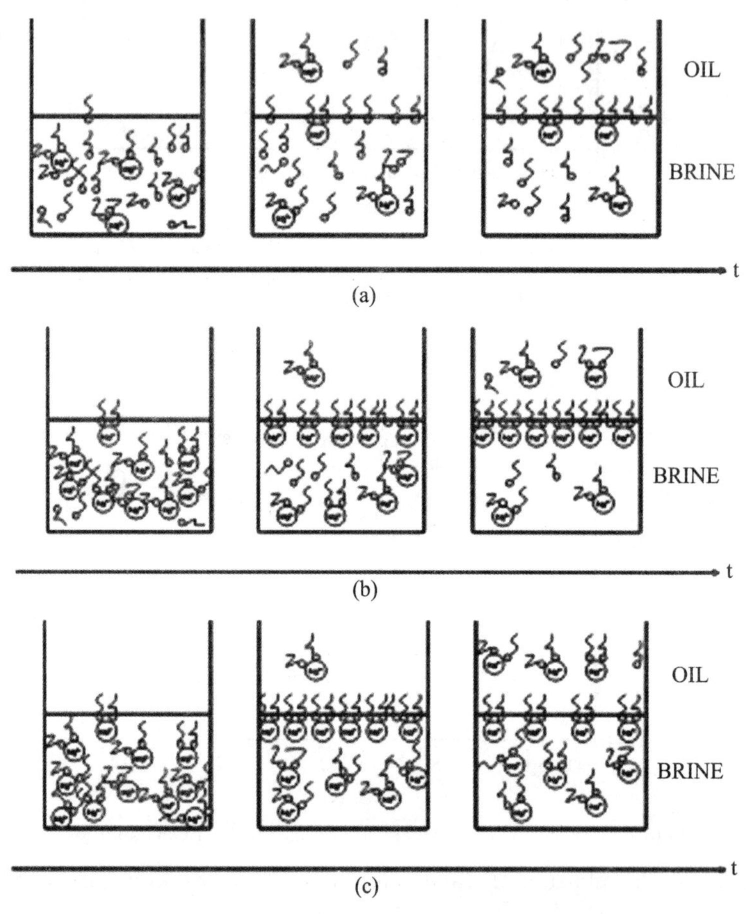

Figure 9 Proposed mechanism for the efect of electrolyte on dynamic IFT.

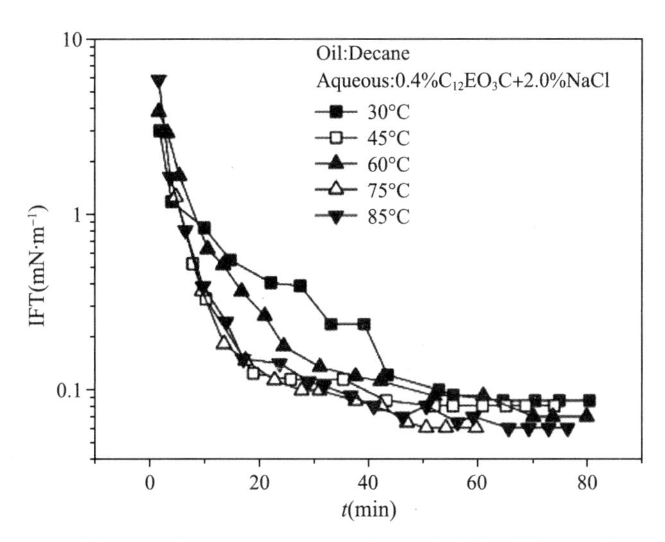

Figure 10 Efect of the temperature on dynamic IFT of the NaCl system.

systems again and the stable value of IFT reaches 10^{-4} mN/m order of magnitude. However, for the 0.4 mass% $C_{12}EO_3C$ system, both the critical concentration of $CaCl_2$ and $MgCl_2$ is 0.04 mass%. The time of appearing transient minimum value of dynamic IFT becomes shorter with the increase of the $CaCl_2$/ $MgCl_2$ concentration in general.

Figure 7 illuminates the stable value of dynamic IFT as a function of the $CaCl_2$/ $MgCl_2$ concentration against di erent $C_{12}EO_3C$ concentrations. It can be observed that the stable value decreases intensively by additional divalent ions. An optimum concentration could be seen in both $CaCl_2$ and $MgCl_2$ cases at di erent $C_{12}EO_3C$ concentrations. Besides, the addition of $MgCl_2$ corresponds to a more sensitive range of the $C_{12}EO_3C$ concentration to achieve ultralow IFT. In other words, the addition of $MgCl_2$ has a stronger tendency to achieve ultralow IFT than that of $CaCl_2$.

3.3.3 Efect of the Electrolyte on hydrophilic-lipophilic Ability.

The addition of di erent electrolytes has a distinct efect on the hydrophilic-lipophilic balance of surfactant solutions. Figure 8 shows the efect of the divalent ion on n_{min} of $C_{12}EO_3C$ solution. In this study, it can be observed that n_{min} increases with the addition of divalent ion, because the partitioning of the surfactant in the oil phase increases as the salinity increases, as shown in Figure 8. We can see from the results that the addition of 0.05 mass% $CaCl_2$ and $MgCl_2$ will improve n_{min} of 1 mass% $C_{12}EO_3C$ + 2.0 mass% NaCl solution from 9 to 11 and 10, respectively. It is worth noting that the addition of $MgCl_2$ has a stronger tendency to achieve ultralow IFT than that of $CaCl_2$ at n_{min}, while the addition of $CaCl_2$ has a stronger tendency to partition surfactant molecules to the oil phase.

3.3.4 Proposed Mechanism for the Efect of the Electrolyte on IFT.

The five major experimental findings related to the interfacial behaviors of $C_{12}EO_3C$ brine solution are as follows:(1) The NaCl concentration has little efect on the IFT. (2) For $CaCl_2/MgCl_2$ systems, beyond a particular divalent ion concentration, the transient values of IFTs drop quickly and pass through an apparent minimum in dynamic IFT curve, followed by a gradual increase to an equilibrium value. (3) For a given $C_{12}EO_3C$ concentration, under a given temperature, the stable value of IFT passes through a minimum with an increasing $CaCl_2/MgCl_2$ concentration. (4) The critical concentration for dynamic IFT coincided with the optimum electrolyte concentration for the stable value of IFT, and this critical concentration is related to the surfactant concentration. (5) The addition of $MgCl_2$ has a stronger tendency to achieve ultralow IFT than that of $CaCl_2$ at n_{min}, while the addition of $CaCl_2$ has a stronger tendency to partition surfactant molecules to the oil phase.

It is well-known that the addition of electrolyte has at least two tendencies on the efect of IFT. One is to partition the surfactant into the oil phase. The other is to decrease the charge repulsive force between interfacial surfactant molecules and result in an increase of the interfacial surfactant amount, which always leads to the reduction of the IFT. Therefore, the experimental result in this paper cannot be explained only by the interfacial adsorption model in Figure 3, and the variation of the hydrophilic-lipophilic balance of the surfactant by adding electrolyte must be taken into account.

Then, according to the experimental results, for Mg^{2+} as an example, we propose the mechanism responsible for the efect of divalent ions on dynamic IFT of $C_{12}EO_3C$ solutions, as shown in Figure 9.

At a lower $MgCl_2$ concentration, for a given oil phase, the binding of the Mg^{2+} ion to $C_{12}EO_3C$ increases the coverage ratio of the surfactant at the interface, which causes a decrease in the IFT. Therefore, the transient values of IFTs decrease gradually to the stable value, and the dynamic IFT curves show a "L" shape, which is schematically illustrated in Figure 9a.

As one increases the concentration of $MgCl_2$, more Mg^{2+} ions bind to interfacial $C_{12}EO_3C$ molecules and reach their maximum value at the critical concentration. As a result, the stable value of dynamic IFT reaches its minimum, as shown in Figure 9b.

Beyond the critical concentration, the binding of Mg^{2+} ions to $C_{12}EO_3C$ will make the obvious enhanced tendency of the

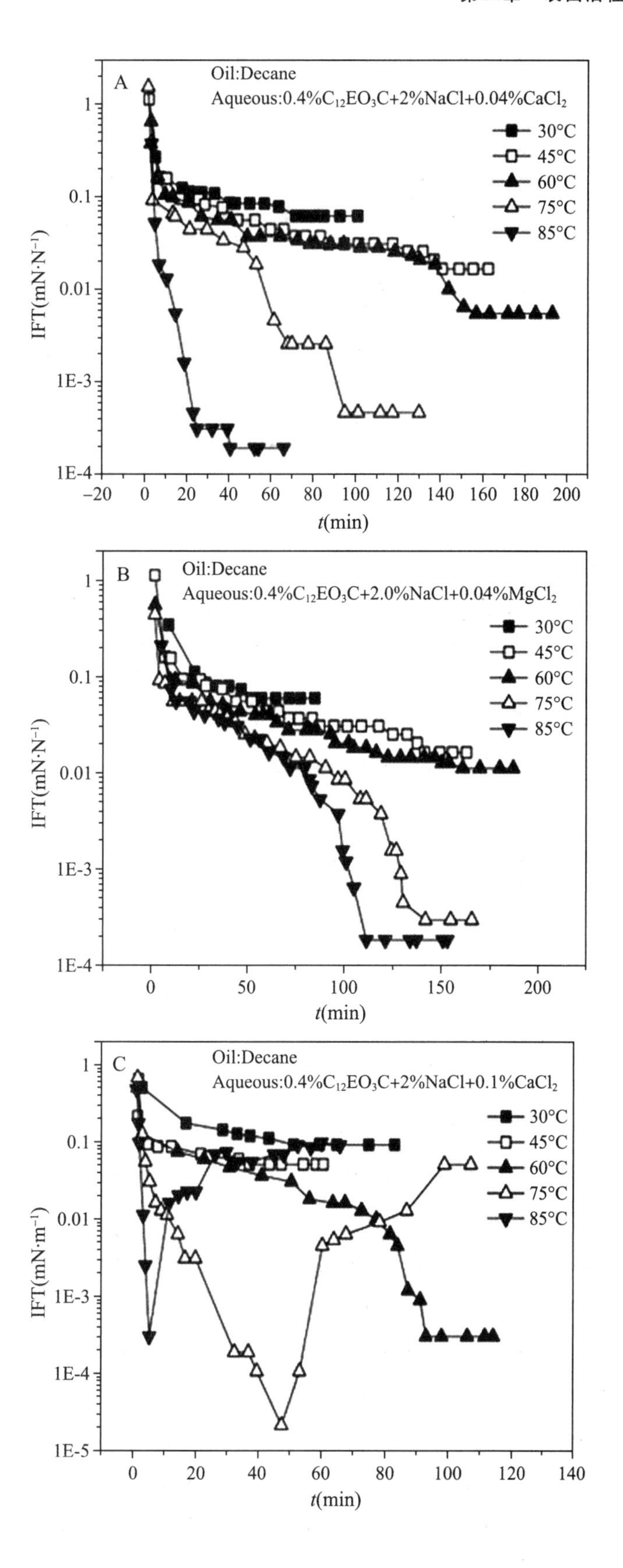
A
Oil:Decane
Aqueous:0.4%C12EO3C+2%NaCl+0.04%CaCl2
30°C
45°C
60°C
75°C
85°C
IFT(mN·N-1)
t(min)
B
Oil:Decane
Aqueous:0.4%C12EO3C+2.0%NaCl+0.04%MgCl2
30°C
45°C
60°C
75°C
85°C
IFT(mN·N-1)
t(min)
C
Oil:Decane
Aqueous:0.4%C12EO3C+2%NaCl+0.1%CaCl2
30°C
45°C
60°C
75°C
85°C
IFT(mN·m-1)
t(min)

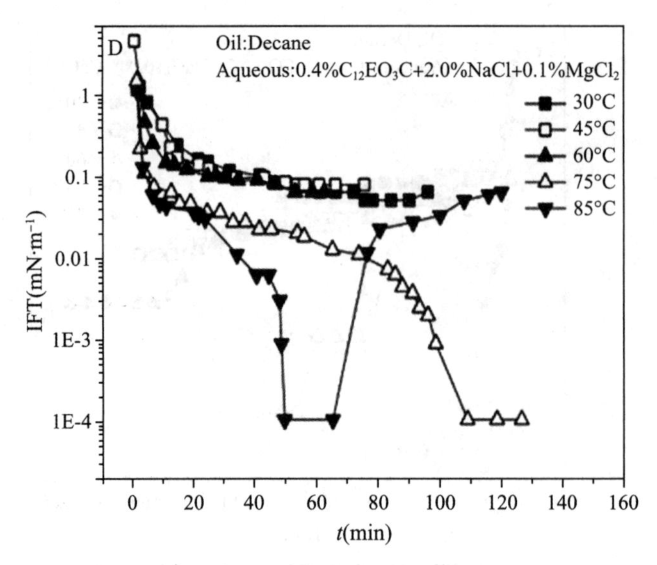

Figure 11 Efect of the temperature on dynamic IFT of the divalent ion system.

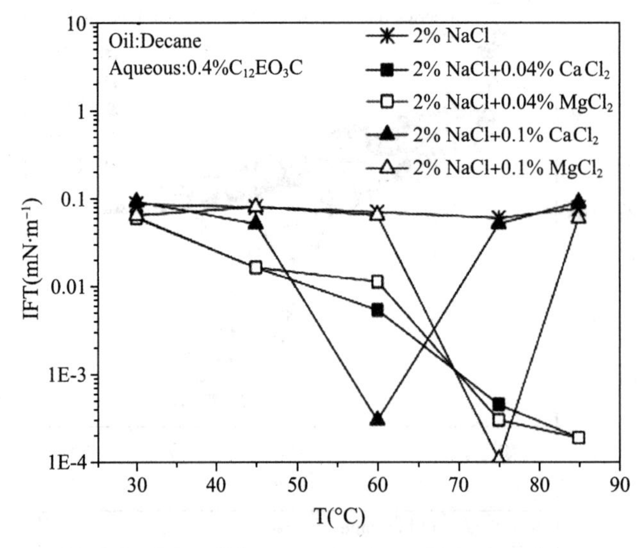

Figure 12 Efect of the temperature on the stable value of dynamic IFT.

oil solubility of the surfactant. Therefore, the binding of Mg^{2+} increases with interfacial aging, which results in the decrease of IFT first, and then the desorption of interfacial surfactant molecules into the oil phase leads to the increase of IFT. As a result, the dynamic IFT shows a "V" shape and the stable value becomes higher than that in an optimum electrolyte concentration, as shown Figure 9c.

The efect of Ca^{2+} ions on dynamic IFT of $C_{12}EO_3C$ solutions is similar to that of Mg^{2+}. However, the extent of binding of Ca^{2+} is lower than that of Mg^{2+} because of the di erence of the ion radius, which leads to the stronger tendency of adding $MgCl_2$ upon achieving ultralow IFT than that of $CaCl_2$. On the other hand, Ca^{2+} shows stronger ability to partition surfactant molecules into the oil phase, as shown in Figure

8, which results in the critical concentration of Ca^{2+} systems being lower than that of Mg^{2+} systems at a low bulk $C_{12}EO_3C$ concentration. Moreover, the "V" shape of the dynamic IFT curve is determined by both binding and partition mechanisms. Therefore, the appropriate ratio of the surfactant/ divalent ion concentration is very important in controlling dynamic IFT behaviors, which makes the "V" shape appear easily at the 0.4 mass% $C_{12}EO_3C$ concentration.

3.4 Efect of the Temperature on IFT.

Generally, the partitioning of non-ionic surfactant in the oil phase will be enhanced as the temperature increases because of the destruction of the hydrogen bonds between polyoxyethylene groups and water molecules. In this part, we have investigated the dynamic IFT at di erent temperatures for 0.4 mass% $C_{12}EO_3C$ and various electrolyte solutions, which can further verify our hypothesis.

Figure 10 shows the dynamic IFT as a function of interfacial age at di erent temperatures from 30 to 85℃ against the solution of 0.4 mass% $C_{12}EO_3C$ and 2 mass% NaCl. We can see from the results that the temperature has little efect on the IFT in the NaCl system, which indicates good water solubility of sodium salt.

Because there exists polyoxyethylene groups in $C_{12}EO_3C$ molecules, the partitioning of the surfactant in the oil phase increases as the temperature increases. Therefore, if our provided mechanism is reasonable, increasing the temperature is similar to increasing the concentration of the divalention. Figure 11 shows the dynamic IFT as a function of interfacial age at di erent temperatures from 30 to 85℃ against the solution of 0.4 mass% $C_{12}EO_3C$ and two concentrations of divalent ion. The temperature has a significant efect on the IFT in $CaCl_2/MgCl_2$ systems. The IFT decreases to an ultralow value with the increased temperature at 0.04 mass% divalent ions. In the case of 0.1 mass% divalent ions, "V" shapes of dynamic IFTs appear at 75 and 85℃ for $CaCl_2$ and $MgCl_2$ systems, respectively. These experimental results are similar to those of electrolytes and can be explained well by the mechanism provided in Figure 9.

Figure 12 illustrates the stable value of dynamic IFTs as a function of interfacial age at di erent temperatures from 30 to 85℃ against the solution of 0.4 mass% $C_{12}EO_3C$ and di erent electrolytes. It can be observed that the stable value decreases gradually for 0.04 mass% divalent ions because the surfactant tends to partition into the aqueous phase even at the highest experimental temperature (85℃). At the same time, the stable value passed through a minimum in the case of 0.1 mass% divalent ions at 60 and 75℃ for $CaCl_2$ and $MgCl_2$ systems, respectively. The efect of the temperature on the stable value of IFT also corresponded to that of the divalent ion concentration.

4 CONCLUSION

The influence of counterions on dynamic IFTs of fatty alcohol polyoxyethylene carboxylate $C_{12}EO_3C$ against alkanes has been studied. The efect of the temperature on IFT has been investigated. On the basis of our experimental results, the follow conclusions can be drawn: (1) The stable value of IFT decreases with an increasing bulk concentration for $C_{12}EO_3C$ against decane, and the n_{min} value of 1.0% surfactant at 2.0% NaCl is 9. (2) The NaCl concentration has little efect on the IFT, and dynamic IFT shows a "L" shape. However, with the increasing $CaCl_2$ or $MgCl_2$ concentrations, dynamic IFT passes through a minimum at a particular salt concentration ("V" shape). Moreover, the stable value of IFT also passes through a minimum at the same salt concentration. (3) The presence of the "V" shape dynamic IFT is related to the $C_{12}EO_3C$ concentration, and an appropriate ratio of the surfactant/divalent ion concentration will benefit the lowering of IFT and "V" shape curve. (4) The addition of the divalent ion has a distinct efect on the hydrophilic-lipophilic balance of surfactant solutions. The addition of 0.05% $CaCl_2$ and $MgCl_2$ will improve n_{min} of $C_{12}EO_3C$ solution from 9 to 11 and 10, respectively. (5) $MgCl_2$ has a stronger tendency to achieve ultralow IFT than that of $CaCl_2$ at n_{min}, while the addition of $CaCl_2$ has a stronger tendency to partition surfactant molecules to the oil phase. (6) For $C_{12}EO_3C$ solution, ultralow IFT could be achieved and "V" shape IFT will appear by improving the temperature because of the enhancement of the oil solubility of the surfactant. (7) An interfacial model combining two mechanisms, partitioning the surfactant into the oil phase and decreasing the charge repulsive force between interfacial surfactant molecules, of the divalent ion responsible for the efect of the electrolyte on dynamic IFT has been provided. All experimental results above can be explained well by this model.

AUTHOR INFORMATION

Corresponding Author

* Telephone: 86-10-82543587. Fax: 86-10-62554670. E-mail: luyiqiao@hotmail.com (L. Z.); zhaosui@mail.ipc.ac.cn (S. Z.).

Notes

The authors declare no competing financial interest.

ACKNOWLEDGMENTS

The authors thank financial support from the National Science and Technology Major Project (2011ZX05011-004) of China.

REFERENCES

[1] Shah, D. O. Fundamental aspects of surfactant-polymer flooding process. In Enhanced Oil Recovery; Fayers, F. J. , Ed. ; Elsevier: Amsterdam, The Netherlands, 1981:1-42.

[2] Lai, C. C. ; Chen, K. M. Colloids Surf. , A 2008,320:6-10.

[3] Well, J. K. ; Stirton, A. J. ; Bistlire, R. G. J. Am. Oil Chem. Soc. 1959, 36:241-247.

[4] Witthayapanyanon, A. ; Acosta, E. J. ; Harwell, J. H. ; Sabatini, D. A. J. Surfactants Deterg. 2006,9:331-339.

[5] Zeng, J. X. ; Ge, J. J. ; Zhang, G. C. ; Liu, H. T. ; Wang, D. F. ; Zhao, N. J. Dispersion Sci. Technol. 2010,31:307-313.

[6] Aoudia, M. ; Al-Shibli, M. N. ; Al-Kasimi, L. H. ; Al-Maamari, R. ; Al-Bemani, A. J. Surfactants Deterg. 2006,9:287-293.

[7] Aoudia, M. ; Al-Maamari, R. S. ; Nabipour, M. ; Al-Bemani, A. S. Energy Fuels 2010,24:3655-3660.

[8] Aoudia, M. ; Al-Harthi, Z. ; Al-Maamari, R. S. ; Lee, C. ; Berger, P. J. Surfactants Deterg. 2010,13:233-242.

[9] Sha, O. ; Zhang, W. ; Lu, R. Tenside, Surfactants, Deterg. 2008,45:82-86.

[10] Al-Sabagh, A. M. ; Azzam, E. M. S. ; Mahmoud, S. A. ; Saleh, N. E. A. J. Surfactants Deterg. 2007,10:3-8.

[11] Padget, J. C. Coat. Technol. 1994,66:89-105.

[12] Patist, A. ; Axleberd, T. ; Shah, D. O. J. Colloid Interface Sci. 1998,208: 259-265.

[13] Rosen, M. J. Surfactants and Interfacial Phenomena; Wiley Publishers: New York, 1989:207-239.

[14] Chan, K. S. ; Shah, D. O. J. Dispersion Sci. Technol. 1980,1:55-95.

[15] Zhang, L. ; Luo, L. ; Zhao, S. ; Yu, J. Y. J. Colloid Interface Sci. 2002, 251:166-171.

[16] Chattopadhyay, A. K. ; Ghaicha, L. ; Oh, S. G. ; Shah, D. O. J. Phys. Chem. 1992,96:6509-6513.

[17] Yan, H. ; Guo, X. L. Langmuir 2011,27:5762-5771.

[18] Zhao, R. H. ; Zhang, L. ; Zhang, L. ; Zhao, S. ; Yu, J. Y. Energy Fuels 2010,24:5048-5052.

[19] Zhang, L. ; Luo, L. ; Zhao, S. ; Yu, J. Y. Ultra low IFT and interfacial dilational properties related to enhanced oil recovery. In Petroleum Science and Technology Research Advances; Montclaire, K. L. , Eds. ; Nova Science Publishers: New York, 2008:81-139.

[20] Cayias, J. L. ; Schechter, R. S. ; Wade, W. H. J. Colloid Interface Sci. 1977,59:31-38.

[21] Cash, L. ; Cayias, J. L. ; Fournier, G. ; Macallister, D. ; Schares, T. ; Schechter, R. S. ; Wade, W. H. J. Colloid Interface Sci. 1977,59:39-44.

[22] Kim, D. H. ; Oh, S. G. ; Cho, C. G. J. Colloid Polymer Sci. 2001,279: 39-45.

[23] Pandey, S. ; Bagwe, R. P. ; Shah, D. O. J. Colloid Interface Sci. 2003, 267:160-166. 3129

Dynamic interfacial tensions between offshore crude oil and enhanced oil recovery surfactants

Song Xinwang[1,2] Zhao Ronghua[3] Cao Xulong[2] Zhang Jichao[2]
Zhang Lei[3] Zhang Lu[3] Zhao Sui[3]

[1] China University of Petroleum (Beijing), Beijing, P. R. China

[2] Geological Scientific Research Institute of Shengli Oilfield Co. Ltd, SINOPEC, Dongying, Shandong, P. R. China

[3] Technical Institute of Physics and Chemistry, Chinese Academy of Sciences, Beijing, P. R. China

Graphichl abstract:

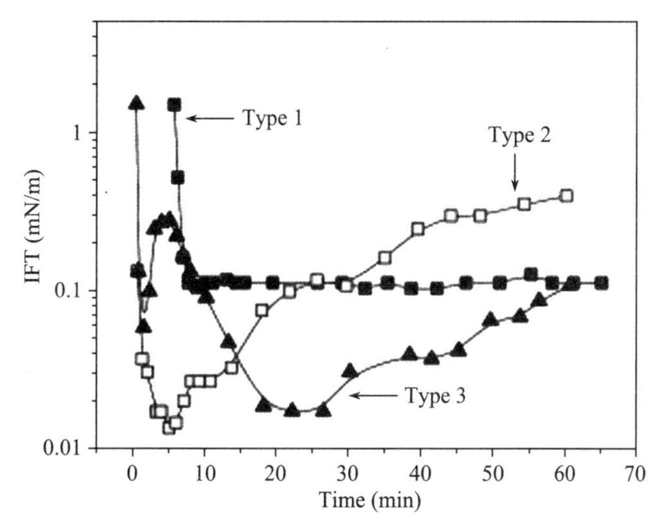

The dynamic interfacial tensions between offshore crude oil and different types of EOR surfactants were investigated by spinning drop method at 65 degree under the condition of weak base (pH、8) in the present work. Effects of surfactant concentration on the dynamic interfacial tension were investigated. In the presence of weak alkali, the active components of crude oil can react at the interface to produce surface-active species in situ. The interactions among the added surfactants, the petroleum active components and in situ produced surface-active species together determine the dynamic interfacial tension behaviors, whose curves show"L","V"and"W"shapes, respectively. Surfactant type and concentration play the crucial roles on dynamic interfacial tension behaviors.

Keywords: Crude oil, dynamic interfacial tension, enhanced oil recovery, surfactant

1 Introduction

It is well recognized that crude oil plays an important role in providing the energy supply of the world among various sources of energy. In the so-called primary oil recovery process (natural flow), the yield is generally only 10%—20% of the original oil in place (OOIP). By application of secondary oil recovery method, such as water flooding, the yield can be increased to 20%—30% OOIP. At the end of water flooding, almost 65%—70% of OOIP is left in the reservoirs, which is believed to be in the form of ganglia and trapped in the pore structure of the rock by capillary forces. In order to recovery additional oil by a chemical flooding process, the capillary number, which determines the microscopic displacement efficiency of oil, should be increased by 3 to 4 orders of magnitude through reducing the interfacial tension (IFT) of oil ganglia from its value of 20—30 mN/m to 10^{-3} mN/m.[1]

Producing ultra-low IFT is one of the most important mechanisms for oil recovery with respect to chemical flood-ing, such as alkaline water flooding, surfactant flooding, surfactant-polymer flooding and alkali-surfactant-polymer flooding. It has long been known that IFT is reduced between an acidic crude oil and an alkaline aqueous phase, due to the producing of in situ surfactant (ionized acid).[2,3] Preformed surfactants have been added to crude oil/alkali systems for several reasons: the resulting IFT is lower than that obtained with alkali or surfactant alone; the salinity at which the optimal phase behavior and the lowest IFT occur is increased; the ionized acid may be lost by adsorption in the soil; and the oil has a low acid content.[4,5] By using an appropriate surfactant system, the IFT between crude oil and chemical flooding solutions can be reduced to lower than 10^{-2} mN/m. The predominant type of surfactant investigated for enhanced oil recovery (EOR) has been sul-fonated hydrocarbons (petroleum sulfonates and synthetic sulfonates).[3,6]

It is well known that acid present in the crude oil reacts with the alkaline solution to produce in situ surfactant (ionized acid) which lowers the oil-water IFT. The ionized acid is very surface active and has a trend to partition into the aqueous phase. The ionized acid may form soap with sodium ions present in the aqueous phase at a high ionic strength. This soap has a trend to partition into the oil phase. The remove of the ionic acids from interface causes the dynamic IFT. A wide range of views exists in the litera-ture on the importance of dynamic IFT in crude oil reser-voirs. The significance of the dynamic IFT minimum for reservoir situations has been discussed by Rubin and Radke[7] and de Zabala and Radke.[8] They suggested that the IFT minimum for acidic crude oils gives the lowest achievable reservoir equilibrium value. Cambridge et al.[9] suggested that dynamic IFT is related to the kinetics of the underlying chemical process. Taylor et al.[10] showed that oil recovered through the surfactant-enhanced alkaline flooding of linear Berea sandstone cores correlates better with the minimum dynamic IFT than with equilibrium IFT values. They reported that an initial low IFT value is most important, while the effect of further changes in IFT with time appears to be relatively unimportant in the core flooding process examined in their work. Thus it is important to examine the factors affecting dynamic IFT.

The mechanisms responsible for dynamic IFT become more complicated when EOR surfactants are added into alkaline solutions. The interactions among surfactants, pet-roleum acids and ionized acid will leads to very different dynamic IFT behaviors. In this paper, the dynamic IFTs between offshore crude oil and different types of EOR surfactants were measured under the condition of weak base (pH 、8). The three main kinds of dynamic IFT curves have been concluded and the possible mechanisms have been provided.

2 EXPERIMENTAL

2.1 Materials

In this study, four kinds of surfactants provide by Shengli Oilfield of China were used. The structures and abbreviations of these surfactants were listed in Table 1.

The offshore crude oil and formation brine sample were col-lected from Shengli Oilfield (Dongying, China). The density of crude oil is 0.9055 g/mL at 65.0℃. All of the aqueous solutions used in this study were prepared with formation brine. The compositions of the formation brine were listed in Table 2. The pH values of formation brine and 0.2 wt% surfactant solutions were listed in Table 3.

2.2 Apparatus and Methods

The spinning drop technique was employed to measure dynamic IFTs. The standard spinning-drop tensiometer had been modified by the addition of video equipment and an interface to a personal computer.[11] The computer had been fitted with a special video board and a menu-driven image enhancement and analysis program. The video board can "capture" a droplet image for immediate analy-sis. Analysis usually consists of measurement of drop length and drop width.

The volumetric ratio of water to oil in the spinning drop tensiometer is about 200. Samples were assumed to be equilibrated when measured values of IFT remained unchanged for half an hour. All experiments were per-formed at 65.0±0.5℃.

3 RESULTS AND DISCUSSION

Reduction of IFT is one of the most commonly mea-sured properties of surfactants in solution, which depends directly on the replacement of molecules of solvent at the interface by molecules of surfactant. When the surfactant molecules replace water and oil molecules of the original oil-water interface, the interaction across the interface is now between the hydrophilic group of the surfactant and water molecules on one side of the interface and between the hydrophobic group of the surfactant and oil molecules on the other side of the interface. Since these interactions are much stronger than the original interaction between the highly dissimilar oil molecules and water molecules, the IFT across the interface is significantly reduced. We can deduce from above that the interfacial surfactant concentration and the surfactant molecular size will be the two most important factors in dominating reduction of IFT.

Table 1 Structures and abbreviations of EOR surfactants

EOR surfactants	Abbreviation
Shengli petroleum sulfonate	SLPS
Polyoxyethylene-anionic amphoteric surfactant	SLNA
Nonionic-anionic amphoteric polymeric surfactant	SLPNA
Oligomeric anionic surfactant	SLOA

Table 2 Composition analysis of the formation brine

Components	Cl^-	SO_4^{2-}	HCO^{3-}	Na^++K^+	Ca^{2+}	Mg^{2+}	Total dissolved solids
Concentration (mg/L)	16 279	443	337	9 760	205	474	27 498

It is well recognized that a proper balance between hydrophilic and lipophilic character in the surfactant is essential for significant surface/interfacial activity. In general, good surface or IFT reduction is shown only by those surfactants that have an appreciable, but limited, solubility in the system under the conditions of use. However, the interfacial surfactant concentration is the necessary condition, not the sufficient, for obtaining low or ultralow IFT. The molecular size of surfactant plays a crucial role in reduction of IFT. Therefore, even if the hydrophilic-lipophilic nature of surfactant is optimized, the IFT value cannot reach ultralow resulted from the insertion of solvent molecules into loose interfacial adsorp-tion film form the oil side. Similarly, solvent molecules can easily insert into loose interfacial adsorption film form the water side when the size of hydrophobic part is bigger than that of hydrophilic part. Only when the size of hydro-phobic part matches that of hydrophilic part, ultralow IFT will appear at optimized hydrophilic-lipophilic balance.[3,12-14]

Unfortunately, the suitable structure of surfactant for EOR cannot be industrialized due to expensive economic cost except for petroleum sulfonates. However, petroleum sulfonates have been limited by its precipitating with bivalent cations. Therefore, the studies of synergisms among added surfactants, petroleum surface-active compo-nents and the formation brine are very important in theory and application.[15-17]

Table 3 pH of the EOR surfactant solutions prepared by formation brine

Solutions	pH
0.2 wt% SLPS	8.09
0.2 wt% SLNA	8.41
0.2 wt% SLPNA	8.30
0.2 wt% SLOA	8.22

3.1 Dynamic Interfacial Tension Between n-Decane and EOR Surfactants

The dynamic IFTs between EOR surfactant solutions and n-decane are shown in Figure 1. It is clear to see that SLPS has the strongest surface activity and the stable value of dynamic IFT for 0.2% SLPS and decane system is in 0.01 mN/m order of magnitude. However, the other three kinds of surfactant, SLNA, SLPAN, and SLOA, can only produce IFT values of 1 mN/m order of magnitude. The dynamic IFTs curve of SLPS in Figure 1 is the most common and most typical characteristic curve: the IFT decreases and then reaches a plateau with time, which curves show typically L-shaped. As known, once the aqueous phase and the oil phase contact, surfactant molecules in the aqueous phase begin to assemble onto interface. As a result, interfacial con-centration increases and the IFT values descend along with time passed. At the same time, interfacial surfactant mole-cules will partition into oil phase due to concentration gradi-ent. In this system, SLPS molecules have a relatively stronger tendency to partition from interface into oil phase. Therefore, the interfacial surfactant concentration will increase monoto-nously and arrive at plateau value till partition equilibrium among the interface and the bulk phases appear.

3.2 Dynamic Interfacial Tension Between Crude Oil and SLPS

The dynamic IFTs between crude oil and SLPS solutions as a function of time have been plotted in Figure 2. We can see from Figure 2 that dynamic IFT curves of the different SLPS concentrations have similar characteristics: drop quickly firstly and then pass through an unobvious mini-mum. At last, the dynamic IFT value has a slight increase until it reaches the platform value. Basically, the dynamic IFT characteristic curves of SLPS are of the L-shaped. The emergency of an unobvious minimum indicates that the in situ generation of surface-active species by the reac-tion of the active component in the crude oil and the alkali will reduce the IFT within a short time under the condition of weak base. When in situ generation of surface-active spe-cies transfers to the aqueous phase, the IFT value increases slightly. On the other hand, the IFT value between SLPS and crude oil is higher than that of n-decane due to the hin-drance of surface-active species to the adsorption of SLPS with high interfacial activity. Anyway, for SLPS solutions, the high interfacial activity of petroleum sulfonate plays a major role, while the synergism between alkaline-active components in crude oil has little effect.

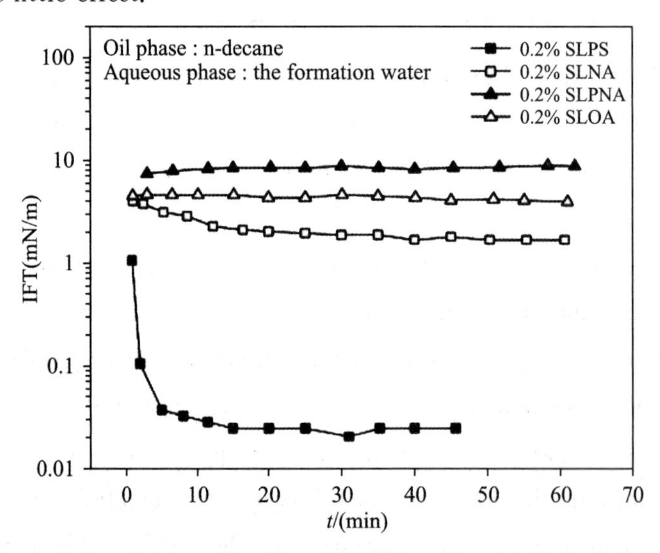

Fig. 1 Dynamic interfacial tensions between EOR surfactant solutions and n-decane.

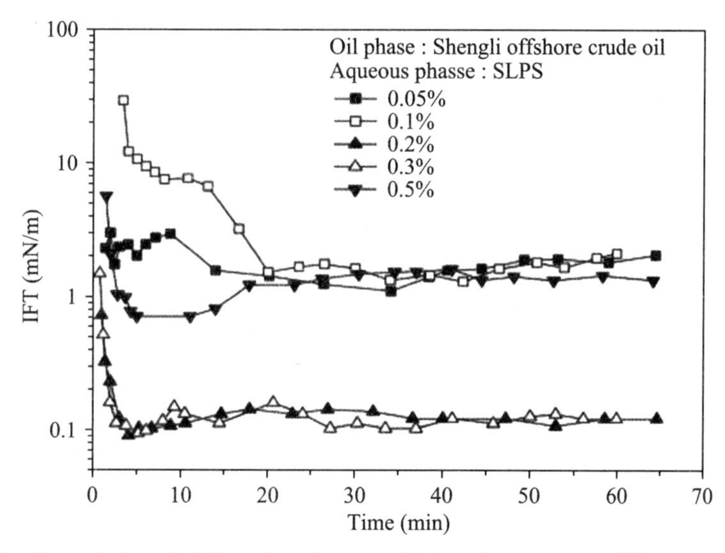

Fig. 2 Dynamic interfacial tensions between SLPS solutions and offshore crude oil.

3.3 Dynamic Interfacial Tension Between Crude Oiland SLNA

The dynamic IFTs between SLNA solutions and off-shore crude oil as a function of time are given in Figure 3. Unlike SLPS, Figure 3 shows clearly that dynamic IFT values of SLNA solutions pass through a minimum and increase to the stable value against offshore crude oil, especially in the case of high concentration. It is indicates that there exists obvious synergism between SLNA and alkali-active components in crude oil. Moreover, the higher the surfactant concentration is, the shorter the time needed for dynamic IFT to reach its minimum.

Figure 3 shows another type of dynamic IFT curve, which shows V-shaped behavior. For SLNA and offshore system, the surfactants and the in situ generation of surface-active species by the reaction of the active compo-nents in the crude oil and alkali work together to reduce the IFT values in short time. As time goes on, when interfacial surface-active species has transferred from the interface into the aqueous phase, the synergism becomes weak and IFT increases. As can be seen from Figure 3, the surfactant concentration can influence the transient IFT significantly. At low surfactant concentration, the synergism among sur-factant and the surface-active species was the strongest and the value of dynamic IFT is lowest. With an increase in the surfactant concentration, the synergism has been weakened and the IFT increases. In addition, an increase of the sur-factant concentration would lead to a faster diffusion exchange, which results in reducing the time for the transi-ent IFT minimum. The above results indicate the syner-gism among SLNA and acidic components in crude oil, and their ionized acids is controlled by their ratio and the extreme increase of interfacial surfactant concentration may destroy the synergism.[18]

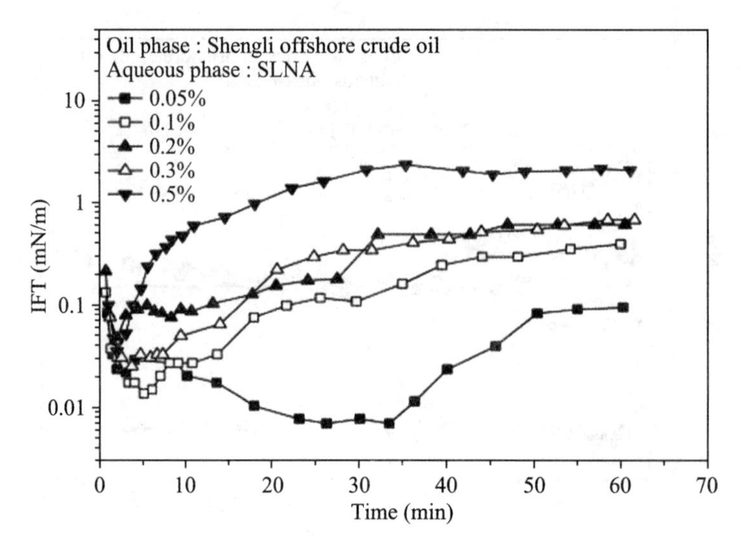

Fig. 3 Dynamic interfacial tensions between SLNA solutions and offshore crude oil.

3.4 Dynamic Interfacial Tension Between Crude Oil and SLPNA and SLOA

Figures 4 and 5 show the dynamic IFT between offshore crude oil and SLPNA or SLOA, respectively. It can be seen that the transient values are in 0.01 mN/m order of magni-tude and the stable values kept about the order of magni-tude 0.1 mN/m, which indicates that it has obvious synergism existed among surfactants, acidic components in offshore crude oil, and the in situ generated surface-active species for SLPNA or SLOA system.

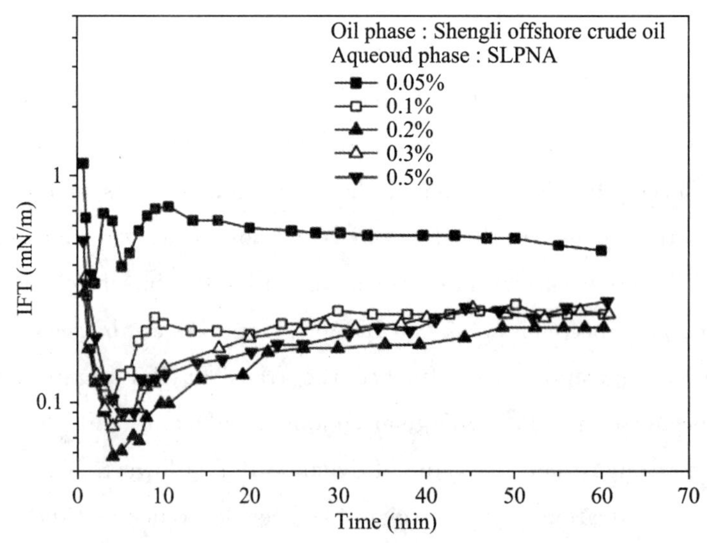

Fig. 4 Dynamic interfacial tensions between SLPNA solutions and offshore crude oil.

It is very interesting to point out that the dynamic IFT of SLPNA and SLOA at low concentration passes through a minimum quickly (<5 minutes), increases to a certain value, and then re-decreases or passes through another minimum. Thus, the dynamic IFT curves show W-shaped behavior. Such phenomena are believed to be due to the complexity of organic acidic components in crude oil. Crude oils of different provenance are all-different in com-position and contain a wide variety of acidic components. The carboxylic acids with different molecular weights and structures present in crude oil are complicated, which have not been fully identified. But so far, it is well known that their number of carbon atoms falls in the range from C_8 to C_{26} and their molecular weights are in the range 200 to 700. In the previous studies,[19] we have found that the acid fractions with lower average molecular weight (Mn＜500) have considerable aliphatics in the R side chain and show stronger interfacial activities and dynamic IFT behavior with alkaline solutions. The fractions with the higher average molecular weight (Mn＞500) have considerable aromaticity in the R side chain and show lower interfacial activities with alkaline solutions. For low concentration of SLPNA and SLOA, the diffusion rate of surfactants from the aqueous phase to the interface is slow, in a short time, which has little effect on the interface. Here the interfacial ionized acid produced by the reaction of low molecular weight organic acids with alkali adsorp-tion at interface is the main factor that affects the reduction of IFT. When the ionized acid transfers from the interface into the aqueous phase, IFT goes up. With interface aging, the added surfactants begin to adsorb on the interface, as a result the IFT decreases again. In addition, the possibility of high molecular weight organic acidic components being extracted to the interface by alkali and the synergism among all interfacial active species can also reduce IFT. With the ratio of added surfactant to high molecular weight organic acids changing, its synergism reaches at the optimum and the dynamic IFT passes through another minimum.

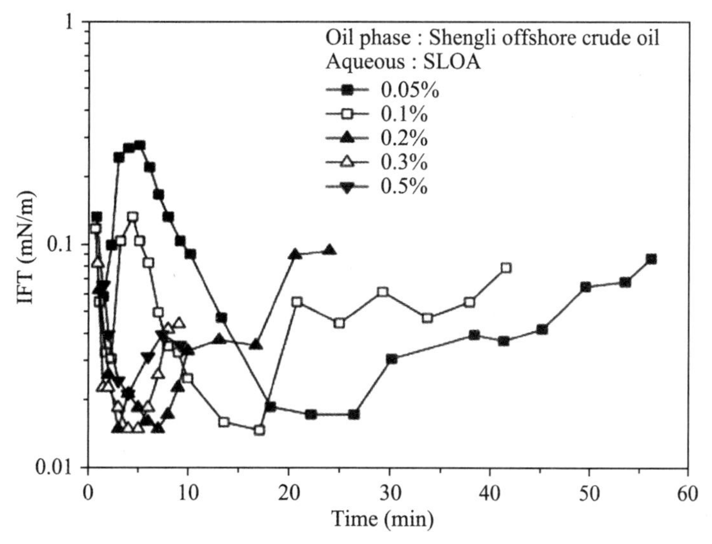

Fig. 5 Dynamic interfacial tensions between SLOA solutions and offshore crude oil.

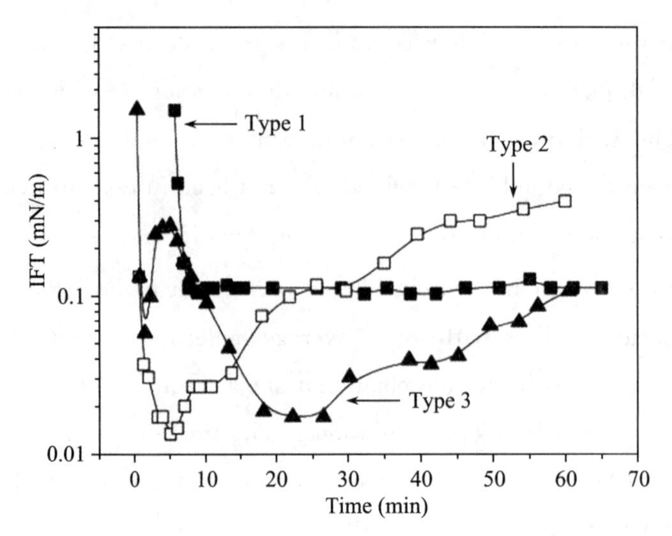

Fig. 6 Three main types of dynamic interfacial tension curves between EOR surfactant solutions and offshore crude oil.

4 Conclusions

The dynamic interfacial tensions between offshore crude oil and different types of EOR surfactants were investigated under the condition of weak base (pH 、8) in the present work. The interfacial active components in the crude oil can react with alkali at the interface to produce surface-active species in situ. The interactions among the added sur-factants, the crude oil components and in situ produced surface-active species together determine the dynamic IFT behaviors. The three main kinds of dynamic IFT curves have been concluded and the possible mechanisms have been provided as shown in Figure 6.

1. In case of SLPS, the added surfactant with high inter-facial activity plays a major role on the IFT. Therefore the IFT value decreases and then reaches a plateau with time, which curves show typically L-shape.

2. For SLNA, the synergism between the added surfac-tant and the ionized acid produced by the reaction of low molecular weight acids mainly determines the dynamic IFT behavior. The IFT passes through a mini-mum quickly and increases to a stable value, which curves show V-shape.

3. When the synergism among the added surfactant, the different petroleum acids and their ionized acid all have obvious effect, the dynamic IFT passes through more than two minima, which curves show W-shaped beha-vior. Such phenomena can be abserved at low concen-tration of SLPNA and SLOA.

References

[1] Shah, D. O. (1981) In Fundamental Aspects of surfactant-polymer Flooding Process (ENHANCED OIL RECOV-ERY), edited by F. J. Fayers; Amsterdam, the Netherlands: Elsevier:1-42.

[2] McCaffery, F. G. (1976) J. Can. Pet. Technol., 15:71-74.

[3] Zhang, L., Luo, L., Zhao, S., and Yu, J. Y. (2008) In Ultralow Interfacial Tension and Interfacial Dilational Properties Related to Enhanced Oil Recovery (Petroleum Science and Technology Research Advances), edited by M. L. Montclaire; New York: Nova Science:81-139.

[4] Taylor, K. C., Hawkins, B. F., and Islam, M. R. (1990) J. Can. Pet. Technol., 29: 50-55.

[5] Taylor, K. C. and Nasr-El-Din, H. A. (1996) Colloids Surf. A, 108: 49-72.

[6] Hernandez, C. (2003) In Surfactantfor Enhanced Oil Recovery Processes (Advances in Incremental Petroleum Production), edited by I. Lakatos; Akade'miai Kiado':Budapest:53-69.

[7] Rubin, E. and Radke, C. J. (1980) Chem. Eng. Sci., 35:1129-1138.

[8] deZabala, E. F. and Radke, C. J. (1986) Soc. Pet. Eng. Res. Eng., 1: 29-43.

[9] Cambridge, V. J., Wolcott, J. M. and Constant W. D. (1989) Chem. Eng. Commun., 84:97-111.

[10] Taylor, K. C., Hawkins, B. F., and Islam, M. R. (1990) J. Can. Pet. Technol., 29:50-55.

[11] Zhao, R. H., Zhang, L., Zhang, L., Zhao, S., and Yu. (2010)J. Y. Energy Fuel, 24:5048-5052.

[12] Doe, P. H., El-Emary, M., Schechter, R. S., and Wade, W. H. (1977) J. Am. Oil Chemists Soc., 54:570-577.

[13] Doe, P. H., El-Emary, M., Schechter, R. S., and Wade, W. H. (1978) J. Am. Oil Chemists Soc., 55:505-512.

[14] Doe, P. H., El-Emary, M., Schechter, R. S., and Wade, W. H. (1978) J. Am. Oil Chemists Soc., 55:513-520.

[15] Zhang, L., Luo, L., Zhao, S., and Yu, J. (2002) J. Colloid Interface Sci., 249:187-193.

[16] Zhang, L., Luo, L., Zhao, S., and Yu, J. (2002) J. Colloid Interface Sci., 251:166-171.

[17] Zhang, L., Luo, L., Zhao, S., and Yu, J. (2003) J. Colloid Interface Sci., 260:398-403.

[18] Chu, Y. P., Gong, Y., Tan, X. L., Zhang, L., Zhao, S., An, J. Y., and Yu, J. Y. (2004) J. Colloid Interface Sci., 276:182-187.

[19] Zhang, L., Luo, L., Zhao, S., Xu, Z. C., An, J. Y., and Yu, J. Y. (2004) J. Petro. Sci. Eng., 41:189-198.

稳态荧光探针法研究对-烷基-苄基聚氧乙烯醚羧酸甜菜碱的聚集行为

董林芳[1]　徐志成[1]　曹绪龙[2]　王其伟[2]　张　磊[1]　张　路[1]　赵　濉[1]

（1. 中国科学院理化技术研究所，北京　100190；

2. 中国石化胜利油田分公司 地质科学研究院，山东　东营　257015）

摘要：利用荧光探针法和表面张力法测定了一类疏水基中含有苯基的新型甜菜碱两性离子表面活性剂对-烷基-苄基聚氧乙烯醚羧酸甜菜碱（ABECB）的临界胶团浓度（cmc）、胶团微极性和表面张力（γ_{cmc}）。研究结果表明，荧光探针（芘）法可用来测定这类表面活性剂的临界胶团浓度（cmc），且测定结果与表面张力法（吊片法）接近；ABECB具有较低的cmc和γ_{cmc}值，表明此类表面活性剂具有优良的表面活性；胶团的微极性随着疏水链长的增大而略微减小，氧乙烯（EO）单元数的增大对ABECB胶团核内的微极性影响不明显。

关键词：对-烷基-苄基聚氧乙烯醚羧酸甜菜碱；稳态荧光探针；临界胶团浓度；胶团微极性；表面张力

文章编号：1674-0475(2013)03-0215-07

中图分类号：O647　**文献标识码**：A

表面活性剂在三次采油过程中是一个不可缺少的重要组分，针对三次采油用表面活性剂的主要性能要求，设计和制备甜菜碱两性离子表面活性剂具有十分重要的意义。普通表面活性剂作为驱油剂在高温、高矿化度下驱油效果差，同时由于含碱，会对地层及油井带来巨大伤害，而烷基酚聚氧乙烯醚羧酸甜菜碱型两性离子表面活性剂能够避免这些问题[1]。但是烷基酚聚氧乙烯醚类化合物由于分子内存在酚氧键，所以在生物降解过程中会产生具有生物毒性的烷基酚[2,3]，限制了该类表面活性剂在实际生活和工业生产中的应用[4,5]。对-烷基-苄基聚氧乙烯醚羧酸甜菜碱（ABECB）分子以苄氧键代替酚氧键，可在根本上解决降解过程中的生物毒性问题，对环境更加友好。

表面活性剂的许多性质与其在水溶液中的聚集行为密切相关，因此研究各种表面活性剂在水溶液中的聚集行为具有重要意义。从20世纪70年代发展并广泛应用的荧光探针法具有操作简单、对体系无特殊要求、用量少、对体系的干扰小等优点，已经广泛应用于cmc及聚集数等的测定[6-9]。本文通过稳态荧光探针法研究了6种自制的ABECB

的临界胶团浓度和微极性，讨论了 ABECB 分子结构对其聚集行为的影响，希望通过本文的研究加深对这类表面活性剂胶团微观结构的认识。

1 实验部分

1.1 实验仪器和试剂

HitachiF-4600 型荧光分光光度计(日本日立株式会社)；Dataphysics DCAT21 动态接触角/表界面张力测量仪，德国 Dataphysics 公司；THX-05 低温恒温循环器，宁波天恒仪器厂；Millipore Simplicity@UV 超纯水机，美国 Millipore 公司。实验温度均为(30±0.1)℃。

探针芘为 Alfa Aesar 公司产品，使用前未经进一步纯化，将其配制成浓度为 1.0×10^{-4} mol/L 的无水甲醇溶液备用；水为纯水；6 种 ABECB 均由本实验室合成[10,11]，其结构经 FT-IR、^{1}HNMR、ESIMS-确定，结果列于表 1。其结构及编号如图 1 所示。

表 1 ABECB 的物理参数 Physical parameter of ABECB

Surfactants	ESI-MS m/z		^{1}HNMR (400 MHz, $CDCl_3$, δ)	FT-IR(γ/cm^{-1})
	$[M+H]^+$	[M+Na]		
C_8BE_3CB	438.5	460.5	0.85—0.89(t,3H),1.26—1.29(m,10H),1.56—1.60(m,2H),2.56—2.60(t,2H),3.17(s,6H),3.60—3.65(m,8H),3.76(s,2H),3.89(s,4H),4.48(s,2H),7.14—7.16(d,2H),7.22—7.24(d,2H)	3020,2925,2855,1630,1464,1396,1101,968,894,852,720
$C_{10}BE_3CB$	466.4	488.4	0.86—0.89(t,3H),1.26—1.30(m,14H),1.57—1.60(m,2H),2.57—2.61(t,2H),3.19(s,6H),3.61—3.66(m,8H),3.84(s,2H),3.89(s,4H),4.49(s,2H),7.15—7.17(d,2H),7.22—7.24(d,2H)	3020,2925,2855,1630,1465,1397,1346,1099,850,721
$C_{12}BE_3CB$	494.6	516.6	0.86—0.89(t,3H),1.25—1.30(m,18H),1.56—1.58(m,2H),2.57—2.60(t,2H),3.18(s,6H),3.60—3.63(m,8H),3.84(s,2H),3.88(s,4H),4.48(s,2H),7.15—7.17(d,2H),7.22—7.24(d,2H)	3020, 2924, 2854, 1630, 1464, 1400, 1243,1095,851,720
$C_{12}BE_1CB$	406.6	428.6	0.86—0.90(t,3H),1.26—1.30(m,18H),1.58—1.61(m,2H),2.58—2.61(t,2H,3.31(s,6H),3.85(s,2H),3.91(s,2H),3.94(s,2H),4.50(s,2H),7.14—7.16(d,2H),7.18—7.21(d,2H)	3020,2924,2853,1630,1466,1399,1341,1120,852,721

(续表)

Surfactants	ESI-MS m/z		^{1}HNMR (400 MHz, $CDCl_3$, δ)	FT-IR(γ/cm^{-1})
	[M+H]$^+$	[M+Na]		
$C_{12}BE_2CB$	450.5	472.5	0.86—0.89 (t, 3H), 1.25—1.30 (m, 18H), 1.57—1.61(m, 2H), 2.57—2.61 (t, 2H), 3.27(s, 6H), 3.61(s, 2H), 3.66 (s, 2H), 3.85(s, 2H), 3.92(s, 4H), 4.49 (s, 2H), 7.14—7.16(d, 2H), 7.19—7.21 (d, 2H)	3020, 2924, 2854, 1628, 1465, 1398, 1246, 1103, 893, 852, 721
$C_{12}BE_4CB$	538.6	560.6	0.86—0.90 (t, 3H), 1.25—1.30 (m, 18H), 1.56—1.60(m, 2H), 2.56—2.60 (t, 2H), 3.18(s, 6H), 3.62—3.64(m, 12H), 3.76(s, 2H), 3.85—3.87(m, 4H), 4.50(s, 2H), 7.14—7.16(d, 2H), 7.23—7.25(d, 2H)	3020, 2924, 2854, 1630, 1465, 1398, 1346, 1103, 851, 722

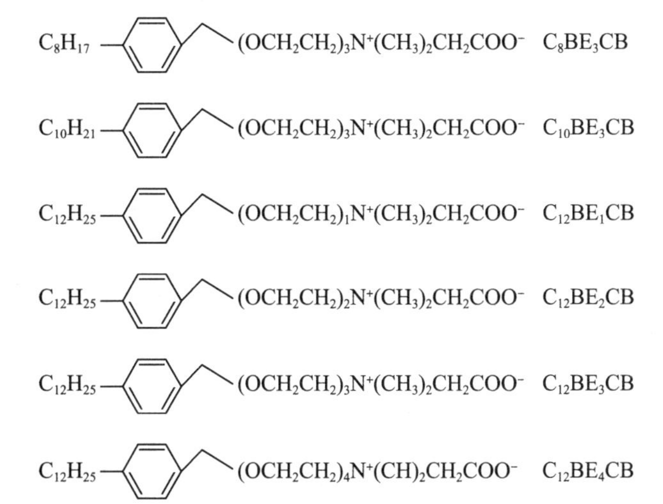

图 1 ABECB 的结构式 Structures of ABECB

1.2 临界胶团浓度和胶团微极性的测定

取 5 μL 芘的甲醇溶液(1.0×10^{-4} mol/L)放入一系列 5 mL 的容量瓶中，通 N_2 将甲醇吹干，依次加入 5 mL 不同浓度的 ABECB 水溶液，放入超声浴槽中分散 2 h，测定芘的荧光发射光谱，所用激发波长为 335 nm，狭缝宽度：EX：5.0 nm，EM：2.5 nm，激发电压为 700 V。

1.3 表面张力的测定

用纯水分别配制一系列不同浓度的 ABECB 水溶液，放置过夜。测量前将待测液倾入测量皿中，设定温度下恒温静置 30 min 后用 Wilhelmy-plate 法测定其表面张力。

2 结果与讨论

单体芘在 335 nm 处激发后，其在溶液中的荧光发射光谱有 5 个振动峰，峰Ⅰ(λ=373 nm)与峰Ⅲ(λ=383 nm)的荧光强度之比(I_1/I_3)强烈地取决于溶剂的极性[12]。

芘的 I_1/I_3 值随 ABECB 水溶液浓度的变化如图 2 所示，当溶液中 ABECB 的浓度增大到一定值时曲线峰型发生突变。芘荧光特性的突变表明其所处环境极性的变化，即开始形成胶团，因此，第一个突变点对应于表面活性剂的 cmc 值[13]。

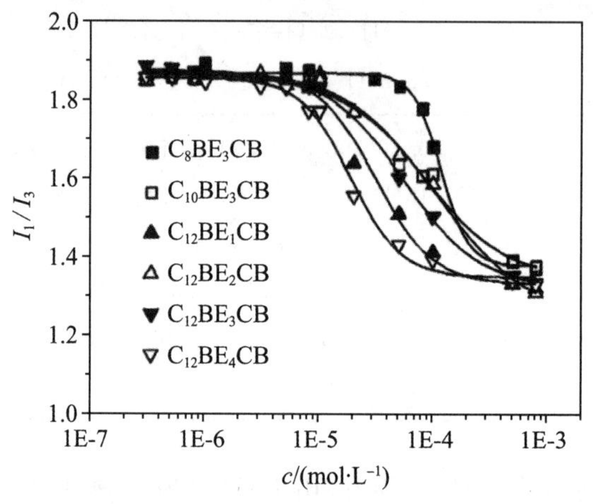

图 2 芘 I_1/I_3 值与表面活性剂浓度的关系

Fig. 2 Relationship between I_1/I_3 and the concentration of different surfactants

由图 2 可以看出，当表面活性剂浓度很稀(小于 cmc)时，I_1/I_3 值基本不变，这是因为此时表面活性剂分子本身以单体形式存在于水溶液中，浓度的增加对改变分散于水中的芘所处的环境没有多大影响；当表面活性剂浓度高于 cmc 后，表面活性剂分子开始形成胶束，极端疏水的芘分子则从水环境转移至疏水环境，即开始增溶于胶束栅栏层疏水微区中，I_1/I_3 值发生突变，随着表面活性剂浓度的进一步增大，I_1/I_3 值逐渐减小，直至到达第二突变点[14]。从图 2 中我们还可以发现，ABECB 的荧光强度之比 I_1/I_3 值的降低趋势较缓慢，各条曲线的第一和第二突变点间的跨度均有一个数量级多，这是因为刚形成的预胶束结构不完善，水分子仍能自由出入并在其周围大量存在，随着 ABECB 浓度不断增大，预胶束结构不断完善，使周围水分子进入的机会减小，致使内核极性也不断减小，直至结构完整的胶束形成时为止[15]。用稳态荧光探针法测得的 ABECB 的 cmc 列于表 2。

表 2 ABECB 的临界胶束浓度和表面张力

Tab. 2 cmc and γ_{cmc} of the synthesized surfactants

Surfactant	C_8BE_3CB	$C_{10}BE_3CB$	$C_{12}BE_3CB$	$C_{12}BE_1CB$	$C_{12}BE_2CB$	$C_{12}BE_4CB$
cmc^a (10^{-6} mol/L)	86.56	20.10	6.96	7.64	7.31	5.99
cmc^b (10^{-6} mol/L)	112.18	15.12	6.21	6.11	6.10	5.27
γ_{cmc} b(mN/m)	27.9	28.4	29.9	31.6	31.5	34.9

a. 由探针荧光法测得；b. 由 Wilhelmy-plate 法测得。实验温度均为(30±0.1)℃。

图 3 为 ABECB 的表面张力曲线，表面张力法测得的 cmc 及 γ_{cmc} 也列于表 2。从表 2 可以看出，芘探针荧光法可用来测定 ABECB 的 cmc，测定结果与表面张力法接近。实验得到了较低的 cmc 和 γ_{cmc} 值，说明此类表面活性剂具有优良的表面活性。

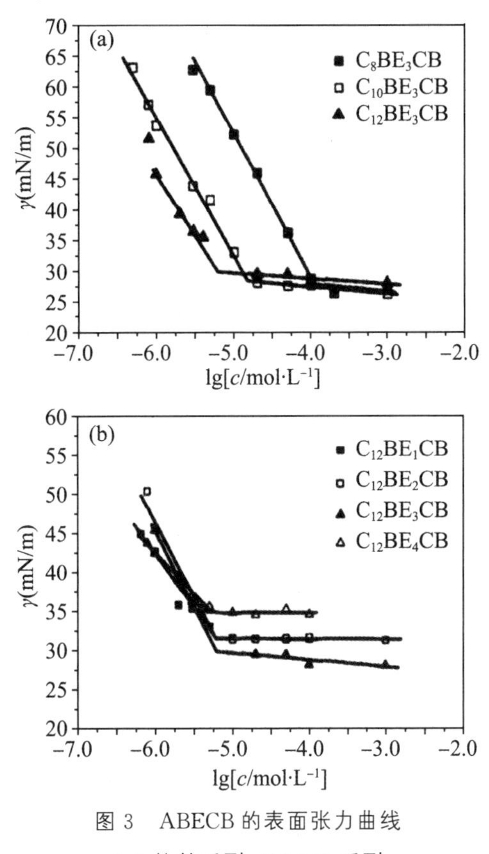

图 3 ABECB 的表面张力曲线

(a) 烷基系列；(b) EO 系列

Fig. 3 Surface tension curves of ABECB surfactants

(a) alkyl group, (b) EO group

本文还考察了探针分子芘增溶在胶团微环境的微极性。芘的荧光发射光谱的 I_1/I_3 值被广泛用作衡量芘所处环境的极性，表 3 列出了 ABECB 在第二个平台时的 I_1/I_3 值(30±0.1)℃。

表 3 ABECB 的微极性(I_1/I_3)

Tab. 3 Micropolarity (I_1/I_3) of the synthesized surfactants

Surfactant	C_8BE_3CB	$C_{10}BE_3CB$	$C_{12}BE_3CB$	$C_{12}BE_1CB$	$C_{12}BE_2CB$	$C_{12}BE_4CB$
I_1/I_3	1.39	1.38	1.34	1.33	1.33	1.34

从表 3 可以看出，代表芘探针分子所处环境极性的 I_1/I_3 值随着 ABECB 分子的疏水链中碳数的增加而减小，EO 单元数的增大对 ABECB 胶团核内的微极性影响不明显。因为芘是一种稠环烃，所以在 ABECB 的胶束中优先增溶在栅栏层，随着 ABECB 疏水链链长的增加，疏水基相互作用增强，使得胶团内表面活性剂分子排列更加紧密，因而渗透

到栅栏层中水分子数量相应减少,导致芘周围环境的极性减小[9]。而EO单元数的增大对ABECB胶团核内的微极性影响不明显,一方面是因为氧乙烯单元的增加导致水合作用增强,因而渗透到栅栏层中的水分子数量相应增多,这会使得荧光探针芘分子所处的环境极性增大;另一方面,当分子中EO数目逐渐增加时,氧乙烯单元的体积增大会使其成卷曲缠绕状态,也会阻碍水分子向栅栏层中的渗透,而胶束排列的紧密程度并没有降低,因而芘所处位置的微观极性变化也不明显[16]。

3 结论

荧光探针(芘)法可用来测定ABECB表面活性剂的临界胶团浓度(cmc),且测定结果与表面张力法(吊片法)接近。表面活性测试表明,该类表面活性剂具有优良的表面活性。胶团的微极性随着疏水链长的增大而略微减小,EO单元数的增大对ABECB胶团核内的微极性影响不明显。

参考文献

[1] 孙文彬,李应成,袁明,等.烷基酚聚氧乙烯醚羧酸盐型甜菜碱及其制备方法[P].中国专利,102276489.2011-12-14.

Sun W B, Li Y C, Yuan M, et al. Alkyl-phenol polyoxyethylene ether carboxybetaine preparation method[P]. Chinapatent,102276489. 2011-12-14.

[2] Cevdet U,Mesude I,Ayse E,et al. The bioaccumulation of nonylphenol and its adverse effect on the liver of rainbow trout [J]. EnvironmentalResearch,2003,92:262-270.

[3] Jorge E L R,Isabelle S A,Clifford P R,et al. Analysis ofoctyl-and nonylphenol and their ethoxylates in water and sediments by liquid chromatography/tandem mass spectrometry [J]. AnalyticalChemistry,2003,75(18):4811-4817.

[4] Catherine A H, Eduarda M S, Afsaneh J, et al. Nonylphenol affects gonadotropin levels in the pituitary gland and plasma of female rainbow trout [J]. EnvironmentalScienceTechnology,2001,35(14):2909-2916.

[5] Charles A S,James B W,R L Blessing,et al. Measuring the biodegradability of nonylphenol ether carboxylates,octylphenol ether carboxylates,and nonylphenol [J]. Chemosphere,1999,38(9):2029-2039.

[6] Thomas J K. Radiation-induced reaction in organized assemblies [J]. Chem. Rev. ,1980,80(4):283-299.

[7] 王琳,王东贤,宫清涛,等.稳态荧光探针研究支链烷基苯磺酸钠的聚集行为[J].感光科学与光化学,2004,22(1):20-27.

Wang L, Wang D X, Gong Q T, et al. The study of aggregation properties of sodium alkylbenzene sulfonate by steady-state fluorescence [J]. Photographic Scienceand Photochemistry, 2004, 22(1): 20-27.

[8] 方云, 刘雪锋, 夏咏梅, 等. 稳态荧光探针法测定临界胶束聚集数[J]. 物理化学学报, 2001, 17(9): 828-831.

Fang Y, Liu X F, Xia Y M, et al. Determination of critical micellar aggregation numbers by steady-state fluo-rescence probe method [J]. ActaPhys. Chim. Sin., 2001, 17(9): 828-831.

[9] 王显光, 严峰, 张春荣, 等. 稳态荧光探针法研究烷基苄基均质聚氧乙烯醚丙烷磺酸钠的聚集行为[J]. 感光科学与光化学, 2007, 25(1): 32-40.

Wang X G, Yan F, Zhang C R, et al. The study of aggregation properties of sodium alkyl-benzyl homogeneous polyoxyethylenated propane sulfonate by steady-state fluorescence [J]. PhotographicScience and Photochemis-try, 2007, 25(1): 32-40.

[10] 董林芳, 徐志成, 靳志强, 等. 对-烷基-苄基聚氧乙烯醚羧酸甜菜碱的合成与表面活性[J]. 精细化工, 2012, 29(12): 1163-1166

Dong L F, Xu Z C, Jin Z Q, et al. Synthesis and surface properties of p-(n-alkyl)-benzyl polyoxyethylene ether carboxybetaine [J]. FineChemicals, 2012, 29(12): 1163-1166.

[11] 董林芳, 徐志成, 赵荣华, 等. 对-(月桂基)苄基聚氧乙烯醚羧酸甜菜碱的合成与表面活性[J]. 日用化学工业, 2013, 43(1): 40-31.

Dong L F, Xu Z C, Zhao R H, et al. Synthesis and surface properties ofp-(n-lauryl)-benzyl polyoxyethylene ether carboxybetaine [J]. ChinaSurfactant Detergent & Cosmetics, 2013, 43(1): 40-31.

[12] Kalyanasundaram L, Thomas J K. Environmental effects on vibronic band intensives in pyrene monomer fluorescence and their application in studies of micellar systems [J]. J. Am. Chem. Soc., 1977, 99: 2039-2044.

[13] Kim J H, Domach M M, Tilton R D. Effect of electrolytes on the pyrene solubilization capacity of dodecyl sulfate micelles [J]. Langmuir, 2000, 16: 10037-10043.

[14] 严峰, 王显光, 曹绪龙, 等. 荧光法测定 N-(α-烷苯氧基)十四酰基牛磺酸钠的临界胶束浓度[J]. 感光科学与光化学, 2007, 25(2): 115-122.

Yan F, Wang X G, Cao X L, et al. CMC determination of sodium N-(α-Alkylphenoxy)-tetradecanoyltaurate by fluorescence method [J]. Photographic Science and Photochemistry, 2007, 25(2): 115-122.

[15] 姜永才, 叶建平, 吴世康. 表面活性剂溶液预胶束的形成及其聚集数的测定

[J]. 化学学报,1992,50(11):1080-1084.

Jiang Y C, Ye J P, Wu S K. Premicelle formation in surfactant solution and measurement of its average aggre-gation number [J]. Acta Chimica Sinica, 1992, 50 (11):1080-1084.

[16] 祝荣先,靳志强,张路,等.稳态荧光探针研究 Guerbet 醇聚氧乙烯醚羧酸钠的聚集行为[J].感光科学与光化学,2006,24(2):110-117.

Zhu R X, Jin Z Q, Zhang L, et al. The study of aggregation properties of sodium guerbet tetradecyl polyoxy-ethylene ether carboxylate by steady-state fluorescence [J]. Photographic Science and Photochemistry, 2006, 24(2):110-117.

The study of aggregation properties of p-(n-Alkyl)-benzyl polyoxyethylene ether carboxybetaine by steady-state fluorescence

Dong Linfang[1] Xu Zhicheng[1] Cao Xulong[2] Wang Qiwei[2]
Zhang Lei[1] Zhang Lu[1] Zhao Sui[1]

(1. Technicalln stitute of Physics and Chemistry, Chinese Academy of Sciences, Beijing 100190, P. R. China;
2. Geological Scientific Research Institute, Shengli Oilfield Company of SINOPEC, Dongying 257015, Shandong, P. R. China)

Abstract: The values of cmc, micelle micropolarity and γ_{cmc} of p-(n-alkyl)-benzyl polyoxyethylene ether carboxybetaine (ABECB) have been measured by fluorescence and Wilhelmy-plate method. It indicated that probe (pyrene) fluorescence emission spectra method could be used to determine the critical micelle concentration (cmc) of p-(n-alkyl)-benzyl polyoxyethylene ether carboxybetaine, and the results are compatible with that of Wilhelmy-plate method. The values of cmc and γ_{cmc} show that these surfactants have high surface activity. The micellar micropola-rity was observed to decrease, as the change of carbon atom number of the substituted hydro-phobic chain of benzene ring, while the effect of increasing polyoxyethylene number on micellar micropolarity is little.

Keywords: p-(n-alkyl)-benzyl polyoxyethylene ether carboxybetaine; probe fluorescence emis-sion spectra; critical micelle concentration; micellar micropolarity; surface tension

磺基甜菜碱型两性离子表面活性剂的相行为研究

于 涛[1] 杨 柳*[1] 曹绪龙[2] 潘斌林[2] 丁 伟[1] 张 微[1]
邢欣欣[1] 李金红[1]

(1. 东北石油大学化学化工学院石油与天然气化工省重点实验室,黑龙江大庆 163318;
2. 中国石化胜利油田有限公司地质科学研究院,山东东营 257015)

摘要:采用自制的4种磺基甜菜碱,运用多种方法对4种磺基甜菜碱/短链醇/正癸烷/NaCl/水形成的微乳液体系相行为进行了研究,并考察了温度、磺基甜菜碱的分子结构、短链醇浓度及其分子结构等对微乳液相行为的影响。实验表明:温度越高,中相微乳液形成的中相体积越大;SB9体系形成中相微乳液时,所需要的最小w(醇)为2%,最大醇宽为12%;SB12体系形成中相微乳液时所需要的最小w(醇)为4%,最大醇宽8%;随着磺基甜菜碱烷基碳数的增加,微乳区面积增大,增溶能力降低;最佳增容参数SP^*和表面活性剂在油相和水相的平均溶解度$S^{O.W}$均随短链醇碳链的增加而增加;平衡界面膜上的表面活性剂和醇在整个微乳液体系中所占的质量分数CS,CA,短链醇平衡界面膜所占的质量比AS,短链醇在油水相中的平均溶解度$A^{O.W}$,均随短链醇碳链的增加而减小;w(醇)的增加使得微乳液体系发生由WinsorⅠ→WinsorⅢ→WinsorⅡ型的相态变化。

关键词:磺基甜菜碱微乳液鱼状相图 Winsor 相图
中图分类号:O648 文献标识码:A 文章编号:1006-7906(2013)05-0061-05

Phase behavior research of sulfobetain zwitterionic surfactant

Yu Tao[1] Yang Liu[1] Cao Xulong[2] Pan Binlin[2] Ding Wei[1]
Zhang Wei[1] Xing Xinxin[1] Li Jinhong[1]

(1. Provincial Key Laboratory of Oil and Gas Chemical Technology, Chemistry and Chemical Engineering College, Northeast Petroleum University, Daqing 163318, China;
2. Research Institute of Geological Sciences, Sinopec Shengli Oilfield Company, Dongying 257015, China)

Abstract: Using four kinds of sulfobetain, the phase behavior of the sulfobetain/n-propanol/n-decane/NaCl/water microemulsion systems is studied by different methods, and the effects of temperature, molecular structure of the sulfobetaine surfactants, the concen-tration of short chain alcohols and its molecular structure on these microemulsion properties are discussed. The results show that the higher the temperature, the larger the phase microemulsion formation in phase volume is. The minimum alcohol needed concentration is 2% and the biggest alcohol width is 12% when SB9 system formed in phase microemulsion, the minimum alcohol concentration is 4% and the biggest alcohol wide is 8% when SB12 system formed in phase microemulsion. With the alkyl carbon numbers of sulfobetaine surfactant molecule increasing, the area of microemulsion region increases but the solubilization capacity decreases. With the length of carbon chain of short chain alcohol increasesing, the optimum compatibilization parameters(SP^*) and average solubility of surfactant in the oil and water phase (S^{ow}) increase, the mass fraction of surfactant on the balance interfacial film(C_S), the mass fraction of al-cohol in the microemulsion system(C_A), the mass fraction of short chain alcohol balance interfacial film (A^S) and the average solubil-ity of short chain alcohol in oil-water phase($A^{o,w}$) decrease. Along with the increase of alcohol mass fraction, the microemulsion sys-tem is transformed from Winsor Ⅰ to Winsor Ⅲ and then to Winsor Ⅱ.

Keywords: Sulfobetain; Microemulsion; Fishlike phase diagrams; Winsor phase diagrams

甜菜碱型两性离子表面活性分子结构中同时含有阴离子基团和阳离子基团，分子呈电中性，有良好的耐盐性，较低的刺激性，较好的生物降解性[1-2]。

磺基甜菜碱型两性离子表面活性剂是一种性能优良的两性离子表面活性剂，是两性离子表面活性剂中研究较早、用量较大的品种之一。美国专利早在 1977 年就报道了磺基甜菜碱两性表面活性剂在三次采油中的应用[3]。该两性表面活性剂在水溶液中电离生成的两性离子对金属离子具有螯合作用，抗盐能力强，耐多价阳离子的性能好，可应用于高温高盐的恶劣油藏。

微乳液是 2 种互不相溶的液体在表面活性剂和醇的作用下形成的热力学稳定、各向同性和澄清透明的分散体系[4]。微乳液一般由表面活性剂、助表面活性剂、油和水(或盐溶液)等组成，有超增溶能力[5-9]。在三次采油中，若以中相微乳液(与过剩水相和过剩油相达到三相平衡时的微乳液)作为驱油体系，其驱油效率的提高潜力最大[10]。

本文利用 Winsor 相图法和 ε-β“鱼状”相图法，研究了自制的 4 种磺基甜菜碱在不同条件下形成的微乳液体系的相行为，并考察了温度、磺基甜菜碱的分子结构、短链醇浓度及其分子结构等对微乳液相行为的影响。这对深入研究磺基甜菜碱表面活性剂结构与性能的关系及其作用原理具有重要的意义,同时也为4种新型磺基甜菜碱的结构表征和实际应用提供了重要的参考。

1 实验部分

1.1 仪器和试剂

N,N-二甲基-N(2-羟基-3-对壬基苯氧基)乙基铵磺基甜菜碱,代号:SB9-1;N,N-二甲基-N(2-羟基-3-壬基苯氧基)乙基铵磺基甜菜碱,代号:SB9-2;N,N-二甲基-N(2-羟基-3-对十二苯氧基)乙基铵磺基甜菜碱,代号:SB12-1;N,N-二甲基-N(2-羟基-3-十二苯氧基)乙基铵磺基甜菜碱,代号:SB12-2;以上4种磺基甜菜碱均按文献方法[11]实验室自制。氯化钠(天津市大茂化学试剂厂);正丙醇、正丁醇、正戊醇和正癸烷(天津市科密欧化学试剂有限公司),均为分析纯;实验用水为石英亚沸二次蒸馏水。

1.2 实验过程

1.2.1 Winnsor相图法

以蒸馏水为溶剂,配制质量分数2.9%的表面活性剂溶液。在若干个带刻度的具塞试管中加入5 mL质量分数为20%的NaCl水溶液、5 mL表面活性剂溶液和5 mL正癸烷以及相应量的正丙醇(表面活性剂与助表面活性剂的物质的量比为1∶3)。充分摇匀后,分别在25,35和45℃的恒温水浴中静置24 h,直至各相体积不再变化时记录总体积和各相体积,计算出各相所占的体积分数,绘制相体积分数随醇浓度变化的相态图即醇度扫描图。

1.2.2 ε-β“鱼状”相图法

以蒸馏水为溶剂,配制浓度不同的表面活性剂溶液。向若干个带刻度的具塞试管中,分别加入5 mL表面活性剂溶液、5 mL质量分数为20%的NaCl溶液、5 mL正癸烷以及相应量的正丙醇(表面活性剂与助表面活性剂的物质的量比为1∶3),充分摇匀后,在40℃恒温水浴中静置24 h,记录达到相平衡时的体积,计算各相所占体积分数,绘制ε-β“鱼状”相图。

2 结果与讨论

2.1 温度对中相微乳液相行为的影响

考察了不同温度下,SB9-1/正丙醇/正癸烷/NaCl/水体系和SB9-2/正丙醇/正癸烷/NaCl/水体系的中相微乳液相行为,见图1。

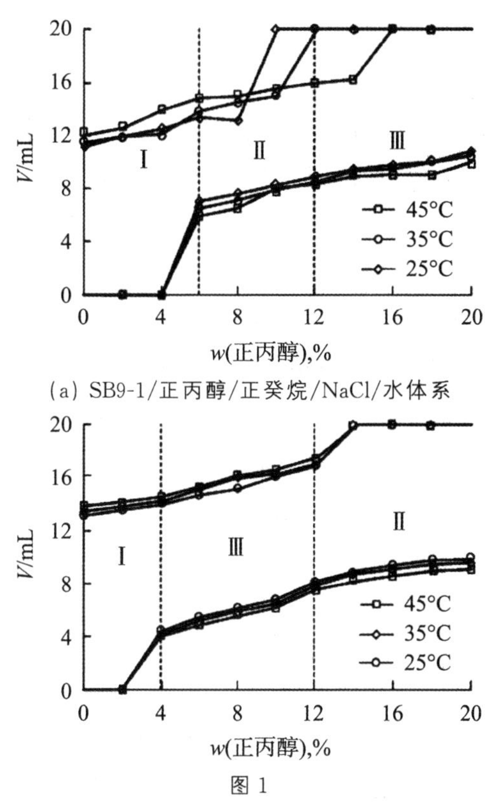

(a) SB9-1/正丙醇/正癸烷/NaCl/水体系

图 1

(b) SB9-2/正丙醇/正癸烷/NaCl/水体系图

1 不同温度下,2 种体系的醇度扫描图

由图 1 可见:在所考察的温度范围内,SB9-1/正丙醇/正癸烷/NaCl/水体系和 SB9-2/正丙醇/正癸烷/NaCl/水体系都形成了中相微乳液;随着温度的升高,形成微乳液中相体积增大,更有利于形成中相微乳液。这是因为对于离子型表面活性剂,温度升高,调节了表面活性剂体系的亲水亲油性,使得表面活性剂分子运动加强,相互碰撞、聚并几率增加[12];亲水基与水的亲和作用减弱,亲油基与油的亲和作用提高,最终使得表面活性剂与正丙醇分子之间缔合作用增强,微乳液更易于形成。这也说明正丙醇从连续相进入界面层的能量变化是由正丙醇的分子形态和温度共同决定的。

2.2 短链醇质量分数对微乳液相行为的影响

短链醇质量分数对微乳液相行为的影响见图 2。

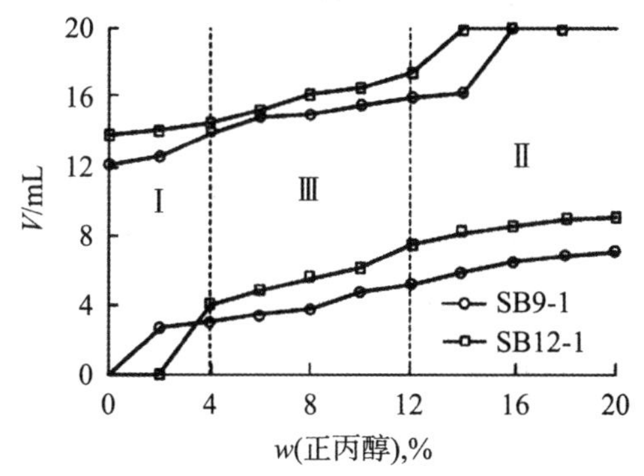

图 2　磺基甜菜碱/正丙醇/正癸烷/NaCl/水体系在 45℃下的醇度扫描图

由图 2 可见:在 45℃条件下,SB9-1 微乳液体系在 w(正丙醇)为 2%时开始形成中相微乳液(即 WinsorⅢ型),w(正丙醇)为 14%时中相微乳液消失;SB12-1 微乳液体系在 w(正丙醇)为 4%时开始形成中相微乳液,w(正丙醇)为 12%时中相微乳液消失。下相微乳液(即 WinsorⅠ型)液滴的界面膜上的 w(正丙醇),随着界面膜上的 w(正丙醇)的增加而增加,界面膜的曲率变小,微乳液液滴间发生聚结,对油的增溶量增大,使得微乳液区与水之间产生密度差,微乳液富集相从下相中分离出来,出现微乳液、剩余油相、水相三相平衡的 WinsorⅢ型微乳液。此时,界面膜就扩散到中相微乳液中。当 w(正丙醇)继续增大时,界面膜向油相凸出,中相微乳液消失,出现上相微乳液和剩余水相两相平衡的上相微乳液(即 WinsorⅡ型)。这一现象也可以用临界排列参数理论[13]解释。正丙醇的加入使得表面活性剂亲水基的占有面积降低,调节了微乳液体系的亲水亲油性,从而改变了离子型表面活性剂界面膜的自发弯曲特性,使得临界排列参数 P 变大。因此,微乳液的类型随 w(正丙醇)的增加发生 WinsorⅠ→WinsorⅢ→WinsorⅡ型的转变。

2.3 磺基甜菜碱结构对微乳液相行为的影响

磺基甜菜碱结构对微乳液相行为的影响见图 3。

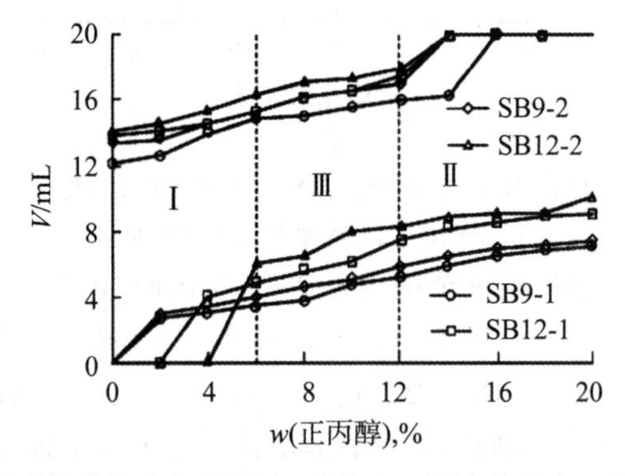

图 3 不同结构的磺基甜菜碱/正丙醇/正癸烷/NaCl/水体系在 45℃下的醇度扫描图

由图 3 可见:当 w(正丙醇)为 2%时,SB9-1、SB9-2 体系开始形成中相微乳液;w(正丙醇)为 4%时,SB12-1 体系开始形成中相微乳液;w(正丙醇)为 6%时,SB12-2 体系开始形成中相微乳液。磺基甜菜碱的碳链越长,开始形成中相微乳液所需的 w(正丙醇)越大。这可能是由于随着磺基甜菜碱分子亲油基长度的增加,使得临界排列参数 P 增大,从而使得磺基甜菜碱界面膜易于自发弯向油相而形成微乳液;此外,磺基甜菜碱亲油基长度的增加,使得形成中相微乳液的聚集数增大,增溶量增加。

在配制磺基甜菜碱微乳液体系时,油和水的体积比为 1∶1。对于 WinsorⅢ型体系而言,若剩余油相和剩余水相的体积相等,中相微乳液中的油和水的体积一定相等,通常将这一状态称为最佳状态[14]。通常 WinsorⅢ型体系的油/水界面张力可达到超低,而在最佳状态时,油/水界面张力达到最佳[10-13],而且对水和油都有较高的增溶能力。因此,磺基甜菜碱可以在三次采油中达到提高原油采收率的目的[15]。磺基甜菜碱/正丙醇/正癸烷/NaCl/水体系形成中相微乳液的醇含量范围和醇宽见表 1。

表 1 磺基甜菜碱/正丙醇/正癸烷/NaCl/水体系形成中相微乳液的醇含量范围和醇宽

项目	磺基甜菜碱			
	SB9-1	SB9-2	SB12-1	SB12-2
w(正丙醇),%	2～14	2～12	4～12	6～12
醇宽,%	12	10	8	6

由表 1 可见:SB9 体系形成中相微乳液时,所需要的最小 w(正丙醇)为 2%,最大醇宽为 12%;SB12 体系形成中相微乳液时所需要的最小 w(正丙醇)为 4%,最大醇宽 8%。

2.4 短链醇结构对中相微乳液相行为的影响

短链醇结构对中相微乳液相行为的影响见图 4。图中,ε 表示短链醇占整个体系的质量分数,β 表示表面活性剂占整个体系的质量分数。

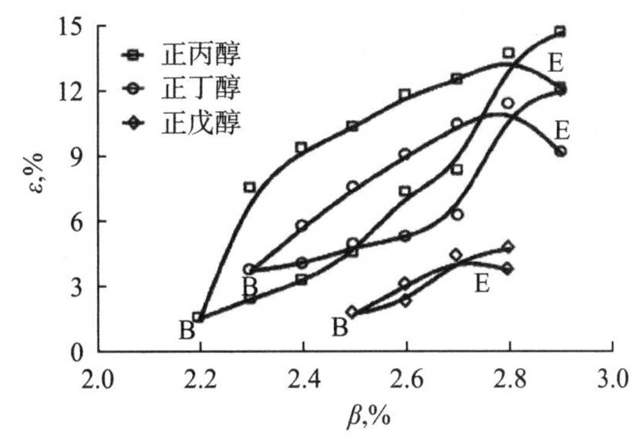

图 4 SB9-1/短链醇/正癸烷/NaCl/水体系在形成中相微乳液时的 ε-β 鱼状相图

由图 4 可见:ε-β 鱼状相图的鱼头 B 点表示中相微乳液刚开始形成,鱼尾 E 点表示中相微乳液刚好消失,此时等量的水和油被增溶形成单相微乳液;随着醇质量分数 ε 的增加,保持表面活性剂质量分数不变,微乳液体系发生 WinsorⅠ→WinsorⅢ→WinsorⅡ型的转变。

对 SB9-1/短链醇/正癸烷/NaCl/水形成的微乳液体系,微乳液体系的变量如下。

1) 油在水、油两组分中的质量分数:

$$a=\frac{m_o}{m_o+m_w}$$

式中,m_o,m_w 分别为油在油、水两相中的质量。

2) 表面活性剂在总体系中的质量分数:

$$\beta=\frac{m_s}{m_s+m_A+m_o+m_w}$$

式中,m_s,m_A 分别为表面活性剂和短链醇的质量。

3）短链醇在总体系中的质量分数：

$$\varepsilon=\frac{m_A}{m_S+m_A+m_o+m_w}$$

4）最佳增溶参数为每克表面活性剂所增溶的油和水的总量,可通过鱼尾点计算：

$$SP^*=\frac{1-\beta_E-\varepsilon_E}{\beta_E}$$

5）由平衡界面膜中表面活性剂和助表面活性剂醇在总组分中所占的质量分数,可得到由表面活性剂和助表面活性剂组成的界面膜中醇的质量分数：

$$A^S=\frac{\varepsilon_i}{\beta_i+\varepsilon_i}$$

6）表面活性剂和醇在油相和水相的平均溶解度：

$$S^{O,W}=\beta_B, A^{O,W}=\varepsilon_B$$

7）平衡界面膜上的表面活性剂和醇在整个微乳液体系中所占的质量分数：

$$C_S=\beta_E-\frac{aS^{O,W}(1-\varepsilon_E-\beta_E)}{1-S^{O,W}-A^{O,W}}, C_A=\varepsilon_E-\frac{aA^{O,W}(1-\varepsilon_E-\beta_E)}{1-S^{O,W}-A^{O,W}}$$

计算结果列于表 2。

表 2　短链醇对 SB9-1/正癸烷/NaCl/水微乳液体系影响的物化参数

短链醇	物化参数									
	β_B	ε_B	β_E	ε_E	C_S	C_A	$S^{O.W}$	$A^{O.W}$	A^S	SP*
正丙醇	0.022	0.015	0.028	0.137	0.008 92	0.124 0	0.022	0.015	0.933	29.8
正丁醇	0.023	0.037	0.028	0.114	0.007 01	0.080 2	0.023	0.013	0.920	30.6
正戊醇	0.025	0.017	0.027	0.043	0.002 73	0.026 5	0.025	0.010	0.907	34.4

由表 2 可见:短链醇平衡界面膜所占的质量比 A^S 与短链醇在油水相中的平均溶解度 $A^{O,W}$,均随着短链醇碳链的增加而减小。短链醇作为助表面活性剂进入界面层后,由于醇碳链的增长使其疏水性增强,要使界面膜达到亲水亲油平衡,所需短链醇量较少;随着短链醇链长的增加,最佳增溶参数 SP* 增大,微乳液体系的增溶能力增加。短链醇碳链越长,穿透界面膜的能力越大,使得表面活性剂分子较易进入界面膜并增加界面膜的亲油性,所以需要较少的醇量来平衡界面膜。微乳液体系增溶等量的水和油所需要的表面活性剂的量也减少,微乳液体系的增溶能力增强[17]。

3　结论

1）磺基甜菜碱 SB9-1 体系,在所考察的 25,35,45℃3 个温度下均能形成中相微乳液。温度越高,形成微乳液的醇宽越宽。

2）SB9 体系形成中相微乳液时,所需要的最小 w(正丙醇)为 2%,最大醇宽为 12%;SB12 体系形成中相微乳液时所需要的最小 w(正丙醇)为 4%,最大醇宽 8%。

3）磺基甜菜碱碳链的增加，使体系形成中相微乳液所需的 w(醇)增大，醇宽变小；而短链醇碳链的增加，使中相微乳液中相区域减小。中相区域由大到小顺序为：正丙醇，正丁醇，正戊醇。

4）随短链醇碳链的增加，SP^*，$A^{O,W}$ 的值增加，C_S，C_A，A^S，$A^{O,W}$ 的值均减小。

参考文献

[1] 徐进云，郑帼，葛启，等. 十八烷基甜菜碱的合成与应用性能[J]. 纺织学报，2005，26(1)：22-24.

[2] 王军，周晓微. 十二烷基二甲基羟丙基磺基甜菜碱的合成工艺优化[J]. 陕西科技大学学报，2007，25(2)：86-90.

[3] MADDOX J J，TATE J F. Surfactantoil recovery process usable in high temperature formations having high concen-trations of polyvalentions：US，4008165[P]，1977.

[4] SCHULMAN J H，STOECKENIUS W，PRINCE L M. Mechanism of formation and structure of microemulsions by electron microscopy[J]. J Phys Chem，1959(63)：1677-1680.

[5] 王桂香，韩恩山，许寒. 微乳液的理论研究进展[J]. 化学工程师，2007，21(12)：31-33.

[6] 李方，李干佐，房伟，等. 阴阳离子表面活性剂复配体系的中相微乳液研究[J]. 化学学报，1996，54(1)：1-6.

[7] 李干佐，郑立强，徐桂英，等. 中相微乳液的形成和特性：Ⅳ. 表面活性剂复配及醇和油的影响[J]. 科学通报，1993，38(22)：2042-2044.

[8] 周雅文，张高勇，王红霞. 汽油微乳液拟三元相图及电导率研究[J]. 日用化学工业，2004，34(4)：211-214.

[9] 张强，汪晓东，金日光. 油酸/氨水-醇-汽油-水微乳液体系拟三元相图的分析[J]. 北京化工大学学报：自然科学版，2001，28(2)：37-39.

[10] 肖进新，赵振国. 表面活性剂应用原理[M]. 北京：中国轻工业出版社，2003.

[11] 丁伟，李淑杰，于涛，等. 新型磺基甜菜碱的合成与性能[J]. 化学工业与工程，2012，29(1)：26-29.

[12] CROSS J. Anionic surfactants：Analytical chemistry [M]. New York：Marcel Dekker，1998：45.

[13] 赵国玺，朱瑶. 表面活性剂作用原理[M]. 北京：中国轻工业出版社，2003.

[14] 丁伟，张志伟，李钟，等. 不同链长烷基芳基磺酸盐形成微乳液的性质[J]. 高等学校化学学报，2012，33 (2)：395-399.

[15] 于涛,李钟,丁伟,等. 十四烷基芳基磺酸形成的分子有序组合体[J]. 物理化学学报,2010,26(2):317-323.

[16] 吴雨彤. 水油比对微乳液相行为及物化性质的影响研究[D]. 济南:山东师范大学,2011.

二乙烯三胺-长链脂肪酸体系的界面剪切流变性质

杨　勇[1]　李　静[2]　曹绪龙[1]　张继超[1]　张　磊[2]　张　路[2]　赵　濉[2]

（1. 中国石化胜利油田分公司地质科学研究院，山东东营　257015；
2. 中国科学院理化技术研究所，山东北京　100190）

摘要：利用双锥法研究了长链脂肪酸模拟油-二乙烯三胺溶液体系的界面剪切流变性质，考察了界面老化时间、应变幅度和剪切频率对界面剪切流变参数的影响，探讨了脂肪酸的疏水链长和饱和度与界面膜性能的关系。结果表明，在脂肪酸模拟油-二乙烯三胺溶液体系中，脂肪酸与有机碱二乙烯三胺反应，生成阴离子型表面活性剂脂肪酸皂；脂肪酸皂与二乙烯三胺能够通过静电相互作用形成界面聚集体，与脂肪酸皂混合吸附，形成一定强度的界面膜。随脂肪酸疏水链长增加，体系界面膜排列更为紧密，剪切复合模量、弹性模量和黏性模量均随之而增大；脂肪酸疏水链中引入双键，分子尺寸增加，体系界面膜变得疏松，剪切复合模量、弹性模量和黏性模量均明显降低。

关键词：二乙烯三胺；脂肪酸；界面；剪切流变；煤油

中图分类号：O647　**文献标识码**：A

Interfacial shear rheological properties of diethylenetriamine-long chain fatty acids system

Yang Yong[1]　Li Jing[2]　Cao Xulong[1]　Zhang Jichao[1]
Zhang Lei[2]　Zhang Lu[2]　Zhao Sui[2]

（1. Geologicaland Scientific Research Institute of Shengli Oilfield Co. Ltd，SINOPEC，Dongying 257015，China；
2. TechnicalInstitute of Physics and Chemistry，Chinese Academy of Sciences，Beijing 100190，China）

Abstract：The interfacial shear rheological properties of fatty acids model oil-diethylenetriamine solution system were studied by biconical method. The effects of aging time，strain amplitude and shear frequency on interfacial shear rheological data of the system were investigated. The relationships between the length and saturation of alkyl chain in fatty acid and the interfacial shear rheological properties were also expounded. The experimental results showed that the soaps produced by the reaction of fatty acid and organic alkali formed the interfacial aggregates with diethylenetriamine through electrostatic interaction. Then the mixed adsorption of the aggregates and soap molecules formed the interfacial film with certain strength on oil-water interface. The mixed adsorption film will become more compact with the increase of alkyl chain length in fatty acid molecule, resulting in the enhancement of interfacial shear complex moduli, elastic moduli and viscous moduli. On the other hand, the introduction of the double bond to hydrophobic chain of fatty acid molecule will increase the molecular size, leading to the looser adsorption film, therefore, the interfacial shear rheological data of the system will decrease obviously.

Keywords:diethylenetriamine;fatty acid;interface;shear rheology;kerosene

化学驱是中国注水开发油田提高采收率的重要手段,碱-表面活性剂-聚合物(ASP)三元复合驱是一种大幅度提高采收率的驱油技术。ASP体系中加入的碱与石油中的酸性物质发生反应,所生成的表面活性物质与加入的表面活性剂产生协同作用,增大界面活性,可以减少表面活性剂的用量。经矿藏实验证实,该技术对于酸值高的原油是一种很有效的提高石油采收率的方法[1-2]。但是无机碱导致结垢严重,给注采系统带来困难。使用有机碱替代无机碱,既能充分的利用原油中的石油酸,又有望解决结垢等问题。因此,开展高效有机碱代替三元复合驱中无机碱的研究对提高原油采收率具有非常重要的意义[3-4]。

胜利油田原油具有酸值高的特点,在化学驱采油中,迫切需要寻求一种能够克服结垢的替代碱,从而使石油酸得以充分利用。碱与原油中酸性组分的反应不仅能有效地降低油-水界面张力,同时对于驱油过程中的乳化、油墙形成等过程也有重要影响。目前,关于有机碱对油-水界面张力的影响方面研究较多[5-6],而对于更为重要的界面流变性质则关注较少。

界面流变学是研究界面膜在外力作用下形变的学科,是界面膜性质表征最为直接和有力的手段。根据外力作用形式的不同,界面流变分为扩张流变和剪切流变。若施加的是使界面形状不发生改变而面积发生变化的外力,则发生扩张流变,其反映的主要是界面层及界面附近的微观弛豫过程和分子间相互作用的信息;若施加的是使界面形状发生改变而面积不发生变化的外力,则发生剪切流变,其反映的主要是界面层结构和膜的机械强度的信息[7]。

原油中的主要活性组分是石油酸，其组成、结构十分复杂[8]。笔者以不同链长、不同饱和度的脂肪酸为模型化合物，研究了有机碱二乙烯三胺与脂肪酸在煤油-水界面上形成的吸附膜的剪切性质，有助于深入理解三元复合驱采油过程中的界面现象，优化配方设计。

1　实验部分

1.1 原料及试剂

十二酸，分析纯，北京旭东化工厂产品；十六酸、硬脂酸，分析纯，天津化学试剂六厂产品；油酸，分析纯，西陇化工股份有限公司产品；二乙烯三胺（DETA），化学纯，西陇化工股份有限公司产品；航空燃料，北京化学试剂公司产品，经过柱提纯，室温下与重蒸后的去离子水的界面张力约为 42 mN/m；实验用水为经重蒸后的去离子水，电阻率大于 18 MΩ・cm。

1.2 界面剪切流变实验

利用安东帕 MCR501 界面剪切流变仪和控制应变的实验模式进行界面剪切流变实验。首先固定剪切频率，改变应变幅度，进行线性黏弹区域扫描；然后，在线性区域范围内选择应变幅度，测定固定频率下的动态界面剪切流变；当界面达到平衡后，进行剪切频率扫描。测定温度均为(30.0±0.1)℃，水相为 1×10^{-2} mol/L 的二乙烯三胺（DETA）水溶液，油相为 1×10^{-2} mol/L 不同结构脂肪酸的煤油，简记为二乙烯三胺-脂肪酸体系。界面剪切流变的稳态数据均为油-水界面张力和动态界面剪切流变达到平衡时的数值，此时脂肪酸与二乙烯三胺间的反应已经完全。

2　结果与讨论

2.1 二乙烯三胺-脂肪酸体系(DETA-FA)的界面剪切流变性质

2.1.1 界面剪切流变的线性黏弹区域

在控制应变模式的界面剪切流变实验中，当应变幅度在某一区域时，界面剪切复合模量的数值不随应变幅度变化而变化，这段区域称为线性黏弹区域。为确保实验数据的可靠性和可比性，最终的界面剪切流变数据应在线性范围内测定获得[9]。DETA-不同链长脂肪酸体系的界面剪切复合模量与应变幅度的关系如图 1 所示。由图 1 可见，当应变幅度过低时，DETA-十二酸和 DETA-十六酸体系的复合模量数值不稳定；当应变幅度过大时，DETA-硬脂酸体系的复合模量有降低的趋势，这是过于强烈的剪切外力破坏了界面膜结构的缘故。综合考虑 3 种脂肪酸体系的线性响应范围，选取应变幅度为 1%。

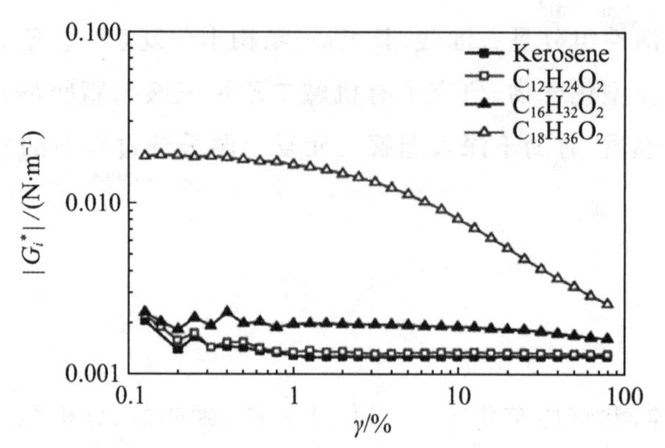

图 1　DETA-不同链长脂肪酸体系的界面剪切复合模量($|G_i^*|$)与应变幅度(γ)的关系

Fig. 1　$|G_i^*|$ vsγ of the systems of DETA-FA with different chain lengths $\omega=1$ Hz

2.1.2 动态界面剪切流变性质

DETA-不同链长脂肪酸体系的界面剪切弹性模量和黏性模量随时间的变化如图 2 所示。由图 2 可见，随着脂肪酸中烷基链长的增大，体系界面剪切弹性模量和黏性模量均增大。采用的 DETA 水溶液的 pH 约为 11，碱性较强，脂肪酸与 DETA 充分反应，生成脂肪酸皂类阴离子表面活性剂。脂肪酸皂的疏水链越长，则其在油-水界面上的吸附量越大，界面分子间的疏水相互作用也越强。十二酸皂和十六酸皂的界面活性较差，无法在界面上形成紧密的吸附膜，因此，相应体系的界面剪切的弹性模量与煤油近似；不过，表面活性剂分子在界面上的吸附会产生如扩散交换等弛豫过程，因此黏性模量略有增强。而硬脂酸皂能够在界面上形成较为紧密的吸附膜，因此该体系的界面弹性模量和黏性模量均随时间逐渐增大，直至达到平衡。

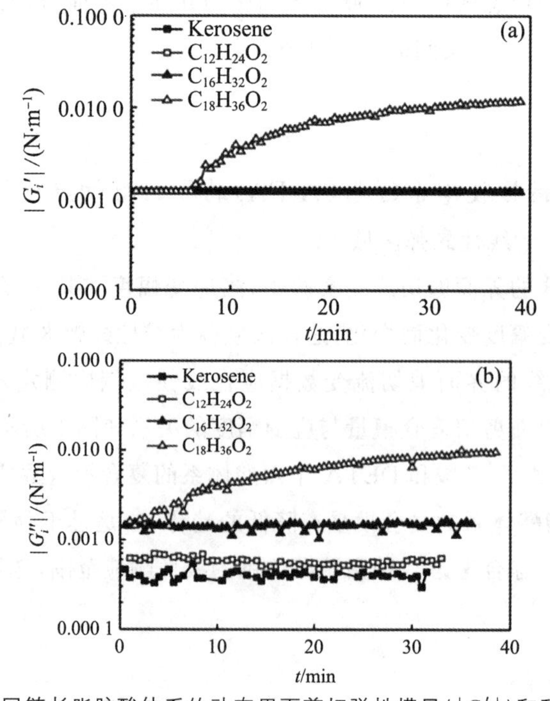

图 2　DETA-不同链长脂肪酸体系的动态界面剪切弹性模量($|G_i'|$)和黏性模量($|G_i''|$)

Fig. 2　$|G_i'|$, $|G_i''|$ of the systems of DETA-FA with different chain lengths $\omega=1$Hz (a) $|G_i'|$; (b) $|G_i''|$

为了深入理解硬脂酸吸附膜的性质，将 DETA-硬脂酸体系的界面剪切弹性模量和黏性模量列在一起，如图 3 所示。由图 3 可见，当时间较短时，界面上硬脂酸分子较少，则模量较低，且界面膜以黏性为主；当吸附硬脂酸分子数目较多时，界面膜以弹性为主。

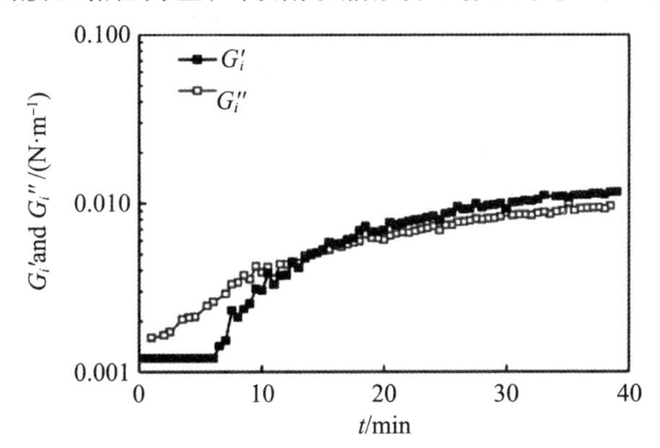

图 3 DETA-硬脂酸体系的动态界面剪切弹性模量($|G_i'|$)和黏性模量($|G_i''|$)

Fig. 3 $|G_i'|$, $|G_i''|$ of DETA-stearic acid system $\omega=1$ Hz

2.1.3 剪切频率对 DETA-不同链长脂肪酸体系界面剪切流变性质的影响

剪切频率是影响界面膜性质的重要参数，通过测定不同剪切频率下的流变参数，可以获得界面膜结构和强度的信息。一般而言，随着剪切频率增加，界面通过各种弛豫过程耗散外力作用的程度不断减弱，界面对抗剪切形变的阻力增加，因此，界面剪切复合模量随频率增大而增大[10]。

剪切频率对 DETA-不同链长脂肪酸体系界面剪切复合模量的影响示于图 4。由图 4 可见，十二酸分子在界面上吸附数量最低，相应体系的界面膜的性质与纯煤油相当，在整个实验范围内，界面对抗剪切形变的阻力主要来自溶剂分子的贡献，因此，其剪切复合模量与剪切频率的双对数曲线为直线，斜率约为 1.8。十六酸分子能形成一定强度的界面膜，在较低的剪切频率范围内，相应体系的界面剪切复合模量约为 10^{-4} N/m，且随频率仅略有升高；当剪切频率高于 0.2 Hz 时，界面膜结构被破坏，剪切复合模量随频率的变化趋势变得与煤油相似。硬脂酸分子形成的界面膜排列最为紧密，DETA-硬脂酸体系在低频下的界面剪切复合模量数值比十六酸体系的界面膜高 2 个数量级；由于硬脂酸形成的界面膜强度较大，需要更为强烈的剪切力才能破坏，因此，剪切复合模量从吸附膜结构控制到溶剂控制的转折点剪切频率(转折频率)约为 3 Hz，远高于十六酸体系形成界面膜的转折频率。

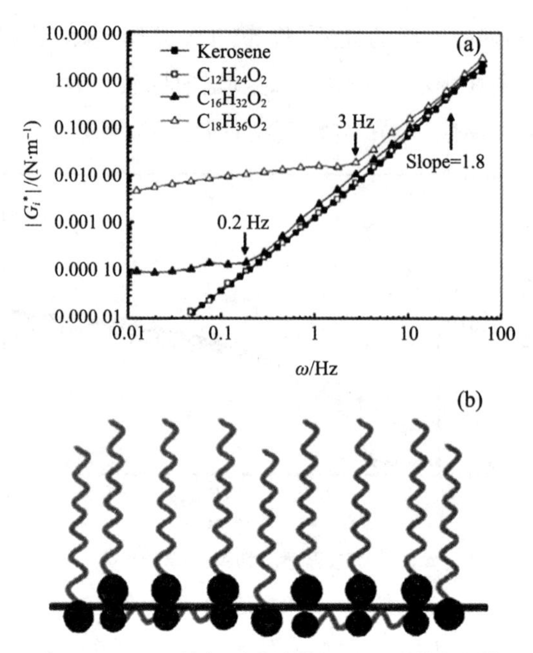

图 4　剪切频率(ω)对 DETA-不同链长脂肪酸体系的界面剪切复合模量($|G_i^*|$)的影响

Fig. 4　The effect of ω on $|G_i^*|$ of the systems of DETA-FA with different chain lengths (a) $|G_i^*|$ vs ω; (b) Schematic diagram

值得注意的是,DETA-硬脂酸体系的界面膜的强度较高,与沥青质形成的界面膜的剪切复合模量相当[11-12]这与 DETA 的特殊结构有关。如图 4(b)所示,DETA 有 3 个正电中心,能促进有机酸皂在界面上吸附,并形成聚集体。硬脂酸皂与聚集体在界面上混合吸附,形成强度较高的界面膜;而链长较短的脂肪酸在界面上吸附量较少,分子间相互作用也较弱,不能形成紧密的混合膜。

DETA-不同链长脂肪酸体系界面剪切弹性模量和黏性模量随频率的变化如图 5 所示。由图 5 可见,十六酸体系和硬脂酸体系的界面吸附膜的弹性模量均略大于黏性模量;随着剪切频率增大,界面膜结构被破坏,弹性模量开始降低;当剪切频率进一步增大时,则体现为纯溶剂的剪切流变特征。因此,适宜剪切频率下的界面剪切流变数据才能反映界面膜的结构特征。

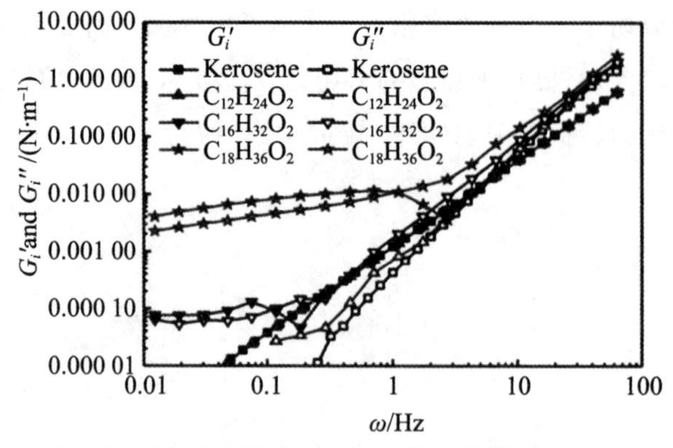

图 5　剪切频率 ω 对 DETA-不同链长脂肪酸体系界面剪切弹性模量($|G_i'|$)和黏性模量($|G_i''|$)的影响

Fig. 5　The effect of ω on $|G_i'|$ and $|G_i''|$ of the systems of DETA-FA with different chain lengths

2.2 DETA-不同饱和度脂肪酸体系的界面剪切流变性质

考察了DETA-不同饱和度脂肪酸体系的界面膜剪切流变性质。油酸体系的界面剪切流变的线性范围较宽，也选取1%作为应变幅度。

DETA-硬脂酸和DETA-油酸体系的界面剪切弹性模量和黏性模量随时间的变化如图6所示。由图6可见，油酸体系的界面剪切弹性模量和黏性模量均远低于硬脂酸体系，表明油酸皂和DETA在界面上无法形成排列紧密的吸附膜。DETA-硬脂酸和DETA-油酸体系的界面剪切复合模量、弹性模量和黏性模量随频率的变化趋势如图7所示。由图7可见，油酸体系界面膜与十六酸体系界面膜较为接近，强度远低于硬脂酸体系界面膜。这是由于烷基链中引入双键，增加了链的柔性，在界面上占据更大的空间，阻碍了分子的紧密排列造成的。相关机理如图8所示。

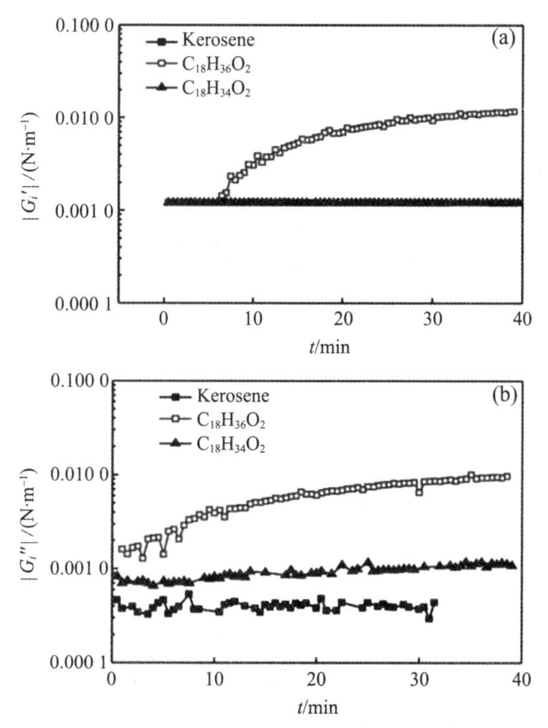

图6　DETA-硬脂酸和DETA-油酸体系的动态界面剪切弹性模量($|G_i'|$)和黏性模量($|G_i''|$)

Fig. 6　$|G_i'|$ and $|G_i''|$ of DETA-oleic acid and DETA-stearic acid systems (a) $|G_i'|$; (b) $|G_i''|$ $\omega=1$ Hz

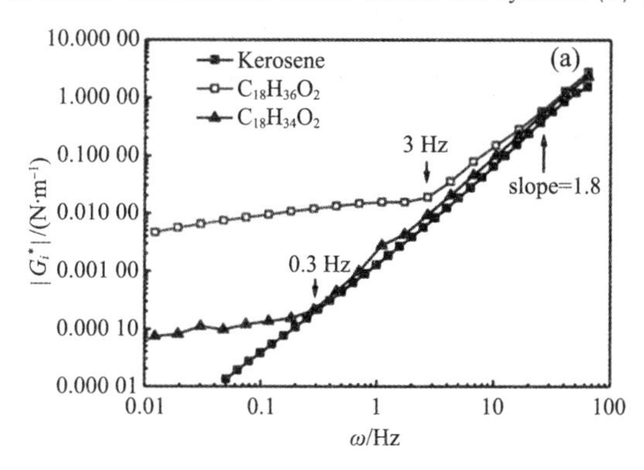

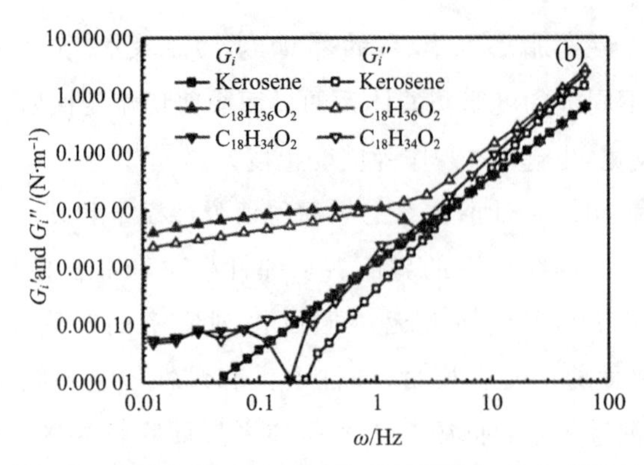

图 7　剪切频率(ω)对 DETA-硬脂酸和 DETA-油酸体系界面剪切复合模量($|G_i^*|$)、弹性和黏性模量($|G_i'|$、$|G_i''|$)的影响

Fig. 7　The effects of ω on $|G_i^*|$, $|G_i'|$ and $|G_i''|$ of DETA-stearic acid and DETA-oleic acid systems (a) $|G_i^*|$; (b) $|G_i'|$, $|G_i''|$

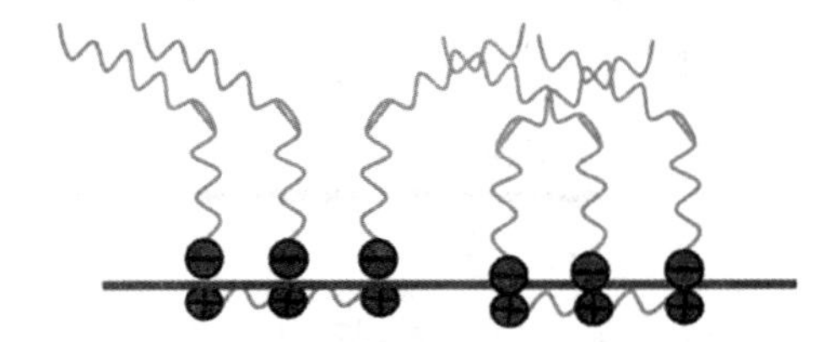

图 8　油酸分子在油-水界面吸附示意图

Fig. 8　Schematic of adsorptions of oleic acid molecules on oil-water interface

3　结论

(1) 在二乙烯三胺溶液-脂肪酸模拟油体系中，脂肪酸与有机碱二乙烯三胺反应，生成阴离子型表面活性剂脂肪酸皂；多正电中心的二乙烯三胺能够通过静电相互作用与脂肪酸皂形成界面聚集体，增加界面膜的强度。

(2) 脂肪酸皂与聚集体在油-水界面上混合吸附，形成一定强度的界面膜。随疏水链长增加，界面上脂肪酸皂分子数目增加，分子间相互作用增强，界面膜排列更为紧密，剪切复合模量、弹性模量和黏性模量均随之而明显增大。

(3) 脂肪酸疏水链中引入双键，分子尺寸增加，空间阻碍效应明显，使得界面膜变得疏松，剪切复合模量、弹性模量和黏性模量均大幅度降低。

(4) 剪切频率是影响界面剪切流变数据的关键因素，高频条件下的界面剪切流变数据主要反映溶剂分子的影响，只有适宜的剪切频率条件下，流变数据才能反映界面膜的结构信息。

参考文献

[1] 俞稼镛,宋万超,李之平,等. 化学复合驱基础及进展[M]. 北京:中国石化出版社,2002.

[2] ZHANG L,LUO L,ZHAO S,et al. Effect of different acidic fractions in crude oil on dynamic interfacial tensions in surfactant/alkali/model oil systems [J]. J Petro Sci Eng,2004,41:189-198.

[3] 王芳,田秀芳,白琰,等. 有机碱/HPAM 二元复合驱体系的研究[J]. 精细石油化工进展,2007,10(8):12-14.(WANG Fang,TIAN Xiufang,BAI Yan,et al. Study on organic base/HPAM binary combination flooding system [J]. Advances in Fine Petrochemicals,2009,10(8):12-18.)

[4] 宗丽平,易泽勇,马秀伟. 应用有机碱改进三元复合驱工艺[J]. 国外油田工程,2007,23(8):1-4. [ZONG Liping,YI Zeyong,MA Xiuwei. The technology improvement of ASP flooding by the application of organic alkali [J]. Foreign Oilfield Engineering,2007,23(8):1-4.]

[5] 石静. 有机碱乙醇胺对胜利原油界面张力的影响[J]. 山东大学学报(工学版),2013,43(3):1-5.[SHI Jing. Effect of ethanolamine on interfacial tensions of Shengli crude oils [J]. Journal of Shandong University (Engineering Science),2013,43(3):1-5.]

[6] 石静.有机碱三元复合驱油体系与胜利原油的协同作用[J]. 石油化工应用,2013,32(16-19):1-5.[SHI Jing. Study on synergisms between organic alkali ASP flooding and Shengli crude oil [J]. Petrochemical Industry Application,2013,32(1):16-19.]

[7] MILLER R,FERRI J K,JAVADI A,et al. Rheology of interfacial layers [J]. Colloid and Polymer Science,2010,288 (9):937-950.

[8] 李明远,吴肇亮. 石油乳状液[M]. 北京:科学出版社,2009:40-68.

[9] KRGEL J,DERHATCH S R,MILLER R. Interfacial shear rheology of protein-surfactant layers [J]. Advances in Colloid and Interface Science,2008,144(1-2):38-53.

[10] GUILLERMIC R M,SAINT-JALMES A. Dynamics of poly-nipam chains in competition with surfactants at liquid interfaces: From thermoresponsive interfacial rheology to foams [J]. Soft Matter,2013,9(4):1344-1353.

[11] FAN Y,SIMON S,SJBLOM J. Interfacial shear rheology of asphaltenes at oil-water interface and its relation to emulsion stability: Influence of concentration,

solvent aromaticity and nonionic surfactant [J]. Colloids and Surfaces A,2010,366(1-3):120-128.

[12] VERRUTO V J,LE R K,KILPATRICK P K. Adsorption and molecular rearrangement of amphoteric species at oil-water interfaces [J]. Journal of Physical Chemistry B,2009,113(42):13788-13799.

CTAB改变油湿性砂岩表面润湿性机理的研究

侯宝峰[1] 王业飞[1] 曹绪龙[2] 张 军[3] 宋新旺[2]
丁名臣[1] 陈五花[1] 黄 勇[1]

(1. 中国石油大学(华东)石油工程学院,山东青岛 266580;
2. 中国石化胜利油田分公司地质科学研究院,山东东营 257015;
3. 中国石油大学(华东)理学院,山东青岛 266580)

摘要:利用红外、原子力显微镜(AFM)、Zeta电位测定、接触角测定及岩心自发渗吸实验等手段研究了阳离子型表面活性剂CTAB(十六烷基三甲基溴化铵)改变油湿性砂岩表面润湿性的机理。结果表明:CTAB改变油湿性砂岩表面润湿性的性能优异。由于静电引力作用,CTAB正电性离子头基与吸附在砂岩表面的原油当中的羧酸基团形成离子对,当CTAB的浓度超过临界胶束浓度(CMC)时,形成的离子对就会从砂岩表面解吸附并增溶于CTAB形成的胶束当中,从而露出干净的水湿表面,砂岩表面因此实现润湿反转。

关键词:油湿性砂岩;润湿性改变;阳离子型表面活性剂

中图分类号:TE357.46 文献标识码:A

三次采油中,表面活性剂之所以能大大提高原油的采收率,主要是由于表面活性剂对油层当中油-水界面、液-固界面的性质影响很大[1]。前者主要是表面活性剂对油水界面张力的影响,后者主要是表面活性剂对岩石表面润湿性的改变[2]。近些年来,在石油开采当中通过表面活性剂改变岩石表面的润湿性发挥着越来越大的作用[3-5],而且研究者发现阳离子表面活性剂在改变油湿性固体表面润湿性方面的性能明显优于阴离子、非离子型表面活性剂[6-7]。国内外很多专家学者对阳离子型表面活性剂改变油湿性碳酸盐岩表面润湿性的机理进行了较多的研究[7-9],而对于阳离子型表面活性剂改变油湿性砂岩表面润湿性机理研究得较少,也不够透彻。本文通过多种手段研究了阳离子型表面活性剂CTAB改变油湿性砂岩表面润湿性的机理,为提高三次采油采收率提供了理论指导。

1 实验部分

1.1 实验材料

石英片(25 mm×25 mm×2 mm),云母片(15 mm×15 mm×0.15 mm),分析纯石英砂,人造砂岩岩心,岩心基本数据如表1所示。

表1 岩心基本数据

Tab.1 Basic data of cores

编号	长度/cm	直径/cm	孔隙度/%	渗透率/10^{-3} μm^2	原始含油饱和度/%
h-1	5.02	2.50	22.50	92.10	70.10
h-2	5.07	2.50	22.30	91.95	70.05
h-3	5.05	2.50	22.40	92.05	70.20

原油(来自胜利胜坨油田,25℃下密度0.862 g/cm^3,黏度61.5 mPa·s,酸值1.423 mg/g),分析纯正庚烷、NaCl、$CaCl_2$ 和 $MgCl_2 \cdot 6H_2O$,地层水离子组成如表2所示。

表2 地层水组成

Tab.2 Composition of formation water

组分	质量浓度/(mg·L^{-1})
Na^+	6 907.64
Ca^{2+}	421.16
Mg^{2+}	223.52
Cl^-	12 521.1

分析纯表面活性剂CTAB和POE(1),化学纯表面活性剂TX-100。

1.2 实验方法

石英片、石英砂及云母片的处理方法:首先将样品置于铬酸洗液中浸泡24 h,再用去离子水将样品表面的残余铬酸清洗掉。然后将样品置于20%正庚烷-原油体系当中,于75℃下浸泡处理5 d。之后用正庚烷清洗样品,直至洗出液无色为止。最后将样品置于干燥器当中干燥1 d,以备后续实验使用[10-11]。

岩心处理方法:首先将新买的岩心干燥,之后将岩心在真空状态下饱和水,测定岩心的孔隙体积,并将岩心置于岩心夹持器中测定岩心的渗透率。随后将岩心饱和油,驱替至少5倍孔隙体积的原油,建立岩心束缚水饱和度。最后将饱和好原油的岩心放在

一个密闭的容器里并浸没于原油当中,将岩心在75℃下老化处理30 d[12-13]。

样品的表面活性剂处理:将上述处理好的石英片、石英砂及云母片置于不同的表面活性剂溶液中浸泡处理24 h。除了Zeta电位测定实验之外,其他实验所用的表面活性剂的质量分数均固定在0.30%。

界面张力测定:实验采用 TX500 旋滴法界面张力仪测定不同溶液与原油在室温下的界面张力。

样品的红外测定:利用 Nicolet 6700 傅立叶红外光谱仪测定样品透光率,波数范围 400～4 000 cm^{-1}[14]。

样品的 AFM 测定:本文利用 Veeco Nanoscope Ⅲ型原子力显微镜研究物质在云母片表面的微观 吸附形貌,于空气中采用敲击模式进行 AFM 扫描。

样品表面的 Zeta 电位测定:Zeta 电位是连续相与附着在分散粒子上的流体稳定层之间的电势差,其对固体表面的性质影响很大。本文采用英国马尔文 Nano-Zeta-Meter Zeta 电位测定仪测定固体表面的 Zeta 电位。

接触角测定:本文利用躺滴法于室温下测定接触角,实验装置如图 1 所示。首先将石英片浸没在表面活性剂溶液当中 24 h,然后利用一弯注射器头将油滴注射到石英片下方,使油滴与石英片接触。为了获得较为准确的接触角值,注射 4 滴油于石英片下方不同的位置,从而可获得接触角的平均值。

由于润湿滞后,实验中令油滴在石英片上吸附 30 min 左右,以使接触角达到平衡[11]。使用高分辨率的数码摄像机将油滴的形状拍摄下来,并利用相关的软件计算得到接触角值。

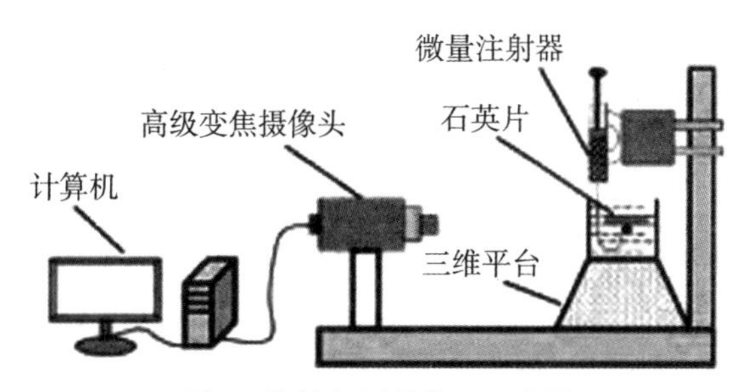

图 1　接触角测量装置示意图

Fig. 1　Schematic diagram for the measurement of contact angle

岩心自发渗吸实验:岩心自发渗吸实验于室温下在 Amott cell 当中进行。实验过程中将岩心垂直立于 Amott cell 当中,并浸没于自吸液,产出油的体积由 Amott cell 上端的刻度读出[13]。岩心在 Amott cell 中的自发渗吸如图 2 所示。

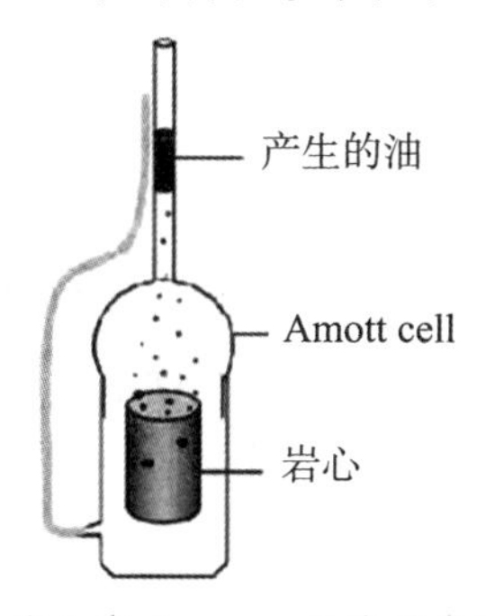

图 2　岩心在 Amott cell 中的自发渗吸

Fig. 2　Schematic diagram for spontaneous imbibition of a core in Amott cell

2 实验结果与讨论

2.1 红外测定

各官能团所处的吸收带如表3所示。

图3是纯石英砂的红外谱图,图3中696 cm^{-1} 和800 cm^{-1} 分别是Si—O键的弯曲振动和伸缩振动导致的吸收峰。在1 100 cm^{-1} 处有一宽而尖锐的吸收峰,其与Si—O键非对称伸缩振动有关。而在3 480 cm^{-1} 处由于O -H键的伸缩振动,有一个很宽的吸收峰。

表3 各官能团的吸收带

Tab. 3 Absorption bands of different functional

吸收带/cm^{-1}	代表官能团
3 700～3 100	O—H伸缩振动带
2 850,2 918	烷基链C—H(—CH_3 和—CH_2)伸缩振动带
1 900～1 680	C═O伸缩振动带
1 640～1 590	共轭及苯环上的 C═C伸缩振动带
1 460～1 376	烷基链C—H弯曲振动带
1 300～1 050	C—O伸缩振动带
1 100～1 000	Si—O非对称伸缩振动带
800	Si—O非对称伸缩振动带
758	苯环取代C—H弯曲振动带
696	Si—O对称弯曲振动带

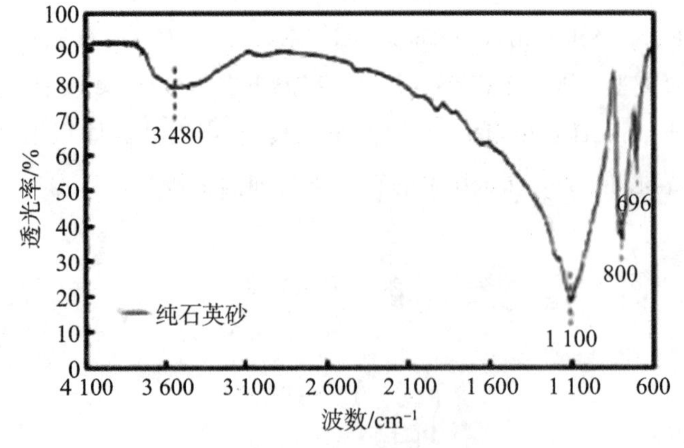

图3 纯石英砂红外谱图

Fig. 3 IR spectrum of pure quartz powder

图4为石英砂经20%正庚烷-原油体系浸泡处理后的红外谱图。图4中758 cm^{-1} 和1 590～1640 cm^{-1} 分别对应着苯环取代C—H弯曲振动和C═C的伸缩振动,1 460

cm^{-1} 和 1 376 cm^{-1} 处的吸收峰与饱和烃的C—H对称变形有关，1 693 cm^{-1} 处的吸收峰归属于C═O键的伸缩振动，2 850 cm^{-1} 和 2 918 cm^{-1} 处的新吸收峰与烷基链C—H（—CH_3 和—CH_2）的伸缩振动是相对应的。通过图3与图4的比较可以看出，石英砂经20%正庚烷-原油体系浸泡处理后，原油当中的很多组分吸附在了石英砂表面。

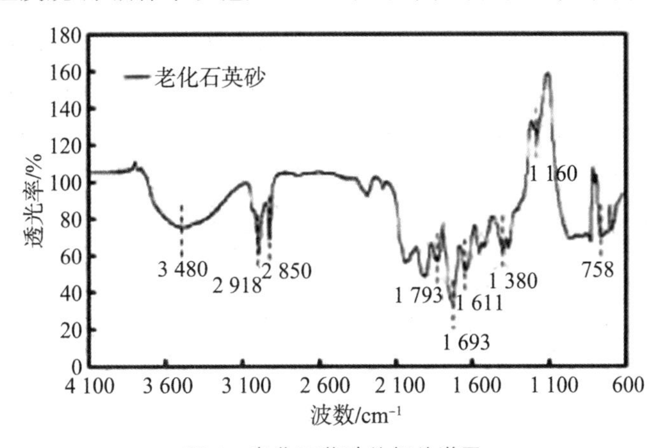

图4　老化石英砂的红外谱图

Fig. 4　IR spectrum of aged quartz powder

图5是老化的石英砂经过CTAB溶液浸泡处理24 h之后的红外谱图。与图4相比，图5中很多吸收峰消失，例如2 918 cm^{-1} 和1 693 cm^{-1} 两处的吸收峰，其分别对应烷烃链当中C—H的伸缩振动和原油当中C═O基团的伸缩振动。可能是由于CTAB与原油当中的羧酸基团形成了离子对，离子对不可逆地从砂岩表面解吸附，并增溶于CTAB形成的胶束当中，从而导致这些基团的消失。这一结果验证了阳离子表面活性剂CTAB改变油湿性砂岩表面润湿性的机理为“离子对机理”[6-7,15]。

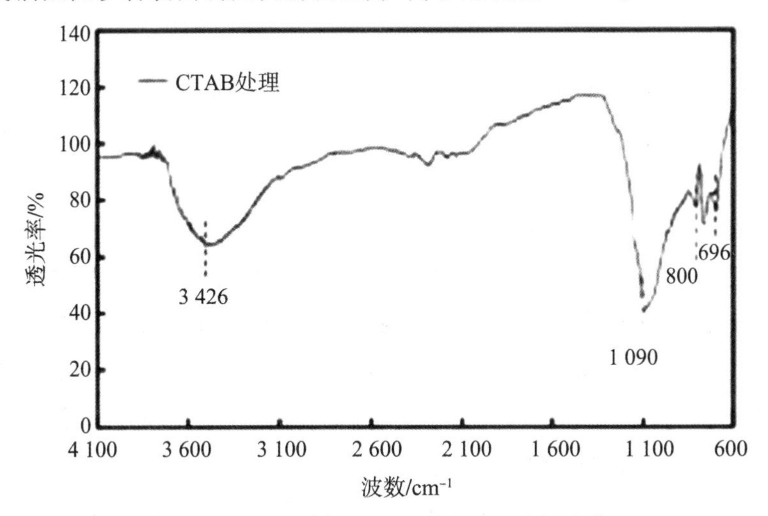

图5　老化石英砂经CTAB处理之后的红外谱图

Fig. 5　IR spectrum of aged quartz powder treated with CTAB

2.2 Zeta 电位测定

图 6 为老化后的石英砂在不同质量分数 CTAB 溶液当中分别浸泡处理 24 h 之后的 Zeta 电位。从图 6 中可以看出,随着 CTAB 质量分数的增加,Zeta 电位先增加后减小,然后趋于平衡。在 CTAB 质量分数比较小的时候(小于 CMC),由于溶液当中未形成胶束,CTAB 与原油中的羧酸基团形成的离子对不能从砂岩表面解吸附,导致随着 CTAB 质量分数增大,Zeta 电位越来越大。当 CTAB 质量分数比较大的时候(大于 CMC),CTAB 与原油中的羧酸基团形成的离子对从砂岩表面解吸附,并增溶于形成的胶束当中,从而导致 Zeta 电位减小。最后随着 CTAB 质量分数继续增大,Zeta 电位趋于平衡。此实验结果与红外测定的结果具有一致性,间接地验证了 CTAB 改变油湿性砂岩表面润湿性的机理为离子对机理的合理性。

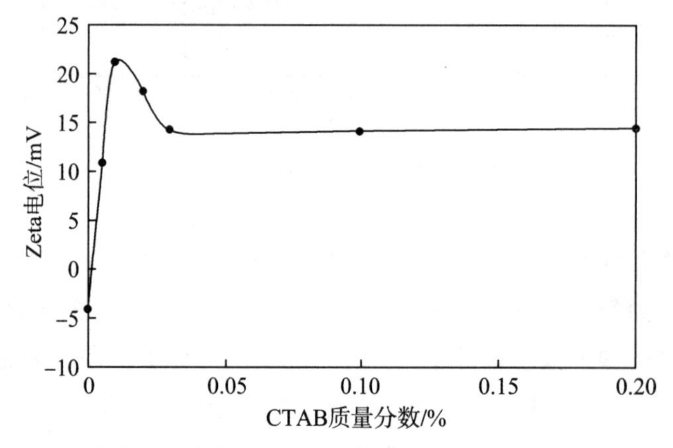

图 6 老化石英砂在不同质量分数 CTAB 溶液中的 Zeta 电位

Fig. 6 Zeta potential of quartz powder treated with CTAB solution of different mass fraction

2.3 AFM 测定

为了更直观地观察 CTAB 对油湿性砂岩表面润湿性的改变,并进一步研究 CTAB 改变润湿性的机 理,本文做了相关的 AFM 测定实验,对不同样品在云母表面的微观吸附形貌进行了扫描。从样片的微观形貌图当中可以获得很多的信息,如固体表面的平均粗糙度,其可以用来表征固体表面润湿性的改变[7,16-17]。

图 7 是不同样品在云母片表面的吸附形貌,其中图的左侧为二维(2D)图,右侧为三维(3D)图。新鲜剥离的云母片表面的平均粗糙度为 0.35 nm,而新鲜剥离的云母片经 20%正庚烷-原油体系浸泡处理后的表面的平均粗糙度为 42.50 nm。通过二者 3D 图的比较,可以看出新鲜剥离云母片经 20%正庚烷-原油体系浸泡处理后,有很多颗粒吸附在了云母片表面,表面变得非常粗糙。图 7(c)是老化云母片经 CTAB 溶液浸泡处理后的表面。通过图 7(b)和图 7(c)中 3D 图的比较,可以看出老化云母片表面经 CTAB 溶液浸泡处理后,许多颗粒消失,而且表面的平均粗糙度比新鲜剥离云母片经 20%正庚烷-原油体系浸泡处理后的表面(图 7(b))的平均粗糙度大大减小,仅为 0.40 nm。这主要是由于 CTAB 与原油当中的羧酸基团形成了离子对,离子对进而从固体表面解吸附导致的。

这一实验结果与红外测定、Zeta 电位测定的结果是一致的，从而进一步确认了 CTAB 改变油湿性砂岩表面润湿性的机理为离子对机理。

(a) 新鲜剥离的云母片表面

(b) 新鲜剥离云母片经20%正庚烷-原油体系浸泡处理后的表面

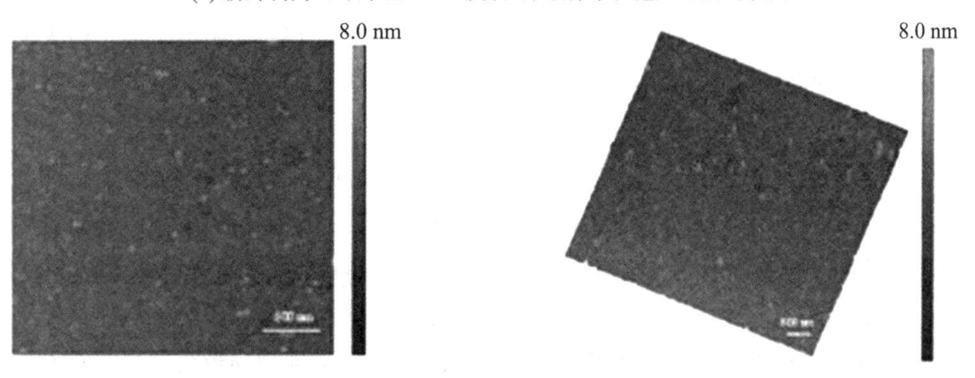

(c) 老化云母片经CTAB溶液浸泡处理后的表面

图 7 不同样品在云母片表面的吸附形貌

Fig. 7 Adsorption morphology of different samples on the surface of mica sheet

2.4 接触角测定

为了进一步验证离子对机理的合理性，本文将 CTAB 与阴离子型表面活性剂 POE (1)、非离子型表面活性剂 TX-100 进行对比，又进一步做了接触角测定实验[7]。图 8 是老化石英片经不同表面活性剂处理后的接触角测定情况。从图 8 可以看出，未处理过的老化石英片上的原始接触角为 140°左右，在 CTAB 处理过的石英片上的接触角为 56°左右,而在 TX-100 和 POE(1)处理过的石英片上的接触角分别为 98°和 110°左右。显然 CTAB 在改变油湿性砂岩表面润湿性方面性能优于阴离子型表面活性剂 POE(1)和非离子型表面活性剂 TX-100。这主要是由于形成离子对的驱动力为静电吸引力,其作用力远远大于范德瓦尔斯力、氢键等作用力。接触角测定结果间接地验证了 CTAB 改变油湿性砂岩表面润湿性的机理为离子对机理。

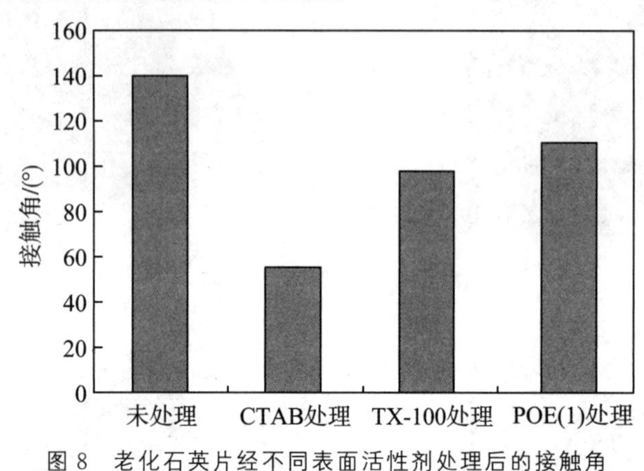

图 8 老化石英片经不同表面活性剂处理后的接触角

Fig. 8 Contact angles of quartz plates treated with three kinds of surfactants

2.5 岩心自发渗吸实验

为了进一步验证接触角实验结果的可靠性,于室温下又进一步做了岩心在上述 3 种类型表面活性剂溶液当中的自发渗吸实验。

经测定 CTAB 溶液、TX-100 溶液和 POE(1)溶液与原油之间的界面张力分别为 0.050 mN/m、0.322 mN/m 和 2.329 mN/m。图 9 是室温下岩心在不同类型的表面活性剂溶液当中的渗吸曲线。从图 9 可以看出,岩心在 CTAB 溶液当中的最终自发渗吸采收率为 12.0%,而岩心在 TX-100 和 POE(1)溶液当中的最终自发渗吸采收率却很低。实验过程中,油滴均从岩心的四周产生,这意味着岩心自发渗吸的过程受毛管力控制。相对于 TX-100 和 POE(1)来讲,CTAB 溶液与原油之间的界面张力较低,但岩心在 CTAB 溶液当中的最终自发渗吸采收率却远远高于岩心在其他 2 种表面活性剂溶液当中的渗吸采收率。这主要是由于 CTAB 改变油湿性砂岩表面润湿性的机理为离子对机理,其改变固体表面润湿性的性能明显优于其他 2 种表面活性剂导致的。岩心自发渗吸实验结果与上述实验结果具有一致性,进一步间接证明了 CTAB 改变油湿性砂岩表面润湿性的机理为离子对机理[15]。

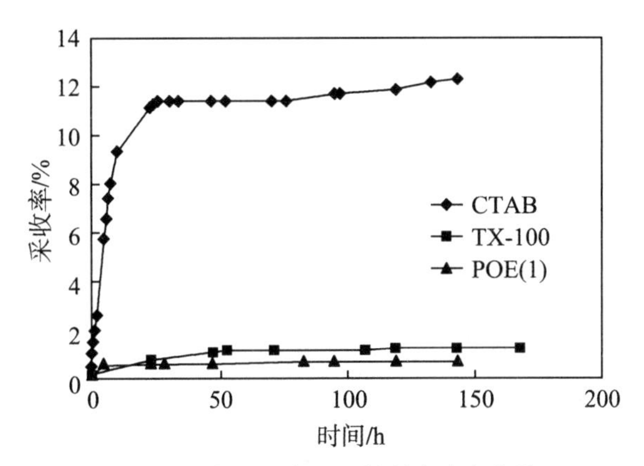

图 9　岩心在不同表面活性剂溶液中的渗吸

Fig. 9　Spontaneous imbibition of cores in different surfactant solutions

3　结论

（1）阳离子型表面活性剂 CTAB 改变油湿性砂岩表面润湿性的机理为离子对机理，即 CTAB 与原油当中的负电性羧酸基团由于静电引力作用而形成离子对，离子对从砂岩表面解吸附并增溶于 CTAB 形成的胶束当中，从而露出干净的水湿表面，砂岩表面因此实现润湿反转。

（2）阳离子型表面活性剂 CTAB 改变油湿性砂岩表面润湿性的能力强于阴离子型表面活性剂 POE(1)和非离子型表面活性剂 TX-100。

参考文献

[1] 蒋平，张贵才，葛际江，等. 润湿反转机理的研究进展[J]. 西安石油大学学报：自然科学版，2007，22(6)：78-84.

JIANG Ping，ZHANG Gui-cai，GE Ji-jiang，et al. Progress in the research of wettability reversal mechanism [J]. Journal of Xi'an Shiyou University：Natural Science Edition，2007，22 (6)：78-84.

[2] Bi Z，Liao W，Qi L. Wettability alteration by CTAB adsorption at surfaces of SiO_2 film or silicagel powder and mimic oil recovery [J]. Applied Surface Science，2004，221(1)：25-31.

[3] Wang Y，Xu H，Yu W，et al. Surfactant induced reservoir

wettability alteration：Recent theoretical and experimental advances in enhanced oil recovery [J]. Petroleum Science，2011，8(4)：463-476.

[4] Leslie Zhang D，Liu S，Puerto M，et al. Wettability alteration and spontaneous imbibition in oil-wet carbonate formations [J]. Journal of Petroleum Science and Engineering，2006，52(1)：213-226.

[5] Mohan K,Gupta R,Mohanty K. Wettability altering secondary oil recovery in carbonate rocks [J] . Energy&Fuels,2011,25(9):3966-3973.

[6] Standnes D C,Austad T. Wettability alteration in chalk: 2. Mechanism for wettability alteration from oil-wet to water-wet using surfactants [J]. Journal of Petroleum Science and Engineering,2000,28(3):123-143.

[7] Jarrahian K,Seiedi O,Sheykhan M,et al. Wettability alteration of carbonate rocks by surfactants: A mechanistic study [J] . Colloids and Surfaces A: Physicochemical and Engineering Aspects,2012,410:1-10.

[8] Gupta R,Mohanty K K. Wettability alteration mechanism for oil recovery from fractured carbonate rocks [J] . Transport in Porous Media,2011,87(2):635-652.

[9] Golabi E, Seyedeyn-Azad F, Ayatollahi S. Chemical induced wettability alteration of carbonate reservoir rocks [J] . Iranian Journal of Chemical Engineering, 2009,6(1):67.

[10] 王业飞,徐怀民,齐自远,等.原油组分对石英表面润湿性的影响与表征方法[J].中国石油大学学报:自然科学版,2012,36 (5):155-159. WANG Ye-fei,XU Huai-min,QI Zi-yuan,et al. Effects of crude fractions on quartz surface wettability and character-ization method [J] . Journal of China University of Petroleum:Edition of Natural Science,2012,36(5):155-159.

[11] Qi Z,Wang Y,He H,et al. Wettabilityalteration of the quartz surface in the presence of metal cations [J] . Energy&Fuels,2013,27(12):7354-7359.

[12] Standnes D C, Nogaret L A D, Chen H L, et al. An evalua-tion of spontaneous imbibition of water into oil-wet carbonate reservoir cores using a nonionic and a cationic surfactant [J] . Energy & Fuels,2002,16 (6):1557-1564.

[13] Strand S, Standnes D C, Austad T. Spontaneous imbibition of aqueous surfactant solutions into neutral to oil-wet car-bonate cores:effects of brine salinity and composition [J]. Energy & Fuels,2003,17 (5):1133-1144.

[14] 王业飞,王所良,徐怀民,等.沥青质与石英表面相互作用及润湿性改变机理[J]. 油气地质与采收率,2011(4):72-74.

WANG Ye-fei, WANG Suo-liang, XU Huai-min, et al. Interaction between asphaltene and quartz surface and mech-anisms of wettability alteration [J]. Petroleum Geology and Recovery Efficiency,2011(4):72-74.

[15] Salehi M,Johnson S J,Liang J T. Mechanisticstudy of wet-tability alteration using surfactants with applications in naturally fractured reservoirs [J]. Langmuir, 2008,24 (24):14099-14107.

[16] Buckley J S,Lord D L. Wettability and morphology of mica surfaces after

exposure to crude oil [J] . Journal of Petroleum Science and Engineering, 2003, 39 (3): 261-273.

[17] Seiedi O, Rahbar M, Nabipour M, et al. Atomic force microscopy (AFM) investigation on the surfactant wettability alteration mechanism of aged mica mineral surfaces [J]. Energy & Fuels, 2010, 25 (1): 183-188.

责任编辑:董　瑾

Study on the mechanism of CTAB changing the wettability of oil wet sandstone

Hou Baofeng[1] Wang Yefei[1] Cao Xulong[2] Zhang Jun[3]
Song Xinwang[2] Ding Mingchen[1] Chen Wuhua[1] Huang Yong[1]

(1. Faculty of Petroleum Engineering, China University of Petroleum (East China), Qingdao 266580, Shandong, China;
2. Re-search Institute of Geology Science, Shengli Oilfield Company of Sinopec, Dongying 257015, Shandong, China;
3. Faculty of Sciences, China University of Petroleum (East China), Qingdao 266580, Shandong, China)

JXSYU 2015 V. 30N. 5p. 95-100

Research and application of high power emission technology of resistivity logging tool for casing wells

Abstract: The mechanism of the wettability change of oil wet sandstone caused by hexadecyl trimethyl ammonium bromide (CTAB) is studied by the different methods including Fourier transform infrared (FTIR), atomic force microscopy (AFM), zeta potential measurement, contact angle measurement and core spontaneous imbibition experiment. The results show that CTAB has excellent performance in changing the wettability of oil-wet sandstone. Due to electrostatic attraction, the ion pairs are formed by the interaction between the positively charged head groups of the cationic surfactants and the carboxylic acid groups of the crude oil, the ion pairs could be desorbed from the sandstone surface and solubilize into the micelles formed by CTAB when the concentration of CTAB is above critical micelle concentration (CMC). Thus, a clean water-wet surface is exposed and the water-wetness of the sandstone surface is im-proved.

Keywords: oil wet sandstone; wettability alteration; cationic surfactant

Hou Baofeng[1], Wang Yefei[1], Cao Xulong[2], Zhang Jun[3], Song Xinwang[2], Ding Mingchen[1], Chen Wuhua[1], Huang Yong[1]
(1. Faculty of Petroleum Engineering, China University of Petroleum (East China), Qingdao 266580, Shandong, China; 2. Re-search Institute of Geology Science, Shengli Oilfield Company of Sinopec, Dongying 257015, Shandong, China; 3. Faculty of Sciences, China University of Petroleum (East China), Qingdao 266580, Shandong, China) JXSYU 2015 V. 30 N. 5 p. 95-100

Research and application of high power emission technology of resistivity logging tool for casing wells.

脂肪醇聚氧乙烯醚丙基磺酸盐的合成与性能研究

王时宇[1*] 曹绪龙[2] 祝仰文[2] 刘 坤[1] 丁 伟[1] 曲广淼[1]

(1.东北石油大学化学化工学院 石油与天然气化工省重点实验室,黑龙江 大庆 163318;

2. 中国石化胜利油田分公司地质科学研究院,山东 东营 257015)

摘要:以脂肪醇聚氧乙烯醚(AEO-5)和3-氯-2-羟基丙磺酸钠为主要反应原料,合成了脂肪醇聚氧乙烯醚丙基磺酸钠(AESO-5)。考察了反应温度、反应时间、原料摩尔比对产率的影响,并对其理化性能进行了测试。结果表明:在3-氯-2-羟基丙磺酸钠与醇醚钠摩尔比为1∶1.5,反应温度为84℃,反应时间为6h的条件下,目的产物收率最高为62.97%。产物的表面张力和临界胶束浓度分别为31.06 mN/m和2.2×10^{-4} mol/L。产物与十二烷基苯磺酸钠(SDBS)、脂肪醇聚氧乙烯醚硫酸钠(AES)相比,耐温抗盐性较好。

关键词:脂肪醇聚氧乙烯醚丙基磺酸钠;合成;表面张力;耐温抗盐

中图分类号:TQ423.3 文献标志码:A 文章编号:0253-4320(2016)01-0071-04

Synthesis and performance of alkyl alcohol polyoxyethylene propyl ether sulfonate

Wang Shiyu[1*] Cao Xulong[2] Zhu Yangwen[2]

Liu Kun[1] Ding Wei[1] Qu Guangmiao[1]

(1. Provincial Key Laboratory of Oil & Gas Chemical Technology, Chemistry and Chemical Engineering College of Northeast Petroleum University, Daqing 163318, China;

2. Gelogical Scientific Research Institute, Shengli Oilfield Branch Company, SINOPEC, Dongying 257015, China)

Abstract: The alkyl alcohol polyoxyethylene propyl ether sulfonate (AESO-5) is synthesized through two steps using fatty ethoxylate (AEO-5) and sodium 3-chloro-2-

hydroxy propanesulfate made by our own laboratory as the main raw materials. The effects of reaction temperature, reaction time and molar ratio of raw materials on the yield of AESO-5 are studied. The structure of AESO-5 is determined by FT-IR. The surface (interfacial) tension and the capacity of emulsifying, temperature resistance, salt tolerance are also tested. The optimum synthesis conditions of AESO-5 are obtained as follows: 84℃ of reaction temperature, 6 hours of reaction time and 1.5 molar ratio of sodium 3-chloro-2-hydroxy propanesulfate to AEO-5. Under the optimum reaction conditions, the yield of AESO-5 can reach up to 62.97%. The surface tension and the critical micelle concentration of AESO-5 are 31.06 $mN \cdot m^{-1}$ and 2.2×10^{-4} $mol \cdot L^{-1}$, respectively. Compared with sodium dodecyl benzene sulfonate (SDBS) and sodium alcohol ether sulphate (AES), the product AESO-5 has better temperature resistance and salt tolerance properties, which can be used to enhance oil recovery in a high temperature and high salinity reservoir.

Keywords: alkyl alcohol polyoxyethylene propyl ether sulfonate; synthesis; surface tension; emulsify; temperature resistance and salt tolerance

根据我国油藏的地质特性,化学驱已成为提高老油田采收率的重要技术之一。其中表面活性剂驱因经济和技术上的优势已得到广泛应用,被认为是最有潜力的一种驱油方式[1-5]。当前应用较多的有阴离子型表面活性剂(如石油磺酸盐、合成烷基苯磺酸盐)和非离子型表面活性剂(如醇醚硫酸盐)。但由于非离子型表面活性剂不耐温,而阴离子型表面活性剂耐盐性能差,使得他们均不能单独应用于高温、高矿化度条件下的油藏。而将非离子表面活性剂和阴离子活性剂复配使用时,往往会出现严重的色谱分离现象[6]。因此,设计一种兼具抗盐与耐温性能的表面活性剂已成为现今该领域探究的热点。

脂肪醇聚氧乙烯醚磺酸盐是一种新型阴-非两性表面活性剂,由于其分子内同时含有氧乙烯基和磺酸基 2 种不同性质的亲水基,因此具有良好的水溶性、抗硬水、抗吸附、耐高温高盐和高效的发泡能力等特点,可以用作高温高矿化度油藏的化学驱油剂[7-12]。

目前合成醇醚磺酸盐主要有 stecker 法、羟乙基磺酸钠法、氯丙基磺酸钠法、丙烷磺内酯法、硫酸酯盐转化法、烯烃加成法等[13]。其中氯丙磺酸钠法副产物少,产物易分离提纯,设备简易,实验工业化的可能性更大,但此方法的相关文献报道较少。笔者根据磺氯丙基磺酸钠法,以脂肪醇聚氧乙烯醚(AEO-5)、3-氯-2-羟基丙磺酸钠为主要反应原料合成醇醚磺酸盐,优化了反应工艺条件并对该表面活性剂的表(界)面张力、乳化和耐温抗盐等性能进行了评价。

1　实验部分

1.1 主要试剂与仪器

3-氯-2-羟基丙磺酸钠(质量分数≥97%),实验室自制(参考文献[14]中所述的方法制备);脂肪醇聚氧乙烯醚(AEO-5),工业级(质量分数≥96%),辽宁省大石桥市宏达化工厂生产;甲苯、氢氧化钠、环己烷、无水乙醇均为分析纯,天津市大茂化学试剂厂生产;脂肪醇聚氧乙烯醚硫酸钠(AES),活性物质量分数为70%,广州市诚壹明化工有限公司生产;十二烷基苯磺酸钠(SDBS),分析纯,上海盛众精细化工有限公司生产;海明1622(质量分数>99%),北京迈瑞达科技有限公司生产。

DF-Ⅱ数显集热式磁力搅拌器,金坛市盛蓝仪器制造有限公司生产;R-201型旋转蒸发仪,上海申胜生物技术有限公司生产;DZF型真空干燥箱,北京市永光明医疗仪器有限公司生产;TENSOR红外光谱仪,布鲁克光谱仪器公司生产;超低界面张力仪SITE100,德国Kruss公司生产。

1.2 脂肪醇聚氧乙烯醚磺酸钠的合成方法

醇醚钠的合成反应:向连有分水器和冷凝管的500 mL四口烧瓶中投入一定摩尔比的AEO-5和氢氧化钠,再向其中加入100 mL环己烷,油浴搅拌加热到80～90℃,利用共沸蒸馏将反应生成的水除去。反应完成后,去除反应瓶中残余的环己烷。

磺化反应:将上述醇醚钠转移至恒压滴定漏斗中,并向反应瓶中投入一定量的甲苯和3-氯-2-羟基丙磺酸钠,升温,逐滴加入醇醚钠,充分搅拌反应后得到粗产物,采用直接两相滴定法测定粗产物中活性物的含量并计算产物的收率[15]。

1.3 产物后处理

所得粗产物杂质主要为未反应的脂肪醇聚氧乙烯醚(AEO-5)和过量的3-氯-2-羟基丙磺酸钠,旋蒸除去粗产物中的甲苯,再将产物用热无水乙醇溶解,趁热滤除无机盐;加入适量的水(Φ=30%体系),用沸程为60～90℃的石油醚多次萃取至石油醚层无色,旋转蒸发掉无水乙醇和水,经无水乙醇重结晶3次,60℃真空干燥后得到纯净的白色粉末固体。

2　结果与讨论

2.1 反应条件优化

2.1.1 物料配比对产物收率的影响

根据化学动力学原理,增大3-氯-2-羟基丙磺酸钠与醇醚钠的摩尔比,有利于产物收率的提高。由于磺化剂以固体粉末形式参与反应,过多的磺化剂会凝结成块而不能很好地参与磺化反应,并且成块的磺化剂会包裹住搅拌器使反应难以进行。因此,研究了原料配比对合成产物收率的影响,结果如图1所示。

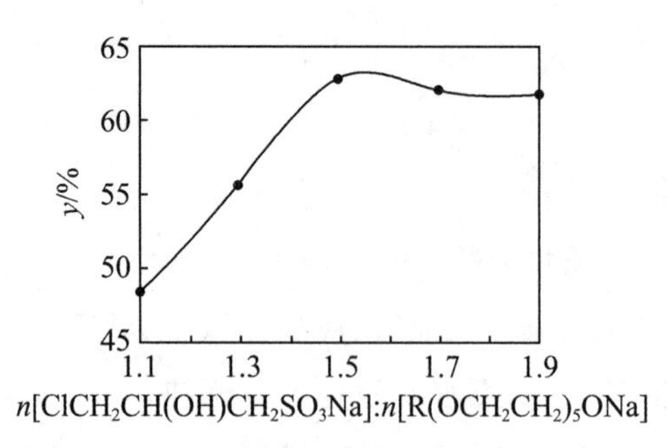

图 1 $n[ClCH_2CH(OH)CH_2SO_3Na]/n[R(OCH_2CH_2)_5ONa]$对合成产物脂肪醇聚氧乙烯醚磺酸钠(5)收率(y)的影响

由图 1 可知,随着 3-氯-2-羟基丙磺酸钠与醇醚钠的摩尔比的增大,产物收率逐渐升高。当 3-氯-2-羟基丙磺酸钠与醇醚钠的摩尔比达到 1.5∶1 时,产物收率最高。继续加大原料摩尔比,产物的收率变化平缓。由此可知,当 3-氯-2-羟基丙磺酸钠与醇醚钠的摩尔比为 1.5∶1 时,2 种物料接触面积已达到饱和状态,因此较适宜的 3-氯-2-羟基丙磺酸钠与醇醚钠的摩尔比为 1.5∶1。

2.1.2 反应温度对产物收率的影响

在 3-氯-3-羟基丙磺酸钠与醇醚钠摩尔比为 1.5∶1,反应时间为 6 h 的反应条件下,温度对产物收率的影响如图 2 所示。

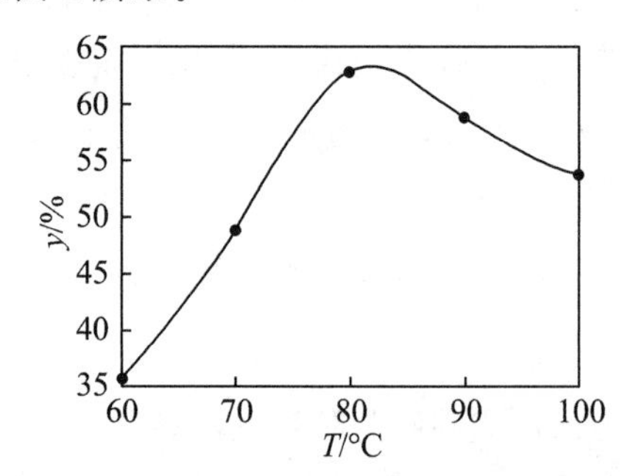

图 2 反应温度对合成产物脂肪醇聚氧乙烯醚磺酸钠(5)收率(y)的影响

由图 2 可知,当反应温度小于 80℃时,产物收率增加很快,这说明温度对产物的收率有较大影响,80℃时产物的收率到最高;之后温度继续升高,目的产物的收率开始下降,原因是反应温度的升高会增加反应物中活化分子的有效碰撞,继而快速提高目的产物的收率,但超过一定温度,副反应逐渐增多,目的产物和副产物相互作用会进一步发生副反应。另外,由于原料浓度降低,副产物浓度升高,根据质量作用定律,此时副反应速率占主导地位,进而影响产物的有效含量。因此,适宜的反应温度为 80～85℃。

2.1.3 反应时间对产物收率的影响

在 n(3-氯-2-羟基丙磺酸钠)/n(醇醚钠)=1.5,反应温度为 80℃的条件下,反应时间对产物收率的影响如图 3 所示。由图 3 可知,目的产物的收率随着反应时间的延长而升高,在 6 h 时达到最高。但反应时间的延长不仅会提高原料转化率,同时也会降低化学反应的选择性并导致副反应的增加,从而使反应收率略有下降。因此,实验最适宜的反应时间为 6 h 左右。

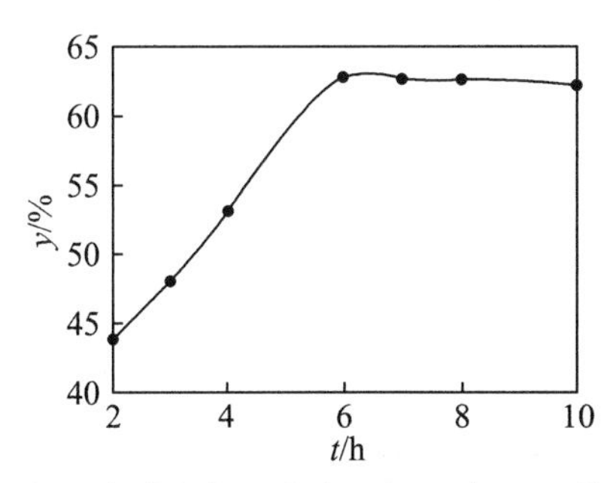

图 3　反应时间对合成产物脂肪醇聚氧乙烯醚(5)收率(y)的影响

2.2 合成产物的红外光谱

产物脂肪醇聚氧乙烯醚磺酸钠的红外光谱图如图 4 所示。其主要吸收峰为:3 358.75 cm^{-1} 为—OH 的伸缩振动,2 938.61 cm^{-1} 为—CH_3 的红外伸缩振动特征峰,2 842.71 cm^{-1} 为—CH_2— 的红外伸缩振动特征峰,729.51 cm^{-1} 为—$(CH_2)_n$— 的摇摆振动峰,在 1 093.26 cm^{-1} 处有尖锐吸收峰,证明存在 C—O—C 醚键,在 1 164.92、1 037.41 cm^{-1} 处有强尖锐吸收峰,为—SO_3Na 的特征峰。确定合成了脂肪醇聚氧乙烯醚磺酸钠。

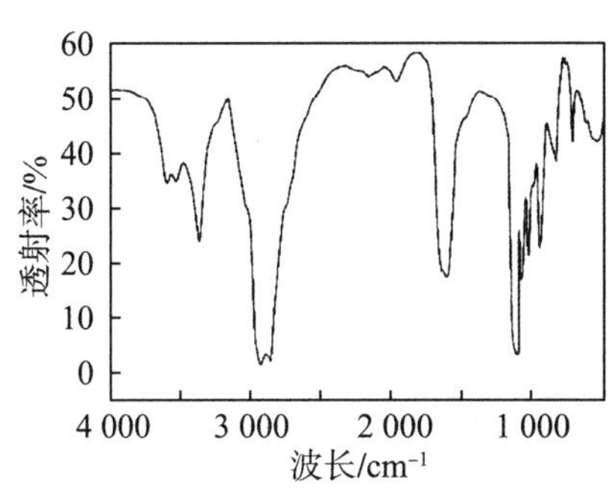

图 4　产物 AESO-5 的红外吸收光谱

2.3 表面张力的测定

以表面张力(γ)对表面活性剂浓度的对数(lg c)作图,如图 5 所示,其中临界胶束浓度取决于斜率突变处的数值。由图 5 可知,表面活性剂水溶液的临界胶束浓度为 2.2×10^{-4} mol/L,临界胶束浓度下的表面张力为 31.06 mN/m。AESO 与 SDBS 的表面性能数值对比如表 1 所示。从表 1 中可以看出,AESO-5 的临界胶束浓度比 SDBS 低 1 个数量级,表面张力相对较低,表现出优异的表面活性。

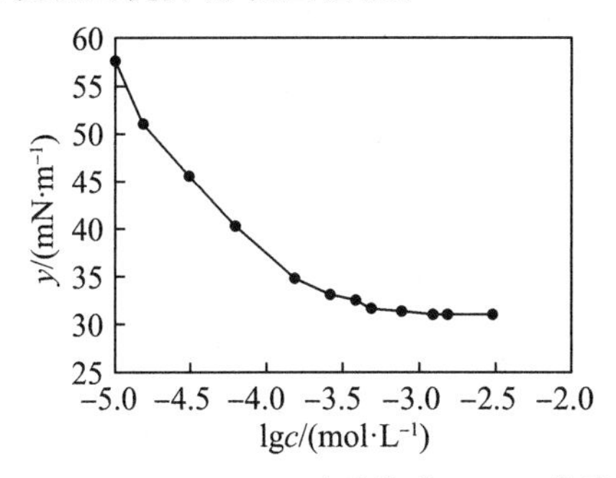

图 5　30℃下 AESO 水溶液的 γ-lg c 曲线

表 1 AESO-5 与 SDBS 表面性能比较

名称	CMC/(mol·L^{-1})	γ_{cmc}/(mN·m^{-1})
AESO	2.2×10^{-4}	31.06
SDBS	14.7×10^{-4}	40.83

2.4 界面张力的测定

在 45℃条件下,质量分数为 0.3%的 AESO-5 溶液与大庆原油间界面张力随时间的变化情况如图 6 所示。由图 6 可知,AESO-5 表面活性剂溶液与原油间的界面张力整体呈下降趋势,表面活性剂与原油接触 30 min 之前,其界面张力迅速下降,当界面张力达到最低值 8.13×10^{-3} mN/m 之后,则变化趋于稳定。

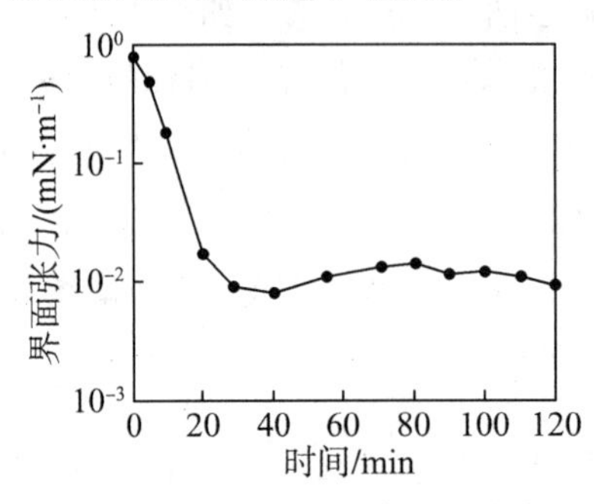

图 6 AESO-5 的动态界面张力

2.5 耐温性能

质量分数为 0.5%的表面活剂 AESO-5 和 AES 水溶液随放置时间的延长,剩余活性物质量分数的变化情况如表 2 和表 3 所示。

表 2 80℃条件下 AESO-5、AES 和 SDBS 的活性物质量分数与水解时间的关系

水解时间/h	AESO-5 活性物质量分数/%	AES 活性物质量分数/%	SDBS 活性物质量分数/%
0	62.9	70.0	93.6
2	62.7	69.6	93.5
8	62.2	68.3	93.2
12	62.1	64.7	93.0
24	61.4	57.4	92.4

表 3 120℃条件下表面活性剂 AESO-5 的活性物质量分数与水解时间的关系

水解时间/h	0	2	4	8	12
活性物质量分数/%	62.8	62.5	61.2	60.8	59.7

由表 2 和表 3 可知,在 80℃条件下,AESO-5 和 SDBS 水溶液中的活性物质量分数下降相对值分别为 1.5%和 1.2%;同等条件,AES 水溶液水解现象比较明显。而在 120℃的条件下,AESO 水溶液的活性物质量分数下降相对值为 3%左右,相比 80℃的测

试条件，活性物质量分数下降幅度较大，表明温度也会对 AESO-5 水溶液的稳定性有影响。但 2 种测试温度下，AESO-5 的有效质量分数变化均不明显，说明产物具有良好的抗盐性能。

2.6 耐盐性能

配制质量分数为 0.5% 的 AES、SDBS、AESO-5 表面活性剂水溶液，向其中加入 NaCl，搅拌溶解，随后置于 40℃ 水浴中保温 24 h，观察溶液是否有新相出现，如表 4 所示。由表 4 可知，当 NaCl 的质量浓度为 50 g/L 时，表面活性剂 SDBS 溶液开始有新相出现，而 AESO-5 和 AES 表面活性剂溶液对应的 NaCl 的质量浓度分别为 150 g/L 和 120 g/L。通常由于无机盐的作用，溶液中反离子量会增多，而这会降低表面活性剂胶束的负电荷密度，使得表面活性剂溶液更容易出现胶束沉淀，使溶液出现新相。但 AESO-5 和 AES 表面活性剂的分子结构中均含有氧乙烯基团，当其与水分子形成氢键时，表面活性剂分子的亲水性增大，在一定程度上阻碍了胶束的凝聚，使得溶液中不易出现新相，耐盐性大大增强。

表 4　氯化钠质量深度对 AESO-5、AES 和 SDBS 耐盐性能的影响

表面活性剂	NaCl 质量浓度/(g·L^{-1})					
	20	50	80	120	150	180
AESO-5	－	－	－	－	＋	＋
SDBS	－	＋	＋	＋	＋	＋
AES	－	－	－	＋	＋	＋

注："-"表示未析出盐，"＋"表示析出盐。

3　结论

(1) 以脂肪醇聚氧乙烯醚(AEO-5)、3-氯-2-羟基丙磺酸钠为主要原料，两步法合成了产物脂肪醇聚氧乙烯醚磺酸钠(AESO-5)。优化工艺条件为：n(3-氯-2-羟基丙磺酸钠)：n(醇醚钠)＝1.5：1，反应温度为 84℃，反应时间为 6 h。产物的收率可达 62.97%。通过红外光谱分析，证实合成了产物 AESO-5。

(2) 性能测定结果表明：脂肪醇聚氧乙烯醚磺酸钠(AESO-5)的表面张力为 31.06 mN/m，临界胶束浓度为 2.2×10^{-4} mol/L，明显优于传统的阴离子表面活性剂(SDBS)。产物 AESO-5 与大庆原油间界面张力最低值能达到 8.13×10^{-3} mN/m。

(3) 耐温抗盐测试表明：脂肪醇聚氧乙烯醚磺酸钠(AESO-5)与十二烷基苯磺酸钠(SDBS)相比，耐盐性更为突出。在 80℃ 和 100℃ 的条件下，相对于脂肪醇聚氧乙烯醚硫酸钠(AES)来说，热稳定性更好。

参考文献

[1] 唐红娇,侯吉瑞,赵凤兰,等.油田用非离子型及阴-非离子型表面活性剂的应用进展[J].油田化学,2011,28(1):115-118.

[2] Iglauer S, Wu Y, Shuler P, et al. New surfactant classes for enhanced oil recovery and their tertiary oil recovery potential [J]. J Pet Sci Eng, 2010 71 (1/2): 23-29.

[3] Ayoub J A, Hutchins R D—Bas F V D, et al. New results improve fracture cleanup characterization and damage mitigation [J]. Spe Production & Operations, 2009, 24 (3): 374-380.

[4] Mohammadi H, Delshad M, Pope G A. Mechanistic modeling of alkaline / surfactant / polymer floods [J] . Spe Reservoir Evaluation & Engineering, 2009, 12 (4):518-527.

[5] Zhang F, Hao J K, Lv W, et al. Design and application of down hole oil production and water injection separation in the same well of high water cut and fault-block oil field [J]. Fault-Block Oil & Gas Field, 2010,17(3):357-359.

[6] 牛瑞霞,宋华,孙双波,等.适用于高矿化度油藏的醇(酚)聚氧烯基醚磺酸盐合成进展[J]. 化工进展, 2013, 32(1):166-173.

[7] Stache H W. Anionic surfactants:Organic chemistry, surfactant science series 56 [M]. New York: Marcel Dekker Press,1996:233.

[8] 张永民,牛金平,李秋小.脂肪醇醚磺酸盐类表面活性剂的合成进展[J].日用化学工业,2008,38(4):253－256.

[9] 肖进新,赵振国.表面活性剂应用原理[M].北京:化学工业出版社,2003.

[10] Gale W W, Saunders R K, Ashcrft TL. Oilrecovery method using a surfactant: US, 3977471[P]. 1976-08-31.

[11] Aoudia M, Al-Harthi Z, Al-maamari R S, etal. Novel alkyl ether sulfonates for high salinity reservoir: Effect of concentration on transient ultralow interfacial tension at the oil-water interface [J]. Journal of Surfactants & Detergents, 2010,13 (3):233-242.

[12] Schmitt K D. Two-tailed surfactants having one aromatic containing tail and their use in chemical water flooding: US, 4545912 [P]. 1985-03-05.

[13] 牛金平,韩亚明.驱油用磺酸盐的工业化生产现状与发展趋势[J].日用化学品科学,2008,11:4-7.

[14] 申云霞,毕彩丰,赵宇,等.2-羟基-3-氯丙磺酸钠的合成及结构表征[J].化工进展,2009,28(12):2218-2220.

[15] 毛培坤编.表面活性剂产品工业分析[M].北京:化学工业出版社,2003:20-22.

辛基酚聚氧乙烯醚磺酸盐界面行为的分子动力学模拟

单晨旭[1]　曹绪龙[2]　祝仰文[2]　刘　坤[1]　曲广淼[1]　吕鹏飞[1]　薛春龙[1]　丁　伟[1]

([1]东北石油大学化学化工学院,黑龙江大庆 163318;

[2]中国石化胜利油田分公司地质科学研究院,山东东营 257015)

摘要:采用分子动力学模拟(MD)的方法在分子层面上考察辛基酚聚氧乙烯醚磺酸盐(OPES)在油-水界面的界 面行为。模拟结果表明:辛基酚聚氧乙烯醚磺酸盐可以大幅降低油-水界面的界面张力,在 OPES 浓度达到饱和浓度时,系统界面张力仅为 3.85 mN·m^{-1};OPES 中磺酸基是主要亲水基团,具有良好的亲水性;温度在 318～373 K 时,界面张力由 24.63 mN·m^{-1} 下降到 17.43 mN·m^{-1},这说明 OPES 具有良好的抗高温性能;当 Na^+ 浓度在 1%～5%的环境下 OPES 性质稳定,界面张力仅有 4.47 mN·m^{-1} 的小幅增加,因此 OPES 具有良好的耐盐性,并且其对 Na^+ 的耐盐性能好于对 Ca^{2+} 的耐盐性。

关键词:辛基酚聚氧乙烯醚磺酸盐;分子动力学模拟;界面张力;抗温;抗盐

中图分类号:O 641　文献标志码:A　文章编号:0438-1157(2016)04-1416-08

Molecular dynamics simulation for interface behavior of octylphenol polyoxyethylene ether sulfonate

Shan Chenxu[1]　Cao Xulong[2]　Zhu Yangwen[2]　Liu Kun[1]

Qu Guangmiao1, Lv Pengfei[1]　Xue Chunlong[1]　Ding Wei[1]

([1] College of Chemistry and Chemical Engineering, Northeast Petroleum University, Daqing 163318, Heilongjiang, China;

[2] Geological Scientific Research Institute, Shengli Oilfield Branch Company, Dongying 257015, Shandong, China)

Abstract: Behaviors of octylphenol polyoxyethylene ether sulfonate (OPES) molecules on the oil-water interface were studied through molecular dynamics simulation (MD). The results showed that OPES could weaken tension of the oil-water interface significantly. The interface tension was only 3.85 mN • m^{-1} at OPES saturation. The sulfonic group in OPES was the main hydrophilic group and had good hydrophilicity. The interface tension declined from 24.63 mN • m^{-1} to 17.43 mN • m^{-1} when temperature increased from 318 K to 373 K, indicating the good high temperature resistance of OPES. OPES maintained the stable properties within 1%～5% Na^{+} concentration with only 4.47 mN • m^{-1} increase of the interface tension. Therefore, OPES had good salt tolerance and could tolerate higher Na^{+} concentration than Ca^{2+}.

Keywords: octylphenol polyoxyethylene ether sulfonate; molecular dynamics simulation; interfacial tension; heat resistance; salt tolerance

引言

在三次采油中,为提高原油采收率,经常利用表面活性剂来降低油水界面张力,目前国内部分油 田综合含水量已高达 90%,单独的阴离子、非离子型表面活性剂已经不能满足当前的采油要求,阴非离子型表面活性剂作为一种同时有非离子及阴离子 表面活性剂优点的两性表面活性剂对于目前日益严 苛的采油环境的适应性更强[1-2]。本文研究的辛基酚 聚氧乙烯醚磺酸盐(OPES)是一种具有优良的乳化、耐温、耐盐性能的阴非两性表面活性剂,它已经作 为分散剂、润湿剂、乳化剂、洗涤剂等被广泛地应 用于石油、日化、纺织等领域[3-5]。

分子动力学模拟主要是利用牛顿力学来模拟分子的运动,从不同状态下的体系抽取样本进行构型积分并以此为基础计算体系的热力学量等宏观性质。从 20 世纪 90 年代后期,人们开始利用计算机模拟研究表面活性剂的性能,它可以将真实环境中的实验现象在分子层面进行解释[6-10]。对液液界面的研究作为分子动力学模拟的重要研究方向之一近年来受到广泛的关注和报道,如 Jang 等[11]利用 MD 模拟了苯磺酸基在不同位置时十六烷基苯磺酸盐的界面张力等界面性能。Wardle 等[12]考察了表面活性剂对无机盐、水和正己醇构成的混合物中钠离子迁移的影响。陈贻建等[13]用 MD 模拟方法对表面活性剂在气-液、固-液、液-液界面的自组装现象进行深刻解释分析。因此利用 MD 方法研究表面活性剂的界面张力、抗温、抗盐等界面性能具有重要意义。国内对于应用分子动力学模拟来研究表面活性剂性能的起步较晚,特别是对具有耐温、耐盐性能的表面活性剂的研究较少,本文通过分子动力学模拟来研究辛基酚聚氧乙烯醚磺酸盐的油-水界面行为、抗温、抗盐性能,可为实际实验提供较为准确的指导。

1 分子动力学模拟的模型选择与模拟方法

20 世纪 80 年代以来,人们相继研发出可以适合不同环境的力场,如 GROMOS、

OPLS、AMBER、CHARMM 等。本文选择 Gromacs[14] 中 GROMOS53a6[15-16] 力场，以辛基酚聚氧乙烯醚磺酸盐为研究对象进行模型构建。

分子的物理化学性质由其分子结构决定，因此合理的分子结构以及准确的原子电荷是模拟准确性的基础保证。首先要对模拟对象用 GAMESS(US)[17-18] 进行结构优化，然后利用 Kollman-Singh 方法计算电荷，另外如果分子内存在对称结构还需进行电荷平均化来保证电荷分配的合理性。由于本文采用联合原子力场，因此还要去除 sp^3 杂化。图 1 为优化后的 OPES 分子结构以及电荷分布，图中绿色小球为碳原子，白色小球为氢原子，红色小球为氧原子，黄色小球为硫原子。

在进行分子动力学模拟之前构建出合理的力场是极为重要的工作[19]。本文通过 Automated Topology Builder (ATB) and repository[20] 生成的 GROMOS[15] 系列力场参数，利用现有的数据库以及量子化学进行计算，同时它可以充分考虑到分子中的对称结构，使其反映出的分子性质及参数更为精确。但 ATB 只能处理原子数小于 40 的分子，对于分子数大于 40 的分子结构需进行拆分。在获取准确的电荷及键参数之后利用 Packmol[21-22] 程序定向排列分子将其堆砌成立方体结构。此外，本文选取的油-水界面需使表面活性剂平均分布在水相两侧，亲水基靠近水相，疏水基靠向油相。图 2 为初始状态下体系截图，其中中间红色部分为水分子，左右两侧蓝色部分为癸烷分子，油水中间即 OPES 分子。

本文中所有体系所堆砌的盒子均为 5 nm×5 nm×17.5 nm 长方体，并在 x、y、z 方向选择周期性边界条件。系综选择 NPT(等粒子等温等压系综)，初始压力为 $1.013\ 25\times10^5$ Pa，水模型使用 SPC[23] (simple point charge)，温度采用 Nose-Hoover[24] 热浴法，压力采用 Parrinello-Rahman[25] 压浴法，由于模拟过程中系统为等压变化，所以本文模拟的所有系统最终压力值均在 $1.008\ 1\times10^5$ ～ $1.017\ 8\times10^5$ Pa 之间。在体系能量最小化后，先进行 100ps 的 NVT 模拟，使体系升温到 300K 并在此温度下产生初速度，再进行 1 ns 的 NPT 模拟使体系密度达到合理状态，再进行 12ns 的 NPT 模拟，控温及控压的弛豫时间为 0.5、4.0ps，积分步长为 2fs，在模拟过程中添加适当的阴阳离子保持体系为电中性。

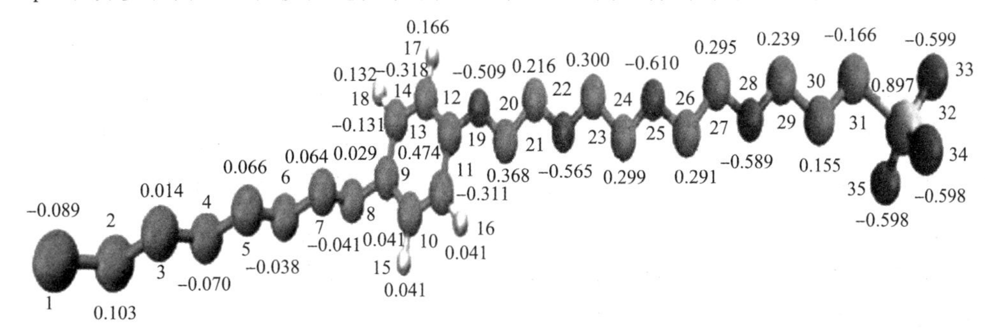

图 1　辛基酚聚氧乙烯醚磺酸盐的分子结构以及电荷分布

Fig. 1　Molecular structures and united atom charge distribution of OPES

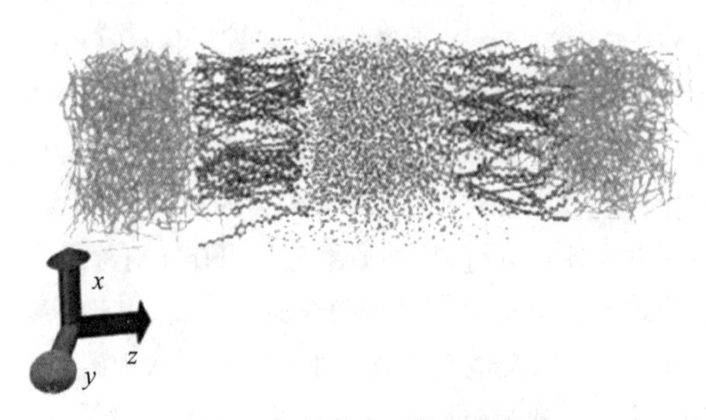

图 2 初始状态下体系截图

Fig. 2 Snapshot of simulation system

2 模拟结果与分析

2.1 界面性能

为考察 OPES 的界面行为及性能进行了 8 组对比模拟实验，分别记作 S20、S50、S80、S100、S120、S140、S160、S180，即通过改变表面活性剂的数量对比各个体系的各相密度、界面宽度、界面张力以及界面聚集形态等界面行为，得出 OPES 表面活性剂浓度对界面行为的影响以及变化趋势等结论。

对比模拟实验均在由 500 个癸烷分子、5 000 个水分子构成的油-水界面以及温度为 318 K 的条件下进行。所有体系在平衡后表面活性剂的亲水基插入水相，亲油基插入油相并且形成非常稳定的界面。当表面活性剂浓度不断增加(从 S20 到 S160)时由于单个表面活性剂分子的占有面积逐渐减少，分子的排列呈由分散变为紧凑的趋势。但当表面活性剂数量增大到 180 时，部分分子开始脱离原来平面，此时表面活性剂浓度已达到饱和状态。这一过程的界面张力变化如表 1 所示。

疏水尾链碳原子序参数(order parameter，S_{CD})可以用来表示疏水尾链的有序性

$$S_{CD}=(3\cos^2\theta-1)/2 \tag{1}$$

S_{CD} 可用式(1)来计算，θ 代表 C_{n-1} 和 C_{n+1} 原子之间向量与界面垂直方向的角度。图 3 所示为上述 8 个体系的序参数曲线，对于每一条序参数曲线都随着碳原子序号的增加序参数逐渐增大，这说明了疏水链末端的碳原子有序性更强。从图中还可以观察到 S20 的曲线几乎水平，这是由于 OPES 在界面的浓度过低其分子可以自由摆动。当表面活性剂的数量从 20 增加到 180 时，SCD 曲线不断上升，这说明随着表面活性剂数量的增加疏水链排列的有序性也在不断增强。并且 S160 和 S180 的序参数曲线相当接近，这表明此时表面活性剂的浓度已经达到界面的饱和浓度。

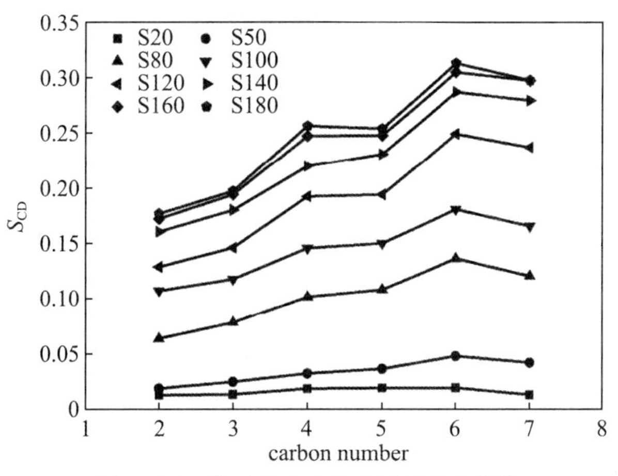

图 3　不同体系的疏水链碳原子的序参数

Fig. 3　Order parameter of hydrophobic chain atoms of different systems

对于表面活性剂来说降低界面张力的能力是考察其性能好坏的重要指标之一，下面通过考察不 同体系的界面张力和界面宽度来进一步说明 OPES 的界面性能，如表 1 所示。对于界面张力可以利用式(2)来计算，其中 L_Z 为盒子高度；P_{xx}、P_{yy}、P_{zz} 分别为 x、y、z 方向的压力。

$$\gamma=\left|\frac{L_z}{2}\left(P_{zz}-\frac{P_{xx}+P_{yy}}{2}\right)\right| \tag{2}$$

从表 1 可以发现表面活性剂数量的增加使得界面张力逐渐下降。其中当 OPES 数量由 20 增加到 80 时，界面张力值较高，这说明当表面活性剂浓度较低时并不能起到很好的降低界面张力的作用；随着表面活性剂数量进一步增加界面张力逐渐下降，当 OPES 数量为 180 时，达到临界饱和状态，此时界面张力仅为 3.85 $mN \cdot m^{-1}$，此变化规律与真实实验规律相同[4]。这同时也说明辛基酚聚氧乙烯醚磺酸盐可以很好地降低界面张力，是一种性能优良的表面活性剂。界面宽度用体系密度图中表面活性剂的密度曲线宽度来表示。随着表面活性剂个数增加界面宽度递增，起初界面宽度增加速度较快是因为界面 OPES 浓度过低并未饱和；当 OPES 数量达到 100～140 时，界面宽度仅有少量缓慢增加这是由于此时界面正在逐渐接近饱和状态。随着 OPES 的数量达到 160 和 180 时，界面宽度增加幅度变大，进一步验证此时界面已达到饱和状态。

表 1　不同表面活性剂浓度下体系界面张力和界面宽度

Table 1　Interfacial tension and width of different surfactant concentration

Number of surfactant	Interfacial tension /$mN \cdot m^{-1}$	Interfacial width/nm
20	52.23	2.97
50	36.94	4.70
80	24.63	4.95
100	20.89	5.08

（续表）

Number of surfactant	Interfacial tension /mN・m^{-1}	Interfacial width/nm
120	19.61	5.11
140	9.96	5.16
160	7.62	5.57
180	3.85	5.80

本文提取体系稳定的S80、S120进行分析，两个体系的各部分密度图如图4所示。在平衡状态下两体系中水的平均密度分别为989.86 kg・m^{-3}、992.43 kg・m^{-3}，与国际温标318 K时水密度990.2 kg・m^{-3}值接近；另外，两体系中癸烷平均密度为711.49 kg・m^{-3}、710.72 kg・m^{-3}，与真实状况下癸烷密度711.2 kg・m^{-3}值接近，这表明模拟体系的模型选择、力场参数都是准确的，可以得出可靠结论。

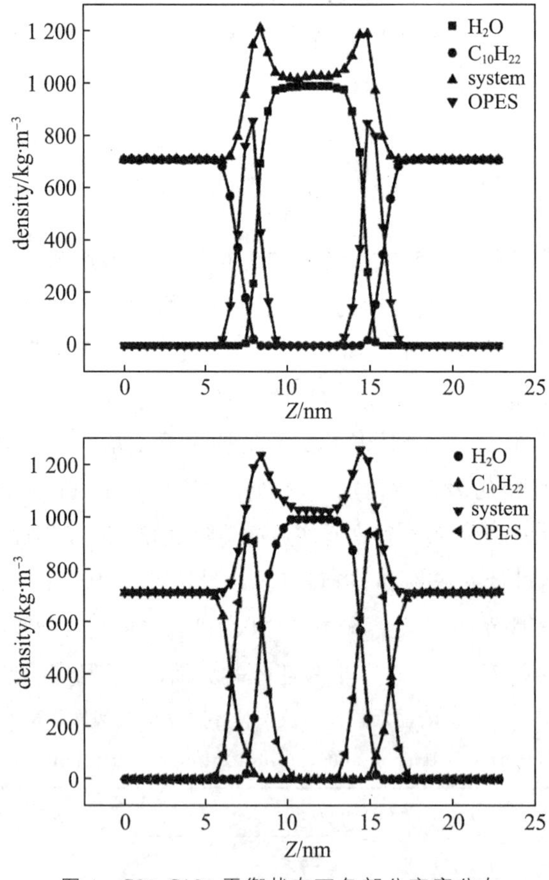

图4 S80、S120平衡状态下各部分密度分布

Fig. 4 Density profile of S80 and S120 simulation system under equilibrium state

在OPES结构中，有两个亲水基团分别为氧乙烯基(OG)、磺酸基(SDMSO)，本文通过径向分布函数(通常指的是给定某个粒子的坐标，其他粒子在空间的分布概率)来对比两者与水之间的作用力。图5为表面活性剂中氧乙烯基和磺酸基与水分子中氢原子的

径向分布函数 $g_{OG\text{-}HW}(r)$、$g_{SDMSO\text{-}HW}(r)$。由图可知，$g_{SDMSO\text{-}HW}(r)$曲线第 1 个峰值出现在 0.306 处，这表明磺酸基中的氧原子与水中的氢原子之间较强的氢键作用形成了第 1 水层；在距离为 0.458 时，出现第 2 个峰值，数值有所下降，这表示逐渐减小的氢键作用形成了第 2 水层；第 3 水层形成在 0.688 处，此时磺酸基与水的作用进一步减弱，但 3 处的峰值均远大于氧乙烯基的峰值。这表明磺酸基与水分子的作用力远高于氧乙烯基，所以磺酸基为辛基酚聚氧乙烯醚磺酸盐结构中的主要亲水基团。

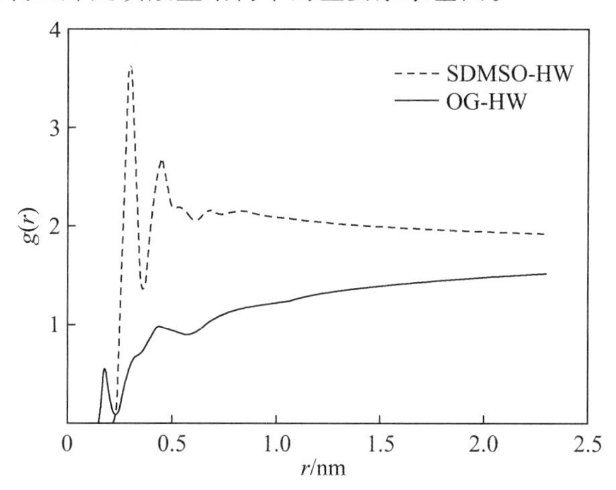

图 5　S80 中表面活性剂中亲水基团与水中氢原子之间的径向分布函数

Fig. 5　RDF between hydrophilic group and hydrogen of water molecule for S80 simulation system

2.2 温度对癸烷＋水＋OPES 体系油-水界面张力的影响

由于采油环境日益严苛，一些表面活性剂在高温条件下与水之间的氢键易断裂，使得其亲水性能大幅降低，因此抗高温性能是考察表面活性剂好坏的重要指标之一。

本文选取 4 组对比模拟实验，保持表面活性剂数量为 80 不变，控制温度分别为 318、343、358、373K，记作 S80T318、S80T343、S80T358、S80T373。

表 2 通过界面张力、表面活性剂与水的氢键数量、势能 3 组数据对比得出表面活性剂的界面张力随着温度升高而降低的结论。数据表明在 4 组模拟实验过程中，界面的宽度并没有发生改变。因此界面张力下降的主要原因是由于 OPES 势能的降低导致分子之间的作用力也随之降低。

表 2　不同温度下各体系的界面性能

Table 2　Interfacial properties of systems at different temperatures

System	Interfacial width/nm	Number of hydrogen bond	Interfacial tension /$mN \cdot m^{-1}$	Potential energy /$kJ \cdot mol^{-1}$
S80T318	5.44	570	24.63	−351 691
S80T343	5.44	549	23.61	−340 130
S80T358	5.43	538	21.28	−333 771
S80T373	5.44	526	17.43	−325 870

另一个值得注意的改变是虽然随着温度的升高 OPES 与水之间的氢键有微量的下降但并没有达到浊点,况且磺酸基具有良好的亲水性,因此,OPES 并没有因为温度升高而失效,反而能提高其在油-水界面的性能。

2.3 盐对癸烷+水+OPES 体系油-水界面张力的影响

大量数据表明,在高盐油藏表面活性剂的化学稳定性易受到影响,其结构可能受到改变或破坏进 而影响石油采收率。石油磺酸盐、烷基苯磺酸盐等表面活性剂在高盐度的环境下极易失去活性[26-27]。因此,表面活性剂是否具有良好的抗盐性能显得尤为重要。对于辛基酚聚氧乙烯醚磺酸盐从结构上来说其具有的磺酸基结构应使其有良好的耐盐性能。

本文选取 5 组对比模拟实验,保持 OPES 数量为 80、温度为 318K 不变,分别向体系内加入 1%、2%、3%、4%、5%的 NaCl 溶液,记作 S80Na1、S80Na2、S80Na3、S80Na4、S80Na5。图 6 所示为 S80Na2 体系在平衡状态下界面状态,其中蓝色小球为 Na^+。Na^+ 几乎全部分散于水相中,在表面活性剂附近的分布很少,因此可以初步确定盐对表面活性剂的影响较小,辛基酚聚氧乙烯醚磺酸盐具有抗盐性。

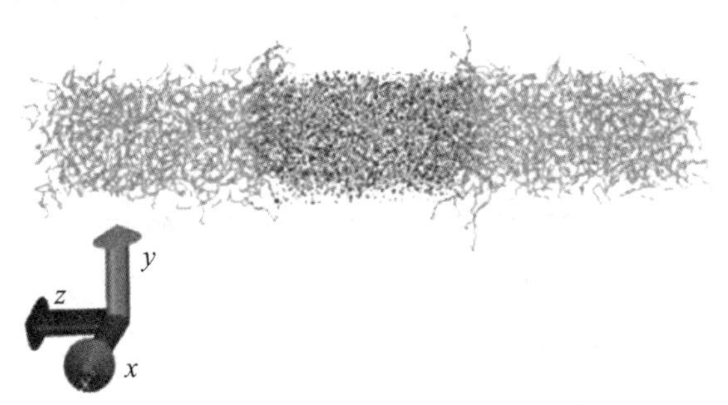

图 6 S80Na2 平衡状态下界面状态

Fig. 6 Interfacial distribution snapshot of S80Na2 simulation system

为进一步确定 OPES 的耐盐性能,可以再通过不同体系平衡状态时相应的界面张力和氢键数量来讨论,相关数据如表 3 所示。

表 3 不同浓度 Na^+ 溶液体系界面张力以及 OPES 与 H_2O 的氢键数量

Table 3 Number of hydrogen bonds between OPES and water and interfacial tension in systems with different concentrations of Na^+

System	Interfacial tension/$mN \cdot m^{-1}$	Number of hydrogen bond
S80Na1	24.71	558
S80Na2	25.02	549
S80Na3	25.99	536
S80Na4	26.17	534
S80Na5	29.18	523

模拟数据显示,随着 NaCl 浓度的升高,表面活性剂在油水界面的界面张力仅有小

幅升高，这是由于体系中不断加入 NaCl 使得 OPES 更加亲油，使得部分表面活性剂分子向油相中跃迁。另外，氢键数量有少量下降这是由于在体系不断添加 Na^+、Cl^- 过程中，替换了水相中的水分子使得水分子数量减少从而影响了氢键数量。

图 7 中显示了在不同 NaCl 浓度的体系中，磺酸基中的氧原子与水中氢原子的径向分布 $g_{OS\text{-}HW}(r)$，可以看出其峰值并没有因为 NaCl 浓度的增加而发生很大改变，这更能说明阳离子并不能对表面活性剂的性能造成影响。

下面同样通过疏水链碳原子序参数的变化来进一步验证 OPES 的抗盐性。从图 8 中可以看出，在同一体系中随着碳原子的增加序参数值增大，这说明越接近疏水尾链的末端的碳原子有序性越好。

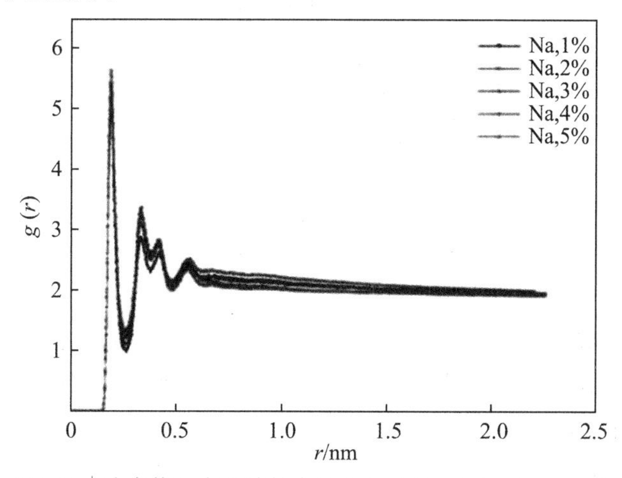

图 7　在不同 Na^+ 浓度体系中磺酸基中的氧原子与水中氢原子的径向分布函数

Fig. 7　RDF between oxygen of sulfonate group and hydrogen of water molecule for different concentrations of Na^+

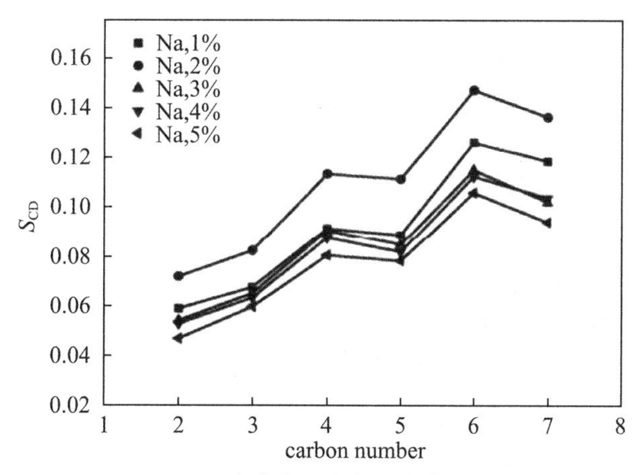

图 8　不同 Na^+ 浓度下疏水尾链碳原子序参数

Fig. 8　Order parameter of hydrophobic chain atoms with different amount of Na^+

同时，对于 NaCl 浓度为 1%、3%、4%、5%的体系，疏水尾链碳原子的序参数并未发生太大改变，NaCl 浓度为 2%时其序参数值还要大于 1%时的序参数值，这说明在浓度为 2%的 NaCl 溶液中疏水链碳原子间的相互作用力最强。

提取 2%NaCl 浓度时体系的疏水尾链碳原子序参数与同浓度的 $CaCl_2$ 体系进行对比,对比结果如图 9 所示。在 $CaCl_2$ 溶液中疏水链碳原子的 SCD 值明显高于无盐溶液以及 2%的 NaCl 溶液中的 SCD 值,因此,在 Ca^{2+} 的环境下碳原子的摆动空间与灵活性要小于在 Na^{+} 的环境中。

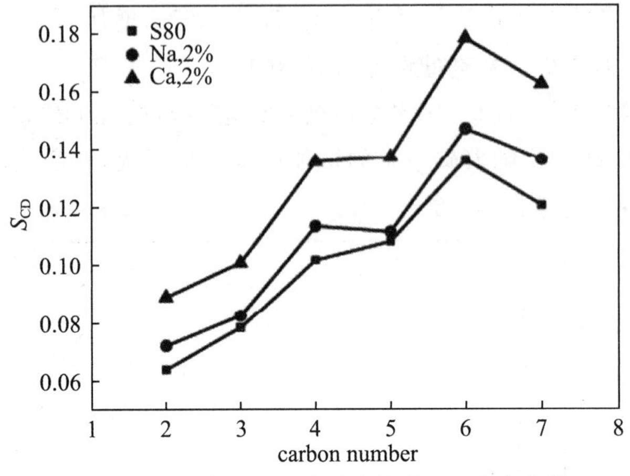

图 9 在不同离子溶液中疏水链碳原子的序参数

Fig. 9 Order parameter of hydrophobic chain atoms with Na^{+} and Ca^{2+}

图 10 中曲线分别代表在 2%NaCl 溶液、2%$CaCl_2$ 溶液中磺酸基中的氧原子与水中氢原子之间的径向分布函数,如图所示两条曲线的峰值并未有太大差别,这说明 OPES 对 Na^{+}、Ca^{2+} 都有很好的抗盐性。

进一步分析 OPES 对 Na^{+}、Ca^{2+} 抗盐性的差别,考察了磺酸基中的氧原子与 Na^{+}、Ca^{2+} 的径向分布函数,如图 11 所示。图中两曲线的峰值出现较大差距,其中 Na^{+} 曲线的峰值明显小于 Ca^{2+} 曲线的峰值,这表明亲水基团与 Na^{+} 的作用较小,也就是说 Na^{+} 对 OPES 的性质影响较小。因此,辛基酚聚氧乙烯醚磺酸盐的抗盐性顺序为 $Na^{+}>Ca^{2+}$,与对序参数所做的分析结论相同。

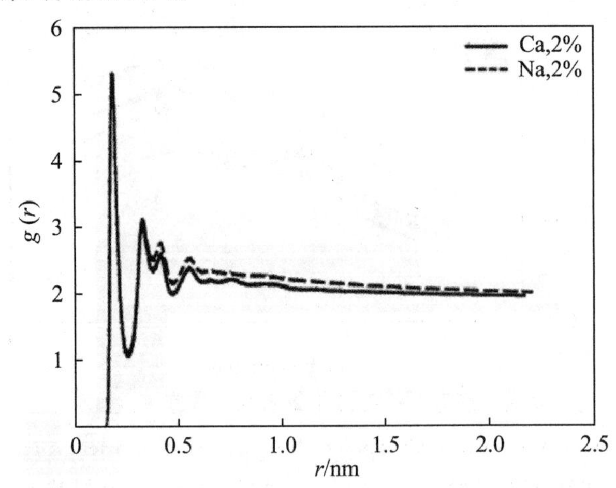

图 10 在不同离子溶液中磺酸基中的氧原子与水中氢原子的径向分布函数

Fig. 10 RDF of interaction between oxygen of sulfonate group and hydrogen of water molecule for solutions with Na^{+} and Ca^{2+}

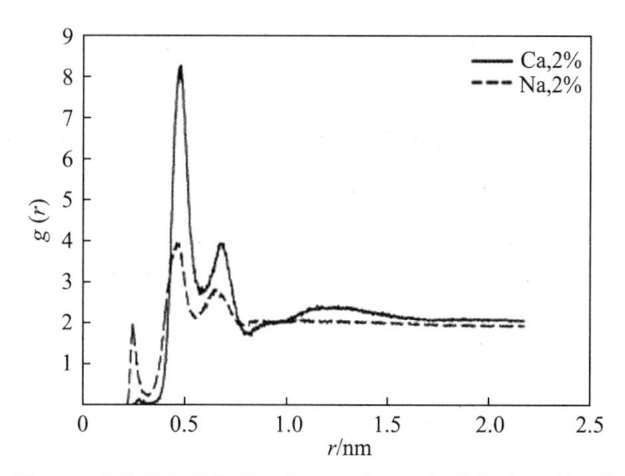

图 11 磺酸基中的氧原子与不同离子之间的径向分布函数

Fig. 11 RDF of interaction between oxygen of sulfonate group and different ions

3 结论

（1）分子动力学模拟可以准确模拟辛基酚聚氧乙烯醚磺酸盐在油-水界面的界面行为及性能。

（2）辛基酚聚氧乙烯醚磺酸盐可以大幅降低油-水界面的界面张力。

（3）辛基酚聚氧乙烯醚磺酸盐中磺酸基为其主要亲水基团且疏水链尾端碳原子有序性较好。

（4）辛基酚聚氧乙烯醚磺酸盐在温度为 318～373 K 时界面张力随温度升高而减小，具有良好的抗高温性能。

（5）辛基酚聚氧乙烯醚磺酸盐在 Na^+ 浓度为 1%～5%的高盐条件下性质稳定，界面张力仅有小幅增加，并且其对 Na^+ 的耐盐性好于对 Ca^{2+} 的耐盐性。

References

[1] 韩巨岩，王文涛，崔昌亿，等. 烷基苯基聚氧乙烯醚磺酸盐 [J]. 日用化学工业，1998，28(6)：60-62. DOI：10. 13218/j. cnki. csdc. 1998. 06. 016.

HAN J Y, WANG W T, CUI C Y, et al. Alkyl phenyl polyoxyethylene ether sulfonate [J]. *China Surfactant Detergent & Cosmetics*, 1998, 28 (6): 60-62. DOI: 10. 13218/j. cnki. csdc. 1998. 06. 016.

[2] 唐红娇，侯吉瑞，赵凤兰，等. 油田用非离子型及阴-非离子型表面活性剂的应用进展 [J]. 油田化学，2011，28(1)：115-118. TANG H J, HOU J R, ZHAO F L, et al. Application progress of nonionic and anionic-nonionic surfactants used in oil field [J]. *Oilfield Chemistry*, 2011, 28 (1): 115-118.

[3] 王业飞，李继勇，赵福麟. 高矿化度条件下应用的表面活性剂驱油体系 [J]. 油

气地质与采收率，2001，8（1）：1-7. DOI：10. 13673/j. cnki. cn37-1359/te. 2001. 01. 019.

WANG Y F，LI J Y，ZHAO F L. Surfactants oil displacement system in high salinity formations [J]. Oil & *Gas Recovery Technology*，2001，8(1)：1-7. DOI ：10. 13673/j. cnki. cn37-1359/te. 2001. 01. 019.

[4] 贺伟东，李瑞冬，葛际江，等. 辛基酚聚氧乙烯醚磺酸盐的合成与界面张力的测定 [J]. 石油化工高等学校学报，2010，23（4）：20-24. DOI：10. 3696/j. issn. 1006-396X. 2010. 04. 005.

HE W D，LI R D，GE J J，et al. The synthesis and interfacial tension determination of octylphenol polyoxyrethylene ether sulfonate [J]. *Journal of Petrochemical Universities*，2010，23（4）：20-24. DOI：10. 3696/j. issn. 1006-396X. 2010. 04. 005.

[5] 张永民，牛金平，李秋小. 壬基酚醚磺酸钠及其与重烷基苯磺酸钠复配物的耐盐性 [J]. 油田化学，2009，26(1)：72-75.

ZHANG Y M，NIU J P，LI Q X. The salt-tolerance of sodium nonyl phenol polyoxyethylene ether sulfonates and their mixture with a heavy alkylbenzene sulfonate [J]. *Oilfield Chemistry*，2009，26(1)：72-75.

[6] 王峰，艾池，曲广淼，等. 十二烷基磺丙基甜菜碱表面活性剂的气液界面聚集行为分子动力学模拟[J]. 计算机与应用化学，2014，31(7)：833-837. DOI：10. 11719/com. app. chem20140715. WANG F，AI C，QU G M，et al. Molecular dynamics simulation on the aggregation behavior of N-dodecyl-N，N-dimethyl-3-ammonio-1-propanesulfonate at air/water interface [J]. *Computers and Applied Chemistry*，2014，31（7）：833-837. DOI：10. 11719/com. app. chem20140715.

[7] 李亚娉. 表面活性剂界面行为和构效关系研究 [D]. 济南：山东大学，2014：1-4.

LI Y P. Interfacial behavior and structure-function relationship of surfactants [D]. Jinan：Shandong University，2014：1-4.

[8] WANG L，HU Y H，SUN W，et al. Molecular dynamics simulation study of the interaction of mixed cationic/anionic surfactants with muscovite [J]. *Applied Surface Science*，2015，327(327) 364-370.

[9] CHEN T，ZHANG G C，JIANG P，et al. Dilational rheology at air/water interface and molecular dynamics simulation research of hydroxyl sulfobetaine surfactant [J]. *Journal of Dispersion Science & Technology*，2014，35（3）：448-455. DOI：10. 1080/01932691. 2013. 785364.

[10] 郑凤仙. 表面活性剂在固液界面及限制空间中的吸附和聚集行为的分子模拟研究 [D]. 北京：北京化工大学，2009.

ZHENG F X. Molecular simulation study on adsorption and aggregation behaviors of surfactants in a solid liquid interface and confined space [D]. Beijing: Beijing University of Chemical Technology, 2009.

[11] Jang S S, LIN S T, MAITI P K, et al. Molecular dynamics study of a surfactant-mediated decane-water interface: effect of molecular architecture of alkyl benzene sulfonate [J]. *The Journal of Physical Chemistry B*, 2004, 108 (32): 12130-12140. DOI: 10. 1021/jp048773n.

[12] WARDLE K E, HENDERSON D J, ROWLEY R L. Molecular dynamics simulation of surfactant effects on ion transport through a liquid-liquid interface between partially miscible liquids [J]. *Fluid Phase Equilibria*, 2005, 233 (1): 96-102. DOI:10. 1016/j. fluid. 2005. 03. 033.

[13] 陈贻建,苑世领,徐桂英. 表面活性剂界面自组装的分子动力学模拟 [J]. 化学通报, 2004, 67 (11): 813-820. DOI:10. 14159/j. cnki. 0441-3776. 2004. 11. 005.

CHEN Y J, YUAN S L, XU G Y. Molecular dynamics simulations for interfacial self-assembly of surfactant [J]. *Chemistry*, 2004, 67 (11): 813-820. DOI:10. 14159/ j. cnki. 0441-3776. 2004. 11. 005.

[14] 丁伟,李思琦,宋晓伟. 基于分子动力学模拟的表面活性剂力场界面的构建及分析 [J]. 化学通报,2014,77(10):973-973.

DING W, LI S Q, SONG X W. Building and analysis of interface of surfactant's force field based on molecular dynamics simulation [J]. *Chemistry*, 2014, 77 (10): 973-973.

[15] OOSTENBRINK C, VILLA A, MARK A E, et al. A biomolecular force field based on the free enthalpy of hydration and solvation: the GROMOS force-field parameter sets 53A5 and 53A6 [J]. *Journal of Computational Chemistry*, 2004, 25 (13): 1656-1676. DOI: 10. 1002/jcc. 20090.

[16] CHRIS O, THEREZA A S, NICO F A, et al. Validation of the 53A6 GROMOS force field [J]. *European Biophysics Journal*, 2005, 34 (4): 273-284. DOI:10. 1016/S0732-8893(01)00362-5.

[17] PERRA M J, WEBER S H. Web-based job submission interface for the GAMESS computational chemistry program [J]. *Journal of Chemical Education*, 2014, 91 (12): 2206-2208.

[18] SCHMIDT M W, BALDRIDGE K K, BOATZ J A, et al, General atomic and molecular electronic structure system [J]. *Journal of Computational Chemistry*, 1993, 11 (14)1347-1363. DOI:10. 1002/ jcc. 540141112.

[19] 丁伟,高翔. 表面活性剂 GROMOS53a6 力场参数文件的构建[J]. 精细石油化

工进展,2013,14(3):1-4.

DING W,GAO X. Method of generating GROMOS53a6 forcefield topology file for surfactant [J]. *Advances in Fine Petrochemicals*,2013,14(3):1-4.

[20] MALDE A K, ZOU L, BREEZE M, et al. An automated force field topology builder(ATB) and repository: version 1.0 [J]. *American Chemical Society*, 2011,7(12):4062-4037. DOI:10.1021/ct200196m.

[21] MARTíNEZ L, ANDRADE R, BIRGIN E G, et al. PACKMOL:a package for building initial configurations for molecular dynamics simulations. [J]. *Journal of Computational Chemistry*,2009,30(13):2157-2164. DOI:10.1002/jcc.21224.

[22] MARTíNEZ J M, MARTíNEZ L. Packing optimization for automated generation of complex system's initial configurations for molecular dynamics and docking [J]. *Journal of Computational Chemistry*, 2003,24(7):819-825. DOI:10.1002/jcc.10216.

[23] LOMBARDERO M, MARTLN C, Jorge S, et al. An integral equation study of a simple point charge model of water [J]. *Journal of Chemical Physics*, 1999,110(2):1148-1153. DOI:10.1063/1.478156.

[24] TUCKERMAN M E, MARTYNA G J. Comment on simple reversible molecular dynamics algorithms for Nose-Hoover chain dynamics [J]. *Journal of Chemical Physics*,1999,111(7):3313.

[25] MANFRED H U. Comments on a continuum-related Parrinello-Rahman molecular dynamics formulation [J]. *Journal of Elasticity*, 2013,113(1):93-112. DOI:10.1007/s10659-012-9412-3.

[26] 王翀,王瑞,丁伟,等.2,5-二甲基十四烷基苯磺酸钠异构体的合成及其EACN值的测定[J].精细石油化工,2009,26(9):8-11.

WANG C, WANG R, DING W, et al. Synthesis of isomeric sodium 2,5-dimethylte tradecyl benzene sulfonates and determination of the EACN value [J]. *Speciality Petrochemicals*, 2009, 26 (9): 8-11.

[27] 丁伟,宿雅彬,张春辉,等.支链异构十五烷基间二甲苯磺酸钠溶液表面性质及其溶剂行为[J].应用化学,2009,26(9):1023-1026.

DING W, SU Y B, ZHANG C H, et al. Surface properties of branch in isomers of pentadecylm-xylene sulfonates solution and solvent behavior [J]. *Chinese Journal of Applied Chemistry*,2009,26(9):1023-1026.

Mechanisms of enhanced oil recovery by surfactant-induced wettability alteration

Hou Baofeng[1]　Wang Yefei[1]　Cao Xulong[2]　Zhang Jun[3]
Song Xinwang[2]　Ding Mingchen[1]　Chen Wuhua[1]

([1] School of Petroleum Engineering, China University of Petroleum (East China), Qingdao, China
[2] Geoscience Research Institute of Shengli Oilfield Company, SINOPEC, Dongying, China
[3] School of Science, China University of Petroleum (East China), Qingdao, China)

Abstract: Different measurements were conducted to study the mechanisms of enhanced oil recovery (EOR) by surfactant-induced wettability alteration in this paper. The adhesion work could be reduced by the surfactant-induced wettability alteration from oil-wet conditions to water-wet conditions. Surfactant-induced wettability alteration has a great effect on the relative permeabilities of oil and water. The relative permeability of the oil phase increases with the increase of the water-wetness of the solid surface. Seepage laws of oil and water are greatly affected by surfactant-induced wettability alteration. Water flows forward along the pore wall in the water-wet rocks and moves forward along the center of the pores in the oil-wet rocks during the surfactant flooding. For the intermediate-wet system, water uniformly moves forward and the contact angle between the oil-water interface and the pore surface is close to 90°. The direction of capillary force is consistent with the direction of waterflooding for the water-wet surface. While for the oil-wet surface, the capillary force direction is opposite to the waterflooding direction. The highest oil recovery by waterflooding is obtained at close to neutral wetting conditions and the minimal oil recovery occurs under oil-wet conditions.

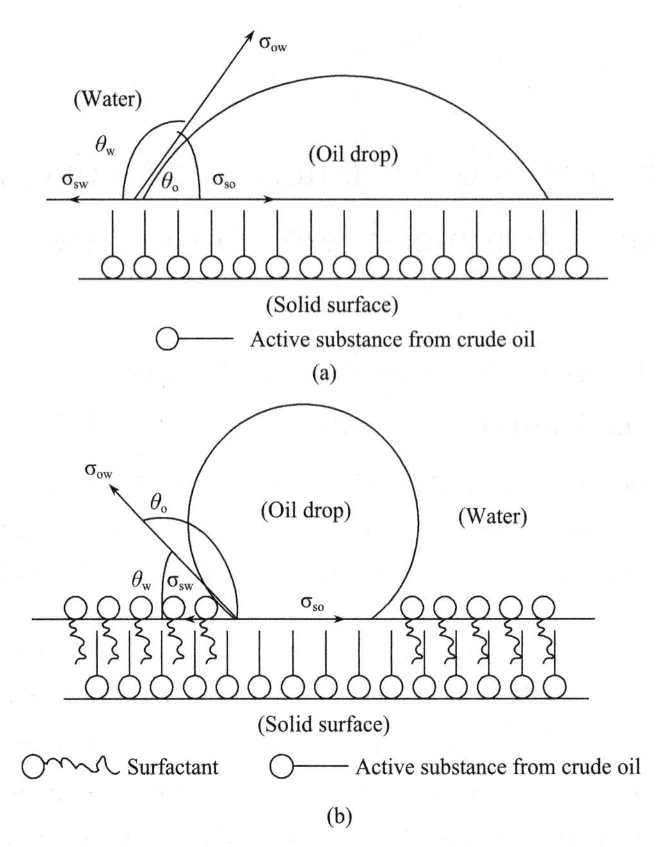

Keywords: EOR mechanisms; Wettability alteration; Surfactants; Adhesion work; Relative permeability; Seepage laws; Capillary force

1 Introduction

Oil recoveries are very low for most of the world's oil fields and various chemical reagents are employed to enhance oil recovery. Surfactants have played a huge role during the oil exploration in the past few years. For surfactants, the EOR mechanisms include: lowering interfacial tension, wettability alteration, emulsifying, increasing surface charge density and so on. Among these mechanisms, surfactant-induced wettability alteration plays an important role during the field development. In recent years, surfactants have been greatly used to alter the reservoir wettability to improve oil recovery[1-10].

Reservoir wettability has a great impact on the distribution of residual oil saturation, capillary forces, particle migration, oil recovery, the microscopic distribution and the relative permeabilities of oil and water. Oil recovery could be enhanced with the changes of the above parameters by surfactant-induced wettability alteration. Great changes will take place for adhesion work after the wettability alteration of solid surface by surfactants. Generally, the more water-wet the solid surface, the smaller the adhesion work. Oil droplets could more easily strip from the more water-wet surface. Reservoir wettability has a great impact on relative permeabilities of oil and water. Anderson et al.[11] studied the effect of reservoir wettability on relative permeabilities of oil and water and found that the relative permeability of the oil phase gradually decreases and the relative permeability of the aqueous phase increases with increasing the oil-wetness of the solid surface. Reservoir wettability has also a great effect on the seepage laws of oil and water during the surfactant flooding. Jamaloei et al.[12] studied the microscopic displacement mechanisms of surfactant flooding in oil-wet and water-wet porous media and some useful experimental phenomenon were observed by them. In the water-wet rock, water moves forward along the pore wall and capillary force is the driving force of waterflooding. While in the lipophilic rock, it is difficult to contact with the pore surface and to flow on the pore surface for the water phase. Water flows along the center of the pore and capillary force is the resistance of waterflooding in the oil-wet rocks. A variety of measurements were used to investigate the influence of surfactant-induced wettability alteration on the above mentioned parameters in this paper.

Oil recovery could be greatly affected with the changes of the above mentioned physical parameters by surfactant-induced wettability alteration. Much more attention has been paid to the study of the relationship between reservoir wettability and oil recovery by waterflooding in recent years [13-20]. Wang et al.[8] systematically summarized the relationship between waterflooding recovery and reservoir wettability. Nearly neutral wetting rock samples could have the maximum oil recovery by waterflooding, while the minimal oil recovery could be obtained in the oil-wet rock samples. Tweheyo[20] also investigated the influence of reservoir wettability on oil production characteristics. They concluded that the highest oil recovery by waterflooding was obtained in the neutral wetting systems and the lowest oil recovery occurred under the oil-wet conditions. The relationship between reservoir wettability and oil recovery by surfactant flooding were studied and were analyzed theoretically in this paper.

The purpose of this paper is to study the effect of surfactant-induced wettability alteration on the various physical parameters and to analyze the mechanisms of enhanced oil recovery by surfactant-induced wettability alteration through different measurements. The changes of the adhesion work, capillary forces, oil recovery, the relative permeabilities and the seepage laws of oil and water by surfactant-induced wettability alteration were investigated in the present study.

2 EXPERIMENTAL

2.1 Materials

2.1.1 Formation Water

In this paper, the employed formation water was obtained from Shengtuo Oilfield, China. The ionic composition of the formation water is shown in Table 1.

2.1.2 Surfactants

The used surfactants were all chemically pure grade in the present study and the full names of the surfactants are shown in Table 2. The concentrations of the surfactants POE(1), GD70 and DTAB were fixed at 0.035 wt%, 0.05 wt% and 0.06 wt%, respectively.

2.1.3 Crude Oil

The crude oil used in this paper was also obtained from Shengtuo Oilfield, China. The basic physical properties of the crude oil were determined at room temperature (20℃) in this work (Table 3).

2.1.4 Cores

The cylindrical cores were cemented by epoxy and were originally water-wet in the present study. Basic physical parameters of the cores are shown in Table 4.

2.2 Methods

2.2.1 Interfacial Tension (IFT) Measurement

The interfacial tension between oil and water was determined using spinning drop method at room temperature in the present paper. The IFT values for surfactant solutions against crude oil are shown in Table 5.

2.2.2 Contact Angle Measurement

Sessile drop method was utilized for the contact angle measurement in this work and this measurement was conducted at room temperature. During the process of the contact angle measurement, one quartz plate was first immersed in the surfactant solution and was then fixed to the container. After that, four parallel oil drops were injected out from the curved syringe to the lower surface of the quartz plate. The four parallel oil drops were left on the solid surface for 30 minutes due to wetting hysteresis[21]. The shape of the oil droplets could be recorded using a high-resolution camera and the contact angles can be calculated by the related software.

2.2.3 Core Handling

The water phase permeabilities of the cores were first determined and then the cores were saturated with crude oil. After that, these oil-saturated cores were placed in a closed container and were immersed in the crude oil at 75℃ for 14 days[22]. The used oil-wet cores in this study were obtained in this way.

2.2.4 Relative Permeability Measurement

The oil-wet cores were first prepared before the relative permeability measurement and the relative permeability measurement was performed at room temperature. The surfactant solution was injected into the cylindrical oil-wet core at a rate of 0.10 mL/min until the instantaneous water cut was up to 99.0%. The oil production was recorded before the water breakthrough. Furthermore, the cumulative oil production, the cumulative fluid production and the pressure difference between the two ends of the rock sample were recorded at the moment of the water breakthrough. The experimental data were recorded frequently at beginning of the water breakthrough.

2.2.5 Microscopic Displacement

The glass-etched model was used to study the microscopic characteristics of oil displacement by different surfactants in this paper. The schematic diagram of the microscopic displacement is shown in Figure 1. The microscopic model was saturated with crude oil and the irreducible water saturation was established. The oil-wet model was prepared before the microscopic displacement. Then, the surfactant solution was injected into the oil-wet model at a rate of 0.002 mL/min until no oil was produced[23]. The seepage laws of oil and water were observed during the process of the microscopic displacement.

3 RESULTS AND DISCUSSION

3.1 Decrease Of Adhesion Work

A series of physical parameters will be changed with the wettability alteration of the solid surface induced by different surfactants. In order to study the EOR mechanisms of wettability alteration of the solid surface caused by different surfactants, the effect of surfactant-induced wettability alteration of the solid surface on the adhesion work was studied in this work. The contact angle, θ, is measured through the water phase in this paper. The adhesion work of different systems could be calculated by the following equation:

$$W=\sigma(1-\cos\theta) \tag{1}$$

where W is the adhesion work, σ is the interfacial tension between oil and water and θ is the contact angle between the solid surface and the oil-water interface.

Table 6 shows the adhesion work of different systems treated with the surfactants.

From Table 6 we can see that the adhesion work decreases with the decrease of the contact angle, i. e. adhesion work will be greatly reduced when the water-wetness of the solid surface is enhanced by different surfactants. Due to the decrease of adhesion work, oil displacement efficiency could be greatly increased.

Figure 2 shows the schematic diagram of the force analysis of oil droplets on the solid surface. Hydrophobic chains will face toward aqueous solution when active substances of crude oil are adsorbed on the solid surface, rendering the solid surface oil-wet. Oil droplets will spread on the solid surface when placed on an oil-wet surface (Figure 2(a)). The force conditions of oil droplets could be analyzed through the Young's equation of the equilibrium state (Equation 2) when the surfactants are injected into the solution.

$$\sigma_{so}=\sigma_{sw}+\sigma_{ow}\cos\theta_w \tag{2}$$

where σ_{so} is the interfacial tension between oil and the solid surface, σ_{sw} is the interfacial tension between water and the solid surface, σ_{ow} is the interfacial tension between oil and water and θ_w is the contact angle between the solid surface and the oil-water interface (Figure 2). The interfacial tension σ_{ow} will decrease when the surfactants are adsorbed on the oil-water interface. The surfactants could be adsorbed on the surface of the film formed by the active substances of crude oil by hydrophobic interaction, making the interfacial tension σ_{sw} lower. The solid-oil interfacial tension (σ_{so}) remains substantially unchanged during this process. The value of cos θ_w will increase to meet the balance of Young's equation, making the contact angle θ_w smaller and rendering θ_o larger. Thus, the oil droplets will be curled up and could be easily taken to improve the efficiency of water flooding (Figure 2(b)).

The whole process of peeling off from the solid surface for the oil droplets could be divided into three stages. Stage 1: The contact area between oil droplets and the solid surface is rapidly reduced after the injection of the surfactants into the water. Stage 2: The surfactants penetrate and diffuse along the oil-solid interface during this stage, which is the longest stage during the spontaneous detachment of oil drops from the solid surface. Stage 3: The contact area between oil droplets and the solid surface decreases rapidly and oil droplets are stretched by the influence of buoyancy (expressed as "necking"), making the oil droplets peel off from the solid surface[24-25].

3.2 Changes In Relative Permeabilities Of Oil And Water

The originally oil-wet cores were used to study the effect of reservoir wettability on the relative permeabilities of oil and water in the present study. The relative permeabilities of oil and water were measured in the originally oil-wet synthetic cores.

The injected surfactants could be adsorbed on the core surface, making the core surface show different wettabilities (Figure 3).

Figure 3 shows the typical curves of the relative permeabilities of oil and water under different wettability conditions. The originally oil-wet surface was changed to be a water-wet surface (63.9°) after the injection of the surfactant POE(1) into the core (Figure 3(a)). From Figure 3(a) we can see that the values of the irreducible water saturation S_{wc} and the residual oil saturation Sor were 29.0% and 30.0%, respectively. Furthermore, the water saturation corresponding to the intersection of the relative permeability curves of the aqueous phase and the oil phase was greater than 50% (Figure 3(a)). The intersection is called "isotonic point", hereinafter the same. Oil recovery of this system can be calculated by the following equation:

$$R=\frac{(S_{oi}-S_{or})}{S_{oi}} \tag{3}$$

where S_{oi} is the original oil saturation and S_{or} is the residual oil saturation. The oil recovery of this system was calculated to be 57.75% of original oil in place (OOIP) from Equation 3.

From Figure 3(b) we can see that the originally oil-wet surface was changed to be a nearly intermediate-wet surface (85.0°) after the surfactant flooding using DTAB. As shown in Figure 3(b), the water saturation corresponding to the isotonic point was also greater than 50% and the corresponding oil recovery was 73.33% of OOIP. Figure 3(c) shows that the water saturation corresponding to the isotonic point was lower than 50% and the range of two-phase flow of oil and water was smaller compared to the systems using POE(1) and DTAB. For the oil-wet core, the relative permeability of the oil phase was smaller compared to the water-wet and intermediate-wet cores under the same water saturation (Figure 3(c)). For this system, less oil was produced (S_{or} =50.0%) and the oil recovery was as low as 37.50% of OOIP. The above results indicate that the highest waterflooding recovery occurs at nearly intermediate-wet conditions.

Reservoir wettability has a great effect on relative permeabilities of oil and water. The relative permeability of the oil phase decreases with the increase of the oil-wetness of the solid surface. In the water-wet system, water is usually distributed in the small gaps or on the inner wall of the rock, making the flow resistance of oil smaller. For the oil-wet rock, water is distributed neither in the small gaps nor on the inner wall of the rock, but in the pores with a form of water droplets. Due to the complexity of the pore structure, these water droplets will encounter the resistance when meeting the narrow pore throat, producing Jamin effect in this way. The Jamin effect hinders the flow of the oil phase and lowers the relative permeability of the oil phase[11].

3.3 Effect Of Wettability On Seepage Of Oil And Water

Surfactant induced wettability alteration has a great impact on the flow of oil and water in porous media. Microscopic displacement was utilized to study the seepage rules of oil and water under different wettability conditions caused by different surfactants in this paper.

The originally oil-wet model surface was changed to be a water-wet surface using POE(1) at 0.035 wt% (Figure 4(a)). As can be seen from Figure 4(a), the water flowed forward along the solid wall during the surfactant flooding (as shown in red circles). While in the oil-wet surface, the water flowed forward along the center of the pore (as shown in red circles), producing film-like residual oil in this way during the experiment (Figure 4(b)). As shown in Figure 4(c), the water uniformly moved forward and the contact angle was close to 90° when the model surface was changed to be intermediate-wet, as shown in red circles. The above experiment rules could be presented by the following diagram (Figure 5).

Crude oil displacement mechanisms vary with the change of rock wettability. For the water-wet rock, water is the wetting phase and flows forward along the pore wall during the water flooding (Figure 5(a)). Due to capillary forces, water automatically enters into the small pores and crude oil is displaced in the water-wet rock. While in the large pores of the water-wet rock, the water proceeds along the pore wall and the water film is first formed. The water film then gradually thickens and occupies the pore channels, making crude oil be expelled. For the oil-wet rock, water is the non-wetting phase and capillary forces are the flooding resistance for the water flooding. Due to the presence of the adhesion force at oil-rock interface during the waterflooding, it is difficult to flow forward along the pore surface for the water (Figure 5(b)). Furthermore, the water viscosity is lower than that of the oil, producing water fingering along the center of the pores. The fingering strengthens when the drive pressure increases. The varying thickness films of crude oil will be left on the pore walls after the waterflooding (Figure 5(b)). As shown in Figure 5(c), the water uniformly moves forward and the contact angle between the oil-water interface and the pore surface is close to 90° for the intermediate-wet system.

3.4 Changes In Capillary Force

Capillary forces will change with the alteration of reservoir wettability. In order to study the mechanisms of enhanced oil recovery by surfactant-induced wettability changes, the magnitude and the direction of capillary forces under different wettability conditions were discussed in this paper. The capillary force could be calculated using the following equation:

$$P_c = \frac{2\sigma\cos\theta}{r} \tag{4}$$

where P_c is the capillary force, σ is the interfacial tension between oil and water, θ is the contact angle between the solid surface and the oil-water interface and r is the capillary radius.

Table 7 shows the capillary forces of different systems. From Table 7 we can see that the capillary force for the system using GD70 is -0.258×10^6 mN/m^2, which represents that the direction of the capillary force is opposite to the water flooding direction and the capillary force is the water flooding resistance. While for the system using DTAB and POE(1), the capillary forces are 0.043×10^6 and 0.219×10^6 mN/m^2, respectively.

As can be seen from Table 7, reservoir wettability has a great impact on capillary forces. The effect of reservoir wettability on capillary force direction is shown in Figure 6. As depicted in Figure 6(a), the liquid level rises due to the presence of the capillary force.

Water is the wetting phase and the direction of capillary force is consistent with that of waterflooding for the water-wet surface (Figure 6(a)). While the capillary force direction is opposite to the waterflooding direction for the oil-wet surface (Figure 6 (b)). Because of the resistance caused by the capillary force, the liquid level drops (Figure 6(b)) (26) .

3.5 Effect Of Wettability On Oil Recovery By Waterflooding

Wettability alteration induced by surfactants has a great impact on water flooding recovery. Effect of surfactant-induced wettability alteration on oil recovery by waterflooding and its theoretical analysis were studied in the present paper.

Figure 7 shows the oil recoveries by waterflooding under different wettability conditions caused by the above surfactants. From Figure 7 we can see that the intermediate-wet rock could have the highest oil recovery by waterflooding and the minimal oil recovery occurred at oil-wet conditions. The results are in good agreement with the relevant literature and can be analyzed with the capillary number equation and the pore-doublet model. The capillary number can be calculated by the following equation:

$$N_c = \frac{\mu_d V_d}{\sigma\cos\theta} \tag{5}$$

where N_c is the capillary number, μ_d is the viscosity of displacing fluid, V_d is the displacement speed, σ is the oil-water interfacial tension and θ is the contact angle. As we all know, the greater the capillary number is, the lower the residual oil saturation of the waterflooding is. As can be seen from Equation 5, the maximum capillary number occurs when the contact angle is close to 90°, i. e. the minimum residual oil saturation of waterflooding happens under the intermediate-wet conditions.

The pore structure of oil reservoir could be assumed to be a pore-doublet model ($r_2 < r_1$) (Figure 8)[27]. If the capillary surface is water-wet, the capillary force will be the driving force of water flooding. For the fluid in the capillaries, the flow speed in the small capillary is greater than that in the bigger capillary when the waterflooding speed is too low. The leading edge of oil and water in the small capillary will first reach the outlet and part of crude oil in the big capillary will be blocked by the water which first reaches the outlet. Thus, some residual oil will be left in the big capillary, reducing the efficiency of water flooding. Due to the presence of viscous forces, the moving speed of oil-water interface in the big capillary is greater than that in the small capillary when the waterflooding speed is high enough. The residual oil will be left in the small capillaries in this case, lowering the oil recovery.

If the capillary surface is oil-wet, viscous forces and capillary forces are not conducive to the flow of oil-water interface in the small capillary. In this case, the residual oil will be left in the small capillaries regardless of the size of water flooding speed.

If the capillary surface is intermediate-wet, the displacement fluid will uniformly move forward in the capillaries and the sweep efficiency will be greatly increased, thereby improving oil recovery.

4 CONCLUSIONS

Different measurements were conducted to investigate the effect of surfactant-induced wettability alteration on the various physical parameters of the reservoir system and the EOR mechanisms were confirmed in this work. The following conclusions could be drawn:

Adhesion work will be greatly reduced with the increase of the water-wetness of the solid surface caused by different surfactants. Thus, oil drops could easily strip from the solid surface.

If the water saturation remains constant, the relative permeability of the oil phase will decrease and the relative permeability of the water phase will increase with the increase of the oil-wetness of the solid surface. For the water-wet surface, the water saturation corresponding to the isotonic point is greater than 50%. While for the oil-wet surface, the water saturation corresponding to the isotonic point is smaller than 50%.

Water is the wetting phase and flows forward along the pore wall in the water-wet rocks. While in the oil-wet rocks, water is the non-wetting phase and moves forward along the center of the pores. The water uniformly moves forward and the contact angle between the oil-water interface and the pore surface is close to 90° for the intermediate-wet system.

Wettability of the solid surface has a great influence on the magnitude and the direction of capillary force. Water is the wetting phase and the direction of capillary force is consistent with the direction of waterflooding for the water-wet surface. While the capillary force direction is opposite to the waterflooding direction for the oil-wet surface.

The intermediate-wet rock has the highest oil recovery by waterflooding and the minimal oil recovery occurs under the oil-wet conditions.

NOMENCLATURE

K	rock permeability ($10^{-3}\ \mu m^2$)
PV	pore volume (mL)
S_{oi}	initial oil saturation (fraction, %)
wt%	weight %
IFT	interfacial tension (mN/m)
OOIP	original oil in place

ACKNOWLEDGEMENTS

The authors are grateful to the financial support by the Program for Changjiang Scholars and Innovative Research Team in University (IRT1294) and the Fundamental Research Funds for the Central Universities (13CX05019A & 15CX02006A).

References

[1] Hirasaki, G. J.; Miller, C. A.; Puerto, M. Recent advances in surfactant EOR. In SPE Annual Technical Conference and Exhibition, edited by G. J. Hirasaki; Denver, Colorado: Society of Petroleum Engineers, 2008:20-54.

[2] Standnes, D. C. ; Austad, T. Wettability alteration in chalk: 2. Mechanism for wettability alteration from oil-wet to water-wet using surfactants. Journal of Petroleum Science and Engineering 2000,28:123-143.

[3] Standnes, D. C. ; Nogaret, L. A. ; Chen, H.-L. ; Austad, T. An evaluation of spontaneous imbibition of water into oil-wet carbonate reservoir cores using a nonionic and a cationic surfactant. Energy&Fuels 2002,16:1557-1564.

[4] Golabi, E. ; Azad, F. S. ; Branch, O. Experimental study of wettability alteration of limestone rock from oil-wet to water-wet using various surfactants. In SPE Heavy Oil Conference, edited by E. Golabi; Calgary, Alberta, Canada: Society of Petroleum Engineers,2012:12-14.

[5] Robin, M. Interfacial phenomena: Reservoir wettability in oil recovery. Oil&Gas Science and Technology 2001,56:55-62.

[6] Zhou, X. ; Morrow, N. ; Ma, S. Interrelationship of wettability, initial water saturation, aging time, and oil recovery by spontaneous imbibition and waterflooding. SPE Journal 2000,5:199-207.

[7] Salehi, M. ; Johnson, S. J. ; Liang, J.-T. Mechanistic study of wettability alteration using surfactants with applications in naturally fractured reservoirs. Langmuir 2008,24:14099-14107.

[8] Wang, Y. ; Xu, H. ; Yu, W. ; Bai, B. ; Song, X. ; Zhang, J. Surfactant induced reservoir wettability alteration: Recent theoretical and experimental advances in enhanced oil recovery. Petroleum Science 2011,8:463-476.

[9] Wu, Y. ; Shuler, P. J. ; Blanco, M. ; Tang, Y. ; Goddard, W. A. An experimental study of wetting behavior and surfactant EOR in carbonates with model compounds. SPE Journal 2008,13:26-34.

[10] Rostami Ravari, R. ; Strand, S. ; Austad, T. Combined surfactant-enhanced gravity drainage (SEGD) of oil and the wettability alteration in carbonates: the effect of rock permeability and interfacial tension (IFT). Energy & Fuels 2011, 25: 2083-2088.

[11] Anderson, W. G. Wettability literature survey part 5: the effects of wettability on relative permeability. Journal of Petroleum Technology 1987,39,1:453-451,468.

[12] Jamaloei, B. Y. ; Kharrat, R. Analysis of microscopic displacement mechanisms of dilute surfactant flooding in oil-wet and water-wet porous media. Transport in porous media 2010,81:1-19.

[13] Jadhunandan, P. ; Morrow, N. R. Effect of wettability on waterflood

recovery for crude-oil/brine/rock systems. SPE reservoir engineering 1995,10:40-46.

[14]Haugen, Å. ; Fernø, M. A. ; Bull, ø. ; Graue, A. Wettability impacts on oil displacement in large fractured carbonate blocks. Energy & Fuels 2010,24:3020-3027.

[15] Morrow, N. R. Wettability and its effect on oil recovery. Journal of Petroleum Technology 1990, 42,1:476-471,484.

[16]Kennedy, H. T. ; Burja, E. O. ; Boykin, R. S. An investigation of the effects of wettability on oil recovery by water flooding. The Journal of Physical Chemistry 1955,59:867-869.

[17] Anderson, W. G. Wettability literature survey-part 6: the effects of wettability on waterflooding. Journal of Petroleum Technology 1987, 39, 1: 605-601,622.

[18] Zhou, X. ; Morrow, N. R. ; Ma, S. Interrelationship of wettability, initial water saturation, aging time, and oil recovery by spontaneous imbibition and waterflooding. SPE Journal 2000,5:199-207.

[19] Agbalaka, C. C. ; Dandekar, A. Y. ; Patil, S. L. ; Khataniar, S. ; Hemsath, J. The effect of wettability on oil recovery: a review. In SPE Asia Pacific Oil and Gas Conference and Exhibition, edited by C. C. Agbalaka; Perth, Australia: Society of Petroleum Engineers, 2008:34-46.

[20] Tweheyo, M. ; Holt, T. ; Torsæter, O. An experimental study of the relationship between wettability and oil production characteristics. Journal of Petroleum Science and Engineering 1999,24:179-188.

[21] Qi, Z. ; Wang, Y. ; He, H. ; Li, D. ; Xu, X. Wettability alteration of the quartz surface in the presence of metal cations. Energy & Fuels 2013,27:7354-7359.

[22] Kathel, P. ; Mohanty, K. Wettability Alteration in a Tight Oil Reservoir. Energy & Fuels 2013,27:6460-6468.

[23] Pei, H. ; Zhang, G. ; Ge, J. ; Jin, L. ; Liu, X. Analysis of microscopic displacement mechanisms of alkaline flooding for enhanced heavy-oil recovery. Energy & Fuels 2011,25:4423-4429.

[24] Kolev, V. ; Kochijashky, I. ; Danov, K. ; Kralchevsky, P. ; Broze, G. ; Mehreteab, A. Spontaneous detachment of oil drops from solid substrates: governing factors. Journal of colloid and interface science 2003,257:357-363.

[25] Verma, S. ; Kumar, V. Relationship between oil-water interfacial tension and oily soil removal in mixed surfactant systems. Journal of colloid and interface science 1998,207:1-10.

[26] Anderson, W. G. Wettability literature survey-part 4: Effects of wettability

on capillary pressure. Journal of Petroleum Technology 1987,39,1:283-281,300.

[27] Lundström, T. S.; Gustavsson, L. H.; Jēkabsons, N.; Jakovics, A. Wetting dynamics in multiscale porous media. Porous pore-doublet model, experiment and theory. AIChE journal 2008,54:372-380.

Table 1 Ionic composition of the formation water

component	concentration, mg/L
Na^+	6 907.5
Cl^-	12 521.6
Ca^{2+}	421.5
Mg^{2+}	223.5

Table 2 Full names of the used surfactants

surfactant	surfactant type	full name
POE(1)	anionic surfactant	Sodium lauryl monoether sulfate
GD70	nonionic surfactant	Alkylpolyglycosides
DTAB	cationic surfactant	Dodecyl trimethyl ammonium bromide

Table 3 Basic parameters of the oil phase

production place	viscosity, mPa·s	density, g/cm^3	acid number, mg of KOH/g
Shengtuo Oilfield	62.0	0.860	1.420

Table 4 Physical parameters of the used cores

core number	PV, mL	porosity, %	S_{oi}, %	k, 10^{-3} μm^2
1	5.10	21.7	73.86	44.5
2	5.12	22.5	77.95	48.6
3	5.36	22.4	74.45	46.7
4	5.65	22.1	74.88	47.9
5	5.18	20.8	75.34	43.2
6	5.25	23.0	78.89	45.0

Table 5 Interfacial tension between oil and water

surfactant	concentration, wt%	IFT, mN/m
POE(1)	0.035	2.492
DTAB	0.06	2.487
GD70	0.05	2.493

Table 6 Adhesion work of different surfactant systems

surfactant	surfactant concentration, wt%	σ, mN/m	θ, °	adhesion work, mN/m
GD70	0.05	2.493	121.2	3.784
DTAB	0.06	2.487	85.0	2.270
POE(1)	0.035	2.492	63.9	1.396

Table 7 Capillary forces of different systems

surfactant	σ, mN/m	θ, °	r, μm	capillary force, 10^6 mN/m^2
GD70	2.493	121.2	10.0	−0.258
DTAB	2.487	85.0	10.0	0.043
POE(1)	2.492	63.9	10.0	0.219

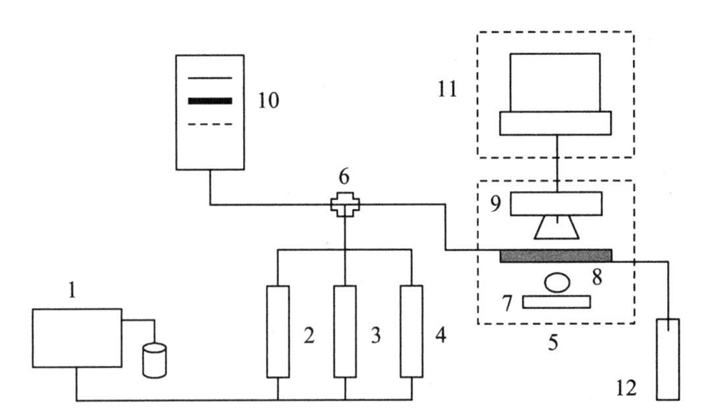

Figure 1 Schematic diagram of microscopic displacement 1-Advection pump 2-Formation water 3-Crude oil 4-Surfactant solutions 5-Oven 6-Pressure sensor 7-Light 8-Microscopic model 9-Video camera 10-Pressure acquisition system 11-Image acquisition system 12-Outlet

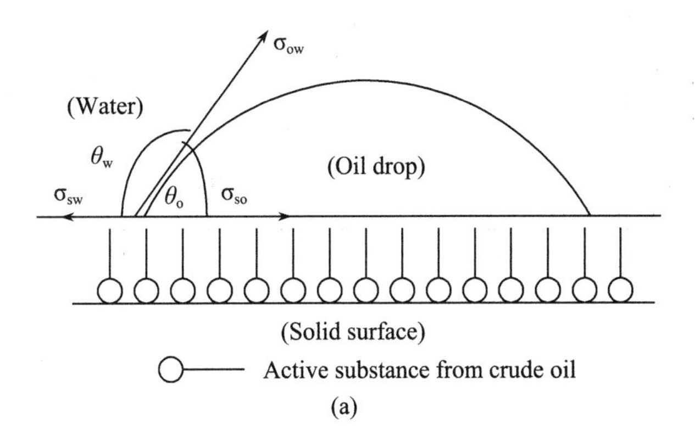

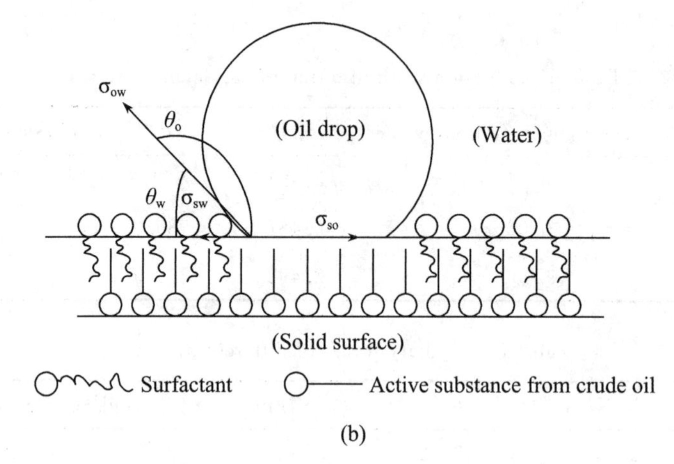

(b)

Figure 2 Schematic diagram of the force analysis of oil droplets

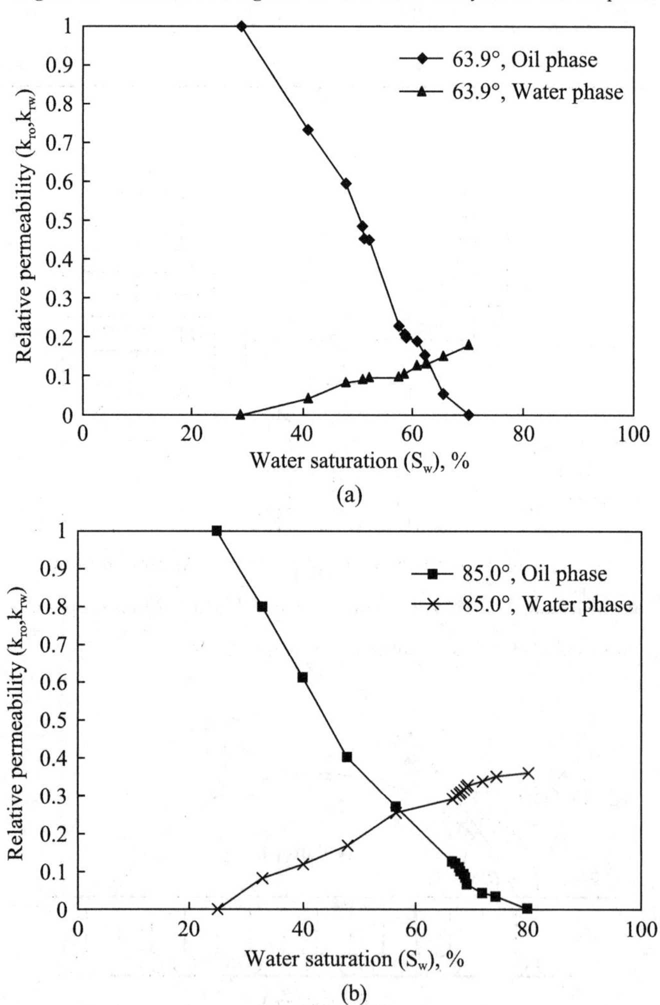

(a)

(b)

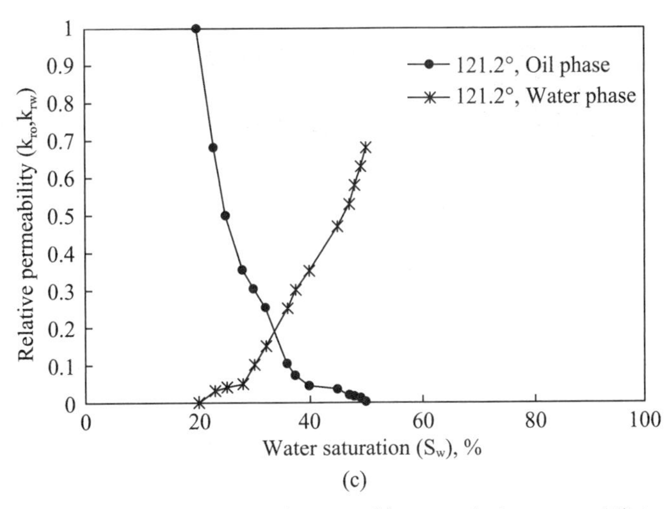

(c)

Figure 3　Effect of wettability on oil/water relative permeabilities

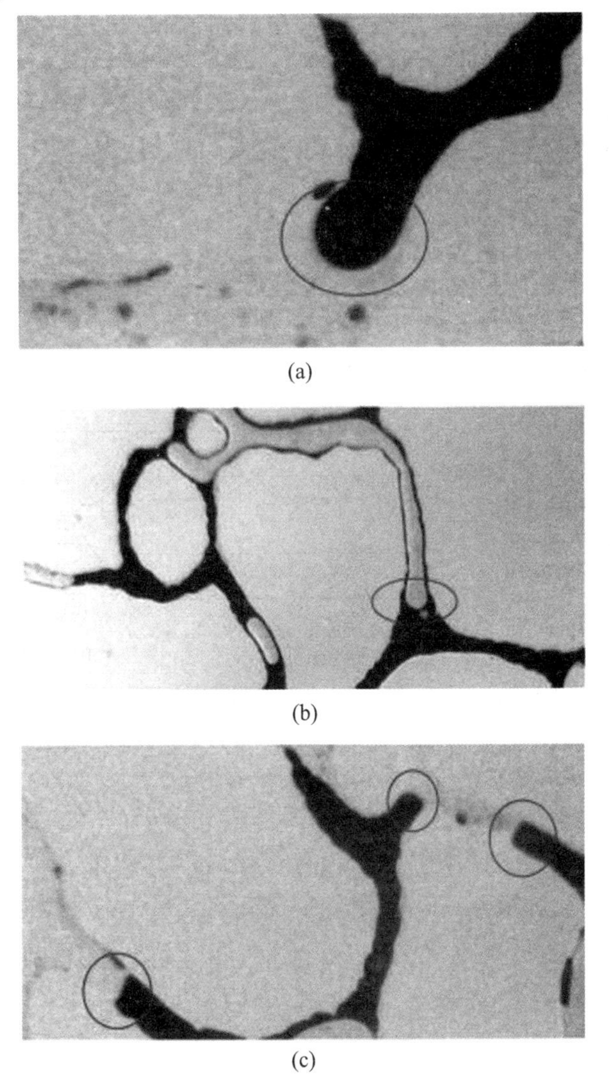

(a)

(b)

(c)

Figure 4　Effect of wettability on oil and water seepage

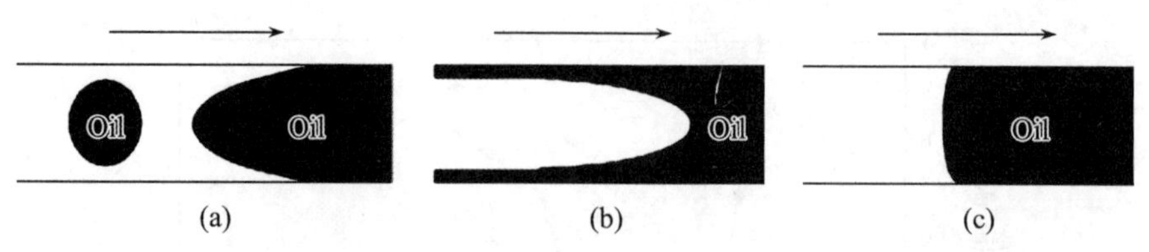

Figure 5 Schematic diagram of seepage rules of oil and water under different conditions wettability

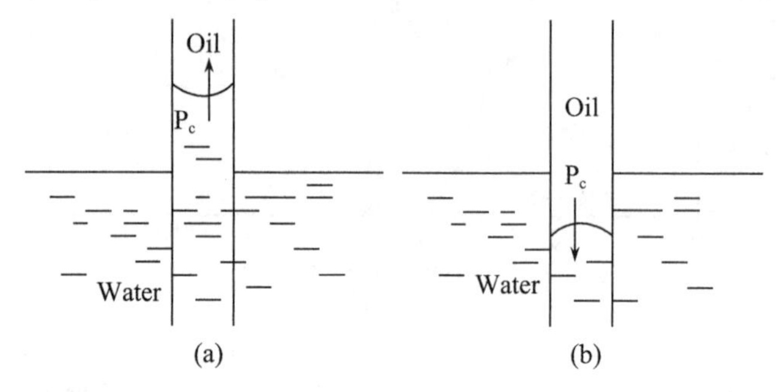

Figure 6 Schematic diagram of capillary forces under different wettability conditions

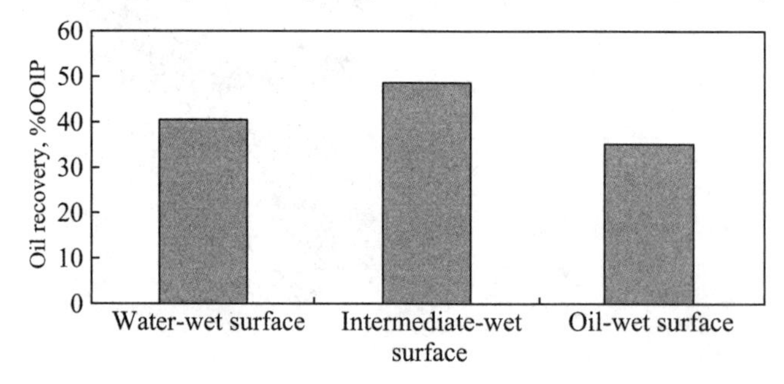

Figure 7 Oil recoveries by waterflooding under different wettability conditions

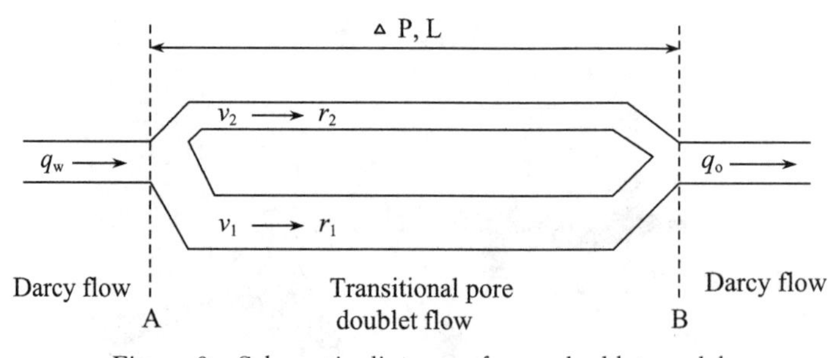

Figure 8 Schematic diagram of pore-doublet model

胜利原油活性组分对原油-甜菜碱溶液体系油-水界面张力的影响

曹加花[1,2] 曹绪龙[3] 宋新旺[3] 徐志成[1] 张 磊[1] 张 路[1,4] 赵 濉[1]

(1. 中国科学院理化技术研究所,北京 100190;

2. 中国科学院大学,北京 100049;

3. 中国石化胜利油田分公司勘探开发研究院,山东东营 257000;

4. 武汉理工大学化学化工与生命科学学院,湖北武汉 430070)

摘要:采用传统的柱色谱四组分分离方法(SARA)将胜利孤岛原油分离得到沥青质、饱和分、芳香分和胶质,采用碱醇液法萃取原油得到酸性组分。测定了正构烷烃、煤油以及原油活性组分模拟油与 2 种不同疏水结构的甜菜碱溶液组成的体系的油-水界面张力。结果表明,在原油活性组分模拟油-甜菜碱溶液体系中,直链甜菜碱由于疏水基团较小,与原油活性组分尤其是酸性组分和胶质发生正协同效应的混合吸附,使油-水界面上表面活性剂分子的含量增加,界面膜的排布更紧密,导致油-水界面张力降低;支链甜菜碱由于具有较大尺寸的疏水基团,煤油中少量的活性物质即可将油-水界面张力降至超低($<10^{-3}$ mN/m),而原油活性组分的加入,则使界面上表面活性剂分子的排布被破坏,削弱了界面膜原有的紧密性,导致油-水界面张力大幅度升高。

关键词:胜利原油;原油活性组分;酸性组分;甜菜碱;界面张力

中图分类号:O647 文献标识码:A

Effect of shengli crude oil active fractions on interfacial tensions of crude oil-betaine solution system

Cao Jiahua[1,2] Cao Xulong[3] Song Xinwang[3] Xu Zhicheng[1]
Zhang Lei[1] Zhang Lu[1,4] Zhao Sui[1]

(1. Technical Institute of Physics and Chemistry, Chinese Academy of Sciences, Beijing 100190, China;
2. University of Chinese Academy of Sciences, Beijing 100049, China;
3. Exploration & Development Research Institute of Shengli Oilfield Co. Ltd, SINOPEC, Dongying 257000, China;
4. School of Chemistry, Chemical Engineering and Life Sciences, Wuhan University of Technology, Wuhan 430070, China)

Abstract: The Shengli crude oil active fractions of asphaltene, saturate, aromatic and resin were obtained by traditional SARA method, and the acid fractions were extracted by NaOH/EtOH mixed solution. The oil-water interfacial tensions (IFT) of the systems composed of n-alkanes, kerosene and model oils containing crude oil fractions, respectively, with two different betaine solutions were studied. The experiment results showed that for the system of crude oil-linear betaine, a small quantity of crude oil active fractions, especially acidic fraction and resin, could decrease the IFT by forming a mixed adsorption film, which enhanced dynamic behavior and the interfacial films to some degree. For the system of crude oil-branched betaine with a larger size of hydrophobic part, absolutely opposite tendency occurred, because the addition of crude oil active fractions could relatively affect the arrangement of surfactant molecules on the interface and weaken the compactness of interfacial films.

Keywords: Shengli crude oil; crude oil active fractions; acid fractions; betaine; interfacial tensions

实现油-水超低界面张力($<10^{-3}$ mN/m)是化学驱提高原油采收率的关键。常规油藏所用表面活性剂如石油磺酸盐等可以将油-水界面张力降至超低,但它们在苛刻条件下会失去活性[1-4]。因此,研究新型表面活性剂与原油体系之间的相互作用对于提高原油采收率至关重要。

原油中存在的天然活性组分会影响油-水界面张力,但原油性质非常复杂,一般只能根据性质将其分离成不同的组分[5-7]。饱和分、芳香分和沥青质的界面活性较低，对界面性质的影响较小，酸性组分和胶质是主要的界面活性物质，影响表面活性剂分子在油相、水相以及界面的分布，对于降低油-水界面张力具有重要的贡献[8-11]。

两性离子表面活性剂界面活性强、毒性低、生物降解性能好，而甜菜碱作为一种两性表面活性剂，既含有阴离子亲水基又含有阳离子亲水基，对高矿化度的油藏具有较好的适应性，而且对碱的依赖性较小，因而在采油中得到广泛应用[12-15]。尽管有关原油界面张力的研究已经得到广泛报道，但系统研究原油活性组分与甜菜碱作用机理仍然十分缺乏。

笔者采用传统的柱色谱四组分分离方法(SARA)分离胜利油田孤岛中一区原油，分别得到沥青质、饱和分、芳香分和胶质，采用碱醇液法萃取原油得到酸性组分，系统考察了各原油活性组分与 2 种不同结构甜菜碱之间的界面性质，加深了对原油活性组分与甜菜碱在界面上相互作用机制的认识。

1　实验部分

1.1 样品及试剂

胜利油田孤岛中一区原油，酸值 2.98 mg KOH/g，油藏温度 70℃，胜利油田提供。烷基磺基甜菜碱(ASB)和芳烷基磺基甜菜碱(BSB)，纯度＞97％，均由中国石油勘探开发研究院提供，具体分子结构见图 1。用二次蒸馏水配制(电阻率＞18.2 mΩ·cm)表面活性剂溶液，所有样品均在 1％NaCl 盐度下测试，且样品溶液呈中性。煤油，购于北京化学试剂公司，经过硅胶柱提纯，在温度 20℃下与重蒸后去离子水之间的界面张力约为 40 mN/m。实验中所用试剂均为分析纯。

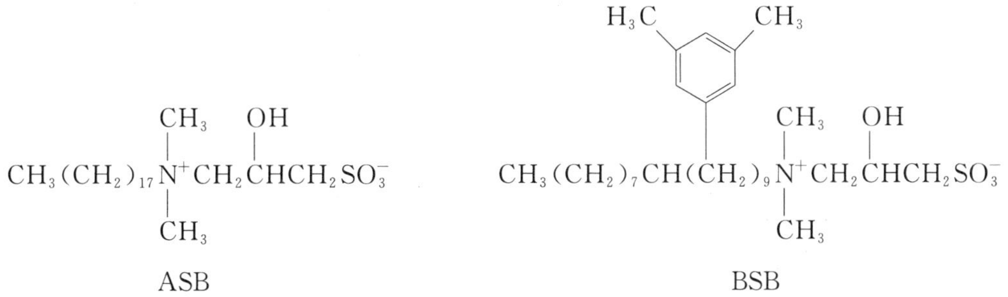

$$CH_3(CH_2)_{17}N^+(CH_3)_2CH_2CH(OH)CH_2SO_3^-$$

ASB

$$CH_3(CH_2)_7CH(C_6H_3(CH_3)_2)(CH_2)_9N^+(CH_3)_2CH_2CH(OH)CH_2SO_3^-$$

BSB

图 1　烷基磺基甜菜碱(ASB)和芳烷基磺基甜菜碱(BSB)的分子结构

Fig. 1　Molecular structures of ASB and BSB

1.2 原油活性组分的分离[6-8]

将原油用正庚烷溶解，过滤后的固体为沥青质，蒸干后的滤液为去沥青质油；去沥青质油经柱色谱分离依次得到饱和分、芳香分和胶质；同时，将原油用环己烷稀释后，通过碱醇液法萃取得到酸性组分。

1.3 原油活性组分模拟油的制备

将原油分离得到的5种活性组分用煤油稀释,分别得到不同质量分数的酸性组分、胶质、沥青质、饱和分和芳香分模拟油。

1.4 实验原理和方法

采用美国彪维工业公司TX500C界面张力仪测定体系的油-水界面张力。油/水体积比约为1/200,转速均为5 000 r/min,实验温度(70.0±0.5)℃。当油-水界面张力(γ)较高时,即油滴长/宽比(L/D)小于4,可以采用式(1)计算γ;当γ较低时,即油滴L/D大于4,可以采用式(2)计算γ。

$$\gamma = 2.741\ 56 \times \exp(-3)\frac{(\rho_h - \rho_d)\omega^2}{C} \qquad L/D < 4 \tag{1}$$

$$\gamma = 3.426\ 94 \times 10^{-7}(\rho_h - \rho_d)\omega^2 D^5 \qquad L/D \geqslant 4 \tag{2}$$

式(1)、(2)中,ρ_h、ρ_d分别为水相和油相密度,mg/L;ω为转速,r/min;C为与油滴长宽比相关的系数。

2 结果与讨论

2.1 原油活性组分的分离结果

2.1.1 原油各活性组分的含量

通过庚烷沉淀及柱层析得到沥青质、饱和分、芳香分及胶质4种组分的质量分数分别为53.6%、20.9%、20.9%和4.6%;采用醇碱液萃取获得酸性组分占原油的1.54%。

2.1.2 原油各组分的元素分析

采用ElementarVarioEL(Germany)元素分析仪测定了原油中各组分所含元素的质量分数,结果列于表1。$n(H)/n(C)$是表征石油分子结构的重要指标,从表1可以看出,沥青质的$n(H)/n(C)$最低,这可能是因为沥青质中含有较多芳环及杂原子;饱和分的$n(H)/n(C)$最高,接近2/1,且不含N元素,说明饱和分极性最低,主要由一些饱和烷烃组成;芳香分和胶质的极性强于饱和分,所以二者N元素的质量分数相对较高。

表1 原油四组分的元素组成

Table 1 The mass fractions of elements in four Fractions of crude oil

Fraction	$w(N)/\%$	$w(C)/\%$	$w(H)/\%$	$n(H)/n(C)$
Saturate	—	85.32	13.80	1.94
Aromatic	0.46	83.04	11.58	1.67
Resin	0.99	79.41	10.79	1.63
Asphaltene	1.09	78.30	9.07	1.39

2.2 正构烷烃-甜菜碱溶液体系的油-水界面张力

表面活性剂是界面活性较强的化合物,在油-水体系中,表面活性剂分子会向油-水界面处扩散，吸附在界面上，顶替界面上的水分子；随着界面上表面活性剂分子数目的增加，油-水界面张力会不断降低。当表面活性剂分子在油-水界面上的吸附达到饱和时，界面张力将不再变化，达到平衡状态。

表面活性剂的结构是影响油-水界面张力最为关键的因素。结构决定了分子尺寸、亲水亲油平衡、吸附-脱附位垒等控制油-水界面张力的各种机理。对于特定的油-水体系，表面活性剂的加入会使油相与水相的不相容性减弱，当表面活性剂分子的亲水头基与水分子之间的相互作用十分接近于表面活性剂分子的亲油尾基与油分子之间的相互作用时，表面活性剂分子可以紧密地吸附在界面上直至饱和，如果分子尺寸适宜，可以实现超低油-水界面张力（$<10^{-3}$ mN/m）[2]。

甜菜碱作为一种特殊的两性离子表面活性剂而被广泛应用于油田中。图 2 为正癸烷与不同质量分数的甜菜碱溶液组成的体系的油-水界面张力随时间的变化。从图 2 可以看出，该体系的油-水界面张力随时间的变化曲线均呈现最常见的“L”型[6]，即随着表面活性剂体相浓度的增加，表面活性剂分子不断向油-水界面处扩散，界面处的表面活性剂分子数目逐渐增加直至饱和。其中，正癸烷-ASB 体系的油-水界面张力在较短时间内略有降低，达到 10^{-1} mN/m 数量级，而正癸烷-BSB 体系的油-水界面张力则能降低至 10^{-2} mN/m 数量级。这是由于甜菜碱的较大的亲水头基平铺在界面上，导致亲水一侧所占面积相对于亲油一侧更大[11]，并非最佳尺寸分配，因而不能实现超低界面张力（$<10^{-3}$ mN/m）。但是 BSB 分子存在支链结构，具有较大的疏水基团，在一定程度上增大了 BSB 在油相一侧的面积，缩小了亲水部分与疏水部分在界面所占面积的差异，因而 BSB 分子可以相对较紧密地排列在界面上，体现出更强的降低油-水界面张力能力。因此，BSB 体系的界面张力比 ASB 体系低一个数量级。

为了更好地研究 2 种表面活性剂降低界面张力的作用机理，笔者测定了一系列不同链长正构烷烃与 ASB 和 BSB 组成的体系的油-水界面张力，结果示于图 3。从图 3 可以看出，油-水界面张力随着体系中正构烷烃碳数的增加有一个最低值，此最低值对应的正构烷烃碳数称为 n_{min} 值。n_{min} 值反映了表面活性剂的亲水亲油平衡能力，n_{min} 值越大，则表面活性剂的油溶性越强[1-2]。ASB 的 n_{min} 值约为 10，而 BSB 的 n_{min} 值约为 9，这是由于支链化增强了表面活性剂的亲水性的结果。另外，在整个烷烃链长范围内，BSB 体系的油-水界面张力总是低于相应 ASB 体系油-水界面张力一个数量级，进一步说明 BSB 分子较强的降低油-水界面张力能力来源于较大的疏水基团，与亲水亲油平衡无关。

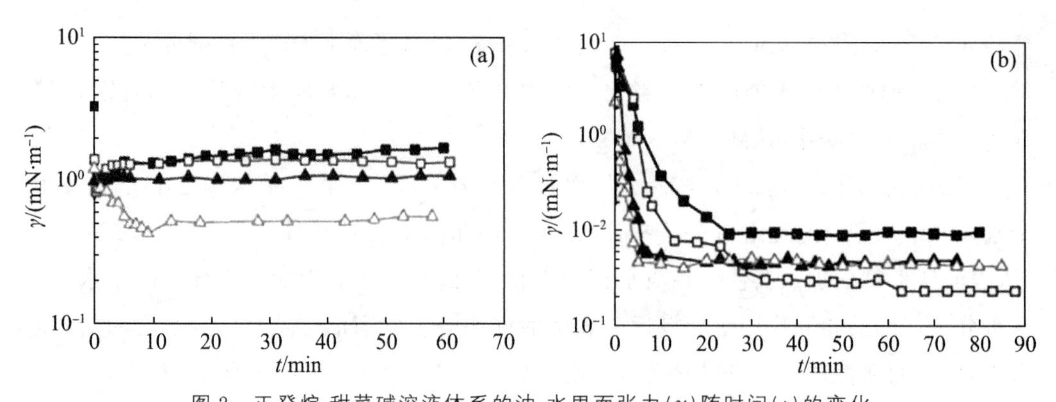

图 2 正癸烷-甜菜碱溶液体系的油-水界面张力(γ)随时间(t)的变化

Fig. 2 The oil-water interfacial tension(γ) of decane-betaine solutions systems vs time(t)

w(NaCl)=1%;T=70℃

Betaine:(a) ASB;(b) BSB

w(Betaine)/%:■ 0.005;□ 0.01;▲ 0.05;△ 0.1

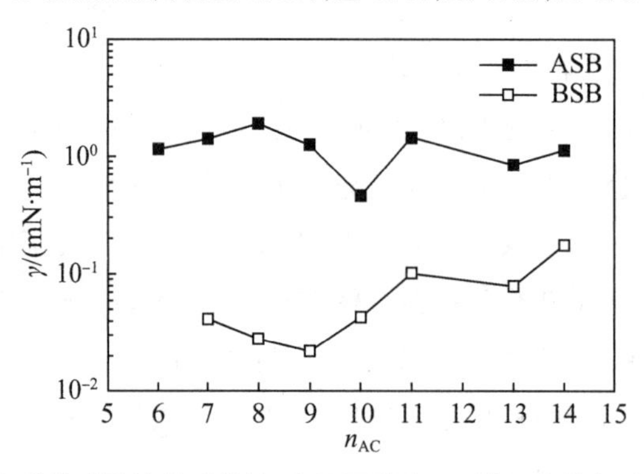

图 3 正构烷烃-甜菜碱溶液体系的油-水界面张力(γ)随正构烷烃碳数(n_{AC})的变化

Fig. 3 The oil-water interfacial tension(γ) of n-alkanes-betaine

Solution system vs alkane carbon number (n_{AC})

w(Betaine)=0.1%;w(NaCl)=1%;T=70℃

2.3 煤油-甜菜碱溶液体系的油-水界面张力

将原油组分溶解在煤油中配制成模拟油,是研究原油-表面活性剂溶液体系界面张力机理的常用模拟手段。值得注意的是,常温下正癸烷-纯水体系的油-水界面张力约为50 mN/m,而本实验中所用煤油-纯水体系的油-水界面张力仅为40 mN/m,说明重蒸、过柱后的煤油仍存在一些无法除去的活性物质。基于以上实验结果,笔者测定了煤油与2种甜菜碱溶液组成的体系的油-水界面张力,结果示于图4。

从图4可以看出,煤油-甜菜碱溶液体系的油-水界面张力明显低于正癸烷-甜菜碱溶液体系的,并且煤油-ASB体系的油-水界面张力有一个明显的最低值,之后开始缓慢回升,最终达到平衡状态,使油-水界面张力曲线呈现“V”型[16]变化。

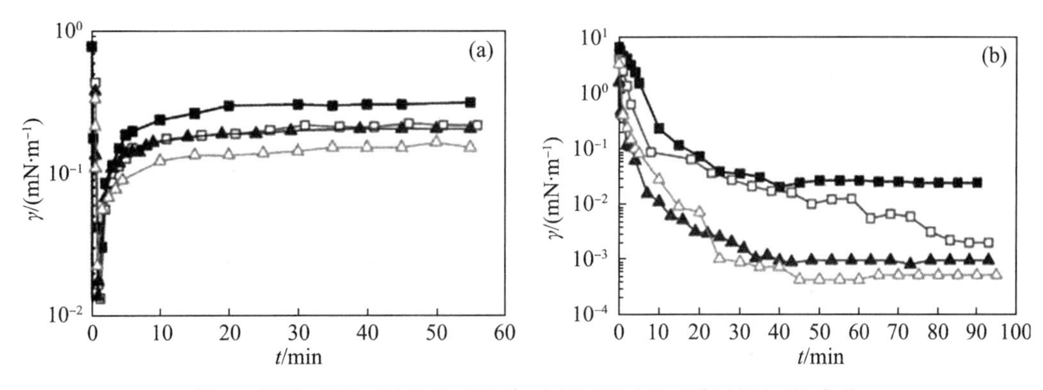

图 4 煤油-甜菜碱溶液体系的油-水界面张力(γ)随时间(t)的变化

Fig. 4 The oil-water interfacia ltension(γ) of kerosene-betaine solutions systems vs time(t)

w(NaCl)=1%;T=70℃

Betaine:(a) ASB;(b) BSB

w(Betaine)/%:■ 0.005;□ 0.01;▲ 0.05;△ 0.1

当甜菜碱分子吸附到界面上时,亲水基团指向水相,亲油基团指向油相;随着时间变化,发生重排,整个亲水基接近平铺在界面上。水相中的 ASB 分子和油相中的活性物质均向界面处扩散,此时 ASB 分子的亲水部分会直立地吸附在界面上,与活性物质形成一层较紧密的混合吸附膜,使界面张力快速达到最低值;随着时间的推移,界面上 ASB 分子的亲水基团会逐渐由直立状态变为平铺状态,从而不断顶替界面上的活性物质,使紧密的界面吸附膜被破坏,导致界面张力回升。煤油-BSB 体系的界面张力随时间的变化曲线仍然呈现"L"型,说明 BSB 分子由于疏水基团较大,亲水基团在界面上处于直立状态或者平铺状态对分子尺寸的影响均不大,因而界面上活性分子的数目会逐渐增加至饱和。总之,煤油中存在的少量活性物质,可以在一定程度上与外加表面活性剂发挥协同作用[17-18],降低体系的油-水界面张力,而且,这种协同效应在 BSB 体系中体现得更为明显。

2.4 原油组分模拟油-甜菜碱溶液体系的油-水界面张力

2.4.1 酸性组分模拟油

在原油中,酸性组分的表面活性最强,对界面性质的影响十分关键。不同质量分数的酸性组分模拟油-甜菜碱溶液体系的油-水界面张力随时间的变化示于图 5。从图 5 可以看出,随着酸性组分质量分数的增加,ASB 体系的油-水界面张力整体有所降低,最低值可以降到 10^{-3} mN/m 数量级,油-水界面张力随时间的变化曲线仍然呈现"V"型。此外,酸性组分的质量分数越高,油-水界面张力达到最低值所需时间越长,尤其是在质量分数 2%时最为明显。由于酸性组分与 ASB 在界面上的吸附存在一定的竞争,延长了界面上活性分子达到最大吸附量所需要的时间。对于 BSB 体系,酸性组分的加入反而增加了油-水界面张力,说明一方面酸性组分会破坏煤油-BSB 体系界面膜原有的紧密性,使酸性组分与表面活性剂的混合吸附体现为负协同效应;另一方面,酸性组分会影响油相的极性,改变表面活性剂分子在油相、水相和界面上的分配,减少了界面上单位面积的表面活性剂分子数目,使油-水界面张力升高,具体机理如图 6 所示。

2.4.2 胶质模拟油

胶质-甜菜碱溶液体系的油-水界面张力与酸性组分-甜菜碱溶液体系的十分相似,因为酸性组分主要存在于胶质中,二者在组成和性质上基本一致。不同质量分数的胶质模拟油与甜菜碱溶液组成的体系的油-水界面张力随时间的变化示于图 7。从图 7 可见,对于 ASB 体系,随着胶质质量分数的增加,油-水界面张力最低值可以从 10^{-2} mN/m 数量级降低至 10^{-3} mN/m,而且随着胶质质量分数的增大,达到最低值所需要的时间也有所延长;对于 BSB 体系,胶质的加入增加了油-水界面张力,而且界面张力达到平衡所需时间随着胶质质量分数的增加而相应地缩短。

2.4.3 沥青质模拟油

沥青质相对分子质量较大,界面活性相对胶质和酸性组分较低,与表面活性剂形成混合吸附的趋势较弱。不同质量分数的沥青质模拟油与甜菜碱溶液组成的体系的油-水界面张力随时间的变化如图 8 所示。由图 8 可知,对 ASB 体系,沥青质对油-水界面张力的影响较小;对 BSB 体系,沥青质在质量分数低于 0.1%时,对油-水界面张力影响较小,但在质量分数大于 0.1%时,却可以大幅度增加油-水界面张力。尽管沥青质分子尺寸较大,但当其质量分数达到 0.1%时,仍可以与 BSB 表面活性剂在界面上发生混合吸附,降低界面上表面活性剂分子的数目,使油-水界面张力不同幅度地升高。

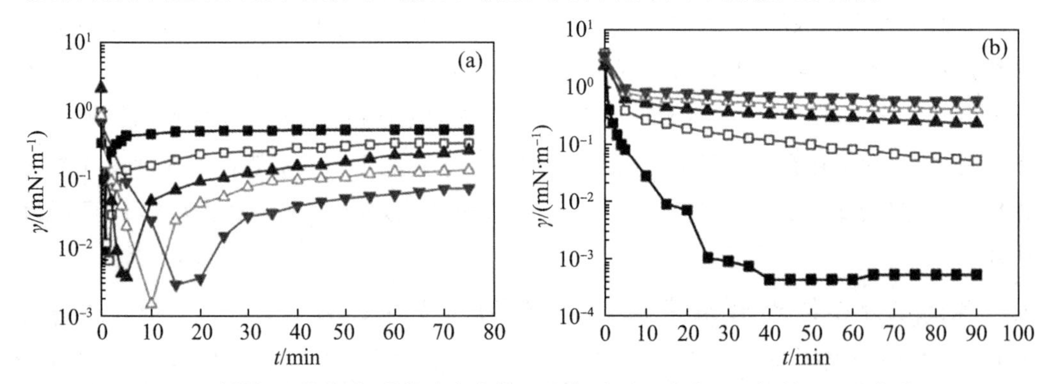

图 5 酸性组分模拟油-甜菜碱溶液体系的油-水界面张力(γ)随时间(t)的变化

Fig. 5 The oil-water interfacial tension(γ) of the system for acidic fraction Model oil and betaine solution vs time(t)

w(Betaine)=1%;w(NaCl)=1%;T=70℃

Betaine:(a) ASB;(b) BSB

w(Acidicfractions)/%:■ 0;□ 0.1;▲ 0.5;△ 1.0;▼ 2.0

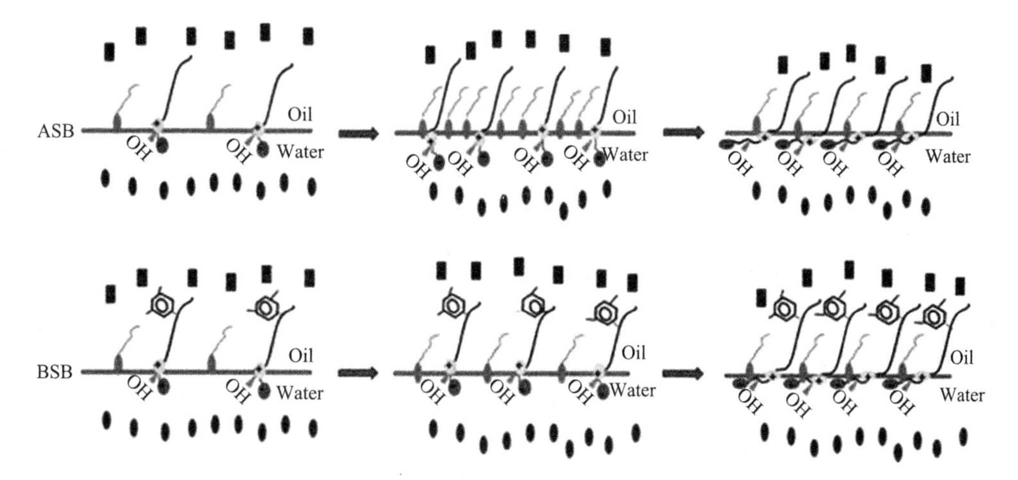

图 6 原油酸性组分与 ASB 和 BSB 在油-水界面上排布的示意图

Fig. 6 Arrangement schematic of ASB or BSB and acidic fractions at oil-water interface

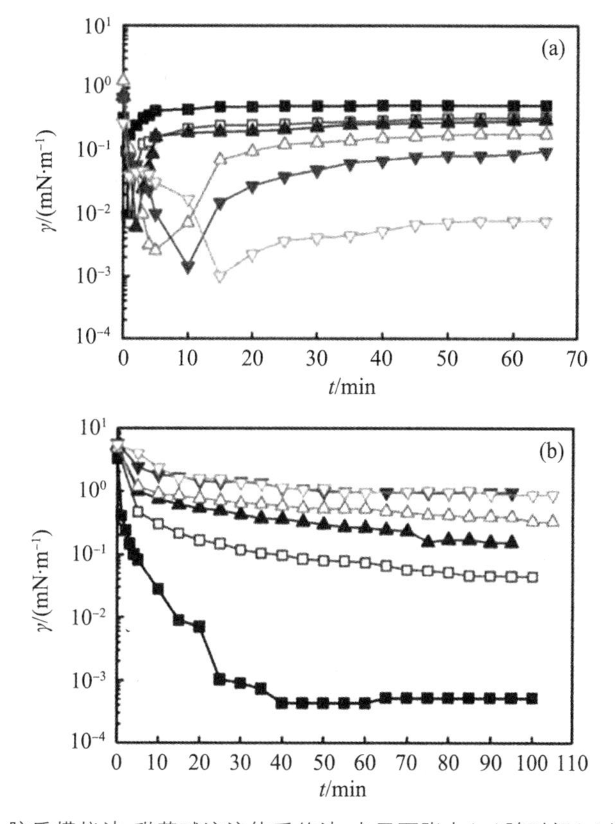

图 7 胶质模拟油-甜菜碱溶液体系的油-水界面张力(γ)随时间(t)的变化

Fig. 7 Theoil-water interfacial tension(γ) of the system for resin model oil and betaine solutions vs time(t)

w(Betaine)=1%;w(NaCl)=1%;T=70℃

Betaine:(a) ASB;(b) BSB

w(Resin)/%:■ 0;□ 0.1;▲ 0.5;△ 1.0;▼ 5.0;▲ 10

2.4.4 不同原油活性组分模拟油-甜菜碱溶液体系油-水界面张力的对比

原油活性组分模拟油-甜菜碱溶液体系的油-水界面张力随活性组分质量分数的变化示于图 9。从图 9 可以看出,就原油活性组分模拟油-ASB 体系而言,活性较低的沥青质、饱和分和芳香分对油-水界面张力的影响较小,均随着活性组分质量分数的增加略有降低,对体系界面膜的排布影响不大;而活性较高的酸性组分和胶质对油-水界面张力的影响比较显著,在质量分数大于 1%时可以将油-水界面张力降低至 10^{-2} mN/m 数量级,此时,酸性组分和胶质与 ASB 分子在界面上发挥了较好的正协同效应。就原油活性组分模拟油-BSB 体系而言,少量活性物质即可将 BSB 体系的界面张力降至 10^{-2} mN/m 以下,此时表面活性剂与活性物质的协同效应最为明显,特别是当加入少量的原油酸性组分和胶质后,体系油-水界面张力均能明显增加。这不仅说明不同组分因活性不同对体系界面性质的影响不同,同时也证明了表面活性剂的尺寸匹配效应对油-水界面张力有关键性的影响。

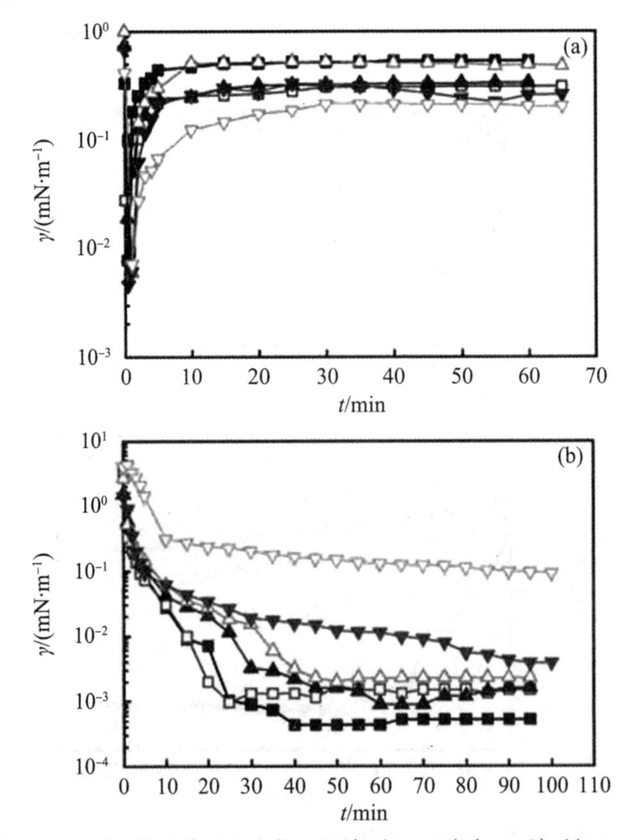

图 8 沥青质模拟油-甜菜碱溶液体系的油-水界面张力(γ)随时间(t)的变化

Fig. 8 The oil-water interfacial tension(γ) of the system for asphaltene model oil and betaine solution vs time(t)

w(Betaine)=1%;w(NaCl)=1%;T=70℃

Betaine:(a) ASB;(b) BSB

w(Asphaltene)/%:■ 0;□ 0.01;▲ 0.05;△ 0.1;▼ 0.5;▲ 1.0

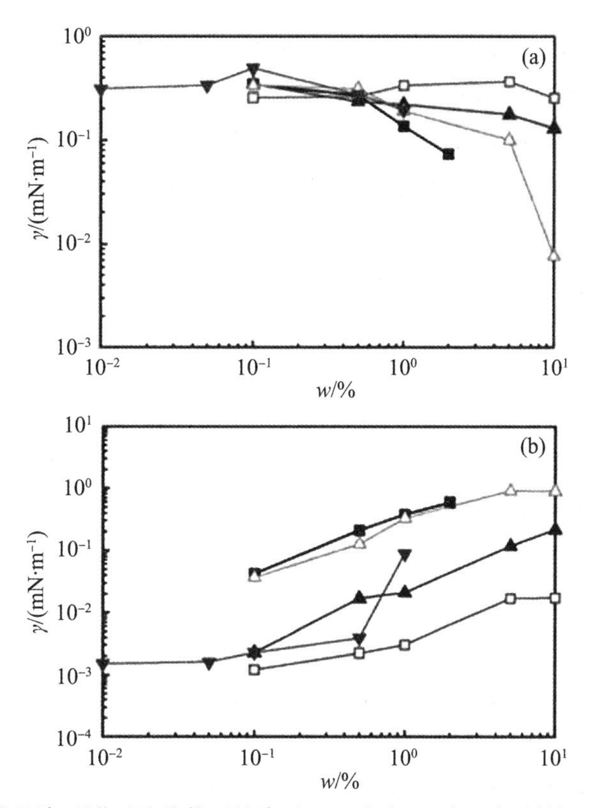

图 9　原油活性组分模拟油-甜菜碱溶液体系的油-水界面张力(γ)随原油活性组分质量分数(w)的变化

Fig. 9　Theoil-water interfacia ltension(γ) of the system for active fraction containing model Oil and betaine solution vs mass fraction(w) of active fractions

w(Betaine)＝1%；w(NaCl)＝1%；T＝70℃

Betaine：(a) ASB；(b) BSB

■ Acidicfractions；□ Saturate；▲ Aromatic；△ Resin；▼ Asphaltene

5 种不同原油组分模拟油与甜菜碱溶液组成的体系的油-水界面张力不同；同时，两类不同分子结构甜菜碱类表面活性剂对体系油-水界面张力的影响也不同。直链的 ASB 分子可以与原油组分发生正协同效应的混合吸附，降低油-水界面张力；支链的 BSB 分子可以与原油组分发生负协同效应的混合吸附，增加油-水界面张力。因此，油相的组成、表面活性剂的结构以及二者的协同作用共同影响着体系的油-水界面张力。

3　结论

（1）在正构烷烃-甜菜碱溶液体系中，甜菜碱类表面活性剂的亲水基团更倾向于平铺在油-水界面上。直链的 ASB 由于疏水基团较小，疏水部分在界面上所占的面积小于亲水部分，油-水界面张力较高；而支链的 BSB 由于疏水基团较大，可以相对增大界面上疏水一侧的面积，油-水界面张力相对较低。

（2）在煤油-甜菜碱溶液体系中，煤油中含有少量的活性物质，能与 ASB 分子发生协同作用，在较短的时间内将油-水界面张力最低值降至超低水平（$<10^{-3}$ mN/m），随着

ASB 分子在界面上由直立变为平铺状态，界面张力开始回升，而 BSB 分子由于疏水基团较大，油-水界面张力不存在最低值现象。

(3) 在原油活性组分模拟油-甜菜碱溶液体系中，原油活性组分中的酸性组分和胶质，界面活性较强，可以与 ASB 分子发生正协同效应的混合吸附，增加界面膜的紧密程度，降低油-水界面张力；而它们与 BSB 分子发生的混合吸附破坏界面膜原有的紧密性，降低界面上表面活性剂分子的数目，使油-水界面张力呈增加趋势。

参考文献

[1] CHAN K ,SHAH D O. Adsorption of surface-active agents from aqueous solution: Hydrophbic substrates [J]. Journal of Dispersion Science and Technology, 1981,2(1):53-66.

[2] ZHANG L, LUO L, ZHAO S, et al. Ultra Low Interfacial Tension and Interfacial Dilational Properties Related to Enhanced Oil Recovery [M]. Petroleum Science and Technology Research Advance. MONTCLAIRE K L , ed. New York: Nova Science Publishers. 2008:81-139.

[3] SHAH D O. Fundamental Aspects of surfactant-polymer Flooding Process [M]. Enhanced Oil Recovery. FAYERS F J,ed. Amsterdam:Elsevier. 1981:1-42.

[4] BERA A, OJHA K, MANDAL A, et al. Interfacial tension and phase behavior of surfactant-brine-oil system [J]. Colloids Surf A,2011,383(1-3):114-119.

[5] ZHAO R H,HUANG H Y,WANG H Y,et al. Effect of organic additives and crude oil fractions on interfacial tensions of alkyl benzene sulfonates [J]. Journal of Dispersion Science and Technology,2013,34(5):623-631.

[6] ZHU Y W,ZHAO R H,JIN Z Q ,et al. Influence of crude oil fractions on interfacial tensions of alkylbenzene sulfonate solutions [J]. Energy Fuel, 2013,27(8): 4648-4653.

[7] 罗澜,赵濉,张路,等. 大庆原油活性组分的分离、分析及界面活性. 油田化学, 2000,17(2):156-158. (LUO lan,ZHAO Sui,ZHANG Lu,et al. Separation, structural analysis and interfacial activity of acidic components in Daqing crude oil [J]. Oilfield Chemistry, 2000,17(2):156-158.)

[8] 赵荣华,黄海耀,董林芳,等. 胜利原油活性组分对其与烷基苯磺酸盐溶液组成的体系油-水界面张力的影响. 石油学报(石油加工), 2012, 28(5): 827-833. (ZHAO Ronghua, HUANG Haiyao , DONG Linfang, et al. Effect of Shengli crude oil fractions on oil-water interfacial tensions of the s ystem of alkylbenzene sulfonate solutions and active fraction[J]. Acta Petrolei Sinica (Petroleum Processing Section),

2012,28(5):827-833.)

[9] POTEAU S ,ARGILLIER J F ,LANGEVIN D ,et al. Influence of pH on stability and dynamic properties of as phaltenes and other amphiphilic molecules at the oil-water interface [J]. Energy Fuels,2005,19(4):1337-1341.

[10] ZHANG L,LUO L,ZHAO S,et al. Effect of different acidic fractions in crude oil on dynamic interfacial tensions in surfactant/alkali/model oil s ystems [J]. Journal of Petroleum Science and Engineering 2004,41 (1-3):189-198.

[11] ZHOU Z H,ZHANG Q,LIU Y,et al. Effect of fatty acids on interfacial tensions of novel sulphobetaines solutions[J]. Energy Fuels,2014,28(2):1020-1027.

[12] ZHANG L, LUO L, ZHAO S, et al. Studies of s ynergism/antagonism for lowering dynamic interfacial tensions in surfactant/alkali/acidic oil s ystems Part 3: Synergism/antagonism in surfactant/alkali/acidic model s ystems [J]. J Colloid Interf Sci , 2003 , 260 (2) : 398-403.

[13] 陈玉萍,丁伟,于涛,等.新型甜菜碱型两性离子表面活性剂界面行为的分子动力学模拟[J].石油学报(石油加工),2012,28(4):592-598.(CHENG Yuping,DING Wei,YU Tao,et al. Molecular dynamics simulation for interface behavior of betaine surfactants in aqueous solution [J]. Acta Petrolei Sinica (Petroleum Processing Section) ,2012 ,28(4) :592-598.)

[14] 李瑞冬,仇珍珠,葛际江,等.羧基甜菜碱-烷醇酰胺复配体系界面张力研究[J].精细石油化工,2012,29 (4):8-12.(LI Ruidong,QIU Zhenzhu,GEJijiang,et al. Studies on interfacial tension for carboxyl betaine/ alkanolamide compound s ystem [J]. Speciality Petrochemicals,2012 ,29(4):8-12.)

[15] 刘凤静,靳志强,张磊,等.甜菜碱与胜利原油间界面张力的影响因素研究.石油化工高等学校学报,2013,26(4):57-61.(LIU Fengjing,JIN Zhiqiang,ZHANG Lei,et al. Factors effecting interfacial tension between betaine and Shengli crude oil [J]. Journal of Petrochemical Universities,2013 ,26(4):57-61.)

[16] SONG X W,ZHAO R H,CAO X L,et al. Dynamic interfacial tensions between offshore crude oil and enhanced oil recovery surfactants [J]. Journal of Dispersion Science and Technology ,2013 ,34(2) :234-239.

[17] ZHU Y W , SONG X W, LUO L , et al. Studies of synergism/antagonism for lowering interfacial tensions in alkyl benzene sulfonate mixtures[J]. Journal of Dispersion Science and Technology ,2009,30(7) :1015-1019.

[18] 楚艳苹,罗澜,张路,等.不同体系中酯与表面活性剂协同效应机理研究[J].物理化学学报,2004,20(7):776-779.(CHU Yanping,LUO Lan,ZHANG Lu,et al. Studies of s ynergism between sodium alkyl benzene sulfonate and methyl oleate in different s ystems [J]. Acta Phys-Chim Sin,2004,20(7) :776-779.)

Surfactant-induced wettability alteration of oil-wet sandstone surface: mechanisms and its effect on oil recovery

Hou Baofeng[1] Wang Yefei[1] Cao Xulong[2] Zhang Jun[3] Song Xinwang[2]
Ding Mingchen[1] Chen Wuhua[1]

Abstract: Different analytical methods were utilized to investigate the mechanisms for wettability alteration of oil-wet sandstone surfaces induced by different surfactants and the effect of reservoir wettability on oil recovery. The cationic surfactant cetyltrimethylammonium bromide (CTAB) is more effective than the nonionic surfactant octylphenol ethoxylate (TX-100) and the anionic surfactant sodium laureth sulfate (POE(1)) in altering the wettability of oil-wet sandstone surfaces. The cationic surfactant CTAB was able to desorb negatively charged carboxylates of crude oil from the solid surface in an irreversible way by the formation of ion pairs. For the nonionic surfactant TX-100 and the anionic surfactant POE(1), the wettability of oil-wet sandstone surfaces is changed by the adsorption of surfactants on the solid surface. The different surfactants were added into water to vary the core surface wettability, while maintaining a constant interfacial tension. The more water-wet core showed a higher oil recovery by sponta-neous imbibition. The neutral wetting micromodel showed the highest oil recovery by waterflooding and the oil-wet model showed the maximum residual oil saturation among all the models.

Keywords: Reservoir wettability O Oil recovery O Surfactant O Wettability alteration O Mechanism

List of symbols:

k	Rock permeability ($10^{-3}\mu m^2$)
S_{oi}	Initial oil saturation (fraction, %)
wt%	Weight %

Abbreviations:

CMC	Critical micelle concentration
IFT	Interfacial tension (mN/m)
OOIP	Original oil in place
PV	Pore volume (mL)

Introduction

Wettability is an important indicator of reservoir physical properties, which has a great effect on oil recovery during the process of enhanced oil recovery (EOR). Many studies have investigated the mechanisms for wettability alteration of oil-wet sandstone surfaces induced by different surfac-tants and the relationship between reservoir wettability and oil recovery by spontaneous imbibition and waterflooding in the past few years[1-9].

Recently, more attention has been paid to the study of mechanisms for wettability alteration of oil-wet solid sur-faces using various surfactants[10-15]. Standnes and Austad[12] studied the mechanisms of wettability alter-ation of the chalk surface induced by various surfactants. They proposed that carboxylic acid groups from crude oil and cationic head groups of cationic surfactants, e. g. cetyltrimethylammonium bromide (CTAB), could interact to form ion pairs as a result of electrostatic attraction. The ion pairs could be desorbed from the solid surface and then solubilized into micelles. Thus, the water wetness of the solid surface is improved. As a result of the adsorption of anionic surfactants on the oil-wet surface, a water-wet bilayer is formed, making the solid surface water-wet.

For these reasons, Salehi et al.[11] proposed that anionic surfactants should be more effective than cationic surfactants in altering the wettability of the oil-wet solid surface. To test the hypothesis, spontaneous imbibition studies were conducted using different surfactants at room temperature. They found that the ultimate oil recovery could be much higher if the mechanism for the wettability alteration of the oil-wet solid surface is ion pair formation. During the process of wettability alteration of oil-wet surfaces, the ion pairs can be formed by the used anionic surfactants and the positively charged basic substances from the crude oil. The results were in good agreement with the above hypothesis. Several experimental means were used to further verify the above hypothesis and to study the mechanisms for the wettability alteration of oil-wet sandstone surfaces caused by different surfactants in this paper.

Determination of the effect of reservoir wettability on oil recovery is a long-standing problem during the process of oil exploration. Zhou et al. [3] studied the relationship between the reservoir wettability and oil recovery by spontaneous imbibition and found that oil recovery by spontaneous imbibition increases with the increase of water-wetness of the solid surface. It is intuitively believed that the more water-wet the reservoir is, the higher the waterflooding recovery is. However, an increasing number of experimental tests show that the minimum residual oil saturation could occur under intermediate wetting or weakly water-wet conditions. Jadhunandan and Morrow[16] studied the influence of reservoir wettability on oil recovery by waterflooding for the crude oil/brine/rock (COBR) systems. They found that the maximum oil recovery by waterflooding occurred at close to neutral wettability for the COBR systems studied. Morrow[4] investigated the effect of reservoir wettability on water-flooding recovery through different instruments and found that the highest oil recoveries were obtained for the neutralwet systems.

Microscopic displacement is an important experimental approach during the study of oil displacement mechanisms, which is also used to study the wettability and its effect on oil recovery. Morrow et al. [17] studied the influence of reservoir wettability on the residual oil saturation in a glass-etched micromodel and found that the residual oil saturation of the weakly water-wet system was lower than that of the strongly water-wet system. Microscopic displacement was used to analyze the relationship between the wettability of the solid surface and the ultimate oil recovery by waterflooding in this paper.

The purpose of this paper was to study the mechanisms for wettability alteration of oil-wet sandstone surfaces induced by different surfactants and the effect of reservoir wettability on oil recovery. Artificial cores cemented by epoxy resin were used to simulate the sandstone surface. Furthermore, microscopic displacement was used to further analyze the relationship between reservoir wettability and oil recovery by waterflooding. Contact angle measure-ments, scanning electron microscopy (SEM), and gravi-metric analysis were utilized to study the mechanisms of wettability alteration of oil-wet solid surfaces induced by different surfactants in the present study.

Experimental Procedures

Oil Phase Materials

The crude oil used in these studies was obtained from the Shengli oilfield in the

Yellow River delta in the north of Shandong province; however, the crude oil used in gener-ating the results in Fig. 2 was produced from the Shengtuo oilfield, Dongying, China. The basic parameters of the crude oil determined at room temperature (r. t., 20℃) are given in Table 1.

Aqueous Phase Materials

Analytical grade reagents, namely, NaCl, $CaCl_2$, and $MgCl_2 \cdot 6H_2O$, were all purchased from Tianjin Bodi Chemical Co., Ltd., China. Deionized water was used to prepare different solutions and the composition of the simulated formation water used in this study is shown in Table 2.

Surfactants

For the surfactants, CTAB, DTAB, SDS, and POE(1) were analytical grade and the other used surfactants were chemically pure grade. All the surfactants used in the present study were obtained from Sinopharm Chemical Reagent Co., Ltd. The full names and abbreviations of the employed surfactants in this study are shown in Table 3.

Table 1 Basic properties of oil sample

Production site of oil sample	Viscosity (mPa · s)	Density (g/cm^3)	Acid number (mg KOH/g)
Shengli oilfield	62.0	0.860	1.420
Shengtuo oilfield	50.5	0.706	2.825

Table 2 Composition of the simulated formation water

Component	Concentration (mg/L)
Na^+	6 907.5
Cl^-	12.521.6
Ca^{2+}	421.5
Mg^{2+}	223.5

Surface Tension Measurements

Surface tension measurements were conducted to determine the critical micelle concentration (CMC) of various surfactant solutions using a DCAT21 surface tensiometer (Dataphysics, Germany). The Wilhelmy plate method was used to measure the surface tension of the surfactant solutions at different concentrations at

20℃[18]. The concentration corresponding to the inflection point in the surface tension isotherm was taken as the CMC.

Core Handling

The core samples were obtained from the China University of Petroleum (Beijing). The quartz sand was first cemented with epoxy resin and then was dried to obtain the core sample. The basic parameters of the employed cores are shown in Table 4. After the deter-mination of basic parameters, the cores were saturated with crude oil and the irreducible water saturation was established. Then these oil-saturated cores were placed in a closed jar and were immersed in the crude oil at 75℃ for 14 days[19].

IFT Measurement

The interfacial tension (IFT) was determined using a TX500C spinning drop tensiometer (Kino Industry Co, USA). The values of the IFT between the surfactant solu-tions and crude oil are shown in Table 5.

Contact Angle Measurement

The contact angles were determined by the sessile drop method using a JC2000D contact angle measurement apparatus (Shanghai Zhongchen Digital Technology Co., Ltd., China) at room temperature[20]. Quartz plates were used to simulate the sandstone surface during the contact angle measurement. Before the contact angle measurement, one quartz plate was immersed in the surfactant solution for 24 h. Several oil drops were first injected out onto the surface of a quartz plate. As a result of the presence of wetting hysteresis, the oil drops must be left on the solid surface for 0.5 h to reach equilibrium. The shapes of the oil droplets were recorded using a high-resolution camera. The contact angles were calculated by the instrument software using the protractor method[21].

Table 3 Basic information of the employed surfactants

Surfactant type	Full name	Abbreviation	CMC (20℃) (%)
Cationic	Hexadecyltrimethylammonium bromide	CTAB	0.034
Cationic	Dodecyltrimethylammonium bromide	DTAB	0.444
Nonionic	Polyoxyethylene octyl phenyl ether	TX-100	0.012
Nonionic	Alkylpolyglycosides	GD70	0.053
Anionic	Sodium lauryl monoether sulfate	POE(1)	0.013
Anionic	Sodium dodecyl sulfate	SDS	0.231

Table 4 Core parameters

Core number	Diameter (cm)	Length (cm)	PV (mL)	Porosity (%)	S_{oi}(%)	k ($10^{-3}\mu m^2$)
X-14	2.50	5.00	5.12	21.9	73.83	44.0
X-23	2.50	5.00	5.13	22.8	77.97	48.3
X-37	2.50	5.00	5.35	22.5	74.39	48.5
X-28	2.50	5.00	5.61	21.9	74.87	47.5
X-2	2.50	5.00	5.14	20.5	75.29	42.0
X-26	2.50	5.00	5.20	22.8	78.85	44.0
X-6	2.50	5.00	5.75	22.6	69.22	41.5
h-50	2.50	5.00	6.08	25.9	68.42	43.0
h-48	2.50	5.00	5.89	26.1	68.76	46.0
X-30	2.50	5.00	5.44	22.2	75.74	45.8
X-34	2.50	5.00	5.30	22.6	76.79	44.5

Table 5 IFT between surfactants and crude oil

Surfactant type	Surfactant concentration (wt%)	IFT (mN/m)
POE(1)	0.035	2.492
DTAB	0.06	2.487
GD70	0.05	2.493
POE(1)	0.02	4.109
TX-100	0.04	4.050
GD70	0.045	3.927
SDS	0.01	2.098
POE(1)	0.05	2.007

Imbibition Studies

The oil-saturated cores were placed vertically in Amott cells and immersed in the surfactant solutions. The volume of produced oil was recorded as a function of time. The imbibition studies were conducted at ambient temperature (20℃) in this study[22].

Microscopic Displacement

The glass-etched micromodel (2.0 × 2.0 cm) was used to investigate the relationship between the solid surface wet-tability and oil recovery by waterflooding, the distribution of residual oil, and the surfactant flooding mechanisms. Figure 1 shows a schematic diagram of the microscopic displacement process. The microscopic model was firstly vacuumed and was then saturated with formation water. The irreducible water saturation was then obtained by displacing with five pore volumes (PV) of oil. After saturation, the micromodel was aged with crude oil for 14 days. The microscopic displacement was then conducted at a slow rate of 0. 002 mL/min. The whole process of the microscopic flooding test could be visualized through a digital video camera [17, 23, 24]. The experi-ments were carried out at room temperature.

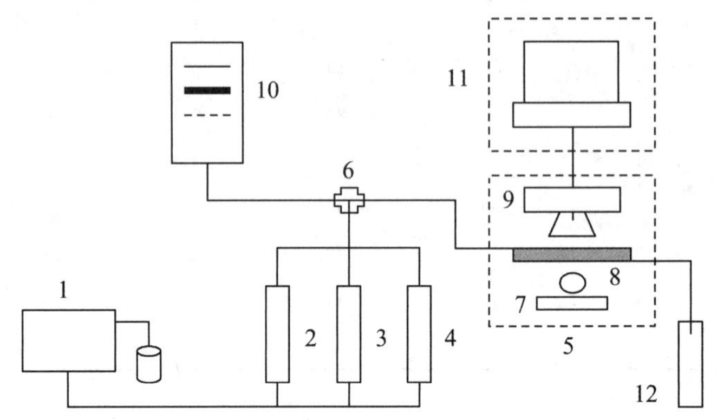

Fig. 1 Schematic diagram of microscopic displacement. 1 Advection pump, 2 simulated formation water, 3 oil, 4 surfactant solution, 5 thermostatic device, 6 pressure transducer, 7 light source, 8 micro-scopic model, 9 digital video camera, 10 pressure data acquisition system, 11 image acquisition system, 12 outlet

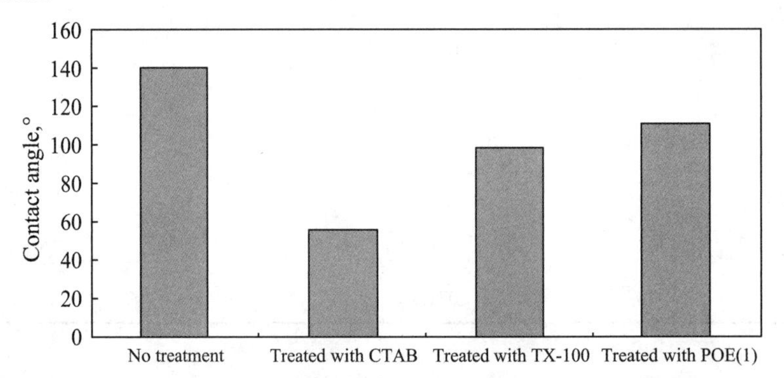

Fig. 2 Results of surfactant treatment of the contact angle measure-ments of crude oil on treated core

Scanning Electron Microscopy (SEM)

The distribution of asphaltenes adsorbed on the core surface was determined using an S-4800 field emission scanning electron microscope (Hitachi, Japan). To study mechanisms the wettability alteration of oil-wet sandstone surfaces using different surfactants, the aged core surfaces before and after surfactant treatment were scanned by SEM [25-29].

Determination of Asphaltene Mass

Gravimetric analysis was used to determine the mass of asphaltenes adsorbed on the core surface. The net mass (m_1) of a single-necked flask was recorded and the employed core was placed in the extraction device. Crude oil components such as wax and gum adsorbed on the core surface were first removed by refluxing with a mixture of n-heptane and ethanol, and then the asphaltenes were removed by refluxing with toluene. Residual asphaltenes were left after the solvent toluene was removed under vacuum by a rotary evaporator. The flask was then placed in a desiccator for 24 h and was dried to constant weight. The mass (m_2) of the single-necked flask together with residual asphaltenes was recorded. Thus, the mass of asphaltenes ($\triangle m$) can be calculated from the difference between m_2 and m_1.

Results and Discussion

Mechanisms for Surfactant-Induced Wettability Alteration of Solid Surface

To study mechanisms of wettability changes of oil-wet sandstone surfaces induced by different surfactants and to verify the hypothesis proposed by Standnes and Austad, contact angle measurements, SEM, and gravimetric anal-ysis were conducted in this study.

(a) (b)

(c) (d)

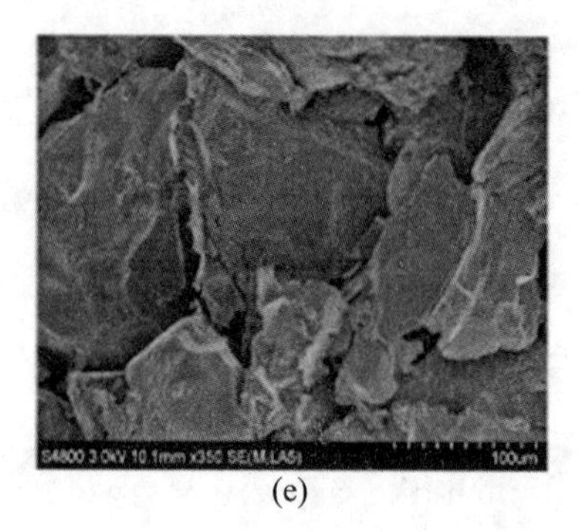

(e)

Fig. 3 SEM micrographs of sandstone core samples. a Clean core surface, b aged core surface, c aged core surface treated with TX-100, d aged core surface treated with POE(1), eaged core surface treated with CTAB

Figure 2 shows the results of the contact angle measurements. The surfactant concentrations were all fixed at 0.30 wt% and the contact angles were measured at ambient temperature. The crude oil used in this section was from the Shengtuo oilfield. As shown in Fig. 2, the contact angle of the solid surface without surfactant treatment is about 140.0°. The contact angles using TX-100 and POE(1) are about 95.0° and 115.0°, respectively[28]. The contact angle is only 57.0° after the core surface was treated with CTAB. From this we concluded that the cationic surfactant CTAB is more effective than the nonionic surfactant TX-100 and the anionic surfactant POE(1) in changing the wettability of the oil-wet sandstone surface. The results are in good agreement with the hypothesis proposed by Standnes and Austad.

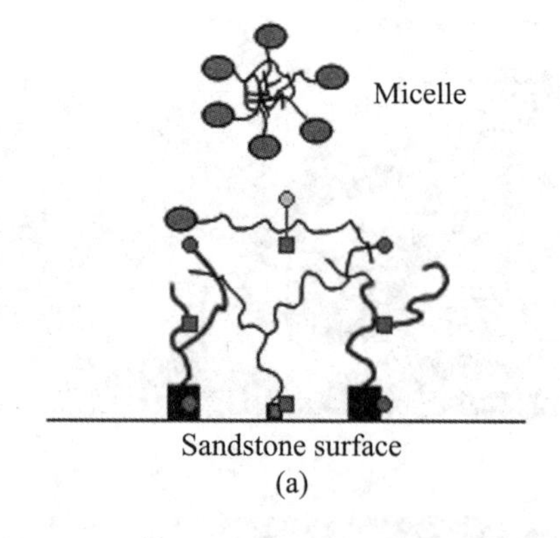

(a)

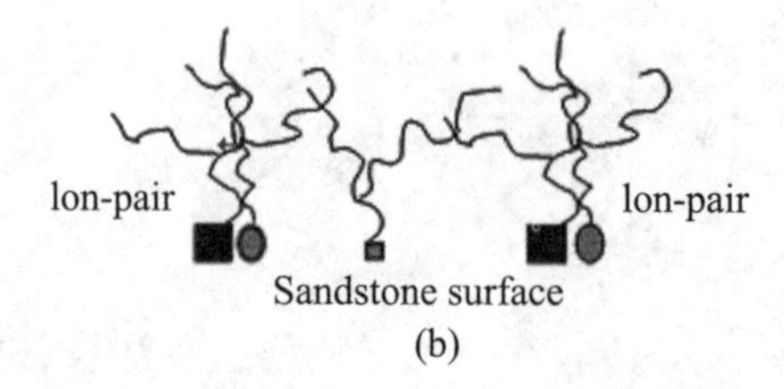

(b)

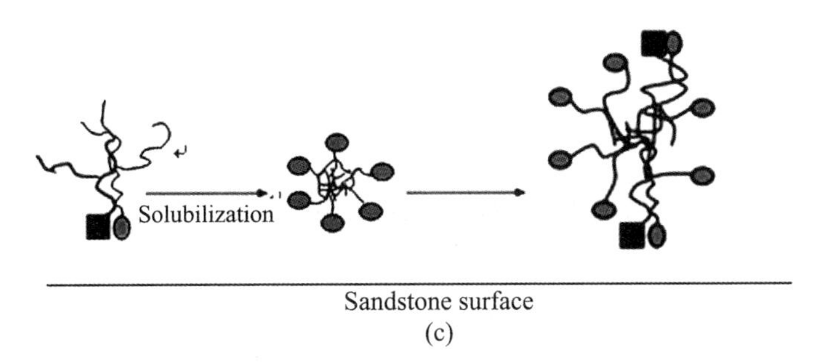

Fig. 4 Mechanism for wettability alteration of oil-wet sandstone surfaces induced by the cationic surfactant CTAB. *Large squares* symbolize negatively charged carboxylic acid groups (—COO—); small *squares* symbolize other *polar* components; *ovals* symbolize cationic quaternary ammonium groups ($—N^+(CH_3)_3$). a Before the formation of ion pairs, b formation of ion pairs, c solubilization of ion pairs into micelles

Generally, sandstone is negatively charged under neutral conditions and contains some clay minerals and a large quantity of aluminosilicate. While crude oil contains large amounts of polar substances, such as resin and asphaltenes, which could be adsorbed on the negatively charged sand-stone surface by polar interactions, surface precipitation, acid/base interactions and ion-binding, etc. [30]

Figure 3 shows SEM micrographs of different samples including the clean core surface and the aged core surfaces with or without surfactant treatment. Figures 4 and 5 show the schematic diagrams of mechanisms for wettability alteration of oil-wet sandstone surfaces caused by different surfactants [31]. As can be seen from Fig. 3a, some solid particles are very small, whereas some solid particles are much larger for the clean core surface. And the surfaces of larger solid particles are relatively smooth. From Fig. 3b we can see that a large number of asphaltenes were adsorbed on the core surface and the core surface became much rougher after the clean core surface was aged with crude oil for 14 days. Figure 3c, d, and e show SEM images of the aged core surfaces treated with TX-100, POE(1), and CTAB, respectively. As shown in Fig. 3c and d, a lot of asphaltenes were still adsorbed on the core surfaces and the core surfaces were very rough after the solid surfaces were treated with TX-100 and POE(1); whereas, owing to the formation of ion pairs, the amount of asphaltenes adsorbed on the core surface was greatly reduced and the core surface became much smoother after the aged core surface was treated with CTAB (Fig. 3e), making the surface of the core water-wet. The experi-mental phenomenon is consistent with the hypothesis proposed above, i. e., cationic surfactants should be more effective than anionic surfactants in altering the wettability of oil-wet solid surface owing to the formation of ion pairs. Thus, mechanisms for wettability alteration of oil-wet sandstone surfaces using different surfactants are nicely revealed by the SEM studies.

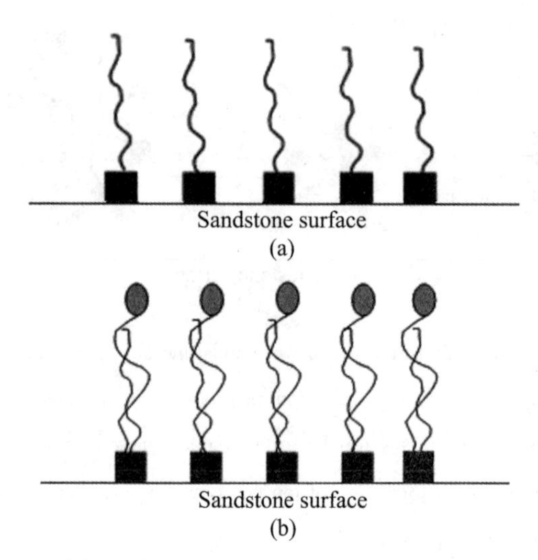

Fig. 5 Mechanism for wettability alteration of oil-wet sandstone surfaces caused by the nonionic surfactant TX-100 and the anionic surfactant POE(1). Ellipses polyoxyethylene groups or polyoxyethylene sulfonate groups, squares some polar components of crude oil. a Adsorption of polar components on the solid surface, b formation of the bilayer caused by hydrophobic interaction

After the spontaneous imbibition, as a result of differences in the wettability alteration capability for the studied surfactants, different amounts of asphaltenes were adsor-bed on the core surfaces. In the present study, gravimetric analysis was used to determine differences in the mass of asphaltenes residue. The mass of each core (2.50 ×5.00 cm) used in the gravimetric analysis was about 50.0 g. After the spontaneous imbibition of CTAB at 0.30 wt% into the oil-wet core, 0.055 g of asphaltenes remained adsorbed on the core surface. While using TX-100 and POE(1) at 0.30 wt% during the spontaneous imbibition, 0.110 and 0.105 g of residual asphaltenes remained adsorbed on the core surfaces, respectively. After contact with the cationic surfactants, the oil-wet sandstone surface is changed to be water-wet by the desorption of ion pairs, exposing a clean water-wet surface. When the anionic surfactants and nonionic surfactants are adsorbed on the oil-wet sandstone surface, the wettability of the solid surface is changed. In this case, the components of the oil phase and the adsorbed surfactants remain on the solid surface. The observations are consistent with the results of the SEM analysis and the mechanisms of wettability changes of the oil-wet solid surface using different sur-factants are indirectly confirmed by the gravimetric analysis.

Effect of Surfactant-Induced Wettability Alteration on Oil Recovery

The spontaneous imbibition recoveries vary with the change of the wettability of the solid surface due to the formation of ion-pairs between the cationic surfactants and the negatively charged carboxylic acid groups from the crude oil. Because of the wettability alteration by the adsorption of the nonionic surfactants and anionic surfactants on the solid surface, the spontaneous imbibition recoveries are also changed.

Figure 6 shows the oil recoveries by spontaneous imbibition under different wettability conditions regulated by different surfactants at room temperature (20℃). About 55.0% of the original oil in place (OOIP) was recovered from the core using POE(1) at 0.035 wt%, whereas less oil was produced from the core using DTAB at 0.06 wt% and no oil was recovered from the core using GD70 at 0.05 wt%. From Table 5 we can see that the IFT values for the three types of surfactants against crude oil are substantially equal. The contact angles between oil-water interface and the sandstone surface treated with POE(1), DTAB, and GD70 are 63.9°, 85.0° and 121.2°, respectively. It is obvious that a conclusion could be drawn from these results, i. e., the stronger hydrophilicity sand-stone surface has a higher imbibition recovery[3].

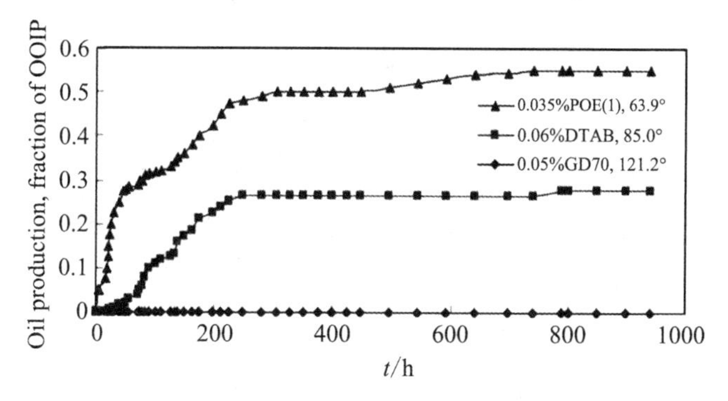

Fig. 6 Spontaneous imbibition recoveries under different wettability conditions adjusted by different surfactants

Figure 7 shows the oil recoveries by spontaneous imbibition under different wettabilities adjusted by differ-ent surfactants at room temperature (20℃). IFT values for the surfactants against crude oil are also basically equal here. Compared to the core using POE(1), more oil was recovered from the core using SDS. The contact angles for the system using SDS and POE(1) are 38.1°and 64.1°, respectively. The trends presented in Fig. 7 are the same as the ones shown in Fig. 6 and are consistent with the previous conclusion, i. e., the stronger hydrophilicity sandstone surface has a higher imbibition recovery.

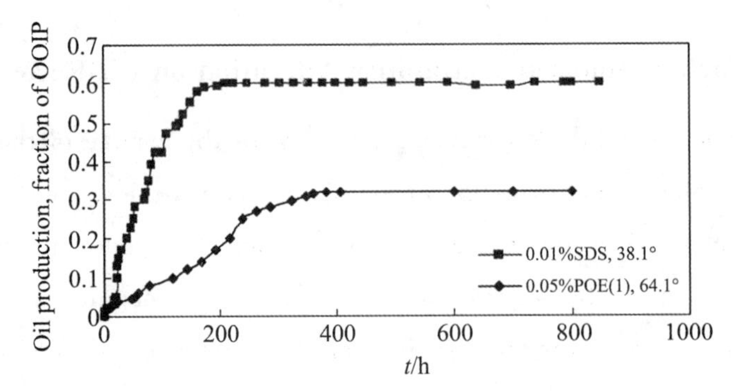

Fig. 7　Spontaneous imbibition of different surfactants into oil-wet cores at different wettabilities

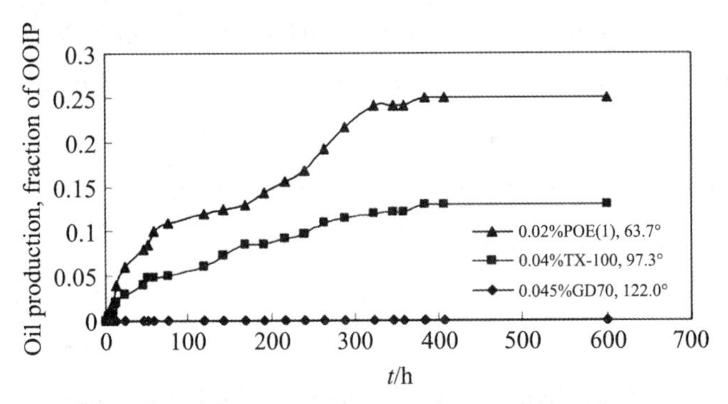

Fig. 8　Effect of surfactant-induced wettability alteration on oil recovery by spontaneous imbibition

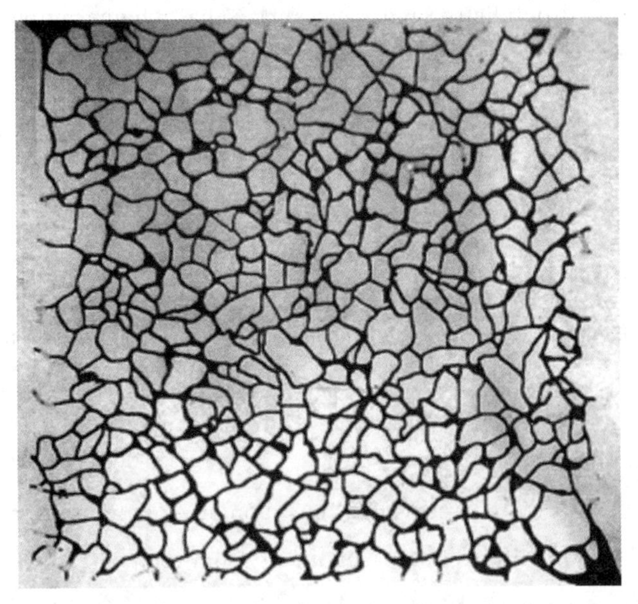

Fig. 9　Glass-etched micromodel saturated with crude oil

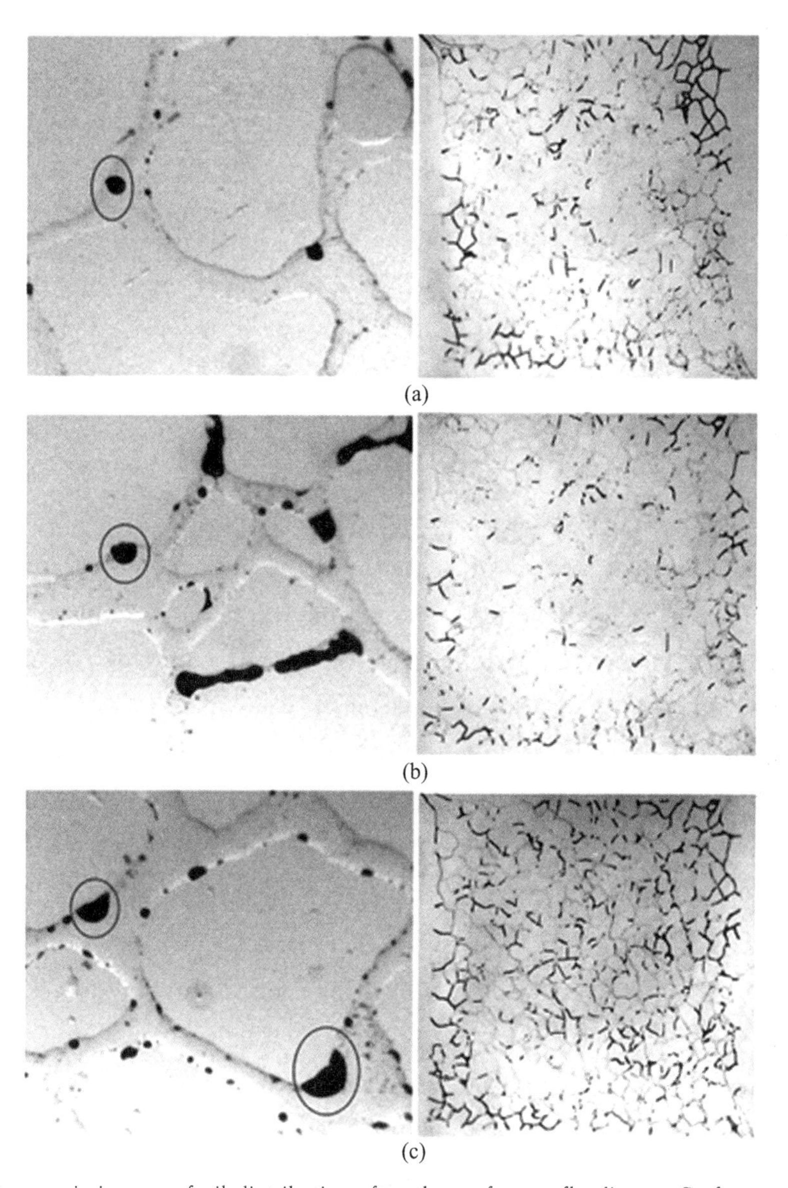

Fig. 10　Microscopic images of oil distribution after the surfactant flooding. a Surfactant, POE(1); concentration, 0.035 wt%, contact angle, 63.9°. b Surfactant, DTAB; concentration, 0.06 wt%; contact angle, 85.0°. c Surfactant, GD70; concentration, 0.05 wt%; contact angle, 121.2°

Figure 8 also shows the oil recoveries by spontaneous imbibition under different wettabilities induced by different surfactants at room temperature (20℃). The curves in Fig. 8 show the same trend as those in the above two fig-ures and the same conclusion could also be drawn from Fig. 8. In summary, the results of the spontaneous imbibi-tion tests are in good agreement with the relevant literature.

Effect of Wettability on Oil Recovery by Waterflooding

Microscopic displacement was used to study the influence of wettability on oil recovery by waterflooding. Figure 9 shows the glass-etched micromodel saturated with crude oil. The upper left corner of the glass-etched micromodel is the inlet and the lower right corner of the micromodel is the outlet, hereinafter the same. The same glass-etched micromodel was used in all the displacement experiments. The sessile drop method was used to measure the contact angles and the quartz plates were utilized to simulate the micromodel surface during the contact angle determination.

From Fig. 10, one can see that different amounts of residual oil were left in the micromodel after three types of surfactant solutions were injected at a slow rate of 0.002 mL/min. The pictures on the left show the partially enlarged images and the entire micromodel images are shown on the right. In the partially enlarged images, the typical oil droplets were marked with circles. From the pictures on the left, one can see that the shape of oil dro-plets was changed and wettabilities of the oil-wet model surfaces were altered after the surfactant flooding.

The original contact angle of the micromodel surface after being aged with crude oil is 140.0°. The contact angles for the system using POE(1), DTAB, and GD70 are 63.9°, 85.0°, and 121.2°, respectively. From the micro-model images in Fig. 10, one sees that the system using DTAB had the minimum residual oil saturation after the surfactant flooding (Fig. 10b), whereas the minimum oil recovery occurred under oil-wet conditions (Fig. 10c). As for the system using POE(1), its oil recovery lay between another two systems using DTAB and GD70 (Fig. 10a). From this measurement, we conclude that the maximum oil recovery by waterflooding occurs under near-neutral wetting conditions and the strongly oil-wet system has the minimal oil recovery. The results are in good agreement with the literature.

Conclusions

Various laboratory methods were utilized to confirm mechanisms for wettability alteration of oil-wet sandstone surfaces induced by surfactants studied in this paper. Spontaneous imbibition tests and microscopic displacement experiments were used to study the influence of reservoir wettability on oil recovery by spontaneous imbibition and waterflooding, respectively. The following conclusions can be drawn:

The cationic surfactant CTAB is able to desorb negatively charged carboxylates of the crude oil in an irreversible way by the formation of ion pairs. CTAB is more effective than the nonionic surfactant TX-100 and the anionic surfactant POE(1) in altering the wettability of oil-wet sandstone surfaces.

The wettability alteration of oil-wet sandstone surfaces induced by the nonionic surfactant TX-100 and the anionic surfactant POE(1) is due to the adsorption of surfactant molecules on the solid surface.

The stronger hydrophilicity sandstone surface has a higher imbibition recovery.

The maximum oil recovery by waterflooding occurs at near intermediate wetting conditions and the oil-wet micromodel shows the lowest oil recoveries by aterflooding.

Acknowledgments Financial support by the Program for Changjiang Scholars and Innovative Research Team in University (IRT1294) and the Fundamental Research Funds for the Central Universities (13CX05019A and 15CX02006A) are all gratefully acknowledged.

References

[1] Kennedy HT, Burja EO, Boykin RS (1955) An investigation of the effects of wettability on oil recovery by water flooding. J Phys Chem 59: 867-869. doi: 10.1021/j150531a015.

[2] Wang Y, Xu H, Yu W, Bai B, Song X, Zhang J (2011) Surfactant induced reservoir wettability alteration: recent theoretical and experimental advances in enhanced oil recovery. Pet Sci 8:463-476. doi:10.1007/s12182-011-0164-7.

[3] Zhou X, Morrow N, Ma S (2000) Interrelationship of wettability, initial water saturation, aging time, and oil recovery by sponta-neous imbibition and waterflooding. SPE J 5:199-207.

[4] Morrow NR (1990) Wettability and its effect on oil recovery. J Pet Technol 42:1476-1484.

[5] Anderson WG (1987) Wettability literature survey-part 6: the effects of wettability on waterflooding. J Pet Technol 39:1605-1622.

[6] Jackson MD, Valvatne PH, Blunt MJ (2005) Prediction of wettability variation within an oil/water transition zone and its impact on production. SPE J 10: 185-195.

[7] Han D, Yuan H, Wang H, Dong F (2006) The effect of wettability on oil recovery by alkaline/surfactant/polymer flooding. In: SPE Annual Technical Conference and Exhibition. Society of Petro-leum Engineers. 10.2118/102564-MS.

[8] Tweheyo M, Holt T, Torsæter O (1999) An experimental study of the relationship between wettability and oil production charac-teristics. J Pet Sci Eng 24:

179-188. doi:10.1016/S0920-4105(99) 00041-8.

[9] Graue A, Bognø T, Baldwin B, Spinler E (2001) Wettability effects on oil-recovery mechanisms in fractured reservoirs. SPE Reserv Eval Eng 4:455-466.

[10] Jarrahian K, Seiedi O, Sheykhan M, Sefti MV, Ayatollahi S (2012) Wettability alteration of carbonate rocks by surfactants: a mechanistic study. Coll Surf A 410:1-10. doi:10.1016/j.colsurfa. 2012.06.007

[11] Salehi M, Johnson SJ, Liang JT (2008) Mechanistic study of wettability alteration using surfactants with applications in naturally fractured reservoirs. Langmuir 24:14099-14107. doi:10. 1021/la802464u

[12] Standnes DC, Austad T (2000) Wettability alteration in chalk: 2. Mechanism for wettability alteration from oil-wet to water-wet using surfactants. J Pet Sci Eng 28:123-143. doi:10.1016/S0920-4105(00)00084-X

[13] Gupta R, Mohanty K (2011) Wettability alteration mechanism for oil recovery from fractured carbonate rocks. Transp Porous Media 87: 635-652. doi: 10.1007/s11242-010-9706-5

[14] Seiedi O, Rahbar M, Nabipour M, Emadi MA, Ghatee MH, Ayatollahi S (2010) Atomic force microscopy (AFM) investigation on the surfactant wettability alteration mechanism of aged mica mineral surfaces. Energy Fuel 25: 183-188. doi: 10.1021/ ef100699t

[15] Golabi E, Azad FS, Branch O (2012) Experimental study of wettability alteration of limestone rock from oil-wet to water-wet using various surfactants. In: SPE Heavy Oil Conference. Society of Petroleum Engineers, pp 12-14.

[16] Jadhunandan P, Morrow N (1995) Effect of wettability on waterflood recovery for crude-oil/brine/rock systems. SPE Reserv Eng 10:40-46.

[17] Morrow NR, Lim HT, Ward JS (1986) Effect of crude-oil-in-duced wettability changes on oil recovery. SPE Form Eval 1:89-103.

[18] Biswas S, Dubreil L, Marion D (2001) Interfacial behavior of wheat puroindolines: study of adsorption at the air-water inter-face from surface tension measurement using Wilhelmy plate method. J Colloid Interface Sci 244:245-253. doi: 10.1006/jcis. 2001.7940

[19] Standnes DC, Nogaret LA, Chen HL, Austad T (2002) An evaluation of spontaneous imbibition of water into oil-wet carbonate reservoir cores using a nonionic and a cationic surfactant. Energy Fuel 16:1557-1564. doi:10.1021/ef0201127

[20] Qi Z, Wang Y, He H, Li D, Xu X (2013) Wettability alteration of the quartz surface in the presence of metal cations. Energy Fuel 27:7354-7359. doi: 10.

1021/ef401928c

[21] Du W, Wu Y (2007) Comparison of hypsometry and goniometry in contact angle measurement. J Text Res 28:29-32.

[22] Strand S, Standnes DC, Austad T (2003) Spontaneous imbibition of aqueous surfactant solutions into neutral to oil-wet carbonate cores: effects of brine salinity and composition. Energy Fuel 17:1133-1144. doi:10.1021/ef030051s

[23] Donaldson EC, Thomas RD (1971) Microscopic observations of oil displacement in water-wet and oil-wet systems. In: Fall Meeting of the Society of Petroleum Engineers of AIME. Society of Petroleum Engineers. 10.2118/3555-MS

[24] Jamaloei BY, Kharrat R (2010) Analysis of microscopic dis-placement mechanisms of dilute surfactant flooding in oil-wet and water-wet porous media. Transp Porous Media 81:1-19. doi:10.1007/s11242-009-9382-5

[25] Robin M (2001) Interfacial phenomena: reservoir wettability in oil recovery. Oil Gas Sci Technol 56:55-62. doi:10.2516/ogst: 2001007

[26] Polson EJ, Buckman JO, Bowen D, Todd AC, Gow MM, Cuth-bert SJ (2010) An environmental-scanning-electron-microscope investigation into the effect of biofilm on the wettability of quartz. SPE J 15:223-227.

[27] Robin M, Combes R, Rosenberg E (1999) Cryo-SEM and ESEM: new techniques to investigate phase interactions within reservoir rocks. In: SPE annual technical conference. Society of Petroleum Engineers. 10.2118/56829-MS

[28] Combes R, Robin M, Blavier G, A dan M, Degreve F (1998) Visualization of imbibition in porous media by environmental scanning electron microscopy: application to reservoir rocks. J Pet Sci Eng 20:133-139. doi:10.1016/S0920-4105(98)00012-6

[29] Robin M, Rosenberg E, Fassi Fihri O (1995) Wettability studies at the pore level: a new approach by use of Cryo-SEM. SPE Formation Eval 10:22-29.

[30] Hou B, Wang Y, Huang Y (2015) Mechanistic study of wetta-bility alteration of oil-wet sandstone surface using different sur-factants. Appl Surf Sci 330: 56-64. doi:10.1016/j.apsusc.2014,12:185.

[31] Buckley J, Liu Y, Monsterleet S (1998) Mechanisms of wetting alteration by crude oils. SPE J 3:54-61.

Hou Baofeng is a doctor from the China University of Petroleum (East China), Qingdao, China. He is engaged in the research of oil field chemistry and enhanced oil recovery.

Wang Yefei is a teacher at the China University of Petroleum (East China), Qingdao, China. He is engaged in the research of oil field chemistry and enhanced oil recovery.

Cao Xulong is a senior engineer at the Geoscience Research Institute of Shengli Oilfield Company, SINOPEC, Dongying, China. He is also engaged in the research of oil field chemistry and enhanced oil recovery.

Zhang Jun is a teacher at the China University of Petroleum (East China), Qingdao, China. He is engaged in the research of materials science and enhanced oil recovery.

Song Xinwang is a senior engineer at the Geoscience Research Institute of Shengli Oilfield Company, SINOPEC, Dongying, China. He is also engaged in the research of oil field chemistry and enhanced oil recovery.

Ding Mingchen is a teacher at the China University of Petroleum (East China), Qingdao, China. He is engaged in the research of oil field chemistry and enhanced oil recovery.

Chen Wuhua is a teacher at the China University of Petroleum (East China), Qingdao, China. He is engaged in the research of oil field chemistry and enhanced oil recovery.

Efect of oleic acid on the dynamic interfacial tensions of surfactant solutions

Liu Miao Cao Xulong Zhu Yangwen Tong Ying Zhang Lei
Zhang Lu Zhao Sui

Technical Institute of Physics and Chemistry, Chinese Academy of Sciences, Beijing 100190, People's Republic of China

University of Chinese Academy of Sciences, Beijing 100049, People's Republic of China

Exploration and Development Research Institute of Shengli Oilfield Company, Limited, Sinopec, Dongying, Shandong 257015, People's Republic of China

Research Institute of Petroleum Engineering, Sinopec, Beijing 100101, People's Republic of China

Abstract: The dynamic interfacial tensions (IFTs) of three di erent types of surfactant solutions, cationic surfactant hexadecyl trimethylammonium (CTAB), anionic surfactant sodium dodecyl benzenesulfonate (SDBS), and non-ionic surfactant Triton X-100 (TX-100), against model oil containing oleic acid (OA) have been investigated by a spinning drop interfacial tensiometer. The influences of the OA concentration and addition of alkali have been studied. The experimental results show that OA added to the oil phase has a significant efect on interfacial behaviors of surfactant solutions. An obvious synergistic efect on reducing IFT can be observed for low-concentration cationic surfactant CTAB solution because the tight mixed adsorption films will be formed through electrostatic attraction between CTAB and OA molecules. However, the synergistic efect will disappear at a high surfactant concentration by the formation of mixed micelles. For the anionic surfactant SDBS system, the mixed adsorption of SDBS and OA molecules leads to a limited decrease of equilibrium IFT as a result of electrostatic repulsion. The hydrophilic group EO chain of TX-100 extends into the aqueous phase and the interfacial film becomes tight with an increasing bulk concentration. As a result, OA shows little efect on the equilibrium IFTs at a high TX-100 concentration. The alkali in the aqueous phase will react with OA in the oil and produce anionic surfactant sodium oleate (SO), which destroys the synergistic efect between CTAB and OA molecules. With an increase of the OA concentration, the dynamic characteristic of IFT has been enhanced and the transient IFTs decrease obviously because more SO molecules have been produced.

1 Introduction

There are still two-thirds of crude oil left in the reservoir after the traditional flooding strategy, primary and secondary oil recovery, which rely on the energy of the oil reservoir itself.[1,2] Considering the current situation, tertiary oil recovery employing the chemical substance surfactant, polymer, and alkali to enhanced oil recovery is a significant and promising work for energy supply for the next few years.[3] The studies of laboratory experiments and theoreticals show that residual oil trapped in the reservoir could be displaced by increasing the capillary number, Nc, which determines the microscopic displacement e ciency of oil.[4,5] Reducing the oil/water interfacial tension (IFT) is one of the most pronounced ways to improve the Nc value. The e ciency of oil recovery will increase strikingly if the IFT value between the displacing aqueous phase and crude oil attains an ultralow value ($<10^{-2}$ mN/m).[6]

In tertiary oil recovery, numerous investigations in both experimental and theoretical characteristics have indicated that alkali flooding is an efective way for enhanced oil recovery of acidic crude oil reservoirs because the reaction of alkaline with the organic acids in crude oil will lead to lowering IFT.[7] On the one and, under alkaline conditions, crude oil emulsions are formed for the adsorption of natural surfactants, which contain arboxylic acid groups.[8,9] These acids ionize, and the generation of surface-active species called in situ surfactants (ionized acids) at the interface results in a remarkable decrease in IFTs. The ionized acid is very surface-active and has a trend to partition into the aqueous phase.[10] The ionized acids may form soaps with sodium ions present in the aqueous phase at a high ionic strength, which has a trend to partition into the oil phase. The removal of the ionic acids from the interface causes the dynamic IFT behavior.[11] Although organic acid components in crude oil are very complex, previous works have shown that the behaviors of simple model systems using a pure hydrocarbon oil acidified with oil-soluble carboxylic acid are closely representative of systems involving real crude oil, particularly with respect to the dynamic character of the IFT observed.[12,13] Zhao et al.[14] and Zhou et al.[15] have studied the influences of alkyl chain lengths of acids on the IFTs in alkyl benzene sulfonate and sulfobetaine systems, respectively. Zhu et al.[16] and Cao et al.[17] have investigated the effect of acidic fractions in crude oil on the IFTs of di erent surfactant solutions. In general, these studies have well-illustrated that fatty acids or acidic fractions influence the IFT through two ways: for the high IFT system, surfactant molecules pack loosely at the interface and organic acid additives may adsorb onto the interfacial spaces among surfactant molecules to form a mixed adsorption film, which will result in the decrease of IFT and show synergism; on the contrary, for the low IFT system, surfactants pack closely at interface and then the acidic additives will compete with surfactant molecules and hardly adsorb onto the interface. In this kind of condition, the organic acid components influence the partition of the surfactant among the oil, water, and interface by changing the equivalent alkane carbon number (EACN) values of model oils, which results in the variation of the IFT.

Many attempts have been made to reduce equilibrium and dynamic IFTs in acidic model oil/alkali/surfactant systems in the past few decades because it is usually not su cient to reach ultralow IFT using surfactant alone. However, their studies mainly with respect to the anionic surfactant/alkali/model oil system[18-22] generally hold the idea that the alkali reacts with the acidic fraction and forms the ionized surfaceactive species, which has a synergistic efect with anionic surfactant molecules with the existence of proper organic acids.[21,22] Therefore, the mechanism responsible for reducing IFT by di erent types of surfactants with alkali solutions against acidic model oil needs to be explored.

In this paper, oleic acid (OA) was employed as the representative of the organic acid component. The dynamic IFTs of the acidic model oil against di erent types of aqueous surfactant systems with and without alkali have been investigated systematically. Moreover, the mechanisms responsible for IFT reduction of di erent types of surfactant solutions against acidic model oil have been provided. The studies in this paper are helpful for understanding interfacial interactions among surfactants, acids, and soaps.

2 EXPERIMENTAL SECTION

2.1 Materials

The cationic surfactant used in this paper is hexadecyl trimethylammonium bromide (CTAB), with a purity of 99+mol %, purchased from Xiamen Pioneer Technology, Inc., Xiamen, China. The anionic and non-ionic surfactants are sodium dodecyl benzenesulfonate (SDBS, with a purity of 99+mol %, purchased from Jinke Company, Tianjin, China) and Triton X-100 (TX-100, with a purity of 99+mol %, purchased from Xilong Company, Guangdong, China), respectively. All inorganic reagents used were analytical-grade. Distilled water was used in the preparation of the aqueous phase. The alkanes with a purity of 99+ mol % (GC) were purchased from Aladdin Industrial Corporation, Shanghai, China. OA, cis-9-octadece-noic acid, with a purity of 99+ mol %, was purchased from Beijing Chemical Works.

2.2 Apparatus and Methods.

Dynamic IFTs were measured by the spinning drop technique with Texas-500C (CNG USA Co.). The standard spinning drop tensiometer had been modified by the addition of video equipment and an interface to a personal computer. The computer had been fitted with a special video board and a menudriven image enhancemenand analysis program. The video board can "capture" a droplet image for immediate analysis. Analyses alway include the measurement of the drop length and drop width.[23] The aqueous surfactant as the bulk phase was injected into the glass tube, and about 1 μL of model oil as the oil phase was put into the middle of the tube. All of the solutions in this paper were prepared by distilled water, and the experimental temperature was 30 ±0.5℃. In all experiments, the measurements of IFTs were at a rotating velocity of 5 000 r/min.

The volumetric ratio of water/oil in the spinning drop tensiometer is about 200.[24] IFTs were assumed to be equilibrated when measured values of IFT remained nearly unchanged for half an hour.[25]

2.3 Quantum Chemical Simulation.

Charge distributions of CTAB, SDBS, OA, and sodium oleate (SO) have been calculated by Gaussion 09 for Windows. Both the three-dimensional molecular geometry optimizations and molecular charge distribution calculations were performed at the B3LYP level using the 6-31G * basic set.[26] The efect of solvent water on the four molecules has been taken into account in the process. The charge is calculated by means of COSMO. Results are plotted at Table 1.

Table 1 Grouped Atomic Partial Charges for Interfacial Active Species

molecule	head group
OA	−0.104
SO	−0.874
SDBS	−1.032
CTAB	+0.874

3 RESULTS AND DISCUSSION

3.1 IFTs of Surfactant Solutions against n-Alkanes.

It is well-known that oil and water have opposite polarity. The greater the dissimilarity in their nature, the greater the IFT value between them.[14] The surfactant is a kind of amphiphilic molecule, and a small amount of it can decrease IFTs between the oil phase and bulk phase considerably.[17] Reduction of IFT is one of the most commonly measured properties of surfactants in solutions.[27] It directly depends upon the replacement of molecules of solvent at the interface by molecules of surfactant.[23] To reach an ultralow IFT ($<10^{-2}$ mN/m) demands that the nature of the material on both sides of the interface must be very similar.[14] In this paper, the IFTs of di erent types of aqueous surfactants against n-alkanes with and without OA have been investigated systematically. It must be pointed out that OA is also surface-active because it has a hydrophobic group alkyl chain and a hydrophilic group carboxyl.

3.1.1 IFTs of Surfactant Solutions against n-Decane.

The dynamic IFTs of the most simply surfactant/n-decane systems have been shown in panels A, B, and C of Figure 1 corresponding to cationic surfactant CTAB, anionic surfactant SDBS, and non-ionic surfactant TX-100, respectively. The similarity of these three pictures in Figure 1 is that the higher the surfactant concentrations, the lower the dynamic IFTs. The equilibrium IFTs of these three aqueous surfactants are displayed in Figure 1D. TX-100 solutions have the lowest IFT values (about 1 mN/m) at a high concentration because non-ionic surfactant molecules without electrostatic repulsion can arrange more tightly at the interface. On the contrary, SDBS solution shows the highest IFT because it has the strongest electrostatic repulsion (see Table 1).

3.1.2 IFTs of Surfactant Solutions against n-Alkanes.

To measure the IFTs of aqueous surfactants against n-alkanes with di erent alkane carbon numbers (ACNs) is very useful to apparently recognize the mechanism responsible for the reduction of IFT because the variations of ACN reflect the partitioning of surfactant molecules among the oil, aqueous phase, and interface.[5,24]

It appears in Figure 2A that the dynamic IFT of 0.05% aqueous TX-100 against n-alkane and the dynamic curves show the most typical "L" shape. That means that there is a continuous decrease of dynamic IFT to the equilibrium value for aqueous surfactants against di erent alkanes. With the gradual di usion of surfactant molecules from the bulk phase to the interface and the replacement of interfacial solvent molecules, the IFT values decrease continuously to the equilibrium. Figure 2B shows the equilibrium IFTs of three di erent aqueous surfactants against n-alkanes. It is very necessary to point out that all of the equilibrium IFTs in di erent types of aqueous surfactants are higher than 1mN/m and the lowest IFT value is the non-ionic aqueous surfactant TX-100 in these three lines. TX-100 has the most efective active interface, and SDBS has the least, under the given conditions. The experimental results can be explained by the electrostatic interaction: ionic surfactants have an electrostatic repulsion between the same charged surfactant molecules, and this efect makes surfactant molecules never packed closely at the interface. SDBS molecules have a more severely electro-static repulsion than CTAB molecules (the quantity of electric charge of SDBS and CTAB is -1.032 and 0.874, respectively). TX-100 has an excellent active interface because of its non-ionic nature.

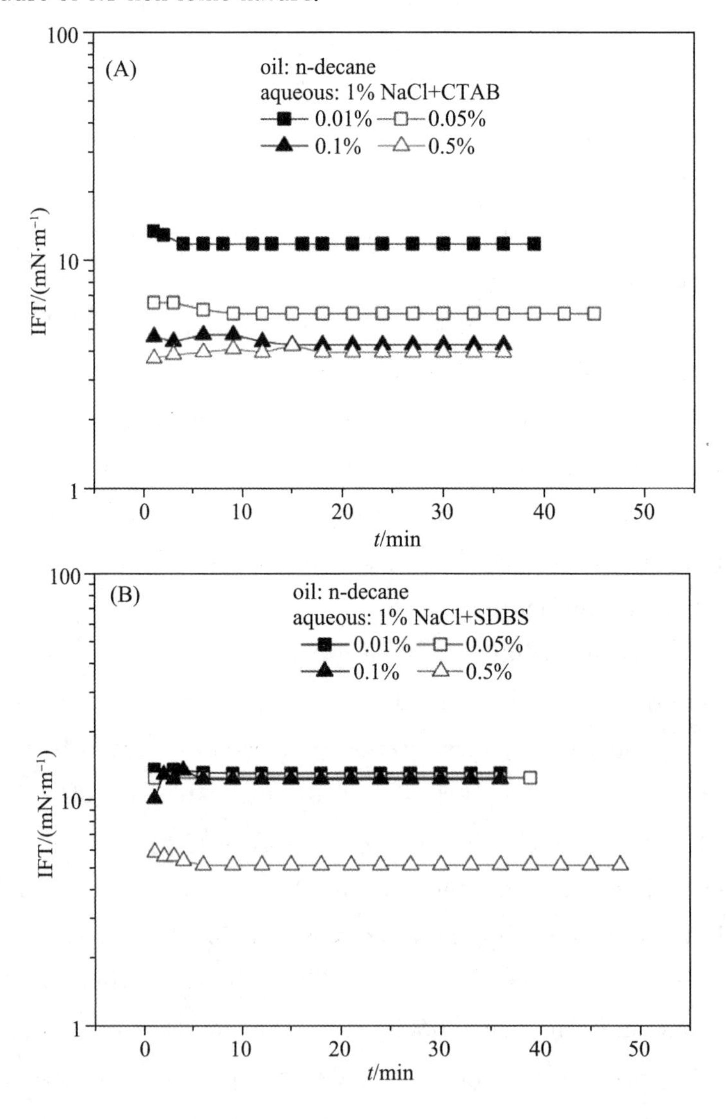

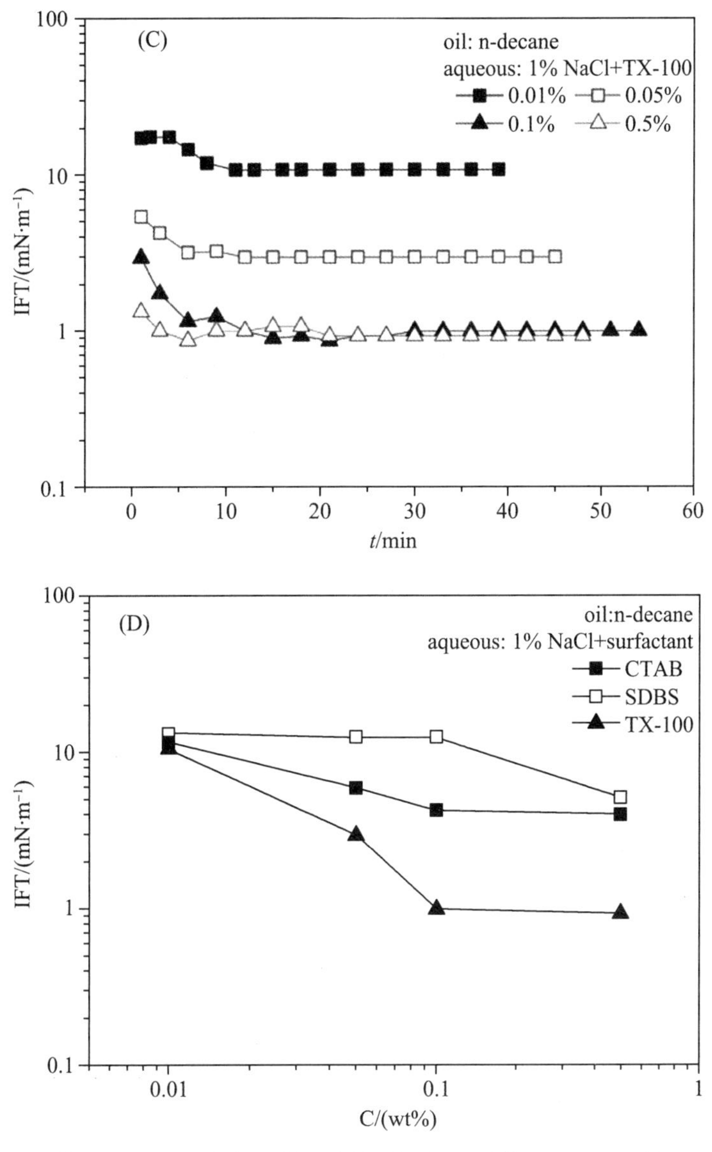

Figure 1　Dynamic IFTs of aqueous (A) CTAB, (B) SDBS and (C) TX-100 against n-decane and (D) equilibrium IFTs of aqueous surfactant.

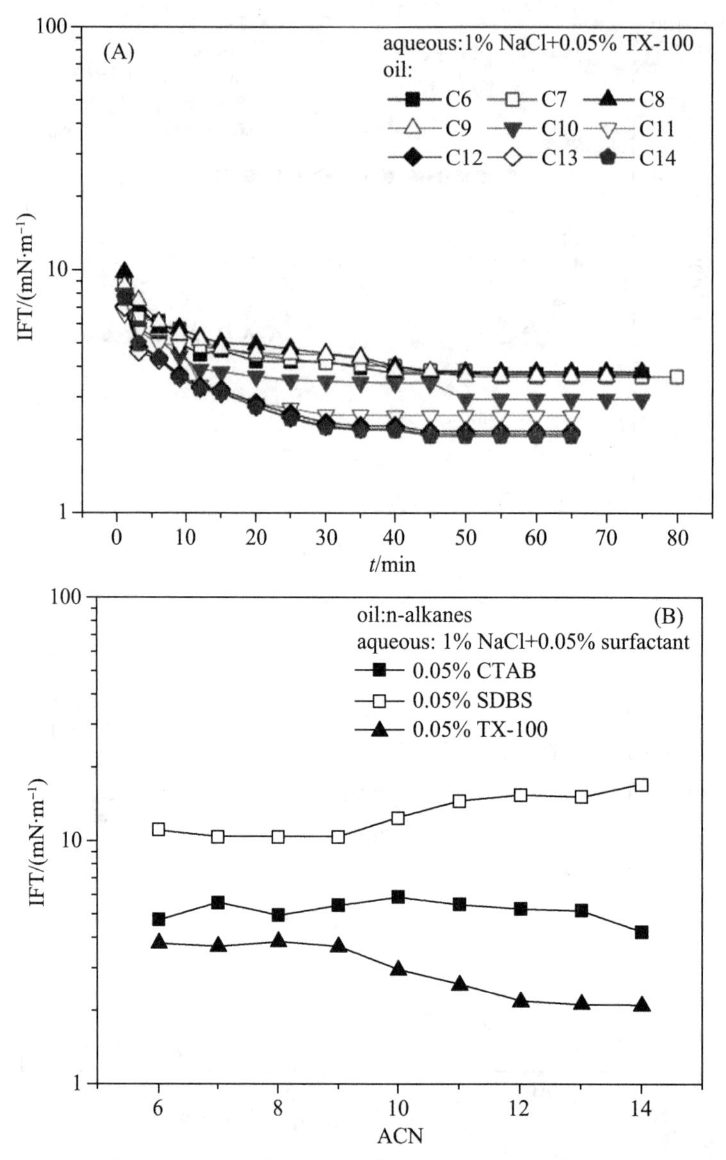

Figure 2 (A) Dynamic IFTs of aqueous TX-100 against n-alkane and (B) equilibrium IFTs of aqueous surfactants against n-alkanes.

Moreover, there is no obvious minimum IFT value in Figure 2B. The reason is that these three surfactants are strongly water-soluble and the change of the hydrophilic lipophilic balance (HLB) caused by the variation of the alkyl carbon number has a weak efect on the IFT value.

3.2 IFTs of Surfactant Solutions against Acidic Model Oils.

3.2.1 Efect of the Surfactant Concentration on the IFTs between Acidic Model Oils and Surfactant Solutions.

Long alkyl carbon organic carboxylic acids (alkyl carbon number more than 6) have a structure similar to the surfactant, with a hydrocarbon chain tail on one side and a polarity head On the other side; therefore, they have a strong tendency to Adsorb onto the interface. 14In this part, the efect of the surfactant concentration on the dynamic IFTs between the bulk phase and n-decane containing 15 mM OA has been investigated, and the results have been plotted in Figure 3.

For most common surfactant solutions, with the increase of the bulk concentration below the critical micelle concentration (cmc), the amount of surfactant molecules at the water/oil interface increases accordingly and the IFTs decrease as a result. Therefore, the dynamic IFT curves show a "L" type, and equilibrium IFTs decrease with the increase of the surfactant molecule concentration. However, this may be reverse when OA has been added to the oil. For cationic aqueous surfactant CTAB, results appear that 0.01 and 0.05 mass% CTAB have low IFT values of 0.01 mN/m order of magnitude against 15 mM OA/decane model oil, lower than 0.1 and 0.5 mass% aqueous CTAB correspondingly in Figure 3A. For the anionic surfactant SDBS system, the equilibrium IFTs of the aqueous surfactant against 15 mM OA/decane model oil decrease with the increase of the surfactant concentration in the range of 0.01—0.1 mass% and the IFT increases slightly at 0.5 mass%. Moreover, the characteristic dynamic behaviors can be observed for CTAB and SDBS solutions, especially at high concentrations. On the other hand, it seems that the surfactant concentration has little efect on both the dynamic behavior and equilibrium IFT values of TX-100 solutions.

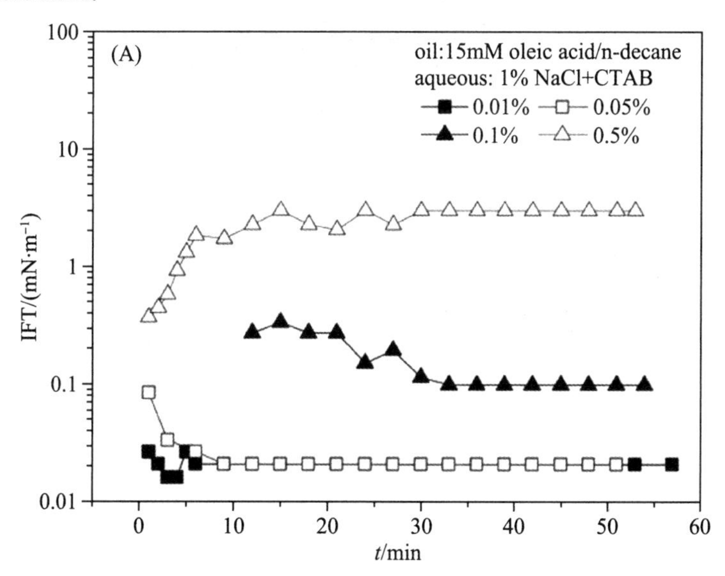

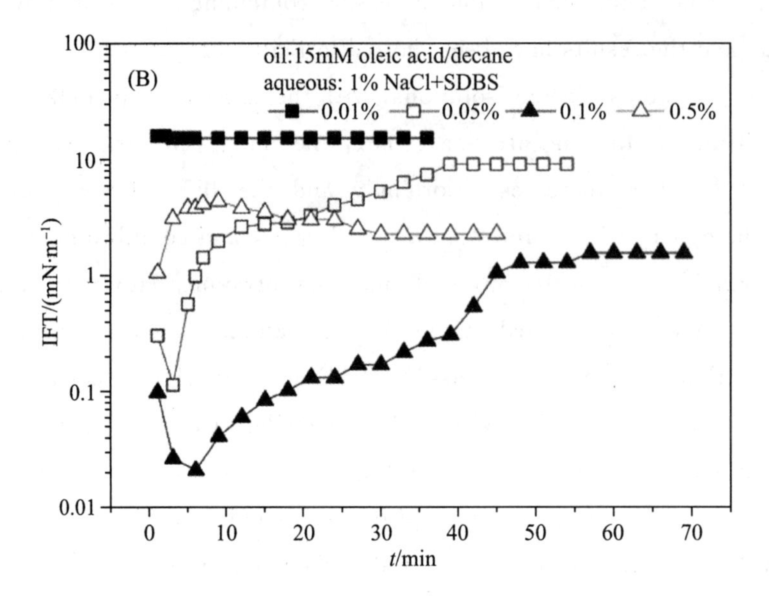
(B)
oil:15mM oleic acid/decane
aqueous: 1% NaCl+SDBS
0.01%
0.05%
0.1%
0.5%
IFT/(mN·m⁻¹)
t/min

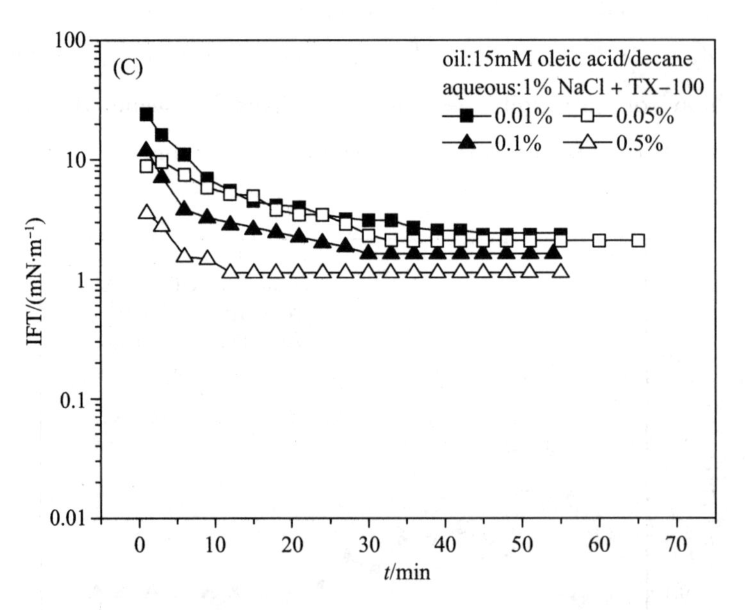
(C)
oil:15mM oleic acid/decane
aqueous:1% NaCl + TX-100
0.01%
0.05%
0.1%
0.5%
IFT/(mN·m⁻¹)
t/min

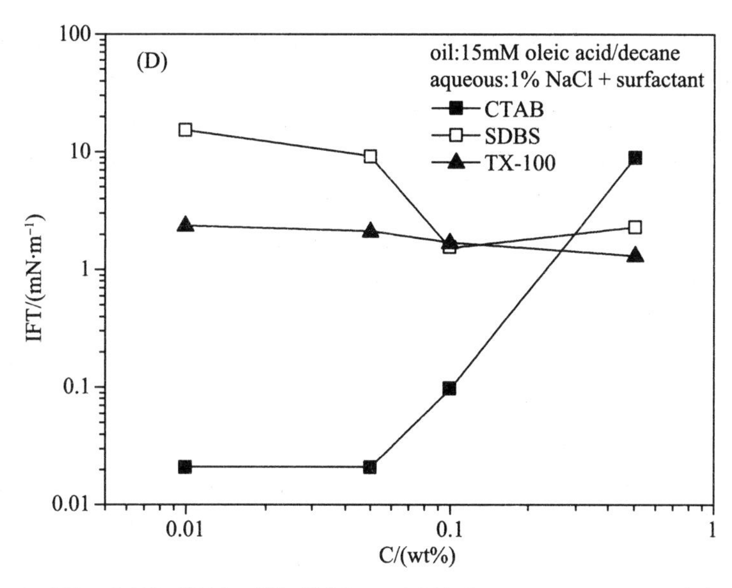

Figure 3 Dynamic IFTs of (A) CTAB, (B) SDBS, and (C) TX-100 solutions and (D) their equilibrium IFTs against model oil containing 15 mM OA.

By comparison of Figures 1D and 3D, one can see that the equilibrium IFTs of SDBS solutions decrease slightly at high concentrations when OA has been added to the oil. On the other hand, similar results can be found for TX-100 solutions at low concentrations. The most obvious decrease of IFT appears for CTAB solutions with the existence of OA.

Generally speaking, the OA molecules and surfactant molecules will adsorb onto the interface independent from the oil and aqueous phase, respectively, which leads to the formation of a tight mixed film at a short time and the decrease of dynamic IFTs under optimized concentrations. However, with the increase of interfacial surfactant molecules and the formation of mixed micelles in solutions, the synergism between OA and surfactant molecules will be destroyed with time and dynamic IFTs pass through a minimum. For non-ionic surfactant TX-100, no minimum can be observed in dynamic IFT curves for all experimental concentrations because the long EO chain of TX-100 lying flat at the interface weakens the synergism efect.

The possible molecule arrangements of di erent surfactants at interfaces are plotted in Figure 4. Because the OA molecules have an electrostatic attraction efect with CTAB molecules, the mixed adsorption film packs closely and appears with low IFTs of 0. 01 mN/m order of magnitude, as diagrammatically shown in the left of Figure 4A. However, for a high surfactant concentration of 0. 5 mass%, the formation of mixed micelles by CTAB and OA molecules leads to the obvious increase of equilibrium IFT, which is consistent with the "V" shape of dynamic IFT curves.

For the anionic surfactant SDBS system, large spaces appear at the interface as a result of electrostatic repulsion among SDBS molecules, even at a high bulk concentration. Therefore, the mixed adsorption of SDBS and OA molecules leads to the decrease of equilibrium IFT. Moreover, the OA and SDBS molecules adsorb simultaneously onto the interface from the oil and aqueous phase, respectively, when a new interface is produced. As a result, a tight interfacial film gradually formed by the mixed adsorption of OA and SDBS, which leads to the reduction of IFT with time. However, more SDBS molecules accumulate at the interface with the aging time, and most OA molecules have been displaced as a result of electrostatic repulsion. The adsorption film becomes looser and a "V" shape of the dynamic IFT curve appears at bulk concentrations after cmc, as shown in Figures 3B and 4B.

For non-ionic surfactant TX-100, equilibrium IFTs decrease obviously at low concentrations as a result of the formation of the mixed film. However, the efect of OA on dynamic and equilibrium IFTs is very slight at a high concentration. The primary reason is that OA molecules can mix/adsorb at the interface and decrease IFT under a low non-ionic surfactant concentration because there are interfacial spaces under this condition. When the TX-100 concentration becomes higher, the hydrophilic group EO chain extends into the aqueous phase and the interfacial film becomes tight. As a result, OA molecules hardly adsorb competitively at the interface, as shown in Figure 4C.

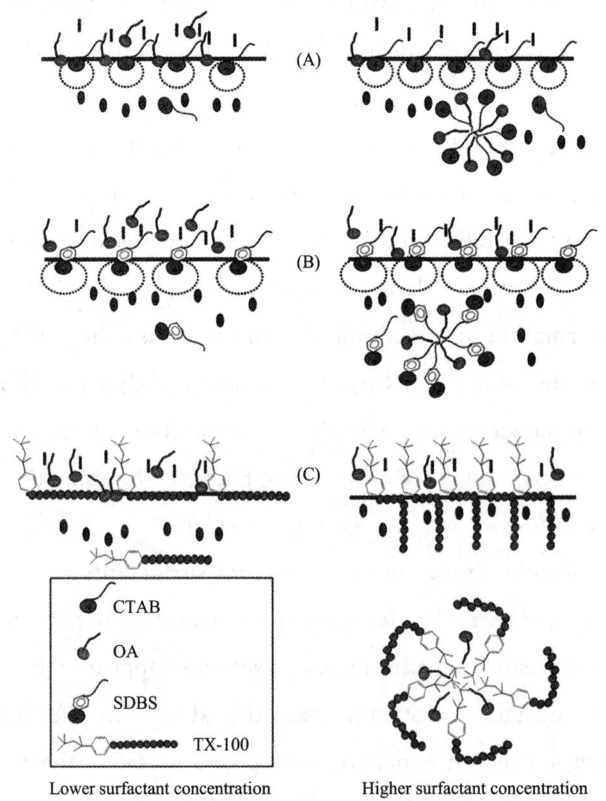

Figure 4 Possible molecule arrangements at the interface for aqueous (A) CTAB, (B) SDBS, and (C) TX-100 against 15 mM OA model oil.

As we discussed above, the dynamic IFT behaviors of surfactant OA systems strongly depend upon the mixed adsorptions of two surface-active molecules. Therefore, the amount of OA in model oil plays a crucial role in controlling IFT. It could be expected that the di usion and competitive adsorption of the OA molecule will be enhanced with an increase of the OA bulk concentration. Moreover, the efect of mixed micelles on the amount of the OA molecule at the interface becomes weaker. The dynamic IFTs of CTAB, SDBS, and TX-100 solutions and their equilibrium IFTs against model oil containing 150 mM OA are plotted in Figure 5. We can clearly see from Figures 3 and 5 that the equilibrium IFTs decrease obviously for all three surfactants and the dynamic behaviors are weakened for CTAB and SDBS systems when 150 mM OA has been added to the oil, which can ensure that our provided mechanism is responsible for IFT behaviors.

3.2.2 Efect of OA on the IFTs between n-Alkanes and Surfactant Solutions.

According to the former studies about IFTs in fatty acid benzenesulfonate systems, fatty acid will a ect IFT through two ways: one is the formation of a mixed Adsorption film, and the other is to change the partitioning of surfactant molecules among the water, oil, and interface.[17]

To investigate the influence of OA on the partitioning of the surfactant, the IFTs of n-alkanes containing OA against di erent types of aqueous surfactants have been studied systematically and plotted in Figure 6.

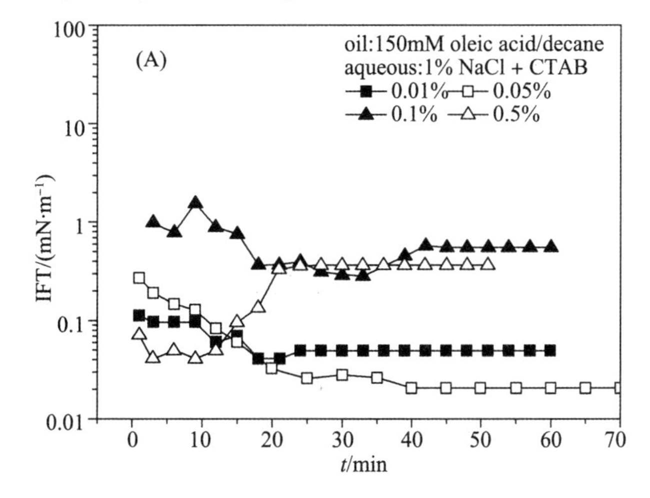

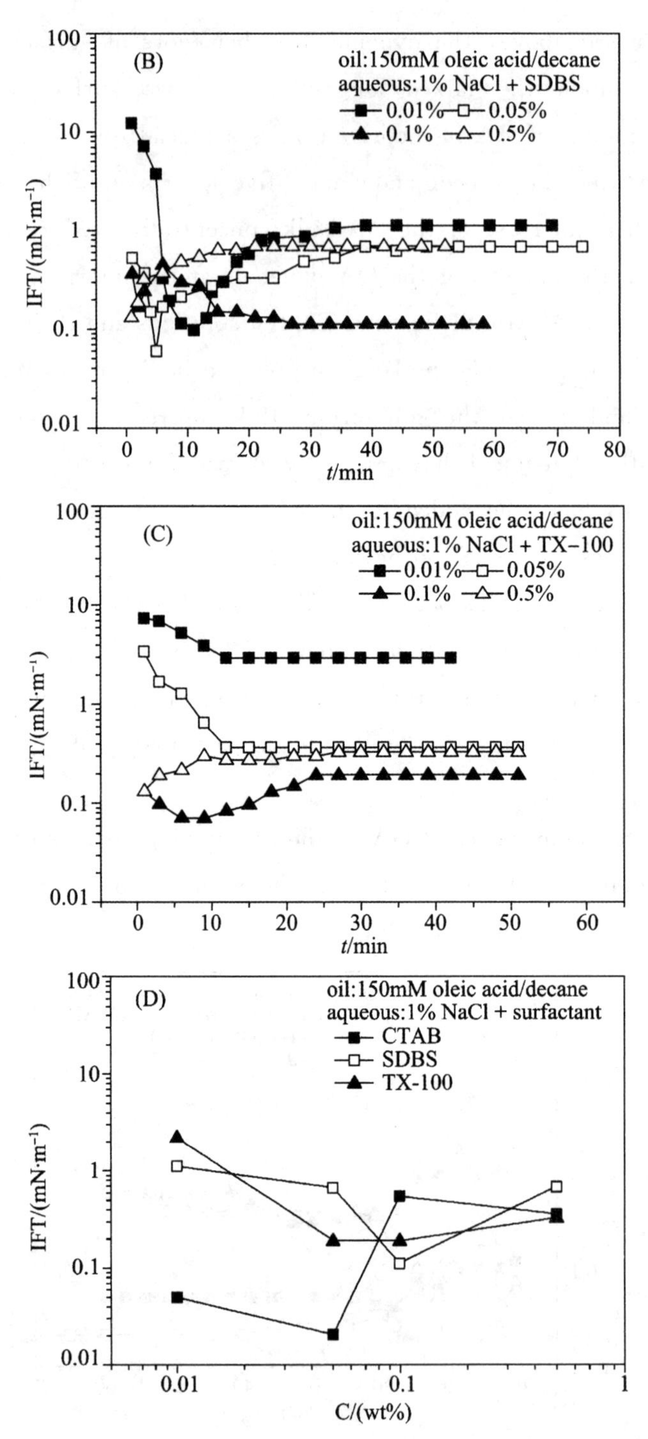

Figure 5 Dynamic IFTs of (A) CTAB, (B) SDBS, and (C) TX-100 solutions and (D) their equilibrium IFTs against model oil containing 150 mM OA.

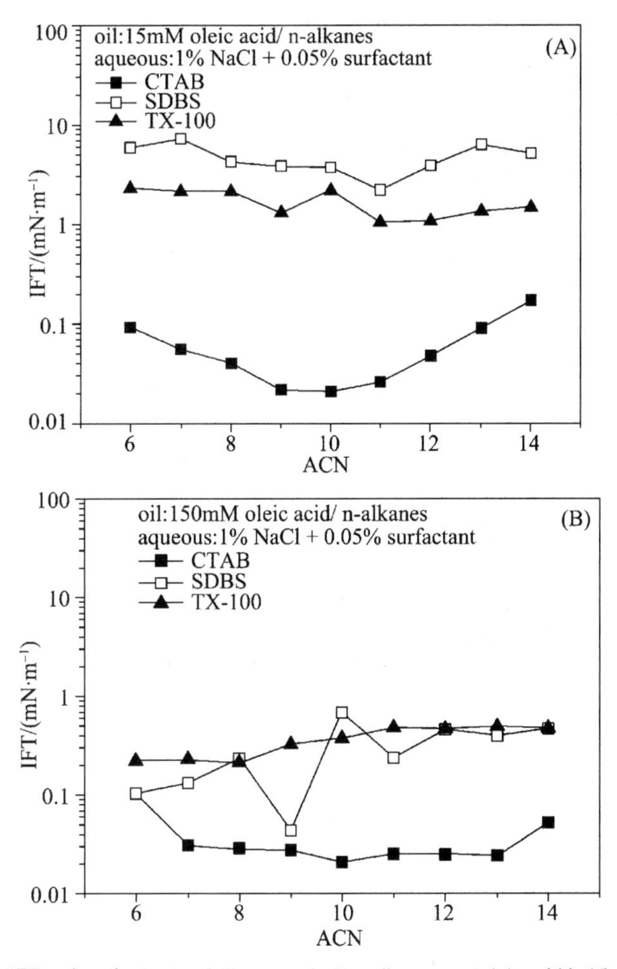

Figure 6 Equilibrium IFTs of surfactant solutions against n-alkane containing (A) 15 mM OA and (B) 150 mM OA.

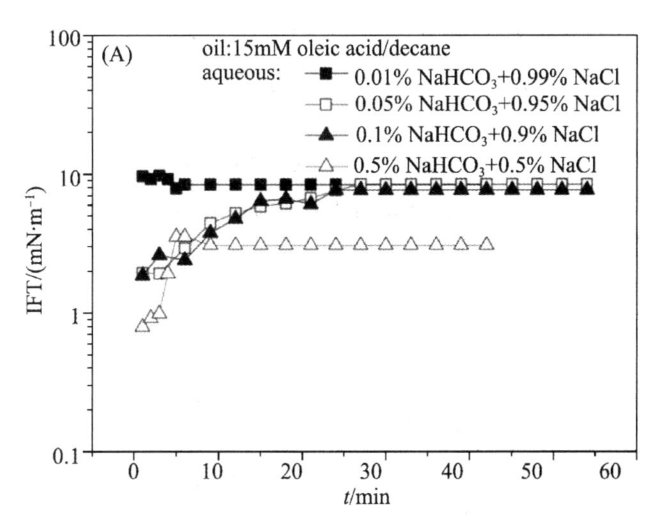

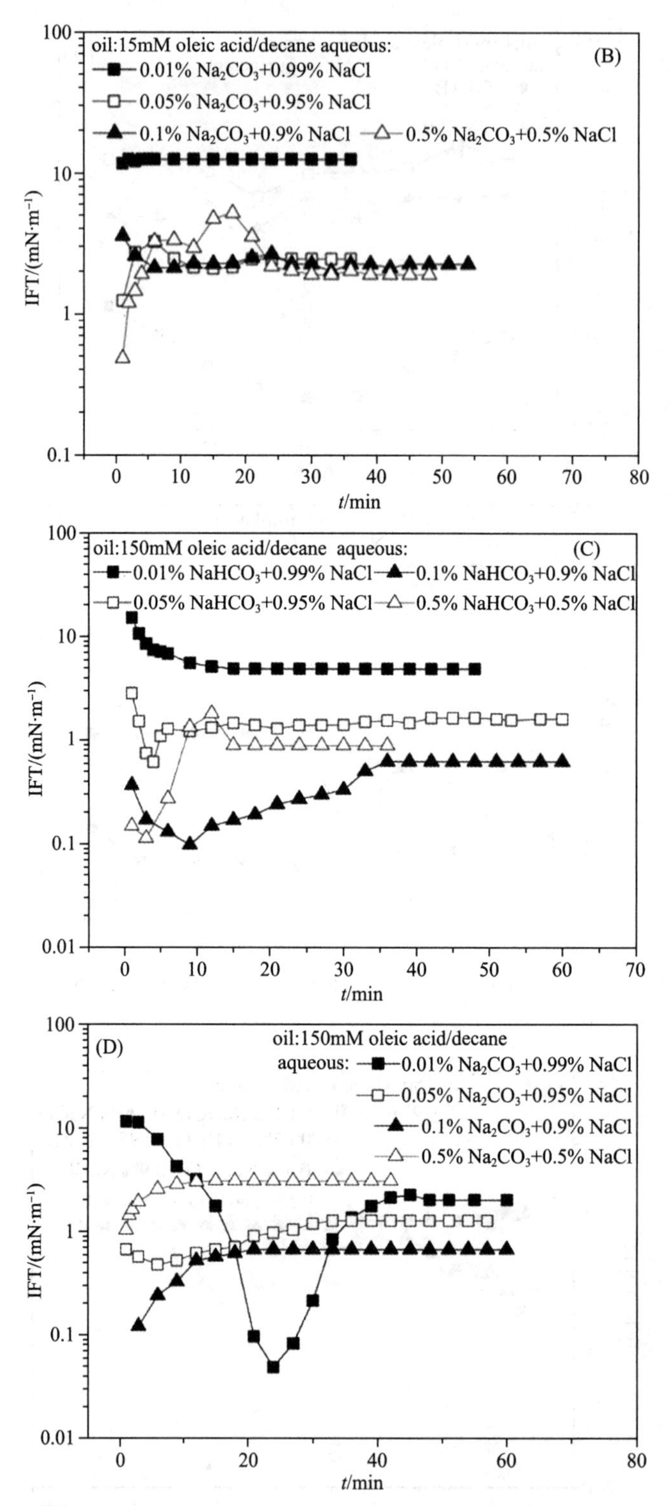

Figure 7 Dynamic IFTs of sodium bicarbonate and sodium carbonate solutions against model oils containing (A and C) 15 mM OA and (B and D) 150 mM OA, respectively.

Surfactant solution may produce a minimum IFT when measured against a series of n-alkanes with di erent alkyl carbon lengths. The number of alkane carbons for the minimum IFT is called n_{min} of this surfactant solution.[3,5] For a surfactant or surfactant mixture, the n_{min} value can quantitatively represent its HLB; the higher its n_{min} value, the lower its HLB.[24] Moreover, a minimum value may appear when the IFTs of hydrocarbon, model oil, or crude oil are measured against a series of aqueous surfactants with varying n_{min} values. The n_{min} value corresponding to the minimum IFT is called the EACN. Similar to the n_{min} concept, the EACN value can quantitatively represent the influence of oil molecules on the exchange of surfactant molecules between the oil phase and the interface. The higher the EACN value, the weaker the trend for a surfactant to partition into the oil.[3,5]

We have proven in previous studies that the IFT versus ACN curve will move evenly under dierent conditions if the HLB of the surfactant controls the properties of the adsorption film. On the other hand, the IFTs of surfactant solutions will decrease against almost all hydrocarbons if the mechanism of mixed adsorption determines the nature of the film.[6,11,14,18] As shown in Figures 2 and 6, all IFTs of SDBS, CTAB, and TX-100 solutions against C_6—C_{14} decrease when 15 mM OA has been added to hydrocarbons, which ensures that mixed adsorption plays the crucial role in reducing IFT. However, from Figure 6A, ACN shows an obvious efect on IFTs of CTAB solutions against 15 mM OA model oil and decane model oil has the lowest IFT value. This demonstrates that the chain length of the oil molecule will a ect the ratio of CTAB and OA at the interface and, as a result, change the IFTs. OA plays a more important role in the synergism of mixed adsorption with the increase of the OA concentration. Therefore, as shown in Figure 6B, the IFTs decrease from 1 to 0. 1 mN/m order of magnitude for SDBS and TX-100 when the OA concentration improves to 150 mM. Moreover, the ACN dependence of IFTs for CTAB solutions is also weakened.

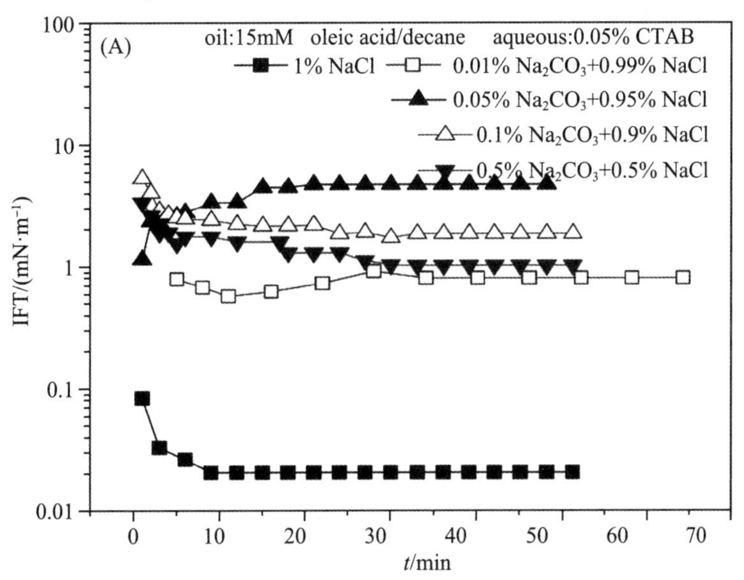

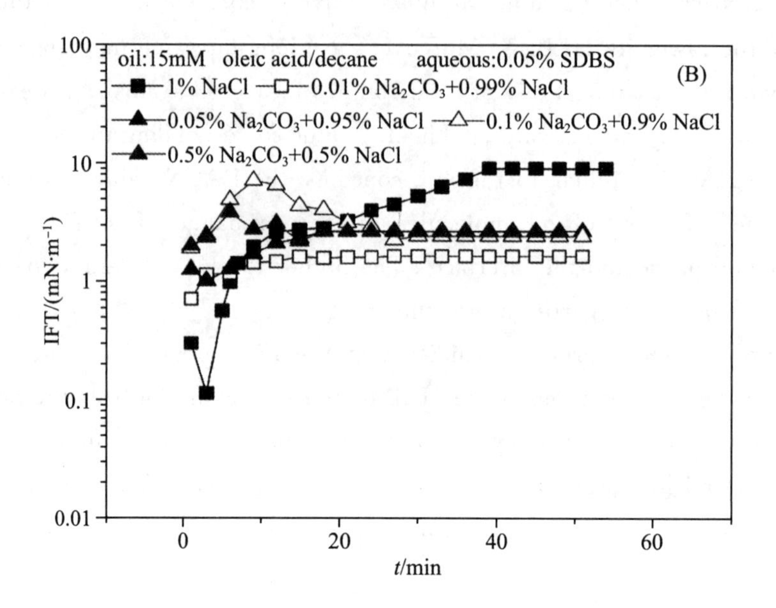
oil:15mM oleic acid/decane aqueous:0.05% SDBS
(B)
1% NaCl
0.01% Na_2CO_3+0.99% NaCl
0.05% Na_2CO_3+0.95% NaCl
0.1% Na_2CO_3+0.9% NaCl
0.5% Na_2CO_3+0.5% NaCl
IFT/(mN·m^{-1})
t/min

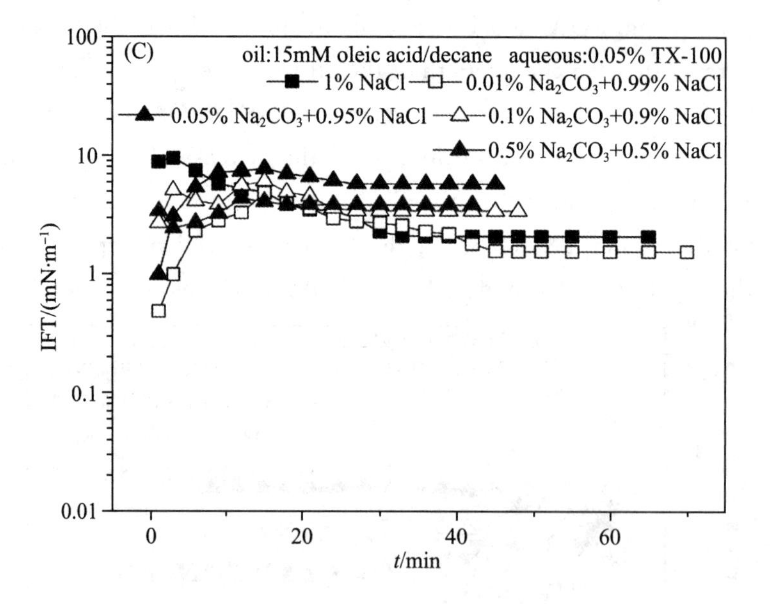
(C)
oil:15mM oleic acid/decane aqueous:0.05% TX-100
1% NaCl
0.01% Na_2CO_3+0.99% NaCl
0.05% Na_2CO_3+0.95% NaCl
0.1% Na_2CO_3+0.9% NaCl
0.5% Na_2CO_3+0.5% NaCl
IFT/(mN·m^{-1})
t/min

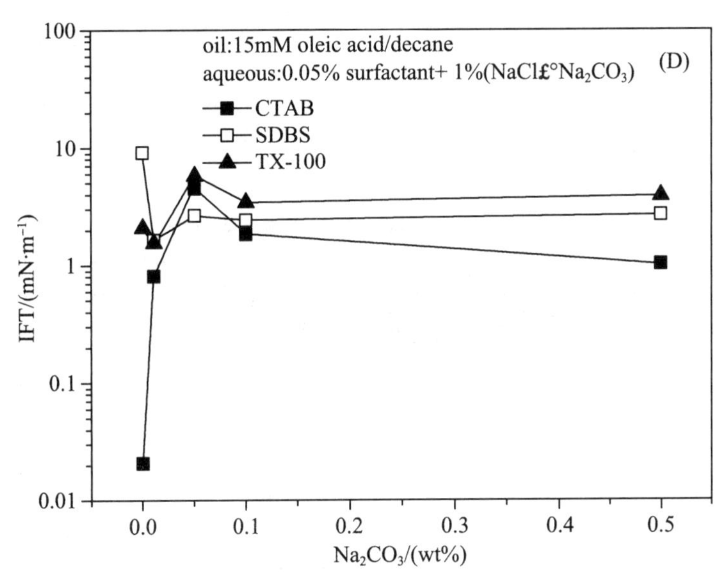

Figure 8 Dynamic IFTs of (A) CTAB, (B) SDBS, and (C) TX-100 and (D) equilibrium IFTs of the aqueous surfactant containing sodium carbonate against model oil containing 15 mM OA.

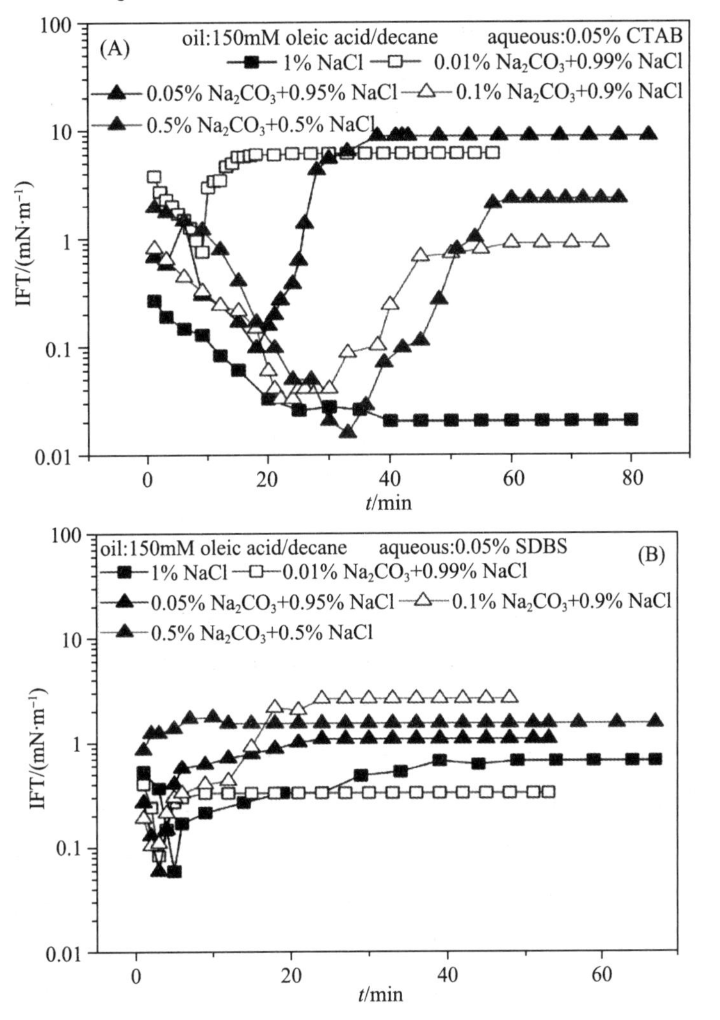

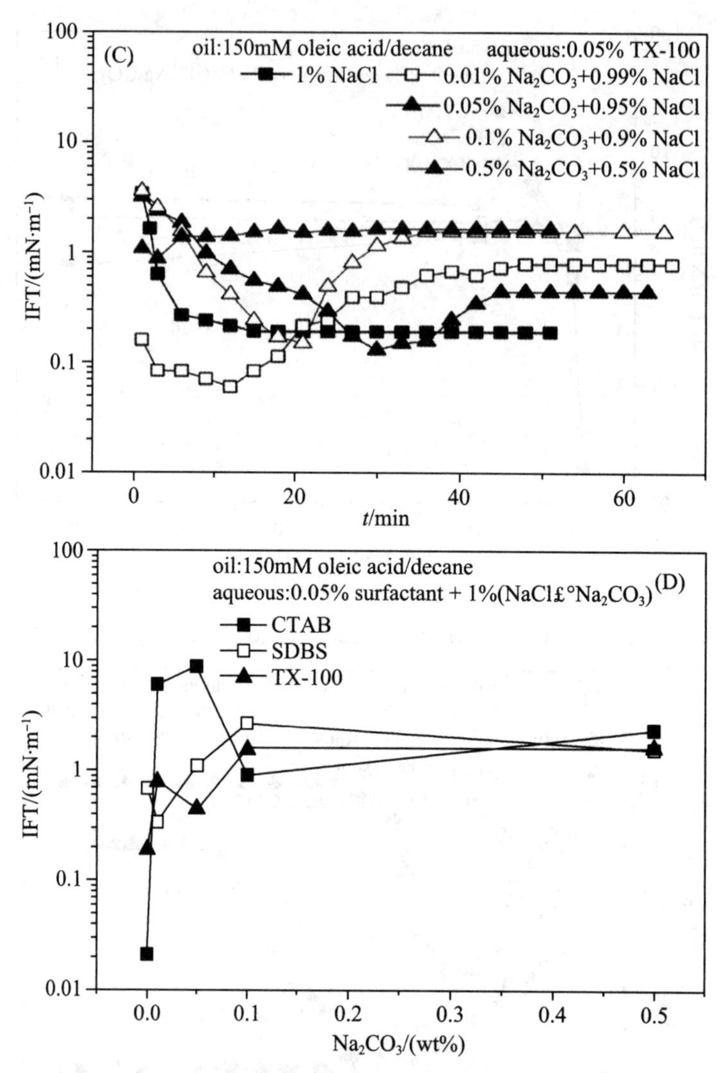

Figure 9 Dynamic IFTs of (A) CTAB, (B) SDBS, and (C) TX-100 and (D) equilibrium IFTs of the aqueous surfactant containing sodium carbonate against model oil containing 150 mM OA.

It is interesting that, for SDBS, the IFT shows a minimum for ACN=9. As we discussed above, the ratio of OA and SDBS molecules at the interface plays a crucial role in determining IFT. With the increase of ACN, the interfacial SDBS concentration decreases. The optimized ratio may appear when the oil is nonane, which results in the minimum for ACN=9.

3.3 IFTs of Acidic Model Oils against Alkaline Solutions.

3.3.1 Efect of the Alkali Concentration on the IFTs of Acidic Model Oils.

Generally, adding an alkali into many flooding systems can enhance the oil recovery e ciency significantly.[27] The injected alkali reacting with acid components, such as petroleum acids, in the reservoir crude oil can generate a kind of surface-active species called in situ soap, which has an excellent ability to reduce the IFT through a synergistic efect together with the added surfactant dramatically.[20,28-30] It is meaningful to investigate the dynamic IFTs between alkaline sloutions and n-decane model oils containing OA. The dynamic IFTs of sodium bicarbonate and sodium carbonate solutions against model oils containing di erent concentrations of OA at a total salinity of 1% have been plotted in Figure 7.

The dynamic IFT curves vary with the aging time when the OA concentrations are 15 and 150 mM in decane. When the concentrations of sodium bicarbonate and sodium carbonate are both 0.01 mass% and the OA concentration is 15 mM OA, the dynamic IFT values were high and nearly keep constant with aging time. With the increase of the alkali concentration, the dynamic IFT curves show a "V" shape. At the same time, the transient minimum IFT and equilibrium values both decrease to 1 mN/m order of magnitude. Moreover, the IFTs reach 0.1 mN/m order of magnitude when the OA concentration increases to 150 mM.

When OA molecules contact with aqueous sodium carbonate or sodium bicarbonate at once, SO will be produced immediately. SO is a kind of surface-active species, which has the ability to reduce the IFT. Then, desorption of SO from the interface to the aqueous phase will lead to dynamic IFT behaviors, such as a "V" curve.

It needs to be noticed that the transient minimum IFT value cannot reach the ultralow value ($< 10^{-2}$ mN/m) for aqueous alkaline against n-decane model oil containing both 15 and 150 mM OA, which means SO has a limited ability to reduce the IFT.

3.3.2 Efect of the Alkali Concentration on the IFTs between Acidic Model Oils and the Aqueous Surfactant.

To investigate the mechanism responsible for systems containing surfactants, OA, and SO in reducing IFT, the dynamic IFTs of aqueous CTAB, SDBS, and TX-100 containing sodium carbonate against OA model oils have been plotted in Figures 8 and 9.

We can learn from Figure 9A that the dynamic and equilibrium IFTs will increase from 0.01 mN/m order of magnitude to 1 mN/m for CTAB solution after the addition of alkali. The reason is that oil-soluble surface-active molecule OA can adsorb from the oil side to the interface and form tight mixed films to show synergism in reducing IFT. However, the anionic surfactant SO prefers to form non-active aggregates with CTAB molecules in the aqueous phase, which results in the obvious increase of IFT.

For anionic aqueous surfactant SDBS, IFTs decrease slightly with the addition of alkali, which indicates that the mixed adsorption film of SDBS and SO may be tighter than that of the SDBS and OA mixed film, as shown in Figure 8B. On the other hand, from Figure 8C, it seems that the mixed film of TX-100 and SO may be looser than that of the TX-100 and OA mixed film. Therefore, the IFTs increase slightly with the addition of alkali for 0.05% TX-100 solution.

The dynamic IFTs are controlled by the surface-active species at the interface. For 150 mM OA model oil, as shown in Figure 9, the dynamic characteristic of IFT has been enhanced and the transient IFTs decrease obviously because more SO molecules have been produced. However, the equilibrium IFTs show little change with the increase of the OA concentration in the oil at a high alkali concentration, which can be attributed to the little ratio of oil to aqueous phase.

4 CONCLUSION

This paper demonstrates the mechanism responsible for dynamic IFT of acidic model oil against three di erent types of surfactant solutions. According to the above studies, the following conclusions can be obtained: (1) With the addition of a small amount of OA to the oil, interfacial properties of CTAB solutions have drastic changes. An obvious synergistic efect on reducing IFT can be observed because the cationic surfactant CTAB and OA molecules form tight mixed adsorption films at the interface through electrostatic attraction. However, the synergistic efect will be destroyed at a high surfactant concentration by the formation of mixed micelles. (2) For the anionic surfactant SDBS system, the mixed adsorption of SDBS and OA molecules leads to a limited decrease of equilibrium IFT as a result of electrostatic repulsion. The formation of mixed micelles may be responsible for the "V" shape of dynamic IFT at bulk concentrations after cmc. (3) For nonionic surfactant TX-100, the hydrophilic group EO chain extends into the aqueous phase and the interfacial film becomes tight with an increasing bulk concentration. As a result, OA shows little efect on the equilibrium IFTs at a high TX-100 concentration. (4) With the addition of alkali to the anionic aqueous surfactant, SO will be produced. SO prefers to form non-active aggregates with CTAB molecules in the aqueous phase, which results in the obvious increase of IFT. The mixed adsorption film of SDBS and SO may be tighter than that of the SDBS and OA mixed film, and IFTs decrease slightly with the addition of alkali. On the other hand, the IFTs increase slightly with the addition of alkali for 0.05% TX-100 solution. (5) The dynamic IFTs are also controlled by the surface-active species in the oil in the presence of alkali. With an increase of the OA concentration, the dynamic characteristic of IFT has been enhanced and the transient IFTs decrease obviously because more SO molecules have been produced.

AUTHOR INFORMATION

Corresponding Authors

* Telephone: 86-10-82543587. Fax: 86-10-62554670. E-mail: luyiqiao@hotmail.com.

* Telephone: 86-10-82543587. Fax: 86-10-62554670. E-mail: zhaosui@mail.ipc.ac.cn.

ORCID

Lu Zhang: 0000-0001-8355-900X

Notes

Theauthorsdeclareno competing financial interest.

ACKNOWLEDGMENTS

The authors are thankful for the financial support from the National Science and Technology Major Project (2016ZX05011-003) and the National Natural Science Foundation of China (21703269).

References

[1] Shah, D. O. Fundamental aspects of surfactant-polymer flooding process. In Enhanced Oil Recovery; Fayers, F. J., Ed.; Elsevier: Amsterdam, Netherlands, 1981:1-42.

[2] Wilson, L. A., Jr. Physico-chemical environment of petroleum reservoirs in relation to oil recovery systems. In Improved Oil Recovery by Surfactant and Polymer Flooding; Shah, D. O., Schechter, R. S., Eds.; Academic Press: New York, 1977: 1-26. DOI: 10.1016/B978-0-12-641750-0.50005-0.

[3] Cayias, J. L.; Schechter, R. S.; Wade, W. H. Modeling Crude Oils for Low Interfacial Tension. SPEJ, Soc. Pet. Eng. J. 1976,16(6):351 -357.

[4] Fainerman, V. B.; Lucassen-Reynders, E. H. Adsorption of single and mixed ionic surfactants at fluid interfaces. Adv. Colloid Interface Sci. 2002,96(1-3): 295-323.

[5] Chan, K. S.; Shah, D. O. Molecular mechanism for achieving ultra low interfacial-tension minimum in a petroleum sulfonate-oil-brine system. J. Dispersion Sci. Technol. 1980,1(1):55-95.

[6] Zhang, L.; Luo, L.; Zhao, S.; Yu, J.-Y. Ultra low interfacial tension and interfacial dilational properties related to enhanced oil recovery. Pet. Sci. Res. Prog. 2008:81-139.

[7] Acevedo, S.; Ranaudo, M. A.; Escobar, G.; Gutierrez, X. Dynamic interfacial tension measurement of heavy crude oil-alkaline systems-The role of the counterion in the aqueous phase. Fuel 1999, 78 (3):309-317.

[8] Jennings, H. Y.; Johnson, C. E.; McAuliffe, C. D. Caustic waterflooding process for heavy oils. JPT, J. Pet. Technol. 1974,26(DEC):1344-1352.

[9] Zhao, Z.; Liu, F.; Qiao, W.; Li, Z.; Cheng, L. Novel alkyl methylnaphthalene sulfonate surfactants: A good candidate for enhanced oil recovery. Fuel 2006,85(12-13):1815-1820.

[10] Sharma, H.; Dufour, S.; Arachchilage, G.; Weerasooriya, U.; Pope, G. A.; Mohanty, K. Alternative alkalis for ASP flooding in anhydrite containing oil reservoirs. Fuel 2015,140:407-420.

[11] Zhang, L.; Luo, L.; Zhao, S.; Yu, J. Y. Studies of synergism/antagonism for lowering dynamic interfacial tensions in surfactant/ alkali/acidic oil systems Part 1: Synergism/antagonism in surfactant/ model oil systems. J. Colloid Interface Sci. 2002, 249(1):187-193.

[12] Touhami, Y.; Hornof, V.; Neale, G. H. Dynamic interfacial tension behavior of acidified oil surfactant-enhanced alkaline systems 1. Experimental studies. Colloids Surf., A 1998,132(1):61-74.

[13] Amaya, J.; Rana, D.; Hornof, V. Dynamic interfacial tension behavior of water/oil systems containing in situ-formed surfactants. J. Solution Chem. 2002,31 (2):139-148.

[14] Zhao, R.-h.; Huang, H.-y.; Wang, H.-y.; Zhang, J.-c.; Zhang, L.; Zhang, L.; Zhao, S. Effect of Organic Additives and Crude Oil Fractions on Interfacial Tensions of Alkylbenzene Sulfonates. J. Dispersion Sci. Technol. 2013,34 (5):623-631.

[15] Zhou, Z.-H.; Zhang, Q.; Liu, Y.; Wang, H.-Z.; Cai, H.-Y.; Zhang, F.; Tian, M.-Z.; Liu, Z.-Y.; Zhang, L.; Zhang, L. Effect of Fatty Acids on Interfacial Tensions of Novel Sulfobetaines Solutions. Energy Fuels 2014,28(2):1020-1027.

[16] Zhu, Y.-w.; Zhao, R.-h.; Jin, Z.-q.; Zhang, L.; Zhang, L.; Luo, L.; Zhao, S. Influence of Crude Oil Fractions on Interfacial Tension of Alkylbenzene Sulfonate Solutions. Energy Fuels 2013,27(8):4648-4653.

[17] Cao, J.-H.; Zhou, Z.-H.; Xu, Z.-C.; Zhang, Q.; Li, S.-H.; Cui, H.-B.; Zhang, L.; Zhang, L. Synergism/Antagonism between Crude Oil Fractions and Novel Betaine Solutions in Reducing Interfacial Tension. Energy Fuels 2016,30 (2):924-932.

[18] Zhang, L.; Luo, L.; Zhao, S.; Xu, Z.-C.; An, J.-Y.; Yu, J.-Y. Effect of different acidic fractions in crude oil on dynamic interfacial tensions in surfactant/ alkali/model oil systems. J. Pet. Sci. Eng. 2004,41:189-198.

[19] Zhang, L.; Luo, L.; Zhao, S.; Yang, B.; Yu, J. Studies of synergism/antagonism for lowering dynamic interfacial tensions in surfactant/alkali/acidic oil systems Part 3: Synergism/antagonism in surfactant/alkali/acidic model systems. J. Colloid Interface Sci. 2003,260:398-403.

[20] Chu, Y.-P.; Gong, Y.; Tan, X.-L.; Zhang, L.; Zhao, S.; An, J.-Y.; Yu, J.-Y. Studies of synergism for lowering dynamic interfacial tension in sodium α-(n-alkyl) naphthalene sulfonate/alkali/acidic oil systems. J. Colloid Interface Sci. 2004,276:182-187.

[21] Zhang, H.; Dong, M.; Zhao, S. Experimental Study of the Interaction between NaOH, Surfactant, and Polymer in Reducing Court Heavy Oil/Brine Interfacial Tension. Energy Fuels 2012,26(6):3644-3650.

[22] Hadji, M.; Al-Rubkhi, A.; Al-Maamari, R. S.; Aoudia, M. Surfactant (in situ)-Surfactant (Synthetic) Interaction in Na_2CO_3/ Surfactant/Acidic Oil Systems for Enhanced Oil Recovery: Its Contribution to Dynamic Interfacial Tension Behavior. J. Surfactants Deterg. 2015,18(5):761-771.

[23] Liu, Z.; Zhang, L.; Cao, X.; Song, X.; Jin, Z.; Zhang, L.; Zhao, S. Effect of Electrolytes on Interfacial Tensions of Alkyl Ether Carboxylate Solutions. Energy Fuels 2013,27(6):3122-3129.

[24] Zhao, R.-h.; Zhang, L.; Zhang, L.; Zhao, S.; Yu, J.-y. Effect of the hydrophilic-lipophilic Ability on Dynamic Interfacial Tensions of Alkylbenzene Sulfonates. Energy Fuels 2010,24:5048-5052.

[25] Guo, K.; Li, H.; Yu, Z. In-situ heavy and extra-heavy oil recovery: A review. Fuel 2016,185:886-902.

[26] Wang, C.; Cao, X.-L.; Guo, L.-L.; Xu, Z.-C.; Zhang, L.; Gong, Q.-T.; Zhang, L.; Zhao, S. Effect of adsorption of catanionic surfactant mixtures on wettability of quartz surface. Colloids Surf., A 2016,509:564-573.

[27] Yuan, F.-Q.; Cheng, Y.-Q.; Wang, H.-Y.; Xu, Z.-C.; Zhang, L.; Zhang, L.; Zhao, S. Effect of organic alkali on interfacial tensions of surfactant solutions against crude oils. Colloids Surf., A 2015,470:171-178.

[28] Bai, Y.; Shang, X.; Zhao, X.; Xiong, C.; Wang, Z. Effects of a Novel Organic Alkali on the Interfacial Tension and Emulsification Behaviors Between Crude Oil and Water. J. Dispersion Sci. Technol. 2014,35(8):1126-1134.

[29] Demirbas, A.; Alsulami, H. E.; Hassanein, W. S. Utilization of Surfactant Flooding Processes for Enhanced Oil Recovery (EOR). Pet. Sci. Technol. 2015,33(12):1331-1339.

[30] Chen, L.; Zhang, G.; Ge, J.; Jiang, P.; Tang, J.; Liu, Y. Research of the heavy oil displacement mechanism by using alkaline/surfactant flooding system. Colloids Surf., A 2013,434:63-71.

Interfacial dilational properties of polyether demulsifiers: eect ofx branching

Zhou Wangwang[a] Cao Xulong[b] Guo Lanlei[b] Zhang Lei[c]
Zhu Yan[a,**] Zhang Lu[a,c,*]

[a] School of Chemistry, Chemical Engineering and Life Sciences, Wuhan University of Technology, Wuhan 430070, PR China

[b] Exploration & Development Research Institute of Shengli Oilfield Co. Ltd, SINOPEC, Dongying 257015, Shandong, PR China

[c] Technical Institute of Physics and Chemistry, Chinese Academy of Sciences, Beijing 100190, PR China

Graphical Abstract: The linear and branched demulsifiers act well for diluted crude oils with higher and lower viscosities respectively.

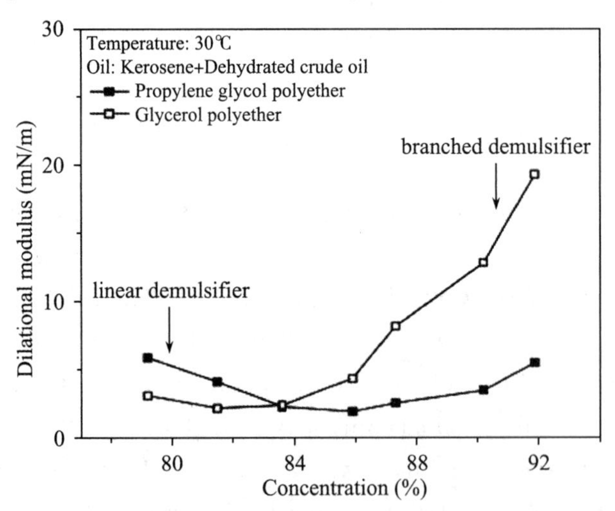

Keywords: Dilational rheological Polyether demulsifi ers Dehydrated crude oil Demulsifi cation

Abstract: The dilational rheology of two di erent types of polyether demulsifiers solutions, linear propylene glycol polyether and branched glycerol polyether, at the water-crude oil interface have been investigated by dilational rheological measurements. This paper research that except the concentration of 10 ppm, linear propylene glycol polyether appears higher moduli than those of branched polyether because branched one has shorter chain at the same molecular weight, which will result in the faster di usion. However, at optimized concentration, the modulus of branched one is higher than that of linear one, which can be attributed to the stronger interactions between branched chains at interface. The linear demulsifier propylene glycol polyether shows stronger ability of reducing modulus for diluted crude oil with higher viscosity. On the contrary, branched demulsifier glycerol polyether acts well when concentrations of diluted crude oils are lower. The reason is that linear demulsifier has the stronger ability of adsorption and branched one shows the stronger ability of replacement of natural active fractions.

1 Introduction

It is well known that the crude oil emulsions contain a large number of natural surfactants, which will form a directional monolayer film at the interface to stable dispersive droplet, so that the crude oil emulsions have a higher stability[1-4]. However, a large amount of crude oil emulsions will increase the storage and transportation load and reduce the service life of the pipeline. Therefore, the chemical emulsions breaking method using demulsifiers have emerged[5,6]. The decisive factor in the emulsion breaking is to change the properties of the in-terfacial film, in other words, to reduce the strength of the film[7-10].

The measurements of interfacial dilational properties are very close to the actual process of droplet coalescences and can be used to reflect the strength of the film[11-15]. Feng et al.[16] studied the interfacial rheology of demulsifiers with di erent structures and concluded that interfacial dilational viscoelasticity could represent the interfacial behavior of demulsifiers with an increase of bulk concentration. Demulsifiers with di erent structures have inconsistent disruption of the original interfacial film. Wang et al.[17] considered the interfacial viscoelastic properties of the linear and the branched demulsifiers against Iranian crude oil using a small oscillatory method and a relaxation method. It was concluded that both demulsifiers could sig-nificantly reduce the dilational viscoelasticity of the crude oil-water interface containing active component. At the same time, as the con-centrations of the demulsifiers increased, the dilation modulus passes a maximum value. Tao et al.[18] researched the relationships between oil-water interfacial films of 12 polyether demulsifiers and the demulsification efects. The results showed that the elastic modulus is a key factor in controlling the dehydration rate. When the demulsifier reduces the elastic modulus to a certain extent (less than 5 mN/m), the emulsion showed a fairly high dehydration rate. Xie et al.[19] studied dilational rheological properties of nine polyether demulsifiers adsorbed to the oil-water interface using a small periodic oscillation method. The results showed that the concentration is an important factor to control the dilational properties of the adsorption layers. For all of these demulsifiers, increasing their concentrations, the dilational viscoelasticity will occur at a maximum.

In this paper, the efects of polyether demulsifiers with straight-chain and branched-chain structures have been studied. The study aims to expound the role of branching in controlling interfacial properties and demulsification of polyether demulsifiers, which is very useful for understanding the mechanisms responsible for the stability and demulsification of crude oil emulsions.

2 Theory

The interfacial tension (IFT) will periodically change with the periodic disturbance of the disturbed interface. Dilational modulus[20] is defined as the ratio of the change in IFT γ to the change in interface area (A), that is,

$$\varepsilon=\frac{d\gamma}{d\ln A}$$

θ is defined as phase angle. For the viscoelastic interface, the phase angle is calculated from the di erence between the periodic change of the interface area and the response of interfacial tension.

$$\theta=\arctan\left\{\frac{\omega\eta_d}{\varepsilon_d}\right\}$$

ω is defined as the frequency of the interface area change, $\omega\eta_d$ represents the viscous part of the modulus, and ε_d represents the elastic part of the modulus. Therefore, the dilational modulus can be written as:

$$\varepsilon=\varepsilon d + i\omega\eta_d$$

3 Experiment methods

3.1 Materials

Propylene glycol polyether and glycerol polyether were synthesized in our laboratory, with a linear and branched molecular structure respectively. Both polyether demulsifiers are two-block polymers with the molecular weight about 8 000 and mole ratio of EO : PO=1 : 3 estimated by the feed ratio. The values of the cmc, and CP (the cloud point) of 1% aqueous are listed in Table 1. Kerosene used in this paper was purified through glass chromatography column containing silica gel layer and the IFT was about 40 mN/m against pure water. Dehydrated crude oil was exploited from Shengli Oil field the properties of crude oil were listed in Table 2. All of the aqueous solutions used in this study were prepared with formation brine. Demulsifier was first dissolved in alcohol to 1% and then diluted to wanted concentrations by formation brine. The composition of the formation brine was listed in Table 3. Viscosities of di erent concentration dehydrated crude oils diluted by kerosene were listed in Table 4.

Table 1 The cmc and CP of polyether demulsifiers

Demulsifier	cmc mol/L	CP℃
propylene glycol polyether	4.5×10^{-7}	8
glycerol polyether	2.2×10^{-6}	11

3.2 IFT measurements

J2000BW spinning drop interfacial tensiometer (DT Shanghai CO.) was used to measure the dynamic IFTs. The interfacial tension meter is controlled by a menu startup program that connects the computer to a video capture device to observe changes in droplet morphology and to connect a heating sleeve control system temperature[21,22]. The glass tube is filled with an oil phase and an aqueous phase. The aqueous phase envelops the oil phase. The aqueous phase is approximately 0.27 mL of demulsifier solution, and the oil phase is injected with approximately 1 μl. By measuring the change of the length and width of the oil droplet, one can calculate the interfacial tension value of the system. Samples were assumed to be equilibrated when measured values of IFT remained unchanged for half an hour. In all cases, the measurements of the IFT were performed at a rotating velocity 5 000 rpm and the standard deviation did not exceed 5%[23].

3.3 Dilational rheological experiments

In this study, the interfacial dilational viscoelasticity meter JMP2000A (Powereach Ltd., Shanghai, China) was employed. The dehydrated crude oil was dilute to a series of concentration using kerosene as model oil. The oil phase is the dehydrated crude oil and kerosene, and the water phase is demulisifiers with di erent concentrations.

To probe the dilational viscoelasticity of the oil-water interfacial film, first in the Langmuir tank, 90 mL of the water phase and 90 mL of the oil phase are injected[24]. The oil phase is in the upper layer and the water phase is in the lower layer. After a certain period of pre-equilibrium, the interface region is expanded and compressed by a sinusoidal oscillating period (0.005—0.1 Hz) through the horizontal sliding of the barrier. The accuracy of the dilational modulus is ±2 mN/m.

4 Results and discussion

4.1 Equilibrium interfacial tensions

The measurements of equilibrium IFT versus bulk concentration isotherms of polyether demulsifiers at kerosene-water interface are presented in Fig. 1. One can see from Fig. 1 that the equilibrium IFTs of two di erent polyether demulsifiers are both decreasing during the experiment concentrations and can produce IFT of 0.1mN/m order of magnitude at low bulk concentrations, which indicates both demulsifiers have higher surface-activity and can easily adsorb onto the interface to modify the nature of film. Moreover, the IFT values of propylene glycol polyether are lower than those of glycerol polyether in general.

Table 2 Density and viscosity of the dehydrated crude oil

Temperature (℃)	50	55	60	65	70	75	80
Density(g/cm[3])	0.950 3	0.948 2	0.944 9	0.942 9	0.939 5	0.937 0	0.934 4
Viscosity (mPa s)	1 590	1 190	708	510	350	271	209

Table 3 Composition analysis of the formation brine

Components	Ca^{2+}	Mg^{2+}	SO_4^{2-}	Mg^{2+}	HCO_3^-	$Na^+ + K^+$	Cl^-
Concentration	784.93	252.15	553.31	252.15	421.27	4 360.62	8 191.50

Table 4 Viscosity of kerosene diluted Dehydrated crude oil

Dehydrated crude oil (mass%)	79.2	81.5	83.6	85.9	87.3	90.2	91.9
Viscosity (mPa・s)	209	271	350	510	708	1 190	1 590

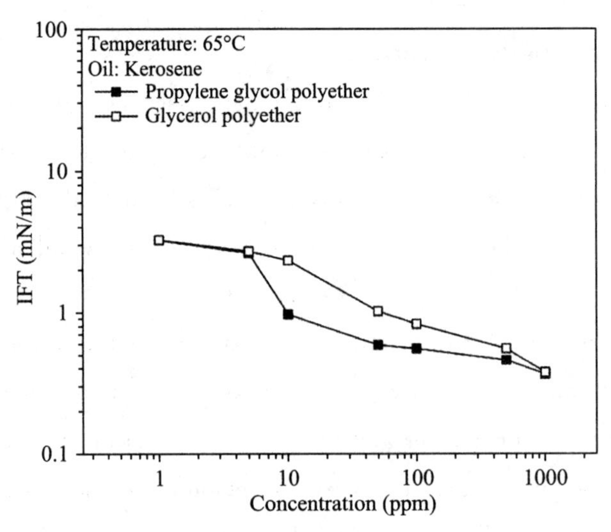

Fig. 1 Equilibrium IFTs as a function of concentration for propylene glycol polyether and glycerol polyether in the aqueous.

The reason is that the larger molecular size of branched glycerol polyether will result in the looser arrangement of adsorbed molecules and the increase of IFT.

4.2 Interfacial dilational properties

4.2.1 Influence of oscillating frequency on interfacial dilational properties of demulsifiers

Oscillating frequency is an essential factor to a ect the dilational viscoelastic parameters. To study the frequency dependence of the polyether demulsifier, the measured frequency ranged from 0.005 to 0.1 Hz. The frequency dependences of demulsifiers propylene glycol polyether and glycerol polyether in the range of experiment concentrations are presented in Figs. 2 and 3, respectively.

It can be observed from Figs. 1 and 2 that the dilational modulus of propylene glycol polyether and glycerol polyether at kerosene-water interface all increase gradually with the increase of oscillating frequency. When the rate of expansion of the barrier is very slow, the interfacial film has enough time to deal with the deformation of the area. Therefore, the resistance of compression-expansion is small during relaxation in the layer. In this situation, the IFT changes little. However, the interfacial film has no time to deal with the deformation of the area and turns out to be insoluble film and resulting in a higher dilational modulus at higher frequency. In addition, the phase angle of propylene glycol polyether and glycerol polyether at kerosene-water interface all reduce gradually as the frequency of oscillating increases and can always keep it at a smaller value (lower than 25°). All these indicate that the interfacial film is predominantly elastic per se. This is due to the fact that the polyether demulsifiers have long hydrophilic (EO) and hydrophobic (PO) chains, which may partly flat at the interface and lead to the slower di usion-exchange process and the stronger molecular interaction.

Referring to the literature, the $\log|\varepsilon|$-$\log\omega$ curve of the demulsifier was found to be quasi-linear, indicating that the relaxation process near the interface adsorption layer was higher than the highest frequency (0.1 Hz) of this experiment. Therefore, the slope value of this curve can quantitatively characterize the relationship between frequency and modulus and reflect the viscoelasticity of the adsorbed membrane[25,26].

Fig. 4 shows the tendency about the slope of $\log|\varepsilon|$-$\log\omega$ with increasing demulsifier concentration at the water-kerosene interface. During experiment concentrations, it demonstrates that the slopes of $\log|\varepsilon|$ versus $\log\omega$ decrease gradually for both demulsifiers. It is known that the lower the slope value, the more elastic the interfacial film is[27]. As bulk concentration with increases, the adsorption film formed by demulsifier molecules becomes tighter, which enhances the interfacial interactions and improves the elastic nature of film. Moreover, the branching of hydrophilic EO chain of glycerol polyether benefits the molecular interactions at the interface and makes the slopes of glycerol polyether be lower than those of propylene glycol polyether. It is very interesting to notice that linear propylene glycol polyether can form tighter adsorption film because of its smaller molecular size and provide lower IFT as shown in Fig. 1. However, the strength of film mainly controlled by the molecular interactions and the rigidity of branched chains can strengthen the elasticity of the film.

4.2.2 Influence of bulk concentration on interfacial dilational properties of demulsifiers

Bulk concentration is a key factor when chemical agents are applied in practice. For interfacial dilational modulus, there are generally two di erent trends of variations with an increase of demulsifier concentration. The increase in the amount of adsorption can enhance the interfacial interaction and improve the dilational modulus. On the contrary, it also can enhance the exchange of demulsifier molecules between the interface and the bulk, which will reduce the interfacial tension change. So there can be a point where the optimal concentration reaches the maximum value of the dilational modulus. The di usion of demulsifier molecules is mainly determined by their own properties.

Fig. 5 shows the influence of bulk concentration on dilational modulus and phase angle for demulsifiers propylene glycol polyether and glycerol polyether at the water-kerosene interface under an oscillating frequency of 0.1 Hz. It can be seen that the interfacial dilational moduli of both demulsifier increase at first and then decrease, and there is a maximum at 10 ppm for two demulsifiers. The maximum of mod-ulus of propylene glycol polyether is slightly lower than that of glycerol polyether, which indicates the stronger intermolecular interactions of branched polyether. However, at other bulk concentration, linear propylene glycol polyether appears higher moduli than those of branched polyether because branched one has shorter chain at the same molecular weight, which will result in the faster di usion and reduce modulus. Moreover, during all experimental concentration ranges, the phase angles of two demulsifiers keep low values from 0 to 10°, which indicated a dominant elastic nature of film, because oscillating frequency of 0.1 Hz is higher than the characteristic time of di usion-exchange process for employed demulsifiers[28]. It also must be pointed out that the maximum value of two demulsifiers is lower than 10 mN/m, which means the low intensity of demulsifier films. Therefore, the adsorption of polyether demulsifier molecules at water-crude oil interface can replace natural surface-active molecules and destroy the rigid interfacial film.

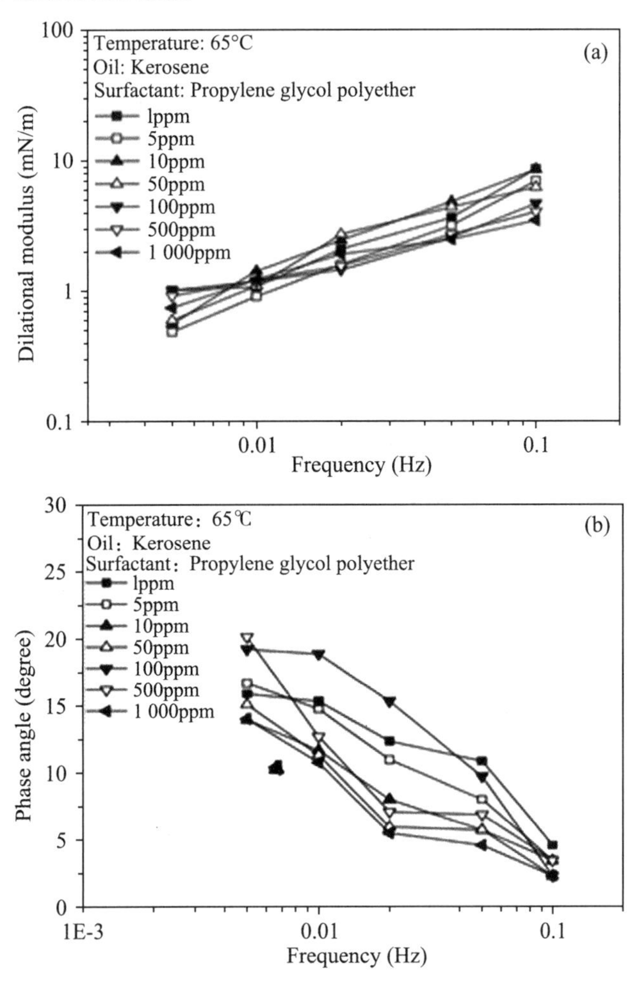

Fig. 2 Influence of the oscillating frequency on (a) dilational modulus and (b) phase angle of propylene glycol polyether at water-kerosene interface.

4.2.3 Influence of oscillating frequency on interfacial dilational properties of diluted crude oils

We know from the literature that the dilational modulus of crude oil sample was highly dependent on crude oil concentrations and its solvent. Aske et al.[29] found when the crude oils were diluted by di erent concentrations of heptane/toluene solvents, the elastic moduli of crude oils varied from 0.5 to 32.9 mN/m. Feng et al.[16] found that the dilational modulus of dilute crude oil at 0.1 Hz frequency were in the range of 3—23 mN/m, and both the storage and loss modulus passed through maximum values within the dilute crude oil concentration ranges.

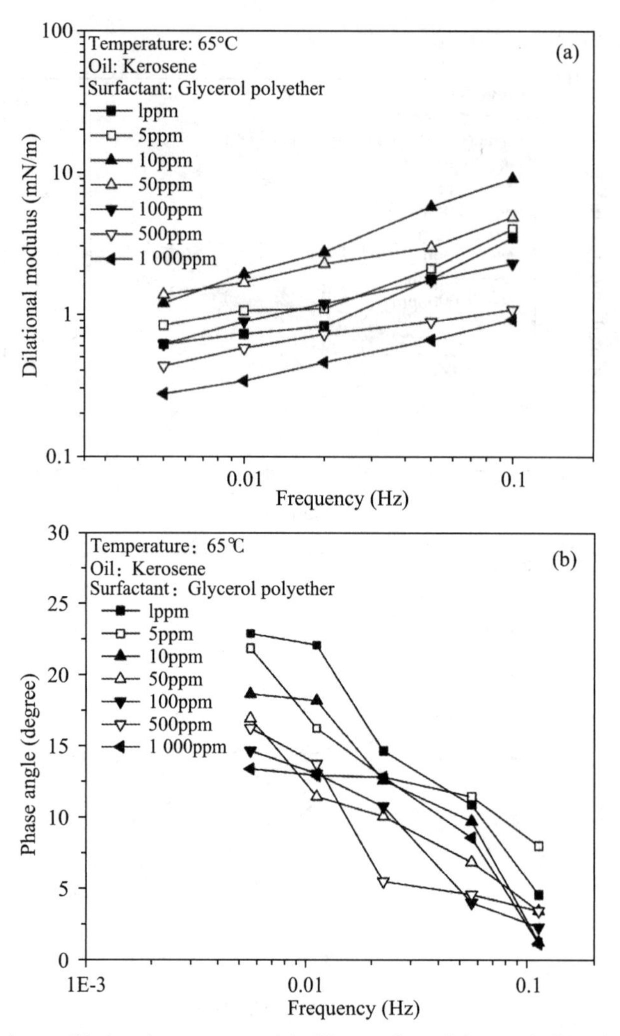

Fig. 3 Influence of the oscillating frquency on (a) dilational modulus and (b) phase angle of glycerol polyether at water-kerosene interface

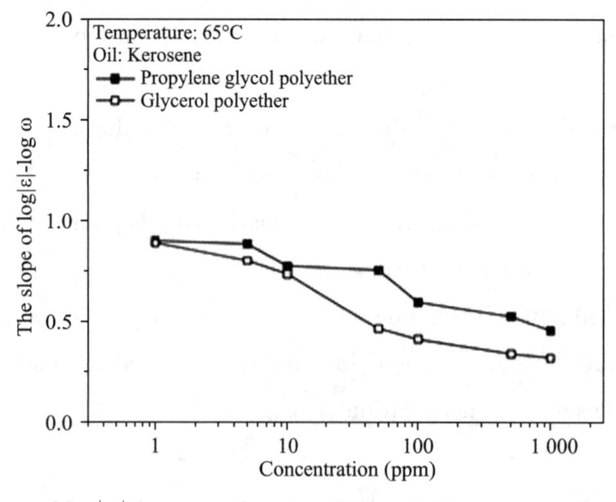

Fig. 4 The slopes of log|ε|-logω as a function of concentration at water-kerosene interfaces.

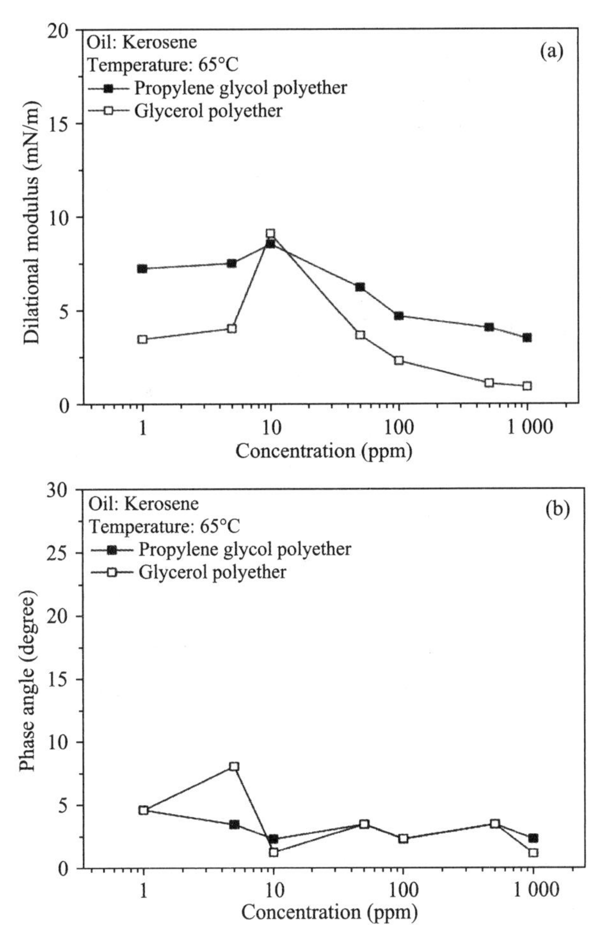

Fig. 5　Influence of bulk concentration on (a) dilational modulus and (b) phase angle of propylene glycol polyether and glycerol polyether at frequency of 0.1 Hz.

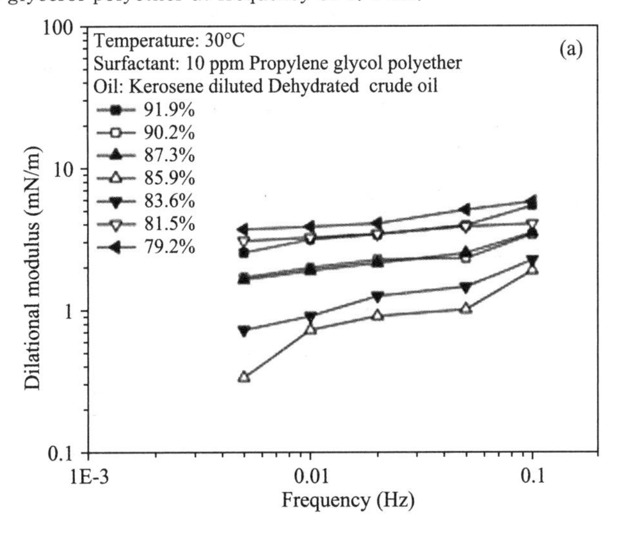

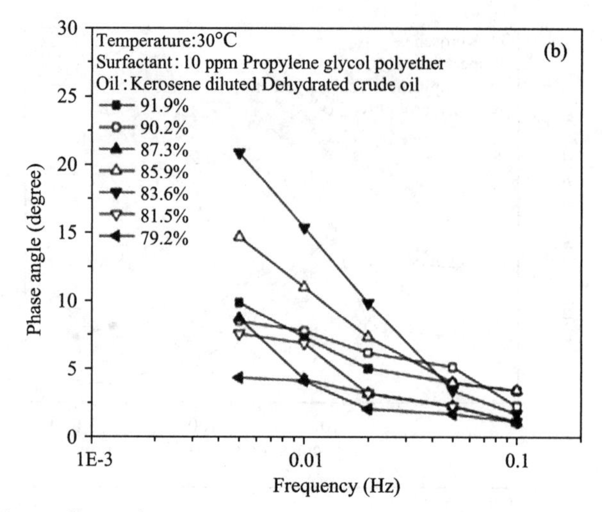

Fig. 6 Influence of the oscillating frequency on (a) dilational modulus and (b) phase angle of propylene glycol polyether at water-diluted crude oil interface.

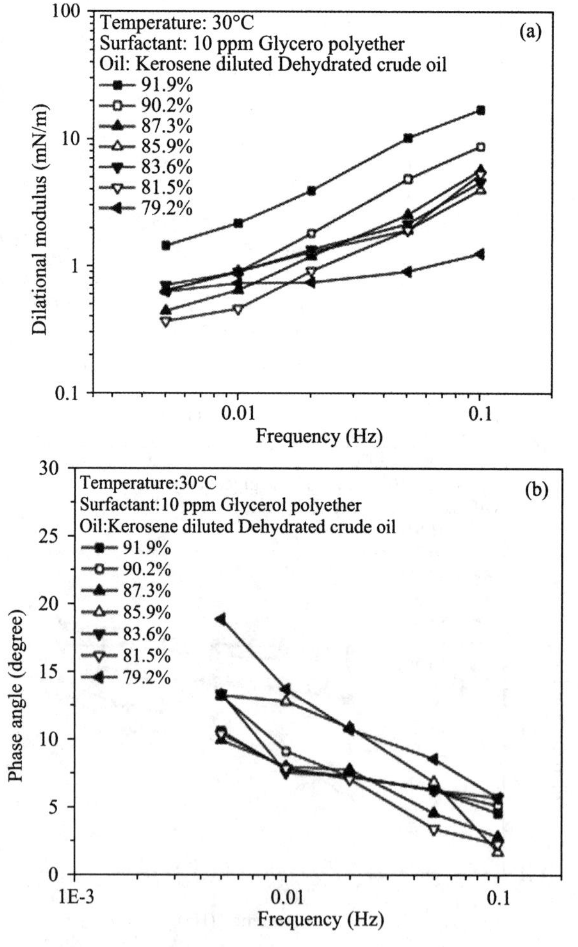

Fig. 7 Influence of the oscillating frequency on (a) dilational modulus and (b) phase angle of glycerol polyether at water-diluted crude oil interface.

We have studied the dilational properties of propylene glycol polyether and glycerol polyether at di erent diluted crude oil-water interfaces. Figs. 6 and 7 illustrate the influence of oscillating frequency on dilational moduli of 91. 9%, 90. 2%, 87. 3%, 85. 9%, 83. 6%, 81. 5% and 79. 2% dehydrated crude oil. Dehydrated crude oil at di erent temperatures(50℃, 55℃, 60℃, 65℃, 70℃, 75℃, 80℃) and diluted crude oil with di erent concentrations at 30℃ (91. 9%, 90. 2%, 87. 3%, 85. 9%, 83. 6%, 81. 5%, 79. 2%) have the same viscosities. As the oscil-lating frequency increases, the variations of dilational modulus and phase angle are similar to those of demulsifiers because the natural surface-active fractions in crude oils play the same role as surfactant. We don't discuss them in detail for brevity.

Fig. 8 illustrates that the influences of propylene glycol polyether and glycerol polyether on interfacial dilational modulus and phase angle of di erent diluted crude oil at 0. 1 Hz. It is very interesting that the linear demulsifier propylene glycol polyether shows stronger ability of reducing modulus and destroying crude oil film when concentrations of diluted crude oils are higher. On the other hand, branched demulsifier glycerol polyether acts well when concentrations of diluted crude oils are lower. It can be speculated that polyether molecules can substitute the natural active molecules of crude oil, which is mainly due to the stronger ability of adsorption.

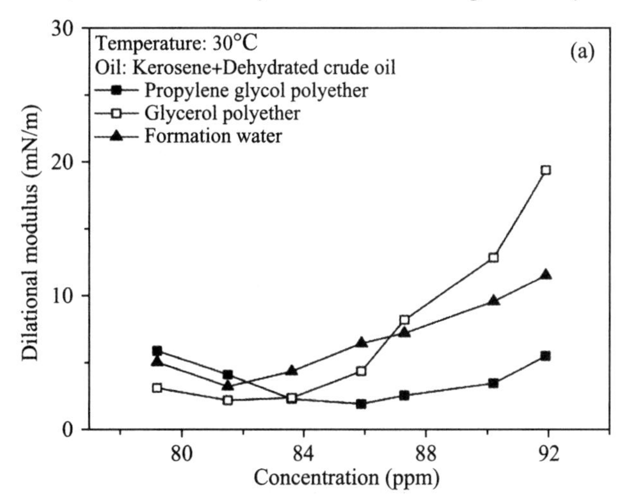

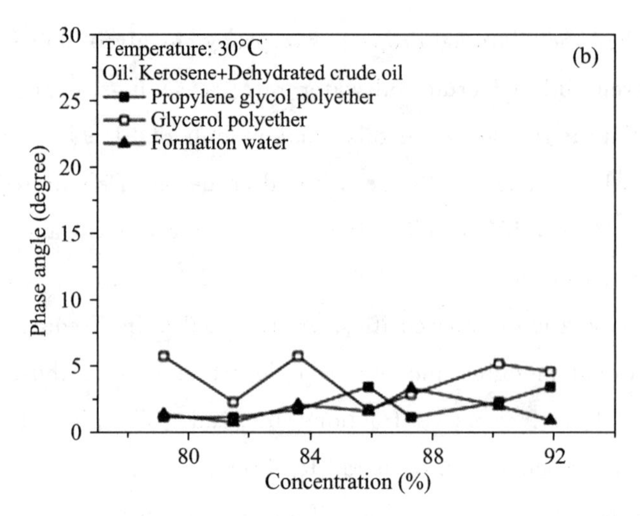

Fig. 8 Influence of demulsifiers on (a) dilational modulus and (b) phase angle of di erent diluted crude oil at 0.1Hz.

As we discussed above, linear demulsifier has the stronger ability of adsorption and branched one shows the stronger ability of replacement of natural active fractions. For rigid films of higher crude oil concentration systems, the ability of adsorption determines the efectiveness of demulsifier. On the other hand, for flexible films of lower crude oil concentration systems, the ability of replacement controls the e ciency of demulsifier.

5 Conclusion

This paper studies the mechanism responsible for interfacial behaviors of crude oil and two di erent types of demulsifier solutions. According to the experimental results, the following conclusions can be obtained:

(1) During all the range of experimental concentrations, the IFT values of propylene glycol polyether are lower than those of glycerol polyether because the larger molecular size of branched glycerol polyether will result in the looser arrangement of adsorbed molecules.

(2) Except the concentration of 10 ppm, linear propylene glycol polyether appears higher moduli than those of branched polyether because branched one has shorter chain at the same molecular weight, which will result in the faster di usion and reduce modulus. However, at optimized concentration, the modulus of branched one is higher than that of linear one, which can be attributed to the stronger interactions between branched chains at interface.

(3) The linear demulsifier propylene glycol polyether shows stronger ability of reducing

modulus and destroying crude oil film for high viscosity system. On the other hand, branched demulsifier glycerol polyether acts well when concentrations of diluted crude oils are lower. The reason is that linear demulsifier has the stronger ability of adsorption and branched one shows the stronger ability of replacement of natural active fractions.

Acknowledgments

The authors thank financial support from the National Science & Technology Major Project (2016ZX05011-003) and the National Natural Science Foundation (21703269) of China.

References

[1] A. Bhardwaj, S. Hartland, Studies on the build up of interfacial film at the crude oil/water interface, J. Dispers. Sci. Technol. 1998(19):465-473.

[2] T. Sun, L. Zhang, Y. Wang, B. Peng, S. Zhao, M. Li, J. Yu, Influence of demulsifiers with di erent structure on dilational viscoelasticity of interfacial film containing surface active fractions from crude oil, J. Colloid Interface Sci. 2002(255): 241-247.

[3] M. H. Ese, J. Sjo¨Blom, H. Førdedal, O. Urdahl, H. P. Rønningsen, Ageing of inter-facially active components and its efect on emulsion stability as studied by means of high voltage dielectric spectroscopy measurements, Colloids Surf. APhysicochem. Eng. Asp. 1997(123-124):225-232.

[4] Y. Y. Wang, L. Zhang, T. L. Sun, H. B. Fang, S. Zhao, J. Y. Yu, Influences of demulsi-fiers with di erent structures on dilational properties of decane-water interface by interfacial tension relaxation methods, Acta Phys.-Chim. Sin. 2003 (153):S337-S338.

[5] H. B. Fang, L. Zhang, L. Luo, S. Zhao, J. Y. Yu, Z. C. Xu, J. Y. An, A. Ottva, H. T. Tien, A study of thin liquid films as related to the stability of crude oil emulsions, J. Colloid Interface Sci. 2001(238):177-182.

[6] C. Cao, J. Lei, L. Zhang, F.-P. Du, Equilibrium and dynamic interfacial properties of protein/ionic-liquid-tyype surfactant solutions at the decane/water interface, Langmuir 2014(30):13744-13753.

[7] P. D. Berger, C. Hsu, J. P. Arendell, Designing and selecting demulsifiers for optimum field performance on the basis of production fluid characteristics, Spe Prod. Eng. 1988(3):522-526.

[8] Y. H. Kim, D. T. Wasan, Efect of demulsifier partitioning on the

destabilization of water-in-oil emulsions, Ind. Eng. Chem. Res. 1996(35):1141-1149.

[9] X. Cao, Y. Li, S. Jiang, H. Sun, A. Cagna, L. Dou, A study of dilational rheological properties of polymers at interfaces, J. Colloid Interface Sci. 2004(270):295-298.

[10] C. Wang, X.-L. Cao, L.-L. Guo, Z.-C. Xu, L. Zhang, Q.-T. Gong, L. Zhang, S. Zhao, Efect of molecular structure of catanionic surfactant mixtures on their interfacial properties, Colloids Surf. 2016(509):601-612.

[11] Y. Wang, Y. Dai, L. Zhang, L. Luo, S. Zhao, M. Li, E. Wang, J. Yu, Hydrophobically modified associating polyacrylamide solutions: relaxation processes and dilational properties at oil-water interface, Macromolecules 2004(37):2930-2937.

[12] B. S. Murray, B. Cattin, E. S. ler, Z. O. Sonmez, Response of adsorbed protein films to rapid expansion, Langmuir 2002(18):9476-9484.

[13] E. P. Kalogianni, E. M. Varka, T. D. Karapantsios, M. Kostoglou, E. Santini, L. Liggieri, F. Ravera, A multi-probe non-intrusive electrical technique for monitoring emulsi-fication of hexane-in-water with the emulsifier C10E5 soluble in both phases, Colloids Surf. A-Physicochem. Eng. Asp. 2010(354):353-363.

[14] J. Maldonado-Valderrama, J. M. R. Patino, Interfacial rheology of protein-surfactant.

[15] L. Dong, Z. Li, X. Cao, X. Song, L. Zhang, Z. Xu, L. Zhang, S. Zhao, Dilational rheological properties of p-(n-alkyl)-benzyl polyoxyethylene ether carboxybetaine at water-decane interface, J. Dispers. Sci. Technol. 2014(36):430-440.

[16] J. Feng, H.-B. Fang, H. Zong, L. Zhang, X.-P. Liu, L. Zhang, S. Zhao, J.-Y. Yu, Efect of demulsifiers on dilatational properties of crude oil-water interfaces, J. Dispers. Sci. Technol. 2012(33):24-31.

[17] Y. Wang, L. Zhang, T. Sun, S. Zhao, J. Yu, A study of interfacial dilational properties of two di erent structure demulsifiers at oil-water interfaces, J. Colloid Interface Sci. 2004(270):163-170.

[18] J. Tao, P. Shi, S. Fang, K. Li, H. Zhang, M. Duan, Efect of rheology properties of oil/water interface on demulsification of crude oil emulsions, Ind. Eng. Chem. Res. 2015(54):4851-4860.

[19] Y. J. Xie, F. Yan, J. X. Li, Interfacial Dilational properties of novel crosslinking phenol-amine resin block polyether demulsifiers at decane-water interfaces, Appl. Mech. Mater. 2012(148-149):202-205.

[20] R. Miller, G. Loglio, U. Tesei, K. H. Schano, Surface relaxations as a tool for studying dynamic interfacial behavior, Adv. Colloid Interface Sci. 1991(37):73-96.

[21] D.-D. Zheng, Z.-H. Zhou, Q. Zhang, L. Zhang, Y. Zhu, L. Zhang, Efect

of inorganic alkalis on interfacial tensions of novel betaine solutions against crude oil, J. Pet. Sci. Eng. 2017(152):602-610.

[22] W.-X. SiTu, H.-M. Lu, C.-Y. Ruan, L. Zhang, Y. Zhu, L. Zhang, Efect of polymer on dynamic interfacial tensions of sulfobetaine solutions, Colloids Surf. A Physicochem. Eng. Asp. 2017(533):231-240.

[23] Vonnegut, Rotating bubble method for the determination of surface and interB.-facial tensions, Rev. Sci. Instrum. 1942(13):6-9.

[24] Z. H. Zhou, L. Zhang, Z. C. Xu, L. Zhang, S. Zhao, J. Y. Yu, Surface dilational prop-erties and foam properties of novel benzene sulfonate surfactants, J. Dispers. Sci. Technol. 2010(32):95-101.

[25] Q. Zhang, Z.-H. Zhou, L.-F. Dong, H.-Z. Wang, H.-Y. Cai, F. Zhang, L. Zhang, L. Zhang, S. Zhao, Dilational rheological properties of sulphobetaines at the wa-ter-decane interface: efect of hydrophobic group, Colloids Surf. A Physicochem. Eng. Asp. 2014(455):97-103.

[26] Z.-H. Zhou, D.-S. Ma, Q. Zhang, H.-Z. Wang, L. Zhang, H.-x. Luan, Y. Zhu, L. Zhang, Surface dilational rheology of betaine surfactants: efect of molecular structures, Colloids Surf. A Physicochem. Eng. Asp. 2018(538):739-747.

[27] M. V. D. Tempel, E. H. Lucassen-Reynders, Relaxation processes at fluid interfaces, Adv. Colloid Interface Sci. 1983(18):281-301.

[28] H. Wang, Y. Gong, W. Lu, B. Chen, H. Wang, Y. Gong, W. Lu, B. Chen, Relaxation processes at fluid interfaces, Appl. Surf. Sci. 2008 (254): 3380-3384.

[29] N. Aske, R. Orr, J. Sj blom, Dilatational elasticity moduli of water-crude oil interfaces using the oscillating Pendant Drop, J. Dispers. Sci. Technol. 2002(23): 809-825.

The effect of demulsifier on the stability of liquid droplets: A study of micro-force balance

Liu Miao[a,c] Cao Xulong[b] Zhu Yangwen[b] Guo Zhaoyang[a]
Zhang Lei[a] Zhang Lu[a,*] Zhao Sui[a,*]

[a] Technical Institute of Physics and Chemistry, Chinese Academy of Sciences, Beijing 100190, People's Republic of China

[b] Exploration and Development Research Institute of Shengli Oilfield Company, Limited Sinopec, Dongying, Shandong 257015, People's Republic of China

[c] University of Chinese Academy of Sciences, Beijing 100049, People's Republic of China

Abstract: An original device was developed by our research group and used for quantitatively and qualitatively studying the stability mechanism of two liquid droplets in another liquid phase. The squeeze time can be employed to detect the stability of liquid droplets. The effect of water soluble emulsifier TX-100 and oil soluble one Span-80 on the squeeze times of both aqueous and kerosene droplets have been investigated. Moreover, the influences of three polyether demulsifiers with different structures, linear PEL, branched PEB and star-type PES, on the stability of liquid droplets have also been studied. Based on experimental results, the hydrophilic head group ethoxylation (EO) chain of TX-100 nearly flats at interface and forms compact interfacial film closed to water phase, which makes kerosene droplets stable at aqueous. On the other hand, two water droplets contain TX-100 will coalesce at once when they contact each other. For oil soluble emulsifier Span-80, the compactness of two sides of interfacial film leads to both stable water droplets and oil droplets. The molecular size of demulsifier may play the crucial role in controlling droplets coalescence: PEL with smaller size can easily destroy aqueous droplets in oil; PEB shows strong effects on both kerosene and aqueous droplets for it has a moderate molecular size; PES with larger size shows strong ability to destroy oil droplets in aqueous but little effects on aqueous droplets in oil.

Keywords: Demulsifier Droplet Coalesce Ethylene oxide Squeeze time

1 Introduction

To study the stability of emulsions is a meaningful work on theoretical research[1,2] and practical application[3,4] because emulsions have widely applications in industry[5], bioengineering[6], medicine[7], agriculture[8], fine chemical[9] and so on. However, emulsion is a kind of highly dispersed[10], unstable system which contains liquid droplets dispersed in another immiscible liquid phase[2]. Thermodynamically, liquid droplets in the unstable system can coalesce with each other readily[11,12] and eventually undergo an apparent flocculation and rapid phase separation[13,14]. To expound the mechanisms responsible for the stability of emulsions is of both theoretically[15,16] and practically[17] importance.

The stability, means the resistance to the coalescence of dispersed droplets in emulsions[2], depends on several factors. It includes two op-positely interactions[18]: attractive interaction makes drops easily coa-lescence like van der Waals forces[19], the emulsions unstable; the other one repulsive interactions leads to dispersed drops in emulsion mechanically prevent coalescence (resistant each other) like electrical or steric barrier[20], the emulsions stable. Therefore, to systematically study the stability of droplets is of great significance in elucidating the mechanism of emulsion stabilization.

Over the past several years, researchers studied the stability of emulsions by optical ways to analysis the size distribution of emulsion drops[3,21-24]. Denzil[22] studied the bridging behavior of two kind of emulsions with respect to physical properties of particle by confocal laser scanning microscope, the emulsions appear three morphology: sparingly covered droplets, bridge clusters of droplets and fully covered droplets. Di Shu investigated the efficiency of water removal from the water-in-oil emulsions by differential scanning calorimetry analysis, the chemical demulsification was more effective than centrifugation in water removal of small droplets[25]. However, these macroscopic or microcosmic[26,27] optical approaches merely express quantitative results of the stability of emulsions.

To resolve this challenging problem, lots of researchers devoted themselves to qualitatively evaluate the coalescence of two emulsion drops by atomic force microscopy (AFM)[13,14,28-30]. Chen Shi[13] utilized AFM technique quantitatively study the interactions between water-in oil emulsion droplets with interfacial adsorbed asphaltenes and found that the stability was obviously influenced by the asphaltene concentration, aging time, contact time and solvent type. Hang Jin[29] employed AFM to quantify the forces caused by the space hindrance of F68 copolymers both in bulk phase and oil/F68 aqueous interface. AFM results by Raymond[30] show there is a close link between static surface forces, interfacial deformation and hydrodynamic drainage. These factors greatly govern dynamic droplets interactions.

However, it must be pointed out that both of the optical ways of quantitative analysis and force measurement methods of qualitative assessment can't capture the liquid droplets coalesce process and correspondingly force curves at the same time. Moreover, AFM method needs preparation period and complex operation process (hydrophobicity of cantilever and the obtained of spring constant of the cantilever etc.). Our research group developed an original, conveniently device[31] (core components contains the force measurement Dynamic Contact Angle Meter Tensiometer 21 (DCAT21) and picture captured device high frequency digital camera Digital 3.0) to study the stability of emulsion droplets that could appear microdroplets squeeze, coalesce process and the corresponding real-time dynamic force curves. We use this novel device to investigate the stability of emulsion droplets, such as two kerosene droplets in aqueous contains different structure demulsifiers and two aqueous droplets contains demulsifiers in oil phase.

2 Experimental section

2.1 Materials

In this work, Triton X-100 (TX-100) with purity of 95% was purchased from Xilong Chemical Company, Guangdong, China. Span-80 was purchased from Beijing North Century Commerce Company, Beijing, China, and the hydrophilic-lipophilic balance value is 4.3. The structures of demulsifiers synthesized in our laboratory are listed in Table 1. The kerosene used in this work was obtained from Beijing Chemical Co. Ltd., with an interfacial tension value of 40mN/m against pure water. All of the solutions were prepared by distilled water.

2.2 Apparatus

The interaction forces between two droplets were measured by the restructured DCAT21, the core components of DCAT including these two parts: high-precision electro dynamically compensated weighting system, the precision is 10^{-5} g, with automatic calibrating function; software-controlled motorized height positioning (the accuracy is 0.1 μm) of temperature controllable sample vessel with variable speed. The DCAT21 was reformed by these ways[31] and vividly appears in Fig. 1: introduced up and down droplets sample injection apparatus (stainless steel needles) with external diameter of 0.91 mm and liquid droplets diameter of 1.1±0.05 mm; a micrometer was introduced to makes up and down droplets alignment; high frequency digital camera Digital 3.0 was introduced to "capture" the process of two liquid droplets drainage, coalescence, etc. All of the test temperature is 25℃.

2.3 Methods

Some essential preparation works need to be done before measure-ments: at first, pull testing aqueous into glass sample cell. Then, formed up and down droplets at the point of needles through Injekt-F syringe made by Germany BRAUN (inside diameter is 10 μm) and Longer syringe pump TJ-2A made by Baoding Longer Precision Pump Co., Ltd. It is vital that up and down droplets must be in align through the adjusting of micrometer (adjustment accuracy is 10 μm) under the sample cell. To make the oil and water have a full contact and interfacial active molecules in the system have an equilibrium adsorption at the interface. With the start of the tests, the software in DCAT21 can reflect the up droplets bearing conditions and the values in weight-position curves automatically calibrate at 0, it means that the horizontal axis shows the relative height of up droplet and the vertical axis shows the force bearing conditions of up needle and up droplet. Down droplets gradually approach to up droplets at a low speed 0.01 mm/s. In addition, the position was set as 0 artificially at the point that two droplets contact each other in the following weight-position curves. The force value is negative when the up droplet bearing an upward force, positive weight value when bearing a downward force correspondingly. It is necessary to state that complete experiment processes include two liquid droplets gradually close, contact, extrusion each other, and finally coalescence will be observed. One experiment repeated 3 times and the squeezing time averaged.

3 Results and discussion

Generally speaking, the vital factors affecting the stability of two liquid droplets contain liquid film drainage, interfacial charge, steric barrier and the mechanical strength of interfacial films[2,32-34]. In this paper, the studied systems only contain nonionic surfactants, so the role of interfacial charge is not taken into account.

3.1 Effect of demulsifiers on the coalescence of two liquid droplets stabilized by TX-100

3.1.1 Effect of TX-100 on the stability of two liquid droplets

Our former experimental results[31] appear that kerosene droplets can stably exist at nonionic surfactant TX-100 aqueous compared with anionic surfactant SDBS and cationic surfactant CTAB. Consequently, the test of kerosene droplets in 1×10^{-4} mol/L TX-100 solution and aqueous droplets containing 1×10^{-4} mol/L TX-100 at kerosene were taken as blank test group. And the test dynamic forces curves have been plotted in Fig. 2.

Table 1 Structures of demulsifiers.

Demulsifier code	PO : EO	Structural formula	Molecular weight
PEL	3 : 1	$(EO)_m(PO)_x(EO)_n$—$OCH_2CH_2CH_2O$—$(EO)_n(PO)_x(EO)_m$	15 000
PEB	3 : 1		15 000
PES	1 : 1		15 000

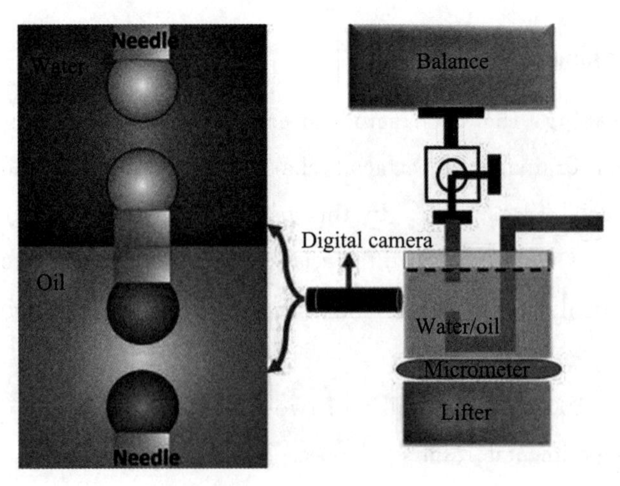

Fig. 1 The reformed apparatus.

With the start of the test, two liquid droplets keeping a certain distance (≥1. 5 mm) gradually get closed and contact. Then, two liquid droplets squeeze each other or coalescence. Comparing Fig. 2 (A) and (B), two kerosene droplets in TX-100 aqueous need to squeeze a relatively long time (about 102s) to coalesce. However, two water droplets in kerosene easily coalesce after contact. This phenomenon can be attributed mainly to the property of interfacial film and may well be depicted by Fig. 3.

The hydrophilic head group ethoxylation (EO) chain nearly flats at interface, hydrophobic group extend into kerosene. When the continuous phase is TX-100 aqueous and inner phase is kerosene, interfacial film closed to water phase has a strong mechanical strength for the close-packed long EO chain, which prevents kerosene droplets from coalescing. Turnover the outer phase and inner phase, the stability of liquid droplets also has been changed. Hydrophobic groups pack loosely at interfacial film nearly to oil phase, as a result, two water droplets are unstable and easily coalesce. There is a common thing in the above two pictures: forces curves appear hops when two liquid droplets coalesce. It is ascribed to the up-steel bear pulling forces from down droplet at coalesce instant.

3. 1. 2 Effect of demulsifiers on the stability of two liquid droplets

This part mainly discusses the effect of three different demulsifiers (branched chain PEB, linear chain PEL, star type PES) on the stability of two liquid droplets. Dynamic force curves of two cases (oil droplets in water and water droplets in oil) of PES appear in Fig. 4 (A—D) as typical representation, and all squeeze times are listed in Table 2.

Firstly, Table 2 reveals that the squeeze times of kerosene droplets in water decrease with the increase of demulsifier concentration, and the squeeze time of aqueous containing 100 ppm PES, PEB and PEL are 38 s, 65 s and 63 s respectively. It is obvious all three demulsifiers can weaken the strength of interfacial film and enhance the coalescence of two kerosene droplets. Moreover, PES shows the strongest ability to destroy the film.

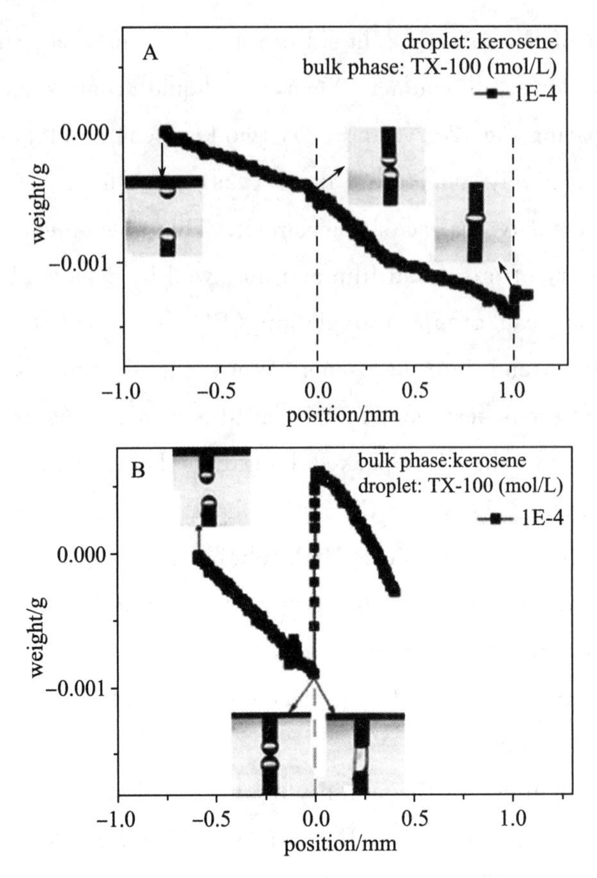

Fig. 2 Dynamic force curves of (A) kerosene droplets in TX-100 aqueous and (B) water droplets containing TX-100 in kerosene.

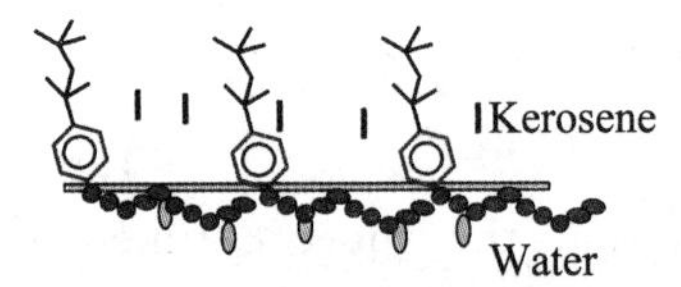

Fig. 3 The arrangement of TX-100 molecules at oil-water interface.

As shown in Fig. 5, the hydrophilic groups of PEB and PEL pack closely at interface, which leads to smaller occupied volumes at interface. On the other hand, the molecule volume of PES is relatively bigger, and hydrophilic groups EO chains are in the outer and flat at interface, hydrophobic groups PO chains and benzene ring extend to oil phase. The "similar to sphere" arrangement way makes PES pack loosely at interface, as a result, more emulsifier molecules will be replaced by PES molecules than those by PEB or PEL molecules.

Besides, all of the water droplets contacted with each other at kerosene will coalesce immediately no matter which one or what concentration of demulsifier in aqueous. The interfacial films near oil phase are unstable because hydrophobic groups of TX-100 and these two demulsifiers pack loosely at interface.

Table 2 The squeezing times (second) of two liquid droplets stabilized by TX-100.

Demuisifier concentration/ppm	Water droplets in kerosene squeezing time/s			Kerosene droplets in water squeezing time/s		
	PES	PEB	PEL	PES	PEB	PEL
0	0	0	0	102	102	102
1	0	0	0	88	87	88
10	0	0	0	84	83	81
100	0	0	0	38	65	63

3. 2 Effectof demulsifiers on the coalescence of two liquid droplets stabilized by Span-80

After having a systematical investigation about liquid droplets stabilized by water-soluble emulsifier TX-100, this part will study the liquid droplets stabilized by oil-soluble surfactant Span-80. Our former study appears when the concentration of Span-80 in kerosene increases to 1×10^{-3} mol/L[31], two water droplets need to be squeezed a long time to coalesce. Consequently, oil phase in this part is 1×10^{-3} mol/L Span-80 in kerosene and water phase is different concentrations demulsifier aqueous. All of the test results of squeeze times of liquid droplets are listed in Table 3.

3. 2. 1 Effect of Span-80 on the stability of two liquid droplets

From Table 3 and Fig. 6 (A), one can find that squeeze time of water droplets in kerosene is 33 s, which is similar to that of kerosene droplets in water 29 s. It illustrates that Span-80 film has a similar tightness on the two sides of interfacial film. Fig. 6 (B) can well explain this result: the hydrophobic tails of long alkyl chain orient into oil phase and hydrophilic group hydroxy and carbonyl groups extend in water and have strong hydrogen bonding interactions with water, which provide hindrance for coalescence at both sides of film.

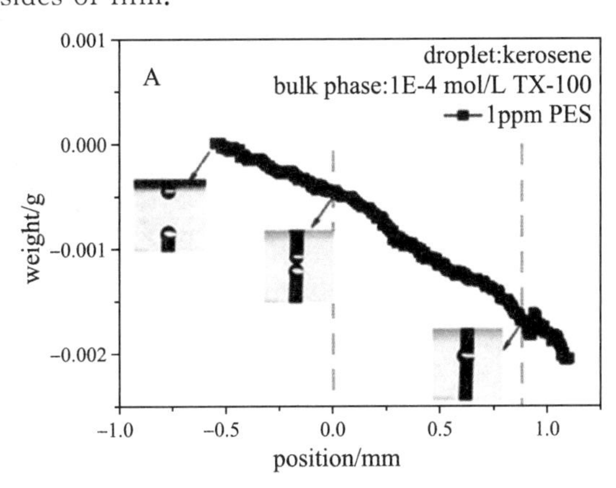

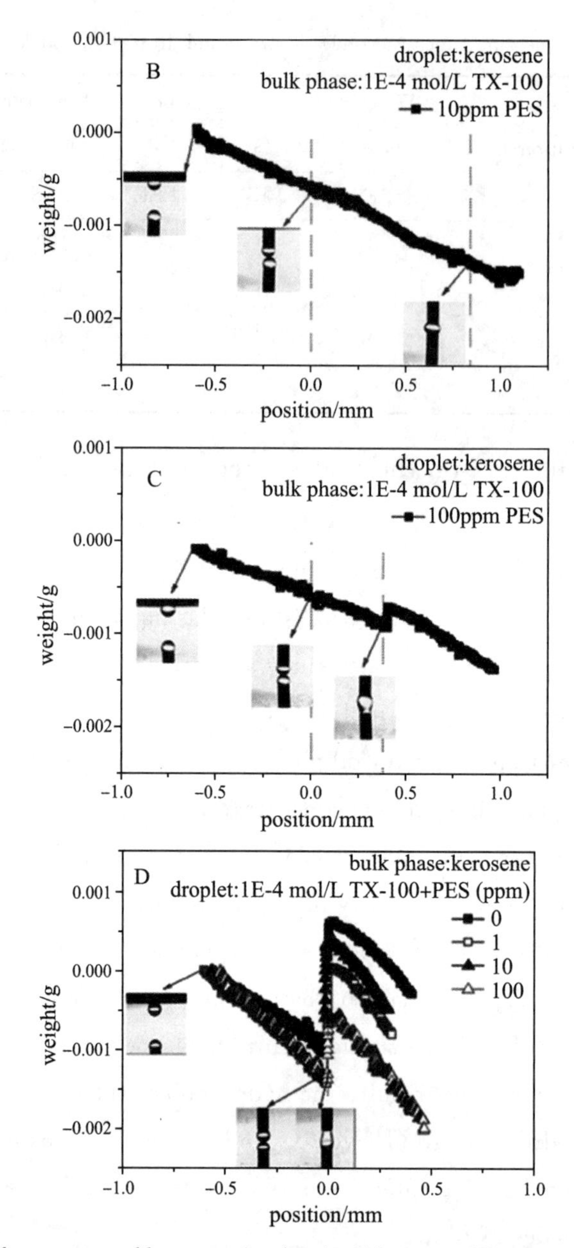

Fig. 4　The dynamic force curves of kerosene droplets in (A) 1ppm, (B) 10ppm, (C)100ppm PES and TX-100 aqueous and (D) water droplets contains PES and TX-100 in kerosene.

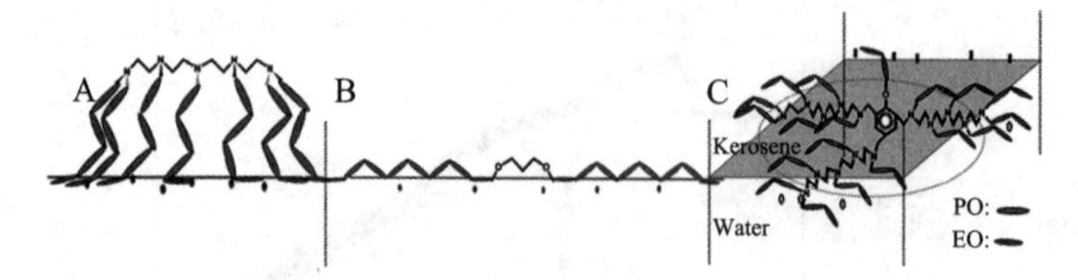

Fig. 5　The arrangements of (A) PEB, (B) PEL and (C) PES at oil-water interface.

Table 3 The squeezing times (second) of two liquid droplets stabilized by Span-80.

Demuisifier concentration/ppm	Water droplets in kerosene squeezing time/s			Kerosene droplets in water squeezing time/s		
	PES	PEB	PEL	PES	PEB	PEL
0	33	33	33	29	29	29
1	27	6	21	19	33	30
10	18	4	27	12	19	48
100	36	3	1	1	5	66

3.2.2 Effect of demulsifier on the stability of two liquid droplets

After adding demulsifiers, stability of liquid droplets has changed significantly. Moreover, demulsifiers with different structures have contrary effects on the properties of interfacial film, and final appearance is the change of stability of liquid droplets. When the aqueous contains 100 ppm PES, squeeze time of water droplets in oil changes little, but that of oil droplets in water aqueous shortened to 1s. As we discussed above, PES can easily destroy Span 80 film from the water side. However, when the bulk is kerosene containing Span 80, PES molecules may form mixed adsorption film and the squeeze time decreases firstly and then increases. On the contrary, for linear chain PEL, with the increase of concentration from 0 to 100 ppm, squeeze time of water droplets in oil decreases to 1s and that of oil droplets in aqueous increases to 66s respectively. Linear PEL has weaker ability to displace Span 80 molecules because of its smaller size, but it may show a good adsorption ability. It seems that the molecular size plays the crucial role in controlling droplets coalescence: the demulsifier with larger size shows strong ability to destroy oil droplets in aqueous and the smaller one trends to destroy aqueous droplets in oil. Branched demulsifier PEB has the moderate molecular size, therefore, PEB show strong effects on both kerosene and aqueous droplets. These results may indicate that the curvature of interface will affect the arrangement of surfactant molecules, and the same demulsifier may have contrary functions for different types of emulsions.

4 Conclusions

After having a systematical investigation about two systems (one contains water-soluble emulsifier TX-100, another contains oil-soluble emulsifier Span-80), the following conclusions could be obtained:

(1) Thehydrophilic headgroupethoxylation (EO) chain of watersoluble TX-100

nearly flats at interface and hydrophobic group extends into oil phase. Interfacial film closed to water phase has a pronounced steric barrier, which makes kerosene droplets stable at aqueous.

(2) Because the long hydrophobic tail of Span-80 oil soluble orients into oil phase and hydrophilic group hydroxy and carbonyl groups extend in water, the compactness of both sides of Span-80 interfacial film leads to stability of both water droplets and oil droplets.

(3) The molecular size may play the crucial role in controlling droplets coalescence: the demulsifier PEL has weaker ability to displace emulsifier molecules because of its smaller size. Therefore, PEL can easily destroy aqueous droplets in kerosene. On the contrary, PES with larger size shows strong ability to destroy oil droplets in aqueous and has little effect on aqueous droplets in oil. PEB shows strong effects on both kerosene and aqueous droplets for it has a moderate molecular size.

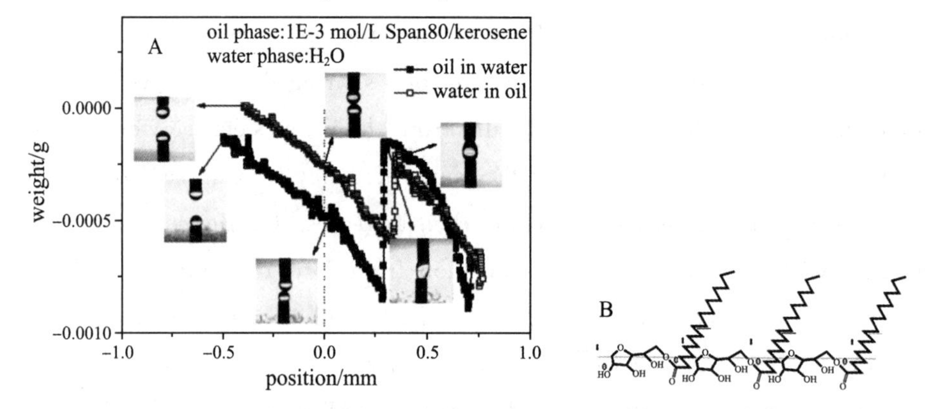

Fig. 6 Force curves of liquid droplets containing Span 80 (A) and (B) the arrangement of Span-80 molecules at interface.

Acknowledgments

The authors thank financial support from the National Science & Technology Major Project (2016ZX05011-003) and National Natural Science Foundation of China (21703269).

References

[1] R. Aveyard, B. P. Binks, J. H. Clint, Emulsions stabilised solely by colloidal particles, Adv. Colloid Interf. Sci,2003(100):503-546.

[2] M. J. Rosen, J. T. Kunjappu, Emulsification by surfactants, Surfactants and Interfacial Phenomena, John Wiley & Sons, Inc. 2012:336-367.

[3] T. N. Hunter, R. J. Pugh, G. V. Franks, G. J. Jameson, The role of

particles in stabilising foams and emulsions, Adv. Colloid Interf. Sci, 2008(137): 57-81.

[4] E. Dickinson, Food emulsions and foams: stabilization by particles, Curr. Opin. Col-loid Interface Sci,2010(15):40-49.

[5] L. D. Zarzar, V. Sresht, E. M. Sletten, J. A. Kalow, D. Blankschtein, T. M. Swager, Dynamically reconfigurable complex emulsions via tunable interfacial tensions, Nature,2015(518):520-524.

[6] F. Ye, M. Miao, B. Jiang, O. H. Campanella, Z. Jin, T. Zhang, Elucidation of stabilizing oil-in-water Pickering emulsion with different modified maize starch-based nanoparticles, Food Chem,2017(229):152-158.

[7] G. Sun, F. Qi, J. Wu, G. Ma, T. Ngai, Preparation of uniform particle-stabilized emulsions using SPG membrane emulsification, Langmuir, 2014 (30): 7052-7056.

[8] V. Temenouga, T. Charitidis, M. Avgidou, P. D. Karayannakidis, M. Dimopoulou, E. P. Kalogianni, C. Panayiotou, C. Ritzoulis, Novel emulsifiers as products from internal Maillard reactions in okra hydrocolloid mucilage, Food Hydrocoll,2016(52):972-981.

[9] H. Yan, X. Chen, H. Song, J. Li, Y. Feng, Z. Shi, X. Wang, Q. Lin, Synthesis of bacterial cellulose and bacterial cellulose nanocrystals for their applications in the stabilization of olive oil pickering emulsion, Food Hydrocoll,2017(72):127-135.

[10] L.-Y. Chu, A. S. Utada, R. K. Shah, J.-W. Kim, D. A. Weitz, Controllable monodisperse multiple emulsions, Angew. Chem. Int. Ed, 2007(46): 8970-8974.

[11] L. Zhang, A. Mikhailovskaya, P. Yazhgur, F. Muller, F. Cousin, D. Langevin, N. Wang, A. Salonen, Precipitating sodium dodecyl sulfate to create ultrastable and stimulable foams, Angew. Chem. Int. Ed,2015(54):9533-9536.

[12] Y. Li, J. Zou, B. P. Das, M. Tsianou, C. Cheng, Well-defined amphiphilic double-brush copolymers and their performance as emulsion surfactants, Macromolecules,2012(45):4623-4629.

[13] C. Shi, L. Zhang, L. Xie, X. Lu, Q. Liu, J. He, C. A. Mantilla, F. G. A. Van den Berg, H. Zeng, Surface interaction of water-in-oil emulsion droplets with interfacially active asphaltenes, Langmuir,2017(33):1265-1274.

[14] H. J. Lockie, R. Manica, G. W. Stevens, F. Grieser, D. Y. C. Chan, R. R. Dagastine, Precision AFM measurements of dynamic interactions between deformable drops in aqueous surfactant and surfactant-free solutions, Langmuir, 2011 (27):

2676-2685.

[15] B. P. Binks, R. Murakami, Phase inversion of particle-stabilized materials from foams to dry water, Nat. Mater,2006(5):865-869.

[16] J. Liu, Y. Zhao, S. Ren, Molecular dynamics simulation of self-aggregation of asphaltenes at an oil/water interface: formation and destruction of the asphaltene protective film, Energy Fuel,2015(29):1233-1242.

[17] L. Ma, Y. Chen, Y. Liu, M. Chen, B. Yang, B. Zhang, Y. Ding, Investigation on the per-formance of a block polyether demulsifier based on polysiloxane for the treatment of aged oil, Energy Fuel,2017(31):8886-8895.

[18] C. Shi, L. Zhang, L. Xie, X. Lu, Q. Liu, C. A. Mantilla, F. G. A. van den Berg, H. Zeng, Inter-action mechanism of oil-in-water emulsions with asphaltenes determined using droplet probe AFM, Langmuir,2016(32):2302-2310.

[19] J. Boyd, P. Sherman, C. Parkinsonl, Factors affecting emulsion stability, and HLB con-cept, J. Colloid Interface Sci,1972(41):359-370.

[20] A. Sanfeld, A. Steinchen, Emulsions stability, from dilute to dense emulsions -role of drops deformation, Adv. Colloid Interf. Sci,2008(140):1-65.

[21] M. Moradi, V. Alvarado, S. Huzurbazar, Effect of salinity on water-in-crude oil emul-sion: evaluation through drop-size distribution proxy, Energy Fuel,2011(25):260-268.

[22] D. S. Frost, J. J. Schoepf, E. M. Nofen, L. L. Dai, Understanding droplet bridging in ionicliquid-based Pickering emulsions, J. Colloid Interface Sci,2012(383):103-109.

[23] A. R. Tehrani-Bagha, Cationic gemini surfactant with cleavable spacer: emulsion sta-bility, Colloids Surf. A Physicochem. Eng. Asp,2016(508):79-84.

[24] D. Pradilla, J. Ramirez, F. Zanetti, O. Alvarez, Demulsifier performance and dehydra-tion mechanisms in Colombian heavy crude oil emulsions, Energy Fuel,2017(31):10369-10377.

[25] D. Shu, Y. Chi, J. Liu, Q. Huang, Characterization of water-in-oil emulsion and upgrading of asphalt with supercritical water treatment, Energy Fuel,2017(31):1468-1477.

[26] A. R. Studart, H. C. Shum, D. A. Weitz, Arrested coalescence of particle-coated droplets into nonspherical supracolloidal structures, J. Phys. Chem. B,2009(113):3914-3919.

[27] A. B. Pawar, M. Caggioni, R. Ergun, R. W. Hartel, P. T. Spicer, Arrested coalescence in Pickering emulsions, Soft Matter,2011(7):7710-7716.

[28] C. Shi, B. Yan, L. Xie, L. Zhang, J. Wang, A. Takahara, H. Zeng, Long-range hydrophilic attraction between water and polyelectrolyte surfaces in oil, Angew. Chem. Int. Ed,2016(55):15017-15021.

[29] H. Jin, W. Wang, H. Chang, Y. Shen, Z. Yu, Y. Tian, Y. Yu, J. Gong, Effects of salt-controlled self-assembly of triblock copolymers F68 on interaction forces between oil drops in aqueous solution, Langmuir,2017(33):14548-14555.

[30] R.R. Dagastine, R. Manica, S.L. Carnie, D.Y.C. Chan, G.W. Stevens, F. Grieser, Dynamic forces between two deformable oil droplets in water, Science, 2006(313):210-213.

[31] L. Guo, J. Li, Y. Zhu, B. Ma, Z. Xu, W. Wang, L. Zhang, L. Zhang, Interaction forces be-tween simulated mini liquid droplets of emulsions, Chem. J. Chin. Univ,2016(37):361-366.

[32] J. Tang, P. J. Quinlan, K. C. Tam, Stimuli-responsive Pickering emulsions: recent ad-vances and potential applications, Soft Matter, 2015 (11): 3512-3529.

[33] B.M. Jose, T. Cubaud, Droplet arrangement and coalescence in diverging/converging microchannels, Microfluid. Nanofluid,2012(12):687-696.

[34] P. Tchoukov, F. Yang, Z. Xu, T. Dabros, J. Czarnecki, J. Sjoblom, Role of asphaltenes in stabilizing thin liquid emulsion films, Langmuir, 2014 (30): 3024-3033.

Structure-activity relationship of anionic-nonionic surfactant for reducing interfacial tension of crude oil

Sheng Songsong[a] Cao Xulong[b] Zhu Yangwen[b]
Jin Zhiqiang[c] Zhang Lei[c] Zhu Yan[a,*] Zhang Lu[ac,**]

[a] School of Chemistry, Chemical Engineering and Life Sciences, Wuhan University of Technology, Wuhan 430070, PR China

[b] Exploration and Development Research Institute of Shengli Oilfield Company, Limited Sinopec, Dongying, Shandong 257015, PR China

[c] Key Laboratory of Photochemical Conversion and Optoelectronic Materials, Technical Institute of Physics and Chemistry, Chinese Academy of Sciences, Beijing 100190, PR China

Abstract: In this article, we have investigated interfacial tensions (IFTs) of anionic-nonionic surfactant fatty alcohol polyoxyethanol carboxylate (C_nEO_mC) and their mixed solutions against n-alkanes, crude oil and model oils containing crude oil fractions, such as saturates, aromatics, resins and asphaltenes. The experimental results show that EO group plays a crucial role in reducing IFT by adjusting the area occupied for C_nEO_mC molecules at the interface and only surfactant with optimum EO number produces lower IFTs against n-decane, named size compatibility. The hydrophilic-lipophilic balance is another important factor for reducing IFT. Ultralow IFT value can only be achieved when both hydrophilic-lipophilic balance and size compatibility are met simultaneously. Fortunately, one can adjust the hydrophilic-lipophilic balance and keep size compatibility simultaneously by mixing C_nEO_mC with different EO numbers to produce ultralow IFTs for both hydrocarbons and crude oil. Moreover, the spaces with different sizes occur at the mixed adsorption film. As a result, we can observe synergism between model oils containing resins, aromatics or asphaltenes during wide concentrations and mixed solution of $C_{14}EO_3C$ and $C_{14}EO_7C$ with mass fraction ratio of 1 : 1. This may be another reason for the ultralow IFT values of mixed solutions against crude oil. The mechanism provided in this paper is very important for the design of chemical flooding at high salt reservoir conditions.

Keywords: nionic-nonionic surfactant Interfacial tension Crude oil fractions Crude oil Structure-activity relationship

1 Introduction

As a strategic energy source, crude oil plays an important role in the national economic development and social stability of a country. However, nearly two-thirds of the oil remains in the oil reservoir after primary (relying on the original reservoir energy) and secondary (through water supplement energy) oil recovery. Therefore, it's very necessary to carry out further enhanced oil recovery (EOR). Surfactants employed in EOR can function by reducing oil/water interfacial tension (IFT)[1-4], emulsification with crude oil[5] and controlling the wetta-bility of reservoir[6]. It is well-known that IFT is an important indicator in EOR, which can be dramatically reduced by using appropriate surfactants. If the IFT between crude oil and flooding aqueous can be reduced to an ultralow value ($<10^{-2}$ mN/m), the efficiency of oil recovery will be remarkably increased. Thus, researchers are committed to seeking a highly efficient, low cost and environmentally friendly way for EOR.

Generally, traditional anionic surfactants such as petroleum sulfo-nates and heavy alkylbenzene sulfonates are widely used in EOR because of their wide range of sources and low cost[7-10]. However, they are not suitable for high temperature and high salt reservoir conditions because of their high Krafft point and poor salt solubility. In recent years, attention of salt-tolerant surfactant systems has been mostly focused on[11-15]. Therefore, it is urgent to find surfactants applied to high salt reservoir conditions.

At present, there are two types of surfactants that have been recognized for their application in high-salt reservoirs: anionic-nonionic surfactants and zwitterionic surfactants. Though zwitterionic surfactants usually have excellent properties, the high cost of production makes it difficult to apply them in oilfield yielding. Anionic-nonionic surfactants are prepared primarily by introducing a certain amount of nonionic chain segments into the anionic surfactant hydrophilic head group [10]. Due to its two hydrophilic groups in the molecule, it has the adva-ntages of both nonionic and anionic surfactants, namely good salt resistance to decomposition and dispersion, and good compatibility[16-19].

Witthayapanyanon et al. [18] found that the introduction of additional PO and EO groups in the extended surfactant yielded lower IFT and lower optimum salinity, both of which are desirable in most formulations. Fatty alcohol polyoxyethanol carboxylate

(AEC) is one type of Anionic-nonionic surfactants. Two hydrophilic groups (—CH_2CH_2O— and —CH_2COO—) are contained in the chemical structure of AEC, making it display greater surface activity than the traditional surfactants, received much attention of researchers. However, the fundamental mech-anisms responsible for lowering IFT between crude oil and AEC solution under high-salinity circumstances have not been sufficient until now. Therefore, understanding the structure-property relationship of anionic-nonionic surfactants plays a crucial role in many industrial applications.

In addition to the type and concentration of surfactant, active components in crude oil also play an extremely important role in reducing the IFT value[20-28]. The composition of crude oil is greatly compli-cated, including resins, asphaltenes, saturates and aromatics. Zhu et al. [25] discussed the mechanism of affecting IFT between crude oil fractions and alkylbenzene sulfonate solutions. Zhou et al. [29] reported the effect of crude oil fractions on IFTs of sulfobetaine solutions. Cao et al. [30] also introduced synergism/antagonism between crude oil and betaine solutions in reducing IFT. These articles revealed that the addition of active substances may result in the destruction of the hydrophilic-hydrophobic balance, which relatively affects the original arran-gement of surfactant molecules and weakens the compactness of interfacial films. Asphaltenes is generally insoluble in pentane, hexane, heptane, but soluble in benzene, capable of forming a stable interface film at the oil-water interface[20,21,26]. Saturates and aromatics, accounting for a large proportion of crude oil, are unable to form an interfacial mixed film because of low activity[22,24]. Resins contain the main interfacial active substance, petroleum acid molecules, greatly affecting the distribution of surfactant molecules between oil phase and water phase [27,31,32].

In this work, we mainly aim to expound the structure-activity rela-tionship of AEC for reducing IFTs. For this purpose, the resins, asphaltenes, saturates and aromatics from Gudong crude oil were separated. The IFTs of AEC with different chain lengths and ethylene oxide (EO) groups, against n-alkanes, crude oil or model oils containing crude oil fractions were measured by a spinning drop tensiometer. This work will be helpful to understand the mechanism responsible for IFT reduction by synergism between crude oil fractions and AEC for EOR.

2 Experimental section

2.1 Materials

Fatty alcohol polyoxyethanol carboxylates $RCH_2(EO)_mCH_2COONa$ employed in this paper were synthesized by our laboratory. The structures and abbreviations are listed in Table 1. The purities of $C_{14}EO_3C$, $C_{14}EO_5C$, $C_{14}EO_7C$, $C_{14}EO_9C$, $C_{16}EO_3C$ and $C_{18}EO_3C$ checked by elemental analysis and ^1H nuclear magnetic resonance (NMR) spectroscopy are above 95 mol%.

Gudong crude oil with a density of 0.929 6 g/cm^3 at 65℃ was collected from Shengli oilfield in China. The compositions of Gudong formation brine are shown in Table 2. Kerosene used in the research was purified by silica gel column chromatography (100-200 mesh), and its IFT against pure water reached a stable value of 40 mN/m. Double-distilled water (resistivity of >18.2MΩ · cm) was used in the preparation of formation brine and surfactant solutions with the approximate pH value of 7.0.

Table 1　Structures and abbreviations of fatty alcohol polyoxyethanol carboxylate.

Abbreviations	R	m
$C_{14}EO_3C$	$CH_3(CH_2)_{12}$-	3
$C_{14}EO_5C$	$CH_3(CH_2)_{12}$-	5
$C_{14}EO_7C$	$CH_3(CH_2)_{12}$-	7
$C_{14}EO_9C$	$CH_3(CH_2)_{12}$-	9
$C_{16}EO_3C$	$CH_3(CH_2)_{14}$-	3
$C_{18}EO_3C$	$CH_3(CH_2)_{16}$-	3

Table 2　Composition analysis of the formation brine.

Ca^{2+} (mg/L)	Mg^{2+} (mg/L)	Fe^{3+} (mg/L)	K^+ (mg/L)	Na^+ (mg/L)	Cl^- (mg/L)	HCO_3^- (mg/L)	SO_4^{2-} (mg/L)	TDS (mg/L)
549.8	57.94	4.73	62.31	4 480.3	6 252.11	540.2	12.37	10,954

2.2 Separation of crude oil fractions

Gudong crude oil was separated into four components (resins, asphaltenes, saturates, and aromatics) according to SARA analysis. The fractions obtained according to SARA are shown in Table 3. The specific separation method of crude oil components is described in detail elsewhere[25,26]. The four components obtained are separately distributed in kerosene to form model oils.

The elemental mass fraction of the four components in the crude oil was determined by Elementar Vario EL (Germany). The analysis results are shown in Table 4.

It's recognized that the molar ratio of the elements hydrogen and carbon [n(H)/n(C)] is an important indicator for characterizing the structure of petroleum molecules. On the basis of the results of the elemental analysis experiments in Table 3, it can be seen that the n(H)/n(C) of saturated fraction is the highest among the four active components, indicating that it has the highest saturation, resulting in poor polarity, mainly composed of some saturated alkanes. Asphaltene has the lowest n(H)/n(C), explaining that asphaltene has the lowest saturation, which may be caused by aromatic rings and heteroatoms. What's more, the mass fractions of nitrogen in asphaltenes is relatively high, manifesting the polarity of asphaltenes is comparatively strong. The n(H)/n(C) of the aromatics is lower than the saturates probably because it contains the aromatic ring.

2.3 Apparatus and methods

The IFTs were measured by a Texas-500C spinning drop interfacial tensiometer (CNG Co.). The surfactant solution as the outer phase was injected into the glass tube, and about 1 μL of oil phase as an inner phase was put into the middle of the tube. In all cases, the measure-ments of the IFT are at a rotating velocity of 5 000 rpm. All experiments were performed at the formation temperature of oilfield of 65.0±0.5℃. The measurement error of the IFT value is lower than ±5% when IFT value is lower than 1 mN/m.

3 Results and discussion

3.1 IFTs between n-alkanes and anionic-nonionic surfactant solutions

Surfactant adsorbs at the oil-water interface, replacing solvent molecules, thereby greatly reducing the IFT. The better the displacement ef-fect, the lower the IFT value. In other words, the more surfactant molecules adsorbed on the interface, the lower the IFT value; at the same time, the hydrophilic portion and the hydrophobic portion are similar in size. The surfactant capable of achieving ultralow IFT must have a relatively balanced hydrophilic-hydrophobic ability and larger molecular size.

Table 3 Mass fraction of crude oil fractions.

Component	Saturates(wt%)	Aromatics(wt%)	Resins(wt%)	Asphaltenes(wt%)
Mass fraction	19.9	63.8	12.2	4.1

Table 4 Mass fraction of elements in each fraction of Gudong crude oil.

Component	N(wt%)	C(wt%)	H(wt%)	$n(H)/n(C)$
Saturate		87.33	12.54	1.72
Aromatic	1.42	86.00	10.43	1.46
Resin	1.48	81.30	10.48	1.55
Asphaltene	1.59	85.97	8.92	1.25

Fatty alcohol polyoxyethanol carboxylate (AEC) is one type of anionic-nonionic surfactant that can produce ultralow IFT under high salinity conditions. The effect of the surfactant concentration on the equilibrium IFTs between AEC surfactants and n-decane is plotted in Fig. 1.

Fig. 1(A) shows that the equilibrium IFT values for these four surfactants with the same carbon chain length decrease as the bulk concen-tration increases. However, the reduction of IFT for $C_{14}EO_5C$ and $C_{14}EO_7C$ is always greater than $C_{14}EO_3C$ and $C_{14}EO_9C$, which indicates that EO group plays a crucial role in reducing IFT by adjusting the area occupied for AEC molecule at interface and only optimum EO number will produce lower IFT.

Fig. 1(B) shows the effect of alkyl chain length on the equilibrium IFTs. Unlike $C_{14}EO_3C$, the equilibrium IFTs of $C_{16}EO_3C$ and $C_{18}EO_3C$ pass through minimum at lower bulk concentration and increase with an increase of concentration after of 0.05%, this may be attributed to the poor solubility for $C_{16}EO_3C$ and $C_{18}EO_3C$ at high concentrations. Moreover, with the increase of alkyl chain length from C_{14} to C_{18}, the IFT values continuously increase at 0.4%, which indicates the hydrophilic-lipophilic balance is another important factor.

In general, the distribution of surfactants in n-alkane decreases as the carbon number of alkanes increases. This is because the intermolecular cohesive energy increases when the chain length of oil molecule increases. Therefore, for an aqueous solution of a surfactant with a fixed concentration, the IFT with a series of n-alkane at a certain salt concentration can be determined, and it may be found that the IFT with a specific alkane carbon number is the lowest. The alkane carbon number (ACN) for the minimum IFT is called n_{min} of the surfactant. The n_{min} value of a surfactant reflects the relative distribution ability of the surfactant in the aqueous phase (or oil phase) and can be used as a quantitative characterization method for its hydrophilic-hydrophobic balance (HLB) ability: the larger the n_{min} value, the weaker the hydrophilicity of the surfactant and the stronger the hydrophobicity. In other words, the greater the n_{min}

value of the surfactant, the greater the carbon number of the oil molecules that match it[1,7].

Fig. 2 shows the equilibrium IFT values as a function of the alkane carbon number (ACN). As the results show the n_{min} values of $C_{14}EO_3C$, $C_{14}EO_5C$, $C_{14}EO_7C$, $C_{14}EO_9C$ are about 12, 11, 10 and 10 respectively. In other words, with the number of EO in the molecule increasing, n_{min} value gradually decreases, indicating that the hydrophilicity of the surfactant is enhanced[33]. Particularly, the IFT of $C_{14}EO_3C$ against n-alkanes under optimal hydrophilic-hydrophobic balance conditions can be reduced to 10^{-2} mN/m. Fig. 2(B) shows the effect of alkyl chain length on equilibrium IFT values as a function of the ACN. The n_{min} value of $C_{14}EO_3C$, $C_{16}EO_3C$ is both about 12. There is little change in n_{min} with increasing the number of carbons from C_{14} to C_{16}, because $C_{14}EO_3C$ has strong hydrophobicity enough and an increase of alkyl chain length only shows little effect on n_{min}. Moreover, the high IFT values of $C_{18}EO_3C$ for all n-alkanes caused by the bad solubility in aqueous solution make it impossible to determine n_{min} value of $C_{18}EO_3C$. It is also worth noting that the IFT of $C_{16}EO_3C$ can reach 10^{-3} mN/m order of magnitude at the optimal conditions.

Based on the experimental results above, we can find that the differ-ence in n_{min} between these surfactants is small, while disparity of IFT values at n_{min} is large, which illustrates the hydrophilic-lipophilic balance is not the sole factor for the difference of IFT values. Under the condition of optimum hydrophilic-hydrophobic balance, the longer EO group may partly lies flat at the interface and enlarges the space between adsorbed molecules by steric effect. As a result, IFTs at n_{min} for $C_{16}EO_3C$ and $C_{18}EO_3C$ are lower than the others. On the other hand, the interfacial amount of surfactant is less when ACN of the oil is much lower than n_{min}. Therefore, longer EO group improves the occupied area of surfactant molecules at interface and caused lower IFT values for C_6 to C_{10}.

3.2 IFTs between crude oil and anionic-nonionic surfactant solutions

Fig. 3 displays the effect of surfactants on the equilibrium IFT values of Gudong crude oil as a function of the concentration. Fig. 3(A) shows that IFTs decrease with an increase of surfactants concentration. It's very interesting that the equilibrium IFTs of $C_{14}EO_3C$ are obvious higher than the others. Fig. 3(B) displays the same trend of $C_{16}EO_3C$, while $C_{18}EO_3C$ is different from them for its poor solubility. Comparing Figs. 2 and 3, the trends of variations of IFTs with an increase of EO number against crude oil are the same as those against hydrocarbons with low ACN, which indicates that $C_{14}EO_3C$ is more oil soluble and far away from optimum hydrophilic-hydrophobic balance. Importantly, it's difficult to achieve ultralow IFT against crude oil for all surfactants.

Based on the above experimental results, $C_{14}EO_3C$ with size compa-tibility and $C_{14}EO_7C$ with better hydrophilic ability were selected and compounded at a mass ratio of 1 : 1. The experimental results of IFT be-tween the mixed solutions and Gudong crude oil are shown in Fig. 4. It's surprising to find that the IFT between the mixed solutions and Gudong crude oil reaches 10^{-3} mN/m order of magnitude at the total concentration of 0. 4%. We can assume that, at this compounding ratio, with the concentration increasing, the amounts of $C_{14}EO_3C$ and $C_{14}EO_7C$ molecules adsorbed to the interface increase. At the same time, surfactant molecules can also be closely arranged at the interface at the total concentration of 0. 4%.

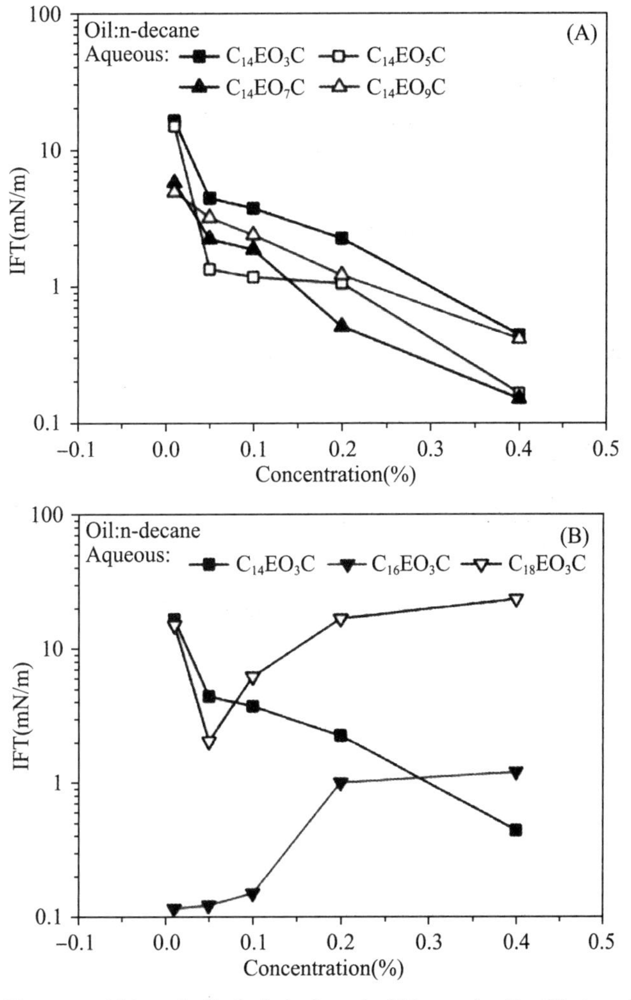

Fig. 1 Effects of EO group (A) and alkyl chain length (B) on the Equilibrium IFTs between AEC solutions and n-decane.

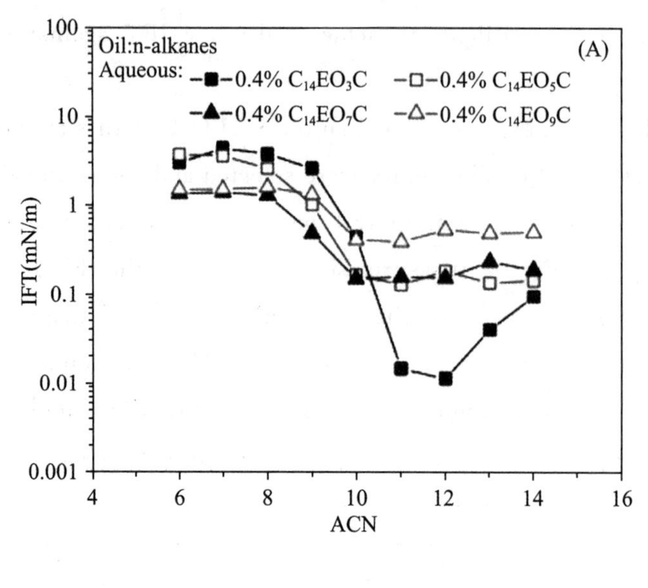

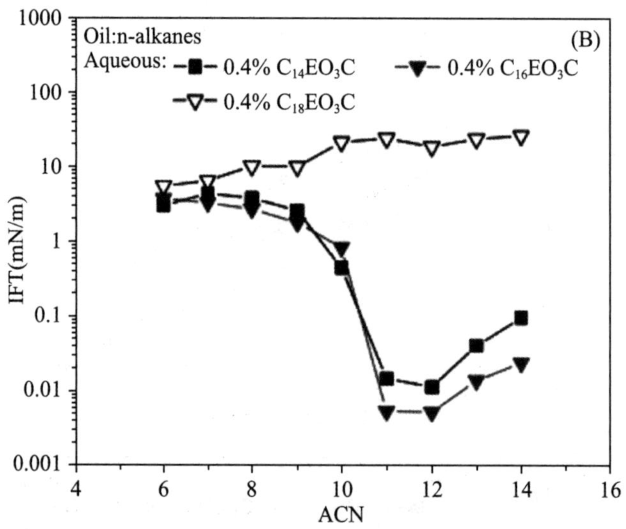

Fig. 2 Equilibrium IFTs between surfactant solutions and n-alkanes with different carbon numbers.

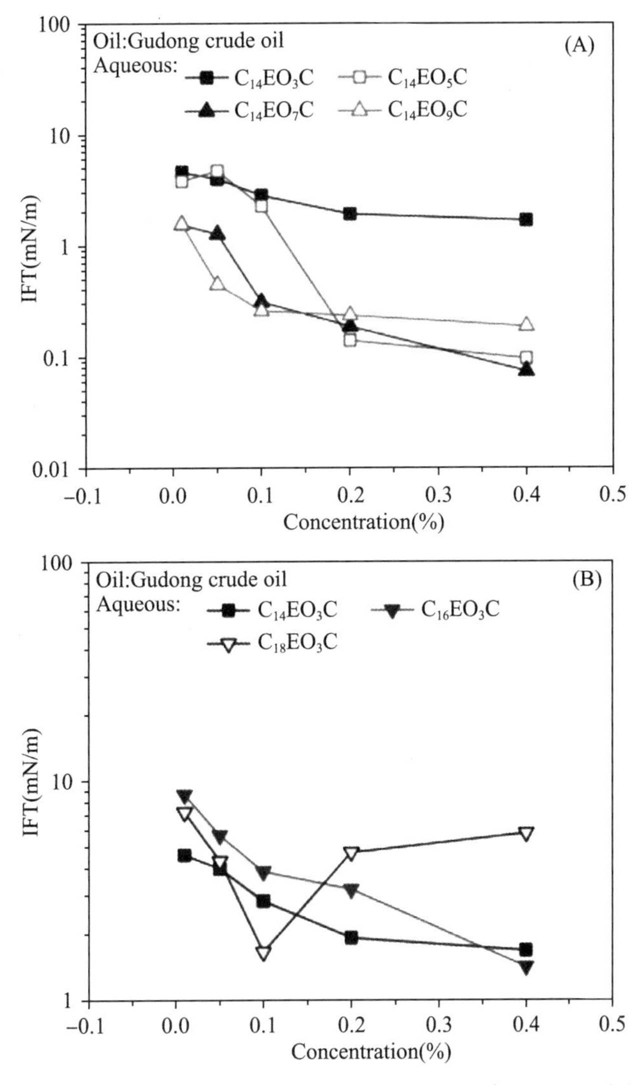

Fig. 3 Equilibrium IFTs between surfactants with different weight fractions and Gudong crude oil.

To better understand the differences in IFTs between separate and combined system, $C_{14}EO_3C$, $C_{14}EO_7C$ and the compound system are sorted out for further explanation. Based on the early studies, a constant concentration of 0.4% is selected for $C_{14}EO_3C$, $C_{14}EO_7C$ and the compound system. It can be seen from the Fig. 5 that the n_{min} value of compound system is 10 and IFT at n_{min} can reach ultralow value. Namely, for the compound system, not only the hydrophilicity is appropriate, but also the EO groups are optimum, which indicates that it has the advantages of $C_{14}EO_3C$ and $C_{14}EO_7C$ at the same time for reducing IFT against Gudong crude oil.

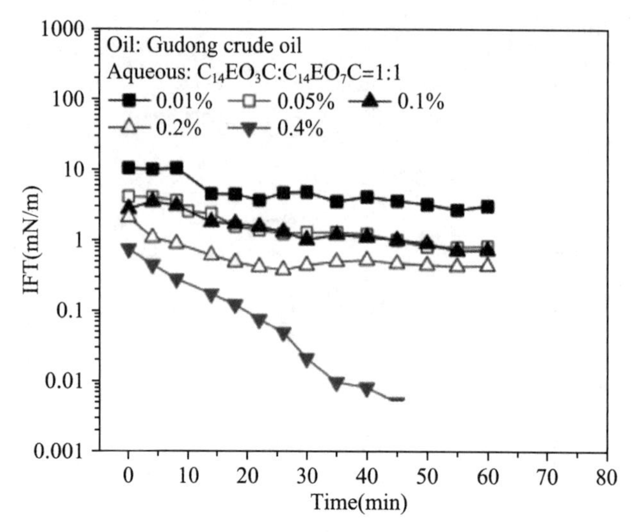

Fig. 4 Dynamic IFTs between mixed solutions with different weight fractions against Gudong crude oil.

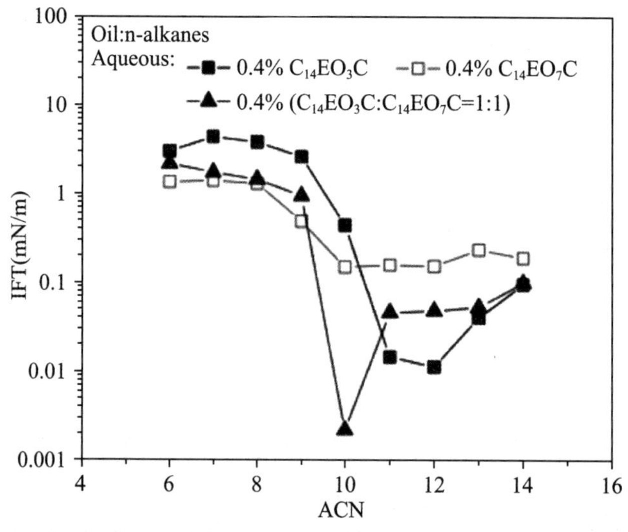

Fig. 5 Equilibrium IFTs between mixed surfactant solution and n-alkanes with different carbon numbers.

3.3 IFTs between the model oil containing crude oil fractions and anionic-nonionic surfactant solutions

3.3.1 Model oil of resins

Dissolving crude oil components in kerosene to form model oil is a common simulation method for studying the IFT mechanism of crude oil-surfactant solution system. Resins in crude oil play an important role in reducing IFT, for most petroleum acids present in the resins. After the addition of the resins, the petroleum acid molecules adsorb to the interface and interact with the surfactant to form a dense adsorption film. There are two major factors affecting the characteristics of the mixed adsorption layer, one is the interaction between acidic molecules and interfacial surfactants, and the other is the arrangement of interfa-cial surfactants molecules.

As shown in Fig. 6, the effect of Gudong resins on the dynamic IFTs of $C_{14}EO_3C$, $C_{14}EO_7C$ and the mixed solutions has been investigated. It is worth noting that IFTs of pure water against n-decane is about 50 mN/m at room temperature, while pure water against kerosene used in this experiment is only 40mN/m, indicating that there are still some unremoving active substances in the kerosene after steaming and the column[30]. Therefore, the dynamic IFTs of surfactant solutions against kerosene have also been measured and plotted in Fig. 6 as blanks.

Three completely same trends can be observed from Fig. 6 that the dynamic IFT curve of $C_{14}EO_3C$, $C_{14}EO_7C$ and the mixed solutions all remain "L" shape, which is the most typical characteristic curve re-ported in the literature[30,34]. Generally, the surfactants are dissolved in the aqueous phase. When the aqueous phase is in contact with the oil phase, the surfactant molecules are adsorbed from the aqueous phase to the interface, and the surfactant starts to enrich at the interface as time passes by. Since the surfactant is an amphiphilic molecule, it has a certain oil solubility. Thus, surfactant molecules at the interface begin to transfer to the oil phase. Finally, the surfactant reaches equilibrium in the aqueous phase, interface and oil phase. Therefore, for most surfac-tant solutions, the dynamic IFT curves will show "L" shape.

Fig. 7 shows the equilibrium IFTs of $C_{14}EO_3C$, $C_{14}EO_7C$ and the mixed solutions as a function of concentration of Gudong resins against model oils. We can see clearly that for both $C_{14}EO_3$ and the mixed solutions, the equilibrium IFTs decrease dramatically to 10^{-3} mN/m order of mag-nitude at the concentration of 0.01% and increase slightly to 10^{-1} mN/m order of magnitude with the increase of resins concentration. On the other hand, the equilibrium IFTs of $C_{14}EO_7C$ solutions change little when the resin concentration is $<0.1\%$ while decrease sharply to ultra low value at 1% resin concentration.

Based on the experimental results above, we assume that a small amount of resins present in kerosene plays a valid synergistic effect with $C_{14}EO_3C$, which can mixed adsorb at the interface and form com-pact film, as a result, IFTs decrease to an ultralow level. However, as the concentration of the resins increases, some $C_{14}EO_3C$ molecules will be replaced by petroleum acid molecules with low surface activity and IFTs become to increase. On the other hand, for $C_{14}EO_7C$ system, adsorp-tion films are loose due to steric effect from long EO group. Therefore, low concentration resins show little effect on IFTs. As the mass fraction of resins in the kerosene increases, a dense adsorption film is formed on the interface, which attributes an obvious decrease in IFT. The IFT curve of the compound system is similar to that of $C_{14}EO_3C$, which indicates $C_{14}EO_3C$ plays a major role in controlling interfacial arrangement.

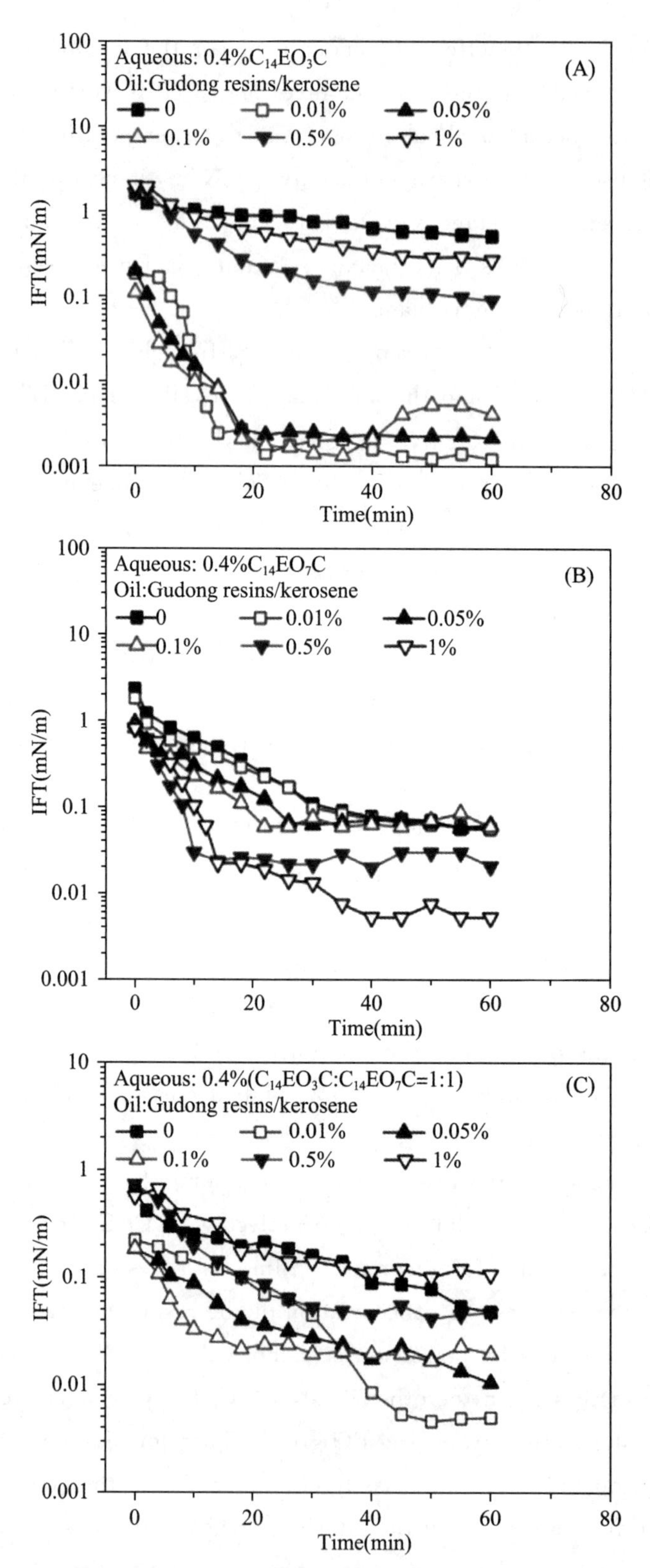

Fig. 6 Dynamic IFTs between (A) $C_{14}EO_3C$, (B) $C_{14}EO_7C$ or (C) mixed solutions against model oil containing Gudong resins with different weight fractions.

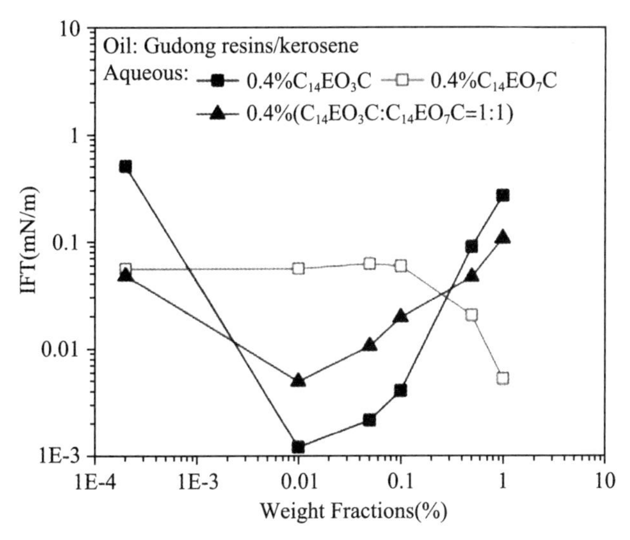

Fig. 7 Equilibrium IFTs between $C_{14}EO_3C$, $C_{14}EO_7C$ or mixed solutions against model oil containing Gudong resins with different weight fractions.

In a word, anionic-nonionic surfactants with short EO group show synergism for reducing IFT with resins at optimum concentrations, while those with long EO group show synergism at higher resins concentrations.

3.3.2 Comparisons of different model oils

The variations of equilibrium IFT values between $C_{14}EO_3C$, $C_{14}EO_7C$ or the compound system against model oils containing four crude oil fractions are plotted in Fig. 8. The possible mechanisms are schemati-cally plotted in Fig. 9. For $C_{14}EO_3C$ system, little effect exists in saturates for its poor activity due to weaker polarity; while other three components can reduce IFTs to 10^{-3} mN/m order of magnitude at an optimum concentration. This phenomenon indicates that $C_{14}EO_3C$ synergizes with a small amount of active substances, and are closely arranged at the interface to form a dense adsorption film. However, when the concentration of the active substance increases, the antagonism occurs as we discussed above.

For $C_{14}EO_7C$ system, only resins with high concentration can reduce IFTs to 10^{-3} mN/m order of magnitude and little effect exists in other three components with lower activity.

Surprisingly, for the mixed solutions, the obvious synergism can be observed for model oils containing aromatics and asphaltenes when fraction concentrations are higher than 0.05%. Moreover, ultralow IFT can be reached during wide aromatics and asphaltenes concentrations. As shown in Fig. 9, we can propose that for $C_{14}EO_3C$ system, surfactant molecules are closely packed on the interface. Therefore, only crude oil fractions with optimum concentrations show synergism with $C_{14}EO_3C$. On the contrary, $C_{14}EO_7C$ molecules arrange loosely at the interface and synergism can hardly be produced. However, for mixed solutions, the spaces with different sizes occur at the interface, as a result, we can observe synergism for model oils containing resins, aromatics or asphaltenes during wide concentrations. This may be another reason for the ultralow IFT values of mixed solutions against crude oil.

4 Conclusion

The Structure-activity relationship of anionic-nonionic surfactant fatty alcohol polyoxyethanol carboxylate C_nEO_mC for reducing interfa-cial tension against crude oil has been investigated by a spinning drop interfacial tensiometer. On the basis of the work above, the following conclusions can be obtained:

(1) EO group plays a crucial role in reducing IFT by adjusting the area occupied for C_nEO_mC molecules at the interface and only optimum EO number will produce lower IFTs against n-decane, as a result of size compatibility.

(2) The hydrophilic-lipophilic balance is another important factor for reducing IFT. The hydrophilic ability of the surfactant is enhanced with an increase of EO number.

(3) Ultralow IFT value can only be achieved when both hydrophilic-lipophilic balance and size compatibility are met simultaneously. $C_{14}EO_3C$ and $C_{16}EO_3C$ can produce ultralow IFT values against hydrocarbons with optimum ACN.

(4) By mixing $C_{14}EO_3C$ and $C_{14}EO_7C$, one can adjust the hydrophilic-lipophilic balance and keep size compatibility simultaneously. Therefore, mixed solution of $C_{14}EO_3C$ and $C_{14}EO_7C$ with mass fraction ratio of 1 : 1 can reduce the IFTs of both n-decane and crude oil to ultralow values.

(5) For anionic-nonionic surfactants with short EO group, only crude oil fractions with optimum concentrations show synergism. On the contrary, those with long EO group arrange loosely at the interface and synergism can hardly be produced. However, for mixed solutions, we can observe synergism for model oils containing resins, aromatics or asphaltenes during wide concentra-tions.

Credit authorship contribution statement

Song-Song Sheng: Investigation, Data curation, Writing-original draft. Xu-Long Cao: Conceptualization, Funding acquisition. Yang-Wen Zhu: Conceptualization, Funding acquisition. Zhi-Qiang Jin: Resources. Lei Zhang: Visualization. Yan Zhu: Conceptualization, Writing-review & editing. Lu Zhang: Writing-review & editing, Supervision.

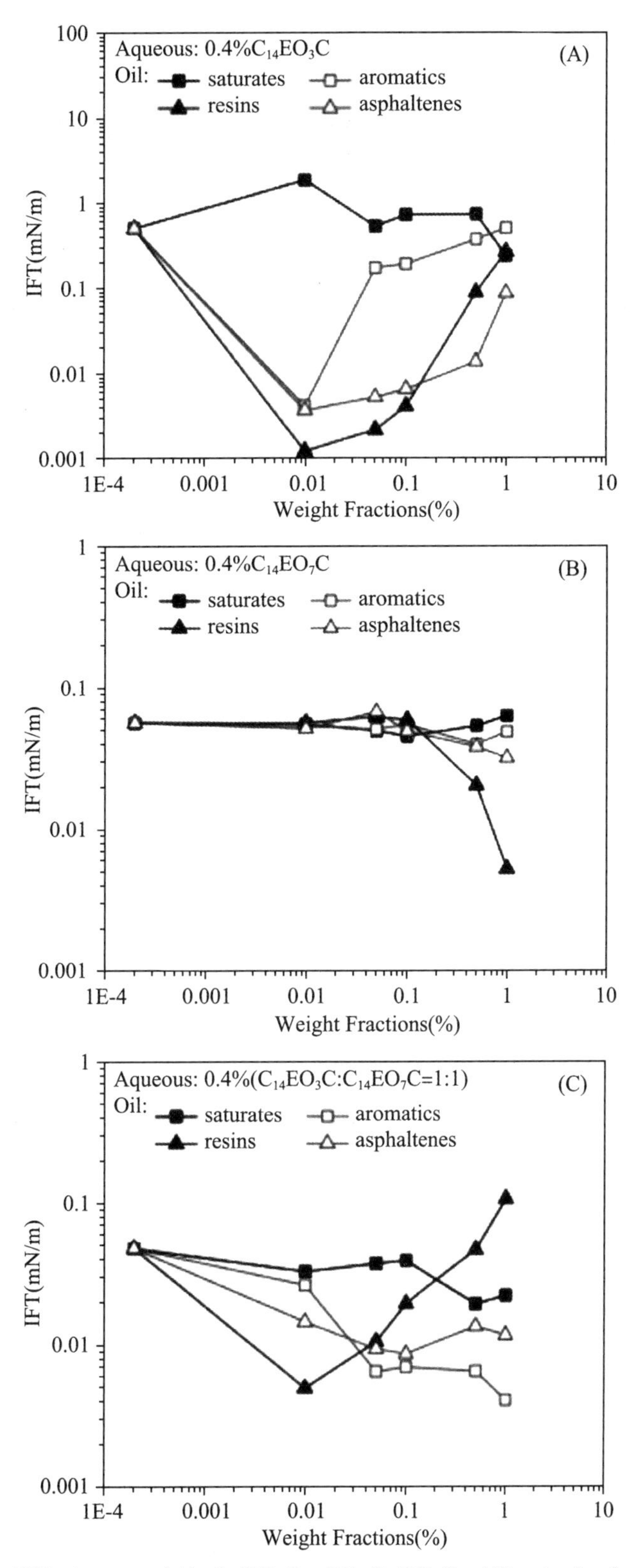

Fig. 8　Equilibrium IFTs between (A) $C_{14}EO_3C$, (B) $C_{14}EO_7C$, (C) mixed solutions against model oils containing different fractions as a function of concentration.

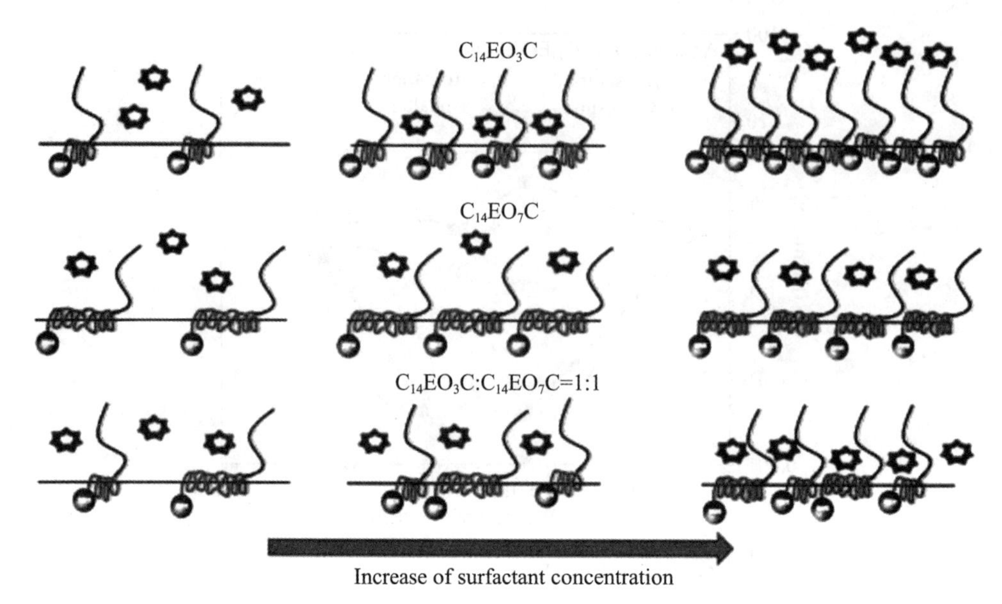

Fig. 9 Schematic mixed adsorption model of surfactant and aromatics or asphaltenes molecules at interface.

Declaration of competing interest

We declare that we have no financial and personal relationships with other people or organizations that can inappropriately influence our work, there is no professional or other personal interest of any nature or kind in any product, service and/or company that could be construed as influencing the position presented in, or the review of, the manuscript entitled.

Acknowledgments

The authors thank financial support from the National Science & Technology Major Project of China (2016ZX05011-003) and National Natural Science Foundation of China (21703269).

References

[1] L. Cash, J. L. Cayias, G. Fournier, D. Macallister, T. Schares, R. S. Schechter, W. H. Wade, Application of low interfacial-tension scaling rules to binary hydrocarbon mixtures, J. Colloid Interface Sci,1977(59):39-44.

[2] N. Pal, N. Saxena, A. Mandal, Studies on the physicochemical properties of synthe-sized tailor-made gemini surfactants for application in enhanced oil recovery, J. Mol. Liq,2018(258):211-224.

[3] P. Pillai, A. Kumar, A. Mandal, Mechanistic studies of enhanced oil recovery by imidazolium-based ionic liquids as novel surfactants, J. Ind. Eng. Chem, 2018(63):262-274.

[4] A. Bera, K. Ojha, T. Kumar, A. Mandal, Water solubilization capacity, interfacial com-positions and thermodynamic parameters of anionic and cationic microemulsions, Colloids Surf. A Physicochem. Eng. Asp,2012(404):70-77.

[5] A. Samanta, A. Bera, K. Ojha, A. Mandal, Effects of alkali, salts, and surfactant on rhe-ological behavior of partially hydrolyzed polyacrylamide solutions, J. Mol. Liq,2018(266):147-159.

[6] N. Pal, N. Saxena, K. V. D. Laxmi, A. Mandal, Interfacial behaviour, wettability alter-ation and emulsification characteristics of a novel surfactant: implications for en-hanced oil recovery, Chem. Eng. Sci,2018(187):200-212.

[7] P. H. Doe, W. H. Wade, R. S. Schechter, Alkyl benzene sulfonates for producing low interfacial-tensions between hydrocarbons and water, J. Colloid Interface Sci,1977(59):525-531.

[8] L. A. Wilson, Physico-chemical environment of petroleum reservoirs in relation to oil recovery systems, in: D. O. Shah, R. S. Schechter (Eds.), Improved Oil Recovery by Sur-factant and Polymer Flooding, Academic Press,1977:1-26.

[9] M. Mulqueen, D. Blankschtein, Theoretical and experimental investigation of the equilibrium oil-water interfacial tensions of solutions containing surfactant mix-tures, Langmuir,2002(18):365-376.

[10] R.-h. Zhao, L. Zhang, L. Zhang, S. Zhao, J.-y. Yu, Effect of hydrophilic-lipophilic ability on dynamic interfacial tensions of alkylbenzene sulfonates, Energy Fuel,2010(24):5048-5052.

[11] A. M. Al-Sabagh, E. M. S. Azzam, S. A. Mahmoud, N. E. A. Saleh, Synthesis of ethoxylated alkyl sulfosuccinate surfactants and the investigation of mixed solutions, J. Surfac-tant Deterg,2006(10):3-8.

[12] C.-C. Lai, K.-M. Chen, Preparation and surface activity of polyoxyethylene-carboxylated modified gemini surfactants, Colloids Surf. A Physicochem. Eng. Asp,2008(320):6-10.

[13] O. Sha, W. Zhang, R. Lu, Synthesis and properties evaluation of nonionic-anionic sur-factants suitable for enhanced oil recovery using sea water, Tenside Surfactant Deterg,2008(45):82-86.

[14] M. Aoudia, R. S. Al-Maamari, M. Nabipour, A. S. Al-Bemani, S. Ayatollahi, Laboratory study of alkyl ether sulfonates for improved oil recovery in

high-salinity carbonate reservoirs: a case study, Energy Fuel,2010(24):3655-3660.

[15] Y. M. Zhang, J. P. Niu, Q. X. Li, Synthesis and properties evaluation of sodium fatty al-cohol polyoxyethylene ether sulfonate, Tenside Surfactant Deterg,2010(47):34-39.

[16] R. K. Mahajan, D. Nandni, Micellization and phase behavior of binary mixtures of an-ionic and nonionic surfactants in aqueous media, Ind. Eng. Chem. Res, 2012(51):3338-3349.

[17] K. C. Taylor, H. A. Nasr-El-Din, The effect of synthetic surfactants on the interfacial be-haviour of crude oil/alkali/polymer systems, Colloids Surf. A Physicochem. Eng. Asp,1996(108):49-72.

[18] A. Witthayapanyanon, E. J. Acosta, J. H. Harwell, D. A. Sabatini, Formulation of ultralow interfacial tension systems using extended surfactants, J. Surfactant Deterg,2006(9):331-339.

[19] N. Pal, K. Samanta, A. Mandal, A novel family of non-ionic gemini surfactants derived from sunflower oil: synthesis, characterization and physicochemical evaluation, J. Mol. Liq,2019(275):638-653.

[20] P. Bouriat, N. El Kerri, A. Graciaa, J. Lachaise, Properties of a two-dimensional asphaltene network at the water-cyclohexane interface deduced from dynamic tensiometry, Langmuir,2004(20):7459-7464.

[21] S. Poteau, J. F. Argillier, D. Langevin, F. Pincet, E. Perez, Influence of pH on stability and dynamic properties of asphaltenes and other amphiphilic molecules at the oil-water interface, Energy Fuel,2005(19):1337-1341.

[22] Q. Shi, D. J. Hou, K. H. Chung, C. M. Xu, S. Q. Zhao, Y. H. Zhang, Characterization of het-eroatom compounds in a crude oil and its saturates, aromatics, resins, and asphaltenes (SARA) and non-basic nitrogen fractions analyzed by negative-ion electrospray ionization Fourier transform ion cyclotron resonance mass spectrometry, Energy Fuel,2010(24):2545-2553.

[23] A. Samanta, K. Ojha, A. Mandal, Interactions between acidic crude oil and alkali and their effects on enhanced oil recovery, Energy Fuel,2011(25):1642-1649.

[24] L. He, X. G. Li, G. Z. Wu, F. Lin, H. Sui, Distribution of saturates, aromatics, resins, and asphaltenes fractions in the bituminous layer of Athabasca oil sands, Energy Fuel,2013(27):4677-4683.

[25] Y.-w. Zhu, R.-h. Zhao, Z.-q. Jin, L. Zhang, L. Zhang, L. Luo, S. Zhao, Influence of crude oil fractions on interfacial tensions of alkylbenzene sulfonate solutions, Energy Fuel,2013(27):4648-4653.

[26] S.-S. Hu, L. Zhang, X.-L. Cao, L.-L. Guo, Y.-W. Zhu, L. Zhang, S. Zhao, Influence of crude oil components on interfacial dilational properties of hydrophobically modified polyacrylamide, Energy Fuel,2015(29):1564-1573.

[27] L. Zhang, L. Luo, S. Zhao, Z. C. X., J. Y. A., J. Y. Yu, Effect of different acidic fractions in crude oil on dynamic interfacial tensions in surfactant/alkali/model oil systems, J. Pet. Sci. Eng,2004(41):189-198.

[28] D. Arla, A. Sinquin, T. Palermo, C. Hurtevent, A. Graciaa, C. Dicharry, Influence of pH and water content on the type and stability of acidic crude oil emulsions, Energy Fuel,2007(21):1337-1342.

[29] Z.-H. Zhou, Q. Zhang, Y. Liu, H.-Z. Wang, H.-Y. Cai, F. Zhang, M.-Z. Tian, L. Zhang, L Zhang, Effect of crude oil fractions on interfacial tensions of novel sulphobetaine solutions, J. Dispers. Sci. Technol,2016(37):1178-1185.

[30] J.-H. Cao, Z.-H. Zhou, Z.-C. Xu, Q. Zhang, S.-H. Li, H.-B. Cui, L. Zhang, L. Zhang, The synergism/antagonism between crude oil fractions and novel betaines solutions in reducing interfacial tension, Energy Fuel,2016(30):924-932.

[31] L. Zhang, L. Luo, S. Zhao, J. Y. Yu, Studies of synergism/antagonism for lowering dynamic interfacial tensions in surfactant/alkali/acidic oil systems part 3: synergism/ antagonism in surfactant/alkali/acidic model systems, J. Colloid Interface Sci,2003(260):398-403.

[32] X. Yang, V. J. Verruto, P. K. Kilpatrick, Dynamic asphaltene-resin exchange at the oil/ water Interface: time-dependent W/O emulsion stability for asphaltene/resin model oils, Energy Fuel,2007(21):1343-1349.

[33] M. Liu, H. Fang, Z. Jin, Z. Xu, L. Zhang, L. Zhang, S. Zhao, Interfacial tensions of oxyethylated fatty acid methyl Ester solutions against crude oil, J. Surfactant Deterg,2017(20):961-967.

[34] X.-W. Song, R.-h. Zhao, X.-L. Cao, J.-C. Zhang, L. Zhang, L. Zhang, S. Zhao, Dynamic interfacial tensions between offshore crude oil and enhanced oil recovery surfactants, J. Dispers. Sci. Technol,2013(34):234-239.

Performance of a good-emulsification-oriented surfactant-polymer system in emulsifying and recovering heavy oil

Wang Yefei[1,2] Li Zongyang[1,2,3] Ding Mingchen[1,2] Yu Qun[3]
Zhong Dong[1,2] Cao Xulong[3]

[1] Key Laboratory of Unconventional Oil & Gas Development (China University of Petroleum (East China)), Ministry of Education, Qingdao, China

[2] School of Petroleum Engineering, China University of Petroleum (East China), Qingdao, China

[3] Exploration and Development Research Institute of Shengli Oilfield, Dongying, China

Abstract: Considering the significant importance of emulsification for heavy oil recovery, a good-emulsification-oriented (GEO) (rather than the conventional ultralow interfa-cial tension (IFT) oriented) surfactant-polymer (SP) system was employed to remove heavy oil. Specifically, the emulsification behavior and viscosity-reduction property of heavy oil in porous media were first studied by flood tests. Then, such a GEO SP was used to displace heavy oil for recovery measurement. It is found that first, as migration increases in porous media, the in situ emulsified oil droplets evolve into smaller and more uniform droplets (without coalescence), thus, exhibit selfadjustment ability to match with, and plug, the pores. An increase in surfactant content and seepage velocity also makes the emulsion's particles smaller and more uniformly distributed. Second, such oil-in-water emulsification indeed helps to reduce heavy oil viscosity in porous media and there is an optimal surfactant content for such viscosity reduction: The lower the water cut, the greater the optimal surfactant content, the worse the viscosity-reduction effect. Third, on its own, the GEO surfactant performs very poorly in recovering heavy oil and is even worse than the single polymer; therefore, such a surfactant should be jointly used with a polymer in an SP system instead of on its own, but an excessive surfactant content in the SP seems to hinder the recovery, while the large slug size tends to contribute significantly to an improved oil recovery.

Correspondence: Mingchen Ding, Key Laboratory of Unconventional Oil & Gas Development (China University of Petroleum (East China)), Ministry of Education, Qingdao 266580, China. Email: Dingmc@upc.edu.cn

Funding information: National Science and Technology Major Project of China,

Grant/Award Number: 2016ZX05058-003-003; National Natural Science Foundation of China,Grant/Award Number: 51504275; Fundamental Research Funds for the Central Universities, Grant/Award Number: 17CX02076

1 INTRODUCTION

As the reserves of conventional light oil begin to dwindle, heavy oil is becoming more and more important for meeting the increasing demand for energy. Heavy oil is usually classed as an unconventional oil resource whose viscosity in reservoirs typically ranges from 50 to 50 000 mPa s.[1] The high viscosity of heavy oil causes poor fluidity which is the key issue restricting its development. The high mobility ratio of water to heavy oil leads to only about 5%—10% of the original oil being recovered in the primary water flooding stage[2]: the poor sweep efficiency caused by the adverse mobility ratio of water and oil is responsible for such low incremen-tal recovery. To reduce the viscosity of heavy oil, and hence enhance its fluidity, thermal recovery methods are usually adopted to enhance the heavy oil recovery rate (considering the great sensitivity of oil viscosity to temperature). Thus, methods such as steam huff and puff injection, steam flooding, in situ combustion, and steam-assisted gravity drainage have all been used.[3-6] These thermal techniques have been successful in certain reservoirs (those with thick pay zones and absence of bottom water). However, they cannot be em-ployed in deep, thin reservoirs due to the severe heat loss during injection. Moreover, consuming a large amount of fuel to generate steam is sometimes not economical or environmentally friendly. Therefore, nonthermal enhanced oil recovery (EOR) methods are required to remove the remaining oil after water flooding.

In recent years, a great deal of research has been under-taken to improve heavy oil recovery using chemical flooding methods, including alkali flooding, alkali-surfactant (AS) flooding, and surfactant-polymer (SP) flooding.[7-9] Adding alkali is cheap and can significantly contribute to water-oil interfacial tension (IFT) reduction, oil-in-water (O/W), or water-in-oil (W/O) emulsification when recovering heavy oil.[10-14] However, the resulting high consumption of alkali and severe scaling problems[15] can restrict the application of alkaline chemical systems. Hence, alkali flooding has not yet been used industrially. In this case, chemical systems based on alkalis may become increasingly more important for the EOR of heavy oil, especially in the context of SP systems. In fact, SP systems have been much researched and applied in the recovery of conventional light oil, taking advantage of the well-known ultralow IFT, emulsification, wettability alternation, and water-oil mobility improvement mechanisms, etc. Following on from the research into chemical flooding in light-oil reservoirs, the IFT behavior of heavy oil-chemical systems has been much investigated, and ultralow IFT has also been suggested for heavy oil recovery applications by some researchers (as it is for light oil).[10-14] Nevertheless, a progressive understanding of the heavy oil recovery mecha-nism is gradually formed such that the role of the emulsification mechanism should be more emphasized in extending the sweep volume[16,17] and recovering heavy oil compared to conventional ultralow IFT. It is believed that, as the heavy oil is very vicious, the poor sweep efficiency (caused by the consequent adverse water-oil mobility ratio) is the primary problem needing to be resolved by SP flooding instead of the conventional displacement efficiency (for light oil).[2,9,18-28]

By chance, the emulsion mechanism bestows dual func-tions in enhancing sweep efficiency. First, it reduces the viscosity of the heavy oil (by forming an O/W emulsion), making it easier to flow from a deep reservoir to the pro-ducing wells.[29,30] Second, it improves the water-oil mobil-ity ratio (by both reducing the viscosity of the heavy oil and plugging the water channels with droplets of emulsi-fied oil), thus, contributing to sweep volume enlargement.

These become highly significant when recovering heavy oil. After a systematic investigation of the contempora-neous situation with respect to heavy oil-chemical flooding, Pei[2] concluded that it is more important to expand the sweep volume than improve the displacement efficiency for heavy oil recovery. This is because of the lower sweep efficiency in the previous water flooding step with heavy oil (compared to light oil). By conducting a series of chem-ical flooding tests on heavy oil, Zhang[18] and Zhou[19] also recognized that the enhanced recovery of heavy oil is mainly due to the plugging of the water channels and not the lowering of the IFT. This is obviously different from light-oil EOR processes where IFT reduction seems to play the dominant role.[20,21] Furthermore, studies have suggested that O/W or W/O emulsions do indeed plug water-flooded channels, which results in improved volumetric sweeping of reservoirs.[22-24] McAuliffe[25,26] found that water flows into less permeable regions, leading to extended sweep volumes when an emulsion enters into the more permeable regions and plug them by the Jamin effect. Yu[27,28] found that a fine-emulsion system exhibits good plugging and sweep-volume-enlargement behavior. More recently, Ding[9] performed a microscopic and macroscopic modeling study of SP flooding of heavy oil reservoirs and found that a good-emulsification-oriented (GEO) SP system clearly outperformed conventional ultralow IFT systems when re-covering heavy oil. This directly verifies the significance of extending the sweep volume using a GEO SP system (instead of a conventional ultralow IFT-oriented system).

If we accept the importance of emulsification in ex-tending sweep volume and recovering heavy oil during SP flooding (according to previous reports), then the detailed emulsification behavior of heavy oil in porous media (in-stead of common bottles), the oil recovery performance, and their influencing factors naturally become the next concerns before applications are devised. However, these issues are in need of clarification as the dominant role of emulsification (over ultralow IFT) in recovering heavy oil has only just been recently recognized[9,18,19] and the existing studies on emulsification behavior have been conducted mostly via bottle tests[31-33] rather than porous media. Therefore, a GEO SP system (that can form O/W emulsions with excellent stability but have non-ultra-low IFT with heavy oil) was adopted in this paper to displace heavy oil (instead of the convention ultralow IFT system with relatively bad emulsion stability).[9] A series of heavy oil in situ emulsification and displacing tests in porous media were designed and conducted to study the emulsification behavior, recovery, and their influencing factors. The results obtained in this research are intended to contribute to the optimization and control of such valuable and prom-ising GEO SP flooding agents in the context of heavy oil recovery.

2 EXPERIMENTAL SECTION

2.1 Fluids and chemicals

The heavy oil used was collected from the Shengli oil field, Dongying, China. The density and viscosity of the oil were measured and found to be 969.2 kg/m^3 (70℃, 0.101 MPa) and 657.0 mPa s (70℃, 0.101 MPa), respectively. The GEO surfactant used is an anionic and nonionic compound variant. The polymer used to prepare the SP system is a partially hydrolyzed polyacrylamide (HPAM) with a molecular weight of 2.0×10^7. The SP system used consists of 0.1%—1.0% surfactant and 0.18% HPAM.

The measured equilibrium IFTs between the SP systems used (0.1% S, 0.3% S, 0.6% S, and 1.0% + 0.18% P) and heavy oil were all at a non-ultra-low level (8.9×10^{-1} mN/m, 3.0×10^{-1} mN/m, 1.8×10^{-1} mN/m and 1.1×10^{-1} mN/m).

The viscosity of the systems was measured at about 33.5 mPa • s (at a shearing rate of 7.34 s^{-1}). The emulsification behavior of this GEO SP system has been thoroughly investigated via bottle tests (including emulsifying capacity and emulsion stability) in a previous paper,[9] and it indeed exhibits an excellent ability to stabilize heavy oil emulsions. The salinity of the simulated formation water is 9.754 g/L (Table 1).

2.2 Heavy oil emulsification tests in sand packs

Emulsification is the key to extending the sweep volume (via the Jamin effect of the emulsified oil droplets), reduc-ing oil viscosity (by the formation of an O/W emulsion), and finally enhancing heavy oil recovery. In this part, rather than the commonly employed bottle test method,[31-33] a series of flood tests were designed and carried out using sand packs. During the tests, liquid samples (containing emulsified oil droplets) produced from sand packs were observed microscopically using a digital microscope; then, images of those emulsified oil droplets were obtained, and finally, statistical analysis based on these images was conducted using ImageJ software for particle size measurement. Thus, we were able to judge the morphologies and size distributions of the emulsions present.

The sand packs used for the flood tests had diameters of 2.5 cm, lengths of 10 and 80 cm, and a permeability of—1 300 mD. Figure 1 shows a schematic representation of the layout of the flood test apparatus. After the models were prepared and their permeability measured, they were saturated with oil, aged for 24 hours, and then subjected to primary water flooding (to a high water cut of 98.0%). Three different SP injection schemes were then implemented:

1. A flood test using a long sand pack (80 cm; permeability—1 300 mD) in which three sampling points were distributed along the axial direction. After water flooding, a large volume of the SP system (1.0 pore volume (PV), 0.3% S+0.18% P) was continuously injected into the model at a rate of 0.25 mL/min. When the injection volume of SP reached 0.3 PV, 0.5 PV, and 1.0 PV, respectively, the liquid samples (one or two drops per sample) were collected from the sampling points and outlet of the model directly onto a slide for online and immediate observation and analysis by a digital microscope and Image J software. This experiment allows the emulsification of the heavy oil to be determined and the effect of migration distance on the emulsion could also be deduced.

2. A series of flood tests using short sand packs (10 cm; perme-ability—1 300 mD) and SP systems with different surfactant contents. Following the previous water flooding step, a limited volume (0.3 PV) of the SP systems was used with 0.1%, 0.3%, 0.6%, and 1.0% surfactant content (injected at a rate of 0.25 mL/min into three separate short sand packs). After the injection of a 0.3 PV SP system, the liquid produced from the outlets of the models was sampled and used for further microscopic observation and particle size analysis. Thus, the effect of varying the surfactant content on the heavy oil emulsification process could be derived.

3. A series of flood tests using a constant SP system (0.3% S+0.18% P) injected at different rates (0.25, 0.5, 1.0, and 1.5 mL/min, equivalent to seepage velocities of 2.0, 4.0, 8.0, and 12.0 m/d) into three short sand packs (10 cm; perme-ability—

1 300 mD). As before, after the injection of 0.3 PV SP system using varied injection rates, the liquid samples were collected from the outlets of the models for particle size analysis. Then, by comparing the morphologies and size distributions of the emulsions produced in these experiments, the effect of seepage velocity on the heavy oil emul-sion could subsequently be determined.

All of the abovementioned experiments were carried out at 70℃ which is the reservoir temperature of the Shengli oilfield.

Table 1 The composition of the formation water in the Shengli oilfield

Content(mg/L)	Ion							
	Cl^-	SO_4^{2-}	CO_3^{2-}	HCO_3^-	$Na^+ + k^+$	Ca^{2+}	Mg^{2+}	Total
	5 423.0	0.0	0.0	656.0	3 414.0	193.0	68.0	9 754.0

2.3 In situ viscosity measurements

The formation of O/W emulsion helps to decrease the viscos-ity of the heavy oil, enhance its fluidity, improve water-oil mobility, increase sweep efficiency, and, ultimately, increase oil recovery. In order to determine the ability of such a GEO surfactant to improve the mobility of the heavy oil as a result of viscosity reduction, heavy oil and surfactant were simultaneously injected into sand packs (10 cm; permeability *c.* 1 300 mD) at a total injection rate of 0.5 mL/min (at 70℃). Then, the stable value of the injection pressure was recorded for in situ viscosity calculation of the oil and water mixtures according to Darcy's law. Furthermore, tests using a variety of surfactant content (0.0%, 0.1%, 0.2%, 0.3%, 0.4%, 0.5%, and 0.6%) and water cut (80%, 70%, 60%, 50%, and 40%) were conducted to investigate the effects of these factors on the reduction in the viscosity of the heavy oil resulting from emulsification.

It is worth pointing out that *only* the surfactant was employed in these experiments; that is, no polymer was added to these systems. This is because the viscosity of the polymer solution varies with the surfactant content and also decreases during the injection process due to shearing degradation. This will affect the measured equilibrium injection pressure and, therefore, the in situ viscosity calculated for the mixture.

2.4 Heavy oil recovery tests

The EOR performance of the GEO SP system and the factors affecting it were finally investigated via another series of flood tests using medium-sized sand packs (30 cm; permeability—1 300 mD). A total of three experimental schemes were then implemented involving flood tests employing:

1. Three different chemical systems: 0.3% S, 0.18% P, and 0.3% S+0.18% P;

2. SP systems with different surfactant contents: 0.1%, 0.3%, 0.6%, and 1.0% (+0.18% P);

3. Different SP slug sizes: 0.1, 0.3, 0.7, and 1.0 PV (for a constant 0.3% S + 0.18% P system).

These experiments thus allow the effect of system type, surfactant content, and slug size on the recovery of heavy oil to be found. The specific experimental procedure used is as follows. First, the wet-packed sand pack was flooded with water to measure its permeability. Then, the temperature of the system was increased to 70℃ after which crude oil was pumped into the model to displace the water and allow oil saturation. Thereafter, the model was flooded with water until the amount of oil produced became negligible (water cut ⩾ 98%). Then, the target chemical solution was injected to displace the remaining oil. This was followed by postwa-ter flooding until the volume of oil gathered in the collector became negligible.

3 EXPERIMENTAL RESULTS AND DISCUSSION

3.1 Emulsification behavior of heavy oil in porous media

The emulsification of the heavy oil is the key to reducing the oil's viscosity and the plugging of the water channels by the emulsified oil droplets which, in turn, enlarges the sweep volume. The behavior of heavy oil (with respect to emulsification capacity, emulsion stability, and emulsion size distribution, etc) has been extensively studied using conventional bottle tests.[31-33] However, its behavior in porous media is different from that in the bottle tests due to the different emulsion-generation mechanisms involved. To be more specific, the shearing action acting on the water-oil interface and snap-off effect occurring at the throats of the pores are thought to be the main mechanisms responsible for emulsion formation in porous media.[34] In contrast, emulsification in bottle tests mainly occurs because of the simple shaking or stirring action employed and the subsequent mixing of the water and oil. As a result, chemical additives behave differently in bottles and porous media. For example, it has been found that an ultralow IFT SP system can give excellent emulsification results using heavy oil in bottle tests, but produce virtually no emulsion in porous media (where it was more inclined to remove oil in the form of "oil wires" instead of an emulsion).[9]

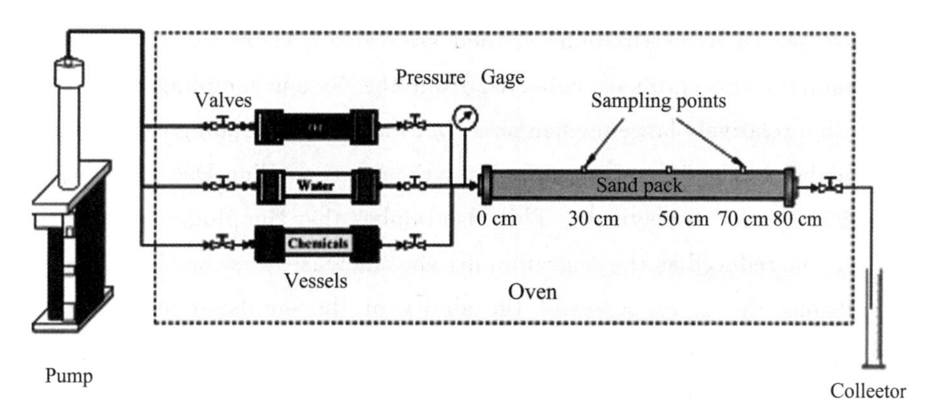

Figure 1 Schematic diagram of the apparatus used in the experiments

Naturally, different emulsification mechanisms may also lead to different factors affecting the emulsification performance. Therefore, in order to more accurately judge the emul-sification behavior of the heavy oil in porous media, attention was paid to the morphologies of the heavy oil emulsions in the sand packs and their size distributions. In addition, the effects of migration, surfactant content, and seepage velocity were also considered.

3.1.1 Effect of migration in porous media

A flood test was first carried out in a long sand pack with three sampling points distributed along its length. The liquid samples collected from these points and the outlet of the model were observed microscopically at different SP injection volumes of 0.3, 0.5, and 1.0 PV (Figure 2).

Figure 2 illustrates that by the time 0.3 PV SP had been injected into the sand bag, some of the heavy oil emulsion had started to appear in the front part of the model (0—50 cm region). At this stage, the chemicals had not reached the back of the model (50—80 cm) which is still in the pure water flooding state so there is almost no emulsion present. Comparing the images obtained from these two parts of the model, it is apparent that pure water alone is hardly able to emulsify the heavy oil at all (at the end of that flooding stage), but the SP system employed is indeed able to do so.

As the injection of the SP continued, the number of emul-sion particles observed clearly became larger. After the whole 1.0 PV of the SP had been injected, the emulsified oil in the 0—30 cm part had migrated to the rear parts of the model so almost no emulsion was found in this part. In the 30—80 cm part, many well-formed heavy oil emulsions were clearly present. This indicates that more and more emulsions were formed with increased migration distance (by the increasing amount of snap-off time occurring with the pores). Coalescence of the emulsion and resulting decrease in number of emulsion particles, as claimed in some existing results,[35,36] did not appear to happen.

Figure 2 also reveals that the emulsion particles got smaller as the migration distance increased. The distributions of their sizes also became more uniform (Figure 3). For example, the emulsion collected from the 50 cm sampling point contained particles with a relatively large median size (D50) and corresponding to 54.7 μm. As it mitigated further to the 70 and 80 cm points, the corresponding D50 values de-creased to 43.0 and 39.0 μm, respectively. This also implies that the plugging capacity of the emulsion may be reduced as the migration distance increases (as the larger the particles in the emulsion, the more powerful the ability of the emulsion to plug the water channels.[34,37] Fortunately, the decrease in the size of the emulsion particles gradually slows down as the migration distance increases. For instance, the median size decreases by 11.7 μm as the emulsion moves from the 50 to 70 cm point (from 54.7 to 43.0 μm)—the corresponding decrease as it moves from 70 to 80 cm is only 4.0 μm. Additionally, from the emulsion-size-distribution curves in Figure 3, it can be seen that, with increased migration, such curves become narrower and higher, indicating the increas-ingly concentrated and more uniform size distribution of emulsions. The size and size-distribution variation observed as the heavy oil emulsion migrated through the sand pack indicates it has the ability to undergo "self-adjustment" to match with, and plug, the pores in the porous media.

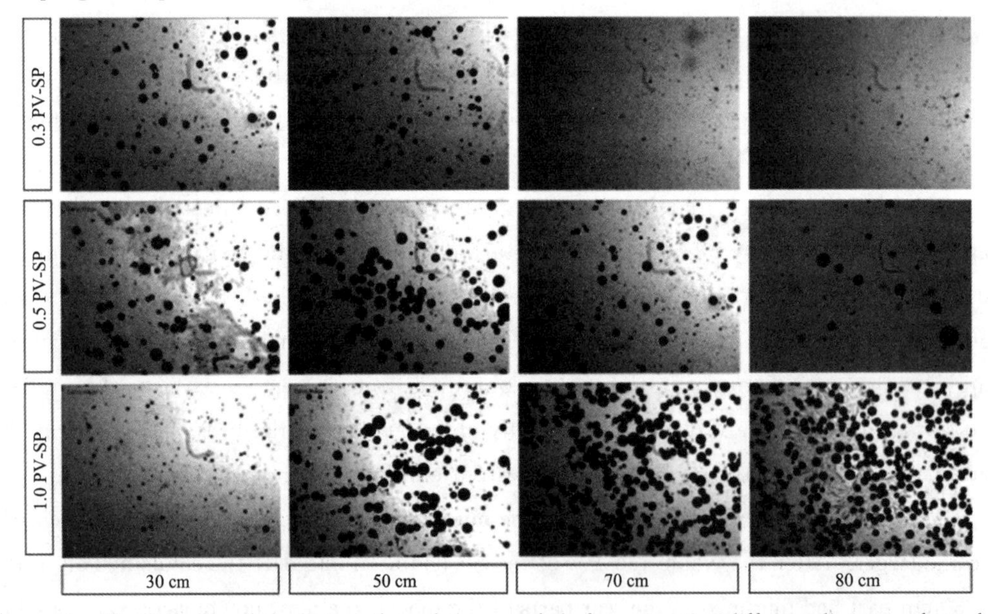

Figure 2 Morphologies of the heavy oil emulsion sampled after migrating different distances through a sand pack (30, 50, 70, and 80 cm) and injecting different amounts of the SP system (0.3, 0.5, and 1.0 PV)

The appearance of a large number of emulsified oil drop-lets in the sand packs (as shown in Figure 2) shows that the heavy oil was readily emulsified even with a limited migration of the SP system (ie, a range of 0—80 cm). In other words, once this GEO SP system is injected into a reservoir, it will work efficiently in emulsifying heavy oil even in the near-wellbore regions. Then, those well-formed emulsions will migrate to the depths of the reservoir, evolve into smaller but more nu-merous oil droplets with better uniformity, trigger and help the generation of new emulsion particles,[9] simultaneously, plug the existing water channels, and contribute to oil recovery.

3.1.2 Effect of surfactant content

The size of the emulsion particles is the key to plugging the water-flooded channels. As observed above, as the emulsion migrates it selfadjusts to fit the pore structure. However, apart from migration distance, there are other factors that may affect the size of the particles in the heavy oil emulsion including surfactant content and seepage velocity. These factors will, therefore, also affect the plugging capacity of the emulsion and need to be investigated.

Figure 4 illustrates the microscopic morphologies and size distributions of emulsions produced using different amounts of surfactant (0.1%, 0.3%, 0.6%, and 1.0%). It is apparent that no obvious signs of an emulsion appear in the sample collected when the surfactant content is 0.1%. It seems that too low a surfactant content cannot endow the SP system with the emulsifying ability to handle the heavy oil. As the sur-factant content is increased to 0.3%, more of it is available for adsorption at the oil-water interface to form and stabilize emulsion particles. Therefore, a large number of large emul-sified oil droplets appear at this concentration resulting in a D50 value of 62.0 μm. To determine whether even higher surfactant contents facilitate a better level of heavy oil emul-sification, surfactant contents of 0.6% and 1.0% were also used. According to Figure 4, the increased surfactant content increases the uniformity of the distribution of the particle sizes in the emulsion and reduces the median size of the oil droplets present (giving smaller D50 values of 45.5 μm and 42.0 μm at surfactant contents of 0.6% and 1.0%). Considering the fact that larger emulsion particles result in a more powerful plugging effect in the water channels,[34,37] increasing the surfactant content from 0.3% to 1.0% may be detrimental to the SP system's plugging capacity. Therefore, too high a surfactant content may be unfavorable and hence undesirable in this respect.

3.1.3 Effect of seepage velocity

Different injection rates (0.25, 0.5, 1.0, and 1.5 mL/min) were used to produce different seepage velocities (2. 0, 4. 0, 8. 0, and 12. 0 m/d) in the sand packs. The morphologies and distributions of the particles produced in the resulting emulsions were then determined (Figure 5). As the velocity increases, the oil droplets in the emulsions clearly tend to decrease in size. At first, the decrease is quite rapid (D50 falls from 62. 0 to 46. 3 μm as the velocity increases from 2. 0 to 4. 0 m/d). Itthen decreases more slowly, showing signs of stabilizing at higher injection rates (D50 falls to 45. 7 and 38. 6 μm when the seepage velocities are 8. 0—12. 0 m/d). At the same time, the oil droplets also become more uniform in size (the distribution curve becomes narrower).

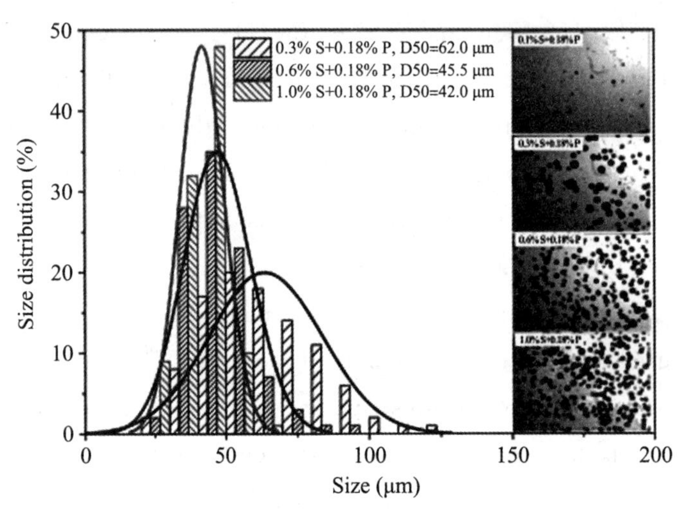

Figure 3 The distributions of the sizes of the particles in the emulsion after it had migrated different distances through the sand pack (50, 70, and 80 cm). The data shown correspond to the injection of 1. 0 PV of the 0. 3% S + 0. 18% P system (injected at a rate of 0. 25 mL/min)

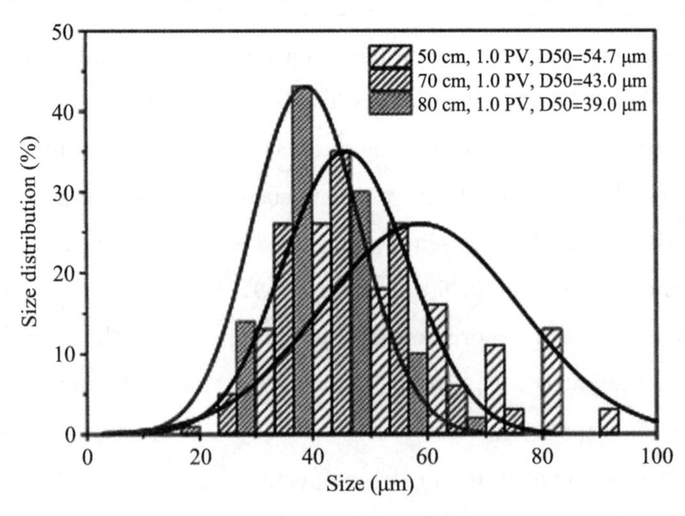

Figure 4 Morphologies and size distributions of emulsion particles produced using different surfactant contents (0. 1%, 0. 3%, 0. 6%, and 1. 0% injected at a rate of 0. 25 mL/min)

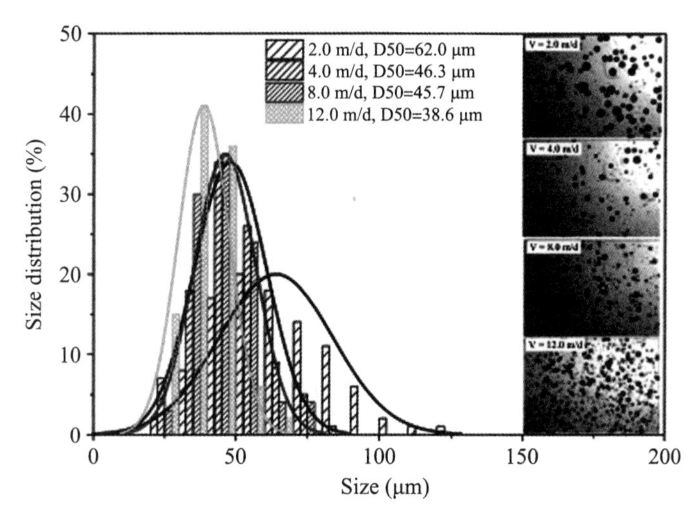

Figure 5 Morphologies and size distributions of emulsion particles produced using different seepage velocities (2.0, 4.0, 8.0, and 12.0 m/d). The SP system used in each case consisted of 0.3% S+ 0.18% P

It may be that larger flow rates lead to stronger shearing and tensile effects being experienced by the oil droplets via the pore throats, and this enhanced breaking effect leads to smaller particles forming in the emulsion: Whatever the explanation, this phenomenon needs to be kept in mind. High injection rates can accelerate oil production and thus help produce more oil in a certain amount of time (making production more economical). However, the resulting high seepage velocity will reduce the sizes of the emulsified oil droplets produced and potentially weaken their ability to plug the pores in the reservoir. Li also reported a phenomenon whereby the plugging capacity of the emulsion decreased with increased seepage velocity.[38]

3.2 Reduction in the viscosity of heavy oil due to emulsification

Reducing the viscosity of heavy oil (by forming an O/W emulsion) and hence enhancing its fluidity is one of the main mechanisms by which such GEO SP system function.[39] The reduction in viscosity achieved is closely related to the emulsification behavior of the heavy oil in porous media and is also influenced by many other factors (eg, surfactant content). In this section, different surfactant solutions were injected into sand packs, the equilibrium injection pressures were measured, and in situ viscosities were calculated according to Darcy's law. At the same time, various oil/water ratios were used (2∶8, 3∶7, 4∶6, 5∶5, and 6∶4) as realized by separately controlling the same injection rate of oil and water, to simulate, on the one hand, different timings of water cut in water-flooded reservoirs (80%, 70%, 60%, 50%, and 40%, respectively). On the other hand, these measures also can reflect the flooding processes in zones with different water cuts in real reservoirs because the development stages of different water cuts (or oil/water ratio) will be experienced in the process of real reservoir development, and the water cuts (or oil/water ratio) duringwater flooding at different positions of a reservoir may also be different due to the varied concentration of the remaining oil.

Figure 6 shows plots of the dynamic injection pressures measured as a function of the injection volume for two cases (water cuts of 80% and 40%). The pressure drop across a sand pack increases at first as the oil and water are simultaneously injected. Then, it tends to stabilize as a steady state is approached. The pressure drops measured when the water cut is lower (eg, 40%) are clearly greater than those when the water cut is higher (eg, 80%). This indicates that reducing theviscosity of heavy oil (via, eg, emulsification) is more necessary in low water cut development stages of a reservoir (or in low water cut areas of a reservoir). The steady state pressure-drop values obtained at the end of oil-water injection processes shown in Figure 6 can be used to calculate the in situ viscosities of the oil-water mixtures. The corresponding viscosity-reduction rates can then be calculated as well, as shown in Figure 7.

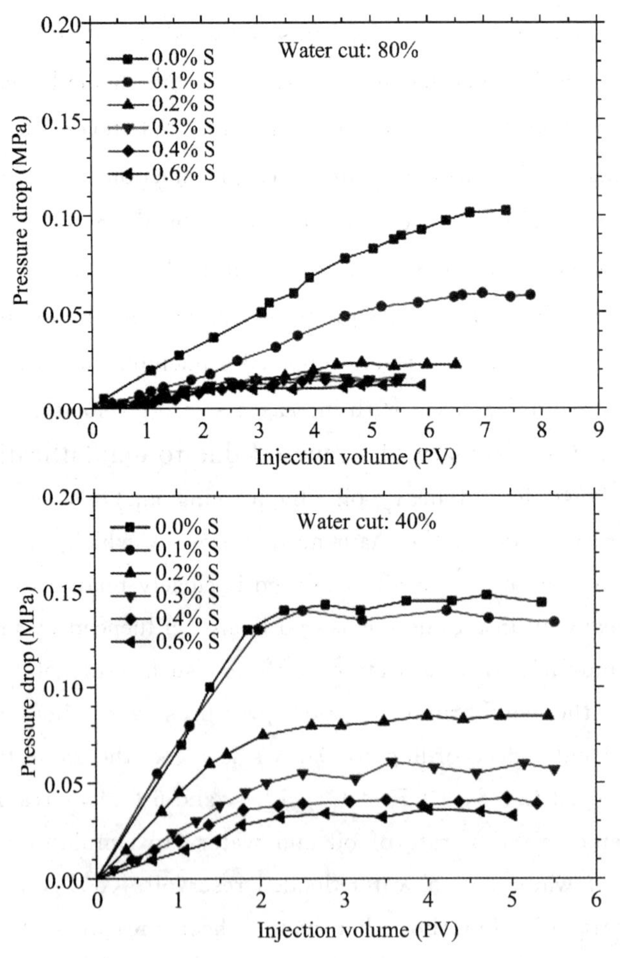

Figure 6 The pressure drops measured during the coinjection of oil and water for different water cuts and surfactant contents

Figure 7 demonstrates that the in situ viscosities of the oil-water mixtures decrease rapidly at first as the surfactant content is increased. Then, the decrease in viscosity slows down and gradually stabilizes. The addition of this GEO surfactant is therefore clearly able to reduce the viscosity of the heavy oil-water mixtures employed. The reduction in the size of the particles in the O/W emulsion and increase in uniformity of their size distributions achieved by increasing the surfactant content (as shown in Figure 4), therefore, contribute to reducing the viscosity of the heavy oil (via the formation of the O/W emulsion). However, there appears to be an optimal surfactant contentwhen it comes to reducing the viscosity of the heavy oil. The 80% water cut data were taken as an example. The in situ viscosities are 81.2, 46.9, 18.3, 12.7, 11.1, and 9.5 mPa s when the surfactant contents are 0.0%, 0.1%, 0.2%, 0.3%, 0.4%, and 0.6%, respectively. The corresponding viscosity-reduction rates are 0.0%, 42.1%, 77.4%, 84.3%, 86.3%, and 88.2%, respectively. Therefore, the optimal surfactant content to reduce the viscosity of the oil seems to be about 0.2%. Above this optimal value, even though the emulsified oil droplets do become smaller and more uniform in size (see Figure 4), the benefits gained do not allow further significant reduction in oil viscosity (as shown in Figure 7). In other words, although we can reduce the viscosity of the heavy oil by increasing surfactant content (because of the reduced size and increased uniformity of the emulsion), beyond the optimal surfactant concentration the additional changes in the properties of the emulsion are less efficient in further reducing the viscosity of the oil.

In addition, comparing the in situ viscosities calculated under different water cut conditions, it can also be seen that the lower the water cut, the greater the optimal surfactant content required to reduce the viscosity of the heavy oil, and the worse the viscosity-reduction effect even with suchoptimal content.

3.3 Heavy oil recovery

3.3.1 Relative performance of S, P, and SP systems

To investigate the details of the roles played by the surfactant and polymer components in displacing heavy oil, flood tests were carried out using three different chemical systems: 0.3% S, 0.18% P, and 0.3% S+0.18% P. To help eliminate measurement errors, longer sand packs (30 cm; permeability *c.* 1 300 mD) were used for these experiments with larger pore volumes and saturated oil volumes. The measured oil recovery factors and pressure drops recorded during S,P, and SPflooding are presented in Figures 8 and 9.

Although the surfactant on its own has a significant viscosity-reduction capacity according to the simultaneous injection results shown in Figure 7, it is very disappointing to see that it yields a very poor 2.1% increase in oil recovery factor (Figure 8). This reflects a fact that it is difficult for the surfactant on its own to remove and emulsify the initially nonflowing heavy oil after water flooding. Only when the oil is already flowing with the surfactant solution (as in our simultaneous oil and water injection experiments) can the surfactant solution effectively emulsify the heavy oil due to the flow-induced mixing effect between the oil and surfactant solution. In other words, before attempting to use surfactant solution to emulsify heavy oil in porous media, we first need to make them flow and mix together. Then, emulsification can occur through the snap-off effect caused by the pores. Actually, the limited pressure increase associated with 0.3% S injection (shown in Figure 9) also suggests that poor activation and emulsification of the heavy oil occur. This is because such behavior would result in a high resistance to flow and hence pressure drop.

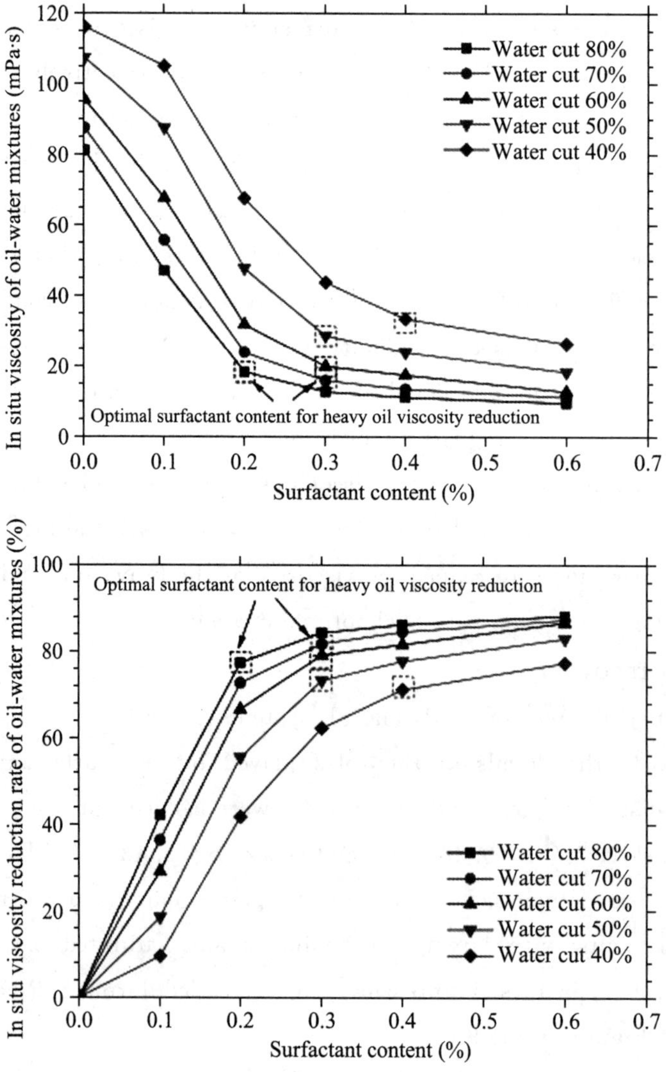

Figure 7 The insituviscosities of the oil-water mixtures and corresponding viscosity-reduction rates plotted as a function of the surfactant concentration

Figure 8 illustrates that a much better 9. 2% increase in recovery factor was obtained when 0. 18% P was injected (compared to that achieved using S-only flooding). Clearly, therefore, the polymer outperforms the surfactant when used on its own to recover heavy oil. This again confirms that the improvement in water-oil mobility and subsequent enlargement in sweep volume (by the viscous polymer solution) is very important when it comes to activating and removing heavy oil. The high injection pressure realized during P flooding (Figure 9) also illustrates the greater ability of the polymer to extend sweep efficiency and remove heavy oil.

Finally, when the surfactant and polymer were used together in the form of an SP system (0. 3% S+0. 18% P), the highest incremental recovery factor of 15. 6% was obtained. The excellent performance of the SP system can be explained in terms of oil displacement characteristics of the surfactant and polymer described above. When the SP system is first injected, the polymer component starts to improve the mobility of the initially static oil and enlarges the sweep volume. The flowing oil then gradually mixes with the SP system and the surfactant component begins to emulsify the heavy oil. Once emulsified oil droplets are formed, they will, in turn, help the polymer component to extend the sweep volume by plugging the water channels (via the Jamin effect). As a result, the oil flow will improve further, which will generate more emulsions. In addition to their contribution to plugging, the emulsion formed also helps to reduce the viscosity of the heavy oil which makes it more mobile (see Figure 7). This may be the main reason for the lower flooding pressure observed in the SP system compared to that associated with the single polymer shown in Figure 9. Ultimately, it is the synergism between the polymer and surfactant that makes the SP system the most potent system for recovering heavy oil. Put briefly, without the assistance of the polymer to extend the sweep volume, the GEO surfactant flooding is very poor and should not be applied on its own to recover heavy oil; therefore, such a surfactant should be jointly used with a polymer, letting the oil first polymerize, then, when emulsified by the surfactant, this, in turn, reduces its viscosity and extending sweep volume by emulsification.

3. 3. 2 Effect of surfactant content on SP flooding

Surfactant-polymer systems with higher concentrations of surfactant will make a greater contribution to the emulsification of the heavy oil. The particles in the emulsion will be smaller and more uniformly distributed in size (Figure 4), and this helps to reduce the viscosity of the heavy oil (Figure 7). However, oil droplets that are too small may not be desirable when it comes to plugging the water-flooded channels.

Therefore, we need to determine whether it is worthwhile increasingthe surfactant content of an SP system (or not) from the point of view of the contribution such an action will make to the recovery of heavy oil. To this end, the oil recovery factors measured as a result of SP flooding using systems with different surfactant contents (0.1%, 0.3%, 0.6%, and 1.0%) are shown in Figure 10.

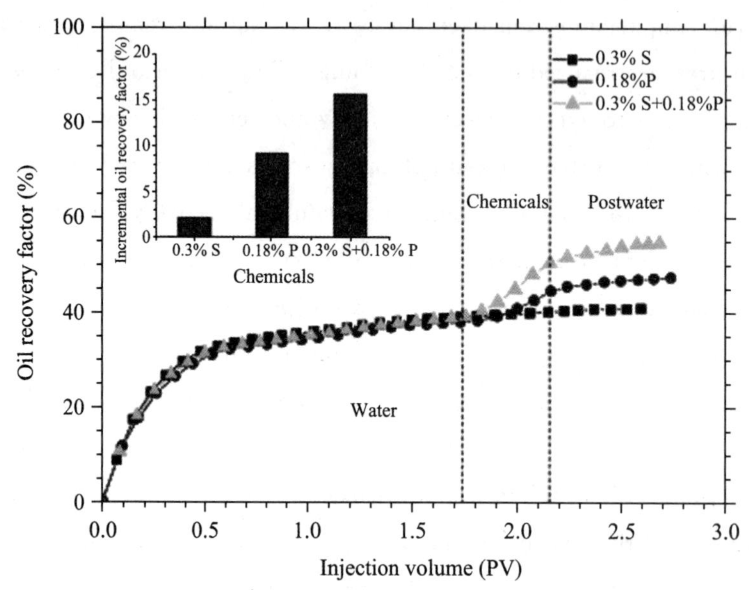

Figure 8 Oil recovery factors obtained using different chemical flooding processes: S, P, and SP

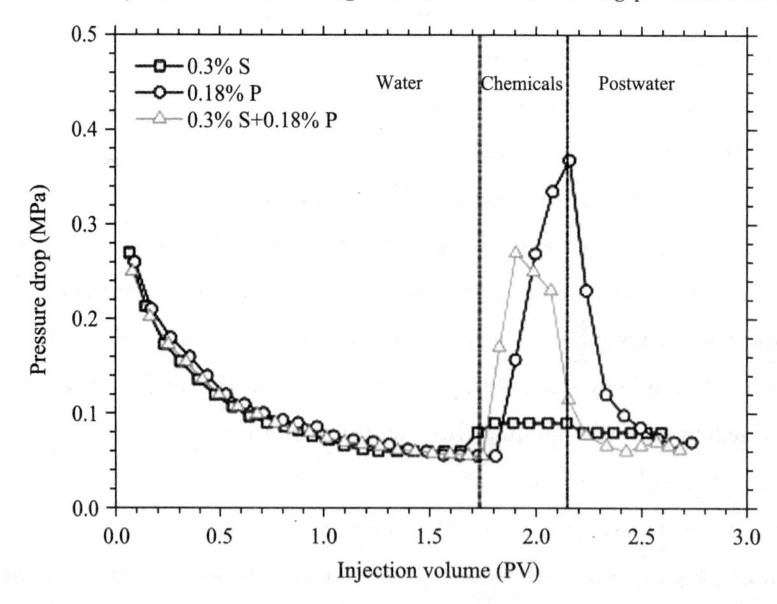

Figure 9 Pressure drops recorded during the different chemical flooding processes: S, P, and SP

The SP system with the lowest surfactant content (0.1%) is hardly able to emulsify the heavy oil (see Figure 4) as there is not enough surfactant adsorbed onto the oil-water interface to stabilize the emulsion. The high injection pressure shown in

Figure 11 also suggests that this SP system provides the worst performance in terms of reducing the viscosity of the heavy oil via emulsification. As a result, it produces the lowest incremental oil recovery rate obtained (10.5%).

Increasing the S content to 0.3% allows the heavy oil to be well emulsified in the porous media (Figure 4). Together with the effect of the polymer on viscosity, this SP system (0.3% S+0.18% P) therefore has a good ability to removeheavy oil (recovery is increased by 15.6%). It also causes the injection pressure being a little lower than that of the 0.1% S+0.18% P system, which indicates it has a strong emulsifying capacity and results in a good reduction in the viscosity of the heavy oil.

Further increasing the surfactant content of the SP system to 0.6% and 1.0% results in the incremental oil recovery factors decreasing to 11.0% and 12.0%, respectively: According to the viscosity curves shown in Figure 7, a surfactant content of 0.6% should reduce the viscosity of the oil very well as a result of O/W emulsification. The low injection pressure of the corresponding SP system (Figure 11) also suggests this is the case. However, the emulsion formed will consist of particles that are smaller (as observed in Figure 4) and this will subsequently weaken the capacity of the emulsion to plug the water-flooded channels which makes the SP flooding process less effective. Thus, too high a surfactant content is not desirable for such GEO SP systems when recovering heavy oil.

3.3.3 Effect of slug size on SP flooding

The effect of slug size on oil recovery is one of the main concerns in chemical flooding. As a result, many related studies have been carried out, especially in relation to the EOR of light oil using chemical systems. Typically, a slug size of about 0.3—0.6 PV is usually recommended for recovering light oil, or heavy oil of relatively low viscosity.[40,41] However, the optimum slug size required with GEO SP systems used to remove "ordinary" heavy oil (ie, of relatively high viscosity like the 657.0 mPa·s samples used in this research) is still not clear. In this section, SP slugs of different sizes (0.1, 0.3, 0.7, and 1.0 PV) were injected into sand packs to displace the heavy oil after water flooding. Figure 12 shows the oil recovery factors thus determined.

Figure 12 shows that the incremental recovery factor increases significantly as the slug size is increased (slug sizes of 0.1, 0.3, 0.7, and 1.0 PV yielding incremental recovery factors of 6.9%, 15.6%, 25.7%, and 29.7%, respectively). The incremental recovery factor clearly does not increase linearly with slug size. Instead, the rate at which it increases slows down with increasing slug size, but the increase in recovery factor is still considerable when the size is enlarged from 0.7 to 1.0 PV. Therefore, a large slug size approaching 1.0 PV may be the optimal dosage to employ when carrying out SP flooding using this GEO system. This is very different to the optimal size recommended for light-oil flooding processes, that is, 0.3—0.6 PV (beyond which further increases in slug size only bring slight increases in incremental recovery factor).[40,41]

The high viscosity and poor fluidity of heavy oil may be responsible for significantly different optimal slug sizes being required during chemical flooding. More specifically, heavy oil is just too viscous to flow after water flooding, so pure water is hardly able to build up a high enough flow resistance (and pressure in the water-flooded channels) to force the subsequent water flow to reach the regions initially saturated by the heavy oil, and hence to remove it. When the SP system is injected, it is able to establish high resistance in such channels due to its higher viscosity compared to pure water. Thus, the displacing agents that follow will migrate to unswept regions and displace the remaining oil there. In fact, this is one of the main mechanisms by which SP flooding has been recognized to act during conventional light-oil flooding.

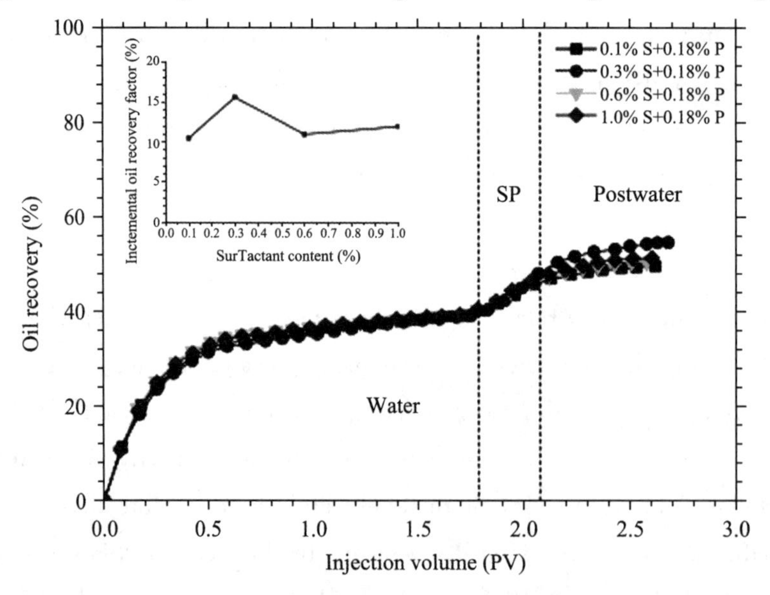

Figure 10 Oil recovery factors achieved by SP flooding with systems having different surfactant contents (0.1%, 0.3%, 0.6%, and 1.0%)

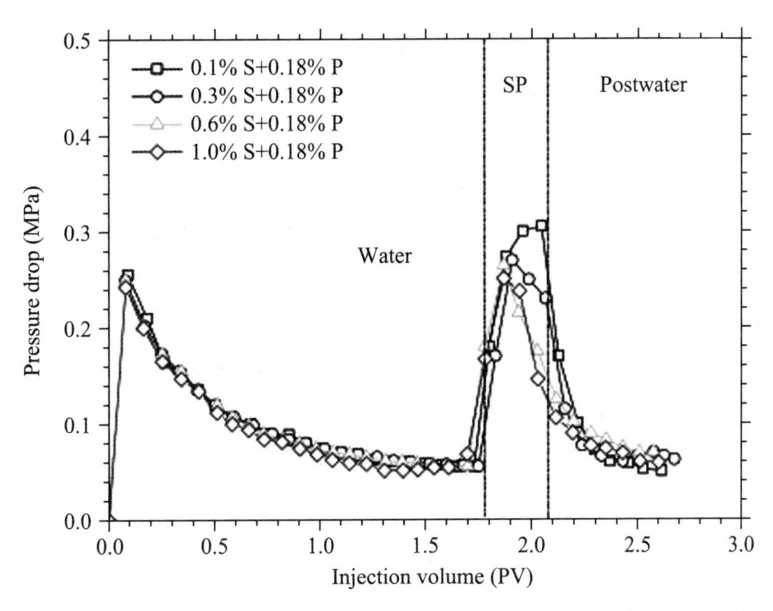

Figure 11 Pressure drops measured during SP flooding with systems having different surfactant contents (0.1%, 0.3%, 0.6%, and 1.0%)

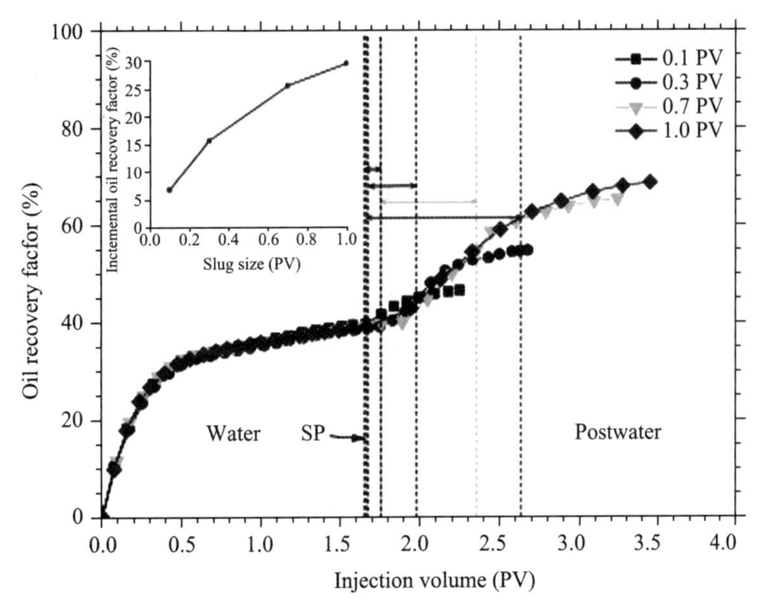

Figure 12 Oil recovery factors achieved by carrying out SP flooding using slugs of different sizes (0.1, 0.3, 0.5, and 1.0 PV)

Unfortunately, high polymer concentrations are often not allowed during heavy oil SP flooding because of limitations imposed by injectivity and economy. As a result, the viscosities of SP systems are far lower than that of heavy oil (in this research, the viscosities of the SP system and heavy oil are about 33.5 and 657.0 mPa · s, respectively, so the water-oil mobility ratio is close to 20). Therefore, the ability of the SP to increase flow resistance in the water-flooded channels is relatively limited so that the enlargement of the sweep volume is also limited. Because of this, increasing the SP slug size to establish larger resistance in the water channels becomes significantly important for extending the sweep volume and recovering heavy oil. Increases in injection pressure as larger slugs are used are clearly observed in Figure 13, indicating that higher flow resistance is built up using larger SP slug sizes. As a result, the incremental recovery factor grows significantlywith increased slug size, as shown in Figure 12, but for light-oil flood processes, it is significantly less viscous, so a much lower resistance is needed in the water-flooded channels to make the subsequent displacing agents migrate to the unswept regions (and hence remove the light oil located there). Therefore, the flow resistance established by an SP system with a relatively modest viscosity (a few dozen mPa can meet such requirements (as the SP and light oil have similar viscosities). As a result, a large slug of SP is not needed to build up high flow resistance in the water-flooded channels when recovering light oil.

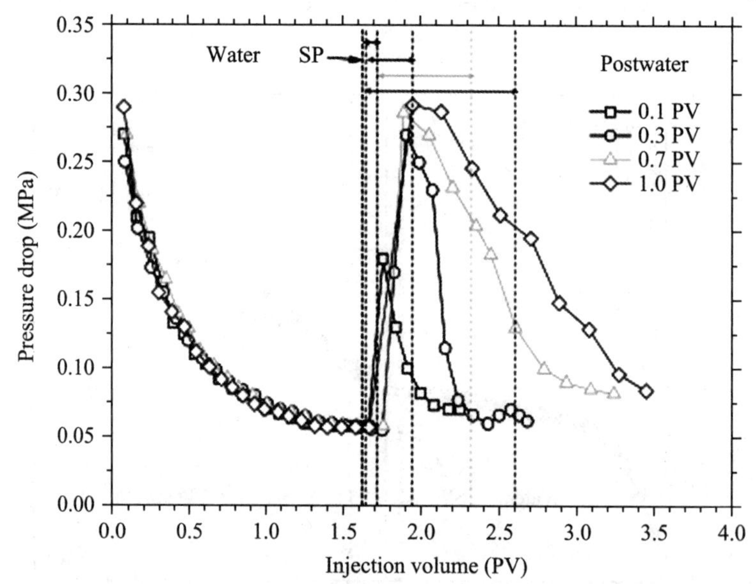

Figure 13 Pressure drops measured during SP flooding experiments using slugs of different sizes (0.1, 0.3, 0.5, and 1.0 PV)

4 CONCLUSIONS

1. Heavy oil could indeed be emulsified in situ by the GEO SP system in porous media. The resulting O/W emulsion undergoes "self-adjustment" when migrating in porous media, evolving into a smaller emulsion with better uniformity to match with, and plug, the pores. Increasing the surfactant content and seepage velocity also results in smaller and more uniform emulsions.

2. Such oil-in-water in situ emulsification indeed helps to reduce heavy oil viscosity

in porous media: There is an optimal surfactant content for such viscosity reduction and the lower the water cut, the greater the optimal surfactant content, and the poorer the viscosity-reduction effect.

3. On its own, the GEO surfactant performs poorly in recovering heavy oil and is even worse than the polymer on its own; therefore, such a surfactant should be jointly used with a polymer.

4. For such GEO SP flooding, too high a surfactant content may impair heavy oil recovery, while a large slug size contributes significantly to the recovery.

ACKNOWLEDGMENTS

The authors thank the National Science and Technology Major Project Fund of China (Grant no. 2016ZX05058-003-003), the National Natural Science Foundation of China (Grant no. 51504275), and the Fundamental Research Funds for the Central Universities (Grant no. 17CX02076) for their financial support.

References

[1] Mai A, Kantzas A. Heavy oil waterflooding: effects of flow rate and oil viscosity. *J Can Pet Technol*. 2009,48:42-51.

[2] Pei HH, Zhang GC, Ge JJ, Liu HQ, Wang Y, Wang C. Advances in enhanced ordinary heavy oil recovery by chemical flood. *Oilfield Chem*. 2010,27(3):350-356.

[3] He CG, Xu AZ, Fan ZF, Zhao L, Shan FC, Bo B. An integrated model for productivity prediction of cyclic steam stimulation with horizontal well. *Energy Sci Eng*. 2019,7:962-973.

[4] Ahmadi MA, Masoumi M, Askarinezhad R. Evolving smart model to predict the combustion front velocity for in situ combustion. *Energy Technol*. 2015,3(2):128-135.

[5] Barillas JL, Dutra TV, Mata W. Reservoir and operational parameters influence in SAGD process. *J Petrol Sci Eng*. 2006,54:34-42.

[6] Zhao DW, Wang J, Gates ID. Optimized solvent-aided steamflooding strategy for recovery of thin heavy oil reservoirs. *Fuel*. 2013,112:50-59.

[7] Thomas S, Ali SM, Scoular JR, Verkoczy B. Chemical methods for heavy oil recovery. *J Can Pet Technol*. 2001,40(3):56-61.

[8] Dong M, Ma S, Liu Q. Enhanced heavy oil recovery through interfacial instability: a study of chemical flooding for Brintnell heavy oil. *Fuel*. 2009,88(6):

1049-1056.

[9] Ding MC, Wang YF, Li ZY, Zhong D, Yuan FQ, Zhu YW. The role of IFT and emulsification in recovering heavy oil during S/SP flooding. *J Ind Eng Chem*. 2019,77:198-208.

[10] Pei HH, Zhang GC, Ge JJ, Jin LC, Liu XL. Analysis of microscopic displacement mechanisms of alkaline flooding for enhanced heavy-oil recovery. *Energy Fuels*. 2011,25(10):4423-4429.

[11] Pei HH, Zhang GC, Ge JJ, Tang MG, Zheng YF. Comparative effectiveness of alkaline flooding and alkaline-surfactant flooding for improved heavy-oil recovery. *Energy Fuels*. 2012,26(5):2911-2919.

[12] Chen LF, Zhang GC, Ge JJ, Jiang P, Tang JY, Liu YL. Researchof the heavy oil displacement mechanism by using alkaline/surfactant flooding system. *Colloids Surf A*. 2013,343:63-71.

[13] Rahul S, Ramgopal V, Pankaj T. Influence of emulsification, interfacial tension, wettability alteration and saponification on residual oil recovery by alkali flooding. *J Ind Eng Chem*. 2018,59:286-296.

[14] Dong MZ, Liu Q, Li AF. Displacement mechanisms of enhanced heavy oil recovery by alkaline flooding in a micromodel. *Particuology*. 2012,10(3):298-305.

[15] Krumrine PH, Mayer EH, Brock GF. Scale formation during alkaline flooding. *J Pet Technol*. 1985,37(08):1466-1474.

[16] Pang SS, Pu WF, Xie JY, Chu YJ, Wang CY, Shen C. Investigation into the properties of water-in-heavy oil emulsion and its role in enhanced oil recovery during water flooding. *J Pet Sci Eng*. 2019,177:798-807.

[17] Liu ZY, Li YQ, Luan HX, Gao WB, Guo Y, Chen YH. Pore scale and macroscopic visual displacement of oil-in-water emulsions for enhanced oil recovery. *Chem Eng Sci*. 2019,197:404-414.

[18] Zhang HY, Chen GY, Dong MZ, Zhao SQ, Liang ZW. Evaluation of different factors on enhanced oil recovery of heavy oil using different alkali solutions. *Energy Fuels*. 2016,30(5): 3860-3869.

[19] Zhou XD, Dong MZ, Maini B. The dominant mechanism of enhanced heavy oil recovery by chemical flooding in a two-dimensional physical model. *Fuel*. 2013, 108:261-268.

[20] Li HB. Oil-displacement system with ultra-low interfacial tension for reservoirs of different permeability under different wettability on rock surface. *Acta Petrolei Sinica*. 2008,29(4):573-576.

[21] Xu DR, Bai BJ, Wu HR, et al. Mechanisms of imbibition enhanced oil recovery in low permeability reservoirs: effect of IFT reduction and wettability alteration. *Fuel*. 2019,244:110-119.

[22] Dranchuk PM, Scott JD, Flock DL. Effect of the addition of certain chemicals on oil recovery during water flooding. *J Can Pet Technol*. 1974,13(3):27-36.

[23] Bryan J, Kantzas A. Investigation in to the processes responsible for heavy oil recovery by alkali-surfactant flooding. *SPE*. 2008,113993.

[24] Bryan J, Kantzas A. Potential for alkali-surfactant flooding in heavy oil reservoirs through oil-in-water emulsification. *J Can Pet Technol*. 2009,48(2):37-46.

[25] McAuliffe CD. Crude-oil-water emulsions to improve fluid flow in an oil reservoir. *J Petrol Technol*. 1973,25(6)721-726.

[26] McAuliffe CD. Oil-in-water emulsions and their flow properties in porous media. *J Petrol Technol*. 1973,25(6):727-733.

[27] Yu L, Dong MZ, Ding BX, Yuan YG. Emulsification of heavy crude oil in brine and its plugging performance in porous media. *Chem Eng Sci*. 2018, 178:335-347.

[28] Yu L, Dong MZ, Ding BX, Yuan YG. Experimental study on the effect of interfacial tension on the conformance control of oil-in-water emulsions in heterogeneous oil sands reservoirs. *Chem Eng Sci*. 2018,189:165-178.

[29] Malkin AY, Zadymova NM, Skvortsova ZN, Traskine VY, Kulichikhin VG. Formation of concentrated emulsions in heavy oil. *Colloids Surf A*. 2016,504:343-349.

[30] Zadymova NM, Skvortsova ZN, Traskin VY, et al. Heavy oil as an emulsion: composition, structure, and rheological properties. *Colloid J*. 2016,78(6):735-746.

[31] Prathibha P, Rohit KS, Ranvijay S, Eswaran P, Ajay M. Effect of synthesized lysine-grafted silica nanoparticle on surfactant stabilized O/W emulsion stability: application in enhanced oil recovery. *J Pet Sci Eng*. 2019,177:861-871.

[32] Pei HH, Zhan S, Zhang GC, et al. Experimental study of nanopar-ticle and surfactant stabilized emulsion flooding to enhance heavy oil recovery. *J Petrol Sci Eng*. 2018,163:476-483.

[33] Tian SD, Gao W, Liu YJ, Kang WL. Study on the stability of heavy crude oil-in-water emulsions stabilized by two different hydrophobic amphiphilic polymers. *Colloids Surf A*. 2019,572:299-306.

[34] Symonds R, Farouq A, Thomas S. A laboratory study of caustic flooding for two Alberta crude oils. *J Can Pet Techol*. 1991,30(1):44-49.

[35] Zhou YZ, Wang DM, Wang ZP, Cao R. The formation and viscoelasticity of pore-throat scale emulsion in porous media. *Pet Explor Dev*. 2017,44(1):110-115.

[36] Ma YF, Zhao FL, Wang X, Hou JR. Dynamic performance evaluation of polymer-surfactant at 30 m long sand-packed model. *Oilfield Chem*. 2017,34(2):300-304.

[37] Jennings HY, Johnson CE, Mcaliliffe CD. A caustic water flooding process for heavy oils. *J Pet Technol*. 1974,26(12):1344-1352.

[38] Liu ZY, Li YQ, Zhuang YT, Chen YH, Ma RC. Comparison of a surfactant/polymer solution and its emulsion flowing in a porous medium. *Pet Sci Bull*. 2017,4(527):535.

[39] Yang S, Xu GL, Liu P, Lun ZM, Sun JF, Qin XJ. Experimental study on chemical viscosity-reducing compound flooding for EOR of heavy oil reservoir. *Pet Geol Recov Effic*. 2018,25(5):80-85.

[40] Zhu XM. (Sheep non-alkali composite wood blocks in Dagang oilfield flooding injection parameter optimization indoor experimental study.) Daqing, China: Northeast Petroleum University,2017:40-43.

[41] Shu HD, Chen HQ, Xiao L, et al. Binary compound flooding in Henan Shuanghe oilfield reservoirs with high temperature above 90℃. *Oilfield Chem*. 2017, 34(1):149-153.

How to cite this article: Wang Y, Li Z, Ding M, Yu Q, Zhong D, Cao X. Performance of a goodemulsification-oriented surfactant-polymer system in emulsifying and recovering heavy oil. *Energy Sci Eng*. 2020,8:353-365.

王丽娟

曹绪龙
化学驱油论文集
（下册）

曹绪龙　编著

中国海洋大学出版社
·青岛·

图书在版编目(CIP)数据

曹绪龙化学驱油论文集 / 曹绪龙编著. — 青岛 ：中国海洋大学出版社，2023.8

ISBN 978-7-5670-3597-3

Ⅰ. ①曹…　Ⅱ. ①曹…　Ⅲ. ①石油 — 开采— 文集　Ⅳ. ①TE357.46-53

中国国家版本馆 CIP 数据核字(2023)第 158756 号

CAO XULONG HUAXUE QUYOU LUNWENJI(XIACE)

曹绪龙化学驱油论文集(下册)

出版发行	中国海洋大学出版社
社　　址	青岛市香港东路 23 号　　**邮政编码**　266071
网　　址	http://pub.ouc.edu.cn
出 版 人	刘文菁
责任编辑	矫恒鹏
电　　话	0532-85902349
电子信箱	2586345806@qq.com
印　　制	青岛国彩印刷股份有限公司
版　　次	2023 年 8 月第 1 版
印　　次	2023 年 8 月第 1 次印刷
成品尺寸	185 mm×260 mm
印　　张	43
字　　数	941 千
印　　数	1～1000
定　　价	598.00 元(上下册)
订购电话	0532-82032573(传真)

发现印装质量问题，请致电 0532-58700166，由印刷厂负责调换。

前　言

提高老油田采收率和新区有效动用是增加经济可采储量和原油产量的主要途径。胜利油田是我国重要的石油工业基地，是一个地质条件复杂、含油层系多、储层种类多、物性变化大、油藏类型多的复式含油盆地，被誉为“石油地质大观园”。不论从油藏类型、油藏条件，还是提高采收率方法来看，依据胜利油藏为对象形成的方法，在全国都具有很好的推广价值。

在石油开采技术不断发展的今天，化学驱作为提高采收率的重要方式，在油田的生产中发挥着越来越重要的作用。胜利油田在20世纪60年代开展化学驱研究，90年代开展矿场试验，经过耐温抗盐聚合物驱理论、耐温抗盐聚合物加合增效理论的研究和发展，建立了高温高盐油藏化学驱理论，形成了聚合物驱、无碱二元复合驱和非均相复合驱技术系列，为油田可持续高质量发展做出了贡献，同时积累了丰富的攻关实践经验。

为进一步推动提高采收率技术的创新与发展，也为了给该领域的科研工作者提供借鉴、学习的机会，本文集选载了曹绪龙及其科研团队著述的154篇文章，以胜利油田为研究对象，全面介绍了胜利油田在化学驱提高采收率方面取得的研究成果。

本文集共分为聚合物及其相互作用、表面活性剂及其相互作用、驱油体系及驱油方法、化学剂浓度分析、驱油实验与数值模拟、矿场应用、论文综述、其他等八个部分，其中，聚合物及其相互作用论文34篇、表面活性剂及其相互作用论文39篇、驱油体系及驱油方法论文38篇、化学剂浓度分析论文3篇、驱油实验与数值模拟论文13篇、矿场应用论文7篇、论文综述3篇、其它文章17篇，内容涉及基础理论研究、驱油剂研发、驱油体系设计、数值模拟研究、矿场试验等诸多方面，是一本专业性强、涉及学科多并具有强烈胜利特色的科技书籍。该书的出版对石油行业的科技工作者和高校从事化学驱研究的师生具有较好的参考和借鉴价值，同时也将对进一步推动我国油田提高采收率发挥积极作用。

本论文集是参与胜利油田化学驱油技术研究的广大科技工作者的集体智慧的结晶，在此向他们表示衷心的感谢！同时也向在本文集编写过程中提供支持与帮助的同志表示谢意。

曹绪龙

2023年6月12日

目录

第一章　聚合物及其相互作用

第二章 表面活性剂及其相互作用

第三章 驱油体系及驱油方法

第四章 化学剂浓度分析

第五章 驱油实验与数值模拟

第六章 矿场应用

第七章 论文综述

第八章 其他

第三章　驱油体系及驱油方法

孤东油砂共驱体系的碱耗

曹绪龙

（胜利油田地质研究院开发试验室）

提要：本文采用静态实验方法，讨论了反应时间、矿物、碱浓度、表面活性剂及固液比对孤东油砂静态碱耗的影响，并给出了描述孤东油砂碱耗的定量模型。结果表明，用引入固液比影响参数后的K-B模型来描述孤东油砂的静态碱耗，能得到良好的线性关系。固液比对不同矿物的碱耗影响，可用固液比影响参数来描述。加入配伍性良好的表面活性剂可在一定程度上减小碱耗量。

碱耗作为注碱采油和共驱采油的重要参数已为人们研究的很多。碱耗模型在文献中已有论述[1-4]。但不同组分的油砂有不同的碱耗模型，目前对碱耗反应的主要机理也没有统一的认识，这就给碱耗预测带来了困难。我们针对孤东油田共驱采油的情况，对孤东油田试7井的油砂碱耗进行了调查，同时对共驱体系中加入表面活性剂对碱耗的影响作了探讨。

碱耗反应是相当复杂的，主要包括(ⅰ)快速的离子交换反应，(ⅱ)可逆的吸附反应，(ⅲ)长期的岩石化学反应，(ⅵ)原油中有机酸与碱的反应。其中(ⅵ)是碱水驱油的有益反应，在整个碱耗中占的比例很小，故在一般现场方案设计中可以忽略。由于不同组分的碱耗反应速度不同[1,2]，不同的油砂由不同的组分构成，把每种组分的碱耗都表示出来，并进行各组分的碱耗计算是很困难的。在实际应用中亦没有必要分清各部分的碱耗量大小，为了使问题简化，我们采用K-B模型来表征整个反应过程[3]，并对此模型做了改进。实验发现，最初二小时的碱耗量占整个碱耗量的80％以上。所以，我们取2小时左右的碱耗量进行对比，并通过求得的其他参数，找出了共驱体系与碱驱体系在不同碱浓度、不同固液比及不同矿物组分时碱耗量的差别。

一、实验方法和试剂

1 方法

取一反应器，加入一定量的矿物样品和100 ml碱液(硅酸钠溶液，下同)或碱-表面活性剂混合液，恒温于60±2℃烘箱内。在不同时刻取样，用ZD-2型酸度计测定瞬时PH。

2 样品与试剂

石英砂:80 目,山东省招远县石英厂生产。蒙脱土:含蒙脱石 86.3%、伊利石 9.1%、石英 4.6%,山东省潍坊膨润土矿生产。油砂:将孤东试 7 井 Ng 上 5^{2+3} 层位油砂用 1∶1 苯-乙醇溶液清洗后备用。油砂中的黏土总量为 5.3%,蒙脱石为 14.7%,伊利石为 40.0%,高岭土为 39.1%,绿泥石为 6.2%。硅酸钠:分析纯,上海第四试剂厂生产。表面活性剂 OP:上海助剂厂生产。石油磺酸盐:青岛化工研究所合成。

二、模型的建立

碱与砂的反应过程是个复杂的真实表面反应和化学反应的综合过程,一般包括下面几步:(ⅰ)本体溶液中的离子向真实表面的扩散;(ⅱ)溶液中的离子与真实表面上离子的交换反应;(ⅲ)溶液中的离子在固体表面活性中心的吸附反应;(ⅳ)固体表面物质在溶液中的溶解作用;(ⅴ)溶液中的离子与岩石表面物质的化学反应。一般认为,反应(ⅱ)、(ⅲ)速度较快,主要影响碱耗的初始阶段;反应(ⅴ)是一个相对缓慢的长期过程,虽然其碱耗量相对较小,但由于现场试验中碱水滞留于油藏中的时间较长,所以也不可忽视。

考虑到以上反应过程,碱耗速度方程可表示为

$$r_T = \sum K_j C^j \tag{1}$$

不同岩石碱耗的主要反应不甚相同,在忽略次要作用后可以得到下面几个有明确物理意义的反应方程:

(ⅰ)一级反应动力学方程[1-2]

$$r = KC \tag{2}$$

(ⅱ)K-B 反应动力学方程[8]

$$\ln(C_0/C) - B(C_0 - C)/C_0 = Kt \tag{3}$$

(ⅲ)吸附-离子交换动力学方程[5]

$$C = KC_r\{1-\exp(-(K_1+K_2)t/Cr)\}C_0/(C_0K_1+K_2) \tag{4}$$

(ⅳ)Krumrine 溶解方程

$$C = C_\theta - (C_\theta - C_0)\exp - Kt \tag{5}$$

我们把实验数据分别用这几种模型在 PC-1500A 微机上处理后,认为用 K-B 模型描述孤东油砂与碱的反应过程较合适。同时,考虑到固液比的影响,在 K-B 模型中引入了固液比影响参数,即得到方程:

$$\ln(C_0 - C) - B(C_0 - C)/C_0 = \alpha K_0 t/\beta \tag{6}$$

这里我们定义固液比影响参数 α:

$$\alpha = \beta K_1/K_0 \tag{7}$$

$$\beta = W\text{o}/W_0 \tag{8}$$

为两固液比之比。以上公式中各符号的意义请参见本文末符号意义表。

三、模型的验证

用石英砂、蒙脱石和孤东油砂的实验数据对上面提出的模型进行了验证，发现用该模型表征反应过程时，数据点基本落在理论线上（图 1、2），相关系数在 0.922 5～0.999 9 之间（表 1—8），大于显著性水平取 0.01 时所需要的 0.874，说明用该方程来表征孤东油砂反应过程是合理的。

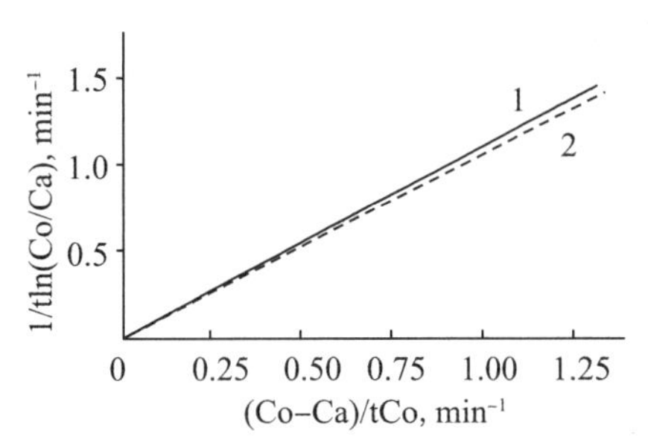

图 1　0.5%(wt)碱液与石英砂的反应

1. 固液比为 3∶20；2. 固液比为 3∶10

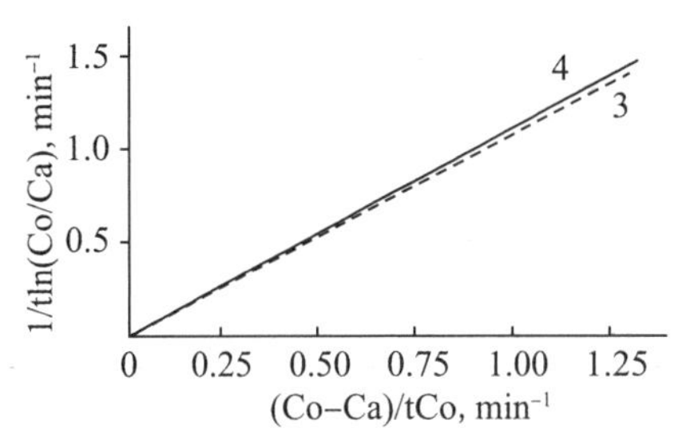

图 2　0.25%碱液与孤东油沙的反应

3. 固液比为 3∶20；4. 固液比为 3∶10

表 1　固液比为 3∶20 时蒙脱石与 0.5% Na_2SiO_3 的静态反应

反应时间 t(min)	0	8	19	48	130	210	340	420	620
pH	12.07	11.42	11.27	11.12	11.08	11.05	10.88	10.90	10.80
1/tln(Co/C)		11.23	5.82	2.73	1.05	0.67	0.48	0.38	0.28
(Co-C)/tCo		5.82	2.66	1.11	0.41	0.26	0.17	0.13	0.09
γ		0.998 0							

表 2　固液比为 3∶10 时蒙脱石与 0.5% Na_2SiO_3 的静态反应

反应时间 t(min)	0	9	52	130	210	340	420	510	620	765
pH	12.07	11.82	11.08	10.80	10.48	10.22	10.14	9.96	9.80	9.50
1/tln(Co/C)		3.84	2.63	2.12	1.05	0.75	0.63	0.57	0.51	0.46
(Co-C)/tCo		2.92	1.04	0.46	0.28	0.17	0.14	0.12	0.10	0.08
γ		0.922 5								

表 3 固液比为 3∶20 时石英砂与 0.5%Na_2SiO_3 的静态反应

反应时间 *t*(min)	0	7	21	61	130	200	310	450	765
pH	12.10	12.03	12.03	12.00	11.92	11.95	11.91	11.90	11.90
1/tln(Co/C)		1.38	0.46	0.23	0.19	0.10	0.08	0.06	0.04
(Co-C)/tCo		1.26	0.42	0.20	0.1u	0.09	0.07	0.05	0.03
γ		0.999 2							

表 4 固液比为 3∶10 时石英砂与 0.5%Na_2SiO_3 的静态反应

反应时间 *t*(min)	0	7	25	62	135	200	450	745
pH	12.1	12.05	12.03	12.00	11.97	11.90	11.92	11.92
1/tln(Co/C)		0.99	0.39	0.22	0.13	0.14	0.06	0.03
(Co-C)/tCo		0.93	0.36	0.20	0.11	0.11	0.05	0.03
γ	0.999 9							

表 5 固液比为 3∶20 时孤东油砂与 0.1%Na_2SiO_3 的静态反应

反应时间 *t*(min)	0	10	20	46	84	120	190	250	330	420	750
pH	11.46	11.35	11.32	11.32	11.30	11.30	11.28	11.29	11.29	11.28	11.25
1/tln(Co/C)		1.52	0.967	0.420	0.263	0.184	0.131	0.094	0.071	0.059	0.039
(Co-C)/tCo		1.34	0.827	0.359	0.220	0.154	0.107	0.078	0.059	0.049	0.031
γ	0.999 9										

表 6 固液比为 3∶10 时孤东油砂与 0.1%Na_2SiO_3 的静态反应

反应时间 *t*(min)	0	10	20	45	84	120	330 •	420	510	620	750
pH	11.46	11.35	11.32	11.30	11.28	11.28	11.27	11.28	11.25	11.25	11.25
1/tln(Co/C)		1.52	0.97	0.49	0.30	0.21	0.08	0.059	0.057	0.047	0.039
(Co-C)/tCo		1.34	0.83	0.41	0.24	0.17	0.06	0.049	0.045	0.037	0.031
γ											

同时,我们发现,改变矿物和固液比时,模型形式不改变,但引入的固液比参数值 α 发生了变化。因此,模型中 α 值不仅表明了固液比对不同矿物碱耗的影响,而且也表明了对同种矿物碱耗的影响。图 3 绘出了固液比影响参数 α 和与固液比之比值 β 之间的关系。其中 a 线是反应速度与固液比变化正比线,b 线表示反应速度与固液比无关,即不论固液比怎样变,反应速度都恒定。Ⅰ区表示反应速度增加大于固液比增加,Ⅱ区表示反应速度增加低于固液比增加。根据实际绘出的 α-1/β 图,可估计出某已知固液比时的碱耗速度,从而可通过室内实验来判断现场固液比时的碱耗速度。

四、实验结果与讨论

1 固液比对静态碱耗的影响

蒙脱石，石英砂和孤东试7井油砂在固液比分别为3∶20和3∶10时与碱反应的试验结果如表1-6所示。数据用K-B模型处理后的结果如图1、2、4、5所示。从这些数据可以看出，固液比的变化引起了碱耗量的变化。但引入α后的K-B模型与实验数据有良好的相符性。

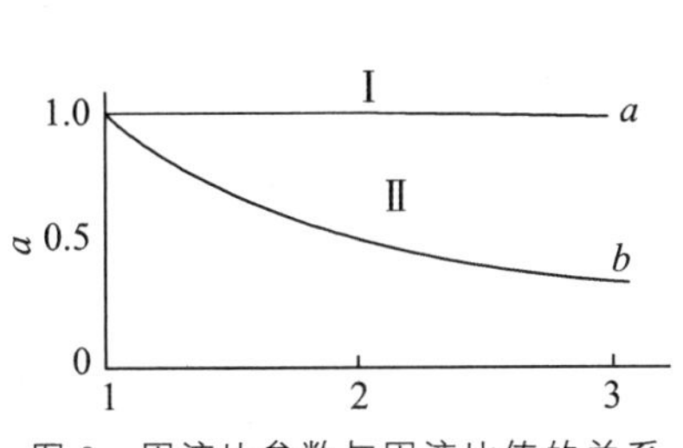

图3 固液比参数与固液比值的关系

a. Ko=βKi；b. Ko=Ki

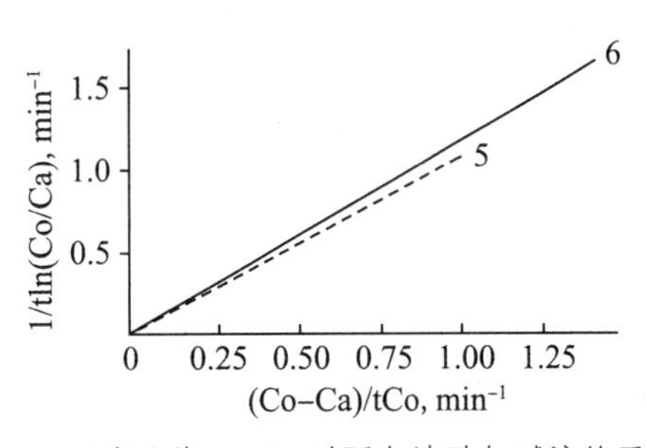

图4 固液比为3∶20时孤东油砂与碱液的反应

5. 碱浓度为1%(wt)；6. 碱液度为0.1%(wt)

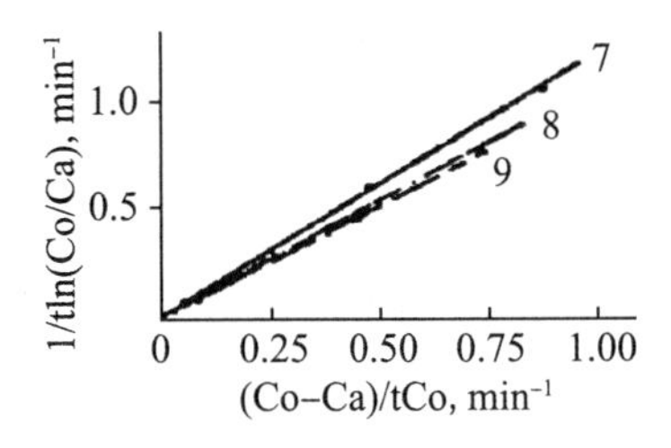

图5 含0.2%OP的诚度与孤东油砂的反应

7.0.1%(wt)碱液，固液比3∶20；8.0.5%(wt)碱液，固液比3∶10；9.0.5%(wt)碱液，固液比3∶20；·为实验点，——为理论线

表7 不同固液比时的蒙脱石碱耗参数(碱浓度C=0.5%)

固液比	反映速度常数	阻抑系数	α	碱耗量(meq/100 g)
3∶20	0.269	1.92	1.00	7.03
3∶10	0.692	1.20	1.286	3.83

* 计算α时以固液比为3∶20时为准，下同。

(ⅰ)蒙脱石在130 min时，固液比大，碱耗量亦大(表1、2)，固液比增大，反应阻抑系数B变小，反应速度常数K增大(表7)。求出的固液比为3∶10的α大于以固液比3∶20为准时的α值，在α-1/β图上落在Ⅰ区。这表明，随着固液比的增大，反应速度常数增大。所以，不能把较小固液比时的反应速度常数用于较大固液比的情况。

(ⅱ)石英砂取 60 min 时两种不同固液比的实际碱耗量数据进行对比发现,石英砂在同一时刻的实际碱耗量相近,阻抑系数 B 变化亦不大,反应速度常数虽然随着固液比增加而增加,但远小于蒙脱石的增加倍数(表 8)。求出固液比为 3∶10 时的 α 值比定为标准固液比 3∶20 时的 α 值小,在 α-$1/\beta$ 图上落在Ⅱ区内。多次试验的结果表明,石英砂的 α-$1/\beta$ 线很接近图 3b 线,故对石英砂的碱耗可近似地用小固液比时的速度常数代替较大固液比时的速度常数。

(ⅲ)孤东油砂取 120 min 时的实际碳耗量数据进行对比发现(表 5、6)固液比增大时,碱耗量增加,但增加程度远小于蒙脱石的情况。两种固液比时的反应阻抑系数 B 相近。反应速度常数增加程度也远较蒙脱石的小(表 9)。求出固液比为 3∶10 时的 α 值比定为标准固液比 3∶20 时的值小,在 α-$1/\beta$ 图上落在Ⅱ区,且接近于图 3b 线。因此,可用室内求得的反应速度常数代替现场情况。从图 2 可看出,用引入 α 值后的 K-B 模型描述孤东油砂碱耗是合理的。

表 8　不同固液比时的石英砂碱耗参数(碱波度 C=0.5%)

固液比	反应速度常数	阻抑系数	α	碱耗量(meq/100 g)
3∶20	7.54×10^{-3}	1.09	1.00	1.73
3∶10	1.18×10^{-2}	1.05	0.782	0.867

表 9　不同固液比时孤东油砂的碱耗参数(碱法度 C=0.1%)

固液比	反应速度常数	阻抑系数 B	碱耗量(meq/100 g)	
3∶20	7.99×10^{-3}	1.14	1.000	0.587
3∶10	1.06×10^{-2}	1.14	0.663	0.323

从上面三种矿物的实际碱耗量、反应速度常数,固液比影响参数和阻抑系数对比可以得出,蒙脱石不仅实际碱耗量大,而且受固液比的影响亦较大,而孤东油砂和石英砂受固液比的影响较小。

2　浓度的影响

用 0.1%(wt)、0.5%(wt)、1.0%(wt)的硅酸钠分别与孤东油砂静态反应,结果如表 5、10、11 所示。图 4 给出了实验数据经 K-B 模型处理的结果。图 5 给出了加入 OP 后碱浓度对碱耗的影响。结果表明,不论是否加入表面活性剂,随着反应物初始浓度的增加,碱耗量都增加。用 K-B 模型处理都能得到良好的线性关系,这些结果与文献中论述的其他油砂的结果一致[1-2]。

表 10　固液比为 3∶20 时感东油砂与 0.5% Na_2SiO_3 的静态反应

反应时间 t(min)	0	9	18	46	84	120	190	330	510	750
pH	12.12	12.05	12.00	12.02	12.00	12.00	11.98	12.00	11.98	12.00
1/tln(Co/C)		1.08	0.92	0.30	0.20	0.14	0.10	0.05	0.04	0.02
(Co-Ca)/tCo		0.99	0.81	0.27	0.17	0.12	0.09	0.04	0.03	0.02
γ		0.999 5								
碱耗量(meg/100 g)		2.13								

表 11　固液比为 3∶20 时孤东油砂与 1% Na_2SiO_3 的静态反应

反应时间 t(min)	0	9	17	44	85	120	250	420	750
pH	12.35	12.30	12.30	12.30	12.28	12.25	12.28	12.26	12.22
1/tln(Co/C)		0.768	0.406	0.157	0.114	0.115	0.038	0.030	0.024
(Co-Ca)/tCo		0.725	0.384	0.148	0.105	0.103	0.036	0.027	0.021
γ		0.999 9							
碱耗量(meg/100 g)		4.4							

3　表面活性剂对碱耗的影响

碱-表面活性剂共驱是目前三次采油的方向之一。碱和表面活性剂的相互补充,使得共驱可行性优于二者中任何一种单一方法。为此,我们在配伍性实验之后,考察了表面活性剂 OP、石油磺酸盐对孤东油砂、石英砂和蒙脱石碱耗的影响。试验结果如表 12、13 和 14 所示。结果表明,在固液比为 3∶10、碱浓度为 0.5%时,加入 OP 后孤东油砂碱耗量有一定程度的减小,反应速度常数亦有减小,说明 OP 对抑制碱耗有一定的作用。图 6 给出了表面活性剂 OP 对引入 α 值后的 K-B 模型的影响。从图中看出,加入 OP 后仍能得到良好的线性关系,即孤东油砂共驱体系的碱耗仍可用修正后的 K-B 模型来描述。图 7、8 分别给出了当固液比为 3∶20,碱浓度分别为 0.1%(wt)和 1.0%(wt)时,加入 0.2%(wt)和 1%(wt)的 OP 和石油磺酸盐后。孤东油砂的碱耗结果。从结果可看出,石油磺酸盐对碱耗的抑制作用比 OP 略好(表 14)。据根这一实验结果,我们认为共驱体系中应优先考虑使用石油磺酸盐。

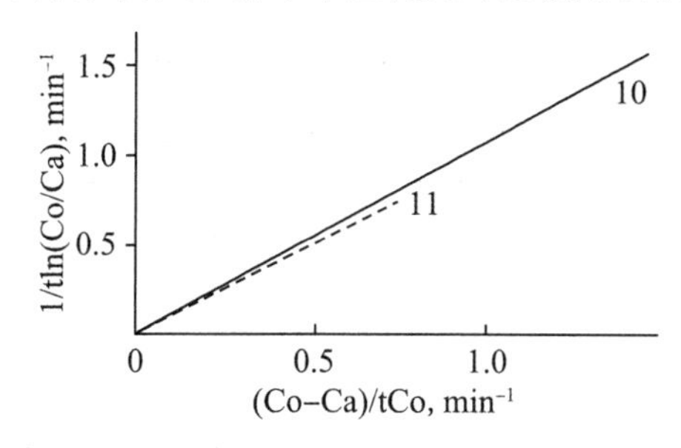

图 6　表面活性剂 OP 对孤东油砂碱耗的影响

10. 未加入表面活性剂;11. 加入表面活性剂 OP0.2%

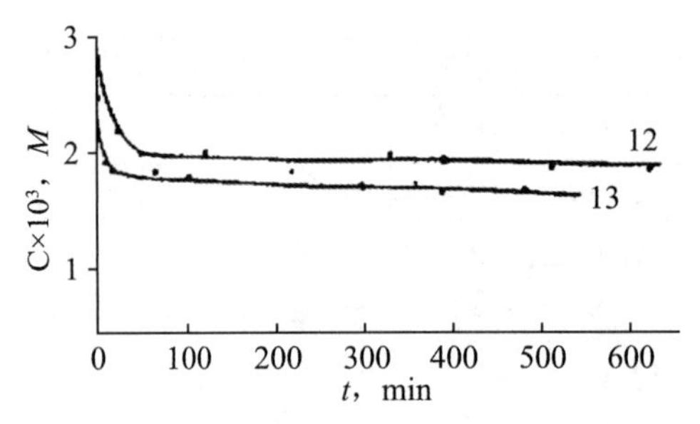

图 7 不同表面活性剂对孤东油砂碱耗的影响

碱浓度为 0.1%(wt) 12. 0.2%OP;13. 0.2%(wt)石油磺酸盐

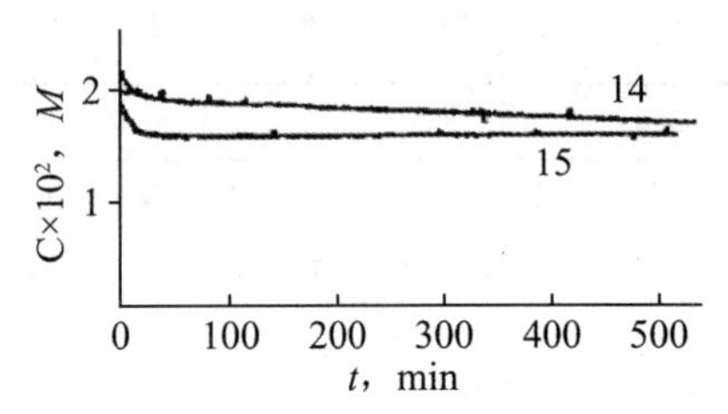

图 8 表面活注剂对孤东油砂碱耗的影响

碱浓度为 1%(wt);14. 1%OP;15. 1%(wt)石油磺酸盐

表 12 表面活性剂 OP(添加 0.2%)对孤东油砂碱耗的影响

反应时间(min)		pH		1/tln(Co/Ca)		(Co/Ca)/tC0		γ	
未加 OP	添加 OP	未加 OP	添加 OP	未加 OP	添加 OP	未加 OP	添加 OP	未加 OP	添加 OP
0	0	12.12	12.10					0.998	0.999 4
9	9	12.03	12.05	1.38	0.768	1.25	0.725		
18	17	12.00	12.00	0.92	0.813	0.81	0.726		
44	43	12.03	12.02	0.28	0.257	0.26	0.235		
85	85	12.00	12.00	0.20	0.163	0.17	0.145		
120	120	11.95	11.98	0.20	0.138	0.16	0.121		
250	250	12.00	12.00	0.066	0.055	0.058	0.049		
330		12.00		0.050		0.044			
420	420	11.98	12.00	0.046	0.033	0.039	0.029		
510	510	11.93	11.98	0.052	0.033	0.042	0.028		
750	750	12.00	12.00	0.022	0.018	0.019	0.017		

表 13 OP 对孤东油砂碱耗参数的影响

	阻抑系数	反应速度常数	碱耗量
加入 0.2%OP	1.09	3.45×10^{-3}	0.867
未加 OP	1.13	5.44×10^{-3}	1.07

表 14　OP 和石油磺酸盐对孤东油砂碱耗参数的影响

表面活性剂种类	碱浓度为 1%		碱浓度为 0.1%	
	反应速度常数	碱耗量	反应速度常数	碱耗量
OP	0.85×10^{-2}	3.26	2.63×10^{-2}	0.727
石油磺酸盐	1.00×10^{-2}	2.53	4.87×10^{-3}	0.620

五、结论

（1）无论是否加入表面活性剂，孤东油砂的碱耗过程均可用修正后的 K-B 模型来描述，得到的模拟结果与实验结果有良好的相符性。

（2）固液比对碱耗有一定的影响，这种影响可用 α 值来描述，由 α-$1/\beta$ 变化趋势，可估计同种矿物在某固液比时的反应速度常数。

（3）不同初始浓度的碱同孤东油砂的反应都可用引入 α 后的 K-B 模型来表征，不论是否加入表面活性剂，随着初始碱浓度的增加，碱耗量都增加。

（4）加入表面活性剂能在一定程度上减小碱耗，石油磺酸盐抑制碱耗的能力较 OP 的强。

符号意义表

B：阻抑系数

Ce：SiO_2 溶解平衡浓度

Cr^*：碱耗最大浓度

K_2：逆向反应速度常数

K_j：j 反应速变常数

C：任一时刻的碱浓度

C_0：初始碱浓度

K_1：正向反应速度常数

K_i：任一固液比时反应速度常数

K_0：定为标准周液比时反应速度常数

r_T：碱耗反应总速度

W_i：任一固液比

α：固液比影响参数

β：两固液比之比值

t：反应时间

W_0：定为标准的固液比

a_j：j 反应级数

γ：相关系数

酸谢王宝谕、俞进桥和陈能范同志对本工作给予了大力支持,在此一并致谢。

参考文献

[1] 刘伟成,等.油层中主要矿物组分的静态碱耗动力学模型.中国科学院大连化物所,1987 年 6 月.

[2] 姜炳南,等.胜利油田孤岛油砂静态碱耗动力学模型及其参数的确定.中国科学院大连化物所,1986 年 12 月.

[3] 姜炳南,等.碱水驱油碱耗动力学模型与应用,中国科学院大连化物所.1984 年 10 月.

[4] Krumrine,P. H. :SPE12670.

[5] 赵学庄.化学反应动力学原理(上册),北京:高等教育出版社,1984.

[6] 黄莉,杨普华.油田化学,1987,4(1):257.

[7] IdenkaNovosad;SPE10605.

(1988 年 6 月 7 日收到)

Caustic consumption of caustic/surfactant coflooding systems on gudong reservoir sand

Cao Xulong(Research insritute of geology,Shengli oilfield)

Abstract: The effects on caustic consumption of caustic/surfactant coflooding systems on Gudong reservoir sand, montmorillonite, and quartz under static experimental conditions are investi-gated in relation to caustic concentration, reactiontime, surfactant added, and solid-to-liquid ratio. The results obtained are simulated on a computer by various kinetic models. A good linear relationship is observed in description of the caustic consumption on Gudong reservoir sand by using K-B model with a parameter introduced which characterizes the influence of solid-to-liquid ratio on caustic consumption on minerals of different kind. A fully compatible surfactant in the coflooding systems decreases the caustic consumption in certain extent.

碱-助表面活性剂-聚合物体系相性质的研究

曹绪龙

（胜利石油管理局地质科学研究院）

提要：本文综述了采用乳化筛选试验法制作活性图，对碱-助表面活性剂-聚合物（APS）驱油体系相性质所作的研究。讨论了活性剂、碱对体系相性质的影响及相性质与界面张力、驱油效率之间的关系。在碱-原油体系中加入低浓度表面活性剂可使活性区加宽，体系最佳含盐量提高。过量碱在体系中起电解质作用。一价阳离子对相性质的影响次序为 $K^+ > Na^+ > NH_4^+$。最佳体系的界面张力可达到很低值。

主题词：碱-助表面活性剂-聚合物油驱体系 三次采油 相特性 界面张力 驱油效率 乳化试验

在碱中加入表面活性剂驱油这一概念的提出可追溯到50年代。对碱-助表面活性剂体系的研究则始于80年代中期。Nelson[1]、Martin[2]等人借助表面活性剂驱油的研究方法对助表面活性剂增效碱驱进行了研究。Nelson等人[1]在考虑碱-助表面活性剂驱油的流度控制问题时提出了碱-助表面活性剂-聚合物（ASP）共驱。80年代后期Terra能源公司的Clark等人[3]在美国怀俄明州克卢县的西基尔油田开展了ASP矿场试验。Lin等人[4]在加拿大阿尔伯达省Grand Forks Lower Mannville D油层开展了ASP研究。我们于1988年结合孤东油田小井距情况开展了ASP研究。从国内外研究结果看，在驱油效率方面ASP法比单一碱驱、聚合物驱优越，可与表面活性剂驱相媲美。ASP法已成为三次采油中当前颇受关注的一种方法。

目前ASP驱的研究内容包括：① 碱、聚合物和助表面活性剂的配伍性；② 动界面张力；③ 三种化学剂的吸附损耗；④ 相性质研究；⑤ 轴向或径向岩芯驱替试验等。通过这些试验选出适用的聚合物、碱和表面活性剂，最后优选出配方。由于ASP驱集碱驱（A）、表面活性剂驱（S）和聚合物驱（P）于一体，在研究中对各单一驱的特性均需加以考虑，例如，表面活性剂驱中的最佳含盐量（最佳盐需求量）、聚合物驱中的盐敏效应，碱驱中的原油酸值等等。过去的基本认识是，在碱中加入表面活性剂是为了弥补碱与石油酸作用生成的石油皂量的不足。最近提出的观点[1]是，加入助表面活性剂可提高体系的盐需求量，过量碱则起电解度的作用，使形成碱-原油低界面张力需要低碱浓度、为应付碱在地层中消耗需要高碱浓度的矛盾获得解决。相性质研究不仅能给出上述最佳盐需求量,也可对碱、表面活性剂的类型和浓度进行筛选,给出不考虑岩石吸附的最佳配方。因此,表面活性剂驱油中的相性质研究,对于 ASP 体系也是非常重要的。本文将讨论 ASP 体系的相性质。

相性质的实验研究方法

对于给定的油藏,虽然温度、原油性质一定,但 ASP 驱油体系相性质的研究中变量仍很多,如碱类型、碱浓度,表面活性剂类型、浓度,含盐量,油水比等,采用一般的相图是难以进行研究的。为此,总结出了利用“活性图”研究相性质的方法[1]。常用的活性图给出给定表面活性剂浓度和碱浓度下的活性区与含盐量、原油体积分数之间的关系。

制作活性图所用的实验方法为乳化筛选试验法。取若干支具塞管,加入定量的已知碱浓 度和表面活性剂浓度的溶液及原油,分别加入定量的不同浓度盐溶液,在油藏温度下静置若 干分钟,在振荡器中振荡若干小时,在空气恒温箱中在油藏温度下放置几天,然后观察样品 管中体系的相行为,确定欠佳区-最佳区和过佳区-最佳区分界处的含盐量及原油体积分数,在活性图上绘出活性区。不过,活性图上虽然绘出了区域边界线,并不表示体系会发生由活 性区到非活性区的突变。另外,这样确定活性图有一定的随意性。

聚合物对 ASP 体系的相性质影响甚微[5],在研究 ASP 体系相性质时一般可不予考虑。

相性质判定指标

Nelson 等人用 Gulf Coast 油藏原油作乳化筛选试验时规定了最佳区、欠佳区、过佳区的判定指标。以后 Saleem 等人[5]绘制委内委拉原油-碱-助表面活性剂体系的活性图,Martin 等人[2]研究碱-盐水-表面活性剂-烷烃或原油体系的相性质,都采用了 Nelson 判定指标。

Nelson 判定指标可叙述如下。过佳区应满足下列条件:① 形成油包水型粘稠乳状液;② 振荡时呈黑色,油相润湿管壁;振荡时发亮,停止振荡后光亮消失。将样品管倒置,油相和水相分别呈细流状流下。最佳区的特征是:① 形成色淡且稳定的乳状液,振荡时显银白色光泽,无未增溶的油和水或只有痕量;② 形成颜色最亮最稳定的乳状液,无未增溶的油和水;或③ 形成颜色淡且稳定的乳状液,有痕量未增溶油。欠佳区的现象是:① 振荡时颜色明亮,振荡停止后慢慢退色,将样品管倒置时水相快速户下流,油相则形成细流并滴出(在这种情况下通常形成最稳定的泡沫);② 振荡时颜色明亮,停止振荡则立即退色,有一部分油被增溶,将样品管倒置时水相快速流下,油相被拉长而形成小油滴残留下来;③ 振荡时呈黑色,油分散成较大的球形油滴。以上判定指标是否适用于一切原油与 AS 体系,需进行考察。

碱对相性质的影响

Martin 等人研究了碳酸钠、氢氧化钠、氢氧化钾、氨水和偏硅酸钠对表面活性剂相性质的影响。在碱浓度 0.1%～0.2%范围内任何碱的加入都使表面活性剂-原油或烃体系的最佳含盐量下降。碱浓度愈大,则最佳含盐量下降愈多。

图 1 所绘为正十四烷-合成石油磺酸盐体系中加入 Na_2CO_3 对最佳含盐量的影响。在给定 NaCl 浓度下增加 Na_2CO_3 浓度,相态按 Winsor Ⅱ(－)→Winsor Ⅲ→Winsor Ⅱ(＋)方向转化。这表明与石油酸反应后过量的碱起电解质的作用。在图 2 中随着所用各种钠碱浓度的增大,体系的最佳含盐量减小而且与碱中钠离子量呈线性关系,与钠平衡的阴离子对相性质影响甚微。这表明不论钠离子来源于那一种碱,对减小体系的最佳含盐量是同样有效的。体系的相性质还与加入的一价阳离子的类型有关,阴离子对相性质的影响力大小可排列为:$K^+ > Na^+ > NH_4^+$,加入钾离子可使最佳含盐量大幅度下降,而加入铵离子则影响很小。Martin 等人对此作作了如下解释。钾离子的半径比钠离子大,离子半径愈大则对相性质的影响愈大;铵离子因水合能力较强,作为电解质的能力下降。不过,只要 ASP 体系配方选择得当,不论使用那种碱,在驱油时均能达到相同的效力。碱的选择主要依赖于价格、产地、油中石油酸转化为石油皂所需要的 pH、油藏温度及配制水组成。

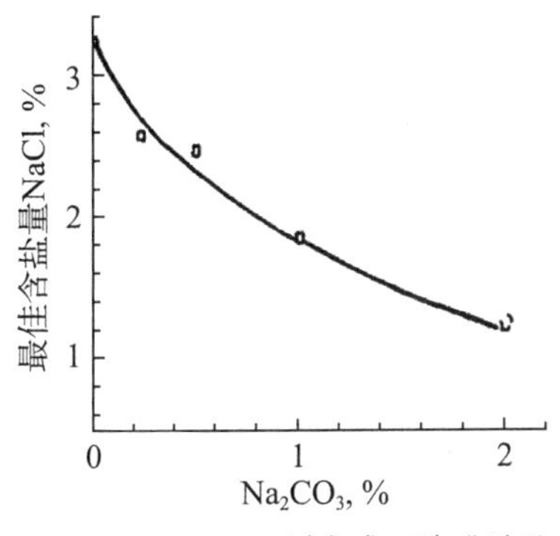

图 1 Na_2CO_3-2.0%合成石油磺酸盐 914-22-正十四烷体系的最佳含盐量与 Na_2CO_3 浓度的关系(75.6℃)

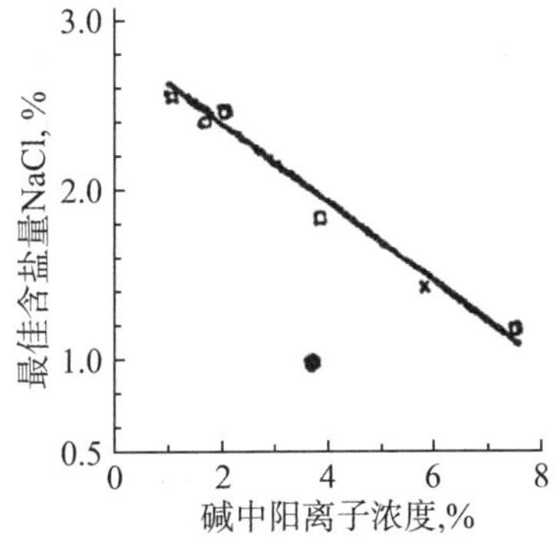

图 2 碱-2.0%合成石油磺盐酸 914-22-正十四烷体系的最佳含盐量与碱中阳离子浓度的关系(75.6℃)

□,Na_2CO_3(0.25%,0.5%,1.2%);○,$Na_2O\cdot(SiO_2)_{3.2}$(1.0%);△,NH_4OH(1.0%);●,KOH(0.1%)

助表面活性剂对相性质的影响

Nelson 等人和 Saleem 等人研究了助表面活性剂对碱-原油体系相性质的影响。图 3 是 1.55% Na_2SiO_3-Gulf Coast 原油(A-O)体系和 1.55%Na_2SiO_3-0.1%Neodol25-3S-Gulf Coast 原油(A-S-O)体系 75.6℃时的活性图。图 4 是 NaOH-委内瑞拉原油(A-O)体系和 NaOH-0.2%Triton X-100-委内瑞拉原油(A-S-O)体系 50℃时的活性图。从这

两幅图看出,仅有碱存在时活性区很窄,为了弥补油藏岩石碱耗、碱扩散及水的稀释而增加碱浓度,会使最佳驱替液很快变成非最佳体系。加入助表面活性剂后体系的最佳盐需求量增大,活性区大大展宽,在碱浓度增加时体系仍会落到活性区内。因此,在碱-原油体系中加入低浓度助表面活性剂,可解决为维持一定驱替速度需要高碱浓度和为获得超低界面张力需要低碱浓度之间的矛盾。

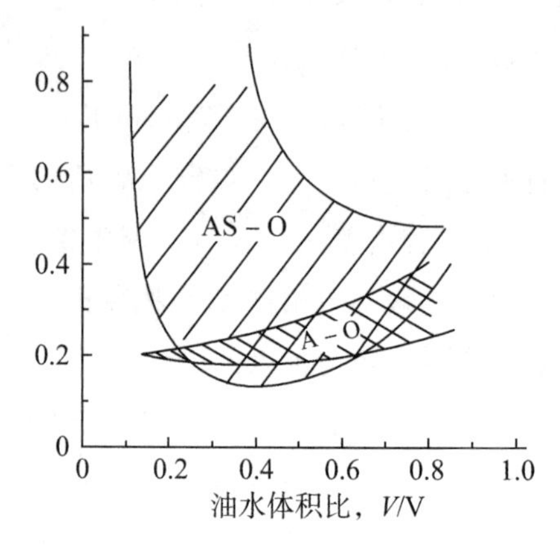

图 3 1.55% $Na_2O \cdot SiO_2$ 与 Gulf Coast 原油体系"(A-O)和
1.55%$Na_2O \cdot SiO_2$+0.1%Neodol25-3S 与 Gulf Coast 原油体系(ASO)的活性图

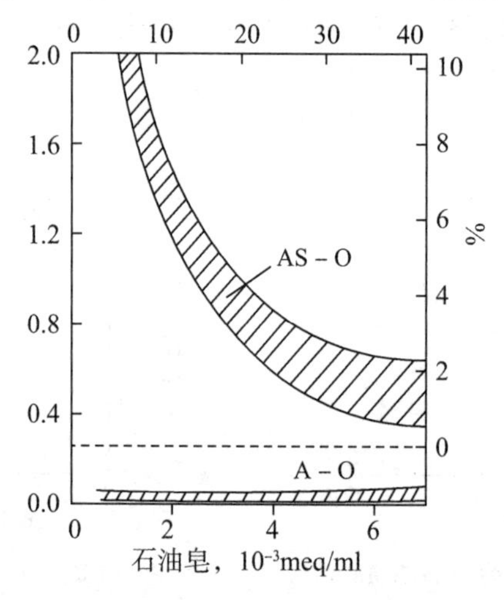

图 4 NaOH 与委内瑞拉原油体系(A-O)和 NaOH+0.2%TritonX100
与委内瑞拉原油体系的活性图一代表总 Na^+ 量中来自 1.55%$Na_2O \cdot SiO_2$ 的 Na^+ 量

在筛选助表面活性剂时应考虑其水解稳定性和油藏温度。烯烃磺酸盐耐水解,盐需求量低于同碳数的烷基磺酸盐,形成的体系活性区对残余油饱和度的敏感性低,有利于实际应用。不同类型表面活性剂对活性区的影响有待进一步研究。

相性质与驱油效率的关系

Nelson70 年代研究表面活性剂驱油体系的相性质时曾指出,要获得最大的驱油效率,设计表面活性剂段塞时应使表面活性剂-原油体系尽可能处在 WinsorⅢ 型状态,后来他又在不改变表面活性剂和碱浓度而改变 NaCl 浓度的条件下作了贝雷岩芯驱替试验。当段塞中 NaCl 浓度按 0.5%，1.5%，3.28%，6.0%次序增加时，驱出油量先增加后下降。当段塞中 NaCl 含量为 0.5%时，在活性图上体系处于欠佳区，在驱油试验中显示欠佳区驱油特征即采出液颜色深，表面活性剂突破早。段塞中 NaCl 含量为 6.0%时体系处于强过佳区内，采出液清澈，表面活性剂突破晚。NaCl 含量为 1.5%时体系处于活性区下边缘处，NaCl 含量为 3.28%时体系处于活性区内，在这两种情况下驱出油量大致相同，均高于前两种情况。这一结果说明，处于活性区内和活性区下边缘的体系驱出油量最高，见图 5。ASP 配方应满足这一条件并在驱替过程中尽可能长时间地维持这一条件。欠佳区和最佳区边缘上的体系，其驱替前缘为水所稀释后便进入活性区范围内。这一结果与表面活性剂驱试验结果一致。Lin 等人研究石油磺酸盐-NaOH-Grand Forks Lower Mann-ville D 油层原油体系时，通过轴向和径向岩芯驱替试验得到了同一结果。

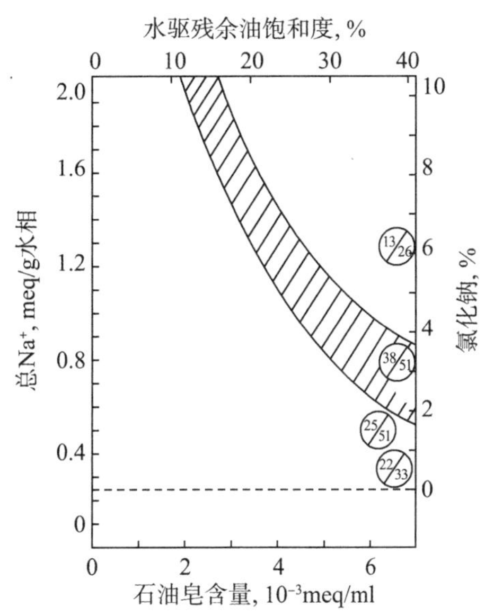

图 5　1.55% $Na_2O \cdot SiO_2$ + 0.2% Neodol 25-3S 段塞在活性图上的位置与水驱残油采出率的关系圆圈内斜线左、右的数学分别为驱至 1 PV 和 2 PV 时采出水驱残油的百分数
——代表总 Na^+ 量中来自1.55% $Na_2O \cdot SiO_2$ 的 Na^+ 量

界面张力与相性质的关系

这是一个有待研究的重要问题。Taylor 等人在研究大卫·劳埃德米斯特原油与 Na_2CO_3-Neodol 25-3S 体系时发现体系的动界面张力随表面活性剂浓度而定。通过与轴向岩芯驱替试验结果的对比，发现体系的初始动界面张力与采收率有关，而平衡界面张力一般不影响采收率。Lin 等人发现，在碱体系中加入表面活性剂可使 Grand Forks 原油与碱溶液之间的界面张力进一步降低。只加少量表面活性剂便可使界面张力降至 mN/m 范围。而原油与表面活性剂溶液之间的界面张力比原油与碱-表面活性剂混合溶

液之间的界面张力高几个数量阶。这表明碱与表面活性剂之间有协同效应。

目前尚未见到关于相性质与界面张力之间关系的论文发表。不过可以肯定，中相乳状液与油、水之间的界面张力小于上相乳状液与盐水之间或下相乳状液与油之间的界面张力，否则处于活性区的体系不可能驱出最多量的油。

结论

借助乳化筛选试验可以制作对分析相性质非常有用的活性图。ASP 体系中过量的碱起电解质的作用，碱中一价阳离子对相性质的影响力大小次序为 $K^+ > Na^+ > NH_4^+$。在碱-原油体系中加入助表面活性剂能使活性区大大展宽，体系盐需求量增加。从活性区下边缘附近至活性区内的 ASP 配方驱油效率最佳。初始动界面张力与驱油关系密切而平衡界面张力影响甚微，在这方面有待作进一步的工作。

致谢作者衷心感谢俞进桥高工在搜集和整理资料过程中给予的指导与协助。

参考文献

[1] Nelson, R. C., Lawson, J. B., Thigpen, D. R., and Stegemeier, G. L.: SPE/DOE12672.

[2] Martin, F. D. and Oxley, J. C.: SPE13575.

[3] Clark, S. R. and Smith, S. M.: SPE 17538.

[4] Lin, F. F. J., Besserer, G. J., and Pitts, M. J.: J. Can. Petrol. Technol., 1987, 26: 54-65.

[5] Saleem, S. M. and Hernanden, A.: J. Surface Sci. Technol., 1987, 3: 1-10.

[6] Nelson, R. C. and Pope, G. A.: Soc. Petrol. Engrs J., Oct. 1987: 325-338.

[7] Taylor, K. C., Hawkins, B. F., Islam, M. R.: "Dynamic Interfacial Tension in Surfactant Enhanced Alkaline Flooding", Paper of Petroleum Society of CIM, 89: 40-44.

[8] Nelson, R. C.: Soc. Petrol. Engrs J., 1982, 22, 2: 259-270.

(1990 年 10 月 4 日收到，1990 年 12 月 14 日修改，1991 年 5 月 8 日第 2 次修改)

The phase behavior of alkaline-cosurfactant -polymer flooding systems: a review

Cao Xu-Long(Research institute of geology, Shengli oil fields)

Abstract: This review article summarized comprehensively the recent situation of phase behavior investigations for alkaline-cosurfactant-polymer (ASP) flooding systems. The following topics are covered: historical aspect, experimental methods for laboratory investigations, criteria for determing phase behavior, phase behavior as influenced by alkaline and surfactant, relations of phase behavior to flooding coefficiency and of interfacial tension to phase behavior.

Keywords: Alkaline-Cosufactant-Polymer (ASP) Flooding Systems, Tertiary Oil Recovery, Phase Behavior, Interfacial Tenston, Flooding Efficiency, Emulsification Test

复合驱油体系与孤东油田馆 5^{2+3} 层原油间的界面张力

曹绪龙 薛怀艳 李秀兰 王宝瑜 袁是高

(胜利石油局地质研究院开发试验室)

提要:本文报导了用于孤东油田馆 5^{2+3} 层复合驱油试验的 ASP 体系与原油间界面张力的研究结果。在 ASP 复合驱油体系中,碱 A 和表面活性剂 S 在降低界面张力上有协同效应,聚合物 P 的影响很小。在单一 A 和 A+S 体系中最佳碱浓度为 0.2%和 1.5%(Na_2CO_3)。1.5%$NaCO_3$+0.4%OP-10 体系有最佳界面特性。

主题词:碱+表面活性剂驱油体系 碱+表面活性剂+聚合物驱油体系 胜利孤东原油 界面张力特性 动态 界面张力

与原油间的界面张力是复合驱油体系的一项重要性质,国外文献已有不少报导[1-3]。本文工作研究了孤东馆 5^{2+3} 层原油与复合驱油体系之间界面张力的特性,为确定矿场驱油体系提供依据。

1 仪器与试剂

德国 Krass 公司制造的 Site 04 型旋转滴界面张力计,用于测定动态界面张力,无锡石油仪器厂制造的 DQ-1 型悬滴界面张力计,用于测定静态界面张力。

碱 Na_2CO_3 和 $Na_2SiO_3 \cdot 9H_2O$,AR 级试剂。

非离子型表面活性剂 OP-10、Js、T80、平平加,上海助剂厂产品。阴离子型表面活性剂 AS、ABS,含量 25%-28%,天津合成洗涤剂厂产品,石油璜酸盐(PS),含量 40%,玉门油田产品,Ty-5,含量 50%,河北辛集石油化工厂产品。

聚丙烯酰胺 3530S,美国 Dfizer 公司产品,水解度 29.7%。

实验原油为模拟油,由 10∶1 的孤东馆 5^{2+3} 层原油与过滤煤油混合而成,68℃时粘度 40.5 mPa· s、密度 0.905 g/cm^3,酸值约 3 mgKOH/g。

过滤黄河水,总矿化度 730 mg/L,pH=8.72,钙镁离子含量 75 mg/L。

2 实验结果与讨论

所有试验均在 68℃下进行。溶液浓度以有效物含量为基准,非离子表面活性剂的含量取为 100%。

2.1 碱与原油间界面张力特性

Na_2CO_3、Na_2SiO_3 与孤东馆 5^{2+3} 层原油间的界面张力 IFT 随碱浓度 C_A 而变化的规律是相似的，随 C_A 的上升起初 IFT 下降，达到最低值后又升高，IFT 达到最低值时对应的 C_A 分别为 0.2%和 0.3%，见图 1。这一现象可以用 Rudin 等人的观点[3]加以解释：

当石油酸与碱生成的石油皂量与剩余石油酸量达到某一比例时，界面张力出现最低值。

因此，凡影响这一比例的因素如原油酸值、pH 值、含盐量、外加表面活性剂等都影响界面张力用悬清法测定碱与原油闻的界面张力特性。

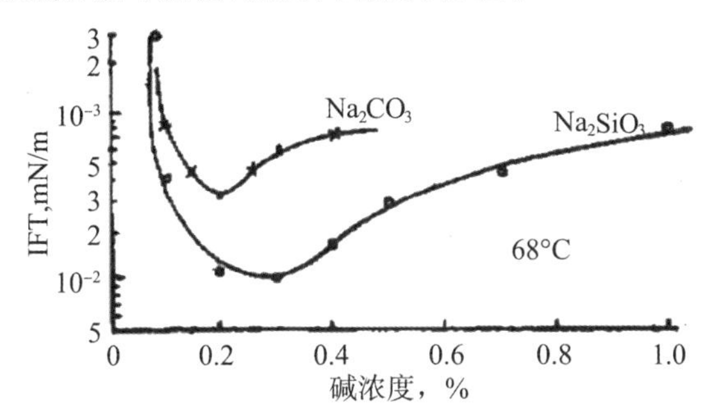

图 1 碱/原油界面张力 IFT 随碱浓度 C_A 的变化界面张力用悬清法测定

2.2 表面活性剂与原油间的界面张力

表 1 列出了 7 种表面活性剂 0.5%溶液与原油间的 IFT 值，OP-10 与 PS 的界面张力最低。

表 1 7 种表面活性剂 0.5%溶液与原油间的界面张力值(mN/m)(68℃)

活性剂	Js	T80	ABS	AS	平平加	OP-10	PS
IFT 值	7.074	9.409	4.589	4.258	3.444	0.513	0.364

2.3 碱+表面活性剂与原油间的动态界面张力

图 2 给出由 1.0%Na_2CO_3 分别与 0.1%OP-10、T80、AS、Ty-5 构成的 A+S 体系与原油间的动态界面张力曲线。由图 2 与图 1 以及表 1 的比较可知，仅加入 0.1%表面活性剂就使 1.0% Na_2CO_3 与原油间的界面张力下降 1-2 个数量级，达到 10^{-2} ～ 10^{-3} mN/m 范围。就降低界面张力来说，碱与表面活性剂之间有某种明显的协同作用。这一结果与文献报导[1]相似。由图 2 还可看到，不同表面活性剂与碱的复配体系与原油间的动态界面张力相差不大，OP-10 和 T80 略好。动态界面张力在前 10 分钟迅速下降，此后下降缓慢。这一现象对驱油有利，但与文献报导[2]有所不同(该文献报导的实验结果是：时间短时 IFT；随时间延长 IFT 增大)。

目前还没有准确描述 A+S/原油动态界面张力特性的模型。我们根据该体系中表面活性剂的吸附/脱附特性对所得结果作如下解释，A+S 与原油接触后碱和外加表面活性剂由水相向油相扩散并生成石油皂，石油皂和外加表菌活性剂在油水界面不断吸附，使界面张力迅速下降；当吸附达到一定程度后开始发生脱附并向水相扩散，这使界面张 力下降速度减慢。

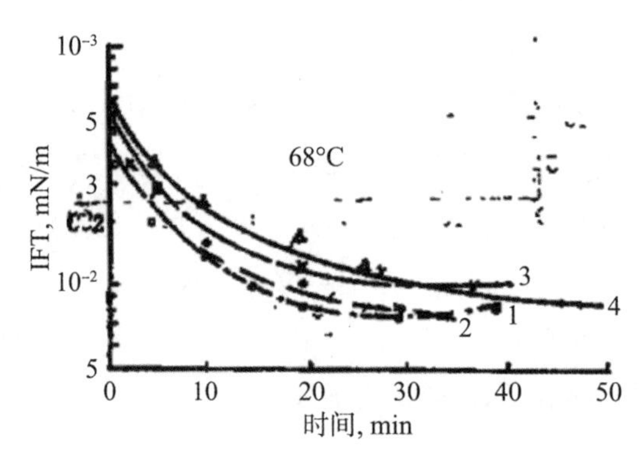

图2 1.0%Na_2CO_3+0.1%S/原油间动态界面张力曲线

1(●),OP-10,2(○),T80,3(×),AS,4(▲),Ty-5

2.4 碱浓度、表面活性剂浓度对碱+表面活性剂与原油间动态界面张力的影响

图3给出0.4%OP-10与不同浓度Na_2CO_3,构成的A+S体系的动态界面张力曲线。碱浓度越低,动态IFT初期下降越快,后期回升也越快,即达到的低IFT时间窗口越宽。碱浓度0.5%时该时间窗口仅几分钟,碱浓度2.0%时IFT下降缓慢且数值较大。Na_2CO_3,浓度为1.5%时,IFT初期下降较快,所达到的数值低,低IFT窗口宽。在A+S/原油体系中同样有最佳碱浓度存在。

图4和图5分别给出碱浓度为1.0%和1.5%而OP-10浓度不同时的动态界面张力曲线。由这两幅图可以看出:① 所有A+S体系与原油的初期动态IFT值相近,约为4×10^{-2} mN/m。② OP-10浓度由0增加到0.1%时IFT值显著下降,再由0.1%增加到0.4%时下降幅度减小。③ Na_2CO_3,浓度为1.0%时,25~35分钟后IFT回升,而Na_2CO_3浓度为1.5%时50分钟后IFT仍未见回升。④ 不同Na_2CO_3浓度下每一OP-10浓度所对应的IFT值在最初15分钟内相差不大,15分钟后差别变得很明显,1.5% Na_2CO_3的低IFT时间窗口要宽得多。而不含表面活性剂的1.0%Na_2CO_3/原油体系的界面张力特性比1.5%Na_2CO_3/原油体系要好。根据所达到的低IFT值和低IFT时间窗口宽度,界面张力特性最佳的A+S体系是1.5%Na_2CO_3+0.4%OP-10。

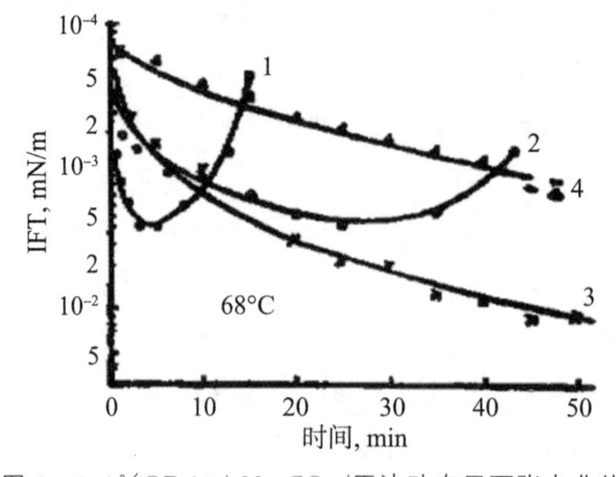

图3 0.4%OP-10+Na_2CO_3/原油动态界面张力曲线

Na_2CO_3浓度:1(○),0.5%,2(●),1.0%,3(×),1.5%,4(△),2.0%

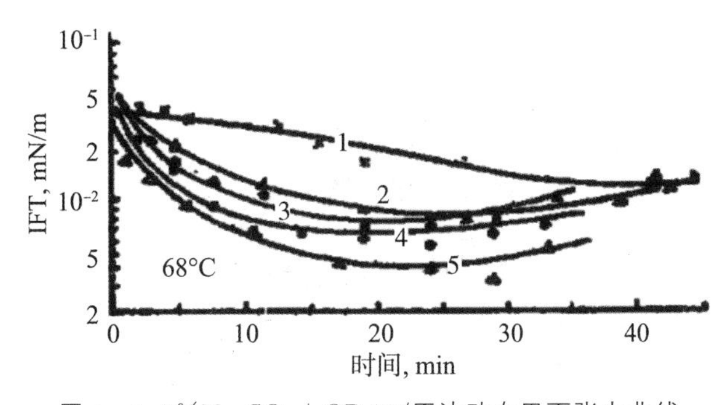

图 4　1.0%Na_2CO_3＋OP-10/原油动态界面张力曲线

OP-10 浓度：1(×)，0%，2(▲)，0.1%；3(●)，0.2%，4(○)，0.3%，5(△)，0.4%

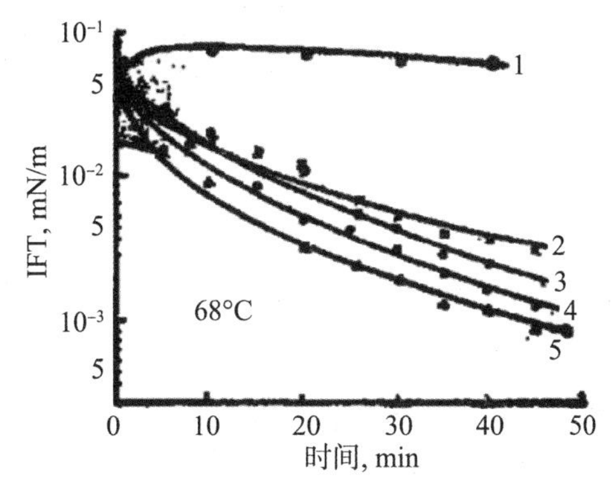

图 5　1.5%Na_2CO_3＋OP-10/原油动态界面张力曲线

OP-10 浓度：1(□).0%，2(×)，0.1%；3(O)，0.2%，4(●)，0.3%；5(△)：0.4%

2.5 碱＋表面活性剂＋聚合物体系与原油间的动态界面张力

Na_2CO_3＋0.4%OP-10＋3530S 体系/原油动态界面张力曲线见图 6。这一组曲线与图 3 所示 Na_2CO_3＋0.4%OP-10 体系/原油动态界面张力曲线相似，所不同的是 0.1%聚合物 3530S 的加入使所达到的低 IFT 值略有下降，达到低 IFT 的时间略有后移。

3 结论

实验研究了孤东馆 5^{2+3} 层原油与单一的碱、表面活性剂、碱＋表面活性剂、碱＋表面活性剂＋聚合物之间的界面张力特性。A/原油体系最低 IFT 值对应的碱液度为 0.2%Na_2CO_3 和 0.3%$Na_2SiO_3 \cdot 9H_2O$。

OP-10 和 PS(0.5%)与原油间的 IFT 值最低。Na_2CO_3＋OP-10/原油界面张力特性与 A、S 的浓度有关，A 与 S 之间有某种协同效应。1.5%Na_2CO_3＋0.4%OP-10 为最佳体系，加入 0.1%P 对此体系界面张力特性的影响不大。因此，采用复合驱油方式 A＋S 和 A＋S＋P 将产生相似的界面张力特性。

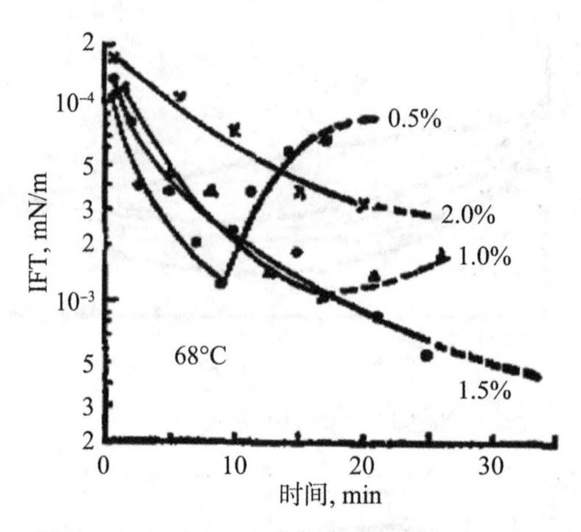

图 6 不同浓度 Na_2CO_3 +0.4%OP-10+0.1%3530S 体系与原油间的动态界面张力曲线虚线表示发生了乳化

参考文献

[1] Liu, F. F. J. J. Canad. Petrol. Technol., 1987, 11-12:54-65.

[2] Taylor, K. C. et al. t. Petrol. Soc. of CIM:Paper No. B9:40-44.

[3] Rudin, J, and Wasou, D. T., SPE21027.

[4] Borwankar, R. P. and Wasan, D. Tg. AIChem J, 1986, 32(3):455-466.

[5] Borwaukar, R. P. and. Wasan, D. T. AIChem J, 1986, 32(3):467-47B.

(1992 年 7 月 8 日收到, 1998 年 4 月 29 日修改)

Dynamical interfaclal tensions between crude oil and alkline/surfactant and alkaline/surfactant/polymer flooding solutions for strata $g5^{2+3}$ of gudong oil field

Cao Xulong　Xue Huaiyan　Li Xinlan　Wang Baoyu　Yuan Shigao

(Department of Oilfield Development, Research Institute of Geology, Shengli Pefrolenm Admlristratipe Bureau)

Abstract: Nonionic surfactant OP-10(S) is selected out from seven surfactants of varions type and alkaline Na_2CO_3, (A)is used in this work as a part of a program on combined chemical flood in strata $G5^{2+3}$ of Gudong Oil Field, Shengli. The interfacial tensions(IFTs) between crude oil(O) and A solutions are lowered by 1～2 magnitudes by introducing a little(0.1%)S. The minimum IFTs in O/A and O/A+S systems are observed at A concentrations 0.2% and 1.5%, respectively. The optimum IFT characteristics are given by 1.5%A+0.4%S solution. 0.1% polyacrylamide added to A+S solutions does not influence much on their IFT characteristics.

Keywords: Alkaline/Surfactant Flooding Solutions, Alkaline/Surfactang/ Polymer Flooding Solutions, Gudong Crude Oil (Shengli), Interfacial Tension Behavior, Dynamical Interfacial Tensions

Tween80 表面活性剂复合驱油体系研究

李干佐　林元　王秀文　舒延凌　张淑珍　毛宏志

(山东大学化学院)

曹绪龙　李克彬　王宝瑜

(胜利石油管理局地质科学研究院开发试验室)

摘要:针对胜利孤东油田的实际情况,通过相态试验、动态界面张力测定、吸附与模型驱油试验,研究了 Tween80/碱/部分水解聚丙烯酰胺复合驱油体系的性能,提出了复合驱油体系配方:0.3%Tween80+1%碱(Na_2CO_3 : $NaHCO_3$ =1 : 1)+0.1%PHPAM(3530S)。研究结果表明,本文提出的复合驱油体系能与胜利孤东油田原油形成中相乳状液,动态界面张力为 10^{-4} mN/m 数量级,是一种较好的驱油体系。

主题词:非离子表面活性剂　三元复合驱油体系　提高原油采收率　配方研究　室内性能评价　胜利孤东油田

为提高原油采收率,国内外开展了表面活性剂/碱/聚合物复合驱油体系(ASP)的研究。试验表明,若表面活性剂选择恰当,ASP 体系的驱油效率相当于 WinsorⅢ型微乳液体系,但表面活性剂的用量却仅为微乳液体系的 1/3 [1]。我们针对胜利孤东油田试验区的实际情况,开展了非离子表面活性剂/碱/聚合物复合驱油体系的研究,现将研究结果简介于下。

1 实验部分

1.1 试剂

Tween80,旅顺化工厂产品。NaCl、Na_2CO_3、$NaHCO_3$ 均为化学纯。部分水解聚丙烯酰胺(3530S),美国 Pfizer 公司产品,分子量 1 615 万,水解度 29.2%。

1.2 实验方法

1.2.1 相态研究

研究方法与文献[2]相同。

1.2.2 动态界面张力测定

用美国产 Texas-500 型旋滴界面张力仪，在 68℃下测定水相与原油界面形成后不同时间的界面张力，测定方法和计算公式见文献[3]。

1.2.3 静态吸附试验

将油砂和试液按固液比 1∶5 的比例置于封闭锥形瓶中，在 68℃下连续搅拌。每隔 1 h 取样 1 次。将所取试样在 3 000 r/min 下离心 30 min。取上清液用硫氰酸钴铵分光光度法测定 Tween80 的含量。以酚酞和甲基橙为指示剂，用标准 HCl 溶液滴定，根据滴定结果求算试样中 Na_2CO_3 和 $NaHCO_3$ 的含量。

1.2.4 模型驱油试验

将孤东油田试验区油砂用 1∶1 的醇-苯混合液洗净并填满长 20 cm、管径 2.5 cm 的直型管，制成驱油模型，控制其空气渗透率为 2～3 μm^2。用模拟地层水（总矿化度 3 403 mg/L）驱替至模型产出液含水与试验区相同即 98%左右，依次注入 0.2 PV 碱＋表面活性剂段塞、0.2 PV 碱＋表面活性剂＋聚合物段塞、0.2 PV 聚合物段塞，最后转注黄河水（总矿化度 730 mg/L）至产出液含水 98%。驱替速度 20 mL/h，试验温度 68℃。实验用模拟油为孤东油田试验区脱气原油与煤油按 8∶1 配成的混合油，68℃时的粘度为 40.47 mPa·s，密度为 0.905 g/cm^3。

表 1　ASP-原油在不同盐度下的相行为

盐浓度(%)	0	0.05	0.1	0.2	0.4	0.6	0.8	1.0	1.5
				油水比 2∶8					
上相体积(mL)	0.2	0.1	0.1	0.3	0.2	0.4	0.4	2.2	2.2
中相体积(mL)	3.0	2.8	2.7	2.4	2.2	2.2	2.2		
下相体积(mL)	6.8	7.2	7.4	7.6	7.8	7.8	7.8	8.0	8.0
相态	Ⅲ	Ⅲ	Ⅲ	Ⅲ	Ⅲ	Ⅲ	Ⅲ	Ⅱ	Ⅱ
				油水比 3∶7					
上相体积(mL)	0.2	0.2	0.2	0.3	0.2	0.2	0.3	3.4	3.4
中相体积(mL)	5.9	4.8	3.6	3.3	3.2	3.1	3.0		
下相体积(mL)	4.3	5.4	6.6	6.7	6.8	6.9	6.8	6.8	6.8
相态	Ⅲ	Ⅲ	Ⅲ	Ⅲ	Ⅲ	Ⅲ	Ⅲ	Ⅱ	Ⅱ
				油水比 4∶6					
上相体积(mL)	0.3	0.1	0.2	0.2	0.4	0.2	0.4	0.4	1.2
中相体积(mL)	8.2	5.6	5.2	4.6	4.3	4.0	4.0	4.0	3.0
下相体积(mL)	1.8	4.6	4.8	5.4	5.7	6.0	6.0	6.0	6.0
相态	Ⅲ	Ⅲ	Ⅲ	Ⅲ	Ⅲ	Ⅲ	Ⅲ		

(续表)

盐浓度(%)	0	0.05	0.1	0.2	0.4	0.6	0.8	1.0	1.5
油水比5∶5									
上相体积(mL)	0.4	0.2	0.3	0.4	0.2	0.2	0.2	0.2	0.2
中相体积(mL)	9.0	6.3	5.9	5.6	5.4	5.2	5.0	5.0	5.0
下相体积(mL)	0.8	3.7	4.1	4.4	4.6	4.8	5.0	5.0	5.0
相态	Ⅲ	Ⅲ	Ⅲ	Ⅲ	Ⅲ	Ⅲ	Ⅲ		

2 结果与讨论

2.1 ASP驱油配方及ASP-原油体系的相态

我们在ASP-原油体系相态研究中发现，乳状液呈棕色时，为中相乳状液，且剩余油相与剩余水相的体积相当，中相体积越大则界面张力越低。因此，我们按照WinsorⅢ型微乳液的指标筛选ASP驱油体系配方。

通过正交试验，确定ASP复合驱油体系配方为0.3%Tween80+1%复碱(Na_2CO_3+$NaHCO_3$，1∶1)+0.1%3530S。对上述配方的ASP与原油形成的体系进行盐度扫描，所得结果列于表1。在盐度0%～0.8%的条件下，体系均能形成稳定的棕色中相乳状液，其中的原油含量为90%左右。中相体积随盐度增大而减小，随油水比增大而增大。

2.2 驱油体系与原油间的动态界面张力

动态界面张力是指油水界面达平衡前的界面张力。动态界面张力的最低值IFT_{min}可衡量表面活性剂的界面活性。IFT_{min}值越低，界面活性越高；达到IFT_{min}所需的时间越短，越有利于驱油。Rudin等人认为[4]，在三次采油中动态界面张力比平衡界面张力更为重要。

2.2.1 驱油体系中各组分对IFT_{min}的影响

以B代表0.4%NaCl，S代表0.3%Tween80，A代表1%复碱，P代表0.1%3530S。不含复碱的体系B、B+P、B+S、B+S+P，IFT_{min}均大于2 mN/m。在盐水中加入复碱(B+A)，IFT_{min}降至8.62×10^{-2} mN/m，再加入Tween80(B+A+S)，IFT_{min}进一步急剧减小。降至2.01×10^{-5} mN/m。这是由于孤东原油酸值高，约为3 mg KOH/g原油，与复碱作用时生成石油酸皂，非离子表面活性剂与石油酸皂在降低界面张力方面具有明显的协同效应。在B+A+S中加入3530S(B+A+S+P)，IFT_{min}由2.01×10^{-5} mN/m升至4.61×10^{-4} mN/m。这说明3530S的存在不利于降低油水界面张力，这是具有一定普遍性的现象，见表2。但3530S能提高驱油体系的粘度。有利于增大波及系数降低表面活性剂的吸附。

表 2　3530S 对 0.3%Tween80+0.4%NaCl+复碱体系与原油间的界面张力的影响(88℃)

3530S 浓度(%)	不同复碱浓度(%)下的界面张力(mN/m)					
	0.2	0.5	0.7	1.0	1.2	1.5
0	2.71×10^{-3}	4.56×10^{-4}	5.46×10^{-5}	5.06×10^{-5}	2.16×10^{-5}	1.46×10^{-6}
0.1	2.76×10^{-3}	2.18×10^{-3}	9.64×10^{-4}	7.89×10^{-4}	2.43×10^{-4}	2.42×10^{-4}

2.2.2 表面活性剂与复碱浓度对 IFT_{min} 的影响

在 ASP 体系中,复碱与表面活性剂对界面张力的影响最大。各种复碱与表面活性剂浓度下的 IFT_{min} 列于表 3。在 Tween80 浓度固定时,IFT_{min} 随复碱浓度的增加先降后升,以复碱浓度为 1%和 1.2%时最低。原油与水的界面张力决定于未解离石油酸与已解离石油酸阴离子在油水两相的比值[4]。未解离石油酸分布在界面的油侧,解离石油酸根分布在界面的水侧。体系中复碱浓度较低时,复碱浓度增加有利于石油酸转变为石油酸皂并进入界面的水侧,使 IFT_{min} 降低;复碱浓度过大时,油侧中的石油酸消耗过多,未解离石油酸与解离石油酸根在油水两侧的比例失调,此时 IFT_{min} 随复碱浓度增加而增大。在复碱浓度固定时,Tween80 浓度增加使 IFT_{min} 增大,但却使 ASP-原油体系达到 IFT_{min} 所需的时间减少,见图 1。

表 3　Tween80、复碱浓度对 IFT_{min} 的影响·(68℃)

Tween80 浓度(%)	不同复碱浓度(%)下的 IFT_{min}(mN/m)					
	0.2	0.5	0.7	1.0	1.2	1.5
0.1	1.38×10^{-3}	7.42×10^{-4}	5.52×10^{-4}	5.45×10^{-4}	6.21×10^{-5}	1.54×10^{-4}
0.2	2.55×10^{-3}	1.44×10^{-3}	9.73×10^{-4}	6.96×10^{-4}	1.55×10^{-4}	4.48×10^{-4}
0.3	4.98×10^{-3}	1.68×10^{-3}	1.05×10^{-3}	7.03×10^{-4}	2.18×10^{-4}	8.00×10^{-4}
0.4	4.95×10^{-3}	2.72×10^{-3}	1.72×10^{-8}	8.42×10^{-4}	4.53×10^{-4}	9.58×10^{-4}

·驱油体系中含 0.1%3530S、0.4%Naa。

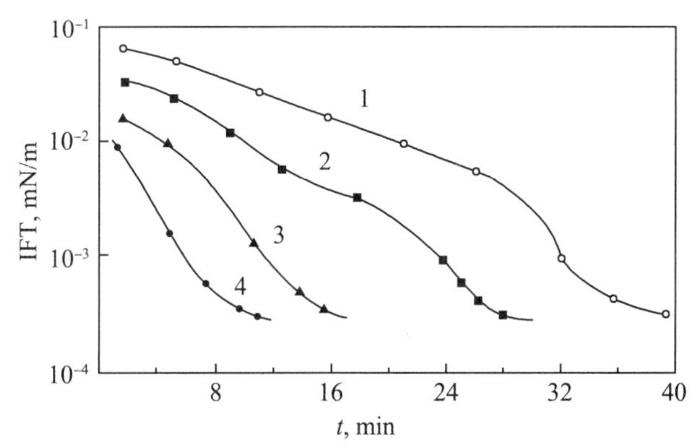

图 1　不同 Tween80 浓度时的油水界面张力与时间的关系

Tween80 浓度:1—0.1%,2—0.2%,3—0.3%.4—0.4%

2.2.3 Na_2CO_3与 $NaHCO_3$ 的配比对 IFT_{min} 的影响

碱水驱油失败的主要原因是碱水的 pH 值过高(>11～12),易与地层岩石反应,碱大量消耗后,低油水界面张力不再能维持。为此 French 等人[5]在复合驱中使用了 pH 值较低的缓冲碱。我们在 ASP 中选择了由 Na_2CO_3 和 $NaHCO_3$ 组成的复配碱。在含 0.1%3530S、0.3%Tween80,总碱浓度 1%的体系中,Na_2CO_3 : $NaHCO_3$ 为 1∶0、2∶1、1∶1、1∶2、0∶1 时,IFT_{min} 分别为 3.47×10^{-4}、4.42×10^{-4}、4.68×10^{-4}、1.76×10^{-3}、4.62×10^{-1} mN/m。复配碱中不含 Na_2CO_3 时,pH 值为 8.5,碱值过低,不利于与石油酸反应,IFT_{min} 达不到超低值。复配碱中不含 $NaHCO_3$ 时,pH 值高达 11,虽然 IFT_{min} 较低,但不适合现场使用。综合考虑各种因素,选择了 Na_2CO_3 : $NaHCO_3$ = 1∶1 的复碱,此时体系既有较低的 IFT_{min} 值,又具有较好的缓冲作用。

2.3 Tween80 的吸附损失与碱耗

2.3.1 Tween80 的吸附损失

对于 Tween80 溶液-油砂体系,Tween80 的浓度由 0.05%增至 0.45%时,吸附量由 1.50 mg/g 油砂增至 5.70 mg/g 油砂。浓度超过 0.35%后,吸附量随浓度变化曲线趋于平台状。0.3%Tween80 溶液的吸附量为 5.47 mg/g 油砂。由于聚合物能提高体系粘度,降低 Tween80 的扩散速度,且碱、聚合物与 Tween80 在油砂上存在竞争吸附,在 0.3% Tween80 溶液中分别加入 1% 复碱、0.1% 3530S、1% 复碱 + 0.1% 3530S 后,Tween80 的吸附量下降至 5.14、4.59、4.35 mg/g 油砂。

2.3.2 碱的损耗

碱与油砂接触时发生化学反应和吸附作用,所引起的碱的总损耗称为碱耗。ASP 配方体系含有 Tween80 和 3530S,其碱耗小于单纯碱溶液的碱耗,见表 4。

表 4 ASP 配方体系和相应复碱溶液与油砂作用后的碱耗

碱	碱耗(mg/g 油砂)	
	ASP 配方体系	1%复碱溶液
Na_2CO_3	5.25	5.67
$NaHCO_3$	1.83	2.83
合计	7.08	8.50

2.4 模型驱油试验结果

模型驱油试验结果见表 5,驱油动态曲线见图 2。从表 5 可见 Tween80 浓度为 0.4%时,每克表面活性剂的采出油量为 72.6 mL,采油成本偏高;浓度为 0.25%时,ASP 采出油量仅占水驱残余油的 37.4%,不利于油田的开发;浓度为 0.3%时,每克表面活性剂的采出油量为 127 mL 左右,驱油效果较好。驱油动态曲线(图 2)表明,注入水量为 2.5 PV 时转注化学剂,至 3.0 PV 时即有油墙形成,此时产出液中含油量增加数十倍,含水量由 98%下降到 55%。

表5 ASP体系模型驱油试验结果

模型编号	Tween 80浓度(%)	原始含油饱和度(%)	转注时驱油参数				试验结束时驱油参数			每克活性剂采油量(mL)
			注入量(PV)	含水量(%)	采出程度(%)	含油饱和度(%)	注入量(PV)	总采收率(%)	含油饱和度(%)	
92-山大-14	0.4	74.3	2.542	98.1	51.6	36.0	4.115	72.6	19.5	72.6
92-山大-3	0.3	73.5	2.597	99.1	48.6	37.8	4.163	69.7	21.3	126.9
92-山大-10	0.3	74.3	2.357	97.3	50.0	37.1	4.525	71.1	20.5	127.6
92-山大-15	0.3	73.0	3.495	97.7	53.9	34.5	4.679	74.4	17.2	115.4
92-山大-16	0.3	72.2	2.730	98.6	50.2	35.9	4.331	72.6	19.0	124.0
92-山大-12	0.25	74.1	3.761	98.8	52.3	35.3	4.481	70.1	21.2	122.5

* 所用ASP驱替体系中复碱浓度1%.3530S浓度0.1%。92-山大-15为锥形管,其余模型为直型管。

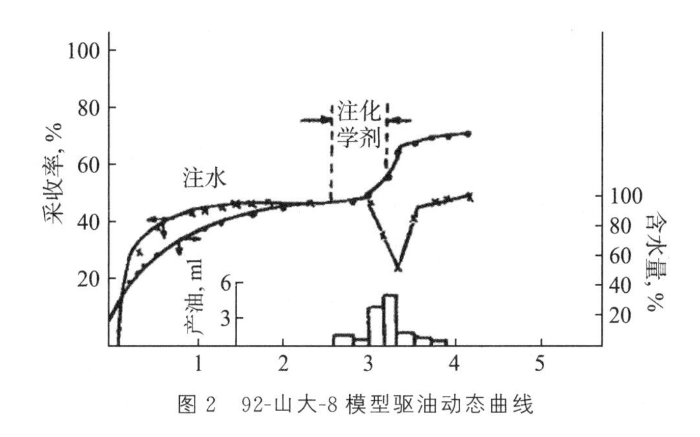

图2 92-山大-8模型驱油动态曲线

3 结论

胜利油田孤东试验区采用非离子表面活性剂Tween80进行碱/表面活性剂/聚合物复合驱油是可行的。配方为1%复碱+0.3%Tween80+0.1%3530S的复合驱油体系能与试验区原油形成中相乳状液,油水界面张力低达10^{-4} mN/m,表面活性剂的静态吸附损耗量与碱耗量符合一般规定要求,驱油效果良好。

参考文献

[1] Schular PJ, Lerner R M, Kuchnc D L. SPE/DOE 14934, 1986.
[2] 李干佐,林元,等.油田化学,1994,11(1):61-65.
[3] 李干佐,宋淑娥,等.化学物理学报,1991,4(4):296.
[4] Rudin J,Wansan D T. SPE 21027.1991.
[5] French T R, Burchfield T E. SPE/DOE 20238,1990.

A new asp-flooding solution of nonionic surfactant tween-80 for the pilot test at gudong oil field in shengli

Li Ganzuo Lin Yuan Wang Xiuwen Shu Yanling Zhang Shuzhen
Mao Hongzhi

(College of Chena. try, Shendomg Unixrnty, Jinrmt(uty)

Cao Xulong Li Kebin Wang Baoyu

(Resexoch Instibte of Ceuogy, Shengli Petrolean Alnanistralue Dremi)

Abstract: A new alkaline/surfactant/polymer flooding solution for Gudong pilot test is formulated on the basis of experimentally determined optimum composition variables and consists of 1. 0% $Na_2CO_3 + NaHCO_3$ (1 : 1, 0. 3% Tween-80, 0. 1% Pfizer's 3530S, and 0. 4% NaCl. The characteristics of the experimental ASP-solution are as follows: the transient interfacial tension between the solution and the crude(acid value—3mg KOH per gram oil)— 10^{-4} mN/m, the adsorption of Tween-80 from the solution—4. 35 mg/g oily sand, the total alkaline consumption—7. 08 mg/g oily sand. In laboratory core tests with the ASP-solution over 45% residual oil after water flooding is recovered.

Keywords: Nonionic Surfactant, ASP-Floofing Solution, Enhanced Oil Recovery, Formulation, Laboratory Performance Evaluation, Gudong Oul Field in Shengli

三元复合驱油体系中化学剂在孤东油砂上的吸附损耗

王宝瑜 曹绪龙 崔晓红 王其伟

(胜利石油管理局地质科学研究院开发试验室)

摘要:用静态和动态吸附方法研究了三元复合驱油体系中各个化学剂在地层温度(68℃)下的吸附损耗。复合体系与单一体系相比,Na_2CO_3和OP-10的吸附量较小,而3530S的较大。在复合体系中,Na_2CO_3、OP-10、3530S的静吸附量分别为1.33、2.4、0.215 mg/g油砂,动吸附量分别为0.67、0.24、0.055 mg/g油砂。

主题词:非离子表面活性剂 碳酸钠 部分水解聚丙烯酰胺 吸附 三元复合驱油体系 油砂 胜利孤东油田

化学剂的吸附损耗是影响化学驱油成败的关键因素之一。有关单一化学剂在油砂上的吸 附损耗,例如各种表面活性剂的吸附(滞留),已有多篇报导[1-3]。孤东油田小井距试验区馆 5^{2+3} 层三元复合驱油先导性试验[4],是在个别井含水100%,中心井含水99%,采出程度53.3%这一苛刻条件下开展的。在研究复合驱油体系配方过程中我们研究了化学剂的吸附损耗问题。本文报导三元复合驱油体系中各化学剂在孤东油砂上的静、动态吸附实验结果。

1 实验方法与实验体系

1.1 吸附实验

静态吸附:将三元复合驱油剂溶液与孤东油砂按等比例(v/w)置于带塞三角瓶中,混匀,置于68℃恒温箱中,定期进行振荡,48小时后取样离心分离,取上层清液作化学剂浓度分析。由初始和平衡浓度求出各该化学剂的吸附量。

动态吸附:在1.5 cm×50 cm玻璃管中充填定量油砂,抽空,饱和地层水,求出孔隙体积。于68℃注入一定PV复合驱油剂溶液,用2 PV黄河水驱替。收集排出液,测定化学剂浓度。由注入和采出液中化学剂浓度求出各该化学剂的吸附量。

1.2 化学剂、油砂与水

表面活性剂OP-10,浊点~75℃,滨州化工厂产品;Na_2CO_3,AR试剂;聚合物

3530S,有效物含量 90.89%,分子量 1 571 万,水解度 29.65%,美国 Pfizer 公司产品。

油砂取自孤东油田,为长石细砂岩,泥质含量 6.4%,其中蒙脱石 23.7%,伊利石 32.6%,高岭石 36.8%,绿泥石 6.37%。经等体积比酒精/苯清洗去油。

过滤黄河水,总矿化度 730 mg/L,离子组成如下(mg/L):$K^+ + Na^+$ 148,Mg^{2+} 34,Ca^{2+} 41,Cl^- 203,SO_4^{2-} 122,HCO_3^- 182。

1.3 化学剂浓度分析

表面活性剂 OP-10 浓度采用硫氰酸钴比色法,用 722 型和 UV-930 型分光光度计在 320 nm 处测定[5]。聚合物 3530S 浓度采用碘-淀粉比色法,用分光光度计在 590 nm 处测定[6]。Na_2CO_3 浓度采用酸碱滴定法测定。

2 结果与讨论

2.1 静态吸附

2.1.1 表面活性剂 OP-10

单一和复合化学剂溶液中 OP-10 的静态吸附实验结果见图 1。在实验浓度范围内,单一溶液中 OP-10 的吸附量与溶液初始浓度基本上呈线性关系,吸附量随初始浓度的增加而增加。OP-10 初始浓度由 2 000 mg/L 增加到 7 000 mg/L 时,其吸附量由 2.0 mg/g 砂增加到 6.7 mg/g 砂。当 OP-10 初始浓度小于 5 000 mg/L 时,液相中的 OP-10 几乎全被油砂吸附。在复合溶液中 OP-10 吸附量也随 OP-10 初始浓度增加而增加。初始浓度小于 2 000 mg/L 时,复合溶液中 OP-10 吸附量与单一溶液接近;大于 2 000 mg/L 后,复合体系中 OP-10 吸附量随浓度增加而增加的趋势变得十分缓和,OP-10 初始浓度由 2 000 mg/L 增加到 7 000 mg/L 时,其吸附量由 2.0 mg/g 砂只增加到 2.9 mg/g 砂。之所以产生这一结果,是由于一方面碱提高了砂岩表面负电性,另一方面聚合物使液相变稠,导致 OP-10 扩散速度降低,并在岩石表面与 OP-10 发生竞争吸附。

2.1.2 碱 Na_2CO_3[7]

图 2 给出单一和复合溶液中 Na_2CO_3 静态吸附实验结果。在两种溶液中 Na_2CO_3 的吸附量均随 Na_2CO_3 初始浓度增加而增加,只是在单一溶液中吸附量的增加幅度要大一些。例如,当 Na_2CO_3 初始浓度为 1.5%时,单一溶液中 Na_2CO_3 的吸附量为 1.55 mg/g 砂,而复合溶液中为 1.33 mg/g 砂。在本实验条件下,碱的吸附量包括了离子交换、岩石矿物溶解、矿物转型、沉淀等反应中碱的消耗量。在复合溶液中粘滞阻力较单一溶液大,离子扩散速率较小,各种反应的速率减慢,故 Na_2CO_3 的吸附量较单一溶液中为小。

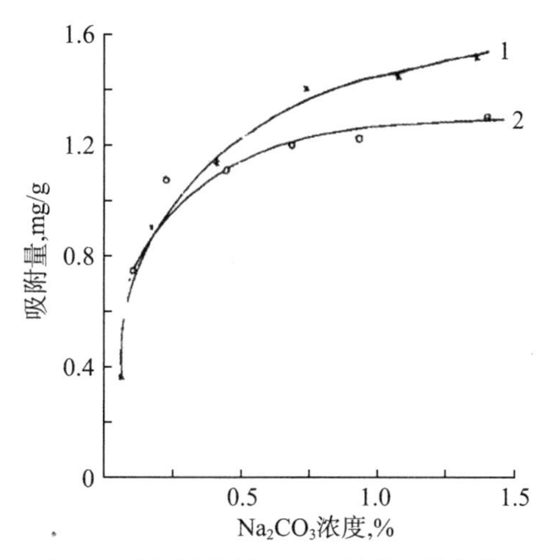

图 1 表面活性剂 OP-10 的静吸附曲线

1-从单一溶液吸附;2-从 ASP 复合体系吸附,复合体系含0.1%3530S 和 1.5%Na_2CO_3

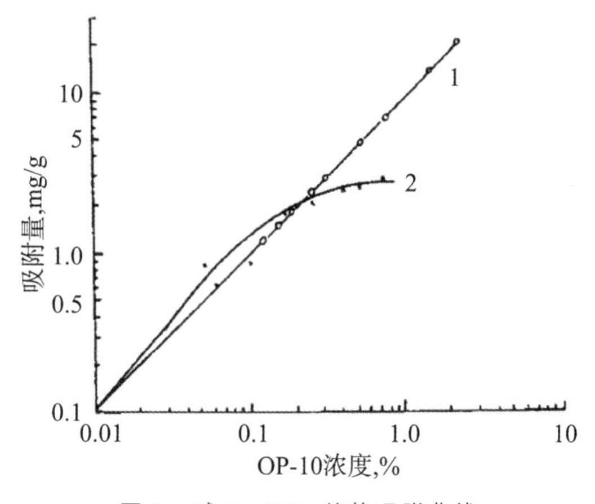

图 2 碱 Na_2CO_3 的静吸附曲线

1-从单一溶液吸附;2-从 ASP 复合体系吸附,复合体系含 0.1%3530S 和 0.4%OP-10

2.1.3 聚合物 3530S

聚合物在油藏中的损耗包括两个方面。一为聚合物分子在岩石表面的吸附,二为岩石孔隙结构空间对聚合物分子线团的机械捕集。静态实验测定值主要为表面吸附量,动态实验测定值为二者综合结果。

图 3 给出单一和复合溶液中聚合物 3530S 静态吸附实验结果。在复合溶液中 3530S 吸附量随初始浓度增加而增加,但增加幅度很小,浓度大于 1 000 mg/L 后增加幅度变得很小。在实验浓度范围内 3530S 的最大吸附量为 0.05 mg/g 砂。在复合溶液中,初始浓度小于 1 000 mg/L 时 3530S 的吸附量变化较小,大于 1 000 mg/L 后变化较大;在全部实验浓度范围内,复合溶液中 3530S 的吸附量均比单一溶液中的吸附量大得多。如初始浓度为 1 000 mg/L 时单一溶液中 3530S 的吸附量为 0.05 mg/g 砂,而复合溶液中吸附量为 0.215 mg/g 砂。

为了查明产生这一现象的原因,我们又测定了 Na_2CO_3 浓度不同但 3530S 浓度相同的二元复合溶液中3530S的静态吸附量,其结果见图4。当Na_2CO_3浓度低于0.3%时,3530S吸附量随Na_2CO_3浓度的增加变化不大,浓度超过0.3%后3530S吸附量随Na_2CO_3浓度增加而迅速增加。这里有两个因素影响实验结果。一是pH增加时岩石表面负电性增大,使聚合物吸附量略有减少;二是Na_2CO_3作为一种盐,浓度增加时压缩聚合物线团和岩石表面扩散双电层,减小二者间电性排斥,使吸附量增加。从图4所示实验结果看,后一因素起控制作用。

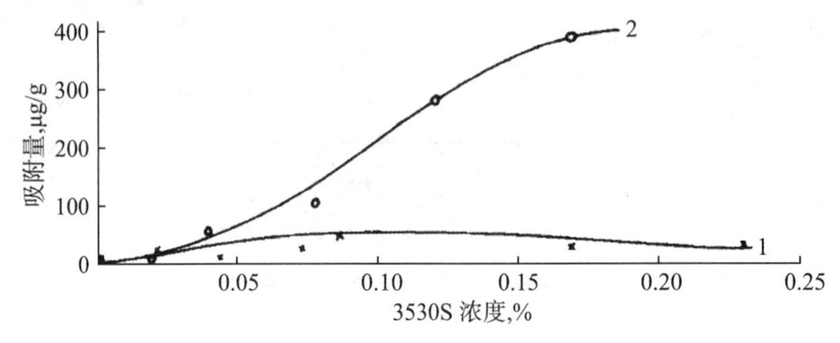

图3　聚合物3530S的静吸附曲线

1-从单一溶液吸附;2-从ASP复合体系吸附,复合体系含0.4%OP-10和1.5%Na_2CO_3

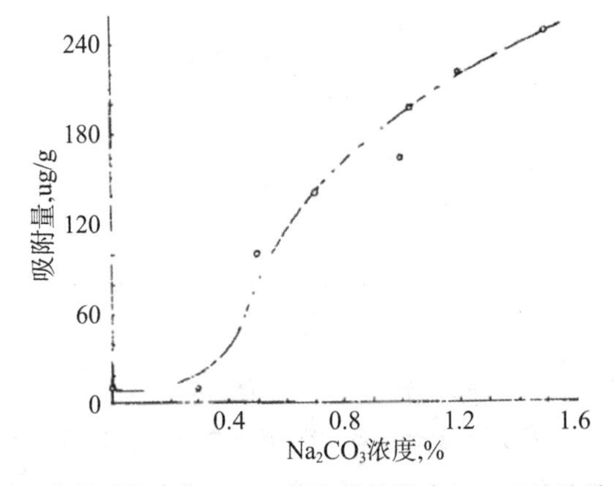

图4　Na_2CO_3含量对聚合物3530S静吸附的影响(3530S溶液浓度为0.1%)

2.2 动态吸附(滞留)

单一表面活性剂溶液和复合化学剂溶液以段塞式注入时,化学剂的动态吸附(滞留)损耗实验结果见表1。单一溶液中OP-10的动态吸附损耗量为0.57 mg/g砂,平行的两次试验中OP-10采出量为注入量的11.5%和14.5%。三元复合化学剂溶液中OP-10的动态吸附损耗量为0.24 mg/g砂,平行的两次试验中OP-10采出量为注入量的18.4%和25.8%。显而易见,在动态实验条件下复合溶液中OP-10的吸附损耗也低于单一溶液中的吸附损耗。这一点有利于复合驱油体系,因为复合驱油体系中表面活性剂的用量低于单一表面活性剂驱油体系。

从表1结果还可看出,在实验条件下复合溶液中Na_2CO_3和3530S的动态吸附(滞留)量分别为0.67 mg/g砂和0.055 mg/g砂。这些数值在复合驱中均可被接受。

表1 88℃时化学剂在孤东油砂上的动态吸附实验结果*

体系	化学剂	实验编号	注入段塞尺寸(PV)	注入化学剂量(mg)	采出体积(mL)	采出化学剂量(mg)	砂量(g)	吸附量(mg/g)	
								实测值	平均值
复合体系	OP－10	91－11－2	0.395	18.20	49	4.7	60	0.23	0.24
		91－12－1	0.418	17.65	48	3.25	60	0.24	
	Na_2CO_3	91－11－2	0.395	125.1	49	89.1	60	0.60	0.67
		91－12－1	0.418	121.3	48	77.1	60	0.74	
	3530S	91－11－2	0.395	7.20	48	3.84	60	0.056	0.055
		91－12－1	0.418	6.98	48	3.79	60	0.053	
单一体系	OP－10	91－12－2	0.475	41.64	64	6.02	60	0.59	0.57
		91－12－3	0.419	36.67	46	4.32	60	0.54	

* 注入速度为 10 mL/h,复合体系含 1.5%Na_2CO_3、0.4%OP-10、0.1%3530S。

3 结论

(1) 与单一溶液的静态吸附量相比,复合溶液中 Na_2CO_3 和 OP-10 在孤东油砂上的静态吸附量较小,而 3530S 的静态吸附量则较大。

(2) 复合溶液中 OP-10 的动态吸附量较单一溶液中为小。

(3) 在含 1.5%Na_2CO_3、0.4%OP-10 和 0.1%3530S 的复合溶液中,此三种化学剂的静态吸附量分别为 1.33、2.4、0.215 mg/g 砂,动态吸附量分别为 0.67、0.24、0.055 mg/g 砂。

参考文献

[1] Bea J H et al. Scc. Petrol Engrs J. Oct 1977:353-357.

[2] Trogus F J ct al. J Petrol Technol. June 1979:769-778.

[3] Glover C J et al. Soc Petrol Engrs J. June 1979:183-193.

[4] 曹绪龙,王宝瑜,等. 油田化学. 1993,10(2):159-163.

[5] 毛培坤,等. 表面活性剂分析. 无锡轻工学院,1984.6.

[6] API Recommented Practices for Evaluation of Polymers Used in Enhanced Oil Recovery Operations,June 1990.

[7] 曹绪龙. 油田化学. 1989,6(2):139-146.

The losses of alkaline, surfactant and polymer in tricomponent chemical flooding solution measured as adsorption on reservoir sand of gudong oil field

Wang Baoyu Cao Xulong Cui Xiachong Wang Qiwei

(Research Institute of Geology, Shengli Petroleum Administrative Bureau)

Abstract: The 48 hrs static adsorption on reservoir sand of Gudong Oil Field, Shengli, from A(alkaline)/S(surfactant),/P(polymer) flooding solutions for Na_2CO_3 and OP-10 is lower and for HPAM 3530S is higher than that from single solutions of these chemicals. The dynamic adsorption of OP-10 from a ASP solution in sandpack flow tests is lower than its static adsorption. The static adsorption of Na_2CO_3, OP-10 and 3530S amounts to 1.33, 2.4 and 0.215 mg/g sand, respectively, and the dynamic adsorption(and retention, or loss) of these chemicals in sandpack flow tests—to 0.67, 0.24 and 0.055 mg/g sand, respectively(the ASP solution contains 1.5% A, 0.4% S and 0.1% P). All measurements are performed at reservoir temperature 68℃.

Keywords: Nonionic Surfactants, Sodium Carbonate, Partially Hydrolyzed Polyacrylamide, Adsorption, ASP Flooding System, Reservoir sand, Gudong Oil Field in Shengli

ASP体系各组分配伍性研究

房会春　曹绪龙　王宝瑜　李向良

（胜利石油管理局地质科学研究院）

摘要：本工作从兼溶（分相）、粘度、热稳定性等方面考察了ASP驱油体系的配伍性，结果表明碱Na_2CO_3、表面活性剂OP-10、聚合物HPAM-3530S的配伍性好，复合溶液的热稳定性优于单一3530S溶液。

主题词：ASP复合驱油体系　组分相容性　粘度　耐温性

本文报导供胜利孤东油田小井距复合驱试验使用的ASP驱油体系中各组分配伍性实验研究结果。

1　实验部分

1.1 实验药剂和实验用水

碱：Na_2CO_3，AR试剂。表面活性剂：OP-10，浊点-75℃，滨州化工厂产品；石油磺酸盐（PS），平均当量426，有效含量40%，玉门炼油厂出品。聚合物HPAM：353QS，分子量1 571万，水解度29.05%，有效含量90.87%；3430S，分子量1 412万，有效含量88.73%，美国Pfizer公司产品。配液用水：过滤黄河水，离子组成如下（mg/l）：$K^+ + Na^+$ 148，Mg^{2+} 34，Ca^{2+} 41，Cl^- 203，SO_4^{2-} 122，HCO_3^- 182，TDS730。

1.2 实验方法

兼溶（分相）试验：观察溶液在68℃静置48小时后或加入某药剂时的分相及沉淀。

热稳定性测定：聚合物（P）溶液和碱、表面活性剂聚合物（ASP）复合溶液抽真空后在68℃恒温静置，定期测定溶液粘度变化。

2　实验结果与讨论

2.1 ASP体系各组分的兼溶性

（1）室温条件组分兼溶性不好的ASP体系可能产生沉淀，在注水管道中发生结垢，堵塞地层。兼溶试验结果表明，用蒸馏水配制的1.5%Na_2CO_3+0.1%3530S复合溶液是稳定的，既不沉淀也不分相。用黄河水配制的该复合溶液有微量沉淀（$CaCO_3$/$MgCO_3$），加入0.4%OP-10后沉淀量不变，也不分层，如加入0.4%PS，则体系立即变浑浊,沉淀量增加。其原因主要是石油磺酸盐在黄河水中溶解性较差,而且可能生成石油磺酸钙(镁)沉淀。因此,Na_2CO_3+OP-10+3530S 体系的组分兼溶性较好,在以下工作中选作实验体系。

(2) 加温条件 OP-10 的浊点为 75℃,如溶液矿化度升高,则浊点下降。当 ASP 体系中 OP-10 的浊点低于地层温度时,OP-10 析出,三种化学剂不再能同时推进。在用矿化度为 4 550 mg/L 的模拟地层水配制的 1.5%Na_2CO_3+0.4%OP-10+0.1%3530S 复合溶液中,OP-10 的浊点为 58℃,只比地层温度高 2℃。为了提高 ASP 复合驱油体系的浊点,还应加入一些阴离子表面活性剂,以确保 OP-10 在地层内不析出。

2.2 单一和复合溶液的粘度

(1) 聚合物 HPAM 3530S 和 3430S 在过滤黄河水中的溶液粘度随溶液浓度的变化见图 1。这两种聚合物溶液的粘度随溶液浓度的增大而迅速增大,0.1%溶液的粘度是 0.01%溶液粘度的 10～12 倍。3530S 的增粘能力优于 3430S。

(2) 0.1%3530S+OP-10 复合溶液的粘度随 OP-10 浓度的增大而稍有增大,见图 1。0.1%3530S+0.4%OP-10 复合溶液的粘度为 12.17 mPa·s,比单一 0.1%3530S 溶液(粘度 11.72 mPA·s)增加 3.8%。这一结果表明 OP-10 有益于 ASP 体系的增粘。

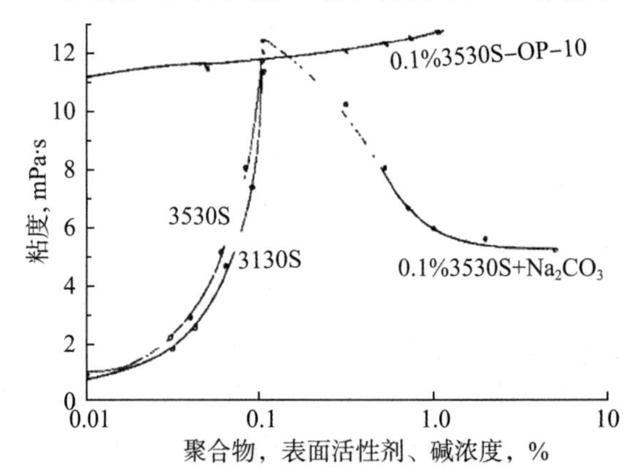

图 1 聚合物 3530S 和 3430S 溶液粘度随溶液浓度的变化及表面活性剂 OP-10 和碱 Na_2CO_3 对0.1%3530S 溶液粘度的影响

溶液用过滤黄河水配制,粘度用 NDJ-1 型旋转粘度计测定,零号转子,30 r/min

(3) 0.1%3530S+Na_2CO_3 在过滤黄河水中的复合溶液的粘度随 Na_2CO_3 浓度的增大先上升而后大幅度下降,见图 1。前一阶段粘度上升是由于少量 Na_2CO_3 的加入使 HPAM 进一步水解,聚合物线团扩张。后一阶段粘度大幅度下降是由于 Na_2CO_3 的加入使溶液含盐量增大,对 HPAM 产生了盐敏效应。1.5%Na_2CO_3+0.1%353QS 溶液的粘度比 1.0%Na_2CO_3+0.1%3530S 溶液的粘度下降 50.4%。

(4) 用过滤黄河水配制的0.1%3530S溶液和1.5%Na_2CO_3+0.4%OP-10+0.1%3530S复合溶液在不同温度下的流变曲线见图2。两种溶液的粘度均随温度升高和剪切速率增大而降低。ASP复合溶液的粘度随剪切速率的下降速率较单一聚合物溶液慢。在温度68℃下,当剪切速率由0.1 s^{-1}增大到30 s^{-1}时,0.1%3530S溶液表观粘度下降72%,而1.5%Na_2CO_3+0.4%OP-10+0.1%3530S复合溶液表观粘度只下降21%。68℃时这两种溶液的零剪切粘度(牛顿粘度)分别为36 mPa·s和5.2 mPa·s。

2.3 单一和复合溶液的热稳定性

0.1%3530S溶液和1.5%Na_2CO_3+0.4%OP-10+0.1%3530S复合溶液的粘度随68℃热老化时间的变化见图3。单一聚合物溶液在受热条件下枯度随时间而下降,20天后渐趋稳定,这时粘度为初始值的50.9%,50天后基本稳定,粘度为初始值的35.9%。ASP复合溶液的初始粘度为5.61 mPa·s,在受热条件下20天内粘度不但不下降反而略有上升,这可能与聚合物溶解过程不断完善和水解过程不断进行有关。受热20天时粘度值为7.27 mPa·s,此后粘度渐趋稳定。受热50天时粘度值下降到6.07 mPa·s。比初始值上升7.57%。这一结果说明,ASP复合溶液虽然初始粘度低于单一聚合物溶液,但热稳定性优于单一聚合物溶液。

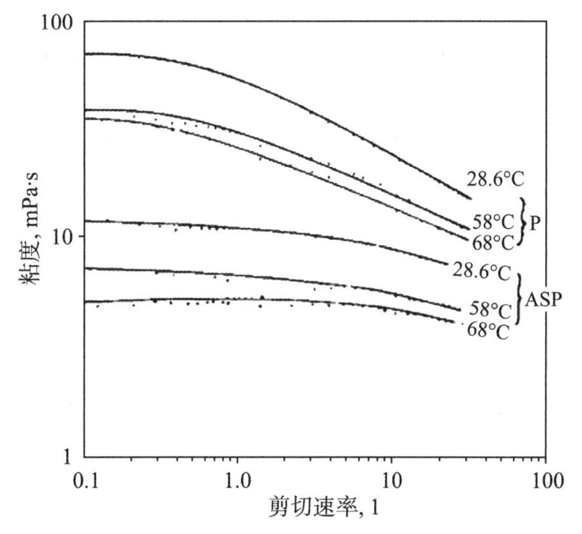

图2　0.1%3530S溶液(P)与0.1%3530S/0.4%OP-10/0.1%Na_2CO_3复合体系(ASP)在不同温度下的流变曲线

溶液用过滤黄河水配制,粘度用RV-20(CV-100)流变仪测定

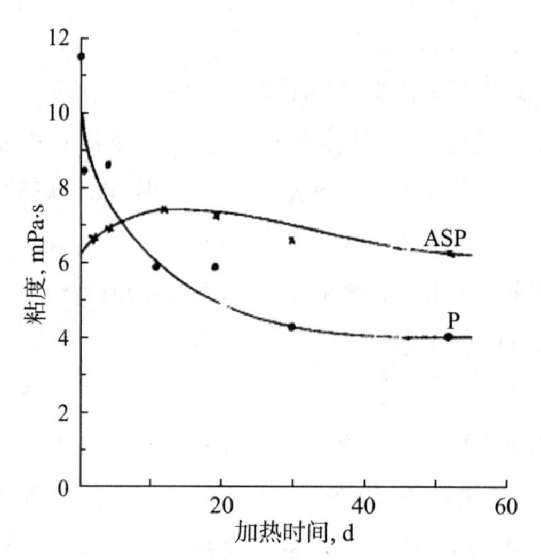

图 3 0.1%3530S 溶液(P)与 0.1%353QS/0.4%OP-10/0.1%Na_2CO_3 复合体系(ASP)的粘度随热处理时间的变化

溶液用过滤黄河水配制;热处理条件:抽真空,68℃;粘度用 NDJ-1 型旋转粘度计测定,零号转子,30 r/min

3 结论

Na_2CO_3＋3530S 与 OP-10 的兼溶性优于与 PS 的兼溶性。在过滤黄河水中 Na_2CO_3＋OP-10＋3530S 体系不产生沉淀,抗剪切能力和热稳定性均优于单一的 3530S 溶液。加入 OP-10 使 3530S 溶液的粘度稍有增大,加入 Na_2CO_3 的量大时 3530S 溶液的粘度下降。

参考文献

[1] 赵福麟. EOR 原理[D]. 东营:石油大学(华东),1985.9.

[2] 俞进桥,王宝瑜. 碱-聚合物-表面性剂复配注入剂的实验室研究[D]. 东营:胜利石油管理局地质科学研究院,1990.4.

The compatibility of components in asp systems

Fang Huichun, Cao Xulong, Wang Baoyu, Li XiangLiang

(Resench Institute of Geology, Shengli Petroleum Administrative Bureau)

Abstract: The compatibility of components in ASP-flooding systems was investigated through miscibility test, rheologic measurement and thermal aging test. It is shown that in Yellow River water Na_2CO_3, OP-10 and polymer 3530S are readily miscible and compatible, and their mixed solution is more shear resistant and more thermally stable in aging test at 68℃ than single 3530S solution does.

Keywords: Alkaline/Surfactant/Polymer Flooding Systems, Compatibility of Components, Viscosity, thermal stability

胜坨油田二区提高采收率方法室内实验研究

薛怀艳 李秀兰 曹绪龙 王宝瑜

(胜利石油管理局地质科学研究院)

薛怀艳,李秀兰,曹绪龙,王宝瑜. 胜坨油田二区提高采收率方法室内实验研究·油气采收率技术,1996,3(1):33-40.

摘要:为寻找适应于胜坨油藏条件的提高采收率的有效方法,在实验室内开展了聚合物驱、复配驱、聚合物交联技术及堵驱结合等方法的实验研究。本文着重叙述了各种驱油体系的筛选及其基本性能实验,以及各驱油体系的驱油效率实验,并对实验结果进行了讨论与研究。

主题词:驱油效率 驱油体系 配伍性 流变性 界面张力 采收率

0 引言

胜坨油田为高渗透油藏,油层非均质性严重,自1966年投入注水开发,至今采出程度为27.4%,综合含水92.1%,已进入特高含水期,因此,寻找行之有效的提高采收率的方法已成为当务之急。该项研究针对胜坨油田二区沙二段油藏特性,选取聚合物驱、复合驱、聚合物交联技术以及堵驱结合等多种三采方法进行室内实验研究,以改善油层非均质性,降低油水粘度比,增大波及体积,降低油水界面张力,从而达到降低含水,提高原油采收率的目的。

1 油层基本性质

胜坨油田二区沙二段油藏深度为1 900～2 400 m,采出程度为26.28%,平均含水92.1%,油层温度75C,层内非均质较严重,有效渗透率0.530 μm^2,岩石孔隙度为26%～35%,泥质含量<5%,原始含油饱和度65%,地面原油密度0.86 g/cm^3,地面原油粘度61.5 mPa·s,原油酸值1.423 mg KOH/g,水型为$CaCl_2$,总矿化度17 487 mg/L。

2 实验条件、试剂及仪器

实验温度:75℃。

主要试剂

碱:NaOH、$NaHCO_3$、Na_2CO_3、Na_2SiO_3,均为分析纯。

活性剂:OP-10、OP-30、CY-1、滨州 P.S.、T80,均为工业级产品。

聚合物:3530S(法国 SNF 公司)、MO-4000HSF(日本三菱化学公司)等。

堵剂:粘土胶(胜利采油厂工艺所)、交联聚合物(胜利油田地质院研制)。

实验仪器:旋转滴界面张力仪(德国 KRUSS SITEO 4 型)、DV-II 型布氏数字粘度计、流程驱油设备、RV-20 流变仪(德国哈克公司)。

驱油模型:采用双管合注分采方式进行驱油实验,模型双管分别长 20 cm、内径 2.0 cm。

3 实验结果及分析

3.1 驱油体系的筛选及其基本性能实验

3.1.1 碱体系

选用四种常用碱 NaOH、$NaHCO_3$、Na_2CO_3、Na_2SiO_3 进行实验。在胜坨地层水中 Ca^{2+} + Mg^{2+} 含量高达 300 mg/L 的情况下,用地层水配制以上碱溶液均产生大量沉淀,碱浓度加大,沉淀增多、碱损耗太大。所以,碱不适合在这种地层水下使用。

3.1.2 活性剂体系

选用非离子活性剂 OP-10、OP-30、T80,阴离子活性剂 CY-1、滨州 P·S 进行实验。

(1) 溶解性实验 OP-10、OP-30 溶解快、溶液清澈、无沉淀,T80 溶液微混,CY-1 需加热才能溶解,其溶液上浮有少量乳黄色不溶物质,滨州 P·S·溶液上浮一层未磺化油。故非离子活性剂在胜坨地层水中的溶解性比阴离子活性剂好。

(2) 配伍性实验 CY-1 与 OP-10 复配后,其溶解性得到改善,溶液清亮,无沉淀,无不溶物质,体系浊点升高,单纯 OP-10 体系浊点仅为 58℃,加入 CY-1 后,体系浊点提高到 82℃,高于油藏温度,从而克服了单纯 OP-10 因浊点低而析出的问题。

(3)界面张力实验 单一活性剂溶液与模拟油间界面张力均为 0.01~0.1 mN/m,其中石油磺酸盐 CY-1 最低。此外,CY-1 与 OP-10 复配后,体系与模拟油间的界面张力较低,为 0.01 mN/m。

3.1.3 碱+活性剂体系

用胜坨注入水配制碱+活性剂溶液,尽管该体系界面张力较低,可低达 0.001 mN/m,但在胜坨水质 Ca^{2+} +Mg^{2+} 含量较高的条件下,含碱的体系都产生大量的沉淀,使碱耗过大,易造成地层伤害。

表1 聚合物性能实验

聚合物名称	有效含量(%)	水解度(%)	分子量($\times10^6$)	粘度(mPa·s)
MO-4000HSF(日本)	91.18	29.1	20.18	14.8
3530S(法国)	90.89	29.7	15.7	9.42
PDA-3000(日本)	90.34	30.7	18	11
白银公司产品(国产)	41.48	30.9	9.16	5.5

3.1.4 聚合物体系

在同等实验条件下,对国内外十多种聚合物进行分子量测定,粘度-浓度曲线测定,水解度、过滤比、热稳定性等多种性能实验,表1列出了四种聚合物主要实验结果。

实验中,选用分子量最大、粘度最高的MO-4000HSF进行流变性实验,分别在50℃、60℃、75℃下测定浓度为0.1%的MO-4000HSF溶液的粘度随剪切速率变化的曲线(见图1)。

3.1.5 活性剂与聚合物复配体系

聚合物选用分子量较大的MO-4000HSF,活性剂选用OP-10、CY-1,进行以下实验。

(1) 配伍性实验 用胜坨模拟水配制0.1% MO-4000HSF溶液及0.1% MO-4000HSF+0.5%OP-10+0.5%CY-1复配溶液,体系粘度分别为6.9 mPa·s和5.8 mPa·s,可见复配体系粘度低于单纯聚合物体系粘度;用孤东污水配制0.1%MO-4000HSF溶液,粘度为14.7 mPa·s其$Ca^{2+}+Mg^{2+}$含量为102 mg/L。可见胜坨水$Ca^{2+}+Mg^{2+}$含量高(300 mg/L),对聚合物粘度影响较大。

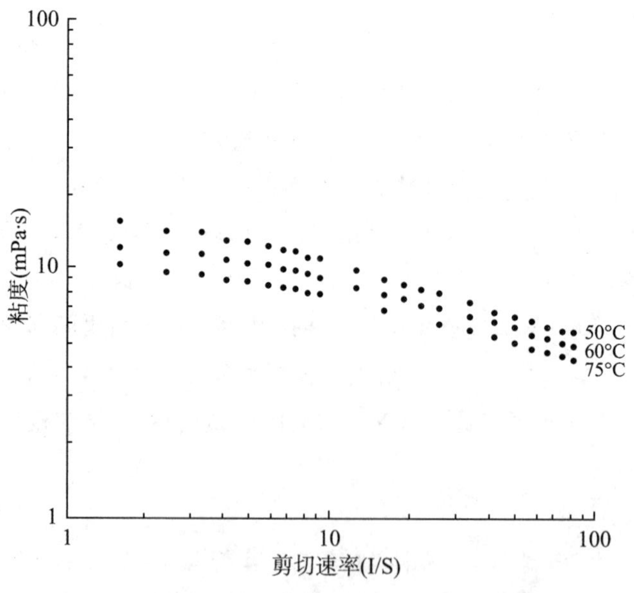

图1 浓度0.1%MO-4000HSF流变性实验曲线

(2) 热稳定性实验 将存有0.1%MO-4000HSF溶液及0.1%MO-4000+0.5%OP-10+0.5%CY-1溶液的容器抽去空气,放入75℃的保温箱中,定期观测其粘度变化

(见图 2),由图 2 所示,复配体系的初始粘度虽低于单纯聚合物体系,但 3 d 后,复配体系粘度大于单纯聚合物体系,30 d 后,复配体系粘度为 4.2 mPa·s,而单独 MO-4000HSF 体系粘度仅为 2.7 mPa·s。可见复配体系的热稳定性优于聚合物体系。

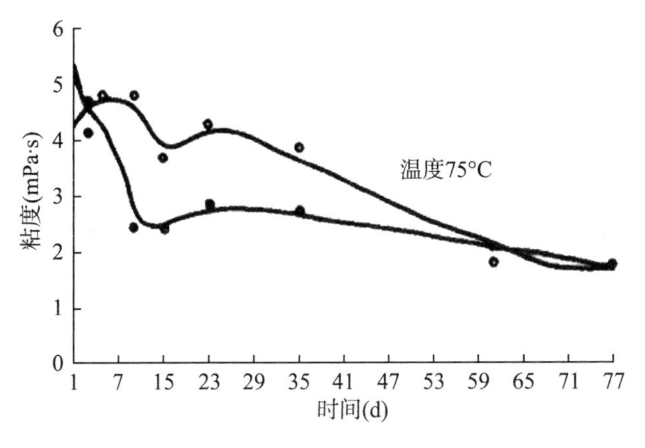

● 0.1%KMO-4000HSF

○ 0.1%MO-4000HSF+0.5%XOP-10+0.5%CY-1

图 2　热稳定实验曲线

(3) 界面张力实验　聚合物与活性剂复配体系的界面张力如图 3 所示,活性剂与聚合物复配后,油水界面张力比用单纯活性剂体系时的界面张力低,低界面张力时间窗口也较宽,这里可以说明 MO-4000HSF 与 OP-10、CY-1 复配后在降低界面张力方面有较好的协同效应。

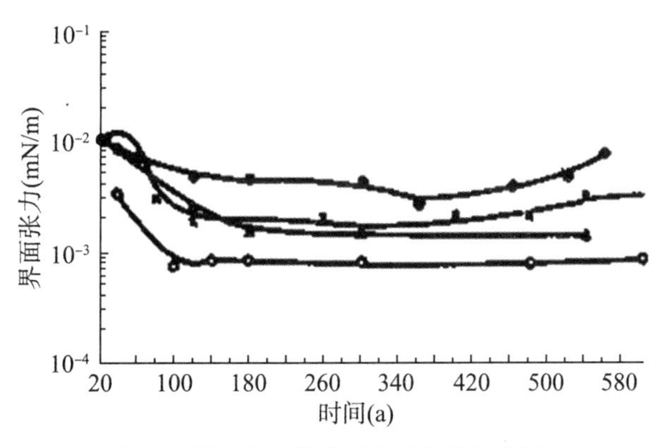

图 3　聚合物+活性剂体系与胜坨原油间动界面张力实验曲线

(4) 流变性实验　图 4 是复配体系流变性实验曲线,分别在 50℃、60℃、75℃下测定该复配体系的粘度随剪切速率的变化。与图 1 比较,聚合物体系初始粘度大于复配体系,随着剪切速率的增加,粘度下降的速率也稍大于复配体系,当转速达 100 r/min 时,两个体系的粘度基本相等。

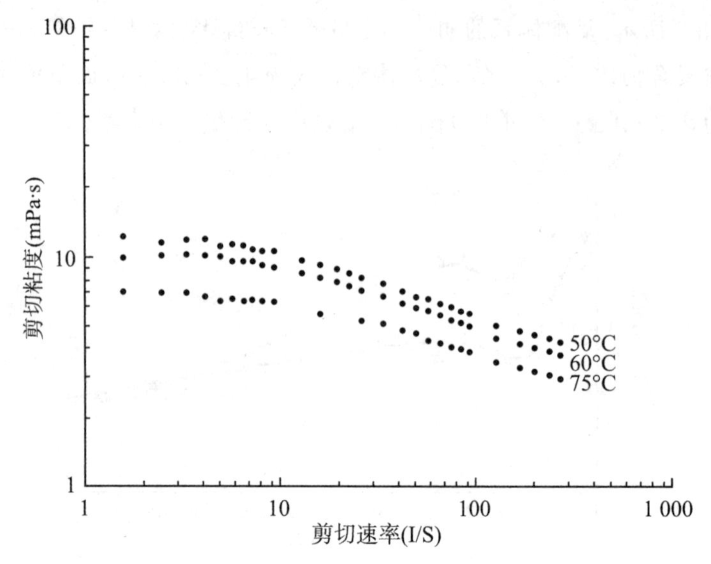

图 4 0.1%MO-4000HSF+0.5%CY-1+0.5%OP-10 复配体系流变性实验曲线

3.1.6 堵、驱结合体系

选用胜坨油田使用的堵水剂，按现场比例 1∶13，用模拟水配制，放入 70 C 烘箱，放置 5～6 h 后成冻胶状。

把成冻胶状的堵剂放入配好的 0.1%MO-4000HSF+0.5%CY-1+0.5%OP-10 溶液中，搅拌 2 h 后观察，大部分不溶解，过滤掉不溶物后测体系的粘度和界面张力。其粘度由加堵剂前的 5.8 mPa·s 增至 55.7 mPa·s 体系界面张力实验结果见图 5，单纯堵剂界面张力较大，加入复配体系中对界面张力没有影响，因此该堵剂与复配体系协同性较好。堵剂加复配液的热稳定性实验结果表明，当堵剂浓度低于 5%时，体系不易形成冻胶状，当堵剂浓度为 5%时，成胶时间为 10 h 左右。

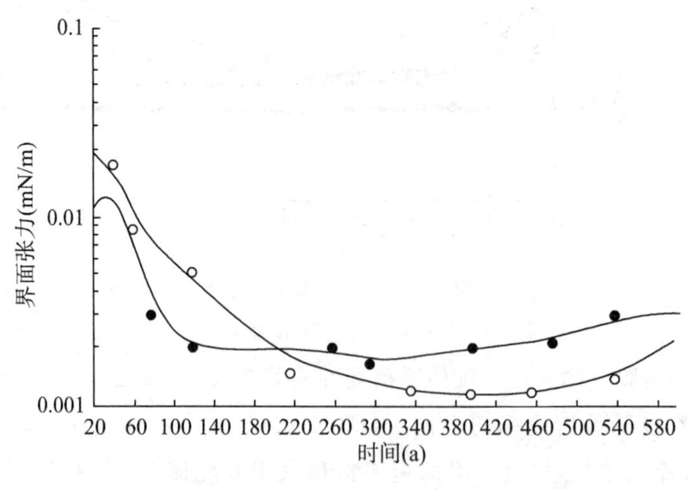

• 0.1%MO+1.0%OP-10 ○ 1.0%OP+0.1%MO+10-DJ

图 5 堵剂对聚合物+活性剂体系动界面张力的影响

3.2 驱油效率实验

根据第一部分实验结果，选出四个体系进行驱油实验：① 聚合物体系；② 聚合物+活性剂复配体系；③ 聚合物交联体系；④ 堵驱结合体系。通过驱油效率实验，可以比较直观地对比分析驱油体系的驱油过程及效益，了解各体系参数对驱油效率的影响，以寻找出提高采收率的有效可行的方法。

表2 胜坨注入水驱油实验模型参数及结果

模型号	K(μm²)	岩心孔隙体积(cm³)	S_o(%)	无水期采收率(%)	η(%)
95-23	5	27.9	71.7	20	55
95-24	1.5	25.3	67.2		50.6
合 23～24		53.2	69.5		51.9
95-35	5	26.7	71.2	14.7	55.8
95-36	0.5	24.9	62.2		3.2
合 35～36		51.6	66.9		32.2

注：K 为渗透率；S_o 为原始含油饱和度；η 为最终采收率。

3.2.1 水驱

根据胜坨油藏非均质性严重的特点，选择了两组模型，渗透率级差分别为 10∶3，10∶1，模型参数及实验结果见表 2。由表 2 可见，渗透率级差小时，高渗透管和低渗透管最终水驱采收率差别不大，分别为 55.0%、50.6%；当渗透率级差大时，两管最终采收率差异较大，分别为 55.8%、3.2%，低渗透管出油较少。说明在渗透率级差大的地层中，单纯用水驱的采收率是较低的。

3.2.2 聚合物驱

选用聚合物 MO-4000HSF 进行驱油实验，进行了聚合物段塞注入量的筛选，聚合物浓度的筛选，并用改变渗透率级差的方式考察它对驱油实验的影响。主要模型参数及实验结果见表 3。由表 3 可见：① 该聚合物体系提高采收率幅度较大，最大可比水驱提高 27.6%，这是由于聚合物溶液粘度大于注入水粘度 6～9 倍，改善了油水粘度比，也降低了地层的水相渗透率。聚合物驱模型的采收率、含水变化典型曲线见图 6。② 在同等条件下，采收率提高幅度随注入聚合物段塞量的增加而增大。进行了注入量为 0.2、0.3、0.4 PV 的三组实验，提高采收率分别为 19.5%、20.3%、27.6%。当注入量为 0.4 PV 时采收率增值与注剂利用率的乘积最大。③ 当浓度分别为 0.1%、0.15%时，可以看出，浓度大，提高采收率效果更好，因为浓度增大使体系粘度增加，驱油效率也就提高。④ 对渗透率级差分别为 10∶3，10∶1 的两组模型，注入聚合物的量和浓度都相同，提高采收率分别为 27.6%、27%，非常接近。可见该体系能有效地适应非均质性地层，除了使高渗透模型产油外，也使低渗透模型大量产油，所以采收率大幅度上升。

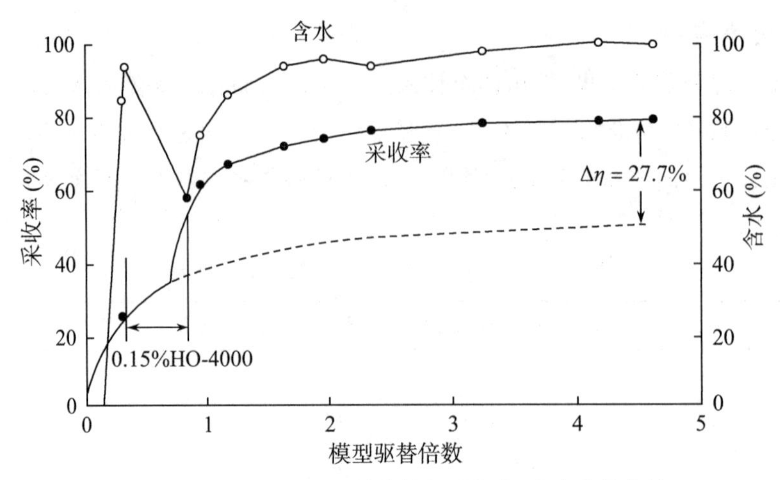

图 6 95-3、95-4 模型驱替倍数与采收率、含水变化曲线

表 3 聚合物体系驱油实验参数及主要结果

模型号	K (μm^2)	岩心孔隙体积 (cm^3)	S_o (%)	转注时含水率 (%)	注入量 (PV)	η	$\Delta\eta$ (%)	U (m^3/t)	$\Delta\eta\times U$	体系配方
95-25	5	27.3	73.3	92		67				
95-26	1.5	25.9	65.6			76.5				0.15%MO-4000HSF
合 25～26		53.2	69.6		0.2	71.4	19.5	452.4	88.2	
95-29	5	26.1	65.1	96		74.1				
95-30	1.5	26.5	56.6			70.2				0.15%MO-4000HSF
合 29～30		52.6	60.8		0.3	72.2	20.3	274.3	55.7	
95-3	5	29.8	73.8			76.4				
95-4	1.5	26.7	67.4			82.2				0.15%MO-4000HSF
合 3～4		56.5	70.8		0.4	79.5	27.6	325.7	89.9	
95-37	5	27.6	70.7	92		66.7				
95-38	0.5	25.6	62.5			50.0				0.15%MO-4000HSF
合 37～38		53.2	66.7		0.4	59.2	27.0	300.2	81.1	
95-21	5	26.2	72.5	96		62.1				
95-22	1.5	25.8	69.8			66.7				0.1%MO-4000HSF
合 21～22		52	71.2		0.4	64.3	12.4	220.7	27.4	

注：$\Delta\eta$——提高的采收率，U——化学剂利用率。

3.2.2 聚合物与活性剂复配体系驱

用胜坨注入水配制 MO-4000HSF＋0.5%OP-10＋0.5%CY-1 复配体系进行驱油实验，模型参数及实验结果见表 4。实验表明：① 该复配体系能大幅度提高采收率，最大比水驱提高 37.7%，该复配体系不仅能有效地改善油水流度比，增大波及体积，还能有效地发挥活性剂降低油水界面张力的作用，因而使采收率明显提高。图 7 给出了复配驱模型采收率、含水变化典型曲线。② 进行渗透级差分别为 10∶3、10∶1 两组实验。注复配体系后，使高渗透模型继续出油，低渗透模型也大量出油，两组实验最终采收率差别不大，分别为 69.3%、69.5%。与水驱结果相比分别提高了 17.4%、37.3%。可见，渗透率级差愈大，提高采收率效果愈明显，因此该复配体系能有效地改善油藏非均质性，也适合于低渗透油藏。③ 改变复配体系中 MO-4000HSF 的浓度进行实验，其结果见表 3，MO-4000HSF 浓度为 0.15%的复配体系的驱油效果明显优于浓度为 0.1%的复配体系。

对聚合物体系与复配体系对比进行驱油实验，结果此较见表 5。由表 5 看出，在实验条件相同的前提下，复配体系提高采收率的幅度大于聚合物体系，但化学剂利用率却远远低于聚合物体系，采收率增值与化学剂利用率的乘积也小于聚合物体系，因此，若从经济方面来看聚合物体系优于复配体系。

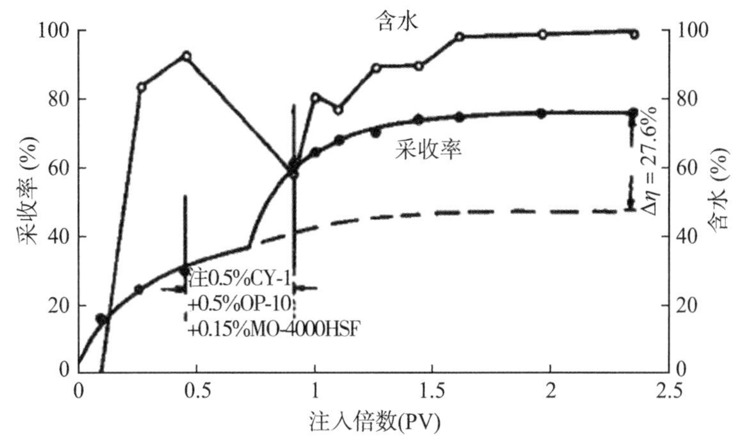

图 7　95-31、95-32 模型驱替倍数与采收率、含水变化曲线

表 4　复配体系驱油实验参数及主要结果

模型号	K (μm^2)	孔隙体积 (cm^3)	S_o (%)	转注时含水(%)	注入量 (PV)	η (%)	$\Delta\eta$ (%)	U (m^3/t)	$\Delta\eta\times U$	体系配方
95-5	5	29.5	67.8	96		72				0.5%CY-1+0.5%
95-6	1.5	26	67.3			66.3				OP-10+0.1%
合 5～6		55.5	67.6		0.4	69.3	17.4	26.7	4.6	MO-4000HSF
95-43	5	27.8	66.5	92		67.0				0.5%CY-1+0.5%
95-44	1.5	24.4	45.1			73.6				OP-10+0.1%
合 43～44		52.8	56.5		0.4	69.5	37.3	47.9	17.8	MO-4000HSF
95-31	5	25.8	67.8	93		73.9				0.5%CY-1+0.5%
95-32	1.5	25.6	60.5			85.3				OP-10+0.1%
合 31～32		51.4	64.2		0.4	79.5	27.5	38.4	10.6	MO-4000HSF

表 5 复配体系与聚合物体系驱油实验结果对比

模型号	K (μm^2)	孔隙体积 (cm^3)	S_o	转注时含水(%)	注剂粘度 (mPa·s)	注入量 (PV)	η (%)	$\triangle\eta$ (%)	U (m^3/t)	$\triangle\eta\times U$	体系配方
95-21	5	26.2	72.5	96	6.96		62.1				0.1%
95-22	1.5	25.8	69.8				66.7				MO-4000HSF
合 21～22		52	71.2			0.4	64.3	12.4	220.7	27.4	
95-5	5	29.5	67.8	96	5.8		72				0.5%CY-1+0.5%
95-6	1.5	26	67.3				66.3				OP-10+0.1%
合 5～6		55.5	67.6			0.4	69.3	17.4	26.7	4.6	MO-4000HSF

3.2.4 交联聚合物体系驱

将聚合物与交联剂配好后放置于 75 C 烘箱中,等交联 5～6 d 再注入岩心模型中进行驱油实验,实验模型参数及主要实验结果见表 6。由表 6 可知,渗透率级差分别为 10∶3、10∶1 的两组实验,最终采收率分别为 66.0%、62.2%,与水驱相比分别提高 14.1%、30.0%,可见对非均质油藏效果较好。

表 6 交联聚合物体系驱油实验参数及实验结果

模型号	K (μm^2)	岩心孔隙体积 (cm^3)	S_o	注剂粘度 (mPa·s)	注入量 (PV)	η (%)	$\Delta\eta$ (%)	备注
95-33	5	27.6	65.2	30 000		64.4		
95-34	1.5	25.7	58.4			68.0		配制第 4 天注,第 8 天驱
合 33～34		53.3	62.9		0.2	66.0	14.1	
95-45	5	26.6	75.2	93		65		
95-46	0.5	24.6	65.0	5 000	58.8			配制第 5 天驱
合 45～46		51.2	70.3		0.2	62.2	30.0	

3.2.5 堵、驱结合体系

用胜坨注入水配制浓度为 5%的堵水剂注入模型,放置 12 h 后分别进行水驱、聚合物驱及复合驱,实验模型参数及主要实验结果见表 7。实验表明:① 对渗透率级差分别为 10∶3、10∶1。两组模型。注入堵剂 0.2 PV,放置 8 h 后,进行水驱,对渗透率级差小的模型.提高采收率效果不明显,对级差大的模型,驱油效率大幅度提高。二种情况提高采收率分别为 1.4%、33%,可见,该堵剂适用于渗透率级差较大的油藏。② 分别注入堵剂 0.2 PV、0.4 PV 于渗透率级差为 10∶3 的两组模型中,提高采收率效果均不明显。③ 对渗透率级差为 10∶3 的模型,注入浓度为 5%的堵剂 0.2 PV,放置 8h 后,分别进行 a)注 0.1%MO-4000HSF0.4 PV,其结果提高采收率为 15.7%,效果好于单纯聚合物驱;b)注 0.5%CY-1+0.5%OP-10+0.1%MO-4000HSF0.4 PV,提高采收率效果不如单纯复合驱效果,这主要是由于实验室模型的局限性较大,所以很难模拟出堵驱效果。

表 7 堵驱结合体系驱油实验参数及实验结果

模型号	K (μm^2)	岩心孔隙体积 (cm^3)	S_o	转注时含水(%)	注入量 (PV)	η (%)	$\Delta\eta$ (%)	实验配方及方法
95-19	5	25.4	74.8	93		48.4		注浓度5%的堵剂 0.2 PV,8 h后水驱
95-20	1.5	24.9	68.3			58.8		
合 19~20		50.3	71.6		0.2	53.3	1.4	
95-7	5	27.8	68.3	96		68.4		注浓度5%的堵剂 0.2 PV,8 h后水驱
95-8	1.5	26.3	76.1			45.0		
合 7~8		54.1	72.1		0.4	56.4	4.5	
95-41	5	27.7	72.2	93		65.5		注浓度5%的堵剂 0.2 PV,8 h后水驱
95-42	0.5	24	54.2			64.6		
合 41~42		51.7	63.8		0.2	65.2	33	
95-9	5	27.7	72.2	96		73.0		注浓度5%的堵剂 0.2 PV.8 h后注 0.1%MO0.4 PV
95-10	1.5	26.1	65.1			64.7		
合 9~10		53.8	68.8		0.2+0.4	67.6	15.7	
95-11	5	29	72.4	96		67.6		注浓度5%的堵剂 0.2 PV.8 h后注 0.5%CY+0.5%OP+0.1%MO 0.4 PV
95-12	1.5	24.3	65.8			62.5		
合 11~12		53.3	69.4		0.2+0.4	65.4	13.5	

4 结论

(1) 胜坨油田的水质因 $Ca^{2+}+Mg^{2+}$ 含量过高,遇碱易产生沉淀,碱耗过高,故采用无碱驱油体系为宜。

(2) 复配体系 0.1%MO-4000HSF+0.5%OP-10+0.5%CY-1 各成分间的配伍性强,既能改善油水流度比(复配体系的粘度为注入水粘度的 6 倍),又能有效降低界面张力(低达 0.001 mN/m),热稳定性也优于单一聚合物体系,能大幅度地提高原油采收率,最高比水驱提高 37.3%。

(3) 0.15%MO-4000HSF 聚合物体系的增粘效果好(为注入水粘度的 9 倍),配伍性强,可提高采收率达 27.6%,在经济上优于复配体系。

(4) 堵水剂能增加驱油体系的粘度,对体系界面张力无影响,配伍性能良好。

(5) 堵驱结合及交联聚合物驱适用于非均质性较严重的油藏。但由于实验室模型的局限性,与现场差距较大,少数实验效果不理想,还有待于进一步研究。

致谢:参加本项工作的还有曾胜文、宋新旺、崔晓红、江小芳、刘异男、刘坤、王其伟、赵华、施小乐、田志铭等。还得到胜利采油厂工艺研究所解通成同志的大力协助,在此一并致谢。

参考文献

[1] LW 拉里著,何生厚,等,译. 化学和热力采油工艺与原理[M]. 济南:山东科学技术出版社,1992.

[2] 赵福麟. EOR 原理[M]. 东营:石油大学出版社,1991.

[3] 沈娟华,张以根,曹绪龙,等. 孤岛油田小井距实验区复合驱油先导实验研究[M]. 北京:石油工业出版社,1995.

[4] Nelson R C, Lawson J B. Cosurfactant-enhanced alkaline flooding [J]. SPE,12672.

[5] Frank F J Lin, George B J. Laboratory evaluations of crosslinked and alkaline-polymer-surfactant flood[J]. Jcpt,1987;28(6):54-65.

[6] Saleem S M. Enhanced oil recovery of acidic crudes by caustic-cosurfactant-polymer flooding. J. Surf. Sci-Tech,1987;3(1):1-10.

本文编辑　陈朝书

Xue Huaiyan, Li Xiulan, Cao Xulong and Wang Baoyu. Laboratory research on EOR techniques in the second block of Shengtuo oil field. OGRT, 1996, 3(1): 33-40.

To determine effective EOR methods suitable for Shengtuo oil field, laboratary research has been conducted on polymer flooding, combination flooding, crosslinked polymer flooding, and the combination of plugging with displacement. This paper stresses the screening of various oil displacement systems and tests on their basic properties, it also discusses the experiment results.

Key words: oil displacement efficiency, oil displacement system, compatibility, rheological property, interfacial tension, oil recovery

ASP 复合驱油体系瞬时界面张力的研究*

牟建海　李干佐　李　英

（山东大学教育部胶体与界面化学开放实验室，济南，250100）

曹绪龙　曾胜文

（胜利石油管理局地质科学研究院，东营，257001）

摘要：以胜利油田孤岛试验区原油为油相，用正交试验筛选了碱/天然混合羧酸盐/聚合物驱油体系，讨论了各组分对 ASP 复合驱油体系油水瞬时界面张力的影响，并探讨了各组分间的相互作用机理及其在油水界面的吸附机理。

关键词：碱/表面活性剂/聚合物复合驱油体系，瞬时界面张力，天然混合羧酸盐

分类号：O647.2

鉴于我国多数油井产出液中含水率高达 90%，开展三次采油研究是十分诱人的课题。表面活性剂在三次采油中的应用已受到重视[1-3]。从化学驱油看，最佳驱油体系是碱/表面活性剂/聚合物组成的 ASP 复合驱。尽管已有许多有关 ASP 驱油体系的报道[4-8]，对体系各组分的作用也进行较为深入的研究[6-9]，但由于驱油体系的成本较高，限制了其在实际生产中的应用。我们从油脂下脚料中制备的天然混合羧酸盐价格低廉，与石油皂间的协同效应显著。针对胜利油田孤岛试验区原油，用正交试验设计方法筛选 ASP 驱油配方，可使油水最小界面张力达到 1.0×10^{-3} mN/m；室内模拟驱油试验表明，该体系能提高采收率 26.2%。本文研究了两个 ASP 驱油体系与原油间的瞬时界面张力（IFT），考察了各组分的影响。

1　实验部分

1.1 仪器与药品

Mettler AE 200 电子自动天平（美国）；Texas-500 旋滴界面张力仪（美国）；阿贝折射仪（上海）；比重计。

胜利油田孤岛试验区原油密度 0.906 g/mL，酸值 0.52 mg KOH/g 原油，地层温度 70℃，地层水矿化度 6 800 mg/L；自制天然混合羧酸盐 SDC，烷基羧酸含量 63.1%，水溶性良好；辅助表面活性剂为市售烷基醇酰胺 ANN，有效含量 97%；按 $w(Na_2CO_3)$/

$w(NHCO_3)=1$ 配制复碱;部分水解的聚丙烯酰胺 HPAM(3530s)由 Pfizer 公司提供,固含量 91%,平均分子量 1.615×10^7,水解度 26%。

1.2 实验方法

用旋滴界面张力仪测定瞬时界面张力。水相为驱油体系,油相为原油,观察油滴形状变化,根据 Prince 等[10,11]的方法计算界面张力。

2 结果与讨论

2.1 ASP 驱油体系与原油间的化学作用

一般认为,碱驱时原油中的酸会迁移到界面上,与水相中的碱反应生成表面活性物质石油皂[12]。石油皂主要处于油滴的水侧,可吸附在油/水界面,也可有少量迁移到油相中,而石油酸则大部分存在于油。吸附、脱附速率引起的瞬时界面张力行为会受到某些因素的影响,如原油的酸值、温度、水相中离子强度、碱浓度、表面活性剂浓度及油相与水相中的粘度等。

原油中有机酸 HA 在油相及水相中的分配平衡为:$HA_o \rightleftharpoons HA_w$,一般情况下,分配系数 $K_D=c(HA_o)/c(HA_w)\gg1$,HA 在水相中很少。水相中的 HA_w 可发生解离,生成石油皂:

$$HA_w+OH^- \rightleftharpoons A_W^- +H_2O \tag{1}$$

Rudin 等[13]指出,未解离的酸和解离的酸根离子在两相中的分配系数 $K_D'=c(A_W^-)/c(HA_w)$ 控制着界面张力,而 K_D' 受 pH 值控制。由式(1)可知,pH 值增大有助于 A_W^- 的生成,因而吸附在界面上的 A_W^- 增多,K_D' 值增大。当该值趋向于 1 时,界面张力值最低。

外加表面活性剂 SDC 是一种天然混合羧酸盐,也可吸附在界面上,与石油皂形成混合胶束,产生复配增效作用,使 A_W^- 量增加。HPAM 的加入会使水相粘度增大,影响各个组成分子的扩散系数,改变原有的吸附脱附化学平衡。HPAM 也可在油水界面吸附,使表面活性物质的吸附减少,从而影响界面张力。

对于胜利油田孤岛试验区原油,用正交试验设计法得到两个驱油体系配方,其 Alkali,SDC,HPAM,ANN 浓度分别为 1.0%,0.5%,0.1%,0(体系Ⅰ)和 1.0%,0.5%,0.1%,0.05%(体系Ⅱ)。

2.2 复碱浓度对界面张力的影响

图 1 是两个驱油体系于不同碱浓度时的瞬时界面张力图。由图 1 可见,当碱浓度由 0.5%变化到 1.5%时,IFT 随时间的延长先减小,达到最小值后又有增大的趋势,这是由于开始时表面活性物质在油水界面上的吸附速率大于脱附速率,因而将在界面上聚集,引起界面张力下降。但随着界面上活性物质浓度的增大,其脱附速率增加,当 $c(A_W^-)>c(HA_w)$(即 $K_D'>1$)时,界面上的活性物质因脱附而减少,界面张力又变大。只有在吸附与脱附速率相等时,才会出现最低界面张力的情况。最佳碱浓度可使体系的最小界面

张力 IFT_{min} 值最低(图 2),当体系碱浓度为 1.0%时,IFT_{min} 有最低值,分别约为 2.2×10^{-3} mN·m^{-1} 和 9.0×10^{-4} mN·m^{-1}。出现这种现象的原因是碱浓度升高时油水界面的 pH 值增大,A_W^- 浓度增加,使 IFT 达到最小值;但碱浓度较大时,A_W^- 的脱附速率也将增加,而且表面活性剂也在界面吸附,使界面上的表面活性物质浓度很高,脱附速率增加,故 IFT 值又变大。加入辅助表面活性剂 ANN 使界面张力值变小,可归于 ANN 与 SDC之间有良好的复配增效作用。

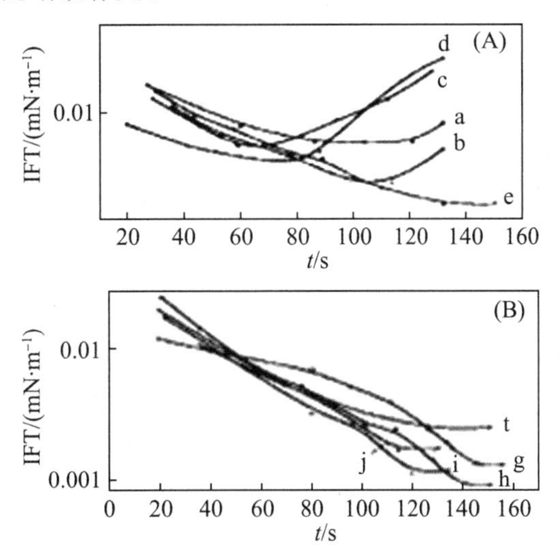

Fig. 1 Effect of alkali concentration on IFT-t behavior

cS =0.5%,cP=0.1%. cA:a. 0.5%;b. 0.8%;c. 1.0%;d. 1.2%;e. 1.5%; f. 0.5%; g. 0.8%;h. 1.0%;i. 1.2%;j. 1.5%. (A) no ANN; (B) c(ANN)=0.05%.

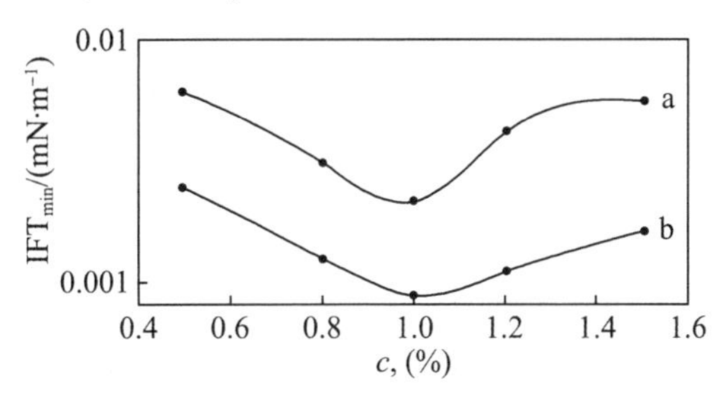

Fig. 2 Effect of alkali concentration on IFT_{min}

cS=0.5%,cP=0.1%. a. No ANN;b. c(ANN)=0.05%

2.3 表面活性剂浓度对界面张力的影响

图 3 为碱浓度 1.0%时不同浓度表面活性剂的瞬时界面张力变化情况。

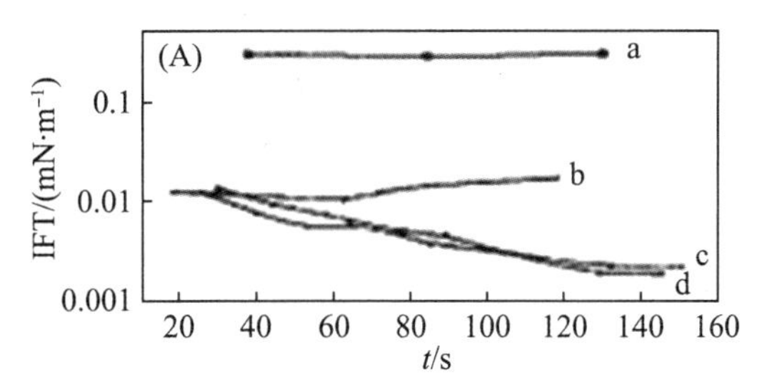

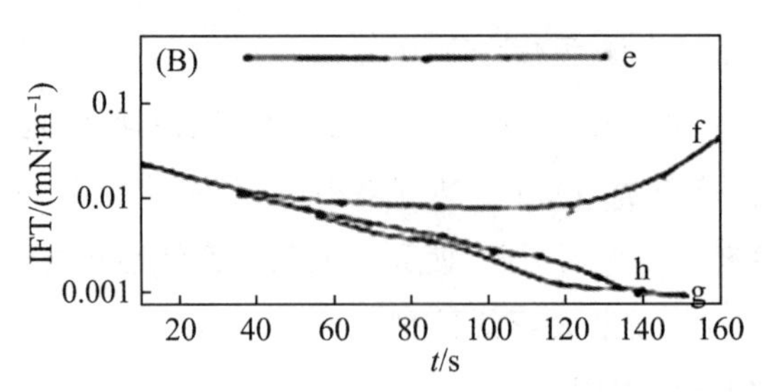

Fig. 3 Effect of surfactant(SDC) concentration on IFT-t behavior

cA=1.0%,cP=0.1%,cS:a. 0.1%;b. 0.3%;c. 0.5%;d. 0.6%;e. 0.1%;

f. 0.3%;g. 5%;h. 0.6%. (A) no ANN; (B) c(ANN)=0.05%.

由图 3 可见,随表面活性剂浓度增大(0.1%~0.6%),IFT 的动态性增强。浓度较低时(<0.3%),界面张力几乎无动态性,且最小值较高(>1.0×10^{-2} mN·m^{-1});浓度增大,界面张力随时间先减小而后增大,但达到最小界面张力值的时间并无一定规律,说明表面活性剂与其他物质的作用机理比较复杂。由最小界面张力 IFT_{min} 与表面活性剂浓度的关系(图 4)可知,当表面活性剂浓度达到 0.5%时,IFT_{min} 的变化不很明显。由于浓度升高会引起脱附速率增大,最终会使 IFT_{min} 增大。由图 3 还可看出,表面活性剂浓度增大,维持最小界面张力值的时间延长。这可能归于浓度增加使油水界面吸附和脱附的平衡很难被破坏,同时表面活性剂与石油皂的相互作用也显著增强。

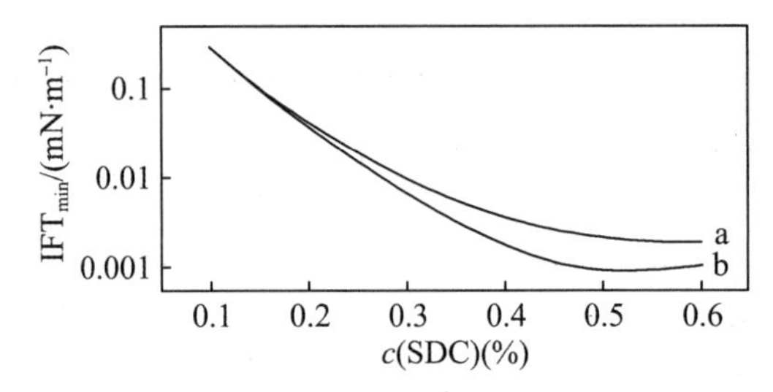

Fig. 4 Effect of cSDC on IFT_{min} cA=0.1%,cP=0.1%.

a. no ANN;b. c(ANN)=0.05%.

2.4 HPAM 对驱油体系界面张力的影响

碱浓度较低时,聚合物水解,静电荷升高,高分子链伸展,粘度增大;而高碱浓度时,电解质屏蔽效应会降低聚合物的水解度,从而抵消水解对粘度的影响,使粘度几乎保持恒定。

据此选择聚合物浓度为 0.1%,测试其对驱油体系界面张力的影响(图 5)。由图 5 可见,无 HPAM 时,两体系的界面张力分别在 90 s 和 120 s 左右达到最低值,但很快又升高;而加入 HPAM 时,界面张力均在 140 s 后才达到最小值,且能维持一段时间。由于加入 HPAM 使体系粘度增大,影响了物质传递,特别是抵制石油皂和 SDC 向界面的迁移,从而推迟了达到 IFT_{min} 的时间。同理,到最小界面张力后,粘度增大使物质由界面向本体相扩散的速率也变慢,故保持 IFT_{min} 有一定时间。粘度对最小界面张力也有影响。从图 5a,图 5b 两条曲线可知,在不加 HPAM 时,IFT_{min} 为 7.0×10^{-3} mN/m,而加入 HPAM 后则降到 2.2×10^{-3} mN/m；当存在 ANN 时，加入 HPAM 后的 IFT_{min} 也由 3.5×10^{-3} mN/m 下降至 9.0×10^{-4} mN/m（图 5 曲线 c，d）。这是由于粘度增大使表面活性物质在界面上排列更加紧密的缘故。

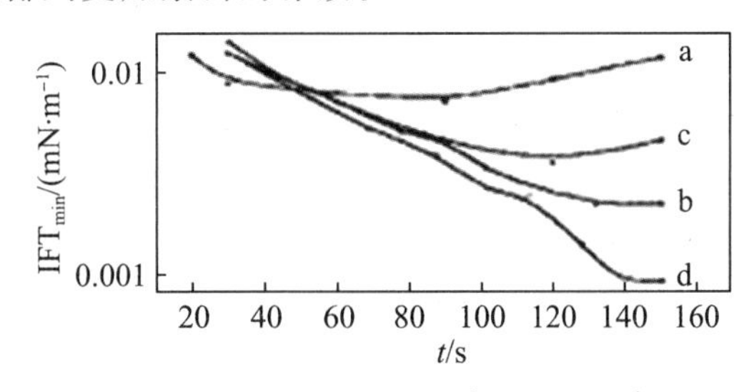

Fig. 5　Effect of HPAM on IFT-t behavior cA＝1.0％，cS＝0.5％. a. no HPAM and no ANN；b. no ANN，cP＝0.1％；c. c(ANN)＝0.05％，cP＝0；d. c(ANN)＝0.05％，cP＝0.1％.

2.5 盐度对体系最小界面张力的影响

加入不同浓度的 NaCl 溶液，考察其对最小界面张力的影响（图 6），发现盐度增加，使 IFT_{min} 的值变小。这是由于体系的离子强度变大，压缩界面的扩散双电层，使表面活性剂排列更紧密。据此也可用提高离子强度的方法来降低界面张力。另外，碱浓度在一定范围内增大可降低 IFT_{min} 值，这也与离子强度的增大有关。由于该驱油体系将注入地层中，因此考察体系的界面张力时应保持其在地层水的盐度范围内。

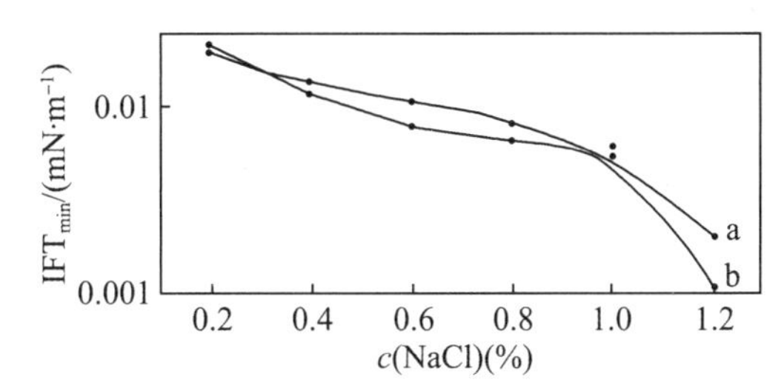

Fig. 6　Effect of c(NaCl) on IFT_{min} a. System Ⅰ；b. System Ⅱ.

2.6 辅助表面活性剂的作用

辅助表面活性剂 ANN 与 SDC 的复配增效作用十分显著。由瞬时界面张力情况可见，加入 ANN 后，各体系达到最小界面张力的时间均有所延长，这可归于表面活性剂间的协同效应，使两者在界面分配达到一定比例时才能到达 IFT_{min}，两种表面活性剂结构不同，到达平衡的时间就会延长，图 2，4，6 结果表明，ANN 的加入使体系的 IFT_{min} 值明显下降。图 7 是两体系的瞬时界面张力变化情况，体系Ⅱ的 IFT 值均小于体系Ⅰ的，表明 SDC 与 ANN 存在明显的协同效应。

2.7 ASP 驱油体系在油水界面吸附机理初探

水相中的物质向油水界面层的传递会极大地影响 IFT 值。根据 Hunsel-Joos 方程[14]：

$$IFT(t)=IFT_e+RT\Gamma^2(c/4Dt)^{1/2}/c_0 \tag{2}$$

体系的 IFT(t)与 $t^{-1/2}$ 的关系为直线时，吸附过程为扩散控制，其中，平衡吸附量 Γ 可由 Gibbs 吸附方程 $\Gamma=-c(dV/dc)/RT$ 得到。由图 7 获得的两个驱油体系达到最小界面张力之前的 IFT(t)～$t^{-1/2}$ 曲线(图 8)可见,达到 IFT_{min} 之前,油水界面吸附尚未平衡时的瞬时界面张力与 $t^{-1/2}$ 近似为直线关系,各组分的混合吸附过程主要由扩散控制,根据直线的斜率和截距,可求出体系Ⅰ,Ⅱ中 SDC 向油水界面扩散的表观扩散系数 D 分别为 1.5×10^{-21},1.9×10^{-19} m^2/s;平衡界面张力 IFT_e[15] 分别为 1.2×10^{-3},5×10^{-2} mN/m;Γ 值分别为 5.5×10^{-9},1.8×10^{-9} mol/m^2(实验得出约 156 s 时吸附达到平衡)。达到 IFT_{min} 后,由于脱附现象十分突出,因而机理也变得相当复杂,可能是由扩散和吸附脱附共同控制,应该用混合动力吸附模型[16]论述。

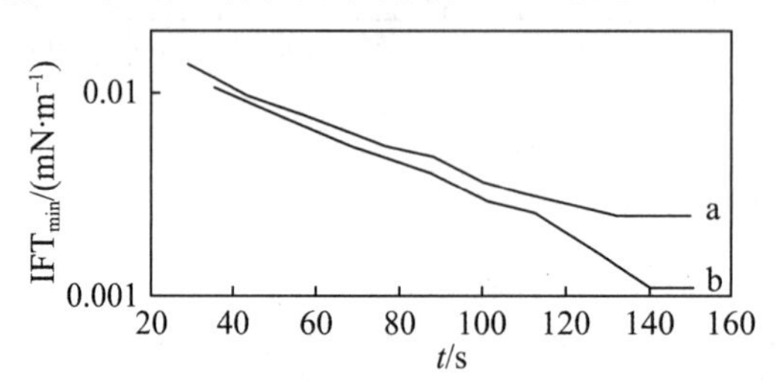

Fig. 7 Transient IFT behavior of systems Ⅰ and Ⅱ

a. System Ⅰ; b. System Ⅱ.

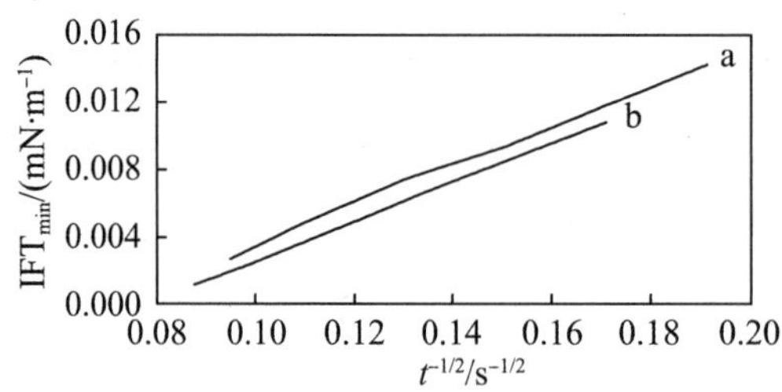

Fig. 8 IFTvs. $t^{-1/2}$ behavior of systems Ⅰ and Ⅱ approaching IFT_{min}

a. Syst em Ⅰ; b. Sys tem Ⅱ.

参考文献

[1] Hao Jingcheng(郝京诚),Liang Fangzhen(梁芳珍), Liu Shuhai(刘树海) et al. Chem. J. Chinese Universities(高等学校化学学报),1998,19(1):111.

[2] Shen Jijing(沈吉静), Zhao Zhenguo(赵振国), Ma Jiming(马季铭). Chem. J. Chinese Universities(高等学校化学学报),1997,18(9):1527.

[3] Cai Lintao(蔡林涛), Yao Shibing(姚士冰), Zhou Shaomin(周绍民). Chem. J. Chinese Universities(高等学校化学学报),1997,18(2): 269.

[4] Wang D. M., Zhang Z. H., Cheng J. C.. SPE Reservoir Eng., 1997, 12(4):229.

[5] Nilsson S., Loh ne A., Veggeland K.. Colloids Surf. (A), 1997,127:241.

[6] Maldal T., Gilje E.. SPE Reservoir Eval. Eng., 1998,1:161.

[7] Chen T. L., Song Z. Y., Fan Y. . S PE Reservoir Eval. Eng., 1998,1:24.

[8] Wang Hongzhuang(王红庄), Yang Puhua(杨普华), Zhu Huaijiang(朱怀江). Oil field Chem. (油田化学), 1997,14(1):62.

[9] Dan Yi (淡宜), Wang Qi(王琪). Chem. J. Chinese Universities(高等学校化学学报),1997,18(5):818.

[10] Li GanZuo (李干佐), Guo Rong (郭荣). Theory and Application of Microemulsion (微乳液理论及其应用), Beijing: Oil Industry Press,1995:18.

[11] Li Fang(李方), Li Ganzuo(李干佐), Fang Wei(房伟), et al. Acta Chimica Sinica(化学学报),1996,54(1):1.

[12] Nasr-El-Din H. A., Taylor K. C.. Colloids Surf. (A),1993,75:169.

[13] Rudin J., Wasan D. T.. SPE21027,1991.

[14] Van Hunsel J., Joos P.. Colloid. Polym. Sci., 1989,267:1026.

[15] Frisch H. L., Mysels K. J.. J. Phys. Chem., 1983,87:3988.

[16] Chang C. H., Franses E. I.. Colloids Surf. (A),1995,100:1.

Transient interfacial tension of crude oil/alkali/surfactant/polymer flooding systems

Mu Jianhai　Li Ganzuo *　Li Ying

(Key Lab. of Colloid and Interf ace Chemistry, Shandong University, Jinan 250100,China)

Cao Xulong　Zeng Shengwen

(Institute of Geological Science, Shengli Oil Field, Dongying 257001, China)

Abstract: The effect of each component of two alkali /surfact ant /polymer flooding systems which use a low price natural mixed carboxylate as the major surfactant on transient and minimum interfacial tension is discussed in this paper. It is found that each component of the ASP flooding systems has an optimum concentration making the minimum interfacial tension on O/W interface lowest: alkali 1.0%, surfactant 0.5%, polymer 0.1%. We also probe into the mechanism of interaction in each component and their adsorption on O/W interface initially.

Keywords: ASP-flooding systems, Transient interfacial tension, Natural mixed carboxylate

(Ed.:Y,X)

孤岛西区三元复合驱体系色谱分离效应研究

隋希华 曹绪龙 王得顺 王红艳 祝仰文 曾胜文

(胜利石油管理局地质科学研究院)

摘要:以孤岛西区复合驱体系为研究对象,借助流动实验和 HPLC 分析方法,对胜利油田孤岛西区三元复合驱体系中各有效组分(Na_2CO_3 + HPAM + PS + BES)在地层油砂上的吸附情况及色谱分离效应进行了研究。实验表明,所有有效组分在地下运移过程中未发生显著的分离,不会对驱油配方中各组分的协同效应、驱油体系的整体性能造成明显影响。文章不仅对孤岛西区复合驱现场,也必将为同类驱油技术的矿场实施提供有益的理论和试验依据。

关键词:复合驱;色谱分离效应;孤岛西区;胜利油田

中图分类号:TE357.46 文献标识码:A 文章编号:1007-2152(2000)04-0001-03

0 引言

三元复合驱配方借助各组分间有效的协同效应,可显著提高体系粘度、降低油水界面张力,从而提高驱油体系的波及体积和驱油效率。三次采油工程在胜利油田的不断实施和扩大,对三元复合驱提高采收率技术提出了更高要求。由于该体系所含化学组分多、驱油机理复杂,因而配方的调整和优化,只能通过加强对体系的理化特性和现场实际的研究才能得以解决。色谱分离效应作为多组分共存时的主要问题,对驱油体系的性能有着重要影响,针对这一问题的研究目前尚未见有关报道。为此本文结合孤岛西区复合驱现场实际,通过流动实验对乳化液的实时监测,开展了这方面的研究工作。

1 实验部分

1.1 实验仪器

流动实验装置由 GJB-1B 高压计量泵,北京兴达技术开发公司 LP-OSC 无脉冲泵,ZK-82A 型真空干燥箱,海安中乔电器厂双联自动恒温箱等组成。

高效液相色谱系统由 Waters515 恒流高压色谱泵,Waters7725i 手动进样阀,Waters2487UV-Vis 检测器及大连江申色谱工作站组成。

1.2 实验材料

法国 SNF 公司产 3530S 型聚丙烯酰胺(固含量 90%,滤过比 1.10,相对分子量 1.8～1.9×10^7);

京大公司产活性剂 BES(固含量 52%);

恒业公司产 PS(固含量 45%);

淄博化学试剂厂生产的 Na_2CO_3、HCl(AR);

孤岛西区 Ng4 层油砂;

模拟原油(70℃条件下粘度为 45 mPa·s);

用来填充油砂的不锈钢管柱(长 30 cm,内径 2.5 cm,渗透率 1.5 μm^2)。

1.3 实验步骤

1.3.1 驱油模型的制作及驱油实验

以 ZK-82A 型真空干燥箱对驱油模型抽空饱和水 1 h,70℃恒温条件下以 0.3 mL/min 流速对模型进行饱和油处理,见油后以流速 0.5 mL/min 饱和至 2.0 PV 结束。

以 0.23 mL/min 的流速将模型水驱至含水 92%后,转三元复合驱,注入倍数为 0.3 PV,最后转水驱(至含水 98%以上结束)。对采出液按采出时间顺序均匀取样,用于对各组分浓度进行分析。

1.3.2 各组分浓度的 HPLC 法分析

(1) 3530S 的分析。

采用中科院兰州化物所合成的 Diol 化学键合固定相,通过紫外扫描、流动相优化等手段对 ASP 体系中的聚合物浓度分析方法进行探索[1,2,3]。确定了如下最佳分析方案:流动相 0.2 mol/L NaH_2PO_4 水溶液,流速 1.0 mL/min, 200 nm×0.5 AUFS 条件下检测。

(2) BES、PS 的分析。

表面活性剂 BES、PS 在驱油过程中起不同作用,二者具有良好的协同效应。其中 BES 为棕黄色透明液体,是一种特殊阴离子与非离子表面活性剂的混合物。PS 呈黑色固体状,均易溶于水,通过对检测波长、流动相体系、交换离子浓度及流速的优化确定了最佳分析方案[4]。

BES 采用 SAX 阴离子交换填料为固定相,0.25 mol/L NaH_2PO_4/CH_3OH(70/30, V/V)溶液,流速 1.0 mL/min,在 235 nm×0.5 AUFS 条件下检测。

PS 采用 SAX 阴离子交换填料与 ODS 填料串联使用作为固定相,0.2 mol/L NaH_2PO_4/CH_3OH(90/10,V/V)溶液,流速 1.0 mL/min,在 270 nm×0.5 AUFS 下检测。

(3) Na_2CO_3 的分析。

采用酸碱滴定法进行分析[5]。0.05 mol/L HCl 水溶液为滴定剂,以酚酞及甲基橙、靛红的混合物为指示剂进行滴定。

2 结果与讨论

以驱油剂注入PV数为横坐标，采出液中各组分浓度为纵坐标作图，得到孤岛西区复合驱体系中各组分的色谱分离曲线(见图1)。

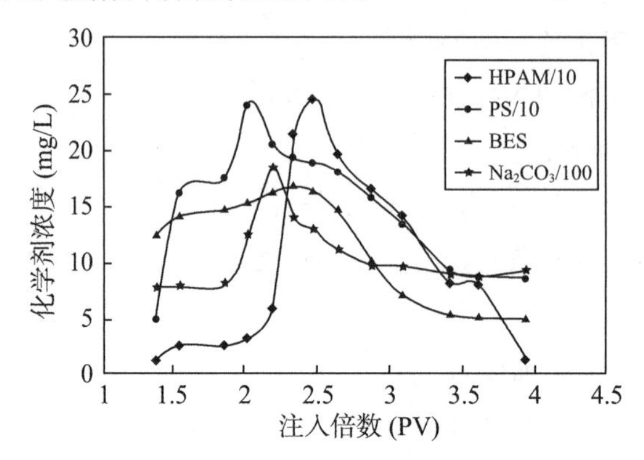

图1 孤岛西区复合驱色谱分离

2.1 色谱分离曲线特征

在注入倍数约1.3 PV时，化学剂浓度开始被检出。之后四种化学剂浓度逐渐升高，并于2.0～2.5 PV之间相继达到最大值。大约在3.4 PV时，各组分浓度均降低至较低水平。而根据化学剂实际开始注入时间(0.3 PV)推算，其浓度达到峰值的对应时间为1.3 PV。

2.2 色谱分离效应曲线的解释

本实验所用孤岛西区油砂孔隙度31.1%，泥质含量2%～3%。3530S分子进入油砂内部后，分子量较高的分子因尺寸较大，在运移过程中只能通过油砂中的大孔隙，不会进入油砂中的小孔隙及微孔，因而这部分分子将以最快速度通过而被最先检出；分子量较小的3530S分子因其尺寸较小，在通过地层油砂中所有孔隙(尤其是小孔隙)时必将有较大的滞留。因此3530S在色谱分离曲线中出峰时间最早，浓度上升最快，且最先得到最大值(见图1)。

活性剂BES、PS分子量大大低于3530S，其分子在驱油过程中与油砂存在的主要作用力与3530S分子有所不同，主要表现为吸附-脱附和分子扩散作用。如图1所示，BES分子量较低，且分子结构为阴离子与非离子活性剂混合物，因而与地层油砂间的吸附作用相对较弱，脱附成为主导作用力，从而导致出峰时间早，浓度由峰值下降最快；PS因分子量较大且分子含芳环等非极性结构，与地层油砂存在较大吸附作用力，因此这种活性剂出峰时间晚，滞留时间最长，且色谱分离曲线脱尾严重。这表明在吸附-脱附平衡作用中吸附占有主导地位。

Na_2CO_3在油砂的吸附-脱附作用力介于两种活性剂之间，出峰及浓度变化情况均介于二者之间。

2.3 色谱分离效应对体系性能的分析

如图2所示，由于在不同时间内，四种化学剂的采出液浓度比符合超低界面张力等值图中最低化学剂浓度的要求。因此，尽管体系存在分离作用，但不会影响驱油体系降低界面张力的性能①。由此可以推断，本三元复合驱油体系在地层油砂中运移的过程中，其降低界面张力和增加粘度的性能将不会受到影响。驱油实验证明，采用本方法较水驱可提高采收率达 22.4%，这也间接印证了本实验得出的结论。据此我们认为，此类配方驱油效果良好，配方设计合理，驱油性能稳定，可以在同类油藏的矿场实施中进行扩大和推广。

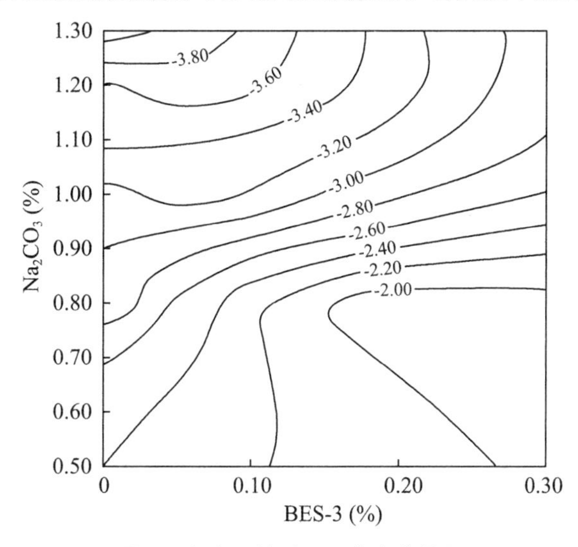

图 2　复合驱体系界面张力等值图

3　结论

（1）色谱分离实验表明，已应用于现场的三元复合驱体系，在地层油砂上的运移过程中，3530S、Na_2CO_3、BES、PS 几种化学剂在 2.0～2.5 PV 的较窄范围内依次出现浓度的最大值，但色谱分离程度不大。

（2）聚合物 3530S 在油砂上的滞留主要取决于体积排阻作用，不可及孔隙体积对小分子量分子滞留有较大影响，活性剂及 Na_2CO_3 的滞留取决于吸附-脱附作用的影响。

（3）体系界面张力等值图表明，孤岛西区复合 驱体系在油砂上的色谱分离效应不会对其驱油性能造成影响。

参考文献

[1] Beazley P R. Anal . Chem. , 1985, 57: 2098.

[2] charfeh S C et al. J. Chromatographia, 1986, 366: 343.

[3] Allison J O et al . SPE, 1985, 13589.

[4] Engelhadt H et al. Chromatographia, 1989, 27(11/12): 535.

[5] 油气田水分析方法. 中华人民共和国石油天然气行业标准，SY 5523-92.

本文编辑　高　岩

信　息

阿拉伯国家天然气工业迅猛发展

据阿拉伯石油公司预测,1999—2005年阿拉伯国家天然气工业的投资将超过300亿美元。投资大幅度增加的主要原因是涌现出话多新的天然气项目,尤其是利比亚向意大利出口天然气项目和卡塔尔北方气田向阿曼、阿联酋和巴基斯坦出口天然气的项目。此外,一旦解除对伊拉克的制裁,将另需50多亿美元的投资。

阿拉伯国家加强天然气项目的开发,一是面向欧洲、远东和东南亚等出口市场;二是满足中东和北非地区各国迅速增长的国内天然气需求。考虑到北非和中东已计划的新项目,预计到2002—2003年度,阿拉伯国家的天然气出口将超过$1\ 000\times10^8\ m^3$。

1985—1998年阿拉伯国家天然气消费由109万桶油当量/日增加到256万桶当量/日,占能源总消费的41.4%。预计到2010年能源总需求将上升到857万桶油当量/日。其中天然气需求为470万桶油当量/日。

推动阿拉伯国家天然气工业迅猛发展的首要原因是该地区天然气探明储量丰富($33.648\times10^{12}\ m^3$,占世界的22%),并且不断有新的天然气田发现。其次,中东和北非国家扩大天然气出口,国内能源利用中尽可实施以气代油的战略,并通过私有化以及与国际石油公司合资等方式向国内外石油公司开放天然气市场。

张小兰　摘自《石油天然气信息》

Oil & gas recovery technology

Abstracts: Sui Xihua, Cao Xulong, Wang Deshun et al. Study on chromatographic fractionation effect of three-component compound flooding system in western area of Gudao. OGRT, 2000, 7(4): 1-3.

Taking compound flooding system in western area of Gudao in Shengli oil field as research object, adsorption condition and chromatographic fractionation effect of each effective component (Na_2CO_3 + HPAM + PS + BES) on formation oil sand are studied with the aid of flow test and HPLC analysis method. The experiment shows that obvious separation of all effective components didn't appear during subsurface migration, and it can't make obvious influence on synergetic effect of each component in displacement formula and the whole performance of displacement system. Useful theory and experiment basis may be provided for the compound flooding fields in western area of Gudao and the same type of fields.

Key words: compound flooding, chromatographic fractionation effect, western area of Gudao, Shengli oil field

Cui Xiaohong, Cao Xulong, Zhang Yigen et al. Formula Study on cross-linked polymer flooding in Gudong 2 area. OGRT, 2000, 7(4): 4-8.

According to the reservoir and underground fluid properties of Gudong 2 area, injected cross-linked polymer system fitting to Gudong 2 area has been developed successfully. This system has suitable cross-linking time, good flowability, high strength and good viscosified property. Gel time and gel strength may be controlled through adjusting the concentration of polymer. The experiment shows that the system may be used for profile control and displacement simultaneously. Lowest gel concentration, gel time, viscosity, strength and stability and so on are all evaluated in this paper. The influence of the factors such as oil sand, shear and concentration on gel and stability has been studied throughly, and the optimum formula is ascertained.

Keywords: cross-linked polymer, cross-linker, stability, gel strength, gel time, Gudong 2 area

Mao Weirong, Gao Li. Steam stimulation recovery of heavy oil reservoir of thin-bed loose sandstone with edge and bottom water. OGRT, 2000, 7(4): 9-12.

Thermal recovery area of steam stimulation in Gudao oil field is a heavy oil reservoir of thin-bed loose sandstone with edge and bottom water, and water invasion is the main factor of influencing its production efect. On the basis of the study of production characteristics of huf and puff that is influenced by edge and bottom water and invasion of injected water, effective measures of raising fluid production at edge, high temperature water shutoff and heightening perforation bottom bound are taken to restrain the harmful efects of water invasion by maximum limit with the use of numerical simulation results, physical simulation results and actual field data. So production effect of huff and puf is improved. Yearly production of viscous oil by thermal recovery in Gudao oil field is above 50×10^4 t and has stable for 5 years, and oil recovery factor is up to 33.5%.

Key words: Gudao oil field, steam stimulation, edge and bottom water, water invasion, recovery characteristic, development effect

Gao Hui, Zhang Lihua, Lang Zhaoxin. Study of numerical simulation for recovery of heavy oil reservoir using CO_2 stimulation. OGRT, 2000, 7(4): 13-15.

孤东二区交联聚合物驱配方研究

崔晓红　曹绪龙　张以根　刘　坤　宋新旺　曾胜文　窦立霞

（胜利石油管理局地质科学研究院）

摘要：针对孤东二区油藏及地下流体特点，成功地研制了适于孤东二区注入的交联聚合物体系。该体系交联时间适宜、流动性好、强度较高、增粘性好。通过调整体系浓度，可以实现对成胶时间和成胶强度的控制。实验表明，该体系可同时用于调剖及驱替。文章对该体系的最低成胶浓度、成胶时间、粘度、强度、稳定性等性能进行了评价。深入研究了油砂、剪切、浓度等因素对成胶、稳定性等方面的影响，并确定了最佳配方。

关键词：交联聚合物；交联剂；稳定性；凝胶强度；交联时间；孤东二区

中图分类号：TE357.46　文献标识码：A　文章编号：1007-2152(2000)04-0004-05

0　引言

交联聚合物常用于调整油层的非均质性。根据使用的交联剂的不同，交联聚合物可以分为三类。第一种为以 Cr(Ⅲ)交联技术为代表的无机交联体系；第二种为以醛或醛与酚的混合物为交联剂生成树脂类凝胶的有机交联体系；第三种为柠檬酸铝交联技术，包括低浓度时形成的 CDG 胶（即胶态分散体系）。近年来，菲利浦（PHILLIPS）公司等推出的有机锆交联技术也属于无机交联体系的一种。其中，Cr(Ⅲ)交联技术由于毒性低，现场使用方便而倍受重视，是国内外近年来使用较多的一种。

孤东二区是一个地层渗透率高，非均质性严重，较为封闭的交联聚合物驱试验区。在本项研究中，选择 Cr(Ⅲ)交联技术。根据交联反应机理，进行了交联剂的研制。在交联剂中加入添加剂以加强交联聚合物的稳定性，研究了油砂等因素对交联聚合物性能的影响，目的在于为矿场推荐合理的交联聚合物驱配方。

1　材料与实验仪器

1.1 实验材料

聚合物 P_1：相对分子质量 1 200 万～1 500 万，水解度 8%～12%，有效含量 30%±3%，胜利油田井下聚合物厂生产。

聚合物 P_2:相对分子质量 1 200 万～1 500 万,水解度 8%～12%,有效含量 30%±3%,齐鲁石化聚合物厂生产。

交联剂:Cr^{3+}-L1、Cr^{3+}-L2、Cr^{3+}-L3,均为墨绿色均匀液体,可与水混溶,pH 值为 2～6,自配研制品。

地层油砂:经 3∶1 乙醇苯洗净的目的层油砂。

1.2 实验条件、仪器及方法

1.2.1 实验条件

实验温度:65℃。

实验用水:胜利盐水,矿化度为 5727 mg/L,$Ca^{2+}+Mg^{2+}=108$ mg/L。

1.2.2 仪器

RV-20 型流变仪:德国 HAAK 公司生产,用于 7.2 s^{-1} 条件下测定凝胶粘度。

DV-Ⅲ型布氏粘度计:美国 BOOKFIELD 公司生产,用于聚合物、交联剂、凝胶粘度测定。

TGU 仪:TIRCO 公司生产,用于衡量交联聚合物溶液成胶前后强度变化。

101-1A 型电热鼓风恒温烘箱。

1.2.3 凝胶强度分类法

按马拉松公司分类法判断生成凝胶的强度等级[1],交联聚合物由不交联到形成会鸣叫的刚性凝胶可分成 A～J 十类。其中,本研究中所能观察到的 A～G 类分别为:A:不交联;B:高流动度的凝胶;C:流动凝胶;D:中等流动度凝胶;E:刚能流动的凝胶;F :高变形的非流动胶;G:中等变形的非流动胶。

2 交联剂的筛选

2.1 交联反应机理

Cr(Ⅲ)与聚合物的交联反应如(1)式所示[2,3]:

$$\mathrm{Cr(III)(L)n+m(P-CO_2^-)} \xrightleftharpoons{Keq} \mathrm{Cr(III)(P-CO_2^-)m(L)n-m+(m-n)L} \quad m\geqslant 2 \qquad (1)$$

式中,L 为 Cr(Ⅲ)的配位体,$P-CO_2^-$ 也是配位体,是聚合物中的聚丙烯酸基团。

由上述反应机理可知,生成交联聚合物的反应是一个配位体交换的络合平衡反应。根据平衡常数 Keq 的计算公式,该反应的反应速度主要取决于配位体交换的速度,也就是 L 从 Cr(Ⅲ)上解离的速度。在 pH 值为 4～7 时,Cr(Ⅲ)在水溶液中以简单无机盐的形式存在,如 L 为 H_2O 或 OH^- 时,成胶速度很快;反之,当 L 为$(OAc)^-$时,在同一 pH 值条件下,聚合物成胶速度会慢得多[4]。也就是说,L 是影响交联反应速度的主要因素[5]。上述络合反应中,L 的络合能力不能太强,否则,难以生成交联聚合物。为控制反应时间,L 的络合能力应适中。因此,L 一般是有机羧酸根。上述反应中加入助剂,也可以改善生成交联聚合物的成胶速度和稳定性等。

2.2 孤东二区所用交联体系的确定

依据上述反应机理，室内合成了三种有机酸铬交联剂，在孤东二区条件下，分别与聚合物 P_1 进行试管交联试验。实验表明，Cr-L3 成胶快、破胶快，需要使用的交联剂、聚合物浓度也过大；Cr-L1、Cr-L2 则取得了满意的交联效果(表 1)，决定选择 Cr-L1、Cr-L2 做进一步筛选。

表 1 交联剂形成交联体系的成胶强度及稳定时间

P_1＋交联剂	2 000＋600 (mg/L)	2 000＋1 000 (mg/L)	2 000＋2 000 (mg/L)	2 000＋4 000 (mg/L)
Cr-L1	C，40 d	C，50 d	D，64 d	—
Cr-L2	C，36 d	C，49 d	D，60 d	—
Cr-L3	不成胶	不成胶	不成胶	C，3 d

注：表中 C，40 d 等为成胶情况。

在 Cr-L1，Cr-L2 与 P_1、P_2 两种聚合物形成交联体系中加入助剂 A 后发现，助剂 A 可以延长各交联体系的稳定时间，尤其对 P_2-Cr-L1 体系有较好的稳定作用，稳定时间可延长至 200 d 左右。由于 P_2-Cr-L1＋A 体系(以下简称 P_2-Cr-L1 体系)的浓度适用范围大，形成的凝胶稳定性好，因而选择该体系作为孤东二区的注入体系进行主要性能评价。

3 P_2-Cr-L1 体系性能评价、影响因素分析及驱油配方确定

3.1 P_2-Cr-L1 体系主要性能评价

性能评价内容为：体系最低成胶浓度测定；交联时间的测定；交联聚合物粘度、强度、稳定性等。

3.1.1 成胶最低浓度测定

无砂条件下不考虑体系由于地下稀释、吸附、滞留等因素造成的化学剂浓度下降，P_2 聚合物的最低成胶浓度为 150 mg/L，Cr-L1 浓度为 400 mg/L，成胶粘度 12.2 mPa·s，相当于交联前的 9 倍，相当于同等水质条件下浓度为 1 200 mg/L 聚合物的粘度，稳定时间为 35 d。

3.1.2 成胶时间测定

实验中，由 1 000～4 000 mg/L 的 P_2 和 600～4 000 mg/L 的 Cr-L1 组成的交联体系在油藏温度和静态状况下成胶时间最长可达 4.7 h。考虑到现场施工是动态条件，地面污水温度低于试验温度及地层水稀释等原因，现场成胶时间将大大延长。

3.1.3 交联体系粘度、强度、稳定性测定

(1) 交联体系粘度测定。

测定了由不同浓度 P_2、Cr-L1 组成的 16 个配方在 65℃放置 24 h 所形成交联聚合物的粘度(表 2)。由实验不难看出，聚合物一旦交联，其粘度远远超过单纯聚合物溶液粘

度(表 3)。

表 2 交联聚合物粘度测定

聚合物浓度(mg/L)	交联剂浓度(mg/L)	凝胶强度	粘度(mPa·s)
1 000	600	C	812
1 200	600	C*	1 004
1 500	600	C*	1 617
1 800	800	D*	2 196
2 000	1 000	D*	2 608
2 500	1 000	E	4 750
3 000	1 000	E	4 761

注:* 为典型配方。

表 3 P_2 聚合物粘度测定

浓度(mg/ L)	1 000	2 000	3 000	4 000	6 000
粘度(mPa·s)	8.9	36	107	419	1 220

(2) 交联前后强度对比。

交联聚合物的凝胶强度可以根据凝胶强度分类法进行划分。表 2 中研究的交联聚合物成胶后强度分别达到了 C～E 级。为了更好地反映成胶后体系强度的变化,测定了 P_2 和 P_2-Cr-L1 体系通过 TGU 仪的压力和时间。实验表明,P_2-Cr-L1 体系通过 TGU 仪比 P_2 困难得多,所需的压力为 P_2 的十倍,时间增长了近两倍(表 4)。压力和流过时间在一定程度上反映了成胶强度,说明成胶后强度大幅度提高。

表 4 凝胶强度的测定

P_2(mg/L)+交联剂(mg/L)	压力(MPa)	流过时间(s)
2 000+0	0.015	16.82
2 000+2 000	0.15	44.59

(3) 交联体系的热稳定性。

图 1 是三个配方在 65℃下粘度随时间的变化。它们显示了同一规律,即在成胶初期粘度由初始值上升到一最高值,然后微降到一“平衡”粘度,并在较长时间保持平稳。在 264 d 的放置中,三个交联体系的粘度损失最大也未超过 30%。在进行试管实验时,一系列配方的稳定时间也超过了一年。

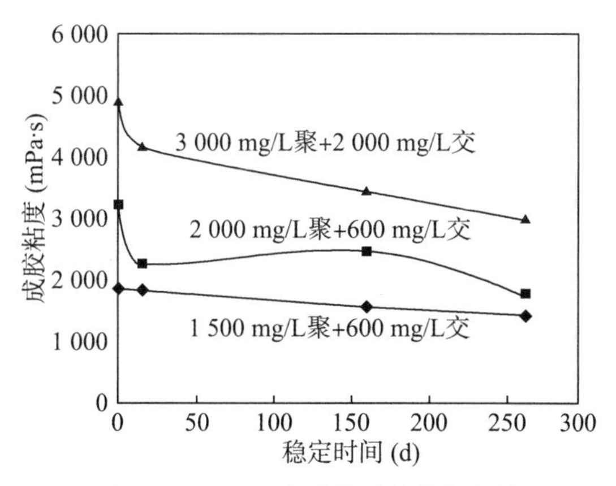

图 1　P_2-Cr-L1 交联体系的热稳定性

3.2 影响因素分析

该部分研究包括聚合物浓度和交联剂浓度的筛选，地层油砂、炮眼剪切对交联体系的影响，聚合物和交联剂成胶适宜浓度的筛选等。

3.2.1 影响成胶时间的因素

（1）聚合物、交联剂浓度对成胶时间的影响。

图 2 给出了交联剂浓度和聚合物浓度对成胶时间的影响。结果表明，当聚合物浓度一定时，成胶时间随交联剂浓度增大而缩短；同理，当交联剂浓度一定时，成胶时间随聚合物浓度增大而缩短。

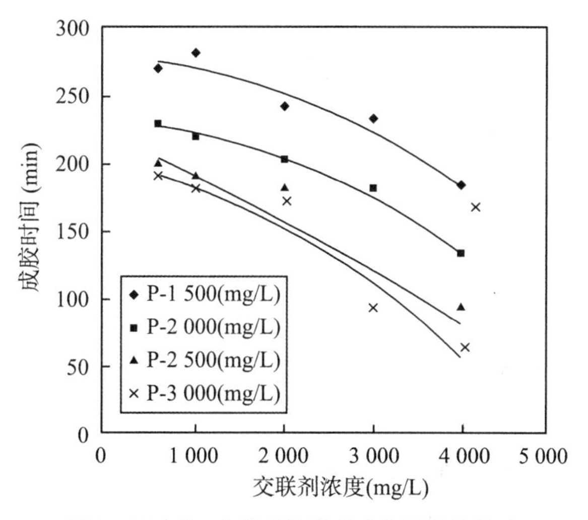

图 2　聚合物、交联剂浓度对成胶时间的影响

（2）温度对成胶时间的影响。

图 3 给出了 P_2（2000 mg/L）＋Cr-L1（600 mg/L）与 P_2（2 000 mg/L）＋Cr-L1（1 000 mg/L）两个配方成胶时间受温度的影响情况，可见温度越高，成胶越快。

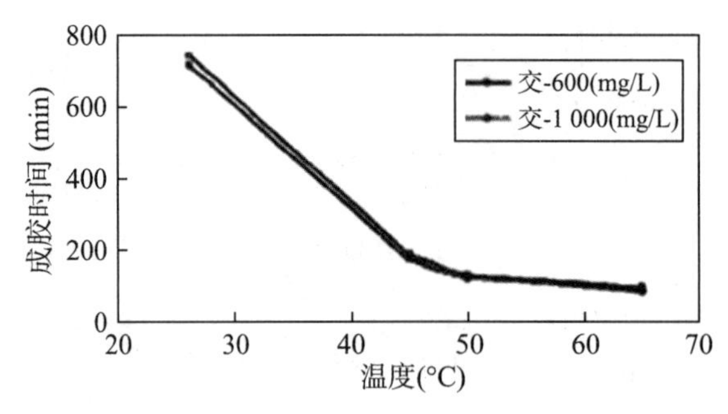

图 3　温度对成胶时间的影响(℃)

3.2.2 地层油砂的影响

影响交联的因素除聚合物、交联剂本身固有性质外，还有环境因素。在地层驱替过程中，交联聚合物在成胶前后一直与油砂接触，这会引起交联聚合物体系浓度、温度、pH 值等多方面的变化。所以必须考察地层油砂对交联体系的影响。

据分析，地层油砂对交联体系的影响基于以下两个反应，即：

$$\mathrm{Cr(L)n} + \mathrm{mClay} \text{——} \mathrm{Cr(Clay)m} + \mathrm{nL} \tag{2}$$

Clay 指的是粘土矿物，粘土吸附对交联聚合物中 Cr^{3+} 造成的浓度损失很小。而当 pH 值大于某一临界值时，将有以下反应：

$$\mathrm{Cr(L)n} + 3\mathrm{OH}^- \text{——} \mathrm{Cr(OH)_3}\uparrow + \mathrm{nL} \tag{3}$$

pH 值增大的主要原因是储层中的碳酸钙(方解石)和菱铁矿($FeCO_3$)等矿物被溶解所致。pH 值增大将促进 $Cr(OH)_3$ 沉淀生成，使 Cr^{3+} 浓度降低。上述两个反应都会造成 Cr^{3+} 的损失，影响稳定性。一般认为，反应(2)发生在与油砂接触的初期，反应(3)发生在驱替过程中，为此开展了以下研究。

(1) 交联剂在油砂上的吸附。

根据研究结果，聚合物在油砂上的吸附可以忽略，因此只研究交联剂在油砂上的吸附。研究了固液比为 1∶10 时，配方 P_2(2 000 mg/L)+Cr-L1(600 mg/L)在 65℃下放置 2 d 成胶、吸附情况。实验表明，该体系在油砂存在条件下能形成满意的凝胶，Cr-L1 的平衡浓度为 518 mg/L，在溶液中的保留率大于等于 85%。

(2) 油砂对交联体系稳定性的影响。

用表 2 中四个典型配方分别做无砂、加砂情况下的强度、稳定性对比考察，固液比为 1∶10。经过一年多观察发现，添加油砂后，除交联聚合物的凝胶强度略有下降外，未见破胶，说明稳定性很好。

综上所述，对于 P_2-Cr-L1 交联体系来说，油砂仅会造成凝胶强度略微降低，但不会影响交联的进行以及生成交联聚合物的稳定性。

3.2.3 剪切对 P_2-Cr-L1 体系交联的影响

在现场注入过程中，聚合物和交联剂溶液都是在井口混合，经炮眼剪切后进入地层。因此，需要考察经炮眼剪切后的成胶能力。

（1）低剪切速率的交联情况。

室温（约16℃）条件下，P_2（2 000 mg/L）+Cr-L1（1 000 mg/L）和P_2（2 000 mg/L）+Cr-L1（2 000 mg/L）两个配方成胶前后经模拟炮眼剪切的实验，剪切速率7 000～8 000 s^{-1}（矿场实际炮眼剪切速率一般为3 500 s^{-1}左右）。未成胶的两个配方经炮眼剪切后，室温下约2 d成胶，凝胶强度与未经剪切的基本一致。已成胶的配方经炮眼剪切后变稀，但是在室温下放置1 d后又成胶，凝胶强度与剪切前相比为同一级别。

（2）高剪切速率下的交联情况。

由于现场剪切情况较为复杂，为了较好地反映交联体系受剪切的程度，以70 000～90 000 s^{-1}的较高剪切速率来进一步模拟现场条件。考察配方为P_2（2 000 mg/L）+Cr-L1（1 000 mg/L），未成胶、已成胶的配方经剪切后，在65℃放置2 d的交联情况见表5。

表5 剪切后的成胶情况

剪切类型	成胶时间	凝胶强度	粘度（mPa·s）	剪切速率（s^{-1}）
未剪切	2 h	D	2 608	0
成胶前	4 h	D	1 796	93 200
成胶后	/	D	1 458	76 800

可看出在较高的剪切速率下，未成胶、已成胶的交联体系经剪切之后，都能成胶，但粘度都有一定的损失。虽成胶后受剪切影响较大，强度仍为同一级别。

3.3 驱油配方确定

3.3.1 P_2-Cr-L1体系成胶的适宜浓度范围

图4为不同浓度P_2和Cr-L1形成的交联聚合物强度等值图，交联聚合物的强度受Cr-L1浓度的影响不大，却随P_2浓度的增加而增大。由此图看出，形成C～D级凝胶的P_2范围是1 000～2 000 mg/L，Cr-L1浓度范围是600～4 000 mg/L。

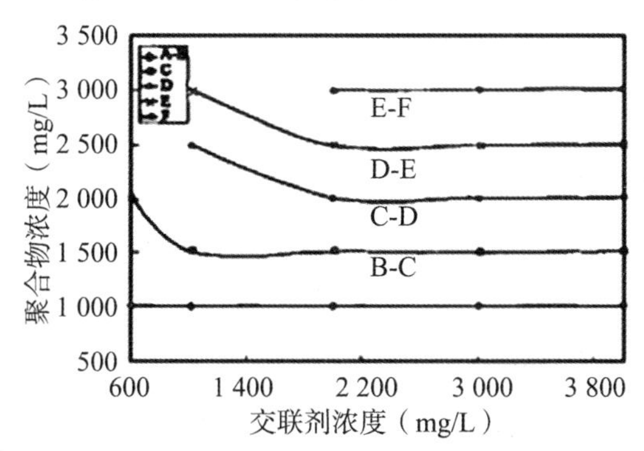

图4 交联聚合物强度与聚合物、交联剂浓度的关系—凝胶强度等值线

图5是P_2浓度一定时Cr-L1浓度对成胶粘度的影响曲线。此图表明，对于每一个P_2浓度，对应着一个与其匹配的Cr-L1浓度范围。在此范围内，Cr-L1浓度增大，凝胶粘度增大。我们称其为该P_2（聚合物）浓度下形成稳定交联的Cr-L1（交联剂）浓度范围。

因为普通的聚丙烯酰胺分子的结构为链状结构。在与Cr(Ⅲ)生成交联聚合物后，由Cr(Ⅲ)将一个个链状结构连接成网状结构，使粘度增加。由于聚合物中的反应基团为$P\text{-}CO_2^-$，对于特定的聚合物来说，一定质量的聚合物中含有一定量的该基团，也只能与一定量的Cr(Ⅲ)反应。因此，一定浓度的聚合物具有相匹配的交联剂浓度范围。

由图5看出，浓度为1 200 mg/L、1 500 mg/L的P_2溶液能形成稳定交联的Cr-L1的浓度范围是600～800 mg/L。浓度为2 000 mg/L的P_2溶液则在考察的Cr-L1浓度范围600～2 000 mg/L内都能形成稳定交联。

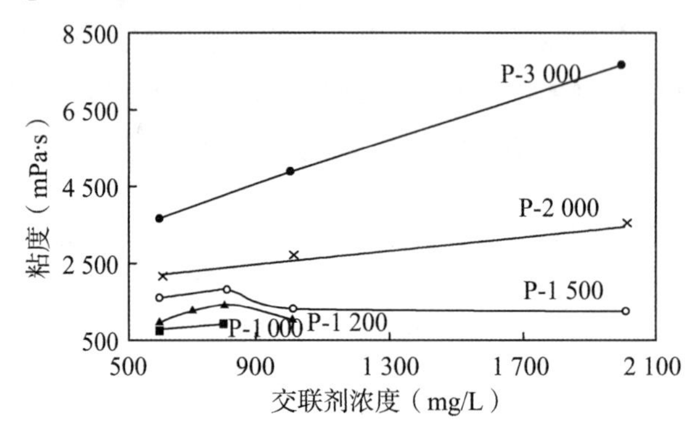

图5　P_2-Cr-L1浓度对交联聚合物粘度的影响曲线

3.3.2 驱油配方确定

综观上述结果，P_2-Cr-L1体系因所使用的P_2和Cr-L1浓度不同可形成B～F级的凝胶。在现场应用中，可以根据现场油藏温度、地层条件，优化选择形成合适强度的交联聚合物配方。考虑孤东二区油藏特点和生产状况，要求交联聚合物既应该有流度控制作用，又应该有调剖作用，故应将凝胶强度选在C～D范围内。推荐的P_2浓度范围是1 000～2 000 mg/L，匹配的Cr-L1浓度范围是600～1 000 mg/L。

4　结论

(1) 以配位络合理论为指导，研制出了交联剂Cr-L1、Cr-L2、Cr-L3。

(2) 筛选确定P_2-Cr-L1体系作为孤东二区交联 聚合物驱注入体系。性能评价结果表明：该体系具有热稳定性好、交联时间适宜、流动性好、强度较高、增粘性好的特点。通过调整体系的浓度，可以控制成胶时间和凝胶强度；具有一定的缓冲能力，现场使用时不用调节pH值；在地层油砂存在下，能够成胶，且稳定时间超过一年；经过炮眼剪切后仍能形成稳定的交联体系。

(3) 根据孤东二区Ng 5层的油藏特点，推荐使用C～D类强度的凝胶，确定配方为：P_2(1 000～2 000 mg/L)＋Cr-L1(600～1 000 mg/L)。

参考文献

[1] Sydansk R D. Acrylamide-Polymer/ Chromium(Ⅲ)-Carboxy-late Gels for

Near Wellbore Martrix Treatments. SPE/DOE20214.

[2] Lockhart T P. Chemical and Sturctural Studies on Cr^{3+}/Poly-acry lamide Gels. SPE 20998.

[3] Lockhart T P et. d. A. New Gelation Technology for In-Depth Placement of Cr^{3+}/Polymer Gels in High-Temerature Reser-voirs. SPE/ DOE 24194.

[4] Lockhart T P. Slow-Gelling Cr^{3+}/Polyacrylamide Solutions fo Reservoir Profile Modification:Dependenceof the Gelation Time on pH . JAPS, (43):1527-1532.

[5] Paola Albonico. Effective Gelation-Delaying Additives for Cr^{3+}/ Polymer Gels. SPE 25221.

本文编辑 高 岩

信 息

陆棚海域遥感地质研究的新方法

利用遥感方法研究陆棚海域地质结构为时已久,但研究区水深一般不超过20～25 m,在少数情况下最大水深也只能增至50～70 m。也就是说,研究深部地质结构的能力有限,不能满足油气地质晋查工作的需要。近年来,全俄遥感方法研究所根据数字宇航照片提出了新的研究方法,明显提高了陆棚区遥感研究工作的效率。根据该方法研究了库页岛、伯朝拉海及白海陆棚数字航片,发现以RSdPS程序进行计算机处理,有时甚至是目视判读,可获得与相邻陆区相似的不均匀线性体网图,作为解释海底地质构造的基础。以伯朝拉海陆棚为例做了典开型分析,编制了线性体分布图及线性走向玫瑰花图,证实新方法的地质研究海水深度可达100～150 m。作者认为,该项研究是一项突破,但仍属初步成果,一些仍有待改进。

文强 摘自俄《俄国地质学》1999年第6期

Oil & gas recovery technology

Abstracts: Sui Xihua, Cao Xulong, Wang Deshun et al. Study on chromatographic fractionation effect of three-component compound flooding system in western area of Gudao. OGRT, 2000,7(4):1-3.

Taking compound flooding system in western area of Gudao in Shengli oil field as research object, adsorption condition and chromatographic fractionation effect of each effective component (Na_2CO_3 + HPAM + PS + BES) on formation oil sand are studied with the aid of flow test and HPLC analysis method. The experiment shows that obvious separation of all effec-tive components didn't appear during subsurface migration, and it can't make obvious influence on synergetic effect of each component in displacement formula and the whole performance of displacement system. Useful theory and experiment basis may be provided for the compound flooding fields in western area of Gudao and the same type of fields.

Keywords: compound flooding, chromatographic fractionation effect, western area of Gudao, Shengli oil field.

Cui Xiaohong, Cao Xulong, Zhang Yigen et al. Formula Study on cross-linked polymer flooding in Gudong 2 area. OGRT, 2000, 7 (4):4-8.

According to the reservoir and underground fluid properties of Gudong 2 area, injected cross-linked polymer system fitting to Gudong 2 area has been developed successfully. This system has suitable cross-linking time, good flowability, high strength and good viscosified property. Gel time and gel strength may be controlled through adjusting the concentration of polymer. The experiment shows that the system may be used for profile control and displacement simultaneously. Lowest gel concentration, gel time, viscosity, strength and stability and so on are all evaluated in this paper. The influence of the factors such as oil sand, shear and concentration on gel and stability has been studied throughly, and the optimum formula is ascertained.

Keywords: cross-linked polymer, cross-linker, stability, gel strength, gel time, Gudong 2 area

Mao Weirong, Gao Li. Steam stimulation recovery of heavy oil reservoir of thin-bed loose sandstone with edge and bottom water. OGRT, 2000,7(4):9-12.

Thermal recovery area of steam stimulation in Gudao oil field is a heavy oil reservoir of thin-bed loose sandstone with edge and bottom water, and water invasion is the main factor of influencing its production efect. On the basis of the study of production characteristics of huf and puff that is influenced by edge and bottom water and invasion of injected water, effective measures of raising fluid production at edge, high temperature water shutoff and heightening perforation bottom bound are taken to restrain the harmful efects of water invasion by maximum limit with the use of numerical simulation results, physical simulation results and actual field data. So production effect of huff and puf is improved. Yearly pro-duction of viscous oil by thermal recovery in Gudao oil field is above 50×10^4 t and has stable for 5 years, and oil recovery factor is up to 33.5%.

Keywords: Gudao oil field, steam stimulation, edge and bottom water, water invasion, recovery characteristic, development effect.

Gao Hui,Zhang Lihua,Lang Zhaoxin. Study of numerical simulation for recovery of heavy oil reservoir using CO_2 stimulation. OGRT,2000,7(4):13-15.

孤岛油田西区复合驱界面张力研究

曹绪龙　王得顺　李秀兰

(胜利油田有限公司地质科学研究院)

摘要:针对胜利油区孤岛油田西区油藏条件开展了单一碱、单一活性剂+碱体系、复配活性剂+碱体系及碱+活性剂+聚合物体系的界面张力实验。探索了碱浓度、活性剂浓度、复配活性剂及聚合物的加入等对动态界面张力的影响规律;考察了驱油体系经岩石吸附和热老化以后的界面张力特征;研究了界面张力等值图。研究表明,该区复合驱油体系即使活性剂有效浓度在150 mg/L条件下亦可达到3×10^{-3} mN/m的低界面张力;聚合物的加入对体系的界面张力影响不大;体系经油砂吸附和热老化后仍能保持较低的界面张力。

关键词:胜利油区;孤岛油田;复合驱油;界面张力;活性剂

中图分类号:TE357.46　文献标识码:B　文章编号:1009-9603(2001)01-0064-03

引言

注水开发后期,毛细管准数一般在$10^{-6}\sim10^{-7}$范围内。增加毛细管准数将显著地提高原油采收率。理想状态下,毛细管准数增加至10^{-2}时,原油采收率可达到100%。通过降低油水界面张力,可使毛细管准数有3~4个数量级的变化。通常,油水界面张力在20~30 mN/m范围内,使用理想的驱替液可使界面张力降至$10^{-3}\sim10^{-4}$ mN/m超低范围内[1]。驱油剂的波及效率和洗油效率是决定采收率的重要参数。提高洗油效率一般是通过增加毛细管准数来实现的,而降低油水界面张力则是增加毛细管准数的主要途径。在室内研究的基础上,于1992年8月在孤东油田实施了三元复合驱,并取得了良好的降水增油效果,试验区中心井提高采收率13.4%。本工作主要依据孤岛油田西区的油藏条件开展了动态界面张力实验研究。

1　实验

1.1 仪器与试剂

用Texas-500型旋转滴界面张力仪测量动态界面张力;用K12型界面张力仪测量静

态界面张力。

试剂为：Na_2CO_3 AR 级；上海助剂厂表面活性剂 BES-3；新河化工厂表面活性剂 PS(A)、PS(B)、PS(C)；北京勘探院研制品 SH2；美国 OCT 公司表面活性剂 ORS；华南理工大学表面活性剂研制品 Mu。

实验用油为孤岛油田西区 14-XN409 井脱水脱气原油，用煤油过滤配制成 70℃时粘度 70 mPa·s 的模拟油；水为孤七注水站注入水、孤岛油田西区实验区产出水；实验温度为地层温度 70℃。分析结果见表 1。

表 1 孤岛西区实验区水分析表 mg/L

	Ca^{2+}	Mg^{2+}	$Na^{+}+K^{+}$	SO_4^{2-}	HCO_3^{-}	Cl^{-}	TDS
注入水	65	13	2174	9	902	2973	6188
地层水	112	17	2459	17	567	3697	6869

2 实验结果与讨论

2.1 碱与原油间界面张力特征

碱与原油中的有机酸发生反应生成活性剂—石油酸皂，使油水界面张力降低。反应如下：

$$R-\overset{\displaystyle O}{\overset{\|}{C}}-OH + OH^{-} \longrightarrow R-\overset{\displaystyle O}{\overset{\|}{C}}-O^{-} + H_2O$$

在适当的盐度下该活性剂的水溶液可使油水界面张力降低 3～4 个数量级。

从图 1 可以看出，油水体系中加入碱以后，界面张力有了明显的降低，由 10^{1} mN/m 降至 10^{-1}～10^{-2} mN/m。当碱的浓度为 1.0%时界面张力最低。在碱浓度 0.5%～1.0%区间界面张力随碱浓度的增加而降低，在 1.0%～1.5%区间随碱浓度的增加而升高。这主要是由于体系中加入碱后，增加了表面活性剂的阳离子浓度，降低了界面处的双电层，从而增加了活性剂分子的表面活性，导致界面张力的降低[2]。在一定的盐度和 pH 值范围内产生超低界面张力，如果超出该范围，界面张力将增加。

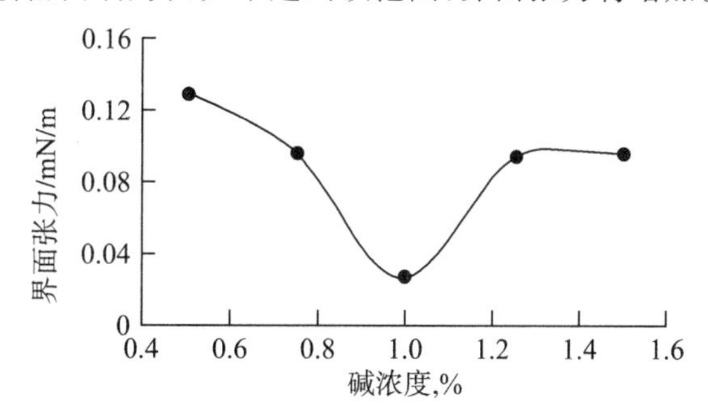

图 1 不同碱浓度与原油界面张力曲线图

2.2 单一活性剂+碱体系与原油间界面张力特征

图 2 给出了活性剂+Na_2CO_3 体系与原油间界面张力曲线。由于活性剂的作用与碱和活性剂协同效应的存在,体系的界面张力达到了 $10^{-2}\sim10^{-3}$ mN/m。对比结果以 BES-3 体系为最好。

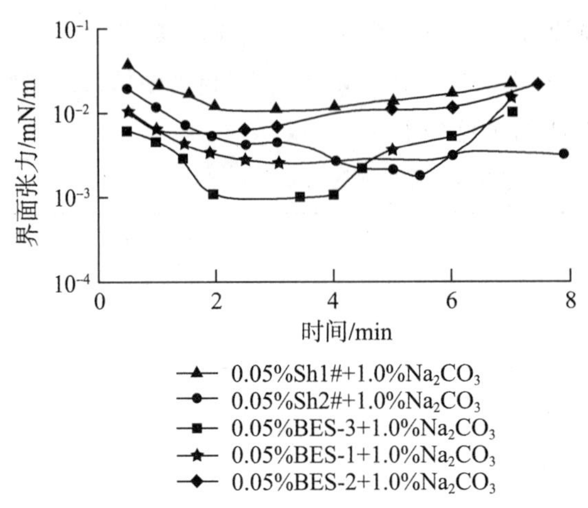

图 2 活性剂+碱与原油界面张力曲线图

2.3 复配活性剂+碱体系与原油间界面张力特征

采用正交实验法设计实验,得出了复配活性剂实验体系与原油间的界面张力实验结果(表 2)。当 PS(A):BES-3=1:2,碱的浓度为 1.0%时,体系界面张力最低且最稳定。

图 3 给出了不同活性剂浓度条件下复配活性剂+碱体系与原油间的界面张力。从表 2 和图 3 的数据中看出,体系的界面张力并没有随着体系中活性剂浓度的增大而降低。实验证明,界面张力随活性剂浓度的变化趋势为:最初时界面张力随着活性剂浓度的增大而降低;当活性剂的浓度增至某一特定值时界面张力达到最低,此时若活性剂的浓度继续增大,体系界面张力亦保持不变,有时会出现上升趋势。原因是当界面与水、油间的相互作用力达到平衡时,活性剂分子则大量聚集于油水界面,体系才能达到超低界面张力。随着体系中活性剂分子继续增多,三相间分子的作用力也随之有不同程度的变化,破坏了这种平衡,界面张力反而出现了上升趋势。

表 2 复配活性剂界面张力实验结果表

PS+BES(%)	Na_2CO_3(%)	PS(A):BES	界面张力(mN/m)	稳定性
0.10	1.0	1:2	3.60×10^{-4}	稳定
0.18	1.0	2:1	1.11×10^{-3}	稳定
0.30	1.0	1:1	3.88×10^{-3}	稳定
0.10	1.2	1:1	2.17×10^{-3}	稳定
0.18	1.2	1:2	6.20×10^{-4}	稳定

（续表）

PS+BES(%)	Na_2CO_3(%)	PS(A)∶BES	界面张力(mN/m)	稳定性
0.30	1.2	2∶1	5.32×10^{-3}	不稳定
0.10	1.5	2∶1	5.22×10^{-3}	不稳定
0.18	1.5	1∶1	2.05×10^{-3}	稳定
0.30	1.5	1∶2	5.60×10^{-3}	不稳定

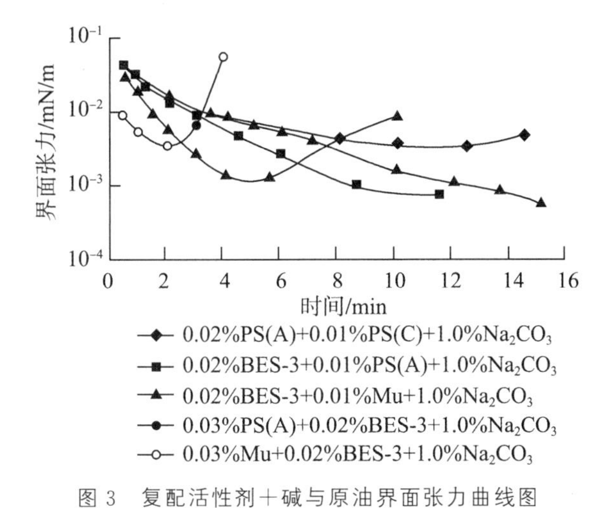

图3 复配活性剂+碱与原油界面张力曲线图

2.4 碱+活性剂+聚合物与原油间界面张力特征

碱、活性剂、聚合物三元复合驱的主要机理是增大波及体积和提高洗油效率。体系中聚合物的作用就是增大波及体积。一般来讲，聚合物的加入对体系的超低界面张力并没有太大的影响，只是延缓了体系达到该超低界面张力的时间，这是由于增加体相粘度减缓了活性剂分子的界面扩散速度所致。有些聚合物中加入了乳化剂、乳液稳定剂等活性剂组分，如复合驱油体系中加入该类聚合物就会引起体系的界面张力进一步降低[3]。研究表明：在孤岛西区油藏条件下体系中加入浓度为0.15%的聚合物3530S后对界面张力的影响并不大。

2.5 界面张力等值线图的编制

在渗流过程中，由于碱同地层原油、水和岩石间的相互作用引起碱耗，活性剂也由于地层岩石的吸附、水的稀释作用使其浓度降低。碱与活性剂浓度的变化必然引起体系界面张力的变化，进而影响驱油效率。因此，研究碱浓度、活性剂浓度与界面张力之间的关系是十分必要的。所谓界面张力等值线图是用等值线的形式直观地描述碱浓度、活性剂浓度与界面张力之间的关系（图4）。当碱浓度大于1%时，活性剂浓度在较宽的范围内界面张力小于1×10^{-3} mN/m；当活性剂浓度小于0.1%时，体系具有较好的界面张力特征。表明该体系在孤岛油田西区油藏条件下有较宽的低张力区，即使在活性剂浓度较低的情况下也能维持较低界面张力。

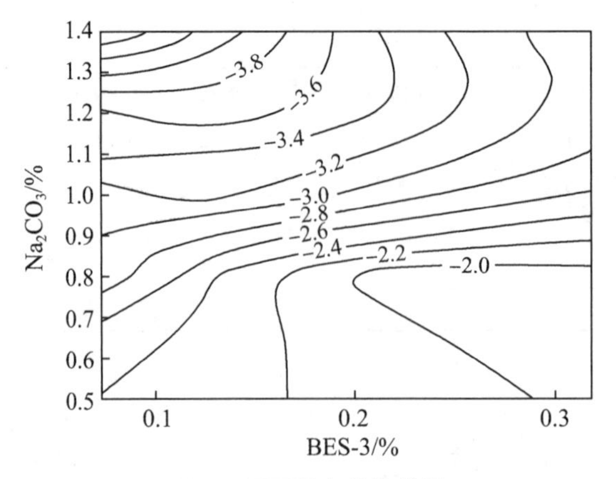

图 4 界面张力等值线图

2.6 地层岩石吸附对界面张力的影响

前已述及,碱和活性剂注入地层以后,由于地层水的稀释和地层岩石的吸附作用使其浓度降低,影响驱油效率。为了考察岩石吸附对体系界面张力的影响,开展了驱油体系经油砂吸附前后的界面张力实验(图 5)。结果说明,推荐体系的界面张力经油砂吸附以后尽管有些变化,但最低界面张力仍能保持在 10^{-4} mN/m。

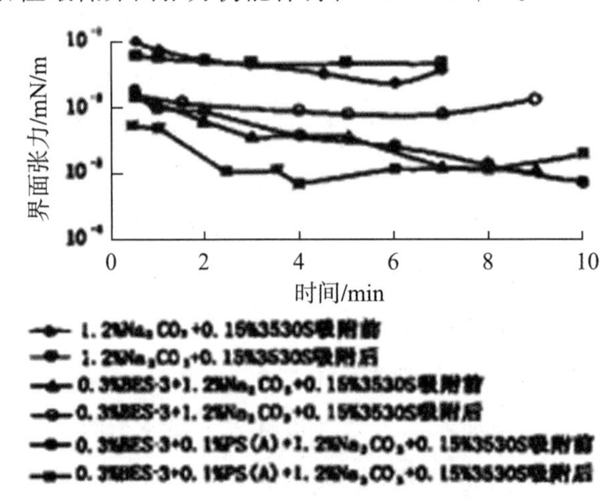

图 5 油砂吸附与体系界面张力关系图

2.7 热老化对体系界面张力的影响

由于地层温度较高,驱油体系在地层中的运移又是一个较长的过程,必然引起活性剂降解、有效浓度降低、界面张力变化,进而影响体系的驱油效率。热老化对体系界面张力的影响研究见图 6。

图 6 说明,0.2%BES−3+0.1%PS+1.2%Na_2CO_3 与 0.2%BES−3+0.1%PS+1.2%Na_2CO_3+0.15%3530S 两个体系在 70℃条件下,随着热老化时间的加长,界面张力基本保持不变,说明该类体系注入油藏以后能保持较强的降低界面张力能力。

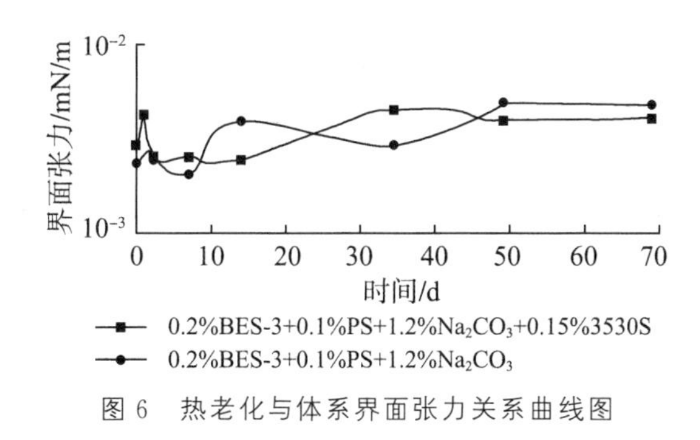

图 6 热老化与体系界面张力关系曲线图

3 结论

复配活性剂体系能更有效地降低油水界面张力,进而满足矿场驱油的需要。

在孤岛油田西区油藏条件下,0.2%BES－3＋0.1%PS＋1.2%Na_2CO_3 体系中加入0.15% HPAM 后,其界面张力变化不大。

在孤岛油田西区油藏条件下,0.2%BES－3＋0.1%PS＋1.2%Na_2CO_3＋0.15% HPAM 体系经油砂吸附和热老化后仍能保持较低的界面张力。

参考文献

[1] 杨承志.化学驱理论与实践[M].北京:石油工业出版社,1996.

[2] 张继芬.提高石油采收率技术[M].北京:石油工业出版社,1997.

[3] Krumrine PHetal. Surfactant Polymer and Alkali Interactions in Chemical Flooding Pocesses. SPE11778.

胜利油区复合驱油体系研究及表面活性剂的作用

陈业泉[1]　曹绪龙[2]　王得顺[2]　周国华[2]　李秀兰[2]

(1. 胜利油田科技处，257001；2. 胜利油田地质科学研究院，山东东营，257015)

摘要：为建立胜利油区复合驱油体系进行了室内实验。介绍了复合驱油的机理及表面活性剂的作用。重点针对孤岛西区进行了复合驱油配方的设计与矿场应用效果评价。

关键词：胜利油田　复合驱　表面活性剂　表面活性剂的应用　配方设计　矿场试验

经过“八五”“九五”攻关，在界面张力、活性剂检测、活性剂吸附损耗、乳化、注入方式等试验的基础上，胜利油田分别于 1992 年 8 月和 1997 年 5 月，实施了国内首例复合驱试验——孤东小井距复合驱油先导试验和复合驱扩大试验——孤岛西区三元复合驱配套试验，并取得了良好的降水增油效果。小井距试验区中心井提高采收率 13.4%，孤岛西区三元复合驱在主体段塞尚未注完的情况下，综合含水已由 95.7%降至 89.4%，累计增油 4.45×10^4 t胜利油区适合三元复合驱的区块群有飞雁滩埕 126 块及孤东二区 Ng5 所代表的区块群，18 个单元，覆盖储量 1.4746×10^8 t，潜力为 2.710×10^7 t。但是根据管理局的统一部署飞雁滩埕 126、孤东二区 Ng5 等 7 个单元在“九五”期间已安排实施聚合物驱。“九五”期间也已经对孤岛西区复合驱试验区和孤东七区西 $Ng5^4—6^1$ 复合驱试验作了安排，“十五”及“十一五”期间将对可能的区块实施工业化推广。十年间总共安排 5 个工业化推广区块，动用地质储量 9.440×10^7 t，提高采收率 13%，增加可采储量 1.227×10^7 t。

1　复合驱油机理及表面活性剂的作用

1.1 复合驱油机理

复合驱是指由两种或两种以上驱油剂组合起来的一种驱替方式，主要通过提高洗油效率和扩大波及体积来提高采收率。提高洗油效率主要通过降低油水界面张力和乳化夹带作用两个方面来实现，扩大波及体积和流度控制主要通过增大驱替液与原油形成的乳化液和聚合物溶液的粘度来实现。

其中表面活性剂主要有以下几个作用：降低油水界面张力、与其他化学剂产生协同效应、促进乳化和改变岩石的润湿性。

1.2 表面活性剂的作用

1.2.1 降低油水界面张力

在决定石油采收率的众多因素中，驱油剂的波及效率和洗油效率是最为重要的两个参数。提高洗油效率一般是通过增加毛细管准数来实现的，而降低油水界面张力则是增加毛细管准数的主要途径。注水开发后期，毛细管准数一般在 10^{-6}～10^{-7}。增加毛细管准数将显著地提高原油采收率，理想状态下毛细管准数增加至 10^{-2} 时，原油采收率可达到 100%。通过降低油水界面张力，可使毛细管准数有 3～4 个数量级的变化。通常油水界面张力在 20～30 mN/m 范围内，理想的表面活性剂可使界面张力降至 10^{-6}～10^{-7} mN/m 的超低范围内[1]。

1.2.2 协同效应

复合驱油体系中，碱与原油中的有机酸反应生成一种新型的表面活性剂——石油酸皂，该活性剂可与体系中人为加入的表面活性剂产生协同效应，使得复配体系的界面张力远低于单一体系的界面张力。这也是复合驱油体系能够大幅度提高原油采收率的主要原因之一。

1.2.3 乳化作用

活性剂在适当的盐度下能与原油形成油包水型乳化液，其粘度远高于原油的粘度，这种高粘度的乳化液既调节了地层流体间的流度比，又为剩余油的剥落和剥落油滴的聚并起到了很大的作用。同时由剥落油滴乳化聚并而形成的高粘度富油带也起到了一个活塞推进的作用。

1.2.4 改变岩石的润湿性

研究结果表明[2]，驱油效率与岩石的润湿性密切相关。油湿表面导致驱油效率差，水湿表面导致驱油效率好。合适的表面活性剂，可以使原油与岩石的润湿接触角变小，降低油滴在岩石表面的粘附功。

2 表面活性剂的应用

2.1 表面活性剂复配体系

由于各区块的油藏条件（油层渗透率、地层温度、地层水矿化度等）各不相同，原油的性质（如芳烃含量、原油酸值、烷烃含量等）也各有差异，因此对某一个具体区块驱油效率很高的配方，对另外的一个区块并不一定好用。另外，针对特定的原油，采用几种表面活性剂复配的方法更容易找到适宜 HLB 值的活性剂体系。因此在复合驱油中，应针对不同的区块或单元筛选出对它适用的活性剂复配体系。在针对某特定区块筛选配方过程中应遵循的原则是对大量的活性剂进行筛选和调配，优选出高效廉价的驱油配方，以便该区块“对症下药”。

2.2 复合驱油配方的设计与性能评价

复合驱油配方主要借助各组分间有效的协同效应来提高波及体积和驱油效率。但是由于体系所含的化学组分多、驱油机理复杂,为达到理想的驱油效果,需要对体系的理化特征进行研究。为保证研究工作的顺利开展,建立了复合驱室内配方设计与性能评价技术,制定了配方设计与性能评价工作流程,见图 1。

该项技术主要包括界面张力试验、活性剂筛选评价、化学剂吸附损耗试验、溶解性试验、配伍性试验、色谱分离试验和物理模拟试验等内容,其技术核心是化学驱油体系的确定、化学剂降低界面张力的性能、复配体系的稳定性、吸附损耗及对界面张力的影响和驱油效率的高低。

根据上述流程开展了大量的室内试验,首先通过界面张力试验、抗钙镁能力试验、乳化试验,对 50 余种表面活性剂进行了筛选评价。研究了不同活性剂间的加和作用及活性剂配比对界面张力的影响。绘制了界面张力等值图,见图 2。

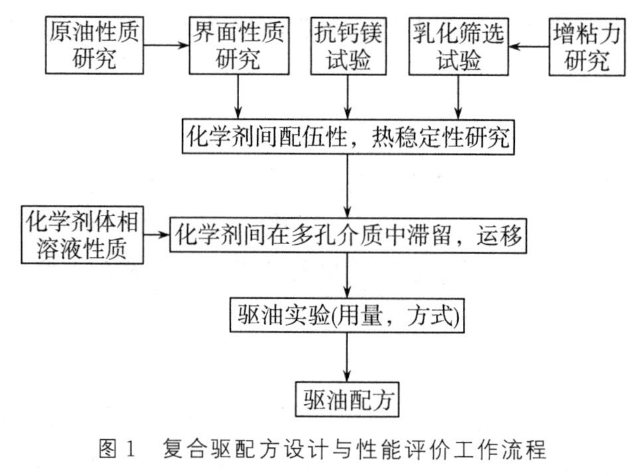

图 1 复合驱配方设计与性能评价工作流程

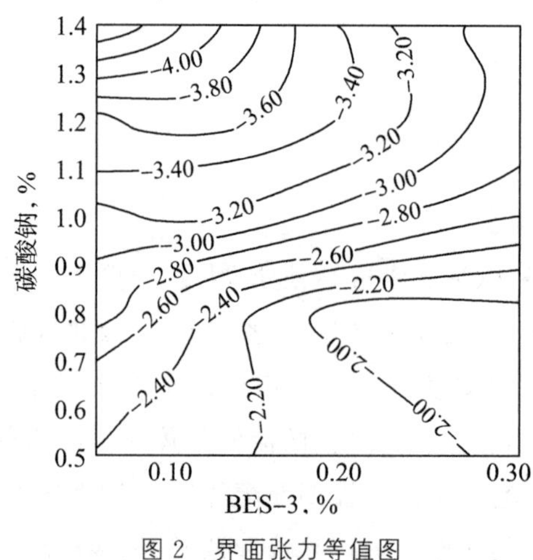

图 2 界面张力等值图

运用该图既可以对单一活性剂的性能进行评价,又可以对复合驱油配方进行评价和优化。

综合上述研究结果初步确定了由国产活性剂复配的孤岛西区复合驱油配方:

1.2%Na_2CO_3+0.2%BES−3+0.1%PS+0.15%3530S

建立了活性剂BES、PS的高效液相色谱浓度分析方法,PS的浓度分析方法最低检出限量为1 mg/L;BES-3的最低检出限量为20 mg/L。并运用建立化学剂浓度分析方法对体系的理化特征进行了全面的评价。

表1 孤岛西区复合驱配方性能指标

配方	0.2% BES-3+0.1% PS+1.2% Na_2CO_3+0.15% 3530S
界面张力/(mN·m^{-1})	7×10^{-4}
活性剂,%	0.03(商品)
提高采收率,%	16.8
采出液状态	无明显乳化现象
每克砂吸附量/mg	3.5(BES−3)

经综合评价认为:① 该配方与目的层原油的界面张力最低达7×10^{-4} mN/m,即使在油砂吸附之后或热老化70 d后,界面张力仍低达10^{-3} mN/m数量级,该配方超低界面张力区域较宽、适用性强;② 该配方原油在高碱浓度时与目的层原油形成比原油粘度高的乳化液,在低碱浓度时形成比原油粘度低的乳化液,因此,宜于注入井附近扩大波及体积,可增加远注入井区域流体的渗流;③ 推荐的配方体系中活性剂吸附损耗量较小,不仅有利于活性剂在地下传播,而且配方中使用活性剂的浓度低;④ 该配方室内提高采收率可达到22.4%(OOIP,Original oil in place),对目的层原油有较好的驱油能力。

采用国产活性剂复配方式具有显著特点,与国内外同类成果相比,其效果更好,价格更低。1996年经国家验收和专家鉴定。认为在配方设计和高效廉价活性剂方面有重大突破,研究成果达到国际领先水平。

2.3 矿场应用

该配方在矿场投入使用以后,取得了良好的降水增油效果,孤岛西区注采动态曲线见图3。

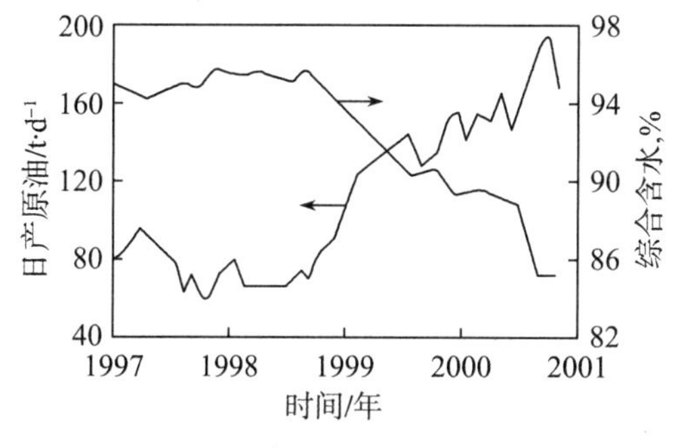

图3 孤岛西区注采动态曲线

在主体段塞尚未注完的情况下,综合含水已由95.7%降至89.4%,累计增油4.45×104 t其中中心评价井含水由98.5%最低降到85.0%,日产油由1.0 t增至14.3 t。

3 结束语

从复合驱油的主要机理和胜利油田复合驱油的实践和现状可以看出,复合驱油技术具有广阔的推广应用前景。但在开发过程中也暴露出一些问题,如仍需进一步解决好高效廉价表面活性剂、结垢防垢问题和采出液的破乳问题。

参考文献

[1] 杨承志.化学驱理论及实践[M].北京:石油工业出版社,1996.
[2] 赵福麟.采油化学[M].东营:石油大学出版社,1994.

Study on asp flooding system for shengli oil field and the effects of surfactant application

Chen Yequan[1] Cao Xulong[2] Wang Deshun[2] Zhou Guohua[2] Li Xiulan[2]

(1. Technical Department in Shengli Oil Field,257001)

(2. Geological Scientific Research Institute,Shengli Oil Field,Shandong Dongying,257015)

Abstract: A series of experiments were conducted for design of alkali-surfactant-polymer(ASP)flooding system in Shengli Oil Field. A formula of ASP flooding and fluid for Western Gudao Oil Field designed and evaluated by field tests. The mechanism of ASP flooding was described also.

Keywords: Shengli oil field; alkali-surfactant-polymer flooding; surfactant; application; formulation; field test

表面活性剂在胜利油田复合驱中的应用研究

周国华　曹绪龙　李秀兰　崔培英　田志铭

（中国石化股份有限公司胜利油田分公司地质科学研究院，东营 257015）

摘要：针对胜利油区孤岛油田西区和孤东七区油藏条件，考察表面活性剂体系的界面张力、稳定性、吸附损耗、驱油效率等指标，给出了碱/聚合物/表面活性剂三元复合驱（ASP）和表面活性剂/聚合物二元复合驱（SP）的配方。孤岛西区现场使用至 2001 年 7 月，已增油 89.7 kt，提高采收率 455％。

关键词：油藏　界面张力　驱油体系　表面活性剂　表面活性剂助剂　胜利油田

由表面活性剂组成的碱/活性剂/聚合物（ASP）三元复合驱、二元复合驱（SP）、泡沫复合驱可用于水驱或聚合驱后的二、三类油藏中。水驱后约有 2/3 的油滞留在油层中，由于毛管力的作用，这部分油滞留在较细的或喉径较窄的毛管孔道中，处于高分散状态，因油水间的界面张力约 30 mN/m，要驱替这部分油，单靠增大压差很难。因驱动所需压差为 9 806.7 kPa，而注水时压差仅为 196.1～392.3 kPa，远小于驱动压差。所以，利用表面活性剂降低界面张力，提高毛管数的特性驱出这部分剩余油是可行又有前途的方法。在孤东小井距三元复合驱油先导试验完成、中心井采收率提高 13.4％效果后[1]，我们开发了适合油田应用的表面活性剂，并研制出适合孤东、孤岛试验区的表面活性剂配方。

1　表面活性剂的筛选依据

用于驱油的表面活性剂主要是阴离子和非离子表面活性剂。阴离子表面活性剂以钠盐为主：烷基苯磺酸盐、烷基磺酸盐、烷基萘盐、羧酸盐、硫酸盐及磷酸盐。大量试验表明，非离子型表面活性剂驱油效果好，因为其在岩石表面的吸附量少，并对地层水中的高价阳离子（Ca^{2+}，Mg^{2+} 等）不敏感。将非离子型和阴离子型表面活性剂复配使用，发挥非离子型表面活性剂乳化作用和阴离子型表面活性剂润湿作用、分散作用，效果更好。针对不同性质的油水，需选择适宜的表面活性剂，下面讨论表面活性剂的筛选依据。

1.1 界面张力与乳液类型的关系

在复合驱体系中，筛选配方的主要手段是相性质实验，从中相乳液体积和含盐需求量方面筛选适宜的碱、表面活性剂及其用量。Healy[2]等证实,当体系处在最佳含盐需求量状态(表面活性剂增溶油和水的能力相同)时,体系形成最低界面张力,此时相态为Winsor Ⅲ型,界面张力与乳液类型的关系见图1。Nelson[3,4]指出,碱/表面活性剂体系形成中相乳液时,驱油效率最高。实践证明,体系界面张力最低时,油水相的增溶参数相等。最佳增溶参数(S)和最低界面张力(γ)的关系为

$$\gamma=\frac{C}{S^2} \tag{1}$$

式中,C为常量,mN/m;γ为界面张力,mN/m。

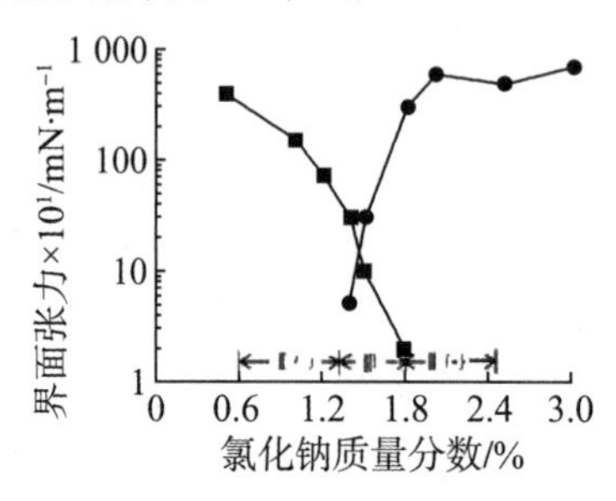

图1 界面张力与乳液类型的关系

Winsor Ⅱ(+)—乳液-水共存;Winsor Ⅱ(-)—乳液-原油共存;Winsor Ⅲ—原油-乳液-水共存

1.2 值与烷焕数的关系

表面活性剂的亲水-亲油平衡值(HLB值)常用作筛选表面活性剂的依据。对特定的油来说,当表面活性剂的HLB值在某一范围内能最大程度地富集于界面上。原油的烷灶数与表面活性剂HLB值的关系见图2[5]。不同原油对应不同烷灶数,从图中找到Winsor-Ⅲ型相应的HLB值范围,即可确定适宜的表面活性剂。

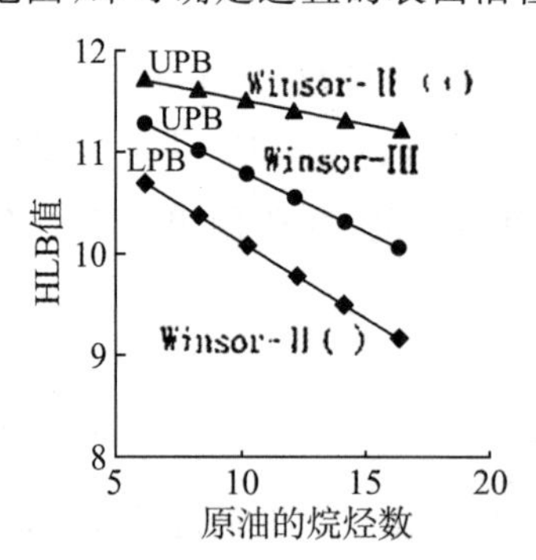

图2 烷怪数与HLB值的关系

LPB—最低含盐需求量;UPB—最高含盐需求量;OPB—最佳含盐需求量

1.3 HLB值的计算及复配活性剂作用

HLB值大,亲水性强;HLB值小,亲油性强。据表面活性剂在水中的溶度,由实践经验得出其HLB值。表面活性剂的HLB值及其在水中的性质见表1[6]。

表1　表面活性剂的HLB值及其在水中的性质

HLB值范围	加入水后的性质
1～4	不分散
3～6	分散不好
6～8	剧烈振荡后成乳色分散体
8～10	稳定乳色分散体
10～13	半透明至透明分散体

Davies将HLB值作为结构因子的总和来处理，把表面活性剂结构分解为一些基团，根据各种基团的HLB数值，即可计算表面活性剂的HLB值：

$$HLB=7+\Sigma(\text{亲水基团数})-\Sigma(\text{亲油基团数}) \tag{2}$$

实验证明，相同成HLB值的复配表面活性剂比单一表面活性剂形成的乳化体系更稳定。复配表面活性剂体系的HLB值计算如下：

$$HLB_{\text{混合}}=f_A\times HLB_A+(1-f_A)\times HLB_B \tag{3}$$

式中，A，B为表面活性剂，f_A 为表面活性剂A的质量分数。

原油是混合体系，不同油组分需不同HLB值的活性剂乳化，达最佳驱油效果。因此，采用复配活性剂满足原油不同组分的需要。根据表面活性剂的驱油机理，表面活性剂应满足油水界面上表面活性高，在地层岩石表面的吸附量尽量少。在选择注水用表面活性剂时，须考虑地层岩石的矿物组成、地层水和注入水的化学组成、地层温度和油藏枯竭程度。

2　三元复合驱的应用

胜利石油管理局1996年进行常规井距复合驱扩大试验，针对孤岛油田西区试验区的油藏条件，进行新型三元复合驱油配方的室内研究，确定了三元复合驱油体系配方。试验区位于孤岛油田西区西北部，试验区面积0.61 km^2。目的层有效厚度16.2 m，孔隙体积316×10^4 m^3，地质储量1 972 kt。原油地下粘度约70 mPa·s，油层温度69℃，孔隙度31.1%，空气渗透率1.52 μm^2。原油酸值1.7 mgKOH/g，产出水矿化度6 864 mg/L，钙镁含量130 mg/L；注入水矿化度6 188 mg/L，钙 镁含量为78 mg/L。

2.1 室内试验

2.1.1 表面活性剂的筛选

考察不同活性剂/碱复合驱油体系的界面张力，其活性剂质量分数为0.05%～0.5%，Na_2CO_3 质量分数为0.5%～1.5%。结果表明，BES、PS、ORS和SH2$^\#$等表面活性剂组成的碱/表面活性剂 体系与孤岛油田原油相比，其界面张力较低。考虑到工业化程度和价格，选用BES、PS进行了复配试验，单一活性剂和复配活性剂体系的界面张力见图3。从图3可看出，复配活性剂体系的界面张力比单一活性剂低，表明两种活性剂间有协同效应。

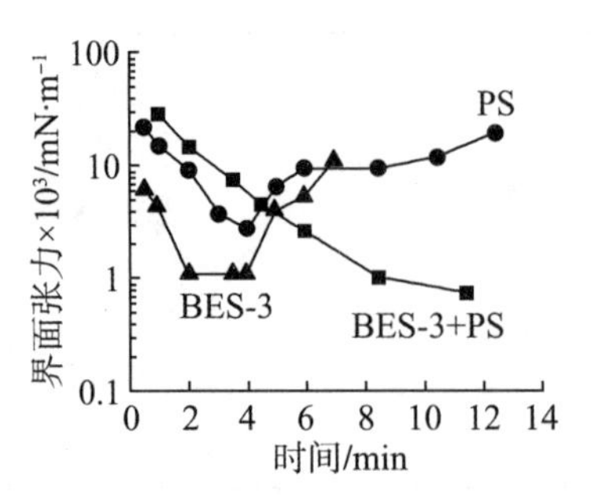

图 3 时间与界面张力的关系

2.1.2 乳化试验

考察复配活性剂 BES-3+PS 在不同碱含量条件下乳液的粘度变化,结果见图 4。从图 4 可见,碱质量分数为 0.4%左右时,乳液粘度为原油粘度的 1/10,而碱质量分数为 1.0%时,乳液粘度为原油的 1.6 倍以上。因此,表面活性剂/碱复合驱油体系在高碱含量时,波及体积增加,而在低碱含量时,渗流阻力减少。

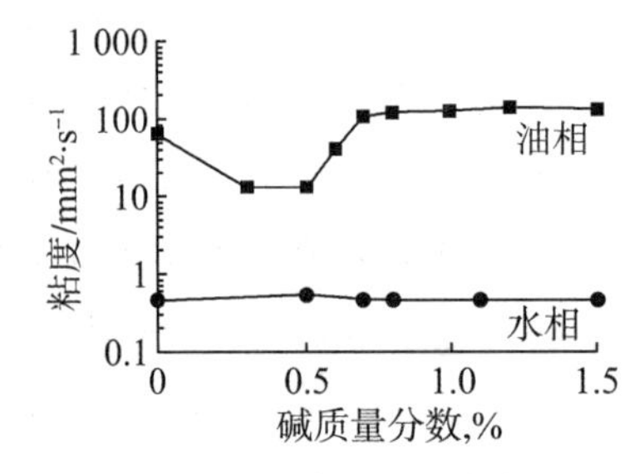

图 4 碱质量分数与乳液粘度的关系

2.1.3 油砂吸附对界面张力的影响

为考察岩石吸附表面活性剂/碱复合体系后界面张力的变化,进行了油砂吸附前后界面张力的试验,结果见图 5。从图 5 可见,吸附后其最低界面张力仍达 10^{-3} mN/m,表明表面活性剂被部分吸附后,表面活性剂/碱复合驱油体系仍保持较好的界面张力。

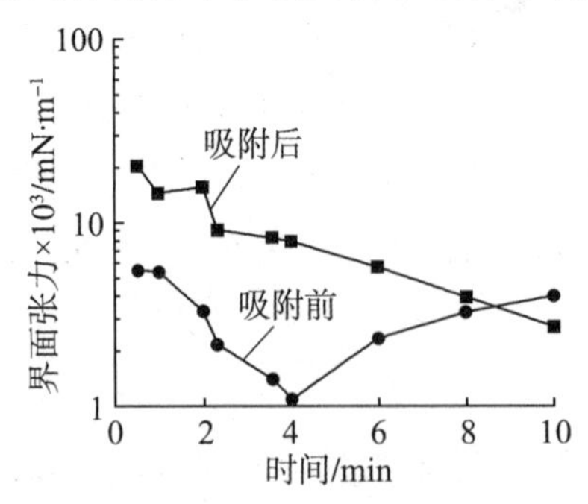

图 5 油砂吸附与界面张力的关系

2.1.4 热老化对界面张力的影响

在 70℃下,考察老化时间对 0.2%BES+0.1%PS+1.2%Na_2CO_3 和 0.2%BES+0.1%PS+1.2%Na_2CO_3+0.15%3530S 聚合物复合驱油体系界面张力的影响,结果见

图6。从图6可见，随老化时间增加，体系界面张力基本不变，说明复合驱油体系注入油藏后，保持低界面张力的能力强。

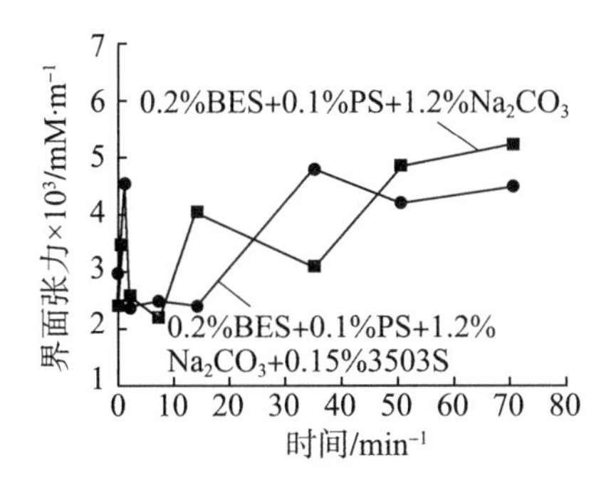

图6 热老化时间与界面张力的关系

2.1.5 单一及复配表面活性剂对驱油效果的影响

单一和复配表面活性剂体系的驱油实验结果见表2。从表2可看出，使用复配活性剂体系驱油效率明显增加，表明表面活性剂复配后，洗油能力增加，采出了一部分难以采出的原油。因此，最佳ASP三元复合驱配方为1.2% Na_2CO_3+0.2% BES+0.1% PS+0.15% 3530S。

表2 不同活性剂体系在孤岛油田上驱油试验结果

试验编号	活性剂质量分数,%		碱质量数,%	聚合物质分数,%	提高采收率,%
	BES	PS			
GD-4	0.3	0	1.2	0.15	17.4
GD-9	0	0.3	1.2	0.15	21.7
GD-8	0.2	0.1	1.2	0.15	25.3

2.2 现场应用效果

孤岛西区试验区自1997年6月水驱转注ASP三元复合驱驱油后，日产油由67 t/d增至142 t/d，综合含水由95.7%降至90.6%。至2001年7月已增油89.7 kt，提高采收率4.55%，增油效果显著。孤岛西区复合驱试验区采油曲线见图7。

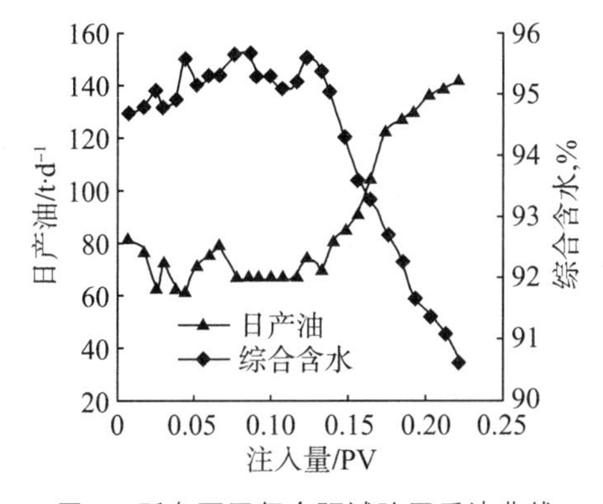

图7 孤岛西区复合驱试验区采油曲线

3 二元复合驱油体系的应用

尽管 ASP 三元复合驱效果好，但碱的加入，使现场工作量加大，注水井结垢严重，产出液处理困难。因此，选用胜利石油管理局设计院生产的石油磺酸盐表面活性剂，进行石油磺酸盐/聚合物二元复合驱油配方研究。

试验区位于孤东七区西南部，油层埋深 1 255～1 333 m，共有 3 个含油小层，油层孔隙度 34%，平均渗透率 1 278×10^{-3} μm^2，渗透率变异系数 0.581，原始含油饱和度 72%，地下原油粘度 45 mPa·s。原始地层水矿化度 3 152 mg/L，目前产出水矿化度 5 315 mg/L，油层温度 70℃，原始地层压力 12.9 MPa，饱和压力 10.2 MPa，试验区采出程度 29.04%。

3.1 配方筛选

3.1.1 石油磺酸盐的筛选

对 20 多种石油磺酸盐的界面张力进行了分析与研究，部分石油磺酸盐的界面张力试验结果见表 3。

表 3 单一石油磺酸盐界面张力试验结果

石油磺酸盐	最低界面张力/mN·m^{-1}	稳定时间/min
0.4% SPS-1 石油磺酸盐	7×10^{-2}	50
0.6% SPS-1 石油磺酸盐	4×10^{-2}	65
0.6% SPS-3 石油磺酸盐	6×10^{-2}	90

由表 3 可看出，SPS-1 石油磺酸盐与模拟油的界面张力最低，其质量分数为 0.6% 时，界面张力仅为 4×10^{-2} mN/m。因此，靠单一活性剂难以获得理想的效果。

3.1.2 复配助剂的筛选

以 SPS-1 石油磺酸盐为主剂，选择助剂与之复配，对复配体系的配伍性、界面张力等进行研究。将石油磺酸盐与助剂的比例定为 3∶1，活性剂的总质量分数由 0.16%增至 0.6%，测定石油磺酸盐/助剂体系的界面张力，其界面张力等值图见图 8。从图 8 可看出，当石油磺酸盐和助剂的质量分数分别为 0.2%～0.45%和 0.05%～0.15%时，其界面张力较低。当石油磺酸盐和助剂的质量分数分别为 0.3%和 0.1%时，界面张力降至 2.95×10^{-3} mN/m。因此，初步确定室内试验配方为 0.3%石油磺酸盐+0.1%助剂。另外，从图中还可看出，石油磺酸盐/助剂体系有较宽的质量分数适用范围，在总质量分数低的情况下，也能与原油达到超低界面张力。

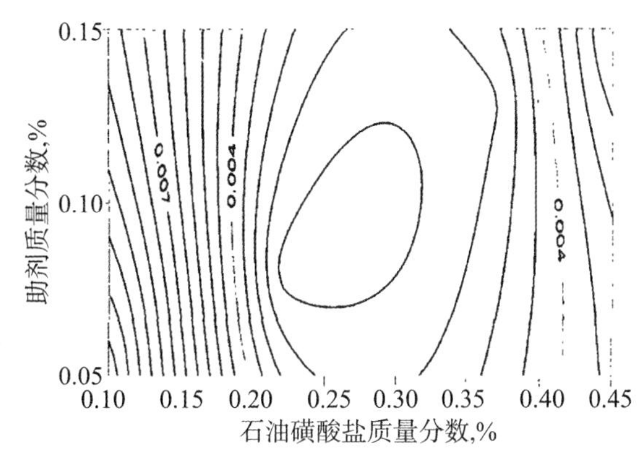

图 8 石油磺酸盐/助剂界面张力等值图

3.2 吸附损耗试验

二元复合驱注入地层以后，活性剂与地层油砂发生复杂的物理化学作用，吸附于岩石表面，影响二元复合驱配方的驱油效率。因此，进行了孤东七区油藏条件下的吸附损耗试验，将不同的复配表面活性剂体系与洗净烘干的油砂以 3∶1 比例混合，在 70℃水浴中振荡 24 h，离心处理后，取清液分析其浓度，采用高效液相色谱法检测复配体系中石油磺酸盐的浓度。不同复配体系的石油磺酸盐吸附损耗试验结果见表 4。

表 4 石油磺酸盐吸附损耗试验结果

配 方	石油磺酸盐浓度/$mg \cdot L^{-1}$		吸附量/$mg \cdot g^{-1}$
	吸附前	吸附后	
0.4% SPS-1	3 228.0	2 727.0	1.503
0.3%SPS-1+0.1%助剂	2 107.3	1 983.0	0.373

从表 4 可看出，石油磺酸盐体系吸附量较小，和助剂复配后其吸附量进一步降低。

3.3 驱油试验

3.3.1 复配表面活性剂质量分数的筛选

固定聚合物质量分数为 0.15%、二元复合驱注入量 0.3 PV，石油磺酸盐和助剂按一定配比复配，考察复配活性剂对提高采收率的影响，结果见表 5。从表 5 可看出，复配活性剂总质量分数从 0.16%到 0.4%，提高采收率从 11.9%提高到 18.1%。因此，最佳复配活性剂总质量分数为 0.4%。

表 5 复配活性剂质量分数筛选

模型编号	配方	提高采收率,%
孤东-11#	0.3%SPS+0.1%助剂	18.1
孤东-4*	0.2%SPS+0.07%助剂	12.8
孤东-24′	0.12%SPS+ 0.04%助剂	11.9

3.3.2 二元复合驱注入量的筛选

固定复配活性剂质量分数为0.4%(0.3%SPS+0.1%助剂),3530S聚合物质量分数0.15%,考察注入量对提高采收率的影响结果,见表6。从表6可看出,注入量为0.4 PV时,提高采收率最大,但注入量为0.3 PV时,综合利用率较高。因此,最佳注入量为0.3 PV。

表6 二元复合驱注入量的筛选

模型编号	注入量/PV	提高采收率/%	综合利用率/%
孤东-12#	0.2	14.0	0.151
孤东-11#	0.3	18.1	0.169
孤东-13"	0.4	20.3	0.159

3.3.3 二元复合驱与聚合物驱的驱油效果

在同等经济条件下,考察了二元复合驱与聚合物驱的驱油效果,结果见表7。从表7可看出,3530S聚合物质量分数0.15%时,注入量为0.54 PV,提高采收率为15.2%;而二元复合驱注入量为0.3 PV,提高采收率18.1%,表明复合二元驱驱油效果比单一聚合物驱好。

表7 复合二元驱与聚合物驱的驱油效果

模型编号	配方	注入量/PV	提高采收率/%
孤东-11*	0.3%SPS+0.1% 助剂+0.15%3530S	0.3	18.1
孤东-18*	0.15%3530S	0.54	15.2

4 结论

(1) ASP三元复合驱在孤岛油田西区应用效果显著。室内试验中,针对目的层油藏,根据界面张力、化学剂吸附损耗、配伍性、乳化等试验,提出的最佳三元复合驱配方为1.2%Na_2CO_3+0.2%BES+0.1%PS+0.15%3530S。现场实施后,含水下降,至2001年7月份已增油89.7 kt,提高采收率4.55%。

(2) 二元复合驱的最佳配方为0.3%SPS-1+0.1%助剂+0.15%3530S,其界面张力为2.95×10^{-3} mN/m,注入量为0.3 PV时,提高采收率可达18.1%。

参考文献

[1] Chenglong W, Baoyu W, Xulong C, et al. Soc Petrol Engrs, 1997: 38321, 605-618.

[2] Healy R N, Reed R L, Stenmark D G. Soc Petrol Engrs, 1976: 5565, 147-160.

[3] Nelson R C. Soc Petrol Engrs, 1982: 8824, 259-270.

[4] Nelson R C. Pope G A. Soc Petrol Engrs，1978：6773，325-328.

[5] Drew Mayers. Surfactant Science and Technology. Second Edition. VCH Publishers Inc，1992.

[6] 赵国玺.表面物理化学[M].北京:北京大学出版社,1984：472-476.

Study of application of surfactant in combinational flooding solution in shengli oil field

Zhou Guohua Cao Xulong Li Xiulan Cui Peiying Tian Zhiming

(Geological Scientific Research Institute of Shengli Oil Field Corporation，SINOPEC，Dongying 257015)

Abstract：According to reservoir conditions of Western Gudao and Gudong in Shengli Oil Field，we examined interfacial tension，stability，adsorptive loss and oil displacement efficiency，etc. The compositions of ASP and SP were given. After their use in Western Gudao，89. 7 kt oil had been produced up till July 2001 and enhanced oil recovery was 4. 55%.

Keywords：reservoir，interfacial tension，oil displacing system，surfactant，cosurfactant，Shengli Oil Field

阴离子表面活性剂与聚丙烯酰胺间的相互作用

曹绪龙* 蒋生祥

(中国科学院兰州化学物理研究所兰州 730000)

孙焕泉 江小芳 李 方

(中国石油化工集团胜利油田有限公司东营)(美国纽约州立大学纽约)

摘要:借助于表面活性剂与聚合物溶液比浓粘度和紫外光谱吸收特征的变化,研究了阴离子表面活性剂月桂酸钠和十二烷基磺酸钠与聚丙烯酰胺(PAM)之间的相互作用。实验结果表明,在中性或弱酸性溶液中聚丙烯酰胺分子可与月桂酸钠发生氢键缔合,形成 PAM-$C_{11}H_{23}COONa$ 聚集体,使 PAM 大分子链上带有大量的电荷,从而使 PAM-$C_{11}H_{23}COONa$ 混合溶液表现出聚电解质的浓度行为。在十二烷基磺酸钠与聚丙烯酰胺形成的体系中,只有当十二烷基磺酸钠的浓度大于临界胶束浓度时,通过十二烷基磺酸钠胶束与 PAM 的离子偶极作用方可观察到溶液粘度增加现象。

关键词 聚丙烯酰胺,表面活性剂,相互作用,聚集体

中图分类号:o641.3 文献标识码:A 文章编号:1000-0518(2002)09-0866-04

复合驱油方法利用碱和表面活性剂在降低界面张力方面的超加合作用而具有提高微观驱油效率和流度控制的双重作用,所以近年来发展迅速并在矿场中得到了一定规模的应用[1,3]。但是体系中表面活性剂和聚合物之间会形成表面活性剂-聚合物的聚集体(复合物)。因此了解聚合物与表面活性剂之间的相互作用对认识复合驱油体系特征有重要意义。有关表面活性剂与聚合物之间相互作用的报道较多,Myers[2]报道了表面活性剂与非离子型和离子型聚合物的相互作用并对形成的机理进行了分析。但是由于材料对形成的聚集体的宏观性质影响较大,本文通过体系比浓粘度的变化研究了 2 种阴离子表面活性剂月桂酸钠($C_{11}H_{23}COONa$)和十二烷基磺酸钠($C_{12}AS$)与聚丙烯酰胺(PAM)的溶液特性。

1 实验部分

试剂和仪器:月桂酸钠($C_{11}H_{23}COONa$,益阳油脂化工厂,$C_{11}H_{23}COOH$ 与 NaOH 中和制得,分析纯);聚丙烯酰胺(PAM),分子量 8×10^6(南中塑料厂);十二烷基磺酸钠

$C_{12}AS$（市售），用乙醇结晶 2 次；去离子水。稀释型乌氏粘度计；UV-260 型紫外-可见分光光度计（日本岛津公司）

2　结果与讨论

2.1 $C_{11}H_{23}COONa$ 对聚合物溶液比浓粘度的影响

根据矿场应用情况，实验在 pH=7 条件下进行。单一聚合物（PAM）水溶液的 $Z_{sp}\times c_p^{-1}$-cp 曲线呈线性关系时显示出非聚电解质的粘度行为，可用哈金斯方程处理。但是当在聚合物溶液中加入不同浓度的月桂酸钠后，聚合物溶液的 $Z_{sp}\times c_p^{-1}$-cp 曲线偏离了线性关系。

由图 1 可见，在 PAM 溶液中加入 $C_{11}H_{23}COONa$ 后，其粘度表现出明显的聚电解质特征，并且电粘效应随 $C_{11}H_{23}COONa$ 浓度的增加而增强。用界面张力法[4]测得 $C_{11}H_{23}COONa$ 临界胶束质量浓度为 6.30 g/L。

通过核磁共振等研究发现，在 $C_{11}H_{23}COONa$ 临界胶束浓度之前，PAM 分子与 $C_{11}H_{23}COONa$ 分子发生缔合作用，且氢键起到了主导作用[5]。缔合过程可用单个表面活性剂分子（S）与聚合物链（P）之间逐级缔合方程来描述：

每步缔合过程受质量作用方程控制并受实验条件（如温度、溶剂、离子强度、pH 等）影响。从本实验结果看，当表面活性剂 $C_{11}H_{23}COONa$ 浓度大于临界胶束浓度时，聚合物溶液的 $\eta_{sp}\times c_p^{-1}$ 随溶液的稀释而显著升高，这主要是因为聚合物与 $C_{11}H_{23}COONa$ 胶团发生缔合作用，是大分子链上电荷数目增多，静电斥力使聚合物分子进一步伸展。

$$
\begin{aligned}
P + S &\xrightarrow{k_1} PS \\
PS + S &\xrightarrow{k_2} PS_2 \\
PS_2 + S &\xrightarrow{k_3} PS_3 \\
&\cdots\cdots \\
PS_{(n-1)} + S &\xrightarrow{k_n} PS_n
\end{aligned}
\qquad (1)
$$

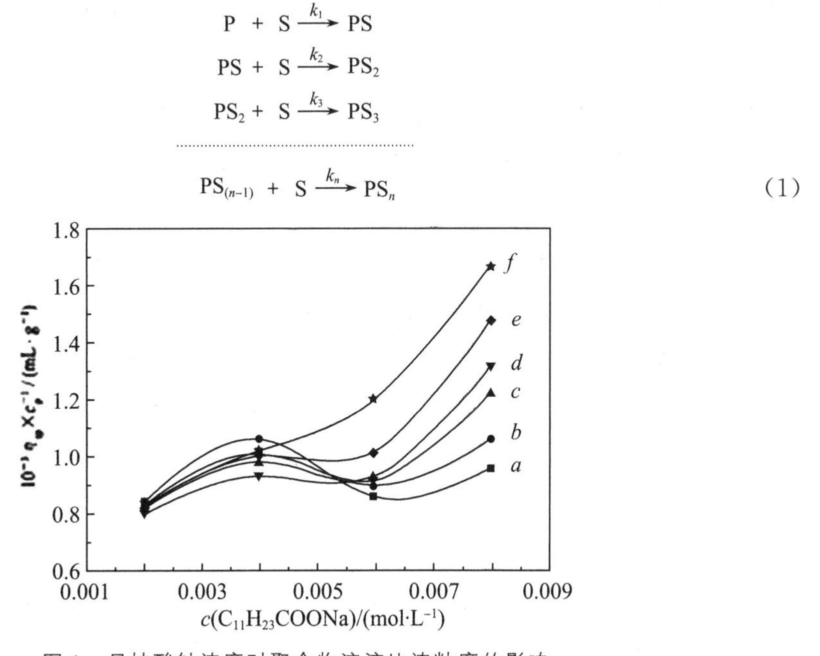

图 1　月桂酸钠浓度对聚合物溶液比浓粘度的影响

Fig. 1　The effect of sodium laurateConcentration on reduced viscosity Of polymer solution

c(PAM)/(mg · L^{-1})：a. 100；b. 150；c. 200；d. 250；e. 320；f. 500

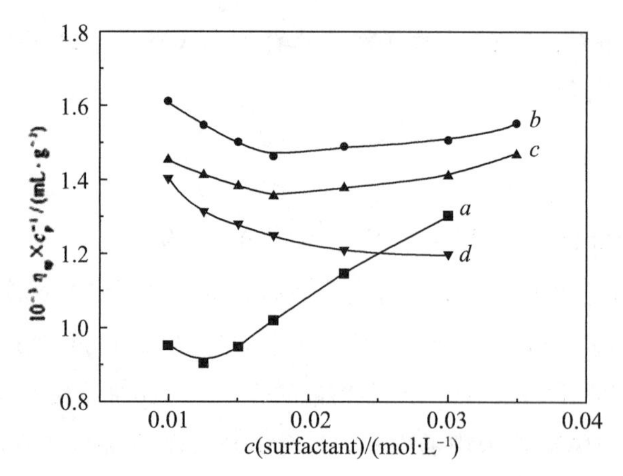

图 2 表面活性剂及浓度对 PAM 溶液比浓粘度的影响

Fig. 2 The effects of surfactants and Their concentration on reduced viscosity Of polymer solution

a. 100 mg/LPAM-$C_{11}H_{23}COONa$；b. 500 mg/LPAM-$C_{11}H_{23}COONa$；

c. 100 mg/LPAM-$C_{12}AS$；d. 500 mg/LPAM-$C_{12}AS$

2.2 十二烷基硫酸盐与 $C_{11}H_{23}COONa$ 对聚合物溶液 $\eta_{sp}\times c_p^{-1}$ 影响的对比

图 2 给出了不同浓度 $C_{12}AS$ 和 $C_{11}H_{23}COONa$ 对 2 种质量浓度聚合物溶液的 $Z_{sp}\times c_p^{-1}$ 影响的实验结果。图中可见，PAM-$C_{11}H_{23}COONa$ 与 PAM-$C_{12}AS$ 体系的粘度变化趋势虽然相同，即当表面活性剂的浓度较小时 PAM 体系的 $Z_{sp}\times c_p^{-1}$ 均随表面活性剂浓度的增大而降低，且只有当表面活性剂的浓度大到一定值后方出现 $Z_{sp}\times c_p^{-1}$ 随表面活性剂浓度增大而升高的现象。但是 2 个体系出现 $Z_{sp}\times c_p^{-1}$ 最低值所对应的表面活性剂浓度却不同，如在 PAM 质量浓度为 100 mg/L 时，$Z_{sp}\times c_p^{-1}$ 最低值所对应的 $C_{11}H_{23}COONa$ 与 $C_{12}AS$ 的浓度分别为 1.25×10^{-2} mol/L 和 1.75×10^{-2} mol/L。这主要是因为 $C_{11}H_{23}COONa$ 与 PAM 之间有氢键作用而 $C_{12}AS$ 与 PAM 之间没有或作用很弱[5]。

由图 2 还可看出，在 PAM-$C_{12}AS$ 体系中，PAM 质量浓度较高时(500 mg/L)，体系的 $Z_{sp}\times c_p^{-1}$ 随 $C_{12}AS$ 浓度的增大而降低。$C_{12}AS$ 的浓度大于 2×10^{-2} mol/L 时变化不大。而在 PAM 质量浓度较低时(100 mg/L)，$C_{12}AS$ 的浓度在 2×10^{-2} mol/L 左右体系的 $Z_{sp}\times c_p^{-1}$ 达到最低值，之后随 $C_{12}AS$ 浓度的升高，PAM 体系的 $Z_{sp}\times c_p^{-1}$ 增大。CMC 实验表明，本文得到的 $C_{12}AS$ 的 CMC 浓度为 2.02×10^{-2} mol/L，因此，上述变化特征是在 CMC 附近发生的。

这一结果表明，PAM 与 $C_{12}AS$ 之间的疏水结很弱，二者之间的相互作用主要是大分子与活性剂胶束之间的缔合。PAM 浓度较大时，大分子链间相互作用使之成线团状，不利于大分子与胶束间的缔合。当 PAM 浓度较低时，大分子与溶剂之间的相互作用促使大分子链伸展，伸展的 PAM 大分子链通过离子-偶极作用与负电荷的 $C_{12}SO_3^{2-}$ 胶束发生缔合，形成 PAM-胶束聚集体，从而导致体系粘度的升高。

2.3 $C_{12}AS$-PAM 体系的紫外吸收光谱特征

不同质量浓度单一 $C_{12}AS$ 水溶液体系和 $C_{12}AS$-PAM 混合溶液体系的吸收光谱表明，2 个体系吸收光谱在紫外范围内基本重合。即使 $C_{12}AS$ 浓度大于其临界胶束浓度时，也未发现红移现象，如在 λ=190.0～230.0 nm 范围内，500 mg/L 的 PAM 溶液的 λ_{max}=203.4 nm，500 mg/L 的 PAM 和 500 mg/L 的 $C_{12}AS$ 混合溶液的 λ_{max} 仍为 203.4 nm。

2.4 $C_{11}H_{23}COONa$-PAM 体系的紫外吸收光谱特征

在不同 pH 条件下单一 PAM 体系、单一 $C_{11}H_{23}COONa$ 和 PAM-$C_{11}H_{23}COONa$ 混合溶液体系最大紫外吸收峰波长结果见表。由表 1 可见，PAM-$C_{11}H_{23}COONa$ 混合溶液体系的 λ_{max} 表现出明显的红移现象，且在 pH 值为 4 左右时，红移现象最为明显。结合 PAM 和 $C_{11}H_{23}COONa$ 的结构特征、本文实验结果及核磁共振的结果[5]，可以判定在 PAM-$C_{11}H_{23}COONa$ 混合溶液体系中有氢键缔合物形成。

在中性或弱酸性水溶液中，羧酸盐同时存在下述平衡：

$$C_{11}H_{23}COONa + H_3O^+ \rightleftharpoons C_{11}H_{23}COOH + H_2O$$
$$\downarrow$$
$$C_{11}H_{23}COO^- + H^+$$

$C_{11}H_{23}COONa$ 作为质子给予体，可以与 PAM 分子上的酰胺基发生氢键缔合。当 pH 值较低时，PAM 分子上的酰胺基发生质子化，反而会削弱 PAM-$C_{11}H_{23}COONa$ 之间形成氢键的能力，而 pH 值较高时，体系中存在较多的 $C_{11}H_{23}COONa$，不易与 PAM 形成氢键缔合物。因此，λ_{max} 的红移现象在 pH 值过高或低时不再发生。该推论亦可对图 2 的实验结果作出合理解释，即在 PAM-$C_{11}H_{23}COONa$ 体系中，以疏水力形成的 $C_{11}H_{23}COOH$-$C_{11}H_{23}COO$ 二聚体可通过氢键结合在 PAM 大分子链上，致使大分子链因带电荷而更加伸展，从而在 $C_{11}H_{23}COONa$ 小于 CMC 值时就出现了电粘效应。随着 $C_{11}H_{23}COONa$ 加入量的增大，体系中 Na^+ 浓度增加，破坏大分子链周围水化膜的作用增加，使大分子链卷曲，溶液粘度降低。当 $C_{11}H_{23}COONa$ 浓度大于 CMC 时，PAM 可与胶束缔合，致使 PAM 大分子的电荷量显著增加，体系粘度增高。

表 1　不同体系条件下混合溶液体系最大紫外吸收峰波长(nm)

Table1　The maximum UV absorption wave-length(nm) of surfactant-polymer solutions

$10^4 c(\text{compounds})/(g \cdot mL^{-1})$	pH 7.0	6.0	4.0	2.5	1.0
PAM 1.25，$C_{11}H_{23}COONa$ 0	195.5	196.0	197.4	197.8	197.8
PAM 0，$C_{11}H_{23}COONa$ 1.25	—	—	—	195.4	199.4
PAM 1.25，$C_{11}H_{23}COONa$ 1.25	197.6	198.0	205.0	199.2	200.0

致谢：在本课题研究过程中得到山东大学李干佐教授、中国科学院兰州化学物理研究所陈立仁研究员的热情指导；山东大学胶体与界面化学研究所徐桂英、沈强同志参加

了部分实验工作并提出了很好的建议,在此一并致谢!

参考文献

[1] Wang Chenglong(王成龙), Cao Xulong(曹绪龙), Wang Baoyu(王宝瑜), et al. Soc Petrol Eng J[J], 1997, SPE, 38321.

[2] Myers D. Surfactant Science and Technology [M]. NewYork: VCH Publishers, INC. 1992:173.

[3] Maurice B, Robert S. Schechter Microemulsion and Related Systems[M]. New York: Marcel DEKKER, INC. 1988:27.

[4] Shen Zhong(沈钟), Wang Guoting(王果庭). Colloidand Surface Chemistry (胶体与表面化学)[M]. Beijing(北京): Chemical Industry Press(化学工业出版社), 1997:443.

[5] Li Fang(李方). Doctoral Dissertation([博士学位论文]). Jinan(济南):College of Chemical Engineering and Chemistry, Shandong University(山东大学化学化工学院), 1997.

On the interaction between polyacrylamide and anionic surfactants

Cao Xulong * , Jiang Shengxiang

(Lanzhou Institute of Chemical Physics, Chinese Academy of Sciences, Lanzhou 730000)

Sun Huanquan, Jiang Xiaofang

(Geological Research Institute of Shengli Oilfield, SINOPEC,Dongying)

Li Fang

(State University of New York at Buffalo, New York, 14260)

Abstract: The interactions between anionic surfactants (sodium laurate ($C_{11}H_{23}COONa$), sodium dodecyl sulf onate($C_{12}AS$)) and polyacrymide have been investigated by measurements of the reduced viscosity and ultraviolet spectra of the polymeric solutions. The results showed that in neutral and weak acid solutions PAM formed an aggregate with $C_{11}H_{23}COONa$ behaved as a polyelectrolyte through hydrogen bonding and there was a large amount of charges in the PAM main chain caused by aggregate. However, PAM-$C_{12}AS$ aggregate was formed through dipolar interactions only in the case when the concentration of $C_{12}AS$ was higher than critical micelle concentratio.

Keywords: polyacrylamide, surfactant, interaction

胜利油区二元复合驱油先导试验驱油体系及方案优化研究

张爱美　曹绪龙　李秀兰　姜颜波

(胜利油田有限公司地质科学研究院，山东 东营 257015)

摘要：首先根据胜利油区的油藏特点和复合驱的特点，在孤东油田七区西南 Ng54-61 层优选出了具有广泛代表性的试验区。然后根据试验区具体的油藏情况，室内筛选出了适合该油藏类型的以石油磺酸盐为主剂的二元复合驱油体系，研究表明，该体系可与模拟油达到超低界面张力状态，其吸附量较小、乳化性能和热稳定性好，驱油效果好。结合室内研究的结果，在建立基于模糊综合评判模型的二元复合驱注采参数优化设计方法的基础上，对化学剂组合方式、配方浓度、段塞尺寸、注入方式、注采速度等参数进行了优化。综合研究认为，该二元复合驱油体系驱油效率高、油藏适应性强，可较大幅度提高采收率。

关键词：胜利油区；二元复合驱；石油磺酸盐；化学驱；先导试验

中图分类号：TE357　文献标识码：A

复合驱是由两种或两种以上驱油剂组合起来的一种驱动方式，它的主要机理是提高洗油效率和增大波及体积[1,2,3]。复合驱油技术研究方面，国内外在八十年代中期开始进行大量室内试验研究，并进行了十余个矿场试验，技术上都获得了成功[4,5]。"八五"期间胜利油田孤东小井距三元复合驱油体系(碱＋活性剂＋聚合物)先导性试验作为我国首例复合驱油先导试验，取得了突破性进展。在中心井含水率达 98.5%，水驱采出程度 54.4%，接近水驱残余油的情况下，实施三元复合驱后，中心井提高原油采收率 13.4%，使总采收率达到 67%[5]。但是三元复合驱也暴露出一些问题，如注入过程中的结垢影响注入的问题、采出液的破乳影响集输的问题、表面活性剂价格昂贵影响经济效益的问题等。为了克服三元复合驱的弊端，继续探索复合驱油技术在胜利油田的应用与发展，有必要开展无碱的二元复合驱油体系(活性剂＋聚合物)研究，同时筛选出适合胜利油区油藏条件的高效廉价的表面活性剂产品。该先导试验是在三元复合驱实践的基础上提出的，并被列为中国石化 2003 年度油气田开发重大先导试验项目。

1 试验区选区

试验区选区是复合驱先导试验方案编制的一个重要环节,它在一定程度上决定了试验的油藏物质基础。根据复合驱的特点和胜利油区已开发复合驱试验单元的经验[5,6],本试验区选区的原则为:① 试验区有代表性,能代表胜利油区主力油田注水油藏的开采规律;② 采用常规开发井网,试验结果具有普遍推广意义;③ 地质情况清楚,油层发育良好,无明显大孔道窜流;④ 井网和注采系统完善,井况良好,中心受效井多;⑤ 储层流体性质适中,油层温度适中。根据以上原则,从油藏本身特点、开发动态情况及与聚合物实施的配套性等方面进行综合研究,确定在孤东油田七区西 $Ng5^4$-6^1 南部开展二元复合驱油先导试验。

试验区油层埋深为 1 261～1 294 m,共发育 5^4、5^5 和 6^1 三个含油小层,含油面积 0.94 km^2,地质储量 277×10^4 t,目的层原始油层温度 68℃,平均渗透率 $1\ 320\times10^{-3}$ μm^2,纵向渗透率变异系数 0.58,地下原油粘度 45 mPa.s,目前地层水矿化度及 $Ca^{2+}+Mg^{2+}$ 含量分别为 8 207 mg/L 和 231 mg/L,平均剩余油饱和度 45.5%。经研究认为,该试验区除目前地层水矿化度和 $Ca^{2+}+Mg^{2+}$ 含量稍高外,其他主要油藏参数皆处于复合驱的有利范围内[1,2,3],进行二元复合驱油先导试验是可行的。试验区设计注入井 10 口,生产井 16 口,中心受效井 6 口,观察井 2 口,取心井 2 口。截至 2002 年底,试验区综合含水 97.4%,采出程度 34.4%,注入压力 11.9 MPa,预测水驱采收率 36.3%。

2 驱油体系室内研究

表面活性剂+聚合物(P)二元复合驱是一个复杂的化学驱油过程。对该驱油体系室内研究主要是对表面活性剂、聚合物类型进行筛选并对它们之间的相互作用、配伍性等进行研究,以期充分发挥表面活性剂降低界面张力的作用,又可发挥聚合物在流度控制、防止或减少化学剂段塞的窜流的作用,更重要的是发挥二者的协同效应并最大限度地消除它们之间的相互干扰(如相分离)。同时,综合考虑其在油层内的损耗、热稳定性影响等因素,探索与试验区油藏相适应的配方体系,初步确定合理的注采参数范围。

2.1 试验条件

试验用油:孤东 7-36-166 井模拟油(地层温度 69℃下,原油粘度 45 mPa.s)。

试验用水:7-36-135 产出水($Ca^{2+}+Mg^{2+}$ 浓度 236 mg/L,总矿化度 8 127 mg/L)、44-1 计量站注入水($Ca^{2+}+Mg^{2+}$ 浓度 282 mg/L,总矿化度 8 956 mg/L)。

2.2 表面活性剂筛选

为减少对外部活性剂市场的依赖、发展胜利油田自己的后续产业,同时考虑到活性剂结构与原油结构的相似性,本先导试验立足以胜利原油为原料合成的表面活性剂(石油磺酸盐)为主剂,这不仅可以降低成本,而且适用性强。室内对胜利不同的石油磺酸盐样品在不同浓度下的界面张力进行了测试,结果 1# 石油磺酸盐(SPS-1)样品界面张力最低,但也仅达到 4×10^{-2} mN/m。于是提出了"助表面活性剂增效石油磺酸盐体系"的研究思路。以 SPS-1 作为主剂,选择不同的助剂与之复配,考察了 26 种复配体系的界面张力,结果 0.3% SPS-1+0.1% 1# 活性剂体系的界面张力最低,可达到 2.95×10^{-3} mN/m,进入了超低界面张力区(图 1)。

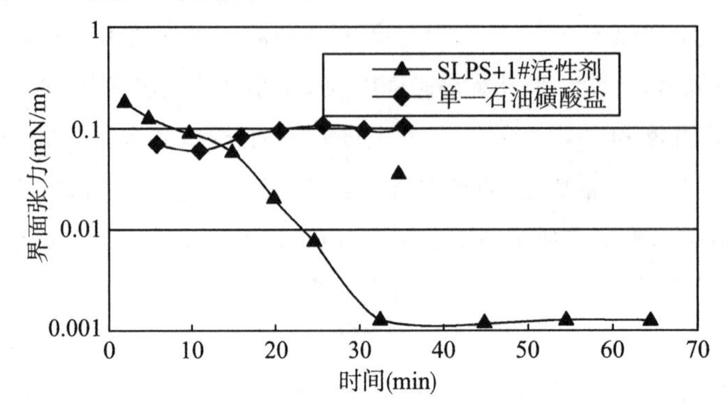

图 1 单一磺酸盐与复配后界面张力比较

2.3 聚合物筛选

针对试验区目的层的油藏条件,对目前市场上的 20 余种聚合物产品进行了性能评价,从而筛选出 10 种性能比较稳定的聚合物,结果认为这 10 种聚合物产品从增粘性上来看,均适合于二元复合驱的需要。

2.4 室内配方筛选

二元复合驱要取得较好的效果,要求体系的界面张力进入超低、配伍性好、吸附量小、热稳定好、驱油效率高。

2.4.1 界面张力试验

将 SPS-1 作为主剂与 1# 活性剂按 3∶1 进行复配,进行了界面张力测定(图 2),结果表明,当 SPS-1 浓度在 0.2%~0.4%范围内,1# 活性剂浓度在 0.05%~0.15%范围内时为最佳活性区。这表明该体系在试验区的油藏条件下有较宽的低张力区,可以满足驱油的需要。

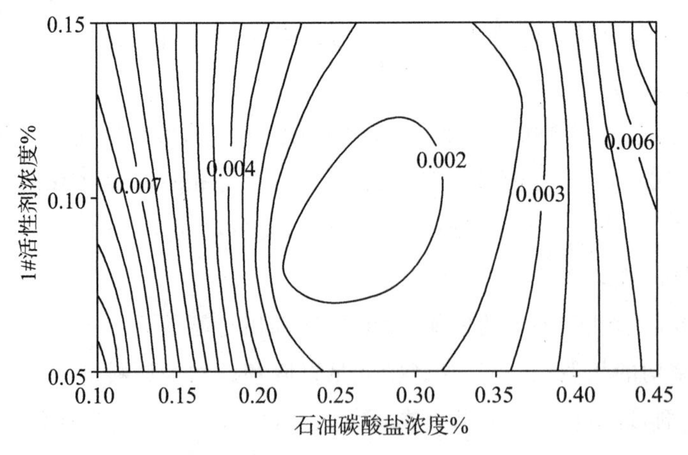

图 2 复配体系界面张力等值图

同时研究了加入聚合物后体系的界面张力情况，结果认为，加入聚合物体系界面张力有所升高，但当活性剂总浓度在0.3%～0.5%范围内时其界面张力仍在10^{-3} mN/m数量级，同时，聚合物是复合驱体系中改善流度比的主要因素，因此加入聚合物是十分必要的。

2.4.2 配伍性试验

当石油磺酸盐体系中加入1#活性剂后，由于增溶作用的存在，溶液的状态有了明显的改善。单一石油磺酸盐体系液面上有明显的分散油，而复配活性剂体系的液面上没有明显的分散油，说明体系配伍性较好。

2.4.3 吸附损耗试验

试验表明，溶液吸附24 h后油水界面张力明显增加。在不改变油砂的情况下，更换两次新溶液，界面张力可恢复到10^{-3} mN/m；改变油砂，而用同一溶液，界面张力继续增大(图3)。

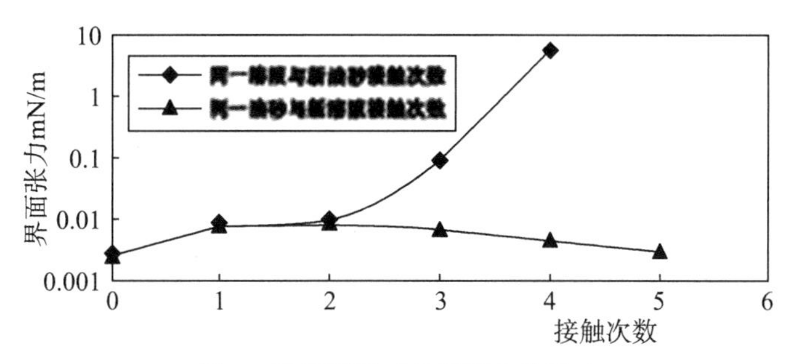

图3　油砂吸附对界面张力的影响

所以该体系存在一定的吸附性，现场注入时一定要确保足够的注入段塞和注入浓度，才能保证复配体系与油水的低界面张力，取得较高的驱油效率。

2.4.4 热稳定性试验

试验表明，复配体系在试验区油藏温度下，粘度及界面张力的热稳定性均较好，能满足现场驱油的需要。

2.4.5 驱油试验

室内模拟试验区的油藏情况，进行驱油试验。分别研究了石油磺酸盐浓度、1#表活剂浓度、聚合物浓度以及不同的化学剂组合方式对采收率的影响。初步推荐驱油配方为：0.3%SPS-1+0.1%1#活性剂+0.15%P，注入段塞0.3 PV。该配方在室内可提高采收率18.1%。

3　方案优化研究

复合驱数值模拟研究采用了改进的SLCHEM软件。注采参数优化是根据正交设计方法[7]和模糊数学原理[8]，在复合驱油藏数值模拟研究的基础上，考虑各项技术、经济指标综合影响，利用基于模糊综合评判模型的优化设计方法对复合驱注采参数进行了优化设计，从而确定出最佳的注采参数。

3.1 配方浓度、段塞尺寸优化

参考室内试验结果,建立二元复合驱注入参数水平取值表(表1),表面活性剂、聚合物浓度以及段塞尺寸均在其合理的取值范围内等间距取四个水平值。

表1 复合驱浓度、段塞优化参数水平取值表

参数 / 水平数	A 表活剂浓度%	B 聚合物浓度%	C 段塞尺寸 PV
1	0.3	0.13	0.25
2	0.4	0.15	0.30
3	0.5	0.17	0.35
4	0.6	0.20	0.40

选用 $L_{16}(4^5)$ 正交表,根据正交设计表可产生16套方案,分别对各方案进行复合驱数值模拟及经济评价,可得到各方案的技术、经济指标。同时,分别对各方案进行模糊综合评判,可得到其综合评判值。表2为正交设计表,它反映了方案中各注采参数的水平数取值及该套方案的综合评判值。从表2中可以看出:① 试验区各注入参数按它们对开发效果影响的敏感程度,依次排序为:聚合物浓度、表活剂浓度、段塞尺寸;② 表活剂、聚合物浓度存在配伍性,该试验区的最优组合为0.4%S+0.17%P;③ 考虑技术和经济综合因素,段塞尺寸存在一个最优值。该试验区的优化结果为0.3 PV。

表2 复合驱注入参数优化正交设计表

方案号	A1	B2	C3	4	5	综合评判值
1	1	1	1	1	1	0.141 1
2	1	2	2	2	2	0.572 0
3	1	3	3	3	3	0.823 5
4	1	4	4	4	4	0.744 7
5	2	1	2	3	4	0.627 9
6	2	2	1	4	3	0.758 0
7	2	3	4	1	2	0.884 5
8	2	4	3	2	1	0.840 0
9	3	1	3	4	2	0.439 0
10	3	2	4	3	1	0.613 2
11	3	3	1	2	4	0.909 1
12	3	4	2	1	3	0.928 3
13	4	1	4	2	3	0.220 4
14	4	2	3	1	4	0.527 3
15	4	3	2	4	1	0.959 5
16	4	4	1	3	2	0.810 4

（续表）

方案号	A1	B2	C3	4	5	综合评判值
I_j	0.570 3	0.654 7	0.654 7	0.381 7	0.388 5	
II_j	0.777 6	0.617 6	0.771 9	0.361 1	0.344 7	
III_j	0.722 4	0.894 2	0.657 5	0.343 9	0.354 7	
IV_j	0.629 4	0.830 9	0.615 7	0.393	0.391 8	
S_j	0.025 8	0.053 9	0.013 6	0.001 4	0.001 7	
较优水平	A2	B3	C2			
因子主次	2	1	3			

3.2 注入方式优化

国内外研究表明，前置牺牲段塞＋主体段塞＋后置保护段塞的注入方式开采效果将优于单一段塞注入方式。其理由在于：增加前置牺牲段塞可减少主段塞中活性剂的吸附，同时提高体系的增溶能力；而设计后置保护段塞可减缓后续水驱的“指进”和“窜流”，以保护主段塞。本方案在保持主体段塞不变的情况下，分别设置了前置保护段塞和后置保护段塞，共设计了四种注入方式，进行优选。从计算结果可以看出（表 3），设置保护段塞后，提高采收率幅度增加，其中以方案 4 提高采收率最高，因此推荐采用方案 4。

表 3　不同的注入方式效果对比表

方案号	注入方式	提高采收率%	折合注聚利用率 t/ t
1	0.3 PV×(0.4%S+0.17%P)	15.76	101.9
2	0.05 PV×0.2%P +0.3 PV ×(0.4%S+0.17%P)	18.35	106.0
3	0.3 PV×(0.4%S+0.17%P) + 0.05 PV×0.15%P	17.23	102.3
4	0.05 PV×0.2%P +0.3 PV×(0.4%S+0.17%P)+0.05 PV×0.15%P	19.72	105.5

3.3 注采速度优化

在优化配方浓度、段塞尺寸的基础上，分别设计了 1 100 m^3/d、1 200 m^3/d、1 300 m^3/d、1 400 m^3/d、1 500 m^3/d 五种注入速度（表 4），结果认为，随注采速度的增大，提高采收率和注聚利用率略有下降，但其开采时间明显缩短。在矿场实施中，考虑到地层注入能力及设备的限制和矿场实验状况，优选注入速度为 1 400 m^3/d。

表 4 不同的注入速度效果对比表

方案号	注入速度	提高采收率 %	折合注聚利用率 t/t
1	1 100	19.97	106.9
2	1 200	19.79	105.9
3	1 300	19.72	105.5
4	1 400	19.69	104.6
5	1 500	19.22	102.9

3.4 推荐注入方案

2003 年 2 月对该先导试验项目进行审查时,行业内专家认为现场实施需考虑以下几方面的影响因素:① 岩石对化学剂吸附比较严重,在驱油过程中化学剂因被岩石吸附使得浓度降低,影响驱油效果;② 岩石吸附对界面张力影响的试验结果也表明,复配体系经岩石吸附后界面张力值明显增大,影响了驱替液的洗油效率;③ 由于本试验区水淹严重,造成化学剂在地下被稀释,使得浓度发生变化影响驱油效果。因此在经济条件允许的情况下要适当增加化学剂的用量。推荐矿场注入方案为采用清水配置母液、污水稀释注入,注入段塞为三段塞注入方式:0.05 PV×0.2%P+0.3 PV×0.45%SPS-1+0.15%1#活性剂+0.17%P)+0.05 PV×0.15%P,注入速度为 0.11 PV/a。

3.5 实施增油效果预测

根据数值模拟预测的复合驱增油规律,并考虑目前试验区的开井率、注采对应率等因素,预测先导试验矿场实施后可提高采收率 12.0%,增产原油 33.3×10^{4} t。

4 结论

试验区能代表胜利油区主力油田注水油藏的开采规律,试验结果具有普遍推广意义。室内筛选出了适合试验区油藏条件的二元复合驱油体系,该体系可与模拟油达到超低界面张力状态,其配伍性好、吸附量较小、热稳定性好。驱油试验表明复合驱体系驱油效果较好。根据正交设计方法和模糊数学原理,在复合驱油藏数值模拟研究的基础上,确定出最佳的注采参数,该先导试验矿场实施后可提高采收率 12.0%。该项技术若能成功,在胜利油区具有巨大的应用前景。

参考文献

[1] 侯吉瑞. 化学驱原理与应用[M]. 北京:石油工业出版社,1998.

[2] 杨承志,韩大伟,等. 化学驱油理论与实践[M]. 北京:石油工业出版社,1998.

[3] 赵福麟. EOR 原理[M]. 北京:石油工业出版社,1991.

[4] Clark S. R. etc. Design and Application of an Alkaline-Surfactant-Polymer Recovery System to the West Kiehl Field, SPE 17538.

［5］宋万超，张以根，等．孤东油田碱-表面活性剂-聚合物复合驱油先导试验效果及动态特点［J］．油气采收率技术，1994，1(2)：51-54．

［6］张贤松．孤岛油田复合驱油扩大试验区试验方案研究和实施［R］．东营：胜利石油管理局，1999．

［7］正交试验设计方法编写组．正交试验设计方法［M］．上海：上海科学技术出版社，1979．

［8］汪诚义．模糊数学引论［M］．北京：北京工业学院出版社，1988．

Abstract: Based on the method used to build GM(1,1) model and its improved method, the paper select two development indexes—annual oil production and water production are selected in the paper as prediction objects and a method based on the grey theory to predict development indexes of oil field is established. On selecting the data range that has the monotony and preprocessing the data properly, we set up the GM(1,1) model to predict development indexes of oil field. After applying the method to Ⅷ and Ⅸ oil sets of Shuanghe oil field, it is proved that the method is applicable for predicting the development indexes.

Keywords: Development Index; Prediction Method; Grey Theory; GM Model

THE STUDY ON SENSITIVITY OF SHA 2 UPPER 4-7 GROUPS IN THE SOUTHERN AREA OF PUCHENG OIL-FIELD/YANG Heshan, ZHONG Jian-hua, WABG Yong, et al. Institute of Earth Resources and Information, University of Petroleum (East China), Dongying, 257061/ Xinjiang Shiyou Xueyuan Xue bao, 2004, 16(3): 32-36.

Abstract: the sensitivity of reservoir, which is discussed here, is actually an outer expression of the inner factors of the reservoir such as the rock's fabric, the integration, the characteristics of the pores and throats. With the help of the sensitivity test on the study groups-Sha 2 upper 4-7 groups in the southern Area of Pucheng Oil-field, a conclusion is arrived at that the low permeable sha 2 upper 4-7 groups are characterized by their own behaviors in the sensitivity as following: non speed sensitivity, a moderate to slightly heavy water sensitivity, a slightly strong to weak alkali sensitivity, an ultimate strong acid sensitivity, a strong to moderate salt sensitivity. What's more, the correlation between the inner factors of the reservoirs and the experimental results on the sensitivity has been analyzed, and it is believed that it is the clay minerals and the characteristics of the pores and throats that afect the sensitivity of the reservoir in the studied area, and ofer some good advice for the protection of the reservoir in development.

Keywords: reservoir; sensitivity; clay minerals; formation damage

STUDIES ON THE SENSITIVITY EXPERIMENTS OF LOW PERMEABILITY RESERVOIR ON PU67-BLOCK SHASANZHONG RESERVOIR OF ZHONGYUAN OIL FIELD/ SONG Han-hua, WANG Youu, JIANG Hou-yan, et al. Petroleum Engineering College, ChangJiang University, Jinzhou Hu Bei 4340231. /Xinjiang Shiyou Xueyuan Xue bao,2004,16(3): 37-39.

Abstract: The studies of the five-sensitivity experiments in low permeability reservoir are presented in this paper, and the experiments are studied on the basis of ShaSanZhong 6-10 reservoir of PU67-Block of ZhongYuan PuCheng oil filed. And the results show that the damage degree of velocity sensitivity is weak in this reservoir, and its damage degree of water sensitivity is middling weak, and at the same time this reservoir is damaged owing to the low acid sensitivity and the low alkali sensitivity.

Keywords: the sensitivity of reservoir; clay mineral; nonclay mineral; the protection of hydrocarbon reservoir

STUDY ON OIL FLOOD SYSTEM AND PLAN OPTIMIZATION IN THE PILOT TEST OF SURFACTANT/POLYMER FLOOD IN SHENGLI OIL FIELD/ ZHANGAi-mei, CAOXu-long, LI Xiu-lan et al. Research Institute of geological Science, ShengLi oiield Company Ltd/Xinjiang Shiyou Xueyuan Xue bao, 2004, 16(3): 40-43.

Abstract: Coupling reservoir characteristic in Shengli oil field with surfactant polymer flood peculiarity, pilot test area is selected in southwest Ng54-61 of District Seven in Gudong oil field. And surfactant polymer flooding system is ascertained in experiment laboratory which is adaptable to this type of reservoirs. The main surfactant is petroleum sulfonate. This system possesses ultralow interfacial tension with the crude oil. Its adsorbance is little. Its emulsibility and thermal stability is fine. And its displacement efficiency is high. By numerical simulation, such parameters as combination manner of chemical agent, surfactant concentration, polymer concentration, slug size, injecting manner and injection rate are optimized. For the surfactant polymer flooding system, its displacement eficiency is high. It has strong reservoir applicability and could enhance oil recovery greatly.

Keywords: Shengli oil field; surfactant polymer flood; petroleum sulfonate; chemical flooding; pilot test

APPLICATION OF HYDRAULIC NUIT ANALYSIS IN HYDRAULIC FRACTUREING OF LOW PERMEABILITY RESERVOIR/GAO Hai-hong, QU Zhan-qing, SHANG Zhao-hui, ZHAO Gang. Petroleum Enginer college of University of Petroleum, DongYing ShanDong 257061 China/Xinjiang Shiyou Xueyuan Xue bao, 2004, 16(3): 44-47.

Abstract: Hydraulic fracturing is effective well stimulation in low permeability reservoir. Zhuang 74 unit of zhuang xi oil field is very low in permeability, and is varied greatly in facies belt and properties of layers. to increasee the efect of the hydraulic and the accuracy of fracture numerical simulation so as to optimize fracturing parameters, the anisotropy must be taken into consideration. This paper applies hydraulic unit analysis to zhuang 74 unit, divides it into four hydraulic units, combines with numerical simulation, optimizes fracturing parameters, gets better result on the spot. This method is very instructive in low permeability.

Keywords: Hydraulic unit; Low permeability reservoir; Hydraulic fracturing; Numerical simulation

STUDY ON THE ACID FRACTURING PIPE STRING OF TAHE OIL FIELD/ MA Wei-rong, PEI Fui-lin, Zhang Bong, et al. Oil Production Technology Research Institute Of Xinjiang Oil Field, Karamay, 834000 China/Xinjiang Shiyou Xueyuan Xue bao, 2004, 16(3): 48-52.

Abstract: The technique of acid fracturing of Tahe oil field Ordovician system has been a dominant crafts among the oil reservoir exploitation after six years eforts. The invalidation causes of pipe string have been analysed in this paper, the main causes have also been found out. The new pipe string adapting to Tahe oil field is designed and proofread. On-the-spot test has been carried out, which acquired better effect. The next ameliorative advice is also put forward in this paper.

Keywords: Tahe oil field; acid fracturing; pipe string; open hole packer; hydraulic anchor

THE APPLICATION OF EMULSIFIED ACID FRACTURING IN TAHE OIL FIELD/JI Chuaniang, YUAN Fei, TAN Fang et al. Engineering supervision Center of Northwest Bureau, SINOPEC, Urumqi, Xinjiang, 830011 China/Xinjiang Shiyou Xueyuan Xue bao, 2004, 16(3): 53-55.

Abstract: According to the property of high temperature and deep well acid fracturing of Tahe oil field Ordovician system carbonate oil deposit, the emulsified acid prescription of the density of 28% HCL has been developed, and every technical parameter has been determined, which has got better efect in field application. Key Words: emulsified acid; property of retard; capability of emulsification; acid fracturing

TECHNICAL COUNTERMEASURES FOR WATER CONTROL AND OIL STABILIZATION IN HIGH WATER-CONTAINING EXPLOITATION PERIOD OF

GRIT FRAME RESERVOIR/LIChen, YANGXiao-li, et al. Oil Recovery Plant three, XinJiang Oiield Company PetroChina, Karamy XinJiang 834007/Xinjiang Shiyou Xueyuan Xue bao, 2004, 16(3): 56-59.

Abstract: For high containing water exploitation period of grit frame reservoir, because of non-uniformity, the onrush of water injected by single orientation, it is very limited for utilizing water injected and the efect of water drive is inadequate. However, it is very difficult to break such adverse phenomenon for ordinary ways of inject water. So, by dint of the knowledge of reservoir exploitation contradiction in different phase, after scientific investigation for potential of remanent oil, and guided by the correlative theory and mechanism of non-uniform reservior and grit frame reservoir exploitation, we adopt many efective on-the-spot work and have brought about continuous stable production of Triassic.

耐温抗盐交联聚合物驱油体系性能评价*

刘　坤　宋新旺　曹绪龙

胜利油田有限公司地质科学研究院

摘要：研究了部分水解聚丙烯酰胺(聚合物)与耐温抗盐交联剂所形成的流动凝胶体系的性能，并进行了物理模拟试验。结果表明，体系对聚合物本身的性能要求宽松，水解度为4%～30%、相对分子质量大于900×10^4的聚合物干粉成胶性能均较好；体系对温度敏感，温度较低时，体系不发生反应，所以有足够的时间配制溶液，且耐高温(90℃)，可用于地层水矿化度为50 000 mg/L的地层。该体系由有机试剂组成，与岩石接触后不产生离子交换、沉淀反应；吸附小，与聚合物以共价键结合，形成的凝胶性能稳定，具有较高的粘弹性，可大幅度提高采收率。

关键词：耐温抗盐交联剂；耐温抗盐；流动凝胶；物理模拟；提高采收率

中图分类号：TE357.431　文献标识码：B　**文章编号**：1009-9603(2004)05-0065-03

引言

在聚合物溶液中加入交联剂使其在地层温度下缓慢交联形成流动凝胶，可以提高单一聚合物的性能[1,2]。研制的耐温抗盐交联剂[3]与聚合物形成高粘弹性流动凝胶，可有效堵塞大孔道，降低驱替相渗透率，提高低压水井的注入压力，实现深部调剖与驱替相结合，不仅可以提高油层驱替介质的粘度，进一步减缓聚合物驱替中的指进和舌进现象，还可扩大波及体积，提高聚合物驱效果。由于该体系具有耐温抗盐的特点，对胜利油区高温高盐油藏提高采收率具有积极意义。

1　驱油机理

耐温抗盐交联体系的交联机理比较复杂，交联方法多种多样，有化学交联，有物理交联，还有链之间的缠结等。主要作用机理是醛类物质与水发生反应生成甲撑二醇，甲撑二醇与酚类物质缩合生成了水杨醇中间体[4]。水杨醇、醛与聚合物直接反应形成热稳定性的凝胶。由于酚中的-OH基很活泼，因此苯酚首先生成邻-羟甲基苯酚，或称之为水杨醇，该产物经进一步缩合反应生成化合物Ⅰ，聚合物中的酰胺基也能与甲撑二醇缩合生成化合物Ⅱ。两者进一步缩合成三维结构体。化合物Ⅰ也可与聚合物中的酰胺基缩合形成类似的三维结构体。

耐温抗盐交联聚合物驱油体系是由交联剂和聚合物组成,在地层条件下发生交联反应,以共价键形式形成三维网状结构,且成胶时间和凝胶的强度可以控制,既保证了成胶的速度,又保证了成胶后的长期稳定性[5-7]。

2 性能评价

2.1 对聚合物的适应性

对不同相对分子质量和水解度的聚合物产品(水解度都为30%,相对分子质量为926×10^4~$2\ 050\times10^4$)的适应性进行了研究,所用水的矿化度为5 727 mg/L,温度为70℃,粘度测定采用RS-150流变仪。在聚合物和交联剂浓度分别为1 000 mg/L时进行试验,试验结果表明,随着聚合物相对分子质量的增加,交联时间缩短,交联后粘度增加(图1)。这是因为随着聚合物相对分子质量的增加,所形成的交联聚合物的交联点增多,交联密度增加,因此交联时间缩短,凝胶粘度增大,但增至一定程度后,反应达到平衡,粘度趋于稳定。同样,在相对分子质量为$1\ 500\times10^4$,水解度为4%~14%,聚合物和交联剂浓度都为1 000 mg/L时进行试验,结果表明,随着水解度的增加,交联时间延长,凝胶粘度减小。达到平衡时,水解度为4%,14%和30%时体系粘度分别为1 400,1 050和660 mPa·s,因为聚合物中与交联剂进行交联反应的主要是酰胺基,水解度越高,可反 应的酰胺基团越少,所以凝胶粘度减小。

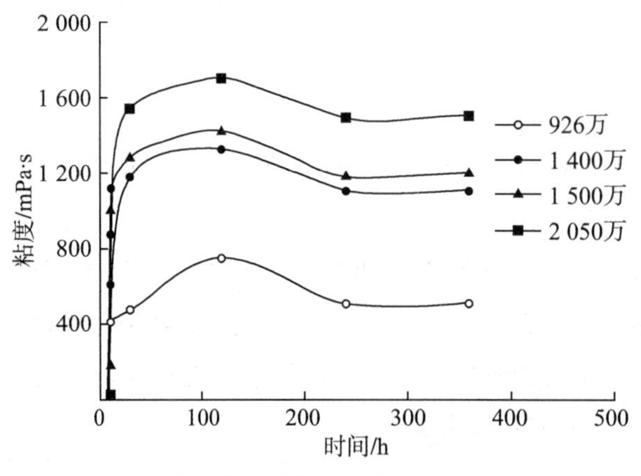

图1 聚合物相对分子质量对体系的影响

2.2 对油藏的适应性

2.2.1 耐高温

在聚合物和交联剂浓度均为1 000 mg/L,60~90℃条件下进行试验,随着温度的升高,反应速度加快。体系对温度非常敏感,温度较低时,体系不成胶或成胶时间很长,所以有足够的时间配制溶液,且可泵性好,可耐90℃高温(图2)。

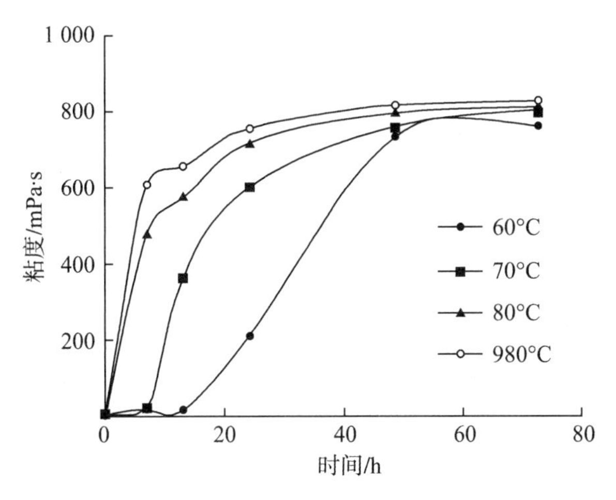

图2　温度对体系的影响

2.2.2 矿化度适用范围

聚合物和交联剂浓度均为1 000 mg/L，在矿化度为700～59 802 mg/L条件下试验。随着矿化度的增高，反应速度加快，凝胶粘度变化不大，体系稳定后粘度为800 mPa·s左右。该有机体系具有一定的抗盐性。

2.2.3 交联剂的吸附情况

许多交联体系在无油砂条件下交联稳定性很好，若添加地层油砂后，由于pH值的改变而不能成胶。所以选用了地层油砂对体系进行试验。试验用聚合物浓度为750和1 500 mg/L两种体系，固液比为1∶5，加油砂与不加油砂进行试验对比，样品混匀后在地层温度下放置，定期测量其粘度变化(图3)。同样配方下，由于吸附的影响，加砂体系交联时间比没添加油砂的体系略长，粘度稍低一些，但仍能形成稳定的凝胶。该体系与岩石接触后不产生离子交换和沉淀反应，吸附小，通过岩层时仍能很好交联。

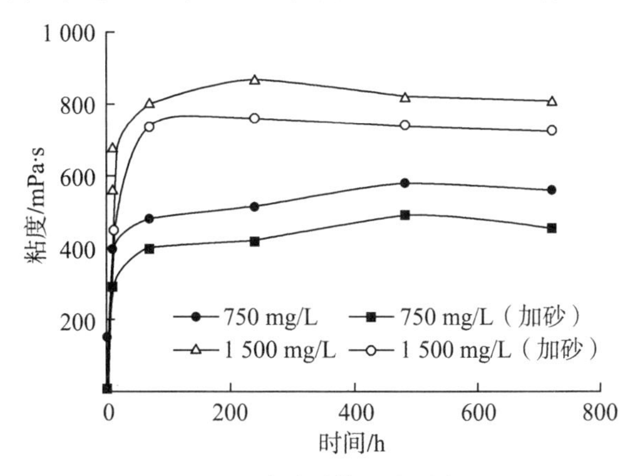

图3　油砂对体系的影响

2.3 体系的粘弹性

定量评价聚合物交联后的特性是比较困难的，由于“爬杆”现象的影响，用一般旋转粘度计测出的数据很不稳定，而且并不能代表体系的真正粘度，只能半定量地判定交联情况。由于这种体系交联后所形成的凝胶不是纯粘性流体,而是一种粘弹性流体,为此采用流变仪对被测样品施加一振荡信号,测量体系交联后粘性和弹性随频率的变化情况。粘性模量和弹性模量可以直观地对比粘性和弹性所占的比例,而且测量简单、快速。实验中对比了交联聚合物驱油体系与单一聚合物体系的粘弹性,单一聚合物体系的粘性模量和弹性模量均为 0.2 Pa 左右,而交联后粘性模量为 0.8 Pa 左右,弹性模量为 2.5 Pa 左右。粘弹性远好于单一聚合物,其中弹性部分在体系中占主导地位,因而体系流动性好,可塑性强。体系在注入时粘度小,首先进入高渗透层,凝胶形成后,由于粘性和弹性的提高,迫使后续注入液流转向进入低渗透区,从而达到封堵高渗区的目的。

2.4 抗剪切能力

凝胶体系中聚合物 3530S 的浓度为 1 500 mg/L,交联剂浓度为 800 mg/L,温度为 70℃。用两种方法考察体系的抗剪切能力:① 先经过 15 000 s^{-1} 的剪切速率剪切后(毛管剪切),再放入保温箱;② 样品放入保温箱成胶后再经过 15 000 s^{-1} 的剪切速率剪切,成胶前进行剪切与不剪切的体系粘度几乎相同;③ 凝胶形成以后再剪切,粘度有所下降,但粘度远大于单一聚合物的粘度。体系注入地层后的高剪切区在炮眼附近,此时体系还没有成胶,所以炮眼剪切不会对体系造成大的影响。

2.5 物理模拟试验

物理模拟试验可以综合反映体系的驱油效果,该试验采用的是 30 cm 长的均质管式模型和三维非均质模型,非均质模型体积为 $500\times500\times100\ mm^3$,对角方向渗透率为 $(3\,000\pm300)\times10^{-3}\ \mu m^2$,两侧渗透率为 $(500\pm200)\times10^{-3}\ \mu m^2$。试验中注入 0.3 倍孔隙体积聚合物和 0.2 倍孔隙体积交联聚合物。结果表明,均质管式模型聚合物驱比水驱提高采收率 11.5%,交联聚合物比水驱提高采收率 19.7%;而三维非均质模型聚合物驱比水驱提高采收率 14%,交联聚合物比水驱提高采收率 34.2%,表明聚合物驱改善非均质效果明显(表 1)。

表 1 均质管式模型与三维非均质模型结果对比

驱油方式	均质管式模型		三维非均质模型	
	最终采出率,%	提高采收率,%	最终采出率,%	提高采收率,%
水驱	56.0		24.0	
聚合物驱	67.5	11.5	38.0	14.0
交联体系	75.7	19.7	58.2	34.2

3 结论

所研制的耐温抗盐交联聚合物驱油体系适应性强,不同水解度和不同相对分子质量的聚合物在一定浓度范围内都能够形成稳定的凝胶,体系的耐温抗盐性能好,粘弹性大,抗剪切能力强,调节油藏非均质性的能力强,可以大幅度提高石油采收率。

参考文献

[1] 杨承志.化学驱提高石油采收率[M].北京:石油工业出版社,1999.

[2] 刘一江,王香增.化学调剖堵水技术[M].北京:石油工业出版社,1999.

[3] 刘坤.耐温抗盐交联聚合物驱油体系研究[J].精细石油化工进展,2003,4(12):1-4.

[4] Moradi-Arahi A, Bjomson C, Doe P H. Thermally stable gels for near-wellbore permeability contrast modifications. SPE Advanced Tech-nology Series,1993,11(1):140-145.

[5] Albonico P, Bartosek M, Lockhart T P et al. New polymer gels for re-ducing water production in high temperature reservoirs. SPE 27609.

[6] Moradi-Arahi. Altering high temperature subteranean formation per-meability. USP 4994194.

[7] Seright R S. Impact of permeability, and lithology on gel perform-ance. SPE/DOE 24190.

编辑 高 岩

Multipore characteristic appears in carbonate fracture-cave reservoirs. The fracture-cave system with strong heterogeneity determined complex oil-water movement and affected the field development level. By analyzing the physical properties of reservoir bodies, reservoir fluid features and karst-cave distribution rule in Tahe carbonate oilfield, the reservoir bodies were classified to ancient near-surface karst reservoir belts and ancient karst channel reservoir belts. Combined with the field practice in Tahe oilfields, the change rule of oil production and water cut and formation energy in Tahe carbonate oilfield was summarized. The distribution features and well spacing of well pattemn of the ancient near-surface karst reservoir belts and ancient karst channel reservoir belts were studied. The possibility of waterflooding development, rational single well production rate and oil recovery rate were discussed.

Key words: fracture-karst reservoir bodies, develop-ment strategy, well pattem, well spacing, development scheme, Tahe oilfield

Zhang Guoping, Xiao Liang, Hu Yanxia et al. Applications of in situ gas generating for enhanced oil recovery to Zhongyuan faulted block oilfield. PGRE, 2004, 11(5): 60-61.

Technical mechanism and laboratory core simulation experiment of in situ gas generating for enhanced oil recovery are introduced. Typical block is analyzed in detail. It is used in the field for 21 well-times. Average water injection pressure reduced 9.5MPa, the average waterflood amount of single well increased 5 321.9 m^3, the cumulative waterflood amount increased 111 759 m^3, and cumulative incremental oil of oilwells is 15 187.49 t. The field experiment shows that the application of this technology of in situ gas generating plays positive role in controlling water, stabilizing oil production and in controlling natural decline in partial block and well group, which is a new technology of EOR with applica-tion prospect.

Key words: in situ gas generating, carbon dioxide, oil displacement by pressure reducing, enhanced oil recovery

Wang Haifeng, Wu Xiaolin, Zhang Guoyin et al. Research progress on the surfactants for ASP flooding in Daqing oilfield. PGRE, 2004, 11(5): 62-64.

Research progress on the surfactants for ASP flooding in Daqing oilfield in recent years is summarized. The researches and applications of surfactants such as petroleum sulfonate, petroleum carboxylate, lignosulfonate, alkyl benzene sulfonate and biosurfactant are introduced respectively. Based on that, the direction of the future research and progress of the surfactants for ASP flooding in Daqing oilfield is pointed out.

Key words: ASP flooding, surfactant, research pro-gress Liu Kun, Song Xinwang, Cao Xulong. Property evaluation of oil displacement system with temperature-salt resisting crosslinked polymer. PGRE, 2004,11(5): 65-67.

The property of flow gel system formed by partially hy-drolyzed polyacrylamide (PHPA)and temperature-salt resisting crosslinker is studied, and then physical simulation experiment is conducted. Result indicates that this system is widely suitable for the polymer proper. The gel-forming property of the polymer dry powder is good when the hvdrolysis degree is 4%～30% and the relative molecular weight is above nine millions. The system is sensitive to the temperature. When the tem-perature was lower, the system had no reaction, so the time was enough for preparing solution and the system could resist the higher temperature (90℃) applying to the stratum with 50 000 mg/L of the formation water sa-linity. This system made up of organic reagents, when it met the rock, there was no ionic exchange and no deposition reaction, and the adsorption was little. When the system combined with the polymer by covalent bond, the gel property was stable with high viscoelasticity, so the recovery efficiency can be enhanced by a big margin.

Key words: temperature-salty resisting crosslinker, temperature-salty resisting, flow gel, physical simula-tion, enhance recovery efficiency

Zhang Aimei. Standard revision and evaluation of resources classification for polymer flooding in Shengli petroliferous province. PGRE, 2004, 11(5): 68-70.

Based on the analysis of polymer flooding reservoirs, combining with laboratory experiment and numerical simulation, classification standard of polymer flooding resources was revised for Shengli petroliferous prov-ince. By studying such factors as crude oil viscosity, formation temperature, air permeability, salinity of formation water, ion concentration and reserve scale, new classification standard is established. And the polymer flooding resources in Shengli petroliferous province were evaluated and classified by the new standard.

胜利石油磺酸盐驱油体系的动态吸附研究*

王红艳[1,2] 曹绪龙[2] 张继超[2] 田志铭[2]

(1. 西南石油学院，新都 610500；2. 中石化胜利油田有限公司地质科学研究院，东营 257015)

摘要：石油磺酸盐表面活性剂驱油时在油藏中的吸附损耗、色谱分离会直接影响复合驱油体系协同效应的发挥，从而影响驱油效果。为此，用动态吸附实验考察了石油磺酸盐表面活性剂浓度对油砂吸附量的影响；考察不同长度填砂模型对胜利石油磺酸盐与助剂1#混合物的色谱分离现象。结果表明，胜利石油磺酸盐浓度在10 g/L以上时，油砂吸附量最大达4.8 mg/g；胜利石油磺酸盐与助剂1#之间存在色谱分离现象，且填砂模型越长，分离越明显，说明现场应用时，只要适当增加注入液浓度，复配协同效应可得到发挥。

关键词：胜利石油磺酸盐 助剂 吸附 色谱分离

石油磺酸盐是原油磺化得到的产品，具有较好的水溶性和耐盐性，是三次采油用量最大的表面活性剂，具有较好的应用前景[1]。为减轻油田对外部化学剂市场依赖，以胜利原油为原料，研制出高效廉价的石油磺酸盐表面活性剂，并初步实现石油磺酸盐产业化，为以石油磺酸盐为主表面活性剂的复合驱的顺利实施奠定了基础。

石油磺酸盐驱油时在油藏中的吸附损耗直接影响驱油效率，研究其在油砂上的吸附特征对复合驱配方和驱油机理的研究具有十分重要的意义[2-3]。由于不同表面活性剂的吸附、脱附能力不同，导致注入流体在渗流过程中的分离。复合驱油体系靠协同作用显著降低油水相界面张力，但色谱分离会影响复配协同作用的发挥，从而影响驱油效果。为此，笔者对孤岛Ng4二元驱中石油磺酸盐与助剂1#在油藏温度条件下的动态吸附规律与色谱分离进行考察，为驱油配方设计及表面活性剂用量计算提供参考。

1 实验部分

1.1 实验材料

吸附剂：石英砂，孤岛反排油砂；胜利石油磺酸盐(SLPS)：胜利油田设计院生产，平均相对分子质量425；助剂1#：胜利京大公司提供，平均相对分子质量400。

1.2 实验方法

动态吸附实验在不同直径及不同长度的填砂模型上进行,孔隙体积 65 mL。注入 0.3 PV 的驱替液,然后转水驱,检测驱替液出口含量,含量为零时结束实验。

2 结果与讨论

2.1 胜利石油磺酸盐最大吸附量研究

实验在内径 0.5 cm,长 2 m 的填砂模型中进行。油砂表面的正电位是引起表面活性剂吸附的主要原因。油砂与溶液接触过程中,表面活性剂与实验水中以及从油砂上交换下来的二价离子相互作用发生沉淀,也是表面活性剂损失的重要原因[4]。石油磺酸盐表面活性剂浓度对油砂吸附量的影响见图 1。

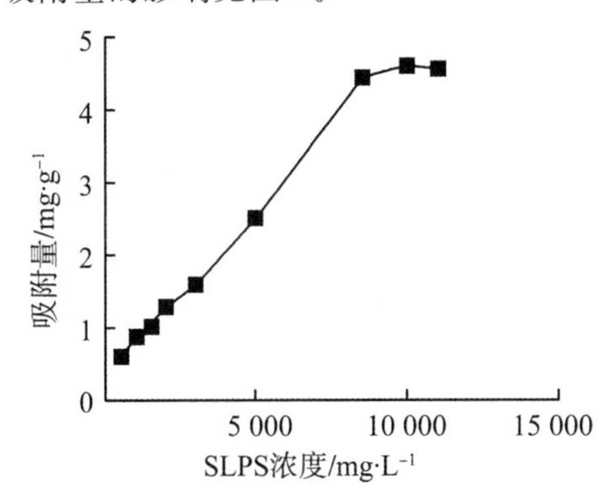

图 1 胜利石油磺酸盐浓度对吸附量的影响

由图 1 可见,当胜利石油磺酸盐浓度在 500~8 500 mg/L 时,SLPS 浓度与油砂吸附量呈线性关系,此时 SLPS 浓度低于其临界胶束浓度。浓度高于 10 000 mg/L 时,吸附量基本不再随浓度的变化而变化,此时吸附量为胜利石油磺酸盐被油砂吸附的最大量(4.8 mg/g)。实验中观察到,达到吸附量最大的时间也取决于注入速度,因此,实验中应取油藏实际注入速度。

2.2 胜利石油磺酸盐的动态吸附研究

图 2 为胜利石油磺酸盐在直径 2.5 cm,长 30 cm 管式模型上的动态吸附曲线。曲线可分为 3 个阶段,Ⅰ阶段以离子扩散为主;Ⅱ阶段以对流为主,是石油磺酸盐产出的主体部分;Ⅲ阶段表现为吸附、脱附,曲线出现了拖尾现象。

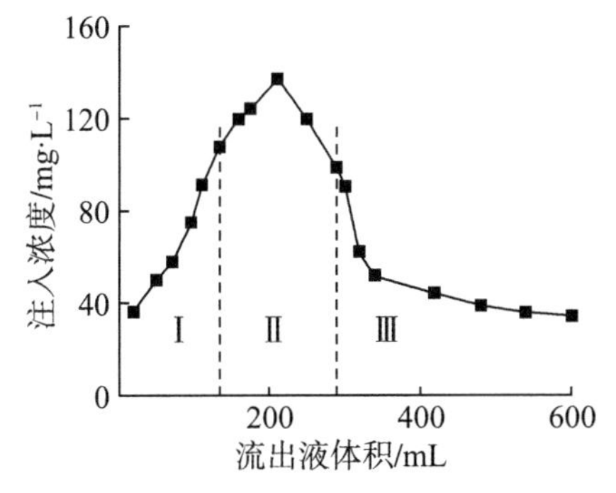

图 2 胜利石油磺酸盐动态吸附曲线

2.3 胜利石油磺酸盐复配体系的色谱分离研究

由于三元复合驱中碱的加入会带来结垢和乳化,给注入和采出液处理带来困难[5],因此胜利油田开展了二元驱现场试验。二元驱中各组分在地层运移过程中表现出的吸附滞留与色谱分离已成为影响复合驱效果的重要因素,所以必须对体系中各组分之间的色谱分离进行研究,以确定其最终注入浓度。为此,分别在内径均为 1 cm,长分别为 50 cm、100 cm 细管填砂模型中注入 0.4%胜利石油磺酸盐与 0.15%助剂 1# 的混合物,考察复配体系的动态吸附规律及色谱分离现象,结果分别见图 3 和图 4。

由图 3 和图 4 均可看出,两种活性剂之间存在一定的色谱分离现象,而且填砂模型越长,色谱分离越明显,在 100 cm 长填砂模型中两种活性剂峰值到达时间相差约为 0.7 PV,说明只要适当增大注入液浓度,两者的复配协同效应可以发挥作用,只是在驱替液前沿由于地下水的稀释与吸附作用的存在,两者的协同作用会受到一定影响。

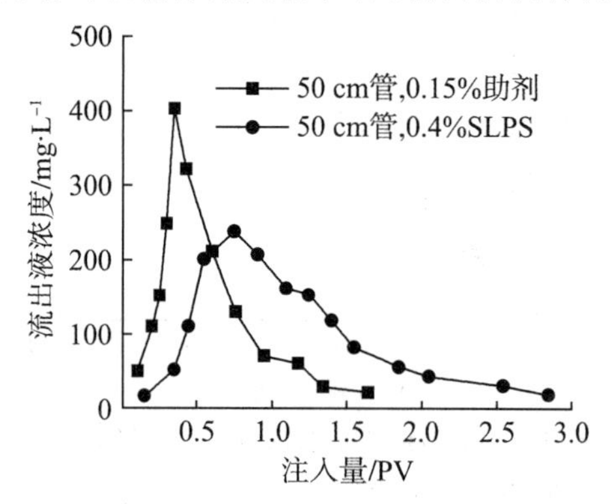

图 3 复配体系在 50 cm 模型上色谱分离现象

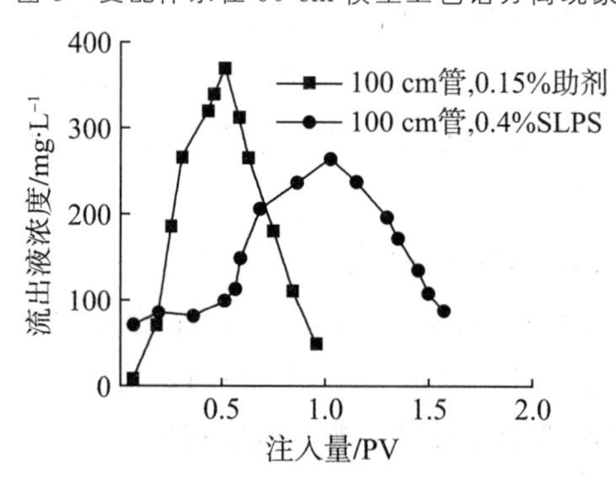

图 4 复配体系在 100 cm 模型上色谱分离现象

3 结论

(1) 胜利石油磺酸盐表面活性剂浓度对油砂吸附量影响的考察结果表明,胜利石油磺酸盐浓度在 10 000 mg/L 以上时,油砂对其吸附量基本稳定,不再随浓度的增加而增加,此时油砂吸附量最大,达 4.8 mg/g。

(2) 由 0.4%胜利石油磺酸盐与 0.15%助剂 1# 组成的混合物进行动态吸附规律及色谱分离现象考察,结果表明胜利石油磺酸盐与助剂 1# 之间存在色谱分离现象,现场应用中只要适当增加注入液浓度，两者的复配协同效应可以得到发挥。

参考文献

[1] 康万利.大庆油田三元复合驱化学剂作用机理研究[M].北京：石油工业出版社，2001：4-13.

[2] 杨承志，韩大匡.化学驱油理论与实践[M].北京：石油工业出版社，1996：208-210.

[3] 俞稼镛，宋万超.化学复合驱基础及进展[M].北京：中国石化出版社，2002：173-178.

[4] 杨普华，杨承志.化学驱提高石油采收率[M].北京：石油工业出版社，1988：152-166.

[5] 赵福麟.EOR原理[M].东营：石油大学出版社，2001：131-142.

Dynamic Adsorption of Flooding System Comprising Shengli Petroleum Sulfonate

Wang Hongyan[1,2]　Cao Xulong[2]　Zhang Jichao[2]　Tian Zhiming[2]

(1. Southwest Petroleum Institute, Xindu 610500;

2. Geological and Scientific Research Institute of Shengli Oilfield, SINOPEC, Dongying 257015)

Abstract: The adsorption and chromatographic separation of Shengli petroleum sulfonate(SLPS) had negative effects on the synergism of compounded flooding system and recovery of oil. The changes of adsorption capacity with the concentration of SLPS were examined by using dynamic adsorption experiment, and chromatographic separation of SLPS and 1# assistant surfactant was also studied. The result showed the maximum adsorption of SLPS was up to 4.8 mg/g at the concentration over 10 g/L, and there was chromatographic separation between SLPS and 1# assistant surfactant, which didn't affect the recovery by increasing the surfactant concentration in flooding liquid.

Key Words: Shengli petroleum sulfonate, assistant surfactant, adsorption, chromatographic separation

复合化学驱油体系吸附滞留与色谱分离研究

王红艳[1,2] 叶仲斌[1] 张继超[2] 曹绪龙[2]

(1,西南石油大学石油工程学院,四川成都 610500;2.中石化胜利油田有限公司地质科学研究院,山东东营 257015)

摘要:主要针对胜利石油磺酸盐与助表面活性剂的二元驱体系中,研究其在油藏条件下,驱油体系的静、动态吸附规律与色谱分离特征。结果表明:胜利石油磺酸盐在低浓度下为线性吸附,最大静态吸附量为4.6 mg/L,加入助表面活性剂可以有效降低其吸附量;石油磺酸盐与助表面活性剂存在一定色谱分离,且二元驱中聚合物的加入加大了这种色谱分离效应。试验证明,只要适当增大注入液中活性剂浓度,二者的复配协同效应可以得到发挥。

关键词:石油磺酸盐;吸附滞留;色谱分离;二元驱

中图分类号:TE357.43 文献标识码:A

引言

石油磺酸盐由于其成本低,具有较好的水溶性和耐盐性,故常作为驱油剂使用[1]。为了减轻对油田外部化学剂市场的依赖,胜利油田已实现了石油磺酸盐的产业化,建成了年产2万吨的生产能力,为以石油磺酸盐为主剂的复合驱的顺利实施奠定了基础。复合驱中各组分在地层运移过程中表现出的吸附滞留与色谱分离已成为影响复合驱效果的重要因素[2-3]。由于不同表面活性剂的吸附、脱附能力不同导致了注人流体在渗流过程中的分离。复合驱油体系是靠超加合作用达到显著降低油水相界面张力的效果的,但色谱分离的存在会影响到超加合效应的发挥,从而影响到驱油效果。

目前0.15%聚合物+0.4%SLPS+0.15%1#的二元复合驱先导试验在孤东七区西已经取得明显的降水增油的效果。本文主要针对胜利石油磺酸盐(SLPS)与助剂1#,研究其在油藏条件下,驱油体系在不同介质上的静、动态吸附规律与色谱分离情况,为驱油配方设计及活性剂用量计算提供理论指导。

1 试验条件与方法

1.1 试验材料

吸附剂：石英砂（30～60 mm，63%；60～80 mm，26%；80～150 mm，11%）；孤岛油砂。

SLPS：胜利设计院生产（平均分子当量为425），活性物含量为34%；挥发份为18%；未磺化油含量为10%；无机盐及杂质为3%。

助剂1#：胜利海峰公司提供（平均分子当量为400），活性物含量为62%。

SLPS与助剂浓度测定方法：HPLC法[4]。色谱柱：SAX阴离子交换填料；流动相：A液甲醇/水＝60/40，B液甲醇/0.2 mol/LNaH_2PO_4（60/40）；流速为1.0 ml/min。

1.2 静态吸附特征研究试验方法

将一定量油砂与表面活性剂溶液混合后，放人水浴振荡器中振荡，取上层清液用一次性0.45 μm微孔过滤膜过滤后用高效液相色谱法测定其剩余浓度，计算吸附量。试验条件：温度70℃；固液比1∶3；振荡时间24 h。

1.3 动态吸附特征及色谱分离研究试验方法

动态吸附试验在直径为1.5 cm、不同长度的模型上进行。注入0.3 PV的驱替液，然后转水驱，检测出口浓度至浓度为0时结束试验。

2 结果与讨论

2.1 SLPS的静态吸附特征研究

2.1.1 低浓度下SLPS的吸附特征

固体表面的正电位是引起活性剂吸附的主要原因，在油砂与溶液接触过程中，表面活性剂与试验水中以及从油砂上交换下来的二价离子相互作用，生成的沉淀也是表面活性剂损失的重要原因[5]。低浓度条件下SLPS的吸附曲线表明：在低浓度条件下，石油磺酸盐的浓度低于其CMC值，以单分子形式在固体表面吸附，未达到饱和，基本呈线性吸附，曲线相关系数0.992。试验表明当浓度高于10 000 mg/L时，吸附量基本不再随浓度而变化。

SLPS在石英砂上的吸附并没有表现出最大吸附特征。这主要是由于SLPS是一个由挥发分、无机盐、未磺化油、活性组份及其他杂质组成的一个混合物，这些组份对SLPS的吸附量及吸附等温线均产生了影响。

2.1.2 吸附时间对SLPS吸附的影响

SLPS吸附量随着吸附时间的增加而增加，但是在大于20 h后增加幅度减小，到35 h后已经基本达到平衡。

2.1.3 固液比对SLPS吸附的影响

理论上固液比值对表面活性剂吸附等温线的形状及吸附量基本没有影响[3]。如图

1 中固液比为 2∶1 和 1∶1 时的吸附等温线。但是从图中固液比为 1∶3 和 1∶10 时的吸附等温线可以看出：随固体数量的增加，SLPS 吸附量减少。这是由于溶液中的胶束排斥作用的结果。

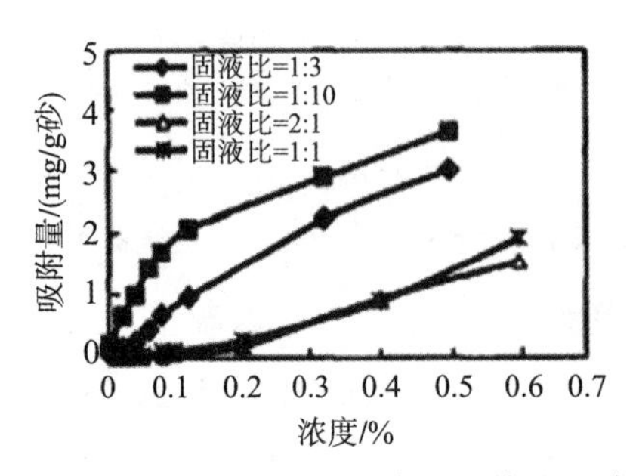

图 1 固液比对 SLPS 吸附等温线的影响

2.1.4 矿化度对 SLPS 吸附的影响

SLPS 将在胜利油田不同区块开展现场试验，因此开展了其在不同矿化度条件下的吸附特征研究。分别用不同矿化度的水配置 SLPS 样品，TDS＝5 000 的水中 Ca^{2+} + Mg^{2+} ＝80 mg/L，TDS＝6 188 的水中 Ca^{2+} + Mg^{2+} ＝78 mg/L，TDS＝8 246 的水中 Ca^{2+} + Mg^{2+} ＝276 mg/L。试验结果见图 2。

由图 2 可见，随着矿化度的增加，SLPS 的吸附量增加。无机电介质增加了表面活性剂在固-液界面上的吸附。这种增加一方面是盐的加入引起临界胶束浓度降低引起的，另一方面水中钙、镁离子与石油磺酸盐的沉淀损失也是吸附量增加的一个原因。

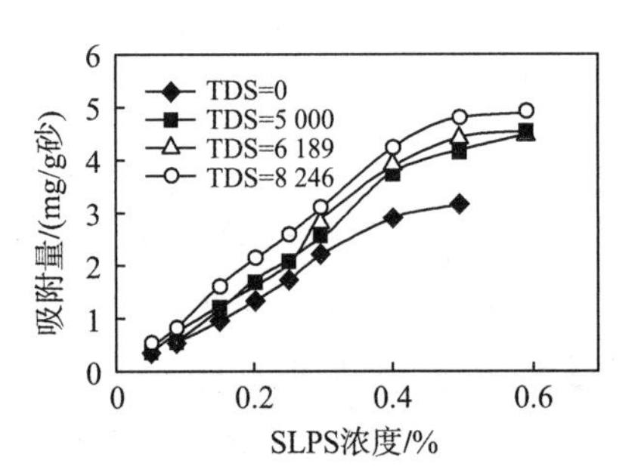

图 2 矿化度对 SLPS 吸附量的影响

2.1.5 助剂对 SLPS 吸附的影响

单一 SLPS 降低界面张力的能力是有限的，助剂 1＃ 的加入可以使油水界面张力达到超低(10^{-3} mN/m)。本文对加入助剂后 SLPS 在孤岛油砂上的吸附等温线进行了试验。从图 3 结果可以看出加入助剂后 SLPS 的吸附量减少。这是由于助剂 1＃ 中的非离子部分优先吸附的结果。

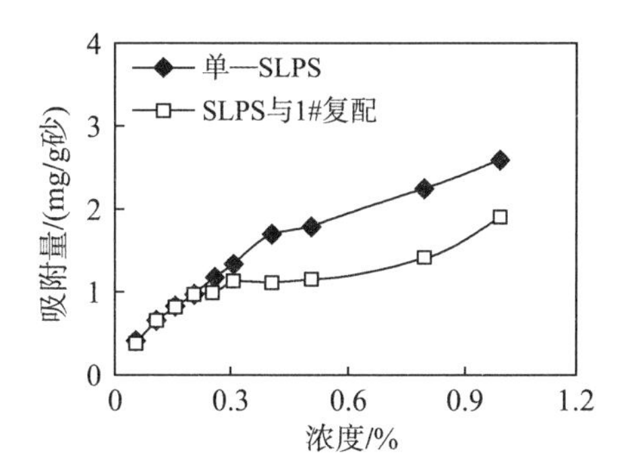

图 3 助剂对 SLPS 吸附等温线的影响

2.2 SLPS 驱油体系色谱分离研究

2.2.1 SLPS 与助剂的色谱分离

试验在不同长度细管模型中进行,分别注入 SLPS 与 1♯ 的混合物,研究其色谱分离情况。图 4 为 100 cm 长管中化学剂的分离情况。可以看出:它们之间存在一定的色谱分离现象,在 100 cm 长模型中两种活性剂峰值到达时间差值约为 0.7 PV,比在 50 cm 长模型中分离现象严重。由室内试验界面张力等值图结果知 SLPS 与助剂 1♯ 总浓度为 0.20 时,其界面张力即可达到超低。由试验结果知适当增大注入液浓度,二者的复配协同效应可以发挥作用,只是在驱替液前沿由于地下水的稀释与吸附作用的存在二者的协同作用会受到一定影响。

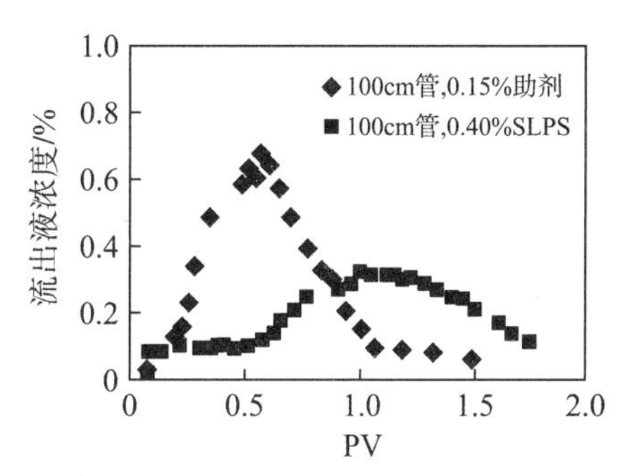

图 4 SLPS 与助剂 1♯ 动态吸附曲线

2.2.2 SLPS 二元驱的色谱分离

由于三元复合驱中碱的加入会带来结垢和乳化,给注入和采出液处理带来困难[6-7]。因此胜利油田开展了二元驱先导试验。二元驱中各组分在地层运移过程中表现出的吸附滞留与色谱分离已成为影响复合驱效果的重要因素。色谱分离试验在 100 cm 的细管模型上进行。注入的二元驱配方为 0.15%HPAM+0.4%SLPS+0.1%1♯。

从试验结果知,配方中三种物质存在色谱分离。HPAM 存在不可及体积,所以最先出来,HPAM 的加入使段塞的波及体积增加,因而加剧了 SLPS 与助剂 1♯ 的色谱效应,两种活性剂峰值到达时间差值约为 0.85 PV。

因此在现场注入时应适当加大活性剂的用量,以使活性剂超加合效应得到发挥。驱油体系的动态吸附规律与色谱分离结果为驱油配方设计及活性剂用量计算提供了理论指导。目前以SLPS为主剂,中石化的重大先导试验在孤东七区西南Ng54-61层开展,试验中活性剂浓度由方案中的0.45%增至0.65%,先导试验取得了明显的降水增油效果。含水由试验前98%降到93%,日产油从34 t增加至95 t。到目前为止,已经累积增油18 027 t。

3 结果与讨论

(1) 通过对SLPS在不同介质、不同条件下的静态吸附试验研究,得到了其吸附特征及影响因素。低浓度下SLPS基本呈线性吸附,曲线相关系数0.991 9。试验表明当浓度高于10 000 mg/L时,吸附量基本不再随浓度而变化,最大吸附量为4.6 mg/L。

(2) 通过对SLPS驱油体系的色谱分离研究发现:体系之间存在色谱分离,在现场应用中适当增大注入液浓度,二者的复配协同效应可以得到发挥。

参考文献

[1] 康万利.大庆油田三元复合驱化学剂作用机理研究[M].北京:石油工业出版社,2001:4-13.

[2] 杨承志,韩大匡.化学驱油理论与实践[M].北京:石油工业出版社,1996:208-210.

[3] 俞稼镛,宋万超.化学复合驱基础及进展[M].北京:中国石化出版社,2002:173-178.

[4] 曹绪龙.复合驱油体系中化学剂的检测方法及相互作用研究[D].北京:中科院物理化学研究所2001级博士论文,2004:35-48.

[5] 杨普华,杨承志.化学驱提高石油采收率[M].北京:石油工业出版社,1988:152-166.

[6] 赵福麟.EOR原理[M].山东东营:石油大学出版社,2001:131-142.

[7] 曹绪龙,孙焕泉,姜颜波,等.孤岛油田西区三元复合驱矿场试验[J].油田化学,2002,19(4):350-353.

(编辑 朱和平)

chuan 610500, China), SUN Lei, LUO Li-qiong, et al. JOURNAL OF SOUTHWEST PETROLEUM INSTITUTE,VOL 28, N0.2, 48-51, 2006(ISSN1000-2634, IN CHINESE)

The effect of gaseous condensate-water on condensate-gas system used to be ignored in conventional condensate-gas system phase behavior test. Condensate gas reservoir exploitation practice in home & abroad indicated that the effect of gaseous con-densate-water on condensate-gas system phase behavior should not be ignored. So to instruct the condensate gas reservoir ex-ploitation more efficiently, we must set up the condensate-gas system PVT phase test and compute the method of the considering gaseous condensate-water. In this paper, we tried to explore and design the condensate-gas system PVT phase test method with abundant condensate-water, compared with the conventional oil/gas two-phase equilibrium system test results, and ana-lyzed how the gaseous water content affect the condensate-gas system.

Key words: condensate gas reservoir; high-temperature & high-pressure; condensate-water; oil/gas/water multi-phase e-quilibrium; PVT phase test

WELL TEST METHOD IN HEAVY OIL THERMAL RE-COVERY WITH CONSIDERATION OF GRAVITY OVERRIDE

HUO jin (Heavy Oil Production Co, PetroChina Xinjiang Oil Field Branch Company, Kalamay 834000, China) JIA Yong-lu, YU Jia, et al. JOURNAL OF SOUTHWEST PETROLEUM IN-STITUTE, VOL. 28, N0. 2, 52-55, 2006 (ISSN1000-2634, 1N CHINESE)

Aimed at Gravity Override in a steam-injection processes, a well test analysis model of wedge-shaped compound reservoir is set up to simulate to slant steam front. The mathematical mod-el considering heat loss is given and solved by Laplace transformation, then stehfest algorithm is used to obtain dismentionless wellbore pressure. The influence of the parameter on the dismentionless wellbore pressure is discussed, such as slant angle, heat loss, and mobility ratio.

Key words: Gravity Override; compound reservoir; welltest analysis; Laplace transformation; stehfest algorithm

DEPLETION EXPLOITATION PERFORMANCE PRE-DICTION FOR THE SATURATION CONDITIONS GAS CONDENSATE RESERVOIR CONTROLLED BY SIN-GLE-WELL

PENG Cai-Zhen (Southwest Petroleum University, Chengdu Sichuan 610500, China), SUN Lei, TANG Yong, et al. JOURNAL OF SOUTHWEST PETROLEUM INSTITUTE, VOL. 28, N0. 2, 56-60, 2006 (ISSN1000-2634, IN CHINESE)

According to the mole conservation on the water and the hy-drocarbon components in reservoir, adopting the K-value theory of pseudo-component division, integrating Darcy's law and the method of the pseudo-component phase equilibrium calculation, the two-dimension three-phase single-well compositional mathe-matical model was presented. The effects of gravity, capillary pressure and interfacial tension, the compressibility of water and rock, the heterogeneity and anisotropy of rock physical proper-ties, were considered. The difference equation for the mathemat-ical model was accomplished by using the way of implicit pressure and explicit saturation(IMPES). The nonlinear equation was linearized by adopting Newton iteration method. The answer of the difference equation group was solved by utilizing linear laxation method and direct solving method. A set of FORTRAN calculation program was written. The result of study was used for the simulation calculation of the saturated conditions condensate gas reservoir controlled by single-well. It put emphasis on eval-uating produced gas velocity affects development performance of stable production fixed number of years and produced degree for the saturation conditions gas condensate reservoir. It provided the foundation for development plan optimization of the saturation conditions gas condensate reservoir and gas well reasonable pro-duce system regulation.

Key words: saturation conditions condensate gas reservoir; numerical model; pseudo-component; history match; exploita-tion gas velocity.

ANALYSIS OF THE RECOVERY DEGREE OF THE UP-WARDS BLOCK OF THE WESTREN BEIYIER LINE IN THE SAERTU OILFIELD

QIAN Jie(Chengdu University of technology, Chengdu Sichuan 610059, China), ZHANC Guo-zheng; YUAN Guan-juen, et al. JOURNAL OF SOUTHWEST PETROLEUM INSTITU-TE, VOL. 28, NO. 2, 61-63, 2006(ISSN1000-2634, IN CHI-NESE)

After polymer-flooding the chief reservoir in the Westemn Beiyier Line in the central Saertu Oilfied, polymer-flooding will be done upwards into Saertu reservoir SI110-SIl110 witch is the second type oil layers where the polymer-flooding is first per-formed. For holding the accurate recovery degree prior to polymer-flooding concerns the result evaluation and application on polymer-flooding in the second type oil layers, on the basis of comparing the advantages and disadvantages by computing the static and performance data in 4 calculating methods, it was concluded that the static calculating method is more rational. We had taken the results calculated with static method as the recov-ery degree ahead of polymer-flooding.

Key words: recovery degree; polymer; upwards going; waterflooding curve

STUDY OF ADSORPTION AND CHROM ATOGRAPHIC SEPARATION OF FLOODING SYSTEM OF SHENGLI PETROLEUM SULFONATE

WANG Hong-yan(Southwest Petroleum University, Chengdu Sichuan 610500,

China), YE Zhong-bin, ZHANG Ji-chao, et al. JOURNAL OF SOUTHWEST PETROLE-UM INSTITUTE, VOL. 28, NO. 2, 64-66, 2006 (ISSN1000-2634, IN CHINESE)

The static and dynamic adsorption law and chromatographic separation character of flooding system of Shengli petroleum sul-fonate(SLPS)and assistant surfactant have been studied. Results show that SLPS is linear adsorption at lower concentration, and maximum adsorption capacity is 4. 6 mg/L. The adsorption capacity has been reduced obviously when adding assistant sur-factant. There is chromatographic separation in some extant when inject SLPS and assistant surfactant together, the effect of chromatographic separation is more obvious when adding HPAM in flooding system. According to the experiments, the cooperative effect will exhibit by increasing the surfactant concentration of the flooding liquid.

Key words: petroleum sulfonate; dynamic adsorption; chromatographic separation; surfactant/polymer flooding

SPECTOPHOTOMETRIC FLOW-INJECTION ANALY-SIS ON TRACEI-IN RESERVOIR WATER

XIAO Xin-feng (Southwest Petroleum University, Cheng du Sichuan 610065, China), ZHANG Xin-shen, LUO Ya-jun, et al. JOURNAL OF SOUTHWEST PETROLEUM INSTITUTE, VOL. 28, NO. 2, 67-70, 2006 (ISSN1000-2634, IN CHINESE)

In the present method, a new, simple and rapid spectopho-tometric flow injection technique for the accuracy and determination of I-in reservoir water has been developed. The method is based on the decrease of absorbance intensity of rhodamine B due to the complexation with[I_2Br]-, I-can be selectively ox-idated to form I_2Br Ce(ⅳ) in acidic medium and I2 reacts with Br" to form[I_2Br]-. At this moment, some changes of the solution's color takes place. It is found useful for sensitivity in CT-MAB media. The absorption peak of the complexation, which is increased linearly by addition of I~, occurs at 585 nm. Optimization of chemical and flow injection variables has been made. Under the optimized conditions, the calibration curve obtained is linear over the range 0. 050 0—1. 00 mg/L, the linear equation is H(mV)=76. 6 c(mg/L)-2. 933 3, r=0. 999 6, the detection limit is 0. 020 mg/L, and the precision(RSD=0. 85%, n=11) is found quite satisfactory. This method has good sensitivity and selectivity. Application of the method to the analysis on I-in reservoir water resulted a good agreement between the expected and found values.

Key words: low injection; I; rhodamine B; spectopho-tometry

RELATION BETWEEN STRUCTURE OF ETHENE-VINYLACETATE COPOLYMERS AND THEIR POUR POINT DEPRESSION

JIANG Qing-zhe(University of Petroleum, Beijing 102249, China), YUE Guo, SONG Zhao-zheng. JOURNAL OF SOUTHWEST PETROLEUM INSTITUTE, VOL. 28, N0. 2, 71-74, 2006(ISSN1000-2634, IN CHINESE)

All kinds of molecule structures of the ethylene-vinyl acetate copolymers(EVA) have been synthesized by solution poly-merization. Their crystallinity, solubility and the effect of lowering pour point of crude have been studied. Results show that on-ly carbon atoms far from polar group of EVA take part in crystallizing. When carbon number of alkyl chain of EVA, which takes part in crystallizing, is about equal to three fourths of average carbon number of wax, the effect of EVA pour point depressants is best. When the average relative molecular mass of EVA is 1.2×10^4, the effect of EVA is best. The distribution of relative molecular mass has little influence.

Key words: ethylene-vinyl acetate copolymer; distribution of relative molecular mass; pour point depression effect

RESEARCH OF THE REACTIVE PERFORMANCE OF N-OCTANE ON DIFFERENT ZEOLITE CATALYSTS

KE Ming(University of Petroleum, Beijing 102249, China), CHEN Yan-guang, HUANG Yong, et al. JOURNAL OF SOUTHWEST PETROLEUM INSTITUTE, VOL. 28,NO. 2,75-79, 2006(ISSN1000-2634, IN CHINESE)

The hydroisomerization and aromatization properties of a series of different zeolite catalysts, primary Ni/HZSM-5, Ni/HFER, Ni/SAPO-11, Ni/Hβand Ni/HY were compared in a micro-reactor by using of n-octane as probing molecular. Results show that Ni/HZSM-5 zeolite with the same pore structures have the approximate product distribution, with the increase of Si/Al ratio, the liquid product yield and hydroisomerization selectivity have been improved. The large pore zeolite have more superior isomeric product selectivity than medium pore and small pore zeolite, the isomeric and aromatic selectivity of Ni/HY catalyst are 41.83% and 3.4% respectively. However, the Ni/HZSM-5 catalyst with a higher Si/Al ratio has higher activity and excellent capability to retard the formation of coke, at the same time it has the optimum isomerization and aromatization performance.

Key words: zeolite; isomerization; aromatization; n-octane

VISCOSITY REDUCTION EFFECTS AND MICROANAL-YSES FOR VISCOUS WATER-IN-OIL CRUDE WITH CHEMICALS OF XINTAN IN SHENGLI OILFLELD

JING Jia-qiang (Southwest Petroleum University, Cheng-du Sichuan 610500, China), LI Ji, YANG Li, et al. JOURNAL OF SOUTHWEST PETROLEUM INSTITUTE, VOL. 28, N0.2, 80-84, 2006(ISSN1000-2634, IN CHINESE)

Viscous water-in-oil crude of Xintan in Shengli Oilfield has high viscosity, poor flowability, high consumption of energy when gathering and transportation at normal temperature. Some measures have to be taken to reduce its viscosity; otherwise its gathering and transportation are difficult. Therefore a kind of phase inversion and viscosity reduction agent VRKD18.

Gemini 表面活性剂对疏水缔合聚丙烯酰胺界面吸附膜扩张流变性质的影响

曹绪龙[1] 张 磊[2] 程建波[3]

(1. 中国石化胜利油田分公司地质科学研究院,东营 257015;
2. 中国科学院理化技术研究所,北京 100190;
3. 烟台大学化学生物理工学院,烟台 264005)

摘要:采用小幅低频振荡和界面张力弛豫技术,考察了疏水缔合水溶性聚丙烯酰胺(HMPAM)在正癸烷-水界面上的扩张黏弹性质,研究了不对称 Gemini 表面活性剂 $C_{12}COONa\text{-}p\text{-}C_9SO_3Na$ 对其界面扩张性质的影响。研究发现,疏水链段的存在,使 HMPAM 在界面层中具有较快的弛豫过程,扩张弹性显示出明显的频率依赖性。表面活性剂分子可以通过疏水相互作用与聚合物的疏水嵌段在界面上形成类似于混合胶束的特殊聚集体。表面活性剂分子与界面聚集体之间存在快速交换过程,可以大大降低聚合物的扩张弹性。同时,聚合物分子链能够削弱表面活性剂分子长烷基链之间的强相互作用,导致混合吸附膜的扩张弹性远低于单独表面活性剂吸附膜。

关键词:扩张弹性;扩张黏性;疏水缔合聚丙烯酰胺;Gemini 表面活性剂;弛豫过程
中图分类号:O647 文献标识码:A 文章编号:0251-0790(2010)05-0998-05

疏水缔合水溶性聚合物因其独特的分子自组装行为和溶液性能而备受关注。通常,在这类聚合物结构中含有少量的疏水基团,它们以侧链或端基的方式连接在水溶性聚合物的主链上,由于疏水缔合作用形成的空间网状结构,增大了聚合物的流体力学体积,因而具有良好的增黏、耐温、抗盐及耐剪切等性能[1-2]。同时,由于此类聚合物的分子链上带有少量疏水基团,对表面活性剂的加入非常敏感,很少量的表面活性剂就能对聚合物溶液的性能造成很大的影响。由于这种物理交联的可逆性使得此类聚合物溶液具有独特的流变性能,因而在三次采油、涂料、化妆品和药品制剂等领域已展示出广泛的应用前景[3-5]。近年来,国内外对疏水缔合共聚物/表面活性剂混合体系的表面张力以及黏度等体相性质研究较多,但对其界面流变性质的研究报道较少[6-12]。本文采用小幅低频振荡和界面张力弛豫技术考察了疏水缔合水溶性聚丙烯酰胺(HMPAM)在正癸烷-水界面的扩张黏弹性质,研究了疏水缔合聚合物与不对称 Gemini 表面活性之间的界面相互作用机理。

1 理论基础

当界面受到周期性压缩和扩张时，界面张力也随之发生周期性变化，扩张模量定义为界面张力变化与相对界面面积变化的比值，即

$$\varepsilon = \mathrm{d}\gamma / \mathrm{dln}A \tag{1}$$

式中，ε 为扩张模量，γ 为界面张力，A 为界面面积。

对于黏弹性界面，界面张力与界面面积的周期性变化之间存在一定的相位差 θ，称为扩张模量的相角，扩张模量可写作复数形式[13]：

$$\varepsilon = \varepsilon_d + \omega_i \eta_d \tag{2}$$

式中，ε_d 为扩张弹性，η_d 为扩张黏度，ω 是界面面积正弦变化的频率。实数部分 ε_d 和虚数部分 $\omega\eta_d$ 分别称作储存模量和损耗模量，分别反映黏弹性表面的弹性部分和黏性部分的贡献。

扩张弹性和扩张黏度分别按下式计算：

$$\varepsilon_d = |\varepsilon| \cos\theta \tag{3}$$

$$\eta_d = (|\varepsilon| / \omega) \sin\theta \tag{4}$$

界面张力弛豫实验是通过对瞬间形变后的界面张力衰减曲线进行 Fourier 变换得到界面扩张黏弹性参数的方法。对于一个存在多种弛豫过程的实际体系，由于弛豫过程具有可加和性，因此衰减曲线可以用几个指数方程之和表示[14]：

$$\Delta\gamma = \sum_{i=1}^{n} \Delta\gamma_i \exp(-\tau_i t) \tag{5}$$

式中，τ_i 为第 i 个过程的特征频率，$\Delta\gamma_i$ 为第 i 个过程的贡献，n 为总过程的个数。因此，用式(5)对界面张力衰减曲线进行拟合，可以得到界面上存在的弛豫过程的个数、每个过程的特征频率(或时间)及各个过程的贡献等信息。

2 实验部分

2.1 样品及试剂

不对称 Gemini 表面活性剂 $C_{12}COONa\text{-p-}C_9SO_3Na$ 由中国科学院理化技术研究所提供[15]；正癸烷为分析纯，天津博迪化学有限公司生产；疏水缔合水溶性聚丙烯酰胺(HMPAM)由中国石化胜利油田分公司地质科学院研究院提供，分子量为 $4\times10^6-5\times10^6$；实验用水为经重蒸后的去离子水，电阻率≥18 MΩ·cm。Gemini 表面活性剂分子结构如下所示。

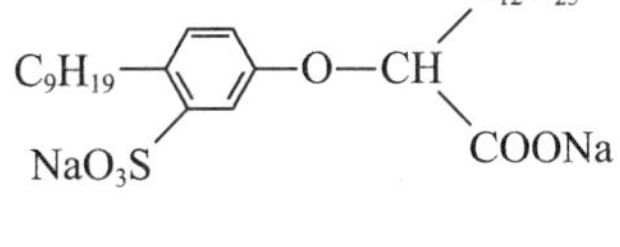

C_9H_{19}—(苯环)—O—CH($C_{12}H_{25}$)(COONa)，苯环上另有 NaO_3S 取代

2.2 实验方法

界面扩张黏弹性用上海中晨JMP2000型界面扩张黏弹性测定仪测定。正癸烷作为油相,水相为用二次蒸馏去离子水配制的不同浓度表面活性剂和HMPAM的溶液。

首先在Langmuir槽中注入90 mL水相,再小心地将50 mL油相铺在水相之上,油相的高度要能浸没整个吊片。在恒温(30℃)条件下预平衡6 h,然后在不同的工作频率(0.005～0.1 Hz)下进行正弦周期振荡实验。平衡1 h后进行弛豫实验,扩张幅度为15%,扩张时间为2 s,实验温度为30℃。

3 结果与讨论

3.1 疏水改性聚丙烯酰胺HMPAM的界面扩张流变性质

疏水改性聚丙烯酰胺分子具有亲水部分和疏水部分,因此表现出一定的界面活性。频率对浓度为1 500 mg/L的HMPAM界面扩张流变性质的影响见图1。从图1可以看出,随扩张频率的增大,HMPAM的扩张弹性逐渐增大,而扩张黏性则随着频率增大产生一个不明显的极大值。从图1还可以看出,除极低频率外(0.005 Hz),HMPAM溶液界面吸附膜的弹性大于黏性。

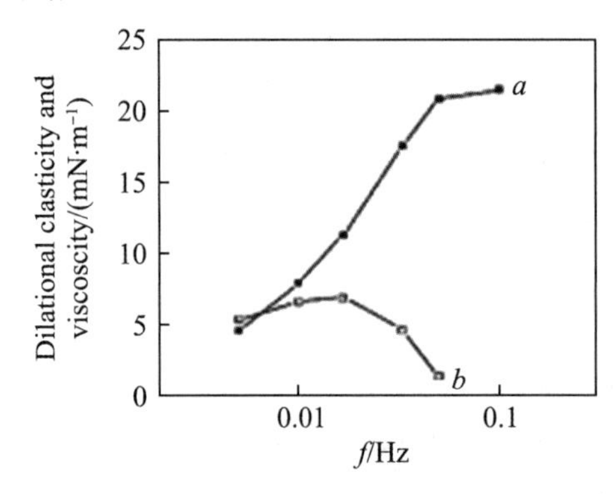

Fig. 1 Frequency dependency of interfacial Dilational elasticity(a) and viscosity(b) for 1 500 mg/L HMPAM

根据Noskov等[16]提出的模型,在大分子表面膜内存在两个微区:一个为紧贴着气相的比较窄的浓度区域,称为“最接近微区”;另一个为链节总浓度很低的形成嵌入到液相的“尾状”和“环状”结构的区域,称为“末端微区”。表面张力的大小主要由“最接近微区”中链节浓度的大小决定。表面扩张弛豫过程是由于将聚合物链拉至表面(链节由“末端微区”进入“最接近微区”)或将聚合物链挤出表面(由“最接近微区”进入“末端微区”)而产生的。聚合物分子链内形变产生的慢弛豫过程和分子链段的扩散交换引起的快弛豫过程是聚合物分子界面现象的基础。

疏水缔合聚合物是由少量疏水基团或疏水链段改性的水溶性高分子,具有表面活性,可以吸附在界面。当HMPAM分子吸附在界面上时,存在于紧靠着油相的“最接近区域”的疏水链段决定了界面张力的大小;而丙烯酰胺链段则以“尾状”或“环状”形态深入到“末端微区”中。疏水链段在界面层中具有相对较快的弛豫过程，因而扩张弹性显示出明显的频率依赖性。

扩张频率对吸附膜扩张弹性的影响存在两种极限情况：扩张频率极低时，界面形变所造成的界面张力梯度在实验时间内可以几乎完全消除；扩张频率极高时，其工作频率已远大于发生在界面上和界面附近的各种弛豫过程的特征频率，界面膜则表现为不溶膜的特性。只有在中间频率范围内，扩张弹性是随着频率降低而减小的，这是因为当界面变形速度较慢时，界面分子有足够的时间通过从界面和体相向新生成界面扩散，修复由界面面积变化产生的界面张力梯度，因而扩张弹性随着形变速度的降低而减小[17]。

扩张黏性随工作频率的变化规律与弹性有所不同，当工作频率远低于各种弛豫过程的特征频率时，实验中对界面膜施加的影响可以通过各种弛豫过程完全消除，扩张黏性接近于零；当工作频率远高于发生在界面上和界面附近的各种弛豫过程的特征频率时，可以看作实验中各种弛豫过程未曾发生，因而扩张黏性也接近于零。由于弛豫过程有其特征频率，当工作频率与某种弛豫过程的特征频率一致时，该弛豫过程对扩张模量的贡献最大。因此，随着工作频率的增大，扩张黏性会出现一个或多个极大值，分别对应于不同弛豫过程的特征频率。

界面上吸附分子的微观弛豫过程可以通过界面张力弛豫法进行研究，获得各种弛豫过程的贡献($\Delta\gamma_i$)及弛豫过程的特征时间 T_i，相关数据列于表 1。从表 1 可以看出，1 500 mg/L HMPAM 聚合物溶液与正癸烷的界面上主要存在特征时间分别为 6.67 和 26.32 s 的弛豫过程，上述过程反映了界面层内不同区域疏水链段的交换以及聚合物分子链在界面层进行构象转换的弛豫过程。

Table1 Interfacial relaxation proceses and their characteristic time for 1 500 mg/L HMPAM polymer

Polymer 1 500 mg/L	$\Delta\gamma_1$(mN/m)	t_1/s	$\Delta\gamma_2$(mN/m)	t_2/s
HMPAM	5.24	6.67	0.59	26.32

3.2 HMPAM 和 $C_{12}COONa$-p-C_9SO_3Na 混合体系的界面扩张性质

图 2 为不同浓度的 Gemini 表面活性剂 $C_{12}CONa$-p-C_9SO_3Na 与 1 500 mg/L 疏水改性聚合物 HMPAM 混合体系界面扩张弹性随工作频率的变化趋势图。从图 2(A)中可以看出，不同浓度的 Gemini 表面活性剂加入疏水改性聚合物 HMPAM 溶液中，降低了疏水改性聚合物 HMPAM 的界面扩张弹性对频率的依赖性，并且随着表面活性剂浓度增大，聚合物溶液的扩张弹性降低。

从图 2(B)中可以看出，低浓度($10^{-6}-10^{-5}$mol/L)时，$C_{12}COONa$-p-C_9SO_3Na 溶液的扩张弹性值较高，且随着工作频率的变化较小。随着体相浓度增大，扩张弹性降低。这是因为浓度较低时，在界面压缩条件下，$C_{12}COONa$-p-C_9SO_3Na 分子的 2 根柔性疏水长链可能发生缠绕和变形，导致吸附分子取向改变和界面分子重排等慢弛豫过程，产生较高的界面扩张弹性；而在高浓度时，分子以扩散交换为主，界面扩张弹性显著降低。

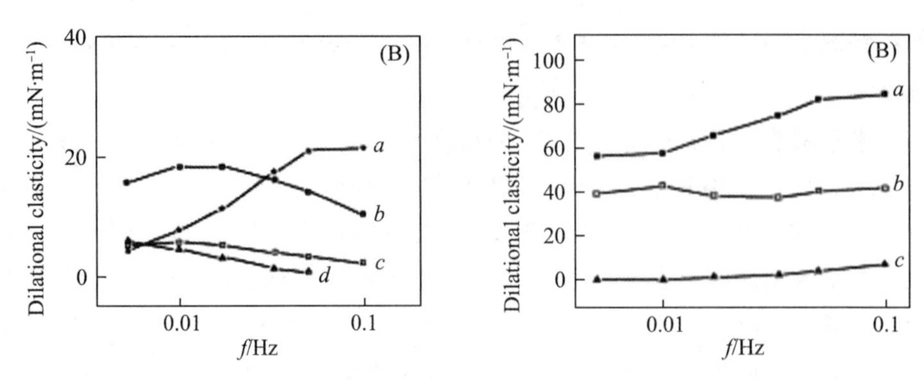

Fig. 2 Interfacial dilational elasticity as a function of frequency for 1 500 mg/L HMPAM at diferent $C_{12}COONa$-p-C_9SO_3Na concentrations(A) and the surfactant on its own solution(B)

添加的 $C_{12}COONa$-p-C_9SO_3Na 对疏水改性聚合物 HMPAM 的界面浓度主要有两个方面的影响:一方面是与聚合物分子在界面上产生竞争吸附,形成混合吸附膜;另一方面是表面活性剂分子与聚合物的疏水嵌段在界面上发生相互作用。上述两个方面都会降低聚合物分子的界面浓度,从而造成聚合物的界面扩张性质的改变。在低表面活性剂浓度时,$C_{12}COONa$-p-C_9SO_3Na 的 2 根柔性的长疏水链相互缠绕交叠,界面分子间的相互作用较强,因而扩张弹性较高。而对于 $C_{12}COONa$-p-C_9SO_3Na 和 HMPAM 的混合体系,表面活性剂和聚合物存在吸附竞争,聚合物疏水嵌段可能插入表面活性剂分子之间,破坏了表面活性剂分子原来的强相互作用,导致扩张弹性较单独表面活性剂界面膜大幅度降低;同时,在低频率条件下,由于混合吸附膜中表面活性剂分子的存在,其扩张弹性又比单独聚合物的高。

随着表面活性剂浓度增大,界面上表面活性剂分子与聚合物分子中的疏水嵌段形成聚集体,这些聚集体联结了不同聚合物分子,从而形成二维空间网络,提供较慢的弛豫过程,降低了扩张弹性对频率的依赖性。同时,界面上还存在单独表面活性剂分子与聚集体之间的快速交换过程,该过程导致混合体系扩张弹性比单独表面活性剂存在时明显降低。

不同浓度 $C_{12}COONa$-p-C_9SO_3Na 的 1 500 mg/L HMPAM 聚合物溶液的界面弛豫过程相关参数如表 2 所示。从表 2 中可以看出,低浓度(10^{-6} moL)条件下,表面活性剂分子在界面上存在三类弛豫过程:最快的弛豫过程特征时间为 1～2 s,对应于表面活性剂分子在界面和体相的扩散交换过程;最慢的弛豫过程特征时间在 10^2 s 数量级,可能对应于界面吸附膜大量分子的重排;中间弛豫过程特征时间约为 20 s,对应于分子疏水支链缠绕等构象变化。随着浓度增大(10^{-5} mol/L),慢弛豫过程的特征时间变短,贡献减小;浓度进一步增大(10^{-4} mol/L),慢弛豫过程消失,界面膜性质由快过程控制。

Table2 Interfacial relaxation proceses and their characteristic time for 1 500 mg/L HMPAM polymer at Diferent $C_{12}COONa\text{-}p\text{-}C_9SO_3Na$ concentrations

$C_{12}COONa\text{-}p\text{-}C_9SO_3Na$ concentration (mol/L)	$\Delta\gamma_1$/(mN/m)	t_1/s	$\Delta\gamma_2$	t_2/s	$\Delta\gamma_3$	t_3/s
1×10^{-6}	3.71	1.58	0.94	25.00	0.90	454.55
1×10^{-5}	5.27	1.39	0.99	18.52	0.64	208.33
1×10^{-4}	1.71	1.92	—	—	—	—
1×10^{-6}+HMPAM	2.50	40.00	1.31	222.22	0.72	1 265.82
1×10^{-5}+HMPAM		Fast process	1.42	66.23	0.42	416.67
1×10^{-4}+HMPAM		Fast process	0.13	41.67	1.58	909.09

对于 $C_{12}CONa\text{-}p\text{-}C_9SO_3Na$ 和 HMPAM 的混合体系，聚合物的存在增加了各个弛豫过程的特征时间，这是由于 Gemini 表面活性剂分子与不同聚合物分子通过疏水作用相互联结造成的. 高表面活性剂浓度时，混合体系弹性大大降低，推测界面上存在表面活性剂分子与聚集体之间的快速交换过程，由于该过程特征弛豫时间太快而无法监测。我们曾利用不同分子尺寸的表面活性剂进行考察，证明只有分子尺寸足够大，才能捕捉到这种快弛豫过程[6]。

图 3 示出了不同浓度的 $C_{12}COONa\text{-}p\text{-}C_9SO_3Na$ 与 1 500 mg/L HMPAM 混合体系界面扩张黏性部分随工作频率的变化趋势图。由图 3 可以看出，不同浓度的 Gemini 表面活性剂加入到 HMPAM 溶液中后，界面扩张黏性随着频率增大而增大，其理论极大值出现在高于实验频率的范围。同时，在低频率条件下，混合体系的扩张黏性比单独聚合物的低；而高频率条件下，单独聚合物的扩张黏性比混合体系的高，表明混合体系界面上确实存在一个较快的弛豫过程。但目前还无法获得其准确的特征弛豫时间。

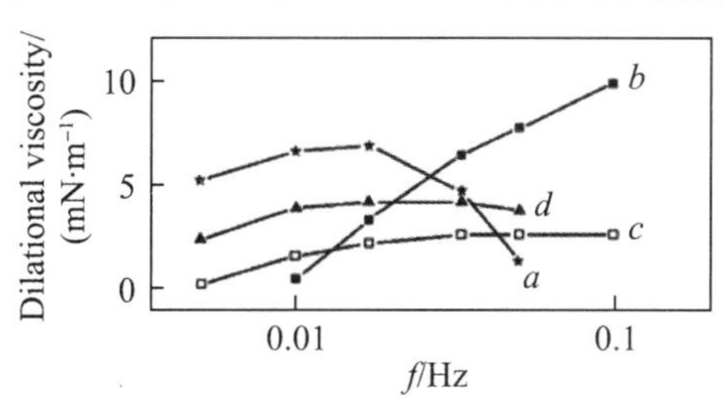

Fig. 3 Interfacial dilational viscosity as a function of frequency for the surfactant and 1 500 mg/L polymer solutions

参考文献

[1] TsitsilianisC, Iliopoulos. I, DucouretG. Macromolecules [J], 2000, 33: 2936-2943.

[2] ShalabyS. W, McCormickC. L, BotlerG. B Water Soluble Polymers Synthesis, Solution Properties and Applications [M], Washington: Am. Chem. Soc. ,1991.

[3] ShawK. G, LeipoldD. P, J. Coa. tTechno. l [J]1985,57:63-72.

[4] SchulzD. N, GlasJ. E, Polymers as Rheology Modifiers [M], Washington: Am. Chem. Soc, 1991.

[5] TaylorK. C, Nasr-E DinH. A, J. Pe. tSc. iEng. [J], 1998, 19:265-280.

[6] WangD. X, LuoL, ZhangL, eta Journal of Dispersion Science and Technology [J], 2007, 28 (5) :725-736.

[7] LuoL, WangD. X, ZhangL, eta. Journal of Dispersion Science and Technology [J], 2007, 28(2):263-269.

[8] WANGDong-Xian(王东贤), LUOLan(罗澜), ZHANGLu(张路), eta. l. ActaPhy. sChmi. Sin.(物理化学学报) [J], 2005, 21(11):1205-1210.

[9] WangY. Y. , DaiY. H. , ZhangL. , eta. l. J. ColoidInter. fSc. i [J], 2004, 280: 76-82.

[10] WangY. Y. , DaiY. H. , ZhangL. , eta. l. Macromolecules [J], 2004, 37: 2930-2937.

[11] CaoX. , LiY, JiangY, eta. l. Journal of Coloid and Interface Science [J], 2004, 270(2):295-298.

[12] DINGWei(丁伟), LIUHai-Yan(刘海燕), YUTao(于涛), eta. l. Chem. J. Chinese Universities(高等学校化学学报) [J], 2008:29(4):868-870.

[13] LucasenJ, GilesD. J. Chem. So. cFaradayTrans1 [J], 1975, 71:217-232.

[14] MuryB. S. , VenturaA. , LalemantC. . ColoidsSur. fA [J], 1998, 143:211-219.

[15] YanF. , HuangY. P. , WangX. G. , eta. l. Journal of Dispersion Science and Technology [J], 2008, 29(5):670-675.

[16] NoskovB. A, AkentievA. V, BilibinA. Yu, eta. l. Adv. Coloid Inter faceSc. i [J], 2003, 104:245-271.

[17] ZHANGChun-Rong(张春荣), SONGXin-Wang(宋新旺), CAOXu-Long(曹绪龙), eta. l. Chem. J. ChineseUniversities(高等学校化学学报) [J], 2007, 28(4):714-718.

Effect of Gemini surfactant on the Interfacial Dilational Properties of Hydrophobicaly Modified Partly Hydrolyzed Polyacrylamide

Cao Xulong[1]　Zhang Lei　Cheng Jianbo[3]

(1. Geological Research Institute of Shengli Oilfield Co. Ltd, SINOPEC, Dongying 257015, China;
2. Technical Institute of Physicsand Chemistry, Chinese Academy of Sciences, Beijing 100190, China;
3. Science and Enginering Colege of Chemistry and Biology, Yantai University Yalntai 264005, China)

Abstract: The dilational viscoelasticproperties of hydrophobicaly modified partly hydrolyzed polyacryl amide in the absence or presence of Gemini surfactant were investigated at the decane water interface by means of longitudinal method and the interfacial tension relaxation method. The dilational elasticity of polymer solution shows strong frequency dependence due to the fast process involving the exchange of hydrophobic micro-domains between the proxmial region and distal region in the interface. Gemini surfactant plays diferent roles in influencing the structure of adsorbed polymer layer at diferent surfactant concentrations.

At low surfactant concentration, the addition of surfactant molecules for med mixed film with polymer resulting in the increase of dilational elasticity at low frequency. At the same time, polymer chain could sharply decrease the dilational elasticity of surfactant film mainly due to the weakening of the strong interactions among longalkyl chains in surfactant molecule. At high surfactant concentration, the adition of surfactant molecules can decrease the dilational elasticity of polymer solution due to the fast process involving in the exchange of surfactant molecules between the interface and the mixed complex formed by surfactant molecules and micro-domains.

Keywords: Dilational elasiticity; Dilational viscosity; Hydrophobically modified partly hydrolyzed polyacry amide(HMPAM); Gemini surfactant; Relaxation process,

多胺修饰β-环糊精与阴离子表面活性剂的相互作用

孙焕泉[1] 刘 敏[2] 曹绪龙[3] 崔晓红[3] 石 静[3] 郭晓轩[2] 陈 湧[2] 刘 育[2]

(1. 中国石化胜利油田分公司,东营 257000;
2. 南开大学化学系,元素有机化学国家重点实验室,天津 300071;3. 胜利油田分公司地质科学研究院,东营 257015)

摘要:采用荧光和紫外-可见光谱滴定法测定了单-[6-(氨基)-6-脱氧]-β-环糊精(NH_2-β-CD)、单-[6-(乙二胺)-6-脱氧]-β-环糊精(DEN-β-CD)、单-[6-(二乙烯三胺)-6-脱氧]-β-环糊精(DETA-β-CD)和单-[6-(三乙烯四胺)-6-脱氧]-β-环糊精(TETA-β-CD)在磷酸缓冲溶液(25℃,pH=7.2)和碳酸缓冲溶液(25℃,pH=10.5)中分别与阴离子表面活性剂辛烷基磺酸钠(OAS)、癸烷基磺酸钠(TDS)和十二烷基磺酸钠(SDS)形成化学计量比为1∶1的超分子配合物的稳定常数。结果表明,中性条件下多胺修饰环糊精键合3种表面活性剂客体的稳定常数可达$2.6\times10^2\sim4.35\times10^4$ L/mol,碱性条件下的键合能力则相对较弱。无论在碱性或中性条件下,主-客体键合能力都随客体分子中碳链的增长而增大。

关键词:环糊精;光谱滴定;阴离子表面活性剂;超分子化学

中图分类号:O621.3;O657 文献标识码:A 文章编号:0251-0790(2011)04-0879-06

环糊精作为第二代超分子主体化合物,具有良好的水溶性和低毒性,能够选择性地与疏水分子形成包结配合物,因此被广泛应用于仿生化学和酶模型研究中[1-4]。近年来,环糊精与表面活性剂形成包结配合物或超分子体系的研究及在食品、化工、农业、化妆品和医药等领域的应用备受关注[5-16]。然而,该研究大多集中于天然环糊精[17],有关修饰环糊精与表面活性剂相互作用的报道很少。相比于天然环糊精,多胺修饰环糊精不但水溶性好,而且带有正电修饰基团,可与阴离子表面活性剂表现出强的键合行为。本文选择了4种多胺修饰β-环糊精,采用荧光和紫外-可见光谱滴定法研究了它们在中性和碱性条件下与几种阴离子表面活性剂的相互作用。

1 实验部分

参照文献[18,19]方法合成单-[6-(氨基)-6-脱氧]-β-环糊精(NH_2-β-CD)、单-[6-(乙二

胺)-6-脱氧]-β-环糊精(EN-β-CD)、单-[6-(二乙烯三胺)-6-脱氧]-β-环糊精(DETA-β-CD)和单-[6-(三乙烯四胺)-6-脱氧]-β-环糊精(TETA-β-CD),结构见 Scheme 1;中性红(Neutralred,NR),分析纯,日本和光公司;酚酞(PP)经乙醇重结晶后使用;辛烷基磺酸钠(OAS)和癸烷基磺酸钠(TDS),分析纯,天津科锐思公司;十二烷基磺酸钠(SDS)经乙醇/水重结晶后使用。用去离子水配制 pH=7.2 的 Na_2HPO_4-NaH_2PO_4(0.025 mol/L)和 pH=10.5 的 Na_2CO_3-$NaHCO_3$ 缓冲溶液(0.025 mol/L)。

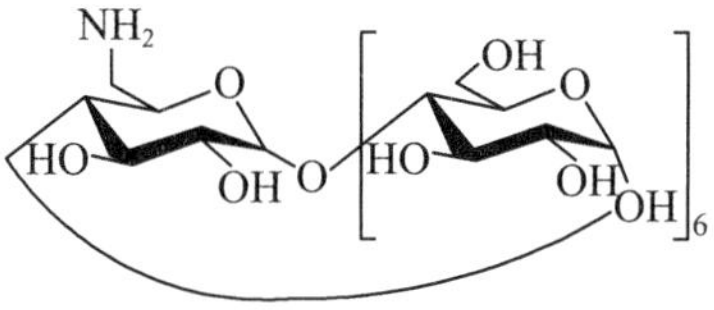

NH_2-β-CD

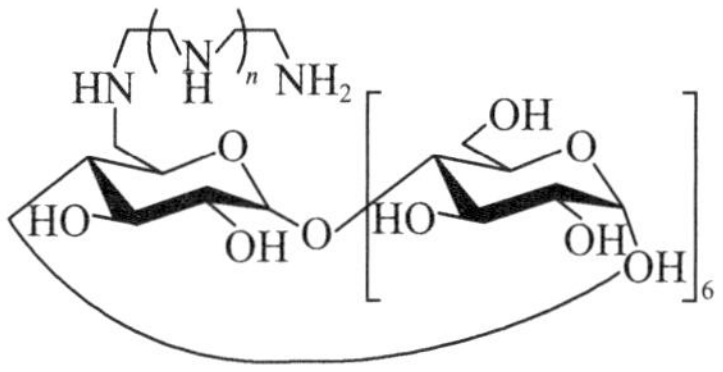

n=0,EN-β-CD; n=1,DETA-β-CD; n=2,TETA-β-CD

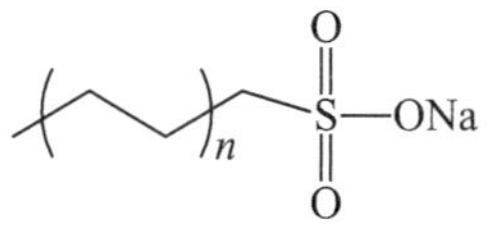

n=3,OAS; n=4,TDS; n=5,SDS

Scheme 1　Structures of polyamine-modified β-cyclodextrins and anionic surfactants

Shimadzu UV-2401PC 型紫外-可见光谱仪,采用普通石英样品池(光程 10 mm),以 PTC-348WI 型恒温器控制温度为(25.0±0.1)℃;Varian Cary Eclipe 荧光光谱仪,采用 10 mm×10 mm×40 mm 石英池,以 Single Cell Peltier Accessory 恒温器控制温度为(25.0±0.1)℃;DDC-307 型电导率仪(上海雷磁仪器厂);QBZY 系列全自动表面张力仪(上海方瑞仪器有限公司)。

2　结果与讨论

2.1 表面活性剂的临界胶束浓度(cmc)

在 25℃下,阴离子表面活性剂辛烷基磺酸钠(OAS)、癸烷基磺酸钠(TDS)和十二烷基磺酸钠(SDS)的临界胶束浓度(cmc)分别为 0.16,0.043 和 0.012 mol/L[20]。实验所用表面活性剂的浓度均在 8.0×10^{-3} mol/L 以下,因此可推断阴离子表面活性剂均以单体形式存在。

2.2 主-客体包结配位的化学计量比

采用等摩尔连续变化法(Job's 方法)测定了主-客体包结配位的化学计量比。固定主体 H 和客体 G 的总摩尔数不变,连续改变两种组分的比例,测量各组混合液的电导率与相同条件下主体和客体自身电导率的差值 Δk,以 Δk 对任一组分的摩尔分数作图,即可得到主-客体包结配位的化学计量比。以 EN-β-CD 和 TDS 的作用为例,由图 1 可见,Job's 曲线的最大值出现在 EN-β-CD 的摩尔分数为 0.5 处,表明主-客体间包结配位的化学计量比为 1∶1,与文献[21,22]结论一致。

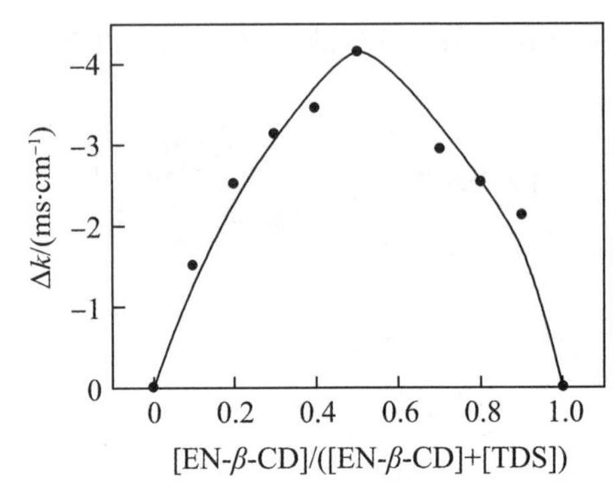

Fig. 1 Job's plot for inclusion complexation of EN-β-CD with TDS

[EN-β-CD]+[TDS]=2.0×10^{-3} mol/L,25℃.

2.3 中性条件下多胺环糊精与表面活性剂键合常数的测定

采用竞争包结法测定了中性条件下多胺修饰 β-环糊精与阴离子表面活性剂包结配位的键合常数。因中性红(NR)与环糊精键合能力适中,且其在环糊精空腔内外荧光较强,因此选择 NR 分子作为荧光探针。在确定了主-客体间包结配位的化学计量比为 1∶1 的条件下,通过观测不同浓度的主体环糊精存在下 NR 荧光光谱的变化 ΔI_f,根据非线性最小二乘法拟合公式[23]可以计算 NR 与 NH_2-β-CD 包结配位的稳定常数。多胺环糊精与 NR 包结配位的稳定常数 K 列于表 1 中。图 2 为在 pH=7.2 的磷酸缓冲溶液中,随 NH_2-β-CD 的加入 NR 荧光光谱的变化及 NR 的分子结构。

Table 1 K values of host-guest inclusion complexation in phosphate buffer solution(pH=7.2)at 25℃

Host	Guest	K/(L · mol^{-1})	Host	Guest	K/(L · mol^{-1})
NH_2-β-CD	NR	331	DETA-β-CD	NR	532
EN-β-CD	NR	812	TETA-β-CD	NR	670

固定主体环糊精和探针 NR 的浓度,向环糊精/中性红体系中加入不同浓度的表面活性剂。图 3 示出了主体 NH_2-β-CD 与 NR 的混合体系中加入 OAS 的荧光光谱。由于 NR 被 OAS 分子挤出环糊精空腔,造成体系的荧光强度降低,竞争过程如下式所示:

$$\text{CD}\cdot\text{NR}+\text{OAS}\rightleftharpoons\text{CD}\cdot\text{OAS}+\text{NR} \tag{1}$$

则包结配位稳定常数(K)可根据下式计算[24]

$$K_s=\frac{[\text{CD}]_0-[\text{CD}]}{[\text{CD}]([\text{OAS}]_0-[\text{CD}]_0+[\text{CD}])} \tag{2}$$

式中,$[\text{CD}]_0$ 为环糊精主体的初始浓度,$[\text{OAS}]_0$ 为表面活性剂客体的初始浓度,[CD]

和[OAS]分别为环糊精主体与表面活性剂客体的平衡浓度，该方程中未知量[CD]可依下式求出：

$$[CD]=\frac{F-F_{NR}}{K(F_{CD-NR}-F)} \tag{3}$$

式中，F_{NR}，$F_{CD\text{-}NR}$ 和 F 分别为 NR 自身的荧光强度、环糊精与 NR 混合体系的荧光强度及环糊精/中性红/表面活性剂同时存在时的荧光强度。采用此方法计算出的环糊精主体与表面活性剂的包结配位稳定常数(K_s)及相应的自由能变化(ΔG^0)列于表 2。

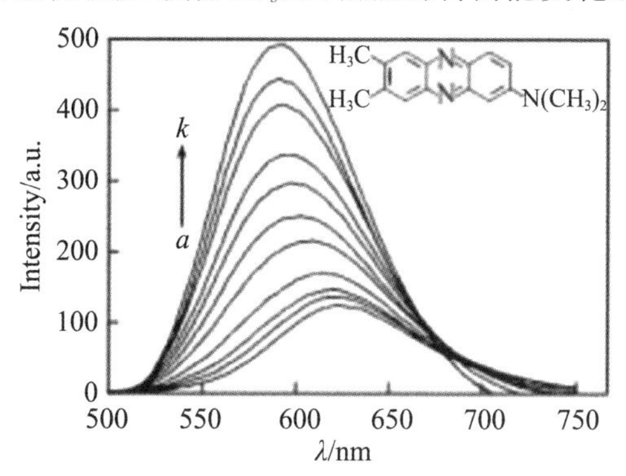

Fig. 2 Fluorescence intensity changes of NR (1.0×10^{-5} mol/L) with addition of NH_2-β-CD(0— 2.5×10^{-3} mol/L from curves a—k) in phos-phate buffer solution (pH=7.2) at 25℃ Inset: the structure of neutral red(NR).

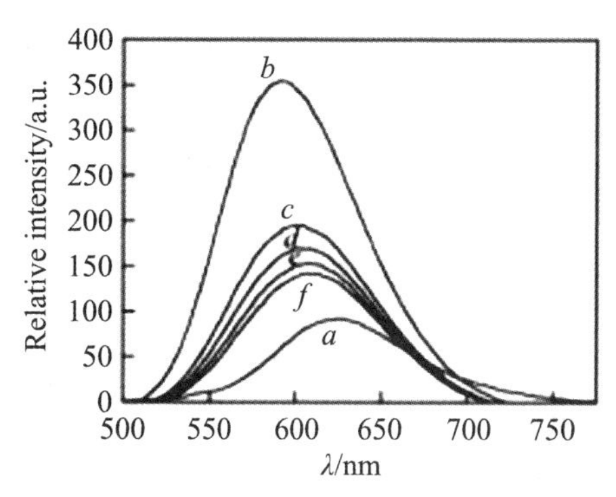

Fig. 3 Effect of NH_2-β-CD and OAS on the fluores-cence spectra of NR in phosphate buffer solu-tion (pH=7.2) at 25℃ a. 1.0×10^{-5} mol/L NR; b. a+2.0×10^{-3} mol/L NH_2-β-CD; c—f. b+OAS(4.5×10^{-4}—4.0×10^{-3} mol/L).

Table 2 Complex stability constants(K_s) and Gibbs free energy change($-\Delta G^0$) for 1∶1 inclusion complexation of surfactant guests with host CDs in phosphate buffer solution(pH=7.2) at 25℃

Host	Guest	$K/(L\cdot mol^{-1})$	$\lg[K/(L\cdot mol^{-1})]$	$-\Delta G^0/(kJ\cdot mol^{-1})$
NH_2-β-CD	OAS	262	2.42	13.80
	TDS	3 731	3.57	20.39
	SDS	9 142	3.96	22.61

(续表)

Host	Guest	K_s/(L·mol^{-1})	[K_s/(L·mol^{-1})]	$-\Delta G^0$/(kJ·mol^{-1})
EN-β-CD	OAS	2 559	3.41	19.45
	TDS	19 200	4.28	24.45
	SDS	43 534	4.64	26.48
DETA-β-CD	OAS	400	2.60	14.85
	TDS	4 723	3.67	20.97
	SDS	13 578	4.13	23.59
TETA-β-CD	OAS	1279	3.11	17.73
	TDS	3 980	3.60	20.55
	SDS	23 929	4.38	24.99

在控制实验中,固定NR的浓度,向溶液中加入不同浓度的表面活性剂客体。NR在620 nm处的荧光强度基本无变化,说明NR与客体表面活性剂之间无相互作用。

2.4 碱性条件下多胺环糊精与表面活性剂键合常数的测定

在碳酸缓冲溶液(pH=10.5)中,以酚酞(PP)为探针,采用竞争包结法测定了多胺环糊精与阴离子表面活性剂包结配位的键合常数[25]。通过观测不同浓度的主体环糊精存在下PP光谱的变化ΔA,根据非线性最小二乘法拟合公式计算出环糊精与PP包结配位的稳定常数(K),结果列于表3。

固定主体环糊精和PP的浓度,向环糊精/PP体系中加入不同浓度的表面活性剂。图4示出了主体NH_2-β-CD与酚酞的混合体系中加入OAS的紫外光谱及探针PP的结构。可见,表面活性剂分子的加入使得PP被逐出环糊精空腔,造成体系的紫外吸收强度升高。

Table 3 K_s values of host-guest inclusion complexation in carbonic acid buffer solution(pH=10.5) at 25℃

Host	Guest	K/(L·mol^{-1})
NH_2-β-CD	PP	10 250
EDA-β-CD	PP	11 030
DETA-β-CD	PP	11 060
TETA-β-CD	PP	14 190

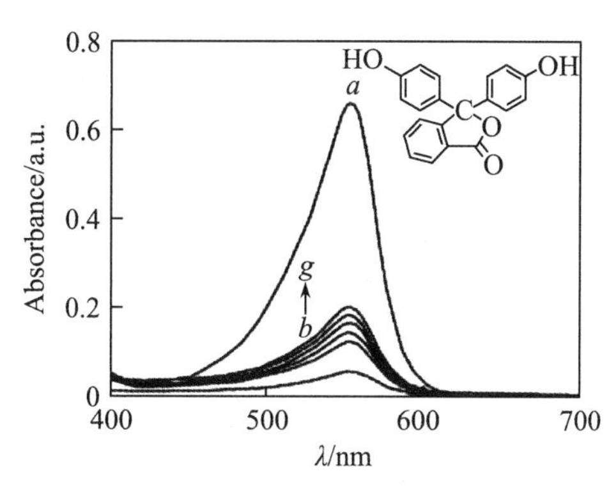

Fig. 4 Effect of NH_2-β-CD and OAS on the UV-Vis spectra of phenolphthalein in carbonic acid buffer solution(pH=10.5)at 25℃

Inset: the structure of phenolphthalein. a. 3.0×10^{-5} mol/L phenolphthalein; b. a+4.5×10^{-4} mol/L NH_2-β-CD; c—g: b+OAS($4.5\times10^{-4}-4.0\times10^{-3}$ mol/L).

在控制实验中,PP的紫外光谱中550 nm处的吸光度并不随OAS的加入而变化。因此,当混合体系中同时存在环糊精主体与PP之间及环糊精主体与表面活性剂客体分子间的平衡时,环糊精主体与表面活性剂客体的配位稳定常数(K_s)可根据以下方程计算:

$$K_s=\frac{c_0-c-(a_0-a)}{c[b_0-(c_0-c)+(a_0-a)]} \tag{4}$$

$$a=a_0-\frac{\Delta A}{\Delta\varepsilon}=\frac{\Delta A_\infty-\Delta A}{\Delta\varepsilon} \tag{5}$$

$$c=\frac{a_0-a}{K\cdot a} \tag{6}$$

式中,a_0,b_0和c_0分别为PP、表面活性剂客体及环糊精主体的初始浓度;a,b和c分别为它们的平衡浓度;ΔA_∞为PP与环糊精形成配合物前后吸光度的差别,即$\Delta A_\infty=\Delta\varepsilon a_0$。测得的环糊精与表面活性 剂客体分子的配位稳定常数(K_s)及相应的自由能变化值($-\Delta G^0$)列于表4。

Table 4 Complex stability constants(K_s)and Gibbs free energy change($-\Delta G^0$)for 1∶1 inclusion complexation of surfactant guests with host CDs in carbonic acid buffer solution(pH=10.5)at 25℃

Host	Guest	K_s/(L·mol^{-1})	lg[K_s/(L·mol^{-1})]	$-\Delta G^0$/(kJ·mol^{-1})
NH_2-β-CD	OAS	152	2.18	12.45
	TDS	1 543	3.19	18.20
	SDS	6 670	3.82	21.83
EN-β-CD	OAS	1 370	3.14	14.66
	TDS	7 360	3.87	22.07
	SDS	10 500	4.02	22.95

(续表)

Host	Guest	$K/(L\cdot mol^{-1})$	$lg[K/(L\cdot mol^{-1})]$	$-\Delta G^0/(kJ\cdot mol^{-1})$
DETA-β-CD	OAS	247	2.39	13.66
	TDS	2 200	3.34	19.8
	SDS	10 700	4.03	23.00
TETA-β-CD	OAS	798	2.90	16.56
	TDS	2 336	3.37	19.23
	SDS	14 189	4.15	23.70

2.5 键合模式

采用二维核磁方法研究了环糊精主体与表面活性剂客体之间的键合模式。图 5(A)为 EN-β-CD 与 OAS 包结配位的 ROESY 波谱图。可见,峰 A 对应于环糊精空腔内的 3 位氢与 OAS 的 1 位甲基氢的 NOE 相关,峰 B 和峰 C 分别对应于环糊精空腔内 5 位和 6 位氢与 OAS 的 8 位氢的 NOE 相关。由此可以推断 OAS 包结进入了环糊精空腔,且其磺酸根阴离子位于环糊精的小口端。其键合模式如图 5(B)所示。在此基础上,进一步用分子动力学模拟的方法对主-客体之间的键合模式进行了确认。结果表明,在包合物稳定存在的条件下,其键合模式与用二维核磁推测的结果是一致的[图 5(C)]。

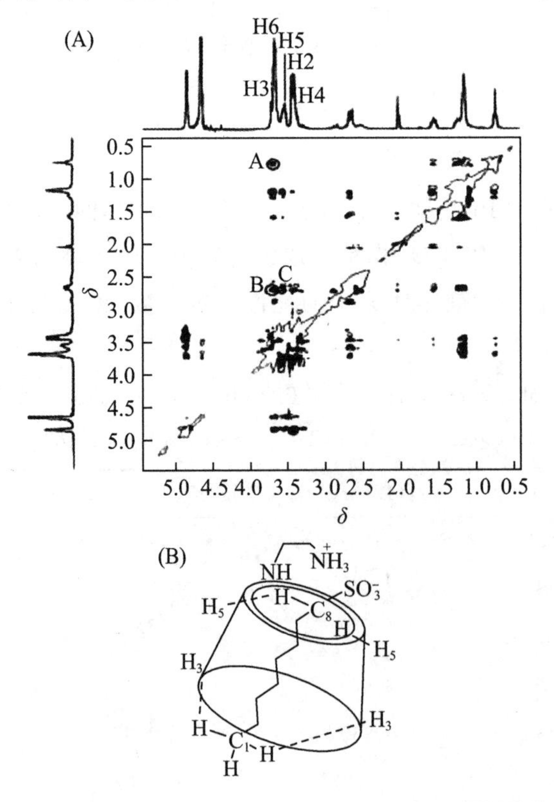

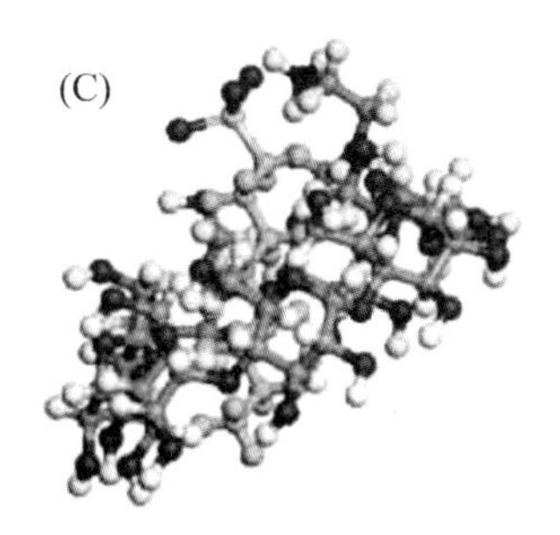

Fig. 5 ROESY spectrum of EN-β-CD/OAS complex in D_2O (A), possible binding mode(B) and Sheleton scheme of complex of EN-β-CD with OAS from MD simulations(C)

2.6 键合能力

环糊精的空腔具有疏水性，在与客体分子包结配位时，主-客体之间疏水性的匹配对主-客体键合常数和选择性具有重要影响。由表2和表4数据可见，对于具有不同长度疏水链的磺酸盐型阴离子表面活性剂客体，与多胺环糊精主体键合能力的强弱顺序为：十二烷基磺酸钠＞癸烷基磺酸钠＞辛烷基磺酸钠，即键合能力随表面活性剂客体疏水链的加长而逐渐升高。例如，在中性条件下，多胺环糊精主体对辛烷基磺酸钠的键合常数为 $2.6\times10^2\sim2.6\times10^3$ L/mol，当表面活性剂客体疏水链增加2个—CH_2—单元后，键合常数升高至 $3.7\times10^3\sim1.9\times10^4$ L/mol。而当表面活性剂客体疏水链再增加2个—CH_2—单元后，键合常数进一步升高至 $9.1\times10^3\sim4.4\times10^4$ L/mol。在碱性条件下，主-客体键合常数也显示出同样的变化规律。这是由于随着表面活性剂客体烷基链长度的增加，其疏水性也随之提高，从而导致表面活性剂客体分子与环糊精空腔之间产生更强的疏水相互作用。此现象表明，主-客体间的疏水相互作用是二者包结配位的主要驱动力。客体分子中的磺酸基与环糊精空腔边缘的羟基及侧臂胺基之间的氢键相互作用也在主-客体间的包结配位过程中起到重要作用。

在比较不同pH值下主-客体键合能力时发现，中性条件下多胺修饰环糊精对表面活性剂客体的键合能力明显高于碱性条件下的键合能力。一种可能的原因是，碱性条件下，多胺环糊精的侧臂会自包结进入环糊精空腔，且多胺侧臂越长，自包结的趋势越明显[26]，这就对表面活性剂客体分子进入环糊精空腔造成了较大的位阻，从而不利于主-客体的键合。在中性条件下，主体环糊精侧臂上的胺基部分质子化并移出环糊精空腔[27,28]，减少了对表面活性剂客体进入环糊精空腔的阻力。另一方面，表面活性剂的磺酸基以磺酸根阴离子形式存在，能够与多胺环糊精上质子化的胺基发生静电相互作用，从而导致中性条件下多胺修饰环糊精对表面活性剂客体更强的键合能力。

3 结论

以紫外-可见光谱、荧光光谱和二维核磁为手段，利用竞争包结法研究了不同pH值条件下几种多胺修饰环糊精与具有不同长度疏水烷基链的磺酸型阴离子表面活性剂之间的键合行为和稳定常数。结果表明，随着表面活性剂疏水烷基链的增长，多胺环糊精与客体的键合能力逐渐增强。而多胺环糊精在中性条件下键合磺酸型阴离子表面活性剂的能力明显强于碱性条件下的键合能力。

参考文献

[1] Bender M. L., Komiyama M.. Cyclodextrin Chemistry [M], Berlin: Springer,1978: 28-33.

[2] Harada A. ; Ed. : Semlyen J. A.. Large Ring Molecules[M],Chichester: Wiley & Sons,1996: 407-432.

[3] Szejtli J., Osa T.. Cyclodextrins,Comprehensive Supramolecular Chemistry [M],Oxford: Pergamon,1996: 3.

[4] Szejtli J.. Cyclodextrins and Their Inclusion Complexes[M], Budapest: Akademiai Kiado, 1982: 204-232.

[5] van Dongen S. F. M., de Hoog H. P. M., Peters R. J. R. W., Nallani M., Nolte R. J. M., van Hest J. C. M.. Chem. Rev., 2009,109: 6212-6274.

[6] Jing B., Chen X., Wang X. D., Yang C. J., Xie Y. Z., Qiu H. Y.. Chem. Eur. J., 2007,13: 9137-9142.

[7] Wang Y. P., Ma N., Wang Z. Q., Zhang X.. Angew. Chem. Int. Ed. [J],2007,46: 2823-2826.

[8] Zeng J. G., Shi K. Y., Zhang Y. Y., Sun X. H., Zhang B. L., Wan P. B., Jiang Y. G., Wang Y. P., Wang Z. Q., Zhang X.. Chem. Commun., 2008: 3753-3755.

[9] Park C., Lee I. H., Song Y., Rhue M., Kim C.. Proc. Natl. Acad. USA, 2006,103: 1199-1203.

[10] Zhou J., Tao F. G., Jiang M.. Langmuir,2007,23: 12791.

[11] Wang C. Z., Gao Q.. Huang J. B.. Langmuir,2003,19: 3757-3761.

[12] Zhang H. C., Shen J., Liu Z. N., Bai Y., An W., Hao A. Y.. Carbohydr. Res., 2009,344: 2028-2035.

[13] Zhang H. C., An W., Liu Z. N., Hao A. Y., Hao J. C., Shen J., Zhao X. H., Sun H. Y., Sun L. Z.. Carbohydr. Res., 2010,345: 87-96.

[14] Wang Y. P., Xu H. P., Zhang X.. Adv. Mater., 2009,21: 2849-2864.

[15] Wan P. B., Jiang Y. G., Wang Y. P., Wang Z. Q., Zhang X.. Chem. Commun., 2008: 5710-5712.

[16] REN Shen-Dong(任申冬),CHEN Dao-Yong(陈道勇),JIANG Ming(江明). Chem. J. Chinese Universities (高等学校化学学报),2010,31 (1): 167-171.

[17] PANG Jin-Yu(庞瑾瑜),XU Gui-Ying(徐桂英),BAI Yan(白燕),ZHAO Tan-

Tao(赵涛涛). Chem. J. Chinese Universities(高等学校化学学报), 2009,30 (4): 735-740.

[18] Liu Y., You C. C., Li B.. Chem. Eur. J., 2001,7: 1281-1288.

[19] Hamasaki K., Ikeda H., Nakamura A., Ueno A., Toda F., Suzzuki I., Osa T.. J. Am. Chem. Soc., 1993,115: 5035-5040.

[20] Rosen M. J.. Surfactants and Interfacial Phenomena[M]. Hoboken: John Wiley & Sons Inc., 2004: 122-123.

[21] Chen X. M., Yang H. Y., He P. S.. Chin. J. Chem. Phys., 2009,22: 541-544.

[22] Eli W. J., Chen W. H., Xue Q. J.. J. Chem. Thermodyn., 1999,31: 1283-1296.

[23] Liu Y., Han B. H., Zhang H. Y.. Curr. Org. Chem., 2004,8: 35-46.

[24] Zhu X. S., Sun J., Wu J.. Talanta, 2007, 72: 237-242.

[25] YOU Chang-Cheng(尤长城), ZHAO Yan-Li(赵彦利), LIU Yu(刘育). Chem. J. Chinese Universities(高等学校化学学报)[J], 2001,22 (2): 218-222.

[26] YOU Chang-Cheng(尤长城), ZHANG Min(张旻), LIU Yu(刘育). Acta Chim. Sinica(化学学报), 2000,58: 338-342.

[27] Alcalde M. A., Gancedo C., Jover A., Carrazana J., Soto V. H., Meijide F., Vázquez T.. J. Phys. Chem. B, 2006,110: 13399-13404.

[28] May B. L., Kean S. D., Easton C. J., Lincoln S. F.. J. Chem. Soc., Perkin Trans. 1, 1997: 3157-3160.

Interactions between polyamine-modified β-cyclodextrins and anionic surfactants

Sun Huanquan[1*] Liu Min[2] Cao Xulong[3] Cui Xiaohong[3] Shi Jing[3]
Guo Xiaoxuan[2] Chen Yong[2] Liu Yu[2*]

(1. Shengli Oilfield Company, SINOPEC, Dongying 257000, China ; 2. State Key Laboratory of Elemento-Organic Chemistry, Department of Chemistry, Nankai University, Tianjin 300071, China ;
3. Geological Scientific Research Institute of Shengli Oilfield Company, SINOPEC, Dongying 257015, China)

Abstract: The binding constants between several polyamine-modified β-cyclodextrins, i. e., mono(6-amino-6-deoxy)-β-cyclodextrin(NH_2-β-CD), mono[6-(ethylenediamino)-6-deoxy]-β-cyclodextrin (EN-β-CD), mono[6-(diethylenetriamino)-6-deoxy]-β-cyclodextrin (DETA-β-CD), mono[6-(triethylene tetraamino)-6-deoxy]-β-cyclodextrin(TETA-β-CD) and three anionic surfactants ($C_n H_{2n+1} SO_3 Na$: n= 8, OAS ; n=10, TDS ; n=12, SDS) in phosphate buffer solution(25℃, pH=7.2) and carbonic acid buffer solution(25℃, pH=10.5) were measured by the methods of UV-Vis and fluorescence spectroscopy titration using phenol-phthalein and neutral red as spectral probes. The 1 : 1 stoichiometry of hosts and guests was validated by Job's experiments, and the inclusion modes were determined by 2D NMR spectroscpy. The obtained results show that the binding constants between polyamine-modified β-cyclodextrins and anion surfactants were in the region of 2.6×10^2—4.35×10^4 L/mol in neutral condition, whereas in basic condition the binding would become weaker (1.5×10^2—1.42×10^4 L/mol). In both of two pH conditions, the binding ability became much stron-ger with increasing the carbon number of anionic surfactants. The different binding abilities can be derived from the cooperative effect of hydrophobic, electrostatic and hydrogen bonding interactions.

Keywords: Cyclodextrin; Spectroscopy titration; Anionic surfactant; Supramolecular chemistry

(Ed. : H,J,K)

不同电荷基团修饰环糊精与离子型表面活性剂的键合行为研究

孙焕泉[a]　刘　敏[b]　张瀛溟[b]　曹绪龙[c]　崔晓红[c]
石　静[c]　陈　湧[b]　刘　育[*,b]

(a 中国石化胜利油田分公司 东营 257000)
(b 南开大学化学系元素有机化学国家重点实验室 天津 300071)
(c 中国石化胜利油田分公司地质科学研究院 东营 257015)

摘要：采用紫外-可见光谱和荧光光谱滴定的方法测定了单(6-脱氧-6-苯胺)-β-环糊精(1)、单(6-脱氧-6-乙二胺)-β-环糊精(2)和单[6-氧-(4-苯甲酸)]-β-环糊精(3)在磷酸缓冲溶液中分别与阴离子表面活性剂十二烷基硫酸钠(SDS)和阳离子表面活性剂十二烷基三甲基溴化铵(DTAC)包结配位的稳定常数，并通过二维核磁等手段研究了主-客体之间的键合模式。结果表明，在环糊精的小口端修饰带有不同电荷的取代基，引入静电相互作用的识别位点，能够有效地改变环糊精对于离子型表面活性剂的键合能力，从而实现主体环糊精对于客体分子的选择性识别。

关键词 环糊精；表面活性剂；光谱滴定

Binding behaviours of different charge groups modified cyclodextrins with ionic surfactants

Sun Huanquan[a] Liu Min[b] Zhang Yingming[b] Cao Xulongc Cui Xiaohong[c]
Shi Jing[c] Chen Yong[b] Liu Yu[*,b]

([a] Shengli Oilfield Company, SINOPEC, Dongying 257000)
([b] State Key Laboratory of Elemento—Organic Chemistry, Department of Chemistry, Nankai University, Tianjin 300071)
([c] Geological Scientific Research Institute of Shengli Oilfield Company, SINO PEC, Dongying 257015)

Abstract: The complex stability constants for the inclusion complexation of mono-(6-anilino-6-deoxy)-β-cyclodextrin (1), mono-(6-ethylenediamino-6-deoxy)-β-cyclodextrin (2) and mono-[6-O-(4-carboxyl-phenyl)]-β-cyclodextrin (3) with anionic surfactant sodium dodecyl sulfate (SDS) and cationic surfactant dodecyl trimethyl ammonium bromine (DTAC) were determined by means of fluorescence and UV-Vis titrations in phosphate buffer solution, and their binding modes were investigated by 2D NMR spectroscopy. These results indicate that the introduction of different charge groups on the primary face of cyclodextrins can effectively change the binding abilities of the modified cyclodextrins towards ionic surfactants, and a high selectivity is achieved through the electrostatic interaction between host and guest compounds.

Keywords: cyclodextrin; surfactant; spectral titration

环糊精(cyclodextrins)是淀粉经环糊精葡萄糖基转移酶催化降解得到的、由 *D*-吡喃葡萄糖通过 α-1,4-糖苷键首尾相连形成的半天然化合物. 最常见的 α-,β-和 γ-环糊精分别由 6,7 和 8 个葡萄糖单元构成,环糊精及其衍生物的分子识别与组装是超分子化学的一个重要分支[1-3]。另一方面,表面活性剂因其在工业生产和生活等方面的大量应用,一直受到人们的关注[4]。因此,环糊精与表面活性剂相互作用的研究成为当前超分子化学的一个热点领域[5-8]。目前基于环糊精和表面活性剂的研究主要集中在天然环糊精上,化学修饰环糊精和表面活性剂键合行为的研究报道并不多见[9]。经化学修饰后的环糊精相比于天然环糊精显示出更高的分子键合能力和选择性。利用环糊精侧臂上的修饰基团与包结在环糊精空腔内的离子型表面活性剂头基之间的静电相互作用，可以扩展主体环糊精的分子选择性和对客体的识别能力[10]。在目前的研究工作中，我们合成了三种不同电荷基团修饰的β-环糊精衍生物，并系统研究了它们与不同类型的离子型表面活性剂之间的键合能力和选择性，并从疏水和静电相互作用等方面讨论了修饰环糊精侧臂上的不同取代基对于形成超分子包结配合物的影响，进而推断出上述三种主体环糊精与客体表面活性剂在水溶液中的构型(Scheme 1)。

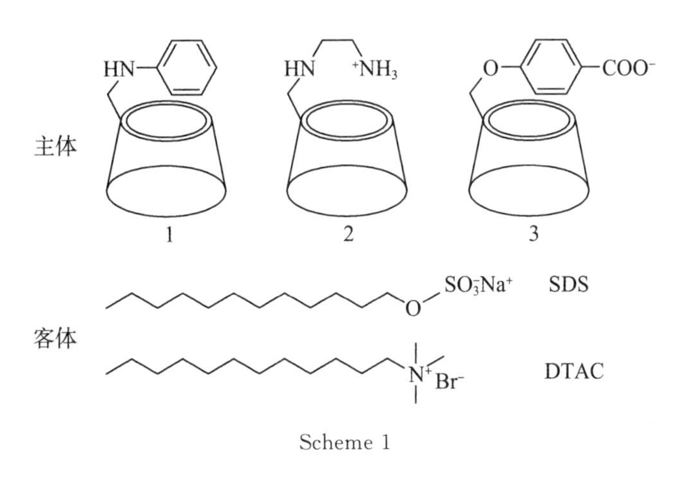

Scheme 1

1 实验部分

1.1 试剂与仪器

β-环糊精购于日本 Wako，重结晶三次后真空干燥(95℃，12 h)以备用。单(6-脱氧-6-苯胺)-β-环糊精(1)[11]、单(6-脱氧-6-乙二胺)-β-环糊精(2)[12]和单[6-氧-(4-苯甲酸)]-β-环糊精(3)[13]均按文献方法合成。罗丹明 B(RhB)、十二烷基硫酸钠(SDS)均为分析纯试剂，购于天津科锐思试剂公司。十二烷基三甲基溴化铵(DTAC)购于百灵威科技有限公司。

Shimadzu UV-2401PC 紫外-可见光谱仪，采用普通石英样品池(光程 10 mm)，以 PTC-348 WI 恒温器控制温度为(25.0±0.1)℃；Varian Cary Eclipe 荧光光谱仪，采用 10 mm×10 mm×40 mm 石英池，光谱测定以 Single Cell Peltier Accessory 恒温器控制温度为(25.0±0.1)℃；Varian Mercury400 核磁谱仪；pH 7.2 的 Na_2HPO_4-NaH_2PO_4 缓冲溶液(0.1 mol/L)以去离子水配制。在 pH=7.2 条件下，1 为中性分子[14]，2 中的氨基质子化以阳离子形式存在[15]，3 以阴离子形式存在[16]。

1.2 方法

采用紫外-可见光谱滴定法，固定含有生色团的主体浓度，其紫外吸收强度会随着客体表面活性剂浓度改变而改变，通过这种光谱变化来计算主-客体包结配位的稳定常数。在实验中，首先通过等物质的量连续变化法(Job's 方法)测定主一客体包结配位的化学计量比，然后根据非线性最小二乘法拟合公式计算包结配位的稳定常数。实验中所采用的表面活性剂浓度均在 5.0×10^{-4} mol/L 以下，由此可以推断在实验中客体表面活性剂均以单体形式存在[17]；对于主体 2，用 RhB 作为光谱探针，用竞争包结的方法测定主一客体包结配位常数[18]，具体方法是先通过荧光光谱滴定法测定竞争试剂 RhB 与主体 2 形成配合物的稳定常数(K)，然后固定 RhB 和主体环糊精的浓度，改变客体表面活性剂浓度，通过 RhB 在 582 nm 处的荧光强度变化，计算出主体环糊精与表面活性剂形成包结配合物的稳定常数。

2 结果与讨论

2.1 主-客体包结配位的化学计量比

在 pH 7.2 的磷酸缓冲溶液中固定主体 H 和客体 G 的总浓度(5.0×10^{-5} mol/L)不变，连续改变两组分的比例，测量各组混合液的吸光度与相同条件下主体自身吸光度的差值 ΔA，以 ΔA 对任一组分的摩尔分数作图，可以得到主-客体包结配位的化学计量比。以主体 1 和 SDS 的作用为例，由图 1 可以看出 Job's 曲线的最大值出现在主体 1 的摩尔分数为 0.5 处，表明主-客体间包结配位的化学计量比为 1∶1，这与文献报道的结果一致[19]。

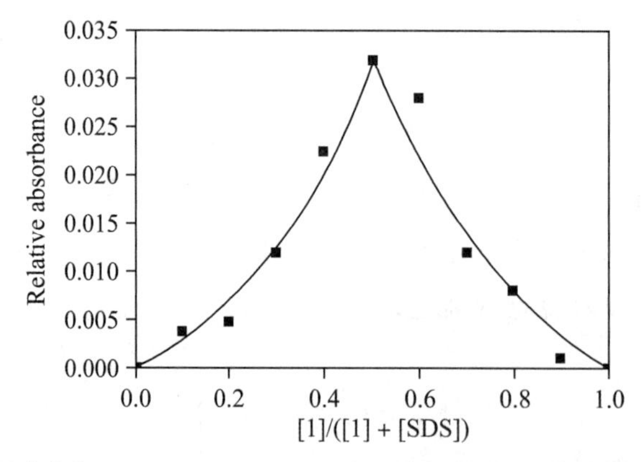

图 1　磷酸缓冲溶液(0.1 mol/L，pH 7.2)中主体 1 与 SDS 包结配位的 Job's 曲线
(242 nm，[1]+[SDS]=5.0×10^{-5} mol/L)

Figure 1　Job's plot of 1/SDS system in NaH_2PO_4-Na_2HPO_4 buffer solution
(0.1 mol/L，pH 7.2) (242 nm，[1]+[SDS]=5.0×10^{-5} mol/L)

2.3 环糊精主体与表面活性剂客体稳定常数的测定

2.3.1 紫外光谱研究

我们通过紫外光谱滴定实验定量地研究了主体环糊精 1 和 2 与客体表面活性剂之间的键合行为。随着客体表面活性剂 DTAC 的加入，主体环糊精 1 在 242 nm 处的紫外吸收强度逐渐上升(如图 2 所示)。

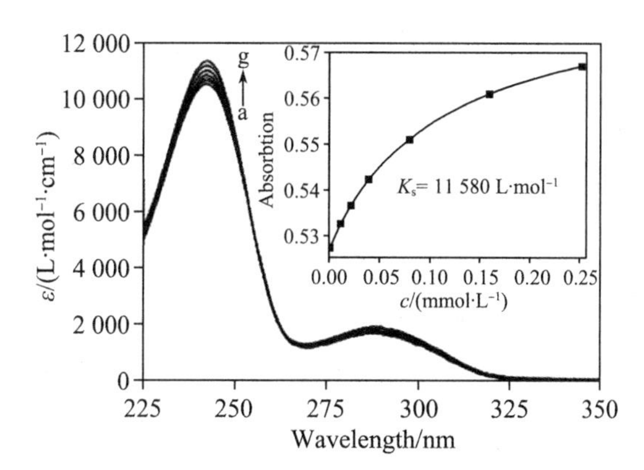

图 2 磷酸缓冲溶液(0.1 mol/L，pH 7.2)中主体 1 随 DTAC(0～2.4×10^{-4} mol/L)的不断加入其紫外光谱的变化

Figure 2 UV/Vis spectral changes of 1 (5.0×10^{-5} mol/L) upon addition of DTAC (0～2.4×10^{-4} mol/L) in $NaH_2PO_4-Na_2HPO_4$ buffer solution (0.1 mol/L, pH 7.2) at 25℃

Inset: Least-squares curve-fitting analyses of the absorption intensity (at 242 nm) to calculate the complex stability constant (Ks) of 1/DTAC system. a: 5.0× 10^{-5} mol/L host 1; b: a+1.0×10^{-5} mol/L DTAC; c: a+2.0×10^{-5} mol/L DTAC; d: a+4.0×10^{-5} mol/L DTAC; e: a+8.0×10^{-5} mol/L DTAC; f: a + 1.6×10^{-4} mol/L DTAC; g: a+2.4×10^{-4} mol/L DTAC

通过 Job's 曲线可得主体环糊精与客体在水溶液中形成化学计量比为 1∶1 的配合物,根据非线性最小二乘法拟合公式(1)可以计算出主体 1 与 DTAC 包结配位的稳定常数(K_s),如图 2 插图所示。

其中$[H]_0$，$[G]_0$，ΔA，$\Delta\varepsilon$ 分别为主体的初始浓度、客体的初始浓度、加入客体后主体的吸光度变化值及加入客体后主体的摩尔吸光系数变化值。采用此方法测得的环糊精与表面活性剂客体分子的配位稳定常数(K_s)及相应的自由能变化($\Delta G°$)，如表 1 所示[20,21]。主体环糊精 1 和 DTAC 的键合常数(K_s)以及相应的 Gibbs 自由能变($-\Delta G°$)分别为 11 580 L·mol^{-1} 和 23.19 kJ·mol^{-1}，主体 2 和 DTAC 键合的键合常数(K_s)以及相应的 Gibbs 自由能变($-\Delta G°$)分别为 8 450 L·mol^{-1}和 22.41 kJ·mol^{-1}，然而主体 3 同客体 DTAC 键合的键合常数(K_s)，以及相应的 Gibbs 自由能变($-\Delta G°$)增加到 33 640 L·mol^{-1} 和 25.84 kJ·mol^{-1}。同主体 1,2 相比，主体 3 对于阳离子表面活性剂的扩展键合能力是由于主体 3 的苯甲酸基团在 pH=7.2 的磷酸缓冲溶液中带有负电荷，引入了静电吸引的识别位点，提高了主-客体之间的键合配位能力。

2.3.2 荧光光谱研究

由于主体环糊精 2 不含有生色团，本文采用竞争包结法测定了主体 2 与客体包结配位的稳定常数。因为 RhB 与环糊精键合能力适中，且其在环糊精空腔内外荧光较强，因此选择 RhB 分子作为荧光探针。

通过荧光光谱滴定法测得竞争试剂 RhB 与主体 2 形成配合物的稳定常数(K)为 7

379 L·mol^{-1}(图 3A)。固定主体环糊精和探针 RhB 的浓度分别为 2.0×10^{-4} 和 1.0×10^{-6} mol/L(二者键合 95%以上),向环糊精/RhB 体系中加入不同浓度的表面活性剂。图 3B 给出了主体 2 与 RhB 的混合体系中加入 DTAC 的荧光光谱,则主-客体的键合常数 K_s 可通过文献方法[22]计算。采用此方法获得的环糊精主体与表面活性剂的包结配位稳定常数(K_s)和相应的自由能变化(ΔG°),如表 1 所示。在控制实验中,固定 RhB 的浓度,向溶液中加入一定浓度的表面活性剂客体。RhB 在 582 nm 处的荧光强度基本无变化,说明在一定浓度范围内 RhB 与客体表面活性剂之间没有相互作用。

$$\Delta A=\frac{\Delta\varepsilon([\mathrm{H}]_0+[\mathrm{G}]_0+K_s)-\sqrt{(\Delta\varepsilon)^2([\mathrm{H}]_0+[\mathrm{G}]_0+K_s)^2-4(\Delta\varepsilon)^2[\mathrm{H}]_0[\mathrm{G}]_0}}{2} \tag{1}$$

表 1　磷酸缓冲溶液(0.1 mol/L, pH 7.2)中环糊精 1～3 与表面活性剂客体分子 1∶1 键合时的配位稳定常数(K_s)和 Gibbs 自由能变化($-\Delta G_\circ$)

Table 1　Complex stability constant (K_s) and Gibbs free energy change ($-\Delta G_\circ$) for 1∶1 inclusion complexation of host compounds 1～3 with surfactant guests in $NaH_2PO_4-Na_2HPO_4$ buffer solution (0.1 mol/L, pH 7.2) at 25℃

Guests	Host	K_s/(L·mol^{-1})	$-\Delta G^\circ$/(kJ·mol^{-1})	lg K_s
DTAC	1	11 580	23.19	4.06
	2	8 450	22.41	3.93
	3	33 640	25.84	4.53
SDS	1	11 180	23.88	4.05
	2	46 270	26.63	4.67
	3	31 780	25.70	4.50

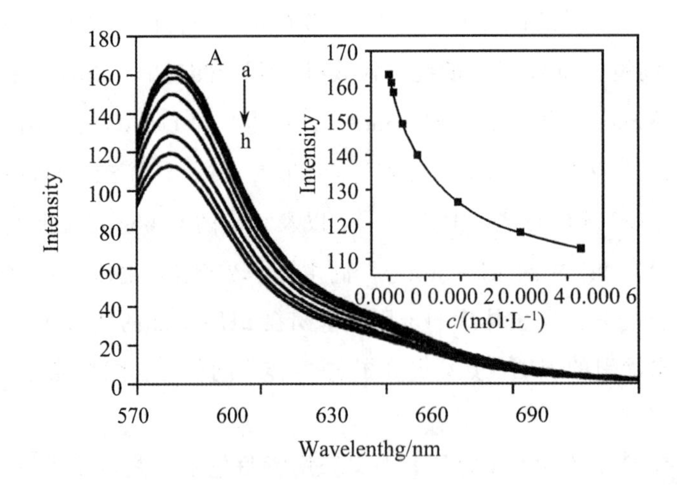

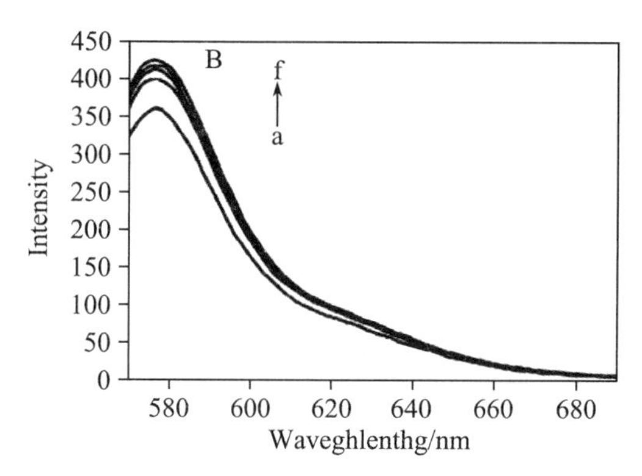

图 3　(A)磷酸缓冲溶液(0.1 mol/L,pH=7.2)中 RhB 随主体 2(0～2.5×10^{-4} mol/L)的不断加入其荧光光谱的变化和非线性最小二乘拟合曲线的到主体 2 与 DTAC 的键合常数;(B)当体系中含有主体 2(2.0×10^{-4} mol/L)时,磷酸缓冲溶液(0.1 mol/L,pH 7.2)中 RhB(1.0×10^{-6} mol/L)随 DTAC(0～2.5×10^{-4} mol/L)的不断加入其荧光光谱的变化

Figure3 (A)Fluorescence changes of RhB (1×10^{-6} mol/L) upon addition of 2 (0～2.5×10^{-4} mol/L) in NaH_2PO_4－Na_2HPO_4 buffer solution (pH 7.2) at 25℃; (B) Fluorescence changes of RhB (1.0×10^{-6} mol/L) upon addition of DTAC (0～2.5×10^{-4} mol/L) in the presence of 2 (2.0×10^{-4} mol/L) in NaH_2PO_4－Na_2HPO_4 buffer solution (pH=7.2) at 25℃

Inset: Least－squares curve－fitting analyses of the fluorescence intensity (at 582 nm) to calculate the complex stability constant (K_s) of 2/DTAC system. (A) a: 1.0×10^{-6} mol/L RhB; b: a+2.0×10^{-6} mol/L host 2; c: a+5.0×10^{-6} mol/L host 2; d: a+5.0×10^{-5} mol/L host 2; e: a+1.0×10^{-4} mol/L host 2; f: a+2.0×10^{-4} mol/L host 2; g: a+3.5×10^{-4} mol/L host 2; h: a+5.5×10^{-4} mol/L host 2. (B) a: 2.0×10^{-4} mol/L host 2+1.0×10^{-6} mol/L RhB; b: a+0.5×10^{-4} mol/L DTAC; c: a+1.0× 10^{-4} mol/L DTAC; d: a+1.5×10^{-4} mol/L DTAC; e: a+2.5×10^{-4} mol/L DTAC; f: 1.0×10^{-6} mol/L RhB

2.4 键合模式

本文采用二维核磁方法研究了环糊精主体与表面活性剂客体之间的键合模式。图 4－A1 为主体 2 和客体 SDS 包结配位的 ROESY 谱图,从图中可以看到峰 A,B 对应于环糊精空腔内的 3 位氢与 SDS 的 1 位甲基氢的 NOE 相关,峰 C 对应于环糊精空腔内 5 位和 6 位氢与 SDS 的 11 位氢的 NOE 相关。峰 D 为环糊精空腔与表面活性剂链上的 NOE 相关。由此,可以推断出 SDS 包结进入了环糊精空腔,且其头基位于环糊精的小口端。其键合模式如图 4-B1 所示。

图 4-A2 为主体 2 与 DTAC 包结配位的 ROESY 谱图,从图中可以看到峰 A 对应于环糊精空腔内的 3 位氢与 DTAC 的 1 位甲基氢的 NOE 相关,峰 B 对应于环糊精空腔内 5 位与 DTAC 的 11 位氢的 NOE 相关。峰 C 为环糊精空腔 1 位氢与表面活性剂氮甲基上氢的相关,由此,可以推断 DTAC 包结进入了环糊精空腔,且其头基由于静电排斥弯向环糊精外侧。其键合模式如图 4-B2 所示。

图 4-A3 为主体 1 与 SDS 包结配位的 ROESY 谱图,从图中可以看到峰 A 对应于环糊

精的苯环侧臂上的氢与 SDS 上 1 位甲基氢的 NOE 相关，峰 B 对应于环糊精的苯环侧臂上的氢与 SDS 的链上的氢的 NOE 相关。峰 C 为环糊精空腔氢与表面活性剂链上氢的相关，表明表面活性剂链进入了环糊精空腔，峰 D 为环糊精的苯环侧臂氢与环糊精相关。在本文工作中，主体对于 SDS 的包结并没有使主体 1 的苯环侧臂完全从 β-环糊精的空腔驱逐出来，而是从空腔的中部来到小口端的端口区域，由此推断其键合模式如图 4-B3 所示。

2.5 结果与讨论

环糊精的空腔具有疏水性，在与客体分子包结配位时，主-客体之间疏水作用和范德华力对主-客体之间的包结配位常数和选择性具有重要影响，这两个作用力的强度主要由客体尺寸、形状、疏水性以及与环糊精匹配的程度决定。另外，拥有电荷中心的修饰环糊精衍生物，取代基团可以与合适客体分子产生静电相互作用，使主-客体间在范德华力和疏水相互作用的基础上增加新的识别点，而在特定情况下这种经典作用力可以提升为配合物形成的主要驱动力，从而产生特殊的分子识别效果。胺基修饰环糊精在一定 pH 值范围内可以质子化[15]，从而带有正电荷，由于存在静电相互作用，在某种程度上增强了所形成包结配合物的稳定性。

如图 5 所示，在 pH 7.2 条件下，主体 1 侧链上的氮原子没有被质子化[14]，整体分子呈现中性，对客体 DTAC 和 SDS 键合的驱动力均来源于疏水作用、氢键和脱溶剂化等弱相互作用，因此主体 1 对两种客体分子的选择性并不明显，键合常数均在 10^4 L·mol^{-1} 左右；当键合带有负电的客体 SDS 时，由于主体 2 上的氮原子由于质子化而带上正电荷，主客体之间存在上述弱相互作用的同时，还会产生的静电吸引力，这种协同作用使二者之间的键合能力明显增强，键合常数达到 4.6×10^4 L·mol^{-1}；相反，当键合带有正电荷的客体分子 DTAC 时，由于存在静电排斥作用，与主体 1 相比，键合能力明显减弱。因此，主体 2 对两种客体分子表现出了明显的选择性。由于主体 3 侧链上的羧基没有质子化，即以 COO^- 离子形式存在，当键合带有正电荷的客体分子 DTAC 时，由于静电吸引作用，与主体 1 和 2 相比，键合能力明显增强。这里需要指出的是，主体分子 3 与 SDS 也表现出相对较强的键合，而并未像主体 2 因静电排斥作用而给出弱的键合。从二维核磁实验可以看出，主体 2 的修饰基团乙二胺被客体分子完全顶出环糊精空腔，客体穿过空腔；而主体 3 对于客体分子的包结并没有使主体 3 的苯甲酸修饰基团完全从 β—环糊精的空腔内驱逐出来，而是从空腔的中部来到小口端的端口区域[11]。因此，带有负电荷的主体分子 3 在包结带有正电荷的表面活性剂 DTAC 时正负电荷的相互吸引虽然对于键合起到了促进作用，但另一方面环糊精的空腔不利于包结带有正电荷的客体分子[23,24]。而主体分子 3 在包结带有负电荷的表面活性剂 SDS 时虽然主客体的负电荷之间的静电排斥不利于键合，但环糊精的空腔更倾向于包结带有负电荷的客体分子[24]。基于这两种因素的影响，主体分子 3 虽然带有负电荷，但在包结带有不同电性的表面活性剂时并不能像带有正电荷的主体分子 2 一样体现出好的客体选择性。

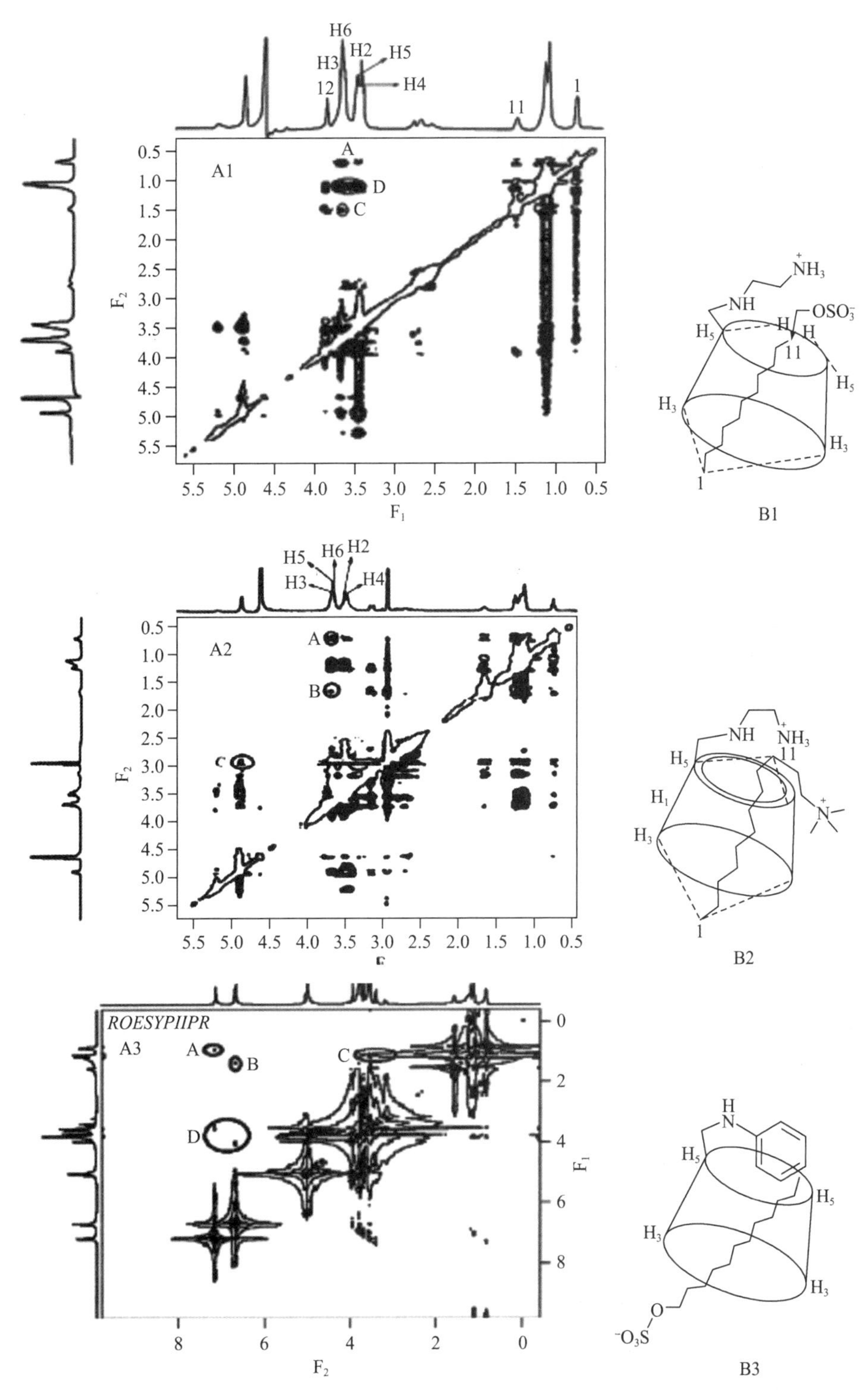

图 4　(A)主体(5.0×10^{-5} mol/L)与客体(5.0×10^{-5} mol/L)的包结配合物在 25℃下的 ROESY 谱图(溶剂:D2O);(B)包结配合物可能的构型

Figure 4　(A) ROESY spectrum of host/guest complex in D2O(5.0×10^{-5} mol/L);(B) possible binding mode

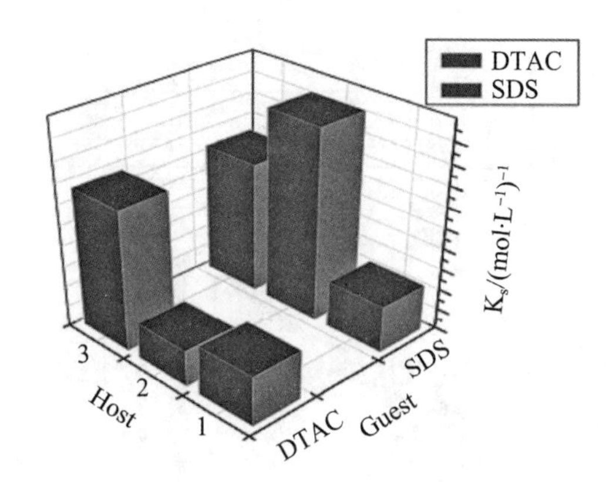

图 5 在 298.15 K 磷酸缓冲溶液中主体环糊精 1～3 与客体分子形成包结配合物的配位稳定常数(K_s)

Figure 5 Complex stability constants(K_s) of inclusion com－plexation of hosts 1～3 with guest molecules in NaH_2PO_4－ Na_2HPO_4 buffer solution(pH 7.2) at 25℃

3 结论

本文通过紫外-可见光谱、荧光光谱和二维核磁等手段,系统研究了 pH7.2 条件下不同电荷基团修饰β-环糊精与离子型表面活性剂之间的键合行为和包结配位的稳定常数。结果表明,通过引入带电荷的取代基,能够产生新的识别位点,从而可以有效地改变环糊精对于带电荷客体的分子选择性。

References

[1] Brewster, M. E.; Loftsson, T. Adv. Drug Delivery Rev. 2007, 59, 645.

[2] Chen, Y.; Liu, Y. Chem. Soc. Rev. 2010, 39, 495.

[3] Bellia, F.; Mendola, L. －D.; Pedone, C.; Rizzarelli, E.; Saviano, M.; Vecchioa, G. Chem. Soc. Rev. 2009, 38, 2756.

[4] Huang, H.; Li, L.; M, Q.; Feng, Y. －Q.; He, Z. －K. Chin. J. Anal. Chem. 2010, 38, 249 (in Chinese).

(黄晖, 李丽, 马乔, 冯钰锜, 何治柯, 分析化学, 2010, 38, 249.)

[5] Zeng, J.-G.; Shi, K.-Y.; Zhang, Y.-Y.; Sun, X.-H.; Zhang, B. －L. Chem. Commun. 2008, 3753.

[6] Zhang, H.-C.; Shen, J.; Liu, Z. －N.; Bai, Y.; An, W.; Hao, A. －Y. Carbohydr. Res. 2009, 344, 2028.

[7] Xing, H.; Lin, C.-X.; Xiao, J.-X. Acta Chim. Sinica 2008, 66, 1382 (in Chinese).

(邢航, 林崇熙, 肖进新, 化学学报, 2008, 66, 1382.)

[8] Sun, D.-Z.; Wang, S.-B.; Wei, X.-L.; Yin, B.-L.; Li, L.-W. Acta

Chim. Sinica 2004, 62, 1247 (in Chinese).

(孙得志，王世兵，魏西莲，尹宝霖，李林尉，化学学报，2004，62，1247.)

[9] Pang, J.-Y.; Xu, G.-Y.; Bai, Y.; Zhao, T.-T. Chem. J. Chin. Univ. 2009, 30, 735 (in Chinese).

(庞瑾瑜，徐桂英，白燕，赵涛涛，高等学校化学学报，2009，30，735.)

[10] Liu, Y.; Cao, R.; Chen, Y.; He, J.-Y. J. Phys. Chem. B 2008, 112, 1445.

[11] Liu, Y.; You, C.-C.; Wada, T.; Inoue, Y. Supramol. Chem. 2000, 12, 299.

[12] Liu, Y.; You, C.-C.; Li, B. Chem. Eur. J. 2001, 7, 1281.

[13] Fan, Z.; Zhao, Y.-L.; Liu, Y. Chin. Sci. Bull. 2003, 48, 1535.

[14] Liu, Y.; You, C.-C.; Kunieda, M.; Nakamura, A.; Wada, T.; Inoue, Y. Supramol. Chem. 2000, 12, 299.

[15] May, B.-L.; Kean, S.-D.; Easton C.-J.; Lincoln, S.-F. J. Chem. Soc. Perkin Trans. 1 1997, 3157.

[16] Rubinson, A. K. J. Phys. Chem. 1984, 88, 148.

[17] Bai, G.-Y.; Wang, J.-B.; Yang, G.-Y.; Han, B.-X.; Yan, H.-K. Acta Chim. Sinica 2000, 58, 1103 (in Chinese).

(白光月，王金本，杨冠英，韩布兴，闫海科，化学学报，2000，58，1103.)

[18] Nau, W.-M.; Ghale G.; Hennig, A.; Hüseyin, B.; Bailey, D.-M. J. Am. Chem. Soc. 2009, 131, 11558.

[19] Chen, X.-M.; Yang, H.-Y.; He, P.-S. Chin. J. Chem. Phys. 2009, 22, 541.

[20] Yang, B.; Yang, L.-J.; Lin, J.; Chen, Y.; Liu, Y. J. Inclusion Phenom. Macrocyclic Chem. 2009, 64, 149.

[21] Chen, Y.; Han, N.; Yang, H.; Liu, Y. Acta Chim. Sinica 2007, 65, 1076 (in Chinese).

(陈湧，韩宁，杨华，刘育，化学学报，2007，65，1076.)

[22] Zhu, X.-S.; Sun, J.; Wu, J. Talanta 2007, 72, 237.

[23] Ooya, T.; Inoue, D.; Choi, H.-S.; Kobayashi, Y.; Loethen, S.; Thompson, D.-H.; Ko, Y.-H.; Kim, K.; Yui, N. Org. Lett. 2006, 8, 3160.

[24] Connors, K.-A. Chem. Rev. 1997, 97, 1325.

(Y1010141 Cheng, F.; Zheng, G.)

新型磺基甜菜碱与聚丙烯酰胺作用的研究*

丁 伟[1] 李金红[1] 曹绪龙[2] 刘 坤[2] 潘斌林[2] 于 涛[1] 邢欣欣[1]
张 微[1] 杨 柳[1]

(1. 东北石油大学化学化工学院石油与天然气化工省重点实验室，黑龙江大庆 163318；
2. 胜利油田地质科学研究院，山东东营 257000)

摘要：合成了一种耐温抗盐表面活性剂，通过红外光谱分析了该表面活性剂的结构。将其与 2 500 万分子量聚丙烯酰胺进行复配，考察了复配体系的表面、界面性能。研究结果表明：所合成的产物为目标产物；磺基甜菜碱表面活性剂的临界胶束浓度(cmc)为 2.19×10^{-3} mol/L，临界胶束浓度下的表面张力(γ_{cmc})为 25.51 mN/m；加入聚合物后临界胶束浓度变为 4.09×10^{-3} mol/L，γ_{cmc} 变为 26.65 mN/m；表面活性剂质量浓度在 0.8～1.5 g/L，可使胜利原油油水间的界面张力达到超低数量级(10^{-3} mN/m)；聚合物的加入有利于乳状液的形成。

关键词：甜菜碱表面活性剂　表面张力　界面张力　乳化性能

中图分类号：TE357.46　文献标识码：A

文章编号：1006-7906(2013)06-0033-03

Research on interaction between novel sulfobetaine and polyacrylamide

Ding Wei[1] Li Jinhong[1] Cao Xulong[2] Liu Kun[2] Pan Binlin[2]
Yu Tao[1] Xing Xinxin[1] Zhang Wei[1] Yang Liu[1]

(1. Provincial Key Laboratory of Oil & Gas Chemical Technology, College of Chemistry & Chemical Engineering, Northeast Petroleum University, Daqing 163318, China; 2. Shengli Oilfield Geologic Science Research Institute, Dongying 257000, China)

Abstract: A heat resisting and salt tolerant surfactant is synthesized, and the structure is analyzed by Infrared (IR). The sulfobetaine surfactant is compounded with 25 million molecular weight of polymer, the surface and interfacial properties of the compound system are investigated. The experimental results indicate that the product is the target one, the critical micelle concentration of sulfobetaine surfactant is 2.19×10^{-3} mol/L, the surface tension(γ_{cmc})is 25.51 mN/m; with the polymer added, the cmc become 4.09×10^{-3} mol/L, and the γ_{cmc} become 26.65 mN/m; the mass concentration of surfactant is in the range of 0.8 g/L to 1.5 g/L, the water/crude oil interfacial tension is arrived ultra low value 10^{-3} mN/m from Shengli oilfield. It is conductive to form the emulsion due to the polymer added.

Keywords: Betaine surfactang; Surface tension; Interfacial tension; Emulsifying performance

甜菜碱两性表面活性剂的分子结构中同时含有阴离子、阳离子 2 种亲水基团，在酸性条件下表现出阳离子表面活性剂的特性，碱性条件下表现出阴离子表面活性剂的特性，使其具有良好的界面活性、耐硬水性及生物降解性[1-3]。目前，甜菜碱表面活性剂在三次采油中已有应用[4-5]，但单独的表面活性剂溶液只能降低界面张力，不能提高注入水的波及体积，不能将原油采收率提高到满意程度；而单独的聚合物溶液只能提高驱替相的黏度，不能明显降低界面张力，因而不能大幅提高采收率。若能综合这二者的优点，形成既降低界面张力又增黏的聚表二元体系，就能达到大幅度提高采收率的目的[6]。本研究合成了一种新型磺基甜菜碱表面活性剂，将其与聚丙烯酰胺复配，考察了复配后体系的表面性能、界面性能及乳化性能。

1 实验部分

1.1 实验药品与仪器

实验药品:新型磺基甜菜碱两性表面活性剂(实验室自制,质量分数大于95%);聚丙烯酰胺(相对分子质量2 500万,大庆炼化公司);液蜡(天津市大茂化学试剂厂,分析纯);试验水(矿化度32.308 g/L);原油(胜利孤岛原油)。

主要仪器:DCAT41接触角/表面张力测定仪(德国DataPhysics公司);TX500C旋滴型界面张力仪(美国盛维公司);FTIR-8400S红外光谱仪(日本岛津公司)。

1.2 实验方法

1.2.1 产物结构表征

用FTIR-8400S红外光谱仪分析提纯样品结构。

1.2.2 表面性能测定

以矿化水为溶剂,配制不同浓度的表面活性剂溶液,聚合物质量浓度1.0 g/L,配制后置于恒温水浴中,采用滴体积法,在(25±0.2)℃下测其表面张力。

1.2.3 界面性能测定

以矿化水为溶剂,配制一系列不同浓度的表面活性剂溶液,聚合物质量浓度1.5 g/L,在85℃条件下,采用旋滴法测定其与胜利孤岛原油的油水界面张力。

1.2.4 乳化性能测定

采用分水时间法[7],以矿化水为溶剂,将合成的磺基甜菜碱表面活性剂配成一系列不同浓度的水溶液,并在每份溶液中加入聚合物,质量浓度1.0 g/L。在85℃条件下,100 mL具塞量筒中,分别加入40 mL液蜡(油相)和40 mL表面活性剂-聚合物水溶液,按紧玻璃塞上下猛烈震动5次,静置1 min;如此重复震动、静置5次,记录分出10 mL水所用时间;重复3次,取平均值。

2 结果与讨论

2.1 磺基甜菜碱表面活性剂结构表征

图1为合成产物红外光谱图。

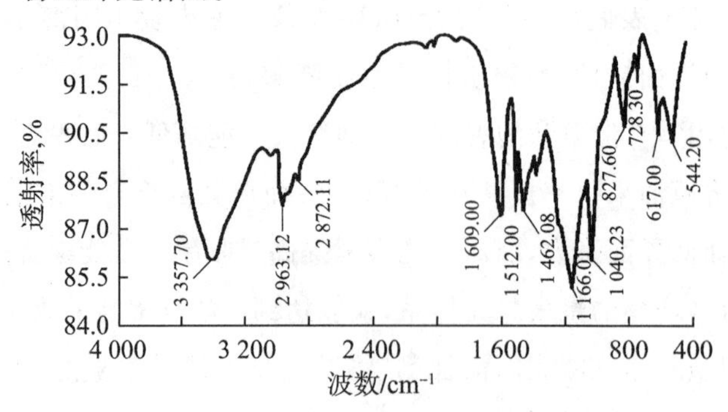

图1 合成产物的红外光谱图

由图 1 可见：3 357.70 cm^{-1} 处为 O—H 的伸缩振动吸收峰，说明有羟基存在；2 963.12 cm^{-1} 处为饱和烷烃 C—H 的伸缩振动吸收峰，2 963.12 cm^{-1} 和 2 872.11 cm^{-1} 是—CH_2—的反对称伸缩振动吸收峰，1 609 cm^{-1} 处是 N—H 弯曲振动吸收峰；1 512 cm^{-1} 中等强度吸收峰是苯环骨架的伸缩振动吸收峰，1 462.08 cm^{-1} 为—CH_2 和—CH_3 的箭式弯曲振动吸收峰；1 166.01 cm^{-1} 处的强度吸收峰为醚键中的 C—O 伸缩振动引起的，1 040.23 cm^{-1} 是—SO_3—伸缩振动吸收峰，728.3 cm^{-1} 为亚甲基的骨架振动。可见所合成产品具备目标产物特征结构。

2.2 表面性能

磺基甜菜碱表面活性剂溶液的表面张力 γ 与浓度 c 之间的关系如图 2 所示。

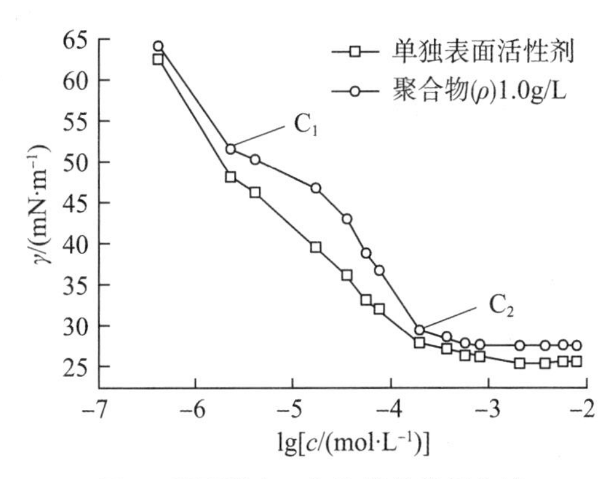

图 2　表面张力 γ 与浓度的关系曲线

由图 2 可见：随着表面活性剂浓度的增大，表面张力逐渐降低，当降低至一定值时不再变化；加入聚合物后体系表面张力有所升高。加入聚合物后曲线上出现 2 个转折 C_1 和 C_2，这是聚合物与磺基甜菜碱表面活性剂相互作用的结果。在 C_1 点表面活性剂分子吸附到聚合物分子链上，随着表面活性剂浓度的增加吸附到聚合物链上的表面活性剂增多，在 C_2 点时达到饱和吸附，表面活性剂浓度继续增加，表面张力基本保持不变。由 γ-lgc 曲线可计算出单独表面活性剂溶液的临界胶束浓度(cmc)为 2.19×10^{-3} mol/L，临界胶束浓度时的表面张力(γ_{cmc})为 25.51 mN/m；加入相对分子质量 2 500 万、质量浓度 1.0 g/L 的聚合物后，临界胶束浓度变为 4.09×10^{-3} mol/L，γ_{cmc} 变为 26.65 mN/m，可见加入聚合物使达临界胶束浓度的表面活性剂浓度增大。

2.3 界面性能

在 85℃条件下，不同浓度磺基甜菜碱表面活性剂、聚丙烯酰胺复配体系与胜利孤岛原油油水间的界面张力随时间变化情况如图 3 所示。

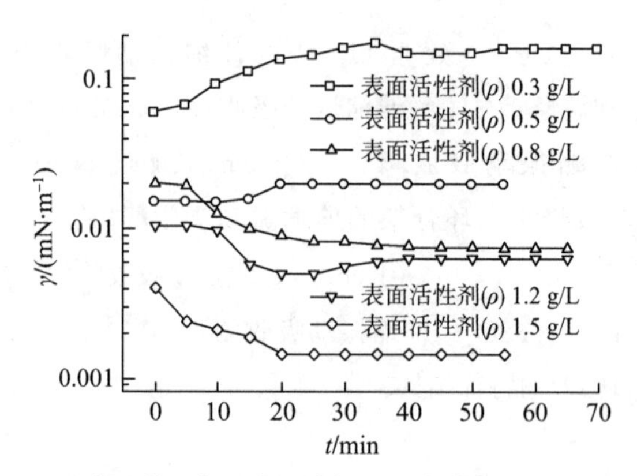

图 3 复配体系与孤岛原油间的界面张力与时间关系

由图 3 可见:表面活性剂浓度增大时,界面张力降低,当表面活性剂质量浓度为 0.8,1.2,1.5 g/L 时,界面张力可达超低数量级(10^{-3} mN/m),界面张力随时间延长而降低,达到稳定后不再变化;当表面活性剂质量浓度为 0.3,0.5 g/L 时,界面张力随时间的延长而增大,达到一定值时不再变化。

2.4 乳化性能

复配体系乳化性能曲线如图 4 所示,以分出 10 mL 水所用时间长短作为乳化性能的评价指标,所需时间越长,乳化性能越好。

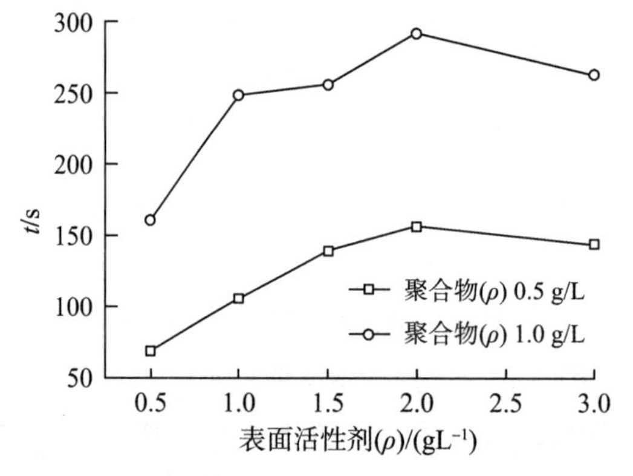

图 4 磺基甜菜碱表面活性剂乳化性能曲线

由图 4 可见:随磺基甜菜碱表面活性剂质量浓度的增大,分出 10 mL 水所需要的时间先增加后减小;当聚合物质量浓度为 0.5 和 1.0 g/L 时,表面活性剂质量浓度为 2.0 g/L 时分出 10 mL 水所需要的时间最长;另外,随着聚合物浓度的增加,分出 10 mL 水所需要的时间延长,由此可知,聚合物的加入有利于乳状液的形成。

3 结论

(1) 合成了甜菜碱两性表面活性剂,经红外光谱确定产物为目标产物。

(2) 磺基甜菜碱两性表面活性剂与聚丙烯酰胺复配后,使得达临界胶束浓度的表面活性剂浓度升高,且聚合物和表面活性剂相互作用使溶液的表面张力比单独表面活性剂溶液的表面张力略高。

(3) 聚丙烯酰胺质量浓度为1.0 g/L,表面活性剂质量浓度为0.8～1.5 g/L,可使胜利油田孤岛原油的油水界面张力达到超低数量级。

(4) 加入聚合物后,分出10 mL水的时间延长,表明聚合物的加入有利于乳状液的形成。

参考文献

[1] 丁伟,李淑杰,于涛,等.新型磺基甜菜碱的合成与性能[J].化学工业与工程,2012,29(1):26-29.

[2] 王军,周晓微.十二烷基二甲基羟丙基磺基甜菜碱的合成工艺优化[J].陕西科技大学学报,2007,25(2):86-90.

[3] MAR AA,PARRA J L. Solubilization of phospholipid bi-layers by C14-alkyl betaine/anionic mixed surfactant sys-tems[J]. Colloid Polym Sci,1995(273): 331-338.

[4] 吴文祥,闫伟,刘春德.磺基甜菜碱BS11的界面特性研究[J].油田化学,2007,24(1): 57-59.

[5] 吴文祥,张武,刘春德.磺基甜菜碱BS11/聚合物复合体系驱油实验研究[J].油田化学,2007,24 (1): 60-62.

[6] 叶仲斌,施雷庭,杨建军,等.新型缔合聚合物与表面活性剂的相互作用[J].西南石油学院学报,2003,25 (1): 55-58.

[7] 于涛,刘华沙,王超群,等.烷基芳基磺酸钠对烷烃的乳化性能[J].应用化学,2011,28(5): 560-564.

聚合物疏水单体与表面活性剂对聚/表二元体系聚集体的作用

季岩峰　曹绪龙　郭兰磊　闵令元　窦丽霞　庞雪君　李　斌

(中国石化胜利油田分公司勘探开发研究院,山东东营 257015)

摘要:将不同疏水单体的两亲聚合物与阴离子表面活性剂十二烷基硫酸钠(SDS)复配得到聚/表二元体系,利用流变学方法、动态光散射和芘荧光探针法,研究疏水单体与SDS对两亲聚合物及其二元体系聚集体的作用规律。结果表明:由于疏水缔合和电性作用,SDS在水溶液中会与两亲聚合物发生相互作用。当SDS质量浓度为0～50 mg/L时,二元体系溶液表观粘度大幅度增大,芘的荧光光谱 I_1/I_3 值明显降低,这说明二元体系溶液表观粘度的增大主要是因为表面活性剂增强了疏水微区间的缔合作用,从而增大了溶液中原有空间网状结构的规模与强度;继续增大SDS的质量浓度,溶液表观粘度开始迅速下降,由于强烈的缔合和静电作用,使得两亲聚合物聚集体发生解离,并与SDS形成混合聚集体,大量小分子表面活性剂的加入,降低了混合聚集体的流体力学半径,导致溶液粘度及粘弹性下降。对比不同疏水基团碳原子数(十二烷基、十六烷基和十八烷基)的影响规律发现,SDS的加入对疏水单体为N-十八烷基丙烯酰胺的两亲聚合物作用最小。这是因为,随着碳原子数的增加,两亲聚合物的疏水基团缔合强度增高,高分子之间排斥作用越弱,聚集体结构更加紧密,SDS便难以进行解离和重组作用。

关键词:两亲聚合物表面活性剂　缔合强度　流变特性　微极性

中国图分类号:TE357.431　文献标识码:A　文章编号:1009-9603(2016)04-0095-07

Influence of hydrophobic groups of polymer and surfactant on aggregation of polymer/surfactant binary system

Ji Yanfeng Cao Xulong Guo Lanlei Min Lingyuan Dou Lixia Pang Xuejun Li Bin

(Research Institute of Exploration and Development, Shengli Oilfield Company, SINOPEC, Dongying City, Shandong Province, 257015, China)

Abstract: A binary system was formed with the addition of the anionic surfactant SDS into amphiphilic polymer solution with different hydrophobic groups. The interactions between hydrophobic groups, SDS and amphiphilic polymer and the binary system were studied by using rheological measurements, dynamic light scattering and pyrene fluorescence probe. The results show that due to the effect of hydrophobic association and electrical property, SDS will interact with amphiphilic polymers in the aqueous solution. At a low concentration of SDS(0-50 mg/L), the apparent viscosity of the binary system solution will have dynamic increase and the fluorescence spectrum of pyrene I_1/I_3 value will be decreased obviously. This suggests that the increasing of apparent viscosity is mainly because the hydrophobic association of micro ranges is enhanced by surfactant, and the size and strength of the original space grid structure are also increased in the solution. When the concentration of SDS is higher, the apparent viscosity of the solution starts to go down rapidly. Strong association and electrostatic action result in the separation of the aggregations of the amphiphilic polymer and the formation of mixed aggre gations. A lot of small-molecule surfactants is added, which makes the dynamic diameter of the mixed aggregations fluid decrease. And viscosity and viscoelasticity of the solution drop. Comparing rules of different number of carbon atoms(dodecyl, hexadecyl and octadecyl) in hydrophobic groups, it is found that the influence of SDS on the amphiphilic polymer with hydrophobic monomer of N-octadecyl acrylamide is minimal. This is because that along with the increasing of the number of carbon atoms, the association strength of the hydrophobic groups in the amphiphilic polymer is higher, and repulsive interactions among high molecules become weaker and the aggregate structure is closer, and it will be harder to be dissociated and recombined by SDS.

Keywords: amphiphilic polymers; surfactant; association intensity; rheological properties; micro polarity

随着中国化工业的高速发展,具有不同特殊物化特性的高分子水溶性聚合物层出不穷。由于部分水解聚丙烯酰胺(HPAM)具有合成工艺简单、成本低廉及较好的溶液增粘能力,被广泛应用在石油化工领域。为了提高 HPAM 的粘弹性和增粘能力,研究人员在高分子主链上引入疏水单体[1-2],当该聚合物质量浓度高于临界聚集浓度时,由于缔合作用,疏水单体在水溶液中发生交联形成具有一定强度的聚集体,使得流体力学半径增大,溶液粘弹性提高。鉴于两亲聚合物独特的流变学特性,目前已被广泛应用在油田化学驱方面[3-8]。在实际的三采过程中,单一聚合物溶液不能完全满足提高采收率的要求,这是因为,聚合物虽可以改善油水流度比,但是不具备降低油水界面张力的能力,很难将原油从吸附岩层上剥离下来,形成稳定的乳状液。因此将表面活性剂与聚合物复配得到聚/表二元体系,以达到改善流度比和降低油水界面张力的目的。目前,对聚/表二元体系已有很多研究。当 HPAM 与不同类型表面活性剂(阳离子型、阴离子型以及非离子型)复配后[9],可改变体系的粘弹性和界面活性。通过调配聚表类型和配比,可以应对不同油藏条件。但两亲聚合物疏水基团缔合强度对表面活性剂与两亲聚合物相互作用的机理没有深入研究。为此,笔者通过研究 3 种具有不同疏水单体的两亲聚合物在水溶液中与十二烷基硫酸钠(SDS)相互作用规律,探讨小分子表面活性剂对高分子聚集体解离、混合重组作用机理,以及疏水基团缔合强度对混合聚集体作用影响规律,以期为两亲聚合物与表面活性剂二元体系的建立和应用提供参考和依据。

1 实验器材与方法

1.1 实验器材

实验室自制不同碳链长度疏水基团的两亲聚合物:N-十二烷基丙烯酰胺/丙烯酰胺/丙烯酸钠三元共聚物(APC12),N-十六烷基丙烯酰胺/丙烯酰胺/丙烯酸钠三元共聚物(APC16),N-十八烷基丙烯酰胺/丙烯酰胺/丙烯酸钠三元共聚物(APC18)。三者水解度均为 30 mol%,疏水单体的物质的量分数均为 1 mol%。实验用阴离子表面活性剂为 SDS,分析纯,国药集团。

实验仪器主要包括:Brookfield DV-II+旋转粘度计、安东帕 MCR301 流变仪、Hitachi F4500 荧光光度仪以及配有 ALV-5000 数字式时间相关器的 ALV/SP-125 激光光散射仪。

1.2 实验方法

溶液流变特性测试 在温度为 45℃、剪切速率为 7.34 s^{-1} 的条件下,利用 Brookfield

DV-II+旋转粘度计测定质量浓度为 1 500 mg/L 的两亲聚合物溶液的表观粘度。在同样条件下，利用安东帕 MCR301 流变仪同轴圆筒系统测定溶液粘弹性能，以确定流变学参数。

聚集体流体力学半径测试　利用 ALV/SP-125 激光光散射仪，在温度为 45℃、波长为 632.8 nm 的条件下，对溶液进行 30°～90°变角测试，利用 Stocks-Einstein 公式[10-11]计算流体力学半径。

聚集体微极性测试　在配制的 100 mL 两亲聚合物溶液中，加入适量的芘乙醇分散液(芘的物质的量浓度为 1.0×10^{-5} mol/L)，声波处理 24 h。在荧光激发波长为 335 nm、测量范围为 350～450 nm[12-13]的条件下，利用 Hitachi F4500 荧光光度仪测定溶液中芘的荧光光谱。

2　聚/表二元体系溶液宏观流变学特性

由于两亲聚合物存在疏水缔合单体，因此其在水溶液中会发生疏水缔合作用并形成聚集体。当溶液浓度较低时，在水溶液中主要是单高分子内缔合，当溶液浓度逐步增大且高于临界聚集浓度时，聚合物高分子从分子内缔合转变为分子间缔合，形成稳定的空间网络结构，宏观表现为溶液粘度较高[14-15]。将同样具有疏水缔合作用的表面活性剂，加入到浓度高于临界聚集浓度的两亲聚合物溶液中，参与聚集体的形成，势必会影响其物化性能。

2.1 表观粘度

两亲聚合物为水溶性高分子聚合物，增粘能力是考察其性能的重要指标。将 SDS 分别加入到质量浓度为 1 500 mg/L 的 3 类两亲聚合物(APC12，APC16 和 APC18)溶液中，研究其表观粘度的变化规律。结果(图 1)表明：随着 SDS 质量浓度的增加，3 类聚/表二元体系溶液的表观粘度均先短暂平衡后急剧增加，分别在 SDS 质量浓度为 125，210 和 230 mg/L 时体系表观粘度达到最大值；随着 SDS 质量浓度的进一步增大，体系表观粘度快速降低。这是因为，SDS 在水溶液中是一种电解质，可以电离成为 1 个阳离子和 1 个阴离子，两亲聚合物与 SDS 之间的相互作用主要为疏水缔合和电性作用。

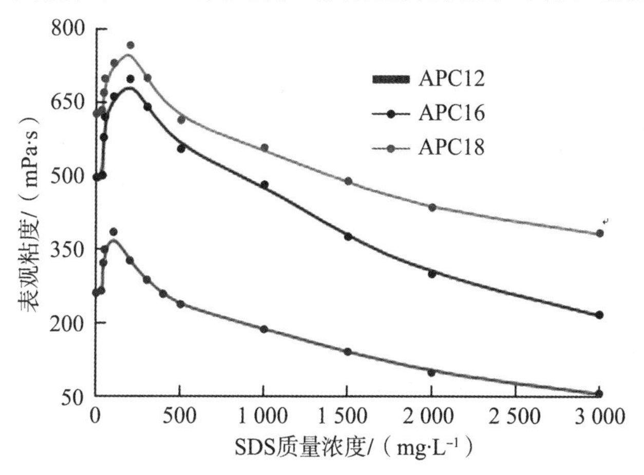

图 1　SDS 质量浓度对二元体系溶液表观粘度的影响

Fig. 1　Influence of SDS concentration on apparent viscosity of AP/SDS mixed solutions

对于质量浓度高于临界聚集浓度(1 000 mg/L)的两亲聚合物,SDS对两亲聚合物溶液表观粘度的影响遵循Biggs3段模型[16]。第1阶段:当溶液中未加入SDS时,两亲聚合物在水溶液中发生以分子间缔合为主、分子内缔合为辅的相互作用,使得溶液内存在大量的疏水微区,两亲聚合物的亲水端基裸露在疏水微区外围,使得高分子发生内向卷曲;当溶液中加入少量的表面活性剂时,其SDS的疏水端基开始与两亲聚合物的疏水端基发生纠缠,形成混合胶束,而亲水端基由于具有小尺寸分子优势,积极取代两亲聚合物的亲水端基进而保护混合胶束的疏水内核,使得高分子链变得相对伸展。同时,由于表面活性剂的电离作用,使得混合胶束表面形成大量电荷,从而压缩分子双电层,导致高分子链收缩。因此,当SDS质量浓度较低(0~50 mg/L)时,体系粘度没有发生明显变化。第2阶段:随着SDS质量浓度的增加,两亲聚合物亲水端基被SDS逐步完全取代,继续加入SDS时,SDS开始与两亲聚合物部分疏水端基形成新的混合聚集体,并促使两亲聚合物进一步发生分子间缔合,形成规模更大、结构更加紧密的空间网络结构,使得体系粘度快速增大。因此,当SDS质量浓度为50~100 mg/L时,体系粘度急剧增大。第3阶段:SDS质量浓度继续增大,溶液中混合聚集体浓度继续增加,每一个聚集体中平均包含的疏水端基数小于2,这意味着原本形成的聚合物空间网络结构发生解离,体系粘度急剧下降,当溶液中聚集体结构被完全解离后,体系粘度达到平衡。

对比不同疏水基团碳原子数的影响规律发现,在同一聚合物质量浓度下,APC18体系粘度最高,APC12体系粘度最低。对于两亲聚合物而言,疏水基团碳链长度不同,使其具有不同的疏水缔合强度,并随着疏水单体碳原子数的增多,疏水缔合能力增强,所形成聚集体结构强度越大,参与分子间缔合的高分子聚合物也越多,增粘能力越强[17]。溶液中加入SDS,APC18体系粘度变化幅度最小(图1)。这是因为,两亲聚合物的疏水基团疏水性越强,电荷之间排斥作用相对越弱,聚集体内部结构更为紧实,相同质量浓度下,SDS很难对聚合物所形成的聚集体进行解离和重组。

2.2 粘弹性能

质量浓度为1 500 mg/L的APC12溶液加入SDS前后的粘弹性测试结果(图2)表明:加入SDS前后APC12溶液的储能模量(G′)总是大于损耗模量(G″),溶液呈现弹性凝胶特性。这主要是缔合结构具有较高弹性所致,这种弹性随着疏水缔合强度的增强而变大[11]。当APC12溶液加入质量浓度为100 mg/L的SDS时,溶液的储能模量和损耗模量均有所增大,说明溶液的粘弹性均有所增大。这是因为SDS的疏水基团促进两亲聚合物缔合结构的形成,增强了聚集体强度。而当APC12溶液加入质量浓度为1 000 mg/L的SDS时,SDS在溶液中形成胶束,解离了两亲聚合物原本形成的聚集体结构,溶液的粘弹性均呈明显下降趋势。

高分子形成的聚集体在不同的剪切条件下会发生变化。两亲聚合物溶液为假塑性流体[17],当剪切速率较低时,聚集体结构不足以被破坏,体系的表观粘度不会发生变化;当剪切速率逐渐增大到一定强度后,聚集体结构发生扭曲、解离,体系的表观粘度降低,宏观表现为剪切变稀性;当剪切速率进一步增大,聚集体结构最终被完全破坏,高分子主链从卷曲态转变为伸展状态,溶液表观粘度不再发生变化。首尾2个剪切过程为流体极限阶段,不能反映流体在剪切过程中溶液的流变学变化性质。分析APC12溶液及其二元体系表观粘度在剪切变稀阶段的表观粘度变化曲线(图3),并利用幂律公式进行拟合,结果(表1)表明:两亲聚合物溶液及其二元体系流变曲线的幂律指数均小于1,且随着疏水链碳原子数的增大而呈现减小的趋势。

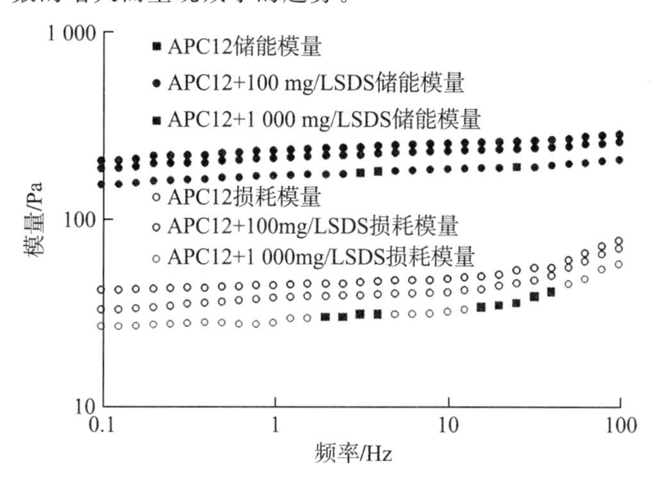

图2 APC12溶液及APC12/SDS二元体系溶液粘弹性

Fig. 2 Viscoelasticity of APC12 and APC12/SDS mixed solutions

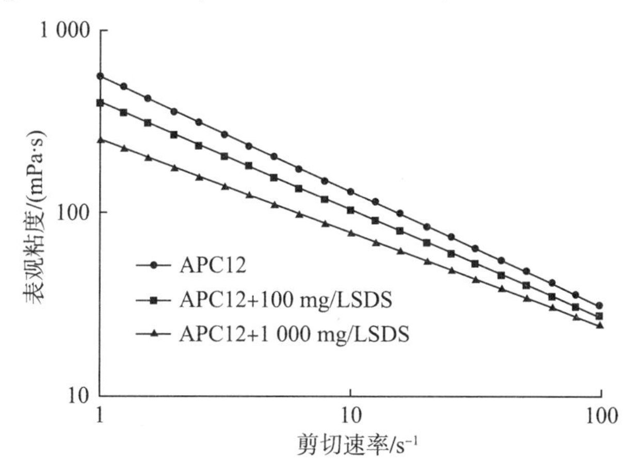

图3 APC12溶液及其二元体系在不同剪切速率下的表观粘度变化

Fig. 3 Apparent viscosity of APC12 and APC12/SDS mixed solutions under different shearing rates

表1 3类两亲聚合物流变曲线幂律公式拟合常数

Table1 Fitting constants of the rheological curves of amphiphilic polymer solutions by power law model

体系组成	lg$K$	幂律指数
1 500 mg/L APC12	2.752	0.341
1 500 mg/L APC12+100 mg/L SDS	3.625	0.234

(续表)

体系组成	lg$K$	幂律指数
1 500 mg/L APC12+1 500 mg/L SDS	2.617	0.426
1 500 mg/L APC16	3.215	0.253
1 500 mg/L APC16+100 mg/L SDS	3.752	0.202
1 500 mg/L APC16+1 500 mg/L SDS	3.023	0.286
1 500 mg/L APC18	3.723	0.201
1 500 mg/L APC18+100 mg/L SDS	4.120	0.195
1 500 mg/L APC18+1 500 mg/L SDS	3.768	0.213

注:K 为稠度系数,$Pa \cdot s^n$。

通过幂律公式可知,稠度系数越大且幂律指数越小,聚合物及其二元体系的增粘能力越强。分析稠度系数和幂律指数可知:对于同一种两亲聚合物,随着 SDS 质量浓度的增加,体系增粘能力呈现先增后减的趋势;两亲聚合物疏水基团缔合强度越大,体系增粘能力越强。当溶液中加入少量的 SDS 时,其非极性疏水端基促进了两亲聚合物在溶液中形成更致密的空间网络结构,两亲聚合物溶液粘弹性及剪切粘度均有所增加;当加入的 SDS 质量浓度过高时,小分子表面活性剂在水溶液中形成胶束,由于胶束具有强的非极性,使得两亲聚合物的疏水基团增溶到胶束中,进而使原本形成的聚集体发生解离,溶液粘弹性能和增粘性能均有明显的下降。对于不同疏水强度的两亲聚合物,疏水基团缔合强度越高,其所形成的聚集体结构性能越强,SDS 小分子较难增溶聚集体中的疏水基团,进而使聚集体发生解离,因此,随着疏水基团烷基链的增长,SDS 对所形成聚集体的影响变小。

3 聚/表二元体系溶液微观作用机理

上述宏观表现必定是聚合物在溶液中微观聚集行为所致。因此,利用荧光光谱和动态光散射,从分子角度分析以上作用规律,进一步探讨两亲聚合物与 SDS 之间的作用机理。

对于高分子聚合物溶液而言,流体的宏观性质是由其在溶液中的聚集形态所决定的,特别是对于具有疏水缔合基团的两亲聚合物,其粘弹性及增粘、抗剪切性质均取决于其在水溶液中形成的高分子聚集体结构。聚集体是因不同分子间弱相互作用通过组装或自组装结合而成的。这种作用力可以是分子间正负电荷的吸引力,也可以是分子间形成新的氢键和配位键,还可以是聚集体之间的特殊相互识别作用,甚至是聚合物中部分功能基团和无机物共结晶而形成的。对于两亲聚合物体系,其聚集体主要是通过非极性疏水基团缔合作用而形成的,属于弱相互作用力。也正是因为相互作用较弱,使得该聚集体在受到高速剪切时容易发生解离。

3.1 聚集体流体力学半径

两亲聚合物在水溶液中可形成聚集体，由于大量高分子发生分子间交联作用，使得聚集体在溶液中存在颗粒特性，发生布朗运动。利用动态光散射法监测聚集体的布朗运动现象，得到聚集体空间尺寸与构象的变化信息，从而研究表面活性剂的加入对两亲聚合物所形成聚集体的解离、混合作用规律。

3类两亲聚合物加入表面活性剂前后所形成聚集体的流体力学半径的变化规律(图4)表明，同一条曲线存在2个峰，一个峰在100 nm左右，另一个峰在2 000 nm左右。这2个峰代表着不同的物理意义，前者主要是高分子内疏水缔合所形成聚集体的流体力学半径，后者为高分子间缔合所形成的。由于SDS的加入对分子内缔合所形成的聚集体流体力学半径影响较小，且当聚合物质量浓度高于临界聚集浓度时，分子间缔合成为主导，因此主要研究分子间缔合峰。APC12，APC16和APC18在水溶液中所形成聚集体的流体力学半径分别为2 982，2 577和1 900 nm，随着疏水基团碳原子数的增加，流体力学半径减小。这主要是因为，对于不同疏水基团的两亲聚合物，疏水基团的疏水性越强，电荷之间排斥作用相对减弱，导致聚集体内部结构更为紧实，表现为聚集体流体力学半径较小。

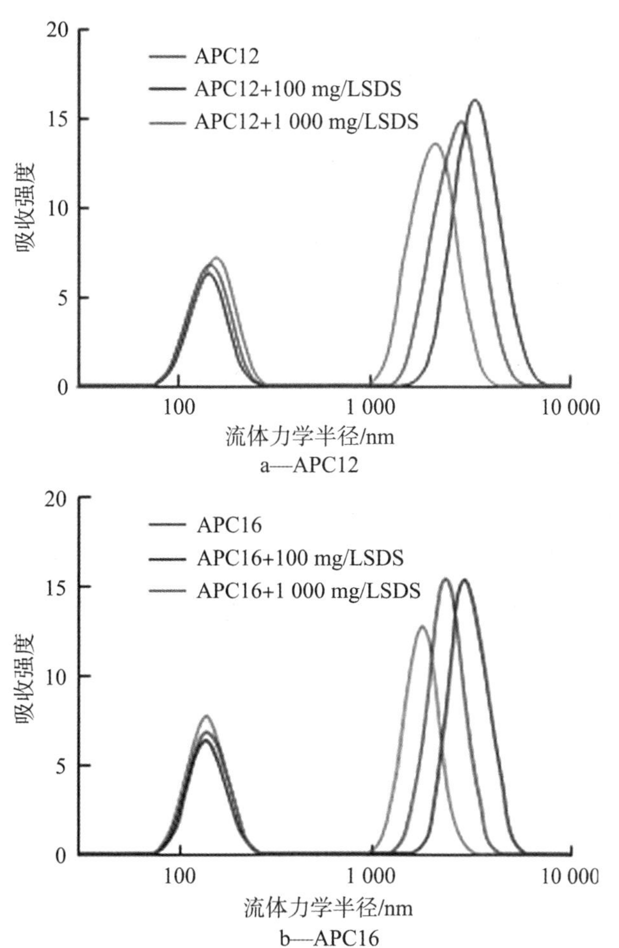

a—APC12

b—APC16

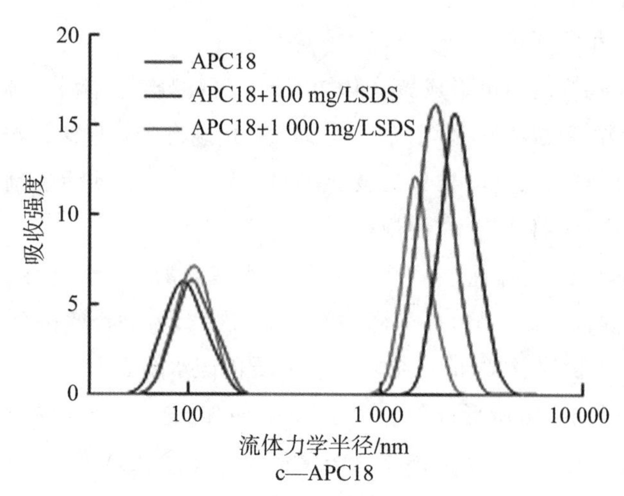

c—APC18

图 4　SDS 对 3 类两亲聚合物溶液中聚集体流体力学半径的影响

Fig. 4　Effect of SDS concentration on apparent hydrodynamic radius in 3 amphiphilic polymer solutions

在两亲聚合物溶液中加入高低 2 种质量浓度的 SDS，当加入 SDS 的质量浓度为 100 mg/L 时，3 类两亲聚合物中聚集体流体力学半径均呈增大趋势。此时，由于疏水基团具有较高的非极性，使得原本游离在外的聚合物疏水基团更容易参与到分子间缔合过程中，导致混合聚集体的流体力学半径增大。当加入的 SDS 质量浓度为 1 000 mg/L 时，SDS 疏水基团与两亲聚合物疏水基团发生强烈的相互作用，由于 SDS 为小分子，空间位阻小，两亲聚合物疏水基团更倾向与 SDS 发生缔合，使得原本形成的聚集体发生解离，导致流体力学半径降低。这也是聚合物溶液表观粘度先降后增的主要原因。

3.2 聚集体微极性

两亲聚合物聚集体的形成主要是极性与非极性基团相互作用的结果。聚集体流体力学半径的变化是其周围极性变化所致。为了更加深入地研究 SDS 对两亲聚合物聚集体的作用机理，使用芘荧光探针法探究聚集体微极性的变化规律。

由于大量非极性疏水基团的存在，两亲聚合物在水溶液中形成非极性微区。当溶液中加入非极性物质时，根据相似相溶原理，非极性物质会进入非极性微区。由于芘具有非极性及荧光特性，其被用于探索两亲聚合物聚集体属性的研究中。荧光光谱可以反映芘探针所在微环境的极性，进而推断出溶液中聚合物所形成聚集体的物化特性[16,18]。从图 5 可以看出，APC12 的芘稳态荧光发射光谱存在 5 个特征吸收峰，其中第 1 振动峰和第 3 振动峰的波长分别为 373 和 385 nm。

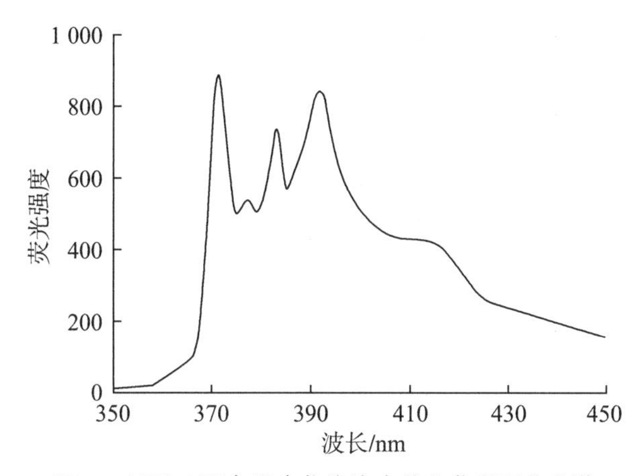

图 5　APC12 两亲聚合物溶液中芘的荧光吸收光谱

Fig. 5　Fluorescence spectra of pyrene in amphiphilic polymer APC12 solutions

第 1 振动峰与第 3 振动峰对应的荧光强度的比值为 I_1/I_3，该值是随着芘所处环境的极性变化而改变的[19]。分析芘的 I_1/I_3 值与溶液 SDS 质量浓度的关系(图 6)可以看出：当 SDS 质量浓度较低时，表面活性剂在水溶液中参与聚集体的形成，由于表面活性剂分子较小，使原本进入聚集体的芘解离到溶液中，I_1/I_3 值呈降低趋势；当 SDS 质量浓度大于 50 mg/L 时，I_1/I_3 值开始增大。这是因为，当 SDS 质量浓度较低时，其与两亲聚合物形成混合胶束，由于电离作用，SDS 在混合胶束表面形成大量电荷，压缩分子双电层，使得高分子聚集体链收缩，增溶了游离在水溶液中的芘，使得 I_1/I_3 值呈增大趋势。当 SDS 的质量浓度进一步增大时，表面活性剂与两亲聚合物的疏水端基形成新的混合聚集体，随着混合聚集体在溶液中浓度的增大，平均每个聚集体包含的疏水端基数逐渐下降，使得两亲聚合物高分子链更加伸展，导致所形成的疏水微区不足以稳定原本在两亲聚合物疏水微区中的芘，使得芘解离到水溶液中，I_1/I_3 值明显上升。当 SDS 质量浓度增至临界聚集浓度时，I_1/I_3 值开始迅速降低，此时体系中 SDS 与两亲聚合物形成稳定的混合聚集体，由于疏水缔合作用，芘大量进入新形成的聚集体中。当表面活性剂分子在聚合物链上的结合达到饱和后，随着 SDS 质量浓度的继续增加，I_1/I_3 值基本保持不变，这可能因为水溶液中已不存在可解离的两亲聚合物聚集体，新增的 SDS 游离在水溶液中，浓度的增加对溶液微极性影响很小[20-21]。对比不同疏水强度的疏水单体，同等浓度时在水溶液中形成的聚集体缔合强度不同。疏水缔合强度高的两亲聚合物溶液中的芘浓度最低，SDS 的解离作用也最小。

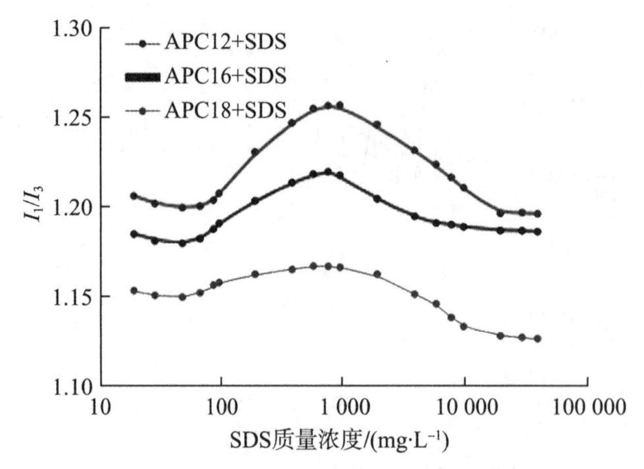

图 6 芘的 I_1/I_3 值与 SDS 质量浓度的关系

Fig. 6 Relationship between SDS concentration and I_1/I_3 of pyrene

4 结论

对于疏水基团不同的两亲聚合物,疏水基团的疏水性越强,电荷之间排斥作用相对越弱,导致聚集体内部结构更为紧密,表现为聚集体流体力学半径较小。当两亲聚合物溶液中加入 SDS 后,SDS 分子与两亲聚合物发生相互作用,当 SDS 质量浓度较低时,非极性疏水基团的存在有助于构建更紧密的两亲聚合物空间网络结构,提高两亲聚合物溶液的粘弹性能及增粘性能;当 SDS 的质量浓度大于 100 mg/L 时,聚/表二元体系中两亲聚合物聚集体开始发生解离,与表面活性剂形成混合聚集体。疏水基团疏水性越弱,SDS 的解离作用越强,对溶液微极性影响越大,越容易形成稳定的混合聚集体。

参考文献

[1] Fernandes L S, Homem-De-Mello P, Lima E C, et al. Rational design of molecularly imprinted polymers for recognition of cannabi-noids: a structure-property relationship study[J]. European Polymer Journal, 2015,71 (2):364-371.

[2] 陈明贵,周智,杨光,等. 两亲聚合物对非均质稠油油藏化学驱的适用性研究[J]. 油气地质与采收率,2015, 22 (6):116-120. Chen Minggui, Zhou Zhi, Yang Guang, et al. Research on feasibili-ty of amphiphilic polymer for chemical flooding in heterogeneous heavy oil reservoir[J]. Petroleum Geology and Recovery Efficiency, 2015,22 (6):116-120.

[3] 曹绪龙,刘坤,韩玉贵,等. 耐温抗盐缔合聚合物的合成及性能评价[J]. 油气地质与采收率,2014,21 (2):10-14.

Cao Xulong,Liu Kun,Han Yugui,et al. Synthesis and properties of heat-tolerance and salt-resistance hydrophobically associating water-soluble polymer[J]. Petroleum Geology and Recovery Efficiency,2014,21 (2):10-14.

[4] Wever D A Z,Picchioni F,Broekhuis A A. Polymers for enhanced oil recovery: A paradigm for structure-property relationship in aqueous solution[J]. Progress in Polymer Science,2011,36(11):1 558-1 628.

[5] 张磊.两亲聚合物/表面活性剂在岩石矿物上的吸附滞留特性研究[D].青岛:中国石油大学(华东),2013.

Zhang Lei. Study on characteristics of adsorption and retention of amphiphilic polymers and surfactants on rocK_s and minerals[D]. Qingdao: China University of Petroleum(East China),2013.

[6] 徐斌.两亲聚合物聚集体调控增强乳化的方法及机理研究[D].青岛:中国石油大学(华东),2013.

Xu Bin. Study on aggregation control of amphiphilic polymers and its mechanism on enhancing emulsification [D]. Qingdao: China University of Petroleum (East China),2013.

[7] 古国华,郭玉,傅洵,等.两亲聚合物用于乳液聚合的研究进展[J].青岛化工学院学报:自然科学版,2001,22 (4):302-306.

Gu Guohua, Guo Yu, Fu Xun,et al. The development of amphiphilic polymers as emusifiers in emulsion polymerization[J]. Journal of Qingdao Institute of Chemical Technology:Science & Technolo-gy Edition,2001,22 (4):302-306.

[8] 康万利,张磊,孟令伟,等.两亲聚合物在岩石矿物上的静态吸附研究[J].应用化工,2012,41 (11):1 865-1 867,1 871.

Kang Wanli,Zhang Lei,Meng Lingwei,et al. The study of static adsorption of the amphiphilic polymer on rocK_s and minerals[J]. Applied Chemical Industry,2012,41 (11):1 865-1 867,1 871.

[9] 付京,宋考平,王志华,等.高分子质量聚合物溶液与二类油层匹配性研究[J].特种油气藏,2015,22 (2):129-132.

Fu Jing, Song Kaoping, Wang Zhihua, et al. Research on compatibility between high molecular polymer solution and type-II oil layer [J]. Special Oil & Gas Reservoirs,2015,22 (2):129-132.

[10] 李方,李干佐,汪汉卿,等.荧光和动态光散射方法研究两性表面活性剂胶束的聚集和相互作用[J].高等学校化学学报,1998,19(7):1 117-1 120.

Li Fang, Li Ganzuo, Wang Hanqing, et al. Studies on the aggregation and interaction of DDAPS micelles with fluorescence and light-scattering methods[J]. Chemical Journal of Chinese Universities,1998,19(7):1 117-1 120.

[11] Zhang Q,Chen Z,Li Z. Simulation of tin penetration in the float glass process(float glass tin penetration)[J]. Applied Thermal Engineering,2011,31(6):1 272-1 278.

[12] Amiji M M. Pyrene fluorescence study of chitosan self-association in aqueous solution[J]. Carbohydrate Polymer,1995,26(3):211-220.

[13] Fisher A, Houzelle M C, Hubert P, et al. Detection of intramolecular associations in hydrophobically modified pectin derivatives using fluorescent probes[J]. Langmuir,1998,14 (16):4 482-4 498.

[14] Kratz K, Hellweg T, Eimer W. Structural changes in PNIPAM microgel particles as seen by SANS,DLS,and EM techniques[J]. Polymer,2001,42 (15):6 631-6 639.

[15] Minatti E, Viville P, Borsali R, et al. Micellar morphological changes promoted by cyclization of PS-b-PI copolymer: DLS and AFM experiments [J]. Macromolecules,2003,36(11):4 125-4 133.

[16] Biggs S,Selb J,Candau F. Effect of surfactant on the solution properties of hydrophobically modified polyacryamide[J]. Langmuir, 1992,3(8):838-847.

[17] Ji Y,Kang W,Liu S,et al. The relationships between rheological rules and cohesive energy of amphiphilic polymers with different hydrophobic groups [J]. Journal of Polymer Research,2015,22 (3):1-7.

[18] Yusa S I, Hashidzume A, Morishima Y. Interpolymer Association of cholesterol pendants linked to a polyelectrolyte as studied by quasielastic light scattering and fluorescence techniques[J]. Langmuir,1999,15(26):8 826-8 831.

[19] 耿同谋,吴文辉. 荧光探针研究 P (AM/NaAA/DiC6AM)在水溶液中的缔合行为[J]. 华东理工大学学报:自然科学版,2006,32 (2):150-154.

Geng Tongmou,Wu Wenhui. Associating behaviors in aqueous solution of twin-tailed hydrophobically associating water-soluble terpolymers P (AM/NaAA/DiC6AM) utilizing fluorescence probe [J]. Journal of East China University of Science and Technology:Natural Science Edition,2006,32(2):150-154.

[20] 耿同谋. P (AM/NaAA/DiACn)与十二烷基硫酸钠的相互作用[J]. 应用化工,2009,38(9):1 286-1 288.

Geng Tongmou. Interactions between P(AM/NaAA/DiACn) and sodium dodecyl sulfate[J]. Applied Chemical Industry,2009,38 (9):1 286-1 288.

[21] 王增林,宋新旺,祝仰文,等. 海上油田二元复合驱提高采收率关键技术——以埕岛油田埕北 1 区西部 Ng4—5 砂层组为例[J]. 油气地质与采收率,2014,21 (2):5-9.

al. Study on key techniques of surfactant-polymer flooding for offshore field-case of sand groups of Ng4-5 in westem Chengbei1 block,Chengdao oilfield[J]. Petroleum Geology and Recovery Efficiency,2014,21 (2):5-9.

编辑 常迎梅

Both-branch amphiphilic polymer oil displacing system: Molecular weight, surfactant interactions and enhanced oil recovery performance

Ji Yanfeng Wang Duanping Cao Xulong Guo Lanlei Zhu Yangwen

Exploration and Development Research Institute, Shengli Oilfield Co. Ltd., Sinopec, Dongying 257000, PR China

highlights: Both-branch amphiphilic polymers were synthesized by means of an aqueous micellar copolymerization. The effect of the molecular weight and surfactant interactions on the rheological properties and oil-water interfacial activity were studied.

The performance of high molecular weight P(AM/BHAM/NaA) shows a high effi ciency in displacing residual oil.

graphical abstract: The influence of micro network structure on macroscopic oil recovery of the both-branch amphiphilic polymers.

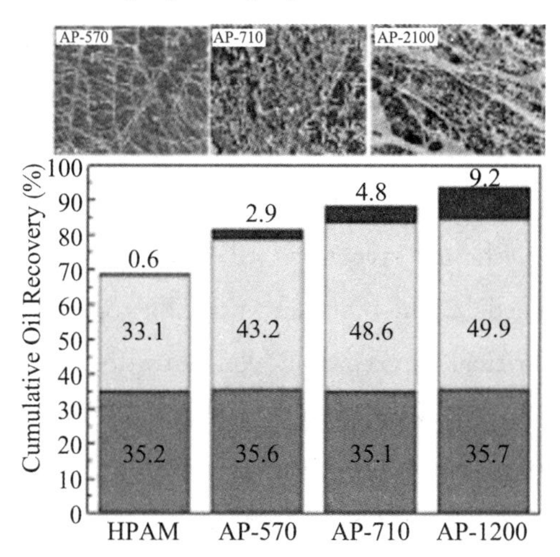

article info: Article history: Received 6 April 2016

Received in revised form 8 September 2016

Accepted 11 September 2016

Available online 12 September 2016

Keywords: Both-branch amphiphilic polymer Polymer/surfactant compound system Aggregation behavior Linear regression method Dynamic interfacial tension Enhanced oil recovery

abstract: Hydrophobically modified polyacrylamidepolymers were synthesized by an aqueous micellar copoly-merization, the N-benzyl-N-n-hexadecyl acrylamide being used as hydrophobic comonomer. The effect of the molecular weight (MW) and surfactant interactions on the rheological properties and oil-water interfacial activity were studied. The results indicate that increasing MW, the critical aggregation concentration of amphiphilic polymer will be decreased, and the network structural strength increased, which was measured by fluorescent probes, DLS and electron microscope. Mixing with the surfactant, sodium dodecyl sulfate (SDS), the solution viscosity increased at first and decreased later on with increasing the ratio of SDS concentration, and the dynamic interfacial tension between oil and water can be reduced by two orders of magnitude compared to the polymer solution only. The result of enhance oil recovery performance of the polymer/surfactant compound system is that it can significantly reduce the residual oil saturation due to the synergistic effect of the increasing viscosity and surface activity. The overall recovery efficiency was raised by 10%~25% OOIP compared to the baseline polymers.

1 Introduction

Amphiphilic polymer is an interesting class of water-soluble polymer that, in principle, combines the properties of polymer with the ability to self-assemble spontaneously, due to the hydrophobic interactions that arise from hydrophobic groups [1,2,3,4]. Such systems have also been described thoroughly by theory, which has predicted the formation of the net structures, depending on the ratio of association [5]. However, the polymer needs a constant concentration to form the net structure. This concentration is called critical aggregation concentration (CAC). Above a critical aggregation concentration, the hydrophobic groups in these polymers tend toward associate in aqueous solution by intermolecular hydrophobic interaction leading to the formation of polymolecular associations. As a consequence, these copolymers exhibit thickening properties equivalent of those observed for higher molecular weight homopolymers. Thereby, these polymers exhibit particular rheological properties in solution, due to the reversible dissociation process of the physical links occurring under shearing [6,7].

The aggregation behavior of amphiphilic polymers in aqueous solutions can be altered by varying the size of molecular weight (MW) and/or modifying solution temperature and/or the addition of surfactant. From the external view of a macromolecule, the molecular weight and polymeric material composition directly affect the rheological properties of polymer solution, such as appar-ent viscosity [8], viscoelasticity [9] and seepage characteristic in porous media[10]. And the interaction of amphiphilic polymer with small-molecule surfactants, for example, sodium dodecyl sulfate (SDS), is of fundamental importance for many industrial applications such as tertiary oil recovery and latex paint technology [11,12]. But in recent paper, there are seldom researches on the effect of the different of polymer's MW on the aggregation behav-ior in the solution about both-branch amphiphilic polymer and the presence of surfactant on the ability of enhanced oil recovery.

In this work, the systematic studies on the aggregation behavior of the both-branch amphiphilic polymer aqueous solutions, acrylamide/N-benzyl-N-n-hexadecyl/sodium acrylate terpolymer-P(AM/BHAM/NaA), are performed in different relative molecular weight and the presence of surfactant in order to obtain further evidence for it could be a high-performance oil displacement system.

2 Experimental section

2.1 Materials

Analytical reagent grade acrylamide (AM), sodium acrylate (NaA), potassium persulfate (KPS), sodium bisulfite (SBS), ben-zyl chloride, n-hexadecylamine, sodium formate, sodium chloride, acryloyl chloride, ethanol, dimethyl sulfoxide (DMSO), sodium dodecyl sulfate (SDS), sodium hydroxide, acetone and allyl bromide were purchased from the Aladdin Chemical Reagent Factory (Shanghai, China). The comonomer, N-benzyl-N-n-hexadecyl acry-lamide (BHAM) [13] were prepared as described throughout the literature, respectively.

2.2 Polymerization

Acrylamide/N-hexadecylacrylamide/sodium acrylate terpolymer-P (AM/BHAM/NaA) were synthesized using a micellar copolymerization route [14]. It should be emphasized that such a micellar process differs strongly from other polymerizations carried out in the presence of a surfactant, i. e. emulsion or microemulsion processes [15]. In this technique use of a sur-factant is necessary to solubilize the hydrophobic monomer into micelles dispersed in water. Each reaction was conducted in a 1 000 mL, three-necked, round-bottomed fiask equipped with a condenser, mechanical stirrer, rubber septum cap, and nitrogen inlet/outlet. The reactor, containing an ionic surfactant, sodium dodecyl sulphate (SDS), and hydrophobic monomer (BHAM) in deionized water, was heated to 50℃ from ambient temperature using a thermostated water bath with continuous stirring under nitrogen until the hydrophobic monomer was solubilized within surfactant micelles and the solution was optically transparent. Nitrogen was also bubbled separately through an aqueous acrylamide solution within a fiask, the acrylamide solution then being transferred into the reaction vessel. The mixture was kept at a constant temperature 50℃ with continuous stirring and under a nitrogen purge for 1～1.5 h to ensure the complete removal of trapped air due to strong foaming arising from the presence of surfactant in the solution. When the mixture was homogeneous, an aqueous solution of $K_2S_2O_8$ was added into the reactor. The reaction was carried out at 50℃ for 1～6 h with purging nitrogen and vigorous stirring. At the end of the reaction either a viscous polymer solution or viscous polymer gel (especially for those with higher BHAM content or low SDS content) was obtained. Both were homogeneous and clear. After cooling the final reaction mixture, the aqueous polymer solutions or gel-like samples were precipitated into a large excess of methanol, methanol being found to be ideal precipitant for this system. The polymer was then washed in a large excess of methanol to remove unreacted monomers and surfactant for a minimum of two days at the ambient temperature. The polymer recovered by filtration was repeatedly washed in methanol under stirring and vacuum, then it was ground into a fine powder using scissors and then a food blender. Finally the methanol was allowed to evaporate off at ambient temperature for a minimum of 48 h prior to drying to constant weight in a vacuum oven at 46℃. By weighing the dry copolymer after each swelling and washing, it was confirmed that a minimum of two days swelling and three times washing using methanol was sufficient to ensure complete removal of unreacted monomers. Conversions were obtained gravimetrically. The overall concentration of monomers was between 1.9～2.5 wt.%, and the concentration of the initiator $K_2S_2O_8$ was 0.3 wt.% relative to the total monomer feed. The concentration of SDS used was 3～90 times its CMC. Candau and coworkers reported [16] that the presence of SDS during the micellar polymerization leads to a reduction in the molecular weight of the copolymers. Therefore, for comparison polyacrylamide homopolymer was also prepared in the presence of surfactant. Our results showed that the molecular weight of P(AM/BHAM/NaA) was reduced slightly upon addition of SDS into the polymerization system, however, this was found not to be very significant. And the chemical formula for P(AM/BHAM/NaA) was shown in Fig. 1.

2.3 Polymer composition analysis

Copolymer composition was determined by ^1H NMR spec-troscopy in deuterium oxide (D_2O) according to literature methods [17,18]. Fig. 2 is an example of a spectrum for a copolymer con-taining BHAM. The hydrophobe content was calculated from the relative integrated area of peak A corresponding to the protons of the terminal CH_3 groups of the alkyl chains compared to that of peak A and B corresponding to the protons of CH groups in the poly-mer backbone. For the present hydrophobic monomer (BHAM) of P(AM/BHAM/NaA), the peak C, D and F represent CH_2 and CH_3 of an alkyl hydrophobic chain, and E is the peak of benzene ring. The ^1H NMR spectroscopy in Fig. 2proved that the target compound was gotten.

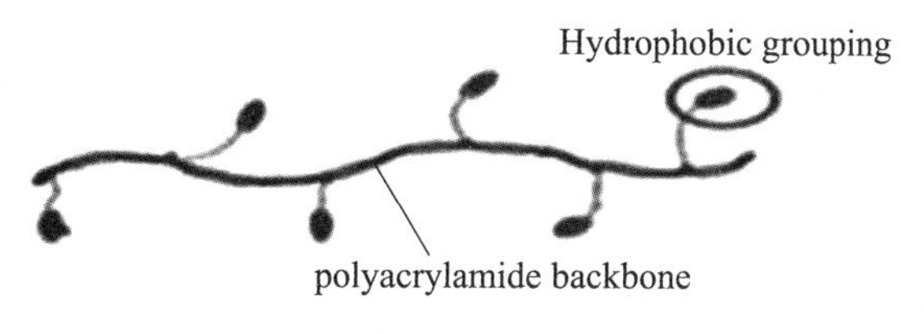

$$\text{-}(CH_2\text{-}CH)_{74}\text{-}(CH_2\text{-}CH)_{1}\text{-}(CH_2\text{-}CH)_{25}\text{-}$$

(side groups: $C{=}O$ / NH_2; $C{=}O$ / $N(CH_2C_6H_5)(C_{16}H_{33})$; $C{=}O$ / ONa)

Fig. 1 The chemical formula for the polymers, P(AM/BHAM/NaA), monomer molar ratio: AM : BHAM : NaA=74 : 1 : 25;.

2.4 Fluorescence spectroscopy experiment

The fiuorescence spectroscopy experiments were performed by using a FluoroMax-4 (HORIBA Jobin Yvon Inc.). A pyrene stock solu-tion (1.0×10^{-3} mol/L) was prepared in methanol. A sample of this solution was introduced into empty vials, and the solvent was evaporated in a vacuum. Afterwards, the vials were filled with 10 mL of the amphiphilic polymer solution, and then gently stirred for 24 h to ensure the incorporation of the molecular probe into polymer hydrophobic micro domains. The final pyrene concentration was 1.0×10^{-6} mol/L. At this low concentration, no excimer band due to the interaction of an active state pyrene with a ground－state pyrene was observed.

2.5 Atomic force microscopy

Atomic force microscopic (AFM) images were taken by a com-mercial Nanoscope

III (Digital Instruments, Santa Barbara, CA) using Si_3N_4 probe to analyze the apparent morphology of the amphiphilic polymer (1 000 mg/L) and inclusion complex.

2.6 Dynamic light scattering

Dynamic light scattering technique is used the intensity fiuctua-tion change of sample to test its particle size. The working principle is base on the variation of the intensity of scattered light, which is the diffusion coefficient of the molecule tested by a doppler frequency shift-D_{app}. Using Stokes-Einstein formula, hydrodynamics radius of particle can be derived.

$$R_{h,app} = k_B T/(6\pi\eta D_{app}) \tag{1}$$

k_B is Boltzmann constant, η is solvent viscosity, and T is absolute temperature.

DLS is gotten by Laser Light Scattering Spectrometer (ALV-5000/E/WINMultiple Tau Digital Correlator). At first, the sample must be centrifugal dust remove, and the test temperature is 40℃.

2.7 Scanning electronic microscopy

SEM photographs were determined on a JMS-6380LV. The polymer solution was frozen by liquid nitrogen. And then were made electrically conductive by coating, in a vacuum with a thin layer of gold (approximately 300 A) for 30 s and at 30 W. The pictures were taken at an excitation voltage of 10 kV and a magnification of 5 000×.

2.8 Dynamic interfacial tension (IFT)

The oil-water interfacial tension (IFT) between aqueous phase and oil phase was measured at 25℃ using a M6500 Spinning Drop Tensiometer manufactured by Grace Instrument Co., USA. Triplicates of each measurement were performed and the mean value of IFT was provided.

3 Results and discussion

3.1 Aggregation behavior and microscopic characteristics of P (AM/BHAM/NaA) different relative molecular mass

A series of defferent molecular weight (MW) of hydrophobically modified polyacrylamide copolymers have been synthesized via the micellar copolymerization procedure. And the association behavior was studied by pyrene fiuorescence. As Fig. 3, all samples were excited at 335 nm, and the emission spectra of pyrene showed vibronic peaks at $\lambda_1 = 372$ nm (strength I_1) and $\lambda_3 = 382$ nm (strength I_3). The vibrational structure of the fiuorescence bands showed sensitivity to the local polarity of the micro environment at the binding sites. The ratio between the fiuorescence intensitiesof peaks I_1 and I_3 were used to evaluate the polarity of the local microenvironment of pyrene.

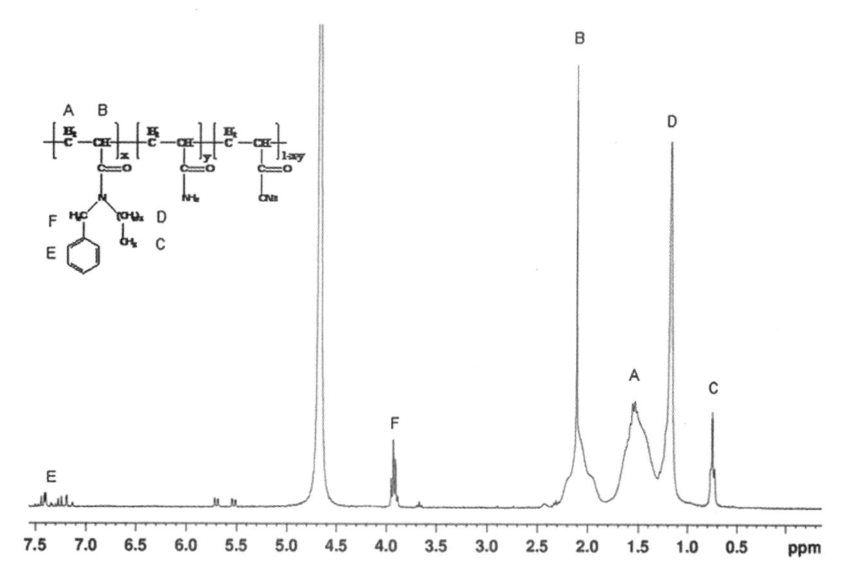

Fig. 2 ^{1}HNMR spectrum of P(AM−NaA−BHAM).

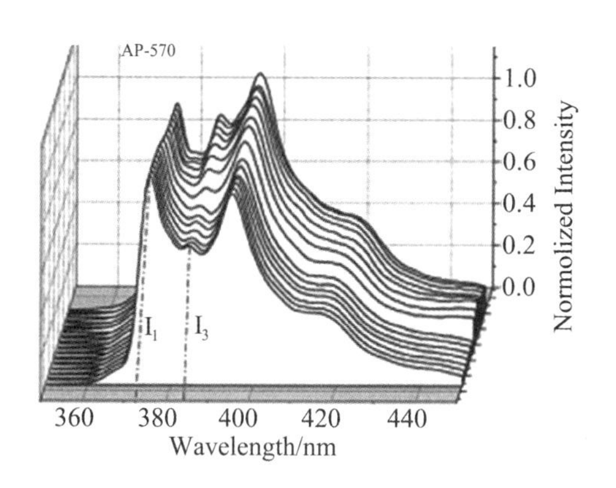

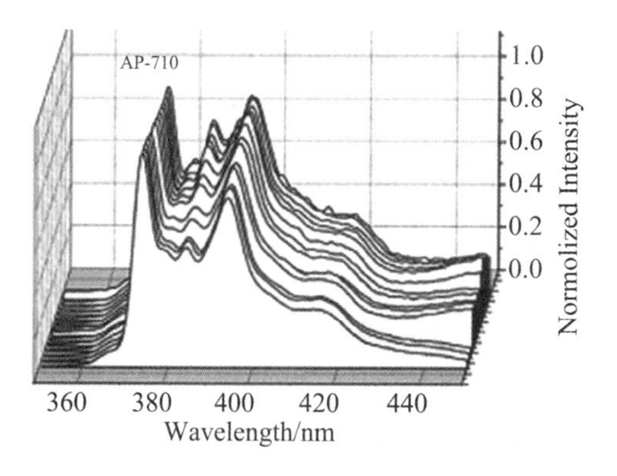

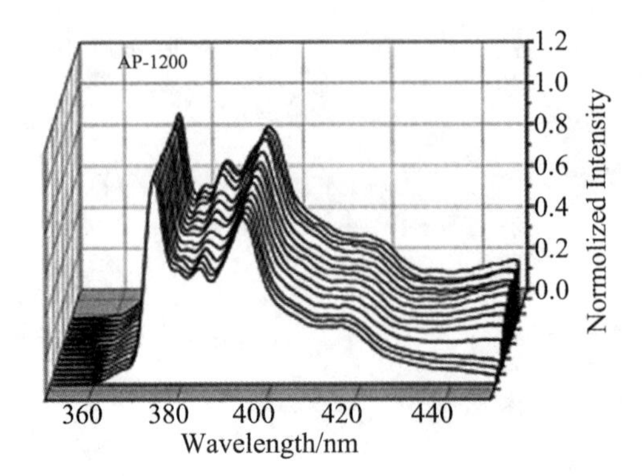

Fig. 3 The Fluorescence emission spectra of pyrene at different polymer concentrations.

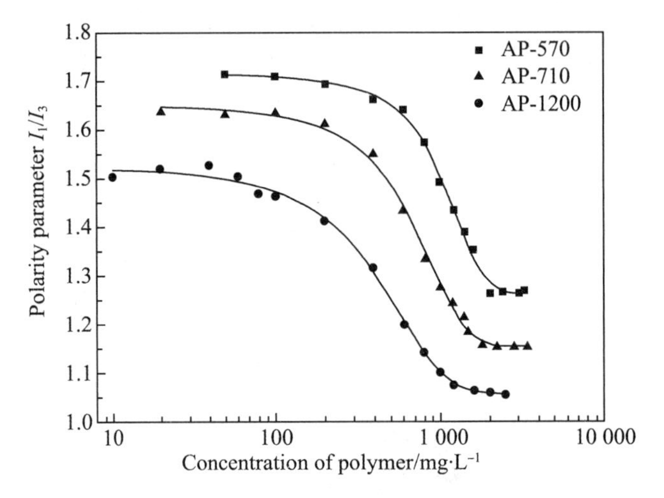

Fig. 4 The curves of I_1/I_3 versus logc for amphiphilic polymer.

The most widely used fiuorescent probe (pyrene) was applied to this work to study the aggregation behavior of amphiphilic polymer. The ratio I_1/I_3 of its emission spectrum can be linearly cor-related to the polarity parameter ET(30) of Dimroth and Eichardt, which corresponds to the most-used polarity scales so far [19]. This study assessed the infiuence of the different relative molecular mass to the aggregation behavior of amphiphilic polymers.

Fig. 4 is the curves of I_1/I_3 versus logc for amphiphilic poly-mer based on Fig. 3. It shows that at a lower concentration of amphiphilic polymer, the value of I_1/I_3 is lower than pure water slightly. And as increasing of the amphiphilic polymer concentra-tion, it remains about the same. This was mainly due to the presence of a small amount of amphiphilic molecules, which bring a small hydrophobic microenvironment. And it cannot provide adequate space for pyrene in the solution. With increasing of the polymer concentration, about the concentrations of AP-1200, AP-710 and AP-570 are over 60 mg/L, 200 mg/L and 400 mg/L respectively, the value of I_1/I_3 will be a sudden change. It turned out that, the pyrene in the hydrophobic microenvironment gradually increased in this system because the hydrophobic groups of amphiphilic polymers have begun to form the association structures. When the con-centration of AP-1200, AP-710 and AP-570 was above 1 000 mg/L, 1 600 mg/L and 2 000 mg/L respectively, the value of I_1/I_3 never changing with increasing the concentration of polymers, which reveals that the amphiphilic polymers have formed the aggregate structures and the hydrophobic pyrene molecules have been com-pletely soluble in a hydrophobic micro environment. While, in low concentration of AP-570 solution, the pyrene is same as pure water, which means it cannot form a hydrophobic region, and I_1/I_3 of AP-570 is higher than AP-1 200 and AP-710. The value of I_1/I_3 is the measurement of micro-polar. The bigger value of I_1/I_3 is, the higher the intensity of polar is, so association formed by amphiphilic polymer is weaker. It indicates that the intensity of association increases as molecular weight.

Aguiar et al. [20] and others point out using the curves of I_1/I_3 versus logc for amphiphilic polymer, the CAC of amphiphilic poly-mer could be determined. Fig. 4 is the curve of Boltzmann function for different amphiphilic polymer concentrations. It was regres-sion fitted by decreasing function of the Boltzman, and the formula shows as follow.

$$y=\frac{A_1-A_2}{1+e^{(x-x_0)/\Delta x}}+A_2 \tag{2}$$

Where y is dependent variable, and it was corresponding with the value of I_1/I_3. x is as independent variable corresponding to con-centration of amphiphilic polymer. A_1 and A_2 are upper and lower limit value of S-shaped curve, and x_0 is the center of the curve.

Fig. 5 indicates the meaning of several parameters, such as $(x_{CAC})_1$ and $(x_{CAC})_2$, which mean the first and second CAC respectively. The value of $(x_{CAC})_1$ is equal as x_0 which could be determined by regression fitting using Formula 2. But the value of $(x_{CAC})_2$ must be determined by analytical calculation. From Fig. 4, $(x_{CAC})_2$ is the point of intersection of $y_3=f(x)$ and $y_2=A_2$. So the equation of y_3 must be determined. For S-shaped Boltzmann curve, the corre-sponding slope of central point is shows as Formula (3).

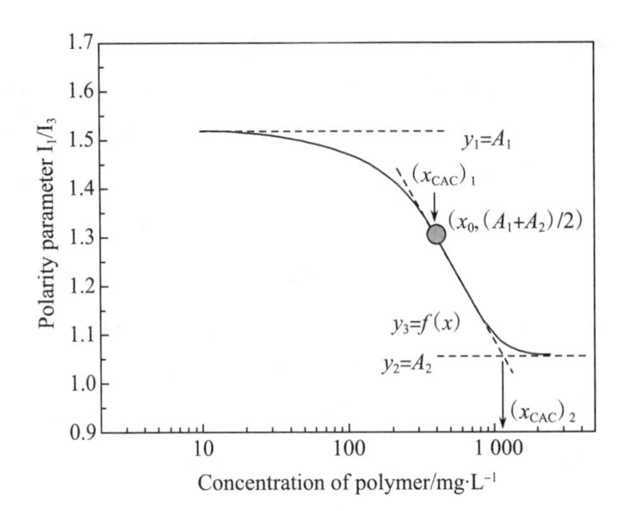

Fig. 5 The curve of Boltzmann function for different amphiphilic polymer concen-trations.

Table 1 The fitting parameter results of different molecular weight for amphiphilic polymer.

	x_0	$\lrcorner x$	R^2	$(x_{CAC})_1$	$(x_{CAC})_2$	$x_0/\Delta x$
AP-570	1015.74	354.21	0.996 5	1 015.74	1 724.16	2.86
AP-710	614.22	310.32	0.996 5	614.22	1 234.86	1.979
AP-1200	255.59	285.41	0.997 1	285.41	826.41	0.895

$$\left(\frac{dy}{dx}\right)_{x=x_0}=\frac{A_2-A_1}{4\Delta x} \tag{3}$$

The equation of y_3 could express as follow.

$$y_3=ax+b \tag{4}$$

The slope a cloud be obtained from Formula(3). And because the formula through the point (x_0, y_0), formula (4)can be transformed into Formula (5),

$$y_3=\frac{A_1+A_2}{2}+\frac{A_1-A_2}{4\Delta x}x_0-\frac{A_1-A_2}{4\Delta x}x \tag{5}$$

The value of $(x_{CAC})_2$ can be confirmed by point of intersection of the Formula $y_3=f(x)$ and $y_2=A_2$.

$$\frac{A_1+A_2}{2}+\frac{A_1-A_2}{4\Delta x}x_0-\frac{A_1-A_2}{4\Delta x}(x_{CAC})_2=A_2 \tag{6}$$

$(x_{CAC})_2$ could be obtained from Formula (7).

$$(x_{CAC})_2=x_0+2\Delta x \tag{7}$$

Fig. 4 was regression fitted by using Formula (2). The results show as Table 1, and the value of $(x_{CAC})_1$ and $(x_{CAC})_2$ list into it. AP-570 has the higher value of CAC, next is AP-710, and the lowest is AP-1200. It indicates that with increasing the relative molec-ular mass of the amphiphilic polymer, the CAC of the polymer decreased. This is because when the relative molecular mass of amphiphilic polymer increases, the number of hydrophobic groups on the polymer chains also increases, and it provides a more colli-sion probability to form a hydrophobic region.

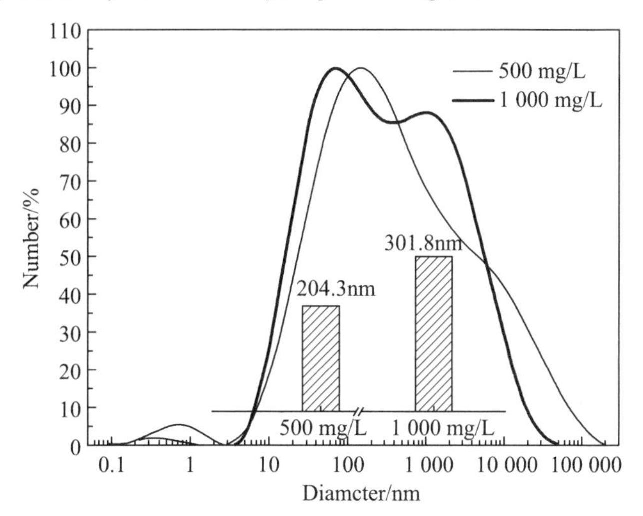

Fig. 6 The curve of apparent hydrodynamic radius distribution of amphiphilic poly-mer by DLS.

Combining the evaluation results of dynamic light scattering and using an example of AP-570 illustrate the relationship of the hydrodynamic radius and concentration. When the concentration of AP-570 is 500 mg/L, only a single peak shows in Fig. 6. It indicates that there is no intermolecular hydrophobic associa-tion to form a network in the solution. And the average apparent radius of fiuid mechanics is 204. 3 nm. There is a multi-peak in the polymer concentration of 1 000 mg/L system, and at this point the average apparent hydrodynamic radius is 301. 8 nm. And in a larger radius position appeared a peak. It suggests that the inter-molecular hydrophobic association was formed. Above the CAC, the hydrophobic groups in these polymers tend to associate in an aqueous solution by intermolecular hydrophobic interaction leading towards the formation of polymolecular associations.

The results of fiuorescent probes and dynamic light scattering show that amphiphilic polymer in an aqueous solution has a critical aggregation concentration. At concentrations above the CAC, inter-molecular hydrophobic association among polymers dominates polymer behavior. The viscosity increases significantly because of the extensive inter-molecular hydrophobic association, which leads to network structures in the solutions. But non-hydrophobic association btween the polymer dominates the behavior while its concentration is less than CAC, so the network structure cannot be formed, which lead to a very low viscosity in amphiphilic polymer solution. At concentration below the CAC, amphiphilic polymer only formed intramolecular association, and when the concentration of it above the CAC, the amphiphilic polymer formed intermolecular association network structure (Fig. 7), at the same time, the larger the molecular weight, the earier to form the struc-ture, and the smaller of the CAC.

Fluorescent probe technology indirectly confirmed the exis-tence of the aggregate structure in the amphiphilic polymer solution. Using the method of microscopic imaging, the morphol-ogy characteristic of aggregation was studied directly. Fig. 8is the atomic force microscope (AFM) of AP-570 in concentration of 500 mg/L, 1 000 mg/L and 1 500 mg/L respectively. The different color of light and shade in the picture indicates different loca-tion of the aggregation. A low-set location shall be dark color. For P(AM/BHAM/NaA) solutions at 500 mg/L, it has a little bulge and low height to the association structure. As concentration has increased, there are growing numbers of bulges. It shows the larger concentration of P(AM/BHAM/NaA), the more aggregation pro-duces, and the structural strength of the aggregation would be stronger.

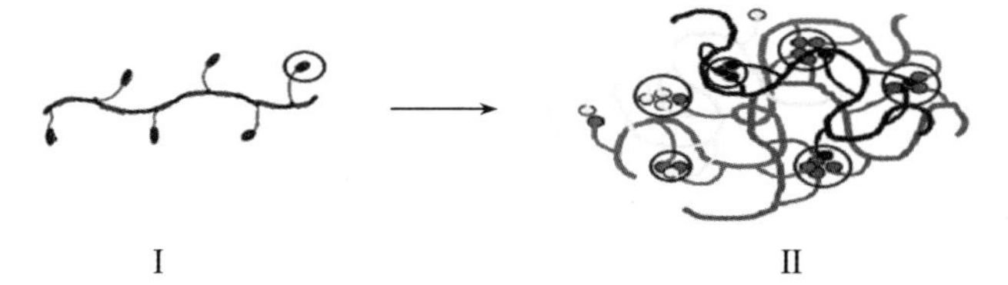

Fig. 7 The schematic diagram of the aggregate formation process of amphiphilic polymer.

Under the same concentation, the polymer network structure is also different with a different molecular weight. Fig. 9 is scanning electron micrograph for the amphiphilic polymer solutions with different MW. At a low MW, the molecular chain is shorter, so it can only form the simple molecular chain entanglement or intramolec-ular association network structure. Along with MW increase, the network structural strength increased and skeleton of network thicker.

It indicates that when the concentration of amphiphilic polymer is lower than the CAC, there is no existence of a large aggregatestructure in the solution. When the concentration reaches a cer-tain line (about 1 000 mg/L), it forms dendritic association structure can be observed clearly in the solution. This is because that above CAC, hydrophobic groups can be spontaneous associating to form supramolecular network structure. Meanwhile, adding a surfac-tant into the solution forms polymer/surfactant compound system. The surfactant also has hydrophobic association ability, and it will infiuence the solution properties.

3.2 The surfactant interaction research

It has been well-recognized that polymers can associate with ionic surfactants, in which the surfactant head groups are located in the proximity of the polymer hydrophilic groups such as OH and $CONH_2$, through forming hydrogen bonds[21,22]. Fig. 10 shows the apparent viscosity of amphiphilic polymer solution in the presence of SDS. The curve pattern showed that the solution viscosity to the ratio of SDS concentration increased at first and decreased later on. As the surfactant is initially added, there is already a substantial quantity of hydrophobe in the solution. It would seem feasible therefore that the surfactant associates directly with those hydrophobic regions of the copolymer solution in a noncoopera-tive process. Similar behavior has been proposed for a variety of systems in which hydrophobic domains exist in solution before the addition of any surfactant [23]. As the concentration of surfactant increases, a point is reached at which there is sufficient for effectively solubilize the hydrophobes, forming mixed micelles, without the assistance of the hydrophilic chains. This corresponds to the onset of the dramatic increases in the viscosity. There is not, as yet, sufficient surfactant to solubilize each region of hydrophobe separately, and so two or more regions, either of the same chain or of others, are incorporated into the same micelle. Obviously, adding a further surfactant to approach its CMC, each hydrophobe region may be solubilized by a single micelle leading to a decrease in the viscosity.

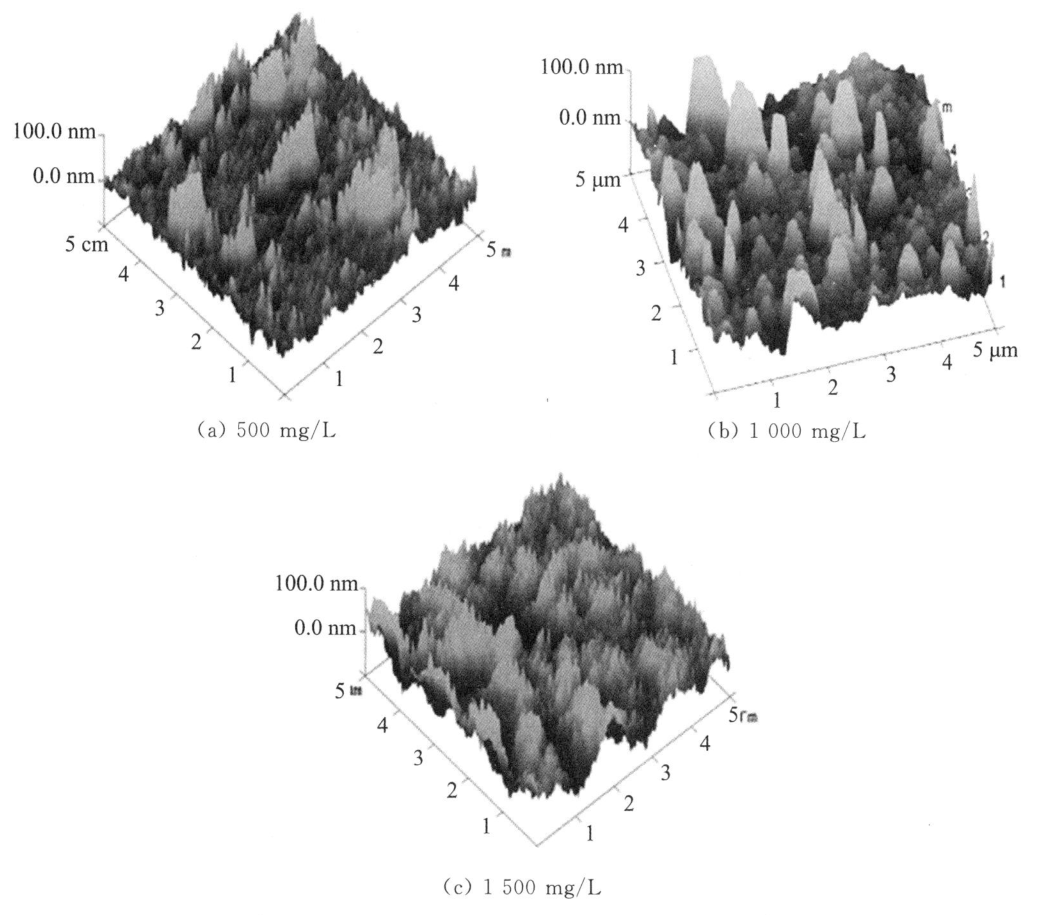

(a) 500 mg/L

(b) 1 000 mg/L

(c) 1 500 mg/L

Fig. 8　The AFM photos of P (AM/BHAM/NaA)-570 at 500 mg/L, 1 000 mg/L and 1 500 mg/L. 1 500 mg/L.

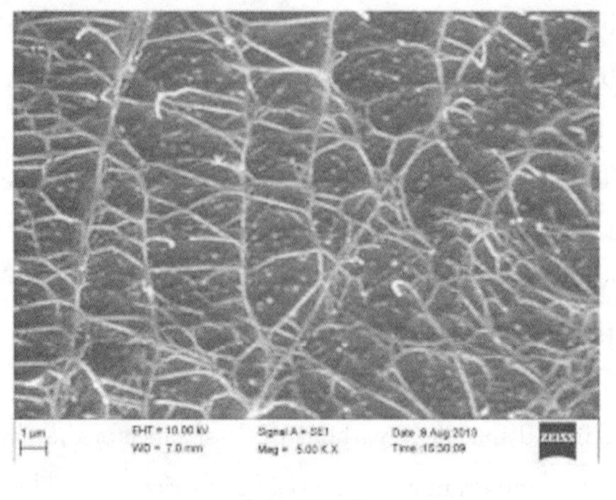

(a) AP-570

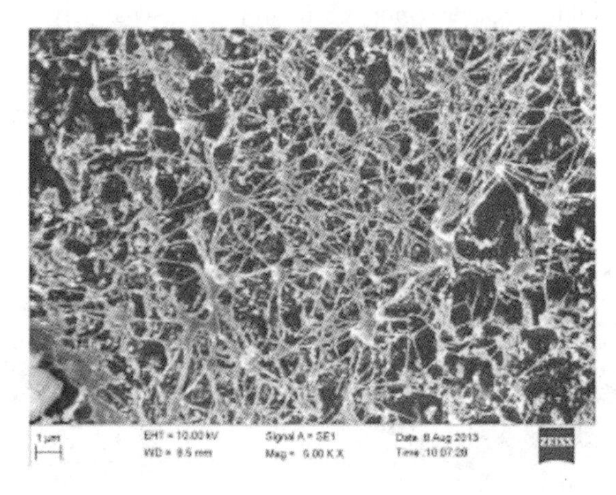

(b) AP-710

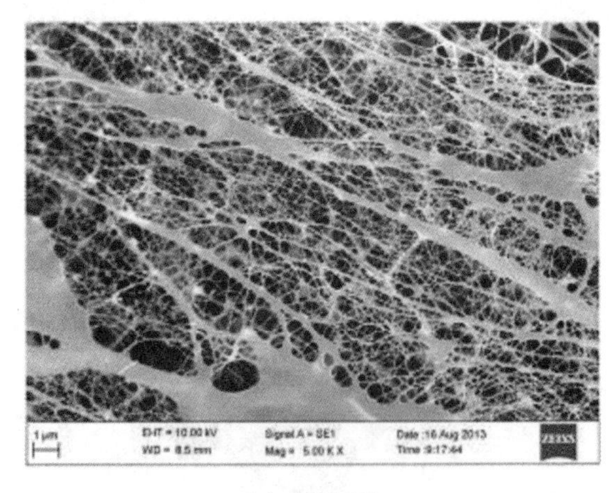

(c) AP-1200

Fig. 9 SEM photos of different molecular weight of P(AM/BHAM/NaA) at 1 000 mg/L. 5 000×.

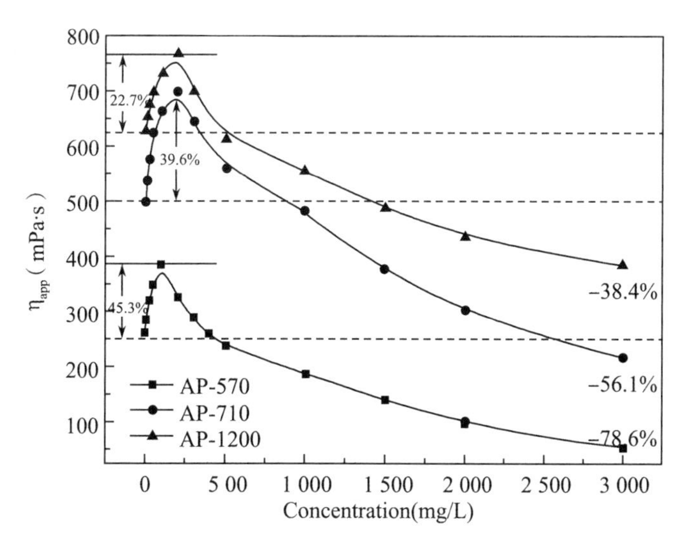

Fig. 10　Infiuence ofSDS concentration on apparent viscosity of AP/SDS mixed solu-tions. The concentration of polymer is 1 500 mg/L, T=25℃.

This kind of interaction between amphiphilic polymer and surfactant not only affects solution viscosity, but also changes the interfacial activity. The dynamic IFT between the polymer solution and the oil from the Shenli Oil Field as a function of time were plotted as shown in Fig. 11. As seen, for the HPAM, the IFT values almost hold constant at approximately 36 mN/m and do not exhibit a noticeable changes with time. On the contrary, the IFT of amphiphilic polymers, AP-1200 AP-710 and AP-570, were dramatically reduced to much lowervalues (33, 30 and 24 mN/m, respectively) in 10 min, after which the curves almost leveled off with time. And the reduction effectiveness of the amphiphilic polyer will be reduced with the increase with molecular weight, because larger average molecular weight of polymer was unfavourable for the adsorption on the surface of oil and water. The polymers is mixed with surfactant to form the polymer/surfactant compound system. The IFT had significantly lowered, and it had the same trends with polyer solution. The reduction of IFT was the major contributor to surfactant molecules, which have two kinds of function in aqueous solution: (1) It has an ability of reduce oil-water interfacial tension; (2) Interacting with amphiphilic polymer, the original aggregation will occur induced dissociation, and the new aggregation will be formed by polymer and surfactant, which has a good performance of reduce O/W tension. With the reduce of MW, this capacity will be stronger.

Generally, the associated polymeric systems with surfactant have two advantages as follows: (1) Enhanced viscoelasticity resulting from the associative networks, which facilitates theim provement of volumetric sweep efficiency and also microscopic displacement efficiency; (2) In the case of the associated HPAM, the EOR surfactant can decrease the interfacial tension between oil phase and water phase by approximately two orders of magnitude, which is expected to be able to boost the microscopicdisplacement efficiency.

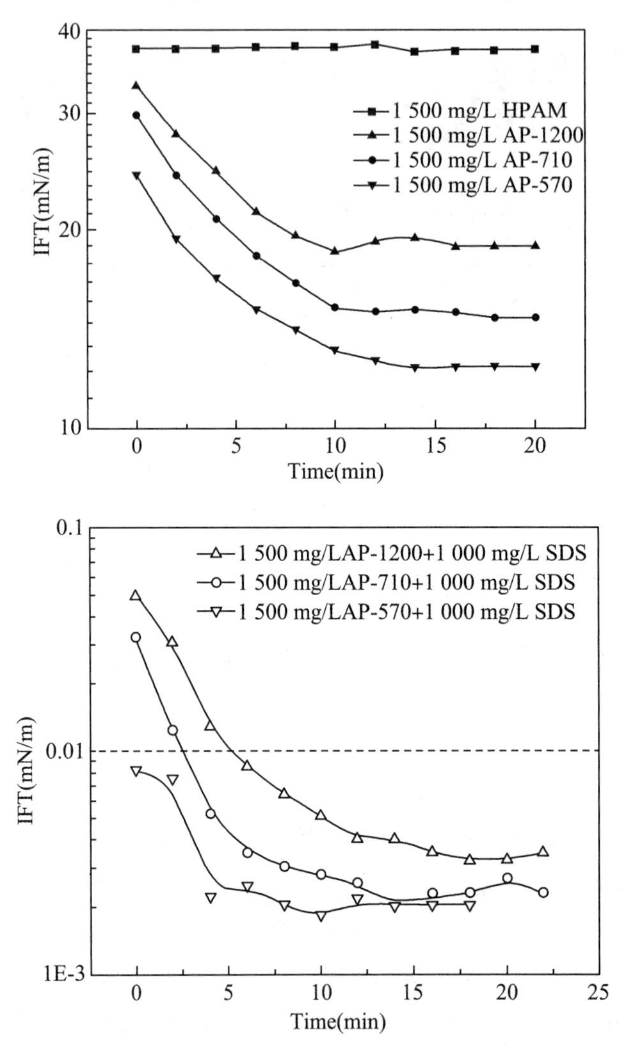

Fig. 11 Dynamical IFT between polymer solutions and crude oil. Description: The dynamic IFT as a function of time was investigated to find the interaction between polymer solution and oil.

3.3 Enhanced heavy oil recovery

The enhance oil recovery performance of the associated polymeric systems were evaluated in sandpack fiooding tests. Prior to polymer fiooding, the water fiooding was carried out as the secondary oil recovery to generate the residual oil saturated porous media. The results show that after injecting 8 PVs, the Water/Oil Ratio (WOR) of the effiuent was nearly 100%, which indicates that the water fiooding method was exhausted. Thus, a polymer slug (1. 0 PV) was then injected as the enhanced oil recovery mode followed by an extended water fiooding (1-3 PVs). Fig. 12presents the results from the oil displacement tests for different molecular weight of P(AM/BHAM/NaA). For this heavy crude oil, only about 35% OOIP can be produced by waterfiooding under the experimen-tal conditions. However, when polymer solutions were injected into the mature reservoirs, the cumulative oil recovery was consid-erably increased. For example, the most widely used EOR polymer, HPAM, raised the oil recovery by more than 33% OOIP. Moreover, it is interesting to observe that the oil recovery efficiency was further improved at different extents by fiooding the amphiphilic polymer systems. Herein, it is worthy to highlight the performance of high molecular weight P(AM/BHAM/NaA), which shows proper propagation in the porous media and also high efficiency in displacing residual oil. Compared to HPAM, 15% OOIP more incremental oil was produced using P(AM/BHAM/NaA). This fact evidenced the potential of these associative systems in the heavy oil recovery process. The preeminent EOR performance of P(AM/BHAM/NaA) is probably attributed to (1) mobility reduction capacity result-ing from the associated and enlargednetworks; and (2) relatively lower oil-water interfacial tension. Regarding molecular weight of amphiphilic polymer, it is found that oil recovery increased with the increase of molecular weight.

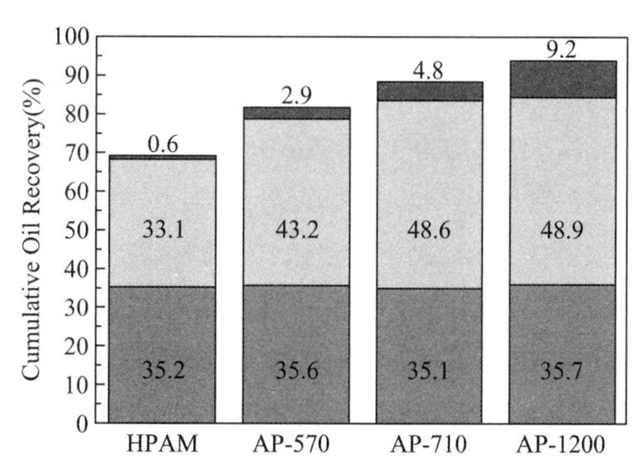

Fig. 12 Cumulative oil recovery in waterfiooding, polymer fiooding, and extended water fiooding.

Description: The efficiency of the polymer systems in displac-ingheavy oil was shown in the figure.

These results including rheological properties, phase behavior, and EOR performance, allow us to conclude that the hydrophobic group associated polymer systems are promising fiooding agents for enhancing heavy oil recovery.

4 Conclusions

The current study evaluated the potential of P(AM/BHAM/NaA)/SDS systems in enhanced heavy oil recovery. The major conclusions can be drawn as follows:

1) The different MW of P(AM/BHAM/NaA) were synthesized by an aqueous micellar copolymerization and was determined by ^{1}H NMR spectroscopy.

2) The effect of the molecular weight on the rheological proper-ties indicated that with increasing the relative molecular mass of the amphiphilic polymer, the CAC of the polymer decreased, which was measured by fiuorescent probes and dynamic light scattering. This is because when the relative molecular mass of amphiphilic polymer increases, the number of hydrophobic groups on the polymer chains also increases, and it provides a more collision probability to form a hydrophobic region.

3) The apparent viscosity of amphiphilic polymer solution in the presence of SDS shows that the solution viscosity to the ratio of SDS concentration increased at first and decreased later on, and the dynamic interfacial tension between oil and polymer solution can be reduced by two orders of magnitude because the disassociation and recombination for surfactant molecules on the aggregation of amphiphilic polyers.

4) The heavy oil recovery was further raised through fiooding of the polymer/surfactant compound system, in which AP-1200 pro-duced 24.4% OOIP more cumulative oil than HPAM as a result of the viscosity control ability and decreased IFT.

Acknowledgments

This research was supported by the National Science and Technology Major Project (No. 2016ZX05011-003), The Certificate of China Postdoctoral Science Foundation (No. 2016M592241).

References

[1] K. Y. Cho, Y. S. Yeom, H. Y. Seo, Y. H. Park, H. N. Jang, K. Y. Baek, H. G. Yoon, Rational design of multiamphiphilic polymer compatibilizers: versatile solubility and hybridization of noncovalently functionalized CNT nanocomposites, ACS Appl. Mater. Interfaces 7, (2015): 9841-9850.

[2] C. Céline, D. S. L. Marli Miriam, C. Christophe, C. Olivier, N. Taco, Structure of pH sensitive-selfassembled amphiphilic di-and triblock copolyelectrolytes: micelles: aggregates and transient networks, Phys. Chem. Chem. Phys, 15 (2013): 3955-3964.

[3] D. Konkolewicz, M. J. Monteiro, S. Perrier, Dendritic and hyperbranched polymers from macromolecular units: elegant approaches to the synthesis of functional polymers, Macromolecules, 44 (2011): 7067-7087.

[4] L. Holmberg, P. Piculell, Oil-continuous microemulsions mixed with an amphiphilic graft copolymer or with the parent homopolymer: polymer-droplet interactions as revealed by phase behavior and light scattering, Colloids Surf. A 250 (2004): 325-336.

[5] A. V. D. And, M. Rubinstein, Hydrophobically modified polyelectrolytes in dilute salt-free solutions, Macromolecules 33 (2000): 8097-8105.

[6] T. S. C. Pai, C. Barner-Kowollik, T. P. Davis, Synthesis of amphiphilic block copolymers based on poly(dimethylsiloxane) via fragmentation chain transfer (RAFT) polymerization, Polymer 45 (2004): 4383-4389.

[7] V. Francisco, R. G. Gilbert, Characterization of branched polysaccharides using multiple-detection size separation techniques,J. Sep. Sci. 33 (2010): 3537-3554.

[8] J. L. Yan, M. H. Yi, W. Feng, Effect of irradiation on the molecular weight, structure and apparent viscosity ofxanthan gum in aqueous solution, Adv. Mater. 239 (2011): 2632-2637.

[9] S. Xiaoyuan, W. Frédéric, K. R. Thomas, Intrinsic viscoelasticity in thin high-molecular-weight polymer films, Phys. Rev. E Stat. Nonlin. Soft Matter Phys. 89 (2014): 062604-062604.

[10] M. Wang, Y. F. Chen, R. Hu, Coupled hydro-mechanical analysis of a dam foundation with thick fiuvial deposits: a case study of the Danba Hydropower Project, Southwestern China, Eur. J. Environ. Civil Eng. 20 (2015): 1-26.

[11] R. Miller,J. B. Li, M. Bree, G. Loglio, A. W. Neumann, H. MÖhwald, Interfacial relaxation of phospholipid layers at a liquid-liquid interface, Thin Solid Films 327 (1998): 224-227.

[12] J. B. Li, G. Kretzschmar, R. Miller, H. MÖ hwald, Viscoelasticity of phospholipid layers at different fiuid interfaces, Colloids Surf. A 149 (1999): 491-497.

[13] W. Xue, I. W. Hamley, Rapid swelling and deswelling of thermoreversible hydrophobically modified poly (N-isopropylacrylamide) hydrogels prepared by freezing polymerization, Polymer 43 (2002): 3069-3072.

[14] I. Lacík, J. Selb, F. Candau, Compositional heterogeneity effects in hydrophobically associating water-soluble polymers prepared by micellar copolymerization, Polymer 36 (1995): 3197-3211.

[15] M. Häger, K. Holmberg, U. Olsson, Synthesis of an amphiphilic polymerperformed in an oil-in-water microemulsion and in a lamellar liquid crystalline phase, Colloids Surf. A 189 (2001): 9-19.

[16] A. Hill, F. Candau, J. Selb, Properties of hydrophobically associating

polyacrylamides: inuence of the method of synthesis, Macromolecules 26 (1993): 4521-4532.

[17] L. Hanykova, NMR and thermodynamic study of phase transition in aqueous solutions of thermoresponsive amphiphilic polymer, Chem. Lett. 41 (2012): 1044-1046.

[18] G. M. Wilmes, D. J. Arnold, K. S. Kawchak, Effect of chain rigidity on blockcopolymer micelle formation and dissolution as observed by 1H NMR spectroscopy,J. Polym. Res. 18 (2011): 1787-1797.

[19] C. Damas, M. Adibnejad, A. Benjelloun, A. Brembilla, M. C. Carré, M. L. Viriot, Fluorescent probes for detection of amphiphilic polymer hydrophobic microdomains: a comparative study between pyrene and molecular rotors, Colloid Polym. Sci. 275 (1997): 364-371.

[20] J. Aguiar, P. Carpena,J. A. Molina-Bolívar, On the determination of the critical micelle concentration by the pyrene 1:3 ratio method,J. Colloid Interface Sci. 258 (2003): 116-122.

[21] I. Iliopoulos, T. K. Wang, R. Audebert, Viscometric evidence of interactions between hydrophobically modified poly (sodium acrylate) and sodium dodecyl sulfate, Langmuir 7 (1991): 617-619.

[22] X. Xin, G. Xu, H. Gong, Interaction between sodium oleate and partially hydrolyzed polyacrylamide: a rheological study, Colloids Surf. A 326 (2008): 1-9.

[23] F. M. Winnik, H. Ringsdorf,J. Venzmer, Interaction of surfactants with hydrophobically-modified poly(N-isopropylacrylamides). 2. Fluorescence label studies, Langmuir 7 (1991): 905-911.

三次采油用小分子自组装超分子体系驱油性能

徐 辉 曹绪龙 孙秀芝 李 彬 李海涛 石 静

(中国石化胜利油田分公司勘探开发研究院,山东东营 257015)

摘要:胜利油区三类油藏由于温度和矿化度较高,常规聚丙烯酰胺驱油剂无法在该类油藏条件下取得较好的驱油效果。为此,利用自制的超分子主剂和辅剂,研制耐温抗盐非聚丙烯酰胺类新型超分子体系,并对其基本性能、驱油效果及微观聚集形态进行分析。研究结果表明:超分子体系利用小分子之间的自组装,能够形成和常规聚丙烯酰胺一样致密的网络聚集体;在胜利油区三类油藏条件下,超分子体系质量分数为0.1%~0.25%时相对于常规聚丙烯酰胺,粘度提高1倍以上;不除氧条件下,超分子体系在30 d内粘度保持稳定;超分子体系质量分数为0.15%时的单管物理模拟实验能提高采收率18%以上;因此,超分子体系是一种非常有前景且能适用于胜利油区高温高盐油藏的新型驱油体系。

关键词:超分子体系 高温高盐油藏 增粘性 驱油效果 微观聚集形态

中图分类号:TE357.43 文献标识码:A 文章编号:1009-9603(2017)02-0080-05

Study on oil displacement performance of self-assembled small-molecule supramolecular system for EOR

Xu Hui, Cao Xulong, Sun Xiuzhi, Li Bin, Li Haitao, Shi Jing

(Research Institute of Exploration and Development, Shengli Oilfield Company, SINOPEC, Dongying City, Shandong Province, 257015, China)

Abstract: Due to high temperature and high salinity in type-Ⅲ reservoirs of Shengli oilfield, conventional polyacrylamide oil displacement agent cannot effectively enhance oil recovery. Therefore, a new type of non polyacrylamide supramolecular system with temperature resistance and salt tolerance was developed with self-made supramolecular main agent and assis-tant agent. And the basic performance, oil displacement efficiency and microscopic accumulation morphology were charac-terized. The result shows that the supramolecular system can form dense network aggregate like HPAM by self-assembly between small molecules. Under the condition of type-Ⅲ reservoirs in Shengli oilfield, the viscosity of the supramolecular system increases by more than one time with mass concentration of 0.1%~0.25% compared with HPAM. Without oxygen re-moval, the viscosity of the supramolecular system keeps stable within 30 days. Single pipe physical simulation may enhance oil recovery by over 18% when the mass concentration of supramolecular system is 0.15%. So the supramolecular system is a very promising new type of oil displacement, suitable for high temperature and high salinity reservoirs in Shengli oilfield.

Key words: supramolecular system; high temperature and high salinity reservoir; viscosifying ability; oil displacement effi-ciency; microscopic accumulation morphology

DOI:10.13673/j.cnki.cn37-1359/te.2017.02.013

根据进行化学驱的难易程度,将胜利油区发育的油藏分为3类。一类和二类油藏由于温度和矿化度相对较低(温度小于等于80℃,矿化度小于等于20 000 mg/L,钙镁离子质量浓度小于等于400 mg/L),现场聚合物驱效果显著,截至2015年,已累积增油量超过2 000×10^4 t。三类油藏由于温度和矿化度都较高(温度大于80℃,矿化度大于20 000 mg/L,钙镁离子质量浓度大于400 mg/L),常规聚丙烯酰胺类[1]驱油剂在该类油藏条件下的增粘性较差,且高温降解严重,无法取得较好的驱油效果。因此,研发新型耐温抗盐驱油剂成为解决高温高盐油藏提高采收率的关键。

目前,新型耐温抗盐驱油剂的研发主要是通过对常规聚丙烯酰胺进行改性,在合成过程中引入耐温抗盐基团[2]、刚性的环状结构[3]或者疏水缔合单体[4-5]来提高驱油聚合物的增粘性和耐温抗盐性。但由于改性后的聚合物主体仍然是聚丙烯酰胺,因此无法彻底解决聚合物在高温高盐油藏条件下分子链卷曲严重以及易降解[6]的问题。为此,笔者提出不再对常规聚丙烯酰胺进行改性,而是以超分子工程学及超分子化学理论为基础[7],研制以小分子为主体,利用小分子间自组装而形成的新型超分子体系,并在胜利油区三类油藏条件下,对这种新型超分子体系的增粘性、耐温性、抗盐性、长期热稳定性以及驱油效果进行评价,最后利用冷冻蚀刻和扫描电镜联用的方式对体系在水溶液中的微观聚集形态进行表征,分析超分子体系具有良好增粘性及驱油效果的原因。

1 实验器材与方法

1.1 实验器材

实验试剂 PNTS01(自制的小分子量超分子主剂)、PNTS02(自制的小分子量超分子辅剂)、无水乙醇(分析纯)、常规聚丙烯酰胺(水解度为24%,相对分子质量为2×10^7)和石油磺酸盐阴离子表面活性剂。

实验用水 蒸馏水。总矿化度为 5 727 mg/L（Ca^{2+} 与 Mg^{2+} 质量浓度之和为 108 mg/L)属于胜利油区一类油藏模拟水，总矿化度为 19 334 mg/L(Ca^{2+} 与 Mg^{2+} 质量浓度之和为 514 mg/L)属于胜利油区二类油藏模拟水，总矿化度为 32 868 mg/L(Ca^{2+} 与 Mg^{2+} 质量浓度之和为 874 mg/L)属于胜利油区三类油藏模拟水。

实验仪器：三口烧瓶、回流冷凝管、温度计、旋转蒸发仪、恒温油浴锅、磁力搅拌器及转子、安东帕 MCR301 流变仪、室内物理模拟试验评价装置、德国蔡司扫描电镜(型号 EVO18 Special Edition)；英国 QUORUM 公司 PP3000T 冷冻传输制备系统。

1.2 实验方法

超分子体系的研制 在三口烧瓶中加入一定量的 PNTS01 超分子主剂，将乙醇加入到带有磁力搅拌子、回流冷凝管和温度计的三口烧瓶中，在一定温度条件下搅拌，采用滴液管滴加 PNTS02 超分子辅剂，反应 24 h。在一定温度和压力下减压蒸馏，除去溶剂乙醇，得到 PNTS 超分子体系。

超分子体系基本性能评价 利用胜利油区不同类型的模拟水配制成不同质量分数的超分子体系，在实验温度为 85℃的条件下，测试体系的增粘性，并进一步考察体系的耐温性、抗盐性和长期热稳定性。

超分子体系驱油效果评价 首先对岩心饱和油，然后水驱至含水率为 95%，接着注入孔隙体积倍数为 0.3 的胜利油区三类油藏模拟水配制的超分子体系，最后进行后续水驱至含水率为 100%。实验温度为 85℃，注入水矿化度为 32 868 mg/L；岩心模型为单管石英砂填充岩心，长度为 30 cm，直径为 2.5 cm；岩心渗透率为 $1\ 500\times10^{-3}\ \mu m^2$，孔隙度为 34.7%；注入段塞的孔隙体积倍数为 0.3；超分子体系的注入速度为 0.23 mL/min，质量分数为 0.15%。

超分子体系微观聚集形态表征用微量注射器取少量蒸馏水配制质量分数为 0.15% 的超分子体系放在样品台上，将样品台放在传输杆上并固定住样品，放入液氮里冷冻 30 s 以上，接着将样品台放入 PP3000T 冷冻传输制备系统的样品制备室，进行升华和喷金，最后用传输杆将样品台放入扫描电镜中观察体系的微观聚集形态。

2 实验结果与分析

2.1 超分子体系基本性能评价

2.1.1 增粘性

用胜利油区三类油藏模拟水配制成质量分数为 0.1%～0.25%的超分子体系和常规聚丙烯酰胺溶液，测定相同质量分数条件下 2 种体系的粘度并进行对比。由图 1 可以看出：虽然超分子体系为小分子，但是通过小分子之间的自组装后仍然能够表现出和相对分子质量在千万级别的聚丙烯酰胺一样的增粘性[8]；而且在胜利油区三类油藏模拟水的配制条件下，质量分数为 0.1%～0.25%的超分子体系在相同质量分数条件下的粘度相对于现场所用的常规聚丙烯酰胺增加 1 倍以上，因此超分子体系具有在胜利油区三类油藏条件下增大驱油剂波及体积的潜力。

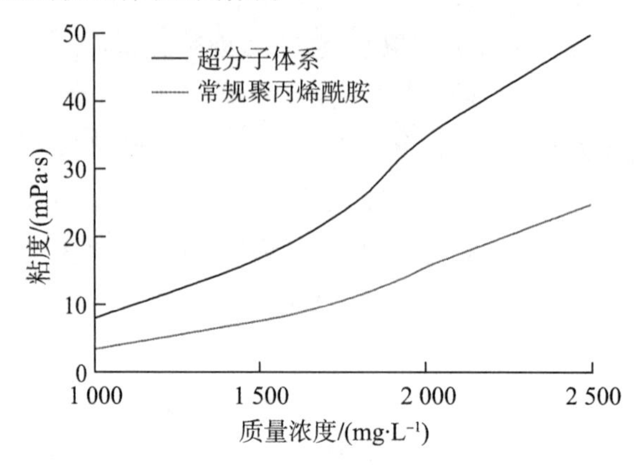

图 1　超分子体系和常规聚丙烯酰胺在胜利油区三类油藏条件下增粘性对比

Fig. 1　Comparison of viscosifying ability between supramolecular system and common HPAM under condition of type-Ⅲ reservoirs in Shengli oilfield

2.1.2 耐温性

用胜利油区三类油藏模拟水配制成质量分数为 0.15%的超分子体系,测试温度为 55～85℃条件下该体系的粘度,并与相同条件下常规聚丙烯酰胺的粘度进行对比。由图 2 可以看出:超分子体系和常规聚丙烯酰胺的粘度都随着温度的增加而降低,其原因为温度升高使超分子体系的分子运动速度加快,减弱体系的自组装能力,粘度随之降低,同时温度升高使常规聚丙烯酰胺分子链自由度增加,分子链间缠绕能力减弱,粘度也降低。但超分子体系和常规聚丙烯酰胺随温度的升高,粘度降低的程度不同,超分子体系在低温时随温度的升高,粘度降低的幅度较大;而当温度高于 75℃后,随温度的升高,粘度降低的幅度减小。常规聚丙烯酰胺随着温度的升高,粘度平缓下降,温度每升高 10℃,粘度降低幅度基本一致。但由于超分子体系初始粘度值远高于常规聚丙烯酰胺,因此温度超过 85℃,质量分数为 0.15%的超分子体系的粘度仍是常规聚丙烯酰胺的 2 倍以上。

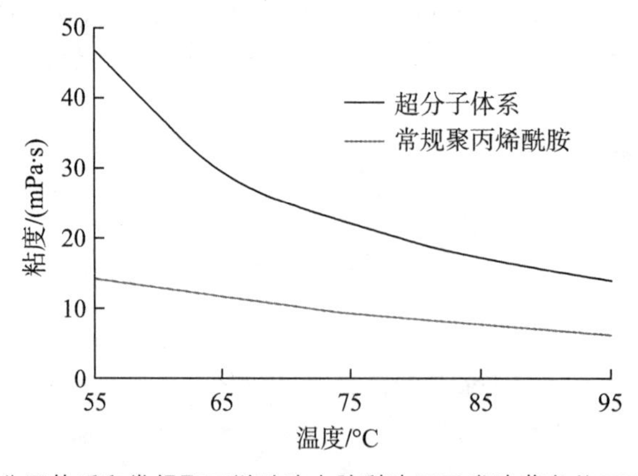

图 2　超分子体系和常规聚丙烯酰胺在胜利油区三类油藏条件下耐温性对比

Fig. 2　Comparison of temperature resistance performance between supramolecular system and commonHPAM under condition of type-Ⅲ reservoirs in Shengli oilfield

2.1.3 抗盐性

分别用胜利油区一类、二类和三类油藏模拟水配制成质量分数为 0.15%的超分子体系，测试温度为 85℃条件下超分子体系的粘度，并与相同条件下所用常规聚丙烯酰胺的粘度进行对比。由图 3 可以看出：常规聚丙烯酰胺随着矿化度的增加，粘度不断降低，矿化度由 5 727 mg/L 升至 32 868 mg/L，粘度由 24.3 mPa·s 降低至 7.6 mPa·s，降低幅度近 70%；而对于超分子体系，随着矿化度的增加，粘度基本不变，具有良好的抗盐能力。这是因为随着矿化度的升高和钙镁离子的增加，常规聚丙烯酰胺中的羧酸根中的负电荷会被水中的阳离子屏蔽，导致羧酸根之间的排斥力减弱，分子链产生卷曲，粘度降低，而超分子体系的粘度是通过小分子间的自组装形成，且不含有对盐敏感的基团，因此随着矿化度的增大，体系粘度基本不变，表明超分子体系更适用于矿化度高的油藏。

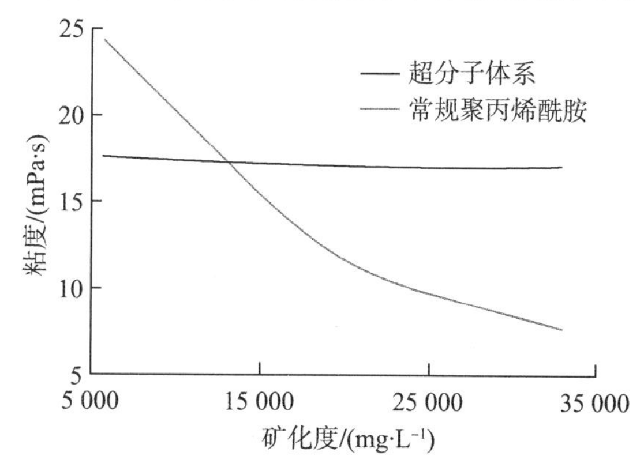

图 3　超分子体系和常规聚丙烯酰胺在胜利油区不同矿化度油藏条件下粘度对比

Fig. 3　Comparison of viscosity between supramolecular system and common HPAM under condition of reservoirs of different salinity in Shengli oilfield

2.1.4 长期热稳定性

目前，实验室对常规聚丙烯酰胺长期热稳定性的评价，需要首先对聚丙烯酰胺溶液进行除氧，然后再密闭封存，放入具有一定温度的烘箱中。因为在不除氧的条件下，水中的氧自由基很快使聚丙烯酰胺断链[9-10]，在较短的时间内会造成聚丙烯酰胺溶液的粘度大幅降低。该实验是在不除氧的条件下，用胜利油区三类油藏模拟水配制成质量分数为 0.2%的超分子体系，然后直接放入温度为 85℃的烘箱中，每隔一定时间测定一次超分子体系的粘度，观察该体系的长期热稳定性，并与现场所用常规聚丙烯酰胺溶液长期热稳定性进行对比。

由图 4 可知：在不除氧的条件下，常规聚丙烯酰胺在高温和高矿化度油藏条件下，一方面溶液中的氧自由基会使聚丙烯酰胺断链；另一方面，酰胺根会发生水解产生羧酸根，羧酸根再与水中的 Ca^{2+}，Mg^{2+} 结合产生沉淀，导致在 4 d 之内溶液粘度降低幅度超过 55%，15 d 之内溶液粘度降低幅度超过 85%。而超分子体系是通过小分子的自组装而形成，且不含对盐敏感的基团，因此溶液中氧自由基不会使超分子体系断链，水中的 Ca^{2+}，Mg^{2+} 也不会和超分子体系相结合产生沉淀，因此超分子体系在 30 d 之内粘度基本保持稳定,具有良好的长期热稳定性。

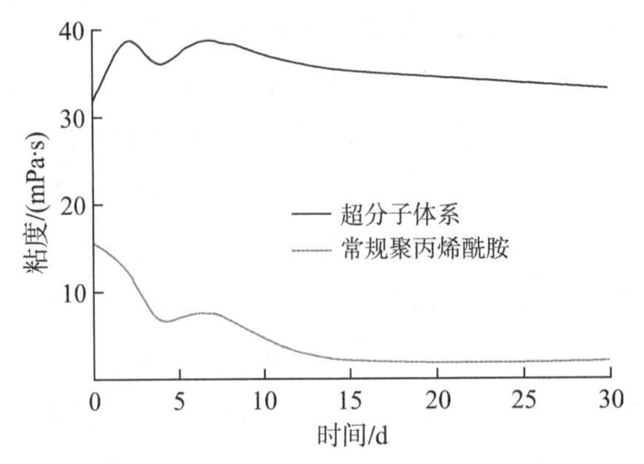

图 4 超分子体系和聚丙烯酰胺在胜利油区三类油藏条件下长期热稳定性对比

Fig. 4 Comparison of long-term thermal stability between supramolecular system and polyacrylamide under condition of type-Ⅲ reservoirs in Shengli oilfield

2.2 超分子体系驱油效果评价

在温度为 85℃、配制水矿化度为 32 868 mg/L 的实验条件下,通过填砂管单管物理模拟实验,得到超分子体系的驱替曲线(图 5)。由图 5 可以看出:在水驱采收率达到 56%,采出液含水率超过 95%时,注入超分子体系,能够使注入压力增加;在后续水驱时,注入压力在一段时间保持稳定,然后慢慢降低,超分子体系能够产生较高的阻力系数和残余阻力系数,同时产生较深的含水率降低漏斗;最终注入超分子体系后再进行后续水驱采收率达到 74.65%,超分子体系提高采收率 18%以上。

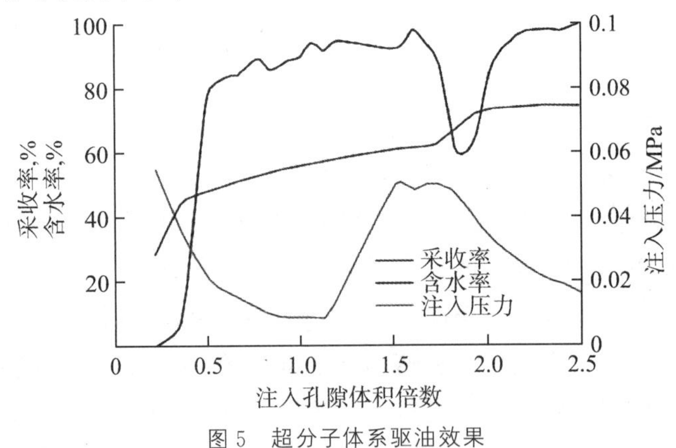

图 5 超分子体系驱油效果

Fig. 5 Oil displacement effect of supramolecular system

2.3 超分子体系微观聚集形态表征

通过对超分子体系在水溶液中微观聚集形态的表征来解释其在胜利油区三类油藏条件下具有高增粘性和良好驱油效果的原因。目前对溶液微观聚集形态的研究主要有原子力显微镜法[11]、透射电镜法[12]和扫描电镜的方法[13]。采用冷冻蚀刻和扫描电镜联用的方式研究超分子体系的结构,通过液氮快速冷冻超分子体系,使溶液瞬间固化,锁定溶液的水化分子形态,再通过扫描电镜对固化后的结构进行观察,可以最大限度的观察该体系在水溶液中真实的分布情况，比直接用扫描电镜观测溶液有明显的优势。

由图6可知：在同样放大5 000倍的条件下，通过扫描电镜观察到的超分子体系在水溶液中呈现和常规聚丙烯酰胺一样的网络聚集体；对比发现超分子体系网络结构的边长为300 nm～2 μm，同时，在中间出现了更致密的小分子聚集体，这种小分子聚集体是由于小分子间氢键的相互作用而形成的(图6a)；常规聚丙烯酰胺网络结构的边长为1～2 μm，其网络结构是通过大分子链之间相互缠绕和分子链之间阴离子的相互排斥而形成的，没有明显的小分子聚集体(图6b)。油田现场常用的石油磺酸盐表面活性剂主要由石油磺酸盐小分子组成，由于石油磺酸盐小分子之间不能形成自组装，因此，在石油磺酸盐表面活性剂的微观聚集形态(图6c)中未出现网络聚集体，而是呈分散的簇状结构，这种结构无法起到增大聚合物溶液粘度的作用。实验结果表明，超分子体系虽然是由小分子组成，但是在水溶液中通过小分子之间的自组装，能够形成致密的空间网络结构，依靠这种结构，该体系能够形成良好的增粘性，且具有良好的驱油效果。

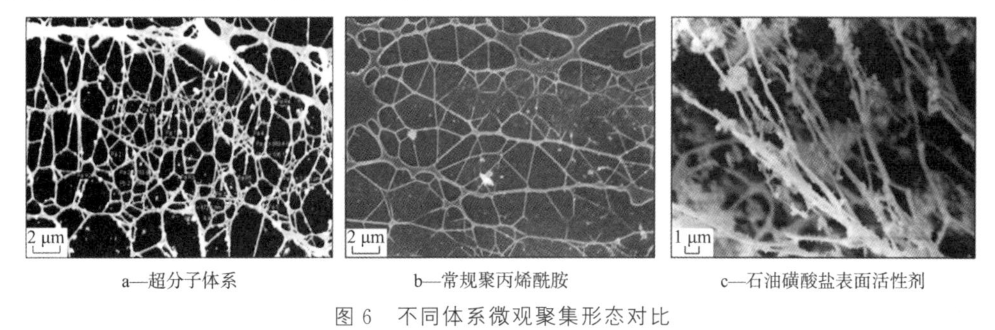

a—超分子体系　b—常规聚丙烯酰胺　c—石油磺酸盐表面活性剂

图6 不同体系微观聚集形态对比

Fig. 6 Comparison of microscopic accumulation morphology between different systems

3 结论

对新型超分子体系的基本性能、驱油效果及微观聚集形态进行分析，结果表明，通过小分子的自组装形成的超分子体系在水溶液中能够形成和高分子一样致密的网络聚集体，可以用于驱油。超分子体系克服了常规聚丙烯酰胺类驱油剂在高温高盐油藏条件下增粘性和耐温抗盐性较差的不足，还兼具在高温高盐油藏条件下长期热稳定性好的优点，是一种非常有前景的且适用于高温高盐油藏的新型驱油体系。目前超分子体系相对于常规聚丙烯酰胺成本高50%左右，下一步需要优化合成条件，降低体系的合成成本，使超分子体系在经济上能够达到在油田现场使用的要求。

参考文献

[1] 付京，宋考平，王志华，等. 高分子质量聚合物溶液与二类油层匹配性研究[J]. 特种油气藏，2015，22 (2)：129-132.

Fu Jing, Song Kaoping, Wang Zhihua, et al. Research on compati-bility between high molecular polymer solution and type-II oil layer[J]. Special Oil & Gas Reservoirs, 2015, 22 (2): 129-132.

[2] 韩玉贵,王秋霞,宋新旺,等.水溶性两性共聚物的表征及其溶液性质[J].石油化工,2007,36(5):507-512.

Han Yugui,Wang Qiuxia,Song Xinwang,et al. Characterization and solution properties of a water-soluble amphoteric copolymer [J]. Petrochemical Technology, 2007, 36 (5): 507-512.

[3] 钟景兴,陈煜,谭惠民.AM/NVP 二元共聚物的溶液性能[J].高分子材料科学与工程,2005,21 (4):220-223.

Zhong Jingxing, Chen Yu, Tan Huimin. Study on solution proper-ties of AM/NVP copolymer[J]. Polymer Materials Science and En-gineering,2005,21 (4):220-223.

[4] 马先平,卢祥国,张德富,等.疏水缔合聚合物渗流特性及其影响因素[J].大庆石油地质与开发,2015,34(3):102-107.

Ma Xianping,Lu Xiangguo,Zhang Defu,et al. Seepage character-istics and its influencing factors of hydrophobic associating poly-mer [J]. Petroleum Geology & Oilfield Development in Daqing,2015,34(3):102-107.

[5] 季岩峰,曹绪龙,郭兰磊,等.聚合物疏水单体与表面活性剂对聚/表二元体系聚集体的作用[J].油气地质与采收率,2016,23(4):95-101.

Ji Yanfeng,Cao Xulong,Guo Lanlei,et al. Influence of hydropho-bic groups of polymer and surfactant on aggregation of polymer/surfactant binary system[J]. Petroleum Geology and Recovery Ef-ficiency,2016,23(4):95-101.

[6] 赵林,姜汉桥,李俊键,等.基于流线的聚合物驱热降解数值模拟[J].油气地质与采收率,2016,23(6):76-81.

Zhao Lin, Jiang Hanqiao, Li Junjian, et al. Numerical simulation study of thermal degradation in polymer flooding based on stream-lines[J]. Petroleum Geology and Recovery Efficiency,2016,23(6):76-81.

[7] Gennady V Oshovsky, David N Reinhoudt, Willem Verboom. Su-pramolecular chemistry in water[J]. Angewandte Chemie Interna-tional Edition,2007,46(14):2 366-2 393.

[8] 王德民,程杰成,杨清彦.粘弹性聚合物溶液能够提高岩心的微观驱油效率[J].石油学报,2000,21(5):45-51.

Wang Demin, Cheng Jiecheng, Yang Qingyan. Viscous-elastic polymer can increase micro-scale displacement efficiency in cores [J]. Acta Petrolei Sinica,2000,21 (5):45-51.

[9] 杨怀军,罗平亚.污水降解聚合物因素分析及控制方法[J].油田化学,2005,22 (2):158-162.

Yang Huaijun,Luo Pingya. Factors for and control of polymer deg-radation in recycled produced water solutions[J]. Oilfield Chemis-try,2005,22 (2):158-162.

[10] 詹亚力,郭绍辉,闫光绪.部分水解聚丙烯酰胺降解研究进展[J].高分子通报,

2004,4 (2):70-74.

Zhan Yali, Guo Shaohui, Yan Guangxu. Study on degradation of partially hydrolyzed polyacrylamide[J]. Polymer Bulletin,2004,4(2):70-74.

[11] 张瑞,叶仲斌,罗平亚.原子力显微镜在聚合物溶液结构研究中的应用[J].电子显微学报,2010,29(5):475-481.

Zhang Rui, Ye Zhongbin, Luo Pingya. The atomic force microsco-py study on the microstructure of the polymer solution[J]. Journal of Chinese Electron Microscopy Society, 2010,29(5):475-481.

[12] 侯吉瑞,刘中春,张淑芬,等.碱对聚丙烯酰胺的分子形态及其流变性的影响[J].物理化学学报,2013,19(3):256-259

Hou Jirui,Liu Zhongchun,Zhang Shufen,et al. Effect of alkali on molecular configuration of polymer and its rheologic behavior[J]. Acta Physico-Chimica Sinica,2013,19(3):256-259.

[13] 冯玉军,罗传秋,罗平亚,等.疏水缔合水溶性聚丙烯酰胺的溶液结构的研究[J].石油学报:石油加工,2001,17 (6):39-44.

Feng Yujun, Luo Chuanqiu, Luo Pingya, et al. Study on character-ization of microstructure of hydrophobically associating water-sol-uble polymer in aqueous media by scanning electron microscopy and environmental scanning electron microscopy [J]. Acta Petrolei Sinica:Petroleum Processing Section,2001,17 (6):39-44.

编辑 王 星

环糊精二聚体与双支化两亲聚合物包合体系的构筑

季岩峰　曹绪龙　王端平　郭兰磊　孙业恒　闵令元

(中国石化胜利油田分公司勘探开发研究院,山东东营 257015)

摘要:将实验室自制的环糊精二聚体(66βCDsu)与双支化两亲聚合物(P(AM/BHAM/NaA))加入到水溶液中,发生包合作用构筑包合体系。研究该体系的增黏及溶液流变学特性,并采用荧光光谱仪、扫描电镜及差示扫描量热仪等研究包合体在溶液中的包合机制及结构形态。结果表明,由于P(AM/BHAM/NaA)中1个疏水单体中存在2个疏水基团(正十六烷基和苄基),因此环糊精与疏水单体BHAM最大摩尔包合比为2∶1,完全包合后溶液中没有游离态疏水基团,因此溶液不存在临界聚集浓度(CAC)。当环糊精与BHAM摩尔包合比为1∶1,体系存在明显的CAC,这是由于环糊精首先包合双支化疏水单体中的苄基形成包合体,而正十六烷基依然存在于水溶液中。当P(AM/BHAM/NaA)浓度为800 mg/L时,该体系中存在2种聚集方式,一种是疏水基团的疏水缔合;另一种是包合作用。通过扫描电镜证明了不同体系的微观聚集形态。

关键词:环糊精二聚体;双支化两亲聚合物;包合体系;流变学特性;微观聚集形态

中图分类号:O636.1+2　**文献标识码**:A　**文章编号**:1000-7555(2017)07-0045-08

环糊精(CD)是直链淀粉在环糊精葡萄糖基转移酶(由芽孢杆菌发酵而成)作用下生成的一系列通过α-1,4糖苷键相互连接的环形低聚糖的总称,通常每个环糊精分子含有6～12个D-(+)-吡喃葡萄糖单元[1-3]。由于其葡萄糖单元数目不同,空间尺寸也有所不同。目前,常见的环糊精是分别含有6、7和8个葡萄糖单元的α-、β-和γ-环糊精。该类聚合物具有“疏水空腔”和“亲水外壁”[4-6],因此能与很多具有疏水-亲水结构的化合物发生包合作用,体系具有独特的流变学特性,进而被广泛地研究。Kobayashi等[7]采用荧光和诱导圆二色谱研究了环糊精与芘的包结配位作用,发现芘分子不能进入α和β-环糊精疏水空腔形成包合物,而7-环糊精与芘可以发生1∶1、1∶2、2∶1和2∶2的包合。Gosselet等人[8],制备了疏水改性N,N-二甲基丙烯酰胺/甲基丙烯酸羟乙酯共聚物与环糊精/环氧丙烷共聚物的包合作用,包合后体系黏数是包合前N,N-二甲基丙烯酰胺/甲基丙烯酸羟乙酯共聚物黏度的十倍,并证明了环糊精包合客体物质可以形成一种稳定的包合物,这也为环糊精超分子科学提供了理论和实验的可能性。

本文,在实验室合成环糊精二聚体和双支化两亲聚合物基础上,研究了 2 种聚合物在水溶液中的包合机理。由于多支化两亲聚合物的 1 个疏水单体中含有 2 个疏水基团,因此该类包合体系在不同的包合比条件下,在水溶液中存在不同的聚集行为,对溶液黏度具有不同的影响。

1　实验部分

1.1 主要原料

环糊精二聚体(66βCDsu):N,N-bis(6A-deoxy-6A-β-cyclodextrin) succinamide 合成方案遵循文献[9],其合成路线见 Fig. 1。

双支化两亲聚合物-P(AM/BHAM/NaA):N-苄基,N-十六烷基丙烯酰胺三元共聚物为实验室自制(分子式如 Fig. 2 所示),其水解度均为 25%(摩尔分数),疏水单体含量均为 1%(摩尔分数)。

1.2 制备方法

将一定质量的 P(AM/BHAM/NaA)颗粒缓慢分散在蒸馏水中,高速搅拌 10 min,使 P(AM/BHAM/NaA)颗粒能够很好地分散在水中而不团聚。当 P(AM/BHAM/NaA)颗粒出现膨胀时,降低搅拌速度,并将其放入 45℃水浴锅中 4 h。随后停止搅拌,将 P(AM/BHAM/NaA)溶液放入 45℃的恒温箱中老化 24 h。按照比例取一定质量的 66βCDsu,加入老化完成的 P(AM/BHAM/NaA)溶液中。在 45℃水浴锅中低速搅拌 4 h 使 66βCDsu 完全溶解于 P(AM/BHAM/NaA)溶液中。将得到的溶液放入 45℃的恒温箱中老化 48 h。

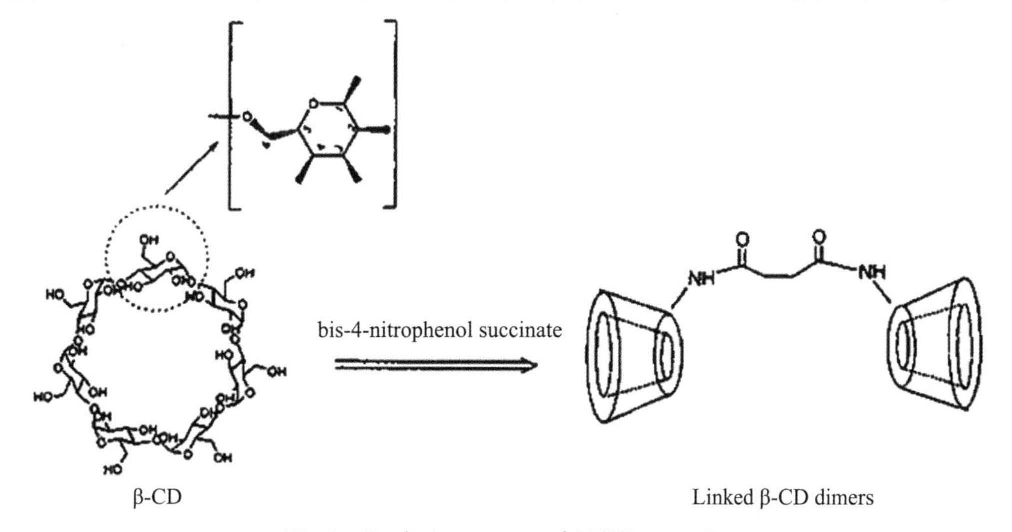

Fig. 1　Synthetic processes of 66βCDsu reaction

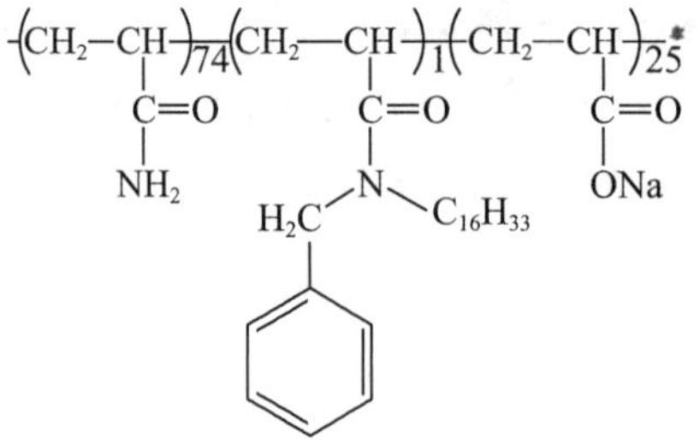

Fig. 2 Molecular formula of P(AM/BHAM/NaA)

1.3 测试与表征

1.3.1 聚合物表观黏度测定:利用 RW200 Digital 搅拌器(IKA,德国)搅拌已制备好的P(AM/BHAM/NaA)及其包合体系,转数设定为 150 r/min。4~6 h 以后,将溶解好的聚合物放入 FYL-YS-138 L1 恒温箱中,45℃恒温熟化 24 h。采用 DV-Ⅱ黏度计(Brookfield,美国)测量 P(AM/BHAM/NaA)溶液的表观黏度,转子转速为 6 r/min。

1.3.2 荧光探针法:在聚合物溶液中加入适量芘的乙醇溶液,芘的最终浓度为 1.0×10^{-6} mol/L。将溶液密封采用超声波对芘增溶后 45℃并于恒温静置 24 h。采用 Hitachi F4500 荧光光度计记录芘的荧光光谱。激发波长 335 nm,发射光谱记录范围为 350~450 nm。芘荧光光谱有 5 个振动峰,其中第 1 振动峰与第 3 振动峰强度比 I_1/I_3 可以反映芘所处微环境的微极性。

1.3.3 流变性测定:流变法是研究 P(AM/BHAM/NaA)及其包合体系溶液性质的重要工具。使用 MCR 301 流变仪椎板系统(Anton-Paar,奥地利)(量杯半径 14.466 mm,转子半径 13.330 mm)测定聚合物溶液的稳态流变以及动态黏弹性能,剪切速率范围为 0.01~100 s^{-1};对于动态黏弹性测试,首先进行振幅扫描以确定线性平台区(Linear Viscoelastic Region,LVR),在线性平台区范围内进行频率扫描,频率为 0.01~100 Hz,测试温度为 45℃。

1.3.4 DSC 测定:准确称取 8 mg(精确至 0.001 mg)样品于铝盒中,用空铝盒为参比。采用差示扫描量热(DSC)仪(TA,美国)分别对 P(AM/BHAM/NaA),66βCDsu、包合体系进行测定,绘制 DSC 曲线。扫描速度为 10℃/min,扫描范围为 30~800℃。

1.3.5 SEM 测定:利用扫描电子显微镜(JMS-6380LV,日本电子),用微量注射器取 P(AM/BHAM/NaA)溶液至承载物上,移至液氮上预冷冻,预冷冻后将容器浸入液氮冷冻 2~3 min,再迅速转移至真空冷冻干燥机,冷冻抽真空干燥 48 h,取出喷金镀膜,将含有 P(AM/BHAM/NaA)的膜放在扫描电镜下观察分子线团的形貌。

2 结果与讨论

2.1 包合体系浓度对体系黏度的影响

作为一种水溶性聚合物体系,表观黏度是重要物化特性。为了系统研究不同两亲聚合物的种类对包合体系包合作用的影响,绘制了 P(AM/BHAM/NaA)在包合前后体系的黏浓曲线。如 Fig. 3 所示，当 P(AM/BHAM/NaA)浓度高于 800 mg/L 时，其溶液黏度显著增加。这主要是 P(AM/BHAM/NaA)的疏水基团可以发生缔合作用从而在水溶液中形成疏水微区。随着 P(AM/BHAM/NaA)浓度的增加，在溶液中将发生分子内缔合、分子间缔合并最终在溶液中形成具有一定黏弹性的空间网络结构。当 P(AM/BHAM/NaA)浓度较低时，其在溶液中以分子内缔合为主。随着浓度的升高，P(AM/BHAM/NaA)以分子间缔合为主，超分子聚集体主要存在溶液中，P(AM/BHAM/NaA)在溶液中形成空间网络结构，溶液表观黏度显著增加，黏度显著提高的转折点浓度就是 P(AM/BHAM/NaA)的临界聚集浓度(CAC)[10,11]。

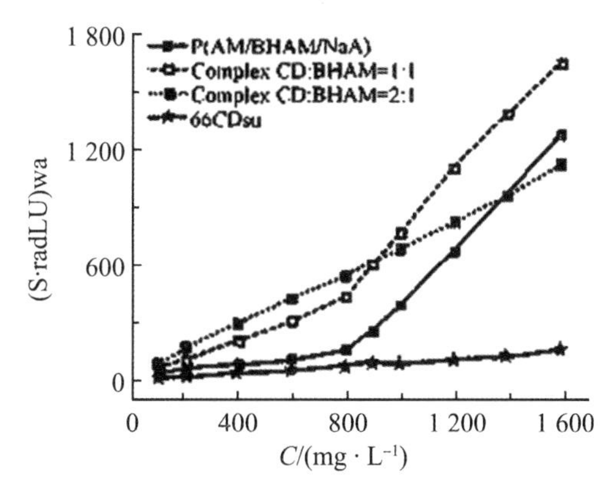

Fig. 3　Effect of polymer concentration on the apparent viscosity of amphiphilic polymers and their mixture with 66pCDsu (pH =7, 4S 6 r/min)

当向 P(AM/BHAM/NaA)溶液中加入一定摩尔比的 66βCDsu 时，环糊精二聚体在水溶液中与 P(AM/BHAM/NaA)中的疏水基团发生包合作用。从而在 P(AM/BHAM/NaA)浓度较低时(低于 CAC 浓度)，便可以形成高分子空间网络结构。从而达到低浓高黏特性。如 Fig. 3 所示，在 P(AM/BHAM/NaA)溶液中加入 66βCDsu 时，体系在低浓环境下就具有较高的表观黏度。环糊精二聚体与 P(AM/BHAM/NaA)中疏水基团 BHAM 摩尔比为 2∶1 时，P(AM/BHAM/NaA)溶液中不再有明显的 CAC 浓度。这是因为，此时 P(AM/BHAM/NaA)中的疏水基团全部与环糊精发生包合作用，溶液中未包合的疏水基团很少，很难发生分子间疏水缔合作用，因此，不具有 CAC 浓度转折点。但当环糊精与 BHAM 摩尔比为 1∶1 时，即 n(CD)∶n(BHAM)=1∶1 时，由于 BHAM 含有 2 个疏水基即正十六烷基和苄基，在水溶液依然存在大量未发生包合作用的疏水基团。当 P(AM/BHAM/NaA)浓度高于 CAC 浓度时，未被包合的疏水基团会发生分子间的疏水缔合作用，从而与包合作用发生协同，增大溶液的表观黏度。

2.2 包合比对体系黏度的影响

上文证明了包合前后，聚合物浓度对溶液黏度的影响。结果发现，P(AM/BHAM/NaA)与 66βCDsu 的包合比对溶液黏度具有较大的影响。下面讨论包合比的大小对其溶液表观黏度的影响规律。为了排除疏水基团分子间缔合作用导致的溶液黏度升高，本实验使用 P(AM/BHAM/NaA)的浓度为 600 mg/L。

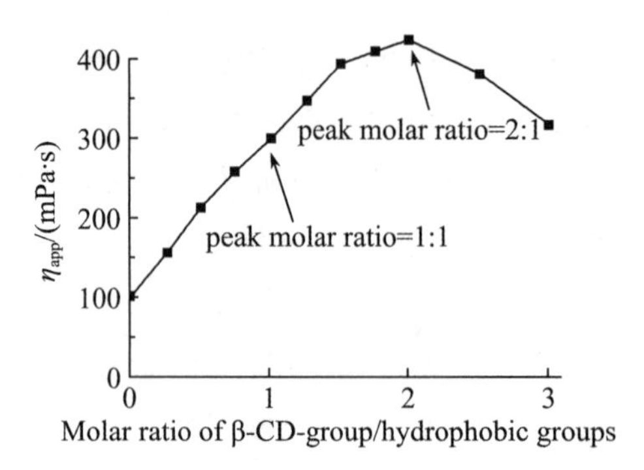

Fig. 4 Apparent viscosity as a function of mole ratio of CD groups in 66βCDsu aqueous solutions to hydrophobic group in amphiphilic polymer (pH=7,45℃,6 r/min, C (amphiphilic polymer) =600 mg/L)

Fig. 4 所示,当n(CD)∶n(BHAM)=2∶1 时,P(AM/BHAM/NaA)的包合溶液出现了黏度最大值。这是因为 P(AM/BHAM/NaA)的结构中存在 2 个疏水基团(正十六烷基和苄基),并且 1 个疏水基团被 1 个环糊精所包合。由于大量 66βCDsu 的存在,使得 P(AM/BHAM/NaA)所有疏水基团与环糊精发生包合作用,进而形成流体力学半径更大、结构更加稳定的包合体系,溶液黏度升到最高值。正是由于此时溶液中已不存在游离的疏水基团,不能发生疏水基团分子间的缔合作用,P(AM/BHAM/NaA)溶液中 CAC 消失。当n(CD)∶n(BHAM)=1∶1 时,CAC 出现,这是主要是该包合体系中疏水基团没有完全被环糊精包合,网络结构上依然存在裸露在水溶液中的疏水基团。当溶液中 P(AM/BHAM/NaA)浓度升高至 CAC 以上,疏水基团在溶液中又会发生分子间疏水缔合,从而在包合作用基础上有机结合了疏水缔合作用,进而使得体系表观黏度更高(如 Fig. 3 所示)。Fig. 5 为双支化两亲聚合物与环糊精二聚体在包合比为 1∶1 时,在溶液中聚集行为的结构示意图。

2.3 包合体系的稳态流变特性

Fig. 6 是 P(AM/BHAM/NaA)及其包合体系溶液黏度随剪切速率的变化曲线。可以看到,3 种体系的流变曲线的形状相同,均出现了第一牛顿区。随着剪切速率的增加,3 种体系黏度均表现出剪切变稀的性质。对于 P(AM/BHAM/NaA)及其包合体系溶液,在不同的剪切条件下,疏水基团间的缔合结构以及分子链间的缠结会产生不同的变化,如 Fig. 6 所示,在较低的剪切速率下,稳态剪切不足以破坏疏水基团形成的缔合结构,导致体系的黏度不会发生变化,溶液中形成的依然是分子链间的缠结和分子间缔合的网络结构;但当达到一定的剪切速率后,分子链发生变形,缔合结构被破坏,聚合物主链也会逐渐形成一定得取向,促进了分子内的缔合,表明剪切过程中分子间的解缠和解缔合作用明显大于分子间的缔合速度,表现出黏度随剪切速率逐渐降低的现象;继续增加剪切速率,缔合结构完全被破坏,聚合物主链完全伸展,使体系的黏度不再随剪切速率变化,此时表现出明显的剪切变稀行为。

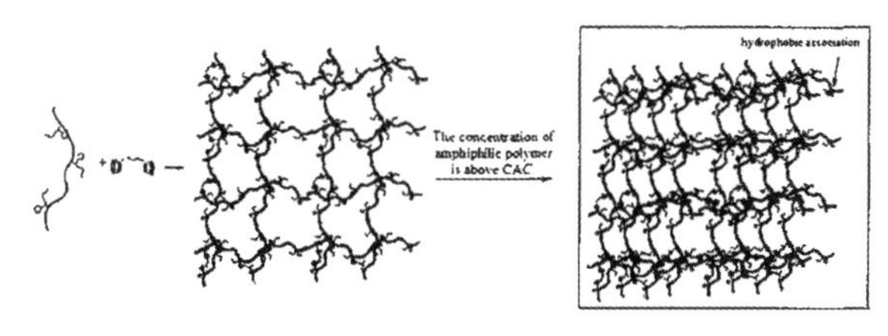

Fig. 5 Association model of P(AM/BHAM/NaA) inclusion complex networks

(mole ratio of β-CD to BHAM is 1∶1)

对流变曲线利用 Carreau 数学模型进行拟合[12]，Carreau 模型的表达式见式(1)，Fig. 6 中实线即为利用 Carreau 模型拟合后的曲线，其相关系数均达到 0.9 以上。

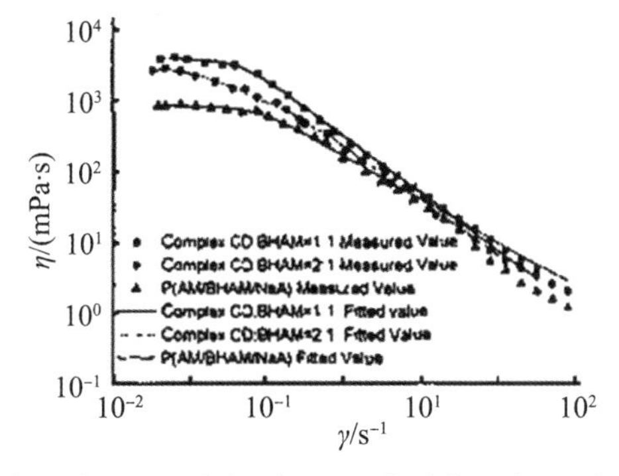

Fig. 6 Shear rheological curves of the three amphiphilic polymers' solutions t=45℃,

the concentration of amphiphilic polymer is 1000 mg/L

$$\eta=\eta_\infty+\frac{\eta_0-\eta_\infty}{[1+(\lambda\gamma)^2]^{(n-1)/2}} \tag{1}$$

式中，η——剪切黏度，mPa·s；η_0——零剪切黏度，mPa·s；η_∞——极限剪切黏度，mPa·s；n——幂率指数；λ——时间常数。零剪切黏度和极限剪切黏度是定值，与剪切速率无关，两亲聚合物溶液黏度介于二者之间，即 $\eta_0>\eta>\eta_\infty$，零剪切黏度是指高分子溶液在临界剪切速率之前是牛顿流体，其黏度是一定值，很小的剪切力无法超越分子链与链之间形成的稳定的结构强度，结构依然保持稳定。零剪切黏度的大小与聚合物的浓度、相对分子质量、盐浓度等有关。

Tab. 1 Shear rheology fitting parameters of P(AM/BHAM/NaA) and its inclusion complexes

Type	a/(mPa·s)	b/min	c(L)	Correlation
P(AM/BHAM/NaA)	856.3	0.452 1	0.923 2	0.981 2
Complex (CD∶BHAM=1∶1)	1 531.4	0.424 9	0.963 5	0.969 8
Complex (CD∶BHAM=2∶1)	2 436.3	0.414 6	0.952 5	0.961 5

由 Tab. 1 可以看到，当环糊精单体与 BHAM 摩尔比为 2∶1 时，包合体系(n(CD)∶n(BHAM)=2∶1)的零剪切黏度最高。这主要是因为，由于环糊精对两亲聚合物中疏水基团进行了完全包合，形成网络结构。这种网络结构流体力学体系要大于由疏水缔合作用所形成的聚集体结构，且形成的结构性强，从而具有更高的零剪切黏度。

2.4 包合体系的动态黏弹特性

不同的包合体系在水溶液中黏弹性有所不同。Fig. 7 为不同包合体系的储能模量和损耗模量，G′和 G″分别代表了聚合物溶液的弹性和黏性。流变学测量表明所有的包合体系溶液均为弹性主导。

当 n(CD) ∶ n(BHAM)＝2 ∶ 1 且 P(AM-BHAM/NaA)浓度为 600 mg/L 时，包合体系储能模量最大。这是因为 CD 包合了所有的正十六烷基和苄基，网络结构大于其他包合体系，黏度最大。在这种条件下，高浓度的 P(AM/BHAM/NaA)包合体系溶液也不能产生分子间疏水缔合，聚合物网络结构体系受限制。但是，当包合比为 1 ∶ 1 时，包合溶液中含有大量的未被包合的疏水基团。在溶液中存在 2 种形式的聚集行为。

Fig. 7 所示，当 P(AM/BHAM/NaA)浓度为 1 200 mg/ L，其储能模量和损耗模量均最高。机理将在下文分析。

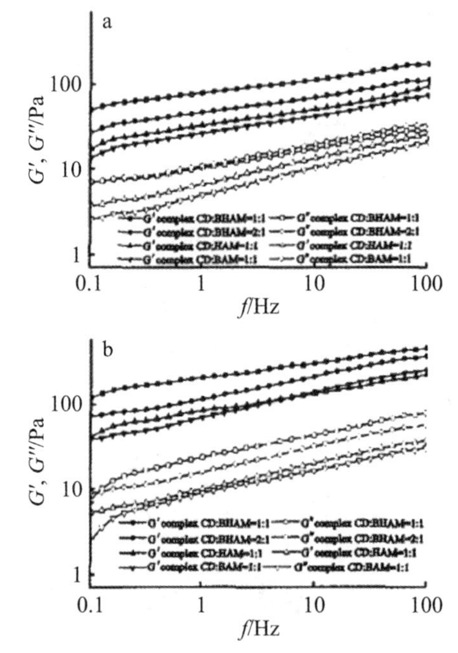

Fig. 7 Storage modulus G' and loss modulus G'' of polymer inclusion solutions a: concentration of amphiphilic polymers is 600 mg/L; b: concentration of amphiphilic polymers is 1 400 mg/L

以上研究表明了包合作用对包合体系宏观增黏性能以及微观结构黏弹性能的影响。下文使用 DSC 从热力学角度来证明环糊精与两亲聚合物中的疏水基团发生了包合作用，并采用紫外光谱法计算了最大包合比，用 AFM 观察了不同包合体系溶液的微观形貌特征。

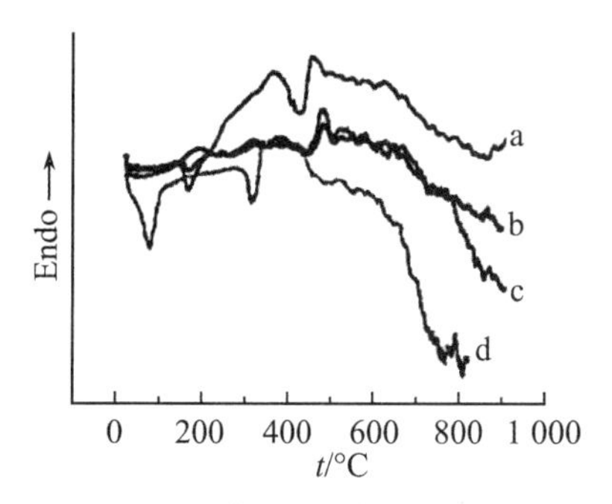

Fig. 8　DSC diagrams of 66βCDsu, P（AM/BHAM/NaA）and its inclusion complexes a：P（AM/BHAM/NaA）；b：inclusion complex（CD∶BHAM=1∶1）；c：inclusion complex（CD∶BHAM= 2∶1）；d：66CDsu

2.5 包合作用的热分析

Fig. 8 为环糊精二聚体、两亲聚合物及其包合体系的差示扫描量热图谱。在 DSC 图谱中，P(AM/BHAM/NaA)在 160.82℃出现 1 个尖锐的吸热峰；66βCDsu 的 DSC 曲线表明 β 环糊精在 160.82℃并没有吸收峰，94.7℃出现的吸热峰说明 β-CD 含有少量水存在。在包合后体系的 DSC 图谱中，吸热峰明显降低，但峰的位置基本没有发生漂移；在包合体系(n(CD)∶n(BHAM)=2∶1)的 DSC 图谱中，在 160.82℃并未出现吸收峰，且吸收峰结构与 P(AM/BHAM/NaA)和 66βCDsu 均有所不同，这说明完全包合体系已经形成。包合体系(n(CD)∶n(BHAM)=1∶1)的 DSC 图谱中，在 160.82℃的位置出现微弱的吸收峰，这是 P(AM/BHAM/NaA)中疏水基团 BHAM 所产生的特征吸收峰，此时说明体系中依然有裸露在环糊精以外的疏水基团的存在，此时的包合体系并没有发生完全包合。

包合体系(n(CD)∶n(BHAM)=1∶1)的 DSC 图谱中，在 160.82℃的位置出现微弱的吸收峰，这是 P(AM/BHAM/NaA)中疏水基团 BHAM 所产生的特征吸收峰，此时说明体系中依然有裸露在环糊精以外的疏水基团的存在，此时的包合体系并没有发生完全包合。通过 DSC 研究，可以证明包合体系的形成，但包合比的大小只能粗略的比较，因此本文使用紫外光谱法结合 Job's 方法和 Benesi-Hildebrand 方程进一步研究 P(AM/BHAM/NaA)和 66βCDsu 的包合机理。

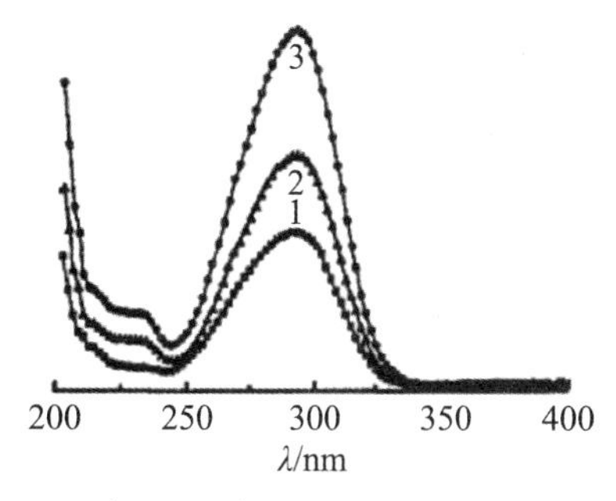

Fig. 9　Absorption spectra of P(AM/BHAM/NaA) and its inclusion complex (1) 6.0×10^{-5} mol^{-1} BHAM, (2) 6.0×10^{-5} mol^{-1} β-CD and 6.0×10^{-5} mol^{-1} BHAM, and (3) the mixture of 1.2×10^{-4} mol^{-1} β-CD and 6.0×10^{-5} mol^{-1} BHAM

2.6 包合体系的包合比研究

Fig. 9 显示 P(AM/BHAM/NaA)在有无 66βCDsu 时的紫外光谱。P(AM/A-β-CD/NaA)本身没有共轴基团不存在紫外吸收。在 290 nm 处的吸收峰是苯环紫外吸收峰。当苯环完全被环糊精包合后,290 nm 处吸收强度显著地增加。和说明苯环正好进入环糊精空腔。当苯环进入环糊精空腔时,环糊精大量的电子云增强了疏水基团的电子云密度,从而增加了苯环的共轭效应,并且最大吸收波长变得更长。P(AM/BHAM/NaA)的吸收强度由于加了 66βCDsu 变得更强。等吸收点在 290 nm 处被发现,这说明 P(AM/BHAM/NaA)和 66βCDsu 存在明显的相互作用。使用 Job's 公式计算包合比[13]。结果表明[β-CD]/([Benzene]+[β-CD])的最大值为 2/3,这说明包合比为 2∶1。这与以上的是实验结果相符。

分析疏水基团和环糊精空腔相对大小可以提供包合体系的结构信息并且明确包合体系的包合作用[14]。环糊精疏水空腔内径为 0.68 nm,而苯环的直径为 0.67～0.68 nm,这正好与环糊精空腔相匹配。因此 P(AM/BHAM/NaA)的疏水基团可进入环糊精空腔形成稳定的包合体系。因为正十六烷基的轴直径为 0.45 nm,柔顺的正十六烷基可以进入环糊精空腔,从而形成包合体系。然而环糊精的疏水内腔容积为 0.262 nm^3,正十六烷基和苯环的体积总和大于 0.3 nm^3,因此,这 2 个疏水基团不能同时进入环糊精疏水空腔,形成包合结构。

环糊精对疏水单体的包合作用,主要是与疏水单体中的疏水基团(苯环和十六烷基)发生包合。环糊精将首先与包合常数大的疏水基团,形成稳定的包合体系。由于正十六烷基没有紫外吸收特性,因此通过紫外光谱无法测试其与环糊精的包合常数。于是本文在同样合成条件下,制备了单支化 N-苄基/丙烯酰胺/丙烯酸钠三元共聚物 P(AM/BAM/NaA)来研究正十六烷基疏水基团存在与不存在时对体系紫外吸收的影响,进而得到 P(AM/BAM/NaA)和 P(AM/BHAM/NaA)的包合常数。固定 2 种聚合物浓度不变,配置不同 β-CD 浓度(最大浓度为环糊精与疏水单体比为 1∶1)的混合溶液,测定 P(AM/BAM/NaA)和 P(AM/BHAM/NaA)加入前后吸光度的变化值 ΔA[15]。根据 Benesi-Hildebrand 方程[15]。计算环糊精对 2 种疏水单体聚合物的包合常数:

$$\frac{1}{\Delta A}=\frac{1}{K_S\Delta e[G]_0[H]_0}+\frac{1}{\Delta e[G]_0} \tag{2}$$

式中,ΔA——P(AM/BHAM/NaA)和 P(AM/BAM/NaA)加入 β-CD 前后吸光度的变化值;Ks——包合常数;Δe——P(AM/BHAM/NaA)和 P(AM/BAM/NaA)加入 β-CD 前后的摩尔吸光系数差值;$[G]_0$——P(AM/BHAM/NaA)和 P(AM/BAM/NaA)的初始浓度;$[H]_0$——β-CD 的初始浓度。以 $1/\Delta A$(Y 轴)对 $1/[H]_0$(X 轴)绘制回归曲线,所得直线的截距与斜率的比值为 Ks。通过计算可得,β-CD 对 P(AM/BHAM/NaA)和 P(AM/BAM/NaA)的包合常数分别为 93.4 L/mol 和 92.8 L/mol。实验表明,双支化两亲聚合物与单支化两亲聚合物包合常数相差很小,也就是正十六烷基的加入对体系包合常数的影响很小。即苄基中的苯环先进入环糊精疏水空腔，形成稳定的包合体系。当双支化 P(AM/BHAM/NaA)与环糊精包合比为 1∶1 时，裸露在外的疏水基团多为正十六烷基。

2.7 包合体系的微观聚集形态

由于极性－非极性相互作用，正十六烷基不能稳定、舒展地存在于水溶液中。正十六烷基链内和链间依然会发生疏水缔合作用。本文使用荧光芘探针，研究了 P(AM/BHAM/NaA)及其包合体系在 2 个激发波长特征带的 I_1/I_3 Dimroth 和 Eichardt 的 $E_T(30)$极性参数[17,18]，见 Fig. 10。用以研究不同包合比对不同包合体系相互作用的影响。

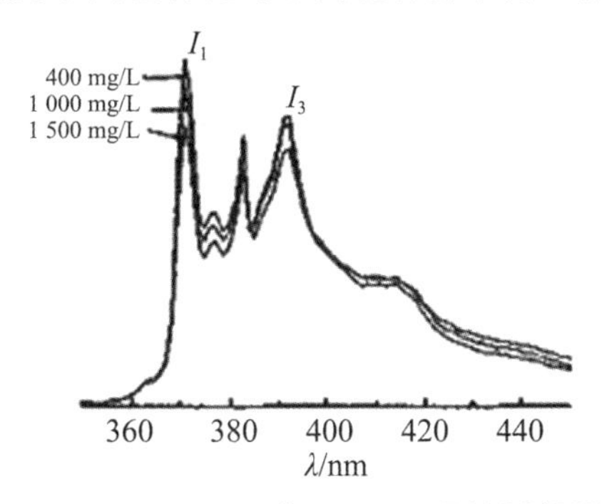

Fig. 10　Fluorescence emission spectra of pyrene in P(AM/BHAM/ NaA) solutions

由于极性很低，芘存在高的分配系数因此优先进入有疏水基团组成的微极性区域，从而在水溶液中只有很小的浓度。I_1/I_3 的大小反映了芘溶解在水中和微区中的大小，I_1/I_3 的降低与芘的微环境的改变相关联。当 P(AM/BHAM/NaA)浓度很低时，由于疏水缔合作用所形成的微区很小，因此进入疏水微区的芘较少，I_1/I_3 变化不大。当聚合物浓度增加的时候，芘进入微区的量急剧增加。导致 I_1/I_3 比值迅速降低。当 P(AM/BHAM/NaA)浓度超过其 CAC 浓度时，大量芘进入疏水微区。此时 I_1/I_3 的比值代表着溶液的极性微环境。

聚合物和包合体系的结构对局部极性环境有很大的影响。由 Fig. 11 可知，P(AM/BHAM/NaA)及其 1∶1 包合体系依然存在 CAC 值。这是因为分子间的疏水缔合作用的存在。2∶1 包合体系荧光谱线下降相对缓慢，这是由于所有的疏水基团被 CD 所包合，从而很难形成微极性区。

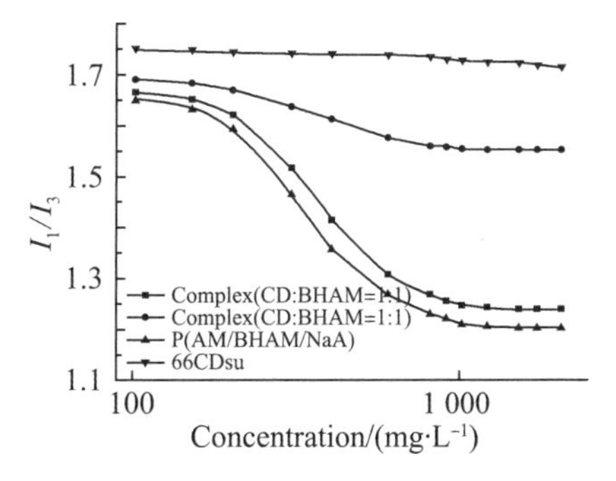

Fig. 11　Variation of the ratio I_1/I_3 with the amphiphilic polymer concentration

如果芘被环糊精所包合,那么同样也会导致 I_1/I_3 的比值的下降。因此将芘的甲醇溶液加入到不同浓度的 66βCDsu 溶液中。然后使用相同的方法在相同的环境下进行试验。Fig. 11 表示 I_1/I_3 并没有明显的降低。说明了 P(AM/BHAM/NaA)包合体系(n(CD)∶n(BHAM)=1∶1)中 I_1/I_3 的降低都是由于疏水基团的原因。

2.8 包合体系的微观形貌

通过荧光探针法发现了包合作用于疏水缔合作用所形成的高分子空间网络结构的不同。本文使用扫描电镜更直观地研究不同结合方式所构筑的网络结构。Fig. 12a 是 P(AM/BHAM/NaA)及其包合体系所形成的高分子网络结构的扫描电镜(SEM)图片,图中颜色的明暗表示结构的不同。对于 P(AM/BHAM/NaA)溶液,浓度越高凸起越来越多,说明溶液中由于疏水缔合作用所形成的缔合结构越多。由于包合作用是环糊精单体对疏水单体的包合,正是由于包合作用,该体系在 SEM 图片中不能体现出由于分子间蔬水缔合作用形成的聚集体的形貌,如 Fig. 12b 为包合体系(CD∶BHAM=2∶1)。然而,当包合比为 1∶1 时,溶液中高分子此时存在 2 种结合方式,疏水缔合和包合作用。因此如 Fig. 12c,出现较 P(AM/BHAM/NaA)溶液少的聚集体形貌。从分子形貌特征角度证明了此时溶液中聚集方式,并与荧光探针法所得的结论相同。

Fig. 12 AFM images of amphiphilic polymer and inclusion complexes solutions (the concentration of amphiphilic polymer: 900 mg/L) (a): 1000 mg/L P(AM/BHAM/NaA); (b): CD∶BHAM=2∶1; (c): CD∶BHAM=1∶1

3 结论

环糊精二聚体与双支化两亲聚合物在水溶液中发生包合作用构筑了包合体系。采用流变学研究其网络结构的宏观流变特性,同时采用荧光光谱、显微照片等手段研究其微观的侧链分子间的包合作用。探明包合机理及包合体系的微观形貌和缔合结构。

(1) 通过黏度法,研究了 P(AM/BHAM/NaA)与 β-CD 聚合物最大摩尔包合比为 2∶1。并使用紫外-可见分光光度法以及荧光法证明了包合比及包合常数,确定最佳包合类型。

(2) 增黏实验及微观形貌特征表明,当 66βCDsu 和 P(AM/BHAM/NaA)按照 n(CD)∶n(BHAM)=1∶1 包合时,该体系中存在 2 种缔合方式,一种是疏水基团的疏水缔合;另一种是包合作用。这 2 种作用在水溶液中发生协同,从而该体系具有较强的增黏能力。

(3) 采用流变学研究包合体系网络结构的宏观流变特性；使用微极性试验和DSC证明了P(AM/BHAM/NaA)疏水分子能够顺利地进入到β-CD空腔，从而说明P(AM/BHAM/NaA)与β-CD聚合物间形成了稳定的包合结构。采用AFM方法从微观角度观察了不同包合比条件下体系的形貌特征。

参考文献

[1] 胡杰，曹顺生，吉海燕. 环糊精在高分子材料制备中的作用[J]. 高分子材料科学与工程，2009，25(3)：150-153.

Hu J，Cao S S，Ji H Y. Influence of cyclodextrins on the synthesis of polymer materials[J]. Polymer Materials Science & Engineering，2009，25(3)：150-153.

[2] 童林荟. 环糊精化学：基础与应用[M]. 北京：科学出版社，2001.

[3] Klyamkin A A，Topchieva I N，Zubov V P. Monomolecular films of pluronic —cyclodextrin inclusion complexes at the water-gas interface[J]，Colloid Polym. Sci.，1995，273：520-523.

[4] 张敏，宋洁，葛正浩，等. β-环糊精与聚丁二酸丁二醇酯(PBS)包合物对PBS结晶性的影响[J]. 高分子材料科学与工程，2012，28(5)：96-99.

Zhang M，Song J，Ge Z H，et al. Curing behaviors and properties of waterborne hyperbranched polymers and waterborne epoxy resin systems [J]. Polymer Materials Science & Engineering，2012，28(5)：96-99.

[5] Szejtli J. Introduction and general overview of cyclodextrin chemistry. [J]. Chem. Rev.，1998，98：1 743-1754.

[6] Yamashita Y，Fujisawa K，Inoue M，et al. Evaluation of dendrimer conjugate with glucuronylglucosyl-β-cyclodextrin for treatment of transthyretin-related fami lialamyloi doti c polyneuropathy[J]. Asian J. Pharm. Sci.，2015：DOI：0. 1016/1. ajps. 2015. 11. 072.

[7] Fanali S. Separation of optical isomers by capillary zone electrophoresis based on host-guest complexation with cyckxlextrins [J]. J. Chromat 曙 r. A，1989，474：441-446

[8] Gosselet N M，Beucler F，Renard E，et al. Association of hydrophobically modified poly（N，N-dimethylacrylamide hydroxyethylmethacrylate） with water soluble β-cyclodextrin polymers[J]. Colloids Surf.，A，1999，155：177-188.

[9] Pham D T，Clements P，Easton C J. Dimerisation and complexation of 6-(4，-t-butylphenylamino） naphthalene-2-sulphonate by β-cyclodextrin and linked p-cyclodextrin dimers [J]. Supramol. Chem.，2009，21：510-519.

[10] Jain N, Trabelsi S, Guillot S, et al. Critical aggregation concentration in mixed solutions of anionic polyelectrolytes and cationic surfactants[J]. Langmuir, 2004, 20: 8496-8503.

[11] Busche B J, Tonelli A E, Balik C M. Cbmpatibilization of polystyrene/poly (dimethyl siloxane) solutions with star polymers containing a γ-cyclodextrin core and polystyrene arms[J]. Polymer, 2010, 51: 454-462.

[12] Carreau P J, Macdonald I F, Bird R B. A nonlinear viscoelastic model for polymer solutions and melts-II [J]. Chem. Eng. Sci., 1968, 23: 90b911.

[13] lamanaka B T, Teixeira A A, Teixeira A R R, et al. Further development of the general rule on correlation between host-guest ratio and topology of polymorphic inclusion compounds and their crystallization temperatures [J]. Acta Crystallogr, 2006, 62: 221.

[14] Chen G, Jiang M. Cyclodextrin-based inclusion complexation bridging supramolecular chemistry and macromolecular self-assembly [J]. Chem. Soc. Rev., 2011, 40: 2254-2266.

[15] Khouri S J, Abdel-Rahim I A, Shamaileh E M. A thermodynamic study of a-, β-, and γ-cyclodextrin-complexed m-methyl red in alkaline solutions[J]. J. Incl. Phenom. Macro., 2013, 77: 105112.

[16] Pickering P J, Chaudhuri J B. Equilibrium and kinetic studies of the enantioselective complexation of (dl)-phenylalanine with copper (II) N-decyl-(1)-hydroxyproline[J]. Chem. Eng. Sci., 1997, 52: 377-386.

[17] Tatikolov A S, Costa SMB. Medium effects on the isomerization of an anionic polymethine dye [J]. Chem. Phys. Lett., 2007, 440: 73-78.

[18] Bosch E, Roses M, Herodes K, et al. Solute-solvent and solventsolvent interactions in binary solvent mixtures. 2. Effect of temperature on the αEt(30) polarity parameter of dipolar hydrogen bond acceptor-hydrogen bond donor mixtures [J]. J, Phys. Org. Chem., 1996, 9: 403-410.

Inclusion mechanism of the cyclodextrin dimers and double branched amphiphilic polymer

Ji Yanfeng　Cao Xulong　Wang Duanping　Guo Lanlei　Sun Yeheng　Min Lingyuan

(Exploration and Development Research Institute, Shengli Oilfield Co. Ltd., SINOPEC, Dongying 257015, China)

Abstract: The cyclodextrin dimers 66βCDsu mixed with double branched amphiphilic polymer P (acrylamide/N-benzyl-N-n-hexadecyl acrylamide/sodium acrylate) (P (AM/BHAM/NaA)) in the solution to form the inclusive system. The viscosifying capacity and rheological properties of the inclusive system were studied and through fluorescence spectrophotometer, SEM and DSC, the inclusion mechanism and structure of inclusive systems were studied to establish the intrinsic relationships between microstructure and rheological properties. The results show that the viscosity of inclusion complex solution significantly increases under the concentrations of amphiphilic polymers which are less than the critical aggregation concentration (CAC). The maximum viscosity for multi-sticker amphiphilic polymer complexes is got at the mole ratio of CD/BHAM 2 : 1. The UV Vis absorbance verifies that the mole ratio of the inclusion complex between CD and BHAM is 2 : 1 stoichiometry. Because of the inclusion association between the host and guest polymers, the solution of inclusion complex has much higher viscoelasticity even under the low amphiphilic polymer concentration. When the mole ratio of CD to BHAM is 1 : 1, the CAC of the inclusion complex solution still remains. Furthermore, above the CAC, two types of associations, inclusion association and intermolecular hydrophobic association, can occur in the complex solution. Using the Carreau model, the rheological curve was fitted and it indicates that the inclusive system (n(CD) : n(BHAM) = 2 : 1) has the highest zero-shear rate viscosity, because the complete inclusion make the system form an intensity network structure. And these interactions were also verified by SEM.

Keywords: cyclodextrin dimer; double branched amphiphilic polymer; inclusive systems; rheological properties; microscopic structures

新型物理交联凝胶体系性能特点及调驱能力研究

徐 辉 曹绪龙 石 静 孙秀芝 李海涛

中石化胜利油田分公司勘探开发研究院

摘要:针对目前化学交联凝胶和体膨体颗粒类堵剂存在的地下交联不易控制及和孔喉尺寸较难匹配的不足,通过在聚丙烯酰胺侧链上引入含有18个碳的季铵盐疏水性功能基团,设计合成了以分子间物理交联为主的新型调驱体系,并对体系的分子结构特点,在水溶液中的微观聚集形态和性能进行了表征和评价。研究结果表明,物理交联调驱体系疏水性功能基团在水溶液中自发聚集,产生物理交联,形成致密的网络聚集体,聚集体的网络尺寸为200～400 nm,相对于化学交联凝胶,具有更好的增黏性、黏弹性和抗剪切性,同时通过调整疏水性功能基团的含量,可调整体系的成胶时间。在地层存在大孔道的条件下,注入未交联的颗粒,颗粒能在岩心中产生物理交联,有效封堵高渗带。物理交联体系分子间的作用力没有化学交联的共价键强,体系在地层中有一定的运移能力,因此物理交联体系具有调驱性能,是一种比较有前景的新型高效堵调体系。

关键词:物理交联 高黏弹性 网络聚集体 抗剪切性 调驱性能

DOI:10.3969/j.issn.1007－3426.2018.01.013

Performance characteristic and profile modification research of new type physical cross-link gel system

Xu Hui Cao Xulong Shi Jing Sun Xiuzhi Li Haitao

Exploration and Development Institute, Sinopec Shengli Oil Field Branch, Dong ying, Shandong, China

Abstract: In view of the current shortages of chemical cross-linking gel and bulk particle plugging agent, which are difficult to control gelling and match the formation pore size. By introducing quaternary ammonium salt hydrophobic functional group it has eighteen atoms of carbon to polyacrylamide side chain, a new type of profile modification gel system basing on inter molecular physical cross-link is designed. The characteristic of molecular structure, the micro-aggregation morphology and property in aqueous solution are characterized and evaluated. The results show that the hydrophobic functional group of physical cross-linking gel system aggregated spontaneously in aqueous solution, produced physical crosslinking, formed dense network aggregate, and the aggregate network size is between 200～400 nm. It is qualified with better viscidity, viscoelasticity and shear resistance compared with chemical cross-linking gel, at the same time by adjusting the content of hydrophobic functional group, formation colloid time can be adjusted. The non cross-linked particle is injected into strata where there is big pore, the particle can produce physical crosslinking in the core, which can plug the high permeability zone effectively. The intermolecular force of physical crosslink is weaker than covalent bond of chemical crosslink, so the system is qualified with certain ability of migration in the formation. The physical crosslink system has the properties of profile control and flooding. It is a new type of promising efficient plugging and profile control system.

Keywords: physical cross-link, high viscoelasticity, network aggregate, shear resistance, profile modification performance

油田出水是油田开发过程中普遍存在的问题[1]，对高渗条带的调控技术在高含水油田控水稳产措施中占有重要地位，随着高含水及高温高盐油藏水驱问题的日益复杂，对该领域技术要求越来越高，现场应用比较成熟的调控体系主要有地下化学交联体系和体膨体颗粒类调堵体系[2-4]。地下化学交联体系存在地下成胶不易控制及污染环境的不足[5]，而体膨体颗粒存在颗粒尺寸与地层孔隙尺寸配伍性的问题[6]。

新型物理交联调驱体系 TDJ 是一种含有特殊 R 基团(含有 18 个碳的季铵盐单体)的功能聚合物，它利用超分子化学和结构流体的理论，通过疏水性功能基团 R 在水溶液中相互聚集产生的氢键和范德华力，使聚丙烯酰胺长分子链之间在水溶液中相互作用形成非化学交联的超分子网络聚集体。本研究对这种调驱体系的分子结构特点、微观聚集形态和性能进行表征和评价，并和常规的化学交联凝胶体系进行比较，同时对体系的调驱能力进行评价和分析，希望能够对含高渗带的非均质性较强的老油田改善水驱开发效果，进一步提高采收率提供借鉴。

1 实验部分

1.1 物理交联体系结构设计与合成

通过在聚丙烯酰胺侧链上引入具有较强疏水性的 R 功能基团,单体之间相互聚集,使聚丙烯酰胺大分子链之间产生物理交联,形成物理交联体系。通过调整 R 单体加量,可合成具有不同交联时间的凝胶体系。式(1)和式(2)分别为物理交联体系的结构式和聚合反应方程式。

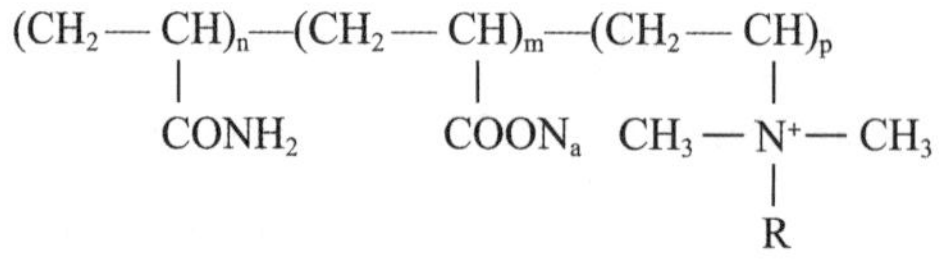

$$\underset{\underset{CONH_2}{|}}{(CH_2-CH)_n}-\underset{\underset{COON_a}{|}}{(CH_2-CH)_m}-\underset{\underset{CH_3-\underset{\underset{R}{|}}{N^+}-CH_3}{|}}{(CH_2-CH)_p} \tag{1}$$

$$(n+m)\underset{\underset{CONH_2}{|}}{CH_2{=}CH}+P\,\underset{\underset{CH_3-\underset{\underset{R}{|}}{N^+}-CH_3}{|}}{CH_2{=}CH}+mNaOH\longrightarrow$$

$$\underset{\underset{CONH_2}{|}}{(CH_2-CH)_n}-\underset{\underset{COONa}{|}}{(CH_2-CH)_m}-\underset{\underset{CH_3-\underset{\underset{R}{|}}{N^+}-CH_3}{|}}{(CH_2-CH)_p}+mNH_3 \tag{2}$$

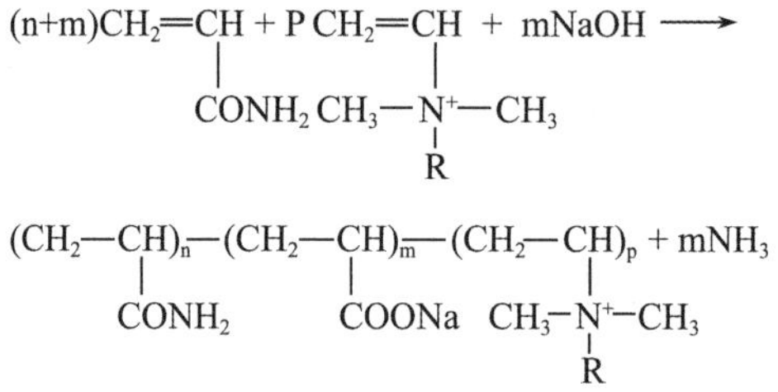

具体合成步骤为:

(1) 在 1 000 mL 烧杯中加入去离子水 717.39 g、在搅拌状态下依次加入乙二胺四乙酸二钠 0.1 g、甲酸钠 0.01 g、甲基丙烯酸二甲基氨基乙酯 0.05 g,丙烯酰胺 250 g、含有 R 功能单体的季铵盐单体(自制,质量分数分别为 1.0%、1.5%和 2.0%)12.5 g,充分搅匀。

(2) 向步骤(1)的溶液中通入高纯 N_2 60 min,加入质量分数为 0.05%的引发剂过硫酸钾—亚硫酸氢钠和偶氮二异丁脒盐酸盐(AIBA)进行聚合,在 0～15℃条件下绝热聚合 6 h,得到 TDJ 胶体。

(3) 将步骤(2)制备的 TDJ 胶体剪成粒径约 1～2 mm 的颗粒,然后加入胶体质量 3%的 NaOH 水溶液于 100℃水解 3 h,然后在 100℃的鼓风干燥箱中进行干燥。

(4) 经粉碎得到白色颗粒,颗粒的大小为亚微米级,功能单体质量分数分别约为 1.0%、1.5%和 2.0%的 TDJ 凝胶颗粒。

1.2 物理交联凝胶体系性能评价

1.2.1 实验条件

实验用水:总矿化度为 9 675 mg/L,$\rho(Ca^{2+}+Mg^{2+})$为 311 mg/L 的胜利油田孤东采油厂污水;实验温度:75℃。

实验仪器:Antonpar MCR301 流变仪,电子天平(±0.01 g 和±0.000 1 g),数显搅拌器,磁力搅拌器,磁力转子,500 mL 烧杯,玻璃棒,烘箱,物理模拟调驱模型。

物理交联凝胶配制:将合成的 TDJ 凝胶干粉用油田污水配制成一定浓度的溶液,静置在 75℃烘箱中一定时间，体系分子间自发形成物理交联凝胶。

化学交联凝胶配制：将常规驱油聚合物溶于水，配制成一定浓度的溶液，加入酚醛树脂交联剂，放入 75℃烘箱中，发生化学交联，形成化学交联凝胶。

1.2.2 物理交联体系性能特点评价

(1) 成胶时间研究：取含不同功能单体含量的物理交联体系，在油藏温度下，溶于油田污水中，考察体系的成胶时间。

(2) 增黏性、黏弹性和抗剪切性评价：固定功能单体质量分数为 1.0%的 TDJ 体系，通过流变仪测试体系的增黏性、黏弹性；通过剪切流变曲线，考察体系的抗剪切性能；最后通过原子力显微镜，观测体系通过岩心剪切前后的微观聚集状态。

(3) 注入方式对调驱性能影响评价：利用双管模型（长 30 cm、内径 2.5 cm、渗透率级差 7.5 μm^2 ∶ 1.5 μm^2），先水驱，接着以不同方式注入质量分数为 0.3%、功能单体质量分数为 1.0%的物理交联体系 TDJ；最后进行后续水驱，考察注入方式对体系调驱性能的影响。

2 结果与讨论

2.1 物理交联体系成胶时间研究

分别取功能单体 R 质量分数为 1.0%、1.5%和 2.0%的 TDJ 颗粒，加入水中搅拌使其分散，配成质量分数为 0.5%的物理凝胶体系，然后放入 75℃烘箱中，每隔一定时间，测试体系的黏度，结果如图 1 所示。

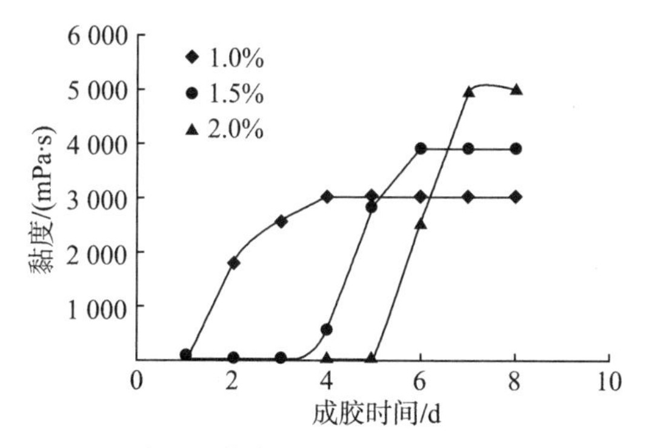

图 1 不同功能单体含量下 TDJ 黏度随时间变化曲线

Figure 1 Curves of viscosity of TDJ with different functional monomer content versus time

由图 1 可知，随着功能单体含量的增加，体系成胶的时间越来越长，且成胶后的黏度越来越高。这主要是因为功能单体为疏水型单体，含量的增加导致体系的疏水性增强，完全溶于水的时间进一步增加，由于分子链之间的物理交联作用在体系完全溶于水之后才能形成，因此，成胶时间随单体含量的增加逐渐变长；同时，由于功能单体含量的增加导致体系中疏水基团浓度增加，高浓度的疏水基团在盐溶液中更倾向于聚集在一起，从而导致聚丙烯酰胺长分子链在水溶液中聚集程度增大，因此成胶后的黏度进一步升高。

2.2 物理和化学交联调驱体系增黏性和黏弹性对比

高黏度是调驱体系具有良好的调驱性能的前提,因此,对调驱体系的增黏性和黏弹性进行评价,在质量分数为0.2%~0.8%的条件下,测试并对比物理交联体系TDJ和化学交联凝胶体系的黏度(化学交联体系聚合物和交联剂总浓度和物理交联体系相同,聚合物和交联剂质量浓度比为2∶1)。在固定扫描频率为1 Hz,应变为20%的条件下,测试不同质量浓度的物理交联体系的复合黏弹模量,结果如图2和图3所示。

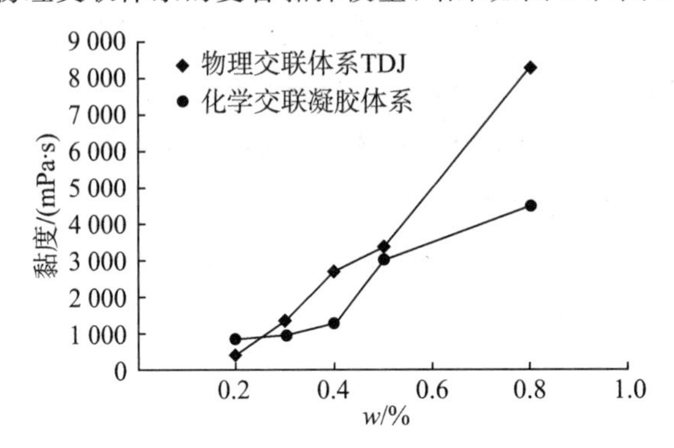

图2 物理交联体系TDJ和化学交联凝胶体系增黏性对比

Figure 2 increasing viscosity contrast of TDJ and chemical crosslinking system

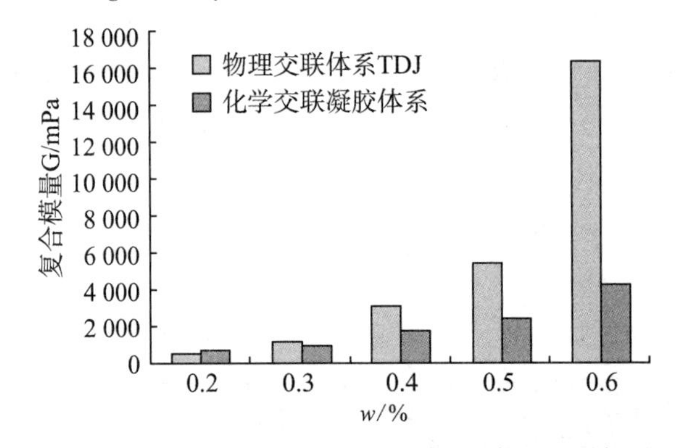

图3 物理交联体系TDJ和化学交联凝胶体系黏弹性对比

Figure 3 Vscoelastic contrast of physical crosslinking system TDJ and chemical crosslinking system

由测试结果可知,两种凝胶体系的黏度和黏弹性均随着浓度的增加而增加,在质量分数低于0.3%时,体系的功能基团较少,分子间物理交联作用较弱,体系黏度和复合模量低于化学交联凝胶体系,随着浓度的升高,物理交联体系交联强度不断增加,黏度和复合模量相对于化学交联凝胶体系具有一定优势,且浓度越高,优势越明显。而高浓度条件下TDJ体系所具有的高黏弹性能充分排驱地层流体,有效占据地下高渗层孔隙,在地层中形成“段塞”,产生较强的封堵。

2.3 物理交联调驱体系抗剪切性评价

调驱体系具备较好的抗剪切性能是体系能够产生长期有效封堵的必要条件,因此,需对体系的抗剪切性进行评价。选择质量分数为0.8%的物理交联体系TDJ和化学交联凝胶体系，在 0.001～800 s^{-1} 的剪切速率条件下，考察体系的剪切流变曲线，对比在不同剪切速率下两种体系的黏度，考察两者的抗剪切性，结果如图 4 所示。

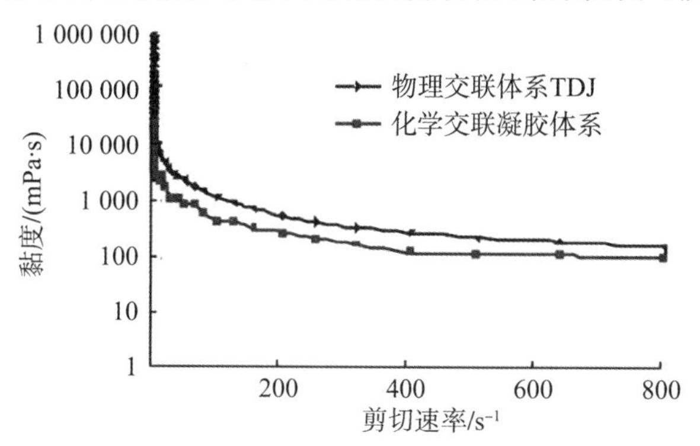

图 4 物理交联体系和化学交联凝胶体系剪切流变曲线对比

Figure 4 Shear rheological curve contrast of physical crosslinking system and chemical crosslinking system

由图 4 可知，在剪切速率较低的条件下，物理交联体系的黏度高于化学交联体系，随着剪切速率的增加，物理交联体系 TDJ 和化学交联凝胶体系黏度都有一定幅度的下降，但即使在剪切速率达到 800 s^{-1} 时，物理交联体系的黏度仍然高于化学交联体系。实验结果表明，无论是低剪切速率条件还是高剪切速率条件，物理交联体系都具有良好的抗剪切性能，这也进一步证明物理交联后的整体强度不弱于化学交联体系。

2.4 物理和化学交联调驱体系微观聚集形态对比

用原子力显微镜法[7-8]，对物理和化学交联体系通过渗透率为 3 μm^2 岩心前后的微观聚集形态进行研究，考察通过岩心剪切前后两者微观聚集形态的对比。所测的体系质量分数为 0.5%，功能单体质量分数为 1.0%，结果如图 5 和图 6 所示。

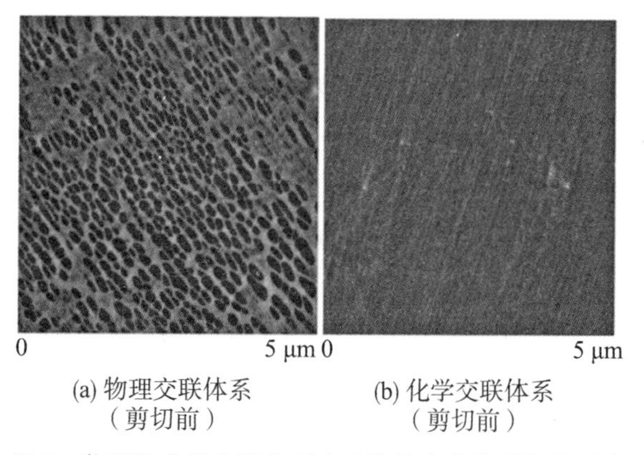

(a) 物理交联体系（剪切前） (b) 化学交联体系（剪切前）

图 5 物理和化学交联体系在水溶液中的微观聚集形态

Figure 5 Micro-aggregation morphology of physical and chemical crosslinking system in aqueous solution

由图 5 可知，在原始状态下，能够清楚地看到物理交联体系在水溶液中形成非常致密的网络聚集体，可以测量出聚集体网络的边长在 200～400 nm 之间。化学交联体系是分子之间通过共价键产生的化学交联，在水溶液中呈现纤维状结构，结构同样比较致密。

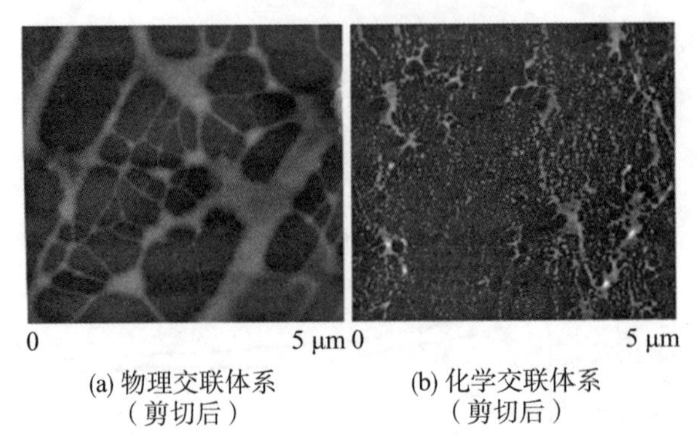

图 6 物理和化学交联体系岩心剪切后在水溶液中的微观聚集形态

Figure 6 Micro-aggregation morphology of physical and chemical crosslinking system in aqueous solution after core shearing

由图 6 可知,经过岩心剪切后,物理交联体系仍能够形成交联的网状结构,但空间中的网络结构与剪切前的相比变得较为稀疏,表明通过岩心剪切后,黏度出现了一定程度的降低,但整体结构没有被破坏。化学交联体系经过剪切后,体系由剪切前的呈整体结构的纤维状分散成一块一块的簇状结构,表明化学交联体系经过岩心剪切后,体系黏度不仅大幅度降低,同时结构整体也出现较严重的破坏。

2.5 物理交联体系注入方式和调驱能力研究

由于物理交联体系在水溶液中交联后黏度大幅度升高,如果体系在物理交联之后再注入,注入过程中会对低渗透带产生较大程度的污染,因此需优化体系的注入方式。

注入方式 1:首先把亚微米级的 TDJ 颗粒在水溶液中充分溶解,使体系在水溶液中先产生物理交联,然后注入 0.3 PV 的 TDJ 体系,TDJ 注完直接进行后续水驱,考察后续水驱之后高渗和低渗管液量的变化。

注入方式 2:首先把亚微米级的 TDJ 颗粒分散在水中,加入一定浓度的聚合物使其悬浮,在未产生物理交联之前,立即注入 0.3 PV 至双管岩心中,然后注入一定量的水推动体系往前运移一定距离,等待一段时间,使体系在岩心中产生物理交联,再进行后续水驱。

由图 7 和图 8 可知,采用先进行物理交联再注入岩心的注入方式。由于交联后黏度较大,体系对低渗管产生了较大的污染,在注入过程中,低渗管进入了较多的交联体系,再进行后续水驱时,高渗管突破压力更小,高渗管首先突破。因此,采用注入方式 1,体系没有对高渗管产生较好的封堵。采用先注入未交联的 TDJ 颗粒,使 TDJ 颗粒在岩心中产生物理交联,再进行后续水驱时,由于注入之前体系黏度较低,体系更容易进入高渗管,在高渗管内物理成胶之后,再进行后续水驱,高低管发生液流转向。同时,由于物理交联后分子间缔合产生的氢键和范德华力的强度低于化学交联的共价键,后续水驱后,体系能够继续向前运移。因此,后续水驱高渗管并没有完全堵死,注入物理交联体系能够更有效地同时提高高渗管和低渗管的采收率。由于体系的交联时间和交联强度可通过调节功能单体的含量进行控制,因此现场可以根据实际情况选择不同成胶时间和成胶强度的物理交联体系进行注入。

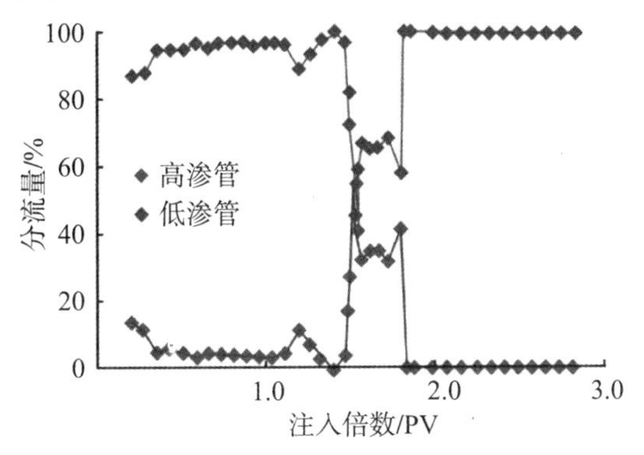

图7　注入方式1后续水驱分流变化

Figure 7　Follow-up water flow variation with injection mode1

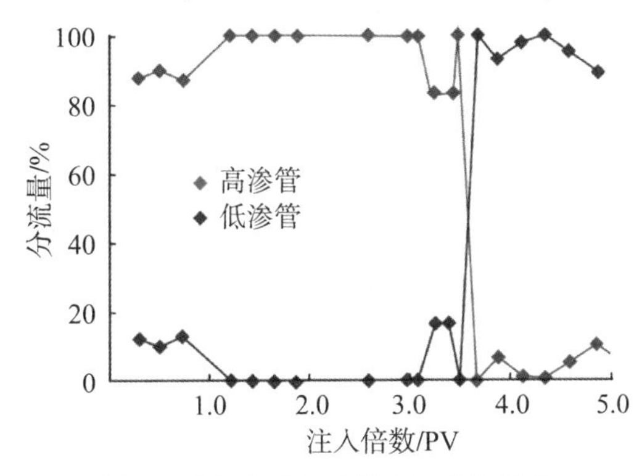

图8　注入方式2后续水驱分流变化

Figure 8　Follow-up water flow variation with injection mode2

3　结论

(1) 利用超分子化学和结构流体的理论,以丙烯酰胺为主体,加入含有18个碳的季铵盐单体,通过过硫酸钾—亚硫酸氢钠和偶氮二异丁脒盐酸盐(AIBA)引发剂进行自由基聚合,合成了以分子间物理交联为主的凝胶体系,并对体系的性能特点进行了评价。结果表明,该体系能够在水溶液中形成非常致密的网络聚集体,具有高黏度、高黏弹性和抗剪切性能。

(2) 物理交联体系相对于化学交联凝胶体系有两个最大的优势:一是可以在水溶液中自发形成凝胶,同时可以通过调整疏水性R单体的加量调节成胶时间和成胶强度;二是由于物理交联体系是通过功能单体相互聚集产生的氢键和范德华力进行增黏,即使经过岩心剪切之后,体系仍可重新聚集,使长分子链重新产生聚集,具有良好的结构恢复能力。

(3) 物理交联体系能够产生有效封堵关键在于选择正确的注入方式。采用先注入未交联的TDJ颗粒,使TDJ颗粒在岩心中产生物理交联的注入方式,能够减少体系对低渗带的污染,有效封堵高渗带。同时,物理交联体系分子间的作用力并没有化学交联的共价键强,因此体系在地层中可以运移,调驱能力更均衡。

参考文献

[1] 何长,李平,汪正勇,等. 大孔道的表现特征及调剖对策[J]. 石油钻采工艺,2000,22(5):63-66.

[2] 王平美,韩明,白宝君. 用于油层深部处理的胶态分散凝胶[J]. 油气采收率技术,1997,4(3):77-80.

[3] 胡书勇,张烈辉,余华洁,等. 油层大孔道调堵技术的发展及其展望[J]. 钻采工艺,2006,29(6) :117-120.

[4] 付美龙,罗跃,何建华,等. 聚丙烯酰胺凝胶在裂缝孔隙双重介质中的封堵性能[J]. 油气地质与采收率,2008,15(3):70-72.

[5] 戴彩丽,张贵才,赵福麟. 影响醛冻胶成冻因素的研究[J]. 油田化学,2001,18(1):24-26.

[6] 李东旭,侯吉瑞,赵凤兰,等. 预交联颗粒调堵性能及粒径与孔隙匹配研究[J]. 石油化工高等学校学报,2010,23(2) :25-28.

[7] 张瑞,秦妮,彭林,等. 注入速度对疏水缔合聚合物剪切后恢复性能的影响[J]. 石油学报,2013,34(1):122-127.

[8] 姚同玉,刘庆纲,刘卫东,等. 部分水解聚丙烯酰胺的结构形貌研究[J]. 石油学报,2005,26(5) :81-84.

编辑 冯学军

Interaction between polymer and anionic/nonionic surfactants and its mechanism of enhanced oil recovery

Wang Yefei, School of Petroleum Engineering, China University of Petroleum (East China), Qingdao, China

Hou Baofeng, School of Petroleum Engineering, Yangtze Un iversity, Wuhan, China

Cao Xulong, Geoscience Research Institute of Shengli Oilfield Company, SINOPEC, Dongying, China

Zhang Jun, School of Science, China University of Petroleum (East China), Qingdao, China

Song Xinwang, Geoscience Research Institute of Shengli Oilfield Company, SINOPEC, Dongying, China

Ding Mingchen, School of Petroleum Engineering, China University of Petroleum (East China), Qingdao, China

Chen Wuhua, School of Petroleum Engineering, China Universityof Petroleum (East China), Qingdao, Chinaof Petroleum Engineering, Yangtze University, No. 111, University Road, Caidian 430100, Wuhan, P. R. China. E-mail: hbf370283@163.com

Abstract: Various experimental methods were used to investigate interaction between polymer and anionic/nonionic surfactants and mechanisms of enhanced oil recovery by anionic/nonionic surfactants in the present paper. The complex surfactant molecules are adsorbed in the mixed micelles or aggregates formed by the hydrophobic association of hydrophobic groups of polymers, making the surfactant molecules at oil-water interface reduce and the value of interfacial tension between oil and water increase. A dense spatial network structure is formed by the interaction between the mixed aggregates and hydrophobic groups of the polymer molecular chains, making the hydrodynamic volume of the aggregates and the viscosity of the polymer solution increase. Due to the formation of the mixed adsorption layer at oil and water interface by synergistic effect, ultra-low interfacial tension (~2.0×10^{-3} mN/m) can be achieved between the novel surfactant system and the oil samples in this paper. Due to hydrophobic interaction, wettability alteration of oil-wet surface was induced by the adsorption of the surfactant system on the solid surface. Moreover, the studied surfactant system had a certain degree of spontaneous emulsification ability (D50= 25.04 μm) and was well emulsified with crude oil after the mechanical oscillation (D50= 4.27 μm).

GRAPHICAL ABSTRACT

Interaction between surfactants and polymers (a) Hydrophobic interaction (b) Hydrogen bond

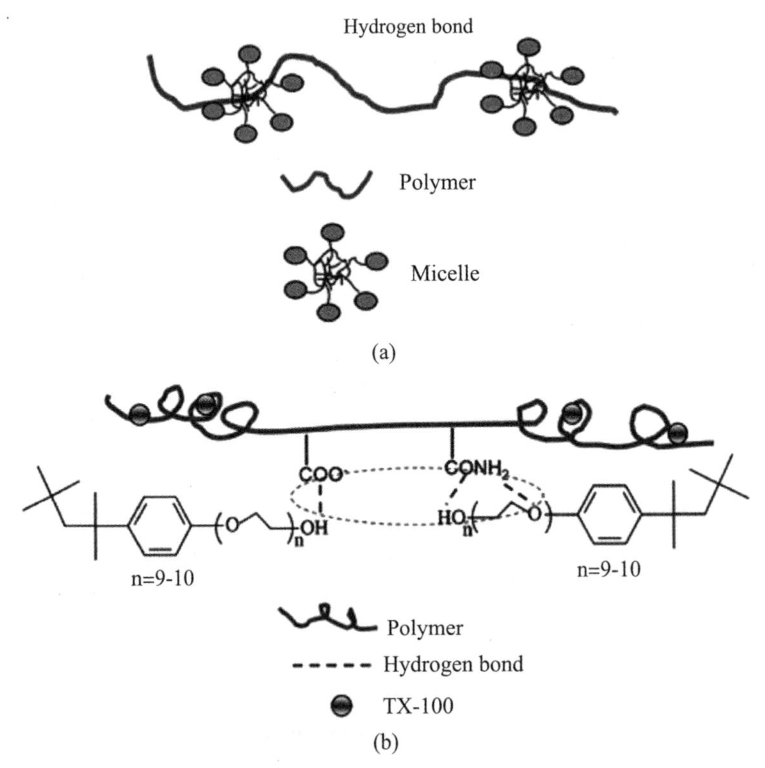

In this study, there were some major interactions existing between surfactants and polymers, such as hydrophobic interaction and hydrogen bond. Due to hydrophobic interaction, surfactant micelles would associate with polymer chains, which can be described as "string of beads model" (Fig. (a)). In addition, hydrogen bond exists between polymers and nonionic surfactants TX-100(Fig. (b)). The complex surfactant molecules are adsorbed in the mixed micelles or aggregates formed by the hydrophobic association of hydrophobic groups of polymers when the surfactant is added into water, making the surfactant molecules at oil-water interface reduce. A dense spatial network structure is formed by the interaction between the mixed micelles or aggregates and hydrophobic groups of the polymer molecular chains, making the hydrodynamic volume of the aggregates and the viseosity of the polymer solution increase.

Keywords: anionic/nonionic surfactants, emulsification, interaction, mechanism of enhanced oil recovery, wettability alteration

1 Introduction

Oil recoveries of oilfield around the world are very low. Many effective methods have been used to enhance oil recoveries of oilfields, including chemical flooding, microbial flooding, thermal recovery and so on.[1-4] Among these methods, chemical flooding plays a significant role during oil exploitation. Alkali/Surfactant/Polymer (ASP) flooding is a very important method of chemical flooding, for which there are some problems during the large-scale industrial application.[5-7] Alkali has been extensively used in ASP flooding, producing multivalent ions precipitate and mineral dissolution.[8] The viscosity of Surfactant/Polymer (SP) system is significantly higher than that of the ASP system under the same conditions. Ultra-low interfacial tension and higher oil displacement efficiency can also be achieved for SP flooding. In recent years, more attention has been paid to the study of the surfactants used in SP flooding.[9,10]

There are many interactions between surfactants and polymers, including electrostatic attraction, hydrophobic interactions, hydrogen bonding and so on.[11-21] Due to the presence of multiple interactions, surfactants have a certain impact on the viscosity of the polymer solution. Jime'nez-Regalado[22] studied the effect of surfactants on the viscoelastic behavior of the polymer solution using steady-flow and oscillatory rheological experiments. They found that the high concentrations of surfactants led to the decrease of the viscosity of the polymer solution and the surfactants made hydrophobically modified polyacrylamides lose their associative properties. In addition, polymer has a certain influence on the interfacial activity of surfactants.

Oil recovery is closely related to the following three factors: interface activity, wettability alteration and emulsification. Interfacial activity should be good for one excellent performance of surfactant system, i. e. the surfactant system has a strong ability to reduce oil-water interfacial tension. Adhesion work for oil droplets on the solid surface reduces and oil recovery increases with the decrease of oil-water interfacial tension. Surfactant-induced wettability alteration is one of mechanisms for enhanced oil recovery by surfactants. Wettability of oil-wet surface can be altered by various surfactants and the corresponding oil recovery will be changed. Hou et al.[23] investigated the relationship between surfactant-induced wettability alteration and oil recovery by spontaneous imbibition. They found that the imbibition recoveries vary with the wettability changes of the oil-wet surface caused by various surfactants.[24,25] During the surfactant flooding, emulsifying of crude oil has a great impact on oil recovery.[26] Oil recovery can be greatly enhanced by the carrying effect and Jamin effect induced by emulsions.[26,27]

The purpose of this paper is to investigate interaction between polymer and anionic/nonionic surfactants and to analyze mechanisms of enhanced oil recovery by the anionic/nonionic surfactants using various experimental methods.

2 Experimental

2.1 *Materials*

2.1.1 *Aqueous phase*

For the reagents, petroleum ether, n-heptane, NaCl, $CaCl_2$ and $MgCl_2 \cdot 6H_2O$ are all analytical-grade, which were purchased from Tianjin Bodi Chemical Co., Ltd., China. The ionic composition of the aqueous phase is shown in Table 1.

2.1.2 *Oil phase*

Basic properties of oil samples used in this paper are shown in Table 2 and the oil sample E was employed in the other parts of this work except Sec. 3.2.1.

2.1.3 *Surfactants and polymers*

For the surfactants, petroleum sulfonate ($RAr\text{-}SO_3Na$, R: alkane chains, Ar: aryl group) was produced from Shengli Oilfield and Triton X-100 (p-Octyl polyethylene glycol phenyl ether) was purchased from Sinopharm Chemical Reagent Co., Ltd. In this paper, the abbreviations of petroleum sulfonate and Triton X-100 are "SLPS" and "TX-100", respectively. In the mixed system, the mass ratio of the anionic surfactant SLPS and the nonionic surfactant TX-100 was 1 : 1. In addition, the total concentration of SLPS plus TX-100 was 0.30 wt% in the present study.

For the polymers, the solid content and the degree of hydrolysis were 91.37% and 32.48%, respectively. Furthermore, the average relative molecular mass of the employed polymer was 1.462×10^6. The polymers used in the present study were industrial-grade and were obtained from Shandong Bao Mo Biological Chemical Co., Ltd., China. The chemical composition of the used polymer in the present study is polyacrylamide. The product number of the used polymer is CA-75, which is an amphipathic copolymer.

2.1.4 *Solid surface*

In this study, quartz plates were used for the contact angle measurement. The smooth quartz plates used in the present study have dimensions of 20×20×2 mm, which are naturally occurring silicate minerals composed of silicon dioxide (SiO_2).

2.2 *Methods*

2.2.1 *Preparation of oil-wet solid surface*

The clean quartz plate was first immersed in the crude oil and then the system was placed in an oven at 70℃ for 15 days. After that, the oil-soaked quartz plate was washed with n-heptane until the washing liquid was colorless. Then the quartz plate was dried for the subsequent experiments.[28]

2.2.2 *Contact angle measurement*

Sessile drop method was used to measure the contact angles between the oil-water interface and the quartz surface at 70℃ in the present paper. Before the contact angle measurement, the surfactant solution was first poured into a glass container and then an oil-wet quartz plate was placed in the glass container (Figure 1). After that, four oil droplets were injected out onto the lower surface of the quartz plate to get the average value of the contact angles. The video camera was used to obtain the shape of the oil droplets throughout the experiment and the experiment duration was 30 minutes. According to the shape of the oil droplets, the values of the contact angles were calculated using the relevant software.[29]

2.2.3 *Determination of emulsifying ability*

Surfactant solutions (20 mL) and crude oil (20 mL) were mixed and were added into a test tube. The test tube was then placed in a water bath at 70℃ for 10 minutes and was shocked every 2 minutes. After that, the test tube was allowed to stand for several hours. The median particle size of the emulsion droplets was determined by the laser particle size analyzer (mastersizer 3000, Xingtai, China).

3 Results and Discussion

3.1 *Interaction between anionic/nonionic surfactants and polymer*

Due to interaction between surfactants and polymers, the viscosity of polymer solution and interfacial tension between oil and water can be changed. Interaction between the employed anionic/nonionic surfactants and polymer was analyzed and investigated in this study.

3.1.1 *Effect of interaction on the viscosity of polymer solution*

In this study, there were some major interactions existing between the studied surfactants and polymers, such as hydrophobic interaction and hydrogen bond (Figure 2). Due to hydrophobic interaction, surfactant micelles associate with polymer chains, which can be described as "string of beads model".[30,31] In addition, hydrogen bond exists between polymers and nonionic surfactants (TX-100) (Figure 2 (b)).

The Brookfield DV-Ⅱ viscometer was used to determine the viscosity of the solution at 70℃ in this paper. Figure 3 shows that the viscosity of polymer solutions increased after the mixed surfactants were added into the solutions. Due to hydrophobic interaction, polymers associate with surfactant micelles and the number of charges for the polymer chain increases when adding surfactants into the polymer solution (Figure 3). Due to the enhanced electrostatic repulsion, the polymer chain is further extended, making the structure viscosity of the polymer increase. In addition, The polymer used in this study is an amphiphilic polymer. Due to the intramolecular and intermolecular association of the polymer, many hydrophobic microdomains appear in the solution when the surfactant is not added into water. The complex surfactant molecules are adsorbed in the mixed micelles or aggregates formed by the hydrophobic association of hydrophobic groups of polymers when the surfactant is added into water. A dense spatial network structure is formed by the interaction between the mixed micelles or aggregates and hydrophobic groups of the polymer molecular chains, making the hydrodynamic volume of the aggregates and the viscosity of the polymer solution increase.[32]

3.1.2 *Effect of polymer on interfacial tension between oil and surfactant system*

Figure 4 shows the effect of polymer on the interfacial tension between oil and surfactant system. From Figure 4 one can see that the values of interfacial tension increased after the polymers were added into the surfactant solution. When the polymer is added into the surfactant solution, part of surfactants originally adsorbed at oil-water interface associate with the polymer in the bulk phase due to hydrophobic interactions and hydrogen bonding (Figure 5), i.e. the complex surfactant molecules are adsorbed in the mixed micelles or aggregates formed by the hydrophobic association of hydrophobic groups of the amphiphilic polymers. Thus, the number of surfactant molecules at oil-water interface decreases, resulting in the increase of the value of interfacial tension.

3.2 *Mechanisms of enhanced oil recovery by anionic/nonionic surfactants*

Compared with single surfactants, anionic/nonionic surfactants have a stronger ability to enhance oil recovery. In general, there are three main mechanisms for enhanced oil recovery by surfactants, including: reduction of interfacial tension between oil and water, wettability alteration and emulsification. Mechanisms of enhanced oil recovery by the studied anionic/nonionic surfactants and the related properties were investigated in the present study.

3.2.1 *Interfacial activity of surfactant system*

Spinning drop method was used to determine the interfacial tension between crude oil and surfactant solutions at 70℃ in the present study, hereinafter the same. Figure 6 shows the values of the interfacial tension between the surfactant complex system (SLPS/TX-100) and various oil samples. The surfactant system had a high interfacial activity. From Figure 6 one can see that as the surfactant concentration increased, the values of the interfacial tension were getting lower and lower and finally reached equilibrium. Furthermore, the equilibrium values were maintained within the range of ultra-low interfacial tension. Synergistic effect exists in the two kinds of surfactants of the composite system. A mixed adsorption layer is formed at oil-water interface, making the value of interfacial tension between oil and water be lower. Thus, the oil recoveries are greatly enhanced.

3.2.2 *Ability to alter the wettability of rock surface*

Figure 7 shows the results of contact angle measurement. Figure 7 shows that the contact angles of quartz plates treated with the surfactant system were lower than that of quartz plates treated with the other two surfactants (SLPS or TX-100) at different concentrations. In the mixed system (SLPS/TX-100), a very good synergistic effect exists between the two surfactants SLPS and TX-100, making the performance of the surfactant system better than that of the single surfactant. Figure 8 shows the schematic diagram of mechanisms for wettability alteration of oil-wet sandstone surface caused by the surfactant system. The solid surface was initially oil-wet before the contact angle measurement in this study (Figure 8(a)). After the mixed surfactants were added into the solution, hydrophobic chains of the surfactants and the hydrocarbon chains of crude oil associated with each other due to hydrophobic interaction, making the solid surface water-wet.[28,33]

3.2.3 *Emulsifying properties of surfactant system*

Figure 9 shows the emulsification phenomenon of oil and water. From Figure 9 (b) we can see that there was a certain ability of spontaneous emulsification for the surfactant system and crude oil. After the mechanical oscillation, the complex system was well emulsified with crude oil (Figure 9 (c)).

Figs. 10 and 11 show the microscopic pictures of emulsions and the size distribution data of emulsion droplets formed by different methods, respectively. Figure 10 shows that the diameter of the emulsion droplets for the spontaneous emulsification was larger than that for the mechanical emulsification. The size of emulsion droplets prepared by mechanical emulsification was relatively concentrated. Moreover, the number of emulsion droplets for the mechanical emulsification was larger than that for the spontaneous emulsification (Figure 10). The D50 particle size of the emulsion droplets formed by spontaneous emulsification and mechanical emulsification are 25.04 and 4.27 μm, respectively (Figure 11).

4 Conclusions

Interaction between polymer and anionic/nonionic surfactants and mechanisms of enhanced oil recovery by the anionic/nonionic surfactants were investigated using a variety of experimental methods in this paper. The following conclusions can be drawn:

The complex surfactant molecules are adsorbed in the mixed micelles or aggregates formed by the hydrophobic association of hydrophobic groups of polymers when the surfactant is added into water, making the surfactant molecules at oil-water interface reduce and the value of interfacial tension between oil and water increase.

A dense spatial network structure is formed by the interaction between the mixed micelles or aggregates and hydrophobic groups of the polymer molecular chains, making the hydrodynamic volume of the aggregates and the viscosity of the polymer solution increase.

Due to the formation of the mixed adsorption layer at oil and water interface by synergistic effect, ultra-low interfacial tension ($\sim 2.0 \times 10^{-3}$ mN/m) can be achieved between the novel surfactant system and the oil samples in this paper.

The ability to change the wettability of oil-wet surface for the studied complex surfactants was stronger than that for the single surfactants in the present study. Due to hydrophobic forces, the oil-wet sandstone surface was changed towards water-wet surface caused by the adsorption of the surfactant system on the solid surface.

The studied surfactant system had a certain degree of spontaneous emulsification ability (D50 = 25.04 μm) and could be well emulsified with crude oil after the mechanical oscillation (D50=4.27 μm).

Nomenclature

k=core permeability (10^{-3} μm^2)

PV=pore volume (mL)

S_{oi}=initial oil saturation (fraction, %)

$wt\%$=weight%

IFT=interfacial tension (mN/m)

Acknowledgments

Financial support by the Program for Changjiang Scholars and Innovative Research Team in University (IRT1294), the Fundamental Research Funds for the Central Universities (13CX05019A &15CX02006A), National Natural Science Foundation of China (Youth Fund)(51704036) are all gratefully acknowledged.

References

[1] Terry, R. E. Enhanced Oil Recovery. In *Encyclopedia of Physical Science and Technology*; Robert A. Meyers Ed.; Academic Press: New York, 2001; Vol. 18: 503-518,

[2] Sedaghat, M. H.; Ghazanfari, M. H.; Parvazdavani, M.; Morshedi, S. *ASME J. Energy Resour. Technol.* 2013, 135(3): 32901.

[3] Kamath, K.; Yan, S. *ASME J. Energy Resour. Technol.* 1981, 103(4): 285-290.

[4] Yassin, M. R.; Ayatollahi, S.; Rostami, B.; Hassani, K.; Taghikhani, V. *ASME J. Energy Resour. Technol.* 2015, 137(1): 12905.

[5] Shen, P.; Wang, J.; Yuan, S.; Zhong, T.; Jia, X. *SPE J.* 2009, 14(2): 237-244.

[6] Carrero, E.; Queipo, N. V.; Pintos, S.; Zerpa, L. E. *J. Pet. Sci. Eng.* 2007, 58(1): 30-42.

[7] Hou, J.; Liu, Z.; Zhang, S.; Yang, J. J. *Pet. Sci. Eng.* 2005, 47(3): 219-235.

[8] Bortolotti, V.; Gottardi, G.; Macini, P.; Srisuriyachai, F. EUROPEC/EAGE Conference and Exhibition, Society of Petroleum Engineers, No. SPE 121832, Amsterdam, The Netherlands, Jun. 8-11, 2009.

[9] Samanta, A.; Ojha, K.; Sarkar, A.; Mandal, A. Adv. Pet. Explor. Dev. 2011, 2(1): 13-18.

[10] Gogarty, W. B. *J. Pet. Technol*. 1978, 30(8): 1089-1101.

[11] Hongyan, W.; Xulong, C.; Jichao, Z.; Aimei, Z. *J. Pet. Sci. Eng*. 2009, 65(1): 45-50.

[12] Arai, H.; Murata, M.; Shinoda, K. *J. Colloid Interface Sci*. 1971, 37(1): 223-227.

[13] Wang, G.; Olofsson, G. *J. Phys. Chem. B* 1998, 102(46): 9276-9283.

[14] Carlsson, A.; Karlstroem, G.; Lindman, B. J. *Phys. Chem*. 1989, 93(9): 3673-3677.

[15] Thalberg, K.; Lindman, B. *J. Phys. Chem*. 1989, 93(4): 1478-1483.

[16] Hansson, P.; Lindman, B. *Curr. Opin. Colloid Interface Sci*. 1996, 1(5): 604-613.

[17] Goddard, E. *J. Colloid Interface Sci*. 2002, 256(1): 228-235.

[18] Anthony, O.; Zana, R. *Langmuir* 1996, 12(8): 1967-1975.

[19] Mészáros, R.; Thompson, L.; Bos, M.; Varga, I.; Gilányi, T. *Langmuir* 2003, 19(3): 609-615.

[20] Goddard, E. D. *Colloid Surf*. 1986, 19(2): 255-300.

[21] Patel, U.; Dharaiya, N.; Parikh, J.; Aswal, V. K.; Bahadur, P. *Colloid Surf. A-Physicochem. Eng. Asp*. 2015, 481: 100-107.

[22] Jiménez-Regalado, E.; Selb, J.; Candau, F. *Langmuir* 2000, 16(23): 8611-8621.

[23] Hou, B. F.; Wang, Y. F.; Huang, Y. *J. Dispersion Sci. Technol*. 2015, 36(9): 1264-1273.

[24] Wang, Y. F.; Xu, H. M.; Yu, W. Z.; Bai, B.; Song, X.; Zhang, J. *Pet. Sci*. 2011, 8(4): 463-476.

[25] Zhou, X. M.; Morrow, N. R.; Ma, S. X. *SPE J*. 2000, 5(2): 199-207.

[26] Liu, Q.; Dong, M.; Ma, S.; Tu, Y. *Colloid Surf. A-Physicochem. Eng. Asp*. 2007, 293(1): 63-71.

[27] Li, J.; Qu, Z.; Kong, L. *Pet. Explor. Dev*. 1999, 26: 93-94.

[28] Hou, B. F.; Wang, Y. F.; Huang, Y. *Appl. Surf. Sci*. 2015, 330: 56-64.

[29] Qi, Z.; Wang, Y.; He, H.; Li, D.; Xu, X. *Energy Fuels* 2013, 27(12): 7354-7359.

[30] Groot, R. D. *Langmuir* 2000, 16(19): 7493-7502.

[31] Goddard, E. *J. Am. Oil Chem. Soc*. 1994, 71(1): 1-16.

[32] Kang, W.; Xu, B.; Wang Y., Lia, Y.; Shana, X.; Anb, F.; Liu, J.

Colloid Surf. A-Physicochem. Eng. Asp. 2011, 384(1): 555-560.

[33] Salehi, M.; Johnson, S. J.; Liang, J.-T. *Langmuir* 2008, 24(24): 14099-14107.

Table 1. Composition of the simulated formation water.

component	concentration, mg/L
Na^{+}	6 907.5
Cl^{-}	12 521.0
Ca^{2+}	421.7
Mg^{2+}	223.6

Table 2. Basic properties of oil samples.

Oil sample number	production place of oil sample	viscosity, mPa・s	density, g/cm^3	acid number, mg of KOH/g
A	Shengli Oilfield	62.5	0.892	1.423
B	Shengli Oilfield	64.1	0.867	1.439
C	Shengli Oilfield	63.6	0.859	1.440
D	Shengli Oilfield	61.2	0.871	1.419
E	Shengli Oilfield	61.7	0.859	1.415

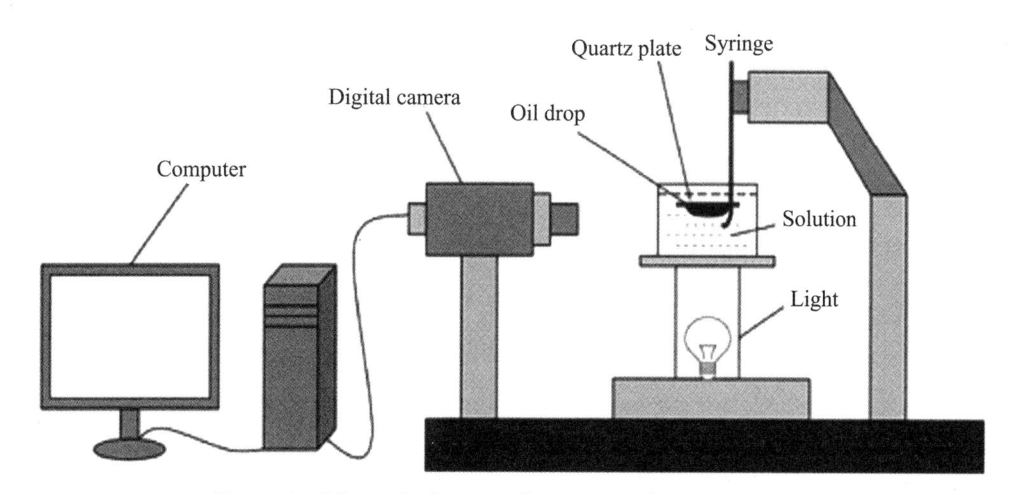

Figure 1. Schematic diagram of contact angle measurement.

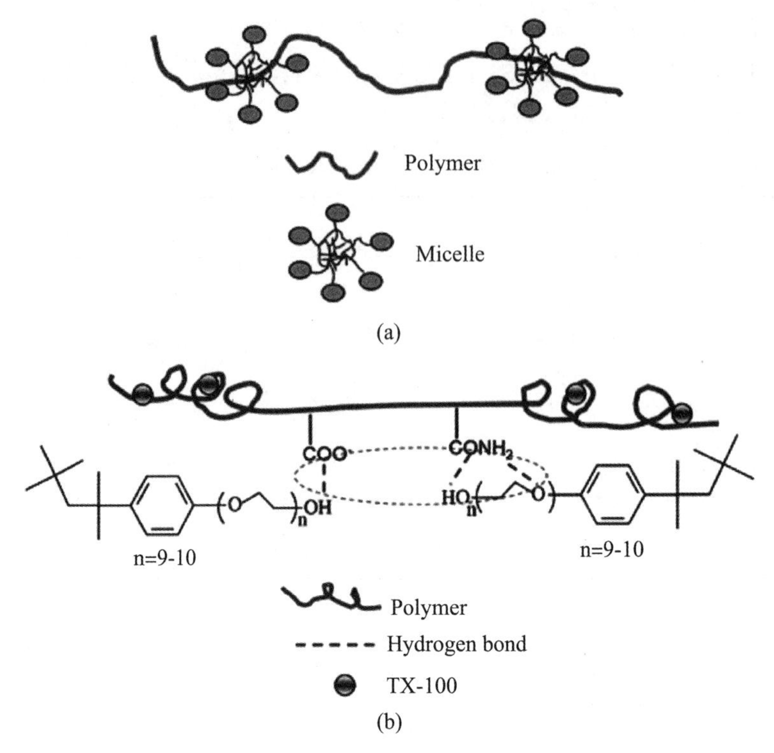

Figure 2. Interaction between surfactants and polymers (a) Hydrophobic interaction (b) Hydrogen bond.

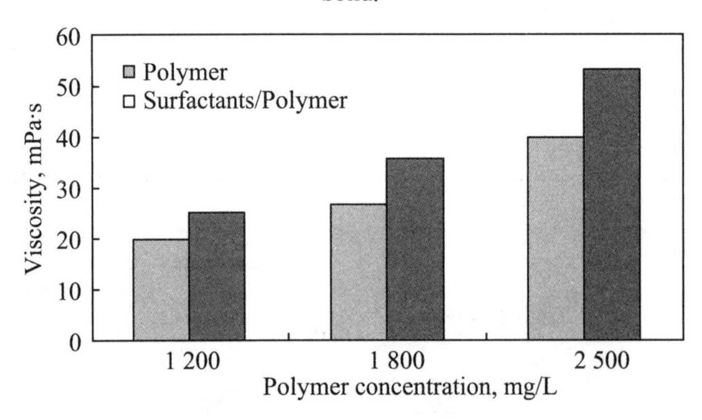

Figure 3. Viscosities of different systems.

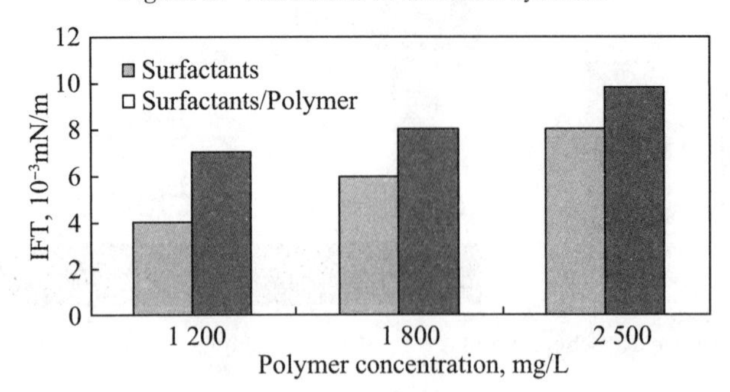

Figure 4. Interfacial tension between different systems and crude oil.

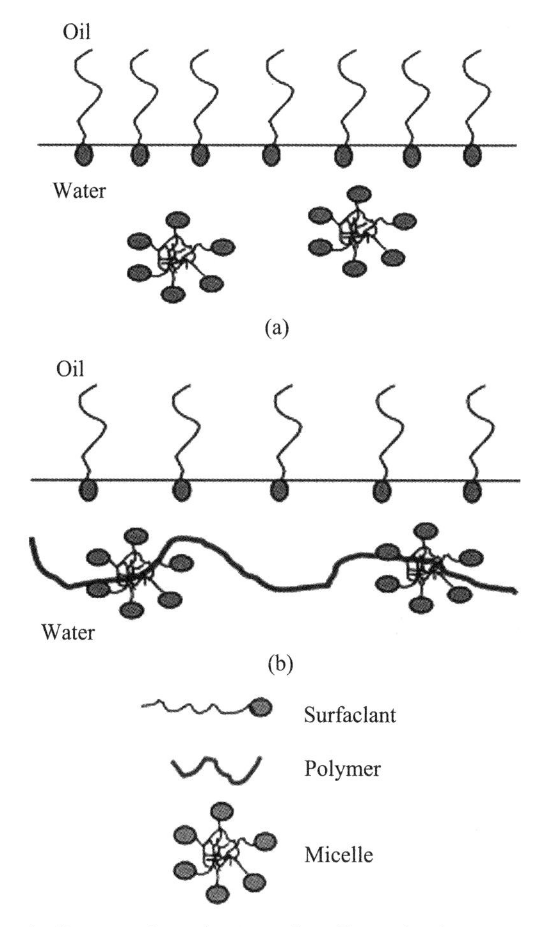

Figure 5. Schematic diagram of mechanisms for effect of polymer on interfacial tension

(a) No polymer (b) Adding polymer.

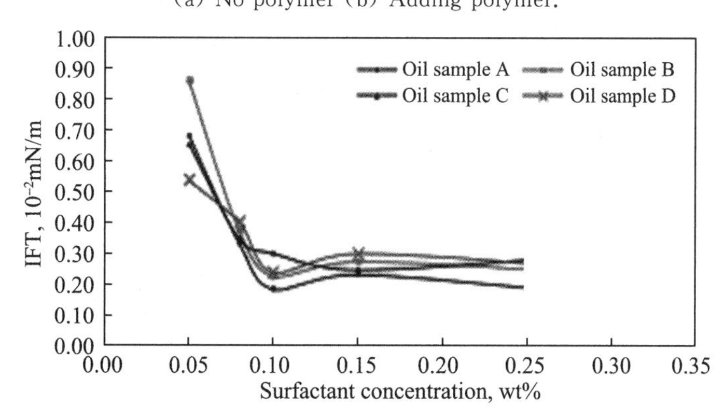

Figure 6. Interfacial tension between different concentrations of surfactant complex system and various oil samples.

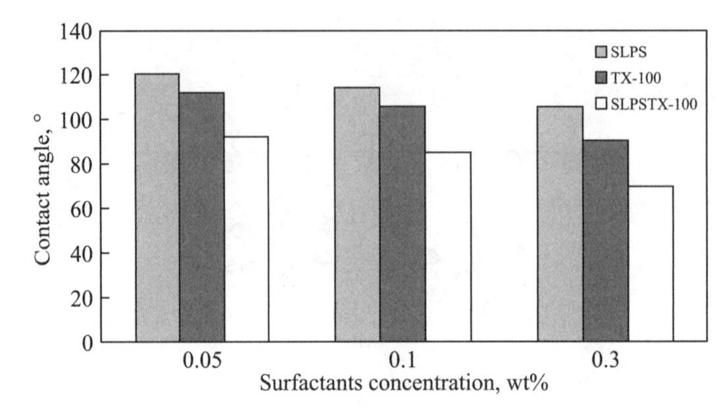

Figure 7. Contact angles of oil-wet quartz plates treated with various systems at different concentrations.

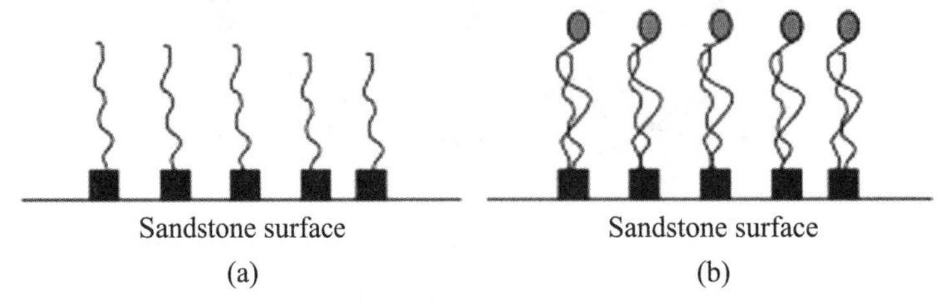

Figure 8. Mechanisms for wettability alteration of oil-wet sandstone surface caused by the surfactant complex system. Ellipses: hydrophilic groups of the surfactants; Squares: polar components of crude oil. (a) Adsorption of polar components on the sandstone surface (b) Formation of the bilayer structure.

Figure 9. Emulsification phenomenon (a) No surfactants (b) Adding surfactants and spontaneous emulsification (c) Adding surfactants and mechanical emulsification.

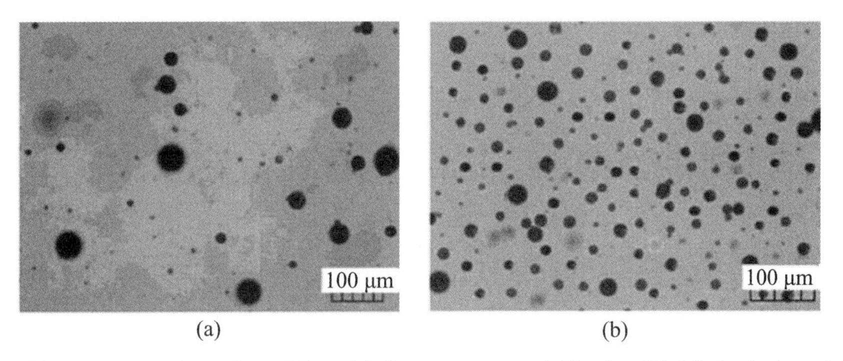

Figure 10. Microscopic pictures of emulsions (a) Spontaneous emulsification (b) Mechanical emulsification.

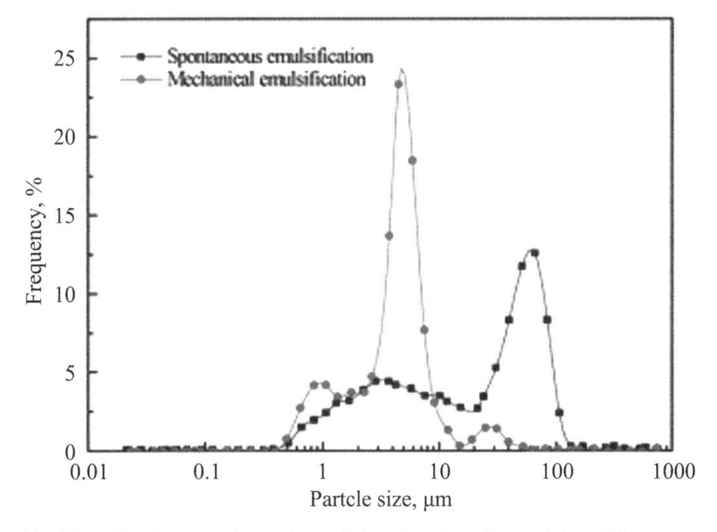

Figure 11. Size distribution data of emulsion droplets formed by different methods.

胜利油区海上油田二元复合驱油体系优选及参数设计

赵方剑[1,2] 曹绪龙[3] 祝仰文[2] 侯 健[1] 孙秀芝[2] 郭淑凤[2] 苏海波[2]

(1. 中国石油大学(华东)石油工程学院,山东 青岛 266580;

2. 中国石化胜利油田分公司 勘探开发研究院,山东 东营 257015;

3. 中国石化胜利油田分公司,山东 东营 257001)

摘要:胜利油区海上油田水驱开发产量呈现逐年下降的趋势,急需转换为二元复合驱方式进行提速增效开发。二元复合驱作为大幅度提高原油采收率和采油速度的有效方法,已经在陆上油田大规模推广应用。但是,由于油藏条件存在差异,海上油田无法照搬陆上油田方案。按照海上油田开展二元复合驱技术要求,通过对多个驱油剂产品的溶解性、热稳定性、抗盐性等关键指标进行优选,得到由聚合物C10与复配的高效活性剂组成的配伍性和稳定性良好的二元复合驱油体系。结合室内物理模拟实验和数值模拟研究,对二元复合驱油技术的注入黏度、注入速度以及段塞尺寸等注入参数进行了优化设计。研究结果表明,该二元复合驱油体系注入0.4 PV之后,采收率提高29.2%,相对于同段塞尺寸单一聚合物驱增加11.3%。数值模拟预测结果显示,在最佳黏度比为0.5、注入速度为0.07 PV/a、最佳注入段塞尺寸为0.42 PV的条件下,与水驱相比二元复合驱可提高采出程度11.6%,最大幅度提高海上油田采油速度和原油采收率。

关键词:二元复合驱;提高采收率;注入参数;采油速度;海上油田

中图分类号:TE357.46 文献标识码:A

Injection parameters optimization of binary combination flooding system in offshore oil field, Shengli oil province

Zhao Fangjian[1,2] Cao Xulong[3] Zhu Yangwen[2] Hou Jian[1] Sun Xiuzhi[2] Guo Shufeng[2] Su Haibo[2]

(1. School of Petroleum Engineering, China University of Petroleum (East China), Qingdao City, Shandong Province, 266580, China;

2. Exploration and Development Research Institute, Shengli Oilfield Company, SINOPEC, Dongying City, Shandong Province, 257015, China;

3. Shengli Oilfield Company, SINOPEC, Dongying City, Shandong Province, 257001, China)

Abstract: The production rate through water flooding in offshore oil field, Shengli oil province is decreasing year by year, it is urgent to convert water flooding to binary combination flooding to accelerate the development and improve producing efficiency. As an effective method to dramatically improve oil recovery and oil recovery rate, binary combination flooding technology has been widely used in onshore oil fields. However, due to the difference of reservoir conditions, offshore oil field cannot copy the reservoir development program from onshore oil fields. According to the technical requirements of binary combination flooding in offshore oil field, Shengli oil province, the key factors such as solubility, thermal stability and salt resistance of oil displacement agents are evaluated, and the selected polymer C10 and high-efficiency compound surfactant form the binary combination flooding system with excellent performance of compatibility and stability. The injection viscosi-ty, injection rate and plug size are optimized based on physical simulation and numerical simulation. The results show that after the injection volume of 0.4 PV, the EOR of binary compound flooding system reaches 29.2%, which is 11.3% higher than that of single polymer flooding system with the same plug size. Under the condition of the optimal viscosity ratio of 0.5, the injection rate of 0.07 PV/a and the optimal injection plug size of 0.42 PV, numerical simulation predicted that the EOR of binary compound flooding is 11.6% higher than that of water flooding, which can obtain the maximum increase of the oil recovery rate and oil recovery in offshore oil field.

Keywords: binary combination flooding; EOR; injection parameters; oil recovery rate; offshore oil field

胜利油区埕岛油田经过多年的开发，形成了中心平台加卫星平台的生产模式，随着单元综合调整工作的到位，新井逐年减少，水驱开发产量呈现逐年下降的趋势。埕岛油田自 2016 年综合含水率超过 80%后，液量快速增长，逐渐接近海工处理液量能力上限，导致油田开采成本快速上涨，效益变差。预计 2023 年后开采成本将高于 50 美元/bbl，中心平台设计寿命期内水驱采出程度不足 40%。因此，埕岛油田急需转换开发方式，进一步提高采收率和采油速度。

分析发现，二元复合驱可大幅度提高油藏采油速度和采收率，且已经在胜利油区陆上油田开展了大规模工业化推广应用，在实施过程中最高采油速度可达 2.9%，平均提高采收率可达 12%[1-6]，成为增油稳产的重要技术支撑，累积增油量超过 $3\ 000\times10^4$ t，获得了良好的经济效益和社会效益。将海上油田与陆上油田二元复合驱单元的油藏条件进行类比发现，二者的地下原油黏度、空气渗透率、地层温度以及地层水矿化度等指标均相当，且海上油藏的单井产液量、注入量、综合含水率及采出程度等动态条件要优于陆上已经实施二元复合驱的单元，具备实施该技术的有利条件[7-9]。但是，直接将陆上油田二元复合驱技术应用在海上油田存在 3 个方面的难题：① 海上平台空间狭小，要求二元复合驱油体系必须采用短流程的注入工艺。② 由于历史上曾经采用注入海水开发方式，储层残留的海水形成了高矿化度油藏环境，要求二元复合驱油体系必须具有良好的抗盐性能。③ 受海工工程和开发成本限制，海上油田通常采用稀井网布井方式，导致注入流体在地层中运移时间相对较长，因此，要求二元复合驱油体系具有良好的长效稳定性。针对这 3 个方面的难题，研发了高效长效的二元复合驱油体系，并结合室内物理模拟实验和数值模拟设计了海上油藏二元复合驱方案，以期为以胜利油区埕岛油田为代表的海上油田的规模化应用提供技术依据。

1　二元复合驱油体系优选

1.1 长效聚合物优选

为了实现海上平台短流程的注入工艺，需要大幅缩短聚合物在溶解环节的时间，以确保经过短流程配注的聚合物进入地层能达到性能指标要求。海上稀井网的布井方式决定聚合物经过多孔介质在地层中运移时，受剪切、吸附、热降解等因素影响的时间更长，再加上注入海水带来的高矿化度环境，需要聚合物产品具有长效稳定性和耐盐性[10-12]。

1.1.1 聚合物溶解时间评价

针对海上油田特殊的技术要求，选取 3 大类聚合物样品开展物化性能评价优选(表 1)。样品 C1 和 C2 属于陆上油田化学驱单元普遍使用的特性黏数约为 2 300 mL/g 的常规聚丙烯酰胺聚合物；样品 C3 和 C8 属于特性黏数为 2 900 mL/g 左右的超高相对分子质量聚合物，因此其增黏能力高于 C1 和 C2，但由于该聚合物在水中水力学体积较大，导致溶解时间最长，均在 60 min 以上；样品 C9 和 C10 属于引入 AMPS 耐盐功能单体的聚合物，特性黏数约为 2 300 mL/g，其增黏性和溶解性均高于其他 2 类聚合物，这是因为 C9 和 C10 这 2 个聚合物含有强极性磺酸基团，增加了高分子骨架电荷数，在提高聚合物增黏性的同时增加了聚合物骨架在水中的分散性能，提升了其溶解性能及溶解速度。因此，样品 C9 和 C10 为能够满足快速溶解要求的聚合物。

表 1 不同聚合物样品基本物化性能评价

Table 1 Evaluation of basic physical and chemical properties of polymers

样品编号	固含量(%)	不溶物含量(%)	水解度(%)	滤过比	残余单体(%)	表观黏度(mPa·s)	特性黏数(mL/g)	溶解时间(min)
C1	91.20	0.060	23.2	1.03	0.02	13.7	2 263	57
C2	90.13	0.024	22.9	1.05	0.04	14.7	2 354	56
C3	89.55	0.042	21.6	1.09	0.02	16.1	2 930	65
C8	90.01	0.012	21	1.00	0.01	15.9	2 997	69
C9	89.66	0.028	17.6	1.08	0.03	16.3	2 290	40
C10	89.16	0.048	13.6	1.15	0.03	17.2	2 380	45
标准	≥89.0	≤0.2	≤24	≤2	≤0.1	≥12.5	≥2 200	

1.1.2 聚合物抗盐性能评价

选取海上平台产出水（矿化度约为 9 000 mg/L）配制质量浓度为 5 000 mg/L 的聚合物母液，在 500 r/min 转速下搅拌 2 h，放置 24 h 后继续将母液分别稀释至 1 000，1 500，2 000，2 500 及 3 000 mg/L 进行黏度测试。从测试结果（图 1）可以看出，在不同质量浓度条件下，引入 AMPS 耐盐功能单体的样品 C9 和 C10 聚合物的黏度更高，在质量浓度为 2 000 mg/L 条件下，黏度均达到 40 mPa·s 以上，表现出良好的抗盐性。

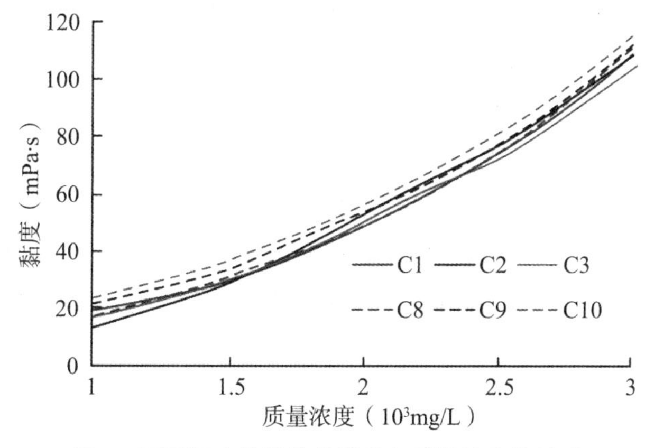

图 1 不同聚合物溶液的黏度与质量浓度的关系

Fig. 1 Relationship between viscosity and concentration of different polymers

1.1.3 聚合物长期稳定性评价

海上油田比陆上油田的注采井距大，经过地层中的长距离运移，聚合物溶液分子与油藏孔隙介质之间作用时间加长，吸附量大大增加的同时聚合物溶液在油藏中的老化时间增加，从而加剧黏度损失，因此，评价聚合物吸附及黏度的长期稳定性十分必要。为评价聚合物产品抗吸附性，配制质量浓度为 3 000 mg/L 的 6 种聚合物溶液样品，通过 30～40 目石英砂模拟地层注入条件，在 30℃条件下放于恒温水浴振荡器振荡 24 h，然后取出样品，离心机分离后，测定吸附后聚丙烯酰胺溶液黏度并计算静吸附黏度保留率[13-15]。将配制好的聚合物溶液进行高纯氮密闭封装，放入温度为 65℃的烘箱内，测定聚合物样品放置 90 d 后的表观黏度保留率，评价其热稳定性(表 2)。

实验结果表明，6 种聚合物样品均表现出良好的抗吸附性，其中，C9 和 C10 这 2 种聚合物由于含有强极性 AMPS 单体，使其高分子骨架具有更好的热稳定性及抗吸附性，90 d 后黏度保留率超过 90%，其中，C10 聚合物增黏性能和稳定性能最好。

表 2　不同聚合物样品吸附及热稳定性测试结果

Table 2　Test results of polymer adsorption and thermal stability

样品编号	静吸附黏度保留率(%)	90 d 后黏度保留率(%)
C1	92.1	88.5
C2	93.6	89.1
C3	92.3	87.6
C8	94.5	85.4
C9	91.8	90.2
C10	94.5	90.8

综合上述所有的评价结果，优选样品 C10 为海上驱油用聚合物。

1.2 高效表面活性剂体系复配

采用单一结构的表面活性剂往往很难在较宽的浓度范围内获得超低的界面张力，而磺酸盐类活性剂具有与原油分子结构相似的疏水基团，对原油具有较好的适应性。为了确保表面活性剂在地层残留海水环境中达到 10^{-3} mN/m 数量级的超低界面张力，提高洗油效率，根据界面受力及界面压力分析，通过调整亲油基、亲水基类型、EO 数改善抗钙镁能力并提高活性，同时引入氧乙烯调整分子尺寸，再辅以适宜的助剂，最终形成高效驱油用阴-非两性表面活性剂 S4。在此基础上选择与海上原油相似度高的胜利石油磺酸盐(SLPS)与阴-非两性表面活性剂 S4 进行复配，并对体系性能进行评价。在总质量分数为 0.4%的条件下，测定了不同质量浓度配比时 SLPS+S4 复配体系的界面张力，由图 2 可以看出，复配体系在 2∶1～1∶3 范围内均可获得超低界面张力[16-18]。

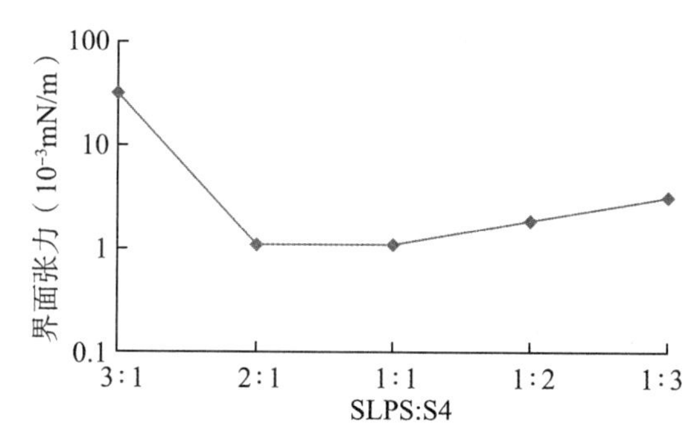

图 2　SLPS＋S4 复配体系不同质量浓度配比时的界面张力

Fig. 2　Concentration ratio of SLPS＋S4 compound system

对 SLPS＋S4 复配体系的超低界面张力的总质量分数窗口进一步测试发现(图 3)，在总质量分数为 0.1%～0.6%时均可以达到超低界面张力，表明复配体系的总质量分数窗口较宽。

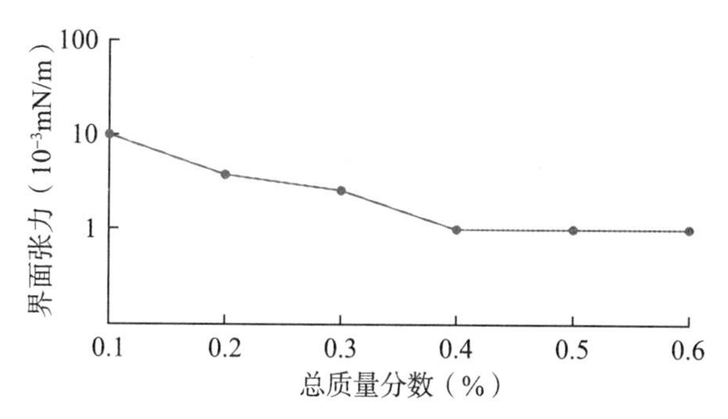

图 3　SLPS＋S4 复配体系总质量分数窗口曲线

Fig. 3　Total mass fraction of SLPS＋S4 compound system

1.3 二元复合驱油体系设计

为了确保聚合物和表面活性剂性能的发挥，在二元复合驱油体系设计过程中，两者的配伍性成为重要的评价指标，配伍性较好的驱油体系可以在较短时间内达到各项性能指标，并呈现良好的长期稳定性。

1.3.1 聚合物与表面活性剂相互作用

聚合物 C10 与 SLPS＋S4 复配体系组成二元复合驱油体系，进一步研究聚合物与表面活性剂的相互作用。利用产出水分别配制质量分数为 0.30%，0.25%，0.20%的聚合物溶液，然后与 SLPS＋S4 复配体系混合搅拌，总质量分数达 0.40%，测定混合体系的黏度及对目的层原油的界面张力。由实验结果可以看出(表 3)，C10 聚合物在不同质量分数条件下，加入 0.2%SLPS＋0.2%S4 复配体系之后，体系黏度均升高，而且都能在 100 min 内达到超低界面张力，这说明 SLPS＋S4 复配体系与聚合物 C10 具有良好的配伍性。

表 3 不同质量分数聚合物与表面活性剂相互作用测试结果

Table 3 Test results of interaction between surfactant and polymer with various mass fraction

二元复合驱油体系	黏度(mPa·s)	界面张力(10^{-3} mN/m)	达到超低界面张力时间(min)
0.30%C10+0.2%SLPS+0.2%S4	120	5.0	100
0.25%C10+0.2%SLPS+0.2%S4	97	3.3	80
0.20%C10+0.2%SLPS+0.2%S4	61	2.6	60

1.3.2 热稳定性评价

为考察二元复合驱油体系在残留海水环境和较长时间地层运移条件下的稳定性，对其进行了热稳定性实验，在 30 d 内以 5 d 为间隔分 6 次测试了二元复合驱油体系的界面张力和黏度保留率。由其测试结果(表 4)可以看出，30 d 内二元复合驱油体系界面张力仍然保持超低，黏度保留率达 100%以上，能够满足长效热稳定性要求。

表 4 二元复合驱油体系热稳定性测试结果

Table 4 Test results of thermal stability of polymer/surfactant flooding system

测试次数	界面张力(10^{-3} mN/m)	黏度保留率(%)
1	5.0	100
2	5.2	101
3	5.1	101
4	6.0	101
5	7.1	101
6	9.2	101

2 二元复合驱油体系注入参数优化设计

胜利油区海上油田采用的海上平台一般设计使用寿命是 15 a，海底管线需要定期进行寿命评估，因此，在实施二元复合驱的过程中，要充分考虑在设备的使用寿命期内最大幅度地提高原油采收率，需要对体系的注入黏度、注入速度以及注入段塞尺寸等参数进行优化设计，以达到最佳经济效益。

2.1 注入黏度

二元复合驱油体系驱替过程中，驱替前缘油水总流度为

$$\lambda_{ow}=\frac{KK_{rw}}{\mu_w}+\frac{KK_{ro}}{\mu_0} \tag{1}$$

作为驱替相的二元复合驱油体系，其流度为

$$\lambda_{sp}=\frac{KK_{rw}}{R_k\mu_{sp}} \tag{2}$$

当二元复合驱油体系流度等于其驱替前缘油水总流度,即 $\lambda_{sp}=\lambda_{ow}$ 时,为理想条件下的最优流度。为了确保海上油田在海工设施寿命期内达到最大采收率,二元复合驱油体系应该达到最高的合理流度比,考虑相对渗透率不变的条件下,利用数值模拟的手段,得到了海上二元复合驱油体系黏度比与提高采收率关系曲线(图 4)。当黏度比大于 0.5 时,二元复合驱油体系提高采收率曲线上升幅度开始放缓,因此选取黏度比为 0.5。埕岛油田地下原油黏度为 30~70 mPa·s,考虑沿程剪切黏度损失及地层黏土、岩石对聚合物的吸附作用,有效驱替注入黏度应大于 35 mPa·s。

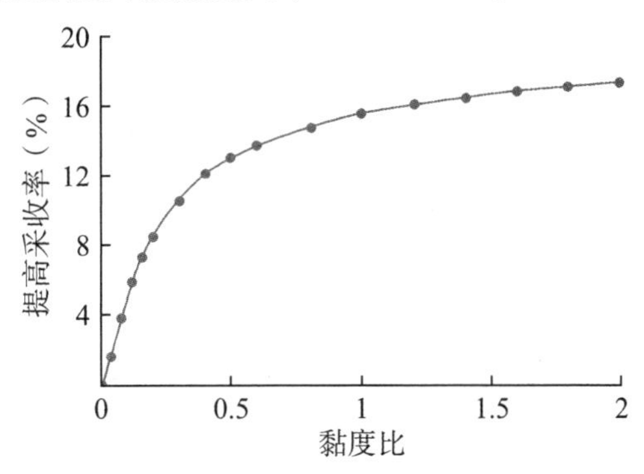

图 4 黏度比与提高采收率关系曲线

Fig. 4 Relationship between viscosity ratio and enhanced oil recovery

2.2 注入速度

二元复合驱油体系注入速度太快容易发生窜聚现象,而注入速度过小则无法形成有效的注入压力,影响增油效果。因此,二元复合驱油体系要在考虑注入能力和增油效果的情况下优选合理的注入速度。随着二元复合驱油体系的注入,由于驱替相黏度的增加,导致渗流阻力增大,注入压力呈现上升趋势,吸水指数呈现下降趋势。借鉴陆上化学驱项目动态变化规律,建立海上油田单井注入二元复合驱油体系最大注入能力计算模型,其表达式为

$$Q_{i\max}=(p_i-p_{th}-\Delta p_q)\times(I_{iw}-\Delta I_w)\times H_i \tag{3}$$

$$\Delta p_q=c\,\frac{Q_i}{H_i\lambda_{sp}}\ln\frac{r}{r_w} \tag{4}$$

按照吸水指数下降值分别选取 50%,30%,10%,利用计算模型核算了注入井的单井最大注聚能力,对应 0.06,0.07 及 0.08 PV/a 共 3 种注入速度,分别预测了相应注入速度条件下采收率提高幅度,从图 5 可以看出,注入速度超过 0.07 PV/a 后,提高采收率幅度减小,考虑到干线注入压力并借鉴陆上二元复合驱单元吸水指数下降值的经验,优选二元复合驱油体系的注入速度为 0.07 PV/a[19-20]。

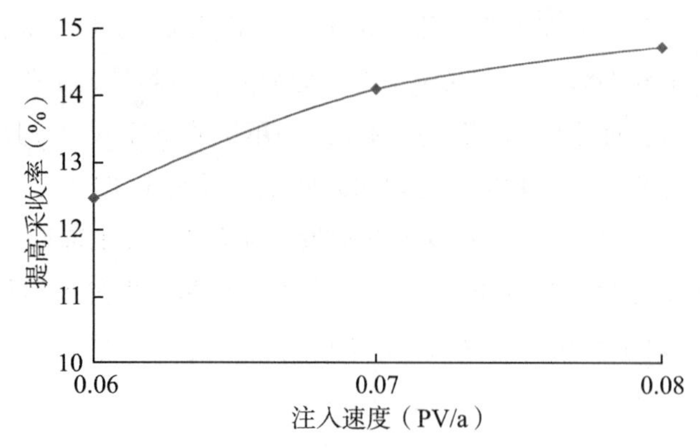

图5 不同注入速度下提高采收率曲线

Fig. 5 Relationship between injection rate and enhanced oil recovery

2.3 注入段塞尺寸

海上平台和设备到达年限后需要进行安全评估及延寿改造，为了实现海上油田大幅度提高采收率的目标，必须在延长二元复合驱油体系注入时间的同时也要考虑经济效益。按照0.07 PV/a的注入速度，分别设计了注入4，5，6，7，8 a共5种注入方案，综合评价其增油效果和经济效益(图6)。随着二元复合驱油体系注入段塞尺寸的增加，阶段采出程度呈现不断上升的趋势，但是通过计算增油量与折算聚合物注入量比值(折算吨聚增油量)发现，经济效益随着注入时间增加而不断下降，当注入段塞尺寸为0.42 PV时达到平衡，因此选择0.42 PV作为最佳注入段塞尺寸。

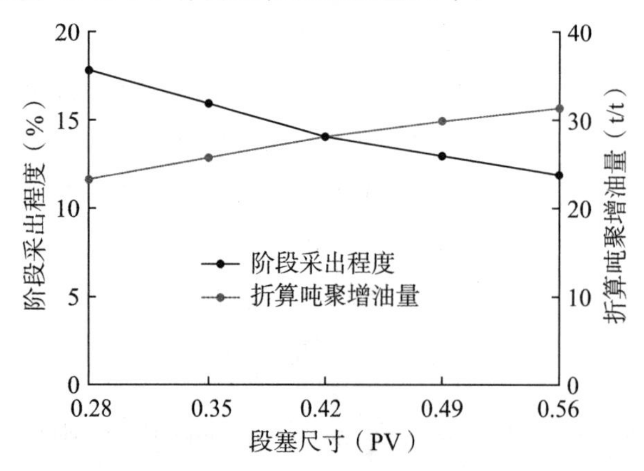

图6 不同注入段塞尺寸下阶段采出程度与折算吨聚增油量曲线

Fig. 6 Comparison between oil recovery and oil increment per ton polymer under different injection plug sizes

3 应用效果评价

3.1 实验室物理模拟评价结果

通过在实验室模拟海上油藏配注条件和地层条件，对优选的二元复合驱油体系进行增油效果评价：实验温度设定为65℃，采用海上平台产出水对驱油体系进行配制，设定填砂管的渗透率级差为1∶3，分别注入段塞尺寸为0.4 PV的单一聚合物驱油体系和二元复合驱油体系进行对比，结果显示，二元复合驱油体系提高采收率达到了 29.2%，相对于同尺寸段塞单一聚合物驱的 17.9%增加 11.3%。

3.2 油藏数值模拟结果预测

采用数值模拟的研究手段对水驱与二元复合驱油体系的效果进行了预测对比(图7)，在数值模拟软件中设定二元复合驱油体系中聚合物质量浓度为 3 000 mg/L，表面活性剂质量浓度为 4 000 mg/L，注入段塞尺寸为 0.42 PV，注入速度为 0.07 PV/a，预测结果显示，15 a 末水驱采出程度为 38.4%，二元复合驱采出程度达 50.0%，后者较前者的采出程度提高 11.6%，预计累积增油量为 140×10^4 t。

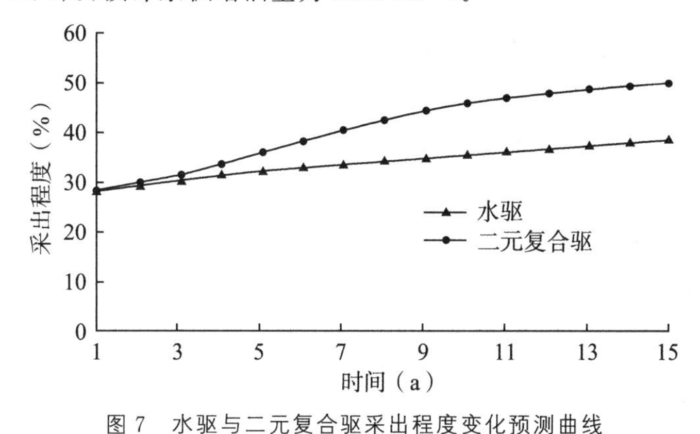

图 7 水驱与二元复合驱采出程度变化预测曲线

Fig. 7 Prediction curves of oil recovery by water flooding and polymer/surfactant flooding

4 结论

胜利油区海上油田开展二元复合驱的油藏条件与陆上油田存在差异，因此，不能照搬陆上油田的二元复合驱油体系及参数。通过室内物理模拟实验和数值模拟对二元复合驱油体系进行了优选，筛选出溶解时间短、抗盐性和热稳定性较好的聚合物 C10，并复配出超低界面张力、质量分数窗口宽的表面活性剂，将二者结合最终得到配伍性和稳定性良好的海上油田二元复合驱油体系 C10＋SLPS＋S4。通过物理模拟驱油实验，该二元复合驱油体系注入段塞尺寸为 0.4 PV 之后，提高采收率 29.2%，相对于同尺寸段塞单一聚合物驱增加 11.3%。为了最大幅度提高海上油田采油速度和原油采收率，优化设计海上油田二元复合驱最佳黏度比为 0.5，优选合理的注入速度为 0.07 PV/a，最佳注入段塞尺寸为 0.42 PV，采用数值模拟预测可以提高采出程度 11.6%。

符号解释：

c ——矿场统计回归得到的系数；

H_i——注入井射孔厚度，m；

I_{iw}——注入井吸水指数，$m^3/(MPa\cdot d)$；

K——绝对渗透率，mD；

K_{ro}——油相相对渗透率；

K_{rw}——水相相对渗透率；

p_i——注入井地层静压,MPa；

p_{th}——注入井启动压力,MPa；

Q_i——单井日注入量,m^3/d；

$Q_{i\max}$——单井注入二元复合驱油体系最大日注入能力,m^3/d；

r——二元复合驱油体系驱替前缘半径,m；

r_w——井筒半径,m；

R_k——阻力系数；

ΔI_w——吸水指数下降值,$m^3/(MPa \cdot d)$；

Δp_q——注入井压力上升值,MPa；

λ_{ow}——驱替前缘油水总流度,mD/(mPa·s)；

λ_{sp}——二元复合驱油体系流度,mD/(mPa ·s)；

μ_o——油相的黏度,mPa·s；

μ_{sp}——二元复合驱油体系黏度,mPa·s；

μ_w——水相的黏度,mPa·s。

参考文献

[1] 赵方剑.胜利油田化学驱提高采收率技术研究进展[J].当代石油石化,2016,24(10):19-22.

ZHAO Fangjian. Research progress of chemical flooding enhanced oil recovery technologies in Shengli Oilfieldb[J]. Petroleum & Pet-rochemical Today, 2016, 24(10):19-22.

[2] 石静,曹绪龙,王红艳,等.胜利油田高温高盐稠油油藏复合驱技术[J].特种油气藏,2018,25(4):129-133.

SHI Jing,CAO Xulong,WANG Hongyan,et al. Combination flood-ing technology used in high-temperature, high-salinity heavy oil reservoirs of Shengli Oilfield[J]. Special Oil & Gas Reservoirs, 2018,25(4):129-133.

[3] 明玉坤.基于响应曲面法的二元复合驱注采参数优化方法——以孤东油田七区西 Ng41-51 区块为例[J].油气地质与采收率,2017,24(3):91-97.

MING Yukun. Optimization method of the injection-production parameters for SP flooding based on the response surface method-ology-A case study of Ng41-51 submember in the west of the 7th block of Gudong oilfield[J]. Petroleum Geology and Recovery Effi-ciency,2017,24 (3):91-97.

[4] 田冀,刘晨,张金庆,等.海上油田平台扩容潜力优化模型的建立及应用[J].断块油气田,2018,25(2):204-207.

TIAN Ji, LIU Chen, ZHANG Jinqing, et al. Establishment and ap-plication of optimized model for expansion potential of offshore oil platform[J]. Fault-Block Oil and Gas Field,2018,25(2):204-207.

[5] 李丹.高低质量浓度段塞组合聚合物驱提高采收率实验[J].大庆石油地质与开发,2018,37(6):121-124.

LI Dan. Experiment of the polymer flooding EOR in the combina-tion of high and low-mass-concentration slugs [J]. Petroleum Ge-ology & Oilfield Development in Daqing,2018,37(6):121-124.

[6] 刘毅,程诗睿,胡子龙,等.微生物复合降黏技术提高稠油底水油藏采收率研究及应用[J].油气藏评价与开发,2017,7 (2):70-73.

LIU Yi, CHENG Shirui, HU Zilong, et al. Research and applica-tion of the composite microbial viscosity reduction technology to improve the recovery of heavy oil reservoir[J]. Reservoir Evalua-tion and Development,2017,7 (2):70-73.

[7] 刘义刚,卢琼,王江红,等.锦州 9-3 油田二元复合驱提高采收率研究[J].油气地质与采收率,2009,16(4):68-70,73.

LIU Yigang, LU Qiong, WANG Jianghong, et al. Research on EOR by binary combination flooding in Jinzhou9-3 Oilfield[J]. Petro-leum Geology and Recovery Efficiency,2009,16(4):68-70,73.

[8] 丁玉娟.胜利油区配注污水水质对化学驱效果的影响[J].油气地质与采收率,2014,21 (3):66-69.

DING Yujuan. Influence of injection water quality on performance of chemical flooding in Shengli oilfield[J] . Petroleum Geology and Recovery Efficiency,2014,21 (3):66-69.

[9] 王锦林,吴慎渠,王晓超,等.渤海 S 油田注聚井注入压力界限潜力[J].大庆石油地质与开发,2017,36(1):109-113.

WANG Jinlin, WU Shenqu, WANG Xiaochao, et al. Potential of the polymer injection pressure limit in Bohai S Oilfield [J]. Petro-leum Geology & Oilfield Development in Daqing,2017,36(1):109-113.

[10] 刘义刚,卢琼,史锋刚.锦州油田无碱二元复合驱实验研究[J].海洋石油,2010,30(1):48-52.

LIU Yigang,LU Qiong,SHI Fenggang. The alkali-free SP flooding experiment in Jinzhou Oilfield[J]. Offshore Oil,2010,30(1):48-52.

[11] LIU Y,HOU J,LIU L,et al. An inversion method of relative per-meability curves in polymer flooding considering physical proper-ties of polymer [J]. SPE Journal,2018,23(5):1 929-1 943.

[12] 祝仰文,刘歌,曹绪龙,等.阴离子聚丙烯酰胺/三乙醇胺超分子体系的表征及性能[J].高等学校化学学报,2016,37(9):1 728-1 732.

ZHU Yangwen, LIU Ge, CAO Xulong, et al. Characterization and property of sulfonated polyacrylamide/triethanolamine supramo-lecular system [J]. Chemical Journal of Chinese Universities,2016,37(9):1 728-1 732.

[13] 李宗阳,王业飞,曹绪龙,等.新型耐温抗盐聚合物驱油体系设计评价及应用[J].油气地质与采收率,2019,26(2):106-112.

LI Zongyang,WANG Yefei,CAO Xulong,et al. Design evaluation and application of a novel temperature-resistant and salt-tolerant polymer flooding system [J]. Petroleum Geology and Recovery Effi-ciency,2019,26(2):106-112

[14] 杜庆军,侯健,徐耀东,等.聚合物驱后剩余油分布成因模式研究[J].西南石油大学学报:自然科学版,2010,32(3):107-111.

DU Qingjun,HOU Jian,XU Yaodong,et al. Study on genetic mod-el of remaining oil distribution after polymer flooding[J]. Journal of Southwest Petroleum University: Science & Technology Edi-tion,2010,32(3):107-111.

[15] 周凤军,李金宜,瞿朝朝,等.海上厚层油藏早期聚合物驱剩余油分布特征实验研究[J].油气地质与采收率,2017,24 (6): 92-96.

ZHOU Fengjun, LI Jinyi, QU Zhaozhao, et al. Laboratory core ex-periments of remaining oil distribution at the early polymer flood-ing in offshore thick reservoirs[J]. Petroleum Geology and Recov-ery Efficiency,2017,24 (6):92-96.

[16] 张瑶,付美龙,侯宝峰,等.三次采油高温高盐油藏用表面活性剂的研究进展[J].当代化工,2019,48(2):350-353.

ZHANG Yao,FU Meilong,HOU Baofeng,et al. Research progress of surfactants for tertiary recovery of high temperature and salinity reservoirs[J]. Contemporary Chemical Industry,2019,48 (2):350-353.

[17] 王斌,董俊艳,王敏,等.阴非离子表面活性剂在高温高盐油藏的研究与应用[J].化学工程与装备,2018,13(11):144-145,125.

WANG Bin, DONG Junyan, WANG Min, et al. Research and ap-plication of anionic nonionic surfactant in high temperature and high salinity reservoirs [J]. Chemical Engineering & Equipment, 2018,13(11):144-145,125.

[18] 赵健慧,赵冀,周代余,等.高温高盐油藏驱油用表面活性剂性能评价[J].新疆石油地质,2013,34(6):680-683.

ZHAO Jianhui, ZHAO Ji, ZHOU Daiyu, et al. Performance evalua-tion of surfactant for EOR in high temperature and high salinity reservoirs [J]. Xinjiang Petroleum Geology,2013,34(6):680-683.

[19] 于金彪. 油藏数值模拟历史拟合分析方法[J]. 油气地质与采收率,2017,24(3):66-70.

YU Jinbiao. History matching analysis method on reservoir numer-ical simulation [J]. Petroleum Geology and Recovery Efficiency, 2017,24 (3):66-70.

[20] 安志杰.海南3断块二元驱技术研究与应用[J].石油钻采工艺,2017,39(3):370-374.

AN Zhijie. Research on S / P combinational flooding technology and its application in Fault Block Hainan3[J]. Oil Drilling & Pro-duction Technology, 2017, 39(3):370-374.

编辑 裴 磊

基于核磁共振技术的黏弹性颗粒驱油剂流动特征

孙焕泉[1]　徐　龙[2]*　王卫东[2]　曹绪龙[1]　姜祖明[1]　祝仰文[1]　李亚军[2]
宫厚健[2]　董明哲[2]

(1. 中国石油化工集团胜利油田勘探开发研究院,东营 257015;
2. 中国石油大学(华东)石油工程学院,青岛 266580)

摘要:首次通过核磁共振技术和岩心驱替实验相结合的方法,对黏弹性颗粒(B-PPG)驱油剂在岩心中的流动特征进行深入研究。驱替结果表明:B-PPG在岩心孔隙介质中以重复堵塞-变形-通过的方式运移。核磁共振横向弛豫时间(T_2)图谱明确了B-PPG在孔隙介质中的调剖作用,即B-PPG首先运移进入大孔隙,然后对其进行封堵,促使液流转向较小孔隙。核磁共振图像直观揭示了B-PPG驱油剂在岩心孔隙中的流动分布状态,B-PPG驱油剂驱替前缘均匀推进,能够进入水驱波及不到的区域,具有显著的调整微观非均质性、增大波及系数的能力。渗透率、B-PPG浓度及注入速率对B-PPG驱油剂的流动特征具有显著影响。增大渗透率能够提高B-PPG在孔隙介质中注入性;增大浓度或降低注入速率有利于B-PPG堵塞大孔隙,液流转向较小孔隙,提高波及系数。

关键词:核磁共振;黏弹性颗粒;T_2图谱;运移方式;流动特征

中图法分类号:TE357;文献标志码:A

Flow characteristics of viscoelastic particle flooding agent based on nuclear magnetic resonance technology

Sun Huanquan[1] Xu Long[2*] Wang Weidong[2] Cao Xulong[1]
Jiang Zuming[1] Zhu Yangwen[1] Li Yajun[2]
Gong Houjian[2] Dong Mingzhe[2]

(1. Exploration and Development Research Institute of Shengli Oilfield, China Petroleum & Chemical Corporation, Dongying 257015, China;
2. School Petroleum Engineering, China University of Petroleum (East China), Qingdao 266580, China)

Abstract: The flow characteristics of viscoelastic particle (B-PPG) oil displacing agent in the core were firstly in depth studied by a combination of nuclear magnetic resonance technology and core displacement experiments. The displacement results show that B-PPG migrates in the porous medium by repeated blocking-deformation-passing. The nuclear magnetic resonance transverse relaxation time (T_2) spectrum clearly clarifies the profile-controlling effect of B-PPG in the pore medium, that is, B-PPG first enters the large pores, then blocks them, and promotes the fluid flow to the smaller pores. The nuclear magnetic resonance image visually reveals the flow distribution of B-PPG oil displacing agent in the core pores. The B-PPG oil displacing agent advances uniformly in the displacement front, can enter the area which cannot be reached by water flooding, and has the ability to significantly adjust the micro heterogeneity and in-crease the spread coefficient. Permeability, B-PPG concentration and injection rate have significant effects on the flow characteristics of B-PPG oil displacement agents. Increasing the permeability can improve the injectability of B-PPG in the porous medium. Increasing the concentration or decreasing the injection rate helps B-PPG to block large pores and the fluid flow turns to smaller pores, which im-proves the spread coefficient.

Keywords: nuclear magnetic resonance; viscoelastic particles; T_2 spectrum; migration model; flow characteristics

目前,大多数油田已进入后续高含水开发阶段,具有产量递减、含水上升快、含水高、聚合物后大孔道等特点,且地下剩余油高度分散,多年的综合治理使得层内、层间非均质性突出,开发形势非常严峻[1],因此,如何实现注入流体在油层深部的转向成为提高油田采收率的主要目标。研究发现,与普通体膨型预交联凝胶颗粒相比[2],带有支化链的新型预交联黏弹性颗粒(branched preformed particle gel,B-PPG)兼有水溶性高分子和凝胶的双重特征,在水中可形成黏稠液体和黏弹性颗粒共存的分散体系,能够满足油田非均质性油层提高采收率的要求[3]。在胜利油田先导性试验中,B-PPG 驱油剂取得良好的控水增油效果[4],但其在地层中的流动特征及渗流机理尚缺乏系统的实验研究。近年来,尽管人们通过填砂管模型、微观平板模型及数学模型等不同的方法对 B-PPG 流动规律做了大量的研究工作[5-7],但由于这些研究方法得到的结果主要以定性、经验关系式为主,不能准确反映 B-PPG 流体在真实地层中的运移规律,对 B-PPG 在孔隙介质中的流动特征和渗流机理缺乏直接有力的证明,不能准确阐述清楚 B-PPG 体系的流动特征,从而制约了该项技术的推广和应用。为此,通过核磁共振(NMR)技术与岩心驱替实验相结合的方法,对 B-PPG 在岩心孔隙介质中运移及流动特征分析的同时,又表征了流体在岩心内部的流动分布状态。对于 B-PPG 驱油剂在油田的推广、应用及改进具有重要的理论意义和应用价值。

1 实验部分

1.1 实验材料及仪器

实验所用材料包括 B-PPG、$MnCl_2$ 和模拟地层水。B-PPG 分散液由 B-PPG 干粉和模拟地层水混合配制,其中,B-PPG 干粉颗粒为 60～120 目,粒度中值(D_{50})为 460.2 μm;模拟地层水由 NaCl、$CaCl_2$、$MgCl_2 \cdot 6H_2O$ 和蒸馏水配制,矿化度为 20 921 mg/L。

核磁共振-驱替装置由核磁共振分析仪和模拟油藏驱替系统组成,如图 1 所示。其中,核磁共振仪为低磁场 MacroMR12-110H-I 型分析仪;填砂管为无磁性聚醚醚酮材质,内径为 25 mm、长度为 60 mm,石英砂目数为 60～100 目。

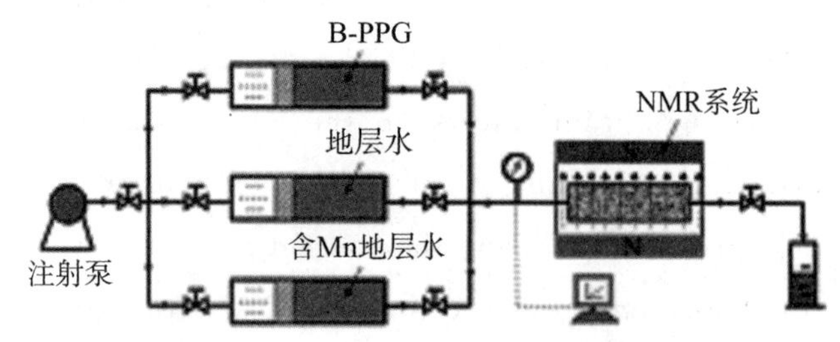

图 1 核磁共振-驱替装置示意图

Fig. 1 Schematic diagram of NMR-displacement device

1.2 实验方法

1.2.1 核磁共振测试

核磁共振技术只采集流体信号，通过抑制弛豫时间的方法来区分不同流体，在油田开发中可用于分析岩石孔隙体积、喉道大小及流体分布[8-9]。其中，横向弛豫时间(T_2)谱反映孔隙内比表面的大小，T_2 与孔隙尺寸成正比，信号幅度与孔隙内液体体积成正比；核磁共振成像模块通过在目标物体上施加 3 个相互垂直的、可控的线性梯度磁场实现信号的空间定位，可采集岩心内部剖面图像，直观表征流体流动特征及分布状态。

研究主要涉及水驱和 B-PPG 驱流体的信号区分。由于水和 B-PPG 流体的弛豫时间有重叠部分，导致两者的信号不能明确区分。研究表明：$MnCl_2$ 中 Mn^{2+} 可加速水中 H 质子的弛豫衰减，使得弛豫时间不再有重叠，使用质量分数为 0.5%的 $MnCl_2$ 水溶液代替水进行驱替，可有效区分水驱与 B-PPG 驱流体的 T_2 谱和图像[10]。由于 $MnCl_2$ 会影响 B-PPG 的黏弹性，不适合用于配制 B-PPG 分散液。因此，采用质量分数为 0.5%的 $MnCl_2$ 水溶液进行水驱实验。

1.2.2 驱替实验

石英砂装填填砂管，测试填砂管物性，进行核磁共振 T_2 测试，获取干燥填砂管岩心的 T_2 图谱；抽真空加压，饱和模拟地层水，进行 T_2 图谱测试；低流速注入含 $MnCl_2$ 模拟地层水驱替模拟地层水，至少 2 PV(PV 为孔隙体积)，充分置换模拟地层水，以屏蔽岩心中的水信号；以恒定流速相继进行含 $MnCl_2$ 模拟地层水驱、B-PPG 驱和后续含 $MnCl_2$ 模拟地层水驱，每个阶段均驱替至压力稳定，然后进行下一流体驱替，驱替过程中进行 T_2 与成像测试，获取不同驱替阶段时流体在岩心中的 T_2 图谱和剖面图像。实验在 82℃条件下进行，全程记录岩心两端压力差变化。

2 实验结果及分析

2.1 B-PPG 在岩心中的流动规律

图 2 所示为岩心(渗透率为 1.7 μm^2)入口端压力随注入孔隙体积(PV)数的变化。水驱后随着 B-PPG(2 500 mg/L)的注入，压力快速上升，当注入 2.2 PV 的 B-PPG 后，压力曲线开始呈现出锯齿状波动“平台”。后续水驱时，压力迅速下降并逐渐稳定，稳定压力大于初始水驱压力。通过以上现象可以看出：注入 B-PPG 时，黏弹性颗粒在岩心孔喉处堆积堵塞水流通道，导致注入端形成憋压，压力逐渐升高。当注入端压力升高到一定程度时，黏弹性颗粒在高压力梯度下，发生形变通过孔喉，暂时性堵塞被破坏，注入水继续通过造成压力下降。黏弹性颗粒通过孔喉后，在水的推动下继续向岩心深部运移直到在下个孔喉处再次发生堵塞，导致憋压，压力梯度增大，颗粒再次发生形变通过孔喉，压力下降。所以 B-PPG 在岩心孔隙介质中的运移方式为：堵塞→变形→通过→运移过程的重复发生，因此，压力变化呈现出锯齿状波动现象。

转后续水驱后，由于大量 B-PPG 被冲出，没有后续颗粒的进入，导致注入端压力迅速下降。但是,由于孔隙介质对 B-PPG 的吸附滞留作用,在岩心中还残存有一些颗粒,这些颗粒在水的推动下继续以堵塞-变形-通过的方式在孔喉处运移。因此,在后续水驱的过程中也会有压力曲线波动的现象。

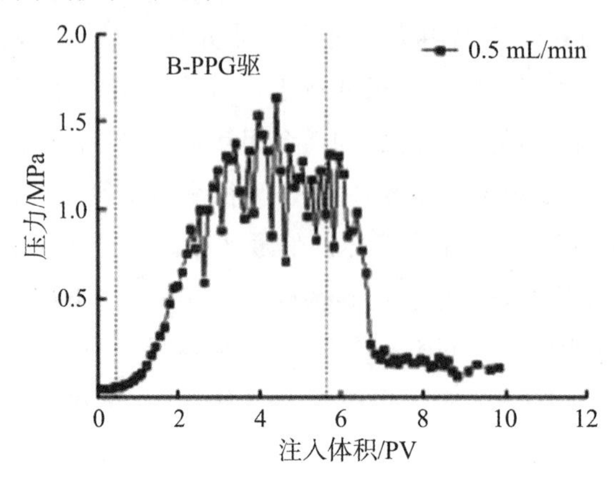

图 2 岩心入口端压力随累计注入 PV 数的变化

Fig. 2 The change of the inlet pressure of the core with the accumulated PV number

图 3 所示为不同因素对 B-PPG 流动规律的影响。从图 3(a)可以看出:岩心透率越小,压力升高速度越快,压力“稳定”平台越高,后续水驱的稳定压力也越高,并且波动越剧烈。渗透率越小,孔隙介质等效孔喉半径越小,黏弹性颗粒通过孔喉时所需要的变形量越大,即促使颗粒发生变形的压力梯度也越大。渗透率为 3 μm^2 和 5 μm^2 的稳定压力相近,说明当渗透率在一定范围内时,或者孔喉半径在一定范围内时,颗粒通过孔喉所需要的压力梯度相近,即 B-PPG 对这一范围内的渗透率适应性相近。

图 3(b)所示为 B-PPG 浓度的影响。由图 3(b)可以看出,B-PPG 的注入浓度越高,压力上升越快,稳定压力越高。随着 B-PPG 浓度的增大,单位体积流体内黏弹性颗粒在孔喉注入端的堆积增多,发生堵塞-变形-通过形式的颗粒的数量也增多,造成注入端压力也相应升高。B-PPG 浓度增高会增加颗粒在孔隙介质中的吸附滞留量,后续水驱压力也随之增大。当浓度增大到吸附滞留量达到饱和后,后续水驱压力也趋于一致。对于 1 000 mg/L 的 B-PPG 分散液,注入 5.5 PV 仍没有达到稳定压力,说明 B-PPG 在此浓度时并未起到有效封堵吼道的效果。

由达西公式可知:流动压力与流体注入速率成正相关关系,如图 3(c)所示。同时,随着注入速率的增大,单位时间内进入孔隙介质中的黏弹性颗粒的数量也增多,在孔喉处以堵塞-变形-通过形式运移的颗粒数量增多,这就需要更大的压力梯度来满足增多的黏弹性颗粒在孔隙介质中运移的需要。综合以上两种因素,增大 B-PPG 流体注入速率时需要提供更大的泵压支撑。

2.2 B-PPG 在岩心中的分布状态及规律

图 4(a)为模拟地层水流动过程中的 T_2 图谱。可以看出,随着地层水的注入,岩心中的流体信号幅度越来越高,1.5 PV 后趋于稳定,表明流体在岩心内达到饱和。在整个

水驱过程中，T_2 图谱的分布范围基本不变，信号幅度峰顶点对应的 T_2 也基本不变（约为621 ms），表明注入水基本都在相同尺寸范围内的孔隙中流动，没有波及更小尺寸的孔隙。

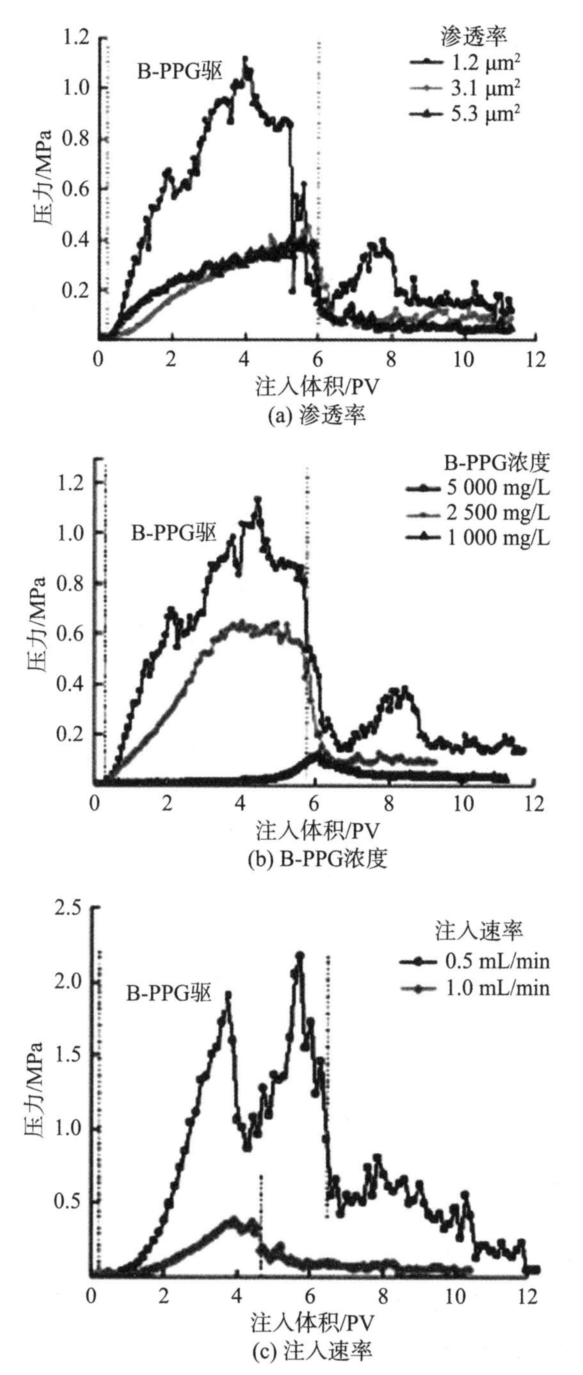

图3 不同渗透率、B-PPG 浓度和注入速率时，岩心入口端压力随累计注入 PV 数的变化

Fig. 3 The pressure of the core during flooding at different permeabilities B-PPG concentrations and injection rates

图 4(b)为 B-PPG(2 500 mg/L)流动阶段的 T_2 图谱。从图 4(b)可以看出,在注入 3.0 PVB-PPG 后信号幅度趋于稳定,岩心内 B-PPG 流体达到饱和,与由图 2 流体压差稳定 PV 数结果一致,表明此时 B-PPG 在岩心中的吸附滞留达到饱和。从图 4 中还可以看出,随着 B-PPG 的不断注入,T_2 图谱范围逐渐向较小值方向延伸。在注入 0.25 PV 时,曲线峰顶点对应的 T_2 为 572 ms;当注入 3.0 PV 时,曲线峰顶点对应的 T_2 为 454 ms,再继续注入 B-PPG,曲线峰顶点对应的 T_2 变化不大。T_2 越小,表明流体进入的孔隙尺寸越小。即在 B-PPG 驱替过程中,B-PPG 除了进入到较大尺寸的孔隙外,也逐渐进入到较小尺寸的孔隙,表明 B-PPG 流动过程中出现了液流转向现象:B-PPG 在封堵大尺寸的孔隙后,后续颗粒逐渐转向进入小尺寸孔隙。微观上 B-PPG 进入低渗区域,宏观上增大了驱油剂的波及面积。

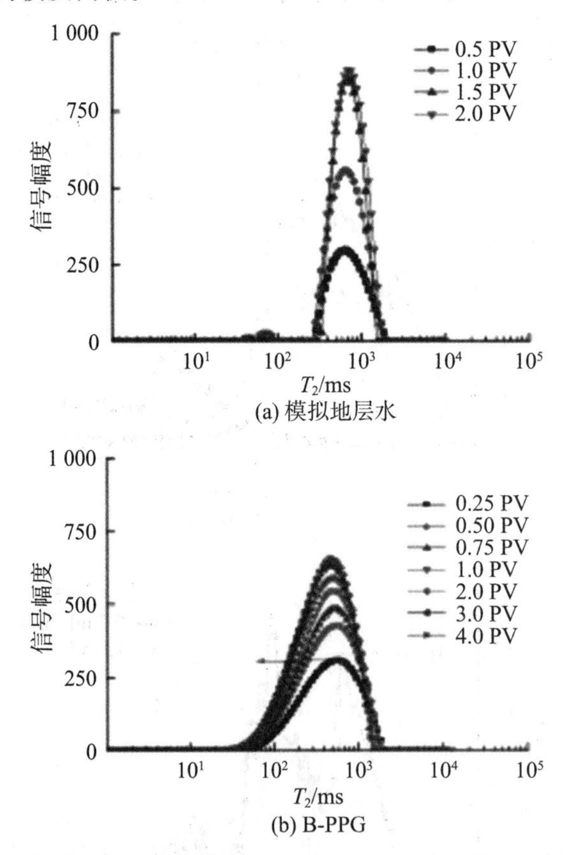

图 4 模拟地层水和 B-PPG 动过程中的核磁共振 T_2 图谱变化(渗透率为 1～2 μm^2)

Fig. 4 NMR T_2 curves of formation and B-PPG during flow(permeability is 1～2 μm^2)

对水驱和 B-PPG 驱过程进行核磁成像,得到流体在岩心中的饱和度变化,如图 5 所示。从图 5 可以看出,地层水从入口端均匀进入岩心后,在后续的流动中逐渐形成一个水流通道。后续地层水主要沿水流通道流动,很少波及其他区域。而 B-PPG 在岩心中的驱替前缘像活塞一样均匀推进,没有出现明显的流动通道。在注入 3.0 PV 后,B-PPG 可以波及岩心中 90%以上的区域,进入水驱不能波及的孔隙。与水驱效果相比,B-PPG

表现出较强的波及能力。

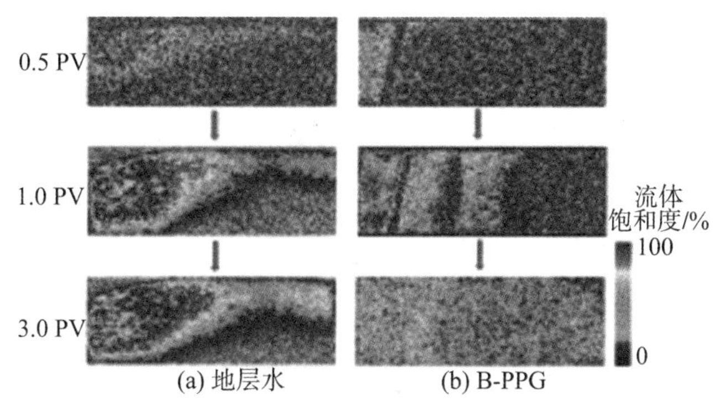

图 5 地层水和 B-PPG 在岩心中的流体饱和度变化

Fig. 5 Changes in flow saturation of formation water and B-PPG in cores

图 6 所示为不同条件时，B-PPG 在岩心中稳定流动阶段的 T_2 图谱。从图 6(a)可以看出，较低渗透率岩心中 B-PPG 流体的 T_2 图谱信号峰分布范围更偏向于较小值，且峰顶点对应的 T_2 更小。渗透率越小，孔隙的平均尺寸越小，B-PPG 可以进入渗透率为 1.3 μm^2 地层中的小尺寸孔隙中。从图 6 可以看出，较高渗透率岩心中的总信号幅度更大，这是因为渗透率较高的地层孔隙度相对较大，B-PPG 流体的饱和量越大，所以获得的信号量更多。

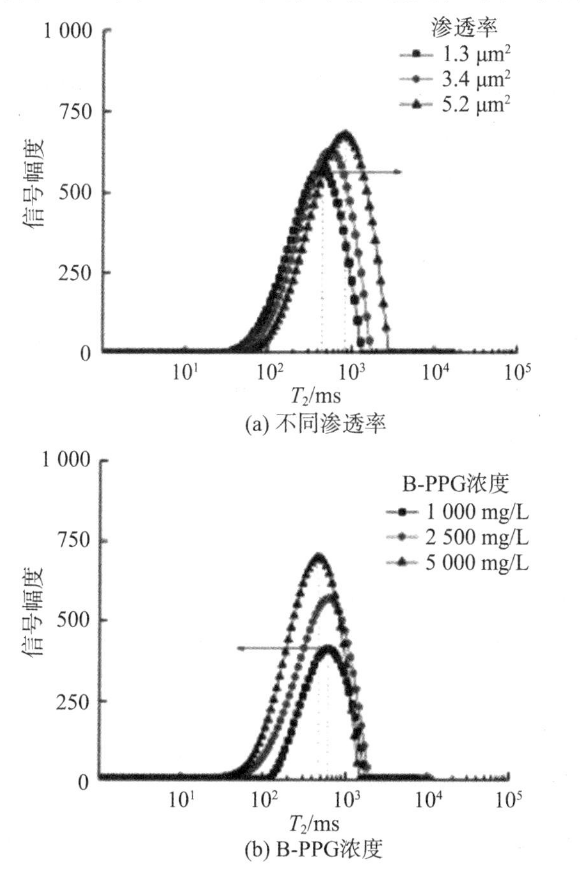

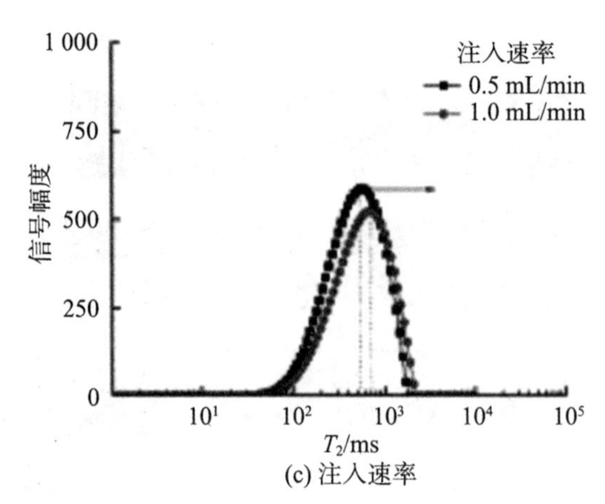

(c) 注入速率

图 6　不同渗透率、B-PPG 浓度和注入速率时，B-PPG 在岩心中稳定流动阶段的 T_2 图谱变化

Fig. 6　NMR T_2 curves of B-PPG during steady flow at different permeabilities, B-PPG concentrations and injection rates

从图 6(b)可以看出，较高 B-PPG 浓度的 T_2 图谱分布范围更偏向于较小值，且信号峰峰顶点对应的 T_2 更小；较高浓度 B-PPG 在岩心中的总信号幅度也更大。浓度越高，单位注入体积的流体中 B-PPG 含量越多，黏弹性颗粒间的交联作用越明显，增强了颗粒封堵高渗通道及液流转向的能力，能进入小尺寸隙中的颗粒更多，因此，测得的信号更强。

从图 6(c)可以看出，低注入速率时的 T_2 图谱分布范围更偏向于较小值，信号峰顶点对应的 T_2 更小，较低流速时岩心中的总信号幅度也更大，表明：在实验的注入流速范围内，较低的流速更有助于 B-PPG 对大孔道进行封堵，液流转向小尺寸孔隙，提高波及系数。

不同渗透率、B-PPG 浓度和注入速率时，流体在岩心中的饱和度变化如图 7 所示。分别比较图 7(a)与图 7(b)～图 7(d)可知，在实验的参数范围内，B-PPG 流体可波及岩心中绝大部分区域。渗透率越大、浓度越高或注入速率越低时，B-PPG 在岩心中的饱和度越高，说明：增大渗透率有助于 B-PPG 颗粒在岩心中的注入性，增大浓度或降低注入速率有利于 B-PPG 堵塞大孔道、液流转向较小尺寸孔隙，以上均可提高 B-PPG 在微观非均质区域的波及系数。

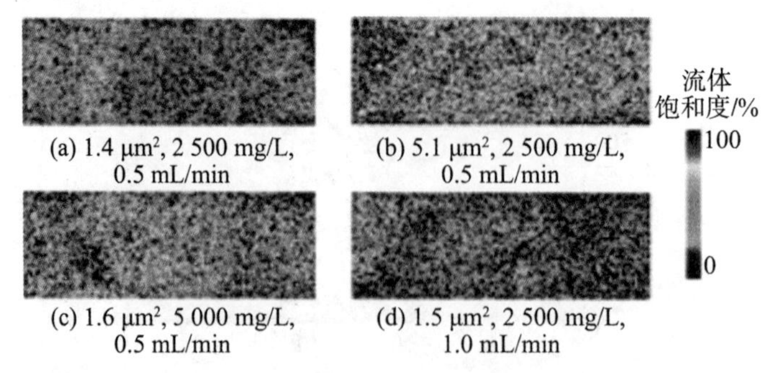

图 7　不同渗透率、B-PPG 浓度和注入速率时，流体在岩心中的饱和度变化

Fig. 7　Changes in flow saturation of B-PPG in cores at different permeabilities, B-PPG concentrations and injection rates

3　结论

通过核磁共振技术与岩心驱替实验相结合的方法系统研究了 B-PPG 在岩心中的流动特征，得出如下结论。

（1）B-PPG 在孔隙介质中以重复堵塞-变形-通过的方式运移。

（2）核磁共振 T_2 图谱明确了 B-PPG 在孔隙介质中的调剖作用。微观上 B-PPG 进入低渗孔隙，宏观上增大驱油剂的波及面积。

（3）核磁共图像清晰呈现出 B-PPG 在岩心中的流动分布状态。B-PPG 在岩心中的驱替前缘均匀推进，没有出现明显的流动通道，能够进入水驱波及不到的孔隙，表现出优异的波及能力。

（4）渗透率、B-PPG 浓度及注入速率对 B-PPG 驱油剂的流动特征具有明显影响。增大渗透率有助于 B-PPG 颗粒在孔隙介质中的注入性；增大浓度或降低注入速率有利于提高 B-PPG 的调剖能力。

参考文献

[1] 张莉.胜利油田聚合物驱油技术经济潜力分析[J].石油勘探与开发，2007，34(1)：79-82.

Zhang Li. Analysis of polymer flooding technology and economic po-tential in Shengli Oilfield [J]. Petroleum Exploration and Develop-ment，2007，34 (1)：79-82.

[2] 吴应川，白宝君，赵化廷，等.影响预交联凝胶颗粒性能的因素分析[J].油气地质与采收率，2005，12 (4)：55-57，86.

Wu Yingchuan，Bai Baojun，Zhao Huating，et al. Analysis of fac-tors affecting the properties of pre-crosslinked gel particles [J]. Pe-troleum Geology and Recovery Efficiency，2005，12 (4)：55-57，86.

[3] Yu L，Sang Q，Dong M. Enhanced oil recovery ability of branched preformed particle gel in heterogeneous reservoirs [J]. Oil & Gas Science and Technology-Revue d'IFP Energies nouvelles，2018，73：65.

[4] 崔晓红.新型非均相复合驱油方法 [J].石油学报，2011，32 (1)：122-126.

Cui Xiaohong. A study on the heterogeneous combination flooding system [J]. Acta Petrolei Sinica，2011，32 (1)：122-126.

[5] 于龙，宫厚健，李亚军，等.非均质油层黏弹性凝胶颗粒提高采收率机理研究[J].科学技术与工程，2014，14 (17)：59-63.

Yu Long，Gong Houjian，Li Yajun，et al. Study on the mechanism of enhanced recovery of viscoelastic gel particles in heterogeneous reservoirs [J]. Science Technology and Engineering，2014，14 (17)：59-63.

[6] Sang Q, Li Y, Yu L, et al. Enhanced oil recovery by branched-preformed particle gel injection in parallel-sandpack models [J]. Fuel,2014,136: 295-306.

[7] Zhou K,Hou J,Sun Q,et al. An efficient LBM-DEM simulation method for suspensions of deformable preformed particle gels[J]. Chemical Engineering Science, 2017,167: 288-296.

[8] 狄勤丰,张景楠,华帅,等.聚合物-弱凝胶调驱核磁共振可视化实验[J].石油勘探与开放,2017,44 (2): 270-274.

Di Qinfeng, Zhang Jingnan, Hua Shuai, et al. Visualization experi-ments on polymer-weak gelprofile control and displacement by NMR technique [J]. Petroleum Exploration and Development,2017, 44 (2): 270-274.

[9] 高辉,程媛,王小军,等.基于核磁共振驱替技术的超低渗透砂岩水驱油微观机理实验[J].地球物理学进展,2015,30 (5): 2157-2163.

Gao Hui,Cheng Yuan,Wang Xiaojun,et al. Experimental of mi-croscopic water displacement mechanism based on NMR displace-ment technology in ultra-low permeability sandstone [J]. Progress in Geophysics,2015,30 (5): 2157-2163.

[10] Cheng Y,Di Q,Gu C,et al. Visualization study on fluid distri-bution and end effects in core flow experiments with low-field MRI method [J]. Journal of Hydrodynamics,2015,27 (2): 187-194.

第四章　化学剂浓度分析

氧化铝包裹硅胶核-壳型色谱填料的制备及正相色谱性能研究

曹绪龙[1]　祝仰文[1]　严　兰[1]　郭　勇[2]　梁晓静*[2]

(1. 中国石化胜利油田分公司地质科学研究院，东营 257015；
2. 中国科学院兰州化学物理研究所西北特色植物资源化学重点实验室，兰州 730000)

摘要：采用层层自组装(LBL-SA)技术制备了新型的氧化铝包裹硅胶(Al_2O_3/SiO_2)核-壳型色谱填料。对该填料进行了光电子能谱表征、扫描电镜表征、比表面孔径表征，结果表明 Al_2O_3 被成功的组装到了 SiO_2 表面。考察了 Al_2O_3/SiO_2 填料在正相色谱条件下对碱性、中性和酸性化合物色谱分离行为并与 SiO_2 填料进行了比较性研究。结果表明，由于氧化铝表面的 Lewis 酸性位点的存在，相比较于 SiO_2 填料而言，Al_2O_3/SiO_2 在分离碱性化合物和中性化合物时具有很大的优势，但是也是由于这些 Lewis 酸性位点，造成酸性化合物在 Al_2O_3/SiO_2 柱上具有很长的保留时间和严重拖尾的峰形，流动相中少量乙酸的加入可以有效地掩盖 Al_2O_3 表面的 Lewis 酸性位点，减弱其对酸性化合物的吸附。

关键词：正相色谱；氧化铝包裹硅胶核-壳型色谱填料；硅胶；氧化铝；路易斯酸性位点

中图分类号：O652.6；0657.7　文献标识码：A

文章编号：1000-0720(2015)02-0134-05

金属氧化物(氧化锆、氧化钛、氧化铝等)具有优异的表面性质，除了被用于固相微萃取填料[1]外，这些金属氧化物很早就已被用作 HPLC 柱填料，并对其色谱性能进行了深入研究。相对于氧化锆、氧化钛而言，氧化铝是两性氧化物，根据流动相的 pH 不同既可以表现出阴离子的交换性能又可以表现出阳离子的交换性能[2]。有关氧化铝的离子色谱性能已有较多研究[3-7]。氧化铝填料除了被用于离子色谱外，还可以被于正相色谱[8-9]。从氧化铝的等电点(pI=7)和硅胶的等电点(pI=3)的对比可以看出，硅胶表面比氧化铝表面的酸性要强[9]。氧化铝的表面化学性质要比硅胶复杂的多，它的表面存在一系列的对被分析物产生保留的作用位点，其中包括 Bröbsted 酸性位点、Lewis 酸性和碱性位点，但是 Bröbsted 酸性位点的数量较少，并且 Lewis 酸性和碱性位点的酸碱性也较弱[8]。在氧化铝表面,未饱和的Al(Ⅲ)表现为Lewis酸性位点,可以和能提供孤对电子的化合物产生Lewis酸碱相互作用,相反地,硅胶表面则不存在这种配体交换性质。在现代HPLC发展的早期,氧化铝就已作为一种性质优异的固定相而被用于正相色谱。但是氧化铝填料至今并没有得到广泛的应用。造成这种现象的主要原因是,像其他金属氧化物一样,氧化铝很难制备成像硅胶一样具有优良物理结构(颗粒大小及分布、孔径大小及分布、比表面积等)的多孔微球。

为了改善金属氧化物的物理结构,敦惠娟等[10]利用LBL-SA技术在色谱用硅胶微球表层组装了多层纳米氧化锆颗粒,制备了ZrO_2/SiO_2核-壳型色谱填料,刘芹等[11]采用相同技术制备了SiO_2/SiO_2核-壳型填料。在此基础上,本研究以聚苯乙烯磺酸钠为自组装介质[12,13],采用层层自组装技术制备出Al_2O_3/SiO_2核-壳型色谱填料,研究Al_2O_3/SiO_2在正相色谱条件下对不同性质化合物的分离性能,并且与硅胶进行比较性研究。

1 实验部分

1.1 仪器与试剂

Agilent 1100 Series高效液相色谱仪;VG ESCALAB 210型光电子能谱仪(美国ThermoFisher公司);ASAP2010型比表面孔度分析仪(美国Micrometritics公司);JSM-6701FE型扫描电子显微镜(日本电子光学);675213-100型高压气动装柱机(北京分析仪器厂)。

异丙醇铝(>98%)和聚苯乙烯磺酸钠(PSS,平均MW 70 000)购于Acros公司;球形硅胶由本实验室制备,平均粒径7 μm,比表面积47.2 m^2/g,孔体积0.36 m^3/g,孔径30.9 nm,使用前酸化,蒸馏水洗涤,120℃真空干燥,色谱表征所用的试剂均为分析纯。

1.2 Al_2O_3/SiO_2核-壳型色谱填料的制备

(1) 参照文献[14,15],以异丙醇铝为原料,以冰乙酸为催化剂,以乙酰丙酮为稳定剂,在乙醇溶液中制备出淡蓝色透明的氧化铝溶胶。

(2) 将10 g硅胶加入到1 mg/mL的PSS水溶液中,超声20 min、水洗、干燥。

(3) 将吸附有PSS的硅胶加入到100 mL纳米氧化铝溶胶液中,超声20 min、搅3 h、水洗、干燥。

(4) 重复上述(1)(2)步骤10次后,将填料放入预先加热到600℃的马福炉内煅烧4 h,即可得到Al_2O_3/SiO_2核-壳型色谱填料。

1.3 色谱条件

Al_2O_3/SiO_2核-壳型色谱填料和SiO_2填料以CCl_4为匀浆液,正己烷为顶替液,在40 MPa压力下装入不锈钢柱(150×4.6 mm I.D.)中。色谱操作均在室温下进行;进样量20 μL;流速1.0 mL/min;检测波长为254 nm;以正己烷和乙醇/正己烷混合液为流动相。分析样品用流动相溶解,色谱柱的死时间(t_0)由溶剂峰确定。

2 结果与讨论

2.1 Al_2O_3/SiO_2 核-壳型色谱填料表征

2.1.1 光电子能谱(XPS)表征

对 Al_2O_3/SiO_2 进行了 XPS 表征来测定其表面的氧化铝的存在及含量。由 XPS 结果可以得到该填料表面的元素含量分别为 O(68.6%),C(9.3%),Si(8.2%),Al(11.9%),由此可以计算出该填料的表面氧化物含量分别为 SiO_2(22.51%),Al_2O_3(17.75%)。

2.1.2 扫描电镜(SEM)表征

对 Al_2O_3/SiO_2 填料的表面形貌采用 SEM 进行了考察(如图 1 所示)。图 1 为单个 Al_2O_3/SiO_2 球的 SEM 图(放大 12 000 倍),可以看出组装后的填料仍然保持着很好的球形结构和非常均匀的表面形貌,说明 Al_2O_3 纳米颗粒在硅胶表面的组装并没有破坏硅胶优异的物理结构。

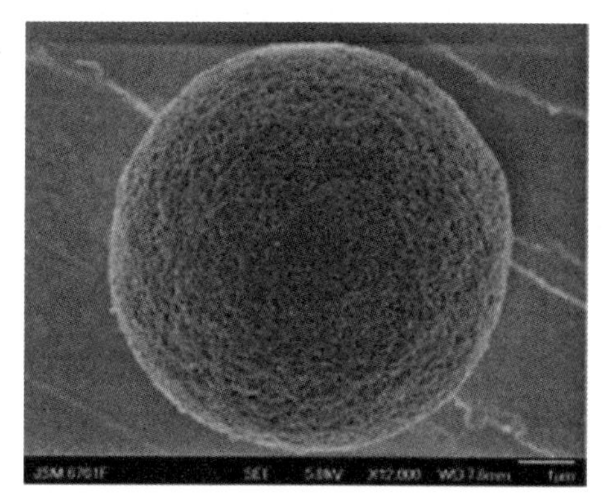

图 1 Al_2O_3/SiO_2 核-壳型色谱填料的 SEM 的图

Fig. 1 The SEM image of Al_2O_3/SiO_2

2.1.3 比表面积、孔体积、孔径表征

采用氮吸附脱附法测定的 Al_2O_3/SiO_2 核-壳型色谱填料的比表面积为 51.1 m^2/g、孔体积为 0.27 cm^3/g 以及孔径为 21.3 nm。相对于作为内核 SiO_2 而言,该核-壳型填料的比表面积增加了 3.9 m^2/g,孔体积减小了 0.09 cm^3/g,孔径减小 9.6 nm。

2.2 对碱性化合物的分离

文献[9,10]指出,在正相色谱条件下,相对于硅胶而言,氧化铝表面表现为碱性,所以碱性化合物可以在氧化铝上得到很好的分离,保留时间短,峰形对称。图 2A 和图 2B 为以乙醇/正己烷(10∶90,V/V)为流动相,四种碱性化合物分别在 Al_2O_3/SiO_2 和 SiO_2 柱上的色谱分离图。表 1 给出了碱性化合物在 Al_2O_3/SiO_2 柱和 SiO_2 柱上的色谱参数。从图 2 和表 1 可以看出,相比较于 SiO_2 柱而言,碱性化合物可以在 Al_2O_3/SiO_2 柱上得到很好的分离并且峰形也更为对称。该现象很好的证明了氧化铝表面的非酸性特点。此外,从图 2 和表 1 还可以看出,相对于 SiO_2 柱而言,邻硝基苯胺、间硝基苯胺、对硝基苯胺在 Al_2O_3/SiO_2 柱上具有更长的保留时间和更大的分离度。这是因为这些苯胺类化合物上的硝基可以吸引苯环上的 π-电子,使得硝基表现出很强的 Lewis 碱性。和在 SiO_2 柱上相比,这些硝基苯胺类化合物不仅可和 Al_2O_3/SiO_2 表面上的羟基产生氢键相互作用，还可以和 Al_2O_3/SiO_2 表面上 Lewis 酸性位点产生 Lewis 酸碱相互作用，两种作用力的叠加使得硝基苯胺类化合物在 Al_2O_3/SiO_2 柱上具有更强的保留和更大的分离度。

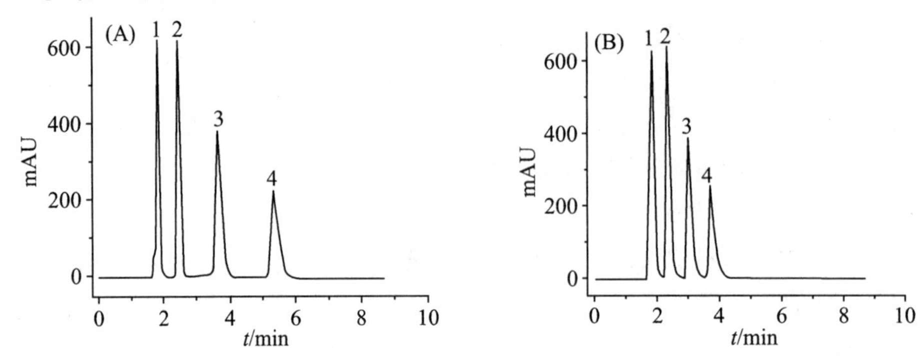

图 2　对碱性化合物的色谱分离图

Fig. 2　Separation of basic compounds

1-N，N一二甲基苯胺，2-邻硝基苯胺，3-间硝基苯胺，4-对硝基苯胺，流动相：*V*（乙醇）：*V*（正己烷）＝10∶90；色谱条件：Al_2O_3/ SiO_2柱(A)；SiO_2柱(B)

表 1　碱性化合物在 Al_2O_3/SiO_2 柱和 SiO_2 柱上的色谱参数

Tab. 1　Chromatographic parameters of basic compounds on Al_2O_3/SiO_2 and SiO_2 columns

化合物	Al_2O_3/SiO_2				SiO_2			
	柱效/(N/m)	拖尾因子(T)	保留因子(k)	分离度(Rs)	柱效/(N/m)	拖尾因子(T)	保留因子(k)	分离度(Rs)
N,N-二甲基苯胺	11 580	1.18	0.096	—	4 720	0.92	0.076	—
邻硝基苯胺	17 726	1.22	0.465	5.0	10 613	1.41	0.372	1.9
间硝基苯胺	11 320	1.30	1.18	2.4	11 853	1.39	0.777	2.6
对硝基苯胺	14 480	1.30	2.203	6.0	11 300	1.38	1.192	2.1

2.3 对中性化合物的分离

图 3 为苯、萘、芴、苯甲醚、荧蒽、屈 6 种中性化合物分别在 Al_2O_3/SiO_2 柱和 SiO_2 柱上的色谱分离图。从图 3A 可以看出 6 种中性化合物可以仅用正己烷为流动相就能在 Al_2O_3/SiO_2 柱得到很好的分离。由于这些中性化合物都具有 π-电子体系，使得这些中性化合物都表现出一定的 Lewis 碱性，因此可以和 Al_2O_3/SiO_2 表面的 Lewis 酸性位点产生 Lewis 酸碱相互作用而使其在填料表面得到一定的保留，并且这些中性化合物的洗脱顺序和它们的 π-电子体系大小相一致，π-电子体系越大，在 Al_2O_3/SiO_2 的保留就越强，反之亦然。从图 3B 可以看出，这些中性化合物在 SiO_2 柱上几乎没有保留，这是因为硅胶填料表面基本不存在 Lewis 酸性位点，没有能和这些中性化合物产生相互作用的位点。和图 3A 相比，苯甲醚在 SiO_2 柱上的出峰顺序发生了变化，这是因为相比较于其他中性化合物，苯甲醚上的氧原子可以和硅胶表面的硅羟基产生氢键相互作用，因此延长了苯甲醚在 SiO_2 柱上的保留时间。通过图 3A 还可以看出这些中性化合物在 Al_2O_3/SiO_2 的保留时间都比

较短，这是因为 Al_2O_3 表面的 Lewis 酸性位点所表现出的 Lewis 酸性较弱，并且对 Al_2O_3/SiO_2 的 XPS 表征结果也表明 Al_2O_3/SiO_2 表面还有一定量的 SiO_2 没有被 Al_2O_3 所覆盖。

2.4 对酸性化合物的分离

考察了苯酚、间苯二酚、间苯三酚 3 种酸性化合物以乙醇/正己烷(10∶90，V/V)为流动相，分别在 Al_2O_3/SiO_2 柱和 SiO_2 柱上的色谱分离(如图 4A 和图 4B 所示)。从图 4A 可以看出，酸性化合物在 Al_2O_3/SiO_2 柱具有很强的吸附，造成酸性化合物的保留时间长，色谱峰形拖尾。相反的，酸性化合物在 SiO_2 柱上则有很好的分离，保留时间短，色谱峰形对称。造成这种现象的原因是，和 SiO_2 填料相比，Al_2O_3 的表面存在很多 Lewis 酸性位点，这些 Lewis 酸性位点对表现为 Lewis 碱性的酸性化合物具有很强的保留。流动相中少量乙酸的加入可以很大程度上改善酸性化合物在 Al_2O_3/SiO_2 柱上的分离(图 4Aa)，这是因为少量乙酸(表现为 Lewis 碱性)的加入可以有效地掩盖 Al_2O_3 表面的 Lewis 酸性位点，减弱其对酸性化合物的吸附。

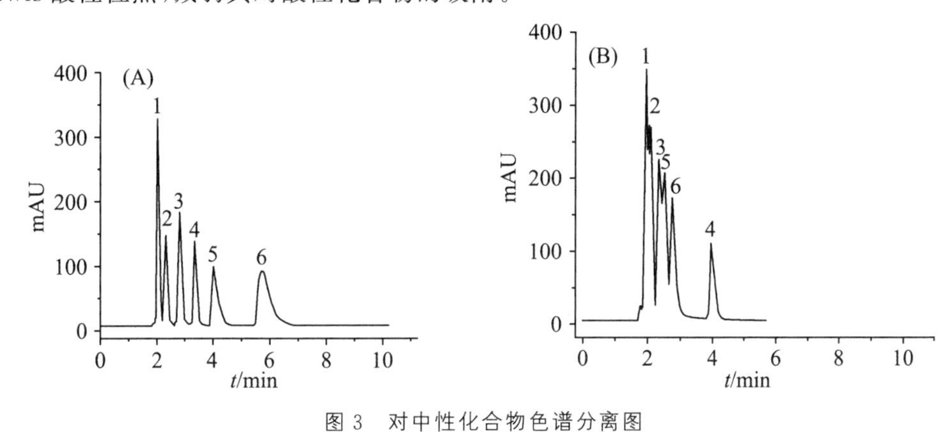

图 3 对中性化合物色谱分离图

Fig. 3 Separation of neutral compounds

1-苯，2-萘，3-芴，4-苯甲醚，5-荧蒽，6-屈；流动相：正己烷；色谱条件：Al_2O_3/SiO_2 柱(A)；SiO_2 柱(B)

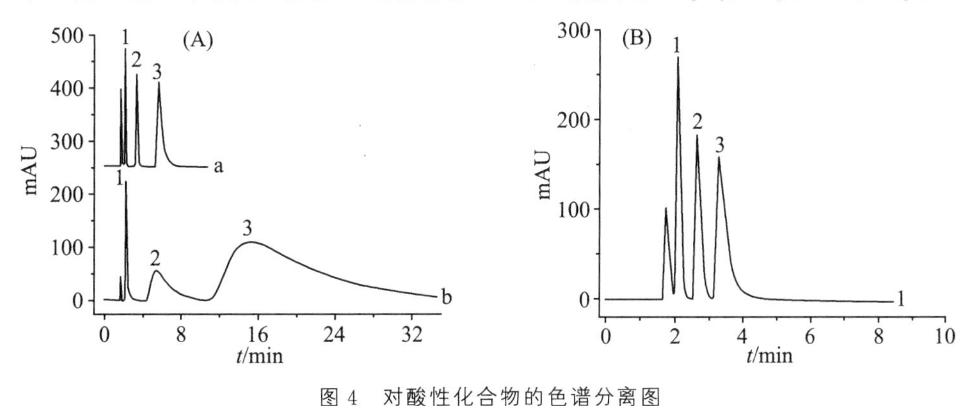

图 4 对酸性化合物的色谱分离图

Fig. 4 Separationofacidiccompounds

1-苯酚，2-间苯二酚，3-间苯三酚；流动相：V(乙醇)∶V(正己烷)=10∶90(a)，

V(乙醇)∶V(正己烷)∶V(乙酸)=10∶90∶0.1(b)；色谱条件：Al_2O_3/SiO_2 柱(A)；SiO_2 柱(B)

3 结论

考察了新型 Al_2O_3/SiO_2 核-壳型色谱填料在正相色谱条件下对碱性、中性和酸性化合物的色谱分离行为并与 SiO_2 填料进行了比较性研究。由于氧化铝表面的 Lewis 酸性位点的存在，相比较于 SiO_2 填料而言，Al_2O_3/SiO_2 核-壳型色谱填料在分离碱性化合物和中性化合物时具有很大的优势，但也是由于这些 Lewis 酸性位点，造成酸性化合物在 Al_2O_3/SiO_2 柱上具有很长的保留时间和严重拖尾的峰形，流动相中竞争 Lewis 碱的加入可以有效地改善酸性化合物在 Al_2O_3/SiO_2 柱上的分离。

参考文献

[1] 隋丽丽，隋崴崴，肖晓杏．分析试验室，2013，32(8)：11.

[2] Blackwell J A，Carr P W．Anal Chem，1992，64(8)：853.

[3] Wirth H J，Eriksson K O，Holt P，et al．J Chromatogr，1993，646(1)：129.

[4] Amphlett C B．Inorganic Ion Exchangers．Elsevier，Amsterdam，1964：84.

[5] Michal J．Inorganic Chromatographic Analysis．New York：Van NostrandReinhole，1970：75.

[6] Fuller M J．Chromatogr Rev，1971，14(1)：45.

[7] Clearfield A．Inorganic Ion Exchange Materials．Boca Raton，FL：CRC Press，1982.

[8] Schmitt G L，Pietrzyk D J．Anal Chem，1985，57(12)：2247.

[9] Kurganov A，Trüdinger U，Isaeva T，et al．Chromatogr，1996：42(3-4)：217.

[10] Dun H J，Zhang W Q，Wei Y，et al．Anal Chem，2004，76：5016.

[11] 刘　芹，张红丽，张志欣，等．分析试验室，2011，30(9)：1.

[12] Liang X J，Chen Q S，Liu X，et al．J Chromatogr A，2008，1182：197.

[13] Liang X J，Wang S，Niu J G，et al．J Chromatogr A，2009，1216：3054.

[14] 黄西朝，祝保林．分析试验室，2012，31(7)：83.

[15] 黄西朝，祝保林．分析试验室，2014，33(3)：357.

Study on the preparation and normal phase chromatographic performance of core-shell type chromatographic packing of alumina coated silica

Cao Xulong[1], Zhu Yangwen[1], Yan Lan[1], Guo Yong[2] and Liang Xiaojing*[2] (1. Geological and Scientific Research Institute of Shengli Oilfield, Sinopec, Dongying 257015; 2. Key Laboratory of Chemistry of Northwestern Plant Resources, Lanzhou Institute of Chemical Physics, Chinese Academy of Sciences, Lanzhou 730000) , Fenxi Shiyanshi, 2015, 34 (2): 134-138.

Abstract: A new stationary phase Al_2O_3/SiO_2 has been prepared by layer-by-layer (LbL) self-assembly technique and characterized by XPS, SEM and surface analysis. Chromatographic properties of SiO_2 and a new core-shell stationary phase Al_2O_3/SiO_2 have been investigated in normal phase mode in the separation of basic, neutral and acidic compounds. In contrast to SiO_2, the chromatographic behavior of the Al_2O_3/SiO_2 revealed the basic property of alumina. Therefore, the separation of basic compounds on Al_2O_3/SiO_2 seems very promising with more symmetrical peaks. Compared with SiO_2, Al_2O_3/SiO_2 also has a distinct advantage in the separation of neutral compounds, but due to the presence of Lewis acid sites on the surface of Al_2O_3, Lewis bases such as hydroxybenzenes adsorb very strongly on it. Lewis acidic sites on Al_2O_3 can be effectively concealed by added little of acetic acid in the mobile phase.

Keywords: Normal phase chromatography; Al_2O_3/SiO_2 core-shell chromatographic support; Silica; Alumina; Lewis acidic site

王丽娟

复合驱注采液中活性剂PS浓度的高效液相色谱分析方法研究

曹绪龙[1]，隋希华[1]，施晓乐[1]，江小芳[1]，王贺振[1]，蒋生祥[2]，刘　霞[2]

(1. 胜利石油管理局地质科学研究院，山东东营　257015；

2. 中国科学院兰州化学物理研究所，甘肃兰州　730000)

摘要：合成了强阴离子交换及硅胶基质ODS键合高效液相色谱(HPLC)柱填料。针对胜利油田复合(碱＋表面活性剂＋聚合物)驱油工程实施的需要，建立了适合于注采液中表面活性剂PS的HPLC分析方法。该方法对PS标准样品的最小检出限为0.4 mg/L，线性范围为50 mg/L～1 000 mg/L，回收率为95.7%～99.8%。实践证明，该方法的建立为油田的复合驱油配方设计、注入方案调整、产品质量监控及驱油机理研究等提供了有力的技术支持。

关键词：高效液相色谱法；表面活性剂PS；复合驱油；注入-采出液；油田

中图分类号：O658　文献标识码：A　文章编号：1000-8713(2001)02-0164-03

1　前言

在胜利油田孤岛西区三元复合驱油配方中，PS作为辅助活性剂，起着降低活性剂BES在油藏表面的无效吸附、提高驱油配方效率的重要作用。在该区块三元复合驱油方案实施过程中，无论是驱油配方的设计、活性剂的吸附损耗测定、驱油剂之间的协同效应研究，还是现场驱油试验研究等，均需对活性剂PS的浓度进行准确快速的分析。胜利油田孤岛西区注入的活性剂PS为易溶于水的棕黑色固体。因商业保密原因，其具体成分和理化性质等信息无法确定，这给其分析方法的建立带来了很大的困难。为此本文在大量实验基础上，建立并完善了适用于注采液中PS浓度分析的双柱HPLC分析方法。实验表明，该方法灵敏度高、准确性好、所得数据具有良好的重现性，对PS商品及其在油田注采液中的浓度可进行准确快速的分析，完全能满足现场应用的要求。国内外尚未见此类研究的报道。

2　实验部分

仪器：HPLC 系统由 Waters 515 泵，Waters 2487 检测器，Waters Millennium[32] 色谱管理系统组成。色谱柱为 2 根 4.6 mm i.d. ×50 mm 不锈钢管，填料为 ODS-C_{18} 硅胶键合相和硅基键合强阴离子交换剂。

试剂：Na_2CO_3，NaCl，$MgCl_2$，$CaCl_2$，CH_3OH，CH_3-CHOH-CH_3，CH_3I 均为分析纯试剂；水经二次蒸馏处理；中科院兰州化学物理研究所提供的堆积硅胶（粒径 5 μm，比表面积 110 m^2/g，平均孔径 6.70 nm）及十八烷基甲基二氯硅烷；山东盖县化工厂提供的胺丙基三乙氧基硅烷（分析纯）；胜利油田提供的 PS 标样。

3　结果与讨论

市售阴离子交换填料中，树脂型填料在流动相中溶胀严重，渗透性及耐压性差；硅胶基质填料交换容量小，难以对矿场浓度大、杂质含量高的 PS 样品进行准确分析。为此本文根据 PS 样品化学结构（可能含有弱离子性及非离子性化合物）未知、矿场样品组成复杂等特点，参考文献[1-3]合成了专用的固定相。

3.1 色谱柱填料的合成

3.1.1 ODS 柱填料

（1）键合反应　十八烷基甲基二氯硅烷与硅胶表面的羟基发生化学反应，可得化学键合相：

$$\text{(-OH)}_2 + C_{18}H_{37}(CH_3)SiCl_2 \longrightarrow \text{(-O)}_2Si(CH_3)(C_{18}H_{37})$$

（2）钝化反应　经过化学键合后，由于空间位阻等原因，仍有少量的羟基残留在硅球表面，需进行钝化反应：

$$\text{-OH},\ \text{(-O)}_2Si(CH_3)(C_{18}H_{37}),\ \text{-OH} + NH[Si(CH_3)_3]_2 \longrightarrow \text{-O-}Si(CH_3)_3,\ \text{(-O)}_2Si(CH_3)(C_{18}H_{37}),\ \text{-O-}Si(CH_3)_3$$

3.1.2 强阴离子交换填料

（1）键合反应　氨丙基三乙氧基硅烷与硅球表面上的羟基发生化学反应，可得到化学键合相：

$$\text{(-OH)}_2 + (C_2H_5O)SiC_3H_6NH_2 \longrightarrow \text{(-O)}_2Si(OC_2H_5)(C_3H_6NH_2)$$

(2) 钝化反应：

$$\text{—OH},\ \text{—O},\ OC_2H_5,\ Si,\ \text{—O},\ C_3H_6NH_2,\ \text{—OH} + NH[Si(CH_3)_3]_2 \longrightarrow \text{—O—Si(CH}_3)_3,\ \text{—O},\ OC_2H_5,\ Si,\ \text{—O},\ C_3H_6NH_2,\ \text{—O—Si(CH}_3)_3$$

(3) 季铵化反应　上述的键合相与碘甲烷在异丙醇溶剂中进行季铵化反应，得到了强阴离子交换填料：

$$\text{—O—Si(CH}_3)_3,\ \text{—O},\ OC_2H_5,\ Si,\ \text{—O},\ C_3H_6NH_2,\ \text{—O—Si(CH}_3)_3 \quad CH_3I \longrightarrow \text{—O—Si(CH}_3)_3,\ \text{—O},\ OC_2H_5,\ Si,\ \text{—O},\ CH_2\text{—}CH_2\text{—}CH_2\text{—}N^+(CH_3)_2I^-,\ \text{—O—Si(CH}_3)_3$$

3.2 色谱条件的选择

3.2.1 检测波长的选择

对 PS 标样进行紫外扫描发现，PS 在 255 nm～274 nm 之间有一个吸收稳定区，且这一波长下的共存杂质 BES、聚丙烯酰胺和无机盐等吸收很弱，故选定检测波长为 270 nm。

3.2.2 色谱柱的选择

(1) 强阳离子交换柱。

因无法得知活性剂 PS 具体的化学成分，我们在最初的实验中采用磺酸型的强阳离子交换柱对 PS 的分离进行探索，发现无论流动相中甲醇与水的比例怎样改变，或是否含盐，PS 在强阳离子交换柱上几乎不滞留。这一结果表明，PS 不属于阳离子型表面活性剂。

(2) 强阴离子交换柱。

采用合成的强阴离子交换填料对 PS 进行分析。用 V(甲醇)：V(水)＝70：30 的溶液作流动相，当流速为 1.0 mL/min 时，PS 出两个色谱峰，其中第 1 个峰不滞留(图 1)。

固定其他实验条件(270 nm 检测，流速 1.0 mL/min)，改变流动相中甲醇与水的配比时，第 1 个峰滞留值几乎不变化，第 2 个峰的滞留值随流动相中甲醇的比例增大而减小，结果见图 2。若保持流动相中甲醇与水的体积比为 70：30，当加入 NaCl 后，第 1 个峰的滞留值不随 NaCl 的浓度大小而变化，第 2 个峰的保留时间将随 NaCl 浓度的增大而减小(见图 3)。

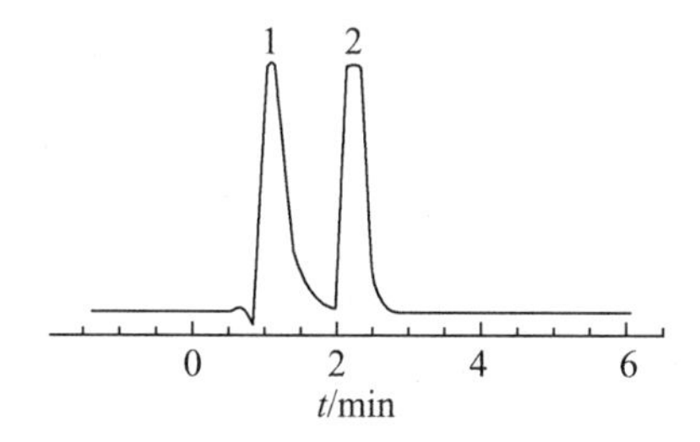

图 1　PS 在强阴离子交换柱上的色谱图

Fig. 1　Chromatogram of PS on ion exchanger

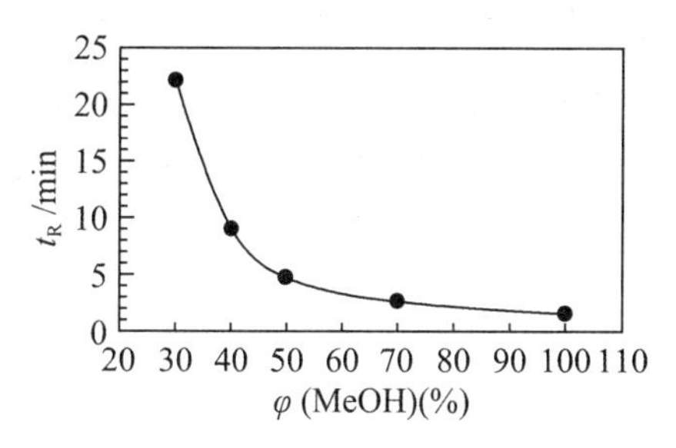

图 2　甲醇的体积分数对图 1 中第 2 个峰保留值的影响

Fig. 2　Effect of methanol volume fraction on t_R of second peak in Fig. 1

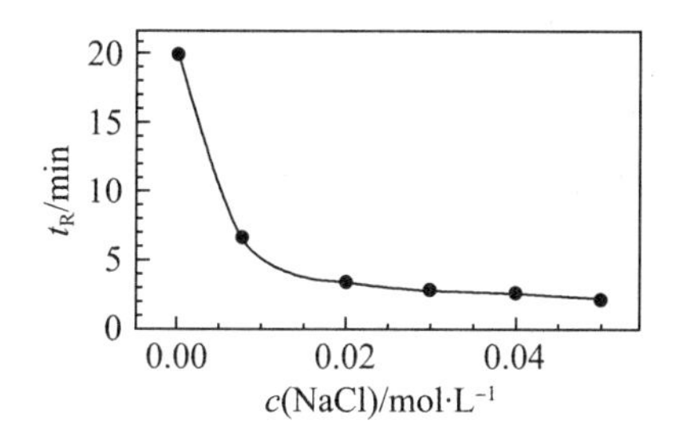

图 3　NaCl 浓度对图 1 中第 2 个峰滞留的影响

Fig. 3　Effect of NaCl conc. on t_R of second peak in Fig. 1

(3) ODS 硅基键合柱。

PS 在阳离子交换柱上不滞留，表明它不属于阳离子型化合物，其在阴离子交换柱上滞留的部分应为阴离子化合物，不滞留的组分应属非离子型化合物。我们采用 V(甲醇)：V(水)=70：30 的溶液为流动相，在流速 1.0 mL/min，UV 270 nm 条件下检测，在自己合成的 ODS 柱上对驱油剂 PS 进行分离后得到两个峰(见图 4)，第 1 个峰不滞留，第 2 个峰有滞留，且滞留值随流动相中甲醇比例的减小而增大，具有明显的反相色谱特征(图 5)。第 1 个峰的滞留随甲醇体积分数的增加而稍有增大，可能是由于甲醇比例增大导致离子型的化合物在流动相中溶解度下降所致。

(4) ODS 柱-SAX 柱联用。

上述单柱分离驱油剂 PS 的色谱结果表明：PS 由非离子型组分和阴离子型组分组成，且上述分析条件下，无法对其进行定性定量分析。

我们尝试采用反相 ODS 柱和强阴离子交换 SAX 柱串联对 PS 进行色谱分析,收到了良好的效果:在 ODS 柱上滞留的组分在 SAX 柱上不滞留,在 ODS 柱上不滞留的组分在 SAX 柱上滞留,从而达到活性剂 PS 在双柱串联体系上被滞留的目的。结果表明,选择流动相为 V(甲醇)∶V(水)=50∶50 的溶液,流速 1.0 mL/min,紫外检测波长 270 nm,采用 ODS 柱和 SAX 柱串联方式,PS 样品被压缩为一 个对称的色谱峰,这对定量工作和现场应用无疑具 有重要意义(图 6)。

3.3 共存物的干扰情况

进行现场分析时,PS 的共存物有表面活性剂 BES、聚丙烯酰胺、Na_2CO_3、无机盐和微量油。其中微量油通过萃取和过滤可去除;一部分共存物在所选的检测波长 270 nm 处无响应;其他共存杂质与表面活性剂 PS 获得了完全的分离,对定量分析不产生干扰。实验表明在进样量为 10 μL 时,共存的无机盐不会对 PS 保留值产生显著影响(见图 7)。

3.4 线性与回收率

以厂家提供的 PS 作标准样品(含量视为 100%),配制一系列不同浓度的标准溶液进行色谱分析,以峰高定量。用 2 g/L 的 BES,12 g/L 的 Na_2CO_3,0.3 g/L 的聚丙烯酰胺和矿化度为 7 g/L 的水与不同量的 PS 配制成一系列样品进行测定,对建立的 PS 分析方法的线性和回收率进行了考察。实验表明,当 PS 的质量浓度为 50~1 000 mg/L 时,以峰高(mm)为 Y,PS 的质量浓度(mg/L)为 X,则 PS 的线性方程 $Y=0.030\,6X-0.517\,3$,$r^2=0.999\,8$。方法的线性范围宽,回收率达 95.7%~99.8%,具有较好的可靠性(见表 1)

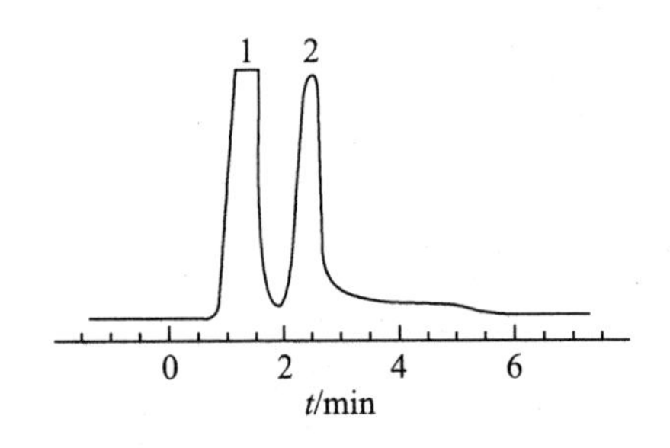

图 4 PS 在 ODS 柱上的色谱图

Fig. 4 Chromatogram of PS on ODS packing

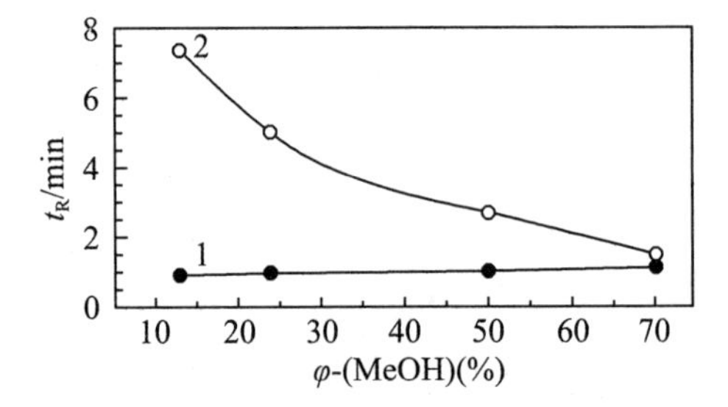

图 5 甲醇浓度对 PS 滞留的影响

Fig. 5 Effect of methanol conc. on t_R of PS

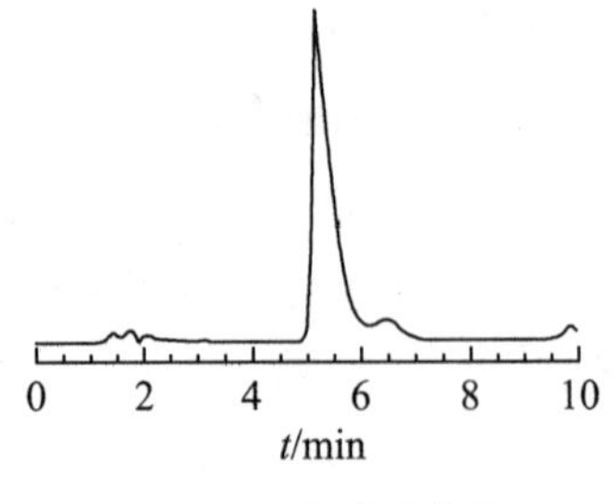

图 6 PS 色谱分离图

Fig. 6 Chromatogram of PS the peak 1 in Fig. 4;2. the peak 2 in Fig. 4.

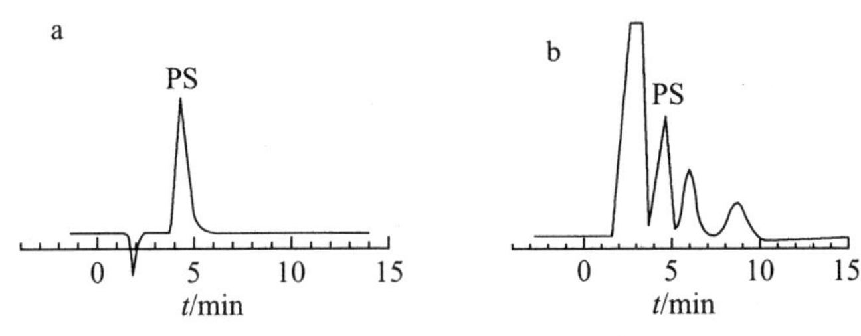

图 7 PS与杂质共存时的色谱图

Fig. 7 Chromatograms of PS with impuritiesa. 1 g/L PS;

b. 1 g/L PS+2 g/L BES+12 g/L Na_2CO_3+0.3/L HPAM+7 g/L(TDS) water.

表 1 PS回收率测定结果

Table1 Testing results of recovery rate of PS

No.	Added(mg/L)	Found(mg/L)	Recovery(%)
1	100	95.7	95.7
2	500	498.3	99.7
3	1 000	998.3	99.8

3.5 建立的方法在实际中的应用

目前胜利油田年注入PS近300 t,有多条供货渠道。采用本文建立的方法,对产品一律按"到厂家抽检-合格后进货-现场注入地层前抽检-合格后注入"的程序检测,由色谱峰数目、峰高、保留时间确定其浓度和质量,任一环节不合格均不能进货,为胜利油田三次采油工程提供了有力的技术保障。

采用本方法对注入、采出液中的PS浓度进行分析,可为其在地层中的吸附损耗及驱油机理研究提供准确的佐证,对驱油剂在地层大孔道中的窜流现象进行准确验证。结合室内驱油试验和表面张力情况,对复合驱油配方中的PS浓度进行分析,可为复合驱油配方及时提供数据和依据。

参考文献

[1] Engelhardt H, O rth P. J Liq Chromatogr, 1987, 10(8&9):1999-2022.

[2] Majors R E. J Chromatogr Sci, 1974, 12:767-778.

[3] Englhardt H, Czok M, Schultz R, et al. J Chromatogr, 1988, 458:79-92.

Study on analysis of concentration of surfactant PS in injected-produced liquor used in ASP flooding with HPLC

Cao Xulong[1], Sui Xihua[1], Shi Xiaole[1], Jiang Xiaofang[1], Wang Hezhen[1], Jiang Shengxiang[2], Liu Xia[2]

(1. Research Institute of Geological Science, Shengli Petroleum Administrative Bureau, Dongying 257015, China;
2. Lanzhou Institute of Chemical Physics, The Chinese Academy of Sciences, Lanzhou 730000, China)

Abstract: Ion exchanger and ODS bonded silica gel for HPLC packing were prepared. An HPLC method for the analysis of surfactant PS concentration in injected-produced liquor has been established to meet the need of A SP flooding developed in Gudao West Block of Shengli Oilfield, with a minimum detectable limit of 0.4 mg/L, a linear range of 50 mg/L～1 000 mg/L and recoveries of 95.7%～99.8%. This method has provided great technical support in a variety of fields to the design of the prescription for ASP flooding, the regulation of injection measurement, the quality control of products and the study on the mechanism of oil flooding etc.

Key words: high performance liquid chromatography; surfactant PS; ASP flooding; injected-produced liquor; Oilfield

王丽娟

离子交换色谱测定原油中的单双石油磺酸盐

赵 亮[2]，曹绪龙[3]，王红艳[3]，刘 霞[1]，蒋生祥[1]

（1. 中国科学院兰州化学物理研究所 甘肃省天然药物重点实验室，甘肃兰州 730000；
2. 中国科学院研究生院，北京 100039；
3. 中国石化胜利油田有限公司地质科学研究院，山东东营 257015）

摘要：建立了一种新的离子交换高效液相色谱在线测定原油中单双石油磺酸盐的方法，通过使用六通阀切换，在一根强阴离子交换柱上完成在线纯化和分离。原油样品经过二氯甲烷-甲醇（体积比 60∶40）溶剂稀释后进入 HPLC 测定。原油中单石油磺酸盐的检出限为 10.0 μg/g，双石油磺酸盐的检出限为 12.5 μg/g。2-萘磺酸钠和萘-1，5-二磺酸钠分别在 4.8～120.0 mg/L 和 4.0～100.1 mg/L 范围内有良好的线性。该方法准确、重复性好，可以满足原油中石油磺酸盐的分离测定。

关键词：离子交换色谱；单石油磺酸盐；双石油磺酸盐；在线纯化；原油

中图分类号：O657.75；TQ423.34 文献标识码：A

文章编号：1004-4957(2007)04-0496-04

Determination of petroleum monosulfonates and petroleum disulfonates in crude oil by anion-exchange chromatography

Zhao Liang[1,2], Cao Xulong[3], Wang Hongyan[3], Liu Xia[1], Jiang Shengxiang[1]

(1. Key Laboratory for Natural Medicine of Gansu Province, Lanzhou Institute of Chemical Physics, the Chinese Academy of Sciences, Lanzhou 730000, China;
2. The Graduate School of Chinese Academy of Sciences, Beijing 100039, China;
3. Geological and Scientific Research Institute Shengli Oil field, SINOPEC, Dongying 257015, China)

Abstract: A method was developed for the determination of petroleum monosulfonates

(PMS) and disul-fonates (PDS) in crude oil by ion-exchange chromatography. The analytical procedure consisted of on-line purification and anion-exchange separation using a SAX column connected with a six-ports witching valve. The crude oil sample was detected by HPLC after diluting with dichloromethane-methanol(60∶40 by volume). The limits of quantification for PMS and PDS in crude oil were 10.0 μg/g and 12.5 μg/g, respectively. The linear ranges of the calibration curves for sodium 2-naphthalene sulfonate and sodium 1, 5-naphthalene disulfonate in crude oil were 4.8～120.0 mg/L and 4.0～100.1 mg/L, respectively. This method has been successfully applied to the determination of PMS and PDS incrude oil samples.

Keywords: Anion-exchange chromatography; Petroleummonosulfonates; Petroleum disulfonates; On-line purification; Crude oil

石油磺酸盐是几种驱油剂中比较重要的一种，广泛应用于油田的驱油生产中。为了详细了解石油磺酸盐驱油剂在地下的地层层析过程以及考察单石油磺酸盐和双石油磺酸盐的驱油效果，需要对地下采出水、原油、岩芯中的单双石油磺酸盐分别进行测定。长期以来，原油中石油磺酸盐的测定是一大难点，国内外对原油中石油磺酸盐的测定一直没有较为简便的方法。大多数情况下，采用先将石油磺酸盐从大量原油中萃取出来，采用化学滴定法测定原油中石油磺酸盐含量。然而，该法操作繁琐、准确度不高，在原油中石油磺酸盐含量较低的情况下，测定误差较大。同时也不能将单双石油磺酸盐分别测定。目前，已经报道的研究工作有水样和微乳液中的单双石油磺酸盐含量测定的高效液相色谱分析方法[1-5]、质量法[6]、电导法[7]、膜渗析分离-重量法[8]、分步滴定法[9]、红外光谱法[10]。本文建 立了原油中微量单双石油磺酸盐离子交换高效液相色谱在线分离测定方法，利用石油磺酸盐分子带负电基团的性质，在强阴离子交换柱上将其在线纯化后分离，既能有效消除原油的干扰，又能同时分离检测单石油磺酸盐和双石油磺酸盐，该法操作简便、测定准确，可以用于实际样品分析。

1 实验部分

1.1 仪器与试剂

1 100 系列高效液相色谱仪(美国安捷伦公司)；LC-10 ATvp 高效液相色谱泵(岛津公司)；HW-2000 型 2.01 版色谱工作站(南京千谱软件公司)；六通切换阀 K502(中国科学院上海原子核研究所科学仪器厂)；色谱分析柱 SAX50×4.6 mm i. d.（兰州化学物理研究所分离分析研究室合成装填，与杜邦公司商品化 Zorbax SAX 色谱柱相同）。2-萘磺酸钠(98%)和萘-1,5-二磺酸钠(98.5%)对照品购于上海试剂三厂；二氯甲烷、甲醇均为分析纯试剂(天津试剂二厂)，经重蒸后使用；磷酸二氢钠为分析纯试剂(北京化学试剂公司)，分析用水为去离子水。

1.2 在线纯化高效液相色谱装置

在线纯化高效液相色谱装置如图1所示。图中虚线箭头所示流路为原油样品在线纯化过程,此时,原油中石油磺酸盐被保留在分析柱上,干扰物被洗脱出来。随后将切换阀转动至粗体箭头所示流路,完成单双石油磺酸盐的分离与测定。

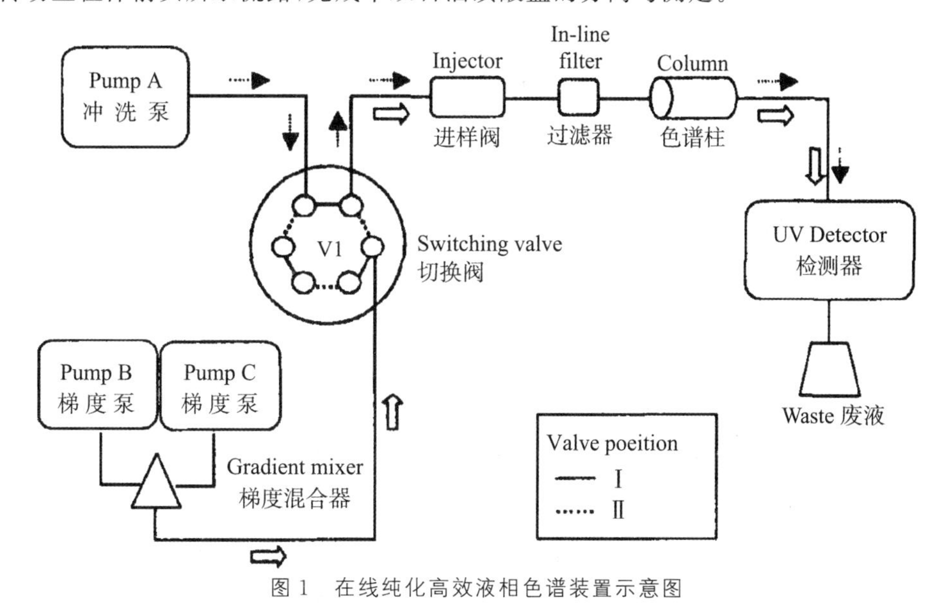

图1　在线纯化高效液相色谱装置示意图

Fig. 1 Schematic diagram of the on-line purification HPLC system

箭头所示由阀切换控制的流路:虚线箭头表示纯化流路,阀切换至位置Ⅰ;

粗体箭头表示分析流路,阀切换至位置Ⅱ(the arrows indicate the direction of the flow depending on the valve position: dashed line, V1=Ⅰ(clean-up), bold line V1=Ⅱ(analysis))

1.3 色谱条件

色谱柱:强阴离子交换柱 SAX 150 mm×4.6 mm i.d.;流动相:A. 二氯甲烷;B. 甲醇-水(体积比60∶40);C. 甲醇-0.3 mol/L 磷酸二氢钠水溶液(体积比60∶40)。样品进入系统以后,以梯度洗脱方式进行:0~10 min, 100% A; 10~11 min100% B; 11~20 min, 100% B;20~26 min, 100% C。流速为:1 mL/min 检测波长为278 nm。

1.4 标准溶液与样品制备

1.4.1 标准溶液

由于石油磺酸盐是由原油馏分经磺化而制成,是一种成分相当复杂的混合物,没有成分完全符合的标准品用于定量分析。文献[4]中曾介绍用石油磺酸盐分离纯化制成标准品,但该方法所制得的单双石油磺酸盐作为标准品纯度不高,且其纯度仍然需要2-萘磺酸钠和萘-1,5-二磺酸钠标定。国内外目前也没有好的方法得到石油磺酸盐标准品。所以,本实验直接采用比较容易得到的2-萘磺酸钠为对照品测定单石油磺酸盐,用萘-1,5-二磺酸钠为对照品测定双石油磺酸盐。

准确称量对照品2-萘磺酸钠(含量98%)30.6 mg 和萘-1,5-二磺酸钠(含量98.5%)

25.4 mg 用甲醇溶于 25 mL 容量瓶中,定容至刻度并用甲醇依次稀释成系列浓度,备用。于 5 个 10 mL 容量瓶中,分别称取空白原油 1.0 g,加入二元溶解溶剂二氯甲烷-甲醇(体积比 60∶40)7 mL,超声溶解 5 min 待原油完全溶解后,用移液管吸取上述系列浓度的标准溶液 1 mL,分别加入 5 个溶解原油的容量瓶中,超声 5 min,定容至刻度,配制成质量浓度为 4.8～120.0 mg/L(2-萘磺酸钠)和 4.0～100.1 mg/L(萘-1,5-二磺酸钠)的标准溶液,用 0.45 μm、φ13 mm 尼龙注射过滤膜过滤,进样。

1.4.2 样品

称量 1.0 g 待测原油样品于 10 mL 容量瓶中,用 7 mL 二氯甲烷-甲醇(体积比 60∶40)溶解,超声溶解 5 min,定容至刻度。用 0.45 μm、φ13 mm 尼龙注射过滤膜过滤,滤液进样分析。

2 结果与讨论

2.1 色谱柱选择

石油磺酸盐是由石油炼油厂的芳烃馏份油经磺化反应而成,其成分十分复杂,不仅分子中芳香骨架不相同,而且磺酸基团的数目也不相同,有单、双磺酸盐。另外,原油也是成分非常复杂的混合物,存在大量的干扰物质。利用石油磺酸盐分子中带负电的磺酸根离子性质,本实验选用了强阴离子交换柱作为离子交换色谱分析柱,使石油磺酸盐被吸附在强阴离子交换柱而原油中的干扰物不被吸附首先洗脱,通过切换阀改变流动相组成,在强阴离子交换柱上将石油磺酸盐进一步分离成单磺酸盐和双磺酸盐。实验表明,结果良好。

2.2 色谱条件优化

2.2.1 流动相 A 的选择

由于原油中石油磺酸盐含量较少,为了保证检测到较低浓度的石油磺酸盐,待测原油样品应尽可能减小稀释倍数,致使大量原油进入分析色谱柱,因而需要选择一种对原油有良好洗脱能力的溶剂,作为洗脱原油的流动相。经过实验选定二氯甲烷为流动相 A,不仅对滞留柱上的原油洗脱完全,而且洗脱时间短。

2.2.2 甲醇含量的影响

以胜利石油磺酸盐为测定样品,考察了流动相中甲醇含量(50%～80%)对单双石油磺酸盐的分离度(Rs)的影响,实验结果表明,流动相中甲醇含量为 60%比较适宜。

2.2.3 磷酸二氢钠缓冲液浓度的影响

在甲醇含量为 60%的条件下,考察了流动相中磷酸二氢钠的浓度(0.1～1.0 mol/L)对单双石油磺酸盐分离度的影响,结果表明磷酸二氢钠缓冲溶液的浓度用 0.3 mol/L 为佳。

2.2.4 检测波长的选择

待测样品中可能存在芳香羧酸盐,对石油磺酸盐的分析有一定的干扰,为了最大限度地避免这种干扰,考察了石油磺酸盐、芳香羧酸盐、对照品和原油的紫外吸收光谱,结果发现,选择检测波长为 278 nm 可以有效地消除芳香羧酸盐的干扰。

2.3 标准曲线

取“1.4”项下的标准原油溶液，每个浓度测定3次，取平均峰面积对标准品质量浓度作图，得到标准曲线方程。2-萘-磺酸钠在4.8～120.0 mg/L范围内有良好的线性，回归方程为：$A=3.0\times10^{3}\rho+5.0\times10^{4}$，r=0.997 4；萘-1,5-二磺酸钠在4.0～100.1 mg/L范围内有良好的线性，回归方程为：$A=3.4\times10^{3}\rho+9.5\times10^{3}$，r=0.998 1。

2.4 精密度

取对照品标准原油溶液进样测定，1天内重复测定5次，以峰面积计算日内精密度，2-萘磺酸钠、萘-1,5-二磺酸钠的相对标准偏差(RSD)为3.1%，萘-1,5-二磺酸钠相对标准偏差为3.9%。对上述样品在6 d内分别测定，日间的相对标准偏差分别为3.4%和4.6%。

2.5 回收率与检出限

取已知浓度样品溶液，分别加入一定量的2-萘-磺酸钠和萘-1,5-二磺酸钠对照品，进行测定并计算回收率(n=5)，2-萘-磺酸钠回收率分别为96%，98%，101%，萘-1,5-二磺酸钠回收率分别为96%，97%，96%。将已知含量胜利石油磺酸盐按“1.4”项制备方法配制成原油溶液，测得原油中单石油磺酸盐的检出限为10.0 μg/g，双石油磺酸盐的检出限为12.5 μg/g(S/N=3)。

2.6 对照品适用性验证

取胜利工业用石油磺酸盐产品(SLPS-20031017)依照样品制备项配制成固形物质量浓度为16.6、41.6、416.0、1 660.0 mg/L的溶液，采用文献[4]的高效液相色谱方法用2-萘磺酸钠和萘-1,5磺酸钠为对照品进行分析测定，分别计算单石油磺酸盐和双石油磺酸盐在石油磺酸盐产品(SLPS-20031017)中的百分含量，从而计算得到胜利石油磺酸盐产品中总的石油磺酸盐的含量。同时，以经典滴定方法测定4个不同浓度样品中的总石油磺酸盐百分含量。测定结果见表1。比较两种方法测定的结果显示，采用文献[4]的高效液相色谱方法用2-萘磺酸钠和萘-1,5-二磺酸钠作为对照品测定的结果与经典滴定法基本一致，说明在无法得到石油磺酸盐标准品的情况下，采用2-萘-磺酸钠和萘-1,5-二磺酸钠作为对照品来定量测定石油磺酸盐是可行的。

表1 经典滴定方法和液相色谱方法测定石油磺酸盐含量

Table1 Determination results of petroleum sulfonates(PS) by conventional titration and HPLC methods

Sample	Titration method		HPLC method	
$\rho/(\mathrm{mg\cdot L^{-1}})$	PS content w/%	Average w/%	PS content w/%	Average w/%
16.6	38.0	38.7	34.4	34.3
41.6	41.4		33.0	
416.0	36.0		34.5	
1 660.0	39.2		35.4	

2.7 原油样品分

按照本文建立的分析方法，对含石油磺酸盐的原油样品进行测定，色谱结果见图 2。A 为空白原油色谱图；B 为含 479.8 μg/g 2-萘-磺酸钠和 400.3 μg/g 萘-1,5-二磺酸钠的对照品原油色谱图；C 为含有 106.5 μg/g 单石油磺酸盐和 89.3 μg/g 双石油磺酸盐原油样品色谱图。

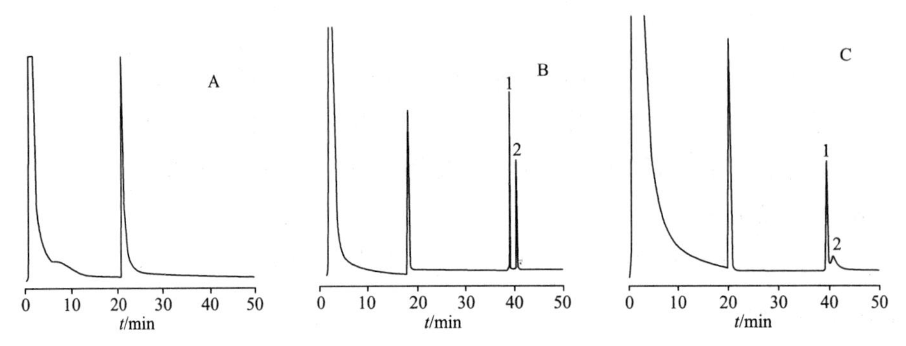

图 2 空白原油以及对照品和石油磺酸盐在原油中的色谱分离图

Fig. 2 Chromatogram of blank crude oil, crude oil spiked with naphthalene sulfonate, and petroleum sulfonates in crude oil samples A. 空白原油(blank crude oil); B. 对照品原油(spiked sample of naphthalene sulfonate, peak 1. 479.8 μg/g sodium 2-naphthalene sulfonate; peak 2. 400.3 μg/g sodium 1, 5-naphthalene disulfonate); C. 石油磺酸盐(petroleum sulfonates in crude oil sample, peak 1. 106.5 μg/g PMS(petroleum monosulfonates); peak 2. 89.3 μg/g PDS(petroleum disulfonates))

参考文献

[1] BEAR G R, LAWLEY C W, RIDDLE RM. Separation of sulfonate and carboxylate mixtures by ion-exchange high-per-Formance liquid chromatography[J]. J Chromatogr, A, 1984, 302:65-78.

[2] BEARGR. Universal detection and quantitation of surfactants by high-performance liquid chromatography by means of the Evaporative light-scatering detector[J]. J Chromatogr, A, 1988, 459:91-107.

[3] BEARG R. Surfactant characterization by reversed-phase ion pair chromatography[J]. J Chromatogr, A, 1986, 371:387-402.

[4] 刘友邦，赵亮，陈立仁，等. 高效液相色谱法快速分析驱油用石油磺酸盐[J]. 油田化学，1987, 4(4):305-311.

[5] 蒋生祥，陈立仁，赵明方，等. 多维高效液相色谱法分析石油中的微量石油磺酸盐[J]. 分析化学，1992, 20(6):677-679.

[6] 蒋怀远，饶福焕，蒋宝源. 驱油用石油磺酸盐成分分析[J]. 油田化学，1985, 2(1):75-82.

[7] 陆彬,李之平.电导法测定石油磺酸盐的混合临界胶束浓度[J].油田化学,1991,8(2):172-174.

[8] 史俊,李谦定,董秀丽.克拉玛依炼油厂石油磺酸盐组成分析及表面活性研究[J].石油与天然气化工,2000,29(5):247-249.

[9] 吴一慧,黄宏度,王尤富,等.混合溶液中石油羧酸盐和磺酸盐含量的测定[J].油田化学,2000,17(3):281-284.

[10] 朱莉,王利平,秦昉.红外光谱法测定石油磺酸盐含量[J].无锡轻工业大学学报,1997,16(2):76-78.

王丽娟

第五章　驱油实验与数值模拟

孤东馆 5^{2+3} 层油藏三元复合驱油体系双截锥体模型驱替试验

曹绪龙　王宝瑜　李可彬　刘异男　袁是高

（胜利石油管理局地质科学研究院开发试验室）

摘要：在模拟孤东小井距试验区水线的双截锥体模型上，使用孤东馆 5^{2+3} 层原油和地层水，系统研究了三元复合驱油体系段塞尺寸、段塞中表面活性剂浓度、复合注入方式、驱替速度对采收率提高量的影响。由研究结果确定的化学剂总量为 0.4 PV，段塞中表面活性剂浓度为 0.4%，注入方式为三段骞式：0.2 PV(A＋S)/0.2 PV(A＋S＋P)/0.2 PV(P)。

主题词：三元复合驱油体系　驱油效率　室内模型驱油试验　驱替方式　三次采油　孤东油田（胜利）

孤东小井距试验区于 1987 年 12 月开发，到 1989 年底中心井含水已达 98%。之后，为维持试验区与孤东七区油藏的平衡，在注采比非常接近的条件下继续水驱。目前中心井含水已达 99%以上，采出程度高达 53.3%，部分采油井如试 12 井含水达 100%。该试验区位于孤东油田西北部，其馆 5^{2+3} 层纵向矛盾很小，变异系数仅为 0.33，平面矛盾较大，水驱试验初期、后期动态及监测资料和水驱拟合结果表明，试 11 井与试 12 井、试 5 井与试 6 井之间有大孔道。由于井距仅 50 m，水驱波及体积已很大，剩余油饱和度约为 37%，因此采用聚合物驱油效果不会理想。碱-表面活性剂-聚合物(ASP)　复合驱油技术在国内外室内和现场试验中驱油效果良好[1-8]，如能应用于孤东小井距试验区，风险可能较小。

在三元复合驱油体系研究中我们对十余种表面活性剂和三种碱进行了筛选，通过配伍性、界面张力、相态和化学剂吸附损耗等研究提出了由 1.5%Na_2CO_3、0.1%OP-10 和 0.1%3530S 组成的复合驱油体系。本文研究该体系的注入段塞尺寸、表面活性剂浓度、注入方式、注入速度等因素对室内模型驱油效率的影响，以便为现场方案设计提供依据。

1　实验方法及材料

1.1 人工岩心（双截锥体模型）

该模型外形为基底连在一起的两个截头圆锥体，长 20 cm，两端直径 2 cm，中部（两锥体共用基底）直径 6 cm。模型用振荡填砂法制成，渗透率控制在 2.5-3.0 μm^2 范围。

此模型模拟试验区储油层内的水线。

1.2 驱替试验程序

① 称量模型干重(g);② 将模型抽空 1 小时,饱和试验区地层水;③ 称量模型湿重(g),求出模型孔隙体积(ml);④ 将模型置于 88 ℃保温箱中,恒温 1 小时后饱和试验区馆 5°层模拟油,求出含油饱和度 Soi,圆用地层水驱至含水 98%,求出残余油饱和度 S。,⑧ 按预定方式顺序注入化学剂;⑦ 再用黄河水驱至含水 08%,求出最终采收率和采收率提高量。

1.3 实验药剂

碳酸钠;分析纯;OP-10,含量～100%,浊点 75～85 ℃,上海助剂厂,聚丙烯酰胺 3530S,分子量 1 571 万,水解度 25%～30%,有效含量～91%,美国 Pfizer 公司产。

1.4 实验用水

黄河水;总矿化度 732 mg/l;试验区地层水,总矿化度 3 402 mg/l。

1.5 实验用油

孤东馆 5^{2+3} 层脱气脱水原油与过滤煤油的混合物,68 ℃时粘度 40.47 mPa.s,酸值 2.98 mgKOH/g,密度 0.905 g/cm^3。

2　实验结果与讨论

2.1 注入化学剂段塞尺寸(注入量)对驱油效率的影响

图 1 给出注入由 1.5%Na_2CO_3、0.4%OP-10 和 0.1%3530S 组成的三元复合驱眷液 0.2、0.3、0.4 和 0,5 PV 的驱替试验结果。由图可见,采收率提高量 $\Delta\eta$ 随化学剂段塞尺寸增大而增大,注入 0.5 PV 时达到 16%(OOIP),但注入单位重量化学剂(碱+表面活性剂+聚合物)驱出的油量 A 却随段塞尺寸增大而下降。根据因子 $\Delta\eta\times A$ 可权衡 $\Delta\eta$ 和 A 两项指标,化学剂段塞尺寸以 0.4 PV 为宜。

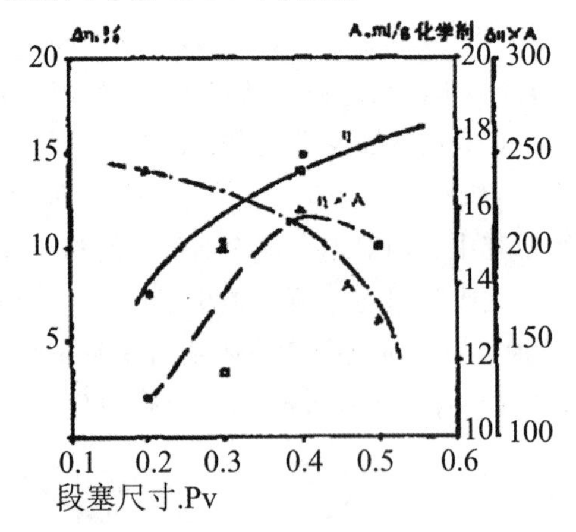

图 1　1.5%Na_2CO_3+0.4%OP-10+0.1%3530S 混合

段塞尺寸(注入量)对驱油效果的影响:采收率

提高量 $\Delta\eta$、单位重量化学剂驱出的油量 A 及 $\Delta\eta\times A$ 与段塞尺寸的关系(注入速度 1.28 m/d)

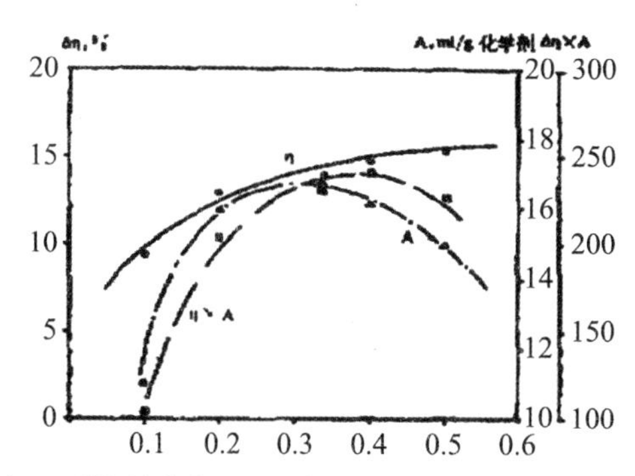

图 2 混合段塞中表面活性剂浓度 Cs 驱油效果的影响：Aη、A 及 Δη×A 与 Cs 的关系

2.2 段塞中表面活性剂浓度对驱油效率的影响

考虑到表面活性剂的吸附损耗，维持混合驱替液中碱和聚合物浓度不变、驱着条件不变，考察了表面活性剂浓度对驱油效率的影响，结果见图 2。可以看出，随表面活性剂浓度 C,的增加，采收率提高量 Δη 起初迅速增加，以后增幅减小。这说明表面活性剂浓度低时，吸附造成的影响较大。注入每克化学剂驱出的油量 A 则随 Cs 的增加经过一最大值。综合考虑 Aη 和 A 两项指标，选择 OP-10 的浓度为 0.35%(参见图 2 中 Δn×A-Cg 曲线)。

2.3 化学剂注入方式对驱油效率的影响

表 1 列出了五种注入方式的驱替试验结果。其中 91-9-Z11 号模型转注化学剂时含水较高，与其他四个模型的可比性略差。但 91-g-Z9 号模型的注入方式〔0.2 PV(A+S)/0.2 PV(A+S+P)/0.2 PV(P)〕提高采收率幅度之大是显而易见的，最终采收率高达 76.67%，比转注化学剂时的 57.6%提高 18.97%，比水驱外延至 4 PV 时的预测值提高 15.2%，采出残余油的 44.90%。

表 1 复合化学剂注入方式对室内模型驱油效果的影响

模型编号	孔隙体积(mL)	So(%)	无水期采收率(%)	转注化学剂时		注入方式*	总采收率(%)	采收率提高**(%)	驱出残余油(%)
				采收率(%)	含水(%)				
91-9-Z1	108.3	78.1	86.6	57.6	96.4	0.4 PV(A+S+P)	74.10	14.9	33.9
91-9-Z9	112.3	72.1	27.2	57.7	97.9	0.2 PV(A+S)/0.2 PV(A+S+P)/0.2 PV(P)	76.67	15.2	44.9
91-9-Z10	110.3	71.8	31.9	56.5	97.5	0.2 PV(A+S)/0.2 PV(0.07%P)/0.2 PV(A+S+P)/0.2 PV(0.03%P)	70.89	12.9	33.14

(续表)

模型编号	孔隙体积(mL)	So(%)	无水期采收率(%)	转注化学剂时		注入方式*	总采收率(%)	采收率提高**(%)	驱出残余油(%)
				采收率(%)	含水(%)				
91-9-Z11	111.0	72.1	37.88	59.38	99.0	0.2 PV(P)/0.2 PV(0.07%P)/0.2 PV(0.03%P)/0.4 PV(A+S)	74.88	12.4	38.2
91-9-Z13	117.3	69.9	38.05	60.98	97.6	0.4 PV(A+S)/0,3 PV(P)/0.2 PV(0.05%P)	76.22	11.7	39.1

*体系中化学剂浓度未注明的,A为1.5%,S为0.4%,P为0.1%。

**相对于水驱空白试验值的提高量。

91-9-Zg号模型的注入方式,既造成了流度控制梯度,又发挥了某些段塞的优势。1.5%Na_2CO_3+0.4%OP-10、1.5%Na_2CO_3+0.4%OP-10+0.1%3530S和0.1%3530S三个段塞的粘度(68℃)依次增大,由0.496到6.6再到14.3 mPa·s。与混合段塞Na_2CO_3+OP-10+3530S相比,Na_2CO_3+OP-10段塞的增溶能力较强(在68℃时为Na_2CO_3+OP-10+3530S段塞的1.76倍),3530S段塞的流度控制能力较强。

2.4 注入速度对驱油效率的影响

提高注入速度可改变毛细管数从而提高采收率,但会增大孔隙中的剪切速率(7而减小聚合物的流度控制能力。表2列出了三种注入速度的驱油试验结果。其中1.28 m/d模拟瓢东小井距试验区开发初期水前缘平均推进速度,2.0和2.7 m/d分别为该区目前可能的水前缘推进上下限。表2结果说明采用2.7 m/d的注入速度时驱油效果最佳,最终采收率比水驱提高17.64%。由于试验区平面矛盾较大,在现场试验中确定注入速度时,除本文结果外还应考虑其他有关的因素。

表2 注入速度对三段塞*驱油效果的影响

模型编号	孔隙体积(mL)	So(%)	无水期采收率(%)	转注化学剂时		注入速度		最终采收率(%)	水驱最终采收率(%)	采收率提高**(%)	驱出残余油(%)
				采收率(%)	含水(%)	mL/h	m/d				
91-9-Z9	112.3	72.1	27.2	57.7	97.9	30	1.28	76.67	61.47	15.2	44.9
91-9-Z16	107.7	71.5	36.9	54.3	97.7	45	2.00	71.95	56.02	15.93	39.49
91-9-Z17	108.9	70.7	33.5	55.1	96.8	60	2.70	75.71	58.07	17.64	45.95

*驱替方式。0.2 PV(1.5%A+0.4%S)/0.2 PV(1.56A+0.4%S+0,1%P)/0.2 PV(0.1%P)。

**相对于水驱最终采收率的提高量。

2.5 典型模型的动态曲线

图3给出91-9-Zg号模型的采收车、驱出液含水量和注入化学剂前后的产油量变化

曲线。注入化学剂使采收率提高 15.2%，主入化学剂后水驱 9.8 PV 时驱出液含水由近 98%下降至约 50%，此时出现较厚油墙，瞬时产油量最大。继续水驱时产油量下降，含水上升，但产油仍可延续约 1 PV。这一点与聚合物驱的情况不同，表明三元复合体系中碱和表面活性剂起明显的作用。

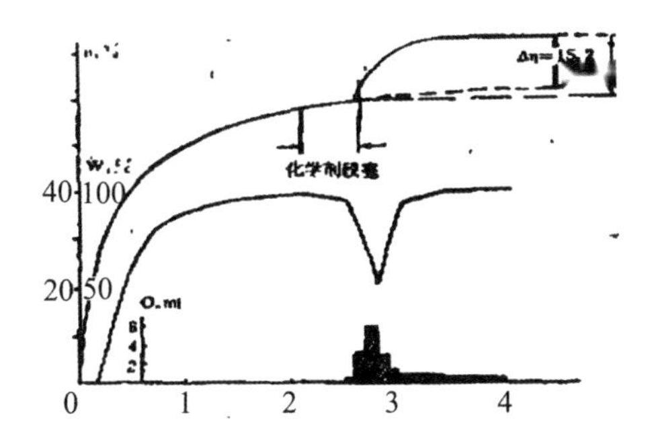

图 3　91-9-Z9 模型驱替试验中采收率 η、含水率 W、产油量 O 随注入量的变化(注入速度 1.28 m/d)

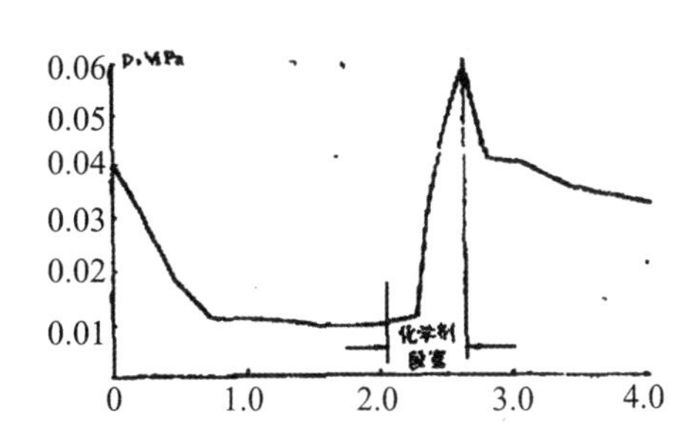

图 4　91-9-Z9 模型注入压力随注入量的变化(注入速度 1.28 m/d)

图 4 给出驱替过程中该模型的注入压力变化曲线。注入化学剂后，碱和表面活性剂的乳化作用及聚合物的增粘作用使注入压力迅速上升，最高达到 0.03 MPa，比开始注水时的压力(即饱和油见油时的压力 0.04 MPa)增加约 50%，比转注化学剂时的压力增加约 5 倍。水驱至 4 PV 时注入压力为 0.03 MPa，比开始注水时减小 15%，比转注化学剂时增加约 2.5 倍。考虑到试验区油藏平均渗透率较大(2.5～3.0 μm^2)，根据驱替试验结果，预计现场注入化学剂过程中不会出现注入困难。

2.6 经济评价

根据室内实验结果估算(三元复合驱油的数值模拟技术尚不完备)，采用 ASP 法采出 1 吨油的化学剂费用为 272.9 元，这个结果非常令人满意。三元复合驱油效率高，应用范围广泛，且可缩短水驱时间，因此现场实施时风险性相对较小。

3　主要结论

孤东小井距试验区注入化学剂段塞尺寸以 0.4 PV 为宜。段塞中表面活性剂浓度对驱油效果影响较大，以 0.35%为佳。最佳注入方式为：0.2 PV(1.5%Na_2CO_3+0.4% OP-10)/0.2 PV(1.5%Na_2CO_3+0.4%OP-10+0.1%3530S)/0.2 PV(0.1%3530S)。注入速度对驱油效率有明显影响。室内驱替试验中不存在注入困难。注入速度为 1.28 m/d 时，采用最佳注入方式可提高采收率 15.2%，驱出残余油 44.90%。注入化学剂后水驱至 0.8 PV 时出现最厚油墙，此时含水由 98%降至 50%。

致谢　作者衷心感谢胜利石油管理局地质科学研究院允许发表此论文，本研究工作得到了沈娟华副院长，姚远勤总工理师、屈智坚和俞进桥高级工程师的支持与协助，特此表示深切谢忱。

参考文献

[1] French, T. R. and Burchfinld T. E.: SPE20238.

[2] Clark, S. R., Pitts, M. J., and Smitb, S. M.: SPE17583.

[3] Doll, T. E.: SPE17801.

[4] Saleem, S. H., J. Surface Sci. Teclnol., 1987, 3: 1-10.

[5] Lin, F. F. J. and Besserer. G. I., J. Canad. Petrnl. Technol., 1987(11/12): 54-65.

[6] 俞进桥,王宝瑜.陆相石油地质.1990,12:53-67.

[7] API RP63 "Recommended Practices for Evaluation of Polymers Used in Enlnced Oil Recovery Operations", American Petroleum institute, Ist ed., June 1,1990.

[8] Meyers, J. J., Pitts, M. J., and Kon Wyatt; SPE/DOE 24144.

微焦点 X 射线计算机层析(CMT)及其在石油研究领域的应用

李玉彬　李向良　张奎祥　曹绪龙

(胜利石油管理局地质科学研究院,山东东营,257015)

摘要:工业微焦点 CT 的 X 射线源焦点尺寸小于 5 微米,成象分辨率明显优于常规焦点 CT。在石油勘探开发研究中,微焦点 CT 在许多技术领域进行了应用,取得了崭新的成果:(1) 确定岩心的基本物理参数—计算和描绘密度、孔隙度、流体饱和度分布;(2) 描述岩心的微观特征一直观描述岩石内部孔隙结构和流体分布特征;(3) 特殊油藏岩心分析评价—观察碳酸岩、火成岩等内部一些常规技术无法获得的特征;(4) 岩心驱替和提高采收率分析一重建油水在岩心内部的宏观分布状态;(5) 其他方面的应用——岩心筛选、地层伤害评价、岩石力学评价等。

关键词:微焦点 X　射线　CT　石油

X-ray computed micro-tomography(CMT) and its application to petroleum research

Li Yubin　Li Xiangliang　Zhang Kuixiang　Cao Xulong

(Geological Scientifc Research Institute of Shengli Petroleum Administrative Bureau, Dongying, Shangdong, 257015)

Abstract: The image resolution of industrial X-ray computed microtomography (CMT) with focal spot of X-ray source less than 5 micrometer is quite better than that of conventional focal spot CT. In petro-leum exploration and development research, computed microtomography was used to many technological areas with revolutionary achievements: (1) determination of the basic petrophysical. properties of cores-calculating and illustrating distribution of buck density, porosity and fluid saturation; (2) imaging of microscopic porous of rock—intuitive describing micro-structure of porous and oil-water distribution in porous; (3) evaluation of special lithological reservoir cores—investigating characteristics of cabonatite, mogatic rock, etc., that can not be done by conventional methods; (4) analysis of core flooding and enhanced oil recov-ery—reconstructing macroscopical distribution of oil and water in cores; (5) application of other aspects: screening core, evaluation of formation damage, evaluating rock mechanical properties of cores, and so on.

Key Words: microfocus X-ray CT petroleum

1　CT 技术在石油领域应用的简介

自从 Hounsfield 工程师开发了第一台 X 射线 CT 系统以来,随着计算机工业的飞速发展,CT 机性能不断提高,石油工程师应用医用 CT 研究岩石结构和流体分布也不断深入。CT 渗透到石油研究中的各个领域,CT 在岩石基本物理性质、特殊岩心性质、岩心地质特征描述、岩心驱替和提高采收率研究等众多方面起到了重要作用。

1979 年,Bellaire 研究中心的 Vinegar 利用医用 CT 完成第一个岩心扫描结果。随后近二十年来,在石油领域应用 CT 的研究工作涉及到岩石物理的各个方面,特别在岩心动态驱替特征方面的研究相当细致和深入。如 Wang 等(1984)描述了岩心驱替中使用 CT 扫描观测原油驱替[4];Wellington 和 Vinegar(1985) 利用 CT 研究泡沫作为 CO_2 流度控制剂的益处[5]。Withjack(1988)通过 CT 确定岩心性质,指导混相驱试验,同时描述和测量相对渗透率[6]。Garg 等(1996)利用 CT 扫描和神经网络技术描述水驱残余油的分布[2]。

由于医用 CT 的 X 射线源能量和系统空间分辨率较低,CT 在研究一直处于一种宏观描述的水平,也限制了岩心地质特征描述领域的发展。随着近年来高分辨率微焦点 X 射线 CT 技术的进步,研究成果的水平明显提高。如 Jasti 等(1993)介绍了微焦点 X 射线 CT 系统对孔隙结构和孔隙中的流体进行成象[3]。Coles 等(1998)利用同步加速器微焦点 X 射线 CT 研究流体在孔隙中的运移[1]。

1997 年底,胜利油田地质科学研究院从美国 BIR 公司引进了带有微焦点 X 射线管的 ACTIS 工业 CT 系统,该系统为岩心分析研制,我们重点在岩心微观成象和微观驱替方面进行了探索性研究,完成图象的分辨率达到了国外同期水平。本文介绍 ACTIS 微焦点工业 CT 系统,以及在石油研究领域的应用的实例,包括常规焦点 CT 能够完成的研究,也介绍常规 CT 无法完成的岩心地质特征和微观驱替特征描述。

2　ACTIS CT 系统介绍

BIR ACTIS 工业 CT 系统采用物体旋转的三代扫描方式(示意图参见文献 8),这套工业 CT 系统的特点表现在：系统配有两套 X 射线源，一个是 PANTAK 320 kV 小焦点源，一个是 FeinFocus 225 kV 微焦点源。FeinFocus 微焦点源的靶点尺寸可以达到微米，对特征尺寸真正实现精确测量，它对直径为 4.8 mm 的物体成象，空间分辨率达到 10 μm。

系统的检测系统采用了 Toshiba 三视野图象增强器，可以实现真正的体积层析（VCT），一次扫描可以重建 100 个切片，提高了检测效率但降低了空间分辨率。

系统中有两个机械转台，立式转台用于扫描静态物体，卧式转台专门设计用于对岩心流动试验装置进行扫描。立式转台更靠近 X 射线源，有利于提高微焦点 CT 的空间分辨率系统的控制和数据采集系统采用奔腾芯片处理机，内置高速阵列处理板进行图象重建。每幅图象扫描时间为 18 秒～5 分钟，重建时间 5 秒～1 分钟。VCT 的功能，重建 100 幅 256×256 的图象需要 6 分钟。图象象素矩阵能达到 1 024×1 024，显示达到 4 096 阶灰度。

3　微焦点 CT 在石油研究领域的应用

利用微焦点 CT 进行岩心分析的费用较高，然而 CT 能提供岩心样品中密度、孔隙度、饱和度的分布，而常规测量方法只能给出整个岩心样品的平均值。另外，就岩石地质结构特征而言，如碳酸岩裂缝油藏岩心的孔隙结构，采用常规分析方法压汞毛管压力曲线、打磨剖光照相、铸体薄片显微照相显然存在缺陷。压汞法是建立在平行毛管束理论的基础上，它不适于裂缝和孔洞类的岩心分析；剖光照相和铸体薄片仅能观察到有限的几个剖面，无法认识岩心内部的空间分布规律；而且需要二次机械破坏，它损害了岩心固有的组织结构。

利用 CT 进行岩心分析还有以下优点：① 岩心驱替试验的同时进行扫描，测量带有围压条件下的参数；② 可以无损检测一些特殊的密闭样品；③ 弥补常规方法无法进行分析的不足；④ 观测孔隙度、饱和度的分布；⑤ 定量检测孔隙通道或裂缝；⑥ 测量不规则样品的参数。而高分辨率微焦点 CT 图象空间分辨率达到微米级，明显提高了岩心分析水平。

3.1 利用 CT 确定油层基本物理参数

岩心的基本物理参数——体积密度、孔隙度、饱和度是油田勘探和开发中最基本的物性参数，但利用 CT 进行岩心分析具有常规试验无法相比得优越性。

微焦点 CT 对岩心进行扫描后，计算密度的方法有单能扫描法、双能扫描法、线性插值法；确定孔隙度的方法有单次扫描法、两次扫描法、测井解释法、图象分析法；确定饱和度的方法有单能两次扫描确定含水率、双能扫描确定含水率、单能扫描确定两相流体饱和度、单能扫描确定三相流体饱和度、双能扫描确定三相流体饱和度、图象分析法确定流体饱和度（详细阐述参见文献 7）。

微焦点 CT 试验得到 21 块岩心的平均密度（包括 CT 插值法，CT 单能扫描干样法，CT 单能扫描饱和水样法）与煤油法测定值的平均百分绝对误差不到 1.2%。21 块岩心饱和煤油法、氦孔隙计法、CT干样单次扫描法、CT完全饱和水单次扫描法、CT两次扫描法孔隙度测定值表明常规测试方法，如饱和煤油法与氦孔隙计法之间的误差为8.53%，而CT两次扫描法与饱和煤油法之间的误差为7.81%。岩心两种方法获得的相对渗透率曲线的对比，饱和度变化是一致的。CT试验与常规测试结果的对比说明利用微焦点CT确定岩心的基本物理参数不仅在理论上可行，而且测试结果也相当准确。

3.2 岩石微观特征描述

储层岩石的宏观渗流特性决定于孔隙的微观结构，如孔隙的形状、大小、分布和连通性等。微焦点CT对孔隙和微裂缝的微观描述是其他技术无法比拟的，弥补了常规方法的许多不足。

砂岩

从微焦点CT图象上能，直接观察到粒间孔隙的形状、大小、分布，以及颗粒的形状、大小、分布和密度，可以对砂岩岩心进行评价。高斜73井岩心CT切片主要由0.1～0.3 mm的颗粒组成，分选较好，同常规粒度分析结果是一致的。孔喉宽度从几微米到200多微米，最大孔喉远大于压汞测得的结果(72 μm)，我们认为压汞平行毛管束理论假设有一定局限性。二值化处理后，测得平均配位数3.4，平均孔喉比4.43，平均面孔率27.8，最大孔隙等面积圆半径145 μm。

特殊岩性岩心

图1　WP1井岩心微焦点CT扫描切片

埕北30井白云质灰岩，清楚出分辨灰质部分(暗色)和白云岩晶体部分(亮色)，晶体棱角清晰可辩。灰质和晶体之间裂缝宽度约0.01 mm，晶间孔隙宽度为0.05～0.3 mm。单10-斜检1井硫松砂砾岩，颗粒分选差，砾石大小均匀，一般在2 mm～4 mm，大颗粒之间充填物细，含油孔隙直径小，存在0.02～0.1 mm的裂缝，大颗粒内也存在微细裂缝。图1为王平1井岩心CT图象，左边为2.5 cm岩心的CT图象，而右边上、下两图分别取岩心不同部位的微观图象。右上图为层理部分，微裂缝0.01～0.02 mm；右下图为孔洞区域，孔隙直径0.01～0.4 mm，颗粒直径0.02～0.8 mm。

3.3 微观油水分布和运移

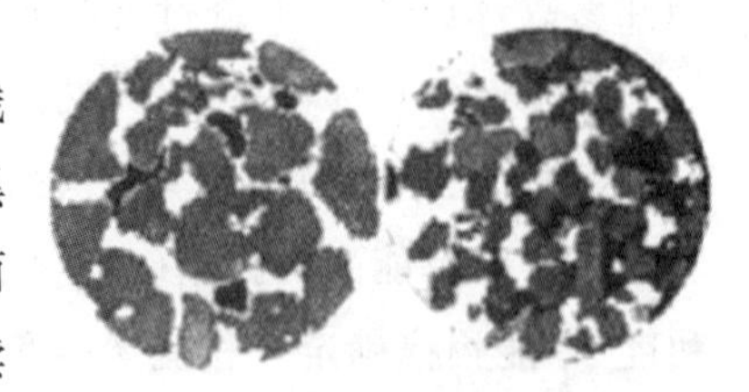

图2　微观油水残余油分布CT切片

图2(左)为粒径0.5～1 mm洗净油砂充填模型残余油状态下的CT扫描切片。图象重建域约5 mm，黑色的是油，暗色的是颗粒，白色的是水。测得切片的面孔率为33.7%，此时含油饱和度为14.8%。水呈连续相，孤立的油滴在孔隙中间分布，表明岩心润湿性亲水。

图2(右)为0.2 mm油砂充填模型残余油状态下的微观CT切片，重建域2.5 mm，

面孔率为37.6%，经过3小时0.001～0.005 ml/min 3%KI溶液驱替后残余油饱和度为30.7%。残余油大部分为连续相，同左图中颗粒大的试验相比，残余油饱和度明显高。

3.4 岩心地质特征描述

岩性是油藏描述的重要组成部分，微焦点CT确定岩石类型、层序、层理、孔隙类型、沉积结 构、岩性非均质性、渗透率变化等具有突出的优势，能清晰反映的是岩心三维空间结构的变化。医用CT也能完成这些工作，但只能对某些特征进行近似。以下CT岩心分析实例是常规方法难以完成的，绝大多数也是常规焦点CT无法获得的。

3.4.1 均质砂岩

某些用肉眼观察岩心表面十分均匀，没有任何差别，然而CT扫描切片图象却不同，可以看到颗粒密度不同，以及密度和孔隙度的分布。表1是依据图象对岩心的分析结果：

表1 四种不同类型砂岩CT扫描成象分析结果

井号	样品号	渗透率 $10^3\ \mu m^2$	煤油法孔隙度	CT密度	CT孔隙度	胶结类型	颗粒分选性	颗粒磨圆度	最大颗粒	最大孔喉
3-4-J111	255		31.5	1.88	29.5	接触胶结	好	好	0.25	0.15
孤北341	44	52.6	15.3	2.29	16.4	孔隙胶结	中	中	0.35	0.25
河130	33	6.0	6.0	2.59	5.1	基质胶结			0.20	0.10
坨76	9		16.7	2.07	17.1	孔隙胶结	差	较好	0.90	0.50

3.4.2 特殊类型岩心性质

裂缝和微裂缝分布

描述裂缝非常困难，有人曾经采用几十个参数描述裂缝，近年来随着分形几何技术的发展使得描述裂缝变得相对有可能。但是无论数学模型如何先进，都必须建立在物理模型的基础上。常规方法仅能描述裂缝在一个或几个平面的分布，而无法构造裂缝在立体空间内的展布。图3是滨675井岩心CT图象重建的三维裂缝和孔洞的分布，能直观获得裂缝的宽度、方位、体积等参数。

砂砾岩的储集性

在砂砾岩油藏中，存在几毫米到几十厘米直径的砾石。通常的观点认为砾石是一个不含任何储集和渗流空间的实心体。通过微焦点CT扫描发现部分砾石存在储集和渗流空间，显然从前的观点就存在认识上的错误。埕913井岩心的CT图象一侧为破碎的砾石，而单从岩心外表观察无法发现这种现象。CT结果表明整个岩心的孔隙度是0.12，除去砾石的基质孔隙度是0.16，这个概念其他任何方法无法获得的。

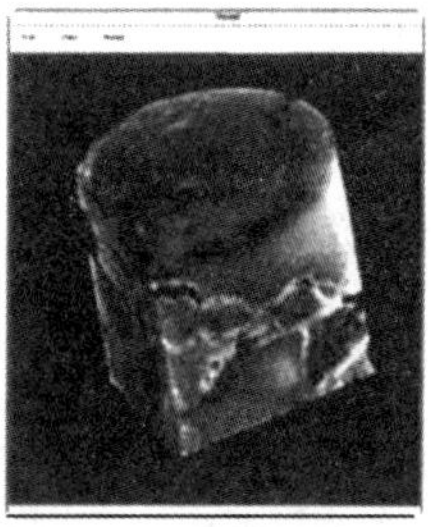

图3 滨675井岩心重建三维裂缝和孔洞状态

疏松砂岩制作筛选

疏松砂岩岩心的制作作为一项成熟技术一直在各个油田应用,它是经过煮蜡、钻取、包裹、清洗等过程后,进行常规岩心分析和特殊岩心分析,其制作是否合格常被提出质疑。利用微焦点 CT 能清楚地观察到一些岩心存在密度上的不均匀,有的还存在明显裂缝,这些特征的存在导致严重的后果,常规岩心分析和特殊岩心分析的不准确。利用微焦点 CT 可以快速方便的对疏松岩心进行筛选,将合格样品用于其他分析试验。

层理判断

XYS9 井岩心的微焦点 CT 图象,所示的层理现象是在岩心表面用肉眼无法观察到的。从外表看,该岩心致密,表面光滑,岩性一致,进行岩石类型划分往往认为它不属于沉积岩。从 CT 图象上观察 到具有水平层理,并且沿层理方向产生裂缝。

压缩系数异常的解释

埕北 303 井片麻岩岩心的压缩系数明显不同于该层其他岩心。随压力升高,压缩系数降低缓慢或没有变化;而一般情况下压缩系数随压力升高明显降低。这种现象在一段时期内没有得到令人满意的解释。而由 CT 图象能观察到岩心内部具有高密度颗粒,难以压缩,它对裂缝产生支撑。

孔洞连通性

岩心孔洞是良好的储集空间,但好的储集未必是好的渗流空间,它还取决于这些储集空间的连 通状态。微焦点 CT 观察到商 742 井火成岩岩心中的孔洞成簇分布,中间缺少良好的渗流通道,酸化压裂作业使这些成簇的孔洞形成渗流通道,取得良好效果的机理即在于此。

岩心污染

钻井过程会造成岩心污染,被污染后的岩心也会造成孔隙度、渗透率和饱和度等分析数据的不 准确。防止这种分析数据的不准确,比较有效的方法就是利用 CT 筛选岩心,选用未被污染的部分进行取样分析,确保分析数据的准确可靠。微焦点 CT 图象观察到丰 128 井岩心越靠近岩心外表污染越严重,取样时我们应该尽量避开污染过的外表层。

3.5 油水驱替动态特征描述

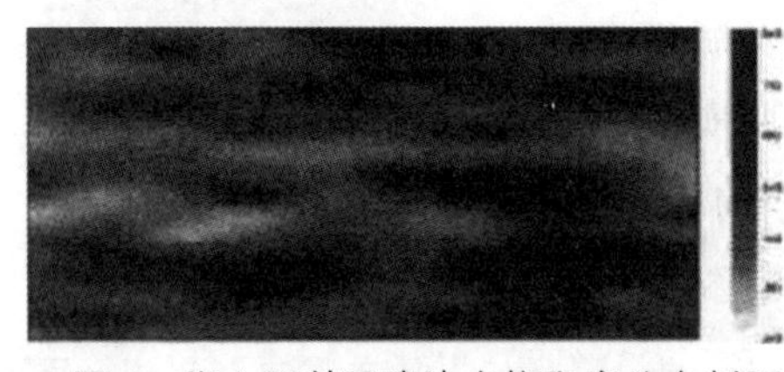

图 4 岩心驱替重建油水饱和度分布剖面

这项研究在常规焦点 CT 上也能很好的完成,在微焦点 CT 上没有明显优势。图 6 是 5 cm 直径砂岩岩心水驱油试验轴向剖面,试验首先完全饱和 5%KI 溶液,然后饱和粘度为 48 mPas 的机油建立束缚水状态,然后以 10 ml/min 的速度泵入 KI 溶液进行驱替。在岩心完全饱和油和完全饱和水时,分别在对应的位置扫描 9 个切片,在动态驱替过程中在对应位置连续扫描。图 6 是驱替到残余油饱和度时油水饱和度大小分布的轴向重建剖面,由完全饱和水、完全饱和油、残余油状态下对应 9 个位置的 27 个扫描切片的象素 矩阵 512×512 变换成 36×36,利用单能扫描确定两相流体饱和度的方法计算出剖面上饱和的值。图象反映了水高速冲刷下残余油饱和度分布特征,水明显沿水流通道窜进。

4　结束语

在石油科研领域,应用微焦点 CT 成功地对岩心进行了扫描;直观地描述岩石特性,如孔隙度分布、密度变化、裂缝、孔洞、层理、岩心伤害等;能准确计算岩心的基本物理参数——体积密度、孔隙度、饱和度;并且成功地对特殊油藏岩心进行了综合描述分析,将精细油藏描述推向更高的层次。

利用小块样品进行微观成象,真实反映了孔隙介质的微观结构,它能对油田地质特征更准确地 进行了定量描述,对孔隙和微裂缝的描述是其他技术无法完成的。完成了对油水微观驱替实验的扫描,图象结果真实反映孔隙介质的网络结构特征,再现了真实润湿状态下的油水运动规律和驱替特征。这项技术对进行岩石微观描述、可采储量核算、驱油机理研究等方面意义重大,并且蕴藏着巨大的发展潜力,尤其是在油田发展后期微观剩余油分布和提高采收率方面的研究。

同时微焦点 CT 已成功应用于岩石力学、特殊样品密度分析、精密元件无损检测、生物医学等 许多方面,尽管包括岩心分析在内许多研究是初步的,但是相信微焦点 CT 在微观世界的探索是没有止境的。

参考文献

[1] E. Coles, R D. Hazlet, P. Spame, W. E. Soll, E. L Murgge, K W. Jones. Pore level imaging of fhuid transgport using syn-chrotron *X*-ray microlamography. JPSE, 19(1998),53-63.

[2] A. Garg, A. R Kovsack, M Nicavesh, L M Castanicr, T. W. Pstrek, CT scan ad naml ndwork tedmology far canstructian of detailed distibution of residaal oll saturation during waterfloading SPE Westem rgianal meting, Anchaage Alaska,1996, 695-710, SPE35737.

[3] J. K. Jasi, Gerald Jeslon, Lee feldkamp. Miauscxpic inaging of porous media with *X*-ray computer tomography. SPEFE, 1993,189-193,SPE20495.

[4] S. Y. Wang, Seyda Aynal, F. S. Caselang, and Carl C. Gryte, 1984. Reoonsnuction of oil sthmaton disribunian histaries during immisible liquid-liquid displacement by computer-assisted tomography. AIChE, 30(4): 642-646.

[5] S. L. Wellington. H. J. Vinegar,1987, *X*-ray computerized tomography. JPT,1987,885-898,SPE16983.

[6] E M Wihgax, Compued tomogophy sudies of 3D misible diglsemat htavoir in a labonday fve-soa modl:634 ammual techmical conference and edhibition of the society of petroleum engineers, Houston, 1988, 435-447,SPE18096.

[7] 李玉彬,等.利用计算机层析(CT)确定岩心的基本物理参数[J].石油勘探与开发,1992.12:92-96.

[8] 李玉彬,等.微焦点X射线计算机层析(CT)及其在无损检测中的应用[J].无损检测,1999,21(12):549-552.

利用体积CT法研究聚合物驱中流体饱和度分布

曹绪龙[1]　李玉彬[2]　孙焕泉[2]　付　静[3]　盛　强[2]

（1. 中国科学院兰州化学物理研究所　甘肃兰州　730000；
2. 胜利油田有限公司地质科学研究院　山东东营　257015；
3. 石油大学(华东)　山东东营　257062）

摘要：对CT图像进行了X射线硬化处理、X射线漂移校正、图像矩阵变换及体积CT位置校正，研究出用体积CT对岩心流体饱和度进行测量的技术。利用该技术对聚合物驱过程中油水饱和度分布进行了测量，结果表明，该技术适合于测量聚合物驱过程中油水饱和度分布。尽管试验中使用了均质岩心，在水驱阶段形成了明显的水窜通道，窜流通道上的油 饱和度明显低于其他区域，微观波及的不均匀性非常明显。注入聚合物段塞后，有效地富集了水驱之后的剩余油，利用聚合物驱，除了能改善波及体积外，还成功地观察到水驱之后进一步提高洗油效率的现象。

关键词：体积CT技术；聚合物驱；岩心；含水饱和度；含油饱和度；试验研究
中图分类号：T E357.431　文献标识码：A

Determination of fluid saturation in polymer flooding by volume-CT method

Cao Xulong[1]　Li Yubin[2]　Sun Huanquan[2]　Fu Jing[3]　Sheng Qiang[2]

(1. Lanzhou Institute of Chemical Physics, Chinese Academy of Sciences, Lanzhou 730000, China)
2. Geosciences Research Institute of Shengli Oil Field Company Limited, Dongying 257001, China;
3. Petroleum University, Dongying 257015, China)

Abstract: A method for determining oil and water saturations in cores by volume-CT scanner was developed on the basis of X-ray hardening treatment, X-ray drifting correction, martix transformation of imagines and position correction of CT images with volume-CT technology. It comes to the conclusion that this volume-CT technology can be used to measure the oil and water saturations in bothar tificial and natural cores during and after polymer flooding. Experiments indicated that there was a fingering phenomenon in the stage of water flooding, and heterogeneous water sweep channels existed clearly during water flooding. Oil saturation in water channels was obvious lower than that of other areas. The sweep volume was enlarged, and the residual oil was enriched efectively. The oil recovery was enhanced, after polymer slug was injected into the core. The application of volume-CT technology to core flooding shows that polymer flooding not only plays a very important role in improvement of sweep volume, but also enhances oil recovery, in comparison to water flooding.

Key words: volume-CT technology; polymer flooding; cores; water saturation; oil saturation; experiment

CT 技术被国内外广泛用于研究多孔介质中的渗流特点已有 20 多年，在国外已经形成了一套完备的试 验体系[1]。近年来，CT 技术与油藏工程的新理论和新方法不断结合，推动了渗流力学的技术进步[2]。在三次采油提高采收率方面，CT 的应用对于发现新的渗流机理和规律起到了重要作用，Dana George Wreath 利用 CT 技术测量了岩心中油的饱和度[3]，Seetharman Ganapathy 利用 CT 技术探索了岩心在聚合物驱条件下油饱和度的变化[4]，Mengwu Wang 将 CT 技术用于测量岩心饱和油及聚合物驱替过程流体饱和度分布[5]。由于岩心制作材料、射线硬化、图像处理、定位等存在一些问题，在他们对驱替过程中的流体饱和度分布的实验中均未取得满意的结果。在国内，尽管石油科技人员在这方面进行了不懈的努力[6,7]，但有关利用 CT 技术研究多孔介质渗流方面的报道甚少。

本次试验中使用的体积 CT 与医用 CT 多次扫描三维重建技术不同，它借助了一套具有体积层析功能的微焦点工业 CT 系统，一次扫描最多可以重建 100 个切片，从而完成一个三维立体的重建，扫描速度明显高于切片 CT，保证了岩心驱替不停泵连续测试。

1 CT 扫描聚合物驱替试验

1.1 仪器设备

美国 BIR 公司生产的 ACTIS 微焦点工业 CT 机，分辨率 10 μm 、最大射线电压 320 kV。美国 TER-RATEK 公司生产的 CORESCAN 岩心驱替装置包括 ISCO 高精度注入泵、油水计量系统等。

岩心及实验条件：选取两个不同的岩心进行试验，试验 1 采用人造岩心，渗透率为 684×10^{-3} μm^2，孔隙度 27.8%，直径为 2.5 cm，长度为 5.7 cm。试验 2 采用天然岩心，渗透率为 $98.2\times10^{-3}\ \mu m^2$，孔隙度 25.5%，直径为 2.5 cm，长度为 6.5 cm。

1.2 试验步骤

试验过程同常规岩心驱替试验类似，首先将实验岩心模型抽真空饱和水，再用油驱水建立束缚水状态，然后注水驱替至含水率为 98%，接下来注入 0.2 PV 聚合物溶液段塞，最后进行后续水驱至产出液含水率达 98%后结束试验。试验与常规聚合物驱替试验相比，不同点是在试验过程中不断地进行扫描。优点在于无须出口计量设备，利用 CT 即可完成内部计量，并且能获得岩心内部饱和度分布数据；无须对驱替系统的死体积进行测量。

在聚合物驱替试验中，根据试验目的的不同，有选择的在各流动相中加入衰减溶质。本次试验中，为了有效观察聚合物驱替过程中油水饱和度的分布，在水和聚合物溶液中加入 Nal，并使水和聚合物溶液的射线衰减系数一致。

2　利用 CT 图像计算岩心饱和度

将岩心从 i、j、k 3 个方向上划分为若干网格体（图 1），CT 扫描所测得的就是每个网格体上的 X 射线线性衰减系数。在实际测试中，为了便于应用，将衰减系数转换成 CT 数，每个网络的 CT，数用 CT^{ijk} 表示。当岩心用油水两相饱和时（聚合物和水看成同一相），每个单元网格的 CT 数表示如下

$$CT_b^{ijk}=(1-\varphi^{ijk})CT_r^{ijk}+\varphi^{ijk}(S_w^{ijk}CT_w+S_o^{ijk}CT_o) \tag{1}$$

式中，CT_b^{ijk} 为油水驱替过程中每个网格的 CT 数；ijk 为每个网格的孔隙度；S_o^{ijk} 和 S_w^{ijk} 分别为每个网格的油、水饱和度；CT_r、CT_o 和 CT_w 分别为岩石骨架、油和水的 CT 数。

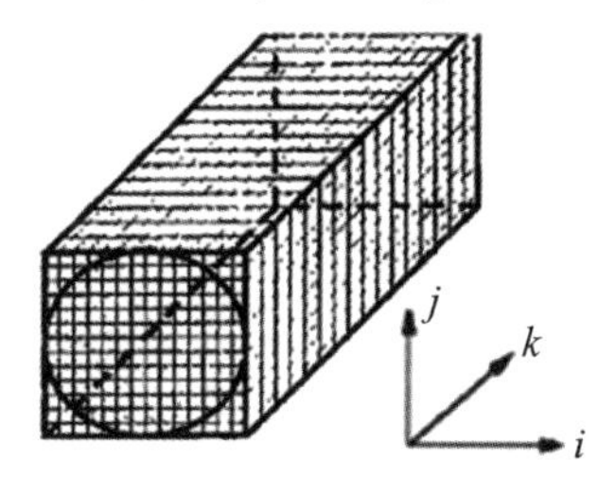

图 1　岩心网格划分示意图

Fig. 1　The grid schematic drawing of cores

对完全饱和水或完全饱和油的岩心进行扫描，其 CT 数表达式如下

$$CT_{bw}^{ijk}=(1-\varphi^{ijk})CT_r^{ijk}+\varphi^{ijk}CT_w \tag{2}$$

$$CT_{bo}^{ijk}=(1-\varphi^{ijk})CT_r^{ijk}+\varphi^{ijk}CT_o \tag{3}$$

式中，CT_{bw}^{ijk} 和 CT_{bo}^{ijk} 分别是完全饱和水、完全饱和油时每个网格的 CT 数。

由式(1)、式(2)和式(3)得到每个网格的油水饱和度为

$$S_{bo}^{ijk}=\frac{CT_b^{ijk}-CT_{bw}^{ijk}}{CT_{bo}^{ijk}-CT_{bw}^{ijk}} \tag{4}$$

$$S_{bw}^{ijk}=1-S_{bo}^{ijk} \tag{5}$$

利用 CT 扫描获得完全饱和油水时的 CT 数分布 CT_{bw}^{ijk}、CT_{bo}^{ijk}，在驱替过程中用扫描技术获得不同时刻的 CT 数 CT_{b}^{ijk}，即可计算出不同时刻饱和度的分布。

3　CT 数字图像处理

扫描得到 CT 图像后，单纯利用这些图像进行三维或二维轴向切片重建，无法反映出驱替过程中油水的动态变化。要反映油水驱替动态，必须对 CT 图像进行一系列的处理，依次为射线硬化处理、图像矩阵变换、射线漂移处理和体积 CT 位置校正等，然后才能进 行饱和度分布的计算。

3.1 射线硬化处理

X 光机发射的射线是连续的多色光谱。射线穿过物体时，低能部分首先被物体表面吸收，造成一种物体表面吸收射线能力强的假象，这种现象就是射线硬化。这种图像的特点是边缘灰度大、中心灰度低，不同衰减系数的物体引起灰度变化的范围也不同。如果不进行射线硬化处理，得到的饱和度分布图像也是边缘的饱和度值高，中间的值低。

CT 系统本身附带有射线硬化处理模块，但这些模块不适用于油水驱替试验。因为油水驱替是一个动态过程，饱和油水的岩心本身的衰减系数在不断变化，无法构造一个统一的方程来消除驱替过程中的射线硬化现象。惟一可行的方法就是利用计算程序对图像数据进行有效处理。经过几年来的反复实践，目前是利用切片 CT 数的平均值创建一个反硬化图像(边缘灰度小、中心大)来消除射线硬化现象。

3.2 图像矩阵变换

对于多孔介质中某一质点 P，以其为质心的体积为 ΔV，其孔隙体积为 ΔV_P，孔隙度为

$$=\frac{\Delta V_P}{\Delta V} \tag{6}$$

绘制与 ΔV 的关系曲线，当 ΔV 缩小到某个体积 ΔV_0 附近时，开始剧烈振荡。当 ΔV 在孔隙上时，=1，当 ΔV 在骨架颗粒上时，=0。

对于饱和度的计算也是如此。CT 扫描获得的图像的矩阵一般为 512×512，对于一个 25 mm 直径的岩心，每个像素仅代表 0.05 mm，如果简单地以 CT 像素矩阵为网格，就会使计算的饱和度分布数值大幅度波动。进行矩阵变换的另外一个原因就是图像噪音，去除噪音的方法之一就是多点数据平均法。实践表明，将网格变换成 16×16，对于细砂岩、中砂岩均能获得满意的结果。

3.3 *X* 射线漂移校正

由于驱替试验和扫描成像是一个漫长的过程，在试验过程中外界条件的变化，如电网电压、外界磁场、自然光线等，导致相同材料的物质 CT 数略有差别。图 2 为标定材料在试验过程中 CT 数的变化，可明显地看出电网电压对 CT 数据有影响，但在午间和晚间 CT 数相对平稳。利用图 2 给出的数据即可对不同时间的 CT 图像数据进行校正。

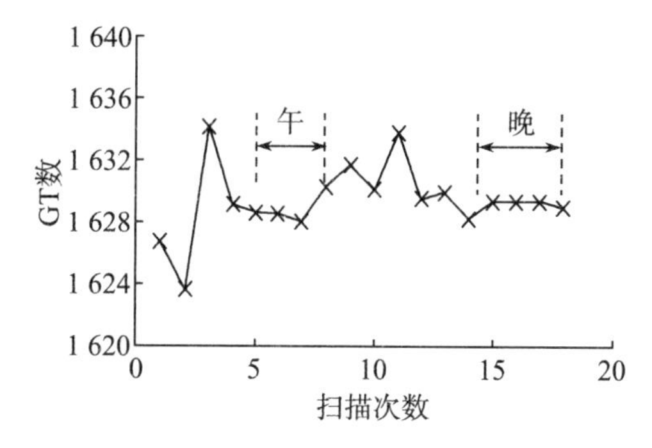

图 2 标定材料在不同扫描时刻的平均 CT 数值

Fig. 2 The average CT unmber for correcting material during different stage

3.4 体积 CT 位置校正

体积 CT 应用锥形 X 射线束，射线的边缘和中间强度有一定差别，中心射线束的强度最大，两边最小。另外，体积 CT 除了中心切片真正再现的是同一个切片，其他切片实际上是再现多个椭圆体的叠加。这两种情况导致了岩心的 k 方向(图 1)各个切片 CT 数的平均值略有差别(图 3)。为了消除这种差别，须利用与饱和岩心衰减接近的均匀柱状材料进行标定，然后对不同位置的切片进行简单校正及运算。

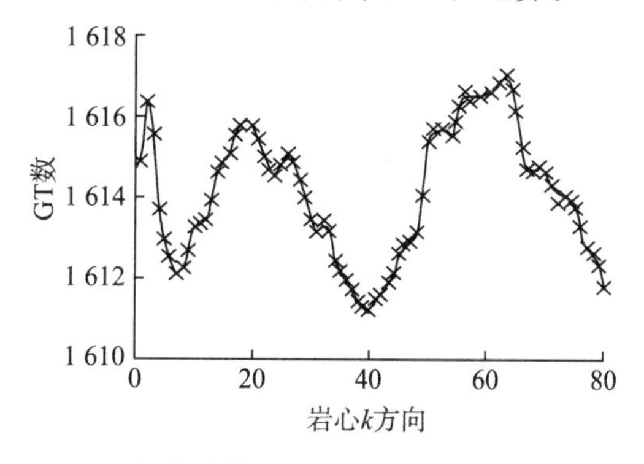

图 3 标定材料 k 方向切片 CT 数的平均值

Fig. 3 The average CT number in K direction for correcting material

4 试验结果与讨论

利用体积 CT 扫描能够计算驱替过程中油、水饱和度的分布、水驱剩余油分布状态、聚合物段塞油富集状态以及后续水驱残余油状态，同时也能非常容易地获得采收率曲线和注聚合物提高采收率的幅度。

结果表明，试验 1 的水驱采收率为原始地质储量的 60%，注入聚合物段塞后，采收率提高 16%，最终采收率达到原始地质储量的 76%。试验 2 的水驱采收率为原始地质储量的 56%，注聚合物提高采收率 15%，最终采收率达到原始地质储量 71%。利用分离器计量试验 1 和 2 的最终采收率分别为 77%和 69%，这表明对 CT 图像处理后计算的采收率数据是可靠的。同时，利用 CT 技术分析驱替过程流体饱和度对认识流体饱和度有重要的意义。

图 4 给出了试验 1 完成的轴向切片的 CT 成像处理结果。图中,1～7 注入水体积倍数分别为 0、0.36、0.83、2.58、2.84、3.21、4.55 。1 为原始含油饱和度,水驱开始阶段;5 为注入聚合物溶液段塞阶段。从中可以看出,随着水的不断注入,岩心内部含油饱和度不断减少。在水驱阶段,即使对均匀人造岩心,水驱波及仍不均匀,甚至存在着一定的指进现象。例如,在水驱至 0.36 PV 时,岩心中上部含油饱和度下降很快,而其他部位含油饱和度下降较慢。在水驱结束时,岩心中剩余含油饱和度很不均匀,尽管平均含油饱和度约为 40%,但有些部位(如岩心中部偏出口处)含油饱和度高达 60%以上,而水驱首先冲洗到的岩心中部偏上位置含油饱和度不足 20%。由于水易沿已形成的通道驱进,故仅用水驱方式很难再驱动这部分剩余油。在聚合物驱阶段,呈现出与水驱不同的 3 个特征。① 改善了水驱指进状况,出现了近似于活塞式的驱动,使水驱后剩余油得到驱动,如注聚合物后使岩心入口端含油饱和度由水驱后约 50%下降至 20%左右。② 聚合物驱使水驱后的剩余油饱和度进一步降低,如在注聚合物后岩心中上部含油饱和度由约 20%下降至 10%左右,岩心其他部位含油饱和度也有不同程度的降低。此结果支持 MENG WU WANG 得出的聚合物驱降低水驱剩余油饱和度的结果[5]。③ 驱替介质波及体积进一步扩大。例如在注聚合物前,岩心中下部靠出口处,水驱后波及不充分,而注聚合物后该部位剩余含油饱和度由 60%降至 30%以下。上述现象表明,注聚合物不权有利于波及体积的增加及流度的控制,提高了洗油效率,而且还有利于水驱后含油饱和度的进一步降低。但是有关此类实验还需要非均质模型的进一步验证。

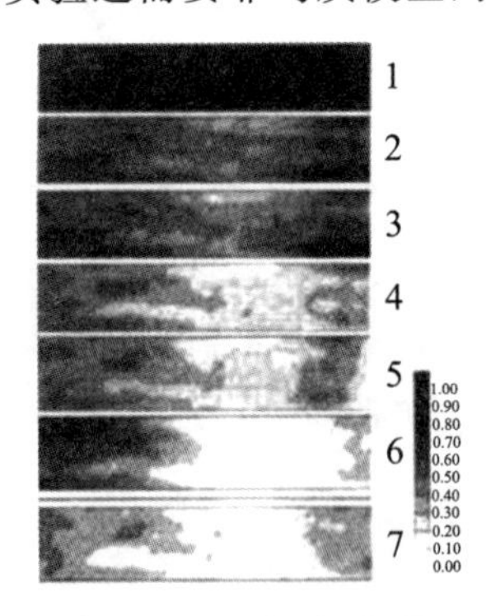

图 4 不同驱替时刻岩心内部饱和度分布

Fig. 4 Oil and water saturation in core determined by volume CT during water and polymer flooding

5 结论

在进行了 X 射线硬化处理、X 射线漂移校正、图像矩阵变换、体积 CT 位置校正等基础上,提出了用体积 CT 对岩心流体饱和度进行测量的技术。该技术对水驱、聚合物驱过程中油水饱和度分布的测量是可行的。

利用体积CT成功再现了聚合物驱替各个阶段的饱和度分布状况，绘制出采收率曲线。试验结果说明，聚合物驱有效地提高了波及的均匀性，并降低了水驱剩余油饱和度，为研究聚合物驱替机理提供了可行的手段。

参考文献

[1] Welling ton S L, Vinegar H J. X-Ray computerized tomography[C]. SPE 16983, 1987:885-898.

[2] Grag A, Kovscek A R, et al. CT scan and neural net work technology for construction of detailed distribution of residu-al oil saturation during waterflooding [C]. SPE 35737, 1996:695-710.

[3] Dana George Wreath. A study of polymer flooding and residual oil saturation [D]. Austin, the University of Texas, 1989:35-49.

[4] Seetharaman Ganapathy. Simulation of heterogeneous sand-stone experiments characterized using CT scanning[D]. Austin, the University of Texas, 1993:87-95.

[5] Mengwu Wang. Labo ratory investig ation of factors afecting residual oil saturation by poly merflooding[D]. Austin, the University of Texas, 1995:9-21.

[6] 赵碧华. 用CT扫描技术观测岩心中液流特性[J]. 石油学报, 1993, 13(1): 91-96.

[7] 李玉彬. 微焦点X射线计算机层析(CMT)及其在石油研究领域的应用[J]. CT理论与应用研究, 2000, 19(3):35-40.

编辑　孟伟铭

聚硅材料改善低渗透油藏注水效果实验

张继超　曹绪龙　汤战宏　马宝东　张书栋

胜利油田有限公司地质科学研究院

摘要:胜利油区低渗透油藏普遍存在注水压力高、吸水能力差、开发效果差等问题。应用纳米技术,开展了低渗透 砂岩油田提高注水能力实验。聚硅纳米材料对岩心的润湿性影响和增注实验研究发现,采用合适的聚硅纳米材料 处理岩心,能够改变岩石润湿性,可以较大幅度地提高岩心的渗透率,改善注水效果。

关键词:聚硅材料;润湿性;渗透率;增注;低渗透油藏

中图分类号:TE348　**文献标识码**:B　**文章编号**:1009-9603(2003)04-0059-02

引言

聚硅材料以二氧化硅为主要成分,是二氧化硅化学改性产品,憎水亲油能力极强,为颗粒或白色粉末状无机非金属纳米材料,具有无毒、无味、无污染等特性,离散颗粒尺寸一般为 10～500 nm[1]。应用范围广泛,但目前国内油田应用的报道较少。

具正电性的聚硅纳米颗粒进入岩石孔隙后,吸附于岩石孔隙内表面,其强憎水性可以降低注入水在孔隙中的流动阻力,避免粘土颗粒的水化膨胀,因而可以降低注入压力,起到一定的增注作用[2]。

1　物理特征

采用透射电镜对聚硅样品进行颗粒粒径测定, 用 0 号柴油进行聚硅样品的分散性实验。由于制备工艺的差别,不同的聚硅材料颗粒将发生不同程度的团聚,从实验结果看,进口样品分散性优于国产样品(表 1)。

表 1　聚硅样品分散性能

样品	颗粒粒径/nm	分散性	备注
101	5～100	分散性好,充分搅拌形成稳定的悬浊液,容器底部基本无未分散颗粒	进口样品
727	35～45	分散性略差,充分搅拌后容器底部有少量团状或片状颗粒	国产样品

（续表）

样品	颗粒粒径/nm	分散性	备注
829	30～50	分散性差，充分搅拌后容器底部有较多团状或片状颗粒	国产样品
808	10～25	分散性较好，充分搅拌后形成较稳定的悬浊液，容器底部颗粒较少	国产样品

2 对岩石润湿性的影响

实验选用渤南油田沙河街组天然岩心，岩心长度为2.5 cm，空气渗透率为0～50×$10^{-3}\mu m^2$，模拟油粘度为1.4 mPa·s，实验温度60℃，聚硅材料用0号柴油分散，浓度为1.5 g/L。采用自吸法测定岩心的润湿性[3]（表2）。

表2 聚硅材料对岩石润湿性影响

岩心编号	岩心直径/cm	孔隙度/%	水相渗透率/$10^{-3}\mu m^2$	处理剂驱替倍数	水湿指数	油湿指数	相对润湿指数	润湿性评价	聚硅材料
7	6.18	17.2	0.75		0.039	0	∞	亲水	
1	5.51	15.9	0.20	3	0.107	0.135	0.79	亲油	101
11	6.87	18.6	1.92	0.5	1.38	0.088	(1.57)	亲水	101
13	7.65	18.2	1.67	1	0.286	0.292	0.98	中性	101
8	6.74	17.3	0.16	3	0.022	0.222	0.1	亲油	727
10	7.46	16.2	0.62	2	0.143	0.380	0.38	亲油	727
12	6.38	16.1	1.24	1	0.112	0.157	0.71	亲油	727

实验结果表明，聚硅材料对改变岩石润湿性有一定的作用。从聚硅材料的分散液处理量分析，聚硅材料分散液用量小于1倍孔隙体积时效果变差，但对改变岩石润湿性仍有一定作用；聚硅材料分散液用量小于0.5倍孔隙体积时，其岩石润湿性变化最小（带括号数据）；聚硅材料分散液用量大于1倍孔隙体积时，岩石润湿性由亲水变化为亲油，其憎水作用可减小注入水在孔隙中的流动阻力，并有利于对残余油的驱替。

3 增注实验

3.1 实验条件

实验选用渤南油田沙河街组天然岩心，空气渗透率为30×10^{-3}～$50\times10^{-3}\mu m^2$，实验温度60℃，采用恒流法进行驱替实验[4]。聚硅材料用0号柴油分散，浓度为1.5 g/L。

3.2 实验方法

首先将岩心抽空，并饱和浓度为3%的KCl溶液，在0.5 mL/min条件下用柴油驱替一定体积的水至压力稳定，建立束缚水。用浓度为3%的KCl溶液低速驱替一定体积的油至压力稳定，建立残余油。选定一驱替速度，用KCl溶液测定岩心在此条件下的水相渗透率。然后向岩心注入10倍孔隙体积聚硅纳米材料分散液,60℃下放置12小时。再采用0.5 mL/min的驱替速度,用浓度为3%的KCl溶液驱替至压力稳定,测定岩心的水相渗透率。

3.3 实验结果

实验发现,聚硅材料对降低岩心注入压力,提高岩心的渗透率具有一定效果。国产样品727,829,808可将岩心渗透率提高25%～66%(表3)。

表3 处理前后岩心水相渗透率

岩心	聚硅样品	处理前水相渗透率/10^{-3} μm^2	处理后水相渗透率/10^{-3} μm^2	岩心渗透率保留率/%	备注
75-3	101	4.737	3.981	84	岩心端面有纳米颗粒沉积,厚度为1～2 mm
	101	4.737	2.753	58	
	101	4.737	5.104	108	清洗岩心端面
90	101	3.830	3.953	103	清洗岩心端面
206	727	5.589	6.972	125	清洗岩心端面
83	829	2.759	4.582	166	清洗岩心端面
14	808	3.761	4.702	125	清洗岩心端面

聚硅颗粒具有很强的活性,进入岩心后极易吸附在岩心内部的孔壁上,其强憎水性使岩石表面水膜中的水被部分排出,水膜变薄,有效孔径变大,渗透性有所改善。同时岩石润湿性的改变,使注入水与岩石表面的水膜不再亲和,水的流动阻力减小,表现为注入压力降低。聚硅材料的增注效果是与渗透率大小相关的。如果渗透率过大,则水滴通过孔隙时不会形成水膜,也就不会造成阻力,就没有应用聚硅材料改善润湿性的必要。如果渗透率过小,则聚硅颗粒不能进入孔隙中,增注效果同样不能得到充分发挥。

国外聚硅样品在柴油中的分散性较好,但在注入过程中会沉积或堵塞于岩心的端面或近端面,形成一种膏状沉积物,厚度为1～2 mm。国产聚硅样品虽然分散性稍差,其悬浊液静置一段时间后,大颗粒团聚后沉于容器底部,未随柴油注人岩心,对岩心端面的堵塞相对小些,但岩心端面仍有沉积层形成。

采用聚硅材料处理岩心后,由于聚硅材料颗粒粒径分布不均匀或在柴油中分散性差,较大颗粒和未分散开的粒团沉积于岩心表面,或在岩心人口端堵塞流动孔喉,导致注入压力上升,渗透率下降。清洗或反冲洗岩心端面后,渗透率可以得到明显的改善,并高于初始渗透率。所以聚硅材料处理地层后,必须对注入井的渗流端面进行清洗或反冲洗,并大排量驱替一定体积的注入水,以减少近井地带聚硅颗粒的密度,提高增注效果。

4 结论与建议

聚硅材料能够改变岩石润湿性,对改善低渗透油田的增注效果具有一定作用。

聚硅材料的注入工艺条件、作业程序的选择对聚硅材料增注效果影响较大,特别是岩心端面的清洗程度影响增注效果。

建议进一步开展聚硅材料增注机理、对低渗透油藏区块的适应性、注入工艺优化等方面研究。

参考文献

[1] 徐国财,张立德. 纳米复合材料[M]. 北京:化学工业出版社,2002.

[2] Ju Binshan, Dai Shugao, Luan Zhian et al. A study of wettability and permeability change caused by adsorption of nanometer structured polysilicon on the surface of porous media. SPE 77938.

[3] 秦积舜,李爱芬. 油层物理学[M]. 东营:石油大学出版社,2001.

[4] 孙良田. 油层物理实验[M]. 北京:石油工业出版社,1992.

reservoir flow units in Matouzhuang oilfield has close relationship with sedimentary microfacies, using space distribution of the flow units may analyse its producing reserves status and forecast the distribution of the remaining oil.

Key words: flow unit, sequence, reservoir architectural structure, Matouzhuang oilfield

Zhao Xiaodong, Shang Zhaohui, Sui Qingguo et al. Application of technique of plugging removal near-wellbore zone and profile control far-wellbore zone in middle-high permeability reservoirs of Zhuangxi. PGRE, 2003, 10(4):57-58.

Heterogeneity is serious in middle-high permeability reservoirs of Zhuangxi oilfield. Because the reinjecting wastewater doesn't qualify for a long time, near-well-bore zones of water injection wells are seriously polluted. If only acidizing measure being taken, the in-creased injection water will flow easily along relatively high permeability layers in non-uniform formation, waterflood effect for mid-low permeability oil-bearing for-mations is bad. For this type of water injection wells, a field operating technology of plugging removal near-wellbore zone and profile control far-wellbore zone is proposed in this paper. This technique has been operated eight well-times, corresponding oil wells aren't watered and good results of increment oil are achieved.

Key words: acidizing, increased injection, water plugging, operating technology,

Zhuangxi oilfield

Zhang Jichao, Cao Xulong, Tang Zhanhong et al. Experiment of polysilicic material for improving waterflooding effect in low permeability oil reservoir. PGRE, 2003, 10(4):59-60.

High in jected water pressure, poor water absorbing capacity, poor development effect and some other questions generally appear in low permeability oil reservoir of Shengli petroliferous province. Experiments of enhancing injection capacity in low-permeability sand-stone reservoir are conducted by nanometer technology. Research on the influence of polysilicic nanometer material on core wettability and augmented injection found that the core treated by polysilicic nanometer material with proper size may change rock wettability, enhance core permeability by a big margin and improve water-flooding effect.

Key words: polysilicic material, wettability, permea-bility, augmented injection, low-permeability oil reservoir

Geng Hongzhang, Pan Juling, Zhou Kaixue. Experimental study on paraffin control by magnetic treatment of crude oil. PGRE, 2003, 10(4):61-63.

In order to study the influence of the effect of the paraffin control by strong magnetic treatment of crude oil, the magnetic treatment conditions are selected based on the orthogonal experiment design method. Optimization parameters of magnetic treatment on degassed crude of Gudao are obtained, that is 200mT of magnetic induction density, four peak, 70℃ of magnetic treat ment temperature; and the influencing factors of paraffin control are analyzed through measuring the cooling curves of crude oil before and after treating with strong magnetic field. The result indicates that the magnetic induction density and magnetic treatment temperature are the important factors to the effectiveness of paraffin control. But the effectiveness of paraffin control is not ideal with the increasing of magnetic induction density.

Key words: magnetic treatment, paraffin control, orthogonal experiment design

Wu Yuntong, Lin Yonghong, Tang Zhanhong et al. Formation damage by injected water quality in He 141 block. PGRE, 2003, 10(4):64-66.

He 141 block is a type of oil reservoir with low porosity and low permeability. Owing to its low formation energy, it needs to be developed by water flooding. Rule and degree of Es, formation damage by different qualities of injected water are studied in this paper. It gives recommendatory criteria of water quality and suggestion of treating injected water.

Key words: formation damage, clay mineral, sensitivity, permeability retention rate, suspended matter, median grain diameter, bacteria

Shang Tonglin. Fine filtration technology of oilfield produced water. PGRE, 2003,10(4):67-68.

Fine filteration technology of the produced water during the oilfield production is introduced. Application of microfiltration treatment of fiber, filtering element and diatomaceous earth dynamic film to oilfield produced water is summarized. And various filter media are evaluated on economics and technology. The research shows that at the initial period, most filters can purify the water to 1～3 mg/L suspended solid content at 1～2 μm median grain diameter, and 5～8 mg/L oil content. The problem of backwashing in the filtration media influenced the operation life of filters. There fore, the paper points out the development direction for fine filtration technology of oilfield produced water.

Key words: low permeability, microfiltration, suspen-ded solid content, grain diameter, oil content

Tan Heqing, Peng Cuncang, Li Wenhuaet al. Ap-plication of cased-hole formation resistivity log in monitoring distribution of remaining oil. PGRE, 2003, 10(4):69-70.

编辑 高 岩

生物聚合物黄胞胶驱油研究

刘 坤　宋新旺　曹绪龙　祝仰文

胜利油田有限公司地质科学研究院

摘要:为了探索高温高盐油藏提高采收率的方法,拓展聚合物驱的应用领域,对三种生物聚合物黄胞胶进行了增粘性能、耐温性能和抗盐性能的初步筛选。建立了生物聚合物的评价方法,确认了生物聚合物黄胞胶的应用条件,提出了改进建议。结果表明:生物聚合物黄胞胶适合于地层温度小于80℃,地层水矿化度大于20 000 mg/L的油藏。

关键词:生物聚合物;黄胞胶;应用性能;评价方法

中图分类号:TE357.46　文献标识码:B　文章编号:1009-9603(2003)05-0068-03

引言

目前,胜利油区聚合物驱规模进一步扩大,聚合物驱已经成为老油田稳产的主要措施。根据胜利油田“十五”三次采油规划的安排和有限公司的统一部署,须进一步扩大聚合物驱的应用领域,在Ⅰ、Ⅱ类油藏取得突出效果的基础上,加大高温高盐油藏聚合物驱提高采收率的研究力度。但由于聚丙烯酰胺的耐温抗盐性差,技术受到限制,一般只适用于地层温度小于80℃,原油粘度小于70 mPa·s,地层水矿化度小于10 000 mg/L的单元[1];生物聚合物与合成聚合物相比具有在高盐度下仍保持高粘度,溶解性好,不会发生沉淀和絮凝;溶液在剪切力作用下发生剪切变稀,表观粘度下降,但分子结构不发生改变,当剪切力减小或消失时,粘度可以恢复的特性[2,3]。但是,常规生物聚合物的价格较高,过滤性能差以及生物稳定性较差,限制了应用规模。

1　试验

1.1 仪器设备

DV-Ⅲ型布氏粘度计,筛网粘度计,抽空装置,RS-150流变仪,物理模拟试验装置等。

1.2 试验条件

生物聚合物样品Flocon 4800CT的有效含量为7.0%;Xanthangum的有效含量为

7.2%;ZB1的有效含量为2.98%。试验温度为50,60,70,80和90℃。样品浓度为500,750,900,1 000和1 200 mg/L。试验用水有清水、胜坨地层水、临盘地层水、埕岛地层水、东辛地层水,矿化度分别为618,17 000,28 317,31 549和59 802 mg/L。试验用油为胜坨油田坨二站外输原油。

1.3 发酵液浓度测定

称取发酵液50 g左右,在搅拌下加入2倍体积的乙醇中,即出现沉淀,沉淀用滤布滤出,用适量丙酮洗涤一次,将沉淀滤出拧干、破碎,置于50℃恒温干燥箱中烘至恒重,准确称重。首先将其配置成5 000 mg/L的母液,无茵搅拌器搅拌2分钟,然后稀释到所需浓度,无茵搅拌器搅拌1分钟、磁力搅拌10分钟。

2 生物聚合物黄胞胶的筛选

2.1 增粘性能

用矿化度为59 802 mg/L的水,分别将三种生物聚合物配制成500,750,900,1 000,1 200,1 500和1 800 mg/L的溶液,在温度80℃条件下,测定其粘度(图1)。在同一温度、同一矿化度条件下,生物聚合物黄胞胶溶液的粘度随浓度的增加而增大,相同浓度的三种生物聚合物样品中ZB1和Flocon 4800CT粘度相对较高,Xanthangum粘度相对较低。

2.2 粘度与温度的关系

用矿化度为59 802 mg/L的水,分别将三种生物聚合物配制成1 000 mg/L的聚合物溶液,在50,60,70,80和90℃下,分别测定其粘度(图2)。温度较低时,生物聚合物具有很好的增粘能力,温度越高,增粘能力越差。在同一浓度、同一矿化度条件下,相同温度时的粘度比较三种生物聚合物样品中ZB1和Flocon 4800CT粘度相对较高,Xanthangum粘度相对较低。

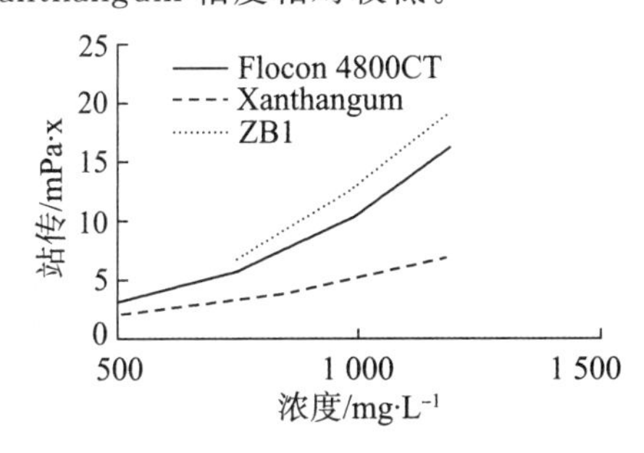

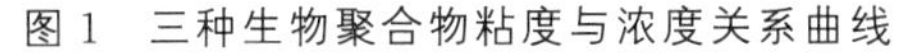
图1 三种生物聚合物粘度与浓度关系曲线

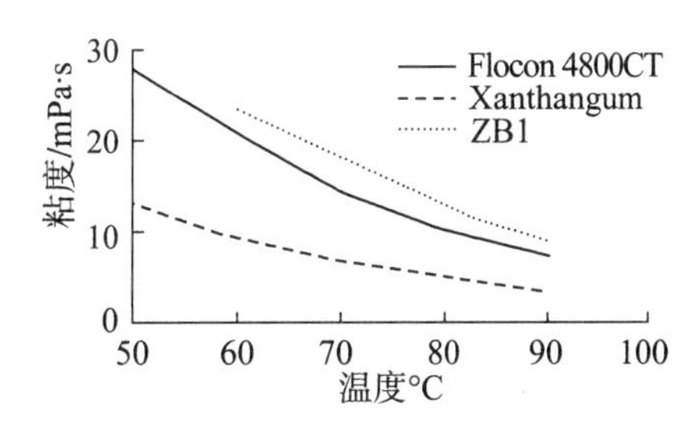

图2 三种生物聚合物粘度与温度关系曲线

2.3 粘度与盐水矿化度的关系

从聚合物分子构象来看,聚丙烯酰胺呈柔性链,盐敏性很强,耐盐性差;而生物聚合物黄胞胶呈螺旋刚性链,耐盐性很好。试验中分别用五种矿化度的水,将生物聚合物配制成1 000 mg/L的溶液,在温度80℃下测定其粘度(图3)。在地层温度下,生物聚合物溶液的粘度随盐度变化不大,对盐度不敏感,并且随盐度增加略有增大。

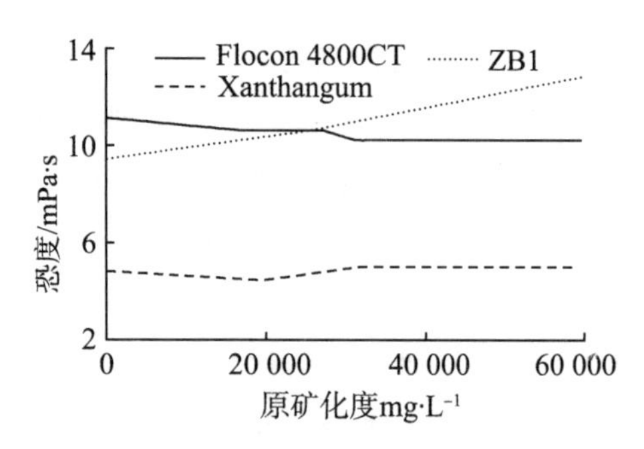

图 3 三种生物聚合物粘度与矿化度关系曲线

据初步评价,ZB1 的增粘能力、耐温性和耐盐性与 Flocon 4800CT 基本相当。由于国产样品运输方便。从性能、价格和运输等综合考虑、初步筛选 ZBI 作进一步应用性能评价并进行驱油效率试验,确定生物聚合物的应用条件。

3 应用性能评价

3.1 抗机械剪切性能

当聚合物溶液发生形变和流动时,它所承受的剪切应力或拉伸应力(或两者组合)增加到足以使聚合物分子断裂时,聚合物将出现机械降解。聚合物溶液在通过炮眼和进入地层时所承受的应力作用最大,为了模拟这种剪切作用,将 ZB1 黄胞胶溶液通过加压进行毛管剪切,剪切后用布氏粘度计在 6 转/min 下测粘度。在剪切速率为 10^5 l/s 时生物聚合物 ZB1 粘度保留率为 98%(放置 1 小时),说明生物聚合物经剪切后放置一段时间,粘度能基本恢复。在注入流程设计中,对于聚丙烯酰胺应尽量避免或降低机械剪切。而对于黄胞胶,为破坏其细菌细胞残骸及微凝胶,则需较高的机械剪切。因此在黄胞胶注入流程设计中,应考虑加两道 1 mm 孔板进行剪切,提高注入能力。

3.2 过滤性能

用 0.2 μm 的膜过滤矿化度为 19 481 mg/L 的水,配制成 1 000 mg/L 的聚合物溶液,生物聚合物 ZB1 的过滤性能较差,用与聚丙烯酰胺同样的方法测定过滤比,基本上测不出。在试验中,用各种方法试图改善生物聚合物溶液的过滤性能,如母液配制时,延长无菌搅拌器高速剪切的时间,只是使溶液粘度降低,而过滤性能没有改善;较长时间的低速剪切水合 也没有改善过滤性能。生物聚合物的过滤性能差,导致溶液的注入能力差,主要是由于有不完全溶解聚合物的微小聚集体、固有的细菌细胞残骸和发酵过程中残余的蛋白质物质所致。其直径为 0.3~0.5 mm,长度为 0.7~2.0 mm。除去这些不理想的物质的方法有筒式和硅藻土过滤法、粘土絮凝法以及化学和酶沉降法。

3.3 溶液的长期稳定性

长期稳定性通常是指聚合物溶液在地下油藏岩石孔隙中,能够保持其粘度而不发生降解的性质。聚合物溶液进入地层后,由注入井到生产井往往需要几个月甚至几年的时间,在这期间会受到许多因素的影响导致粘度的损失,而聚合物溶液能否维持较高的粘度是影响驱油效果的重要因素之一,因此,研究聚合物溶液在地层条件下的稳定性是很有必要的。

造成聚合物溶液粘度下降的原因主要是高分子的降解，包括机械降解、化学降解和生物降解。机械降解主要发生在注入设备和近井地带的高速流区，造成分子链的断裂；而化学降解和生物降解则多发生在地层深部。生物聚合物在经过注入设备和近井地带时的剪切不足以造成其机械降解，主要表现为细菌降解和热氧降解 4。所以筛选合适的粘度稳定剂是非常重要的。用矿化度为 19 481 mg/L 的水按以上提供的方法配制生物聚合物溶液，溶液稀释以及加入添加剂后，测定粘度，分装后，无氧条件下，抽空、充氮反复几次，然后烧结密封。暴氧条件下，装入样品后，直接烧结密封。放入 80℃烘箱中，定期测定其粘度的变化(图 4)。生物聚合物 ZB1 在高温下的长期热稳定性较差。在基本不含氧的条件下，90 天的粘度保留率为 62.4%；在暴氧条件下，90 天的粘度保留率为 10.4%；添加稳定剂后，90 天的粘度保留率为 62.8%。所以，在现场实施时，要加入适量的粘度稳定剂。

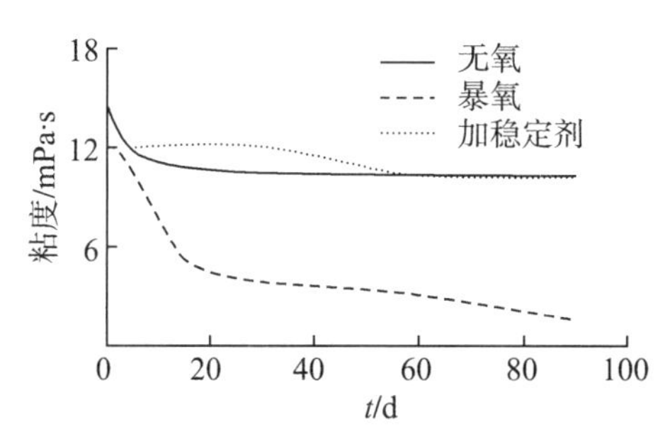

图 4 黄胞胶的热稳定性

3.4 驱油效率试验

驱油效率试验是驱油过程的综合反映，通过驱油效率试验，可以比较直观地对比分析聚合物驱油过程的开发特征及效益，为编制开发方案提供依据。

试验用石英砂，填充管式模型，单管模型长 30 cm，截面积 4.9 cm^2。试验温度为 80℃，用坨二站脱水脱气原油配制模拟油，80℃下粘度为 25.27 mPa·s。用不同矿化度的盐水，在同等经济条件下注入黄胞胶和 3530S 聚合物。试验步骤：① 模型抽空饱和水，用称重法测出孔隙体积；② 模拟油驱水，求出油饱和度；③ 地层水进行水驱，至含水 95%时注入 0.3 PV 一定浓度的黄胞胶和 3530S 聚合物，然后水驱至含水 98%以上，结束试验(表 1)。结果表明，在价格相同的条件下，水的矿化度小于 20 000 mg/L 时，聚丙烯酰胺比黄胞胶效果要好；当水的矿化度大于 20 000 mg/L 时，黄胞胶比聚丙烯酰胺效果要好。所以，黄胞胶更适合用于矿化度高的油藏。

表 1 驱油试验模型参数及试验结果

总矿化度/mg·L^{-1}	渗透率/μm^2	化学剂	水驱采收率%	最终采收率%	提高采收率%
5 727	1.52	ZB1S	70.3	82.9	12.6
5 727	1.38	3530S	63.3	81.0	17.7
19 481	1.37	ZB1S	68.2	81.5	13.3

(续表)

总矿化度/mg·L^{-1}	渗透率/μm^2	化学剂	水驱采收率%	最终采收率%	提高采收率%
19 481	1.44	3530S	69.4	82.4	13.0
59 802	1.40	ZB1S	65	77.9	12.9
59 802	1.48	3530S	65.2	75.7	10.5

4 结论

在高温、高矿化度条件下,生物聚合物黄胞胶具有粘度较高,稳定性较强,抗剪切性能较好;一般适合于矿化度大于 20 000 mg/L 的油藏。在目前条件下,如果将生物聚合物黄胞胶应用于油藏,建议在过滤性能方面做一些改进,并添加合适的杀菌剂,提高它的热稳定性。

参考文献

[1] 梁凤来.高温三次采油用黄原胶的稳定剂[J].油田化学,1990,7(2),198-202.

[2] 赵颖,张华,朱金才.黄胞胶溶液在油田开发中的应用[J].断块油气田,1997,4(4):9-14.

[3] 张柏英,孙景民,康恒.黄胞胶驱油现场试验效果分析[J].钻采工艺,1999,22(2):70-71.

[4] 张柏英,孙景民,康恒.胜利油田孤东七区油井黄胞胶驱油应用效果分析[J].石油与天然气化工,1999,28(1):49-52.

编辑 高 岩

He Yongming, Li Aifen, Tao Jun et al. A new productivity prediction equation of vertical wells in plastic fluid type heavy oil reservoirs. PGRE, 2003, 10(5):57-58.

Heavy oil always has non-Newton rheological behavior in the formation. In order to study its influence on productivity of vertical wells, we assume fluid as Bingham one. Moving radius and moving boundary pressure are derived based on its startup pressure gradient. Productivity of vertical wells is derived on the above three parameters. Compared with other equations which being considered without the above 3, influence of sone parameter, especially startup pressure gradient, on the productivity of vertical wells is analyzed in the new equation. Calculation correlation shows that non-New-ton rheological behavior of the heavy oil has certain influence on productivity prediction results.

Key words: vertical well, heavy oil, startup pressure gradient, productivity prediction, moving radius

Zeng Liufang, Lu Yunzhi, Li Linxiang. Research on remaining oil distribution rule in extra-high water cut stage of Gudong oilfield. PGRE, 2003, 10(5):59-61.

Influencing factors of remaining oil distribution in extra-high water cut stage of Gudong oilfield are studied u-sing a large number of data of production, log and testing. It points out that injetion-production correlation, macroscopic channel, cumulative injection pore volume and so on are the main influencing factors of remaining oil distribution. The distribution rule is summarized as "two-section, three-layer, four-area". That is to say, the remaining oil is enriched in interlayer low-permea-bility section and low watered out section influenced by interlayer; formations with large difference between interlamination and poor physical property, low watered out layer with low injection pore volume and poor pro-ducing or nonproducing thin layer; the position of bad dynamic injection-production correlation in plane, lag area influenced by interwell percolating properties and macroscopic channel, bypassing area and largely rug-ged microstructure high position. In addition, the degree of knowing the oil-bearing formation is also a non-neglectful potential. The main methods of tapping the potential of remaining oil in old oilfield are discussed, which have direct significance in field application.

Key words: extra-high water cut stage, remaining oil distribution, influencing factor, Gudong oilfield

Liu Hua, Sheng Ruyan. Recovery method of attic remaining gas in Es, gas reservoir with edge water in Sheng 1 area of Shengtuo oilfield. PGRE, 2003, 10(5):62-63.

Initial unstable working system had resulted in locally watered out in Es, gas reservoir with edge water of Shengtuo oilfield. On the basis of comprehensive analysis, potential-tapping program is put forward using horizontal well to exploit the remaining gas. Develop-ment results of various well patterns are studied using extended Black Oil software SimBest Ⅱ. The research indicates that side-track horizontal well with horizontal length equal 250 meters is optimum, which can provide successful experiences for similar type of reservoirs.

Key words: water-drive gas reservoir, remaining gas, edge-water fingering, horizontal well, side-track-well, numerical simulation

Geng Zhaohua, Wang Jihua, Hao Zhenxian et al. Technologies of EOR in late stage of high water cut in Wenzhong oilfield. PGRE, 2003, 10(5):64-65.

Geologic characteristics of complicated fault-block oil reservoir in Wenzhong area are introduced. Application and effect of some technologies of double target direc-tional well, sidetracking, updating-water injector, well pattern recombination, adjusting displacement by profile control of water injection wells are discussed. Producing degree by waterflooding rises from 55.2% to 65.4%, oil recovery efficiency from 43.03% to 46.59% and recoverable reserves increased by 69×10^4 t. It gives a new way to keep sustained and high-efficiency exploitation for the complicated fault-block oil field.

Key words: complicated fault-block oil reservoir, high water cut stage, directional well, sidetracking well, recovery efficiency

Wu Yongchao, Zhang Huazhen. Experimental study of low-alkaline ASP ternary combination flooding. PGRE, 2003, 10(5):66-67.

For the particularity of ASP combination flooding, the concentration changes of alkaline and polymer have certain influence on viscosity and interfacial tension of the system at the given surfactant concentration. Optimum range of alkaline concentration in ASP system is selected with oil displacement efficiency experiments. The result shows that for the ASP combination flooding system used in this test, while considering the ultralow interfacial tension, the viscosity of the system must be considered, and its ultralow interfacial tension is not the essential condition of enhancing oil recovery.

Key words: low alkaline, interfacial tension, viscosi-ty, ternary combination flooding

Liu Kun, Song Xinwang, Cao Xulong et al. Oil displacement research by biopolymer xanthan gum. PGRE, 2003. 10(5):68-70.

In order to study on enhancing oil recovery in the high temperature and high brine reservoir, and in order to widen application range of polymerflooding, the viscosi-fying, heat-resisting and salt-resisting behaviors of three biopolymer xanthan gums have been screened primari-ly. The evaluation method of biopolymer is estab-lished. The applying conditions of biopolymer xanthan gum are determnined. The improving suggestion is put forward. The result showed that biopolymer xanthan gum is suitable to the reservoirs with formation temper-ature less than 80℃ and formation water salinity higher than 20,000 mg/L.

Key words: biopolymer xanthan gum, applying behavior, evaluation method

Song Kaili, Zhang Junqing, Liu Changfu. Present application situation and improved measure of sucker rod pump at Shengli petroliferous province. PGRE, 2003, 10(5):71-72.

Quite a number of sucker rod pumps in Shengli petro-liferous province have been troubled by the problems of low pump efficiency, short turnaround and so on. The causes of these problems are analyzed from the aspects of coordination between the capacity of fluidsupply and flowing back, influence of the pumping parameters on the oil pump eficiency, influence of eccentrically wearing between sucker rod and tubing on lifting fluid, the pump structure and its application. On this basis, some improved countermeasure and suggestions are presented.

Key words: sucker rod pump, present situation, influencing factor, advance, suggestion, Shengli petrolif-erous province

Xiong Youming, Guo Shisheng, Liu Jing. Study on A3 well completion method in Pinghu oilfield of East China Sea. PGRE, 2003, 10(5):73-76.

According to the properties of Pinghu oilfield in East China Sea, the well completion method of A3 horizontal well is determined as initial well completion by slot-ted liner and later sand control by metal fibre screen pipe. For the well completion method by slotted liner, well completion parameters, single slotted liner length and pipe string structure in well completion are opti-mized. The software of HWCS developed by ourselves, has been used to predict the productivity of slotted liner well completion. After A3 well completion, oil produc-tion is about 1 200 m^3/d without sand production, which tallies with the prediction results. It shows that the prediction is 100% correct.

Key words: horizontal well, well completion method, slotted liner, metal fibre sand control screen pipe, sand control, productivity of well completion, Pinghu oil-field of East China Sea

Liu Yuanliang, Jiao Qiaoping, Li Jun et al. Re-search on feasibility of perforating and fracturing matching technology in Daqingzi well exploration area of Jilin oilfield. PGRE, 2003, 10(5):77-78.

Aiming at the low successful ratio of fracturing, low productivity after fracturing and low fracture scale in Daqingzi well exploration area, the matching technology between perforating and fracturing is studied. The influence of perforating density, bore diameter, perforating depth and orientation on fracturing is analyzed. Through the demonstration indoor and theoretic calculation, the perforating parameters suitable to Daqingzi well exploration area are given.

Key words: perforating, fracturing, parameter optimi-zation, match technology, Daqingzi well exploration area

Wang Dengqing, Sun Xiuzhao, Xie Jinchuan et al. Development of double-layer water injection sand screening pipe and its application in Chengdao oil-field. PGRE,

2003，10(5)：79-80.

Chengdao oilfield of ShengLi has faced the waterflood-ing problem since developed on a large scale in 1995. The difficult points of offshore oilfield lie in serious sand flow and short-term validity of chemical sand con-trol. The traditional single-string waterflood or double-strings waterflood cant satisfy the needs of sand control and waterflood at the same time. The single well opera-tion expenses are high. Aim at the current situations of well conversion in Chengdao oilfield, we develop the integrated string with zonal injection and sand control. This paper introduced the structure, principle, technology and characteristic of the double-layer waterflood and sand screen pipe and its complete technology. It has solved the difficult problems of multistage zonal injecting and sand control of Chengdao oilfield, and it will have wider application prospects.

Key words: water injection, sand screen, develop-ment, application, Chengdao oilfield

Exact solutions for nonlinear transient flow model including a quadratic gradient term

Cao Xulong(曹绪龙)[1,2], Tong Dengke(同登科)[3], Wang Ruihe(王瑞和)[3]

(1. Institute of Chemical Physics, Chinese Academy of Sciences, Lanzhou 730 000, P. R. China;

2. Geological Science Research Institute, Shengli Oilfield Co. Ltd, Dongying, Shandong 257000, P. R. China;

3. Department of Applied Mathematics, University of Petroleum, Dongying, Shandong 257061, P.R. China)

(Communicated by ZHANG Hong_qing)

Abstract: The models of the nonlinear radial flow for the infinite and finite reservoirs including a quadratic gradient term were presented. The exact solution was given in real space for flow equation including quadratic gradiet term for both constant rate and constant pressure production cases in an infinite system by using generalized Weber transform. Analytical solutions for flow equation including quadratic gradient term were also obtained by using the Hankel transform for a finite circular reservoir case. Both closed and constant pressure outer boundary conditions are considered. Moreover, both constant rate and constant pressure inner boundary conditions are considered. The diference between the nonlinear pressure solution and linear pressure solution is analyzed. The diference may be reached about 8% in the long time. The efect of the quadratic gradient term in the large time well test is considered.

Key words: nonlinear flow; integral transform; analytical solution; well test analysis

Chinese Library Classification: TE312 Document code: A

2000 Mathematics Subject Classification: 83C15; 35Q35; 34A25

Introduction

In the certain operations, the small pressure gradients may cause significant error of the predicted pore pressure, such as hydraulic fracturing, large-drawdown flows, drill-stem test and large-pressures pulse testing. However, classical pressure transient models are assuming small compressibility or small pressure gradient, and the nonlinear quadratic gradient term is neglected[1]. In order to describe the effect of the quadratic gradient term, Odeh and Babu[2] built the nonlinear pressure transient model considering the effect of the quadratic gradient term. They discover that the absolute change in well bore pressure is different for injection and pumping conditions, unlike what is predicted by the linear solutions. Finjord and Aadnoy[3] presented steady state and approximate psudosteady state solution. Wang and Dusseault[4] used the same technique to predict pore pressure around boreholes. Chakrabarty et al.[5] showed a noticeable effect of the nonlinear term for the constant discharge case with high injection rate and small reservoir transmissibility. Tong Dengke[6-8] discussed the flow problem of fluid in double porous media including the effects of the quadratic-gradient term. But the exact solution in real space has not given, let alone the constant-pressure producing and finite formations. Real reservoirs are finite. Thus, it is necessary that the pressure transient model including the nonlinear quadratic gradient term and well test theory be perfected.

In this paper, we give the exact solution in real space for flow equation including quadratic gradient term for both constant-rate and constant pressure production cases from in an infinite system by using generalized Weber transform. Analytical solutions for flow equation including quadratic gradient term are also obtained by using the Hankel transform for a finite circular reservoir case. Both closed and constant pressure outer boundary conditions are considered. Moreover, both constant rate and constant pressure inner boundary conditions are considered. The nonlinear transient pressure behavior characteristic of fluids through porous media including a quadratic gradient term is analyzed. The sensitivity of the system response to the nonlinear parameter and outer boundary are also examined in detail. Conventional well test model is a special case of the nonlinear well test model with a quadratic gradient term(that is, $\alpha = 0$).

1 The Nonlinear Pressure Transient Analysis Model Considering the Effect of the Quadratic Gradient Term

The following assumptions are made in deriving the mathematical model considered in the present study:

1) The porous medium has a uniform thickness h, and the radial flow takes place around well bore with the well penetrating the entire formation thickness;

2) The porous medium is homogeneous and isotropic;

3) Porosity and permeability are constant;

4) Fluid compressibility is constant, and fluid viscosity is constant.

Analytical pressure-transient problem for single-phase radial flow of a slightly compressible fluid into a well of the cylindrical-symmetry reservoir center with constant rate or constant pressure production is described as follows:

$$\frac{\partial 2 PD}{\partial r2 D}+\frac{1}{rD}\frac{\partial PD}{\partial rD}-\alpha\left(\frac{\partial PD}{\partial rD}\right)^2=\frac{\partial PD}{\partial TD}, \tag{1}$$

with the initial condition given by

$$p_D|_{t_D=0}=0. \tag{2}$$

The inner boundary condition is as follows:

$$\left.\frac{\partial p_D}{\partial r_D}\right|_{r_D=1}=-1. \tag{3}$$

or

$$p_D|_{r_D=1}=1 \tag{4}$$

Considering well bore storage, the inner boundary condition becomes

$$C_D\left.\frac{\partial p_D}{\partial t_D}\right|_{r_D=1}-\left.\frac{\partial p_D}{\partial r_D}\right|_{r_D=1}=1 \tag{5}$$

The outer boundary conditions are as follows:

$$\left.\frac{\partial p_D}{\partial r_D}\right|_{r_D=R}=0 \tag{6}$$

or

$$p_D|_{r_D=R}=0 \tag{7}$$

or

$$\lim_{r_D\to\infty} p_D=0. \tag{8}$$

For the constant rate production, dimensionless pressure and dimensionless compressibility are defined as

$$p_D=\frac{2\pi kh(p_i-p)}{\mu q},\alpha=\frac{q\mu c}{2\pi kh}$$

For the constant rate production, dimensionless pressure and dimensionless compressibility are defined as

$$p_D=\frac{(p_i-p)}{p_i-p_w},\alpha=c(p_i-p_w)$$

The rest are defined as

$$r_D=\frac{r}{r_w},t_D=\frac{kt}{c\phi\mu r_w^2},C_D=\frac{C}{2\pi\phi c_t hr_w^2},$$

where p_w is the well bore pressure. k, are the porosity and permeability respectively. μ is the fluid viscosity, c is fluid compressible coefficient, p_i is the initial pressure.

The six typical initial value and boundary value problems for the nonlinear flow equation with the effect of quadratic gradient term:

1) The flow Problem Ⅰ are made of Eqs. (1), (2), (3) and (5) for the constant-rate production with closed outer boundary condition;

2) The flow Problem Ⅱ are made of Eqs. (1), (2), (3) and (6) for the constant-rate production with finite constant_pressure outer boundary condition;

3) The flow Problem Ⅲ are made of Eqs. (1), (2), (4) and (5) for the constant-pressure production with closed outer boundary condition;

4) The flow Problem Ⅳ are made of Eqs. (1), (2), (4) and (6) for the constant-pressure production with a finite constant_pressure outer boundary condition;

5) The flow Problem Ⅴ are made of Eqs. (1), (2), (3) and (7) for the constant-rate production with an infinite outer boundary condition;

6) The flow Problem Ⅵ are made of Eqs. (1), (2), (4) and (7) for the constant-pressure production with an infinite outer boundary condition.

2 Exact Solutions for the Flow Problem of Fluid Through the Porous Media Considering the Effect of the Quadratic Gradient Term

2.1 The exact solution for the flow Model Ⅰ

Because the partial differential equation in Model Ⅰ is a nonlinear differential equation, the equation can not be solved. Introducing transform

$$p_D = -\alpha^{-1} \ln x \tag{8}$$

Model Ⅰ may be simplified as

$$\frac{\partial^2 x}{\partial r_D^2} + \frac{1}{r_D}\frac{\partial x}{\partial r_D} = \frac{\partial x}{\partial t_D} \tag{9}$$

$$x\big|_{t_D=0} = 1 \tag{10}$$

$$\left(\frac{\partial x}{\partial r_D} - \alpha x\right)\bigg|_{r_D=1} = 0 \tag{11}$$

$$\frac{\partial x}{\partial r_D}\bigg|_{r_D=R} = 0. \tag{12}$$

The generalized Hankel transform[5] is defined as

$$\tilde{x}(s_n, t_D) = \int_1^{r_{De}} r_D B(s_n r_D) x(r_D, t_D) \mathrm{d}r_D \tag{13}$$

where

$$B(s_n r_D)=[s_n J_1(s_n)+\alpha J_0(s_n)]Y_0(s_n r_D)-[s_n Y_1(s_n)+\alpha Y_0(s_n)]J_0(s_n r_D) \tag{14}$$

where s_n are the n_th positive root of the following equation:

$$[sJ_1(s)+\alpha J_0(s)]Y_1(sr_{De})-[sY_1(s)+\alpha Y_0(s)]J_1(sr_{De})=0 \tag{15}$$

Applying the Hankel transform to Eqs. (9)～(12)yield

$$\frac{\partial\tilde{x}}{\partial t_D}=-s_n^2\tilde{x},\tilde{x}|_{t_D=0}=\frac{2\alpha}{\pi s_n^2}$$

The above differential equation is solved, and one has

$$\tilde{x}=\left(\frac{2\alpha}{\pi s_n^2}\right)\exp[-s_n^2 t_D] \tag{16}$$

The application of the inversion of the Hankel transformation to Eq. (16)yields

$$x(r_D,t_D)=\pi\alpha\sum_{n=1}^{\infty}\frac{\exp[-s_n^2 t_D]B(s_n r_D)J_1^2(s_n r_{De})}{\{[s_n J_1(s_n)+\alpha J_0(s_n)]^2-(s_n^2+\alpha^2)J_1^2(s_n r_{De})\}} \tag{17}$$

Inserting (17) into (8), yields

$$p_D(r_D,t_D)=-\frac{1}{\alpha}\ln\left[\pi\alpha\sum_{n=1}^{\infty}\frac{\exp[-s_n^2 t_D]B(s_n r_D)J_1^2(s_n r_{De})}{\{[s_n J_1(s_n)+\alpha J_0(s_n)]^2-(s_n^2+\alpha^2)J_1^2(s_n r_{De})\}}\right] \tag{18}$$

The pressure solution of well bore($r_D=1$)can be simplified as

$$p_{wD}(t_D)=-\frac{1}{\alpha}\ln\left[\pi\alpha\sum_{n=1}^{\infty}\frac{\exp[-s_n^2 t_D]J_1^2(s_n r_{De})}{\{[s_n J_1(s_n)+\alpha J_0(s_n)]^2-(s_n^2+\alpha^2)J_1^2(s_n r_{De})\}}\right] \tag{19}$$

The inverse solution in Laplace transform for Model Ⅰ with the wellbore storage can be written as

$$p_D(r_D,t_D)=-\frac{1}{\alpha}\ln\{1-\alpha L^{-1}[K_0(r_D\sqrt{s}I_1(r_{De}\sqrt{s})+I_0(r_D\sqrt{s})K_1(r_{De}\sqrt{s})]\times[s(C_D s+\alpha)\lambda_1+s\sqrt{s}\lambda_2]^{-1}\}, \tag{20}$$

where $\lambda_1=I_0(\sqrt{s})K_1(r_{De}\sqrt{s})+K_0(\sqrt{s})I_1(r_{De}\sqrt{s})$, $\lambda_2=K_1(\sqrt{s})I_1(r_{De}\sqrt{s})-I_1(\sqrt{s})K_1(r_{De}\sqrt{s})$.

2.2 The exact solutions for Models Ⅱ～Model Ⅵ

The other solutions are obtained by using the Hankel transform and Weber transform, and listed in Table 1.

Table 1 Exact solutions for nonlinear transient flow model including a quadratic gradient term

Type	The pressure solutions in real space	s_n satisfied equation
Ⅱ	$p_D(r_D,t_D) = -\frac{1}{\alpha}\ln\left[\frac{\alpha\ln r_D+1}{\alpha\ln r_{De}+1}+\frac{\pi\alpha(1-\alpha\ln r_{De})}{(\alpha\ln r_{De}+1)}\times \sum_{n=1}^{\infty}\frac{\exp[-s_n^2 t_D]B_1(s_n r_D)J_0^2(s_n r_{De})}{[s_n J_1(s_n)+\alpha J_0(s_n)]^2-(s_n^2+\alpha^2)J_0^2(s_n r_{De})\}}\right]$	$B_1(s_n r_D)=[s_n J_1(s_n)+\alpha J_0(s_n)]Y_0(s_n r_D)-[s_n Y_1(s_n)+\alpha Y_0(s_n)]J_0(s_n r_D)$, s_n is the n th positive root of the following equation $B_1(sr_{De})=0$.
Ⅲ	$p_D(r_D,t_D) = -\frac{1}{\alpha}\ln\left[e^{-\alpha}+\pi(e^{-\alpha}-1)\times\sum_{n=1}^{\infty}\frac{\exp[-s_n^2 t_D]B_2(s_n r_D)J_1^2(s_n r_{De})}{\{J_0^2(s_n)+J_1^2(s_n r_{De})\}}\right]$	$B_2(s_n r_D)=Y_0(s_n)J_0(s_n r_D)-J_0(s_n)Y_0(s_n r_D)$, s_n is the n th positive root of the following equation $Y_0(s)J_1(sr_{De})-J_0(s)Y_1(sr_{De})=0$.
Ⅳ	$p_D(r_D,t_D) = -\frac{1}{\alpha}\ln\left[e^{-\alpha}+\frac{1-e^{-\alpha}}{\ln r_{De}}\ln r_D+\pi(e^{-\alpha}-1)\sum_{n=1}^{\infty}\frac{\exp[-s_n^2 t_D]B_3(s_n r_D)J_0^2(s_n r_{De})}{\{[J_0^2(s_n)-J_0^2(s_n r_{De})\}}\right]$	$B_3(s_n r_D)=Y_0(s_n)J_0(s_n r_D)-J_0(s_n)Y_0(s_n r_D)$, s_n is the n th positive root of the following equation $B_3(s_n r_{De})=0$.
Ⅴ	$p_D(r_D,t_D) = -\frac{1}{\alpha}\ln\left[1-\frac{2\alpha}{\pi}\int_0^{\infty}\times\frac{(1-\exp[-u^2 t_D])[\alpha\varphi_{0,0}(r_D,1,u)+u\varphi_{0,1}(r_D,1,u)]du}{u\{[\alpha J_0(u)+uJ_1(u)]^2+[\alpha Y_0(u)+uY_1(u)]^2\}}\right]$	$\varphi_{m,n}(x,y,\lambda)=Y_m(x\lambda)J_n(y\lambda)-J_m(x\lambda)Y_n(y\lambda)$, where $J_m(x)$, $Y_n(x)$ is the first type and the second type Bessel function, respectively.
Ⅵ	$p_D(r_D,t_D) = -\frac{1}{\alpha}\ln\left[e^{-\alpha}+\frac{2(e^{-\alpha}-1)}{\pi}\times\int_0^{\infty}\frac{\varphi_{0,0}(1,r_D,s)\exp[-s^2 t_D]ds}{[J_0^2(s)+Y_0^2(s)]s}\right]$	$\varphi_{0,0}(1,r_D,s)=J_0(sr_D)Y_0(s)-Y_0(sr_D)J_0(s)$

2.3 The discussion of the solution for Model Ⅴ 1) The solution in Laplace space

By using transform (8) and applying Laplace transform to Model Ⅴ, yield

$$\frac{\partial^2\bar{x}}{\partial r_D^2}+\frac{1}{r_D}\frac{\partial\bar{x}}{\partial r_D}=s\bar{x},\ \left(\frac{\partial\bar{x}}{\partial r_D}-\alpha\bar{x}\right)\Big|_{r_D=1}=0,\ \lim_{t_D\to\infty}\bar{x}=1/s$$

The solution in the Laplace space is obtained

$$\bar{x}(r_D,s)=\frac{1}{s}-\frac{\alpha K_0(\sqrt{s}r_D)}{s[\alpha K_0(\sqrt{s})+\sqrt{s}K_1(\sqrt{s})]} \tag{21}$$

Substituting the Laplace inversion of Eq. (21) into Eq. (8) yields

$$p_D(r_D,t_D)=-\frac{1}{\alpha}\ln\left[1-\alpha L^{-1}\left\{\frac{K_0(\sqrt{s}r_D)}{s[\alpha K_0(\sqrt{s})+\sqrt{s}K_1(\sqrt{s})]}\right\}\right] \tag{22}$$

The inverse solution of the Laplace Transform with wellbore storage is written as

$$p_D(r_D,t_D)=-\frac{1}{\alpha}\ln\left[1-\alpha L^{-1}\left\{\frac{K_0(\sqrt{s}r_D)}{s[(C_D s+\alpha)K_0(\sqrt{s})+\sqrt{s}K_1(\sqrt{s})]}\right\}\right] \tag{23}$$

2) The approximate solution for the short-time

As $s \to \infty$, the modified Bessel function $K_v(z)$ can be approximated as

$$K_v(z) \approx \sqrt{\frac{\pi}{2z}} e^{-z} \tag{24}$$

Using Eq. (24), Eq. (21) can be reduced to the following form:

$$\bar{x}(r_D, s) = \frac{1}{s} - \frac{\alpha \exp[-(r_D - 1)(\sqrt{s})]}{\sqrt{r_D}(\alpha + \sqrt{s})s}. \tag{25}$$

Inverting Eq. (25) by using Laplace transform tables, one obtains, after simplifying,

$$x(r_D, t_D) = 1 - \frac{1}{\sqrt{r_D}}\left[\operatorname{erfc}\left(\frac{r_D - 1}{2\sqrt{t_D}}\right) - \exp[\alpha^2 t_D + \alpha(t_D - 1)]\operatorname{erfc}\left(\frac{r_D - 1}{2\sqrt{t_D}} + \alpha\sqrt{t_D}\right)\right] \tag{26}$$

Substituting Eq. (26) into Eq. (8), the short-time approximate solution can be obtained

$$p_D(r_D, t_D) = -\frac{1}{\alpha}\ln\left[1 - \frac{1}{\sqrt{r_D}}\operatorname{erfc}\left(\frac{r_D - 1}{2\sqrt{t_D}}\right) + \frac{1}{\sqrt{r_D}}\exp[\alpha^2 t_D + \alpha(r_D - 1)]\operatorname{erfc}\left(\frac{r_D - 1}{2\sqrt{t_D}} + \alpha\sqrt{t_D}\right)\right] \tag{27}$$

3) The large-time approximate solution

For $t_D \to \infty$, such that $s \to 0$, one obtains

$$K_0(s) = -\left(\ln\frac{s}{2} + \gamma\right), K_1(s) = \frac{1}{s} \tag{28}$$

Substituting Eq. (26) into Eq. (21), after simplifying

$$\bar{x}(r_D, s) = \frac{1}{s}\left[1 + \alpha\left(\ln\frac{\sqrt{s}\, r_D}{2} + \gamma\right)\left(1 + \alpha\left(\ln\frac{\sqrt{s}}{2} + \gamma\right)\right) + O(\alpha^2)\right] \tag{29}$$

Inverting Eq. (29) by using Laplace transform tables, one obtains

$$x(r_D, t_D) = 1 - \frac{\alpha}{2}\ln\frac{4t_D}{Cr_D^2} + \frac{\alpha^2}{4}\left(\ln\frac{4t_D}{Cr_D^2}\right)^2 - \frac{\alpha^2\pi^2}{24} \tag{30}$$

where $C = e^{\gamma}$, $\gamma = 0.577\ 2$ is Euler constant.

The large-time approximate solution can be written as

$$p_D(r_D, t_D) = -\frac{1}{\alpha}\ln\left[1 - \frac{\alpha}{2}\ln\frac{4t_D}{Cr_D^2} + \frac{\alpha^2}{4}\left(\ln\frac{4t_D}{Cr_D^2}\right)^2 - \frac{\alpha^2\pi^2}{24}\right] \tag{31}$$

3　The Pressure Behavior of the Nonlinear Flow Model with the Quadratic Gradient Term

In the present analysis, we have considered the nonlinear pressure distribution in an infinite reservoir during constant-rate production. Fig. 1 demonstrates the variation of nonlinear dimensionless well bore-pressure with time for different values of CD, namely, 0 and 1 000. The nonlinear solutions are characterized by two values of α.

From Fig. 1 it can be seen that irrespective of the effect of the dimensionless well bore storage, the nonlinear and linear solutions show very small differences at small time. However, the difference increases with time if a pressure value is controlled by the magnitude of α. From Fig. 1, the difference between the nonlinear solution of $\alpha = 10^{-4}$ and the nonlinear solution of $= 10^{-2}$ at $t_D = 10^9$ is 9%.

In order to quantify the difference between the linear and nonlinear pressure solutions, we define the following term:

$$\varepsilon = 1 - \frac{p_{Dnl}}{p_{Dl}} \tag{32}$$

where p_{Dnl} and p_{Dl} are dimensionless nonlinear and linear solutions, respectively, for any given set of boundary conditions. From (32) it can be seen that the greater the deviation of the term "ε" from zero, the more is the difference between the linear and nonlinear pressure solutions.

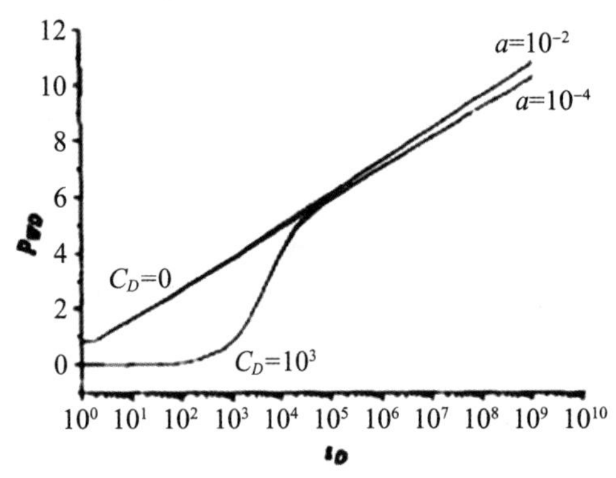

Fig. 1 Semi_lg plots of pressure versus time depending on α infinite outer boundary

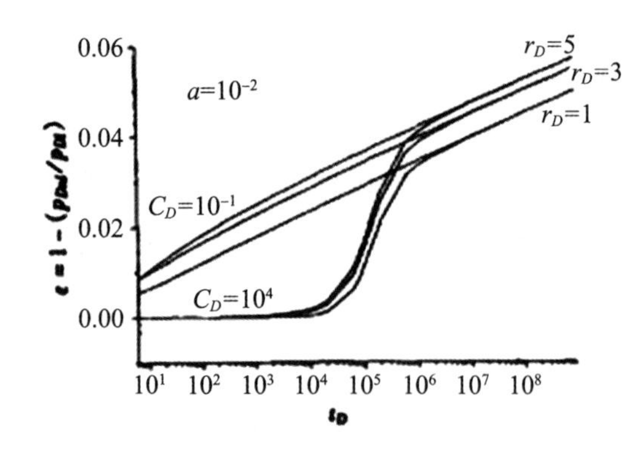

Fig. 2 Error in nonlinear solution at different radii infinite outer boundar

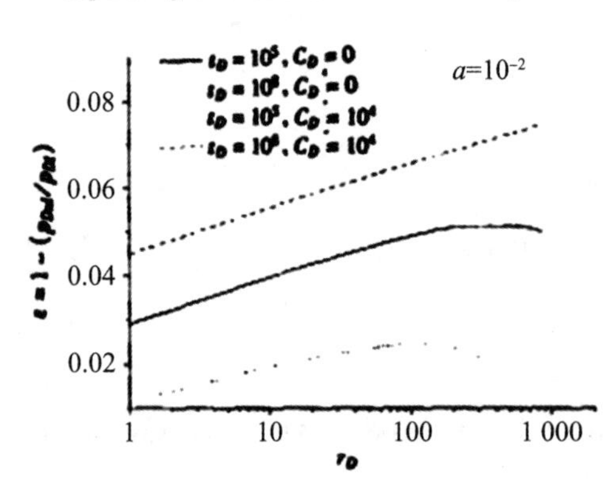

Fig. 3 Error in linear pressure solution versus radius at different times: infinite medium

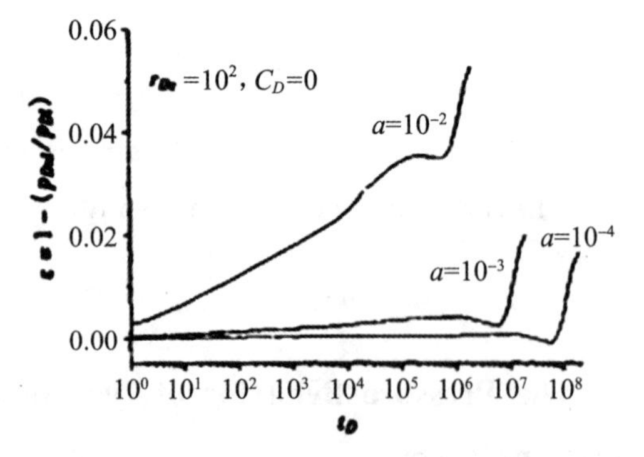

Fig. 4 Error in linear pressure solution versus time at different α: finite closed formation

Figure 2 exhibits the temporal variation of the error at different radii close to the well bore, for two different values well bore storage. At early times, the error is smaller at different radi closer to the well bore, for larger magnitudes of well bore storage. The error incrases with time. Fig. 3 shows that magnitude of the error at any radial distance would depend not only on the values of α and C_D, but also on time. The smaller the times(e. g. , at $t_D = 10^5$), the smaller error is for the values of α and C_D. At larger times(e. g. , at $t_D = 10^8$), the spatial distribution of error is not affected by the value of C_D. At any time, the spatial distribution of the error would increase with increasing distance from the well bore. At large distances, however, the error tends to flaten out after reaching its maximum value. The maximum error would probably be of the order of 10%.

During the constant-rate production, the error is affected by the values of α, r_{De} and C_D. At any time the variation of the error between nonlinear and linear pressure solutions are affected by the value of α. At early times, the error is smaller at the well bore. With increasing times, the error increases. The larger is the value of α, the more is the error(e. g. , for $\alpha = 10^{-2}$, the error reaches about 6%). As α is small, the error may be neglected (such as $\alpha = 10^{-4}$). From Fig. 5 it can be seen that the error is not nearly affected by the magnitudes of C_D at early times. With increasing times, the effect of C_D is more and more. The effect time of the error is earlier and larger as the magnitude of C_D is small. It can be seen from the Fig. 6 that the error distributions are affected by the values of r_{De} at any given time. The larger is the value of r_{De}, the greater is the error. Especially, the error is clearer at larger times for the value of r_{De}.

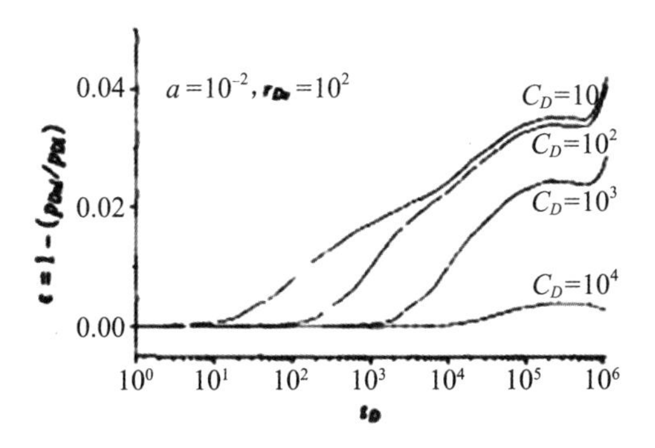

Fig. 5 Error in linear pressure solution versus time at different C_D: finite closed formation

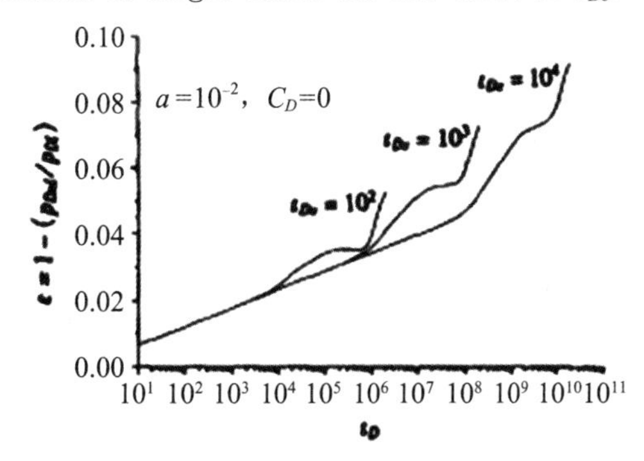

Fig. 6 Error in linear pressure solution versus time at different r_{De}: finite closed formation

References

[1] TONG Dengke, Ge Jiali. An exact solution for unsteady seepage flow through

fractal reservoir [J]. Acta Mechanica Sinica, 1998, 30(5):621-626. (in Chinese)

[2] Odeh A S, Babu D K. Comprising of solutions for the nonlinear and linearized diffusion equations [J]. SPE Res Eng, 1998, 3(4):1202-1206.

[3] Finjord J, Aadony B S. Effects of quadratic gradient term in steady tate and quasi teady tate solu-tions for reservoir pressure [J]. SPE Form Eval, 1989, 4 (3): 413-417.

[4] Wang Y, Dusseault M B. The effect of quadratic gradient terms on the borehole solution in poro-elastic media [J]. Water Resour Res, 1991, 27 (12): 3215-3223.

[5] Chakrabarty C, Farouq Ali S M, Tortike W S. Analytical solutions for radial pressure distribution in-cluding the effects of the quadratic_gradient term [J]. Water Resour Res, 1993, 29(4):1171-1177.

[6] TONG Deng_ke, CAI Lang_lang, CHEN Qin_lei. Flow analysis of fluid in double porous media in-cluding the effects of the quadratic _ gradient term [J]. Engineering Mechanics, 2002, 19(3):99-103. (in Cinese)

[7] TONG Deng _ ke, CAI Lang _ lang. Dynamical characteristics of double porosity double permeability model including the effects of the quadratic_gradient term [J]. Chinese Journal of Computational Physics, 2002, 19(2):177-182. (in Chinese)

[8] TONG Deng_ke, LIU Min_ge. Non_linear flow fractal analysis on reservoir with double_media [J]. Journal of the Univeristy of Petroleum, 2003, 27(2):59-62. (in Cinese)

考虑二次梯度项影响的非线性不稳定渗流问题的精确解释

曹绪龙[1,2]，同登科[3]，王瑞和[3]

（1. 中国科学院兰州化学物理研究所，甘肃兰州 730000；
2. 胜利油田有限公司地质科学研究院，山东东营 257000；
3. 石油大学应用数学系，山东东营 257061）

（张鸿庆推荐）

摘要：考虑了二次梯度项影响的非线性径向流动问题的无限大地层和有界地层渗流模型。在井底定流量和定压生产时，对无限大地层及有界地层（包括封闭和定压地层）六种情况，利用广义 We-ber 变换和广义 Hankel 变换求得了实空间的解析解，分析了非线性压力解与线性压力解的差异，发现在晚时段其差异可达 8% 以上。因此在试井长时要考虑二次梯度项的影响。

关键词：非线性流动；积分变换；解析解；试井分析

中图分类号：TE312　文献标识码：A

引　言

在某些操作过程中，很小的压力梯度可能引起预测孔隙压力分布方面的显著错误，例如水力压裂，大的压降流动，试井，钻杆测试和大的压降脉冲测试，特别地，高灵敏度的压力测量装置必定能检测到如此小的压力波动。在流动系统中，忽略这个波动是由于探测线性化近似，而不是由于异常。而以往的理论大多是假设小的压缩率或小的压力梯度，忽略非线性二次梯度项[1]。为了描述二次梯度项的影响，欧德（Odeh）和巴布（Babu）在流动方程中保留二次梯度项[2]，建立了考虑二次梯度项影响的非线性压力不稳定试井模型. 发现非线性解注入并压力变化与生产井压力变化不同，不像线性解预测的那样。Finjord 和 Aadnoy 用这种非线性方法解决稳态和拟稳态条件的油藏压力[3]。Wang 和 Dusseault 在弹性多孔介质方面用一种相似的方法[4]。Charabarty 的研究表明[5]，非线性项在具有高注入量和小油藏传导率定流量生产时有显著影响。同登科用数值方法[6][7][8]，讨论了考虑二次梯度项影响的双重介质流动问题，但他们都没有给实空间的精确解，更没有讨论定压生产和有界地层。实际油藏是有限的，因此有必要进一步完善考虑二次梯度项影响的非线性渗流模型及其试井理论。

利用 Weber 变换和 Hankel 变换给出了定流量和定压生产时，无限大地层，有界封闭和有界定压地层考虑二次梯度项影响的实空间精确解，讨论了非线性压力动态特征、非线性参数和边界对压力动态的影响，传统模型为非线性参数 $a=0$ 时的特例。

1　考虑二次梯度项影响的非线性压力不稳定分析模型

假设：1）储层厚度一致为 h，且产层厚度全部打开，流体径向流入井内（或从井眼将流体 注入地层）；2）孔隙介质均一，各向同性；3）孔隙度和渗透率是常数（与压力无关）；4）流体的压缩系数是常值，流体粘度是常数。在储层中心一口井定产量或定压生产时，考虑二次梯度项影响的弱可压缩液体圆柱对称问题可描述如下

$$\frac{\partial^2 p_D}{\partial r_D^2}+\frac{1}{r_D}\frac{\partial p_D}{\partial r_D}-a\left(\frac{\partial p_D}{\partial r_D}\right)^2=\frac{\partial pD}{\partial t_D}。\tag{1}$$

初始条件

$$p_D|_{t_D=0}=0,\tag{2}$$

内边界条件

$$\left.\frac{\partial p_D}{\partial r_D}\right|_{r_D=1}=1,\tag{3}$$

$$p_D|_{r_D=1}=0,\tag{4}$$

考虑井眼储集时，内边界条件为

$$C_D\left.\frac{\partial p_D}{\partial t_D}\right|_{r_D=1}-\left.\frac{\partial p_D}{\partial r_D}\right|_{r_D=1}=1$$

外边界条件为

$$\left.\frac{\partial p_D}{\partial r_D}\right|_{r_D=r_{De}}=0,\tag{5}$$

$$p_D|_{r_D=r_{De}}=0,\tag{6}$$

$$\lim_{r_D\to\infty}p_D=0。\tag{7}$$

对于定量生产时，无因次压力无因次压缩系数定义为：

$$p_D=\frac{2\pi kh(p_i-p)}{\mu q},\alpha=\frac{q\mu c}{2\pi kh}。$$

对于定压生产时

$$p_D=\frac{(p_i-p)}{p_i-p_w},\alpha=c(p_i-p_w)。$$

其他无因次量定义为

$$r_D=\frac{r}{r_w},t_D=\frac{kt}{c\phi\mu r_w^2},C_D=\frac{c}{2\pi\phi c_t h r_w^2}$$

其中，P_w 为定压生产时井底压力 4，©分别为孔隙度和渗透率”为液体粘度，C 为流体压缩系数，P，是初始地层压力。

考虑二次梯度项影响非线性渗流问题的 6 个典型初边值问题：1）定产量生产有界封闭地层问题，式(1)，(2)，(3)，(5)，简称为问题Ⅰ；2）定产量生产有界定压地层问题，式(1)，(2)，(3)，(6)，简称为问Ⅱ；3）定压生产有界封闭地层问题，式(1)，(2)，(4)，(5)，简称为问题皿；4）定产量生产有界封闭地层问题，式(1)，(2)，(4)，(6)，简称为问题 W；5）定产量生产无限大地层问题，式(1)，(2)，(3)，(7)，简称为问题 V；6）定压生产无限大地层问题，式(1)，(2)，(4)，(7)，简称为问题Ⅵ。

2　考虑二次梯度项影响的不稳定渗流问题的精确解

2.1 模型 I 的精确解

模型中的微分方程是非线性方程，无法直接求解，作变换 pD=－a^{-1}ln$x$，　(8)

模型 I 可化为

$$\frac{\partial^2 x}{\partial r_D^2}+\frac{1}{r_D}\frac{\partial x}{\partial r_D}=\frac{\partial x}{\partial t_D},\tag{9}$$

$$x|_{t_D=0}=1。\tag{10}$$

$$\left(\frac{\partial x}{\partial r_D}-ax\right)\Big|_{r_D=1}=0,\tag{11}$$

$$\frac{\partial x}{\partial r_D}\Big|_{r_D=r_{De}}=0。\tag{12}$$

作广义 Hankel 变换[5]

$$\tilde{x}(x_n,t_D)=\int_1^{r_{De}}r_D B(s_n r_D)x(r_D,t_D)\mathrm{d}r_D,\tag{13}$$

其中，

$$B(s_n r_D)=[s_n\mathrm{J}_1(s_n)+\alpha\mathrm{J}_0(s_n)]\mathrm{Y}_0(s_n r_D)-[s_n\mathrm{Y}_1(s_n)+\alpha\mathrm{Y}_0(s_n)]\mathrm{J}_0(s_n r_D)\tag{14}$$

Sn 是方程

$$[s\mathrm{J}_1(s)+\alpha\mathrm{J}_0(s)]\mathrm{Y}_1(sr_{De})-[s\mathrm{Y}_1(s)+\alpha\mathrm{Y}_0(s)]\mathrm{J}_1(sr_{De})=0\tag{15}$$

的第 n 个根。

对方程(9)－(12)关于 r_D 应用广义 Hankel 变换得

$$\frac{\partial\tilde{x}}{\partial t_D}=-s_n^2\tilde{x},\tilde{x}|_{t_D=0}=\frac{2\alpha}{\pi s_n^2}$$

求解得

$$\tilde{x}=\left(\frac{2\alpha}{\pi s_n^2}\right)\exp[-s_n^2 t_D]\tag{16}$$

对(16)作 Hankel 逆变换得

$$x(r_D,t_D)=\pi\alpha\sum_{n=1}^{\infty}\frac{\exp[-s_n^2t_D]B(s_nr_D)\mathrm{J}_1^2(s_nr_{De})}{\{[s_n\mathrm{J}_1(s_n)+\alpha\mathrm{J}_0(s_n)]^2-(s_n^2+\alpha^2)\mathrm{J}_1^2(s_nr_{De})\}} \tag{17}$$

将式(17)代入式(8)得

$$p_D(r_D,t_D)=-\frac{1}{\alpha}\ln\left[\pi\alpha\sum_{n=1}^{\infty}\frac{\exp[-s_n^2t_D]B(s_nr_D)\mathrm{J}_1^2(s_nr_{De})}{\{[s_n\mathrm{J}_1(s_n)+\alpha\mathrm{J}_0(s_n)]^2-(s_n^2+\alpha^2)\mathrm{J}_1^2(s_nr_{De})\}}\right] \tag{18}$$

井壁处的压降(rD=1)

$$p_{wD}(t_D)=-\frac{1}{\alpha}\ln\left[\pi\alpha\sum_{n=1}^{\infty}\frac{\exp[-s_n^2t_D]\mathrm{J}_1^2(s_nr_{De})}{\{[s_n\mathrm{J}_1(s_n)+\alpha\mathrm{J}_0(s_n)]^2-(s_n^2+\alpha^2)\mathrm{J}_1^2(s_nr_{De})\}}\right] \tag{19}$$

模型Ⅰ考虑井筒储集的拉氏空间反演解为

$$p_D(r_D,t_D)=-\frac{1}{\alpha}\ln\{1-\alpha L^{-1}[\mathrm{K}_0(r_D\sqrt{s})\mathrm{I}_1(r_{De}\sqrt{s})+I_0(r_D\sqrt{s})\mathrm{K}_1(r_{De}\sqrt{s})]\times$$
$$[s(C_Ds+\alpha)\lambda_1+s\sqrt{s}\lambda_2]^{-1}\}, \tag{20}$$

其中,

$$\lambda_1=\mathrm{I}_0(\sqrt{s})\mathrm{K}_1(r_{De}\sqrt{s})+\mathrm{K}_0(\sqrt{s})I_1(r_{De}\sqrt{s}),\lambda_2=\mathrm{K}_1(\sqrt{s})I_1(r_{De}\sqrt{s})-I_1(\sqrt{s})\mathrm{K}_1(r_{De}\sqrt{s})。$$

2.2 模型Ⅱ~模型Ⅵ的精确解

利用 Hankel 变换和 Weber 可得其他情况的解,列于表1。

表1 考虑二次梯度项影响的不稳定渗流问题的精确解

类型	实空间压力解	s_n 满足的方程
Ⅱ	$p_D(r_D,t_D)=-\frac{1}{a}\ln\left[\frac{a\ln r_D+1}{a\ \ln r_{De}+1}+\frac{\pi a(1-a\ \ln r_{De})}{(a\ \ln r_{De}+1)}\times\sum_{n=1}^{\infty}\frac{\exp[-s_n^2t_D]B_1(s_nr_D)\mathrm{J}_0^2(s_nr_{De})}{\{[s_n\mathrm{J}_1(sn)+\alpha\mathrm{J}_0(s_n)]^2-(s_n^2+\alpha^2)\mathrm{J}_0^2(s_nr_{De})\}}\right]$	$B_1(s_nr_D)=[s_n\mathrm{J}_1(s_n)+a\mathrm{J}_0(s_n)]\mathrm{Y}_0(s_nr_D)-[s_n\mathrm{Y}_1(s_n)+a\mathrm{Y}_0(s_n)]\mathrm{J}_0(s_nr_D)$ s_n 是方程 $B_1(sr_{De})=0$ 的第 n 个根。
Ⅲ	$p_D(r_D,t_D)=-\frac{1}{a}\ln\left[e^{-a}+\pi(e^{-a}-1)\times\sum_{n=1}^{\infty}\frac{\exp[-s_n^2t_D]B_2(s_nr_D)\mathrm{J}_1^2(s_nr_{De})}{\{\mathrm{J}_0^2(s_n)+\mathrm{J}_1^2(s_nr_{De})\}}\right]$	$B_2(s_nr_D)=\mathrm{Y}_0(s_n)\mathrm{J}_0(s_nr_D)-\mathrm{J}_0(s_n)\mathrm{Y}_0(s_nr_D)$ s_n 是方程 $\mathrm{Y}_0(s)\mathrm{J}_1(sr_{De})-\mathrm{J}_0(s)Y_1(sr_{De})=0$ 的第 n 个根。
Ⅳ	$p_D(r_D,t_D)=-\frac{1}{a}\ln\left[e^{-d}+\frac{1-e^{-d}}{\ln r_{De}}\ln r_D+\pi(e^{-a}-1)\times\sum_{n=1}^{\infty}\frac{\exp[-u^2t_D]B_3(s_nr_D)\mathrm{J}_0^2(s_nr_{De})}{\{J_0^2(s_n)-\mathrm{J}_0^2(s_nr_{De})\}}\right]$	$B_3(s_nr_D)=\mathrm{Y}_0(s_n)J_0(s_nr_D)-\mathrm{J}_0(s_n)Y_0(s_nr_D)$ s_n 是方程 $B_3(sr_{De})=0$ 的第 n 个根。

（续表）

类型	实空间压力解	s_n 满足的方程
V	$p_D(r_D,t_D)=-\frac{1}{a}\ln\left[1-\frac{2a}{\pi}\int_0^m\times\frac{(1-\exp[-u^2t_D][a\varphi_{0,0}(r_D,1,u)+u\varphi_{0,1}(r_D,I,u)]du}{u\{[aJ_0(u)+uJ_1(u)]^2+[aY_0(u)+uY_1(u)]^2\}}\right]$	$\varphi_{m,n}(x,y,\lambda)=Y_m(x\lambda)J_n(y\lambda)-J_m(x\lambda)Y_n(y\lambda)$，$J_m(x)$，$Y_n(x)$分别为一类第二类 Besal 函数。
Ⅵ	$p_D(r_D,t_D)=-\frac{1}{a}\ln\left[e^{-a}+\frac{2e^{-a}-1)}{\pi}\times\int_0^\infty\frac{\varphi_{0,0}(1,r_D,s)\exp[-s^2t_D]ds}{[J_0^2(s)+Y_0^2(s)]s}\right]$	$\varphi_{0,0}(1,r_D,s)=J_0(sr_D)Y_0(s)-Y_0(sr_D)J_0(s)$

2.3 对模型 V 的解的讨论

1）拉氏空间的解

对模型 V 经变换(8)后关于 t_D 作 Laplace 变换

$$\frac{\partial^2\bar{x}}{\partial r_D^2}+\frac{l}{r_D}\frac{\partial\bar{x}}{\partial r_D}=s\bar{x},\left(\frac{\partial\bar{x}}{\partial r_D}-a\bar{x}\right)\bigg|_{r_D=1}=0,\lim_{t_D\to\infty}\bar{x}=1/s。$$

由此可得拉氏空间的解为

$$\bar{x}(r_D,s)=\frac{1}{s}-\frac{a\mathrm{K}_0(\sqrt{s}r_D)}{s[a\mathrm{K}_0(\sqrt{s})+\sqrt{s}\mathrm{K}_1(\sqrt{s})]}。\tag{21}$$

代回变换(8)得模型 V 的拉氏空间的反演解为

$$p_D(r_D,t_D)=-\frac{1}{a}\ln\left(I-aL^{-1}\left\{\frac{\mathrm{K}_0(\sqrt{s}r_D)}{s[a\mathrm{K}_0(\sqrt{s})+\sqrt{s}\mathrm{K}_1(\sqrt{s})]}\right\}\right)。\tag{22}$$

考虑井眼储集的拉氏空间的反演解为

$$p_D(r_D,t_D)=-\frac{1}{a}\ln\left(1-aL^{-1}\left\{\frac{\mathrm{K}_0(\sqrt{s}r_D)}{s[(C_Ds+a)\mathrm{K}_0(\sqrt{s})+\sqrt{s}K_1(\sqrt{s})]}\right\}\right)\tag{23}$$

2）短时渐近解

当 S∞时，此时的贝塞尔函数可近似为

$$\mathrm{K}_v(z)\approx\sqrt{\frac{\pi}{2z}}e^{-z}\tag{24}$$

利用式(24)可将式(21)近似为

$$\bar{x}(r_D,s)=\frac{1}{s}-\frac{\alpha\exp[-(r_D-1)(\sqrt{s})]}{\sqrt{r_D}(a+\sqrt{s})s}。\tag{25}$$

对式(25)进行拉氏反演得

$$x(r_D,t_D)=1-\frac{1}{\sqrt{r_D}}\left[\mathrm{erfc}\left(\frac{r_D-1}{2\sqrt{t_D}}\right)-\exp[a^2t_D+a(t_D-1)]\mathrm{erfc}\left(\frac{r_D-1}{2\sqrt{t_D}}+a\sqrt{t_D}\right)\right]\tag{26}$$

那么短时渐近解为

$$p_D(r_D,t_D)=-\frac{1}{a}\ln\left[1-\frac{1}{\sqrt{r_D}}\mathrm{erfc}\left(\frac{r_D-1}{2\sqrt{t_D}}\right)+\frac{1}{\sqrt{r_D}}\exp[a^2t_D+a(t_D-1)]\mathrm{erfc}\left(\frac{r_D-1}{2\sqrt{t_D}}+a\sqrt{t_D}\right)\right] \quad (27)$$

3) 长时渐近解

当 $t_D\infty$时,s0 此时有

$$\mathrm{K}_0(s)=-\left(\ln\frac{s}{2}+\lambda\gamma\right),\mathrm{K}_1(s)=\frac{1}{s}。 \quad (28)$$

利用式(26)可将式(21)得

$$\bar{x}(r_D,s)=\frac{1}{s}\left[1+a\left(\frac{\sqrt{s}r_D}{2}+\gamma\right)\left(1+a\left(\frac{\sqrt{s}}{2}+\gamma\right)\right)+O(a^2)\right] \quad (29)$$

对(29)作拉氏反演得

$$x(r_D,t_D)=1-\frac{a}{2}\ln\frac{4t_D}{Cr_D^2}+\frac{a^2}{4}\left(\ln\frac{4t_D}{Cr_D^2}\right)^2-\frac{a^2\pi^2}{24} \quad (30)$$

其中,C=eγ,γ=0.577 2,是欧拉常数.则长时渐近解为

$$p_D(r_D,t_D)=-\frac{1}{a}\left[1-\frac{a}{2}\ln\frac{4t_D}{Cr_D^2}+\frac{a^2}{4}\left(\ln\frac{4t_D}{Cr_D^2}\right)^2-\frac{a^2\pi^2}{24}\right] \quad (31)$$

3　压力曲线动态特征

考虑常量生产,无限大地层情况,图 1 是线性和非线性无因次井眼压力解在 $C_D=0$,1 000 两种不同的值随时间的变化,非线性解被 a=0.01,0.000 1 两个值所刻化。

从图 1 可以看出,考虑或不考虑无因次井储系数,线性解和非线性解之间的差异很小,线性解与 $a=0.000\ 1$ 时非线性解几乎完全重合。但是,随着时间的增加,线性解与非线性解间的差异由 a 的值控制,从图 1 可知,线性解和非线性解的误差在 $t_D=10^9$ 时可达 9%,而在早时却没有多少差别.

为了定量的描述线性解与非线性解间的差异,我们定义

$$\varepsilon=1-\frac{P_{Dnl}}{P_{Dt}} \quad (32)$$

其中,P_{Dnl} 加和 P_{Dl} 以分别为无因次非线性和线性解,从式(32)知 ε 与 0 的偏离越大,线性解与非线性解的差异越大。

图 2 展现的是井眼附近对于两个不同的井储系数 C_D 值误差的瞬时变化。在初时段半径越接近井眼和井储系数越大,误差越小。随着时间的增加,误差增加。为了研究在较大的半径和较长的误差动态,我们观察图 3,发现在任何径向距离的误差的大小不仅依赖于 α 和 C_D 值,而且还依赖时间,时间越短(如 $t_D=10^5$),对任何 α 和 C_D 值误差越小。在较长的时间(如 $t_D=10^8$),误差的空间分布不受 C_D 的影响。从图 3 中注意到在

任何给定的时间，误差的空间分布随着距井眼的径向距离的增加而增加直到最大值达到，然而当最大值达到后误差趋向于平坦或下降，如果我们感兴趣的是较长时间和距井眼较距离处的压力，误差可能很大，甚至可达10%左右。

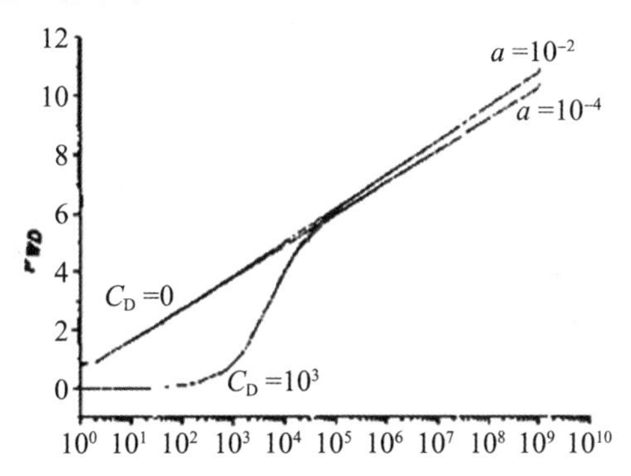

图1　无限大地层线性与非线性压力解比较

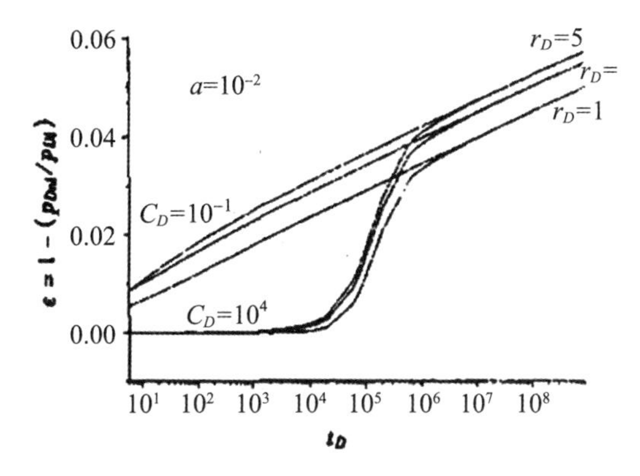

图2　在不同的时间无限大地层的压力解的误差的变化

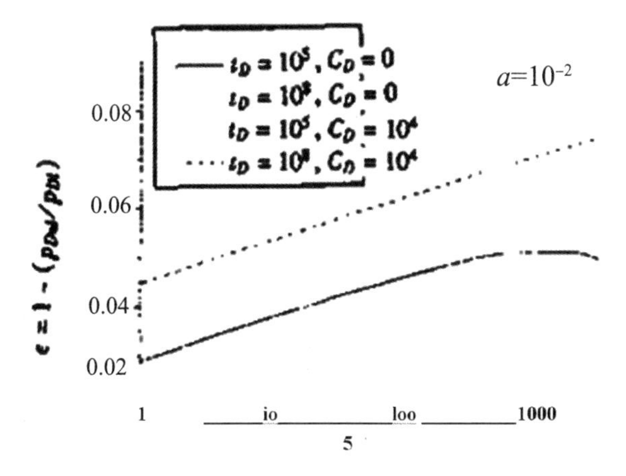

图3　在不同的半径无限大地层压力解的误差变化

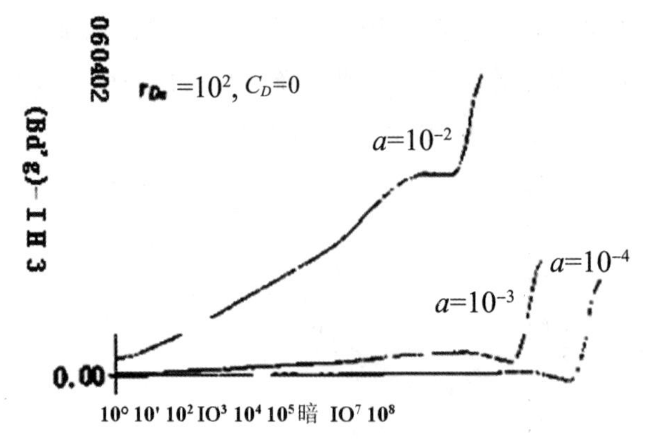

图 4　有界封闭系统误差依赖 Q 的变化

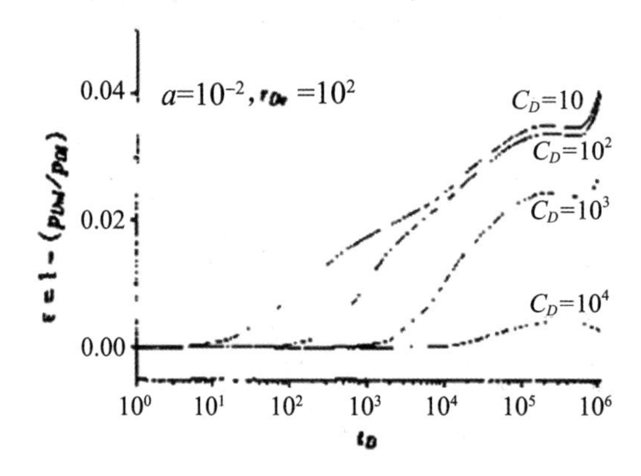

图 5　有界封闭系统误差依赖 tp 的变化

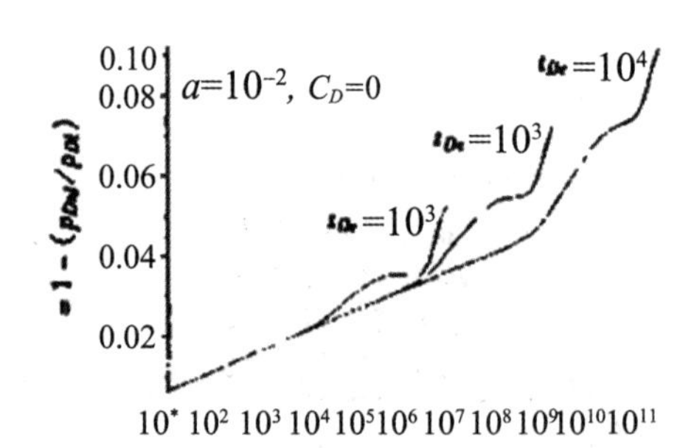

图 6　有界封闭系统误差依赖 r 每的变化

对于定量生产、有界封闭系统，误差受到 α，r_{De}，C_D 的影响。在图 4 中，α 在整个流动过程中都影响着线性解与非线性解之间的误差的变化，初始阶段影响较小，随着时间的增加，对误差的影响越来越大，α 越大影响越大(当 $\alpha=10^{-2}$，影响可达 6%左右)，对于较小的 α，影响可忽略不计(当 $\alpha=10^{-4}$ 几乎没有影响)。从图 5 中可看出，初始阶段，误差几乎不受 C_D 的影响，随着时间的增加，C_D 的影响越来越明显，C_D 越小，对误差的影响时间越早且影响越大，C_D 越大，对误差影响的越晚，且不超过 0.5%，此时可忽略其影响。在图 6 中，r_{De} 在整个流动过程中对误差都有影响，r_{De} 越大，影响越大，特别对于较大的时间值，影响更明显。

参考文献

[1] 同登科,葛家理. 分形油藏不稳定渗流问题的精确解[J]. 力学学报,1998,30(5):621-626.

[2] Odeh A S, Babu D K. Comprising of solutions for the nonlinear and linearized difiiision equations[J]. *SPE Res Eng*, 1998,3(4):1202-1206.

[3] Fiixjord J, Aadony B S. Effects of quadratic gradient term in steady-state and quasi-steady-state solutions for reservoir pressure[J]. *SPE Form Eval*, 1989,4(3):413-417.

[4] Wang Y, Dusseault M B. The effect of quadratic gradient terms on the borehole solution in poroelastic media [J]. *Water Resour Res*, 1991, 27(12):3215-3223.

[5] Chakrabarty C, Farouq Ali S M, Tortike W S. Analytical solutions for radial pressure distribution including the effects of the quadratic-gradient term [J]. *Water Resour Res*, 1993, 29(4):1171-1177.

[6] 同登科,蔡郎郎,陈钦雷. 考虑二次梯度项影响的双重介质流动分析[J]. I程力学, 2002, 19(3):99-103.

[7] 同登科,蔡郎郎. 考虑二次梯度项影响的双渗模型的动态特征口[J]. 计算物理,2002, 19(2):177-J82.

[8] 同登科,刘敏鸽. 双重介质分形油藏非线性流动分析[J]. 石油大学学报, 2003, 27(2):59-62.

Exact solutions for nonlinear transient flow model Including a quadratic gradient term

Cao Xulong[1,2], Tong Dengke[3], Wang Ruihe[3]

(1. Institute of Chemical Physics, Chinese Academy of Sciences, Lanzhou 730000, P. R. China;

2. Geological Science Research Institute, Shengli Oilfield Co. Ltd, Dongying, Shangdong 257000, P. R. China;

3. Department of Applied Mathematics, University of Petroleum, Dongying, Shandong 257061, F. R. China)

Abstract: The models of the nonlinear radial flow for the infinite and finite reservoirs including a quadratic gradient term were presented. The exact solution was given in real space for flow equation including quadratic gradiet term for both constant-rate and constant pressure production cases in an infinite system by using generalized Weber transfbnn. Analytical solutions for flow equation including quadratic gradient term were also obtained by using the Hankel transform for a finite circular reservoir case. Both closed and constant pressure outer boundary conditions are considered. Moreover, both constant rate and constant pressure inner boundary conditions are considered. The difference between the nonlinear pressure solution and linear pressure solution is analyzed. The difference may be reached about 8% in the long time. The effect of the quadratic gradient term in the large time well test is considered.

Key words: nonlinear flow; integral transform; analytical solution; well test analysis

水溶性高分子弱凝胶体系凝胶化过程的 Monte Carlo 模拟

杨健茂[2],曹绪龙[1],张坤玲[2],宋新旺[1],邱 枫[2],许元泽[2]

(1. 中国石化胜利油田有限公司地质科学研究院,东营 257015;
2. 复旦大学高分子科学系,聚合物分子工程教育部重点实验室,上海 200433)

摘要:为研究弱凝胶的形成过程,并把高分子弱凝胶用于三次采油,采用三维 Monte Carlo 模拟了高分子溶液凝胶化过程。模拟预测了凝胶化开始的时间,得到了凝胶化过程中分子量分布的演化规律和胶团生长的三维图像。发现生成溶胶与凝胶团的歧化过程,初始聚合物的浓度对能否形成凝胶至关重要,低于临界浓度不能形成凝胶。模拟了凝胶化速度和聚合物浓度以及交联剂浓度的关系,并与粘度随凝胶化时间变化的实验结果进行比较,结果表明,聚合物浓度较高时,浓度对交联反应的影响减弱,这一趋势与实验结果相一致。

关键词:Monte Carlo 模拟;弱凝胶;凝胶化;三次采油

中图分类号:O631 **文献标识码**:A **文章编号**:025-10790(2006)03-0579-04

高分子弱凝胶是由高分子溶液进行一定程度的交联形成的,虽然是流体,但表现出强烈的粘弹性和高粘滞性。国内外学者对弱凝胶的凝胶化过程都有一定的研究[2]。弱凝胶已应用于石油开采、生物医药和食品工业等多个领域,为了提高弱凝胶在大规模矿场三次采油的采收率,亟待研究凝胶化过程的影响因素,以便在合理的数学模型指导下取得最大的经济效益。

从空间结构上看,高分子凝胶化是一种随机的、非均一的发展过程,Monte Carlo 模拟是对凝胶的生成过程进行研究的一种简便可行的方法。Chen 和 Chiu[3]发展的 Monte Carlo 模拟方法,可用于具有复杂化学反应机理的交联反应体系。Nosaka 等[4]建立了可以描述三维空间中有引发剂、单体和交联剂存在,通过自由基聚合机理发生交联化反应的模型。Wen 等[5,6]建立了既考虑到化学反应动力学常数,又考虑到高分子的空间分布的 Monte Carlo 方法。还有一些模型[7,8]也部分考虑了一些关键速率过程的影响。上述模型中,均未考虑单体和高分子的扩散和空间受限效应,秦原等[9]利用协同运动算法进行了二维模拟,本文发展的 Monte Carlo 方法则可以明确考虑这两种效应,并利用三维 Monte Carlo 凝胶化模拟模型,对实用体系的模型参数、物料参数与工艺条件之间的关系及工程预测的可靠性进行了探索,预测了弱凝胶体系凝胶化所需时间与空间分布,以期对弱凝胶应用于三次采油提供理论指导。

1 Monte Carlo 模拟

1.1 模拟方法

采用的模拟链运动和交联的离散格子模型,可以明确地考虑交联反应中链构型、浓度分布及凝胶团的空间形状等时空演化。其基本思想是:高分子链由一系列链段组成,在空间作随机运动,如果任意两个链段之间的距离小于一个特定的尺度(通常为 1~2 个链段长度)时,则两者以某种概率 k 发生交联。所用参数及其物理意义:N 为三维空间中高分子链的数目,对应于实验中高分子溶液的浓度;m 为一个高分子链中链段数;k 可称为反应概率参数,对应于扩散速率与交联剂反应常数的比值,反映发生反应所需要的碰撞次数;t 为所用的时间步数,对应于一定长度的实际时间段。假设格子模型满足周期性边界条件,在一次计算中高分子链段的扩散速率不变,链段间碰撞发生交联反应的几率不变,反应空间大小不变。

1.2 胶团尺寸的三维发展和高分子浓度的影响

模拟发现,在高分子链的浓度太低时发生分子间交联的概率极低,不能发生凝胶化,只有当高分子链浓度达到一定水平时,体系在交联剂的作用下才能发生交联(图 1)。其中初始高分子链的数目是 $N=1\ 000$,链长 $m=100$,体系的尺寸 $L_x=L_y=L_z=100$,因此高分子的链段浓度 $c=mN/(L_x L_y L_z)=0.1$。这里假定每一个链段占据一个单元格子,链段浓度与实测高分子浓度之间存在某种比例关系。

在交联反应的初期,生成的高分子集团的分子量大小相差不多,最大和次大的集团个数也比较多,且空间分布均匀。在反应接近凝胶点时,分子量最大的高分子集团只有一个,并且其中的链段在空间中的分布不均匀。在达到凝胶点时,分子量最大的凝胶团贯穿整个体系。相反,当高分子浓度较低时(c=0.004),则在模拟时间内分子量最大的高分子集团的增长非常缓慢,难以达到凝胶化所必需的尺度(如图 2 所示)。应该指出,这两种情况下,链段碰撞并反应的几率相同,也就是交联剂浓度一样,后者不能发生凝胶化是因为分子间交联的概率远远低于分子内交联。因此,采用这个模型可以预测发生凝胶化的临界高分子浓度。

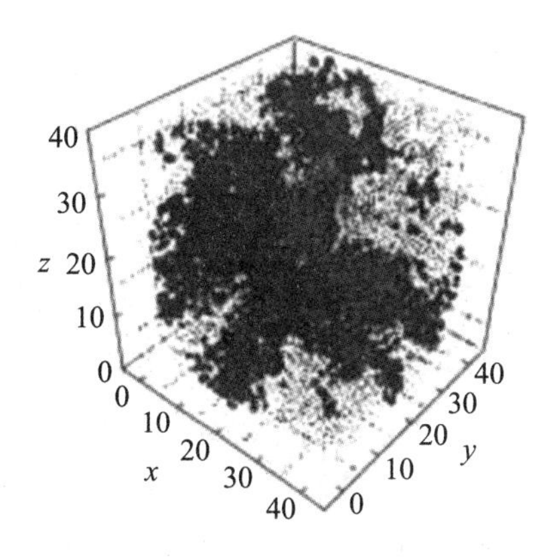

Fig. 1 An example of gelation

The middle dark sites stand for the largest percolated molecular group, the most dark for the second largest for the light dark, the rest mole-cules. $N=1\ 000$, $L_x=L_y=L_z=100$, $c=0.1$, $t=500\ 00$.

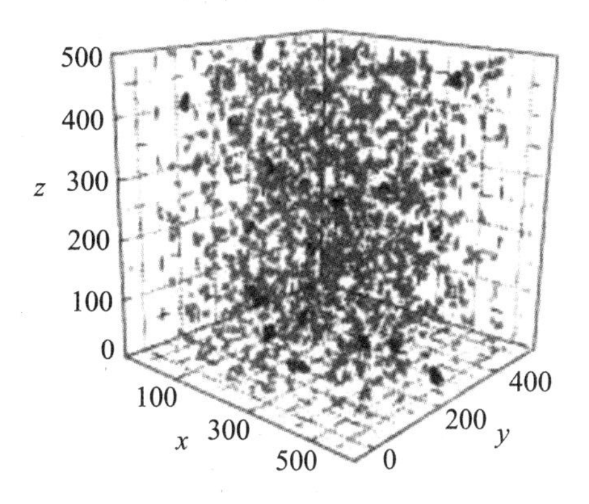

Fig. 2 An example of system without gelation

$N=1\ 000$, $L_x=L_y=L_z=500$, $c=0.004$, $t=500\ 000$(the final state of simulation). The dark sites stand for the two largest molec-ular groups for the gray, the rest molecules.

2 实验部分

ARES 流变仪(同轴圆筒):美国 TA Instruments 公司;A utosizer 4700 激光光散射仪:英国 Malvern 公司。部分水解聚丙烯酰胺 HPAM,M 4000,重均分子量 3.013×10^7,胜利油田提供。间苯二酚,上海化学试剂五联化工厂生产;酚醛交联剂,自制。

按照 PAM/交联剂/间苯二酚/盐水的不同配方配制溶液 200 mL,分别放置在磨口锥形瓶中,置于 80℃的烘箱内使其交联,而后在不同的时间取出定量样品,冷却后进行粘弹性的测定,取点的方法由实验进程而定。通氮气保持溶液体系中无氧,流变实验中样品的测试温度均为 25℃。采用动态激光光散射法(DLS)测定溶液中聚丙烯酰胺的分子形态[10]。

3　理论和实验的比较

3.1 模拟与动态光散射谱的比较

样品中 HPAM 和交联剂(苯酚和甲醛)的质量浓度为 1 000 mg/L,间苯二酚的质量浓度为 100 mg/L,在 80℃ 交联,在每一个时间点取出的凝胶液用标准盐水稀释至 50 mg/L,过滤灰尘后在 25℃对其进行光散射测定,结果见图 3。图 4 是 Monte Carlo 模拟结果,参数为 $L_x=L_y=L_z=100$,$t=500\ 000$,$n=1\ 000$,$m=100$,$k=100$。分子量分布随时间的演化统计结果也示于图 4,在交联反应的初期,体系中的分子量分布比较窄,逐步变宽;分子量增长到一定程度时,分子量分布出现双峰,即体系分为两个集团,其中一个分子量很大,但集团内高分子数量很少,对应于凝胶集团,另一个平均分子量比较小,但集团内高分子数量较多,对应于溶胶集团。这是高分子交联中典型的歧化现象。这一模拟结果与动态光散射的结果一致。由动态光散射图(图 3)可以看到,粒子半径从一个峰逐渐变为两个峰,这两个峰的距离随时间增大。

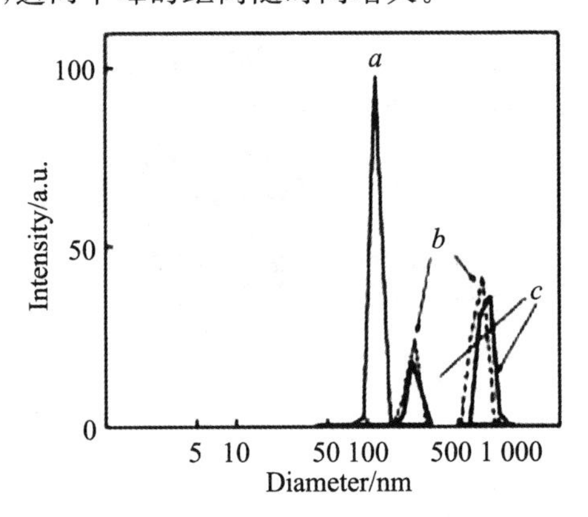

Fig. 3　DLS of sample HPAM(50 mg/L)

Reaction time a. 0 h, b. 3 h, c. 5 h.

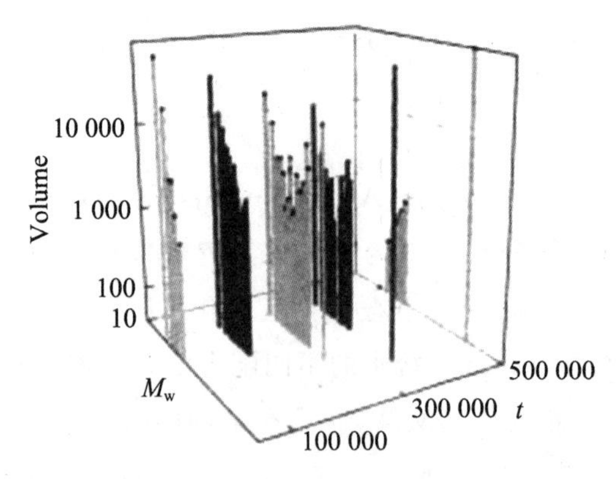

Fig. 4　The evolvement of mass molecular weight distribution at a higher concentration

$L_x=L_y=L_z=100$, $t=500\ 000$, $N=1\ 000$, $m=100$.

3.2 交联剂的浓度对粘度的影响

为检验 Monte Carlo 模拟方法的可靠性，利用凝胶点以前聚合物溶液的粘度基本和其重均分子量成正比的关系，把模拟结果与流变测量结果作对比。部分水解聚丙烯酰胺在不同交联剂浓度下的粘度随时间的变化如图 5 所示。交联剂的质量浓度为 500，1 000，1 500 mg/L。在 Monte Carlo 模拟中通过调节 k 参数，可以近似模拟交联剂的浓度效应。交联剂的浓度越大，交联反应的速率越快，相应的 k 值就越小。假设 k 值和交联剂浓度成反比，得到了重均分子量随时间变化的理论预测曲线，经过归一化处理，如图 5 中所示，与粘度曲线转折点符合甚佳。可见 Monte Carlo 模拟准确地预言了凝胶化开始的时间和交联剂浓度的关系。

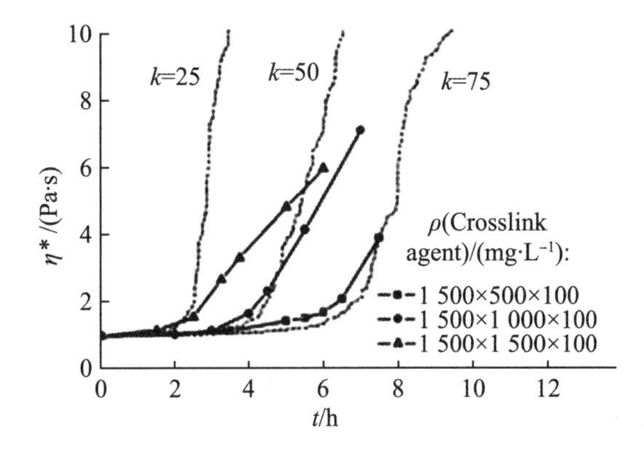

Fig. 5 Comparison between simulation and exper-iments at different concen trations of crosslink agent ρ(HPAM)=1 500 mg/L, the dashed lines are the Monte Carlo smiulation results.

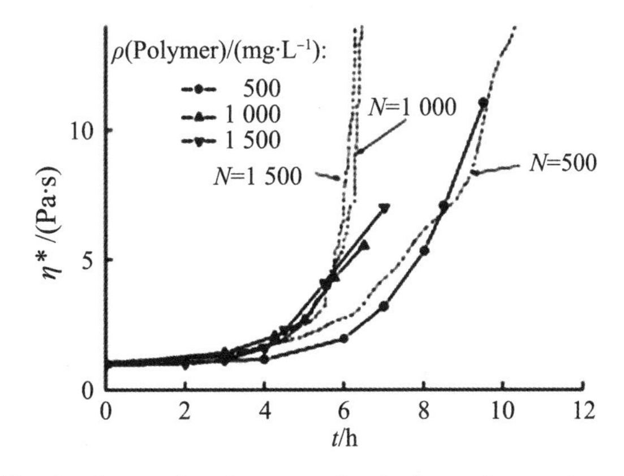

Fig. 6 Comparison between simulation and exper-iment at different concen trations of polymer The dashed lines are the Monte Carlo simulationr-esults.

3.3 聚丙烯酰胺浓度对粘度的影响

将 Monte Carlo 模拟重均分子量随时间变化的结果与聚丙烯酰胺浓度对粘度的影响结果的实验值进行对比,见图 6。模拟结果和实验结果基本符合,特别是在聚丙烯酰胺浓度相对较高的区域,聚合物浓度对交联反应影响不大,与实验结果一致。定量预测涉及到模型中的基本假设,即每个单元空间中高分子链段实际占有的比例为浓度 c。按图 6 中 $N=1\ 000$,$m=100$,$L_x=L_y=L_z=100$,$c=0.1$,相应于实验中 PAM 分子量 3.013×10^7,实测质量浓度 1 000 mg/L 来计算,每个链段分子量为 3.013×10^5,每单元格中高分子重量占据 1%,即浓度转换因子 c/c_{exp} 为 100。对可交联的最低的聚丙烯酰胺浓度的模拟预测值和其他的物理量的模拟值与实测值的定量比较,往往与链段即单元格的选取有关,还有待于进一步优化算法和选取恰当的转换因子。

参考文献

[1] JI Shu-Ling(纪淑玲),PENG Bo(彭勃),LIN Me-iQin(林梅钦) et at. A cta Polym erica Sinica(高分子学报)[J], 2000, (1):65-68.

[2] WANG K-eLiang(王克亮), YANG Jing(杨靖),ZHEN Jing(甄静). Petroleum Geo logy and Research in Daqing(大庆石油地质与开发)[J], 1999, 18(2): 38-40.

[3] Chen Y. C., Ch iu W. Y.. Macrom olecu les[J], 2000, 33(18):6672-6684.

[4] N osaka M., Takasu M., Katoh K.. J. Chem. Phys. [J], 2001, 115(24): 11333-11338.

[5] Wen M., Scriven L. E., M c Cormick A. V.. Macrom olecu les[J], 2003, 36(11):4140-4150.

[6] Wen M., Scriven L. E., M c Cormick A. V.. Macrom olecu les[J], 2003, 36(11):4151-4159.

[7] Boots H., Pandy R. B.. Polym. Bul (Berlin)[J], 1984, 11:415.

[8] Bowman C. N., Peppas N. A.. J. Polym. Sc, Polym. Chem. [J], 199 29 (11):1575-1583.

[9] QIN Yuan(秦原), LIU Hong-Lai(刘洪来), HU Ying(胡英). Chem. J. Chinese U niversities(高等学校化学学报)[J], 2002, 23(7):1447-1449.

[10] Klu cker R., Mun ch J. P., Schosseler F.. Macrom olecu les[J], 1997, 30: 3839-3848.

Monte carlo simulation of the gelation process of the aqueous polymer weak gel

Yang Jianmao[2], Cao XuLong[1], Zhang Kunling[2],
Song XinWang[1], Qiu Feng[2], Xu Yuanze[2*]

(1. Research Institute of Geological Sciences, ShengliO iield, SINOPEC, Dongying 257015, China;
2. Deparmtent of Macromolecular Science, Key Labora tory of Molecular Eng ineering of Poylmers of Ministry of Education, Fudan University, Shanghai 200433, China)

Abstract: To study the gelation process of weak gel and apply weak gel for the industrial practice of tertiary oil recovery, three dmiensional Monte Carlo simulations were carried out to obta in the grow th of average molecular weights and gel grow th miages. It is found that the initial polymer concentration is critical for the gel for mation. The dismutation process of solgel groups is clearly observed. The relationship bewteen the speed of gelation and the concentration of crosslink agentw as also smiulated. The predicted dependency of the gel tmieupon the concentration of crosslink agent is in good agreement with the expermiental data of viscosity versustmie. Also, the polymer concentration effects agree well with viscosity data. The smiulation correctly predicts that the concen tration effect on crosslink reaction will be weaker at the higher polymer concentration.

Key words: Monte Carlo smiulation; Weak gel; Gelation; Tertiary oil recovery

(Ed. : Y, Z)

用三维非均质模型研究聚合物分布规律*

祝仰文[1] 曹绪龙[1] 宋新旺[1] 张战敏[2] 韩玉贵[1]

(1. 胜利油田地质科学研究院,山东东营 257015;

2. 华北油田公司采油工艺研究院,河北任丘 062552)

摘要:为研究聚合物驱后地下滞留聚合物的分布规律及再利用的可行性,利用50 cm×50 cm×10 cm 三维模型设计了中间没有隔板和有2个隔板的非均质模型,在饱和实际原油的基础上模拟层间(有隔层)及层内(没有隔层)2种情况下,聚合物驱油后各层聚合物采出总量及吸附滞留量,研究聚合物的吸附滞留规律。结果表明,注入聚合物总量的60%以上吸附滞留在地下,并且中、高渗透层吸附滞留量占90%以上,从分布规律来看,由于后续流体的冲刷,从注入井到采出井,滞留浓度逐渐增加,由于地下吸附滞留量较大,可以进一步开展地下吸附滞留聚合物再利用的可行性研究。

关键词:三维非均质模型;聚合物驱后;吸附滞留;分布规律

中图分类号:TE357.46 **文献标识码**:A

聚合物驱油技术作为一项成熟技术在胜利油田和大庆油田得到了大规模的推广应用,已经成为降水增油的主要技术手段。聚合物驱油过程中由于吸附和机械捕集而使一部分聚合物分子滞留在地下多孔介质中[1],这部分的量有多大,是否有利用的价 值,一直存在争议,关于地下聚合物的滞留、分布及再利用的研究报导也较少。试验利用50 cm×50 cm×10 cm 三维模型设计了中间没有隔板和有2个隔板的非均质模型,研究了在有原油存在条件下聚合物采出总量及滞留量分布规律,为聚合物驱后聚合物溶液再利用提供了指导。

1 实验装置及实验方法

1.1 主要试剂

聚合物(MO-4000),日本三菱公司生产,固相含量91%,水解度24.7%,相对分子量约2 200万。

1.2 实验装置

该装置由驱动系统、加热保温系统、模型、压力控制系统、采出液收集系统等部分组成。

模型长 L 为 50 cm，宽 W 为 50 cm，厚 H 为 10 cm，采用非均质添砂模型，根据不同粒径的石英砂混合填充来达到所需的渗透率要求。根据胜利油田油藏变异系数设计渗透率分别为 3 600×10^{-3} μm^2、1 200×10^{-3} μm^2、350×10^{-3} μm^2 的三维非均质模型，平均渗透率为 1 700×10^{-3} μm^2。为了模拟层内及层间的情况，开展了 2 组实验：一组模拟层内非均质条件，3 层之间没有隔板，层间存在相互渗流；另一组实验模拟层间非均质条件，3 层之间有 2 个隔板，层间不存在相互渗流。模型平放，不考虑重力作用，从左注入，自动分配注入量，见图 1。

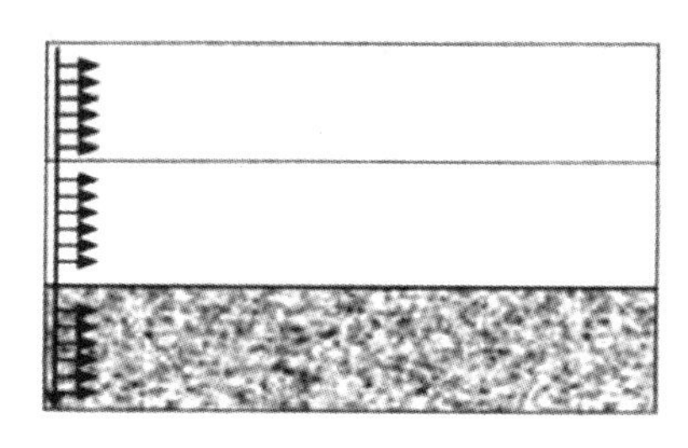

图 1 三维非均质模型示意图

1.3 实验方法

饱和用水矿化度 19 334 mg/L，离子组成见表 1，实验温度 70℃。

模拟油用胜坨油田外输原油配制，温度 70℃时模拟油黏度约为 50 mPa・s。

表 1 地层水离子组成 mg/L

TDS	Cl^-	HCO_3^-	Na^+	Ca^{2+}	Mg^{2+}
19 334	11 373	533	6 914	412	102

聚合物浓度 1 500 mg/L，温度 70℃，转速 6 r/min 条件下测量黏度为 15 mPa・s。

用模拟水饱和模型，模拟油驱替饱和地层水，束缚水饱和度约为 20% 30%。实验驱替速度为 1. 0 mL/min，含水 92%，转注 0.3 PV 聚合物，之后水驱 98%以上，结束实验。

2 产出及滞留聚合物计算方法

2.1 产出及吸附滞留聚合物总量计算

实验过程中，注入聚合物后，从模型出口收集采出液，油水分离后，按照中华人民共和国石油天然气行业标准《驱油用丙烯酰胺类聚合物性能测定（SY/T5862-93)》推荐的溶液中聚丙烯酰胺含量的测定方法进行产出液聚合物浓度分析测定。

首先根据分液率曲线计算得到高、中、低 3 层每层的注入量，然后根据产出液聚合物浓度，应用公式(1)计算聚合物质量，从而计算单层模型采出量、吸附滞留量及模型总的采出量及吸附滞留总量

$$M=\sum C_i V_i \tag{1}$$

式中，M 为聚合物质量，mg；C_i 为第 i 号样聚合物浓度，mg/l；V_i 为第 i 号样体积，l。

2.2 聚合物吸附滞留量及分布测定

实验结束后,拆开模型,将模型均匀分成 24 个网格,每层取 8 个点,在每个网格内取相等量的砂样,取样点示意图见图 2。取出的砂样按照中华人民共和国石油天然气行业标准《驱油用丙烯酰胺类聚合物性能测定(SY/T5862-93)》推荐的聚丙烯酰胺静态吸附量测定方法进行聚合物滞留量测定,分析每个网格中聚合物吸附滞留量。对数据进行归一化处理,绘制等值图。

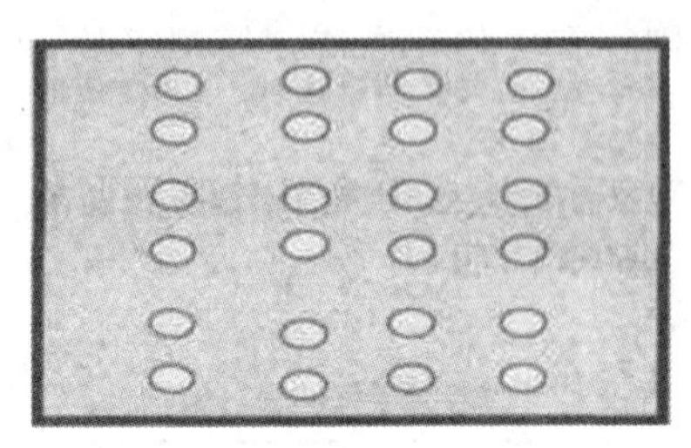

图 2　模型取样点示意图

3　实验结果及分析

3.1 层间非均质模型聚合物吸附滞留量

模型共注入聚合物 0.3 PV,模型孔隙体积 10 L,总量 4 500 mg。根据分液率曲线(图 3),高、中、低渗透层聚合物注入量分别为 3 028.5 mg、1 170.0 mg、301.5 mg,高、中、低 3 个渗透层聚合物注入量分别占聚合物总注入量的 67.3%、26.0%和 6.7%,可见聚合物大部分进入高渗透层。

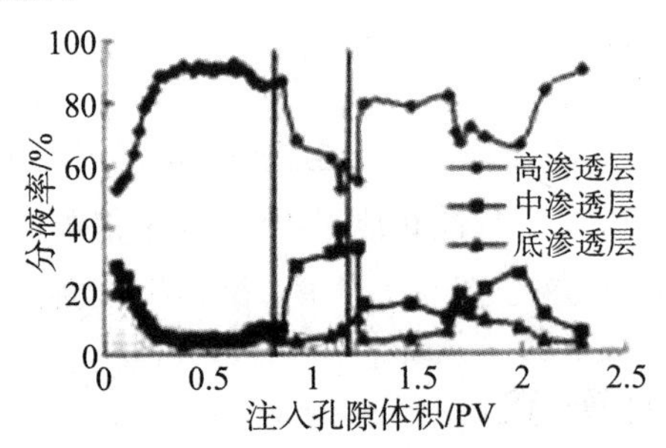

图 3　注入孔隙体积与分液率的关系

采用公式(1)的计算方法,可以得到聚合物驱层间非均质模型实验单层聚合物吸附滞留量,结果见表 2。高、中、低 3 个渗透层聚合物注入量分别占总注入量的 67.3%、26.0%、6.7%,采出量分别占聚合物总采出量的 84.2%、15.0%和 0.8%。研究表明[2,3],不同分子量的聚合物与不同渗透率的地层具有一定的配伍关系,当孔隙喉道半径与聚合物分子线团回旋半径之比大于 5 时聚合物可以通过且不产生堵塞。从实验结果来看,高渗透层聚合物进的多,采出的也多,这主要是由于注入的聚合物分子量较高,达到 2 200 万,分子水动力学尺寸较大,与中高渗透层配伍性好。

表 2 层间非均质模型聚合物采出及吸附滞留量

渗透层	注入量/mg	采出量/mg	吸附滞留量/mg	吸附滞留率/%
高渗透层	3 028.5	1 455.9	1 572.6	52.0
中渗透层	1 170.0	259.4	910.6	77.8
低渗透层	301.5	12.8	288.7	95.8
合计	4 500.0	1 728.1	2 771.9	61.6

高渗透层聚合物吸附滞留率为 52.0%，中渗透层聚合物吸附滞留率为 77.8%，低渗透层聚合物吸附滞留率为 95.8%，3 层平均吸附滞留率为 61.6%，具有利用价值。

从 3 层滞留量比例来看，高渗透层占 56.7%，中渗透层占 32.9%，低渗透层占 10.4%，聚合物吸附滞留主要在中、高渗透层，达到 90%，因此聚合物驱后开展地下吸附滞留聚合物的再利用研究，如何利用中、高渗透层中的聚合物是研究的重点。

3.2 非均质模型聚合物产出液变化规律

聚合物驱层间非均质模型实验注入孔隙体积与产出液聚合物浓度关系曲线见图 4。在注聚期间，就有聚合物在高渗透层产出，且在注聚合物驱结束后，产出液聚合物浓度达到峰值 750 mg/L 左右，而此时中、低渗透层产出液聚合物浓度很低，表明聚合物进入高渗透层相对较多，中渗透层产出液在注入 1.4 PV 时，开始有聚合物产出，至注入 2.0 PV 左右浓度达到最高，整个实验过程中，低渗透层产出液聚合物浓度一直很低，聚合物进入低渗透层相对较少。主要是由于聚合物存在不可及孔隙体积效应[4]，高渗透层孔隙喉道半径大，在岩心中渗流过程中，聚合物沿高渗透层中孔隙半径较大的通道渗流，并且流动速度较快，因而高渗透层采出浓度高、采出量大。

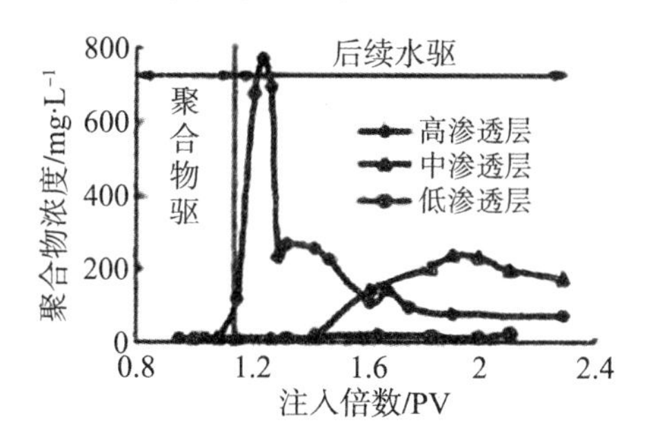

图 4 注入孔隙体积与产出液聚合物浓度关系

3.3 模型内聚合物吸附滞留量分布规律

非均质模型聚合物吸附滞留量分布(数据经归一化处理)见图 5。高渗透层吸附滞留量相对较高，其次为中渗透层，低渗透层吸附滞留量最少。总体来看，由进口至出口聚合物吸附滞留量逐渐增加，并且越靠近采出井，吸附滞留量越高，低渗透层由于聚合物注入少，采出的更少，吸附滞留率高达 95%。从图 5 中可见，聚合物改善了层间非均质性，差异减小，吸水均匀。

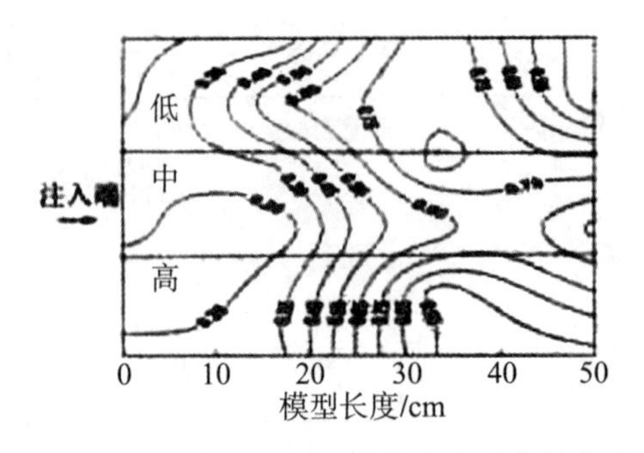

图5 层间非均质模型聚合物浓度分布等值图

3.4 层内非均质模型聚合物实验

层内非均质模型与层间实验结果类似,聚合物吸附滞留总量达到67.3%(由于存在层间窜流,无法计算单层吸附滞留量)。

4 结论和建议

(1) 聚合物驱过程中如果地层不存在大孔道窜流,聚合物驱后在地层中的吸附滞留总量将超过60%,其主要原因是目前矿场驱油的聚合物主要是部分水解聚丙烯酰胺,其主链上的—$CONH_2$ 基团水解后形成—COONa,电离后—COO^- 带负电,而地层岩石一般带正电,造成吸附滞留量较大。

(2) 从分布规律来看,由于后续流体的冲刷,从注入井到采出井,滞留浓度逐渐增加;聚合物驱吸附滞留主要在中、高渗透层,达到总注入量的50%以上,说明聚合物驱过程中聚合物主要进入渗透率比较高的部位,起到了调整非均质性作用,但是也加剧了后续水驱的矛盾。

(3) 聚合物驱后地下滞留大量的聚合物,继续开展聚合物再利用研究十分必要,但是由于热老化及后续水驱等作用,聚合物将进一步降解和采出,因此建议矿场应该及早开展聚合物再利用研究。

参考文献

[1] 赵福麟. 油田化学[M]. 山东东营:石油大学出版社, 2000-07:102-108.

[2] 卢祥国, 高振环. 聚合物分子量与岩心渗透率配伍性——孔隙喉道半径与聚合物分子线团回旋半径比[J]. 油田化学, 1996, 13(1):72-75.

[3] 高振环, 卢祥国, 陈静惠. 高分子量聚合物的分子量对岩心渗透率的适应性研究[J]. 油田化学, 1994, 11(1):73-76.

[4] 孙焕泉, 张以根, 曹绪龙. 聚合物驱油技术 [M]. 山东东营:石油大学出版社, 2002-04:72-77.

编辑 景 暖

Key words: powerful ant-isloughing agent clay dispe-rsion; suppress; antisloughing

QIAO Meng-zhan, Ruifeng Chem ica l Industry L miited Corp. , Jidong O ilfield, Tang shan 063200, Hebe, i China Research and application of LG vegetable gum additiv CAI J-ihua, WU Xiao-ming, YANG Qian yun, GU Sui ODPT, 2006, 28(6): 31-34.

Abstract: Using widely distributed and cheap ild vegetable gum. we developed a LG vegetable gum additive. The paper in troduced the physical and chemical performance of LG gum and its efects in fresh wate, rsalt water and complex salt water drilling fluid systems first he results of soak test, shale expansion test and shales'rolled recovery showed the good rejection capability and an ti collapsing. Result of lubrication tests howed that reduced rate of lubrication factor of 4% LG vegetable gum in slurry was 57.9%. It can resist 100℃. Then the active mechanism of viscosifying action, loss control and lubrication of LG gum was elaborated. Finally, the author introduced its application performance in no-dig and coal field drilling engi neering. The lab test and field application have shown that LG gum is a good viscosity increase, rfluid loss additive, saltresisting agent, anti-caving agent and lubricant, also because it comes from the leaves of camphor section plan, t it has a prom ising future.

Keywords: LG vegetable gum; fluid loss contro; l rejection capability; lubrication; no-dig; coal field drilling

CAI J-ihua, Engineering Faculty of China University of Geoscience Wuhan 430074, Hube, China

Study on residual oil quantitative distribution in abnormal high pressure reservoir by static and dynamic method. HUANG Hu, i FAN Yuping, SUN Lixu. ODPT, 2006, 28(6): 35-38.

Abstract: On the basis of microstructure, sedmientary microfac ie sand water out logging interpretation processing, the fine reservoir description and residua l oil quantitative distribution in low permeability reservoir with abnormal high pressure were studied with 3D geologicmodel construction and numerical smiulation software. The distribution area and quantity of residual oil were determined and comprehensivead justing measurese stablished. The study result shows that it is important to study residual oil distribution with the integrated study of geologic model construction and numerical simulation.

Key words: microstruc ture; sedmientary microfacies; water out; geologic model; numeric smiulation; residual oil

HUANG Hui, China University of Petroleum, Changping 102249, Beijing, China

Research on polymer distributing law by the three-dimensional anisotropic mode

ZHU Yang-wen, CAO Xu-long, SONG X in-wang, ZHANG Zhan-min, HAN Yu-gu. i ODPT, 2006, 28(6):39-41.

Abstract:In order to study polymerd istributing law after polymer flooding and the feasibility of polymer recycle, two three-dmiensional anisotropic models of 50×50 ×10 cm were designed. One of the models had a clap board to simulate the polymer distribution under multilayers condition, and one model had no clapboard to smiulate polymer distribution under a single oil-bearing layer. The rule of polymer adsorption post polymer flooding was studied by analyzing the gross polymer production volume and absorption volume. The results show that more than 60% of total injecting polymer has been adsorbed, where more than 90% of to taladsorption is held up by high permeability or medium permeability layer and the adsorption content gradually increases from injecting well to picking well. The research results provide favorable evidence for recycling the adsorbed polymer.

Key words: three-dimensional an isotropic mode; lpost polymer flooding; adsorption;distributing law

ZHU Yan wen, Geological Research Institute of Shengli Oilfield Company Limited, Dongying 257015, Shandong, China Study about on-line curing technology of hydro-phobically associating polymer AP-P4 for flooding.

GUO Yong-jun, FENG Ru-sen, MA Jun-long, SHU Zheng, NIU Hong-bin. ODPT, 2006, 28(6):42-45.

Abstract:Due to narrow and small space and lmiited bearing capacity on offshore platform, a new polymer dissolution technology, on-line curing, is put forward that dissolution velocity can be greatly accelerated by increa sing specific surface area of polymer swellingpar ticles. A ccording to the theory of enforced swelling-particles cutting, a set of on-line curing experi mental installation has been designed and manufactured. The influence of different kinds of factor such as size of filter screen, tmies of cyclic filtrating, pre-swelling time of AP-P4, flow rate of pump, mode of combining filter screen and static mixer and concentration of AP-P4, have been studied in field expermients. Expermiental results shows that on-line curing technology can make AP-P4 completely dissolve in pipe line with in 35min while abolishing curing-tank thoroughly. It has been proved that the on-line curing technology can smiplify process of preparing polymer shorten dissolution time greatly and decrease bearing capacity of platform efectively and it is feasible for offshore polymer flooding.

Key words:hydrophobically associating polymer;polymer flooding;on-line curing;

of shore platform

GUO Yong-jun, State Key Lab of Oil and Gas Reservoir Geology and Exploitation, Southwest Petroleum University, Chengdu 610500, Sichuan, China

Application of microbial injection for oil recovery enhancemen CHENG Chang-ru, ZHANG Shu-qin, YAN Yun-gu, i LI Yu-ping, XU W e-isheng, X ING L-iguo. ODPT, 2006, 28(6):46-48

Abstract: To solve the oil production increase of Dagang Oilfield, several TOR (tertiary oil recovery) technologies have been studied to miprove the oil recovery rate. Microbialen-hanced oil recovery is a rapidly developing te chnology in past several years. Based on the bioticcharacte ristics of microbe, the injection crafts is very important for microbial enhanced oil recovery. To guarantee the injection concentration of microbe, a series of research have been conducted, such as the develop.

润湿性对油水渗流特性的影响

宋新旺[1,2],张立娟[1,3],曹绪龙[2],侯吉瑞[1,3],岳湘安[1,3]

(1. 石油工程教育部重点实验室(中国石油大学),北京昌平 102249;

2. 中国石化胜利油田分公司地质科学研究院,山东东营 257015;

3. 中国石油大学提高采收率研究中心,北京昌平 102249)

摘要:用亲水露头砂制备水测渗透率分别为0.030、0.197、0.508 μm^2 的3组人造柱状岩心,每组岩心分为表面亲水(不处理)、中性润湿(经甲基硅油处理)、亲油(经胜利原油煤油混合油处理)3类,其表面润湿性经切片测定动态接触角确认。分别用矿化度19.4 g/L的模拟胜利地层水和模拟油测定岩心渗透率。相同渗透率岩心的水测渗透率,岩心表面亲水时最高,中性润湿时次之,亲油时最低,变化幅度随渗透率的增大而减小,最大达600倍;油测渗透率则在表面中性润湿时为最高,亲油时次之,亲水时最低。岩心的水驱采收率,以岩心表面中性润湿时为最高,表面亲水时大幅下降,表面亲油时最低,但与表面亲水时相比,降幅不大。认为油润湿是低渗透油藏注水压力高的原因之一,将油藏润湿性转为中性,可提高水(及化学驱油剂)的注入能力。表3参8。

关键词:砂岩油藏;岩心;润湿性;水相渗透率;油相渗透率;水驱;采收率;低渗透油藏

中图分类号:TE311:T E341:O647.5 **文献标识码**:A

润湿性是储层的一项重要物性,对开发效果有显著的影响[1-6]。Morrow等[6]通过多组岩心实验发现,原油/盐水/砂岩体系处于弱水湿或中性润湿状态时,驱油效率最高。鄢捷年等[7]的室内研究结果也表明,随水湿性逐渐减弱,驱油效率呈上升趋势,弱水湿岩样的驱油效率最高。刘中云等[8]评价了各种润湿类型岩石的水驱效果,结果也显示,弱亲水岩石的水驱采收率最高,强亲油岩石水驱采收率最低。本文在前人工作的基础上,分别将高、中、低三种渗透率岩心的润湿性由亲水转变为亲油和中性润湿,研究润湿性对油水在不同渗透率岩心中的渗流特性及采收率的影响。

1 实验

1.1 实验仪器与材料

SCAT表面/界面张力仪(德国Haake公司),RS-600流变仪(德国Haake公司),2PB00C型平流泵(北京卫星制造厂,流量范围0.01～5 mL/min),岩心夹持器(规格为

φ2.5 cm×15 cm),恒温箱、高压中间容器等。

甲基硅油;模拟胜利地层水,含 Na^+ 7 001.4 mg/L,Ca^{2+} 412.0 mg/L,Mg^{2+} 101.9 mg/L,Cl^- 11 839.2 mg/L,矿化度 19 354.5 mg/L;模拟油(胜利地面脱气原油与煤油混合油,体积比分别为 1∶5 和 1∶9,黏度分别为 2.044 mPa·s 和 1.048 mPa·s)。亲水性露头砂人造圆柱状岩心,共计 36 块,分为 3 组,水测渗透率 K_w 分别为 30×10^{-3}、197×10^{-3}、508×10^{-3} μm^2,简称低渗(30)、中渗(197)、高渗(508)岩心。同一渗透率值的岩心取自同一块方岩心。实验温度 80℃。

1.2 实验方法

(1) 将岩心润湿性由亲水变为中性润湿:将亲水圆柱状岩心在 10%硅油中浸泡 24 小时后烘干,切片取样测量地层水的动态接触角以确定润湿性。

(2) 将岩心润湿性由亲水变为亲油:将亲水圆柱状岩心在模拟油(胜利原油与煤油体积比 9∶1)中浸泡 24 小时后烘干,切片取样测量地层水的动态接触角以确定润湿性。

(3) 油水渗流及水驱油实验步骤:① 取三块具有相同渗透率、润湿性的岩心,抽真空;② 第一块饱和模拟地层水进行水渗流实验,注入速度从低到高,待压力稳定后,换下一个流速,记录不同注入速度下的压力动态及采液量;③ 第二块饱和模拟油(胜利原油与煤油体积比 1∶5)进行油渗流实验,注入速度从低到高,待压力稳定后,换下一个流速,记录不同注入速度下的压力动态及采液量;④ 第三块饱和模拟地层水,模拟油驱水至束缚水饱和度,油驱水至含水 100%,测量原始含油饱和度、水驱油采液量、采油量,计算采收率。

2 结果与讨论

2.1 润湿性对水渗流的影响

岩石的润湿性可根据水在固体表面的接触角 θ 来划分,一般当 $\theta<75°$ 时为水润湿或亲水、当 $75°<\theta<105°$ 时为中性润湿、当 $\theta>105°$ 时为油润湿或亲油。表 1 给出几种不同渗透率亲水岩心处理后的动态接触角。可见岩心处理后由亲水变为中性润湿和亲油。对比不同润湿条件下的水测渗透率值可知,当岩心的润湿性由亲水向中性和亲油转变时,水通过岩心的渗流阻力大幅增加。当岩心渗透率为 30×10^{-3}、197×10^{-3}、508×10^{-3} μm^2 的亲水岩心转变为中性润湿性后,渗透率分别降至 12.91×10^{-3}、62.68×10^{-3}、83.41×10^{-3} μm^2,相差 2.32、3.14、6.09 倍,低渗亲水岩心转为亲油润湿后,水的渗流阻力相差 600 倍。分析认为这是低渗油藏注水压力高的重要原因之一。当岩心的润湿性转为亲油时,水通过岩心孔隙的过程中需克服毛管力做功,渗透率越低,孔隙平均尺度越小,毛管阻力越大,使渗流阻力越高。当岩心由亲水转为中性时,水渗流过程中的毛管阻力基本为零,但需克服来自壁面的黏附功,黏附功与流通面积成正比,渗透率越高,水的流通面积越大,渗流所需克服的黏附功越高,因而渗流阻力也相对较高。对于中性润湿岩心,高渗透岩心的渗流阻力最高,中渗次之,低渗最低。

2.2 润湿性对油渗流的影响

表2给出了不同润湿条件下的油测渗透率。可以看出,当岩心的润湿性由亲水向中性和亲油转变时,不同渗透率岩心的油测渗透率值均先增加后减少,中性润湿时的最高。岩心渗透率为30×10^{-3}、197×10^{-3}、508×10^{-3} μm^2的亲水岩心转变为中性润湿性后,油测渗透率由0.76×10^{-3}、1.15×10^{-3}、1.38×10^{-3} μm^2分别增加至9.53×10^{-3}、7.20×10^{-3}、18.78×10^{-3} μm^2,相差12.54、6.26、13.60倍。由此实验结果可知,不论岩心渗透率高或低,中性润湿条件最有利于油的流动。

表1 不同润湿性条件下的水测渗透率

岩心分类	岩心编号	直径/cm	长度/cm	孔隙度/%	润湿性	前进角/°	后退角/°	水测渗透率 $K_1/10^3$ μm^2	K_w/K_1
低渗(30)	A-1	2.53	9.70	23.91	亲水	0	0	30.00*	1.00
	A-2	2.52	9.42	24.14	中性	82.38	65.68	12.91	2.32
	A-3	2.53	10.42	24.97	亲油	109.36	107.65	0.05	600.00
中渗(197)	A-4	2.47	10.09	27.42	亲水	0	0	197.00*	1.00
	A-5	2.52	9.41	29.77	中性	76.45	72.60	62.68	3.14
	A-6	2.53	9.52	31.68	亲油	112.37	107.90	1.94	101.55
高渗(508)	A-7	2.29	8.29	39.69	亲水	10	9.00	508.00*	1.00
	A-8	2.51	7.25	35.66	中性	95.06	79.01	83.41	6.09
	A-9	2.51	8.35	41.63	亲油	118.32	115.43	9.59	52.97

* $K_1=K_w$。

表2 不同润湿性条件下的油测渗透率

岩心分类	岩心编号	直径/cm	长度/cm	孔隙度/%	润湿性	水测渗透率 $K_w/10^3$ μm^2	油测渗透率 $K_2/10^3$ μm^2	K_2/K_o
低渗(30)	B-1	2.53	10.02	23.91	亲水	30.00	0.76*	1.00
	B-2	2.52	9.72	24.14	中性		9.53	12.54
	B-3	2.51	7.02	24.97	亲油		2.04	2.68
中渗(197)	B-4	2.45	10.09	27.42	亲水	197.00	1.15*	1.00
	B-5	2.52	9.37	29.77	中性	7.20	6.26	
	B-6	2.50	7.23	31.68	亲油		3.03	2.64
高渗(508)	B-7	2.37	8.48	39.69	亲水	508.00	1.38*	1.00
	B-8	2.53	6.95	35.66	中性		18.78	13.60
	B-9	2.52	9.26	41.63	亲油		4.10	2.97

* $K_2=K_o$。

表 3 不同润湿性条件下的水驱采收率

岩心分类	岩心编号	直径/cm	长度/cm	孔隙度/%	润湿性	水测渗透率 $K_w/10^3\ \mu m^2$	水驱采收率 $E/\%$	$E\text{-}E_w/\%$
低渗(30)	C-1	2.52	7.43	23.91	亲水	30.00	41.30*	0.00
	C-2	2.52	7.25	24.14	中性		45.46	4.16
	C-3	2.51	9.19	24.97	亲油		40.22	−1.08
中渗(197)	C-4	2.52	8.60	27.42	亲水	197.00	50.93*	0.00
	C-5	2.49	8.93	29.77	中性		53.98	3.05
	C-6	2.51	8.43	31.68	亲油		49.90	−1.03
高渗(508)	C-7	2.51	9.50	39.69	亲水	508.00	62.75*	0.002.51
	C-8	2.51	8.50	35.66	中性		66.67	3.92
	C-9	2.53	9.19	41.63	亲油		62.22	−0.53

* $E=E_w$。

2.3 润湿性对水驱采收率的影响

表 3 为不同渗透率、不同润湿性条件下的水驱采收率结果。可以看出，在不同润湿性条件下，水驱采收率随着渗透率的增加而增加；对于不同渗透率的岩心，中性润湿条件下的水驱采收率最高，亲水条件下次之，亲油条件下最低。岩心渗透率为 30×10^{-3}、197×10^{-3}、$508\times10^{-3}\ \mu m^2$ 的亲水岩心转变为中性润湿性后，水驱采收率分别提高 4.16 %、3.05 %、3.92 %。这说明对于不同渗透率的岩心，将岩心的润湿性由亲水和亲油转向中性润湿条件均有利于增加油的流动能力，增加水的流动阻力，从而降低流度比，改善水驱效果。不同润湿性岩心中的油水渗流和水驱采收率的实验结果还表明，在长期注水油藏的化学驱中，其岩石的润湿性往往亲油性较弱或为中性润湿条件，润湿性由亲油转向中性润湿是提高采收率的一个重要机理。此外，在使用化学 剂驱油提高采收率时，要特别注意避免油藏的润湿性转为亲油性。

3 结论

岩石润湿性对油水渗流有显著的影响，当由亲水向中性和亲油转变时，水渗流阻力大幅增加，油渗流阻力减少；对于低渗透油藏，油性润湿是注水压力高的原因之一，建议将其润湿性转为中性润湿条件以提高水的注入能力；若长期注水油藏转向化学驱，润湿性由亲油转向中性润湿是提高采收率的重要机理之一。

参考文献

[1] 任晓娟，刘宁，曲志浩，等. 改变低渗透砂岩亲水性油气层润湿性对其相渗透率的影响 [J]. 石油勘探与开发，2005，32 (3):123-124/134.

[2] Jadhunandan P, M orrow N R. Effect of w et tabilit y on waterflood recovery for crude oil/brine/rock syst em [Z]. SPE 22 597, 1991.

[3] 黄春，汤志强，蒋官澄. 润湿性对悬东 29 块稠油油藏注水采收率的影响[J]. 石油钻采工艺，1999，21(3):92-94.

[4] 蒋明煊. 油藏岩石润湿性对采收率的影响[J]. 油气采收率技术，1995，2 (3): 25-31.

[5] 王凤清，姚同玉，李继山. 润湿性反转剂的微观渗流机理[J]. 石油钻采工艺，2006，28(2):40-42.

[6] M orrow N R. Wett ability and its effect on oil recovery [J]. JPet rol Technol，1990，(12):1476-1484.

[7] 鄢捷年. 油藏岩石润湿性对注水过程中驱油效率的影响[J]. 石油大学学报(自然科学版)，1998，22(2):43-46.

[8] 刘中云，曾庆辉，唐周怀，等. 润湿性对采收率及相对渗透率的影响[J]. 石油与天然气地质，2000，21(2):148-150.

Effects of wettability on oil and water flow through porous media

Song Xinwang[1,2], Zhang Lijuan[1,3], Cao Xulong[2], Hou Jirui[1,3], Yue Xiangan[1,3]

(1. Mi nist ry of Ed ucation Key Laboratory of Pet roleum Engineering in China Un iversity of Petroleum, Changping, Beijing 102249, PR of China;
2. Research Inst it ut e of Geological Science, S hengli Oi lfield Branch Company, S inopec, Dongying, S handong 257015, PR of China;
3. Enhanced Oi l Recovery Research Center, China Univer si ty of Petroleum, Changping, Beijing 102249, P R of China)

Abstract: Lots of artificial cores of water permeability, K_w, 0.030, 0.197, and 0.508 μm^2 are prepared from an outcrop sand and are treated partly with a methyl silicone oil and partly with a crude/kerosene mixture to make their surface neutrally wetting or oil wetting with the remained cores keeping water wetting surfaces. The surface wettability of cores is ascertained by determining dynamic contact angles on cut sections. The permeability of the cores is measured to a flooding water of TSD= 19.4 g/L and a simulation oil. The water phase permeability is arranged in a decreasing order: water wetting > neutrally wetting > oil wetting at equal K_w value and the changes in permeability increases with increasing K_w value, giving a largest value of 600 times. The oil phase permeability of cores at equal K_w v alue is ar ranged in a decreasing order: n eutrally wetting > oil wetting > water wetting. The highest oil recovery by waterflood is observed in neutrally wetting cores, a significant decrease in oil recovery—in water wetting cores and the lowest, a little droped, oil recovery—in oil wetting cores. It is considered that oil wettability of rock surfaces is one of the important reasons for raised water injection pressure at low permeability sandstone oil reservoirs and conversion of reservoir rock wettability to neutrally wetting is a way of enhacing injectivity of water (and chemical flooding fluids).

Key words: sandstone oil reservoirs; cores; wettability; water phase permeability; oil phase permeability; water flood; oil recovery; low permeability reservoirs

油藏润湿性对采收率影响的实验研究

宋新旺[1]，程浩然[2]，曹绪龙[1]，侯吉瑞[3*]

（1. 中国石化胜利油田地质科学研究院，山东东营 257015；2. 清华大学机械系，北京 100084；
3. 中国石油大学采收率中心，北京 102249）

摘要：为了研究润湿性对油藏水驱采收率的影响，实验选用了低、中、高三种不同渗透率的露头砂岩心，利用甲基硅油、模拟油等化学剂将其润湿性由亲水状态改为中性和亲油状态，通过岩心驱替实验，研究了不同润湿性对水驱油采收率的影响。实验表明，岩心的润湿性由亲水和亲油状态转向中性润湿后，最大限度降低了多孔介质中毛管渗流阻力，从而提高了水驱效果。本实验条件下，建议将水驱后期油藏的润湿性由亲水或亲油状态转为中性润湿，则有利于水驱效率的大幅度提高。

关键词：润湿性；采收率；渗流阻力；波及效率

中图分类号：TE312　**文献标识码**：A　doi：10. 3696/ j. issn. 1006-396X. 2009. 04. 013

The effects of wettability on oil recovery efficiency

Song Xinwang[1], Cheng Haoran[2], Cao Xulong[1], Hou Jirui[3*]

(1. Research Institute of Geological Science of S hengli Oi l Field, Sinopec, Dongy ing Shandong 257015, P. R. China;

2. Department of Mechanical Engineering, Tsinghua University, Beijing 100084, P. R. China;

3. The Center of EOR, China University of Petroleum, Beijing 102249, P. R. China)

Received 12 December 2008; revised 18 June 2009; a ccepted 17 September 2009

Abstract: For achieving a better understanding of the effects of wettability on the oil recovery, the core displacement experiments were carried out. By using outcrop man-made cores with low, middle and high permeability, and converting core's wet tability from water-wet and oil-wet to intermediate wet, the oil recovery efficiency after water flooding on the cores was studied. The results show that converting core's wettability from water-wet and oil-wet to intermediate wet is favorable to decreasing. maximumly the capillary force as a resistance force in reserv oir pores, and enhancing water displacement effects. So the suggestion that the wettability of water-wet and oil-wet should be changed into intermediate wet on the experimental condition.

Key words: Wettability; Oil recovery efficiency; Seepage resistance; Sweep efficiency

* Corresponding author. Tel.: +86-10-89731663; fax: +86-10-89734612; e-mail: houjirui@126. com

当水驱油时,地层原油采收率或驱油效果在很大程度上与水对地层岩石的润湿性有关[1-3],而原油/岩石/盐水体系润湿性是一个复杂问题,不仅与原油组分、岩矿组分、地层水性质等密切相关,而且与流体接触时间的长短密切相关[4-8]。在油藏复杂的多孔介质中,因润湿性和界面张力差异而在不同润湿性的毛管中产生不同的毛管力,不仅影响油水在多孔介质中的渗流,也将进一步影响最终的采收率。为了进一步揭示润湿性对水驱油藏采收率的影响,本文开展了不同润湿性条件下岩心驱替实验。

1 实验部分

1.1 模拟地层水

实验所用地层水为胜利地质院提供的地层水配制的模拟地层水,其中 Mg^{2+},Ca^{2+},Na^{+},Cl^{-} 离子质量浓度分别为 101.874, 412.000, 7 001.436, 11 839.241 mg/L;总矿化度为 19 334 mg/L。

1.2 模拟地层油

由于实验中所用的胜利原油为地面的脱气脱水原油,原油粘度较高。而在实际油藏条件下,原油粘度介于 1.9~2.5 mPa·s,所以实验中使用模拟油,用胜利原油和煤油来制备。

1.3 实验岩心

实验所用岩心为实验室制备的人造岩心,以气测渗透率为参考分成高渗、中渗、低渗3种,润湿性为强亲水的。

将含质量分数10%硅油的煤油溶液饱和进入强亲水岩心,并于油藏温度下浸泡12 h,保持温度,连续以低速(0.05~0.1 mL/min)长时间非匀速注入地层水以驱除硅油,48 h后观察连续2 h内无油珠出现为止,干燥保存。

中性润湿:将含质量分数10%硅油的煤油溶液饱和进入强亲水岩心,并于油藏温度下浸泡 12 h,保持温度,连续以低速(0.05～0.1 mL/min)长时间非匀速注入地层水以驱除硅油,48 h 后观察连续 2 h 内无油珠出现为止,干燥保存。

强亲油性:将胜利原油与煤油的体积比为 9∶1 的模拟油饱和进入强亲水岩心,并于油藏温度下保持 24 h,保持温度,连续以低速(0.05～0.1 mL/min)长时间非匀速注入地层水,以驱除多余油相,至 48 h 后观察 2 h 内无油珠出现为止,可直接用于实验。岩心的基础数据见表 1。

表 1 实验所用岩心基础的数据

Table 1 Basal data of cores used in experiments

润湿性类别	渗透率类别	岩心编号	长度/m	直径/m	渗透率/md	孔隙体积/mL
强亲水	高渗	ML3-a11	0.095 02	0.025 08	594	13.742
	中渗	MB4-b11	0.085 98	0.025 20	183	13.706
	低渗	SWC1-c11	0.074 52	0.025 12	32	11.176
中等润湿	高渗	M L3-a12	0.085 03	0.025 10	601	12.309
	中渗	M B4-b12	0.089 34	0.024 94	191	12.886
	低渗	SWC1-c12	0.072 52	0.025 20	35	11.241
强亲油	高渗	M L3-a13	0.091 92	0.025 06	597	14.877
	中渗	M B4-b13	0.084 32	0.025 14	214	13.605
	低渗	SWC1-c13	0.074 32	0.025 26	35	11.225

1.4 实验流程

渗流实验仪器包括高压恒流泵、温箱、岩心夹持器等,实验流程图见图 1。

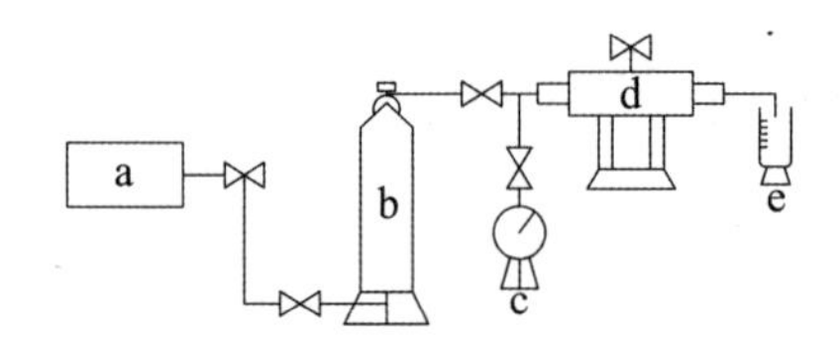

Fig. 1 Flow diagram of experimental process

图 1 渗流实验流程示意图

a 驱替泵;b 中间容器;c 测压传感器;d 岩心夹持器;e 产出液收集量筒

1.5 实验条件

实验模拟胜利油田的油藏温度(80℃),岩心的围压控制为 3.5 MPa 左右。

1.6 实验步骤

(1) 用电子天平称取各岩心的干重,记录数据。

(2) 对岩心抽真空,然后饱和模拟地层水,取出称取湿重,记录数据。

(3) 检查设备是否完好,高压恒流泵冲液,调节各级阀门。

(4) 中间容器中加入模拟地层水(或模拟油)。

(5) 岩心放入岩心夹持器中,加围压至 3.5 MPa。

(6) 打开温箱温控开关,调节温度至 80℃,恒温 24 h。

(7) 饱和模拟油在 3 PV 以上,并且出口无水产出。

(8) 打开高压恒流泵排液,以 0.2 mL/min 的流速注入地层水,驱替岩心,直至不再出油为止,在该过程中要记录压力、出油量。

(9) 整理设备,实验完成。

2 结果与讨论

不同润湿性高、中、低渗岩心的水驱油采收率结果见表 2,强亲水、中等润湿、强亲油性高、中、低渗的岩心采收率与累计注入量(以岩心孔隙体积倍数 PV 计)的关系比较分别见图 2,3,4。

表 2 水驱油采收率实验结果

Table 2 Experimental results of water-flooding recovery

润湿性类别	渗透率类别	岩心编号	饱和油/mL	出油量/mL	注入体积/mL	PV	采收率,%
强亲水	高渗	M L3-a11	10.2	6.4	60	4.708 837	62.745 1
	中渗	M B4-b11	10.8	5.5	68	4.961 331	50.925 9
	低渗	SWC1-c11	9.2	3.8	92	8.231 926	41.304 3
中等润湿	高渗	M L3-a12	9.9	6.6	88	7.149 240	66.666 7
	中渗	M B4-b12	11.3	6.6	120	8.820 287	57.894 7
	低渗	SWC1-c12	9.9	4.5	104	9.251 846	45.454 5
强亲油	高渗	M L3-a13	13.5	8.4	120	8.066 142	62.222 2
	中渗	M B4-b13	11.4	6.1	100	7.201 498	53.982 3
	低渗	SWC1-c13	9.2	3.7	108	10.562 350	40.217 4

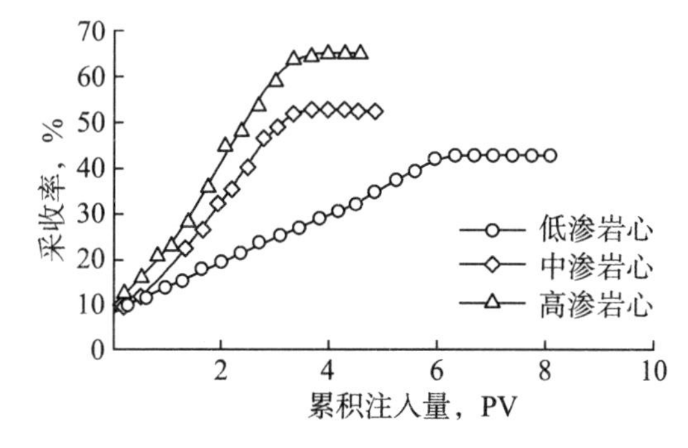

Fig. 2 Comparison of relationship curves of recovery efficiency and PV on strong hydrophilic high, medium and low-permeability core

图 2 强亲水高、中、低渗岩心采收率与 PV 关系比较

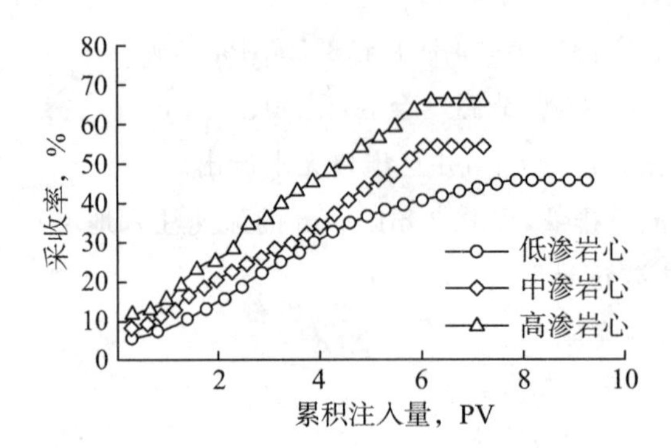

Fig. 3 Comparison of relationship curves of recovery efficiency and PV on medium-wet high, medium and low-permeability core

图 3 中润湿高、中、低渗岩心采收率与 PV 关系比较

由以上结果可以看出,不同润湿性的高、中、低渗岩心采收率的变化趋势基本上是一致的,都是逐渐增加,而后稳定在某一值。其中强亲水高渗岩心的采收率最高为62.74%,其次是中渗岩心的采收率基本上稳定在 50.92%,低渗岩心的采收率最低,为41.3%;中等润湿的高渗岩心的采收率稳定在 66.67%,中渗岩心的采收率稳定在53.98%,低渗岩心的采收率稳定在 45.45%;强亲油的高渗岩心采收率稳定在 62.22%,中渗岩心的采收率稳定在 53.98%,低渗岩心的采收率稳定在 40.21%。

图 5 为不同润湿性岩心采收率的对比。由图 5 可以得到,中等润湿条件的高、中、低渗岩心的最终水驱采收率高于强亲水的高、中、低渗岩心,而强亲水的高、中、低渗岩心的最终水驱采收率高于强亲油的高、中、低渗岩心。

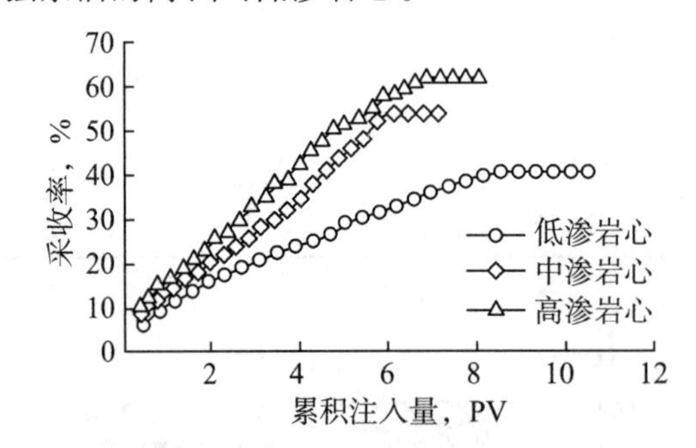

Fig. 4 Comparison of relationship curves of recovery efficiency and PV on lipophilic high, medium and low-permeability core

图 4 强亲油高、中、低渗岩心采收率与 PV 关系比较

亲水的油藏也有利于水驱采收率的提高,但幅度不如中等润湿油藏大。尽管在亲水岩石孔隙中毛管力为驱油的动力,但由于孔隙结构的复杂性及驱替过程中的润湿反转,动态水驱油过程中,多孔介质中的原油已被分割成大小不等的油滴或孤岛油,这些不连续的油滴或孤岛在流经孔喉时产生贾敏效应,形成渗流阻力,且孔喉越小、孔隙结构越复杂,产生的渗流阻力越大。强亲水低渗透岩心中,虽然毛管力使注入水更有利于进入较

小尺度的孔隙,但因粘土吸水膨胀,使孔隙结构更为复杂,因此水驱效率最低。

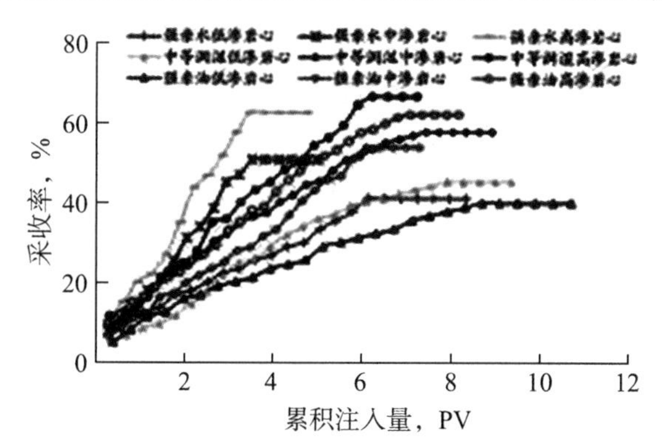

Fig. 5　Comparison of relationship curves of recovery efficiency and PV on high, medium and low-permeability cores with different wettability

图 5　不同润湿性高、中、低渗岩心采收率与 PV 的关系比较

对于强亲油岩心,因复杂孔隙结构中广泛分散的油/水界面产生的毛管力使得渗流阻力方向混乱,而且油膜影响最终采收率,因而使得强亲油岩心的采收率低于强亲水岩心的采收率。强亲油油藏在开采过程中不要过多的采取水驱方法来开采,这样不利于最终采收率的提高。

而中等润湿条件的岩心孔隙内,因孔隙尺度差异造成的分散残余油,在静止状态呈现大小不一的油柱,尽管流动过程中会产生润湿滞后,但相对强亲水与强亲油的岩心相比,产生的贾敏效应较小,更加利于油在孔喉中的流动,因此提高了水的波及效率和宏观驱油效率,从而提高了最终的采收率。

根据实验结果,在实际油藏的开采过程中,中等润湿条件油藏的水驱采收率最高,鉴于此点,可以采取一系列措施来改变油藏的润湿性,使其变为中等润湿,以提高最终水驱采收率。

参考文献

[1] 任晓娟，刘宁，曲志浩，等. 改变低渗透砂岩亲水性油气层润湿性对其相渗透率的影响[J]. 石油勘探与开发，2005，32(3):123-124;134.

[2] Jadhunandan P，Morrow N R. Effect of wettability on waterflood recovery for crude oil/brine/rock sy stem [J]. SPE 22597，1991.

[3] 舒小彬，刘建成，韩传见，等. 储层岩石润湿性对开发的影响 [J]. 内蒙古石油化工，2004，6:135-136.

[4] 韩翻珍，高芒来，刘永斌. MD 膜驱剂与烷基季铵盐复配改性粘土的润湿性[J]. 石油化工高等学校学报，2008，21(1):21-24;29.

[5] 王凤清，姚同玉，李继山. 润湿性反转剂的微观渗流机理 [J]. 石油钻采工艺，2006，28(2):40-43 .

[6] Morrow N R. We ttability and its effect on oil recovery [J]. JPT，1990(12):1476-1484 .

[7] 鄢捷年. 油藏岩石润湿性对注水过程中驱油效率的影响 [J]. 石油大学学报:自然科学版，1998，22(2):43-46.

[8] 刘中云，曾庆辉，唐周怀，等. 润湿性对采收率及相对渗透率的影响 [J]. 石油与天然气地质，2000，21(2):148-150.

(Ed.: Y Y L, Z)

非均相复合驱非连续相渗流特征及提高驱油效率机制

侯　健[1,2]，吴德君[1,2]，韦　贝[1,2]，周　康[1,2]，巩　亮[3]，曹绪龙[4]，郭兰磊[5]

（1. 非常规油气开发教育部重点实验室（中国石油大学（华东）），山东青岛 266580；

2. 中国石油大学（华东）石油工程学院，山东青岛 266580；

3. 中国石油大学（华东）新能源学院，山东青岛 266580；

4. 中国石化胜利油田分公司，山东东营 257015；

5. 中国石化胜利油田分公司勘探开发研究院，山东东营 257015）

摘要：非均相复合驱是聚合物驱后油藏大幅度提高采收率的重要方法。微流控实验及室内双管岩心流动实验均表 明非均相复合驱具有非连续相渗流特征，表现为压力一定条件下微通道内流速仍大幅波动、不同渗透率双管岩心中分流量持续交替变化。综合考虑凝胶颗粒弹性变形特征、聚合物增黏与非牛顿特性、表面活性剂降低界面张力作用及润湿改变等化学剂作用机制，建立非均相复合驱微观渗流格子玻尔兹曼模拟方法。基于微观数值模拟结果，揭示非均相复合驱提高驱油效率机制，颗粒暂堵升压后变形运移、固液增阻导致液流转向、驱替液增黏抑制黏性指进、强化启动微观剩余油，进而提高微观波及系数及微观洗油效率。

关键词：非均相复合驱；非连续相渗流；凝胶颗粒；微观渗流模拟

中图分类号：TE 312　文献标志码：A

引用格式：侯健，吴德君，韦贝，等. 非均相复合驱非连续相渗流特征及提高驱油效率机制[J]. 中国石油大学学报（自然科学版），2019，43(5)：128 鄄 135.

Hou Jian，Wu Dejun，Wei Bei，et al. Percolation characteristics of discontinuous phase and mechanisms of improving oil displacement efficiency in heterogeneous composite flooding [J]. Journal of China University of Petroleum（Edition of Natural Science），2019，43(5)：128-135.

Percolation characteristics of discontinuous phase and mechanisms of improving oil displacement efficiency in heterogeneous composite flooding

Hou Jian[1,2], Wu Dejun[1,2], Wei Bei[1,2], Zhou Kang[1,2], Gong Liang[3], Cao Xulong[4], Guo Lanlei[5]

(1. Key Laboratory of Unconventional Oil & Gas Development(China University of Petroleum(East China)), Ministry of Education, Qingdao 266580, China;
2. School of Petroleum Engineering in China University of Petroleum(East China), Qingdao 266580, China;
3. College of New Energy in China University of Petroleum (East China), Qingdao 266580, China;
4. Shengli Oilfield Branch Company, SINOPEC, Dongying 257015, China;
5. Shengli Oilfield Exploration and Development Research Institute, SINOPEC, Dongying 257015, China)

Abstract: Heterogeneous composite flooding is an important method for greatly improving oil recovery after polymer flooding. The microfluidic and double-sandpacks displacement experiments show that the heterogeneous composite flooding appeared with percolation characteristics of discontinuous phase, which can be reflected by a large fluctuation of flow velocity in the microchannels and the continuous alternating change of fractional flow in sandpacks with different permeability. In this study a microscopic percolation lattice Boltzmann method for heterogeneous composite flooding was established in which the elastic deformation characteristics of gel particles, the viscosity increase and non-Newtonian properties of polymers and the interfacial tension reduction and wettability altering mechanism of surfactants were considered and simulated. Based on the simulation results, the mechanisms of improving oil displacement efficiency via heterogeneous composite flooding were revealed. Firstly temporary blockage of flow pathways by gel particles can lead to pressure increase, and then the particles can remigrate with their deformation under high pressure. Secondly, the particle-liquid interactions can increase flow resistance and conse-quently change the flow direction. Meanwhile, the polymer in the flooding fluid can increase the fluid's viscosity and inhibit viscous fingering. At last, the composite fluid can also move up of the remaining oil in microscopic state. The combination of These mechanisms can improve both the microcosmic sweep coefficient and microcosmic oil displacement efficiency.

Keywords: heterogeneous composite flooding; discontinuous phase flow; gel particles; microscopic percolation simulation

聚合物驱后油藏仍有约一半或以上的原油滞留在地下未被采出,须进一步探索新的提高采收率方法[1]。与水驱相比,聚合物驱后油藏开发面临着更大的难题:一是储层非均质更严重,注入流体易窜流;二是可动剩余油更分散,原油采出困难,现有驱油体系难以采出。针对以上问题,在驱油介质中引入可变形运移的软固体冶凝胶颗粒,与聚合物和表面活性剂溶液互配加大调驱力度,构建起非均相复合驱油体系,达到大幅度提高聚合物驱后油藏采收率的效果[2]。目前,非均相复合驱技术已在胜利油田得到推广应用[3-4],在孤岛、孤东及胜坨等特高含水主力油田共推广单元 17 个,提高采收率 8.3%,增油降水效果显著。由于凝胶颗粒的加入,非均相复合驱过程表现为非连续相流动。驱油体系中由于聚合物和表面活性剂的存在,渗流还具有非牛顿流体流动、油水界面张力降低、润湿性改变等特性。笔者采用微流控及双管岩心流动实验揭示非均相复合驱非连续相渗流特征,综合考虑凝胶颗粒、聚合物及表面活性剂等特性,建立非均相复合驱微观渗流格子玻尔兹曼模拟方法,阐释非均相复合驱提高驱油效率机制,为非均相复合驱大幅度提高采收率提供理论支持。

1　非连续相渗流特征

非均相复合驱油体系中包含连续相(含水、聚合物、表面活性剂等组分)和非连续相(凝胶颗粒)。凝胶颗粒吸水膨胀后成为可变形的"软固体冶,在渗流过程中运移至孔隙喉道可产生堵塞,随着喉道两端压差增大,颗粒能变形通过喉道,继续向深部运移。凝胶颗粒运移、滞留堵塞、变形通过、再运移的渗流现象,使非均相复合驱流动表现出非连续相渗流特征。

1.1 微流控实验

建立凝胶颗粒悬浮液微观渗流实验系统,其中微流体进样系统通过外接钢瓶洁净气体作为压力输出源,依次经过精密调压阀、气压控制泵、储液瓶以及精度流量控制阀,最后连接芯片传感器以控制驱替压力。实验可施加压力为 0～0.7 MPa,微通道喉道长度为 300 μm,宽度为 24 μm,如图 1 所示。

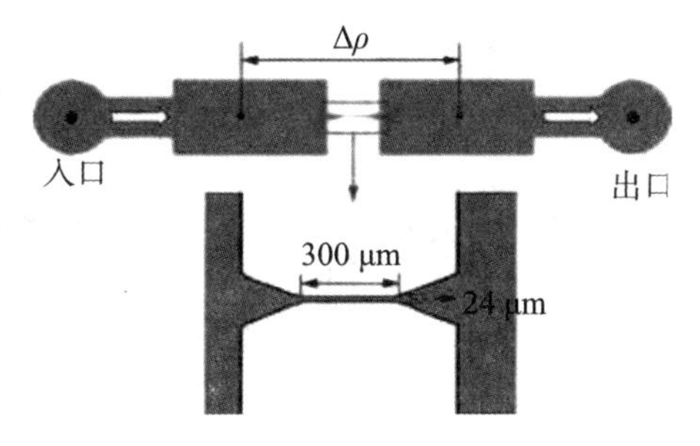

图 1　微通道模型示意图

Fig. 1　Sketch map of microchannel model

凝胶颗粒流动处于周期性的堵塞和变形通过状态,导致流体流速持续大幅度波动变化,图 2 为微通道流速变化曲线。流动开始阶段,凝胶颗粒进入微通道发生堵塞;随着微通道两端压差增加至 0.02～0.021 MPa,颗粒变形通过微通道,流速大幅上升;后续颗粒进入微通道后再次发生堵塞,流速迅速下降到较低水平;继续增加微通道两端压差到 0.049 2

～0.049 5 MPa 时，颗粒又可变形通过，流速上升。如此往复，整个过程中流速不断波动，表现出非连续相流动特征。当微通道两端压差达到约 0.181 MPa 时，通道流速逐渐上升，说明颗粒可连续通过。

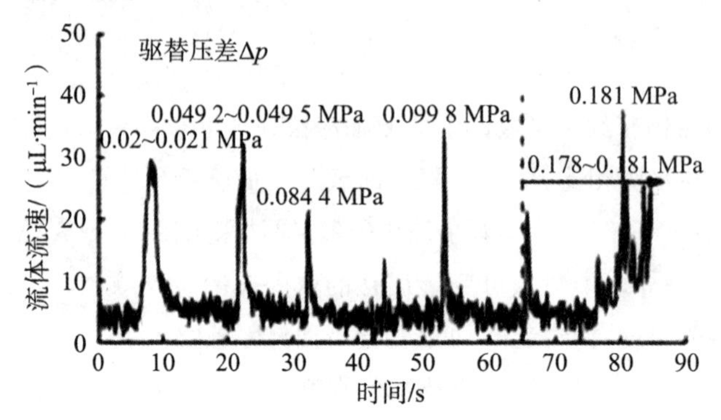

图 2　微通道流速变化曲线

Fig. 2　Flow rate curve in microchannel model

1.2 双管岩心流动实验

双管岩心流动实验装置包括注入系统、温度、压力测量控制系统和采集系统[5]。模型渗透率级差为 6，实验温度为 70℃，注入速度为 1 mL/min。实验过程中，依次注入 1 V_P(V_P 为孔隙体积)水、0.6 VP 质量浓度为 2 000 mg/L 的化学剂溶液、1.4 V_P 的后续水，记录双管分流量变化。

两种化学剂溶液注入双管前后分流量变化曲线如图 3 所示。图 3(a)表明：聚合物溶液注入前，高低渗管分流量比例稳定在约 9：1；聚合物溶液开始注入后，高渗管分流量有一定程度下降，低渗管分流量上升；聚合物溶液注入结束后，分流量恢复到初始水平。图 3(b)表明：凝胶颗粒溶液注入前高低渗管分流量稳定在 87：13；凝胶颗粒溶液注入时高渗管的分流量迅速下降，低渗管的分流量迅速上升，并且两管的分流量交替波动；凝胶溶液注入结束后，高低渗管的分流量仍保持交替波动。与聚合物溶液注入相比，凝胶颗粒的不断堵塞和运移引起高低渗管分流量更大幅度的波动变化，且在后续注水过程中仍持续波动，表现出非连续相流动特征。

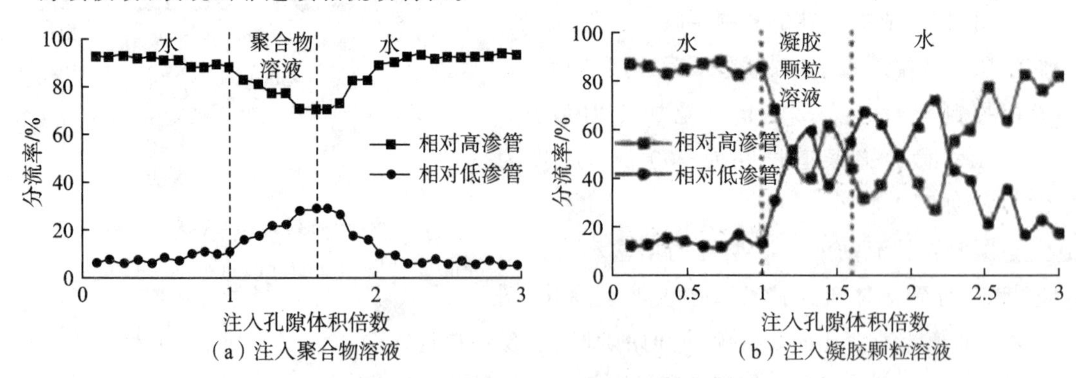

图 3　双管岩心中分流量变化曲线

Fig. 3　Fractional flow curves in two rock cores

2　微观渗流模拟方法

非均相复合驱微观渗流特性应考虑凝胶颗粒、聚合物及表面活性剂特性。凝胶颗粒的流动除具有刚性颗粒流特征[6]外，还具有弹性变形特征[7]；聚合物具有流体增黏、非牛顿流体流动等特征；表面活性剂具有油水界面张力降低、岩石润湿性改变等特征。选用格子玻尔兹曼方法(LBM) 开展非均相复合驱微观渗流模拟，该方法在微观渗流模拟方面优势明显，其多相流模拟过程无需额外追踪油水界面，且可利用反弹格式简洁处理流体与壁面的作用[8]。模型采用 Shan-Chen 多相流 LBM 模型模拟油水流动[9]，主要化学剂机制表征方法如图 4 所示，另外利用主动溶质法模拟表面活性剂传质扩散过程[10]，通过调节松弛时间模拟聚合物增黏作用。

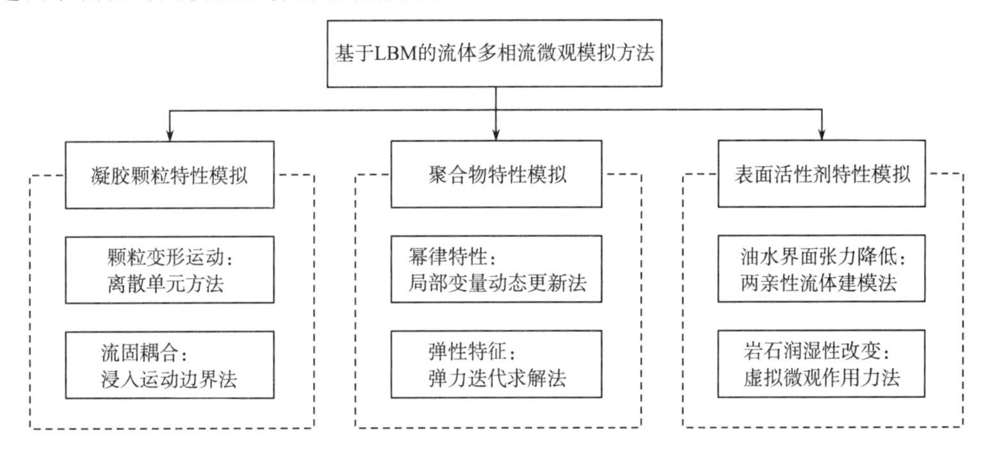

图 4　非均相复合驱微观渗流模拟化学剂机制表征方法

Fig. 4　Mechanism characterization of chemical agents in microscopic seepage simulation of heterogeneous composite flooding

2.1 凝胶颗粒特性模拟

借鉴浸入边界法[11]的基本思想，将固体边界离散为一系列虚拟边界点，则凝胶颗粒原始曲线边界即可通过各边界离散点相连围成的多边形近似表示。凝胶颗粒变形及运移的模拟相当于各边界离散点相对位置变化和运移的模拟，如图 5 所示。

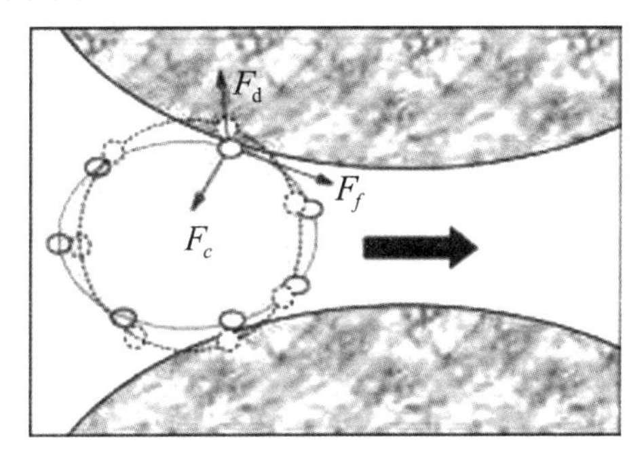

图 5　凝胶颗粒弹性变形模拟方法示意图

Fig. 5　Schematic diagram of gel particle elastic deformation

根据牛顿第二定律,考虑各边界离散点受到的流固作用力 F_f、固固接触作用力 F_c 和形变恢复力 F_d,即可更新各边界离散点的速度和位置,其表达式为

$$ma=F_f+F_c+F_d, \tag{1}$$

$$V=V_{old}+a\Delta t, x=x_{old}+v\Delta t。 \tag{2}$$

式中,m 为边界离散点的质量,其值为凝胶颗粒质量与边界离散点个数的比值,kg;a 为边界离散点的加速度,m/s^2;Δt 为时间步长,s;v_{old} 和 v 分别为边界离散点更新前后的速度,m/s;x_{old} 和 x 分别为边界离散 点更新前后的位置。

流固作用力 F_f 采用考虑附加碰撞项的格子玻尔兹曼方法(LBM)[12]根据动量定理进行模拟计算,其表达式为

$$F_f=\frac{(\Delta x)^2}{\Delta t}\sum_{k=1}^{n}B_{sk}\sum_{i=0}^{8}\Omega_i^s e_i, \tag{3}$$

$$\Omega_i^s=[f_{-i}(x,t)-f_{eq\text{-}i}(\rho,u)]-[f_i(x,t)-f^{eqi}(\rho,v_p)]。 \tag{4}$$

式中,驻 x 为格子间距,m;e_i 为格子离散速度,m/s,赘$_i^s$ 为 LBM 模拟方法中反映凝胶颗粒对流场影响的附加碰撞项;B_s 为赘的权重函数;$f_i(x,t)$为 t 时刻处于位置 x 处的粒子速度分布函数;$f^{eqi}(x,t)$为局部平衡态速度分布函数;u 和 v_p 分别为流体和颗粒在位置 x 处的速度,m/s。

固固接触作用力 F_c 采用软球模型离散单元法(DEM)[13]进行模拟计算 ,其表达式为

$$F_c=F_n n+F_t t, \tag{5}$$

$$F_n=k_n\delta_n+c_n\frac{d\delta_n}{dt}, F_t=k_t\delta_t+c_t=\frac{d\delta_t}{dt}。 \tag{4}$$

式中,F_n 和 F_t 分别为颗粒间的法向作用力和切向作用力,n;n 和 t 分别为单位法向向量和单位切向向量;k_n 和 k_t 分别为法向和切向接触刚度,N/m;c_n 和 c_t 分别为法向和切向阻尼系数,$n/(\mathrm{m}\cdot\mathrm{s}^{-1})$;$\delta_n$ 和 δ_t 分别为法向重叠量和切向位移,m。

形变恢复力 F_d 反映凝胶颗粒弹性变形后促使 其恢复至初始形状的作用力[14],其表达式为

$$F_d=F_b n+F_s t, \tag{7}$$

$$F_b=E_b(\beta-\beta_0), F_s=E_s(I_{ab}-I_{ab0})。 \tag{8}$$

式中,F_b 和 F_s 分别为凝胶颗粒变形后离散边界点受到的弯曲形变恢复力和拉伸形变恢复力,n;茁和茁 0 分别表示凝胶颗粒变形前后相邻 3 个边界离散点之间的向量夹角,(°);I_{ab} 和 I_{ab0} 分别表示凝胶颗粒变形前后相邻两个边界离散点之间的距离,m;E_b 和 E_s 分别为凝胶颗粒边界弯曲刚度和拉伸刚度。

2.2 表面活性剂特性模拟

表面活性剂是两亲性分子,在油水界面中极性的亲水基易与水相结合,非极性的亲油基易与油相结合,表现出在界面聚集的现象,进而降低了界面张力。参考两亲性流体

建模方法[15]，在 LBM 中将表面活性剂分子团抽象等价为具有一定浓度的偶极子，同时引进一个方向矢量演化偶极子平均排列方向，如图 6 所示。

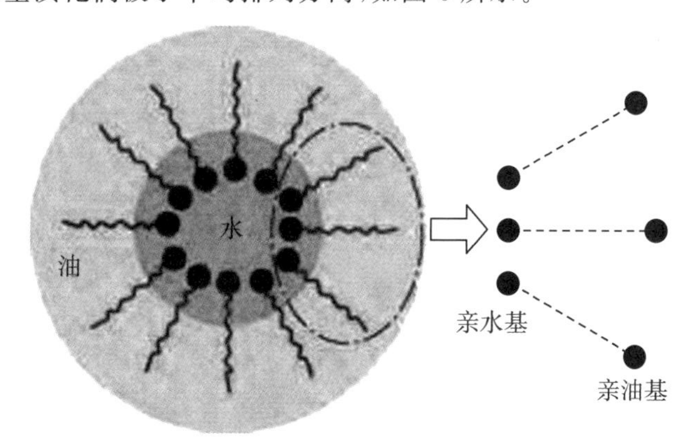

图 6 偶极子等价模型

Fig. 6 Dipole equivalent model

偶极子两端电性不同，分别代表了亲水基及亲油基，其可与油水间产生吸引力或排斥力，力的作用通过牛顿第二定律体现。t 时刻处于位置 x 处，油或水所受表面活性剂分子的作用力 $F_{\sigma s}$ 为

$$F_{\sigma s}=-2\rho_\sigma(x,t)G_{c,\sigma}\sum_{i\neq 0}\mathrm{d}(x+\mathrm{e}_i\Delta t,t)\cdot\theta_i\rho_s(x+\mathrm{e}_i\Delta t,t)w_i,\tag{9}$$

其中，

$$\theta_i=I-D\frac{\mathrm{e}_i\mathrm{e}_i}{\mathrm{e}_i^2}.$$

式中，$G_{c,\sigma s}$ 为油或水与表面活性剂间的作用强度；ρ_σ 为油或水的密度，kg/m^3；ρ_s 为表面活性剂密度，kg/m^3；d 为偶极子排列方向；w_i 为权重系数；I 为单位张量；D 为格子维数。

表面活性剂受油水作用力 $F_{s\sigma}$ 为

$$F_{s\sigma}=2\rho_s(x,t)\mathrm{d}(x,t)\sum_\sigma G_{c,\sigma s}\sum_{i\neq 0}\theta_i\rho_\sigma(x+\mathrm{e}_i\Delta t,t)w_i.\tag{10}$$

表面活性剂间的作用力 F_{ss} 为

$$F_{ss}=-\frac{4D}{c^2}G_{c,ss}\rho_s(x,t)\sum_{i\neq 0}\rho_s(x,t)\{\mathrm{d}(x+\mathrm{e}_i\Delta t,t)\mathrm{d}(x,t):\theta_i\mathrm{e}_i+[\mathrm{d}(x+\mathrm{e}_i\Delta t,t)\mathrm{d}(x,t)+d(x,t)d(x+\mathrm{e}_i\Delta t,t)]\cdot\mathrm{e}_i\}.\tag{11}$$

式中，$G_{c,ss}$ 为表面活性剂组分间的作用力强度；c 为格子速度，m/s；"："为双点积运算符，双点积的出现是由于相邻格点表面活性剂均为具有矢量方向的偶极子，需要利用张量运算将力场分配到不同方向。

表面活性剂在壁面的吸附浓度与壁面吸附强度有关，为此引入一个表面活性剂与壁面的微观作用力 $F_{ads,s}$ 以控制表面活性剂在壁面的吸附[16]：

$$F_{ads,s}=-G_{ads,s}\rho_s(x,t)\sum w_i s(x+\mathrm{e}_i\Delta t,t)\mathrm{e}_i.\tag{12}$$

式中,$G_{ads,s}$ 为表面活性剂与固壁的作用强度;s 为表示是否为固壁的指示函数。

表面活性剂会在岩石表面上吸附并影响岩石的润湿性,由于表面活性剂的吸附浓度与表面活性剂-固壁作用强度 $G_{ads,s}$ 成正比,且水-固的界面张力与水-固壁作用强度 $G_{ads,w}$ 成正比,可通过建立 $G_{ads,w}$ 与 $G_{ads,s}$ 的关系等价改变流固作用力实现润湿性改变[17],

$$G_{ads,w}=G_{ads,w0}+k_1 C_{swall}\ln(1+k_2 G_{ads,s}). \tag{13}$$

式中,$G_{ads,w0}$ 为初始的水与壁面的作用强度系数;C_{swall} 为壁面表活剂质量浓度,g/L;k_1 和 k_2 为系数,可用来调节润湿性改变幅度。

2.3 聚合物特性模拟

LBM 中的流体黏度主要与松弛时间有关,考虑幂律流体特性时需要令松弛时间随局部剪切速率变化[18],进行局部动态参数更新,松弛时间子可表示为

$$\tau=3\frac{K}{\rho}\gamma^{n-1}+\frac{1}{2}. \tag{14}$$

式中,K 为稠度系数;n 为流态指数;γ 为剪切速率,m/s。

考虑黏弹性流体的弹性时可先根据流体本构方程及速度场计算出弹力,然后耦合到 LBM 模型中,通过改变速度场反映弹力的影响。速度场改变后又可重新计算弹力,如此反复迭代求解。以线性 Max-well 黏弹流体为例,弹力 F_{el} 由应力张量的散度计算[19],表达式为

$$F_{el}=\nabla\cdot T=\frac{\eta}{\tau_{el}}\int_{-\infty}^{t}\mathrm{e}^{\left(-\frac{t-t^1}{\tau_{el}}\right)}\nabla^2 u(x,t')\mathrm{d}t'. \tag{15}$$

其中,

$$\tau_{el}=\eta/E.$$

式中,T 为应力张量,n;浊为聚合物表观黏度,mPa·s;μ 为流体速度,m/s;τ_{el} 为记忆时间,s;E 为弹性模量,Pa;t'为时间积分变量,s。

3 提高驱油效率机制

从扩大微观波及系数和提高微观洗油效率两个方面,基于微观渗流模拟方法揭示非均相复合驱油体系提高驱油效率机制。一方面,非均相复合驱油体系中可变形运移的“软固体冶凝胶颗粒可引起流动增阻,迫使后续驱替液体转变流动方向,同时聚合物使驱替液黏度增加有效抑制黏性指进,从而达到扩大微观波及系数的作用;另一方面,表面活性剂的加入降低了油水界面张力,增加了分散油滴形变能力,同时改变了岩石壁面润湿性,导致油滴在壁面上的黏附功大大降低,从而提高了微观洗油效率。

3.1 颗粒暂堵升压后变形运移

为研究凝胶颗粒在孔喉中的变形运移过程,建立多喉道模拟模型,如图 7 所示。模拟区域长为 1 300 μm,最宽处为 200 μm,最窄处为 40 μm。模拟过程中实时统计模型两

端的驱替压差变化。

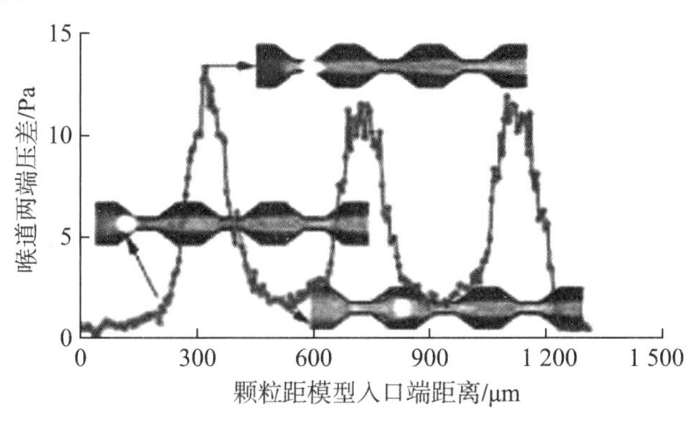

图 7　凝胶颗粒通过孔喉时的压差变化

Fig. 7　Pressure difference variation when gel particles pass through throats

可以看出，当凝胶颗粒在水流携带作用下运移至喉道入口处时，由于压差较小，凝胶颗粒将暂时滞留堵塞，从而导致模型两端驱替压差升高；随着驱替压差的不断升高并达到凝胶颗粒变形通过所需的临界值后，凝胶颗粒开始变形并逐步向喉道内部运移；在壁面摩擦和颗粒变形作用的影响下，颗粒在喉道运移过程中，模型两端的驱替压差维持在较高水平；而当凝胶颗粒进一步运移通过喉道出口端时，模型两端驱替压差开始下降，凝胶颗粒也将逐渐恢复至其原始形状，并在流体携带作用下继续向前运移，直至到达下一喉道时再次重复这一过程[20]。因此，凝胶颗粒在多喉道模型中具有颗粒暂堵、升压后变形运移的微观渗流机制。

3.2 固液增阻导致液流转向

为研究凝胶颗粒变形运移对流体流动的影响，建立多孔介质模型，如图 8 所示。模拟区域长度为 340 μm，宽度为 100 μm。模拟并统计凝胶颗粒溶液流动过程中多孔介质两端的压差变化，并将其与纯流体流动时的压差比值定义为无因次流动增阻倍数[21]。

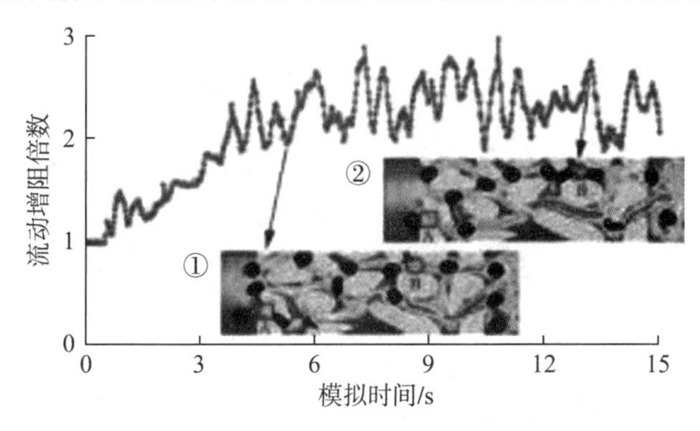

图 8　多孔介质中凝胶颗粒溶液运移模拟结果

Fig. 8　Simulation results of gel particle dispersion system migrating in porous media

可以看出，随着凝胶颗粒溶液的注入，流动增阻倍数逐渐增加，直至达到一相对较高值后呈现持续波动状态。图 8 中①表明，由于凝胶颗粒滞留堵塞及变形运移引起的阻碍作用,A 通道流体流量较小,而 B 通道由于没有凝胶颗粒影响,成为主要的液流方向之一。图 8 中②表明,由于 A 通道中凝胶颗粒已变形通过,颗粒阻碍作用消失,因此 A 通道成为主要液流方向,而此时 B 通道由于入口处颗粒堆积堵塞,流体难以流动通过,不再是主要液流方向。

由于凝胶颗粒在多孔介质中不断滞留堵塞、变形运移,导致各喉道流动阻交替变化,进而引起不同通道内流体流速交替变化,即固液增阻导致液流转向的微观渗流机制。

3.3 驱替液增黏抑制黏性指进

在毛细管数一定的条件下,一维均质多孔介质模型驱替过程黏性指进模拟结果如图 9 所示,其中无量纲界面长度为驱替前缘界面长度与多孔介质宽度之比。随着油水黏度比的增加,分形维数及无量纲界面长度增加,即黏性指进的自相似性提高,指状形状更细长,指进更明显,驱替突破时间更早。

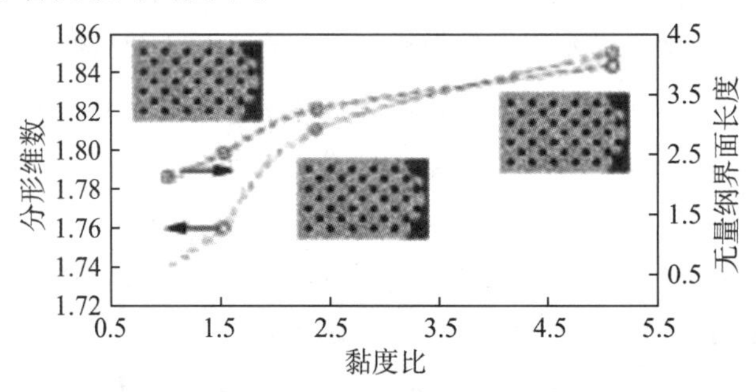

图 9 油水黏度比对黏性指进的影响

Fig. 9 Effect of viscosity ratio on viscous fingering in porous media

在驱替压差一定的条件下,二维随机多孔介质模型(模型尺寸为 2 mm×2 mm)中水驱、聚合物驱及非均相复合驱黏性指进现象对比如图 10 所示。

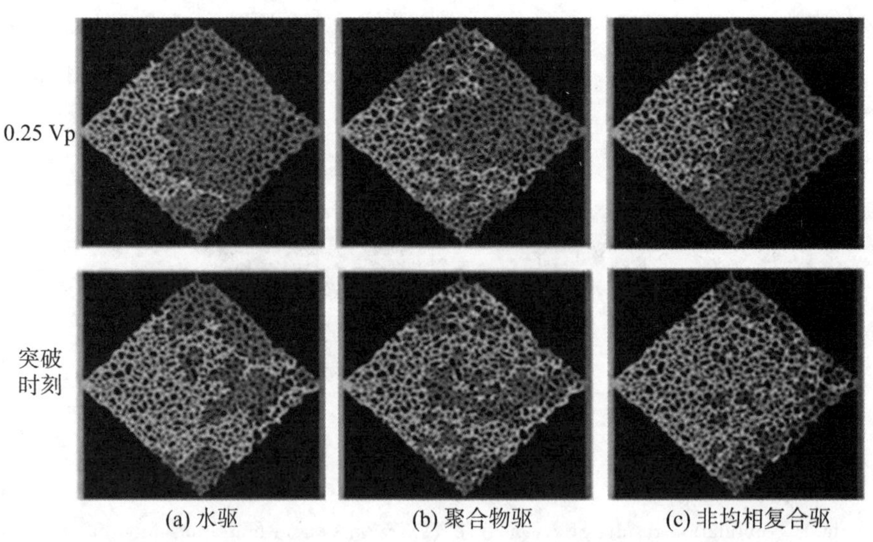

图 10 水驱、聚合物驱及非均相复合驱二维驱替前缘对比

Fig. 10 Comparison of oil-water interface of two-dimensional displacement

在水驱过程中驱替前缘不均匀前进，指进现象明显，突破时刻模型角隅处及主流线中心残存大量剩余油；聚合物驱时驱替前缘推进更均匀一些，突破时驱替液波及范围更大；非均相复合驱时主流线中心的绝大部分油都能被驱替出来，角隅处的油得到有效动用，驱替前缘较其他方案最均匀，波及面积最大，驱替效果最佳。

非均相复合体系中由于聚合物加入增加了水相黏度，降低了油水黏度比，同时黏弹性颗粒起到了液流转向作用，因此可有效抑制复合驱黏性指进并扩大微观波及系数。

3.4 强化启动微观剩余油

水驱或聚合物驱后微观剩余油主要包括孔内分散型、壁面吸附型和角隅型剩余油等类型[22]。主要以赋存量较大的孔内分散剩余油及壁面吸附型剩余油为研究对象，讨论了非均相复合驱强化启动微观剩余油的作用。

图 11 为相同剪切作用强度下孔内分散油滴在不同表面活性剂质量分数下的最终形态。初始油滴形状为圆形，剪切流开始后，液滴在表面活性剂的影响下开始旋转、变形，逐渐从圆形变为椭圆或长条状，如图 11(a)、(b)所示。如果最大形变状态时剪切作用不足以卡断冶液滴，则液滴在马兰戈尼力及毛管力下开始回缩，直到剪切力与马兰戈尼力、毛管力达到平衡。如果最大形变达到破裂界限，则油滴会进一步分散为子液滴，如图 11(c)所示。总体上随着表面活性剂质量分数的增加，油滴更易变形甚至破裂，这主要是由于油水界面张力降低导致了毛细管数增大，大变形及小液滴形态也使油滴能更容易通过孔喉。

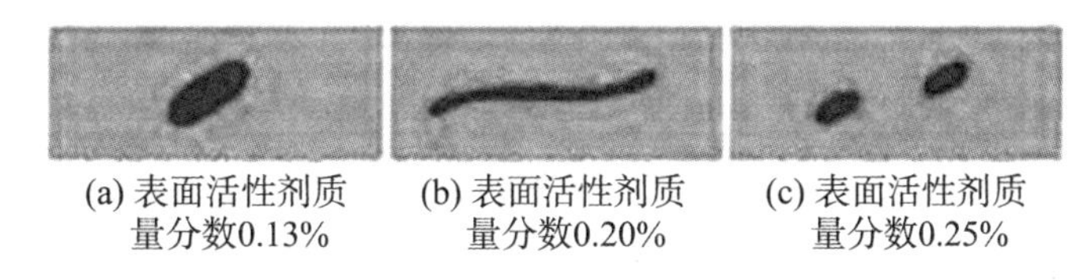

(a) 表面活性剂质量分数0.13%
(b) 表面活性剂质量分数0.20%
(c) 表面活性剂质量分数0.25%

图 11　剪切流中含不同表面活性剂质量分数的油滴形状

Fig. 11　Droplet shapes in shear flow with different surfactant mass fraction

设置岩石壁面原始状态为水湿，进行定压力梯度驱动模拟。模拟过程中微观数据模型中的表面活性剂质量分数与实际表面活性剂质量分数存在转换关系，可根据油水界面张力降低程度确定。不同表面活性剂质量分数下壁面附着剩余油的启动模拟结果如图 12 所示，其中无量纲启动压力梯度为剩余油启动压力梯度与界面张力之比[16]。随着表面活性剂质量分数的增加，界面张力降低，使原来不能从壁面剥离的油滴逐渐脱离下来，且表面活性剂质量分数越高，剥离程度越高；同时由于吸附导致了润湿性改变，吸附浓度越大，润湿性改变效果越好，油滴启动压力梯度降低，剥离程度大幅度增加。水湿情况下壁面附着剩余油所需要的启动压力梯度小，油滴可以完全剥离，而油湿情况下油滴的启动压力梯度大，且会有油膜吸附在壁面上而不能完全剥离。总体上非均相复合体系中表面活性剂降低了油水界面张力，改变了岩石壁面润湿性，使油滴在壁面上的黏附功大大降低，在低界面张力及润湿反转的双重作用下可使壁面剩余油启动压力降低 90%以上，从而有效促进壁面吸附型剩余油的动用。

非均相复合驱过程中表面活性剂可增大孔内分散油滴的形变,促进油滴破裂,使油滴更容易通过孔喉,从而有效减小贾敏效应;同时可大幅度降低油滴在壁面的黏附功,降低壁面吸附型剩余油启动压力,最终有效提高微观洗油效率。

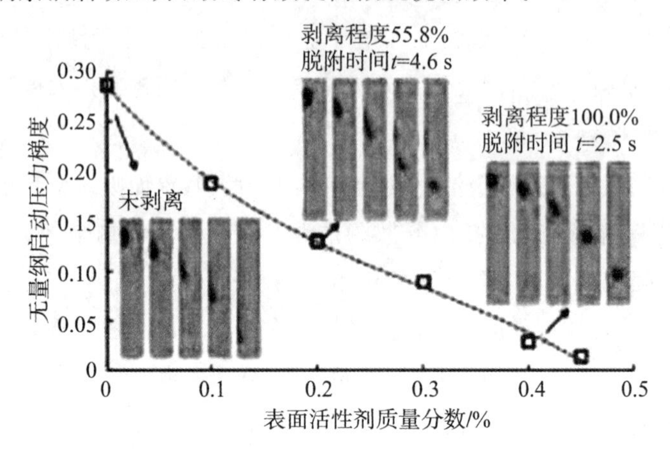

图12 壁面剩余油启动压力梯度随表面活性剂质量分数变化关系

Fig. 12 Starting pressure gradient of oil droplet on wall changing with surfactant mass fraction

4 结论

(1) 微通道两端压差一定条件下凝胶溶液流速仍大幅度波动,非均质双管岩心模型中分流量持续交替变化,均表明非均相复合驱体系流动具有非连续相渗流特征。

(2) 非均相复合驱通过扩大微观波及系数及提高微观洗油效率两方面提高了驱油效率,基于微观数值模拟揭示了非均相复合驱颗粒暂堵升压后变形运移、固液增阻导致液流转向、抑制黏性指进并扩大微观波及系数、强化启动微观剩余油等作用机制。

参考文献

[1] 赵福麟，王业飞，戴彩丽，等. 聚合物驱后提高采收率技术研究[J]. 中国石油大学学报(自然科学版)，2006，30(1):86-89.

ZHAO Fulin, WANG Yefei, DAI Caili, et al. Techniques of enhanced oil recovery after polymer flooding[J]. Jour-nal of China University of Petroleum(Edition of Natural Science), 2006, 30(1):86-89.

[2] 曹绪龙. 非均相复合驱油体系设计与性能评价[J]. 石油学报(石油加工)，2013，29(1):115-121.

CAO Xulong. Design and performance valuation on the heterogeneous combination flooding systems [J]. Acta Pe-trolei Sinica (Petroleum Processing Section), 2013, 29(1):115-121.

[3] 张莉，刘慧卿，陈晓彦. 非均相复合驱封堵调剖性能及矿场试验[J]. 东北石油大学学报，2014，38(1):63-68.

ZHANG Li，LIU Huiqing，CHEN Xiaoyan. Pilot test of PPG/polymer/surfactant flooding after polymer flooding[J]. Journal of Northeast Petroleum University，2014，38(1):63-68.

[4] 孙焕泉. 聚合物驱后井网调整与非均相复合驱先导试验方案及矿场应用:以孤岛油田中一区 Ng3 单元为例[J]. 油气地质与采收率，2014，21(2):1-4.

SUN Huanquan. Application of pilot test for well pattern adjusting heterogeneous combination flooding after polymer flooding: case of Zhongyiqu Ng3 block，Gudao Oilfield[J]. Oil & Gas Recovery Technology，2014，21(2):1-4.

[5] WUD，ZHOUK，ANZ，et al. Experimental study on the matching relationship between PPG size and reservoir heterogeneity[R]. SPE 193709，2018.

[6] ZHOUK，HOUJ，SUNQ，et al. A study on particle suspension flow and permeability impairment in porous media using LBM—DEM—IMB simulation method [J]. Transport in Porous Media，2018，124:681-698.

[7] ZHOUK，HOUJ，WUD，et al. Study on flow diversion caused by deformable preformed particle gel[R]. SPE 192965 2018.

[8] LIUH，KANGQ，LEONARDI C R，et al. Multiphase lattice Boltzmann simulations for porous media applica-tions[J]. Computational Geosciences，2016，20(4):777-805.

[9] SHAN X，CHENH. Lattice Boltzmann model for simulating flows with multiple phases and components[J]. Physical Review E，1993，47(3):1815.

[10] SUKOP M，Jr THORNE D T. Lattice Boltzmann modeling[M]. Berlin: Springer，2006.

[11] PESKIN C S. Flow patterns around heart valves: a numerical method[J]. Journal of Computational Physics，1972，10(2):252-271.

[12] NOBLE D R，TORCZYNSKI J R. A lattice-Boltzmann method for partially saturated computational cells[J]. International Journal of Modern Physics C，1998，9(8):1189-1201.

[13] CUNDALL P A，STRACK O D L. A discrete numerical model for granular assemblies[J]. Geotechnique，1979，29(1):47-65.

[14] GROSS M，KRüGER T，ARNIK F. Fluctuations and diffusion in sheared athermal suspensions of deformable particles[J]. Europhysics Letters，2015，108(6):68006.

[15] CHEN H，BOGHOSIAN B M，COVENEY P V，et al. A ternary lattice

Boltzmann model for amphiphilic fluids[J]. Proceedings of the Royal Society of London (Series A: Mathematical, Physical and Engineering Sciences), 2000, 456 (2000):2043-2057.

[16] WEI B, HOU J, SUKOP M C, et al. Pore scale study of amphiphilic fluids flow using the Lattice Boltzmann model[J]. International Journal of Heat and Mass Transfer, 2019, 139:725-735.

[17] WEI B, HOU J, WU D, et al. Pore scale simulation of surfactant flooding by lattice Boltzmann method[R]. SPE 193660 2018.

[18] SULLIVAN S P, GLADDEN L F, JOHNS M L. Simulation of power-law fluid flow through porous media using lattice Boltzmann techniques[J]. Journal of Non-Newto-nian Fluid Mechanics, 2006, 133(2/3):91-98.

[19] ISPOLATOV I, GRANT M. Lattice Boltzmann method for viscoelastic fluids[J]. Physical Review E, 2002, 65(5):056704.

[20] ZHOU K, HOU J, SUNQ, et al. An efficient LBM-DEM simulation method for suspensions of deformable preformed particle gels [J]. Chemical Engineering Science, 2017, 167:288-296.

[21] ZHOU K, HOU J, SUN Q, et al. Study on the flow resistance of the dispersion system of deformable preformed particle gel in porous media using LBM-DEM-IMB method[J]. Journal of Dispersion Science and Technology, 2019, 40(10): 1523-1530.

[22] 余义常，徐怀民，高兴军，等. 海相碎屑岩储层不同尺度微观剩余油分布及赋存状态:以哈得逊油田东河砂岩为例[J]. 石油学报，2018，39(12):1397-1409.

YU Yichang, XU Huaimin, GAO Xingjun, et al. Distribution and occurrence status of microscopic remaining oil at different scales in marine clastic reservoirs:a case study of Donghe sandstone in Hadeson Oilfield[J]. Acta Petrolei Sinica, 2018, 39 (12):1397-1409.

(编辑 李志芬)

化学驱粘性指进微观渗流模拟研究

于　群[1]，王惠宇[2,3,*]，曹绪龙[1]，郭兰磊[1]，韦　贝[2,3]，石　静[1]
（1. 中国石化胜利油田分公司勘探开发研究院，山东东营 257015）
（2. 非常规油气开发教育部重点实验室（中国石油大学（华东）），山东青岛 266580）
（3. 中国石油大学（华东）石油工程学院，山东青岛 266580）

摘要：化学驱是提高原油采收率的重要方法，化学剂的加入使得地下流体粘度界面张力发生变化，粘性指进演变特征更为复杂。基于格子玻尔兹曼伪势多相流模型，综合考虑油水两相间相互作用、聚合物增粘与非牛顿特性、表面活性剂传质扩散作用及降低界面张力机理，建立了化学驱微观渗流格子玻尔兹曼模拟方法。基于数值模拟方法了研究二维单通道内粘性指进发展及演变特征，结果表明：粘性指进主要受毛细管数及粘度比影响，毛细管数越大、粘度比越高指进现象越明；多孔介质亲油性越强，粘性指进越明显；非牛顿流体在通道内粘度分布不均，剪切变稀聚合物可促进粘性指进发展；同时重力作用可导致粘性指进前缘纵向的分布不均。

关键词：粘性指进；化学驱；格子玻尔兹曼方法

粘性指进是由粘度较小的流体驱替粘度较大的流体时产生的一种界面不稳定现象，它广泛存在于自然界、工业生产和生物医疗等领域的多相渗流之中，由于两相界面的非均匀推进像"手指"一样，故称为指进[1]。在油田开发过程中，需要注入其他流体将原油驱替出来，粘性指进现象使得大量原油由于未被波及而残存在地下。化学驱实施过程中矿场通过在驱替液中加入聚合物、表面活性剂等化学剂提高采收率，聚合物可以增大驱替相粘度起到了抑制粘性指进作用；但表面活性剂降低了油水界面张力又促进了指进的发展[2]。化学剂的加入使得粘度及界面张力等性质发生改变，因此基于微观模拟方法研究粘度界面张力变化作用机制对于深化化学驱微观渗流机理认识、改善油气田开发效果具有重要意义。

目前常用的孔隙尺度微观渗流模拟方法主要包括孔隙网络模型、格子玻尔兹曼方法（LBM）及传统的计算流体动力学（CFD）方法。其中 LBM 作为一种粒子类方法在模拟多相流中无需追踪界面且能高效处理复杂多孔介质边界，在渗流模拟中具有极大优势。Chin 等利用 LBM 伪势多相流模型研究了不同粘度比下二维通道内的粘性指进现象[3]，Kang 等随后进一步研究了二维通道内毛细管数及润湿性对粘性指进的影响[4],Dong 等则考虑了重力并且研究了二维简化多孔介质内的粘性指进现象[5]。Huang 等利用 LBM 颜色多相流模型研究了非混相两相流在二维随机多孔介质内的流动,并基于模拟结果绘制了三种驱替状态的相图[6]。雷体蔓等模拟了化学反应中混溶流体在微通道中的粘性指进现象,重点研究指进的形态位置随化学反应速率和稳态浓度参数的变化[7]。笔者选用 LBM 伪势多相流模型,研究二维单通道内粘性指进产生、发展及演变特征,讨论粘度比、毛细管数、润湿性、流体非牛顿性等特征等对粘性指进的影响及其作用机制。

1 化学驱格子玻尔兹曼模拟方法

1.1 LBM 单相流模型

LBM 是离散 Boltzmann 方程而来的数值模拟方法,离散过程时通过 BGK 近似可得到经典的 LBGK 模型[8]。模型主要包括离散速度模型、平衡分布函数及演化方程三部分。在 LBGK 模型中,通过引入一个离散的速度分布函数 $f_i(x,t)$来演化流体运动,其满足以下演化方程:

$$f_i(x+\mathrm{e}_i\Delta t,t+\Delta t)=f_i(x,t)-\frac{1}{\tau}(f_i(x,t)-f_i^{eq}(x,t)), \tag{1}$$

式中,i 表示离散速度方向,e_i 为离散速度,$f_i^{eq}(x,t)$为平衡分布函数,X 为空间位置,T 为松弛时间,为满足低马赫数的要求,松弛时间一般要大于 0.5,否则会出现数值不稳定,离散速度 e_i 依赖离散速度模型,在此选用 $D2Q9$ 模型,则 e_i 取值为[9]:

$$[\mathrm{e}_0,\mathrm{e}_1,\mathrm{e}_2,\mathrm{e}_3,\mathrm{e}_4,\mathrm{e}_5,\mathrm{e}_6,\mathrm{e}_7,\mathrm{e}_8]=c\begin{bmatrix}0 & 1 & 0 & -1 & 0 & 1 & -1 & -1 & 1\\ 0 & 0 & 1 & 0 & -1 & 1 & 1 & -1 & -1\end{bmatrix}, \tag{2}$$

平衡分布函数 $f_i^{eq}(x,t)$计算公式为:

$$f_i^{eq}(x,t)=w_i\rho\left[1+\frac{e_i\cdot u}{c_8^2}+\frac{(e_i\cdot u)^2}{2c_8^4}-\frac{u^2}{2c_8^2}\right]. \tag{3}$$

式中,ω_i 代表每个方向的权重系数,c_s 为格子声速,ρ、u 为宏观密度和宏观速度。对于 D2Q9 模型 $w_0=\frac{4}{9},w_{1-4}=\frac{1}{9},w_{5-8}=\frac{1}{36}$.

密度及速度可以通过下式计算:

$$\rho=\sum_i f_i \tag{4}$$

$$u=\frac{1}{\rho}\sum_i f_i\mathrm{e}_i \tag{5}$$

上述 LBM 方案可以通过多尺度展开还原为 N-S 方程,揭示了微观 LBM 和宏观方程间的联系,此时流体运动粘度为 $v=c_\varepsilon^2\left(\tau-\frac{\Delta t}{2}\right)$。

1.2 油水两相流伪势模型

选用 LBM 伪势模型中的多组分多相流模拟油水流动[10-11]，以模型两相流中，系统存在油水 2 个不同的组分，其分布函数表示为 $f_i^\sigma(x,t)$或 $f_i^\sigma(x,t)$，或每个组分都遵循单相流的 LBGK 模型。组分间受力由两部分组成，一部分来自与不同组分间粒子相互排斥作用，另一部分则为与壁面的相互作用。

不同组分间的排斥力计算公式为[12]：

$$F_c(x,t)=-\rho_\sigma(x,t)G_c\sum_i w_i\rho_\sigma(x+\mathrm{e}_i\Delta t,t)\mathrm{e}_i, \tag{6}$$

式中，Gc 为控制不同组分间作用强度的系数，可以调节油水界面张力。

组分受到壁面作用力为：

$$F_{ads,\sigma}(x,t)=-G_{ads,\sigma\rho\sigma}(x,t)\sum_i w_i s(x+\mathrm{e}_i\Delta t,t)\mathrm{e}_i, \tag{7}$$

式中，$G_{ads,\sigma}$ 为控制壁面与组分间作用强度的系数，可以调整岩石润湿性。

多相流中计算平衡分布函数时需要每个组分的平衡速度计算，受力后的平衡速度 u^{eq} 用牛顿第二定律

更新：

$$u^{ep}=u'+\frac{\tau_\sigma(F_{e,\sigma}+F_{ads,\sigma})}{\rho_\sigma}, \tag{8}$$

式中，u'为复合速度，由每个组分的微观及宏观参数加权而来，定义为，

$$u'=\sum_\sigma\left(\sum_i\frac{f_i^\sigma\mathrm{e}_i}{\tau_\sigma}\right)/\left(\sum_i\frac{\rho_i^\sigma}{\tau_\sigma}\right) \tag{9}$$

1.3 LBM 非牛顿流体模型

在幂律流体的 LBM 模型中，粘度应随局部剪切速率变化[13]：

$$v=\frac{K}{\rho}\gamma^{n-1}=\frac{1}{3}\left(\tau-\frac{1}{2}\right)\Rightarrow\tau=3\frac{K}{\rho}\gamma^{n-1}+\frac{1}{2}。 \tag{10}$$

式中，K 为稠度系数，Y 为剪切速率，n 为流态指数，$0<n<1$ 为假塑性流体，$n>1$ 为膨胀性流体。

在伪势多相流模型中，溶解密度可能较小甚至接近 0，导致此时式(9)中分子趋近于，模型稳定性可能受到考验。可以牛顿流体粘度为基准消去密度项：

$$\tau=\gamma^{n-1}\left(\tau_0-\frac{1}{2}\right)+\frac{1}{2}。 \tag{11}$$

式中，T_0 为牛顿流体的基准松弛时间。

1.4 含表面活性剂的 LBM 模型

在含表面活性剂 LBM 模型中，引入一个表面活性剂分布函数 g，其与溶剂共用流场速度，单相流的模型演化方程、离散速度模型及平衡分布函数式仍然适用，但是一般可以选择更少的离散速度方向(如 D2Q5)和更简单的平衡分布函数：

$$g_i^{eq}(x,t)=w_iC\left(1+\frac{e_i\cdot u}{c_8^2}\right), \tag{12}$$

式中,g_i 为溶质的速度分布函数,$C=\sum_i g_i$ 为溶质浓度.其还原的宏观方程为对流扩散方程,$D_\delta=c_\delta^2\left(\tau_\delta-\frac{\Delta t}{2}\right)$ 为其扩散系数。同时如若要体现表面活性剂浓度对油水界面张力的影响,可以引入偶极子模型表征表面活性剂分子两亲性特征,从而令油水界面张力随表面活性剂浓度升高而降低[14]。

2 计算模型

计算模型采用格子单位,选用大小 400×66 的二维通道为模拟区域,模拟开始前通道内充满油相,从入口处注入驱替相且采用速度边界条件,出口采用充分发展边界,上下边界采用固壁反弹格式[15]。基础参数方案选取各组分松弛时间为 1.0,各相的主相密度为 1.0,溶解相密度为 0.01,$G_c=2.0$,$G_{ads,1}=-G_{ads,2}$。除非特别说明,在进行其他参数的影响因素分析时润湿性取中性润湿 90°,非牛顿特性不予考虑。定义油水粘度比为 $M=\frac{\mu_o}{\mu_w}=\frac{\rho_o v_o}{\rho_w v_w}=\frac{\rho_0(\tau_0-0.5)}{\rho_w(\tau_w-0.5)}$,其中 O、W 代表油、水组分,可通过改变油水的松弛时间调节粘度比;定义毛细管数为 $Ca=\frac{u\mu_w}{\sigma_{ow}}=\frac{u\rho_w(\tau_w-0.5)}{3\sigma_{ow}}$,式中 σ_{ow} 油水界面张力,可通过调节流体速度及表面活性剂浓度调节毛细管数。单通道内的粘性指进通过指状长度 T 及宽度 W 描述,定义如图 1 所示。

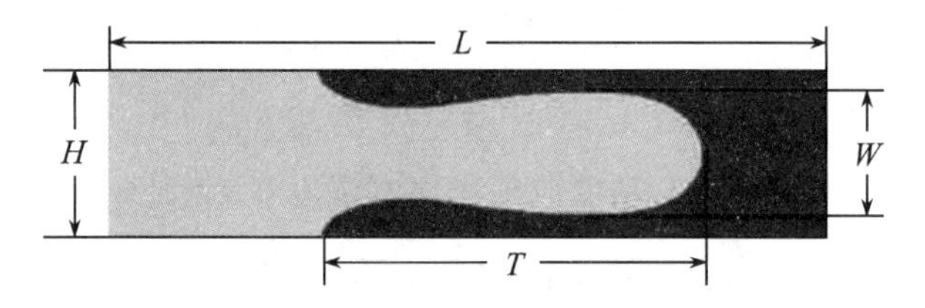

图 1 二维单通道内的粘性指进示意图

3 影响因素分析

3.1 毛细管数的影响

单通道内毛细管数对粘性指进影响如图 2 所示,图中给出了粘度比为 1.0 及 5.12 时不同毛细管数下的油水界面演化过程。模拟结果表明,在其他参数一定时毛细管数越大指进越明显,两相界面在毛细管数较小时为稳定的抛物线状,随着毛细管数的增加,尖端逐渐顶起,当毛细管数达到一定界限,形成了明显的指状。同时在大粘度比下指进现象更容易发生,如图 2(b)中毛细管数为 0.018 6 时已可明显观察到粘性指进现象,但图 2(a)中毛细管数为 0.028 时仍未发生指进。

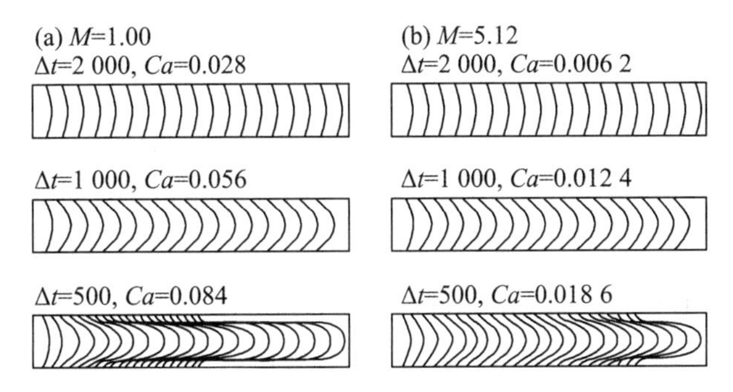

图 2　单通道内毛细管数对粘性指进的影响

利用无量纲数 W/H 及 T/L 分别表征单通道内粘性指进的指状宽度及长度，粘度比为 1 时的不同毛细管数下的指状形态演化统计如图 3 所示。统计结果表明，随着毛细管数的升高指状宽度逐渐减小，指状长度逐渐升高，即指状会越来越细长，同时指进现象发展初期和后期形状变化较为缓慢，而在毛细管数 0.05～0.1 范围内变化十分明显。

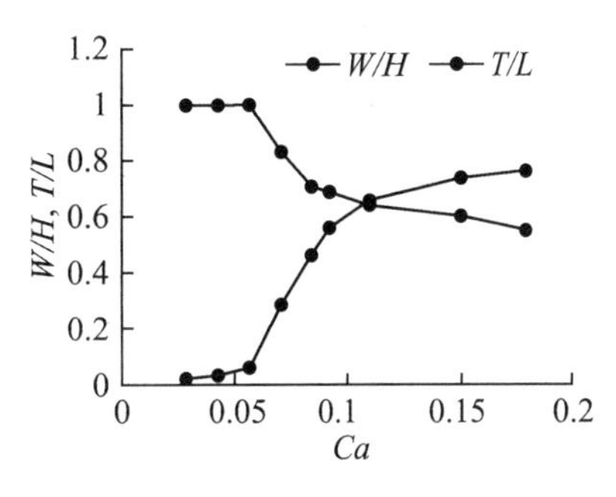

图 3　单通道内毛细管数对指状形状的影响

3.2 粘度比的影响

单通道内粘度比对粘性指进影响如图 4 所示，模拟结果表明，在其他参数一定的情况下，粘度比越大指进越明显，两相界面随粘度比的变化与随毛管数变化一致即随着粘度比的增加，界面尖端逐渐顶起，当粘度比到达一定界限，形成明显的指状。

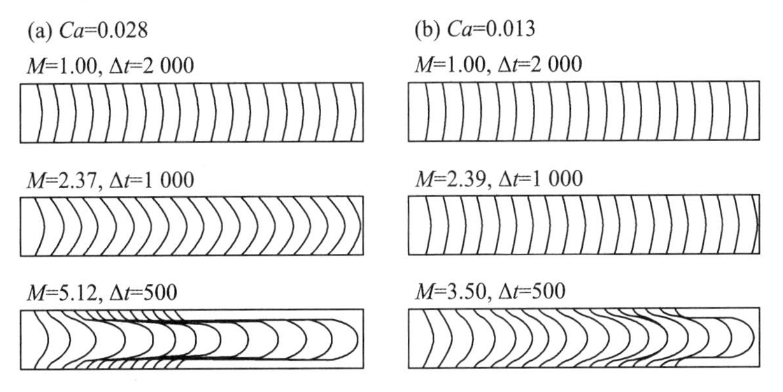

图 4　单通道内粘度比对粘性指进的影响

毛细管数为 0.028 时的不同粘度比下的指状形态演化统计如图 5 所示。统计结果表明，随着粘度比的增加指状宽度逐渐减小，指状长度逐渐升高，且升高速度越来越快，使得指状越来越细长。

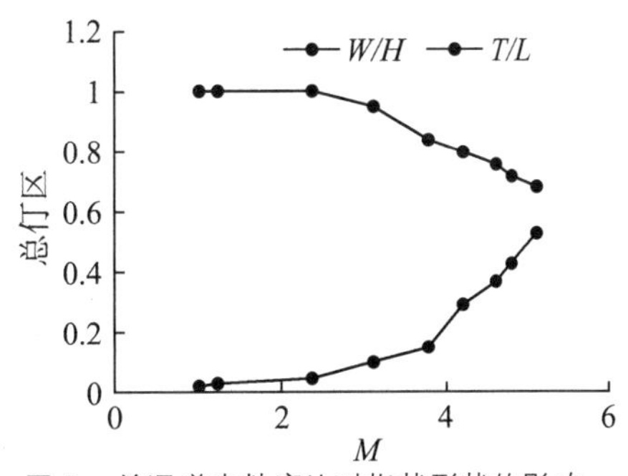

图 5　单通道内粘度比对指状形状的影响

3.3 润湿性的影响

通过改变油相接触角，得到了不同润湿性对粘性指进的影响规律，如图 6 所示，同时给出了 T

=6 000 时不同润湿性下的前缘对比。随着油相润湿角增大即管壁越来越亲水,指进现象得到了抑制,这是由于在水湿情况下驱替液更容易沿管壁流动从而使得通道中心的指进现象得到缓解;而油湿时水相显然会更倾向于沿管壁中心流动,从而进一步加剧了指进现象。同时对比图 6(a)及图 6(b)小粘度比下润湿性对指进现象影响更为明显,这是由于粘性指进的主控因素是粘度比及毛细管数,润湿性影响占次要因素。

粘度比为 1、毛细管数为 0.1 时不同润湿性情况下的指状状态发展如图 7 所示,统计表明指状宽度在不同润湿性下基本一致,随着润湿性的改变指进长度逐渐下降,即在油相润湿角增加的过程中指状变的越来越短。

3.4 非牛顿特性影响

幂律指数对粘性指进影响如图 8 所示,(a)、(b)、(c)分别为相界面演化过程、$T=10\ 000$ 时的流体粘度分布、相同时间步下的指进前缘相界面形态对比。假塑性流体粘度表现出孔隙中间高壁面附近低的分布,膨胀性流体粘度分布则相反,这是由于孔隙中心虽然速度大但是剪切速率小,而边界处虽然速度小但是剪切速率大,导致流体在不同幂律指数下呈现不同的粘度分布。由于幂律流体在流动过程根据幂律指数不同而粘度发生了变化,呈现出幂律指数越小,粘性指进越明显的趋势,且膨胀性流体表现出中间细两边粗的指状,这可能与粘度部分不均有关。

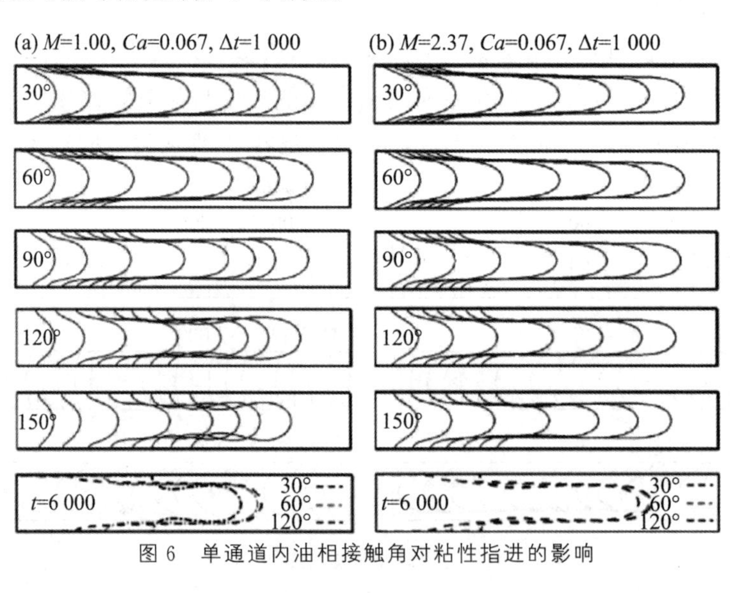

图 6　单通道内油相接触角对粘性指进的影响

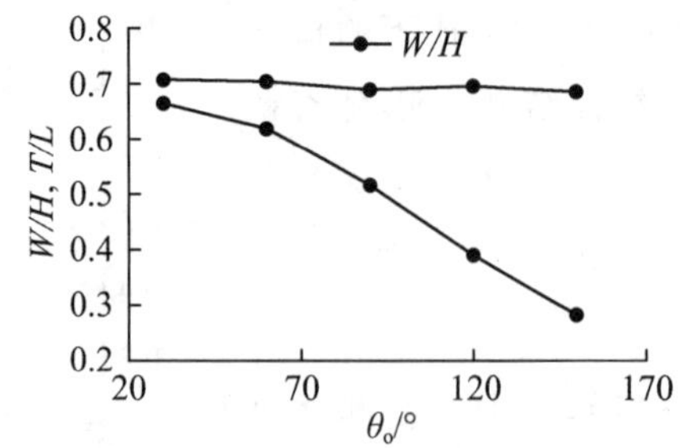

图 7　单通道内油相接触角对指状形状的影响

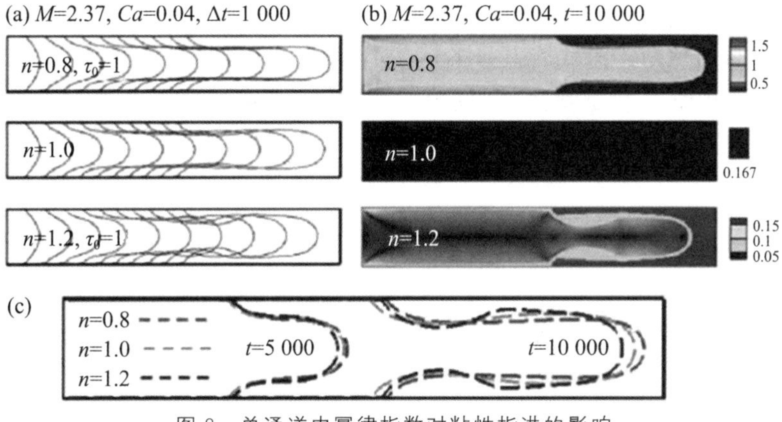

图 8　单通道内幂律指数对粘性指进的影响

3.5 重力的影响

重力作用通过在模型中添加一个体力实现，可通过调节重力加速度模拟不同密度比及邦德数下的粘性指进现象，模拟结果如图 9 所示。结果表明随着重力加速度增大即邦德数的增大，油水界面向重力方向移动，使得油水分布在纵向上也产生了不均匀现象，这在多孔介质流动模拟中会造成驱替液在纵向上的突破时间不一。同样，重力差异在粘度比越小的情况下越明显，这是由于重力差异的呈现需要一定时间，而粘度比过大时指进发展较为迅速造成的。

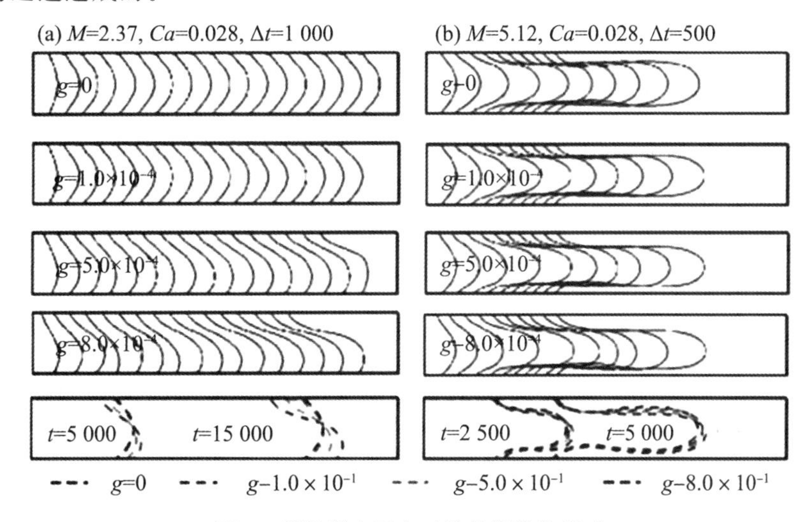

图 9　单通道内重力对粘性指进的影响

4　结论

（1）粘性指进调控机制研究表明，毛细管数越大、粘度比越高、多孔介质越油湿则指进现象越明显。粘性指进主控因素主要为粘度比和毛细管数，润湿性为次要影响因素。

（2）假塑性流体在多孔介质中的粘度呈现孔隙中心高而固壁边界附近低的分布，膨胀性流体粘度呈现相反分布，且聚合物的剪切变稀特性可促进粘性指进，而重力主要影响纵向驱替界面的分布。

(3) 聚合物的增粘效果降低了油水粘度比,可以抑制粘性指进;表面活性剂降低了油水界面张力,增加了毛细管数,促进了指进发展,可通过降低流动速度来抑制粘性指进。

参考文献

[1] 苏玉亮, 吴春新张祥吉. 非活塞驱替条件下粘性指进的表征及影响因素[J]. 特种油气藏,2009, 16(5):51-53.

[2] 刘哲宇, 李宜强, 冷润熙, 刘振平, 陈鑫. HEJAZI Hossein. 孔隙结构对砾岩油藏聚表二元复合驱提高采收率的影响[J]. 石油勘探与开发,2020, 47(1):129-139.

[3] Chin J. Lattice Boltzmann simulation, of the flow of binary immiscible fluids with different viscosities using the Shan-Chen microscopic interaction model [J]. Philosophical Transactions of the Royal Society of London. Series A: Mathematical, Physical and Engineering Sciences, 2002, 3.60(17.92):547-558.

[4] Kang Q, Zhang D, Chen S. Immiscible displacement in a channel: simulations of fingering in two dimensions[J]. Advances in Water Resources, 2004, 27(1):13-22.

[5] Dong B, Yan Y Y, Li W Z. LBM simulation of viscous fingering phenomenon in immiscible displacement of two fluids in. porous media[J] Transport in Porous Media, 2011, 88(2):293-314.

[6] Huang H, Huang J, Lu X. Study of immiscible displacements in porous media using a color-gradientbased multiphase lattice Boltzmann method[J]. Computers & Fluids, 2014, 93:164-172.

[7] 雷体蔓, 孟旭辉, 郭照立. 微通道内反应流体粘性指进的数值研究[J]. 计算物理 2016, 33(01):30-38.

[8] Bhatnagar P L, Gross E P, Krook M. A model for collision processes in gases. I. Small amplitude processes in charged and neutral one-component systems[J]. Physical review, 1954, 94(3):511.

[9] Qian Y H, D'Humières D, Lallemand P. Lattice BGK models for Navier-Stokes equation[J]. EPL(Europhysics Letters), 1992, 17(6):479.

[10] Shan X, Chen H. Lattice Boltzmann model for simulating flows with multiple phases and components[J]. Physical Review E, 1993, 47(3):1815.

[11] Shan X, Chen H. Simulation of nonideal gases and liquid-gas phase transitions by the lattice Boltzmann equation [J]. Physical Review E, 1994, 49(4):2941.

[12] Chen S, Doolen G D. Lattice boltzmann method for fluid flows[J]. Annual Review of Fluid Mechanics, 1998, 30(1):329-364.

[13] Sullivan S P, Gladden L F, Johns M L. Simulation of power-law fluid flow through porous media using lattice Boltzmann techniques[J]. Journal of Non-newtonian Fluid Mechanics 2006, 133(2-3):91-98.

[14] 侯健，吴德君，韦贝，周康，巩亮，曹绪龙，郭兰磊. 非均相复合驱非连续相渗流特征及提高驱油效率机制[J]. 中国石油大学学报(自然科学版)，2019，43(5):128-135.

[15] 何雅玲，王勇，李庆. 格子 Boltzmann 方法的理论及应用[M]. 北京:科学出版社,2008.

Micro-simulation study of the viscous fingering in chemical flooding

Yu Qun[1], Wang Huiyu[2,3], Cao Xulong[1], Guo Lanlei[1], Wei Bei[2,3], Shi Jing[1]

(1. Shengli Oilfield Exploration and Development Research Institute, Dongying 257015, China)

(2. Key Laboratory of Unconventional Oil & Gas Development(China University of Petroleum(East China)), Ministry of Education, Qingdao 266580, China)

(3. School of Petroleum Engineering in China University of Petroleum(East China), Qingdao 266580, China)

Abstract: Chemical flooding is an important method to enhance oil recovery. The addition of chemical agents changes the viscosity and interfacial tension of injected fluids, and the evolution characteristics of viscous fingering are more complicated. Based on the Lattice Boltzmann pseudo-potential multiphase model, we established a numerical simulator at pore scale for chemical flooding, considering the oil-water interactions, polymer thickening and non-Newtonian characteristics, surfactant mass transfer and diffusion, and the mechanism of reducing interfacial tension. Based on the numerical simulation method, the characteristics of the generation, development and evolution of viscous fingering in a two-dimensional single channel are studied. The results show that the viscosity fingering is mainly affected by the capillary number and viscosity ratio. The stronger the lipophilicity of the porous medium, the more obvious the viscous fingering; the non-Newtonian fluid has an uneven viscosity distribution in the channel, and the shear thinning polymer can promote the development of viscous fingering; at the same time, the gravity contributes to uneven vertical distribution of viscous fingering.

Keywords: viscous fingering; chemical flooding; lattice Boltzmann method

第六章　矿场应用

孤东油田小井距注水示踪剂的选择及现场实施

王宝瑜　曹绪龙

（胜利石油管理局地质科学研究院）

提要　本文介绍了4种可同时使用的注水示踪剂的室内选择及其在孤东油田小井距试验区的应用情况。试验表明，硫氰酸氨、碘化钾、亚硝酸钠和钼酸铵是可同时使用的注水示踪剂。利用中心井的示踪剂采出曲线粗略地计算了示踪剂在地层中的损耗，并引用Maghsood等人的理论初步分析了中心井与4口注水井之间油层渗透率的纵向分布情况。

为了更好地开发孤东油田，研究孤东油田注水开发的全过程及用化学驱进行三次采油的效果，决定在孤东油田七区西北部开展小井距的注水开发试验。根据试验要求，决定在注入水中添加示踪剂，以便摸清注入水在地下的流动情况，准确地判断油井见水时的来水方向及水线推进速度。根据油井示踪剂的采出曲线，可以了解试验区的渗透率分布状况，这对于注水开发及化学驱提高采收率的实施都是很重要的。

大庆油田和胜利油田在六十年代曾进行过注水示踪剂的室内筛选和现场实施，先后对硫氰酸铵、硝酸铵、碘化钾、甲醛、萤光素和苯酚等十几种化学药剂进行过室内试验，胜利油田胜坨一区、堤东油田先后进行过单井一种注水示踪剂（硫氰酸铵）的现场试验。大庆油田曾在现场同时使用三种注水示踪剂：硫氰酸铵、硝酸铵、碘化钾。本次注水开发试验要求同时在四口注水并注入四种示踪剂，并在同一口生产井中检测。室内和现场试验证明，硫氰酸铵、亚硝酸钠，碘化钾和钼酸铵是能同时使用的四种注水示踪剂。这为油田注水开发中运用多种注水示踪剂提供了依据。

一、室内试验

注水示踪剂需要满足下列条件；① 地层及地层水中含量极少；② 溶于水而不溶于油；③ 不与油层、原油及地下水起化学反应；④ 注入弛层后能与水一起流动，被岩石的吸附量很小；⑤ 价格便宜，货源广；⑥ 分析方法简单可推，灵敏度高；⑦ 如果同时使用几种示踪剂，还要求示踪剂彼此间无干扰。

1. 示踪剂间的干扰试验

根据孤东小井距试验的具体情况，要求选用的四种示踪剂混合时不产生沉淀，分析时彼此间没有干扰。我们将硫氰酸铵、碘化钾、亚硝酸钠和钼酸铵四种示踪剂按一定的浓度比混合在一起，观察溶液的变化，然后用比色分析法分析其中一种示踪剂的浓度，并与原始浓度对比。试验结果见表1。这四种示踪剂按一定浓度比混合时，肉眼观察到的溶液透明，无沉淀产生。无论分析那一种示踪剂，分析误差除个别值外均小于规定范围(相对误差<10%) 所以这四种物质的分析彼此间没有干扰

表1 示踪剂的干扰试验数据表

示踪剂	干扰物	浓度比	示踪剂浓度(mg/L)	示踪剂浓度测定值(mg/L)	相对误差(%)
硫氰酸铵	碘化钾、亚硝酸钠、钼酸铵	1∶5∶1∶20	0.071	0.071	0
		1 ∶10∶2∶35	0.071	0.075	5.6
		1∶15∶3∶70	0.071	0.070	−1.4
碘化钾	硫氰酸铵、亚硝酸钠、钼酸铵	5∶1∶1∶40	0.185	0.163	−11.9
		10∶2∶2∶40	0.370	0.385	4.1
亚硝酸钠	硫氰酸铵、碘化钾、钼酸铵	3∶1∶5∶20	0.058	0.060	3.4
		3∶2∶10∶40	0.058	0.060	3.4
		3∶4∶15∶80	0.058	0.057	−1.7
钼酸铵	硫氰酸铵、碘化钾、亚硝酸钠	40∶1∶5∶1	0.040	0.038	−5.0
		40∶2∶10∶2	0.040	0.040	0
		40∶4∶15∶3	0.040	0.037	−7.5

① 本文中四种示踪迹的浓度分别以硫氰酸铵、I^-、NO^{2-}、和 Mo 的浓度(mg/L)表示。

② 浓度比是指示踪迹浓度:干扰物浓度(按“干扰物”栏顺序)。

2. 示踪剂在油砂上的吸附试验

为了了解示踪剂在地层的吸附情况，为确定示踪剂的注入浓度提供依据，我们在室内进行了示踪剂的静态吸附试验。采用孤东油田试7井(上馆陶组 5^{2+3} 层)的油砂，用溶剂洗净后烘干，称20 g放入干净的三角瓶中，分别加入40 ml硫氰酸铵(100 mg/l)、碘化钾(100 mg/l)、亚硝酸钠(100 mg/l)、钼酸铵(700 mg/l)溶液，使溶液与油砂混合均匀，放入60℃烘箱中。定期取样分析示踪剂溶液的浓度，结果见图1。图1中的四种物质，除硫氰酸铵的吸附量稍大以外，其余三种的吸附量都较小，均能满足现场的需要。

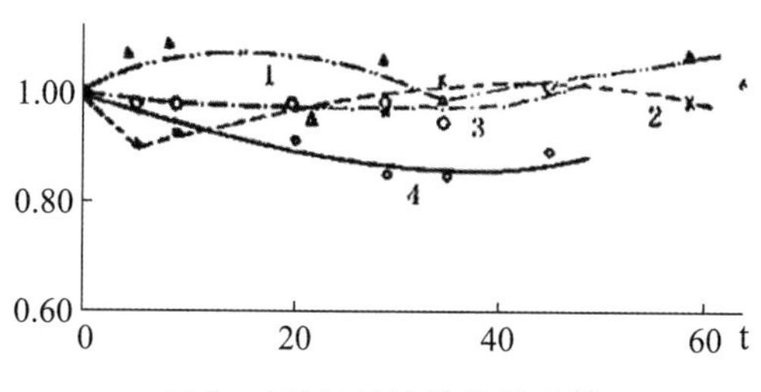

图 1 4 种示踪迹的吸附曲线

1-$NaNO_2$; 2-KI; 3-钼酸铵; 4-NH_4CNS

3. 油砂中示踪剂的检测

称取上述试 7 井的油砂 50 g 于三角瓶中，并加入 100 ml 蒸馏水，在 60℃烘箱中放置 3 天，然后取样分析，结果见表 2。表 2 中的数据说明，孤东小井距试验区 5^{2+3} 层油砂中含 CNS^-、I^-、NO^{2-}、Mo 的量都很低，如果现场实施时注入的示踪剂浓度较大，则可以不考虑油砂中所含的示踪剂成份。

表 2 油砂中示踪剂的含量

踪示剂	NH_4CNS	I^-	NO_2^-	Mo
每克油砂中示踪剂含量(μg)	0.009 5	0.004 6	0.006 5	0.009 0

研究地层水与示踪剂相互作用的主要任务，是考察地层水与示踪剂混合后是否产生沉淀以及地层水中是否含有示踪剂。

试验用的地层水样为孤东 2-20-56 井(上馆陶 6^1 层)水，总矿化度为 15 517 mg/L (其中含 Mg^{2+} 17 mg/L，Ca^{2+} 485 mg/L，SO_4^{2-} 8 mg/L，HCO_3^- 256 mg/L)，pH=7.64。表 3 列出了 NH_4CNS、I^-、NO^{2-}、和 Mo 在地层水中的含量。6^1 层地层水中含有一定量的 I^- 和 NO^{2-}，不含 CNS^- 和 Mo。将 700 mg/L 硫氰酸氨、100 mg/L 碘化钾、100 mg/L 亚硝酸钠和 100 mg/L 钼酸铵水溶液与地层水按 1∶1(体积)混合时，用肉眼观察不到有沉淀产生。

表 3 孤东 2-20-56 井水中示踪剂的含量

示踪剂	NH_4CNS	I^-	NO_2^-	Mo
含量(mg/l)	0	0.110	0.109	0

以上室内试验表明，硫氰酸铵、碘化钾、亚硝酸钠和钼酸铵是可以同时使用的注水示踪剂，它们具有易溶于水(钼酸铵需加热)，价格较便宜、货源广、分析方法简单且灵敏度较高等优点。

二、现场试验

1. 试验区简况

孤东油田小井距试验区位于七区西北部，井按 50 米井距、五点法面积井网加平衡井方式布，由 13 口井组成试验井组，包括中心井——试 7 井为主的 9 口采油井，4 口注水井(试 3、试 6、试 8 和试 11 井)。试验层是七区的主力油层——馆陶组 5^{2+3} 层，平均有效厚度为 10.8 米，该层在试验区内大片连通，油层渗透率高，倾角为 0.5°左右，孔隙度为 34.2%。试验区于 1987 年 12 月 22 日投入注水开发。

2. 示踪剂的注入

1987 年 12 月 24 日，试 3、试 6，试 8 和试 11 井分别投入 NH_4CNS、$NaNO_2$、钼酸铵和 KI(见图 2)，12 月 31 日注完示踪剂，具体数据见表 4。

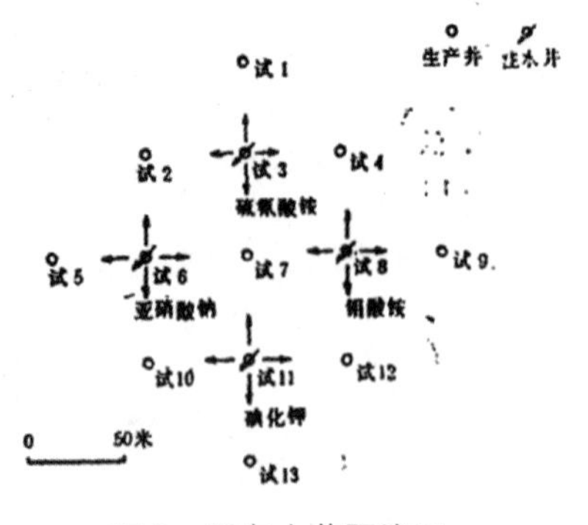

图 2 孤东小井距注示踪剂示意图

表 4 注入示踪剂数据表(注入原位为馆 5^{2+3})

注水井号	注入日期	注入时间(b)	示踪剂			注水量 (m^3)	平均注入浓度(%)
			名称	浓度(%)	注入量(m^3)		
试 3	1987.12.24-29	121	NH_4CNS	11.5	2.22	131.77	0.190
试 6	1987.12.24-29	113	$NaNO_2$	11.5	2.14	156.7	0.155
试 8	1987.12.24-31	162.5	钼酸铵	2.54	5.91	157.22	0.092
试 11	1987.12.24-27	65	KI	11.5	2.45	50.1	0.537

3. 示踪迹的检测

试驱区注水开发 29 天后，试 12 井首先见水。示踪剂检测结果表明，这些水是先后来自试 11、试 8 井的注入水。然后试 5 井见水，同时见到示踪剂。1988 年 1 月 30 日，试 7 井见水，示踪剂检测发现该井 2 月 1 日同时见到四口注水井的来水，验证了数值模拟的计算结果，即四口注水井的含水饱和度、压力梯度和水线推进速度在试 7 井周围是均匀的。同时也达到了小井距试验区配产配注的预期效果。试 9、4、13 井也先后见水，并相应见到各注水井的示踪剂。试 10 井于 3 月 31 日见水，但未见到试 6、11 井的示踪剂 $NaNO_2$ 和 KI。试 1、2 井未投产。具体数据见表 5。

表 5 油井见水及初见示踪剂数据表*

见水井号	见水情况			见示踪剂情况			来水方向			
	见水日期	无水采油天数	初含水(%)	日期	示踪剂	切见浓度(mg/L)	井号	注水天数	注水强度(m^3/d·m)	水线推进速度(m/d)
试 12	1988.1.20	29	7.6	1988.1.20	KI	0.56	试 11	29	3.33	1.33
				1988.1.25	钼酸铵	1.46	试 8	34	2.99	1.63
试 5	1988.1.29	38	2.3	1988.1.30	$NaNO_2$	0.24	试 6	38	3.05	1.36
试 7	1988.1.30	39	1.0	1988.2.1	KI	1.20	试 11	39	3.45	1.43
					NH_4CNS	1.53	试 3	39	2.00	1.17
					$NaNO_2$	0.54	试 6	39	3.06	1.06

（续表）

见水井号	见水情况			见示踪剂情况			来水方向			
	见水日期	无水采油天数	初含水（%）	日期	示踪剂	切见浓度（mg/L）	井号	注水天数	注水强度（m^3/d·m）	水线推进速度(m/d)
					钼酸铵	26.4	试 8	39	3.03	1.51
试 9	1988.3.6	75	1.5	1988.3.6	钼酸铵	9.38	试 8	75	3.197	0.529
试 4	1988.3.29	98	0.4	1988.3.31	NH_4CNS	0.35	试 3	100	1.990	0.553
				1988.3.29	钼酸铵	2.8	试 8	98	3.56	0.55
试 13	1988.2.4	44	20.0	1988.2.7	KI	5.8	试 11	44	3.489	1.093

* 各井的投产日期均为 1987.12.22。

根据生产井中示踪剂浓度的检测结果，作出反映示踪剂浓度与注水时间关系的示踪剂采出曲线。此曲线是示踪剂通过地层时扩散和驱动、吸附和解吸过程的综合反映。

4. 结果与讨论

(1) 在一多层体系的油藏中，示踪剂采出曲线反映的是各层采出示踪剂的总和。在注水速度恒定的情况下，示踪剂到达的时间和浓度是各层孔隙度、渗透率和层厚度的函数。Maghsood 等人建立的流体通过非均质油藏时的互混理论及其对五点法井网示踪剂采出曲线的数学处理方法，很好地解决了通过示踪剂采出曲线来了解油藏非均质的问题。该理论认为，示踪剂在油层的各个小层中形成了不规则的移动，从小层中穿过的示踪剂能够形成互混区；对于每个互混状态，采出曲线上都相应有一个较明显的峰。通过模拟每个峰即可分出油层中的小层。根据 Maghsood 的数学模型[1]，图 3 给出了适用于交错行列布井的 C_D-PV 图，其中 C_D 为无因次浓度。从图中看出，当井距小或 a 值大时，曲线比较宽。图 4 给出了在不同井网条件下，示踪剂穿过油层的模拟曲线[1,2]。孤东试 7 井的示踪剂采出曲线具有相似的特点。

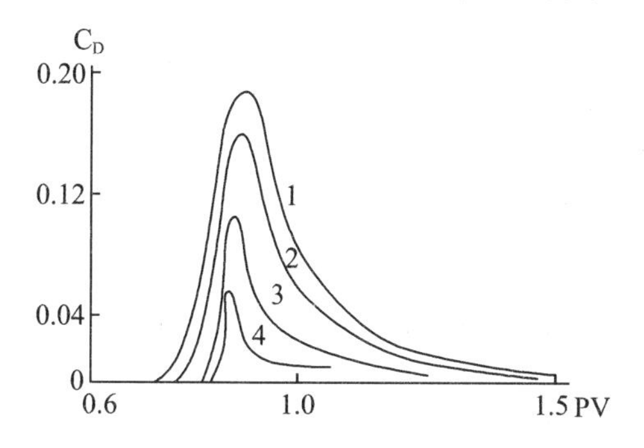

图 3 非均质油藏中交错行列布井示踪剂采出曲线[1]

a/a：1-500；2-1 000；3-S000；4-S0000

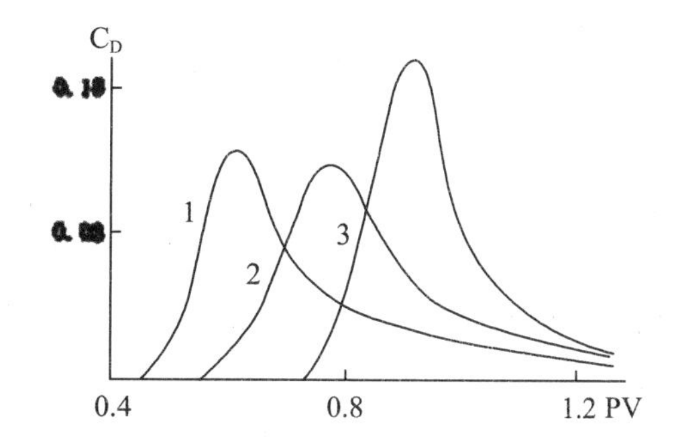

图 4 a/a=500 时不同井网的 zjs 踪剂采出曲线[1]

1-线性驱；2-五点法井网；3-交错行列布井线性驱

我们应用 Maghsood 等人提出的理论对孤东小井距示踪剂采出曲线（图 5）作了如下分析。

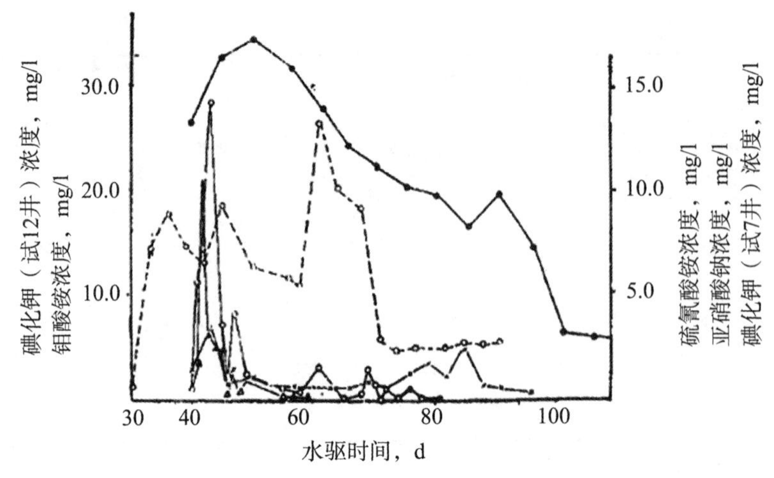

图 5 试 7 井和试 12 井示踪剂采出曲线

—·—钼酸铵(试 7 井);—×—硫氰酸铵(试 7 井);

—△—亚硝酸钠(试 7 井);—○—碘化钾(试 7 井);

—○——碘化钾(试 12 井)

① 从平面分布看,试 12 井最早见水(表 5)。见水的同时见到碘化钾,而且碘化钾的浓度一直较高(图 5)。此外,该井含水上升很快,这说明在试 11 与试 12 井之间有一个高渗透率通道。试 10 井见水最晚,没有见到 KI,说明试 11 与试 10 井之间的渗透率低。试 5 井见水的同时见到 $NaNO_2$,见水后含水上升很快(含水上升速度为每天 0.94%),但试 10 井没有检测到 $NaNO_2$,这表明试 6 与试 5 井之间有一高渗透带,试 6 与试 10 井间是低渗透带。试 10 井没有见到试 11 和试 6 井两个方向的示踪剂,表明示踪剂向试 10 井方向的流动量非常小。又因试 10 井见水最晚,因此,试 10 与试 6、试 10 与试 11 井两个方向的渗透性差。

② 根据 Maghsood 的流线互混理论和试 7 井的 4 条示踪剂采出曲线,可以分析层内的渗透率纵向分布。硫氰酸铵的采出曲线共出现三个明显的峰,分别在注示踪剂后第 43、79 和 85 天出现。据此可初步认为,试 3 与试 7 井间 Ng 5^{2+3} 层的渗透率大致可分为三个极差,其中第 43 天出现的峰代表从较大渗透带中穿过的示踪剂。由于三个峰高和峰面积相差不大,故三个级差的厚度相近。$NaNO_2$ 的采出曲线除了一个主峰外,不存在具有较大影响的峰,这表明试 6 与试 7 井间的层内没有明显的透率率级差。碘化钾的采出曲线除了第 42 天出现的主峰外,第 50 天出现的峰亦较明显,其余 4 个则不太明显,同时第 50 天出现的峰与主峰相比小得多,故试 11 与试 7 井间除了有一个发育较好、厚度较大的高渗透带外,还有一较薄的低渗透带。钼酸铵的采出曲线有两个峰值,分别在第 50 天和 90 天出现,说明试 8 与试 7 井间有两个不同的渗透带。由于第一个峰大且持续时间长,第二个峰相对较小,所以高渗透带较厚,较低渗透带较薄。

(2) 我们根据这 5 条曲线认为,这四种示踪剂在地层中的扩散速度小。硫氰酸铵和钼酸铵的初见浓度很高,1 天以后就达到示踪剂浓度的高峰;碘化钾和 $NaNO_2$ 扩散速度稍大,2～3 天后达到浓度高峰。由于注示踪剂和注水基本是同时进行的(相差两天),生产井见水的同时也见到了示踪剂,这说明示踪剂的滞后现象不严重,能与注入水一起流动。

(3) 根据试7井四条示踪剂采出曲线中浓度的第一峰值及示踪剂注入的初始浓度,可以估算出相当于示踪剂在地层中每移动1米所消耗的浓度(mg/L):硫氰酸铵18.93;碘化钾26.55;亚硝酸钠10.11;钼酸铵3.9。相比而言,碘化钾和硫氰酸铵的消耗较大,而钼酸铵的消耗小。这些数据可供确定注水示踪剂初始浓度参考。

三、结论

(1) 室内和现场试验结果表明,硫氰酸铵、碘化钾、亚硝酸钠和钼酸铵在地层水和注入水中含量很低时,是可以同时使用的四种注水示踪剂。

(2) 硫氰酸铵、钼酸铵、碘化钾和亚硝酸钠在地层中的扩散速度均小,能满足注水示踪剂的要求。

(3) 用示踪剂采出曲线进行的粗略计算结果表明,硫氰酸铵、钼酸铵、碘化钾和亚硝酸钠四种示踪剂在地层中的消耗浓度以碘化钾和硫氰酸铵较大,钼酸铵最小。

(4) 应用Maghsood等人提出的流线互溶理论对试7井示踪剂采出曲线进行的分析表明,试6井与试5井和试11井与12井之间有一高渗透带。

参考文献

[1] Maghsood et al: "Analysis of Well-to-Well Tracer Flow to Determine Reservoir Laye-ring", *J. Petrol TechnoL*, Oct. 1984, 36, No4, 1753-1762.

[2] Brigham, W · E. et al. "Tracer Testing for Reservoir Description", SPE 14102.

[3] 吴先承,王解忠,等. 孤东油田馆陶组油层小井距注水开发全过程及提高原油采收率现场试验总体设计. 胜利油田地质研究院,1987年8月,未发表资料.

[4] 屈智坚. 孤东油田小井距馆 5^{2+3} 层试验方案. 胜利油田地质研究院,1987年7月,未发表资料.

Selecting and testing tracers for flooding water flow in close well spacing district in gudong oil field

Wang Baoyu Cao Xulong

(Research Institute of Geology, Shengli Oil Field)

Abstract

In this paper the results of laboratory screening test, of the first practical application at a close well spacing district in Gtidong, Shengli Oil Field, and of analysis of four flooding water tracers injected simultaneously. These are ammonium thiocyanate, potassium iodide, sodium nitrite, and ammoniutn molybdenate. The losses of these tracers in reserviors are roughly estimated from the curves of each tracer production in the central well. The vertical distribution of permeabilities of the reserviors between the central and the four peripheral injecting wells is analysed preliminaryly in the light of the theoretical concepts published.

孤岛油田西区三元复合驱矿场试验

曹绪龙[1],孙焕泉[2],姜颜波[2],张贤松[2],郭兰磊[2]

(1. 中国科学院兰州化学物理研究所,甘肃兰州 730000;
2. 中国石化胜利油田公司地质科学研究院,山东东营 257015)

摘要:孤岛油田西区三元复合驱矿场扩大试验区包括6口注水井和13口采油井。介绍了试验区油藏地质特征。所设计的超低界面张力三元复合驱油溶液(主段塞)配方为:1.2%Na_2CO_3+0.3%复配表面活性剂(阴离子表面活性剂BES+木质素磺酸盐PS)+0.15%聚合物3530S。从1997年5月开始实施化学剂注入,实际注入情况如下:前置段塞(0.20%聚合物溶液)0.097 PV;主段塞0.309 PV;后置段塞(0.15%聚合物溶液)0.05 PV,2001年11月转后续水驱。本文介绍注完化学剂时的试验效果,包括:全区油井综合含水由94.7%降至84.5%,日产油量由82 t升至194 t,累计增油10.42×10^4 t,提高采收率5.27%,预计最终提高采收率12.04%;注水井纵向各层吸水均匀化,注水井流动系数和流度下降;注水利用率提高,每采出1 t原油的耗水量由17.9 t降至10.4 t。从受效方向、油层构造、沉积微相、窜流等4个方面分析了各油井增油减水效果不同的原因。图3表3参7。

关键词:三元复合驱(ASP驱);矿场扩大试验;胜利孤岛油田;驱替效果;提高采收率;效果分析

中图分类号:TE357.46 文献标识码:A

碱/表面活性剂/聚合物(ASP)三元复合驱是上世纪80年代发展起来的三次采油新方法,它既能提高波及体积又能提高驱油效率,从而大大提高原油采收率。1992年8月在孤东油田开展了小井距三元复合驱试验[1],取得了良好的降水增油效果,试验区中心井提高采收率13.4%。为使三元复合驱油技术尽快进入矿场工业化应用,在孤岛油田西区开展了常规井距、注水开发井网的扩大试验。

1 试验区概况

1.1 地质特征

试验区位于孤岛背斜构造的西翼,构造比较简单,平缓,东高西低,构造高差约

20 m,地层倾角 0.5°～1.5°,油层埋深 1 190～1 310 m。试验区目的层砂层组为上馆陶组 4^2～4^4,为一套正韵律的曲流河沉积,油层分布广、厚度大、渗透率高,层内非均质严重。岩性以粉细砂岩为主,油层胶结疏松,成岩性差。纵向上可分 3 个小层,4^4 层厚度大,连通性好,分布广;4^2 层呈条带状分布,4^3 层仅在局部发育。

试验区含油面积 0.61 km^2,有效厚度 16.2 m,平均渗透率 1.52 μm^2,孔隙度 32%。

原始油层温度 69℃,油层压力 12.3 MPa,地下原油粘度约 70 mPa·s,原油酸值 1.7 mg KOH/g,属低凝重芳烃原油,具有密度大、粘度高、凝固点低、含蜡量少、含硫量低等特点。地层水为 $NaHCO_3$ 型,地层水矿化度 6 864 mg/L,Ca^{2+} + Mg^{2+} 含量 143 mg/L;回注污水矿化度 6 188 mg/L,Ca^{2+} + Mg^{2+} 含量 78 mg/L。

1.2 开发概况

试验区内共有注入井 6 口,生产井 13 口,注采井距 212 m,以生产井 5-142 和 6-121 为中心井,由 6 个井组组成长方五点井网(图 1)。

试验前全区日产油 82 t/d,综合含水 94.7%,采出程度 28.5%,预测最终水驱采收率 31.0%。

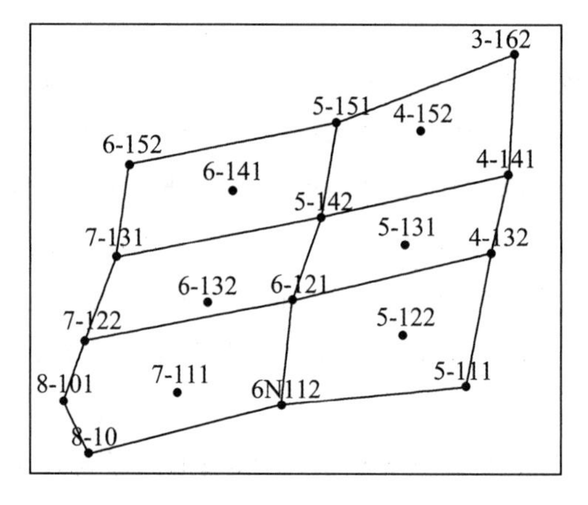

图 1 试验区井位示意图

●—水井;○—油井

2 试验注入简况

2.1 复合驱方案设计

复合驱油体系的配制费用是制约其发展的一个重要因素,设计的关键在于兼顾驱油效果和经济效益。室内配方研究结果表明,表面活性剂复配有助于界面张力的降低和低界面张力范围的拓宽,可以达到廉价高效的目的。适用于试验目的层的三元复合驱油体系为:1.2% Na_2CO_3 + 0.3%复配表面活性剂 + 0.15%聚合物 3530S,复配的两种表面活性剂均为阴离子型,一种为 BES,一种为木质素磺酸盐。该体系与目的层原油界面张力最低达 7×10^{-4} mN/m,即使在油砂吸附或热老化 70 天后,界面张力仍低于 10^{-3} mN/m,且体系色谱分离效应不明显[2],室内驱油实验提高采收率高达 22.4%

(OOIP)[3]。

通过数值模拟对化学剂注入浓度、段塞大小、注入方式等参数进行了优化，结合油藏综合研究，最终确定矿场采用三段塞注入方式[4]：

第一段塞为聚合物前缘段塞，0.05 PV×2 000 mg/L，注入聚合物干粉 316 t，溶液 15.8×10^4 m^3。

第二段塞为聚合驱主体段塞，0.3 PV，注入溶液 94.8×10^4 m^3，其中：碱浓度 1.2%，用量 11 376 t；复配表面活性剂浓度 0.3%，用量 2 844 t；聚合物浓度 1 700 mg/L，用量 1 612 t。

第三段塞为聚合物后置段塞，0.05 PV×1 500 mg/L，注入聚合物干粉 237 t，溶液 15.8×10^4 m^3。

2.2 方案的矿场实施

试验区 1997 年 5 月开始注聚合物前缘段塞，1998 年 5 月开始注三元复合驱主体段塞，2001 年 5 月转注第三段塞。由于矿场复合驱地面设备投产滞后，第一段塞的注入时间延长，共注入聚合物溶液 0.097 PV，总用量超过方案设计，主段塞注入 0.309 PV，第三段塞按设计方案注入 0.05 PV。到 2001 年 11 月，已累积注入聚合物干粉 2 936 t，复配表面活性剂 2 652 t、碱 10 403 t，将转入后续水驱。

3 试验效果

3.1 油井含水下降，产油量增加

在前缘聚合物段塞注入期间，全区含水逐渐平稳，在主段塞注入 0.03 PV 后生产井陆续见效，综合含水开始大幅度下降，日产油大幅度上升。与试验前相比，全区综合含水由 94.7% 下降到 84.5%，下降了 10.2%，日产油由 82 t/d 上升到 194 t/d，增加 1.36 倍(图 2)[5]。

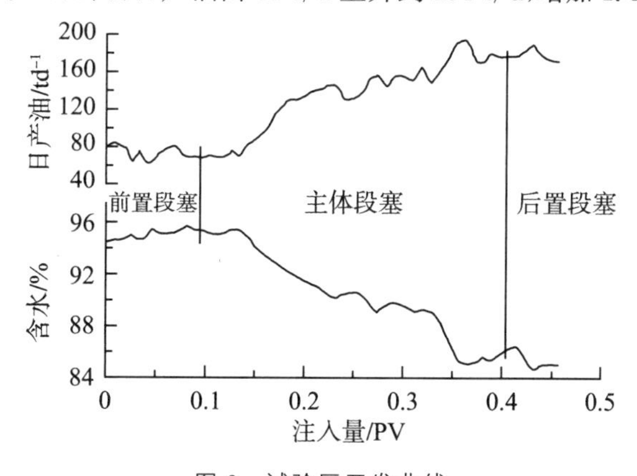

图 2 试验区开发曲线

全区 13 口油井全部见到明显的降水增油效果(表 1)，其中效果好的一类井 6 口(表 1 中 1～6 号)，平均单井增油 12 331 t；二类井 4 口(表 1 中 7～10 号)，平均单井增油 6 661 t；三类井 3 口(表 1 中 11～13 号)，平均单井增油只有 1 196 t，效果较差，目前已失效。

表1 油井单井见效分析表*

序号	井号	见效时间年月	含水%			增油量 t
			见效前	最低	目前	
1	5-111	1 999.3	95.0	64.0	81.4	16 213
2	3-162	1 998.6	95.9	83.2	91.5	14 769
3	6-121	1 998.6	94.0	81.8	83.2	14 058
4	6-152	2 000.2	93.9	67.6	71.9	11 567
5	7-131	2 000.4	96.5	79.8	81.0	9 202
6	8-101	1 998.12	95.8	82.3	87.2	8 174
7	4-141	1 998.10	92.4	82.1	88.8	7 251
8	8-10	1 998.11	95.6	88.9	88.9	6 811
9	7-122	1 999.2	97.2	82.1	87.2	6 748
10	5-151	1 999.1	96.6	81.2	95.8	5 835
11	6N112	2 000.8	90.5	81.0	92.1	1 512
12	5-142	1 998.9	97.4	90.9	关井	1 495
13	4-132	2 000.11	96.9	88.6	96.7	581

*统计截止日期2001年11月。

到2001年11月,试验区已累积增油 10.42×10^4 t,提高采收率5.29%。预计最终可增油 23.7×10^4 t,提高采收率12.04 %。

3.2 油层纵向矛盾得到调整,驱油效率明显提高

注入化学剂段塞后,驱替相的粘度提高,油层中渗流阻力增加,流动系数和流度明显变小(表2)。

表2 试井解释成果表

井号	流动系数/$\mu m^2\cdot m(mPa\cdot s)^{-1}$		流度/$\mu m^2(mPa\cdot s)^{-1}$	
	试验前	试验期间	试验前	试验期间
5-122	0.26	0.08	0.007	0.006
6-141	0.21	0.08	0.014	0.005
7-111	0.12	0.09	0.005	0.004

吸水剖面测试结果表明:注入井纵向非匀质性有所改善,原来吸水量大的层吸水量下降,而吸水量少的层吸水量上升,层间差异减小。例如7-111井(表3),试验前 4^3 层吸水最好,4^4 层吸水最差,每米相对吸水量相差2.1倍;试验后 4^3 层吸水受到有效的遏制,4^4 层吸水量明显增加,层间吸水量趋于均匀。这说明化学剂溶液首先进入高渗透层并发生吸附和捕集,渗流阻力增加,转向进入吸水差的层,扩大了波及体积,改善了层间动用状况[6]。

表 3 7-111 井吸水剖面变化对比表*

层位	全层相对吸水量%		平均每米相对吸水量%		层间比值	
	A	B	A	B	A	B
4^2	37.6	25.8	3.4	2.3	2.4	1.0
4^3	48.8	33.2	4.4	3.0	3.1	1.3
4^4	13.6	41.0	1.4	4.3	1.0	1.9

* A:试验前;B:试验期间。

碳氧比能谱测井(C/O)是利用元素碳和氧对快中子的非弹性散射截面的差别及放出的伽马射线能量的差别确定剩余油饱和度的测井方法。矿场 C/O 资料表明:试验期间驱油效率明显提高。注采井间的监测井 5-检 142 井测试结果表明,注入化学驱油剂后,试验层含水饱和度开始随时间增加而下降,含油饱和度逐渐增加。在化学驱油剂的作用下剩余油逐渐富集并向油井推进。生产井产出的原油初馏点降低,300℃馏份增加[7]。

3.3 改善了开发效果,提高了注水利用率

试验区驱替特征曲线(图 3)明显向产油轴偏移,开发形势变好。特别是注入主体段塞后这种偏移更加明显,这说明与单一聚合物驱相比,三元复合驱不仅扩大了波及体积,而且提高了驱油效率,提高采收率的幅度更高。

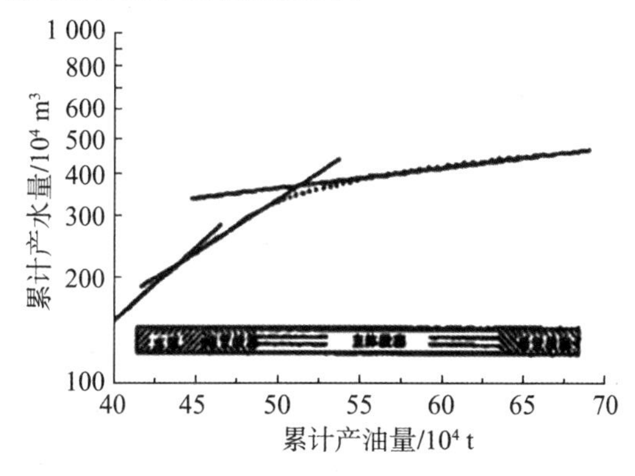

图 3 试验区驱替特征曲线

试验前油井平均含水 94.7%,每采 1 t 原油耗水 17.9 t。试验期间含水下降,产油量逐渐增加,耗水量降低。试验中化学剂注入期间全区阶段累产油 20.88×10^4 t,累产水 216.9×10^4 t,每采 1 t 原油耗水 10.4 t,比试验前减少 7.5 t。据此测算,试验区少产水 156.6×10^4 t,相当于 0.5 PV,注水利用率提高,注水量减少。

4 见效状况分析

由表 1 可以看出,虽然油井见效率 100%,但见效程度差别较大。根据油藏描述的结果,结合油水井动态特点,通过对单井见效特点进行分析,认为影响效果的因素主要有以下几个方面。

(1) 受效方向 中心井受效方向多,效果好。中心井 6-121 含水由 94.0%下降到

81.8%,下降了12.2%,日产油由7.0 t/d上升到23.6 t/d,上升了2.4倍,已累积增油14 058 t。目前含水仍在下降,日产油仍在上升,效果明显。

(2) 油层构造 试验区构造简单,东高西低,处于高部位的油井见效相对较早,效果好。东部高部位的油井平均单井增油是西部的1.7倍。

从产出液中化学剂的浓度分析可以看出,东部高部位井产出液中平均碱浓度655 mg/L,聚合物浓度610 mg/L,其中4-141井产出液中碱浓度达1 176 mg/L;西部井产出液中平均碱浓度314 mg/L,聚合物浓度285 mg/L,远远低于东部井。构造的高低决定注入液的流向,从而影响油井的增油减水效。

(3) 沉积微相 4-132井位于试验区东部高部位,生产层位为4^3、4^4,直到2000年11月才见到降水增油的效果。其原因主要是该井目的层处于天然堤微相,油层发育较差,与注入井的连通稍差(图4)。

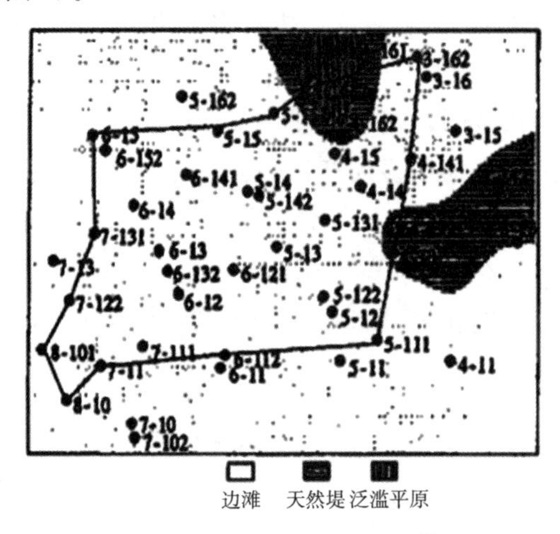

图4 试验区沉积微相图(4^{42})

(4) 窜流 层内窜流影响注入流体的驱替方向,导致油井见效的差异。中心井5-142位于主河道,渗透率较高,试验初期效果明显,但1999年4月产出液中就见到了化学剂,2000年2月由于产出化学剂浓度太高而关井。多井干扰试井测试结果表明,注入井6-141与5-142间存在严重的大孔道,导致化学剂窜流。而同井组的6-152和7-131两口井见效差,当5-142井关井后,这两口井含水开始大幅下降。

参考文献

[1] 王宝瑜,曹绪龙,王其伟,等. 孤东小井距三元复合驱现场试验采出液相态变化及组分浓度测定[J]. 油田化学,1994,11(4):327-330.

[2] 隋希华,曹绪龙,王得顺,等. 孤岛西区三元复合驱体系色谱分离效应研究[J]. 油气采收率技术,2000,7(4):1-3.

[3] 曹绪龙,王得顺,李秀兰. 孤岛油田西区复合驱界面张力研究[J]. 油气地质与

采收率，2001，8(1):64-66.

[4] 王成龙，胡重浩. 胜利油田孤东小井距试验区先导性试验研究[A]. 载:冈秦麟. 化学驱油论文集(下册). 石油工业出版社，1998:1-14.

[5] 李华斌，李洪富，杨振宇，等. 大庆油田萨中西部三元复合驱矿场试验研究[J]. 油气采收率技术，1999，6(2):15-19.

[6] Buell R S, Kazemi H, Poettmann F H. Analyzing injectivity of polymer solutions with the hall plot[Z]. SPE 16 963, 1987.

[7] 廖广志，牛金刚，王刚，等. 大庆油田三元复合驱矿场试验效果评价方法分析[J]. 石油勘探与开发，1998，25(6):44-46.

Enlarged field test on ASP-flood at east district of gudao oil field

Cao Xulong[1], Sun Huanquan[2], Jiang Yanbo[2], Zhang Xiansong[2], Guo Lanlei[2]

(1. Lanzhou Institute of Chemical Physics, Chinese Academy of Sciences, Lanzhou, Gansu 730000, China;

2. Geological Scientific Research Institute, Shengli Oilfield Company, Sinopec, Dongying, Shandong 257015, China)

Abstract: There are 6 injection and 13 production wells in the APS-flooding test area of east district of Gudao oil field, Shengli. The geologic characteristics of the reservoir in the test area are described. The ASP-flooding solution (main slug) designed is composed of 1.2% Na_2CO_3, 0.3% combinational surfactants (anionic surfactant BES+lignosulfonate PS), and 0.15% polymer 3530S and possesses ultralow interfacial tension with the crude oil. The injection of the chemicals has been started in may 1997 according to a modified scheme as follows: 0.20% polymer solution as pad slug of size 0.097 PV; main ASP slug of size 0.309 PV; 0.15% polymer solution as postpad slug of size 0.05 PV; and successive flooding water started in Nov 2001. The test results at the end of chemical injection are presented: in the whole test area the combained water cut decreases from 94.7% to 84.5%, the oil production increases from 82 t to 194 t per day, the incremental oil production reaches 1.04×10^5 t accumulatively, the enhancement in oil recovery obtained accounts 5.27% and that predicted by UTCHEM at the end of the field test will be of 12. 04%; the watertake of reservoir layers in injection wells becomes more even and the flow coefficient and the fluidity in injection wells are lowered; the ratio of produced oil to produced water is decreased from 17. 9 to 10. 4. The differences in chemical flood response of some production wells are analysed in terms of flooding directions, layer construction, sedimentary microfacies, and existance of channellings.

Keywords: alkaline/surfactant/polymer(ASP) flood; enlargedfield test; Gudao oil field in Shengli; flood response; enhancement in oil recovery; results analysis

Development and application of dilute surfactant-polymer flooding system for Shengli oilfield

Wang Hongyan*, Cao Xulong[1], Zhang Jichao[2], Zhang Aimei[1]

Geological and Scientific Research Institute, Shengli Oilfield, SINOPEC, Dongying 257015, PR China

articleinfo

Article history: Received 4 April 2007

Accepted 25 December 2008

Keywords:

surfactant-polymer flooding Shengli petroleum sulfonate interfacial tension dynamic behavior pilot fi eld trial

abstract

A dilute surfactant-polymer(S-P) flooding system has been designed and developed for Gudong oilfi eld with Shengli petroleum sulfonate (SLPS) as the primary ingredient. The dynamic behavior of the system and the interactions of the system components have been investigated through various methods, including DPD molecular modeling technology and dynamic interfacial-tension analysis. The results have shown a signifi cant synergistic effect between sulfonate and nonionic surfactant. The interfacial tension (IFT) and its time to reach equilibrium could be dramatically decreased, suggesting a fast diffusion-adsorption characteristic of ionic surfactants as well as the high surface activity of nonionic surfactants. The S-P flooding formulation was finalized as 0. 3% (w/w) SLPS + 0. 1% (w/w) 1 # + 0. 15% polyacrylamide (PAM), in which 1 #, the secondary surfactant, is able to enhance the interfacial activity of SLPS and the flooding effi ciency of the system. The S-P flooding system manifests a wide range of IFT with the lowest value of 2.95×10^{-3} mN/m. The pilot fi eld trial of the system has exhibited outstanding performance to improve oil production and reduce water cut since the first injection of the main slug of the S-P flooding system started in June 2004. Until July 2008, the accumulated oil-production had risen by 17.8×10^4 tons, with the oilrecovery increase by 6.4%. The field trial provides useful information for the further large-scale application of the SP system in Shengli oilfi eld.

1. Introduction

After being developed for more than thirty years, Shengli oilfi eld becomes increasingly expensive to exploit due to high water cut in the main oilfi eld reservoir. Meanwhile, it is getting extremely diffi cult to explore new oil reservoirs. Therefore, it is of great importance to improve oil production of current oilfields by using new technologies.

Alkaline-surfactant-polymer(ASP) flooding invented in 1980s has been regarded as a potential enhanced-oil-recovery (EOR) technology which is more powerful than polymer flooding. Extensive studies on ASP technology have been carried out in the U. S., Germany, and the North Sea(Hernandez et al., 2003; Carrero et al., 2007). In China, a number of bench-scale and on-site experiments on ASP technology were carried out in Daqing and Shengli oilfi elds, which had indicated satisfactory capability of ASP systems to increase oil production(Baoyu et al., 1994; Xulong et al., 2002; Kang, 2001). Meanwhile, disadvantages identified during the implementation process of the ASP flooding technology, such as, severe scaling in the injection lines and strong emulsification of the produced fluid, (Hongyan et al., 2005; Zhang and Xiao, 2007) also limited its further application in the field.

In order to overcome the drawbacks associated with ASP flooding, alkali-free surfactant-polymer(SP) flooding technology was developed, and its application in Shengli oilfield has been extensively investigated by authors' research institute. The pilot field trial of our S-P flooding system in southwest Gudong 7 th region $Ng5^4$-6^1 was the first field experiment of the S-P flooding technology in China. The objective of the work focused on demonstrating the feasibility of the S-P flooding technique for further enhancement of oil recovery after polymer flooding. Moreover, the field trials provided useful information on the S-P flooding technology, such as, the formulation design and optimization. This could eventually lead to wider implementation of the S-P flooding system in Shengli Oilfield.

T. Austad and I. Fjelde had previously indicated that significant improvements can be obtained by coinjecting surfactant and polymer at a rather low chemical concentration. Furthermore, the key factor in selecting chemicals is to avoid S-P complex form ation in order to still maintain a very low IFT at low surfactant concentration(Austad et al., 1994). Among all the efforts to selecting good chemicals for S-P flooding, petroleum sulfonate, a good oil displacement agent, has gained more and more attention as used in chemical enhanced oil recovery technology. Lu Zhang et al. studied the effect of different acidic fractions in crude oil on dynamic interfacial tensions in surfactant/alkali/model oil systems (vander Bogaert and Joos, 1980), A study by H. S. Al-Hashim had focused on the adsorption and precipitation behaviour of petroleum sulfonates on Saudi Arabian limestone (Zhang et al., 2004). F. E. DeBons and L. E. Whittington compared performance of the petroleum sulfonate with lignin in Berea sandstone cores (Al-Hashim et al., 1988). All these studies were limited on the laboratory development stage, none petroleum sulfonate-polymer pilot test have been reported.

Using Shengli petroleum sulfonate(SLPS) as the primary component, this effort studied the effect of the secondary surfactant and polymer in the formulation, as well as their capability to enhance the overall oil-recovery performance.

2. Surfactant design for the S-P flooding system

2.1. Experiment

2.1.1. Reagents

Dodecyl polyoxyethylene polyoxypropylene ether, $C_{12}H_{25}(EO)_4(PO)_5H$ (LS45) and $C_{12}H_{25}(EO)_5(PO)_4H$ (LS54), (both purchased from Henkel company, Germany) are colorless, viscous liquid with purity greater than 99.95%; sodium dodecyl benzene sulfonate (SDBS), sodium hexadecyl sulfonate(AS), sodium dodecyl sulfonate(SLS), and nonylphenol polyethylene oxide ether (TX, $n=8-9$) are all of analytical purity grade, purchased from Shanghai reagent company, China; Shengli petroleum sulfonate (SLPS) contains 31.4% active ingredient(Table 1).

2.1.2. Experimental methods

Droplet volume method was utilized to measure dynamic surface tension; TX-500 C spin drop apparatus from Bowing Industry Corporation, USA, was used to measure IFT. Viscosity of the polymers was examined using DVIII viscometer in reservoir conditions. Chromatographic separation tests were conducted for the surfactant flooding system under reservoir conditions. The tests were performed in a tube model with inner diameter of 1.5 cm and length of 50 cm. In the tests, unconsolidated model porous media composed of fine silica sands of different mesh was used and its permeability was about $1.5\times10^{-3}\ \mu m^2$ which was close to the reservoir conditions.

The tube was prevacuumized and then saturated by water before injection of the surfactant fl ooding fluid of 0. 3 PV. Afterwards, the tube was flooded by water until the outgoing concentration of surfactants became zero.

In oil displacement test, the formation water and re-injection water were prepared at salinity of 4 876 mg/L and 6 188 mg/L, respectively. The oil mixture was formulated by kerosene and dehydrated crude oil from Shengli 16-011 well to simulate underground crude oil at 50 mPa s viscosity. Testing temperature was 70℃. The tests were performed in a core with inner diameter of 2. 5 cm and length of 30 cm. The heterogeneousness of the formationwas simulated by dual-core system of different permeability: 1 500$\times 10^{-3}$ μm^2 and 4 500$\times 10^{-3}$ μm^2. The core-flooding procedure is as follow:

Vacuum-pump test core for 2 h, then saturated with reservoir water. Afterwards the core was flushed with (1) oil until a state cor-responding to Swi was reached. (2) Water until a state corresponding to Sor was reached. (3) A slug of surfactant formula until water cut to 100%. Injection speed was 0. 23 mL/min.

Table1 List of chemicals

No.	Name	Company	Purity grade
1	Dodecyl polyoxyethylene polyoxypropylene ether $C_{12}H_{25}(EO)_4(PO)_5H$ (LS45) and $C_{12}H_{25}(EO)_5(PO)_4H$ (LS54)	Henkel company, Germany	Greater than 99. 95%
2	Sodium dodecyl benzene sulfonate(SDBS);	Shanghai reagent	Analytical
3	Sodium hexadecyl sulfonate(AS);	company, China	purity grade
4	Sodium dodecyl sulfonate(SLS);		
5	Nonylphenol polyethylene oxide ether(TX, n =8−9)		
6	Shengli petroleum sulfonate(SLPS)	Shengli zhongsheng company, China	31. 4% active ingredient

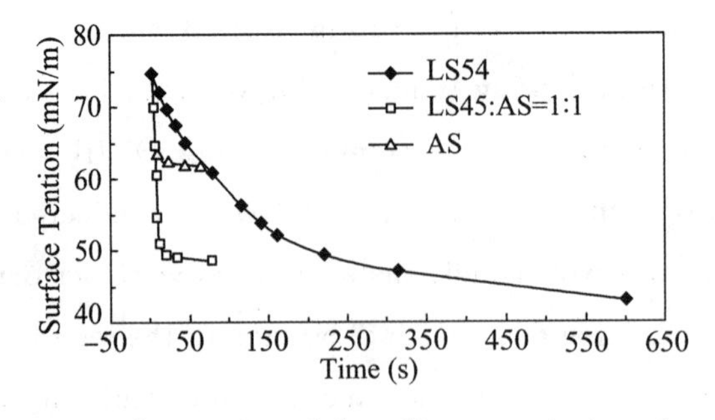

Fig. 1 Dynamic surface tension of the sulfonate-nonionic surfactant system.

2.2. Results and discussion

2.2.1. Dynamic surface tension of the sulfonate-nonionic surfactant system

SLPS was been selected as the primary surfactant of the S-P flooding system for the pilot field trial in Gudong oilfi eld due to its compatibility with oil reservoir as SLPS is produced directly from Shengli crude oil. Meanwhile Employing SLPS as the main component of the S-P flooding formula also lowers the reliance on the outside chemical sources. As an anionicsurfactant, SLPSshows strong electricrep ulsion among the polar heads. A variety of hydrophobic chains structures is expected as SLPS is produced from Shengli crude oil. This results in a loose arrangement of chains in the interfacial membrane and thus the low interfacial activity. As a consequence, SLPS itself is incapable to decrease the oil/water IFT to an extremely low level of 10^{-3} mN/m. Nevertheless, previous studies had indicated that co-adsorption of different types of surfactants in oil/water interface could generate interfacial membranes with tight and ordered arrangement of molecules due to weaker steric and electric interactions in the system(Salager and Mongan, 1979; Myers, 2009; Sheng and Wang, 2001). In this case, the oil/water IFT can be further reduced as a result of the synergistic effects between different surfactants.

The dynamic surface tension of AS-LS54 was measured, as shown in Fig. 1, to investigate the activating behavior of the sulfonate-nonionic surfactant system and the synergistic interactions between the two types of surfactants.

Fig. 1 illustrates that the surface tension for AS equilibrates immediately in water while it takes a much longer time for LS54. The equilibrium surface tension are determined to be 62 mN/m for AS and 42 mN/m for LS54, respectively, which is 20 mN/m lower than that for AS. In contrast, the surface tension of AS-LS54 (in1:1ratio) reaches the steady state as effi cient as AS, and possesses a low value similar to that for LS54. Since AS-LS54 shows augmentation of diffusion coeffi cient as well as surface activity compared to the single surfactant system, sulfonate-nonionic surfactant has been determined to be the basic formulation of the S-P flooding system.

Table2 IFT(mN/m) for sulfonate and sulfonate-nonionic surfactant systems

Oil phase	Surfactant						
	SDBS	SDBS: LS45 =9:1	SDBS: TX-100=8:2	SDBS:TW-80=8:2	SDBS:AES =9:1	SLS	TX
1-Octane	1.209	/	0.528	0.983	/	>3	0.500
Toluene:1-Octane=1:2	0.456	0.505	0.142	0.321	0.418	2.460	/
Toluene	0.333	/	0.227	0.082	/	/	>3

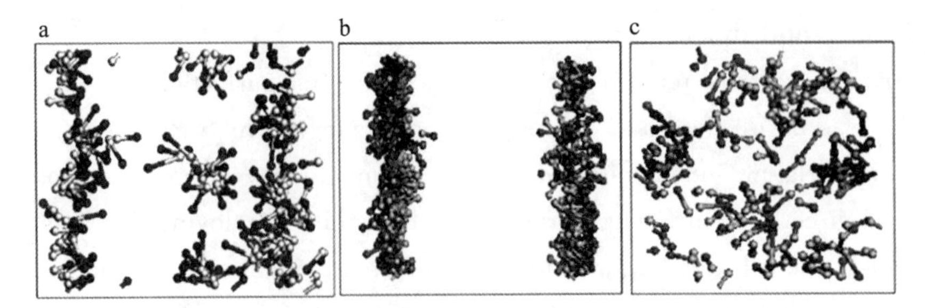

Fig. 2 Arrangement of sulfonate and sulfonate-nonionic surfactant in the interface through molecular modeling. (a) SDBS array in the interface. Red and yellow colors represent the head and tail of SDBS respectively. (b), (c) SDBS-TX array in the interface. Pink and blue colors represent the head and tail of SDBS respectively, while red and green colors correspond to the head and tail of TX respectively. (For interpretation of the references to color in this fi gure legend, the reader is referred to the web version of this article.)

2.2.2. Synergistic effect between sulfonate and nonionic surfactant

Structure-function relationships have been investigated for combinations of sulfonate and various nonionic surfactants through measurement of oil/water IFT. Total concentration of each surfactant system studied is 0.05% by weight in 0.7% NaCl brine.

Table 2 demonstrates a remarkable decrease of 1-octane/water IFT upon the addition of a small amount of nonionic surfactant (TX-100 or Tween-80) in SDBS, suggesting a certain synergistic effect between SDBS and nonionic surfactants. With toluene as the oil phase, combination of SDBS and Tween gives the lowest oil/water IFT among all the surfactant systems. It has been found that the oil/water IFT gradually increases with the content of saturated alkanes in oil phase. Lower oil/water IFT by introducing toluene into saturated alkanes shows that structure similarity between surfactant hydrophobic chains and oil phase molecules favors the IFT reduction. Clearly, the most prominent synergistic effect for SDBS comes from the combination of nonionic surfactants that contain aromatic rings in the hydrophobic chains. Since the ratio of aromatic hydrocarbon and alkane in Gudong crude oil is near 1∶2 the combination of the right molecule chain and the number of EO with SLPS gain the lowest interfacial tension in surfactant-polymer flooding system.

2.2.3. Molecular modeling of the synergistic effect between sulfonate and nonionic surfactants

Dissipative Particle Dynamics(DPD), one of molecular modeling techniques, was used in this study to simulate the synergistic effect between sulfonate and nonionic surfactants(Dong et al., 2004). DPD originally proposed by Hoogerbrugge and

Koelman in 1992 is a stateof-the-art mesoscale simulation method for the study of complex fluids, such as polymeric or colloidal suspensions. In DPD, molecular cluster in complex fl uid system is denoted as 'bead', which is taken to be an effective soft sphere that acts as a center of mass. Based on Newton's motion equation, each bead interacts with the remaining beads through soft potentials, subjected to dissipative and fluctuating forces. In current simulation, each surfactant molecule was represented by two beads, hydrophilic head bead and hydrophobic tail bead, linked by elastic springs with elastic constant to be 4. 0 KT. Simulationwas carried out in a $20 \times 10 \times 10$ simulation box at density of 3. 0 using 20,000 time steps and a time step interval of 0. 05. The mass of bead and system temperature were all set to be 1. 0 DPD unit.

Table 3 IFT of SLPS and its combination with various surfactants

Number	Surfactant system	IFT (mN/m)
1	0. 3%SLPS	7.62×10^{-2}
2	0. 3%SLPS−01+0. 1%JDQ-1	8.61×10^{-3}
3	0 3%SLPS−01+01%JDQ-2	565×10^{-3}
4	0. 3%SLPS−01+0. 1%1#	2.95×10^{-3}
5	0. 3%SLPS−01+0. 1%4#	6.03×10^{-3}
6	0 3%SLPS−01+01%T1501	981×10^{-3}
7	0. 3%SLPS−01+0. 1%T1402	6.00×10^{-3}
8	0. 3%SLPS−01+0. 1%4-02	5.10×10^{-3}

Simulation(Fig. 2a) shows loose arrangement of SDBS in the interface with cavities that can not be inserted by other free SDBS molecules no matter how large the concentration of SDBS is. However, the TX clusters canenterthecavities(Fig. 2b, c) because of theweaker repulsion between the nonelectric polar head of TX and electric polar head of SDBS than that between two electric polar heads of SDBS. Therefore, combination of TX with SDBS offers a synergistic effect to tremendously diminish the IFT by signifi cantly increasing the surfactant density in the interface.

2. 2. 4. Surfactant formulation design for the S-P flooding system in Gudong oilfield

The crude viscosity was determined to be 45 mPa s and the reservoir temperature to be 68℃ in the southwest Gudong 7 th oilfi eld $Ng5^4$-6^1. The estimated salinity is 6188 mg/L for injection water and 8207 mg/L for produced water. The bivalent ion (Ca^{2+} and Mg^{2+}) in injection water is 189 mg/L.

Based on aforementioned synergistic studies and reservoir conditions, SLPS was formulated as the primary ingredient together with a variety of complementary surfactants of different types and structures. Table 3 shows the oil/water IFT for these formulations. The formulation with the lowest IFT (2.95×10^{-3} mN/m) in Table 3 corresponds to the combination of SLPS and the secondary surfactant, 1#, which is a nonionic surfactant with TX-100 as the basic ingredient. Furthermore, the dynamic oil/ water IFT for both SLPS and 1# has been illustrated in Fig. 3. The results suggests that the IFT is able to reach the ultra-low level at 10^{-3} mN/m as the total surfactant concentrations are in the range of 0.15%~0.65%.

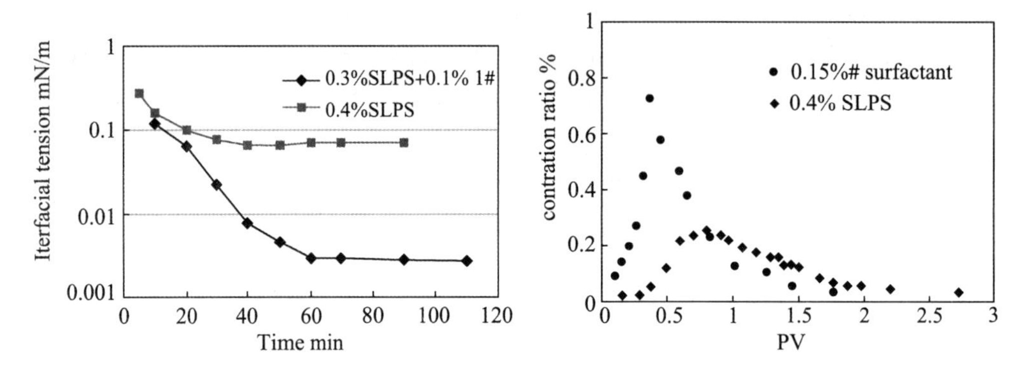

Fig. 3. Dynamic IFT of the surfactant fl ooding system. Fig. 4. Dynamic adsorption of SLPS and 1#.

2.2.5. Chromatographic separation of surfactants in the S-P flooding system in Gudong oilfi eld

An important consideration of S-P flooding system is to avoid possible chromatographic separation of surfactants, which occurs during the movement of the flooding chemicals in the oilfi eld formation since the flooding system is composed of surfactants of various structures. Undoubtedly, chromatographic separation would dramatically decrease the flooding efficacy and oil recovery (Austad et al., 1994; Wang et al., 2005; Sui Xihua et al., 2000).

The dynamic adsorption of the injected SLPS-1# mixture in Fig. 4 suggests that there exists a chromatographic separation phenomenon between SLPS and 1#. The time difference of 0.5 PV between the outgoing concentration peaks of these two surfactants was observed. To decrease the possible surfactant adsorption and chromatographic separation in the oilfield formation, it has been suggested to increase the total concentration of the injected surfactants to be higher than the threshold value of 0.15%. As noted before, the IFT for SLPS-1# is able to achieve the super low level (10^{-3} mN/m) as long as the total concentration of two surfactants is above 0.15% and below 0.65%. Synergism between SLPS and 1# might be weakened at the front edge of the flooding fluid because of the dilution effect by underground water and the adsorption of surfactants inthe formation. Based on above studies, it is recommended to start with 0.45% SLPS+0.15% 1# in the field trial.

3. Polymer design for the S-P flooding system

3.1. Polymer viscosity

Oil recovery can be improved through the increase of oil wash capacity by surfactant, and further promoted through the expansion of swept volume by polymer (Cao et al., 2002). As a result, it is recommended to incorporate polymer into the surfactant flooding system to maximize oil recovery. Four polymers have been carefully selected for the S-P flooding system based on the oil reco-very effects in the previous field applications. Viscosity of the polymers was examined using DVIII viscometer in reservoir conditions, i. e., brine salinity of 6 188 mg/L and temperature of 68℃. The experiment shows high viscosity for all of the four polymers at concentration of 1 500 mg/L, as shown in Fig. 5.

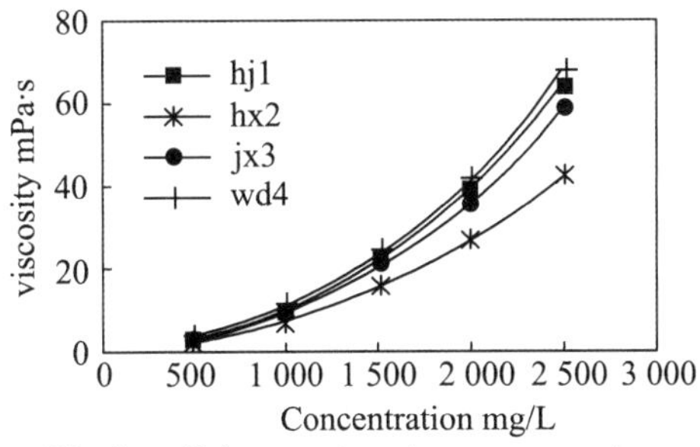

Fig. 5. Polymer viscosity-concentration relationship.

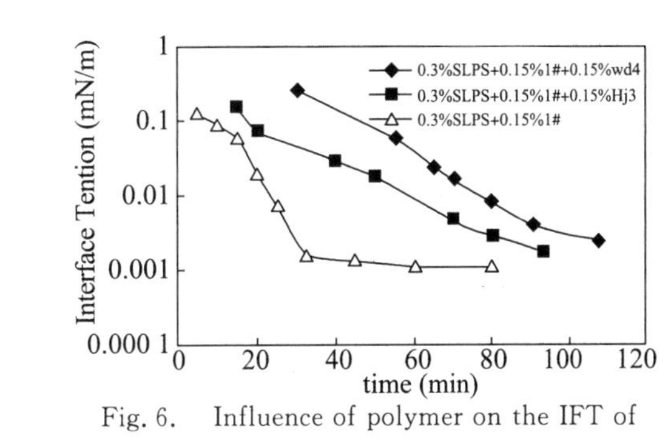

Fig. 6. Influence of polymer on the IFT of surfactants in the fl ooding system.

3.2. Influence of polymer on the interfacial tension

Various polymers (0.15%) have been added into the surfactant flooding system consisting of 0.3% SLPS and 0.1% 1#, and their infl uence on the IFT has been evaluated accordingly. Because of the elevation of system viscosity upon the addition of polymers, the diffusion of surfactant from water phase towards oil/water interface slows down, extending the time for IFT to reach the super low level (Fig. 6). Nevertheless, there is no difference on the order of the lowest IFT for both the S-P flooding system and polymer-free surfactant system, indicating that addition of polymer does not affect surfactants' ability to reduce the oil/water IFT.

3.3. Oil displacement test

Oil displacement tests were performed to investigate the EOR performance of different S-P flooding systems under reservoir conditions: actual formation temperature, pressure, permeability, and the degree of oil saturation. The results are shown in Table 4.

It has been found that an oil recovery enhancement of 18.1% can be achieved by injecting 0.3% SLPS+0.1% 1#+0.15% hj1 of 0.3 PV. Tests also showed that S-P flood out performed polymer flood under the same core conditions and economical cost. Considering the adsorption consumption and costs, the polymer concentration is suggested to be 0.15%~0.2% in the S-P flooding system.

4. Application of the S-P flooding system in the pilot fi eld trial in Gudong oilfi eld

The pilot experiment of the S-P flooding system was carried out in the southeast zone of west Gudong 7th oilfi eld $Ng5^4$-6^1, which is characteristic of oil containing area of 0.94 km^2, and oil reservoir of 277.5×10^4 t in depth of 1261-1294 m. The target zone consists of three oil containing strata (54, 55, and 61) with approximately 34% porosity.

Table 4 EOR comparison between S-P flooding and polymer flooding

No.	Formulation	Injection (PV)	OOIP(%)
Model-11#	0.3%SLPS+0.1%1#+0.15%P	0.3	18.1
Model-18#	0.15%P	0.54	15.2
Model-6#	0.15%P	0.3	11.7

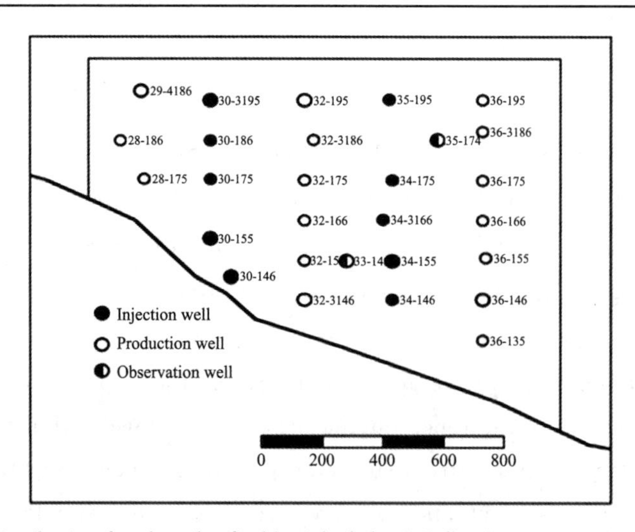

Fig. 7. Well distribution for the pilot fi eld trial of the S-P flooding system in Gudong oilfi eld.

The field trial involved twenty six wells, including sixteen production wells, ten injection wells, and three observation wells, as shown in Fig. 7. The water cut of the field was determined to be 98.2%, and the oil recovery to be 34.4% before the trial. In order to alleviate "fingering" and "crossflow" of the flooding system in the formation, the three-slug injection methodology was established in the field trial. The first slug was the polymer pre-protection slug of 2 000 mg/L polymer solution of 0.05 PV. The second slug was the main slug of 0.3 PV solution consisting of 1 700 mg/L polymer, 0.45% SLPS, and 0.15% 1# and the third one was the polymer post-protection slug of 1 500 mg/L polymer solution of 0.05 PV.

The pilot field trial has showed significant water-cut reduction and oil production enhancement since the injection of the main slug of the S-P flooding system in June 2004. Most current field data show that the water cut has been continuously decreased by 13%, i. e. from 98.2% in year 2004 to 85.2% in year 2007. Until July 2008, the oil production had increased dramatically by 159 t/day, i. e. from 34 t/day to 193 t/day (Fig. 8), The single well's oil production has reached to 2.0×10^4 t and the cumulative oil production had risen by 11.5×10^4 t with the oil recovery enhancement of 4.15%. Fourteen out of sixteen production wells in the field trial have manifested water-cut reduction and oil-production improvement at various degrees. The increase rates from field trials are clearly higher than that of exclusive polymer flooding.

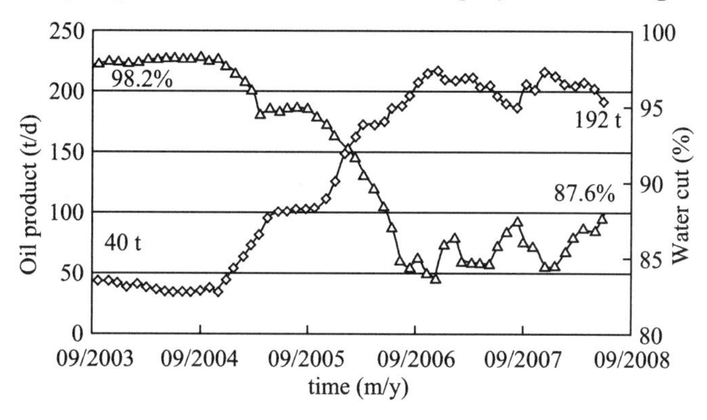

Fig. 8.　Oil production and water cut in the pilot fi eld trial of the S-P flooding system in Gudong oilfi eld.

5. Conclusions

The SP formulation in this work exhibits the efficient diffusion-adsorption properties of ionic surfactants as well as the high surface activity of nonionic surfactants. An excellent synergistic effect has been obtained between the primary surfactant (SLPS) and secondary nonionic surfactant.

The finalized S-P flooding formulation used in the pilot field trial is capable to reduce the IFT to 2.95×10^{-3} mN/m, and improve the oil recovery by 18.1% in laboratory oil displacement tests. Since the injection of themainslug of theS-P flooding system in June 2004, the field trial has demonstrated tremendous decrease of water cut and enhancement of oil production. It has been reported that the accumulative oil-production increase had reached 17. 8×10^4 t in July 2008. The increase rate of oil recovery and decrease rate of water cut from S-P flooding are clearly higher than those using single polymer flooding.

Current studies of surfactant dynamic activities and synergistic effects for the S-P flooding system will provide theoretical guidance for the future design and development of combination flooding systems. Meanwhile, the success of the pilot field trial of the S-P flooding system in Gudong oilfield will build up a solid foundation for the further large-scale application of the system.

References

[1] Al-Hashim, H. S. , Celik, M. S. , Oskay, M. M. , Al-Yousef, H. Y. , 1988. Adsorption and precipitation behaviour of petroleum sulfonates from Saudi Arabian limestone. J. Pet. Sci. Eng. 335-344 October.

[2] Austad, T. , Fjelde, I. , Veggeland, K. , Taugbol , K. , 1994. Physicochemical principles of low tension polymer flood. J. Pet. Sci. Eng. 10, 255-269.

[3] Baoyu, Wang, Xulong, Cao, Qiwei, Wang, Shengwen, Zeng, Xiaohong, Cui, 1994. Phase transition and composition concentration of the produced fl uid in the alkaline-surfactant-polymer fl ooding fi eld trial in Gudong oilfi eld with short-well spacing. Oilfield Chem. 11 (4), 327-330.

[4] Cao, Xulong, Jiang, Shengxiang, Sun, Huanquan, Jiang, Xiaofang, Li, Fang, 2002. Interactions of polymers and surfactants. Appl. Chem. 19 (9), 866-869.

[5] Carrero, Enrique, Queipo, Nestor V. , Pintos, Salvador, Zerpa, Luis E. , 2007. Global sensitivity analysis of alkali-surfactant-polymer enhanced oil recovery processes. J. Pet. Sci. Eng. 58(1-2), 30-42 August.

[6] Dong, F. L. , Li, Y. , Zhang, P. , 2004. Mesoscopic simulation study on the orientation of surfactants adsorbed at the liquid/liquid interface. Chem. Phys. Lett 399, 215.

[7] Hernandez, Clara, Chacon, Larry J. , Anselmi, Lorenzo, 2003. ASP system design for an offshore application in La Salina Field, Lake Maracaibo. SPE Reserv. Evalu. Eng. 6, 147-156 June.

[8] Hongyan, Wang, Benyan, Zhang, Jichao, Zhang, Wenli, Tu, Lenyan, Zhang, Wen, Zhang, 2005. Scale inhibitors for the alkaline-surfactant-polymer flooding system in Zhengli oilfield. Oilfield Chem. 22(3), 252-254.

[9] Kang, Wanli, 2001. Mechanism of the Alkaline-Surfactant-Polymer Flooding Chemicals in Daqing Oilfield, vol. 1. Petroleum Industry Publishers, Inc., Beijing, pp. 4-15.

[10] Myers, Drew, 2009. Surfactant Science and Technology. VCH Publishers, Inc., New York, pp. 6-24.

[11] Salager, J. L., Mongan, J. C., 1979. Optimum formulation of surfactant/water/oil system for minimum interfacial tension or phase behavior. Soc. Pet. Eng. J. 107-115 April.

[12] Sheng, Zhong, Wang, Guoting, 2001. Colloid and Surface Chemistry [M]. Chemical Industry Publishers, Inc., Beijing, pp. 347-351.

[13] Sui, Xihua, Cao, Xulong, Wang, Deshun, Wang, Hongyan, 2000. Chromatographic separation effects of the alkaline-surfactant-polymer flooding system in the west Gudao oilfield. Oil Gas Recov. Technol. 7(4), 1-3.

[14] van der Bogaert, R., Joos, P., 1980. Diffusion-controlled adsorption kinetics for a mixture of surface active agents at the solution-air interface. J. Phys. Chem. 84, 190-194.

[15] Wang, Hongyan, Cao, Xulong, Zhang, Jichao, Tian, Zhiming, 2005. Dynamic adsorption of the Shengli petroleum sulfonate flooding system. Dev. Specialty Petrochem. 6 (3), 28-29.

[16] Xulong, Cao, Huanquan, Sun, Yanbo, Jiang, Xiansong, Zhang, Lanlei, Guo, 2002. Combination flooding field trial in the west Gudao oilfi eld. Oilfi eld Chem. 19(4), 350-353.

[17] Zhang, Luhong, Xiao, Hong, 2007. Optimal design of a novel oil-water separator for raw oil produced from ASP flooding. J. Pet. Sci. Eng. April.

[18] Zhang, Lu., Luo, Lan, Zhao, Sui, 2004. Effect of different acidic fractions in crude oil on dynamic interfacial tensions in surfactant/alkali/model oil systems. J. Pet. Sci. Eng. 189-198 January.

The study and pilot on heterogeneous combination flooding system for high recovery percent of reservoirs after polymer flooding

Cao Xulong, President, RIPED, Sinopec, China
Guo Lanlei, Sinopec, China
Wang Hongyan, Sinopec, China
Wei Cuihua, Sinopec, China
Liu Yu, Sinopec, China

Abstract

The reservoir conditions of ZYQ Ng_3 at Gudao oilfield are characterized as high water cut, high recovery percent and severe heterogeneity. To counteract the challenging reservoir conditions, branched preformed particle gel (B-PPG) was designed and synthesized which has dual features of in-depth profile control and displacement. A heterogeneous combination flooding system was developed that consists of B-PPG, polymer and surfactant. Significant synergistic effect was found among the chemicals in the system. Compared with polymer flooding, heterogeneous combination flooding exhibits higher sweep efficiency and displacement efficiency. It could meet the demands for reservoirs after polymer flooding.

The pilot was conducted with 15 injectors and 10 producers. The geological reserve was 1.23 million tons. After water flooding, polymer flooding and subsequent water flooding, the composite water cut and oil recovery before the pilot were 98.2% and 52.3%, respectively. After applying the heterogeneous combination flooding system in November 2011, daily oil production increased from 3.3 to 79 ton/day. The water cut decreased to 81.3%, with the maximum drop of 16.9%. The enhanced oil recovery was 6.62% and is estimated to be 8.5%. The ultimate recovery could reach 63.6% OOIP.

1. Introduction

With the popularization of polymer flooding technology, the problem of further increasing oil recovery after polymer flooding is concerned. After polymer flooding, remaining movable oil in reservoirs is still up to 50%, therefore, developing a new flooding technology to extract the residual oil is very essential[1-3]. Polymer flooding started to implement in Zhongyiqu Ng3 block, Gudao oil plant of Shengli oilfield in 1994[4], and finished by the end of 2006. By analyzing the data of sealed coring wells in Zhongyiqu Ng3 block, it is found that after polymer flooding, reservoir heterogeneity gets serious and the residual oil becomes more dispersed, so the existing flooding technology is invalid for after-polymer flooding reservoirs. Aimed at these problems, a new chemical flooding system, Heterogeneous Phase Combination Flooding system (HPCF) consisting of B-PPG, polymer and surfactant is designed. B-PPG has not only a cross-linked network but also linear branched chains, so it does not dissolve completely in water or saline solutions, and the mixture is a heterogeneous aqueous solution, that is, a B-PPG suspension. The cross-linked network provides an excellent temperature tolerance mechanism, elastic deformation, and anti-shearing properties, while the water-soluble linear branched chains provide high viscosity and suspension properties. On one hand, B-PPG can interact with polymer to increase viscoelasticity of the system and expand swept volume; on the other, surfactant plays a role in reducing the oil-water interfacial tension and improving the displacement efficiency. Then the HPCF formula is optimized and pilot test is carried out in after-polymer flooding reservoirs in Zhongyiqu Ng3 block, Gudao oil plant of Shengli oilfield, which has achieved remarkable results.

2. Experimental

2.1 Test Material

Surfactant: 30% petroleum sulfonate (SLPS) and 50% nonionic surfactant (GD-3).

HPAM: white powder, viscosity-average molecular weight was about 2.6×10^7 with solid content of 88.2%.

B-PPG: white powder, self-prepared with different particle size.

2.2 Experiment Conditions

Most experiments were conducted under type I reservoir condition of Shengli Oilfield without special mentioned, brine salinity was 6 666 mg/L, $Ca^{2+}+Mg^{2+}=$

129 mg/L. Crude oil was from 11×3 009 well in Zhongyiqu Ng3 block, Gudao oil plant of Shengli oilfield, with viscosity of 51.6 MPa·s at 70℃.

2.3 Equipments

TEXAS 500C spinning drop interfacial tensiometer was used to measure IFT between chemical solution and crude oil.

HAAKE 600 Rheometer was used to measure viscoelasticity parameters such as viscosity, phase angle and storage modulus (G') of test fluid.

Core displacement equipment was used for blocking, fluid dispersion, and displacement test.

2.4 Experimental methods

2.4.1 Rheological measurements

The viscoelasticity of B-PPG suspensions were investigated by oscillatory measurements on HAAKE MARS Rheometer. Samples were squeezed between stainless steel parallel plates and subjected to oscillating rotational deformations with a frequency of 1 Hz and constant stress of 0.1 Pa. Steady-state shear viscosity measurements were performed using the same instrument with a shear rate of 7.34 s^{-1}.

2.4.2 Oil-Water Interfacial Tension measurements

0.4wt% surfactant solution was prepared by brine with salinity of 6 666 mg/L and oilwater interfacial tension was measured by TEXAS 500C spinning drop interfacial tensiometer at 70℃.

2.4.3 Fluid Diversion experiments

Core model: ϕ2.54 cm×30 cm dual sand packed tube models, permeability was 1.0 μm^2:3.0 μm^2 measured by gas. After injecting water, 1 500 mg/L B-PPG+1 500 mg/L HPAM was injected and the change of fluid output the two tubes was recorded.

2.4.4 Displacement experiments

Core model: ϕ2.54 cm × 30 cm dual sand packed tube models, permeability was 1.0 μm^2:3.0 μm^2 measured by gas. After injecting water, crude oil from Zhongyiqu Ng3 block, Gudao oil plant was injected to core models and aged for 24 h before use. Firstly, water was injected until the water-cut was 98%; then 0.3 PV HPAM solution was injected, followed with water flooding until overall water cut was 98%; finally 0.3 PV HPCF system was injected followed with water flooding until overall water cut was 98%.

3. Formula design of hpcf system

3.1 Performance Evaluation of B-PPG

3.1.1 Viscoelasticity Property

Good viscoelastic property of B-PPG is important for ensuring that the displacement system can expand swept volume and migrate. Eight B-PPGs with different viscoelasticities were compared and the result is listed in Table 1. B-PPG-1 and B-PPG-2 had no viscoelasticity and B-PPG-3～B-PPG-8 had good viscoelasticities. For example, storage modulus (G') of B-PPG-6 was 5.42 Pa, and phase angle (δ) was 33.2°, indicating that the B-PPG-6 shows a good viscoelastic property.

Table1　The viscoelastic parameters of B-PPG

sample	G'/Pa	G''/Pa	δ/°
B-PPG-1	—	0.396	90.0
B-PPG-2	—	0.314	90.0
B-PPG-3	1.846	1.736	43.2
B-PPG-4	0.513	0.895	60.2
B-PPG-5	0.461	0.815	60.5
B-PPG-6	5.42	3.55	33.2
B-PPG-7	0.956	1.306	55.4
B-PPG-8	0.186	0.616	73.2

3.1.2 Profile Control Property

A parallel connection of two sandpacked cores with the permeability ratio of 3∶1 was used to study fractional flow of B-PPG after polymer flooding, and the fractional flow curve of B-PPG suspension in porous media is shown in Figure 1. It can be seen that, the production ratio of the higher permeability tube and lower permeability tube before B-PPG flooding was 95.8%∶4.2%. It turned to 2%∶98% after B-PPG suspension was injected, which was obvious "fluid diversion". Also, the fluctuation in fractional flow curve proved that the migration of B-PPG particles through the porous media was a dynamic process of plugging and deforming to pass through pore throats at the same time.

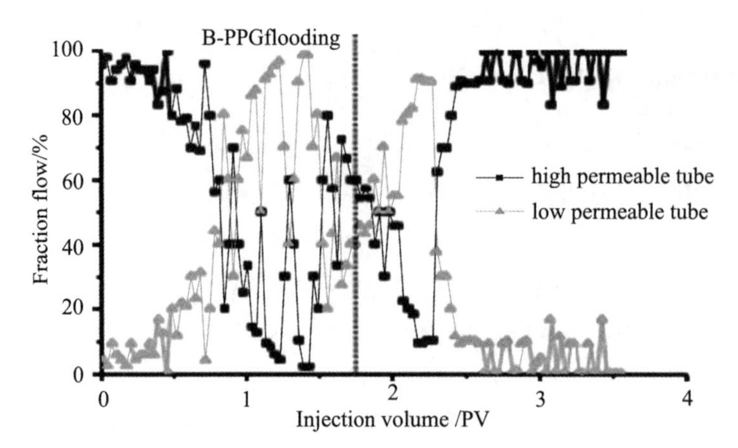

Fig. 1 The fractional flow curve of the two sandpacked tube during B-PPG flooding

3. 2 Selection of Surfactant

The ability of reducing interfacial tension of oil and water to ultra-low is an important criterion for selection of surfactants. The interfacial activity of petroleum sulfonate SLPS could be improved by using auxiliary surfactant with appropriate molecular structure and the oil-water interfacial tension could be greatly reduced by their synergistic effects[8]. It was found that the interfacial tension could reach an ultra-low value when SLPS and GD-3 were compounded in a wide range.

At the ratio of 1 : 1, the compounded system could reduce interfacial tension to ultralow with the concentration of 0. 1% ~ 0. 6%. Considering the adsorption and dilution by formation water, the surfactant in HPCF system formula was confirmed as 0. 2%SLPS+0. 2%GD-3. 0

3. 3 Interaction between B-PPG and Polymer

In HPCF system, polymer has the ability of mobility control and suspending B-PPG, so the interaction between B-PPG and polymer were investigated. The viscoelastic properties of different system are listed in table 2.

Table 2 The viscoelastic properties of different system(70℃, TDS=6 666 mg/L)

Sample	concentration/mgL^{-1}	η/mPa s	G′/Pa	G″/Pa	δ/°
HPAM	1 500	23. 7	0. 087	0. 23	69. 5
B-PPG-6	1 500	6. 7	0. 35	0. 2	29. 6
HPAM+B−PPG−6	1 500+1 500	68. 3	1. 015	0. 613	31. 1

In table 2, for HPAM G″ was larger than G′ and δ was 69. 5°, indicating viscosity dominated the system. However, for B-PPG-6, G″ was smaller than G′ and δ was

29.6°, indicating elasticity dominated the system. When HPAM and B-PPG were compounded at the ratio of 1 : 1 with the concentration of 3 000 mg/L, apparent viscosity and viscoelasticity of the compounded system increased greatly, exhibiting excellent viscoelastic character.

3.4 Results of Displacement Tests

The result of oil displacement experiment by HPCF is shown in Figure 2. It can be seen that the ultimate recovery was 79.9% and its oil recovery increment was 13.6% after polymer flooding.

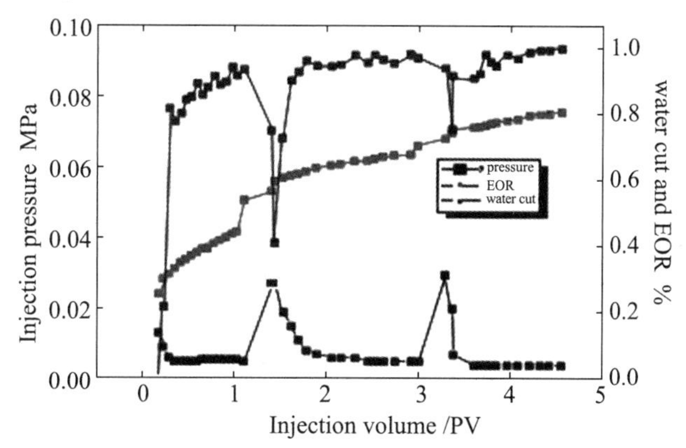

Figure 2 Displacement curve of HPCF system after polymer flooding

4. Filed application of hpcf system

Pilot test area is located in the southeast of Zhongyiqu Ng3 block, Gudao oil plant, with total oil-bearing area of 0.275 km^2 and geological reserves of 1.23 million tons. For the test area, permeability variation coefficient is 0.538, porosity is 33% and the air permeability is 1.5～2.5 μm^2; formation oil viscosity is 46.3 mPa · s, formation water salinity is 5 923 mg/L and original formation temperature is 70℃.

After water flooding, polymer flooding and subsequent water flooding, the comprehensive water cut and recovery percent of the pilot test area is 98.2% and 52.3%, respectively. HPCF system was injected in November 2011. The averageinjection concentrationof polymer and B-PPG were both 1 296 mg/L, and the viscosity of injection fluid at wellhead was 50 mPa · s; the average injection concentrationof petroleum sulfonate and auxiliary surfactant were both 0.2%. By December 2015, project design injection was completed.

Great success has been achieved after HPCF was implemented, and effects were as follows:

(1) Water cut decreasing and oil increasing obviously.

The composite water cut decreased from 98.2% to 81.3% with the max decrease rate of 16.9%; and the daily oil production increased from 3.3 t/d to 79 t/d, as shown in Figure 3. The project had enhanced oil recovery by 6.62%, and the oil recovery was expected to be improved by 8.5% with the ultimate recovery factor of 63.6%.

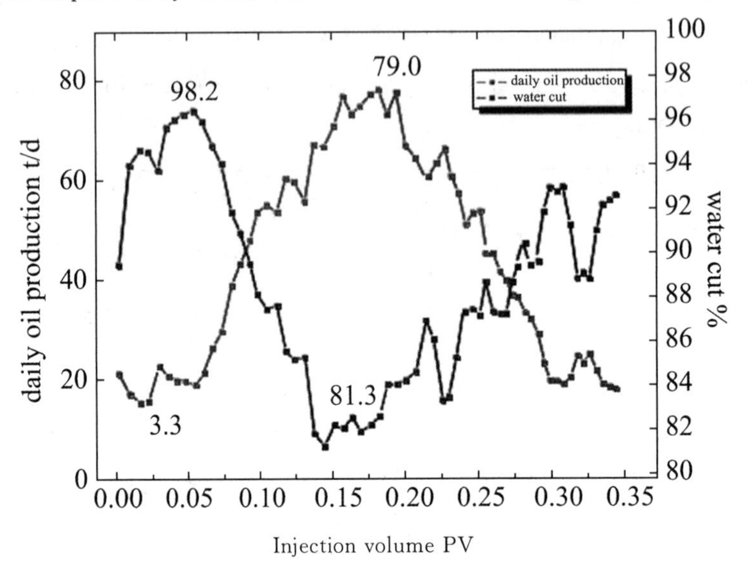

Figure 3 Theproduction curveof HPCF system in Zhongyiqu Ng3 block, Gudao oil plant

(2) Oil pressure of injection well increasing obviously.

The average injection pressure of injection well in pilot test area was 7.2 MPa before HPCF implemented. After HPCF was carried out, the injection pressure increased greatly, with the highest value of 11.5 MPa. This is because that B-PPG has strong plug ability, leading to formation seepage resistance increasing and injection pressure rising.

(3) Vertical injection profile improving obviously.

It can be seen from continuous monitoring data of injection profile (table 3) that the vertical injection profile was changing. Before HPCF implemented, the Ng33 layer mainly absorbed water, while 8 months later, the relative water intake capacity of Ng33 layer decreased to 47.69% and that of Ng34 layer increased to 52.32%, indicating that the vertical injection profile is improved. However, 5 months later the relative water intake capacity of Ng33 layer increased to 64.33% and that of Ng34 layer decreased to 35.67%, which due to the migration of B-PPG in reservoir. Thus, the continuous monitoring data of injection profile also proved that the migration of B-PPG through reservoir was a dynamic process of plugging and deforming to pass through pore throats at the same time, resulting in double effects of profile control and displacement.

Table 3 Thestatistical data of injection profile for 11-311

Date	Relative water intake capacity%	
	$Ng3^3$	$Ng3^4$
2011.3.22	81.53	18.47
2011.11.28	47.69	52.32
2012.3.7	40.98	59.02
2012.5.15	64.33	35.67
2012.11.13	44.76	55.24
2013.3.6	66.25	33.75

5. CONCLUSION

(1) HPCF system consisting of B-PPG, polymer and surfactant was designed. These chemical agents had synergistic effect and HPCF system could further increase oil recovery by 13.6% after polymer flooding.

(2) Pilot test of HPCF was carried out in after-polymer flooding reservoirs in Zhongyiqu Ng3 block, Gudao oil plant of Shengli oilfield, which has achieved remarkable results: oil pressure of injection well increased obviously; vertical injection profile improved obviously; water cut decreased and oil increased obviously, which proved that increasing oil recovery by HPCF technology in after-polymer flooding reservoirs was successful and effective.

第七章　论文综述

pH 调控蠕虫状胶束研究进展

陈维玉[1]，曹绪龙[2]，祝仰文[2]，曲广淼[2]，丁　伟[1]

（1. 东北石油大学化学化工学院，黑龙江大庆　163318；

2. 中国石化胜利油田分公司地质科学研究院，山东东营　257015）

摘要：介绍了近年来 pH 响应型蠕虫状胶束的研究进展，包括 pH 响应型阳离子表面活性剂体系、pH 响应型阴离子表面活性剂体系、CO_2 响应型表面活性剂体系以及 pH 响应型两性表面活性剂体系，并对其发展前景进行了展望。pH 响应型蠕虫状胶束具有易于控制、过程可逆、实现方便等一系列优点。新型耐温抗盐 pH 响应型蠕虫状胶束能满足驱油剂的黏度要求，具有很好的降低界面张力的能力，而且还可以减少在地层的吸附损失，在提高采收率领域更具有研究价值。

关键词：表面活性剂　pH 响应型蠕虫状胶束　黏弹性

表面活性剂在水溶液中的浓度达到一定值后，可以通过自身缔合发生自组装行为，形成囊泡、液晶、球状胶束、棒状胶束、蠕虫状胶束[1-2]甚至巨胶束[3-4]等有序结构。促使自组装发生的动力不是共价键或离子键，而是范德华力、疏水作用、氢键作用以及静电作用等[5-6]。

蠕虫状胶束被广泛应用于催化化学[7]、柔性材料制备[8]、生物医药[9]、护肤品[10-11]及油田化学[12]等领域，尤其是 pH 响应型的蠕虫状胶束，在油田自转向酸化液方面已经取得了一定的应用成果[13]。智能型蠕虫状胶束是近年来的研究热点。智能型蠕虫状胶束是指在构成表面活性剂的分子中引入具有环境刺激响应的功能基团，刺激因素包括浓度、温度、紫外光、添加剂、pH 等[14-18]。其中 pH 是一种易于控制的环境因素，若能通过改变体系 pH 而实现蠕虫状胶束的制备，并使该过程可逆，将具有十分重要的意义[19-20]。

pH 响应型蠕虫状胶束体系中含有 pH 刺激响应基团，主要包括—COOH、—NH_2、—ArOH、—OPO_3H_2 和—PO_3H_2，这些基团与酸或碱发生作用，导致表面活性剂分子的 HLB 值或反离子的结构发生变化，使蠕虫状胶束实现破坏—重筑的可控转变[21-22]，宏观表现主要是黏弹性发生改变，微观方面主要是分子的聚集状态发生改变。近年来随着各国相关领域研究者的不懈努力，在通过 pH 调控制备蠕虫状胶束方面取得了很大进展。

1 pH 响应型阳离子表面活性剂体系

目前研究最多的阳离子表面活性剂蠕虫状胶束体系，主要是由氯化十六烷基吡啶及十六烷基三甲基溴化铵(CTAB)、十六烷基三甲基氯化铵(CTAC)、十八烷基三甲基氯化铵(OTAC)等季铵盐参与构筑。

Lin 等[23]采用 CTAB 和邻苯二甲酸(PPA)制备了一种 pH 响应型蠕虫状胶束体系。研究结果表明，pH 为 3.90～5.35 时，黏度随 pH 的增加而降低。pH 的增大导致 PPA 的质子化程度降低，使体系在柱状胶束和虫状胶束之间转变，pH 对体系的影响如图 1 所 示。Verma 等[24]研制了由 CTAB 和邻氨基苯甲酸(AA)构筑的蠕虫状胶束体系。pH 的变化主要改变了 AA 的净电荷数，导致其与 CTAB 的作用程度有所不同，使胶束的长度和缠绕程度发生改变。CTAB 与邻苯二甲酸氢钾也可制备 pH 响应的蠕虫状胶束体系，pH 的变化使表面活性剂分子的聚集状态发生改变，具体过程为球状胶束—短棒状胶束—虫状胶束。Lin 等[25]采用 CTAB 和 n-癸基磷酸(DPA)制备了自组装体系，鉴于 DPA 有 2 个 pK_a 值，根据中和值 α[26-28]的不同，借助透射电镜即可观察到蠕虫状胶束的存在。

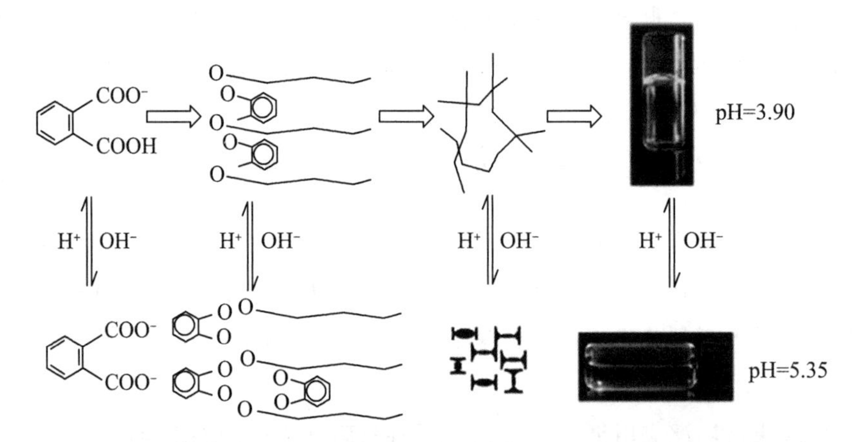

图 1 随着 pH 的改变，体系从虫状胶束到短柱状胶束的转变

Chu 等[29]采用不溶的长链叔胺与马来酸构建了一种蠕虫状胶束。pH 为 6.20～7.29 时，溶液显示了从牛顿流体到黏弹溶液的可逆变化。由于季铵头基的部分质子化，在适当的 pH 下，蠕虫状胶束的长链相互缠结，形成网状结构，具有很高的黏弹特性。长链叔胺 UC_{22}AMPM 与马来酸体系的分子结构如图 2 所示。

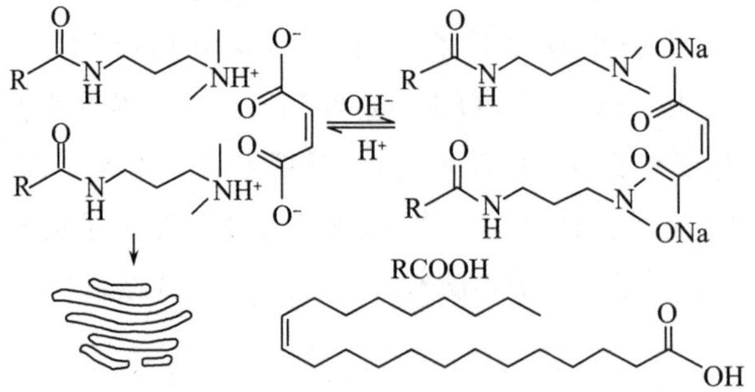

图 2 UC_{22}AMPM 与马来酸体系的分子结构

在pH响应型蠕虫胶束的构筑过程中，当pH发生变化时，有机酸与碱生成有机酸盐，有机酸盐的水溶性发生变化，能够压缩表面活性剂头基面积，排列更紧密，从而改变临界堆积参数值，促进胶束增长。

2　pH响应型阴离子表面活性剂体系

目前阴离子蠕虫状胶束体系方面的报道较少，该类体系主要是基于油酸钠(NaOA)、氨基酸、胆酸等构筑而成[30]，具有价格低廉、来源广泛、表面活性高、易降解等优点，是一种绿色环保的阴离子表面活性剂。

Lu等[31]采用NaOA和氯化钠构建了pH响应型可逆蠕虫状胶束体系。研究结果表明，pH为9.43时，表现出牛顿流体特征；pH为9.53～10.18时，溶液体系具有黏弹性；pH大于10.18时，出现牛顿平台，体系呈剪切变稀特性，证明有蠕虫状胶束的生成。张世鑫等[32]采用NaOA与盐酸三乙胺构建了一种随pH可逆的蠕虫状胶束体系，发现在很窄的pH范围内，黏度可由3 mPa·s增至27 649 mPa·s，具有典型的剪切变稀现象，这是蠕虫状胶束特有的性质。

Baccile等[33]通过光学显微镜和红外傅立叶光谱仪等研究了蠕虫状胶束的形成及胶束间的相互作用。采用了经发酵得到的酸化后的槐糖脂，因为分子结构中含有多个—COOH，随着—COOH离子化程度的不同，可以形成不同pH的体系。Stefan等[34]采用小角中子散射(SANS)、小角X射线散射(SAXS)、动态光散射(DLS)等分析手段，对由胆汁盐、磷脂、辛酸、辛酸甘油酯及NaCl构建的蠕虫状胶束体系进行了研究。当pH为2～4时，体系从蠕虫状胶束转变为囊泡。该结果对脂肪消化释放短链脂肪酸这一过程具有指导意义，该研究成果有望应用于药物运输领域。Han等[35]采用芥酸钠(NaOEr)制备蠕虫状胶束，并对体系的相行为和流变行为进行了对比研究。研究结果表明有机助溶盐的加入可以降低NaOEr的Krafft点的温度，有利于蠕虫胶束的构筑。

Leana等[36]构筑了具有生物适应性的蠕虫状胶束体系，采用具有pH响应性的苯丙氨酸的基团取代胆酸中的羟基，得到2种衍生物构筑的胶束体系。随着pH的改变，可实现从球状结构至蠕虫状胶束的转变，高浓度形成凝胶。

当表面活性剂本身对pH具有响应特性时，体系中加入无机盐对胶束的聚集方式有所影响。根据Israelachvili[37]理论，加入的无机盐离子可以压缩表面活性剂头基的扩散双电层和水化层，屏蔽头基之间的静电斥力，减小头基占据面积，从而提高临界堆积参数值，促进蠕虫状胶束的线性增长。另一方面无机盐的加入使得表面活性剂的临界胶束浓度大幅降低，使胶束的聚集数增加，从而促进胶束的生长。

3　CO_2响应型表面活性剂体系

以pH作为触发剂制备蠕虫状胶束，方法简单、便捷，但也有严重的缺点，首先调节

pH 所应用的酸或碱必须等物质的量比,其次每次循环后生成的无机盐副产物都存在于体系中,这些盐的存在会影响虫状胶束的构筑,甚至破坏蠕虫状胶束的结构,而且这些化学原料的成本较高,三废处理的环境成本和经济成本也较高。采用 CO_2 气体作为 pH 调节剂则可以避免以上缺点。CO_2 气体具有生物相容性,易于再生。有研究者采用超临界 CO_2 气体构筑了蠕虫状胶束[38-39]。

Zhang 等[40]制备了一种受 CO_2 气体调控的可逆型阴离子 N,N,N’,N’-四甲基丙二胺(TMPDA)蠕虫状胶束体系。CO_2 气体可以导致叔胺的质子化,使体系形成类似双子结构的蠕虫状胶束;当 CO_2 被移出体系后,体系恢复为水样的流体状态。

Zhang 等[41]研究了由单一组分构建的蠕虫状胶束体系,避免多组分构筑的蠕虫状胶束体系经过多孔材质时发生色谱分离效应。通过透射电镜观察发现,体系在球形结构、囊泡、蠕虫状胶束、网状结构之间进行转变。该小组还研究了一种由长链叔胺制备的 CO_2 响应型蠕虫状胶束体系,随着 CO_2 气体的通入和移除,体系由均匀透明的黏弹溶液变成黏度较低的乳浊液。

4 pH 响应型两性表面活性剂体系

两性表面活性剂 C_nDMAO[42-43]在 pH 为 7 时分子的净电荷数为零,表面活性剂头基之间静电斥力微弱,正是由于微弱的静电斥力,在无任何添加剂的情况下自发组装形成蠕虫状胶束。Rathman 等[44]发现中和值 α 为 0.5 时,红外光谱中 CH_2 和 CNO 弯曲带消失,CH 伸缩频率最小化,表明了分子间氢键的形成,此时蠕虫状胶束的长度达到最大。通过改变体系的 pH,能够实现油烯基二甲基氨基氧化物从蠕虫状胶束到囊泡的形貌控制。

Brinchi 等[45]合成了 N-十二烷氧基苯基 N,N-二甲基氧化铵(pDoAo),并由 pDoAo 和 CH_3SO_3Na 制备了蠕虫状胶束体系。采用动态流变仪对该物质的黏度进行了测定,胶束的长度和黏度随 pH 的变化增大。Ghosh 等[46]采用氨基酸类的两性表面活性剂和阴离子表面活性剂十二烷基磺酸钠构筑了蠕虫状胶束体系,并利用荧光探针技术对分子在磷酸盐缓冲溶液中的聚集方式进行了说明。

Chen 等[47]通过两性表面活性剂和马来酸酐构建了一种用于提高油田采收率的蠕虫状胶束体系。研究结果表明,当表面活性剂的质量浓度为 2%、pH 为 8 时,体系的黏度值高达 4.5×10^5 mPa · s。驱油效率对比试验结果表明,质量分数为 0.4%的该两性表面活性剂体系的 pH 为 6~8 时,驱油效率远大于水驱和聚驱。

5 结语

pH 响应型蠕虫状胶束作为智能胶束的一个分支,具有构筑简单、可循环多次等优点,尤其是 CO_2 气体响应型蠕虫状胶束,更具有环境友好、构筑成本低廉的优点。越来越多的研究者采用新型 Gemini 表面活性剂构筑蠕虫状胶束，但是如何引入合适的响应分子，分子中 pH 响应基团的数量、分子的空间位阻效应对蠕虫状胶束的构筑有何影响，构筑蠕虫状胶束时是否有规律可循等问题会给研究者带来挑战。采用分子模拟和实验相结合的方法可以更好地研究影响蠕虫状胶束构筑的因素。在油田化学提高采收率领域，新型的 pH 调控的耐温抗盐型阴离子表面活性剂蠕虫状胶束体系具有很好的实际应用价值，响应条件易于实现，能满足聚驱的黏度要求，具有表面活性剂降低界面张力的性能，能有效提高采收率。pH 响应型阴离子蠕虫胶束具有更好的发展前景，可以减少地层的吸附损失，保持胶束的浓度和黏度，是一个颇具价值的研究方向。

参考文献

[1] Oelschlaeger C, Willenbacher N. Mixed wormlike micelles of cationic surfactants: effect of the cosurfactant chain length on the bending elasticity and rheological properties[J]. Colloids and Surfaces A: Physicochemical and Engineering Aspects, 2012,406:31-37.

[2] Lehn J M. Toward self-organization and complex matter[J]. Science, 2002, 295:2400-2403.

[3] Rodrigues R K, Ito T H, Sabadini E, et al. Thermal-stability of mixed giant micelles of alkyltrimethylammonium surfactants and salicylate[J]. Journal of Colloid and Interface Science, 2011, 364:407-412.

[4] Cates M E, Fielding S M J. Rheology of giant micelles[J]. Advances in Physics, 2006, 55:799-879.

[5] Lowik D W P M, Leunissen E H P, Heuvel M van den, et al. Stimulus responsive peptide based materials Chem[J]. Soc Rev, 2010, 39:3394-3412.

[6] Pisarcik M, Polakovicova M, Pupak M. Biodegradable gemini surfactants. Correlation of area per surfactant molecule with surfactant structure[J]. Journal of Colloid and Interface Science, 2009, 329:153-159.

[7] Tokarev I, Minko S. Multiresponsive hierarchically structured membranes: new, challenging, biomimetic materials for bio-sensors, controlled release, biochemical gates, and nanoreactors[J]. Adv Mater, 2009, 21:241-247.

[8] Bajpai A K, Shukla S K, Bhanu S, et al. Responsive polymers in controlled drug delivery[J]. Prog Polym Sci,2008,33:1088-1118.

[9] Lee K Y, Mooney D J. Hydrogels for tissue engineering[J]. Chem Rev, 2001, 101(7):1869-1879.

[10] Pucci A, Bizzarri R, Ruggeri G. Polymer composites with smart optical properties[J]. Soft Matter, 2011, 7:3689-3700.

[11] Yoon B, Huh J, Ito H. Smart self-adjustment of surface mi-celles of an amphiphilic block copolymer to nanoscopic pattern boundaries[J]. Adv Mater, 2007, 19:3342-3348.

[12] Morvan M, Degre'G, Leng J, et al. New viscoelastic fluid for chemical EOR [J]. SPE 121675,2009.

[13] Zeiler C, Alleman D, Qu Q. Use of riscoelastic surfactant-based diverting agents for acid stimulation[J]. SPE 90062, 2004.

[14] Stanway R, Sproston J L, El-Wahed A K. Applications of electro-rheological fluids in vibration control: a survey[J]. Smart Mater Struct, 1996, 5:464-482.

[15] Tsuchiya K, Orihara Y, Kondo Y. Control of viscoelasticity using redox reaction[J]. Am Chem Soc, 2004, 126:12282-12283.

[16] Sakai H, Orihara Y, Kodashima H. Photoinduced reversible change of fluid viscosity[J]. Am Chem Soc, 2005, 127:13454-13455.

[17] Lin Y, Cheng X, Qiao Y. Creation of photo-modulated multi-state and multi-scale molecular assemblies via bina-ry-state molecular switch[J]. Soft Matter, 2010, 6:902-908.

[18] Eastoe J, Vesperinas A. Self-assembly of light-sensitive surfactants[J]. Soft Matter, 2005, 1:338-347.

[19] Brinchi L, Germani R, Profio P D. Viscoelastic solutions Formed by worm-like micelles of amine oxide surfactant[J]. Colloid Interface Sci, 2010, 346:100-106.

[20] Ali M, Jha M, Das S K. Hydrogen-bond-induced micro-structural transition of ionic micelles in the presence of neutral naphthols: pH dependent morphology and location of surface activity[J]. J Phys Chem B, 2009, 113:15563-15571.

[21] Graf G, Drescher S, Meister A. Self-assembled bolaamphi-phile fibers have intermediate properties between crystalline nanofibers and wormlike micelles: formation of viscoelastic hydrogels switchable by changes in pH and salinity[J]. J Phys Chem B, 2011, 115:10478-10487.

[22] Ghosh S, Khatua D, Dey J. Interaction between zwitterionic and anionic surfactants: spontaneous formation of zwitanionic vesicles[J]. Langmuir, 2011, 27: 5184-5192.

[23] Lin Y Y, Xue H N, Huang J B. Microstructures and rheologi-cal dynamics of viscoelastic solutions in a catanionic surfac-tant system[J]. Journal of Colloid and

Interface Science, 2009, 330(2):449-455.

[24] Verma G, Aswal V K, Hassan P. pH-responsive self-assembly in an aqueous mixture of surfactant and hydrophobic amino acid mimic[J]. Soft Matter, 2009, 5(15):2919-2927.

[25] Lin Y Y, Han X, Cheng X H. pH-regulated molecular self-assemblies in a cationic-anionic surfactant system: from a"1-2"surfactant pair to a"1-1"surfactant pair [J]. Langmuir, 2008, 24:13918-13924.

[26] Ikeda S, Tsunoda M, Maeda H. The effects of ionization on micelle size of dimethyldodecylamine oxide[J]. Colloid Inter-face Sci, 1979, 70:448-455.

[27] Kaimoto H, Shoho K, Sasaki S. Aggregation numbers of dodecyldimethylamine oxide micelles in salt solutions[J]. J Phys Chem, 1994, 98:10243-10248.

[28] Maeda H, Kakehashi R. Effects of protonation on the thermo-dynamic properties of alkyl dimethylamine oxides[J]. Adv Colloid Interface Sci, 2000, 88:275-293.

[29] Chu Z L, Feng Y J. pH-switchable wormlike micelles[J]. Chemical Communications, 2010, 46(47): 9028-9030.

[30] Travaglini L, D'Annibale A, Gregorio di M H. Between pep-tides and bile acids: self-assembly of phenylalanine substi-tuted cholic acids[J]. Journal of Physical Chemistry B, 2013, 117(31): 9248-9257.

[31] Lu H, Shi Q, Huang Z, et al. pH-responsive anionic worm-like micelle based on sodium oleate induced by NaCl[J]. J Phys Chem B, 2014, 118(43): 12511-12517.

[32]Zhang S X, Cheng L J, Wei S B. Effect of pH on the rheolog-ical properties of wormlike micelles formed by sodium oleate/triethylamine hydrochloride[J]. Fine Chemicals, 2014, 31(8):969-972.

[33] Baccile N, Babonneau F, Jestin J. Unusual, pH-induced, self-assembly of sophorolipid biosurfactants[J]. ACS Nano, 2012, 6(6):4763-4776.

[34] Salentinig S, Phan S, Darwish T A. pH-responsive micelles based on caprylic acid[J]. Langmuir, 2014, 30(2): 57296-57303.

[35] Han Y X, Feng Y J, Sun H Q. The effect of tail length of anionic surfactant on rheological behaviors of wormlike micel-lar solutions[J]. Journal of Physical Chemistry B, 2011, 21(115): 6893-6902.

[36] Travaglini L, Da A, Gregorio di M C. Between peptides and bile acids: self-

assembly of phenylalanine substituted cholic acids[J]. Journal of Physical Chemistry B, 2013, 117(31):9248-9257.

[37] Israelachvili J N, Mitchell D J, Ninham B W J. Theory of self-assembly of hydrocarbon amphiphiles into micelles and bilayers[J]. J Chem Soc, Faraday Trans 1, 1976, 72:1525-1568.

[38] Yan Q, Zhou R, Fu C. CO_2-responsive polymeric vesicles that breathe[J]. Angew Chem Int Ed, 2011, 50:4923-4927.

[39] Han D, Tong X, Boissière O. Two-way CO_2-switchable triblock copolymer hydrogels[J]. Macro Molecules, 2012, 45(18):7740-7745.

[40] Zhang Y, Feng Y, Wang Y, et al. CO^2-switchable viscoe-lastic fluids based on a pseudogemini surfactant[J]. Lang-muir, 2013, 29(13):4187-4192.

[41] Zhang Y, Feng Y, Wang J. CO_2-switchable wormlike mi-celles[J]. Chem Commun, 2013, 49:4902-4904.

[42] Maeda H, Tanaka S, Ono Y. Reversible micelle-vesicle conversion of oleyldimethylamine oxide by pH changes [J]. J Phys Chem B, 2006, 110, 12451-12458.

[43] Kawasaki H, Souda M, Tanaka S. Reversible vesicle forma-tion by changing pH[J]. J Phys Chem B, 2002, 106:1524-1527.

[44] Rathman J F, Christian S D. Determination of surfactant activities in micellar solutions of dimethyldodecyl amine oxide[J]. Langmuir, 1990, 6(2):391-395.

[45] Brinchi L, Germani R, Profio P D. Viscoelastic solutions formed by worm-like micelles of amine oxide surfactant[J]. Colloid Interface Sci, 2010, 346:100-106.

[46] Ghosh S, Khatua D, Dey J. Interaction between zwitterionic and anionic surfactants: spontaneous formation of zwitanionic vesicles[J]. Langmuir, 2011, 9(27):5184-5192.

[47] Chen I, Yegin C, Zhang M, et al. Use of pH-responsive amphiphilic systems as displacement fluids in enhanced oil recovery[J]. SPE Journal, 2014, 19(6):1035-1046.

Progress of research on pH-responsive wormlike micelles

Chen Weiyu[1], Cao Xulong[2], Zhu Yangwen[2], Qu Guangmiao[2], Ding Wei[1]

(1. Chemistry and Chemical Engineering Institute of Northeast Petroleum University, Daqing, Heilongjiang 163318;

2. Geoscience Research Institute of Sinopec Shengli Oilfield Company, Dongying, Shandong 257015)

Abstract: The progress on the research on pH-responsive wormlike micelles including pH-responsive cationic surfactant system, pH-responsive anionic surfactant system, CO_2-responsive surfactant system and pH-responsive ampholytic surfactant system in recent years is introduced, and the future development of pH-responsive wormlike micelles is discussed in this paper. pH-responsive wormlike micelles have several advantages like being easy to control, being reversible in process, and convenient application, etc. New heat resistant and salt tolerant pH-responsive wormlike micelles can satisfy the requirement for the viscosity of oil displacing agent, are capable of reducing the interfacial tension well enough, and can reduce formation adsorption. They have the value for studying how to enhance oil recovery.

Key words: surfactant; pH-responsive wormlike micelle; visco-elasticity

索尔维在中国的香兰素生产设施投产

索尔维公司于 2015 年 11 月 23 日宣布，在中国江苏省镇江市的香兰素生产设施业已投产，从而提高了其香兰素的生产能力 40%。新的生产预计将帮助满足亚洲市场快速增长的需求，增强了索尔维公司在中国和亚洲的地位道。索尔维公司的香味性能部门也在美国路易斯安那州巴吞鲁日和法国 Saint-Fons 拥有香兰素设施。索尔维公司是世界上最大的香兰素生产商。

Chemical Week, 2015-11-24

胜利油田 CO_2 驱油技术现状及下步研究方向

曹绪龙,吕广忠,王 杰,张 东,任 敏

(中国石化胜利油田分公司,山东东营 257000)

摘要:CO_2 驱是提高低渗透油藏采收率和减少温室气体排放双赢的主要技术。针对胜利油田低渗透油藏 CO_2 驱面临的混相难、易气窜、波及系数低等技术难题,采用物理模拟和数值模拟相结合的方法,明确了超前注 CO_2 混相驱的开发机理,形成了特低渗透油藏的超前注 CO_2 混相驱开发技术,现场应用后增产效果明显,单井日产油增加了5倍。提出降低混相压力的原理和技术思路,研发了降低混相压力体系,降幅可达22%。分析胜利油田 CO_2 驱规模应用面临的挑战及对策,提出了深化 CO_2 驱提高石油采收率的相态理论、研发低成本扩大 CO_2 驱波及体积技术、发展 CO_2 非完全混相驱、气窜通道描述与预警等 CO_2 驱的发展方向,为油田实现 CO_2 驱规模应用提供技术支撑。

关键词:超前注气;特低渗透;降低混相压力;CO_2 混相驱;研究方向

中图分类号:TE357.42 文献标志码:A

Present situation and further research direction of CO_2 flooding technology in Shengli Oilfield

Cao Xulong, Lyu Guangzhong, Wang Jie, Zhang Dong, Ren Min

(Sinopec Shengli Oilfield Company, Dongying, Shandong 257000, China)

Abstract: CO_2 flooding is effective for enhancing the oil recovery in low permeability reservoir and reducing the greenhouse gas emissions. In order to solve the technical problems of difficult miscible phase, easy gas channelling and low sweep coefficient for CO_2 flooding in low permeability reservoir in Shengli Oilfield. By the combination of physical and numerical simulation, the development mechanism of the CO_2 injection miscible flooding long in advance is clarified, and the comprehensive techniques for extra low permeability reservoir is formed. After field application, the stimulation effect is obvious, the daily production of oil per well increase by 5 times. The principle and technical idea of reducing the miscibility pressure are put forward, and the system of reducing the miscible pressure system is developed, which can make the pressure decrease by up to 22%. The challenge and countermeasure faced by scale application of CO_2 flooding in Shengli Oilfield are analyzed, and the development directions of CO_2 flooding are proposed, such as deepening the phase state theory of oil recovery enhanced by CO_2 flooding, developing CO_2 flooding technology with expanded sweep volume at low cost, developing incomplete CO_2 miscible flooding, and description and early warning of gas channeling. All these provide technical support for oil field to realize scale application of CO_2 flooding.

Key words: advanced gas injection, ultra-low permeability reservoir, reduce miscibility pressure, CO_2 miscible flooding, development direction

胜利油田中低渗油藏(渗透率<$100\times10^{-3}\ \mu m^2$)资源丰富,开发潜力大。截至目前,探明地质储量 12.67×10^4 t,控制储量 3.02×10^4 t。其中,未动用探明储量 3.72×10^4 t,受埋藏深、物性差、丰度低等影响,常规水驱开发难以动用,亟需单控储量相对较高的有效开发方式;已动用储量单井产液能力低(日产液 8.8 t),采收率低(18.9%),亟需寻求新的能量补充方式和提高采收率方法。

作为一种优选的驱油剂,在 130℃、30 MPa 油藏条件下 CO_2 黏度低(0.05 mPa·s),是同条件下水黏度的 1/5,同时,CO_2 与边界层间内摩擦力极小,且不存在新边界层形成的问题,注入能力强,是补充低渗储层能量的良好介质。超临界 CO_2 和原油混相后,可降低界面张力,克服贾敏效应,有效动用小孔喉原油,大幅度提高驱油效率[1-2]。因此,CO_2 驱是提高低渗油藏动用率和采收率的重要技术。

国外自 20 世纪 50 年代开始,在 CO_2 驱开发理论和矿场应用等方面进行了大量研究工作[3],CO_2 驱成为重要的提高采收率方法,技术相对成熟,已经规模化应用。1952 年,WHORTON 等[4]获得了第一项利用 CO_2 采油的专利权;1958 年,壳牌首先在 Permian 盆地试验注 CO_2 驱油[5];1972 年,首个商业项目在美国德州 Kelly-Snyder 油田实施[6]。20 世纪 70 年代以来,国外 CO_2 驱提高采收率技术得到了快速的发展,美国和前苏联等国家都进行了大量的 CO_2 驱工业性试验,取得了显著的经济效益,采收率可以提高 10%~25%[7-8]。得益于减税政策的支持,20 世纪 80 年代,美国开发了多个 CO_2 气田,并建成了长距离 CO_2 输送管道,为油田开展 CO_2 驱油提供了稳定气源,CO_2 驱油技术得到进一步推广和发展[9]。

国内自 20 世纪 60 年代开始开展 CO_2 驱油理论与技术研究[10]。1963 年,大庆油田首先进行了 CO_2 提高石油采收率的方法探索,并于 1969 年在葡 I4-7 层和萨南东部过渡带进行了矿场试验,形成了对 CO_2 驱油方法可行性的初步认识。受 CO_2 气源不足、CO_2 驱矿场规模小、气窜和腐蚀等问题影响,2000 年之前,CO_2 驱油技术一直发展缓慢。自"十一五"以来,国家和中国石化、中国石油等各大油气公司高度重视 CO_2 捕集、驱油和埋存技术研发,针对中国陆相油藏的原油特点和储集层特征,相继设立了多个不同层位的 CO_2 驱油与封存的相关研发项目,包括国家重点基础研究发展计划(973 计划)、国家高技术研究发展技术(863 计划)、国家重大科技专项以及各大油气公司设立的重大支撑配套项目 。经过多年攻关,基本形成了适合我国陆相沉积油藏的 CO_2 驱油理论、油藏工程优化设计、注采输出工艺和 CO_2 循环注入等系列技术。同时,中国石油吉林油田、大庆油田、长庆油田,中国石化华东油气分公司、胜利油田、中原油田和延长油田等开展了大量 CO_2 驱先导试验和推广应用。

对胜利油田 CO_2 驱油技术发展历程、CO_2 驱技术的最新进展与 CO_2 驱油矿场试验区的实施情况及应用效果进行系统阐述,并针对 CO_2 驱技术发展所存在的问题以及矿场试验取得的经验和教训,提出了 CO_2 驱技术下一步的发展方向,为形成一套适合我国陆相沉积油藏特征的 CO_2 驱开发理论和技术体系,扩大 CO_2 驱油应用规模提供借鉴和参考。

1 胜利油田 CO_2 驱提高采收率技术 的发展历程

自 20 世纪 60 年代以来,胜利油田持续开展 CO_2 驱提高采收率技术攻关,先后经历了室内实验研究、关键技术攻关和先导试验/扩大试验三个阶段,初步形成具有胜利特色的 CO_2 驱配套技术。

1) 第一阶段:CO_2 驱室内研究阶段(1967—1995 年)

1967 年 4 月,胜利油田开展了 CO_2 驱提高稠油采 收率的室内实验研究工作,实验结果表明:15.1 MPa 条件下,CO_2 可使孤岛原油黏度降低 91%;先注 0.1 PV 的 CO_2 再注水,可使无水期采收率提高 9.5%;若将同体积的 CO_2 与水分散交替注入,则最终采收率将提高 17.4 %。

1968 年,针对滨南平方王油田油层灰质含量高、渗透率低的特点,为使 CO_2 资源就地利用,改善该地区的注水效果,开展了注碳酸水的室内实验。将平方王油田的岩心用 CO_2 浓度为 2.6%的碳酸水经 150 PV 冲刷后,使水相渗透率提高了 3.2 倍,实验表明,碳酸水驱油的采收率可增加 5%。

1978 年,开始 CO_2 混相驱试验研究,并列入石油化学工业部"六五"科技攻关项目。引进了长观察窗 PVT 仪及 RUSKA 混相仪 ,开展了 CO_2^- 甲烷体系相态、CO_2 对原油的混相萃取实验和混相特征研究,初步探讨了 CO_2 驱油机理,为 CO_2 驱油技术的发展奠定了基础。

2) 第二阶段:CO_2驱关键技术攻关阶段(1996—2006 年)

为提高 CO_2 吞吐和 CO_2 驱的成功率,利用有限的 CO_2 资源获得最大的经济效益,

开展了 CO_2 吞吐和 CO_2 驱提高采收率机理、CO_2 近混相驱室内实验研究，制定了筛选条件，攻关 CO_2 驱油藏工程优化设计技术、气窜控制技术和防腐防垢等技术，为 CO_2 矿场试验奠定了基础。

1998 年 9 月，垦利油田垦 153-斜 2 井为胜利油田第一口注 CO_2 吞吐试验井，注入 65 t 液态 CO_2。注气前，日产液为 9.2 t，日产油为 5.1 t；注气后，日产液为 23.6 t，日产油为 10.3 t，累积增油 5 781 t。自 1998 年开始，在东辛、桩西、滨南和纯梁等采油厂进行了 CO_2 单井吞吐增油技术试验，平均单井增产原油 500 t 以上。

2001 年，依托《大芦湖油田樊 124 块二氧化碳驱技术研究》项目，首次全面、系统的进行了 CO_2 混相驱室内实验研究，包括相态特征研究、长细管驱替实验研究、长岩心物理模拟实验研究、CO_2 混相驱过程中沥青质沉淀析出的条件和石蜡析出温度的变化规律的研究。樊 124 块 CO_2 驱实验结果表明：樊 124 地层油与 CO_2 有较好的互溶性，在 29 MPa 条件下，1 t 原油可溶解 326 m^3 标准状况的 CO_2，原油体积膨胀了 46.6%，原油黏度下降了 36.7%，最小混相压力为 25.9 MPa，对比初期气水交替注入的方式，采收率提高了 16%以上。

3）第三阶段：CO_2 驱先导试验/扩大试验阶段（2007—2019 年）

2007 年，依托中国石化科技项目《低渗透油藏 CO_2 驱提高采收率先导试验》在纯梁采油厂高 89-1 块开展 CO_2 驱先导试验，部署 10 注 14 采的 CO_2 驱五点法井网，采用纯 CO_2 气体连续稳定注入的方式进行开发，采收率预计可由 8.9%提高到 26.1%，提高了 17.2%。截至 2019 年底，累积注入 31×10^4 t 的 CO_2，累积增油 8.9×10^4 t，区块采出程度为 15.7%，中心井区为 18.4%（提高了 9.5%）。通过技术集成与创新，初步形成 CO_2 驱适应性评价体系和评价方法、CO_2 驱室内实验技术、CO_2 驱油藏工程优化设计技术、CO_2 驱注采工艺、地面工程技术和燃煤电厂烟气 CO_2 捕集纯化处理技术等，建成国内外首个工业化规模燃煤电厂 烟气 CO_2 捕集、驱油与地下封存的全流程示范工程，实现 CO_2 减排与提高石油采收率的双重目标。

在高 89-1 块开展 CO_2 驱先导试验成功的基础上，陆续在高 89-1 块、高 891 块、高 899 块、樊 142-10 块、商 853 块和桩 23 块等建立 CO_2 驱油开发单元，覆盖地质储量 730 $\times10^4$ t，注气井 24 口，生产井 60 口，预计平均提高采收率 10%以上，产量占比逐年增大，成为胜利油田低渗透油藏在低油价下稳产、上产的重要保障。

2　特低渗透油藏超前注 CO_2 混相驱技术

2.1 超前注 CO_2 混相驱开发技术机理

2.1.1 注气增能机理

物理模拟和数值模拟表明：随着超前 CO_2 注入量增加，地层压力逐步增加，混相能力和范围增加，驱替类型由“非混相驱”变为“混相驱”。随着压力增加，CO_2 在原油中的溶解度增加（图 1），原油体积系数增加，储存能量增加了液体内的动能；当地层压力下降后,CO_2在原油中的溶解度减少,CO_2 从原油中分离出来,液体内产生气体驱动力,维持地层压力。

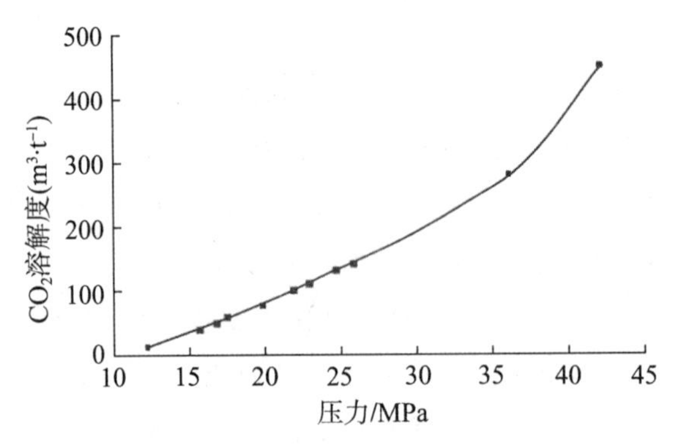

图 1 CO_2 溶解度和压力关系曲线

Fig. 1 Relation between CO_2 solubility and pressure

2.1.2 传质增效机理

在一定油藏条件下,CO_2 处于超临界状态,密度近于液体,黏度近于气体,扩散系数是液体的 100 倍。一方面,通过 CO_2 与原油的传质,改善了原油的性质,可以大幅度降低原油黏度和界面张力,改善宏观波及效率;另一方面,CO_2 通过扩散进入较小孔隙后,溶胀作用可使小孔隙中的原油得到有效的动用,进而改善微观波及效率。长岩心驱油实验表明(图 2),不同压力保持水平下,随着压力和采收率增加,气窜出现得越晚。

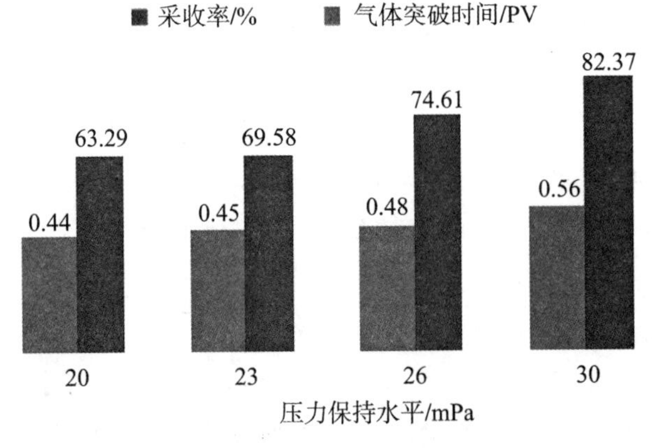

图 2 压力保持水平与采收率和气体突破时间对比

Fig. 2 Comparison of pressure level with recovery ratio and gas breakthrough time

CO_2 与原油接触时间越长,传质作用越强,压力和组分分布越均匀,合理的注入速度使混相压力前缘和组分前缘达到最优匹配,可充分发挥传质作用,最大幅度降低界面张力、原油黏度,增加驱油效率。利用油藏数值模拟技术,研究了不同注入速度条件下压力、CO_2 组分和黏度分布规律(图 3),可以看出,相同注入量下,注入速度过大致使压力场、组分场、黏度场分布不均衡,会造成 CO_2 过早气窜,影响最终采收率。

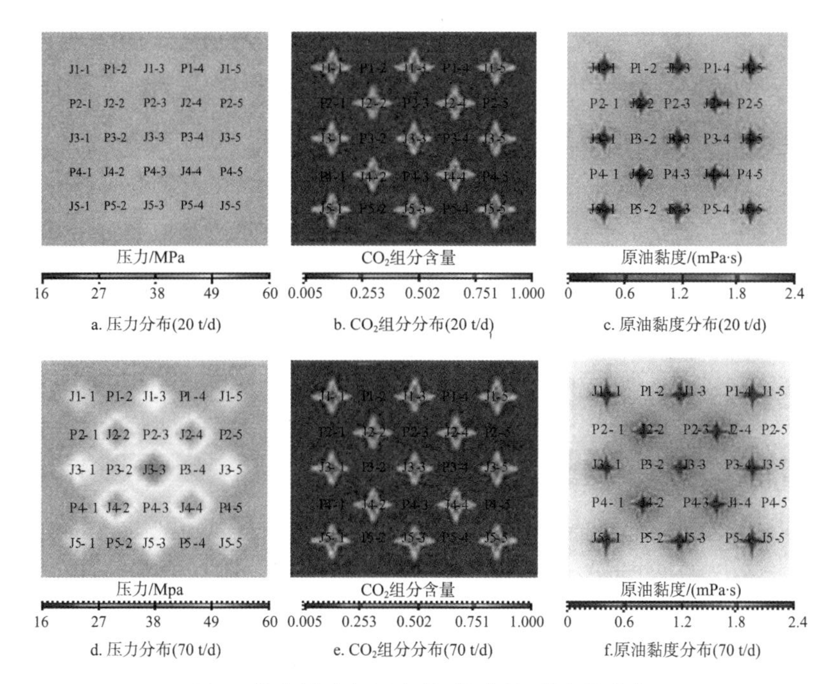

图 3 注气速度与压力场、组分场、黏度场分布

Fig. 3 Distribution of gas injection speed and pressure field, composition field, viscosity field

2.2 超前注 CO_2 优化设计技术

超前注 CO_2 优化设计指通过混相能力(地层压力/最小混相压力)、超前注入速度和注采方式优化,实现压力(水平、分布和前缘)和 CO_2(注入量、分布和前缘)的适配,使 CO_2 均衡驱替,达到提高经济效益和采收率的目标。

2.2.1 混相能力优化

通过建立井组模型,综合考虑技术和经济因素,以采收率和换油率(增油量/累积注气量)为综合评价指标,优化了不同渗透率下的合理混相能力。数值模拟结果表明:同一渗透率条件下,随着压力恢复水平的升高,采收率逐渐增加;当混相能力达到 1 附近时,随着混相能力增大,采收率提高幅度变缓,换油率则呈现先升高后降低的趋势;当混相能力为 1~1.1 时,存在拐点(图 4)。

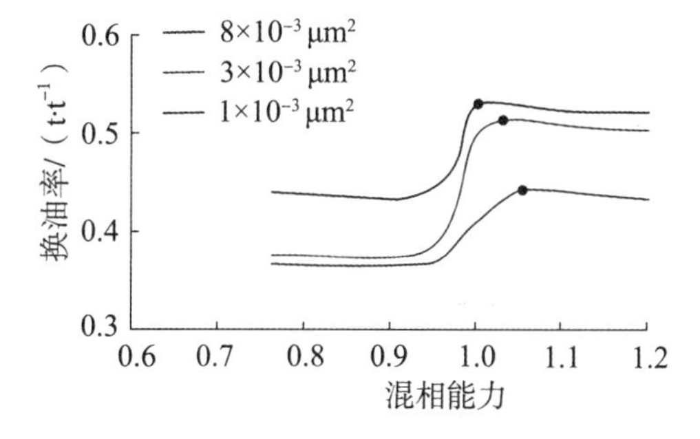

图 4 换油率随混相能力变化曲线

Fig. 4 Variation curve of oil exchange rate with miscible ability

2.2.2 注入速度优化

室内研究表明,CO_2 通过扩散和弥散等传质作用,可大幅度提高宏观驱油效率和波及效率。CO_2 传质作用受两个因素影响,即 CO_2 组分前缘和最小混相压力前缘。当注入速度较慢时,CO_2 组分前缘滞后于最小混相压力前缘,CO_2 与地层原油接触较少;随着注入速度增快,CO_2 组分前缘与最小混相压力前缘之间的距离逐渐减少,CO_2 与地层中更多原油发生作用;继续提高注入速度,CO_2 组分前缘超过最小混相压力前缘,最小混相压力前缘的 CO_2 越多,发生传质作用的 CO_2 就越少。

实际生产时,油田在前期具有较高的采油速度,因此,从经济角度分析,注气速度越快,恢复压力所需时间越短,越有利于尽快收回投资。然而,较高的注气速度也存在两方面的问题:① CO_2 组分前缘超过最小混相压力前缘,发生传质作用的 CO_2 就越少,CO_2 传质增效作用难以发挥;② 高注气速度会使储层存在被压裂开的风险。

图 5 为不同渗透率和油层厚度下合理注气速度图版。研究结果表明:渗透率越高,油层厚度越大,合理注气速度越快。基于注气速度图版设计方案时,可以根据渗透率和油层厚度确定合理注气速度。樊 142-7-斜 4 井组渗透率为 $1.2\times10^{-3}\ \mu m^2$,油藏厚度为 9.2 m,可以确定初期合理注气速度为 15 t/d。

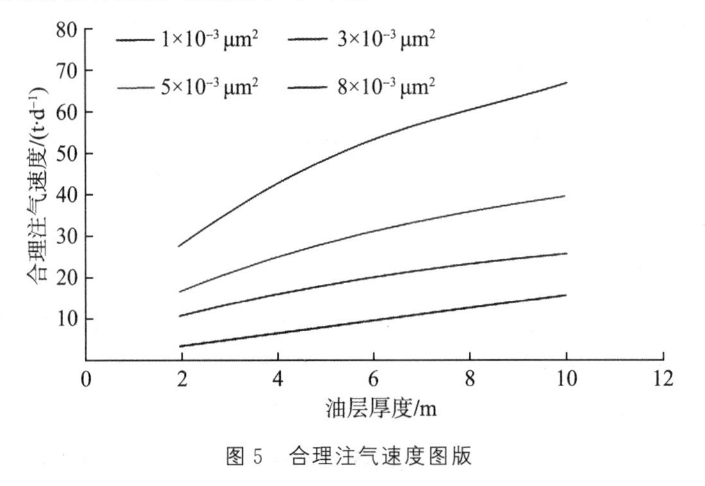

图 5 合理注气速度图版

Fig. 5 Plate of reasonable gas injection speed

2.2.3 注采方式优化

通过长岩心驱替实验,对脉冲注入、恒速注入、交替注采等注采方式进行了优化。结果表明:注入速度一致时,交替注采方式驱油效率最高,脉冲注入次之,恒速注入最低。这是由于三种注入方式下,CO_2 在岩心中扩散时间不同。以恒速注采为基准,交替注采增加了 CO_2 气体与岩心孔隙流体的接触时间,从而使得扩散时间变长,驱油效率得到了提高。而脉冲注采由于注入速度的增加,导致 CO_2 气体与岩心孔隙流体的接触时间变短,扩散时间变短,使得 CO_2 气体未经充分扩散便被后续注入气体向出口端推 进,驱油效率降低。因此,对于超前注 CO_2 混相驱,最佳注采方式是交替注采方式,其次是脉冲注入方式。

2.3 超前注 CO_2 混相驱矿场应用

樊 142-7-斜 4 井组位于正理庄油田樊 142-10 块(图 6),东北部发育坝砂,西南部发育滩砂,含油面积为 0.94 km^2,地质储量为 32.6×10^4 t,渗透率为 1.2×10^{-3} μm^2,注气井 1 口,油井 6 口,注采井距为 243～676 m。方案设计地层压力保持水平为混相压力的 1.3 倍,采用连续注入方式,设计 CO_2 注入速度分别为 15、20、25、30 t/d,预计提高采收率 14%。

2013 年 6 月开始注气,注气前地层压力为 17 MPa, 6 口油井关井恢复地层压力,注气速度为 15～30 t/d。截至 2016 年底,CO_2 累积注入量为 1.9×10^4 t,油井地层压力恢复至 33.7 MPa,地层压力与混相压力的比值为 1.07。

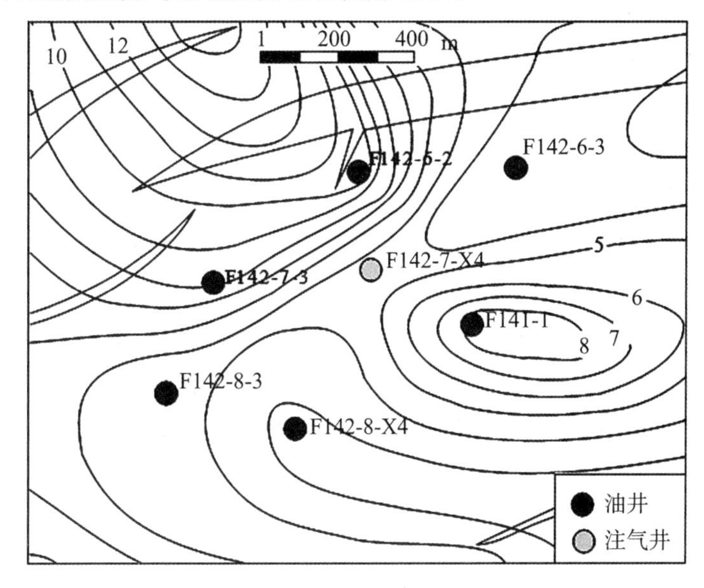

图 6　樊 142-7-斜 4 井组井位

Fig. 6　Well location of F142-7-X4

井组注气后,持续进行了地层压力监测、井流物组分分析,实时分析地层压力、组分分布及前缘推进情况,确定油井的开井时机,根据压力、组分分布及前缘情况进行油气井动态调控,确保井组注 CO_2 的开发效果。2016 年 11 月以来,樊 141-1 井、樊 142-6-3 井和樊 142-6-2 井先后开井,三口井皆自喷生产,产量为 5～6 t/d,远大于注气前产量(泵抽 1 t/d)。截至 2019 年 12 月,井组 CO_2 累积注入量为 3.9×10^4 t,累积增油量为 0.7×10^4 t,阶段累积注入量为 1.9×10^4 t,每注入 1 t CO_2,原油增产 0.37 t。

3　降低混相压力技术

3.1 降低最小混相压力的机理

在 CO_2 与原油接触混相过程中,驱替前缘处的原油与 CO_2 混合部分的黏度下降显著(图 7),对于原油具有极强的抽提性和溶解性,此混合部分的出现对原油驱替效果的影响十分显著。

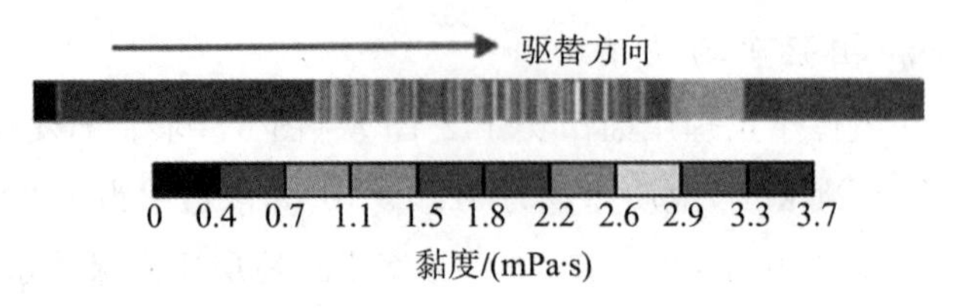

图 7 细管实验中黏度变化场

Fig. 7 Change field of viscosity in thin tube experiment

当油藏最小混相压力过高时,可借助加入化学剂,促使 CO_2 与原油混合部分更易产生原油与 CO_2 混相,从而显著改善 CO_2 非混相驱油效果,达到降低最小混相压力和提高采收率的目的。此化学剂应具有与 CO_2 流体互溶的能力,通过改变 CO_2 流体热力学性质,增强 CO_2 混溶原油的能力,当油藏最小混相压力过高时,原本无法形成混相的原油与 CO_2 体系,通过加入化学剂后形成混相或达到混相驱替的效果。

3.2 降低最小混相压力体系研究

3.2.1 增效剂优选

增效剂之所以能够增强 CO_2 与原油间的混相程度,是因为加入的化学剂能够通过改变 CO_2 的密度、极性等物理性质,从而改变二者的相平衡,即增效剂能够增大 CO_2 中的原油组分。根据超临界状态下,CO_2 的溶解特性以及地层驱油的实际情况,综合考虑增效剂的极性、水性、沸点、凝点、稳定性及毒性等因素,优选出增效剂 DJY13。加入增效剂后,平衡体系气相中 C_3-C_{15} 的含量增加了 3 倍(图 8),表明增效剂能大幅度提高 CO_2 的抽提能力。

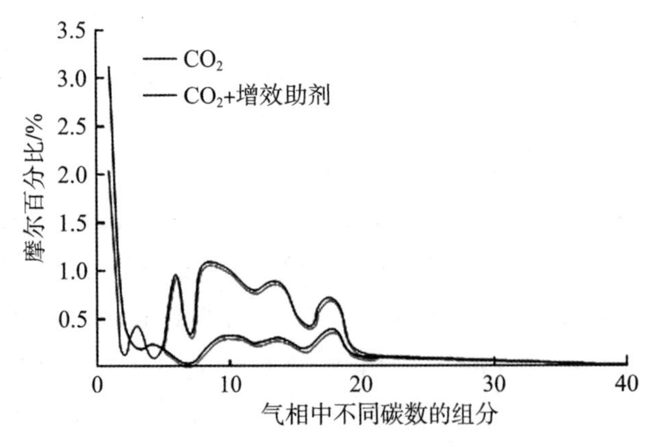

图 8 平衡体系中气相组分变化曲线

Fig. 8 Variation curve of gas phase composition in balance system

3.2.2 增溶剂

CO_2 与原油在没有混相时,维持原始的热力学相平衡状态。加入增溶剂后,原有平衡的分子间相互作用,发生改变,直至新的分子间相互作用,达到新的热力学相平衡稳定状态。这使 CO_2 与原油的非混相平衡状态更趋向于混相平衡状态,从而达到促进 CO_2 与原油混相的目的。

增溶剂分子包含亲 CO_2 基团与亲油基团，两种基团分别与 CO_2 和原油分子接触。加入增溶剂前，气液两相中 CO_2 与原油直接作用，维持一个热力学相平衡状态；加入增溶剂后，由于增溶剂与 CO_2 和原油溶解的能力不同，增溶剂更易将 CO_2 分子“拖拽”至原油。因此，第三组分分子作为纽带增强了 CO_2 与原油分子间相互作用，达到新的热力学相平衡状态，使 CO_2 与原油的非混相平衡状态更趋向于混相，达到促进 CO_2 与原油混相的目的，这就是增溶剂的微观作用机理。为了促进 CO_2 与原油的混相或混溶，增溶剂可根据分子的极性及链长设计，达到改善 CO_2 与原油混相条件的目的。合成的增溶剂应包含苯环基团、酯基、长碳链等，并且是具有一定对称程度的分子，其基本结构如下：

O
‖
R_1—C—O—R_2

其中，R_1 和 R_2 分别为长度不等的直碳链；苯环基团的作用是作为分子主体骨架，促进增溶剂与原油混溶；酯基与 CO_2 结构类似，可促进增溶剂与 CO_2 相互作用；而不同长度的碳链可以调节增溶剂与两相的混溶能力，即可通过调整碳链长度，定向地将分子设计为增溶剂；通过调节分子极性，还可促进分子与 CO_2 相互混溶的能力。

基于以上原理，设计合成了增溶剂 S6，图 9 为加入不同增溶剂后，原油中 CO_2 的溶解情况。结果表明：增溶剂 S6 具有最佳的增溶效果，在油藏温度、压力范围内，CO_2 在原油中的溶解能力增强了 1.5 倍。

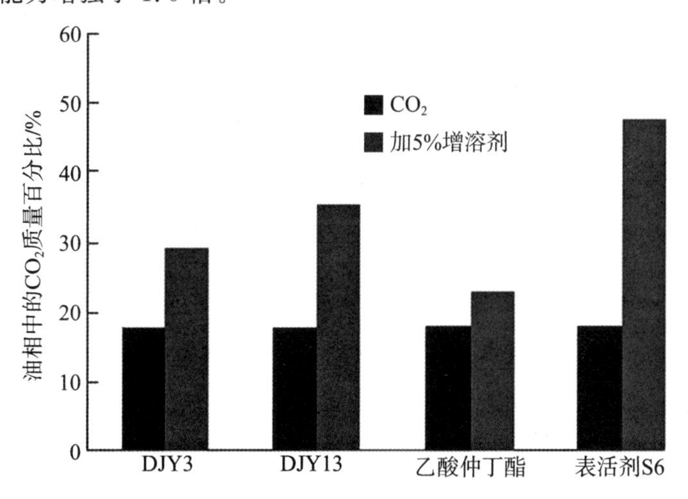

图 9 平衡体系中油相中 CO_2 质量百分比

Fig. 9 CO_2 mass percent in oil phase of balance system

3.2.3 降低混相压力体系研制

在 CO_2 与原油体系中加入化学助剂可有效地改善 CO_2 与原油间的互溶度，增强 CO_2 非混相驱油效果，达到降低最小混相压力和提高采收率的目的。然而单一的化学助剂往往仅具备增效和增溶两种作用中的一个，因此，需要将增效剂和增溶剂进行复配，通过调整配比来实现增效和增溶两种性能之间权重大小的可调控，使复配体系达到兼顾增效和增溶的作用。

通过长细管实验研究了增效剂 DYJ13 和增溶剂 S6 不同配比时对混相压力的影响。结果表明:当 DYJ13∶S6 为 3∶7 时,混相压力降低幅度最大,混相压力可以由 31.65 MPa 下降到 24.6 MPa,降低幅度达到 22%。

3.3 降低混相压力体系的矿场应用

利用室内物理模拟实验,对降低混相压力体系的注入时机、注入浓度、注入方式和段塞尺寸等进行了优化设计(图 10)。结果表明:最佳注入时机是先注入降混相压力体系再注 CO_2,注入浓度为 25%,注入方式为段塞注入,段塞尺寸为 0.05 PV。

基于以上优化结果,在高 891-12 井组进行了矿场试验。2019 年 4 月,在高 891-12 井注降混相压力体系 100 t,周围 3 口井产量增加了 50%,有效期为 4～6 个月。

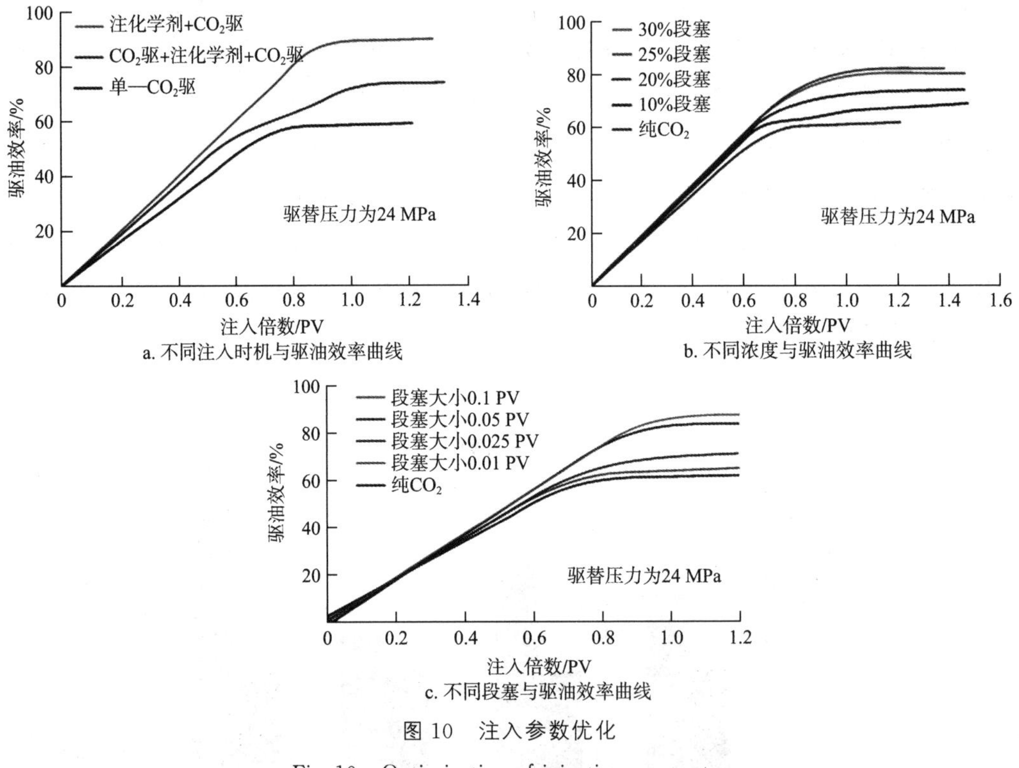

图 10 注入参数优化

Fig. 10 Optimization of injection parameter

4 面临的挑战及下步方向

4.1 CO_2 驱规模应用面临的挑战

与国外 CO_2 驱相比,胜利油田 CO_2 驱面临三个挑战:① 油藏条件和原油性质决定了 CO_2 与原油混相压力高。国外实施 CO_2 驱的区块原油性质好,混相压力一般低于 12 MPa;胜利油田的原油重质组分含量高、黏度大、密度高、温度高,混相压力一般在 30 MPa 左右。② 油藏非均质性强,CO_2 驱易气窜。国外 CO_2 驱主要用于水驱效果较好的中低渗油藏,水驱后转 CO_2 气水交替驱,利于控制气窜;胜利油田主要用于水驱无法正常开发的低渗透、特低渗透油藏，衰竭开采后能量补充差，非均质强，易气窜，同时采取连续注气、间歇注气，不利于气窜的控制。③ 低成本气源匮乏、运输成本高。以美国为例，CO_2 主要来自天然 CO_2 气田，成本低于 200 元/t，输送方式以管道输送为主，运输成本为 0.06 元/(t·km)。胜利油田的 CO_2 气源则是由高碳天然气分离(450～650 元/t)或燃煤电厂捕集处理(500～850 元/t)，CO_2 输送方式采用罐车，运输成本为 1 元/(t·km)。

4.2 CO_2 驱开发对策

转变发展方式，降低源头成本，是实现 CO_2 驱效益开发的关键问题。应对三个挑战的对策是实现三个转变：一是转观念，聚焦混相驱开发，研究由满足混相压力转向降低混压，通过 CO_2 与化学剂相结合发展化学增效，CO_2 驱油技术由原来的单一注气提升地层压力改为超前注气和化学剂降低混相压力并重的复合手段；二是转思路，拓展 CO_2 驱领域，由提高特低渗油藏采收率转向提高中低渗油藏采收率和提高特低渗油藏动用率并重；三是转方式，注重节约气源成本，由用天然气源转向找工业尾气，基于天然气源不足的状况，转向与齐鲁石化第二化肥厂、青州宇信钙业、山东联盟化工等 CO_2 排放企业积极合作，获得廉价 CO_2 工业尾气，在利用 CO_2 的同时减排，实现经济效益和社会效益双赢。

4.3 下一步研究方向

4.3.1 深化 CO_2 驱提高石油采收率的相态理论

胜利油田原油类型和组成相对复杂，开展 CO_2 混相驱过程中的 CO_2^- 地层油体系的相态特征及其影响因素研究，建立 CO_2 与复杂烃类物质构成的多组分体系相态及相态表征方法，分析 CO_2 混相采油过程中地层油物理化学性质及其与压力和温度的敏感性、轻组分抽提与重组分沉积特点等，平衡相态过程中的相态理论、动态过程中的相态理论、多孔介质中(微观尺度)的相态理论和完善适合胜利油田原油与 CO_2 的状态方程的建立和发展，为 CO_2 驱油藏数值模拟提供技术支撑。

4.3.2 低成本扩大 CO_2 驱波及体积技术

CO_2 驱油是一种有效提高采收率的方法[11-16]，理论上驱油效率接近 100%，但是，CO_2 相对原油，具有低黏度、低密度与储层的非均质性特征，黏性指进和重力超覆等问题大大降低了 CO_2 的波及系数，实际油藏的 CO_2 驱提高采收率仅维持在 7%～20%。如何扩大波及体积大幅度提高石油采收率是 CO_2 驱面临的一大挑战。

低成本扩大 CO_2 驱波及体积技术是通过向油层注入低流度物质占据 CO_2 流动通道或者通过改变 CO_2 注采工作制度使流线发生改变，从而起到抑制 CO_2 窜流的作用。WAG(水气交替)注入技术、CO_2 泡沫调驱技术和 CO_2 增稠技术是典型的扩大气驱波及体积的技术。以 CO_2 水气交替为例：一方面，CO_2 与水形成贾敏效应，改变液流方向，提高波及体积；另一方面，有效地控制了驱替流体的流度，实现了注 CO_2 提高微观驱油效率和注水提高宏观波及系数的有机结合，进而提高了原油采收率。

4.3.3 CO_2 非完全混相驱研究的发展

传统 CO_2 驱混相理论认为,地层压力低于最小混相压力的油藏都属于非混相驱。但是,与中高渗透油藏不同,低渗透油藏注采井间压力变化较大,注入井底附近压力一般大于 40 MPa,远大于最小混相压力;而在生产井底附近的压力为 15 MPa 左右,又远小于最小混相压力,这就意味着注入井附近为混相驱,生产井附近为非混相驱。因此,低渗透油藏压力空间变化对 CO_2 混相状态产生的影响不能忽视,用单一的混相或非混相定性描述 CO_2 驱替过程,不能完全反映混相状态的分布,具有较大局限性。

CO_2 非完全混相驱目前尚处于攻关阶段[17-19],应重点加强含水和微纳米孔隙对混相状态的影响,CO_2 非混相驱替机理,优化设计方案,非混相驱替三相的相渗曲线响应特征、非混相驱开发特征曲线及见效特征,CO_2 突进,窜流规律及改善波及效率方法研究,进一步提高 CO_2 驱开发效果。

4.3.4 气窜通道识别、描述及预警技术研究

受储层非均质、重力超覆和不合理的注采制度等影响,气窜是 CO_2 驱过程中普遍存在的生产问题。气窜通道的识别和描述主要借助于昂贵的现场测试,缺乏一套基于生产动态资料分析的气窜通道高效识别和描述方法。同时,目前研究仅仅是针对气窜通道形成后的识别和描述,无法在气窜通道形成前对气窜进行预警。针对不同油藏地质条件及开发阶段气窜通道的形成与演化规律所表现出的生产动态响应,需要建立大量基础模型,利用数值模拟分析不同地质及开发条件下气窜类型、气窜通道位置、气窜通道形态等来明确对应的生产动态响应特征,结合理论分析和实验研究结果,形成大数据学习知识库,采用大数据深度学习算法,建立气窜通道识别、描述及预测模型,实现气窜形成前的预测及气窜形成后的识别,指导现场生产。

5 结论

1) 经过多年的探索和攻关配套,胜利油田 CO_2 驱提高采收率技术的发展历程可划分为室内实验研究、关键技术攻关和先导试验/扩大试验三个阶段,形成具有胜利特色的 CO_2 驱配套技术。

2) 深化了超前注 CO_2 驱注气增能、传质增效的开发机理,形成特低渗油藏超前注 CO_2 开发技术,制定了技术实施界限,应用在樊 142-7-斜 4 井组现场增产效果明显,单井日产油增加了 5 倍以上。

3) 提出了降低混相压力的原理和技术思路,研发了强化 CO_2 对原油组分抽提能力的增效剂和增强 CO_2 溶解能力的增溶剂,建立了降低混相压力体系,室内实验表明最大混相压力降幅可达 22%,在高 891-12 井组矿场应用初见成效。

4) 针对 CO_2 驱规模化应用面临的挑战和技术瓶颈,提出了改善 CO_2 驱开发效果的技术对策和下一步发展方向,研究成果对于加快推进胜利油田 CO_2 驱工业化具有较好的指导意义,对于同类油藏开展 CO_2 驱油与封存项目具有借鉴意义。

参考文献

［1］史云清，贾英，潘伟义，等. 低渗致密气藏注超临界 CO_2 驱替机理[J]. 石油与天然气地质，2017，38(3)：610-616.

SHI Y Q，JIA Y，PAN W Y，et al. Mechanism of supercritical CO_2 flooding in low-permeability tight gas reservoirs[J]. Oil & Gas Geology，2017，38(3)：610-616.

［2］张本艳，周立娟，何学文，等. 鄂尔多斯盆地渭北油田长 3 储层注 CO_2 室内研究[J]. 石油地质与工程，2018，32(3)：87-90.

ZHANG B Y，ZHOU L J，HE X W，et al. A laboratory study on CO_2 injection of Chang 3 reservoir of Weibei oilfield in Ordos basin[J]. Petroleum Geology & Engineering，2018，32(3)：87-90.

［3］British Petroleum Company. BP statistical review of world energy[R]. London：BP Company，2018.

［4］WHORTON L P，BROWNSCOMBE E R，DYES A B. Method for producing oil by means of carbon disoxide[P]. US 2623596，1952.

［5］杨勇，吕广忠，张东，等. 胜利油田特低渗透油藏 CO_2 驱技术研究与实践[J]. 油气地质与采收率，2020，27(1)：1-9.

YANG Y，LYU G Z，ZHANG D，et al. Research and application of CO_2 flooding technology in extra-low permeability reservoirs of Shengli Oilfield[J]. Petroleum Geology and Recovery Efficiency，2020，27(1)：1-9.

［6］LANGSTON M V，HOADLEY S F，YOUNG D N. Definitive CO_2 flooding response in the SACROC unit[C]//paper SPE-17321-MS presented at the SPE Enhanced Oil Recovery Symposium，16-21 April 1988，Tulsa，Oklahoma，USA.

［7］江怀友，沈平平，卢颖，等. CO_2 提高世界油气资源采收率现状研究[J]. 特种油气藏，2010，17(2)：5-10.

JIANG H Y，SHEN P P，LU Y，et al. Present situation of enhancing hydrocarbon recovery factor by CO_2[J]. Special Oil & Gas Reservoirs，2010，17(2)：5-10.

［8］廖洪. S 油藏注 CO_2 提高采收率技术[D]. 成都：西南石油大学，2017.

LIAO H. The EOR study of CO_2 injection in S reservoir[D]. Chengdu：Southwest Petroleum University，2017.

［9］秦积舜，韩海水，刘晓蕾. 美国 CO_2 驱油技术应用及启示[J]. 石油勘探与开发，2015，42(2)：209-216.

QIN J S，HAN H S，LIU X L. Application and enlightenment of carbon dioxide flooding in the United States of America[J].

Petroleum Exploration and Development，2015，42(2)：209-216.

[10] 贾凯锋，计董超，高金栋，等. 低渗透油藏 CO_2 驱油提高原油采收率研究现状[J]. 非常规油气，2019，6(1)：107-114.

JIA K F，JI D C，GAO J D，et al. The exisiting state of enhanced oil recovery by CO_2 flooding in low permeability reservoirs[J]. Unconventional Oil & Gas，2019，6(1)：107-114.

[11] 王海妹. CO_2 驱油技术适应性分析及在不同类型油藏的应用——以华东油气分公司为例[J]. 石油地质与工程，2018，32(5)：63-65.

WANG H M. Adaptive analysis of CO_2 flooding technology and its application in different types of reservoirs[J]. Petroleum Geology & Engineering，2018，32(5)：63-65.

[12] 商琳琳. 龙虎泡油田高台子致密油层 CO_2 驱实验研究[J]. 石油地质与工程，2018，32(5)：60-62.

SHANG L L. Experimental study on CO_2 flooding of Gaotaizi tight oil layers in Longhupao oilfield[J]. Petroleum Geology & Engineering，2018，32(5)：60-62.

[13] 国殿斌，徐怀民. 深层高压低渗油藏 CO_2 驱室内实验研究——以中原油田胡96 块为例[J]. 石油实验地质，2014，36(1)：102-105.

GUO D B，XU H M. Laboratory experiments of CO_2 flooding in deep-buried high-pressure low-permeability reservoirs：A case study of block Hu96 in Zhongyuan Oilfield[J]. Petroleum Geology & Experiment，2014，36(1)：102-105.

[14] 李剑，段景杰，姚振杰，等. 低渗透油藏水驱后注 CO_2 驱提高采收率影响因素分析[J]. 非常规油气，2017，4(6)：45-52.

LI J，DUAN J J，YAO Z J，et al. Analysis on influence factors of enhanced oil recovery in CO_2 flooding after water flooding in low permeability reservoir[J]. Unconventional Oil & Gas，2017，4(6)：45-52.

[15] 邓瑞健，田巍，李中超，等. 二氧化碳驱动用储层微观界限研究[J]. 特种油气藏，2019，26(3)：133-137.

DENG R J，TIAN W，LI Z C，et al. Microscopic limits of reservoir producing for carbon dioxide flooding[J]. Special Oil & Gas Reservoirs，2019，26(3)：133-137.

[16] 丁妍. 濮城油田低渗高压注水油藏转 CO_2 驱技术及应用[J]. 石油地质与工程，2019，33(6)：73-76.

DING Y. Technology and application of CO_2 flooding in low-permeability and high-pressure water injection reservoirs in Pucheng oilfield[J]. Petroleum Geology & Engineering，2019，33(6)：73-76.

[17] 胡伟，吕成远，王锐，等. 水驱油藏注 CO_2 非混相驱油机理及剩余油分布特征[J]. 油气地质与采收率，2017，24(5):99-105. HU W, LYU C Y, WANG R, et al. Mechanism of CO_2 immiscible flooding and distribution of remaining oil in water drive oil reservoir[J]. Petroleum Geology and Recovery Efficiency, 2017, 24(5) : 99-105.

[18] 高云丛，赵密福，王建波，等. 特低渗油藏 CO_2 非混相驱生产特征与气窜规律[J]. 石油勘探与开发，2014，41(1):79-85. GAO Y C, ZHAO M F, WANG J B, et al. Performance and gas breakthrough during CO_2 immiscible flooding in ultra-low permeability reservoirs[J]. Petroleum Exploration and Development, 2014, 41(1):79-85.

[19] 吕成远，王锐，赵淑霞，等. 低渗透油藏 CO_2 非混相驱替特征曲线研究[J]. 油气地质与采收率，2017，24(5):111-114. LYU C Y, WANG R, ZHAO S X, et al. Study on displacement characteristic curve in CO_2 immiscible flooding for low permeability reservoirs[J]. Petroleum Geology and Recovery Efficiency, 2017, 24(5) : 111-114.

（编辑　余聪）

聚合物驱研究进展及技术展望

曹绪龙,季岩峰,祝仰文,赵方剑

(中国石化胜利油田分公司,山东东营 257000)

摘要:随着我国对石油的需求量不断增加,对油气田的进一步挖潜显得至关重要。化学驱技术是提高采收率的重要技术之一,而聚合物驱作为最主要的化学驱提高采收率方法,在矿场上已经得到了广泛应用,并取得良好的驱油效果。该文通过对聚合物驱的基本原理以及各种驱油用聚合物的发展现状进行综述,对聚合物驱的矿场应用效果进行总结,展望了聚合物驱在高温高盐等苛刻油藏环境下的发展方向。通过综述可以看出,虽然耐温抗盐共聚物、速溶聚合物、两亲聚合物等功能型聚合物已成功研发,但应用于矿场的聚合物类型仍然有限,如何将新型聚合物的研发成果应用于现场提高采收率是重点发展方向。随着不同聚合物类型的研发,对聚合物驱油机理的研究需要继续深入。

关键词:聚合物;驱油原理;研究进展;矿场实验;新型聚合物

中图分类号:TE357 文献标识码:A

Research advance and technology outlook of polymer flooding

Cao Xulong, Ji Yanfeng, Zhu Yangwen, Zhao Fangjian

(Sinopec Shengli Oilfield, Dongying, Shandong 257000, China)

Abstract: With the increasing oil demand in China, it is very important to further tap the potential of oil and gas fields. Chemical flooding technology is one of the important technologies of EOR, and polymer flooding, as the most important method of EOR, has been widely used in the field and achieved good oil displacement effects. Therefore, by summarizing the basic principles of polymer flooding, the development status of various kinds of polymer for oil displacement and the field application effect of polymer flooding, the development direction of polymer flooding in harsh reservoir conditions, such as high temperature and high salt, has been prospected. Through the review, although functional polymers, such as temperature resistant and salt resistant copolymers, instant polymers and amphiphilic polymers, have been successfully developed, the types of polymers used in the field are still limited. How to apply the research and development achievements of new polymers to on-site EOR is the key development direction. With the development of different polymer types, further research on polymer flooding mechanism is needed.

Key words: polymer, oil displacement principle, research advance, field tests, novel polymer

随着工业的发展,世界能源的需求日益增加。相比于太阳能、风能等其他能源,现阶段的化石燃料,特别是石油和天然气,在能源供应方面发挥着更加重要的作用[1]。近年来,我国石油的需求量不断增加。2019 年,我国原油进口量超过 5×10^8 t,同比增长 9.5%,石油对外依赖度上升至 72%[2]。为缓解这种情况,对老油田进一步挖潜以提高原油采收率或通过发现新的油气田来提高油气产量是至关重要的[3]。油田开采一般经历 3 个阶段。通过油层自身能量进行石油开采的方式称之为一次采油。一次采油结束后,将气体或水注入油田储层以保持油层与生产井之间的压力差,这种方式称为二次采油。在一次和二次采油阶段之后,由于油井的含水率不断增加,需要更多的地面设备来分离油井采出的油气和水,导致产油成本高昂,很难以低成本继续采油[4-7]。三次采油是二次采油后进一步提高采收率的重要阶段,利用化学、热能、物理、生物等方法改变储层岩石或驱替液的性质,从而提高波及系数和洗油效率。化学驱提高采收率(CEOR)是提高采收率的重要技术之一,在油田得到广泛应用。有关提高采收率的研究表明,全球 11%提高采收率项目中,有 11%是化学驱。在化学驱提高采收率技术中,聚合物驱所占比例超过 77%,23%是聚合物/表面活性剂二元驱[8]。

聚合物驱通过向水相中加入聚合物,增加水的黏度,降低水油流度比,同时降低水的相对渗透率,实现吸水剖面的调整,提高水相波及体积[9]。聚合物驱已在油田现场应用 50 多年,与其他提高采收率技术相比成本更低[10]。但是在矿场应用过程中也存在一些问题:在高温高矿化度油藏,聚合物分子易发生蜷曲导致流度控制能力减弱;在海上油田聚合物溶解熟化时间长,配聚效率低;在低渗透稠油油藏聚合物注入性差,难以驱替稠油。各种新型聚合物的研发有效地解决了上述问题,促进了聚合物驱在油田的进一步应用[11]。该文首先综述了聚合物驱的基本原理以及各种驱油用聚合物的发展现状,然后对聚合物驱的矿场应用效果进行总结,最后展望了聚合物驱未来的发展趋势。

1 聚合物驱研究进展

1.1 聚合物驱的原理

1.1.1 增加水相黏度,避免横向指进

在非均质油藏中,水驱会导致严重的指进现象,即受层内非均质性及油水黏度差异的影响,注入水的渗流速度远远快于原油,在注水井和采出井间形成优势通道,最终导致水驱后油藏的大部分区域没有被水波及[12]。1964 年,PYE[13] 与 SANDIFORD[14] 提出向水中加入水溶性聚合物能够降低水的流度,进而降低水油流度比,实现活塞式的水驱油方式。Buckly-Leverett 方程表明流度比对含水率存在影响,如式(1),并认为当水油流度比 M 降低到 $M<1$ 时,水驱油为活塞式的驱替,此时平均含水饱和度较大,剩余油的数量也因此减少。

$$f_w=\frac{\frac{k_w}{\mu_w}}{\frac{k_w}{\mu_w}+\frac{k_o}{\mu_o}}=\frac{M}{1+M}=\frac{1}{1+\frac{1}{M}} \tag{1}$$

式中,f_w 为含水率,%;k_w 为水相渗透率,μm^2;μ_w 为水相黏度,Pa·s;k_o 为油相渗透率,μm^2;μ_o 为油相黏度,Pa·s;M 为水油流度比。

1.1.2 提高纵向波及

在纵向上,聚合物先与高渗储层接触,在阻力相对较小的大孔道中缓慢渗流。由于聚合物与孔壁间存在相互吸附作用,使高渗层渗透率减小。此外,聚合物溶液黏度高,渗流过程中存在很大的摩擦阻力,降低了在高渗透层中流动速度,减小聚合物溶液在高渗透层与低渗透层间推进速度之差,调整吸水剖面,从而扩大纵向波及[15]。

1.1.3 黏弹性提高洗油效率

聚合物溶液不同于水,是一种非牛顿流体,非牛顿流体是黏弹性流体,黏弹性流体产生的拉伸变形和黏弹效应,使得孔隙夹缝中不易波及的残余油,在孔道中变成丝絮状态或以活塞式推进,被聚合物溶液驱替出来,从而提高洗油效率[16]。然而,在高温高矿化度下,聚丙烯酰胺和部分水解聚丙烯酰胺易水解,洗油效率大幅降低,此时向主链引入特殊官能团的多元共聚物在高温高矿化度下亦可表现出良好的黏弹性,进一步提高了高温高矿化度油藏中的洗油效率[17]。

1.1.4 胶束作用剥离岩石壁面原油

近年来,有学者提出两亲聚合物和聚合物表面活性剂等新型功能聚合物可通过胶束作用将原油从岩石壁面剥离,从而提高采收率 。聚合物将原油从岩石表面剥离分为 5 个阶段,分别为:表面物理吸附、形成高覆盖层、聚合物预胶束沉积、聚合物胶束覆盖、胶束裹挟原油剥离。聚合物与原油作用后,通过吸附作用在油膜表面形成吸附层,随着吸附作用的进行,聚合物胶束覆盖整个油膜,随后原油进入聚合物胶束的疏水内核,最终聚

合物胶束裹挟原油从岩石壁面剥离，其机理如图1所示[18]。但是该类聚合物目前在矿场应用较少，并且其驱油机理并不适用于传统表面活性剂。

1.1.5 吸附、捕集作用降低岩石渗透率

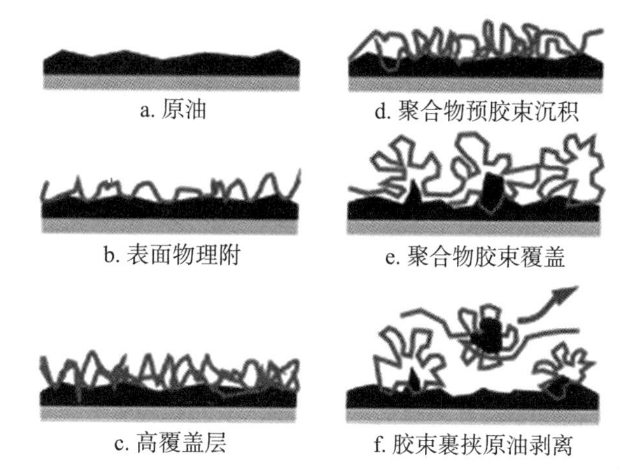

图1　聚合物与原油相互作用机理

Fig. 1　Interaction mechanism between polymer and crude oil

聚合物分子可以被孔隙表面吸附，也可以被狭窄吼道捕集。聚合物降低岩石渗透率的作用可由两个参数表示，分别为阻力系数（RF）和残余阻力系数（RFF）。阻力系数表示聚合物溶液与水相比所增加的流动阻力[19]，如式（2）所示。阻力系数越高，说明聚合物溶液越难在岩石孔隙中渗流。残余阻力系数是初始水驱的流度与注聚后后续水驱的流度之比，表征了聚合物分子由于吸附滞留在岩石中导致渗透率降低的程度[20]，如式（3）所示：

$$RF=\frac{\lambda_{水驱}}{\lambda_{聚合物驱}}=\frac{\Delta P_{聚合物驱}}{\Delta P_{水驱}} \tag{2}$$

$$RFF=\frac{\lambda_{初始水驱}}{\lambda_{后续水驱}}=\frac{\Delta P_{后续水驱}}{\Delta P_{初始水驱}} \tag{3}$$

式中，RF 为阻力系数；RFF 为残余阻力系数；$\lambda_{水驱}$ 为水流度，$\mu m^2\cdot(Pa\cdot s)^{-1}$；$\lambda_{聚合物驱}$ 为聚合物溶液流度，$\mu m^2\cdot(Pa\cdot s)^{-1}$；$\Delta P_{聚合物驱}$ 为聚合物驱压差，MPa；$\Delta P_{水驱}$ 为水驱压差，MPa；$\lambda_{初始水驱}$ 为初始水驱时水相流度，$\mu m^2\cdot(Pa\cdot s)^{-1}$；$\lambda_{后续水驱}$ 为后续水驱时水相流度，$\mu m^2\cdot(Pa\cdot s)^{-1}$；$\Delta P_{初始水驱}$ 为初始水驱压差，MPa；$\Delta P_{后续水驱}$ 为后续水驱压差，MPa。

1.2 驱油用聚合物研究进展

1.2.1 传统驱油用聚合物

传统的聚合物驱主要使用的聚合物有两类：合成聚合物与天然聚合物。部分水解聚丙烯酰胺（HPAM）是应用最广泛的合成聚合物[21]，结构式如图2所示。

部分水解聚丙烯酰胺合成原料易得，合成方式简单，成本低，易于工业化生产，具有良好的增黏性能，是聚合物驱应用最多的聚合物。但是HPAM在多孔介质中受到高温

高矿化度及微生物活动等环境因素的影响，极易发生降解导致黏度损失[22]。HPAM 在高温下易发生水解反应，使水解度增大，羧基含量增加，在高矿化度地层水中，高浓度的二价金属离子(Ca^{2+} 和 Mg^{2+})通过静电作用与羧基结合，减弱高分子链的带电性，从而使 HPAM 分子发生蜷曲甚至出现絮凝[23]。因此，HPAM 不适用于高温高矿化度油藏。

$$-\!\!\left[CH_2-\underset{\underset{NH_2}{|}}{\underset{O=C}{\underset{|}{CH}}}\right]_x\!\!\left[CH_2-\underset{\underset{O^-Na^+}{|}}{\underset{C=O}{\underset{|}{CH}}}\right]_y\!\!-$$

图 2　HPAM 结构式

Fig. 2　HPAM structural formula

另一种应用较为广泛的天然聚合物是黄原胶，它是一种高分子生物多糖，通常由一种叫作黄瘤的细菌通过发酵过程产生，图 3 为黄原胶的结构式[24]。

黄原胶分子在低温下具有双螺旋结构，但随着温度的升高，高温下双螺旋结构改变为无序螺旋[25]。加入少量盐后，由于电荷屏蔽效应，黄原胶溶液黏度会降低。黄原胶比聚丙烯酰胺更适用于高盐条件。一般来说，黄原胶在较高的矿化度和温度下比 HPAM 更稳定，但它们都在高温高矿化度(HTHS)条件下降解。因此，黄原胶也不适用于高温高矿化度储层。

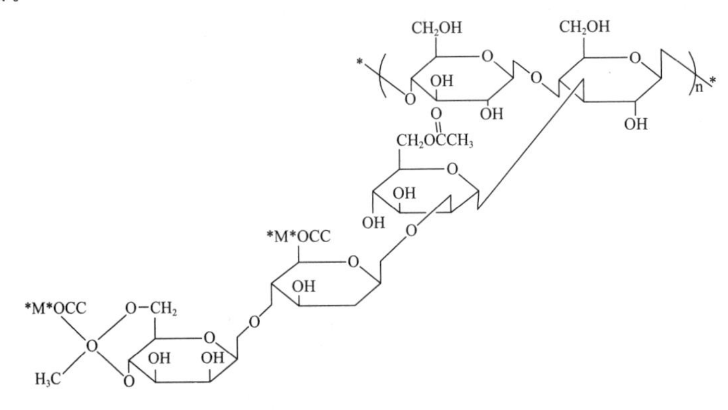

图 3　黄原胶分子结构式

Fig. 3　Molecular structural formula of xanthan gum

1.2.2 耐温抗盐聚合物

近年来，随着油田开发的不断深入，储层条件较好的油藏经过常年的水驱和聚合物驱开发，增油稳产的难度越来越大，高温高矿化度油藏逐渐成为开发的重点。由于常规的 HPAM 并不适用于高温高矿化度油藏，学界针对提高丙烯酰胺类聚合物耐温抗盐性做了大量研究，促进了耐温抗盐聚合物的合成与研发，形成了耐温抗盐单体共聚物、疏水缔合聚合物、温增黏聚合物、纳米颗粒增强聚合物等产品。

(1) 耐温抗盐单体共聚物。

耐温抗盐单体共聚物从聚合物分子的化学组成出发，设计研发不同的耐温抗盐单体,将其与丙烯酰胺、丙烯酸通过自由基聚合合成耐温抗盐共聚物。耐温抗盐单体多选用含 C-N、C-C、C-O 等对温度和矿化度不敏感的强键单体,避免使用酯基等弱键单体[26]。耐温抗盐单体根据带电性质可分为非离子型、阴离子型、阳离子型、两性型、疏水缔合型。

非离子型耐温抗盐单体本身不带电,因此对于矿化度不敏感,这类单体主要包括 N-乙烯基吡咯烷 酮、N,N-二甲基丙烯酰胺等。李新勇等[27]使用非离子耐温抗盐单体 N-丙烯酰吗啉合成了耐温抗盐三元共聚物,结果表明由于吗啉环的引入,共聚物具有优异的耐温抗盐及抗老化性能。

阴离子耐温抗盐单体是使用及研究最广泛的一类单体,目前应用最多的是 2-丙烯酰胺基-2-甲基丙磺酸。这类单体通常含有磺酸基等强极性基团,在高温下抑制酰胺基的水解,同时取代 HPAM 中的部分羧基,降低水解度,减弱羧基与二价金属离子的结合,提高聚合物的抗盐能力。孙群哲等[28]使用苯乙烯磺酸钠作为耐温抗盐单体,制备出磺化聚丙烯酰胺 SPAM,虽然分子量仅有 736 万,但与工业化产品相比耐温抗盐性能有很大改善。近年来,多种新型耐温抗盐单体被研发出来,但所合成的耐温抗盐聚合物仅限于室内研究阶段,矿场应用最广泛的仍为含 AMPS 单体的耐温抗盐聚合物。

阳离子单体一般不能作为驱油用耐温抗盐聚合物的聚合物单体,因为阳离子聚合物极易在地层中吸附,造成黏度损失。同样的,两性离子单体也存在地层吸附问题。但是两性离子单体合成的共聚物具有反聚电解质性质,由于两性离子单体含有数目相等的正负电荷,其溶液黏度反而会随矿化度的增加而增大,是一种前景广阔的耐温抗盐聚合物[29]。但是,两性聚合物容易因为高温下酰胺基水解等原因导致分子链正负电荷数目发生变化,从而失去反聚电解质效应,这是两性聚合物急需解决的问题[30]。

(2) 疏水缔合聚合物。

疏水缔合聚合物(HAWP)是在聚丙烯酰胺分子链中引入少量(一般小于 2%)缔合型单体形成的水溶性聚合物[31]。根据疏水基在分子链上的分布不同,分为接枝型和嵌段型。疏水缔合聚合物存在临界缔合浓度(CAC),即高于 CAC 时,聚合物分子链开始聚集[32],链间疏水基发生缔合,形成空间网状结构,实现增黏性能的提升[33]。

1967 年,DUBIN 和 STRAUSS[34]合成第一种疏水缔合聚合物,是由烷基乙烯基醚与马来酸共聚而成。而疏水缔合聚合物第一次用于提高采收率是由 Landoll Corporation(美国蓝道)在 1985 年疏水缔合聚合物的美国专利中提出。此后研究人员研发出多 种疏水单体,形成了多种类型的疏水缔合聚合物。MAIA 等[35]利用 N,N-二己基丙烯酰胺合成出疏水缔合聚合物,并与工业化产品 PHPA 做比较,发现外加 Na^+ 不会影响疏水缔合聚合物溶液的黏度。马喜平等[36]合成出一种疏水性季铵盐单体 ADMA-16。并以 AM、DMC 与 ADMA-16 为反应单体合成了一种三元疏水缔合聚合物 PADA,具有良好的耐温抗盐性。除此之外,近年来,许多学者研发出一些特殊分子结构的疏水缔合聚合物,具有良好的应用前景。SHARKER 等[37]通过 RAFT 链转移自由基聚合方法合成了17个聚合物臂的核交联聚(N-异丙基丙烯酰胺)星型聚合物,具有温度响应和pH响应特性;刘锐[38]利用自制的亲水多官能度大分子PAMAMF与AM、AA、阳离子疏水单体(DMDCC)共聚合成出超支化缔合聚合物,超支化缔合聚合物相比于普通疏水缔合聚合物具有更强的黏弹性。

(3) 温增黏聚合物(TVP)。

温增黏型聚合物通过在聚合物分子链中引入具有较低的临界溶解温度(LCST)特性的疏水温敏嵌段或接枝形成[39],在高温高矿化度油藏具有良好的应用前景。常见的温敏型单体包括聚(环氧乙烷)-聚(环 氧丙烷)(PEO-PPO)、聚(N-异丙基丙烯酰胺)(PNI-PAM)和基于丙烯酰胺二丙酮(MPAD)的大单体共聚物等[40-41]。

温增黏型聚合物在室温下表现出与常规高分子溶液相似的性质,但当温度高于LCST时,温敏基团依靠物理作用相互缠结形成疏水微区,使得溶液黏度大幅提高[42]。

1994年,HOURDET等[43]利用聚丙烯酸胺修饰PEO温敏基团,合成出温增黏型聚合物,并首次将其 引入油田开发领域。此后,许多学者也开展了适用于高温高矿化度油藏的温增黏聚合物的研发工作。WANG等[44]通过自由基聚合合成出温增黏型聚合物P(AM-MPAD),并研究了其在高温高盐条件下的性能。结果表明外加盐浓度的增大能够降低LCST,使聚合物在较低温度下就可以表现出温增黏特性。郭睿威等[45]对温增黏聚合物聚(N-异丙基丙烯酰胺)(PNIPAm)进行了研究,结果表明随着温敏基团接枝率和链长的增大,温增黏效应逐渐增强。

(4) 纳米颗粒增强聚合物。

纳米颗粒增强聚合物是向聚合物中引入纳米颗粒,如二氧化硅、纳米氧化铝、氧化钛、氧化镍等,通过氢键作用或其他分子间作用力使纳米颗粒与聚合物分子结合,利于纳米颗粒自身的特性,改善聚合物的耐温抗盐性及抗剪切性,是一种应用前景广阔的聚合物驱油体系。ZHENG等[46]将纳米二氧化硅使用硅烷偶联剂十六烷基三甲氧基硅烷进行改性,使纳米二氧化硅表面接枝疏水长链,增强纳米二氧化硅与疏水缔合聚合物分子间的相互作用,从而表现出更优异的增黏性及耐温抗盐性。

1.2.3 特殊类型聚合物

高温高矿化度油藏的开发使耐温抗盐聚合物的研究不断深入。此外,海上油田、稠油油藏等特殊类型油藏的开发也促进了速溶聚合物、两亲聚合物等特殊类型聚合物的合成与研究。

(1) 速溶聚合物。

随着原油需求量的增加和陆上油田开发的深入,近年来海上油田开发数量逐渐增多,聚合物驱油技术在海上油田试验规模逐渐扩大。与陆上油田相比,海上平台空间狭小,难以通过增加熟化罐数量或容积来提高熟化效果。此外,海上油田矿化度高,增大了聚合物溶解难度。聚合物干粉溶解效果差不仅会影响聚合物自身增黏性和驱油效果,而且会造成聚合物溶液注入压力虚高甚至造成注入困难。

针对上述问题，课题组提出研发速溶聚合物用以改善聚合物驱在海上油田的开发效果。目前主要的研究思路有：① 引入亲水的离子型结构单元或适量的其他单体进行共聚改性，减少聚丙烯酰胺链上酰胺基及其氢键的数量；② 在合成过程或后处理过程中加入适量的能与酰胺基产生氢键的低分子物质（如致孔剂），以减少聚丙烯酰胺链间氢键数量，或加入亲水性的表面活性剂即渗透剂，使聚丙烯酰胺颗粒与水接触后在表面形成水膜，降低水的表面张力，促进水向颗粒内的扩散和聚合物的溶解；③ 改变聚合物产品剂型 。采用反相乳液聚合法生产反相乳液产品或水分散聚合法生产水包水乳液。乳液型聚丙烯酰胺的分散相是高浓度的聚合物溶液，分散粒子尺寸很小，溶解时不需要经过溶胀阶段只是浓溶液的稀释。ZHENG 等[47]利用 PDMC 为稳定剂，浓硫酸铵溶液为分散介质，通过双水相聚合法制备了溶解性能良好的疏水缔合型聚合物 HAPAM，该聚合物溶液 20 s 内黏度即可达到峰值。上述方法虽然能够有效提高聚合物的溶解速率，但是合成过程中需要大量有机溶剂及表面活性剂，生产成本高且易对环境造成污染。也有许多学者仅通过优化聚合物合成工艺，研发出具有速溶能力的聚合物。

（2）两亲聚合物。

随着常规油田产量的递减，稠油油田的开采对于保证我国石油能源的供应具有重要意义。但是稠油中含有大量的胶质沥青质，流动性差，难以开采，需要对稠油进行降黏处理。稠油冷采是稠油开发的重点研究方向，即通过加入化学降黏剂来降解原油中的胶质沥青质或使原油乳化成为 O/W 型乳状液从而实现原油的化学降黏。表面活性剂驱是稠油冷采的重要技术，但是对于非均质性强的稠油油藏，表面活性剂容易沿高渗层渗流。许多学者提出将两亲单体引入聚合物中，合成两亲聚合物，从而综合聚合物和表面活性剂的技术优势，促进聚合物驱在稠油油藏的应用。张向峰等[48]采用沉淀法研究了两亲聚合物中高分子量组分（HM）和低分子量组分（LM）各自的流变和乳化性质，结果发现 HM 组分对黏性起主导作用，而 LM 组分具有较强的乳化能力。朱洲等[49]合 成了不同疏水基含量并具有盐增黏特性的丙烯酰胺类甜菜碱两亲聚合物 PAMA，在高矿化度条件下，PAMA 能够形成稳定的 O/W 型乳状液，有利于高矿化度稠油油藏提高采收率。于斌[50]研究了两亲聚合物复配体系的性能，将 S 型两亲聚合物与 L 型两亲聚合物复配，复配体系中的 S 型两亲聚合物的聚醚结构链段与 L 型两亲聚合物的异构烷烃链形成类混合胶束，增强体系的结构强度，实现对两亲聚合物体系增黏性能和乳化性能的调控。

（3）核壳型聚合物。

聚合物注入过程中经历配聚污水及注聚管道内的 Fe、S 腐蚀、井口炮眼剪切，近井地带岩石吸附，实际的地下工作黏度有一定损失，不利于聚合物的深部调驱。本课题组通过分子包埋法，将 HPAM 包埋在外壳中，形成核—壳型聚合物，保护 HPAM 分子，从而实现延时增黏和靶向增黏，提高聚合物在地层深部的流度控制能力。张超[51]评价了核壳型聚合物的延时增黏性和深部封堵能力，结果表明核壳型聚合物在 85℃下老化 90 d 后，黏度保留率可达 92.11%，其受溶液矿化度、pH 的影响，岩心的吸附量均小于聚丙烯酰胺,且 50 cm 长岩心 3 个测压点均检测到压力上升,具有良好的深部运移封堵特性。

驱油用聚合物研发的不断深入归功于不同类型油藏的逐渐开发,未来驱油用聚合物研发重点仍以提高各种类型油藏适应性为主。保证驱油用聚合物具有良好的溶解性、增黏性、耐温耐盐性、抗剪切性、抗吸附性、注入性。同时,也需保证合成聚合物的成本。

2 聚合物驱矿场应用效果

世界上已开展的聚合物驱大部分应用于砂岩油藏,因为聚合物极易吸附在碳酸盐岩油藏的岩石表面。但目前许多聚合物驱项目在碳酸盐岩油藏也逐渐开展。截至目前,全世界范围内 24 个国家开展了 733 个聚合物驱项目,均以陆上砂岩油藏为主,仅有 8 个海上聚合物驱,其中约仅 1/7 为碳酸盐岩油藏[52]。

2.1 常规聚合物驱矿场试验

最初的聚合物驱仅限于室内研究,但许多研究人员报道了聚合物驱的一些现场应用,2012 年 MOE SOE LET 等[53]报道了 Suriname 的 Tambaredjo 油田采用注入 25~40 mPa·s 的聚合物溶液成功采出非均质储层中 500 mPa·s 黏度的原油,进一步说明聚合物驱在稠油开采中具有很大潜力。2001 年,Yabin 报道了大庆油田聚合物驱的矿场应用,大庆油田聚合物驱的矿场应用始于 1996 年,聚合物驱提高采收率比水驱高出 12%。而后,大庆油田在聚合物驱的基础上大足发展了二元或三元复合驱。据报道 2019 年大庆油田结合同心分层注入技术和分质注入技术,油井累计增油将近 126×104 t[54]。

相对于 HPAM,黄原胶在聚合物驱油中的应用报道较少。黄原胶首次应用是在挪威的 Vorhopo-Knesebeck 油田和德国的 Eddesse-Nord 油田[55]。一般来说,黄原胶在较高的盐度和温度下比丙烯酰胺更稳定,但它们都在高温高盐度(HTHS)条件下降解。因此,这类合成生物聚合物不适用于高盐度、高温储层[56]。

2.2 高温高盐聚合物驱矿场试验

进入 20 世纪后,聚合物驱在高温高矿化度油藏的应用受到限制,促进了许多新型聚合物的研发[57]。如对聚丙烯酰胺进行硫化处理、温增黏聚合物(TVPs)、纳米颗粒增强聚合物体系[58]。但目前矿场应用较多的仍为耐温抗盐聚丙烯酰胺类聚合物,如表 1 所示。后两种方式仍处在实验室研究阶段,尚未有矿场试验报道。

胜利油田报道了 2012 年 11 月胜利油田在胜坨油田二区 Ed34 单元开展了梳形抗盐聚合物驱先导试验[59]。试验区含油面积 3.2 km^2,地质储量 421×10^4 t,地下原油黏度 70 mPa·s,油藏温度 70℃,矿化度 18 035 mg/L,钙镁离子含量 678 mg/L,属于典型的高盐高钙镁型油藏。经聚驱后,含水由 96.5%降至 88.5%,下降 8.0 个百分点。日产油量由 86 t 上升至 268 t,增加了 182 t,提高采收率 2.5%。

表1 国内外一些高温高盐油藏聚合物驱矿场试验

Table1 Fieldtest of polymer floodinginsomehightemperatureandhighsalinity reservoirs in Chinaandother countries

国家	油田	开始时间	岩性	聚合物类型	聚合物浓度/(mg·L^{-1})	注入孔隙体积倍数	注入方式	结果评价
印度	Sanand	1985年04月	砂岩	丙烯酰胺	800	0.15	梯式	提高采收率11%,投资回报率28%
德国	Edesse-Nord	1985年10月	砂岩	黄原胶聚合物	800~1 000	0.20	一体	增产5%~7%,经济成功
美国	White Castle	1986年01月	砂岩	聚丙烯酰胺	500	0.40	一体	未评价
德国	V-K	1989年12月	砂岩	黄原胶聚合物	500~600	0.40	梯度	提高采收率6%,经济成功
德国	Ploen-ost	1989年05月	砂岩	合成聚合物	1 000	0.60	梯度	大幅增产,经济成功
美国	Adon Road	1990年10月	砂岩	胶态分散凝胶	600	0.30	梯度	提高采收率18%,0.72美元/桶
中国	双河油田	1994年02月	砂岩	聚丙烯酰胺	800		梯度	提高采收率10%,吨油成本390元/t
中国	胜利油田胜二区	2012年11月	砂岩	聚丙烯酰胺	2 200	0.65	梯式	提高采收率7.3%

将聚合物微交联或者制作成凝胶颗粒可有效提升聚合物耐温耐盐性能。WU等[60]研究了一种超高温超高矿化度聚合物微球,在中国某油田成功应用。在该研究中,采用了微凝胶颗粒(SMG)作为新型高分子材料,可在高温条件下保持稳定,采收率从44%提高至66%。

3 结论

在对前人工作总结的基础上,对聚合物驱未来发展的主要结论和建议如下:

(1) 聚合物驱通过增加水相黏度,降低水油流度比,提高波及系数;通过自身黏弹性对原油产生拉拽作用从而提高洗油效率。但随着多种适用于不同类型油藏聚合物的研发,对聚合物驱油机理的研究需要继续深入。

(2) 聚合物能够适用的油藏条件不断拓宽,TVPs(温增黏聚合物)、疏水缔合聚合物、耐温抗盐共聚物的研发使聚合物能够应用于高温高盐油藏;速溶聚合物的研发提高了海上油田的配聚效率;两亲聚合物的研发将聚合物驱引入稠油油藏。未来适用于多种类型油藏的功能型聚合物的研发是聚合物驱发展的重要方向,但是如何降低功能型聚合物的研发及应用成本也是亟须解决的问题。

(3)目前聚合物驱矿场实验仍以耐温抗盐部分水解聚丙烯酰胺、交联聚合物凝胶、聚合物微球、聚表二元复合体系为主,如何将新型聚合物的研发成果转化为矿场应用的采收率提高是今后聚合物驱的重点发展方向。

参考文献

[1] 钱兴坤,刘朝全,姜学峰,等. 全球石油市场艰难平衡发展风险加大—2019年国内外油气行业发展概述及2020年展望[J]. 国际石油经济,2020,27(1):2-9.

QIAN X K, LIU Z Q, JIANG X F, et al. Overview of the domestic and foreign oil and gas industry development in 2019 and outlook for 2020[J]. International Petroleum Economics, 2020, 27 (1):2-9.

[2] 阿依加马力·艾尼. 高含水油田水驱聚驱后剩余油分布特征研究及提高采收率对策[D]. 青岛:中国石油大学(华东),2017. AINI A. Study on residual oil distribution after water and polymer flooding for high water cut field and the method of improving oil recovery[D]. Qingdao:China University of Petroleum(East China), 2017.

[3] MOGENSENA K, MASALMEHA S. A review of EOR techniques for carbonate reservoirs in challenging geological settings[J]. Journal of Petroleum Science and Engineering, 2020, 195:1-13.

[4] SCHEXNAYDER P, BAUDOIN N, CHIRDON W M. Enhanced oil recovery from denatured algal biomass: Synergy between conventional and emergent fuels[J]. Fuel, 2020.

[5] 王斌,周迅,王敏,等. 三次采油技术在中原油田的应用进展[J]. 油田化学,2020,37(3):552-556.

WANG B, ZHOU X, WANG M, et al. Application of tertiary oil recovery technology in Zhongyuan Oilfield[J]. Oilfield Chemistry, 2020, 37(3): 552-556.

[6] 魏华旭. 三次采油阶段提高采收率的措施[J]. 化学工程与装备,2020,49(10):105.

WEI H X. Measures to enhance oil recovery in tertiary oil recovery stage[J]. Chemical Engineering & Equipment, 2020, 49(10): 105.

[7] 刘玉章. EOR聚合物驱提高采收率技术[M]. 北京:石油工业出版社,2006.

LIU Y Z. EOR Polymer flooding enhanced oil recovery technology[M]. Beijing: Petroleum Industry Press, 2006.

[8] RELLEGADLA S, PRAJAPAT G, AGRAWAL A. Polymers for enhanced oil recovery: Fundamentals and selection criteria [J]. Applied Microbiology &

Biotechnology, 2017, 101(15): 1-16.

[9] 侯吉瑞,陈宇光,吴璇,等. 聚合物表面活性剂溶液微观驱油特征[J]. 油田化学,2020,37(2):292-296.

HOU J R, CHEN Y G, WU X, et al. Microscopic oil displacement characteristics of polymeric surfactant solution[J]. Oilfield Chemistry, 2020, 37(2): 292-296.

[10] 金亚杰. 国外聚合物驱油技术研究及应用现状[J]. 非常规油气,2017,4(1): 116-122.

JIN Y J. Progress in research and application of polymer flooding technology abroad[J]. Unconventional Oil & Gas, 2017, 4(1):116-122.

[11] 崔名喆,张建民,吴春新,等. 渤海油田低渗透油藏黏性指进特性分析[J]. 中国石油勘探,2018,23(5):94-99.

CUI M Z, ZHANG J M, WU C X, et al. Analysis of viscous fingering characteristics of low permeability reservoirs in Bohai oilfield[J]. China Petroleum Exploration, 2018, 23(5):94-99.

[12] ALMANSOUR A O, ALQURAISHI A A, ALHUSSINAN S N, et al. Efficiency pf Enhanced Oil Recovery Using Polymer-ugmented Low Salinity Flooding [J]. Journal of Petroleum Exploration & Production Technology, 2017, 7(1): 1149-1158.

[13] PYE D J. Improved secondary recovery by control of water mobility[J]. Journal of Petroleum Technology, 1964, 16(8):911-916.

[14] SANDIFORD B B. Laboratory and field studies of water floods using polymer solutions to increase oil recoveries[J]. Journal of Petroleum Technology, 1964, 16(8):917-922.

[15] 未志杰,康晓东,何春百,等. 非均质稠油油藏聚合物驱吸液剖面变化规律[J]. 科学技术与工程,2018,18(8): 61-66. WEI Z J, KANG X D, HE C B, et al. Polymer flood injection conformance behavior of heterogeneous heavy-oil reservoirs[J]. Science Technology and Engineering, 2018, 18(8): 61-66.

[16] 孙刚. 利用双极化干涉法研究聚合物与原油的相互作用[J]. 油田化学,2017, 34(2):290-295.

SUN G. Insight into interactions between polymer and crude oil with dual polarization interferometry[J]. Oilfield Chemistry, 2017, 34(2):290-295.

[17] ABEL A, McCORMICK C L. Mechanistic insights into temperature-dependent trithiocarbonate chain-end degradation during the RAFT polymerization of N-arylmethacrylamides[J]. Macromolecules, 2016, 49(2):465-474.

[18] 邢恩浩,陶鑫,金磊,等. 阻力系数和残余阻力系数的测定及影响因素的研究

[J]. 辽宁化工,2016,45(3):284-287.

XING E H, TAO X, JIN L, et al. Determination and influence factors of resistance coefficient and residual resistance coefficient [J]. Liaoning Chemical Industry, 2016, 45(3):284-287.

[19] ZHANG H C, GOU S H, ZHOU L H, et al. Synthesis and properties of betaine hydrophobic modified polymer flooding[J]. Applied Chemical Industry, 2019, 48(11):2627-2631.

[20] KURŞUN I, IPEKOĞLU B, ÇELIK M S, et al. Flocculation and adsorption-desorption mechanism of polymers on albite[J]. Developments in Mineral Processing, 2000, 13:24-30.

[21] 沈天阳,徐兆冉. POSS的合成及其在传统聚合物改性中的研究进展[J]. 化工管理, 2018,33(18):29-30.

SHEN T Y, XU Z R. Synthesis of POSS and its research progress in modification of traditional polymers[J]. Chemical Engineering & Equipment, 2018, 33(18):29-30.

[22] ELHAEIAB R, KHARRATC R, MADANI M. Stability, flocculation, and rheological behavior of silica suspensions-augmented polyacrylamide and the possibility to improve polymer flooding functionality[J]. Journal of Molecular Liquids, 2020.

[23] YOO H S, KIM H S, SUNG W M, et al. An experimental study on retention characteristics under two-phase flow considering oil saturation in polymer flooding[J]. Journal of Industrial and Engineering Chemistry, 2020, 87:120-129.

[24] 胡钰灵, 全红平, 黄志宇. 疏水改性黄原胶XG-C16溶液性能及微观结构探究[J]. 应用化工,2019,48(11):2657-2661. HU Y L, QUAN H P, HUANG Z Y. Study on solution properties and microstructure of hydrophobic modified Xanthan gum—XG-C16[J]. Applied Chemical Industry, 2019, 48(11):2657-2661.

[25] XU L, XU G Y, LIU T, et al. The comparison of rheological properties of aqueous welan gum and xanthan gum solutions[J]. Carbohydrate Polymers, 2013, 92(1):516-522.

[26] 祝仰文. 超高分三元共聚物流变特性及驱油性能[J]. 油气地质与采收率, 2018,25(6):78-83.

ZHU Y W. Study on rheology and oil displacement properties of ultra high molecular weight terpolymer[J]. Petroleum Geology and Recovery Efficiency, 2018, 25(6):78-83.

[27] 李新勇, 罗攀登, 刘坤, 等. 含有吗啉基的耐温抗盐聚合物合成及性能[J]. 当代化工,2020,49(5):838-841.

LI X Y, LUO P D, LIU K, et al. Synthesis and performance evaluation of salt-resistant and heat-tolerant polymer containing morpholine groups[J]. Contemporary Chemical Industry, 2020, 49(5):838-841.

[28] 孙群哲,宋华,李锋,等. 三元驱油用磺化聚丙烯酰胺的合成与性能研究[J]. 化学工业与工程技术,2014,35(3):41-44. SUN Q Z, SONG H, LI F, et al. Study on synthesis and property of terpolymer sulfonated polyacrylamide used in EOR[J]. Journal of Chemical Industry & Engineering, 2014, 35(3):41-44.

[29] 尚克剑. 两性聚合物的合成及应用[J]. 新疆化工,2016,41(1):5-10.

SHANG K J. Synthesis and application of amphoteric polymers[J]. Xinjiang Chemical Industry, 2016,41 (1):5-10.

[30] 伊卓,刘希,方昭,等. 三次采油耐温抗盐聚丙烯酰胺的结构与性能[J]. 石油化工,2015,44(6):770-777.

YIZ, LIU X, FANG Z, et al. Structure and properties of temperature-tolerant and salt-resistant polyacrylamide for tertiary oil recovery [J]. Petrochemical Technology, 2015, 44(6):770-777.

[31] BERRET J F, CALVET D, COLLET A. Fluorocarbon associative polymers [J]. Current Opinion in Colloid & Interface Science, 2003, 8(3) : 296-306.

[32] BAI Y R, SHANG X S, WANG, Z B, et al. Experimental study on hydrophobically associating hydroxyethyl cellulose flooding system for enhanced oil recovery[J]. Energy & Fuels, 2018, 32(6):6713-6725.

[33] 王晓藜. 聚丙烯酰胺类疏水缔合聚合物的合成与表征[D]. 济南:山东大学,2015.

WANG X L. Study on synthesis and characterization of hydrophobically associating acrylamide-based copolymers[D]. Jinan:Shandong University, 2015.

[34] DUBIN P L, STRAUSS U P. Hydrophobic hypercoiling in copolymers of maleic acid and alkyl vinyl ethers[J]. Journal of Physical Chemistry, 1970, 74(14): 2842-2847.

[35] MAIA A M S, BORSALI R, BALABAN R C. Comparison between a polyacrylamide and a hydrophobically modified polyacrylamide flood in a sandstone core [J]. Materials Science and Engineering, 2009, 29(2):505-509.

[36] 马喜平,廖明飞,董江洁,等. 疏水缔合聚合物 PADA 的合成与性能评价[J]. 应用化工, 2020,49(3):669-673.

MA X P, LIAO M F, DONG H J, et al. Synthesis and performance evaluation of hydrophobic association polymer PADA[J]. Applied Chemical Industry, 2020, 49(3): 669-673.

[37] SHARKER K K, TAKESHIMA S, TOYAMA Y, et al. pH-and thermo-

responsive behavior of PNIPAM star containing terminal carboxy groups in aqueous solutions[J]. Polymer, 2020:122735.

[38] 刘锐. 超支化缔合聚合物的制备及驱油性能[J]. 西南石油大学学报(自然科学版),2015,37(2):145-152.

LIU R. Preparation of hyperbranched association polyacrylamide and its oil displacement properties[J]. Journal of Southwest Petroleum University(Science & Technology Edition), 2015, 37(2):145-152.

[39] LI X E, XU Z, YIN H Y, et al. Comparative studies on enhanced oil recovery: Thermoviscosifying polymer versus polyacrylamide[J]. Energy & Fuels, 2017, 31(3):2479-2487.

[40] CHEN Q S, WANG Y, LU Z Y, et al. Thermo-viscosifying polymer used for enhanced oil recovery: Rheological behaviors and core flooding test[J]. Polymer Bulletin, 2013, 70(2): 391-401.

[41] ROY D, BROOKS W L A, SUMERLIN B S. New directions in thermoresponsive polymers[J]. Chemical Society Reviews, 2013, 42(17):7214-7243.

[42] MOGOLLóN J L, YOMDO S, SALAZAR A, et al. Maximizing a mature field value by combining polymer flooding, well interventions, and infill drilling[C]// paper SPE-194652-MS presented at the SPE Oil and Gas India Conference and Exhibition, 9-11 April, 2019, Mumbai, India.

[43] HOURDET D, L'ALLORET F, AUDEBERT R. Reversible thermothickening of aqueous polymer solutions[J]. Polymer, 1994, 35(12):2624-2630.

[44] WANG Y, FENG Y J, WANG B Q, et al. A novel thermoviscosifying water-soluble polymer: Synthesis and aqueous solution properties[J]. Journal of Applied Polymer Science, 2010, 116(6): 3516-3524.

[45] 郭睿威,吴大成,鹿现栋,等. 热缔合接枝物 HPAM-g-PNIPAm 的温敏增稠性能[J]. 石油化工,2003,32(8):690-694.

GUO R W, WU D C, LU X D, et al. Thermothickening properties of graft copolymer HPAM-g-PNIPAm[J]. Petrochemical Technology, 2003, 32(8):690-694.

[46] ZHENG C, CHENG Y, WEI Q, et al. Suspension of surface-modified nano-sio2 in partially hydrolyzed aqueous solution of polyacrylamide for enhanced oil recovery[J]. Colloids & Surfaces A Physicochemical & Engineering Aspects, 2017, 524(1):169-177.

[47] ZHENG C C, HUANG Z H. Self-assembly and regulation of hydrophobic associating polyacrylamide with excellent solubility prepared by aqueous two-phase polymerization[J]. Colloids & Surfaces A Physicochemical & Engineering Aspects,

2018, 555:621-629.

[48] ZHANG X F, YANG H B, WANG P X, et al. Oil-displacement characteristics and EOR mechanism of amphiphilic polymers with two molecular weights[C]// paper SPE-192385-MS presented at the SPE Kingdom of Saudi Arabia Annual Technical Symposium and Exhibition, 23-26 April, 2018, Dammam, Saudi Arabia.

[49] 朱洲,康万利,杨红斌,等. 磺基甜菜碱型两亲聚合物的合成及其流变特性[J]. 石油化工高等学校学报,2017,30(5):32-36. ZHU Z, KANG W L, YANG H B, et al. Synthesis and rheological properties of a sulfobetaine amphiphilic polymer[J]. Journal of Petrochemical Universities, 2017, 30(5):32-36.

[50] 于斌. 驱油两亲聚合物性能调控方法及增效机理研究[D]. 青岛:中国石油大学(华东),2019.

YU B. Study on property control of amphiphilic polymers for oil displacement and its synergistic mechanism [D]. Qingdao: China University of Petroleum (East China), 2019.

[51] 张超. 延时增粘聚合物增粘特性及运移规律研究[D]. 青岛:中国石油大学(华东),2016.

ZHANG C. Study on Thickening Properties and Migration Law of Delayed Viscosity Increasing Polymer [D]. Qingdao: China University of Petroleum (East China), 2016.

[52] JIANG Z Z, ZHU J R. Cationic polyacrylamide: Synthesis and application in sludge dewatering treatment[J]. Asian Journal of Chemistry, 2014, 26(3):629-633.

[53] MOE SOE LET K P, MANICHAND R N, SERIGHT R S. Polymer flooding a 500-cpoil[C]// paper SPE-154567-MS presented at the SPE Improved Oil Recovery Symposium, 14-18 April, 2012, Tulsa, Oklahoma, USA.

[54] 侯巍. 大庆油田三次采油提高采收率技术研究[J]. 化工管理,2019,34(10):222-224.

HOU W. Analysis of influencing factors of polymer flooding and application of stratified injection technology in Daqing Oilfield [J]. Chemical Enterprise Management, 2019, 34(10):222-224.

[55] LI J R. Experimental investigation and simulation of polymer flooding in high temperature high salinity carbonate reservoirs[D]. Abu Dhabi: The Petroleum Institute, 2015.

[56] KAMAL M S, SULTAN A S, AL-MUBAIYEDH U A, et al. Review on polymer flooding: Rheology, adsorption, stability, and field applications of various

polymer systems[J]. Polymer Reviews, 2015, 55(3):491-530.

[57] SHENG J J, LEONHARDT B, AZRI N. Status of polymer-flooding technology[J]. Journal of Canadian Petroleum Technology, 2015, 54(2):116-126.

[58] ZHU D W, WEI L M, WANG B Q, et al. Aqueous hybrids of silica nanoparticles and hydrophobically associating hydrolyzed polyacrylamide used for EOR in high-temperature and high-salinity reservoirs[J], Energies, 2014, 7(6):3858-3871.

[59] 赵方剑,曹绪龙,祝仰文,等. 胜利油区海上油田二元复合驱油体系优选及参数设计[J]. 油气地质与采收率,2020,27(4):133-139.

ZHAO F J, CAO X L, ZHU Y W, et al. Injection parameters optimization of binary combination flooding system in offshore oil field, Shengli oil province[J]. Petroleum Geology and Recovery Efficiency, 2020, 27(4):133-139.

[60] WU X C, YANG Z J, XU H B, et al. Success and lessons learned from polymerflooding a ultra high temperature and ultra high salinity oil reservoir-A case study from West China[C]// paper SPE-179594-MS presented at the SPE Improved Oil Recovery Conference, 11-13 April, 2016, Tulsa, Oklahoma, USA.

(编辑 常燕)

第八章　其他

DP-4 型泡沫剂的研制及其性能评价

陈晓彦[1] 王其伟[1] 曹绪龙[1] 李向良[1] 周国华[1] 张连壁[2]

1. 胜利油田有限公司地质科学研究院;2. 胜利油田有限公司孤东采油厂

摘要:在筛选发泡性能好、稳定性强的 AK 系列活性剂和稳泡能力强的 TY-10 的基础上,采用复配增效原理,研制成 DP-4 型泡沫剂。驱油试验时,对其发泡能力、稳定性与地层水的配伍性及耐油性进行了评价。结果表明,DP-4 型泡沫剂的泡沫性能良好,可提高采收率 18%。

关键词:泡沫剂;泡沫驱;强化泡沫驱;半衰期;采收率

中图分类号:TE357.46 文献标识码:B 文章编号:1009-9603(2003)02-0054-02

引言

泡沫在油藏运移过程中的视粘度远远大于其他驱替介质的粘度[1],而且泡沫粘度随渗透率的增大而增大。泡沫的另一特性是遇水稳定,遇油破灭,因此对于油水具有选择性封堵能力[2]。泡沫应用于采油具有众多的优点,其致命弱点是不稳定性,因此,研制性能优良的泡沫剂是泡沫驱的关键[3,4]。通过借鉴现有的泡沫剂,研制了 DP-4 泡沫剂,并对其性能进行了评价。

1 室内试验

1.1 仪器与材料

采用 Ross-miles 泡沫仪,控温精度为±0.5℃的岩心流动试验装置进行参数测定。试验材料有表 面活性剂 S-80、JFC 、O-20、SA-20、OP-10、OP-12、OP-15、十二烷基苯磺酸钙、甘油单月桂酸酯硫酸钠、FE-6、ABS、SP-169、SDS、ABS、AK、S-6、R-7、BQ-2、TY-10 及国产聚合物。

1.2 试验方法

1.2.1 起泡体积及半衰期的测定

将泡沫剂及表面活性剂用黄河水配成 0.5%的溶液,按 GB/T 13173.6-91《洗涤剂发泡力的测定》测定 30℃、常压下其起泡体积及半衰期。

1.2.2 耐油性试验

将 50 mL 泡沫剂及 10 g 模拟油(原油与煤油比为 1∶1)加入泡沫仪接受器中,测定起泡体积及稳定性。

1.2.3 泡沫驱油试验

用石英砂充填直径 25 mm、长 300 mm 的模型,抽空并饱和水、油,连接到驱替流程中,在 60 ℃、6 MPa 回压条件下,用黄河水驱替岩心至含水 98%,注入 0.3 PV 泡沫剂段塞,记录两端的压差及采出程度。

1.3 主剂筛选

用罗氏泡沫法测定了 14 种泡沫剂的起泡体积、半衰期及其与黄河水的配伍性,结果发现 FE-6、AK 的起泡体积分别为 360,370 mL,半衰期分别为 70,80 min。AK 与黄河水的配伍性也好于其他试剂,所以选择 AK 为主剂。

1.4 助剂筛选

单一成分泡沫剂的泡沫性能很难达到理想的水平,因此,泡沫剂绝大多数为复配物。对泡沫剂起协同作用的物质有不同成分的活性剂、无机盐类、高分子聚合物、醇类。高分子聚合物主要通过增加体系的粘度来增加泡沫膜厚度和膜弹性,而一些活性剂是通过与主剂的不同电性交替排列,增加膜分子密度,减少气体穿透,增加泡沫稳定性。

首先筛选出稳泡性能良好的 CN、T Y-10 两种产品作为泡沫剂的助剂,再将两者与主剂 AK 进行配伍性试验。助剂 CN 与主剂 A K 的浓度比为 10∶时,起泡体积为 400 mL,半衰期为 110 min,其配伍性最好,泡沫性能最强;助剂 T Y-10 与 AK 在不同的浓度下,其泡沫性能发生变化,当 T Y-10 与 AK 之比等于 1∶2 时,起泡体积为 360 mL,半衰期为 350 min,泡沫的综合性能较好。因此,泡沫剂 DP-4 的配方为 30%A K+15%T Y-10+1%CN+64%水。

2 DP-4 型泡沫剂性能评价

2.1 与地层水的配伍性

配制泡沫剂的地层水取自胜坨油田,矿化度为 17 000 mg/L,钙、镁离子含量大于 1 000 mg/L。比较了 DP-4 型泡沫剂及其他几种市售泡沫剂泡沫性能(表 1),可以看出多数产品与地层水的配伍性很差,起泡体积小、半衰期短,而且有沉淀,DP-4 型泡沫剂性能明显好于常规泡沫剂。

表 1 泡沫剂与地层水的配伍性

泡沫剂	起泡体积/mL	半衰期/min	与地层水的配伍性
S-6	280	50	半透明,有白色沉淀物
R-7	270	45	半透明,有白色沉淀物
BQ-2	100	30	浑浊,白色沉淀

（续表）

泡沫剂	起泡体积/mL	半衰期/min	与地层水的配伍性
D P-4	340	665	清亮透明
D P-5	50	10	浑浊，白色沉淀
A BS	340	30	浑浊，白色沉淀

2.2 耐油性

DP-4 型泡沫剂注入油层后，由于油水界面张力远远小于水气表面张力，当三相界面共存时，按界面能趋于减小的规律，活性剂将大量由水气界面转移到油水界面，使水气界面张力升高，破坏了泡沫的稳定性，这也是泡沫遇油易破的主要原因。为了评价 DP-4 泡沫剂的耐油性，在有油存在条件下，对 DP-4 型泡沫剂与其他活性剂进行了泡沫性能试验（表 2），结果表明，DP-4 型泡沫剂与模拟油相遇后，体积减小到 330 mL，消失时间为 60 min，是试验中最好的泡沫剂，说明虽然模拟油对泡沫剂的起泡能力，稳定性影响很大，但可以看出 DP-4 型泡沫剂抗油性效果较好。

表 2　模拟油对泡沫剂稳定性的影响

泡沫剂	泡沫体积/mL	消失时间/min
OP-10	150	11
AEO-9	250	11
ABS	160	10
AK	280	40
S-6	240	20
DP-4	330	60

2.3　驱油效果试验

注泡沫剂的目的是提高原油采收率，聚合物是一种性能良好的驱油剂，可以使泡沫的稳定性更强，将泡沫剂与聚合物复配，可以充分发挥两种体系的优点，起到协同作用（图 1，图 2）。

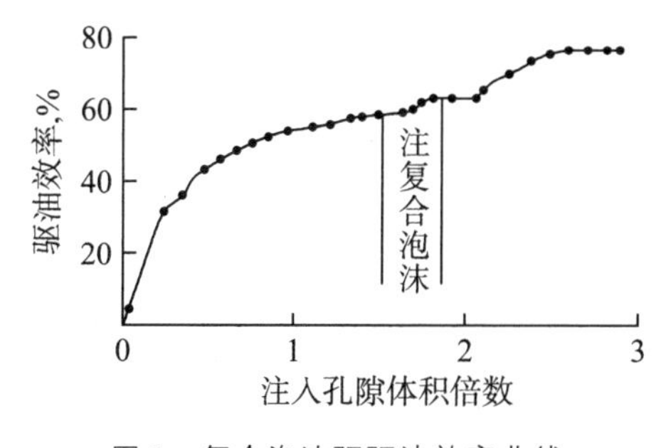

图 1　复合泡沫驱驱油效率曲线

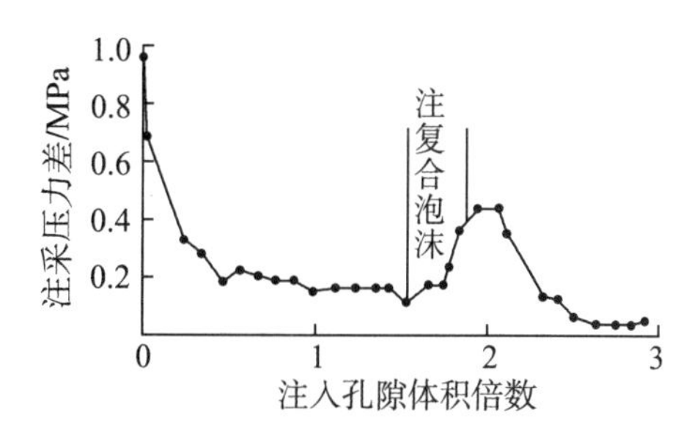

图 2　复合泡沫驱油注采压差曲线

由图 1、图 2 可以看出，当水驱至注入倍数 1.5 PV 时，采出液含水率为 95.8%，按气液比 3∶1 注入 0.3 PV 复合泡沫段塞后转注水，注入压差由 0.16 MPa 上升到 0.43 MPa，采出液含水率由 95.8%下降到最低的 33%，驱油效率从 58.3%提高到 76.5%。说明复合泡沫段塞的注入可大幅提高驱油效率。

3 结论

复配的 DP-4 型泡沫剂具有起泡能力强、半衰期长、耐油性强，与高矿化度地层水的配伍性好等特点。驱油试验表明，复合泡沫驱能够较大幅度地提高采收率，将在三次采油领域起到较大的作用。

参考文献

[1] 佟曼玉. 油田化学. 东营：石油大学出版社，1996.

[2] 周静，谭永生. 稳定泡沫流体的机理研究. 钻采工艺，1999，22(6):21-27.

[3] 麻金海. 胜利金家油田泡沫驱油室内研究. 油田化学，1997，14(2):156-158.

[4] Yang C Z，Hong Y H，Han D K. Analysis and explanation to indus-t rial pilot foam flooding results on the Laojunmiao field in China. SPE 17387.

编辑　高岩

board thickness, radius and location on the law of oil water movement is studied by reservoir numerical simulation. It is demonst rated that the bigger the block board radius is, the longer the effective period and the better development effects are; and the higher the block board location is, the better the development effect is; and the block board thickness hasn't obvious effect on the development effect.

Key words: heavy oil reservoir, bottom water, coning, numerical simulation, block board thickness, block board radius, block board location.

Liu Dating, Zhang Shujuan. Study on the remaining oil distribution of carbonate buried hill reservoir in Renqiu. PGRE, 2003, 10(2):46-47.

According to the comprehensive analysis of reservoir dynamic and static data, it is proposed that the geometric shape of top bed, distribution of interior restraining barrier and developing degree of fractures in buried hill are the main geological factors for controlling the distribution of remaining oil in the reservoir. Through applying various new technologies and methods comprehensively, the above geological factors are described finely. The patterns of remaining oil distribution controlled by various factors are summarized. On the basis of the types of remaining oil distribution, the regions and future direction for tapping the potential of the reservoir are put forward.

Key words: buried hill reservoir, remaining oil distribution, carbonatite, Renqiu oilfield.

Xu Shucheng, Zhang Taibin, Bian Shiju etal. Re-view and stimulation for low productivity & low efficiency exploratory wells. PGRE, 2003, 10(2):48-49.

The reasons of low productivity of some old wells are analyzed, and the stimulation is given by the review of the old wells in recent years. The stimulation treatments have been implemented in selected 22 wells with the result of 15 response producer, 68% of success ratio and 9.5t of average daily incremental oil production for single well. The applying stimulation technique can tap the potential of old wells sufficiently.

Key words: old well review , low productivity & low efficiency well, heavy oil reservoir, stimulation.

Xu Xinli, Yang Ping, You Xingyou. Developing turbidite oil reservoir in high water cut period by horizontal wells—taking Liangjialou oilfield as example. PGRE, 2003, 10(2):50-51.

Middle Es3 turbidite sandstone oil reservoir of Liangjialou has been in high water cut period. Research on fine reservoir description and rule of remaining oil distribut ion shows t hat movable remaining oil has great potential, it enriched in positive micro-tectonics and fracture ridge structural element.

Notable result of high production and non water cut in early period has been achieved by applying horizon-tal wells. Basedn that, optimal design and system management mode of the horizontal wells are summarized.

Key words: turbidite oil reservoir, microtectonics, extra high water cut period, remaining oil, horizontal well, Liang jialou oilfield.

Wang Qinggui, Peng Shangqian. Application of horizontal well technology in Es1 thin oil bearing formations of Ying31 fault block. PGRE, 2003, 10(2):52-53.

The problems of commingled production at high water cut and interlayer interference result in low recovery efficiency in oil reservoirs with thin formations after vertical wells are applied in Ying31 fault block of Dong xin oilfield where oil bearing formations are thin with wide distribution. In order to increaset he produced reserves in thin oil bearing formations and explore the production pattern of remaining oil by horizontal well, pilot experiment is done. The implemented results show that horizontal well technology applied in thin oil bearing formations can improve the seepage flow capacity and development effect of reservoirs.

Key words: horizontal well, thin oil bearing forma-tion, application, Es1, Dongxin oilfield.

Chen Xiaoyan, Wang Qiwei, Cao Xulong et al. Preparation and performance evaluation of DP-4 type of foamagent. PGRE, 2003, 10(2):54-55.

Based on the selection of AK series surfactant with good foaminess and strong foam stability and T Y-10 with good foam stabili ty, DP-4 foaming agent is developed applying the principle of compounding for higher efficiency. Its foaming capacity, stability, compatibility with formation water and oil resistance are evaluated in oil displacement test. The result shows that foaming performance of DP-4 type of foamagent is good, and EOR will increase 18%.

Key words: foamagent, foam flooding, enhanced foam flooding, half life period, recovery efficiency.

Xu Qinglian. Tertiary recovery method for heteroge-neous oil reservoirs. PGRE, 2003, 10(2):56-57.

Based on the areal and interlayer heterogeneous mod-els, different oil displacement modes of waterflood-ing, polymer flooding, ASP flooding, gel flooding as well as gel plus ASP flooding are studied using nu-merical simulation. The results indicate that the pro-file control by gel flooding is the key to exploit the potential in the heterogeneous reservoirs. Gel flooding plus ASP flooding can decrease the oil saturation con-siderably in the low permeable zones of the reservoir, develop the whole reservoir evenly, and then improve the oil recovery obviously.

泡沫封堵能力试验研究

王其伟，曹绪龙，周国华，郭　平，李向良，李雪松

（中国石化胜利油田有限公司地质科学研究院，山东东营　257015）

摘要：借助岩心驱替流程试验，研究了含油模型水驱之后注泡沫、复合泡沫及聚合物驱后注单一泡沫、复合泡沫的泡沫体系封堵能力，结果表明，单一泡沫驱在水驱残余油条件下，难以形成有效的封堵，而复合泡沫体系在各种残余油条件下均能形成有效的封堵调剖能力，证明复合泡沫体系是一种性能良好的调剖驱油剂。

关键词：泡沫；复合泡沫；封堵能力；封堵压差

中图分类号：TE357.46；TE358.3　文献标识码：A

引　言

泡沫体系视粘度高、封堵调剖能力强[1]，封堵能力随渗透率的增大而增大[2]，对油水的封堵具有选择性[3]，因此，泡沫驱已经成为一种很有发展前途的三次采油方法，并且在堵水调剖方面[4]，也有重要的用处。本文通过不同泡沫体系在不同条件下在模型两端所建立的驱替压差及双管模型高低渗管产液百分数曲线，来证明泡沫在不同残余油条件下所具有的堵水调剖能力，为泡沫复合体系在三次采油领域的应用提供可靠的试验依据。

1　试验部分

1.1 主要仪器设备及材料

岩心驱替流程，控温精度±0.5℃；气体质量流量计，控制流量0～30 mL；回压阀，控压0～10 MPa；调压阀，控压精度0.02 MPa；数字压力表，精度0.01 MPa。

模拟地层水：矿化度8 379 mg/L，Ca^{2+}＋Mg^{2+}含量52 mg/L，Na^{+}＋K^{+}含量3 061 mg/L，Cl^{-}含量4 086 mg/L，HCO_3^{-}含量1 085 mg/L，CO_3^{2+}含量95 mg/L，DP-4泡沫剂（自制品），有效含量45％；聚合物3530S，SNFI公司产品，固含量91％，分子量为1 500万。

模拟油：孤岛中一区原油与过滤煤油按9∶1配置的模拟原油；

试验模型：石英砂敲制模型，模型尺寸Φ25 mm×600 mm，渗透率1.5 μm^2；

注入气:不特别指明,注入气体皆为氮气。

1.2 试验方法

泡沫封堵能力试验:将岩心驱替流程连接、安装、调试,将人工敲制的石英砂模型抽空饱和水、饱和油,注入速度 0.5 mL/min,水驱至含水 98%;将聚合物或泡沫体系,注入岩心,压力稳定后,记录各个阶段的模型两端压差。

1.3 试验条件

试验温度:60℃;

回　压:6MPa。

2 结果与讨论

2.1 单一泡沫体系

石英砂充填模型饱和油后,水驱至含水 98%,水驱采出程度为 55%,残余油饱和度为 45%,在水驱残余油条件下注入单一泡沫体系,压差仅从水驱时的 0.05 MPa,提高到泡沫驱的 0.1 MPa,压差提高幅度较小,并且见气早,产气量不稳定。虽然单一泡沫体系在不含油的石英砂模型中泡沫稳定,封堵能力强,根据阻力因子与气液比关系试验,气液比 1∶1 时,阻力因子可达 1 500 倍以上,视粘度为 750 mP·s,模型两端压差为 2.5 MPa。而在含油模型中泡沫稳定性差,不能形成稳定的泡沫体系,说明油对泡沫体系有非常大的影响,也验证了泡沫选择性封堵的特性。

2.2 聚合物驱后注单一泡沫体系

图 1 为饱和油模型水驱至含水 98%后转注聚合物至含水 98%,注泡沫体系的注采压差随 PV 数的变化曲线,水驱聚合物驱后原油采出程度达 72%,剩余油为 28%,从图上可以看出,水驱、聚合物驱之后,单一泡沫体系能够形成有效的封堵,封堵压差从水驱时的 0.1 MPa 上升到聚合物驱的 0.37 MPa,注入泡沫后压差上升到 0.75 MPa,为聚合物浓度 1 800 mg/L 时注采压差的 2 倍,并且随着泡沫注入程度的增加,压差仍会继续增加;试验现象中还发现,注入泡沫之后,压差首先降低然后逐渐增加,但增加的幅度缓慢,并且出口产气量较多;这说明泡沫在残余油 28%时,虽然能够形成泡沫,能够建立起压差,但在此条件下,泡沫体系的稳定性较差。

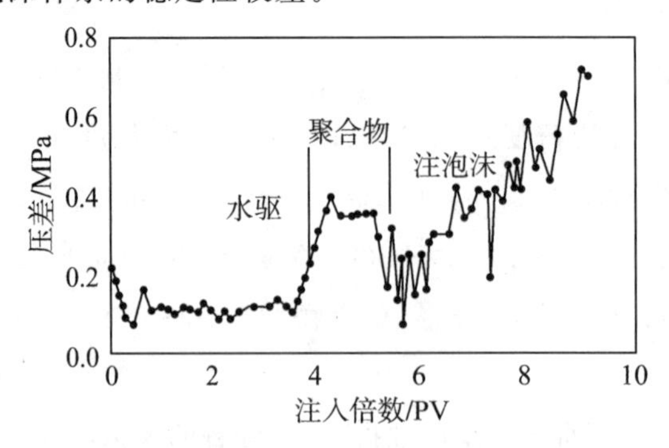

图 1　含油模型水驱、聚合物驱、泡沫驱压差曲线

2.3 复合泡沫体系

图 2 为饱和油模型水驱之后注复合泡沫体系的压差、采出程度、含水随 PV 数综合变化曲线。

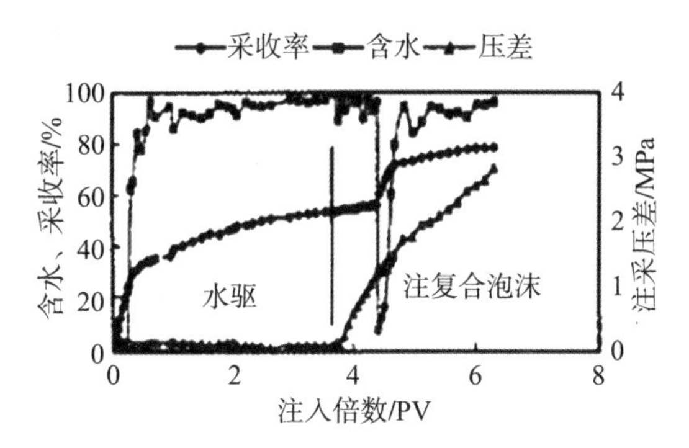

图 2　含油模型水驱泡沫复合驱综合曲线图

从图 2 可以看出，注入复合泡沫段塞后，注采压差上升的速度较快，水驱注采压差为 0.05 MPa，而注入复合泡沫后，注采压差升高到 2.8 MPa，是水驱压差的数十倍，比 1 800 mg/L 聚合物压差高 5 倍以上，并且从压差曲线上升趋势看，泡沫压差仍能够增加，从原油采出程度曲线看，原油采出程度由水驱时的 54%提高到 79%，采出程度提高了 25%，含水曲线有一个较大的漏斗，含水最低下降到 10%，因此认为强化泡沫体在残余油达 46%时，复合泡沫体系稳定，不但封堵调剖能力强，并且有较好的驱油能力。

2.4 聚合物驱后注复合泡沫体系

图 3 为饱和原油模型水驱、聚合物驱后注复合泡沫体系压差随 PV 数变化曲线。

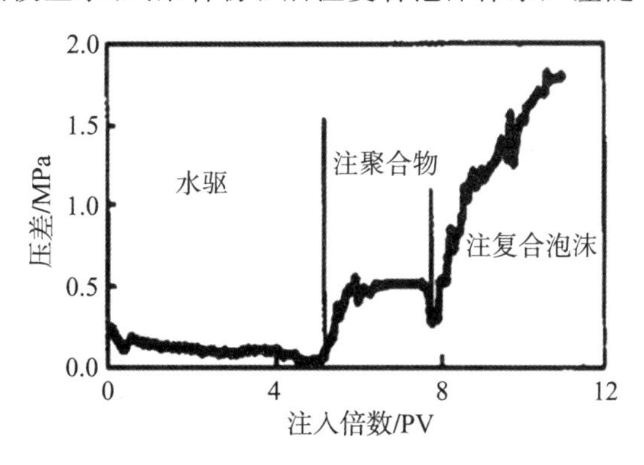

图 3　单管含油模型水驱、聚合物驱、泡沫驱压差曲线

从图 3 可以看出，注采压差由水驱时的 0.05 MPa 上升到聚合物驱的 0.47 MPa。聚合物驱后原油采出程度达 72%，残余油含量为 28%，注入复合泡沫后，注采压差迅速上升到 1.8 MPa，并且随着注入程度的增加，压差仍能提高。试验在注入复合泡沫 0.8 PV 时见气，说明复合泡沫在残余油含量 28%时，有非常好的稳定性，能够形成有效的封堵调剖能力。

2.5 天然气复合泡沫体系

图 4 为水驱之后注天然气复合泡沫体系的压差及采出程度随 PV 数变化的综合曲线。

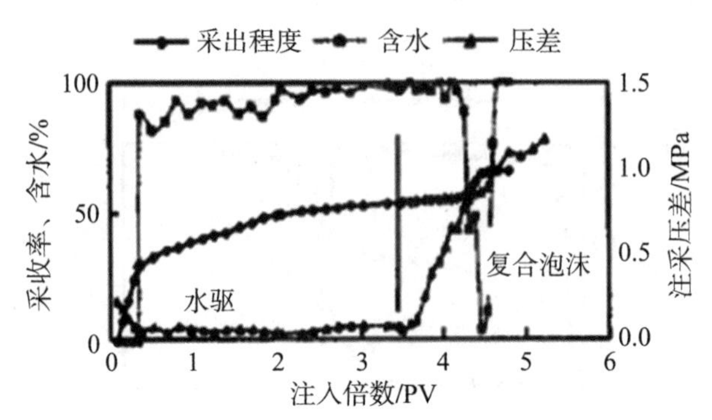

图 4　天然气复合泡沫体系驱油综合曲线图

从图 4 可以看出，水驱结束时采出程度为 52%，残余油含量为 48%，从压差曲线可以看出天然气复合泡沫有良好的封堵调剖能力，压差上升的幅度较大，由注水时的 0.07 MPa，提高到注天然气复合泡沫的 1.17 MPa，上升速度快，并且仍有上升的趋势，从试验现象观察，天然气形成的泡沫稳定，注入倍数达 0.85 PV 时，出口才见到气体产出，也就是说，天然气泡沫在模型中均匀推进，改善了注入流体的指进现象，采出程度也有较大程度的提高，由水驱时的 54%，提高到 69%，提高了 15%，含水一度下降到 5%，因此，可以认为，天然气复合泡沫体系，稳定性强，驱油性能良好。

2.6 双管模型特殊条件下注单一泡沫体系

油藏中经过多年的注水开发，残余油饱和度分布很不均匀，相差很大，室内实验无法真实的模拟现场条件，因此我们模拟了油藏驱油的极限条件，分别制作了两根渗透率不等的模型，高渗管渗透率 2.44 达西，低渗管渗透率 1.01 达西，将低渗管饱和模拟油，高渗管饱和模拟水，并联水驱后注单一泡沫体系，图 5 为产出液随 PV 数变化曲线。

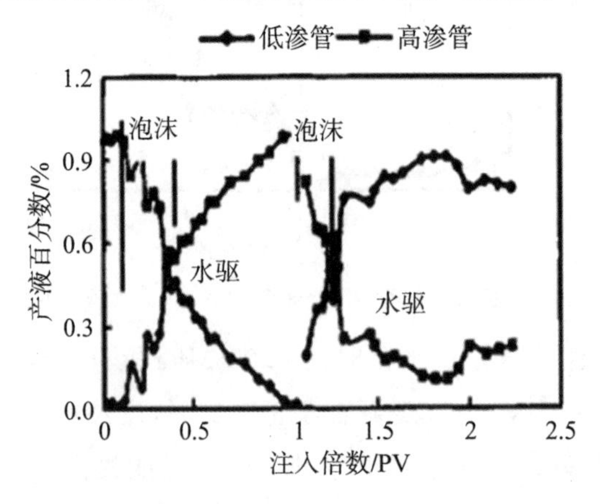

图 5　不同渗透率双管模型注入泡沫段塞产液百分数曲线

从图中曲线可以看出，双管模型注水后，水主要从高渗管产出，高渗管产液占总液量的 98%，注入 0.3 PV 泡沫后，注入压差增大，压差从水驱时的 0.05 MPa 提高到注泡沫剂的 0.4 MPa，压差提高了 8 倍，高渗管产出液减少，低渗管产出液增大，转注水后，压差下降到 0.02 MPa，高渗管产液量恢复，再注入 0.2 PV 泡沫段塞后，压差迅速上升到 0.15 MPa，虽然压差上升的幅度较小，但注入泡沫后低渗管产液量增大，高渗管产液量减少，随着注入泡沫段塞及转注水后，高、低渗管产出液百分数反转，低渗管产液量远远大于高渗管，低渗管产液量占总液量的 90%，高渗管液量占总液量的 10%，转注水后，低渗管液量仍占总液量的 80%，从高低渗管产液比例可以说明泡沫对于油水有很强的选择性，因此，在没有油存在的条件下，单纯泡沫体系也有非常好的封堵性。

3　结论

试验结果表明单一泡沫体系在没有残余油存在时稳定性强，具有良好的封堵能力，但耐油性差，而复合泡沫体系在残余油含量较高时有非常好的封堵调剖能力，可以较大幅度的提高注采压差从而提高注入流体的波及系数，提高驱油效率，同时在试验中可以看到，强化泡沫体系改进了单一泡沫体系遇油不稳定的特点，使泡沫体系在一定残余油存在时，仍有良好的调剖驱油能力，天然气泡沫体系的成功为泡沫驱拓展了新的应用空间，可以大大降低注入氮气泡沫制氮成本。

参考文献

[1] 赵福麟. EOR 原理[M]. 山东东营：石油大学出版社，2001.

[2] Moradi-Araghi A，Johnston E L，Zornes D R，et al. Phillips Petroleum Company Laboratory，Evaluation of surfanctants for CO_2-foam applications at the South Cowden [C]. This paper was presentationat the 1997 SPE International Symposium on Oilfield Chemistry held in Houston，Tex as，February 1997.

[3] 廖广志，李立众，孔繁华，等. 常规泡沫驱油技术[M]. 北京：石油工业出版社，1999.

[4] 宋育贤. 泡沫流体在油田上的应用[J]. 国外油田工程，1997，13(1)：5-8，17.

（编辑　朱和平）

in recent years，large-scale reservoir numerical simulation can be performed at a relatively low cost in Daqing oil field. Firstly，the first PC-Cluster used for reservoir numerical simulation is established in Daqing oil field，and based on this developing environment，the serial black oil simulator is paralleled. Then，this PC-Cluster parallel simulating technique is applied in seven oil production districts of Daqing oil field. The hardware and system software configuration of PC-Clusters established in Daqing oil field is briefly introduced，the idea and method for paralleling the serial applying results is described and presented.

Key words: fine geological study; residual oil; numerical simulation; PC-Cluster; parallel black oil simulator

EXPERIMENTAL STUDY OF FOAM BLOCKING ABILITY

WANG Qi-wei (Shengli Oil Field Corporation, Dongying Shandong 257015, China), CAO Xu-long, ZHOU Guo-hua, et al. JOURNAL OF SOUTHWEST PETROLEUM INS TI-TUTE, VOL. 25, NO. 6, 40-42, 2003 (ISSN1000-2634, IN CHINESE)

In this paper, the blocking ability of several foam flooding systems have been studied, which is foam flooding after oil-bearing model water flooding, single-pure foam flooding after complex foam or polymer flooding, complex foam flooding. Result shows that the single-pure foam flooding cannot form effective block under the condition of residual oil saturation, while complex foam flooding system can form effective block under various residual oil saturation. It can be determined that the complex foam flooding system is an useful profile control agent.

Key words: foam; complex foam; blocking ability; pres-sure difference of blocking

STUDY ON OIL DISPLACEMENT MECHANISMSOF QUA-TERNARY CATIONICS

YAO Tong-yu (Institute of Porous Flow & Fluid Mechanics, Langfang Hebei 065007, China), LIU Fu-hai, LIU Wei-dong. JOURNAL OF SOUTHWEST PETROLEUMINS TI-TUTE, VOL. 25, NO. 6, 43-45, 2003 (ISSN1000-2634, IN CHINESE)

Oil displacement mechanism of quaternary cationics has been studied in several ways: surface and interface tension, wettability, surface electrical behavior. It's thought that there has been low interface tension mechanism, wetting conversion mechanism and surface electrical conversion mechanism while displaced by the surfactants. When the concentration of the surfactants is as high as 600 mg • L^{-1}, the oil recovery increment of 5%—6% is available. It is expectable that cationic surfactants with adsorbance property probably have such mechanisms in oil displacement, and have some potentiality.

Key words: quaternary cationics; low interfacial tension; wetting conversion; surface electrical conversion

SYNTHESIS AND PERFORMANCE OF AN ORGANIC CHROME CROSSLINKING AGENT FOR DEEP PROFILE CONTROL.

LI Yong-tai (Southwest Petroleum Institute, Nanchong Sichuan 637001, China), PENG Zhi-gang, FENG Qian, et al. JOURNAL OF SOUTHWEST PETROLEUM INSTI-TUTE, VOL. 25, NO. 6, 46-48, 2003 (ISSN1000-2634, IN CHINESE)

An organic chrome crosslinking agent of weak gel for deep profile control has been synthesized by using a monose compound as ligand based on its complexing constant of the complexing agent. Results indicate that the crosslinking system has the properties of convenient operation, easy injection and good performance. The contraction between the vertical and plane in formation has been solved effectively and the oil recovery has been enhanced.

Key words: weak gel; profile control; crosslinking agent; enhanced oil recovery

STUDY ON MECHANISM OF WAXY CRUDE OIL STRUC-TURE FORMATION

JING Jia-qiang (Southwest Petroleum Institute, Nanchong Sichuan 637001, China), YANG Li, Q IN Wen-ting, et al.

JOURNAL OF SOUTHWEST PETROLEUMINS TITUTE, VOL. 25, NO. 6, 49-52, 2003(ISSN1000-2634, IN CHINESE)

Based on the infrared spectra of paraffin and asphaltene, the microstructure of simulated wax y oils and crystallog raphic theories, mechanism of structure formation for wax y crude oils has been discussed. Results show that the mechanism is that of the formation of crystalline nucleuses.

油田污水中溶解氧的流动注射分析方法研究

flow injection determination of dissolved oxygen in oilfield produced water

曹绪龙[1,2],蒋生祥[1],隋希华[2],王红艳[2]

Cao Xu-long[1,2],Jiang Shengxiang[1],Sui Xihua[2],Wang Hongyan[2]

(1. 中国科学院兰州化学物理研究所,甘肃兰州 730000;

2. 中国石化胜利油田有限公司地质科学研究院,山东东营 257015)

(1. Lanzhou Institute of Chemical Physics, Chinese Academy of Sciences, Lanzhou 730000, China;

2. Research Institute of Geological Science, Shengli Oilfield, SINOPEC, Dongying 257015, China)

摘要:根据可逆氧化还原反应原理,采用六通阀密闭配样进样方式建立了油田污水中溶解氧含量的流动注射分析方法,优化了反应温度、流动相的pH值及反应管内径等测试条件。采用本实验方法最小检出量为0.025 mg/L,活水中溶解氧的线性范围为0.05～8.45 mg/L,可在1 500 mg/L聚丙烯酰胺(HPAM)存在的条件下进行准确测量。采用离子交换法可以避免杂质的干扰。

关键词:污水;溶解氧;流动注射分析方法

中图分类号:TE357.6 **文献标识码**:A

Abstract: Based on reversible redox theory and prismatic technology, the flow injection analysis method for determination of the mass concent ration of oxygen dissolved in waste water formulated to be injected into the oil field was established. In this method, the sam pling process was performed with a sealed six-port valve. Also, the test conditions such as pH, reaction temperature, inner diameter of reaction tube, etc. were optimized. This novel method has a linear range of 0.05～8.45 mg/L of dissolved oxygen with a minimum detective concent ration of 0.025 mg/L. Even mixed with 1 500 mg/L HPA M in the waste water, the mass concent ration of dissolved oxygen can be measured accurately. The influence of all kinds of impurities can avoided by the ion exchange method.

Key words: produced water; dissolved oxygen; flow injection analysis

在油田开发过程中，尤其在开采后期，往往需要注入大量污水。由于水中溶解的氧含量关系到采油设备和注水管道的腐蚀，因此它是控制注入水水质的重要参数之一。随着化学驱油技术的不断进步，化学驱的实施规模不断扩大[1-3]。但是，在化学驱过程中，配注污水中的溶解氧易与驱油剂发生化学反应，如聚合物的氧化降解等，造成聚合物溶液粘度的损失，导致聚合物驱油效果的降低。此外，由于配注用的污水透明度差，氧化性离子杂质含量高，经典的 Winkler 法、溶解氧测定管法等已不能满足油田污水中溶解氧快速准确分析的需要，因此研究新的污水中的溶解氧含量的准确快速测定方法十分必要。流动注射法 FIA (Flow injection analysis)是近 20 年来发展起来的分析技术，现已应用到多个领域[4]。在本课题中采用此方法，通过六通阀密闭配样、阳离子交换柱去除杂质离子等技术以及优化测定条件等，在体系浑浊、高浓度驱油用聚丙烯酰胺条件下，实现了污水中溶解氧的准确、快速和抗干扰的仪器化测定，可为油田的注采方案设计、配方调整和聚合物粘度稳定的研究提供技术支持。

1 原理

亚甲基兰在一定酸度条件下可被葡萄糖还原为无色；当其遇到溶解氧时又可定量地被氧化为蓝色物质[5]。由于这种兰色物质在紫外光谱 610 nm 的条件下的吸光度与溶解氧含量成正比，因而由此可以得到水中溶解氧的含量。

2 仪器和试剂

2.1 仪器

Waters515 恒流输液泵；Waters600 泵控制器；Waters Lambda-Max 481 紫外可见检测器；德国 Binder 恒温箱；大连江申色谱工作站。

2.2 试剂

葡萄糖 AR，上海第二化学试剂厂生产；亚甲基兰 AR，上海第二化学试剂厂生产；C_2H_5OH、KOH、$FeCl_2$、$FeCl_3$、Na_2CO_3 均为分析纯；实验用水为去离子水；溶解氧试样取自胜利油田现场。

3 实验过程

3.1 测量流程

流程示意见图 1。

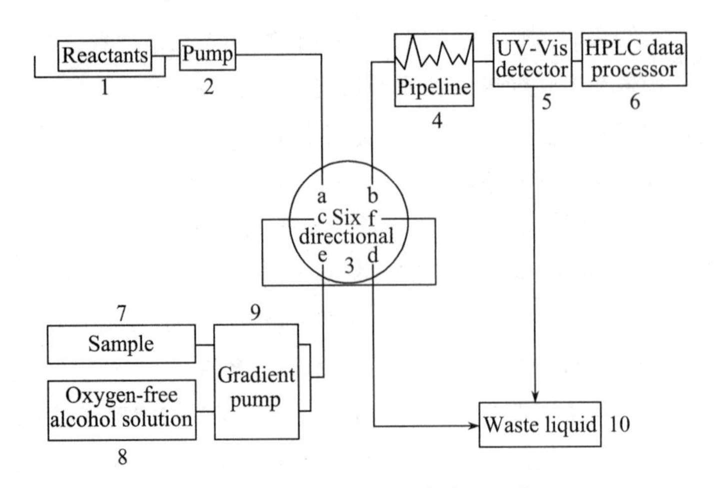

图1　油田污水中溶解氧测定流程示意图

Fig. 1　Schematic diagram for the determination of dissolved oxygen in oilfield produced water

1—Reactants; 2—Pump; 3—Six-directional valve; 4—Pipeline; 5—U V-Vis detector; 6—HPLC dat a processor; 7—Sample; 8—Oxygen-free alcohol solution; 9—Gradient pum p; 10—Waste liquids

3.2 反应液的制备

取500 mL 95%乙醇倾入1 000 mL二颈烧瓶1中,加入亚甲基兰0.062 5 g和葡萄糖0.325 g,此时溶液呈蓝色。加热煮沸,以10%KOH溶液将溶液pH值调至7.2,继续加热回流至溶液变为无色,通N_2保护。测定时该反应液在加热回流过程中由输液泵2泵入测定系统。

3.3 测定过程

进样前,反应液经流路1—2—3(a—b)—4—5—6反应并达到平衡状态。然后,样品经流路7—9—3(e—f—c—d)20 μL定量管,此时切换六通阀3使反应液处于流路1—2—3—(a—c—f—b)—4—5—6状态。反应液与样品在完全隔绝了外界O_2的干扰下在反应管4中进行反应。生成的蓝色物质在紫外波长610 nm下进行检测。

3.4 标准曲线绘制

制作标准曲线时,图1中7为饱和空气的水样品,8为无氧乙醇,由梯度泵9将二者按比例混合后送入20 μL定量管。此时切换六通阀3使反应液处于流路1—2—3—(a—c—f—b)—4—5—6状态,反应液与标准样品在反应管4中进行反应。数据采集完毕后,流路恢复到初始状态。改变泵的流速可溶解氧质量得浓度不同的标准样品。标准样品中溶解氧的质量浓度计算公式为

$$\rho_x = v_w \rho_o / (v_e + v_w)$$

式中,ρ_x为样品中溶解氧的质量浓度,mg/mL;v_w为饱和空气水的流速,mL/min;v_e为无氧乙醇的流速,mL/min;ρ_o为饱和空气水中氧的溶解度,mg/mL。

3.5 样品测定

样品测定时，图1中7为测试样品，梯度泵9将样品泵入锂型强阳离子交换柱，去除干扰物后进入20 μL定量管。此时切换六通阀3使反应液处于流路1—2—3—(a—c—f—b)—4—5—6状态，反应液与样品在反应管4中进行反应。数据采集完毕后，流路恢复到初始状态，然后进行下一个样品的测定。从标准曲线上可以查出样品中溶解氧的含量。

4　影响因素的考察

4.1 分析条件的确定

4.1.1 流动相的pH值

取500 mL 95%乙醇倾入1 000 mL二颈烧瓶中，加入亚甲基兰0.062 5 g和葡萄糖0.325 0 g，此时溶液呈蓝色。加热煮沸，滴加10%KOH溶液调节pH值，考察流动相的pH值对实验结果的影响。可以看出，在500 mL流动相中加入1.0 mL 10%KOH溶液后，反应液回流30 min，即变为无色，此时测得pH值为7.2。如加入10% KOH少于1.0 mL，即使回流1 h，溶液也不退色。如多于1.0 mL，溶液则变为黄色。因此，在本实验方法中，采用流动相溶液pH=7.2作为实验条件。

4.1.2 反应管材质

取500 mL 95%乙醇倾入1 000 mL二颈烧瓶中，加入亚甲基兰0.062 5 g和葡萄糖0.325 g，加热煮沸，以10%KOH溶液将溶液pH值调至7.2，继续加热回流至溶液变为无色，通N_2保护。在实验中，首先采用聚四氟乙烯管路，实验结果的重现性较差。继续考察发现，不进样时反应管内流动相也会变蓝，在反应液中水的比例较大时尤为明显。这表明聚四氟乙烯管有渗漏氧气的可能。分别选用石英毛细管和不锈钢管进行实验，上述现象得以消除。为方便耐用起见，选用不锈钢管作为反应管。

4.1.3 反应管内径

选用不同内径的不锈钢反应管进行实验，其它条件同本文4.1.2节，流速1.0 mL/min，温100 ℃，管长20 m，检测波长610 nm，测定结果见表1。可以看出，减小内径可以有效地减少涡流扩散的影响，有利于峰形的改善。选用内径为20 μm的不锈钢毛细管作为反应管。

表1　反应管路内径与流出峰形的关系

Table 1　Relation between inner diameter of reaction tube and effluent peak shape

Φ[1)]/μm	H/A ratio[2)]/cm^{-1}
20	129
25	104
35	87

1) Inner diameter; 2) Peak height/peak area ratio

4.1.4 反应温度

在氧气与亚甲基兰反应生成性质稳定的次甲基兰的过程中,反应越充分,越有利于检测灵敏度的提高。为此,固定其它实验条件如4.1.3,考察了不同温度对分析灵敏度的影响,结果见表2。可以看出,提高温度,有利于检测灵敏度的提高。选择100℃作为反应温度。

表2 反应温度与检测灵敏度的关系

Table 2 Relation between reaction temperature and sensitivity ofdetection

t/℃	H/cm
20	21
30	39
50	44
60	65
80	84
90	110
100	124

4.1.5 检测波长

对反应生成的蓝色物质进行紫外扫描,结果见表3。可以看出,亚甲基兰-氧在610 nm处存在最大的吸收峰高。选择610 nm作为检测波长。

表3 吸收波长与最大峰高的关系

Table 3 Relation between the absorption wavelength and maximum peak height

λ/nm	H/cm
600	24
610	26
620	23
630	20

4.1.6 载流速率

溶解氧与亚甲基兰化学反应需要一定的时间才能完全,降低流速有利于延长样品在反应管中的停留时间,从而提高检测灵敏度,但由于分子扩散作用的影响,峰形会变得更差。本方法选择载流速率为1.0 mL/min时,峰型较好,分析速度较快。

4.2 干扰物的影响及排除

油田污水成分复杂,特别是各种化学添加剂的加入,使得油田污水的处理更为困难。为了考察污水中各种成分对本方法的干扰,在标准样品中分别加入Fe^{2+}、Fe^{3+}、部分水解聚丙烯酰胺(HPAM)、微量油以及HEDP防垢剂、DP-1杀菌剂、JD-1防垢剂等,测定回收率。结果表明，油田污水中存在的HPAM、微量油以及上述水处理剂对测定结果影响较小，可以忽略。Fe^{2+}、Fe^{3+}的存在对测定结果有较严重影响，必须排除。

4.2.1 干扰物的影响及排除

取500 mL 95%乙醇倾入1000 mL二颈烧瓶中，加入亚甲基兰0.062 5 g、葡萄糖0.325 g、1.0 mL 10%KOH溶液，加热回流25 min后作为反应液，依次测定暴露在空气中的300 mg/L $FeCl_2$、$FeCl_3$水溶液的峰高。结果表明，其峰高分别为标准样品峰高的0.5和3.5倍，说明Fe^{2+}、Fe^{3+}的存在严重干扰溶解氧的测定。

在有氧的条件下，水中铁离子只以Fe^{3+}形式存在。如在进样泵与六通阀之间加装1根50 mm长的锂型强阳离子交换柱，泵入500 mg/L的$FeCl_3$水溶液，经过该柱后产生的峰高则与标准水样完全相同。这表明样品经过锂型强阳离子交换树脂处理后可以完全排除其中铁离子的干扰。

4.3.2 最佳分析条件

综上所述，最佳分析条件为流速1.0 mL/min，反应温度100℃，检测波长610 nm，流动相配比500 mL乙醇中加入0.062 5 g亚甲基兰和0.325 g葡萄糖，流动相的pH=7.2，反应管路内径0.20 mm，管长20 m。除干扰柱为50 mm长的锂型强阳离子交换柱。

5 方法的准确性

5.1 方法的线性范围

按图1所示，分别泵入含有饱和氧的去离子水与流动相，设定不同流速比例，测定溶解氧含量在0.025～8.45 mg/L范围内的峰高。实验结果表明，线性关系很好，实验数据回归后相关系数r=0.994 9，溶解氧含量在0.025 mg/L时峰高大于2倍基线噪音。由此确定本方法线性范围为0.05～8.45 mg/L，最小检出量为0.025 mg/L。

5.2 方法的重复性及可靠性

为了确定方法的重复性，从胜利油田孤岛采油厂孤7注取得实际配聚污水进行了7次重复测定，其中HPAM含量为1 500 mg/L。结果见表4。可以看出，最大相对偏差为5.6%，本方法的重复性较好。取纯净的含不同质量浓度溶解氧的水样，分别用经典的Winkler方法和本方法测定溶解氧含量，结果见表5。可以看出，最大相对偏差为7.60%，因此，采用本方法测定污水的溶解氧含量是可靠的。

表4　方法的重现性

Table 4　Repeatability of method

Sample No.	H/cm	Relative deviation/%
1	5.80	−1.2
2	5.80	−1.20
3	5.80	−1.20

(续表)

Sample No.	H/cm	Relative deviation/%
4	5.90	0.5 1
5	5.90	0.5 1
6	6.20	5.60
7	5.70	−2.90
A verage	5 87	

表 5 方法的可靠性

Table 5 Reliability of method

ρ/mg · l^{-1}		Relative deviation/%
FIA	Winkler	
7.50	7.79	3.72
1.52	1.53	0.65
3.61	3.87	6.72
1.33	1.44	7.60
3.61	3.87	6.72
1.33	1.44	7.60

6 实例

用自制的现场密闭取样容器取孤岛油田含 HPAM 的污水采用本方法测定其中的氧含量,测定条件见本文 4.2.2 节。结果见表 6。可以看出,采用本方法测定所得结果的重复性很好。

表 6 现场样品中溶解氧测定结果

Table 6 Determination for dissolved oxygen in oil field samples

Sample from G udao oilfield	ρ(HPAM)/mg · l^{-1}	ρ/mg · l^{-1}	
		No. 1	No. 2
No. 3 injection allocation station	0	0.03	0.03
Fresh water	1 500	7.84	7.58
No. 5 dist ributing room for water injection	1 516	0.05	0.06
No. 5-142 wellhead	1 487	20.48	21.03

7 结论

采用紫外分光光度法,建立了油田污水中溶解氧测定的流动注射分析方法,优化了测试条件。采用本方法最小检出量为 0.025 mg/L,线性范围为 0.05～8.45 mg/L,有可能应用于油田污水中溶解氧含量的准确快速分析。

参考文献

[1] Li Y, Cao X L. Practice and knowledge of polymer flooding in high temperature and high salinity reservoirs of Sheng li petroliferous area[A]. SPE paper 68697, SPE Asia and Pacific Annual Technical Conference and Exhibition[C]. Kuala Lumpur, Malaysia. 2000.

[2] Wang C L, Cao X L. Design and application of ASP system to the close well spacing, Gudong oilfield[A]. SPE38321, SP E Annual Improving Oil Recovery Conference[C]. Tulsa, Oklahoma. 1997.

[3] 孙焕泉,张以根,曹绪龙. 聚合物驱油技术[M]. 东营:石油大学出版社,2002:280-294.

[4] 方兆伦. 流动注射分析方法[M]. 北京:科学出版社,1999:134-189.

[5] 陈立仁,蒋生祥,刘霞,等. 高效液相色谱基础与实践[M]. 北京:科学出版社,2001:216-236.

SCL-1污水处理剂的研制与应用

王增林[1],马宝东[2],曹绪龙[2]

(1. 石油大学(华东)石油工程学院,山东东营　257061;

2. 胜利油田有限公司地质科学研究院,山东东营　257015)

摘要:在考虑污水处理剂与聚丙烯酰胺配伍性的基础上,通过分子改性研制出了一种适用于油田注聚区采出液处 理的非离子药剂SCL-1。室内研究和现场试验结果表明,该药剂对孤四联注聚合物区采出液具有优良的综合处理能力,而且该药剂能改善污水水质,从而可有效缓解污水对聚丙烯酰胺溶液的降粘作用。

关键词:污水处理;油水分离;破乳;原油脱水;聚合物

中图分类号:TE868　文献标识码:B　文章编号:1001-0890(2004)03-0065-03

随着聚合物驱油规模的不断扩大,该项驱油技术在各类油藏条件下均遇到了不同程度的困难。大量矿场应用显示,采用污水配注聚合物后,由于污水中复杂的化学组分对聚丙烯酰胺的降解作用,导致聚合物溶液在配制和注入过程中粘度损失严重,控制流度能 力变差,进而限制了该项驱油技术在部分油藏的推广应用。各项室内试验和矿场研究表明,注聚合物区配注水水质,尤其是用于配注聚合物溶液的污水水质,是导致聚合物溶液粘度损失的主要因素之一。

目前胜利油田各注聚合物区均沿用常规回注污水处理工艺,外输污水含油量高达200 mg/L,部分污水站外输污水含油量甚至在1 000 mg/L以上。SCL-1污水处理剂的研制过程中,首次引入了水处理剂与聚丙烯酰胺的配伍性研究,通过分子改性技术克服注聚合 物区采出液粘度高、乳化程度深、油水分离阻力大、吸附损耗严重等技术难点。

1　室内研制

1.1 配方设计

聚合物驱由于高分子聚丙烯酰胺的存在,污水粘性增大,油水分离缓慢;三元复合驱不仅含有聚丙烯酰胺,还含有碱和表面活性剂,采出污水粘性大,乳化严重,油水很难靠自然沉降分离1。

污水处理剂的研制以胜利油田孤岛采油厂四号联合站(以下简称“孤四联”)的注聚

合物区采出液为研究对象,以药剂的油水分离、破乳、絮凝和净水等性能为目标,在配方设计中首先充分考虑了药剂与聚丙烯酰胺溶液的配伍性。通过对国内外相关试剂的筛选,初选6种试剂为主要原料,分别对孤四联注聚合物区采出液的油水分离能力进行了初步评价,试验结果见图1。

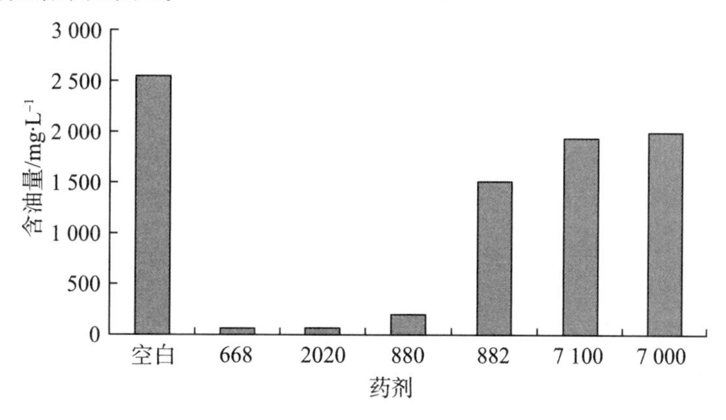

图1 ROMAX系列试剂油水分离性能分析

图1中的数据结果显示,以ROMAX668、2020和880为主要原料的3种药剂在孤四联注聚合物区采出液的处理中显现出了明显的油水分离效果,其中以ROMAX668的油水分离性能最为突出。但是ROMAX668在采出液处理过程中存在油水分离速度较慢,处理后的采出液油水过渡层较厚,过渡层油水乳化严重,油水界面不稳定等方面的不足。

ROMAX 668是非离子改性聚醚类化合物,与阴离子聚丙烯酰胺具有良好配伍性能,在水溶液中对聚丙烯酰胺分子的活性基团和分子结构无任何破坏性影响。该化合物可溶于水,其分子结构经过改性后具有亲油基团。根据相似相溶的原则,在油水混合液中,ROMAX 668的亲油基团能够吸附水中的小油滴,使其聚集为大油泡,从而达到油水分离的目的。ROMAX 668具有很高的溶解度,能够完全溶解于水,它几乎不会进入油相,因此它不会影响分离所得的油的性质。

油水界面上的状态模型以及液滴在界面的寿命显示各类原油乳状液具有不同的稳定性:O/W型乳状液水滴接近空间阻碍很小,较易聚结;W/O型乳状液水滴接近空间阻碍很大,不易聚结;O/W/O型乳状液油滴与油水界面拉近时,空间阻碍作用大,液滴不易破 裂,寿命很长,O/W/O型乳状液较稳定[2-5]。该现象表明,ROMAX 668油水分离剂对水包油型乳状液具有良好的破乳性能,但对于油包水型乳化液的破乳效果较差。

1.2 配方优化

聚合物驱和三元复合驱采出污水难以处理的原因:聚合物使采出污水的粘度增加,表面活性剂使油珠严重乳化,微小油珠很难凝聚,即油珠的粒径小,导致油珠上升速度慢[5-7]。

为改善药剂对聚合物区采出液油包水型乳化液的破乳性能、聚集能力以及对原油的脱水性能,针对聚合物区采出液粘度大,乳化形式复杂多变的特性,通过分子改性的HF

系列药剂对 ROMAX 668 进行性能改进，形成系列药剂，它们对孤四联注聚合物区采出液的油水分离效果见图 2。

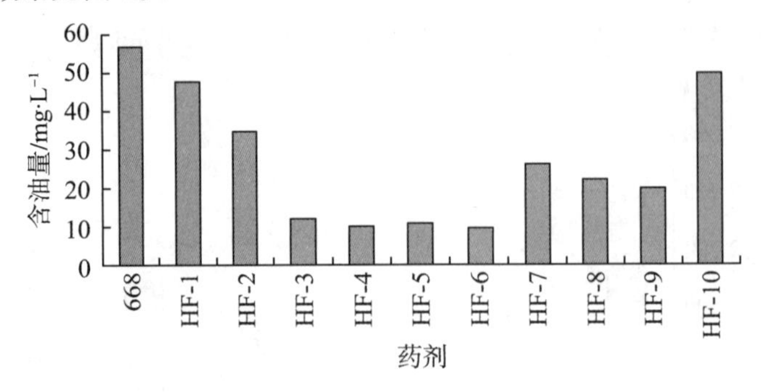

图 2　ROMAX 系列试剂油水分离性能分析

HF 系列药剂是以聚氧丙烯聚氧乙烯醇类非离子表面活性剂为主要成分的活性剂。活性物含量为 60%，以羟值测定的分子量为 1500～2000，溶剂为醇类水溶液，活性物结构式为

$$R\text{-}(C_3H_6O)_m\text{-}(C_2H_4O)_n\text{-}(C_3H_6O)_p\text{-}H$$

HF 系列药剂在污水处理中的主要作用是促进 ROMAX 668 对油包水型乳状液进行破乳以及油滴和悬浮颗粒的聚集。

图 2 所示的 1～10 组试验分别为 HF 系列药剂对 ROMAX 668 油水分离性能的改性试验。试验结果显示，HF-3、HF-4、HF-5 和 HF-6 对 ROMAX 668 油水分离性能的优化结果最为明显。在以上 4 组试验中，孤四联注聚合物区采出液的水层含油量从 2 500 mg/L 迅速降至 10 mg/L 左右。在室内研究的基础上，以 ROMAX 668 和 HF-3、HF-4、HF-5、HF-6 为主要原料，针对孤四联注聚区采出液合成了 SCL-1、SCL-2、SCL-3、SCL-4 四种污水处理剂，

2　性能评价

2.1 污水处理剂与聚丙烯酰胺的配伍性评价

由于清水不足和污水外排困难等方面的原因，目前各油田注聚合物区均采用污水注聚合物溶液 8。通过 SCL 系列污水处理剂在孤四联注聚合物区采出液中对聚丙烯酰胺溶液粘度的影响评价，主要确定其与采出液中的各种组分相互作用后对聚合物的影响，试验结果见表 1。

表 1　SCL 系列污水处理剂对聚合物溶液粘度影响

药剂名称	聚合物溶液粘度/mPa·s				
	0[①]	1[①]	5[①]	10[①]	20[①]
空白	19.9	19.1	19.1	19.2	19.0
SCL-1	20.4	20.2	20.5	20.5	20.5

（续表）

药剂名称	聚合物溶液粘度/mPa·s				
	0①	1①	5①	10①	20①
SCL-2	19.8	19.8	19.8	20.2	20.1
SCL-3	20.6	20.6	21.0	20.9	20.5
SCL-4	19.5	19.4	19.7	19.6	19.1

① 为保温时间 d。

表 1 所示的评价试验结果显示，与不加药剂的空白样品相比，在孤四联注聚合物区采出液中分别加入 20 mg/L 的 SCL 系列污水处理剂后，其配制的聚丙烯酰胺溶液在 70℃条件下放置 20 d 后仍未出现粘度损失现象，说明 SCL 系列污水处理剂与孤四注聚合物区采出液中的各种组分相互作用后对聚丙烯酰胺溶液的粘度无任何不良影响，二者具有良好的配伍性。

2.2 污水处理剂对采出液的油水分离性能评价

确定 SCL 系列污水处理剂对聚丙烯酰胺溶液的粘度无任何不良影响后，首先对药剂的合成进行了中试试验，并对中试产品进行了一系列的性能评价试验。SCL 系列污水处理剂对孤四注聚区采出液油水分离性能评价试验，结果采出液经 SCL-1、SCL-2、SCL-3 和 SCL-4 处理后，污水含油量分别为 10.0、11.8、18.2 和 26.2 mg/L。可见，4 种污水处理剂对孤四联注聚区采出液均具有优良的油水分离性能，其中 SCL-1 污水处理剂的油水分离性能最好，经该药剂处理后采出液下部水层迅速澄清，水体浊度明显下降，水中含油量降至 10.0 mg/L。

2.3 污水处理剂对原油的破乳脱水性能评价

联合站分离出的原油在外输前要经过电脱水器进行脱水处理。因此，为保证电脱水器的正常运行，要求油水分离后的原油含水必须在 30%以下。为此，针对孤四联注聚合物区采出液中的原油，进行了 SCL-1 原油脱水能力的一系列评价试验，结果见表 2。

由表 2 可看出，污水处理剂 SCL-1 对孤四联注聚合物区采出液中的原油具有良好的破乳脱水能力。当加药质量浓度为 10～30 mg/L 时，其破乳脱水能力与 100 mg/L 的现场污水处理剂(XPI-5085B/LGS-2)的破乳脱水效果基本相似，均能从 80 mL 原油中脱出 10.0 mL 的水；当加药质量浓度为 20 mg/L 时，从 80 mL 原油中脱出了 12.5 mL 的水，达到了最高值。

表 3 SCL 系列污水处理剂原油脱水性能评价

药剂名称	药剂质量浓度/mg·L^{-1}	原油体积/mL	脱水量①/mL	脱水率,%
XPI-5085B/LGS-2	100	80	10.0	29.8
	25	80	5.0	14.9
SCL-1	30	80	10.0	29.8
	25	80	11.0	32.7
	20	80	12.5	37.2
	15	80	11.0	32.7
	10	80	10.0	29.8
	5	80	7.0	20.8

① 为 120 min 脱水量。

3 现场应用效果评价

在现场试验期间,主要考察了污水处理剂 SCL-1 对孤四联注聚合物区采出液的油水分离性能,现场检测结果见图 3。

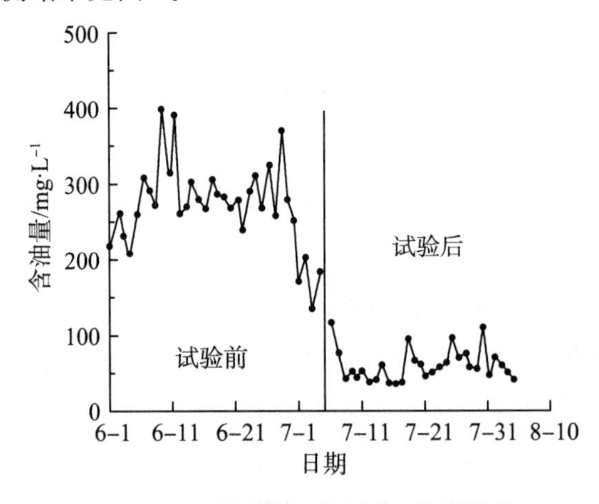

图 3 孤四联外输污水含油量检测结果

图 3 的现场检测结果显示,6 月份孤四联外输污水含油量均在 200 mg/L 以上(月平均值为 290 mg/L),从 7 月 6 日采用 SCL-1 完全替代原破乳剂和预脱水剂对采出液进行化学处理以后,外输污水含油量迅速下降,至 7 月 8 日孤四联外输污水含油量降至 43 mg/L,污水水质明显改善。

现场试验期间,污水处理剂 SCL-1 的加药质量浓度从 20 mg/L 到 10 mg/L 进行了一系列的调整和检测,如图 3 所示,孤四联外输污水含油量均保持在 50 mg/L 左右,最终将 SCL-1 在孤四联注聚合物区采出液处理过程中最经济的质量浓度确定为 11.5 mg/L。

4　结束语

新型污水处理剂 SCL-1 在注聚合物区采出液处理过程中显示出了优良的油水分离和破乳性能，对聚丙烯酰胺溶液的粘度无任何不良影响，克服了注聚合物区采出污水破乳技术的难点，充分体现了复合型药剂在污水处理过程中一剂多能的特点。该药剂在油田注聚合物区采出液化学处理领域具有广阔的应用前景，作者附言：

室内研究和现场应用过程中，地质科学研究院陈晓彦、张本艳、王润芳，长安集团张元成、林春玲以及孤岛采油厂张锡波等科研人员做了大量的基础研究和跟踪分析工作，在此表示诚挚的感谢。同时感谢长安集团研发中心和孤岛采油厂工艺所、科技办以及孤四联的大力协助。

参考文献

[1] 雷乐成，陈琳，张瑞成. 油田废水的 COD 构成分析及生物可降解性研究[J]. 给水排水，2002，28(6)：44-47.

[2] 李绵贵，张万忠，王卫前，等. LBT 系列双功能水处理剂在油田的应用[J]. 油田化学，2001，18(4)：53-55.

[3] 项勇，常斌、大港油田污水处理技术[J]. 石油规划设计，2002，13(3)：17-18.

[4] 牟永春、薛爱琴，李绯. 油田含油污水深度处理技术探讨[J]石油规划设计，2002，13(3)：40-44.

[5] 牟建海. 原油破乳机理研究与破乳剂的发展[J]. 化工科技市场，2002，(4)：26-30.

[6] 刘生福，刘治，蔡彩霞，等. 中原卫城油田原油破乳剂的破乳机理研究[J]. 工艺·试验，2002，(3)：31-33.

[7] 李平、郑晓宇，朱建民. 原油乳状液的稳定与破乳机理研究进展[J]. 精细化工，2001，18(2)：89-93.

[8] 牛金刚，孙钢，高飞. 油田产出污水配制聚合物实验研究[J]大庆石油地质与开发，2001，20(2)：17-19.

Development and application of SCL-1 sewage treatinent additive

Wang Zenglin[1], Ma Baodong[2], Cao Xulong[2]

(1. College of Peroleum Enginering, University of Peroleum (Huadong), Domgying, Shandong, 257061, China;

2. Geological Research Institute of Shengli Oilfield Co. Lld., Dongying, Shandong, 257015. China)

Abstract: SCL-1, the new developed water treatment agent, is a kind of nonionic surfactant, which is suitable for treating the polymer flooding produced liquid. Both lab experiments and field applications showed that it has efficient capability of sewage disposal. In addition, it could improve the wastewater quality and weaken the viscosity reducing of HPAM influenced by wastewater efficiently.

Key words: sewage disposal; oil water distribution; demulsification; crude oil dehydrating; ploymer

低渗透油田增注用 SiO_2 纳米微粒的制备和表征

曹 智[1]，张治军[1]，赵永峰[2]，曹绪龙[3]，张继超[3]

（1. 河南大学特种功能材料重点实验室，河南开封 475001；

2. 胜利油田胜利化工有限公司，山东东营 257099；

3. 胜利油田地质科学院，山东东营 257099）

摘要：用原位表面修饰法制备了低渗透油田增注用 SiO_2 纳米微粒，用透射电镜、X 射线粉末衍射、红外光谱和热分析等手段对其形貌、化学结构和热力学性质进行了研究，考察了其在有机介质中的分散性。结果表明经表面修饰的 SiO_2 纳米微粒在柴油、液体石蜡中有很好的亲油性，岩心驱替试验表明该材料能有效提高水相渗透率。

关键词：油分散性；SiO_2；低渗透油田；纳米微粒

中图分类号：TB383　文献标识码：A　文章编号：1008-1011(2005)01-0032-03

Preparation and characterization of SiO_2 nanoparticles-water Flooding Enhancement Agent of Low Permeability Oil Field

Cao Zhi[1], Zhang Zhijun[1], Zhao Yongfeng[2], Cao Xulong[3], Zhang Jichao[3]

(1. Laboratory for Special Functional Materials, Henan University, Kaifeng 475001, Henan, China;

2. Shengli Chemical Co. of Shengli Oil Field, Dongying 257099, Shandong, China;

3. Academic of Geological of Shengli Oil Field, Dongying 257099, Shandong, China)

Abstract: SiO_2 nanopartic le s were prepared by in-situ m odifica tion method. The morphology, structure, and thermodynamic properties were studied by TEM, XRD,

FT-IR and TG/DTA. The oil dispersibility in organic so lvents was evaluated. The results showed that the silicananoparticles had a good oil dispersibility in diesel oil and parafin oi The core driving expermient proved that the nanoparticles can enhance the water perm eability rate effectively.

Key words:oil dispersion; silica; low permeability pools; nanopartic les

目前,纳米材料在物理、化学、生物、机械、电子等众多领域中正发挥着不可替代的作用[1-3],但在实际应用过程中纳米材料的均匀性、稳定性和分散性一直成为其主要制约因素[4]. SiO_2 纳米微粒是一种优良的新型功能材料,具有良好的热稳定性和抗氧化性。但市售的大部分 SiO_2 纳米微粒在有机介质中的分散性均较差,而将 SiO_2 纳米微粒应用于低渗透油田,增加水驱时的注水量,要求其在柴油中有良好的分散性和稳定性。作者合成的 SiO_2 纳米微粒,在柴油、液体石蜡、邻苯二甲酸二乙酯中分散性良好,在油酸、二甲苯等介质中也有一定的分散性。作者用 TEM, XRD, FT-IR, TG/DTA 等手段对其形貌、化学结构和热力学性质进行了研究,并进行了岩心驱替试验。

1 实验部分

1.1 制备过程

以廉价的 Na_2SiO_3 和质子酸(如 HCH_2SO_4 等)为原料,采用原位表面修饰法在纳米微粒生成过程中加入有机修饰剂,并通过化学键与 SiO_2 表面的—OH 结合,从而能有效地阻止 SiO_2 纳米微粒的自身团聚。在制备过程中要控制质子酸的加入速度,待反应出现沉淀时加入修饰剂,反应终点的 pH 值控制在 9 左右,产物经过滤、洗涤、喷雾干燥得成品。

1.2 结构表征和岩心驱替试验

用 JEM-2010 透射电镜观察产品形貌,用 XP'ert Pro X 射线衍射仪考察其晶型,用 AVATAR 360FT-IR 红外光谱仪和 EXSTA R-6000 型热分析仪分别研究其表面键合结构及热力学性质,岩心驱替试验在胜利油田地质科学院岩心驱替试验装置上进行[5]。

2 结果和讨论

2.1 DNS-1 型 SiO_2 纳米微粒的形貌及结构表征

图 1 是 SiO_2 纳米微粒分散于无水乙醇中的电镜照片,可以看出微粒粒径在 10~20 nm 左右,粒径分布比较均匀。图 2 是 SiO_2 纳米微粒的 XRD 图谱,从图中可见,在 $2\theta=21.2°$ 处出现了一个宽化的衍射峰,根据 X 射线衍射图所得到的衍射角度 θ 和半峰宽 β 值按 Scherrer 方程可以近似计算出 SiO_2 纳米微晶的平均尺寸为 1.5 nm。结合 TEM 结果,表明粒径为 10~20 nm 的二氧化硅纳米微粒是由许多平均粒径为 1.5 nm 的 SiO_2 微晶组成的,在宏观上表现为无定形态。

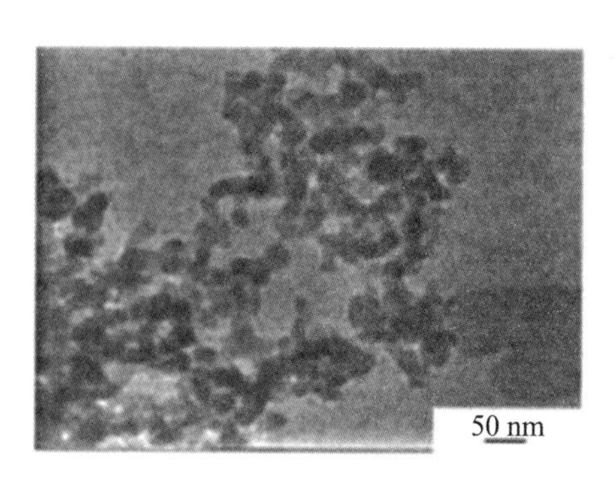

图 1　SiO_2 纳米微粒的电镜照片

Fig. 1　TEM image of silica nanoparticles

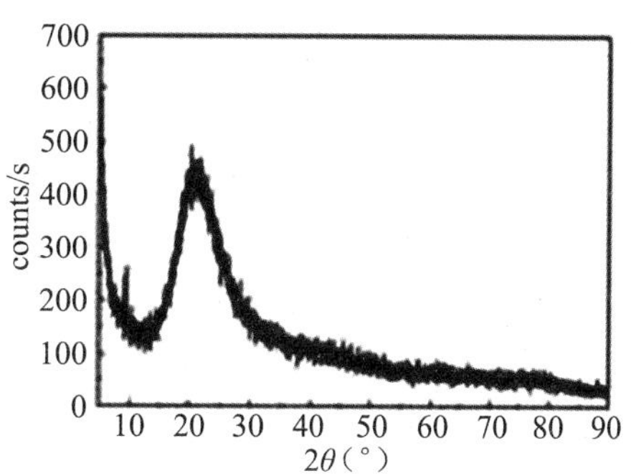

图 2　SiO_2 纳米微粒的 XRD 图谱

Fig. 2　XRD pattern of silica nanoparticles

图 3 是 SiO_2 纳米微粒的 FT-IR 图谱。3 420 cm^{-1} 处可归属为 SiO_2 表面上 O—H 的伸缩振动吸收和结构水的 O—H 的伸缩振动吸收，1 095 cm^{-1} 处为 Si—O—Si 反对称伸缩振动引起的强吸收带，802 cm^{-1}，473 cm^{-1} 处为 Si—O 键的弯曲振动和摇摆振动峰，1 650 cm^{-1} 为表面吸附水 H—O—H 的弯曲振动吸收峰。从图中还能清晰看到，SiO_2 经过修饰后在 2 963 cm^{-1}，2 930 cm^{-1} 处出现了特征吸收峰，可归属为有机碳链上 C—H 键的振动吸收峰，表明 SiO_2 纳米微粒表面修饰上了烷基链。

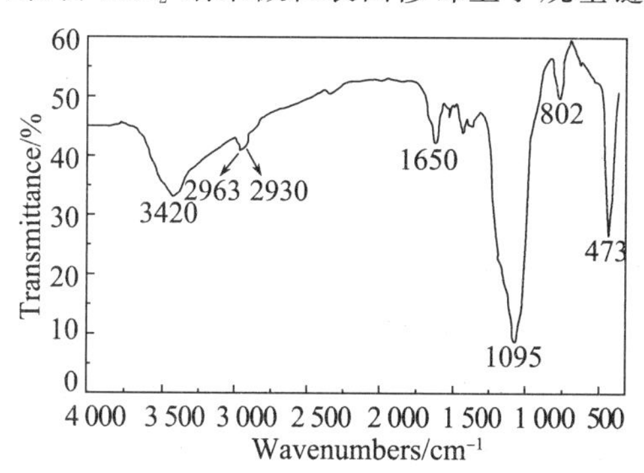

图 3　SiO_2 纳米微粒的红外图谱

Fig. 3　FT-IR pattern of silica nanoparticles

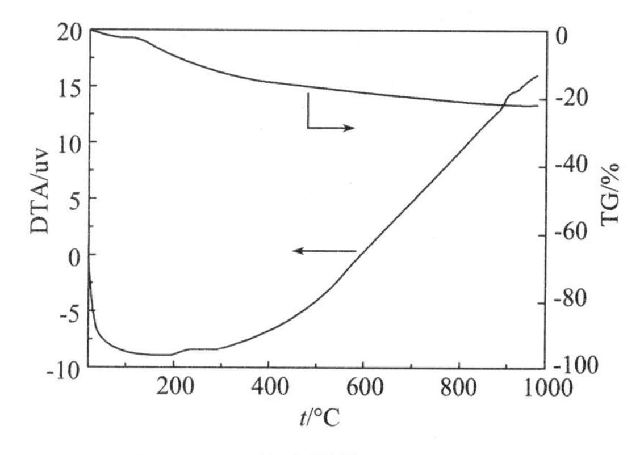

图 4　SiO_2 纳米微粒的热分析图谱

Fig. 4　TG/DAT curves of silica nanoparticles

样品的热分析曲线如图 4 所示，从图中可以看出，失重主要有两个阶段，一个是 100℃之前，失重 2.1%，另一个是 120℃至 400℃之间，失重 12.8%，至 1 000℃时总失重达 21.5%。结合差热曲线分析，120℃前吸热主要是吸附水的蒸发引起的，180℃后开始放热，232℃处的放热峰是有机物燃烧引起的，600℃后继续放热是由于 SiO_2 纳米微粒中的孔隙塌陷造成的微粒团聚引起的[6]。从热分析结果可粗略估计 SiO_2 纳米微粒中 SiO_2 含量约为 78.5%，水分约为 2.1%，有机物含量约为 19.4%。

2.2 DNS-1 型 SiO_2 纳米微粒的性能

2.2.1 DNS-1 型 SiO_2 纳米微粒的分散性

SiO_2 纳米微粒在柴油、液体石蜡、邻苯二甲酸二乙酯中可以稳定分散，形成透明的类真溶液。这是由于 SiO_2 纳米微粒表面键合有亲油性基团，所以具有良好的亲油性。此外，它在油酸、二甲苯中也有一定的分散性，但溶液稍有浑浊，且稳定性较差。

2.2.2 增注性能

低渗透油田储集层的特点是孔隙喉道细小，渗透性差，流体渗流速度缓慢，油井产量很低，表现为水井注不进，油井难以采出。这就要求注水开发时采取相应措施[7]。DNS-1 型 SiO_2 纳米微粒是专门针对低渗透油田设计的增注材料，它自身粒径小，不会阻塞地层孔道，在地层吸附后，改变岩石表面性质，能有效提高水相渗透率，增加水的注入量。

按文献[5]进行了 SiO_2 纳米微粒的驱替实验。实验结果表明，岩心经 SiO_2 纳米微粒处理后水相渗透率提高了 20%，说明用 SiO_2 纳米微粒处理低渗透岩心具有明显的增注效果。图 5 是 SiO_2 纳米微粒增注机理示意图。其中图 5(A)是地层孔道原始状态。图 5(B)是未经处理的粘土经注入水的浸泡，膨胀而堵死孔道或使孔道变窄，注入水难以顺利通过。图 5(C)是经处理后的情况，SiO_2 纳米微粒吸附在岩石表面，赶走岩石表面的吸附水，使孔道变宽，注入水顺利流过。SiO_2 纳米微粒的柴油悬乳液注入地层后，易于吸附在岩石和孔隙表面，降低水化膜的厚度或将岩石表面的吸附水驱走，从而增加了孔道的有效半径，而且还能够包覆在粘土表面，阻止注入水的浸入起到防膨胀的作用。还能使由于长期受注入水冲刷而亲水的近地岩石表面变为憎水，降低了注入水的流动阻力有利于注入水的流动，提高了水相渗透率。

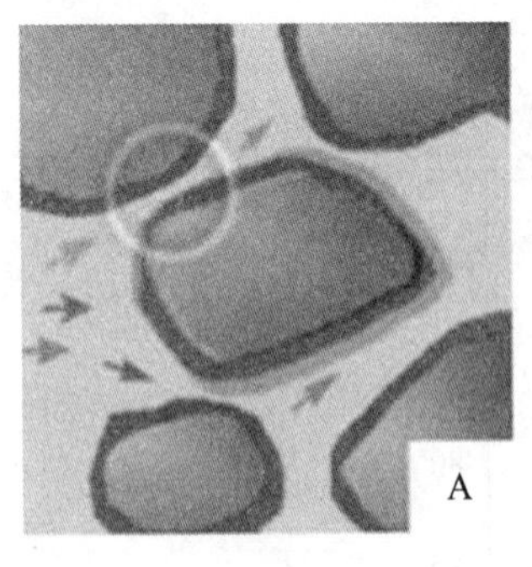

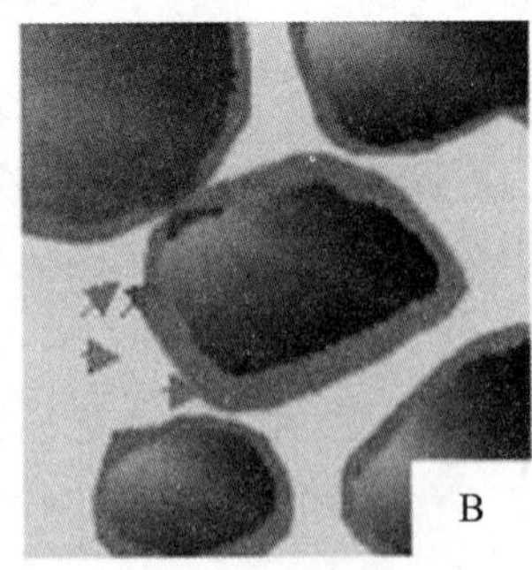

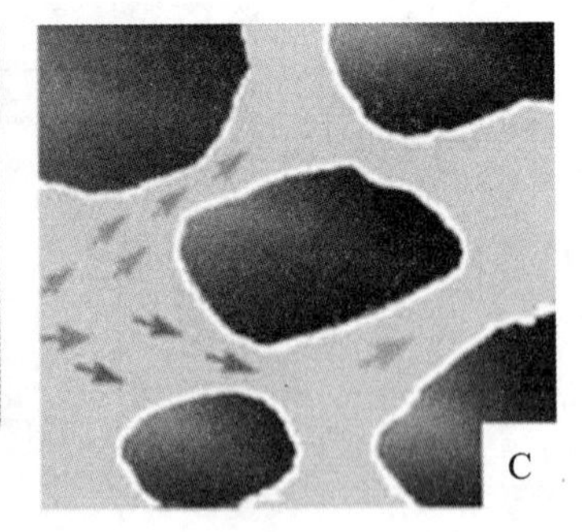

A 原始状态，B 处理前，C 处理后水流情况

图 5 增注机理示意图

Fig. 5 Schem atic mechanism of enhancing water flooding volume

综上所述，经原位表面修饰法制备的 SiO_2 纳米微粒粒径均匀，表面修饰有亲油性烷基链，有机物含量约为19.4%，在柴油、液体石蜡、邻苯二甲酸二乙酯中分散性良好。岩心驱替试验表明，低渗透岩心经 SiO_2 纳米微粒处理后可提高水相渗透率20%，说明用该材料处理水井后可增加水的注入量。SiO_2 纳米微粒有望成为新型的油田采油助剂-低渗透油田增注剂。

致谢：感谢胜利油田地质科学研究院提供岩心驱替试验数据。

参考文献

[1] 张立德，牟季美. 纳米材料和纳米结构[M]. 北京：科学出版社，2002.

[2] Bönnem ann H，R icha rds R M. Nanoscopic metal particles-syn thetic methods and potential applications[J]. Eur J Inorg Chem，2001，(10)：2455-2480.

[3] Hrubesh L W. Aerogel applications[J]. J Non-cryst Solids，1998，(225)：335-342.

[4] 李小红，孙蓉，张治军，等. 以化学键结合的聚合物-无机纳米复合材料[J]. 自然科学进展，2004，14(3)：1-5.

[5]苏咸涛，吕广忠，栾志安，等. 纳米聚硅材料在油田开发中的应用[J]. 石油钻采工艺，2002，24(3)：48-51.

[6] Slobodan B. Thermal changes of amorphous SiO_2 as a consequence of a precipitation temperature[J]. PPSCE，2000，44(2)：133-140.

[7] 何琰，伍友佳，吴念胜. 特低渗透油藏开发技术[J]. 钻采工艺，1999，22(2)：20-24.

交替式注入泡沫复合驱实验研究

周国华[1]，曹绪龙[2]，王其伟[2]，郭　平[2]，李向良[2]

(1. 中国石化股份有限公司石油勘探开发研究院，北京　100083；

2. 中国石化股份有限公司胜利油田分公司地质科学研究院)

摘要：针对孤岛油田中二中 Ng3-4 聚合物驱后油藏条件，通过物模手段研究了交替式注入泡沫复合驱的特征及驱油能力。长细管实验表明，交替式注入泡沫复合驱，低气液比时模型两端的封堵压差上升缓慢，交替周期越大封堵效果越差。非均质模型驱油实验表明，泡沫复合驱主要是通过扩大波及体积来提高采收率，在与聚合物驱相同注入量条件下，其封堵能力是聚合物驱的 2 倍。

关键词：泡沫复合驱；交替式注入；封堵作用；气液比；交替周期；提高采收率

中图分类号：TE357.42　文献标识码：A

引　言

泡沫复合驱体系由表面活性剂、聚合物和气体组成，由于该体系具有选择性封堵作用[1-2]，在多孔介质运移过程中视粘度远远大于其他驱替介质的粘度[3]特点，在提高采收率技术研究中[4-5]越来越受到关注。这项驱替技术是结合化学驱与气驱[6-7]的综合优势，可看作表面活性剂加合聚合物后注气，也可看作注气过程中加入控制流度比的起泡剂和提高泡沫稳定性的聚合物[8]。在非混相驱中，人们发现水气交替注入可以改善注气效果[9]，主要是由于水气交替注入能够降低水油流度比而增加水驱波及体积，能降低水驱过的油层剩余油饱和度而提高驱油效率；同时在重力分异作用下，可以通过气驱扫及正韵律厚油层上部那些水驱扫及不到的油层；能降低气相渗透率而降低气体流度，减缓气窜的发生[10]。

孤岛油田中二中 Ng3-4 注聚单元部分井组窜聚严重、注聚效果差，结合气驱与化学驱物模研究手段，对泡沫复合驱体系交替式注入的动态特征、封堵、驱油能力做了系统研究，为指导现场实施注氮气泡沫复合驱提供了大量的室内实验基础。

1　实验部分

1.1 实验用材料

N 聚合物：法国 SNF3530A；固含量 89.02%，水解度 25.5，滤过比 1.22；发泡剂：

DP-4(室内自制);实验用水:模拟地层水,矿化度 6 227 mg/L;实验用原油:模拟油,60℃粘度 102 mP·s

1.2 实验方法与步骤

将 DP-4 发泡剂、氮气按一定注入方式注入岩芯,由岩芯两端的压差变化确定发泡剂的封堵能力。即发泡后岩芯两端的压差与其水驱压差之比。驱油效率实验,采取双管合注合采,石英砂填充不锈钢管式模型;长岩芯驱替实验将岩芯换成玻璃珠填充长细管模型。

2 实验结果与分析

2.1 封堵能力实验研究

2.1.1 注入动态特征

在研究泡沫交替式注入的动态特征时,设定气液比为 1∶1,即向岩芯中交替注入起泡剂溶液段塞和气体段塞,在岩芯内形成泡沫。实验表明(图 1):

交替式注入产生的阻力因子随着注入量增加呈波动式上升,当注入倍数为 3.4 PV 时达到平衡。分析认为在注入初始阶段,岩芯中经历一个极稀体系的泡沫产生、破灭的过程,所以在初始阶段表现为随注入量增加阻力因子缓慢升高。并且交替式注入产生的泡沫是不连续的,转注气时,封堵压差较高,而转注液时,封堵压差有所下降,因此交替式注入在封堵强度上会出现交错式上升的现象,但是随着注入气量、液量不断增加,产生的泡沫量也逐渐增加,封堵效果呈增强趋势,最终达到一个趋于平衡的状态。

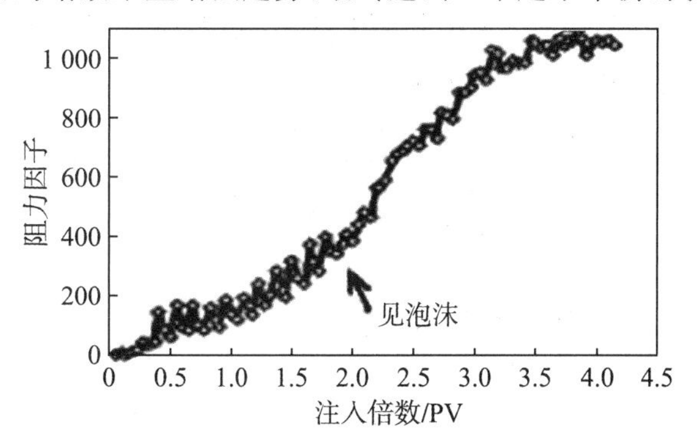

图 1 阻力因子与注入倍数曲线

2.1.2 不同气液比对封堵效果的影响

由于受设备耐压能力的限制,现场注入时无法实现高气液比,甚至气液比 1∶1 也不易控制实验正常进行,只能采取低气液比注入。我们在长岩芯模型上,对低气液比采取交替式注入做了一组对比实验(图 3),选择气液比分别为 0.5∶1,1∶1,注入段塞总量为 0.6PV,交替周期为 6。气液比 0.5∶1 达到最高压差 2 600 kPa 时注入倍数为 0.63 PV;气液比 1∶1 达到最高压差 2 730 kPa 时,注入倍数为 0.51 PV。即气液比为 1∶1 的模型两端压差上升速度快,达到最高压差所需注入倍数小;二者最高压差绝对值差别不大。

转入后续水驱后，封堵压差下降速度初期较快，以后逐渐减缓，实验结束时模型两端压差基本一致；当气液比较低时，气体在模型中扩散速度相对较慢，因而泡沫的封堵压差上升较为滞后。两种气液比最终的封堵效果一致，说明在实验条件范围内这两种气液比对驱替效果影响不大。

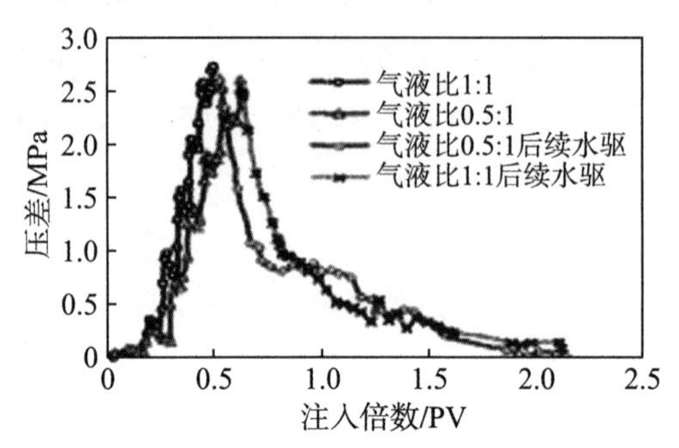

图 2 不同气液比与封堵压差的关系曲线

2.1.3 交替周期对封堵效果的影响

交替周期即为室内实验中设计的交替段塞。实验选择气液比为 1：1，在长细管模型上采用小段塞交替和大段塞交替进行实验(图 3)。结果表明，在注入泡沫段塞总量相等的条件下，采用小段塞交替注入时模型两端产生的最大压差为 2 730 kPa，是大段塞交替注入时最大压差 2 240 kPa 的 1.22 倍，表明采用小段塞交替注入时泡沫的封堵效果更好。这是因为在气液交替注入过程中，气液交替周期直接影响到气液两相的分散程度，缩短气液交替注入周期，有利于加大气液两相的分散程度，在多孔介质运移过程中充分产生泡沫。在后续水驱阶段二者动态特征基本一致，都能在较长时间内保持一定的压差，说明泡沫在多孔介质中的稳定性较好。

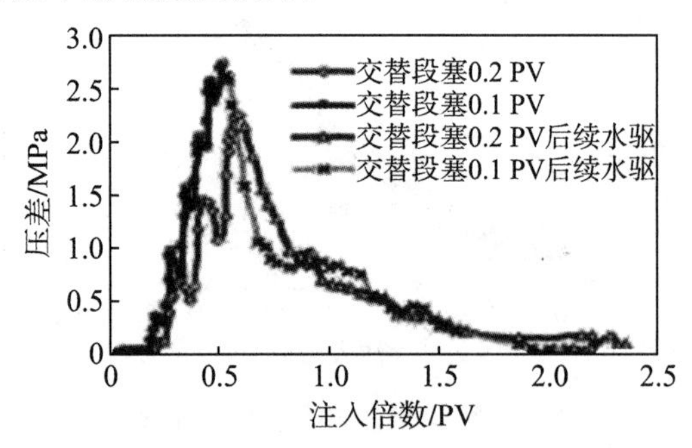

图 3 不同交替周期与压差关系曲线

2.2 驱油能力实验研究

2.2.1 泡沫复合驱交替式注入驱油效率实验

在渗透率级差为 4：1 的双管模型上，采取合注合采，回压 6 MPa，水驱至综合含水为 100%时，注入前置起泡剂与聚合物段塞 0.1 PV，再以气液比 1：1，氮气驱交替周期 0.05 PV、液相驱交替周期 0.05 PV，完成六个气相-液相驱组合后，转注水至结束。

实验结果如图 4、图 5 所示，水驱结束时采出程度 33.5%，在六个组合完成时采出程度 47.1%，实验结束时采出程度 51.7%，比水驱提高采收率 18.2%。在泡沫复合驱交替式进行过程中，压差最高增至 0.38 MPa 比水驱结束时 0.04 MPa 增长了 8 倍，这个封堵压差的增长是相当可观的。在未实施泡沫复合驱前，低渗管产液接近于零，当注入倍数达到 2.26 PV 时，低渗管与高渗管产液量几乎相当。说明泡沫复合驱有很强的封堵调剖能力，并且可以根据需要通过改变注入量来控制其强度，从这一点来说，目前还未见到比这种驱替技术更理想的封堵效果的报道。

2.2.2 相同注入量的聚合物驱驱油效率实验

在渗透率级差为 3.3∶1 的双管模型上，采取合注合采，回压 6 MPa 水驱至综合含水为 100%时，转注聚合物段塞 0.4 PV，后转水驱驱至综合含水为 100%时再转注聚合物段塞 0.3 PV，转注水至结束。

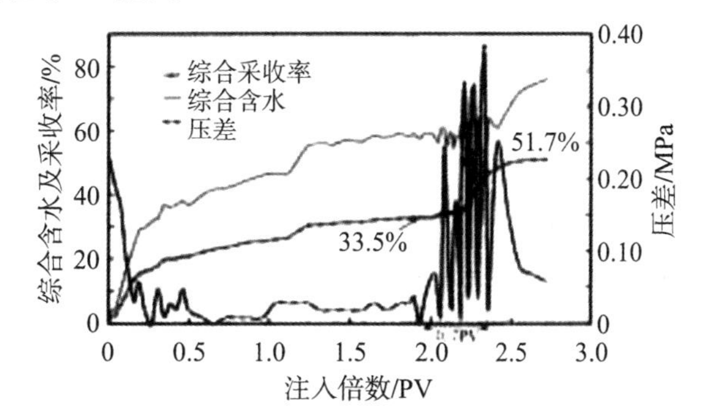

图 4　泡沫复合驱综合采收率、含水及压差曲线

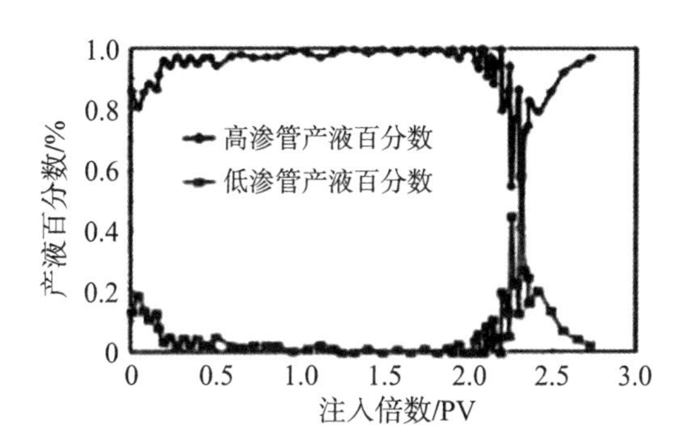

图 5　泡沫复合驱高低渗管产液百分数曲线

由图 6 所示，水驱结束时采出程度 35.2%，注入聚合物第一段塞 0.4 PV 结束时采出程度 45.7%，第二次水驱结束时采出程度 49.2%，第二次注入聚合物段塞 0.3 PV，实验结束时采出程度 50.0%，共计比水驱提高采收率 14.8%。注入聚合物前压差 0.06 PV，第一次注入聚合物压差最高增至 0.19 PV，第二次注入聚合物压差有小幅增长最高达 0.13 PV。由实验结果可明显看到，再次注入聚合物的效果并不显著，明显差于泡沫复合驱。因物模实验是采用管式模型的缘故，故在水驱后的提高采收率两种注入方式差别不大。

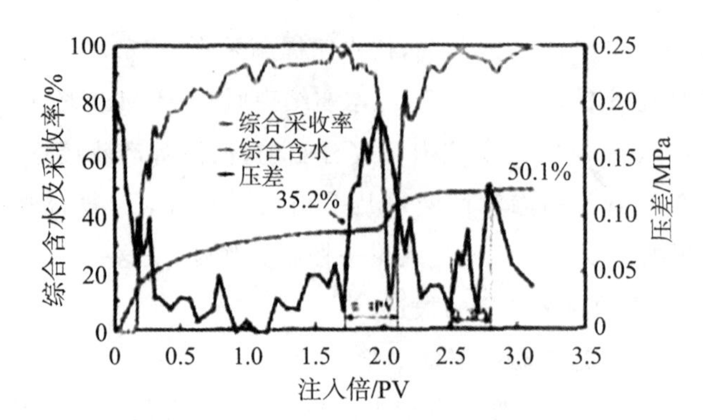

图 6　聚合物驱综合采收率、含水及压差曲线

3　结论

(1) 泡沫驱复合交替式注入产生的压差呈交错式上升,当达到一定的注入量,压差增加趋于平衡。

(2) 在其他实验条件相同的情况下,低气液比时的封堵压差比高气液比时上升缓慢,并且达到最高封堵压差时所需的注入量较大。现场应用时可根据这个特点调整气液比尽量在地层深部产生封堵作用而不影响其驱替效果。

(3) 在其他实验条件相同的情况下,采用大段塞交替注入即交替周期较长,封堵效果略差。

(4) 泡沫复合驱气液交替式注入,主要是通过 扩大波及体积来提高采收率的。室内实验在注入段塞与聚合物驱同等的条件下,水驱后提高采收率略高于聚合物驱 3.4%;随着注入量增加,其封堵压差越大,是相同注入量聚合物驱的两倍。

参考文献

[1] 廖广志,李立众,孔繁华,等. 常规泡沫驱油技术[M]. 北京:石油工业出版社, 1999.

[2] 杨承志,黄琰华,刘彦丽,等. 泡沫驱油过程中起泡剂分配方式的研究[J]. 石油勘探与开发, 1985, 12(3):57-64, 67.

[3] 佟曼玉. 油田化学[M]. 东营:石油大学出版社, 1996.

[4] 贾忠盛, 潘严富, 王滨玉. 大庆油田烃气非混相驱矿场试验[J]. 石油勘探与开发, 1997, (01):39-41, 93.

[5] 赵长久, 麻翠杰, 杨振宇, 等. 超低界面张力泡沫体系驱先导性矿场试验研究[J]. 石油勘探与开发, 2005, 32(1):127-130.

[6] Chiang JinC, SawyalSubirK, CastanierLouisM, et al. Foam as a mobility control agent in steam injection proceses[C]. SPE8912, 1980.

[7] Yaghobi, Hosein. Laboratory investigation of parame Ters affecting CO_2-foam mobility in sand stone at reservoir conditions[C]. SPE29168, 1994.

[8] Robert D S Polymer-enchanced foams[C]. SPE25168, SPE25175, 1994.

[9] Greenwalt W A, VelaSaul, Christian LD, etal. A Field Test of Nitrogen WAG Injectivity[C]. SPE8816, 1982.

[10] 郭万奎. 注气提高采收率技术[M]. 北京:石油工业出版社, 2003.

（编辑 朱和平）

ALTERNATE INJECTION FOAM COMBINATION FLOOD ING EXPERMENT

ZHOU Guo-hua (Research Institute of Petroleum Exploitatio and Development, Petrochina, Beijing100083, China), CAO Xu-long, WANG Qi-wei, eta. JOURNAL OF SOUTHWEST PETROLEUM UNIVERSITY, VOL. 29, NO. 3, 94-96, 2007 (ISSN1000-2634, IN CHINESE)

Abstract: In this paper the character of polym er-surfactant foam combination flooding of alternate injection is studied with physical model test under the reservoir condition of central part of Block2 in central Gudao Oilfield. The long slim tube test shows that at low gas liquid ratio, the blocking pressure diference between inlet and outlet of slim tube increases more slowly and the bigger alternate slug is the worse the block effect. The oil displacement test on heterogeneous porousmodel shows that EOR of polymer-surfactant foam combination flooding mainly depended on increasing sweep volume, at the same injection volume, its blocking ability is twice bigger than polymer flooding.

Keywords: polymer-surfactant foam combination flooding; alternate injection; blocking effect; gas liquid ratio; alter nate slug; EOR

THE EXPERI MENTAL STUDY ON IPN GEL PARTICLESBLANKING-OFF TO FRACTURED CORE

XIEQuan(State Key Laboratory Of Oil and Gas Reservoir Geology Exploitation enginering · Southwest Petroleum University, 610500), PUWan-feng, DINGXun-yi, eta. l JOURNAL OF SOUTHWEST PETROLEUM UNIVERSITY, VOL. 29, NO. 3, 97-99, 2007(ISSN1000-2634, INCHINESE)

Abstract: The authors of the paper investigate IPN, 606, 612 pre-cross linked gel Particles' distribution of blank-off coefficient in fractured core, meanwhile, study these gel Particles' blank-off rate, reserve rate and the gradient of particles' break-through pressure in the fractured core after 4, 10, 30, and 180 hours respectively. The results of the experiment show that gel particles' distribution of blank-off coefficient and their dynamic law in the fractured core can be used to evaluate the particles' migration law and profile control property in the fractured core. It is also identified IPN gel particles illustrate a good blank-off effect. When IPV water is injeceted, the blank-off coeficient of long fractured core is quite steady, which is apparentely higher than that of 606 and 612 gelparticals. The blank-off Rate in the fractured core after 180 days can reach 903. Percent and the reserve rate of blank-off Water efficien-cy99 Percent. This kind of gel can be used as a deep profile control agent in the channeling reservoirs.

Keywords: IPN gel particles; fractured core; blank-off coefficient; blank-of rate; reserve rate; break-through Pressure gradient of the particles; profile adjusting; expermient.

A PROBEON THE RELATED PROBLEMS OF LAYRING POLYMER INJECTION

CHEN Zhang-qing(Changjiang University, Qianjiang Hubei 434023, China), LI Lin, QIAN Yu, eta. l JOURNAL OF SOUTHWEST PETROLEUM UNIVERSITY, VOL. 29, NO. 3, 100-103, 2007(ISSN1000-2634, IN CHINESE)

Abstract: The layring polymer injection is the one of the effecyive ways enhancing oil recovery. At the present, there are 302 wells incentral Saertu development area in Daqing oilfield having been practiced layring injection, which take up 37. 6% of total wells, the way has played a unreplacible role in polymer flooding recovery. But impacted by the multiple factors such as adaptability of reservoirs and layring injection technique, the way needs furtherly improved in practice. Through the analysis to reservoirs production dynamic situation of layring polymer in-jection wells in different polymer injection stages, dynamic change and production, the well-selecting rule and lay-ring time of the layring wells are described, the adaptability and the existing problems of layring injection technique

表面活性剂疏水链长对高温下泡沫稳定性的影响

曹绪龙[2],何秀娟[1],赵国庆[1],宋新旺[2],王其伟[2],曹嫣镔[3],李 英[1]

(1. 山东大学胶体与界面化学教育部重点实验室,济南 250100;
2. 中国石油化学公司胜利油田地质科学研究院,东营 257015;
3. 中国石油化学公司胜利油田采油工艺研究院,东营 257017)

摘要:选用不同疏水链长的α-烯烃磺酸盐(AOS)形成泡沫,分别用泡沫衰减法和泡沫岩芯封堵法测定不同温度下的泡沫稳定性,并采用动态表面张力、界面流变、分子模拟等方法研究了表面活性剂在气/液界面的吸附行为和界面吸附层的性质,分析了高温下泡沫的稳定机制。实验结果表明,在高温下,极性头的“锚定作用”减弱,表面活性剂疏水链难以在气液界面保持以直立状态吸附,疏水链碳数大于20的表面活性剂分子难以分立吸附,其疏水链相互交叉缠绕,增强了泡沫膜的强度,减缓了气体通过液膜的扩散,形成的泡沫在高温下具有较好的稳定性。

关键词:泡沫稳定性;高温泡沫;泡沫封堵;蒸汽驱;疏水链长度

中图分类号:O648.2+4 文献标识码:A 文章编号:0251-0790(2007)11-2106-06

泡沫在日用化工、矿物浮选、食品、石油开采等工业领域中均具有广泛的应用,因此对泡沫性质特别是泡沫稳定性的研究已引起越来越多的关注,关于泡沫的稳定机理研究已有报道[1-5],但大多数的文献报道集中在对常温下泡沫性质的研究方面。实际上,在一些高温的工业过程中,泡沫也具有重要的应用。如稠油开采方式,主要是注蒸汽热采。但是由于储油层的层间、层内均存在非均质性,蒸汽的低密度和低粘度易引起重力超覆和汽窜,产生不均匀的垂直扫油效率,地层中残余油的饱和度高,蒸汽波及系数小,采收率降低,严重影响油田开发的经济效益和矿藏资源的有效利用。多年来的研究结果表明,上述驱油体系的流度问题可以通过注入蒸汽泡沫的形式得到改善。泡沫不仅可以选择性地控制流度,改善注入流体在非均质油藏中的驱替状况,降低储层中高渗透层流体的锥进,还可在注蒸汽热采中调整油层吸汽剖面,封堵蒸汽汽窜通道,提高蒸汽波及系数,提高采收率[6-8]。研究高温下泡沫的稳定机理,开发高温泡沫稳定剂是该技术的关键。α-烯烃磺酸盐(AOS)是耐高温、抗盐性能好的表面活性剂,在较大的温度范围内均具有良好的泡沫稳定性。本文选用不同疏水链长的AOS形成泡沫,分别用泡沫衰减法和泡沫岩芯封堵法测定不同温度下的泡沫稳定性,并采用动态表面张力、界面流变、分子模拟等手段研究 AOS 在气/液界面的吸附行为和界面吸附层的性质,分析了高温下泡沫的稳定机制,这对高温泡沫稳定剂的开发和应用具有指导意义。

1 实验部分

1.1 试 剂

疏水链碳数为 14,16 及 20～24 的 α 烯-烃磺酸盐,分别表示为 AOS_{14},AOS_{16} 和 $AOS_{20\sim24}$,由南风化工集团西安研究所提供。

1.2 泡沫稳定性的测定方法

低于 100℃时,采用传统的泡沫衰减法考察泡沫的稳定性。用直径为 100 μm 的玻璃珠烧结而成的玻璃砂滤板上方预置 40 mL 的测试溶液,以 75 mL/min 的速度从玻璃砂滤板下方注入氮气,形成一定体积的泡沫柱,观察泡沫的衰减,用半衰期即泡沫体积衰减一半所需的时间表征泡沫的稳定性。

采用泡沫岩芯封堵法考察高温下泡沫的稳定性。该方法是根据泡沫在模拟岩心管中运移时建立的压差来衡量泡沫稳定性,如果泡沫运移时在岩心管两端产生明显而稳定的封堵压差,说明泡沫在此条件下有较好的稳定性[9-11],实验装置见 Scheme1。自制模拟岩心管渗透率为 4×10^{-12} m²;测量时 2 号泵以 1 mL/min 的速度注 入泡沫剂溶液,1 号泵以 4 mL/min 的速度注入 纯水,使得混合后泡沫剂的质量分数为 0.35%;调节锅炉加热到设定温度;氮气的注入速度为 0.02 L/min;记录 5 号和 6 号压力传感器的压力随着泡沫注入的变化,压差记为 P,即为封堵压差。实验采用同一根岩心管,在相同的试验条件下,封堵压差越高,泡沫稳定性越好。

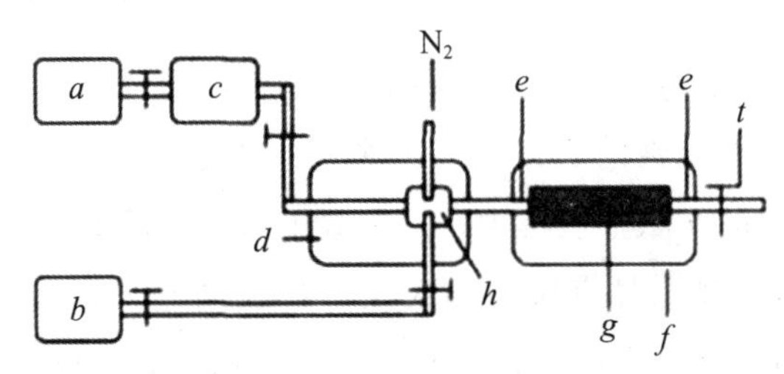

Scheme 1 Schematicdiagram of devices for foam-block determination

a. Pump in jecting pure water b. pump injecting surfactantsolution;

c. boiler; d. insulationcan; e. temperature and presure sensor;

f. heating jacker; g. core pipe; h. container where liquid and gas mixture; i. valve.

1.3 表面张力测定

吊环法测量平衡表面张力,采用 JYW-200B 型自动张力仪(承德试验机有限责任公司)。

采用最大气泡压力法[12,13],利用自组装动态表面张力仪测定体系的动态表面张力,采用 Rosen 模型处理数据,从而得到表面活性剂分子的动态行为参数。

1.4 界面流变

采用法国.IT.Concept公司Tracker全自动液滴界面流变仪。当气泡在扩张压缩条件下变形时,便产生阻止表面变化的阻力,从而产生界面粘性和弹性响应,界面粘弹模量E(mN/m)定义为界面张力与界面面积相对变化的比值:

$$E = d\gamma/(dA/A) = d\gamma/d(\ln A) \tag{1}$$

当界面面积正弦变化时,任何时刻的界面流变都可以用FOURRIER分析,式(1)中,界面面积形变dA/A是输入值,界面张力γ是输出结果,分析结果为界面粘弹模量E,其中实部和虚部分别称为存储模量和损耗模量,分别代表了弹性和粘性部分的贡献[14-16],其表达式如下:

$$E = |E|^{\zeta} = E' + iE'' \tag{2}$$

实验测量时,预先向样品池内加入10 mL待测溶液,恒温放置,然后用注射器缓慢吹起一个体积为5 μL的气泡,待泡沫稳定后添加振荡,振幅为泡沫初始体积的10%,振荡周期分别为3,5,7,9,11和13,s分析施加应力和界面形变对时间的响应。

1.5 分子模拟

采用全原子分子动力学模拟,研究了不同温度下不同碳链长度AOS分子在泡沫双层液膜中的排布情况和动态行为,探讨了常温和高温下泡沫稳定机制。模拟初始构型搭建和计算方法与“SDSNBF液膜模拟”[17,18]中的相同,即将16个AOS分子按照六角状排列成一单层,模拟盒子3个方向均为周期性。之后,把它们放在水层两侧组合成一个体系,应用分子动力学(MD)模拟到平衡。

模拟条件选择PCFF力场,采用NVT系综。通过HooverNose方法保持温度恒定,弛豫时间选择0.2 p^{s}采用Ewald加和方法处理长程静电相互作用;van derWaals相互作用应用Lennard-Jones势能函数表示,截断半径选择1.5 nm[19-22]。模拟执行500 ps的平衡动力学以后,至少1.5 ns的时间过程收集动力学信息。模拟选择1 fs的步幅,采用Cerius2 Ve.l4.6模拟软件。

2 结果与讨论

2.1 温度对泡沫稳定性的影响

采用泡沫衰减法测定的AOS_{14},AOS_{16}和$AOS_{20\sim24}$的泡沫稳定性随温度变化的数据列于表1。

从表1中实验数据可以看出,AOS_{14},AOS_{16}和$AOS_{20\sim24}$形成的泡沫稳定性均随温度的升高而降低;在同一温度下,泡沫的稳定性随着疏水链碳原子数的增加而增加,其中$AOS_{20\sim24}$形成的泡沫稳定性增加非常显著。

Table1 Half-life of foam formed by AOS Afected by temperature*

Temperature/℃	AOS_{14}	Half life/min AOS_{16}	$AOS_{20\sim24}$
60	10.0	26.0	65
70	5.6	9.3	57
80	4.3	5.0	35

* The mass fraction of foam is 0.35%.

图 1(A)是 AOS_{16} 形成的泡沫在 60 和 250 ℃下的封堵曲线。由图 1(A)可以看出，泡沫在 60℃下能建立稳定的压差，但在 250℃高温下几乎不能建立压差，表明此时泡沫已不能稳定存在。图 1(B)是 $AOS_{20\sim24}$ 形成的泡沫在 250 和 300℃下的封堵曲线，可以看出 $AOS_{20\sim24}$ 在 250 和 300℃高温下运移时都建立了稳定的封堵压差，这充分说明 $AOS_{20\sim24}$ 形成的泡沫在高温下仍具有良好的稳定性。

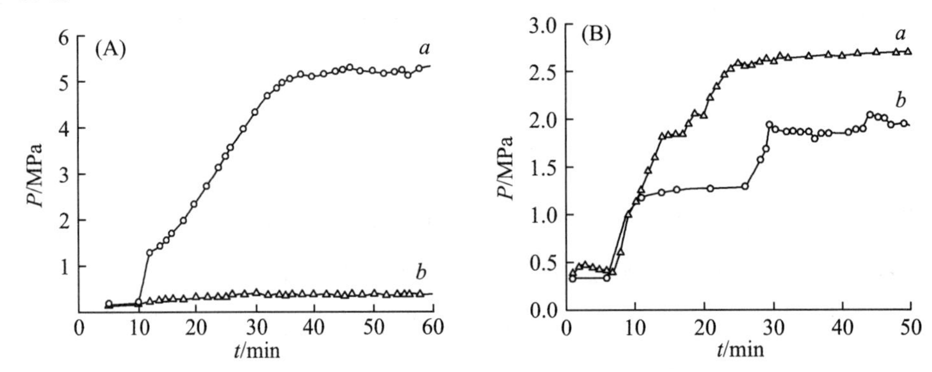

Fig. 1 Pressure diference found by AOS_{16} and $AOS_{20\sim24}$ foam when transported insand-pack pipe at Different temperatures

(A) a. AOS_{16}, 0.35%, 60℃; b. AOS_{16}, 0.35%, 250℃;

(B) a. AOS_{20-24}, 0.35%, 250℃; b. AOS_{20-24}, 0.35%, 300℃.

实验结果表明，泡沫的稳定性随着温度的升高而降低，随着疏水链长度的增加而升高，$AOS_{20\sim24}$ 形成的泡沫在 300 ℃时仍然保持了较高的稳定性。

2.2 疏水链长度改变对表面活性剂在气/液界面吸附行为的影响

经测定，25℃时 AOS_{14}，AOS_{16} 和 $AOS_{20\sim24}$ 的临界胶束浓度分别为 0.1%，0.06%和 0.02%，平衡表面张力值 γ_e 分别为 32.5，33.4 和 43.1 mN/m；随着疏水链长度的增加，临界胶束浓度(cmc)减小，吸附饱和后的平衡表面张力值升高。

泡沫的形成是能量升高的过程，根据 Gibbs 原理，体系总是趋向处于较低的表面能状态，因此低表面张力可使泡沫体系能量降低，有利于泡沫的稳定。毛细管压力与溶液的表面张力成正比，表面张力低时，毛细管压力小，泡沫排液速度慢。

AOS 形成的泡沫稳定性随着疏水链长度的增加而升高，但吸附饱和后的平衡表面张力值却随着疏水链长度的增加而增加，因此可见平衡表面张力不是在高温下泡沫稳定与否的决定性因素。

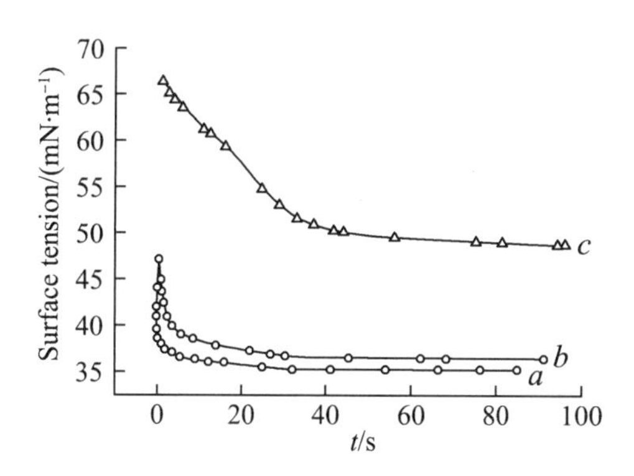

Fig. 2 Dynamic surface tension of AOS_{14}, AOS_{16}, $AOS_{20\sim24}$ solutions at 25℃

Mass fraction:a. AOS_{14}, 0.2%; b. AOS_{16}, 0.12%;. c$AOS_{20\sim24}$, 0.05%.

图2是25℃时AOS_{14},AOS_{16}和$AOS_{20\sim24}$溶液的动态表面张力曲线,实验浓度采用约2倍的cmc。采用Rosen模型处理得到的表面活性剂动态行为参数列于表2。从图2可以看出,AOS_{14}和AOS_{16}到达平衡区表面张力所用的时间差别不大,但$AOS_{20\sim24}$所需时间明显变长,而且平衡区表面张力值$\gamma AOS_{20\sim24} > \gamma AOS_{16} > \gamma AOS_{14}$。

用Rosen模型处理得到的动态行为参数,n值反映了吸附初期($t\rightarrow0$)表面活性剂分子从本体溶液扩散到表面下层的扩散过程,值越小,表面活性剂扩散越快。* t反映了吸附后期($t\rightarrow\infty$)表面活性剂分子从表面下层到溶液表面的吸附过程,t^*越小,吸附势垒越大,表面活性剂分子越不易吸附在溶液表面[9]。

Table2 Dynamic surface tension param eters dealt With Rosen model

Sample	Mass fraction(%)	n	t^*
AOS_{14}	0.20	0.144 56	0.043 90
AOS_{16}	0.12	0.294 24	0.119 57
$AOS_{20\sim24}$	0.05	0.936 33	1.563 43

从实验结果可以看出,AOS疏水链越长,n值越大,AOS分子在体相的运移速度越慢,因此,同等条件下$AOS_{20\sim24}$分子从体相到达界面需要的时间较长;AOS_{14}分子疏水链比较短,分子体积相对较小,在体相内运移速度比较快,n值较小。

表面活性剂由体相向界面吸附是头基的水化能力和疏水链的疏水作用共同作用的结果。AOS_{14},AOS_{16}和$AOS_{20\sim24}$具有相同的极性头,而$AOS_{20\sim24}$疏水链较长,疏水作用强,其吸附在界面上的趋势比较大,一旦吸附,脱附相对比较困难,因此t^*最大[23]。

2.3 界面流变性能

当温度为60℃,表面活性剂质量分数为0.35%时,由AOS_{14},AOS_{16},$AOS_{20\sim24}$形成的单层泡沫膜的粘弹性膜量随泡沫振荡周期的变化结果见图3。

由图3结合式(1)可以看出,当气泡在周期振荡下,膜量的变化主要取决于界面张力的变化。当一个新鲜的气泡膜在体相形成时,表面活性剂分子在界面的吸附和脱附决定了界面膜量的大小。添加振荡时,界面受到扰动后表面活性剂分子迅速修复界面膜,使界面张力达到平衡，膜量值则比较低[24-26]。AOS_{14} 和 AOS_{16} 分子尾链都比较短，分子扩散比较快，自修复能力比较强，这种分子的快速交换使得膜量值较低，膜量值随周期的变化也比较小。而 $AOS_{20\sim24}$ 的泡沫膜在同样的实验条件下，膜量随振荡周期的延长而降低，直至达到 一个稳定值，较长振荡周期时 $AOS_{20\sim24}$ 分子也有足够的时间修复气泡界面，但仍具有较高的膜量值，表现出了不溶性膜的性质，表明其膜量不仅仅由表面活性剂分子在界面上形成的张力梯度决定。

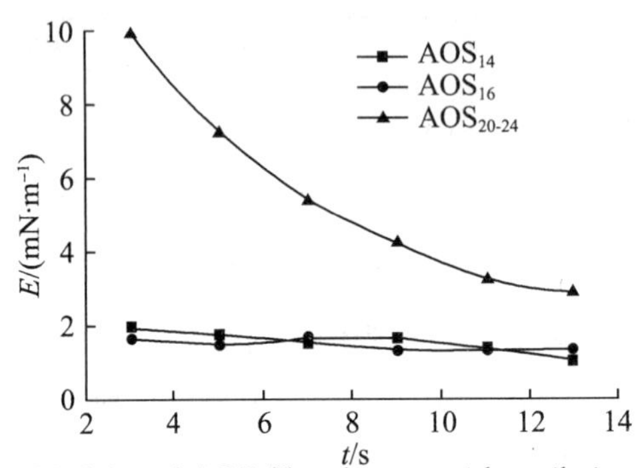

Fig. 3 Modulus of AOS film changes with oscilating period

The mass fraction of surfactant is 0.35%, temperature is 60℃.

图 4 是 AOS 形成的泡沫膜弹性膜量随振荡周期的变化。由图 4 可以看出，AOS_{14} 和 AOS_{16} 形成的界面膜的弹性膜量随振荡周期变化很小；$AOS_{20\sim24}$ 形成的界面膜随振荡周期的延长而减小，然后达到一 稳定值，而且其稳定膜量值比 AOS_{14} 和 AOS_{16} 形成的界面膜的弹性膜量值稍有升高，表明界面膜的弹性 不是高温稳定性的决定因素。

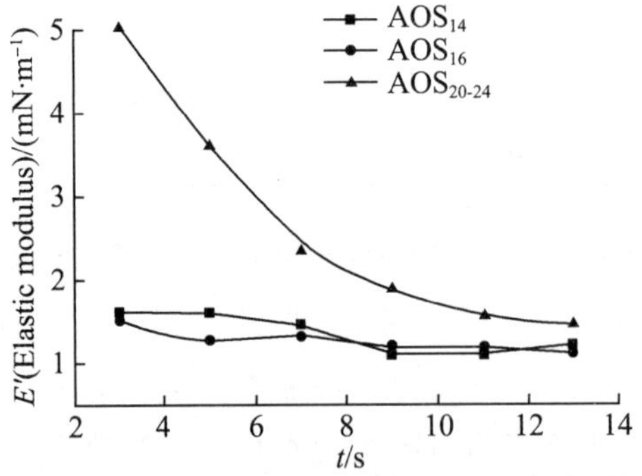

Fig. 4 Elastic modulus of AOS film changes with Oscilating period

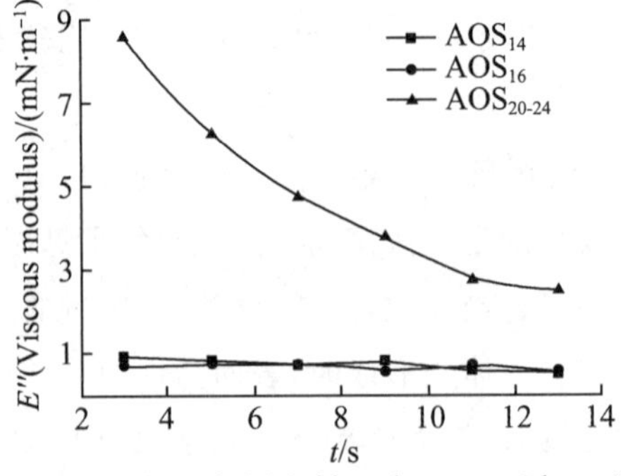

Fig. 5 Viscous modulus of AOS film changes with oscilating period

界面粘度是指界面分子层内的粘度，主要是表面活性剂分子在其表面单分子层内的亲水基间的相互作用及水化作用产生的。良好的起泡剂在吸附层内必须有较强的相互作用，同时亲水基团要有较强的水化能力。前者使膜有较高的机械强度，后者可提高液膜的表面粘度。

由图 5 可以看出，AOS_{14} 和 AOS_{16} 形成的界面膜的粘性膜量随振荡周期几乎未变化，$AOS_{20\sim24}$ 形成的界面膜随振荡周期的延长而减小，然后达到一稳定值，稳定膜量值与 AOS_{14} 和 AOS_{16} 的相比，粘性膜量升高 4～5 倍，高的界面粘性使 $AOS_{20\sim24}$ 形成的泡沫抵抗外界干扰的能力增强，提高了泡沫的稳定性。

从分子结构上分析，AOS_{14}，AOS_{16} 和 $AOS_{20\sim24}$ 的亲水基相同，分子结构的差别只是疏水链长度的不同。疏水链长度增加可能带来的变化有：(1) 疏水链变长，分子间的疏水相互作用增强，导致粘性膜量增加；(2) 表面活性剂在界面排布时，头基的水化作用对分子的构象有"锚定作用"，保持了疏水链的独立存在，但当疏水链超长时，疏水链开始发生弯曲缠绕，这种空间的交叉使得界面膜粘性膜量值突升，膜强度增大。

2.4 分子模拟结果

2.4.1 表面活性剂疏水链间的相互作用随疏水链长度的变化

采用分子动力学模拟分析范德华力随疏水链长度的变化可以看出，分子间的范德华相互作用随疏水链长度的线性变化为

$$VDW=-1.870\ 22+0.638\ 86^{n}$$

式中，n 为表面活性剂疏水链中碳原子个数，结果见图 6。

界面流变实验表明，$AOS_{20\sim24}$ 形成的界面膜的粘弹性膜量与 AOS_{14} 和 AOS_{16} 的相比，弹性膜量稳定值稍有升高，粘性膜量平衡值突升 4～5 倍，而范德华力随表面活性剂疏水链长呈线性变化。对比两个结果可以看出，界面膜量的变化不仅仅与分子间疏水作用力有关，还与表面活性剂分子在界面上排布的构象变化有关。

2.4.2 表面活性剂分子界面排布构象—疏水链卷曲程度

本文采用参数 R 来表示疏水链的弯曲程度，$R=L/L_0$，L_0 表示表面活性剂分子疏水链最伸展时的长度；L 表示表面活性剂分子疏水链在一定条件下平衡时的长度。

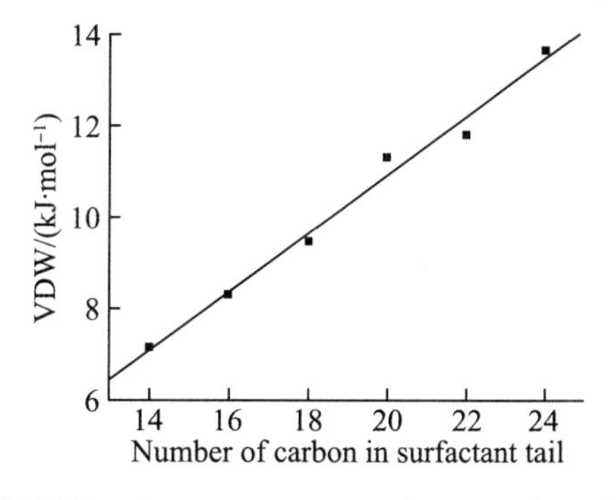

Fig. 6 Plot of VDW pair energy versus carbon Number of surfactant tail

模拟结果表明,温度升高使得疏水链变得弯曲,如 AOS_{14} 疏水链平均长度在 298 K 时为 1.473 nm,而在 473 K 为 1.331 nm,疏水链长度均差为 0.142 nm。高温下,AOS_{14} 和 $AOS_{20\sim24}$ 疏水链卷曲程度的变化列于表 3。从表 3 可以看出,$AOS_{20\sim24}$ 的疏水碳链比 AOS_{14} 的疏水碳链弯曲程度要大,这种更大程度的弯曲使得 $AOS_{20\sim24}$ 的长疏水链相互缠绕交错(图 7)。由图 7 可见,空间的相互作用使得界面膜的强度增大,界面流变的膜量升高,另外疏水链相互交叉还减缓了气体通过液膜的扩散,因此,超长疏水链(碳数>20)的 AOS 形成的泡沫在高温下有较好的稳定性。

Table3 Value of R for AOS with diferent hydrophobic chain length at 473 K

Length/nm	L_0	L	$R=L/L_0$
AOS_{14}	1.798 3	1.330 5	0.074 3
AOS_{20}	2.546 1	1.693 7	0.066 5

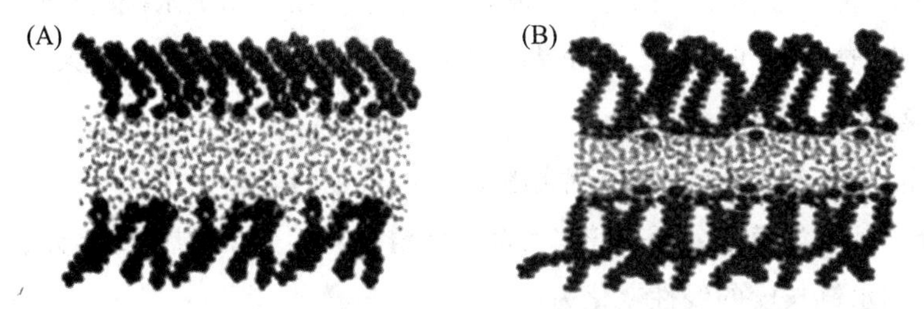

Fig. 7 Conformation of AOS in AOS_{14} (A) and AOS_{20} (B) foam film

For visual clarity the sodium ions are drawn as Vander Walss pheres, and water molecules are line stye and the other atoms are drawn by CPK style. Three cells are repeated in the x-cordinate in order to show the interface clearly.

参考文献

[1] Ashok Bhakta EliRuckenestein. Adv. Colloid Interface Sc. i[J], 1997, 70: 1-124.

[2] HAN Guo-Bin(韩国彬), WU Jin-Tian(吴金添), XU Xiao-Ming(徐晓明). Chem. J. Chinese Universities(高等学校化学学报[J], 2001. 22(7):1177-1180.

[3] Jennifer Hanwright, James Zhou. Langmui[J], 2005, 21(11):4912-4920.

[4] Cosmia Stubenrauch, Reinhard Miller. J Phys Chem. B[J], 2004, 108(20): 6412-6421.

[5] Cosmia Stubenrauch, Khristo Khristov. J Coloid Interface Sci. [J], 200 286 (2):710-718.

[6] Cheng Lin-Song, Liang Ling. J Petroleum Sci& Engineering[J], 2004, 41 (3):199-212.

[7] BergeronV. Coloids& Surfaces A[J], 1997, 123/124:609-622.

[8] WEI Yan, Clarence A Miller Coloids& Surfaces A[J], 2006, 282/283:348-359.

[9] Li Ying, Zhang Peng, Zhao Guo-Qing. Coloids& SurfaceA[J], 2006, 272:124-129.

[10] VidrineW. K., WilsonB. C. S. Coloids& Surfaces A[J], 2000, 175:277-289.

[11] FerguiQ, BertinH. J Petroleum Sci& Enginering. [J], 1998 20:9-29.

[12] EastoeJ, DaltonJ S. Adv Colloid and Interface Sci[J], 2000, 85(2):103-144.

[13] FainermanV. B, KazakovV. N. Colloids& Surfaces A[J], 2004, 250:97-102.

[14] Cao Xu-Long, Li Yang. J Coloid Interface Sci[J], 2004, 270(2):295-298.

[15] Ben jamins J, CagnaA, Lucasen-ReynderF. H. Colloids& Surfaces A[J], 1996, 114:245-254.

[16] SoninA. A, BonfilonA, LangevinD. J. Coloid Interface Sci[J], 1994, 162(2):323-330.

[17] FernandoB, JordiF, Langmui[J], 2004, 20(12):5127-5137.

[18] JangS. S, GoddardW. A. J. PhysChem. B[J], 2006, 110(15):7992-8001.

[19] PaiY. H, ChenL. J. J Chem. Eng. Data[J], 1998, 43(4):665-667.

[20] Tsierkezos N. G, Molinou. I E. J Chem Eng Data[J], 1998 43(6):989-993.

[21] CalvoE, BrocosP, BravoR, eta. l. J Chem Eng Data[J], 1998, 43(1):105-111.

[22] SastryN. V, ValandM. K. J Chem Thermodyn[J], 1998 30(8):929-938.

[23] SabrinaH Myrick, Elias IFranses Coloids& SurfacesA [J], 1998, 143:503-515.

[24] KoelschP, MotschmanH. Langmui[J], 2005 21(14):6265-6269.

[25] Cosmia Stubenrauch, Reinhard Miler J Phys Chem B[J], 2004, 108 (20):6412-6421.

[26] MartinA Bos, Ton van Vliet Adv Coloid Interface Sci[J], 2001, 91(3):437-471.

Effect of hydrophobic chain length of surfactants on foam stability at high temperature

Cao Xulong[2], He Xiujuan[1], Zhao Guoqing[1], Song Xinwang, Wang Qiwei[2], Cao Yanbin[3], Li Ying[1]

(1. Key Laboratory for Coloid& Interface Chemistry of State Ministry of Education, Shan dong University Jinan 250100, China;

2. Geological Scientific Research Institute, Shengli Oilfield, Dongying 257015, china

3. Oil Production Technology Institute Shengli Oilfield Dongying 257017, China)

Abstract: The foam stability stabilized by α-olefin sulfonate(AOS) with diferent hydrophobic chain lengths at temperature below 100℃ and above 200℃ was investigated by foam column decay method and foam block in sand-pack pipe, respectively. The results show that the foam stability increases with the increase of hydro-phobic chain length. Dynamic surface tension, interfacial rheometer and molecular smiulation were used to study the foam stablity me chanism at a high temperature. From the expermient resulst, we can see that when the hydrophobic chain length was longer than 20 carbon atoms, the hydration of surfacrtant head could not sup-port the hydrophobic chain to keep vertical or unatached, and the hydrophobic chains cross-link with each other, which increase the strength of the foam film, decrease the diffusion of the gas through the film and make the foam stable at a high temperature. Keywords Foam stability; High temperature foam; Foam block; Steam flooding; Hydrophobic chain length. (Ed: V, I)

泡沫加二元复合体系提高采收率技术试验研究

王其伟
中国石油大学(华东)化学化工学院,山东东营　257062
胜利油田分公司地质科学研究院,山东东营　25701 5)
郑经堂
(中国石油大学(华东)化学化工学院,山东东营　257062
曹绪龙,郭　平,李向良,王军志
(胜利油田分公司地质科学研究院,山东东营　257015)

[摘要]通过物理模拟试验研究了聚合物驱油体系、泡沫复合驱油体系、二元复合驱油体系及泡沫与二元的复合体系提高采收率的能力、驱替过程中产出液含水变化及注采压差变化。试验证明,泡沫加二元复合驱油体系集合了泡沫及二元驱油体系的特点,具有强调及强洗的双重作用,泡沫的高视粘度及选择性封堵提高了驱油体系的波及面积,二元体系的高界面活性提高了驱油效率,减少了油藏的残余油的存在,使泡沫体系更稳定,多种组份相互作用,使驱油效果最大化;注入 0.3 PV 复合段塞,综合 采收率提高 33.7%,提高采收率效果明显好于单纯的聚合物驱、二元驱及泡沫复合驱,在注入相同段塞条件下能够增加采收率 4%～13%。

[关键词]泡沫驱;二元驱;采收率;含水;二元加泡沫复合驱油体系

[中图分类号]TE357.46　[文献标识码]A　[文章编号]1000-9752(2008)03-0134-04

胜利油区主力油田逐步进入高含水、特高含水期,稳产难度大,开发矛盾日渐突出,而勘探新增储 量的难度也越来越大,成本越来越高。进一步提高已探明、已开发储量的采收率已经成为十分迫切的工作,提高高含水油田采收率技术越来越受到人们的关注。三次采油目前较为成熟的技术是聚合物驱及二元驱[1]。二元驱在现场应用中取得了良好的效果,是一项很有前途的提高采收率的技术[2]。而泡沫驱由于其良好的封堵性能及对油水选择性的特点,已被许多石油科技工作者所看好[3]。泡沫体系具有强调[4]特点,与二元体系具有强洗特点相结合,可形成具有强调强洗的泡沫加二元的复合泡沫体系。笔者通过试验研究验证了泡沫加二元复合体系提高采收率的能力及特点,试图为高含水油田提高开发效果及聚合物驱后进一步提高采收率提供一种新的方法。

1 试验部分

1.1 主要仪器设备、材料及试验模型

1）主要仪器设备 泡沫物理模拟驱油流程：控温范围 0～150℃，控温精度±0.5℃；气体质量流量控制器，耐压范围 0～30 MP a，控制流量 0～30 mL/min；回压阀：控压 0～30 MPa；调压阀：控压 0～30 MPa，控压精度±0.01 MPa；数字压力表：压力范围 0～20 MPa，精度±0.01 MPa。

2）材料 模拟地层水：矿化度 6 379 mg/L，$Ca^{2+}+Mg^{2+}$ 含量 52 mg/L，$Na^{+}+K^{+}$ 含量 2 061 mg/L，Cl^{-} 含量 3 086 mg/L，HCO_3^{-} 含量 1 085 mg/L，CO_3^{2-} 含量 95 mg/L；泡沫剂：为 DP-4 工业品，有效含量 36%；聚合物 3 530S：SN FI 公司产品，固含量 91%，分子量 1 500 万；模拟油：河口埕东油田原油与过滤煤油按 9∶1 配置的模拟油，原油粘度 47 mPa·s；注入气体：高纯氮气；SLPS 磺酸盐：胜利石油磺酸盐，有效含量 35%，中胜环保化工有限公司生产；表面活性剂 CEA-1：有效含量 35%，东营广贸化工科技有限公司生产。

3）试验模型 用清洁分筛后的河道砂敲制模型，模型尺寸 25 mm×800 mm，高渗模型渗透率为 3.90 μm^2，低渗模型渗透率为 1.32 μm^2，渗透率级差接近 3∶1。

1.2 试验方法

进行泡沫流程驱油试验时，将人工敲制的不同渗透率的填砂模型抽空饱和水、饱和油，并联接入流程，水驱至含水 98%；注入一定孔隙体积倍数的化学剂溶液段塞后转水驱，至产出液含水 98%后，试验结束，记录各个试验阶段的注入压力、产出液含水及采出程度变化情况。

1.3 试验条件

试验温度 60℃，回压 6 MPa，注入速度 0.5 mL/min，原油粘度 47 mPa·s，气液比 1∶2，泡沫剂浓度 0.5%，注入段塞 0.3 PV。

2 结果与讨论

2.1 泡沫加二元复合驱油体系的建立

泡沫驱油体系具有较好的封堵调剖能力，并且对于油水具有选择性封堵的特点[5]。泡沫剂作为一种表面活性物质，同时具有降低油水界面张力的作用。但是，通常泡沫体系降低油水界面张力的能力较差，达不到超低界面张力[6]，即使达到超低界面张力，由于油水界面张力低，地层残余油在泡沫液膜铺展，使泡沫迅速破灭，使泡沫体系不稳定[7]。因此，泡沫驱油体系最重要的作用在于增大驱替介质的波及面积，提高波及系数[8,9]。

二元复合驱是继聚合物驱后在现场应用中较为成功的三次采油方法。由于二元体系能够使油水界面张力达到 10^{-3} mN/m 的超低水平，增大了驱替液的毛管数，能够很好地提高驱油体系的洗油效率，提高采收率。但是，由于二元复合驱提高波及面积的能力

较差，对于非均质较为严重的油藏及聚合物驱后的油藏，使用受到一定程度的限制。

将泡沫驱的选择性封堵及较强的调剖能力与二元复合驱的强洗油能力有机地结合在一起，能够有效地发挥2种驱油方式的特点，形成一种新的驱油方式。

前期的试验研究结果认为，0.5%的DP-4泡沫剂与1 800 mg/L的HPAM聚合物形成的泡沫复合体系具有较好的泡沫性能及良好的提高采收率的能力[10]：起泡体积360 mL，半衰期135 min，界面张力0.285 mN/m，提高采收率21.5%。

二元体系的配方为：0.3%的表面活性剂（0.1%CEA－1＋0.2%石油磺酸盐）＋1 800 mg/L的3 530S聚合物溶液，界面张力为5.6×10^{-3} mN/m，60℃模拟地层水配制溶液粘度为21.5 mPa·s。化学驱的优化注入段塞为0.3 PV[11]，根据数模计算结果，泡沫加二元复合驱油体系，首先注入0.15 PV的泡沫复合段塞改善油藏的非均质性、提高波及面积，再注入0.15 PV的二元段塞提高驱油效率，这种注入方式提高采收率的效果好于其他方式。

2.2 泡沫加二元复合驱油体系提高采收率

将制作好的渗透率分别为3.90 μm^2、1.32 μm^2的填砂模型接入泡沫物理模拟驱油流程，分别抽空饱和水，将抽饱和水后的模型以1.0 mL/min的速度注入1.5 PV的模拟油，计量驱出的模拟水体积，作为饱和油量。将模型并联，以0.5 mL/min的速度进行水驱，水驱至含水98%时，以同样的速度注入0.15 PV的泡沫复合段塞及0.15 PV的二元段塞，然后转水驱。图1为双管非均质模型泡沫加二元复合驱提高采收率的综合曲线，水驱采收率为39.0%，注入0.3 PV的泡沫加二元复合段塞后采收率提高到72.7%，双管非均质模型综合采收率提高了33.7%。从采收率变化曲线可见，注入复合段塞后，综合采收率在水驱的基础上有一个明显的上升台级，并且在上升的过程中有一个不光滑的波动，这是由于2个不同的驱油体系提高采收率的机理不同造成的。

图2为双管非均质模型泡沫加二元复合驱综合含水变化曲线，水驱至综合含水98%时注入泡沫加 二元复合段塞，综合含水下降到51.4%，与单纯聚合物驱、二元驱及泡沫复合驱不同的是综合含水有2个下降的漏斗。总体含水下降漏斗宽于其他驱油方式。

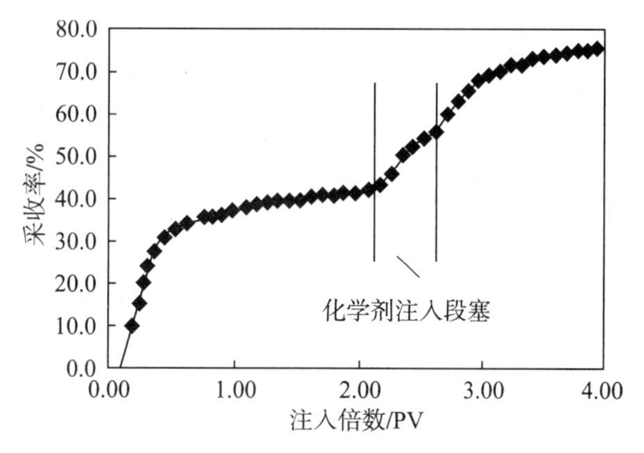

图1 泡沫加二元复合驱提高采收率综合曲线

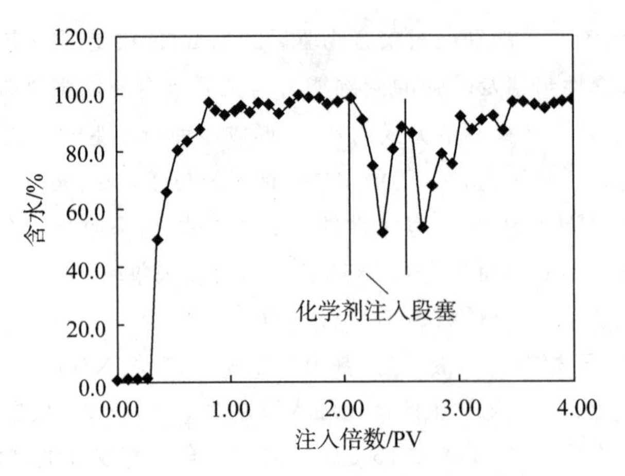

图 2　泡沫加二元复合驱综合含水变化曲线

图 3 为双管非均质模型泡沫加二元复合驱注采压差变化曲线。从曲线上可见，饱和油模型水驱时，随着高渗模型残余油饱和度的降低，注采压差迅速下降，使低渗模型非均质性更加严重，低渗模型的剩余油更加难于驱动。注入复合段塞后，注采压差迅速上升，恢复到原始启动压差，使难于驱动的低渗模型剩余油恢复流动，转注水后，注采压差开始下降，但是下降速度低于原始模型水驱时的下降速度，且水驱时的残余压差较大，能够继续维持低渗模型的驱动。

图 4 为双管非均质模型泡沫加二元复合驱高、低渗模型产液变化曲线，从曲线变化可以看出，饱和油模型并联水驱时，随着高渗模型残余油饱和度的降低，注采压差下降，高渗模型产液量上升，低渗模型产液量下降，高、低渗模型产液量更加不均匀。注入复合段塞后，随着注采压差上升，低渗模型产液量上升，高渗模型产液量下降，高、低渗模型产液量趋于均匀，模型的非均质性得到调整。转注水后，随注采压差下降，高、低渗模型产液量逐渐变大，驱替效果变差。

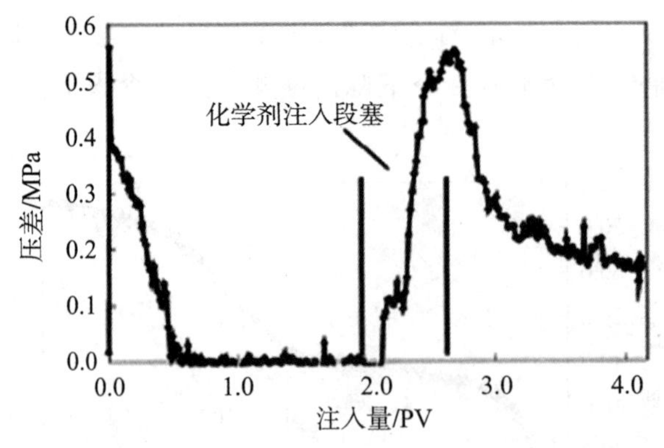

图 3　双管非均质模型泡沫加二元复合驱注采压差变化曲线

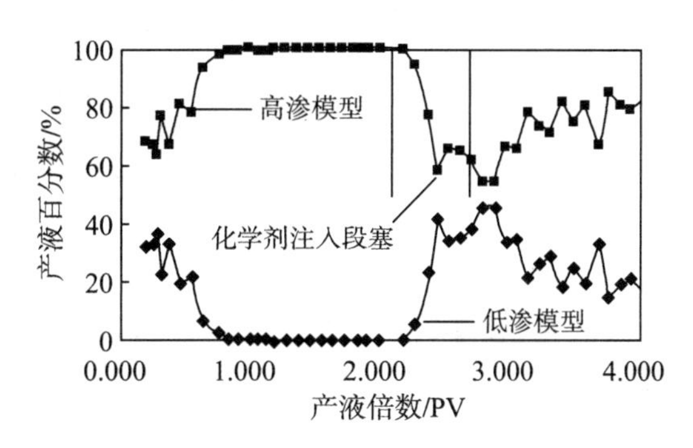

图4 双管非均质模型泡沫加二元复合驱高、低渗模型产液变化曲线

2.3 不同驱油体系提高采收率效果对比

为了对比泡沫加二元复合驱提高采收率的效果，按照同样的试验方法，制作渗透率近似的填砂模型，注入相同大小的段塞，聚合物浓度为 1 800 mg/L，泡沫剂浓度为 0.5%，试验温度为60℃，二元驱化学剂(0.1%CEA－1＋0.2%石油磺酸盐)总浓度为 0.3%，分别进行非均质双管模型水驱后聚合物驱、二元驱及泡沫复合驱，试验结果见表1。

表1 不同驱油体系提高采收率效果对比

	水驱采收率/%	最终采收率/%	采收率提高值/%	最低含水/%	最高压差/MPa
聚合物驱	38.0	58.7	20.7	44.3	0.579
二元驱	32	56.8	24.8	34.3	0.285
泡沫复合驱	34.1	63.8	29.7	51.3	0.540
泡沫加二元复合驱	39.0	72.7	33.7	51.4	0.545

将水驱后聚合物驱、二元驱、泡沫复合驱及泡沫加二元复合驱综合采收率曲线画在同一坐标上，如图5所示。从图5中的曲线可以看出，几种驱油方式变化规律近似，泡沫加二元复合驱提高采收率的值明显高于其他方式。

3 结语

试验证明，泡沫加二元复合驱油体系由于将泡沫的选择性封堵及增加注入流体波及面积的能力与二元体系降低油水界面张力、提高驱油效率的能力结合在一起，其降水增油提高采收率的能力优于聚合物驱、泡沫复合驱及二元驱；虽然降水幅度及提高注采压差的能力与其他方式相比，没有太多的优势，但含水下降的有效期较长，注入相同尺寸的化学剂溶液段塞，与其他注入方式相比提高采收率增加4%～13%。该项研究将为高含水及非均质严重的油藏提高采收率提供了一种新的选择。

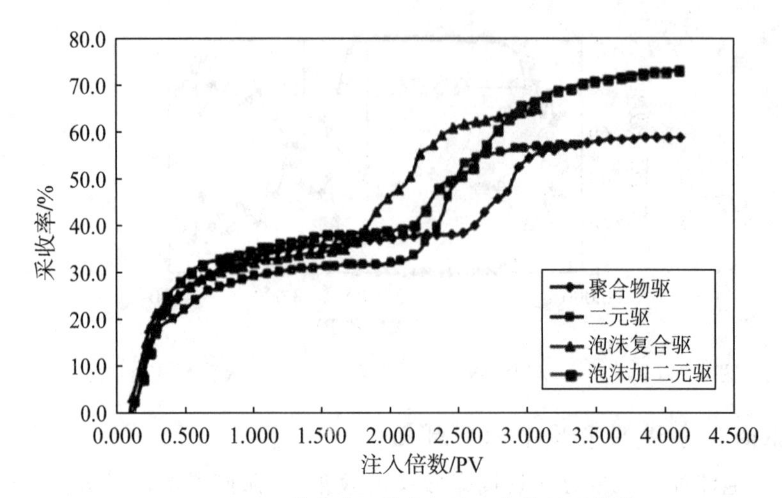

图 5 不同驱替方式采收率与注入倍数关系曲线

参考文献

[1] 王德民. 大庆油田"三元""二元""一元"驱油研究[J]. 大庆石油地质与开发, 2003, 22(3):1-9.

[2] 王红艳, 叶仲斌, 张继超. 复合化学驱油体系吸附滞留与色谱分离研究[J]. 西南石油学院学报, 2006, 28(2):64-66.

[3] 王其伟, 周国华, 郭平, 等. 泡沫封堵能力试验研究[J]. 西南石油学院学报, 2003, 25(6):40-42.

[4] 徐婷, 李秀生, 张学洪, 等. 聚合物驱后提高原油采收率平行管试验研究[J]. 石油勘探与开发, 2004, 31(6):98-100.

[5] 赵长久, 麻翠杰, 杨振宇, 等. 超低界面张力泡沫体系驱先导性矿场试验研究[J]. 石油勘探与开发, 2005, 32(1):127-130.

[6] 周国华, 宋新旺, 王其伟, 等. 泡沫复合驱在胜利油田的应用[J]. 石油勘探与开发, 2006, 33 (3):369-373.

[7] MU Jianhai, LI Ganzuo, LI Ying. An experimental study on alkaline/ surfactant/ polymer flooding systems using natural mixed carbox-ylate[J]. Chinese Journal of Chemical Engineering, 2001, 9(2):162-166.

[8] 宫俊峰, 曹嫣镔, 唐培忠, 等. 高温复合泡沫体系提高胜利油田稠油热采开发效果[J]. 石油勘探与开发, 2006, 33(2):212-216.

[9] 张俊, 周自武, 王伟胜, 等. 葡北油田气水交替驱提高采收率矿场试验研究[J]. 石油勘探与开发, 2004, 31(6):85-87.

[10] 王其伟, 宋新旺, 周国华, 等. 聚合物驱后泡沫驱提高采收率技术研究[J]. 江汉石油学院学报, 2004, 26(1):105-107.

[11] 张贤松，王其伟. 聚合物强化泡沫复合驱油体系试验研究[J]. 石油天然气学报，2006，28(2):137-138.

[编辑]苏开科

king validate time was as high as 86 days, its water cut decreased obviously. It provides a new way for heavy oil production in low product ion oilfield in an economic way.

Key words: deep layer; heavy oil reservoir; gas in jection; viscosity reduction and production increase; feasibility

Research of A Simplified Field Operation and High Performance Oil Base Drilling Fluid XU Ming-biao (College of Petroleum Engineering, Yangtze University Jingzhou 434023, Hubei, China)

ZHANG Chun-yang (CNOOC Department of Oilfield Development and Production, Beijing, 100010, China)

GONG Chun-wu (College of Chemistry and Environmental Engineering, Yangtze University Jingzhou 434023, Hubei, China)

ZENGjing, HUANG Hong-xi (College of Petroleum Engineering, Yangtze University Jingzhou 434023, Hubei, China)

Abstract: By means of laboratory experiments, the result showed that the oil base drilling fluid has an excellent ability of rheological property, stronge lectrical stability, simplified field operation; after 150℃, 16 hours heated and rolled, emulsion breaking voltage accessed or exceeded 600 volt, HTHP filter loss was less than 5 mL; for another side, this oil base mud had excellentanti-polluted ability such as sea water and drill cuttings. Besides this drilling fluid system, density and the oilwater ratio were adjustable, it was beneficial to be constructed in fields.

Key words: simplified; field operation; high performance; oil based rilling fluid

Evaluation on Performance Test of the Operation Fluid with Low Formation Damage

LIU Wei (College of Energy Resources, Chengdu University of Technology, Chengdu 610051, Sichuan, China; Department of Exploration Project Management, Sheng li Oil field Co. Lt d. , SINOP EC, Dong ying 257001, Shan dong, China) Wang Xinhai, Wang Zhongde(Key Laboratory of E xp lorat ion Technol ogies for Oil and Gas Reservoirs(Yang tz e University), Ministry of Educati on, Jing zhou 434023, Hubei, China) Ruan Yan(Department of Production Operation, Tar im Oilfield Company, CNPC, Korla 841 000, Xinjiang, China)

Abstract: in consideration of the charact eristics of Jishan sands tone reservoirs, a formulae system of reservoir protection fluid with low formation damage was developed from the aspects of swell preventi on, emulsification resistance and low surface tension, experiments of reducing flow back pressure, high temperature, oil emulsification resistance and core damage were carried out. The results how s that the reservoir protection fluid is put into well completion fluid and well flushing fluid in low permeability sandst one reservoirs, by which clay mineral swelling, emulsion created by using well completion fluid can be effectively prevented, flow back pressure of completion fluid is obviously reduced. It can be used for preventing water sensitivity and water locking without formation damage.

Key words: formation damage; formation damage prevention; reservoir protecti on fluid; performance test; evaluation

Physical Simulation Test on Hydraulic Fracture Expansion in Horizontal Wells

SHI Ming-yi, JIN Yan, CHEN Mian (MOE Key Laboratory of Petroleum Engineering in China University of Petroleum, Beijing 102249, China)

LOU Yi-shan (College of Petroleum Engineering, Yangtze University, Jingzhou 434023, Hubei, China)

Abstract: Experimental study on hydraulic fracture geometry initiated from horizontal wells was important for studying mechanics of horizontal well fracturing. According to experimental requirement, the larg e-size triaxial simulator was reconstructed in the rock mechanics research center in China University of Petroleum. Two groups of horizontal well hydraulic fracturing experiments were conducted. The first group was observed for the influence of horizontal well azimuth on fracturing initiation and expansion, the second group was aimed to under stand the fracturing initiation and propagation when the ratio of two horizontal stresses changed from 1.5 to 2.0. Laboratory observation indicates that the expansion near horizontal wells is not perpendicular to the direction of minimum horizontal stress, and is random under the condition of small difference between the two horizontal stresses, and the ratio of the two horizontal stresses also affects hydraulic fracture geometry near wellbore.

Key words: hydraulic fracturing; horizontal well; physical simulation; fracture; fracture initiation Experimental Study on Enhanced Oil Recovery with Foam and Dual Component Complex System

Wang Qiwei (College of Chemistry and Chemical Engineering, China University of Petroleum, Dongying 257062, Shandong, China; Research Institute of Geological Science, Shengli Oilfield, SINOPEC, Dongying 257015, Shandong, China)

ZHENG Jing-tang (College of Chemistry and Chemical Engineering, China University of Petroleum, Dongying 257062, Shandong, China)

CAO Xu-long, GUO Ping, LI Xiang-liang, WANG Jun-zhi(Research Institute of Geological Science, Shengli Oilfield, SINOPEC, Dongying 257015, Shandong, China)

Abstract: Physical simulation experiments were conducted on enhanced oil recovery(EOR) capacities of polymer flooding system, foam complex flooding system, dual component flooding system, foam and dual component complex systems as well as on the water content changes in produced fluid and differen tial pressure changes during the flooding. Experiments show that foam and dual component complex flooding system has the characteristics of foam and dual component flooding systems. The complex system has dual action of strong profile control and strong flus-hing, the sweep efficiency of the complex flooding system is raised by using apparent viscosity and selected shut-off of the foam, displacement efficiency is improved by high interfacial activity of the dual component system, also the remaining oil in the reservoir is reduced, by which the foam system is more steady, multi-components acteachother, thus maximum oil displacement efficiency is obtained. 33.7% of integrated oil recovery percent is raised by injecting 0.3 PV complex slug. The effect of enhanced oil recovery is better than that of polymer flooding, dual component flooding and foam complex flooding, under the condition of same slug injection, 4% to 13% of oil recovery percent can be raised.

Key words: foam flooding; dual-component flooding; oil recovery rate; water content; dual component and foam complex flooding

Synthesis and its Performance Evaluation on Low-temperature Oxidization Viscosity Reducer for Heavy Oils

LIU De-xin(College of Petroleum Engineering, China University of Petroleum, Dongying 257061, Shandong, China)

CHEN Lian-tao (Downhole Operation Company, Shengli Oilfield Company, SINOPEC, Dongying 257077, Shandong, China)

DING Jun-tao(Gaosheng Oil Production Plant, Liaohe Oilfield Company, CNPC, Panjin 124125, Liaoning, China)

CAO Peng-yun(Bitumen Co. Ltd, CNPC, Panjin 124125, Liaoning, China)

Abstract: The cat aly zed oxidative performance of(Bi_2O_3) $(acac)_2$(LTO-CB) for Heavy oil was in vest igated according to the technology of low-temperat ure oxidizati on viscosi ty-reducing. (Bi_2O_3) $(acac)_2$ was an organic compound synthesi sed with the metal oxide of IIA and acac. Parameters, in cluding reaction time, catalyst concent ration and temperature that affected the change in viscosity-reducing were as well investi gated. The experimental results show that the effect of viscosity-reducing for Heavy oil with LTO-CB is related with concentration, hydrogen proton donor and reaction duration. At low-temperature condition LTO-CB can effectively reduce the viscosity by oxidative degradation of asphal tenes in heavy oil, and the rate of viscosity reducing reaches 50 percent for heavy oils with dif-Ⅸ.

三次采油中泡沫的性能及矿场应用

王其伟[1,2],郑经堂[1],曹绪龙[2],郭 平[2],李向良[2]

(1. 中国石油大学化学化工学院,山东东营 257061;

2. 胜利油田分公司地质科学研究院,山东东营 257015)

摘要:在室内试验装置上考察了泡沫流体在多孔介质中的性能。结果表明:泡沫流体流动阻力较大,其视粘度超过各单一组分粘度;泡沫体系的视粘度随渗透率的增大而增大;在多孔介质中流动时,泡沫压差分布均匀,不会形成端面堵塞;泡沫对于油水具有选择性封堵能力,在残余油含量较低时,封堵能力强,流速低,残余油含量高时,泡沫不稳定,封堵性能差;泡沫流体 良好的驱油效果,可以提高采收率20%以上。现场应用证明,泡沫能够显著提高注入压力,改善油层吸水剖面,提高油井产量,降低含水率。

关键词:泡沫;复合泡沫驱;封堵能力;封堵压差;三次采油

中图分类号:TE357.29 文献标识码:A

Foam capacityin tertiary oil recovery and aplication inpilot

Wang Qiwei[1,2], Zheng Jingtang[1], Cao Xulong,

Guo Ping, Li Xiangliang[2]

(1. Colege of Chemistry and Chemical Enginering in China University of Petroleum, Dongying 257061,

Shandong Provinc, e China;

2. Geological Scientific Research Institute of Shengli Oilfield Branch Company,

Dongying 257015 Shandong Provinc, e China)

Abstract: The foam capacity in pore medium was investigated on in-house lab setup. The results show that foam has high flowing resistant, and its apparent viscocity is larger than any single component viscocity. The apparent viscocity of foam system increases with the permeability increasing. As a result of unifo distribution of foam presure diference, end jam can't form. The block of foam has selectivity too iland water. With low-contentre mainingol foam has porstability and weak block capacity. The foam system can enhance oil recovery by more than20%. The pilot test results show that the foam can increase in jection presure enormously, improve the injection profil, eenhance oil production and decrease water cut.

Key words: foam; combination foam flooding; block capacity; block presure diference tertiary oil recovery

泡沫驱是一种用泡沫作为驱油介质的三次采油方法,它能够极大地提高注入流体的视粘度,增加波及面积,在一些油田的矿场试验中取得了良好的效果[1-2]。泡沫是一种假塑性流体[3],随着化学工业的发展和泡沫体系性能的增强,为其在三次采油的应用拓展了广阔的领域。

1 试验

1.1 主要仪器设备及材料

岩心驱替物理模拟装置:美国 Temco 公司生产,控温精度±0.5℃;气体质量流量控制计的控制流量为 0~30 mL/min;回压阀的控压为 0~10 MPa;调压阀的控压精度为 0.01 MPa;数字压力表精度为 0.01 MPa。

ST700 混相仪:法国 VINCI 公司产,控温精度±0.5℃;气体质量流量控制计的控制流量为 0~100 mL/min;回压阀的控压范围为 0~35 MPa,控压精度为 0.01 MPa;调压阀的调压范围为 0~35 MPa,调压精度为 0.01 MPa;数字传感器的精度为 0.01 MPa。ST700 混相仪长细管模型为管式人工填充玻璃珠模型,长 12 m,内径 6 mm,渗透率 16 μm^2,带有 7 个测压点,每 2 m 分布一测压孔,与压力传感器相连,模型孔隙度 35%,模型出口接可视观察窗及回压控制阀,自动采集数据。

泡沫剂:为自制 DP-4 产品,pH 值为 7~8,密度 1.0~1.01 kg/L,有效含量 36%。

模拟地层水:矿化度 8 379 mg/L,$Ca^{2+}+Mg^{2+}$ 含量 52 mg/L,$Na^{+}+K^{+}$ 含量 3 061 mg/L,Cl 含量 4 086 mg/L,HCO_3^- 含量 1 085 mg/L,CO_3^{2-} 含量 95 mg/L。

聚合物 3530S:SNF 公司产品,固体含量 91%,分子量 1.5×10^7,水解度 25%。

模拟油:孤岛中二区原油与过滤煤油按 9∶1 配置,原油粘度为 102 mPa·s 。

试验模型:采用石英砂填充模型,模型尺寸为 Y25 mm×600 mm。

注入气为氮气。

1.2 试验方法

阻力因子测定。将岩心驱替流程连接、安装、调试，将石英砂填充模型抽空饱和水，注入模拟地层水测定水驱时的基础压差，注入聚合物或泡沫，待压力稳定后测定注入压差，计算阻力因子：$R=\Delta p/\Delta p_0$。式中，Δp 为起泡剂的流动压差，MPa，Δp_0 为对应流量下的基础压差，MPa。

驱油试验。将人工填制的不同渗透率的石英砂模型抽空饱和水、饱和油，并联接入流程，水驱至含水 98%；注入一定孔隙体积（V_p）倍数的泡沫段塞后转水驱，至含水 98%后，试验结束，记录各个试验阶段的注入压力、产出液含水率及采出程度变化情况。

1.3 试验条件

试验温度为 60℃，回压为 6 MPa，注入速度 0.5 mL/min，若不特别注明，气液比均为 1，泡沫剂浓度为 0.5%。

2 结果分析

2.1 泡沫体系封堵性能

图 1 为聚合物及泡沫体系阻力因子曲线。石英砂填充模型空气渗透率为 1.50 μm^2。曲线 1 为 2 500 mg/L 3530S 聚合物的阻力因子曲线，聚合物体系最大阻力因子稳定值为 151；曲线 2 为泡沫体系的阻力因子曲线，驱替稳定时阻力因子最大值为 1 560。泡沫体系的阻力因子比单纯聚合物的阻力因子大得多。在此温度压力下，模拟水的粘度为 0.5 mPa·s，若将体系的阻力因子折合成视粘度，泡沫体系的视粘度将远远大于聚合物的粘度，这是由于聚合物溶液虽然体系粘度大，但是溶液是均匀的流体，在孔隙介质流动时，流动阻力只与溶液的粘度有关，而泡沫流体在孔隙介质中运动时，流动阻力一方面受流体的粘度影响，同时由于气泡在孔隙介质的孔喉处需要一定力的作用，发生变形才能通过，附加了额外的阻力，因此在微孔介质中，泡沫体系具有较大的流动阻力和较好的封堵能力，能够有效地提高注入流体的波及面积。泡沫的视粘度远远大于组成泡沫体系的单一的气体、活性剂溶液及聚合物溶液的粘度。

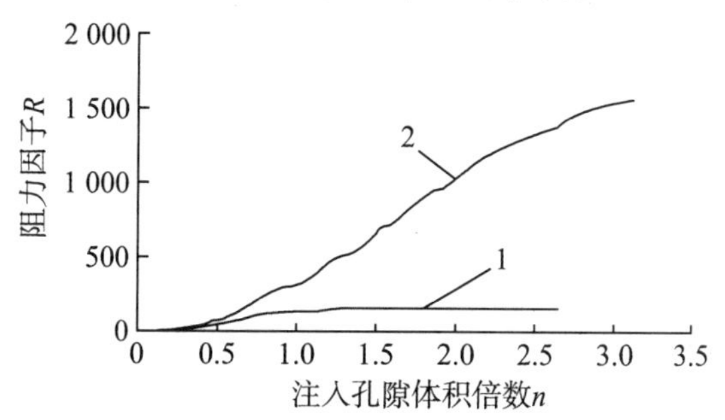

图 1 聚合物及泡沫体系阻力因子曲线

2.2 泡沫体系阻力因子与模型渗透率的关系

常规流体如水及高分子聚合物溶液在多孔介质中运移时，其粘度是相对稳定的，随着渗透率的增大,流动压差减小,符合流体的运动规律,可以用达西定律表示:$Q=kA\Delta p/(\mu L)$,对于一定尺寸的模型,面积 A 和长度 L 为确定值,同一条件下的流体粘度 μ 也是固定的,压差 Δp 一定,流量 Q 与渗透率 k 呈线性关系。对于泡沫流体,其流动特征不符合达西定律,流量 Q 与渗透率 k 呈非线性关系。

泡沫阻力因子与渗透率的关系是泡沫所独有的特点,泡沫的封堵能力随渗透率的增大而增大,在油藏中表现为对高渗层封堵的选择性[4]。图 2 为泡沫阻力因子与渗透率关系曲线。

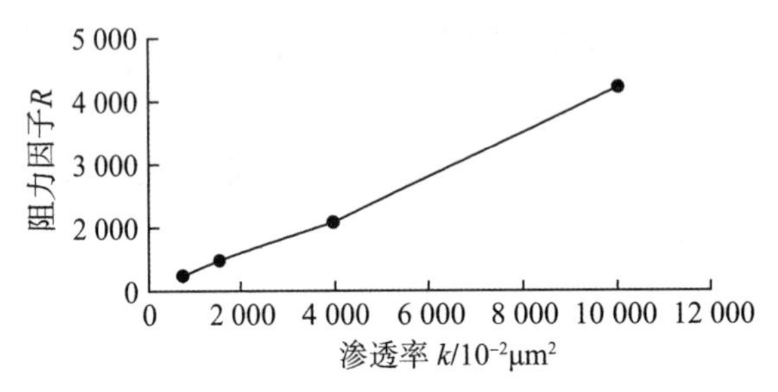

图 2 渗透率与阻力因子关系曲线

李和全等人[5]用数学模型的方式描述了泡沫流体的粘度特性,模型综合考虑了孔隙结构、泡沫结构、表活剂的浓度、液相和气相饱和度以及流速等的影响。结果表明,岩石绝对渗透率与泡沫表观粘度呈线性关系,并且表观粘度随渗透率增大而增大,这与试验中得出的阻力因子与渗透率呈正比的线性关系相吻合。

2.3 泡沫形成的压差与气液比的关系

将氮气和泡沫剂溶液按设计比例混合注入长细管模型,同时测定模型两端的注入压差及各测压点之间的压差,将 4 种气液比模型两端形成的压差与模型注入孔隙体积倍数作图,结果见图 3。

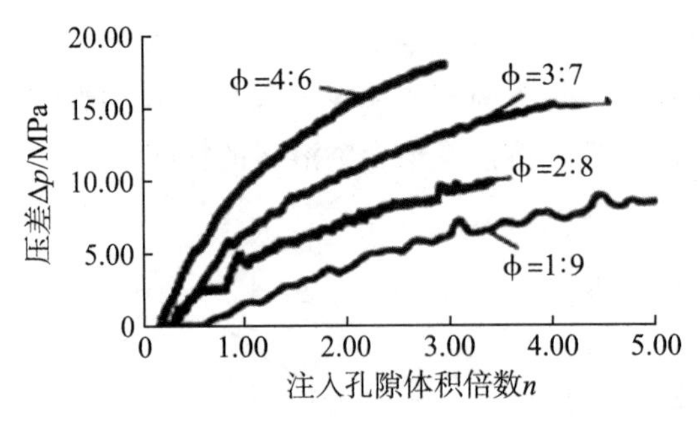

图 3 不同气液比(ϕ)长细管模型两端压差随注入孔隙体积的变化曲线

从图 3 可以看出,模型两端的压差随注入量的增大而增大,气液比越大,注入相同孔隙体积倍数时形成的压差越大。这是由于在相同注入倍数时,气液比越大,单位时间内生成的泡沫越多,封堵能力越强。同时泡沫是一种流体,气泡受到外力作用会发生变形通过孔隙介质,在一定渗透率的孔隙介质中,气液比一定的泡沫体系随着注入量的增大,模型两端的压差逐渐增大。当增大到一个稳定值后,两端的压差不再随注入量的增大而增大，不同气液比的泡沫体系有不同的稳定值，随着气液比的增大，产生泡沫的速度变快，形成稳定泡沫的时间变短，泡沫的稳定性增大。

从图中不同气液比泡沫体系形成的压差可以看出，随着气液比的增大，初始起压时间缩短。由气液比 1∶9 时注入 0.61 VP 泡沫开始起压，减少到气液比 4∶6 时 0.13 VP 开始起压，开始起压时间由 195 min 缩减到 60 min，在一定气液比范围内泡沫体系的气液比较大时，单位时间内产生的泡沫量较多，只有当泡沫在孔隙介质中累积到一定的孔隙体积倍数，在孔隙的横截面上形成一定稳定的泡沫层后，才能表现出有效的封堵能力。

2.4 泡沫体系在孔隙介质中的压力分布

泡沫体系在孔隙介质中如何分布，是否会在近井附近形成端面堵塞，在油藏深部能否形成有效的封堵调剖作用，是长期以来人们关心的问题，长细管试验重点解决泡沫在孔隙介质的分布及运移问题。试验证明，泡沫体系不会形成端面堵塞。图 4 为气液比 4∶6 时泡沫体系相邻测压点之间的压差与注入量的关系曲线。从图 4 可以看出，泡沫进入孔隙介质，首先在模型的进口端形成封堵，随着注入量的增大，不断向深部运移，孔隙介质前端的压差不再增大，封堵压差成均匀方式向后运动，与泡沫体系的运移方向相同，进口与第 1 测压点，第 1 与第 2，第 2 与第 3，第 3 与第 4，第 4 与第 5，第 5 与第 6 测压点之间的稳定压差分别为 0.053，0.54，0.42，0.32，0.31，0.36 MPa，除进口与第一测压点压差低外，其他各点等距离之间的压差相差较小，说明泡沫体系在孔隙介质中分布较为均匀。泡沫的进口与第一个测压点的压差远远低于孔隙介质内部形成的压差，分析原因，是由于气液混合注入时，在进口端气液还没有完全混合均匀，不能形成均匀的泡沫，因此其封堵压力低于内部，随着气液向深部运移，气液逐渐混合形成泡沫，封堵压差达到稳定状态，泡沫在模型孔隙中封堵压力分布均匀，压力沿着流体的流动方向由高到低线性分布。

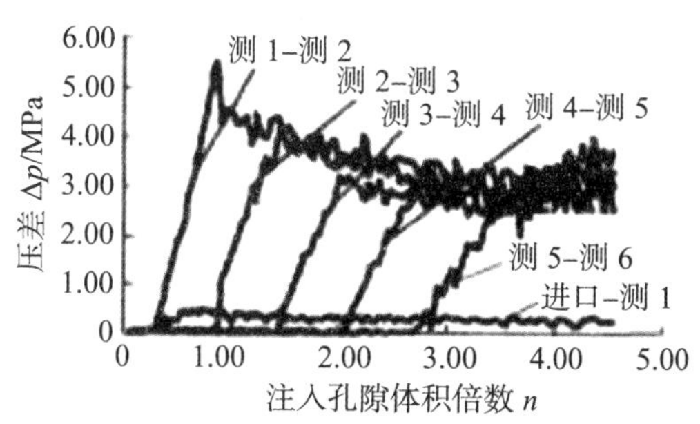

图 4　长细管模型相邻测压点之间压差随注入体积的变化曲线

2.5 泡沫体系在孔隙介质中的稳定性

液膜强度是决定泡沫稳定性的主导因素，表面活性剂溶液形成的泡沫液膜厚度越大，强度越高，排液期越长，越能抵抗外部影响造成的形变，泡沫的稳定性越好。与处于平衡状态的泡沫相比，在多孔介质中流动或静止的泡沫所处的环境更为复杂。处于平衡状态的泡沫形成后，进行的是自然的排液过程，泡沫在排液过程中受到的外力影响小，液膜中的液体在重力作用下排出,使得液膜变薄,失去弹性,与同时发生的气体的扩散一起导致了泡沫的破裂。多孔介质中泡沫的稳定性因素要复杂得多,除了上述影响泡沫稳定性的因素外,泡沫在运动过程中受到的冲击、挤压、摩擦同样影响到泡沫的稳定性,研究泡沫在多孔介质的动态及静态稳定性,对于泡沫能否应用于油藏驱油具有重要意义。

将充满泡沫的长细管模型放置,继续测定不同测压点的压力,从长细管进出口两端的压差曲线可以看出,在长细管中放置 3 500 min,泡沫体系具有较好的稳定性,进出口两端的压差下降幅度不大,仍然具有较好的封堵能力,从图 5 长细管模型各测压点的压力变化曲线可以看出,泡沫体系在孔隙介质中静止放置时,除进口与第一测压点的压力值很快重合外,各测压点压力基本稳定,呈缓慢减小趋势,并由前向后运移,压力缓慢下降。长细管模型中除进口与第一测压点之间的压差外,其他各测压点间的压差变化幅度较小,总体上,泡沫在没有残余油存在的长细管多孔介质模型中性能稳定,在没有外力作用下基本处于静止状态,泡沫破灭及运移速度缓慢。

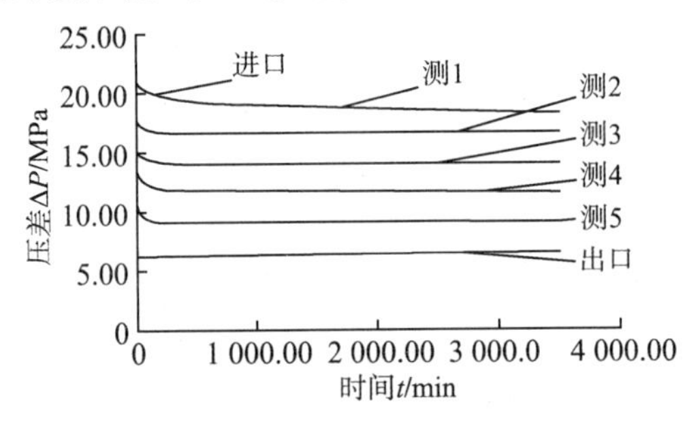

图 5　长细管模型各测压点的压力变化

2.6 泡沫体系在孔隙介质中水驱时的变化规律

在长细管模型中静止放置 3 500 min 后,以注泡沫相同的速度注水,从长细管模型各测压点压力变化曲线(图 6)可以看出,泡沫从前向后均匀运移,长细管模型的前端压力减小,进口与第一测压点的压力线重合,后端测压点的压力值先上升后下降。泡沫段塞在后续水的压力驱动下呈波浪式向后运动。从图 6 可以看出,其进出口压差维持较长的时间,注入 10 VP 水后,泡沫体系仍然具有较好的封堵压差,泡沫体系在孔隙介质中有良好的残余封堵能力。

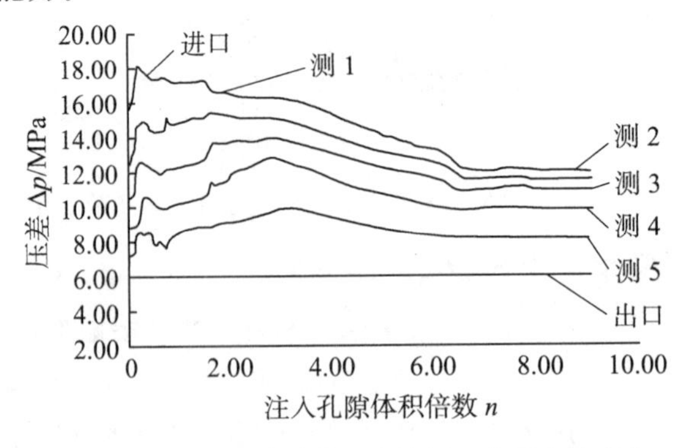

图 6　后续水驱时泡沫体系各个测压点压力的变化

2.7 泡沫体系对于油水的选择性调剖能力

油藏中经过多年的注水开发，残余油饱和度分布很不均匀，相差很大[6]，室内试验无法真实地模拟现场条件，因此模拟了油藏驱油的极限条件，分别制作了两根渗透率不等的模型，高渗管渗透率 2.44 μm^2，低渗管渗透率 1.01 μm^2，将低渗管饱和模拟油，高渗管饱和模拟水，将高、低渗模型管并联，水驱后注单一泡沫体系，图 7 为产出液随注入孔隙体积倍数的变化曲线。

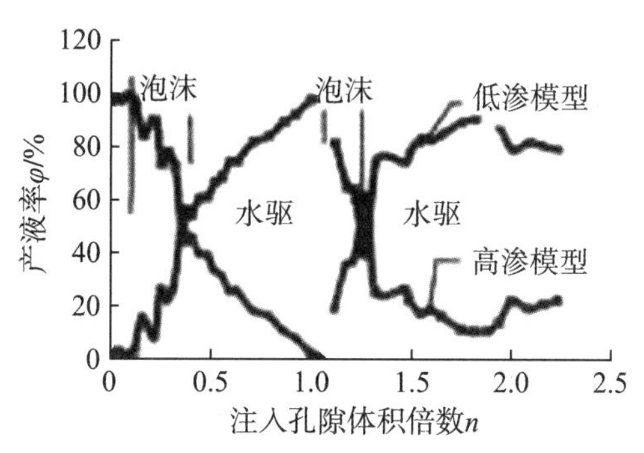

图 7　不同含油模型泡沫驱产出液量曲线

从图 7 可以看出，双管模型注水后，水主要从高渗管产出，高渗管产液占总液量的 98%；注入 0.3 VP 泡沫后，注入压差增大，压差从水驱时的 0.05 MPa 提高到注泡沫剂的 0.4 MPa，高渗管产出液减少，低渗管产出液增大，转注水后，压差下降到 0.02 MPa，高渗管产液量恢复，再注入 0.2 VP 泡沫段塞后，压差迅速上升到 0.15 MPa，虽然压差上升的幅度较小，但注入泡沫后低渗管产液量增大到总液量的 90%，高渗管产液量减少至总液量的 10%，高、低渗管产出液百分数反转。转注水后，低渗管液量仍占总液量的 80%，在泡沫的作用下，低渗模型产液量远远高于高渗模型，这一试验有力地说明了泡沫体系对于油水的选择性封堵特性。产生这一特殊现象的主要原因是，高渗模型不含油，泡沫形成有效的封堵，而低渗模型由于残余油的影响，泡沫不稳定，不能形成有效的封堵[7]，因此，导致低渗模型的产液量高于高渗模型，从孤岛中二区 28-8 井单井注泡沫试验的吸水剖面，可以证明泡沫体系这一优良特性。注泡沫前 4^2 层吸水 100%，4^4 层吸水 0，注入泡沫体系后，4^2 层吸水变为 44%，4^4 层吸水 56%，说明泡沫体系具有良好的选择性封堵能力。

2.8 泡沫体系良好的驱油能力

泡沫体系良好的驱油特性也是泡沫驱的一大特点，但是单一泡沫体系由于受到残余油的影响，稳定性较差，驱油效果不理想[8]。强化泡沫体系是在泡沫体系中加入定量的聚合物，聚合物的加入可以提高泡沫体系粘度，增强泡沫的稳定性[9-10]，同时可以减少泡沫剂在油藏中的吸附损耗。强化泡沫体系具 有泡沫体系及聚合物的双重优点。

试验模拟地层条件下油藏非均质的特点，制作不同渗透率的双管模型，高渗管渗透率 2.39 μm^2，低渗管渗透率 0.99 μm^2，双管并连，回压为 6 MPa，注入速度为 0.5 mL/min，水驱至综合含水98%,注入0.1 VP强化泡沫溶液(1 800 mg/L 3530S+0.5% DP-4泡沫剂)前置段塞,再注入0.3 VP强化泡沫段塞(1 800 mg/L 3530S+0.5% DP-4泡沫剂与氮气混合式注入),气液比为0.5∶1,水驱至含水98%。图8为强化泡沫驱综合采收率曲线,图9为高低渗模型产液曲线,表1为高低渗模型不同阶段采出程度数据。

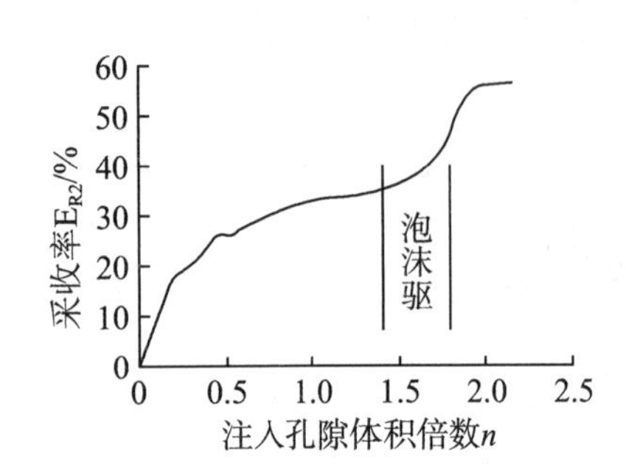

图8 强化泡沫体系综合采收率曲线

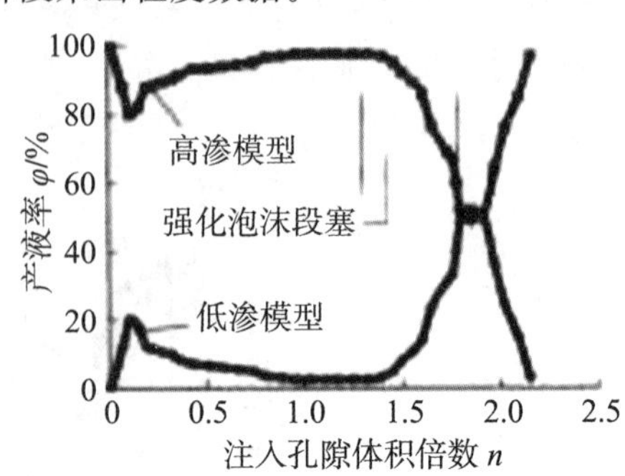

图9 强化泡沫高、低渗透率管产液曲线

表1 非均质模型强化泡沫驱采出程度数据

模型	水驱采收率 E_{R1}/%	泡沫驱采收率 E_{R2}/%	泡沫驱采收率提高值 ΔE_R/%
低渗管	17.8	40.63	22.83
高渗管	52.13	72.13	20.0

试验数据显示,水驱至综合含水98%时,高、低渗模型水驱综合采收率为34.8%,转注0.1 VP的泡沫前置段塞及0.3 VP强化泡沫段塞后继续水驱至综合含水98%时,综合采收率为56.3%,与水驱相比提高采收率21.5%。低渗模型泡沫驱最终采收率提高了22.83%,高渗模型提高采收率20.0%,低渗模型提高采收率高于高渗模型,这是因为低渗模型水驱后剩余油含量较高,高渗模型剩余油潜力较小[11]。试验证明泡沫体系对于低渗层的驱油能力优于其他方法。

从高低渗透率模型产液曲线上可以看出,注入泡沫段塞后,高渗管产液量减少,低渗管产液量增加,一度高、低渗管产液量相等,证明了泡沫体系良好的调剖能力[12]。转水驱后,高低渗管产液量逐渐恢复到原来的水平。

3 现场试验结果

在孤岛油田中二区中部28-8井,开展了注强化泡沫试验研究,试验设计注入前置段塞10 d,0.18%聚合物+1.5%泡沫剂,主段塞170 d,0.18%聚合物+0.75%泡沫剂+氮气,气液比1∶1,注入氮气130.9×10^4 m^3,聚合物28.5 t,泡沫剂118 t。由于试验过程中出现的气窜等问题,试验方案作了适当调整,气液比降为3∶7。试验取得了较好的效果,注入井的注入压力由5.5 MPa提高到7.2 MPa,注水剖面得到了明显改善,主力吸水层4^2层由试验前的吸水量100%下降到试验后的43%,而其他两个试 验前不吸水的油层由吸水量接近0提高到57%。注入强化泡沫明显改善了油藏的非均质性和吸水剖面。13口受效井,有10

口不同程度见效，井组产油量由试验前的70 t/d提高到155 t/d，日增油85 t，综合含水率由试验前的93.8%下降到88.1%，下降5.7%，其中29-3井的含水率由95%下降到42.5%，下降52.5%，降水效果非常明显。单井试验说明，强化泡沫驱能够较好地改善油层非均质性，提高油井产量，改善水驱效果，是一种很有前途的三次采油方法。

4　结论

(1) 泡沫体系在多孔介质中具有较高视粘度，其视粘度超过组成体系的气体、泡沫剂溶液粘度，甚至高于聚合物溶液粘度；泡沫体系的视粘度具有随渗透率的增大而增大的特性。

(2) 在一定范围内泡沫封堵能力随气液比的增大而增大，气液比大时泡沫形成封堵的时间缩短。

(3) 泡沫在多孔介质中形成的压差分布较为均匀，各测压点的压力从进口至出口逐渐下降；泡沫不会形成端面堵塞。

(4) 泡沫体系在油藏运移时对于油水具有选择性的驱动特性，在残余油含量较低时，封堵能力强，流动度低，残余油含量高时，泡沫不稳定，封堵性差，甚至形不成有效的封堵。

(5) 泡沫具有较好地提高采收率的能力，并且能够提高低渗层即残余油较多的区域的采收率。

参考文献

[1] 廖广志，李立众，孔繁华，等. 常规泡沫驱油技术[M]. 北京：石油工业出版社，1999：126-130.

[2] 张思富，廖广志，张彦庆，等. 大庆油田泡沫复合驱油先导性矿场试验[J]. 石油学报，2001，22(1)：49-53.

ZHANG Si-fu，LIAO Guang-zh，iZHANG Yan-qing，eta.l ASP-foam pilottest of Daqing oilfield[J]. ActaPetrolel Sinic，a2001，22(1)：49-53.

[3] 宋育贤. 泡沫流体在油田上的应用[J]. 国外油田工程，1997，13(1)：5-8.

SONG Yu-xian. Foam liquid applying in oilfield[J]. Foreign Oil field Enginering，1997，13(1)：5-8.

[4] 王其伟，郭平，周国华，等. 泡沫体系封堵性能影响因素研究[J]. 特种油气藏，2003(3)：79-81.

WANGQwe，iGUOPing，ZHOU Guo-hu，a eta. lThe r searchofefectfactoroffoam block ability[J]. Special Oil& GasReservoir，s 2003(3)：79-81.

[5] 李和全，李淑红，吴波，等. 一个气液两相泡沫驱模型[J]. 大庆石油学院学报，

1999，23(3)：15-18.

LIHe-quan，LIShu-hong，WU Bo，eta. lA modelof gas-liquidfoam flooding[J]. Journal of Daqing Petroleum Institu，e 1999，23(3)：15-18.

[6] 赵福麟，张贵才，周洪涛，等. 二次采油与三次采油的结合技术及其进展[J]. 石油学报，200 22(5)：38-42.

ZHAOFu-lin， ZHANG Guca， i ZHOUHong-tao， et al. The combinationtechniqueofsecondaryoilrecoverywith tertiaryoilrecovery and itsprogress [J]. ActaPetrolel Sinic，a 2001，22(5)：38-42.

[7] 王其伟，周国华，郭平，等. 泡沫封堵能力试验研究[J]. 西南石油学院学报，2003，25(6)：40-42.

WANGQi-we，i ZHOU Guo-hu，a GUO Ping，eta. lE permiental study offoam blockingability[J]. Journal of South west Pet roleum Institue 200 25(6)：40-42.

[8] 刘中春，侯吉瑞，岳湘安，等. 泡沫复合驱微观驱油特性分析[J]. 石油大学学报：自然科学版，2003，27(1)：49-53.

LIU Zhong-chun， HOU Ji-ru， i YUEXiang-an， eta. lM cro-visualanalysisonoildisplacementinAlkaline-surfac-tant-ploymerfoam floding[J]. Journal of the University of Petroleum，China(Edition of Natural Science)，2003，27(1)：49-53.

[9] 赵晓东. 泡沫稳定性综述[J]. 钻井液与完井液，1992，9 (1)：7-14.

ZHAO Xiao-dong. Summarize of foam stability [J]. Dril ing Fluid& CompletionFlu 1992，9 (1)：7-14.

[10] 佟曼玉. 油田化学[M]. 东营：石油大学出版社，1996：236-240.

[11] SUDARSHITA Regismond，FRANCOISEM Winnk GODDARDE Desmond. Stabilizationofaqueous foams by polymer/sur fact ants ystemefectofsurfactantchain length[J]. Coloidsand Surfaces A，1998，141(12)：165-171.

[12] PACELLILJZitha. Foam drainageinporousmedia[J]. Transportin Porous Medi，a 2003，52：1-16.

[13] SCHWARTZLW，ROY R V. A mathematical foran expandingfoam[J]. Journal of Coloid and Interface Scienc，e2003，264(8)：237-249.

(编辑 刘为清)

Ultra-stable aqueous foam stabilized by water-soluble alkyl acrylate crosspolymer

Lv Weiqin[a], Li Ying[a,*], Li Yaping[a], Zhang Sen[a], Deng Quanhua[a], Yang Yong[b], Cao Xulong[b], Wang Qiwei[b]

[a] Key Laboratory of Colloid and Interface Chemistry of State Education Ministry, Shandong University, 27 South Shanda Road, Jinan, Shandong 250100, PR China

[b] Geological Scientific Research Institute, Shengli Oilfi eld, 3 Liaocheng Road, Dongying, Shandong 257100, PR China

highlights

The foam formed from HMPAA solution was ultra-stable. The foam films characteristics were measured by FT-IR. The mechanism of the foam stability was revealed by diverse techniques.

graphicalabstract

Ultra-stable aqueous foam stabilized by water-soluble alkyl acrylate crosspolymer (HMPAA) was introduced. The HMPAA molecules adsorbed at the gas/water interface could interact with each other to form network structure through the hydrophobic force, and the interface could be covered very well. Besides, huge amount of water molecules was found to be strapped in the foam fi lm and would not be drained out. There fore the gas permeability of HMPAA foam film was low, and the coalescence of bubbles in the foam was postponed.

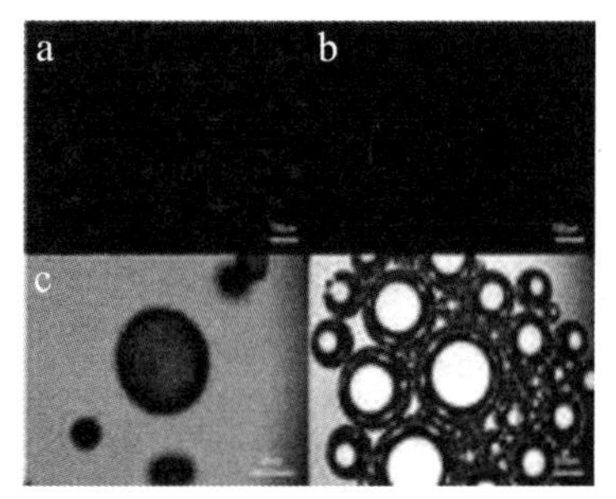

abstract

Aqueous foam solely stabilized by a kind of hydrophobic modified water-soluble polymer, alkyl acrylate crosspolymer (HMPAA), was found to be extraordinary stable, no matter in static state or under disturbance, even if CO_2 was used as gas agent. The high water-holding capacity of HMPAA foam film demonstrated by Foam Scan and FT-IR measurement was in accordance with the low gas transmission through the foam film, which was detected by FT-IR, too. Fluorescence Microscope, TEM and Molecular Dynamic(MD) simulation were used to get information about the adsorption and array behavior of HMPAA on the gas/water interface and in foam film, it was found that the comb polymer molecules adsorbed on the interface clustered to form plat network, which covered the interface very well like a "shell". By combining all these results, the mechanism of ultra high foam stability of HMPAA was revealed, and a novel approach to achieve long-term foam aqueous foam was proposed.

Keywords: Ultra-stable foam Water-soluble polymer Fluorescence microscopy FT-IR CO_2 foam

1. Introduction

Aqueous foam is a system that gas dispersed in liquid which has been widely used in daily life and industrial applications, such as extinguishing fires, mineral floatation, enhanced oil recovery, food industry, personal care products, etc.[1-3]. In many applications, the foam is required to have long-term stability. This is an ongoing technical challenge because foam is a multiphase dispersed system with highly developed interface. This makes the foam system is thermodynamically and kinetically unstable. After the foam is generated, the thinning of foam film caused by liquid drainage, coalescence of the foam bubbles would be aggravated by continuous diffusion of gas through the foam film, the rupture of foam films occur constantly, and both would be strengthened under disturbance[4-11]. How to dramatically increase the stability of the aqueous foam is always a huge challenge for not only theoretical research but also application technology.

In typical industrial applications of aqueous foam, for example, foam flooding in EOR[12], Nitrogen, air and CO_2 are usually used as the gas agents. In recent years, the underground injection of greenhouse gas for storage or displacement attracted much more attention[13], and CO_2 foam flooding in oil recovery which uses captured CO_2

from fl ue gas is one of the most impressive techniques[14-17]. However, aqueous foam using CO_2 as foaming gas is much more difficult to be stabilized as compared with that of N_2 and air. There fore, developing technologies that can be used to improve the stability of CO_2 foam is of technical and application value.

Surfactant is one of the most typical type of the aqueous foam stabilizers[5,11,18]. However, the stability of foam formed from surfactant solution is not always satisfied, especially under harsh conditions, such as high temperature, high salt concentration, or CO_2 being used as gas agent. Recently, ultra-stable aqueous foam stabilized by polymer rods or solid particle with or without surfactant has been reported[19-28]. For example, Alargova and co-workers[23] produced extraordinary stable foams using hydrophobic polymer microrods, which interacted with each other like dense thick "hairy" forming rigid intertwined protective shells around the bubbles to keep bubbles stable. Wege[27] reported foams and emulsions with extraordinary stability by using hydrophobic cellulose microparticles formed in situ with a liquid-liquid dispersion technique. However, the use of unsoluble polymer rods or particles could be limited in the production applications.

In this paper, alkyl acrylate crosspolymer HMPAA[29-31] was used as aqueous foam stabilizer. It was found out that the foam formed from HMPAA solution has extremely high static and dynamic stability, not only for foams using N_2 and air asgas agent, but also for that using CO_2. The drainage of foam and the change of foam film following drainage process were measured by FoamScan and FT-IR. The transmission of gas through foam film was also determined by FT-IR. The behavior of polymer molecules in bulk phase and foam film was investigated by fluorescence microscopy, TEM and molecular simulation. Through all the results, the mechanism of ultra stability of foam was revealed, which represented a novel approach to achieve long-term foam stability.

2. Materials and methods

2.1 Materials

The alkyl acrylate crosspolymer HMPAA used in the paper was introduced in detail in our previous study[30], of which the ingredient ratio of hydrophobic modified segments is 10%, the average molecular weight is about 100, 000. The aqueous solutions of HMPAA were prepared by following procedure: a certain amount of polymer was added in water, the solution was kept stirred using magnetic stirrer at 30 ±0.1℃ for 24 h before used in further measurement.

Anionic surfactant Sodium dodecyl sulfate(SDS, purity>99%), was bought from

Sigma Aldrich. The fluorescent indicator Rhodamine B(purity>99%) was purchased from Aladdin. Freshly distilled water (twice distilled) was used in all solution preparations.

2.2. Methods

2.2.1. The measurement of static foam stability

The FoamScan device(TECLIS, France) was utilized to monitor foam properties (foamability, foam stability and drainage)[32,33]. The bubble sizes can be analyzed with the cell size analysis(CSA) function(TECLIS, France) which allows for a visualization of the bubbles coalescing process. In our experiments, an initial liquid volume of Vs= 60 ml was foamed by sparking N_2 through a porous disk(pore sizes=40−100 山 m) at a constant gas flow rate 100 ml/min. The total foam volume reached to 150 ml, the changing of the volume of solution and the liquid fraction in the foam column were measured by three electrodes. The electrodes were named as the first, the second and the third electrode from the bottom, middle and top of the foam column respectively. Pictures of the foamswere recorded using the cell size analysis(CSA) camera. The CSA software from TECLIS was used to analyze the bubble size. The mean diameter and mean area of the bubbles the size distribution and other parameters could be obtained from the analysis.

2.2.2. Dynamic stability of foam

The dynamic foam stability was determined using rotordisturbing method[5]. The foam column was formed and contained in a transparent glass bucket with interlayer, which is connected with water bath, the temperature was kept at 30±0.1℃. The apparent viscosity of the foam was measured right after being formed, with a Digital Viscometer NDJ-8 S with # 2 cylinder rotor. The rotation speed of the rotor was kept at 6 rpm and the shearing stress was fixed in the determination. The apparent foam viscosity under disturbing was constantly recorded until the foam column collapsed. The data were recorded when reproducible values were observed.

2.2.3. Fluorescence microscope observation

Rhodamine B(0.0002 wt%) was used as the fluorescence label of the polymer chain. For bulk phase observing, the HMPAA solution was dropped in a silica square cell. For foam film observing, appropriate volume of HMPAA solution was deposited on a wire mesh with the grid radius of 0.1 mm to form thin film. One drop of Rhodamine B solution(0.0002 wt%) was dropped on the mesh. Then the fluorescence microscope image was taken with Olympus BX53 Microscope(Olympus, Japan). Green filter was utilized for imaging.

2.2.4. Transmission Electron Microscope(TEM)

Appropriate volume of HMPAA solution was deposited on a 5 mm×5 mm copper grid to form thin foam film. One drop of phosphotungstic acid solution was dropped, and quickly dried under IR light. The JEM-100 CXII(100 kV) was used.

2.2.5. Molecular dynamics simulation

A reasonable double-layer film model was prepared for the simulation of the wet foam films. A65×20×105 grid containing 1 0 polymer molecules which contain 25 repeating units, 6000 water molecules were used. The details were introduced in our early study[5,11].

The layer was first minimized by smart minimize algorithm with 50,000 steps to avoid the possible molecule overlap and make the confi gurations more reasonable. After a 2 ns MD equilibration period, at least a 1 ns MD production was run to obtain the dynamic information, the result of which was used for analysis.

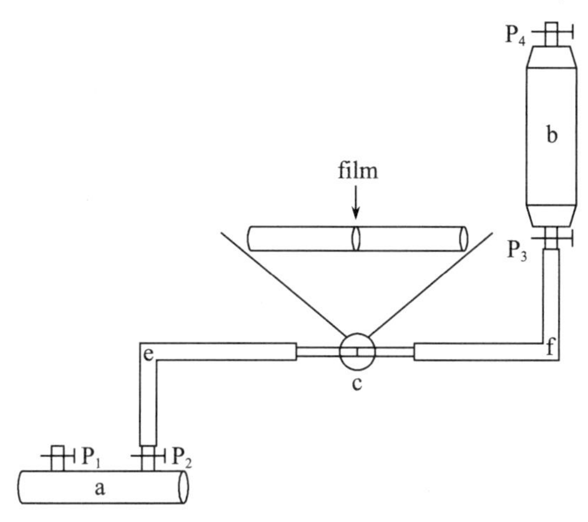

Fig. 1. The schematic of the film permeability measurement equipment. a is gas cell, b is glass container, c is quartz tube, e and f are silicone tube, p1-p4 are pistons. During the experiment the gas cell is always in the test chamber.

2.2.6. Detection of foam film characteristics by Fourier Transforms Infrared Spectroscopy(FT-IR)

Nicolet IS5 (Thermo Fisher Scientific) was used to measure the FT-IR spectroscopy in the frequency range of 4000-400 cm^{-1}, at a resolution of 1 cm^{-1} with a total of 1s6 scans. A quartz tube(90 mm long, 8 mm diameter) was placed in the test chamber. A vertical foam film was formed inside the tube by blowing HMPAA solution through a capillary and was scanned constantly as a function of time.

A setup as shown in Fig. 1 was used in measurement of the gas permeability of the foam film. A sealed gas cell with pistons full of nitrogen was put into the test chamber and scanned as background. A quartz tube(φ8 mm, 90 mm long), inside of which a vertical foam film was formed in advance, was connected with the glass column through silicone tube on one side, and connected with a glass container which was full of CO_2 on the other side. Turned on the piston 2 on the class cell first, and then open the piston 3, the spectroscope was scanned constantly as a function of time.

3. Results and discussion

3.1. The static foam stability

The foamability and foam stability of HMPAA was determined using gas flow method by FoamScan. The time needed to get equal volume of foam was recorded to represent the foamability. There is no signifi cant difference between the time needed to get 150 ml foam from 60 ml 0.1 wt% HMPAA and 0.3 wt% SDS, as shown in Table 1. The foam evolution was monitored as a function of time to refl ect the foam stability, the half lifetime t_{half} was also listed in Table 1. According to the results, the t_{half} of the N_2 foam stabilized by HMPAA could be about 90 h, which is ten times longer than that stabilized by SDS. For CO_2 foam, the t_{half} of the foam formed from HAPAA solution was about 8 times as that of SDS. And the lifetime of foam lamellar film formed from 0.1 wt% HMPAA solutions in air were much longer than that from 0.3 wt% SDS which was 110 and 50 min respectively[11]. It was obvious that the aqueous foam stabilized by HMPAA is much more stable than that stabilized by SDS.

The critical associative concentration(CAC) of HMPAA in aqueous solution is determined to be about 0.08 wt%, which was shown in Fig. 2. It can be seen that even the CO_2 foam formed from much diluted HMPAA solution demonstrated very good stability, the thalf could be hours. The bulk phase viscosity of HMPAA solutions showed a linear increase with the increase of concentration above CAC. The half lifetime of the CO_2 foam increased with the increase of HMPAA concentration above CAC, too. But the t_{half} curve became flatten when the HMPAA concentration is higher than 0.12 wt.%, which did not coincide with the constant increase tendency of the bulk phase viscosity as a function of concentration, especially when the viscosity is higher than 0.2 Pa s. The high stability of the foam formed from HMPAA solution is not solely linked to the increase of the bulk phase viscosity.

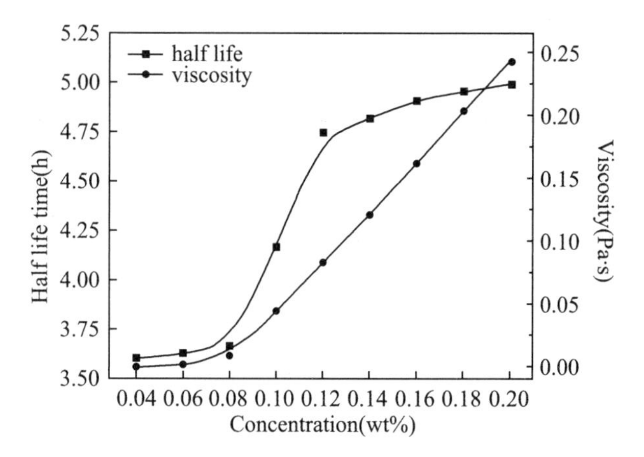

Fig. 2. Variation of the thalf(■) of CO_2 foam formed from HMPAA and solution HMPAA solution viscosity(●) as a function of concentration of

3. 2. The dynamic foam stability

The measurement of apparent viscosity of the foam column as a function of time is shown in Fig. 3. The end of the curve indicated the collapse of the foam column, which refl ects the dynamic foam stability under disturbing. In SDS foam, the apparent viscosity slightly increased at the initial period, which was induced by the adhension effect aroused from the thinning of foam film following gravitational drainage, and then decreased because of the collapse of the foam. The apparent viscosity of the foam formed from HMPAA solution was much higher than that from SDS solution. There was a increase of the foam viscosity coursed by drainage initially, too. But the apparent viscosity of HMPAA foam still maintain high even under shear for hours, the lifetime of HMPAA foam is almost 20 times longer that of SDS foam under the same disturbing condition, which showed solidly that the HMPAA foam is outstandingly dynamically stable.

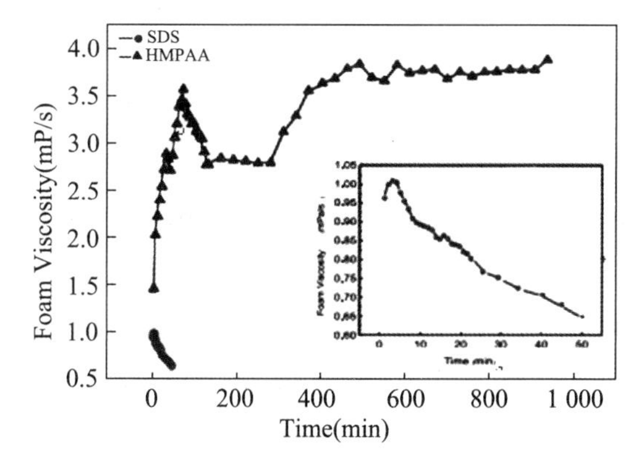

Fig. 3. Variation of dynamic apparent viscosity of foams formed from SDS 0. 3 wt%(●) and HMPAA 0. 1wt%(▲) under rotation as a function of time. The inset shows the curve of the SDS foam in detail.

Table 1 The foamability and half lifetime(t_{half}) of the foams formed from surfactant or polymer solution using N_2 or CO_2 as gas agent. T=30±0.1℃.

Concentration of foam stabilizer	Time for forming 150 ml foam/s	t_{half} of N_2 foam/h	t_{half} of CO_2 foam/h	Lifetime of lamellar foam film in air/min
0.3 wt%SDS	42	10	0.5	50
0.1 wt% HMPAA	49	90	4	110

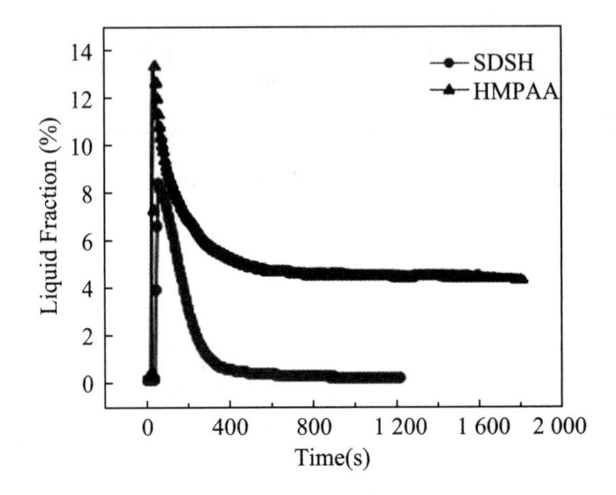

Fig. 4. The variation of liquid fraction of the foams as a function of time: formed from SDS 0.3 wt%(●) and HMPAA 0.1 wt%(▲).

3.3. Determination about the foam drainage and coalescence of bubbles

The time evolution of the liquid fraction of the foam column was monitored directly after the foam was generated to give out information about drainage. The measurement data of the third electrode was shown in Fig. 4. The peak value of the curve represents the initial liquid fraction of the fresh foam, which are 13.5% and 7.5% for HMPAA and SDS foam respectively. This indicates that the HMPAA foam has greater water-carrying capability. During the drainage process the liquid fraction decreased until it reaches a constant level, The finial liquid fraction refl ect the water-holding capacity, which is 5% for HMPAA foam and almost 0 for SDS foam. There fore there is very little amount of water remained in the SDS foam film after complete drainage. It is very clear that both the water-carrying capability and the water-holding capacity of the HMPAA foam is high. These phenomena provide important clue for why the HMPAA foam has high static and dynamic stability.

In the above experiment, the difference of foams bubble sizes for the two foam systems attracted our attention. According to images in Fig. 5, the initial bubbles size of the HMPAA foam is bigger than that of SDS. After 20 min, the bubble size of SDS foam increases about 10 times than initial, while the bubbles size of HMPAA foam just changed slightly. This suggests that the coalescence of bubbles in HMPAA foam is effi ciently retarded. It is necessary to point out that the bubble size of the foam formed from HMPAA solution at very low concentration also increases distinctly, which shows that the coalescence of bubbles correlated with the adsorption amount of polymer molecules at gas/water interface.

3.4. The adsorption and array behavior of HMPAA on gas/water interface and in foam film

Fig. 6(a) is the fluorescence microscopy image of bulk phase, it was found that the polymer molecules was distributed uniformly in the bulk phase solution. Fig. 6(b) is the fl uorescence microscopy image of the same solution except for bubbles being put into the solution. It was clear that HMPAA molecules had the tendency to be adsorbed at the air/water interface. The thickness of the air/water interfacial layer was smaller than the size of the polymer aggregations in bulk phase. This can be due to that polymer opens its structure when it is adsorbed at the interface driven by the hydrophobic side chains. The bubbles in HMPAA solution were covered very well by HMPAA molecules at the air-water interface which led to the decrease of the light transmittance in the foam, as shown in Fig. 6(c). However, in SDS system there is no light transmittance reduction across the bubbles.

Fig. 7 showed the microscope, fluorescence microscope and TEM images of foam films formed from 0.05 wt% HMPAA solution and 0.1 wt% HMPAA solution. According to Fig. 6, the polymer chains curled up and present as an aggregation state at low concentration(<CAC), the coverage degree of interface by polymer molecule was low, large proportion of the film area was uncovered. At higher concentration(>CAC), the polymer chains got stretching and connect with each other forming a network structure. The interfacial coverage degree became high. The air-water interfacial dilational modulus of 0.1 wt% HMPAA solutions was determined to be 28.64 mN/m, which is surely higher than normal surfactants. The results agreed very well with that in Figs. 2 and 5 i.e. full coverage of interface in foam film by the polymer molecules was important to maintain high foam stability.

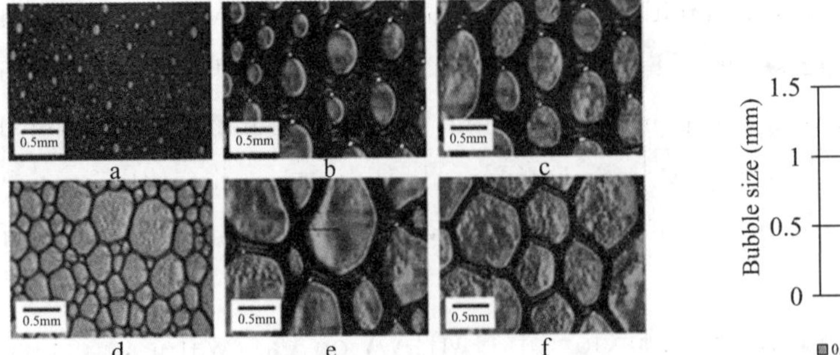

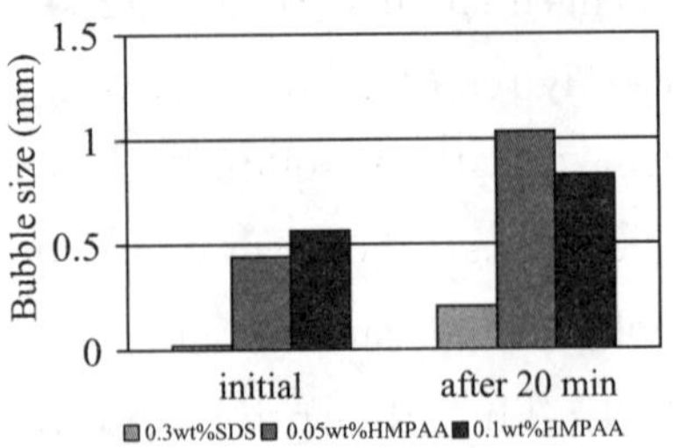

Fig. 5. The images of foams formed from different solutions: 0. 3 wt% SDS(a,d), 0. 05 wt% HMPAA(b,e), 0. 1 wt% HMPAA(c,f). (a-c) Images were taken immediately after the foam was formed. (d-f) Images were taken after 20 min, and the histogram of the bubble size was shown in the right of the images.

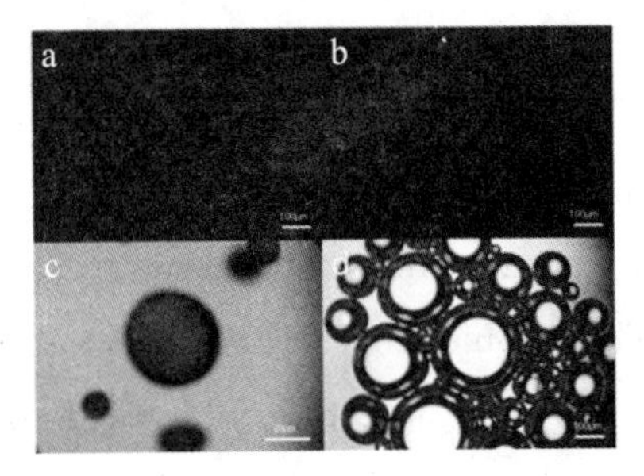

Fig. 6. (a) The fluorescence microscopy image of bulk phase solution of HMPAA. (b) The fluor escence microscopy image of bubbles in HMPAA solution. (c) The optical microscope image of bubbles in HMPAA solution. (d) The optical microscope image of bubbles in SDS solution.

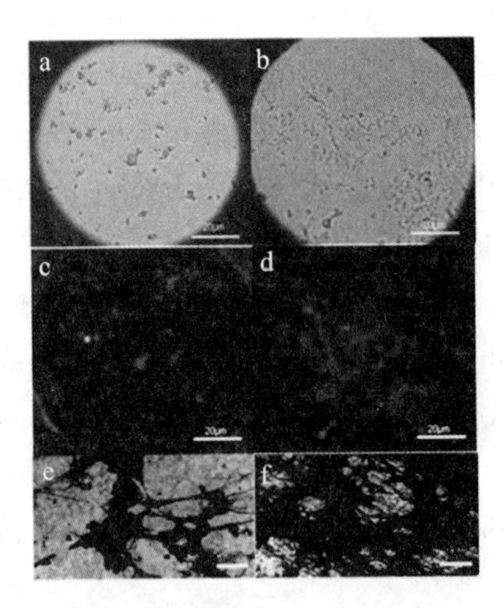

Fig. 7. The images of foam films: microscope(a,b), fluorescence microscope(c, d), and TEM (e, f). (a, c and e) is of film formed from 0. 05 wt% HMPAA solution; (b, d and f) is of film formed from 0. 1 wt% HMPAA solution.

In order to get detailed information about the HMPAA molecular behaviour at the gas/water interface, molecular dynamics(MD) was used to simulate the foam film, and one of the snapshots was shown in Fig. 8. It was shown that, in the foam film, the hydrophobic side of HMPAA stretched out into the gas phase, the main chain docked beneath the gas/water interface and draw close to each other clustering, thus explained how the protective "shell" structure around the bubbles in Fig. 6(b and c) could be formed. This simulation result agrees very well with the experimental observation, and more systematic works are under going.

3.5. Characteristics of the foam film detected by FT-IR

The FT-IR spectroscopes of foam film formed from SDS and HMPAA solution were shown in Fig. 9. The absorption band around 1640 cm^{-1} and the region from 3200 to 3600 cm^{-1} all correspond to vibration modes of OH. The peak around 2350 cm^{-1} corresponds to the characteristic absorption of C-O. As shown in Fig. 9(a and b), all the absorption peak belong to water molecules got weakened quickly for SDS foam film, which means that the amount of water in the foam film decreases quickly along the drainage process. While all the absorption peaks belonging to water molecules do not change so much for the HMPAA foam film, and the light transmittance was very low. These indicated that the water content of HMPAA foam film was very high. This is consistent with previous observation that water-holding capacity was notably good. Therefore plenty of water mole cules was entrapped in the foam film and could not be drained out.

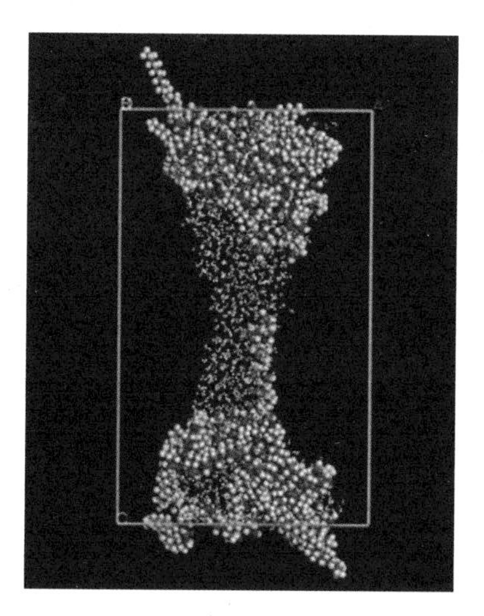

Fig. 8. Snapshots of the equilibrium confi guration of HMPAA foam film in MD simulation. The atoms drawn as vander Waals spheres are shown as small colored sphere: C, gray; H, white; O, red.

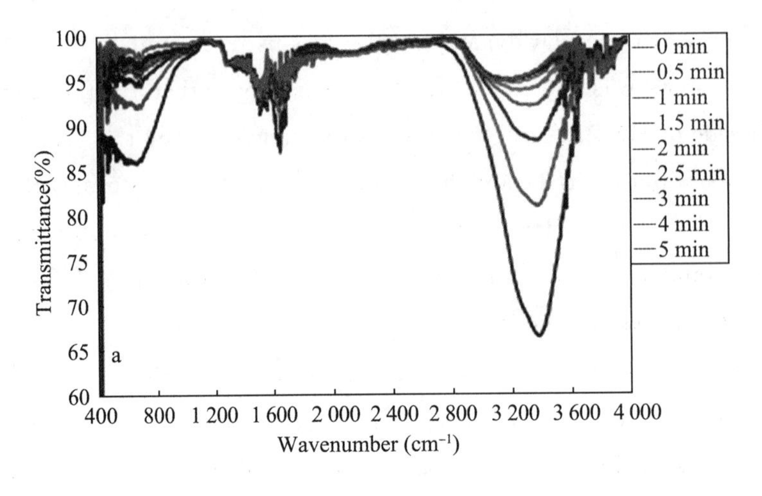

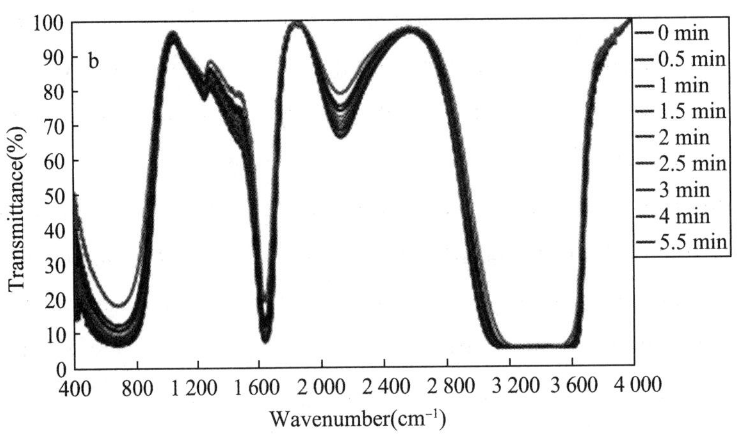

Fig. 9.　FT-IR spectra of foam film formed from(a) SDS and (b) HMPAA solutions.

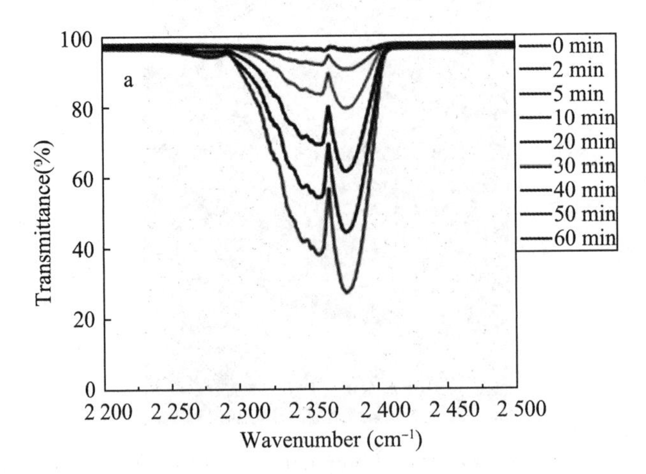

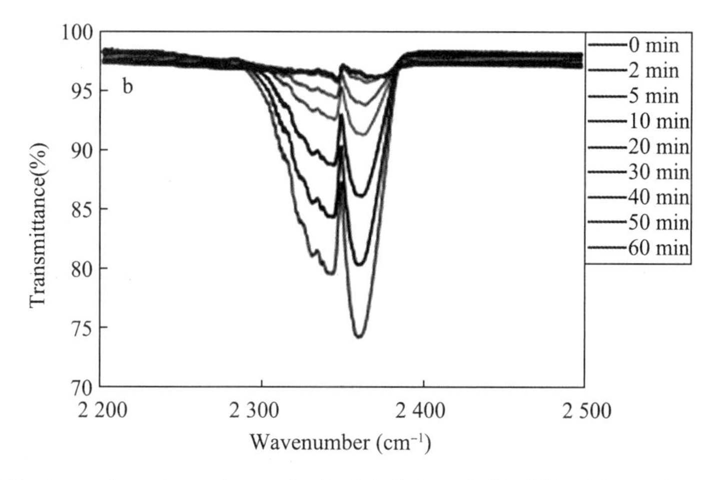

Fig. 10.　CO_2 absorption peaks in (a) SDS and (b) HMPAA condition.

The gas permeability of foam film was detected by FT-IR, the results was shown in Fig. 10 (a and b). The CO_2 transmitted across the foam film decreased because of the light adsorption of CO_2 molecules. It can be seen that the gas permeability of HMPAA foam film was much lower than that of SDS. This agrees very well with that shown in Figs. 7 and 8 and further confirms that the gas/water interface was well covered by interacted polymer chains. In this case the thickness of interfacial layer is too high to allow the gas transmission efficiently. The interfacial layer of surfactant does not have such effect, so the gas could easily transmit through the foam film, and the coalescence of foam bubbles was easily to happen. The FT-IR results correspond perfectly with the experimental results mentioned above.

4. Conclusion

Ultra-stable aqueous foam was obtained using comb watersoluble polymer HMPAA as foam stabilizer. The influences of the polymer behavior and foam film properties on the foam stability have been discussed. It was found that the HAPAA molecules can be adsorbed onto air-water interface and interact with each other to form network structure through the hydrophobic and hydrogen bond interaction. The gas-water interface is therefore covered with the structured network. As a result, the thickness of the foam film of HMPAA was observed to be high and huge amount ofwater molecules were also trapped and hard to be drained out. Therefore the gas permeability of HMPAA foam film became low whichmakes and the coalescence of bubbles in the foam was postponed. According to above, the aqueous foam stabilized by HMPAA becomes extremely stable, even under disturbing or using carbon dioxide as gas agent. We believe that our mechanisms on how HMPAA helps to stabilize foam can help to design a system with supper stable foam.

Acknowledgments

The funding from National Science Fund of China (no. 21173134) and National Municipal Science and Technology Project (no. 2008ZX05011-002) is gratefully acknowledged. Thanks Dr. M. Tang (P& G) for the helpful discussion during the revise course of this work.

References

[1] O. Paulson, R. J. Pugh, Flotation of inherently hydrophobic particles in aqueous solutions of inorganic electrolytes, Langmuir 12(1996) 4808-4813.

[2] B. P. Binks, Particles as surfactants—similarities and differences, Curr. Opin. Colloid Interface Sci. 7 (2002) 21-41.

[3] R. Farajzadeh, A. Andrianov, P. L. Zitha, Investigation ofimmiscible and miscible foam for enhancing oil recovery, J. Ind. Eng. Chem. Res. 49 (2010) 1910-1919.

[4] X. Li, S. I. Karakashev, G. M. Evans, Effect of environmental humidity on static foam stability, Langmuir 28(2012) 4060-4068.

[5] X. Hu, Y. Li, X. He, C. Li, Z. Li, C. Cao, X. Xin, P. Somasundaran, Structure behavior property relationship study of surfactants as foam stabilizers explored by experimen taland molecular simulationapproaches, J. Phys. Chem. B116 (2012) 160-167.

[6] W. M. Jacobi, K. E. Woodcock, C. S. Grove, Theoretical investigation of foam drainage, Ind. Eng. Chem. 48(1956) 2046-2051.

[7] W. Yang, X. Yang, Molecular dynamics study of the influence of calciumions on foam stability, J. Phys. Chem. B114(2010) 10066-10074.

[8] S. Jun, D. D. Pelot, A. L. Yarin, Foam consolidation and drainage, Langmuir 28(2012) 5323-5330.

[9] R. M. Muruganatha, N. Krastev, R. Müller, H. J. Foam, Films stabilized with dodecyl maltoside. 2. Film stability and gas permeability, Langmuir 22(2006) 7981-7985.

[10] L. Wang, R. H. Yoon, Effect of pH and NaCl concentration on the stability of surfactant-free foam films, Langmuir 25(2008)294-297.

[11] C. Li, Y. Li, R. Yuan, et al., Study of the microcharacter of ultrastable aqueous foam stabilized by a kind offl exible connecting bipolar-headed surfactant with

existence of magnesium ion, Langmuir 29(2013) 5418-5427.

[12] L. L. Schramm, Foams: Fundamentals and Applications in The Petroleum Industry, American Chemical Society, Washington, DC, 1994.

[13] S. Holloway, Underground sequestration of carbon dioxide—a viable greenhouse gas mitigation option, Energy 30(2005) 2318-2333.

[14] M. Blunt, F. J. Fayers, F. M. Orr, Carbon dioxide in enhanced oil recovery, Energy Convers. Manage. 34(1993) 1197-1204.

[15] V. G. Reidenbach, P. C. Harris, Y. N. Lee, Rheological study offoam fracturing fluids using nitrogen and carbon dioxide, SPE Prod. Eng. 1(1986) 31-41.

[16] D. X. Du, A. N. Beni, R. Farajzadeh, Effect of water solubility on carbon dioxide foam flow in porous media: an X-ray computed tomography study, Ind. Eng. Chem. Res. 47(2008) 6298-6306.

[17] P. L. Bondor, Applications of carbon dioxide in enhanced oil recovery, Energy Convers. Manage. 33(1992) 579-586.

[18] K. Golemanov, N. D. Denkov, S. Tcholakova, Surfactant mixtures for control of bubble surface mobility in foam studies, Langmuir 24(2008)9956-9961.

[19] U. T. Gonzenbach, A. R. Studart, E. Tervoort, Stabilization offoamswith inorganic colloidal particles, Langmuir 22(2006) 10983-10988.

[20] U. T. Gonzenbach, A. R. Studart, E. Tervoort, Macr oporous ceramics from partic lest abilized wet foams, J. Am. Chem. Soc. 90(2007) 16-22.

[21] B. P. Binks, T. S. Horozov, Aqueous foams stabilized solely by silica nanoparticles, Angew. Chem. Int. Ed. 117(2005)3788-3791.

[22] U. T. Gonzenbach, A. R. Studart, E. Tervoort, Ultrastable particle-stabilized foams, Angew. Chem. Int. Ed. 45(2006)3526-3530.

[23] R. G. Alargova, D. S. Warhadpande, V. N. Paunov, Foam superstabilization by polymer microrods, Langmuir 20(2004) 10371-10374.

[24] A. L. Fameau, A. Saint-Jalmes, F. Cousin, Smart foams: switching reversibly between ultrastable and unstable foams, Angew. Chem. Int. Ed. 50(2011) 8264-8269.

[25] R. Petkova, S. Tcholakova, N. D. Denkov, Foaming and foam stability for mixed polymer-surfactant solutions: effects of surfactant type and polymer charge, Langmuir 28(2012) 4996-5009.

[26] N. Kristen-Hochrein, A. Laschewsky, R. Miller, Stability of foam fi lms of oppositely charged polyelectrolyte/surfactant mixtures: effect of isoelectric point, J. Phys. Chem. B 115(2011) 14475-14483.

[27] H. A. Wege, S. Kim, V. N. Paunov, Long-term stabilization offoams and emulsions with in-situ formed microparticles from hydrophobic cellulose, Langmuir 24 (2008) 9245-9253.

[28] Z. G. Cui, Y. Z. Cui, C. F. Cui, Aqueous foams stabilized by in situ surface activation of $CaCO_3$ nanoparticlesviaadsorptionofanionic surfactant, Langmuir 26(2010) 12567-12574.

[29] O. E. Philippova, D. Hourdet, R. Audebert, pH-responsive gels of hydrophobically modifi ed poly(acrylic acid), Macromolecules 30(1997) 8278-8285.

[30] Q. Li, R. Yuan, Y. Li, Study on the molecular behavior of hydrophobically modified poly (acrylic acid) in aqueous solution and its emulsion-stabilizing capacity, J. Appl. Polym. Sci. 128(2013) 206-215.

[31] K. T. Wang, I. Iliopoulos, R. Audebert, Viscometric behaviour ofhydrophobically modified poly(sodium acrylate), Polym. Bull. 20(1988)577-582.

[32] E. Carey, C. Stubenrauch, Properties of aqueous foams stabilized by dodecyltrimetylammonium bromide, J. Colloid. Interface Sci. 333(2009) 619-627.

[33] E. Carey, C. Stubenrauch, A Disjoining pressure study of foams fi lms stabilized by mixtures of nonionic(C_{12} DMPO) and an ionic surfactant(C_{12} TAB), J. Colloid Interface Sci. 343(2010) 314-323.

Molecular array behavior and synergistic effect of sodium alcohol ether sulphate and carboxyl betaine/sulfobetaine in foam film under high salt conditions

Sun Yange[a], Li Yaping[a], Li Chunxiu[a], Zhang Dianrui[a], Cao Xulong[b], Song Xinwang[b], Wang Qiwei[b], Li Ying[a,*]

[a] Key Laboratory of Colloid and Interface Chemistry of State Education Ministry, Shandong University, 27 South Shanda Road, Jinan, Shandong 250100, PR China

[b] Geological Scientifi c Research Institute, Shengli Oilfi eld, 3 Liaocheng Road, Dongying, Shandong 257100, PR China

highlights

The foam stability of the mixed surfactant systems under high salinity was studied. The detailed molecular array behavior of the surfactant mixtures was investigated by MD. How the inorganic cations affect the foam stability of the mixed surfactants was discussed.

graphicalabstract

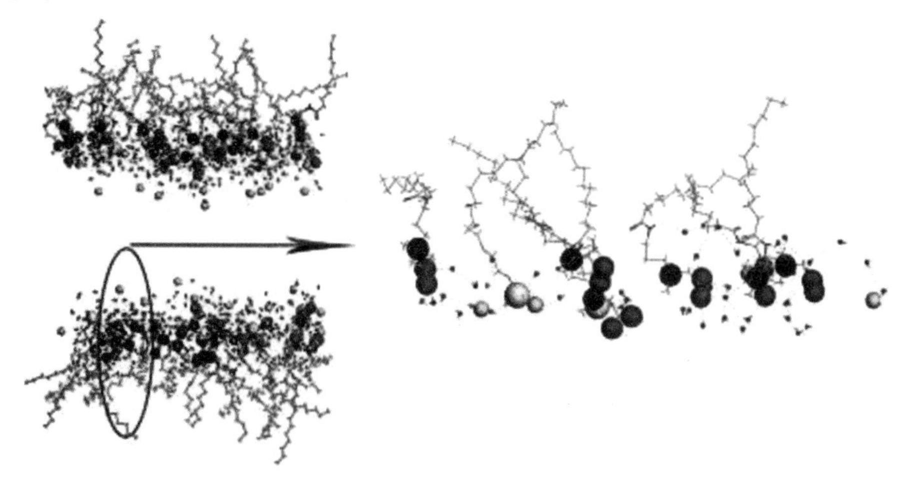

abstract

In this paper, the molecular array behavior of the mixed systems of anionic-nonionic surfactant sodium alcohol ether sulphate(AES) and zwitteroinic surfactants carboxyl betaine(CAB) or sulfobetaine(DSB) at the gas/water interface in foam films was investigated by molecular dynamics simulation approach, which helped to understand in microscopic view how salts effected the foam stability of the mixed surfactants and their synergistic effect. It was found that the multivalent inorganic cations could be bounded to the negatively charged head groups of the surfactants and participated in the formation of the interface layer directly, not only infl uenced the interfacial adsorption behavior of the surfactant molecules, but also changed the state of the water molecules in the foam films. Foam decay method was utilized to determine the foam stability experimentally. The experimental results agreed very well with the simulation results. The knowledge about how the microscopic character of the surfactant mixtures at Y. Sun et al. /Colloids and Surfaces A: Physicochem. Eng. Aspects 480(2015) 138-148139 the interface with multivalent cations coexisting effect the foam properties could provide useful guidance for the application of surfactants under high salinity condition, such as the formula design for foam flooding system used in offshore enhanced oil recovery (EOR).

Keywords: Mixed surfactants Synergistic effect Molecular dynamics simulation Multivalent inorganic cations

1. Introduction

Surfactants is a kind of typical functional substance having a lot of performances, including solubilization, emulsifi cation, wetting, dispersing, foaming, etc. , and has been generally used in many industrial fields, such as food, daily chemical, oil recovery, paper making, textile dyeing and finishing, medicine industry, etc. [1-7]. Considerable studies have shown that the mixed surfactant systems often have better performance than mono surfactant systems from both physicochemical perspective and the point of view of application, and the study about the synergistic effects of mixed surfactants is drawing more and more attentions[8-11].

Most of the literatures about the synergistic effect of the mixed surfactants focused on the mixture of nonionic/ionic surfactants and amphoteric/ionic surfactants[12-19]. It has been reported that the co-adsorption of non-ionic/ionicsurfactants or amphoteric/ionic surfactants on the gas/water or oil/water interfaces could improve the interfacial activity signifi cantly by reducing the electrical repulsion between the charged head groups and increasing the interfacial density of the adsorbed molecules[20,21]. Among the diverse mixed surfactant systems, the mixed systems of amphoteric and ionic surfactants attracted our attention, because of the great potential for applications under high salinity conditions[22-25].

It is well-known that the array behavior of the surfactant molecules at interfaces critically determined their performance, so the detailed information about the interfacial adsorption behavior of the surfactant molecules is the key for determining the structure-performance relationship of surfactants. Due to the limit of methods researching molecules at the interfaces, study about the array behaviors of the mixed surfactants was relatively lacked, and the microscopic mechanism about the synergistic effect of mixed surfactants was still not clear[13,14,18]. Besides, it has been found that the effect of inorganic salts on the nature of surfactants is remarkable[13, 21], but few references has reported about the specific interfacial array behavior of surfactant molecules under high salinity conditions. The short of knowledge about the synergism regularity of the mixed surfactants under high salinity condition results in the lack of the necessary theoretical basis and guidance for the usage in the practical application.

In this paper, the foam stability of the mixed systems of anionic-nonionic surfactant sodium alcohol ether sulphate(AES) and zwitterionic surfactants carboxyl betaine(CAB) or sulfobetaine (DSB) was investigated. The impact of inorganic salts and the component proportion of the surfactants on the foam properties have been explored. It was found that the synergistic effect of the mixed surfactant systems was further strengthened under high salinity conditions, achieving better foam stability and even higher interfacial activity. Thereby the two mixed systems were testifi ed to be good candidates for being used as foam stabilizer under high salinity conditions, for example, as foam flooding agent in offshore oil production, which could increase the oil recovery by improving sweep efficiency and oil washing efficiency[26,27]. In order to find out the mechanism, molecular dynamics(MD) simulation was used to investigate the array behavior of the mixed surfactants in the foam films. The simulation results showed that the multivalent inorganic cations, such as Mg^{2+} and Ca^{2+}, participated in the formation of the interface layer directly. The variation of the strength of the interaction between Mg^{2+}, Ca^{2+} and the different negatively charged groups of the surfactants was found to be the key factor that resulted in the different array behavior of the mixed surfactants, thereby affected the microstate of the molecules in the foam films and the foam properties.

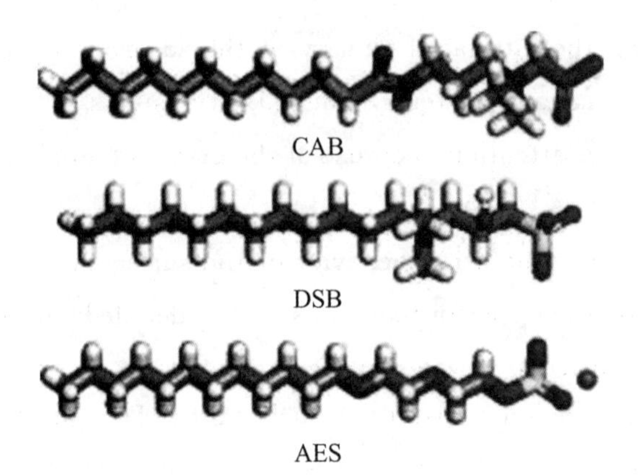

Scheme 1. Molecular structures of the surfactants used in the paper. Atoms or ions are labeled as follows: Na^+, purple; S, yellow; O, red; N, mazarine; C, gray; H, white. (For interpretation of the references to color in this fi gure legend, the reader is referred to the web version of this article.)

2. Materials and methods

2.1. Materials

Alkyl polyoxyethylene ether sodium sulfate(AES) (70% solid content, the rest component is water) and Cocoamidopropyl Betaine(CAB) (>99.0% purity) was provided by the Geological Scientific Research Institute of Shengli Oilfi eld Co. of SINOPEC and used as given. Sulfobetaine(DSB) was synthesized and purified by Dr. J. Chen in Jinling Petrochem. Corp., SINOPEC. The chemical structures of the surfactants were shown in Scheme 1. NaCl(A. R.) and $MgCl_2 \cdot 6H_2O$(A. R.) were purchased from Tianjin Guangcheng Chemical Corp. $CaCl_2$(A. R.) was purchased from Tianjin Fengchuan Chemical Corp. Dodecane(A. R.) was purchased from Tianjin Miou Chemical Corp. Sea water was collected from the Bohai sea, with a total salinity of 32082 ppm(Na^+:10,638 ppm; Ca^{2+}:398 ppm; Mg^{2+}: 1042 ppm).

2.2. Methods

2.2.1. The measurement of oil/ water interfacial tension

The oil/ water interfacial tension was measured using the TX-500C dynamic interfacial tension meter at 50±0.1℃ with a rotating speed of 6000 r/min. The solution was injected into the tube and kept rotated for about 5 min to get pre-equilibrium before 0.5 wL oil was injected. Dodecane was used as oil phase in this determination.

2.2.2. The measurement of static foam stability

The static foam stability was determined using foam decay method, as described in

Refs.[28,29]. The temperature was kept at 50±0.1℃. Foam column was formed by sparging N_2 into 50 mL surfactant solution through a porous disk (pore sizes = 40—100 wm) at a constant gas flow rate 100 mL/min, until the total foam volume got to 150 mL, the half-life time $t_{1/2}$ of the foam column was recorded to reflect the foam stability.

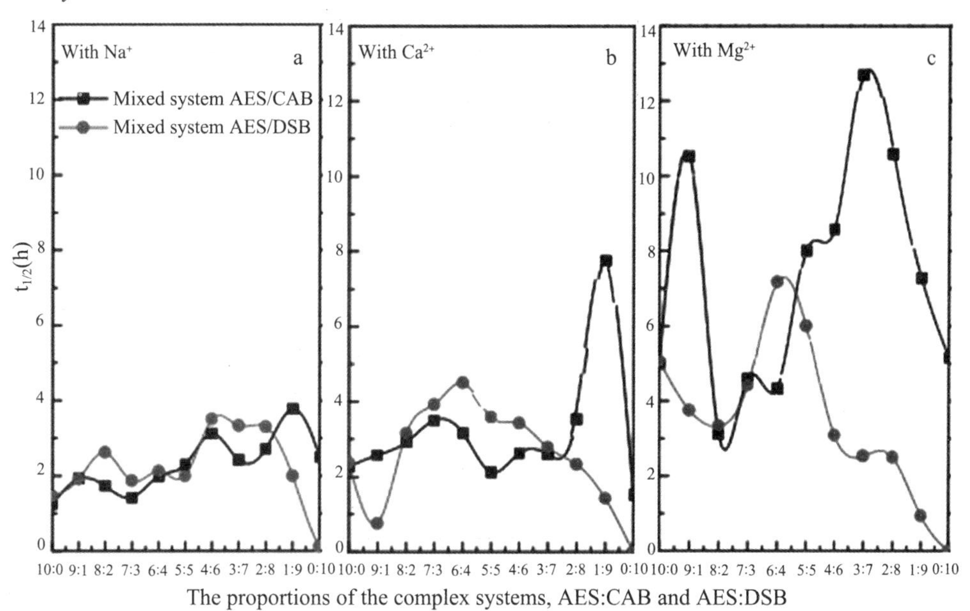

Fig. 1. Half life time, $t_{1/2}$, of the foams formed from mixed systems with different proportion and different salt concentration obtained by foam decay method under 50 ℃. (a) containing 10 g/L Na^+; (b) containing 0.4 g/L Ca^{2+}; (c) containing 1 g/L Mg^{2+}.

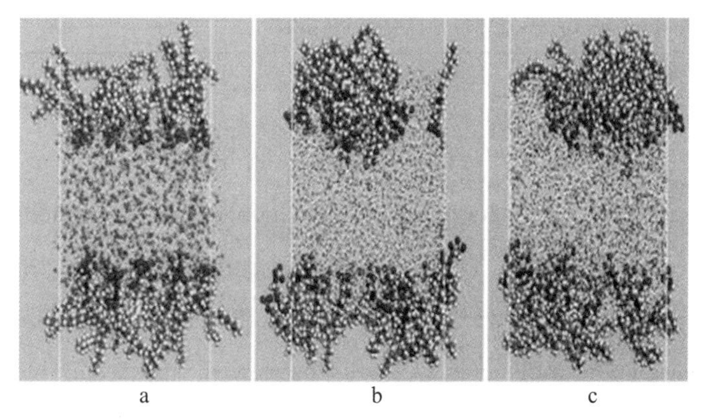

Fig. 2. The illustrative diagrams of AES/CAB foam films gave by molecular dynamics simulation. Molecul ranumber proportion of AES and CAB is 3:22. (a) withNa^+ (n_{Na+} = 480); (b) with Ca^{2+} ($n_{Ca^{2+}}$ = 16); (c) with Mg^{2+} ($n_{Mg^{2+}}$ = 64). Atoms or ions are labeled as follows: Na^+, purple; Ca^{2+}, green; Mg^{2+}, blue; S, yellow; O, red; N, mazarine; C, gray; H, white. (For interpretation of the references to color in this fi gure legend, the reader is referred to the web version of this article.)

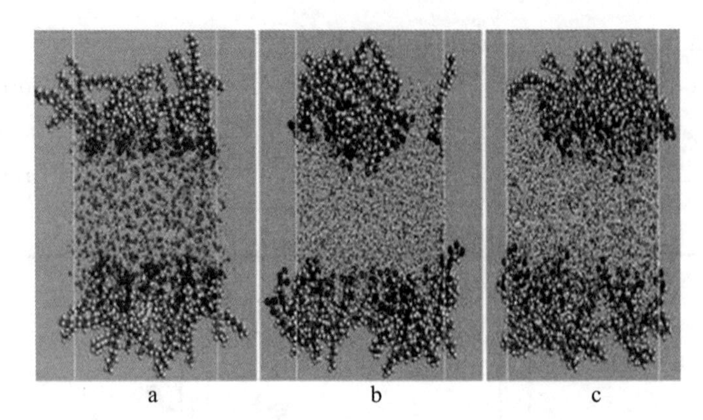

Fig. 3. The illustrative diagrams of AES/DSB foam films gave by molecular dynamics simulation containing different cations. (a) Na^+ ($n_{Na^+}=480$); (b) Ca^{2+} ($n_{Ca^{2+}}=16$); (c) Mg^{2+} ($n_{Mg^{2+}}=64$). Molecular number proportion of AES and DSB is 12:13. Atoms or ions are labeled as follows: Na^+, purple; Ca^{2+}, green; Mg^{2+}, blue; S, yellow; O, red; N, mazarine; C, gray; H, white. (For interpretation of the references to color in this figure legend, the reader is referred to the web version of this article.)

Table 1 Details of simulated systems.

System		a=b(Å)	Lx (Å)	Ly(Å)	Lz(Å)	n_{water}	n_{Na^+}	$n_{Mg^{2+}}$	$n_{Ca^{2+}}$	n_{Cl^-}
	With NaCl	10	43.30	50.00	113.50	1650	480			480
AES/CAB	With $MgCl_2$				103.11			64		128
	With $CaCl_2$				104.50				16	32
	With NaCl	10	43.30	50.00	112.27	1650	480			480
AES/DSB	With $MgCl_2$				103.11			64		128
	With $CaCl_2$				103.27				16	32

2.2.3. Molecular dynamics simulation

The details of the molecular model and the simulation method used in this paper were described in Refs. [30,31]. As andwich-like double-layer film model was used for simulating the foam films[32]. The charges and potentials of all the atoms in the surfactant molecules were assigned based on the calculation using the compass force field[33,34]. 25 surfactant molecules were disposed with space suitable for hexagonal close packing to form a surfactant monolayer in a simulation box imposed to periodic boundary conditions in all three spatial directions at first, the size of the simulation box refers to the maximum adsorption area data of surfactants used in previous simulation and experimental results[30,31]. The surface adsorption area per molecule for AE_3S, CAB and DSB is presumed to be 75Å², 85Å², 85Å². A 25Å thick slab of the water phase(the number of water molecules is 1650) using the flexible SPC/E model with the same cell parameters as surfactants cell was set[35]. Right numbers of sodium ions as counterions of surfactants were added randomly into the water box, and the additional inorganic saltions were added to simulate the proportion of inorganicions contained in the sea water. Two surfactant monolayers were placed on opposite sides of the water phase with hydrophilic head groups of surfactants inserted to the water phase. Details of the simulating box are listed in Table 1.

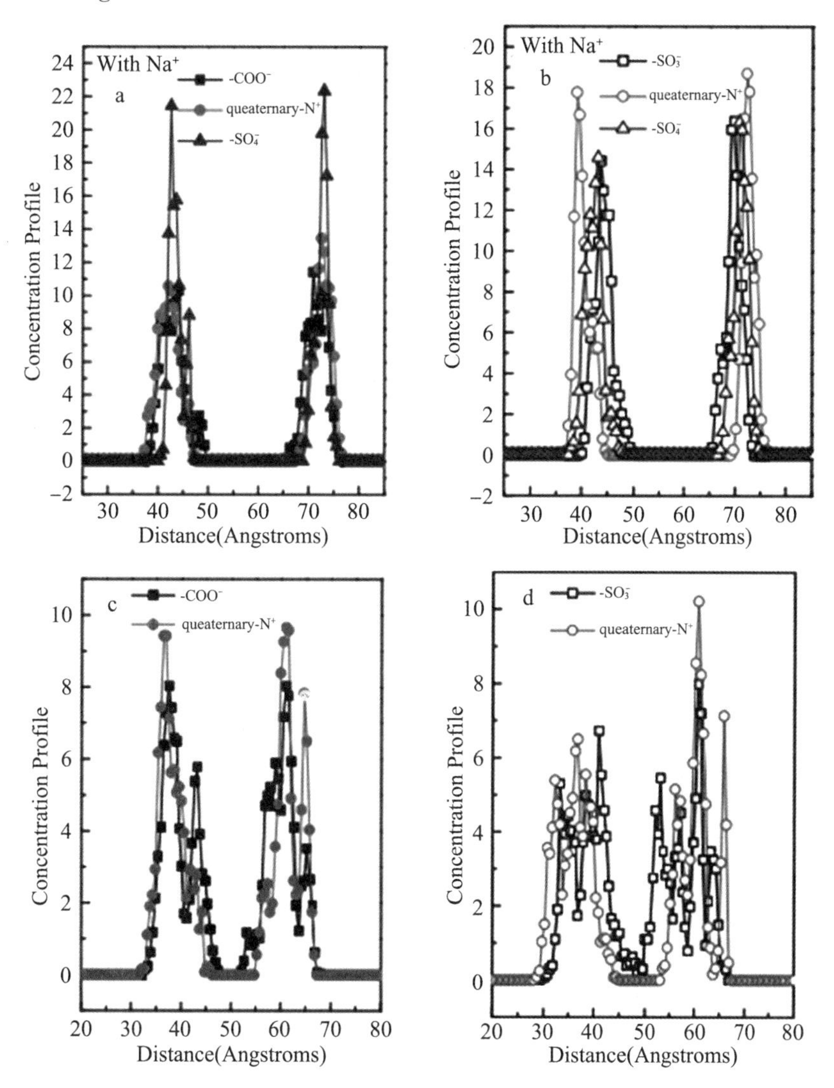

Fig. 4. The concentration distribution of the head groups of surfactant molecules in different foam films. (a) AES/CAB mixed system; (b) AES/DSB mixed system; (c) unitary CAB system; (d) unitary DSB system. (—■—), COO-of CAB; (—●—), quaternary ammonium ion—N^+ of CAB; (—□—)—SO_3^-; of DSB; (—○—), quaternary ammoniumion—N^+ of DSB; (—▲—or—△—), —SO_4^- of AES.

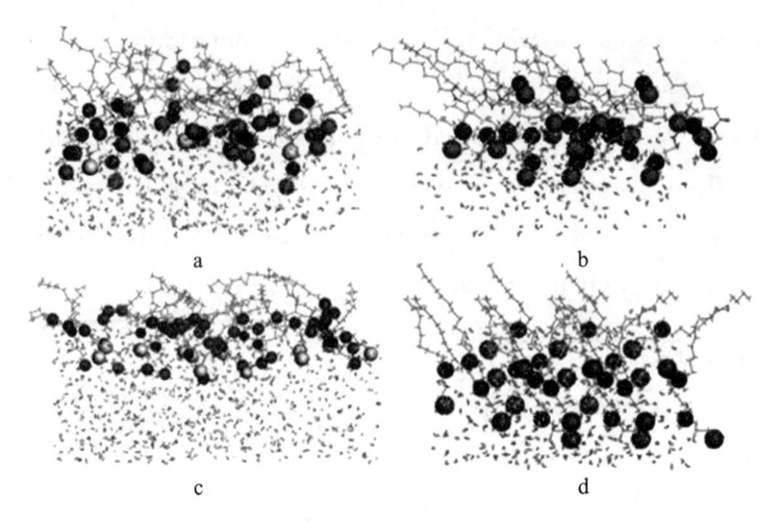

Fig. 5. The detail illustrative diagrams of the surfactant molecules in foam film with Na^+. (a) AES/CAB mixed system; (b) unitary CAB system; (c) AES/ DSB mixed system; (d) unitary DSB system. Atoms labeling the head groups are highlighted as follows: S(—SO_3^-), green; S(—SO_4^-), yellow; O(—EO), red; N(quaternary ammoniumion—N^+), mazarine; C(—COO^-), purple. (For interpretation of the references to color in this figure legend, the reader is referred to the web version of this article.)

The total energy of the foam film system was given as the combination of valence terms and nonbond interactions, and the summation of energies are listed in Eq. (1):

$$E = E_{bonds} + E_{angles} + E_{dihedrals} + E_{cross} + E_{VDW} + E_{elec} \quad (1)$$

After the full explicit atom model was constructed, molecular dynamics simulation was conducted to explore the interfacial behavior of the surfactants on foam fi lms, in which the NVT ensemble was carried out with a time step of 0. 001 ps. The temperature was controlled using a Hoover-Nose thermostat[36,37] with a relaxation time of 0. 2 ps. The simulations were performed at T = 323 K, being the same as experimental conditions. For the long-range electrostatic potential statistics, the Ewald summation method was used[38]. After the dynamics simulation ran 2 ns to get equilibrium period, another 1 ns production run was conducted for confi rmation and analyzation.

3. Results and discussion

3.1. Effect of salts on the foam stability of the mixed surfactant systems

Half life time of the foams formed from solutions of AES/CAB and AES/DSB mixtures with different mass ratio and different saltconcentration was determined using foam decay method under 50±0.1℃, as shown in Fig. 1. The total concentration ofthe mixed surfactants was kept at 0. 3 wt%. The concentration of the Na^{+}, Ca^{2+}, and Mg^{2+} referenced the proportion of the cations in the sea water.

In situations the foam was formed from solutions containing 10 g/L sodium chloride, as shown in Fig. 1 a, the foam stability for both the AES/CAB and AES/DSB mixed systems were higher than unitary system in very wide proportion range, which implied that the synergistic effect do exist in the mixed systems. When the solutions containing 0. 4 g/ L calcium chloride, as shown in Fig. 1 b, the stability of the CAB and DSB foam both decreased abruptly with Ca^{2+} coexisting, while the foam stability of all the AES/CAB and AES/DSB mixed systems were higher than DSB or CAB unitary system. The AES/CAB mixed system with the mass ratio AES:CAB=1:9 achieved the best foam stability with Ca^{2+} coexisting, the $t_{1/2}$ of which got as three times long as that of the AES system. The synergistic effect was more obvious when the solutions containing 1 g/L magnesium chloride, as shown in Fig. 1 c. It has been reported in one of our works that aqueous foam formed from solution of anionic-nonionic surfactant AE_3 S with Mg^{2+} coexisting could be very stable, which was driven from twoaspects: one is the favorable arrangement of surfactant molecules on the gas/water interface, and the other is the increase of capacity of the foam films for resolutelyholding water molecules deduced by dipolar pair formed by the head groups of AE_3 S and the hydrated Mg^{2+} viaintermolecular coactions, both related to the presence of magnesiumions[30]. The foam stability of the AES/CAB mixed system was further increased comparing with AES unitary system in very large mass ratio range with the existence of Mg^{2+}, the $t_{1/2}$ of the mixed system with the mass ratioAES : CAB=3 : 7 even get to 13 h, which was as three times long as that of AES foam. The foam stability of the AES/DSB mixed system was also higher than DSB unitary system in the whole mass ratio range, and was higher than AES unitary system in the mass ratio range AES : DSB= 6 : 4-4 : 6.

In order to probe the mechanisms of the distinguished effect of multivalent inorganic cations on foam stability of the mixed surfactants systems, molecular dynamics simulation of foam films were carried out to investigate the molecular behavior of the surfactants with Mg^{2+} or Ca^{2+} coexisting.

3. 2. The effect of inorganic cations on the interfacial array behavior of surfactant molecules of the AES/CAB and AES/DSB mixtures

In the dynamics molecular simulation, the molecular number proportion ratio of AES and CAB was chosen to be 3 : 22(approximately corresponding to the mass ratio 1 : 9) for AES/CAB system, and the molecular number proportion ratio of AES and

DSB was set to be 12 : 13(approximately corresponding to the mass ratio 6 : 4) for AES/CAB system. The two mixed systems both presented the optimum foam stability under the chosen mass ratio.

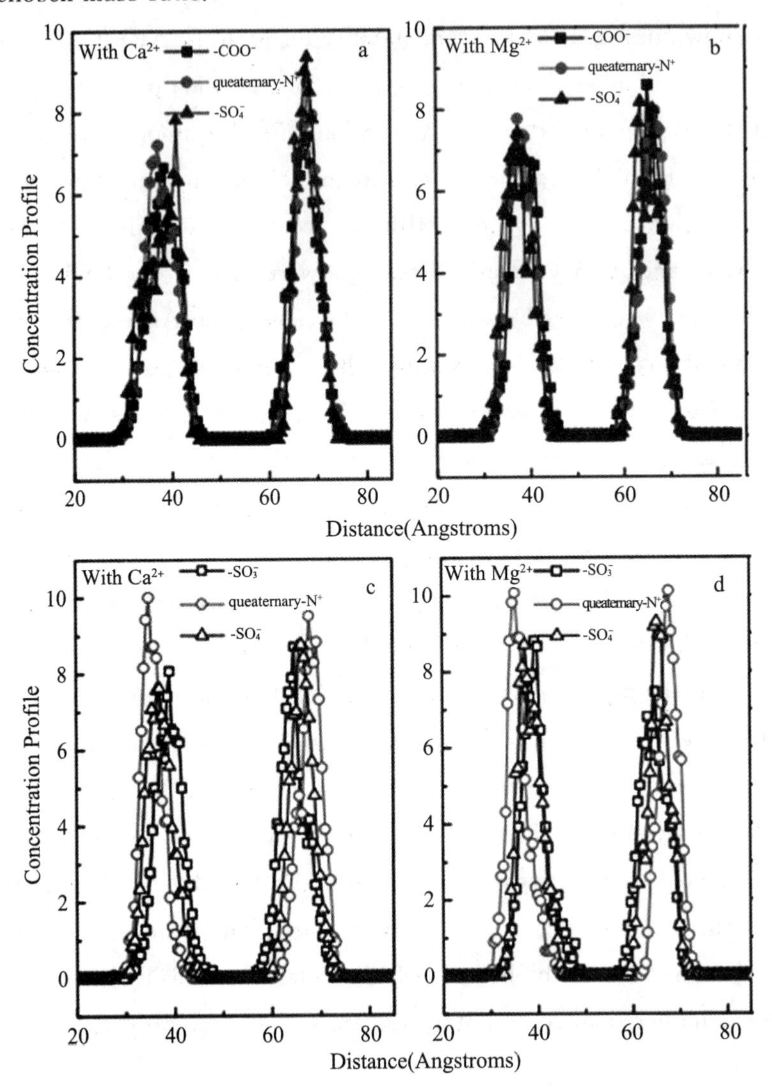

Fig. 6. The concentration distribution of head groups of surfactant molecules with multivalent cations coexisting. (a and b) The AES/CAB mixed system with Ca^{2+} and Mg^{2+} coexisting respectively. (—■—) —COO^- of CAB; (—●—), quaternary ammoniumion—N^+ of CAB; (—▲—), SO_4^- of AES. (c and d) The AES/DSB mixed system, (—□—),—SO_3^- of DSB; (—○—), quaternary ammoniumion N^+ of DSB; (—△—),—SO_4^- of AES.

Figs. 2 and 3 showed the illustrative diagrams of foam films given by molecular dynamics simulation for the AES/CAB and AES/DSB mixed systems respectively. The detailed information about the position distribution of the head groups of surfactants and the inorganicions was analyzed to quantitatively describe the array behavior of the molecules in foam films, and the interactions between the charged surfactant head groups and the cations were investigated precisely, as shown and discussed below.

3.2.1. The molecular array behavior of the mixed surfactants in foam film

As it was shown in Figs. 2 and 3, the surfactant molecules adsorbed at the air/water interface got more clustered obviously with Mg^{2+} and Ca^{2+} coexisting comparing with that with Na^{+}, and part of the Mg^{2+} and Ca^{2+} participated in the molecular aggregation in the interface layer, which indicated that the multivalent cations might acted as "bridge", not only reduced the electrostatic repulsion of the charged head groups, but also gathered the surfactants together, and the amounts of the adsorbed surfactant molecules at the interface could be increased. In order to describe the array state of the molecules in the foam films, the equilibrium distributions of all the molecules were analyzed.

Fig. 4 showed the variation of the particle density of head groups of CAB and DSB and AES as a function of distance on the perpendicular direction of the foam film interface. It was found that in the unitary CAB (Fig. 4 c) and DSB (Fig. 4 d) system, the positive charged and negative charged part of the head groups were arranged in a staggered state, which was reasonable because the electric repulsion interaction would be avoided in this array manner. While in the foam fi lms of the AES/CAB mixed systems, the profile of the distribution of the positive charged and negative charged part of the head groups of CAB were almost coincided. The detailed illustrative diagram of the foam films of the AES/CAB and AES/DSB mixed systems were shown inFig. 5. It could be clearly seen that the—SO_4^- groups of AES inserted more deeply into the water phase than—COO^- groups of CAB (Figs. 4 a and 5a), and the EO (—CH_2CH_2O—) groups of AES located at the same depth localization and separated the charged groups of CAB, as shown in Fig. 4 a. The peak of the distribution profi le of the head groups of DSB molecules in the AES/DSB mixed system got narrower (Fig. 4 b) than the DAB unitary system, too. The above information confi rmed the co-adsorption state of the mixed surfactants and responded the good synergistic effect of the AES/CAB and AES/DSB mixed systems, which coincides with the experimental results given by Fig. 1 a.

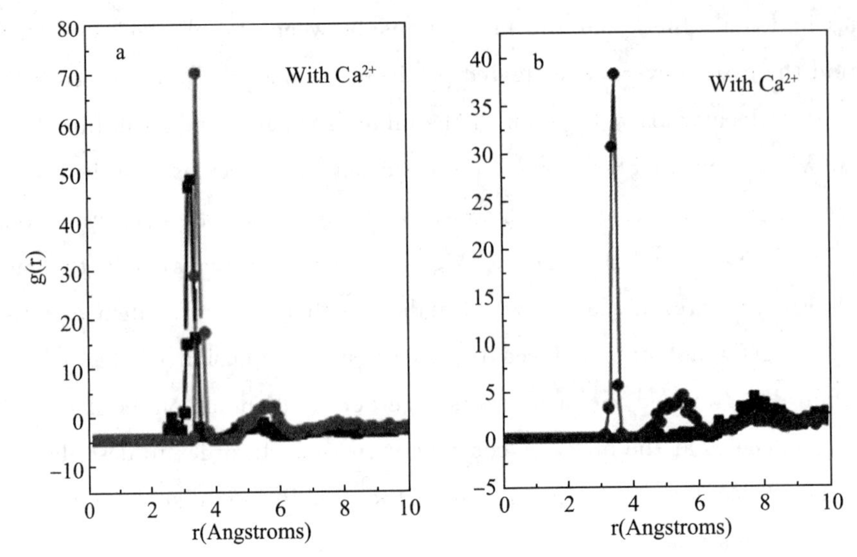

Fig. 7. Radial distribution function of Ca^{2+} ions around head groups of surfactants. (a) (—■—), Ca^{2+} around—COO^- of CAB in AES/CAB system; (—●—), Ca^{2+} around—SO_3^- of DSB in AES/DSB system; (b) (—■—), Ca^{2+} around SO_4^- of AES in AES/CAB system; (—●—), Ca^{2+} around—SO_4^- of AES in AES/DSB system.

Fig. 6 showed the variation of the particle density profile of the head groups of surfactant molecules in the foam films of the AES/CAB and AES/DSB mixed systems with Mg^{2+} and Ca^{2+} coexisting. Comparing Fig. 6 with Fig. 4 (a and b), it could be found that the peak of the distribution profi le of head groups of AES and CAB overlapped and the thickness of the interface layer got larger with Mg^{2+} and Ca^{2+} coexisting than that with Na^+ (Fig. 6 a and b). For AES/DSB system(Fig. 6 c and d), the thickness of the interface layer got larger with Mg^{2+} and Ca^{2+} coexisting comparing with Na^+ coexisting, too. The widening of the interface layer in the foam films could prompt the enhancement of the capacity of the foam films for holding water molecules, and would surely benefit the increase of the foam stability.

3.2.2. The interaction between the head groups of surfactants and the counterions

As shown in Figs. 2 and 3, Mg^{2+} and Ca^{2+} ions entered into the subsurface area of the surfactant head groups, and participated in the formation of the interface layer in the foam films directly. The RDF of multivalent inorganic cations to the head groups of surfactant molecular were calculated according to Eq. (2) and the results were shown in Figs. 7 and 8:

$$g_{ab}(r)=\frac{\Delta(N_{ab}(r))}{\rho_b \Delta V(r)} \tag{2}$$

where ρb is the bulk density of the b atom, Δ(Nab(r)) is the average number of type b atoms at the distance between r and r+dr from type a atom, ΔV(r) denotes the shell volume, gab (r) is simply obtained by computing the average local density of type batom around a type a atom in a specific shell volume element.

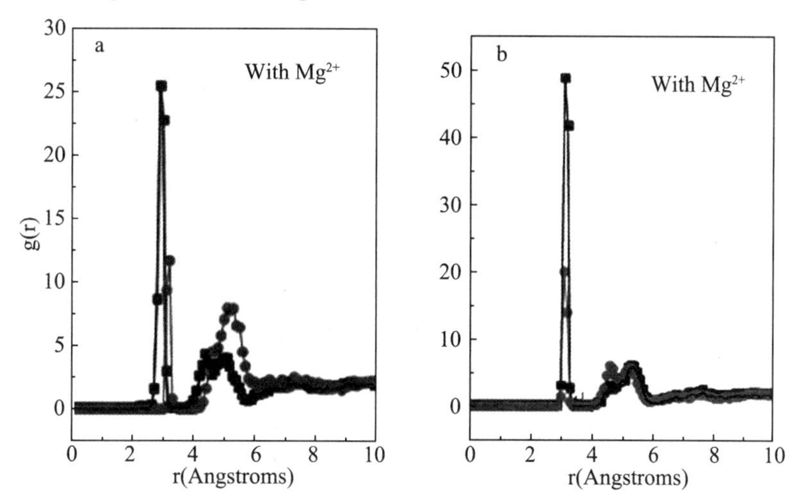

Fig. 8. Radial distribution function of Mg^{2+} ions around head groups of surfactants.

(a) (—■—), Mg^{2+} around—COO^- of CAB in AES/CAB system;

(—●—), Mg^{2+} around—SO_3^- of DSB in AES/DSB system;

(b) (—■—), Mg^{2+} around—SO_4^- of AES in AES/CAB system;

(—●—), Mg^{2+} around—SO_4^- of AES in AES/DSB system.

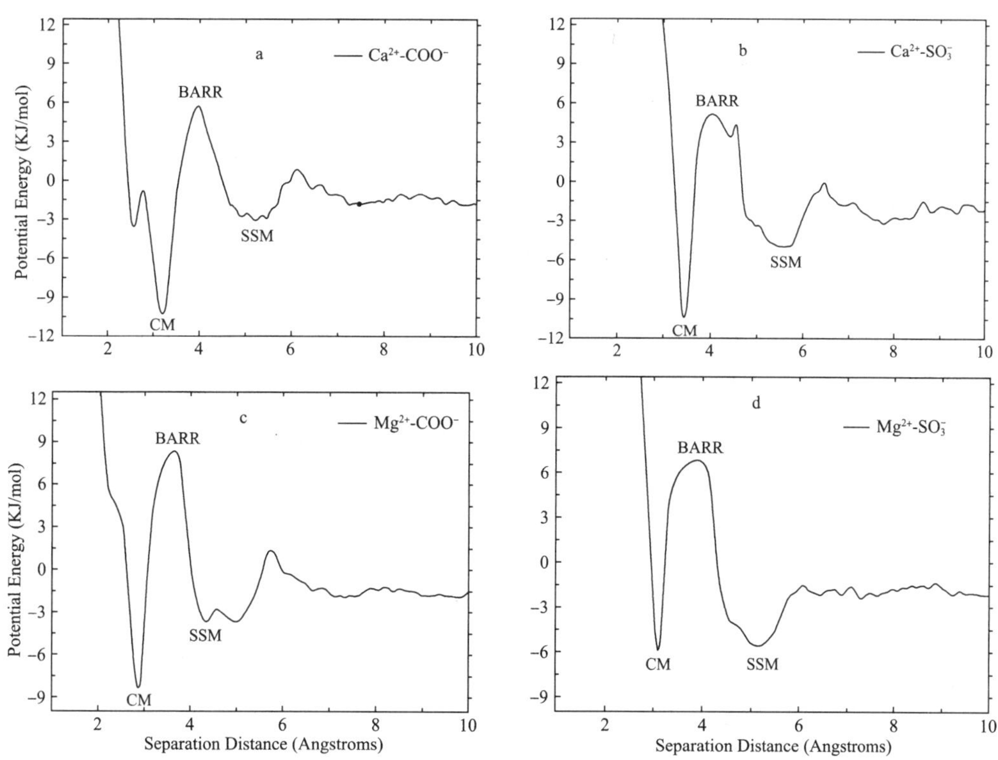

Fig. 9. Pair potential of mean force between head groups and ions. (a and c) AES/CAB system; (b and d) AES/DSB system.

As it was shown in Figs. 7 and 8, the existence of Mg^{2+} and Ca^{2+} ions within distance no more than 5Å from the head groups of the surfactants was assured, which indicates the definite interaction between them. It is interesting to point out that Ca^{2+} could be found beside the—SO_3^- of DSB and—SO_4^- of AES in AES/DSB system, while it could only be found beside the—COO^- group of CAB in AES/ CAB system, no Ca^{2+} was found beside the—SO_4^- of AES in AES/DSB system, which means the interaction between Ca^{2+} and—COO^- is stronger than that between Ca^{2+} and—SO_3^-. Mg^{2+} could be found in the distance nearer than 5A° from the negatively charged head groups of AES, CAB and DSB, which means that Mg^{2+} can be bounded to the head groups of AES, CABand DSB by theelectrostatic interaction.

The energy profile between the cations and the surfactant head groups was determined by calculating the potential of mean force(PMF) , using the equation $E(r) = k_B T \ln g(r)$[30,39], where k_B is Boltzmann's constant and T is the simulation temperature. Fig. 9 showed the PMFs of the different head groups of surfactants and cations derived from the correlation g(r). The detailed results, such as the minimum contact free energy (CM), the minimum solvent-separated free energy(SSM), and the dissociation energy barrier(BARR) for the formation and separation of the dipolar pairs formed by the anionic group of surfactant and the cations are listed in Table 2. $\Delta E^+ = E_{BARR} - E_{SSM}$ is the binding energy barrier and $\Delta E^- = E_{BARR} - E_{CM}$ is the dissociation energy barrier. Parameter K was defi ned as the ratio of ΔE^+ and ΔE^- to reflect the relative tendency of binding and dissociation. $K<1$ means binding is easier happening and dissociation is harder processing, while it is the opposite for $K>1$[30].

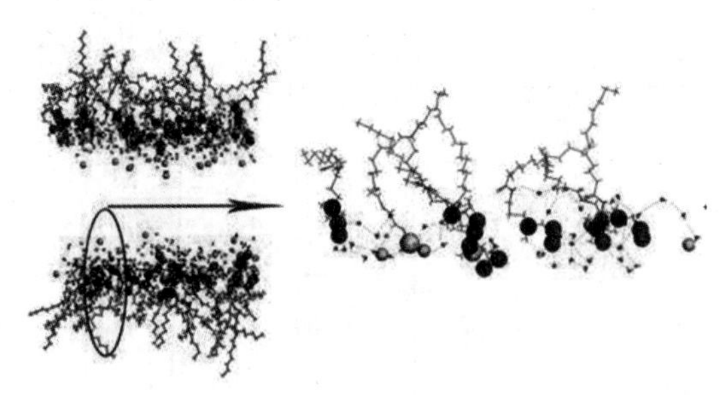

Fig. 10. Co-hydration of counterions and surfactants head groups. The atom coloring scheme is: C, gray; H, white; S, yellow; divalent ion, blue; oxygen atom of water in the first hydration shell by the surfactants head groups, red; water hydrated by the divalent counterions, pink; water shar ed by the counterions and surfactant first hydration shell, green. (For interpretation of the references to color in this figure legend, the reader is referred to the web version of this article.)

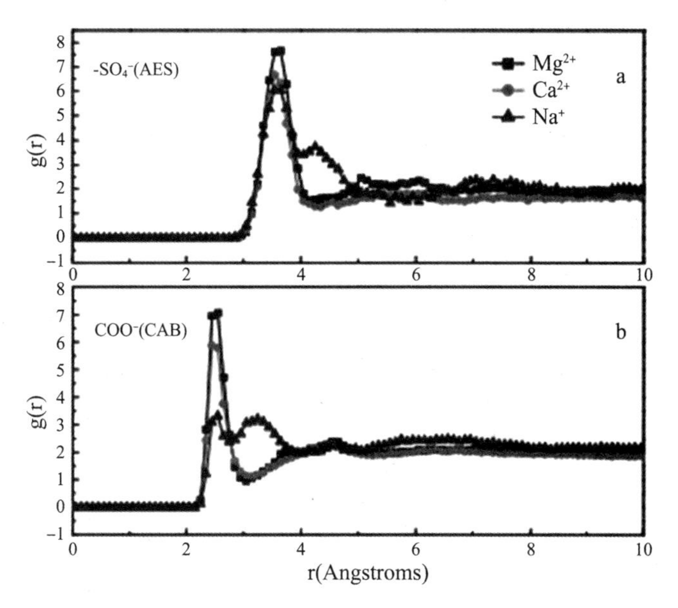

Fig. 11. Radial distribution function of water molecules around the head groups—SO_4^- of AES(a) and around——COO^- of CAB (b) with different inorganic cations coexisting in the AES/CAB system. (—■—), Mg^{2+}; (—●—), Ca^{2+}; (—▲—), Na^+.

According to Fig. 9 and Table 2, the order of the values of the K parameter is K(—COO^-—Ca^{2+} in AES/CAB system)<K(—SO_3^-—Ca^{2+} in AES/DSB system) and K(—COO^-—Mg^{2+} in AES/CAB system)<K(—SO_3^-—Mg^{2+} in AES/DSB system), which means the interaction degree between Ca^{2+}, Mg^{2+} and—COO^- of CAB are both stronger than SO_3^- of DSB.

3.2.3. Mechanism about the effect of the cations on the foam film properties

According to the above discussion, divalent cations interacted strongly with negatively charged head groups of the surfactant, penetrating into their hydration shells, constructed dipolar pair and shared hydration water molecules with each other, as shown in Fig. 10. As a result, the distribution intensity of the water molecules in the hydration shell of the surfactant head groups increased for the mixed systems, as shown in Figs. 11 and 12. The amount of water molecules trapped inside the interface layer of the AES/DSB and AES/CAB mixed systems with Mg^{2+} or Ca^{2+} coexisting was larger, corresponding to the enhancement of the hydration force between the two interface layers of the foam film, which defi nitely enhanced the stability of foam fi lms. It could be concluded that the foam was stabilized by the surfactants and multivalent cations together, and benefited a lot from the increased hydration force, which agreed with the results reported before[30].

The radial distribution function of water molecules around inorganic cations was analyzed, and the results were shown in Fig. 13. The interaction degree between water molecules and Mg^{2+} was the strongest, and then the Ca^{2+} followed by the Na^{+}, which explained the strong and general positive effect of the multivalent cations on the synergism of the mixed surfactants on foam stability, as shown in Fig. 1. For the AES/CAB and AES/DSB mixed systems, the different interaction degree between—COO^{-} of CAB and—SO_3^{-} of DSB with Mg^{2+} or Ca^{2+} (Table 2) determined the equilibrium distribution of the multivalent cations around the head groups, which resulted in the different degree for the effect on the foam stability.

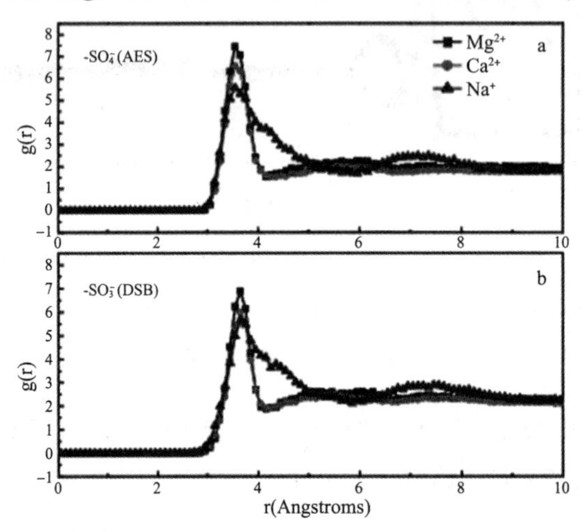

Fig. 12. Radial distribution function of water molecules around the head groups—SO_4^{-} of AES(a) and around—SO_3^{-} of DSB(b) with different inorganic cations coexisting in the AES/DSB system. (—■—), Mg^{2+}; (—●—), Ca^{2+}; (—▲—), Na^{+}.

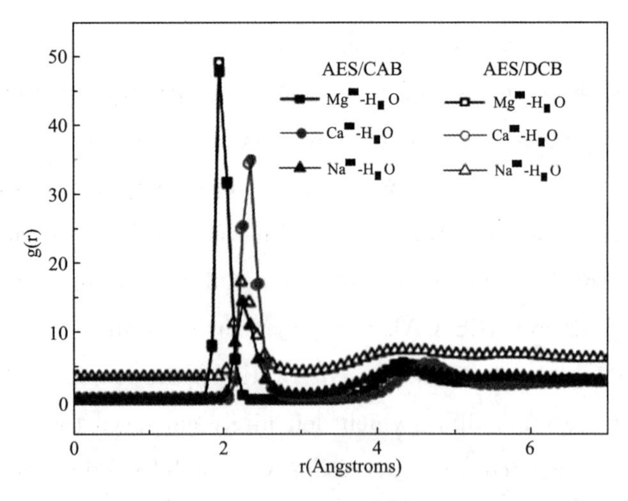

Fig. 13. Radial distribution function of water molecules around inorganic cations. Solid symbol denotes the AES/CAB system and open symbol denotes AES/DSB system.

Table 2 Values of the SSM, CM and the energy barriers of binding-dissociation process of cationic ions and negative charged groups of CAB, DSB and AES.

Complex system	Dipolar pair	CM	BARR	SSM	ΔE^{+}	ΔE^{-}	K
AES/CAB	$-COO^{-}-Ca^{2+}$	−10.32	5.80	−3.00	8.8	16.12	0.546
AES/DSB The energy unit is kJ/mol	$-SO_3^{-}-Ca^{2+}$	−10.59	5.23	−5.45	10.68	15.82	0.675
AES/CAB	$-COO^{-}-Mg^{2+}$	−8.39	8.43	−3.74	12.17	16.82	0.724
AES/DSB The energy unit is kJ/mol	$-SO_3^{-}-Mg^{2+}$	−5.74	6.94	−5.58	12.52	12.68	0.987

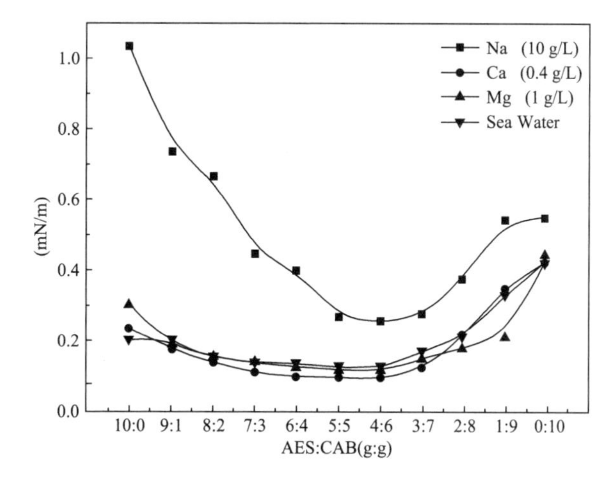

Fig. 14. Variation of the oil/water interfacial tension of the solutions of the AES/CAB mixed systems containing different inorganic salts as a function of the mass proportion under 50±0.1℃.

3.3. Effect of inorganic salts on the oil/water interfacial activity of the AES/CAB binary system

The oil/water interfacial tension of AES/CAB mixed system with different inorganic salts and mass ratios were measured under 50±0.1℃, the results were shown in Fig. 14. It has been found that the oil/water interfacial activity of the AES/CAB mixed system abruptly increased with multivalent cations coexisting, the interface tension reduced more than an order of magnitude in a very wide mass ratio range, including sea water was used as themedium. The results confi rmed that the mixed of surfactant molecules and the multivalent cations not only reduced the electrostatic repulsion of head groups, but also increased the amounts of the adsorbed surfactant molecules at the interface, as illustrated in Figs. 2 and 3. Considering the foam stability and the oil/ water interfacial activity of the AES/CAB mixed system were further strengthened under high salt concentrations, especially with multivalent cations coexisting, it was sure a good candidate system being used as foaming agent with low oil/water interface tension under high salt conditions, which could be used, for example, in offshore foam flooding.

4. Conclusion

In this paper, the synergistic effects between nonionic-ionic surfactant sodium alcohol ether sulphate (AES) and zwitterionic surfactants carboxyl betaine (CAB) or sulfobetaine (DSB) with multivalent cations ions coexisting were investigated. The experimental results showed that some of the mixed surfactant systems with Mg^{2+} or Ca^{2+} coexisting have high foam stability. The microscopic descriptions of the foam films were given by molecular dynamics simulation. It was found that the multivalent cations could interact with the negatively charged head groups of surfactants and took part in the formation of the interface layer directly, not only infl uenced the array behavior ofsurfactant molecules, but also changed the microstate of the water molecules trapped in the foam film, thus further affected the foam property. The formation of the dipolar pairs of the negative charged head groups of surfactants and hydrated Mg^{2+} or Ca^{2+} increased the capacity of foam films for resolutely holding of water molecules, doing favor for the increase of the foam stability. In addition, the existence of the strong interaction between multivalent cations and the head groups of surfactants contributed to the reducing of the electrostatic repulsion between the charged head groups, and enabled the multivalent cations acting as "bridge" which combined the surfactant molecules together, thus increased the number of the adsorpted surfactants at the gas/water and oil/water interface.

The knowledge got by investigating the microscopic character of the mixtures of surfactants at interfaces could provide useful guidance for the design and application of compound surfactant systems under the condition of high salinity, such as low tension foam flooding system used in offshore EOR.

Acknowledgments

The funding from National Science Fund of China (nos. 21173134 and 21473103) and National Municipal Science and Technology Project (no. 2008 ZX05011-002) is gratefully acknowledged.

References

[1] P. D. T. Huibers, D. O. Shah, Evidence for synergism in nonionic surfactant

mixtures: enhancementofsolubilizationinwater-inoil microemulsions, Langmuir 13 (1997) 5762-5765.

[2] J. Weiss, D. Julian McClements, Mass transport phenomena in-oilinwater emulsions containing surfactant micelles: solubilization, Langmuir 16 (2000) 5879-5883.

[3] B. -H. Chen, C. A. Miller, Rates of solubilization of triolein/fatty acid mixtures by nonionic surfactant solutions, Langmuir14(1998) 31-41.

[4] H. Hirata, A. Ohira, N. Iimura, Measurements of the Krafft point of surfactant molecular complexes: insights into the intricacies of "solubilization", Langmuir 12(1996) 6044-6052.

[5] K. Kundu, B. K. Pau, Effect ofpolyoxyethylene type nonionic surfactant and polar lipophilic oil on solubilization of mixed surfactant microemulsion systems, J. Chem. Eng. Data 58(2013) 2668-2676.

[6] J. Allouche, E. Tyrode, V. Sadtler, L. Choplin, J.-L. Salager, Single-and two-step emulsifi cation to prepare a persistent multiple emulsion with a surfactantpolymer mixture, Ind. Eng. Chem. Res. 42(2003) 3982 -3988 .

[7] S. Tcholakova, N. D. Denkov, T. Danner, Roleofsurfactanttypeandconcentration for the mean drop size during emulsifi cation in turbulent flow, Langmuir 20(2004) 7444-7458.

[8] F. A. Siddiqui, E. I. Franses, Surface tension and adsorption synergism for solutions of binary surfactants, Ind. Eng. Chem. Res. 35 (1 996) 3223 -3232.

[9] M. Mulqueen, K. J. Stebe, D. Blankschtein, Dynamic interfacial adsorption in aqueous surfactant mixtures: theoretical study, Langmuir 17(2001) 5196-5207.

[10] A. Shiloach, D. Blankschtein, Prediction of critical micelle concentrations and synergism of binary surfactant mixtures containing zwitterionic surfactants, Langmuir 13(1997) 3968-3981.

[11] N. Poorgholami-Bejarpasi, M. Hashemianzadeh, S. M. Mousavi-khoshdel, B. Sohrabi, Role of interaction energies in the behavior of mixed surfactant systems: a lattice Monte Carlo simulation, Langmuir 26(17) (2010) 13786-13796.

[12] S. Lu, P. Somasundaran, Tunable synergism/antagonism in a mixed nonionic/anionic surfactant layer at the solid/liquid interface, Langmuir 24 (2008) 3874-3879.

[13] A. Bera, A. Mandal, B. B. Guha, Synergistic effect of surfactant and salt mixture on interfacial tension reduction between crude oil and water in enhanced oil recovery, J. Chem. Eng. Data 59(2014) 89 -96.

[14]S. K. Mehta, S. Chaudhary, R. Kumar, K. K. Bhasin, Facile solubilization of organochalcogen compounds in mixed micelle formation ofbinary and ternary cationic-nonionic surfactant mixtures, J. Phys. Chem. B113(2009)7188-7193.

[15] H. Matsubara, T. Nakano, T. Matsuda, T. Takiue, M. Aratono, Effect of preferential adsorption on the synergism of a homologous cationic surfactant mixture, Langmuir 22(2006) 2511-2515.

[16] M. A. Mir, N. Gull, J. M. Khan, R. H. Khan, A. A. Dar, G. M. Rather, Interaction of bovine serum albumin with cationic single chain+ nonionic and cationic gemini+ nonionic binary surfactant mixtures, J. Phys. Chem. B14(2010)3197-3204.

[17] X. Wang, R. Wang, Y. Zheng, L. Sun, L. Yu, J. Jiao, R. Wang, Interaction between zwitterionic surface activity ionic liquid and anionic surfactant: Na^{+}-driven wormlike micelles, J. Phys. Chem. B117(2013) 1886-1895.

[18] I. Svanedal, G. Persson, M. Norgren, H. Edlund, Interactions in mixed micellar systems ofan amphoteric chelating surfactant and ionic surfactants, Langmuir 30(2014) 1250-1256.

[19] L.-S. Hao, Y.-T. Deng, L.-S. Zhou, H. Ye, Y.-Q. Nan, P. Hu, Mixed micellization and the dissociated margules model for cationic/anionic surfactant systems, J. Phys. Chem. B116(2012) 5213-5225.

[20] G. Suryanarayana, P. Ghosh, Adsorption and coalescence in-mixedsurfactant systems: air-water interface, Ind. Eng. Chem. Res. 49 (2010) 1711-1724.

[21] Y. R. Suradkar, S. S. Bhagwat, CMC determination of an odd carbon chain surfactant (C13 E20) mixed with other surfactants using a spectrophotometric technique, J. Chem. Eng. Data 51(2006) 2026-2031.

[22] T. Jimbo, A. Tanioka, N. Minoura, Characterization of an amphoteric-charged layer grafted to the pore surface of a porous membrane, Langmuir 14(1998) 7112-7118.

[23] H. Matsumoto, Y. Koyama, A. Tanioka, Characterization of novel weak amphoteric charged membranes using 志 potential-measurements: effect of dipolarion structure, Langmuir 17(2001) 3375-3381.

[24] N. Schelero, G. Hedicke, P. Linse, R. V. Klitzing, Effects of counterions and Coions on foam films stabilized by anionic dodecyl sulfate, J. Phys. Chem. B 114 (2010) 15523-15529.

[25] H. Zhang, C. A. Miller, P. R. Garrett, K. H. Raney, Defoaming effect ofcalcium soap, J. Colloid Interface Sci. 279(2004) 539-547.

[26] L. L. Schramm, Foams: Fundamentals and Applications in The Petroleum Industry, American Chemical Society, Washington, DC, 1994.

[27] T. Zhu, D. O. Ogbe, S. Khataniar, Improving the foam performance for mobility control an improved sweep effi ciency in gas flooding, Ind. Eng. Chem. 43 (2004)4413-4421.

[28] S. A. Koehler, S. Hilgenfeldt, H. A. Stone, A generalized view of foam drainage: experiment and theory, Langmuir 16(2000) 6327-6341.

[29] A. Bhattacharyya, F. Monroy, D. Langevin, J. F. Argillier, Surface rheology and foam stability of mixed surfactant-polyelectrolyte solutions, Langmuir 16 (2000) 8727-8732.

[30] C. Li, Y. Li, R. Yuan, W. Lv, Study of the microcharacter of ultrastable aqueous foam stabilized by a kind of fl exible connecting bipolarheaded surfactant with existence of magnesium ion, Langmuir 29(2013) 5418-5427.

[31] X. Hu, Y. Li, Structure-behavior-property relationship study of surfactants as foam stabilizers explored by experimental and molecular simulation approaches, J. Phys. Chem. B 116(2012) 160-167.

[32] S. S. Jang, W. A. Goddard III, Structures and properties of Newton black films characterized using molecular dynamics simulations, J. Phys. Chem. B 110 (2006) 7992-8001.

[33] H. Sun, P. Ren, J. R. Fried, The COMPASS force field: parameterization and validation for phosphazenes, Comput. Theor. Polym. Sci. 8(1998) 229.

[34] H. Sun, COMPASS: an ab initio-force field optimized for condensed-phase applications-overview with details on alkane and benzene compounds, J. Phys. Chem. B102(1998) 7338.

[35] H. J. C. Berendsen, J. R. Grigera, T. P. Straatsma, Effect of support pretreatments on carbon-supported iron particles, J. Phys. Chem. 91(1987) 6269.

[36] S. Nose, A unifi ed formulation ofthe constant temperature molecular dynamics methods,J. Chem. Phys. 81(1984) 511-519.

[37] W. G. Hoover, Canonical dynamics: equilibrium phase-space distributions, Phys. Rev. A 31(1985) 1695-1697.

[38] P. Ewald, Die Berechnung Optischer und Elektrostatischer Gitterpotentiale, Ann. Phys. 369(1921) 253-287.

[39] H. Yan, X.-L. Guo, S.-L. Yuan, C.-B. Liu, Molecular dynamics study ofthe effect of calcium ions on the monolayer ofSDC and SDSn surfactants at the vapor/liquid interface, Langmuir 27(2011) 5762-5771.

不同含油饱和度时泡沫的稳定性及调驱机理研究

曹绪龙[1] 马汉卿[2] 赵修太[2] 王增宝[2] 陈文雪[2] 陈泽华[2]

(中石化胜利油田勘探开发研究院[1];中国石油大学(华东)石油工程学院[2],青岛 266580)

摘要:利用搅拌起泡法研究了含油饱和度对于泡沫稳定性的关系,利用微观玻璃模型研究了不同含油饱和度时泡沫的调驱机理,利用双管并联填砂岩心模型评价了不同含油饱和度时的调剖效果。结果表明:含油饱和度对泡沫稳定性有明显影响,含油饱和度0.05～0.2区间内,泡沫稳定性缓慢降低,含油饱和度0.2～0.4区间内,泡沫稳定性快速降低,含油饱和度0.4～0.6以上区间内泡沫稳定性缓慢降至较低水平;微观玻璃模型驱替实验表明,含油饱和度在0.4～0.6区间内泡沫不能稳定存在,泡沫依靠泡沫剂溶液的洗油作用提高采收率,0.05～0.2区间内泡沫稳定性较好,主要依靠气泡的调堵作用提高采收率,0.2～0.4区间内调堵作用和洗油作用协同;含油饱和度对于泡沫的调剖效果有显著影响,在实验条件下,含油饱和度在0.1～ 0.3区间内,驱替压差可达1.2 MPa,泡沫调剖作用显著,在0.3～0.6区间内,驱替压差逐渐降至0.3 MPa,调剖效果不理想,含油饱和度大于0.6,驱替压差低于0.2 MPa,泡沫失去调剖作用。

关键词:泡沫 稳定性 调驱机理 含油饱和度

中图法分类号:TE357.46; 文献标志码:A

泡沫体系的稳定性是影响泡沫驱效果的核心因素,在油藏条件下能否形成泡沫及泡沫稳定时间直接影响泡沫驱的实施效果[1-3]。泡沫破坏的机理主要有液膜的排液[4]和气体透过液膜的扩散[5],油相的存在使得起泡剂分子由气-水表面向油-水界面迁移,对泡沫的稳定性产生显著的影响,地层含油饱越高,泡沫越不稳定,液膜易聚并消泡[6-9]。因此泡沫在含油饱和度较高地层与含油饱和度较低的地层的驱油机理会存在显著差别。利用搅拌法起泡实验,评价了泡沫体系与油相接触后的稳定性,利用微观玻璃模型研究了不同含油饱和度下泡沫的调驱机理,利用双管并联填砂岩心模型评价了不同含油饱和度时泡沫的调驱效果。

1 实验部分

1.1 材料与仪器

起泡剂为胜利油田提供的低张力起泡剂XR-1,溶液质量浓度为0.5%;模拟地层水矿化度11 400 mg/L(离子组成 Na^+ :4 500 mg/L,Ca^{2+} :500 mg/L,Mg^{2+} :

100 mg/L,Cl^- ：5 300 mg/L,HCO_3^- ：400 mg/L)；模拟原油由胜利油田孤东原油配制(体积比原油：煤油＝1：1.5),60℃下黏度为 26 mPa·s;发泡气体为氮气。

岩心驱替实验装置包括泡沫发生器、气体流量计、高压恒流泵、中间容器、填砂模型管、高温高压模型夹持器、压力表,恒温箱等。

1.2 实验方法

1.2.1 含油情况下泡沫的稳定性实验

使用模拟地层水配置质量浓度为 5%的起泡剂溶液,60℃下使用搅拌法起泡(转速 12 000 r/min,时间 5 min),然后加入一定量的模拟原油配置成含油饱和度 0.05 至 0.6 不等的泡沫体系,用半衰期法测量体系的稳定性。

1.2.2 并联填砂岩心模型驱替实验

将并联填砂岩心模型抽真空,后饱和模拟地层水后再饱和模拟地层原油,60℃下老化 24 h 后用气液比 1：1 的泡沫驱,记录驱替过程中高、低渗填砂岩心模型中的采出程度、分流量、压力数据。岩心参数见表 1。

表 1　并联填砂岩心模型的参数

Table 1　Diameter of parallel sand pack

玻璃模型	高渗岩心	低渗岩心
渗透率/($\times10^{-3}$ μm^2)	8 942	2 473
孔隙度/%	40.1	32.3

1.2.3 微观玻璃模型驱替实验

将微观玻璃模型抽真空后饱和水然后饱和油,利用高温高压夹持器进行气液比为 1：1 的泡沫驱。

2　结果与讨论

2.1 泡沫体系对油相的稳定性实验

泡沫体系半衰期与含油饱和度的关系曲线如图 1 所示。

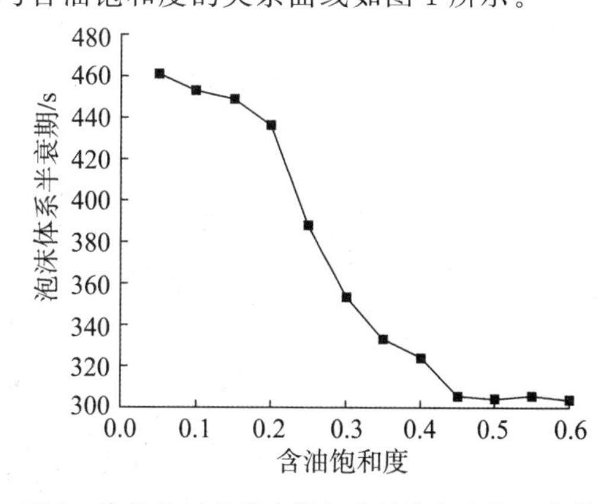

图 1　泡沫体系的半衰期与含油饱和度关系曲线

Fig. 1　Curve of foam half-time period and oil saturation

由图 1 可知,随含油饱和度的增高,泡沫的稳定性下降[10],且油相的存在对泡沫体系的稳定性影响显著,含油饱和度 0.6 时的半衰期比含油饱和度 0.05 时的半衰期降低约 40%。含油饱和度在 0.05～0.2 区间内,泡沫的半衰期缓慢降低,这是由于泡沫体系内含油较少时,转移到油-水界面的泡沫剂分子比较少,吸附在气-水界面上剩余的表面活性剂足以使泡沫保持相对稳定的状态,故半衰期下降不明显;含油饱和度在 0.2～0.4 区间内,泡沫的半衰期快速降低,在这一区间内随着含油饱和度的增加,大量泡沫剂分子吸附油水界面[11],剩余吸附在气-水表面的泡沫剂浓度不足以维持液膜稳定所需要的低张力,故液膜大量破裂聚并,泡沫消泡剧烈,稳定性迅速下降;含油饱和度在 0.4～0.6 区间内,泡沫剂在油-水界面和气-水界面的吸附达到平衡,泡沫的半衰期趋于稳定。

2.2 不同含油饱和度时泡沫的调剖机理研究

根据节 2.1 中泡沫稳定性变化趋势的不同而确定三个含油饱和度区间:0.05～0.2、0.2～0.4 以及 0.4～0.6,使用微观玻璃模型进行不同含油饱和度下泡沫的调剖机理研究,含油饱和度采用数格法确定。

2.2.1 含油饱和度 0.05～0.2 区间泡沫的调驱机理

这一区间内泡沫的稳定性较好,泡沫容易再生,有利于气泡在高渗孔道内多排累积,对高渗大孔道形成封堵(图 2 红圈部分),调整地层吸水剖面,使后续的泡沫更多的进入到低渗孔道中进行驱油(图 2 中蓝色箭头),被洗出的油滴沿着气泡与孔隙壁运移(图 2 中白色箭头),此时体系主要依靠气泡的调堵作用使液流转向提高波及系数来提高采收率,而体系中稀表面活性剂溶液在乳化剥离膜状剩余油和盲端剩余油的时候能发挥作用。

2.2.2 含油饱和度 0.2～0.4 区间泡沫的调驱机理

由图 3 可知,含油饱和度在 0.2～0.4 区间内泡沫的半衰期开始快速下降,这意味着在地层中随着含油饱和度的升高,液膜破裂聚并成的大气泡在地层中越来越多,泡沫的平均泡径增大。狭长的大气泡会沿着优势渗流通道窜走[12],但仍然有一部分泡径适宜的气泡(图 3 红圈内)封堵在大孔道入口起到调堵作用。在此区间内,泡沫体系即通过泡沫剂溶液的洗油作用驱油,又通过气泡的调堵作用使得液流转向提高采收率,二者协同作用,相辅相成。

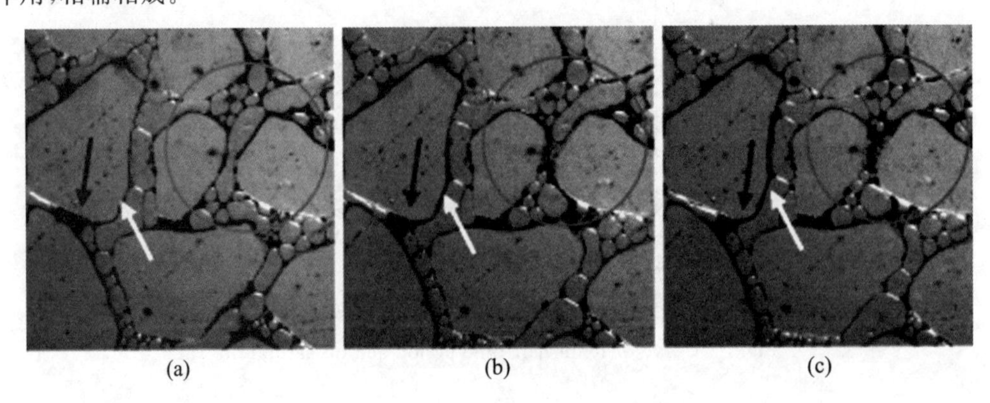

(a) (b) (c)

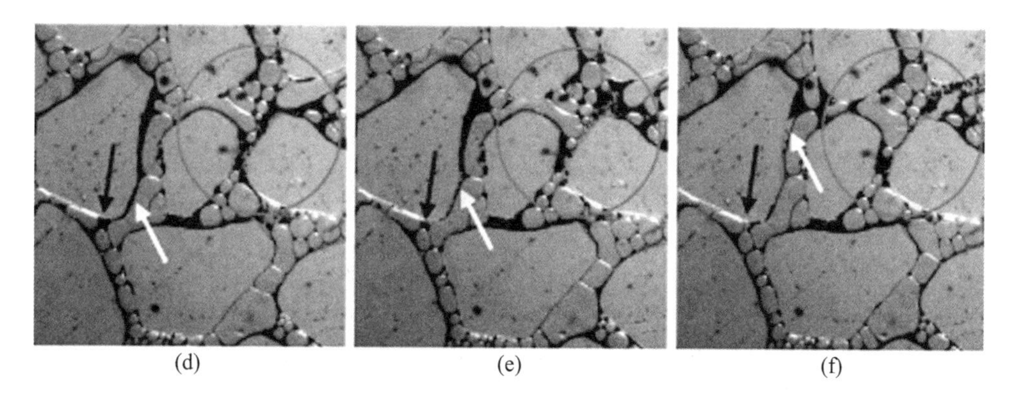

图 2　含油饱和度 0.05～0.2 时泡沫调驱机理

Fig. 2　Mechanics on foam profile control between oil saturation 0.05～0.

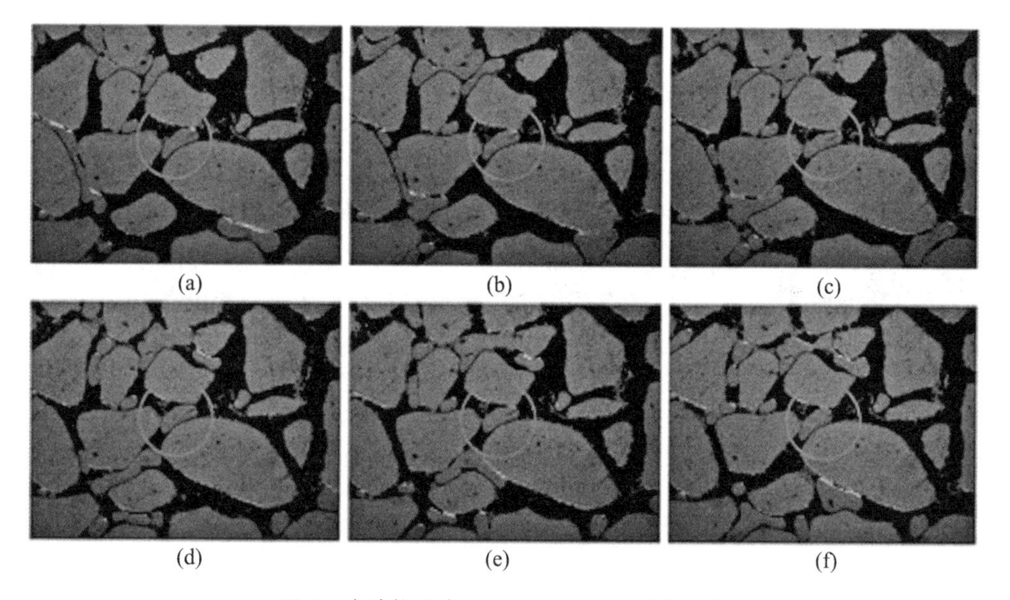

图 3　含油饱和度 0.2～0.4 时泡沫调驱机理

Fig. 3　Mechanics on foam profile control between oil saturation 0.2～0.4

2.2.3 含油饱和度 0.4～0.6 区间泡沫的调驱机理

在地层含油饱和度在 0.4～0.6 区间时，由于地层中未被驱替的原油过多，泡沫剂分子大量向油-水界面转移，导致泡沫前缘非常不稳定，再生能力降低，小气泡极易聚并成大气泡，大气泡不易在孔隙中形成封堵，故运移阻力较小，易从泡沫体系中分离出来先于稀表面活性剂体系进入油相单独运移(图 4 红色箭头)，气相易沿优势孔道发生突进，后续的泡沫剂溶液作为表面活性剂体系继续驱替原油(图 4 中白色箭头)，在较高的含油饱和度下，泡沫剂溶液的驱油机理对于提高采收率起到关键作用。

2.3 不同含油饱和度时泡沫的调驱效果评价

泡沫对高渗地层的封堵作用是靠贾敏效应的叠加来实现的，实现贾敏效应的叠加要求泡沫泡径不能远大于孔道尺寸(即形成连片状的大气泡)，否则将会发生气窜使得泡沫失去调剖效果,而油相的存在对泡沫的稳定性和平均泡径有显著的影响[13-15],故采用并联填砂岩心模型评价了高渗管中含油饱和度 0.05～0.6 时泡沫对于非均质地层的调剖效果,如图 5 所示。

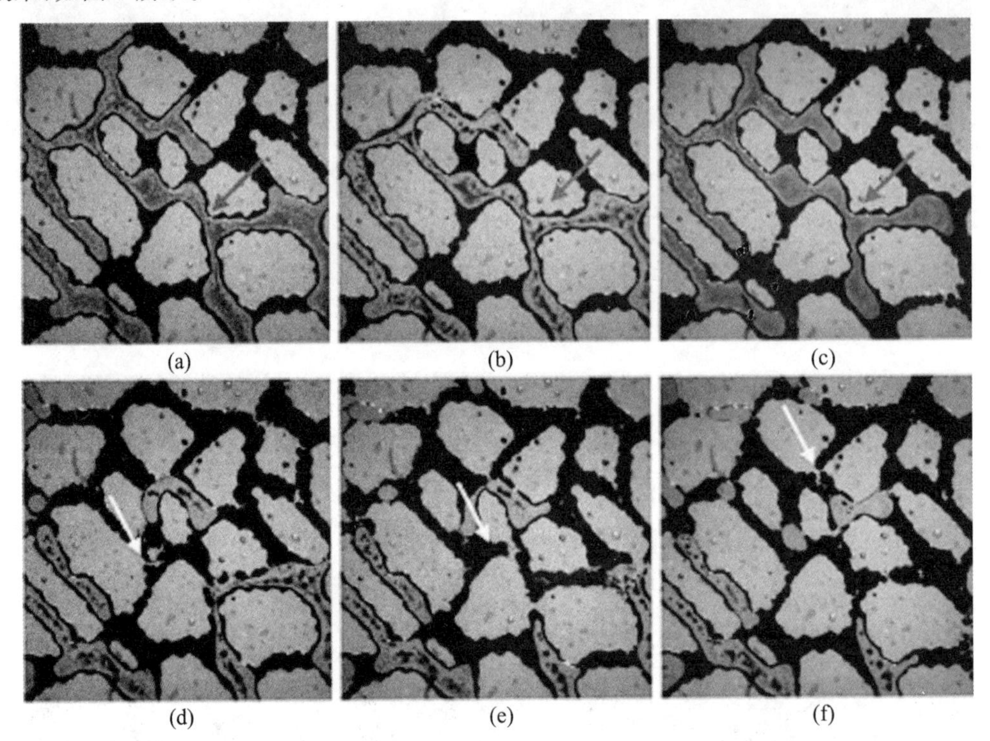

图 4　含油饱和度 0.4～0.6 时泡沫调驱机理

Fig. 4　Mechanics on foam profile control between oil saturation 0.4～0.6

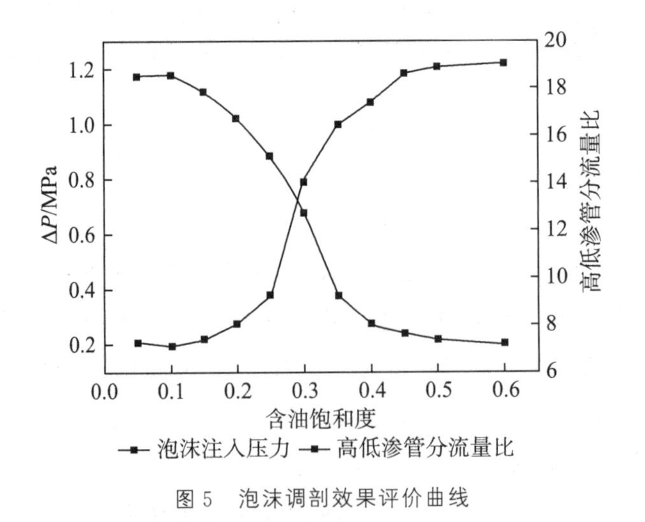

图 5　泡沫调剖效果评价曲线

Fig. 5　Evaluation curve foam profile control effect

由图 5 可知,驱替压差 ΔP 随含油饱和度的升高而逐渐降低,高渗岩心与低渗岩心分流量之比随含油饱和度的升高而逐渐升高。含油饱和度在 0.1～0.30 区间内,驱替压差高最高可达 1.2 MPa 左右，高低渗管分流量比可达 6，调堵作用显著；含油饱和度在 0.30～0.60 区间内，驱替压差逐渐降低至 0.3 MPa 左右，高渗管和低渗管的分流量比高于 12(即高渗管产液 12 mL，低渗管仅产液 1 mL)，泡沫在高渗管中的调剖作用不理想；含油饱和度高于 0.6 时，驱替压差降至 0.2 MPa 以下，泡沫基本无调剖作用。

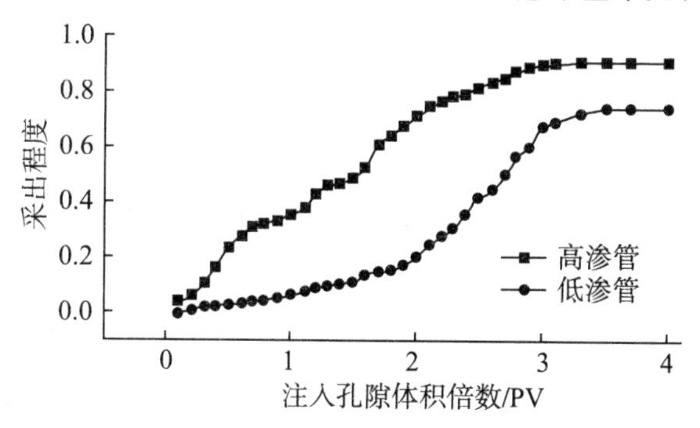

图 6　岩心采出程度与注入孔隙体积倍数曲线

Fig. 6　Curve of recovery efficiency and PV value

并联岩心模型采出程度与注入孔隙体积倍数变化曲线如图 6 所示。

高渗模型采出程度先快速增加至含油饱和度 0.30 左右，后缓慢增加至含油饱和度 0.1 左右后达到稳定；低渗模型采出程度先缓慢增加至含油饱和度为 0.3 左右后快速增加至含油饱和度 0.1 以下达到稳定。

综上，对于低张力泡沫剂 XR-1 泡沫体系，含油饱和度在 0～0.1 区间内，低渗岩心采出程度高，且趋于平稳，调剖效果显著；含油饱和度 0.1～0.30 区间内，低渗岩心采出程度迅速增加，调剖效果明显；含油饱和度 0.30～0.6 区间内，低渗岩心采出程度缓慢增加，调剖效果不理想；含油饱和度高于 0.6 时，低渗岩心采出程度基本不变，泡沫不具有调剖作用。

3　结论

(1) 油相的存在对泡沫稳定性有明显影响，含油饱和度 0.05～0.2 区间内，泡沫稳定性缓慢降低，含油饱和度 0.2～0.4 区间内，泡沫稳定性快速降低，含油饱和度 0.4～0.6 以上区间内泡沫稳定性缓慢降至较低水平。

(2) 微观玻璃模型驱替实验表明，含油饱和度在 0.4～0.6 区间内泡沫不能稳定存在，泡沫依靠泡沫剂溶液的洗油作用提高采收率，0.05～0.2 区间内泡沫稳定性较好，主要依靠气泡的调堵作用提高采收率，0.2～0.4 区间内调堵作用和洗油作用协同作用。

(3) 并联岩心管实验表明含油饱和度对泡沫的调剖效果影响显著，随着含油饱和度的逐渐升高，泡沫的调剖效果逐渐变差，最终丧失调剖作用。在实验条件下，含油饱和度在 0.1～0.3 区间内，驱替压差可达 1.2 MPa，泡沫调剖作用显著，在 0.3～0.6 区间内，驱替压差逐渐降至 0.3 MPa，调剖效果不理想，含油饱和度大于 0.6，驱替压差低于 0.2 MPa，泡沫失去调剖作用。

参考文献

[1] 郭程飞,李华斌,吴忠正,等. 起泡剂在不同驱油方式下的驱油效果. 油田化学,2014; 31(4) : 534-537.

Guo C F, Li H B, Wu Z Z, et al. Effect of foaming agent under dif-ferent displacement method. Oilfield Chemistry,2014; 31(4):535-537.

[2] 马丽萍,黎晓茸,谭俊领. 空气泡沫流体垂直管注入特征实验研究. 科学技术与工程,2015; 15(19):136-138.

Ma L P,Li X R,Tan J L. Experimental study on the air-foam fluid characteristics in vertical tube. Science Technology and Engineering, 2015; 15(19): 136-138.

[3] 孙乾,李兆敏,李松岩,等. SiO_2 纳米颗粒的泡沫体系驱油性能研究. 中国石油大学学报(自然科学版),2014; 38(4):124-131

Sun Q,Li Z M,Li S Y,et al. Oil displacement performance of stabi-lized foam system by SiO_2 nanoparticles. Journal of China University of Petroleum(Edition of Natural Science),2014; 38(4),124-131.

[4] 王玮,李楷,宫敬,等. 分散体系中液滴间力学行为. 科学通报,2015; 60(24):2272-2281.

Wang W, Li K, Gong J, et al. Recent progess of drop deformation and force behavior in disperse systems. Science China, 2015; 60(4):2271-2281.

[5] 郭丽梅,王亚丹,管保山,等. 泡沫压裂液稳定剂及衰变机理研究. 精细石油化工,2014; 31(5):9-13.

Guo L M,Wang Y D,Guan B S,et al. Study on stability and decay mechanism of foam fracturing fluid. Speciality Petrochemicals,2014; 31(5):9-13.

[6] 王琦,习海玲,左言军. 泡沫性能评价方法及稳定性影响因素综述. 化学工业与工程技术,2007; 28(2):25-30.

Wang Q,Xi H L,Zuo Y J. Review on measurement techniques of performance and influence factors of Stability for foam. Journal of Chemical Industry & Engineering, 2007,28(2):25-30.

[7] 贾新刚,燕永利,屈撑囤,等. 原油对水相泡沫稳定作用机理研究进展. 日用化学工业,2010; 40(1):54-59.

Jia X G,Yan Y L,Qu C T,et al. Progress in study of effect of crude oil on stability of aqueous foam. China Surfactant Detergent & Cos-metics,2010; 40(1):54-59.

[8] 燕永利,王瑶,山城,等. 油滴作用下的气泡液膜破裂机理研究进展. 日用化学工业,2013;43(4):303-308.

Yan Y L, Wang Y, Shan C, et al. Progress in research work with re-spect to the mechanism foe foam rupture by action of oil droplets. China Surfactant Detergent & Cosmetics, 2013; 43(4):303-308.

[9] 李兆敏,王鹏,李松岩,等. 纳米颗粒二氧化碳泡沫稳定性的研究进展. 西南石油大学学报(自然科学版),2014; 36(4):155-161.

Li Z M, Wang P, Li S Y, et al. Advances of researches on improving the stability of CO_2 foams by nanoparticles. Journal of Southwest University of Petroleum(Science & Technology Edition), 2014; 36(4):155-161.

[10] 端祥刚,侯吉瑞,李实,等. 耐油起泡剂的研究现状及发展趋势. 石油化工, 2013; 42(8):935-940.

Duan X G, Hou J R, Li S, et al. Progress and future trends in research of oil resistant foaming agent. Petrochemical Technology, 2013; 42(8):935-940.

[11] 马魁菊,邹若华,陈舟圣,等. 泡沫剂辅助蒸汽与孤岛稠油相互作用研究. 应用化工,2014; 43(1):24-27.

Ma K J, Zou R H, Chen Z S, et al. Study on the interaction of frot-her assisted steam flooding and Gudao viscous crude oil. Applied Chemical Industry, 2014; 43(1): 24-27.

[12] 梁于文,熊运斌,高海涛,等. 强非均质性油藏空气泡沫调驱先导实验——以胡状集油田胡 12 断块为例. 油气地质与采收率, 2009; 16(5):69-71.

Liang Y W, Xiong Y B, Gao H T, et al. Pilot test of air foam drive of serious heterogeneous reservoirs-taking Hu12 fault block in Huzhuangji oilfield as an example. PGRE, 2009; 16(5):69-71.

[13] 陈淮,吴树森. 双液泡沫的粒径分布及其影响因素研究. 上海师范大学学报(自然科学版),1996; 25(1) : 59-64.

Chen H, Wu S S. Study on partical radius distribution and the influ-ential factors of biliquid foam. Journal of Shanghai Teachers Univer-sity(Natural Sciences), 1996; 25(1):59-64.

[14] 黄浩,李华斌,牛忠晓,等. 高压下泡沫性能参数的研究. 科学技术与工程, 2013; 13(3):694-696.

Huang H, Li H B, Niu Z X, et al. Experimental study on bearing characteristics under high Pressure. Science Technology and Engi-neering, 2013; 13(3):694-696.

[15] 华子东. 洼 16 块非均质油藏泡沫复合驱油技术试验. 石油地质与工程,2010; 24(5):121-127.

Hua Z D. Experiment on Wa16 Block heterogeneous reservoir foam combination flooding technology. Petroleum Geology and Engineer-ing, 2010; 24(5):121-127.

Foam stability and mechanism on profile control under different oil saturation

Cao Xulong[1], Ma Hanqing[2], Zhao Xiutai[2], Wang Zengbao[2], Chen Wenxue[2], Chen Zehua[2]

(Sinopec Shengli Oilfield Exploration and Development Research Institute[1]; UPC Institute of Petroleum Engineering[2] Qingdao 266580, P. R. China)

[Abstract] The foam stability and oil saturation was studied with stirring method, the mechanism on profile con-trol was studied with micro model, and the profile control effect was studied with parallel san pack. Results show that, oil saturation has an obvious influence on foam stability, in 0.05~0.2 interval, stability decreases slowly, in 0.2~0.4 interval, stability decreases sharply, finally in 0.4~0.6 interval stability reaches a stable level. The en-hancement of recovery relies on surfactant system swapping in high oil saturation in oil saturation 0.4~0.6, while in low saturation 0.05 ~0.2 relies on water shut-off of Jamin effect. In the condition of this paper, the flood pres-sure can be 1.2 MPa at oil saturation 0.1~0.3, profile control effect works well, and pressure decreases to 0.3 MPa at oil saturation 0.3~0.6, the profile control effect is not obvious, finally when oil saturation reaches 0.6 and higher, profile control effect decreases to none.

[Key words] foam stability profile control & displacement mechanism oil saturation

Effect of dynamic interfacial dilational properties on the foam stability of catanionic surfactant mixtures in the presence of oil

Wang Ce[a], Zhao Li[a], Xu Baocai[a,*], Cao Xulong[b], Guo Lanlei[b], Zhang Lei[c], Zhang Lu[c,*], Zhao Sui[c]

[a] School of Food and Chemical Engineering, Beijing Technology and Business University, Beijing, 100048, PR China

[b] Exploration & Development Research Institute of Shengli Oilfield Co. Ltd, SINOPEC, Dongying 257015, Shandong, PR China

[c] Technical Institute of Physics and Chemistry, Chinese Academy of Sciences, Beijing, 100190, PR China

GRAPHICAL ABSTRACT

The mixing ratio of anionic/ cationic surfactants and the hydrophobic chain length of the surfactants infl uence the foam stability in the presence of oil through the interfacial properties. The fast generation of a tightly packed interfacial layer is in favor of the foamability and foam stability in the presence of oil.

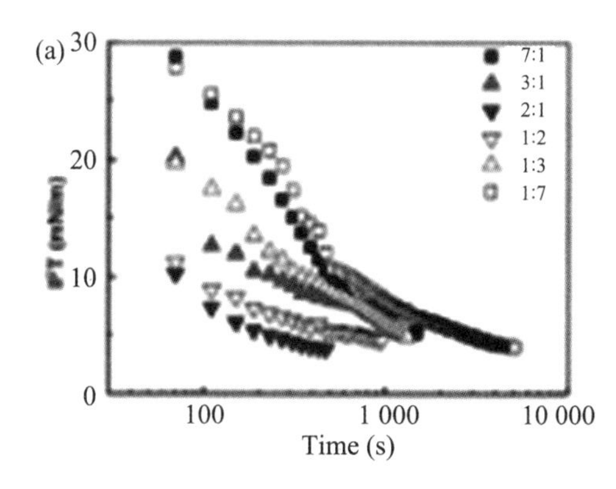

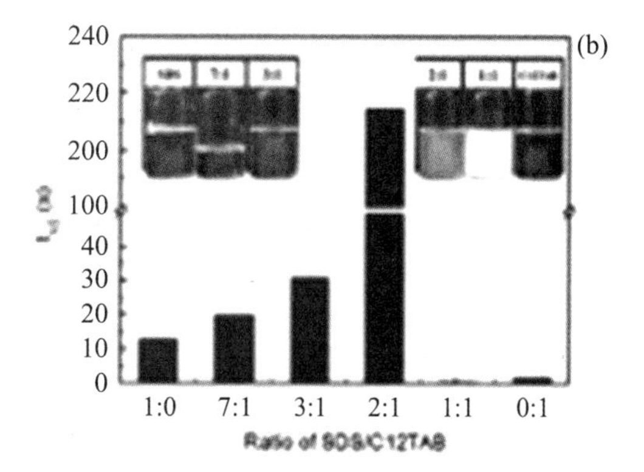

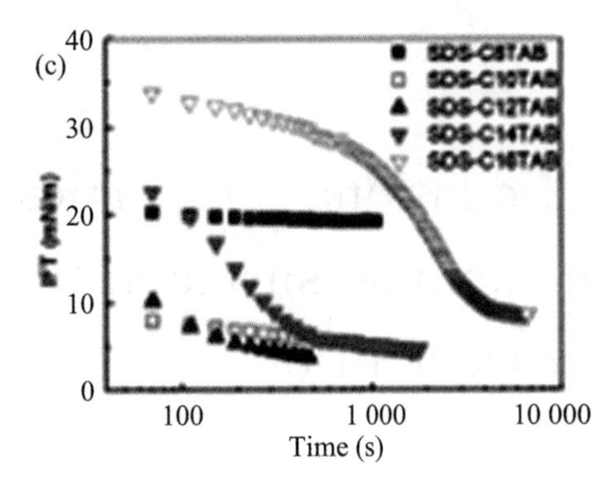

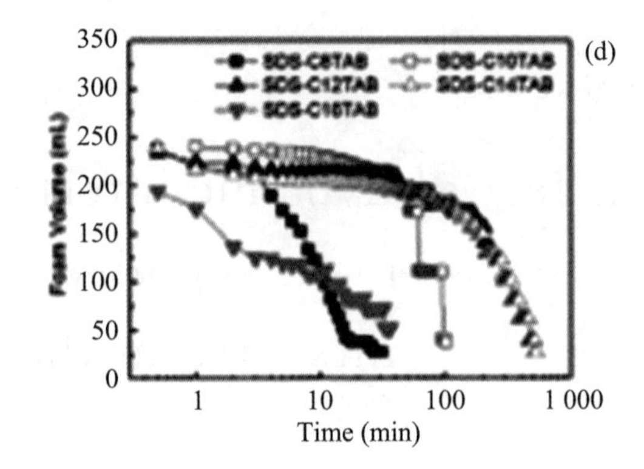

ABSTRACT

The dynamic interfacial dilational properties of catanionic surfactant mixtures and the relationship with foam stability in the presence of oil are studied in the present work. The dynamic interfacial dilational properties of catanionic mixtures at aqueous oil interface have been studied by means of oscillating the drop profile method. The mixing ratio of anionic/cationic surfactants and the hydrophobic chain length of the surfactants influence the foam stability in the presence of oil through the interfacial dilational properties. Over a wide range of mixing ratio of SDS/C12 TAB, interfacial tension and dilational modulus of the catanionic mixtures reaches similar equilibrium values. But as the mixing ratio approaches 1 : 1, comparable anionic and cationic surfactants areadsorbed at interface. This significantly speeds up the dynamic adsorption progress, which benefi ts the fast foaming progress. In addition, the increase of the alkyl chain length of CnTAB in SDS-CnTAB(n = 8, 10, 12, 14, 16) mixtures enhances the equilibrium modulus, but slows down the dynamic adsorption progress. Thus the foamability and stability decreases when surfactants have overlong hydrophobic chains. Overall, the fast generation of a tightly packed interfacial layer is in favor of the foamability and foam stability in the presence of oil.

Keywords: Catanionic surfactant mixtures　Interfacial tension　Dilational modulus　Foam stability　Oil-resistance

1. Introduction

As mixtures of discrete gas bubbles and continuous liquid films, foams have shown complex properties and been applied in various industrial fields, such as enhanced oil recovery, food industry, cosmetics, etc[1-3]. In many processes, the stability of foams in the presence of oil is desired[4-7]. In our previous study, catanionic surfactant mixtures were proposed as a type of foaming agent and have shown remarkable advantage in enhancing the foam stability in the presence of oil[8].

Catanionic surfactant mixtures are systems consisting of both anionic and cationic surfactants which have drown a wide interest[9-22]. As a result of the intensive electrostatic attraction of the hydrophilic groups with opposite charge, cationic and anionic surfactant monomers self-assemble into various aggregates in bulk solution, such as micelles and vesicles[16, 23, 24]. At interface, catanionic surfactant mixtures are closely packed and dramatically decrease the surface tension and critical micelle concentration (cmc)[25-27]. The adsorption layers of catanionic mixtures also show strong viscoelasticity and confer high disjoining pressures when two interfaces are approaching each other when forming a thin liquid film[22]. This is in favor of the stability of foam films[18, 28-31]. Therefore, catanionic surfactant mixtures are promising systems in enhancing foamability and foam stability[9,22,27].

Foaming ability of the surfactant solution may be correlated with the equilibrium surface tension or dynamic surface tension, owing to the different foaming methods. For foam generated by blowing gas into solution, the surfactant molecules have not enough time to diff use and adsorb at the air/ water interface to reach the equilibrium, therefore the dynamic surface tension of the solution becomes the key factor determining the foaming ability[32]. For the foaming progress in the presence of oil, the gas flow will disperse the oil phase and large amount of fresh aqueous-oil and gas-aqueous interfaces are formed and foaming agents would adsorb at both kinds of interfaces. In foam films, as an oil drop approaches gas-water interface, a thin aqueous film, which is called pseudoemulsion film, is formed between gas-water and oil-water interfaces. It has been widely accepted that the stability of pseudoemulsion film is the key factor to the foam stability in the presence of oil[33-36]. For oilfree foams, the foams of catanionic mix-tures have shownsignifi cantly stability and the life time is obviously longer than the foam life in the presence of oil in the same condition[22,27]. That indicates the properties of gas-water interface meet the requirement of foams with suffi cient stability. Thus, instead of gas-water interface, the oil drops in the foam liquid films should be the dominant factor that infl uences the foam stability. Therefore, it is crucial to study the interfacial properties of oil-water interface. But most of the studies about catanionic mixtures focus on gas-aqueous interface and oil-free foams, the oil-aqueous interfacial properties of catanionic surfactants are scarcely reported (especially the dynamic interfacial properties), which, however, is worth further study[37-40].

In our previous work, the equilibrium interfacial dilational rheological properties of catanionic mixtures at aqueous interface were tested. The interfacial dilational properties are slightly dependent on the mixing ratio. But the dilational modulus shows obvious increase with the increase of electronic charge density of the headgroups and the hydrophobic chain length of the surfactants[41]. The dynamic dilational properties can provide more accurate information about the adsorption progress of interfacial layers, which have been comprehensively investigated in recent years due to the commercialized instruments[42-44]. But the dynamic interfacial dilational properties of catanionic mixtures have been rarely reported and require a systematic research.

In the present work, to further research the dynamic interfacial properties of catanionic mixtures and the relationship with foam stability in the presence of oil, the dynamic dilational properties of catanionic mixtures at aqueous-oil interface have been investigated by means of oscillating drop method at low frequency (0.1 Hz). The influence of the mixing ratios and the hydrophobic chain length of cationic and anionic mixtures on the dynamic interfacial dilational properties and foam stability are also studied.

2. Experimental methods

2.1. Materials

The anionic surfactant used in this article is sodium dodecyl sulfate (SDS, purchased from Xilong company, Guangdong, China) that have been recrystallized from ethanol. The cationic surfactants are octyltrimethylammonium bromide(C8 TAB, purchased from J & K), decyltrimethylammonium bromide (C10 TAB, purchased from J & K), dodecyltrimethylammonium bromide(C12 TAB, purchased from Hushi Co. Shanghai, China), tetradecyl trimethylammonium bromide(C14 TAB, purchased from J & K), cety ltrimethylammonium bromide (C16 TAB, purchased from Jinke Fine Chemical Institude, Tianjin, China) that have been recrystallized from acetone. The purity of the surfactants has been verified by surface tension measurements. Catanionic surfactant mixtures are mixed at various ratios(7 : 1, 3 : 1, 2 : 1, 1 : 1, 1 : 2, 1 : 3 and 1 : 7). The solutions were prepared with ultrapure water with resistivity of 18.2 MΩ · cm. Kerosene was applied as the model oil, which was purified by silica gel column chromatography(100-200 meshes) in this research.

2.2. Foam production and characterization in the presence of oil

The oil-resistance of foams is characterized by evaluating the stability of foams generated from the mixture of kerosene and aqueous solution[34,45]. To be more specific, kerosene is dispersed in the surfactant solution using ultrasonic wave and the mass fraction of kerosene is 1%. After that, the oil-aqueous mixture is poured into the column with water bath and heated to the test temperature 50℃. The foams are generated by blowing N_2 at a flow rate of 10 mL/s through a porous glass filter at the bottom of the glass column where the foaming agent under investigation is placed. Around 200 mL foams are formed after 20 s flow. At the same time, oil drops are carried with aqueous solution and kept in foam films. To characterize the foam stability in the presence of oil, the evolution of the foam volume is recorded and the evolution curves of the foam volume of diff erent systems are compared. Generally, the time when the foams volume decreases to the 1/2 of the initial value, $t_{1/2}$, is recorded to characterize the stability of foams[46].

2.3. Interfacial dilational rheological experiment

To study the interfacial rheology properties, interfacial dilational rheological experiment was performed using an oscillating drop tensiometer (OCA 20, DataPhysics, Germany). The main elements of the method are a dosing system, a light source, CCD camera, an oscillator, and a needle for drop formation. The interface is created by injecting an aqueous drop (volume: 8 μL) from a gas-tight syringe into kerosene and the drop is attached to a stainless steel needle. A computer-controlled periodical oscillation of the drop area(A) is performed at chosen amplitude (ΔA/A, 10%) at a frequency of 0.1 Hz. The image of the drop is recorded by CCD camera and digitized to allow the analysis of its shape by fitting the Laplace equation to its coordinates. The results of these harmonic relaxation experiments were analyzed using a Fourier transformation[47,48]. A typical diagram of the area oscillation and IFT response along with time is displayed in Fig. 1.

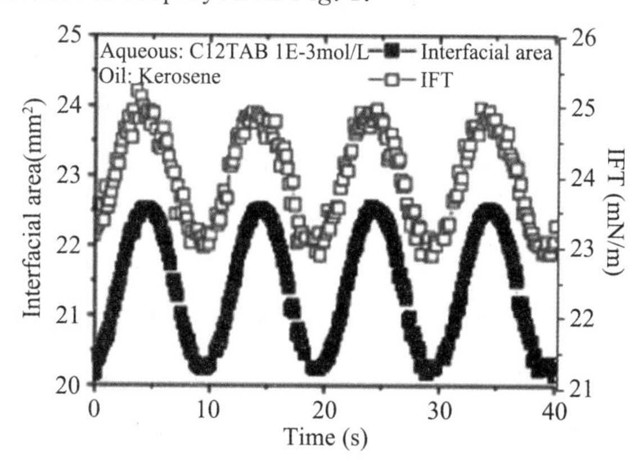

Fig. 1 Sinusoidal variations of interfacial area and IFT of an oscillating aqueous drop in kerosene at the frequency of 0.1 Hz at 30℃.

The dilational modulus is defi ned by Gibbs in the following expression:

$$\varepsilon = \frac{d\gamma}{d\ln A} \tag{1}$$

where ε is the dilational modulus, γ is interfacial tension (IFT), and A is the interfacial area. Dilational modulus depicts IFT response with the changes of area. According to the defi nition of dilational modulus, it refl ects the ability to resist the deformation of the interface. An interface with high dilational modulus shows better stability. Therefore, it benefi ts the stability of the pseudoemulsion film.

3. Results and discussion

3.1. Infl uence of mixing ratio of anionic/cationic surfactant on interface and foam properties

In order to study how the ratio of anionic/cationic surfactant influences the foam stability in the presence of oil through the dynamic interfacial properties, SDS and C12 TAB are taken as model surfactant and mixed at various ratios. The dynamic interfacial properties and foam stability in the presence of oil are tested.

3.1.1. Dynamic interfacial tension and dilational modulus of SDS-C12TAB mixtures at diferent ratios

Dynamic interfacial dilational rheology measurement has provided a possible way to study the adsorption progress and the conformational behavior of interfacial active agents at an interface. It has been used instudying the interfacial behaviors of surfactants, protein, etc[43,44,49]. Therefore, this method was employed in this work to investigate the possible adsorption process of catanionic surfactant mixtures.

The dynamic IFT and dilational modulus of SDS-C12 TAB mixtures of diff erent ratios at aqueous-oil interface are shown in Fig. 2. Generally, the amount of surfactants adsorbed at interface increases with time, which leads to the decrease of IFT and the increase of dilational modulus. Both of the parameters finally reach constants when the interfacial adsorption becomes equilibrium[44]. Take the SDS-C12 TAB mixtures at 7∶1 as an example, at low bulk concentration (below 1×10^{-5} mol/L), IFT decreases from the initial value to the equili-brium in a small extent and dilational modulus also increases slightly because of the small amount adsorption of surfactants. As the increase of concentration, the initial IFT begins to decrease, and equilibrium IFT can decrease to a much lower value. Meanwhile, a signifi cant increase in dilational modulus can be observed at high concentration.

It is worthy noting that IFT and dilational modulus reaches equilibrium at diff erent speed when SDS and C12 TAB are mixed at different ratios. The initial and equilibrium IFT of SDS-C12 TAB at the con-centration of 5×10^{-5} mol/L and 1×10^{-4} mol/L are shown in Fig. 3. For instance, at the concentration of 5×10^{-5} mol/L, in the case of the 7 ∶ 1 and 1 ∶ 7 systems, with either SDS or C12 TAB obviously excess in the solution, the initial IFT of both systems is over 30 mN/m and the dy-namic decrease of IFT is relatively slower. As the ratio of SDS and C12 TAB approaches 1 ∶ 1, the initial IFT decreases fast and IFT reaches equilibrium in short time. In the case of 3 ∶ 1 and 1 ∶ 3, the initial IFT is around 25 mN/m, while the initial IFT decreases below 20 mN/m when SDS and C12 TAB are mixed at 2 ∶ 1, 1 ∶ 2 and 1 ∶ 1.

Similarly, the dynamic dilational modulus also shows a remarkable dependence on the ratio of SDS and C12 TAB. As shown in Fig. 2(b, n), at high concentration(5×10^{-5} mol/L), dilational modulus increases gradually from around 10 mN/m to nearly 45 mN/ m when SDS and C12 TAB are mixed at 7 ∶ 1 and 1 ∶ 7. However, as the ratio of SDS and C12 TAB is close to 1 ∶ 1, dilational modulus reaches a plateau around 45 mN/m in very short time.

In spite of the signifi cant diff erence in the dynamic IFT and dila-tional modulus between SDS-C12 TAB mixtures of various ratios, the equilibrium values coincide well with each other (1×10^{-4} mol/L). Fauser H reported the surface adsorption of SDS-C12 TAB mixtures is governed by the stoichiometric formation of catanionic surfactant complexes and independent of the mixing ratio[27]. At 5×10^{-5} mol/L, there are not enough catanionic complexes to generate a saturated compact interfacial layer. The concentration of catanionic complexes increases with the mixing ratio approaches to 1 ∶ 1. Therefore, the equilibrium IFTs decreases accordingly. When concentration is high enough(1×10^{-4} mol/L), there are sufficient catanionic complexes to form a saturated layer at any mixing ratio. Therefore, equilibrium IFTs and dilational modulus keeps constant regardless the diff erence of the mixing ratios of SDS/C12 TAB, which could speculate that the cata-nionic complexes at interface share a similar composition.

For catanionic mixtures, both anionic and cationic surfactant will diff use from the bulk to interface once a new interface is generated. The initial composition of cationic and anionic surfactants at interface is closely related to the mixing ratio in the bulk. Therefore, for mixture systems with one of the components signifi cantly excess(like 7 ∶ 1 or 1 ∶ 7), there would be a dominant surfactant at the interface. Although a small amount of surfactant with opposite charge neutralizes the elec-trostatic repulsion, most of the dominant surfactant molecules are still under the repulsion from each other and cannot pack closely at inter-face during the initial stage. Thus a high IFT is observed at the begin-ning. During the dynamic progress, the large amount of excessive sur-factant will attract the opposite one from the solution through electrostatic attraction. As the interfacial concentrations of anionic and cationic surfactants get balanced gradually, electrostatic repulsion between the same surfactant is replaced by the electrostatic attraction from molecules with opposite charge. Molecules of catanionic surfac-tants are tightly packed at interface. Therefore, IFT decreases and di-lational modulus increases gradually until the interfacial adsorption layer reaches equilibrium over a long time. However, when SDS and C12 TAB are mixed in similar concentration, SDS and C12 TAB would be adsorbed at interface almost equally at the initial stage. Thus the electrostatic attraction becomes dominant from the beginning and a closely packed aggregation is generated quickly. As a result, the initial IFT is quite low and very close to the equilibrium value. So, the systemsof 2 : 1 , 1 : 2 and 1 : 1 reach equilibrium in short time.

3.1.2. Influence of ratio of SDS/C12 TAB on the foam stability in the presence of oil

Catanionic surfactant mixtures have been demonstrated a promising foam agent in enhancing the foam stability in the presence of oil in our previous report[8]. In this study, the influence of the mole ratio of anionic/cationic surfactant on foam stability is investigated. Anionic surfactant SDS and cationic surfactant C12 TAB are mixed at different ratios, 1 : 0, 7 : 1, 3 : 1, 2 : 1, 1 : 1 and 0 : 1, and the total concentration of surfactants is 0.015 mol/L. Kerosene, as model oil, was added to the surfactant solutions before foaming. The oiltolerance of the foams generated by these catanionic agents is evaluated and the $t_{1/2}$ of the foams is shown in Fig. 4(a).

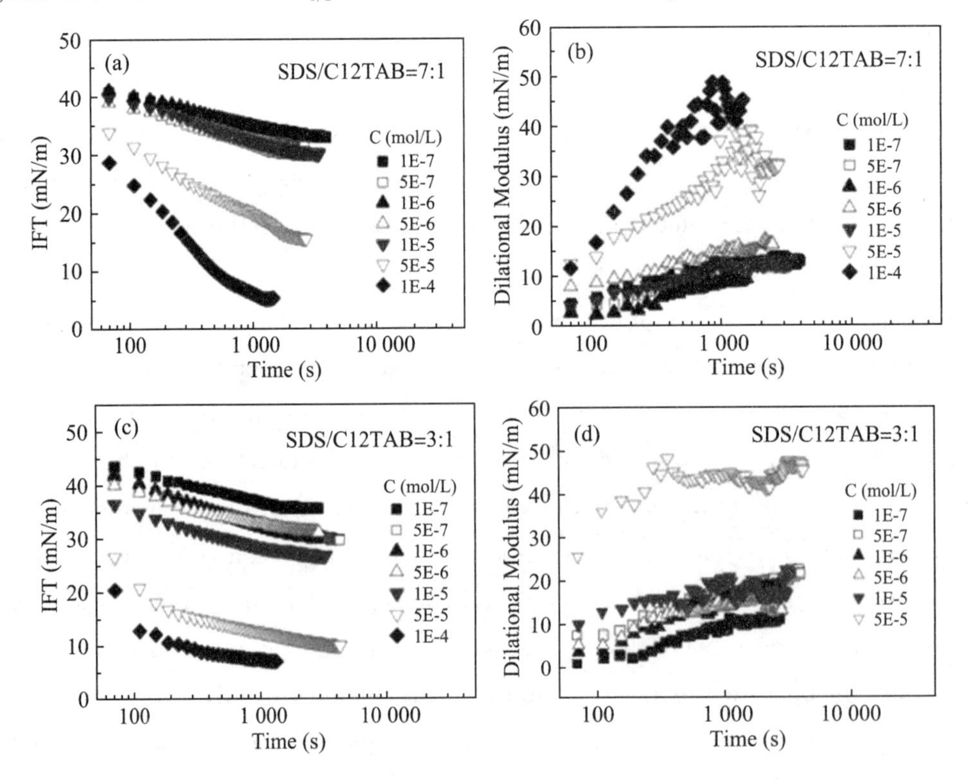

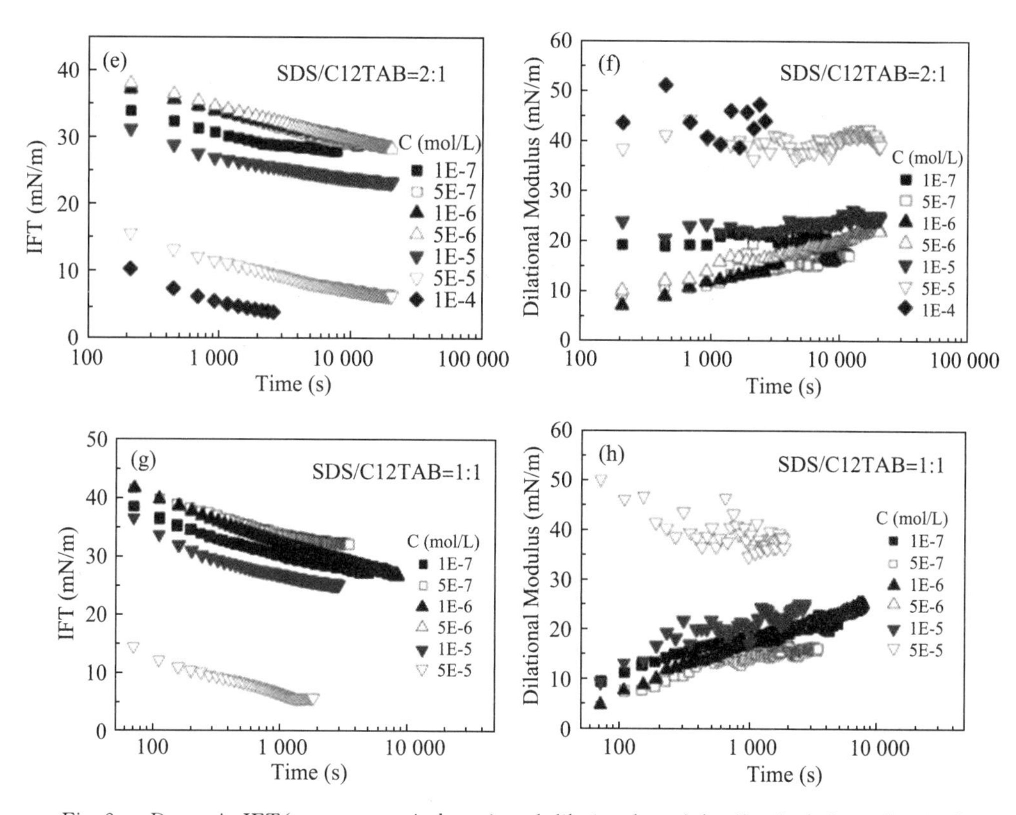

Fig. 2. Dynamic IFT(a, c, e, g, i, k, m) and dilational modulus(b, d, f, h, j, l, n) of SDS-C12 TAB mixtures at diff erent ratios at 30℃. (Oil: Kerosene).

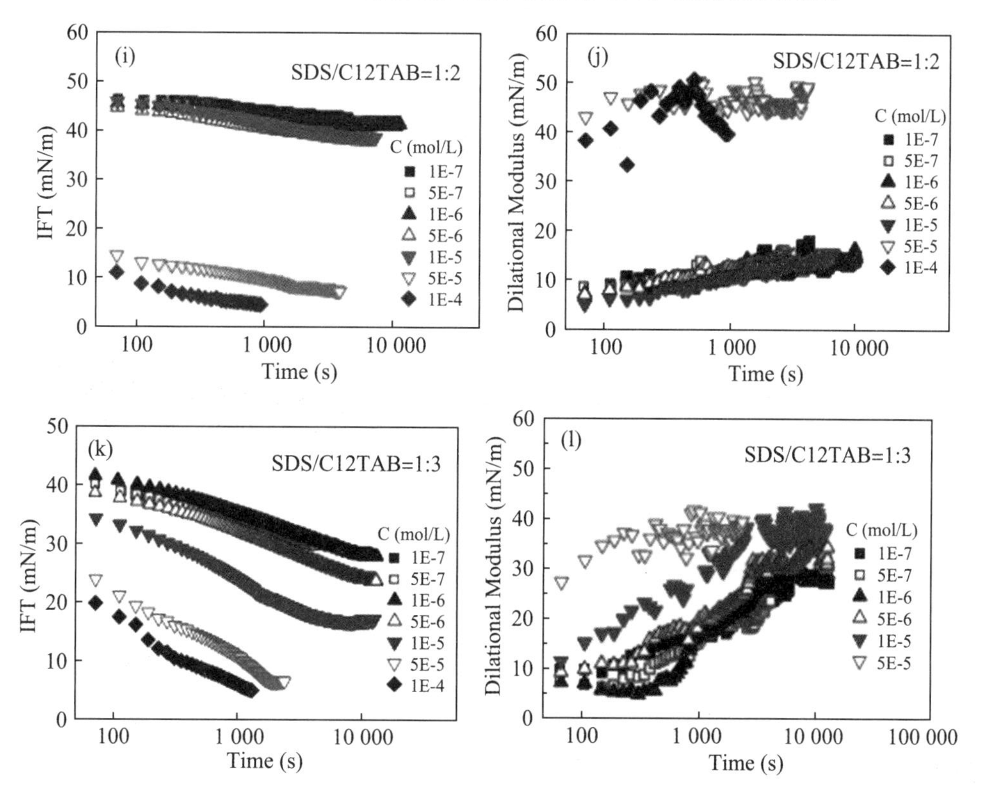

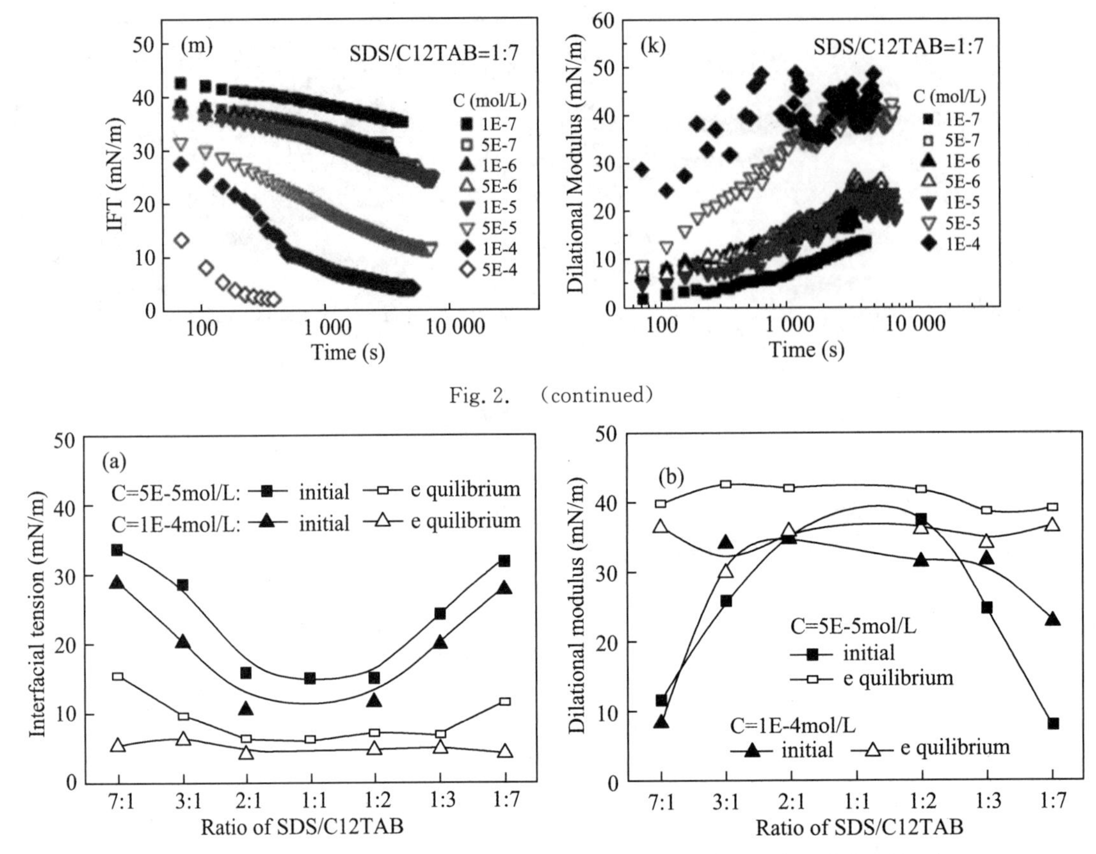

Fig. 2. (continued)

Fig. 3. Initial and equilibrium values of IFT(a) and dilational modulus(b) of SDS-C12 TAB mixtures at 30℃.

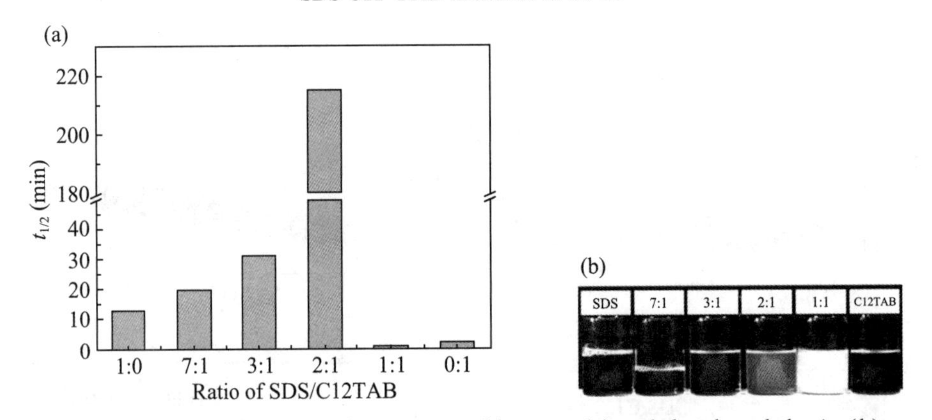

Fig. 4. The $t_{1/2}$ of foams in the presence of kerosene(a) and the phase behavior(b) for catanionic mixtures of SDS-C12 TAB at 30℃. (C=0.015 mol/L).

As the ratio of SDS to C12 TAB decreases from 7 : 1 to 2 : 1, the foam stability in the presence of oil increases gradually and the foam stability reaches the maximum at the ratio of 2 : 1. However, in the case of the equimolar mixture of SDS-C12 TAB, both foamability and oil-tolerance are reduced. According to the ternary phase diagram for C12 TAB/SDS/H_2O[26], precipitation takes place in this condition which is shown in Fig. 4(b). Thus, the actual concentration of surfactants in the solution is signifi cantly decreased. As a result, the foam properties for the equimolar mixtures of SDS-C12 TAB decline dramatically.

In terms of the interfacial dilational viscoelasticity, the equilibrium dilational modulus of SDS-C12 TAB mixtures at diff erent ratios are all around 45 mN/m, which are well above the values of SDS and C12 TAB, respectively[50]. Although the equilibrium interfacial properties of SDS-C12 TAB are similar at different ratios, the foam stability shows obvious dependence on the ratio of SDS/C12 TAB. Since the foams studied in this work are generated by blowing N_2 into solution, which is a fast progress, the possible reason might lie in the dynamic interfacial properties. Therefore, it requires a fast adsorption of foaming agents at the interface. As the ratio of SDS/C12 TAB approaching to 1 : 1, the dy-namic adsorption process reaches the tightly packed state at a fast speed, which meets the requirement of the fast foaming process.

3. 2. Influence of the hydrophobic chain length of surfactant on interface and foam properties

Apart from the mixing ratio of anionic/cationic surfactant, the length of hydrophobic chains of the surfactants also aff ects the infl uence of the length of hydrophobic chains on the dynamic interfacial properties and foam stability, SDS is mixed with C8TAB, C10 TAB, C12 TAB, C14 TAB, and C16 TAB at 2 : 1 respectively. The interfacial dilational rheological experiment and foam stability measurement in the presence of oil are performed.

3. 2. 1. Dynamic IFT and dilational modulus of SDS-CnTAB mixtures

For catanionic mixtures, the interaction between anionic and cationic surfactants includes not only the electrostatic attraction between the hydrophilic heads, but also the hydrophobic interaction between the alkyl chains[28,51,52]. As the increase of the length of hydrophobic tails of cationic surfactant, the hydrophobic interaction between anionic and cationic surfactants increases, which leads a closer pack of catanionic mixtures at the interface. Thus the equilibrium IFT decreases and the plateau of dilational modulus increases with the growth of the alkyl chains[41].

The dynamic IFT and dilational modulus of SDS-CnTAB mixtures at aqueous-oil interface are displayed in Fig. 5. It can be seen clearly that, for SDS-C8TAB and SDS-C10 TAB at high concentration, the IFTs reach low values from the initial stage and maintain constant during the test, which indicates the diff usion adsorption progress of surfactants is very fast. The dilational modulus also keeps constant with time. Because the hydrophobic chains of cationic surfactants are small and molecules have high movability, when an interface appears, the small molecules can be transferred quickly from bulk to the interface. Therefore, IFT reach equilibrium quickly. Even for SDS-C12 TAB (Fig. 2(e,f)), the initial IFT and dilational modulus are quite close to the equilibrium values because of the fast diff usion exchange from bulk to interface.

Compared with the above systems, the IFTs arrive at a plateau much slower for SDS-C14 TAB and SDS-C16TAB. For example, when the total concentration is 1×10^{-4} mol/L, the initial IFT of SDS-C8 TAB is 20 mN/m and keeps almost constant, while an obvious decrease from over 30 mN/m to 5 mN/m can be observed in the dynamic IFT curve of SDS-C16 TAB. The dilational modulus of SDS-C8 TAB mixture at high concentration(1×10^{-4} mol/L) keeps almost constant around the initial value (~20 mN/m) while the dilational modulus of SDS-C16 TAB increases from 15 mN/m to 45 mN/m gradually. It could speculate that the diff usion-adsorption progress gets slower with the increase of the alkyl chain length. Thus it takes longer time for the catanionic mixtures to reach equilibrium.

3.2.2. Influence of hydrophobic chain length on the foam properties in the presence of oil

Fig. 6 illustrates the foam volume as a function of time in the presence of oil for SDS-CnTAB mixtures at 50℃. Foam stability increases as the hydrocarbon chain length of cationic surfactants get closer to that of the anionic surfactant(SDS, n=12) and the most stable foams are generated by mixture of SDS-C12 TAB and SDS-C14 TAB. Previous reports have pointed out that a highly elastic interface can oppose the local depletion of surfactant in the interface more effectively, and therefore, surfactant solutions with high interfacial modulus values are expected to show higher foam stability[32,53,54]. For SDS-C8TAB and SDS-C10 TAB systems, even though the dilational modulus arrive plateau in very short time, the value keeps in low constant. Thus, the foams show poor stability. As the increase of alkyl chains of cationic surfactant, the hydrophobic interaction between the alkyl chains are enhanced which contributes to tightly packed interfacial layers and stable films[51,55]. Therefore, foam stability increases with the length of the hydrophobic chains of cationic surfactants when $n<12$.

However, the long alkyl chain makes the molecular diffusion, orientation and rearrangement more difficult, so it takes longer time for these surfactants to cover a new interface. Even though SDS-C16 TAB mixture can reach high interfacial dilational modulus eventually, there is no suffi cient time for the surfactants to generate a closely packed adsorption layer at the oil-aqueous and gas-aqueous interfaces during the fast foaming process. As a result, the SDS-C16 TAB mixture shows lower foaming ability, and the foam volume decreases faster than other catanionic mixtures.

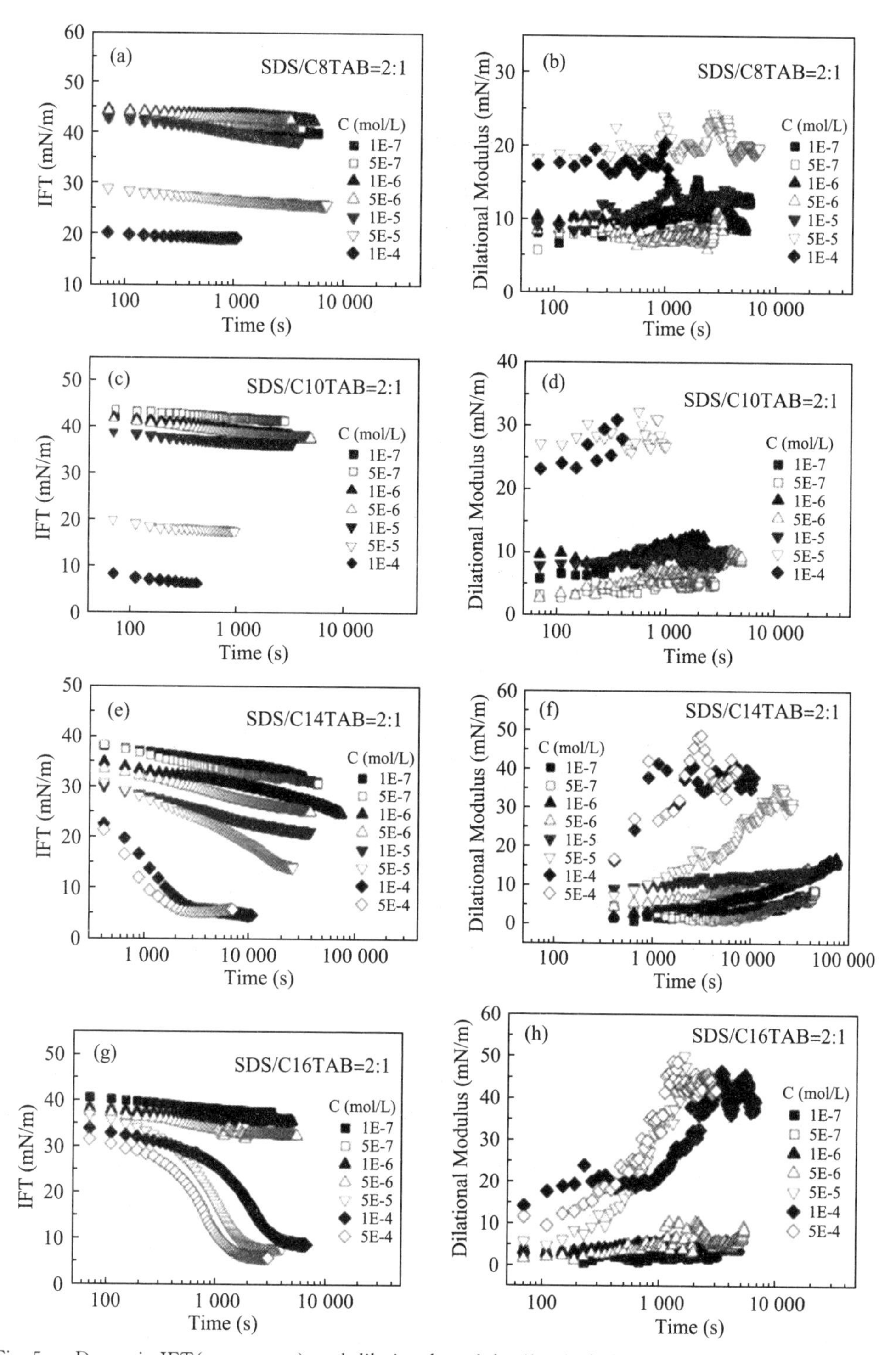

Fig. 5. Dynamic IFT(a, c, e, g) and dilational modulus(b, d, f, h) of SDS-CnTAB(n=8, 10, 14, 16) mixtures at the ratio of 2∶1 at 30℃.

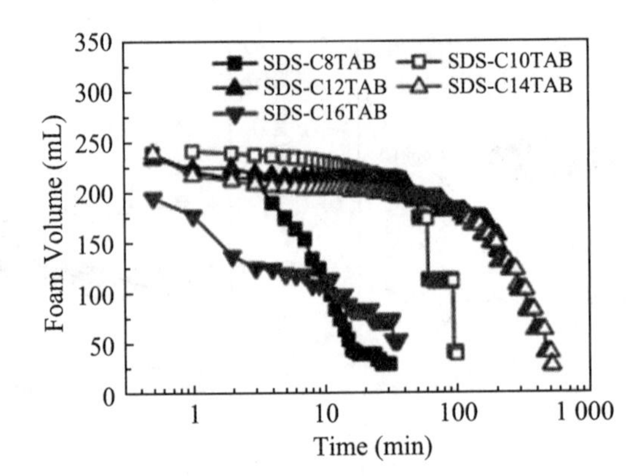

Fig. 6. The evolution of foam volume in the presence of kerosene for catanionic mixtures of SDS-CnTAB at the ratio of 2 : 1.

Fig. 7. TEM image of SDS-C14 TAB at the ratio of 2 : 1 at 45℃, and the total concentration is 0.015 mol/L.

It is worth noting that in the case of SDS-C14 TAB, foam has shown outstanding stability, even though the dynamic interfacial progress is much slower than SDS-C12 TAB. The possible reason is that the viscosity of SDS-C14 TAB is obviously higher than other systems. Packed vesicles of SDS-C14 TAB in the solution have been observed by TEM as shown in Fig. 7 which increases the viscosity. It helps to reduce the drainage of foams and thus provides suffi cient time for the adsorption and rearrangement of catanionic mixtures at interface. Therefore, it is still possible for SDS-C14 TAB to form a tightly packed adsorption layer and to improve the foam stability.

4. Conclusions

The dynamic interfacial properties of catanionic surfactant mixtures at aqueous-oil interface and their foam stability in the presence of oil have been studied in the present work. The experimental results show that the mixing ratio and the alkyl chain length of the surfactants will influence the foam stability in the presence of oil through the interfacial dilational properties.

Although SDS-C12 TAB mixtures generate interfacial layer with similar equilibrium IFT and dilational modulus over a wide range mixing ratio, the dynamic adsorption progress shows noteworthy dependence on ratio. As the ratio of SDS/C12 TAB approaches to 1 : 1, comparative SDS and C12 TAB diff use from the bulk to interface when a new inter-face appears. A tightly packed layer with high modulus is formed in very short time which is in favor of the generation of foaming progress except the 1 : 1 system, in which precipitate generates and the surfac-tants concentration decreases.

In addition, as the increase of the chain length of CnTAB in SDSCnTAB mixtures, the hydrophobic interaction among the molecules at interface is strengthened and increases the equilibrium modulus. On the other hand, the increase of alky chain length declines the movability of the molecules. SDS-C16 TAB mixtures adsorb at interface so slowly that the catanionic mixtures cannot cover the interface during the foaming progress. Thus the foamability and stability decreases when n excesses 14. In the condition of n=14, packed vesicles are formed and increases the viscosity of the liquid. The drainage slows down and provides more time for the adsorption, and therefore, the foams have also shown outstanding stability.

Acknowledgments

The authors thank financial support from the National Science & Technology Major Project(2016 ZX05011-003) and National Natural Science Foundation(21703269) of China.

References

[1] R. Farajzadeh, A. Andrianov, P. L. J. Zitha, Investigation of immiscible and miscible foam for enhancing oil recovery, Ind. Eng. Chem. Res, 49(2010): 1910-1919.

[2] A. Andrianov, R. Farajzadeh, M. M. Nick, M. Talanana, P. L. J. Zitha, Immiscible foam for enhancing oil recovery: bulk and porous media experiments, Ind. Eng. Chem. Res, 51(2012): 2214-2226.

[3] S. Rouimi, C. Schorsch, C. Valentini, S. Vaslin, Foam stability and interfacial properties of milk protein-surfactant systems, Food Hydrocolloids, 19 (2005): 467-478.

[4] E. S. Basheva, D. Ganchev, N. D. Denkov, K. Kasuga, N. Satoh, K. Tsujii, Role of betaine as foam booster in the presence of silicone oil drops, Langmuir, 16(2000): 1000-1013.

[5] L. Arnaudov, N. D. Denkov, I. Surcheva, P. Durbut, G. Broze, A. Mehreteab, Eff ect of oily additives on foamability and foam stability. 1 . Role of interfacial properties, Langmuir, 17(2001): 6999-7010.

[6] A. K. Vikingstad, M. G. Aarra, A. Skauge, Eff ect of surfactant structure on foam-oil interactions-comparing fluorinated surfactant and alpha olefin sulfonate in static foam tests, Colloid Surf. A-Physicochem. Eng. Asp, 279(2006): 105-112.

[7] R. Farajzadeh, A. Andrianov, R. Krastev, G. J. Hirasaki, W. R. Rossen, Foam-oil interaction in porous media: implications for foam assisted enhanced oil recovery, Adv. Colloid Interface Sci, 183(2012): 1-13.

[8] C. Wang, H. B. Fang, Q. T. Gong, Z. C. Xu, Z. Y. Liu, L. Zhang, L. Zhang, S. Zhao, Roles of catanionic surfactant mixtures on the stability of foams in the presence of oil, Energy Fuels, 30(2016): 6355-6364.

[9] J. Eastoe, J. Dalton, P. Rogueda, D. Sharpe, J. F. Dong, J. R. P. Webster, Interfacial properties of a catanionic surfactant, Langmuir, 12(1996): 2706-2711.

[10] L. L. Brasher, E. W. Kaler, A small-angle neutron scattering (SANS) contrast variation investigation of aggregate composition in catanionic surfactant mixtures, Langmuir, 12(1996): 6270-6276.

[11] X. Li, G. Zhao, The physico-chemical properties of the catanionic surfactants mixture, Acta Phys. -Chim. Sin, 8(1992): 191-196.

[12] K. L. Herrington, E. W. Kaler, D. D. Miller, J. A. Zasadzinski, S. Chiruvolu, Phase-behavior of aqueous mixtures of dodecyltrimethylammonium bromide (DTAB) and sodium dodecyl-sulfate(SDS), J. Phys. Chem, 97(1993): 13792-13802.

[13] D. J. Iampietro, L. L. Brasher, E. W. Kaler, A. Stradner, O. Glatter, Direct analysis of SANS and SAXS measurements of catanionic surfactant mixtures by fourier transformation, J. Phys. Chem. B, 102(1998): 3105-3113.

[14] T. Zemb, M. Dubois, B. Deme, T. Gulik-Krzywicki, Self-assembly of flat nanodiscs in free catanionic surfactant solutions, Science, 283(1999): 816-819.

[15] T. Bramer, M. Paulsson, K. Edwards, K. Edsman, Catanionic drug-surfactant mixtures: phase behavior and sustained release from gels, Pharm. Res. 20(2003).

[16] H. Q. Yin, Z. K. Zhou, J. B. Huang, R. Zheng, Y. Y. Zhang, Temperature-induced micelle to vesicle transition in the sodium dodecylsulfate/dodecyltriethylammonium bromide system, Angew. Chem.-Int. Ed, 42 (2003): 2188-2191.

[17] T. Bramer, N. Dew, K. Edsman, Catanionic mixtures involving a drug: a rather general concept that can be utilized for prolonged drug release from gels, J. Pharm. Sci. 95(2006): 769-780.

[18] Y. Wang, C. M. Pereira, E. F. Marques, R. O. Brito, E. S. Ferreira, F. Silva, Catanionic surfactant fi lms at the air-water interface, Thin Solid Films 515 (2006): 2031-2037.

[19] B. M. D. O'Driscoll, E. A. Nickels, K. J. Edler, Formation of robust, free-standing nanostructured membranes from catanionic surfactant mixtures and hydrophilic polymers, Chem. Comm. (2007): 1068-1070.

[20] L. Zhao, J. Liu, L. Zhang, Y. Gao, Z. Zhang, Y. Luan, Self-assembly properties, aggregation behavior and prospective application for sustained drug delivery of a drug-participating catanionic system, Int. J. Pharm. 452(2013): 108-115.

[21] L. Chiappisi, H. Yalcinkaya, V. K. Gopalakrishnan, M. Gradzielski, T. Zemb, Catanionic surfactant systems-thermodynamic and structural conditions revisited, Colloid. Polym. Sci. 293(2015): 3131-3143.

[22] D. Varade, D. Carriere, L. R. Arriaga, A. L. Fameau, E. Rio, D. Langevin, Drenckhan, On the origin of the stability of foams made from catanionicW. surfactant mixtures, Small 7(2011): 6557-6570.

[23] H. Q. Yin, J. B. Huang, Y. Q. Gao, H. L. Fu, Temperature-controlled vesicle aggregation in the mixed system of sodium n-dodecyl sulfate/n-dodecyltributylammonium bromide, Langmuir 21 (2005): 2656-2659.

[24] E. W. Kaler, A. K. Murthy, B. E. Rodriguez, J. A. N. Zasadzinski, Spontaneous vesicle formation in aqueous mixtures of single-tailed surfactants, Science 245(1989): 1371-1374.

[25] E. W. Kaler, K. L. Herrington, A. K. Murthy, J. A. N. Zasadzinski, Phase-behavior and structures of mixtures of anionic and cationic surfactants, J. Phys. Chem. 96(1992): 6698-6707.

[26] G. Kume, M. Gallotti, G. Nunes, Review on anionic/cationic surfactant mixtures, J. Surfactants Deterg. 11(2008): 1-11.

[27] H. Fauser, M. Uhlig, R. Miller, R. von Klitzing, Surface adsorption of oppositely charged SDS:C(12) TAB mixtures and the relation to foam film formation

and stability, J. Phys. Chem. B 119(2015): 12877-12886.

[28] T. Gilanyi, R. Meszaros, I. Varga, Phase transition in the adsorbed layer of catanionic surfactants at the air/solution interface, Langmuir 16(2000): 3200-3205.

[29] A. Stocco, D. Carriere, M. Cottat, D. Langevin, Interfacial behavior of catanionic surfactants, Langmuir 26(2010): 10663-10669.

[30] T. H. Chou, Y. S. Lin, W. T. Li, C. H. Chang, Phase behavior and morphology of equimolar mixed cationic-anionic surfactant monolayers at the air/water interface: isotherm and brewster angle microscopy analysis, J. Colloid Interface Sci. 321(2008): 384-392.

[31] B. Sohrabi, H. Gharibi, B. Tajik, S. Javadian, M. Hashemianzadeh, Molecular interactions of cationic and anionic surfactants in mixed monolayers and aggregates, J. Phys. Chem. B, 112(2008): 14869-14876.

[32] F. Yan, L. Zhang, R.-H. Zhao, H.-Y. Huang, L.-F. Dong, L. Zhang, S. Zhao, J.-Y. Yu, Surface dilational rheological and foam properties of aromatic side chained Nacyltaurate amphiphiles, Colloid Surf. A-Physicochem. Eng. Asp. 396 (2012): 317-327.

[33] A. Hadjiiski, S. Tcholakova, N. D. Denkov, P. Durbut, G. Broze, A. Mehreteab, Effect of oily additives on foamability and foam stability. 2. Entry barriers, Langmuir 17(2001): 7011-7021.

[34] L. Sun, W. Pu, J. Xin, P. Wei, B. Wang, Y. Li, C. Yuan, High temperature and oil tolerance of surfactant foam/polymer-surfactant foam, RSC Adv. 5 (2015): 23410-23418.

[35] K. Osei-Bonsu, N. Shokri, P. Grassia, Foam stability in the presence and absence of hydrocarbons: from bubble-to bulk-scale, Colloid Surf. A-Physicochem. Eng. Asp. 481(2015): 514-526.

[36] F. F. Gao, H. Yan, Q. W. Wang, S. L. Yuan, Mechanism of foam destruction by antifoams: a molecular dynamics study, Phys. Chem. Chem. Phys. 16 (2014): 17231-17237.

[37] J. X. Zhao, W. S. Zou, Foams stabilized by mixed cationic gemini/anionic conventional surfactants, Colloid. Polym. Sci. 291(2013): 1471-1478.

[38] V. B. Fainerman, E. V. Aksenenko, S. V. Lylyk, M. Lotfi, R. Miller, Adsorption of proteins at the solution/air interface infl uenced by added nonionic surfactants at very low concentrations for both components. 3. dilational surface rheology, J. Phys. Chem. B 119(2015): 3768-3775.

[39] C. Cao, J. M. Lei, L. Zhang, F. P. Du, Equilibrium and dynamic

interfacial properties of protein/ionic-liquid-type surfactant solutions at the decane/water interface, Langmuir, 30(2014): 13744-13753.

[40] P. A. Yazhgur, B. A. Noskov, L. Liggieri, S. Y. Lin, G. Loglio, R. Miller, F. Ravera, Dynamic properties of mixed nanoparticle/surfactant adsorption layers, Small, 9(2013): 3305-3314.

[41] C. Wang, X.-L. Cao, L.-L. Guo, Z.-C. Xu, L. Zhang, Q.-T. Gong, L. Zhang, S. Zhao, Eff ect of molecular structure of catanionic surfactant mixtures on their interfacial properties, Colloid Surf. A-Physicochem. Eng. Asp. 509(2016): 601-612.

[42] Y. P. Huang, L. Zhang, L. Luo, L. Zhang, S. Zhao, J. Y. Yu, Dilational properties of hydroxy-substituted sodium alkyl benzene sulfonate at surface and water-decane interface, Acta Phys.-Chim. Sin. 23(2007): 12-15.

[43] B. A. Noskov, D. O. Grigoriev, A. V. Latnikova, S. Y. Lin, G. Loglio, R. Miller, Impact of globule unfolding on dilational viscoelasticity of beta-lactoglobulin adsorption layers, J. Phys. Chem. B 113(2009): 13398-13404.

[44] X.-L. Cao, J. Feng, L.-L. Guo, Y.-w. Zhu, L. Zhang, L. Zhang, L. Luo, S. Zhao, Dynamic surface dilational properties of anionic gemini surfactants with polyoxyethylene spacers, Colloid Surf. A-Physicochem. Eng. Asp. 490(2016): 41-48.

[45] R. Singh, K. K. Mohanty, Synergy between nanoparticles and surfactants in stabilizing foams for oil recovery, Energy Fuels 29(2015): 467-479.

[46] G. Zhao, C. L. Dai, D. L. Wen, J. C. Fang, Stability mechanism of a novel three-phase foam by adding dispersed particle gel, Colloid Surf. A-Physicochem. Eng. Asp. 497(2016): 214-224.

[47] R. Miller, V. B. Fainerman, A. V. Makievski, J. Kragel, D. O. Grigoriev, V. N. Kazakov, Sinyachenko, Dynamics of protein and mixed protein/surfactant adsorptionO. V. layers at the water/fluid interface, Adv. Colloid Interface Sci. 86(2000): 39-82.

[48] V. B. Fainerman, E. V. Aksenenko, S. V. Lylyk, A. V. Makievski, F. Ravera, J. T. Petkov, J. Yorke, R. Miller, Adsorption layer characteristics of tritons surfactants 3. Dilational visco-elasticity, Colloid Surf. A-Physicochem. Eng. Asp. 33 4(2009): 16-21.

[49] K. Fang, G. Zou, P. He, Dynamic viscoelastic properties of spread monostearin monolayer in the presence of glycine, J. Colloid Interface Sci. 266(2003): 407-414.

[50] C. Wang, X.-L. Cao, L.-L. Quo, Z.-C. Xu, L. Zhang, Q.-T. Gong, L.

Zhang, S. Zhao, Eff ect of molecular structure of catanionic surfactant mixtures on their interfacial properties, Colloid Surf. A-Physicochem. Eng. Asp. 509(2016): 601-612.

[51] S. Y. Shiao, V. Chhabra, A. Patist, M. L. Free, P. D. T. Huibers, A. Gregory, S. Patel, Shah, Chain length compatibility eff ects in mixed surfactant systems for D. O. technological applications, Adv. Colloid Interface Sci. 74(1998): 1-29.

[52] M. K. Sharma, D. O. Shah, W. E. Brigham, Correlation of chain length compatibility and surface properties of mixed foaming agents with fl uid displacement effi ciency and eff ective air mobility in porous media, Ind. Eng. Chem. Res. 23 (1984): 213-220.

[53] L. K. Shrestha, E. Saito, R. G. Shrestha, H. Kato, Y. Takase, K. Aramaki, Foam stabilized by dispersed surfactant solid and lamellar liquid crystal in aqueous systems of diglycerol fatty acid esters, Colloid Surf. A-Physicochem. Eng. Asp. 293(2007): 262-271.

[54] E. Santini, L. Liggieri, L. Sacca, D. Clausse, F. Ravera, Interfacial rheology of span 80 adsorbed layers at paraffin oil-water interface and correlation with the corresponding emulsion properties, Colloid Surf. A-Physicochem. Eng. Asp. 309 (2007): 270-279.

[55] A. Patist, V. Chhabra, R. Pagidipati, R. Shah, D. O. Shah, Eff ect of chain length compatibility on micellar stability in sodium dodecyl sulfate/alkyltrimethylammonium bromide solutions, Langmuir 13(1997): 432-434.

Conined structures and selective mass transport of organic liquids in graphene nanochannels

Jiao Shuping,[†] Zhou Ke,[†] Wu Mingmao,[‡] Li Chun,[‡]
Cao Xulong,[§] Zhang Lu,[*,‖] Xu Zhiping[*,†]

[†]Applied Mechanics Laboratory, Department of Engineering Mechanics, and Center for Nano and Micro Mechanics, and

[‡]Department of Chemistry, Tsinghua University, Beijing 100084, China

[§]Exploration & Development Research Institute of Shengli Oilfield Co. Ltd, SINOPEC, Dongying 257015, Shandong, China

[‖]Technical Institute of Physics and Chemistry, Chinese Academy of Sciences, Beijing 100190, China

[*]S Supporting Information

Abstract: Selective transport of liquids is an important process in the energy and environment industry. The

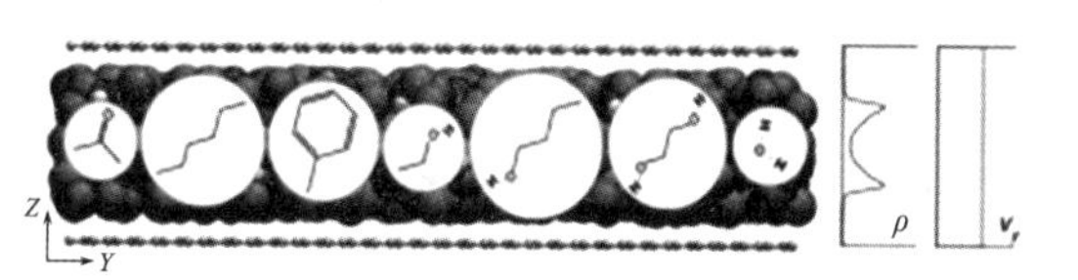

increased energy consumption and the demands of clean water and fossil fuels have urged the development of high performance membrane technologies. Nanoscale channels with the critical size for molecular sieving and atomistically smooth walls for significant boundary slippage are highly promising to balance the trade off between permeability and selectivity. In this work, we explore the molecular structures and dynamics of organic solvents and water, which are confined within nanoscale two-dimensional galleries between graphene or graphene oxide sheets. Molecular dynamics simulation results show that the layered order and significant interfacial slippage are universal for all molecular liquids, leading to notable flow enhancement for channels with a width of few nanometers, in the order of ethylene glycol > butanol > ethanol > hexane > toluene > water > acetone. The extracted dependence of permeability, selectivity on the channel width, and properties of molecular liquids clarify the underlying mechanisms of selective mass transport in nanofluidics, which help to understand and control the filtration and separation processes of molecular liquids. The performance of graphene oxide membranes for permeation and filtration is finally discussed based on the calculated flow resistance for pressure driven flow or molecular diffusivity for diffusive flow, as well as the solubility and wettability of membranes.

INTRODUCTION

Membrane-based technologies have been widely applied in industrial filtration and separation processes including chemical, pharmaceutical separation, and water purifi cation, where permeability and selectivity are the two key figures of merits measuring the performance.[1,2] Recent developments in nanomaterial synthesis and nanofluidic device development have promoted active research in selective mass transport in materials embedding low-dimensional nanochannels. As an example, a graphene oxide (GO) membrane is considered as a promising material for nanofiltration applications as the two-dimensional(2D) interlayer gallery features a typical width ranging from 0.6 to a few nanometers, which can be tuned through swelling in solution, interlayer cross-linking, or intercalation,[3,4] allowing both passive and active controls of the size-sieving process at the molecular level. The ultralow wall friction off ered by the atomically smooth surfaces of pristine sp[2] regions in the GO sheets further facilitates fast performance of GO membranes.[5,6] Moreover, their excellent mechanical and chemical stabilities in organic and even strong acidic, alkaline, or oxidative solvents offer additional merits for Recent studies on the molecular structures and selective fluidic transport processes in GO membranes have been mainly focused on water,[5,8-10] as driven by the increasing demand of clean water around the world, whereas the behaviors of organic solvents under nanoconfinement have been less explored, partly due to the complexity in their molecular structures. However, the separation processing of organic solvents is no doubt of critical importance in the energy and environment industry and should be discussed timely considering the rising energy consumption.[1] A few recent work concluded with diverse results in probing the permeability and selectivity of organic solvents through pervaporation, diffusion, and nano-filtration through GO membranes.[7,11,12] Specifically, Nair et al. reported that the GO membrane is almost impermeable for gases such as He and organic molecule steam (ethanol, hexane) but allows unimpeded transport of the water vapor.[8] Measurements of reduced GO (rGO) membranes and ultrathin GO membranes demonstrate that the permeabilities of several organic solvents including ethanol, isopropanol, nonpolar solvents such as toluene and hexane. 12-14 The GO methanol, acetonitrile, acetone, butanol, and hexane are inversely proportional to the bulk shear viscosity η.[7,13]

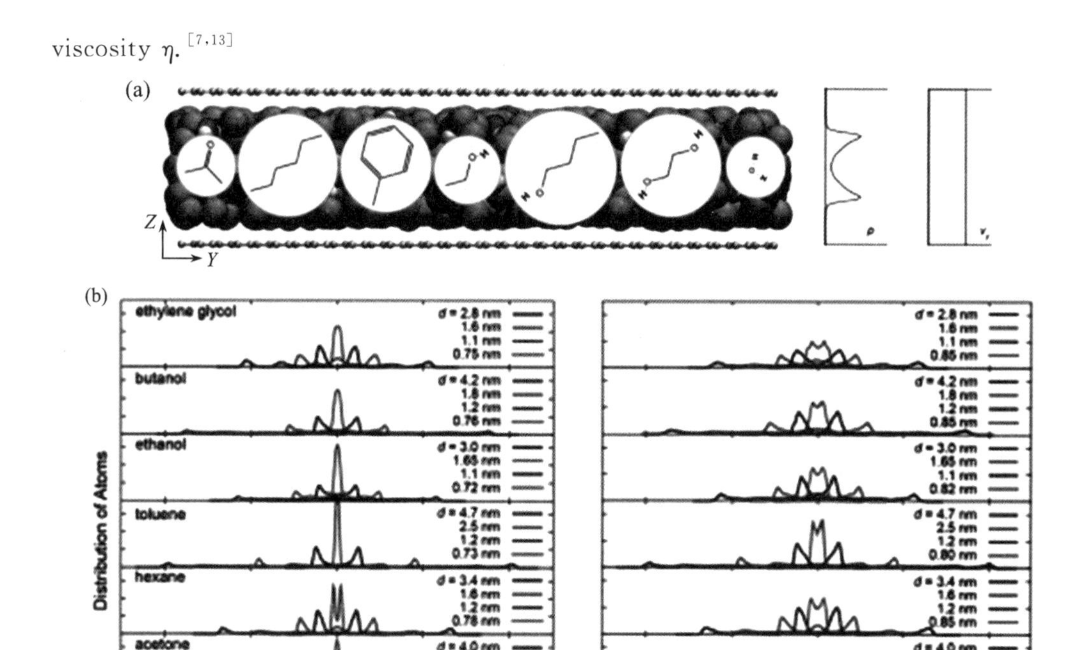

Figure 1.　Molecular structures of organic liquids that are confined in graphene and GO nanochannels. (a) Simulation snapshots and illustrative plots of the atom density distribution(ρ) and flow velocity(v_y) profiles. (b) Distribution of atoms in the liquids along the width of nanochannels.

The permeability of GO membranes is determined by both the flow resistance and solubility of solvents within the nanochannels. Experimental studies reported that the GO membrane can swell in water and selected organic solutions (such as acetone, ethanol, butanol, and methanol) but not for nonpolar solvents such as toluene and hexane.[12-14] The GO membranes swell in a humid environment or water, with the interlayer distance increasing up to 6-7 nm,[4] and the capillary force was concluded to be one of the driving force for fast water transport in graphene channels.[8] GO membranes soaked in organic solvents also swell with an increased spacing, which is reported for acetone(0.98[13], 1.29[12] nm), n-hexane(0.78[13], 0.88[12] nm), ethanol (1.66[13], 1.58[12] nm), butanol(0.92[13] nm), and toluene(0.84[12] nm). On the other hand, the flow resistance in a nanochannel includes contributions from the wall friction and viscous dissipation in the liquid. It was argued that the lipophilicity of graphene to hydrocarbons could lead to a nonslip nature of organic molecule flow between GO layers.[13] These facts lead to concerns on the microscopic origin of the permeability measured in experiments.[7,12,13]

To understand the performance of GO membranes in the selective transport of molecular liquids, it is necessary to explore the process at the molecular scale by considering organic liquids and water as a set of liquids with varying properties including the viscosity and molecule-wall interaction. The microscopic origin of molecular flow in the nanochannels and its universality and peculiarity among different types of liquids lay the ground for the design of membranes or devices with well-balanced trade-off between permeability and selectivity. Relevant discussions, however, have not been made in the literature, although quite a few experimental studies were reported as mentioned earlier. In contrast to bulk properties such as solubility that can be experimentally measured, direct assessment of flow resistance in nanochannels embedded the membranes acquires singlechannel measurements, which is yet technically challengcarrieding.[15-18] Molecular dynamice(MD) simulations are thus carried out to probe the molecular structures and flow of the molecular liquids under nanoconfinement. We find that the interfacial slippage and layered order are not unique for water but universal for all molecular liquids under exploration, which include both polar and nonpolar species. The permeability and selectivity of graphene nanochannels are then analyzed based on the simulation results. Combining with the reported solubility of solvents and measured wettability, we explain the underlying mechanisms of recent experimental measurements of the filtration and separation performance for GO membranes.

RESULTS AND DISCUSSION

Molecular Structures of Nanoconined Organic Liquids. In this work, we explored six organic molecules acetone, hexane, toluene, ethanol, butanol, and ethylene glycol, in addition to water that was explored in the previous work.[9] Molecular structures and the distribution of atoms confined within the graphene nanochannels are shown in Figure 1, which indicatesthat thes patial arrangement of the molecules is highly dependent on the confinement. For very strong confinement with d = 0.64 − 0.8 nm, monolayer water structures with an ordered quasi-square hydrogen-bonding(H-Bond) network form at room temperature, which was reported to be responsible for the fast water diff usion in the graphene nanochannels.[19] This long-range translational order is lost, however, as d increases because of the formation of cross-layer H-bonds between the water bilayer, trilayer, or thicker ones.[19] This in-plane order of the organic liquids explored in this study is generally less prominent than water except for ethylene glycol where ordered H-bond structures are also identified, which could be attributed to the lower polarity of organic liquids compared to the water molecules with relatively stronger H-bonds.

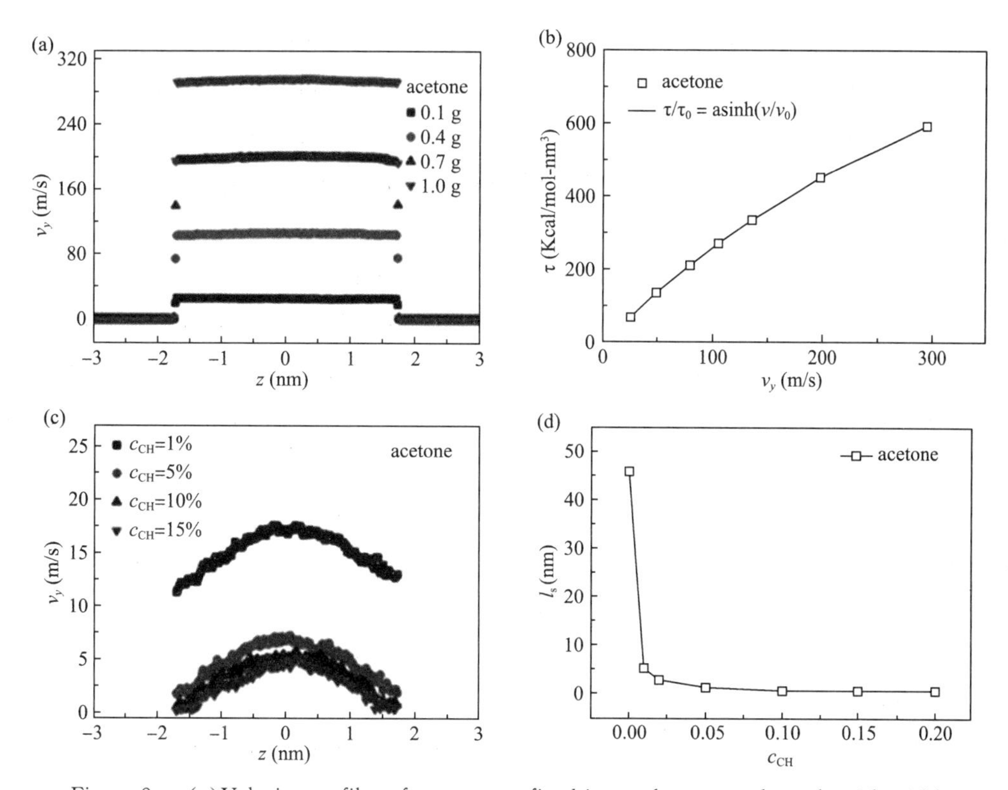

Figure 2. (a)Velocity profiles of acetone confined in graphene nanochannels with width d=4.0 nm, which is driven to flow in the NEMD simulations, where a pressure gradient g=0.686 MPa/nm is applied. (b)Interfacial shear strength between graphene and acetone plotted as a function oftheflow velocity v_y measured at the solid-liquid interface. (c)Velocity profiles of acetone flow in the GO nanochannel with the ratio of oxidation c_{OH} ranging from 0 to 20%. (d)Interfacial slip lengths measured for acetone confined within the GO nanochannels. The error bars indicatethe standard deviations of data obtained from simulations with four initial configurations generated from simulating annealing simulations.

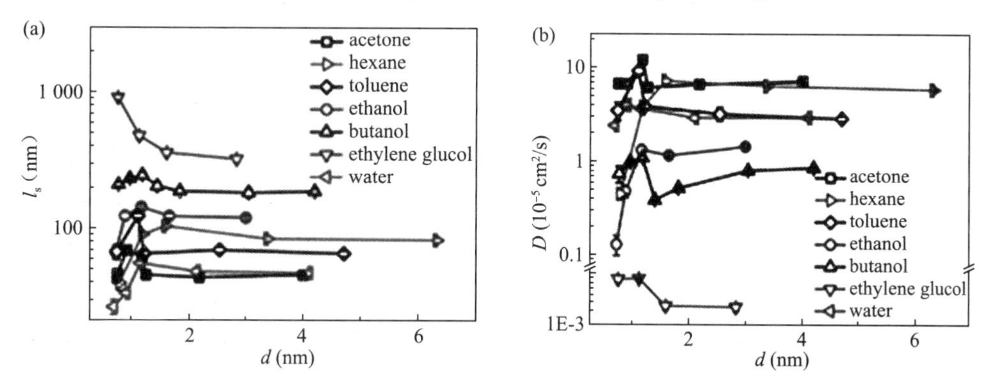

Figure 3. (a)Slip lengthsand(b) coeffi cients of self-diffusion calculated for the molecular liquids confined in graphene nanochannels with widthd, where the thinnest channel contains only molecular monolayers. The error bars indicate the standard deviations of data obtained from MD simulations with four diff erent initial configurations.

As d increases to 1.0～1.2 nm, multilayered structures of molecular liquids emerge for acetone, hexane, toluene, ethanol, butanol and ethylene glycol(up to a bilayer), and water(up to a trilayer beyond 1.15 nm). The layered order appearing for all of the molecular liquids universally is resulted from nanoscale spatial confinement, irrespective to the nature of intermolecular interactions. In even wider channels, there are two distinct layers near the graphitic walls, with uniformly distributed atoms in the central regions. The similarity in the layered order between organic liquids and water under nanoconfinement implies that the unconventional flow characteristics reported for water in atomistically smooth graphene nanochannels may also apply for the organic liquids.

As the graphene sheets are oxidized, the H-bonds within the intercalated water layers are perturbed by the oxygen containing functional groups that could form H-bonds with the water molecules. The translational order of monolayer water thus is destructed but the layering order is preserved as a result of the nanoconfinement. The later argument holds also for the organic liquids.

Interfacial Slip and Flow Enhancement. To probe the nature of flow, we first quantify the interfacial friction between molecular liquids and the graphitic walls. Our nonequilibrium MD(NEMD) simulation results show that the boundary slippageis significant for graphene nanochannels, with notable velocity jump near the wall(see Figure 2a for d=4 nm). It is thus diffi cult to extract the slip length I_s from the velocity profile across the channel through its definition $I_s = v_y/(dv_y/dz)$, where the liquid is driven to flow by a fictitious body force. Instead, we calculate the interfacial shear stress τ between the liquid and walls as a function of flow velocity v_y measured at the solid-liquid interface. The relation between τ and v_y can be fitted using an inverse hyperbolic sine function $\tau/\tau_0 = a\ \sinh(v_y/v_0)$ according to the transition state theory (Figure 2b).[20,21] The fitting parametres τ_0 and v_0 can be used to calculate the frictional coeffi cient $\lambda=\tau_0/v_0$ and slip length $I_s=\eta/\lambda$(Table S1), where the bulk viscosity η is used here. On the other hand, one could also quantify the boundary flow resistance through equilibrium MD(EMD) simulations by evaluating the self-correlation function of interfacial shear stress, where λ is calculated following the Green-Kubo formalism.[22] We confirm the consistency between NEMD and EMD predictions of I_s, approving the use of bulk viscosity values for the liquid confined between graphene walls, which is not obvious considering the fact that the viscosity of the liquid usually increases with the strength of nanoconfinement. To further clarify this issue, we calculate the liquid viscosity by considering nanoconfinement, which shows that the values obtained for the graphene and GO nanochannels with d = 4 nmare closeto the bulk values except for ethylene glycol that maybeattributedto the extended molecular configuration and ordered H-bond network (Figure S1, Table S2). The predicted values of I_s and λ are significant under strong spatial confinement because of the appearance of ordered molecular structures at the liquid-wall interfaces (Figures 3a and 4a, b), in consistency with the results reported in recent studies, [23] with which one can then evaluate the flow enhancement factor compared to nonslip flows, $\varepsilon = 1 + 6\, I_s/d$, for molecular liquids confined in graphene nanochannels. [9] Without loss of generality, we will then use the results obtained for the graphene and GO nanochannels with width d = 1.15 nm that corresponds to the formation of bilayer structures of organic liquids and the water trilayer and d = 4 nm as a wide-channel model that can be experimentally accessed at the single-channel level. [15-18]

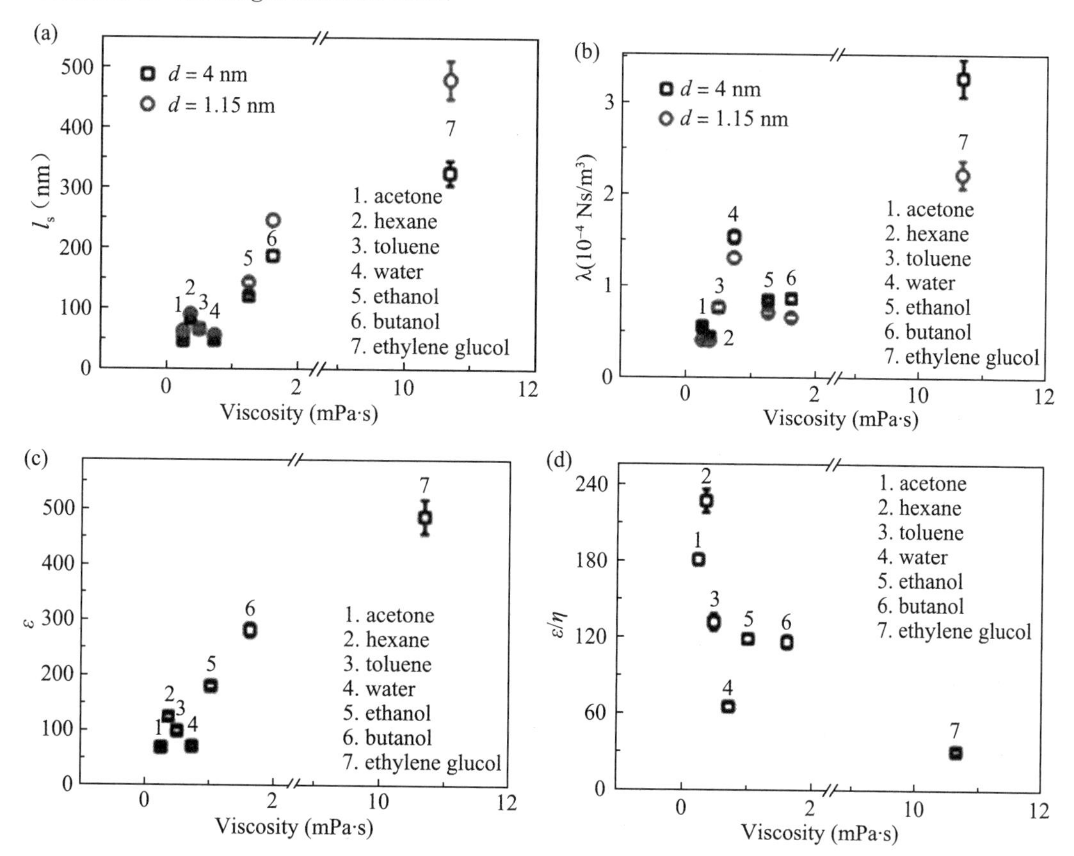

Figure 4.　(a) Slip lengths, (b) interfacial friction coeffi cients, (c) flow enhancement factor, and(d) value of η/e that measures the contrast in the flow resistance of the organic liquids confined in graphene nanochannels with width d = 1.15 nm(a-d) and 4 nm(c,d). The error bars indicate standard deviations of data obtained from simulations with four different initial configurations of the molecular liquids.

The values of l_s and λ for the molecular liquids with d=1.15 and 4 nm summarized in Figure 4a, b demonstrate large slip lengths ranging from tens to hundreds of nanometers. The slip length l_s displays almost the samed dependence except for ethylene glycol, demonstrating a combined effect from molecular ordering and spatial confinement in the channels. The value of l_s increases with d as the layering order becomes less distinct and then declines with d, converging to a constant(Figure 3a). For ethylene glycol with high viscosity(Figure S1), the value of l_s decreases with d monotonically. From the definition $l_s = \eta/\lambda$ we find that, in contrast to the viscosity and frictional coeffi cients that are the intrinsic properties of the liquid and the liquid-wall interface, respectively, the value of l_s is a phenomenological quantification of the interfacial flow slippage. Ethanol and butanol exhibit more significant slippage than water because of their higher viscosity and lower wall friction. These two factors endow the nanoconfined liquids almostflat velocity profiles across the channel, compared to the amplitude of velocity jumps at the liquid-wall interfaces(Figure 2a), in stark contrast to the feature of nonslip viscous flow. Ethylene glycol, although bearing high wall friction(twice as highas water), still exhibits significant slippage thanks to its high viscosity. For acetone, hexane, and toluene, the low interfacial friction coeffi cients and viscosities appear as two competitivefactors, resulting in comparable slippage with that of water. These results obtained for wide graphene nanochannels demonstrate the universal features within the polar or nonpolar molecular liquids that are clearly not unique for water. In addition, water has no advantage in slippage over organic liquids as their interaction with the graphene wall is comparable(Table S3). Specifically, the interfacial energy densities calculated are 16.7, 18.7, 23.67, 16.8, 20.8, and 22.0 kcal mol^{-1} nm^{-2} for the organic liquids(acetone, hexane, toluene, ethanol, butanol, and ethylene glycol, respectively), and the value is 10.8 kcal mol^{-1} nm^{-2} for water, which is lower but still on the same order. These findings are contrary to the argument of lipophilicity induced nonslip that was proposed in the recent work.[13] The universal feature of flow enhancement is also highlighted in Figure 4c through the value of e for nanochannels with d = 4 nm, where the flow enhancement of water confined within graphene nanochannels(e=70) is significantly lower than the values of hexane, toluene, ethanol, butanol, and ethylene glycol(e=125, 100, 182, 283, and 489, respectively) but comparable with the value e=69 for acetone. Interestingly, the flow enhancement factors of organic molecules display almost same ordering as the shear viscosity, except for hexane and toluene, where the viscosity of toluene is higher than hexane but the enhancement of toluene is less significant, which may be attributed to the relatively higher accessible surface of toluene in contact with the wall consideringthe geometry of molecules.

Considering the enhancement with respect to the nonslip prediction of molecular liquid flow, the flow rate can be expressedas $Q = \varepsilon\Delta Pd^3/12\eta L$, where ΔP is the pressure difference applied across the nanochannel with length L(Figure 4c,d). The flow enhancement can be quantified by a factor $\varepsilon = 1 + 6\ I_s/d = 1 + 6\ \eta/\lambda d$, and consequently, the permeability through nanochannels is determined by not only the shear viscosity of liquids but also the liquid-wall interaction that is characterized by the interfacial friction coeffi cient λ, which contradicts with the $1/\eta$ dependence. For nonslip flow($I_s \ll d$), we have $Q = \Delta Pd^3/12\eta L \sim 1/\eta$, and the relation between the flow rate and viscosity can be fitted with a line with intercept at the origin,[7,12,13] whereas for the flow with significant slippage($I_s \gg d$), we have $Q = \Delta Pd^2/2\lambda L \sim 1/\lambda$ that is not directly relevant with the viscosity.

The significant interfacial slippage of nanoconfined flow can be suppressed by the presence of oxygen containing groups in GO.[10] Our simulation results(Figure 2c) indicate that the interfacial slippage and velocity across the channel are significantly reduced because of the presence of oxygen-rich functional groups in graphene, even at a low oxidation level of $c_{OH} = 5\%$. From the simulation results(Figure S2), we find that the molecular layers adhered to the wall have a finite velocity. However, this conclusion may be limited by the nanometer size of models which cannot capture the characteristic length scale of GO sheets with highly inhomogeneous atomic structures. The distributed oxidized regions could hinder the continuous flow of molecular layers, making them immobile as reported in the experiments for water and hydrocarbons confined in the oxidized silicon nanochan nels.[24-26] Our simulation results show that the slip length declines remarkably in the GO nanochannels(Figure 2c,d). The value of l_s is reduced by 92.3% to 3.62 nm for acetone confined between channels with width d= 4 nm and low concentration functionalization(c=~1%), similarly for other molecular liquids. As a result, the flow enhancement factor is reduced to 6.4 and 1.6 for c=~1 and ~20%, respectively. Convective liquid flow though GO membranes would thus be significantly impeded by the oxidized patches in GO sheets, and the percolated pristine regions may constitute the major transport pathway. Recent experimental studies[7,12,13] on organic liquid flow through the GO membranes show no evidence of flow enhancement, whereas the permeance is inversely proportional to viscosity.[7,12,13] This fact may be attributed to the breakdown of interfacial slippage by oxygencontaining functional groups on the graphene surface.

In addition to pressure-driven liquid flow, pervaporation and concentration-driven diffusion were also explored in recent experiment studies, where the self-diff usion of molecular liquids is the fundamental molecular process.[7,11,12] We calculate the coeffi cients of self-diffusion for the molecular liquids using MD simulations as well(Figure 3b), which can be used to measure the eff ective sizes of solutes through the Renkin equation.[12] We find that the general feature of l_s-d dependence(Figure 3a) is echoed in the D-d relation, although the order between the values for molecules with high diff usivity is not distinct. Specifically, for acetone, hexane, toluene, ethanol, butanol, and water, D increases with d for very narrow channels and then decreases and converges, whereas for ethylene glycol, the diff usivity is very low and decreases with d. The results also show that, remarkably, liquids with larger slip lengths are usually less diff usive (except for hexane). Consequently, the permeability measured from the pressure-driven flow and diff usion experiments should be analyzed separately by specifying the underlyingmechanisms. Selectivity. To understand the mechanism of selective fluidic transport in graphene and GO nanochannels, we need to quantify the critical channel width for the molecules to travel. From EMD simulations, we find the value of d corresponding to the monolayer liquids is d=~0.7—0.78 nm for the organic liquids(Table S4) and d=0.65 nm for water. Considering the fact that the value of d for dry GO membranes is only 0.65 nm,[4] rejection of organic molecules but efficient transport of water could be activated, which explains the reported fact that dry GO membranes are impermeable for organic vapors(e. g., ethanol, hexane steam) but transparent for the water vapor.[8]

The contrast in the critical channel width for various molecular liquids under strong confinement could be utilized to separate water and organic liquids by size exclusion, as the value of d for swollen membranes in water or organic solvents is in the range between 0.65 and~1.4 nm(water) and~1.9 nm(organic solvents).[4,8,13,27] For nanochannels wider than 1.2 nm that can accommodate all of the molecular liquids under investigation, the selectivity between the pressure-driven molecular flow could be achieved by the contrast in the combined factor 1/λ or λ or l_s/η for the nonslip flow in GO nanochannels. It was reported that the thickness of GO membranes is optimized to be 1 μm by considering the uniformness and permeability, where the average percentage of holes in the GO sheet is as small as 2%.[12] In these membranes with thousands of GO layers, organic molecules with more extended molecular structures (e. g., long alkane chains) and high wall affi nity (e. g., toluene) may block thetransport pathway at narrow channels. In previous experiments, both water and organic molecules could pass through and the flow is proportional with $1/\eta$ as the case in macro nonslip flow, whereas in the case of ultrathin GO membranes with thickness <8 nm, similar dependencies on η are also reported.[13] This is because that wide channels with $d>\sim1$ nm could form in the transport pathway by the presence of pore or slits and different membrane preparation processes.[13] This fact could be attributed to the complex microstructures of the GO membranes,[14] with a two-scale feature(~1 nm sheet and ~100 nm lamellae) and a composite transport pathway consisting of 2D galleries, slits, and one-dimensional nano-pores. In a recent work, it was pointed out that the fluidic transport in the 2D channels decays exponentially with the thickness of membranes, whereas the cross-layer transport demonstrates a linear dependence.[13] The combination of these two contributions to the overall flux and the diff erent behaviors of the confined molecular liquids as discussed in this work explain the turnover of contrast between organic liquids and water.[13] On the other hand, the mechanism of selective transport in diff usive flow is diff erent. The size-sieving eff ect demonstrated in concentration-driven diff usion can be captured by the Renkin equation through the ratio between d and the eff ective size of molecules.[12]

It should be remarked that the measured permeability reported in the literature depends not only on the flow resistance or diff usivity in nanochannels but also the solubility,[28,29] which can be quantified by the Hansen solubility parameters $\delta=[\delta_d, \delta_p, \delta_h]$. The dispersion, polar, and Hbonding components δ_d, δ_p, and δ_h measure the polarity and interaction strength of the solvent, and a membrane swells in solvents that are close in the δ-space.[30,31] The data collected and summarized in Figure 5 and Table S5 show that the diff erence in δdbetween the solvents under investigation is much smaller than that in δ_p and δ_h.[31] Accordingly, the solvents can be divided into two clusters according to the distance in the subspace $[\delta_p, \delta_h]$.[32] Although the Hansen solubility parameters of GO membranes are unknown, these results explain recent findings that GO membranes do not swell in solvents with low polarity, such as hexane, toluene,[13,14] benzene, and naphthalene. However, the increased interlayer distance between GO sheets in allyl alcohol and methanol could accommodate these molecules. The solubility is closely related to the measured membrane permeability. Although hexane and toluene are the fastest permeating molecules in the graphene channels based on our simulation results, they cannot expand the interlayer gallery of GO membranes for a high permeability that can be measured experimentally.[13,14] This fact leads to the conclusion that permeation through highly laminated ultrathin GO membranes may not be controlled by interlayer transport.[13] To further quantify the affi nity of GO membranes for molecular liquids explored in this work, the liquid contact angles are measured(see Supporting Information, Figure S3 and Table S6). The results show that wettability is not directly correlated with the solubility and the changes in d-spacings of swollen GO membranes, indicating the existence of an energy barrier for solvent molecules to enter the membranes and for the interlayer gallery to expand, which acquires further studies.

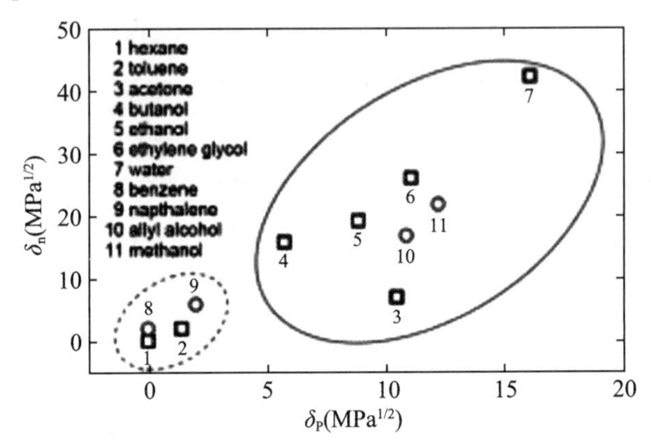

Figure 5. Hansen solubility parameters [δ_p, δ_h](the squares are the datareported inprevious experiments, whereas the circles are those reported only in this work), which can be divided into two clusters by measuringthe subspace distance. GO membranes swell in solvents belonging to the higher cluster but not in those clustered in the lower one.

CONCLUSIONS

In brief, we explore the molecular structures and flow characteristics of organic solvents and water, which are confined within the nanoscale 2D galleries between graphene and GO sheets. MD simulations were performed, showing that the layered order and interfacial slippage under nanoconfinement are universal for all molecular liquids, which result in significantflowenhancement that follows the order of ethylene glycol>butanol>ethanol>hexane>toluene>water>acetone. The dependence of flow resistance and molecular diff usivity onthe channel width was calculated, demonstrating distinct correlation with structural ordering of molecular liquids in the nanochannels and chemistry of the walls. The permeability and selectivity of nanoconfined liquid transport were then discussed in the regimes of pressure-driven flow and diffusion. From the simulation results, we conclude that the process of selective mass transport across a GO membrane is defined by flow resistance in the embedded transport pathway, as well as the solubility and wettability that can be experimentally measured. These results and understandings could help for the understanding and design of molecular processes in filtration and separation of molecular liquids.

MODELSANDMETHODS

Molecular Structures. All of the MD simulations are performed in three dimensions, with water or organic molecules confined between two 2. 7 × 14 nm graphene or GO sheets, as illustrated in Figures 1a and S4. We study graphene and GO nanochannels with width dranging from 0. 6 to 4 nm, and the number of molecules N ranges from 100 to 1000. Periodic boundary conditions (PBCs) are applied in all directions. The molecular models of GO sheets proposed in the literature consist of hydroxyl and epoxy groups on the basal plane and carbonyl groups at defective sites and edges.[33] For the surface functional groups on graphene, the hydroxyl groups are reported to be able to stay enriched in the long-living quasiequilibrium state,[33] and a typical fraction of hydroxyl species relative to the amount of carbon atoms in GO is~20%.[34] Considering these experimental evidences, we construct in this work hydroxyl-functionalized graphene with $c=n_{OH}/n_C=0-20\%$. Here, n_{OH} and n_C are the numbers of hydroxyl groups and carbon atoms, respectively. The distribution of hydroxyl groups is sampled randomly on both sides of the sheet.

MD Simulations. We perform MD simulations using the largescale atomic/molecular massively parallel simulator (LAMMPS).[35] The all-atom optimized potentials for liquid simulations (OPLSAA)[36] are used for the graphene and GO sheets. The SPC/E model[37] of water and united atom optimized potentials for liquid simulations(OPLS-UA) of organic molecules are used in our study, which were widely adopted in the literature as they predict reasonable mass densities, viscosities, and relatively low computational cost compared to other models with more complex forms of potential functions.[37-41] Long-range Coulomb interactions are computed by using the particle-particle particle-mesh algorithm.[42] The interaction between carbon atoms (in graphene and GO) and oxygen atoms (in water) is modeled through the 12-6 Lennard-Jones(L-J) potential function with parameters $\varepsilon_{C\text{-}O}=4.063$ meV and $\sigma_{C\text{-}O}=0.319$ nm, which predict a water contact angle(WCA) of $\theta_{c,G}=98.4°$ for graphene, in consistency with the value measured experimentally.[9,43] The WCA $\theta_{c,GO}$ for GO is lower than $\theta_{c,G}$ and decreases with the concentration c of oxygen-rich functional groups. For a typical value of c=20% for GO, our simulation results is $\theta_{c,GO}=26.8°$, which is also close to recent experimental reports.[44] The 12-6 L-J interactions between mixed-type atom pairs are evaluated using the Lorentz-Berthelot mixing rules. The key parameters used in our model are listed in Table S3.[45]

The molecular structures of organic liquids are obtained through a simulating annealing process followed by 1 ns thermal equilibration at 300 K by using the Berendsen thermostat with a damping time constant of 100 fs. This process is particularly important for longchain organic molecules such as butanol and ethylene glycol. Product simulations of 10 ns are carried out.after equilibration. In all of the simulations, we consider both full and partial(by fixing a few carbon atoms) planar constraints of carbon atoms in the GO sheets that do not lead to notable difference for the conclusions we draw in this work.

NEMD Simulations. In our NEMD simulations, the Poiseuille flow is driven by applying a constant gravity to all of the atoms in the molecular liquids. It usually takes a few hundred picoseconds to reach the steady state where the external force is balanced with wall friction, and the simulations are continued for ten more nanoseconds to collect data. The interfacial shear stress is calculated from the external force as $\tau = Nma/2A$, where N is the number of liquid molecules, m is the mass of a liquid molecule, a is the gravity applied, and A is the interfacial area.[21] It should be noted that this NEMD approach has the limitation that the unweighted force applied to all of the atoms in the molecular liquids is not of physical significance, although the computational cost could be saved by utilizing the PBCs on the flow direction. A more realistic setup is to model a channel connected to two reservoirs where a pressure difference is applied.[46] To assess the reliability of our NEMD setup, we compare the slip length for water flow between graphene sheets and conclude that comparable results are obtained from these two methods (Figure S2).

Interfacial Friction Coeicients. As a parameter that quantifies energy dissipation at the liquid-wall interface, the interfacial friction coefficient λ can be calculated via the Green-Kubo relation of fluctuating interfacial forces F between atoms in the liquid and wall in the EMD simulations, that is

$$\lambda \frac{1}{Ak_{a}T}\int_{0}^{\infty} F(t)F(0)\,\mathrm{d}t \qquad (1)$$

Here, k_B is the Boltzmann constant and T is the temperature. We calculate the autocorrelation function(ACF) of F(t) with binned data set of 20 ps in our 10 ns long EMD simulations. It should be pointed out that there is a well documented difficulty to obtain the Green-Kubo relation via EMD simulations because of the finite-size eff ect often leading to vanishing friction coefficients in a very long time simulation.[21,22] The integration of ACF in eq 1 should thus be done within a reasonable cutoff time t_c to resolve this issue. A widely adopted recipe is to use the time corresponding to the onset of a plateau in the integration as t_c.[21,22] In our analysis, the value of t_c is several picoseconds for the organic liquids and water.

Viscosities of Molecular Liquids. The bulk viscosity of organic liquid η can be calculated via the Green-Kubo relation through the ACF of the fluctuating pressure tensor P at thermal equilibrium, that is

$$\eta_{\alpha\beta}=\frac{V}{k_BT}\int_0^{\infty}P_{\alpha\beta}(t)P_{\alpha\beta}(0)\ \mathrm{d}t \tag{2}$$

where $\alpha,\beta=$ x, y, z($\alpha\neq\beta$), and the bulk viscosity is calculated as $\eta=\frac{1}{3}(\eta_{xy}+\eta_{yz}+\eta_{xz})$. The ACFs are calculated by binning the data from simulations into records of 10—500 ps in our 10 ns long EMD simulations to assure the convergence of results.

The viscosity of nanoconfined liquids is extracted from the flow velocity profiles in GO nanochannels, where the boundary slippage is absent, through the relation $Q=\Delta Pd^3/12\eta L$. The value of η obtained from the measured flow rate Q and the driven force applied $\Delta P/L$ decreases with the interlayer spacing d, and the d dependence is similar for all of the molecular liquids under investigation.

Difusion Coeicients of Molecular Liquids. The molecular diffusion coefficient D is calculated from the correlation function of center-of-mass positions r_i of the liquid molecules through the mean-square distance in thermal equilibrium, by using the Einstein's relation

$$D=\lim_{t\to\infty}|r(t)-r(0)|^2/2d_it \tag{3}$$

where $d_i=2$ is the dimension of space where the molecules diff use, t is the time span of simulations that is typically a few nanoseconds, and ... is the thermodynamic ensemble average. Time averaging is performed for 1 500 or more time series in each MD run, which is under nanoconfinement displays a spatial variation.[47,48] We verified this by calculating the local in-plane diff usion coeffi cient,[49] D(z), acrossthe graphene channel, using acetone as an example. The results show that diff usion near the wall is suppressed because of the presence of distinct molecular layers adhered to the wall (Figure S5). As the focus of our work is on the permeability of molecular liquids confined in nanochannels, the spatially averaged in-plane diff usivity

ASSOCIATEDCONTENT

A list of parameters used in the text; notes on the forcefieldparameters and liquid contact angle measurements, slip lengths obtained from EMD and NEMD; mass densityand shear viscosity of molecular liquids; a list of the force field parameters;

width of graphene nanochannels that intercalate monolayer molecular liquids; Hansen's solubility parameters; contact angles andsurface tension of molecule liquids; shear viscosities for bulk and nanoconfined molecular liquids; density andvelocityprofiles obtained from reservoir and forcing models; snapshots of liquid droplets on the GO and rGO membranes; a 3D snapshot of simulated system; and spatial variation of local in-plane diff usion coeffi cients(PDF)

ACKNOWLEDGMENTS

This work was supported by the National Natural Science Foundation of China through grant no. 11472150, the National Science & Technology Major Project (2016ZX05011-003), and the Tsinghua University Initiative Scientific Research Program through grant no. 2014z22074. The computation was performed on the Explorer 100 cluster system at Tsinghua National Laboratory for Information Science and Technology.[50]

REFERENCES

[1] Lively, R. P.; Sholl, D. S. From Water to Organics in Membrane Separations. Nat. Mater. 2017, 16: 276-279.

[2] Sammells, A. F.; Mundschau, M. V. Membrane Technology in the Chemical Industry; Wiley, 2006.

[3] Xu, Z. Graphene Oxides in Filtration and Separation Applications. In Graphene Oxide; Gao, W., Ed.; Springer, 2015: 129-147.

[4] Zheng, S.; Tu, Q.; Urban, J. J.; Li, S.; Mi, B. Swelling of Graphene Oxide Membranes in Aqueous Solution: Characterization of Interlayer Spacing and Insight into Water Transport Mechanisms. ACS Nano, 2017, 11: 6440-6450.

[5] Mi, B. Graphene Oxide Membranes for Ionic and Molecular Sieving. Science 2014, 343: 740-742.

[6] Jiao, S.; Xu, Z. Non-Continuum Intercalated Water Diffusion Explains Fast Permeation through Graphene Oxide Membranes. ACS Nano 2017, 11: 11152-11161.

[7] Huang, L.; Chen, J.; Gao, T.; Zhang, M.; Li, Y.; Dai, L.; Qu, L.; Shi, G. Reduced Graphene Oxide Membranes for Ultrafast Organic Solvent Nanofiltration. Adv. Mater. 2016, 28: 8669-8674.

[8] Nair, R. R.; Wu, H. A.; Jayaram, P. N.; Grigorieva, I. V.; Geim, A. K. Unimpeded Permeation of Water through Helium-Leak-Tight Graphene-Based

Membranes. Science 2012, 335: 442-444.

[9] Wei, N. ; Peng, X. ; Xu, Z. Breakdown of Fast Water Transport in Graphene Oxides. Phys. Rev. E: Stat. , Nonlinear, Soft Matter Phys. 2014, 89: 012113.

[10] Wei, N. ; Peng, X. ; Xu, Z. Understanding Water Permeation in Graphene Oxide Membranes. ACS Appl. Mater. Interfaces 2014, 6: 5877 -5883.

[11] Huang, K. ; Liu, G. ; Lou, Y. ; Dong, Z. ; Shen, J. ; Jin, W. A Graphene Oxide Membrane with Highly Selective Molecular Separation of Aqueous Organic Solution. Angew. Chem. , Int. Ed. 2014, 53: 6929-6932.

[12] Huang, L. ; Li, Y. ; Zhou, Q. ; Yuan, W. ; Shi, G. Graphene Oxide Membranes with Tunable Semipermeability in Organic Solvents. Adv. Mater. 2015, 27: 3797-3802.

[13] Yang, Q. ; Su, Y. ; Chi, C. ; Cherian, C. T. ; Huang, K. ; Kravets, V. G. ; Wang, F. C. ; Zhang, J. C. ; Pratt, A. ; Grigorenko, A. N. ; Guinea, F. ; Geim, A. K. ; Nair, R. R. Ultrathin Graphene-Based Membrane with Precise Molecular Sieving and Ultrafast Solvent Permeation. Nat. Mater. 2017, 16: 1198-1202.

[14] Putz, K. W. ; Compton, O. C. ; Segar, C. ; An, Z. ; Nguyen, S. B. T. ; Brinson, L. C. Evolution of Order During Vacuum-Assisted Self-Assembly of Graphene Oxide Paper and Associated Polymer Nanocomposites. ACS Nano 2011, 5: 6601-6609.

[15] Xie, Q. ; Alibakhshi, M. A. ; Jiao, S. ; Xu, Z. ; Hempel, M. ; Kong, J. ; Park, H. G. ; Duan, C. Fast Water Transport in Graphene Nanofluidic Channels. Nat. Nanotechnol. 2018, 13: 238-245.

[16] Radha, B. ; Esfandiar, A. ; Wang, F. C. ; Rooney, A. P. ; Gopinadhan, K. ; Keerthi, A. ; Mishchenko, A. ; Janardanan, A. ; Blake, P. ; Fumagalli, L. ; Lozada-Hidalgo, M. ; Garaj, S. ; Haigh, S. J. ; Grigorieva, I. V. ; Wu, H. A. ; Geim, A. K. Molecular Transportthrough Capillaries Made with Atomic-Scale Precision. Nature 2016, 538: 222-225.

[17] Esfandiar, A. ; Radha, B. ; Wang, F. C. ; Yang, Q. ; Hu, S. ; Garaj, S. ; Nair, R. R. ; Geim, A. K. ; Gopinadhan, K. Size effect in ion transport through angstrom-scale slits. Science 2017, 358: 511-513.

[18] Jung, W. ; Kim, J. ; Kim, S. ; Park, H. G. ; Jung, Y. ; Han, C. -S. A Novel Fabrication of 3. 6 nm High Graphene Nanochannels for Ultrafast Ion Transport. Adv. Mater. 2017, 29: 1605854.

[19] Boukhvalov, D. W. ; Katsnelson, M. I. ; Son, Y. -W. Origin of Anomalous

Water Permeation through Graphene Oxide Membrane. Nano Lett. 2013, 13: 3930-3935.

[20] Yang, F. Slip Boundary Condition for Viscous Flow over Solid Surfaces. Chem. Eng. Commun. 2009, 197: 544-550.

[21] Xiong, W.; Liu, J. Z.; Ma, M.; Xu, Z.; Sheridan, J.; Zheng, Q. Strain Engineering Water Transport in Graphene Nanochannels. Phys. Rev. E: Stat., Nonlinear, Soft Matter Phys. 2011, 84: 056329.

[22] Bocquet, L.; Barrat, J.-L. Hydrodynamic Boundary Conditions, Correlation Functions, and Kubo Relations for Confined Fluids. Phys. Rev. E: Stat. Phys., Plasmas, Fluids, Relat. Interdiscip. Top. 1994, 49: 3079-3092.

[23] Dai, H.; Liu, S.; Zhao, M.; Xu, Z.; Yang, X. Interfacial Friction of Ethanol-Water Mixtures in Graphene Pores. Microfluid. Nanofluidics 2016, 20, 141.

[24] Gruener, S.; Hofmann, T.; Wallacher, D.; Kityk, A. V.; Huber, P. Capillary Rise of Water in Hydrophilic Nanopores. Phys. Rev. E: Stat., Nonlinear, Soft Matter Phys. 2009, 79, 067301.

[25] Gruener, S.; Huber, P. Spontaneous Imbibition Dynamics of an n-Alkane in Nanopores: Evidence of Meniscus Freezing and Monolayer Sticking. Phys. Rev. Lett. 2009, 103, 174501.

[26] Gruener, S.; Wallacher, D.; Greulich, S.; Busch, M.; Huber, P. Hydraulic Transport across Hydrophilic and Hydrophobic Nanopores: Flow Experiments with Waterand n-Hexane. Phys. Rev. E 2016, 93, 013102.

[27] Lerf, A.; Buchsteiner, A.; Pieper, J.; Sch ttl, S.; Dekany, I.; Szabo, T.; Boehm, H. P. Hydration Behavior and Dynamics of Water Molecules in Graphite Oxide. J. Phys. Chem. Solids 2006, 67: 1106-1110.

[28] Marchetti, P.; Solomon, M. F. J.; Szekely, G.; Livingston, A. G. Molecular Separation with Organic Solvent Nanofiltration: A Critical Review. Chem. Rev. 2014, 114: 10735-10806.

[29] Karan, S.; Jiang, Z.; Livingston, A. G. Sub-10 nm Polyamide Nanofilms with Ultrafast Solvent Transport for Molecular Separation. Science 2015, 348: 1347-1351.

[30] Hansen, C. M. Hansen Solubility Parameters: A User's Handbook; CRC Press, 2002.

[31] Drioli, E.; Giorno, L.; Fontananova, E. Comprehensive Membrane Science and Engineering; Elsevier, 2017.

[32] Park, S.; An, J.; Jung, I.; Piner, R. D.; An, S. J.; Li, X.;

Velamakanni, A.; Ruoff, R. S. Colloidal Suspensions of Highly Reduced Graphene Oxide in a Wide Variety of Organic Solvents. Nano Lett. 2009, 9: 1593-1597.

[33] Kim, S.; Zhou, S.; Hu, Y.; Acik, M.; Chabal, Y. J.; Berger, C.; de Heer, W.; Bongiorno, A.; Riedo, E. Room-Temperature Metastability of Multilayer Graphene Oxide Films. Nat. Mater. 2012, 11: 544-549.

[34] Dreyer, D. R.; Park, S.; Bielawski, C. W.; Ruoff, R. S. The Chemistry of Graphene Oxide. Chem. Soc. Rev. 2010, 39: 228-240.

[35] Plimpton, S. Fast Parallel Algorithms for Short-Range Molecular Dynamics. J. Comput. Phys. 1995, 117: 1-19.

[36] Jorgensen, W. L.; Maxwell, D. S.; Tirado-Rives, J. Development and Testing of the OPLS All-Atom Force Field on Conformational Energetics and Properties of Organic Liquids. J. Am. Chem. Soc. 1996, 118: 11225-11236.

[37] Berendsen, H. J. C.; Grigera, J. R.; Straatsma, T. P. The Missing Term in Effective Pair Potentials. J. Phys. Chem. 1987, 91: 6269- 6271.

[38] Svishchev, I. M.; Kusalik, P. G.; Wang, J.; Boyd, R. J. Polarizable Point-Charge Model for Water: Results under Normal and Extreme Conditions. J. Chem. Phys. 1996, 105: 4742-4750.

[39] B ez, L. A.; Clancy, P. Existence of a Density Maximum in Extended Simple Point Charge Water. J. Chem. Phys. 1994, 101: 9837-9840.

[40] Wu, Y.; Tepper, H. L.; Voth, G. A. Flexible Simple Point-Charge Water Model with Improved Liquid-State Properties. J. Chem. Phys. 2006, 124: 024503.

[41] Giovambattista, N.; Rossky, P. J.; Debenedetti, P. G. Phase Transitions Induced by Nanoconfinement in Liquid Water. Phys. Rev. Lett. 2009: 102, 050603.

[42] Hockney, R. W.; Eastwood, J. W. Computer Simulation Using Particles; Taylor & Francis, 1989.

[43] Rafiee, J.; Mi, X.; Gullapalli, H.; Thomas, A. V.; Yavari, F.; Shi, Y.; Ajayan, P. M.; Koratkar, N. A. Wetting Transparency of Graphene. Nat. Mater. 2012, 11, 217-222.

[44] Huang, H.; Song, Z.; Wei, N.; Shi, L.; Mao, Y.; Ying, Y.; Sun, L.; Xu, Z.; Peng, X. Ultrafast Viscous Water Flow through Nanostrand-Channelled Graphene Oxide Membranes. Nat. Commun. 2013, 4, 2979.

[45] Shih, C.-J.; Lin, S.; Sharma, R.; Strano, M. S.; Blankschtein, D. Understanding the pH-Dependent Behavior of Graphene Oxide Aqueous Solutions: A Comparative Experimental and Molecular Dynamics Simulation Study. Langmuir 2011, 28: 235-241.

[46] Thomas, J. A.; McGaughey, A. J. H. Water Flow in Carbon Nanotubes: Transition to Subcontinuum Transport. Phys. Rev. Lett. 2009, 102, 184502.

[47] Kusmin, A.; Gruener, S.; Henschel, A.; Holderer, O.; Allgaier, J.; Richter, D.; Huber, P. Evidence of a Sticky Boundary Layer in Nanochannels: A Neutron Spin Echo Study of n-Hexatriacontane and Poly(Ethylene Oxide) Confined in Porous Silicon. J. Phys. Chem. Lett. 2010, 1: 3116-3121.

[48] Hofmann, T.; Wallacher, D.; Mayorova, M.; Zorn, R.; Frick, B.; Huber, P. Molecular Dynamics of n-Hexane: A Quasi-Elastic Neutron Scattering Study on the Bulk and Spatially Nanochannel-Confined Liquid. J. Chem. Phys. 2012, 136, 124505.

[49] Castrill n, S. R.-V.; Giovambattista, N.; Aksay, I. A.; Debenedetti, P. G. Effect of Surface Polarity on the Structure andDynamics of Water in Nanoscale Confinement. J. Phys. Chem. B 2009, 113: 1438-1446.

[50] Zhang, W.; Lin, J.; Xu, W.; Fu, H.; Yang, G. SCStore: Managing Scientific Computing Packages for Hybrid System with Containers. Tsinghua Sci. Technol. 2017, 22: 675-681.

Research paper molecular dynamics simulation of thickening mechanism of supercritical CO_2 thickener

Xue Ping[a], Shi Jing[b], Cao Xulong[b], Yuan Shiling[a,*]

[a] Department of Chemistry, University of Shandong University, 27 Shanda Nanlu, Jinan 250100, China

[b] Shengli Oil Field Exploration and Development Research Institute, 257000, China

abstract

Due to small shear viscosity of supercritical carbon dioxide, oil drilling leads to viscous fi ngering. Addition of polymer as thickener in supercritical CO_2 has been used for improving the sweep efficiency of flooding. The simulated shear viscosity of supercritical CO_2 systems increased with the addition of the polymer. The groups of the polymer have different ability to bind the supercritical CO_2 molecules. The binding effect can mainly be attributed to space grid structure of polymer to reduce movement of CO_2 molecules, in which the van der Waals interaction was the main constraint interaction.

Keywords: Molecular dynamics　Thickener　Supercritical carbon dioxide fluid　Viscosity

1. Introduction

In petroleum industry, a large amount of residual oil still remains in the petroleum reservoir after primary and secondary procedures. To increase the product, the enhanced oil recovery[1,2] process is necessary. Supercritical carbon dioxide[3-5] flooding, as an enhanced recovery process method, is one of the best choices due to its low cost, non-toxicity and non-flammability. The critical temperature and critical pressure of supercritical carbon dioxide is 304. 1 K and 7. 38 MPa, respectively, which is lower than the temperature and pressure in the petroleum reservoir. Therefore, carbon dioxide can be transformed into supercritical liquid when it is put into the reservoir[6,7].

This favourable condition is a good news for the application of this method, since it can save the energy of changing CO_2 from gaseous to supercritical state. Although this method has enormous advantages, some technical challenges still need to be overcome.

The main technical diffi culty is based on the mobility and viscosity[8,9] of supercritical carbon dioxide. However, the low viscosity of supercritical carbon dioxide[10], 10^{-2} mPa · s, can lead to the sweep effi ciency induction and pass through high permeability zone in flooding[11,12] and relatively low sand concentration in fracturing[13]. Zuhair et al. investigated dense CO_2 oil displacement experiment toward fractured carbonate reservoir[14]. Their results indicated that the effect of CO_2 thickening agents added is superior to pure carbon dioxide flooding. In practical application, dense CO_2 also is used for fracturing operation. By this means, the deliverability after fracturing is even better than pure CO_2 fracturing[15]. According to these discussions, it is obvious that thickening of supercritical carbon dioxide is an effi cient way to improve the sweep effi ciency of CO_2 flooding in EOR.

The viscosity of polymer in supercritical CO_2 fluid is very low and extremely limited[16,17] which restricts their use in being thickeners to control fluid mobility in CO_2 flooding[5,18]. Substantial efforts have been devoted to find good supercritical CO_2 thickener. The thickener of CO_2 fluid mainly is polymer, and the research about it has undergone several stages of development. The earliest polymer thickener, silicon-base polymers[19] and fluoropolymers[20,21], have been made in the design of CO_2^- philes to modify solvent properties of supercritical CO_2. However, both of them are expensive, and fluoropolymers can cause environmental issues[22]. All of those are unsuitable to apply on a large scale. After that, the polymer thickener focused on the high molecular weight. Since the polymer with high molecular weight is diffi cult to dissolve in supercritical carbon dioxide fluid, another cosolvent is commonly required to be added into the polymer system. Thus, some relatively low molecular polymers are employed to improve CO_2 viscosity, but polar bond of the polymer reduced their solubility in CO_2 system. Therefore, plenty of cosolvents still need to be added. Recently, the new trend is to design new easy soluble thickener to increase the viscosity of CO_2 system.

To find better soluble thickener, the study about the thickening mechanism of supercritical CO_2 thickener is necessary, because the mechanism is helpful for designing and using of thickener in CO_2 flooding. Molecular dynamics (MD) simulations in studying the mechanism have more excellent performance than traditional experimental method at the molecular level.

Senapati et al. used MD simulation to investigate the structural properties of reserve micelles in supercritical carbon dioxide, and found that the properties of polyether and perfl uoropolyether surfactant are in good agreement with the experimental data[23]. In the study of Saharay et al., an ab Initio MD simulation was used to explore the microscopic details on the intra-and intermolecular structure and dynamics of supercritical carbon dioxide with pressure[24]. In the study of Qin et al., the structural and dynamical properties of supercritical carbon dioxide fluid with the hydroxylated and silylated amorphous silica surfaces were studied by MD simulation. And vdW interaction potential between supercritical carbon dioxide molecules and silica surface was also calculated[25]. In the work of Vaz et al., diffusion coefficients and structural properties of ketones in supercritical carbon dioxide fl uid at infi-nite dilution were discussed[26]. These researches proved that MD simulation is an efficient tool to study the supercritical carbon dioxide system at the molecular level.

In this paper, we studied the thickener, poly (vinyl acetate-covinyl ether) (PVAEE), by using the MD simulations. The microstructure is observed, and the viscosity of thickener is calculated in order to have some insight to the microscopic information and properties of the thickener. Structure and dynamical properties of solvation shell around the groups of polymer are characterized via diffusion coeffi cient, relaxation time and the potential of mean force. We explained thickening mechanism of CO_2 thickener, which can be used to guide researcher to design and screening thickener in the future.

2. Simulation details and methods

The MD simulations were performed by using the GROMACS 4. 6. 7 package[27,28]. The AMBER 03 all-atom force field[29] was adopted to calculate all potential energies which include bond stretching, angle bending, torsion and nonbonded interactions. And the nonbonded interaction between atoms was described by the long-range electrostatic interaction and the short-range vander Waals (vdW) interaction. The electrostatic interaction was calculated by Coulombic equation and the vdW interaction was represented by 12-6 Lennard-Jones function[30]. Materials Studio software was used to construct polymer PVAEE(Fig. 1) which contain 34 mol% vinyl ethyl ether[31]. Degree of polymerization of each chain was N=50. In the simulation, PVAEE was chosen as target molecule to study the dynamic behavior on the solvent of supercritical carbon dioxide. Three systems(system A, system B and system C) was defined. System A was filled with supercritical carbon dioxide liquid. System B was added polymer PVAEE with the content of 1. 19 wt% and in supercritical carbon dioxide liquid. System C was added polymer PVAEE with the content of 2.35 wt% and filled with supercritical carbon dioxide liquid.

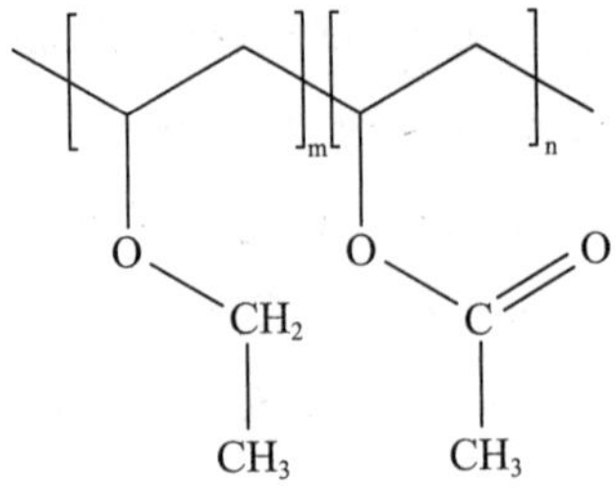

Fig. 1. Structure formula of ester group and ether group in polymer PVAEE.

The simulation cell (10 nm × 10 nm × 10 nm) was constructed with periodic boundary conditions[32] in all directions. The Particle Mesh Ewald (PME) method[33] was employed to treat the long-range electrostatic interactions and the cutoff radius of nonbonding interaction was 1.2 nm. The total initial confi gurations were conducted by energy minimization using steepest descent method. Then all the simulations were conducted in NVT ensemble for 30 ns. The v-rescale thermostat algorithm[34] was used for temperature coupling, controlling temperature at 308 K[35-38] for all simulations. The step length was 2 fs. And the LINCS algorithm[39] was used to constraint bond lengths. Energy fluctuation of system and the potential of mean force of PVAEE was used to determine balance of the system. After 30 ns NVT ensemble, we continued 500 ps NVT ensemble to acquire the MD trajectories. Throughout the simulation an interval of 1.0 ps was applied to collect the trajectories for further analysis. MD trajectories were visualized using VMD 1.9.1.

3. Results and discussion

3.1. The viscosity of supercritical carbon dioxide fluid

CO_2 injection is considered one of the most effective enhanced oil recovery processes applicable to light and medium oil reservoirs. Moreover, the viscosity of CO_2 fluid in oil was the most important factor. Therefore, to improve the sweep efficiency of CO_2 flooding in EOR processes, it is necessary to know the viscosity property of CO_2-crude oil mixtures at reservoir conditions. Consequently, the next will focus on the viscosity firstly in the simulation.

The viscosity of supercritical carbon dioxide fluid can be obtained from transverse-current autocorrelation function[40-42] which was used to examine the behavior of correlations formed from the amplitudes of spontaneous fluctuations for plane waves. More detailed formula derivations are given in Supporting Information.

Supplementary data associated with this article can be found, in the online version, at https: //doi. org/10. 1016/j. cplett. 2018. 07. 006.

We employed this method and Amber all-atom model to calculate the viscosity. Long-wave length fluctuations decay exponentially with a decay constant $1/\tau_H = \mu k^2/\rho$, where μ is the shear viscosity, k is the fl uctuation's wave vector, and $\rho$is the density of system. In this approach, the shear viscosity can be determined by a decay constant, of which the parameter k and μ are known already. The transverse-current autocorrelation function $C_\perp(kx, t)$ decay as $C_\perp(kx,t)$ $e^{-}(\mu k_x^2/\rho)^t$, if k is small enough and t is large enough. Thus, by fitting $C_\perp(kx, t)$ to an exponential decay, the shear viscosity can be extracted from $C_\perp(kx,t)$. Then, the shear viscosity can be calculated from the decay constant. But, the simulated values of μ exhibited some dependence on k which was the smallest values. For a large enough system, it is possible to correct the values of l. This can be expressed as $\mu = \mu_\infty + ak^2$. Thus, the data was used to make curve fitting where the k→0 limit taken to obtain μ_∞. The detailed reckoning formula was showed in S2.

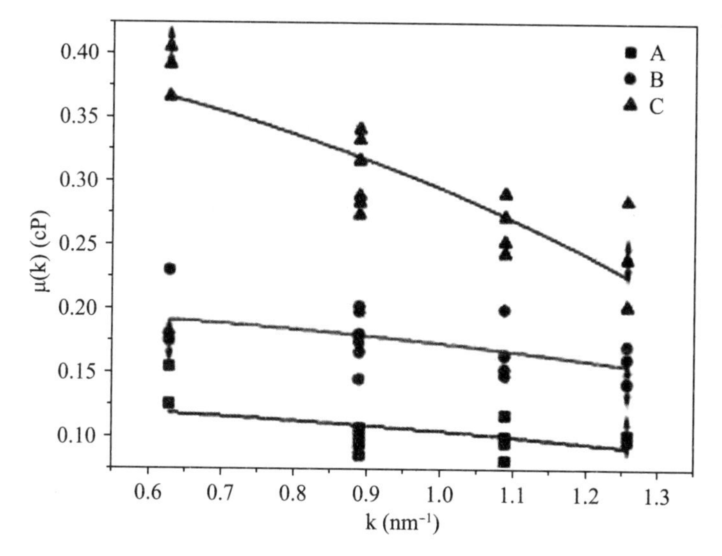

Fig. 2. Viscosities of liquid carbon dioxide of three different systems. The squares are calculated from system A; the circles are calculated from system B; the triangles are calculated from system C.

For all three systems, 500 ps MD trajectories were used to obtain the shear viscosity. The temperature was set at 308 K and the density was set to the liquid saturation density of 0. 854 g/cm^3[35]. Using Eq. (11) in Supporting Information, the viscosity are plotted as a function of wave vector of fl uctuation(k) in Fig. 2, and the values of l_1 and parameter a, are listed in Table 1. The viscosities can be gotten from carve fi tting as k goes to zero. Compared with Palmer's work, in which the data of viscosity is about half of the experiment data, we selected Amber 0 3 force field instead of the Murthy-Singer-McDonald force field. According to extrapolation method from Fig. 2, the simulated viscosity of pure supercritical carbon dioxide liquid is 0. 1268 cP, and it is similar to the experimental value of supercritical carbon dioxide, 0. 08 cP.[43] The corresponding values have been listed in Table 1. From the statistical calculation of the data, the viscosity increases with increasing percentage concentration of poly PVAEE. Moreover, compared with the viscosity of pure supercritical carbon dioxide liquid, system C is treble in viscosity. We will explore the mechanism of fluid viscosity further in a later section.

3. 2. Aggregation structure and balance of system

System B, as a typical simulation system, was used to carry out specific research. After 30 ns simulation, the system has been balanced. Please see the Supporting Information S1. The final confi gurations were provided in Fig. 3. We noticed that there are many mesh structure in the simulated box. The chains of polymers cross each other. In fact, the supercritical CO_2 molecules were constrained in the mesh structure through the non-bond interaction between the polymer and CO_2 molecules. The polymer chains can cross in the CO_2 flooding, resulting into whole moving system. That constrained interaction can increase the viscosity of CO_2 flooding.

3. 3. Molecular dynamics of supercritical carbon dioxide

3. 3. 1. Diffusion property

Mean square displacement(MSD)[44] of supercritical carbon dioxide with different distance range of polymer has a linear relation with time evolution. Compared with in the solvent layer of ether group of polymer, the MSD of CO_2 in the solvent layer of ester group of polymer has a relatively lower slope. Slope of curves can be gotten by linear fitting. Then, diffusion coeffi cient D of CO_2 can be calculated according to the slope of curves. The diffusion constant using the Einstein relation[44] was defi ned as:

$$D=\frac{1}{2dN}\lim_{t\to\infty}\frac{d}{dt}\sum_{i=1}^{N}\langle[\overrightarrow{r_{i(t)}}-\overrightarrow{r_{i(t)}}]^2\rangle$$

where N is the number of target molecular, d is the number of dimensionality and d=3, $\overrightarrow{r_{i(t)}}$ and $\overrightarrow{r_{i(0)}}$ are the ith particle' s coordination at time t and time 0, $\frac{d}{dt}\sum_{i=1}^{N}\langle[\overrightarrow{r_{i(t)}}-\overrightarrow{r_{i(t)}}]^2\rangle$ is the straight slope of curve, as showed in Fig. 4. The diffusion coeffi cient of CO_2 in the solvent layer of ester group, ether group and solvent system are $12.94\pm0.05*10^{-5}$ cm^2/s, $13.90\pm1.32^{*}10^{-5}$ cm^2/s, $17.83\pm0.40^{*}10^{-5}$ cm^2/s, respectively.

It is clear from these data that the diffusion coeffi cient of CO_2 near ester group or ether group is less than that in the system without addition of polymer. The result indicates that the polymer has the ability to bind CO_2 molecules. The decrease of diffusion coefficient also can demonstrate thickening ability of the polymer. Another result also should be mentioned. The diffusion coeffi cient of CO_2 around ester group is slightly lower than that around ether group of the polymer, which means the different binding ability at each part of the copolymer. The difference may make copolymer as useful thicker because we can achieve the desired thickening effect through using different comonomers.

Table 1　The values of μ_∞ and a in fitting curve equation of transverse-current autocorrelation-function in the equation $\mu=\mu_\infty+ak^2$, and percentage concentrations by weight of poly PVAEE and the viscosity of supercritical carbon dioxide liquid.

System	Parameters		C(wt%)	η(cP) of simulation	η(cP) of experiment
	μ_∞	a			
A	0.126 8	−0.022	0	0.126 8	0.08
B	0.203 7	−0.03	1.19	0.203 7	
C	0.413 5	−0.118	2.35	0.413 5	

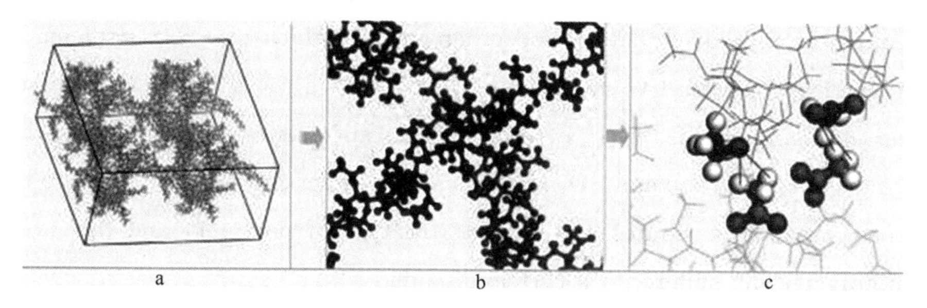

Fig. 3.　(a) A mesh structure of polymer PVAEE in supercritical carbon dioxide fluid. In order to exhibit the mesh structure clearly, a supercell(20 nm×20 nm×10 nm) was constructed. (b)、(c) The cross structure between polymer chains. In(b), PVAEE polymers are represented by different colors. In(c), red balls, O; write balls, H; blue and pink balls, C, respectively. (For interpretation of the references to colour in this figure legend, the reader is referred to the web version of this article.)

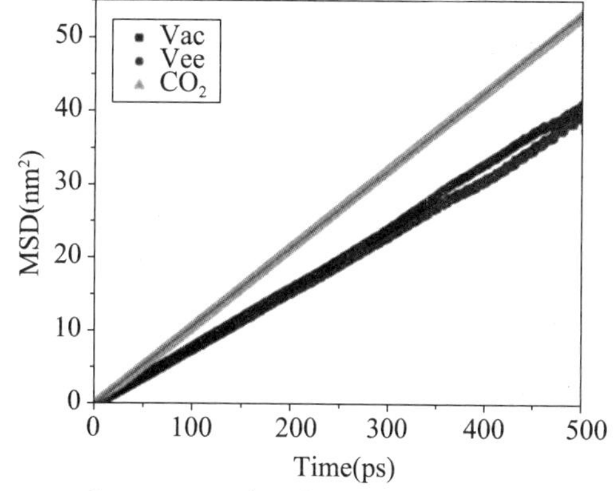

Fig. 4　MSD-time curves of supercritical carbon dioxide near polymer groups in system B.

3.3.2. Residence time dynamics

The residence time of supercritical carbon dioxide near ether group and ester group can well describe the interaction between polymer group and supercritical carbon dioxide, and the longer residence time shows the stronger binding capacities of polymer group to supercritical carbon dioxide. The residence time is determined by percentage of supercritical carbon dioxide within initialthickness range in one time step, which can be characterized by the time correlation function $C_r(t)$[45,46]:

$$C_r(t)=\frac{1}{N_w}\sum_{j=1}^{N_w}\frac{\langle P_{Rj}(0)P_{Rj}(t)\rangle}{\langle P_{Rj}(0)^2\rangle}$$

where P_{Rj} is a binary operator that equals to 1 when the jth supercritical carbon dioxide remains in the initial selected range at time t, otherwise that equals to 0, as show in Fig. 5. N_w is the number of supercritical carbon dioxide in the initial selected range. $< >$ is ensemble average.

In the simulation, CO_2 molecules in the range of 0.4 nm around the group were selected according to RDF(S3), as shown in Fig. 5. The residence time of CO_2 around ester group is longer than that around ether group, which means a stronger binding effect of ester group with CO_2 molecules. The result is the same as that obtained by analyzing diffusion coeffi cient of CO_2. The analysis above indicates that the combination of ester group and CO_2 molecules is stronger and more effective to increase the viscosity, which attributes to the grid structure of polymer, and the interaction between polymer and supercritical carbon dioxide.

3.4. The potential of mean force

To compare the adsorption capacity of ester and ether groups along polymer chain with CO_2 molecules, the potential of mean force(PMF) was used. In our simulation, the PMF was calculated by the RDF through the equation:

$$E(r)=-k_B T\ln(r)$$

where k_B is Boltzmann constant, T is the simulation temperature, and g(r) represents the radial distribution function between the groups and CO_2 molecules. Fig. 7 indicates the energy barrier to be overcome when the external supercritical carbon dioxide is close to the group of polymer. And the CO_2-ester group and the CO_2-ether group are showed respectively, the following instructions are made: (1) The contact minimum (CM) is at about 0.4 nm, which indicates the distance that the polymer directly contacts with CO_2 molecules. (2) The second minimum which appeared at about 0.85 nm is the second solvent-separated minimum(SSM). And it represent that the second solvent layer contacts with CO_2 molecules. The corresponding energy values of

CM and SSM determined the binding stability of CO_2 and polymer in the first solvent layer and in the second solvent layer, respectively. (3) A higher energy barrier between CM and SSM is the barrier of solvent layer (BS). It corresponds that the carbon dioxide from the second solvent layer into the direct contact with the group of polymer demanded to overcome the energy barrier. With increasing distance between CO_2 and group of polymer, PMF tend to be zero at infinite.

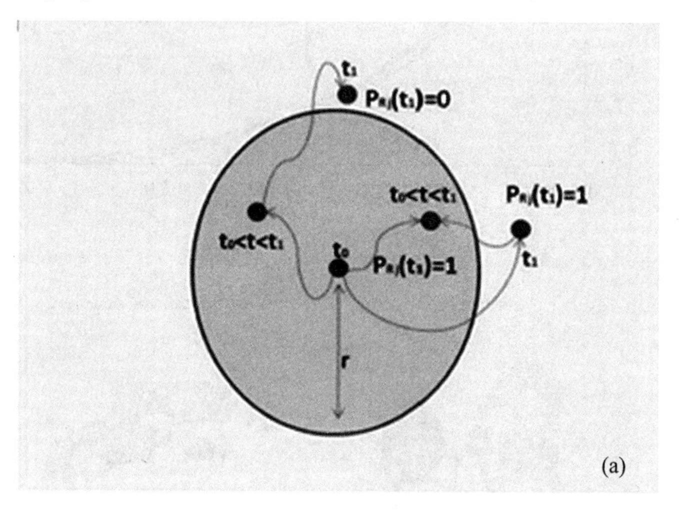

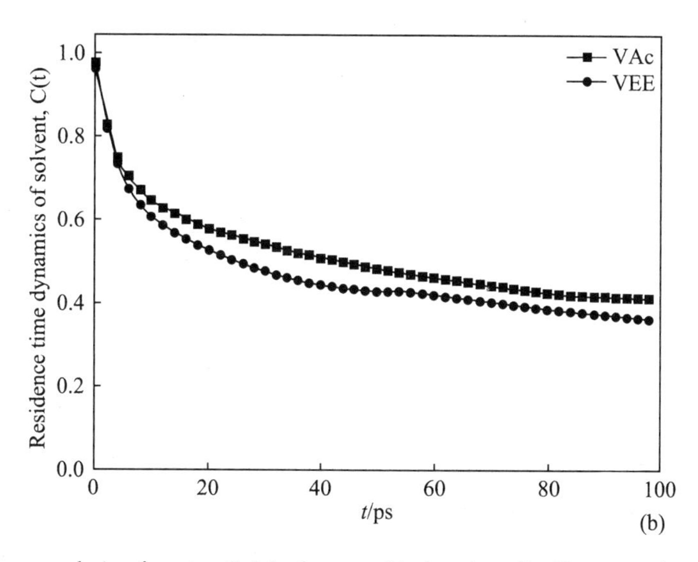

Fig. 5　Time correlation function $C_r(t)$ of supercritical carbon dioxide near polymer group (a) Illustration of definition of P_{Rj}, $P_{Rj}(t)=1$ when supercritical carbon dioxide molecule in the first shell at t_0 and t_1. The location of CO_2 molecule during t_0建 t_1 does not count. (b) Survival time correlation functions of CO_2 near groups of polymer PVAEE.

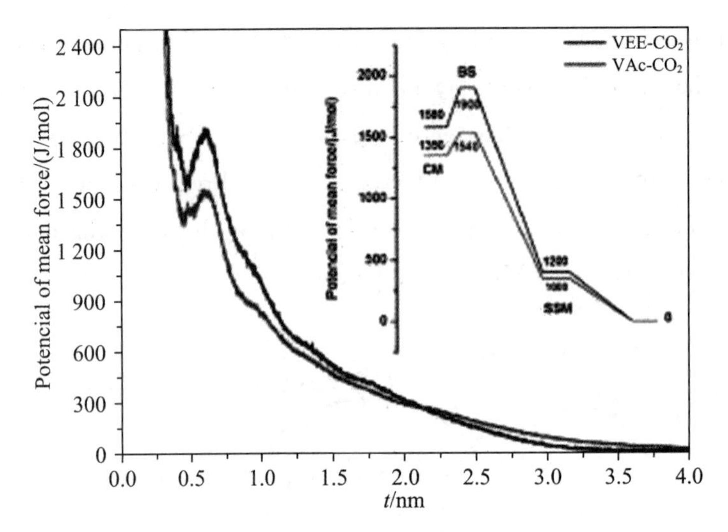

Fig. 6 Potential of mean force between CO_2 and polymer groups;

Inset on right top contains only minimum and maximum points of energy for clarity.

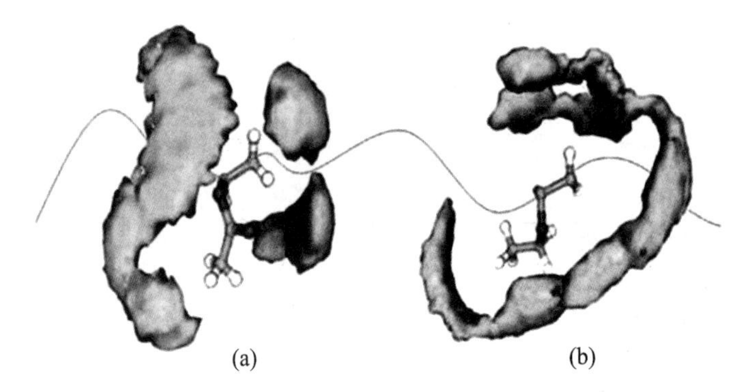

Fig. 7 Spatial distribution functions of supercritical carbon dioxide molecules around the ester group(a) and ether group(b).

The binding energy barrier between the CO_2 and polymer group depends on BS and SSM, $\Delta E^+ = E_{BS} - E_{SSM}$; while the dissociation energy barrier between CO_2 and group of polymer can be determined by BS and CM, $\Delta E^- = E_{BS} - E_{SSM}$. Table 2 showed the binding and dissociation energies between polymer group and CO_2. As shown in Fig. 6 and Table 2, the following conclusions are made: (1) The solvent layer between ester group and CO_2 is more stable than that between the ether group and CO_2, because energy in the corresponding solvent layer is 1350 J/mol which is smaller than that between eater group and CO_2; (2) The binding energy barrier between ester group and CO_2 ($\Delta E^+ = 540$ J/mol) is lower than that between ether group and CO_2 ($\Delta E^+ = 700$ J/mol), the energy barrier is required to be low, which means that combination of ester group and CO_2 is more easily.

Meanwhile, the dissociation energy between ether group and CO_2 ($\Delta E^- = 1450$ J/mol) are greater than that between ester group and CO_2 ($\Delta E^- = 1020$ J/mol). In other words, the supercritical carbon dioxide are easy to combine with ether group. And the ester group is also more easily to dissociate.

3. 5. Spatial distribution function

The spatial distribution functions (SDFs)[47] of CO_2 molecule can give vivid description of the distributions of supercritical carbon dioxide molecules around the ether group and ester group. In Fig. 7(a), polymer side chains are surrounded by a big ribbon, which is called the first solvation shell. And another smaller ribbon spreads around carbon-oxygen double bond. It shows that the O atom in the carbonyl group would enhance the dissolution of supercritical carbon dioxide with polymer. A small oval-shaped ribbon distribute at carbon hydrogen bonds of polymer backbone. This result is mainly due to the existence of weak hydrogen bond between the O atom in CO_2 and the H atom of the polymer chain. In Fig. 7(b), one half annulated ribbon surrounded ether group. This ribbon also represents the first solvation shell of CO_2 molecules. The appearance of solvation shell at ester and ether group can be attributed to the interaction between CO_2 and these groups, which also is one of the proofs that the CO_2 can be bound by the polymer.

Table 2 Binding and dissociation energies barrier between polymer group and CO_2.

Polymer group	ΔE^+/(J mol^{-1})	ΔE^-/(J mol^{-1})
VEE	700	320
VAc	540	190

3. 6. Interactions of polymer groups in supercritical carbon dioxide fluid

As the discussion above, the viscosity of CO_2 flooding attributed to the interaction between the polymer group and CO_2 molecules. This interaction belongs to the non-bonded energies including electrostatic interaction and VDW interaction, showed in Fig. 8. We noted that the VDW interaction energy is more than the electrostatic interaction between CO_2 molecules and polymer groups. We think the VDW interaction energy between CO_2 molecules and group of polymer is dominant driving force, which is important for improving the viscosity of carbon dioxide and stability of polymer in the fluid.

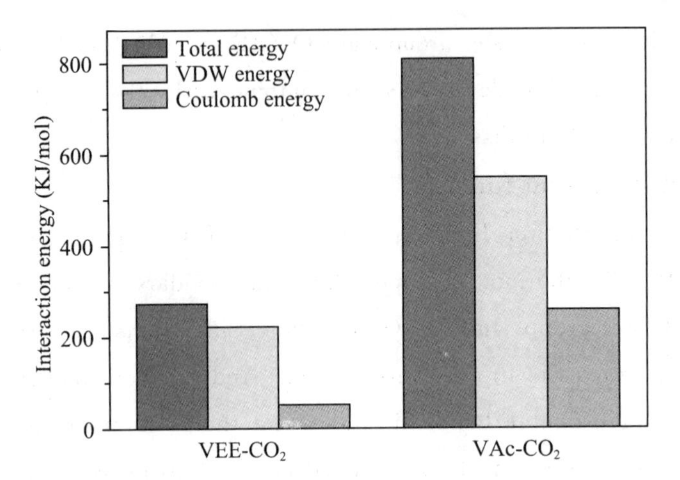

Fig. 8 The interaction energy of supercritical carbon dioxide and ether or ester group including total energy, VDW energy and coulomb energy All the data are averaged using the last 500 ps MD simulations.

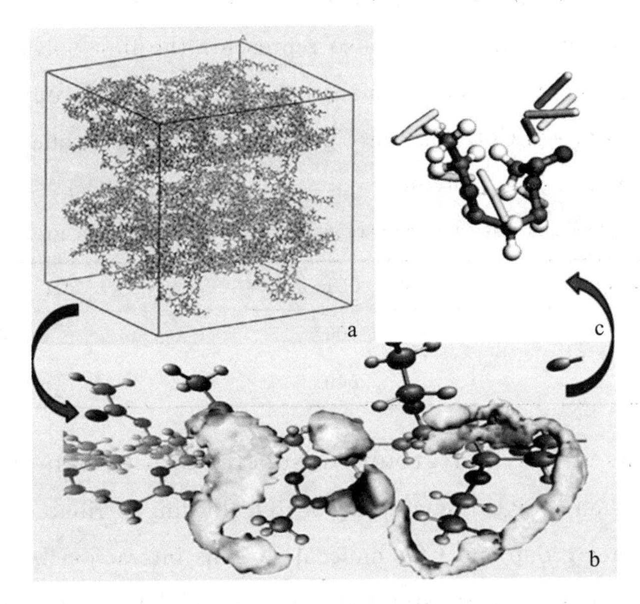

Fig. 9 Illustration of polymers in CO_2 fluid and CO_2 molecules located around the group of polymer.

To sum up, in supercritical carbon dioxide fluid, the structure of polymers has been changed into space grid structure, showed in Fig. 9. That structure decreased the ability of CO_2 molecules mobility. Besides, the appearance of solvation shell at ester and ether group can be attributed to the interaction between CO_2 and these groups. The dominating interaction is VDW interaction. All of these proofs that the CO_2 can be bound by the polymer.

4. Conclusions

The MD simulations were performed to calculate viscosity of supercritical liquid and investigate the mechanism of supercritical CO_2 thickener. The studied results showed that the addition of the random polymer PVAEE observably increases the viscosity of supercritical CO_2. The more the concentration of PVAEE increases, the higher the viscosity is. It is agreement with the experiment. These results suggest that this polymer can be considered as one of the effective thicker. According to the diffusion coeffi cient of supercritical CO_2 at the system in the present or absent of the polymer PVAEE, we can know that the addition of polymer can decrease the diffusion of supercritical CO_2. It indicates that the polymer can interact with CO_2 molecules so as to constraint their movement in the mesh structure of polymers. Another interesting result from the diffusion coefficient analysis is that the ester group has higher ability to bind CO_2 molecule than ether group, which is the same conclusion as that from the residence time of CO_2 molecule. This means that the difference represents very desirable during the future of molecular design for the supercritical CO_2 thickener. The simulated models help us to further understand the dynamical process of thickener. The results of this work may provide a forward step on understanding the thickening mechanism and providing a reference to select better thickeners.

Acknowledgement

We gratefully appreciate the financial support from the National Science Foundation of China (No. 21573130) and National Science and Technology Major Project(2016ZX05011-003).

References

[1] L. W. Lake, Enhanced Oil Recovery, Prentice Hall, Englewood Cliffs, NJ, 1989.

[2] F. Orr, J. Taber, Science 224(1984): 563.

[3] Y. Gu, S. Zhang, Y. She, J. Polym. Res. 20(2013): 61.

[4] J. M. DeSimone, Science 297(2002): 799.

[5] S. Cummings, D. Xing, R. Enick, S. Rogers, R. Heenan, I. Grillo, J. Eastoe, SoftMatter 8(2012) 7044.

[6] S. Wang, Q. Feng, M. Zha, F. Javadpour, Q. Hu, Energy Fuels 32(2018)

169.

[7] S. Wang, Q. Feng, F. Javadpour, T. Xia, Z. Li, Int. J. Coal Geol. (2015) 9: 147-148.

[8] D. Chakravarthy, V. Muralidharan, E. Putra, D. Hidayati, D. S. Schechter, Mitigating oil bypassed in fractured cores during CO_2 flooding using WAG and polymer gel injections, in: SPE/DOE Symposium on Improved Oil Recovery, Society of Petroleum Engineers, Tulsa, Oklahoma, USA, 2006.

[9] H. Liu, F. Wang, J. Zhang, S. Meng, Y. Duan, Pet. Explor. Dev. 41 (2014) 513.

[10] E. Heidaryan, T. Hatami, M. Rahimi, J. Moghadasi, J. Supercrit. Fluids 56(2011) 144.

[11] Q. Feng, S. Wang, G. Gao, C. Li, J. Petrol. Sci. Eng. 75(2010) 13.

[12] S. Wang, Q. Feng, X. Han, PLoS One 8(2014), e83536.

[13] Z. Hamdi, M. B. Awang, B. Moradi, Low temperature carbon dioxide injection in high temperature oil reservoirs, Int. Pet. Technol. Conf. (2014).

[14] Al Yousef, Study of COZ. 2 Mobility Control in Heterogeneous Media Using CO_2 Thickening Agents, Texas A& M University, 2012.

[15] S. Luk, M. Apshkrum, Appetite 12(1996) 241.

[16] Beckman, Cheminform 35(2004) 1885 E. J..

[17] J. Eastoe, S. Gold, S. Rogers, P. Wyatt, D. C. Steytler, A. Gurgel, R. K. Heenan, X. Fan, E. J. Beckman, R. M. Enick, Angew. Chem. 118(2006) 3757.

[18] S. Cummings, R. Enick, S. Rogers, R. Heenan, J. Eastoe, Biochimie 94 (2012) 94.

[19] R. Mertsch, B. Wolf, Macromolecules 27(1994) 3289.

[20] J. Eastoe, B. M. H. Cazelles, D. C. Steytler, J. D. Holmes, A. R. Pitt, T. J. Wear, R. K. Heenan, Langmuir 13(1997) 6980.

[21] Z. Huang, C. Shi, J. Xu, S. Kilic, R. M. Enick, E. J. Beckman, Macromolecules 33(2000) 5437.

[22] E. Beckman, Chem. Commun. (2004) 1885.

[23] S. Senapati, M. L. Berkowitz, J. Phys. Chem. B 107(2003) 12906.

[24] M. Saharay, S. Balasubramanian, J. Phys. Chem. B 111(2007) 387.

[25] Y. Qin, X. Yang, Y. Zhu, J. Ping, J. Phys. Chem. C 112(2008) 12815.

[26] R. V. Vaz, J. R. B. Gomes, C. M. Silva, J. Supercrit. Fluids 107 (2016) 630.

[27] H. J. C. Berendsen, D. van der Spoel, R. van Drunen, Comput. Phys.

Commun. 91(1995) 43.

[28] D. Van Der Spoel, E. Lindahl, B. Hess, G. Groenhof, A. E. Mark, H. J. Berendsen, J. Comput. Chem. 26(2005)1701.

[29] J. Wang, R. M. Wolf, J. W. Caldwell, P. A. Kollman, D. A. Case, J. Comput. Chem. 25(2004) 1157.

[30] N. Du, R. Song, H. Zhang, J. Sun, S. Yuan, R. Zhang, W. Hou, Colloids Surf., A 509(2016) 195.

[31] D. Hu, S. Sun, P. Yuan, L. Zhao, T. Liu, J. Phys. Chem. B 119(2015) 3194.

[32] I.-C. Yeh, G. Hummer, J. Phys. Chem. B 108(2004) 15873.

[33] U. Essmann, L. Perera, M. L. Berkowitz, T. Darden, H. Lee, L. G. Pedersen, J. Chem. Phys. 103(1995) 8577.

[34] G. Bussi, D. Donadio, M. Parrinello, J. Chem. Phys. 126(2007), 014101.

[35] D. Hu, S. Sun, P.-Q. Yuan, L. Zhao, T. Liu, J. Phys. Chem. B 119 (2015)12490.

[36] M. Nobakht, S. Moghadam, Y. Gu, Energy Fuels 21(2007) 3469.

[37] Y. Song, N. Zhu, Y. Zhao, Y. Liu, L. Jiang, T. Wang, Phys. Fluids 25 (2013), 053301.

[38] X. Wang, Y. Gu, Ind. Eng. Chem. Res. 50(2011) 2388.

[39] V. E. Petrenko, M. L. Antipova, D. L. Gurina, Russ. J. Phys. Chem. A 89(2015) 411.

[40] B. J. Palmer, Phys. Rev. E 49(1994) 359.

[41] H. Zhong, S. Lai, J. Wang, W. Qiu, H.-D. Lüdemann, L. Chen, J. Chem. Eng. Data 60(2015) 2188.

[42] B. Hess, J. Chem. Phys. 116(2002) 209.

[43] M. S. Zabaloy, V. R. Vasquez, E. A. Macedo, J. Supercrit. Fluids 36 (2005) 106 .

[44] N. F. A. van der Vegt, Macromolecules 33(2000) 3153.

[45] Q. Shao, Y. He, A. D. White, S. Jiang, J. Phys. Chem. B 114 (2010) 16625.

[46] J. C. Hower, Y. He, M. T. Bernards, S. Jiang, J. Chem. Phys. 125(2006) 214704.

[47] J. C. Shelley, M. Sprik, M. L. Klein, Langmuir 9(1993) 916.